BASALTIC VOLCANISM ON THE TERRESTRIAL PLANETS

Eruption of basaltic volcano Stromboli in September, 1971. Photographer: Charles A. Wood.

BASALTIC VOLCANISM ON THE TERRESTRIAL PLANETS

By members of the

BASALTIC VOLCANISM STUDY PROJECT

Sponsored by

The Lunar and Planetary Institute
Houston, Texas 77058

Pergamon Press

New York • Oxford • Toronto • Sydney • Frankfurt • Paris

Pergamon Press Offices:

U.S.A.	Pergamon Press Inc., Maxwell House, Fairview Park, Elmsford, New York 10523, U.S.A.
U.K.	Pergamon Press Ltd., Headington Hill Hall, Oxford OX3 0BW, England
CANADA	Pergamon Press Canada Ltd., Suite 104, 150 Consumers Road, Willowdale, Ontario M2J 1P9, Canada
AUSTRALIA	Pergamon Press (Aust.) Pty. Ltd., P.O. Box 544, Potts Point, NSW 2011, Australia
FRANCE	Pergamon Press SARL, 24 rue des Ecoles, 75240 Paris, Cedex 05, France
FEDERAL REPUBLIC OF GERMANY	Pergamon Press GmbH, Hammerweg 6 6242 Kronberg/Taunus, Federal Republic of Germany

---

This book may be cited as:
Basaltic Volcanism Study Project (1981)
Basaltic Volcanism on the Terrestrial Planets. Pergamon Press, Inc., New York. 1286 pp.
In all cases the author should be cited as Basaltic Volcanism Study Project.

Typeset by the Lunar and Planetary Institute, printed by Publishers Production International, and bound by Arnold's Bindery in the United States of America.

Library of Congress Cataloging in Publication Data.

Main entry under title:

Basaltic volcanism on the terrestrial planets.

Includes bibliographies and indexes.
1. Basalt. 2. Volcanism. 3. Planetary geology.
I. Basaltic Volcanism Study Project.
QE462.B3B27 1981 552'.26 81-17800
ISBN 0-08-028086-2 AACR2

Basaltic Volcanism Study Project

1976—1979

Project Leaders

T. R. McGetchin
LPI Director 1977–79

R. O. Pepin
LPI Director 1974–77

R. J. Phillips
LPI Director 1979–

Team Leaders

J. B. Adams, Team 2
K. C. Burke, Team 6
W. K. Hartmann, Team 8
J. W. Head, III, Team 5
W. M. Kaula, Team 10

G. E. Lofgren, Team 1
S. C. Solomon, Team 9
G. W. Wetherill, Team 7
J. A. Wood, Jr., Team 4*
P. J.Wyllie, Team 3

Project Administrator

P. H. Jones

Project Support Staff

R. B. Merrill
R. Ridings
P. Robertson

A. Sweet-Jackson
H. Thorson
K. Wohletz

*M. N. Toksöz headed Team 4 from 1976 to 1977.

Project Participants†

5-76 John B. Adams
University of Washington

10-76 Isidore Adler
University of Maryland

10-76 Thomas H. Ahrens
California Institute of Technology

10-77 Claude J. Allègre
Institute de Physique du Globe

10-76 Edward Anders
Enrico Fermi Institute

5-76 Donald L. Anderson
California Institute of Technology

5-76 James R. Arnold
University of California, San Diego

5-76 A. E. Bence
State University of New York

1979 Michael J. Bielefeld
Computer Science Corporation

5-76 Karl Blasius
Science Applications, Inc.

5-76 Christopher Brooks
University of Montreal

5-76 Wilfred Bryan
Woods Hole Oceanographic Institute

1979 W. Roger Buck
Massachusetts Institute of Technology

10-76 Kevin Burke
State Univertsity of New York at Albany

10-76 Michael Carr
U.S. Geological Survey, Flagstaff

10-76 Patrick Cassen
NASA Ames Research Center

5-76 Clark R. Chapman
Planetary Science Institute

5-76 Michael M. Charette Resigned 1977
University of Washington

1979 Guy Consolmagno
Harvard-Smithsonian

5-76 S. E. DeLong
State University of New York at Albany

5-76 Michael R. Dence
Department of Energy, Mines & Resources, Ottawa

5-76 John F. Dewey
State University of New York at Albany

1979 Jimmy Diaz
U. S. Geological Survey, Flagstaff

5-76 Colin H. Donaldson
University of Manchester

5-76 Michael B. Duke
NASA Johnson Space Center

7-77 Michael Dungan
Lunar and Planetary Institute

5-76 Peter Eberhardt
University of Bern

10-76 Fraser P. Fanale
Jet Propulsion Laboratory

5-76 Ronald Greeley
University of Santa Clara/NASA Ames

5-76 David H. Green
Australian National University

7-77 John C. Green
University of Minnesota

1979 Richard A. F. Grieve
Department of Energy, Mines & Resources, Ottawa

5-76 John Guest
University of London Observatory

5-76 Stephen E. Haggerty
University of Massachusetts

5-76 Stanley R. Hart
Massachusetts Institute of Technology

5-76 William K. Hartmann
Planetary Science Institute

5-76 Larry A. Haskin
NASA Johnson Space Center

5-76 James W. Head
Brown University

5-76 Albert T. K. Hsui
Massachusetts Institute of Technology

5-76 Anthony J. Irving
Lunar and Planetary Institute

5-76 Torrence V. Johnson
Jet Propulsion Laboratory

5-76 Kenneth Jones
Jet Propulsion Laboratory

5-76 William M. Kaula
University of California

5-76 Susan E. Kesson
Australian National University

5-76 William S. F. Kidd
State University of New York at Albany

10-76 John Lewis
Massachusetts Institute of Technology

10-76 Peter W. Lipman
U. S. Geological Survey, Hawaii

5-76 Gary E. Lofgren
NASA Johnson Space Center

10-76 Alexander R. McBirney Resigned 1977
University of Oregon

5-76 Thomas B. McCord
Massachusetts Institute of Technology

5-76 Thomas McGetchin
Los Alamos Scientific Laboratory

5-76 Russell B. Merrill
Lunar and Planetary Institute

1979 Albert E. Metzger
Jet Propulsion Laboratory

10-76 John Minear
NASA Johnson Space Center

5-76 Peter Molnar
Massachusetts of Technology

1979 John W. Morgan
U. S. Geological Survey, Reston

8-80 Peter J. Mouginis-Mark
Brown University

10-76 V. Rama Murthy
University of Minnesota

10-76 Anthony J. Naldrett
University of Toronto

5-76 Michael J. O'Hara Resigned 1977 Assoc. Mem. 1978
University of Edinburgh

8-80 E. Marc Parmentier
State University of New York at Albany

5-76 James J. Papike
State University of New York at Stony Brook

10-76 Robert O. Pepin
Lunar and Planetary Institute-University of Minnesota

1979 Roger J. Phillips
Lunar and Planetary Institute

5-76 Carle Pieters
Massachusetts Institute of Technology

6-79 Dean C. Presnall
University of Texas at Dallas

5-76 Arch M. Reid
University of Cape Town

5-76 Raymond T. Reynolds
NASA Ames Research Center

5-76 J. Michael Rhodes
NASA Johnson Space Center

5-76 E. A. Ringwood
Australian National University

7-77 Ernest Schonfeld
NASA Johnson Space Center

5-76 Peter H. Schultz
Lunar and Planetary Institute

5-76 Eugene M. Shoemaker
California Institute of Technology

5-76 Raymond Siever
Harvard University

5-76 Norman H. Sleep
Northwestern University

5-76 Larry Soderblom
U. S. Geological Survey

5-76 Sean C. Solomon
Massachusetts Institute of Technology

7-77 R. S. J. Sparks
University of Rhode Island

7-77 Edward Stolper
Harvard University

5-76 David W. Strangway
University of Toronto

5-76 Robert G. Strom
University of Arizona

5-76 Paul E. Tapponier
University of Montpellier

7-77 S. Ross Taylor
Australian National University

5-76 Fouad Tera
California Institute of Technology

5-76 M. Nafi Toksöz Resigned 1977
Massachusetts Institute of Technology

5-76 Donald L. Turcotte
Cornell University

5-76 Kark Turekian
Yale University

10-76 Thomas Usselman
University of New York at Buffalo

6-78 David T. Vaniman
Los Alamos Scientific Laboratory

5-76 Randall Van Schmus
University of Kansas

5-76 David Walker
Harvard University

7-77 George R. L. Walker
Imperial College

10-76 Heinrich Wänke
Max-Plank Institut, Mainz

5-76 Stuart Weidenschilling
Carnegie Institution

5-76 George Wetherill
Carnegie Institution

6-78 James Whitford-Stark
Brown University

7-77 Donald Wise
University of Massachusetts

5-76 Charles A. Wood
Brown University

5-76 John A. Wood
Smithsonian Astrophysical Observatory

1979 Alex Woronow
University of Arizona

5-76 Peter J. Wyllie
University of Chicago

---

†*Dates preceding names are those on which participants joined the Project. Institutions are those with which participants were affiliated longest while working on the Project.*

Thomas R. McGetchin, 1936—1979

The purpose of the Basaltic Volcanism Study Project was to bring together a number of scientists with diverse backgrounds to attack a problem and process that was common to the terrestrial planets and which represented a significant part of planetary evolution. It is entirely fitting that we dedicate this book, the culmination of the Basaltic Volcanism project, to Thomas R. McGetchin. Not only did Tom contribute significantly to the project through his leadership, but the major characteristics of the endeavor represented fundamental cornerstones of his life—a goal of understanding volcanism, an involvement with space exploration and the return of scientific data from other planets, a comparison of these data to cast new light on our understanding of the Earth, and a fundamental belief in people and their ability to work together.

Tom's love of volcanoes was legendary. Volcanoes are so awesome that it is easy to understand Tom's excitement, but exploring a volcano with Tom was an experience that touched every intellectual and emotional fiber of one's being. Tom's unbridled enthusiasm was an infectious combination of scientific curiosity and emotional awareness. He simply wanted to know "What makes volcanoes work?" He talked in simple terms, such as understanding the "magma plumbing system," but thought in complex ways about the physics of fluids and the ascent and emplacement of magma. His breadth of knowledge and interests dictated an incredible scientific assault on volcanoes—in addition to field mapping and sampling everything in sight for geochemical and petrologic analysis, Tom made seismological, acoustical, and infrared observations, and employed high-speed photography for ballistic analysis of pyroclastic eruptions. In many disciplines there is a gap, and sometimes even an abyss, between observation and theory, and between qualitative and quantitative approaches. In a quiet, enthusiastic, and yet non-judgmental way, Tom tried to bridge these gaps in volcanology. One sometimes wondered whether Tom was even *aware* of these basic differences in approach and philosophy. But then, in the middle of one of his slide presentations would appear, accompanied by a big grin, a photograph of some volcanic outcrop—with nothing but a slide rule for scale!

But a major part of Tom's view of volcanoes was emotional and this was one of the most significant gifts he gave to scientists who had been taught that feeling has no place in science. There is, unfortunately, no way to describe this experience. But it had to do with picking up a xenolith and pausing to realize that you were actually *holding* a piece of the Earth's mantle, or standing on the flanks of an erupting volcano that was simultaneously assaulting every sense one possessed. It was, perhaps, best communicated by Tom when he would look up at the sky or into the volcano and simply say "geeee-zuss!"

The idea of the Basaltic Volcanism project had great appeal for Tom. He had a keen awareness of new ideas and the future. He thus found exciting the prospect of synthesizing the results from planetary exploration and applying them to understanding planetary volcanism as a process. Tom was one of those rare individuals who was comfortable with both what he did know and what he didn't know. His firm desire was that this book would contain at least as many questions as it did answers.

From Tom's view, however, all these things were secondary to the importance of human relationships. Tom had a beautiful, unassuming respect for people and their ideas. He believed that everyone had something to offer, and that the challenge was to find it or to help *them* find it. He also believed in what might be called scientific synergism. If scientists with different backgrounds could be brought together to work on a problem, the resulting contribution would surely be more significant than the sum of the individual contributions. A corollary was that each individual would be enriched by the experience of intellectual and emotional interaction with colleagues from different backgrounds. To Tom, the Basaltic Volcanism project was a grand opportunity for exploring and developing, and watching develop, human relationships of all kinds.

While in Hawaii a few months before his death, Tom took time to pause and reflect on what he felt were the important things in life. As was often the case, his philosophy was almost indistinguishable from his daily life, and thus his words seem strangely applicable both to the spirit of this endeavor and to life in general:

> *"Shaping your stone means quietly doing your job, as well as you can. Your identity will soon be lost to history but your stone, if well shaped and polished will fit into the structure we call civilization and hold its weight, as time sweeps past us and others build upon us. History is full of greed, horror and the worst in mankind—but humanness is built of well shaped loving lives. What we do matters and if there is beauty in the world it is because many quiet souls have shaped their stones well and the cathedral of life is beautiful after all."*

Contents

CHAPTER 1

PETROLOGY AND CHEMISTRY OF TERRESTRIAL, LUNAR AND METEORITIC BASALTS

CHAPTER 2

REMOTE SENSING OF BASALTS IN THE SOLAR SYSTEM

CHAPTER 3

EXPERIMENTAL PETROLOGY OF BASALTS AND THEIR SOURCE ROCKS

CHAPTER 4

GEOPHYSICAL AND COSMOCHEMICAL CONSTRAINTS ON PROPERTIES OF MANTLES OF THE TERRESTRIAL PLANETS

CHAPTER 5

DISTRIBUTION AND MORPHOLOGY OF BASALT DEPOSITS ON PLANETS

CHAPTER 6

TECTONICS OF BASALTIC VOLCANISM

CHAPTER 7

RADIOGENIC AND STABLE ISOTOPES, RADIOMETRIC CHRONOLOGY, AND BASALTIC VOLCANISM

CHAPTER 8

CHRONOLOGY OF PLANETARY VOLCANISM BY COMPARATIVE STUDIES OF PLANETARY CRATERING

CHAPTER 9

THERMAL HISTORIES OF THE TERRESTRIAL PLANETS

CHAPTER 10

IMPLICATIONS OF BASALTIC VOLCANISM FOR THE EVOLUTION OF PLANETARY BODIES

READER'S GUIDE

The reader is presumed to be familiar with terms in the *Glossary of Geology* (R. L. Bates and J. A. Jackson, eds., American Geological Institute, second edition, 1980). Jargon has been eliminated wherever possible, and is defined where its use was inescapable, as for example in discussions of lunar rocks. Boldface page numbers in the subject index denote pages on which terms are defined or illustrated. Standard references for meteorite names are the *Catalog of Meteorites* (M. H. Hey, 1966, British Museum), the *Appendix to the Catalog of Meteorites* (R. Hutchison, A. W. R. Bevan, and J. M. Hall, 1977, British Museum), the *Catalog of Antarctic Meteorites, 1977-1978* (U. B. Marvin and B. Mason, eds., 1980, Smithsonian Insitution Press), and *Antarctic Meteorite Descriptions 1976-1977-1978-1979* (R. Score, C. M. Schwarz, T. V. V. King, B. Mason, D. D. Bogard, and E. M. Gabel, 1981, NASA Johnson Space Center Curatorial Branch Publication 54, JSC No. 17076).

Team 1 has collected a great deal of new descriptive and analytical data for the reference suites which were inappropriate for inclusion in Chapter 1 but which constitute one of the most comprehensive data bases available. These data are reproduced in an appendix which will be found on microfiche cards mounted in a pocket at the rear of the book.

MICROFICHE APPENDIX

Table of Contents

*Roman numerals represent fiche card numbers in this Table. Note that the cards themselves are marked with arabic numerals.

Introduction

Basalts have erupted on earth throughout known geologic history, and on the moon for more than a billion years before it became thermally quiescent 2.5 to 3 billion years ago. From the earth today to the small asteroid Vesta 4.5 billion years in the past, on Mars, Venus, and Mercury, the generation of basalts has spanned the history of the solar system and an enormous range of planetary mass. Basaltic volcanism is a fundamental process, the expression of stages of partial melting in the evolution of terrestrial planets.

The study of the process and products of basaltic volcanism is to first order the study of the generation and migration of magma within planets, its eruption and distribution on planetary surfaces, and the petrological, chemical and chronological characteristics of basaltic rocks. From this perspective, basalts could be viewed in isolation from other basic planetary processes. But in a broader sense, in the full range of its expression in the inner solar system, basaltic volcanism is not separable from earlier stages in the evolution of planets. Planetary accretion, the nature and initial distribution of heat sources, global differentiation and the subsequent physical and chemical states of planetary mantles, and tectonic responses to stress are all intrinsic elements of the total basaltic volcanism problem. Thus the variety of disciplines involved in a comprehensive effort to understand the process through time and across the suite of terrestrial planets must be very wide, ranging from the petrology, geochemistry, and radiometric chronology of terrestrial, lunar and meteoritic basalts to the remote characterization of unsampled planetary surfaces by astronomical techniques and spacecraft instruments; from the cosmochemistry, physics, and geophysics of primitive nebular matter, planetary formation, and internal thermal history to the geology of past and present tectonic motions and volcanic landforms.

The abundance and ubiquity of basalts, their role as probes of chemical compositions and physical conditions in planetary mantles that are hidden from direct observation by depth and time, and their relevance to basic questions of planetary origin and evolution indicate that basaltic volcanism is a geologic phenomenon of particular importance. Remote-sensing data and, for the earth, moon and meteorites, actual samples of basalt, now allow the process to be examined under the different conditions existing on all the terrestrial planets with degrees of perspective that were impossible when only one planet was accessible to study. However, there have existed barriers to the development of the study of basaltic volcanism through comparative planetology. The terrestrial, lunar and planetary research communities historically have been compartmentalized, drawing their support from separate agencies and from program offices with different objectives. Gaining access to primary research materials in these communities has often been difficult for "outside" scientists. During the past decade, the outpouring of new data has been so large and has affected so many different disciplines that few individuals have been able to maintain the perspective and expertise necessary for a true overview of the subject.

All of these aspects of basaltic volcanism—the vitality of interest in the subject, its significance and generality in the comparative planetology of the inner solar system, its complex and multi-disciplinary data base, the breadth of relevant theory, and technical impediments to individual scientific mastery—suggested that progress in understanding the problems and implications of basaltic volcanism, planet by planet and as a cross-cutting process in planetary evolution, could be significantly advanced by a comprehensive study, structured as an interdisciplinary, interactive project involving scientists in all associated research areas. Such an effort, the Basaltic Volcanism Study Project, was launched in 1976 by the Lunar and Planetary Institute (then the Lunar Science Institute) with financial support from the National Aeronautics and Space Administration.* This book is the final report of that project.

*NASA Contract Nos. NSR-09-051-001 and NASW-3389.

The goals, interactive scientific structure, and organizational framework of the project were defined initially by the Lunar and Planetary Science Council, a standing advisory committee of the Institute's parent consortium, the Universities Space Research Association, and by the LPI Director and scientific staff. The principal objectives of the project were to review the current states of knowledge in all disciplines pertaining to basaltic volcanism on the terrestrial planets, to identify gaps and to catalyze new research initiatives in areas where data or theory were absent or sparse, and to publish the overviews of basaltic volcanism, the unifying concepts, and the new insights generated by the scientific interactions of the project membership. Ten study areas, each chartered to focus on a particular aspect of basaltic volcanism as a planetary process, comprised the scientific backbone of the project and collectively defined its scope. These areas, each addressed by a study team, were:

(1) the petrology and chemistry of terrestrial, lunar, and meteoritic basalts;
(2) the compositions of planetary basalts as determined by remote-sensing techniques;
(3) the compositions of planetary mantles as inferred from experimental petrology of basalts;
(4) geophysical and cosmochemical constraints on properties of mantles of the terrestrial planets;
(5) the distribution and morphology of basalt deposits on planetary surfaces;
(6) tectonics of basaltic volcanism;
(7) radiogenic and stable isotopes, radiometric chronology, and basaltic volcanism;
(8) chronology of planetary volcanism by comparative studies of planetary cratering;
(9) thermal histories of the terrestrial planets; and
(10) implications of basaltic volcanism for the evolution of planetary bodies.

A detailed description of the charter, role, and scope of each of these study areas is given in the "Overview" section.

Leaders for the study teams were nominated and enlisted by the Council and the LPI Director. Overall scientific and administrative responsibility for the project was vested in a steering committee composed of the team leaders, representatives of the Council, and the Institute Director, with the Director serving as leader of the project. Between four and twelve members, each filling a particular role in the intradisciplinary work of the team, were chosen for each team by its leader in consultation with the steering committee, on the bases of expertise and current research interests. Approximately 80 participants were active at any one time during the three years of regular meetings. A total of 105 individual scientists were ultimately involved in one phase or another of the project. Their names and affiliations are listed under "Project Participants."

The Basaltic Volcanism Study Project was above all else a pilot exercise in intensive communication, mutual education, and synthesis. It presented the dual challenges of requiring contributions from a very wide range of scientific disciplines, and organized collaboration across these disciplines as well as across planets. Focused meetings of single teams, and broader plenary sessions and conferences involving all teams together, formed the heart of the project. Each team held a series of workshops during the study, usually individually but occasionally with one or more other teams. The workshop environment sparked the dynamic and creative interactions which became the hallmark of the overall project. Individuals whose previous communications with each other had often been limited to exchanges at conferences and in their written work, met face-to-face at intervals over a three-year period to expound and discuss their sometimes differing theories, and to listen. In the process, some rough edges were smoothed, and extreme viewpoints tended to move toward more central ground. These occasionally electric meetings ultimately produced improved consenses and comprehensive summaries of the state of knowledge in each research area. Critical gaps and questions were identified for later exploration by team members and by their various research communities. In later stages of the project, the final written product of each team's activities was outlined, debated, and prepared for publication as one of the chapters in this book. Three interteam meetings and conferences were convened at yearly intervals to foster cross-disciplinary exchanges among the study groups, to provide the basis for synthesis of the work of all groups, and to review and integrate early drafts of each team's contribution to the final report of the project. The steering committee met periodically to review progress and to provide a forum for scientific and logistic discussion and planning among the project leaders.

Apart from the more formal team workshops, interteam sessions and the scientific conferences that were part of the interteam activities, and meetings of the steering committee, other channels of communication were developed among teams and with the membership as a whole. Observers from other teams were encouraged to attend team workshops in areas of particular scientific relevance to the work of their own study groups. Discussions at each team workshop were summarized by the leader and distributed to all project members in a newsletter circulated periodically by the Lunar and Planetary Institute. This newsletter contained schedules, locations and agendas for all project meetings, current team membership rosters, and lists of scientific publications related to the project which had been produced by participants in the course of the study (the BV Contribution Series). Annotated bibliographies of literature considered by the team leaders to contain particularly pertinent background information, data, and summary reviews in their teams' areas, which would therefore be informative for all participants, were also circulated to the full membership via the newsletter.

Publications and reports from the project, in addition to this volume, include articles in scientific journals and collections of summaries and extended abstracts of talks given at conferences held during the first two interteam meetings. Independent publication of research stemming from the project was encouraged. Many such papers have appeared in the literature, authored by individual team members, by groups of members within particular teams, and collaboratively by members of different teams. A current list of titles in this BV Contribution Series, reprints of articles, and copies of the interteam report and abstract volumes are available from the LPI Library.

The Basaltic Volcanism Study Project was sponsored, logistically organized, and administered by the Lunar and Planetary Institute under the aegis of an Institute proposal which was approved and funded by NASA. Travel and other project expenses for members were paid from these funds. This direct support of participants was one of the principal reasons for the success of the study, because it guaranteed high attendance at the core of interactive meetings around which the project was built. Responsibility for general administrative and financial management was assumed by the LPI Director, and by the Project Administrator who authorized and disbursed funds, arranged travel and accommodations for meetings and workshops, organized and supported the logistics of interteam and steering committee meetings, prepared and distributed the newsletter, and edited the interteam conference report and abstract volumes. This book was produced as camera-ready copy by the LPI Publications and Production Services Offices for publication by Pergamon Press on a non-profit basis. Costs of the editorial, graphic arts, and typesetting services provided by the Institute were funded directly by NASA under the contract for the study. For this phase of the project, the steering committee was reconstituted in late 1978 as an editorial committee and advisory board on editorial policy to the study teams and to the LPI production staff. Each member was responsible for obtaining scientific reviews of one or two chapters, and for working with the team leaders on revisions and to reconcile overlaps, gaps, and discrepancies. Subcommittees or individual members of the editorial committee wrote the "Overview" and provided guidelines for chapter format and indexes. The Managing Editors and Manuscript Coordinator planned and directed the actual production of the book.

It remains to acknowledge special individual contributions, levels of effort, and overall support that were crucial to the definition, initiation, progress, and products of the study project. A. J. Calio of NASA Headquarters played a key role in stimulating the Council and the LPI Director to think about the scientific potential of a pilot study in comparative planetology, and in the negotiations leading to its approval. Members of the Lunar and Planetary Science Council at the time the project was formulated were D. W. Strangway (chairman), D. L. Anderson, C. R. Chapman, L. A. Haskin, J. W. Head, W. M. Kaula, J. J. Papike, and J. A. Wood. The original prospectus for the proposed basalt study was written by J. A. Wood, J. W. Head, and M. Nafi Toksöz. An important public description of the intent, structure and goals of the project at an early stage was published by J. A. Wood [(1977) Basaltic Volcanism in Terrestrial Planets. *Geotimes* **22**, 26-29)]. The Project was supported initially by NASA's Lunar and Planetary Programs Office, headed by E. A. Flinn; continued funding and strong scientific encouragement were provided for the balance of the study by W. L. Quaide of the Planetary Geochemistry and Geophysics Office, Solar System Exploration Division. The topic of this study fell within the realm of the lunar and planetary geosciences, for which the Johnson Space Center has responsibility within NASA. The close scientific cooperation and extensive material assistance from JSC, including financial support for all JSC participants, were provided by the Planetary and Earth Sciences Division, headed by L. A. Haskin during the early stages of the project and thereafter by M. B. Duke. T. R. McGetchin, in addition to his contributions to Team 5, led the project with infectious enthusiasm during his tenure as LPI Director from 1977 to 1979. All sections of this book were reviewed by scientists external to the project. Thanks and appreciation are extended to all sixty of them, and especially to D. E. Wilhelms, the only reviewer to read the entire manuscript in draft form. The final product reflects the thoroughness and dedication of the LPI staff in preparing the volume for publication, and in coping with the multitude of editorial and production complexities inherent in a work of this length, with its many authors, reviewers, and editors.

This book was the principal goal of the study project from its inception. It was intended to reflect three progressive stages in the study of basalts by the project. The first was to review what is known, to identify what is unknown, and to ask the leading questions in each of the ten broad research areas represented by the teams. Important gaps in the various data bases would then be filled by experimental or theoretical research carried out by team members. The final objective was to interrelate and unify the principal concepts of basaltic volcanism across both disciplines and planets. All three stages are represented in this volume, but to varying extents. First and foremost, the book stands as a comprehensive review of current knowledge about basaltic volcanism. New data generated by project members are incorporated into several chapters, particularly in the report by Team 1 on the petrology and chemistry of the basalt reference suites. Although the short duration of the project's active phase constrained both the opportunities for experimental and theoretical initiatives and the ability of the teams to utilize these data in developing the directions to be taken in their work under the direct auspices of the project, the data now exist to provide raw material and fresh impetus for the next generation of research. For example, primitive basaltic compositions identified by Team 1 could not be studied experimentally by Team 3, since the two groups worked simultaneously rather than sequentially, but these compositions are now available for future investigations by experimental petrologists. Much the same can be said about interrelating and synthesizing the data, theory and concepts in each of the study areas to develop and focus the major interdisciplinary themes of basaltic volcanism as a general planetary process. Meaningful interaction among the study groups had to grow from the collective intradisciplinary data bases and overviews constructed by the teams. The project reached a level of maturity appropriate to broad interdisciplinary studies as its active phase came to an end. Even so, the degree of integration among disciplines in this volume is unprecedented in previous books of separately-authored chapters in the earth and planetary sciences. But this is just the first step. A comprehensive synthesis of the work of all the teams is the principal challenge offered to readers of this book. Many of the potential interdisciplinary linkages across the topical chapters remain to be forged. A question raised in one chapter may be answered in another. A firm conclusion reached from the point of view of a particular discipline may differ from an explicit or implicit conclusion on the same subject in another chapter. It is from these conflicts among disciplinary perceptions of the problems and their resolutions that the next steps of critical research and simplifying insights will emerge.

The direct relationships of the various sections of this book to basaltic volcanism vary, as they must in any comparative study. While some chapters, for example those on petrology and morphology, focus specifically on the subject, others dwell on processes, techniques, and the general aspects of planetary modeling that are germane to the temporal and spatial settings of basaltic volcanism. In all, the chapters provide a nearly exhaustive framework for scholarly pursuit of the topic. The authors have written for readers who are acquainted with the fundamentals of geology, geochemistry and geophysics, but who are not necessarily working in these fields. Several teams have included detailed introductions to the research areas of their chapters. A particular strength of the book thus lies in its ability to serve not only as a catalyst for future studies, but also as a textbook and review and reference work for advanced undergraduate students, graduate students, and professionals alike.

Final perspective on the contributions of *Basaltic Volcanism on the Terrestrial Planets* to the understanding of basaltic volcanism as a fundamental planetary process, to the scientific broadening of those working in the diverse areas of the field and the education of those interested in joining it, and to the stimulation of new research addressing the unanswered questions is left, as it must be, to the reader and to time. Some of the right questions have been asked; a few have been answered, others may have been answered. That basaltic volcanism is an interweaving of many disciplinary and planetary threads is clear, but only part of the fabric is visible, and its pattern is far from revealed.

Robert O. Pepin
Minneapolis

Pamela H. Jones
Houston

Overview

Basalt is a dark, igneous rock characterized by small grain sizes (less than about 1 mm) and containing roughly equal proportions of plagioclase feldspar and calcium-rich pyroxene, with less than 20% of other minerals (typically olivine, calcium-poor pyroxene, and iron-titanium oxides). The name is derived indirectly from the Greek *basanos*, meaning touchstone (which is an entirely different rock: a dark, flinty stone used to test gold and silver alloys). Basalt is by far the commonest igneous rock on the surface of the solid Earth, and occurs abundantly on other terrestrial planets.

Basaltic volcanism appears to be the most accessible of the major processes in planetary thermal and chemical evolution. Hence, in several aspects, basaltic volcanism furnishes the greatest prospect of making comparative planetology a meaningful science.

The purpose of this overview is to give the reader a grasp of the general organization of the book. We also try to define the most common terms and to explain the most basic concepts. This overview attempts to be a précis, leaving interpretive generalizations to concluding sections within the chapters and to Chapter 10. Numbers of the sections being described are given in the margin.

Chapter 1 1.1

Petrology and chemistry of terrestrial, lunar and meteoritic basalts refines the definition of "basalt" and furnishes a data base of rock chemistry, mineralogy and petrography.

While the concept of basaltic volcanism is simple—it is the extrusion of partial melt from a relatively primitive mantle—in practice, the variety of circumstances in which basaltic volcanism occurs and the variety of rocks which result require careful examination and description of a diverse set of lavas. This diversity is such that the only rigid quantitative measure of a basalt adopted is arbitrary: that the SiO_2 content be less than ~54 wt.%. But nearly all of the basaltic samples described in the chapter can be characterized as dark rocks containing less than 52% SiO_2, more than 6% Al_2O_3 and CaO, and less than 12% MgO. They contain plagioclase and pyroxene as major phases, and olivine is often present as well.

1.2

That the most coherent data base exists for lunar mare basalts is a consequence of the systematic analyses of samples returned by the Apollo project. The main drive of the work preparatory to writing Chapter 1 was to carry out comparably systematic analyses of terrestrial and meteoritic samples. The bulk of the text in Chapter 1 consists of descriptions of eleven rock suites: their geologic settings; mineral chemistries; petrographies; major, trace and isotope chemistries; and petrogenesis. Six of these suites are terrestrial. Two suites are comprised partly or entirely of ultramafic rocks; i.e., rocks having higher MgO and lower Al_2O_3 contents than basalts proper. One of these ultramafic suites represents the oldest terrestrial volcanism known, which occurred more than 2.5 b.y. ago during the Archean era, while the other consists of ultramafic xenoliths, which are nearest to being products of the upper mantle that are complementary to basalts. Finally, two suites are lunar and one is meteoritic.

Detailed analytical data and descriptions of the rocks described in section 1.2 are tabulated in an appendix on microfiche cards in a pocket at the back of the book. A Table of Contents for the microfiche appendix is included in the Reader's Guide.

1.2.1

Archean volcanism included the entire range of compositions which have erupted in more recent times. In addition there are ultramafic lavas which are nearly unique to the Archean. These lavas, called komatiites, have MgO contents of 12–33% and characteristic "spinifex" textures: bladed or needle-like olivine and pyroxene crystals. Komatiites occur in sequences of basalts within Archean greenstone belts (so called because many of the rocks are now green as a result of hydrous metamorphism), which occur in all old cratons. Komatiites are widely considered to signify higher mantle magma temperatures—200°C higher or more—than occur now, and higher degrees of partial melting in source regions: perhaps more than 30% higher. How much the greater degree of partial melting contributed to the greater diversity of compositions erupted during the Archean is much debated. The reference suite for komatiites is from the Canadian craton.

1.2.2

Flood basalts (plateau basalts) are broad, flat, large-volume, low viscosity flows which have erupted along fissures in the continental crust. These basalt plateaus constitute very large terrestrial features and are considered important enough to treat in two sections: pre-Tertiary (section 1.2.2) and Tertiary (section 1.2.3) flood basalts.

Comparisons are made among four pre-Tertiary (older than 60 m.y.) basalt plateaus: the Paraná Basin, which is early Cretaceous (119–149 m.y. ago); the Karroo Province, which is mainly Jurassic (166–206 m.y.); the Siberian Platform, which is Permian to Triassic (216–248 m.y.); and the Lake Superior Basin (Keweenawan), which is Late Precambrian (1100–1140 m.y.). The reference suite is from the Lake Superior Basin.

The average thickness of the flood basalt accumulations is about one kilometer, though they are much thicker locally, and their original areas were on the order of 2×10^6 km^2. Individual flows are typically 10–30 m thick. The flood lavas were erupted in zones of tectonic tension, and filled broad, subsiding basins. Compositionally, the plateaus are dominated by somewhat "evolved" tholeiites, transitional basalts, and basaltic andesites, but more primitive high-alumina olivine tholeiites are common in the older plateaus.

1.2.3

Tertiary flood basalts have been produced by the voluminous volcanism associated with the continental divergence which has been occurring over the last 100 m.y. They differ from the earlier Phanerozoic flood basalts mainly in that they are more accessible. The areas selected for comparison are the Columbia River Province, the Deccan Traps, and the Brito-Arctic Province. The reference suite is from the Columbia Plateau.

1.2.4

Continental rift volcanism occurs in association with extensional tectonics, such as characterize the Rio Grande Rift from which a suite was selected for this study. In the Rio Grande region, voluminous tholeiitic basalts occur along the rift axis. On the rift flanks, volcanism is more silicic and alkaline. In some rifts, alkaline basalts predominate even on the main axis.

1.2.5

Ocean-floor basalts are by far the most abundant rocks on the Earth's surface. They appear to be erupted largely at ocean ridges, and are predominantly tholeiitic. They are remarkably similar in composition throughout the world, except near the intersection of a ridge with an ocean island feature (e.g., Iceland). Compositional ranges are 49–51% SiO_2, 14–17% Al_2O_3, 12–5% MgO, and 0.01–0.3% K_2O. These "normal" ocean-floor basalts are depleted in light rare-earth elements (REE) and in incompatible lithophiles, such as K, Ba, Rb and Sr. Other ocean-floor basalts have compositions which approach those of ocean island basalts.

1.2.6

Oceanic intraplate volcanism, characterized mainly by oceanic island basalts, is small in volume compared with ocean ridge volcanism. The Hawaiian island chain is the archetype for ocean intraplate volcanism; however, oceanic intraplate volcanism also forms seamounts and isolated islands. A reference suite from Hawaii is described. Ocean island basalts are predominantly tholeiitic in character, but cover a much wider range of compositions than ocean-floor basalts and grade into alkali olivine basalts. They have higher K_2O, REE, and incompatible element contents than ocean floor basalts.

1.2.7

Island-arc basaltic rocks differ from other terrestrial basalts in that they constitute only a minor portion of the volcanism in their tectonic settings—about 20% volumetrically—with the bulk being andesitic volcanism. Furthermore, the basalt compositions are generally quartz-normative, with SiO_2 varying from 48 to more than 53%, Al_2O_3 16–20%, and TiO_2 less than 1%. MgO contents are variable but usually too low relative to FeO ($Mg/Mg + Fe \sim 0.5$) for the basalt liquids to have been in equilibrium with typical mantle olivines. The most systematic variation with tectonic setting is positive correlation of K_2O with distance from the trench, and hence with depth to the subduction zone. A suite of rocks from New Britain is examined in detail, and there is a comparative discussion of other island-arc basalts.

1.2.8

The basaltic meteorites consist essentially of two types of meteorites, eucrites and shergottites. In addition, basaltic constituents are found in two other composite types of meteorites, howardites and mesosiderites. Eucrites are very old, 4.54 b.y., and appear to have been formed by partial melting of a primitive (i.e., chondrite-like) source material. The eucrites are severely depleted in both volatiles and siderophiles, apparently having differentiated in an environment in which free iron was present and the oxygen fugacity was very low. Shergottites, however, may be relatively young—0.65 b.y. old—and appear to have formed in a volatile-rich environment.

1.2.9

Lunar mare basalts erupted some 3.15–3.95 billion years ago, covering about 17% of the lunar surface, but constituting less than 1% of the total volume of the lunar crust. Compositionally, they differ from terrestrial basalts in having extremely low alkali, lower SiO_2 and Al_2O_3, higher FeO and in some cases higher TiO_2 contents. The absence of Fe_2O_3, the dearth of alkalies, the greater enrichments in refractory rare earths, the occurrence of Cr^{2+}, and various other features indicate that these basalts were produced as partial melts in a source region of low oxygen fugacity and low volatile content.

1.2.10

Lunar highland basalts are known to have existed almost by inference alone, because virtually all highland samples have been subjected to appreciable impact brecciation, melting and metamorphism. Nevertheless, there are a few highland samples which are sufficiently undisturbed that basaltic textures can still be identified (Intermediate-K Fra Mauro basalts), and there are others which are thought to be basaltic because their compositions plot close to that of the olivine-plagioclase cotectic in the olivine-silica-anorthite system (Low-K Fra Mauro basalts). Both highland basalt types differ compositionally from mare basalts in having higher Al_2O_3 and lower TiO_2, FeO, and MgO contents.

1.2.11

Ultramafic xenoliths are ultramafic rocks which have been brought to the Earth's surface relatively undisturbed by reactions with their host magmas and which represent our best samples of the residue left by the partial melting events which generated the basalts. Spinel lherzolites dominate the reference suite, which also contains garnet lherzolites and one garnet-spinel harzburgite. Data presented here suggest that these xenoliths cannot be interpreted as simple partial melting residues, but rather provide evidence for depletion and enrichment of lithophile elements within the mantle by many events involving processes of partial melting and fluid metasomatism.

1.3

Planetary comparisons are made of major and trace element chemistries, mineralogies, petrographies, oxygen fugacities and volatile contents.

1.3.1

Major and trace element chemistry of planetary interiors can be constrained by studies of erupted basalts. The simplest measure of primitiveness is the $Mg/(Mg + Fe^{2+})$ ratio, which is abbreviated throughout this volume as Mg′. Differences in major element chemistries can be examined systematically in terms of this ratio. More striking are trends in trace elements. While some trace element patterns are diagnostic—such as the Eu deficiency relative to other REE, which indicates that plagioclase had been removed from the source material prior to the generation of the mare basalts—others have multiple interpretations. It is inferred from the trace element data, as well as from radioisotopic data (see Chapter 7), that the mantle has developed appreciable heterogeneity.

1.3.2

Silicate mineralogy is devoted to a statistical analysis of mineral samples selected from the ten basaltic suites described in section 1.2. The samples selected include 1271 pyroxenes, 274 olivines, and 705 feldspars. The parameter $X_{Fe} = Fe^{2+}/(Fe^{2+} + Mg)$ is emphasized; three cations are plotted against it for the olivines, and six for the pyroxenes. In addition, mean compositions are given for each mineral and suite. Special emphasis is placed on those characteristics of silicate chemistry which reflect planetary constraints on the host basalts.

1.3.3

Comparative petrography and cooling rates provides a classification of textures and inferred cooling rates. Several factors affect the formation of textures, including nucleation, growth, and cooling rates; composition and flowage; and crystal settling. These factors interact in a complex manner which makes it difficult to assign unique cooling rates to a particular texture or group of basalts.

1.3.4

Temperatures and gas fugacities examines oxygen-buffering equilibria, such as those involving fayalite-magnetite-quartz (FMQ) and iron-wustite (IW), to determine the temperature (T) and oxygen fugacity (f_{O_2}) which governed during crystallization of samples drawn from the reference suites. In general, terrestrial basalts fall along an f_{O_2} vs. T line defined by FMQ buffering, although local variations occur. Oxygen fugacities in lunar basalts were lower by factors $\gtrsim 10^{-3}$ for the same temperatures, so that assemblages usually equilibrated below the IW line. Meteorites are much less systematic, with f_{O_2} values that range from earth-like for Shergotty to moon-like for some equilibrated chondrites.

1.3.5

Volatiles in terrestrial basalts consist primarily (>99.9%) of H_2O, CO_2 and SO_2. However, at the very low oxygen fugacities which characterize lunar and some meteoritic basalts, H_2S and CO become relatively important species. The extent of magma degassing is sensitive to and diagnostic of the conditions under which lavas erupt and crystallize.

1.4.1

Processes that affect basaltic magma composition include several aspects of the generation, transport crystallization, mixing, and separation of magmas. To each process, there is a range of responses of the magma composition, depending upon whether equilibrium or kinetic relationships control. Important processes include partial melting, fractional crystallization, gravity and flow differentiation, magma mixing, assimilation, and possibly vapor transport. Identification of the processes which have been dominant in the evolution of a particular basalt suite requires comprehensive study.

1.4.2

Primary magmas are characterized and some major types are scrutinized. A substantial body of evidence documents the existence of primary komatiite and picritic magmas. Undoubtedly, some basalts are derivatives of these more "primitive" melts through extensive differentiation. However, if as appears to be the case, primary melts such as komatiites and picrites are produced as consequences not of single, extensive melting events, but of sequential melting episodes in which previously melted and depleted mantle is re-melted, then it follows that primary basaltic magmas are necessary precursors to the production of these more "primitive" melts. If correct, this argument points to a spectrum of primary magma compositions. Thus, basalts may be both primary and derivative, and both types may erupt within a single magmatic province.

Chapter 2

2.1

Remote sensing of basalts in the solar system extends our knowledge of basaltic compositions as far as possible beyond those samples which are actually in our hands. The principal techniques discussed are gamma-ray and X-ray spectrometry from lunar orbiting spacecraft, visible and near-infrared spectral reflectance from the Earth, multispectral mapping, and remote-controlled *in situ* measurements, such as X-ray spectrometry by the Viking Lander. The greatest difficulty in inferring volcanic compositions from remote-sensing data is the interference caused by the effects of surface alteration processes (including impact-generated thermal metamorphism, hydration, and oxidation) and by processes which can redistribute surface material (such as aeolian transport or impact ejection).

2.2

The principal inferences, planet-by-planet, follow:

Mercury: Basalts probably exist in the plains units, with TiO_2 contents lower than those of basalts in the lunar maria.

Venus: The least-known planet, with the sole surface composition information being a gamma-radiation level comparable to those of terrestrial basalts observed at two Venera landing sites, and an appreciably higher level observed at the third site.

Earth: Remote-sensing methods (such as airborne gamma-ray surveys and multispectral imaging), although well developed for some geologic applications, rarely have been used to derive new information about basalts.

Moon: Remote-sensing techniques, both orbital and earth-based, have been applied most extensively to the Moon. From orbiter gamma-ray and X-ray measurements, there appear to be positive correlations of Al, Si and negative correlations of Fe, Mg with topographic elevation. An unusual feature is correlation of high contents of the radiogenic elements K, U, Th with the western hemisphere maria, Imbrium and Procellarum. Earth-based reflectance spectrophotometry indicates that there are appreciable variations in the TiO_2 contents of mare surfaces, ranging from less than 2% to more than 6%, with values less than 3% being most common. The positive correlation of TiO_2 content with age does not appear to be as systematic as was suggested by the returned samples. Other spectral features, which cannot be so readily interpreted in terms of compositions, suggest that basalt types exist which differ appreciably from those at the landing sites.

Mars: The darker areas appear to be predominantly basaltic; however, the surface dust layer, while possibly derived from basalt, has been altered significantly.

Asteroids: Vesta is unique in having a reflectance spectrum which strongly resembles the spectra of basaltic achondrite meteorites.

Galilean Satellites: There is no spectral evidence for the existence of basalt on the surfaces of these bodies.

Chapter 3

3.1

Experimental petrology of basalts and their source rocks discusses the use of experimental data to infer the compositions and conditions of the sources of basalts, the experimental techniques themselves, and the resulting interpretations of terrestrial, lunar and meteoritic basalts.

In principle, given a source composition inferred from xenoliths and nonpetrological considerations (such as those discussed in Chapter 4), experimental techniques can discover the ranges of pressures and temperatures within which partial melting, and hence differentiation, will occur and the compositions of the resulting products. Another experimental variable is the volatile content of the source region or of the magma. Most lunar and terrestrial basaltic compositions cannot be derived directly from primitive mantle compositions; rather, they have been generated in some region which itself was the product of an earlier differentiation event, or they have experienced significant alteration since leaving the source region through, for example, interactions with other rocks encountered during ascent. Hence, interpretations of experimental results must also consider more complex constraints, such as the tectonic associations of basaltic samples or geotherms inferred by geophysical modelling, and proceed step-by-step.

3.2

In order for experimental results to be used with confidence, precautions must be taken to insure that the equilibrium which would persist on geologic time scales is duplicated in the shorter experimental time scale. This section is devoted to discussions of experimental techniques, of experimental problems which can affect interpretation of results, and of the documentation of techniques and procedures.

3.3-3.5

The principal inferences concerning terrestrial basalts from experimental petrology are that (1) most basalts contain much more information about near-surface fractionation and emplacement conditions than about mantle conditions; (2) the few relatively primary basalts originated in a source region which contained olivine and pyroxenes, some aluminous mineral, and Mg/Fe > 1; (3) volatiles are much less important in the generation of basaltic magmas than of more siliceous magmas; and (4) the temperatures of magmas range from 1000°C to 1600°C, with most being near 1200°C. Lunar basalts also are products of evolved sources; those of the Fra Mauro basalts having been much closer to the surface and more plagioclase-rich than those of the mare basalts. Eucritic meteorites come closest to being products of a primitive source, and good arguments can be made for that source's having been chondritic in composition.

3.6 Recasting model chemical compositions for mantles of the terrestrial planets in terms of mineralogies predicted by experimentally-determined phase relations indicates that the mantle of Venus probably is similar to that of the Earth (80–400 km phase assemblage: olivine + diopside + enstatite + garnet), while the mantles of Mercury and Mars are more likely to contain spinel than orthopyroxene at equivalent pressures (depths > 210 km on Mars). Experimental studies of these model mantle compositions offer opportunities to predict compositions of liquids which could be derived from them by partial melting, along with the differentiation processes most likely to have modified those liquids subsequently. Most basalts on planetary surfaces, including those most likely to be obtained by unmanned samplers, probably have been modified by low-pressure equilibration events. Even if primary liquid compositions cannot be found among samples returned from other planets, meaningful constraints can be placed on mineralogies and compositions of magma source regions by integrating the "inverse approach" outlined in the experimental flow chart of Fig. 3.1.4 with other information and deductions concerning planetary geology and evolutionary history.

4.1 **Chapter 4**

Geophysical and cosmochemical constraints on the mantles of terrestrial planets discusses nonpetrological constraints on the source regions of basaltic volcanism.

4.2 Mantle compositions must correspond to rocks with mass densities which are consistent with the known geophysical properties of the planets. These properties are the planets' mean densities (known for all the planets), moment-of-inertia factors (known for Earth, Moon and Mars), and seismic velocity profiles (known for Earth and Moon).

4.3 Not every possible combination of chemical elements must be considered in order to infer the compositions of planetary mantles. The non-volatile elements are present in the sun and in meteorites in a characteristic abundance pattern, and it can be safely assumed that the terrestrial planets are similarly composed. However, there must be second-order chemical differences among the planets, since their mean densities (corrected for temperature and pressure effects) are not identical. Current theories of planet formation involve processes (such as condensation from a hot vapor) that would lead to chemical differences of the sort needed to rationalize the observed density differences. A variety of mantle compositions result from these theories.

4.4

To test the mantle compositions of section 4.3 against the geophysical constraints of section 4.2, each of the former (a chemical model) must be translated into a planetary density profile (geophysical model). This requires knowledge or estimates of (a) the present-day temperature profiles of the planets; (b) the stable mineral assemblages for each model composition at the various temperatures and pressures in its planet; and (c) the equations-of-state for planetary materials (effects of pressure and temperature on density).

4.5

Most of the chemical models of section 4.3 produce density profiles that agree reasonably well with the geophysical constraints of section 4.2, or which can be made to agree by adjusting their compositions slightly. These models define ranges of plausible mantle compositions for the terrestrial planets. These compositions do not differ greatly from that of the Earth's mantle: they consist predominantly of olivine and pyroxene, with the Mg/(Mg + Fe) molar ratio varying directly with core size, from over 0.90 in Mercury to less than 0.80 in Mars and the Moon. The most anomalous mantle probably is Mercury's, which may contain the bulk of the planet's Al_2O_3 in the form of spinel, because of SiO_2 deficiency.

Chapter 5

5.1

Distribution and morphology of basalt deposits on planets examines the forms of basaltic volcanism and attempts to derive constraints on the chemistry of the lava and on the physics of the emplacement process. As with other chapters, the initial emphasis is on principles inferred from volcanic activity on Earth, as a basis for extrapolation to other planets.

5.2

Rheological properties are the most fundamental in determining morphologies of basaltic lava flows. At temperatures above their liquidus temperatures, silicate melts behave as Newtonian fluids with viscosities which range from 10^0 to 10^5 poises and which are correlated positively with silicate content of the melt. At lower temperatures, such as those typical of terrestrial flows, melt behavior is non-Newtonian. Other factors which influence rheological properties of melts include dissolved gas contents, degree of undercooling, and degree of crystallinity. In addition to rheological factors, flow morphology is influenced by discharge rate, by topography, by preexisting tectonic patterns and by presence of ground or surface water.

5.3

Terrestrial volcanoes are commonly classified either by morphology or by eruptive style. This section presents a detailed discussion of the morphologies of terrestrial volcanic products, including subaerial and subaqueous extrusive environments, calderas, and subsurface (intrusive) activity, in order to establish a foundation for subsequent interpretations of volcanic morphologies seen on other planets.

5.4

Lunar volcanic deposits, among which the lunar maria are the largest and most striking examples, are characterized by broad, very gently sloping flows, reflecting low magma viscosity due to low silica and volatile contents. The great extent of some Imbrium flows has led to the inference of high discharge rates.

5.5

The surface of Mercury is less well-known than that of Earth's moon, so it is more difficult to identify volcanic deposits unambiguously. Possible mercurian volcanic landforms include the smooth plains which occur within and around basins, the intercrater plains, domes, possible vents, and sharp albedo contrasts between crater floors and walls reminiscent of lunar basalt-filled craters.

5.6 Mars possesses a rich, diverse array of volcanic features that range widely in size, in distribution across the planet's surface, and in age. Martian volcanic landforms are characterized, including the large central-vent volcanoes and plains units.

5.7 The analysis of venusian volcanism is based upon interpretations of low-resolution radar images of the planet's surface. Possible volcanic features include the Beta Regio feature, which is topographically similar to large shield constructs on other planets, and relatively smooth areas which may be volcanic plains like those on the Moon, Mars and Mercury.

5.8 The Galilean satellite Io is volcanically active. Many meteorites are igneous rocks. This section reviews the evidence for basaltic and nonbasaltic volcanism on the smaller bodies of the solar system.

6.1

Chapter 6

Tectonics of basaltic volcanism examines the structural surroundings of volcanic occurrences.

6.2 On Earth, basaltic eruption is closely linked to the plate structure which dominates the tectonics. Most of the basalt now erupting is produced at mid-ocean ridges, and considerable amounts also erupt at convergent plate boundaries in Andean and island arcs. Minor quantities of basalt are associated with the third kind of plate boundary, the transform fault. Much smaller, but substantial, terrestrial basaltic eruptions occur away from plate boundaries in isolated areas (hot spots) and in intracontinental rifts. Tectonic environments of basaltic volcanism are reviewed throughout the terrestrial geologic record. Ophiolite complexes provide the best evidence of basaltic activity at plate boundaries and at hot spots during the latter half of the geologic record, while continental volcanism is best recorded in ancient rifts and in those few ancient flood basalt provinces which have escaped erosion.

6.3

6.4 Basaltic eruption seems to have been associated with plate boundaries during the Archean as well, when higher heat flow would have led to greater lengths of fast-spreading ridges and subducting trenches than exist today. Evidence of this activity is preserved in slivers of oceanic and island arc rocks found in greenstone belts, and in the form of the continental lithosphere, which appears to have formed largely by accretion of island arc igneous rocks emplaced above down-going slabs of oceanic material.

6.5 Mars, Mercury and Earth's moon currently have stable global lithospheres, while Io clearly has a very active lithosphere. Our knowledge of venusian tectonics suffers from limited surface observations, but the absence of ocean rise features implies that the venusian lithosphere is not now experiencing plate tectonics similar to those of the Earth. Subduction may now be prevented by a much thicker crust.

6.6

Theories for the initiation of magma migration are adaptations of models for flow in porous media. After magma flows have coalesced, buoyancy effects become important.

Chapter 7

7.1

Isotopic and trace element studies of basaltic volcanism discusses the use of isotopes and minor constituents to infer the chronology and constraints on the evolution of basaltic materials. The chapter begins by describing the basic principles of isotopic and trace element techniques, and then describes and interprets the isotopic and trace element characteristics of meteorites, lunar rocks, and terrestrial rocks.

The most familiar use of isotopes is to infer a crystallization age: the time when radiogenic atoms became fixed in a crystal matrix and decay products began to accumulate systematically in quantities proportional to parent element contents. Such an age, however, is a particular instance of a more general property of radiogenic isotope systems: they constrain the time from which the rock system must have been closed to exchanges with external systems. The rock systems to which such constraints may apply can be quite large, such as the source regions of continental and ocean ridge basalts and even entire planets. Nonradiogenic isotopes are also valuable indicators of physical conditions or of different source materials: e.g., the oxygen isotope ratios are related to temperature of formation. Radiogenic, nonradiogenic and trace elements help constrain the history of basaltic materials, such as the percent of partial melting, the prior modification of the source region from some "primary" composition, and contamination by interaction of the magma with other rocks or magmas in its ascent.

7.2

Some basaltic achondrites, including the eucrites, appear to be among the oldest rocks in the solar system, approximately 4.54 b.y. in age, which indicates that volcanism occurred very early in some small bodies. However, a few achondrites have apparent crystallization ages as young as 1.3 b.y. The oxygen isotope data on basaltic achondrites fall close to the terrestrial $\delta^{17}O$ vs. $\delta^{18}O$ fractionation line, but with some enrichment in ^{16}O.

7.3

The age of the Moon inferred from the "model" Rb-Sr age of all rocks is about 4.5 b.y. The oxygen isotopic data fall in a fairly tight cluster on the terrestrial fractionation line, allowing the Moon to come from the same cosmogenic sources as the Earth. The bulk of the lunar crust differentiated more than 4.4 b.y. ago, as indicated by the highland rock isochron. The source regions of mare basalts also became separated from highland rocks at least that early. However, crystallization ages older than 4.0 b.y. are rare, apparently because of resetting by major impacts.

All dated mare basalts fall in the range 4.0–3.1 b.y. Trace element data, such as the rare earths, indicate that the mare basalts were formed by small percentages of partial melting of a low-oxygen-fugacity source region. They also require either that the source was already differentiated, or that the magma interacted significantly with cumulate rocks during its ascent.

7.4

The Earth presents an appreciably more complex picture, which can be inferred from Sr, Nd, and Pb radiogenic isotope systematics. If core formation was accompanied by U-Pb fractionation, as is probable, then it occurred more than 4.4 b.y. ago. However, substantial mantle heterogeneities appear to have persisted throughout the Earth's evolution, with variations in Rb/Sr, U/Pb Sm/Nd ratios of 2–3. The source regions of continental basalts appear to be similar to chondritic meteorites in trace element and in radiogenic isotope contents, and to have evolved only slightly since the Archean. However, the source regions of oceanic abyssal and ridge basalts appear to be 50–80% depleted in large ion lithophiles (Ba, U, Rb, Cs, etc.), suggesting either that they are differentiated from material from which continental crust had been derived previously, or that more enriched mantle regions are not accessible to ocean rises. Examination of ultramafic rocks, such as lherzolites, requires that they have lost lithophilic elements as well as volatiles. The presence of primordial ^{3}He and other inert gases in oceanic tholeiites indicates that the Earth is not yet completely outgassed. The presence of CO_2 and H_2O in the mantle is indicated by stable isotope data. These oxygen and carbon data may also indicate initial contamination in some cases. Clustering of rock chronological dates suggests that volcanism has occurred episodically in the Earth.

8.1

Chapter 8

8.2-8.3

Cratering chronology of volcanic planetary evolution uses the densities and morphologies of impact craters to estimate the ages of various unsampled planetary surface provinces, interpreted as volcanic. Crater diameter distributions, obtained by a variety of techniques, exist for 53 planetary provinces. Results indicate reasonable correlations among radiometric ages of rocks collected at a site, crater densities in the region of the rock samples, and crater morphometry in the dated provinces. "Reasonable" implies a factor of 2 or less uncertainty in definition of crater density for mare-scale provinces.

8.4

These data are analyzed by correlating dated provinces on the Earth and Moon with observed crater densities and impact rates on those planets. This correlation of crater density and surface age in the Earth/Moon system, where ages range from less than 1 b.y. to about 4 b.y., reveals that the meteoroid flux has decreased from a rate which, prior to 4 b.y. ago, was more than ten times as great as the present flux. The lunar cratering rate is regarded as known within a factor of 2 or 3 for most of the history of the solar system.

8.5

Extension of these ideas to other planets yields greater uncertainties in age because of uncertainty in the crater production rates among the different planets. The dynamics of small-body interactions with a planet affect impact probability and therefore affect crater production rates calculated for the planets.

8.6

Curves of predicted crater density vs. crater retention age are plotted for the Moon, Mercury, Mars, Venus and Earth. From these, estimated ages are derived and listed for the 53 planetary provinces. Results suggest a long history of declining volcanism on the Moon and Mercury. Substantial volcanism also occurred several b.y. ago on Mars, but probably persisted into the last third of martian history, and may still occur intermittently. Thus volcanism persisted longer on Mars than on the Moon or Mercury. Recent radar observations indicate densely and sparsely cratered regions on Venus, with some evidence of basaltic volcanism, but present data are inadequate to constrain the volcanic chronology. Only the oldest shields on Earth are substantially cratered, consistent with the Earth's erosive and volcanic activity. 8.7

8.8

More precise dating of smaller provinces could be obtained if the origins of small craters were better understood. Some techniques using smaller craters are compared, but further work in this area is needed to resolve controversial issues.

8.9

Calculated extrusion rates associated with filling of various lunar basins suggest that lunar volcanism in the "premare" period, in regions now classified as uplands, probably exceeded extrusion rates during mare formation. Many of the "premare" volcanic units are modified or nearly obliterated by subsequent cratering.

8.10

Volcanism probably was an important process during the era of intense early cratering, but has not been fully appreciated as such. The data imply early heating of the planets, with subsequent cooling rates and consequent durations of volcanic activity being correlated with planetary mass. Refinement of knowledge of impact and crater production rates throughout the solar system would allow important advances in dating the processes of planetary evolution.

Chapter 9 9.1

Thermal histories of the terrestrial planets concerns the mathematical modelling of the evolution of the planetary energy fluxes underlying basaltic volcanism. The fluid dynamical concepts and techniques and the relevant physical properties are discussed before treating each planet in turn.

9.3

The applicable formulations are the energy and momentum (Navier-Stokes) equations of thermal convection. Use of plausible physical values in these equations indicates that, whereas a planetary mantle is viscous enough to make inertial effects negligible (i.e., the Prandtl number is very high), the likely heat sources and large scale, together with the low thermal conductivity, make it overwhelmingly probable that nonlinear convection dominates heat transfer (i.e., the Rayleigh number is well above critical). Because the full integration of the relevant equations is well beyond current computer capacities, simplifications must be adopted in order to model planetary thermal evolution. One of the most promising simplifications is based on an assumed power-law relationship between the heat delivered by convection and the Rayleigh number for a planetary mantle.

9.4

The principal uncertainty about physical properties affecting planetary thermal evolution is the magnitude of the various possible heat sources in the early history of the planet. The radioactive sources which are important in the long run, U, Th, and K, would require several hundred m.y. to heat up a cold planet to the point that significant amounts of convection and magmatism would occur. For those planets which have differentiated an appreciable core, the resultant release of gravitational energy was sufficient to provide heat for crustal differentiation and other very early volcanic and tectonic activity. For other planets (such as the Moon and the eucrite parent body), evidence of early crustal differentiation implies an exotic energy source: accretional, electromagnetic, or short-lived radioactive. The accretional energy acquired by growth from any plausible planetesimal population is sufficient for such early heating of a lunar-sized or larger body, but for smaller bodies is almost as speculative as the alternatives: (1) ohmic dissipation arising from induction by an extremely active (approximately 10^7 times the present flux) solar wind; and (2) decay of ^{26}Al, as has been evidenced by ^{26}Mg excesses in a few Allende refractory inclusions. Another energy source available only to a close satellite which has much lower tidal Q than its planet is tidal dissipation: the flexing of the satellite due to noncircularity or (less plausibly) to tilt of its orbit; this heat source probably dominates in Io and may have been important for several other outer planets' satellites.

Other physical properties which affect subsequent planetary evolution involve the equation-of-state: how do pressure and temperature affect the viscosity and the density (including phase transition effects)? In a convecting planet, are temperature gradients modified from an adiabat toward a constant-viscosity curve?

9.5

Among the planets, the Earth has been the most examined and debated. It has adequate energy sources, and is large enough that primordial heat retained from infall and core accretion may still be appreciable, but radioactive decay is probably the major heat source today. The Earth's mantle undoubtedly has been convecting since its early history. Major problems involving the thermal and compositional evolution of the Earth are (1) the relative contributions of primordial and radioactive heat; (2) the degree of episodicity of the convection, and the related question of whether the plate tectonic pattern of the last 200 m.y. is atypical; (3) the above-mentioned equation-of-state effects on temperature and viscosity; (4) the development of heterogeneity in the mantle, and its implications for volatiles and lithophiles (including radioactive heat sources); and (5) the thermal regime of the core: energy sources for the geodynamo and mechanisms of heat transfer between core and mantle. A problem which is particularly pertinent to basaltic volcanism is the relationship of magmatism to convection and to the plate tectonic structure, including the correlation of volcanism with convective up- and down-streams, and how magma penetrates the lithosphere.

The Moon's evolution has gone much further toward a quiescent conclusion, mainly because the Moon is smaller than the Earth, so it was affected by less energy of formation, negligible energy release upon core differentiation (if it ever occurred), and faster cooling because of its larger surface area/mass ratio. The apparent balance of expansion and contraction in the Moon's surface requires heating of the center to offset cooling of the outer layer. Other conditions required are the maintenance of sufficient heat at intermediate depths for mare basalt source melting 3–4 b.y. ago and the dominance of a compressive regime in the outer parts, shutting off volcanism after this epoch. Models satisfying these conditions entail appreciable early upward differentiation heat sources in the outer two or three hundred kilometers, and achieve a present state somewhat like that indicated by the seismological data: a 900 km radius asthenosphere inside a thick lithosphere. The major problem in lunar evolution is explaining the source of the field which impressed remanent magnetism 3.5 b.y. ago.

Mars is intermediate between Earth and the Moon in the extent of its evolution as well as its size. Mars is more oxidized than either of the other planets, which affects the extent to which a core could differentiate. The predominantly extensional tectonics may imply that formation of the martian core was delayed.

The core of Mercury almost certainly differentiated rather early, since surface evidence of the major planetary expansion which would have resulted from core formation is no longer preserved. The predominantly contractive evolution which has prevailed since core formation would have stifled volcanism.

The early evolution of Venus may have resembled that of the Earth during the Archean. Since then, the higher temperatures and the absence of water may have caused subduction to cease, resulting in the accumulation of a much thicker basalt layer.

Significant amounts of volcanism on smaller bodies, such as Vesta and the eucrite parent body, appear to depend on the poorly known early heat sources which were discussed above. The greatest problem is to account for the source of the heat required to produce those few basaltic achondrites which appear to have formed within the last 2 b.y.

Chapter 10

10.1

Implications of basaltic volcanism for planetary evolution reviews the major problems of planetary evolution and attempts to draw conclusions concerning constraints and their consequences for basaltic volcanism. The solar nebula and formation of the terrestrial planets are discussed, since these processes necessarily are pertinent to the differences in planetary bulk compositions which are evidenced in the basalts, as well as to the primordial heating required to satisfy isotopic and modelling constraints on early differentiation and thermal evolution. Problems are most severe for the Moon, whose peculiar bulk composition and early differentiation of a thick crust appear to require special circumstances.

10.3

The subsequent compositional and thermal evolutions of the planets are discussed separately, because their characters are so distinct.

The main problem of terrestrial evolution relevant to basaltic volcanism appears to be the development of heterogeneities within the mantle, which have been evolving for at least 1.5–2.0 b.y. The source regions of oceanic ridge basalts appear to be significantly depleted in some lithophilic isotopes relative to the sources of ocean island and continental basalts. Inference as to whether these source regions are separated radially, as well as laterally within the mantle, depends upon seismological evidence and upon thermomechanical considerations, both of which are uncertain. In any case, a large amount of basaltic material must have been returned to the mantle throughout the Earth's history by subduction, which raises questions as to the manner and place of its deposition: the extent of remixing, exchange of material between upper and lower mantle, etc.

The Moon has had a simpler history, starting from a massive crustal differentiation very early and cooling ever since, with the waning phases of basaltic volcanism occurring some 3.0 b.y.

ago. The evolution of Mars has been more prolonged, perhaps because core formation was delayed. Much of the martian surface is consistent with basaltic flows subsequently modified by erosive processes. The evolution of Mercury has been simpler: a general cooling and contraction since the pulse of heat from core formation. Whether the mercurian surface contains appreciable basalt depends upon the bulk SiO_2 abundance. Venus has no features which resemble mid-ocean ridges. There is only a single mode in its distribution of elevations, which suggests that plate tectonics has ceased because more basalt has been retained at shallow depths than has been retained on Earth, which has led to a modern venusian surface which is much more dominated by continent-like material than the Earth's surface.

10.4

Finally, an attempt is made to draw systematic inferences and to isolate the major remaining problems of basaltic volcanism. The evident major determinant of overall evolutionary pattern is size: the larger the planet, the more slowly it cools, and hence the longer it preserves an active tectonic and differentiating regime. The degree of return of basaltic material to the mantle by lithospheric subduction is greater in a large planet, probably influenced both by the larger gravity and by a greater proportion of the denser eclogite phase being in the lithosphere. Bulk composition also influences evolution, as is evidenced by differences in the histories of Mercury and Mars. Differences in volatile evolution and in the consequent atmospheric temperature have led Venus to thermal and crustal evolutionary paths very different from those of the Earth.

The basalts resulting from the evolution of terrestrial planets can be categorized as belonging to two types: primordial and evolutionary. The former are associated with initial large scale crustal differentiation, and now are accessible only for the eucrite parent body and, in a much transmuted form, the Moon. Evolutionary basalts arise from the later heating of mantles; hence their source depths are correlated inversely with planet size, while their abundances are correlated directly.

The chapter and the book end with a discussion of the outstanding problems and directions for further research, ranging from extensions of the data base to better models for small body heating and differentiation.

Overleaf:

The juxtaposition of textures typical of basalts is shown in this photomicrograph: A subophitic pyroxene encloses plagioclase laths, while immediately adjacent plagioclase laths form a network enclosing finely crystalline matrix in an intersertal texture. A dendritic intergrowth of plagioclase and pyroxene is located in the left central part of the field of view. Photo courtesy of G. Lofgren.

1. Petrology and chemistry of terrestrial, lunar and meteoritic basalts

Gary E. Lofgren, Team Leader
—*NASA Johnson Space Center*
Houston, Texas

A. E. Bence
—*Exxon Production Research*
Houston, Texas

Michael B. Duke
—*NASA Johnson Space Center*
Houston, Texas

Michael A. Dungan
—*Department of Geological Sciences*
Southern Methodist University

John C. Green
—*Department of Geology*
University of Minnesota at Duluth

Stephen E. Haggerty
—*Department of Geography and Geology*
University of Massachusetts

L. A. Haskin
—*Department of Earth and Planetary Sciences*
Washington University

Anthony J. Irving
—*Department of Geological Sciences*
University of Washington

Peter W. Lipman
—*U. S. Geological Survey*
Denver, Colorado

Anthony J. Naldrett
—*Department of Geology*
University of Toronto

James J. Papike
—*Department of Earth and Space Sciences*
State University of New York at Stony Brook

Arch M. Reid
—*Department of Geology*
University of Cape Town

J. Michael Rhodes
—*Department of Geography and Geology*
University of Massachusetts

S. Ross Taylor
—*Research School of Earth Science*
Australian National University

David T. Vaniman
—*Los Alamos Scientific Laboratory*
Los Alamos, New Mexico

1.1 INTRODUCTION

The theme of this book is the study of basaltic volcanism as a stage in planetary evolution: to use the eruption of lava from the interior of a planet as evidence of the thermal and chemical processes of the planet. All lavas extruded on a planet's surface can provide information about events in the interior. From a petrologic point of view, however, it is the more primitive lavas, which rise to the surface of a planet with minimal changes, that can provide the most information about the interior. In general, these more primitive lavas are basalts and so in Chapter 1 our sampling has concentrated on basaltic rocks.

Basalts are dark-colored, relatively fine grained volcanic rocks composed mostly of plagioclase and pyroxene, with lesser amounts of iron oxides and, in many cases, olivine. Chemically, basalts are characterized by Fe, Ca and Mg contents commonly in the range 5–15 wt.% oxide and by SiO_2 contents generally in the range 38–53 wt.%. This is a very general working definition because, for example, many lunar mare basalts and meteoritic basalts have FeO contents greater than 20 wt.%.

Three criteria commonly have been used to define basalts: SiO_2 content, color index and plagioclase composition. As more analyses have become available, the composition of plagioclase has been increasingly recognized as an unreliable criterion. Plagioclase is usually zoned, so it is difficult to obtain an analysis of a crystal's bulk composition. Andesites, which by some definitions should have plagioclase compositions of less than An_{50}, quite commonly have much more calcic plagioclase compositions. Color index is difficult to use as a criterion because basalts usually contain a high percentage of uncrystallized material which makes an accurate determination of color index difficult. The SiO_2 content used to discriminate between basalts and andesites is generally 53 wt.%. Many lavas that are useful indicators of their source regions, however, would not be classed as basalts by this criterion. Thus, we have relaxed this restriction and have included lavas in the reference suites with SiO_2 contents as great as 58%. At the low-SiO_2 end of the continuum of observed lava compositions, basalts grade into rocks such as picrites, komatiites and kimberlites. Here we have included a suite of Archean komatiites and basalts, as well as a suite of mantle-derived peridotite nodules.

It is our task in this chapter to characterize and compare the chemistry and petrology of the solar system basalts of which we have samples: from the Earth, the Moon, and the basaltic meteorites. In order to compare basaltic rocks from different planets, it is necessary to have a data base of uniformly high quality. Such a data base does exist for lunar basalts because of intensive study during the last decade. Unfortunately there has been no comparable data base for basalts from Earth or the meteorites. Team 1 had the objectives of selecting a representative suite of basalts from the Earth and meteorites and of producing a data set for these rocks which would be comparable to the lunar data base. The time and resources available have constrained our choice to no more than a total of 100–150 terrestrial samples. While it is not possible in 150 samples to characterize all of the petrologic processes on Earth by which basalts have been derived, such a sample set should provide a first-order characterization of terrestrial basalts.

The choice of suites of basalts representative of the Moon and the basaltic meteorites is restricted by the number of available samples (i.e., lunar samples returned by the Apollo and Luna missions and meteorites that reside in accessible collections). For the Earth, the task of selection is more difficult. It seems most appropriate to select basalts primarily within a tectonic framework but also with regard to the temporal aspects. Basaltic volcanism occurs both at plate boundaries and within plates, so suites have been chosen from both kinds of provinces. In order to restrict individual terrestrial suites to a reasonable number of samples, it generally was necessary to select samples from larger collections which had already been sufficiently well-characterized that intelligent choices could be made. This requirement severely, and perhaps unduly, limits the choices. Several nongeologic requirements have also imposed restrictions. Thus, it was necessary to select the suites so that the analytical work could be completed within the time limits of the project. Suites were, for the most part, selected from one general locality rather than from many, widespread localities. The ocean island suite, for example, is from Hawaii. Whereas this may limit the variety of basalts in the suites, it does allow the selection of petrologically significant samples from a well-characterized environment.

Cenozoic basalts dominate the suites, but older (e.g., Precambrian) suites have been included in order to assess changes in the character of basaltic volcanism which may have occurred with time.

The reference suites are described and compared with other basalts of similar types in section 2 of this chapter. For the most part, summary data and diagrams are presented in the text; a complete tabulation of the new data collected on the reference suites is contained in the Appendix (microfiche packet). The Archean suite, selected and described by A. J. Naldrett and I. E. M. Smith, contains many komatiites from Munro Township, Canada, which are compared to Fe-tholeiites and high-alumina basalts from elsewhere in the Abitibi greenstone belt. Komatiites are, with few exceptions (e. g., Echeverria, 1979), unique to the Precambrian but their presence in this suite is felt to be justified because of their importance as possible primary magmas.

Continental flood (or plateau) basalts are split into two suites based on age. The pre-Tertiary plateau basalts are represented by a suite of Keweenawan lavas from Minnesota which was selected and described by J. C. Green. The suite consists mostly of olivine tholeiites and weakly alkaline basalts, but a few basaltic andesites and andesites are included. The Tertiary plateau basalts are represented by a suite from the Columbia River Province selected by T. L. Wright and described by S. E. Haggerty and A. J. Irving. This is the smallest suite, with only ten samples, and was selected to represent the range of observed composition types. As is true of most suites, the number of samples of a particular type does not necessarily represent its volumetric abundance.

A suite of basaltic samples from the Taos volcanic field in the Rio Grande Rift was chosen and described by P. W. Lipman and M. A. Dungan. The suite is predominantly olivine tholeiite with some alkalic basalts and andesites. The suite does not include the highly undersaturated basalts commonly associated with continental rift volcanism because they do not occur in this part of the rift. Mildly undersaturated basalts are found in the southern parts of the rift and outside of the rift on either side, but are not directly associated with the Taos Plateau volcanic field.

The ocean floor suite comprises basalts which presumably erupted at an ocean ridge spreading center. The samples were chosen by J. M. Rhodes with the cooperation of W. B. Bryan, W. G. Melson and S. R. Hart. In contrast to other suites, the samples were selected from different localities (from the FAMOUS area, 28°N on the Mid-Atlantic Ridge, from the Galapagos and Juan de Fuca Ridges and from the East Pacific Rise) and represent a range from primitive to evolved basalts.

Oceanic, intraplate volcanism is represented by a suite from the island of Hawaii, which was selected by the staff of the Hawaiian Volcano Observatory to represent basaltic volcanism on that island. The suite, described by A. E. Bence, is dominated by tholeiites from Kilauea and Mauna Loa volcanoes, but also includes alkali olivine basalts, hawaiites, ankaramites and trachytes from Hualalai, Mauna Kea and Kohala volcanoes.

Island arc basalts are volumetrically abundant in areas of active plate subduction. A typical suite from New Britain, Papua-New Guinea, was selected and described by R. W. Johnson and S. R. Taylor. These basalts are closely associated with more silicic volcanic rocks (basaltic andesites, andesites and dacites) which are not included in the suite.

The extraterrestrial suites include the meteoritic basalts and the lunar sample collection. Basaltic material occurs in the basaltic achondrite meteorites as monomict breccias (eucrites) or as eucritic clasts in polymict breccias (howardites). A second group of basaltic meteorites, the shergottites, are chemically and mineralogically distinct from the more abundant basaltic achondrites. A. M. Reid and M. B. Duke selected six meteorites for detailed study: Juvinas, Pasamonte, Stannern, Nuevo Laredo, Aioun el Atrouss, and Shergotty. In light of the already extensive data available for some of the meteorites, and because of the small size of the fragments in the sample suite, complete chemical and mineralogical studies were not attempted for each sample.

The lunar suite is divided into two groups, the mare and the highland basalts. The mare basalt suite includes samples from all the Apollo landing sites, except 16, and from the Luna 16 and 24 sites. The emphasis in the selection of this suite by J. J. Papike was to cover the range of chemical compositions observed in mare basalts. Whereas the mare basalts are clearly volcanic rocks, the highland basalts may not be. Most samples returned from the lunar highlands are either breccias or crystallized impact-produced melts of these breccias (melt rocks), although a small number of endogenic volcanic rocks (KREEP basalts) have been recognized. The highland breccias are pieces of the fragmented early lunar crust which most likely did contain basalts. Many of the breccias and melt rocks have basalt-like compositions and several are monomict. Since these breccias and melt rocks are the only available representatives of possible premare lunar volcanism, we felt it was important to include them, so a suite has been selected and described by D. T. Vaniman.

The final suite to be discussed does not consist of basalts, but of terrestrial mantle peridotites which could be solid residues complementary to the basaltic liquids. A. J. Irving has selected and described a suite of ultramafic nodules (including both spinel and garnet lherzolites) from alkalic basalts and kimberlites. Eight lherzolites from Kilbourne Hole, New Mexico, U.S.A., represent some of the variety from a single locality. Nodules from eight other sites around the world provide a small sampling of the regional variation in bulk chemistry and mineralogy.

Following a systematic description of each of the suites, they are compared one to another from a planetary perspective. Basalts are partial melts of planetary interiors, and provide information on the chemistry and mineralogy of their source regions. Thus comparison of differences in the compositions of basalt from different planets allows us to compare some aspects of planetary interiors. These conclusions, in turn, can be used to constrain models for accretionary histories of the planets and for the paths of mantle evolution in the terrestrial planets.

There appear to be major interplanetary differences in mantle compositions with respect to Fe, Ti, Al, Na, Cr, Ni, and in refractory to volatile element ratios. The $Mg/(Mg + Fe^{2+})$ ratio (denoted elsewhere in this chapter as Mg') is distinctly different for different planetary mantles; lunar basalts have Mg' values as high as 0.6, compared to terrestrial basalts, 0.7 to 0.9, and basaltic meteorites, up to 0.5. These differences must reflect inherent differences in the compositions of the source regions on the different planets, as discussed in section 1.3.1.

A detailed comparison of the silicate mineralogy on the three sampled planetary bodies, treated in section 1.3.2, further illustrates differences in planetary compositions and goes on to establish constraints on the temperatures, pressures, and oxygen fugacities of the source regions. The valence state of Fe in pyroxene shows the Moon to have had melts which were significantly more reduced during crystallization than similar terrestrial melts. The more calcic feldspars on the Moon and in basaltic meteorites, compared with those on the Earth, indicates the lower Na content of the Moon and meteorites.

The textures of basalts and their contribution to understanding the histories and cooling environments of basalts are discussed in section 1.3.3. The interpretation of basalt textures has not yet reached the same level of sophistication as have the discussions of chemistry and mineralogy, but one significant fact emerges from studies of comparative petrography. The basaltic magmas which reach the surface of the Earth, as evidenced by the nature of their phenocrysts, have experienced a more complex formational and migration history than have basalts on either the Moon or the eucrite parent body(ies).

In the last two sections (1.3.4 and 1.3.5) on planetary comparisons, the variations in the volatile contents (hydrogen, carbon, and sulfur gas species) and the gas fugacities are discussed. The presence of water on the Earth has resulted in chemical differentiation trends in basaltic magmas which are markedly different from those observed on the other bodies sampled. The oxygen fugacity during basalt formation and crystallization on the terrestrial planets is known with a fair degree of confidence. Generally, terrestrial basalts crystallize within the magnetite stability field, whereas lunar basalts form in the stability field of metallic iron. Mars may crystallize volcanic rocks in the stability field of hematite. Current speculation suggests that basalts on Venus form in the wüstite field and, on Mercury, in the metallic iron field. Meteorites present conflicting evidence for environments ranging from moon-like for the eucritic basalts and to earth-like for the shergottites.

Sections 1.4.1 and 1.4.2 concentrate on the important issue of how we can identify primary magmas—those liquids that can tell us the most about the interiors of planets—and the processes which may modify those important liquids before they reach the surface. Finally, in section 1.5 we summarize the major conclusions of this petrological and chemical examination of the planets.

1.2 SURVEY OF MAJOR BASALT TYPES

1.2.1 MAFIC AND ULTRAMAFIC VOLCANISM DURING THE ARCHEAN

1.2.1.1 Introduction

Volcanic rocks constitute a significant proportion of supracrustal rocks deposited during the Archean, in many areas considerably more than 50% of the total (e.g., Goodwin *et al.*, 1972). The proportions of mafic to intermediate to felsic volcanics vary among different volcanic belts. In an idealized Archean volcanic succession, basaltic rocks predominate in the lower stratigraphic units, whereas the upper units are made up of rock suites in which intermediate and acid rocks predominate. Significant changes in the compositions of basalts making up these units accompany changes in relative proportions of the different rock types. Although preservation of such sequences is typically fragmentary, well-developed sequences have been described from the Abitibi greenstone belt of northern Canada, where Goodwin (1977) estimates the relative proportions of rock types making up the whole succession to be ultramafic rocks 1%, basalt 55%, andesite 33%, dacite 7%, and rhyolite 4%.

Typically, these are metamorphosed to greenschist and locally to amphibolite facies, although lower metamorphic grades are not uncommon and basaltic rocks containing primary igneous minerals have been described from several areas. Because metamorphism is so widespread in Archean volcanic terrains, classification schemes based solely on present mineral contents are impractical. Similarly, because of the probability of redistribution of some large-ion-lithophile elements, notably K, Rb and Ba, during metamorphism (Hart, 1969; Hart *et al.*, 1970; Brooks *et al.*, 1970; Jahn and Shih, 1974; Jahn and Condie, 1975), schemes based on normative mineralogies and alkali contents must be treated with caution. Here we follow a classification scheme established by current usage, which subdivides Archean basaltic and ultramafic rocks according to their MgO and FeO contents and to the abundance of selected refractory lithophile elements (Ti, Y, Zr, Sc, REE), which are generally accepted to be fairly immobile during weathering and metamorphism. Compositionally, Archean volcanic rocks resemble their Phanerozoic counterparts in many respects, but the extrusion of komatiite lavas containing up to 33 wt.% MgO and crystallizing with characteristic "spinifex" textures appears to have been unique to the Archean. [Komatiite lavas have been extruded subsequent to the Archean, but the liquid compositions have been less magnesian (< 22% MgO) than those of Archean komatiites.] Komatiites constitute as much as 10 volume percent of the total volcanic succession in some greenstone belts where the stratigraphy is reasonably well understood (Fig. 1.2.1.1).

Considerable discussion has developed in the literature over what constitutes and what does not constitute a komatiite (Nesbitt and Sun, 1976; Arndt *et al.*, 1977; Naldrett and Turner, 1977; Sun and Nesbitt, 1978). This discussion stems partly from the difficulty of distinguishing basalts that are differentiates of more magnesian komatiite magmas from those that are not. In this section we restrict our discussion of komatiites to those rocks containing more than 12% MgO, treating these first and then following with a discussion of the different types of less magnesian Archean rocks.

1.2.1.2 Description of principal rock types

Komatiites

Definition and terminology. The term "komatiite" comes from the Komati River area of the Barberton Mountainland, South Africa, where Viljoen and Viljoen (1969) first described these rocks as lavas. They used textural and chemical criteria to distinguish komatiites from other ultramafic rocks. Their chemical criteria were refined by Brooks and Hart (1974) who suggested that komatiites were noncumulate rocks with $MgO > 9$ wt.%, $K_2O < 0.5\%$, $TiO_2 < 0.9\%$ and $CaO/Al_2O_3 > 1$. It is now appreciated that these criteria are too restrictive. Arndt *et al.* (1977) have proposed an extended definition in which komatiites are regarded as a suite of cumulate and noncumulate rocks. In the noncumulate rocks, (1) MgO ranges from > 30 to < 8 wt.%; (2) $Mg/Mg + Fe$ ratios are generally high, especially when compared with rocks from other magma series having similar Al_2O_3 contents; (3) TiO_2 contents are low, in almost all cases < 1 wt.%; (4) Ni and Cr always are high (usually in excess of 100 ppm and 140 ppm, respectively); (5) the CaO/Al_2O_3 ratio usually ranges between 0.8 and 1.1, although in some samples it is as low as 0.6 and in certain areas, such as the Barberton Mountainland, some rocks have ratios as high as 2; (6) skeletal crystals including bladed (olivine spinifex or spinifex texture) or hollow (hopper) olivines are developed in liquids containing > 20 wt.% MgO and hollow, needle-like pyroxene crystals (pyroxene spinifex texture) occur in liquids containing 12–20 wt.% MgO.

Here we refer to noncumulates with MgO > 20 wt.% as peridotitic komatiites and to similar rocks with MgO < 20 wt.% as basaltic komatiites. However, it is difficult to distinguish basaltic rocks that are members of the komatiite suite, as evidenced by their close association with komatiite lava flows (see description of Fred's Flow below), from basalts that have no spatial association and probably no genetic link with komatiites on the basis solely of chemical or textural criteria. Therefore, we refer to rocks with 12–8.5 wt.% MgO as magnesian basalts. We recognize that some of these basalts actually are differentiates of peridotitic komatiite magmas and hence belong to the komatiite suite, while others may be primary mantle-derived magmas and more properly should be referred to as tholeiites. The nomenclature that has been used for komatiitic rocks in the past is compared with that used here in Table 1.2.1.1. In addition, Table 1.2.1.1 contains a suggested nomenclature for cumulate rocks which are not discussed in detail in this study.

Field relations. Field relations have been described in many papers including Viljoen and Viljoen (1969), Pyke *et al.* (1973), Lewis and Williams (1973), Bickle *et al.* (1975), Arndt *et al.* (1977), and Lajoie and Gélinas (1978).

Table 1.2.1.1 Komatiite nomenclature.

MgO content (wt. %)	Cumulate or Noncumulate	This Study	Arndt *et al.* (1977)	Naldrett & Turner (1977)	Sun & Nesbitt (1978)	Viljoen & Viljoen (1969)
>44	c	komatiitic dunite	—	dunite	—	dunite
30-44	c	komatiitic peridotite	—	peridotite	—	peridotite
30-35	nc	peridotitic komatiite	peridotitic komatiite	spinifex peridotite	STPK	peridotitic komatiite
20-30	c	komatiitic peridotite	—	pyroxene peridotite	—	peridotite
20-30	nc	peridotitic komatiite	peridotitic komatiite	spinifex pyroxene peridotite	STPK	peridotitic komatiite
12-20	c	komatiitic pyroxenite	—	pyroxenite	—	pyroxenite
12-20	nc	basaltic komatiite	pyroxenitic komatiite	pyroxenite -flow type	STB	basaltic komatiite geluk and (Badplaas types)
10-12	nc	magnesian† basalt	basaltic komatiite	magnesian basalt	B	basaltic komatiite (Barberton type)
8.5-10	nc	magnesian† basalt	basaltic komatiite	basalt (komatiitic series)	B	basaltic komatiite (Barberton type)
<8.5	nc	basalt†	basaltic komatiite	basalt (komatiitic series)		basaltic komatiite (Barberton type)

† With these rocks it is very difficult to tell on the basis of petrography or chemistry whether they are of komatiitic affiliation — proximity in the field indicates that some examples are.

STPK = Spinifex-textured peridotitic komatiite.

STB = Spinifex-textured basalt.

B = Basalt.

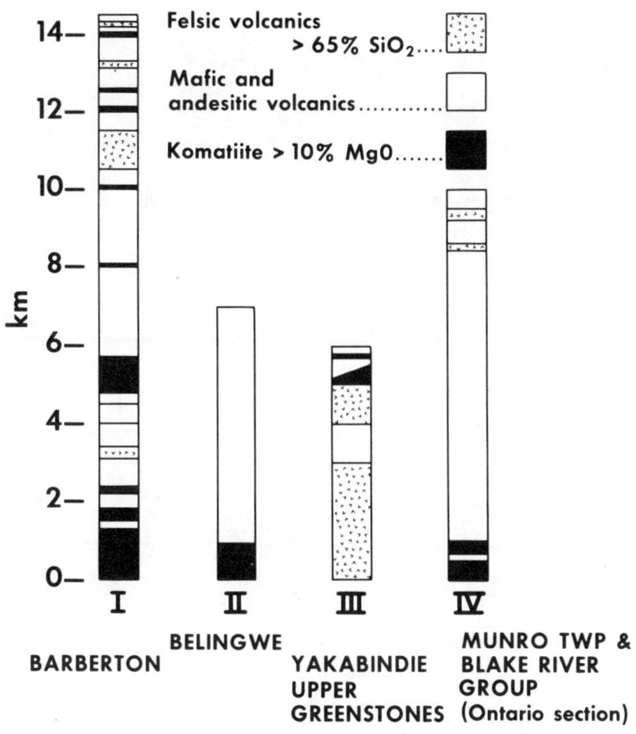

1.2.1.1 Comparison of proportions of types of volcanic rocks in some well-studied Archean greenstone belts.

Komatiites are found within sequences of basalts within Archean greenstone belts. Where a given belt is characterized by stratigraphic cycles in which the volcanism progresses upwards from mafic through intermediate to felsic compositions, komatiites tend to occur near the base of the basaltic succession (e.g., Blake River Group, Ontario, Goodwin *et al.*, 1972; Onverwacht formation, Barberton Mountainland, Viljoen and Viljoen, 1969; Fig. 1.2.1.1). Where this orderly sequence is lacking, komatiites may occur anywhere within the basaltic sequences, such as in the Norseman-Coolgardie and Yakabindie areas of Western Australia (Gemuts and Theron, 1975; Naldrett and Turner, 1977).

Within a given sequence of komatiitic rocks, the most ultramafic lavas tend to occur at the base and to pass upwards into successively less magnesian lavas. An example of this is shown in Fig. 1.2.1.2, which illustrates two Mg-rich to Mg-poor cycles of komatiitic volcanism, one overlying the other in Munro Township, Ontario, and both overlying a sequence of picritic and Fe-rich tholeiitic rocks. The stratigraphic positions of rocks in the reference suite collected expressly for this study are indicated in this figure. In discussing chemical variation within the reference suite, particularly trace element variation, it will be emphasized whether the samples belong to the underlying picrite-tholeiite association, or to the lower or upper komatiite sequence.

Figure 1.2.1.3 illustrates variations in the form of peridotitic komatiite lava flows; the variations are attributed by Arndt *et al.* (1977) to different degrees of settling of phenocrysts suspended in the magma. In the

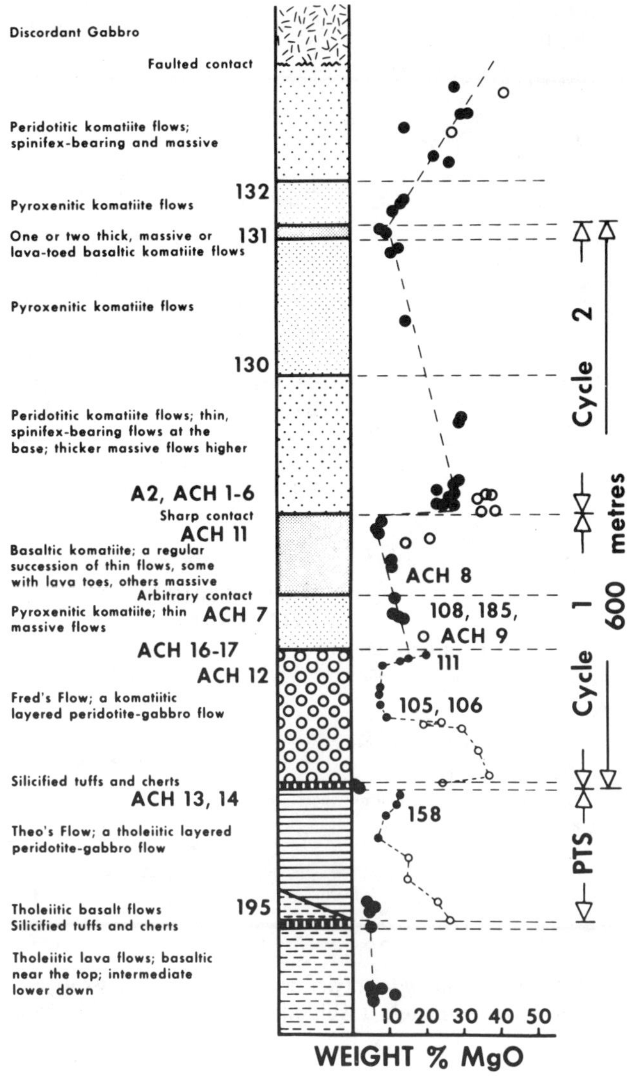

1.2.1.2 Stratigraphic section in northern Munro Township. Tholeiitic lavas forming the base of the section are overlain by a chemically distinct suite, the picrite-tholeiite series (PTS) which in turn is overlain by peridotitic and basaltic komatiitic rocks of cycle 1 (PKLC and BLC). These are overlain by peridotitic and basaltic komatiites of cycle 2 (PKUC and BUC). The MgO contents of samples throughout the succession are indicated, with open circles referring to cumulate rocks, and dark circles to non-cumulates thought to be representative of liquid compositions. Stratigraphic positions of samples used in this study are indicated. Figure is adapted from Arndt *et al.* (1977).

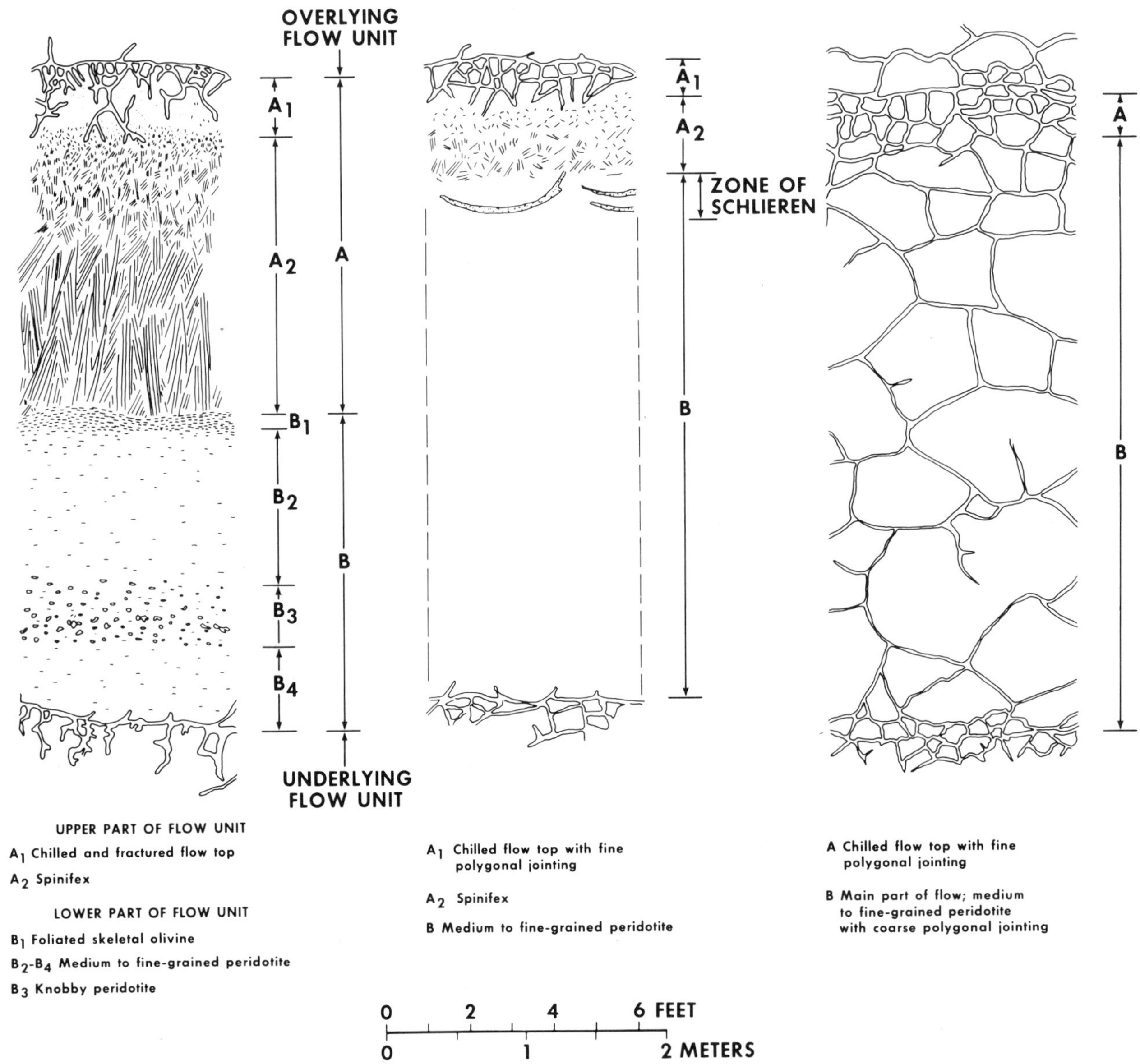

1.2.1.3 Three types of peridotitic komatiite flow ranging from a very well settled flow on the left through a partially settled flow to a completely unsettled flow on the right. Figure from Arndt *et al.* (1977).

most completely settled form, the flows are characterized by a fine-grained "quench" cap (A_1 zone) grading downwards with increase in grain size to a "spinifex" zone (A_2) in which the fabric is dominated by parallel olivine blades oriented subperpendicularly to the plane of the flow. This is underlain by several cumulus zones passing downwards from the B_1 zone of cumulus skeletal olivine crystals to the B_2 zone of closely packed, equant crystals to the B_3 zone in which the cumulus olivine becomes progressively less closely packed, grading into a chilled base (the B_4 zone). In the least settled form, equant olivine grains with skeletal overgrowths occur evenly dispersed throughout the flow, which is characterized by a pronounced system of polygonal joints interpreted as shrinkage cracks. The peridotitic komatiite flows vary from a few tens of centimeters to 20 meters in thickness and from a few meters to several kilometers in lateral extent.

Basaltic komatiite flows exhibit similar variability, although the spinifex zone consists either of close-packed, acicular augite interspersed with rare bladed olivine or, in less magnesian variants, of elongate parallel needles of augite with no bladed olivine. The degree of settling is less pronounced; cumulus zones, although sometimes present, are less common than in peridotitic komatiites and the majority of the flows exhibit no crystal settling. Basaltic flows vary from less than 1 m to more than 10 m in thickness. Their lateral extents range from a few meters up to the scale of an outcrop (~ 100 m), but it is not known how extensive individual flows are beyond this.

Some komatiitic flows exhibit fractional crystallization. One such flow, "Fred's Flow", in Munro Township (Arndt *et al.*, 1977; Arndt, 1978; Fig. 1.2.1.4) is characterized by a spinifex zone grading downwards from olivine to pyroxene spinifex and by a basal zone of cumulus olivine passing into an augite (+ minor orthopyroxene) cumulate and then into quartz gabbro.

Petrography and mineralogy. The petrography of komatiite flows of differing compositions is summarized in Table 1.2.1.2 (adapted from Arndt *et al.*, 1977). Some points require stressing. Olivine and minor amounts of chromite are the only nonskeletal phases in peridotic komatiites. Olivine, clinopyroxene and minor orthopyroxene occur as nonskeletal phases in basaltic komatiites. Magnesian basalts of komatiitic affiliation contain only very rare skeletal (but not bladed) olivine, but may carry plagioclase and clinopyroxene phenocrysts.

Selected olivine, pyroxene and chromite analyses are given in Table 1.2.1.3. Overall, olivine compositions range from Fo_{95} to Fo_{82}, although relatively few compositions have been documented in the Fo_{80-85} range. Those Fo_{80-85} olivines that have been analyzed come from pyroxenitic komatiites (Naldrett and Mason, 1968). Little zoning has been detected across cumulus olivine grains, although a range of 4–7 mole percent is commonly observed among spot analyses from a given sample, and the tendency for olivine to be altered around the margins may obscure zoning. Significant zoning, with Fo increasing inwards from 89 at the margin to 94 at the center, has been documented in blades from spinifex-textured rocks. The composition of olivine at the centers of spinifex blades is fairly close to that in the underlying cumulus zone, although in cases where they have been studied closely it appears that the cores of spinifex blades may average 2 mole percent lower in forsterite content than the cumulus olivine. There is also a suggestion (Arndt *et al.*, 1977) that the composition of olivine becomes less forsteritic as one moves towards stratigraphically-lower parts of the spinifex zone—an observation which is consistent with downward crystallization coupled with a small degree of fractionation through the spinifex zone. Olivines are remarkably rich

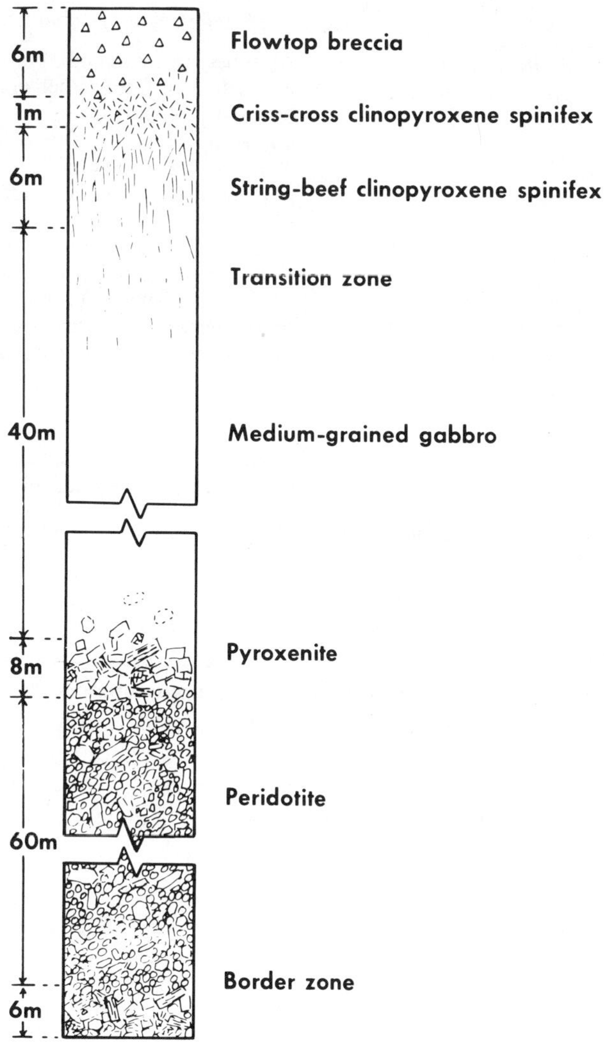

1.2.1.4 Typical section through a differentiated peridotitic komatiite flow (Fred's Flow). Figure from Arndt *et al.* (1977).

in Cr_2O_3 (0.1–0.4 wt.%) and differ in this respect from all other terrestrial olivines except those observed as inclusions in diamonds and in garnet lherzolites (Meyer, 1975).

Numerous data are available on skeletal pyroxenes in komatiitic lavas, which range from 34–50 mole percent enstatite (Fig. 1.2.1.5a). These pyroxenes are remarkable for their high (6–9 wt.%) Al_2O_3 contents (Fig. 1.2.1.5b and c). They fall outside the normal ranges of terrestrial and lunar pyroxenes because of their high overall Al_2O_3 and the high proportion of $Al(Z = 4)$ to $Al(Z = 6)$. Data on cumulus pyroxenes in komatiites are relatively sparse, but indicate that cumulus pyroxenes are normal with respect to both absolute

Table 1.2.1.2 Petrographic characteristics of komatiites and related basalts.†

Peridotitic komatiite	All types are composed of olivine grains and minor chrome spinel in a matrix of fine-grained clinopyroxene and devitrified glass. In cumulates olivine grains are close-packed, solid, roughly equant and comprise 60 – 80% of the rock; in spinifex texture olivine forms large skeletal platy grains (35 – 60%); and in spinifex-free, non-cumulate rock, olivine may be equant or skeletal (45 – 70%).	MgO > 20%
Basaltic komatiite	1. Equant solid, or platy skeletal grains of olivine (0 – 35%) in fine-grained matrix of clinopyroxene and devitrified glass; or 2. skeletal subcalcic clinopyroxene needles in devitrified glass groundmass; cumulates consist of closely-packed equant grains of pyroxene and minor olivine. Plagioclase may or may not be present.	MgO between 12 and 20%
Magnesian basalt	Little or no olivine in the groundmass; plagioclase instead. Clinopyroxene and plagioclase form spinifex texture, "graphic" intergrowths, or normal subophitic texture.	MgO 8.5 – 12%

†Adapted from Arndt *et al.* 1977.

and tetrahedrally coordinated Al. The unusual compositions of the skeletal pyroxenes are attributed to disequilibrium crystallization consequent upon supercooling.

Many acicular pyroxene needles consist of augite which encloses a core of either calcic-clinobronzite (magnesian pigeonite) or, more commonly, hydrous alteration after clinobronzite. One example of this (Naldrett and Mason, 1968) is shown by the points connected by a line in Fig. 1.2.1.5a. Sometimes even more complex zoning develops in these needles (Arndt, 1977).

Chromite is generally rich in Cr_2O_3 (~ 50 wt.%), fairly low in Al_2O_3 (12–15 wt.%) and moderately high in MgO (11–14 wt.%). No difference has been observed

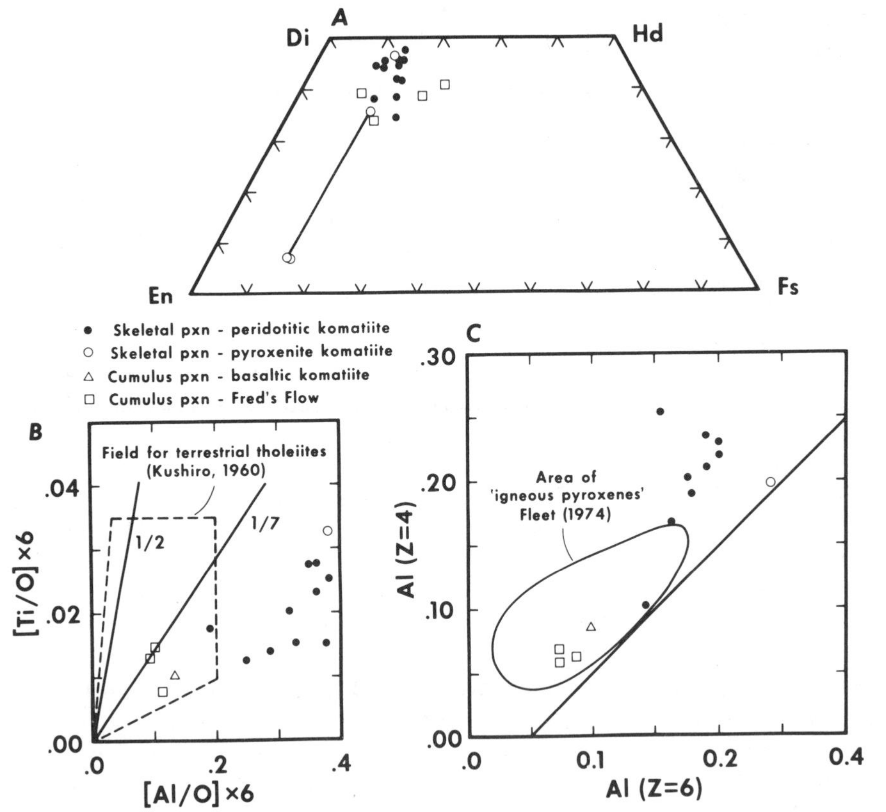

1.2.1.5 **a.** Pyroxene quadrilateral showing compositions of pyroxenes from komatiite flows. **b.** The same samples on a Ti/O vs. Al/O plot. **c.** The same samples on an Al(Z = 4) vs. Al(Z = 6) plot.

Table 1.2.1.3 Microprobe analyses of komatiite minerals.

Olivine	1 SA1157	2 SA1160	3 SA2059	4 SA2060	5 SA2048	6 SA2047	
SiO_2	41.1	40.9	40.8	40.8	40.5	40.8	
$FeO^{\dagger}$	8.2	8.1	8.1	7.4	11.4	6.6	
MgO	50.2	50.2	51.0	51.6	48.1	52.6	
CaO	0.25	nd	0.26	nd	0.27	nd	
Cr_2O_3	0.23	0.26	0.29	0.18	0.33	0.14	
NiO	nd	nd	0.39	0.32	0.40	0.55	
Fo	0.916	0.917	0.916	0.926	0.888	0.934	
Pyroxene	1 SA1162	2 SA2041	3 SA2051	4 P9118	5 P9235	6 Mun12	7 FF6
SiO_2	48.8	47.7	47.9	49.5	49.1	51.6	51.0
Al_2O_3	8.5	8.2	7.8	5.7	8.8	3.10	2.19
TiO_2	nd	0.95	0.96	0.44	1.17	0.35	0.49
$FeO^{\dagger}$	7.1	8.4	8.3	8.5	7.4	9.5	13.2
MnO	nd	0.04	0.07	nd	nd	0.21	0.24
MgO	14.3	13.2	12.4	17.0	12.9	17.7	13.8
Cr_2O_3	0.47	0.39	0.38	nd	nd	nd	nd
CaO	21.1	20.2	21.6	18.6	20.3	17.2	18.9
Na_2O	nd	0.20	0.17	nd	nd	0.13	nd
Wo	45.3	44.8	47.6	38.0	46.1	34.9	39.0
En	42.8	40.7	38.1	48.3	40.5	50.0	39.7
Fs	11.9	14.5	14.3	13.7	13.3	15.0	21.3
Chromite	1 SA1162	2 SA2041	3 SA2044	4 SA2054	5 SA336a	6 SA336b	
MgO	11.7	13.5	11.8	3.76	12.3	13.4	
* FeO	16.7	13.4	16.5	28.3	14.9	13.8	
MnO	0.30	0.32	0.47	0.96	0.52	nd	
* Fe_2O_3	3.92	5.26	7.92	14.0	7.70	9.01	
Al_2O_3	15.0	12.8	14.4	13.8	14.6	13.9	
Cr_2O_3	53.0	54.9	50.3	39.5	48.9	49.4	
TiO_2	0.31	0.34	0.35	0.69	0.39	0.41	

$FeO^{\dagger}$ Total Fe calculated as FeO.
* Fe in chromites was apportioned between Fe^{2+} and Fe^{3+}, assuming a stoichiometric formula for spinel and that TiO_2 substitutes as ulvospinel.
Key to samples (see Fig. 1.2.1.3 for zone identities). *Olivine:* 1 to 4 — all from same peridotitic komatiite flow (Arndt *et al.*, 1977); 1 — skeletal olivine from A_1 zone; 2 is a bladed olivine from A_2 zone; 3 and 4 — equant olivine from B_2 zone; 5 — Bladed olivine from A_2 zone (Arndt *et al.*, 1977); 6 — Equant olivine from B_2 zone (Arndt *et al.*, 1977). *Pyroxene:* 1 to 4 — skeletal cpx needles from A_2 zone of peridotitic komatiite flows (Pyke *et al.*, 1973; Arndt *et al.*, 1977); 5 — skeletal cpx needle from groundmass of pyroxenitic komatiite flow (Arndt, 1975); 6 — subhedral prismatic cpx from subophitic basaltic komatiite lava (Arndt, 1975). *Chromite:* 1 — Equant euhedral grain from B_2 zone, peridotitic komatiite flow (Arndt *et al.*, 1977); 2 — Equant euhedral grain from A_2 zone, peridotitic komatiite flow (Arndt *et al.*, 1977); 3 — Equant euhedral grain from A_2 zone, peridotitic komatiite flow (Arndt *et al.*, 1977); 4 — Equant euhedral grain from B_2 zone, peridotitic komatiite flow (Arndt *et al.*, 1977); 5 — Skeletal grain from A_2 zone, peridotitic komatiite flow; 6 — Equant euhedral grain from A_2 zone of same flow as 5.

between the compositions of equant and skeletal grains, although analyses of the latter are rare because of their small size.

Other ultramafic rocks

Naldrett and Mason (1968) and Arndt *et al.* (1977) draw attention to a suite of Mg-rich lavas which they call picrites (implying compositions which are Mg-rich, but not due to the presence of olivine phenocrysts) and to olivine-rich intrusions in the Abitibi belt which do not fulfill many of the criteria for komatiites. Specifically, despite aphyric rocks associated with the lavas and intrusions containing up to 15% MgO, spinifex textures are absent, and the aphyric rocks contain higher TiO_2, have a lower Mg′-value for a given Al_2O_3, and have distinctive REE patterns. The cumulate rocks also display many of these chemical characteristics. The layered intrusions differ from komatiitic layered intrusions in containing a much higher proportion of augite pyroxenite to both peridotite and gabbro. Chemically, the gabbroic differentiates of these intrusions resemble the Fe-rich basalts of Munro Township which are discussed later and to which they may be related.

Basaltic rocks

Many Archean greenstone terrains display a stratigraphic succession of volcanic rock types from compositions which are primitive in terms of their trace element contents (particularly REE abundance patterns) to overlying units made up of rocks which are evolved in terms of such geochemical parameters (e.g., Goodwin, 1977; Hawkesworth and O'Nions, 1977; Hallberg *et al.*, 1976; Gélinas *et al.*, 1977). In an idealized succession, lower units of a volcanic pile are made up of predominantly basaltic submarine lavas associated with komatiites, cherts and turbidites. While they contain a significant proportion of basalt (commonly rich in Al_2O_3), overlying units also contain more intermediate to felsic rocks, together with abundant pyroclastic and epiclastic material. A primary division of Archean basaltic rocks can be recognized which separates "primitive" basalts, whose compositions indicate that they were derived from unmodified mantle, from geochemically evolved basaltic rocks which are members of suites dominated by intermediate to acid rocks for which multistage petrogenetic models have been proposed (see later discussion). In this section, the primitive basalts are divided into magnesian basalt (MgO 12–8.5%), basalt (MgO $<$ 8.5%, Σ FeO $<$ 12%), and Fe-rich basalt (ΣFeO $>$ 12%). The geochemically evolved basalts, which occur in association with and are chemically transitional into the andesite suites, are referred to here as Al-rich basalts. A fifth, and comparatively rare, Archean basaltic rock type comprises rocks with high alkali contents and in which there is good mineralogical evidence that the high alkali contents are primary magmatic features; these rocks are referred to as alkali-rich basalts.

Magnesian basalt. As defined here, magnesian basalt contains between 12 and 8.5% MgO, and thus includes rocks which have been referred to by other workers as high-Mg basalts (e.g., Hallberg *et al.*, 1976a). Magnesian basalts are essentially clinopyroxene-plagioclase basalts containing accessory oxides with or without olivine. The fabric is usually equigranular, and olivine, which occurs in some samples, is the only phase reported as phenocrysts. Their mineralogy and chemical compositions are intermediate between those of basaltic komatiite and the basalt described below. Magnesian basalt is found both as a component of komatiite associations and in Archean greenstone terrains spatially separate from komatiites. Although discussed here as a group, magnesian basalts probably originate in several different ways (see following discussion). However, we regard magnesian basalt as an important rock type, many occurrences of which represent primary magma from which other Archean basalt types may have been derived.

Magnesian basalts are characteristically low in alkalies, TiO_2 and Al_2O_3, with Ca/Al ratios < 1 and high Ni ($>$ 100 ppm) and Cr ($>$ 500 ppm) abundances. Chondrite-normalized REE abundances lie in the range 5 to 15 and patterns are typically flat or show LREE depletion.

Basalt. Archean greenstone belts typically include a wide range of basaltic volcanic rocks which are essentially intermediate between magnesian basalts and the Fe-rich basalts described below. Such basalt is found in a variety of rock associations, for example, interbedded with magnesian basalt and Fe-rich basalt in the Abitibi greenstone belt, and with acid tuff and rhyolite as well as with pods and stringers of dunite and peridotite in Western Australia (Hallberg *et al.*, 1976a). Most basaltic rocks making up Archean greenstone terrains are of this type.

Unaltered basalts of this type are made up of clinopyroxene and plagioclase with accessory opaque oxides with or without rare olivine (almost invariably altered) or quartz. As with magnesian basalts, fabrics are equigranular, with olivine phenocrysts observed in only occasional samples. Their compositions are essentially basaltic ($SiO_2 < 53\%$) with MgO = 8.5–4%, ΣFeO $<$ 12%, Ca/Al ratios < 1, low total alkali contents (typically < 3 in little-altered specimens) and TiO_2 contents in the range 0.5–1.5%. Refractory lithophile and large-ion-lithophile element abundances are typically higher than in the more Mg-rich basalts described earlier. Chondrite-normalized REE abundances lie within the range 10–20 and patterns are typically flat or show a small degree of LREE enrichment ($La_N/Sm_N < 1.5$).

Fe-rich basalts. Fe-rich basalts, defined by Pearce and Birkett (1974) as basaltic rocks in which ΣFeO (anhydrous basis) exceeds 12%, are important components of some Archean volcanic sequences (e.g., Goodwin, 1977; Pearce and Birkett, 1974; Naldrett and Turner, 1977) and have been described as minor components of others, including komatiite associations (Arndt *et al.*, 1977). In this study, it has become clear that Fe-rich basalts associated closely with komatiites in Munro Township differ in several important respects from those which occur much higher in the stratigraphic sequence. Fabrics are equigranular and phenocrysts are absent. The primary minerals in these Fe-rich basalts are essentially plagioclase, clinopyroxene and opaque oxides with or without accessory quartz. A distinctive feature of many of these rocks is the abundance of oxides which occur either as ubiquitous small equant grains or as feathery skeletal crystals.

Apart from high FeO contents (12–20%), Fe-rich basalts typically show high TiO_2 (1.5–4%), P_2O_5, Zr and alkalies relative to other basaltic rocks. However, TiO_2 contents of two Fe-rich basalt flows described from the Abitibi greenstone belt by Pearce and Birkett (1974) are comparatively low (0.7–1%). REE abundance patterns typically are flat or slightly LREE-depleted, although LREE-enriched Fe-rich basalts have been described in association with komatiites by Arth *et al.* (1977). There is an overall increase in chondrite-normalized REE abundance within the range 15–40, which is correlated with increase in the FeO/MgO ratio. Coupled with this is the development of a small positive Eu anomaly.

Al-rich basalts. Archean volcanic suites, which are dominated by rocks that are intermediate to acid in terms of SiO_2 content, have been reported from Western Australia (Hallberg *et al.*, 1976b); southern Africa (Davis and Condie, 1977; Hawkesworth and O'Nions, 1977) and Superior Province, Canada (Goodwin, 1977). Basaltic rocks which form minor components of most of these suites are referred to as Al-rich basalts in recognition of the fact that Al_2O_3 contents typically (although not always) are greater than 15%; the associated rocks are andesitic with 16–20% Al_2O_3.

Petrographic data for Al-rich basalts are not widely available, possibly because of the pervasive alteration which is so common among Archean volcanic rocks. Unaltered Al-rich basalts from the Abitibi belt are porphyritic rocks containing phenocrysts of clinopyroxene and plagioclase in a groundmass of feldspar and pyroxene. This mineralogy appears to be characteristic of basaltic and andesitic members of evolved Archean volcanic suites (Hallberg *et al.*, 1976a), although more commonly such rocks are altered (Hawkesworth and O'Nions, 1977).

Al-rich Archean basalts commonly (but not invariably) have Al_2O_3 contents in the range 15–22%, have variable to high CaO (7–11%), ΣFeO in the range 4–10%, and moderate to low TiO_2 (0.9–1.4) and MgO (4–7%). Total alkali contents may be high (up to 7%), but because of the problem of secondary alteration, these contents may not have any primary significance. Large-ion-lithophile element and total REE contents are comparatively high. Chondrite-normalized REE patterns typically show pronounced light-heavy fractionation ($La_N/Yb_N > 3$) and low-heavy REE abundances.

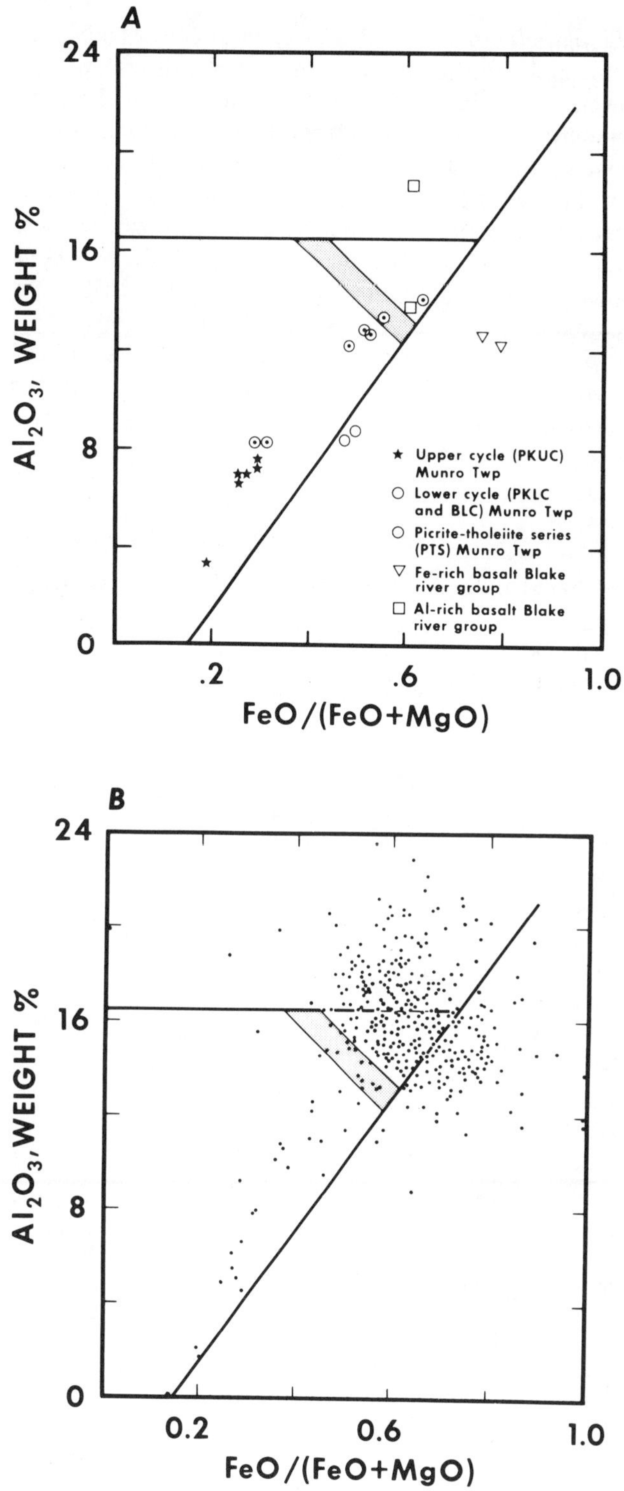

1.2.1.6 Al_2O_3 vs. FeO/(FeO + MgO) plots, where FeO represents total iron. **a.** Rocks of the reference suite analyzed for this study. **b.** Approximately 800 analyses from the Blake River group and immediately underlying rocks, Abitibi greenstone belt, Ontario (Naldrett and Goodwin, 1977). See text for explanation.

Alkali-rich basalts. Alkali basalts are not widely documented from Archean greenstone terrains. Reported occurrences of rocks termed hawaiite or mugearite in Superior Province greenstone belts (Goodwin, 1977) essentially are artifacts of alkali-element mobility and of the limitations of normative classification schemes. However, there is one group of Archean volcanic rocks, represented by the Timiskaming Group in the Kirkland Lake area of the Abitibi greenstone belt described by Cooke and Moorhouse (1969), which have mineralogies and chemical compositions consistent with crystallization from an alkali-rich magma. The Timiskaming Group consists of lava flows interbedded with conglomerate, sandstone and argillite; volcanic rock types include a variety of compositions described as basalt, leucite trachyte and mafic trachyte.

Timiskaming Group basaltic rocks consist of augite, biotite, plagioclase, leucite and their alteration products with less common K-feldspar and hornblende and accessory magnetite and apatite (Cooke and Moorhouse, 1969). Chemical data for these rocks are restricted to major element analyses presented by Cooke and Moorhouse (1969). The basaltic rocks of the suite (which includes compositions with up to 58% SiO_2) are characterized by high alkali contents (up to 11%), K_2O/Na_2O ratios greater than 1, relatively low MgO, CaO and TiO_2 and high ΣFeO and Fe_2O_3/FeO ratios; Al_2O_3 contents are high and variable (up to 24%).

Table 1.2.1.4 Major and trace element data for the Archean Basalt Reference Suite.

wt. %	Method	ACH-1	ACH-2	ACH-3	ACH-4	ACH-5	ACH-6	ACH-7	ACH-8	ACH-9	ACH-10
SiO_2	1	42.64	43.25	39.68	42.81	42.22	42.83	50.45	48.37	48.13	49.06
TiO_2	1	0.31	0.34	0.14	0.33	0.30	0.31	0.61	0.61	0.60	1.95
Al_2O_3	1	6.50	7.23	3.09	7.02	6.40	6.61	12.14	12.86	12.21	12.20
Fe_2O_3	1	2.80	1.40	4.38	1.30	2.78	2.69	0.47	1.01	0.64	3.47
FeO		6.75	8.77	4.05	8.93	6.94	7.35	10.27	10.03	10.61	12.00
MnO	1	0.15	0.18	0.13	0.17	0.15	0.18	0.18	0.20	0.20	0.25
MgO	1	25.74	24.06	33.89	24.28	26.35	26.08	9.38	8.96	10.07	4.92
CaO	1	6.14	6.92	1.75	6.67	5.94	6.31	10.09	10.68	10.23	6.40
Na_2O		0.28	0.77	0.04	0.65	0.52	0.49	2.27	2.28	2.28	3.75
K_2O	1	0.07	0.13	0.01	0.13	0.09	0.15	0.02	0.05	0.03	0.70
P_2O_5	1	0.03	0.03	0.02	0.04	0.02	0.02	0.04	0.07	0.06	0.15
H_2O^+		8.13	5.36	11.34	5.16	6.95	7.36	3.07	3.79	4.56	3.75
H_2O^-											
CO_2		0.13	0.14	0.17	0.12	0.14	0.11	0.22	0.19	0.09	0.06
Total		99.67	98.58	98.69	97.59	98.80	100.49	99.21	99.22	99.71	98.66
Mg'†		0.846	0.826	0.894	0.826	0.847	0.841	0.635	0.619	0.641	0.392
ppm											
Cs	2	0.5	0.29	0.10	0.24	0.21	0.4	0.4		0.04	0.90
Rb	1	2.5	2.0	0.4	3.2	1.9	3.0	0.8	1.2	0.9	11
Ba	2										
Sr	1	12	5	4.7	35	25	33	19	76	45	62
Th	2			0.13	0.09	0.11	0.11	0.11	0.08	0.15	0.34
Hf	2	0.47	0.51	0.21	0.49	0.48	0.43	1.06	1.05	1.05	2.83
Zr	1										
Y	1	7.4	8.4	4.4	7.9	7.3	7.0	14	14	14	34
La	2	0.37	0.37	0.16	0.36	0.32	0.29	1.17	1.38	1.17	5.77
Ce	2	1.40	1.17	0.50	1.35		1.1	3.43	3.89	3.47	15.5
Sm	2	0.63	0.62	0.38	0.61	0.58	0.53	1.39	1.46	1.35	3.91
Eu	2	0.23	0.23	0.11	0.26	0.24	0.21	0.40	0.53	0.45	1.27
Tb	2	0.17	0.19	0.10	0.16	0.15	0.16	0.35	0.37	0.36	0.96
Yb	2	0.90	0.90	0.38	0.85	0.80	0.75	1.51	1.61	1.48	3.95
Lu	2	0.14	0.15	0.061	0.15	0.11	0.12	0.28	0.25	0.23	0.59
Sc	2	24.7	25.3	11.9	24.6	23.1	21.1	41.0	43.1	41.8	41.0
Cr	2	2640	2600	1690	2600	2530	2370	1030	500	750	37
Ni	2	1290	1150	2130	1170	1380	1270	280	105	150	—
Pb	1	2	3	1	N.D.	4	3	4	5	3	6
Ga	1	8	8	5	8	8	9	14	15	15	20

†Fe_2O_3/FeO set to 0.1.
Note: Method 1 – XRF, Method 2 – INAA.
N.D. = Not Determined.

1.2.1.3 Major element chemistry

Analyses of the sample suite collected for this study, in which most of the rock types discussed above are represented, are given in Table 1.2.1.4. These are augmented by representative major element analyses from the literature in Tables 1.2.1.5a and b. The compositional ranges of Archean komatiites and basalts are illustrated in Figs. 1.2.1.6–8.

$FeO/(FeO + MgO)$ vs. Al_2O_3 variation

Figure 1.2.1.6a illustrates the variation of $FeO/(FeO + MgO)$ with Al_2O_3 for all analyses presented in this study, whereas Fig. 1.2.1.6b illustrates the same data for approximately 800 basalts from the Blake River Group, Abitibi belt, Ontario, Canada (Naldrett and Goodwin, 1977). These authors have shown that what at first sight appears to be a shot-gun scatter resolves into a diagram which is useful for discriminating certain major basaltic types. For the Blake River succession Fe-rich basalts characteristic of one stratigraphic interval are found to plot to the right of the sloping line, whereas Al-rich basalts characteristic of another interval mostly plot above the 16.5 wt.% Al_2O_3 line. Within these stratigraphic intervals, rocks with transitional chemistry are present but are relatively uncommon. Peridotitic and basaltic komatiites are responsible for the string of points above the line trending down towards low values of $FeO/(FeO + MgO)$.

Table 1.2.1.4 *(Continued)*

wt %	Method	ACH-11	ACH-12	ACH-13	ACH-14	ACH-15	ACH-16	ACH-17	ACH-18	ACH-19
SiO_2	1	48.96	50.21	47.04	46.38	50.26	43.20	43.36	46.70	50.28
TiO_2	1	0.76	0.59	1.00	0.97	0.70	0.43	0.43	2.56	1.11
Al_2O_3	1	13.32	11.70	8.24	8.00	13.03	7.72	7.51	11.89	17.83
Fe_2O_3	1	1.43	0.93	1.50	1.24	2.98	2.37	2.22	5.49	0.87
FeO		10.63	9.97	11.70	11.82	5.28	8.16	7.67	11.68	8.12
MnO	1	0.21	0.18	0.28	0.26	0.17	0.19	0.17	0.25	0.20
MgO	1	7.03	11.53	13.38	14.19	5.12	22.45	23.16	4.42	5.60
CaO	1	8.66	9.21	10.43	9.26	16.04	6.57	6.33	8.26	4.34
Na_2O		3.14	1.46	1.30	1.09	0.23	0.88	0.68	3.43	5.02
K_2O	1	0.24	0.16	0.10	0.07	0.13	0.10	0.09	0.09	0.56
P_2O_5	1	0.06	0.04	0.08	0.08	0.32	0.06	0.04	0.23	0.19
H_2O^+		3.84	2.77	3.20	4.48	4.41	5.70	7.63	2.90	4.73
H_2O^-										
CO_2		0.09	0.30	0.16	0.23	0.57	0.14	0.07	0.06	0.23
Total		99.37	99.05	98.41	98.07	99.24	97.97	99.36	97.96	99.68
Mg'†		0.539	0.678	0.670	0.685	0.560	0.812	0.826	0.345	0.555
ppm										
Cs	2	0.14	0.15	0.07			0.15	0.09	0.34	0.86
Rb	1	6.1	3.7	0.8	0.7	1.5	2.0	1.3	2.9	10
Ba	2	60								
Sr	1	88	43	69	89	262	16	12	133	69
Th	2	0.13		0.36	0.15	1.89			0.47	1.40
Hf	2	1.46	0.96	1.91	1.74	2.68	0.78	0.83	3.14	3.29
Zr	1					85				103
Y	1	18	13	17	16	19	11	10	32	19
La	2	2.10	1.31	4.18	3.42	18.0	1.19	1.20	5.24	12.9
Ce	2	5.83	4.05	12.7	10.3	41.0	3.64	3.70	15.5	29.0
Sm	2	1.91	1.34	2.99	2.62	4.27	1.03	1.04	3.91	3.21
Eu	2	0.63	0.49	0.99	0.86	1.25	0.34	0.32	1.55	1.14
Tb	2	0.46	0.40	0.62	0.58	0.46	0.32	0.30	0.97	0.57
Yb	2	1.97	1.50	1.80	1.56	2.08	1.10	1.22	3.68	1.90
Lu	2	0.33	0.26	0.26	0.23	0.31	0.19	0.18	0.59	0.31
Sc	2	41.9	39.9	31.6	29.1	24.0	25.4	24.7	47.4	24.0
Cr	2	106	1180	1460	1320	317	2470	2450	14	129
Ni	2	60	330	580	530	109	1180	1250	—	90
Pb	1	4	N.D.	5	5	6	4	4	6	4
Ga	1	16	14	13	15	19	10	10	22	17

Table 1.2.1.5a Selected major element analyses of komatiitic rocks.[(a)]

	1	2	3	4	5	6	7	8	9	10	11	12
SiO_2	45.8	45.6	42.8	45.0	45.2	49.0	50.4	49.5	48.6	52.2	53.1	50.7
Al_2O_3	7.30	7.35	4.52	5.84	3.66	9.73	11.3	12.5	12.2	12.3	14.4	14.4
TiO_2	0.30	0.27	0.22	0.28	0.20	0.52	0.65	0.45	0.66	0.59	0.83	0.73
FeO[(b)]	11.2	11.7	11.5	9.65	11.0	11.4	9.91	10.7	11.1	11.6	10.9	12.4
MnO	0.21	0.21	0.24	0.18	0.22	0.24	0.19	0.21	0.24	0.18	0.18	0.23
MgO	26.1	25.9	33.6	31.9	32.2	17.8	14.4	13.2	12.2	10.6	7.77	6.92
CaO	7.64	7.62	5.73	5.68	5.28	8.71	10.5	10.4	11.5	11.5	11.2	13.5
Na_2O	0.69	0.83	0.29	0.52	0.44	1.18	2.34	2.01	1.70	0.58	1.83	1.77
K_2O	0.10	0.06	0.04	0.06	0.17	0.06	0.00	0.01	0.17	0.29	0.02	0.37
Cr_2O_3	0.24	—	0.34	—	—	0.31	0.23	—	—	0.02	0.02	—
Ni	—	0.1336	0.1930	—	—	0.0582	0.0645	—	—	0.0094	—	—
P_2O_5	—	—	—	0.02	0.02	—	—	—	—	—	—	0.08
Total	99.58	99.67	99.47	99.13	98.39	99.01	99.98	98.98	98.37	99.87	100.25	101.10
Mg′	0.822	0.814	0.853	0.868	0.853	0.756	0.742	0.710	0.685	0.644	0.586	0.665

(a) Recalculated to eliminate volatiles or loss on ignition. (b) Total Fe calculated as FeO. **1.** Average of spinifex zone through flow in Munro Township (Arndt *et al.,* 1977), Table 6. **2.** SA 2048 of spinifex-textured sample, Munro Township (Arndt *et al.,* 1977), Table 7, No. 4. **3.** Olivine spinifex-textured sample, Yakabindie area (Naldrett and Turner, 1977), Table 1, No. 106. **4.** Olivine spinifex-textured sample, Scotia mine (Sun and Nesbitt, 1977), Table 1A No. 331/307. **5.** Olivine spinifex-textured sample, Barberton (Sun and Nesbitt, 1977), Table 1A, No. 49J. **6.** Basaltic komatiite, Yakabindie (Naldrett and Turner, 1977), Table 1, No. 55. **7.** Olivine-rich basaltic komatiite, Munro Township (Arndt *et al.,* 1977), Table 7, No. P9-132. **8.** Pyroxene-rich basaltic komatiite, Munro Township (Arndt *et al.,* 1977), Table 7, No. P9-185. **9.** Basaltic komatiite. **10.** Magnesian basalt, Munro Township (Arndt *et al.,* 1977), Table 7, No. P9-108. **11.** Basalt, Munro Township (Arth *et al.,* 1977), Table 1, No. 131. **12.** Basalt, Scotia mine (Sun and Nesbitt, 1977), Table 1A, No. 830/450.

Table 1.2.1.5b Selected major element analyses representative of Archean Basalts.

	1	2	3	4	5	6	7	8	9	10	11	12	13	14
SiO_2	49.29	51.17	51.33	47.65	51.5	49.76	47.10	50.45	50.97	47.64	47.18	48.55	46.54	50.86
TiO_2	0.65	0.64	0.62	0.64	1.18	0.70	0.93	1.47	2.58	0.87	0.49	0.60	0.55	1.07
Al_2O_3	13.45	11.01	12.02	14.69	13.99	14.30	14.45	14.80	12.29	16.46	13.04	18.02	15.24	21.02
Fe_2O_3	—	2.53	3.16	2.13	1.94	—	2.23	—	—	3.12	1.70	1.74	5.57	1.43
FeO	12.33†	8.05	7.58	9.15	10.10	9.43†	12.66	15.18†	17.70	7.17	4.40	6.89	3.31	7.01
MnO	0.21	0.16	0.18	0.18	0.18	0.24	0.22	0.25	0.29	0.20	0.10	0.13	0.16	0.17
MgO	11.58	11.49	10.51	8.38	6.85	6.03	8.15	4.93	3.44	6.36	8.63	7.12	3.81	1.09
CaO	8.58	8.50	8.27	11.90	10.98	15.43	13.43	8.50	7.64	13.08	7.74	10.55	7.14	2.56
Na_2O	2.00	1.72	1.92	1.42	1.82	2.52	2.00	2.58	2.31	2.15	3.42	1.61	3.71	0.65
K_2O	0.44	0.04	0.62	0.06	0.19	0.17	0.23	0.21	0.30	0.15	0.17	0.51	5.38	9.64
P_2O_5	0.10	0.11	0.12	0.06	0.05	0.07	0.19	—	—	0.42	0.22	0.95	0.35	0.37
Loss	1.36	3.95	3.54	2.16	1.77	4.68	—	—	—	—	12.80	3.5	7.80	3.84
Total	99.99	99.37	99.87	98.42	100.36	103.33	101.59	98.37	97.52	97.62	99.89	100.17	99.56	100.31

†Total Fe calculated as FeO.

1. Amphibolitic metavolcanic (magnesian basalt), Lawlers greenstone belt, W. Australia, Sun and Nesbitt (1978), Table 1A. No. 10. **2.** Magnesian basalt, Meekatharra, W. Australia, Hallberg, *et al.* (1976a), Table 1, No. 8. **3.** Magnesian basalt, Meekatharra, W. Australia, Hallberg, *et al.* (1976a), Table 1, No. 9. **4.** Archean basalt, southern Africa, Hawkesworth and O'Nions (1977), Table 2A, No. 5. **5.** Archean basalt, southern Africa, Hawkesworth and O'Nions (1977), Table 2A, No, 10. **6.** Archean basalt, Scotia, W. Australia, Sun and Nesbitt (1978), Table 1A, No. 20. **7.** Basalt sample iron-rich flow, Abitibi belt, Canada, Pearce and Birkett (1974), Table 1, No. 1. **8.** Fe-rich basalt, Lawlers, W. Australia, Naldrett and Turner (1977), Table 1, No. 520. **9.** Fe-rich basalt, Lawlers, W. Australia, Naldrett and Turner (1977), Table 1, No. 513. **10.** Al-rich basalt, Abitibi belt, Canada. Unpublished data (no. S25). **11.** Al-rich basalt, southern Africa, Hawkesworth and O'Nions (1977), Table 2A, No. 3. **12.** Al-rich basalt, southern Africa, Hawkesworth and O'Nions (1977), Table 2A, No. 7. **13.** Leucite tephrite, Timiskaming Group, Canada, Cooke and Moorehouse (1969), Table 1, No. 1. **14.** Leucite trachyte, Timiskaming Group, Canada, Cooke and Moorehouse (1969), Table 1, No. 3.

Basalts not characterized by these extremes of chemistry plot within the triangle bounded by the two lines and the stippled zone. Arndt *et al.* (1977) and Naldrett and Turner (1977) have used the same diagram to illustrate the difference between closely-associated Fe-rich basaltic and komatiitic rocks in Ontario and Western Australia, respectively. Komatiites lie above the sloping line, Fe-rich basalts below. Arndt *et al.* (1977) point out that the distinction applies to cumulates as well as to liquids. It should be stressed that not all rocks plotting above the line are komatiitic. Many basalts, indeed most ocean ridge basalts, lie above the line. However, the plot is particularly useful to distinguish komatiites from suites of Mg-rich rocks which are not believed to be komatiitic but with which komatiites are often associated. Some komatiites, apparently having Al_2O_3-depleted compositions (Nesbitt and Sun, 1976), plot below the line, but these are comparatively rare. It is seen from Fig. 1.2.1.6a that the suite collected for this study fulfills the criteria outlined in earlier studies. One of the most important points illustrated by this diagram is that all Archean mafic and ultramafic volcanics belong to a continuum. Certain members are very common and others are missing, but the continuum becomes apparent when many samples are studied.

FeO–MgO–(Na_2O + K_2O). The FMA diagram should be treated with caution in dealing with Archean rocks because of the possibility of alkali element mobility. In Fig. 1.2.1.7 peridotitic and pyroxenitic komatiites, magnesian basalts, basalts and iron-rich basalts form a trend on the right-hand side of the diagram reflecting variations in FeO/MgO within an essentially low alkali suite. High-Al basalts that contain more alkalies and have FeO/MgO ratios around 1 plot further away from the Fe-Mg side than other basalts and, when grouped with the Al-rich andesites with which they are closely associated in the field, define a trend across the center of the diagram. The alkali-rich Timiskaming basaltic rocks with FeO/MgO ratios close to 1 and high K_2O contents plot toward the left-hand side of the diagram.

TiO_2 and MnO. TiO_2 increases with declining Mg′ (Fig. 1.2.1.8a). Komatiites are characterized by values ranging from about 0.3 wt.% in the peridotitic variants to 0.7–0.9 wt.% in basaltic members of this suite. An upper limit of 1 wt.% has been used as a discriminant for basalts related to this suite by Naldrett and Cabri (1976). Fe-rich basalts are comparatively rich in TiO_2 (~ 1.5 wt.%). The picrites from Munro Township are rich in TiO_2 (~ 1 wt.%) in comparison with komatiites of the same Mg′ (~ 1.6 wt.%). A less marked, but nevertheless distinct difference, is apparent between peridotitic komatiites of the upper cycle and lower cycle, with the former containing slightly less TiO_2 for the same Mg′. MnO (Fig. 1.2.1.8b) shows little variation with Mg′, being between 0.17 and 0.20 wt.% in most komatiites, magnesian basalts, basalts and Al-rich basalts but being somewhat higher (0.2–0.26) in Fe-rich basalts.

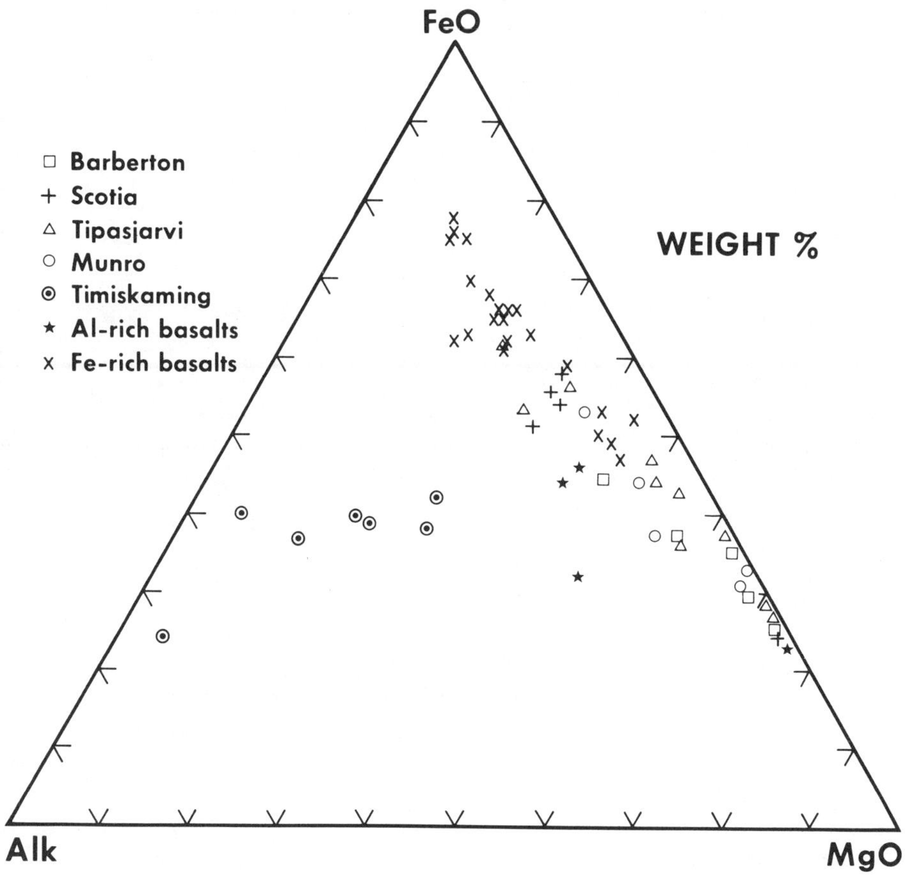

1.2.1.7 Diagram illustrating variation in FeO (total iron), MgO and Na_2O + K_2O. If andesites closely associated with the Al-rich basalts had also been plotted on the diagram, they would have defined a trend very similar to that of the Timiskaming, alkalic basalts.

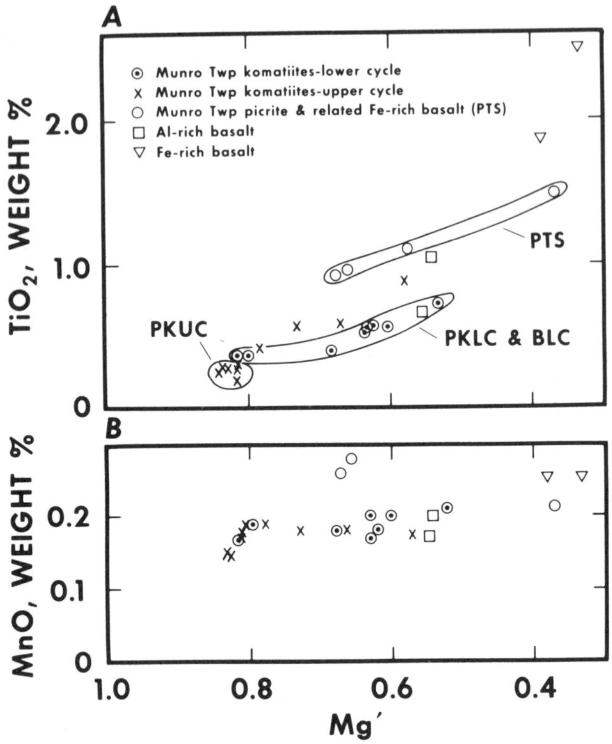

1.2.1.8 **a.** TiO_2 variation in Munro Township and Blake River group rocks expressed as a function of Mg′. Note that rocks of the picrite-tholeiite series, the komatiite lower cycle and the komatiite upper cycle have progressively lower concentrations of TiO_2 for a given Mg′, attributed to their derivation from progressively more depleted mantle. The picrite-tholeiite series and the komatiite lower cycle both show a trend of increasing TiO_2 with decreasing Mg′, attributed to progressive differentiation within the series or cycle. **b.** MnO variation with Mg′ for the same rocks as in **a.**

1.2.1.4 Variations in some trace elements

In discussing trace element data for Archean basaltic rocks, the main problem is uneven coverage. The following discussions are based on data from our reference suite and from Nesbitt and Sun (1976), Hermann *et al.* (1976), Hawkesworth and O'Nions (1977), Naldrett and Turner (1977), Sun and Nesbitt (1978) and Hallberg *et al.* (1976b). In particular, high-Al basalts are poorly represented. Data from the reference suite are given in Table 1.2.1.4 and representative analyses from the literature are presented in Table 1.2.1.6.

Ni, Co, Cr and Sc

The variation of Ni with Mg′ in rocks believed to represent liquid compositions from a variety of locations is shown in Fig. 1.2.1.9. The Ni contents decrease from values of 1500–1700 ppm in rocks with Mg′ >0.85 to less than 150 ppm in those with Mg′ < 0.6. Figure 1.2.1.9a illustrates the behavior of Ni in komatiitic liquids, showing that values cluster within 150 ppm of the mean for any given Mg′. Values obtained in this study for Munro Township agree well with values for similar rocks from throughout the world. The picritic (nonkomatiitic) rocks underlying the komatiitic sequences in Munro Township stand out in Fig. 1.2.1.9a as being anomalously rich in Ni. Figure 1.2.1.9b illustrates behavior over the basaltic range. Given the scatter of the data in this figure, the Fe-rich and Al-rich basalts do not have unusual Ni values compared with basalts which have similar Mg′ and which are associated with

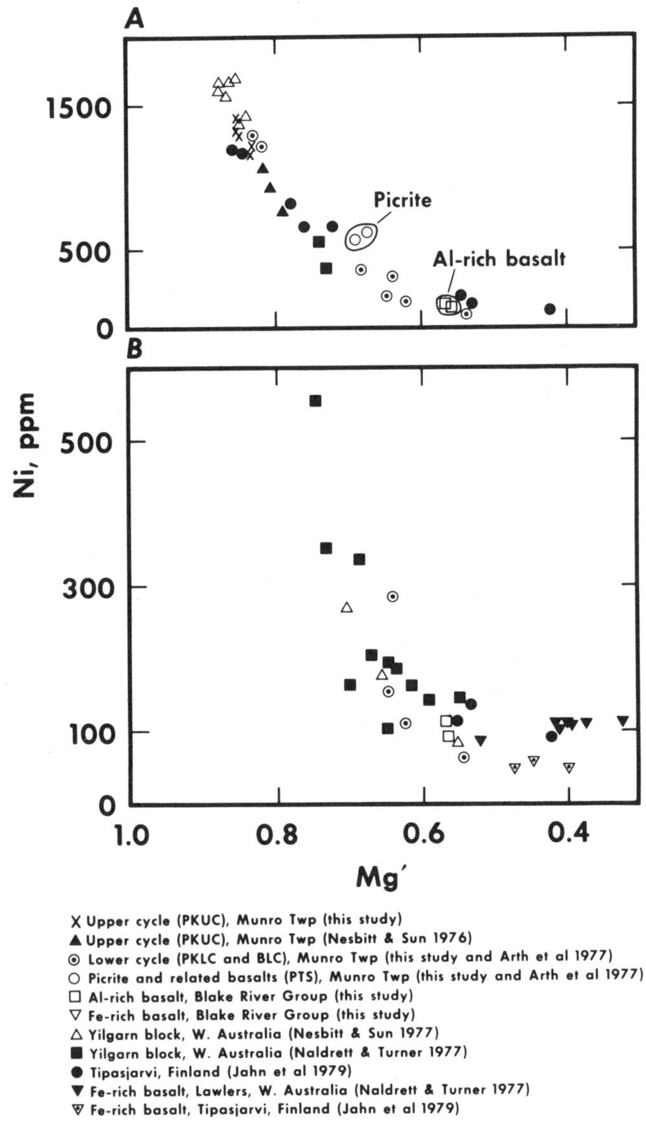

1.2.1.9 Variation of Ni contents with Mg′ for Archean komatiites (A) and basalts (B) from different localities.

komatiites. No cobalt data were obtained for our reference suite, but data drawn from the literature (Fig. 1.2.1.10) indicate that cobalt content varies from just over 100 ppm in rocks with high Mg′ to 60–80 ppm in those with low Mg′. No data are available for the Al-rich basalts, but no other rock types stand out as being anomalous. Chromium data (Fig. 1.2.1.11a) for rocks with high Mg′ exhibit a marked scatter, with the values obtained for our Munro Township reference suite being approximately 600 ppm lower than those reported for similar rocks obtained from the same localities by Nesbitt and Sun (1976). We suspect a problem of analysis here. For Mg′ < 0.8, Cr values decrease sharply with decreasing Mg′. It can be seen from Fig. 1.2.1.11b that Al-rich basalts have Cr contents which are very similar to those of other basalts having the same Mg′. Fe-rich basalts are low in Cr (generally < 100 ppm) but, when viewed in the light of their Mg′ and the scatter in the data, they cannot be said to be anomalously low; rather, they fit on the extension of any curve fitted to the remainder of the data.

Scandium (Fig. 1.2.1.12) increases from 20–25 ppm in peridotitic komatiites to 40–45 ppm in magnesian basalts and basalts. The Fe tholeiites which overlie the Munro series have Sc values which are similar to those of most other basalts. However, the picrites from Munro Township are lower in Sc than basaltic komatiites of similar Mg′ and, when viewed along with the

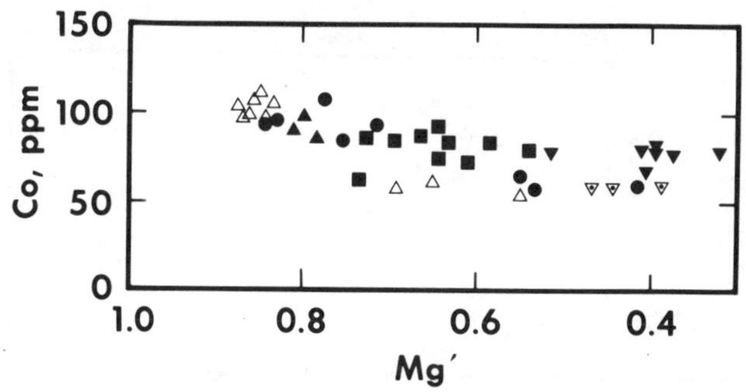

1.2.1.10 Variation of Co with Mg′ for Archean komatiites and basalts from different localities. See Fig. 1.2.1.9 for key.

related Fe-enriched basalts from Munro, display a trend of sharply decreasing Sc with decreasing Mg′. The two Al-rich basalts analyzed for this study have < 25 ppm Sc, approximately half that expected for normal basalts of the same Mg′.

Sr, Rb, Ba and Cs

Variations in Sr, Ba and Cs are illustrated in Fig. 1.2.1.13. Rb is not illustrated because the data are highly scattered (Arth *et al.*, 1977), probably as a result of alteration. Cs also is highly variable, so we place little confidence in the possibility that these data are representative of the original rock compositions. Sr and Ba

Table 1.2.1.6 Rare-earth element abundances for Archean basaltic rocks (in ppm).

		Tipasjarvi, E. Finland (Jahn *et al.*, 1979)								Scotia, W. Australia					
		S847	S831	S828	S818	S833	S829	S830	S824	SD4/193	331/307	SD13/764	1170/20	830/450	830/420
La	0.315	0.340	0.895	2.40	2.17	1.06	1.518	1.887	2.705	0.520	0.626	1.53	3.01	2.66	2.18
Ce	0.813	1.63	3.02	7.36	7.16	2.864	4.31	5.33	7.26	1.54	1.80	4.44	8.99	6.60	5.63
Nd	0.597	1.96	2.56	6.15	6.17	2.326	3.67	4.41	5.37	1.34	1.61	3.55	7.60	4.81	4.20
Sm	0.192	0.762	1.05	2.15	2.22	0.762	1.245	1.470	1.800	0.480	0.586	1.16	2.49	1.52	1.34
Eu	0.0722	0.262	0.386	0.788	0.861	0.263	0.440	0.499	0.656	0.211	0.202	0.403	0.853	0.594	0.501
Cd	0.259	1.08	1.61	2.95	3.02	1.015	—	1.89	2.54	0.692	0.847	1.57	3.45	2.04	1.84
Dy	0.325	1.35	2.06	3.72	3.83	1.253	1.98	2.36	3.24	0.840	1.03	1.84	3.89	2.44	2.16
Er	0.213	0.858	1.27	2.43	2.44	1.811	1.28	1.52	2.20	0.550	0.668	1.13	2.45	1.56	1.37
Yb	0.208	0.861	1.30	2.44	2.40	0.805	1.245	1.435	2.29	0.550	0.677	1.05	2.28	1.47	1.30
Lu	0.0323	0.137	—	0.364	0.367	0.130	0.176	0.201	0.352	—	—	—	—	—	—
MgO (wt.%)	—	27.2	15.9	7.9	6.95	26.7	21.6	15.9	5.39	31.9	31.2	12.2	7.40	6.92	6.03
Mg′	—	0.843	0.751	0.550	0.534	0.831	0.773	0.718	0.418	0.868	0.861	0.688	0.541	0.559	0.562

S847	Meta-peridotitic komatiite (Jahn *et al.*, 1980).	S824	Meta-basalt (Jahn *et al.*, 1980).
S831	Meta-basaltic komatiite (Jahn *et al.*, 1980).	SD4/193	Spinifex-textured peridotitic komatiite (Sun and Nesbitt, 1978).
S828	Meta-basalt (Jahn *et al.*, 1980).		
S818	Meta-basalt (Jahn *et al.*, 1980).	331/307	Ultramafic tuff (Sun and Nesbitt, 1978).
S833	Meta-peridotitic komatiite with micro-spinifex texture (Jahn *et al.*, 1980).	SD13/764	Amphibolitic basalt (Sun and Nesbitt, 1978).
		1170/20	Amphibolitic basalt (Sun and Nesbitt, 1978).
S829	Meta-peridotitic komatiite (Jahn *et al.*, 1980).	830/450	Amphibolitic basalt (Sun and Nesbitt, 1978).
S830	Meta-basaltic komatiite (Jahn *et al.*, 1980).	830/420	Amphibolitic basalt (Sun and Nesbitt, 1978).

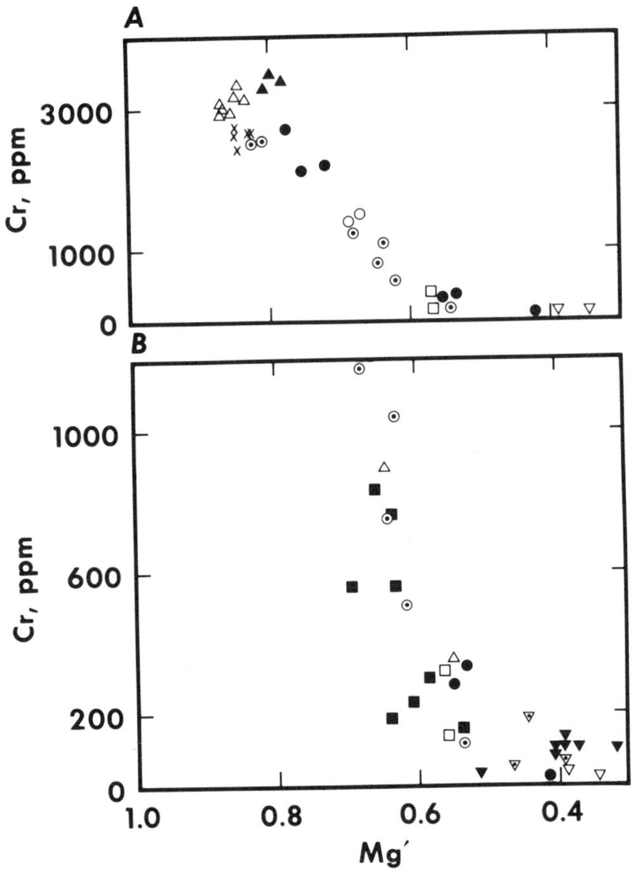

1.2.1.11 Variation of Cr with Mg′ for Archean komatiites (A) and basalts (B) from different localities. See Fig. 1.2.1.9 for key.

increase with decreasing Mg′; however, the Ba values are relatively scattered, which may also be due to alteration.

Data for these elements in basaltic rocks are extremely limited; therefore, we have drawn upon unpublished data from Superior Province in discussing their variations. All of these elements show increases in abundance from magnesian basalt through Archean basalt to iron-rich basalt. The andesitic suite, of which high-Al basalts form a part, typically is higher in Rb, Sr and Ba than the other basalts.

Hf, Zr, Th, Y, Nb and V

Data for these elements, drawn from our reference suite and from the literature (largely Nesbitt and Sun, 1976; Jahn *et al.*, 1979), are summarized in Figs. 1.2.1.14–16. The Hf data (Fig. 1.2.1.14a) are taken solely from this study and, in general, indicate an increase with decreasing Mg′. However, considering first the Munro rocks, the picrites have values which are distinctly higher for a given Mg′ (≈ 1.8 ppm) than values for the lower sequence of komatiites (increasing from ~ 0.8–1.5 ppm) which, in turn, are distinctly higher than values for the upper sequence of komatiites (< 0.5 ppm). The Al-rich and Fe-rich basalts are all richer in Hf, with values of approximately 3 ppm. Similar behavior is exhibited by Y, with the picrites containing 5–10 ppm more Y than komatiites of the lower cycle with the same Mg′ and the peridotitic komatiites of the lower cycle containing 5–8 ppm more Y than those of the upper cycle. The Fe-rich basalts are high in Y, but the Al-rich basalts do not show the same enrichment in Y that they show in Hf.

Zr increases (Fig. 1.2.1.15) steadily with decreasing Mg′ from approximately 15 ppm in peridotitic komatiites to over 40 ppm in basalts. The Al-rich basalts are particularly rich in Zr, having values around 90 ppm.

Data for Th are restricted to those obtained in this study. The komatiites and related basalts from Munro Township contain less than 0.2 ppm Th, which shows little variation with Mg′. The two Munro picrite analyses do not agree well, but one is distinctly higher (0.36 ppm) than the values obtained for the komatiite. The Fe-rich basalts contain approximately 0.4 ppm. The Al-rich basalts are particularly rich in Th, containing 1.4–1.9 ppm. Data on Nb (Fig. 1.2.1.14d) from the literature indicate a progressive increase with decreasing Mg′ within the komatiite-magnesian basalt-basalt association from 0.1–1.5 ppm Nb in peridotitic komatiites to approximately 3.5 ppm Nb in basalt.

Data for V (Fig. 1.2.1.16) are drawn solely from the literature. V shows a progressive increase with decreas-

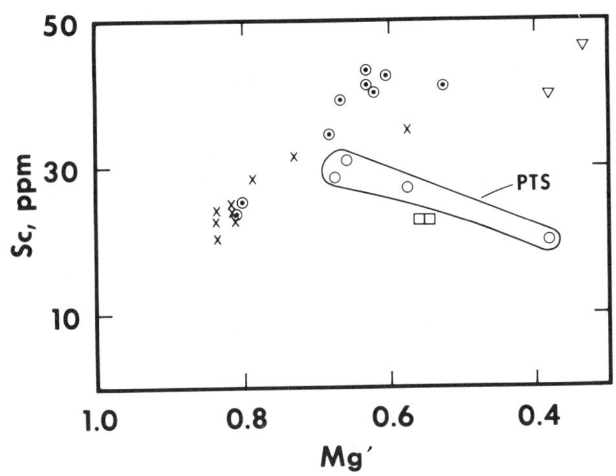

1.2.1.12 Variation of Sc with Mg′ for rocks from Munro Township and the Blake River group. Note the difference in behavior between the komatiitic rocks and those from the picrite-tholeiite series (PTS). See Fig. 1.2.1.8 for key.

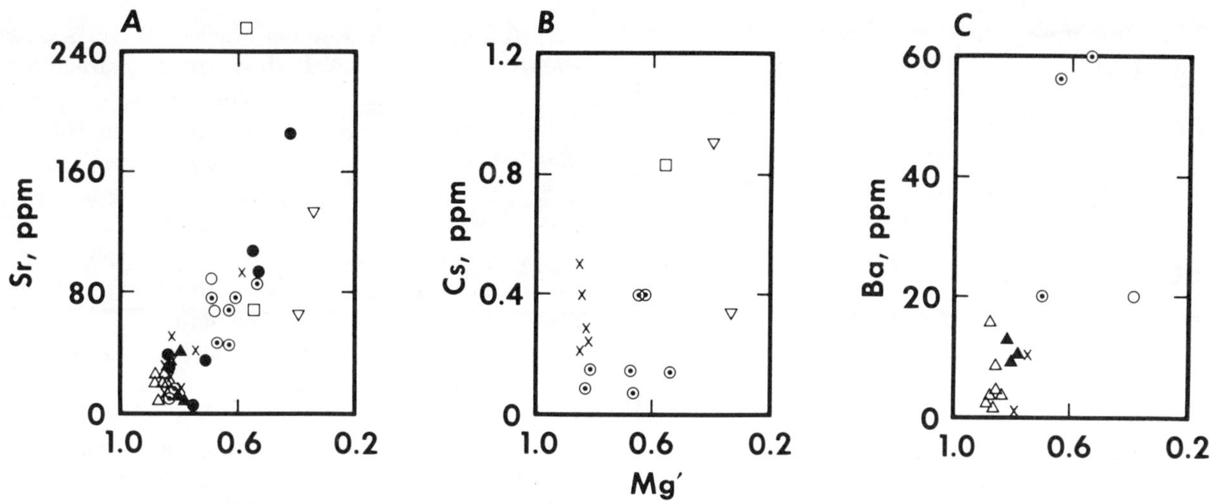

1.2.1.13 Variations of Sr, Cs and Ba with Mg′ for Archean komatiites and basalts from different localities. See Fig. 1.2.1.9 for key.

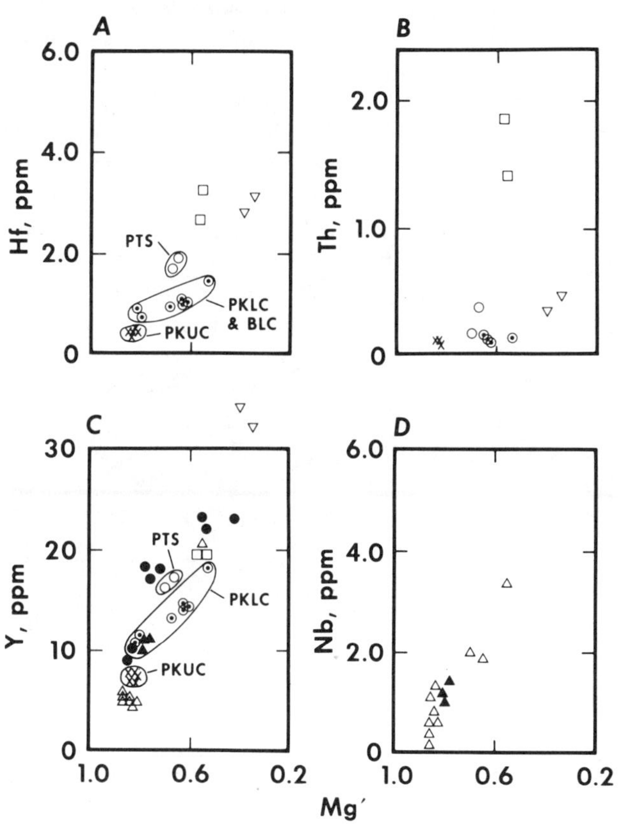

1.2.1.14 Variations of Hf, Th, Y and Nb with Mg′. Note that Hf and Y show behavior similar to TiO_2 in that both decrease for rocks of similar Mg progressively from the picrite tholeiite series (PTS) through the komatiite lower cycle (PKLC and BLC) to the peridotitic komatiites of the upper cycle (PKUC). See Fig. 1.2.1.9 key.

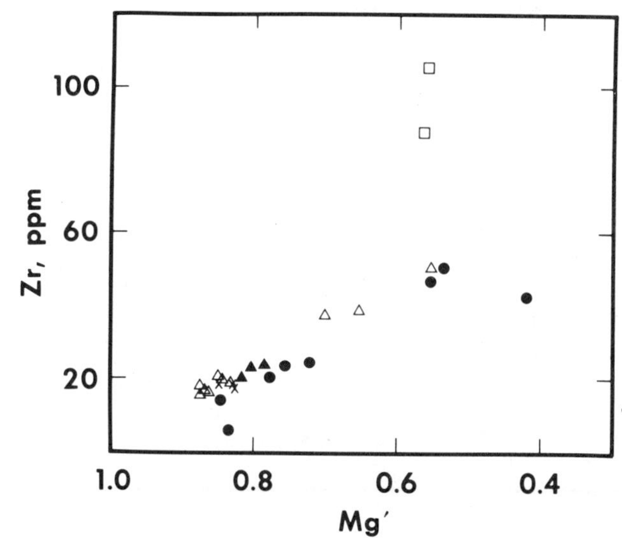

1.2.1.15 Variation of Zr with Mg′ for Archean komatiites and basalts from different localities. See Fig. 1.2.1.9 for key.

ing Mg′ from approximately 15 ppm in the most magnesian peridotitic komatiites to over 300 ppm in basalt. No data are available for Al- or Fe-rich basalts.

Rare-earth elements (REE)

Rare earth patterns for peridotitic and basaltic komatiites, magnesian basalts and basalts analyzed in this study are shown in Fig. 1.2.1.17. These patterns are compared with our data (plus data for two samples from Arth *et al.*, 1977) on the Munro picrite-tholeiite

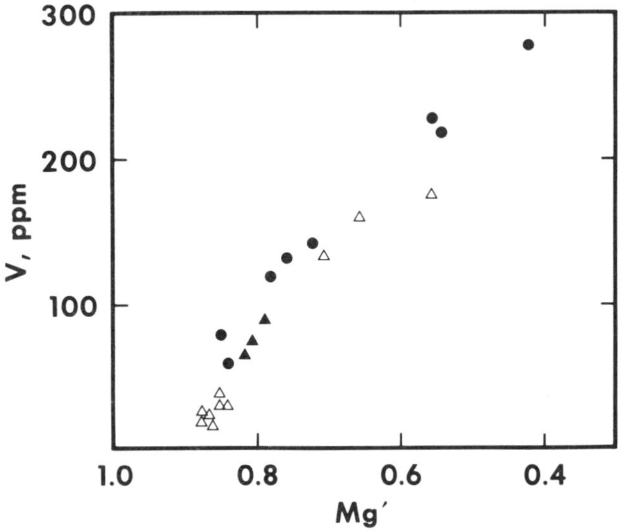

1.2.1.16 Variation of V with Mg′ for Archean komatiites and basalts from different localities. See Fig. 1.2.1.9 for key.

series, and our data on Fe-rich and Al-rich basalts in Fig. 1.2.1.18.

Some of the key points are summarized in Fig. 1.2.1.19 where our data are augmented with those of Arth *et al.* (1977) and Sun and Nesbitt (1978). These key points are:

1. The peridotitic komatiites of the upper cycle of Munro Township (PKUC) are characterized by low overall REE abundances ($Sm_N \cong 3$; where the subscript "N" denotes the chondrite-normalized abundance) and unusually low abundances of Ce_N to Sm_N (about 0.5). The Sm_N and Yb_N abundances for these rocks are distinctly lower than for other rocks studied.

2. The basaltic rocks of the upper cycle do not show the same characteristics as the peridotitic komatiites.

3. All rocks of the lower komatiitic cycle (PKLC and BLC) in Munro Township show similar characteristics in terms of element ratios, with Ce_N to Sm_N ratios mostly varying between 0.6–0.8 and Sm_N to Yb_N ratios mostly between 0.9–1.0. There is a progressive increase in Sm_N in rocks of this series with decrease in Mg′.

4. The Fe-rich basalts of the Blake River group are characterized by flat chondrite-normalized patterns, with both samples analyzed for this study containing close to 20 times chondritic abundances.

5. The Munro picrite-tholeiite association (PTS), which previously has been thought to be related to the Fe-basalts of the Blake River group, is different from these rocks in having Sm_N/Yb_N ratios of 1.8–2.1.

6. The Al-rich basalts have high Ce_N/Sm_N and Sm_N/Yb_N ratios of about 2, although reference to Fig. 1.2.1.18 indicates that the REE profiles flatten out for elements heavier than Gd.

A comparison with other data on komatiites and associated basalts reported in the literature indicates that the Munro data are representative. Jahn *et al.* (1980) distinguish a light REE depleted series of komatiites in the Tipasjarvi area of Finland, in addition to a series with a flat profile and another with a slightly enriched light REE profile. The degree of light REE depletion of the Munro peridotitic komatiites of the upper cycle, as evidenced by their low Ce_N/Sm_N ratios, is greater than that reported elsewhere, even for rocks with higher Mg′. Sun and Nesbitt (1978) point out that very magnesian (31 wt.% MgO) spinifex textured peridotitic komatiites from the Yakabindie area have lower heavy REE abundances than those of the Munro upper cycle, while their La abundances are twice as high. On a world-wide basis, it would seem that, for komatiites and associated basalts, the overall REE abundances form a continuum, increasing with decreasing Mg′, but that the Ce_N/Sm_N ratios, while characteristic for a given set

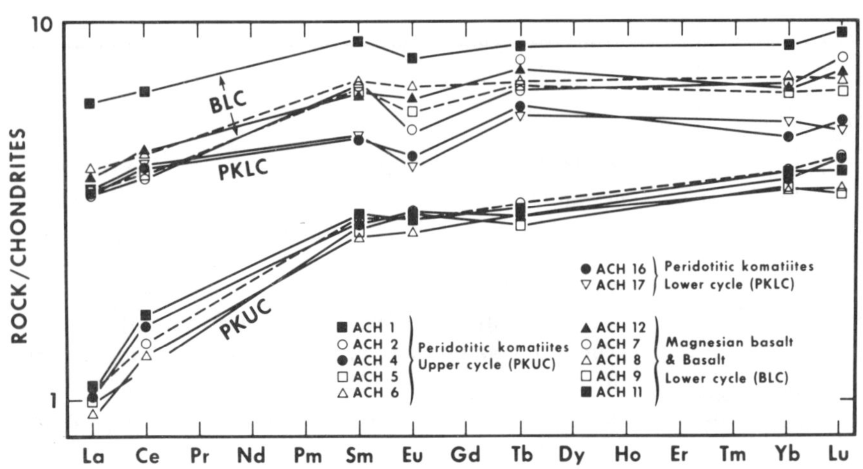

1.2.1.17 Chondrite-normalized rare-earth element patterns obtained in this study for peridotitic komatiites of the upper cycle and peridotitic komatiites and basalts of the lower cycle from Munro Township.

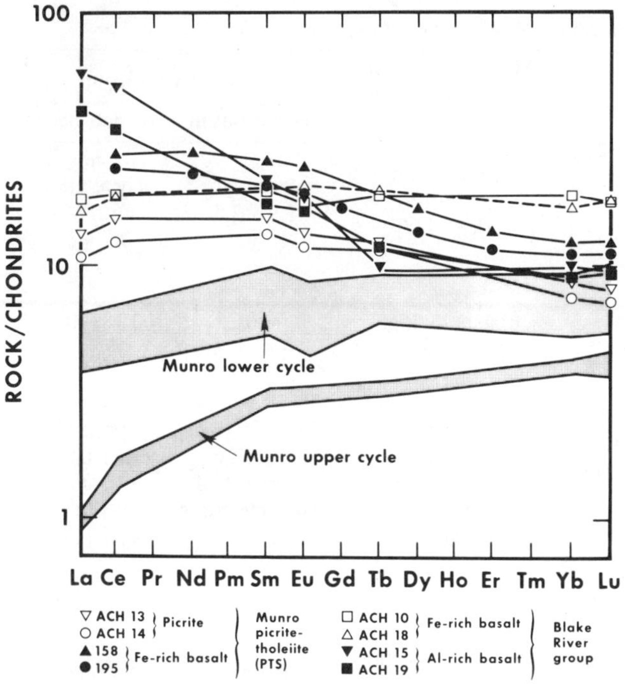

1.2.1.18 Comparison of chondrite-normalized rare-earth element patterns of rocks from the Munro Township picrite-tholeiite series (PTS) and of Al-rich and Fe-rich basalts of the Blake River group with data presented in Fig. 1.2.1.17 (stippled).

of rocks in a given area, also form a continuum that has little relationship to Mg′.

The two patterns reported for Fe-rich basalts in this study have shapes similar to those for similar rocks elsewhere, although abundances ranging from 20–40 times chondritic have been found for a Fe-rich basalt suite from the Blake River Group (I. E. M. Smith, pers. comm., 1980).

The data available for Al-rich basalts are extremely limited, although the pattern of high Ce_N/Y_N ratios (typically > 3) found in our samples is characteristic of other areas (e.g., Taylor and Hallberg, 1977; Hawkesworth and O'Nions, 1977). No clear correlation between total abundance and other compositional features has yet been noted for these rocks.

Summary of trace and minor element data

Trace and minor element data are summarized in Table 1.2.1.7. In general, Ni, Co and Cr show positive correlations with Mg′, while TiO_2, Sr, Ba, Hf, Zr, Y, Nb, V, Ga and REE show negative correlations. Data for Cs and Rb are too scattered to display any correlations, probably due to mobility during alteration.

In the three cycles in Munro Township, certain parameters show a marked change between cycles (excluding the basaltic komatiites of the upper cycle that do not seem to be related to the peridotitic komatiites) which is distinct from the variation within the cycle. These parameters include TiO_2, Hf, Y, REE and the Ce_N/Lu_N ratio. Sc content decreases in the picrite-tholeiite series with decreasing Mg′, contrasting strongly with its behavior in the two komatiite cycles where it increases as Mg′ decreases from 0.85–0.65, below which the Sc content remains relatively constant.

The Fe-rich basalts fall on extensions of the komatiite trend, except for being very high in TiO_2, Hf and Y (no data are available for Ba, Zr, Nb or V). The Al-rich basalts differ from basalts of similar Mg′ in their low Sc and high Sr (only one analysis), Hf, Zr, Th and Ce_N/La_N ratios (no data are available for Ba, Nb or V).

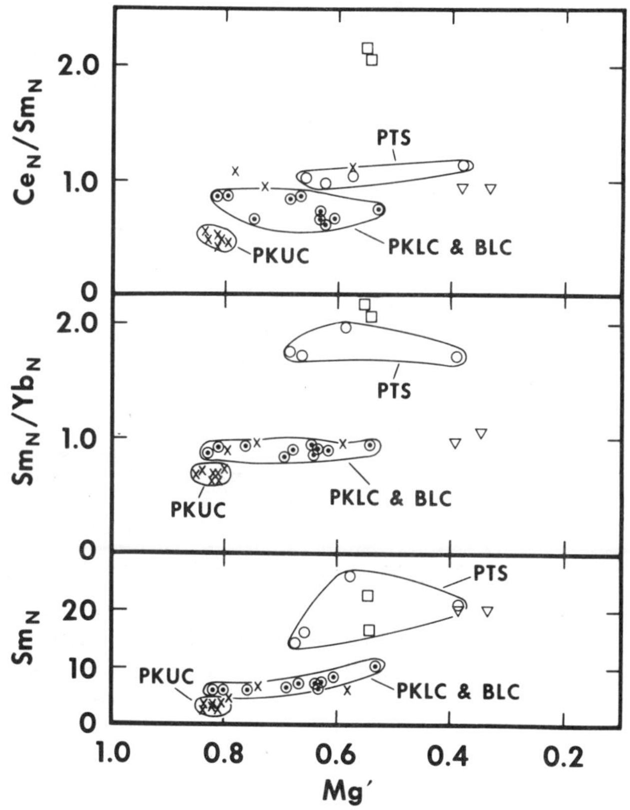

1.2.1.19 Variation of chondrite-normalized Sm abundance and REE ratios with Mg′, illustrating the progressive decrease in overall rare-earth concentration and Ce_N/Sm_N and Sm_N/Yb_N ratios, from the picrite-tholeiite series (PTS) through rocks of the komatiite lower cycle (PKLC and BLC) to peridotitic rocks of the komatiite upper cycle (PKUC) in Munro Township. These decreases parallel those of TiO_2, Hf and Y, illustrated in Figs. 1.2.1.8, 1.2.1.14 and 1.2.1.15. See Fig. 1.2.1.8 for key.

Table 1.2.1.7 Summary of minor and trace element data.

	Komatiites General	Komatiites Munro Township Upper Cycle	Komatiites Munro Township Lower Cycle	Basalt Munro Twp.	Munro Township Picrite-Tholeiite Association	Blake River Group Fe-rich Basalt	Blake River Group Al-rich Basalt
TiO_2	-Mg′	-Mg′ 0.2 – 0.9	-Mg′ 0.4 – 0.8	0.8 – 1.5	higher than komatiites 0.9 – 1.5	very high 1.9 – 2.5	no correlation with Mg′ 0.7 – 1.1
MnO	around 0.15 in rocks with highest Mg′ rising rapidly to around 0.18 and showing little variation with decreasing Mg′ thereafter			~ 0.20	picrites unusually high ~ 0.27	~ 0.25	0.18 – 0.20
Ni	+Mg′ – 1500 – 1700 ppm in rocks with Mg′ > 0.85 decreasing progressively to 300 – 500 ppm in rocks with Mg′ = 0.7			+Mg variable but 100 – 200 ppm depending on Mg′	picrites slightly higher than komatiites on basis of Mg′ (500 ppm instead of 300 ppm)	50 – 100 ppm but falls on extension of komatiite curve	similar to other basalts with similar Mg′
Co	+Mg′ decreasing from 105 ppm to 75 ppm			60 – 90 ppm	—	—	—
Cr	+Mg′ decreasing from 3000 ppm at Mg′ ~ 0.85 to 400 ppm at Mg′ ~ 0.65			100 – 300 ppm fall on extension of komatiite trend	Picrites ~ 1400 ppm — fall on komatiitic curve. No data on basalt.	little relation to Mg′ < 100 ppm	120 – 320 ppm — similar to other basalts with similar Mg′
Sc	-Mg	-Mg′ 20 – 35 ppm	-Mg′ 25 – 40 ppm	40 ppm	+Mg′ 30 ppm in picrite 20 ppm in basalt	40 – 50 ppm	low ~ 24 ppm
Sr	scattered but -Mg′ no distinctions apparent between different cycles at Munro			scattered 80 – 190 ppm	-Mg′ picrites ~ 40 ppm basalt ~ 70 ppm	fall on possible extensions of komatiitic Mg′ trend	70 to 260 ppm
Ba	-Mg′ — very scattered data			—	—	—	—
Cs	very scattered data — no generalizations apparent						

1.2.1.5 Petrogenesis

Four of the most important questions concerning the genesis of the Archean basalts and komatiites are: (1) To what extent are each of the processes, fractional crystallization and mantle partial melting, responsible for the diversification within the komatiite suite?; (2) What is the relationship between komatiites and basalts?; (3) What is the origin of the vast amount of (highly fractionated) Fe-rich basalt?; and (4) What is the origin of the Al-rich basalt-andesite suite?

With respect to the first question, opinions are very divergent. Viljoen and Viljoen (1969) suggested that fractional crystallization was important. Herrmann *et al.* (1976) and Sun and Nesbitt (1978) suggested that variable degrees of mantle melting, melting of depleted mantle and original heterogeneities in the mantle accounted for much of the variation. Arndt *et al.* (1977) and Naldrett and Turner (1977) favored both fractional crystallization and variable partial melting, as did Arth *et al.* (1977).

As stated previously, variations in major element chemistry are consistent with either fractionation of olivine followed by olivine + pyroxene or the progressive melting of a lherzolitic mantle. In Munro Township the presence of Fred's Flow, the komatiite flow that has differentiated to produce a quartz gabbro containing 7.5 wt.% MgO from an initial liquid containing 20% MgO, provides strong evidence for at least some fractional crystallization. The major element composition of the gabbroic differentiate is identical to those of many of the basaltic flows found in Munro Township. Also in Munro, the occurrence of volcanic cycles in which the lavas become successively less magnesian

Table 1.2.1.7 *(Continued)*

	Komatiites Munro Township: General	Komatiites Munro Township: Upper Cycle	Komatiites Munro Township: Lower Cycle	Basalt Munro Twp.	Munro Township Picrite-Tholeiite Association	Blake River Group: Fe-rich Basalt	Blake River Group: Al-rich Basalt
Hf		peridotites < 0.5 ppm	-Mg′ peridotites 0.75 – 1 ppm	—	picrites high in Hf — ~ 1.8 ppm in contrast to komatiites of similar Mg′ with ~ 1 ppm	high ~ 3ppm	high on basis of Mg′ ~ 3 ppm; twice as high as basalts associated with komatiites that have same Mg′
Zr	-Mg′ — no data to distinguish between areas			40 – 50 ppm	—	—	very high 85 – 105 ppm in comparison with basalts of same Mg′ with 45 – 50 ppm
Th	no correlation with Mg′ approximately 0.1 ppm			as for komatiites	possibly (1 out of 2 analyses) higher than komatiites	approximately 0.4 ppm — distinctly higher than extension of komatiite trend	very high 1.3 – 1.9 in comparison with < 0.2 in komatiites of same composition
Y	-Mg′	7 – 8 ppm	-Mg′ 10 – 13 ppm	-Mg′ 13 – 18 ppm	picrites high in γ(16 – 18 ppm) in contrast to komatiites of similar Mg′	very high 30 – 35 ppm	~ 19 ppm similar to komatiitic basalts in Munro Twp.
Nb	-Mg′ data scattered but show reasonable correlation			as for komatiites	—	—	—
V	-Mg′ — good correlation ~ 115 at Mg′ > 0.85 to ~ 300 ppm at Mg′ = 0.55			as for komatiites	—	—	—
Ga	generally decreasing with increasing Mg′ in all rocks						
Pb	very slight decrease with increasing Mg′, but no good trend						
REE	-Mg′ variable Ce_N/Sm_N $Sm_N/Yb_N \sim 1$	peridotites have low $\simeq$ $Ce_N/Sm_N \simeq 0.5$ and $Sm_N/Yb_N \simeq 0.75$ basalts with ratios $\simeq 1$	-Mg′ Ce_N/Sm_N 0.7 Sm_N/Ce_N 1 true for all rocks	-Mg′— generally flat patterns	-Mg′ HREE depleted $Ce_N/Sm_N \simeq 1$ $Sm_N/Yb_N \simeq 2$	-Mg′ flat patterns	variable abundance but LREE enriched Ce_N/Sm_N and $Sm_N/Yb_N \simeq 2$

Note: -Mg′ or + Mg′ indicates negative or positive correlation with Mg′, respectively.

upwards within a given cycle is difficult to explain by a sequence of melts derived by progressive melting of a mantle source, but is the logical result of a near-surface, fractionating magma chamber's being tapped at intervals to give rise to surface extrusions.

On the other hand, the REE patterns appear to be inconsistent with the possibility that many of the rocks are related by fractional crystallization. Hermann *et al.* (1976) and Sun and Nesbitt (1978) point this out as do Arth *et al.* (1977). The last authors have suggested that the source of many of the komatiites in Munro Township may be progressive melting in a single zone of the mantle, possibly a rising diapir. They suggest that the initial partial melt was a member of the picrite-tholeiite series discussed in this study. They further suggest that the light REE-enriched patterns of these magmas (discussed later) are consistent with a small degree of melting (< 10%) of a source in which garnet remains behind in the residuum. Subsequent melting of this zone, now somewhat depleted but containing a proportion of the original melt, could produce a moderately depleted magma which then could fractionate to produce a series of magmas with moderate light REE depletion. A very magnesian, highly light REE-depleted melt would be produced by yet a third stage of melting. Langmuir *et al.* (1977) emphasized the need to retain some of the initial melt within the source region to mix with the products of further melting in order to avoid the very complete elimination of highly incompatible elements from the source. They presented a vivid description of this process for which they coined the term **dynamic melting**.

Following up on the work of Arth *et al.* (1977), Whitford and Arndt (1978) have studied REE in Fred's Flow, which immediately overlies the tholeiitic sills and flows referred to above. They have shown it to have a moderately light REE-depleted pattern, identical with that which we report for the lower komatiite cycle. Progressive differentiates of this magma are believed to be recorded in spinifex zones within the flow, and to show a negative correlation of total REE abundance with Mg′. REE data obtained in this study agree with those of Arth *et al.* (1977) and of Whitford and Arndt (1978) and indicate that Fred's Flow is overlain by a series of pyroxenitic and basaltic komatiite lavas which show major element and REE chemistries very similar to those of the successive differentiates of the flow and which could well be derived from it or from a common differentiating magma source. The base of the new cycle (cycle 2 in Fig. 1.2.1.2) is marked by a series of very magnesian, highly REE-depleted lavas (sample A_2 in Table 1.2.1.4), consistent with a third phase of melting of the same source as the tholeiites and Fred's Flow. Our REE data on the picrite-tholeiite series supplement those of Arth *et al.* (1977) and lend support to the suggestion that it is the initial melt which was derived from the source before the magma which was responsible for Fred's Flow and for the remainder of the lower komatiite cycle was produced.

The behavior of TiO_2, Hf and Y, which decrease discontinuously in rocks of similar Mg′ in successively higher cycles at Munro, also supports the concept of a series of melts derived from a source or sources showing progressively greater degrees of depletion. Our data for Hf (Fig. 1.2.1.14a) are some of the most convincing evidence in this regard; these data also show a gradual increase within the Munro komatiite lower cycle (PKLC and BLC), which is suggestive of a trend resulting from fractional crystallization. Similar trends are shown for TiO_2 and Y in both the picrite-tholeiite series and the komatiite lower cycle (Figs. 1.2.1.8a and 1.2.1.14c, respectively).

Our conclusion from Munro Township is therefore that the periodic partial removal of magmas produced by the progressive melting of a single portion of mantle is responsible for much of the diversity exhibited by the rocks, but that these batches of magma have fractionally crystallized to produce additional diversity. TiO_2, Hf and Y are among those elements most suited to sort this problem out.

In contrast to Munro, some areas (e.g., Barberton, Scotia, Tipasjarvi) have rocks which are equally or even more MgO-rich and yet have significantly less light REE-depletion. Langmuir *et al.* (1977) have shown that the amount of melt retained in the source can have a pronounced effect on the degree to which light REE become depleted during the progressive melting described above. The Munro rocks may be the result of more complete melt extraction than rocks of other areas or, alternatively, the differences may be functions of the La/Sm ratios in the sources.

The sample which shows the least relative light REE depletion is 49J from Barberton; the La/Sm ratio in this sample is slightly greater than 1 while the MgO content is 32.2 wt.% (Table 1.2.1.4). Some of the Barberton rocks, including 49J, are also characterized by very high CaO/Al_2O_3 ratios. Green (1975) has suggested that both the high CaO/Al_2O_3 ratio and the REE pattern are results of the selective removal of garnet during magma ascent.

Partial melting itself poses a problem in komatiite genesis. Green (1975) indicates that komatiite liquids, which contain as much as 32% MgO, represent 70% melting of a pyrolitic mantle. Even if one regards the mantle as somewhat more refractory than pyrolite (equivalent, for example, to lherzolite xenoliths from basalts) single-stage melting would require in excess of 50% melting to produce a komatiite liquid. Mantle melting of the type envisioned here requires that the magma and residual olivine crystals remain in physical contact, i.e., that the olivine remains suspended in the melt. Arndt (1977) has considered this problem in the light of settling rates which can reasonably be expected for olivine, and has concluded that the proportion of liquid in an ascending diapir is unlikely to exceed 30–40%. In Chapter 6 it is concluded, from the ability of olivine to deform in response to stress, that the proportion of liquid will be considerably less than 30–40%. The progressive melting of a volume of mantle, as has been proposed for Munro, avoids the necessity for high degrees of melting, because a very much lower degree of melting is required to produce a magnesian melt from a depleted mantle composed largely of residual olivine than would be required to generate that melt from an undepleted mantle.

Turning to the second problem, that of the relationship between komatiites and other Archean volcanics, the Archean basaltic rocks described here include a wide variety of compositions, which implies a number of different petrogenetic processes. A primary division of the basaltic rocks can be made on the basis of REE patterns. Thus, Archean basalts with $La_N/Yb_N < 2$ are considered to have resulted from direct melting of an unfractionated mantle source, while those with La_N/Yb_N ratios > 2 are derived from a modified mantle source (see discussion below). Some of the magnesian basalts, basalts and iron-rich basalts are thought to represent magmas which originated by partial melting of unmodified mantle and which in some cases underwent subsequent fractionation before eruption.

Primary magmas which have high MgO contents (> 20%) appear to be restricted to the Archean. Perhaps this restriction reflects higher degrees of melting which were made possible by higher rates of heat production at that time. However, relatively high-MgO (12–20%), low-K_2O tholeiites which have been reported from the Tertiary of Greenland and eastern Canada (Drever and Johnson, 1966; Clarke, 1970) and from Gorgona Island off the coast of Colombia (Gansser *et al.*, 1979) are possible modern analogues of the Archean basaltic komatiites. Clarke (1970) has suggested that parental magmas with MgO contents of 18–20% may have given rise to a suite of high-magnesium basaltic rocks by differentiation. It has been estimated that such magmas would require a high degree of melting (30%) and temperatures in excess of 1400°C (Nesbitt and Sun, 1976).

In some studies of Archean basaltic rocks, the regular pattern of element variations has led to the suggestion that the basalts are related to magnesian basalts by fractional crystallization, but in other studies there is a distinct compositional gap between the two basalt types (e.g., Hallberg and Williams, 1972) which has been taken to indicate distinct magma sources. Nesbitt and Sun (1976) suggest that the large volumes of relatively low-MgO basalt are probably derived from primary melts which had MgO contents of about 12% and which were produced at lower temperatures and by lower amounts of partial melting than the melts which differentiated to produce the pyroxenitic komatiites and their associated differentiates.

Clearly, physical conditions during the Archean were such that partial melting of primary mantle was capable of producing melts with a wide range of MgO contents. Thus basaltic komatiites may represent a magma type which differentiated to give magnesian basalt and basalt, and some magnesian basalts may represent primary magmas from which basalts were produced. Chemical compositions and, particularly, the REE contents of basaltic rocks in Archean terrains are consistent with the concept of a variety of primary magmas undergoing varying degrees of fractionation.

With respect to the third major question, the chemical compositions of Fe-rich basalts in Archean terrains are in general consistent with a model of fractional crystallization from relatively Mg-rich parental magmas. In this respect, they may be compared with basaltic members of the Fe-rich basalt-plagiogranite series (Coleman and Peterman, 1975) found in association with Phanerozoic tholeiitic basalt. However, three problems arise in connection with this model:

1. Fe-basalt is unusually abundant in some Archean greenstone belts and is not matched by complementary cumulate rocks.
2. Limited experimental work (Wells *et al.*, 1977) and petrographic observations indicate that plagioclase is the liquidus phase in these rocks, which is not consistent with the evidence for extensive fractionation of MgO with respect to FeO.
3. Petrographic observations show that clinopyroxene is a near-liquidus ferromagnesian phase in Fe-basalts. Although quantitative modelling indicates that clinopyroxene-dominated crystal fractionation is feasible from the point of view of major element compositions, the comparatively high Sc contents of Fe-rich basalts cannot be reconciled with the known affinity of Sc for clinopyroxene.

An alternative, that iron-rich magma was produced directly from Archean mantle, is not compatible with currently accepted constraints on the Mg/Fe ratios of liquids in equilibrium with likely mantle residual phases, but may have been feasible if the oxygen fugacity was drastically different from that which exists at present. This possibility remains to be evaluated.

Considering the fourth question, high-Al basalts in Archean volcanic terrains typically are minor components of an association which includes intermediate (andesites) and acid (dacites and rhyolites) members. The contrast, between the compositions of magnesian through to Fe-rich basalts on the one hand and the compositions of high-Al basalts and associated rocks on the other, defines a geochemical discontinuity of major importance in Archean volcanic sequences. This discontinuity is defined by the following features:

1. Increase in the total abundance and variability of Si, Al, alkalies and the large-ion-lithophile elements (Rb, Ba, Sr) in the high-Al basalt association.

2. Similar contents of transition elements, such as Ni and Cr, but markedly lower Sc contents (typically 20–30 ppm) in high-Al basalts relative to tholeiitic basalts (Sc contents of 30–50 ppm).

3. The high-Al basalts and associated rocks display marked light/heavy REE fractionation. Heavy REE are lower than in tholeiitic Archean basalts, light REE abundances overlap and are higher.

4. Archean tholeiitic basalts display an iron enrichment trend on the FMA diagram, high-Al basalts and associated rocks do not.

The two models which have been advanced to explain the geochemical discontinuity between tholeiitic Archean basalt and the overlying high-Al basalt association differ in the degree of significance attached to the discontinuity. Hawkesworth and O'Nions (1977) suggest that the change can be accounted for by a change in the nature of the phases which were fractionating from an essentially homogeneous parent magma. Fractionation of Fe-rich garnet is invoked to explain the Si-enrichment, lack of Fe-enrichment and REE fractionation of the high-Al basalt association. The main objection to this model is the high degree of garnet fractionation required to generate the observed REE fractionation (Williams and Smith, 1979) and the difficulty of deriving basaltic compositions with this degree of fractionation.

An alternative suggestion (Smith *et al.*, 1977) stresses the significance of the geochemical discontinuity between tholeiitic Archean basalts and the high-Al basalt association. According to the model, the magnitude of the discontinuity requires the existence of geochemically distinct sources which give rise to tholeiitic basalts and to high-Al basalts, respectively. An analogy may be drawn between the geochemically-evolved high-Al basalt association and the rocks of modern arc-trench type volcanoes (Taylor and Hallberg, 1977). The implication of this analogy is that interaction with crustal material, possibly through a subduction process, gave rise to chemically-modified mantle which then partially melted to yield Al-rich Archean basalt.

The alkali-rich Archean basalts described by Cooke and Moorehouse (1969) are not well enough known at present to allow speculation as to their origins. However, they resemble the highly potassic lavas found in some modern volcanic arcs (e.g., the Roman province, Italy) in their major element chemistry.

1.2.1.6 References

Arndt N. T. (1975) Ultramafic rocks of Munro Township and their volcanic setting. Unpub. Ph.D. Thesis, Univ. of Toronto.

Arndt N. T. (1977) Ultrabasic magmas and high-degree melting of the mantle. *Contrib. Mineral. Petrol.* **64**, 205–221.

Arndt N. T., Naldrett A. J. and Pyke D. R. (1977) Komatiitic and iron-rich tholeiitic lavas of Munro Township, northeast Ontario. *J. Petrol.* **18**, 319–369.

Arth J. G., Arndt N. T. and Naldrett A. J. (1977) Genesis of Archean komatiites from Munro Township, Ontario: Trace element evidence. *Geology,* **5**, 590–594.

Bickle M. S., Martin A. and Nisbett E. G. (1975) Basaltic and peridotitic komatiites and stromatolites above a basal unconformity in the Belingwe greenstone belt, Rhodesia. *Earth Planet. Sci. Lett.* **27**, 155–162.

Brooks C. and Hart S. R. (1974) On the significance of komatiite. *Geology* **2**, 107–110.

Brooks C., Hart S. R., Krogh T. E. and Davis G. L. (1970) Initial $^{87}Sr/^{86}Sr$ ratios of regionally sampled metavolcanics from the Canadian Shield. *Carnegie Inst. Wash., Yearb.* **68**, 425–426.

Clarke D. B. (1970) Tertiary basalts of Baffin Bay: possible primary magma from the mantle. *Contrib. Mineral. Petrol.* **25**, 203–224.

Coleman R. G. and Peterman Z. E. (1975) Oceanic plagiogranite. *J. Geophys. Res.* **80**, 1099–1108.

Cooke D. L. and Moorhouse W. W. (1969) Timiskaming volcanism in the Kirkland Lake area, Ontario, Canada. *Can. J. Earth. Sci.* **6**, 117–132.

Davis P. and Condie K. C. (1977) Geochemistry of Archean greenstone belts in western Kenya. *Geochim. Cosmochim. Acta* **41**, 271–277.

Drever H. I. and Johnson R. (1966) A natural high-lime silicate liquid more basic than basalt. *J. Petrol.* **7**, 414–420.

Gansser A., Dietrich V. J. and Cameron W. E. (1979) Palaeogene komatiites from Gorgona Island. *Nature* **278**, 545–546.

Gelinas L., Brooks C., Perrault G., Carignan R., Trudel P. and Cote R. (1977) Chemostratigraphic divisions within the Abitibi volcanic belt, Rouyn Noranda District, Quebec. In *Volcanic Regimes in Canada* (W. R. Barager, ed.), p. 265–295. Geol. Assoc. Can. Spec. Publ., 16.

Gemuts I. and Theron A. (1975) The stratigraphy and mineralization of the Archean between Norseman and Coolgardie, Western Australia. In *Ore Deposits of Australia and Papua-New Guinea*, p. 66–74. Monogr. Aust. Inst. Min. Met. **5**.

Goodwin A. M. (1977) Archean volcanism in Superior province, Canadian shield. *Geol. Assoc. Canada Spec. Paper* **16**, 205–241.

Goodwin A. M., Ambrose J. W., Ayres L. D., Clifford P. M., Currie K. L., Ermanovics I. M., Fahrig W. F., Gibb R. A., Hall D. H., Innes M. J. S., Irvine T. N., MacLaren A. S., Norris A. W. and Pettijohn F. J. (1972) The Superior province. In *Variations in Tectonic Styles in Canada*, (R. A. Price and R. J. W. Douglas, eds.), p. 527–624. Geol. Assoc. Canada Spec. Paper. **11**.

Green D. H. (1975) Genesis of Archean peridotitic magmas and constraints on Archean geothermal gradients and tectonics. *Geology* **3**, 15–18.

Hallberg J. A. and Williams D. A. C. (1972) Archean mafic and ultramafic rock associations in the eastern gold fields region, Western Australia. *Earth Planet. Sci. Lett.* **15**, 191–200.

Hallberg J. A., Carter D. N. and West K. N. (1976a) Archean volcanism and sedimentation near Meekatharra, Western Australia. *Precambrian Res.* **3**, 577–595.

Hallberg J. A., Johnson C. and Bye S. M. (1976b) The Archean Mada igneous complex, Western Australia. *Precambrian Res.* **3**, 111–136.

Hart S. R. (1969) K, Rb, Cs contents and K/Rb and K/Cs ratios of fresh and altered submarine basalts. *Earth Planet. Sci. Lett.* **6**, 295–303.

Hart S. R., Brooks C., Krogh T. E., Davis G. L. and Nava D. (1970) Ancient and modern volcanic rocks: a trace element model. *Earth Planet. Sci. Lett.* **10**, 17–28.

Hawkesworth C. J. and O'Nions R. K. (1977) The petrogenesis of some Archean volcanic rocks from southern Africa. *J. Petrol.* **18**, 487–520.

Herrmann A. G., Blanchard D. P., Haskin L. A., Jacobs J. W., Knade D., Korotev R. L. and Brannon J. C. (1976) Major, minor, and trace element compositions of peridotitic and basaltic komatiites from the Precambrian crust of southern Africa. *Contrib. Mineral. Petrol.* **59**, 1–12.

Jahn B. M. and Condie K. C. (1975) On the age of Rhodesian greenstone belts. *Contrib. Mineral. Petrol.* **57**, 317–330.

Jahn B. M. and Shih C. (1974) On the age of the Onverwacht Group, Swaziland Sequence, South Africa. *Geochim. Cosmichim. Acta* **38**, 873–885.

Jahn B. M., Auvray B., Blais S., Capdevila R., Cornichet J., Vidal F. and Hameurt J. (1980) Trace element geochemistry and petrogenesis of Finnish greenstone belts. *J. Petrol.* **21**, 201–244.

Lajoie J. and Gélinas L. (1978) Emplacement of Archean peridotitic komatiites in La Motte Township, Quebec. *Can. J. Earth Sci.* **15**, 672–677.

Langmuir C. H., Bender J. F., Bence A. E., Hanson G. N. and Taylor S. R. (1977) Petrogenesis of basalts from the FAMOUS area: Mid-Atlantic ridge. *Earth Planet. Sci. Lett.* **36**, 133–156.

Lewis J. D. and Williams I. R. (1973) The petrology of an ultramafic lava near Murphy Well, eastern Goldfields, Western Australia. *W. Aust. Geol. Surv. Ann. Rept. 1972*, 60–68.

Meyer H. O. A. (1975) Chromium and the genesis of diamond. *Geochim. Cosmochim. Acta* **39**, 929–936.

Naldrett A. J. and Cabri L. J. (1976) Ultramafic and related rocks: their classification and genesis with special reference to the concentration of nickel sulfides and platinum group elements. *Econ. Geol.* **71**, 1131–1158.

Naldrett A. J. and Goodwin A. M. (1977) Volcanic rocks of the Blake River Group, Abitibi greenstone belt, Ontario, and their sulfur content. *Can. J. Earth Sci.* **14**, 539–550.

Naldrett A. J. and Mason G. D. (1968) Contrasting Archean ultramafic igneous bodies in Dundonald and Clergue Townships, Ontario. *Can. J. Earth Sci.* **5**, 111–143.

Naldrett A. J. and Turner A. R. (1977) The geology and petrogenesis of a greenstone belt and related nickel sulfide mineralisation at Yakabindie, Western Australia. *Precambrian Res.* **5**, 43–103.

Nesbitt R. W. and Sun S-S. (1976) Geochemistry of Archean spinifex textured peridotites and magnesian and low magnesian tholeiites. *Earth Planet. Sci. Lett.* **31**, 433–453.

Pearce T. H. and Birkett T. (1974) Archean metavolcanic rocks from Thackeray Township, Ontario. *Can. Mineral.* **12**, 509–519.

Pyke D. R., Naldrett A. J. and Eckstrand A. R. (1973) Archean ultramafic flows in Munro Township, Ontario. *Geol. Soc. Amer. Bull.* **84**, 955–978.

Smith I. E. M., Goodwin A. M. and Wooden J. (1977) Magma evolution in Archean volcanic piles (abstract). *Abstracts with Program*, Geol. Soc. Amer. **9**, 1181.

Sun S-S. and Nesbitt R. W. (1978) Petrogenesis of Archean ultrabasic and basic volcanics: evidence from rare earth elements. *Contrib. Mineral. Petrol.* **65**, 301–325.

Taylor S. R. and Hallberg J. A. (1977) Rare earth elements in the Marda calc-alkaline suite: an Archean geochemical analogue of Andean-type volcanism. *Geochim. Cosmochim. Acta* **41**, 1125–1130.

Viljoen M. J. and Viljoen R. P. (1969) The geology and geochemistry of the lower ultramafic unit of the Onverwacht Group and a proposed new class of igneous rock. *Upper Mantle Project Spec. Publs.* **2**, 221–244. Geol Soc. S. Africa.

Wells G., Grant A., Pearce T. H. and Roeder P. L. (1977) Archean pillow lavas from the Pickle Crow and Noranda areas. In *Proc. Geotrav. Conf., 1977, Univ. Toronto*, p. 117–118. Univ. of Toronto.

Williams J. and Smith I. E. M. (1979) Geochemical evidence for paired arcs in the Permian volcanics of southern New Zealand. *Contrib. Mineral. Petrol.* **68**, 285–291.

Whitford D. F. and Arndt N. T. (1978) Rare earth elements in a thick layered komatiite lava flow. *Earth Planet. Sci. Lett.* **41**, 188–196.

1.2.2 PRE-TERTIARY CONTINENTAL FLOOD BASALTS

1.2.2.1 Introduction

Geologists have long recognized the existence of very large accumulations of basaltic lava in different parts of the Earth's continents which Sir Archibald Giekie called **plateau basalts**. ". . . There have been periods in the Earth's history when the crust was rent into innumerable fissures over areas thousands of square miles in extent, and when the molten rock instead of issuing, as it does in most modern volcanoes, in narrow streams from a central elevated cone, welled out from these vents or from numerous small vents along their course, and flooded enormous tracts of country without forming any mountain or conspicuous volcanic cone. . ." (Giekie, 1903). Apart from the ocean-floor ridges and plains these are the largest volcanic features on Earth and, like the oceanic basalts, they were produced in a tensional environment. Typified by the Columbia River Province and the Deccan Traps, these features are characterized by their great areal extents, large volumes, immense sizes of some individual eruptive units, associations with dike swarms, subaerial eruptions on continental foundations, and dominantly tholeiitic (subalkaline) basaltic compositions. H. S. Washington (1922) provided an early summary, pointing out further similarities in petrography, composition, and physical character, as well as the wide range in ages of plateau basalts from Keweenawan (late Precambrian) to recent. Tyrrell (1937) followed with another general survey, preferring the term **flood basalt** to plateau basalt since these accumulations generally filled broad, subsiding basins surrounded by uplands; only long after cessation of volcanism did they become plateaus. Tyrrell also outlined the major characteristics mentioned above, and pointed out the common association with abundant diabase sills and other mafic intrusions, the scarcity of central volcanic edifices, the occasional but minor occurrence of pyroclastic debris, and the association of interflow fluvial, lacustrine, or aeolian sediments and rare paleosols indicative of periods of volcanic quiescence. Kuno (1969) provided a more recent review, concentrating on petrochemistry. He pointed out the occurrence of alkali-olivine and high-alumina basalts as well as (quartz) tholeiites in lava plateaus and their intergradational chemistries. Kuno also indicated that despite their large volumes, relatively uniform chemistries, and fissure modes of eruption, most plateau lavas are considerably differentiated or evolved with respect to our ideas of primitive mantle melt compositions.

Some of the largest of the world's flood basalt accumulations are of considerable antiquity, ranging from late Precambrian (Keweenawan, Lake Superior district) to Cretaceous. Besides the Keweenawan basalts, these include the Siberian Platform traps, the Karroo lavas and dolerites of southern Africa, and the traps of the Paraná basin in southeastern South America (Table 1.2.2.1), all of which are discussed below. Cenozoic flood basalts are discussed in another section (1.2.3) of this chapter.

The major basalt accumulations with which we are here concerned grade into somewhat smaller examples of continental flood volcanism, such as the Triassic-Jurassic Newark basalts of eastern North America, the Late Precambrian Coppermine lavas of the Northwest Territories, Canada, and the Jurassic dolerites of Antarctica and Tasmania, among many others.

It is possible that very extensive flood basalt provinces of much greater age, now completely eroded away, once were fed by the many great dike swarms (e.g., Matatchewan, Abitibi, Franklin, Sudbury, etc.) of the Precambrian Canadian Shield (Baragar, 1977).

1.2.2.2 Geological setting

Tectonic environment

The major basalt plateaus discussed here and those pre-Tertiary flood basalts of lesser volume are associated with continental rift environments. Of those discussed here, the older ones (late Precambrian to Permo-Triassic) were contained within their respective continents and no new ocean basin was produced during their formation, whereas, in the case of younger plateaus, rifting was associated with the actual separation of Africa from its neighboring land masses. Some close analogs of Tertiary and Quaternary age are found in Iceland, where the plateau is a product of entirely oceanic rifting, and in the associated Brito-Arctic (Thulean) province, which began as a continental rift.

The Keweenawan lavas of the Lake Superior district, along with their associated mafic intrusions including the Duluth Complex, constitute the crustal rock expression of the great midcontinent gravity high (Fig. 1.2.2.1). This gravity high, which is the largest geophysical feature of central North America, extends southwestward from Lake Superior into Kansas, and an eastern limb bends south through central Michigan into Kentucky. Its interpretation as an abortive continental rift has been extensively discussed by King and

Zietz (1971), White (1972), Chase and Gilmer (1973), and Wold and Hinze (1981), after early mention by Smith *et al.* (1966). The Keweenawan lavas lie on an older, stabilized, low-relief crust of Archean and lower Proterozoic rocks, locally with strong angular unconformity. Several dike swarms cut lower parts of the sequence and cut older rocks beyond the present limits of the lavas (Green, 1977), but most of the volcanic accumulation and its assumed feeder system are buried beneath younger continental Precambrian and Pleistocene sediments and Lake Superior. Both geological and geophysical evidence suggest that the Keweenawan lavas were erupted in several (as many as eight) separate accumulations, each on the order of a few kilometers thick, as the intensity of rifting changed in time and space along the midcontinent rift (White, 1972; Green,

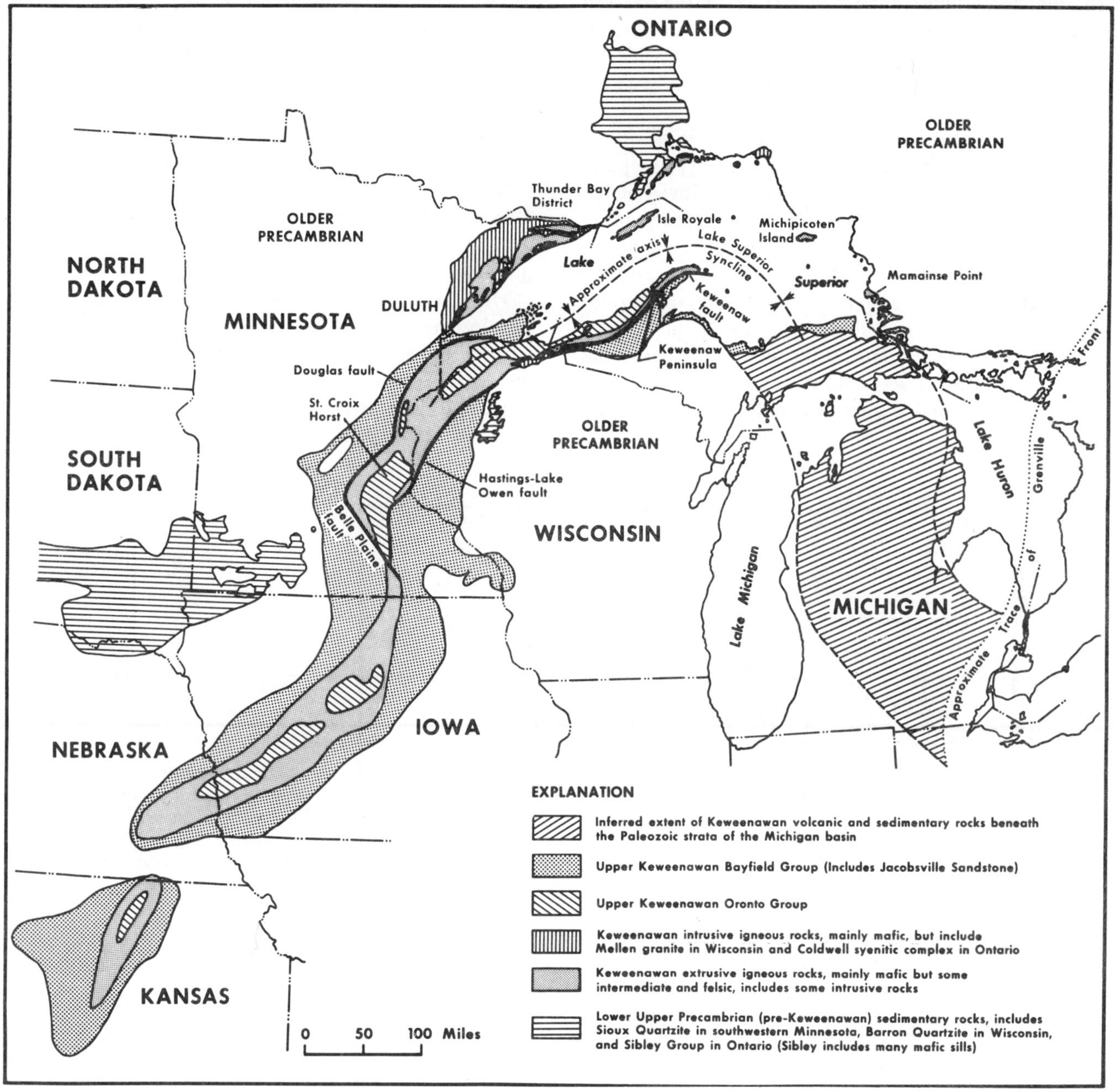

1.2.2.1 Generalized map of Late Precambrian rocks of the midcontinent gravity high of North America. Modified from Craddock (1972), Fig. V-2.

1977). Each separate accumulation subsided centrally and has been partly eroded away. The overall tectonic history has been outlined by Craddock (1972).

Deep seismic studies (Smith *et al.*, 1966) show that, in the central part of Lake Superior, the Mohorovicic discontinuity is anomalously deep (50–55 km), that rocks with seismic velocities appropriate to granitic basement appear to be absent, and that the mafic igneous rocks give an efficient coupling for seismic waves to the subcontinental mantle. In contrast, on the western shore of Lake Superior in Minnesota the seismic experiments show unusually thin crust (25–30 km), which is known from geologic evidence to contain granitic basement beneath the Keweenawan rocks.

The traps of the Siberian Platform, which are of late Permian to early Jurassic age, are found principally in the great Tunguska basin ("syneclise" in Soviet literature) east of the Yenesei River (Fig. 1.2.2.2). The lavas are accompanied by even more voluminous sills, laccoliths, and other shallow intrusions, and lie on and are interbedded with abundant basaltic tuffs and agglomerates. According to Kulikov *et al.* (1972), the Triassic of the West Siberian Plate (at the western border of the Siberian Platform) was characterized by a major rift system which trended approximately north-south and which contained multiple graben structures; the flood basalts and the associated terrigenous sediments and volcanic debris filled the grabens and extended over the entire area as well as over the Tunguska syneclise to the east. More flood basalts across the Yenesei-Katanga depression in the Taimyr area to the north are the same age and apparently part of the same great Triassic magmatic episode (Makarenko, 1974; Milanovskiy, 1976), but they became folded before the start of the Jurassic, unlike the Tunguska basin.

According to Bochkarev and Rudkevich (1976), three successive systems of grabens developed during the early to mid-Mesozoic in the West Siberian Plate, each associated with arching, and with associated downwarps migrating from north to south. The orientations of the Triassic troughs were influenced by the folded structures of the eroded Paleozoic basement. General subsidence followed the magmatism in the late Jurassic and Cretaceous, with eventual marine transgression, but according to Bochkarev and Rudkevich (1976) the subsidence was not caused by the graben development and basalt magmatism.

To the east of the West Siberian rift system, the widespread Triassic trap volcanism was preceded by a major period of subsidence; at the center of the

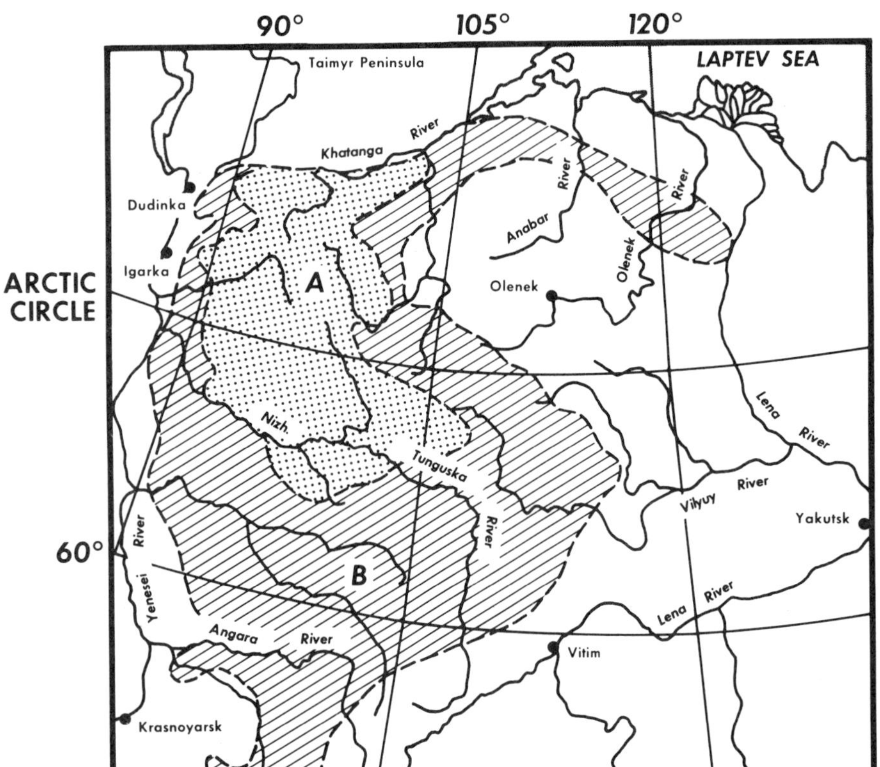

1.2.2.2 Location of Siberian Platform traps. A. area of Permo-Triassic plateau lavas; B. region of abundant Permo-Triassic mafic intrusions in older strata. After Masaitis *et al.* (1966).

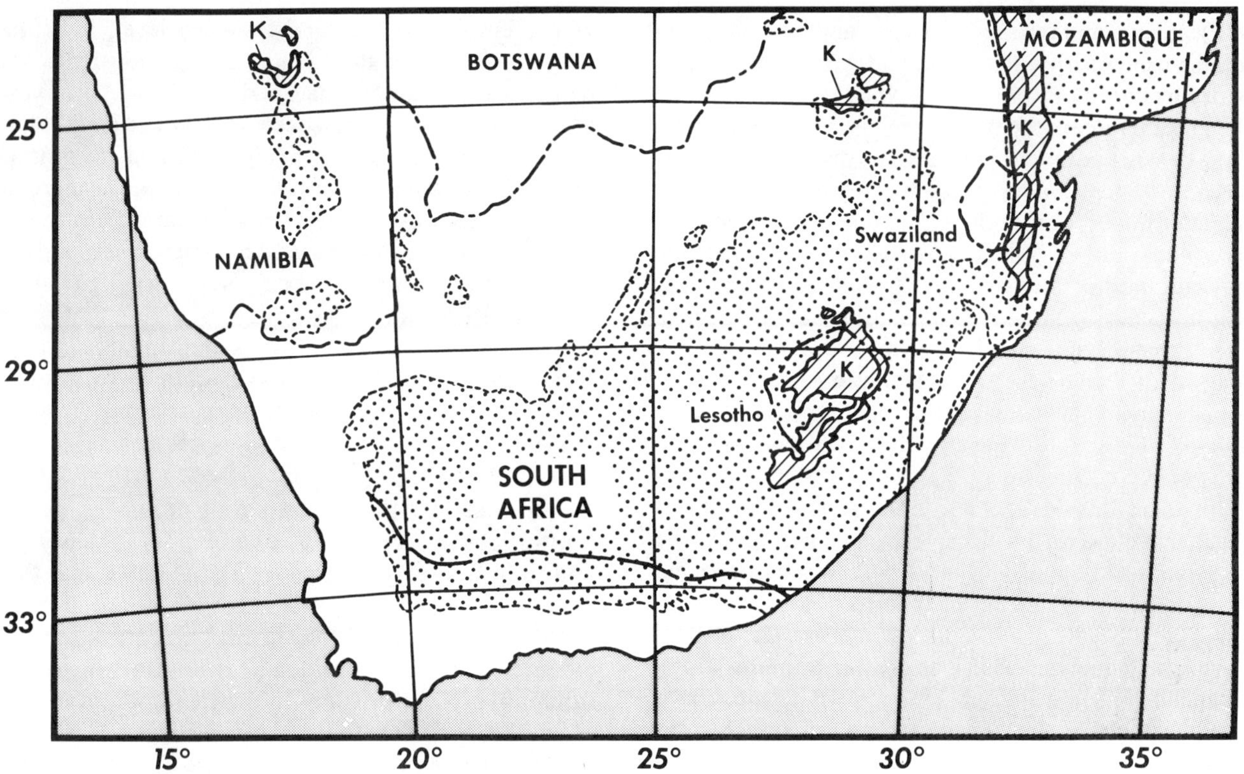

1.2.2.3 Location of Karroo (Jurassic-Cretaceous) plateau lavas in southern Africa (K). Larger patterned area shows extent of sedimentary rocks of Karroo System. Long-dash line shows southern limit of diabase sills in Karroo strata. After Walker and Poldervaart (1949).

Tunguska basin the basement rocks (pre-Permian) lie at a depth of 9 km (Milanovskiy, 1976).

A complex pattern, in space and time, of grabens, horsts, and "junction zones" related to the Siberian Platform magmatism has been envisaged by other workers. Kuznetsov and Naumov (1975), for instance, recognized an older phase (283–240 m.y.b.p.) of tholeiitic magmatism along a network of "seams" in and around the Tunguska basin, which were produced during a period of gentle arching. According to these authors, this phase was followed by related intrusive rocks (240–200 m.y.b.p.) in the central part of the province. Subsidence followed during the Jurassic, producing another peak of tholeiitic and more alkaline activity at 188–170 m.y.b.p., which was concentrated along the southeastern part of the Platform.

Seismic studies of the Siberian Platform (Pritula *et al.*, 1974) show a "normal" crustal thickness of about 35–40 km along the southeast flank and a slightly thinner crust (about 30 km) under the Tunguska basin, with some deep faults producing stepped changes in crustal thickness. These studies also show that the granitic layer has been attenuated where the crust is thinner under the basin.

The Karroo province and the Paraná basalts are both associated with the tensional regime related to the breakup of Gondwanaland during the Jurassic and Cretaceous periods. The Karroo basalts (Fig. 1.2.2.3), like those of the Siberian Platform, were accompanied by great volumes of roughly horizontal dolerite sills of similar composition that penetrated the underlying continental sedimentary strata of Permian and Triassic age. These clastic rocks indicate a long period of prevolcanic subsidence which was broad and gentle over large areas (e.g., much of South Africa), but which elsewhere (e.g., Zimbabwe, Mozambique) was greatly intensified in troughs apparently localized over orogenic belts in the Precambrian basement (Cox, 1970). In some areas, such as southeast of the major Lesotho basalt region (Stormberg lavas, Drakensberg basalts) and in the Tuli syncline, vertical dikes (apparently feeders) are very abundant, and Du Toit (1929) and Gibson (1966) have pointed out the association of an intense north-south dike swarm with a major eastward-dipping monoclinal flexure to the northeast in the Lebombo area. Dips reach 45° in this monocline where the thickest exposed section of lavas is found. Du Toit (1929) recognized the contemporaneity of dike intrusion, basalt extrusion, and flexuring, and the tensional tectonics involved. The Jurassic age of the Karroo magmatism coincides with the separation of southeastern Africa from Antarctica (Dietz and Holden, 1970).

The volcanic rocks are unevenly distributed across southern Africa, as a result of uneven original thickness and erosion. Especially thick sections occur along the western border of Mozambique (Nunanetsi syncline, Lebombo monocline) where 6000–9000 m of lavas are preserved. Except for the 1500 m of horizontal basalts in the Stormberg lavas of Lesotho resting on the relatively stable Karroo shelf and the South African craton, the other principal accumulations are found in basins or troughs which, following the Karroo sedimentation, were largely controlled by basement structures (Cox, 1970); this control is generally analogous to the control of the Permo-Triassic West Siberian rifting by the recently consolidated and eroded Paleozoic orogenic basement. The lava basins or troughs commonly show normal faults with similar dip directions; Cox (1970) interpreted these as results of crustal thinning along weakened trends which were put under tension by a broad convective flow of mantle asthenosphere, not necessarily with localized upwelling at each site of crustal extension.

The Paraná Basin lavas (Serra Geral Formation) were erupted onto a broad area of post-Upper Triassic aeolian sand (Botucatu Sandstone), some of which also reappears as interbeds between the flood lavas (Fig. 1.2.2.4). The upper Paleozoic-Triassic continental sediments underlying the lavas in turn overlie the crystalline rocks of the Brazilian Shield, in a sequence much like that in southern Africa. The basal lavas in the center of the basin now are far below sea level, indicating major posteruptive subsidence. Upper Cretaceous nonmarine sedimentary strata overlie the lavas. Thousands of dikes, some of which exceed 100 km in length, have been found cutting surrounding rocks and are assumed to have been feeders. Many diabase sills are also present locally, intruding the underlying formations, but extrusive rocks clearly dominate. Radiometric ages indicate that eruption occurred over a period ranging from the late Jurassic to mid-Cretaceous with peak activity in the early Cretaceous (Cordani and Vandoros, 1967) just as the South Atlantic Ocean was beginning to open up. Some of the dike swarms (such as those along the northern coast of São Paulo) have a northeasterly trend parallel to the coast, but others trend northwesterly.

In most of these flood basalt provinces, general crustal subsidence took place contemporaneously with and following eruption, covering the lavas with thick continental sediments; in some, subsidence also preceded the initiation of volcanism. This contrasts with a common model of continental rifting in which uplift (arching) precedes rifting. In the Lebombo monocline of southeast Africa, the major subsidence occurred to the east of at least one principal zone of fissures, just as on the Tertiary Columbia Plateau of North America, where one of the principal source fissure zones, the Grande Ronde dike swarm, is well to the east of the center of subsidence. However, in general, the basalt plateaus themselves and their cover prevent us from determining if subsidence was actually centered over the major feeding fissures which supplied magma from the mantle.

Table 1.2.2.1 Physical aspects of the four Pre-Tertiary basalt plateaus.

Basalt Plateau	Age	Maximum and Average Thickness	Present and Possible Original Area	Approximate Volume	Flow Thickness
Paraná Basin	Late Jurassic to early Cretaceous 119 - 149 m.y.	1500-1800 m ave. 650 m	1,200,000 km^2 orig. 2,000,000 km^2?	650,000 km^3	ave. 50 max. >100m
Karroo Province	Late Triassic to early Jurassic 166 - 206 m.y.	8000-9000 m ave. 1000 m?	140,000 km^2 orig. 2,000,000 km^2?	?	ave. 10m (Lesotho) Sills to 1000 m
Siberian Platform	Early Permian to Triassic (most early Triassic) 216-248 m.y.	3500 m ave. 1000 m	>1,500,000 km^2	575,000 km^3 (911,000 km^3 including intrusions)	ave. 30 m
Lake Superior Basin (Keweenawan)	Late Precambrian 1100-1120 m.y. (−1200?)	8000-12,000 m ave. 5000 m? (several basins)	100,000 km^2† orig. 125,000 km^2?	>300,000 km^3 †	3-30 m ave. 25 max. 400m

†In Lake Superior area only; much greater with rest of mafic rocks of the Mid-continent Gravity High.

Size

Many of the principal physical aspects of the major pre-Tertiary basalt plateaus are listed in Table 1.2.2.1.

Basalts are exposed over very large and roughly equal areas in the Siberian Platform, Karroo, and Paraná Basins, and over a somewhat smaller area in the Keweenawan of North America. According to Lebedev (1964), the Siberian Platform traps underlie an area exceeding 1.5×10^6 km^2; this includes the broad area where associated intrusions cut the underlying strata beyond the lava exposures (Fig. 1.2.2.2). Lur'ye and Masaitis (1966) calculated a total volume of 911,000 km^3 for intrusive, extrusive, and pyroclastic rocks; of this total, 574,000 km^3 are volcanic. The fragmental facies is found in thicknesses as great as 700–800 m (average 350 m), and the basaltic lavas have accumulated to thicknesses of up to 2 km (average 1 km) in the Tunguska basin and of 1–2 km in the Taimyr area (Lur'ye and Masaitis, 1966; Milanovskiy, 1976). Over the whole of middle northern Eurasia, early Mesozoic trap magmatism is estimated to have produced a volume of 2×10^6 km^3 and to have affected an area of about 4×10^6 km^2 (Milanovskiy, 1976).

The Paraná Basin, about 1800 km in length by 800 km wide (Fig. 1.2.2.4), has been explored extensively for petroleum and much stratigraphic information has been gained (Cordani and Vandoros, 1967). The greatest portion of the basaltic plateau lies in southwestern Brazil, with smaller areas in Uruguay, Paraguay and Argentina. Maack (1952) has estimated that the basalts now remaining cover 1.2×10^6 km^2, but abundant dikes and basalts outliers suggest that the original area may have approached 2×10^6 km^2. Drilling and geophysical exploration have established the existence of two areas of maximum basalt thickness: to the north in the Paraná Basin proper (up to 1800 m), and to the south in

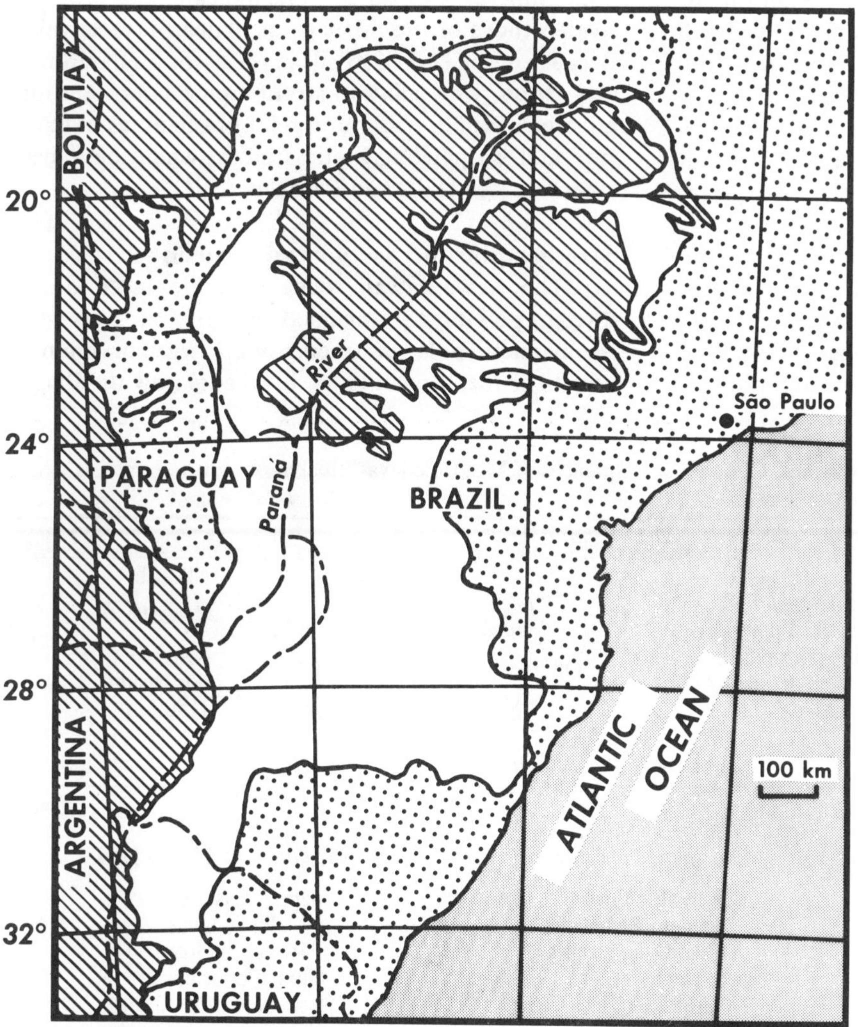

1.2.2.4 Location of Cretaceous Paraná traps. Ruled area: post-volcanic sedimentary strata; blank area: Paraná plateau lavas; patterned area: older rocks. After Cordani and Vandoros (1967).

the Uruguay River area (up to 1200 m) (Beurlen, 1970). Leinz *et al.* (1966) estimated the average basin-wide thickness to be 650 m, with a total volume of 650,000 km^3.

Mafic rocks of the Karroo System, including sills as well as flood basalts, may have originally extended over as much as 2×10^6 km^2 of southern and southeastern Africa (Cox, 1970). The lavas are now exposed over much smaller areas as a result of erosion and of burial by younger Mesozoic sediments. Sills cut the older Permian and Triassic rocks over a wide area to the west (Fig. 1.2.2.3). The major areas of plateau lavas are in Lesotho (formerly Basutoland; Cox and Hornung, 1966), northeastern South Africa, eastern Botswana, and western Mozambique. The thickest development of lavas is in the Nuanetsi-Lebombo area where at least 8000 m of lavas (including considerable rhyolite) are preserved; about 1500 m of basalts constitute the flat-laying sequence in Lesotho (Table 1.2.2.2). No volumetric estimates are offered because of the discontinuous nature of exposures and the unknown amount beneath Cretaceous cover in Mozambique east of the Lebombo monocline, but the total must be very large.

The Keweenawan basalts have the greatest uncertainty in volume, not because of their greater age, but because of their occurrence in several different overlapping plateaus (now basins), each mostly covered by Lake Superior (White, 1972; Green, 1977; Green, 1981) and not well delineated (Fig. 1.2.2.5). For most of these Keweenawan basins, only a single random(?) cross section is exposed, although the youngest basin, containing the Portage Lake volcanics, is stratigraphically well-documented because of intensive diamond drilling and mining development for native copper (White, 1960, 1968; Huber, 1973). However, the base of this sequence is cut out by a major thrust fault. Most of these plateau units have central thicknesses of 2.5–7 km (Table 1.2.2.3). Geophysical modelling of the area south of eastern Lake Superior (Oray *et al.*, 1973) gives similar thicknesses. The lavas now underlie an area of about 100,000 km^2 in the Lake Superior district (including under the lake); their former extent, as suggested by peripheral dike swarms, sills, and stratigraphic/ structural inferences, was probably roughly 125,000 km^2. White (1960) estimated a volume of over 400,000 km^3 but, because the mafic Keweenawan igneous rocks continue in the subsurface for at least 1000 km to the south-southwest from western Lake Superior (the mid-continent gravity high) and for at least several hundred kilometers south of eastern Lake Superior, the original areas and volumes were certainly much greater than the Lake Superior exposures alone indicate, and probably rivaled those of the other great plateau lava terrains. Because of the lack of fossils in the associated sedimentary strata, Keweenawan stratigraphic interpretations must be based on structural relations in areas of exposure and on paleomagnetic studies, which have been extensive (see Halls and Pesonen, 1981).

Stratigraphy and associated intrusions

The Keweenawan rocks were emplaced during three or possibly four magnetic polarity intervals while the magnetic pole was tracing out a large curving path known as the "Logan loop" in what is now the Pacific

Table 1.2.2.2 Stratigraphic table, Karroo province. Generalized sequences of Mesozoic lavas (thicknesses given in meters where known).†

(a) Central part of continent

North *Victoria Falls (Zambesi Valley)*	*Tuli Syncline (Limpopo Valley)*	*Transvaal (Springbok Flats)*	South *Lesotho*
Olivine-poor basalts (1000 m)	Olivine-poor basalts Olivine-rich basalts (total 400 m)	Olivine-poor basalts (400 m)	Olivine-poor basalts (1500 m)

(b) Eastern part of continent

North *Lupata (Zambesi Valley)*		*Nuanetsi, Sabi and Northern Lebombo*	South *Southern Lebombo (Swaziland and Mozambique)*
Phonolites, rhyolites, trachytes	Lupata Series	Rhyolite Group Olivine-poor basalts Olivine rich basalts Nephelinites (total 6700m)	Upper Basalts and alkalic lavas Rhyolite Group Olivine-poor basalts (estimated *c.* 9000 m)
Rhyolite Basalt	Karroo		

†After Cox (1970).

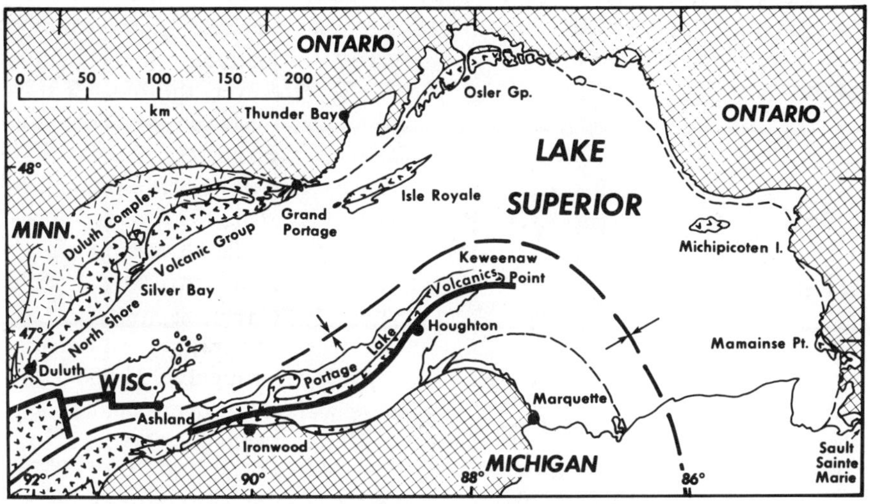

1.2.2.5 Generalized geologic map of the Keweenawan rocks of the Lake Superior region (after Green, 1977). (Courtesy of the Geological Association of Canada.)

Ocean (Palmer, 1970; Robertson and Fahrig, 1971; Books, 1972; Robertson, 1973). Although relationships at Mamainse Point, Ontario, could be interpreted as showing four polarity intervals, or two intervals with fault repetition (R, N, R, N), igneous rocks elsewhere around the Lake Superior basin fit well into a paleomagnetic framework of a lower normal polarity interval (thin basal succession in Michigan only), one major reversed interval, and a major upper normal interval. Correlation from one part of the Lake Superior area to another is so far possible only through these paleomagnetic field reversals, with the notable exception of Isle Royale in northwestern Lake Superior. Here the mapped lava sequence duplicates that of the Portage Lake Volcanics 80 km across the Lake Superior syncline on the Keweenaw Penisula (Huber, 1973). As mentioned above, the Keweenawan plateau lavas occur in several separate but somewhat overlapping accumulations (Green, 1977; Fig. 1.2.2.6), each containing a few hundreds of flows. The stratigraphy in most of these accumulations is well known to the level of individual flows or flow groups (e.g., White, 1968; Sandberg, 1938; Green, 1972; Annells, 1973, 1974; Hubbard, 1975; McIlwaine and Wallace, 1976; Fig. 1.2.2.7).

Table 1.2.2.3 Thicknesses of individual plateaus of the Keweenawan province.†

Plateau, with subunits	Described thickness, km	Magnetic polarity
Keweenaw Point — Isle Royale	2.5-5.2	Normal
Keweenaw Point	2.4-5.2	
Isle Royale	3.1	
Chengwatana	6.0	Normal
North Shore normal	3.7-7.1	Normal
Mamainse — Michipicoten	4.1-5.1	Normal
Mamainse Point	2.6	
Michipicoten Island		
Mamainse Pt. Fm.	1.0	
Michipicoten I. Fm.	1.5	
Mamainse reversed	2.5	Reversed
Mamainse Point	1.2	
Alona Bay	1.3	
Osler	2.75	Reversed
Ironwood — Grand Portage — Nopeming	3.0-6.3	Reversed
Powdermill Group	2.4-5.1	
Grand Portage — Hovland	2.7-3.0	
Ely's Peak	0.4	
Siemens Creek	0.12	Normal

†After Green (1977, 1981).

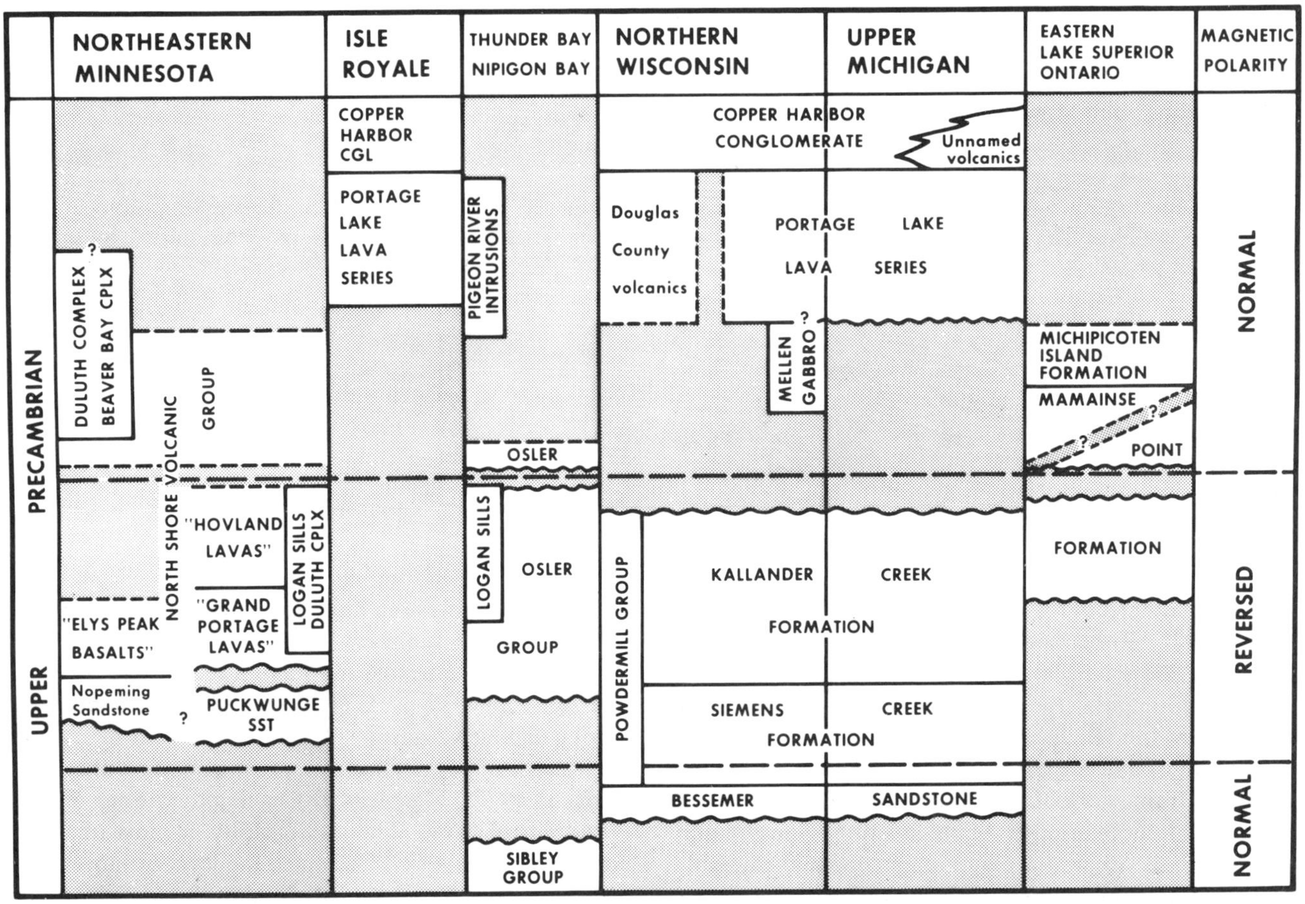

1.2.2.6 Stratigraphic diagram for Keweenawan igneous rocks, Lake Superior area (Green, 1977). (Courtesy of the Geological Association of Canada.)

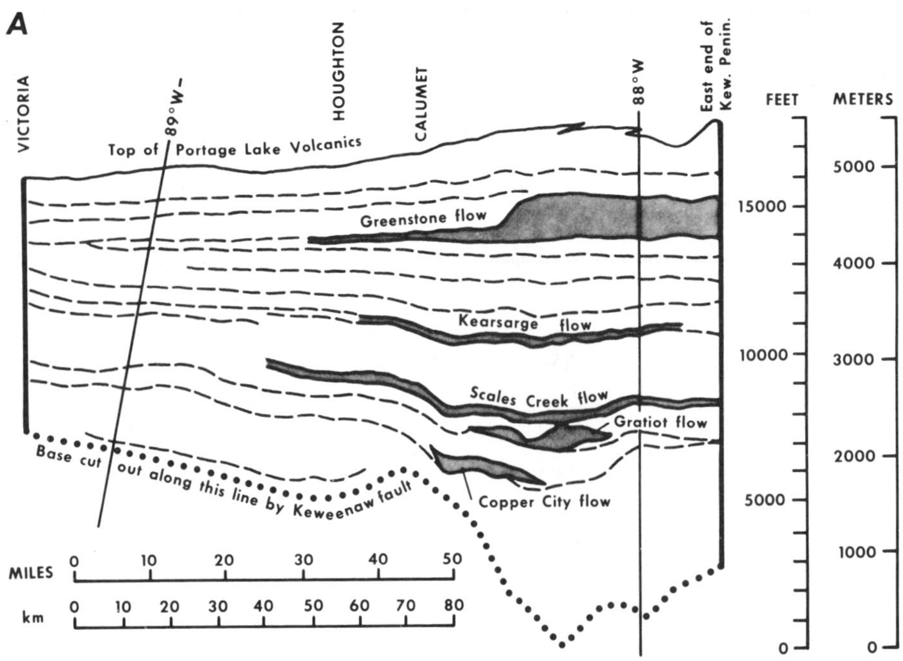

1.2.2.7 Stratigraphy of two Keweenawan lava plateaus. **a**. Portage Lake Volcanics, modified after White (1960). **b**. North Shore Volcanic Group, from Green (1972). (Courtesy of the Minnesota Geological Survey.) These lavas are all older than those in **a**.

Only two large central volcanic complexes have been identified (Michipicoten Island, Ontario, and Porcupine Mountains, Michigan), consisting largely of intermediate and felsic rocks, but others may be present in Minnesota. The two identified stratovolcanoes occur at the top of their respective plateau sequences, but the abundant felsic lavas in Minnesota have intermediate stratigraphic position. In both Michigan and Minnesota there is a gross progression from more alkaline basalts and basaltic andesites in the lower part of the section (reversed polarity plateaus) to more olivine-tholeiitic in the upper normal-polarity plateaus, but there is considerable mixing and alteration of lithologies. In several local sequences of tholeiitic basalts, cyclic compositional variations have been found (McIlwaine and Wallace, 1976; Brannon *et al.*, 1979; Rose and Grimes, 1979) in which Fe and Si change in inverse sense. In three widely separated places around the western part of the basin the first lavas to be erupted were a distinctive chemical and petrographic type containing olivine and Mg-rich augite phenocrysts (which settled in some flows) and somewhat elevated abundances of incompatible elements (see KEW-13 in the Keweenawan reference suite). Picritic basalt enriched in olivine is also found near the base of the section on the east side of the lake at Mamainse point. Farther up in the reversed-polarity sequence all around the basin (except for the Osler Group), basaltic andesites and K-rich basalts containing abundant large plagioclase phenocrysts are common. The olivine tholeiites, though present and mixed with other basalt types in most Keweenawan plateaus, are most abundant in the upper normal polarity sequences in Minnesota, Ontario and Michigan.

B

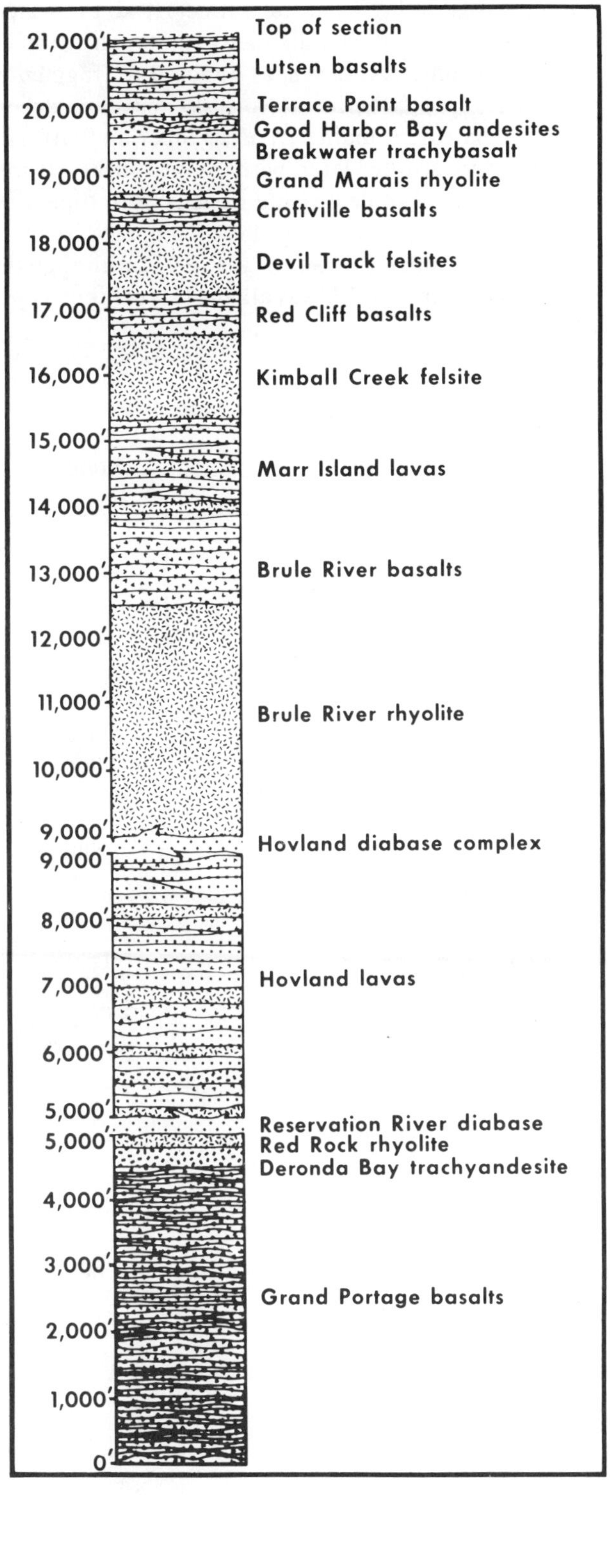

Intrusive rocks are relatively rare in the Keweenawan plateaus that developed on consolidated older basement, except for the great Duluth Complex, one of the world's largest mafic intrusive bodies (Sims and Morey, 1972; Weiblen and Morey, 1980). Consisting of many separate intrusions dominated by troctolite and gabbroic anorthosite but ranging to granite, the Duluth Complex was emplaced generally along the basal contact of the lavas on top of Middle and Lower Precambrian rocks, but some portions were intruded between the older, reversely-polarized and younger, normally-polarized Keweenawan lavas. A large volume of diabase sills ("Logan sills") of reversed polarity intrudes the flat-lying sedimentary Rove Formation of Middle Precambrian age and Sibley Group of pre-Keweenawan late Precambrian age beyond the plateau lavas northwest of Lake Superior. Several diabase sills and other plutons also cut the lavas in the northwest part of Lake Superior in both Minnesota (Green, 1972) and Ontario (McIlwaine and Wallace, 1976), and a few gabbroic and granitic plutons occur in the Keweenawan section in northern Wisconsin.

In contrast to most of the world's basalt plateaus, the Siberian Platform volcanism began with large volumes of basaltic tuffs and breccias, and this pyroclastic activity occasionally reappeared later in the volcanic cycle. Associated with the extrusive rocks is a very large volume of subvolcanic intrusive bodies including dikes, sills, laccoliths, chonoliths and plugs. The sills range from several tens to a few hundreds of meters in thickness, and some of the larger ones are differentiated. These intrusions are concentrated along zones of flexure and particularly around the margins of the Tunguska basin. Because of the important Cu, Ni and Fe ores contained in them, the intrusions ("intrusive traps" in Russian literature) have received much more detailed study than the lavas. Godlevskiy (1959) described four cycles of volcanism and intrusive activity in the Noril'sk region in the northwestern part of the basin. It began with a cycle in the Permian, which was succeeded by three Triassic cycles, all of which can be correlated throughout much of the Platform (Table 1.2.2.4). These cycles are separated by erosion surfaces marking periods of quiescence. Within each cycle lava compositions proceeded from a more felsic to a more mafic composition upwards in the sequence, but nearly all are generally basaltic. In each cycle distinctive units can be traced for many kilometers. Lavas, as opposed to fragmental rocks, were erupted earlier at Noril'sk (Permian, earliest Triassic) than in the main part of the Siberian Platform, and also were erupted later, perhaps into Jurassic time. An overall trend of decreasing basicity with time (and with proximity to the center of the Tunguska basin) is reported by Dmitriev (in Afans'yev, 1975), although the earliest phase at Noril'sk was alkalic (Dyuzhikov *et al.*, 1973).

Table 1.2.2.4 Scheme of development of trap volcanism in Noril'sk Region.†

Volcanic Cycles	Extrusive Facies	Intrusive Facies	Form of Intrusion
III Triassic		Diabasic titanaugite porphyrites	Dikes
	Porphyritic basalts	Gabbro-dolerites	
		Vitrophyric porphyrites	
	Tholeiitic basalts		
	Tuff sandstones		
Break, considerable erosion and weak folding			
II Triassic		Gabbro-diabases with ortho-pyroxene, differentiated (with associated Cu-Ni-Pt ore)	Chono-liths
	Porphyritic (holocrystalline) and vitrophyric basalts	Dolerites and gabbro-dolerites	Sills
	Tholeiitic basalts and diabases	Tholeiitic and ophitic diabases	
	Tuffs and tuffites	Labradorite porphyrites	
Break, weak erosion			
I Triassic	Picrite porphyrites	Vein picrites	Dikes
	Porphyritic (holocrystalline) and vitrophyric basalts	Vitrophyric porphyrites	Sills
		Intersertal diabases	
		Diabasic porphyrites	
		Amphibolized diabases	
	Aphyric basalts		
	Tholeiitic diabases and basalts	Tholeiitic and ophtic dolerites and diabases	Sills
	Two-feldspar porphyritic basalts		
Break, strong erosion			
Permian	Porphyritic hypocrystalline basalts and dolerite-basalts, sometimes with titanaugite	Diabasic porphyrites with titanaugite	Sills
	Spilite-like rocks (sub-alkaline diabases)		
	Vitroclastic, less often lithoclastic basalt tuffs		

†After Godlevskiy (1959).

The Karroo lavas also occur in separate areas with different stratigraphies (Table 1.2.2.2), as described by Cox (1971). It is difficult to correlate units between the major areas of exposure except through general petrographic/chemical criteria which may not be isochronal. In the northern Lebombo monocline and Tuli and Nuanetsi synclines, which are nearly connected, a general progression occurs from olivine-rich basalts (including "limburgites") with occasional nephelinites at the base, up through olivine-poor basalts to rhyolites at the top. To the south along the Lebombo monocline, the main unit is olivine-poor basalt; olivine-rich and alkaline basalts are found at the top, overlying and interbedded with rhyolites (Fig. 1.2.2.8; Cox *et al.*, 1965). Because of relatively poor exposure, detailed stratigraphy has not been determined for these sequences, although several petrographically distinctive groups of basalts within the upper basalts and the upper rhyolitic group can be traced over long distances in the Nuanetsi syncline. In the slightly younger (Cretaceous) Lupata series of Mozambique, rhyolites are found with alkalic differentiates such as trachytes and phonolites, and rare absarokite and shoshonite are described from the Tuli syncline. In the Drakensberg tholeiites of Lesotho the flows, though well exposed, are so uniform as to preclude stratigraphic subdivision (Cox and Hornung, 1966). Associated with the Karroo lavas is a major group of coeval tholeiitic sills which intrude the underlying Karroo sedimentary rocks over a large area of South Africa (Walker and Poldervaart, 1949; LeRoex and Reid, 1978). Several smaller intrusions, some of alkalic nature and including carbonatite, occur in the northeastern areas of Nuanetsi-Sabi and Lupata.

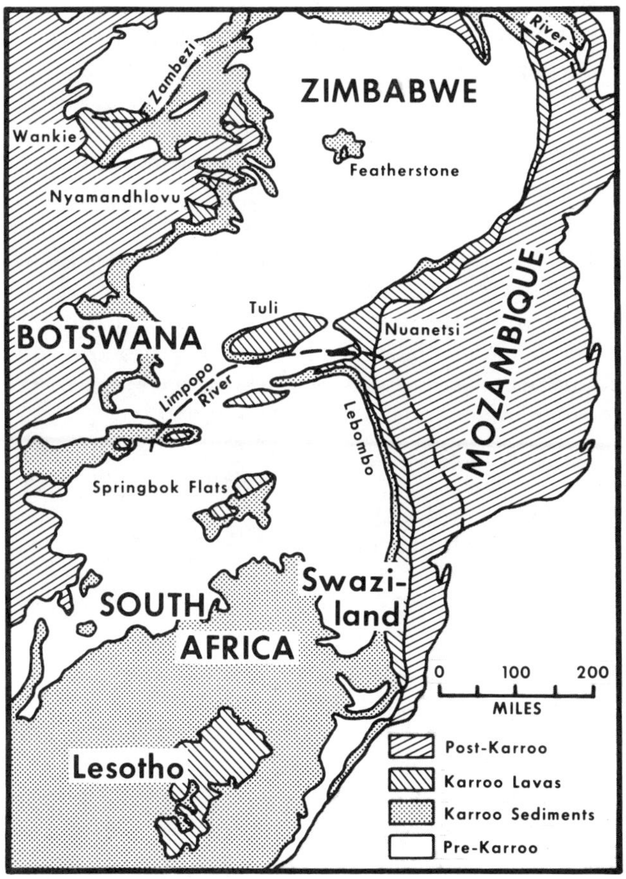

1.2.2.8 Generalized geologic map of northeast Karroo area. After Cox (1971).

Little regional stratigraphic study has been carried out in the Paraná Basin, at least partly because of the general uniformity of character of the lavas and lateral variations in thickness of flows. In some areas thick interflow sandstones can be used to separate volcanic sequences. Using trend-surface analysis of major element chemistry, Ruegg and Amaral (1976) have found four areal geochemical "anomalies" or subregions of different geochemical character. This analysis, however, is geographic, not stratigraphic, and gives little information concerning vertical succession or lateral correlation of individual units. In general, this analysis indicates that the most magnesian basalts are those in areas of the Paraná Basin where the lava sequence is thickest, 1000 m and more according to the isopachs presented by Leinz *et al.* (1966), and thus may imply that the middle and upper part of the section in the deepest portion of the basin is most magnesian.

Many dike swarms, containing thousands of dikes, cut the rocks underlying and surrounding the Paraná traps, and a large number of sills, some over 200 m thick, occur in underlying sedimentary formations, but they are clearly subordinate in volume to the lavas. One sill is estimated to extend over an area of 10 km^2 and a ramified sill system in São Paulo state covers about 900 km^2 (Cordani and Vandoros, 1967). A small alkaline intrusive center is located at Sapukai in eastern Paraguay, west of the erosional edge of the Paraná traps (Palmieri and Arribas, 1975).

Age

These major basalt plateaus reflect continental rifting episodes that persisted over times on the order of 10^6 to 10^7 years. None older than about 1200 m.y. is known; after the Keweenawan (and the contemporaneous Coppermine basalts of northern Canada) a long hiatus of 900–1000 m.y. occurred before rifting again produced large basalt plateaus. That at least five have formed during the past 250 m.y. is surely significant for crustal evolution. Are continents easier to split now than they were in the Precambrian?

The time span for the Keweenawan igneous activity is of interest in outlining the history and character of rifting and as a means of calibrating the paleomagnetic reversals and polar wander curve recorded in the rocks. Most of the rocks, however, have been subjected to some low-grade burial metamorphism and/or hydrothermal alteration. Early K/Ar dates (Goldich *et al.*, 1961) covered a range from about 1.2–1.0 b.y.b.p. Newer K/Ar investigations on Canadian Keweenawan gabbros and diabases (Wanless *et al.*, 1970; Hanson and Malhotra, 1971) give a range of 1300–924 m.y.b.p.; the 1300 m.y.b.p. age is on a reversely-polarized "Logan sill." Rb/Sr dates on Keweenawan igneous rocks range from 1115 m.y.b.p. to about 900 m.y.b.p. (for a summary see Van Schmus *et al.*, 1981). Many of these younger ages, however, are at variance with evidence from paleomagnetic reversals and polar wander. Hanson (1975) has obtained a $^{40}Ar/^{39}Ar$ spectrum age of 1150–1170 m.y.b.p. for Logan sills, but with a considerable, though unspecified, error range. U/Pb studies (Silver and Green, 1972) over nearly the entire stratigraphic span of Keweenawan igneous bodies, both intrusive and extrusive, show a remarkably narrow time span of about 1120–1100 m.y.b.p. for all of the main normal-polarity igneous events and for at least the younger part of the previous reversed-polarity interval, thus fixing the time of field reversal. No age determinations have been reported for flows of the lower-normal magnetic polarity interval during which Keweenawan activity commenced. Thus the U/Pb evidence agrees with the older Rb/Sr ages, but not with the younger ones, and is at variance with many of the published K/Ar dates. These radiometric studies thus suggest a total span of perhaps 80–100 m.y. for Keweenawan rift-related igneous activity, with the bulk of it occurring during the last 20 m.y. of that range.

The Siberian traps are dated paleontologically as beginning in the late Permian, with peak activity in the early Triassic, but lasting into earliest Jurassic. Radiometric age determinations (mostly K/Ar) span the range 172–283 m.y.b.p. with a great concentration between 216 and 248 m.y.b.p. (Kulikov *et al.*, 1972; Feoktistov, 1975; Kuznetsov and Naumov, 1975), essentially in agreement with the stratigraphic evidence.

The Karroo mafic igneous activity succeeded continental deposition of the Karroo system, which ranges in age from upper Carboniferous to lower Jurassic. The youngest lavas interfinger with Cretaceous rocks in Mozambique. Radiometric age determinations on both extrusive and intrusive units, summarized by Cox (1971) range from 209–114 m.y.b.p. by Rb/Sr and K/Ar methods. Of these the great bulk are Jurassic, the early Cretaceous rocks being located mainly at the top of the section at Lupata, Mozambique and in the Kaokoveld basalts of Namibia (Southwest Africa).

The Paraná basalts can be dated stratigraphically as post-Upper Triassic. The overlying Bauru Formation is considered to be upper Cretaceous. Radiometric age determinations, summarized by Cordani and Vandoros (1967), cluster in the period 120–130 m.y.b.p. (mid-Early Cretaceous) with a total range from about 100–150 m.y.b.p. Thus the Paraná volcanism was slightly younger than most of the Karroo activity but coincident with that of Namibia.

1.2.2.3 Petrography and mineralogy

Textures

Textures of the basalts in these four plateaus show strong similarities, ranging from holocrystalline ophitic, subophitic and intergranular to hypocrystalline intersertal. The thickest flows in all areas tend to be ophitic.

The diameter of the clinopyroxene oikocrysts reaches 2–3 cm in the thickest flows of the Keweenawan, but 3–8 mm is more common. Coarse-grained ("pegmatitic," up to 10 mm grain size) schlieren and layers are found in a few of the largest flows. Textural type generally correlates with composition as well: olivine tholeiites are ophitic, whereas quartz tholeiites tend to be aphanitic or fine-grained with intergranular or intersertal textures. In the latter, glass or its devitrified or altered product commonly makes up between 10 and 30% of the rock, and occasionally more. Unaltered glass is common in the Paraná and Karroo basalts, occurs in some Siberian Platform tholeiites, and has also been found in a very few tholeiites and andesites of the Keweenawan (Annells, 1974; Green, 1972).

In the ophitic olivine tholeiites, most but not all of the olivines occur outside the poikilitic augite-plagioclase intergrowths, suggesting that all three minerals crystallized concurrently but that the growing pyroxene preferentially excluded olivines while incorporating small plagioclase laths of comparable size (Annells, 1974). Oxide minerals are late-crystallizing and interstitial in these ophitic rocks. This is supported by experimental data (Green, 1979b).

Porphyritic texture is unusual in the Lesotho basalts except for micro-phenocrysts of plagioclase. A few flows near the base contain up to 12% olivine phenocrysts. In the Lebombo monocline, plagioclase-porphyritic basalts are typical of the tholeiitic series, whereas olivine and augite are more common as phenocrysts in Zimbabwe (Cox *et al.*, 1967). In the Paraná and Lake Superior areas porphyritic rocks are fairly common, although aphyric lavas seem to predominate. Plagioclase is the only common phenocryst phase in the Paraná traps. In the Keweenawan plateaus a few glomeroporphyritic basalts or andesites, as well as normal porphyritic rocks are found in nearly all sequences; they are particularly abundant in the reversed-polarity (older) lavas. Plagioclase is again by far the most common phenocryst mineral, but olivine and/or augite phenocrysts are abundant in some flows near the base of the section in Minnesota, Michigan, and Mamainse Point, Ontario. Plagioclase phenocrysts occasionally attain lengths of 3–5 cm (Hubbard, 1975; Green, 1972). Skeletal olivine and clinopyroxene are found in some "limburgites" with glassy matrices in the northeastern Karroo province (Cox *et al.*, 1965; Vail *et al.*, 1969).

Diktytaxitic cavities are found in many of the lavas and commonly are filled with low-temperature secondary minerals. The finer-grained lavas (generally quartz tholeiites and basaltic andesites) commonly exhibit a laminated structure on weathered surfaces, with alternating subparallel lighter and darker layers 1–3 mm thick that appear to represent different degrees of oxidation. A coarser, cruder banding on a scale of 4–10 cm is found in some Keweenawan ophitic lavas; its origin is obscure.

The textural and structural characteristics of the pre-Tertiary plateau lavas described above suggest that: (a) they were very fluid; (b) perhaps most are aphyric, thus representing slightly superheated magmas and constituting dependable samples (when fresh) of

Table 1.2.2.5 Modal composition ranges of Siberian Platform Lavas.

Rock	No. Samples Represented		Plagioclase	Clinopyroxene	Olivine	Ti-Magnetite	Mesostatis (including glass)
1.	40	Vol. % Range	34.7-56.1	11.5-46.3	2.0-13.2	1.0-5.6	4.0-44.1
		Mean	44	28.7	5.9	3.0	17.6
		Comp. gdmass					
		Range	An_{74-43}	augite, ferroaugite	Fo_{85-35}		
		Mean	60	pigeonite	64		
		Comp. pheno.					
		Range	An_{88-70}				
2.	12	Vol. % Range	27.9-58.8	26.2-60.6	2.9-12.1	1.6-6.9	1.0-25.4
		Mean	44.4	36.5	6.3	6.3	9.9
		Composition					
		Range	An_{68-53}	augite, ferroaugite	Fo_{80-35}		
		Mean	62	salite	62		
3.	4	Vol. % Range	10.5-42.8	0.0-3.9	2.3-8.1		51.2-79.4
		Mean	26.8	1.0	5.6		66.6
		Composition					
		Range	An_{60}	?	Fo_{85}		
4.	8	Vol. % Range	21.8-55.0	24.8-46.1	1.0-1.9	3.3-7.5	5.9-38.5
		Mean	33.9	35.8	1.6	5.7	23.0
		Comp. gdmass					
		Range	An_{64-45}	augite, ferroaugite	?		
		Mean	54				
		Comp. pheno.					
		Range	An_{80}				
5.	2	Vol. % Range	35.4-57.2	37.8-44.7	0.6-1.4	2.3-4.3	2.1-10.2
		Mean	48.3	41.2	1.0	3.5	6.0
		Comp. gdmass					
		Range	An_{62-60}	augite	?		
		Mean	61				
		Comp. pheno.					
		Range	An_{88-82}				
6.	1	Vol. %	8.7		1.1		90.2
		Composition	?		?		
7.	43	Vol. %	43.4	31.4	4.4	3.5	17.3
		Composition	An_{60}	$En_{42}Wo_{42}Fs_{16}$		TiO_2-17.12	
8.	1	Vol. %	48.7	24.3	15.9	3.0	8.1
		Composition	An_{72}	ferroaugite	Fo_{69}		

Table 1.2.2.5 (Continued)

Rock	No. Samples Represented		Plagioclase	Clinopyroxene	Olivine	Ti-Magnetite	Mesostatis (including glass)
9.	1	Vol. %	53.1	29.2	10.0	3.8	3.1
		Composition	An_{77}	ferroaugite	Fo_{67}		
10.	6	Vol. % Range	41.8-56.5	10.1-13.7	3.0-13.1	1.2-4.7	8.2-30.8
		Mean	47.2	24.2	8.9	3.2	16.5
		Composition					
		Range	$An_{77\text{-}64}$	ferroaugite, augite	$Fo_{85\text{-}53}$		
		Mean	71		68		
11.	4	Vol. % Range	20.0-28.3	in aphanitic	2.1-5.2	3.9-9.4	60.2-72.4
		Mean	22.8	groundmass	3.8	5.8	67.4
		Composition					
		Range	An_{80}		$Fo_{60\text{-}57}$		
		Mean			59		
12.	2	Vol. % Range	20.7-24.7	in aphanitic	0.5-0.9	5.9-7.9	66.9-71.1
		Mean	22.7	groundmass	0.7	6.9	69.0
		Composition					
		Range	$An_{81\text{-}80}$		?		
13.	22	Vol. % Range	40-65	20-30		3-10	10-20
		Mean					(1-3 apat.)
		Comp. gdmass					
		Range	$An_{60\text{-}40}$	$En_{12\text{-}36}Wo_{40\text{-}50}Fs_{32\text{-}48}$			
		Comp. phenos.					
		Range	$An_{68\text{-}50}$				
14.	14	Vol. % Range	55-65	15-25		2-3	5-15
		Mean					(2-3 apat.)
		Comp. gdmass					
		Range	$An_{50\text{-}45}$				
		Comp. phenos.					
		Range	$An_{62\text{-}45}$				
15.	22	Vol. % Range	50-60	15-25		3-5	10-20
		Mean					(1-2 apat.)
		Comp. gdmass					
		Range	$An_{60\text{-}38}$	$En_{30\text{-}40}Wo_{35\text{-}44}Fs_{25\text{-}30}$			
		Comp. phenos.					
		Range	(a) $An_{50\text{-}40}$	(corroded)			
		Mean	(b) $An_{68\text{-}38}$				
16.	?	Vol. % Range	41-59	17.26		2-4	13-34
		Composition Range	$An_{70\text{-}56}$	$En_{35\text{-}53}Wo_{39\text{-}46}Fs_{8\text{-}23}$			n 1.52-1.54

1. Olivine basalt	Central part of Tunguska basin:	Dmitriev, 1973, Table 4.
2. Olivine anamezhite	Central part of Tunguska basin:	Dmitriev, 1973, Table 4.
3. Olivine vitrobasalt	Central part of Tunguska basin:	Dmitriev, 1973, Table 4.
4. Basalt	Central part of Tunguska basin:	Dmitriev, 1973, Table 4.
5. Anamezhite	Central part of Tunguska basin:	Dmitriev, 1973, Table 4.
6. Vitrobasalt	Central part of Tunguska basin:	Dmitriev, 1973, Table 4.
7. Anamezhite of Tembenchi River basin	Central part of Tunguska basin:	Dmitriev, 1973, Table 4.
8. Troctolitic basalt	Marginal part of Tunguska basin:	Dmitriev, 1973, Table 2.
9. Olivine basalt with orthopyroxene	Marginal part of Tunguska basin:	Dmitriev, 1973, Table 2.
10. Olivine basalt	Marginal part of Tunguska basin:	Dmitriev, 1973, Table 2.
11. Olivine anamezhite	Marginal part of Tunguska basin:	Dmitriev, 1973, Table 2.
12. Anamezhite	Marginal part of Tunguska basin:	Dmitriev, 1973, Table 2.
13. Basalts with Ti-augite	Noril'sk area, Dyuzrikov et al., 1973	
14. Labradorite-porphyritic basalts	Noril'sk area, Dyuzrikov et al., 1973	
15. "di-feldspathic" porphyritic basalts	Noril'sk area, Dyuzrikov et al., 1973	
16. Tholeiitic basalts	Noril'sk area, Godlevskiy, 1959	

crystal-free magmatic liquids; (c) nearly all of those that are porphyritic reached the plagioclase liquidus field first, though in some transitional basalts and andesites olivine or augite was the first mineral to crystallize.

Modal composition

Published modes of these plateau lavas are not abundant. Some modal analyses of Siberian Platform basalts are given in Table 1.2.2.5 and modes of the Keweenawan lavas of Michipicoten Island and of some Karroo "limburgites" and olivine-rich basalts are presented in Table 1.2.2.6. Modes for samples in the Keweenawan reference suite are found in Table 1.2.2.18.

Modes of these plateau lavas are fairly uniform. The olivine tholeiites typically contain from 8–15% olivine, along with labradorite, augite, interstitial pigeonite, ilmenite and titanomagnetite. Small interstitial clots of residual material may show apatite, K-feldspar, quartz after tridymite or alteration products, as well as overgrowths on the plagioclase and pyroxenes.

Table 1.2.2.6 Modal composition of Karroo and Keweenawan lavas.†

A. Modes of olivine basalts and enstatite-normative "limburgites," Karroo province, Nuanetsi syncline (Cox *et al.*, 1965).

	LM 338	LM 434	(volume %) LM 430	LM 447	KC 32	KC 42(a)	LM 592
olivine + alteration products	5.0	5.7	6.9	9.4	21.2	25.2	26.7
clinopyroxene	46.0	28.7	27.2	35.3	4.0	31.1	16.2
orthopyroxene	—	—	—	—	1.2	—	0.8
plagioclase	44.7	23.3	33.1	35.3	—	—	17.5
opaques	4.6	14.9	6.9	3.0	—	9.6	5.5
glass	—	26.9	25.4	16.9	—	—	28.4
groundmass (undifferentiated)	—	—	—	—	73.5	34.1	—
amygdule material	—	0.7	0.5	—	—	—	5.0

	LM 452	KC 94	(volume %) LM 432	KC 46	LM 428
olivine	14.4	16.7	25.0	29.1	19.4
clinopyroxene	26.0	22.8	24.5	15.7	47.0
opaques	5.2	5.4	10.0	8.9	} 33.6
glass	52.3	53.8	39.2	46.3	
amygdule material	2.1	1.3	1.3	—	—

B. Modes of Keweenawan olivine tholeiites and tholeiite, Michipicoten Island, Ontario (Annells, 1974).

Modal composition of olivine tholeiite lavas

	AK 133	AK 125	AK 144	AK 108
plagioclase	60.6	51.2	57.8	48.9
clinopyroxene	16.4	26.6	24.3	35.8
olivine	19.3	17.6	15.2	8.3
opaques	2.1	1.9	2.2	3.6
others	1.6	2.7	0.5	3.4

(includes chlorite, epidote, zeolite, apatite, etc.)

Mode of South Shore tholeiite

	AK 14
plagioclase	49.9
pyroxene	35.2
opaques	7.3
mesostasis	7.2
others (carbonate, etc.)	0.4

†See also Table 1.2.2.18.

LM 338, dike — Chilembeni Hill, cutting complex; LM 434, flow — Davata Dip; LM 430, probable flow — 2 miles west of Chilembeni; LM 447, dike — Chilembeni Hill, cutting complex; KC 32, dike — Chisume River; KC 42(a), flow — north of Bumburudza River; LM 592, dike — east of Shamandali.

The finer-grained silica-saturated tholeiites and basaltic andesites contain only a few percent or no olivine (typically altered), along with plagioclase, augite, pigeonite, ilmenite and Ti-magnetite. Magnetite appears earlier in the crystallization sequence and commonly is euhedral or subhedral. Interstitial areas may be glassy or may contain apatite, oxides, K-feldspar, quartz and/or alteration products.

The more andesitic lavas, such as in the central volcano of Michipicoten Island in Lake Superior (Annells, 1974), contain phenocrysts of labradorite and two or three pyroxenes, although no orthopyroxene is found in the groundmass.

Partly resorbed phenocrysts, evidently derived from a deep source and not in equilibrium with the low-pressure extrusive liquid, are reported from some localities: Dyuzhikov *et al.* (1973) describe resorbed calcic andesine in basalts which also contain a later generation of labradorite and Vail *et al.* (1969) and Cox *et al.* (1965) report resorbed orthopyroxene in "limburgite" from the Tuli and Nuanetsi synclines.

Whereas most of the Siberian and Keweenawan plateau basalts are holocrystalline, the Paraná and Karroo lavas commonly contain 10–40% glass or more.

Amygdules, diktytaxitic cavities and veins in the lavas are filled with a great variety of secondary minerals including chalcedony, calcite, chlorite, epidote, prehnite, pumpellyite, native copper, and many zeolites, depending largely on depth of burial, proximity to intrusions and rock permeability.

Mineral chemistry

Information on the compositions of the minerals of these plateau lavas is highly variable. The following discussion gives a general impression of characteristic mineral compositions in pre-Tertiary plateau basalts, and Table 1.2.2.5 contains some mineral composition data from the Siberian Platform. Feoktistov (1975) and Nesterenko and Al'mukhamedov (1973) summarize much information on the mineralogy of the associated intrusions. More specific data on the minerals of the Keweenawan North Shore Volcanic Group follow in section 1.2.2.5. Much detailed microprobe mineral chemistry for the lavas of the Keweenawan Portage Lake Volcanics can be found in Scofield (1976), and detailed optical studies of plagioclases of the Paraná intrusive diabase sheets and sills have been published by Ruegg (1972).

Plagioclase. Phenocrysts in the basalts have bytownite cores (An_{80-85}) and labradorite rims (An_{66-50}). Groundmass laths are labradorite and are also zoned, ranging from An_{70} to An_{50} with some andesine rims down to An_{45-40}. The phenocrysts in the andesites are medium labradorite or occasionally calcic andesine. Secondary alteration in some areas has converted original plagioclase to albite, pumpellyite, chlorite, epidote, prehnite, or zeolite (Jolly and Smith, 1972; Scofield, 1976).

Pyroxene. Available analyses show typical augite compositions of roughly Wo_{40} En_{42} Fe_{18} with some outer zones showing increasing Fe content at the expense of both Ca and Mg. Al_2O_3 contents are generally between 2 and 3% in the augite of the Keweenawan Portage Lake Volcanics (Scofield, 1976) and North Shore Volcanic Group. Pigeonite in the Keweenawan basalts is generally Wo_{7-10} En_{52-55} Fs_{37-38}. Orthopyroxene phenocrysts in saturated tholeiites of Michipicoten Island (Annells, 1974) are $Wo_4En_{64}Fs_{32}$.

Olivine typically is altered, but in the few fresh samples reported and studied it is Fo_{72-68} (phenocrysts, Karroo, Lesotho) or Fo_{63} (groundmass, North Shore Volcanics, Keweenawan).

No published data were found for compositions of primary oxide minerals in the lavas; see section 1.2.2.5.

1.2.2.4 Petrochemistry

Major elements

Plateau lavas have long been considered one of the foundations on which the concept of **tholeiitic basalt** has been established, and with very few exceptions the lavas of the four pre-Tertiary plateaus considered here are hypersthene-normative and fall within either the olivine tholeiite or the quartz tholeiite volumes of the Ol-Ne-Di-Qz tetrahedron (Yoder and Tilley, 1962;

Table 1.2.2.7 Normative character of the Pre-Tertiary basalt plateaus.

Plateau	Age	% Normative Character		
		Q	Ol + Hy	Ne
Paraná (50)	Cretaceous	72	28	0
Karroo (107)	Jurassic	57	37	6
Siberia (61)	Permo-Triassic	28	69	3
L. Superior (160)	Late Precambrian	42	51	7

Table 1.2.2.7). Petrographically, as well, these rocks tend to contain the Ca-poor pyroxenes (including subcalcic augite) and the SiO_2-saturated or oversaturated residual material and associated lavas that are characteristic of the tholeiite "family." However, many of the rocks in these plateaus do not satisfy one or more of the other common chemical criteria for tholeiitic rocks. For example, about half of the Paraná basalts and about one-third of the Keweenawan analyses fall on the alkaline side of MacDonald and Katsura's (1964) dividing line for Hawaiian lavas (Fig. 1.2.2.9), as do several from Karroo and Siberia. Irvine and Baragar (1971) and Chayes (1966) have proposed modifications of Yoder and Tilley's (1962) normative classification which distinguish tholeiitic (subalkaline) from alkaline suites.

Silica contents show significant differences among these provinces. The Siberian Platform basalts appear to be the least siliceous, with a range between 46.5 and 51.5% and an average close to 49% for published suite averages and composite samples. The others have averages close to 50% SiO_2, but Karroo lavas have a range of 47–54%, the Keweenawan lavas include several as low as 45%, and both the Keweenawan and Paraná suites contain a few andesites that range up to about 66% SiO_2. Rhyolites are associated with the Karroo basalts in the Nuanetsi-Lebombo area and with all of the Keweenawan plateaus.

Most of these pre-Tertiary plateau basalts have Al_2O_3 contents of about 13.0–15.5% (as in the Tertiary plateaus; see section 1.2.3), but in the Keweenawan, Al-rich olivine tholeiites with 16–18% Al_2O_3 constitute a major proportion of several but not of all plateaus (Fig. 1.2.2.10). These show strong resemblances to island arc tholeiites (see section 1.2.7) and to some ocean-floor basalts (section 1.2.5) in major element chemistry.

Alkali contents are also variable among these plateaus (Fig. 1.2.2.9). Again because only averages rather than individual analyses are available, the Siberian Platform lavas show the narrowest and lowest range (mostly 2.2–3.2% $Na_2O + K_2O$). Significantly higher values, up to 5.8%, are characteristic of some of the earliest lavas. The Karroo lavas have the next-higher alkali contents, most falling between 2.5–4.2% but again with a few more strongly alkaline members. The Paraná traps contain more alkalies, most analyses being in the range of 3.0–5.0% with a mean of 4.2%. Keweenawan lavas show the greatest range, with most analyses falling between 2.3–5.3%.

Magnesia contents of these major plateaus also vary considerably. One hundred sixty published analyses of Keweenawan lavas (SiO_2 < 60%) have a mean of 5.84% MgO; the mean of 44 average compositions of Siberian traps is 7.08%; and the average of 236 Paraná trap analyses is lowest at 5.0%. Eighty-six analyses of

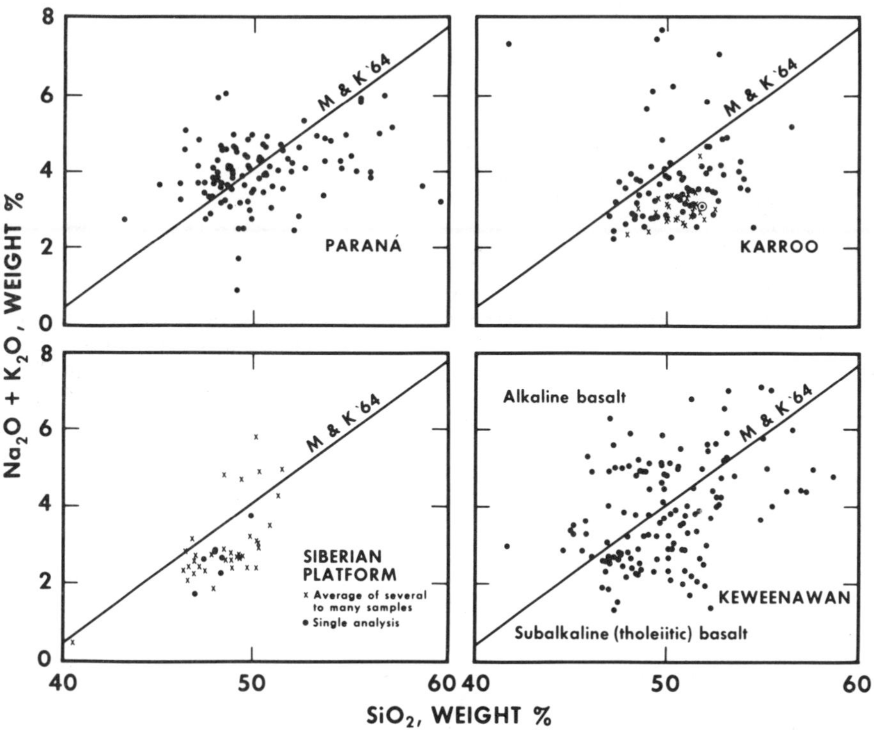

1.2.2.9 $(Na_2O + K_2O)/SiO_2$ diagrams for the four pre-Tertiary basalt plateaus. Diagonal line is MacDonald and Katsura's (1964) discriminator for Hawaiian alkaline/tholeiitic lavas. Karoo lavas: X = Lesotho; ● = Average dolerite.

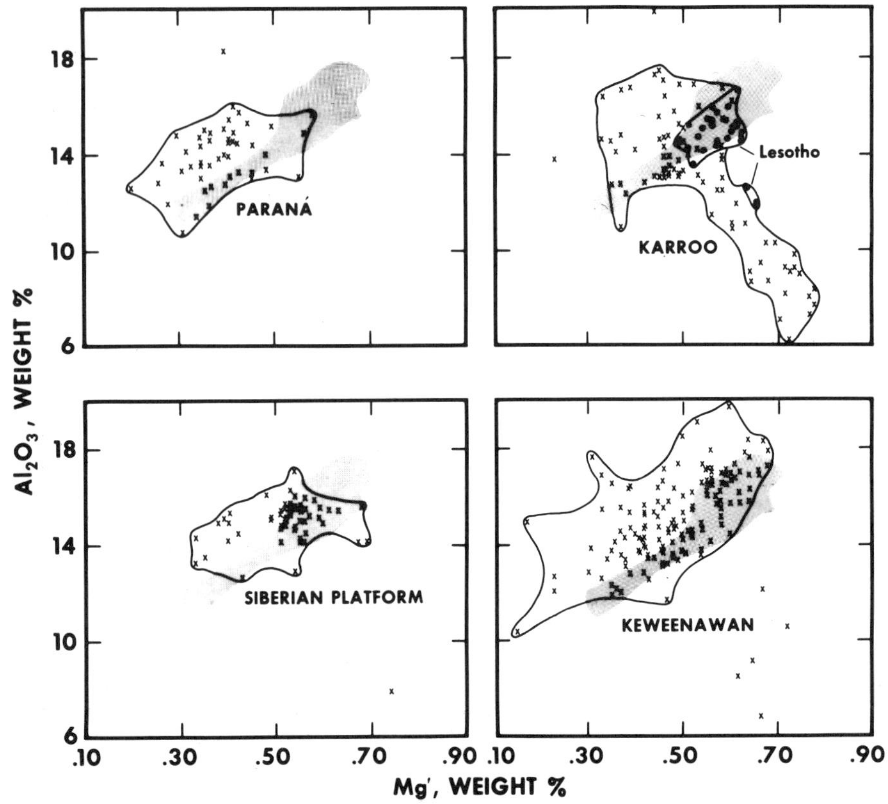

1.2.2.10 Al_2O_3/Mg' diagrams for the four pre-Tertiary basalt plateaus. Shaded area shows composition range of abyssal oceanic glasses (see section 1.2.5). $Mg' =$ atomic $Mg/(Mg + 0.9\ Fe)$. Many of the points for the Siberian Platform represent averages of several flows.

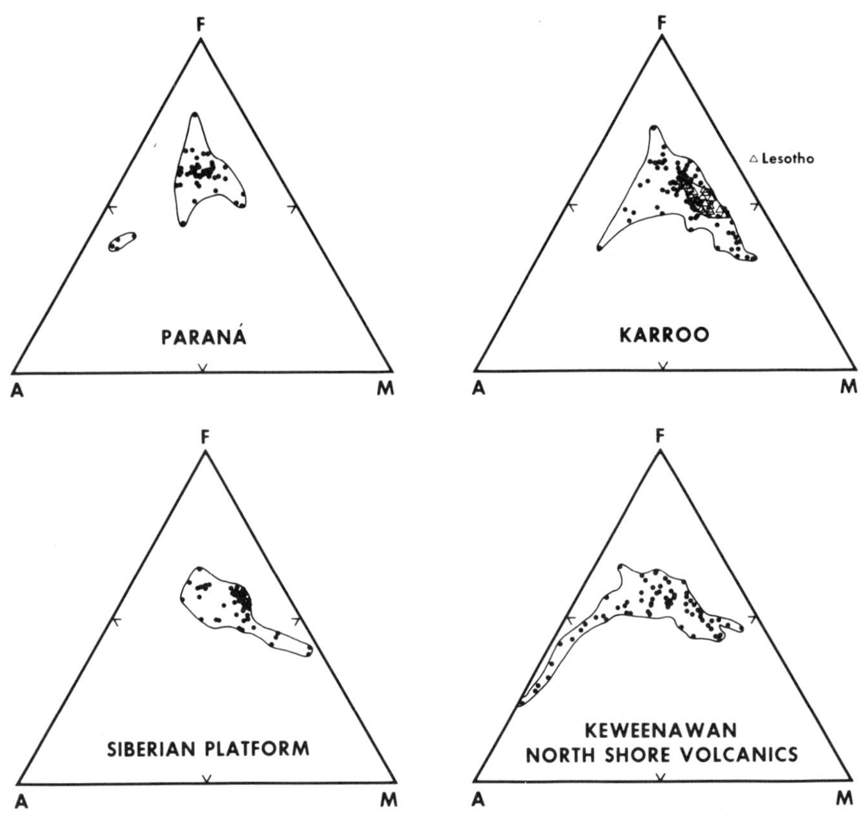

1.2.2.11 AFM diagrams for the pre-Tertiary basalt plateaus. Many of the points for the Siberian Platform are averages of several flows.

Karroo lavas exist, with an average of 7.98% MgO, but this probably is skewed toward high values by unrepresentative sampling of limburgites and similar mafic lavas.

Figure 1.2.2.11 contains AFM diagrams for the four plateaus. They show that most of the lavas are considerably evolved, with more alkalies and higher Fe/Mg ratios than compositions generally considered to be primitive, but that the Keweenawan, Siberian and Karroo provinces all contain some fairly primitive basalts. Analyses which diverge toward the Mg-rich side of the diagram represent rocks which contain excess, accumulated olivine phenocrysts. The picritic limburgites of northern Karroo constitute a distinctive Mg-rich group.

These provinces show a large range in Ti/Mg′ relations, with generally lower TiO_2 at any value of Mg′ than abyssal oceanic glasses (Fig. 1.2.2.12). Again the

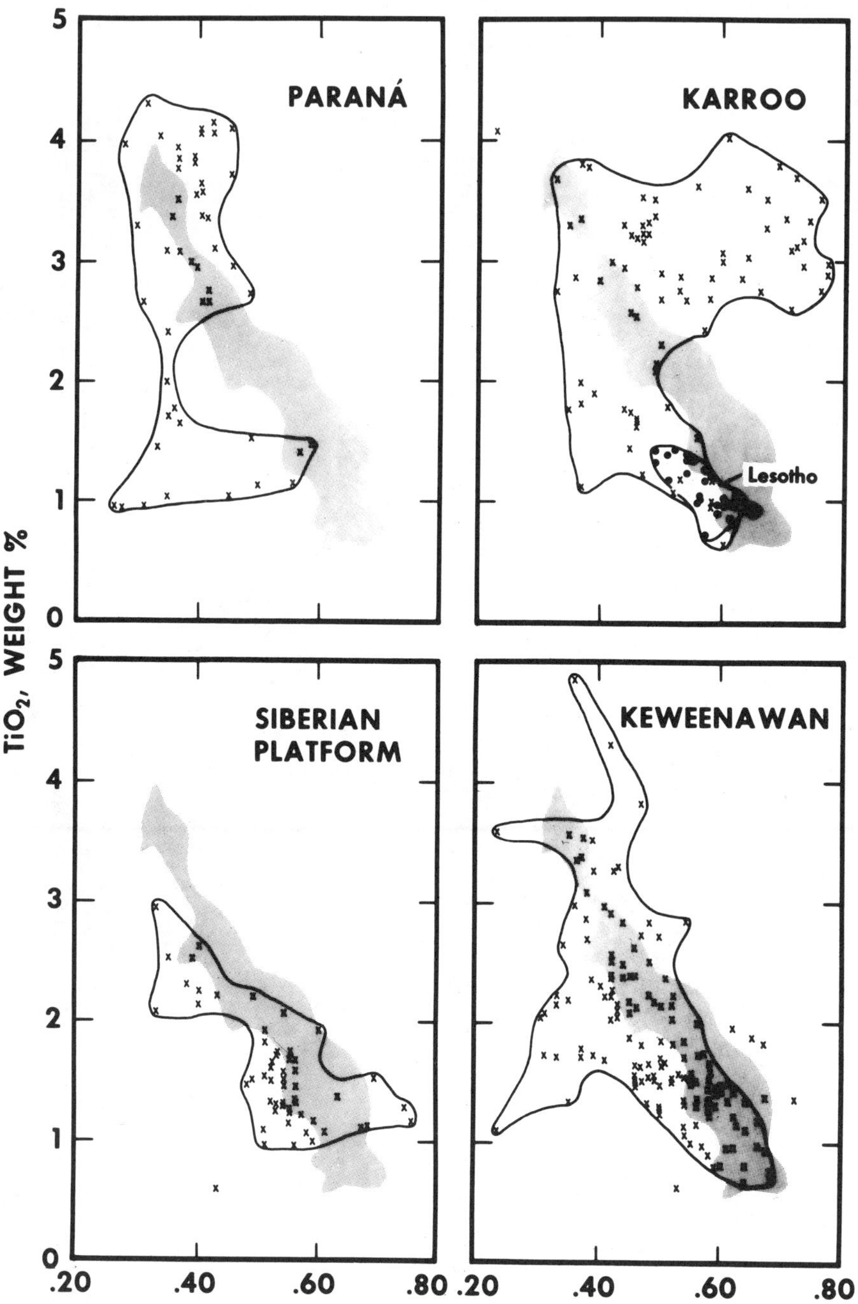

1.2.2.12 TiO_2/Mg′ diagrams for pre-Tertiary basalt plateaus. See caption of Fig. 1.2.2.10 for explanation of shaded area and of Mg′.

Table 1.2.2.8 Representative chemical analyses and cation-equivalent CIPW norms of Keweenawan lavas (SiO_2 <60%).[a]

	1	2	3	4	5	6	7	8	9	10	11	12	13	14	15	16	17
SiO_2	45.71	46.23	46.7	47.1	47.39	47.34	48.90	49.95	50.90	51.6	52.2	52.2	52.4	52.70	53.20	55.42	56.9
TiO_2	1.28	1.42	1.81	1.5	1.53	1.68	1.45	1.19	1.05	2.8	2.3	1.28	2.38	1.76	2.52	2.15	1.81
Al_2O_3	17.47	16.70	16.3	16.1	16.82	16.16	14.10	15.29	14.50	14.1	16.0	13.50	14.0	14.47	15.07	12.29	14.4
Fe_2O_3	7.80	3.99	5.5	5.3	3.09	5.58	7.98	1.12	1.28	4.7	5.7	6.22	5.8	7.44	5.32	9.00	3.3
FeO	3.41	7.27	6.9	6.3	7.91	5.96	5.23	9.05	9.63	8.3	5.5	5.56	7.3	5.55	5.45	3.32	6.3
MnO	0.15	0.17	0.20	0.18	0.17	0.17	0.19	0.22	0.18	0.2	0.1	0.16	0.22	0.24	0.22	0.23	0.25
MgO	6.80	7.78	6.9	7.0	6.65	6.94	6.27	8.38	6.49	3.8	3.4	5.65	4.6	3.70	3.75	2.89	4.1
CaO	10.53	9.75	9.8	10.2	9.87	6.91	9.36	6.29	8.41	7.5	5.2	9.45	7.7	8.01	5.21	3.25	6.1
Na_2O	2.61	2.49	2.4	2.4	2.63	3.75	2.48	3.28	2.80	3.0	4.0	2.22	2.9	3.19	4.00	4.47	3.6
K_2O	0.31	0.29	0.4	0.35	0.48	1.34	0.84	1.27	0.85	0.9	1.7	0.87	1.4	1.14	2.00	2.72	1.4
P_2O_5	0.14	0.16	0.18	0.20	0.18	0.19	0.02	0.19	0.17	0.7	0.6	0.11	0.36	0.25	0.70	0.42	0.48
S	n.d.	0.01	n.d.	n.d.	0.02	0.02	<0.01	n.d.	n.d.	n.d.	n.d.	0.02	n.d.	0.02	0.01	n.d.	n.d.
H_2O+	1.77	2.45 }	3.8	{ 2.4	2.21	3.50	1.64 }	3.40	3.47	{ 1.4	2.0	1.11 }	2.3	0.68	1.95	2.86 }	2.6
H_2O-	1.78	1.41 }		{ 0.81	0.96	0.43	1.76 }			{ 0.7	1.2	1.01 }		0.48	0.41	1.71 }	
CO_2	0.12	0.17	0.2	0.05	0.08	0.09	0.08	0.00	0.08	0.2	0.1	0.18	0.1	0.00	0.10	0.15	0.1
Total	99.88	100.30[c]	101.1	100.0	100.00[c]	100.08[c]	100.30	99.63	99.81	99.9	100.0	99.50	101.5	99.68†	99.91	100.88	101.3
qz									.60	6.25	2.24	5.35	3.72	3.77	2.77	4.04	9.14
or	1.90	1.79	2.45	2.14	2.93	8.25	5.13	7.80	5.22	5.45	10.39	5.29	8.35	6.84	12.13	16.72	8.39
ab	22.96	22.01	21.05	21.02	23.01	29.92	21.68	28.84	24.61	26.01	35.00	19.32	24.77	27.42	34.74	39.34	30.88
an	36.42	35.02	33.71	33.25	33.78	24.27	25.68	24.16	25.44	22.90	21.39	25.00	21.25	22.14	17.71	5.65	19.26
ne						1.70											
di	14.01	11.89	12.86	14.75	12.97	8.45	18.40	5.81	13.93	9.26	1.38	18.73	12.70	14.12	3.83	7.24	7.12
hy	2.12	10.14	11.47	14.22	11.48		19.54	16.48	25.93	20.72	21.49	21.21	21.37	19.14	20.22	19.10	18.91
ol	17.31	13.97	12.09	8.97	10.43	21.37	3.94	12.45									
mt	1.75	1.75	1.98	1.85	1.78	1.84	2.07	1.68	1.80	2.07	1.77	1.85	2.04	2.00	1.69	1.91	1.51
il	2.53	2.82	3.56	2.95	3.00	3.32	2.84	2.35	2.07	5.45	4.52	2.50	4.56	3.40	4.91	4.25	3.49
ap	.32	.37	.41	.45	.41	.43	.05	.43	.39	1.57	1.36	.25	.79	.55	1.57	.95	1.06
Mg′	.56	.60	.54	.56	.55	.56	.50	.62	.54	.38	.39	.50	.42	.37	.42	.33	.47
An(%)	61	61	62	61	59	45	54	46	51	47	38	56	46	45	34	13	38
d.i.[b]	25	24	24	24	26	40	27	37	30	38	48	30	37	38	50	60	48

Norms calculated without H_2O, CO_2 and with $Fe^{3+}/\Sigma Fe = 0.1$.

†Nos. 2 and 5 contain 0.01% Cu; No. 6 contains 0.02% Cu; No. 14 contains 0.05% BaO.

[a]See also Table 1.2.2.19. [b]d.i. = normative qz + or + ab + ne. [c]No's 2 and 5 contain 0.01% Cu; No. 6 contains 0.02% Cu; No. 14 contains 0.05% BaO.

1. Ophitic olivine tholeiite, T-45, North Shore Volcanic Group (NSVG), Green, 1972, p. 309. **2.** "Typical ophite" olivine tholeiite, #3095, Portage Lake Volcanics, (PLV), Keweenaw Point, Broderick, 1935, p. 536. **3.** Olivine tholeiite of Quebec Mine basalts, AK 144, Michipicoten Island, Annells, 1974, p. 128. **4.** Ophitic olivine tholeiite, IR-66, Hill Point Flow, PLV, Isle Royale, Huber, 1973, p. C10. **5.** Olivine tholeiite, ave. of Greenstone Flow, PLV, Keweenaw Point, Broderick, 1935, p. 513. **6.** "Typical glomeroporphyrite" olivine tholeiite, #3103, PLV, Kew. Pt., Broderick, 1935, p. 536. **7.** Tholeiite, #6, Osler Group, McIlwaine and Wallace 1976, p. 16. **8.** Transitional basalt, PP-2, NSVG, Green, 1972, p.309. **9.** Olivine tholeiite, GP-4a, NSVG, Green, 1972, p. 309. **10.** "Andesite," #11 (W-169-993), Kallander Creek Fm., Hubbard, 1975, p. 536. **11.** "Trachyandesite," #9 (W-169-998), Kallander Creek Fm., Hubbard, 1975, p. 536. **12.** Tholeiite, #1, Osler Group, McIlwaine and Wallace, 1976, p. 16. **13.** Quartz tholeiite, AK18, South Shore basalts, Michipicoten I., Annells, 1974, p. 129. **14.** Transitional basalt, #9, NSVG, Sandell and Goldich, 1943, Table 5. **15.** "Fine-grained melaphyre," #3127, PLV, Keweenaw Pt., Broderick, 1935, p. 536. **16.** Trachyandesite, GH-2b, NSVG, Green, 1972, p. 310. **17.** Glassy andesite, AK345, Michipicoten I., Annells, 1974, p. 128.

Karroo tholeiitic limburgites stand out in their combination of high TiO_2 and high Mg′. Of the four provinces, the Keweenawan most resembles the abyssal glasses, with the most primitive members having TiO_2 well under 1% and Mg′ in the upper 60's.

Chemical petrology of Keweenawan lavas has been discussed by several authors. Representative major element analyses are presented in Table 1.2.2.8. Cornwall (1951) and Broderick (1935) studied differentiation within the large "Greenstone" and Kearsarge flows of the Keweenaw Peninsula and between other lavas of the Portage Lake Volcanics. They found only minor crystal settling, but a gradual though modest iron-enrichment trend with constant Si and a steady increase in Fe, Ti, Mn, Cu, P, Na and K combined with later enrichment in Si along with alkalies and incompatible elements to produce residual "pegmatites" and very minor granophyres in the flow centers.

Table 1.2.2.9 Representative chemical analyses, averaged chemical analyses and cation-equivalent CIPW norms of Siberian Platform plateau lavas.[(a)]

	1	2	3	4	5	6	7	8	9	10	11	12
SiO_2	40.57	46.57	46.64	47.44	47.98	48.32	48.49	48.63	49.90	50.2	50.28	50.90
TiO_2	1.14	1.06	1.52	1.66	1.78	1.62	2.41	1.10	2.84	1.9	1.12	1.04
Al_2O_3	7.05	15.25	15.02	14.08	15.11	14.91	14.65	14.50	13.92	15.0	15.17	15.29
Fe_2O_3	5.52	5.46	5.77	4.44	3.60	3.20	4.82	5.09	6.84	4.0	2.66	3.23
FeO	8.56	6.33	6.25	9.08	9.17	9.87	8.75	6.32	6.53	7.4	7.41	6.31
MnO	0.14	0.19	0.20	0.30	0.23	0.22	0.21	0.106	0.15	0.2	0.17	0.16
MgO	19.28	7.87	6.26	8.08	6.62	6.87	4.28	6.70	3.20	8.2	7.20	7.15
CaO	6.33	10.81	10.97	11.56	10.80	10.68	7.33	11.04	9.20	9.6	10.13	10.72
Na_2O	0.29	1.95	2.17	2.37	2.44	2.02	3.19	2.02	3.06	2.2	2.17	2.19
K_2O	0.14	0.11	0.26	0.27	0.40	0.25	1.64	0.58	0.70	0.9	0.89	1.32
P_2O_5	0.065	0.10	0.11	0.17	0.15	0.15	0.68	0.08	0.62	0.4	0.08	0.08
S	n.d.	0.01	n.d.	0.02	—	n.d.	—	—	—	—	—	—
H_2O^+		2.21		0.92	0.85	0.93				—		
H_2O^-	10.82	2.38	4.39	0.08	1.08	0.94	3.23	3.16	3.03	—	2.86	2.08
CO_2		0.08		0.02	0.20	—				—		
TOTAL	99.90	100.38	99.56	100.49	100.41	99.98	99.68	99.34[(c)]	100.02[(c)]	100.0	100.14	100.47
No. of samples represented	2	2	6	1	1	1	22	5	1	4	58	8
qz									3.91		.23	
or	.93	.68	1.61	1.60	2.41	1.51	10.05	3.56	4.27	5.32	5.41	7.93
ab	2.75	17.24	19.29	20.17	21.01	17.42	27.99	17.77	26.70	18.62	18.88	18.84
an	19.67	34.00	32.02	27.13	29.61	31.47	21.58	29.93	22.87	28.40	29.84	28.45
di	12.20	17.62	20.26	24.35	19.89	17.92	10.07	21.89	17.28	13.81	17.51	20.57
hy	30.62	17.87	16.22	4.85	12.64	24.28	15.83	21.45	15.30	25.34	23.99	17.86
ol	28.34	7.93	4.89	15.94	8.42	1.66	5.67	.79		1.99		2.44
mt	2.45	1.89	1.94	2.12	2.03	2.09	2.19	1.83	2.11	1.77	1.62	1.51
il	2.43	2.10	3.03	3.17	3.44	3.14	4.75	2.17	5.56	3.61	2.19	2.01
ap	.16	.23	.25	.37	.33	.33	1.54	.18	1.40	.87	.18	.18
Mg′	.74	.58	.52	.55	.51	.52	.39	.55	.33	.60	.59	.61
An(%)	88	66	62	57	58	64	44	63	46	60	61	60
d.i.[(b)]	3.7	17.9	20.9	21.8	23.4	18.9	38.0	21.3	34.9	23.9	24.5	26.8

Norms calculated without H_2O, CO_2 and with $Fe^{3+}/\Sigma Fe = 0.1$
[(a)]See also Table 1.2.2.13. [(b)]d.i. = normative qz + or + ab.
[(c)]No. 8 incl. 0.02% Cr_2O_3; No. 9 incl. 0.03 Cr_2O_3.
1. "Picrite porphyrite," #7, Noril'sk area, Godlevskiy, 1959, Table 3, p. 13. **2.** Calcite-bearing basalt, #1, Nudimsk Series, Upper Tunguska basin, Dmitriev, 1976, p. 305. **3.** Basalt average, #C-15b, Kochechum formation, lower Tembenchi River, Nesterenko *et al.*, 1964, p. 974. **4.** "Anamezite," Table 3, marginal part of Tunguska syneclise, Dmitriev, 1973, p. 63. **5.** Non-calcite-bearing basalt, #5, Nudimsk series, upper Tunguska basin, Dmitriev, 1976, p. 305. **6.** Basalt, Table 5, central part of Tunguska syneclise, Dmitriev, 1973, p. 65. **7.** Ave. "basalts with titaniferous augite," Ivakinsk suite (first cycle). Noril'sk area, Dyuzhikov *et. al.*, 1973, p. 722. **8.** Tholeiitic basalt, #2, Noril'sk area, Godlevskiy, 1959, Table 3, p. 13. **9.** Vitrophyric basalt, #4, Noril'sk area, Godlevskiy, 1959, Table 3, p. 13. **10.** Weighted ave., basalts of Triassic mesocycle I, Kharayelakh Mtns., Masaitis *et al.*, Table 1, 1966, p. 394. **11.** Ave. basalt, #C-17d, Nadezhdinsk fm., hole 746, Nesterenko *et al.*, 1964, Table 2, p. 974. **12.** Ave. basalt, #C-17h, Nadezhdinsk fm., hole 421, Nesterenko *et al.*, 1964, Table 2, p. 974.

Green (1972), in a description of the North Shore Volcanic Group, noted (a) the dominance of olivine tholeiite and its resemblance to high-alumina basalt and oceanic tholeiite; (b) the complete range in compositions from moderately primitive basalts to rhyolites; (c) a moderate iron-enrichment trend; (d) a general similarity to Icelandic Tertiary plateau lavas and the Thingmuli central volcano; and (e) a somewhat higher K_2O content for several rocks in the basalt-andesite range than is normal for tholeiitic series. Following the experimental results of Green and Ringwood (1967) and others, he concluded that the olivine tholeiite could have been produced by partial melting of pyrolitic mantle at 15–35 km depths, with production of quartz tholeiite at shallower depths by fractional crystallization. He also considered the possibility of partial melting of granitic crust by the large volumes of mafic magma, with or without assimilation, and the potential of mixing two major primitive magma sources, one olivine tholeiite, the other richer in K and incompatible elements, to produce the transitional basalts and andesites.

Phinney (1970) modelled a fractional crystallization scheme using major elements to suggest that the most primitive and Al-rich olivine tholeiite could be parental to several other Keweenawan basalts by fractionally crystallizing plagioclase, olivine, pyroxene, and oxides in the general proportions that make up the anorthositic gabbro of the Duluth Complex.

Annells (1974) proposed olivine tholeiite of mantle origin as the parent liquid for the Michipicoten Island flood basalts. He also proposed that this liquid supplied the heat for melting a large volume of sialic crust, some of which was emplaced directly to form the abundant quartz-porphyry rhyolites, and some of which mixed with the basalt liquid and subsequently differentiated in high-level chambers to produce the andesites and granophyres of the volcano.

Weiblen and Morey (1980) proposed that the variety of rocks in the Duluth Complex could be derived by fractional crystallization of two parental magma types, a high-Al olivine tholeiite and a transitional type rich in both Mg and incompatible elements. However, the lava analysis they selected for the latter cannot be demonstrated to have been a liquid composition because the analyzed rock contains abundant excess augite and olivine phenocrysts accumulated by settling in the flow.

Brannon *et al.* (1979) have concluded, based largely on trace element evidence, that the range in olivine tholeiite compositions in the North Shore volcanics cannot be satisfactorily explained by simple fractional crystallization in a single or renewed chamber or by partial melting of a single, uniform source.

The Siberian Platform lavas are generally rather uniform in composition (Tables 1.2.2.9 and 1.2.2.13), with olivine tholeiite predominant. However, the earliest extrusive phase or volcanic cycle in the Noril'sk area, containing trachybasalts and basaltic andesites, is characterized by much higher amounts of Fe, Na, K, Ti and P and lower MgO than the later cycles (Dyuzhikov *et al.*, 1973). Dimitriev (1973) finds generally more magnesian basalts in the marginal parts of the Tunguska syneclise than in the center and that the crustal chambers which discharged lavas to the surface periodically were replenished with primitive magma from the mantle. Much work has been done on the associated mafic intrusions and their several paths of differentiation (e.g., Nesterenko *et al.*, 1971; Nesterenko and Al'mukhamedov, 1973) which generally lead to silica-oversaturated residua. Ultramafic rocks, including kimberlites (Vasyl'yev and Zolotukhin, 1977), also intrude both the plateau lavas and the surrounding older sedimentary strata.

Major element analyses of representative Karroo lavas are given in Table 1.2.2.10. The great majority are tholeiitic, with quartz-normative or weakly olivine-normative basalts being most common, similar to the voluminous and widespread associated sills and sheets in the underlying Karroo sediments.

Cox and Hornung (1966) concluded that the limited variation in the chemistry of the Lesotho tholeiites has resulted from fractional crystallization at shallow depths. Most of the basalts are aphyric; the few phenocrysts are olivine or plagioclase. Rock compositions plot close to the doubly-saturated plagioclase + olivine surface in the synthetic system Fo-Di-An-Qz at 1 atmosphere, and the few lavas near the base of the section with more olivine phenocrysts (MgO up to 11.8%) can easily be interpreted as being enriched in settled olivine. For the regional Karroo magmatism, Cox (1972) proposed a cyclic history initiated by the rise from deep in the mantle (at least 500 km) of a large body of K-rich picritic magma. At higher levels and to the west and south of this great "plume," a volume of more normal tholeiitic magma was generated which produced the Lesotho, Swaziland, Featherstone, and other olivine-poor tholeiites. The potassic, picritic magmas, rich in incompatible elements, first erupted a small volume of differentiates (e.g., nephelinite, carbonatite) then a large amount of primitive limburgite and olivine-rich basalt. A steady state was subsequently reached, controlled largely by olivine fractionation, in which more ordinary olivine-poor basalts were erupted, followed by a waning Cretaceous stage characterized by the Lupata phonolites and associated alkaline intrusions. Crustal anatexis is called upon to generate the

Table 1.2.2.10 Representative chemical analyses and cation-equivalent CIPW norms of Karroo lavas (SiO_2 < 60%).

	1	2	3	4	5	6	7	8	9	10	11	12	13	14	15
SiO_2	41.46	48.2	48.3	48.8	48.98	49.1	49.4	49.61	50.0	50.05	50.37	50.6	51.2	51.7	52.90
TiO_2	2.62	3.17	1.58	2.54	3.34	2.74	0.93	3.54	1.12	2.54	2.22	2.76	3.18	1.02	2.69
Al_2O_3	13.71	8.52	15.78	14.54	7.74	13.68	14.1	11.28	14.8	12.33	14.16	12.71	12.98	14.1	13.05
Fe_2O_3	9.66	5.20	1.66	3.26	3.75	5.29	5.00	2.13	4.32	3.84	3.03	7.29	4.83	1.50	3.58
FeO	4.36	8.68	9.87	7.82	6.53	8.88	6.17	10.11	7.43	7.69	9.53	4.46	8.96	7.82	8.20
MnO	0.20	0.17	0.16	0.16	0.13	0.15	0.16	0.18	0.17	0.15	0.15	0.14	0.20	0.18	0.14
MgO	7.35	14.07	4.90	5.39	15.52	3.92	7.7	7.67	6.0	6.74	6.19	6.20	5.54	7.7	4.85
CaO	9.63	7.21	11.25	11.17	6.44	8.78	10.43	9.66	10.47	9.18	9.29	9.72	9.40	10.33	8.69
Na_2O	4.67	1.36	2.77	2.27	1.40	2.68	2.26	2.37	2.43	2.45	2.04	2.21	2.58	2.27	2.36
K_2O	2.68	1.56	0.10	0.30	2.44	1.49	0.66	1.83	0.71	1.58	0.47	1.14	0.95	0.60	1.63
P_2O_5	0.47	0.39	0.18	0.28	0.54	0.55	0.11	0.52	0.13	0.36	0.29	0.41	0.38	0.08	0.39
H_2O^+	2.97	1.36	2.73	2.54	2.36	0.93	3.54	0.80	3.85	1.74	1.64	2.41	0.77	1.88	1.35
H_2O^-	0.55	0.77	—	0.96	0.91	0.84	—	0.42	—	0.27	0.75	—	—	—	0.18
CO_2	n.d.	n.d.	n.d.	n.d.	0.00	n.d.	n.d.	0.00	n.d.	0.00	0.00	n.d.	n.d.	n.d.	0.00
TOTAL	100.30	100.66	99.28	100.03	100.08	99.12	100.46	100.12	101.43	99.92	100.13	100.05	100.97	99.18	100.01
qz				2.85		1.01				.31	4.76	3.84	2.92	2.19	6.41
or	4.05	9.36	.61	1.84	14.89	9.05	4.02	10.94	4.30	9.64	2.84	6.90	5.60	3.64	9.78
ab		11.68	24.28	19.90	12.24	23.32	19.73	20.28	21.07	21.39	17.66	19.15	21.79	19.74	20.28
an	8.81	12.72	31.41	29.63	7.88	21.49	27.22	14.90	28.06	18.55	28.74	21.91	20.99	27.25	20.51
lc	9.65														
ne	22.11														
di	31.07	17.16	21.10	21.55	17.42	16.84	20.95	24.91	20.23	21.57	13.80	20.57	19.39	20.37	17.39
hy		29.35	14.48	16.60	27.13	19.06	19.59	14.82	21.37	20.63	25.04	18.90	19.98	23.07	17.47
ol	15.08	10.19	2.66		10.75		4.25	4.17	.31						
mt	2.17	2.18	1.90	1.80	1.65	2.26	1.77	1.96	1.87	1.85	2.02	1.82	2.14	1.52	1.87
il	5.14	6.11	3.11	5.00	6.55	5.35	1.82	6.80	2.18	4.98	4.31	5.37	6.03	1.99	5.19
ap	1.06	.86	.41	.63	1.22	1.24	.25	1.15	.29	.81	.65	.92	.83	.18	.87
Mg′	.53	.68	.46	.50	.76	.36	.59	.56	.51	.54	.50	.53	.45	.62	.46
An(%)	100	52	56	60	39	48	58	42	57	46	62	53	49	58	50
d.i.†	35.8	21.0	24.9	24.6	27.1	33.4	23.8	31.2	25.4	31.3	25.3	29.9	30.3	25.6	36.5

Norms calculated without H_2O, CO_2 and with $Fe^{3+}/\Sigma Fe = 0.1$
†d.i. = normative qz + or + lc + ne.
1. Nepheline basalt, #774, Kloppersfontein, NE Transvaal, Lombaard, 1952. **2.** Porphyritic olivine basalt, #TV43, Tuli syncline, Rhodesia, Vail *et al.*, 1969, p. 409. **3.** Basalt, #S. 13, Swaziland, Cox *et. al.*, 1967, p. 1455. **4.** Augite-plagioclase porphyritic basalt, #TV-22, Tuli area, Vail *et al.*, 1969, p. 410. **5.** "Limburgite," #LM 432, Nuanetsi area, Cox *et al.*, 1965, p. 145. **6.** Aphyric basalt, #TV-29, Tuli area, Vail *et al.*, 1969, p. 411. **7.** Basalt, #B.1, Lesotho, Cox and Hornung, 1966, p. 1421. **8.** Olivine-rich basalt, #LM434, Nuanetsi area, Cox *et al.*, 1965, p. 145. **9.** Basalt, #B.47, Lesotho, Cox and Hornung, 1966, p. 1421. **10.** Basalt, Chikombedzi type, #C868, Nuanetsi area, Cox *et al.*, 1965, p. 145. **11.** Basalt, #DW389, Nuanetsi area, Cox *et al.*, 1965, p. 145. **12.** Basalt, #F.6, Featherstone area, Cox *et al.*, 1967, p. 1454. **13.** Basalt, #Bu.3, Nyamandhlovu area, Cox *et al.*, 1967, p. 1454. **14.** Basalt, #B.72, Lesotho, Cox and Hornung, 1966, p. 1421. **15.** Basalt (Chikwedziwa type), #KC-58, Nuanetsi area, Cox *et al.*, 1965, p. 145.

rhyolitic magmas of Lebombo, although initial Sr isotopic ratios (see below) do not confirm this model; rather, they suggest mixing of mantle- and crust-generated liquids was important. Cox and Jamieson (1974) performed experimental melting studies of some of the Nuanetsi limburgitic and picritic lavas at high pressures. They concluded that these liquids may have been derived from a parental K-rich magma which was in equilibrium with mantle olivine and orthopyroxene at about 12 kb, and which subsequently evolved either by polybaric fractional crystallization of olivine and orthopyroxene or by reaction with harzburgitic wall-rock during its rise to the surface.

The Paraná traps are nearly all quartz-normative tholeiites or transitional basalts and andesites which are relatively poor in Mg and rich in Fe, Ti, and alkalies (Tables 1.2.2.11 and 1.2.2.16). Some dacitic lavas are recorded as well as a few Ne-normative alkali basalts (Ruegg, 1976a). Figure 1.2.2.13 shows the distribution of major elements in analyzed samples. Despite the evolved compositions of these lavas, Sr isotopes (Compston *et al.*, 1968) suggest a uniform mantle source with little or no crustal contamination. This conclusion implies that considerable high-level fractionation must have occurred before eruption (Cordani and Vandoros, 1967; Leterrier *et al.*, 1972; Ruegg, 1975). However, Wilkinson and Binns (1977) stress the possibility that more Fe-rich mantle material may have been a direct source of these voluminous transitional basalts and basaltic andesites.

Trace elements and isotopic ratios

Trace element analyses have been reported for three series of Keweenawan lavas: the Osler Group (McIlwaine and Wallace, 1976), Michipicoten Island (Annells, 1974), and the Portage Lake Volcanics (Keweenaw Point, Michigan: Cornwall and Rose, 1957; Isle Royale: Huber, 1973; see Table 1.2.2.12). Olivine

Table 1.2.2.11 Representative chemical analyses and cation-equivalent CIPW norms of Paraná Basin plateau lavas. (SiO_2 < 60%).

	1	2	3	4	5	6	7	8	9	10	11	12
SiO_2	45.48	46.00	47.80	48.00	48.86	49.10	49.60	50.74	52.02	52.28	55.47	56.41
TiO_2	4.10	3.30	3.00	3.20	3.60	1.00	3.00	3.72	1.46	1.64	1.78	1.46
Al_2O_3	14.31	17.11	12.87	14.42	13.98	15.00	13.97	12.44	15.49	12.53	12.55	13.41
Fe_2O_3[a]	16.60	11.64	14.39	15.37	13.49	13.90	13.60	13.78	10.02	15.50	13.63	12.99
MnO	0.00	0.16	0.18	0.20	0.15	0.20	0.18	0.10	0.13	0.22	0.19	0.11
MgO	5.26	3.57	4.84	2.96	5.12	5.20	3.35	4.22	6.54	4.06	3.48	2.86
CaO	8.24	7.19	9.76	8.32	8.25	10.60	8.26	7.57	10.87	7.64	6.79	6.48
Na_2O	1.13	2.46	2.60	2.23	2.36	1.50	3.35	2.56	1.78	2.04	2.82	2.45
K_2O	0.52	1.24	1.10	1.55	1.17	1.00	1.49	1.57	0.69	2.55	3.04	2.57
P_2O_5	0.95	0.59	0.33	0.78	0.12	0.40	0.43	0.62	0.14	0.19	0.25	0.28
l.o.i.[b]	3.78	6.50	3.00	3.80	3.68	1.60	1.80	2.77	0.94	1.93	1.20	0.92
TOTAL	100.37	99.76	99.87	100.83	100.78	99.50	99.03	100.09	100.08	100.58	101.20	99.94
qz	7.16	2.85		3.93	3.17	2.95	0.50	6.45	5.47	4.06	5.02	10.38
or	3.18	7.86	6.71	9.44	7.12	6.04	9.06	9.53	4.11	15.28	17.97	15.34
ab	9.90	22.32	22.71	19.45	20.57	12.97	29.16	22.26	15.19	17.50	23.86	20.94
an	33.58	34.29	20.85	25.52	24.82	31.91	19.21	18.31	32.52	17.74	12.61	18.18
di	2.46	0.19	22.65	10.45	13.76	16.24	17.14	13.80	17.24	16.53	16.58	10.73
hy	29.46	21.44	16.22	19.46	19.98	23.72	14.83	17.67	19.99	21.61	16.83	17.93
ol			0.74									
mt	2.49	1.81	2.15	2.30	2.01	2.06	2.03	2.05	1.47	2.28	1.98	1.90
il	8.06	6.72	5.88	6.26	7.04	1.94	5.86	7.26	2.80	3.16	3.38	2.80
ap	2.15	1.38	0.74	1.76	0.27	0.89	0.97	1.39	0.31	0.42	0.55	0.62
Mg′	.41	.40	.43	.30	.46	.45	.35	.40	.59	.37	.36	.33
An(%)	77	61	48	57	55	71	40	45	68	50	35	46
d.i.	20.24	33.03	29.42	32.82	30.86	21.95	38.71	38.24	24.78	36.84	46.85	46.66

[a]Total Fe as Fe_2O_3. [b]l.o.i. = loss on ignition. Norms calculated without H_2O, CO_2 and $Fe^{3+}/\Sigma Fe = 0.1$

1. Diabase, São Paulo State, Brazil, Leinz, 1949 (#T3, Ruegg, 1976a, p. 105). **2.** Basalt, São Paulo State, Brazil, #T9, Ruegg, 1976a, p. 105. **3.** Basalt, São Paulo State, Brazil, #T20, Ruegg, 1976a, p. 105. **4.** Basalt, São Paulo State, Brazil, #T27, Ruegg, 1976a, p. 105. **5.** Basalt, Mato Grosso State, Brazil, Almeida, 1946 (#T39, Ruegg, 1976a, p. 105). **6.** Basalt, São Paulo State, Brazil, #T47, Ruegg, 1976a, p. 106. **7.** Basalt, São Paulo State, Brazil, #T59, Ruegg, 1976a, p. 106. **8.** Basalt, Mato Grosso State, Brazil, Almeida, 1946 (#T74, Ruegg, 1976a, p. 106). **9.** Basalt, Uruguay, Walther, 1927 (#T81, Ruegg, 1976a, p. 106). **10.** Basalt, Uruguay, Walther, 1927 (#T83, Ruegg, 1976a, p. 106). **11.** Basalt, Argentina, Walther, 1927 (#T90, Ruegg, 1976a, p. 106). **12.** Basalt, Uruguay, Walther, 1927 (#T94, Ruegg, 1976a, p. 106).

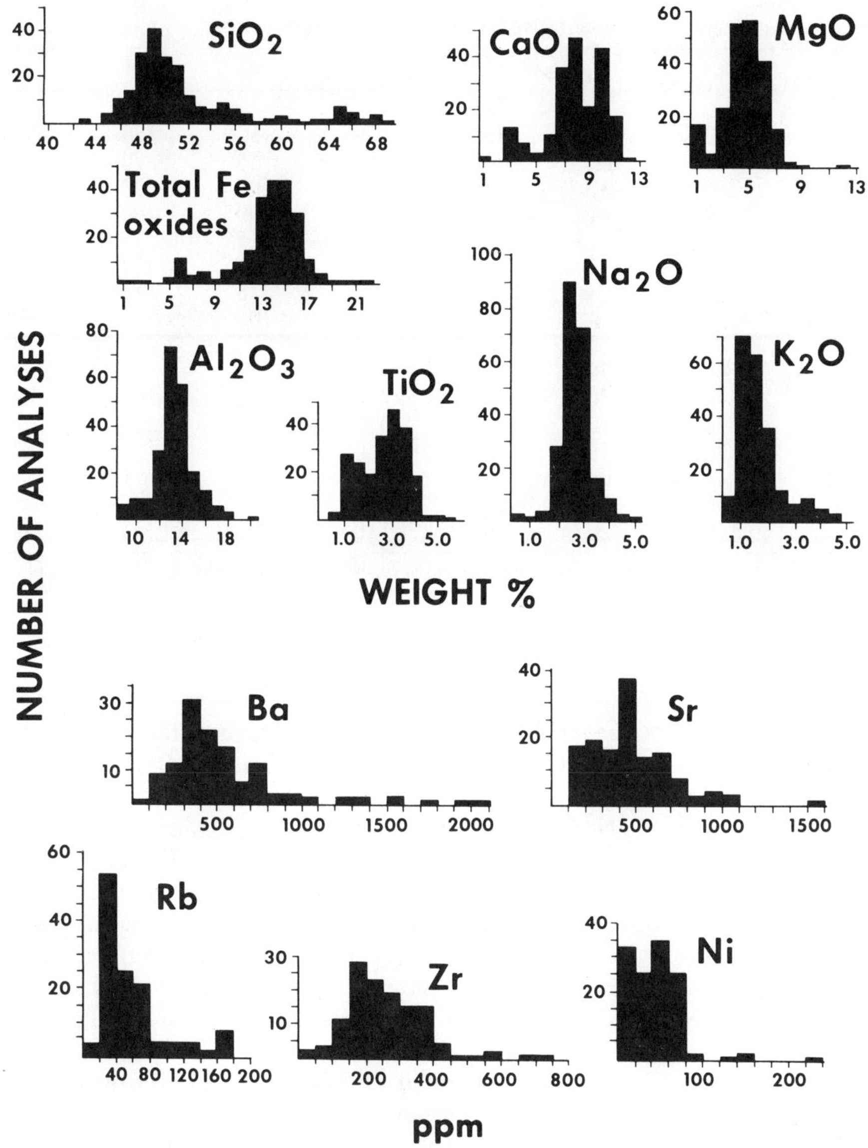

1.2.2.13 Distribution of selected major and trace elements in the Paraná traps, after Ruegg (1976a, b).

tholeiites show ranges of 70–310 ppm Cr, 150–230 ppm Ni, and 70–160 ppm Zr; quartz tholeiites contain 40–200 ppm Cr, 60–140 ppm Ni and 30–260 ppm Zr. The higher Ni and Co values of the Keweenaw Point olivine tholeiites were attributed by Cornwall and Rose (1957) to their above-average olivine contents (Ni = 460 ppm near the base of the very thick Greenstone flow where it is enriched in settled olivine). A major increase in available data is represented by the new analyses presented here for the Keweenawan reference suite from the North Shore Volcanic Group (Tables 1.2.2.22–1.2.2.25).

Rb/Sr relations have been investigated by several workers. Faure *et al.* (1969) found Rb/Sr whole-rock ratios ranging from 0.020 in gabbro to 1.78 in granophyre differentiates of the Duluth Complex and Endion Sill in Duluth, and initial $^{87}Sr/^{86}Sr$ ratios of 0.7055 ± 0.0003 and 0.7046 ± 0.0006 in these two intrusions. They inferred a small amount of crustal contamination. Van Schmus (1971) found a wide range of initial $^{87}Sr/^{86}Sr$ ratios from 0.7055–0.7125 for the Mamainse Point volcanics in Ontario. He outlined four possible interpretations but believed that some samples have been chemically disturbed since crystallization. Wanless and Loveridge determined initial $^{87}Sr/^{86}Sr$ ratios of 0.7068 to 0.7074 ± 0.0010 for the Channel Lake andesite and 0.7109 and 0.711 ± 0.024 for intrusive felsites of the Michipicoten Island central volcano (1972, 1977). Studies by Leeman (1977) of a suite of lavas from the North Shore Volcanic Group found that the mafic rocks have initial $^{87}Sr/^{86}Sr$ ratios of 0.7038 to 0.7059,

Table 1.2.2.12 Trace element contents of representative Keweenawan lavas, in ppm.†

	1	2	3	4	5	6	7	8	9	10
As	< 1	< 1	< 1							
Ba	400	300	150	160	850	380	150	150	150	300
Co	50	50	50	58	n.f.	49	59	50	70	50
Cr	150	100	200	260	37	n.f.	270	200	300	100
Cu	130	150	140	70	41	97	94	200	100	200
Ga	20	20	30					15	15	15
La										50
Nb								15	15	15
Ni	90	110	140	180	n.f.	34	170	200	200	150
Pb	< 10	15	< 10							10
Sc	30	30	40	45	38	49	36	50	50	30
Sr	200	200	100	240	250	240	250	500	300	500
V	250	300	150	320	190	470	320	200	300	200
Y	< 20	< 20	20	39	76	63	40	30	50	50
Yb								3	5	7
Zn	90	100	100	< 500	680	< 500	< 500			
Zr	100	100	80	150	420	260	140	70	150	200

†See also tables 1.2.2.22-25. n.f.: not found. No's **1, 2, 3**: #2, 6, 14 resp., Osler Group, McIlwaine and Wallace, 1976, p. 16. No's **4, 5, 6, 7**: #AK144, AK345, AK14, AK133 resp., Michipicoten I., Annells, 1974, p. 128-129. No's **8, 9, 10**: #IR-66, IR-70A, IR-40 resp., Isle Royale, Huber, 1973, p. C10-C11.

but that some siliceous rocks have undergone disturbances in their Rb/Sr isotopic relations since crystallization. Leeman also found that Pb isotopes imply that material from much older lithosphere (3.99 b.y.), either mantle or crust, contributed to the origins of those liquids. Silver and Green (1972) also reported Pb isotopic evidence for crustal melting or assimilation. Van Schmus *et al.* (1981) in a review of Rb/Sr data in the literature find an initial $^{87}Sr/^{86}Sr$ of 0.7046 ± 0.0005 for a selected group of 27 samples. Sabelin (1979) analyzed 16 lavas from the North Shore Volcanic Group and found an initial $^{87}Sr/^{86}Sr$ of 0.70358 ± 0.00065, consistent with the former value.

Nesterenko *et al.* (1964) have published analyses of trace elements (Cr, V, Ni, Co, Cu, Ba, Sr) in composites covering the whole Siberian Platform plateau lavas (Table 1.2.2.13). Throughout most of the section, Cr ranges between 100 and 150 ppm and Ni between 100 and 120 ppm, but more primitive averages of 260 and 330 ppm (Cr) and 170 and 250 ppm (Ni) were found for two composite samples of the Gudchikhinsk formation. In contrast, they and Dyuzhikov *et al.* (1973) found much lower values of 50 ppm for Cr and 80 ppm for Ni and higher than normal values for Zr, Ba, and Sr in the Ivakinsk formation, the more alkalic early magmatic cycle in the Noril'sk area. Masaitis *et al.* (1966) determined the uranium contents of Siberian platform traps, finding a weighted mean of 0.62 ppm U in basalts and undifferentiated intrusions, and a positive correlation of U with K content. They found this mean value to be similar to that for oceanic crust, higher than that for several other tholeiitic suites but less than that in Hawaiian alkali-olivine basalt. Balashov and Nesterenko (1966) studied rare-earth elements in the Noril'sk basalts (Table 1.2.2.14), finding within each volcanic cycle a decreasing total REE content, a decreasing light/heavy REE fractionation, and a decreasing Ce/Y ratio correlated with decreasing alkali content. By analogy with associated differentiated intrusions, they inferred that these trends imply successive eruptions of increasingly less-fractionated liquids within each cycle.

As was pointed out by Cox *et al.* (1967), the Karroo lavas can be geochemically divided into two major provinces, one in the northeast (Nuanetsi, Tuli, N. Lebombo) characterized by large abundances of incompatible elements, and one to the southwest (Swaziland, Lesotho) in which incompatible elements are relatively depleted (the more typical tholeiites). Trace element contents of representative Karroo lavas are presented in Table 1.2.2.15. The "limburgites" and olivine-rich basalts in the northeast, though Mg-rich, contain enough Si to keep them from being classed as alkaline (nepheline normative) and have unusually large contents of Ba and Sr along with high values of Cr and Ni. The Lesotho tholeiites, on the other hand, range from 160–350 (average 256) ppm Ba, 110–270 (average 190) ppm Sr, and 55–130 (average 85) ppm Zr with moderate to high Cr (170–650, average 317 ppm) and low to moderate Ni (25–350, average 73 ppm; Cox and Hornung, 1966), which suggests that olivine and plagioclase fractionation have been important in their evolution.

Compston *et al.* (1968) and Manton (1968) have analyzed Sr isotopes in a variety of rocks of the Karroo system. Whereas Compston *et al.* (1968) have found relatively uniform initial $^{87}Sr/^{86}Sr$ ratios for tholeiitic Karroo dolerites (0.7045–0.7066, mean 0.7057), Manton has found a wide range of initial ratios in mafic rocks from the Nuanetsi-Lebombo area (0.7042–0.7125) which he explains by crustal contamination. Rhyolites near the top of the southern Lebombo monocline have low initial ratios (0.7042) which indicate mantle derivation

Table 1.2.2.13a Major and trace elements in composite samples of Siberian Platform basalts.[a]

	C-15a	C-15b	C-15c	C-16	C-17a	C-17b	C-17c	C-17d	C-17e	C-17f	C-17g	C-17h	Weighted mean
No.[b]	5	6	7	8	17	25	31	58	18	14	22	8	219
wt.%													
SiO_2	47.06	46.64	46.40	46.50	50.12	49.40	49.42	50.28	46.48	47.24	46.95	50.90	48.70
TiO_2	1.66	1.52	1.38	1.62	2.12	1.50	1.31	1.12	1.24	1.23	1.44	1.04	1.36
Al_2O_3	14.93	15.02	16.35	14.51	14.73	14.13	14.86	15.17	15.28	14.98	13.43	15.29	14.80
Fe_2O_3	5.21	5.77	5.28	5.47	2.56	3.40	3.39	2.66	2.33	3.88	3.36	3.23	3.31
FeO	7.56	6.25	5.87	6.95	8.64	7.56	6.17	7.41	9.34	7.95	6.25	6.31	7.27
MnO	0.20	0.20	0.24	0.19	0.22	0.13	0.18	0.17	0.22	0.20	0.16	0.16	0.18
MgO	6.95	6.26	6.24	6.73	5.32	6.37	7.92	7.20	7.32	6.65	10.62	7.15	7.30
CaO	11.12	10.97	11.15	9.62	7.00	7.84	9.84	10.13	10.01	11.46	9.55	10.72	9.68
Na_2O	2.24	2.17	2.13	1.78	3.50	3.28	2.06	2.17	2.48	2.15	1.98	2.19	2.37
K_2O	0.48	0.26	0.18	0.49	2.30	1.43	0.59	0.89	0.35	0.29	0.58	1.32	0.85
P_2O_3	0.14	0.11	0.08	0.08	0.36	0.14	0.08	0.08	0.09	0.08	0.08	0.08	0.11
rest	2.68	4.39	5.23	6.04	3.02	4.70	4.68	2.86	4.72	3.84	5.26	2.08	4.00
SUM	100.23	99.56	100.58	99.98	99.89	99.88	100.30	100.14	99.86	99.95	99.66	100.47	99.93
ppm													
Cr	120	140	150	130	51	160	260	130	100	120	330	150	160
V	280	240	150	240	68	93	170	170	240	160	210	210	170
Ni	110	120	120	100	54	130	170	120	120	100	250	70	130
Co	33	26	19	31	10	20	16	20	37	23	22	28	25
Cu	50	50	70	140	10	20	20	20	90	42	80	110	43
Ba	280	170	140	160	900	720	350	370	160	150	320	690	400
Sr	540	340	340	280	850	710	360	240	430	270	450	650	430
Ni/Co	33	46	63	32	54	65	101	60	32.5	43.5	13	26	52

[a] From Nesterenko *et al.* (1964).
[b] No. of samples in composite samples.
Basalts: (porphyritic, two-feldspar, tholeiitic and picritic basalts and labradorite andesites). C-15a, b, c — Yambukinsk, Kochechum and Nidym formations; lower Tembenchi River, B. P. Markevich's coll., VAGT; C-16 — upper Severnaya R., V. L. Masaitis' coll., VSEGEI; Spp. C-17a-17h — Noril'sk region, G. V. Nesterenko's coll.; C-17a — Ivakinsk formation, hole 746; C-17b — Syverminsk fm., hole 746; C-17c — Gudchikhinsk fm., hole 746; C-17d — Nadezhdinsk fm., hole 746; C-17e — Morongovsk fm., hole E-18; C-17f — Nadmorongovsk fm., hole E-18; C-17g — Gudchikhinsk fm., hole 421; C-17h — Nadezhdinsk fm., hole 421.

Table 1.2.2.13b Average contents of minor elements, eruptive phase 1, Noril'sk area (ppm).†

	Basalts with titaniferous augite (58)*	Labradorite porphyritic basalts (49)	"Di-felspathic" porphyritic basalts (55)
Cu	90	90	90
Ni	70	80	100
Co	30	50	40
Cr	20	40	90
V	80	120	180
Zr	180	170	210
Sr	310	270	390
Ba	830	610	1020

† From Dyuzhikov *et al.* (1973).
* Number of analyses shown in parentheses.

Table 1.2.2.14 Distribution of the rare earths in the traps of the Siberian Platform (ppm).[a]

	C-17e	C-17d	C-17g	C-17v	C-17b	C-17a	Average
No.[b]	14	18	58	31	25	17	—
La	4.5	5.2	16.5	7.6	19.5	36	13
Ce	11	15	36	14	38	80	29
Pr	1.7	2.7	4.3	2.7	6.6	12	4.3
Nd	7	11	18	10	22	42	16.5
Sm	2.4	3.1	4.3	3.6	5.8	8.5	4.2
Eu	0.8	1.3	1.2	1.0	2.2	1.8	1.3
Gd	2.4	3.6	5.3	3.9	8.0	6.1	4.7
Tb	0.7	0.9	1.6	—	—	—	1.3
Dy	2.9	4.1	5.0	4.1	5.5	2.9	4.3
Ho	0.8	1.2	1.5	1.0	—	—	1.3
Er	1.9	3.0	3.0	2.0	2.7	0.7	2.5
Yb	1.7	2.3	2.4	1.7	2.3	0.4	2.0
Lu	—	0.55	0.60	—	—	—	1.4
Y	18	26	26	19	27	14	.23
$\frac{\Sigma Ce}{\Sigma Y}$	0.95	0.95	1.8	1.1	2.0	6.7	1.7
$RE_2O_3 \cdot 0.10^{-2}$,%	0.65	0.95	1.57	0.85	1.70	2.50	1.27
Thickness[c]	136	308	386	186	144	108	1268

[a] After Balashov and Nesterenko (1966).
[b] Number of samples in composite sample.
[c] Thickness of formation, m.
Noril'sk basalts: cycle II — Nadmorongovsk (C-17e), Morongovsk (C-17d), Nadezhdinsk (C-17g), formations; cycle I — Gudchikhinsk (C-17v), Syverminsk (C-17b), Ivakinsk (C-17a) formations.

(through fractionation of basalt magma?) but rhyolites and subvolcanic granites from Nuanetsi have slightly higher initial ratios of 0.7081–0.7085, suggesting some crustal contamination (Manton, 1968).

Trace element data on Paraná lavas are contained in Table 1.2.2.16 and Fig. 1.2.2.13. This province contains the lowest Ni and Cr and highest Y and Zr and relatively high V compared with other tholeiitic series

Table 1.2.2.15 Trace element contents of representative Karroo lavas.†

	1	2	3	4	5	6	7	8	9	10	11	12	13	14	15
Ba	250	245	250	530	530	1120	100	986	418	786	650	1000	800	400	600
Co	30	40	40				35	52	88	44	35	50	25	30	30
Cr	350	260	550				120				450	1100	320	260	60
Ga	16	14	19				15				22	16	17	19	23
Li	< 10	< 10	< 10				20				< 10	< 10	15	< 10	< 10
Mo	< 3	< 3	< 3				< 3				8	3	3	5	3
Nb	10	15	< 10	< 10	< 10	< 10	15	20	19	25	50	45	40	40	40
Ni	90	55	100	60	75	110	65				250	950	180	190	90
Pb	< 10	< 10	< 10				< 10				< 10	18	< 10	13	10
Rb	8	30	40	35	29	30	< 50	24	1	31	< 50	80	55	< 50	< 50
Sc	50	30	45				50				40	25	40	30	25
Sr	260	150	190	544	558	680	270	1137	446	635	900	1200	700	380	850
V	350	350	300				300				240	240	350	300	300
Y	15	25	15				20				45	20	30	35	30
Zn								128	92	102					
Zr	109	111	107	320	320	395	60	455	161	290	220	320	200	130	220

†Measured in ppm. No's **1, 2, 3**: #B.1, B.47, B.71 resp., basalts, Lesotho, Cox and Hornung, 1966, p. 1422. No's **4, 5**: #Bu.3, Bu.7 basalts, Nyamandhlovu; No. **6**: #F.6 basalt, Featherstone area; No. **7**: #S.13 basalt, Swaziland, all Cox *et al.*, 1967, p. 1454-5. No's **8, 9, 10**: #TV43, 22, 29 resp., Tuli syncline, Vail *et al.*, 1969, p. 409-411. No's **11, 12, 13, 14, 15**: #LM434 olivine-rich basalt; #LM432 limburgite; #C868 basalt; #DW389 basalt; #KC58 basalt; all Cox *et al.*, 1965, p. 147.

Table 1.2.2.16 Concentrations of major and trace elements in Paraná traps.†

	Average	Median	Dispersion	Principal Mode	No. of Analyses
Wt. %					
SiO_2	51.3	49.6	43.1-69.1	48-50	236
TiO_2	2.6	2.4	0.25-5.30	2.5-3.5	"
Al_2O_3	13.4	13.4	8.6-19.8	12-15	"
$FeO + Fe_2O_3$	13.4	13.4	5.1-22.1	13-16	"
MnO	0.17	0.19	0.0-0.49		"
MgO	5.0	4.1	0.9-13.2	4-5	"
CaO	8.0	7.8	0.9-12.4	7.0-9.5	"
Na_2O	2.7	2.4	0.2-4.9	2.5-3.0	"
K_2O	1.7	1.1	0.4-4.5	1.0-2.0	"
ppm					
Ba	535	385	111-2099	300-400	155
Co	43	38	5-99	40-50	109
Cr	136	125	25-500	140-160	109
Cu	149	120	14-999	50-75	155
Ni	48	38	2-350	40-60	155
Rb	54	30	10-195	20-40	155
Sr	429	350	123-1550	400-500	155
V	343	325	27-751	350-400	109
Zn	149	112	10-332	100-125	155
Zr	254	215	52-700	150-200	155

†After Ruegg (1976 a, b).

(Ruegg and Dutra, 1970), again implying an advanced evolutionary stage for these liquids. Compston (1968) found that initial $^{87}Sr/^{86}Sr$ ratios for Serra Geral plagioclases are clustered closely around 0.7057, typical of continental basalts in general but distinctly higher than ratios typical of ocean-ridge basalts.

Wasserburg and DePaolo (1979) have measured Nd isotopes in Siberian Platform basalts. They interpret their results as indicating that these basalts (and the Columbia River and Paraná basalts) were derived from previously undifferentiated, deeper portions of the mantle.

Comparisons and petrogenesis

It has long been recognized that "average" or "typical" continental flood basalts have a distinctive major element composition range relatively high in Si, Fe, Ti and K. Pearce *et al.* (1977) found, for example, a fairly clear statistical distinction between continental basalts (including plateau basalts) and others on an MgO-FeO*-Al_2O_3 diagram. However, the major basalt plateaus actually show a wide range of compositions, overlapping with most of the other principal basalt types. Thus alkali basalts, basaltic andesites, picrite basalts and high-Al basalts all form significant portions of various plateaus which cannot be ignored. In the oldest plateaus (Keweenawan and Siberian), more magnesian basalts which are poorer in incompatible elements play a greater role than they do in younger Mesozoic and Cenozoic flood basalts. In the Lake Superior district, the most abundant single type of lava is a high-Al olivine tholeiite with a moderately primitive range of Mg′ values, which is more like a mid-ocean ridge or Icelandic basalt than a typical Paraná, Deccan or Columbia River lava. Other than this greater proportion of primitive compositions in the older plateaus and the remarkable Mg- and incompatible-element-rich tholeiitic limburgites of Karroo, systematic differences in composition among the world's great plateau basalt provinces are largely masked by the substantial variations observed among basalts within each province. For instance, the several plateaus within the Keweenawan province of the Lake Superior district, although showing much overlap, have distinctly different ranges of Mg′, Al_2O_3, and TiO_2 contents (Green, 1979a).

Other comparisons show that the most characteristic continental flood basalts (as distinct from basaltic andesites) are similar in Al to Hawaiian intra-plate tholeiites but are poorer in Mg, while they are richer in Si, K, Ti and P. They similarly are more evolved than the island arc tholeiites. However, the Keweenawan high-Al olivine tholeiites show strong major element similarities to island arc basalts, Icelandic tholeiites and mid-ocean ridge basalts, though they are higher in Al content than the latter two.

Trace element differences, especially in the rare earths, are also considerable. Most of the plateau basalts have REE abundances which are moderately

*Total Fe calculated as FeO.

enriched over those of ocean-ridge or island arc basalts, and show moderate to strong light/heavy REE fractionation. Again the Lake Superior province shows the most variety (see section 1.2.2.5).

Compared to mid-ocean ridge olivine tholeiites, those of the Keweenawan province are not depleted in light REE; their REE patterns and abundances more nearly resemble those of the FAMOUS area (White and Bryan, 1977), except that the latter do not show the positive Eu anomaly nor the slight depletion in HREE that characterizes the more magnesian Keweenawan tholeiites. The REE patterns and contents of the Keweenawan olivine tholeiites (see Fig. 1.2.2.21) are similar to those of southwestern Iceland and the northern end of the Reykjanes Ridge (Schilling, 1973), but the Keweenawan rocks contain substantially lower concentrations of TiO_2 and P_2O_5. Keweenawan olivine tholeiites have smaller light/heavy REE fractionations than those of Kilauea (intraplate) tholeiites and show positive Eu anomalies which the latter lack. On the other hand, they are more fractionated than island arc tholeiites which also have generally lower REE concentrations (see sections 1.2.3, 1.2.5, 1.2.6 and 1.2.7).

Keweenawan transitional basalts, which are typical in composition of the bulk of the younger plateau basalt provinces, have REE patterns that are generally steeper than those of most Hawaiian and Icelandic tholeiites.

The great range in geochemical characters of plateau basalts is certainly related to the many influences on their origin, perhaps even including the element of geologic time. These factors include such primary aspects as differences in depth and degree of partial melting in the mantle and in homogeneity of mantle sources, as well as evolutionary factors related to the passage of these melts to the surface through thick continental lithosphere. Such secondary influences would include possible zone refining (Langmuir *et al.*, 1977), melting and assimilation, fractional crystallization (Wood, 1978), magma mixing (Rhodes *et al.*, 1979) and at late stages of fractionation, liquid immiscibility (Roedder, 1978; McBirney, 1975). It is likely that each of these processes has played a role in the evolution of some plateau lava. So far, however, very little has been done to model specific evolution/fractionation schemes for pre-Tertiary plateau basalts quantitatively, and most inferences must be drawn from Sr isotopic data and from standard geochemical arguments.

The initial $^{87}Sr/^{86}Sr$ ratios discussed earlier are uniformly higher—though most are only slightly higher—than those of oceanic basalts having similar ages. Although this relationship has been generally interpreted to be the result of assimilation of crustal material, it could also be derived from partial melting of old subcontinental lithospheric mantle which was not depleted in Rb. This is supported in the case of Keweenawan basalts by Pb isotopes (Leeman, 1977). Other, higher initial Sr ratios in intermediate and rhyolitic plateau lavas, such as in the Karroo and Keweenawan plateaus, most probably do result from incorporation of crustal material.

Trace elements everywhere indicate a mantle source undepleted by previous melting events, in contrast with sources postulated for most ocean-ridge basalts. The more primitive high-Al olivine tholeiites appear to have been produced by a relatively high degree (25–30%) of melting of mantle peridotite (Bender *et al.*, 1978; see Chapter 3) which contained abundant spinel and/or Al-rich clinopyroxene. The more widespread transitional basalts and basaltic andesites are more problematical. If they are primary or near-primary basalts (as proposed by Wilkinson and Binns, 1977), they must represent a considerably smaller degree of source melting because of their richness in incompatible elements and low Cr and Ni contents. However, their relatively low Mg/Fe and Ca/Na ratios imply that crustal evolutionary processes, such as crystal fractionation and/or zone refining, have been more important than equilibration with mantle assemblages having high Mg/Fe ratios. If these magmas are indeed close to primary, widespread assumptions as to subcontinental mantle compositions must be greatly modified. Unfortunately, some combination of slow rate of ascent and/or low viscosity of plateau basalt magmas appears to have prevented samples of their mantle source rocks from reaching the surface as xenoliths.

1.2.2.5 The Keweenawan reference suite of the North Shore Volcanic Group

General geology

The North Shore Volcanic Group (NSVG; Goldich *et al.*, 1961) is defined as the Keweenawan lavas and interbedded sediments that are found in northeastern Minnesota adjacent to Lake Superior (Fig. 1.2.2.5). Paleomagnetic investigations (Palmer, 1970; Books, 1972) define a lower reversed polarity and an upper normal-polarity sequence. Green (1972) has provided a general summary of its structure, general character, stratigraphy, and composition, and has concluded (1977) that the reversed and normal sections were parts of two separate lava plateaus that largely overlapped. Excellent exposures are found along the shore of Lake Superior and in some stream valleys, but away from these erosional agents the rocks are mostly covered by glacial drift. The volcanic rocks are intruded by several large sills and by many less regular plutons, especially in the Beaver Bay Complex. The Duluth Gabbro Complex normally underlies the volcanic rocks, but intrudes above their base near Duluth and at the northeastern end.

Preliminary flow-by-flow mapping of nearly all of the shoreline exposures has been published (see Green, 1972, for references) and more detailed mapping by Green correlated with petrochemical studies by Green, Brannon and Haskin, is continuing (Brannon *et al.*, 1979; Green *et al.*, 1980). Intermediate-scale generalized maps are available in Green *et al.* (1977) and a 1:250,000 scale map showing regional relations is also available (Green, 1981).

Structurally, the North Shore Volcanic Group has the form of a large truncated, filled saucer, tilted gently toward Lake Superior. The lavas dip 15° to 20° to the east at the southwest end near Duluth, and about 10° south at the northeast end. At the top of the section, near Tofte and Lutsen, they dip about 10°–12° southeast. The lavas overlie very thin Middle Proterozoic orthoquartzites which in turn overlie Lower Proterozoic (Animikie) shale, slate, and graywacke. The top of the sequence is now covered by Lake Superior, but is thought to have been overlain in part, perhaps along an erosional unconformity, by the Portage Lake Volcanics which constitute a slightly younger lava plateau.

The total stratigraphic section has never been measured at one place (such as by drilling). Cumulative addition of flows mapped in the southwestern limb gives a total of about 9000 meters, including only 340 m in the lower, reversed-polarity unit. The shoreline section in the northeast limb contains a total of about 6400 m, of which the lower 2,750 m has reversed polarity (see Fig. 1.2.2.7).

Original volcanic textures and structures are well preserved, except that most of the original interstitial glass has been altered. Ropy, billowy, "rhino-skin" and rubbly flow tops are common, as are internal structures such as pipe amygdules, vesicle cylinders and segregation veins (see Green, 1972). Nearly all of the flows are

Table 1.2.2.17 Characteristics of major lava types in the North Shore Volcanic Group.[†]

	Olivine tholeiite	Quartz tholeiite-basaltic andesite		Andesite, trachyandesite	Intermediate quartz latite	Quartz latite-rhyolite
Thickness, in m. (normal)	3-18	10-20		10-30		25-185
(max. thick., in m.)	50	>30		75	60	400
Jointing	rough columns in upper ½; sheeting at top	small joints, no good pattern		small joints, irregular and subhorizontal	irregular, blocky, platy near top	platy, small joints, large columns in some thick flows
Top zone	ropy, smooth; round amygdules	amygdular rubble-breccia; stretched amygdules		lobate, wrinkled crust; stretched amygdules	uneven, wrinkled crust or breccia	wrinkled, flow-banded crust, vesicular
Textures	obviously ophitic lower ⅔; amygdaloidal in upper ⅔	aphanitic to fine-grained, pilotaxitic to subophitic; aphyric		aphanitic to fine-grained; porphyritic (plag., augite, magnetite)	aphanitic, porphyritic (plag., ferroaugite, Fe-olivine, magnetite)	aphanitic to felsitic-poikilitic; most porphyritic (K-spar, qtz, |plagioclase, ferroaugite, magnetite)
Mineralogy	olivine, augite, bytownite-labradorite, opaques, zeolites, etc.	augite, labradorite-andesine, opaques, some interstitial glass, K-spar, quartz		andesine, augite, magnetite, some pigeonite, interstitial K-spar	plagioclase, ferroaugite, Fe-olivine, magnetite, K-spar, quartz	as above (ferroaugite rarely preserved)
		Tholeiite	Basaltic andesite			
SiO_2, wt.%	45.7-49.1	50.1-51.1	53.1-55.0	51.4-56.4	60.2-64.9	68.9-75.5
MgO, wt.%	5.3- 8.1	4.7-5.9	2.9-4.3	2.3- 4.5	0.9- 1.9	Tr-0.7
K_2O, wt.%	0.1- 0.5	0.6-0.9	1.9-2.4	0.8- 2.7	2.8- 4.9	3.5-6.2
Other	pipe amygdules at base; thin flow units common	fine oxidation-banding common in lower ½; qtz, agate in amygdules		variable	red-brown; plagioclase commonly altered	red, pink, or gray

[†]After Green (1972).

subaerial, but rounded pillows are found in the basal flow at Duluth and in a flow near the base at Grand Portage (at the northeast end). Pillow-breccia was found at a few localities higher in the section. A few local breccia layers are present, but the interflow sediments, which constitute between 1–3% of the section, are mostly well-rounded epiclastic fluvial sands with very little tuffaceous material (Jirsa, 1978).

Flow thickness is highly variable, ranging from 30 cm to over 100 m. The thinnest flows are generally alkalic basalt and olivine tholeiite, whereas many compositions occur as thick flows. One rhyolite unit is estimated to be 1000 m thick and probably represents a dome. Because of incomplete mapping and gaps in the exposure, no accurate figure for the average flow thickness is available, but a general estimate, based on 456 flows in measured sections, is about 25 meters. Lateral continuity is difficult to establish for many flows because of lack of exposure, but some petrographically distinctive flows or flow groups can be traced as far as 38 km inland from their lakeshore outcrops, establishing the plateau-like structure.

The stratigraphic section has been divided (Green, 1972) into mappable units composed either of thick and distinctive individual flows or groups of similar flows showing distinctive characteristics, much like the stratigraphic units established by Walker (1959) in the Tertiary lavas of eastern Iceland. In fact, the NSVG closely resembles the Icelandic Tertiary plateau lavas both physically and chemically.

Olivine tholeiite with ophitic texture is the most abundant lava type in this Group; these basalts are estimated to make up about 40–45% of the total thickness. They are found scattered throughout the section, though the uppermost 1000 m or so is entirely olivine tholeiite. A few olivine tholeiites contain plagioclase phenocrysts but most are aphyric. Weakly alkaline or transitional basalt, showing relatively coarse grain size, is much less common and may contain phenocrysts of plagioclase. A conspicuous and unique group of basalts at the base of the section contains abundant olivine and augite phenocrysts. Aphanitic to fine-grained saturated tholeiites and basaltic andesites are common; these are generally aphyric. Plagioclase- and augite-porphyritic andesites are less abundant, as are icelandites (intermediate quartz latites). Rhyolite is more common in the NSVG than in the other lava plateaus of the Keweenawan Lake Superior district; it makes up between 15–25% of the stratigraphic section, and occurs as widespread lavas. General characteristics of the various rock types are shown in Table 1.2.2.17.

Table 1.2.2.18 Measured modes of Keweenawan reference suite.[a]

	KEW-1	KEW-2	KEW-3	KEW-4	KEW-5	KEW-6[b]	KEW-7	KEW-8	KEW-9	KEW-10
Quartz										0.4
Plagioclase P									23.0	
Plagioclase G	63.8	40.3	49.0	51.5	51.6	36.6 }	63.2	{ 49.7	26.7	29.6
K feldspar[b]	tr.			tr.			2.9			x
Augite P										24.9 }
Augite G	20.3	23.0	30.7	26.4	27.7	29.0	21.4	24.0 }	32.3	
Pigeonite		tr.	tr.		1.3			2.9		
Orthopyroxene						0.6				
Olivine P									(0.9)	
Olivine G	(9.1)	(17.2)	(12.1)	(9.5)	14.9	16.3	(9.5)	(8.1)	(1.1)	
Hornblende	tr.									
Biotite				tr.						10.9
Magnetite }	3.0	6.1	1.3 }	2.8	1.5 }	1.8	2.5	1.3	{ 10.2	4.5
Ilmenite }		1.7	0.8 }						{ 2.2	3.3
Apatite										0.7
Interstitial residua		3.9	0.1	5.3	2.9	0.2	0.5	6.4	0.5	
Microcrystalline groundmass						15.2				25.5
Cavity fillings:										
Fe-clay, etc.	3.8	7.9	4.5			0.2		7.5	3.1	
Calcite						0.1	tr.			
Zeolite			1.4							

Sample selection

A reference suite of 20 samples from the NSVG has been selected for detailed study. The suite is limited to rocks containing less than 60% SiO_2, although the plateau volcanism involved more siliceous members as well. Selection of sample localities (flows) was a compromise among (a) attempting to show the variety of rock types present in the North Shore Volcanics, (b) choosing aphyric or nearly aphyric rocks that could represent magmatic liquids, and (c) attempting to achieve a balance according to volumetric abundance of the various types. At each locality a sample was chosen from the massive non-amygdaloidal interior of the flow, since petrographic experience had shown such samples to have the greatest chance of retaining fresh primary mineralogy and chemistry. Diktytaxitic cavities, generally filled with chlorophaeite or chlorite, could not be avoided in a few flows. Jolly and Smith (1972) have shown that the massive interiors of Keweenawan flows in Michigan are nearly unaffected chemically by the burial metamorphism to which the NSVG has also been subjected. However, Wood *et al.* (1976) concluded from a chemical study of 70 Tertiary plateau basalts in Iceland that regional zeolite-grade metamorphism had brought about some mobility of Si, Mg, K, Rb, Sr, and light REE. Later, Wood (1978) concluded that significant chemical change had not occurred in those Icelandic lavas showing minimal H_2O and secondary mineralization.

Petrography and stratigraphy

Textures and mineralogies of the reference suite samples are presented in the Appendix (section A-2) and measured modes are given in Table 1.2.2.18. In general, the olivine tholeiites show ophitic texture; sample KEW-6 (KR-20), an olivine tholeiite, has mostly intergranular and radiate texture but it was taken from a more rapidly chilled, finer-grained part of a flow which elsewhere is ophitic. The transitional or weakly alkaline basalts show a variety of textures including subophitic and intergranular, and several of them are porphyritic. The basaltic andesites are generally intergranular or intersertal. Thirteen of the 20 samples are aphyric, three contain fewer than 5% phenocrysts, and the remaining four are significantly porphyritic.

Notes on the locality, geology, and stratigraphy of each sample are in the Appendix, keyed to the stratigraphic scheme of Green (1972).

Mineralogy

Representative microprobe analyses of the principal phases in the suite are presented in the Appendix

Table 1.2.2.18 *(Continued)*

	KEW-11	KEW-12	KEW-13	KEW-14	KEW-15	KEW-16	KEW-17	KEW-18	KEW-19	KEW-20
Quartz								2.4	tr.	1.8
Plagioclase P							48.8	2.4		
Plagioclase G	29.2	45.0	19.2	50.7	34.5	52.1		22.4	38.3	35.6
K feldspar[b]	3.9	x				10.0		40.4	2.8	x
Augite P			34.9					0.4		
Augite G	30.4	22.4	8.5	16.0	32.1	22.2		15.0	8.0	21.0
Pigeonite										
Orthopyroxene										
Olivine P	(3.4)	(13.8)	(11.4)	(4.0)	(7.8)?		(tr.)	(9.1)		
Olivine G										
Hornblende				1.1			0.1		5.7	
Biotite										
Magnetite	4.2	2.7			3.5	6.6	0.4	5.4	4.6	3.9
Ilmenite		1.6	2.5	9.7	0.6					
Apatite	0.2	x		0.5		x		tr.	x	
Interstitial residua	1.1	11.7		4.2	12.8	6.4			26.1	32.2
Microcrystalline groundmass			15.2				50.5			
Cavity fillings:										
Fe-clay, etc.		2.8	8.3	8.8	8.7	2.6	0.2	2.4	9.3	5.5
Calcite										
Zeolite									3.2	

[a]Measured in volume percents.
[b]Too fine-grained for accurate analysis; feldspar probably deficient with respect to pyroxene and olivine.
Note: P = phenocrysts, G = groundmass; x = present, in interstitial residual material; ( ) parens. indicate altered olivine.

and are illustrated in Figs. 1.2.2.14–15. Plagioclase is bytownite (up to An_{83}) to calcic labradorite in the olivine tholeiites, and gradually becomes more sodic through the transitional basalts to An_{32-50} in the andesites (Fig. 1.2.2.14). In the transitional basalts it tends to contain more Or for any particular value of An than in the tholeiites. The plagioclases in KEW-6, a rapidly cooled olivine tholeiite, have higher Or contents and a more limited range of An/Ab ratios than those of the other olivine tholeiites. Phenocrysts generally show strong marginal normal zoning to groundmass compositions. One sample (KEW-3) contains rare disequilib-

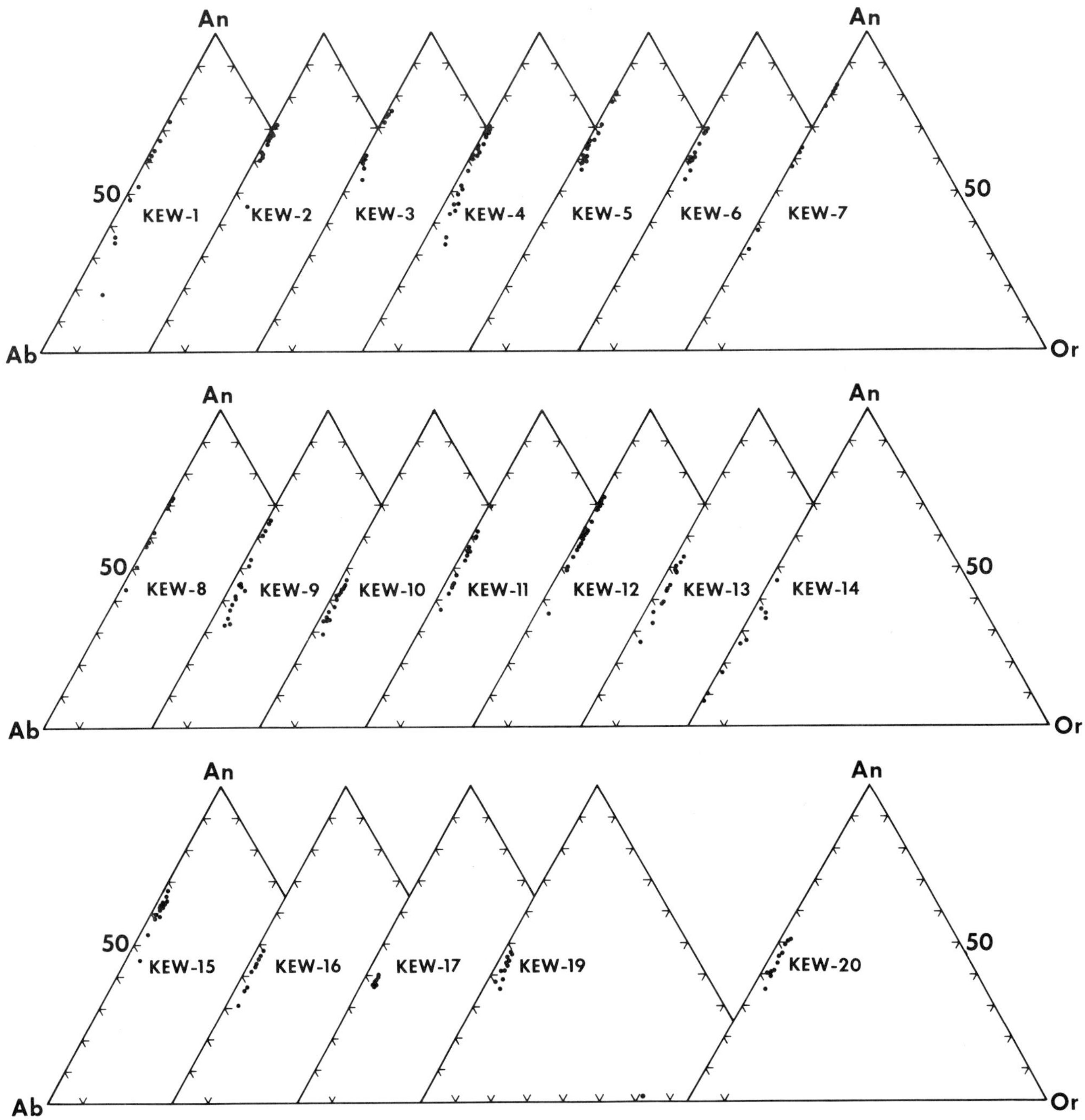

1.2.2.14 Plagioclase compositions for the Keweenawan reference suite, as determined by electron microprobe, in mol. proportions.

rium phenocrysts of An_{60} in a groundmass of $An_{75\text{-}70}$ laths.

Olivine is altered in nearly all samples in which it crystallized; in only two olivine tholeiites of the suite is it fresh and available for analysis (Fo_{68},Fo_{63}). All porphyritic olivine is altered.

Clinopyroxene (Fig. 1.2.2.15) is dominantly augite, being most magnesian in the olivine tholeiites and in the augite-porphyritic transitional basalt KEW-13. The ophitic clinopyroxenes of the olivine tholeiites are zoned to more iron-rich rims. Late, interstitial pigeonite is found in most samples. In the chilled olivine tholeiite KEW-6, the last pyroxene to crystallize was an orthopyroxene, in minute pockets of coarse residual material; otherwise, orthopyroxene is conspicuously absent from the suite. The pyroxenes trace out a compositional trend that is similar to that seen in the Skaergaard intrusion; see also Konda and Green (1974). The augites in the transitional basalts contain slightly more Ca than those in the tholeiites, and pigeonite is generally absent except for KEW-11. The basaltic andesites contain subcalcic augite of intermediate Fe/Mg ratio, with or without pigeonite which in some samples forms the cores of ferroaugite grains.

No primary hydrous minerals are present, except for traces of interstitial brown hornblende in a few quartz tholeiite and andesite flows and in one olivine tholeiite.

Major element chemistry

The general major element character of the suite is illustrated in Figs. 1.2.2.16–18 and Tables 1.2.2.19–20. Nearly all samples are hypersthene-normative, and nearly all fall in the tholeiitic or subalkaline field of the various discriminant schemes, although there are a few which lap over into the alkaline field. Most of the suite is quartz-normative, as are most of the younger plateau basalt formations, but the olivine-normative tholeiites constitute a major portion. The suite as a whole shows a moderately strong iron-enrichment trend which grades into dominant alkali enrichment at the basaltic andesite stage; this trend is very similar to the Thingmuli trend from the Tertiary of Iceland (Carmichael, 1964). Some scatter in the trends suggests that the rocks are not all genetically related by the same process.

The olivine tholeiites (KEW-1–KEW-8) are more coherent than the remainder of the suite. They are high-alumina basalts which show a strong Fe/Mg variation

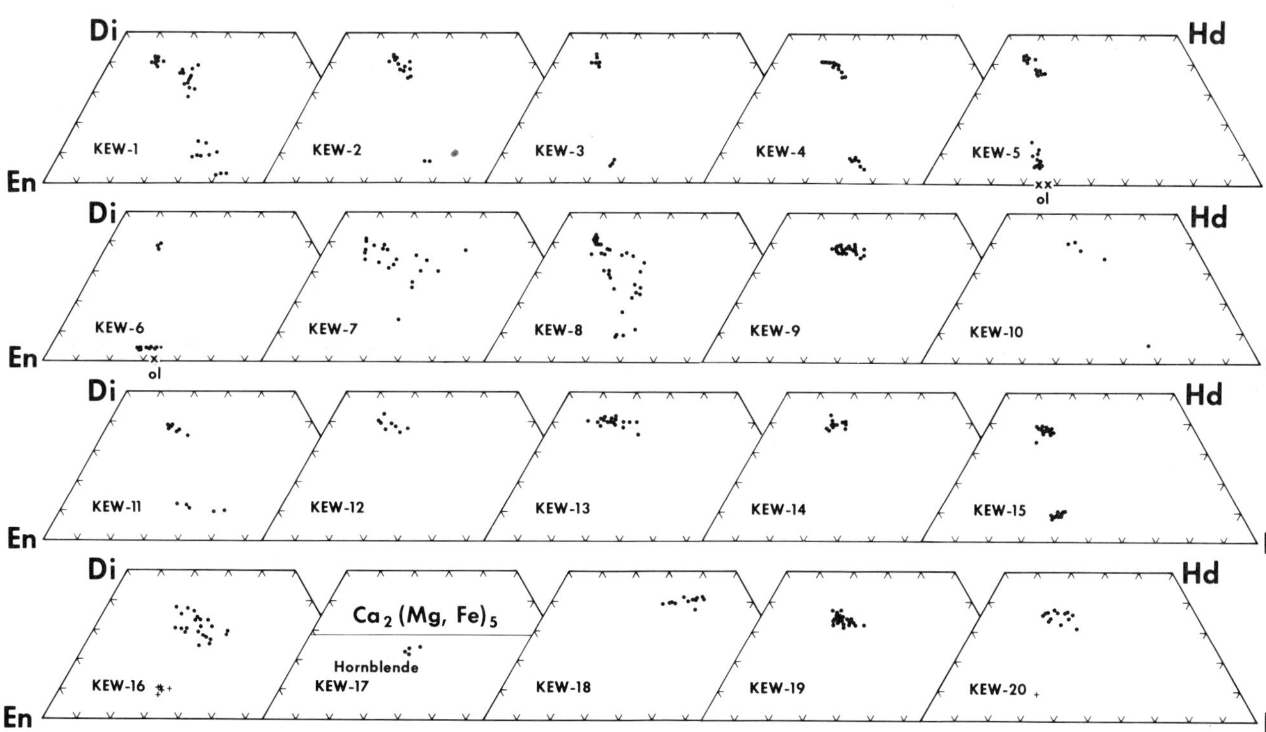

1.2.2.15 Pyroxene and olivine compositions for the Keweenawan reference suite, as determined by electron microprobe. Mol. proportions; x = olivine for KEW-16; + shows compositions of pigeonite in another sample of same flow. Pigeonite altered in KR-20.

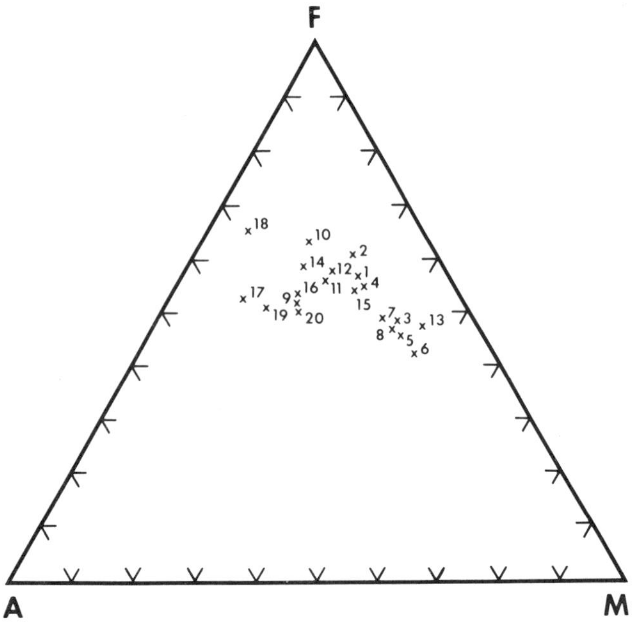

1.2.2.16 AFM diagram for the Keweenawan reference suite.

which resembles the iron-enrichment trend of the Skaergaard intrusion (Wager and Brown, 1967) and other crystal-fractionated tholeiitic bodies. There is no clear correlation of this trend with stratigraphic height in the NSVG, but the samples were collected from localities scattered widely throughout the section and represent various local magma sources. Brannon *et al.* (1979) however, have found a strong correlation of *decreasing* Fe/Mg ratio with stratigraphic height in one section of olivine tholeiites near Knife River that includes three of the flows of this suite.

The most magnesian olivine tholeiites, such as KEW-5 and KEW-6 (Mg′ values 0.62 and 0.66 respectively), which might be considered primitive, are unusually low in such incompatible elements as Ti, K, and P, as well as in the rare earths and other trace elements. Applying the calculations of Roeder and Emslie (1970) to these two aphyric rocks, their MgO and FeO values indicate equilibration with olivine of composition $Fo_{85\text{-}87}$ (KEW-5) and $Fo_{87\text{-}88}$ (KEW-6), which suggests that they could not have lost much olivine by fractionation since leaving their mantle peridotite source. Their major element compositions resemble those of mid-ocean ridge basalts but are even more depleted in Ti and P. They show strong similarities to such other high-alumina basalts as the Warner basalt (Yoder and Tilley, 1962), the chilled margin of the Skaergaard intrusion (Wager and Brown, 1967), and the inferred bulk composition of the Kiglapait intrusion (Morse, 1974) (Table 1.2.2.21). Although physically and chemically the flows resemble Icelandic olivine tholeiites, especially the more primitive series such as at Langjökull, the North Shore samples are somewhat richer in Al and poorer in Ti than those Cenozoic lavas (see section 1.2.3).

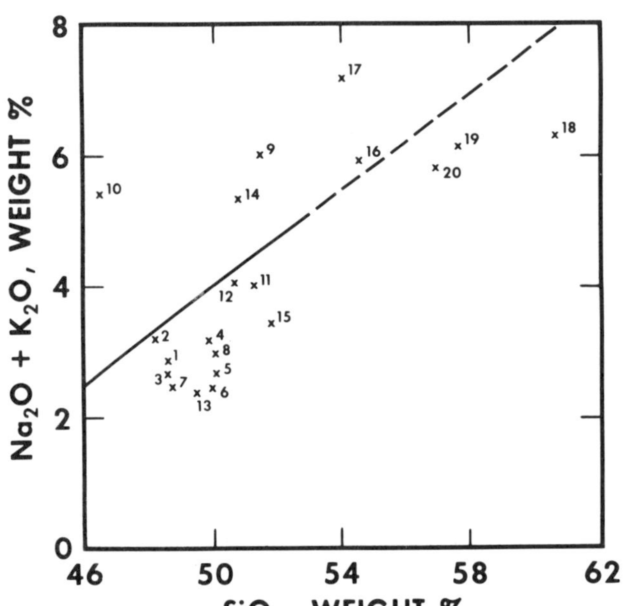

1.2.2.17 $(Na_2O + K_2O)/SiO_2$ diagram for the Keweenawan reference suite. Diagonal line is MacDonald and Katsura's (1964) discriminator for Hawaiian alkaline/tholeiitic lavas.

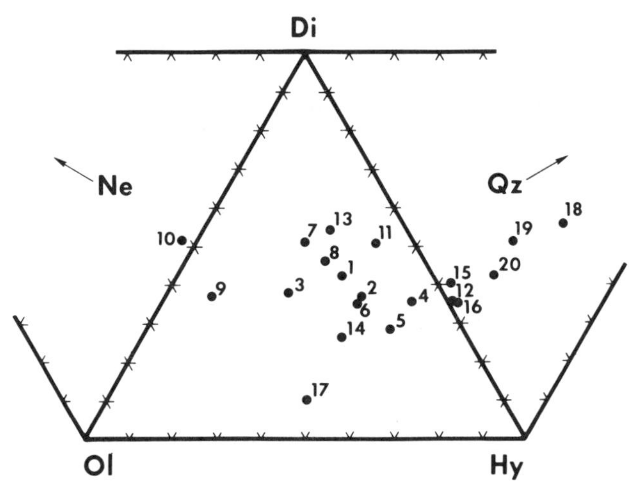

1.2.2.18 Normative Di-Ol-Hy-Qz-Ne diagram for the Keweenawan reference suite, using cation norms (Chayes and Metais, 1964) and a standardized oxidation ratio of $Fe^{+3} = 0.1\ (Fe^{+2} + Fe^{+3})$.

Table 1.2.2.19 Major elements in Keweenawan reference suite.

	KEW-1	KEW-2	KEW-3	KEW-4	KEW-5	KEW-6	KEW-7	KEW-8	KEW-9	KEW-10
Wt.%										
SiO_2	47.69	46.64	47.19	46.98	48.73	48.52	46.87	48.39	50.49	45.90
TiO_2	1.28	2.33	0.95	1.81	0.82	0.72	0.81	0.96	2.05	4.92
Al_2O_3	18.36	15.21	17.04	16.06	18.02	17.57	19.20	15.77	16.24	12.28
Fe_2O_3	6.47	9.49	2.63	5.28	3.68	2.18	5.42	4.45	6.71	3.34
FeO	4.74	4.86	7.69	6.63	5.36	6.10	2.97	5.45	5.09	14.16
MnO	0.17	0.14	0.14	0.17	0.11	0.13	0.12	0.17	0.17	0.22
MgO	5.27	5.68	8.11	6.32	7.77	8.65	5.86	7.81	4.52	4.63
CaO	11.21	9.21	10.76	9.59	10.13	10.83	12.41	10.75	6.75	7.26
Na_2O	2.35	2.56	2.23	2.64	2.39	2.23	2.27	2.44	3.98	3.52
K_2O	0.46	0.54	0.35	0.50	0.21	0.17	0.12	0.42	1.93	1.83
P_2O_5	0.15	0.25	0.13	0.25	0.06	0.03	0.08	0.09	0.24	0.68
S		0.01			0.02	0.01		0.02	0.00	n.d.
Cr_2O_3	(0.03)	0.02	(0.04)	(0.03)	0.02	0.05	(0.06)	0.05	0.01	(0.004)
H_2O+	1.62	3.05	2.55	2.79	3.29	2.25	1.55	3.29	2.04	1.23
H_2O-				1.60			2.08			
CO_2	0.00		0.07	0.16			0.10			0.16
Total	99.77	99.99	99.84	100.78	100.61	99.44	99.86	100.06	100.22	100.13
Mg′	.47	.43	.59	.50	.62	.66	.57	.60	.42	.32

	KEW-11	KEW-12	KEW-13	KEW-14	KEW-15	KEW-16	KEW-17	KEW-18	KEW-19	KEW-20
Wt.%										
SiO_2	50.54	49.54	48.55	49.13	50.19	52.63	53.22	58.50	55.01	55.20
TiO_2	1.49	2.14	1.86	3.19	1.51	2.04	2.05	1.54	1.65	1.61
Al_2O_3	16.59	15.05	9.02	15.14	15.15	13.12	17.55	10.12	14.13	13.51
Fe_2O_3	5.65	10.05	2.63	5.55	5.51	6.23	4.63	8.25	6.90	4.16
FeO	5.47	3.23	10.20	8.30	5.82	6.10	6.18	6.33	2.80	6.43
MnO	0.16	0.17	0.24	0.18	0.15	0.18	0.13	0.30	0.11	0.16
MgO	4.51	5.05	11.69	4.24	5.91	4.51	2.33	1.27	2.88	4.39
CaO	10.06	8.38	11.29	6.17	9.13	5.58	4.79	3.59	5.62	5.53
Na_2O	3.23	2.67	1.62	4.06	2.71	4.18	5.35	2.42	3.79	3.74
K_2O	0.76	1.29	0.71	1.13	0.62	1.54	1.76	3.69	2.07	1.90
P_2O_5	0.23	0.29	0.25	0.40	0.17	0.37	0.51	0.46	0.39	0.22
S		0.01						0.02		0.06
Cr_2O_3	(0.02)	0.03	(0.131)	0.00	(0.02)	(0.01)	(0.001)	0.00	(0.002)	0.01
H_2O^+	1.38	2.53	1.75	2.21	1.92	2.70	1.32	2.50	3.58	2.75
H_2O^-	0.74				1.76	1.35			1.84	
CO_2	0.02		0.02	0.13	0.13	0.07	0.11		0.10	
Total	100.83	100.43	99.83	99.83	100.68	100.60	99.93	98.99	100.87	99.67
Mg′	.43	.42	.62	.39	.49	.41	.29	.14	.36	.43

The next group of samples (KEW-9, 11, 12, and 14) are "transitional" or "Fe-Ti" basalts in the terminology of ocean-floor petrology, or weakly alkaline basalts according to alkali/SiO_2 classifications. They contain higher proportions of K, Ti, P, and generally have lower Mg′ values than the olivine tholeiites but all are hypersthene-normative. They resemble many of the plateau tholeiites of other districts, such as the Columbia River, Deccan, Karroo, and Paraná in major element contents.

KEW-10 is an aphyric rock with very high concentrations of incompatible elements, low Mg′ and normative oligoclase (An_{28}). It has the highest normative Ol content of the suite and low SiO_2. Whether it is Ne- or Hy-normative depends upon the Fe oxidation ratio assumed to have existed during crystallization. It could be called a mugearite. It has been somewhat recrystallized by contact metamorphism by the Duluth Complex, giving rise to modal biotite.

KEW-13 is also relatively rich in K, Ti and P but contains a large amount of magnesian augite and altered olivine phenocrysts; these probably are in excess over the original magma content as they can be seen to be concentrated in the lower parts of some flows. It is an example of a lava variety that was erupted over a large area of western Lake Superior at the initiation of

Table 1.2.2.20 Cation norms of Keweenawan reference suite.†

	KEW-1	KEW-2	KEW-3	KEW-4	KEW-5	KEW-6	KEW-7	KEW-8	KEW-9	KEW-10
Qz										
Or	2.74	3.42	2.07	2.99	1.30	1.23	.73	2.60	11.76	10.80
Ab	20.02	22.77	18.91	22.59	20.82	19.70	19.71	21.41	34.54	29.74
An	38.44	30.21	35.53	30.83	38.94	39.21	42.94	31.93	21.50	12.29
Di	13.74	14.08	13.64	12.12	10.12	13.79	15.58	18.55	9.42	15.60
Hy	14.03	11.37	12.34	16.10	21.05	12.35	9.91	14.20	2.34	.59
Ol	3.46	8.61	9.72	4.71	3.38	9.57	5.08	6.27	12.25	13.51
Mt	3.14	4.19	2.98	3.40	2.64	2.48	2.38	2.90	3.35	5.07
Il	2.45	4.65	1.81	3.48	1.60	1.43	1.58	1.88	3.99	9.33
Ap	.33	.67	.28	.55	.13	.16	.18	.20	.85	1.48
Cr	.03	.03		.04	.03	.08	.09	.07	.02	0.4
An/(An + Ab)	64	56	64	56	64	65	67	59	37	28

	KEW-11	KEW-12	KEW-13	KEW-14	KEW-15	KEW-16	KEW-17	KEW-18	KEW-19	KEW-20
Qz	.09				2.47	1.90		15.30	9.07	4.21
Or	4.50	8.11	4.20	6.85	3.72	9.21	10.42	23.05	12.63	11.87
Ab	27.40	23.68	13.72	35.24	23.26	35.78	45.37	21.70	33.10	33.28
An	28.59	26.23	15.25	20.26	27.73	12.63	13.71	6.41	15.93	15.21
Di	16.54	13.23	31.70	7.32	13.45	10.56	1.17	8.84	8.13	10.67
Hy	14.95	19.39	18.99	12.56	20.59	18.78	12.83	16.24	10.72	17.84
Ol		0.61	6.37	8.05			.26			
Mt	3.12	3.77	3.71	2.20	3.23	3.49	4.65	4.33	4.18	3.14
Il	2.84	4.22	3.53	6.21	2.91	3.92	3.90	3.08	3.23	3.22
Ap	.50	.71	.55	0.90	.38	.82	1.12	1.07	.88	.55
Cr	.03	.05	.19		.03					0.1
An/(An + Ab)	50	51	51	36	53	25	28	22	31	30

†Chayes and Metais, 1964; Fe adjusted to Fe_2O_3/Σ FeO = 0.2 by weight.

Keweenawan volcanism. In this respect it is somewhat analogous to the "limburgites" of the Karroo province but it contains less of the incompatible elements and more CaO than the tholeiitic limburgites.

KEW-15 is a quartz tholeiite which bears a strong physical resemblance to the fine-grained tholeiites of the Northern Ireland and East Iceland Tertiary plateaus.

Table 1.2.2.21 Compositions of Al-rich olivine tholeiites.†

	Cenozoic			Keweenawan		
	Skaergaard	Warner	M.O.R.B.	KR-20	Pigeon Pt.	Edwards I.
SiO_2	48.50	48.14	49.94	49.93	47.4	49.43
TiO_2	1.18	.89	1.51	.74	1.01	.98
Al_2O_3	16.69	18.12	16.69	18.09	18.2	18.57
FeO*	9.84	9.33	8.71	8.31	11.07	8.39
MnO	.16	.17		.13	.16	.16
MgO	8.69	8.93	7.28	8.90	8.46	8.61
CaO	11.48	11.29	11.86	11.14	9.44	11.44
Na_2O	2.39	2.79	2.76	2.29	2.57	1.89
K_2O	.25	.14	.16	.17	.33	.14
P_2O_5	.10	.07	.16	.03	.07	.08
Norm. An (%)	60	60	58	67	63	72
Mg′	.61	.63	.60	.68	.60	.67

†Measured in weight percent. References: Skaergaard — Wager and Brown (1967); Warner — Yoder and Tilley (1962); M.O.R.B. — ave. oceanic tholeiite, Engel et al. (1965); KR-20 — KEW-6, this study; Pigeon Pt. — M.G. Mudrey, Jr., pers. comm. (1979); Edwards I. — Huber (1973).

The last group (KEW-16–KEW-20) contains basaltic andesites, andesites, and trachy-andesites with appreciable K_2O contents (1.5–3.7%). They are either aphyric or plagioclase-phyric (KEW-17). The aphyric ones also chemically resemble many lavas from other plateaus such as the Columbia River, Deccan and Paraná. They generally are quartz-normative, have normative andesine to oligoclase, and variable but low Mg/Fe ratios. KEW-18 is extremely iron-rich with an Mg′ value of 14.

Trace element chemistry

Trace element concentrations in reference suite samples cover a considerable range (Tables 1.2.2.22–

Table 1.2.2.22 Large cations, Keweenawan reference suite.[†]

	KEW-1	KEW-2	KEW-3	KEW-4	KEW-5	KEW-6	KEW-7	KEW-8	KEW-9	KEW-10
ppm										
Cs	0.36	1.0	0.7	2.0				1.2	12	3.7
Rb	1	13	4	10	2	4	9	19	65	45
K_2O	0.46	0.54	0.35	0.50	0.21	0.17	0.12	0.42	1.93	1.83
Sr	243	199	265	284	210	213	268	412	440	430
Ba*		146			51	51		62		378
K/Rb	3820	345	725	415	870	352	111	184	246	338
Rb/Sr	.0041	.065	.015	.035	.0095	.019	.034	.046	.148	.105

	KEW-11	KEW-12	KEW-13	KEW-14	KEW-15	KEW-16	KEW-17	KEW-18	KEW-19	KEW-20
ppm										
Cs	0.3	11	3.8			0.6	2.7	4.9	2.1	1.1
Rb	16	50	46		8	41	54	125	102	19
K_2O	0.76	1.29	0.71	1.13	0.62	1.54	1.76	3.69	2.07	1.90
Sr	259	235	389		184	300	973	157	288	249
Ba*	206	269				740		1130		524
K/Rb	394	214	128		644	312	270	245	169	832
Rb/Sr	.0618	.213	.118		.044	.137	.0555	.796	.354	.076

[†]K as % K_2O. *Ba analyses by isotope dilution mass spectrometry (Sabelin, 1979): all others by XRF at NASA, Houston.

Table 1.2.2.23 Rare-earth elements, Keweenawan reference suite.

	KEW-1	KEW-2	KEW-3	KEW-4	KEW-5	KEW-6	KEW-7	KEW-8	KEW-9	KEW-10
ppm										
La	11.9	15.8	7.82	15.2	4.29	2.94	4.6	5.45	24.1	50.1
Ce	26.5	36.9	17.5	33.5	10.5	6.8	10.4	12.1	54.5	114
Sm	4.02	5.67	2.79	4.80	1.96	1.48	1.99	2.23	7.17	13.7
Eu	1.35	1.79	1.06	1.58	.815	.737	.79	.92	2.01	3.57
Tb	.78	1.23	.57	.83	.38	.34	.53	.45	1.49	2.32
Yb	2.2	3.20	1.70	2.35	1.35	1.15	1.27	1.79	3.70	5.0
Lu	.33	.48	.226	.36	.197	.159	.17	.24	.58	.743

	KEW-11	KEW-12	KEW-13	KEW-14	KEW-15	KEW-16	KEW-17	KEW-18	KEW-19	KEW-20
ppm										
La	17.6	19.4	29.4	49.2	16.1	33.6	68	74.1	45.7	41.5
Ce	40.4	44.4	77.0	111	37.2	72.3	143	190	100	85
Sm	5.60	6.74	8.66	11.2	5.37	7.53	15.0	22.0	8.93	9.97
Eu	1.60	1.91	2.29	3.18	1.48	2.17	4.21	4.95	2.10	2.27
Tb	.96	1.3	.99	1.16	.92	1.41	1.69	3.9	1.02	1.96
Yb	3.23	3.6	2.36	2.3	3.24	3.33	2.9	11.9	2.7	5.5
Lu	.49	.54	.258	.36	.49	.51	.40	1.88	.40	.82

1.2.2.25). The strongly compatible elements Ni and Cr vary from 270 and 390 ppm, respectively, in the primitive olivine tholeiites to almost nothing in the most silicic rocks of the suite (Fig. 1.2.2.20). The strongly incompatible elements La and Th vary by factors of 5.4 and 5.5 times among the olivine tholeiites alone. They vary by factors of 25 and 36 between the most primitive olivine tholeiite and the most trace-element-rich andesite.

Some trace element pairs (e.g., Sm and Hf, Fig. 1.2.2.19) show remarkably tight correlations, not only within the olivine tholeiites but throughout the range of rock types, except in some of the rhyolites (not included in the reference suite). Such strong correlations may imply some genetic relationship among the lavas of the reference suite. This would imply a relationship among all NSVG lavas, as the reference suite is from scattered localities over a wide geographic area and is not part of any single sequence of basalt to rhyolite.

The trace element concentrations follow well-established trends with major element compositions. For example, concentrations of compatible trace ele-

Table 1.2.2.24 Large, high-valency cations, Keweenawan reference suite.

ppm	KEW-1	KEW-2	KEW-3	KEW-4	KEW-5	KEW-6	KEW-7	KEW-8	KEW-9	KEW-10
Th†	1.23	1.83	0.84	1.45	0.44	—	0.34	0.33	3.0	7.4
Zr*	40		69	118			137			400
Hf†	3.3	4.6	2.01	3.65	1.18	1.14	1.3	1.65	5.8	10.7
Nb*	3		5	7			9			34
Ta†	0.65	0.93	0.43	0.84	0.15	—	0.23	0.22	1.4	2.8
Zr/Hf	12.1		34.3	32.3			105			37.4
Zr/Nb	13.3		13.8	16.9			15.2			11.8

ppm	KEW-11	KEW-12	KEW-13	KEW-14	KEW-15	KEW-16	KEW-17	KEW-18	KEW-19	KEW-20
Th†	2.4	2.3	2.36	4.9	2.8	3.6	7.1	10.8	5.8	5.9
Zr*	155		207		150	255	412		263	
Hf†	4.9	5.1	6.3	7.6	4.5	5.9	10.9	19.5	7.2	7.8
Nb*	11		18		8	14	49		14	
Ta†	0.86	0.97	1.6	2.3	0.63	0.8	3.9	3.0	0.9	1.2
Zr/Hf	31.6		32.9		33.3	43.2	37.8		36.5	
Zr/Nb	14.1		11.5		18.8	18.2	8.41		18.8	

†Analyses by INAA. *Analyses by XRF.

Table 1.2.2.25 Ferromagnesian trace elements in Keweenawan reference suite.

ppm	KEW-1	KEW-2	KEW-3	KEW-4	KEW-5	KEW-6	KEW-7	KEW-8	KEW-9	KEW-10
Cr	193	154	292	205	140	357	387	337	100	30
Sc	28.5	28.9	27.7	28.0	22.3	26.0	25.5	32.3	29.3	26.4
Ni	90	170	220	180	270	240	200	140	—	60
Co	41.4	56.8	50.2	48.3	48.2	47.5	45.1	44.4	37.3	47.0
Ni/Co	2.2	3.0	4.4	3.7	5.6	5.1	4.4	3.2		1.3

ppm	KEW-11	KEW-12	KEW-13	KEW-14	KEW-15	KEW-16	KEW-17	KEW-18	KEW-19	KEW-20
Cr	127	207	890	6	124	63	1	6	14	48
Sc	30.5	32.5	34.5	20	30.7	29.2	12.4	27.3	18.4	26.8
Ni	40	90	240	—	110	32	—	—	—	—
Co	38.0	42.4	70.0	41	48.7	37.8	29.3	8.7	26.0	36.4
Ni/Co	1.0	2.1	3.4	—	2.2	0.85	—	—	—	—

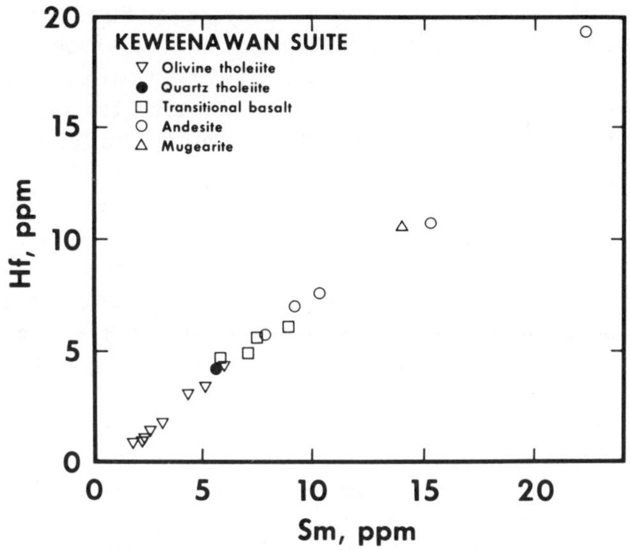

1.2.2.19 Hf/Sm diagram for the Keweenawan reference suite.

ments decrease and those of incompatible trace elements increase as the MgO concentration decreases (Fig. 1.2.2.20). As MgO concentration decreases, REE concentrations increase, the ratio of LREE to HREE increases, and the Eu anomaly shifts from clearly positive to substantially negative (Fig. 1.2.2.21). A few samples (e.g., transitional basalt KEW-14 and andesite KEW-17) deviate from the general trend by showing steeper LREE to HREE slopes.

Qualitatively, the observed trends are those expected for liquids that correspond to residual melts from fractional crystallization. However, Brannon *et al.* (1979) found for 73 adjacent, successive NSVG olivine tholeiites a monotonic trend of composition with stratigraphy in which the youngest flows are the more primitive. Also, they could not account for the major and trace element relationships by simple models for fractional crystallization or partial melting.

A comparison of trace element compositions of the Keweenawan NSVG basalts with those of other basalt groups emphasizes the extent to which their compositions are characteristic of common continental basalts rather than unique to a particular province or age. The observed ranges for Sc, Hf, Th, Cr, Ni, Sr, and Ba substantially overlap those for the Columbia River Plateau basalts, those for the Servilleta basalts of the Taos Plateau, and those for the Archean basalts. They also overlap those for the Siberian Traps (Tables 1.2.2.13, 1.2.2.14), the Paraná basalts (Table 1.2.2.16), the Deccan Traps (section 1.2.3), and the Karroo lavas (Table 1.2.2.15). The NSVG tholeiites show less relative enrichment in LREE than the Deccan or Servilleta basalts, and their patterns roughly parallel those of Columbia River Province basalts. The patterns for individual samples are smoother than those constructed from data published for Siberian Trap basalts (Balashov and Nesterenko, 1966).

Isotope chemistry

Duplicates of seven of the Keweenawan reference suite samples were among 24 from the NSVG analyzed

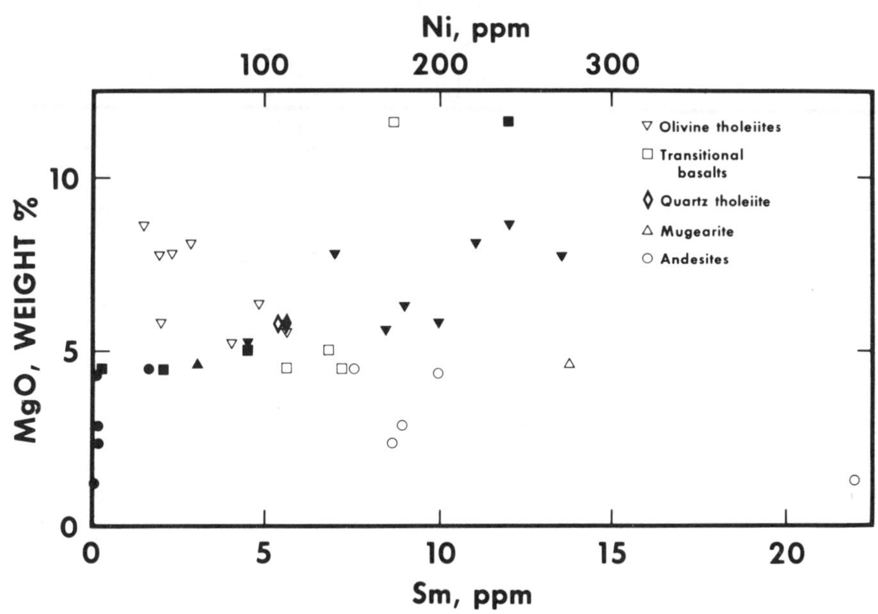

1.2.2.20 Ni and Sm as a function of MgO in the Keweenawan reference suite. Solid symbols are Ni, open symbols are Sm.

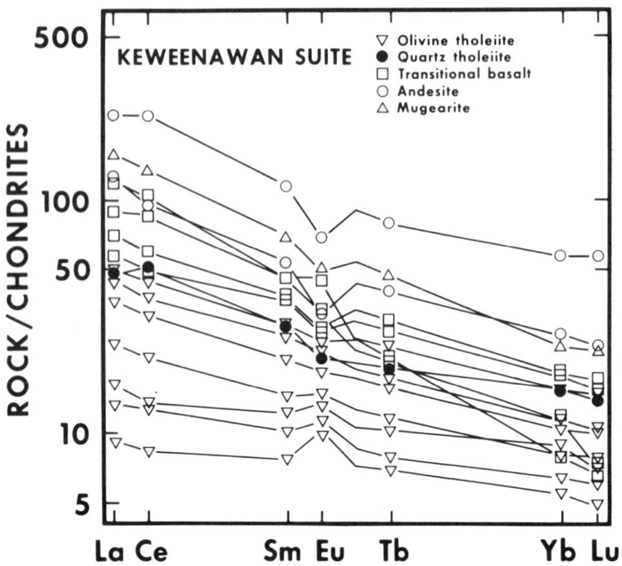

1.2.2.21 Chondrite-normalized rare-earth patterns for the Keweenawan reference suite.

for Rb/Sr and U/Pb relations by Leeman (1977); Table 1.2.2.26. He found the systematics to be complex, with evident disturbance of isotope ratios in many of the felsic samples. Mafic (basaltic and andesitic) samples gave an Rb/Sr isochron age of about 1.12 b.y. in agreement with Silver and Green's (1972) U/Pb results, and initial $^{87}Sr/^{86}Sr$ ratios ranging from 0.7038–0.7059.

The nine NSVG samples he analyzed for Pb isotopes yielded variable $^{206}Pb/^{204}Pb$ ratios at 1.12 b.y. and a "mantle isochron" that suggests the involvement of either mantle (as a source) or lower crust (as a contaminant) about 3.99 ± 0.2 b.y. old in the generation of these liquids.

Sabelin (1979) measured Sr isotopes for twelve samples which are duplicates of the KEW reference suite (Table 1.2.2.26), along with four other samples from the North Shore Volcanic Group. Her samples are different from those studied by Leeman (1977). She also found evidence of some isotopic disturbance but her results are consistent with Leeman's, giving an isochron age of 1.12 ± 0.03 b.y. and an initial $^{87}Sr/^{86}Sr$ of 0.7036 ± 0.00065.

A significant result of these analyses is that the olivine tholeiites (KEW-1–KEW-8) and transitional basalts (KEW-9–KEW-14) clearly show different ranges of values for Rb, K/Rb, and Rb and Sr isotopic ratios, which suggests that they reflect primary values rather than alteration processes.

Petrogenesis

The rocks in the North Shore Volcanic suite cannot all be explained or related by any single petrogenetic model.

Of the rocks in the Keweenawan suite, only the more magnesian olivine tholeiites (KEW-5, 6) have Mg′ values high enough to make them interesting as possible primitive magmas according to most models of

Table 1.2.2.26 Rb/Sr isotopic data for the Keweenawan reference suite.

KEW No.	Rb, ppm	Sr, ppm	K/Rb	$^{87}Rb/^{86}Sr$	$^{87}Sr/^{86}Sr \pm 2\sigma$	Analyst
1	9.14	272	420	0.0973	0.70456	WL
2	11.33	190.5	375	0.172	0.70594 ± 5	TS
3	4.12	316.5	705	0.0377	0.70437	WL
4	10.5	288.4	395	0.106	0.70522	WL
5†	2.32	209.4	678	0.032	0.70399 ± 6	TS
6	2.61	203.2	552	0.037	0.70450 ± 8	TS
7	1.24	251.5	805	0.0143	0.70325	WL
8	15.00	216.9	201	0.200	0.70734 ± 9	TS
9	59.46	393.9	248	0.436	0.71097 ± 7	TS
10	42.43	766.2	378	0.160	0.71116 ± 5	TS
11	19.37	253.1	310	0.221	0.70658 ± 8	TS
12	53.43	234.1	200	0.660	0.71192 ± 8	TS
13	40.43	379.9	176	0.308	0.70796 ± 7	TS
14	—	—	—	—	—	—
15	7.03	196.3	730	0.104	0.70547	WL
16	34.23	357.8	371	0.277	0.70945 ± 5	TS
17	52.4	1238	280	0.123	0.70637	WL
18	129.0	141.0	241	2.645	0.74503 ± 6	TS
19	—	—	—	—	—	—
20	50.14	224.4	304	0.646	0.71407 ± 9	TS

Analysts: TS: Tatiana Sabelin (1979); WL: William P. Leeman (pers. comm., 1978). †Averages of data from duplicate analyses mantle composition. They also have high Cr and Ni concentrations and are aphyric. As mentioned above, KEW-6 would be in equilibrium with olivine of composition Fo_{87-88}, according to Roeder and Emslie (1970), which is possible for a residual mantle phase although most mantle peridotite models have slightly more magnesian olivines. These olivine tholeiite lavas are also depleted in K, Ti and P, suggesting little fractionation, but are not depleted in light REE showing that they did not come from a mantle source depleted by a previous partial-melting event, as is the case for most mid-ocean ridge basalts.

Within the olivine tholeiite group, all incompatible elements, REE and La/Yb increase steadily with Fe/Mg. Because of their low Mg′ values, the more evolved members (e.g., KEW-2) could not have been produced simply by a smaller degree of partial melting of magnesian mantle peridotite. The more magnesian members (e.g., KEW-6) are similar to compositions that have been produced experimentally by a relatively large amount (about 25%) of melting of a spinel peridotite or pyrolite at about 30–40 km depth (Green and Ringwood, 1967). Several more recent experimental studies at high pressures have been carried out on olivine tholeiite compositions generally similar to those of the Keweenawan suite; they are discussed in more detail in Chapter 3. For instance, Bender *et al.* (1978) concluded that an ocean-floor glass from the FAMOUS area could have been derived by about 5% olivine fractionation from a primary partial melt of lherzolite at a depth of about 40 km. These and other experiments show that the abundant olivine-tholeiitic magmas of oceanic and continental environments could be derived by varying amounts of olivine fractionation from partial melts of mantle lherzolite at depths of 35 to 50 km. The amount of melting was sufficient to remove any Al-rich phase such as spinel or clinopyroxene in the source rock, and on trace element considerations Frey *et al.* (1978) concluded that 20–25% melting must have occurred. The compositional range (such as in Fe/Mg ratio) of tholeiitic lavas may be the result either of subsequent crystal fractionation, such as of olivine, en route to the surface, or of partial melting of heterogeneous mantle having a range of Fe/Mg ratios. Isotopic heterogeneity, as in Sr isotopes, has been demonstrated for many mantle sources, and Wilkinson and Binns (1977) and Langmuir and Hanson (1979), among others, also stress the possibility of Fe/Mg heterogeneity in the mantle sources of both continental and oceanic tholeiites. Unfortunately, mantle xenoliths have not been recognized in plateau tholeiites. If mantle heterogeneity is the dominant cause of the compositional variation within the Keweenawan olivine tholeiites, it must be on a small scale with rapid changes and perhaps vertically zoned, since Brannon *et al.* (1979) have found a tight inverse relation between Fe/Mg ratio and stratigraphic height within a sequence of 73 mostly aphyric olivine tholeiite flows which include three from the reference suite. They also concluded that the range of flow compositions could not have been produced either by successive melts of a single homogeneous mantle source, nor by simple fractional crystallization of a single batch of parental magma. A magma mixing model would fit the data but is difficult to test because of the lack of phenocrysts, and also leaves open the origin of the Fe-rich end-member liquid. Processes such as zone refining (Harris, 1957) and dynamic melting (Langmuir *et al.* 1977) could be responsible for much of this variation.

According to Ni/Mg relationships discussed by Sato (1977), four of the Keweenawan reference suite olivine tholeiites (KEW-2,3,4,6) lie along a line of equilibration with mantle olivine, assuming an exchange partition coefficient [(MgO/NiO) magma/(MgO/NiO) olivine = 2.3] and a range of MgO contents for the melts, and KEW-8 has an NiO/MgO ratio close to that predicted for 8% fractionation of olivine from a primary melt containing 11.3% MgO and 400 ppm NiO. Samples KEW-5 and KEW-7 have higher Ni/Mg ratios than Sato predicts for equilibration with mantle olivine, and KEW-1 could have lost both Ni and Mg by ~8% fractionation of olivine from a liquid composition such as KEW-6.

Leeman (1976) modelled possible origins for the McKinney olivine tholeiite of the Snake River Plain, which contains considerably lower Cr and Ni and higher REE concentrations, and higher Cr/Ni and Fe/Mg ratios than the Keweenawan olivine tholeiites but which also has a positive Eu anomaly. He concluded that this lava may represent a primary magma derived from partial melting of an undepleted spinel- or aluminous-clinopyroxene lherzolite at depths of 40–60 km. By analogy, the Keweenawan olivine tholeiites could have a similar origin but involving greater amounts of melting, olivine being the principal mineral contributing to the greater melt volume. Dynamic crystallization experiments by Green (1979b) on KEW-6, the most magnesian composition, reveal that olivine and plagioclase are liquidus minerals at low pressures; crystal fractionation of this liquid would deplete residua in Al and Ni with respect to the other olivine tholeiites analyzed, again implying that KEW-6 was not parental to the other NSVG olivine tholeiites.

Sample KEW-13, a transitional basalt, also has a relatively high Mg′ value (0.62) and could be considered a relatively primitive if not primary basalt of the transitional or weakly alkaline type. However, it probably does not represent a liquid composition because it contains a large proportion of augite and (altered) olivine phenocrysts. Subtracting the augite phenocryst composition, which is highly magnesian, simply increases the Fe/Mg ratio and makes the remaining liquid even less likely to have been a primitive melt. KEW-10, a mugearite, is stratigraphically close to KEW-13 and has a texture somewhat similar to that of the groundmass of KEW-13, but it cannot be simply KEW-13 without its phenocrysts because KEW-10 has much higher Ti, K, P, and lower Mg contents than such a liquid would have. It could have evolved, however, by fractionation from a liquid related to KEW-13.

The other transitional or weakly alkaline basalts (KEW-9–KEW-14) are fairly diverse in composition and probably in origin. KEW-9 contains abundant small plagioclase phenocrysts but has a slight negative Eu anomaly and probably does not contain excess crystals.

The quartz tholeiite, KEW-15, falls close to the general trend of the olivine tholeiites in several but not all chemical parameters. Green and Ringwood (1967) suggest that a liquid of this type could be produced by fractionation of olivine tholeiite at shallow depths (0–15 km).

The andesitic group also has an unclear origin and models are being investigated to explain them. Fractional crystallization of more magnesian tholeiites or transitional basalts is possible but not yet evaluated. Wilkinson and Binns (1977) proposed that the transitional basalts and basaltic andesites so common in plateau provinces could be to a large extent primary melts from Fe-rich mantle. KEW-17 contains abundant large, broken plagioclase phenocrysts which occur in excess over liquid composition according to trace element calculations. KEW-18 is very iron- and LIL-enriched, and its major element composition closely resembles that of an analyzed sample of the differentiated Lester River sill which lies approximately 420 m stratigraphically beneath it (Schwartz and Sandberg, 1940, Table 1, No. 17).

The presence beneath these lavas of the many contemporaneous immense to small intrusive bodies of the Duluth Complex suggests that many of the North Shore lavas could be samples tapped from complex differentiating magma chambers within the upper levels of the crust.

Finally, the possibility of crustal contribution to some of these mafic liquids must be considered. With such a large amount of hot, mafic liquid intruded into the sialic Archean and Proterozoic crust, it is likely that at least some partial melting occurred and thus contributed to parts of the extremely complex plumbing system which must have underlain the Keweenawan lava plateaus. In fact, Rb/Sr and Pb isotope relations discussed above indicate that this has indeed taken place, particularly in some of the more felsic (rhyolitic) liquids. The conclusion of Leeman (1977) that 4-b.y.-old lithosphere was involved in generation of the basalts is most easily interpreted in terms of "stagnant" mantle lithosphere, considering the mafic character of the liquids, but some of the more intermediate and felsic liquids appear to contain a significant amount of crustal material.

1.2.2.6 References

Afanas'yev G. D. (1975) Principal investigations by IGEM in field of geologic petrology. *Internat. Geology Rev.* **17**, 390–410.

Annells R. G. (1973) Proterozoic flood basalts of eastern Lake Superior: the Keweenawan volcanic rocks of the Mamainse Point area, Ontario. *Geol. Surv. Can. Paper 72-10.* 51 pp.

Annells R. G. (1974) Keweenawan volcanic rocks of Michipicoten Island, Lake Superior, Ontario, an eruptive center of Proterozoic age. *Geol. Surv. Can. Bull.* **218**. 141 pp.

Balashov Yu. A. and Nesterenko G. V. (1966) Distributions of the rare earths in the traps of the Siberian Platform. *Geochem. Internat.* **3**, 672–679.

Baragar W. R. A. (1977) Volcanism of the stable crust. In *Volcanic Regimes of Canada* (W. R. A. Baragar, L. C. Coleman and J. M. Hall, eds.), p. 377–405. Geol. Assoc. Canada Spec. Paper 16.

Bender J. F., Hodges F. N. and Bence A. E. (1978) Petrogenesis of basalts from the Project FAMOUS area: experimental study from 0 to 15 kbars. *Earth Planet. Sci. Lett.* **41**, 277–302.

Beurlen K. (1970) *Geologie von Brasilien.* Borntraeger, Berlin. 427 pp.

Bochkarev V. S. and Rudkevich M. Ya. (1976) Early platform stage of development of the West Siberian plate. *Geotectonics* **9**, 181–185.

Books K. G. (1972) Paleomagnetism of some Lake Superior Keweenawan rocks. *U. S. Geol. Survey Prof. Paper 760.* 42 pp.

Brannon J. C., Haskin L. A. and Green J. C. (1979) Compositional differentiation in planetary flood basalts (abstract). In *Lunar and Planetary Science X*, p. 152–154. Lunar and Planetary Institute, Houston.

Broderick T. M. (1935) Differentiation in lavas of the Michigan Keweenawan. *Geol. Soc. Amer. Bull.* **46**, 503–558.

Carmichael I. S. E. (1964) The petrology of Thingmuli, a Tertiary volcano in eastern Iceland. *J. Petrol.* **5**, 435–460.

Chase C. G. and Gilmer T. H. (1973) Precambrian plate tectonics: the Midcontinent Gravity High. *Earth Planet. Sci. Lett.* **21**, 70–78.

Chaudhuri S. and Faure G. (1967) Geochronology of the Keweenawan rocks, White Pine, Michigan. *Econ. Geol.* **62**, 1011–1033.

Chayes F. (1966) Alkaline and subalkaline basalts. *Amer. J. Sci.* **264**, 128–145.

Chayes F. and Metais D. (1964) On the relationship between suites of CIPW and Barth-Niggli norms. In *Carnegie Inst. Wash., Yearb.* **63**, 193–195.

Compston W., McDougall I. and Heier K. S. (1968) Geochemical comparison of the Mesozoic basaltic rocks of Antarctica, South Africa, South America and Tasmania. *Geochem. Cosmochim. Acta* **32**, 129–149.

Cordani U. G. and Vandoros P. (1967) Basaltic rocks of the Paraná basin. In *Problems in Brazilian Gondwana Stratigraphy and Paleontology*, Mar del Plata, 207–231. Int. Symp. Gondwana Stratig. Paleontol., 1st, Mar del Plata. Curitiba, Brazil.

Cornwall H. R. (1951) Differentiation in lavas of the Keweenawan Series and the origin of the copper deposits of Michigan. *Geol. Soc. Amer. Bull.* **62**, 159–201.

Cornwall H. R. and Rose H. J., Jr. (1957) Minor elements in Keweenawan Lavas, Michigan. *Geochim. Cosmochim. Acta* **12**, 209–224.

Cox K. G. (1970) Tectonics and volcanism of the Karroo period and their bearing on the postulated fragmentation of Gondwanaland. In *African Magmatism and Tectonics* (T. N. Clifford and I. G. Gass, eds.), p. 211–235. Oliver and Boyd, Edinburgh.

Cox K. G. (1971) Karroo Lavas and associated igneous rocks of Southern Africa. *Bull. Volc.* **35**, 867–886.

Cox K. G. (1972) The Karroo volcanic cycle. *J. Geol. Soc. London* **128**, 311–336.

Cox K. G. and Hornung G. (1966) The petrology of the Karroo basalts of Basutoland. *Amer. Mineral.* **51**, 1414–1432.

Cox K. G. and Jamieson B. G. (1974) The olivine-rich lavas of Nuanetsi: a study of polybaric magmatic evolution. *J. Petrol.* **15**, 269–301.

Cox K. G., Johnson R. L., Monkman L. J., Stillman C. J., Vail J. R. and Wood D. N. (1965) The geology of the Nuanetsi igneous province. *Phil. Trans. Roy. Soc. Lond.* **A257**, 71–218.

Cox K. G., Macdonald R. and Hornung G. (1967) Geochemical and petrographic provinces in the Karroo basalts of Southern Africa. *Amer. Mineral.* **52**, 1451–1474.

Craddock C. (1972) Regional geologic setting (Late Precambrian). In *Geology of Minnesota: A Centennial Volume* (P. K. Sims and G. B. Morey, eds.), p. 281–291. Minn. Geol. Surv., St. Paul.

Dietz R. S. and Holden J. C. (1970) The breakup of Pangaea. *Sci. Amer.* **223**, 30–41.

Dmitriev Yu. I. (1973) Mezozoiskiy trappoviy volkanizm v tsentre i na peripheriy Tungusskoy syneklisy. *Izvest. Akad. Nauk SSSR, Seri i Geolog. No. 10*, 58–67.

Dmitriev Yu. I. (1976) Trappovyi vulkanizmi genezis mestorozhdeniy islandskogo shpata. Akad. Nauk SSSR. Inst. Geol. *Ocherki Geologicheskoy Petrologii*, 300–308.

DuToit A. L. (1929) The volcanic belt of the Lebombo—a region of tension. *Trans. Roy. Soc. So. Africa* **18**, 189–217.

Dyuzhikov O. A., Lind E. N., Nesterovskiy V. S. and Shestakov Yu. G. (1973) The early phase of effusive traprock magmatism in Noril'sk area. *Internat. Geol. Rev.* **15**, 721–725.

Engel A. E. J., Engel C. G. and Havens R. G. (1965) Chemical characteristics of oceanic basalt and the upper mantle. *Geol. Soc. Amer. Bull.* **76**, 719–734.

Faure G., Chaudhuri S. and Fenton M. D. (1969) Ages of the Duluth Gabbro complex and the Endion sill, Duluth, Minnesota. *J. Geophys. Res.* **74**, 720–725.

Feoktistov G. D., ed. (1975) Mineralogiya trappov yuga Sibirskoy platformy. *Izd. Nauka, Sib. Otd., Novosibirsk.* 87 pp.

Frey F. A., Green D. H. and Roy S. D. (1978) Integrated models of basalt petrogenesis: a study of quartz tholeiites to olivine melilites from South Eastern Australia utilizing geochemical and experimental petrological data. *J. Petrol.* **19**, 463–513.

Geikie A. (1903) *Textbook of Geology*, 4th Ed., Macmillan, London. 1147pp.

Gibson I. L. (1966) Crustal flexures and flood basalts. *Tectonophys.* **3**, 447–456.

Godlevskiy M. N. (1959) Trappyi rudonosnye intruzii Noril'skogo raiona. Gosgeoltekhizdat. Transl. Canada Transl. Bureau.

Goldich S. S., Nier A. O., Baadsgaard H., Hoffman J. H. and Krueger H. W. (1961) The Precambrian geology and geochronology of Minnesota. *Minn. Geol. Surv. Bull.* **41**. 193 pp.

Green D. H. and Ringwood A. E. (1967) The genesis of basaltic magmas. *Contrib. Mineral. Petrol.* **15**, 103–190.

Green J. C. (1972) North Shore Volcanic Group. In *Geology of Minnesota: A Centennial Volume* (P. K. Sims and G. B. Morey, eds.), p. 294–332. Minn. Geol. Survey, St. Paul.

Green J. C. (1977) Keweenawan plateau volcanism in the Lake Superior area. In *Volcanic Regimes of Canada* (W. R. A. Baragar, L. C. Coleman and J. M. Hall, eds.), Geol. Assoc. Canada Spec. Paper 16, 407–422.

Green J. C. (1979a) High-Al olivine tholeiite—a Keweenawan speciality—and the variability of plateau magmatism (abstract). *Abstracts with Program, Geol. Soc. Amer.* **11**, 231.

Green J. C. (1979b) Iron enrichment in experimentally crystallized olivine tholeiite and its application to the differentiation of Keweenawan magmas (abstract). *Abstracts with Program, Geol. Soc. Amer.* **11**, 435.

Green J. C. (1980) Two Harbors Sheet (1:250,000). Minn. Geol. Survey, St. Paul.

Green J. C. (1981) Keweenawan extrusive rocks. In *Tectonic History of the Lake Superior Basin—A Review* (R. J. Wold and W. J. Hinze, eds.), in press.

Green J. C., Haskin L. A. and Brannon J. C. (1980) Geochemistry and stratigraphy of the northeast limb of the Keweenawan North Shore Volcanics, Minnesota (abstract). *EOS (Trans. Amer. Geophys. Union.)*, in press.

Green J. C., Jirsa M. K. and Moss C. M. (1977) *Environmental Geology of the North Shore.* Minn. Geol. Survey, St. Paul. 99 pp.

Halls H. C. and Pesonen L. J. (1981) Paleomagnetism of Keweenawan rocks: A review. In *Tectonic History of the Lake Superior Basin—A Review* (R. J. Wold and W. J. Hinze, eds.), in press.

Hanson G. N. (1975) $^{40}Ar/^{39}Ar$ spectrum ages on Logan intrusions, a Lower Keweenawan flow and mafic dykes in northeastern Minnesota-northwestern Ontario. *Can. J. Earth Sci.* **12**, 821–835.

Hanson G. N. and Malhotra R. (1971) K-Ar ages of mafic dikes and evidence for low grade metamorphism in NE Minnesota. *Geol. Soc. Amer. Bull.* **82**, 1107–1114.

Hubbard H. A. (1975) Lower Keweenawan volcanic rocks of Michigan and Wisconsin. *U. S. Geol. Survey J. of Res.* **3**, 529–541.

Huber K. (1973) The Portage Lake Volcanics (middle Keweenawan) on Isle Royale, Michigan. *U. S. Geol. Surv. Prof. Paper 754-C.* 36 pp.

Irvine T. N. and Baragar W. R. A. (1971) A guide to the chemical classification of the common volcanic rocks. *Can. J. Earth Sci.* **8**, 523–548.

Jirsa M. (1978) Interflow sandstones of the Keweenawan North Shore Volcanic Group (abstract). In *24th Inst. on Lake Superior Geology*, p. 17. Univ. of Wisconsin, Milwaukee.

Jolly W. T. and Smith R. E. (1972) Degradation and metamorphic differentiation of the Keweenawan tholeiitic lavas of northern Michigan, U. S. A. *J. Petrol.* **13**, 273–309.

King E. R. and Zietz I. (1971) Aeromagnetic study of the midcontinent gravity high of the central United States. *Geol. Soc. Amer. Bull.* **82**, 2187–2208.

Konda T. and Green J. C. (1974) Clinopyroxenes from the Keweenawan Lavas of Minnesota. *Amer. Mineral.* **59**, 1190–1197.

Kulikov P. K., Belousov A. P. and Latypov A. A. (1972) West Siberian Triassic rift system. *Geotectonics* **7**, 367–371.

Kuno H. (1969) Plateau basalts. In *The Earth's Crust and Upper Mantle* (P. Hart, ed.), p. 495–501. Amer. Geophys. Union Geophys. Mono. 13.

Kuznetsov M. F. and Naumov V. A. (1975) Trap magmatism and some problems of tectonics in the southeast of the Tunguska syneclise. *Internat. Geol. Rev.* **17**, 781–785.

Langmuir C. H. and Hanson G. N. (1979) MgO/FeO heterogeneity in the mantle sources of oceanic basalts (abstract). *EOS (Trans. Amer. Geophys. Union)* **60**, 408.

Langmuir C. H., Bender J. F., Bence A. E., Hanson G. N. and Taylor S. R. (1977) Petrogenesis of basalts from the FAMOUS area: Mid-Atlantic Ridge. *Earth Planet. Sci. Lett.* **36**, 133–156.

Lebedev A. P. (1964) Correlation of trappean magmatism in the Siberian and some other provinces of the world. *XXII Internat. Geol. Cong., Papers by Soviet Geologists, Problem 7*, p. 14–21. Nauka, Moscow.

Leeman W. P. (1976) Petrogenesis of McKinney (Snake River) olivine tholeiite in light of rare-earth element and Cr/Ni distributions. *Geol. Soc. Amer. Bull.* **87**, 1582–1586.

Leeman W. P. (1977) Pb and Sr isotopic study of Keweenawan lavas and inferred 4 b. y. old lithosphere beneath part of Minnesota (abstract). *Abstracts with Program, Geol. Soc. Amer.* **9**, 1068.

Leinz V., Bartorelli A., Sadowski G. R. and Isotta C. A. L. (1966) Sobre o comportamento espacial do trapp basaltico da bacia do Paraná. *Bol. Soc. Bras. Geol.* **15**, 79–91.

LeRoex A. P. and Reid D. L. (1978) Geochemistry of Karroo dolerite sills in the Calvinia district, western Cape Province, South Africa. *Contrib. Mineral. Petrol.* **66**, 351–360.

Leterrier J., de LaRoche H. and Ruegg N. R. (1972) Composition chimique et parente tholeiitique des roches basaltiques du Bassin du Parana. *C. R. Acad. Sci. Paris. Ser. D.* **274**, 1772–1775.

Lombaard B. V. (1952) Karroo dolerites and lavas. *Trans. Proc. Geol. Soc. So. Africa* **55**, 175–198.

Lur'ye M. L. and Masaitis V. L. (1966) Verchsne-paleozoiskie-nitsnemezozoiskie doleriti i basalti trappovoi formatsii. In *Geologiia Sibirskoi Platformy*, 247–284. Nedra, Moscow.

Maack R. (1952) Die Entwicklung der Gondwana-Schichten Suedbrasiliens und ihre Beziehungen zur Karru-Formation Suedafrikas. *Int. Geol. Congr.* **19**, *Algiers, Sympos. sur les Series de Gondwana*, 339–372.

MacDonald G. A. and Katsura T. (1964) Chemical composition of Hawaiian lavas. *J. Petrol.* **5**, 82–133.

Makarenko G. F. (1974) Trap areas of continents and basalt areas of oceans—areal comparison. Vestnik Moskovskogo Universiteta. *Geologiya.* **29**, 3–30.

Manton W. I. (1968) The origin of associated basic and acid rocks in the Lebombo-Nuanetsi Igneous province, Southern Africa, as implied by strontium isotopes. *J. Petrol.* **9**, 23–39.

Masaitis V. L., Abramovich I. I., Dobin D. A. and Smyslov A. A. (1966) Uranium in Siberian Platform traprocks. *Geochem. Internat.* **3**, 392–405.

McBirney A. R. (1975) Differentiation of the Skaergaard intrusion. *Nature* **253**, 691–694.

McIlwaine W. H. and Wallace H. (1976) Geology of Black Bay Peninsula, District of Thunder Bay, Ontario. *Ont. Div. Mines Geoscience Rept. 133.* 54 pp.

Milanovskiy Ye. Ye. (1976) Rift zones of the geologic past and their associated formations. Report 2. *Internat. Geol. Rev.* **18**, 619–639.

Morse S. A. (1974) Kiglapait magma chemistry and application of a partition model (abstract). In *EOS (Trans. Amer. Geophys. Union)* **55**, 476–477.

Nesterenko G. V. and Al'mukhamedov A. E. (1973) *Geokhimiya differentsirovannykh trappov.* Akad. Nauk SSSR, Sib. Otd., Moscow. 178 pp.

Nesterenko G. V., Avilova N. S. and Smirnova N. P. (1964) Rare elements in the traps of the Siberian platform. *Geochem. Internat.* **5**, 970–976.

Nesterenko G. V., Znamenskiy Ye. B., Al'mukhamedov A. I. and Tsykhanskiy V. D. (1971) Nb, Ta, Zr, and Hf in trap magma differentiation. *Geochem. Internat.* **8**, 725–738.

Oray K., Hinze W. J. and O'Hara N. W. (1973) Gravity and magnetic evidence for the eastern termination of the Lake Superior syncline. *Geol. Soc. Amer. Bull.* **84**, 2763–2780.

Palmer H. C. (1970) Paleomagnetism and correlation of some middle Keweenawan rocks, Lake Superior. *Can. J. Earth Sci.* **7**, 1410–1436.

Palmieri J. H. and Arribas A. (1975) El complejo alcalino potasico de Sapukai (Paraguay oriental). In *II Congreso Ibero-Americano de geologia economica* **4**, 267–300. Dept. Argent. Sci., Buenos Aires.

Pearce T. H., Gorman B. E. and Birkett T. C. (1977) The relationship between major element chemistry and tectonic environment of basic and intermediate volcanic rocks. *Earth Planet. Sci. Lett.* **36**, 121–132.

Phinney W. C. (1970) Chemical relations between Keweenawan lavas and the Duluth complex, Minnesota. *Geol. Soc. Amer. Bull.* **81**, 2487–2496.

Pritula Yu A., Savinskiy K. A., Kogan A. B., Smirnova T. G. and Shtutin I. M. (1974) Structure of terrestrial crust in Tunguska syneclise. *Internat. Geol. Rev.* **16**, 1144–1151.

Rhodes J. M., Dungan M. A., Blanchard D. P. and Long P. E. (1979) Magma mixing at midocean ridges: evidence from basalts drilled near 22° N or the Mid-Atlantic Ridge. *Tectonophys.* **55**, 35–61.

Robertson W. A. (1973) Pole positions from the Mamainse Point lavas and their bearing on a Keweenawan pole path and polarity sequence. *Can. J. Earth Sci.* **10**, 1541–1555.

Robertson W. A. and Fahrig W. F. (1971) The great Logan paleomagnetic Loop—the polar wandering path from Canadian Shield rocks during the Neohelikian Era. *Can. J. Earth Sci.* **8**, 1355–1371.

Roedder E. (1978) Silicate liquid immiscibility in magmas and in the system K_2O-FeO-Al_2O_3-SiO_2: an example of serendipity. *Geochim. Cosmochim. Acta* **42**, 1597–1617.

Roeder P. L. and Emslie R. F. (1970) Olivine-liquid equilibrium. *Contrib. Mineral. Petrol.* **29**, 275–289.

Rose W. I., Jr. and Grimes J. G. (1979) Cyclic compositional changes within flood basalts of the Portage Lake Volcanics, Michigan (abstract). *Abstracts with Program, Geol. Soc. Amer.* **11**, 255.

Ruegg N. R. (1972) Plagioclasios plutonicos e vulcanicos em diabasios da bacia do Paraná. *Revista Brasileira de Geociencias* **2**, 8–17.

Ruegg N. R. (1975) Natureza serial do magmatismo basaltico da bacia do Paraná. *Naturalia* (Fac. Fil Cienc. e Letras de S. José do Rio Preto, Sao Paulo) **1**, 141–152.

Ruegg N. R. (1976a) Characteristicas de distribuicao e teor de elementos principais em rochas basalticas da bacia do Paraná. *Bol. Inst. Geocienc.* **7**, 81–106. Univ. Sao Paulo.

Ruegg N. R. (1976b) Characteristicas de distribuicao e teor de elementos tracos dosados em rochas basalticas da bacia do Paraná. *Naturalia* **2**, 23–45.

Ruegg N. R. and Amaral G. (1976) Variacao regional da composicao quimica das rochas basalticas da bacia do Paraná. *Bol. Inst. Geocienc.*, **7**, 131–147. Univ. Sao Paulo.

Ruegg N. R. and Dutra C. V. (1970) Variation in the content of some trace elements in basaltic rocks from the Paraná basin. *Anais do XXIV Congr. Bras. Geol.* **24**, 219–226. Soc. Bras. Geol., Brasilia.

Sabelin T. (1979) Trace element and Sr isotopic geochemistry of the Keweenawan volcanics of Minnesota. Master's thesis, Univ. of Minn.

Sandberg A. E. (1938) Section across Keweenawan lavas at Duluth, Minnesota. *Geol. Soc. Amer. Bull.* **49**, 795–830.

Sandell E. B. and Goldich S. S. (1943) The rarer metallic constituents of some American igneous rocks. *J. Geol.* **51**, 99–115.

Sato H. (1977) Nickel content of basaltic magmas: identification of primary magmas and a measure of the degree of olivine fractionation. *Lithos* **10**, 113–120.

Schilling J. -G. (1973) Iceland mantle plume: geochemical study of the Reykjanes Ridge. *Nature* **242**, 565–571.

Schwartz G. M. and Sandberg A. E. (1940) Rock series in diabase sills at Duluth, Minnesota. *Geol. Soc. Amer. Bull.* **51**, 1135–1172.

Scofield N. (1976) Mineral chemistry applied to interrelated albitization, pumpellyitization, and native copper distribution in some Portage Lake basalts, Michigan. Ph. D. dissertation, Mich. Tech. Univ. 180 pp.

Silver L. T. and Green J. C. (1972) Time constants for Keweenawan igneous activity (abstract). *Abstracts with Program, Geol. Soc. Amer.* **4**, 665.

Sims P. K. and Morey G. B., eds. (1972) *Geology of Minnesota: A Centennial Volume.* Minn. Geol. Survey, St. Paul. 632 pp.

Smith T. J., Steinhart J. S. and Aldrich L. T. (1966) Lake Superior crustal structure. *J. Geophys. Res.* **71**, 1141–1172.

Tyrrell G. W. (1937) Flood basalts and fissure eruptions. *Bull. Volc.* **1**, 89–111.

Vail J. R., Hornung G. and Cox K. G. (1969) Karroo basalts of the Tuli syncline, Rhodesia. *Bull. Volc.* **33**, 398–418.

Van Schmus W. R. (1971) Rb/Sr age of middle Keweenawan rocks, Mamainse Point and vicinity, Ontario, Canada. *Geol. Soc. Amer. Bull.* **82**, 3221–3226.

Van Schmus W. R., Green J. C. and Halls H. C. (1981) Review of the geochronology of rocks directly associated with the Keweenawan midcontinent rift system. In *Tectonic History of the Lake Superior Basin—A Review* (R. W. Wold and W. J. Hinze, eds.), in press.

Vasyl'yev Yu. R. and Zolotukhin V. V. (1977) Problemy platformennogo magmatizma stat'ya: 3, Ul'traosnovnoy magmatizm platformennykh oblastey. *Geol. Geofiz. Akad. Nauk SSSR, Sib. Otd., No. 2* (206) 33–46.

Wager L. R. and Brown G. M. (1967) *Layered Igneous Rocks.* W. H. Freeman and Co., San Francisco. 588 pp.

Walker F. and Poldervaart A. (1949) Karroo dolerites of the Union of South Africa. *Geol. Soc. Amer. Bull.* **60**, 591–706.

Walker G. P. L. (1959) Geology of the Reydarfjordur area, eastern Iceland. *Quart. J. Geol. Soc. Lond.* **114**, 367–391.

Wanless R. K., Stevens R. D., Lachance B. R. and Delabio R. N. (1970) Age determinations and geological studies, K-Ar isotopic age Report 10. *Geol. Surv. Can. Paper 71–2.*

Washington H. S. (1922) Deccan traps and other plateau basalts. *Geol. Soc. Amer. Bull.* **33**, 765–804.

Wasserburg G. J. and DePaolo D. J. (1979) Nd in flood basalts from the Siberian Platform and models of earth structure (abstract). *EOS (Trans. Amer. Geophys. Union)* **60**, 406.

Weiblen P. W. and Morey G. B. (1980) A summary of the stratigraphy, petrology, and structure of the Duluth Complex. *Amer. J. Sci. Jackson Volume* **280-A**, 88–133.

White W. M. and Bryan W. B. (1977) Sr-isotope, K, Rb, Cs, Sr, Ba and rare-earth geochemistry of basalts from the FAMOUS area. *Geol. Soc. Amer. Bull.* **88**, 571–576.

White W. S. (1960) The Keweenawan lavas of Lake Superior, an example of flood basalts. *Amer. J. Sci. Bradley Vol.* **258-A**, 367–374.

White W. S. (1968) The native-copper deposits of northern Michigan. In *Ore Deposits of the United States, 1933–1967*, **1** (J. D. Ridge, ed.), p. 303–326. Amer. Inst. Mining, Metall., and Petroleum Engineers, N.Y.

White W. S. (1972) Keweenawan flood basalts and continental rifting (abstract). *Abstracts with Program*, Geol. Soc. Amer. **4**, 732–734.

Wilkinson J. F. G. and Binns R. A. (1977) Relatively iron-rich lherzolite xenoliths of the Cr-diopside suite: a guide to the primary nature of anorogenic tholeiitic andesite magmas. *Contrib. Mineral. Petrol.* **65**, 199–212.

Wold R. J. and Hinze W. J., eds. (1981) *Tectonic History of the Lake Superior Basin—A review.* Geol. Soc. Amer., in press.

Wood D. A. (1978) Major and trace element variations in the Tertiary Lavas of eastern Iceland and their significance with respect to the Iceland geochemical anomaly. *J. Petrol.* **19**, 393–436.

Wood D. A., Gibson I. L. and Thompson R. N. (1976) Elemental mobility during zeolite facies metamorphism of the Tertiary basalts of eastern Iceland. *Contrib. Mineral. Petrol.* **55**, 241–254.

Yoder H. S. and Tilley C. E. (1962) Origin of basalt magmas: an experimental study of natural and synthetic rock systems. *J. Petrol.* **3**, 342–532.

1.2.3 TERTIARY CONTINENTAL FLOOD BASALTS

1.2.3.1 Introduction

The previous section (1.2.2) discussed flood basalt volcanism and related intrusives up until the mid-Cretaceous. This type of volcanism continued on the Earth through the Tertiary, producing further enormous thicknesses of flat-lying basalt lavas from widespread centers on a global scale. As with the older lavas, extrusion rates were relatively rapid, the basalts were highly fluid, and lava sheet floods covered large areal expanses on the continents. Among the largest of the continental plateau extrusion complexes are the Columbia River Province in the northwestern United States which is estimated to contain more than 2×10^5 km^3 of basalt, the Deccan Traps in central and northwest India, which currently cover 5×10^5 km^2 and which once may well have been twice as large, and the Brito-Arctic Province which stretches from the British Isles (northern Ireland and northwest Scotland) to the Faeroe Islands, Iceland, Jan Mayen, Greenland, and Baffin Island, an area of approximately 10^6 km^2. It was within the Brito-Arctic volcanic sequences that the concept of basaltic parental liquids was formulated and the distinction between alkali olivine basalts and tholeiites was first recognized.

Although there are still many unresolved problems in the mechanics of flood basalt extrusion, in their depths of origin, in their modes of differentiation, in the degrees of partial melting, and in the magnitude of crustal contamination, it is an established fact that the study of Tertiary basaltic volcanism has contributed much to our understanding of the principles of igneous petrogenesis. Furthermore, because large-volume effusive volcanism appears to be related to the dynamics of lithospheric plate motions, these major volcanic features are of considerable importance in models of crustal evolution.

This section is intended as an overview of the geological setting and geochemistry of Tertiary continental flood basalts. The treatment is in three parts. First, the salient features of the three best known provinces (the Columbia River Province, the Brito-Arctic Province and the Deccan Traps) are described. The Columbia River Province is emphasized, because it has been studied in the greatest detail, and because new major and trace element data have been obtained on a suite of samples for this project. Second, geochemical data for all of these provinces are compared in a synthesis section. Finally, the petrogenesis of these lavas is discussed in the light of the chemical variation patterns, isotopic data and experimental petrologic constraints.

1.2.3.2 Columbia River Province

Geological Setting

The Columbia River Basalt Group comprises a tholeiitic flood basalt sequence having an area of $\sim 2 \times 10^5$ km^2 and an estimated volume of 2×10^5 km^3 (Waters, 1962). These basalts erupted between 17 and 6 m.y. ago (Watkins and Baksi, 1974), with approximately 99% of the volume erupting within a 3 m.y. period shortly after volcanism began (McKee *et al.*, 1977). Lavas were extruded into a basin-like structure that occupied the region between the Cascade Range and the Idaho batholith, extending through northern Oregon, western Idaho and central to eastern Washington (Fig. 1.2.3.1).

Lavas vary in thickness from multiple thin flow units to thick (10–45 m), massive flows that display spectacular vertical and fan colonnades, entablatures, and curvilinear, buckled joint patterns (Schmincke, 1967; see Fig. 5.3.1). Typical flow volumes are 10–20 km^3. However, the Roza and Pomona Members have volumes of several hundred km^3 (Waters, 1961; Schmincke, 1967; Swanson *et al.*, 1975). Lavas erupted from fissure-aligned vent systems, which were tens to over 100 kilometers in length and up to several kilometers wide. Two major dike swarms have been recognized and most lavas have been traced to dike feeders, located mostly on the eastern side of the province (Waters, 1962; Swanson *et al.*, 1975). Average extrusion rates are estimated to have been 10^{-1} to 10^{-2} km^3/day for each linear kilometer of active fissure, and rates as high as 1 km^3/day may have been in effect during the Roza and the Pomona events (Swanson *et al.*, 1975). However, Swanson *et al.* (1975) suggest that, in general, the long-term production rate probably was equivalent to that of historic and present-day activity in Hawaii and Iceland ($\sim 10^{-3}$ km^3/day).

Considerable attention has been focused on stratigraphic correlations of individual formations across this extensive volcanic field, and on the characterization of magma types. The generalized classification by Waters (1961) divided the Columbia River basalt group into an older Picture Gorge and a younger Yakima basalt type, but subsequent studies (e.g., Schmincke, 1967; Swanson and Wright, 1978; Swanson *et al.*,

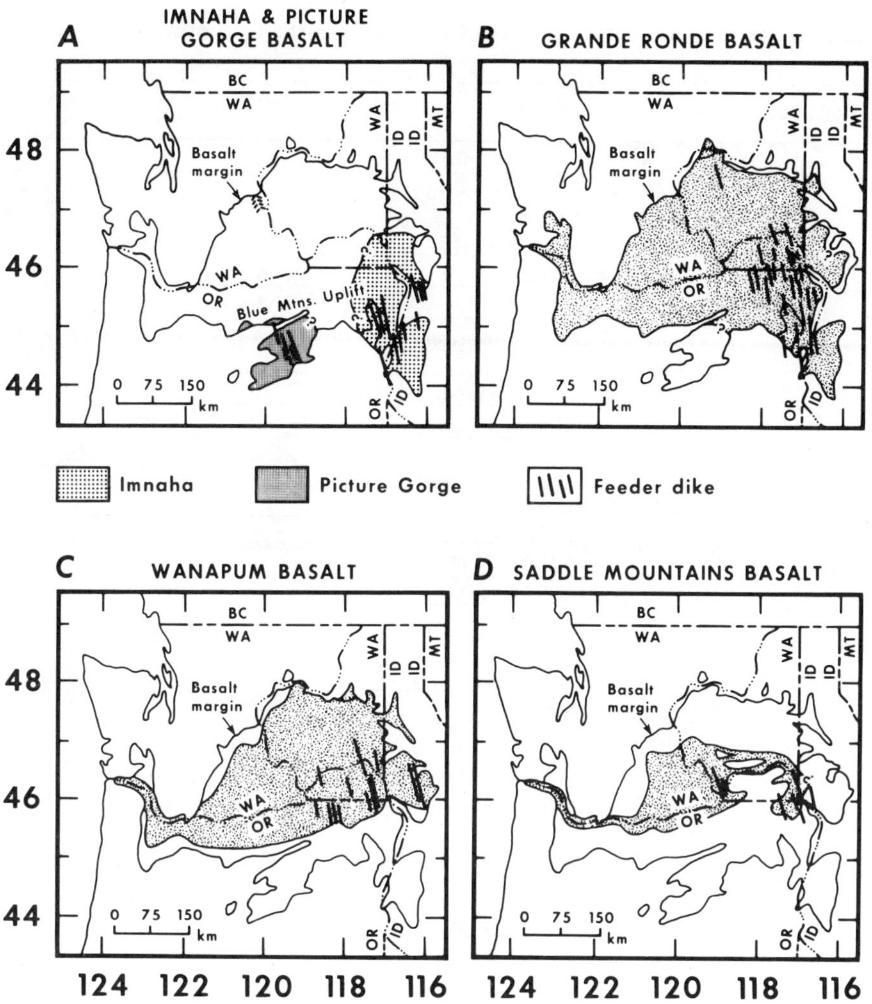

1.2.3.1 Maps showing the distribution of the five formations in the Columbia River Basalt Province and locations of known feeder dikes. After Swanson and Wright (1978).

1979) have separated the sequence into five major formations and many members on the basis of stratigraphy, geochemistry, paleomagnetic polarity and K-Ar age (Table 1.2.3.1). The Imnaha and Picture Gorge Formations (Lower Miocene) are confined to the southeastern and southern regions of the plateaus, respectively. These Formations are nowhere known to be in contact; however, interpretation of magnetostratigraphic data suggests that the Imnaha is older. The Grande Ronde (formerly Lower Yakima) Formation covers most of the Province, and is overlain by the less extensive Wanapum and Saddle Mountains Formations. Flows in the latter Formation were channeled by paleovalleys cut in the Wanapum and Grande Ronde basalts; the larger eruptions (e.g., Pomona, Elephant Mountain) also formed sheet floods on the relatively undissected plateau surface. The Grande Ronde Formation comprises about 80% of the eruptive volume and the Wanapum formation about 15%. The Saddle Mountains Formation, which has the smallest volume (~5%), erupted over the longest period of time (6–12 m.y.).

This variation in volume-time relationships offers a unique opportunity for the study of basaltic magma generation and evolution. Although opinions on petrogenesis vary (e.g., Wright *et al.*, 1973; McDougall, 1976; Swanson and Wright, 1981; Carlson *et al.*, 1981), the apparently insurmountable problem of correlating stratigraphy over large distances has been solved successfully, making this province the best characterized of the Tertiary flood basalt provinces.

Major element chemistry

The dominant rock type in the Columbia River Province is quartz tholeiite, but tholeiitic andesites and olivine tholeiites are also present. Major element data for many lavas have been presented and discussed by Wright *et al.* (1973) and Swanson and Wright (1981), and other significant contributions have been made by

Table 1.2.3.1. Stratigraphic nomenclature, age and magnetic polarity of units within the Columbia River Basalt Province†.

Series		Group	Subgroup	Formation	Member	Age (m.y.)	Magnetic polarity*
MIOCENE	UPPER MIOCENE	COLUMBIA RIVER BASALT GROUP	YAKIMA BASALT SUBGROUP	Saddle Mountains Basalt	Lower Monumental Member	6	N
					/ / /Erosional uncomformity / / / / / / / / / / / / / / /		
					Ice Harbor Member	8.5	N
					basalt of Goose Island		
					basalt of Martindale	8.5	R
					basalt of Basin City	8.5	N
					/ / /Erosional unconformity / / / / / / / / / / / / / / /		
					Buford Member		R
					Elephant Mountain Member	10.5	N,T
					/ / /Erosional unconformity / / / / / / / / / / / / / / /		
					Pomona Member	12	R
					/ / /Erosional uncomformity / / / / / / / / / / / / / / /		
	MIDDLE MIOCENE				Esquatzel Member		N
					/ / /Erosional uncomformity / / / / / / / / / / / / / / /		
					Weissenfels Ridge Member		
					basalt of Slippery Creek		N
					basalt of Lewiston Orchards		N
					Asotin Member		N
					/ / /Local erosional uncomformity/ / / / / / / / / / / / /		
					Wilbur Creek Member		N
					Umatilla Member		N
					/ / /Local erosional uncomformity/ / / / / / / / / / / / /		
					Priest Rapids Member		R
				Wanapum Basalt	Roza Member		T,R
					Frenchman Springs Member		N
					Eckler Mountain Member		
					basalt of Shumaker Creek		N
					basalt of Dodge		N
					basalt of Robinette Mtn.		N
	LOWER MIOCENE			Grande Ronde Basalt		14–16.5	N
							R
				Picture Gorge Basalt		14.6–15.8	N
					basalt of Dayville		
					basalt of Monument Mtn.		
					basalt of Twickenham		R
				—?—			
				Imnaha Basalt			R
							T
							N
							R

†After Swanson *et al.* (1979).
*N = normal polarity, R = reversed polarity, T = transitional polarity.

Hooper (1974), Holden and Hooper (1976) and McDougall (1976). Representative major element analyses of lavas throughout the section are given in Table 1.2.3.2.

The pioneering petrological and chemical studies of Waters (1961) distinguished the Picture Gorge and Imnaha basalts at the base of the section from the overlying basalts of the Yakima Subgroup. Wright *et al.*

Table 1.2.3.2 Representative major element compositions of basalts from the Columbia River Province.†

Formation	*Imnaha Basalt*		*Picture Gorge Basalt*		*Grande Ronde Basalt*			*Wanapum Basalt*					
Member or Type	American Bar	Rock Creek	High-Mg	Low-Mg	High-Mg	Low-Mg	Prineville	Robinette Mountain	Dodge	Shumaker Creek	Frenchman Springs	Roza	Lolo
No. of Analyses	44	68	8	4	13	8	13	9	20	4	8	35	55
SiO_2	51.14	49.53	50.36	51.46	53.78	55.94	51.57	50.01	52.13	54.80	52.29	51.19	50.09
TiO_2	2.24	2.41	1.56	1.79	1.78	2.27	2.71	1.00	1.48	2.46	3.17	3.13	3.15
Al_2O_3	15.06	16.34	15.54	15.39	14.45	14.04	13.87	17.08	15.41	13.86	13.21	14.07	14.31
Σ FeO	13.04	12.38	11.25	12.46	11.35	11.77	12.28	10.01	10.66	13.32	14.38	13.91	13.78
MnO	0.22	0.20	0.20	0.23	0.19	0.19	0.24	0.14	0.19	0.26	0.22	0.23	0.19
MgO	5.07	6.06	6.68	4.86	5.25	3.36	4.44	7.84	5.92	2.84	4.04	4.39	5.18
CaO	9.31	9.15	10.67	9.45	9.07	6.88	8.12	11.01	10.18	6.48	7.90	8.48	8.88
Na_2O	2.58	2.58	2.95	3.29	2.83	3.14	3.36	2.44	3.00	3.18	2.67	2.72	2.57
K_2O	0.91	0.93	0.57	0.74	1.05	1.99	2.02	0.27	0.68	1.87	1.41	1.22	1.07
P_2O_5	0.42	0.41	0.22	0.33	0.28	0.43	1.39	0.19	0.35	0.93	0.71	0.67	0.78
SUM	99.99	100.00	100.00	100.00	100.01	100.01	100.00	99.99	100.00	100.00	100.00	100.01	100.00

Table 1.2.3.2 *(Continued)*

Formation	*Saddle Mountains Basalt*												
Member or Type	Umatilla	Wilbur Creek	Asotin	Slippery Creek	Lewiston Orchards	Esquatzel	Pomona	Elephant Mountain	Buford	Basin City	Martindale	Goose Island	Lower Monumental
No. of Analyses	13	11	6	3	2	12	30	41	8	12	13	8	24
SiO_2	54.70	54.41	50.72	52.12	49.75	54.16	51.88	51.08	54.46	47.45	48.73	47.50	50.44
TiO_2	2.80	1.95	1.45	2.48	2.42	2.82	1.62	3.52	2.17	3.62	3.30	3.79	2.90
Al_2O_3	14.10	14.51	16.23	14.33	15.26	13.84	14.88	13.54	14.29	13.84	13.88	12.50	14.07
Σ FeO	12.63	11.07	9.64	11.64	11.82	12.60	10.55	14.75	11.05	15.22	14.41	17.53	13.78
MnO	0.17	0.21	0.17	0.17	0.21	0.19	0.17	0.20	0.15	0.23	0.20	0.27	0.21
MgO	2.71	4.51	8.19	5.58	7.10	3.91	6.96	4.28	4.85	5.99	5.88	4.41	5.01
CaO	6.14	8.32	10.70	9.64	10.13	7.71	10.67	8.34	8.54	9.71	9.72	8.80	8.67
Na_2O	3.20	2.69	2.22	2.69	2.32	2.66	2.36	2.45	2.75	2.31	2.42	2.44	2.79
K_2O	2.68	1.77	0.51	0.87	0.46	1.70	0.64	1.25	1.39	0.72	0.73	1.23	1.47
P_2O_5	0.88	0.56	0.18	0.49	0.55	0.41	0.25	0.59	0.35	0.91	0.73	1.54	0.66
SUM	100.01	100.00	100.01	100.00	100.02	100.00	99.98	100.00	100.00	100.00	100.00	100.01	100.00

†Data from Swanson and Wright (1981).

Table 1.2.3.3 Samples in the Columbia River Basalt Reference Suite.

Sample Number	Other Sample Number	Chemical Type
CP-1	DSTW 71-137	High-Mg Picture Gorge
CP-2	DSTW 71-31	Rock Creek
CP-3	DSTW 72-104	Robinette Mountain
CP-4	DSTW 72-88	High-Mg Grande Ronde
CP-5	DSTW 72-189	Low-Mg Grande Ronde
CP-6	DSTW 71-105	Roza
CP-7	DSTW 71-125	Umatilla
CP-8	DSTW 71-106	Pomona
CP-9	DSTW 71-116	Elephant Mountain
CP-10	DSTW 71-114	Martindale

(1973) used this initial framework in conjunction with updated stratigraphic studies and a growing body of chemical data to provide a synthesis of the chemical-stratigraphic variations in the province. However, Wright *et al.* (1973, p. 371) stress that . . ."Flows of one or more chemical types may form the *dominant* lithology in a stratigraphic unit, but single flows of the same chemical type may occur in *any* stratigraphic unit."

At least seven chemical types are represented at the base of the stratigraphic column in the Picture Gorge and Imnaha Formations. Their main characteristic is one of a variation in MgO and TiO_2 contents. Wright *et al.* (1973) note that some basalts of the Imnaha Formation are compositionally similar to basalts of the Picture Gorge Formation, but with higher concentrations of TiO_2, and infer this to mean that at least two source regions were tapped.

The Grande Ronde Formation consists of 300 m of flows which show a broad continuum of compositions between high-Mg and low-Mg types (Table 1.2.3.2). A distinctive feature of these lavas is their relatively high SiO_2 content (53–56 wt.%). Here again, some flows in this sequence from the eastern part of the plateau have higher TiO_2 contents than those from farther west, indicating multiple sources within the broad Chief Joseph dike swarm. A sample from Butte Creek was identified by Nathan and Fruchter (1974) as being yet another chemical type (Prineville type). This is a single flow within the Grande Ronde Formation that is distinguished from other flows in the Province by its high P_2O_5, Na_2O and Ba contents and its low TiO_2 content.

The Wanapum Formation has been subdivided into four members (Eckler Mountain, Frenchman Springs, Roza and Priest Rapids) that contain at least seven chemical types. Flows in the upper three members have a range of MgO contents (13.7–15.4 wt.%) similar to that for the Grande Ronde Formation, but have higher FeO and TiO_2 and lower SiO_2 (49–52 wt.%). The

Table 1.2.3.4 Major element compositions of samples in the Columbia River Basalt Reference Suite (values in wt. %).

Sample	CP-1	CP-2	CP-3	CP-4	CP-5	CP-6	CP-7	CP-8	CP-9	CP-10
SiO_2	48.35	48.14	48.71	53.92	55.40	50.17	52.59	50.84	49.98	47.91
TiO_2	1.57	2.71	0.96	1.83	1.96	3.15	2.80	1.55	3.49	3.26
Al_2O_3	15.49	16.19	16.91	14.23	13.67	13.23	13.09	14.71	12.64	13.23
Fe_2O_3	3.26	3.06	2.44	2.27	4.98	3.52	7.25	1.60	2.30	1.84
FeO	8.05	9.56	7.77	9.10	7.16	10.71	5.88	8.75	12.67	12.49
MnO	0.17	0.18	0.20	0.20	0.23	0.22	0.23	0.19	0.22	0.24
MgO	7.03	5.39	7.86	4.25	3.34	4.41	2.67	6.99	4.27	5.80
CaO	9.92	8.66	11.06	8.42	6.95	8.20	6.01	10.48	8.35	9.99
Na_2O	2.76	3.30	2.48	2.92	2.41	2.85	3.18	2.30	2.46	2.38
K_2O	0.51	0.90	0.28	1.35	1.81	1.26	2.52	0.66	1.33	0.90
P_2O_5	0.24	0.36	0.18	0.31	0.32	0.67	0.85	0.25	0.56	0.78
H_2O^+	1.52	0.98	0.40	0.83	1.33	0.85	2.56	0.53	0.67	0.50
CO_2	0.05	0.11	0.04	0.04	0.05	0.05	0.08	0.06	0.04	0.08
SUM	98.92	99.54	99.29	99.67	99.61	99.29	99.71	98.90	98.98	99.40
$\frac{100Mg}{Mg+\Sigma Fe}$	53.3	43.8	58.4	40.5	33.8	36.2	27.7	55.0	34.1	42.2
ΣFeO	10.98	12.31	9.97	11.14	11.64	13.88	12.40	10.19	14.74	14.15

All analyses by X-ray fluorescence, with Na_2O by neutron activation analysis and Fe_2O_3/FeO, H_2O^+ and CO_2 by classical methods (analysts J.M. Rhodes, L.A. Haskin). Data precision 2%, except for SiO_2 and Al_2O_3 (3%).

Robinette Mountain basalt of the Eckler Mountain Member is a relatively "primitive" type, having high MgO (7.8 wt.%) and very low K_2O (0.27 wt.%) and P_2O_5 (0.19 wt.%) contents.

The Saddle Mountain Formation contains many chemically diverse members and types, characterized by wide ranges of MgO (2.7–8.2 wt.%), CaO (6.4–10.8 wt.%) and TiO_2 (1.5–3.6 wt.%) contents. The Umatilla magma type has the lowest MgO (2.69 wt.%) and the Asotin magma type has the highest MgO (8.2 wt.%) content among magma types defined to date for the Columbia River Plateau. The Goose Island type of the Ice Harbor Member has total iron (17.5 wt.% FeO) and TiO_2 (3.8 wt.%) contents which are among the highest reported for terrestrial basalts.

Studies by Hooper (1974) and Holden and Hooper (1976) concentrated on the Imnaha and Grande Ronde basalts in west-central Idaho. They have identified the Rock Creek type at the top of the Imnaha Formation, which contains SiO_2 in the range 48.8–50.5 wt.% and Al_2O_3 >15.2 wt.%, and is a coarse-grained ophitic basalt with abundant plagioclase and olivine phenocrysts. In other respects the west-central Idaho section of the Yakima Subgroup is similar to the section to the west which was described by Wright *et al.* (1973), again except for the occurrence of relatively TiO_2-rich flows in the east.

As part of this study project, ten samples representative of the Columbia River Basalt Province were selected for analysis by T. L. Wright and D. A. Swanson (Table 1.2.3.3). Nine of the samples were described by Wright *et al.* (1973). The remaining sample from Robinette Mountain (Wanapum Basalt Formation) has been included because it has a low TiO_2 (~1 wt.%) content but relatively high concentrations of MgO (~8 wt.%), Al_2O_3 (~17 wt.%) and CaO (~11 wt.%). Petrographic descriptions of these samples are given in Appendix A-3. Major element data are presented in Table 1.2.3.4.

The major element variation for the Columbia River Province is illustrated in Fig. 1.2.3.2, which includes the new data on the reference suite. For the Imnaha, Picture Gorge and Grande Ronde Formations there is a general inverse correlation of SiO_2, K_2O and TiO_2 with MgO. However, samples from the Wanapum

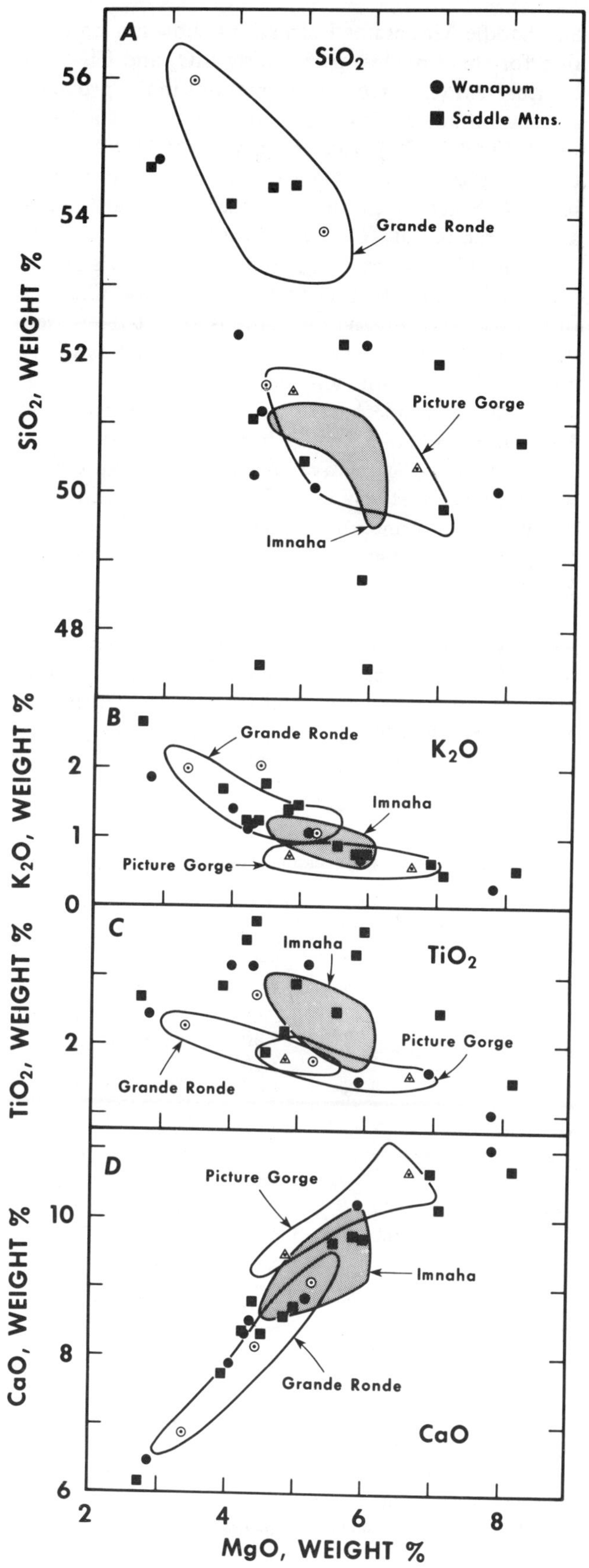

1.2.3.2 Variations in selected major element oxides as functions of MgO for basalts of the Columbia River Province. High-Mg, low-Mg and Prineville types are indicated for the Grande Ronde Formation (⊙), and high-Mg and low-Mg types for the Picture Gorge Formation (▵). Modified from Wright *et al.* (1973) to include data from Swanson and Wright (1981) and Table 1.2.3.3.

and Saddle Mountains Formations show less correlation for these oxides, particularly SiO_2 and TiO_2. In contrast, samples from *all* Formations in the province show a marked positive correlation between CaO and MgO (Fig. 1.2.3.2d), which suggests crystal/liquid control by a phase such as calcic clinopyroxene. Very similar CaO-MgO correlations are observed for the Paraná and Karroo basalts (Cox, 1980, Fig. 2).

The chemical features of the major subdivisions of the Province may be summarized as follows. The Picture Gorge and Imnaha Formations are relatively rich in MgO (4.5–7.1 wt.%) and are characterized by intermediate SiO_2 and low K_2O contents relative to MgO. The Grande Ronde basalts are distinguished by low MgO contents (3.0–5.5 wt.%), but have the highest SiO_2 (up to 56.4 wt.%) relative to MgO. The Wanapum basalts show relatively high TiO_2 and FeO, and relatively low SiO_2 contents over a range of MgO contents comparable to that of the Grande Ronde basalts. Basalts from the Saddle Mountains Formation show the greatest chemical dispersion; these include basalts with the lowest MgO (2.7 wt.%), the highest MgO (8.2 wt.%), and the highest total FeO (17.5 wt.%) contents so far reported; two of the youngest samples (in the Ice Harbor Member) have the lowest SiO_2 (47.5 wt.%), but the highest TiO_2 and P_2O_5 contents relative to MgO.

Trace element chemistry

Trace element data for the basalt types distinguished on the basis of major element composition were reported by Nathan and Fruchter (1974), McDougall (1976) and Swanson and Wright (1978, 1981). Additional data obtained for this project are presented in Table 1.2.3.5. These elements serve to better define the chemical types identified on the basis of major elements.

Data for selected trace elements in a number of the lava types are plotted in Fig. 1.2.3.3. Some elements, such as Sc, Th and Rb, show little systematic variation through the section, whereas others, such as La and Cr, exhibit some stratigraphic correlations. In general, the older flows are depleted in La, but tend to be enriched in Cr relative to younger flows. However, the high-Mg Picture Gorge, Robinette Mountain, Asotin and Pomona types are exceptional in their low La and high Cr abundances, and the Umatilla type conversely is enriched in La and depleted in Cr relative to the other lavas.

Precise nickel data are available for relatively few of the basalt types. Abundances for nearly all analyzed samples fall within the range 8–80 ppm, with one value of 150 ppm (Nathan and Fruchter, 1974; McDougall, 1976; Table 1.2.3.5).

Table 1.2.3.5 Trace element compositions of samples in the Columbia River Basalt Reference Suite (values in ppm).

Sample	CP-1	CP-2	CP-3	CP-4	CP-5	CP-6	CP-7	CP-8	CP-9	CP-10
La	7.03	16.7	6.77	21.9	22.2	26.6	42.7	16.2	32.8	36.2
Ce	17.0	43.9	16.0	48.0	45.6	60.1	90.9	35.3	70.0	82.6
Sm	3.72	7.14	2.72	6.00	5.76	8.58	10.0	4.64	9.36	10.7
Eu	1.37	2.27	1.03	1.86	1.63	2.55	4.10	1.54	2.74	3.23
Tb	0.72	1.40	0.59	1.07	1.01	1.44	1.63	0.88	1.68	1.98
Yb	2.60	3.15	2.20	3.45	2.90	4.05	4.26	2.70	4.55	5.05
Lu	0.38	0.47	0.33	0.51	0.43	0.62	0.65	0.40	0.70	0.74
Y	25	34	20	34	34	41	45	26	49	52
Rb	5.9	20	2.3	31	49	33	47	13	33	13
Cs	—	0.34	—	0.84	1.2	0.4	0.53	—	0.52	—
Sr	274	383	338	316	324	301	263	234	242	250
Ba	280	330	180	560	550	600	3000	260	535	560
Hf	2.35	5.79	1.67	4.90	4.42	5.37	11.4	3.64	7.00	6.99
Th	0.57	2.02	0.36	4.25	5.29	4.18	6.73	2.64	5.77	1.94
Pb	5	5	4	10	13	9	5	8	2	12
Ga	19	26	18	22	22	24	21	20	26	24
Sc	40.1	27.0	37.0	34.4	24.6	36.7	26.5	35.0	31.2	41.0
Cr	188	88.7	156	40.3	5	34.0	2.0	118	17.9	206
Ni	77	150	—	—	—	—	—	—	—	—

Data for rare-earth elements, Cs,Ba,Hf,Th,Sc,Cr and Ni by instrumental neutron activation analysis (analysts M. M. Lindstrom and L. A. Haskin). Data for Y, Rb, Sr, Pb and Ga by X-ray fluorescence (analyst J. M. Rhodes). Data precision 1-3%, except for Y, Rb, Sr, Ba (5%) and Cs, Pb, Ga, Ni (10%).

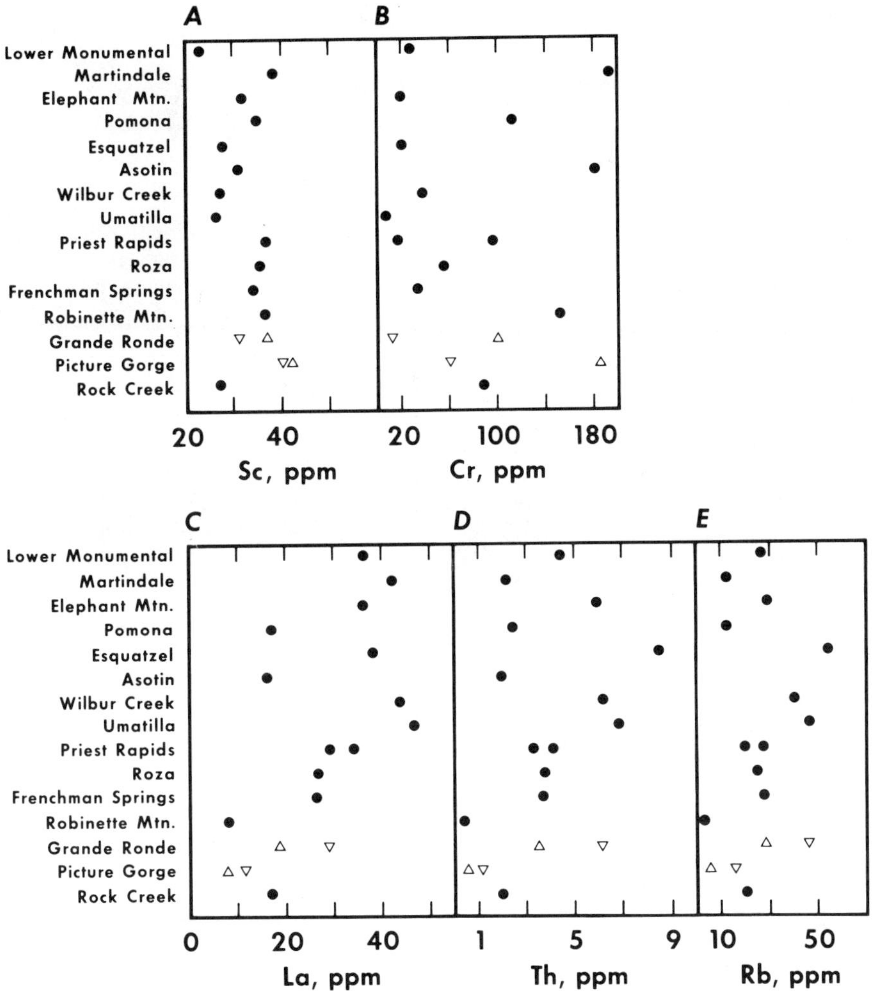

1.2.3.3 Variation in selected trace element abundances for basalt chemical types within the Columbia River Province (arranged stratigraphically). Data (either averages or single analyses) from Table 1.2.3.5, Swanson and Wright (1981) and Nathan and Fruchter (1974). (High-Mg (△) and low-Mg (▽) types are distinguished for the Picture Gorge and Grande Ronde Formations.

Rare-earth element patterns for the Columbia River basalts are shown in Fig. 1.2.3.4. The patterns for most samples from the Grande Ronde, Wanapum and Saddle Mountains Formations fall within three subparallel, overlapping fields, which are displaced to higher abundances in stratigraphic order. All of these samples are light REE-enriched with small negative Eu anomalies. Three of the most MgO-rich samples (the Robinette Mountain, Asotin and Pomona basalts) have much lower REE abundances than the other samples from their respective Formations (Wanapum and Saddle Mountains). In contrast, the very Fe-rich Goose Island basalt is markedly enriched in REE. The older Imnaha and Picture Gorge Formations are represented by only two samples, which have lower La/Sm ratios than samples from the overlying Formations.

Isotope chemistry

Strontium and neodymium isotopic analyses have provided important constraints on petrogenetic models for the Columbia River Province and for continental flood basalts in general. The very large range of variations in both Sr and Nd isotopic compositions in the Columbia River Province is remarkable, and even more noteworthy is the fact that there is a definite stratigraphic correlation for the major formations.

Strontium isotopic data are summarized in Fig. 1.2.3.5. As first pointed out by McDougall (1976), the Imnaha and Picture Gorge basalts have the lowest $^{87}Sr/^{86}Sr$ ratios (0.703–0.704), the Grande Ronde and Wanapum basalts are more radiogenic ($^{87}Sr/^{86}Sr$ $\sim$0.705), and the basalts of the Saddle Mountains Formation display a wide range of even higher $^{87}Sr/^{86}Sr$ ratios. More recent analyses by DePaolo and Wasserburg (1979), Nelson (1980) and Carlson *et al.* (1981) have confirmed McDougall's results and have established a range of values for the Saddle Mountains Formation from $\sim$0.708 (Pomona, Elephant Mountain and Martindale types) to $\sim$0.715 (Esquatzel type).

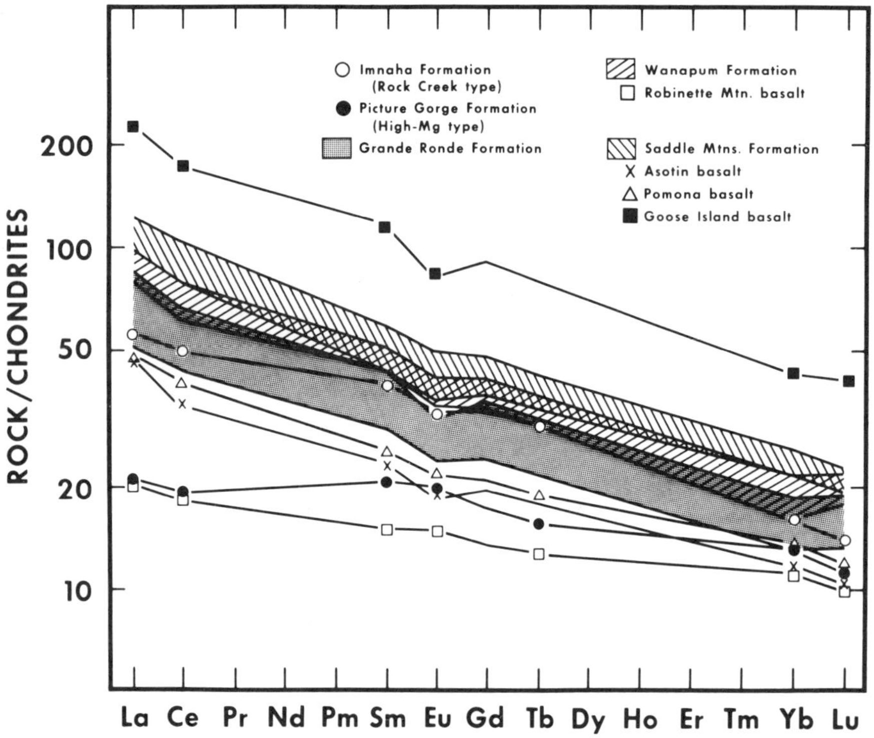

1.2.3.4 Chondrite-normalized rare-earth element distributions for basalts of the Columbia River Province. Data for Grande Ronde Formation, Wanapum Formation (excluding Robinette Mountain Flow), and Saddle Mountains Formation (excluding Asotin, Pomona and Goose Island Types) are shown as ranges. Data from Table 1.2.3.5 and Swanson and Wright (1978, 1981).

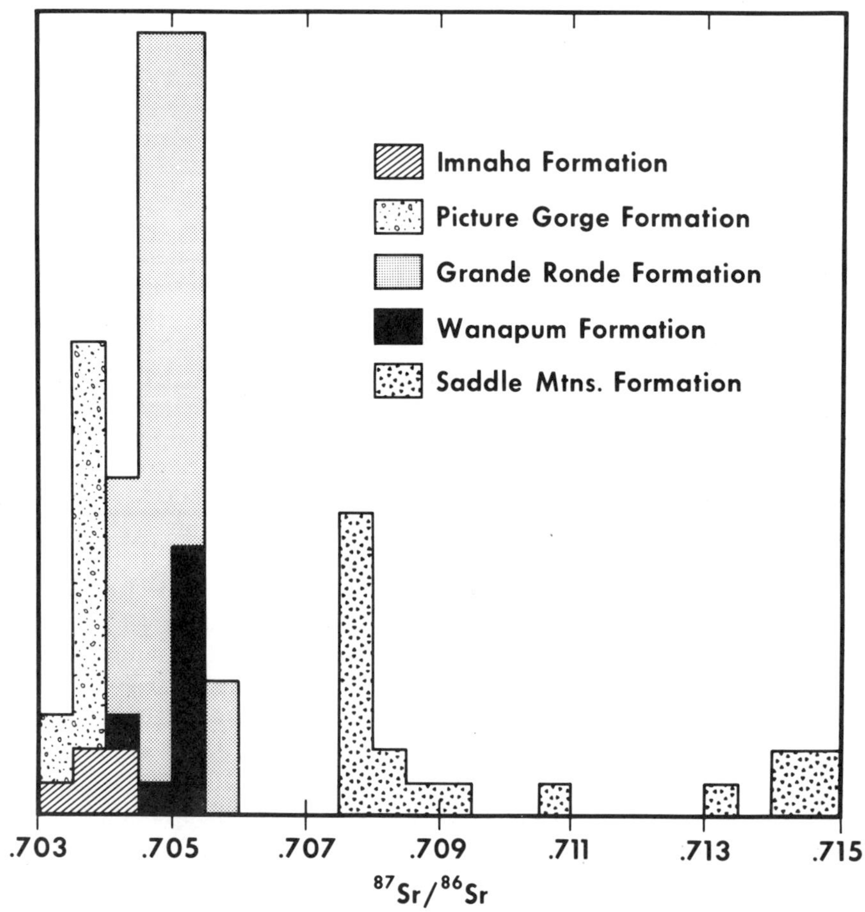

1.2.3.5 Histogram of initial $^{87}Sr/^{86}Sr$ ratios for basalts of the Columbia River Province. Data from Pankhurst and O'Nions (1973), McDougall (1976), DePaolo and Wasserburg (1976b, 1979), Nelson (1980) and Carlson *et al.* (1981).

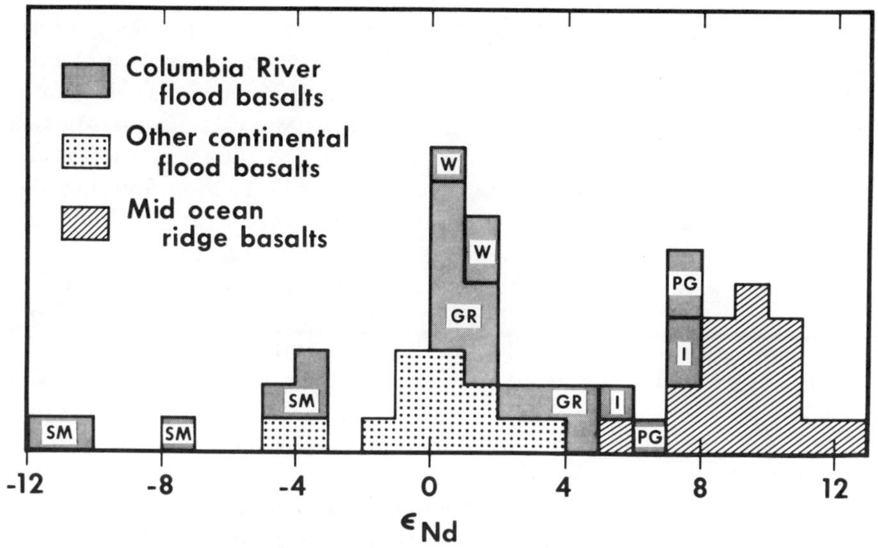

1.2.3.6 Histogram of initial ϵ_{Nd} values for Columbia River and other continental flood basalts compared with those for mid-ocean ridge basalts. For Columbia River samples: I = Imnaha, PG = Picture Gorge, GR = Grande Ronde, W = Wanapum, SM = Saddle Mountains. Other flood basalt samples are from the Deccan, East Greenland, Karroo, Paraná and Siberian Provinces. After DePaolo and Wasserburg (1979), with substantial additional data from Carlson *et al.* (1981).

Within this formation, however, there is no correlation of isotopic ratios with stratigraphic position.

Data for neodymium isotopes confirm the conclusions based on strontium isotopes. In Fig. 1.2.3.6, available data for the Columbia River Province and other continental flood basalt provinces are compared with data for mid-ocean ridge tholeiites. DePaolo and Wasserburg (1976a, 1977, 1979) have emphasized the isotopic distinction between these two major groups of terrestrial basalts, which implies very different geochemical histories for their respective source regions.

Early analyses of samples from the Grande Ronde Formation and from several other flood basalt provinces, which showed ϵ_{Nd} values clustering about zero, led DePaolo and Wasserburg (1976a, b) to suggest relatively primitive, undifferentiated mantle sources for these magmas. However, subsequent work by Carlson *et al.* (1981) has demonstrated that the range in $^{144}Nd/^{143}Nd$ for the Columbia River Province is just as great as that observed in $^{87}Sr/^{86}Sr$ ratios, and that these two isotopic parameters display a remarkable correlation when measured for the same samples (see Fig. 1.2.3.7). Carlson *et al.* (1981) further conclude that the mean Nd isotopic composition of the Grande Ronde Formation (which represents over 80% of the volume of the Columbia River Province) is displaced to values

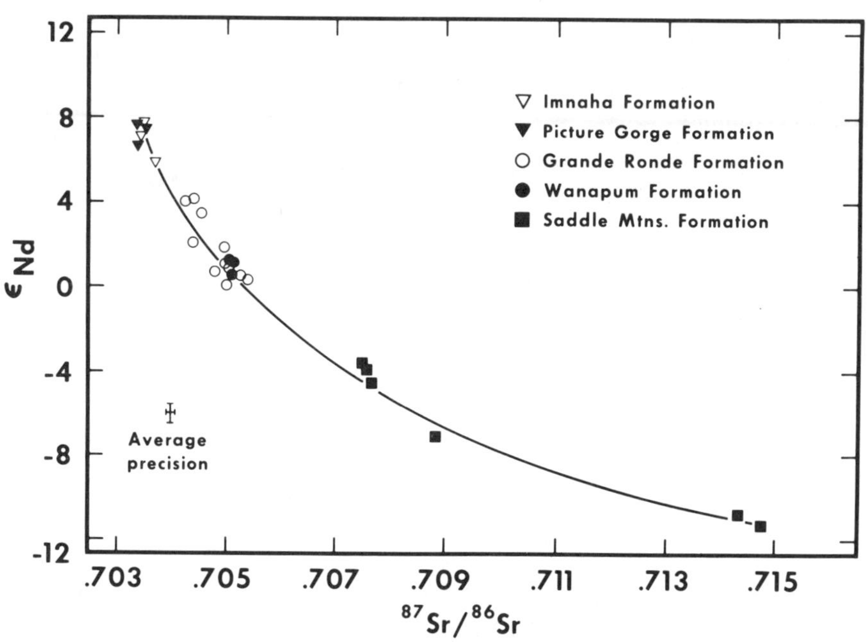

1.2.3.7 Nd-Sr isotopic variation for basalts of the Columbia River Province. After Carlson *et al.* (1981), with additional data from DePaolo and Wasserburg (1976a, b, 1979) and Pankhurst and O'Nions (1973).

of ϵ_{Nd} greater than zero, and that the wide range of isotopic ratios for the Province may largely be a result of contamination of ascending primary magmas by continental crustal material, as also suggested by McDougall (1976).

The petrogenesis of the Columbia River Province and other continental flood basalts is obviously much more complex than once supposed, and realistic models must take these key isotopic results into account as well as the still-growing data base of stratigraphic, petrological, major element and trace element characteristics. Discussion of petrogenetic models is deferred to section 1.2.3.9 after descriptions of other terrestrial plateau basalt provinces.

Tertiary basalts marginal to the Columbia River Plateau

Several other regions of Tertiary tholeiitic basalt volcanism have been recognized in the northwest United States outside of the main areal extent of the Columbia River Plateau. These are not treated in detail here, but some of the more significant studies are mentioned below because of the comparisons that have been made between these lavas and those of the Columbia River Province.

Snavely *et al.* (1968) described a sequence of pillow lavas, water-lain basaltic fragmental debris, and interbedded marine volcaniclastic sedimentary rocks of Early to Middle Miocene age from the Siletz River Volcanics in the central part of the Oregon Coast Range. The older rocks in the sequence are holocrystalline or vitrophyric olivine tholeiites that are more primitive than those of the overlying units, which represent a differentiation sequence from picrites to alkali basalts and mugearites. In a later study of basalts erupted from local vents along the coastal margins of Washington and Oregon, Snavely *et al.* (1973) found a close resemblance in major element, trace element, and isotopic abundances to lavas of the Grande Ronde and Wanapum Formations and the Pomona Member of the Columbia River Province. The location of these vent complexes over 100 km from the nearest known vent system of the Columbia River Province remains a puzzling paleotectonic problem.

The Snake River Plain to the southwest of the Columbia River Plateau lies in a broad, arc-shaped topographic depression that extends from the Yellowstone Plateau across southern Idaho to Oregon. Rhyolitic volcanism began about the same time as basaltic volcanism in the Columbia River Province ($\sim$17 m.y.b.p). Basaltic volcanism on the Snake River Plain is younger than most of the Columbia River volcanism, and is distinct, petrologically and geochemically. The Snake River Clan, as defined by Powers (1960), is relatively low in SiO_2 (45–49 wt.%) and is generally characterized by a decrease in SiO_2 as the Fe/Mg ratio increases. Rocks in this clan are classified as Mg-rich olivine tholeiites to Fe-rich tholeiitic andesites (Powers, 1960; Leeman and Vitaliano, 1976; Stout and Nicholls, 1977). Major and trace element data for the Pleistocene McKinney olivine tholeiite flow, which ranges from 5 to 10 m in thickness and covers an area of $\sim$300 km^2, are given by Leeman (1976) and Leeman and Vitaliano (1976). The low $^{87}Sr/^{86}Sr$ ratio of this flow (0.7066), compared with the more evolved Snake River Plain lavas, led Leeman and Manton (1971) to conclude that crustal contamination has affected the majority of the lavas.

Steens Mountain in southwestern Oregon consists of a 1-km-thick sequence of basalts dated at 15 m.y. which may have accumulated in no more than 50,000 years, and perhaps in as little as 2,000 years (Gunn and Watkins, 1970). The sequence is divided into 8 groups of lavas distinguished primarily by the ratio of pyroxene/plagioclase. Steens Mountain basalts are high alumina (14–23 wt.% Al_2O_3) olivine basalts that are aphyric or coarsely porphyritic. Silica content varies within narrow limits (47.8–50.3 wt.%) and the sequence is best characterized by the variation in Al_2O_3 contents: lavas that are enriched in Al_2O_3 also are enriched in CaO, Sr, Ni, and Cu, and are depleted in TiO_2, ΣFeO, K_2O, MnO, MgO, P_2O_5, Rb and Ba. Data for rare-earth elements in 52 of 70 consecutive flows were given by Helmke and Haskin (1973). Gunn and Watkins (1970) have suggested that the parent magma was an olivine basalt intermediate between high-alumina basalt and olivine tholeiite. The cyclic nature of the chemical variations with time may imply repetitive recharging of shallow magma reservoir(s).

1.2.3.3 British Isles Province

Geological setting

The British Isles Tertiary Province includes the Hebridean Isles of northwest Scotland and the north and northeast of Ireland. Field relationships in these areas have been described by Stewart (1965) and Charlesworth (1963), respectively, but the early classical memoirs by Harker (1904), Bailey *et al.* (1924), and Richey and Thomas (1930) remain useful references.

The province is dominated by plateau basalt sequences, a NW-SE trending dolerite dike swarm, and central igneous intrusions of granite and layered complexes. The type localities of cone sheets are on the islands of Skye and Mull, but sheets are also present on

Rhum, Ardnamurchan and Carlingford. Perhaps the most spectacular development of columnar jointing in any of the plateau basalt provinces on Earth occurs in the Middle Antrim series of Giant's Causeway, where colonnades reach heights of 100 meters or more.

Volcanism began in this province in the late Cretaceous with eruptions around or just before 65–66 m.y. ago, and the major events are considered to have terminated between 60 and 56 m.y. ago. Minor activity may have continued until the Eocene, some 50 m.y. ago (Evans *et al.*, 1973).

The major basalt types that are now recognized were first postulated on the island of Mull by Bailey *et al.* (1924), who described them as a *Plateau Magma-Type* and a *Non-Porphyritic Central Magma-Type*. These magma types now are classed as *alkalic* and *tholeiitic* basalts respectively. The early work set the stage for modern petrogenetic studies which have extended the classification of magma types in the British Tertiary Province to include basaltic hawaiites, mugearites, mildly alkaline and transitional basalts, low-alkali high-calcium olivine tholeiites, and benmoreites (e.g., Ridley, 1971a, b; Holland and Brown, 1972; Thompson *et al.*, 1972; Esson *et al.*, 1975; Beckinsale *et al.*, 1978). However, Tilley and Muir (1962) concluded that all plateau lavas of the Hebridean Tertiary Province are nepheline-normative and that any deviation from this normative classification can be ascribed to pneumatolitic alteration and metasomatic exchange. This was an important consideration in earlier interpretive studies. Precautions were taken in the studies summarized here to insure that samples selected were fresh or were taken well away from major vent complexes, so their compositions should have been less affected by these alteration processes.

Mull

Basaltic lavas on the island of Mull cover an area of ~2 km^2. Approximately half of these basalts are derived from central complexes, and the remainder from fissure-type eruptions around the margins of the island.

The study by Beckinsale *et al.* (1978) has resulted in division of the Mull lava sequence into three units based on $^{87}Sr/^{86}Sr$ ratios. Group 1 comprises an alkaline series (basalt-hawaiite-mugearite) with extremely low initial $^{87}Sr/^{86}Sr$ ratios ($\lesssim$0.7030) and extremely low lithophile element abundances. Group 2 contains hypersthene-normative basalts with more tholeiitic affinities and more evolved rocks of trachytic chemistry with higher initial $^{87}Sr/^{86}Sr$ ratios (0.7055) and lithophile element abundances. Group 3 is less well-defined, has chemical characteristics which lie between those of Groups 1 and 2, with $^{87}Sr/^{86}Sr$ ratios in the range 0.7033–0.7043, and consists largely of aphyric or sparsely olivine-phyric basalts. The major element variation for the Mull lavas is very similar to that for the Skye lavas, and data for these two islands are compared in Fig. 1.2.3.8.

Skye

Lavas on the island of Skye cover approximately 1500 km^2 in five stratigraphically distinct igneous units consisting of alternating sequences of basalts and mugearites, mugearites and trachytes, and basalts overlain by mugearite flows. Individual units range in thickness from 120–500 m, and the total thickness of the plateau series is approximately 2.3 km (Anderson and Dunham, 1966). These lavas lie in a shallow, faulted, oval basin with a long axis trending NW-SE. The southern part of Skye is dominated by the large Cuillin Hills layered mafic-ultramafic complex and by the Red Hills

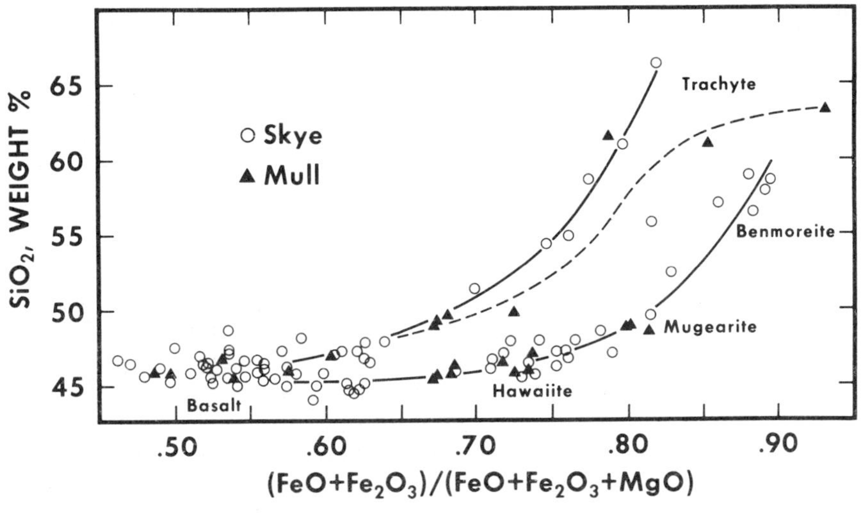

1.2.3.8 Variation in SiO_2 content as a function of $(FeO+Fe_2O_3)/(FeO+Fe_2O_3+MgO)$ for lavas from Mull and Skye. Solid lines emphasize the two trends for evolved lavas from Skye. Dashed line emphasizes the high-SiO_2 group of evolved lavas from Mull. The low-SiO_2 Mull lavas lie on the same trend as those from Skye. After Thompson *et al.* (1972) and Beckinsale *et al.* (1978).

granite-granophyre complex, which are flanked by older plateau basalts.

Studies by Thompson *et al.* (1972), Esson *et al.* (1975), and Mattey *et al.* (1977) have shown that the Skye volcanic rocks include hypersthene- to nepheline-normative basalts, high-Ca and low-alkali olivine tholeiites, basaltic hawaiites, hawaiites, mugearites, benmoreites, and trachytes. Major element data define a continuous evolution series from basalt to either benmoreite or trachyte. These two separate trends are particularly evident on a SiO_2 evolution diagram (Fig. 1.2.3.8). Apart from this Main Skye series Thompson *et al.* (1972) and Esson *et al.* (1975) reported the presence of two aphyric, low-alkali, Ca-rich tholeiitic basalts with compositions distinct from all other Skye lavas, which are interleaved with alkali basalts near the top of the lava pile. Compared with other Skye lavas, these basalts have similar ranges of MgO, total FeO, Al_2O_3 and SiO_2, but have distinctively lower alkali (Na_2O more so than K_2O), P_2O_5, and TiO_2 contents, and higher concentrations of CaO. Esson *et al.* (1975) suggested that these lavas may be related to the parental magmas of the Cuillin Hills complex.

In a companion study of the Skye dike swarms (Mattey *et al.,* 1977), the low-K, high-Ca magma type was found to be present in 70% of the dikes; of the remaining dikes, 15% have compositions equivalent to the hypersthene-normative basalts of the Main Skye lava series. Dike compositions similar to those of the mugearite and benmoreite lava series are rare, and few dikes resembling the nepheline-normative basalts are present. An additional 15% of the dikes appear to represent a magma type which is distinct from both the low-K, high-Ca type and the Main lava series type (hypersthene-normative olivine tholeiites); only one lava is similar to this magma type, and is characterized by a relatively flat REE pattern in contrast with the light REE enriched or depleted patterns that persist in the Main Skye series.

Rhum, Eigg, Canna and Muck

These islands, together with Skye, comprise the Inner Hebrides. Rhum is the most best known, apart from Skye, and is dominated by a "ring fault-controlled cumulate mafic complex" that is considered to have experienced a 2 km vertical movement during emplacement of the body. Granophyres, felsites and explosion breccias are also present (Stewart, 1965). Plateau volcanics rest on Torridonian arkose and Tertiary granophyre on Rhum. Basalt and tuff lie on an eroded surface of Mesozoic sandstones, limestones and shales on Muck. On the island of Eigg, the basalts are extruded onto Cretaceous sediments, and on Canna the lava-sediment contact is below sea level. Numerous basaltic dikes radiate from the central Rhum complex, in addition to the regional NW-SE trending dike swarm that dissects the entire province. Northwest-trending pitchstone dikes are also present on Eigg.

The major volcanic rock types on these islands are members of the basalt-hawaiite-mugearite series, and include felsites and pitchstones (Ridley, 1971a, b). Most of the basalts are nepheline-normative, although some samples show normative hypersthene. The basalts may be generally described as mildly alkaline ranging to transitional. In terms of major and trace element chemistry, the hawaiites and the mugearites are comparable to those on Mull and Skye. The fractionation of basalts to hawaiites on Rhum and Eigg resulted in only a small decrease in total iron and in an increase in K_2O and Na_2O. However, on Muck and Canna the hawaiites are more evolved, being markedly enriched in K_2O and more depleted in total iron.

Ardnamurchan

The Ardnamurchan Peninsula is renowned for the spectacular ring dike and cone sheet complexes that surround the three central intrusions. Alkali olivine basalts predominate, along with local feldspar-phyric flows and one recorded occurrence of mugearite (Stewart, 1965). Tertiary lavas cover an area of approximately 18 km^2 with a total thickness of $\sim$100 meters. These lavas have been strongly affected by intrusive pneumatolysis, so reliable analytical data are unavailable. However, a characterization of the central magma types can be gained from the minor intrusions, as was demonstrated by Mattey *et al.* (1977) for dikes and lavas on the island of Skye.

Holland and Brown (1972) showed that the cone sheets from Ardnamurchan are tholeiitic in character. Ring dikes associated with the three intrusion centers have been studied by Gribble (1974). Analyses of dolerites from each show slight trends in iron-enrichment from intermediate to basic members with a considerable degree of overlap among the trends for all three centers. All these rocks have strong tholeiitic affinities and resemble the less-fractionated cone sheets described by Holland and Brown (1972).

Ireland

The most extensive remnants of Tertiary plateau basalts in the British Isles are found in Antrim and cover $\sim$4000 km^2. Two lateritic horizons separate the sequence into three units. The Upper and Lower Series are typical alkali olivine basalts and the Middle Series is tholeiitic (Patterson, 1951). Trachytes, rhyolites and obsidians are also present. Ages range from 66–61 m.y.

and, based on a mean thickness of 0.5 km, extrusion rates over this 5 m.y. period are estimated to have been 4×10^5 m^3/yr (Evans *et al.*, 1973), a rate that is substantially less than those of Mauna Loa and Kilauea ($\sim 10^{-1}$ km^3/yr; Swanson *et al.*, 1975).

In a study of the quartz tholeiites of the Middle Antrim Series (Giant's Causeway), Patterson and Swaine (1955) reported an initial increase in $\Sigma FeO/(\Sigma FeO + MgO)$ from 0.63–0.65 in adjacent flows at the base of the section; this ratio decreases systematically to 0.57 further up the section and rises again to a maximum of 0.71 at the top of the lava pile. This type of variation occurs in a number of flood basalt provinces and may suggest shallow, pre-eruptive fractionation of a recharged magma reservoir. Schilling and Noe-Nygaard (1974) noted that the earlier flows in this section are light-REE enriched, whereas the later flows are light-REE depleted, a distribution pattern that is also present in the Faeroe Islands. The discontinuity is interpreted to have resulted from a change in the type of volcanicity from *plume* to *oceanic*.

The basalts and tholeiites in the Antrim Province bear some resemblance to those from Skye and Mull. The association of trachytes and the inferred olivine basalt → quartz tholeiite fractionation trend are features common to all these localities.

1.2.3.4 Faeroe Islands Province

The Faeroe Islands, north of Scotland, have a surface area ~1400 km^2 and consist almost entirely of a subaerially extruded lava sequence that is 3 km thick. This sequence has been divided into three series as outlined by Noe-Nygaard (1974), and is dated at 52–60 m.y. (Tarling and Gale, 1968). The Lower Series consists mainly of quartz tholeiite flows, and is overlain successively by a coal seam (containing Eocene pollen) and a ~30 m-thick tuff-agglomerate horizon. The Middle Series is 1350 m thick and is composed of thin, ropy lava flows which are dominantly porphyritic tholeiites interspersed with olivine-phyric flows and rare picrites. The 625-m-thick Upper Series consists of well-defined flows of olivine tholeiites (~10 m thick) with interbedded tuffs.

Volcanism is considered to have started in the west, in close proximity to the Wyville-Thompson ridge, and to have migrated eastwards with time, yielding early quartz tholeiites and late, less-differentiated olivine tholeiites (Noe-Nygaard, 1966, 1974). The coal seam unconformity between the Lower and Middle lava sequences indicates a period of quiescence which was followed by a change in magma composition and the extrusion of picritic basalts. The second unconformity, between the Middle and Upper Series, is marked by a change from thin laminated flows to massive flows. Noe-Nygaard (1974) considers that this event signaled a return to the type of volcanism that had produced the Lower Series quartz tholeiites. Either fissure-aligned vents or flat, shield-type volcanoes are thought to have dominated the period, in association with times of intermittent tension during which dikes and sills were intruded.

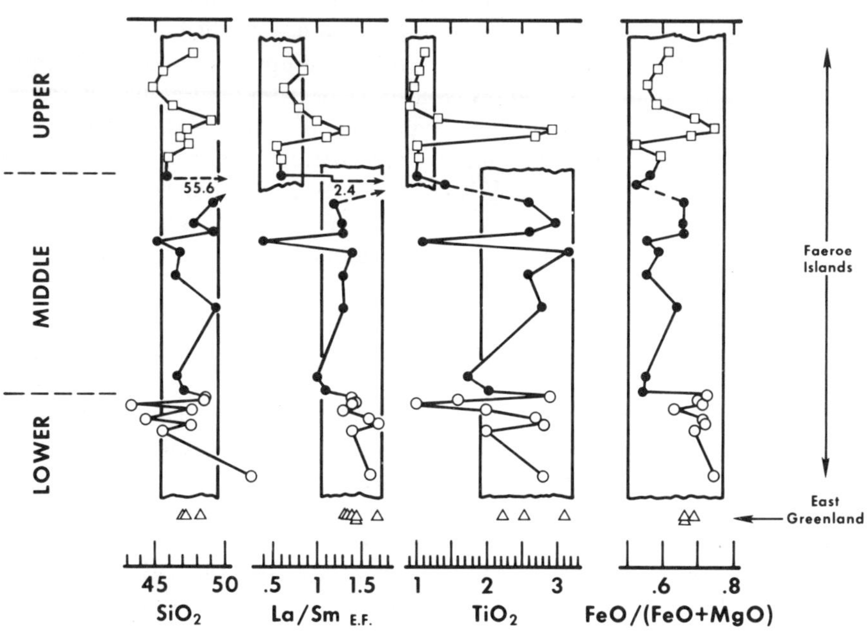

1.2.3.9 Profiles for bulk SiO_2 and TiO_2 contents, chondrite-normalized La/Sm ratio and $\Sigma FeO/(\Sigma FeO+MgO)$ ratio through the 3 km-thick basalt plateau on the Faeroe Islands. Data for Tertiary basalts from Scoresby Sund, East Greenland are shown for comparison. After Schilling and Noe-Nygaard (1974).

The Lower and Middle Series are characterized by high TiO_2 contents (2.6–2.9 wt.%) relative to those of the Upper Series (average 1.0 wt.%TiO_2). The early formed quartz-normative tholeiites have the highest SiO_2 contents (average = 49.5 wt.%), lower Al_2O_3 concentrations, and relatively higher iron and alkali values. These data contrast with those for the Upper Series olivine tholeiites which have lower silica, iron and alkalies, but higher alumina contents.

REE abundances in the 3 km-thick Faeroe sequence show an abrupt change from light-REE enriched to light-REE depleted patterns near the boundary between the Middle and Upper Series (Schilling and Noe-Nygaard, 1974). All flows of the Lower Series, and all of the Middle Series except one, are light-REE enriched with La/Sm enrichment factors >1. The flows in the Upper Series fluctuate between light-REE depleted and light-REE enriched patterns (La/Sm >1), the uppermost flows having light-REE depleted patterns. These REE distributions correlate with TiO_2 contents and with the distinction between quartz and olivine tholeiites of the Lower-Middle and Upper Series, respectively (see Fig. 1.2.3.9). The interpretation of the discontinuity in petrology and in major and trace element chemistry is that the mechanism and the environment of volcanic activity was dominated initially by a plume source, but changed later to oceanic tholeiitic volcanism (Schilling and Noe-Nygaard, 1974).

The light-REE enriched lavas generally have higher $^{87}Sr/^{86}Sr$ ratios (0.7031–0.7045) than the light-REE depleted lavas (0.7026–0.7035), and conversely the former have consistently lower $^{143}Nd/^{144}Nd$ ratios (Gariepy *et al.*, 1981). This correlation, and the existence of one unaltered sample with $^{87}Sr/^{86}Sr$ of 0.7096 and $^{143}Nd/^{144}Nd$ of 0.51205, implies that the light-REE enriched lavas have been affected by crustal contamination.

1.2.3.5 Iceland Province

Geological setting

Iceland has traditionally been regarded as part of the Brito-Arctic Plateau Basalt Province. However, it clearly resembles oceanic crust. Iceland lies along the seismically active Reykjanes-Jan Mayen Ridge system and intersects the transverse aseismic ridges that extend from Greenland to the Faeroe Islands (Pálmason and Saemundsson, 1974).

Iceland is the largest (10^5 km^2) and volcanically the most productive (0.04 km^3 of material/yr and one eruption/5 yrs) landmass astride an active ridge and mantle plume system on Earth (Schilling, 1973; Sigvaldason, 1974). The island is dissected by two major rift zones along which active rifting is taking place, and along which major volcanic events have occurred in the past 4.0 m.y. Postglacial (<10,000 yrs) eruptions have produced approximately 40 km^3 of lava, including the largest single volume (~12 km^3) extruded in historic times which erupted in 1783 from the 25-km-long Laki fissure and which covered an area of 565 km^2 (Noe-Nygaard, 1974; Sigvaldason, 1974). This centrally active province is flanked on the east and west by Tertiary plateau basalts which range in age from 1–16 m.y. (Pálmason and Saemundsson, 1974). Lavas dip gently (1–8°) on either side of the neovolcanic province towards the central region. Dips increase with increasing age, suggesting that a partial warping of the volcanically active central region has taken place in conjunction with tensional break-up (Walker, 1965).

The Tertiary lavas cover approximately 35–40% of the area of Iceland. A sequence on the east coast is at least 12 km in total thickness, with average flow thicknesses of 10 m (Dagley *et al.*, 1967). Continuous sequences on the west coast are more difficult to estimate but are probably in excess of 3 km thick (Bailey and Noe-Nygaard, 1976). The volumetric proportions of rock types in these sequences are as follows: tholeiites 48%, olivine basalts 23%, feldspar-phyric basalts 12%, andesitic lavas 3%, rhyolitic to dacitic lavas 8%, and pyroclastics + sediments 6%.

The basaltic and tholeiitic suites are derived largely from fissure-type eruptions, whereas the intermediate to acid lavas erupted periodically from central volcanoes among the main plateau sequences. Thingmúli is an example of this type of Tertiary volcano in which olivine tholeiites, basaltic andesites, icelandites and pitchstones formed (Carmichael, 1964).

Postglacial volcanic eruptions are derived largely from central volcanoes or from fissure-aligned craters, and the lava types are tholeiites, transitional alkali basalts, alkali olivine basalts, and Fe-Ti rich basalts. Studies by Jakobsson (1972), Sigvaldason (1969, 1974) and Sigurdsson *et al.* (1978) have shown that these basalt types have a regional distribution within the neovolcanic province (Fig. 1.2.3.10). They exhibit correlations between age and fractionation trends within some volcanic centers (Sigvaldason, 1969), and are gradational to basalts of the Mid-Atlantic Ridge in major element, trace element and isotope chemistries (Schilling, 1973; Hart *et al.*, 1973; O'Nions *et al.*, 1976; Schilling *et al.*, 1978).

Neovolcanic Province

Although most of the rock types that have been identified in the oceanic crust are present also in Ice-

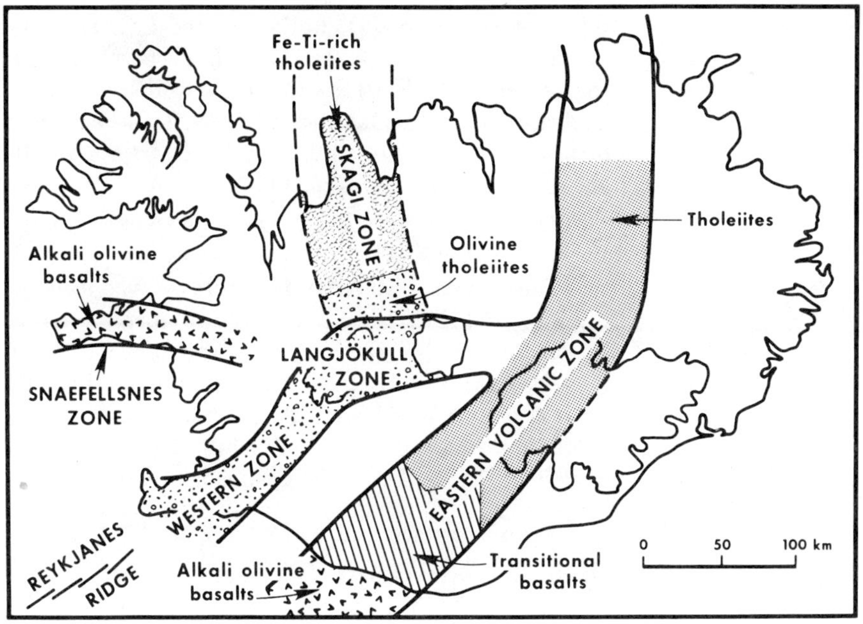

1.2.3.10 Map of Iceland showing the distribution of recent lava types in the neovolcanic zones. Modified from Sigvaldason (1974), with petrologic data from Jakobsson (1972) and Sigurdsson *et al.* (1978)

land, many of them are restricted to Iceland's neovolcanic rift zones and are not present in the Tertiary plateau basalt sequences to the east and west of the spreading center which is active now. The neovolcanic suites may nevertheless, represent end products of an evolutionary volcanic cycle; so they are included in this review for purposes of comparison with the Brito-Arctic Province. Representative major element compositions are given in Table 1.2.3.6.

Convincing demonstrations of elemental and isotopic variations as functions of distance along the Reykjanes Ridge, and away from Iceland, were made by Schilling (1973) and Hart *et al.* (1973). In particular, K_2O, P_2O_5, TiO_2 and $^{87}Sr/^{86}Sr$ decrease systematically with distance away from the margin of Iceland along the ridge axis (Fig. 1.2.3.11). Along a 220-km-long profile on the northward extension of the Reykjanes Ridge, the dominant quartz tholeiitic character of the ridge gives way to olivine tholeiites along the Reykjanes Peninsula (Jakobsson, 1972), to Mg-Al rich basalts in the vicinity of Langjökull, and to Fe-Ti rich basalts in the Skagi area (Sigurdsson *et al.*, 1978; Schilling *et al.*, 1978).

Rock types have a bilateral distribution on either side of the extension of the ridge axis, with alkali olivine basalts occurring in the Snaefellsnes fracture zone and in the southernmost portion of the Eastern Volcanic

Table 1.2.3.6 Average major element compositions of Iceland lavas.[†]

Region Type	Reykjanes Peninsula Olivine Tholeiite	Langjökull Zone Olivine Tholeiite	Skagi Zone Fe-Ti-rich Tholeiite	Askja Mývatn Quartz Tholeiite	Snaefellsnes Zone Alkali Olivine Basalt	Westman Islands Alkali Olivine Basalt	E. Iceland (Tertiary) Olivine Tholeiite
SiO_2	48.01	47.02	47.92	50.26	46.88	46.83	46.73
TiO_2	1.87	1.39	3.30	2.47	2.41	2.38	2.51
Al_2O_3	14.09	14.91	13.65	13.52	15.53	16.12	14.83
Fe_2O_3	2.69	2.22	3.14	3.52	2.30	3.39	6.40
FeO	10.06	9.78	12.16	11.87	9.42	8.69	6.74
MnO	0.21	0.20	0.27	0.24	0.22	0.19	0.21
MgO	8.29	10.45	6.38	5.58	7.51	7.47	6.70
CaO	11.77	11.76	10.51	9.18	11.07	10.26	10.82
Na_2O	2.17	1.89	2.54	2.48	2.90	3.27	2.53
K_2O	0.29	0.10	0.40	0.56	0.71	0.60	0.38
P_2O_5	0.19	0.10	0.33	0.23	0.44	0.32	0.31
H_2O^+	0.28	—	—	0.31	0.28	0.30	1.78
SUM	99.92	99.82	100.60	100.22	99.67	99.82	99.94

[†]Data from Jakobsson (1972), Wood (1976) and Sigurdsson *et al.* (1978).

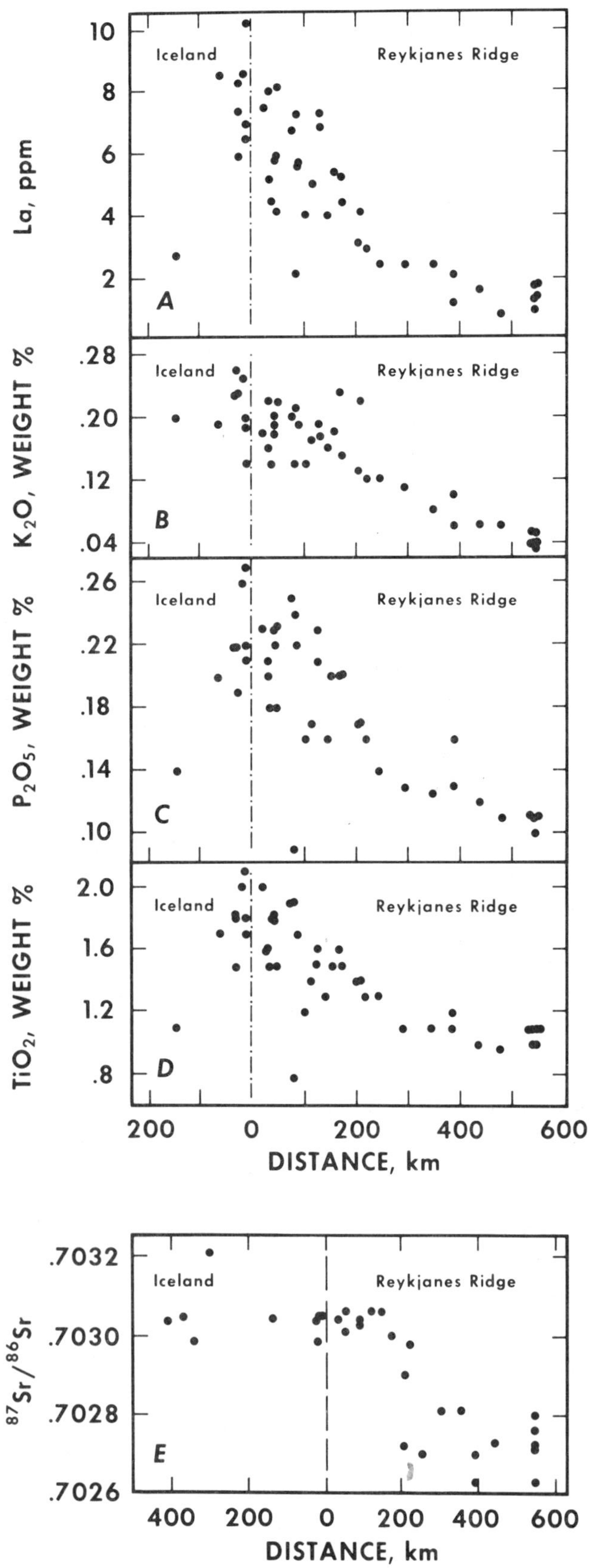

Zone (Jakobsson, 1972). Following the Eastern Volcanic Zone further north, the alkali olivine basalts grade progressively into transitional alkali basalts, and ultimately into tholeiites (Jakobsson, 1972). To the extreme northeast of this region, olivine tholeiites are the dominant rock type (Sigvaldason, 1974).

Considerable attention has been given to comparisons between the lavas of the neovolcanic province and ocean floor basalts (MORB). The more evolved Iceland tholeiites and Fe-Ti rich basalts are distinct from MORBs in having higher Fe, Ti, K and light REE (see Fig. 1.2.3.12a), and lower Al and Mg contents (Brooks and Jakobsson, 1974; Sigvaldason and Steinthorsson, 1974; Sigurdsson *et al.*, 1978; Schilling *et al.*, 1978). Lavas from the Reykjanes Peninsula have relatively flat REE patterns. However, a number of lavas from the Langjökull Zone in central Iceland and from Theistareykir in the northern part of the Eastern Volcanic Zone (O'Nions *et al.*, 1976; Schilling *et al.*, 1978) are light REE-depleted, have $^{87}Sr/^{86}Sr$ ratios of 0.7030, and are otherwise indistinguishable from MORBs on the Reykjanes Ridge.

O'Nions *et al.* (1976) found a strong correlation between $^{87}Sr/^{86}Sr$ and chondrite-normalized Ce/Yb ratios for all Icelandic tholeiites (including some Tertiary examples as much as 16 m.y. old). However, such a correlation does not exist for the alkalic basalts. O'Nions *et al.* concluded that the chemical features of Icelandic basalts are largely reflections of mantle heterogeneities established at least 100–200 m.y. ago.

Alternative models involving mixing of several mantle sources were proposed by Langmuir *et al.* (1978). After an extended evaluation of element and isotope ratio plots, they suggested two separate mixing trends: one beneath Iceland involving an enriched alkali basalt source and a depleted source; the second beneath the Reykjanes Ridge involving a heterogeneous, depleted oceanic source and a more homogeneous, intermediate source feeding the Reykjanes Peninsula and the Reykjanes Ridge within 200 km of Iceland. Langmuir *et al.* were able to rule out magma mixing for Iceland, but did not exclude the possibility of dynamic partial melting of mixed sources. Wood (1979) favored dynamic melting of a depleted source veined by previous magmatic products. Existing data do not permit further discrimination among these various petrogenetic models.

1.2.3.11 Chemical and isotopic variations in recent tholeiitic lavas along a traverse from the Reykjanes Peninsula onto the submarine Reykjanes Ridge. After Schilling (1973) and Hart *et al.* (1973).

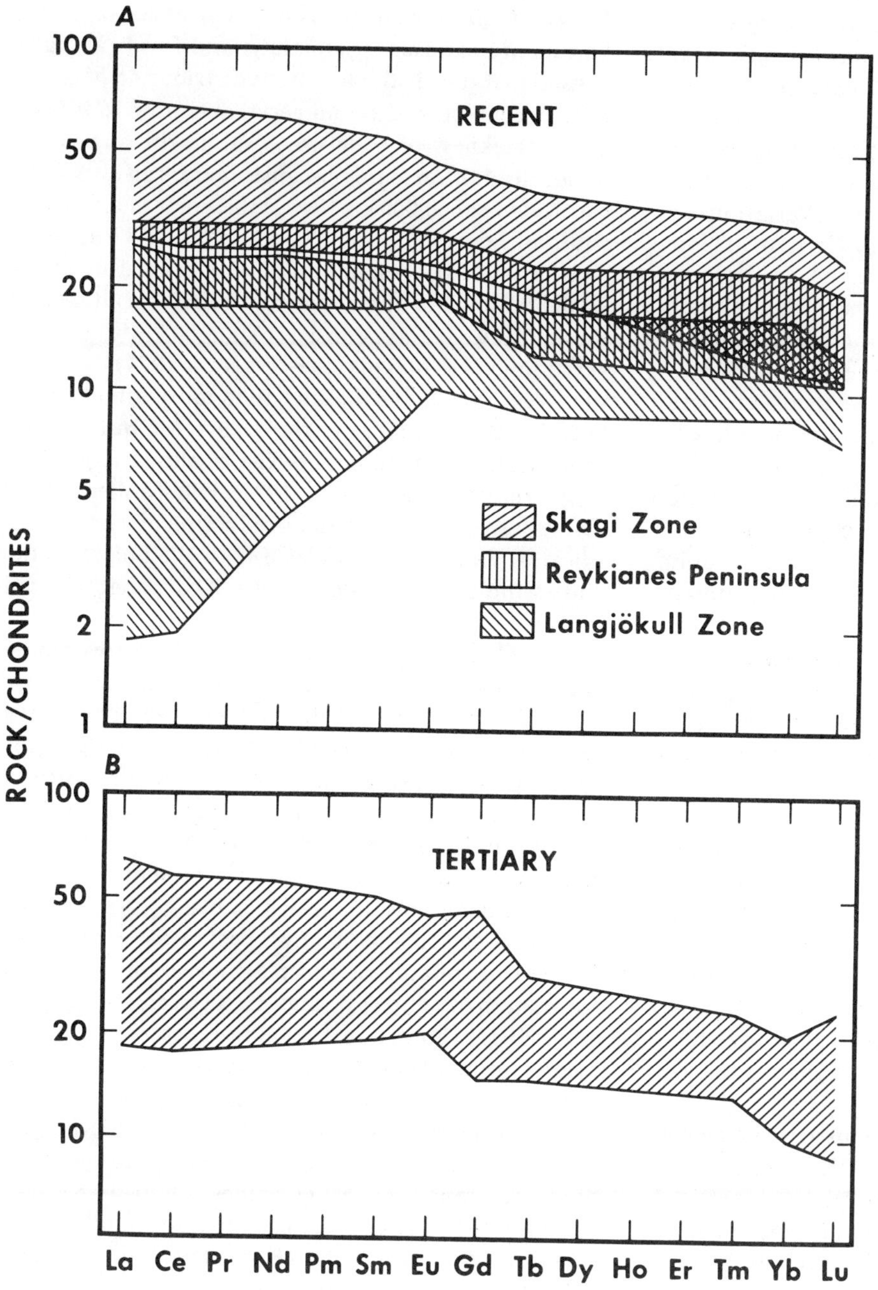

1.2.3.12 Ranges of chondrite-normalized rare-earth element distributions for tholeiitic lavas from Iceland. **a.** Recent tholeiites from three of the neovolcanic zones (Schilling *et al.*, 1978). **b.** Tertiary tholeiites from eastern Iceland (Wood, 1976).

Tertiary Plateau Provinces

A 2 km thick sequence of Tertiary tholeiites from West Iceland was described by Bailey and Noe-Nygaard (1976). These plagioclase-phyric lavas include both olivine-normative and quartz-normative varieties, which are more magnesian than the tholeiitic suites of the Faeroe Islands and East Greenland, and have higher Fe and Mg contents than ocean floor tholeiites. No evidence of a low-pressure fractionation trend towards silicic alkalic lavas, such as occurs at Thingmúli volcano in East Iceland (Carmichael, 1964), has been found. The relatively high abundances of large-ion-lithophile elements (Ti, P, K, Rb, Ba, Sr, Zr) led Bailey and Noe-Nygaard (1976) to propose a plume origin for these lavas.

Wood (1976, 1978) and Wood *et al.* (1976) have discussed major and trace element data for a 12-km-thick sequence of Tertiary lavas in East Iceland (Dagley *et al.*, 1967). For these lavas, it is important to evaluate the effects of low grade, burial metamorphism. Si, Mg, K, Rb, Sr, and the light REE exhibit extensive mobility; however, Ti, P, Zr, Y, Hf, Nb and Ta are thought to have been relatively unaffected by zeolitization, based on high degrees of interelement correlation. Wood *et al.* (1976) suggested that the proposed secular variations in $^{87}Sr/^{86}Sr$ ratios, which range from 0.7036 to 0.7033 in basalts erupted over a 15 m.y. period in Iceland (O'Nions and Pankhurst, 1973), may be seriously compromised by the effects of zeolitization because of the high mobility of Rb and Sr during the alteration process. Because the more refractory elements are less affected by metasomatic transport, they are more likely to reflect the original chemistry of the lava pile.

The rocks range from olivine tholeiites to tholeiitic andesites, and are aphyric, plagioclase-phyric or olivine-phyric. An average composition is given in Table 1.2.3.6. The more primitive lavas are concentrated towards the upper sections of the sequence, with the more evolved rocks being relatively older. This apparent temporal relationship, however, has been interpreted in terms of a spatial variation model in which the dominant, more primitive upper lavas represent the edges of eruptive units (equivalent to shield volcanoes) whereas the lower, more evolved tholeiites are derived from the centers of such units and are, therefore, equivalent to fissure eruptions associated with tensional stresses. The older flows bear a striking resemblance in major and trace element chemistry to the Fe-Ti rich lavas of the Skagi Zone (see Fig. 1.2.3.12), whereas those from the upper section are similar to some of the less evolved Mg-Al rich basalts of the Reykjanes Peninsula. Both rock types are present throughout the East Iceland lava pile.

Major and trace element variations across a single 11-m-thick plagioclase-pyroxene-phyric basalt flow from the East Iceland sequence were reported by Watkins *et al.* (1970), Hart *et al.* (1971) and Lindstrom and Haskin (1981). Whereas some of the observed variations must be ascribed to low grade alteration, Lindstrom and Haskin (1981) conclude that much of the variation (including that observed for the light REE) is the result of short-range segregation of crystals from residual liquid during cooling of this undifferentiated flow.

1.2.3.6 Greenland Province

The absolute extent of the Tertiary volcanic province in Greenland is unknown because of the ice cap. However, thick near-horizontal sequences of plateau basalts are exposed in nunataks and fjords along the coastal margins. Estimates of the distribution of lavas in East Greenland suggest an aerial extent of 16,000 km^2 and a thickness of ~700 m to the north of Scoresby Sund. In addition to this expanse, there is an established basalt cover of 60,000 km^2 with a thickness of 6.5 m to the south of Scoresby Sund (Noe-Nygaard, 1974). On the west coast of Greenland, there is evidence for the erosional removal of approximately 5 m of basalt, and the present-day coverage consists of 40,000 km^2 and a measured thickness of 3 m (Noe-Nygaard, 1974). It is upon the West Greenland basalts that most attention has been focussed, because of the occurrence of large concentrations of metallic Fe-Ni-Co alloys (e.g., Noe-Nygaard, 1974; Pedersen, 1977; Ulff-Møller, 1977). These native alloys formed by reduction of iron-bearing silicates, oxides and sulfides due to the incorporation and assimilation of rafted blocks of carbonaceous sediments into partially consolidated lavas.

The basic volcanic rocks of Greenland are predominantly quartz tholeiites and olivine tholeiites, with subordinate olivine basalts and picrites. These lavas range from massive flows (10–50 m in thickness) to thin aa and pahoehoe flow units that were extruded subaerially. There are few exposures of the underlying bedrock but, where it does crop out, either Archean basement rocks or Cretaceous sediments are present. The age of the Greenland volcanic province has been determined to be between 55 and 66 m.y. (Beckinsale *et al.*, 1970).

A suite of aphyric and plagioclase-phyric samples from south of Scoresby Sund, East Greenland (Fawcett *et al.*, 1973) shows a restricted range in major element compositions (Table 1.2.3.7); the oxidation ratio is relatively high, K_2O contents are relatively low, and TiO_2 contents are relatively high. These tholeiitic magmas have undergone some differentiation to yield the observed range of SiO_2 values (46.0–49.5 wt.%); however, there are no systematic variations observed as functions of stratigraphy (Fawcett *et al.*, 1973). An important sample from this collection is a basaltic glass which contains plagioclase (An_{88}), olivine (Fo_{75}) and minor opaque minerals. Portions of the glass are hydrated, but a major element analysis of unaltered glass is comparable to those of the rock samples.

Rasmussen *et al.* (1976) provided convincing evidence that dredge-hauls obtained from Kap Walløe, along the southeastern continental margin of Greenland, are representative samples (ice-rafted) of the Tertiary plateau basalt province of East Greenland. Of the 31 samples collected, 15 are fresh porphyritic or aphyric olivine tholeiites; the remainder are altered and include

Table 1.2.3.7 Average compositions of basalts from Greenland and Baffin Island.

Locality	Scoresby Sund	Svartenhuk			Baffin Island	
Type	Olivine tholeiite	Quartz tholeiite	Olivine tholeiite	Picritic tholeiite	Olivine tholeiite	Picritic tholeiite
SiO_2	49.0	48.4	46.8	42.8	47.5	44.0
TiO_2	1.9	2.30	1.51	0.97	0.97	0.58
Al_2O_3	13.8	14.2	12.9	7.3	13.8	8.3
Cr_2O_3	—	0.03	0.11	0.24	0.13	0.36
Fe_2O_3	6.2	4.3	2.9	2.7	3.3	2.2
FeO	8.2	8.0	8.0	8.7	6.7	8.8
MnO	0.2	0.20	0.17	0.17	0.17	0.19
MgO	6.6	5.8	11.2	25.7	11.8	26.0
NiO	—	0.01	0.05	0.20	0.04	0.19
CaO	11.1	11.8	11.6	6.9	11.7	7.3
Na_2O	2.3	2.55	2.00	1.08	1.53	0.90
K_2O	0.3	0.38	0.25	0.21	0.10	0.06
P_2O_5	0.2	0.26	0.18	0.12	0.10	0.07
H_2O^+	—	1.03	2.25	2.65	2.20	0.97
SUM	99.8	99.3	99.9	99.7	100.0	99.9

Data from Clarke (1970) and Fawcett *et al.* (1973)

pyroclastics, dolerites, and rhyodacites. In general, the unaltered suite corresponds chemically to the less differentiated sequence studied by Fawcett *et al.* (1973).

The major element compositions of the dominant volcanic rocks from East Greenland are similar to those of rocks from West Greenland in the region of Svartenhuk (Clarke, 1970) and Disko Island (Pedersen, 1977), which suggests that these lavas form a single, continuous province. However, two major differences of considerable interest are present. The first is the association of carbon-contaminated lavas on Disko Island, and the second is the presence of picritic and olivine basalts at Svartenhuk. Furthermore, lavas very similar to the Svartenhuk suite occur on the coast of Baffin Island, Canada, to the west (Clarke, 1970; see Table 1.2.3.7), and also have been dredged from Baffin Bay (Clarke, 1975).

Clarke (1970) found that the variation in major elements (including Ni) for the Svartenhuk and Baffin Island lavas could be accounted for by olivine fractionation of proposed primary picritic magmas (see Fig. 1.2.3.13). However, the two suites differ in their abundances of large-ion-lithophile elements (Clarke, 1970; O'Nions and Clarke, 1972), which suggests differences in their degrees of partial melting. The Svartenhuk lavas are light REE-enriched, and have higher abundances of Ba, Rb, Sr, Y and Zr than the Baffin Island lavas. The REE patterns for the Baffin Island tholeiites range from flat to light REE-depleted and closely resemble patterns for MORBs. The $^{87}Sr/^{86}Sr$ ratios (0.7030–0.7034) are unusually low for continental flood tholeiites, and imply negligible crustal contamination for these particular suites (O'Nions and Clarke, 1972).

Fawcett *et al.*(1973) noted that none of the major element data for East Greenland tholeiites or other basalts corresponds to those of the chilled margins of the Skaergaard intrusive complex. In contrast, there is a striking similarity in TiO_2, K_2O, and P_2O_5 contents between the inferred Skaergaard liquid and the olivine basalts from Svartenhuk (Clarke, 1970), which raises

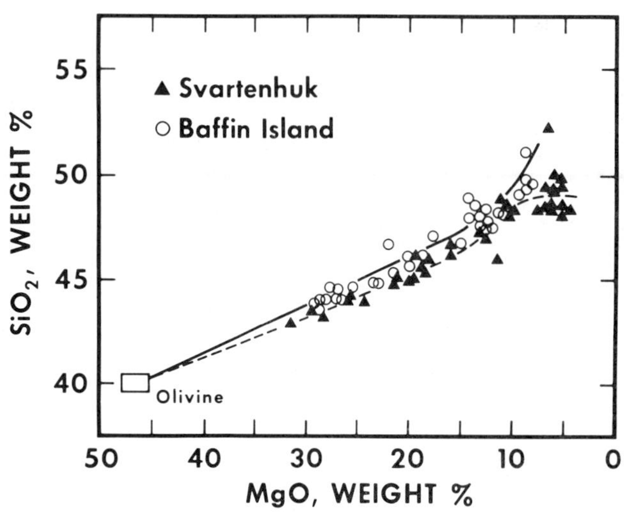

1.2.3.13 Variation in SiO_2 content as a function of MgO content for basalts from Svartenhuk (West Greenland) and Baffin Island (Canada), demonstrating the importance of olivine fractionation in their evolution. After Clarke (1970).

the question of whether the Greenland plateau basalt sequence had one or multiple sources. The major difference is in the high Al_2O_3 content of the Skaergaard chilled margin (17.2 wt.%), and the relatively lower Al_2O_3 concentrations of the East and West Greenland basalt suites (12–14 wt.%). The question of single vs. multiple source regions for the basalts remains unanswered, and their relationship to layered intrusive complexes is equally uncertain.

1.2.3.7 Deccan Traps Province

The Deccan Traps now cover an area of approximately 500,000 km^2 in central and northwest India (Fig. 1.2.3.14). The lava pile varies in thickness from approximately 2000 m along the coastal margin to 100–200 m at its eastern most extent (Sukheswala and Poldervaart, 1958). Although three stratigraphic horizons are recognized, based on intercalated ash and freshwater sedimentary deposits, only two distinct compositional sequences are present. The Lower Traps are uniformly quartz-normative tholeiites that occupy approximately 80% of the east and central regions of the plateau. In contrast, the Upper Traps to the west contain quartz tholeiites, olivine tholeiites, olivine basalts, alkali olivine basalts, nephelinites, carbonatites, ultramafic rocks, intermediate and acid differentiates, and pyroclastics in association with comagmatic gabbros, nepheline syenites, diorites and monzonites (Bose, 1972; Ghose, 1976). The SiO_2-undersaturated suites are broadly aligned along two major rift zones, the Narbada-Son and the Cambay grabens, which are younger than the Trap tholeiites and olivine basalts. Further details of field relations, petrography and other observations are given in Aswathanaraiana (1972).

1.2.3.14 Map showing the distribution of lavas in the Deccan Traps in central and northwest India. After Ghose (1976).

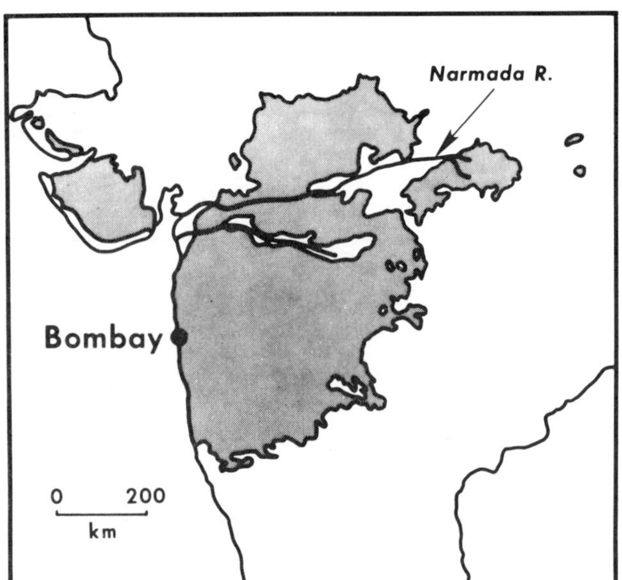

Basaltic volcanism is estimated to have occurred between 65 and 50 m.y. ago (Wellman and McElhinny, 1970; Molnar and Francheteau, 1975), although eruption of the carbonatites and other SiO_2-undersaturated volcanics may have continued until at least 35 m.y. ago (Deans and Powell, 1968). Both fissure and central vents are recognized in the Deccan Traps, with the former type being dominantly responsible for tholeiitic volcanism and the latter for the undersaturated suites. The variations in thickness and in age of the Deccan Traps from east (Lower Traps) to west (Upper Traps) are considered to have resulted from the progressive migration of the magma source over a period of time that was synchronous with the evolution of the Indian Ocean, phases of Himalayan upheaval, and the northward drift of the Indian subcontinent (Ghose, 1976).

Major element compositions of tholeiites and associated differentiated rocks from the Upper and Lower Traps were given by Washington (1922), Sukheswala and Poldervaart (1958) and Bose (1972), and were summarized by Ghose (1976). Representative basalt compositions are listed in Table 1.2.3.8. Both the Lower and Upper Traps are quartz-normative tholeiites that are distinctive in their relatively high TiO_2 contents (~2.5 wt.%). Upper Trap tholeiites are more aluminous than those of the Lower Traps (15.1 wt.% vs. 12.7 wt.%

Table 1.2.3.8 Average major element composition of basalts from the Deccan Traps.†

	Lower Traps (44 samples)	Upper Traps (65 samples)
SiO_2	49.65	48.37
TiO_2	2.66	2.18
Al_2O_3	12.71	15.10
Fe_2O_3	3.70	5.14
FeO	9.93	7.37
MnO	0.21	0.20
MgO	5.31	5.28
CaO	10.19	9.65
Na_2O	2.35	2.91
K_2O	0.64	0.94
P_2O_5	0.31	0.37
H_2O^+	1.93	1.75
H_2O^-	0.57	0.75
SUM	100.16	100.01

†From Ghose (1976).

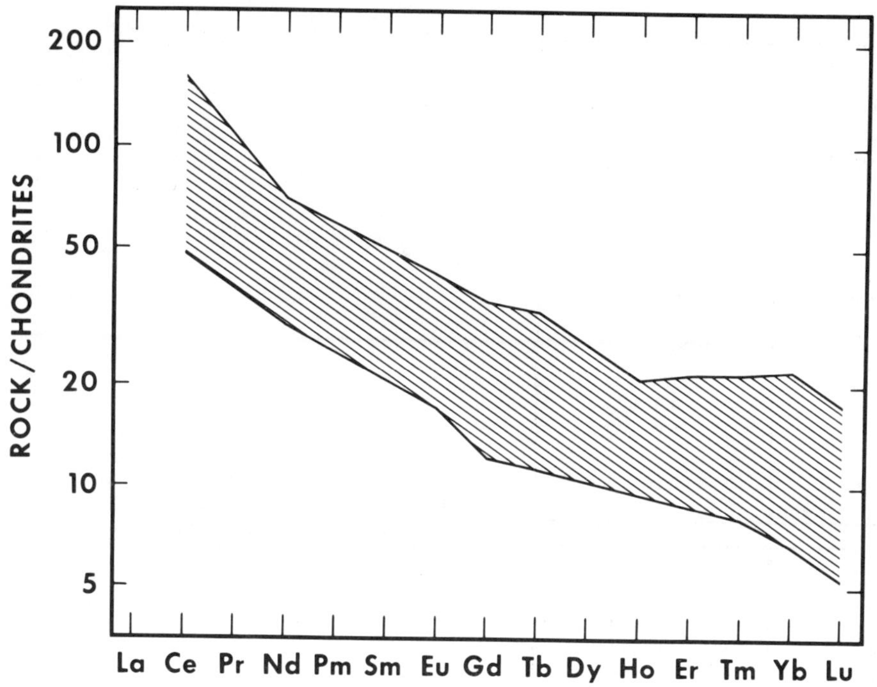

1.2.3.15 Chondrite-normalized rare-earth element distributions for 22 basalts from the Deccan Traps. Data from Frey *et al.* (1968) and Alexander and Gibson (1977).

Al_2O_3) and are also slightly more oxidized (5.14 wt.% vs. 3.70 wt.% Fe_2O_3), but contain less total iron. The basalts show a general increase in total alkalies, in SiO_2/MgO, and in $(Na_2O + K_2O)/CaO$ ratios from east to west, suggesting that the younger Upper Traps became progressively more enriched in alkalies. Normative orthoclase is higher in the Upper flows, whereas total normative pyroxene is greater in the Lower flows (Bose, 1972; Ghose, 1976).

Picritic basalts from near the base of the section in western India have MgO contents as high as 22.4 wt.%. Krishnamurthy and Cox (1977) concluded that these rocks represent liquids with about 16 wt.% MgO which have been enriched in cumulus olivine to a variable extent. Krishnamurthy and Cox further suggested that such picritic magmas, despite their comparative rarity, are parental to the tholeiitic lavas of the province.

Rare-earth element data obtained by Frey *et al.* (1968) and Alexander and Gibson (1977) are illustrated in Fig. 1.2.3.15. The patterns are consistently light REE-enriched with small Eu anomalies, and closely resemble patterns for many of the lavas of the Columbia River Province (cf. Fig. 1.2.3.4). There are further similarities between these two provinces in Sr and Nd isotopic compositions. $^{87}Sr/^{86}Sr$ ratios for ten tholeiite flows analyzed by Alexander and Paul (1977) range from 0.7039–0.7084. Combined Sr and Nd isotopic data for other samples (Macdougall *et al.*, 1981) mostly plot near the mixing curve for the Columbia River basalts (Fig. 1.2.3.7), and provide further evidence in both of these flood basalt provinces for contamination of either the magmas or their depleted mantle sources by continental crustal material.

1.2.3.8 Synthesis of major element variations

In order to compare lavas from the different flood basalt provinces discussed above, a total of 1488 major element analyses were compiled from the literature. In general, analyses obtained by instrumental techniques were preferred wherever possible. An unfortunate aspect of the literature for some provinces is that averages of analyses and variation diagrams are the major sources of data available. Although some data trends were transferable, the data coverage is not uniform among the various provinces and is probably not completely representative for some individual provinces. Despite this, the oxide variation diagrams presented in this section (Figs. 1.2.3.16–20) display a number of linear correlations, population clusters and data dispersions. In general, although there are differences among the provinces in detail, there are also characteristic major element signatures which must reflect common processes in the formation of continental flood basalts during the Tertiary.

A feature of most provinces is the relatively low Mg′ of the most primitive lavas compared with other voluminous terrestrial tholeiites for the ocean floor and oceanic islands (e.g., Fig. 1.2.3.16). Exceptions are the picritic lavas from Greenland and the Deccan Traps,

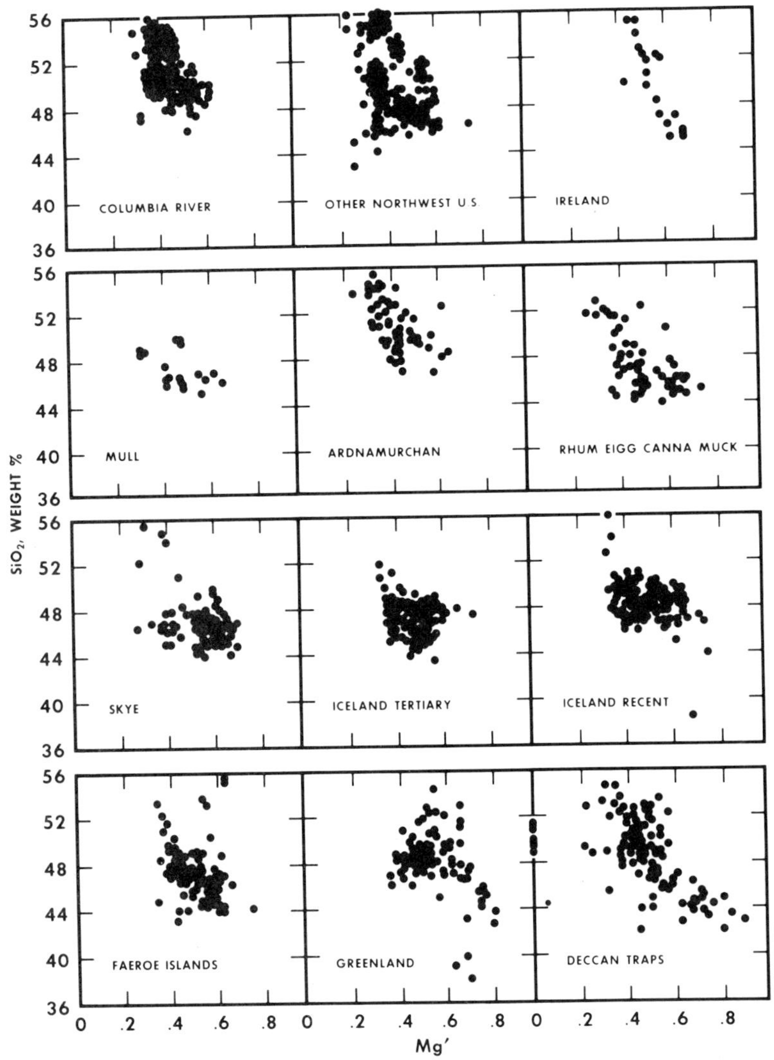

1.2.3.16 Variation in SiO_2 content as a function of Mg/Mg+ΣFe for Tertiary flood basalts.

although the possibility of olivine accumulation in these rocks cannot be ignored. Among pre-Tertiary flood basalts, there is a similar predominance of relatively low-Mg′ basalts (see section 1.2.2), yet more Mg-rich picritic lavas are known in the Lebombo Monocline area of the Karroo Province. The Deccan Traps also display a continuous trend towards the most evolved lavas, although there is considerable scatter in the data. There are differences among the various provinces in the extremes of oxide concentrations, but the overall trends are remarkably similar.

The most alkali-rich lavas occur in the Deccan Traps and some alkali enrichment is present in lavas from Mull, Skye, Iceland and northwest U.S. provinces marginal to the Columbia River Province (Fig. 1.2.3.17). With respect to TiO_2 (Fig. 1.2.3.18), there is a good negative correlation with Mg′ for lavas from Iceland, less distinct correlations exist for the British Isles and Columbia River Provinces, and there is considerable scatter for the Deccan Traps basalts. CaO-Mg′ correlations shown in Fig. 1.2.3.19 are for the most part positive, but some tendency towards a negative relationship (Greenland) or towards scattering (Deccan Traps) is also apparent. Although not illustrated, overall trends of Al_2O_3 contents as functions of Mg′ are mostly positive, with the most aluminous basalts occurring in the northwest U.S. provinces; nevertheless, broadly negative Al_2O_3-Mg′ correlations are exhibited by basalt suites from Mull and the Deccan Traps. Finally, in the AFM plots of Fig. 1.2.3.20, lavas from Skye, Iceland and the Deccan Traps show the most pronounced iron and alkali enrichment trends, whereas the remaining suites are characterized by an absence of significant alkali enrichment.

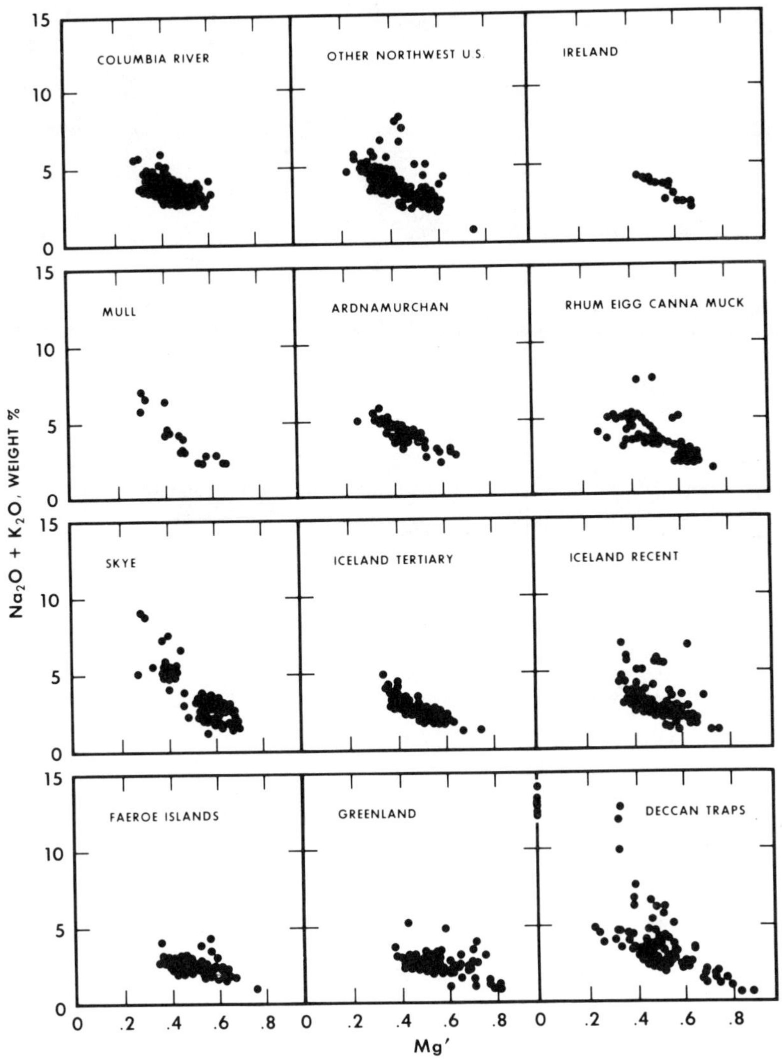

1.2.3.17 Variation in ($Na_2O + K_2O$) contents as a function of $Mg/Mg+\Sigma Fe$ for Tertiary flood basalts.

1.2.3.9 Petrogenesis

The complexity of models which appear to be required to explain the origin and evolution of continental flood basalts has already been mentioned in section 1.2.2. In addition to processes of partial melting, fractional crystallization and possibly magma mixing, which are important for other terrestrial tholeiitic suites, the effects of contamination of flood basalt magmas or their sources by continental crustal material must be evaluated. The occurrence of flood basalts as sequences of flat-lying ponded flows allows important information on eruptive volumes, extrusion rates and stratigraphic correlations of chemistry and petrology to be obtained with greater certainty than for most other terrestrial basalts. Despite this, there is lack of agreement on the relative importance of the processes involved.

A current debate exists as to whether the majority of continental flood basalts are primary or near-primary magmas, or whether they have been derived by fractional crystallization of primary picritic magmas. This type of debate is not unique to flood basalts but has been waged for ocean-floor and other basalts as well (see sections 1.2.5 and 1.4.2). In this regard it should be stated that, although rocks from different flood basalt provinces have many similarities and are conveniently grouped together for description, it is possible that not all flood basalts are created equal.

Discussion in this section includes a summary of key observations on Tertiary continental flood basalts

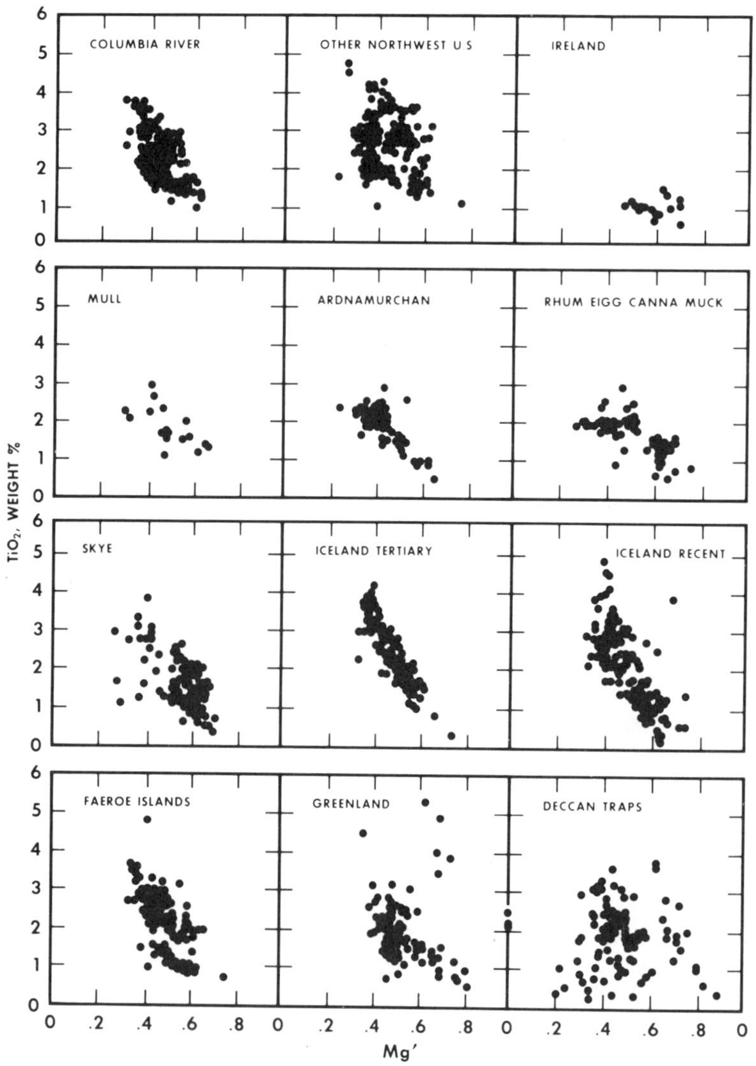

1.2.3.18 Variation in TiO_2 contents as a function of Mg/Mg+ΣFe for Tertiary flood basalts.

and an evaluation of the two major classes of petrogenetic models. Much of this discussion is also relevant to pre-Tertiary flood basalts. The treatment is incomplete, but our intent is to highlight some of the perplexing problems encountered in understanding these lavas. The petrogenesis of basalts from Iceland has already been discussed in section 1.2.3.5.

Models for the origin of flood basalts must account for the following important observations:

1. Individual eruptions are commonly characterized by very large volumes and high extrusion rates.
2. The dominant rock type in most provinces is quartz tholeiite. An exception is the Svartenhuk-Baffin Island suite which is mostly olivine tholeiite and picritic basalt. Minor amounts of picritic basalt also occur in the Faeroe Islands, Deccan and Karroo Provinces. Olivine tholeiites also occur in the upper parts of the Faeroe Islands and Deccan sequences. Alkalic to transitional basalts are dominant in the British Isles Province (except in Ireland) and are also present in the Deccan Traps, but are very rare in the Columbia River Province.
3. Large eruptive units tend to be relatively uniform in composition and generally show little evidence of fractional crystallization or of crystal accumulation chemical trends. This and the absence of grabens or calderas associated with voluminous flows in the Columbia River Province were taken by Swanson and Wright (1981) as evidence for negligible high-level crustal storage of magma. Helz (1980), however, presented evidence for crystal storage of the small-volume Martindale basalts.
4. There is no overall progression from more primitive to more evolved lavas with time. Rather, there usually are various degrees of cyclicity of erupted lava types.
5. Phenocrysts, where present, are plagioclase accompanied by clinopyroxene or olivine or both.

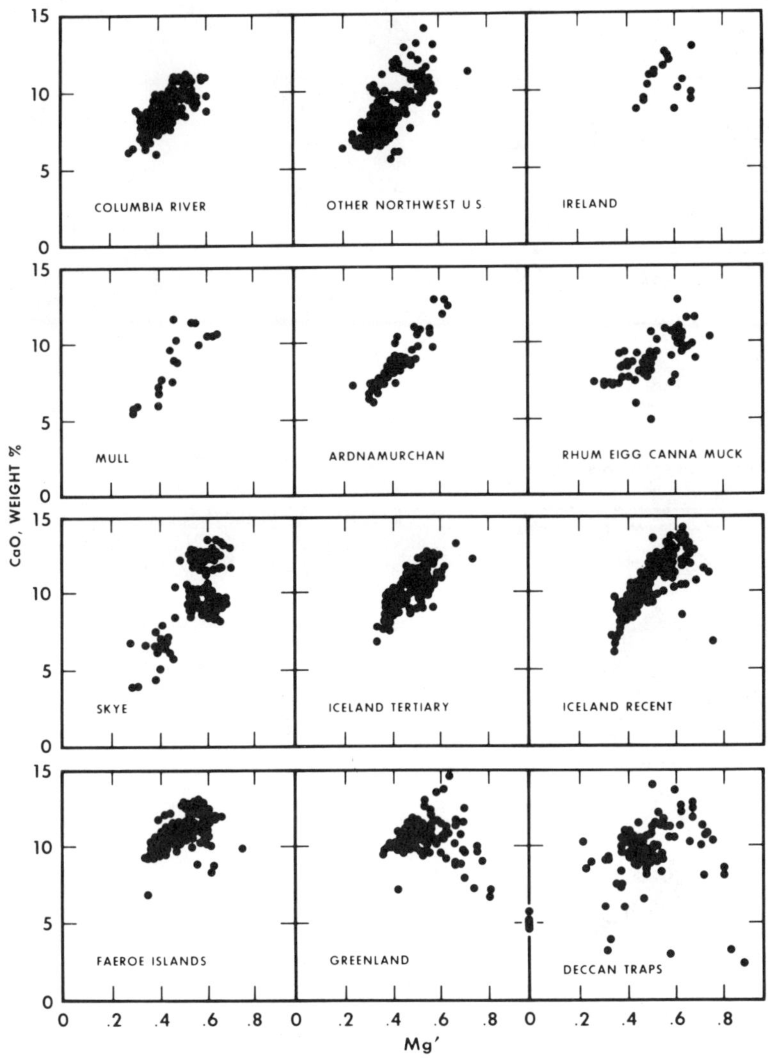

1.2.3.19 Variation in CaO contents as a function of Mg/Mg+ΣFe for Tertiary flood basalts.

There is a notable scarcity of phenocrysts in the lavas of the Columbia River Province. High-pressure xenoliths and megacrysts characteristically are absent from all flood basalt provinces.

6. Where extensive chemical data are available, there is a lack of coherence among major and lithophile trace element abundances. Mixing calculations for *both* major and trace elements are unsuccessful in relating individual Columbia River lava types to each other by fractional crystallization processes (Swanson and Wright, 1981).

7. The most primitive basaltic lavas have Mg′ values which typically are significantly lower than those of other terrestrial tholeiitic lavas [e.g., the maximum Mg/(Mg + ΣFe) observed for the Columbia River Province is 0.60]. If these basalts are primary magmas, then they must have been generated from mantle more Fe-rich than that inferred as the source for ocean-floor and oceanic intraplate tholeiites. The picritic lavas observed in several provinces are more Mg-rich, but whether they are parental to the basalts is debatable in most provinces.

8. The well-defined CaO-MgO correlations in several provinces (Columbia River, Karroo, Paraná) point to the importance of crystal/liquid control by calcic clinopyroxene, during either partial melting or fractional crystallization.

9. Wide ranges in Sr and Nd isotopic ratios are observed in the Columbia River and Deccan Provinces. In the former, there is an irregular trend towards increasing $^{87}Sr/^{86}Sr$ and decreasing $^{143}Nd/^{144}Nd$ with time. For most provinces, $^{87}Sr/^{86}Sr$ values are greater than 0.704; however, lower values are observed for the olivine-rich lavas of the Svartenhuk-Baffin Island suite. These data have been consistently interpreted as indicating crustal contamination, although petrographic evidence of assimilation is apparently lacking.

10. The available isotopic data suggest that the

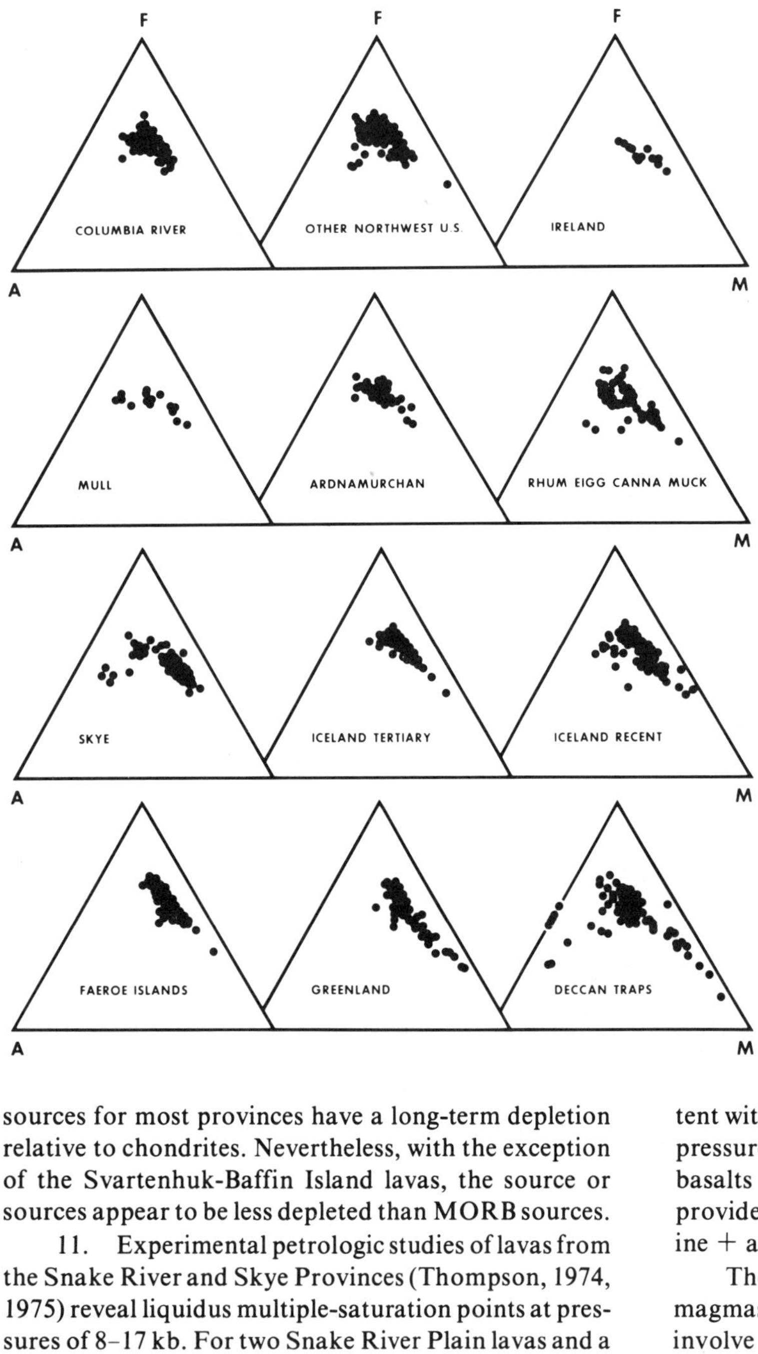

1.2.3.20 AFM plots for Tertiary flood basalts. A = Na_2O+K_2O, F = $FeO+Fe_2O_3$, M = MgO.

sources for most provinces have a long-term depletion relative to chondrites. Nevertheless, with the exception of the Svartenhuk-Baffin Island lavas, the source or sources appear to be less depleted than MORB sources.

11. Experimental petrologic studies of lavas from the Snake River and Skye Provinces (Thompson, 1974, 1975) reveal liquidus multiple-saturation points at pressures of 8–17 kb. For two Snake River Plain lavas and a Skye transitional basalt, the saturating assemblage of olivine + clinopyroxene + plagioclase has been interpreted to have been that which was involved in fractional crystallization of the ascending primary magmas. For a Skye alkali olivine basalt, however, the saturating assemblage is olivine + clinopyroxene + pigeonite at ~17 kb (see Fig. 1.3.26). Thompson (1974) interpreted this as a residual mantle assemblage and suggested that the most primitive Skye alkali olivine basalts are in fact primary magmas. The chemical variation shown by the majority of Skye and Mull lavas is, nevertheless, consistent with fractional crystallization (at both high and low pressures). Experimental studies of the Martindale basalts of the Columbia River Province (Helz, 1980) provide evidence for fractional crystallization of olivine + augite + plagioclase at 3.5 kb.

The debate over primary basalt vs. primary picrite magmas is difficult to resolve because both hypotheses involve assumptions about rock types at depth which are difficult, if not impossible, to test. Proponents of the primary basalt hypothesis (e.g., Thompson, 1975; Swanson and Wright, 1981) are impressed by the total volumes of basalt erupted and the relative compositional uniformity and scarcity of phenocrysts in individual eruptive units, which are taken as evidence of rapid transport from deep (i.e., mantle) reservoirs. The major untestable requirement of this hypothesis is a mantle source region which is more Fe-rich than that inferred for oceanic tholeiites and for many continental and oceanic alkalic basalts.

Samples of the freshest mantle samples available for study, namely peridotite xenoliths in alkalic basalts and kimberlites, are for the most part too Mg-rich to be suitable source materials. A small number of relatively Fe-rich lherzolites do occur in xenolith suites, and Wilkinson and Binns (1977) have proposed these as samples of potential source material for flood basalts. However, there are questions concerning the representativeness of xenoliths as pristine mantle samples (see section 1.2.11), and furthermore there is evidence that at least some relatively Fe-rich lherzolites may be only locally abundant in the upper mantle in the vicinity of pyroxenite dikes produced from ascending magmas (e.g., Irving, 1980).

Other workers (e.g., Clarke, 1970; Cox, 1980) who advocate primary picrite magmas for some or all flood basalt provinces point to the occurrences of picritic lavas (even though they are known to be volumetrically important only in the Svartenhuk and Baffin Island sequences), to some experimental evidence for the expectation of picritic liquids from partial melting of "normal" mantle peridotite (at pressures of $\sim$30kb), and to the magnesian nature of the great majority of peridotite xenoliths which could represent suitable sources. In this hypothesis, the abundant basaltic lavas are to have been derived by extensive, polybaric fractional crystallization of the ascending picritic magmas, which carries the untestable corollary that very large volumes of ultramafic and mafic cumulate rocks must be present beneath flood basalt provinces, perhaps as sill complexes underplated at the crust-mantle boundary (Cox, 1980).

The evidence (from CaO-MgO correlations) for calcic clinopyroxene control in the formation of flood basalts is interpreted differently in each of the two models. Clinopyroxene must be a residual phase in the source region in the primary basalt hypothesis. Swanson and Wright (1981) proposed further that the source was an olivine-poor clinopyroxenite before melting. Such rocks are known among mantle xenolith suites, although they are widely interpreted as high-pressure precipitates from basaltic magmas; nevertheless, the more clinopyroxene-rich xenoliths do possess lower Mg′-values than the typical lherzolites. In the primary picrite hypothesis, clinopyroxene is a major phase in the higher pressure regime of polybaric fractional crystallization. Olivine and plagioclase are also fractionating phases in the model of Cox (1980), although the amount of plagioclase removal must be relatively small to accord with the small Eu anomalies displayed by the basalts.

Most investigators have assumed source regions which are volatile-free. This may not be correct for the subcontinental mantle. Partial melting of peridotite at $\sim$10 kb in the presence of H_2O will produce quartz-normative tholeiitic liquids (e.g., Green, 1973, 1976). If H_2O is involved, then other mobile components must be present as well. This raises the possibility of the operation of metasomatic processes, which have been invoked in the generation of highly alkalic basalts (e.g., Menzies and Murthy, 1980). Metasomatism in the source region could explain some of the incoherence of major element/lithophile element ratios and the overall light REE-enrichment commonly observed.

For either of the models discussed, the isotopic data provide important evidence on the role of continental crustal material. The possibility of assimilation of solid crustal rocks by ascending magmas has been frequently suggested, but bulk chemical and petrographic data do not often support such hypotheses, particularly if the amounts of solids required are large. For the Columbia River Province it is very difficult to generate the extreme Sr and Nd isotopic compositions by assimilation alone. The possibility remains, however, of involvement of lower continental crust in the source region(s), that is, mixing of crust and mantle sources. In such a model H_2O may not be required, and source heterogeneities created by incomplete mixing (coupled possibly with dynamic partial melting) could explain most of the spatial and temporal variability of erupted lavas.

If the common basalts are produced in this way, and erupt with little fractional crystallization, then it is still necessary to explain the picrites and olivine-rich tholeiites of some provinces. The isotopic and REE data for olivine-rich tholeiites from Svartenhuk and Baffin Island imply more primitive source regions which closely resemble sources for ocean-floor basalts. In many respects these rocks are atypical of continental flood basalt provinces. Some of the picritic lavas in other provinces may be accumulative, as are the more magnesian picrites at Svartenhuk and Baffin Island (Clarke, 1970), whereas others may indeed be primary. Nevertheless, for those who contend that primary magmas should be volumetrically abundant, difficulties with the primary picrite hypothesis will remain for most continental flood basalt provinces.

1.2.3.10 References

Alexander P. O. and Gibson J. L. (1977) Rare earth abundances in Deccan Trap basalts. *Lithos* **10**, 143–147.

Alexander P. O. and Paul D. K. (1977) Geochemistry and strontium isotopic composition of basalts from the eastern Deccan volcanic province, India. *Mineral. Mag.* **41**, 165–172.

Anderson F. W. and Dunham K. C. (1966) The geology of northern Skye. *Mem. Geol. Surv. Scotland.* 216 pp.

Aswathanaraiana U., ed. (1972) *International Symposium on Deccan Trap and Other Flood Eruptions, Proceedings, Part I. Bull. Volcanol.* **35**, 509–1177.

Bailey E. B., Clough C. T., Wright W. B., Richey J. E. and Wilson G. V. (1924) Tertiary and post Tertiary geology of Mull, Loch Aline and Oban. *Mem. Geol. Surv. Great Britain.*

Bailey J. C. and Noe-Nygaard A. (1976) Chemistry of Miocene plume tholeiites from northwest Iceland. *Lithos* **9**, 185–201.

Beckinsale R. D., Brooks C. K. and Rex D. C. (1970) K-Ar ages for the Tertiary of east Greenland. *Bull. Geol. Soc. Denmark* **20**, 27–37.

Beckinsale R. D., Pankhurst R. J., Skelhorn R. R. and Walsh J. N. (1978) Geochemistry and petrogenesis of the early Tertiary lava pile of the Isle of Mull, Scotland. *Contrib. Mineral. Petrol.* **66**, 415–427.

Bose M. K. (1972) Deccan basalts. *Lithos* **5**, 131–145.

Brooks C. K. and Jakobsson S. P. (1974) Petrochemistry of the volcanic rocks of the north Atlantic Ridge System. In *Geodynamics of Iceland and the North Atlantic Area*, (L. Kristjansson, ed.), p. 139–154. Reidel, Dordrecht.

Carlson R. W., Lugmair G. W. and Macdougall J. D. (1981) Crustal influence in the generation of continental flood basalts. *Nature* **289**, 160–162.

Carmichael I. S. E. (1964) The petrology of Thingmúli, a Tertiary volcano in eastern Iceland. *J. Petrol.* **5**, 435–460.

Charlesworth J. K. (1963) *Historical Geology of Ireland.* Oliver and Boyd, Edinburgh. 565 pp.

Clarke D. B. (1970) Tertiary basalts from Baffin Bay: possible primary magma from the mantle. *Contrib. Mineral. Petrol.* **25**, 203–224.

Clarke D. B. (1975) Tertiary basalts dredged from Baffin Bay. *Can. J. Earth Sci.* **12**, 1396–1405.

Cox K. G. (1980) A model for flood basalt vulcanism. *J. Petrol.* **21**, 629–650.

Dagley P., Wilson R. L., Ade-Hall J. M., Walker G. P. L., Haggerty S. E., Sigurgeirsson T., Watkins N. D., Smith P. J., Edwards J. and Grasty E. L. (1967) Geomagnetic polarity zones for Iceland lavas. *Nature* **216**, 25–29.

Deans T. and Powell J. L. (1968) Trace elements and strontium isotopes in carbonatites, fluorites and limestones from India and Pakistan. *Nature* **218**, 750–752.

DePaolo D. J. and Wasserburg G. J. (1976a) Nd isotope variations and petrogenetic models. *Geophys. Res. Lett.* **3**, 249–252.

DePaolo D. J. and Wasserburg G. J. (1976b) Inferences about magma sources and mantle structure from variations of $^{143}Nd/^{144}Nd$. *Geophys. Res. Lett.* **3**, 743–746.

DePaolo D. J. and Wasserburg G. J. (1977) The sources of island arcs as indicated by Nd and Sr isotopic studies. *Geophys. Res. Lett.* **4**, 465–468.

DePaolo D. J. and Wasserburg G. J. (1979) Neodymium isotopes in flood basalts from the Siberian Platform and inferences about their mantle sources. *Proc. Nat. Acad. Sci. U.S.A.* **76**, 3056–3060.

Esson J., Dunham A. C. and Thompson R. N. (1975) Low alkali, high calcium olivine tholeiite lavas from the Isle of Skye, Scotland. *J. Petrol.* **16**, 488–497.

Evans A. L., Fitch F. J. and Miller J. A. (1973) Potassium-argon age determination on some British Tertiary igneous rocks. *J. Geol. Soc. Lond.* **129**, 419–443.

Fawcett J. J., Brooks C. K. and Rucklidge J. C. (1973) Chemical petrology of Tertiary flood basalts from the Scoresby Sund area. *Meddel. Om Grønland* **195**, No. 6.

Frey F. A., Haskin M. A., Poetz J. A. and Haskin L. A. (1968) Rare-earth abundances in some basic rocks. *J. Geophys. Res.* **73**, 6085–6089.

Gariepy C., Ludden J. N., Brooks C. and Machado N. (1981) Trace element, Sr- and Nd-isotopic constraints on the genesis of the Faeroe lava pile (abstract). *EOS (Trans. Amer. Geophys. Union)* **62**, 424.

Ghose M. C. (1976) Composition and origin of Deccan basalts. *Lithos* **9**, 65–73.

Green D. H. (1973) Experimental melting studies on a model upper mantle composition at high pressure under water-saturated and water-undersaturated conditions. *Earth Planet. Sci. Lett.* **19**, 37–53.

Green D. H. (1976) Experimental testing of "equilibrium" partial melting of peridotite under water-saturated, high-pressure conditions. *Can. Mineral.* **14**, 255–268.

Gribble C. D. (1974) The dolerites of Ardnamurchan. *Scott. Jour. Geol.* **10**, 71–89.

Gunn B. M. and Watkins N. (1970) Geochemistry of the Steens Mountain Basalt, Oregon. *Geol. Soc. Amer. Bull.* **81**, 1497–1516.

Harker A. (1904) The Tertiary igneous rocks of Skye. *Mem. Geol. Surv. U.K.*.

Hart S. R., Gunn B. M. and Watkins N. D. (1971) Intra-lava variation of alkali elements in Icelandic basalt. *Amer. J. Sci.* **270**, 315–318.

Hart S. R., Schilling J. -G. and Powell J. L. (1973) Basalts from Iceland and along the Reykjanes Ridge: Sr isotope geochemistry. *Nature Phys. Sci.* **246**, 104–107.

Helmke P. A. and Haskin L. A. (1973) Rare-earth elements, Co, Sc and Hf in the Steens Mountain basalts. *Geochim. Cosmochim. Acta* **37**, 1513–1529.

Helz R. T. (1980) Chemical and experimental study of the Ice Harbor Member of the Yakima Basalt Subgroup: Evidence for intracrustal storage and contamination (abstract). *EOS (Trans. Amer. Geophys. Union)* **61**, 68.

Holden G. S. and Hooper P. R. (1976) Petrology and chemistry of a Columbia River Basalt section, Rocky Canyon, West-Central Idaho. *Geol. Soc. Amer. Bull.* **87**, 215–225.

Holland S. G. and Brown G. M. (1972) Hebridean tholeiitic magmas: a geochemical study of the Ardnamurchan cone sheets. *Contrib. Mineral. Petrol.* **37**, 139–160.

Hooper P. R. (1974) Petrology and chemistry of the Rock Creek flow, Columbia River Basalt, Idaho. *Geol. Soc. Amer. Bull.* **85**, 15–26.

Irving A. J. (1980) Geochemistry of composite ultramafic xenoliths in alkalic basalts and implications for magmatic processes in the mantle. *Amer. J. Sci.*, Jackson vol. **280A**, 389–426.

Jakobsson S. P. (1972) Chemistry and distribution pattern of recent basaltic rocks in Iceland. *Lithos* **5**, 365–386.

Krishnamurthy P. and Cox K. G. (1977) Picrite basalts and related lavas from the Deccan Traps of Western India. *Contrib. Mineral. Petrol.* **62**, 53–75.

Langmuir C. H., Vocke R. D., Jr., Hanson G. N. and Hart S. R. (1978) A general mixing equation with applications to Icelandic basalts. *Earth Planet. Sci. Lett.* **37**, 380–392.

Leeman W. P. (1976) Petrogenesis of McKinney (Snake River) olivine tholeiite in light of rare earth and Cr/Ni distributions. *Geol. Soc. Amer. Bull.* **87**, 1582–1586.

Leeman W. P. and Manton W. I. (1971) Strontium isotopic composition of basaltic lavas from the Snake River plain, southern Idaho. *Earth Planet. Sci. Lett.* **11**, 420–434.

Leeman W. P. and Vitaliano C. J. (1976) Petrology of the McKinney Basalt, Snake River Plain, Idaho. *Geol. Soc. Amer. Bull.* **87**, 1777–1792.

Lindstrom M. M. and Haskin L. A. (1981) Compositional inhomogeneities in a single Icelandic tholeiite flow. *Geochim. Cosmochim. Acta* **45**, 15–33.

Macdougall J. D., Lugmair G. W., Mahoney J., Murali A. V. and Gopalan K. (1981) Deccan and Rajmahal Traps, India: further evidence for crustal components in flood basalts (abstract). *EOS (Trans. Amer. Geophys. Union)* **62**, 432.

Mattey D. P., Gibson J. G., Marriner G. F. and Thompson R. N. (1977) The diagnostic geochemistry, abundance, and spatial distribution of high-calcium, low alkali olivine tholeiite dykes in the Lower Tertiary regional swarm of the Isle of Skye, NW Scotland. *Mineral. Mag.* **41**, 273–285.

McDougall I. (1976) Geochemistry and origin of basalt of the Columbia River Group, Oregon and Washington. *Geol. Soc. Amer. Bull.* **87**, 777–792.

McKee E. H., Swanson D. A. and Wright T. L. (1977) Duration and volume of Columbia River basalt volcanism, Washington, Oregon and Idaho (abstract). *Geol. Soc. Amer., Abstracts with Programs* **9**, 463–464.

Menzies M. and Murthy V. R. (1980) Mantle metasomatism as a precursor to the genesis of alkaline magmas—isotopic evidence. *Amer. J. Sci.*, Jackson vol. **280A**, 622–638.

Molnar P. and Francheteau J. (1975) Plate tectonic and palaeomagnetic implications for the age of the Deccan Traps and the magnetic anomaly time scale. *Nature* **255**, 128–130.

Nathan S. and Fruchter J. S. (1974) Geochemical and paleomagnetic stratigraphy of the Picture Gorge and Yakima Basalts (Columbia River Group) in central Oregon. *Geol. Soc. Amer. Bull.* **85**, 63–76.

Nelson D. O. (1980) Strontium isotopic and trace element geochemistry of the Saddle Mountains and Grande Ronde Basalts of the Columbia River Group. Ph.D. Thesis, Oregon State Univ.

Noe-Nygaard A. (1966) Chemical composition of tholeiitic basalts from the Wyville-Thompson Ridge Belt. *Nature* **212**, 272–273.

Noe-Nygaard A. (1974) Cenozoic to Recent volcanism in and around the North Atlantic Basin. In *The Ocean Basins and Margins, Vol. 2. The North*

Atlantic, (A. E. M. Nairn and F. G. Stehli, eds.), p. 391–443. Plenum Press, New York.

Noe-Nygaard A. and Rasmussen J. (1968) Petrology of a 3,000 meter sequence of basaltic lavas in the Faeroe Islands. *Lithos* **1**, 286–304.

O'Nions R. K. and Clarke D. B. (1972) Comparative trace element geochemistry of Tertiary basalts from Baffin Bay. *Earth Planet. Sci. Lett.* **15**, 436–446.

O'Nions P. K. and Pankhurst R. J. (1973) Secular variation in the Sr-isotope compositions of Icelandic volcanic rocks. *Earth Planet. Sci. Lett.* **21**, 13–21.

O'Nions R. K., Pankhurst R. J. and Grönvold K. (1976) Nature and development of basalt magma sources beneath Iceland and the Reykjanes Ridge. *J. Petrol.* **17**, 315–338.

Pálmason G. and Saemundsson K. (1974) Iceland in relation to the Mid-Atlantic Ridge. *Ann. Rev. Earth Planet. Sci.* **2**, 25–50.

Pankhurst R. J. and O'Nions R. K. (1973) Determination of Rb/Sr and $^{87}Sr/^{86}Sr$ ratios of some standard rocks and evaluation of X-ray fluorescence spectrometry in Rb-Sr geochemistry. *Chem. Geol.* **12**, 127–136.

Patterson E. M. (1951) A petrochemical study of the Tertiary lavas of northeast Ireland. *Geochim. Cosmochim. Acta* **2**, 283–299.

Patterson E. M. and Swaine D. J. (1955) A petrochemical study of Tertiary tholeiitic basalts: the middle layer of the Antrim Plateau. *Geochim. Cosmochim. Acta* **8**, 173–181.

Pedersen A. K. (1977) Iron-bearing and related volcanic rocks in the area between Gieseckes Dal and Hammers Dal, North-west Disko. *Rapp. Grønlands Geol. Unders.* **81**, 5–14.

Powers H. A. (1960) A distinctive chemical characteristic of Snake River basalts of Idaho. *U. S. Geol. Survey Prof. Paper* **400**, p. 268.

Rasmussen M. G., Campsie J., Dittmar F. B. and Bailey J. C. (1976) Basalts from the south eastern Greenland continental margin. *Bull. Geol. Soc. Denmark*, **25**, 73–78.

Richey J. E. and Thomas H. H. (1930) The geology of Ardnamurchan, northwest Mull, and Coll. *Mem. Geol. Surv. U. K.*

Ridley W. I. (1971a) The petrology of some volcanic rocks from the British Tertiary Province: the islands of Rhum, Eigg, Canna, and Muck. *Contrib. Mineral. Petrol.* **32**, 251–266.

Ridley W. I. (1971b) Some chemical features of basalts from the British Tertiary Province. *Earth Planet. Sci. Lett.* **11**, 435–439.

Schilling J. G. (1973) Iceland mantle plume: geochemical study of Reykjanes Ridge. *Nature* **242**, 565–571.

Schilling J. G. and Noe-Nygaard A. (1974) Faeroe-Iceland plume: rare-earth evidence. *Earth Planet. Sci. Lett.* **24**, 1–14.

Schilling J. G., Sigurdsson H. and Kingsley R. H. (1978) Skagi and Western neovolcanic zones in Iceland: 2. Geochemical Variations. *J. Geophys. Res.* **83**, 3983–4002.

Schmincke H. U. (1967) Stratigraphy and petrography of four Upper Yakima basalt flows in south-central Washington. *Geol. Soc. Amer. Bull.* **78**, 1385–1422.

Sigurdsson H., Schilling J. G. and Meyer P. S. (1978) Skagi and Langjökull volcanic zones in Iceland: petrology and structure. *J. Geophys. Res.* **83**, 3971–3982.

Sigvaldason G. E. (1969) Chemistry of basalts from the Icelandic rift zone. *Contrib. Mineral. Petrol.* **20**, 357–370.

Sigvaldason G. E. (1974) Basalts from the center of the assumed Icelandic mantle plume. *J. Petrol.* **15**, 497–524.

Sigvaldason G. E. and Steinthorsson S. (1974) Chemistry of tholeiitic basalts from Iceland and their relation to the Kverkfjoll hot spot. In *Geodynamics of Iceland and the North Atlantic Area* (L. Kristjansson, ed.), p. 155–164. Reidel, Dordrecht.

Snavely P. D., Jr., MacLeod N. S. and Wagner J. C. (1968) Tholeiitic and alkali basalts of the Eocene Siletz River Volcanics, Oregon Coast Range. *Amer. J. Sci.* **266**, 464–481.

Snavely P. D., Jr., MacLeod N. S. and Wagner H. C. (1973) Miocene tholeiitic basalts of coastal Oregon and Washington and their relations to coeval basalts of the Columbia River Plateau. *Geol. Soc. Amer. Bull.* **84**, 387–424.

Stewart F. H. (1965) Tertiary igneous activity. In *The Geology of Scotland*, (G. Y. Craigh, ed.), Chapter 13, p. 420–466. Oliver and Boyd, Edinburgh.

Stout M. Z. and Nichols J. (1977) Mineralogy and petrology of Quaternary lavas from the Snake River Plain, Idaho. *Can. J. Earth Sci.* **14**, 2140–2156.

Sukeswala R. N. and Poldervaart A. (1958) Deccan basalts of the Bombay area, India. *Geol. Soc. Amer. Bull.* **69**, 1475–1494.

Swanson D. A. and Wright T. L. (1978) Bedrock geology of the northern Columbia Plateau and adjacent areas. In *The Channeled Scabland*, (V. R. Baker and D. Nummedal, eds.), p. 37–57. NASA Office Space Sci., Washington, D.C.

Swanson D. A. and Wright T. L. (1981) Guide to geologic field trip between Lewiston, Idaho and Kimberly, Oregon, emphasizing the Columbia River Basalt Group. *U. S. Geol. Survey Circular 838*, 1–14.

Swanson D. A., Wright T. L. and Helz R. T. (1975) Linear vent systems and estimated rates of magma production and eruption for the Yakima Basalt on the Columbia Plateau. *Amer. J. Sci.* **275**, 877–905.

Swanson D. A., Wright T. L., Hooper P. R. and Bentley R. D. (1979) Revisions in stratigraphic nomenclature of the Columbia River Basalt Group. *U. S. Geol. Survey Bull. 1457-G*, 59 pp.

Tarling D. H. and Gale N. H. (1968) Isotopic dating and paleomagnetic polarity in the Faeroe Islands. *Nature* **218**, 1043–1044.

Thompson R. N. (1974) Primary basalts and magma genesis I. Skye, northwest Scotland. *Contrib. Mineral. Petrol.* **45**, 317–341.

Thompson R. N. (1975) Primary basalts and magma genesis II. Snake River Plain, Idaho, U.S.A. *Contrib. Mineral. Petrol.* **52**, 213–232.

Thompson R. N., Esson J. and Dunham A. C. (1972) Major element chemical variations in the Eocene lavas of the Isle of Skye, Scotland. *J. Petrol.* **13**, 219–253.

Tilley C. E. and Muir I. D. (1962) The Hebridean plateau magma type. *Edinburgh Geol. Soc. Trans.* **19**, 208–215.

Ulff-Møller F. (1977) Native iron bearing intrusions of the Hammers Dal Complex, Northwest Disko. *Rapp. Grønlands Geol. Unders.* **81**, 15–33.

Walker G. P. L. (1965) Evidence of crustal drift from Icelandic geology. *Phil. Trans. Roy. Soc. Lond.* **A258**, 199–204.

Washington H. S. (1922) Deccan Traps and other plateau basalts. *Geol. Soc. Amer. Bull.* **33**, 765–804.

Waters A. C. (1961) Stratigraphic and lithologic variation in the Columbia River Basalt. *Amer. J. Sci.* **259**, 583–611.

Waters A. C. (1962) Basalt magma types and their tectonic associations: Pacific Northwest of the United States. *Amer. Geophys. Union Geophys. Mon.* **6**, 158–170.

Watkins N. D. and Baksi A. K. (1974) Magnetostratigraphy and oroclinal folding of the Columbia River, Steens, and Owyhee basalts in Oregon, Washington, and Idaho. *Amer. J. Sci.* **274**, 148–189.

Watkins N. D., Gunn B. M. and Coy-yll R. (1970) Major and trace element variations during the initial cooling of an Icelandic lava. *Amer. J. Sci.* **268**, 24–49.

Wellman P. and McElhinny M. W. (1970) K-Ar ages of the Deccan Traps, India. *Nature* **227**, 595–596.

Wilkinson J. F. G. and Binns R. A. (1977) Relatively iron-rich lherzolite xenoliths of the Cr-diopside suite: a guide to the primary nature of anorogenic tholeiitic andesite magmas. *Contrib. Mineral. Petrol.* **65**, 199–212.

Wood D. A. (1976) Spatial and temporal variation in the trace element geochemistry of the Eastern Iceland flood basalt succession. *J. Geophys. Res.* **81**, 4353–4360.

Wood D. A. (1978) Major and trace element variations in the Tertiary lavas of eastern Iceland and their significance with respect to the Iceland geochemical anomaly. *J. Petrol.* **19**, 393–436.

Wood D. A. (1979) Dynamic partial melting: its application to the petrogeneses of basalts erupted in Iceland, the Faeroe Islands, the Isle of Skye (Scotland) and the Troodos Massif (Cyprus). *Geochim. Cosmochim. Acta* **43**, 1031–1046.

Wood D. A., Gibson I. L. and Thompson R. N. (1976) Elemental mobility during zeolite facies metamorphism of the Tertiary basalts of Eastern Iceland. *Contrib. Mineral. Petrol.* **55**, 241–254.

Wright T. L., Grolier M. J. and Swanson D. A. (1973) Chemical variation related to the stratigraphy of the Columbia River basalt. *Geol. Soc. Amer. Bull.* **84**, 371–386.

1.2.4 CONTINENTAL RIFT VOLCANISM

1.2.4.1 Introduction and review

Basaltic volcanism is a common manifestation of extensional tectonics within continental plates (see section 6.3). Conventionally, continental rifting and associated rift volcanism are viewed as processes linked to incipient or abortive continental breakup (East African Rift system and Keweenawan of Lake Superior, respectively). Failed rifts generated during otherwise successful continental fragmentation (eastern U.S. Triassic basins) or volcanic fields stranded at Atlantic-type continental margins (Brito-Arctic Tertiary province) also can be regarded as components of this type of rift setting. However, extensional tectonics and volcanism are not limited to such environments. Structural basins develop within continents as a result of continent-continent collisions (Baikal rift, Rhinegraben). In a setting analogous to the inter-arc basins of oceanic plates, extensional zones may develop behind or within continental margin volcanic arcs (Taupo Volcano-Tectonic Depression, Columbia Plateau).

The Rio Grande rift, from which the reference suite was collected, occurs at the eastern margin of the broad zone of late Cenozoic uplift, extensional faulting and basaltic volcanism that characterizes much of the U.S. Cordillera. Prior to the onset of tensional tectonics in the latest Oligocene, apparently subduction-related calc-alkaline magmatism was prevalent across this same broad zone (Lipman *et al.*, 1972; Coney and Reynolds, 1977). The Rio Grande rift, which was initiated immediately following this widespread volcanic episode, occurs just to the east of the Mogollon-Datil and San Juan volcanic fields (Christiansen and Lipman, 1972). It is interpreted as an example of extension behind a waning magmatic arc, at least in its initial (middle Tertiary) evolutionary phase. The rifting is attributed to a decreased rate of convergence and subduction, leading to back-flow in the asthenosphere (Lipman, 1980). Subsequent uplift, extensional faulting, and volcanism are thought to reflect localized upwelling of this asthenosphere and interaction with the overlying lithosphere. The zone of present rifting was mobilized in late Paleozoic (Uncompahgre) and late Mesozoic-early Tertiary (Laramide) times along roughly north-south trends which are regionally subparallel to and locally reactivated by the Cenozoic activity (Chapin and Seager, 1975; Cordell, 1978). Nonetheless, the role of pre-existing structural trends (both compressional and tensional) in controlling rifting worldwide is controversial (e.g., Mohr, 1978; King and Williams, 1976).

Despite the diversity of the plate tectonic interactions which are thought to give rise to intracontinental rifts, they all lead to generally similar structures and volcanism. The unifying feature appears to be similarities in underlying crust-mantle structure among rifts, regardless of the environment of formation. Although some controversy exists as to whether upwelling mantle splits the continent along conveniently weakened zones or the mantle is forced to rise as continents are pulled apart (see sections 6.3.6–7), geophysical evidence demonstrates that asthenosphere upwelling is a fundamental characteristic of continental rifting. Heat flow, seismic refraction data, electrical conductivity, and gravity studies demonstrate the presence of an elevated lithosphere-asthenosphere boundary, abundant basaltic melt, and crustal attenuation beneath rifts. Most continental rifts are characterized by a positive gravity anomaly over the axial graben (after removing the effect of low density, graben-fill sediments) which is superimposed on a residual negative anomaly over the regional uplift. The regional gravity low is thought to be caused by a negative density contrast resulting from the upward protrusion of low density (and low velocity—7.7 or less vs. 8.1 km/sec) mantle. The local positive anomaly is thought to be due to axial intrusions of basaltic magma into the sialic crust (Girdler, 1976; Cordell, 1978; Ramberg *et al.*, 1978).

The volcanic rocks in continental rifts range from extensive flood basalt provinces (Columbia Plateau, Keweenawan of Lake Superior, Deccan Traps, Paraná Basin, Newer Basalts of southeastern Australia) to sporadically distributed small centers of volcanic activity associated with largely amagmatic graben (Rhinegraben). The Rio Grande rift lacks volcanic rocks along some segments, other segments contain voluminous but short-lived volcanic fields. Compared with the classic East African rift system (including Afar-Red Sea-Gulf of Aden), the Rio Grande rift is relatively lacking in volcanic deposits (Gass, 1970). However, some segments of the East African system are characterized by non-volcanic periods and other segments apparently lack volcanic rocks entirely (King and Williams, 1976). Volcanic periodicity in continental rifts is well documented (Baker *et al.*, 1972). Changes in volcanic style can coincide with periods of either tectonic activity or quiescence. Such changes include: petrology of the dominant magma type, relative proportions of mafic and silicic magmas, rate of magma production and/or mode of eruption. The connection between tectonism and volcanism may reflect changes in the state of the mantle beneath the rift (Baker *et al.*, 1972). However, once rifting has been established, episodicity of faulting and volcanism are likely to be second-order effects. As Gass (1975) emphasizes, an upwelling mantle diapir is unlikely to behave episodically because of constraints imposed by the rate of heat conduction.

The character of volcanism commonly varies transversely across rifts, and flank lavas tend to be more alkaline than those in the axial graben. A wide spectrum of magmas from tholeiite to nephelinite and carbonatite may be represented in the spatial progression away from the rift axis, but basalts of contrasting character can also be erupted in close spatial and temporal association, both within the depression and on the flanks.

The basaltic rocks of the East African rift system (and of several other major rifts) are dominantly alkalic (see Sørensen, 1974). Highly silica-undersaturated lavas are voluminous in some areas, but many of the volcanics within the East African axial zone and within the Afar are hypersthene-normative or tholeiitic in character. Alkalic rocks are also common in volcanic regions associated with extensional tectonics in the western U.S.A., including the southern Rio Grande rift and the Basin-Range Province. However, the Rio Grande is the best defined active rift structure in the western U.S.; the Taos Plateau is the largest and most compositionally diverse young basaltic area in the rift; and other major continental rifts, such as the Ethiopian-Afar section of the African rift system, the Keweenawan Province and most other continental flood basalts, are predominantly tholeiitic. Thus, the study of tholeites from the Taos Plateau volcanic field facilitates comparisons with volcanic rock associations present in continental rifts worldwide.

A major problem of rift volcanism, but one which is beyond the scope of this volume, is the origin of volcanics more silicic than basalt. In some rifts (Barberi *et al.*, 1975), basalt appears to be closely related to more silicic differentiates, and derivation of the silicic magmas by fractional crystallization seems likely. In others, the volume, timing, isotopic, and geochemical relations among evolved magma types are not easily interpretable in terms of descent from any sampled basalt. Do the high heat flow and high basalt production rates of rifts allow substantial amounts of the lower crust to be partially melted in order to produce the silicic magma? This topic deserves additional study.

1.2.4.2 Rio Grande rift

Physiography and structure

The Rio Grande rift of New Mexico and Colorado is characterized by structural features, geophysical anomalies, and volcanism typical of worldwide rifts (Fig. 1.2.4.1). In its northern extent (34–41° N), the axial depression (~50 km wide) is a narrow prong separating the Colorado Plateau on the west from the High Plains on the east. The axial rift valleys constitute a series of discrete basins separated by *en echelon* offsets, which in some cases involve local basement highs and activation of transverse structures which appear to be major crustal flaws of Precambrian ancestry.

The Taos volcanic field occupies the southern San Luis Valley (Fig. 1.2.4.2a, b), a complex asymmetric graben which is bounded by a major fault on the east against the uplifted Sangre de Cristo Mountains. Within the San Luis Valley an axial horst trends north-south. Precambrian basement highs, bounded by transverse structures, are present at both the north and south ends of the San Luis Valley. The other basins of the northern Rio Grande rift (Upper Arkansas River, Espanola, and Albuquerque-Belen) also are structurally complex, asymmetric, and bordered by uplifted Precambrian on at least one side.

South of the Albuquerque-Belen basin, the rift broadens and is expressed by several basins over a width of 150 km. The structural style is reminiscent of the Basin-Range province with which the rift merges. However, the higher heat flow, late Cenozoic volcanism, and gravity anomaly of the southern rift (Ramberg *et al.*, 1978; Cordell, 1978) set it apart from the tectonically less active Basin-Range province of southern Arizona, northern Mexico and western Texas.

Transverse lineaments, which in many places coincide with offsets between rift segments, are also the locus of volcanic centers; both within the axial depression and on the rift flanks. The most notable of these is the Springerville-Raton trend of northern New Mexico. The volcanic centers along this trend have been interpreted as the trace of a hot spot (Suppe *et al.*, 1975), but subsequent dating of basalts by Lipman and Mehnert (1979) indicates that this zone has been active all along its 800 km extent over the last 5 m.y. with no discernible temporal-spatial migration of activity. Thus, these major, pre-existing crustal flaws seem important in localizing volcanism but are not plume tracks.

Ages of rifting and volcanism

The inception of the Rio Grande rift was concurrent with the waning of calc-alkaline volcanism about 27 m.y. ago. The onset of extensional faulting and fundamentally basaltic volcanism is documented by 26–28 m.y. old flows capping fanglomerates, both in the north (San Juan Mountains—Lipman and Mehnert, 1975) and the south (Sierra de las Uvas—Chapin and Seager, 1975).

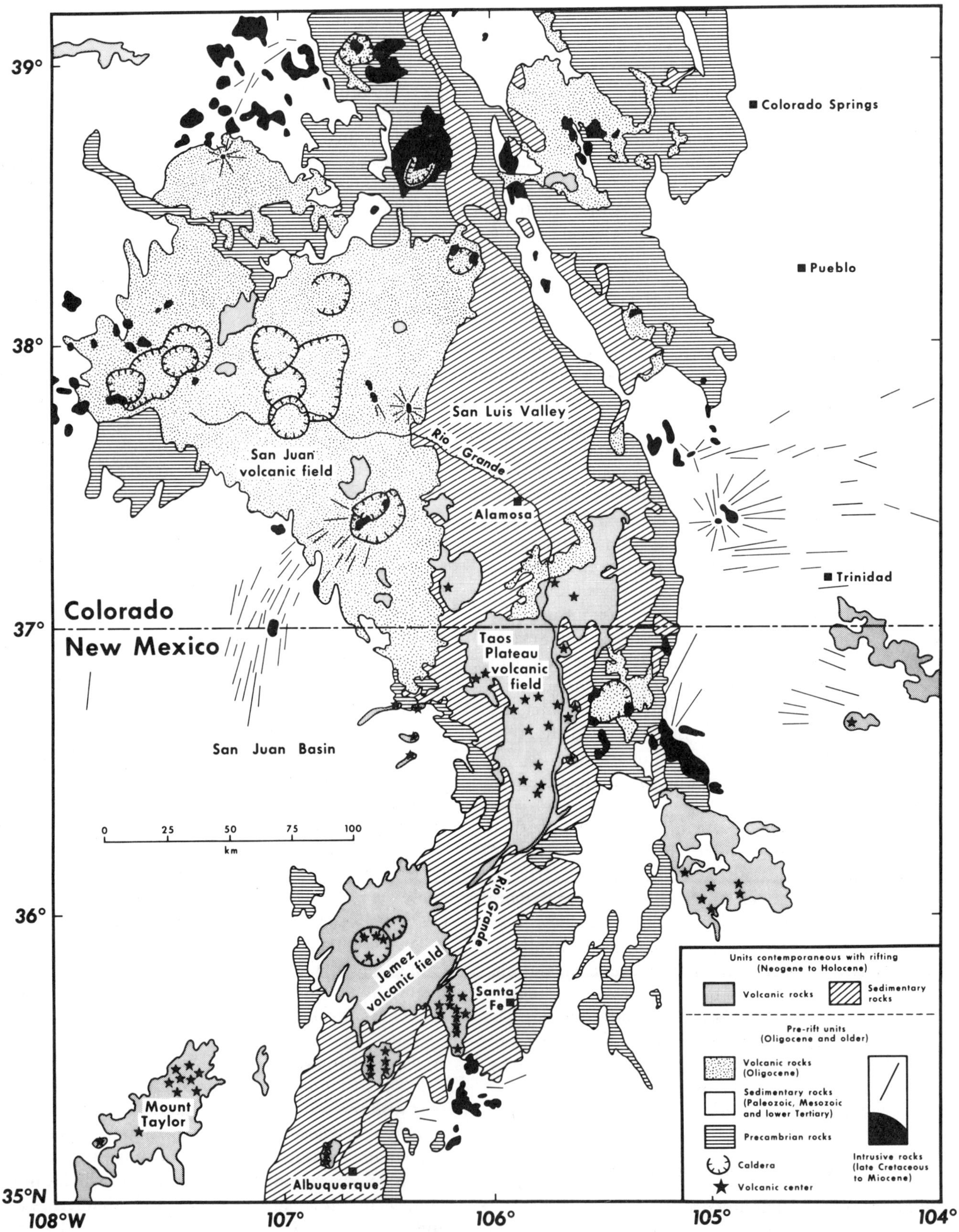

Fig. 1.2.4.1 Generalized geologic map of the northern Rio Grande Rift Region, New Mexico and Colorado. Generalized from Tweto (1978) and Woodward *et al.* (1978).

Andesitic lavas and rhyolitic pyroclastic flows of early-rift age (22–24 m.y.) are present in the Sangre de Cristo Mountains and within the Taos Plateau, exposed on the axial horst (Lipman and Mehnert, 1979). Bachman and Mehnert (1978) have also dated an andesite flow and biotite-bearing pyroclastic deposits in the San Acacia-Joyita Hills area (near Socorro, New Mexico) with early-rift ages (26 m.y.). Rocks and vents of early-rift age are abundant in the Magdelena and Socorro Mountains along the west flank of the rift west of Socorro (Chapin *et al.*, 1978). Compilations of available dates (Lipman and Mehnert, 1975, 1979; Bachman and Mehnert, 1978) demonstrate that magmatic activity has been essentially continuous throughout the history of the rift, but the most active period of basaltic activity has been the last 10 m.y. Within any given area, magmatic activity appears to have been episodic. The relatively short duration of activity within any single volcanic field, compared with the overall time span of rifting, is a characteristic of Rio Grande rift, as well as of many other continental rifts.

Volcanic petrology

With the exception of data generated in this study and by Lipman and co-workers (e.g., Lipman *et al.*, 1978; Lipman and Mehnert, 1975, 1979; Everson and Silver, 1978; Hedge and Lipman, 1972), few integrated studies exist of volcanic rocks for the Rio Grande rift in which isotopic, major and trace element geochemical, and mineral chemical data are available. Thus, igneous events which have accompanied rifting must be compared largely on the basis of major element chemistry.

No broad temporal progression has yet been recognized in the composition of volcanics in the Rio Grande rift, other than the increased proportion of basalts erupted during the last 10 m.y. Rocks which are similar in composition to those erupted during the earliest stages of rifting have erupted throughout history of the rift.

Some significant changes in the compositions of Pliocene and younger volcanics along the axis of the rift correlate with the two-fold structural subdivision of the rift into the narrow northern area and the broader zone south of Socorro. In general, basalts of the northern rift are predominantly tholeiitic, or at least are hypersthene-normative. South of Socorro, hypersthene-normative basalts are present, but nepheline-normative alkalic basalts predominate (Lipman and Mehnert, 1975; Aoki and Kudo, 1976). The Albuquerque-Belen Basin volcanoes range in composition from tholeiitic (San Felipe Mesa and Albuquerque volcanoes) in the north to mildly alkaline in the south (Isleta Volcano and Cat Hills), as documented by Kelley and Kudo (1978). Baldridge (1979a) has shown that the Cerros del Rio field, which occurs on a transverse structural high between the southern Espanola Basin and the Albuquerque-Belen Basin, contains both tholeiites and nepheline-normative basalts.

In the northern Rio Grande rift, basalt compositions vary laterally away from the axial depression. The tholeiitic basalts typical of the axial volcanic fields give way to more alkalic basalts on the flanks. However, basalt types characteristic of either the axis or flanks are found in lesser amounts in the contrasting structural position, where they are closely allied in time and space with the predominant type. As noted above, this is a typical feature of rift volcanism.

The basalts which cap erosion surfaces in the San Juan Mountains, Colorado, are dominantly hypersthene-normative lavas with $SiO_2 > 50\%$ and higher alkalies than the axial tholeiites. Because of the much higher contents of large-ion-lithophile elements compared with the tholeiites, Lipman and Mehnert (1975) called these rocks silicic alkalic basalts. Similarly, in the Ocate volcanic field on the east flank of the Sangre de Cristo Mountains, the dominant magma types are silicic alkalic basalt and alkali-rich hypersthene-normative basaltic andesite. Minor volumes of silicic alkalic basalt are closely associated in space and time with much more voluminous tholeiites in the Taos volcanic field.

With increasing distance from the axis of the rift, the most alkalic basalts present in a given field tend to have progressively higher normative nepheline contents. Examples of flanking volcanic fields with basanites or nephelinites in association with substantially less alkalic lavas are Mt. Taylor (Lipman and Moench, 1972), the adjoining Bandera field (Laughlin *et al.*, 1972) and the Clayton-Raton field (Stormer, 1972, 1978).

In addition to the diverse basaltic types in the various Rio Grande rift volcanic fields, more silicic volcanic rocks are also present, particularly within the latest Cenozoic volcanic fields of the northern rift. Andesite and rhyolite are present in the Mt. Taylor field (Baker and Ridley, 1970), and lavas ranging from basaltic andesite to dacite occur in the Cerros del Rio field (Baldridge, 1979a). The Clayton-Raton field (Stormer, 1972, 1978) also includes dacitic and rhyodacitic lavas in association with more voluminous basalts. Basaltic andesites, in association with silicic alkalic basalts, are important in the Ocate field and the Quaternary Brazos volcanics to the west of the Taos field. Isolated small centers of volcanism in the Albuquerque-Belen Basin (e.g. Los Lunas, Tome Hill, Black Butte—Kelley and Kudo, 1978; Zimmerman and Kudo, 1979) are domi-

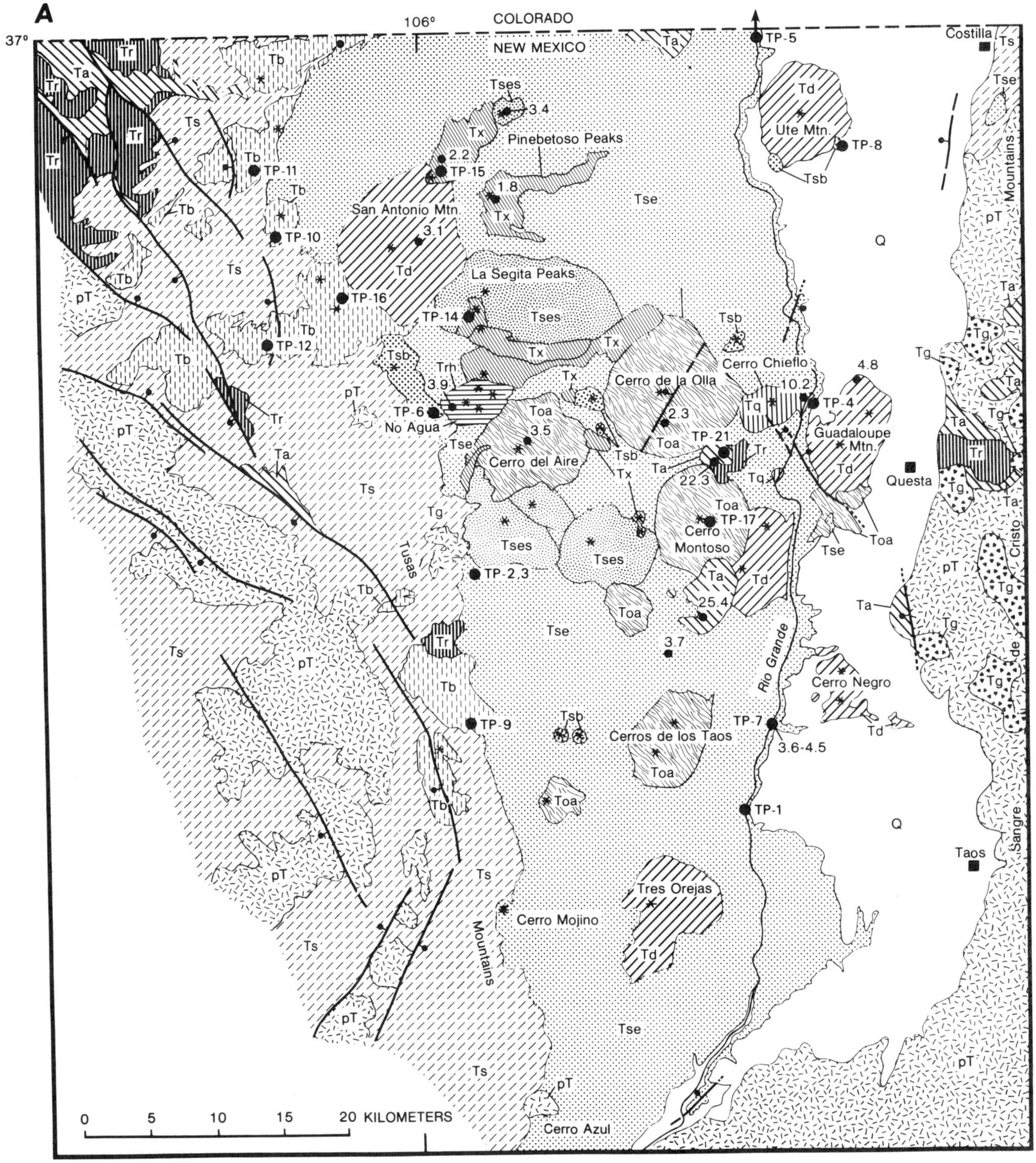

1.2.4.2a Geologic map of the Taos Plateau volcanic field, northern New Mexico, with reference suite sample localities. After Lipman and Mehnert (1979).

nantly basaltic with minor andesite, basaltic andesite and more silicic rocks.

The two most voluminous volcanic fields in the Rio Grande rift, the Jemez and Taos volcanics also contain

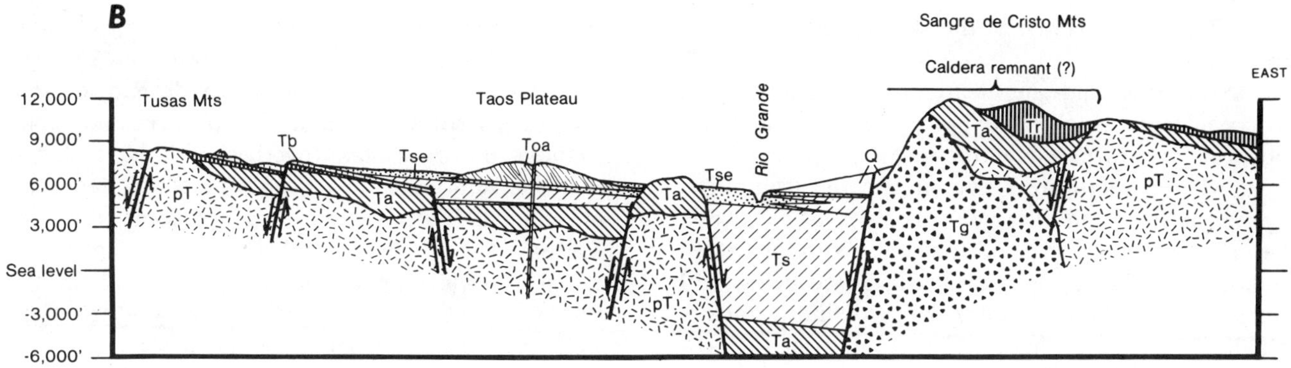

1.2.4.2b Schematic cross section of southern San Luis Valley and Taos Plateau volcanic field. After Lipman and Mehnert (1979).

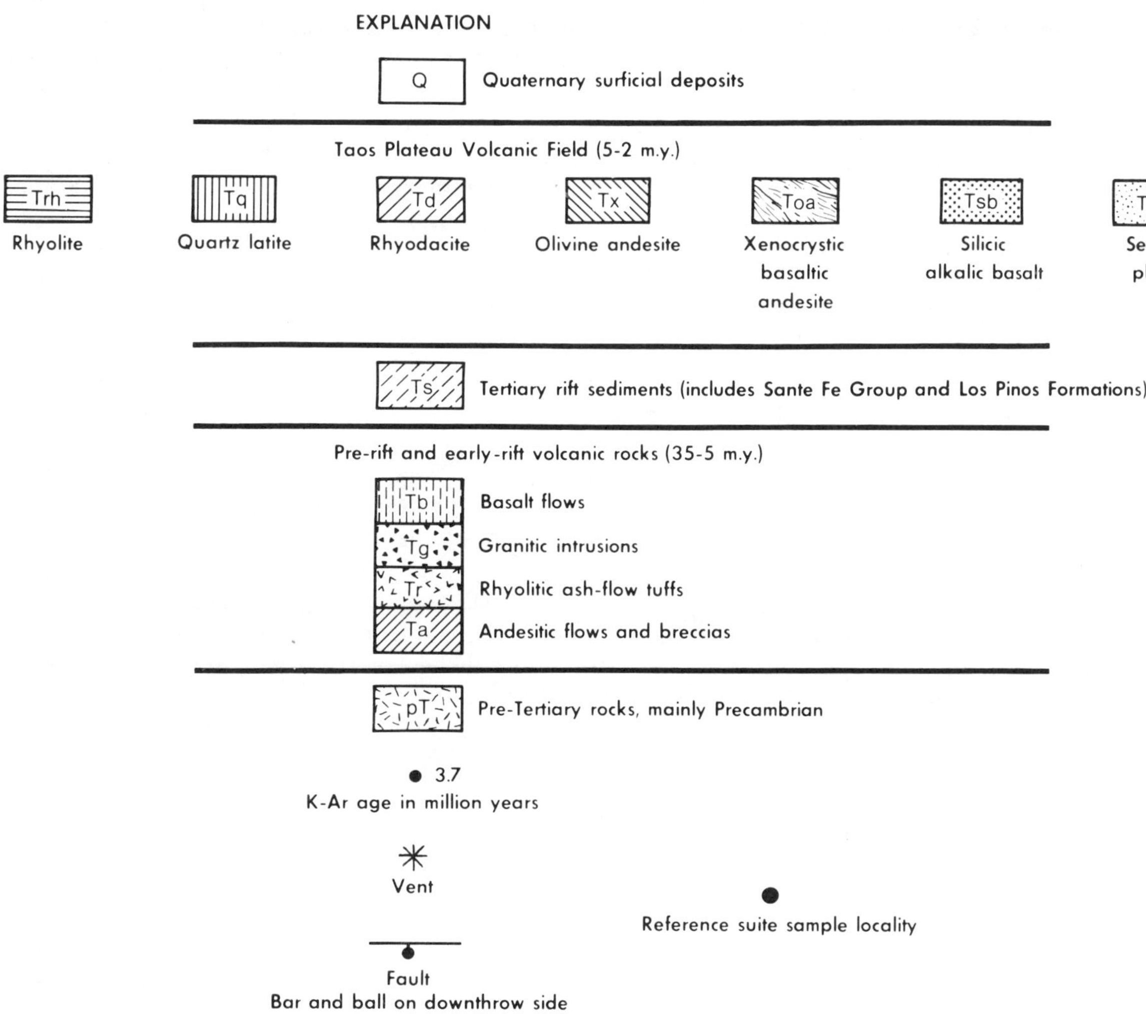

the greatest abundance of silicic volcanics. The Jemez field embodies four cycles, each beginning with basalts (Smith and Bailey, 1968; Smith *et al.*, 1970). The first and fourth phases are bimodal suites, whereas the intervening sequences grade from basalt to highly differentiated lavas. Both tholeiites and silicic alkalic basalts are present, but no nepheline-normative basalts have been recognized, in contrast to the adjacent Cerros del Rio field (Baldridge, 1979a) and the Mt. Taylor complex to the southwest.

None of the andesite, dacite, and/or rhyolite suites has been convincingly related to a parental basalt composition by low-pressure fractionation. In the Taos field, where the data are most complete, the prohibition against deriving andesites from any observed basalt as a parental magma is particularly strong. With the exception of a peraluminous andesite (discussed below), all the silicic rocks (andesite through rhyolite) are calc-alkaline. Normative quartz content increases with silica, although in general, these rocks are relatively alkali-rich. Except in rocks more silicic than 65% SiO_2, hydrous minerals are rare.

1.2.4.3 The Taos Plateau Reference Suite

The reference suite consists of 17 representative samples of broadly basaltic aspect, all from the Taos Plateau volcanic field within the central rift valley and from the adjacent flanking Tusas Mountain (Fig. 1.2.4.1). The suite includes samples of six tholeiitic basalts, three silicic alkalic basalts, two xenocrystic basaltic andesites, and two andesites from within the rift zone, as well as three compositionally diverse basaltic flows from the flanking slope (see Tables 1.2.4.1 and 1.2.4.2, and Appendix section A-4).

The Taos Plateau volcanic field contains over 30 central vent volcanoes ranging in composition from basalt to high-silica rhyolite (Lipman and Mehnert, 1979). They are clustered on the western side of the valley and define an imperfect concentric complex with tholeiitic shields of the Servilleta basalt (TP-1-5 and 14) at the center, rhyodacitic cones with steep slopes (20°) farthest out, and andesitic volcanoes (TP-17) with intermediate slopes (10°) at intermediate positions. A single rhyolite dome complex occurs among the tholeiite shields. Throughout this area there are small cinder cones of silicic alkalic basalt (TP-6-8) and apparently kindred basaltic andesites (TP-15 and 16) containing abundant xenocrysts of quartz and sodic feldspar. Field relations and the similar K-Ar ages for the Servilleta tholeiites and more silicic volcanics indicate that eruption of these diverse rock types has been approximately synchronous over the period of 4.5–2.0 m.y.

None of the reference samples is nepheline-normative nor do any possess the high Cr and Ni contents or high Mg′-values which are expected of basalt which last equilibrated with a mantle residue dominated by Fo_{90} olivine. Either the samples were generated in an unusual source or they underwent differentiation prior to eruption. Present data suggest that unfractionated and/or strongly undersaturated magmas are rare, particularly in the northern rift. An exception is a 25-m.y.-old olivine nephelinite from the Cerros del Rio volcanic field (SE of Valles Caldera at the southern end of the Espanola Basin) which, together with its included peridotite xenoliths, has been studied in detail by Baldridge (1979b). We have included some trace element data for this basalt and for another nephelinite studied by Baldridge in Figs. 1.2.4.4 and 1.2.4.5, for comparative purposes, because these rocks are more alkalic than any samples in the reference suite.

Servilleta basalt

Occurrence. Distinctive olivine tholeiites of the Servilleta basalt (TP-1-5 and TP-14), erupted from shield volcanoes on the western Taos Plateau (Fig. 1.2.4.2a), are the most voluminous and areally most extensive volcanic rock type in the area. These basalts uphold the geomorphic plateau surface and are exposed in the Rio Grande gorge from the Colorado-New Mexico border, where they disappear beneath valley-fill sediments northward, to just north of the Espanola, where a single 2.9-m.y.-old flow overlies Santa Fe Formation (Manley, 1979). The Servilleta lavas occur as vesicular pahoehoe flows which are 1–10 m thick with a median thickness of 2–3 m. The gorge exposures reveal thin fluvial beds between thick packages of thin flows which lack interbedded sediments or soil profiles.

Petrography. Servilleta flows exhibit substantial textural variability within single cooling units, related both to cooling-rate variations as a function of distance from the flow margin and to complex post-eruption differentiation. Rapidly cooled margins are characterized by a hiatal porphyritic texture in which small olivine and plagioclase phenocrysts (1–2 mm) are set in a cryptocrystalline groundmass. Minor octahedra of chromian spinel occur as inclusions in the larger olivine phenocrysts, but no pyroxene phenocrysts have been recognized. Large plagioclase crystals, extensively resorbed, are sparsely present in most samples. Two varieties are observed: equant, euhedral grains with internally resorbed cores and compositions similar to unresorbed plagioclase are the more common, but anhedral grains with more sodic compositions are also present. Scarce, resorbed quartz phenocrysts are visible in many outcrops but are rarely observed in thin section. One xenocryst-charged sample from La Segita Peak (TP-14) contains abundant large, embayed quartz xenocrysts and both types of resorbed plagioclase.

The characteristic texture of Servilleta flow interiors is coarse diktytaxitic. Plagioclase laths (1–4 mm

Table 1.2.4.1 Major element compositions of lavas of the Taos Plateau Volcanic Suite.

	Servilleta Basalts						Tusas Mountains Volcanics			Silicic Alkalic Basalts			Xenocrystic Basaltic Andesites		Andesites	
	TP-1	TP-2	TP-3	TP-4	TP-5	TP-14†	TP-10	TP-11	TP-9	TP-7	TP-6	TP-8	TP-15	TP-16	TP-17	TP-21
SiO_2	51.10	49.56	50.52	51.81	50.83	51.70	53.03	57.85	57.28	52.01	52.55	50.51	57.64	55.35	58.45	57.04
TiO_2	1.19	1.17	1.28	1.21	1.21	1.21	1.15	1.69	1.11	1.55	1.81	1.52	1.36	1.30	1.09	1.24
Al_2O_3	16.27	16.78	15.85	16.45	16.41	15.77	15.59	15.69	17.83	16.94	16.58	16.21	15.77	15.50	16.74	15.50
Fe_2O_3	2.09	3.96	1.65	2.13	1.13	2.92	0.94	2.79	5.00	3.27	3.50	5.57	2.46	1.88	2.20	2.83
FeO	8.84	7.16	9.12	7.72	9.64	7.52	8.46	5.95	4.40	5.87	6.17	4.42	4.62	6.33	4.33	4.18
MnO	0.17	0.15	0.15	0.16	0.17	0.16	0.15	0.13	0.20	0.15	0.14	0.16	0.12	0.14	0.11	0.10
MgO	7.36	7.40	7.15	6.14	7.20	6.94	6.54	2.37	5.60	5.64	4.21	6.40	2.85	4.75	4.12	3.99
CaO	8.95	9.22	9.17	9.00	9.16	8.26	7.84	5.48	4.34	7.76	7.31	8.44	5.71	6.60	5.85	6.02
Na_2O	3.22	3.39	3.16	3.16	3.03	3.13	3.10	3.86	3.47	4.08	4.02	3.55	4.25	3.77	3.81	4.02
K_2O	0.52	0.60	0.52	0.91	0.36	1.04	1.29	2.51	0.56	1.71	2.00	1.93	2.75	2.24	2.73	2.56
P_2O_5	0.20	0.25	0.21	0.26	0.16	0.21	0.20	0.51	0.19	0.35	0.85	0.72	0.72	0.46	0.51	0.53
H_2O+	0.57	—	—	0.65	0.47	0.56	0.62	0.93	1.09	0.85	0.88	0.93	1.26	0.83	1.00	1.01
CO_2	0.10	—	—	0.07	0.05	0.14	0.07	0.07	0.29	0.05	0.06	0.10	0.30	0.07	0.10	0.07
Total	100.58			99.68	99.82	99.57	98.98	99.83	101.36	100.23	100.08	100.46	99.81	100.52	100.04	99.09
*Mg/ $(Mg + Fe^{+2})$	0.576	0.578	0.572	0.558	0.572	0.575	0.582	0.357	0.555	0.559	0.472	0.573	0.453	0.531	0.564	0.540
Q	—	—	—	—	—	—	0.89	8.24	12.09	—	—	—	5.56	1.55	6.45	4.56
Or	3.07	3.55	3.07	5.38	1.13	6.15	7.62	14.83	3.31	10.11	11.82	11.41	16.25	13.24	16.13	15.13
Ab	27.25	28.69	26.74	25.74	26.64	26.49	26.23	32.66	29.36	34.53	34.02	30.04	35.96	31.90	32.24	34.02
An	28.41	28.80	27.53	28.04	30.11	25.91	25.82	18.07	20.41	22.86	21.29	22.60	15.83	21.48	17.78	16.69
Lc	—	—	—	—	—	—	—	—	—	—	—	—	—	—	—	—
Ne	—	—	—	—	—	—	—	—	—	—	—	—	—	—	—	—
Di	12.19	12.72	13.76	12.42	11.84	11.37	10.60	5.16	—	11.16	8.29	12.35	6.88	7.06	6.77	8.29
Hy	16.36	4.19	14.76	19.99	20.12	20.31	24.07	14.03	26.46	2.34	12.38	0.75	12.36	19.51	15.24	14.55
01	8.18	16.98	8.13	1.92	5.16	4.12	—	—	—	13.02	4.35	15.90	—	—	—	—
Mgt	1.55	1.55	1.54	1.40	1.55	1.47	1.35	1.23	1.29	1.28	1.35	1.37	0.99	1.21	0.91	0.98
Ilm	2.26	2.22	1.43	1.30	2.30	2.30	2.18	3.21	1.11	2.94	3.44	2.89	1.58	1.47	1.07	1.36
Ap	0.44	0.55	0.46	0.57	0.35	0.46	0.44	1.11	0.42	0.76	1.86	1.57	1.57	1.01	1.11	1.16
Cor	—	—	—	—	—	—	—	—	4.03	—	—	—	—	—	—	—
An/ (An + Ab)	49.6	48.6	49.2	51.2	52.5	48.0	47.1	34.3	39.6	38.4	37.1	41.5	29.3	38.8	34.2	31.6

†xenocrystic

*normative compositions calculated after recasting $Fe_2O_3/(FeO + Fe_2O_3) = 0.1$

Table 1.2.4.2 Trace element in lavas of the Taos Plateau Volcanic Suite.

	Servilleta Basalts						Tusas Mountains			Silicic Alkalic Basalts			Xenocrystic Basaltic Andesites		Andesites	
	TP-1	TP-2	TP-3	TP-4	TP-5	TP-14†	TP-10	TP-11	TP-9	TP-7	TP-6	TP-8	TP-15	TP-16	TP-17	TP-21
Nb	6.5	8.9	6.0	10.1	5.4	8.8	8.7	22.9	14	20.3	26.8	23.3	29.5	21.7	24.7	21.5
Zr	88	106	85	120	78	105	113	270	95	172	219	150	286	176	210	184
Y	20	19	20	21	20	21	20	31	23	24	31	27	33	25	22	16
Sr	348	434	320	400	273	338	334	567	272	599	1016	1037	907	660	830	1070
Rb	9.1	7.3	7.1	11	3.3	16	22	57	45	26	23	23	48	39	36	36
Ba	300	335	250	440	195	420	450	930	325	625	1275	1270	1850	1025	1000	1360
Sc	23.9	24.9	24.4	24.0	23.5	22.5	21.4	11.5	18.0	23.1	18.6	22.8	13.6	16.9	14.1	13.6
Cr	193	171	163	148	184	199	210	4.0	178	142	32.9	109	28.7	76.4	142	113
Ni	156	66	—	104	136	175	150	—	130	—	—	115	—	95	94	81
La	10.9	13.4	8.76	16.4	7.63	14.8	14.5	41	12.7	22.8	43.1	34.5	55	32.1	42.6	45.7
Ce	26.7	31.2	20.6	36.5	17.4	33.0	32.5	82	24.9	50.4	103	77.2	139	68.2	85.0	94.5
Sm	3.45	3.92	3.21	4.35	3.00	3.71	3.71	8.57	3.69	5.29	10.0	7.32	10.5	6.25	6.65	6.66
Eu	1.16	1.28	1.09	1.35	1.03	1.18	1.19	2.26	1.10	1.58	2.72	2.09	2.70	1.74	1.76	1.84
Tb	0.71	0.67	0.57	0.72	0.62	0.67	0.66	1.10	0.71	0.79	1.17	0.98	1.24	0.86	0.80	0.67
Yb	1.99	1.88	1.90	2.10	1.90	1.96	1.98	2.70	2.33	2.33	2.90	2.55	3.04	2.30	1.92	1.29
Lu	0.30	0.31	0.29	0.32	0.32	0.31	0.32	0.42	0.37	0.40	0.44	0.38	0.46	0.38	0.33	0.20
Th	0.86	0.63	0.76	1.12	0.66	1.84	1.88	5.4	3.03	2.24	1.63	1.82	3.86	3.75	3.93	4.87
Hf	2.45	2.88	2.17	2.99	2.14	2.95	3.06	6.4	2.90	4.41	5.72	4.18	7.47	4.58	5.27	4.83
Pb	6	5	4	6	4	7	7	11	11	9	5	11	15	4	15	16
Ga	20	21	19	19	20	19	19	24	18	21	22	18	20	20	20	21
Th/Hf	0.35	0.22	0.35	0.38	0.31	0.62	0.61	0.84	1.05	0.51	0.29	0.43	0.52	0.82	0.75	1.01
Zr/Y	4.2	5.1	4.3	5.5	3.8	5.1	5.5	8.2	4.0	6.9	6.7	5.5	8.4	7.0	9.3	11.1
Zr/Nb	13.5	11.9	14.2	11.9	14.4	11.9	13.0	11.8	6.8	8.5	8.2	6.4	9.7	8.1	8.5	8.6
Y/Nb	3.2	2.3	3.3	2.2	3.8	2.3	2.4	1.4	1.7	1.24	1.22	1.18	1.15	1.17	0.91	0.77
Rb/Sr	0.025	0.017	0.021	0.028	0.013	0.050	0.065	0.101	0.163	0.043	0.023	0.022	0.052	0.058	0.044	0.033
La/Sm	3.16	3.42	2.73	3.77	2.54	3.77	3.91	4.78	3.44	4.31	4.31	4.71	5.30	5.13	6.41	6.86

long) and equant euhedral to subhedral olivine crystals of varying size (up to 2 mm) are partially enclosed in ophitic clinopyroxene grains. In general, there is a positive correlation between plagioclase and pyroxene grain sizes, such that the grain size variation for these phases becomes seriate rather than hiatal with progressive coarsening. All the Servilleta basalts contain 5–10% of brown, glass-rich mesostasis which occurs mostly in patches close to the polygonal void spaces inherent in the diktytaxitic texture. These areas contain skeletal Fe-Ti oxide, apatite needles, and minor skeletal pyroxene in a matrix of glass which consists of a clear granitic phase and abundant, small (<0.01 mm), spherical blebs of an immiscible dark glassy phase. The contrasts in grain size and in crystal morphology between the mesostasis and the much coarser crystals of the flow interiors are always marked.

Even thin Servilleta flows exhibit complex internal structures involving the development of highly vesicular segregations which occur as cylinders and as horizontal sheets filling fractures (commonly completing a three-dimensional, interconnected framework). The textures of the segregations are quite different from those of their enclosing matrices. The mineralogy of these segregations is the same as that of the adjacent diktytaxitic matrix, but individual phases are more strongly zoned and Fe-Ti oxides are more abundant. The texture is still hiatal in that there is a marked contrast between the coarse crystals and the finer-grained mesostasis, which is largely crystalline, even though the pyroxene, plagioclase, apatite, and Fe-Ti oxides are all acicular and very fine-grained. The characteristic diktytaxitic texture is not well-developed; rather pyroxene and plagioclase are united in a mutual interference or bow-tie arrangement.

Chemistry. The Servilleta basalts are olivine tholeiites with relatively evolved compositions, as indicated by their low Mg′-values of 0.60–0.56 and depleted Cr (<300 ppm) and Ni (<200) contents. The major and trace element compositions of the tholeiitic lavas are surprisingly uniform in light of their wide geographic and temporal distributions. The olivine tholeiites typically have contents of normative hypersthene 15–20%, normative olivine <10%, and normative orthoclase <10%. The normative plagioclase is in the range An_{47-52}. The Servilleta basalts are distinct from all other rock types in the Taos field in terms of their low SiO_2 and total alkalies (Fig. 1.2.4.3), and in terms of trace elements. Of the basalt types present in the Taos Plateau, these contain the lowest abundances of lithophile elements such as Rb, Sr, Ba, K, P, Th, U, Hf and Nb. The Servilleta tholeiites have the highest Zr/Nb ratios (13 ± 1.5 for over 30 samples) and lowest La/Yb ratios (3.2 ± 0.3) of any of the mafic magma types (Figs. 1.2.4.4–5).

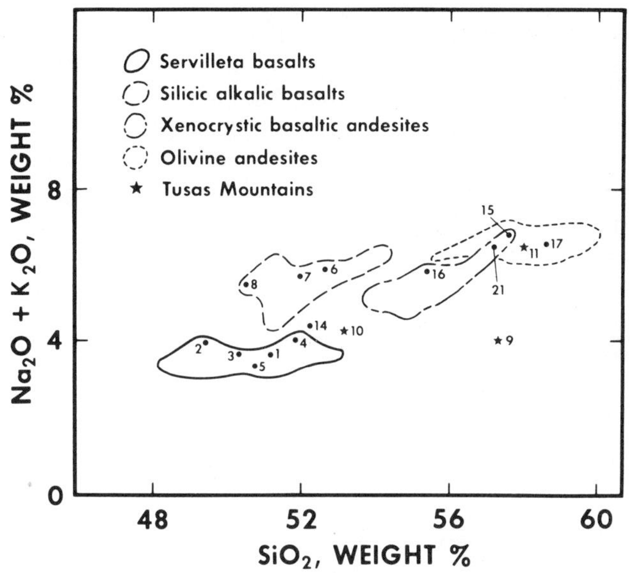

1.2.4.3 Silica-alkalies variation diagram for basaltic lavas of the Rio Grande rift reference suite. Compositional ranges of main basaltic rock types from the Taos Plateau volcanic field are from Lipman and Mehnert (1979, fig. 3).

The vesicular segregations in Servilleta flows, although not represented in the reference suite, are depleted in Cr, Ni and Sr, but enriched in Sc, Th, Rb, Fe, REE, Zr, Nb, Y, Ba, K, Th, and Hf (Haskin *et al.*, 1978). Chondrite-normalized REE patterns of the segregations are similar in slope to those of the adjacent flow interiors but abundances are elevated by a factor of 1.5–2.2, and marked negative Eu anomalies are present. Thus, the segregations appear to have crystallized from late-stage residual liquid (gas-rich) which formerly occupied the voids now present in the diktytaxitic matrix. Kuno (1960) was one of the first to describe these features, which are common to many tholeiitic basalt flows.

Haskin *et al.* (1978) have also found that trace element abundances vary within flows; the fine-grained margins differ significantly from representative interior samples. Analyses of multiple samples from outcrops of single flows (Crossey, 1979) encompass as much as half of the entire compositional range known for the Servilleta (Aoki, 1967; Lipman and Mehnert, 1975). Variations in such elements as K, P, Rb and Sr appear in part to be the result of intraflow heterogeneities which formed by short-range, in-place differentiation. (Reference suite samples, with the exception of TP-3, are from flow interiors.) These intraflow variations probably contribute to the variations in lithophile element abundances seen among rocks which have nearly constant Al_2O_3, ΣFeO, CaO, MgO and Mg/(Mg + Fe), as well as nearly constant phenocryst compositions.

Petrographic evidence also indicates that some contamination of the tholeiites has occurred, although the available isotopic data do not support large-scale assimilation of radiogenic crustal material. The xenocryst-rich Servilleta tholeiite from La Segita Peak (TP-14) is a fine-grained flow-margin sample which contains 5–10% xenocrystic quartz and feldspar. It is relatively enriched in SiO_2, K_2O, and Rb compared with other Servilleta basalts having similar Mg/(Mg + Fe) ratios. It seems likely that these chemical anomalies are related to contamination; however, the concentrations of other elements such as Zr, Nb, Y and the REE are essentially identical to those in less silicic and alkalic samples (e.g., TP-2 and TP-4).

Silicic alkalic basalts

Occurrence. The silicic alkalic basalts (TP-6–8) are minor components of the Taos Plateau field and have a mode of eruption which is different from that of the Servilleta basalts. Individual volcanic centers typically consist of a single cinder cone and one flow. The flows are generally aa rather than pahoehoe and tend to be highly vesicular. Several of the flows are traceable as far as 5–10 km from their vents, but most are less than 5–10 m thick. An exception is a 25 m thick flow in the Rio Grande gorge (TP-7), which may have been ponded by an ancestral Rio Grande. Older silicic alkalic basalts occur in the Tusas Mountains at Los Mogotes volcano (5 m.y.b.p.), just northwest of the Taos field, and throughout the San Juan Mountains (Lipman and Mehnert, 1975), extending back in age to the inception of rifting (27 m.y.b.p.).

The silicic alkalic basalts are predominantly olivine-phyric lavas with very fine-grained, intergranular groundmasses. Chromian spinel octahedra occur rarely as inclusions in the olivine phenocrysts, which is in keeping with the lower Cr contents of the bulk lavas. More evolved flows (TP-6) are multiply saturated (ol + plag + cpx or ol + plag), the crystallization sequence being olivine-plagioclase-clinopyroxene-FeTi oxide, as in the tholeiites.

Sample TP-7, in addition to being chemically and isotopically unique, is petrographicaly unusual. Despite its thickness (25m), the center of the flow is little different texturally from the margin. Both regions are nearly holocrystalline. Olivine phenocrysts, partly altered to iddingsite, and complexly zoned and resorbed plagioclase phenocrysts are set in a nearly equigranular groundmass consisting of stubby plagioclase laths and granular clinopyroxene. The clinopyroxenes are characterized by unusual vermicular textures, suggesting resorption.

Chemistry. The silicic alkalic basalt type is characterized by high alkali (Na_2O = 3.51–4.1%; K_2O = 1.7–2.0%) and high silica (SiO_2 = 50–53%) contents, which result in both normative hypersthene and olivine in

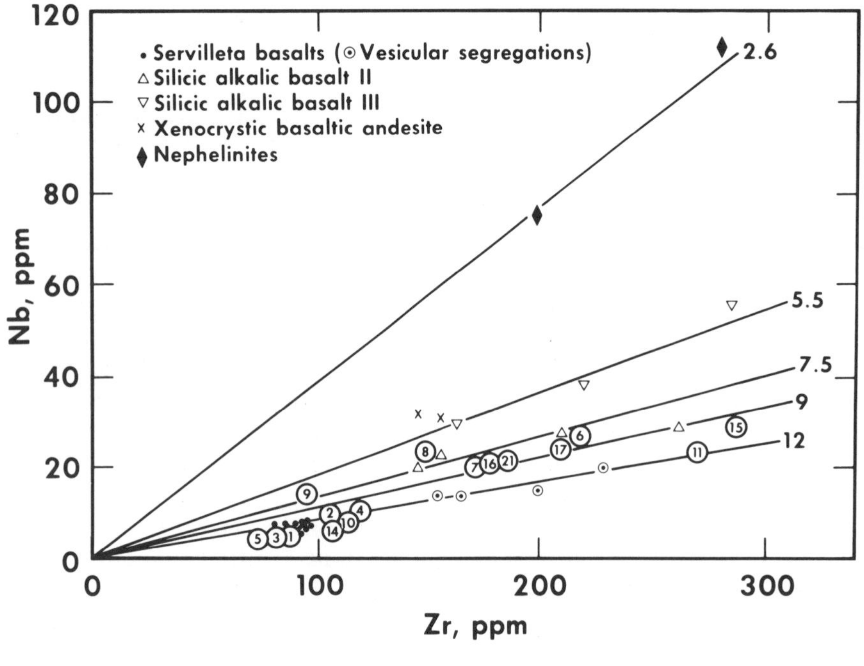

1.2.4.4 Zr-Nb variation diagram for basaltic lavas of the Rio Grande rift. Reference suite lavas are indicated by sample number, all others by the designated symbol. Reference lines are for constant Zr/Nb ratio.

combination with high normative (Ab + Or)/An (Fig. 1.2.4.3). In general, TiO_2 (1.5–1.8%) and P_2O_5 are higher than in the Servilleta tholeiites. These rock have differentiation indices of 41–46 and normative plagioclase compositions of An_{42-37}, which place them on the borderline between basaltic and andesitic fields in some classification schemes (e.g., Irvine and Baragar, 1971). On an alkalies vs. silica variation diagram (Fig. 1.2.4.3), the silici alkalic basalts plot consistently on the alkalic side of the dividing line which was determined empirically for Hawaiian basalts by Macdonald and Katsura (1964). However, the least fractionated members of this suite have Mg′-values which are virtually identical to those of the Servilleta tholeiites.

In addition to major element differences between the two rock types, the silicic alkalic basalts are depleted in Sc, Cr, and Ni and enriched in all lithophile elements relative to the olivine tholeiites. These rocks also differ from the tholeiites in lithophile trace element ratios and rare-earth element patterns. The light rare-earth elements are enriched (Fig. 1.2.4.5b) and the Zr/Y ratio is higher (7.0–5.5 vs. 6.0–3.8), whereas the Zr/Nb (6–9) and Y/Nb (1.2.) ratios are lower (Fig. 1.2.4.4). Elements such as Th, Sr, Ba, Rb and particularly Nb, which tend to be higher in alkalic basalts compared to tholeiites (Pearce and Norry, 1979), are enriched by factors of 2–4. The silicic alkalic basalts resemble the even more silica- and alkali-rich olivine andesites in terms of some trace element abundances and trace element ratios, despite marked differences in major elements. On the basis of presently available data (for about 12 samples), we have defined three subgroups of silicic alkalic basalts (SAB I, SAB II, SAB III), two of which are represented in our reference collection.

The thick silicic alkalic basalt flow (TP-7) at the base of the Dunn Bridge section of the Rio Grande gorge is the lone sample of SAB I in the reference suite and in our larger sample collection. It is chemically intermediate between olivine tholeiites and the common silicic alkalic type on the Taos Plateau (SAB II), particularly with regard to its REE abundances (Fig. 1.2.4.5b). Isotopically, it is distinct from any other flow we have studied and may be a unique flow. Even though the abundances of lithophile elements such as K, Ba, Sr, P and Nb are slightly lower than for other silicic alkalic basalts, ratios such as Zr/Y, Zr/Nb and Y/Nb fall within the characteristic ranges for the rest of the samples (Zr/Y = 6.7 ± 1; Y/Nb = 1.2 ± 0.05), making its general affinity clear.

The main group of silicic alkalic basalts, represented by TP-6 and TP-8, is part of a coherent suite of samples from widely scattered volcanic centers (SAB II). Sample TP-6 is a relatively evolved, apparently multiply-saturated composition containing olivine and plagioclase microphenocrysts. The less-fractionated sample, TP-8, contains only olivine phenocrysts with rare inclusions of chromian spinel. These flows lack the complex intraflow structures and our limited data appear to indicate that the chemical variations so characteristic of the Servilleta tholeiites are absent from these rocks.

The third type (SAB III of Fig. 1.2.4.4; not represented in the reference suite) includes such rocks as an

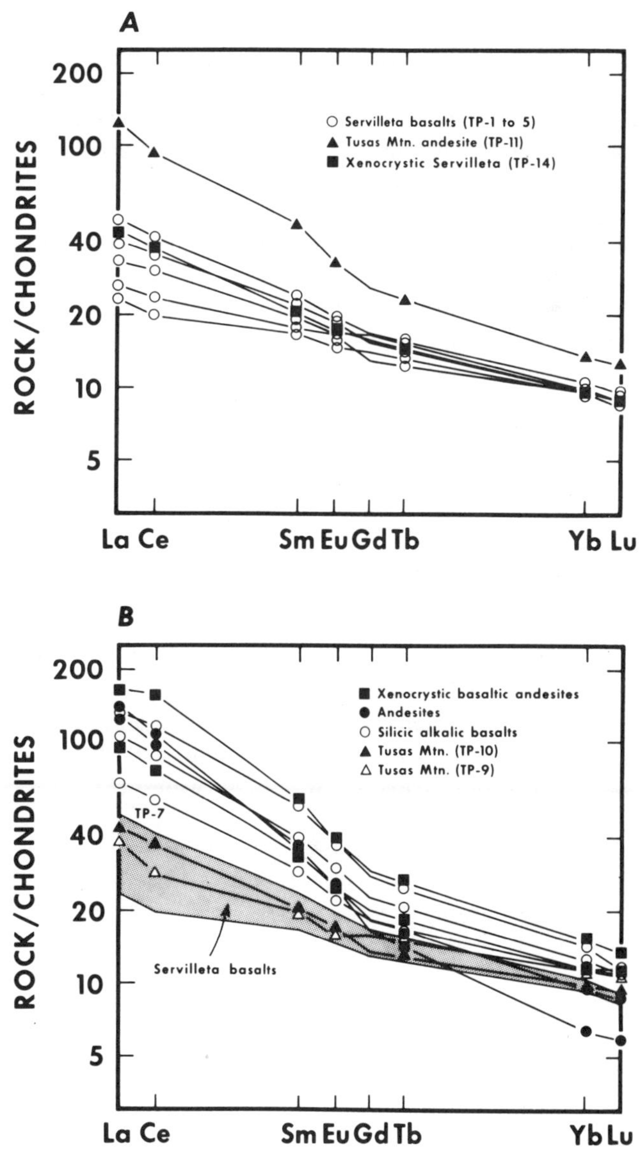

1.2.4.5 Chondrite-normalized rare-earth element abundances. **a.** Servilleta olivine tholeiites plus basalts and andesites of Tusas Mountains. **b.** Silicic alkalic basalts, andesites and xenocrystic basaltic andesites.

alkali-rich basaltic andesite of early rift age which is present within the axial horst of the Taos plateau (>20 m.y.), some basaltic andesites at Los Magotes volcano (5 m.y.) to the north, and several flows in the Brazos volcanics (~250,000 years) to the west of the Tusas Mountains (Lipman and Mehnert, 1975).

Olivine andesites

Occurrence. Over a dozen late-Cenozoic, central-vent volcanoes composed of andesite and rhyodacite occur within the Taos Plateau field. Six of these are dominated by olivine andesites. The reference suite sample (TP-17) representing this group of rocks is from Cerro Montoso. The limited sampling to date (e.g., nine samples from Cerro Montoso) suggests that the lavas exposed at the surfaces of these volcanoes are quite uniform in petrography and composition. However, older flows exposed beneath Servilleta flows in the Rio Grande gorge and samples collected from remnants of apparently older ancestral cones, demonstrate some diversity. These glimpses of slightly older lavas are rare, due to the lack of erosional dissection of the cones, and the parentage of the capping flows is accordingly obscure. Andesite also occurs within the eroded horst block of early rift volcanics exposed near Cerro Montoso, and farther south in the rift (Zimmerman and Kudo, 1979).

Petrography. The andesites contain olivine phenocrysts (Fo_{82-77}), the more magnesian of which enclose abundant octahedra of chromian spinel. Groundmass phases are plagioclase, Fe-Ti oxide, calcic augite and orthopyroxene, which typically are set in a brownish, glassy matrix. Characteristic macroscopic flow banding is expressed by stretched vesicles and parallel laths of groundmass plagioclase. Corroded xenocrystic aggregates of orthopyroxene are sparse but widespread.

Chemistry. The characteristic chemical features of the olivine andesites are high SiO_2 (57–59 wt.%), Mg′-values comparable to those of the Taos Plateau basalts (0.57–0.54), relatively low TiO_2 (1.0–1.2 wt.%), and high alkalies (Fig. 1.2.4.3; Lipman and Mehnert, 1979; Zimmerman and Kudo, 1979). The rocks are calc-alkaline and contain substantial quartz in their norms. Compared with the average andesite of Chayes (1969), these lavas are lower in Al_2O_3 and ΣFeO and higher in K_2O, Na_2O, and MgO.

The trace element systematics of these andesites are distinctly different from those of island-arc mafic magmas (see section 1.2.7; Kay, 1977; Pearce and Norry, 1979). The rare-earth patterns of the Taos andesites are strongly light REE enriched (La/Yb > 22), in accord with other trace element ratios such as high Zr/Y (~8–10) and low Zr/Nb (~6–9). These values are more characteristic of alkali olivine basalts than of andesites. The olivine andesites resemble the silicic alkalic basalts with respect to trace elements, but the andesites are characterized by even higher La/Yb, Zr/Nb and Zr/Y ratios. The olivine andesites also have higher Cr contents (~200–250 ppm), compared with plate-margin andesites and the silicic alkalic basalts.

Xenocrystic basaltic andesites

Occurrence. Basaltic andesite flows, bearing abundant quartz and plagioclase xenocrysts, have erupted from a cluster of vents in the area of San Antonio Mountain – No Agua Peaks, in the NW Taos Plateau. Each vent consists of a single flow and cinder cone, so the basaltic andesites resemble silicic alkalic basalts in eruptive style. Five major vents in this area have been recognized (La Segita peak, Pinebetosa peak, north flank of San Antonio Mountain TP-15, and Cerritos de la Cruz TP-16). The xenocrystic Servilleta tholeiite (TP-14) was collected from a shield on which is superimposed the La Segita xenocrystic basaltic andesite. Xenocryst-bearing basaltic andesites and hawaiites are common elsewhere in the southwestern U.S. (Best and Brimhall, 1974; Stormer, 1972; Hausel and Nash, 1977). The xenocrysts have been interpreted as high-pressure cognate phases (Hausel and Nash, 1977), as phenocrysts associated with the mixing of basalt and porphyritic rhyolite magmas (Lipman *et al.*, 1978), and as accidental material incorporated during the partial assimilation of non-radiogenic lower crustal granulites (Doe *et al.*, 1969). The presence of these xenocrysts in both the Servilleta tholeiites and the chemically distinct basaltic andesites, combined with the close spatial association of the lavas, suggests an origin by contamination.

Petrography. The two xenocrystic basaltic andesites in the reference suite are typical; they contain resorbed xenocrysts and a phenocryst assemblage apparently formed by low-pressure crystallization. TP-16 contains minor xenocrystic orthopyroxene, plus phenocrysts of plagioclase and clinopyroxene, and abundant olivine phenocrysts, which are absent from TP-15.

The plagioclase shapes and zoning patterns in both rocks are complex but generally similar; there are two distinct shape-compositional types of resorbed plagioclase, as well as plagioclase phenocrysts. The two xenocryst types are (1) marginally-corroded grains of oligoclase-andesine and (2) larger crystals which have normally zoned rims overgrown on resorbed interiors.

Chemistry. The two xenocrystic lavas in the reference suite are chemically intermediate between the silicic alkalic basalts and the olivine andesites (Fig. 1.2.4.3). The olivine-phyric lava from Cerritos de la Cruz (TP-16) is similar to silicic alkalic basalts except for its low TiO_2 content. In addition to the predictably higher SiO_2 and alkalies, it has higher La/Sm and Rb/Sr ratios than the silicic alkalic basalts. The relatively magnesian olivine composition (Fo_{80}) and relatively high Mg′-value are typical of moderately fractionated Taos Plateau lavas. The other lava (TP-15) from the adjacent La Segita peak area, lacks olivine phenocrysts and has a substantially more fractionated composition ($Mg' = 0.45$) with higher SiO_2, alkalies, P_2O_5, and other lithophile trace elements (Fig. 1.2.4.5b). Total REE abundances are substantially higher than those in either the silicic alkalic basalts or the andesites, but the La/Sm ratio of this sample (and of TP-16) is intermediate between the two.

Tusas Mountains volcanics

Occurrence. Basaltic lava flows which dip 5–10° east beneath the Taos Plateau volcanic rocks on the west flank of the Rio Grande rift are exposed in the Tusas Mountains (Fig. 1.2.4.2). These flows are interbedded with fluvial and volcaniclastic sediments of the Los Pinos Formation, derived from the San Juan volcanic field to the north and other volcanic areas to the east which now are largely covered by younger deposits within the rift zone. The Tusas volcanics represent an important sample of intermediate rift-age volcanism along the west side of the rift zone. Some Tusas flows closely resemble the younger Servilleta lavas, but others are types not found in the Taos Plateau. The Tusas flows have been mapped in reconnaissance by Butler (1947, 1971), but stratigraphic, petrologic and age relations are poorly known. Similar lava flows along the east margin of the San Juan Mountains to the north range in age from 26–5 m.y.b.p. and include tholeiites, silicic alkalic basalt, xenocrystic basaltic andesite, and olivine andesite (Lipman and Mehnert, 1975). The reference suite includes a basalt (TP-10) and two andesites (TP-9, TP-11) from the Tusas Mountains, but additional sampling has documented a much greater range of compositions.

Petrography. The reference suite basalt (TP-10) is chemically similar to Servilleta basalt; although the characteristic diktytaxitic texture is not developed and its groundmass texture is finer grained. This basalt is olivine-phyric and lacks plagioclase phenocrysts, but contains sparse, resorbed plagioclase xenocrysts. Two Tusas andesites of markedly contrasting petrology represent rock types not present in the Taos Plateau field. Andesite TP-11 is aphyric and exhibits a distinctive hyaloophitic texture, with flow-oriented plagioclase laths and Fe-Ti oxides, sparse olivine altered to iddingsite and equant groundmass pyroxene grains in a brown glass matrix. TP-9 is a peraluminous xenocrystic andesite, which contains olivine microphenocrysts and sparse clinopyroxene phenocrysts in a fine-grained intergranular groundmass.

Chemistry. The limited sampling and chemical analyses reported here and by Lipman and Mehnert (1975) have identified a few basalts identical to Servilleta tholeiites. Other lavas, distinct from typical Servilleta but showing major and trace element similarities to Servilleta, have also been sampled. Basalt TP-10 has many characteristics in common with typical Servilleta basalt but is distinct in its lower CaO and higher SiO_2, K_2O, and Rb contents (Fig. 1.2.4.3). These chemical characteristics are similar to those of the highly contaminated sample TP-14. However, the $^{87}Sr/^{86}Sr$ ratio of this basalt (0.7046) does not suggest that substantial contamination has occurred relative to basalts which show little petrographic or chemical evidence of crustal assimilation.

Andesite TP-11 is an aphyric lava which caps a mesa in the Tusas Mountains. Despite its basaltic aspect in hand specimen, it is intermediate in character (Fig. 1.2.4.3), with a strongly fractionated composition ($Mg' = 0.357$). Compatible elements, such as Mg, Cr, Ni and Sc, are markedly depleted, the light REE are enriched relative to the heavy REE ($La/Sm = 7.78$), and Y is depleted relative to Zr and Nb. Overall, incompatible element concentrations are greater by factors of 1.5–2.5 compared with typical Servilleta basalts.

An even more unusual sample from the Tusas Mountains is TP-9, a quartz-normative peraluminous xenocrystic olivine andesite (Fig. 1.2.4.3) from a relatively young but undated flow (Dorado Basalt of Butler, 1947) in the southern Tusas Mountains. This unit exhibits low, Servilleta-like concentrations of TiO_2, K_2O, Na_2O, Zr, Y, Ba and REE. Light REE concentrations are even lower than in some Servilleta lavas. The $Mg/(Mg + Fe^{2+})$ ratio (Mg′) is lower than those of most Servilleta basalts (0.55 vs. 0.58), but it is high compared with TP-11 which has a similar SiO_2 content. Additional evidence that TP-9 is not simply a fractionation product of a Servilleta-type parent is the lower Zr/Nb ratio (6.8 vs. 11; Fig. 1.2.4.4) and much higher Rb content. The peraluminous character is also anomalous. Despite the high Al_2O_3 and lower CaO contents, which are responsible for its normative corundum, microphenocrysts of clinopyroxene are present along with the typical olivine microphenocrysts, and plagioclase is restricted to the groundmass. Although these phenocrysts are only weakly zoned in terms of Mg/Fe, they have relatively calcic cores ($Wo_{43}En_{43}Fs_{14}$) which are aluminous (4–5 wt.% Al_2O_3) compared to the rims ($Wo_{37}En_{48}Fs_{15}$). High pressure fractionation of aluminous clinopyroxene may have contributed to the unusual chemical character of this rock type, but a parental magma which is unlike any observed basalt would have been required.

Early rift volcanics

Occurrence. High-angle faulting has exposed early rift volcanics (25–21 m.y.b.p.) along an axial horst within the Taos Plateau volcanic field (Figs. 1.2.4.2a, b). The horst exposes dacitic to rhyolitic pyroclastic rocks, andesite and rhyodacite lava flows, and a rhyolite dome. These rocks are broadly similar to some of the younger silicic rocks but differ in detail chemically. The early rift rocks also contain abundant hydrous minerals, whereas the younger rocks are, with very few exceptions, conspicuously devoid of hornblende or biotite.

Petrography. The early rift andesite in the reference suite (TP-21) contains sparse, olivine phenocrysts (altered in part to iddingsite), abundant coarse-grained glomerophyric aggregates of orthopyroxene, and calcic augite and minor plagioclase microphenocrysts in a trachytic groundmass. The pyroxene phenocrysts contain inclusions of Fe-Ti oxides and glass in their cores. Abundant pyroxene phenocrysts, such as the glomerophyric aggregates in TP-21, are absent from younger olivine andesites with similar SiO_2 contents.

Chemistry. The major element compositions of the early rift andesite (TP-21) and of the Cerro Montoso andesites (TP-17) are closely comparable (Fig. 1.2.4.3). The trace elements are also similar, although TP-21 has higher Sr, Ba and light REE.

1.2.4.4 Isotope chemistry

Williams and Murthy (1978, 1979) have determined the Sr and Nd isotopic compositions of four olivine tholeiites, four silicic alkalic basalts, and several more silicic rocks from the Taos Plateau volcanic field (see Appendix A-4 and Fig. 1.2.4.12). The range of $^{87}Sr/^{86}Sr$ ratios in the basalts (0.70369–0.70503) is limited, but analytically significant variations exist. The average ratio of 0.70422 is an acceptable mantle $^{87}Sr/^{86}Sr$ ratio, but it is not so low nor is the range of $^{87}Sr/^{86}Sr$ ratios sufficiently small to rule out crustal contamination of these basalts. The $^{143}Nd/^{144}Nd$ ratios of the tholeiites and of all but one of the silicic alkalic basalts are identical (0.51243 ± 0.00005)($^{143}Nd/^{144}Nd$ ratios are normalized to $^{146}Nd/^{144}Nd = 0.7219$). This ratio is slightly less than the inferred bulk earth ratio of 0.51264 (DePaolo and Wasserburg, 1976; O'Nions *et al.*, 1977) for an undifferentiated chondritic reservoir. Thus the source of most of the Taos Plateau basalts is isotopically homogeneous in Nd, is approximately isotopically homogeneous in Sr, and has been enriched in light REE relative to a primitive chondritic uniform reservoir.

Of the four silicic alkalic basalts, three are isotopically similar, having $^{87}Sr/^{86}Sr$ ratios of 0.70452–0.70503 and $^{143}Nd/^{144}Nd$ ratios of 0.51242–0.51245. The isotopically anomalous sample, TP-7, is distinct in its petrography and trace element chemistry from the other three silicic alkalic basalts. Its low $^{87}Sr/^{86}Sr$ ratio of 0.70369 and high $^{143}Nd/^{144}Nd$ ratio of 0.51255 indicate that it cannot be directly related genetically to the other silicic alkalic basalts. The different isotopic compositions of the two types of silicic alkalic basalts prohibit derivation of one by differentiation of the other. Thus the isotopic differences between TP-7 and the other basalts can be accounted for only by derivation from two different source regions or by mixing.

The isotopic compositions of three olivine andesites are indistinguishable from those of the basalts ($^{87}Sr/^{86}Sr = 0.70451$–$0.70497$; $^{143}Nd/^{144}Nd = 0.51242$–$0.51244$).

1.2.4.5 Mineral chemistry

Olivine

Two types of microprobe analyses were performed. The data summarized as frequency plots in Fig. 1.2.4.6 are three-element analyses which were corrected by the Bence-Albee method. The most magnesian compositions are for phenocryst cores, but the aggregate analyses generally span the range of phenocryst core-rim and groundmass compositions present in the rocks. Eight-element olivine analyses were obtained for representative samples (see Appendix A-4).

The most magnesian olivine phenocrysts ($Fo_{82.5-80.0}$) found in major Taos Plateau rock types occur in varied basalts and andesites with $Mg/(Mg + Fe^{2+})$ values ($Fe^{2+} = 0.9 \times \Sigma Fe$) of 0.58–0.56. The liquidus olivine compositions determined in one-atmosphere, controlled-f_{O_2} experiments (QFM buffer, iron loss minimized by presaturation of Pt loops with Fe) for three of the least-fractionated Taos Plateau lavas (TP-8, TP-10

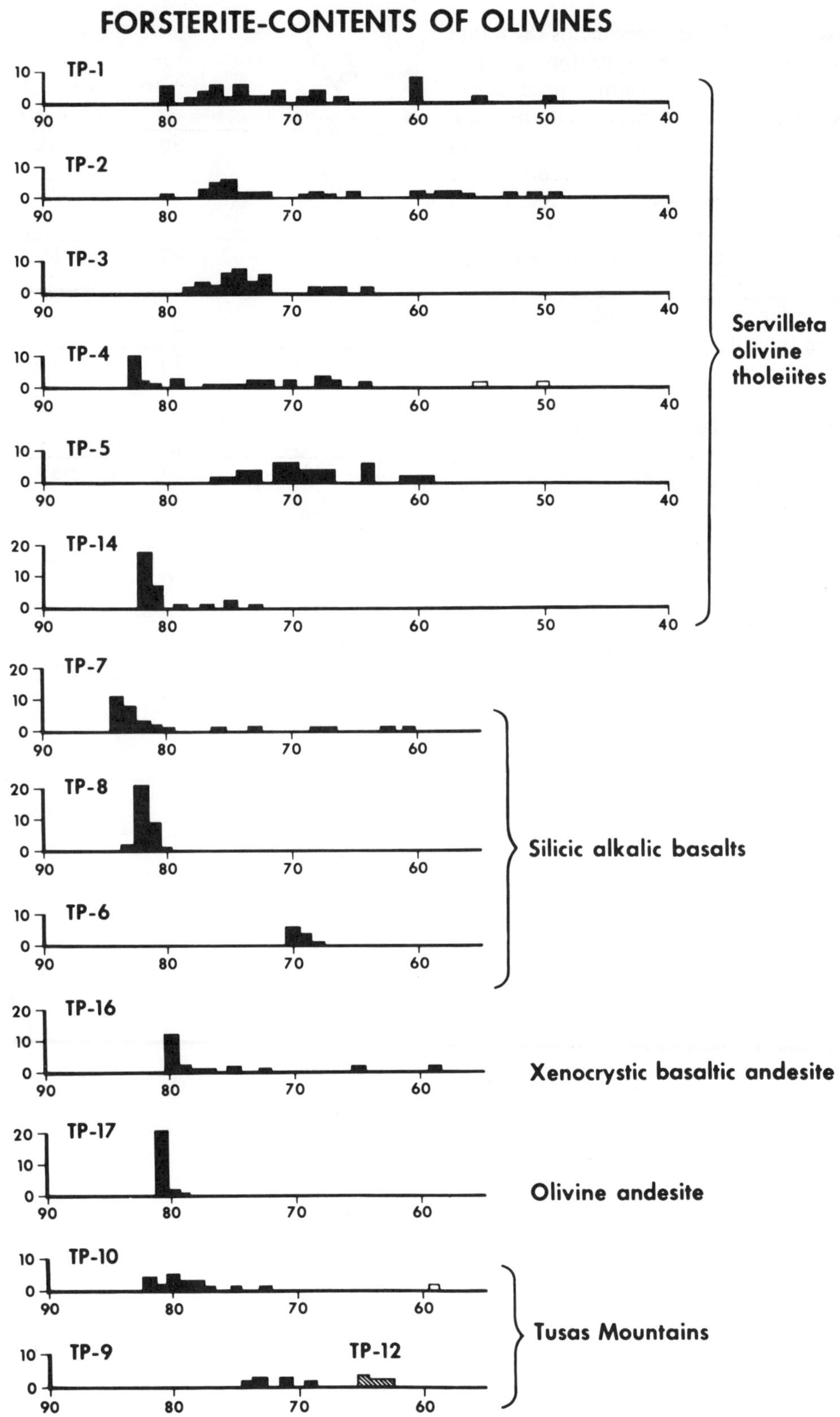

1.2.4.6 Histograms of olivine compositions for the Taos Plateau reference suite (three element analyses for Si, Fe, Mg corrected by Bence-Albee method).

and TP-17) are identical. These experiments show that these rocks are essentially liquid compositions and provide further confirmation of the lack of dependence of olivine-liquid K_Ds (Roeder and Emslie, 1970) on compositional parameters other than Mg/Fe ratio. Many of the analyses of Servilleta tholeiites do not correspond precisely to those of the erupted liquids because of post-emplacement differentiation (Haskin *et al.*, 1978) and because the xenocrystic samples may have been substantially modified by assimilation.

The maximum forsterite content of olivine phenocrysts in Servilleta tholeiites ranges from 82.5–77 (Fig. 1.2.4.6), but Fo content does not correlate with either phenocryst grain size or Mg′-value. Indeed, the sample (TP-5) which is most depleted in lithophile elements contains abundant small phenocrysts of Fo_{77-69}, and the most enriched samples (TP-4 and TP-14) are characterized by large (>2 mm), relatively magnesian phenocrysts ($Fo_{82.5-80}$). TP-4 also has the lowest Mg′-value (0.56) of these samples.

In general, those samples with a homogeneous size population of olivine phenocrysts show a strong maximum in core compositions corresponding to the most magnesian composition in the rock. Samples with rapidly quenched groundmasses lack marked iron-enrichment, whereas coarse-grained diktytaxitic Servilleta tholeiites and Servilleta-type flows show a broad distribution of phenocryst core compositions and contain groundmass olivines ranging up to comparatively iron-rich compositions (Fo_{45} in the vesicular segregations).

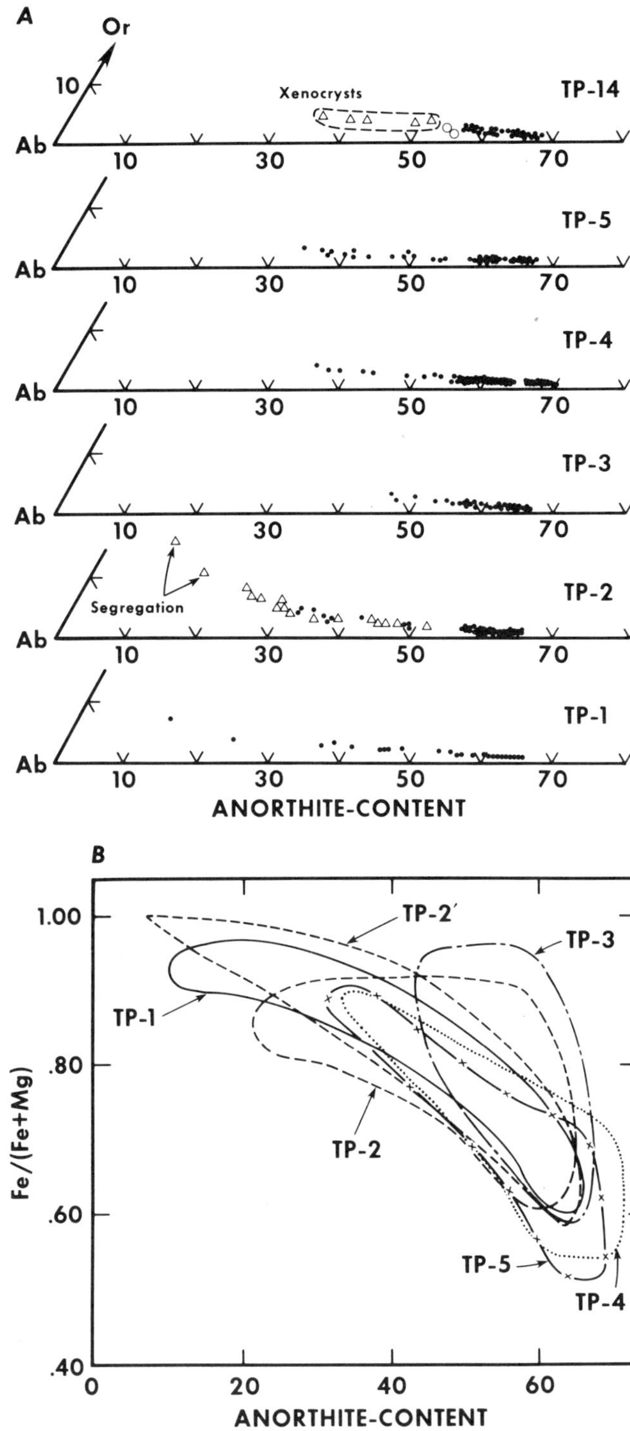

1.2.4.7 Plagioclase compositions in Servilleta basalts. Plagioclase compositions in the early rift andesite (TP-21), the olivine andesite (TP-17) and three rocks from the Tusas Mountains.

Plagioclase

In performing seven-element analyses of cores and rims of plagioclase feldspars care was taken to bracket the range of zoning in each sample. Analyses were also collected in linear core-rim traverses. These data are presented on Ab-An-Or ternary diagrams and in plots of Fe/(Fe + Mg) vs. anorthite content (Fig. 1.2.4.7). In general, Fe/(Fe + Mg) increases with decreasing anorthite component, but the trajectory of the trend and the absolute Fe and Mg concentrations varies substantially among rock types. These variations apparently depend on crystallization history and on bulk lava composition.

The cores of plagioclase phenocrysts in Servilleta tholeiites have maximum An-contents of 72–66 (Fig. 1.2.4.7a). Those rocks with the most anorthitic plagioclase also have the most magnesian olivines. Although the cores are weakly zoned, rims show strong normal zoning. The apparent early Fe/Mg increase at An_{55-70} is due to albite-anorthite oscillatory zoning coupled with a more monotonic increase in Fe/Mg. Plagioclases in relatively rapidly-cooled samples nearer the margins of flows (TP-3, TP-14) continue the trend of strongly increasing Fe/Mg in contrast to those in more slowly cooled rocks which are much more albitic at the same Fe/(Fe + Mg). As with the olivine phenocryst compositions, the maximum anorthite contents of plagioclase phenocrysts in the Servilleta tholeiites are similar to those in other Taos Plateau magma types. The lower K_2O of the olivine tholeiites is, however, reflected in lower Or for a given anorthite content (Figs. 1.2.4.7a, 8a, 9a). The calcic cores of plagioclase phenocrysts in Servilleta lavas have substantially lower Fe/(Fe + Mg) than do the feldspars of any other magma type in the Taos field (Fig. 1.2.4.7b). The apparent cause of this distinction is that the tholeiites were erupted with plagioclase phenocrysts present, whereas plagioclase is confined to the groundmass in most of the other rocks. During rapid cooling subsequent to eruption, plagioclase nucleation is suppressed, and abundant ferromagnesian minerals crystallize prior to the onset of plagioclase crystallization. Comparisons between Servilleta tholeiites and the other lavas indicate that, where plagioclase co-crystallizes with groundmass pyroxene (± olivine), early plagioclase will have comparatively high Fe/(Fe + Mg), and the trend will be one of strong magnesium-depletion in combination with limited An-Ab-Or zoning. This type of trend is well defined by the trends in the andesite and Tusas lavas (Fig. 1.2.4.9b).

Among the silicic alkalic basalts, plagioclase in TP-8 is confined entirely to its groundmass and has the

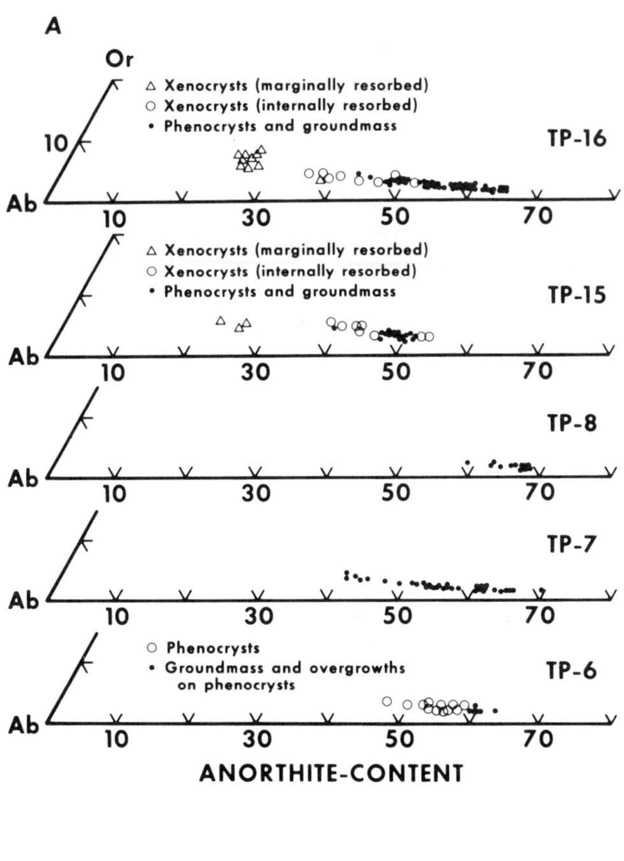

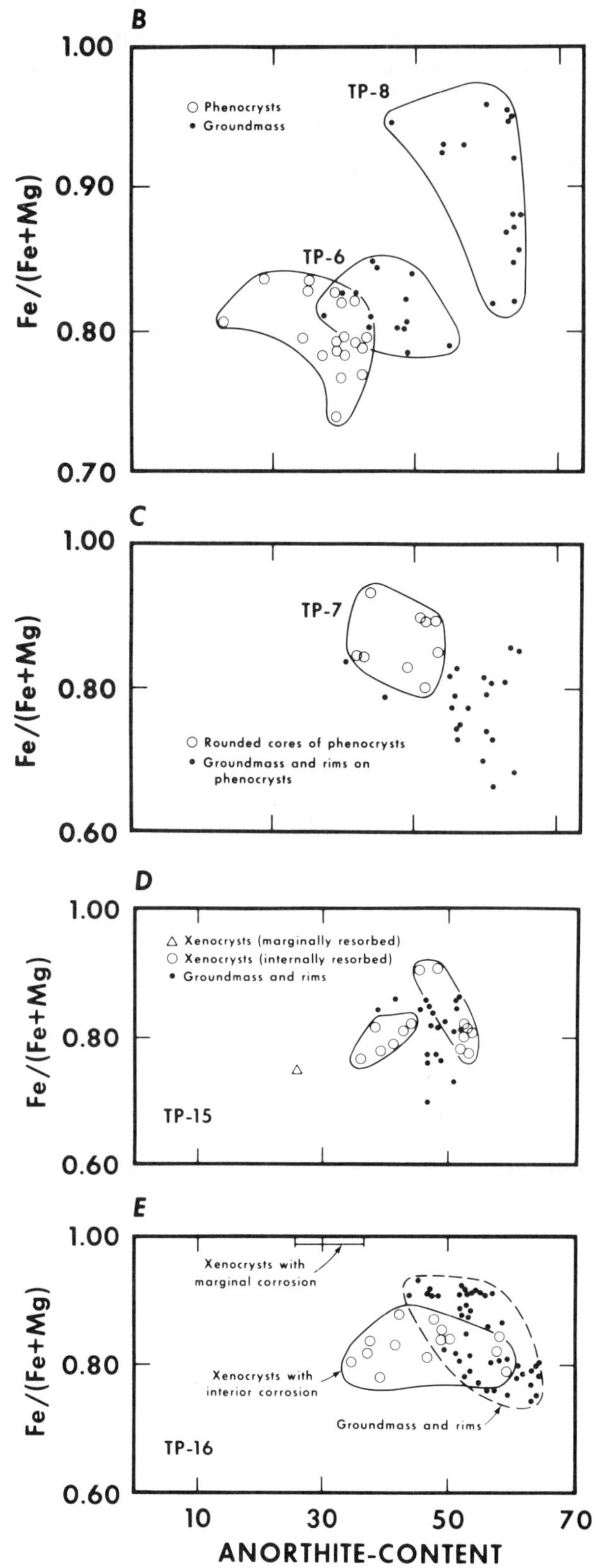

1.2.4.8 Plagioclase compositions in the silicic alkalic basalts and xenocrystic basaltic andesites.

compositional characteristics of plagioclase in the andesites (Fig. 1.2.4.8). Samples TP-6 and TP-7 contain sparse,

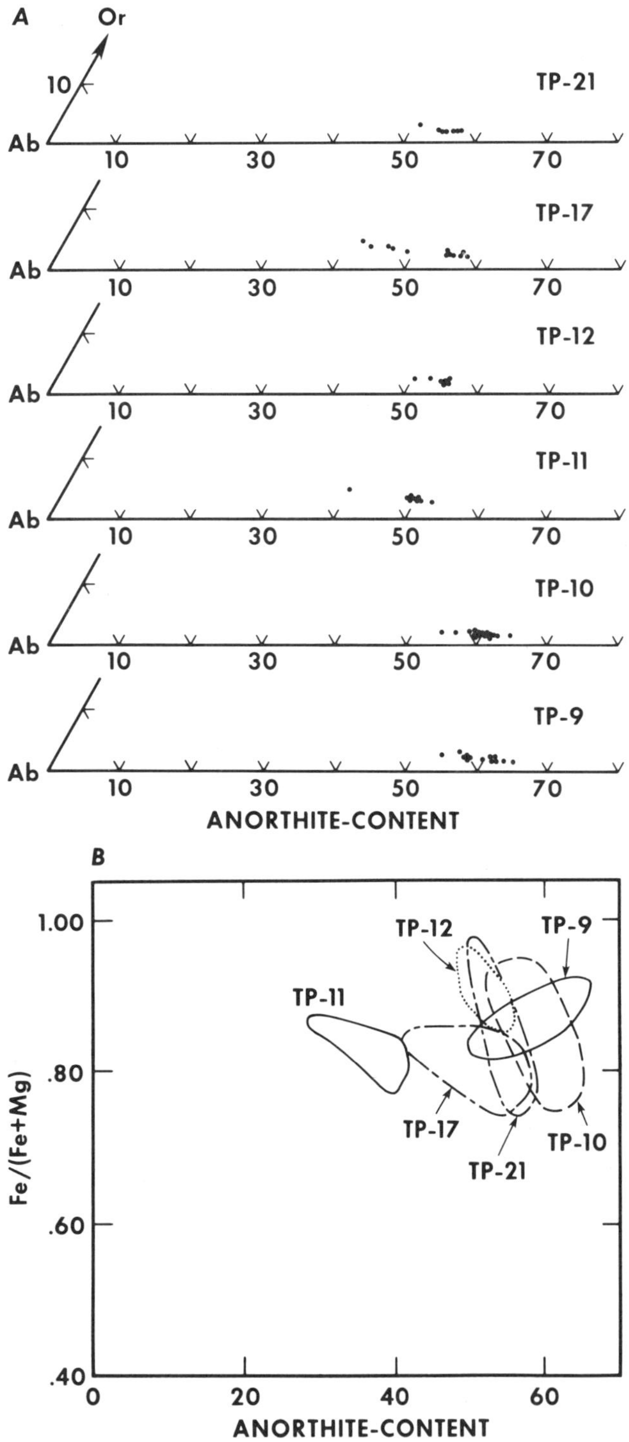

1.2.4.9 Plagioclase compositions in the early rift andesite (TP-21), the olivine andesite (TP-17) and four rocks from the Tusas Mountains.

complexly zoned phenocrysts whose cores are more sodic than the groundmass plagioclase and the overgrowths on the phenocrysts. In TP-7 the phenocrysts are composite, consisting of rounded cores zoned outward from An_{51} to An_{42}. These are mantled by reversely zoned overgrowths which range in composition from An_{56} (adjacent to the core) to An_{64}. The last, outer rim overgrowth is An_{40-42}. The compositions of groundmass laths are equivalent to the reversely zoned overgrowths. Vermicular resorption textures are characteristic of the plagioclase phenocrysts and are present in some laths as well. Cores of the lath-shaped plagioclase phenocrysts in TP-6 are also more calcic than their rims or the groundmass plagioclase. The large ones are weakly zoned from An_{57} to An_{46} and are mantled by overgrowths of An_{62-53}. Although these phenocrysts are not internally resorbed, the cores have rounded corners and are separated from the rims by a sharp hiatus. The origin of the disequilibrium phenocrysts in these rocks remains problematic, but appears to have been different from that of the resorbed xenocrysts present in the Taos Plateau xenocrystic basaltic andesites (also in Fig. 1.2.4.8).

The two petrographic types of xenocrystic plagioclase in the basaltic andesites are distinct compositionally as well as texturally. The marginally resorbed feldspars are oligoclase-andesine. The internally resorbed grains overlap in composition with their unresorbed rims and with the groundmass plagioclase, in terms of ternary feldspar components and of Fe/Mg ratio. This correspondence may be due in part to reequilibration during resorption. The Fe/(Fe + Mg) values for the oligoclase xenocrysts are not very meaningful because the MgO content is below the microprobe detection limit in most grains. Although TP-15 is a much more fractionated rock than TP-16, with plagioclase phenocrysts of An_{52-47} compared with the typical basaltic range of An_{66-50}, the xenocrysts are compositionally very similar in the two samples.

Pyroxenes

Pyroxene occurs in the Servilleta tholeiites as a groundmass phase. Augitic pyroxene, ranging in texture from intergranular to subophitic or ophitic, encloses olivine phenocrysts and plagioclase laths. These grains do not exhibit strong sector zoning or marked Fe/Mg variations in most samples, but the typical range is extended by more strongly zoned grains in the vesicular segregations. However, zoning of titanium and aluminum is significant. The cores of the grains are relatively aluminous compared to the rims and have Ti/Al ratios of 0.2–0.25 (Fig. 1.2.4.10b). Core-rim traverses reveal a consistent trend of decreasing calcium, chromium and aluminum contents coupled with constant or slightly increasing TiO_2. The outermost rims show a marked decrease in TiO_2, probably related to Fe-Ti oxide crystallization. Sample TP-3, which was taken from the lower portion of a flow directly overlying another flow, has a unique texture and trend of pyroxene zoning. The ophitic grains are zoned from typical calcic augite to pigeonite. This trend shows a more pronounced decrease in Ti and Al than the high-calcium pyroxenes. The Tusas Mountain basalt (TP-10) shows compositional trends in its groundmass pyroxenes which are very similar to those of Servilleta basalts (Fig. 1.2.4.11).

Pyroxene composition relations in silicic alkalic basalts vary as a function of bulk composition and cooling history. Sample TP-7, which has very complexly zoned plagioclase phenocrysts, contains two compositional varieties of clinopyroxene. Analyses of a single resorbed grain of very calcic ($Wo_{47}En_{40}Fs_{13}$) and low Ti:Al (1:30) pyroxene are quite distinct from those of more normal groundmass grains, which generally are similar to the typical Servilleta basalt pyroxenes. The peraluminous andesite (TP-9) from the Tusas Mountains also contains two generations of clinopyroxene. Sparse, rounded (but not corroded) phenocrysts are comparatively calcic but have similar or lower Mg/(Mg + Fe) ratios than groundmass pyroxenes (note that groundmass plagioclase also shows reverse Fe/Mg zoning).

The groundmass textures of TP-6 and TP-8 are very fine-grained. Pyroxene was not analyzed. Andesites TP-11 and TP-17 lack pyroxene phenocrysts, but both contain two varieties of groundmass pyroxene. The olivine andesite (TP-17) contains relatively abundant augite and minor low-calcium orthopyroxene. Resorbed orthopyroxene xenocrysts are distinctly more magnesian than the groundmass grains. The groundmass pyroxenes in the substantially more evolved andesite (TP-11) have much lower Mg/(Mg + Fe) ratios. The augites are sub-calcic, and the low-calcium phase probably is pigeonite.

The remaining three andesitic rocks (TP-15, TP-16, and TP-21) all contain phenocrysts of augite and orthopyroxene which are very similar in composition,

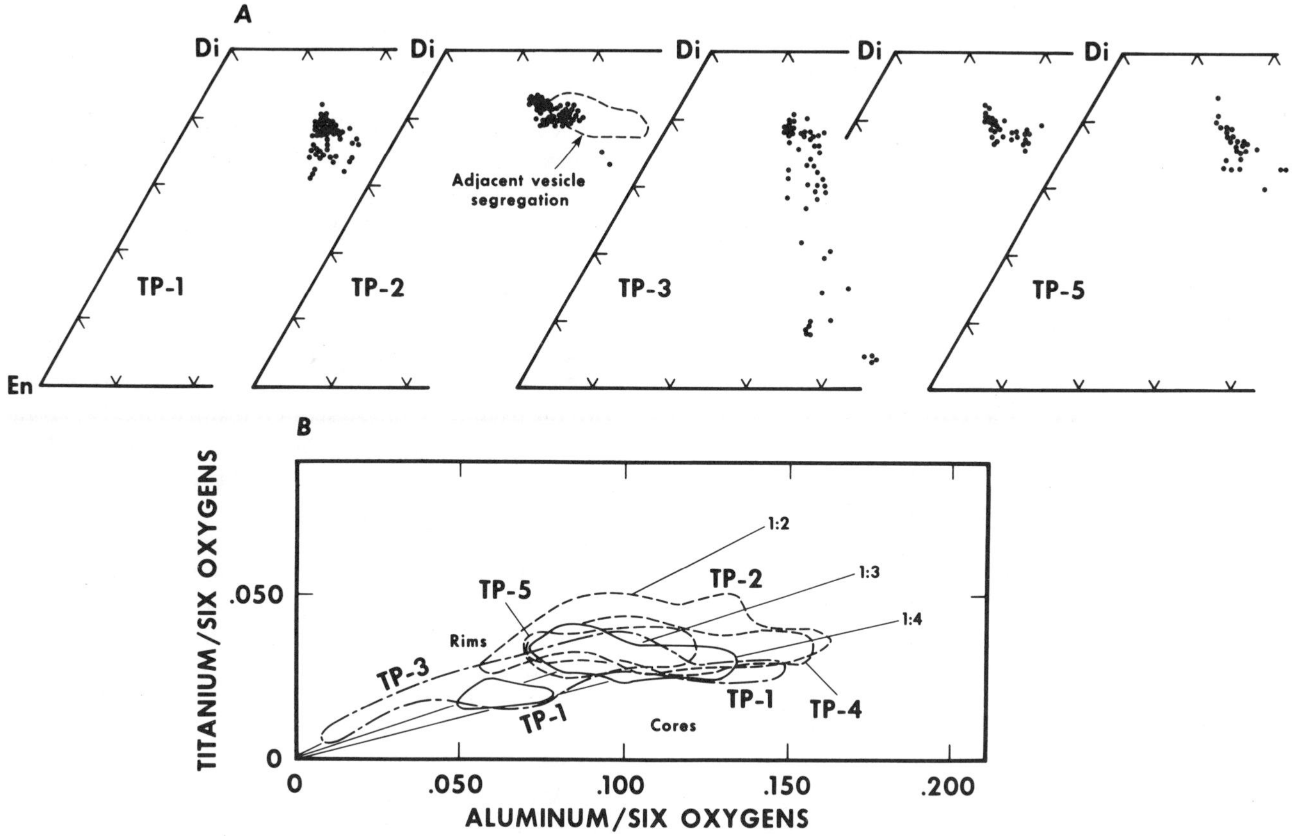

1.2.4.10 Pyroxene compositions in Servilleta basalts (quadrilateral components calculated on the basis of total Fe as FeO).

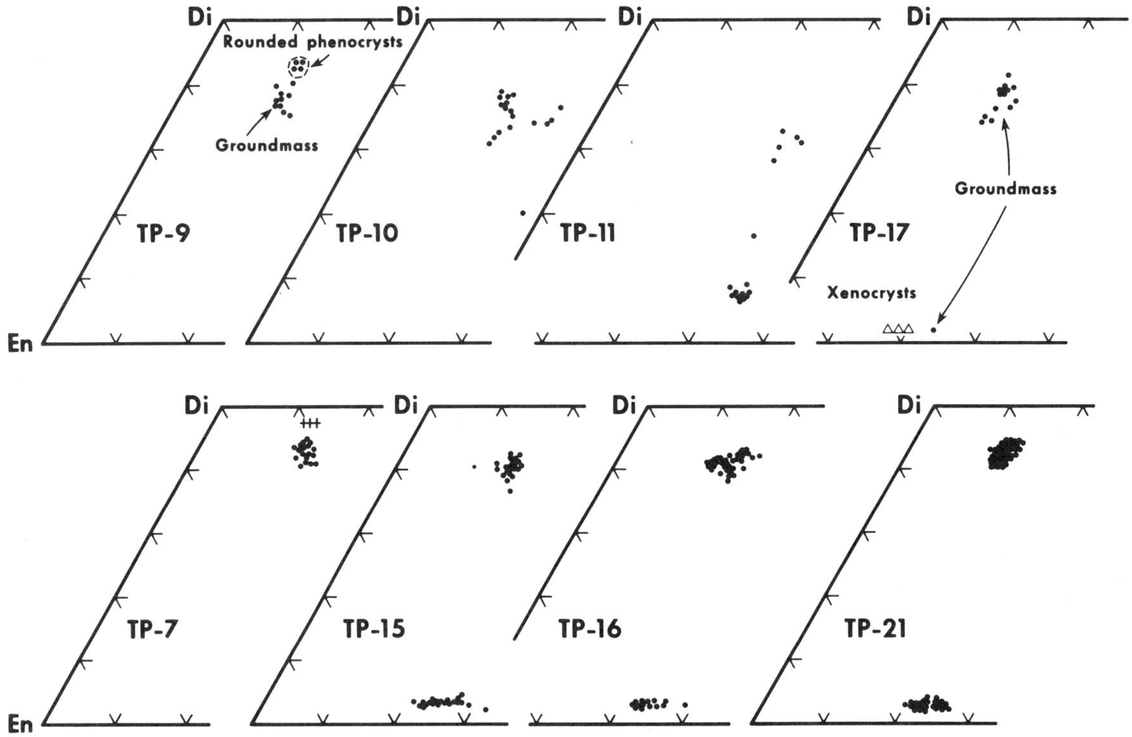

1.2.4.11 Pyroxene compositions in other basalts and andesites of the Taos Plateau reference suite.

in terms of quadrilateral components and Ti:Al ratios. Both TP-21 and TP-15 show strong zoning from low-Al cores to high-Al rims and only minor Fe/Mg zoning. The orthopyroxenes have lower Ti and Al contents than do the coexisting calcic pyroxenes and also have lower Ti:Al ratios. Similar but less pronounced zoning trends are present in TP-16.

1.2.4.6 Petrogenesis

The basaltic volcanic rocks of some rifts are dominantly alkalic, consisting of a spectrum of transitional or slightly undersaturated alkalic basalts, basanites and nephelinites. Volcanism in the northern Rio Grande rift is characterized by abundant tholeiitic basalts, as typified by the Taos Plateau field, but within the rift as a whole a large range of undersaturated rocks is present. This spatial petrologic pattern broadly correlates with geophysically derived models of subrift crust-mantle structure. Beneath the axis of the rift, mantle material is thought to well up and give rise to basaltic magmas that equilibrate with their mantle residues at relatively shallow depths compared to those on the flanks. In the southern Rio Grande rift, the axial geophysical anomalies and structural expression of the rift are indicative of a broader and less effective mantle upwelling, and the axial basalts are more alkalic in nature than those to the north where the axial graben is well defined.

Superimposed on this regional pattern are volcanic fields, such as the Taos Plateau, in which diverse magma types have erupted closely associated in space and time. We have identified three mafic rock types and several more silicic types (rhyodacite-rhyolite). Each of these has erupted from discrete volcanic centers which, with the exception of the scattered silicic alkalic basalt cinder cones, are arranged in a concentric distribution (Servilleta and olivine andesites at the center, rhyodacites at greater distance), suggesting a genetic relationship to a single thermal anomaly for the entire field. In contrast to the chemical diversity among the three mafic rock types, they are very similar isotopically. The Nd isotopic ratios of these three rock types are virtually identical, and very little spread has been documented for the Sr isotopic ratios. Early rift volcanic rocks of basaltic to andesitic composition share these isotopic characteristics.

Available data place several loose constraints on petrogenesis of the Taos Plateau mafic magmas. The diversity of major and trace element geochemistry is not consistent with derivation from a common parent magma. Isotopic evidence suggests that partial melting in a mantle containing long-lived heterogeneities cannot be called upon to produce the observed diversity. The concentric geometry of volcanic rock types implies that the diversity is due to variable conditions of genesis related to a single melting anomaly, but the available data do not uniquely define what these variables may have been.

The uncertainties are due to several factors. All of the magmas are fractionated derivatives of more primitive magmas, unless the mantle beneath the Taos Plateau field is more iron-rich than typical mantle. Thus, primary magmas from which inferences concerning magma generation are ideally derived are not available. Furthermore, potential factors contributing to the observed diversity include such variables as depth of melting, degree of melting, mode of extraction from the source region, presence or absence of volatile components, composition of the volatile phase, and the possibility that the mantle was metasomatically modified by a migratory fluid phase prior to magma generation.

The low abundances of lithophile elements, the large volume of erupted material, and the apparently constant composition of the Servilleta lavas are consistent with generation of magmas in a mantle diapir by relatively large degrees of melting. The much higher concentrations of lithophile elements in the silicic alkalic basalts are suggestive either of smaller degrees of partial melting in a similar source region or of enrichment of these elements to varying degrees by a precursor metasomatic event. In the case of silica-saturated olivine andesites, which are also the most enriched in lithophile elements, infiltration of H_2O-rich vapor is a tempting means for producing the enrichment and for contracting the enstatite phase volume in the source region, thereby giving rise to calc-alkaline magmas (see section 3.3.7). However, we have no internally consistent model for accounting for the characteristics of both the andesites and silicic alkalic basalts as variations of this process.

Variable crustal contamination of a single parental magma type is insufficient to explain the diversity among the three mafic magma types in light of their isotopic similarities. The xenocryst-rich lavas, in which there is abundant petrographic support for contamination, have not been analyzed for both Sr and Nd isotopes. The isotopic data, by themselves, do not rule out minor contamination by lower crustal granulites of all the mafic magma types. Interpretation of the isotopic characteristics of the mafic lavas is best done in light of data gathered on the Taos Plateau rhyodacites, which occur as the outermost volcanoes in the field. Five rhyodacites are similar to each other and to the basalts and andesites in their $^{87}Sr/^{86}Sr$ ratios of 0.70451–0.70497 but differ entirely in $^{143}Nd/^{144}Nd$ ratio from the basalts and andesites. On a plot of $^{143}Nd/^{144}Nd$ vs. $^{87}Sr/^{86}Sr$ (Fig. 1.2.4.12), the combined isotopic data from rhyodacites, andesites, and basalts define a linear array which we interpret as a "mixing" line between mantle derived liquids and lower crustal mafic granulites. In fact, mafic granulites are the only reasonable choice for the low $^{143}Nd/^{144}Nd$ end-member of the mixing line, because they alone have been found to have sufficiently low Rb contents combined with REE patterns similar to those of upper crustal rocks (Heier and Thorensen, 1971; Green *et al.*, 1972). The data rule out contamination by upper crust or by intermediate lower crust. Similar interpretations, involving varying degrees of modification of mantle-derived magmas by assimilation of Precambrian mafic granulites of the lower continental crust, have been proposed (1) on the basis of Sr and Nd isotopic data for the Tertiary volcanic rocks in northwest Scotland (Carter *et al.*, 1978), and (2) from Pb and Sr isotopic data for magmatism in the San Juan volcanic field, Colorado (Lipman *et al.*, 1978).

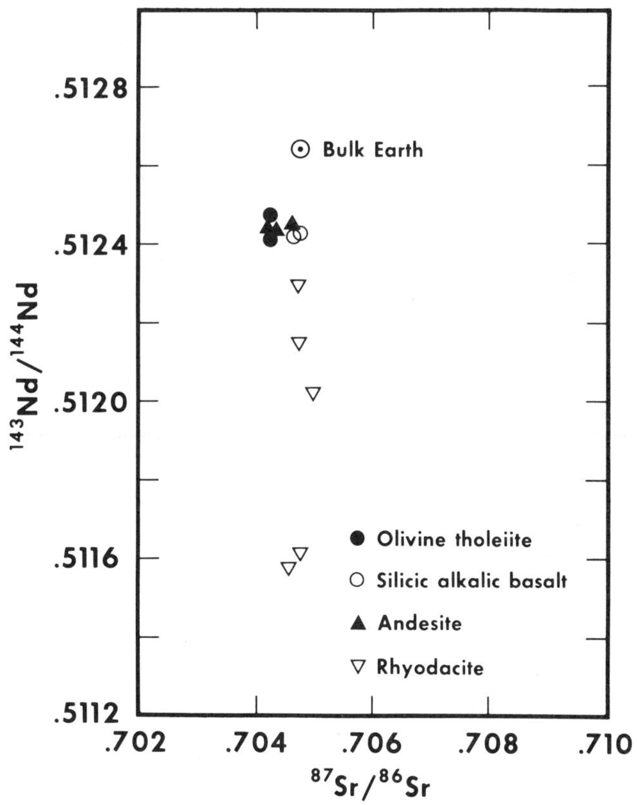

1.2.4.12 Plot of $^{143}Nd/^{144}Nd$ vs. $^{87}Sr/^{86}Sr$ for lavas of the Taos Plateau. Estimated present-day bulk earth composition from O'Nions *et al.*(1979).

For the high $^{143}Nd/^{144}Nd$ end-member, two possibilities are reasonable. One is that the basalts and andesites are one end-member, in which case the mantle-derived liquids tap a slightly light REE-enriched portion of the mantle. The other possibility for this end-member would be a mantle-derived liquid with Nd and Sr isotopic compositions equal to those inferred for a chondritic uniform reservoir. In this case, all young volcanics thus far studied from the Taos Plateau, including the olivine tholeiites, would be mixing products of mafic lower crustal granulite and mantle derived melts.

1.2.4.7 References

Aoki K. I. (1967) Petrography and petrochemistry of latest Pliocene olivine-tholeiites of Taos area, northern New Mexico, U.S.A. *Contrib. Mineral. Petrol.* **14**, 191–203.

Aoki K. I. and Kudo A. M. (1976) Major element variations of late Cenozoic basalts of New Mexico. *New Mexico Geol. Soc. Spec. Publ. 5*, 88–82.

Bachman G. D. and Mehnert H. H. (1978) New K-Ar dates and the late Pliocene to Holocene geomorphic history of the central Rio Grande region, New Mexico. *Geol. Soc. Amer. Bull.* **89**, 283–292.

Baker B. H., Mohr P. A. and Williams L. A. J. (1972) Geology of the Eastern Rift System of Africa. *Geol. Soc. Amer. Spec. Paper 136*.

Baker I. and Ridley W. I. (1970) Field evidence of K, Rb, Sr data bearing on the origin of the Mt. Taylor volcanic field, New Mexico, U.S.A. *Earth Planet. Sci. Lett.* **10**, 106–114.

Baldridge W. S. (1979a) Petrology and petrogenesis of Plio-Pleistocene basaltic rocks from the central Rio Grande Rift, New Mexico and their relation to rift structure. In *Rio Grande Rift: Tectonics and Magmatism* (R. Riecker, ed.), p. 323–353. Amer. Geophys. Union, Washington, D.C.

Baldridge W. S. (1979b) Mafic and ultramafic inclusion suites from the Rio Grande Rift (New Mexico) and their bearing on the composition and thermal state of the lithosphere. *J. Volc. Geotherm. Res.* **6**, 319–351.

Barberi F., Santacroce R., Ferrara G., Treuil M. and Varet J. (1975) A Transitional basalt-pantellerite sequence of fractional crystallization, the Boina Centre (Afar Rift, Ethiopia). *J. Petrol.* **16**, 22–56.

Best M. G. and Brimhall W. H. (1974) Late Cenozoic alkalic basaltic magmas in the western Colorado Plateau and the Basin and Range transition, U.S.A., and their bearing on mantle dynamics. *Geol. Soc. Amer. Bull.* **84**, 3269–3278.

Butler A. P., Jr. (1947) Tertiary and Quaternary geology of the Tusas-Tres Piedras area, New Mexico. Ph.D. dissertation, Harvard University, 188 pp.

Butler A. P., Jr. (1971) Tertiary volcanic stratigraphy of the eastern Tusas Mountains, southwest of the San Luis Valley, Colorado-New Mexico. In *New Mexico Geol. Soc. 22nd Ann. Field Conf. Guidebook, San Luis Basin*, p. 289–300.

Carter S. R., Evensen N. M., Hamilton P. J. and O'Nions R. K. (1978) Neodymium and strontium isotope evidence for crustal contamination of continental volcanics. *Science* **202**, 743–747.

Chapin C. E. and Seager W. R. (1975) Evolution of the Rio Grande Rift in the Socorro and Las Cruces areas. In *New Mexico Geol. Soc. 26th Ann. Field Conf. Guidebook*, p. 297–321.

Chapin C. E., Chamberlin R. M., Osborn G. R., Şanford A. R. and White D. W. (1978) Exploration framework of the Socorro Geothermal area, New Mexico. In *Field Guide to Selected Caldrons and Mining Districts of the Datil-Mogollon Volcanic Field, New Mexico*, p. 131–138. New Mexico Geol. Soc. Spec. Publ. 7.

Chayes F. (1969) The chemical composition of Cenozoic andesite. In *Proc. Andesite Conf.* (A. R. McBirney, ed.), p. 1–11. Dept. Geol. Mineral. Ind. State of Oregon Bull. 65.

Christiansen R. L. and Lipman P. W. (1972) Cenozoic volcanism and plate-tectonic evolution of the Western United States. II Late Cenozoic. *Phil. Trans. Roy. Soc. Lond.* **A271**, 249–284.

Coney P. J. and Reynolds S. J. (1977) Cordilleran benioff zones. *Nature* **270**, 403–406.

Cordell L. (1978) Regional geophysical setting of the Rio Grande rift. *Geol. Soc. Amer. Bull.* **89**, 1073–1090.

Crossey L. J. (1979) Intraflow chemical variations in Servilleta Basalts; Taos Plateau, New Mexico. M. S. thesis, Washington University, St. Louis, Mo.

DePaolo D. J. and Wasserburg G. J. (1976) Inferences about magma sources and mantle structure from variations of $^{143}Nd/^{144}Nd$. *Geophys. Res. Lett.* **3**, 743–746.

Doe B. R., Lipman P. W., Hedge C. E. and Durasawa H. (1969) Primitive and contaminated basalts from the southern Rocky Mountains, U.S.A. *Contrib. Mineral. Petrol.* **21**, 1112–1156.

Everson J. E. and Silver L. T. (1978) Lead systematics of late Cenozoic basalts from the Rio Grande Rift. In *Proc. Internatl. Symp. on the Rio Grande Rift*, p. 36–37. Los Alamos Scientific Laboratory, Los Alamos, New Mexico.

Gass I. G. (1970) The evolution of volcanism in the junction area of the Red Sea, Gulf of Aden and Ethiopian rifts. *Phil. Trans. Roy. Soc. Lond.* **A267**, 369–381.

Gass I. G. (1975) Tectonic and magmatic evolution of the Afro-Arabian dome. In *African Magmatism and Tectonics* (T. N. Clifford and I. G Gass, eds.), p. 285–300. Oliver and Boyd, Edinburgh.

Green T. H., Brunfelt A. O. and Heier K. S. (1972) Rare-earth element distribution and K/Rb ratios in granulites, mangerites and anorthosites, Lofoten-Vesteraalen, Norway. *Geochim. Cosmochim. Acta* **36**, 241–257.

Girdler R. W. (1976) The East African Rift System—Geophysical aspects. In *Geodynamics Progress and Prospects*, (C. L. Drake, ed.), p. 75–80. Amer. Geophys. Union, Washington, D.C.

Haskin L. A., Lindstrom M. E. and Dungan M. A. (1978) Trace element fractionation during crystallization of thin tholeiitic flows (abstract). In *Lunar Science IX*, 468–470. The Lunar Science Institute, Houston.

Hausel W. D. and Nash W. P. (1977) Petrology of Tertiary and Quaternary volcanic rocks, Washington County, southwestern Utah. *Geol. Soc. Amer. Bull.* **88**, 1831–1842.

Hedge C. E. and Lipman P. W. (1972) Upper Cenozoic basalts of the southern Rocky Mountain region: Part 2, Strontium isotopes and the geochemistry of Sr, Rb and Ba. *Abstracts with Program*, Geol. Soc. Amer. **4**, 169.

Heier K. S. and Thorensen K. (1971) Geochemistry of high grade metamorphic rocks, Lofoten-Vesteraalen, North Norway. *Geochim. Cosmochim. Acta* **35**, 89–99.

Irvine T. N. and Baragar W. R. A. (1971) A guide to chemical classification of the common volcanic rocks. *Can. J. Earth Sci.* **8**, 523–548.

Kay R. W. (1977) Geochemical constraint on the origin of Alveton magmas. In *Island Arcs, Deep Sea Trenches and Back-Arc Basins* (M. Talwani and W. C. Pitman, III, eds.), p. 229–242. Amer. Geophys. Union, Washington, D.C.

Kelley V. C. and Kudo A. M. (1978) Volcanoes and related basalts of Albuquerque Basin, New Mexico. *New Mexico Mines and Min. Res. Circular*, 156.

King B. C. and Williams L. A. J. (1976) The East African Rift System. In *Geodynamics Progress and Prospects* (C. L. Drake, ed.), p. 63–74. Amer. Geophys. Union, Washington, D.C.

Kuno H. (1960) High-alumina basalt. *J. Petrol.* **1**, 121–145.

Laughlin A. W., Brookins D. G. and Causey J. D. (1972) Late Cenozoic basalts from the Bandera lava field, Valencia County, New Mexico. *Geol. Soc. Amer. Bull.* **83**, 1543–1551.

Lipman P. W. (1980) Cenozoic volcanism in the western United States: implications for continental tectonics. In *Studies in Geophysics: Continental Tectonics*, p. 161–174. Nat'l. Academy of Sci., Washington, D.C.

Lipman P. W., Doe B. R., Hedge C. E. and Steven T. A. (1978) Petrologic evolution of the San Juan volcanic field, southwestern Colorado: Pb and Sr isotopic evidence. *Geol. Soc. Amer. Bull.* **89**, 59–82.

Lipman P. W. and Mehnert H. H. (1975) Late Cenozoic basaltic volcanism and development of the Rio Grande depression in the southern Rocky Mountains. *Geol. Soc. Amer. Mem.* **144**, 119–154.

Lipman P. W. and Mehnert H. H. (1979) The Taos Plateau volcanic field, northern Rio Grande Rift, New Mexico. In *Rio Grande Rift: Tectonics and Magmatism* (R. E. Rieker, ed.), p. 289–312. Amer. Geophys. Union, Washington, D.C.

Lipman P. W. and Moench R. H. (1972) Basalts of the Mount Taylor volcanic field, New Mexico. *Geol. Soc. Amer. Bull.* **83**, 1335–1344.

Lipman P. W., Prostka H. J. and Christiansen R. L. (1972) Cenozoic volcanism and plate tectonic evolution of the Western United States. I Early and Middle Cenozoic. *Phil. Trans. Roy. Soc. Lond.* **A271**, 217–248.

Macdonald G. A. and Katsura T. (1964) Chemical composition of Hawaiian lavas. *J. Petrol.* **5**, 82–123.

Manley K. (1979) Stratigraphy and structure of the Espanola Basin, Rio Grande Rift, New Mexico. In *Rio Grande Rift: Tectonics and Magmatism,* (R. E. Rieker, ed.), p. 71–86. Amer. Geophys. Union, Washington, D.C.

Mohr P. (1978) Origin and significance of continental rifts. *Proc. Internatl. Symp. on the Rio Grande Rift,* p. 59–62. Los Alamos Scientific Laboratory, Los Alamos, New Mexico.

O'Nions R. K., Hamilton P. J. and Evensen N. M. (1977) Variations in $^{143}Nd/^{144}Nd$ and $^{87}Sr/^{86}Sr$ ratios in oceanic basalts. *Earth Planet. Sci. Lett.* **34**, 13–22.

O'Nions R. K., Carter S. R., Evensen N. M. and Hamilton P. J. (1979) Geochemical and cosmochemical applications of Nd isotope analysis. *Ann. Rev. Earth Planet. Sci.* **7**, 11–38.

Pearce J. A. and Norry M. J. (1979) Petrogenetic implications of Ti, Zn, Y, and Nb variations in volcanic rocks. *Contrib. Mineral. Petrol.* **69**, 33–47.

Ramberg I. B., Cook R. A. and Smithson S. B. (1978) Structure of the Rio Grande rift in southern New Mexico and west Texas based on gravity interpretation. *Geol. Soc. Amer. Bull.* **89**, 107–123.

Roeder P. L. and Emslie P. F. (1970) Olivine-liquid equilibrium. *Contrib. Mineral. Petrol.* **29**, 275–289.

Smith R. L. and Bailey R. A. (1968) Stratigraphy, structure and volcanic evolution of the Jemez Mountains, New Mexico (abstract). In *Cenozoic Volcanism in the Southern Rocky Mountains,* Colorado School of Mines Quarterly **63**, 259–260.

Smith R. L., Bailey R. A. and Ross C. S. (1970) Geologic map of the Jemez Mountains, New Mexico. *U.S.G.S. Misc. Geol. Inv.* Map I-571.

Sorenson H., ed. (1974) *The Alkaline Rocks.* Wiley, London. 622 pp.

Stormer J. C. (1972) Mineralogy and petrology of the Raton-Clayton volcanic field, northeastern New Mexico. *Geol. Soc. Amer. Bull.* **83**, 3299–3322.

Stormer J. C. (1978) Contemporaneous calc-alkaline and basanite-nephelinite suites on the east flank of the Rio Grande depression. *Proc. Internatl. Symp. on the Rio Grande Rift,* p. 59–62. Los Alamos Scientific Laboratory, Los Alamos, New Mexico.

Suppe J., Powell L. and Berry R. (1975) Regional topography, seismicity, Quaternary volcanism, and the present-day tectonics of the western United States. *Amer. J. Sci.* **275A**, 397–436.

Tweto O. (1978) Tectonic map of Rio Grande rift system in Colorado. In *Guidebook to Rio Grande rift in New Mexico and Colorado: N. M. Bur. Mines and Mineral. Res. Circ.* **163**, sheet 1.

Williams S. and Murthy V. R. (1978) Genetic relationships of volcanic rocks from the Taos Plateau, Rio Grande Rift: Rb-Sr and Sm-Nd isotope systematics (abstract). In *1978 International Symposium on the Rio Grande Rift,* p. 100. Los Alamos Scientific Laboratory, Los Alamos, New Mexico.

Williams S. and Murthy V. R. (1979) Sources and genetic relationships of volcanic rocks from the northern Rio Grande Rift: Rb-Sr and Sm-Nd evidence. *EOS (Trans. Amer. Geophys. Union)* **60**, 407.

Woodward L. A., Callender J. F., Seager W. R., Chapin C. E., Gries J. C., Shaffer W. L. and Zilinski R. E. (1978) Geotectonic map of Rio Grande rift region in New Mexico, Chihuahua, and Texas. In *Guidebook to Rio Grande rift in New Mexico and Colorado: N. M. Bur. Mines and Mineral Res. Circ.* **163**, sheet 2.

Zimmerman C. and Kudo A. M. (1979) Geochemistry of andesites and related rocks, Rio Grande Rift, New Mexico. In *Rio Grande Rift: Tectonics and Magmatism* (R. Riecker, ed.), p. 355–381. Amer. Geophys. Union, Washington, D.C.

1.2.5 OCEAN-FLOOR BASALTIC VOLCANISM

1.2.5.1 Introduction

Basalt was first dredged from the ocean floor in 1905 (Wiseman, 1936). Since that time, and particularly in the past 15 years, extensive sampling programs have shown that basaltic lavas with unique chemical characteristics are perhaps the major component of the upper oceanic crust (e.g., Engel *et al.*, 1965; Kay *et al.*, 1970; Bryan *et al.*, 1976). Such basalts have been variously referred to as submarine basalts, sea- or ocean-floor basalts, oceanic basalts or tholeiites, abyssal basalts or tholeiites, and mid-ocean ridge basalts. Since it cannot be shown with certainty that all these basalts have erupted along mid-ocean ridges, and since not all are strictly tholeiitic, the less restrictive term "ocean-floor basalts" will be used here.

Early studies emphasized the unique and consistent chemical composition of ocean-floor basalts, and in particular their tholeiitic characteristics: constant silica values, low potassium abundances, low K/Na ratios, low magmaphile element content, and high K/Rb ratios (e.g., Engel *et al.*, 1965; Kay *et al.*, 1970). More recent studies, benefiting from the spatial and temporal control provided by detailed sampling programs involving drilling, dredging, and the use of submersibles, have shown that chemical and petrological diversity at a single site is frequently as great as that previously identified for ocean-floor basalts in general (e.g., Aumento *et al.*, 1971; Rhodes and Rodgers, 1972; Blanchard *et al.*, 1976; Hekinian *et al.*, 1976; Bryan and Moore, 1977; Bryan, 1979). Furthermore, basalts erupted along topographically "normal" segments of the ridge have different isotopic and trace element characteristics from those erupted along topographic "highs" or platforms associated with islands astride the ridge axis (e.g., Iceland, Azores, Galapagos, Bouvet, Reunion). Bryan *et al.* (1976) designate these two basalt types as Type-I and Type-II, respectively. Although the Type-II basalts have isotopic and trace element affinities with basalts from oceanic islands, they are indistinguishable from Type-I basalts in terms of petrography, mineralogy and major element chemistry. Type-I basalts appear to be less varied than Type-II basalts, which may also be accompanied by transitional tholeiites and alkali basalts (e.g., Muir and Tilley, 1964; Aumento *et al.*, 1971; Johnson, 1979; Wood *et al.*, 1979). Nevertheless, using chemical discriminants, both Type-I and Type-II basalts are readily distinguishable from basalts erupted in other tectonic regimes (Pearce and Cann, 1973; Floyd and Winchester, 1975; Pearce, 1976), with the possible exception of "primitive" tholeiitic basalts from island arcs.

Interest in ocean-floor basalts has been concomitant with the development of the plate tectonics hypothesis. It is now widely agreed that the vast majority of ocean-floor basalts have been erupted and emplaced along the Earth's 65,000 km interconnecting mid-ocean ridge system, and that volcanism along these spreading ridges is the principal mechanism for the formation of the oceanic crust. These basalts are the most voluminous eruptive rocks on Earth and play an important role in the differentiation of the Earth from a relatively primitive mantle to the fractionated crust.

The reference suite

Fifteen samples of ocean-floor basalt and a single rhyodacite have been selected to illustrate the compositional and mineralogical characteristics of ocean-ridge volcanism. Eleven of these are from the Atlantic Ocean and five from the Pacific Ocean. All of the samples are fresh, and many are glassy. Where glass is available, data have been obtained on glass as well as on whole-rock samples. Highly phyric basalts have been avoided in order to minimize compositional effects due to crystal accumulation.

Attempts have been made to include a spectrum of compositions, from "primitive" to evolved for both Type-I and Type-II basalts. Included are: primitive (OF-1, OF-3, OF-5), normal (OF-2, OF-4, OF-6), and evolved (OF-9, OF-10) Type-II basalts from the FAMOUS area of the mid-Atlantic Ridge near 36°N; primitive (OF-7, OF-8) and normal (OF-11) Type-I basalts from the mid-Atlantic Ridge at 28°N; a normal Type-I basalt (OF-16) from the East Pacific Rise; and highly evolved Type-I basalts from the Galapagos Ridge (OF-13, OF-15) and Juan de Fuca Ridge (OF-12). Sample OF-14 is a rhyodacite dredged from the Galapagos Ridge.

In this section, mineralogical and geochemical data obtained on these samples are integrated with data from the literature to provide an overview of ocean-floor basaltic volcanism. Detailed locality information, petrographic descriptions, modal data, normative compositions, and major and trace element abundances for these reference suite samples are given in the Appendix (section A-5).

1.2.5.2 Tectonic setting

Basalts, erupted as thin flows and pillows, have been dredged from the axial crests, median valleys and flanks of the Earth's mid-ocean ridge system, and from the walls of associated fracture zones. Where drilling has penetrated the volcanic basement beneath sedimentary cover, it is found that basaltic flows and pillows, mineralogically and chemically similar to those erupted at spreading ridges, are the prevalent rock type in all ocean basins. These ocean-floor basalts are erupted along the mid-ocean ridges, the constructional margins of the lithospheric plates, in response to mantle upwelling and melting, generating new oceanic crust as the lithospheric plates move apart. Half-spreading rates vary from as little as 1 cm/year on the Mid-Atlantic Ridge near Iceland, to more than 8 cm/year along segments of the East Pacific Rise. Consequently, the present ocean basins are young tectonic features, and ocean-floor basalts must become progressively older away from the mid-ocean ridges: predictions that have been confirmed by the results of the Deep Sea Drilling Project. The oldest ocean-floor basalts described to date are Jurassic in age and were sampled along the western margin of the Atlantic Ocean (Bryan *et al.*, 1977). These basalts are indistinguishable from modern basalts erupted on the Mid-Atlantic Ridge, and must have been erupted shortly after the separation of the Euro-African and North American plates.

Most of the basalts which have been drilled on the ocean floor were probably erupted at mid-ocean ridges. Obvious exceptions are those erupted at seamounts, and other examples of intra-plate volcanism. Chemically, those basalts are readily distinguishable from ocean-floor basalts, and are a minor component of the oceanic crust (see section 1.2.6). A small proportion of ocean-floor basalts may also have been produced along "leaky" fracture zones (Van Andel *et al.*, 1969; Thompson and Melson, 1972). Compositional differences between basalts erupted at spreading ridges, transform faults and aseismic ridges are discussed by Hekinian and Thompson (1976).

Eruption rates are low at mid-ocean ridges, in contrast to oceanic island or "hot spot" volcanism. Moore *et al.* (1974) estimate that about 8600 m^3/km/yr are erupted in the FAMOUS area as compared with about 90,000 m^3/km/yr on Iceland. However, the overall volcanic budget of the mid-ocean ridges is an order of magnitude greater than that of any other form of terrestrial volcanism. Estimates for mid-ocean ridge volcanism vary from as little as 5 km^3/yr to as much as 20 km^3/yr, depending on assumptions concerning the nature of layer 3 of the oceanic crust (Menard, 1967; Anderson, 1974; Schilling *et al.*, 1978). This contrasts with volumes of 1.5 and 1.7 km^3/yr estimated by Schilling *et al.* (1978) for "hot spot" and island arc volcanism, respectively.

Geophysical studies, dredging, drilling, and *in situ* examination by submersibles are providing valuable insights into magmatic processes at mid-ocean ridges. A coherent picture is beginning to emerge concerning the nature and formation of the oceanic crust, a process of profound significance to our understanding of terrestrial differentiation. Detailed geophysical studies (e.g., Talwani *et al.*, 1971; Peterson *et al.*, 1974; Christensen and Salisbury, 1975; Rosendahl *et al.*, 1976; Houtz and Ewing, 1976) have revealed greater complexity in the structure of the oceanic crust than the three-layer model initially proposed by Raitt (1963), indicating the existence of additional sub-layers and greater regional and secular diversity. Table 1.2.5.1 provides a summary of these results.

Layer 2 has been widely correlated with basalt, and where sampled this has proved to be the case. However, the velocities for layer 2A, which correspond closely with velocities determined from downhole logging experiments, are markedly less than those obtained from laboratory measurements on basalts. Kirkpatrick (1979) interprets these results to indicate a high porosity (13–41%) and low density (1.6–2.6) for layer 2A as a consequence of cavities and fracturing in the upper crust. Houtz and Ewing (1976) show that layer 2A thins away from mid-ocean ridges, and is frequently absent in crust older than 20–60 m.y. This is most likely a consequence of cementation accompanying diagenesis as the crust ages.

Layers 2B and 2C have higher seismic velocities, more commensurate with those measured in basaltic rocks and their metamorphosed equivalents. It is widely inferred, by analogy with ophiolite complexes, that

Table 1.2.5.1 Oceanic crustal structure derived from geophysical studies. (Typical values are shown in parentheses.)

Layer	Velocity, km/sec	Thickness, km
1	1.7 — 2.0	<1
2A	2.0 — 4.1 (3.6)	0.0 — 1.5 (0.5)
2B	4.0 — 5.6 (5.2)	0.6 — 1.3 (0.9)
3A	6.5 — 6.9 (6.9)	2.0 — 3.0 (2.5)
3B	7.0 — 7.5	2.0 — 5.0
Mantle	8.1	—

these layers are composed of pillow basalts and sheeted-dike intrusives emplaced along the axes of mid-ocean ridges (Cann, 1974; Christensen and Salisbury, 1975; Dewey and Kidd, 1977). Drilling has shown that thin, rapidly chilled (frequently glassy), extensive flows and lobate "pillow" structures account for a substantial part of the upper 600 m of the crust. Intrusive magmatic units such as dikes and sills appear to be minor in extent. Detailed studies in the FAMOUS area (Moore *et al.*, 1974; Bryan and Moore, 1977; Ballard and Van Andel, 1977) indicate that the extrusive contribution to layer 2 occurs within a very narrow zone (2–3 km) on the floor of the inner rift valley. Recent volcanic activity (< 20,000 yrs) is confined to a narrow zone less than 1 km wide along the central axis of the rift valley. Away from this central zone the basalts become progressively, though not uniformly, older and more fractured because of the pervasive tectonic activity. Faulting is apparently continuous, whereas volcanism is episodic.

The nature of layer 3 is less clearly understood. It is particularly important to recognize how much of layer 3 is generated at mid-ocean ridges as a consequence of intrusive processes and how much is produced by off-ridge processes including volcanism, underplating from the mantle, and intrusion of mobile, serpentinized mantle peridotites. An answer to this question is of fundamental importance: if layer 2 alone is the product of basaltic volcanism and layer 3 is dominated by underplating and hydrated mantle material, then the thermal budget and magnitude of mantle differentiation is drastically different than it would be if the entire 5–8 km thickness of the oceanic crust were a consequence of volcanic processes along a very narrow zone of activity.

There is compelling evidence to suggest that layer 3B increases in thickness away from the mid-ocean ridges (Christensen and Salisbury, 1975), growing at the expense of underlying asthenosphere either by off-ridge intrusions or by underplating of hydrated mantle (Clague and Straley, 1977). Seismic evidence concerning the upper part of layer 3 is less clear. This layer is well developed and is characterized by relatively uniform thicknesses and velocities away from mid-ocean ridges, but its presence at the ridge axes is a subject of debate (Whitmarsh, 1973; Fowler and Mathews, 1974; Rosendahl, 1976).

Petrological arguments, coupled with ophiolite analog studies, suggest that basaltic magma chambers persist at shallow levels in the crust immediately beneath ridge crests, and that the bulk of layer 3 is consequently composed of gabbros and related cumulate rocks which have crystallized from these magma bodies (e.g., Gass, 1968; Moore and Vine, 1971; Cann, 1974; Peterson *et al.*, 1974; Christensen and Salisbury, 1975; Bryan and Moore, 1977; Dewey and Kidd, 1977). Steady-state refilling and mixing of these magma bodies with batches of fresh magma (O'Hara, 1977) would ensure that the magma remains close to multiple-saturation throughout its crystallization history, producing gabbroic rocks, rather than the dunites and troctolites that should dominate crystallization in the absence of continued replenishment (Rhodes *et al.*, 1979; Rhodes and Dungan, 1979).

Geophysical evidence supports the concept of persistent magma chambers at mid-ocean ridge crests (Whitmarsh, 1973; Orcutt *et al.*, 1975; Rosendahl, 1976; Reid *et al.*, 1977). Rosendahl (1976) argues that a wedge-shaped low-velocity zone (magma chamber) containing about 30% melt exists in the crust beneath the axial block of the East Pacific Rise. The width of the magma body at its base is comparable to the width of the axial block (about 20 km), but it becomes very narrow in the upper crust. Undifferentiated layer 3 thickens away from the ridge crest at the expense of the inferred magma chamber; presumably having crystallized from the magma, as suggested by petrological reasoning. Rosendahl (1976) extends his model to other spreading ridges, suggesting that magma chambers may persist beneath other segments of the East Pacific Rise and the Reykjanes Ridge. He argues that magma bodies may be present at the axis of the Mid-Atlantic Ridge, but that, in contrast with those of the East Pacific Rise, they must be very narrow, on the order of 1–2 km. This last suggestion is supported by detailed studies in the FAMOUS area (Moore *et al.*, 1974; Bryan and Moore, 1977; Ballard and Van Andel, 1977).

In summary, a major consequence of these studies is the recognition that the oceanic crust is generated almost entirely by intrusive and extrusive volcanism at mid-ocean ridges as a consequence of lithospheric separation, and that this zone of volcanic activity and crustal formation is extremely narrow, on the order of a few kilometers. Figure 1.2.5.1 attempts to summarize, schematically, current thinking on the nature of the oceanic crust and mechanisms for its generation at the axes of mid-ocean ridges.

1.2.5.3 Petrography and mineralogy

The petrographic characteristics of ocean-floor basalts are established by their bulk compositions and by their emplacement and subsequent cooling histories. A spectrum of textures, ranging from glassy to to fan-spherulitic (variolitic) to intersertal to ophitic to cumulate is observed. Grain sizes are variable, and many of the basalts are porphyritic, including sparsely (< 10%)

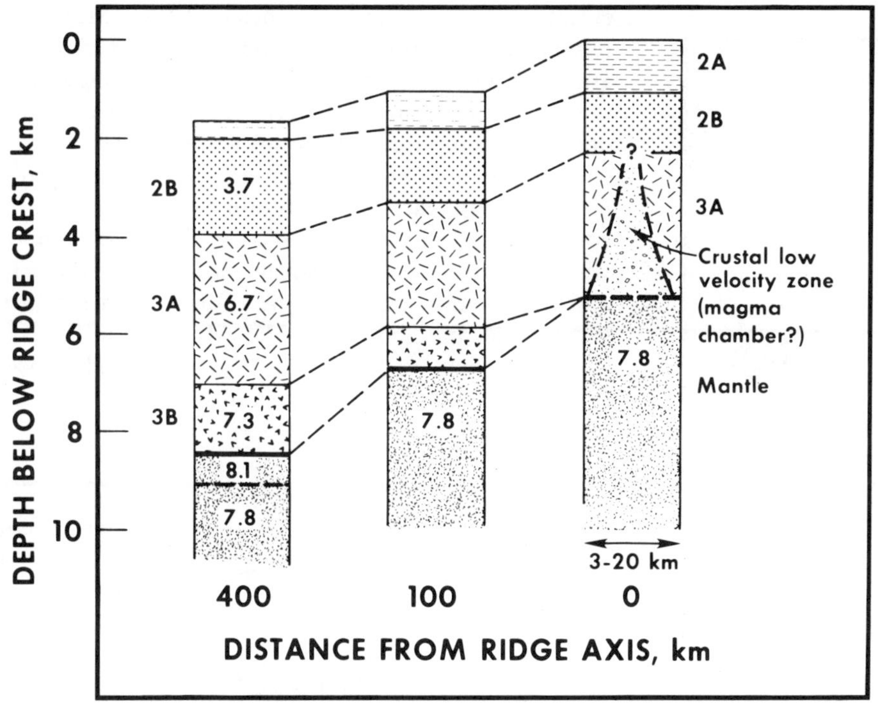

1.2.5.1 Schematic cross-sections of the oceanic crust at a mid-ocean ridge spreading center. Based on interpretations of Peterson *et al.* (1974), Christensen and Salisbury (1975), Houtz and Ewing (1976), and Rosendahl (1976).

and highly phyric (20–30%) variants. Phenocryst assemblages include: olivine (± Mg-Cr spinel); plagioclase plus olivine (± Mg-Cr spinel); and plagioclase plus olivine plus clinopyroxene. Petrographic classifications (Miyashiro *et al.*, 1970; Shido *et al.*, 1971; Bougault and Hekinian, 1974; Hekinian *et al.*, 1976), based upon phenocryst assemblages, recognize olivine tholeiites, plagioclase tholeiites and pyroxene tholeiites. Implicit in this classification is the assumption that the pyroxene tholeiites are evolved derivatives of the other two types, whereas the olivine and plagioclase tholeiites are thought to reflect distinct parental magmas, one which crystallized olivine first, the other plagioclase. Subsequent studies have shown that most phyric basalts are accumulative and that the common crystallization sequence for ocean-floor basalts is olivine, olivine + plagioclase, olivine + plagioclase + clinopyroxene (Bryan *et al.*, 1976; Donaldson and Brown, 1977; Bender *et al.*, 1978; Dungan *et al.*, 1978; Fisk, 1978; Rhodes *et al.*, 1979; Walker *et al.*, 1979).

Olivine in these basalts ranges from skeletal-euhedral forms (in the quench glass of pillow rims) to largely anhedral grains with alteration rims in fine-to-medium grained fan-spherulitic to intersertal textured basalts. Rarely, xenocrystic aggregates are found (Hodges and Papike, 1976). Compositions range from Fo_{91} to Fo_{65} (Fig. 1.2.5.2). Late-stage olivines in alkaline basalts associated with ocean-floor basalts are as iron-rich as Fo_{20}. Increasing Ca and Mn concentrations correlate with increasing Fe; however, high Ca is a consequence of rapid cooling. Annealed olivines from cumulates contain exceedingly low Ca concentrations. Moderate zoning of the olivine phenocrysts is commonplace. Depletion of Mg and enrichment of Ca and Al in the glass surrounding the olivine crystals in the vitrophyres suggest that they grew in place and were not transported with the magma (Rice, 1978). Systematic relationships between olivine core compositions and the bulk chemistry of the rock are noted along the Reykjanes Ridge (Hermes and Schilling, 1976). However, phenocryst cores that are too forsteritic to be in equilibrium with enclosing glass or bulk rock compositions are common in phyric basalts (Rhodes *et al.*, 1979).

Measurement of Fe-Mg exchange between olivines and melt in both natural glassy basalts and experimental studies of ocean-floor basalt compositions give K_D values of about 0.28 (Bryan and Moore, 1977; Bender *et al.*, 1978; Dungan and Rhodes, 1978). Low-pressure crystallization temperatures, as determined directly by crystallization experiments (e.g., Tilley *et al.*, 1972; Bender *et al.*, 1978; Fisk, 1978; Green *et al.*, 1979; Walker *et al.*, 1979), or as calculated from the experimentally calibrated Fe and Mg exchange between olivine and liquid (Roeder and Emslie, 1970), range from about 1270° C for the most primitive samples to about 1150° C for the most evolved samples (Fisk, 1978; Hanson and Langmuir, 1978). Calculated temperatures exceeding this range are most likely to be

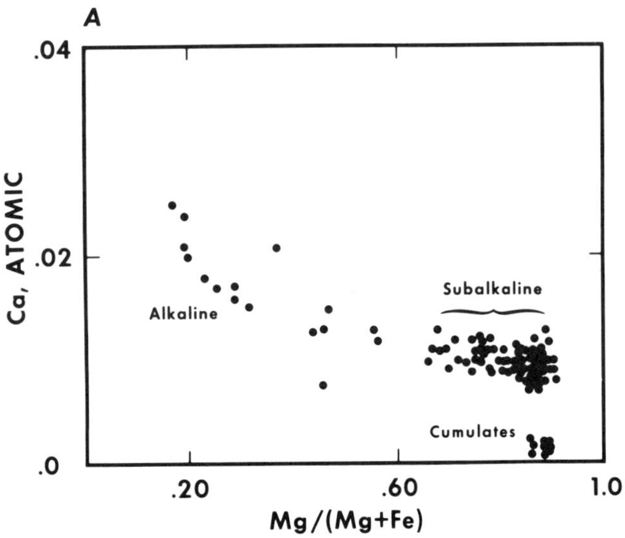

1.2.5.2a. Ca(atoms/6 oxygen) vs. Mg/(Mg + Fe) in 148 ocean-floor basalt olivines. Very low Ca, high Mg/(Mg + Fe) olivines are in cumulates from DSDP leg 334 (Hodges and Papike, 1976).

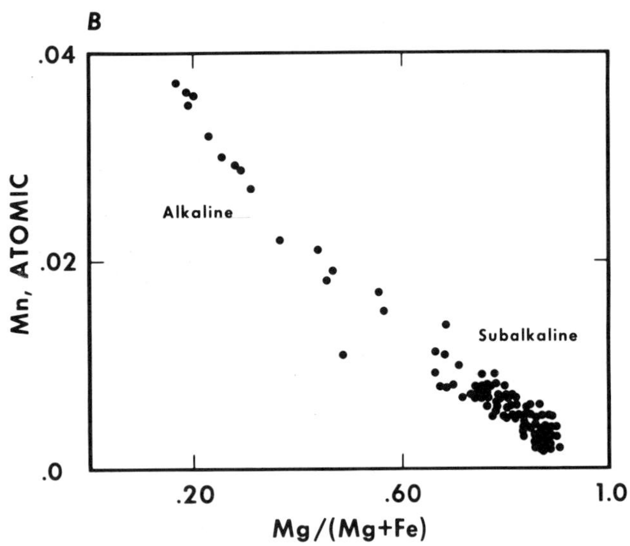

1.2.5.2b. Mn (atoms/6 oxygen) vs. Mg/(Mg + Fe) in 148 ocean-floor basalt olivines.

consequences of nonattainment of equilibrium between the olivine and the bulk rock, as, for example, in the case of refractory megacrysts of olivine (as well as other phases) common in many phyric basalts (Donaldson and Brown, 1977; Rhodes *et al.*, 1979).

Nickel concentrations in ocean-floor basalt olivines range from 1000 to ~ 3000 ppm and may, in the case of vitrophyres, have a concentric zonation with Ni enrichment at the rims (Nabelek, 1979). Measured olivine-glass distribution coefficients (K_D) for Ni are about 10, a value which is in agreement with the experimental results of Hart and Davis (1978) and with values determined empirically from whole-rock data.

Magnesian-chromian spinel, which commonly occurs as octahedra in ocean-floor basalts, precedes or co-crystallizes with high-Mg olivine. The spinel composition is extremely variable from one locality to the next and even within thin sections of individual samples (Sigurdsson and Schilling, 1976; Hodges and Papike, 1976; H. Dick, pers. comm., 1979; Fisk and Bence, 1979; Fig. 1.2.5.3). Experimental studies by Hill and

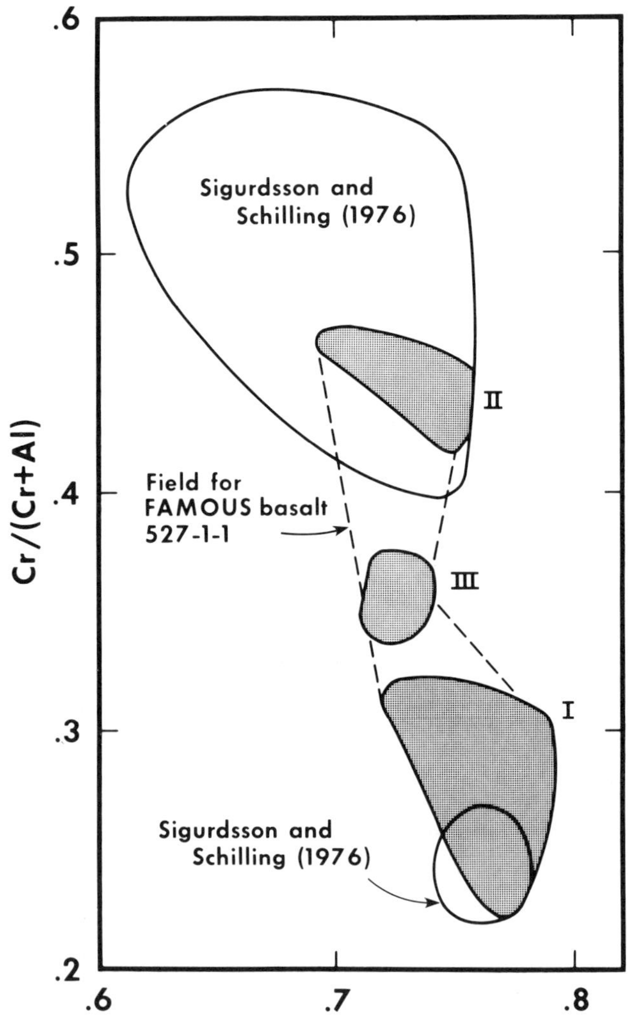

1.2.5.3 Cr/(Cr + Al) (atomic) vs. Mg/(Mg + Fe^{2+}) in ocean-floor basalt spinels. Field for FAMOUS basalt glass 527-1-1 from Fisk and Bence (1979). Fe^{2+} values calculated from charge balance assuming stoichiometry.

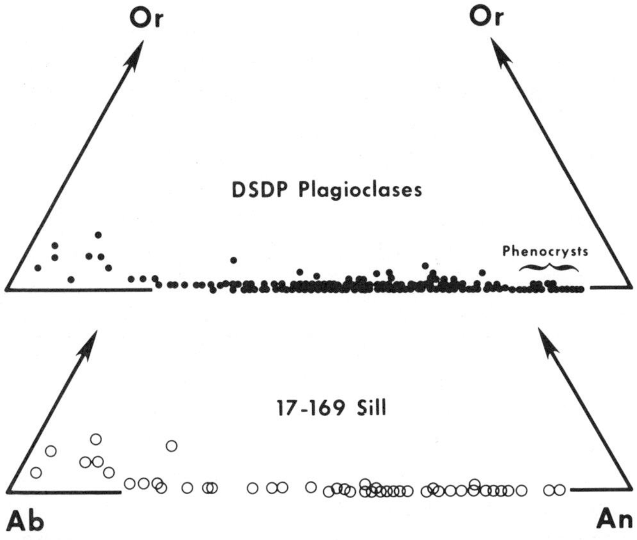

1.2.5.4 Ab-An-Or contents of DSDP ocean-floor basalt plagioclases. Site 169 sill data from Myers *et al.* (1975).

Roeder (1974) and by Fisk and Bence (1979) reveal that the composition of the spinel is highly sensitive to oxygen fugacity. Chrome spinel is the liquidus phase at 1270° C and $\log f_{O_2} = -9.5$ in FAMOUS basalt glass 527-1-1 (Fisk and Bence, 1979). At lower temperatures (1230° C) and $\log f_{O_2} = -7.5$ an intermediate chrome, high alumina spinel co-crystallizes with Fo_{86} olivine.

Plagioclase occurs as delicate swallow-tail microphenocrysts in glasses, as coarse xenocrystic subhedral grains with plagioclase overgrowths in all textural types, as glomeroporphyritic aggregates (frequently with calcic clinopyroxene) in fine-grained intersertal basalts, and as elongate laths with clinopyroxene and glass in the fan-spherulitic basalts. Compositions range from An_{88} in the xenocrystic and phenocrystic plagioclase to $\sim An_{40}$ for interstitial and rim plagioclase compositions (Fig. 1.2.5.4). As was noted for olivine, phenocryst cores that are too anorthitic to be in equilibrium with enclosing glass or with bulk rock compositions are common in phyric basalts (Rhodes *et al.*, 1979). The Or component is usually exceedingly low. However, significant amounts of Fe and Mg are present and may show sectoral control (Bryan, 1974). Variations in Fe, Mg, and Fe/(Fe + Mg) within plagioclases as functions of An content reflect their growth histories. Early forming (high An) plagioclase phenocrysts and xenocrysts are characterized by low Fe and Mg concentrations but by moderately high Fe/(Fe + Mg) ratios. These low concentrations presumably reflect approach to an equilibrium crystallization situation. As the basalt crystallizes, rapid crystal growth precludes crystal-liquid equilibration and higher concentrations of both Fe and Mg are accepted by the plagioclase as it becomes more sodic. By charge balance considerations, the presence of both Fe^{2+} and Fe^{3+} is indicated. Divalent iron is probably present, with Mg, as the component $Ca(Fe,Mg)^{IV}Si_3O_8$ (Bryan, 1974). The progressive decrease in Mg content for compositions more sodic than Ab_{25} may reflect the onset of clinopyroxene crystallization (Mazzullo and Bence, 1976; Bence *et al.*, 1975).

Clinopyroxene is common to all but the most fine-grained and glassy basalts. It occurs as fine, elongate crystals intergrown with plagioclase in the fan-spherulitic basalts, as subophitic anhedral grains in the intersertal basalts, and as ophitic grains enclosing plagioclase and opaques in the ophitic basalts. The vitrophyres, fan-spherulitic basalts, and intersertal basalts occasionally contain glomeroporphyritic aggregates of plagioclase and clinopyroxene. Rarely, phenocrysts, xenocrysts, or xenocrystic aggregates of clinopyroxene are found. The most commonly observed clinopyroxene compositions cluster around $Wo_{30-40}\ En_{50}Fs_{10-15}$ (Fig. 1.2.5.5). Low-calcium compositions are rare and iron-rich compositions occur only in late-stage interstitial pyroxenes. Early crystallizing pyroxenes contain significant amounts of Ti^{4+}, Al, Cr^{3+}, as well as small amounts of Na and Fe^{3+} as components other than quadrilateral components. The most common substitutions occurring in pyroxenes in ocean-floor basalts are: Al for Si in a tetrahedral site combined with Ti or Fe^{3+} substituting for Mg or Fe^{2+} in an octahedral site (Schweitzer *et al.*, 1979). Important components in the first-formed pyroxenes are: $R^{2+}AlSiAlO_6$, $R^{2+}CrSiAlO_6$ and $R^{2+}TiAl_2O_6$. In the late stages $R^{2+}Fe^{3+}SiAlO_6$ becomes an important component. Overall abundances

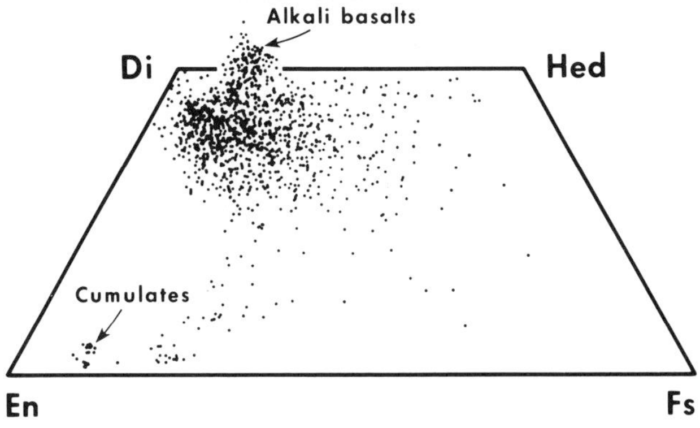

1.2.5.5 Pyroxene data for 1100 ocean-floor basalts and cumulates. Alkali basalt pyroxenes project above the Di-Hed join due to their high contents of $CaAl^{VI}SiAl^{IV}O_6$.

of these "other" components decrease with degree of melt crystallization. Xenocrystic and phenocrystic clinopyroxenes can be frequently distinguished from other pyroxenes in the samples by their high chrome contents and by their low $Mg/(Mg + Fe^{2+})$ ratios (Bence and Bender, 1978).

Skeletal iron-titanium oxides (most commonly titanomagnetite but occasionally ilmenite and only rarely both) are late crystallizing phases in the more slowly cooled basalts. Selected compositions are projected in Fig. 1.2.5.6. Minor components commonly found substituting in these phases are MgO, Al_2O_3, Cr_2O_3 and V_2O_3. The compositions of coexisting Fe-Ti oxides can be used to estimate oxygen fugacities if an equilibration temperature can be independently determined. Mazzullo and Bence (1976) used this technique for Leg 34 Site 319A basalts and estimated f_{O_2}'s of $\sim 10^{-12}$ atmosphere at 1000° C. This estimate may be low as a consequence of halmyrolitic alteration (Ade-Hall *et al.*, 1976).

Haggerty (1976) summarizes the T-f_{O_2} relationships of ocean-floor basalts. Temperatures range from 850–1100° C and f_{O_2} from 10^{-15}–10^{-10} atmospheres and closely follow the quartz-fayalite-magnetite buffer curve.

Amphiboles are rare in ocean-floor basalts. They have been observed in basalts with alkaline affinities at DSDP Sites 169 and 211, and in the cumulates at Site 334, where they are either alteration products or late-stage crystallization products. Late-stage primary amphiboles in a sill at Site 169 in the Central Pacific Basin include Ti-arfvedsonites, ferrorichterites, ferroactinolite and grunerites (Fig. 1.2.5.7) (Myers *et al.*, 1975). Two hornblendes have been noted in hypersthene-normative basalts at DSDP Site 211 in the Indian Ocean (Hekinian, 1974). One of these is probably kaersutite. In the Atlantic, amphiboles are reported at DSDP Site 334 (Cr-pargasites after pyroxene; Hodges and Papike, 1976) and in greenstones from the Mid-Atlantic Ridge at 22° N where they are products of metamorphism (Melson and Van Andel, 1966). Other occurrences have been reported in dredge hauls from fracture zones and from St. Paul's rocks (Melson *et al.*, 1972). Although reported occurrences of amphibole in ocean-floor basalts are few and those of primary amphibole fewer still, detailed documentation of its occurrences and composition is important because of the implications concerning P_{H_2O} in the basalts.

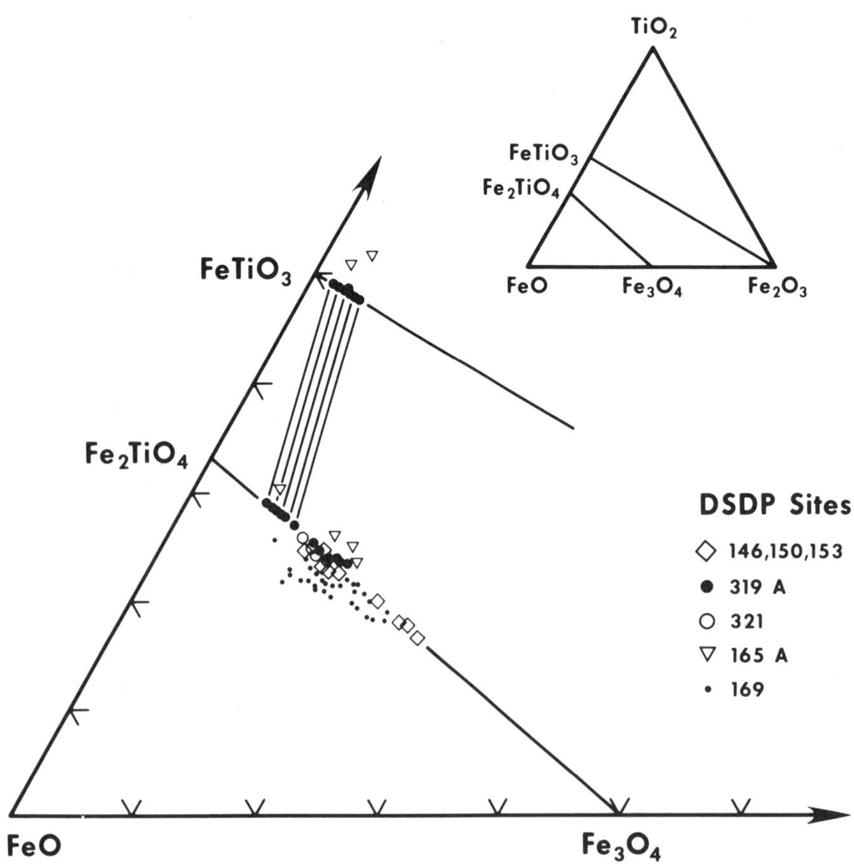

1.2.5.6 Fe-Ti oxide data from seven DSDP sites. Component calculations are described by Mazzullo and Bence (1976).

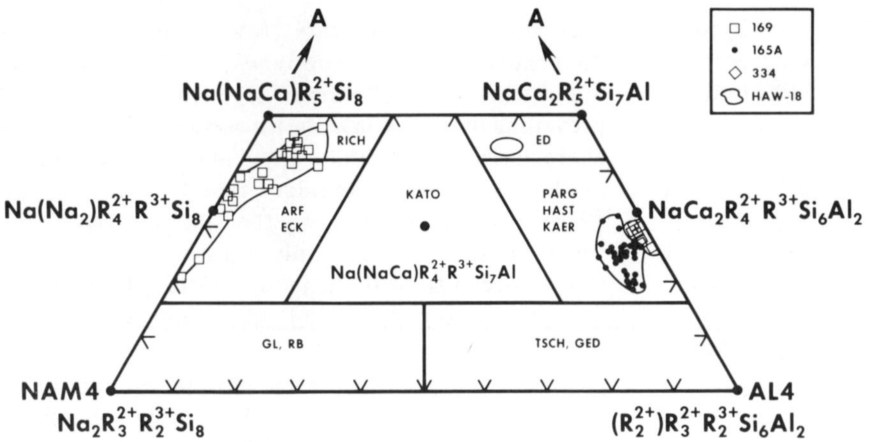

1.2.5.7 Compositions of ocean-floor and Hawaiian amphiboles plotted on "others" diagram of Papike *et al.* (1974). Data from Myers *et al.* (1975), Hodges and Papike (1976), A. E. Bence (pers. comm.) and Table 1.2.6.10.

1.2.5.4 Major element chemistry

Early studies of ocean-floor basalts (e.g., Engel *et al.*, 1965) emphasized their uniformity and unique characteristics, including: high and constant silica values, low potassium content; and low K/Na ratios. Continuing studies over the past fifteen years have shown that, despite broad similarities, differences do exist, reflecting a variety of magmatic processes and the necessity for a heterogeneous mantle source (e.g., Schilling, 1973; Hart *et al.*, 1973; Bryan *et al.*, 1976; Hofmann and Hart, 1978; Bougault *et al.*, 1979). Despite these differences, ocean-floor basalts, as a group, may be readily distinguished from basalts erupted in other tectonic environments (with the possible exception of low-K tholeiites from island arcs) on the basis of major element chemistry (Pearce, 1976). K_2O provides the most effective single discriminant, supported by combinations of TiO_2, SiO_2 and MgO abundances.

Table 1.2.5.2 gives several estimates for average compositions for ocean-floor basalts in general, and for samples from the major ocean basins. There is a tendency for averages obtained from rock analyses to be higher in Al_2O_3 than those calculated from analyses of basaltic glass, which is a consequence of the predominance of plagioclase phenocrysts in many phyric ocean-floor basalts. Higher FeO and TiO_2 contents of Pacific samples, as compared with those from the Atlantic and

Table 1.2.5.2 Average compositions of ocean-floor basalts.

	1 Engel *et al.* (1965)	2 Cann (1971)	3 Melson and Thompson (1971)	4 Pearce (1976)	5 Melson *et al.* (1976)	6 Melson *et al.* (1976)	7 Melson *et al.* (1976)	8 Melson *et al.* (1976)
SiO_2	49.93	49.61	49.21	49.56	50.53	50.68	50.19	50.93
TiO_2	1.51	1.43	1.39	1.42	1.56	1.49	1.77	1.19
Al_2O_3	17.25	16.01	15.81	16.09	15.27	15.60	14.86	15.15
FeO*	8.71	11.49	9.18	10.17	10.46	9.85	11.33	10.32
MnO	0.17	—	0.16	—	—	—	—	—
MgO	7.28	7.84	8.53	7.69	7.47	7.69	7.10	7.69
CaO	11.86	11.32	11.14	11.34	11.49	11.44	11.44	11.84
NaO_2	2.76	2.76	2.71	2.80	2.62	2.66	2.66	2.32
K_2O	0.16	0.22	0.26	0.24	0.16	0.17	0.16	0.14
P_2O_5	0.16	0.14	0.15	—	0.13	0.12	0.14	0.10
Mg′-value	0.62	0.57	0.60	0.60	0.59	0.61	0.55	0.60

1. Average of Atlantic and Pacific ocean tholeiites.
2. Average ocean-floor basalt.
3. Average Mid-Atlantic Ridge basalt.
4. Average for ocean-floor basalts.
5. Average basaltic glass from Atlantic, Pacific and Indian ocean spreading centers.
6. Average basaltic glass from Mid-Atlantic Ridge.
7. Average basaltic glass from East Pacific Rise.
8. Average basaltic glass from Indian Ocean spreading centers.

Indian oceans, are evident in Table 1.2.5.2. This, combined with the greater variability of basalts erupted along the East Pacific Rise relative to those erupted in the Atlantic Ocean (Natland, 1978), suggests that Pacific basalts are more evolved than those in other oceans. This has been widely attributed to differences in spreading rates, although the mechanisms are not clear. There are no discernible differences in major element chemistry between basalts erupted along "normal" segments of mid-ocean ridges (Type-I) and those erupted along topographic highs, adjacent to oceanic islands or platforms (Type-II).

The vast majority of ocean-floor basalts are hypersthene normative: that is, they are tholeiites according to the classification of Yoder and Tilley (1962). Most are olivine tholeiites but, as can be seen from Fig. 1.2.5.8, some are quartz tholeiites, including several from the reference suite (e.g., OF-9, OF-10, OF-12, OF-13, OF-15). The quartz-normative tholeiites are evolved, and are characterized by higher FeO and TiO_2 contents and by correspondingly lower MgO and Mg'-values than are the olivine-normative basalts (Clague and Bunch, 1976; Figs. 1.2.5.11–13). Alkali and transitional basaltic liquids are extremely rare, as is evident from the glass data (Fig. 1.2.5.8), although such rock types are associated with seamounts, aseismic ridges, and some fracture zones. Alkaline and transitional basalts appear to have been associated with tholeiitic volcanism on the Mid-Atlantic Ridge at 45° N for the past several million years (Muir and Tilley, 1965; Aumento *et al.*, 1971; Wood *et al.*, 1979). Although this is a topographically "normal" segment of the ridge, the prevailing tholeiites have the trace element and isotopic characteristics of basalts erupted at topographic highs, or "hot spots" along the ridge (Type-II of Bryan *et al.*, 1976). It is not yet clear whether alkaline and transitional basalts are found only in association with this type of ocean-floor tholeiite, or whether they can also occur with "normal," Type-I ocean-floor basalts.

Figure 1.2.5.9, a plot of normative pyroxene, plagioclase and olivine, shows that most basaltic glass data fall within a well-defined band that follows the olivine-plagioclase cotectic inferred by Shido *et al.* (1971). The whole-rock data are more scattered, but define two lobes, one of olivine accumulation and the other of combined plagioclase-olivine accumulation. To the best of our knowledge, no phenocryst-free basalts plot far from the glass trend, and those with "primitive" compositional characteristics (such as reference suite samples OF-1, OF-3, OF-5, OF-7, OF-8) fall on the olivine side of this trend (Fig. 1.2.5.9). Several conclusions may be drawn from these observations: a) Most ocean-floor basalts lie on or close to the inferred olivine-plagioclase cotectic; b) pyroxene tholeiites are the most evolved, and plot close to the pyroxene-plagioclase join at the end of the cotectic; c) whole-rock samples show compositional evidence for accumulation of either olivine, or plagioclase plus olivine; d) "primitive" basalts plot

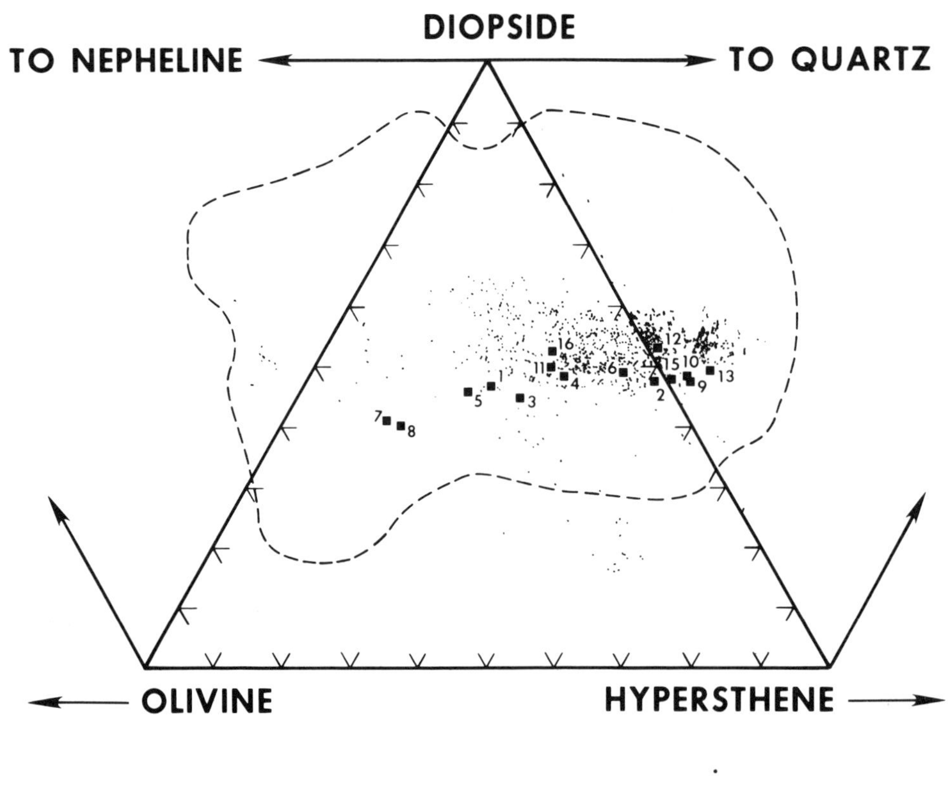

1.2.5.8 Projection of the normative basalt tetrahedron (Yoder and Tilley, 1962) for ocean-floor basalts. Reference suite samples are numbered and plotted as squares; individual basaltic glass analyses are shown as points, and the field for rock analyses is enclosed by a dashed line. Norms have been calculated assuming molar Fe^{3+}/total Fe as $Fe^{2+} = 0.1$.

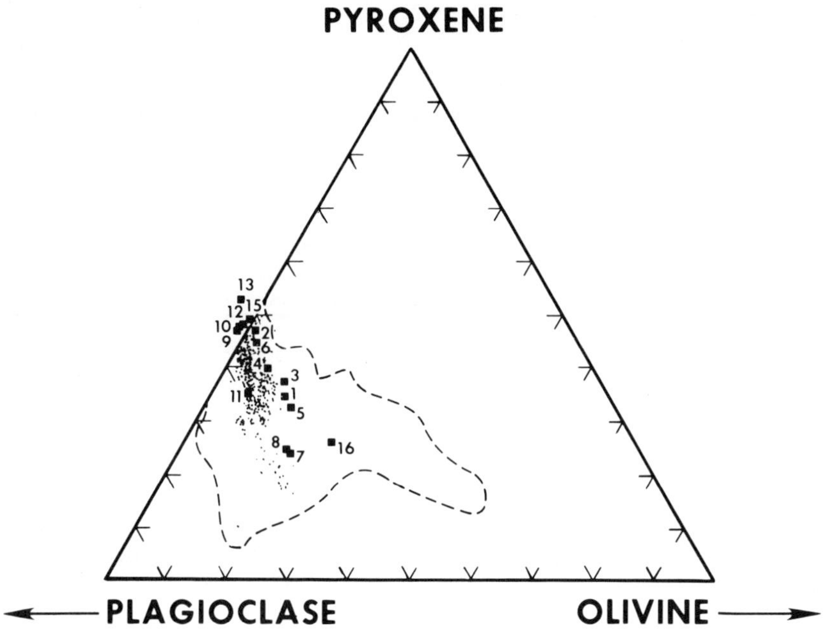

1.2.5.9 Ternary plot of normative olivine-plagioclase-pyroxene for ocean-floor basalts. Symbols are the same as in Fig. 1.2.5.8.

on the olivine side of the inferred cotectic; e) the evidence for "primitive" plagioclase tholeiites (Shido *et al.*, 1971) is equivocal.

Several authors (e.g., Kay *et al.*, 1970; Shido *et al.*, 1971; Bryan *et al.*, 1976; Bender *et al.*, 1978; Rhodes and Dungan, 1979) have pointed out the importance of cotectic crystallization of olivine and plagioclase at low pressures in controlling the bulk chemistry of ocean-floor basalts. This has been discussed with reference to Fig. 1.2.5.9, and is further amplified in Figs. 1.2.5.10 and 1.2.5.12. In Fig. 1.2.5.10, the basaltic glass data and reference suite samples exhibit a covariance in Al_2O_3 and MgO which is indicative of olivine-plagioclase fractionation from "primitive" magmas characterized by high MgO (≃ 10–11%) and Al_2O_3 (≃ 16%) contents. Note that the five "primitive" basalts included in the

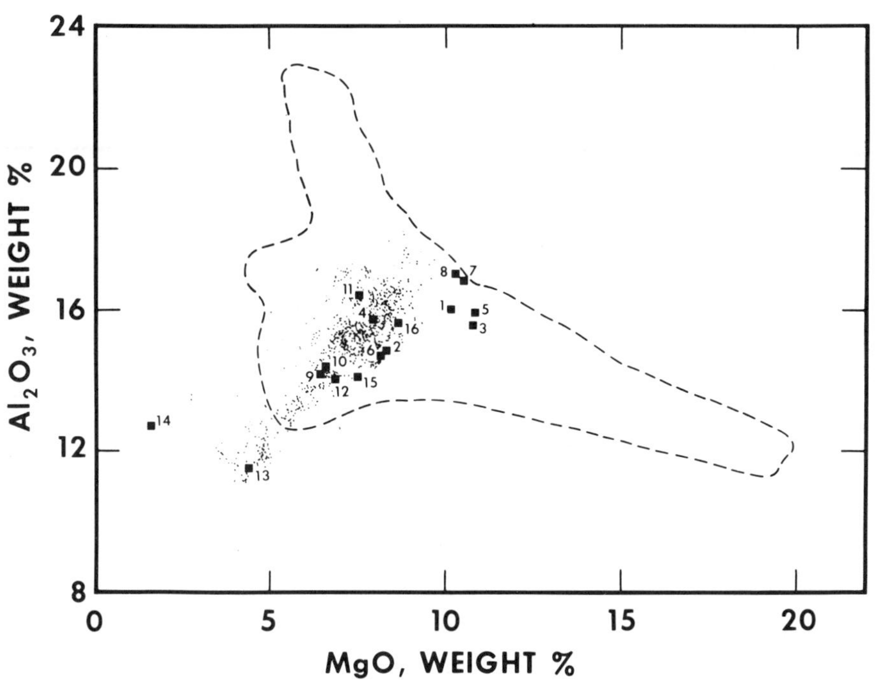

1.2.5.10 MgO-Al_2O_3 relationships for ocean-floor basalts. Symbols are the same as in Fig. 1.2.5.8.

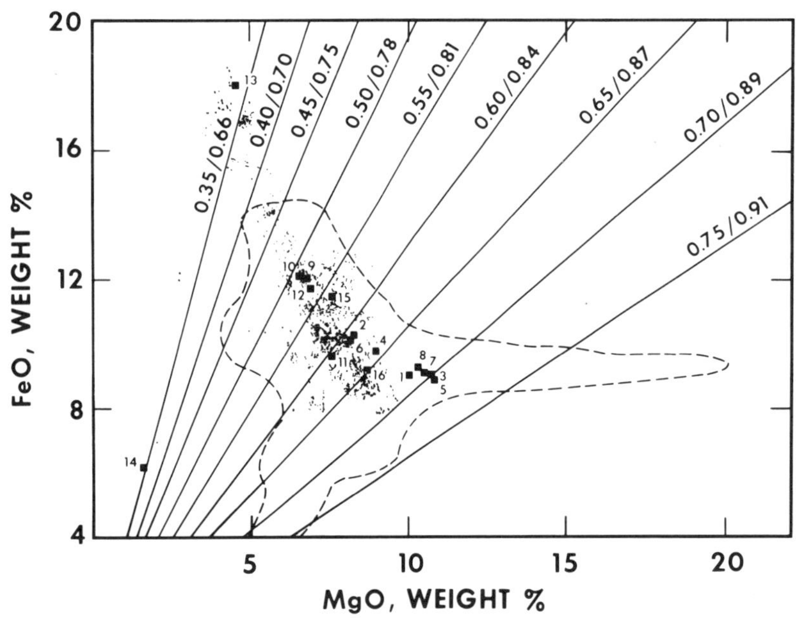

1.2.5.11 MgO-FeO relationships for ocean-floor basalts. Lines of constant Mg′-value are shown together with the forsterite content of equilibrium olivine (calculated assuming K_DFe/Mg = 0.28).

reference suite require a small amount of olivine crystallization in order to gain the plagioclase-olivine fractionation trend illustrated by the glass data. The whole-rock data on this diagram contrast markedly with the glass data, clearly illustrating the influence of olivine, and plagioclase plus olivine, accumulation.

Figures 1.2.5.11 and 1.2.5.12 show pronounced increases in FeO and TiO_2 with decreasing MgO content for both the glass data and for the reference suite samples. The whole-rock data are scattered, as in other diagrams, and again demonstrate the importance of olivine and olivine plus plagioclase accumulation, as well as differentiation, in modifying whole-rock compositions. In Figs. 1.2.5.11–12, the most "primitive" basalts have Mg′-values of about 0.70, and, assuming K_D values $\simeq 0.3$ (Roeder and Emslie, 1970), would crystallize olivine (Fo_{88-90}) prior to joining the trend for the glass data. These trends cannot be produced solely by olivine crystallization, but require co-precipitation of plagioclase and olivine (± clinopyroxene) in order to account for the large increases in FeO and TiO_2. The role of clinopyroxene in the fractionation process is enigmatic. Compositional constraints, in particular a tendency for CaO/Al_2O_3 ratios to decrease with increasing differentiation, indicate that clinopyroxene is an important fractionating phase (Clague and Bunch, 1976). This observation contrasts with clinopyroxene's subordinate role in phenocryst assemblages.

Figure 1.2.5.13 shows that the Mg′-value of ocean-floor basaltic glasses typically is between 0.54 and 0.64. Evolved basalts (e.g., OF-9, OF-10, OF-12, OF-13, OF-15) have lower values, whereas the most "primitive" glass compositions have Mg′-values of about 0.70 (e.g., OF-1, OF-3, OF-5, OF-7, OF-8). Inspection of Figs. 1.2.5.11–12 shows that these "primitive" basalts contain less than 10% FeO and 1% TiO_2. In fact, most "primitive" ocean-floor basaltic magmas identified to date (Bougault and Hekinian, 1974; Frey *et al.*, 1974; Blanchard *et al.*, 1976; Bryan and Moore, 1977; Donaldson and Brown, 1977; Langmuir *et al.*, 1977; Dungan and Rhodes, 1978; Rhodes *et al.*, 1979) have many major element characteristics in common (Table 1.2.5.3). All have high Mg′-values ($\simeq 0.70$), high MgO (10–11%) and CaO (12.0–13.5%) contents, and low FeO (8–9%), TiO_2 (0.5–0.9%) and Na_2O ($\simeq 2.0$%) abundances. That this coherence includes both Type-I and Type-II basalts is evident from the different La/Sm ratios in Table 1.2.5.3.

1.2.5.5 Trace element chemistry

Large, low-valency cations

Early studies of ocean-floor basalts established that most are depleted in large, low-valency cations (e.g., Cs, Rb, K, Ba, Pb, Sr) relative to oceanic island and continental tholeiites (e.g., Engel *et al.*, 1965; Gast 1968; Kay *et al.*, 1970), and that the larger ions are depleted to a greater extent than smaller ions. The consequence of this is that element ratios such as K/Rb, K/Ba, Sr/Rb are characteristically higher for ocean-floor basalts than for oceanic-island basalts or for tholeiites from other tectonic regimes.

It now is recognized that this characteristic depletion in large cation abundances and high small/large

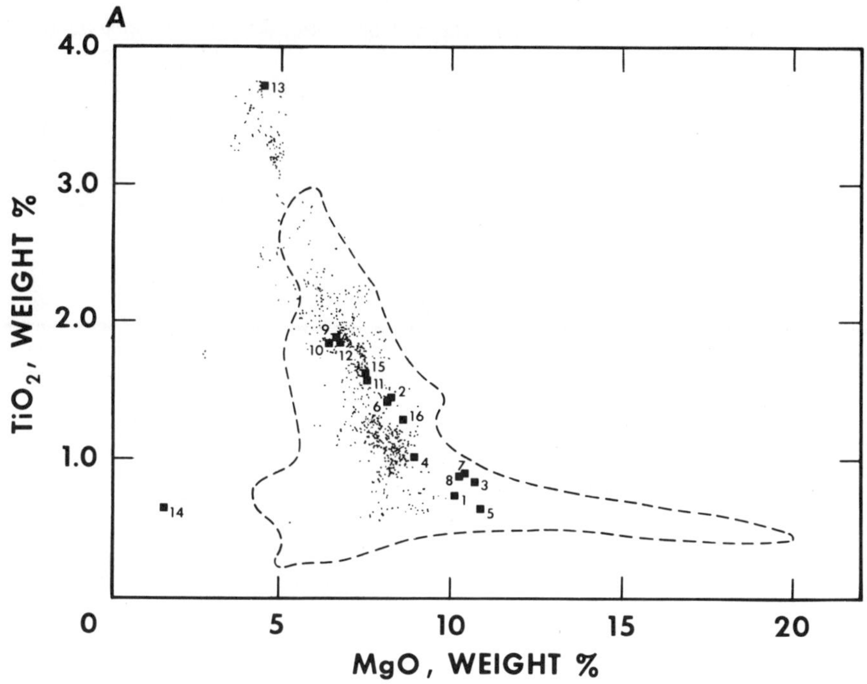

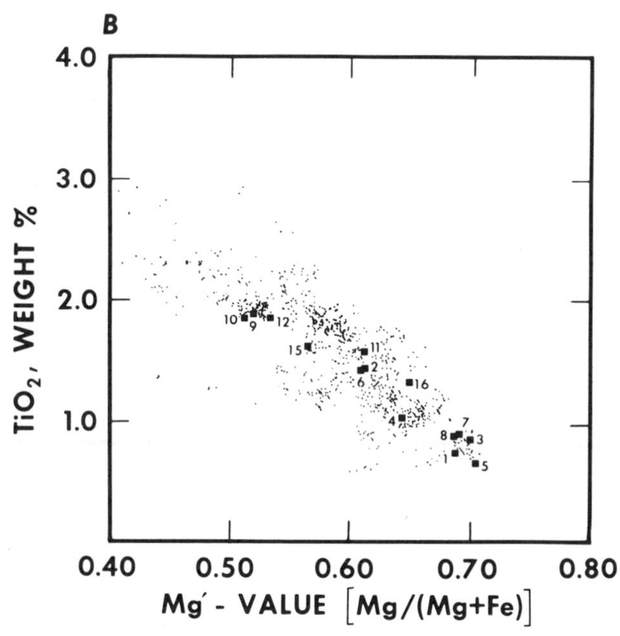

1.2.5.12 **a.** TiO_2 – MgO relationships, and **b.** TiO_2-Mg′-value relationships for ocean-floor basalts. Symbols are the same as in Fig. 1.2.5.8. Note that the basaltic glass field converges towards an apex with Mg′-values of about 0.7 and a TiO_2 content less than 1.0%.

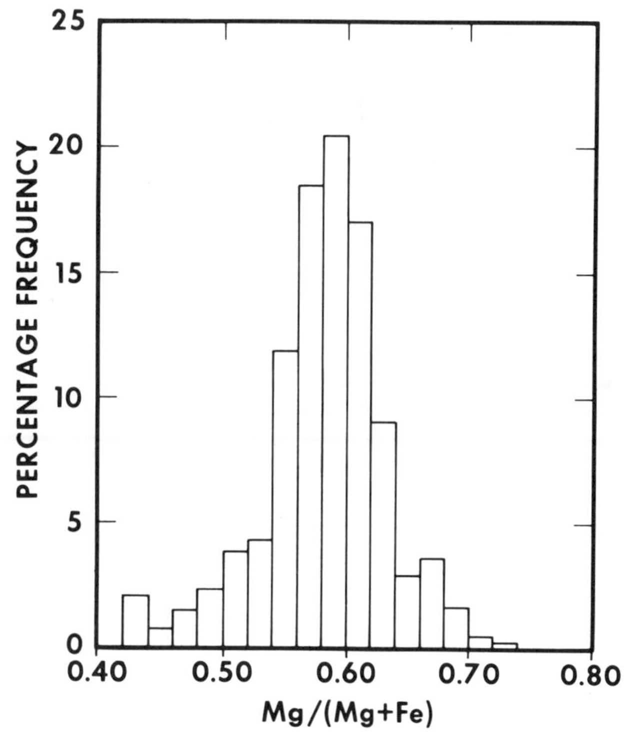

1.2.5.13 Frequency histogram of Mg′-values in ocean-floor basaltic glasses. Data from Frey *et al.* (1974), Melson *et al.* (1976) and Langmuir *et al.* (1977).

cation ratios applies primarily to basalts erupted along "normal" topographic sections of mid-ocean ridges (Type-I basalts of Bryan *et al.*, 1976). Basalts erupted along elevated sections of mid-ocean ridges adjacent to volcanic platforms associated with oceanic islands (e.g., Iceland, Azores, Bouvet, Galapagos) contain higher abundances of large cations and have much lower small/large cation ratios. These "anomalous" basalts (Type-II basalts of Bryan *et al.*, 1976) tend to be more varied than Type-I basalts, with small/large cation ratios that are closely comparable to those found in

Table 1.2.5.3 Chemical characteristics of some potential primary ocean-floor basaltic magmas (recalculated to 100% on an anhydrous basis).

	1	2	3	4	5	6
SiO_2	49.7	49.5	49.1	50.8	50.0	47.25
TiO_2	0.72	0.81	0.62	0.62	0.65	0.79
Al_2O_3	16.4	15.7	16.5	14.4	14.5	13.64
FeO*	7.89	7.45	8.78	6.89	8.26	9.77
MnO	0.12	0.15	0.15	0.10	n.d.	0.14
MgO	10.1	10.0	10.3	12.0	11.7	17.61
CaO	13.2	13.0	12.4	13.6	13.0	9.58
Na_2O	2.00	1.95	1.92	1.43	1.71	0.89
K_2O	0.01	0.17	0.07	0.07	0.03	0.06
P_2O_5	n.d.	0.08	0.06	n.d.	n.d.	0.07
Cr_2O_3	0.07	0.14	0.06	0.10	n.d.	0.20
Mg′-value	0.72	0.73	0.70	0.77	(0.72)	0.78
CaO/Al_2O_3	0.81	0.82	0.75	0.94	0.90	0.70
Ni ppm	320	249	232	—	—	—
Sm ppm	1.60	1.95	1.37	—	—	—
La/Sm	0.63	1.81	1.57	—	—	—

1. Frey *et al.* (1974), 3-14. **2.** Rhodes (unpubl. data), Chain 43 #23,45°N. **3.** Langmuir *et al.* (1977), FAMOUS, 527 1-1. **4.** Donaldson and Brown (1977), Avg. melt inclusion in spinel. **5.** Dungan and Rhodes (1978), interpolated from melt inclusion data. **6.** Elthon (1979), Nt-23, high magnesian Tortuga dike.

oceanic-island tholeiites, or intermediate in value between those of Type-I basalts and oceanic-island tholeiites (e.g., Bryan *et al.*, 1976; Hanson, 1977; White and Schilling, 1978). Typical values for Type-I, Type-II and oceanic-island tholeiites are compared in Table 1.2.5.4. An important observation, brought out in Table 1.2.5.4, is that Sr is the only large cation with similar abundances in both Type-I and Type-II basalts. In fact, a Sr abundance of about 200 ppm serves as a useful discriminant between both Type-I and Type-II ocean-floor tholeiites and tholeiites from other tectonic regimes. Care must be exercised in evaluating the abundance of the large, low-valency cations in ocean-floor basalts, since they are susceptible to seawater alteration. Hart (1971) shows that seawater effects are more pronounced in the order Cs > Rb > K > Ba > Sr.

Most of the large, low-valency cations are incompatible (that is, they are excluded from the lattices of common rock-forming minerals) and have low (< 0.01) crystal/liquid distribution coefficients (Irving, 1978). Consequently, their abundances in basalt should reflect their abundances in the mantle source, the extent of partial melting, and the degree of subsequent differentiation in the magma. If these elements truly are incompatible, as current evidence seems to suggest (e.g., Hanson, 1977; White and Schilling, 1978), then their abundances will not depend on the source mineralogy or on crystallizing phases, so the ratios of these elements

Table 1.2.5.4 Typical large cation abundances for oceanic tholeiites.

	Type I basalts	Type II basalts	Ocean-Island tholeiites
K	1064 ppm	1854 ppm	1600 — 8300 ppm
Rb	1.0 ppm	4.5 ppm	5 — 12 ppm
Ba	12.2 ppm	55 ppm	70 — 200 ppm
Sr	127 ppm	105 ppm	150 — 400 ppm
Cs	13 ppb	58 ppb	—
K/Rb	1046	414	≃ 400
K/Ba	109	34	25 — 40
Sr/Rb	127	23	20 — 70
$^{87}Sr/^{86}Sr$	0.7024 — 0.7030	0.7030 — 0.7035	0.7030 — 0.7050

Data from Hart 1970; O'Nions and Pankhurst, 1974; Hart, 1976; White *et al.*, 1976; White and Bryan, 1977; Hofmann and Hart, 1978 ; Appendix A-5 and section 1.2.6.

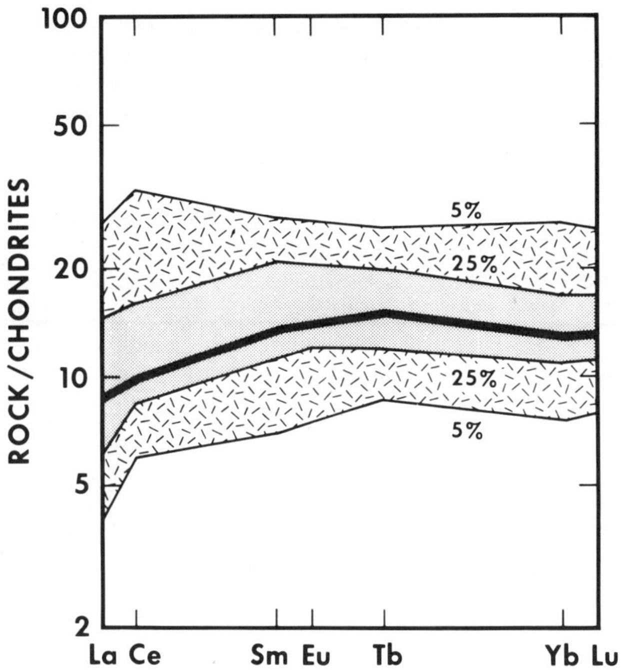

1.2.5.14 Chondrite-normalized rare-earth abundances for ocean-floor basalts, summarizing concentration frequency. Ninety percent of all ocean-floor basalts plot within the outermost contours, 40% within the inner contours, and the mean abundance is shown as a heavy line.

in the basalt should reflect their ratios in the mantle source. An obvious exception is strontium, which will be partitioned into plagioclase either during crystallization or in the peridotitic source if the depth of segregation is shallow enough (Ringwood, 1975). Sun and Hanson (1975) have evaluated the possibility that accessory mantle phases such as phlogopite, amphibole and apatite might serve to fractionate "incompatible" trace element ratios during the melting process, and conclude that although this is a possibility for nephelinitic magmas, it is unlikely in the case of alkaline or tholeiitic magmas. Consequently, it seems clear that the difference in large cation abundances and ratios between "normal" ocean-ridge basalts and oceanic-island tholeiites and alkali basalts reflects a profound difference in their source compositions, a fact which was recognized by Gast (1968) over a decade ago.

Rare-earth elements

Figure 1.2.5.14 summarizes the abundances of rare-earth elements relative to chondritic abundances in ocean-floor basalts. Typical basalts have unfractionated heavy rare-earth (HREE) abundances of about 13–15 times chondrites and are depleted in the light rare-earth elements (LREE), with abundances of about 9–10 times chondrites. As can be seen from Fig. 1.2.5.14, the range is fairly extensive, but most basalts

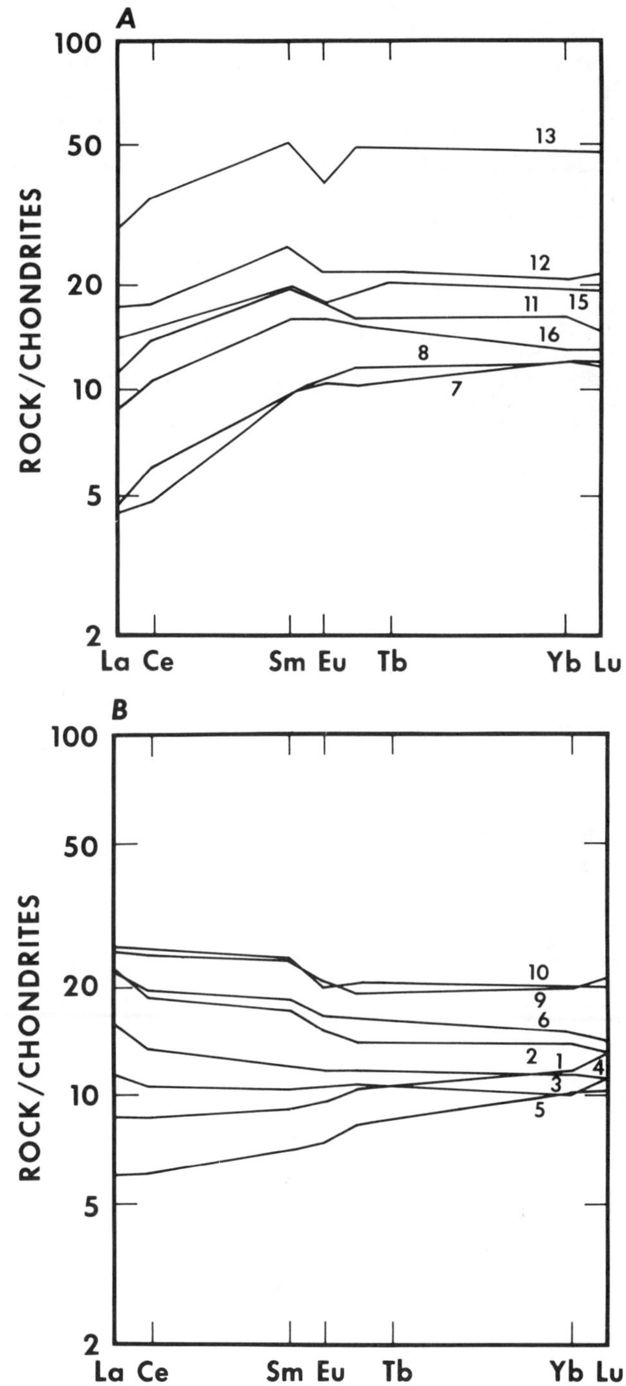

1.2.5.15 Chondrite-normalized rare-earth abundances for the reference suite basalts. **a.** "Normal," Type-I ocean-floor basalts with $[La/Ce]_{E.F.} < 1.0$. **b.** "Anomalous," Type-II ocean-floor basalts with $[La/Ce]_{E.F.} > 1.0$.

(90%) contain less than 30 times chondritic rare-earth abundances, and more than 4 times chondritic LREE and 8 times chondritic HREE abundances. Most of the "primitive" basalts identified to date (Frey *et al.*, 1974; Langmuir *et al.*, 1977; J. M. Rhodes, pers. comm., 1980; this section) have rare-earth concentrations of 10 times chondrites or less (e.g., reference suite samples OF-1, OF-3, OF-5, OF-7, OF-8), whereas extremely differentiated basalts (e.g., reference suite sample OF-13) may contain up to 50 times chondritic REE abundances (Fig. 1.2.5.15a). Crystal fractionation involving the common phenocryst phases of ocean-floor basalts (e.g., olivine, plagioclase, clinopyroxene, spinel), although increasing the REE content of the melt, has little influence on the relative abundances of these elements (Schilling, 1971; Frey *et al.*, 1974), with the consequence that characteristic REE patterns are maintained. However, there is a tendency to develop a negative Eu-anomaly as fractionation proceeds (Kay *et al.*, 1970; Schilling, 1971) because Eu is slightly divalent and behaves like Sr, being partitioned into plagioclase in preference to other REE. It is not an efficient process, so as much as 25% plagioclase must crystallize to produce a perceptible Eu-anomaly. Increases in REE concentrations from 10–50 times chondrites, if accomplished solely by crystal fractionation of a "primitive" melt, imply that as much as 80% of the magma must crystallize to produce the most evolved ocean-floor basalts (Fig. 1.2.5.15).

The two ocean-floor basalt groups identified by Bryan *et al.* (1976) have characteristically different chondrite-normalized rare-earth patterns. The Type-I basalts are depleted in the LREE (Fig. 1.2.5.15a), in keeping with their depletion in large cations, and have chondrite-normalized La/Sm < 1.0. The Type-II basalts, on the other hand, are less depleted in LREE (as is to be expected from their K, Rb and Ba abundances) and tend to have chondrite-normalized La/Sm > 1.0 (Fig. 1.2.5.15b). These differences are also seen in Fig. 1.2.5.16, where there is a tendency for the data to follow two trends, one with "normal" (0.53) chondrite-normalized La/Sm (Schilling, 1971), and the other with chondrite-normalized La/Sm of about 1.0, similar to values for basalts from Mauna Loa (section 1.2.6). Many basalts with intermediate ratios are also observed. The Type-I basalts are mostly from the Pacific and from the Atlantic Ocean south of 30° N, whereas the Type-II basalts come mainly from the Atlantic north of 30° N and from the Galapagos spreading center.

Attempts to model the behavior of the rare-earth elements during partial melting of model mantle compositions (Schilling, 1971; Frey *et al.*, 1974; Schilling, 1975a; White and Schilling, 1978) have led to the following conclusions: 1) Neither garnet nor plagioclase are probable residual components of the source region after melting. This suggests that segregation of the melt occurs at relatively shallow levels in the mantle. 2) If melting is fairly extensive (> 10%), which is generally

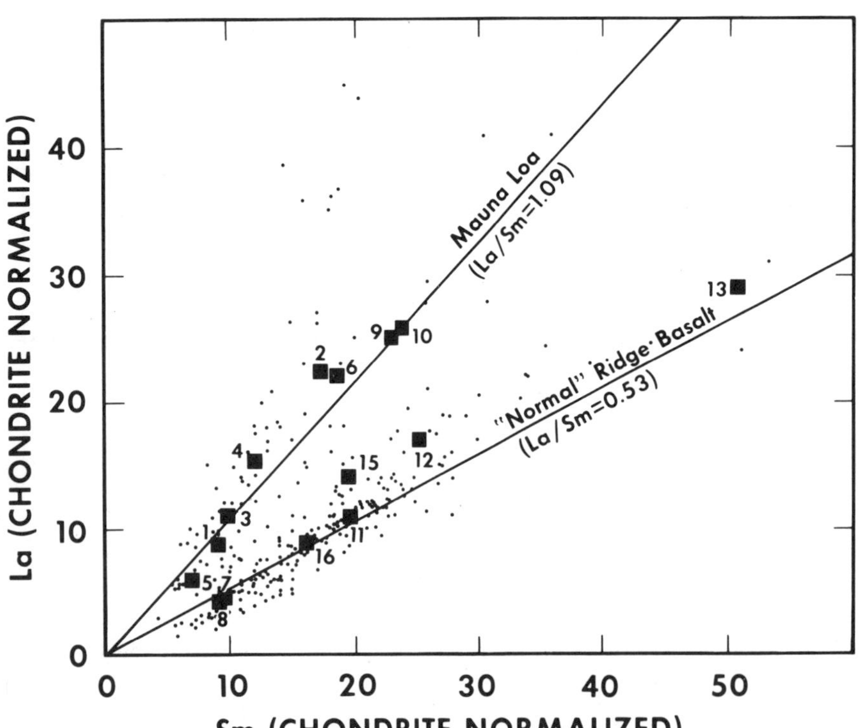

1.2.5.16 Chondrite-normalized La-Sm relationships for ocean-floor basalts. Reference suite samples are numbered and plotted as squares; data from the literature as points. Shown for reference are lines of constant $[La/Sm]_{E.F.}$ ratio for "normal" ocean-floor basalts (Schilling, 1971) and for Mauna Loa (section 1.2.6), an oceanic-island tholeiite.

believed to be the case for tholeiites, there is no effective mechanism to fractionate the REE (and other incompatible elements), so that relative abundances, such as La/Sm, La/Yb etc., observed in the basalt will reflect the ratios of these elements in the source.

Recent studies suggest that not all is so simple. Basalts with markedly different REE characteristics, in such as their La/Sm and La/Yb ratios, have been sampled along one segment of the Mid-Atlantic Ridge at 36° N, from within 1–2 km of each other in the median valley, and even in the same drill hole (Blanchard *et al.*, 1976; Langmuir *et al.*, 1977; Wood *et al.*, 1979), which raises the possibility of multiple mantle sources for spatially and temporally associated magmas. In contrast, drilling along other segments of the Mid-Atlantic Ridge suggests that the mantle source in these segments has remained constant for as long as 35 m.y. (Wood *et al.*, 1979; Rhodes *et al.*, 1979). Langmuir *et al.* (1977), in attempting to explain the crossing rare-earth patterns in FAMOUS area basalts, point out that only the very light REE are truly incompatible, and that ratios commonly regarded as characteristic of the source, such as La/Sm and La/Yb, can be changed by complex melting processes. Consequently, these ratios in the basalt may not be indicative of those in the source, whereas the ratio La/Ce and large cation ratios should be more diagnostic. Viewed in this light, we see that the Type-I basalts all have LREE-depleted patterns, as well as normalized La/Ce < 1.0, whereas Type-II basalts may be LREE-enriched or LREE-depleted, and yet have normalized La/Ce > 1.0 (Fig. 1.2.5.15).

Large, high-valency cations

Most of the high-valency cations commonly measured in ocean-floor basalts (e.g., Th, U, Zr, Hf, Nb, and Ta) tend to be depleted in Type-I ocean-floor basalts relative to Type-II basalts and oceanic-island tholeiites. Measurements of several of these elements in the fresh interiors and altered rims of pillow basalts have shown that their concentrations are only slightly modified, even by extensive alteration. In this respect, they contrast in behavior with most of the large, low-valency cations. Consequently, these elements have been widely used, often in combination with other alteration-resistant elements such as Ti, Y, P and Sr, to discriminate among basalts from different tectonic regimes (e.g., Ridley *et al.*, 1973; Pearce and Cann, 1973; Floyd and Winchester, 1975). However, the similarities between Type-II ocean-floor basalts and those from oceanic islands reduce the utility of these methods.

Ranges of uranium and thorium abundances in unaltered ocean floor basalts are about 0.02–0.43 ppm and 0.2–2.3 ppm, respectively (Tatsumoto *et al.*, 1965; Church and Tatsumoto, 1975; Langmuir *et al.*, 1977; Tatsumoto, 1978; Wood *et al.*, 1979). Th/U ratios in Type-I basalts range from 0.6–3.0, and are lower than those in oceanic-island tholeiites (2.8–6.8). Type-II basalts, on the other hand, have higher Th/U ratios (2.7–5.3), similar to oceanic-island values. Tatsumoto (1978) notes that there is a correlation between La/Sm and Th/U in basalts from the Reykjanes Ridge, which is consistent with the differences between Type-I and -II basalts.

Hafnium abundances in ocean-floor basalts range from 1–8 ppm (e.g., Frey *et al.*, 1974; Rhodes *et al.*, 1976; Blanchard *et al.*, 1976; Rhodes *et al.*, 1978; Wood *et al.*, 1979), averaging about 2.5 ppm. These values are not dissimilar from those reported for some oceanic-island tholeiites (2.6–7.1 ppm; Wood, 1978; section 1.2.6). Similarly, there are no obvious differences in Hf abundances between Type-I and Type-II ocean-floor basalts. Zr/Hf ratios are remarkably constant in ocean-floor basalts and do not differ from oceanic-island ratios. Presumably, these two elements have almost identical distribution coefficients, so they are not readily fractionated relative to one another.

Tantalum occurs in low concentrations in ocean-floor basalts, usually between 0.2 and 1.0 ppm, but occasionally attaining 2–3 ppm in some Type-II basalts (Frey *et al.*, 1974; Rhodes *et al.*, 1976; Bougault *et al.*, 1979; Wood *et al.*, 1979). These values are comparable to those in Icelandic basalts (Wood, 1978). Bougault *et al.* (1979) have emphasized that, unless crystal/liquid distribution coefficients are very low, partial melting processes will modify element ratios, thus obscuring evidence of regional inhomogeneities which may have existed in the mantle source. They suggest that Ta is one of the most "incompatible" elements, and therefore have used the ratio of Ta to other incompatible elements (e.g., Ta/Th) to evaluate mantle inhomogeneity.

Zirconium abundances in ocean-floor basalts (Cann, 1970) range from as low as 32 ppm in "primitive" basalt to as high as 300 ppm in the most highly evolved basalt (Fig. 1.2.5.17). There is no obvious difference in Zr abundance between Type-I and Type-II basalts, nor between ocean-floor tholeiites and those from oceanic islands (Cann, 1970; Pearce and Cann, 1973; Erlank and Kable, 1976; Rhodes *et al.*, 1976; Rhodes *et al.*, 1978; Wood *et al.*, 1979). Niobium, on the other hand, is much more abundant in Type-II basalts, commonly ranging between 4–13 ppm, than in Type-I basalts, where it rarely exceeds 5 ppm except in highly evolved samples (e.g., OF-13, Fig. 1.2.5.17). Basalts at 45° N on the Mid-Atlantic Ridge have exceptionally high Nb abundances (Cann, 1970; Erlank and Kable, 1976;

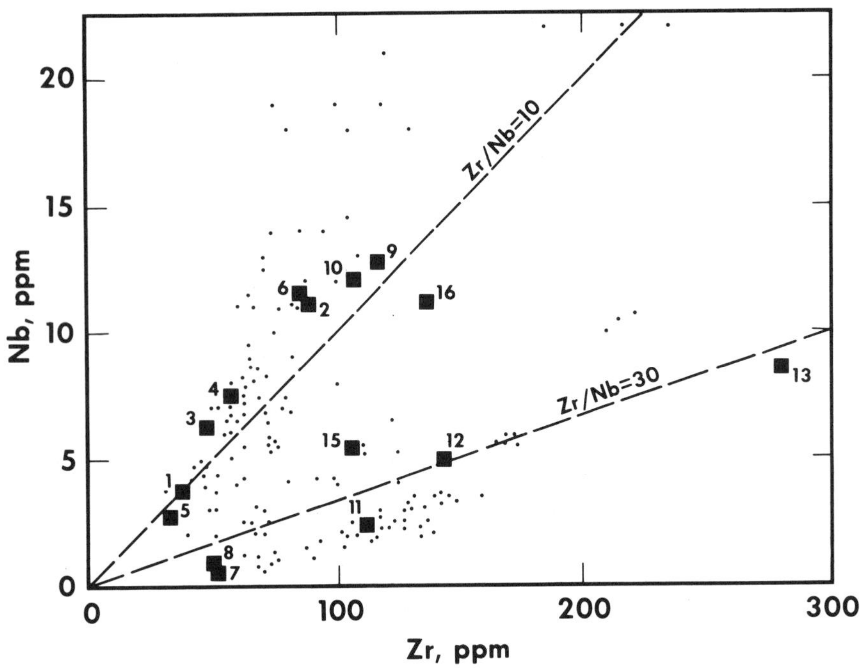

1.2.5.17 Plot of Zr-Nb in ocean-floor basalts. Symbols are the same as those in Fig. 1.2.5.16. Shown for reference are lines of constant Zr/Nb ratio. Type-I basalts have high ratios (>30) whereas Type-II basalts have low ratios (≃10), similar to those of oceanic-island basalts.

Wood *et al.*, 1979). In part, this high abundance reflects the transitional nature of some of the tholeiites sampled in this region (Muir and Tilley, 1964; Aumento *et al.*, 1971), but it also is a characteristic of the tholeiites proper, which contain from 4–11 ppm.

Zr/Nb ratios in Type-I and Type-II ocean-floor basalts are distinctly different. Type-I basalts have high ratios, mostly between 30–55, whereas Type-II basalts are characterized by uniformly low ratios between 6–10 (Cann, 1970; Erlank and Kable, 1976; Rhodes *et al.*, 1976; Rhodes *et al.*, 1978; Appendix A-1.5). As was noted by Erlank and Kable (1976), Zr/Nb ratios in Type-II basalts are comparable to Zr/Nb ratios in oceanic island tholeiites, which commonly range between 6–15. The differences in Zr/Nb ratios in ocean-floor basalts are illustrated in Fig. 1.2.5.17, where two diverging trends are apparent. Plotted Type-I basalts, with low ratios, are mostly from the Pacific Ocean, the Nazca Plate, and from near 22° N on the Mid-Atlantic Ridge. The Type-II basalts are mainly from the Atlantic Ocean, north of 30° N, and are dominated by samples from 45° N and the FAMOUS area.

Although two broad trends are evident in Fig. 1.2.5.17, a range of Zr/Nb values exists from the very high ratios in Type-I basalts from the Mid-Atlantic Ridge at 22° N (Rhodes *et al.*, 1978) to the low values in Type-II basalts from the Mid-Atlantic Ridge at 45° N (Erlank and Kable, 1976; Wood *et al.*, 1979; Rhodes, pers. comm.). Zr/Nb data, together with other high-valency cation ratios, have been used by several authors (e.g., Bougault *et al.*, 1979; Cann *et al.*, 1979; Wood *et al.*, 1979) to explore regional heterogeneities in basalts erupted along the Mid-Atlantic Ridge. It is apparent that regional differences do exist, and that these differences have persisted for long periods, possibly since the opening of the Atlantic.

Ferromagnesian elements

Typical contents of ferromagnesian elements in ocean-floor basalts are as follows: Cr ≃ 270 ppm; V ≃ 250 ppm; Sc ≃ 38 ppm; Ni ≃ 135 ppm; Co ≃ 47 ppm; Cu ≃ 86 ppm; Zn ≃ 85 ppm; and Ga ≃ 17 ppm (Engel *et al.*, 1965; Kay *et al.*, 1970; Melson and Thompson, 1971; Frey *et al.*, 1974; Bougault and Hekinian, 1974; Blanchard *et al.*, 1976; Rhodes *et al.*, 1978; Johnson, 1979; Wood *et al.*, 1979). Some of these elements are determined routinely in trace element studies (e.g., Cr, Sc, Ni, Co) so their abundances and behaviors are well known. Other elements (e.g., V, Cu, Zn, Ga) are rarely measured, so their abundances are known tentatively, at best.

Langmuir *et al.* (1977) noted that the range of transition element abundances in "primitive" ocean basalts is limited, and not dissimilar to ranges of transition element abundances in basalts from other tectonic regimes. This observation contrasts markedly with that for "incompatible" element abundances. Langmuir *et al.* suggest that this limited range implies a mantle that is relatively homogeneous with respect to the transition elements. Crystal/liquid distribution coefficients for the ferro-magnesian minerals indicate that nickel and cobalt will partition into olivine during melting and fractionation processes, and that scandium, chromium and vanadium should enter clinopyroxene (Frey *et al.*, 1974; Irving, 1978). Thus, abundances of these elements should be useful indicators of petrogenetic processes.

Although controversy exists concerning the appropriate olivine/liquid distribution coefficients to use for Ni in basaltic melts (cf. Irving, 1978), there is little doubt that Ni abundances are strongly controlled by olivine fractionation. Ni abundances range from about 300 ppm in "primitive" glassy or aphyric basalts with Mg'-values around 0.70, to as low as 25 ppm in highly evolved basalts. Data for basaltic glasses and aphyric basalts show a well-defined correlation with MgO abundance (Fig. 1.2.5.18), which indicates olivine-controlled crystal fractionation. Data for phyric and holocrystalline basalts are much more scattered, with some picritic samples containing up to 600 ppm Ni (Blanchard *et al.*, 1976) as a consequence of olivine accumulation. Rhodes and Dungan (1979) observe that Ni abundances in many basalts are higher than would be expected from fractional crystallization models. This

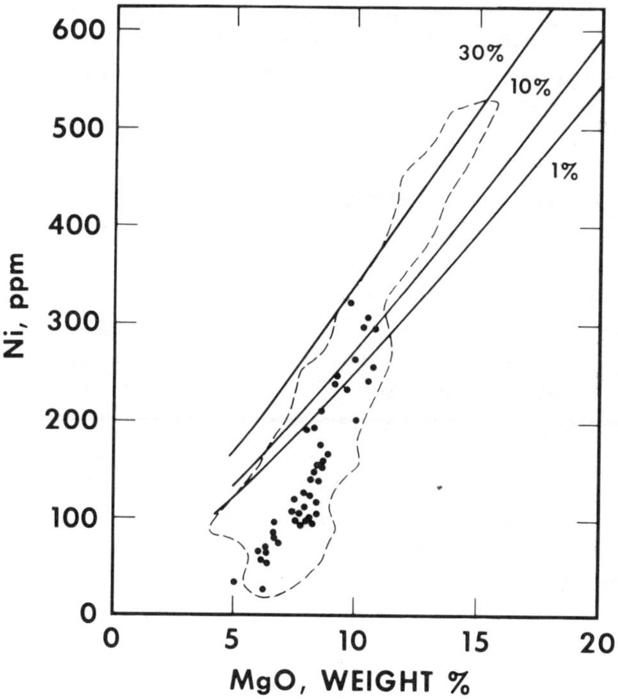

1.2.5.18 MgO-Ni relationships for ocean-floor basalts. Glass and aphyric basalt data are shown as points. The field for phyric or undocumented basalts is enclosed by a dashed line. Note that there are no glass data with Ni values > 300 ppm. Shown for reference are inferred lines for constant percent melting of a model mantle composition (Hart and Davis, 1978).

discrepancy, which is particularly marked in basalts which crystallized olivine alone, was interpreted by Rhodes and Dungan (1979) to result from magma mixing.

Chromium is expected to partition into clinopyroxene; however, its behavior in ocean-floor basalts is very similar to that of Ni. Abundances range from about 700 ppm in "primitive" basaltic glasses with about 10% MgO down to about 100 ppm in evolved, low-MgO basalts. Sample OF-13 in the reference suite, one of the most evolved ocean-floor basalts known, contains only 13 ppm Cr. As with Ni, data for rocks are more scattered. Accumulative basalts may contain from 1000–1700 ppm. These relationships are not functions of pyroxene fractionation, but reflect the simultaneous crystallization of olivine and spinel in ocean-floor basaltic magmas. Langmuir *et al.* (1977) note that, when normalized against chondritic abundances, Cr is depleted relative to other transition elements in ocean-floor and other basalts. However, this is not the case for high-Mg Archean komatiites. Langmuir *et al.* suggest that this implies that residual spinel remains in the mantle after melting has produced most basalts, but that melting which produces komatiites may be sufficiently extensive to incorporate all of the spinel.

Both scandium and vanadium are expected to partition into clinopyroxene, and should be indicators of pyroxene fractionation (Frey *et al.*, 1974). Vanadium varies from about 140 ppm in "primitive" basalts from the FAMOUS area (Langmuir *et al.*, 1977) to between 200 and 400 ppm in typical highly evolved basalts (Frey *et al.*, 1974; Johnson, 1979). There is a tendency for vanadium to increase with increasing TiO_2 content, which suggests that the bulk distribution coefficient is < 1.0, and that the effects of fractionation of plagioclase and olivine overshadow those of pyroxene. A similar behavior pattern must govern scandium. Abundances of Sc in ocean-floor basalts range from 28–49 ppm but often are almost constant, varying by only a few ppm in basalts from a particular locality. Again, there is a tendency to increase slightly with degree of fractionation, which suggests that the bulk distribution coefficient is close to unity. This implies that high-pressure pyroxene fractionation is not an important mechanism for generating the compositional differences observed in associated basalts.

1.2.5.6 Isotope chemistry

The isotopic data for ocean-floor basalts play an important role in understanding mantle-crust evolution. They are discussed thoroughly in Chapter 7, but the major points can be summarized as follows:

1. There is considerable diversity in $^{87}Sr/^{86}Sr$, $^{206}Pb/^{204}Pb$ and $^{143}Nd/^{144}Nd$ ratios in modern ocean-floor basalts, which demonstrates heterogeneity in the mantle source regions.

2. Exchange with seawater can lead to significant increases in $^{87}Sr/^{86}Sr$ in altered or metamorphosed ocean-floor basalts, whereas $^{143}Nd/^{144}Nd$ appears to remain unchanged.

3. Type-I ocean-floor basalts have lower $^{87}Sr/^{86}Sr$ ratios (0.7024–0.7030) than Type-II basalts (0.7030–0.7035), which fall into the lower end of the larger range found for oceanic-island tholeiites (0.7030–0.7050). Along certain segments of mid-ocean ridges, Type-II basalts are transitional between oceanic-island basalts and Type-I basalts. This is illustrated by the Sr-isotopic data for basalts along the Mid-Atlantic Ridge between Iceland and 30° N (Hart *et al.*, 1973; White *et al.*, 1976—see Fig. 1.2.3.11). Data for Pb and Nd isotopes are not yet as plentiful or systematic. Nevertheless, it is clear that $^{143}Nd/^{144}Nd$ ratios tend to be higher in ocean-floor basalts than in oceanic-island basalts (e.g., O'Nions *et al.*, 1977), whereas there is considerable overlap in $^{206}Pb/^{204}Pb$ data (e.g., Sun and Jahn, 1975; Tatsumoto, 1978).

4. All three isotopic systems exhibit distinct depletion in parent/daughter ratios of ocean-floor basalts. This implies, following Gast (1968), that the mantle source for these basalts is depleted in large cations relative to chondritic or bulk-earth model compositions. The covariance of $^{143}Nd/^{144}Nd$ with $^{87}Sr/^{86}Sr$ (e.g., O'Nions *et al.*, 1977) suggest that not only are Type-I basalts derived from a depleted mantle, but so also are basalts showing LREE enrichment such as some Type-II ocean-floor basalts and oceanic-island basalts. Mantle "isochrons" based on the Rb-Sr and Pb-isotope systems indicate that this depletion is long-lived, implying that the mantle source for ocean-floor basalts has persisted for about 1–2 b.y.

1.2.5.7 Petrogenesis

A variety of factors have played major roles in generating the compositional diversity observed in fresh basalts erupted at the mid-ocean ridges. These include: (1) chemical, mineralogical and isotopic heterogeneities in the mantle sources; (2) variable amounts of melting of these sources, and variations in the melting processes; (3) subsequent modification of magma compositions by fractionation and magma mixing both within the mantle and at shallow depths along mid-ocean ridges. The following sections summarize the evidence supporting these processes and evaluate their roles in ocean-floor basalt petrogenesis.

Primitive magmas

The vast volumes and seemingly world-wide compositional uniformity of ocean-floor basalts induced early researchers to propose that they were primary magmas, derived directly from the mantle without significant fractionation or crustal contamination *en route* to the surface (e.g., Engel *et al.*, 1965). It now is widely accepted, following O'Hara (1968), that the majority of these basalts sampled to date are of an evolved nature: that is, they have crystallized from magmas whose compositions have been substantially modified following partial melting and segregation in the mantle. Consequently, it is of crucial importance to an understanding of ocean-floor basalt petrogenesis to identify and characterize "primitive" basalts which may have been derived directly from the mantle without modification. It is these compositions that can provide chemical and mineralogical information about their sources, and on the nature of the evolutionary processes.

The nature of "primitive" or primary ocean-floor basaltic magmas is the subject of continuing and spirited debate (see section 1.4.2). Central to this controversy are the questions of whether *all* ocean-floor basalts are evolved (extensively) from picritic melts containing 16–20% MgO (e.g., O'Hara, 1968; Yoder, 1976; Green *et al.*, 1979; Elthon, 1979; Stolper, 1980), or whether the most "primitive" basalts yet sampled reflect magma compositions that have undergone little or no modification since segregation from their mantle source (e.g., Frey *et al.*, 1974; Blanchard *et al.*, 1976; Bougault and Hekinian, 1977; Langmuir *et al.*, 1977; Bryan, 1979; Presnall *et al.*, 1979; Rhodes *et al.*, 1979).

Inspection of Figs. 1.2.5.10–13 shows that, although most ocean-floor basalts are evolved, basaltic glass data converge toward compositions with MgO contents of about 10% and Mg′-values of about 0.70. The Ni contents of these glasses are around 300 ppm (Fig. 1.2.5.18). These are the most "primitive" ocean-floor basaltic liquids identified to date. There are no glasses with Mg′-values or Ni contents higher than these, and rocks with higher values invariably show signs of olivine accumulation (Blanchard *et al.*, 1976; Rhodes *et al.*, 1979). Examples of such "primitive" basalts have been identified at dispersed sites in the Atlantic Ocean (Rhodes and Rodgers, 1972; Frey *et al.*, 1974; Bougault and Hekinian, 1974; Bryan and Moore, 1977; Langmuir *et al.*, 1977). Typical analyses are given in Table 1.2.5.3. Examples included in the reference suite are OF-1, OF-3, OF-5, OF-7 and OF-8.

Although "primitive" basalts such as these are not common, indirect evidence, including compositions of olivine phenocrysts and melt inclusion data, indicates that such basaltic compositions may be more important in ocean-floor basalt petrogenesis than their apparent volumetric abundances suggest. For example, the maximum forsterite and NiO content commonly observed in ocean-floor basalts (Fo_{90-91} and 0.41%, respectively) is compatible with melts having Mg′-values between 0.70–0.75 and Ni contents between 280–400 ppm. Several authors (e.g., Watson, 1976; Donaldson and Brown, 1977; Dungan and Rhodes, 1978) have attempted to deduce the compositions of primary ocean-floor basaltic magmas from those of melt inclusions in phenocryst phases. Their estimates are remarkably similar, although based upon melt inclusions enclosed in different phenocryst hosts and coming from widely dispersed localities, and compare closely with the "primitive" basalt compositions (Table 1.2.5.3). Thus, the most "primitive" compositions yet identified have many features in common. All are characterized by low TiO_2 (0.5–0.9%) and Na_2O ($\simeq$2.0%) abundances, high CaO (12.0–13.5%) contents, and MgO and FeO contents of about 10 and 8 wt.%, respectively. Although there are close similarities in major element chemistries, there is no such coherence in trace element or isotopic abundances (Table 1.2.5.3). "Primitive" basalts have been identified with the chemical characteristics of both Type-I and Type-II basalts.

These "primitive" ocean-floor basalts have Mg′-values of about 0.7, and crystallize Fo_{90-91} olivine, values that are not inappropriate for melts derived directly from the mantle without substantial modification (e.g., Green, 1970; O'Hara *et al.*, 1975; Ringwood, 1975; Hanson and Langmuir, 1978; Bender *et al.*, 1978). High Ni values of about 300 ppm have also been used as evidence for the primary nature of these basalts (e.g., Kay *et al.*, 1970; Sato, 1977; Hart and Davis, 1978). Figure 1.2.5.18 shows that, if the crystal/liquid distribution coefficients and model mantle compositions of Hart and Davis (1978) are used, the most Ni-rich ocean-floor basaltic glasses could be primary, having resulted from 10–30% partial melting of their mantle sources. If, however, the appropriate olivine/liquid distribution coefficients are lower, as experiments on more magnesium-rich lavas imply (e.g., Arndt, 1977), then the Ni contents of these basalts may be too low for primary melts. The Ni contents of olivine phenocrysts may help resolve this dilemma. For example, although few data are available, the maximum NiO content for olivine phenocrysts in ocean-floor basalts ($\simeq$0.41%) is similar to that of olivines from spinel lherzolite inclusions (Sato, 1977; Nabelek, 1979), evidence that the basaltic olivines may have crystallized from a melt that has undergone little fractionation since segregating from the mantle.

Arguments favoring picritic primary magmas' being parental to ocean-floor basalts rest heavily on experimental constraints and upon the assumption that harzburgites are residues of the partial melting process. It follows that olivine and orthopyroxene should be on the liquidi of primary ocean-floor magmas at pressures corresponding to the depth of segregation (O'Hara, 1968). Some recent experiments support these arguments for primary picritic magmas (e.g., Green *et al.*, 1979; Stolper, 1980), showing that picritic liquids (MgO 13–17%) are saturated with both olivine and orthopyroxene at 10–12 kb, whereas ocean-floor basaltic magmas with lower MgO contents do not have orthopyroxene on their liquidi at any pressure. However, not all experiments support this viewpoint. Bender *et al.* (1978) report orthopyroxene on the liquidus of a primitive basalt from the FAMOUS area between 10–12 kb, and Presnall *et al.* (1979) predict, by analogy with the system $CaO–MgO–Al_2O_3–SiO_2$, that the most primitive ocean-floor basalts will be saturated with olivine and orthopyroxene at about 9 kb. Further discussion of these experimental results is provided in Chapter 3.

The major obstacle to the theory that primary picritic magmas are parental to ocean-floor basalts is their embarrassing absence among samples recovered from the oceans to date. Basaltic glasses with MgO contents greater than 10–11% have not been identified. Rocks with higher values contain accumulated olivine and are far removed from liquid compositions (e.g., Blanchard *et al.*, 1976; Rhodes *et al.*, 1979). Elthon (1979) suggests that dikes in the Tortuga ophiolite complex may be examples of picritic primary magmas from which ocean-floor basalts could be derived (Table 1.2.5.3). He further argues that the bulk composition of the oceanic crust, if represented by typical ophiolite complexes, requires picritic magmas similar to the Tortuga dikes for its formation. This argument is weakened if the oceanic crust is constructed at mid-ocean ridges as a consequence of open-system crystallization (e.g., O'Hara, 1977; Rhodes and Dungan, 1979; Walker *et al.*, 1979) rather than by closed-system crystallization. Furthermore, extrapolation from the specific example of the Tortuga ophiolite to the general case of the oceanic crust may be inappropriate. This is because the Tortuga ophiolite is believed to be a relict of a back-arc basin (Stern, 1980) where, in contrast with larger ocean basins, recycled mantle may give rise to depleted, high-magnesian lavas.

The controversy surrounding the nature of primary ocean-floor basaltic magmas cannot yet be resolved, and probably will be around for some time to come. If picritic magmas are primary and parental to ocean-floor basalts, it appears that all traces of their existence are lost. This loss may be a consequence of magma mixing at crustal levels beneath mid-ocean ridges (O'Hara, 1977; Rhodes *et al.*, 1979; Walker *et al.*, 1979), or the high density of these picritic melts may preclude them from attaining crustal levels (Stolper and Walker, 1980).

Near-surface crystal fractionation

Petrographic and compositional data provide strong evidence for the domination of low pressure crystal fractionation in ocean-floor basalts by olivine and plagioclase. These minerals are the prevailing phenocryst phases (Shido *et al.*, 1971). The compositional effects of plagioclase and olivine fractionation are readily seen in plots of Al_2O_3 with MgO or Mg'-value (Fig. 1.2.5.10). Data for basaltic glasses follow a narrow trend, apparently controlled by the olivine-plagioclase cotectic (Bryan *et al.*, 1976; Bence *et al.*, 1979). Plots for crystalline basalts, on the other hand, are more scattered extending to higher values of MgO and Al_2O_3 (Fig. 1.2.5.10) as a consequence of olivine and plagioclase accumulation.

Experimental melting studies at 1 atm confirm the importance of olivine and plagioclase as early crystallizing phases, and the significance of the olivine-plagioclase cotectic for ocean-floor basalt petrogenesis (Kushiro, 1973; Fujii and Kushiro, 1977; Bender *et al.*, 1978; Dungan *et al.*, 1978; Walker *et al.*, 1979). These studies also show, again supporting the petrographic evidence, that pyroxene begins to crystallize later, anywhere from 40–130° C below the liquidus depending upon bulk composition. Studies by Bender *et al.* (1978) on a "primitive" glass composition from the FAMOUS area showed that olivine crystallized first at 1268° C, was joined by plagioclase at 1235° C; and that cotectic crystallization of olivine and plagioclase (in molar proportions of 1:2) continued down to 1205° C (see Chapter 3). Only after about 50% crystallization were olivine and plagioclase joined by clinopyroxene. Natural glass samples that follow the experimentally determined olivine-plagioclase cotectic, or those more evolved compositions that appear to be saturated with olivine, plagioclase and clinopyroxene, provide strong evidence for crystal fractionation at low pressures. It seems unlikely that any of these basalts could be considered primary magmas.

Trace element abundances are broadly consistent with the petrographic and experimental conclusions. In particular, the rapid depletion of Ni with increasing differentiation (Fig. 1.2.5.18) and the development of negative Eu anomalies in otherwise unfractionated but highly enriched rare-earth patterns (Fig. 1.2.5.15) attest to the importance of olivine and plagioclase fractionation (Kay *et al.*, 1970; Schilling, 1971; Frey *et al.*, 1974). Magmaphile element abundances vary widely in ocean-floor basalts and may increase by a factor of five from the most "primitive" basalts to those with highly evolved compositions (e.g., the Fe-Ti basalts with FeO $>$ 12% and $TiO_2 > 2\%$; Melson *et al.*, 1976). If this increase is due solely to crystal fractionation from these "primitive" compositions, it implies between 40–60% fractionation for the majority of basalts and about 60–80% for the highly evolved basalts. Alternately, different parental magmas may be required to generate the wide spectrum of magmaphile element abundances. These may be derived from different mantle sources, from one source through varying amounts of partial melting, or may arise as consequences of multiple episodes of magma mixing in an open magma chamber (O'Hara, 1977; Rhodes *et al.*, 1979).

Although low-pressure crystal fractionation can be demonstrated in principle, it is an elusive process to follow in detail. Attempts to relate associated magmatic units to one another, to a common parental magma, or to a hypothetical primary magma, solely by crystal fractionation employing the *observed* phenocryst phases are rarely successful. Instead, it appears to be necessary to postulate multiple parental magmas that evolve separately in small, unconnected magma chambers (e.g., Blanchard *et al.*, 1976; Langmuir *et al.*, 1977; Flower *et al.*, 1977; Bryan, 1979). This may result from distinctly different magmaphile element ratios in the basalts, which cannot be produced by crystal fractionation involving the dominant phenocryst phases, and which have been attributed in the Leg 37 and FAMOUS area basalts to heterogeneities in the source or to complex melting processes (e.g., Blanchard *et al.*, 1976; Langmuir *et al.*, 1977).

More commonly, crystal fractionation models relying on a single parental magma fail to account for the observed chemical diversity for two reasons. First, the magmaphile element content of the more evolved basalt is higher than the values predicted from fractionation models based on major element data (e.g., Bryan *et al.*, 1976; Bryan and Moore, 1977; Rhodes *et al.*, 1979). In other words, there is a discrepancy between the extent of fractionation deduced from major element data using the observed phenocryst phases and that inferred from the increase in magmaphile elements (Anderson and Greenland, 1969). The dilemma becomes worse for some basalts that have crystallized only olivine when estimates of crystal fractionation based on magmaphile elements are compared with estimates based on a highly compatible trace element such as Ni (Rhodes and Dungan, 1979).

Second, it is frequently found that, in order to relate associated basaltic units by crystal fractionation, it is necessary to include clinopyroxene as a *major* fractionating component, even though pyroxene is absent or is only a minor component of the observed phenocryst assemblage (Clague and Bunch, 1976; Bryan *et al.*, 1976; Thompson *et al.*, 1976; Bryan, 1979). The inclusion of clinopyroxene stems from a need to increase the magmaphile trace element abundances without drastically modifying the major element chemistry and, more importantly, from the need to lower the CaO/Al_2O_3 ratio with increasing degree of differentiation (Clague and Bunch, 1976; Rhodes *et al.*, 1979). This last point is inescapable: clinopyroxene fractionation must take place. The question is where? In shallow magma chambers as advocated by Bryan (1979); within the mantle as proposed by Bence *et al.* (1979); or in evolved magmas that are subsequently mixed with more "primitive" cogenetic magmas (Rhodes *et al.*, 1979; Walker *et al.*, 1979)? The last two explanations share the common characteristic that, although the petrographic evidence may be lost or obscured, the chemical imprint of clinopyroxene fractionation will remain.

Crystal fractionation within the mantle

In the preceding section it was argued that clinopyroxene fractionation is required to account for the decrease in CaO/Al_2O_3 ratios with increasing differentiation in ocean-floor basalts. Experimental studies at 1 atm (e.g., Kushiro, 1973; Fujii and Kushiro, 1977; Bender *et al.*, 1978; Dungan *et al.*, 1978; Walker *et al.*, 1979) show that, consistent with the petrographic evidence, clinopyroxene crystallizes at lower temperatures than olivine or plagioclase, and therefore is less likely to play a major role in simple low-pressure crystal fractionation. However, clinopyroxene is the liquidus phase for ocean-floor basalt compositions at higher pressures ($\simeq$10 kb) (Kushiro, 1973; Bender *et al.*, 1978; Green *et al.*, 1979; Presnall *et al.*, 1979; Stolper, 1980), so it seems reasonable to suggest that clinopyroxene fractionation under mantle conditions is a viable process in ocean-floor basalt petrogenesis. Donaldson and Brown (1977) describe pyroxene megacrysts which they consider to be unstable relicts of former high-pressure clinopyroxene fractionation. Similarly, Bender *et al.* (1978) note that clinopyroxene microphenocrysts or xenocrysts, which appear to be a minor ($< 1\%$) but prevailing constituent of ocean-floor basalt magmas, may reflect a prior history of high-pressure pyroxene fractionation.

Supporting this suggestion is the observation that these pyroxenes have Fe/Mg ratios which are lower than those of the clinopyroxenes which form first in 1 atm experimental runs or those of groundmass pyroxenes in the same basalts. Although this is an appealing mechanism for producing an imprint of clinopyroxene fractionation in magmas with only olivine or plagioclase as the phenocryst phases, Sc abundances appear to preclude significant high-pressure fractionation. Sc abundances in ocean-floor basalts are remarkably uniform (35–45 ppm), implying that the effective distribution coefficient for Sc is close to unity (Frey *et al.*, 1974); and therefore requiring co-precipitation of olivine and plagioclase as well as clinopyroxene, conditions which would best be explained by low-pressure crystal fractionation. Additional data for Sc, Cr, and V, as well as studies of distribution coefficients for these elements, are needed in order to evaluate these possibilities further.

We have stated earlier that the vast majority of ocean-floor basalts are evolved. O'Hara has argued consistently that this evolution is a substantial process occurring either in the mantle through eclogite fractionation or through extensive olivine fractionation of a picritic magma *en route* to the surface (e.g., O'Hara, 1968; O'Hara, 1973; O'Hara *et al.*, 1975). Eclogite fractionation provides an effective mechanism for increasing LREE relative to HREE, which has been suggested by O'Hara (1973) to explain the change in La/Yb ratios observed by Schilling (1973a) along the Reykjanes Ridge. Frey *et al.* (1974) and O'Nions *et al.* (1976) have explored this process for ocean-floor and Icelandic basalts. O'Nions *et al.* argue that fractionation involving garnet would result in a severe depletion of HREE, with the consequence that a subsequent, highly opportune episode of low-pressure fractionation involving plagioclase would be required to return the HREE to the relatively flat pattern typical of ocean-floor basalts. Frey *et al.* (1974) suggest that garnet plus clinopyroxene fractionation might provide a means for deriving LREE-enriched (Type-II) ocean-floor basalts from "typical" LREE-depleted (Type-I) magmas. However, the extent of fractionation is restricted by almost constant Sc contents. Both Frey *et al.* (1974) and O'Nions *et al.* (1976) point out that the correlation observed between LREE enrichment and Sr-isotope abundances cannot be explained by either high- or low-pressure fractionation, and that these variations are most likely to be source-related. This suggestion receives further support from Nd- and Pb-isotope studies (e.g., O'Nions *et al.*, 1977; Tatsumoto, 1978).

The other suggestion, that basalts result from extensive olivine fractionation of a picritic magma *en route* to the surface from the mantle, has recently been criticized by Hart and Davis (1978). They point out that an ultramafic magma with initial MgO and Ni concentrations of 20% and 600 ppm, respectively, would only contain about 40 ppm Ni by the time olivine crystallization had lowered the MgO content to "normal" values of about 8%. Most ocean-floor basalts have higher Ni abundances than this, particularly those in the 8–10% MgO range. Thus, if the Ni distribution coefficient used by Hart and Davis is applicable, this appears to be an unlikely process for ocean-floor basalt evolution. If, however, lower olivine/liquid Ni distribution coefficients are more appropriate (e.g., Arndt, 1977; Clarke and O'Hara, 1979), this objection to extensive olivine fractionation would not be valid.

Magma mixing

O'Hara (1977) introduced a model involving the fractionation of mixed magmas produced by the repeated injection of successive magma batches into an open magma chamber, in order to account for large differences in trace element abundances seen in otherwise similar basalts. Such a process seems plausible in the tectonic regimes of mid-ocean ridges where magma constantly is being generated and replenished along a narrow zone of volcanic activity. Several subsequent papers (e.g., Donaldson and Brown, 1977; Dungan and Rhodes, 1978; Rhodes *et al.*, 1979; Walker *et al.*, 1979) provide petrographic, experimental, mineralogical and geochemical evidence supporting magma mixing, suggesting that it is a fundamental process accompanying crystal fractionation in ocean-floor basalt petrogenesis. The principal lines of evidence include: (1) olivine and plagioclase phenocrysts that are too refractory to have grown in equilibrium with melts corresponding to the host basalt composition; (2) complex zoning patterns and resorption features in plagioclase phenocrysts indicative of multistage crystallization histories; (3) melt inclusions in phenocrysts that are more "primitive" than the host basalt; (4) chemical evidence for fractionation of clinopyroxene or plagioclase, even though petrographic evidence, corroborated by experimental studies, indicates that olivine is the liquidus phase and the only component that should be involved in the fractionation process.

Rhodes and co-workers (Rhodes *et al.*, 1979; Rhodes and Dungan, 1979) argue that the commonly encountered discrepancies among estimates of the extent of crystal fractionation based on incompatible element abundances, based on major element chemistry or based on a compatible element such as Ni, are inevitable if magma mixing accompanies crystal fractionation. Furthermore, they propose that the apparent requirement that clinopyroxene be a fractionating component, even though it is absent from the phenocryst assemblage, is readily explained by this mechanism. Clinopyroxene fractionation has occurred, but in an evolved, multiply saturated magma prior to injection of and mixing with a more "primitive" magma. The resulting magma mixture may have only olivine on its liquidus, but its bulk composition will bear the imprint of the preceding episode, or episodes, of clinopyroxene fractionation.

Magma mixing explains why basalts that are closely associated spatially frequently cannot be related to a common parental magma by simple crystal fractionation (Rhodes and Dungan, 1979), and removes the necessity of postulating separate batches of parental melts, within narrow geographic and temporal confines, that retain their integrities throughout the melting, emplacement and differentiation processes (e.g., Blanchard *et al.*, 1976; Flower *et al.*, 1977; Bryan, 1979). However, it is unlikely that magma mixing can account for wide differences in incompatible element ratios, as was originally suggested by O'Hara (1977; Pankhurst, 1977). These variations appear to be derived from inhomogeneities in the source (Schilling, 1973; White and Schilling, 1978) or from complex melting processes (Langmuir *et al.*, 1977).

To reconcile the vast volumes of these basalts with their evolved characteristics is a major problem of ocean-floor basalt petrogenesis. Traditional models, involving partial melting followed by crystal fractionation, fail to resolve this discrepancy, since one would expect the relatively "primitive" basalts to be the volumetrically more important. To claim that "primitive" basalts are underplated on the oceanic crust or that they do not make it to the surface is *ad hoc* and avoids the issue. On the other hand, a "steady-state" combination of crystal fractionation and magma mixing provides an effective mechanism for buffering the magma composition (O'Hara, 1977; Rhodes *et al.*, 1979; Walker *et al.*, 1979), the dominant product of which will be moderately evolved basalts. Both "primitive" basalts and highly evolved differentiates should be rare. Observations of ocean-floor basalts confirm this prediction. Furthermore, because of this buffering process, there is no necessity for large magma volumes to exist beneath mid-ocean ridge crests, an observation which is in accord with recent interpretations of geophysical data (Rosendahl *et al.*, 1976) and with recent models for the development of the oceanic crust (Sleep, 1975; Dewey and Kidd, 1977).

1.2.5.8 Partial melting and the mantle source

It is widely accepted that, although crystal fractionation accompanied by magma mixing may produce a broad range of trace and minor element abundances, isotopic ratios will not be modified. It is also improbable that these processes, involving the observed phenocryst phases (i.e., olivine, plagioclase and clinopyroxene), can significantly modify the ratios of elements for which crystal/liquid distribution coefficients are similar and very small (e.g., K/Ba, K/Rb, Rb/Sr, Zr/Nb, La/Ce, La/Sm, La/Yb, etc.). Yet, as is evident from the previous discussions, there is a considerable range of isotopic and trace element ratios among ocean-floor basalts. The trace element differences have been attributed to variable amounts of melting, to mechanics of melting, to compositional differences in the source, and to the residual mineralogy of the source, whereas isotopic differences are attributed to heterogeneity in the mantle source (e.g., Gast, 1968; Schilling, 1971, 1973a, b; Hart *et al.*, 1973; Frey *et al.*, 1974; O'Nions and Pankhurst, 1974; Langmuir *et al.*, 1977; Hanson, 1977; Hofmann and Hart, 1978; Tatsumoto, 1978; White and Schilling, 1978; Bougault *et al.*, 1979; Wood *et al.*, 1979).

Experimental melting studies at high pressures on natural ocean-floor basalts show that they are multiply saturated with olivine and pyroxenes at the liquidus above 8–10 kb (e.g., Kushiro, 1973; Fujii and Kushiro, 1977; Bender *et al.*, 1978; Green *et al.*, 1979; Stolper, 1980). From these experiments, it has been inferred that such pressures reflect the minimum pressures at which a peridotite mantle can melt to produce ocean-floor basaltic magmas. Pressures of 8–10 kb correspond to depths of 25–30 km, within the stability field of spinel peridotite (Ringwood, 1975). Trace element partial melting models are broadly in accord with these estimates, since the most satisfactory solutions for rare-earth element abundances in ocean-floor basalts are obtained by 10–20% melting of a lherzolite source containing olivine, orthopyroxene and clinopyroxene ± spinel (Schilling, 1975a). There is no compelling evidence from these trace element studies that garnet exists in the residual mineralogy, although it is possible that it was present initially but was consumed in the melting process. The status of plagioclase is more equivocal. Sr abundances are less variable than those of the other large cations, and are lower in Type-II basalts than in oceanic-island tholeiites. This raises the possibility of Sr "buffering" by residual plagioclase in the source, although the absence of negative Eu anomalies in "primitive" basalts argues against this interpretation.

An important consequence of trace element partial melting models is the recognition that between 10–30% melting there is little opportunity to alter the ratios of those elements with small crystal/liquid distribution coefficients. This is particularly true when melting exceeds 30%, and for truly incompatible element pairs such as K/Ba, Rb/Sr, La/Ce, Zr/Nb, etc., (Gast, 1968; Schilling, 1975a; Hanson, 1977). Under such circumstances, the ratios of the elements in the basalts probably reflect their ratios in the mantle source. If these assumptions are correct, it implies that the wide range in such element ratios as K/Ba, K/Rb, Rb/Sr, La/Ce, La/Sm, La/Yb in unaltered ocean-floor basalts reflects different source compositions and therefore marked heterogeneities in the mantle.

Strontium isotope ratios, which range from 0.7023–0.7030 in Type-I basalts and from 0.7030–0.7035 in Type-II basalts, confirm the concept of a heterogeneous mantle, unless isotopic equilibrium is not attained during the melting process. O'Nions and Pankhurst (1974) and Flower *et al.* (1975) have challenged the fundamental assumption of local equilibrium during partial melting. They suggest instead that trace element and isotopic equilibrium may not be attained between mineral grains in the mantle, and that partial melting of a single source containing a high Rb/Sr phase (e.g., phlogopite) without attaining isotopic equilibrium could explain the range of trace element and strontium isotopic ratios. Hofmann and Hart (1978) argue on kinetic grounds that, although local isotopic disequilibrium is to be expected in dry crystalline mantle, and therefore in mantle xenoliths, it is unlikely to persist during the partial melting process. This view is supported by the work of Sun and Hanson (1975), who show that although accessory minerals such as phlogopite, apatite, etc., may persist as residual phases during the production of nephelinites by small amounts of melting, they are consumed by the larger amounts of melting required for alkali basalts and tholeiites. Furthermore, recent Pb- and Nd-isotopic data (Sun and Jahn, 1975; O'Nions *et al.*, 1977; Carlson *et al.*, 1978; Tatsumoto, 1978) show variations similar to those in the more abundant strontium isotopic data, further demonstrating the importance of heterogeneous source models for ocean-floor basalts.

In a series of important papers, Schilling has noted progressive REE variations along mid-ocean ridges extending from centers of high volcanic activity as reflected in topographic highs, oceanic platforms and islands. Examples include Iceland (Schilling, 1973a), Afar (Schilling, 1973b), Azores (Schilling, 1975b) and the Galapagos (Schilling *et al.*, 1976). In all of these examples, there is a progressive depletion in the LREE accompanied by a decrease in many incompatible elements (e.g., K, Ba, Rb, Sr, La, TiO_2, P_2O_5, etc.), as one moves from the center of high volcanic activity to "normal" ridge segments. Accompanying these variations in trace element abundances and ratios is a decrease in Sr-isotope ratios from maxima of about 0.7030–0.7035 associated with centers of high volcanic activity, to ratios of about 0.7024–0.7026, characteristic of "normal" ridge segments (Hart *et al.*, 1973; White *et al.*, 1976; White and Schilling, 1978). These variations have been interpreted by Schilling and co-workers in terms of a two-source model, in which one source, a mantle plume rising from deep within the mantle, is the source of high volcanic activity associated with volcanic islands or so-called "hot spots;" and the other, the asthenosphere, is the source of "normal" ocean-ridge volcanic activity. The mantle plume source is enriched in lithophile elements and has high Sr-isotope ratios, whereas the asthenospheric source is depleted in these elements and has lower Sr-isotope ratios. Mixing of these two sources, either before or after melting, then is inferred to be the cause of the observed spread in the isotopic and trace element ratios. Several difficulties or complexities arise from this two-source model:

1) Detailed studies of the FAMOUS area and of DSDP Leg 37 samples (Blanchard *et al.*, 1976; Langmuir *et al.*, 1977) show variations in chondrite-normalized REE patterns ranging from those that are slightly REE-enriched to others that are REE-depleted. Such variations, including crossing REE patterns, have been widely interpreted as reflecting different source chemistries, yet these samples all have similar ratios of highly incompatible elements, such as K/Ba, K/Rb, Rb/Sr and Zr/Nb, which imply derivation from a common source. Thus, it seems that variations in REE abundances may not necessarily reflect different sources, but may possibly also be produced by the mechanics of the melting process (Langmuir *et al.*, 1977). Furthermore, although the FAMOUS area is located on a proposed transition zone and should reflect mixing of material from an asthenosphere source with material from the Azores mantle plume (Schilling, 1975b), the incompatible element ratios are not transitional, but are similar to those of basalts from oceanic islands (Langmuir *et al.*, 1977).

2) A similar problem exists in Iceland. There is a considerable range in REE chemistries, from basalts that are LREE-depleted, similar to ocean-floor basalts, to more abundant, LREE-enriched basalts (O'Nions *et al.*, 1976; Schilling *et al.*, 1978; Wood, 1978—see Fig. 1.2.3.12). Although these basalts all have incompatible element ratios typical of oceanic islands, O'Nions *et al.* (1976) show that the spread in REE ratios is accompanied by large differences in Sr-isotope ratios (0.7029–0.7035), which they interpret as reflecting variable source chemistries beneath Iceland, rather than a single homogeneous plume.

3) There are regions along the Mid-Atlantic Ridge with Type-II basalts that are characterized by high Sr-isotope abundances, high lithophile element abundances, and incompatible element ratios appropriate to oceanic islands. These so-called "anomalous" regions at 45° N and 35° N have been interpreted on geochemical evidence to be additional mantle plumes, even though they lack the morphological features characteristic of other inferred mantle plume centers (White *et al.*, 1976). There is a certain circularity of argument here. Does this mean that all such examples of "anomalous" basalt chemistry reflect sites of incipient mantle plumes, and also that the many oceanic islands, seamounts, or aseismic ridges are or have been centers of mantle plume activity?

4) Attempts to account for trace element and isotopic variations in terms of mixing of a simple two-component mantle source model, although consistent with Sr- and Pb-isotopic data, have not been successful in general for other trace elements (e.g., Sun *et al.*, 1975; O'Nions *et al.*, 1976; Langmuir *et al.*, 1978; White and Schilling, 1978).

In summary, although evidence for large-scale mantle heterogeneity in the source region of ocean-ridge basalts is very compelling, the spatial relationships among these sources are not clearly understood, and it is perhaps premature to embrace the two-source mantle plume model. One fact that does emerge from these studies is that two kinds of basalts are commonly erupted along mid-ocean ridges. The "normal" ocean-floor basalts, corresponding to the Type-I basalts of Bryan *et al.* (1976), are LREE and incompatible element-depleted and are characterized by high K/Ba, K/Rb, and Zr/Nb and low $^{87}Sr/^{86}Sr$ ratios. Basalts of the other kind, corresponding to the Type-II basalts of Bryan *et al.* (1976), commonly are less depleted in REE and incompatible elements and have higher $^{87}Sr/^{86}Sr$ ratios. An important characteristic of this second kind of basalt is that the ratios of truly incompatible elements (e.g., K/Ba, K/Rb, La/Ce, Zr/Nb) are lower than those of "normal" ridge basalts and are indistinguishable from those of oceanic-island tholeiites (Erlank and Kable, 1976; Hanson, 1977). However, it must be emphasized that although the incompatible element ratios are similar to those of the oceanic tholeiites, their major element chemistries and other characteristics, such as Sr contents (Hart, 1976), TiO_2, P_2O_5 content and Zr/Y ratios (Pearce and Cann, 1973), are those of ocean-ridge basalts. Transitional tholeiites and alkali basalts with similar isotopic and incompatible element ratios, may be associated with this second type of basalt, as is the case at 45° N (Aumento, 1968).

It appears that ocean-floor basalts can be derived from two distinct mantle sources, one having compositional characteristics similar to the source region of the tholeiites and alkali basalts of oceanic islands, and the other with those of the mantle beneath "normal" mid-ocean ridge segments. White and Schilling (1978) have attempted to estimate the compositions of two source regions, one the mantle beneath the Azores and the other that beneath "normal" segments of the Mid-Atlantic Ridge, using partial melting calculations involving 10–30% melting of lherzolite. Both sources have relative REE and incompatible element patterns which do not differ significantly from those of the corresponding basalts. Consquently, both sources are depleted in large, low-valency cations (K, Rb, Cs) with respect to refractory elements (Ba, Sr, and REE). However, the Azores source is substantially enriched (8–19 times) in these elements relative to the "normal" ocean ridge source. It also is high in Ba and LREE, and is LREE-enriched relative to chondrites, whereas the "normal" ridge source is characteristically LREE-depleted. Both sources have essentially similar HREE abundances. In many respects, the proposed Azores mantle source is very similar to the source region proposed for the Ross Island basanitoids by Sun and Hanson (1975). Sr- and Pb-isotope studies suggest that these distinct mantle sources may have persisted for about 1–2 b.y. (Sun and Jahn, 1975; Brooks *et al.*, 1976; Tatsumoto, 1978).

1.2.5.9 References

Ade-Hall J. M., Fink L. K. and Johnson H. P. (1976) Petrography of opaque minerals. Leg 34. In *Initial Reports of the Deep Sea Drilling Project* **34**, p. 349–362. U.S. Government Printing Office, Washington, D.C.

Anderson A. T. (1974) Chlorine, sulfur and water in magmas and oceans. *Geol. Soc. Amer. Bull.* **85**, 1485–1492.

Anderson A. T. and Greenland L. P. (1969) Phosphorus fractionation diagram as a quantitative indicator of crystallization differentiation of basaltic liquids. *Geochim. Cosmochim. Acta* **33**, 493–505.

Arndt N. T. (1977) Partitioning of nickel between olivine and ultrabasic and basic komatiite liquids. *Carnegie Inst. Wash., Yearb.* **76**, p. 553–557.

Aumento F. (1968) The Mid-Atlantic Ridge near 45° N. II. Basalts from the area of Confederation Peak. *Can. J. Earth Sci.* **5**, 1–21.

Aumento F., Loncarevic B. D. and Ross D. I. (1971) Hudson Geotraverse: geology of the Mid-Atlantic Ridge at 45° N. *Phil. Trans. Roy. Soc. Lond.* **268**, 623-650.

Ballard R. D. and Van Andel T. J. (1977) Morphology and tectonics of the inner rift valley at 36° 50′ N on the Mid-Atlantic Ridge. *Geol. Soc. Amer. Bull.* **88**, 507–530.

Bence A. E. and Bender J. F. (1978) Clinopyroxenes in mid-ocean ridge basalts: experimental studies (abstract). *EOS (Trans. Amer. Geophys. Union)* **59**, 410.

Bence A. E., Bayliss D. M., Bender J. F. and Grove T. L. (1979) Controls on the major and minor element chemistry of mid-ocean ridge basalts and glasses. In *Deep Drilling Results in the Atlantic Ocean: Ocean Crust* (M. Talwani, C. G. Harrison and D. E. Hayes, eds.), p. 331–341. Maurice Ewing Series 2. Amer. Geophys. Union, Washington, D. C.

Bence A. E., Papike J. J. and Ayuso R. A. (1975) Petrology of submarine basalts from the Central Caribbean: DSDP Leg 15. *J. Geophys. Res.* **80**, 4775–4804.

Bender J. F., Hodges F. N. and Bence A. E. (1978) Petrogenesis of basalts from the Project FAMOUS area: experimental study from 0 to 15 kbars. *Earth Planet. Sci. Lett.* **41**, 277–302.

Blanchard D. P., Rhodes J. M., Dungan M. A., Rodgers K. V., Donaldson C. H., Brannon J. C., Jacobs J. W. and Gibson E. K. (1976) The chemistry and petrology of basalts from Leg 37 of the Deep Sea Drilling Project. *J. Geophys. Res.* **81**, 4231–4246.

Bryan W. B., Frey F. A. and Thompson G. (1977) Oldest Atlantic Seafloor; Mesozoic basalts from Western North Atlantic margin and Eastern North America. *Contrib. Mineral. Petrol.* **64**, p. 223–242.

Bougault H. and Hekinian R. (1974) Rift valley in the Atlantic Ocean near 36° 50′ N: petrology and geochemistry of basaltic rocks. *Earth Planet. Sci. Lett.* **24**, 249–261.

Bougault H., Cambon P., Corre O., Joron J. L. and Treuil M. (1979) Evidence for variability of magmatic processes and upper mantle heterogeneity in the axial region of the Mid-Atlantic Ridge near 22° N and 36° N. *Tectonophys.* **55**, 11–34.

Brooks C., Hart S. R., Hofmann A. L. and James D. E. (1976) Rb-Sr mantle isochrons from oceanic regions. *Earth Planet. Sci. Lett.* **32**, 51–61.

Bryan W. B. (1974) Fe-Mg relationships in sector-zoned submarine basalt plagioclase. *Earth Planet. Sci. Lett.* **24**, 157–165.

Bryan W. B. (1979) Regional variation and petrogenesis of basalt glasses from the FAMOUS area, Mid-Atlantic Ridge. *J. Petrol.* **20**, 293–325.

Bryan W. B. and Moore J. G. (1977) Compositional variations of young basalts in the Mid-Atlantic Ridge rift valley near lat. 36° 49′ N. *Geol. Soc. Amer. Bull.* **88**, 556–570.

Bryan W. B., Thompson G., Frey F. A. and Dickey J. S. (1976) Inferred settings and differentiation in basalts from the Deep Sea Drilling Project. *J. Geophys. Res.* **81**, 4285–4304.

Cann J. R. (1970) Rb, Sr, Y, Zr and Nb in some ocean floor basaltic rocks. *Earth Planet. Sci. Lett.* **10**, 7–11.

Cann J. R. (1971) Major element variations in ocean-floor basalts. *Phil. Trans. Roy. Soc. Lond.* **A268**, p. 495–505.

Cann J. R. (1974) A model for oceanic crustal structure developed. *Geophys. J. Roy. Astron. Soc.* **39**, 169–187.

Cann J. R., Tarney J., Varet J. and Wood D. A. (1979) mantle heterogeneity in the North Atlantic: evidence from Leg 49 geochemistry. In *Initial Reports of the Deep Sea Drilling Project* **49**, p. 841–850. U.S. Government Printing Office, Washington D.C.

Carlson R. W., MacDougall J. D. and Lugmair G. W. (1978) Differential Sm/Nd evolution in oceanic basalts. *Geophys. Res. Lett.* **5**, 229–232.

Christensen N. I. and Salisbury M. H. (1975) Structure and constitution of the lower oceanic crust. *Rev. Geophys. Space Phys.* **13**, 57–86.

Church S. E. and Tatsumoto M. (1975) Lead isotope relations in oceanic ridge basalts from the Juan de Fuca-Gorda Ridge area, N.E. Pacific Ocean. *Contrib. Mineral. Petrol.* **53**, 253–279.

Clague D. A. and Bunch T. E. (1976) Formation of ferrobasalt at East Pacific mid-ocean spreading centers. *J. Geophys. Res.* **81**, 4247–4256.

Clague D. A. and Straley P. F. (1977) Petrologic nature of the oceanic Moho. *Geology* **5**, 133–136.

Clarke D. B. and O'Hara M. J. (1979) Nickel, and the existence of high-MgO liquids in nature. *Earth Planet. Sci. Lett.* **44**, 153–158.

Dewey J. F. and Kidd W. S. F. (1977) Geometry of plate accretion. *Geol. Soc. Amer. Bull.* **88**, 960–968.

Donaldson C. H. and Brown R. W. (1977) Refractory megacrysts and magnesium-rich melt inclusions within spinel in oceanic tholeiites: indicators of magma mixing and parental magma composition. *Earth Planet. Sci. Lett.* **37**, 81–89.

Drake M. J. (1976) Plagioclase-melt equilibria. *Geochim. Cosmochim. Acta* **40**, 457–466.

Dungan M. A. and Rhodes J. M. (1978) Residual glasses and melt inclusions in basalts from D.S.D.P. Legs 45 and 46: evidence for magma mixing. *Contrib. Mineral. Petrol.* **67**, 417–431.

Dungan M. A., Long P. E. and Rhodes J. M. (1978) The petrography, mineral chemistry and one-atmosphere phase relations of basalts from site 395-Leg 45 D.S.D.P. In *Initial Reports of the Deep Sea Drilling Project* **45**, p. 461–478. U.S. Government Printing Office, Washington, D.C.

Elthon D. (1979) High magnesia liquids as the parental magma for ocean floor basalts. *Nature* **278**, 514–518.

Engel A. E., Engel C. G. and Havens R. G. (1965) Chemical characteristics of oceanic basalts and the upper mantle. *Geol. Soc. Amer. Bull.* **76**, 719–734.

Erlank A. J. and Kable E. J. D. (1976) The significance of incompatible elements in mid-Atlantic ridge basalts from 45° N with particular reference to Zr/Nb. *Contrib. Mineral. Petrol.* **54**, 281–291.

Fisk M. R. (1978) Melting relations and mineral chemistry of Iceland and Reykjanes Ridge basalts. Ph.D. dissertation, Grad. School of Oceanography, University of Rhode Island, Kingston, R.I. 254 pp.

Fisk M. R. and Bence A. E. (1979) Experimental studies of spinel crystallization in FAMOUS basalt 527-1-1 (abstract). *EOS (Trans. Amer. Geophys. Union)* **60**, 420.

Flower M. F. J., Robinson P. T., Schmincke H. -U. and Ohnmacht W. (1977) Magma fractionation systems beneath the Mid-Atlantic Ridge at 36°–37° N. *Contrib. Mineral. Petrol.* **64**, 167–195.

Flower M. F. J., Schmincke H. -U. and Thompson R. N. (1975) Phlogopite stability and the $^{87}Sr/^{86}Sr$ step in basalts along the Reykjanes Ridge. *Nature* **254**, 404–406.

Floyd P. A. and Winchester J. A. (1975) Magma type and tectonic setting discrimination using immobile elements. *Earth Planet. Sci. Lett.* **27**, 211–218.

Fowler C. M. R. and Matthews D. H. (1974) A seismic refraction experiment using ocean bottom seismographs and sonobouys on the Mid-Atlantic Ridge near 36° N. *Nature* **249**, 752–754.

Frey F. A., Bryan W. B. and Thompson G. (1974) Atlantic Ocean floor: geochemistry and petrology of basalts from Legs 2 and 3 of the Deep Sea Drilling Project. *J. Geophys. Res.* **79**, 5507–5527.

Fujii T. and Kushiro I. (1977) Melting relations and viscosity of an abyssal tholeiite. *Carnegie Inst. Wash., Yearb.* **76**, 461–465.

Gass I. G. (1968) Is the Troodos massif of Cyprus a fragment of Mesozoic ocean floor? *Nature* **220**, 39–42.

Gast P. W. (1968) Trace element fractionation and the origin of tholeiitic and alkaline magma types. *Geochim. Cosmochim. Acta* **32**, 1057–1086.

Green D. H. (1970) The origin of basaltic and nephelenitic magmas. *Trans. Leicester Lit. Philos. Soc.* **64**, 28–54.

Green D. H., Hibberson W. O. and Jaques A. L. (1979) Petrogenesis of mid-ocean basalts. In *The Earth: Its Origin, Structure and Evolution* (W. H. McElhinny, ed.), p. 265–299. Academic Press, London.

Haggerty S. E. (1976) Opaque mineral oxides in terrestrial igneous rocks. In *Oxide Minerals* vol.3, p. Hg 101–Hg 300. Mineral. Soc. of Amer. Short Course Notes. Washington, D.C.

Hanson G. N. (1977) Geochemical evolution of the sub-oceanic mantle. *J. Geol. Soc. Lond.* **134**, 235–253.

Hanson G. N. and Langmuir C. H. (1978) Modelling of major elements in mantle-melt systems using trace element approaches. *Geochim. Cosmochim. Acta* **42**, 725–742.

Hart S. R. (1970) Sea floor basalts. *Carnegie Inst. Wash., Yearb.* **68**, 403–408.

Hart S. R. (1976) LIL-element geochemistry, Leg 34 basalts. In *Initial Reports of the Deep Sea Drilling Project* **34**, p. 283–288. U.S. Government Printing Office, Washington, D.C.

Hart S. R. and Davis K. E. (1978) Nickel partitioning between olivine and silicate melt. *Earth Planet Sci. Lett.* **40**, 203–219.

Hart S. R., Schilling J. -G. and Powell J. L. (1973) Basalts from Iceland and along the Reykjanes Ridge: Sr isotope geochemistry. *Nature Phys. Sci.* **246**, 104–107.

Hekinian R. (1974) Petrology of the Ninety East Ridge (Indian Ocean) compared to other aseismic ridges. *Contrib. Mineral Petrol.* **43**, 125–147.

Hekinian R. and Thompson G. (1976) Comparative geochemistry of volcanics from rift valleys, transform faults and aseismic ridges. *Contrib. Mineral. Petrol.* **57**, 145–162.

Hekinian R., Moore J. G. and Bryan W. B. (1976) Volcanic rocks and processes of the Mid-Atlantic Ridge rift valley near 36°49′ N. *Contrib. Mineral. Petrol.* **58**, 83–110.

Hermes O. D. and Schilling J. -G. (1976) Olivine from Reykjanes Ridge and Iceland tholeiites and its significance to the two-mantle source model. *Earth Planet. Sci. Lett.* **28**, 345–355.

Hill R. and Roeder P. (1974) The crystallization of spinel from basaltic liquid as a function of oxygen fugacity. *J. Geol.* **82**, 709–729.

Hodges F. N. and Papike J. J. (1976) DSDP Site 334: Magmatic cumulates from oceanic layer 3. *J. Geophys. Res.* **81**, 4135–4151.

Hofmann A. W. and Hart S. R. (1978) An assessment of local and regional isotopic equilibrium in the mantle. *Earth Planet. Sci. Lett.* **38**, 44–62.

Houtz R. and Ewing J. (1976) Upper crustal structure as a function of plate age. *J. Geophys. Res.* **81**, 2490–2498.

Irving A. J. (1978) A review of experimental studies of crystal/liquid trace element partitioning. *Geochim. Cosmochim. Acta* **42**, 743–770.

Johnson J. R. (1979) Transitional basalts and tholeiites from the East Pacific Rise, 9° N. *J. Geophys. Res.* **84**, 1635–1652.

Kay R., Hubbard N. J. and Gast P. W. (1970) Chemical characteristics and origin of oceanic ridge volcanic rocks. *J. Geophys. Res.* **75**, 1585–1613.

Kirkpatrick R. J. (1979) The physical state of the oceanic crust: results of downhole geophysical logging in the Mid-Atlantic Ridge at 23° N. *J. Geophys. Res.* **84**, 178–188.

Kushiro I. (1973) Origin of some magmas in oceanic and circum-oceanic regions. *Tectonophys.* **16**, 211–222.

Langmuir C. H., Bender J. F., Bence A. E. and Hanson G. N. (1977) Petrogenesis of basalts from the FAMOUS area: Mid-Atlantic Ridge. *Earth Planet. Sci. Lett.* **36**, 133–156.

Langmuir C. H., Vocke R. D., Hanson G. N. and Hart S. R. (1978) A general mixing equation with applications to Icelandic basalts. *Earth Planet. Sci. Lett.* **37**, 380–392.

Mazzullo L. and Bence A. E. (1976) Abyssal tholeiites from DSDP Leg 34: The Nazca Plate. *J. Geophys. Res.* **81**, 4327–4351.

Melson W. G. and Thompson G. (1971) Petrology of a transform fault zone and adjacent ridge segments. *Phil. Trans. Roy. Soc. Lond.* **A268**, 423–441.

Melson W. G. and Van Andel T. J. (1966) Metamorphism in the Mid-Atlantic Ridge, 22° N latitude. *Marine Geol.***4**, 165–186.

Melson W. G., Hart S. R. and Thompson G. (1972) St. Paul's Rocks, Equatorial Atlantic: Petrogenesis, radiometric ages, and implications on sea-floor spreading. *Geol. Soc. Amer. Mem. 132*, 241–272.

Melson W. G., Thompson G. and Van Andel T. J. (1968) Volcanism and metamorphism in the Mid-Atlantic Ridge, 22° N latitude. *J. Geophys. Res.* **73**, 5925–5941.

Melson W. G., Vallier T. L., Wright T. L., Byerly G. and Nelen J. (1976) Chemical diversity of abyssal volcanic glass erupted along Pacific, Atlantic, and Indian Ocean sea-floor spreading centers. In *The Geophysics of the Pacific Ocean Basin and Its Margin*, p. 351–367. Amer. Geophys. Union, Washington, D.C.

Menard H. W. (1967) Sea-floor spreading, topography and the second layer. *Science* **157**, 923–924.

Miyashiro A., Shido F. and Ewing M. (1970) Crystallization and differentiation in abyssal tholeiites and gabbros from Mid-Atlantic ridges. *Earth Planet. Sci. Lett.* **7**, 361–365.

Moore J. G., Fleming H. S. and Phillips J. D. (1974) Preliminary model for extrusion and rifting at the axis of the Mid-Atlantic Ridge, 36°48′ North. *Geology* **2**, 437–440.

Moore E. M. and Vine J. F. (1971) The Troodos Massif, Cyprus, and other ophiolities as oceanic crust: evaluation and implications. *Phil. Trans. Roy. Soc. Lond.* **A268**, 443–446.

Muir I. D. and Tilley C. E. (1964) Basalts from the northern part of the rift zone of the Mid-Atlantic Ridge. *J. Petrol.* **5**, 409–434.

Myers C. W., Bence A. E., Papike J. J. and Ayuso R. A. (1975) Petrology of an alkali-olivine basalt sill from site 169 of DSDP Leg 17: The Central Pacific Basin. *J. Geophys. Res.* **80**, 807–822.

Nabalek P. (1979) Nickel partitioning between olivine and liquid in natural basalts (abstract). *EOS (Trans. Amer. Geophys. Union)* **60**, 402.

Natland J. H. (1978) Crystal morphologies in basalts from DSDP Site 395, 23° N, 46° W, Mid-Atlantic Ridge. In *Initial Reports of the Deep Sea Drilling Project* **45**, p. 423–446. U.S. Government Printing Office, Washington, D.C.

O'Hara M. J. (1968) Are ocean floor basalts primary magmas? *Nature* **220**, 683–686.

O'Hara M. J. (1973) Non-primary magmas and dubious mantle plume beneath Iceland. *Nature* **243**, 507–508.

O'Hara M. J. (1977) Geochemical evolution during fractional crystallization of a periodically refilled magma chamber. *Nature* **266**, 503–507.

O'Hara M. J., Saunders M. J. and Mercy E. L. P. (1975) Garnet-peridotite, primary ultrabasic magma and eclogite; interpretation of upper mantle processes in kimberlite. *Phys. Chem. Earth* **9**, 571–604.

O'Nions R. K. and Pankhurst R. J. (1974) Petrogenetic significance of isotope and trace element variations in volcanic rocks from the Mid-Atlantic. *J. Petrol.* **15**, 603–634.

O'Nions R. K., Hamilton P. J. and Evensen N. M. (1977) Variations in $^{143}Nd/^{144}Nd$ and $^{87}Sr/^{86}Sr$ ratios in oceanic basalts. *Earth Planet. Sci. Lett.* **34**, 13–22.

O'Nions R. K., Pankhurst R. J. and Gronvold K. (1976) Nature and development of basalt magma sources beneath Iceland and the Reykjanes Ridge. *J. Petrol.* **17**, 315–338.

Orcutt J., Kennett B., Dorman L. and Prothero W. (1975) Evidence for a low velocity zone underlying a fast-spreading rise crest. *Nature* **265**, 475–476.

Pankhurst R. J. (1977) Open system crystal fractionation and incompatible element variation in basalts. *Nature* **268**, 36–38.

Papike J. J., Cameron K. L. and Baldowin K. R. (1974) Amphiboles and pyroxenes: characterizations of *other* than quadrilateral components and estimates of ferric iron from microprobe data (abstract). *Abstracts with Programs,* Geol. Soc. Amer. **6**, p. 1053–1054.

Pearce J. A. (1976) Statistical analysis of major element patterns in basalt. *J. Petrol.* **17**, 15–43.

Pearce J. A. and Cann J. R. (1973) Tectonic setting of basic volcanic rocks determined using trace element analyses. *Earth Planet. Sci. Lett.* **19**, 290–300.

Peterson J. J., Fox P. J. and Schreiber E. (1974) Newfoundland ophiolites and the geology of the oceanic layer. *Nature* **247**, 194–196.

Presnall D. C., Dixon J. R., O'Donnell T. H. and Dixon S. A. (1979) Generation of mid-ocean ridge tholeiite. *J. Petrol.* **20**, 3–35.

Raitt R. (1963) The crustal rocks. In *The Sea*, v. 3, 85–102. John Wiley, N.Y.

Reid I., Orcutt J. A. and Prothero W. A. (1977) Seismic evidence for a narrow zone of partial melting underlying The East Pacific Rise at 21° N. *Geol. Soc. Amer. Bull.* **88**, 678–682.

Rhodes J. M. and Dungan M. A. (1979) The evolution of ocean floor basaltic magmas. In *Deep Drilling Results in the Atlantic Ocean: Ocean Crust* (M. Talwani, C. G. Harrison and D. E. Hayes, eds.), p. 262–272. Maurice Ewing Series 2. Amer. Geophys. Union, Washington, D.C.

Rhodes J. M. and Rodgers K. V. (1972) Chemical diversity of mid-ocean ridge basalts from a single dredge haul (abstract). *EOS (Trans. Amer. Geophys. Union)* **53**, 1133.

Rhodes J. M., Blanchard D. P., Dungan M. A., Rodgers K. V. and Brannon J. C. (1978) Chemistry of basalts from Leg 45 of the Deep Sea Drilling Project. In *Initial Reports of the Deep Sea Drilling Project* **45**, p. 447–460. U.S. Government Printing Office, Washington, D.C.

Rhodes J. M., Blanchard D. P., Rodgers K. V., Jacobs J. W. and Brannon J. C. (1976) Petrology and chemistry of basalts from the Nazca plate: Part 2—major and trace element chemistry. In *Initial Reports of the Deep Sea Drilling Project.* **34**, p. 239–244. U.S. Government Printing Office, Washington, D.C.

Rhodes J. M., Dungan M. A., Blanchard D. P. and Long P. E. (1979) Magma mixing at mid-ocean ridges: evidence from basalts drilled near 22° N on the Mid-Atlantic Ridge. *Tectonophys.* **55**, 35–62.

Rice S. B. (1978) Diffusion and growth in FAMOUS glasses. M.S. thesis, State University of New York, Stony Brook, N.Y.

Ridley W. I., Rhodes J. M., Reid A., Jakeš P., Shih C., and Bass M. (1973) The petrology of basaltic rocks from the Deep Sea Drilling Project, Leg 6. *J. Petrol.* **15**, 140–159.

Ringwood A. E. (1975) *Composition and Petrology of the Earth's Mantle.* McGraw Hill, N.Y. 618 pp.

Roeder P. L. and Emslie R. F. (1970) Olivine-liquid equilibrium. *Contrib. Mineral Petrol.* **29**, 275–289.

Rosendahl B. R. (1976) Evolution of oceanic crust, 2, constraints, implications and inferences. *J. Geophys. Res.* **81**, 5305–53114.

Rosendahl B. R., Raitt R. W., Dorman L. M., Bibee L. D., Hussong D. M. and Sutton G. H. (1976) Evolution of oceanic crust, 1. A physical model of the East Pacific Rise crest derived from seismic-refraction data. *J. Geophys. Res.* **81**, 5294–5304.

Sato H. (1977) Nickel content of basaltic magmas: identification of primary magmas and a measure of the degree of olivine fractionation. *Lithos* **10**, 113–120.

Schilling J. -G. (1971) Sea-floor evolution: rare-earth evidence. *Phil. Trans. Roy. Soc. Lond.* **A268**, 663–706.

Schilling J. -G. (1973a) Iceland mantle plume: geochemical evidence along Reykjanes Ridge. *Nature* **242**, 565–571.

Schilling J. -G. (1973b) Afar mantle plume: rare earth evidence. *Nature Phys. Sci.* **242**, 2–5.

Schilling J. -G. (1975a) Rare-earth variations across "normal segments" of the Reykjanes Ridge, 60°–53° N, Mid-Atlantic Ridge, 29° S, and East Pacific Rise, 2°–19° S, and evidence on the composition of the underlying low-velocity layer. *J. Geophys. Res.* **80**, 1459–1473.

Schilling J. -G. (1975b) Azores mantle blob: rare earth evidence. *Earth Planet. Sci. Lett.* **25**, 103–115.

Schilling J. -G., Anderson R. N. and Vogt P. (1976) Rare earth, Fe and Ti variations along the Galapagos spreading centre, and their relationship to the Galapagos mantle plume. *Nature* **261**, 108–113.

Schilling J. -G., Unni C. K. and Bender M. L. (1978) Origin of chlorine and bromine in the oceans. *Nature* **273**, 631–636.

Schweitzer E. L., Papike J. J. and Bence A. E. (1979) Statistical analysis of clinopyroxenes from deep sea basalts. *Amer. Mineral.* **64**, 501–513.

Shido F. A., Miyashiro A. and Ewing M. (1971) Crystallization of abyssal tholeiites. *Contrib. Mineral. Petrol.* **31**, 251–266.

Sigurdsson H. and Schilling J. -G. (1976) Spinels in Mid-Atlantic Ridge basalts: chemistry and occurrence. *Earth Planet. Sci. Lett.* **29**, 7–20.

Sleep N. H. (1975) Formation of oceanic crust: some thermal constraints. *J. Geophys. Res.* **80**, 4037–4042.

Stern C. R. (1980) Geochemistry of Chilean Ophiolites: evidence for the compositional evolution of the mantle source of back-arc basin basalts. *J. Geophys. Res.* **85**, 955–966.

Stolper E. (1980) A phase diagram for mid-ocean ridge basalts: preliminary results and implications for petrogenesis. *Contrib. Mineral. Petrol.* **74**, 13–28.

Stolper E. and Walker D. (1980) Melt density and the average composition. *Contrib. Mineral. Petrol.* **74**, 7–12.

Sun S. -S. and Hanson G. N. (1975) Origin of Ross Island basanitoids and limitations upon the heterogeneity of mantle sources for alkali basalts and nephelinites. *Contrib. Mineral. Petrol.* **52**, 77–106.

Sun S. -S. and Jahn B. (1975) Lead and strontium isotopes in post-glacial basalts from Iceland. *Nature* **225**, 527–530.

Sun S. -S., Tatsumoto M. and Schilling J. -G. (1975) Mantle plume mixing along the Reykjanes Ridge axis: lead isotopic evidence. *Science* **190**, 143–147.

Talwani M., Windisch C. and Langseth M. (1971) Reykjanes ridge crest: a detailed geophysical study. *J. Geophys. Res.* **76**, 473–517.

Tatsumoto M. (1978) Isotopic composition of lead in oceanic basalt and its implication to mantle evolution. *Earth Planet. Sci. Lett.* **38**, 63–87.

Tatsumoto M., Hedge C. E. and Engel A. E. J. (1965) Potassium, rubidium, strontium, thorium, uranium, and the ratio strontium-87 to strontium-86 in oceanic tholeiitic basalt. *Science* **180**, 886–888.

Thompson G. and Melson W. G. (1972) The petrology of oceanic crust across fracture zones in the Atlantic Ocean: evidence of a new kind of sea-floor spreading. *J. Geol.* **80**, 526–538.

Thompson G., Bryan W. B., Frey F. A., Dickey J. S. and Sven C. J. (1976) Petrology and geochemistry of basalts from DSDP Leg 34, Nazca plate. In *Initial Reports of the Deep Sea Drilling Project* **34**, 215–226. U.S. Government Printing Office, Washington. D.C.

Tilley C. I., Thompson R. N. and Levenbury P. A. (1972) Melting relations of some oceanic basalts. *Geol. J.* **8**, 59–64.

Van Andel T. J., Phillips J. D. and Von Hermen R. P., (1969) Rifting origin for the Vema fracture in the North Atlantic. *Earth Planet. Sci. Lett.* **5**, 296–300.

Walker D., Shibata T. and DeLong S. E. (1979) Abyssal tholeiites from the Oceanographer Fracture Zone. II Phase equilibria and mixing. *Contrib. Mineral. Petrol.* **70**, 111–125.

Watson E. B. (1976) Glass inclusions as samples of early magmatic liquid: determinative method and application to a South Atlantic basalt. *J. Volc. Geotherm. Res.* **1**, 73–84.

White W. M., Schilling J. -G. and Hart S. R. (1976) Evidence for the Azores mantle from strontium isotope geochemistry of the Central North Atlantic. *Nature* **268**, 659–662.

White W. M. and Bryan W. B. (1977) Sr-isotope, K, Rb, Cs, Sr, Ba and rare-earth geochemistry of basalts from the FAMOUS area. *Geol. Soc. Amer. Bull.* **88**, 571–576.

White W. M. and Schilling J. -G. (1978) The nature and origin of geochemical variation in Mid-Atlantic Ridge basalts from the Central North Atlantic. *Geochim. Cosmochim. Acta* **42**, 1501–1516.

Whitmarsh R. B. (1973) Median valley refraction line, Mid-Atlantic Ridge at 37° N. *Nature* **264**, 297–299.

Wiseman J. D. H. (1936) The petrography and significance of a rock dredged from a depth of 744 fathoms near to Providence Reef, Indian Ocean. *Linnean Soc. Lond. Trans., 2nd Ser.* **19**, 437–443.

Wood D. A. (1978) Major and trace element variations in the Tertiary lavas of Eastern Iceland with respect to the Iceland geochemical anomaly. *J. Petrol.* **19**, 393–436.

Wood D. A., Tarney J., Varet J., Saunders A. D., Bougault H., Joron J. L., Treuil M. and Cann J. R. (1979) Geochemistry of basalts drilled in the North Atlantic by IPOD Leg 49: implications for mantle heterogeneity. *Earth Planet. Sci. Lett.* **43**, 77–97.

Yoder H. S. (1976) *Generation of Basaltic Magma*. National Academy of Sciences, Washington, D. C. 265 pp.

Yoder H. S. and Tilley C. E. (1962) Origin of basalt magmas: an experimental study of natural and synthetic rock systems. *J. Petrol.* **3**, 342–532.

1.2.6 OCEANIC INTRAPLATE VOLCANISM

1.2.6.1 Introduction

Volcanic activity in oceanic regimes is largely associated with plate boundaries. Eruptions at diverging plates (mid-ocean ridges) are responsible for the creation of new oceanic crust, and eruptions at converging plate boundaries (near subduction zones) are related, in part, to crustal consumption. However, significant volcanic activity also occurs in regions far from active plate boundaries. This volcanism, which is associated largely with volcanic islands and seamounts, is distinctly different in style from ocean ridge volcanism. It is a consequence of partial melting which occurs at a localized "hot spot" within the mantle rather than along a rift. Eruption of lava at these restricted localities takes place over long intervals of time and in volumes sufficiently large that a volcanic edifice is built upon the oceanic crust. Production rates ($\Delta v/\Delta t$ km^3/yr) of individual volcanoes are very high; those of the Hawaiian islands range from 0.01–0.2 km^3/yr (Bargar and Jackson 1974; Shaw *et al.* 1980). Although these individual rates are much higher than local Mid-Ocean Ridge Basalt (MORB) extrusion rates, the total annual production rate of all intraplate sites (including those on the continental plates) is much lower than the total MORB extrusion rate. Schilling *et al.* (1978) estimate total intraplate production rates to be ~10% of mid-ocean ridge production rates.

Erupted basalts are dominated by enormous volumes of tholeiites, chemically distinct from mid-ocean ridge tholeiites, which form major portions of the volcanic edifices. These basalts may be oversaturated quartz tholeiites, slightly undersaturated olivine tholeiites, or picrites. In the simplest model, the late stage (post-erosional) activity is dominated by extrusion of alkaline series basalts which form a thin cap over the tholeiitic base (Powers, 1955; MacDonald and Katsura, 1964; Engel *et al.*, 1965). The alkaline basalts cover a wide range of chemical compositions, including alkali olivine basalts, trachybasalts, basanites, nephelinites, and related rocks.

Well-known localities of intraplate volcanism in the modern ocean basins include Tristan da Cunha, Gough, St. Helena, Ascension, the Azores, and the Canary Islands in the Atlantic Ocean; Guadalupe, Hawaii, the Galapagos Islands, and Tahiti in the Pacific Ocean; and Réunion, Rodriguez, and Mauritius in the Indian Ocean. The western Pacific is dotted with numerous seamounts which represent extinct intraplate volcanoes rising above the ocean floor. In the Atlantic, the island of Bermuda is a coral-topped volcanic edifice which formed as a consequence of intraplate volcanism.

It is very difficult, if not impossible, to estimate the proportions of tholeiitic series basalts relative to alkali and nephelinite series basalts in the intraplate regime. Some intraplate volcanoes (Hawaii and Réunion are good examples) are dominated by tholeiitic basalts. Others, for example the Canary Islands, Gough, Tristan da Cunha, Ascension, and Tahiti, appear to be dominated by alkalic rocks.

The Hawaiian Islands, particularly the island of Hawaii, are the classic localities of oceanic intraplate basaltic volcanism. Over the past 150 years, frequent eruptions on Hawaii have provided numerous data on the controls exerted on magma chemistry by the processes of partial melting (Green and Ringwood,1967; Schilling and Winchester, 1969), fractional crystallization (Powers, 1955; Tilley, 1960; Tilley and Scoon, 1961; MacDonald and Katsura, 1964; Murata and Richter, 1966; Schilling and Winchester, 1966, 1969; Green and Ringwood, 1967; Wright and Fiske, 1971; Gunn, 1971; Wright *et al.* 1976; Wright and Okamura, 1977), magma mixing (Wright, 1973; Wright *et al.* 1975), and flow differentiation (Wright and Okamura, 1977). Low- and high-pressure melting experiments on Hawaiian basalt compositions (Yoder and Tilley, 1962; Tilley *et al.* 1965; Green and Ringwood, 1967; Thompson and Tilley, 1976) have confirmed the importance of olivine fractionation at low pressure in the evolution of these basalts. The high-pressure experiments confirmed ideas of source-region mineralogy which had been based largely upon the study of xenoliths in the more alkalic basalts (see, for example, Jackson and Wright, 1970).

Relatively easy access to the frequent eruptions make Hawaii ideal for *in situ* measurement of eruption temperatures (Peck *et al.* 1964; Wright *et al.* 1976), oxygen fugacities (Sato and Wright, 1966), volatile and rare gas contents (Funkhouser *et al.* 1966), and viscosities (Shaw *et al.* 1968; Shaw, 1972). The remoteness of the islands from continental plates and the relative thinness of the underlying oceanic crust reduce problems of contamination; therefore, the chemistries of the basalts tend to reflect the chemistry and mineralogy of the source, the T, P, and f_{O_2} conditions of melting, the degree of melting, and the post-melting fractionation effects.

The objectives of this section are to examine the major and trace element chemistry, mineralogy, and petrography of basalts from the island of Hawaii, to establish how well they represent intraplate ocean island basalts as a group, to compare this group with mid-ocean ridge basalts and to evaluate models for their petrogenesis.

1.2.6.2 Regional tectonic and temporal setting of the Hawaiian chain

The Hawaiian Ridge, together with the Emperor Seamount chain in the northwestern Pacific, forms an elbow-shaped trend of volcanism which has migrated across the Pacific Ocean (see Fig. 5.3.14; Table 1.2.6.1), with the most recent activity occurring in the extreme southeast. Wilson (1963) suggests that the linear Hawaiian chain represents melting which has occurred in an upper mantle source which has remained fixed while the oceanic crust has moved northwestward. Volcanism along the ridge is associated with a series of rifts that are not related to the major east-west fracture zones in the Pacific plate.

The oldest radiometric age yet obtained for volcanoes on the Emperor Seamount chain is a 64.7 ± 1.1 m.y. K-Ar (^{40}Ar-^{39}Ar) age reported by Dalrymple *et al.* (1980) for a Suiko Seamount basalt (DSDP site 433). The northernmost guyot in the chain, Meiji, has an early Maestrichtian (68–70 m.y.) sedimentary age (Worsley, 1973) and a minimum basalt K-Ar age of 61.9 ± 5.0 m.y. (Dalrymple *et al.*,1980). Prior to about 42–43 m.y. ago (the age of volcano #69 at the bend in the Emperor-Hawaiian chain), the motion of the Pacific plate with respect to a fixed hot spot was ~N10° W at a mean rate of about 6 cm/yr (Shaw *et al.*, 1980). At 42–43 m.y. ago, the direction of movement changed to ~N70° W with no change in propagation rate, and, during the last 1.3 m.y., the propagation rate has accelerated to a mean of about 17.8 cm/yr (Shaw *et al.*, 1980).

In the Hawaiian Islands proper, the earliest dated volcanism occurred 4.50–5.62 m.y. ago on Kauai (McDougall, 1964). Additional K-Ar ages (McDougall, 1964) indicate that volcanic activity then progressed

Table 1.2.6.1 Rates of volcanic propagation: Hawaiian-Emperor Chain.

Edifice	Distance to Next Dated Edifice, km	Age, m.y.	Source	Rate, cm/yr
Meiji seamount		72 ± 3	Scholl and Creager (1973)	
Koko seamount	2056	46.4† ± 1.1	Clague and Dalrymple (1973);	8.0
		50.5 ± 3.5	DSDP Scientific Staff (1973c);	
			Bukry (pers. comm.)	
Yuryaku seamount	264	42.3 ± 1.6	Clague *et al.* (1975)	6.4
Kanmu seamount	106	40 ± 3.0	Quoted in Clague and Jarrard (1973)	4.6
Unnamed seamount #63*	589	26.7 ± 0.5	Clague *et al.* (1975)	4.4
Unnamed seamounts #56 and 58*	153	27 − 30†	DSDP Scientific Staff (1973c)	−4.6
Unnamed seamount #57*	72	27.3 ± 0.4	Clague *et al.* (1975)	2.7
Midway Island	168	17.9 ± 0.6	Dalrymple *et al.* (1974)	1.8
Pearl and Hermes Reef	151	20.0 ± 0.5	Clague *et al.* (1975)	−7.2
La Perouse Pinnacles (French Frigate Shoals)	1072	11.7 ± 0.4	Dalrymple *et al.* (1974)	12.9
Necker Island	151	10.0 ± 0.4	Dalrymple *et al.* (1974)	8.9
Nihoa Island	278	7.0 ± 0.3	Dalrymple *et al.* (1974)	9.3
Niihau Island	215	3.0 ± 0.2	Jackson *et al.* (1972)	5.4
Kauai Island	46	4.50 − 5.62†	McDougall (1964)	−1.8
Waianae volcano	145	3.60 ± 0.4	Jackson *et al.* (1972)	7.2
Koolau volcano	35	2.55 ± 0.10	McDougall (1964); Jackson *et al.* (1972); Doell and Dalrymple (1973)	3.3
West Molokai volcano	59	1.84 ± 0.06	McDougall (1964)	8.3
East Molokai volcano	24	1.48 ± 0.04	McDougall (1964)	6.7
West Maui volcano	35	1.30 ± 0.04	McDougall (1964)	19.4
Haleakala volcano	39	0.84 ± 0.03	McDougall (1964)	8.5
Kohala volcano	82	0.42 ± 0.02	McDougall and Swanson (1972)	19.5
Kilauea volcano	100	0		23.8

From Jackson, (1976).

† Where more than one edifice age is given, the underlined age was the one used in the rate calculation, either because it was considered more reliable or because the older of a range of dates is more likely to present the true edifice age.

* Numbered volcanoes refer to the numbering system of Bargar and Jackson (1974).

southeastward to Oahu (Koolau Volcano, 2.55 ± 0.10 m.y.), Molokai (West Molokai Volcano, 1.84 ± 0.06 m.y.; East Molokai Volcano, 1.30 ± 0.04 m.y.), Maui (West Maui Volcano, 1.30 ± 0.04 m.y.; Haleakala Volcano, 0.84 ± 0.03 m.y.). The tholeiitic volcanism at Kohala Volcano on Hawaii has been dated at 0.42 ± 0.02 m.y. (McDougall and Swanson, 1972), but McDougall (1964) inferred from paleomagnetic data that volcanism on Hawaii may have begun as early as 800,000 years ago.

Geophysical data indicate that the oceanic crust directly beneath the Hawaiian chain extends to depths of 15–20 km, in contrast with the 5–6 km depth which is typical for adjacent oceanic crust (Jackson *et al.* 1972). This greater depth is thought to be due to isostatic subsidence, related to the transfer of crustal and mantle materials during volcano growth (Dalrymple *et al.*, 1974).

The island of Hawaii consists of five shield volcanoes (Fig. 1.2.6.1) which are, from oldest to youngest in sequence of peak activity, Kohala, Mauna Kea, Hualalai, Mauna Loa and Kilauea. Distinct tholeiitic, alkalic and nephelinitic lava suites have been recognized (MacDonald and Katsura, 1964; MacDonald, 1968; Jackson, 1976) but rapidly erupted, shield-forming tholeiites comprise more than 99% of the lavas. Smaller volumes of intermittently erupted alkalic and nephelinitic basalts occur as late capping sequences on Kohala, Mauna Kea, and Hualalai volcanoes. The reference suite selected for this study (Appendix A-6, Table 1) includes three alkalic basalts from Kohala, two from Mauna Kea and four from Hualalai. A fifth sample is an ultramafic xenolith from the 1801 flow at Hualalai. Tholeiitic samples in the suite include seven from Mauna Loa and ten from Kilauea.

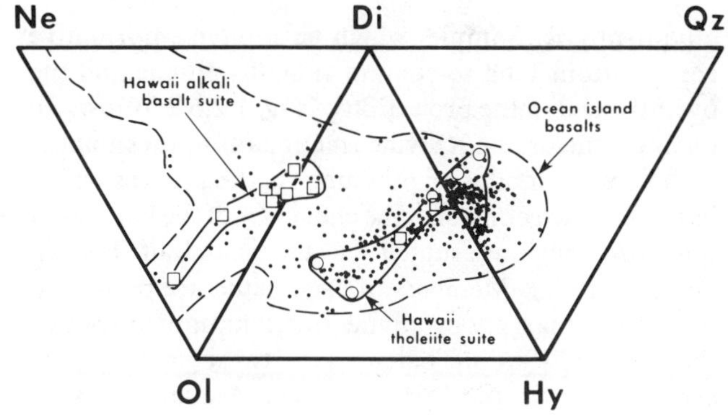

1.2.6.2 Hawaiian basalt normative compositions (dots) projected in the basalt tetrahedron (Yoder and Tilley, 1962), showing fields for the alkalic and tholeiitic basalt members of the Hawaii reference suite and for all ocean island basalts.

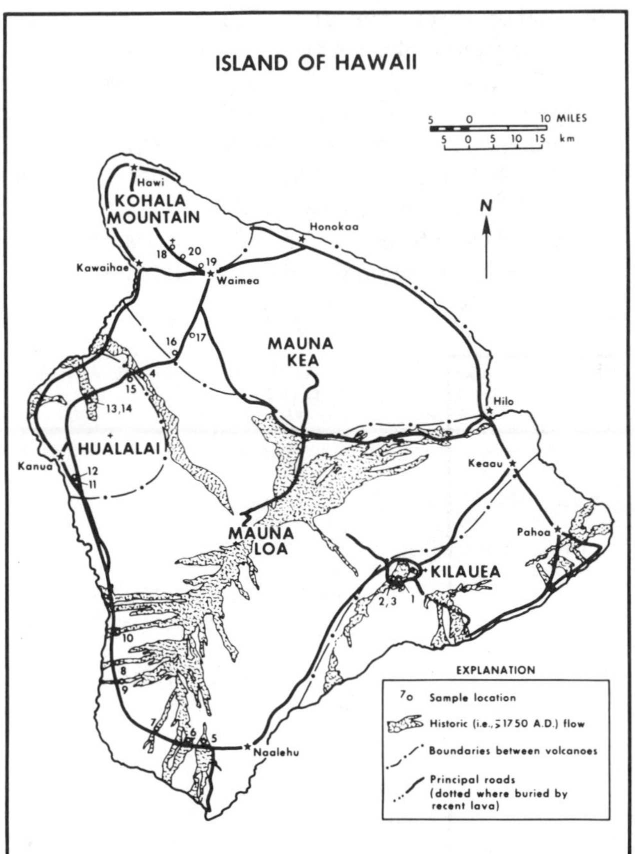

1.2.6.1 Sample location map, Island of Hawaii.

1.2.6.3 Major element chemistry of oceanic intraplate basalts

Major element analyses of approximately 1800 crystalline basalts and basalt glasses from oceanic intraplate eruptive localities have been collected from the literature. Approximately 600 of the rocks analyzed are from the island of Hawaii (Kilauea, 326 analyses; Mauna Ulu, 66; Mauna Kea, 42; Mauna Loa, 117; Kohala, 33; Hualalai, 12). The compositional spectrum of the analyses of basalts from Hawaii has been analyzed collectively and separately for each volcano. Averages for each are reported in Appendix A-6, Table 3.

Most of the analyses of basalts from Hawaii are hypersthene normative (Fig. 1.2.6.2), so the basalts are tholeiites according to the classification of Yoder and Tilley (1962). The majority of these tholeiites are slightly quartz normative and many are also sparsely olivine-phyric. Samples which have olivine-normative compositions tend to contain cumulus olivine and the overall trend in the projection of Fig. 1.2.6.2 follows an olivine control line. Olivine fractionation, presumably under low to moderate pressures, is the process which has dominated control of the chemistry of the batches of tholeiitic magma erupted on the island of Hawaii (Wright, 1971). Members of each batch are related to one another largely by olivine fractionation; however, the chemically unique batches arise through many processes. Lavas in the Kilauea Iki and Makaopuhi lava lakes appear to have experienced three types of crystal fractionation (Wright *et al.*, 1976): (1) gravitational settling of olivine phenocrysts, (2) filter pressing, and (3) downward concentration of augite and plagioclase during convective flow.

The alkali series basalts on Hawaii have compositions which range from slightly hypersthene normative alkali olivine basalts to post-erosional nephelinitic basalts (Fig. 1.2.6.2). MacDonald and Katsura (1964) recognized two distinct types. Alkali basalt series members of the reference suite include a trachyte (5% normative olivine), hawaiite (13%), ankaramite (18%), and an alkali olivine basalt (21%). The projected compositional fields for Hawaii tholeiite and alkali series basalts overlap on Fig. 1.2.6.2. This overlap can also be seen on the variation diagrams presented in Figs. 1.2.6.3–9. In these diagrams, the intraplate, MORB, and abyssal glass fields are outlined, and individual Hawaii basalts are indicated by dots.

The Hawaii tholeiite series basalts are enriched in TiO_2, total iron as FeO, and K_2O and depleted in Al_2O_3 relative to mid-ocean ridge crystalline basalts and basalt glasses. These differences, which characterize the entire intraplate group, are not related to low-pressure fractionation effects, so they must reflect compositional differences between the parent magmas from which the two basalt types (MORB tholeiite and island tholeiite) are derived. These parent liquid compositional differences, in turn, reflect differences in source region chemistry and mineralogy.

Plots of TiO_2 vs. MgO (Fig. 1.2.6.3) and TiO_2 vs. $Mg/(Mg + Fe^{2+})$ (Mg'-value*; Fig. 1.2.6.4) reveal the TiO_2-enriched character of the Hawaii basalts (and of island basalts in general) relative to most ocean ridge glasses and crystalline MORBs. The most primitive crystalline basalts from the island of Hawaii have Mg'-values slightly lower than those of most primitive abyssal glasses (~0.68 as opposed to 0.70–0.72), although their MgO contents are comparable. This is due to the higher total Fe content of the Hawaiian samples (Fig. 1.2.6.5, Table 1.2.6.2). The TiO_2 enrichment of the Hawaiian basalts must be a consequence of some property of mantle source regions that is not restricted to the source regions under Hawaii. Basalts from the Galapagos Islands (McBirney and Williams, 1969), Tahiti (McBirney and Aoki, 1968), St. Helena (Baker, 1969), Mauritius (Shand, 1933), the Canary Islands (Ridley, 1970; Ibarrola, 1969; Muñoz, 1969), Tristan da Cunha (Baker *et al.*, 1964), Bouvet (Baker, 1973), Gough (Le Maitre, 1962), etc., are all enriched in TiO_2 (for comparable Mg'-values) relative to MORBs.

The Hawaii basalt field reflects a very strong olivine (~Fo_{88}) control on liquid compositions within the

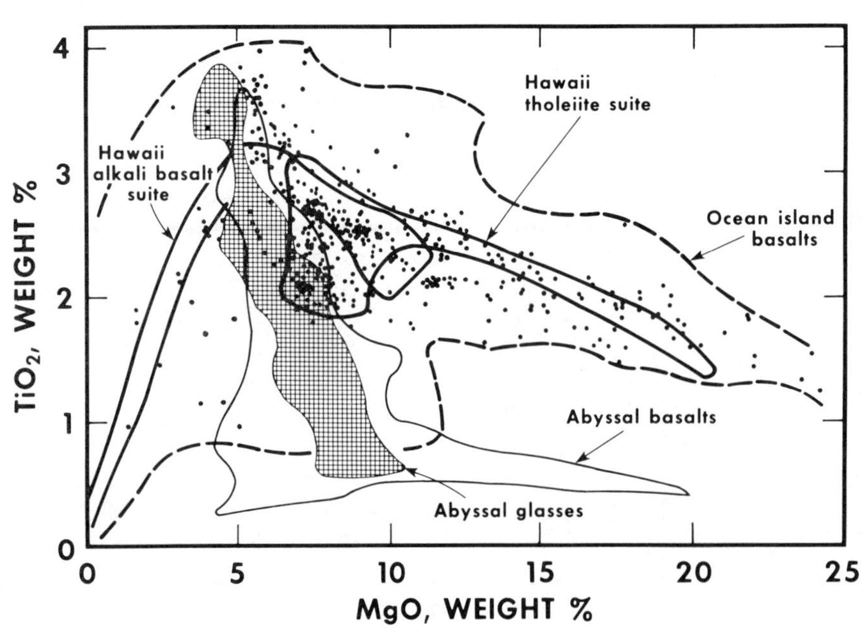

1.2.6.3 TiO_2 vs. MgO variation diagram for ocean island basalts. Fields are shown for crystalline abyssal basalts (MORBs) and for glasses.

Mg' is calculated assuming 10% (atomic) of total iron is Fe^{3+}. Mg'-value is calculated from reported values of FeO when Fe_2O_3 is measured.

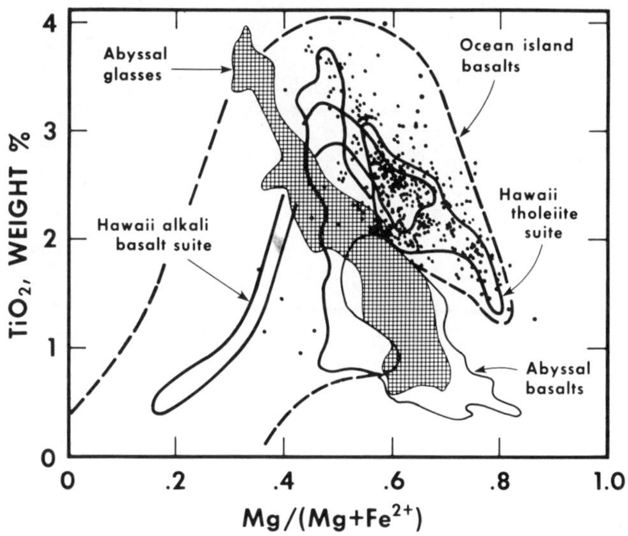

1.2.6.4 TiO_2 vs. $Mg/(Mg + Fe^{2+})$ variation diagram for ocean island basalts. $Mg/(Mg + Fe^{2+})$ calculated assuming that atomic $Fe^{3+}/\Sigma Fe = 0.1$. Fields are shown for crystalline abyssal basalts (MORBs) and for glasses.

olivine primary phase field, which becomes even more pronounced when compositions from individual eruptions are singled out. Residual glasses in the basalt vitrophyres have a relatively restricted compositional range that reflects multiple phase controls.

The abyssal glass field defines the low-pressure liquid-line-of-descent for basaltic liquid which first has been saturated with olivine and plagioclase (at high Mg′-value) and then with olivine plagioclase, and clinopyroxene (at lower Mg′-value) (Bence *et al.*, 1979). Many of the Hawaiian vitrophyres, with the exception of some evolved Mauna Loa basalts, contain only olivine and pyroxene as phenocryst phases so plagioclase control had not yet been imposed upon them when they erupted. A small number of vitrophyres from Mauna Ulu contain phenocrysts of spinel, olivine, clinopyroxene and plagioclase. With four crystalline phases present, these residual liquid compositions were tightly constrained.

Alumina and lime depletion relative to silica in the Hawaii basalts (Figs. 1.2.6.6a, b and 1.2.6.7) is such that the low-pressure liquid-line-of-descent under olivine control projects very near to the point where the liquids are multiply saturated with three crystalline phases (olivine, clinopyroxene, and plagioclase; Figs. 1.2.6.10–11). Slight but systematic differences between the silica contents of Mauna Loa and recent Kilauea summit (also Makaopuhi and Mauna Ulu) lavas are sufficient to change the order of appearance of the crystallizing phases. The low-pressure crystallization sequence of the low-silica Kilauea lavas is olivine followed by plagioclase followed by clinopyroxene. Mauna Loa lavas, on the other hand, crystallize clinopyroxene before plagioclase. The two groups define separate olivine control lines (Figs. 1.2.6.10–11) and, because there is no systematic difference in the Mg′-values of the least-fractionated members of each group, we conclude that the silica differences are inherited from the parental liquids of the basalts. These parental liquids probably fractionated at moderate-to-high pressures in magma

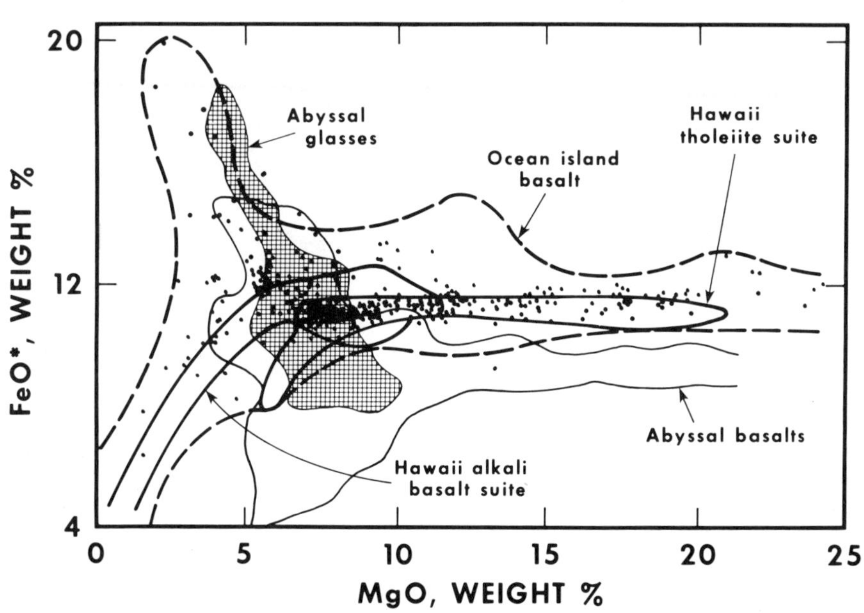

1.2.6.5 Total iron (FeO*) vs. MgO variation diagram for ocean island basalts. Fields for MORB basalts and glasses are shown.

Table 1.2.6.2 Major element compositions and norms of samples in the Hawaii Reference Suite (in wt. %).

	HAW-1	HAW-2	HAW-3	HAW-4	HAW-5	HAW-6	HAW-7	HAW-8	HAW-9
SiO_2	49.28	50.51	50.22	51.75	47.68	51.63	52.12	51.94	52.29
Al_2O_3	12.56	13.45	13.31	13.81	8.95	13.12	13.74	13.81	13.64
Fe_2O_3	3.31	1.78	1.39	2.48	3.67	2.58	2.80	2.57	2.43
FeO	8.20	9.59	9.68	8.35	7.90	8.48	8.38	8.52	8.74
MgO	9.74	7.41	7.28	7.25	21.02	8.53	7.23	7.14	7.90
CaO	10.80	11.18	11.03	10.57	6.99	9.97	10.65	10.64	10.61
Na_2O	2.10	2.28	2.31	2.31	1.48	2.21	2.30	2.31	2.28
K_2O	0.48	0.49	0.47	0.42	0.27	0.33	0.36	0.38	0.37
MnO	0.17	0.17	0.18	0.17	0.16	0.17	0.17	0.17	0.17
TiO_2	2.61	2.63	2.54	2.07	1.40	1.94	2.11	2.10	2.06
P_2O_5	0.27	0.28	0.27	0.24	0.17	0.22	0.23	0.23	0.24
H_2O^t				0.19	0.28		0.23	0.23	0.24
CO_2				0.06	0.06		0.13	0.05	0.08
SUM	99.52	99.77	98.68	99.67	100.03	99.18	100.45	100.09	101.05
Qz		1.37	1.53	2.86		2.83	3.32	3.37	2.64
Or	2.84	2.90	2.78	2.46	1.54	1.95	2.07	2.13	2.07
Ab	17.77	19.29	19.55	19.55	12.52	18.70	19.46	19.55	19.29
An	23.43	25.02	24.56	25.72	16.60	24.91	25.07	25.38	25.13
Di	23.04	23.42	23.24	19.30	12.79	18.86	19.77	19.54	19.54
Hy	20.66	20.09	19.63	22.06	21.40	25.69	21.13	20.82	22.55
Ol	4.01				27.58				
Mt	2.00	2.00	1.95	1.85	1.97	1.93	1.89	1.87	1.87
Il	4.96	5.00	4.82	3.91	2.60	3.69	3.89	3.87	3.82
Ap	0.63	0.65	0.63	0.63	0.42	0.51	0.53	0.58	0.53
Ne									
Mg′	0.636	0.570	0.572	0.592	0.789	0.613	0.573	0.572	0.596

	HAW-10	HAW-11	HAW-12	HAW-15	HAW-16	HAW-17	HAW-18	HAW-19
SiO_2	50.54	46.37	46.88	62.99	50.26	47.33	58.55	46.29
Al_2O_3	12.92	14.18	13.69	18.04	17.33	12.26	17.99	15.20
Fe_2O_3	2.86	4.09	2.41	4.42	4.49	3.18	4.59	2.38
FeO	8.26	8.91	9.86	0	6.38	8.34	2.22	9.16
MgO	8.99	9.47	9.78	0.32	3.89	11.36	1.34	6.11
CaO	9.85	10.33	10.81	0.71	7.05	11.43	2.82	10.90
Na_2O	2.29	2.85	2.47	7.19	4.97	2.16	7.13	2.91
K_2O	0.35	0.93	0.68	4.87	1.91	0.57	2.92	0.77
MnO	0.18	0.19	0.17	0.29	0.22	0.17	0.23	0.18
TiO_2	1.92	2.40	2.03	0.45	2.81	2.40	1.03	3.04
P_2O_5	0.22	0.28	0.24	0.13	0.93	0.29	0.69	0.44
H_2O^t				1.31	0.26		0.43	
CO_2				0.11	0.06		0.03	
SUM	98.38	98.32	99.02	100.83	100.56	99.49	99.97	97.38
Qz	0.80							
Or	2.07	5.38	4.02	28.31	10.99	3.37	17.02	4.55
Ab	19.38	17.42	20.10	57.83	34.60	18.28	56.44	23.70
An	23.94	22.31	24.26	1.60	18.45	22.08	8.17	26.14
Di	19.13	21.20	27.67	0.76	7.17	26.47	0.76	20.61
Hy	26.81					3.27		
Ol		20.75	20.89	4.80	12.80	18.62	8.50	12.97
Mt	1.94	2.22	2.15	0.69	1.81	2.00	1.11	2.02
Il	3.65	4.46	3.86	0.80	5.17	4.56	1.92	5.77
Ap	0.51	0.72	0.56	0.30	2.25	0.67	1.58	1.02
Ne		3.63	0.43	1.63	4.04		2.11	0.50
Mg′	0.624	0.602	0.620	0.161	0.429	0.67	0.310	0.520

Table 1.2.6.2. *(Continued)*

	HAW-20	HAW-21	HAW-22	HAW-23	HAW-24	HAW-25	HAW-26	HAW-27
SiO_2	47.52	50.09	50.92	46.65	50.01	49.63	48.71	50.94
Al_2O_3	15.95	13.37	13.42	9.15	13.07	13.13	11.18	13.53
Fe_2O_3	7.16	1.79	4.77	1.64	2.58	1.72	2.49	2.48
FeO	5.30	9.52	7.87	10.17	9.02	9.77	9.27	8.74
MgO	5.18	7.26	5.47	18.90	8.54	6.78	13.62	7.16
CaO	8.96	11.60	9.53	8.43	11.21	10.31	9.24	11.39
Na_2O	3.56	2.31	2.88	1.60	2.20	2.61	2.15	2.29
K_2O	1.29	0.56	0.76	0.38	0.55	0.64	0.49	0.48
MnO	0.19	0.18	0.18	0.18	0.16	0.19	0.16	0.17
TiO_2	3.29	2.79	3.61	1.90	2.73	3.05	2.42	2.57
P_2O_5	0.64	0.29	0.39	0.20	0.29	0.35	0.26	0.26
H_2O^t	1.16	0.32	0.36	0.25	0.26		0.29	0.32
CO_2	0.05	0.10	0.04	0.06	0.10		0.05	0.15
SUM	100.25	100.18	100.20	99.51	100.72	98.18	100.33	100.48
Qz		0.32	5.13			0.78		1.19
Or	7.33	3.31	4.43	2.25	3.31	3.78	3.13	2.84
Ab	28.12	19.55	24.37	13.54	18.62	22.09	18.19	19.38
An	22.73	24.21	20.88	16.36	23.43	22.22	20.96	24.51
Di	12.83	23.92	17.81	18.28	22.98	21.84	20.02	23.00
Hy		18.88	11.54	11.49	18.87	18.82	19.58	19.20
Ol	13.88			28.91	2.76		8.29	
Mt	2.04	1.95	1.41	2.04	1.98	2.02	2.00	1.92
Il	6.00	5.17	6.67	3.53	5.05	5.79	4.86	4.73
Ap	1.48	0.70	1.00	0.44	0.67	0.81	0.65	0.60
Ne	1.09							
Mg′	0.470	0.570	0.580	0.763	0.603	0.546	0.664	0.572

HAW-1: Tholeiite, Kilauea Volcano, 1921 eruption. Locality of MacDonald and Eaton (1955), Table 9. **HAW-2:** Tholeiite, Kilauea Volcano, September, 1971 eruption. **HAW-3:** Tholeiite, Kilauea Volcano, September, 1974 eruption. **HAW-4:** Tholeiite, Mauna Loa, 1859 eruption. **HAW-5:** Tholeiite, Mauna Loa, 1868 eruption. **HAW-6:** Tholeiite, Mauna Loa, 1887 eruption. **HAW-7:** Tholeiite, Mauna Loa, 1907 eruption. **HAW-8:** Tholeiite, Mauna Loa, 1919 eruption. **HAW-9:** Tholeiite, Mauna Loa, 1926 eruption. **HAW-10:** Tholeiite, Mauna Loa, 1950 eruption. **HAW-11:** Alkali olivine basalt, Hualalai Volcano, prehistoric. **HAW-12:** Alkali olivine basalt, Hualalai Volcano, prehistoric. **HAW-13:** Ultramafic xenolith, Hualalai Volcano, xenolith in 1801 alkali olivine basalt flow. **HAW-14:** Alkali olivine basalt, Hualalai Volcano host for ultramafic xenoliths, 1801 flow. Analysis of 1801 lava in MacDonald (1968), Table 4. **HAW-15:** Trachyte, Hualalai Volcano. **HAW-16:** Hawaiite, Mauna Kea Volcano, west slope. Same locality as C-72, Table 4 of MacDonald and Katsura (1964). **HAW-17:** Ankaramite, Mauna Kea Volcano, west slope. Same locality as C-71, Table 4 of MacDonald and Katsura (1964). **HAW-18:** Trachyte trending to mugearite, Kohala Volcano. Same locality as 62-1, Table 3 of MacDonald and Katsura (1964). **HAW-19:** Hawaiite, Kohala Volcano. Same locality as C-68, Table 3 of MacDonald and Katsura (1964). **HAW-20:** Alkali olivine basalt, Kohala Volcano. Same locality as C-70, Table 3 of MacDonald and Katsura (1964). **HAW-21:** Tholeiite, Kilauea Volcano, summit caldera, 5-31-54. **HAW-22:** Tholeiite, Kilauea Volcano, Puna Rift, Kauelau flow 3-14-55. **HAW-23:** Tholeiite, Kilauea Volcano, Kilauea Iki (IKI-22), 11-18-59, 5 p.m. **HAW-24:** Tholeiite, Kilauea Volcano, Kilauea Iki (IKI-10), 11-29-59, 9 p.m. **HAW-25:** Tholeiite, Kilauea Volcano, Puna Rift (KP-2), 1-15-60, 5:30 p.m. **HAW-26:** Tholeiite, Kilauea Volcano, Puna Rift (KP-25), 2-13-60, 7 p.m. **HAW-27:** Tholeiite, Kilauea Volcano, summit caldera, 11-29-75, 7 a.m.

chamber "holding tanks." The observed silica differences could be generated at that stage through olivine vs. pyroxene controls. The low silica (recent Kilauea, etc.) basalts may have experienced high-pressure fractionation involving pyroxene extraction (see discussion of high-pressure phase relations in section 3.3).

Prehistoric Kilauea lavas have higher silica contents (at comparable MgO concentrations) than do recent Kilauea (and satellite) eruptions (Wright, 1971) and thus are more like the Mauna Loa summit eruptions. Secular changes in Kilauea lava compositions have involved increases in TiO_2, CaO, FeO, K_2O and P_2O_5 and decreases in Al_2O_3 and SiO_2 in the younger lavas (Wright, *et al.*, 1975). Reversals in this overall trend were noted for the 1969–71 Mauna Ulu eruption.

The low overall Al_2O_3 content of tholeiitic basalts from Hawaii, relative to mid-ocean ridge basalts at comparable degrees of differentiation, is the result of a major difference in the Al_2O_3 contents of the primary magmas. Variations in the Al_2O_3 contents of primary magmas can be consequences of (a) chemical variations in the source, (b) variations in the degree of partial

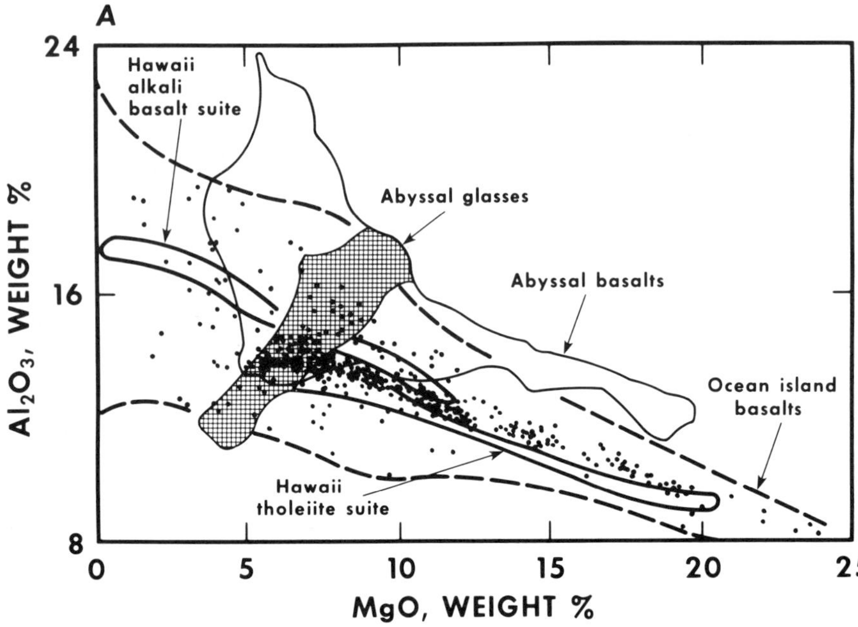

1.2.6.6 a. Al_2O_3 vs. MgO and **b.** $Mg/(Mg + Fe^{2+})$ variation diagrams.

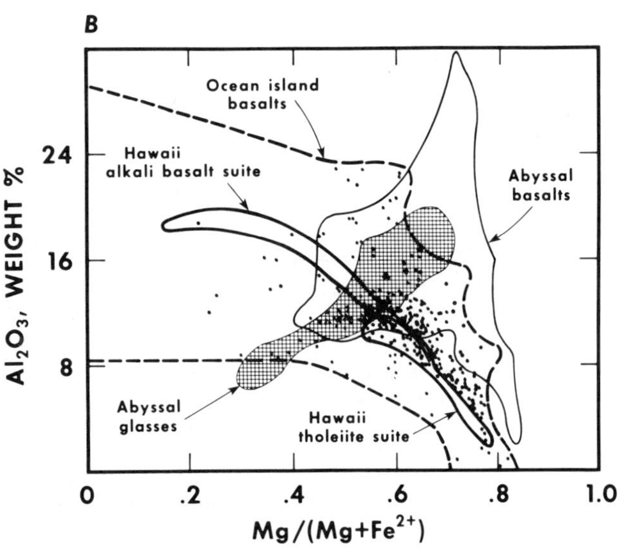

melting, (c) variations in the residual mineralogy of the source, or (d) high-pressure fractionation of an aluminous phase. Experimental investigations at high pressure (Green and Ringwood, 1967) on Hawaiian olivine tholeiite suggest that (c) is unlikely. The relationship of Al_2O_3 content to Mg′-value is in the opposite sense of that which would be expected if a lower degree of partial melting were to have generated the Hawaiian basalts. Hence, explanation (b) is considered unlikely. Variations in TiO_2 and K_2O contents (Figs. 1.2.6.6 and 1.2.6.8) are consistent with different source region chemistry for the island basalts. Consequently, this may be the correct explanation of the lower Al_2O_3 contents of the Hawaii basalts. However, the REE data (see below) suggest that there is residual garnet in the sources of the Hawaiian basalts and of island basalts in general. This may reflect differences in depths of source regions in addition to the recognized compositional differences.

The alkali series volcanics from Hawaii include alkali olivine basalts, hawaiites (andesine-bearing) and mugearites (oligoclase-bearing), which predominate, as well as subordinate trachytes, rhyolites, and ankaramites (olivine or augite cumulates). The most mafic members of this series, like the Hawaii tholeiites, are enriched in TiO_2, FeO, and K_2O, and are depleted in Al_2O_3 relative to abyssal basalts. They form a continuum with the island tholeiites. The more evolved members of the series are enriched in K_2O (Fig. 1.2.6.8) and Na_2O (Fig. 1.2.6.9; Table 1.2.6.2). This series may represent the liquid-line-of-descent followed by a fractionating alkali basalt parent. The relationship between the alkali basalts and the shield-forming tholeiites is yet undefined and is a point of contention.

1.2.6.4 Trace element chemistry

Rare-earth elements

Rare-earth patterns of basalts from oceanic islands are characterized by varying degrees of light rare-earth elements (LREE) enrichment relative to the heavy rare-earth elements (HREE). Island tholeiites, for example those of Hawaii (Schilling and Winchester, 1966) and the Galapagos Islands (McBirney and Williams, 1969;

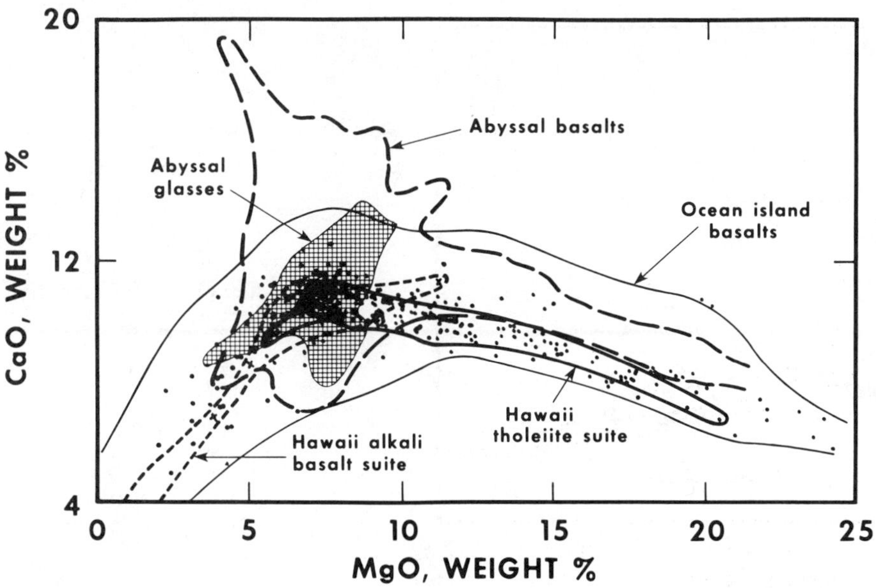

1.2.6.7 CaO vs. MgO variation diagram.

Schilling *et al.*, 1976), though fractionated, have lower LREE abundances (30–50 times chondrites) and lower La_N/Yb_N ratios (2–3) than members of the alkali magma series (Kay and Gast, 1973; White *et al.* 1979). Systematic differences occur within the alkali magma series. Alkali basalts tend to have lower LREE abundances and lower La_N/Yb_N ratios than the associated nephelinites.

The high LREE abundances of ocean island tholeiites relative to MORBs are consequences of the derivation of the island lavas from unfractionated source regions. The even greater enrichment of LREE in the alkali magma series, together with their depletion in HREE (relative to MORBs), is consistent with derivation of these lavas through relatively small degrees of partial melting of a source in which garnet remains in the residuum (e.g., Kay and Gast, 1973).

The Hawaiian lavas have strongly-fractionated rare-earth element (REE) distributions. Tholeiitic basalts from Kilauea Volcano contain LREE in concentrations ranging from ~40–70 times chondritic values and HREE in concentrations from ~6–15 times chondrites (Fig. 1.2.6.12a and Table 1.2.6.3). REE patterns are very slightly concave downward (see also Leeman *et al.*, 1977, 1980), which distinguishes them from high-K MORB tholeiites (see Frey *et al.*, 1974; Bence *et al.*, 1975). Mauna Loa summit lavas (Fig. 1.2.6.12b and

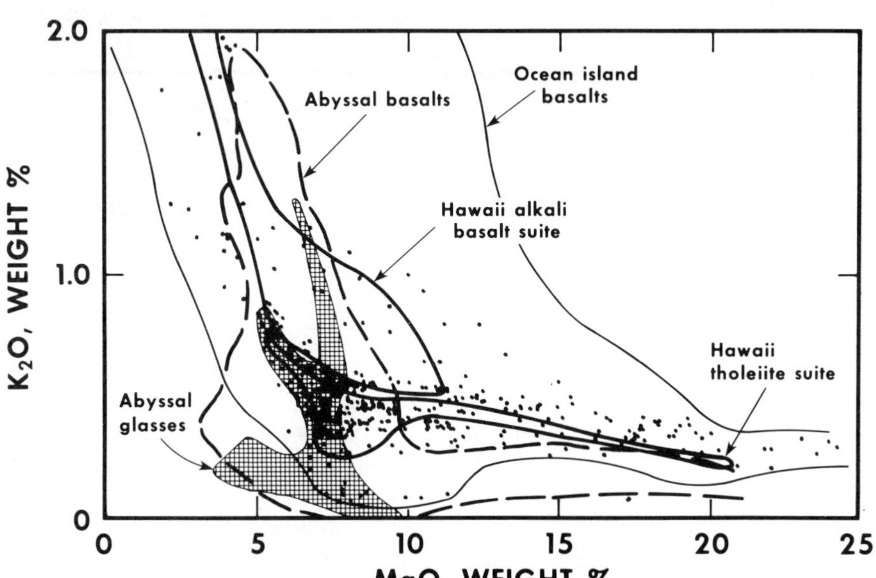

1.2.6.8 K_2O vs. MgO variation diagram.

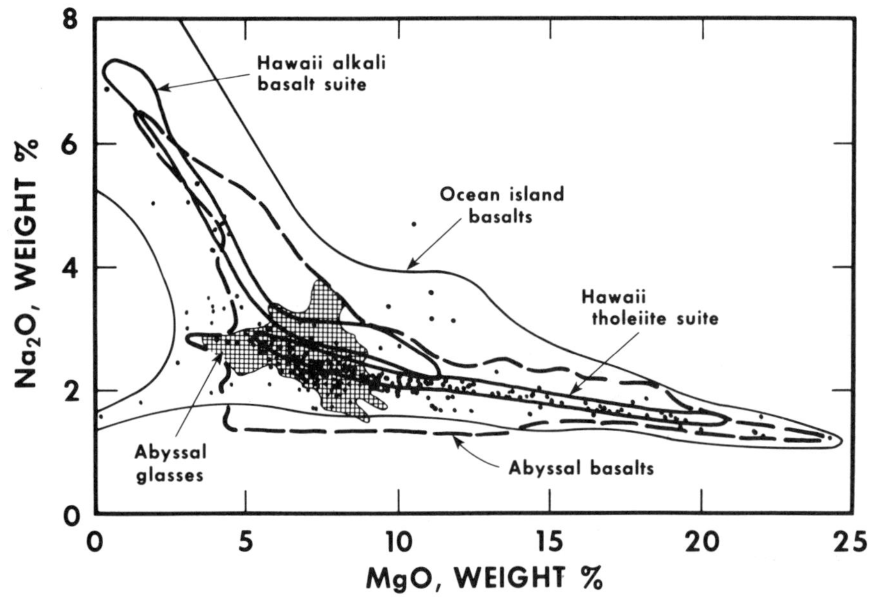

1.2.6.9 Na_2O vs. MgO variation diagram.

Table 1.2.6.4) are similarly fractionated. The reference suite samples from Mauna Loa have a factor of two lower total REE contents than the Kilauea samples. However, REE analyses of Mauna Loa summit and Kilauea lavas reported by Leeman *et al.* (1977, 1980) are, overall, quite similar. Their data and the analyses reported here reveal one pronounced difference between the two localities—the relatively unfractionated nature of the LREE and the pronounced concave downward pattern of the Mauna Loa lavas. These differences, as well as differences in the Co, Cr, and Ni abundances and in K_2O/P_2O_5 ratios, were cited by Leeman *et al.* (1977) as evidence that the Mauna Loa and Kilauea lavas are not related by low- or intermediate-pressure crystal fractionation. This supports a similar conclusion reached by Wright and Fiske (1971). Leeman *et al.* conclude from their data, however, that the lavas could be related by fractionation at high pressures if garnet were involved. Our data, which show that the Mauna Loa lavas are depleted in both LREE and HREE relative to the Kilauea lavas but have similar middle rare-earth abundances, indicate that garnet fractionation operating alone cannot relate the two lava groups. It is clear, however, from the highly fractionated REE that garnet must be involved in the petrogenesis of these lavas, if the source REE were unfractionated. Leeman *et al.* (1977) made this assumption and, using mineral/melt distribution coefficients, calculated that the clinopyroxene/garnet ratios in the sources would have to differ significantly (5.7 for Kilauea and 1.6 for Mauna Loa). They calculated that upper limits on the REE contents of the source rocks are as follows: Kilauea, 0.8–6.1 times chondrites; Mauna Loa, 1.8–13.3 times chondrites.

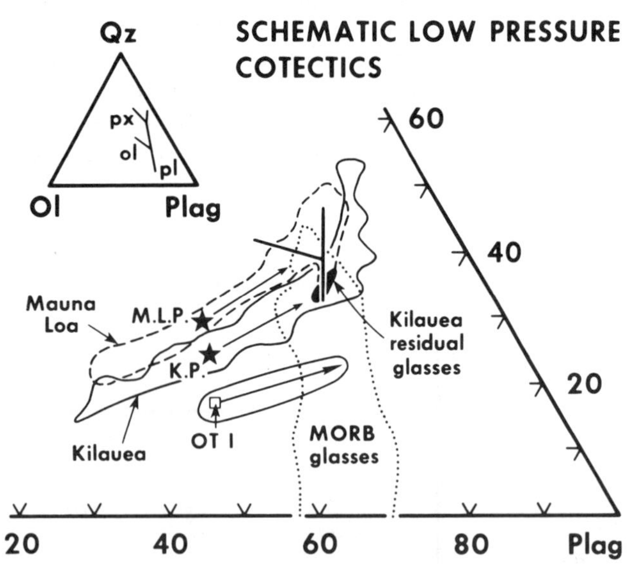

1.2.6.10 Subprojection olivine-plagioclase-quartz, showing fields for Kilauea and Mauna Loa tholeiites, Kilauea residual glasses, and MORB glasses. K.P. and M.L.P. are proposed Kilauea and Mauna Loa parental liquids. OT1 is Kilauea Iki parent proposed by Green and Ringwood (1967). One-atmosphere olivine fractionation paths from the three parental liquids are shown. Ol, cpx and plag fields recalculated from Presnall *et al.* (1978).

Alkaline series basalt REE distribution patterns (Fig. 1.2.6.12c and Table 1.2.6.5) are very similar to the Kilauea tholeiite patterns, although overall REE are enriched and the La/Yb ratios are slightly greater in the alkali basalts. These distributions are consistent with a deep source containing residual garnet (Kay and Gast,

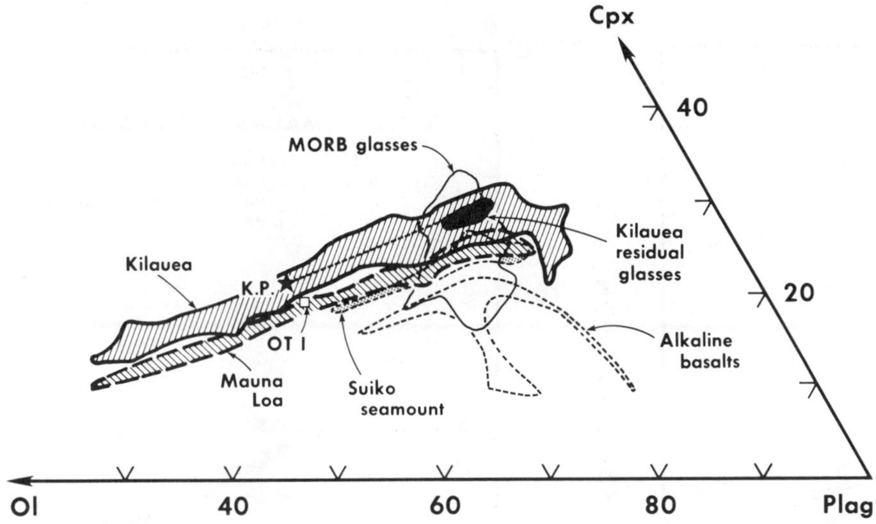

1.2.6.11 Subprojection olivine-plagioclase-clinopyroxene, showing fields for Kilauea and Mauna Loa tholeiites, Hawaii alkaline basalts, Kilauea residual glasses, and MORB glasses. K.P. is proposed Kilauea parental liquid, and dashed line is one-atmosphere olivine fractionation path. OT1 is Kilauea Iki parent proposed by Green and Ringwood (1967). Data for Suiko Seamount from Clague *et al.* (1980).

1973), although the mildly alkalic basalts need not have originated from a source that retains garnet in the residuum (Frey *et al.*, 1978).

Transition metals

The transition metals, which occur in trace abundances (e.g., Cr, Ni, Co), are important indicators of basalt petrogenesis. Nickel, in particular, is a sensitive recorder of olivine fractionation, because of its large mineral/melt partition coefficient (see, for example, Leeman *et al.*, 1977; Hart and Davis, 1978; Nabelek, 1980). The Hawaiian basalts are no exception. Gunn (1971) determined the partitioning behavior of Cr, Ni, and Co in the differentiated Kilauea Iki (1959) lava series and in a Makaopuhi lava lake suite. The nickel data for Kilauea Iki (Fig. 1.2.6.13) reflect pronounced

Table 1.2.6.3 Trace element abundances in Kilauea tholeiites, ppm.†

	HAW-1	HAW-2	HAW-3	HAW-21	HAW-22	HAW-23	HAW-24	HAW-25	HAW-26	HAW-27
La	14.6	13.4	13.0	16.2	21.6	11.5	15.7	18.9	14.9	13.1
Ce	36.5	35.5	33.0	39.5	54.0	27.6	38.3	46.6	36	33
Sm	5.55	6.14	5.82	6.22	8.98	4.50	5.90	6.80	5.80	5.95
Eu	1.92	1.88	1.94	2.02	2.85	1.39	1.99	2.40	1.92	1.91
Tb	0.91	0.90	0.98	0.95	1.39	0.64	0.94	1.11	1.02	0.92
Yb	1.90	1.98	2.05	2.00	3.05	1.40	1.98	2.50	1.95	2.08
Lu	0.27	0.29	0.29	0.29	0.42	0.21	0.29	0.37	0.28	0.30
Sc	30.9	31.3	—	31.7	29.1	25.0	30.3	30.7	27.1	32.0
Cr	630	340	350	330	85	1550	390	260	710	290
Rb	8.4	9.2	9.0	12	13	11	15	13	9	8
Sr	368	371	358	330	396	270	470	390	310	358
Ba	125	150	120	135	200	115	130	170	135	115
Hf	4.47	4.39	4.42	4.65	6.9	3.15	4.45	5.8	4.62	4.28
Zr	(141)	(115)	(162)	(173)	(247)	(118)	(165)	(202)	(159)	(160)
Nb	(24)	(17)	(15)	(19)	(24)	(13)	(18)	(20)	(16)	(15)
Y	22	25	24	25	36	17	24	29	23	25
Th	1.14	1.27	1.00	1.35	1.65	1.08	1.40	1.50	1.25	1.00
Pb	6	5	5	—	5	—	—	—	—	6
Ga	21	21	21	—	25	—	—	—	—	21

†La, Ce, Sm, Eu, Tb, Yb, Lu, Sc, Cr, Ba, Hf and Th by INAA. Remainder by XRF.

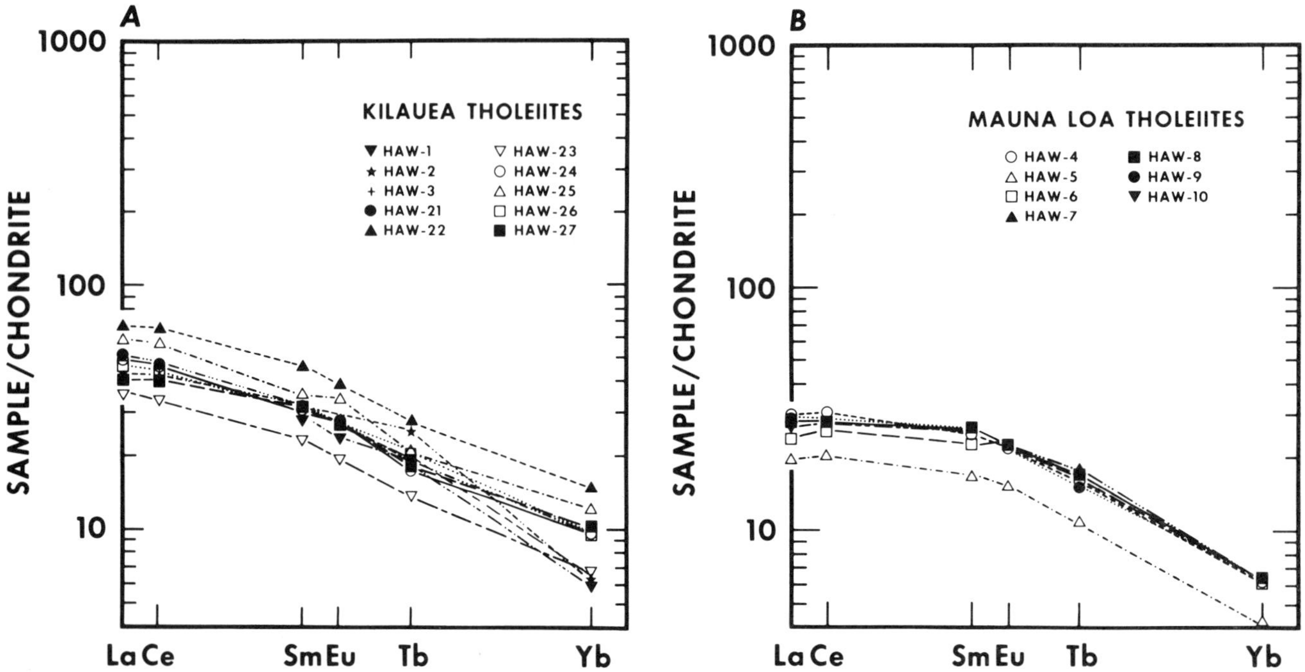

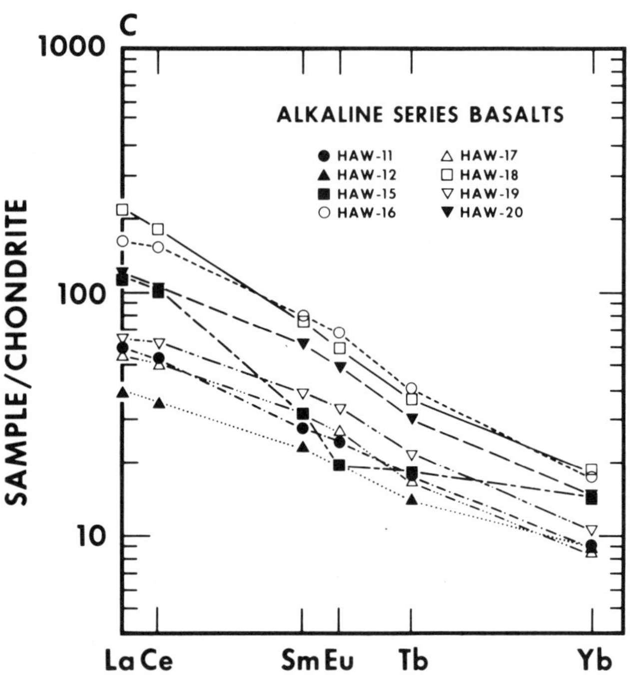

1.2.6.12 Chondrite-normalized rare-earth element abundances for Hawaiian basalts. **a.** Kilauea tholeiites. MORB average from Kay *et al.* (1970); data for high-K MORBs from Frey *et al.* (1974) and Bence *et al.* (1975). **b.** Mauna Loa tholeiites. **c.** Alkaline series basalts.

olivine control. Unfractionated lavas (Mg′-value ~0.7 and lacking cumulate olivine) would have nickel abundances in the vicinity of 400 ppm, which is significantly higher than abundances observed in the most primitive MORB glasses. It is likely that no unfractionated samples of the latter have been sampled and/or analyzed for nickel. Oceanic island alkali series basalts are depleted in Ni relative to island tholeiites and MORBs (Fig. 1.2.6.13).

A note of caution must be sounded regarding the use of Ni (and Co) as indicators of magmatic differentiation. Evidence of sulfide-silicate liquid immiscibility has been noted in the Hawaiian basalts (Desborough *et al.*, 1968) and in mid-ocean ridge basaltic glasses (Czamanske and Moore, 1977). Separation of a sulfide liquid, with its presumed very high partition coefficients for Ni and Co (under conditions of low f_{O_2}), would result in a drastic reduction in the content of these elements in the silicate liquid without a corresponding drop in Mg′-value.

The abundances of chromium in tholeiitic members of the island basalts also correlate with an olivine fractionation path for Mg′-value (Fig. 1.2.6.14). This reflects the co-crystallization of Cr-spinel and olivine in these basalts. The alkaline-basalts, except those alkali olivine basalts which contain cumulus olivine (+ spinel), are markedly depleted in Cr relative to tholeiitic basalts from the oceanic islands and from the mid-ocean ridges. This is a potentially useful discriminator between highly altered tholeiitic and alkaline basalts where their identification by other techniques may be virtually impossible. Unfortunately, high quality Cr analyses are rarely obtained during conventional bulk analyses of basalts.

Scandium, which is preferentially incorporated into clinopyroxene and, thus, should be a useful indicator of high-pressure pyroxene fractionation, has a very small compositional range in island tholeiites (25–35 ppm, Tables 1.2.6.3–4). These abundances are somewhat lower than levels in MORBs (35–50 ppm; see, for

Table 1.2.6.4 Trace element abundances in Mauna Loa tholeiites, ppm.

	HAW-4	HAW-5	HAW-6	HAW-7	HAW-8	HAW-9	HAW-10
La	9.38	6.16	7.58	8.54	8.75	8.81	8.52
Ce	25.0	16.4	21.0	22.8	23.0	23.3	22.5
Sm	4.80	3.30	4.40	4.95	4.98	4.97	4.85
Eu	1.70	1.10	1.60	1.68	1.67	1.64	1.55
Tb	0.81	0.54	0.84	0.90	0.84	0.80	0.86
Yb	2.00	1.35	1.98	2.08	2.08	2.07	2.00
Lu	0.29	0.20	0.29	0.31	0.32	0.30	0.29
Sc	29.5	21.0	29.5	30.4	30.6	30.4	29.0
Cr	350	1405	480	330	320	390	490
Rb	6.9	4.4	4.9	5.2	5.6	5.4	5.2
Sr	344	213	273	297	305	302	294
Ba	—	70	75	75	75	82	72
Hf	3.48	2.38	3.34	3.55	3.60	3.45	3.43
Zr	(129)	(88)	(119)	(131)	(130)	(123)	(121)
Nb	(9)	(6)	(8)	(9)	(9)	(8)	(8)
Y	23	15	23	23	23	23	23
Th	0.56	0.34	0.50	0.53	0.52	0.58	0.55
Pb	<1	4	6	5	5	3	5
Ga	20	14	20	21	22	20	19

Table 1.2.6.5 Trace element abundances in alkaline series basalts from Hualalai, Mauna Kea, and Kohala Volcanoes, ppm.

	HAW-11	HAW-12	HAW-15	HAW-16	HAW-17	HAW-18	HAW-19	HAW-20
	Hualalai			Mauna Kea		Kohala		
La	18.8	12.5	39.9	51.2	17.4	68.4	20.4	38.0
Ce	43.0	29.0	85	125	41.0	146	51.0	85
Sm	5.35	4.38	6.13	15.2	6.15	14.4	7.40	11.8
Eu	1.76	1.43	1.92	4.9	1.92	4.3	2.4	3.5
Tb	0.88	0.69	0.91	1.96	0.80	1.80	1.08	1.49
Yb	1.88	1.88	3.00	3.60	1.75	3.80	2.25	3.08
Lu	0.28	0.26	0.48	0.50	0.25	0.53	0.32	0.42
Sc	27.5	30.1	6.50	9.5	32.3	2.50	28.9	23.2
Cr	447	560	—	2	770	3	200	204
Rb	22	15	108	36	11	67	14	26
Cs	—	—	0.51	0.30	—	0.22	—	—
Sr	500	359	48	1270	459	1320	594	650
Ba	300	200	500	600	175	940	230	340
Hf	3.96	3.00	21.0	11.5	4.20	15.1	5.75	8.5
Zr	(148)	(166)	(905)	(482)	(157)	(681)	(238)	(351)
Nb	(25)	(16)	(127)	(59)	(18)	(88)	(24)	(36)
Y	22	21	23	46	21	41	28	39
Th	1.94	1.20	8.0	4.0	1.4	6.6	1.32	2.9
Pb	7	<1	10	7	6	7	5	5
Ga	19	19	23	24	20	21	23	24

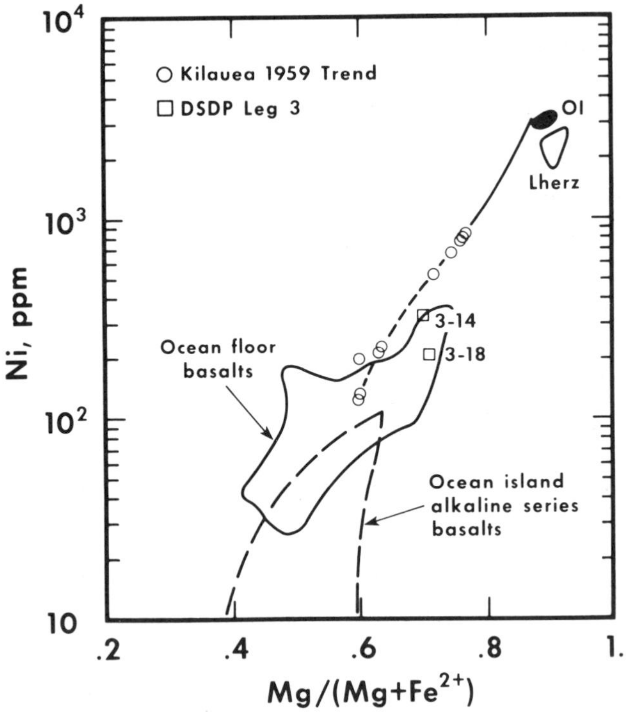

1.2.6.13 Variation of Ni abundance with Mg′ for Kilauea 1959 tholeiites (Gunn, 1971), ocean island alkaline basalts, and ocean floor basalts. Data for primitive MORBs from DSDP Leg 3 from Frey *et al.* (1974). Also plotted are data for bulk spinel lherzolite xenoliths (Lherz) and their olivines (Ol) from Frey and Green (1974).

example, Engel *et al.*, 1965; Frey *et al.*, 1974; Bence *et al.*, 1975; Bence and Taylor, 1977). Those alkaline series basalts which are only slightly undersaturated have abundance levels indistinguishable from those of the island tholeiites (Table 1.2.6.5). Only the most alkaline members of the alkali series lavas have markedly different Sc concentrations ($\lesssim$ 10 ppm). The consistently lower Sc levels in the island tholeiites (and in some alkaline basalts) relative to MORBs are consistent with an island basalt petrogenetic history involving pyroxene fractionation at high pressures, where clinopyroxene is on the liquidus.

Large cations

Large-cation (Ca, Cs, Rb, Ba, K, Pb, Sr) abundances in basalts are controlled by source region abundances, degree of partial melting, residual mineralogy, and crystal fractionation. Their variation within related suites of ocean island basalts is largely a consequence of differentiation through crystal fractionation at low pressures. None is preferentially incorporated into early crystallizing minerals (except Sr, which is incorporated in calcic plagioclase), so all become enriched in residual liquids.

Differences in the abundances and abundance ratios of large-cation elements between ocean island suites are due largely to chemical heterogeneities in upper mantle sources. However, under conditions of low degrees of partial melting, abundances and abundance ratios become extremely sensitive to residual mineralogy.

In general, large cations are enriched in ocean island tholeiites relative to most fresh MORBs.

Potassium content ranges from 0.2 to nearly 1.0 wt.% K_2O (~ 1600–8300 ppm K) in unaltered ocean island tholeiites. Most MORBs and MORB glasses have K_2O contents in the range <0.1 to ~ 0.3 wt.% K_2O ($<$ 800 to ~ 2500 ppm K; Fig. 1.2.6.8; Hart, 1969, 1971). Only the more differentiated or most altered MORB samples have K_2O abundances exceeding ~0.3 wt.%. The rarer MORB high-K tholeiites, which seem to

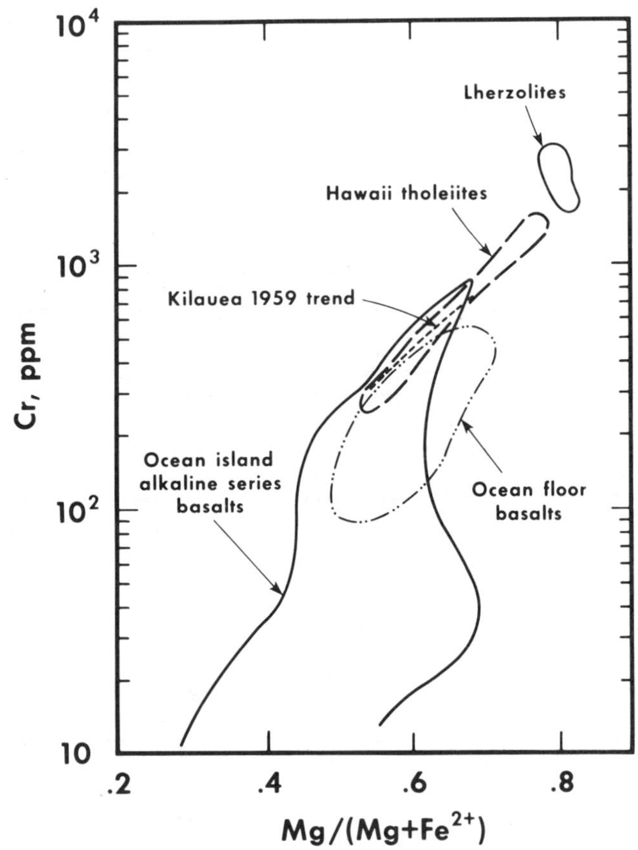

1.2.6.14 Variation of Cr abundance with Mg′ for Hawaii tholeiites, Kilauea 1959 tholeiites (Gunn, 1971), ocean island alkaline basalts, and ocean floor basalts. Also shown are data for spinel lherzolite xenoliths from Frey and Green (1974).

be associated with topographic highs and may result from plume type activity, can have K_2O contents which exceed 0.3 wt.% (see, for example, Frey *et al.*, 1974). Tholeiitic samples in the Hawaiian reference suite (Table 1.2.6.2) have K_2O contents ranging from 0.26–0.75 wt.% (~ 2100–6200 ppm K). In this suite, Kilauea lavas (K_2O = 0.38 to 0.75 wt.%) consistently are enriched in potassium relative to the Mauna Loa summit lavas (K_2O = 0.26–0.42 wt.%).

Ocean island alkali series basalts have K_2O contents ranging from as low as 0.5 wt.% in olivine basalts to as high as ~ 6 wt.% in phonolites and trachytes—the extreme differentiates. Hawaiian alkali basalts have a K_2O range of ~ 0.6–5 wt.% K_2O (Table 1.2.6.2, Fig. 1.2.6.8).

Island tholeiites and typical MORBs have distinctly different Ba and Sr concentrations. Fresh MORB ranges are: Ba = 5 to ≤50 ppm and Sr = 90–200 ppm. Ocean island thoeliite ranges are: Ba = 70–200 ppm and Sr = 150–400 ppm, which are comparable to abundance levels in high-K MORBs (Ba = 80–200 ppm and Sr = 200–300 ppm). Ocean island alkali basalt concentrations are: Ba = 200–1400 ppm and Sr = 400–4000 ppm (with the majority in the range of 500–1200 ppm Sr).

The two tholeiite suites from Hawaii (Mauna Loa summit and Kilauea) can be distinguished on the basis of their Ba and Sr contents and of their K/Ba ratios. Mauna Loa lavas in the reference suite have ranges of 70–82 ppm Ba and 166–330 ppm Sr (Table 1.2.6.4) whereas the Kilauea lavas range from 115–200 ppm Ba and 270–470 ppm Sr (Table 1.2.6.3). Variations in K/Ba ratios (as well as in other incompatible element ratios) are sensitive indicators of mantle heterogeneity (e.g., Erlank and Kable, 1976). Engel *et al.* (1965) calculated the K/Ba ratio for average unaltered MORB to be 121; however, the observed range is from as low as ~ 20 (e.g., Project FAMOUS basalts, Langmuir *et al.*, 1977) to 160 (e.g., DSDP Site 332B, Bence and Taylor, 1977). These sites are located very close to each other on the Mid-Atlantic Ridge, so the observed K/Ba range suggests that the oceanic upper mantle is inhomogeneous with respect to these elements.

Ocean island basalt K/Ba ratios tend to be very low. The average for alkali series basalts is 28 (Engel *et al.*, 1965) while the range for the Hawaii alkali series basalts is 25–80. Hawaiian tholeiite ratios range between 25–40. Mauna Loa summit lavas tend to have higher K/Ba ratios (31–40) relative to Kilauea lavas (25–36; Table 1.2.6.5).

Rubidium, which tends to follow potassium during basalt magmatic processes, is slightly enriched in island tholeiites (~ 5–12 ppm) relative to fresh MORBs (≤5 ppm) (Tables 1.2.6.3–4). Ocean island alkali series basalts have highly variable Rb contents (~ 15–400 ppm). The Hawaii alkali series basalts of the reference suite have Rb abundances in the range 15–100 ppm (Table 1.2.6.5).

Rb-Sr and Sm-Nd isotopic relationships

The isotopic composition of strontium and Rb/Sr ratios have been determined for many ocean island basalts. In general, these lavas have higher $^{87}Sr/^{86}Sr$ ratios than MORBs (0.703–0.707 vs. 0.702–0.703) and variable, but usually higher, Rb/Sr ratios (0.02–2 vs. 0.001–0.02). One exception is Iceland, where the basalts are more like normal ridge tholeiites (Moorbath and Walker, 1965; O'Nions *et al.*, 1976; O'Nions and Grönvold, 1973; O'Nions and Pankhurst, 1973). The higher $^{87}Sr/^{86}Sr$ ratios of the ocean island basalts tend to correlate with higher Rb/Sr ratios, which are properties of their mantle source regions.

The number of ocean island basalts for which Nd isotope ratios ($^{143}Nd/^{144}Nd$) have been measured is relatively small compared with those for which $^{87}Sr/^{86}Sr$ ratios have been measured. O'Nions *et al.* (1977) reported Nd isotope ratios for basalts from Iceland, Hawaii, Ascension, Tristan da Cunha, and Bouvet. The Iceland data reveal differences between Reykjanes Ridge basalts and those erupted on Iceland and differences between post-glacial and Tertiary tholeiites. $^{87}Sr/^{86}Sr$ and $^{143}Nd/^{144}Nd$ ratios for Iceland and Reykjanes Ridge basalts show a covariance with a negative correlation.

Basalts from Hawaii have $^{143}Nd/^{144}Nd$ ratios that range from 0.51288 ± 2 to 0.51307 ± 3 (O'Nions *et al.*, 1977), but there is no correlation with rock type. Furthermore, the negative correlation of Sr- and Nd-isotopic compositions observed for the Iceland-Reykjanes Ridge basalts is not so obvious for Hawaii. The Sr and Nd data do indicate that long-term Rb/Sr and Sm/Nd heterogeneities are present in the sources of the Hawaiian basalts.

MORB Rb/Sr ratios of ~0.01, which are assumed to be identical to those of their source regions, are too low to support the present $^{87}Sr/^{86}Sr$ ratios if we assume a chondritic $^{87}Sr/^{86}Sr$ ratio initially for the Earth's mantle. It appears that the sources have experienced prior melting events which had reduced the Rb/Sr ratio. Ocean island basalts, which have only slightly higher $^{87}Sr/^{86}Sr$ ratios, have Rb/Sr ratios which are higher than those of MORBs by nearly a factor of 3. This ratio is sufficient to have produced an $^{87}Sr/^{86}Sr$ ratio in the islands basalt source region equal to that of basalts derived from a source with a chondritic ratio. It may be concluded, therefore, that sources of the oceanic island basalts are more primitive than sources of the MORBs.

Large, high-valency cations

These elements, which include Th, U, Ce, Zr, Hf, Nb, Ta, Sr, Ti, Mo and W, find it difficult to enter the lattices of common rock-forming silicates because of their high charges. As a consequence, like the large cations, they are preferentially incorporated into the first melt formed during partial melting and are concentrated in the residual liquid during magmatic differentiation.

Ocean island alkali basalts have high concentrations of these elements relative to ocean island tholeiites and MORBs. For example, Zr abundance ranges in the three groups are: ocean island alkali basalts, 200–1000 ppm (e.g., Wager and Mitchell, 1953; Engel *et al.*, 1965; Baker, 1969; Ridley, 1970; Bence *et al.*, 1980; Table 1.2.6.5); ocean island and seamount (high-K) tholeiites, 100–300 ppm (e.g., Wager and Mitchell, 1953; Bence *et al.*, 1980; Leeman *et al.*, 1980; Table 1.2.6.4); MORBs, 15–150 ppm (e.g., Engel *et al.*, 1965; Frey *et al.*, 1974; Ayuso *et al.*, 1976; Bence and Taylor, 1977; Erlank and Kable, 1976; Bryan *et al.*, 1977; Langmuir *et al.*, 1977). Hawaiian tholeiites of the reference suite contain 70–700 ppm Zr (Tables 1.2.6.3–4) and the alkali series reference suite contains 150–900 ppm Zr (Table 1.2.6.5). The differences in zirconium abundances among ocean island tholeiites, MORBs, and ocean island alkali basalts reflect source region chemical differences. MORBs are derivatives of oceanic mantle which is depleted relative to the source regions of oceanic island basalts.

In spite of the large ranges of Zr and Ti abundances in the basalts (see section 1.2.6.3), the Ti/Zr ratios of the three groups are very nearly identical (e.g., Hubbard, 1969).

Relative abundances of niobium (Tables 1.2.6.3–5) are comparable to those of zirconium. Niobium abundances in MORBs range from 1–15 ppm (e.g., section 1.2.5, Bence *et al.*, 1975; Ayuso *et al.*, 1976; Bence and Taylor, 1977; Erlank and Kable, 1976; Mazzullo and Bence, 1976; Langmuir *et al.*, 1977; Rice *et al.*, 1979), in ocean island and seamount tholeiites, 5–25 ppm (e.g., this section, Bence *et al.*, 1980; Cambon *et al.*, 1980), and in ocean island and seamount alkali series basalts, 20–160 ppm (e.g., this section, Ridley, 1970; Erlank and Kable, 1976; Cambon *et al.*, 1980). The Zr/Nb ratios for these associations are less diagnostic of the association than are the Nb abundances. MORB Zr/Nb ratios range from 6–90 and seem to be quite distinctive for each locality. Typical MORBs have Zr/Nb ratios <20. High-K MORBs have ratios as high as 90. Island tholeiites have Zr/Nb ratios ranging from 7–15 and those of island alkaline series basalts range from 5–15. Kilauea tholeiites tend to have slightly lower Zr/Nb ratios (7–10) than Mauna Loa summit lavas (8–15). The similarities among the oceanic island tholeiites, alkali series basalts, and the majority of MORBs indicate that Zr/Nb ratios are similar in their source regions.

The abundances of thorium and uranium in ocean island basalts vary widely depending upon locality and magma type. The overall ranges are, for ocean island tholeiites: Th = 0.3–1.2 ppm, U = 0.1–0.3 ppm (Hamilton, 1965; Bence *et al.*, 1980; Cambon *et al.*, 1980); and for oceanic alkalic series basalts: Th = 1–6 ppm, U = 0.2–0.3 ppm (Hamilton, 1965; Subbarao *et al.*, 1977; Bence *et al.*, 1980; Cambon *et al.*, 1980). Tholeiites from the Hawaii reference suite contain from 0.3–1.7 ppm Th (U was not determined for this suite of samples; Tables 1.2.6.3–4). The Mauna Loa samples are markedly depleted in Th (0.34–0.58 ppm) relative to the Kilauea samples (1.0–1.7 ppm). Alkaline series basalts from Hawaii have Th concentration levels from 1.2–8.0 ppm (Table 1.2.6.5). Fresh MORB abundance levels for Th and U range from 0.1–0.7 ppm and 0.05–0.3 ppm, respectively (e.g., Ayuso *et al.*, 1976; Bence and Taylor, 1977; Langmuir *et al.*, 1977; Subbarao *et al.*, 1977). These abundances are consistent with the depleted characters of MORB source regions discussed previously and with the relatively undepleted characters of ocean island sources.

1.2.6.5 Mineralogy

General

Oceanic island basalts tend to be more complex mineralogically than mid-ocean ridge basalts. The latter characteristically contain olivine (Fo_{80-90}) (±Cr-Al spinel) as the low-pressure liquidus phase, followed by calcic plagioclase ($\leq An_{85}$), subcalcic clinopyroxene, titanomagnetite, and accessory sulfides, silica, etc. (see section 1.2.5 for a detailed discussion of MORB mineralogy). Many MORBs which lack olivine appear to have experienced crystal fractionation. Plagioclase, sometimes concentrated by flow differentiation, appears to be the low-pressure liquidus phase in these compositions. Clinopyroxene is a late-crystallizing phase in the majority of MORBs. However, rare calcic clinopyroxene megacrysts having $Mg/(Mg + Fe^{2+}) \sim 0.9$ in some MORBs may represent an early phase which crystallized under high total pressure or moderate P_{H_2O} conditions (see, for example, Kushiro, 1972; Bence and Bender, 1978; Bender *et al.*, 1978).

Ocean island tholeiites are generally similar mineralogically to MORBs but frequently contain orthopyroxene, in addition to olivine, spinel, plagioclase, clinopyroxene and Fe-Ti oxides, as an important phase.

Hawaiian tholeiites typically have Cr-Al spinel as the liquidus phase. It is closely followed in the paragenetic sequence by olivine (Fo_{88-80}). Plagioclase (An_{70-50}) and clinopyroxene ($Wo_{40-35}En_{50-42}Fs_{10-25}$) occur frequently as rapidly grown phases in the groundmass. Many of the tholeiites contain phenocrysts and microphenocrysts of orthopyroxene ($Wo_{4-5}En_{76-80}Fs_{16-19}$) and clinopyroxene ($Wo_{35-42}En_{46-51}Fs_{9-17}$). A reaction relationship between olivine and orthopyroxene indicates that orthopyroxene follows olivine in the paragenetic sequence. Epitaxial overgrowths of calcic clinopyroxene on orthopyroxene in one sample from Mauna Loa (HȦW-9) indicates that, for this sample, the paragenesis was chromite→olivine→orthopyroxene→(clinopyroxene + plagioclase). (Appendix A-6, Table 1.)

The mineralogies of the alkali magma series and nephelinitic basalts are highly diverse—consequences of their wide ranges of chemical composition. Alkali olivine basalts from Hualalai Volcano (HAW-11 and 12) are olivine (Fo_{85-70}) porphyritic, but also contain clinopyroxene-plagioclase glomerocrysts and clinopyroxene microphenocrysts in a groundmass (variolitic to subophitic) consisting of olivine, plagioclase, clinopyroxene, and titanomagnetite. Chrome spinel, found enclosed within the olivine, was the liquidus phase. The spinel was followed by olivine, clinopyroxene, plagioclase, and titanomagnetite. Alkali olivine basalt from Kohala Volcano is both olivine and plagioclase porphyritic. Its paragenesis was olivine→plagioclase→clinopyroxene→titanomagnetite.

Ankaramite from Mauna Kea is olivine (Fo_{87-56}) porphyritic but also contains microphenocrysts of plagioclase (An_{86}) and clinopyroxene ($Wo_{45}En_{47}Fs_8$) in a fine-grained groundmass of lathy plagioclase, clinopyroxene, and Fe-Ti oxides. Its paragenesis was Cr-Al spinel→olivine→plagioclase→clinopyroxene→Fe-Ti oxides.

Trachytes from Kohala and Hualalai Volcanoes are similar mineralogically, although our Kohala sample is slightly porphyritic (plagioclase and edenite). Both are dominated by fine-grained lathy plagioclase and contain small amounts of edenite, Fe-Ti oxides, ± clinopyroxene.

Spinel

Spinel is ubiquitous in the tholeiitic basalts from Hawaii. It is also present in the alkali olivine and hawaiite members of the alkali series basalts. In both groups spinel is the liquidus or co-liquidus phase with olivine. Its composition is very sensitive to f_{O_2} (e.g., Hill and Roedder, 1974), so spinel is an important indicator of the conditions prevailing during the early stages of Hawaiian basalt crystallization.

Evans and Wright (1972) determined the compositional variations in liquidus chromites from the 1959 Kilauea Iki and 1965 Makaopuhi eruptions at Kilauea Volcano. They noted that Cr_2O_3 variations in rapidly-quenched basaltic pumice correlate positively with MgO and Al_2O_3 and negatively with FeO, Fe_2O_3, TiO_2, and V_2O_3 contents. Furthermore, in these pumices, olivine and spinel appear to have equilibrated. The spinel compositions have re-equilibrated in more slowly cooled and extensively crystallized basalt.

Our studies of spinels in the tholeiitic and alkalic members of the reference suite reveal that there are systematic compositional differences between the two groups (Fig. 1.2.6.15, Table 1.2.6.6). Spinels from both groups are less magnesian than the MORB spinels and have a much larger range of $Mg/(Mg + Fe^{2+})$. The Hawaiian tholeiite spinels have higher Cr_2O_3 (Table 1.2.6.6) and higher $Cr/(Cr + Al + Fe^{3+})$ ratios than

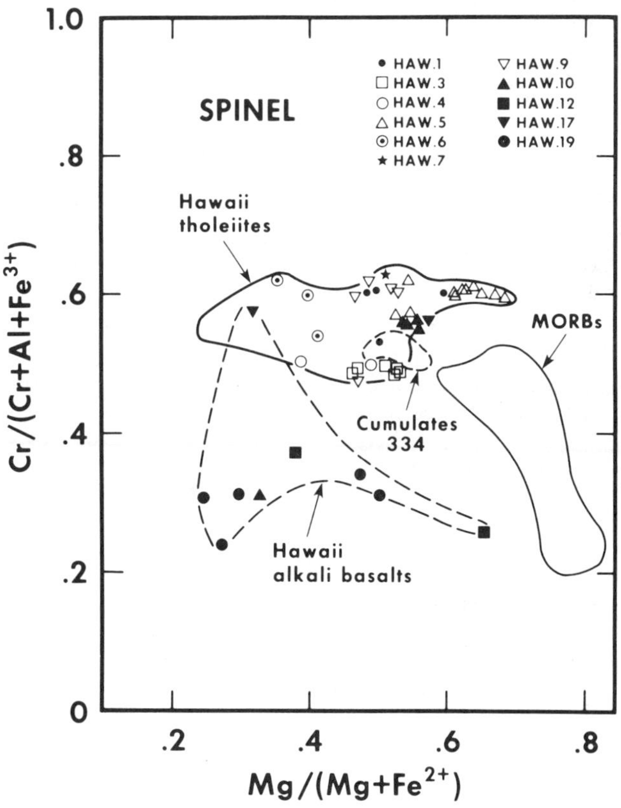

1.2.6.15 $Cr/(Cr + Al + Fe^{3+})$ vs. $Mg/(Mg + Fe^{2+})$ for spinels in ocean island and ocean floor basalts and cumulates. Fe^{3+} was calculated assuming stoichiometric constraints as in Fig. 1.2.6.4.

Table 1.2.6.6 Compositions of Hawaiian spinels.

	1	2	3	4	5	6	7
Al_2O_3	16.1	13.9	14.2	12.4	15.6	14.1	14.2
TiO_2	3.14	3.47	2.05	2.72	1.48	1.62	1.65
"FeO"	29.1	35.3	22.6	27.5	22.1	30.5	22.1
MnO	0.39	0.38	0.27	0.47	0.30	0.43	0.32
MgO	10.3	10.7	14.1	11.7	14.6	10.7	13.2
Cr_2O_3	40.8	36.3	47.3	45.2	47.0	43.0	48.7
SUM	99.8	100.1	100.5	100.0	101.1	100.4	100.2
Al	0.625	0.550	0.535	0.485	0.584	0.553	0.541
Cr	1.060	0.964	1.194	1.187	1.179	1.132	1.244
Fe^{3+}	0.159	0.362	0.186	0.220	0.215	0.300	0.162
Ti	0.078	0.088	0.049	0.068	0.035	0.041	0.040
Fe^{2+}	0.641	0.631	0.418	0.544	0.372	0.549	0.435
Mn	0.011	0.011	0.007	0.013	0.008	0.012	0.009
Mg	0.504	0.534	0.673	0.579	0.690	0.531	0.636

	8	9	10	11	12	13
Al_2O_3	12.9	28.9	19.5	14.7	18.8	5.74
TiO_2	2.16	2.34	2.82	2.38	4.04	18.1
"FeO"	36.2	31.5	40.1	26.3	41.3	58.6
MnO	0.41	0.30	0.55	0.31	0.35	0.43
MgO	8.16	14.2	7.60	12.1	9.66	4.66
Cr_2O_3	40.1	22.0	27.9	44.5	25.8	11.8
SUM	99.9	99.2	98.5	100.3	100.0	99.3
Al	0.523	1.061	0.785	0.565	0.744	0.251
Cr	1.090	0.542	0.753	1.148	0.685	0.347
Fe^{3+}	0.358	0.377	0.598	0.198	0.483	0.536
Ti	0.056	0.055	0.072	0.058	0.102	0.506
Fe^{2+}	0.682	0.443	0.547	0.519	0.697	1.284
Mn	0.012	0.008	0.016	0.009	0.010	0.014
Mg	0.418	0.659	0.387	0.588	0.483	0.258

1. HAW-1 Spinel in rim of olivine megacryst analysis **2** Table 1.2.6.7. **2.** HAW-2 Spinel in olivine microphenocryst analysis **3** Table 1.2.6.7. **3.** HAW-23 Spinel in glass (analysis **6** Table 1.2.6.13). Adj. to olivine (analysis **6** Table 1.2.6.7). **4.** HAW-26 Spinel in olivine. **5.** HAW-5 Spinel in olivine megacryst core (analysis **10** Table 1.2.6.7). **6.** HAW-5 Spinel in olivine megacryst rim (analysis **11** Table 1.2.6.7). **7.** HAW-6 Spinel in olivine megacryst core adjacent to analysis **12** Table 1.2.6.7. **8.** HAW 6 Spinel in olivine megacryst rim adjacent to analysis **13** Table 1.2.6.7. **9.** HAW-12 Spinel in olivine megacryst (analysis **15** Table 1.2.6.7). **10.** HAW-12 Spinel adjacent to olivine megacryst rim (analysis **16** Table 1.2.6.7). **11.** HAW-17 Spinel in olivine megacryst (analysis **19** Table 1.2.6.7). **12.** HAW-19 Spinel in olivine megacryst core (analysis **22** Table 1.2.6.7). **13.** HAW-19 Spinel in olivine megacryst rim (analysis **23** Table 1.2.6.7).

either the alkali basalt or MORB spinels. The contents of Fe^{3+} (as determined by charge balance) are consistently higher in the spinels from the alkali basalts. Titanium is low in the early crystallizing spinels from both basalt series (1.3–3.8 wt.% TiO_2), but is consistently higher in the hawaiite spinels (4–8 wt.% TiO_2).

The spinels within individual tholeiite basalts are highly variable in composition; correlation of compositions with relative times of crystallization is evident. Spinel octahedra which are enclosed in the cores of olivine megacrysts have the highest $Mg/(Mg + Fe^{2+})$ ratios (Fig. 1.2.6.15), MgO, Cr_2O_3 and Al_2O_3 contents. Those included in the olivine rims are more iron-rich and have lower Cr_2O_3 and Al_2O_3 contents.

Early crystallizing spinels (those included in olivine megacryst cores) have the highest contents of Cr_2O_3 (45–48 wt.% Cr_2O_3). The low-pressure crystallization experiments of Hill and Roeder (1974), performed on a Kilauea tholeiite under controlled oxygen fugacity conditions, indicate that the Cr_2O_3 content of the spinel is sensitive to f_{O_2} for $\log f_{O_2} < -3$. High Cr_2O_3 contents are consistent with low f_{O_2} during crystallization. Extrapolation of their experimental data suggests that the high Cr_2O_3, high $Mg/(Mg + Fe^{2+})$ spinels in the Kilauea and

Mauna Loa tholeiites crystallized at $\log f_{O_2} \simeq -10$. This value falls on the QFM buffer curve at 1100° C. Both values are in very close agreement with the direct measurements obtained on the Makaopuhi lava lake by Wright and Okamura (1977).

Olivine

Hawaii tholeiitic olivines have compositions ranging from Fo_{89} to Fo_{70} with a frequency maximum at Fo_{82} and an average of Fo_{84} (Fig. 1.2.6.16). Their Ca, Mn, Al, and Cr ranges (in atoms/formula unit) are as follows: Ca = 0.0057–0.0126 (Fig. 1.2.6.17); Mg = 0.0010–0.0080; Al = 0.0000–0.0060; Cr = 0.0000–0.0060. Representative analyses are reported in Table 1.2.6.2.

The concentrations of nickel in olivine megacrysts and in their host glasses from a Mauna Ulu basalt have been determined using electron microprobe analytical techniques developed specifically for trace element studies (Bence *et al.*, 1981). The olivine megacryst core (Fo_{88}) contains 1900 ± 20 ppm Ni (average of two analytical points, each with 12 replicate analyses) and the surrounding glass (MgO = 7 wt.%) contains 108 ± 10 ppm Ni (average of three analytical points, each with 12 replications). The resulting K_D = 17.6 is in close agreement with the value (19.1) obtained from the Hart and Davis (1978) expression for K_D as a function of the MgO content of the liquid.

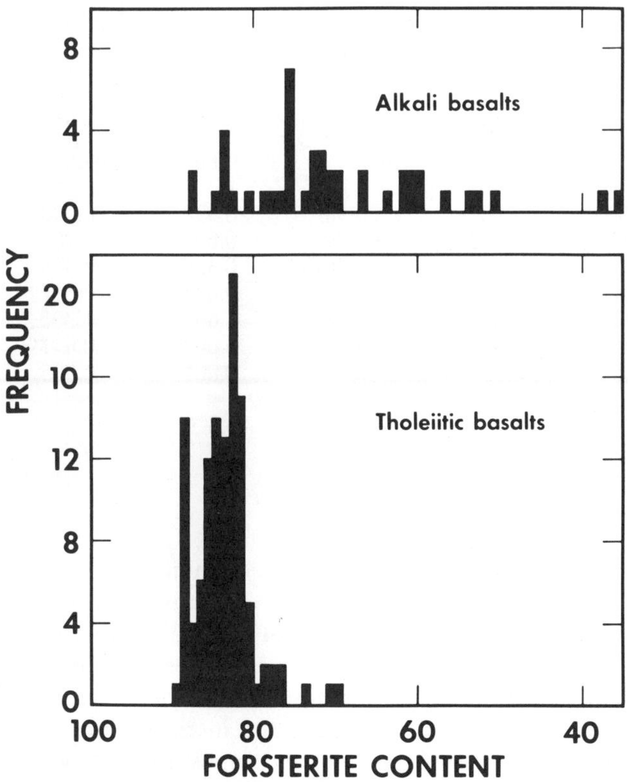

1.2.6.16 Histograms of Hawaiian basalt olivine compositions, based upon 105 analyses of tholeiites and 47 analyses of alkali series basalts.

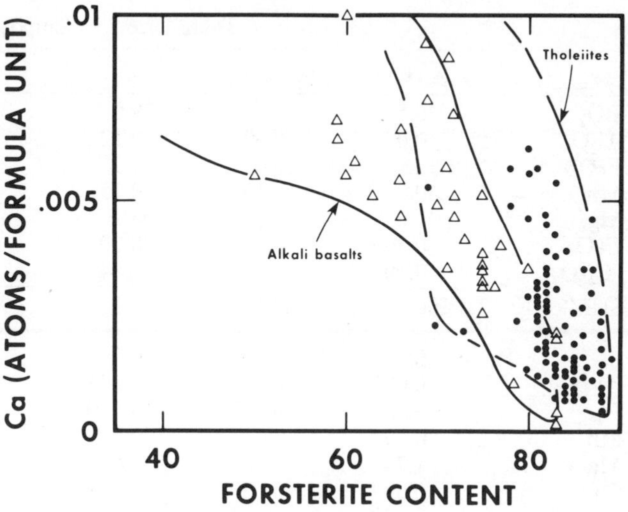

1.2.6.17 Calcium contents of Hawaiian olivines.

Melt inclusions within olivine megacrysts (Table 1.2.6.7, analyses 1 and 6) have evolved compositions; therefore, they are not samples of unfractionated parent magmas.

Olivines in the alkali series basalts have, overall, a larger compositional range (Fo_{87-34}) than those in the tholeiites. An average $Fo_{71.5}$ (Fig. 1.2.6.16) is based upon 47 analyses. Within individual samples, the compositional range is nearly as large (Fo_{83-34} in olivine basalt HAW-12). Systematic differences occur in the minor element contents of the tholeiitic and alkali basalt olivines. The minor element ranges in the alkali basalts (in atoms/formula unit) are: Ca = 0.0075–0.0170; Mn = 0.0040–0.0580; Al = 0.0020–0.0095; Cr = 0.0000–0.0050. Representative analyses are reported in Table 1.2.6.7.

Liquidus temperatures, determined by calibration with experimental studies of olivine-melt equilibria (Roeder and Emslie, 1970; Bender *et al.*, 1978), range from 1272° C in tholeiitic sample HAW-26 to 1317° C in tholeiitic basalt HAW-6. These temperatures are consistent with those observed by direct measurement (Peck *et al.*, 1964; Wright *et al.*, 1976) and by experimentation (Thompson and Tilley, 1976; Fisk and Bence, 1980). The alkali basalts have liquidus temperatures ranging from 1251° C (HAW-12) to 1327° C (HAW-17). Hawaiite HAW-19 has plagioclase preceding olivine as the liquidus phase. In this sample, olivine crystalized at 1196° C.

Table 1.2.6.7 Compositions of Hawaiian olivines.

	1	2	3	4	5	6	7	8
SiO_2	40.5	39.6	39.6	38.3	37.6	39.9	39.0	39.5
Al_2O_3	0.09	0.09	0.12	0.00	0.00	0.00	0.00	0.00
FeO	12.2	15.4	16.7	20.29	28.2	11.7	16.7	14.0
MnO	0.09	0.16	0.15	0.19	0.26	0.07	0.15	0.16
MgO	47.6	44.4	44.2	40.3	34.0	47.8	44.1	46.1
CaO	0.30	0.28	0.30	0.33	0.36	0.28	0.27	0.28
"Cr_2O_3"	0.10	0.16	0.04	0.0	0.00	0.00	0.00	0.02
SUM	100.9	100.1	101.1	99.4	99.9	99.8	100.2	100.1
Si	0.995	0.997	0.991	0.994	1.002	0.990	0.987	0.988
Al	0.003	0.003	0.004	0.000	0.000	0.000	0.000	0.000
Fe	0.251	0.323	0.350	0.440	0.629	0.243	0.353	0.293
Mn	0.002	0.003	0.003	0.004	0.006	0.002	0.003	0.003
Mg	1.742	1.664	1.650	1.557	1.351	1.768	1.663	1.719
Ca	0.008	0.008	0.008	0.009	0.010	0.007	0.007	0.008
Cr	0.002	0.003	0.001	0.000	0.000	0.000	0.000	0.000
Fo	87.3	83.6	82.4	77.8	68.2	87.9	82.5	85.4
Fa	12.7	16.4	17.6	22.2	31.8	12.1	17.5	14.6

	9	10	11	12	13	14	15	16
SiO_2	39.7	39.9	39.1	40.2	38.9	38.8	39.2	37.3
Al_2O_3	0.01	0.05	0.04	0.07	0.07	0.30	0.15	0.29
FeO	13.7	10.8	17.7	11.5	17.7	20.8	15.8	26.0
MnO	0.22	0.09	0.18	0.06	0.21	0.42	0.21	0.52
MgO	45.5	49.0	43.1	48.4	43.0	40.0	44.2	35.1
CaO	0.24	0.25	0.29	0.20	0.26	0.36	0.26	0.38
"Cr_2O_3"	0.00	0.19	0.06	0.25	0.28	0.17	0.17	0.22
SUM	99.4	100.3	100.5	100.7	100.4	100.9	100.0	99.8
Si	0.998	0.982	0.991	0.988	0.987	0.994	0.990	0.993
Al	0.000	0.002	0.001	0.002	0.002	0.009	0.005	0.009
Fe	0.288	0.222	0.375	0.237	0.376	0.446	0.334	0.579
Mn	0.005	0.002	0.004	0.001	0.005	0.009	0.005	0.012
Mg	1.705	1.797	1.628	1.771	1.627	1.528	1.663	1.392
Ca	0.007	0.007	0.008	0.005	0.007	0.010	0.007	0.011
Cr	0.000	0.004	0.001	0.005	0.006	0.003	0.003	0.005
Fo	85.5	89.0	81.3	88.2	81.2	77.4	83.3	70.6
Fa	14.5	11.0	18.7	11.8	18.8	22.6	16.7	29.4

Plagioclase

For the majority of tholeiitic basalts from Hawaii, plagioclase is a relatively minor, late-crystallizing phase (following Cr-spinel, olivine, and clinopyroxene) (Appendix A-6, Tables 1 and 2). Due to the rapid cooling of many of these flows, plagioclase crystallization was inhibited, so its compositional range is restricted ($An_{68}Ab_{31}$–$An_{50}Ab_{49}$; Fig. 1.2.6.18, Table 1.2.6.8). More rarely, plagioclase occurs as phenocrysts or as microphenocrysts, which have more calcic compositions, or in glomerocrysts with clinopyroxene.

Plagioclase is a much more common constituent of alkali series basalts (Appendix A-6, Table 2). Its modal abundance ranges from a low of about 25–35% in alkali olivine basalts to 60–70% in trachytes. The diverse chemistries and crystallization histories of the reference suite alkali basalts are reflected in their plagioclase modal abundances and in the compositional ranges of their plagioclases.

The alkali olivine basalts contain plagioclase phenocrysts (An_{72-58} with olivine phenocrysts) or plagioclase (An_{68-62}) glomerocrysts with calcic clinopyroxenes,

Table 1.2.6.7 (continued)

	17	18	19	20	21	22	23	24
SiO_2	33.8	36.4	40.1	36.2	35.2	38.6	36.6	38.2
Al_2O_3	0.27	0.09	0.06	0.13	0.09	0.04	0.12	0.13
FeO	50.0	34.3	12.5	36.9	37.2	18.8	34.0	22.3
MnO	1.00	1.02	0.14	0.67	2.23	0.28	0.63	0.27
MgO	14.8	28.8	47.4	26.7	24.1	42.0	29.0	38.2
CaO	0.06	0.38	0.29	0.26	1.30	0.31	0.45	0.30
"Cr_2O_3"	0.12	0.04	0.30	0.07	0.06	0.20	0.05	0.06
SUM	100.1	100.9	100.7	100.9	100.2	100.2	100.9	99.5
Si	1.016	0.997	0.990	1.004	1.000	0.987	1.002	0.999
Al	0.009	0.003	0.002	0.004	0.003	0.001	0.004	0.004
Fe	1.258	0.786	0.258	0.856	0.883	0.402	0.778	0.488
Mn	0.026	0.024	0.003	0.016	0.054	0.006	0.015	0.006
Mg	1.633	1.179	1.744	1.104	1.017	1.601	1.183	1.491
Ca	0.002	0.011	0.008	0.008	0.039	0.009	0.013	0.008
Cr	0.003	0.001	0.004	0.002	0.001	0.004	0.001	0.001
Fo	34.5	59.3	87.1	56.3	52.1	79.9	60.3	75.3
Fa	65.5	40.7	12.9	43.7	47.9	20.1	39.7	24.7

"Cr_2O_3" = Total Cr calculated as Cr_2O_3.
1. HAW-1 Megacryst core. **2.** HAW-1 Megacryst rim adjacent to spinel **1** Table 1.2.6.6. **3.** HAW-2 Microphenocryst adjacent to spinel **2** Table 1.2.6.6. **4.** HAW-22 Phenocryst core. **5.** HAW-22 Phenocryst rim. **6.** HAW-23 Phenocryst rim; adjacent to glass analysis **6** Table 1.2.6.13 and spinel analysis **3** Table 1.2.6.6. **7.** HAW-24 Phenocryst with cpx inclusion (analysis **6** Table 1.2.6.9). **8.** HAW-26 Olivine core adjacent to glass inclusion (analysis **9** Table 1.2.6.13). **9.** HAW-4 Megacryst with cpx rim (analysis **9** Table 1.2.6.9). **10.** HAW-5 Megacryst core adjacent to spinel analysis **5** Table 1.2.6.6. **11.** HAW-5 Megacryst rim. **12.** HAW-6 Megacryst core. **13.** HAW-6 Megacryst rim. **14.** HAW-11 Megacryst. **15.** HAW-12 Megacryst adjacent to spinel (analysis **9** Table 1.2.6.6). **16.** HAW-12 Megacryst rim adjacent to spinel (analysis **10** Table 1.2.6.6). **17.** HAW-12 Interstitial. **18.** HAW-16 Interstitial. **19.** HAW-17 Megacryst with spinel inclusion (analysis **11** Table 1.2.6.6). **20.** HAW-17 Megacryst rim. **21.** HAW-18 Interstitial. **22.** HAW-19 Megacryst core. **23.** HAW-19 Megacryst rim. **24.** HAW-20 Megacryst core.

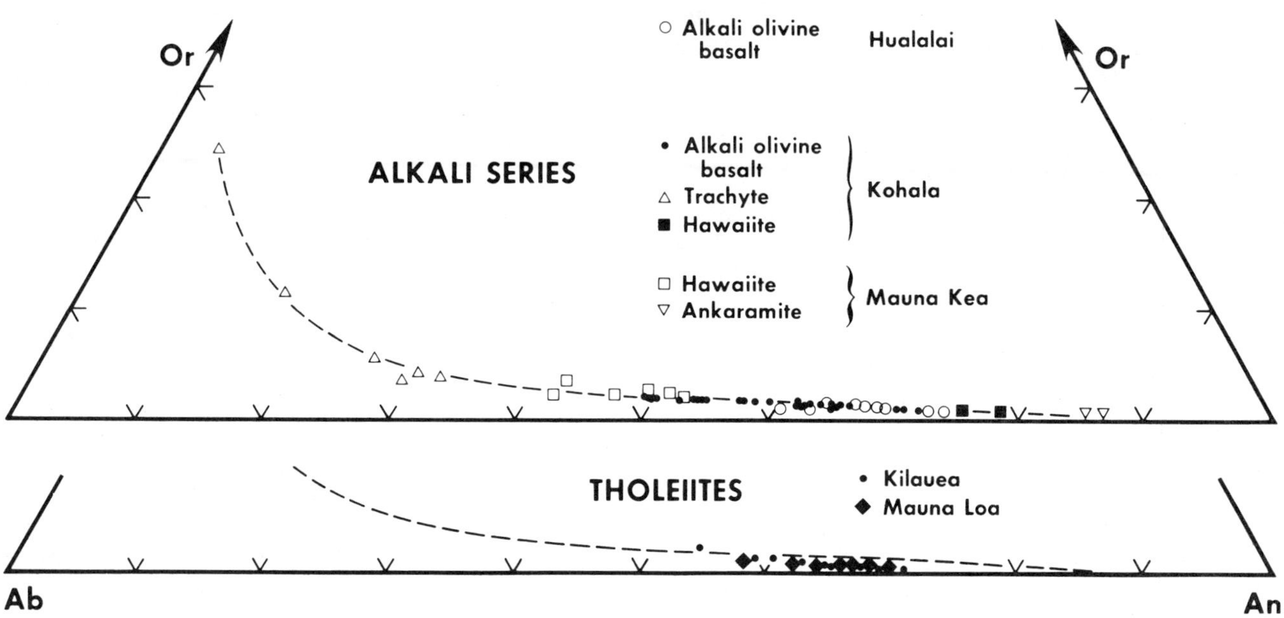

1.2.6.18 Compositions of plagioclase in Hawaiian basalts.

Table 1.2.6.8 Compositions of Hawaiian plagioclases.

	1	2	3	4	5	6	7	8	9	10
SiO_2	51.5	52.7	52.7	51.3	51.5	55.4	46.2	60.6	49.6	54.7
Al_2O_3	29.5	29.4	28.2	30.4	30.0	28.0	33.2	24.5	31.5	27.6
FeO	0.84	0.72	0.91	0.71	1.17	0.73	0.67	0.21	0.82	0.58
MgO	0.09	0.04	0.04	0.12	0.00	0.00	0.07	0.00	0.07	0.03
CaO	13.7	13.2	12.5	14.7	13.0	10.5	17.2	5.92	15.3	11.1
Na_2O	3.46	3.83	4.15	3.37	3.72	5.30	1.67	7.56	2.76	5.05
K_2O	0.13	0.16	0.17	0.07	0.20	0.44	0.06	0.63	0.12	0.35
SUM	99.2	100.1	98.6	100.7	99.6	100.4	99.1	99.5	100.1	99.5
Si	2.366	2.395	2.428	2.327	2.358	2.494	2.150	2.712	2.270	2.490
Al	1.598	1.576	1.532	1.630	1.619	1.490	1.820	1.293	1.699	1.483
Fe	0.032	0.027	0.035	0.027	0.045	0.027	0.026	0.008	0.031	0.022
Mg	0.006	0.003	0.003	0.008	0.000	0.000	0.005	0.000	0.005	0.002
Ca	0.674	0.643	0.617	0.716	0.640	0.506	0.858	0.284	0.750	0.540
Na	0.308	0.338	0.371	0.297	0.330	0.463	0.150	0.656	0.244	0.445
K	0.008	0.009	0.010	0.004	0.012	0.025	0.004	0.036	0.007	0.020
An	68.1	65.0	61.9	70.4	65.2	50.9	84.8	29.1	74.9	53.7
Ab	31.1	34.1	37.2	29.2	33.6	46.6	14.9	67.2	24.4	44.3
Or	0.8	0.9	1.0	0.4	1.22	2.54	0.4	3.7	0.7	2.0

1. HAW-22 Phenocryst core. **2.** HAW-22 Phenocryst rim. **3.** HAW-25 Microphenocryst. **4.** HAW-8 Glomerocryst. **5.** HAW-11 Glomerocryst. **6.** HAW-16 Groundmass. **7.** HAW-17 Phenocryst. **8.** HAW-18 Phenocryst. **9.** HAW-19 Plagioclase in olivine phenocryst. **10.** HAW-20 Phenocryst.

as well as lathy groundmass plagioclases which have the more sodic compositions of the phenocryst and glomerocryst rims (Table 1.2.6.8; Fig. 1.2.6.18).

Hawaiite from Kohala Volcano contains calcic plagioclase phenocrysts ($An_{78}Ab_{22}$) in a groundmass containing lathy, more sodic plagioclases ($An_{53}Ab_{44}Or_3$). Hawaiite from Mauna Kea Volcano is nonporphyritic. Its groundmass contains lathy, sodic plagioclase ($An_{54-39}Ab_{42-58}Or_{1-3}$).

Ankaramite from Mauna Kea contains calcic plagioclase ($An_{86}Ab_{14}$) phenocrysts and more sodic groundmass plagioclases ($An_{73}Ab_{26}Or_1$).

Trachytes from Kohala Volcano are plagioclase-phyric with sodic plagioclase phenocrysts ($An_{30}Ab_{67}Or_3$) in a fine-grained matrix containing alkali feldspar ($An_{16}Ab_{72}Or_{11}$).

The potassium contents of the plagioclases in both basalt series are dependent to a large extent upon the anorthite contents. However, for given anorthite contents, plagioclases in alkali series basalts have consistently higher K contents than those in tholeiitic basalts (Fig. 1.2.6.18).

Crystallization temperatures of the tholeiitic basalts, as determined from the plagioclase geothermometer of Drake (1972), range from 1126°C (HAW-25) to 1207°C (HAW-22). These agree with the extrusion temperatures obtained by direct measurement (1200–1225°C), and are compatible with the slightly higher temperatures obtained from consideration of olivine-melt equilibria.

Temperatures of plagioclase crystallization in the alkali-olivine basalts (1168–1175°C) are consistent with the olivine temperatures. However, the porphyritic hawaiite from Kohala and ankaramite from Mauna Kea both give unreasonably high plagioclase crystallization temperatures (1280°C and 1347°C, respectively). It may be concluded that the plagioclase in these basalts is too calcic to have crystallized from a melt having the same composition as the bulk rock.

Pyroxene

Both clinopyroxene and orthopyroxene are present in the tholeiitic and alkali series basalts from Hawaii. The most common occurrence of clinopyroxene in the tholeiitic basalts is in the fine-grained groundmass, where it preceded or co-crystallized with plagioclase during late-stage rapid cooling, and where the evolved compositions reflect prior removal of olivine. Many of the phyric basalts also contain calcic clinopyroxene megacrysts which crystallized prior to extrusion and were carried along by the host magma. These megacrysts are characterized by high $Mg/(Mg + Fe^{2+})$ ratios (0.85–0.90), and generally have higher Ca contents than groundmass clinopyroxenes in the same samples (Muir and Tilley, 1963). The megacrysts also tend to have lower Al_2O_3 and TiO_2 concentrations than microphenocrysts and groundmass pyroxenes in the same samples. As a consequence the proportion of quadrilateral to "others" components is highest in the megacrysts (Table 1.2.6.9). The high Mg/(Mg + Fe^{2+}) ratios in these megacrysts suggest either that they crystallized from the basalt magma relatively early (presumably at elevated pressures) or that they were accidental xenocrysts—possibly from high-pressure cumulates (cognate) or from the conduit walls through which the magma passed.

Groundmass pyroxenes in the tholeiitic basalts are more iron-rich than the megacrysts and can be significantly more subcalcic (Fig. 1.2.6.19).

Low Ca pyroxene includes both pigeonite and augite. Pigeonite occurs as reaction rims on olivine megacrysts in one of our samples, thus satisfying the classical textural definition of a tholeiite (e.g., Wilkinson, 1968). It also occurs as a late-crystallizing interstitial phase (with calcic clinopyroxene and plagioclase) in HAW-4 from Mauna Loa (Table 1.2.6.9). Orthopyroxene (hypersthene) megacrysts with or without augite megacrysts are also common in the tholeiites. The range of orthopyroxene compositions observed in all of these samples is very small (Fig. 1.2.6.19), which suggests that a single process was responsible for their formation.

Tholeiitic basalts from Suiko Seamount (DSDP Site 433C, Clague *et al.*,1980) contain calcic clinopyroxene phenocrysts and microphenocrysts with compositions ($Wo_{41}En_{51}Fs_8$ to $Wo_{38}En_{50}Fs_{12}$) indistinguishable from those of the megacrysts in Hawaiian tholeiites. Minor element abundances in megacrysts are also very

Table 1.2.6.9 Compositions of Hawaiian pyroxenes.

	1	2	3	4	5	6	7	8	9	10
SiO_2	51.5	52.1	50.0	51.2	48.9	51.6	51.2	52.3	51.8	53.5
Al_2O_3	3.57	3.63	4.31	3.90	5.07	2.81	2.62	2.09	2.17	0.91
TiO_2	0.79	0.99	1.56	1.05	1.68	0.90	1.11	0.83	0.71	0.49
"FeO"	6.21	6.03	7.18	6.82	7.28	5.81	9.19	7.25	9.10	14.8
MnO	0.10	0.10	0.08	0.18	0.13	0.11	0.19	0.15	0.18	0.25
MgO	16.7	16.8	16.4	18.4	16.2	17.1	17.1	17.9	17.3	24.5
CaO	19.7	20.0	18.8	18.0	18.8	20.5	17.8	19.0	16.8	4.03
Na_2O	0.12	0.14	0.19	0.13	0.04	0.07	0.08	0.07	0.24	0.0
Cr_2O_3	0.91	1.16	1.08	0.90	1.02	0.61	0.17	0.12	0.10	0.01
SUM	99.7	101.0	99.6	100.6	99.1	99.5	99.5	99.7	98.4	98.5
FeO	6.21	6.03	6.92	5.62	6.61	5.34	8.45	6.51	8.85	14.8
Fe_2O_3	0.0	0.0	0.30	1.34	0.74	0.53	0.82	0.82	0.28	0.0
Si	1.896	1.892	1.850	1.861	1.818	1.900	1.901	1.921	1.936	1.972
Al^{IV}	0.104	0.108	0.150	0.139	0.182	0.100	0.099	0.079	0.064	0.028
Al^{VI}	0.051	0.047	0.038	0.028	0.047	0.022	0.031	0.023	0.020	0.014
Ti	0.022	0.027	0.043	0.029	0.040	0.025	0.015	0.012	0.031	0.012
Cr	0.026	0.033	0.032	0.026	0.030	0.018	0.005	0.003	0.003	0.000
Fe^{3+}	0.000	0.183	0.008	0.037	0.021	0.015	0.023	0.023	0.008	0.000
Fe^{2+}	0.191	0.000	0.214	0.171	0.206	0.164	0.262	0.200	0.277	0.456
Mn	0.003	0.003	0.003	0.006	0.004	0.003	0.006	0.005	0.006	0.008
Mg	0.916	0.909	0.904	0.994	0.898	0.939	0.944	0.981	0.965	1.345
Ca	0.777	0.778	0.745	0.701	0.751	0.809	0.708	0.748	0.674	0.159
Na	0.009	0.010	0.014	0.009	0.003	0.005	0.006	0.005	0.017	0.000
Wo	41.2	41.6	40.0	37.6	40.5	42.3	37.0	38.8	35.2	8.1
En	48.6	48.6	48.5	53.3	48.4	49.1	49.3	50.9	50.4	68.6
Fs	10.2	9.8	11.5	9.1	11.1	8.6	13.7	10.4	14.5	23.3
Quad	89.6	89.2	85.0	86.1	81.8	90.0	90.1	92.1	93.6	97.2
Others	10.4	10.8	15.0	13.9	18.2	10.0	9.9	7.9	6.4	2.8
	CaAl CaTS	CaMg TAl	CaMg TAl	$CaFe^3$ CaTS	CaMg TAl	CaMg TAl	CaMg TAl	CaMg TAl	Na TAl	Mg TAl

1. HAW-1 Megacryst core. **2.** HAW-1 Megacryst rim. **3.** HAW-1 Groundmass pyroxene. **4.** HAW-3 Microphenocryst. **5.** HAW-21 Microphenocryst; adjacent glass analysis **4** Table 1.2.6.13. **6.** HAW-24 Inclusion in olivine (Analysis **7** Table 1.2.6.7). **7.** HAW-25 Phenocryst. **8.** HAW-26 Phenocryst. **9.** HAW-4 Augite rim on olivine megacryst (Analysis **9** Table 1.2.6.7). **10.** HAW-4 Interstitial pigeonite.

Table 1.2.6.9 (continued)

	11	12	13	14	15	16	17	18	19	20
SiO_2	55.0	54.6	54.4	51.7	51.1	54.1	47.5	51.0	51.0	50.2
Al_2O_3	1.67	2.36	1.79	2.44	3.17	1.66	6.01	3.16	3.74	3.56
TiO_2	0.35	0.37	0.24	0.61	0.67	0.27	2.39	0.70	1.14	1.25
"FeO"	12.7	10.8	11.2	7.76	6.92	10.8	8.03	6.38	5.40	7.90
MnO	0.21	0.17	0.13	0.11	0.12	0.19	0.15	0.16	0.0	0.11
MgO	28.4	26.6	30.0	17.7	17.4	29.4	13.3	16.6	16.6	15.0
CaO	2.30	5.22	2.05	18.4	14.9	1.98	21.5	20.4	22.0	20.6
Na_2O	0.0	0.0	0.0	0.15	0.15	0.0	0.30	0.18	0.22	0.33
Cr_2O_3	0.22	0.35	0.44	0.26	0.95	0.40	0.44	0.59	0.41	0.29
SUM	100.9	99.7	100.3	99.1	99.3	98.8	99.6	99.2	99.5	99.2
FeO	12.2	10.8	9.51	6.61	5.72	9.89	6.26	4.94	3.17	6.34
Fe_2O_3	0.55	0.0	1.86	1.28	1.33	0.98	1.97	1.60	2.48	1.73
Si	1.945	1.942	1.922	1.914	1.886	1.939	1.777	1.887	1.857	1.870
Al^{IV}	0.055	0.058	0.075	0.086	0.114	0.061	0.223	0.113	0.143	0.130
Al^{VI}	0.015	0.004	0.000	0.020	0.023	0.009	0.042	0.025	0.017	0.027
Ti	0.009	0.010	0.006	0.017	0.019	0.007	0.067	0.019	0.031	0.035
Cr	0.006	0.010	0.012	0.008	0.028	0.011	0.013	0.017	0.012	0.009
Fe^{3+}	0.015	0.000	0.049	0.036	0.037	0.027	0.055	0.045	0.068	0.048
Fe^{2+}	0.362	0.321	0.281	0.204	0.177	0.296	0.196	0.153	0.097	0.198
Mn	0.006	0.005	0.004	0.003	0.004	0.006	0.005	0.005	0.000	0.003
Mg	1.500	1.410	1.576	0.973	0.957	1.568	0.739	0.914	0.902	0.835
Ca	0.087	0.199	0.078	0.728	0.745	0.076	0.861	0.809	0.858	0.822
Na	0.000	0.000	0.000	0.011	0.011	0.000	0.022	0.013	0.016	0.024
Wo	4.5	10.3	4.0	38.2	39.7	3.9	47.9	43.2	46.2	44.3
En	77.0	73.1	81.5	51.1	50.9	80.8	41.2	48.7	48.6	45.0
Fs	18.6	16.6	14.5	10.7	9.4	15.3	10.9	8.1	5.2	10.7
Quad	94.5	94.0	92.5	91.4	88.6	93.9	77.7	88.7	85.7	87.0
Others	5.5	6.0	7.5	8.6	11.4	6.1	22.3	11.3	14.3	13.0
	MgAl CaTS	MgAl CaTS	MgAl CaTS	$CaFe^{3}$ CaTS	$CaFe^{3+}$ CaTS	$MgFe^{3}$ CaTS	CaMg TAl	$CaFe^{3+}$ CaTS	$CaFe^{3+}$ CaTS	CaMg TAl

11. HAW-6 Orthopyroxene megacryst core. **12.** HAW-6 Pigeonite rim on olivine. **13.** HAW-7 Orthopyroxene megacryst core. **14.** HAW-7 Augite megacryst core. **15.** HAW-9 Orthopyroxene/augite megacryst (augite). **16.** HAW-9 Orthopyroxene/augite megacryst (opx). **17.** HAW-11 Microphenocryst. **18.** HAW-12 Microphenocryst core. **19.** HAW-17 Megacryst. **20.** HAW-20 Megacryst.

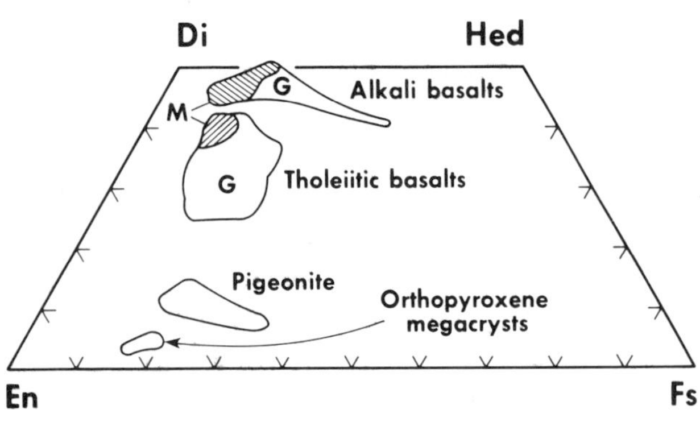

1.2.6.19 Hawaiian basalt pyroxene compositional relationships. M = Megacrysts; G = groundmass.

similar at both sites. Consequently, one petrogenetic model seems capable of explaining both occurrences. The lack of orthopyroxene at the Suiko site suggests that details of the petrogeneses may differ slightly.

The alkaline basalts of Hawaii contain megacrysts with high Ca-Tschermaks and ferri-Tschermaks component contents relative to the tholeiitic megacrysts (Table 1.2.6.9, Fig. 1.2.6.19). The important substitutions appear to be $CaTiAl_2O_6$ and $CaFe^{3+}SiAlO_6$. This is in agreement with the results of the statistical analysis by Schweitzer *et al.* (1979). Groundmass clinopyroxenes are enriched in Fe^{2+} and Ti but tend to be slightly depleted in Ca relative to the megacrysts.

The overlying alkaline basalts at Suiko Seamount (DSDP Sites 433A,B) also contain clinopyroxene megacrysts which are compositionally similar to megacrysts in the alkaline basalts from Hawaii. Higher Wo contents and higher contents of "others" components distinguish these pyroxenes from the tholeiitic megacrysts.

Amphiboles

Amphibole is recognized only in the alkaline basalts from Hawaii. Amphibole in the trachyte (HAW-18) from Kohala Volcano is an edenite (Fig. 1.2.5.7, Table 1.2.6.10). It occurs both as megacrysts and as crystals growing on the walls of vugs. No compositional distinctions between these two occurrences are evident. Trachybasalts collected at DSDP Site 165A, which is located at a topographic high on the Line Islands chain, contain kaersutite that forms epitaxial overgrowths on clinopyroxene phenocrysts. A representative analysis is reported in Table 1.2.6.10.

Mica

Phlogopite is observed as a trace constituent in hawaiite HAW-16. It is reddish brown and has a high TiO_2 content (Table 1.2.6.11). Its composition is comparable to that of phlogopite observed in a San Juan, Colorado, trachyte (Table 1.2.6.11, column 2), although the Mg/(Mg + Fe) ratios are different.

Fe-Ti oxide minerals

Fe-Ti oxides are late-crystallizing phases in all of the Hawaii basalts and occur as interstitial grains,

Table 1.2.6.10 Compositions of Hawaiian and other amphiboles.

	1	2	3	4
SiO_2	49.1	40.1	49.3	49.6
Al_2O_3	5.03	11.6	0.60	0.62
TiO_2	1.35	6.0	0.98	0.35
"FeO"	10.4	12.6	33.73	34.5
MnO	0.47	0.21	0.53	0.55
MgO	16.6	13.0	1.10	3.29
CaO	9.37	11.0	3.26	7.89
Na_2O	4.29	2.59	6.00	0.77
K_2O	1.18	0.94	1.07	0.19
Cr_2O_3	0.05	—	—	—
Σ Oxides	97.8	98.0	96.6	97.7
Si	7.149	5.961	7.954	7.884
^{IV}Al	0.851	2.039	0.046	0.116
^{VI}Al	0.012	0.000	0.068	0.000
Ti	0.148	0.670	0.119	0.042
"Fe^{3+}"	0.000	0.001	0.368	0.111
Fe^{2+}	1.266	1.563	4.185	4.471
Mn	0.058	0.026	0.072	0.074
Mg	3.602	2.876	0.264	0.780
Σ Oct.	5.092	5.137	5.076	5.478
M4 site				
Oct	0.092	0.137	0.076	0.478
Ca	1.462	1.749	0.563	1.345
Na	0.446	0.144	1.361	0.177
Σ M4 site	2.000	2.000	2.000	2.000
A site				
Na	0.765	0.632	0.514	0.060
K	0.219	0.178	0.220	0.039
Σ A site	0.984	0.810	0.734	0.099
Quad	1.60	0	26.6	85.3
"Others"	98.4	100.0	73.4	14.7

Oct. = Octahedrally-coordinated cations

1. HAW-18 megacryst. **2.** Site 165A amphibole rim on clinopyroxene phenocryst. **3.** Site 169 green interstitial amphibole, Myers *et al.*, 1975. **4.** Site 169 colorless interstitial amphibole, Myers *et al.*, 1975.

Table 1.2.6.11 Compositions of phlogopites.

	1	2†
SiO_2	40.3	38.6
Al_2O_3	12.1	14.1
TiO_2	5.48	4.95
"FeO"	9.04	13.2
MnO	0.16	0.20
MgO	20.5	15.6
CaO	0.07	1.06
Na_2O	0.98	0.81
K_2O	8.90	8.68
Cr_2O_3	0.07	—
	97.9	97.2

"FeO" = total Fe calculated as FeO.
† Includes 0.43 BaO and 1.25 F.
1. HAW-16 interstitial grain. **2.** Glassy rhyolite, San Juan, Colorado, Larsen *et al.*, 1937.

together with sodic plagioclase and Fe-rich clinopyroxene. Many tholeiitic basalts contain no Fe-Ti oxide minerals because the liquids were quenched from too high temperatures. One of the tholeiitic basalts (HAW-4, Mauna Loa summit) contains titanomagnetite in the groundmass. An analysis of one grain, included within an iron-rich olivine, is reported in Table 1.2.6.12. This is an unusually high-chrome, high-aluminum titanomagnetite which is quite different from the more typical titanomagnetites that are ubiquitous in the hawaiites and alkali olivine basalts. Similar titanomagnetites (also included within olivine) occur rarely in the alkali basalts (analysis #3, Table 1.2.6.12). The more typical alkali basalt titanomagnetites contain 1–1.5 wt.% Al_2O_3 and very little, if any, Cr_2O_3. No coexisting ilmenite was found in any of our samples; consequently, T-f_{O_2} estimates could not be made.

Residual liquids

Electron microprobe analysis of residual glass in the Hawaiian reference suite basalts reveals a relatively restricted range of compositions (Table 1.2.6.13 and Fig. 1.2.6.20). All but one are slightly quartz normative, which is a consequence of extreme olivine fractionation. Kilauea liquids containing olivine, spinel, clinopyroxene, ± plagioclase define a very small field in the subprojections of the pseudoquaternary (Figs. 1.2.6.10–11). This field appears to define the projection of the quaternary univariant line, where olivine, clinopyroxene, and plagioclase coexist with liquid. Analyses of Mauna Loa residual liquids are unavailable. However, it can be seen from Figs. 1.2.6.10–11 that they will be distinctly more siliceous than the Kilauea residual glasses.

Melt inclusions in olivine megacrysts from the tholeiitic basalts are depleted in both ΣFeO and MgO relative to the residual glasses (Fig. 1.2.6.11, Table 1.2.6.13). Apparently, olivine continued to crystallize from the trapped melt and plated out on the walls of the inclusion. No samples of possible parent magma were detected among these inclusions.

1.2.6.6 Petrogenesis

Parent magma chemistry

Numerous studies of Hawaiian basalt petrogenesis have pointed out the importance of low-pressure olivine

Table 1.2.6.12 Compositions of Hawaiian Fe-Ti oxide minerals.

	1	2	3	4	5	6	7
Al_2O_3	6.51	1.44	4.92	1.51	1.00	1.56	2.85
TiO_2	10.8	22.7	14.5	22.7	22.0	19.9	23.0
"FeO"	57.7	68.3	62.5	67.8	67.5	72.8	62.3
MnO	0.42	0.76	0.60	0.72	1.78	1.50	0.54
MgO	4.65	3.18	4.11	1.84	1.34	1.40	3.63
Cr_2O_3	13.5	1.56	11.1	0.00	0.00	0.01	5.48
SUM	93.6	97.9	97.7	94.6	93.6	97.2	97.8
Al	0.309	0.067	0.226	0.074	0.050	0.076	0.129
Cr	0.430	0.049	0.342	0.000	0.000	0.003	0.167
Fe^{3+}	1.000	0.695	0.929	0.793	0.837	1.138	0.446
Ti	0.327	0.676	0.425	0.705	0.698	0.618	0.665
Fe^{2+}	0.945	1.566	1.109	1.548	1.544	1.374	1.556
Mn	0.014	0.026	0.020	0.025	0.064	0.052	0.018
Mg	0.277	0.188	0.239	0.113	0.084	0.086	0.208

"FeO" = total Fe calculated as FeO.
1. HAW-4 inclusion in olivine (Fo_{73}). **2.** HAW-11 matrix. **3.** HAW-11 inclusion at rim of olivine megacryst. **4.** HAW-16 interstitial. **5.** HAW-18 interstitial. **6.** HAW-18 interstitial. **7.** HAW-19 interstitial.

Table 1.2.6.13 Compositions of Hawaiian glasses.

	1	2	3	4	5	6	7	8	9
SiO_2	52.1	50.3	50.4	49.7	49.6	48.8	48.7	50.4	50.5
Al_2O_3	16.3	13.4	13.5	13.9	13.6	11.9	12.2	13.3	12.9
TiO2	2.90	2.67	2.70	2.89	2.88	2.49	2.50	3.80	2.67
"FeO"	8.97	11.1	11.1	11.0	11.0	11.4	11.4	12.2	9.40
MnO	0.18	0.21	0.21	0.10	0.10	0.11	0.10	0.10	0.10
MgO	2.75	6.98	6.97	6.18	6.24	9.66	9.69	5.20	6.73
CaO	13.5	11.8	11.7	11.1	11.0	10.5	10.8	9.59	12.3
Na_2O	2.53	2.32	2.42	2.39	2.37	1.99	2.05	2.74	2.33
K_2O	0.66	0.58	0.59	0.54	0.57	0.49	0.47	0.75	0.53
Cr_2O_3	0.10	0.09	0.01	0.00	0.01	0.06	0.10	0.00	0.01
SUM	100.0	99.4	99.6	97.8	97.4	97.4	98.0	98.1	97.5

1. HAW-1 glass inclusion in olivine megacryst. **2.** HAW-2 glass adjacent to olivine microphenocryst analysis **3** Table 1.2.6.7 and spinel analysis **2**. **3.** HAW-2 Average glass. **4.** HAW-21 glass adjacent to cpx microphenocryst analysis **5** Table 1.2.6.9. **5.** HAW-21 Average glass. **6.** HAW-23 glass adjacent to olivine analysis **6** Table 1.2.6.7 and spinel analysis **3** Table 1.2.6.6. **7.** HAW-23 Average glass. **8.** HAW-25 Average glass. **9.** HAW-26 inclusion in olivine (analysis **8** Table 1.2.6.7).

fractionation on the compositions of the lavas. Much of this fractionation probably occurs at shallow depths in magma chambers, but fractionation through crystal settling in lava lakes and through flow differentiation at the site of extrusion are important processes as well. This fractionation obscures but, fortunately, does not obliterate all vestiges of the chemistry of the parent liquids.

As noted above, at least two chemically distinct magmas (Mauna Loa and Kilauea) are recognized within the tholeiitic suite on the basis of differing amounts of normative quartz, clinopyroxenes, olivine, and plagioclase (Figs. 1.2.6.10–11). The major distinction between them is in silica content; differences in the major diagnostic oxides (e.g., MgO, FeO, TiO_2, Al_2O_3) are relatively small. These two basalt chemistries may

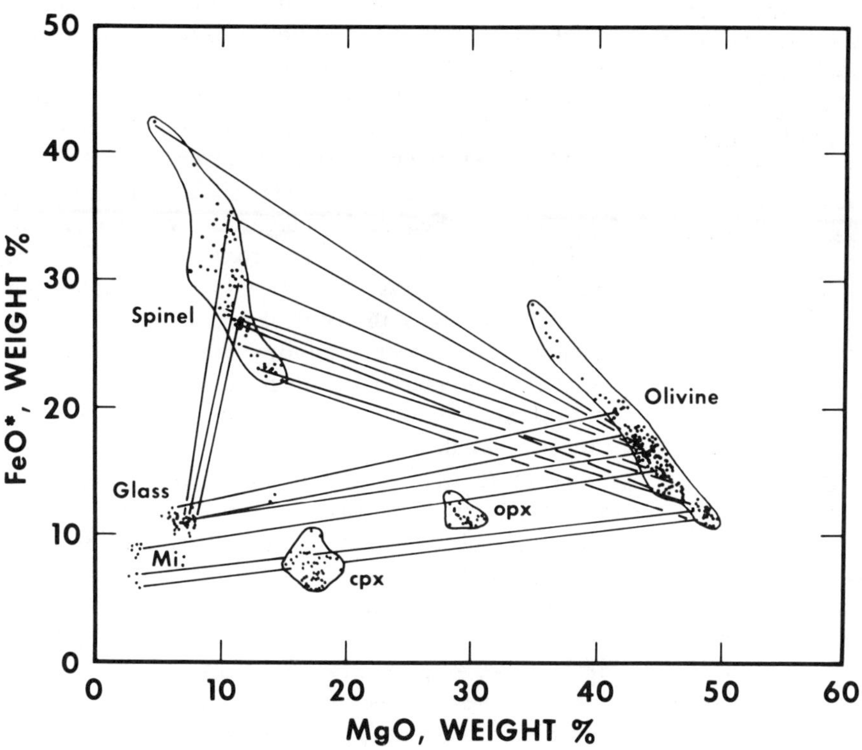

1.2.6.20 Total iron (FeO*) vs. MgO for residual glass and crystal phases in Hawaiian basalts.

be related by fractionation at high pressure. Wright and others (e.g., Wright and Fiske, 1971; Wright *et al.*, 1975) recognize numerous chemically distinct magma batches at Kilauea that cannot be related by low-pressure crystal fractionation but which require complex processes. These various magma batches are quite similar chemically and are less easily distinguished than the Mauna Loa and Kilauea lavas as groups.

Considerations of vitrophyre chemistry and olivine/melt equilibria permit reconstruction of a most probable Kilauea parent magma. Figure 1.2.6.5 shows that, excepting the low MgO basalts, olivine fractionation does not affect the FeO content. Thus the Kilauea parent must have contained ≃11 wt.% total iron as FeO. It is relatively straightforward to calculate Mg′-values and MgO and then to use these values to calculate TiO_2, Al_2O_3, Na_2O, K_2O, and CaO. Silica content is first estimated by difference (assuming values for P_2O_5, H_2O), and then is refined by means of projections, such as that shown in Fig. 1.2.6.10. The calculated Kilauea parent magma has the composition: $Mg/(Mg + Fe^{2+}) \simeq 0.70–0.72$ (in equilibrium with Fo_{89} olivine), MgO = 14.5 wt.%, total iron as FeO = 11 wt.%, TiO_2 = 2.2 wt.%, Al_2O_3 = 11 wt.%, CaO = 9 wt.%, Na_2O = 2 wt.%, and K_2O = 0.4 wt.%. By difference, silica is ~48 wt.%, assuming ~1 wt.% MnO, Cr_2O_3, P_2O_5, H_2O, etc. This composition projects into the primary phase field of olivine in the cpx-ol-plag and qtz-ol-plag subprojections of Figs. 1.2.6.10–11. Natural Hawaiian basalt samples having this composition are olivine cumulates. The Kilauea Iki lava lake parent of MacDonald and Katsura (1961), OTI, has very nearly the correct proportions of normative clinopyroxene, olivine, and plagioclase (Fig. 1.2.6.11), but it is deficient in silica and is not a reasonable parental composition for any of the Hawaiian tholeiites. Mauna Loa lavas require a parent which is slightly more silica-rich (~49–50 wt.% SiO_2) than the calculated Kilauea parent.

Low-pressure liquidus phase relationships

Thompson and Tilley (1976) performed low-pressure melting and crystallization experiments on natural Kilauea basalt compositions—several of which approximate the proposed parent magma (*op. cit.* Table 1, Samples 1959a, 19601). Although their experiments were run in air, and are thus more oxidized than the natural samples, they can still provide information on the maximum temperature of crystallization for a magma of this composition at low pressures. Olivine appeared at 1321°C in 1959a and 1273°C in 19601. These temperatures are comparable to those calculated from the olivine/melt relationships in the reference suite tholeiites. No phase composition data are available for these experiments.

Calculated low-pressure olivine/melt equilibrium relationships for the Mauna Loa parent magma composition reveal that olivine (Fo_{90}) will crystallize (probably following chrome-spinel) at a temperature of 1323°C. Assuming complete olivine fractionation, the residual liquid will evolve along an olivine control line (Figs. 1.2.6.10–11) until it intersects the olivine-clinopyroxene peritectic. This occurs after ~18 wt.% olivine removal. At the peritectic, clinopyroxene ($Wo_{41}En_{48}Fs_{11}$) crystallizes as the olivine attempts to react with the liquid. Continued fractional crystallization, now involving both olivine and clinopyroxene extraction, causes the residual liquid composition to evolve until clinopyroxene and plagioclase ($An_{67}Ab_{32}Or_{<1}$) co-precipitate. Hawaiian basalt glasses, which contain the four crystalline phases, olivine, spinel, clinopyroxene, and plagioclase, define a very small field on the two subprojections of the forsterite-fayalite-silica-anorthite pseudoquaternary (Figs. 1.2.6.10–11). Thus, for Kilauea basaltic liquid compositions, we are able to locate this low-pressure multiple saturation point with considerable accuracy. The temperature calculated from mineral/melt equilibria for compositions in the vicinity of this point is ~1170°C. No similar multiple saturation region for Mauna Loa lavas could be located, due to the lack of samples containing residual glass. Residual MORB glasses have a much larger range of compositions; hence, their region of multiple saturation is larger. Furthermore, the multiple saturation point is the projection of the quaternary univariant line (liquid + olivine + anorthite + clinopyroxene); the effect of lowering the temperature (and changing the composition) is to shift the projection of this line.

A number of crystalline Hawaiian basalt compositions are enriched in the plagioclase component (or depleted in olivine and clinopyroxene) relative to the residual glasses (Fig. 1.2.6.11). Those compositions which fall in the stability field of plagioclase can be explained by olivine + clinopyroxene fractionation under conditions such that plagioclase crystallization is suppressed. Plagioclase accumulation through flotation or through flow differentiation could generate a similar trend. In this instance, the latter is not a reasonable explanation, because the expected trend on the ol-plag-qtz subprojection (Fig. 1.2.6.10) is not observed. Evolved compositions show the effects *both of olivine and of plagioclase removal.*

High-pressure phase relationships

High-pressure melting experiments have not yet been conducted on the proposed Kilauea parent composition. However, Green *et al.* (1967) and Green and Ringwood (1967) conducted experiments on the proposed parent magma composition for the 1959 Kilauea Iki lava lake (MacDonald and Katsura, 1964). The major differences between that composition and the parent proposed here are in silica, alumina, and lime (Table 1.2.6.14). The Kilauea parent proposed here has 48 wt.% SiO_2, 11 wt.% Al_2O_3, and 9 wt.% CaO compared with 46.9 wt.% SiO_2, 13 wt.% Al_2O_3 and 10 wt.% CaO in the proposed Kilauea Iki parent. The major effect of these differences is that the Kilauea Iki parent will fractionate to the olivine-plagioclase cotectic at low pressures. Clinopyroxene will appear after plagioclase. This is contrary to the mineral paragenesis observed in Kilauea basalts. As noted above, the Kilauea Iki composition does not fall along the olivine control line on the olivine-plagioclase-silica subprojection, and hence cannot be the parent magma composition for any Hawaiian tholeiites.

Although the Kilauea Iki "parent" magma composition cannot be used to model low-pressure fractionation effects, its composition is sufficiently close to that of the Kilauea parent proposed here that the high-pressure anhydrous melting experiments of Green and Ringwood (1967) have some relevance. Olivine (Fo_{91}) is the liquidus phase up to about 10 kb where it is replaced by orthopyroxene ($\sim Wo_4En_{88}Fs_8$; $Al_2O_3 \sim 2.0$ wt.%); and it, in turn, is replaced by clinopyroxene at ~ 20 kb.

Compositions of the high-pressure pyroxenes were estimated from partial electron microprobe analyses (Fe, Ca, and Al only) assuming stoichiometric constraints and no Na, Cr, Ti, or Fe^{3+} substitutions. However, as noted above and in Table 1.2.6.9, the "others" components comprise only 5–10% of the pyroxene chemistry. Consequently, generalized comparisons are justified, but should be made with caution.

Orthopyroxene megacrysts in the tholeiitic members of the Hawaii reference suite are much more iron-rich (Fs_{15-20}) than the experimental liquidus orthopyroxenes (Fs_8) of Green and Ringwood (1967). Lower temperature phases contain more iron (Fs_{10-14}), but this still is less than that observed in the natural samples. The discrepancy appears to result from iron loss to the platinum capsules in the high-pressure runs. The effect of iron loss on the liquid compositions is to increase the amount of normative hypersthene and decrease the amount of normative olivine in the liquids. For a given composition this should stabilize orthopyroxene at somewhat lower pressures than if there were no iron loss. We conclude that orthopyroxene ($\sim Wo_4En_{78-80}$ Fs_{16-18}) is the liquidus phase of the Kilauea parent composition at pressures in the vicinity of 10–12 kb.

High-pressure fractionation involving orthopyroxene will drive the liquid composition away from the olivine-silica join of Fig. 1.2.6.10. It follows that the proposed Kilauea parent magma could be derived by high-pressure fractionation (involving orthopyroxene) of a Mauna Loa parent magma, but that the reverse is not true.

At pressures in excess of ~20 kb, calcic-clinopyroxene is on the liquidus of the Kilauea parent magma. Crystal fractionation involving this phase will drive the liquid away from clinopyroxene on Fig. 1.2.6.10. Under these conditions, the major element normative chemistry of a Mauna Loa parent magma could be derived from a Kilauea parent.

The major and trace element characteristics of the Hawaiian alkali basalts are such as to preclude their derivation from a tholeiitic parent (e.g., the proposed Kilauea parent magma) by crystal fractionation at pressure (olivine, orthopyroxene, and clinopyroxene are the possible fractionating phases in the pressure intervals 0–11, 11–20, and >20 kb, respectively). This is contrary to the conclusion of Green and Ringwood (1967, p. 168) that high-pressure fractionation of an olivine tholeiite parent magma (involving extraction of aluminous enstatite) in the depth interval 35–70 km will result in liquids which trend towards alkali olivine basalt. The Green and Ringwood olivine tholeiite parent will fractionate

Table 1.2.6.14 Proposed Hawaiian parent magmas.

	1	2
SiO_2	48.	46.95
Al_2O_3	11.0	13.10
TiO_2	1.35	2.02
"FeO"	11.0	11.0
MnO	—	0.15
MgO	14.5	14.55
CaO	9.0	10.16
Na_2O	2.0	1.73
K_2O	0.4	0.08
CIPW NORM†		
Or	2.4	.48
Ab	16.9	14.6
An	19.9	27.7
Di	19.9	18.2
Hy	15.8	11.7
Ol	17.3	21.3
Mt	2.0	2.0
Ilm	4.2	3.8

$Mg/(Mg+Fe^{2+})$ † 0.72

†Calculated for $Fe^{3+}/\Sigma Fe = 0.1$

1. Proposed Kilauea parent. 2. 1959 Kilauea Iki "parent" (MacDonald & Katsura (1961).

towards alkali olivine basalt at high pressures; however, their choice of a parent magma, as noted above, was inappropriate. A more reasonable parent composition will not fractionate toward alkaline basalts. Similar incompatible element ratios (e.g., K/Ba, K/Rb, Zr/Nb, etc.) and similar strontium isotopic chemistry in the alkaline and tholeiitic basalts from oceanic islands (e.g., Peterman and Hedge, 1971; Hanson, 1977) are consistent with derivation of the two magma series from a common source or from geochemically similar sources by different degrees of partial melting. These sources are enriched in incompatible element abundances with respect to mid-ocean ridge basalt sources. High-pressure crystallization studies of synthetic and natural basalt systems (e.g., Kushiro, 1972; Presnall *et al.*, 1978) reveal that mantle peridotite can give rise to silica undersaturated (alkaline) liquids at total pressures >12 kb (dry) and to silica oversaturated (tholeiitic) liquids at pressures < 12 kb.

1.2.6.7 Source region chemistry and mineralogy

The previous discussions of major and trace element chemistry and of high-pressure equilibria are consistent with a garnet peridotite source. Xenoliths of garnet peridotite occur in post-erosional, highly alkalic basalts of the Honolulu Volcanic Series on Oahu (Jackson and Wright, 1970). Jackson and Wright suggested that the alkaline magmas are derived from a deep source consisting of garnet pyroxenite and garnet lherzolite, and that tholeiitic magmas are derived from a shallower (60–100 km) garnet lherzolite layer. Differences in pressure and in fraction of melting account for the major element chemical differences, while the similar source lithologies account for the similar incompatible trace element ratios. A comparison of major and trace element data for other intraoceanic plate basaltic suites suggest gross geochemical similarities of their sources.

Relative to MORB sources, the island sources have higher total iron, TiO_2, and K_2O. Lower Al_2O_3 contents may be an intrinsic property of these island sources as well; however, garnet retention in the source, which is compatible with the REE data, could also account for this feature.

Incompatible element abundances in the island sources are 2–5 times chondrites (MORB sources are 1–2 times chondrites) and Rb/Sr ratios are ~0.03 as opposed to 0.01 in MORB sources.

1.2.6.8 References

Ayuso R. A., Bence A. E. and Taylor S. R. (1976) Upper Jurassic tholeiitic basalts from DSDP Leg 11. *J. Geophys. Res.* **81**, 4305–4325.

Baker I. (1969) Petrology of the volcanic rocks of St. Helena Island, South Atlantic. *Geol. Soc. Amer. Bull.* **80**, 1283–1310.

Baker P. E. (1973) Islands of the South Pacific. In *The Ocean Basins and Margins. Vol. 1: The South Atlantic*, (A. E. M. Nairn and F. G. Stehli, eds.), p. 493–553. Plenum Press, N.Y.

Baker P. E., Gass I. G., Harris P. G. and LeMaitre R. W. (1964) The volcanological report of the Royal Society expedition to Tristan da Cunha, 1962. *Phil. Trans. Roy. Soc. Lond.* **A256**, 439–578.

Bargar K. E. and Jackson E. D. (1974) Calculated volumes of individual shield volcanoes along the Hawaiian-Emperor chain. *U.S. Geol. Survey Jour. Res.* **2**, 545–550.

Bence A. E. and Bender J. F. (1978) Clinopyroxenes in mid-ocean ridge basalts: experimental studies. *EOS (Trans. Amer. Geophys. Union)* **59**, 410.

Bence A. E. and Taylor S. R. (1977) Petrogenesis of mid-Atlantic Ridge basalts at DSDP Leg 37 Holes 332A and 332B from major and trace element geochemistry. *Initial Reports of the Deep Sea Drilling Project* **37**, p. 705–710. U.S. Govt. Printing Office, Washington, D.C.

Bence A. E., Papike J. J. and Ayuso R. A. (1975) Petrology of submarine basalts from the Central Caribbean: DSDP Leg 15. *J. Geophys. Res.* **80**, 4775–4804.

Bence A. E., Taylor S. R. and Fisk M. (1980) Major and trace element geochemistry of basalts from Ojin, Nintoku, and Suiko seamounts of the Emperor seamount chain: DSDP-IPOD Leg 55, *Initial Reports of the Deep Sea Drilling Project* **55**, p. 599–605. U.S. Govt. Printing Office, Washington, D.C.

Bence A. E., Baylis D., Bender J. F. and Grove T. L. (1979) Controls on the major and minor element chemistry of mid-ocean ridge basalts and glasses. In *Deep Drilling Results in the Atlantic Ocean: Ocean Crust* (M. Talwani, C. G. Harrison, D. E. Hayes, eds.), p. 331–341. Maurice Ewing Series 2. Amer. Geophys. Union, Washington, D.C.

Bence A. E., Brande S., Indelicato G. and Allen F. (1981) Analysis of trace and minor elements in rock-forming minerals using an automated electron probe. *Proc. 8th Int. Conf. X-Ray Optics and Microanalysis*, p. 538–548. Boston.

Bender J. F., Hodges F. N. and Bence A. E. (1978) Petrogenesis of basalts from the project FAMOUS area: experimental study from 0 to 15 kbars. *Earth Planet. Sci. Lett.* **41**, 277–302.

Bryan W. B., Thompson G., Frey F. A., Dickey J. S. and Roy S. (1977) Petrology and geochemistry of basement rocks recovered on Leg 37, DSDP. *Initial Reports of the Deep Sea Drilling Project* **37**, p. 695–703. U.S. Govt. Printing Office, Washington, D.C.

Cambon P., Gordon J. L., Bougalt H. and Treuil M. (1980) Leg 55: Emperor seamounts; trace elements in transitional tholeiites, alkali basalts and hawaiites; mantle homogeneity or heterogeneity and magmatic processes. *Initial Reports of the Deep Sea Drilling Project* **55**, p. 585–597. U.S. Govt. Printing Office, Washington, D.C.

Clague D. A. and Dalrymple G. B. (1973) Age of Kōko seamount, Emperor seamount chain. *Earth Planet. Sci. Lett.* **17**, 411–415.

Clague D. A. and Jarrard R. D. (1973) Tertiary Pacific plate motion deduced from the Hawaiian-Emperor chain. *Geol. Soc. Amer. Bull.* **84**, 1135–1154.

Clague D. A., Dalrymple G. B. and Moberly R. (1975) Petrography and K/Ar ages of dredged volcanic rocks from the western Hawaiian ridge and the southern Emperor seamount chain. *Geol. Soc. Amer. Bull.* **86**, 991–998.

Clague D. E., Fisk M. and Bence A. E. (1980) Mineral chemistry of basalts from Ojin, Ninooku, and Suiko seamounts, Leg 55 DSDP. *Initial Reports of the Deep Sea Drilling Project* **55**, p. 607–637. U.S. Govt. Printing Office, Washington, D.C.

Czamanske G. K. and Moore J. G. (1977) Composition and phase chemistry of sulfide globules in basalt from the mid-Atlantic rift valley near 37° N lat. *Geol. Soc. Amer. Bull.* **88**, 587–599.

Dalrymple G. B., Lanphere M. A. and Clague D. A. (1980) Conventional and $^{40}Ar/^{39}Ar$ K-Ar ages of volcanic rocks from Ojin (site 430), Nintoku (site

432), and Suiko (Site 433) Seamounts and the chronology of volcanic propagation along the Hawaiian-Emperor chain. *Initial Reports of the Deep Sea Drilling Project.* **55**, p. 659–676. U.S. Govt. Printing Office, Washington, D.C.

Dalrymple G. B., Lanphere M. A. and Jackson E. D. (1974) Contributions to the petrography and geochronology of volcanic rocks from the leeward Hawaiian Islands. *Geol. Soc. Amer. Bull.* **5**, 727–738.

Desborough G. A., Anderson A. T. and Wright T. L. (1968) Mineralogy of sulfides from certain Hawaiian basalts. *Econ. Geol.* **63**, 636–644.

Doell R. R. and Dalrymple G. B. (1973) Potassium-argon ages and paleomagnetism of the Waianae and Koolau Volcanic Series, Oahu, Hawaii. *Geol. Soc. Amer. Bull.* **84**, 1217–1242.

Drake M. (1972) The distribution of major and trace elements between plagioclase feldspar and magmatic silicate liquid: an experimental study. Ph.D. dissertation, Univ. of Oregon, Eugene, Oregon.

DSDP Scientific Staff (1973) Deep Sea Drilling Project: Leg 32. *Geotimes* **18** (12), 14–17.

Engel A. E. J., Engel C. G. and Havens R. G. (1965) Chemical characteristics of oceanic basalts and the upper mantle. *Geol. Soc. Amer. Bull.* **76**, 719–734.

Erlank A. J. and Kable E. J. D. (1976) The significance of incompatible elements in mid-Atlantic ridge basalts from 45° N with particular emphasis to Zr/Nb. *Contrib. Mineral. Petrol.* **54**, 281–291.

Evans B. W. and Wright T. L. (1972) Composition of liquidus chromite from the 1959 (Kilauea Iki) and 1965 (Makaopuhi) eruptions of Kilauea Volcano, Hawaii. *Amer. Mineral.* **57**, 217–230.

Fisk M. R. and Bence A. E. (1980) Basalt cotectic boundaries as defined by experimentally determined crystallization paths of Kilauea and Mauna Loa Basalts (abstract). *Abstracts with Program*, Geol. Soc. Amer. **12**, 426.

Frey F. A. and Green D. H. (1974) The mineralogy, geochemistry and origin of lherzolite inclusions in Victorian basanites. *Geochim. Cosmochim. Acta* **38**, 1023–1059.

Frey F. A., Bryan W. B. and Thompson G. (1974) Atlantic Ocean floor: Geochemistry and petrology of basalts from Legs 2 and 3 of the Deep Sea Drilling Project. *J. Geophys. Res.* **79**, 5507–5528.

Frey F. A., Green D. H. and Roy S. D. (1978) Integrated models of basalt petrogenesis: a study of quartz tholeiites to olivine melilitites from south-eastern Australia utilizing geochemical and experimental petrological data. *J. Petrol.* **19**, 463–513.

Funkhouser J. G., Barnes I. L.and Naughton J. J. (1966) Problems in the dating of volcanic rocks by the potassium-argon method. *Bull. Volcanol.* **29**, 709–717.

Green D. H. and Ringwood A. E. (1967) The genesis of basaltic magmas. *Contrib. Mineral. Petrol.* **15**, 103–190.

Green T. H., Green D. H. and Ringwood A. E. (1967) The origin of high alumina basalts and their relationships to quartz tholeiites and alkali basalts. *Earth Planet. Sci. Lett.* **2**, 41–51.

Gunn B. M. (1971) Trace element partitioning during olivine fractionation of Hawaiian basalts. *Chem. Geol.* **8**, 1–13.

Hamilton E. I. (1965) Distribution of some trace elements and the isotopic composition of strontium in Hawaiian lavas. *Nature* **206**, 251–253.

Hanson G. N. (1977) Geochemical evolution of the suboceanic mantle. *J. Geol. Soc.* **134**, 235–253.

Hart S. R. (1969) K, Rb, Cs contents and K/Rb and K/Cs ratios of fresh and altered submarine basalt. *Earth Planet. Sci. Lett.* **6**, 295–303.

Hart S. R. (1971) K, Rb, Cs, Sr, and Ba contents and Sr isotope ratios of ocean floor basalts. *Phil. Trans. Roy. Soc., Lond.* **A268**, 573–587.

Hart S. R. and Davis K. (1978) Nickel partitioning between olivine and silicate melt. *Earth Planet. Sci. Lett.* **40**, 203–219.

Hill R. and Roeder P. (1974) Stability of spinel in basaltic melts. *J. Geol.* **82**, 709–729.

Hubbard N. (1969) A chemical comparison of oceanic ridge, Hawaiian tholeiitic and Hawaiian alkalic basalts. *Earth Planet. Sci. Lett.* **5**, 346–352.

Ibarrola E. (1969) Variation trends in volcanic rocks of the Canary Islands. *Bull. Volc.* **33**, Pt. 3, 729–777.

Jackson E. D. (1976) Linear volcanic chains on the Pacific Plate. *Amer. Geophys. Union, Monograph 19*, p. 319–335.

Jackson E. D. and Wright T. L. (1970) Xenoliths in the Honolulu Volcanic series *J. Petrol.* **11**, 405–430.

Jackson E. D., Silver E. A. and Dalrymple G. B. (1972) Hawaiian-Emperor chain and its relation to Cenozoic circum-Pacific tectonics. *Geol. Soc. Amer. Bull.* **83**, 601–618.

Kay R. and Gast P. W. (1973) The rare earth content and origin of alkali-rich basalts. *J. Geol.* **81**, 653–682.

Kay R., Hubbard N. and Gast P. W. (1970) Chemical characteristics and origin of oceanic ridge volcanic rocks. *J. Geophys. Res.* **75**, 1585–1613.

Kushiro I. (1972) Effect of water on the composition of magmas formed at high pressures. *J. Petrol.* **13**, 311–342.

Langmuir C. H., Bender J. F., Bence A. E., Hanson G. N. and Taylor S. R. (1977) Petrogenesis of basalts from the FAMOUS Area: Mid-Atlantic Ridge. *Earth Planet. Sci. Lett.* **36**, 133–156.

Larsen E. S., Gonyer F. A. and Irving J. (1937) Petrologic results of a study of the minerals from the tertiary volcanic rocks of the San Juan Region, Colorado -6. Biotite. *Amer. Mineral.* **22**, 898–905.

Leeman W. P., Murali A. V., Ma M. -S. and Schmitt R. A. (1977) Mineral constitution of mantle source regions for Hawaiian basalts—rare earth element evidence for mantle heterogeneity. *Oregon Dept. Geol. Mineral. Indus. Bull.* **96**, 169–183.

Leeman W. P., Budahn J. R., Gerlach D. C., Smith D. R. and Powell B. N. (1980) Origin of Hawaiian tholeiites: trace element constraints. *Amer. J. Sci.* **280A**, 794–819.

Le Maitre R. W. (1962) Petrology of volcanic rocks, Gough Island, South Atlantic. *Geol. Soc. Amer. Bull.* **73**, 1309–1340.

Mazzullo L. and Bence A. E. (1976) Abyssal tholeiites from DSDP Leg 34: The Nazca Plate. *J. Geophys. Res.* **81**, 4327–4351.

McBirney A. R. and Aoki K. (1968) Petrology of the Island of Tahiti. *Mem., Geol. Soc. Amer. 116*, p. 523–556.

McBirney A. R. and Williams H. (1969) Geology and petrology of the Galapagos Islands. *Mem., Geol. Soc. Amer. 118*, 197 pp.

MacDonald G. A. (1968) Composition and origin of Hawaiian lavas. *Mem. Geol. Soc. Amer. 116*, p. 477–522.

MacDonald G. A. and Katsura T. (1961) Variations in the lava of the 1959–60 eruption in Kilauea Iki. *Pacific Sci.* **15**, 358–369.

MacDonald G. A. and Katsura T. (1964) Chemical composition of Hawaiian lavas. *J. Petrol.* **5**, 82–133.

McDougall I. (1964) Potassium-argon ages from lavas of the Hawaiian Islands. *Geol. Soc. Amer. Bull. 75*, 107–127.

McDougall I. and Swanson D. A. (1972) Potassium-argon ages of lavas from the Hawi and Polulu volcanic series, Kohala Volcano, Hawaii. *Geol. Soc. Amer. Bull.* **83**, 3731–3738.

Moorbath S. and Walker G. P. L. (1965) Strontium isotope investigations of igneous rocks from Iceland. *Nature* **207**, 837–840.

Muir I. D. and Tilley C. E. (1963) Contributions to the petrology of Hawaiian basalts. II. The tholeiitic basalts of Mauna Loa and Kilauea. *Amer. J. Sci.* **261**, 111–128.

Muñoz M. (1969) Estudio petrológico de las formationes alcalinas de Fuerteventura (Isles Canarias) *Estud. Geol. Inst. Mallada* **25**, 257–310.

Murata K. J. and Richter D. H. (1966) The settling of olivine in Kilauean magma as shown by lavas of the 1959 eruption. *Amer. J. Sci.* **264**, 194–203.

Myers C. W., Bence A. E., Papike J. J., Ayuso R. A. (1975) Petrology of an alkali-olivine basalt sill from Site 169 of DSDP Leg 17: The Central Pacific Basin. *J. Geophys. Res.* **80**, 807–822.

Nabelek P. I. (1980) Nickel partitioning between olivine and liquid in natural basalts: Henry's Law behavior. *Earth Planet. Sci. Lett.* **48**, 293–302.

O'Nions R. K. and Grönvold K. (1973) Petrogenetic relationship of acid and basic rocks in Iceland: Sr isotopes and rare earth elements in late and post glacial volcanics. *Earth Planet. Sci. Lett.* **19**, 397–409.

O'Nions R. K. and Pankhurst R. J. (1973) Secular variations in the Sr isotope composition of Icelandic volcanic rocks. *Earth Planet. Sci. Lett.* **21**, 13–21.

O'Nions R. K., Pankhurst R. J. and Grönvold K. (1976) Nature and development of basalt magma sources beneath Iceland and the Reykjanes Ridge. *J. Petrol.* **17**, 315–338.

O'Nions R. K., Hamilton P. J. and Evensen N. M. (1977) Variations in $^{143}Nd/^{144}Nd$ and $^{87}Sr/^{86}Sr$ ratios in oceanic basalts. *Earth Planet. Sci. Lett.* **34**, 13–22.

Papike J. J., Cameron K. L. and Baldwin K. (1974) Amphiboles and pyroxenes: characterization of other than quadrilateral components and estimates of ferric iron from microprobe data. *Abstracts with Program,* Geol. Soc. Amer. **6**, 1053–1054.

Peck D. L., Moore J. G. and Kojima G. (1964) Temperatures in the crust and melt of Alae lava lake, Hawaii, after the August 1963 eruptions of Kilauea Volcano—a preliminary report in Geological Survey Research 1964. *U.S. Geol. Surv. Prof. Paper 501D,* p. D1–D7.

Peterman Z. E. and Hedge C. E. (1971) Related strontium isotopic and chemical variations in oceanic basalts. *Geol. Soc. Amer. Bull.* **82**, 493–499.

Powers H. A. (1955) Composition and origin of basaltic magma of the Hawaiian Islands. *Geochim. Cosmochim. Acta* **7**, 77–107.

Presnall D. C., Dixon S. A., Dixon J. R., O'Donnell T. H., Brenner N. L., Schrock R. L. and Dycus D. W. (1978) Liquidus phase relations on the join diopside-forsterite-anorthite from 1 atm to 20 kbar: their bearing on the generation and crystallization of basaltic magma. *Contrib. Mineral. Petrol.* **66**, 203–220.

Rice S., Langmuir C. H., Bender J. F., Hanson G. N., Bence A. E. and Taylor S. R. (1979) Basalts from DSDP Sites 417A,D; fractionated melts of a light rare earth depleted source. *Initial Reports of the Deep Sea Drilling Project* **51**, p.1099–1111. U.S. Govt. Printing Office, Washington, D.C.

Ridley W. I. (1970) The petrology of Las Canadas volcanoes, Tenerife, Canary Islands. *Contrib. Mineral. Petrol.* **26**, 124–160.

Roeder P. L. and Emslie R. F. (1970) Olivine-liquid equilibrium. *Contrib. Mineral. Petrol.* **29**, 275–289.

Sato M. and Wright T. L. (1966) Oxygen fugacities directly measured in magmatic gases. *Science* **153**, 1103–1105.

Schilling J. -G. (1973) Iceland mantle plume: geochemical evidence along Reykjanes Ridge. *Nature* **242**, 565.

Schilling J. -G., Anderson R. and Vogt P. (1976) Rare earth, Fe and Ti variations along the Galapagos spreading center and their relation to the Galapagos mantle plume. *Nature* **261**, 108–113.

Schilling J. -G., Unni C. K. and Bender M. L. (1978) The origin of chlorine and bromine in the oceans. *Nature* **273**, 631–636.

Schilling J. -G. and Winchester J. W. (1966) Rare earths in Hawaiian basalts. *Science* **153**, 867–869.

Schilling J. -G. and Winchester J. W. (1969) Rare earth contribution to the origin of Hawaiian lavas. *Contrib. Mineral. Petrol.* **23**, 27–37.

Scholl D. W. and Creager J. S. (1973) Geologic synthesis of Leg 19 (DSDP) results: Far North Pacific, Aleutian ridge, and Bering Sea. *Initial Reports of the Deep Sea Drilling Project* **19**, p. 897–913. U.S. Govt. Printing Office, Washington, D.C.

Schweitzer E. L., Papike J. J. and Bence A. E. (1979) Statistical analysis of clinopyroxenes from deep-sea basalts. *Amer. Mineral.* **64**, 501–513.

Shand S. J. (1933) The lavas of Mauritius. *Quart. J. Geol. Soc. London* **89**, 1–13.

Shaw H. R. (1972) Viscosities of magmatic silicate liquids: an empirical method of prediction. *Amer. J. Sci.* **272**, 870–893.

Shaw H. R., Wright T. L., Peck D. L. and Okamura R. (1968) The viscosity of basaltic magma: an analysis of field measurements in Makaopuhi lava lake, Hawaii. *Amer. J. Sci.* **266**, 225–264.

Shaw H. R., Jackson E. D. and Bargar K. E. (1980) Volcanic periodicity along the Hawaiian-Emperor chain. *Amer. J. Sci.* **280A**, 667–708.

Subbarao K. V., Reddy V. V., Hekinian R. and Chandrasekharam D. (1977) Large ion lithophile elements and Sr and Pb isotopic variation in volcanic rocks from the Indian Ocean. In *Indian Ocean Geology and Biostratigraphy*, (J. R. Heirtzler, H. M. Bolli, T. A. Davies, J. B. Saunders and J. G. Sclater, eds.), p. 259–278. Amer. Geophys. Union, Washington, D.C.

Thompson R. N. and Tilley C. E. (1976) Melting and crystallization relations of Kilauean basalts of Hawaii—The lavas of the 1959–60 Kilauea eruption. *Earth Planet. Sci. Lett.* **5**, 469–477.

Tilley C. E. (1960) Differentiation of Hawaiian basalts; some variants in lava suites of dated Kilauea eruptions. *J. Petrol.* **1**, 47–55.

Tilley C. E. and Scoon G. H. (1961) Differentiation of Hawaiian basalts; trends of Mauna Loa & Kilauean historic magmas. *Amer. J. Sci.* **259**, 60–68.

Tilley C. E., Yoder H. S., Jr. and Schairer J. F. (1965) Melting relations of volcanic tholeiite and alkali rock series. *Carnegie Inst. Wash., Yearb.* **64**, 69–82.

Wager L. R. A. and Mitchell R. L. (1953) Trace elements in a suite of Hawaiian lavas. *Geochim. Cosmochim. Acta* **3**, 217–223.

White W. M., Tapia M. D. M. and Schilling J. -G. (1979) The petrology and geochemistry of the Azores Islands. *Contrib. Mineral. Petrol.* **69**, 201–213.

Wilkinson J. F. G. (1968) The petrography of basaltic rocks. In *Basalts.* (H. Hess and A. Poldervaart, eds.) **1**, p. 163–214. Wiley Interscience, N.Y.

Wilson J. T. (1963) A possible origin of the Hawaiian Islands, *Can. J. Phys.* **41**, 863–870.

Worsley J. R. (1973) Calcareous nanofossils: Leg 19 of the Deep Sea Drilling Project. *Initial Reports of the Deep Sea Drilling Project* **19**, p. 741–750. U.S. Govt. Printing Office, Washington, D.C.

Wright T. L. (1971) Chemistry of Kilauea and Mauna Loa lava in space and time. *U.S. Geol. Survey Prof. Paper 735.* 40 pp.

Wright T. L. (1973) Magma mixing as illustrated by the 1959 eruption, Kilauea Volcano, Hawaii. *Geol. Soc. Amer. Bull.* **84**, 849–858.

Wright T. L. and Fiske R. S. (1971) Origin of the differentiated and hybrid lavas of Kilauea Volcano, Hawaii. *J. Petrol.* **12**, 1–65.

Wright T. L. and Okamura R. T. (1977) Cooling and crystallization of tholeiitic basalt, 1965 Makaopuhi lava lake, Hawaii. *Geol. Survey Prof. Paper 1004.* 78 pp.

Wright T. L., Peck D. L. and Shaw H. R. (1976) Kilauea lava lake: Natural laboratories for study of cooling, crystallization, and differentiation of basaltic magma. *Amer. Geophys. Union Monograph 19*, p. 375–390.

Wright T. L., Swanson D. A. and Duffield W. A. (1975) Chemical compositions of Kilauea east-rift lava, 1968–1971. *J. Petrol.* **16**, 110–113.

Yoder H. S. and Tilley C. E. (1962) Origin of basalt magmas: An experimental study of natural and synthetic rock systems, *J. Petrol.* **3**, 342–532.

1.2.7 ISLAND ARC BASALTS

1.2.7.1 Introduction

Basalts form an important component of "island arc" volcanic suites. The most important distinction of these from other suites of basaltic rocks lies in their tectonic setting, being associated with the great suite of "orogenic" lavas erupted at convergent plate boundaries. Two subsets may be distinguished: "island arc" and "continental margin" volcanic suites. It is not clear that any fundamental genetic difference exists between these two types. There is overlap in all chemical and mineralogical parameters. These voluminous volcanic rocks, which comprise andesites, dacites and rhyolites in addition to basalts (Fig. 1.2.7.1), are second in volumetric importance only to Mid-Ocean Ridge Basalts (MORB) in the terrestrial volcanic environment. This suite of "orogenic" lavas is widely held to be a principal source of new material for continental growth (e.g., Taylor, 1979), and accordingly forms a crucial stage in hypotheses of the overall geochemical evolution of the Earth.

In this section, we are concerned only with the common basaltic members of the suite (Fig. 1.2.7.1). The place of these basaltic members in the overall genetic scheme for the production of the suite is not yet clearly understood. Whether they are parental to the rest of the suite, or are the most mafic members of a continuum of lavas produced by partial melting remains as an open question. They show similarities to the other basalt types discussed in this chapter, but a combination of chemical and isotopic criteria serves to separate them. They are most closely related to basalts from marginal or back-arc basins.

The available petrological, chemical and isotopic data on these rocks are widely scattered and there are few rocks for which a complete set of data is available. For this reason, we have chosen to discuss first the Island Arc Basalt Reference Suite, hereafter referred to as the reference suite, which provides a comprehensive set of petrological, chemical and isotopic data. This information forms a framework within which to fit the less complete data from other areas.

1.2.7.2 Island arc basalt reference suite

Introduction

The New Britain island arc of central Papua New Guinea (Figs. 1.2.7.2–3) has several features in common with other island arcs, and may be taken as a suitable example of an island arc that has produced basaltic magmas.

New Britain has a well-defined Benioff zone and submarine trench; it is therefore regarded as an area where two lithospheric plates have converged, one being consumed beneath the other. In addition, unlike continental-margin volcanic arcs, there is an active marginal basin in the Bismarck Sea to the north, behind the New Britain arc (the nearest continental material is 500 km to the southwest). Several New Britain volcanoes are active or potentially active. Rock types are

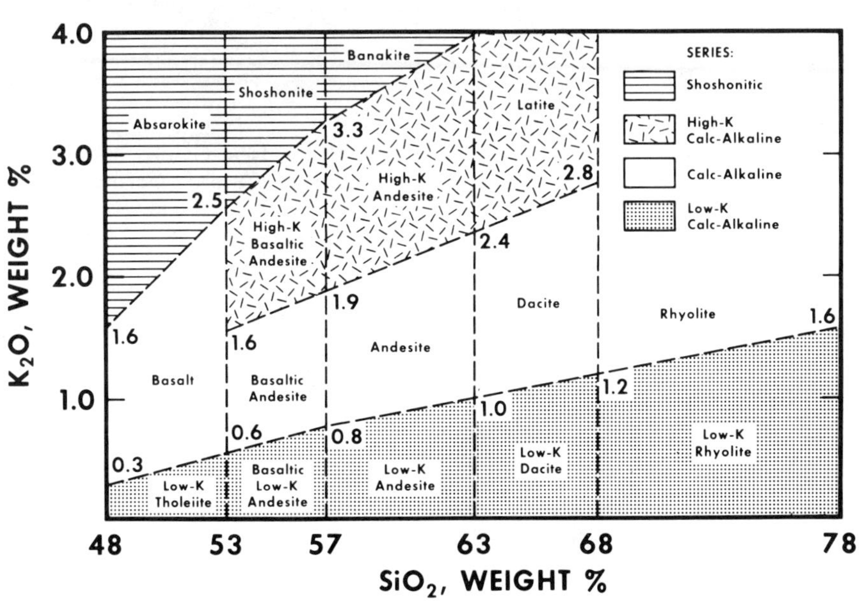

1.2.7.1 Classification scheme for island arc volcanic rocks.

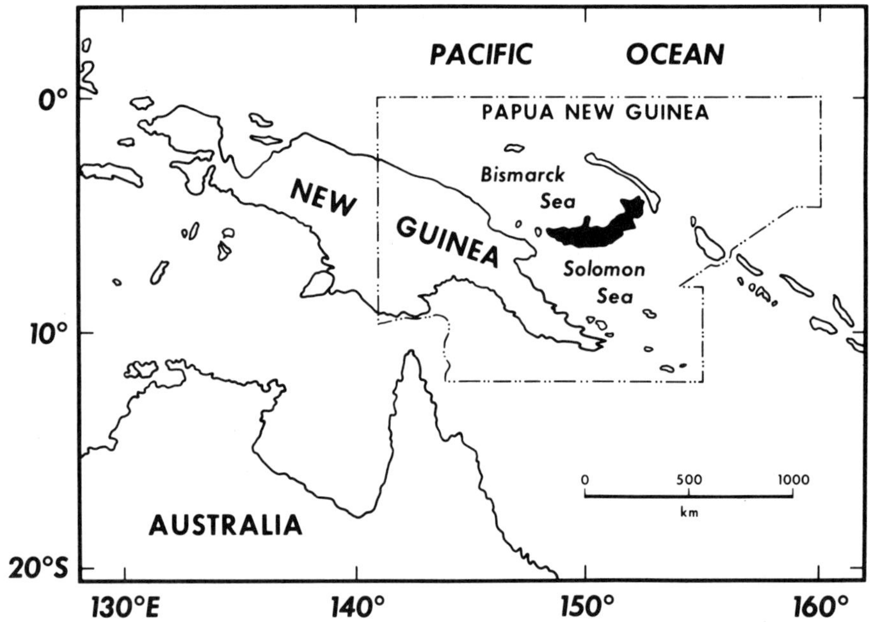

1.2.7.2 Papua New Guinea locality map. New Britain shown solid.

basalt, andesite, dacite, and rhyolite, forming a suite typical of volcanism near convergent plate margins.

The volcanoes of New Britain and its off-shore islands have been comprehensively surveyed by the Australian Bureau of Mineral Resources, in conjunction with the Geological Survey of Papua New Guinea, as part of a wider study of late Cenozoic volcanoes in the Bismarck Sea area. More than 200 rock samples from New Britain volcanoes have been analyzed (Johnson and Chappell, 1979) so a substantial data base exists for the selection of a suite of basaltic rocks. A synthesis of the geology and tectonic setting of the volcanoes and an evaluation of their major element chemistry were given by Johnson (1977).

Tectonic setting

The New Britain Benioff zone dips at about 70°, and is regarded as part of the Solomon Sea plate which

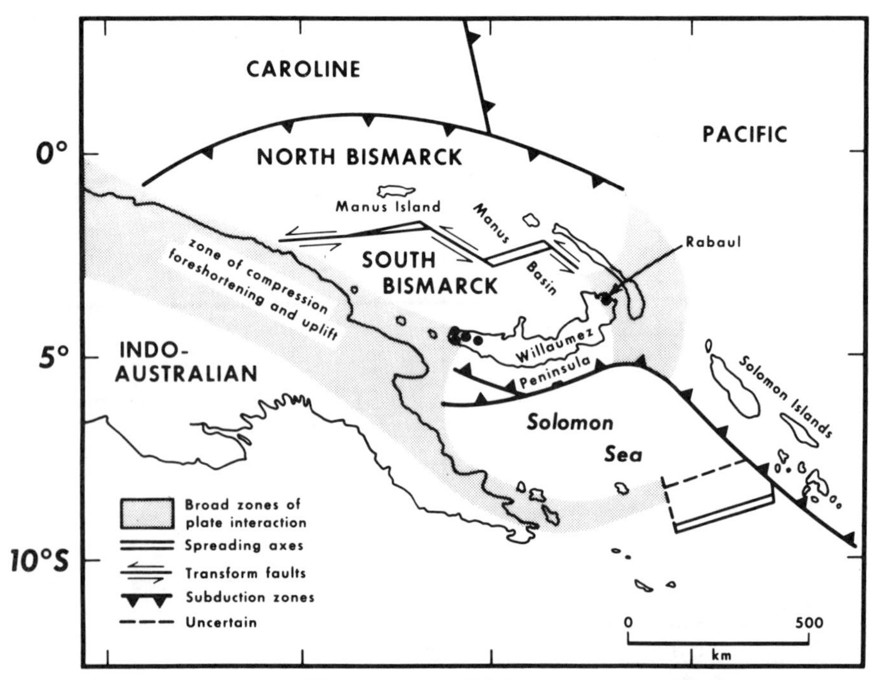

1.2.7.3 Plate configuration in Papua New Guinea. Four minor plates—Caroline, North Bismarck, South Bismarck, and Solomon Sea—are thought to exist between the major Pacific and Indo-Australian plates, but there is relatively little seismicity at the northern margin of the North Bismarck plate and between the Caroline and Pacific plates.

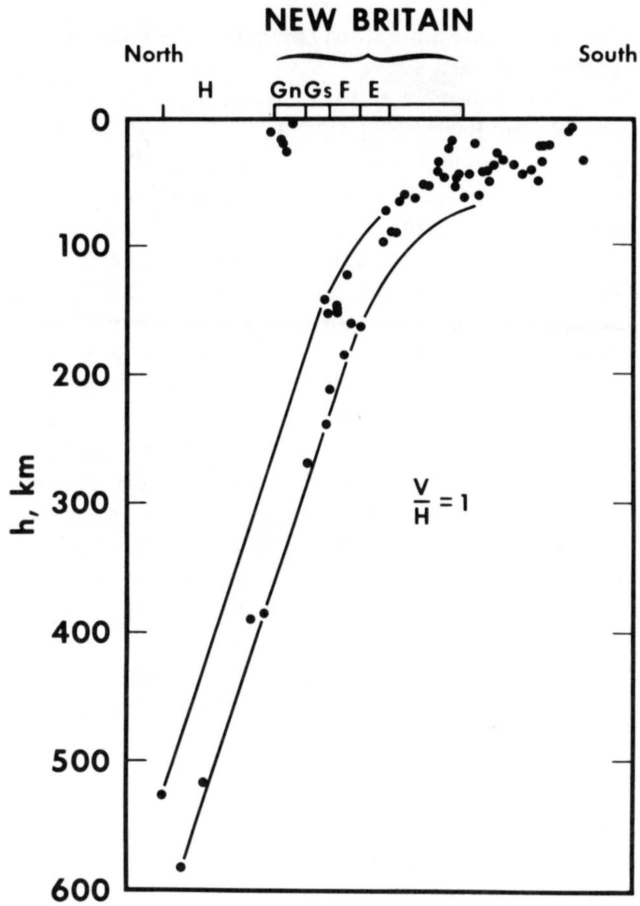

1.2.7.4 Benioff zone beneath central New Britain, showing limits of zones E-to-H defined in Fig. 1.2.7.5 (after Johnson, 1977, Fig. 4).

has been subducted northwards beneath the South Bismarck plate (Figs. 1.2.7.3–4). Both plates are regarded as minor pieces of lithosphere which are trapped, and whose motions are probably governed by the larger Indo-Australian and Pacific plates which, in the Papua New Guinea region, are thought to be converging at 10.7 cm/yr on an azimuth of N255° E (Le Pichon, 1970). Estimates for the present-day rate of convergence of the Solomon Sea and South Bismarck plates are 9.2 cm/yr (azimuth N343° E; Johnson and Molnar, 1972) and 12.0 cm/yr (azimuth N352° E; Curtis, 1973). Krause (1973) argued that the pole of rotation for the South Bismarck/Solomon Sea plate boundary is nearby, and therefore that rates of plate convergence increase along the plate boundary from 6.2 cm/yr in the southwest to 12.4 cm/yr in the northeast. The submarine trench south of New Britain also deepens in this direction to more than 8000 m below sea level.

East of longitude 150° E the submarine trench is a single valley, but to the west it splits into two distinct valleys, the deeper, more southerly one being less than 7000 m deep (Fig. 1.2.7.3). The ridge between the two valleys coincides with a distinct gravity low (Finlayson *et al.*, 1977), and may represent a thrust slice dipping northwards beneath western New Britain (Johnson, 1977).

Late Cenozoic volcanoes on New Britain may be divided into four groups. These include (1) the prominent chain of volcanoes making up Willaumez Peninsula which extends northwards for about 60 km from the central-north coast of New Britain, and (2) the

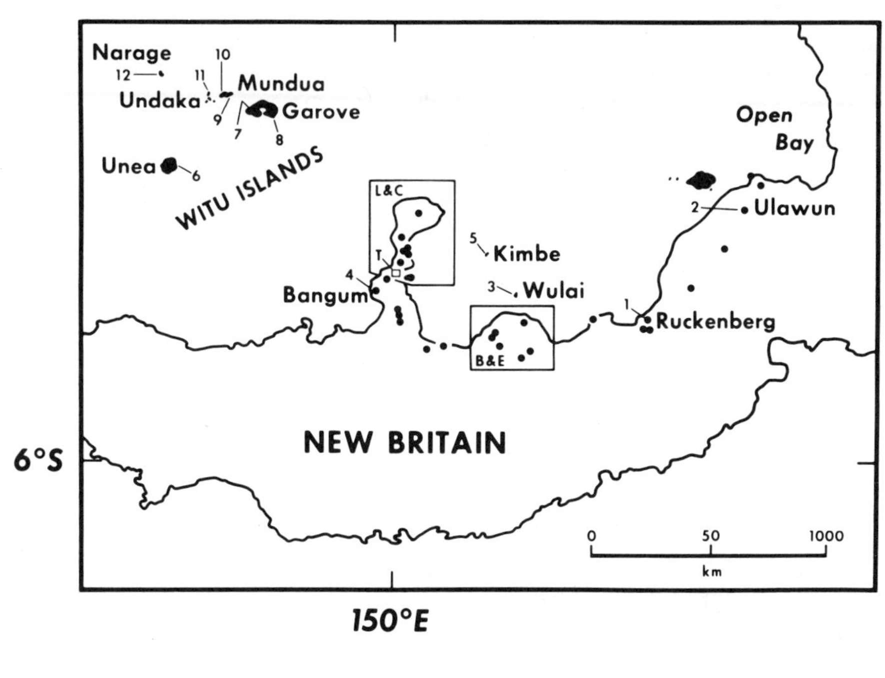

1.2.7.5 Main volcanic centers of central New Britain (solid circles) and Witu Islands, showing localities of Island Arc Basalt Reference Suite samples (numbers 1–12). Island volcanoes shown solid. Rocks from area L & C were described by Lowder and Carmichael (1970) and Lowder (1970); Settlement at Talasea shown by T. Rocks from area B & E were described by Blake and Ewart (1974).

arcuate belt of volcanoes extending eastwards from the southern end of Willaumez Peninsula as far as Open Bay (Fig. 1.2.7.5). A third group, represented by the volcanic complex at Rabaul (including Watom Island) on the extreme northeastern tip of New Britain (Fig. 1.2.7.3; Heming, 1974), appears to have developed independently of those to the southwest, and therefore is not considered to be part of the volcanic province discussed here (Johnson *et al.*, 1978a). Similarly, the fourth group comprising volcanoes at the extreme western end of New Britain (Fig. 1.2.7.2) is also excluded, because they are probably related to the Indo-Australian/South-Bismarck plate boundary. The Witu Islands north of western New Britain constitute another group, and are considered to belong to the same province as the volcanoes of Willaumez Peninsula and the central-north coast of New Britain, because they are part of the progressive pattern of chemical change acoss the New Britain Benioff zone.

The distinctive pattern of the Willaumez Peninsula and central-north coast volcanoes corresponds closely to the features of the submarine trench. The absence of volcanoes immediately west of the southern end of Willaumez Peninsula corresponds with the postulated existence of the thrust slice, which Johnson (1977) speculated may have acted as a supplementary heat sink to the downgoing slab, lowering temperatures in the overlying mantle where primary magmas may otherwise have formed. The extrapolation of the trend of Willaumez Peninsula southwards intersects the New Britain trench at a point where it splits into the two western valleys. The peninsula may overlie a zone where to the west the thrust slice moves beneath the South Bismarck plate more slowly than does the Solomon Sea plate to the east.

The eastern part of the Bismarck Sea is known as the Manus Basin (Fig. 1.2.7.3), and is regarded as an active marginal basin younger than the floor of the western Bismarck Sea. Its inferred spreading directions are northwestwards and southeastwards away from part of the zone of earthquakes that marks the northern margin of the South Bismarck plate (Taylor, 1979). The southwestern edge of the Manus Basin is defined by the Willaumez-Manus Rise which extends from Willaumez Peninsula northwestwards to Manus Island. This rise is thought to represent a tectonic upwarp of the eastern edge of the older seafloor of the Western Bismarck Sea. It is unclear to what extent the back-arc tectonic history has influenced the distribution of volcanoes in New Britain, the Willaumez Peninsula and the Witu Islands (Johnson and Arculus, 1978; Johnson *et al.*, 1979).

Some of the chemistry of the Witu Islands samples (e.g., Ba-REE) resembles that of back-arc basins (see discussion below) and this part of the reference suite appears to be transitional between island arc and back-arc basin volcanic suites.

Sample selection

Twelve basalt samples (SiO_2 less than 53 wt.%) have been selected from the 210 analyzed rocks from the New Britain arc.

In order that intra-arc comparisons of rock compositions can be made, the volcanoes have been divided into four zones, E, F, G and H; and zone G has been subdivided into two groups—G-southern (Gs) and G-northern (Gn)—each of which is represented by far

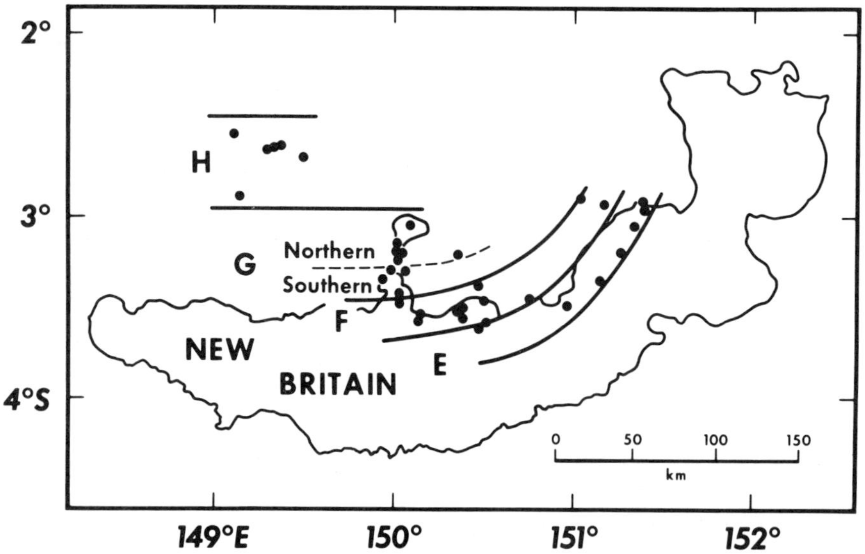

1.2.7.6 Volcanoes (solid circles) of central New Britain and Witu Islands grouped into zones E-to-H (after Johnson, 1977, Fig. 10).

fewer analyzed samples than are zones E, F and H (Johnson, 1977). The zones from E to H overlie progressively deeper parts of the New Britain Benioff zone (Fig. 1.2.7.6).

In common with the analyzed rocks of a number of island arcs, basalts are generally poorly represented in the New Britain data base, compared with andesites and other rocks containing more than 53% SiO_2. Basalts are well represented in zone H, and on Mount Ulawun in zone E, but elsewhere they appear to be quite rare. Zone Gs, for example, is represented by only one basaltic sample—a xenolith from an andesite lava flow.

The twelve basalt samples were chosen so that they covered as uniformly as possible the total range of depths to the Benioff Zone (h values). This procedure favors zone H where the h range is greatest (Fig. 1.2.7.4) and there is a further bias towards zone H because of the significantly larger proportion of basalts found there. The result of this approach is the selection of a wide range of basalt types.

The probability that the twelve rocks are not statistically representative of the true proportions of basalt types in the New Britain arc cannot be overstressed. The wide range of the twelve samples probably includes many of the basalt types found in other arcs, but no alkaline rocks are represented, and none is high in potassium. Some zone-H basalts have compositional affinities with the basalts of marginal basins (Johnson and Arculus, 1978).

Thus the reference suite includes not only examples of "true" island arc basalts, but also representatives of back-arc or marginal basin basalts.

Locality descriptions for the twelve samples are listed in Table 1.2.7.1. Sample IA-12 is oxidized, but is included because it is the only analyzed rock from Narage Island, which is the most northerly of the volcanoes in the arc.

Mineralogy

The twelve basalt samples range from strongly porphyritic to aphyric, and their groundmasses are microcrystalline or glassy. A relatively restricted mineralogy includes olivine, low and high-Ca pyroxenes, plagioclase, spinel, and magnetite (Table 1.2.7.2). No hydrous phases are present. Glomerocrysts are characteristic, and on the basis of microprobe data appear to be mostly cognate accumulations. Groundmass and interstitial glomerocryst glass contains rapidly-grown silicates and oxides with generally skeletal shapes. Four out of the five basalts from zones E to Gn contain orthopyroxene or pigeonite, including the sample from Gn with a Mg′ of 0.69. [$Mg' = Mg/(Mg+Fe^{2+})$, where $Fe^{2+} = 0.9$ total Fe.] In contrast, only one of the samples (IA-7) from zone H contains orthopyroxene. The presence of low-Ca pyroxene is in fact typical of many of the basalts of the New Britain arc.

The restricted number of samples does not permit extended critical comparison of zonal variations in

Table 1.2.7.1 Locality descriptions for twelve New Britain basalts — the Island Arc Basalt Reference Suite.

I A Numbers	BMR Numbers	Johnson & Chappell Numbers	Grid Reference
1	51NG0217F	E2/1	KU721953
2	51NG0065B	E5/11	LV122401
3	51NG0239	F5/2	KV219079
4	51NG2707A	Gs2/1	HQ216093
5	51NG0246	Gn2/1	KV090238
6	48NG0022	H1/1	H485615
7	48NG0036	H2/3	H780851
8	48NG0572	H2/1	H886797
9	48NG0592	H3/6	H691894
10	48NG0605	H3/7	H668905
11	48NG0038A	H3/3	H610905
12	48NG0048A	H3/1	H405987

Field descriptions:
1. Boulder on western flank of Mount Ruckenberg, near Sulu Range. **2.** Lava flow at 920 m above sea level in stream valley running westwards from beneath the east-west escarpment on the southern flank of Mount Ulawun. **3.** Lava flow on headland on northern coast of Wulai Island **4.** Block in lava flow at Cape Schellong on western flank of Mount Bangum. **5.** Boulder on western coast of Kimbe Island. **6.** Boulder on northern side of islet (Duaga Island) off the east-northeastern coast of Unea Island. **7.** Lava flow on headland on northeast side of Papua Bay on western coast of Garove Island. **8.** Block from volcanic rudite deposit at headland (Island Cape) on southeastern coast of Garove Island. **9.** Lava flow on southern coast of Mundua Island. **10.** Lava flow on northern coast at western end of Mundua Island. **11.** Block from volcanic rudite deposit at southeastern end of Undaka Island. **12.** Lava flow on northwestern coast of Narage Island.
Letters of Johnson and Chappell (1979) numbers correspond with zones defined in Fig. 1.2.7.5; BMR = Bureau of Mineral Resources, Australia.

Table 1.2.7.2 Compositional ranges of minerals (phenocryst and groundmass) found in the New Britain Island Arc Basalt Reference Suite.

IA Number	Zone	Mg′	Oliv(Fo)	Cpx			Opx-Pig			Plag(An)	Spinel
				Ca	Mg	Fe	Ca	Mg	Fe		
1	E	0.68	86–79	36–43	49–53	6–11	5–10	69–56	26–34	81–71	TiMt
2	E	0.51	80–69	42–37	46–43	12–20	8–12	58–63	29–30	91–67	CrMt–TiMt
3	F	0.55	76–66	39–42	39–44	13–22	2–4	69–73	25–28	88–78	AlMt
4	G_S	0.59	67	37–46	39–51	10–18	3–9	65–72	24–27	75–69	TiMt
5	G_n	0.69	77–70	36–43	42–48	12–17	3–4	62–64	32–34	70–64	TiMt
6	H	0.49	78–06	21–44	30–52	9–46				85–12	TiMt
7	H	0.51	74–67	27–44	39–53	10–23				90–61	TiMt
8	H	0.64	86–82	32–46	44–51	8–19				84–74	TiMt
9	H	0.68	89–80	40–47	45–51	8–11				84–61	CrAlSp
10	H	0.60	84–81	37–46	45–50	7–18				89–64	TiMt
11	H	0.70	87–80	34–43	40–52	10–23				82–56	CrAlSp
12	H	0.55	86–59	39–45	40–48	11–20				89–59	AlMt

Abbreviations: Oliv(Fo) refers to mole % forsterite; Cpx refers to mole % Ca, Mg and Fe (respectively) in clinopyroxene; Opx-Pig refers to mole % Ca, Mg and Fe (respectively) in orthopyroxene and pigeonite; Plag(An) refers to mole % anorthite; TiMt = titaniferous magnetite; CrMt = chromian magnetite; AlMt = aluminous magnetite; CrAlSp = Chrome-alumina-rich spinel.

A

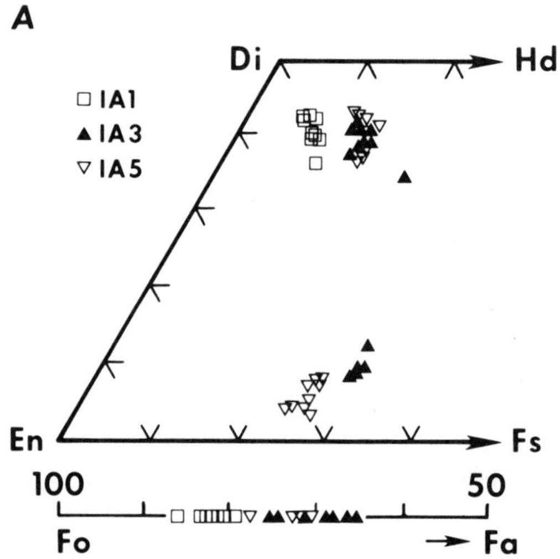

mineralogy. It seems significant, however, that orthopyroxene is rare in the basalts of the Witu Islands (Zone H) and only present in basalts of relatively low Mg′.

The compositional variations of the analyzed minerals are given in Table 1.2.7.2. Figures 1.2.7.7a and b show projected compositions where dramatic or significant ranges occur. Olivine is present in all of the samples, typically as phenocrysts. In some samples groundmass olivine is also present, but in sample IA-4, olivine is found only as a microscopic intergrowth with plagioclase. Olivine compositions do not always correspond to the appropriate Mg′ of the bulk rock using the empirical Roeder-Emslie (1970) distribution coefficient (see section 1.4.2 and Chapter 3). For example, the olivine phenocrysts in sample IA-12 are more forsteritic

B

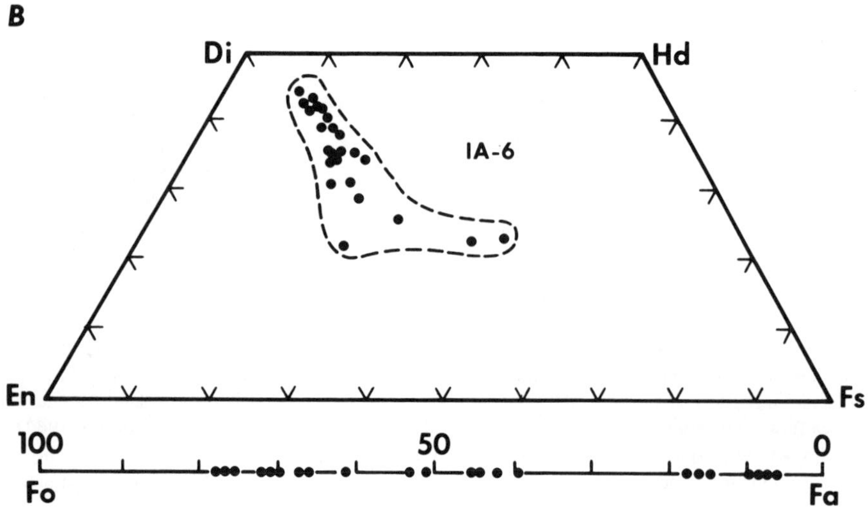

1.2.7.7 Pyroxene and olivine compositions projected in terms of Ca-Mg-Fe.

than would be expected and the converse is true for the aphyric sample IA-5 where the olivine is Mg-poor. A remarkable range (Fo_{78}–Fo_6) of olivine compositions is present in sample IA-6 (see also Fig. 1.2.7.7b; the fayalitic compositions are for groundmass crystallites). In the same rock, there is an extensive range of calcic to subcalcic-augite compositions, and alkali feldspar is present. No orthopyroxene has been detected.

Except in sample IA-6, most clinopyroxenes have a restricted range in Ca, Mg, and Fe, but typically are zoned in Al and Ti from low Al-Ti cores (1.5 wt.% Al_2O_3, 0.3 wt.% TiO_2) to high Al-Ti rims (7 wt.% Al_2O_3, 2 wt.% TiO_2). Oscillatory and sector-zoned phenocrysts are prominent in some zone-H basalts. The presence or absence of orthopyroxene or pigeonite (or both) is not a simple function of Mg′, or FeO or CaO contents. For example, sample IA-4 could be considered more evolved (in terms of Mg′) than sample IA-5, yet IA-4 contains only orthopyroxene whereas IA-5 contains both orthopyroxene and pigeonite. The presence of primary (not inverted pigeonite) orthopyroxene in the low-Mg′ rock could be attributed to a relatively low temperature of crystallization in comparison with IA-3 and IA-5. Low-Ca pyroxenes also show restricted Mg-Fe compositional variation.

Plagioclase is oscillatory zoned, and some phenocrysts are riddled with inclusions (clinopyroxene and glass), or are resorbed. Anorthite-rich phenocrysts characteristic of island arc (and some MOR basalts) are present. However, plagioclase in the samples of zones Gs and Gn is significantly An-poor relative to the normative plagioclase content and overall mineralogy of the bulk rocks. Inconsistencies in terms of unexpectedly An-rich phenocrysts in otherwise relatively evolved basalts are observed in zone-H basalts. Some glomerocryst plagioclase in sample IA-10 is compositionally distinct (An_{89-84}) from phenocryst plagioclase (An_{80-64}), although the ferromagnesian silicates are similar in composition.

The oxides are predominantly titaniferous magnetites. They are present as phenocrysts, inclusions, and in the groundmasses of most samples. Cr-Al-rich spinels are found as inclusions in olivine, clinopyroxene, and plagioclase of samples IA-9 and IA-11. Systematically higher Cr/(Cr+Al) ratios of spinels included in clinopyroxene (0.47–0.58) compared to olivine (0.40–0.46) have been detected in IA-9. High Al contents of magnetite in samples IA-4 and IA-12 (5–9 wt.% Al_2O_3) may be the result of either crystallization at high (10 kb) pressure (Osborne and Watson, 1977), or at relatively high $a^{melt}_{Al_2O_3}$, for sample IA-12 in particular. Projected compositions are presented in Fig. 1.2.7.8 for spinels and magnetite for zone-H basalts.

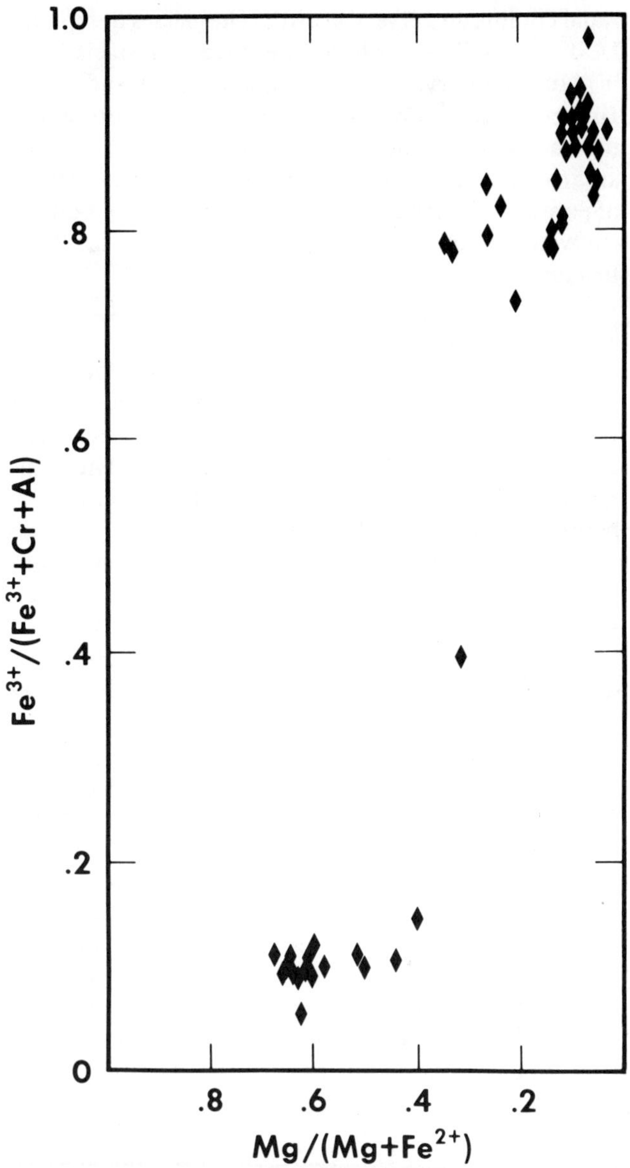

1.2.7.8 Compositions of spinel in zone H basalts. Magnetite plots at top right-hand corner and Cr-Al-rich spinels near the base.

The groundmass and glomerocryst glasses are typically SiO_2-rich, ranging from high-SiO_2 andesite to dacite and rhyolite. In sample IA-4, rhyolite groundmass glass with 7.1 wt.% K_2O and 5.5 wt.% Na_2O is present, and feathery, rapidly-grown plagioclase is distributed throughout it. High-SiO_2 andesitic to dacitic glass is found in the groundmass and in a glomerocryst of sample IA-7, and dacitic to rhyolitic glass with variable K/Na ratio is present in sample IA-10. Feathery plagioclase and stacked-arrowhead magnetite indicate rapid cooling and crystal growth. The glasses in samples IA-4 and IA-7 appear to be late-stage mesostasis.

In summary, a simple mineralogy is indicated. However, the compositional variations observed, together with evidence of disequilibrium crystallization, are inadequately understood in terms of the variations of parameters such as pressure, temperature and different volatile fugacities (H_2O, H_2, O_2, CO_2) for New Britain.

Major element chemistry

The data for these elements are given in Table 1.2.7.3. As noted above, an attempt was made to provide a wide spectrum of compositions. The mean SiO_2 content is 52 ± 1.7 wt.% (range, 47.5–53.1%). The highest value slightly exceeds the nominal division of 53% between basalts and basaltic andesites.

There is a wider spread in Al_2O_3 content, from 14.4–19.5% (mean = 17.3 ± 1.6%). All limits quoted with means are 1 sigma values. TiO_2 values are low and typical of island arc rocks, ranging from 0.30–1.43% (mean value of 0.83 ± 0.32%). The rocks are highly oxidized, which is typical of many island arc rocks. A fixed value of 0.90 for the ratio Fe^{2+}/ total Fe has been used to calculate Mg/(Mg+Fe^{2+}) values (Brooks, 1976). MgO values and Mg′-values fall into two groups. Samples IA-1, IA-5, IA-9 and IA-11 comprise a high-Mg group, with a tight clustering of Mg′-values (0.69–0.71), whereas values for the other samples range from 0.51–0.65. Total-iron is relatively constant, averaging 7.44 ± 1.1%. Representatives of both "tholeiitic" and "high-Al" basalts, as defined by Kuno (1960) using Al_2O_3 - (Na_2O + K_2O) relationships, are found in this suite; samples IA-2, IA-3, IA-6, IA-7, IA-9, IA-10 and IA-12 are high-Al basalts, and also are members of the low MgO-Ni-Cr group. The major element chemistry for the basalt suite shows mild correlations with depth to the Benioff zone (h), particularly for K, Na, Ti and P which show an increase with h.

Trace element chemistry

Large cations. Data for Cs, Rb, Ba, Pb, and Sr, and for the major elements K, Ca and Na are given in Table 1.2.7.4. K/Rb ratios are high, averaging 583 ± 64, excluding sample IA-12. The total range in values is lower than that observed in fresh MORB samples (800–3000; Hart *et al.*, 1972; Tarney *et al.*, 1977; Gill, 1976b; Hawkins, 1977). Both K (0.32 ± 0.18%) and Rb (5.7 ± 3.2 ppm) are much higher than typical MORB values, but are generally similar to those in samples from marginal basins and in some "plume" MORB (Tarney *et al.*, 1977; Sun *et al.*, 1979; Saunders and Tarney, 1979; Weaver *et al.*, 1979). Cesium (mean 0.14 ppm) has the same characteristics as K and Rb. Barium is notably higher in the island arc suite (63 ± 30 ppm) than in MORB samples (10–20 ppm), but is closer to the values in marginal basin rocks (Tarney *et al.*, 1977; Hart *et al.*, 1972; Hawkins, 1977). K/Ba ratios (50 ± 15) are less uniform than are the K/Rb values.

Rare earth elements. These elements (Table 1.2.7.5; Figs. 1.2.7.9–10) have a wide range of concentrations from about 3–20 times chondritic abundances, except in sample IA-12. With the exception of sample IA-12, all the patterns are subparallel to those of chondrites [La_c/Yb_c = 1.21 ± 0.54 (c = chondrite normalized)].

Table 1.2.7.3 Major elements in the New Britain Island Arc Basalt Reference Suite.

	IA-1	IA-2	IA-3	IA-4	IA-5	IA-6	IA-7	IA-8	IA-9	IA-10	IA-11	IA-12
SiO_2	53.1	52.32	52.5	50.9	49.2	50.4	52.6	49.7	49.9	51.7	49.2	47.5
TiO_2	0.30	0.76	0.49	0.52	0.52	0.80	1.43	0.85	1.01	0.99	1.15	0.95
Al_2O_3	14.7	18.66	17.9	15.6	15.3	19.5	16.3	16.5	16.6	16.9	14.4	18.4
Fe_2O_3	1.91	3.22	4.05	5.20	2.55	2.70	1.80	1.55	2.30	2.95	1.40	6.20
FeO	6.90	6.05	5.45	4.85	6.70	6.60	8.45	7.65	6.30	5.75	8.10	3.80
MnO	0.16	0.17	0.16	0.18	0.18	0.16	0.20	0.18	0.17	0.18	0.18	0.19
MgO	9.20	4.69	5.60	6.90	10.1	4.45	5.30	7.95	9.15	6.45	10.8	5.80
CaO	10.9	10.60	10.2	11.7	13.0	11.2	9.80	12.9	11.1	10.9	11.4	12.6
Na_2O	1.58	2.48	2.15	1.69	1.51	2.65	3.00	2.00	2.40	2.50	2.15	2.30
K_2O	0.22	0.30	0.31	0.61	0.17	0.56	0.40	0.20	0.21	0.74	0.17	0.67
P_2O_5	0.03	0.08	0.08	0.11	0.06	0.15	0.16	0.08	0.10	0.18	0.11	0.37
H_2O+	0.30	0.33	0.41	0.82	0.60	0.54	0.04	0.01	0.17	0.35	0.28	0.37
H_2O-	0.22	0.07	0.16	0.48	0.23	0.18	0.20	0.33	0.33	0.15	0.20	0.39
CO_2	0.04	0.05	<0.05	0.05	<0.05	0.05	0.05	0.05	<0.05	0.05	0.05	0.05
rest	0.14	0.13	0.16	0.18	0.19	0.16	0.14	0.13	0.18	0.16	0.21	0.20
SUM	99.70	99.91	99.62	99.80	100.31	100.10	99.87	100.08	99.92	99.95	99.80	99.79
Mg/Mg + Fe	0.68	0.51	0.55	0.59	0.69	0.49	0.51	0.64	0.68	0.60	0.70	0.55

Table 1.2.7.4 Large cations in the New Britain Island Arc Basalt Reference Suite.

	Method	IA-1	IA-2	IA-3	IA-4	IA-5	IA-6	IA-7	IA-8	IA-9	IA-10	IA-11	IA-12
Cs	1	0.07	0.3	0.09	0.16	—	0.18	0.16	0.08	0.06	0.16	—	—
Rb	2	2.6	4.6	4.2	8.6	—	8.6	6.2	3.0	3.0	12.6	2.4	6.6
K%	3	0.18	0.25	0.25	0.51	0.14	0.47	0.33	0.17	0.17	0.62	0.14	0.56
Ba	1	40	66	86	132	33	77	62	37	31	76	38	81
Pb	1	2.6	3.3	1.9	3.4	1.4	3.1	2.5	1.1	1.4	3.2	1.9	2.9
Sr	2	167	297	403	560	228	520	213	211	189	346	187	730
Ca%	—	7.79	7.58	7.29	8.36	9.29	8.00	7.00	9.22	7.93	7.79	8.15	9.00
Na%	—	1.17	1.84	1.60	1.25	1.12	1.97	2.23	1.48	1.78	1.85	1.60	1.71
K/Rb	—	692	534	605	594	—	679	531	564	567	488	575	847
Rb/Cs	—	37	15	47	54	—	48	39	38	50	79	—	—
Rb/Sr	—	0.016	0.015	0.010	0.015	0.005	0.017	0.02	0.014	0.016	0.036	0.013	0.009
Sr/Eu	—	642	594	876	788	430	611	157	310	222	384	191	518
K/Ba	—	45	37	30	39	42	61	53	46	55	81	36	69

Note: Data in ppm except where indicated in wt.%.
Method 1 = SSMS; 2 = XRF, 3 = AAS.

Table 1.2.7.5 Rare-earth elements in the New Britain Island Arc Basalt Reference Suite.

	Method	IA-1	IA-2	IA-3	IA-4	IA-5	IA-6	IA-7	IA-8	IA-9	IA-10	IA-11	IA-12
La	1	0.74	1.37	2.10	4.76	1.62	5.22	5.26	2.89	2.72	5.16	3.97	21.6
Ce	1	2.05	3.7	5.39	11.6	4.49	13.5	14.5	7.53	7.97	13.4	12.1	49.8
Pr	1	0.31	0.55	0.73	1.55	0.68	1.82	2.26	1.10	1.25	1.87	1.88	6.30
Nd	1	1.68	3.09	3.75	7.63	3.72	9.00	12.3	5.75	6.70	9.65	9.33	26.1
Sm	1	0.60	1.06	1.16	1.97	1.21	2.48	3.83	1.83	2.18	2.64	2.70	4.85
Eu	1	0.26	0.50	0.46	0.71	0.53	0.85	1.36	0.68	0.85	0.90	0.98	1.41
Gd	1	0.86	1.49	1.36	2.28	1.70	2.64	4.60	2.39	2.92	2.98	3.23	3.77
Tb	1	0.16	0.28	0.24	0.34	0.32	0.47	0.81	0.41	0.52	0.53	0.59	0.53
Dy	1	1.01	1.83	1.59	2.06	2.04	3.13	5.31	2.90	3.36	3.50	3.88	3.21
Ho	1	0.24	0.42	0.33	0.45	0.47	0.70	1.22	0.66	0.76	0.78	0.84	0.65
Er	1	0.73	1.25	1.06	1.34	1.41	2.06	3.62	1.94	2.26	2.26	2.43	1.79
Tm	3	0.11	0.17	0.15	0.18	0.20	0.30	0.51	0.26	0.31	0.32	0.32	0.24
Yb	1	0.74	1.21	1.04	1.27	1.40	2.07	3.52	1.81	2.14	2.19	2.22	1.69
Lu	3	0.11	0.19	0.16	0.20	0.22	0.31	0.55	0.28	0.33	0.34	0.34	0.26
SUM	—	9.6	17.1	18.7	36.3	20.0	44.6	59.7	30.4	34.3	46.5	44.8	122
Y	—	6.0 (2)	11 (1)	11.5 (1)	13 (1)	13 (2)	18 (1)	30 (2)	17 (2)	21 (2)	21 (2)	25 (1)	18 (2)
La/Yb	—	1.0	1.1	2.0	3.8	1.2	2.5	1.5	1.6	1.3	2.4	1.8	12.8
Eu/Eu*	—	1.1	1.2	1.1	1.0	1.1	1.0	1.0	1.0	1.0	1.0	1.0	1.0

Note: Data in ppm. Method 1 = SSMS; 2 = XRF; 3 = interpolated value. Eu* is the calculated Eu abundance for zero Eu anomaly.

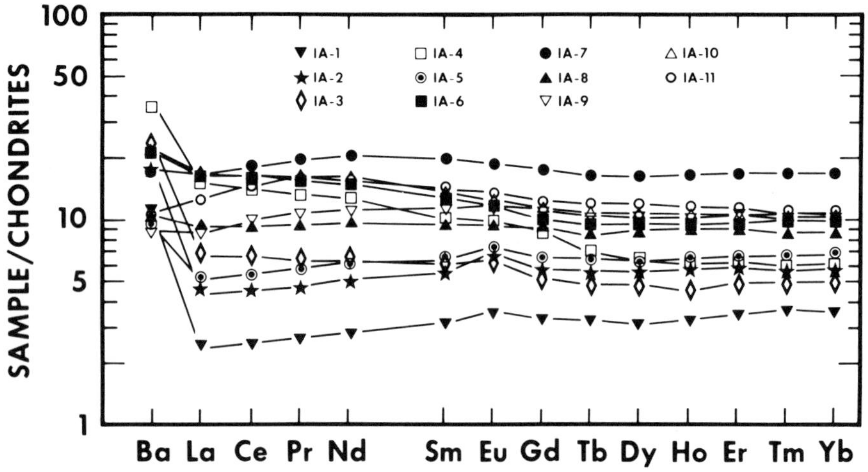

1.2.7.9 Chondrite-normalized REE and Ba data for the Island Arc Basalt Reference Suite.

Sample IA-12 has a strongly enriched light REE pattern (Fig. 1.2.7.10). There is a general increase in ΣREE with increasing depth to the Benioff zone. Those samples nearest to the trench (IA-1, -2, -3, -5) have slight positive Eu anomalies (about 10%), but the other samples show no Eu enrichment or depletion. Although these REE patterns are similar to MORB REE patterns (e.g., Langmuir *et al.*, 1977), they are in general less depleted in LREE. They are within the range reported for marginal basin basalts (e.g., Tarney *et al.*, 1977), but differ from REE patterns in oceanic alkali basalts, except for sample IA-12. There is a distinctive relationship between chondrite normalized Ba and chondrite normalized REE patterns: for those samples nearest the trench (IA-1–6), Ba_c is enriched relative to $LREE_c$; for samples IA-7–11, further from the trench, $Ba_c \approx LREE_c$; while for sample IA-12, furthest from the trench, Ba is depleted relative to $LREE_c$. The enrichment of Ba_c relative to REE_c close to the trench appears to be a distinctive feature of island arc basalts. MORB and marginal basin basalts (e.g., Hawkins, 1977) show either Ba_c falling smoothly on the extension of the REE_c pattern or a pronounced depletion in Ba_c relative to La_c.

The large, high-valence cation contents (Th, U; Table 1.2.7.6) and Th/U ratios (mean 1.70 ± 0.67) are low compared with upper crustal values (typically 3.8). The exception again is sample IA-12, which has a high Th/U ratio of 4.8. The mean Th and U values for MORB (Th: 0.13 ppm, U: 0.6 ppm) in general are much lower than for the island arc basalt suite. The Zr/Hf ratio is nearly constant (33.6 ± 3.9) and close to the chondritic value of 33. Nb has a wide range (0.28–3.9 ppm) and no well-developed correlation with Zr or Ti.

Ferromagnesian trace elements. Two distinct groups are apparent in the data from the suite (Table 1.2.7.7). Samples IA-1, IA-5, IA-9 and IA-11 have high Mg concentrations (mean : 5.92 ± 0.7%), have Mg/(Mg + Fe) ratios of 0.70 ± 0.01, and have high mean Cr (396 ± 104 ppm) and Ni (139 ± 46 ppm) contents relative to most island arc rocks.

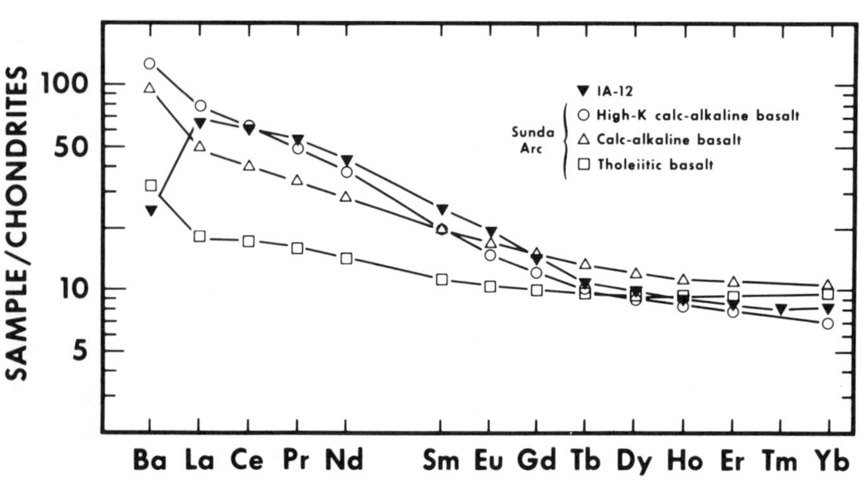

1.2.7.10 Chondrite-normalized REE and Ba data for reference suite sample No. 12 and for basalts from the Sunda arc (Whitford *et al.*, 1979).

Table 1.2.7.6 Large high valence cations in New Britain Island Arc Basalt Reference Suite.

	Method	IA-1	IA-2	IA-3	IA-4	IA-5	IA-6	IA-7	IA-8	IA-9	IA-10	IA-11	IA-12
Th	1	0.07	0.11	0.20	0.35	0.13	0.52	0.59	0.21	0.25	0.71	0.29	1.61
U	1	0.07	0.55	0.15	0.22	0.07	0.32	0.27	0.13	0.11	0.35	0.11	0.33
Zr	1	15.7	19 (1)	21.4	26	20	44	86	34	45	57	64	63
Hf	1	0.50	0.61	0.67	0.80	0.50	1.42	2.43	1.11	1.47	1.76	1.75	1.49
Sn	1	0.23	—	0.28	0.38	0.32	—	1.0	—	0.57	0.62	0.87	—
Nb	1	1.4	0.28	0.43	0.59	0.37	1.75	3.90	1.54	1.77	2.39	3.89	1.33
Ti%	—	0.18	0.46	0.29	0.31	0.31	0.48	0.86	0.51	0.61	0.59	0.69	0.57
Mo	1	0.41	—	0.40	0.50	0.21	—	0.51	—	0.25	0.43	0.37	—
Th/U	—	1.0	0.2	1.3	1.6	1.9	1.6	2.2	1.6	2.3	2.0	2.6	4.8
Zr/Hf	—	31	31	32	32	40	31	35	31	30	32	36	42
Zr/Nb	—	11.3	62	49	44	55	25	22	22	25	24	17	48

Note: Data in ppm except where indicated in wt.%.
Method 1 — SSMS.

Table 1.2.7.7 Ferromagnesian elements in the New Britain Island Arc Basalt Reference Suite.

	Method	IA-1	IA-2	IA-3	IA-4	IA-5	IA-6	IA-7	IA-8	IA-9	IA-10	IA-11	IA-12
Cr	2	283	—	59	69	386	15	38	96	381	59	535	20
V	2	188	290	266	269	258	241	294	261	225	246	230	276
Sc	2	42	33	35	42	42	25	32	37	35	31	33	23
Ni	2	88	21	30	34	119	18	24	61	152	50	196	31
Cu	2	97	81	159	68	127	77	50	60	73	128	86	116
Fe%	—	6.70	10.6	7.07	7.41	6.99	7.02	7.82	7.03	6.51	6.53	8.28	7.29
Mn	—	1200	1300	1200	1400	1400	1200	1500	1400	1700	1400	1400	1500
Zn	2	68	81	70	67	65	70	89	67	68	66	71	68
Mg%	—	5.55	2.83	3.38	4.16	6.09	2.68	3.20	4.80	5.52	3.89	6.51	3.50
Ga		11.0	17.0	15.5	13.5	12.5	17.0	17.5	14.0	15.0	15.5	15.5	18.0
V/Ni		3.5	13.8	8.9	7.9	2.2	13.4	12.3	4.5	1.5	4.9	1.2	8.9
Cr/V		1.5	—	0.22	0.26	1.5	0.06	0.13	0.37	1.7	0.24	2.3	0.072

Note: Data in ppm except where indicated in wt.%. Method 2 – XRF.

Although basalts high in Mg, Cr and Ni are not common in island arc suites, they represent a possibly important chemical type in that they may not have differentiated significantly since they last were in equilibrium with mantle peridotite. These high Mg, Ni and Cr contents contrast sharply with the abundances in the remainder of the suite, in which Mg content averages 3.56 ± 0.71%, Mg′-values range from 0.51–0.65, and both Cr (mean 51 ppm) and Ni (mean 34 ppm) have the low values typical of island arc basalts. There is a general tendency for Ni and Cr to increase with increasing depth to the Benioff zone, but the data base is small.

Vanadium (mean 254 ± 30 ppm) contents are uniform throughout the suite, in contrast with the Mg, Ni and Cr abundances. This results in two clusters of the V/Ni and Cr/V ratios. The average V/Ni ratios for the two groups are 6.9 and 2.1, and the average Cr/V ratios are 0.19 and 1.75, respectively. Scandium is also uniform in the reference suite samples, averaging 33 ± 5 ppm.

Isotope chemistry

Strontium-isotope ratios for twelve samples from the reference suite are listed in Table 1.2.7.8. The methods of analysis were modified only slightly from those reported by Page and Johnson (1974).

$^{87}Sr/^{86}Sr$ values range from 0.70311–0.70361, and have a mean value of 0.70328 ± 0.00015. The small differences in values suggest an isotopically similar source for the basalts. The mean $^{87}Sr/^{86}Sr$ value is somewhat lower than that of 0.7036 noted for 24 rocks of different compositions from a wider range of volcanoes at the southern margin of the Bismarck Sea (Peterman *et al.*, 1970; Page and Johnson, 1974). The $^{87}Sr/^{86}Sr$ values given here include the lowest ones reported for the Papua New Guinea region, but nevertheless are within the range measured in basalts from other island arcs (see Kay *et al.*, 1978).

Correlations between $^{87}Sr/^{86}Sr$ values and various trace element abundances and ratios (e.g., Sr, Rb, Rb/Sr, K/Rb, K) are not apparent. There is some indication that $^{87}Sr/^{86}Sr$ increases with La/Yb ratio, but there is no relationship between the Sr-isotopic ratios and depths to the Benioff zone. Also, intra-volcano differences are nearly as great as the range of ratios for the entire suite.

Nd-isotope ratios have been determined for a number of samples from New Britain, including two basalts from this suite (IA-4 and IA-11, Table 1.2.7.9; DePaolo and Johnson, 1979). The samples have very uniform $^{143}Nd/^{144}Nd$ values that correspond with the similarly uniform $^{87}Sr/^{86}Sr$ values. The Nd ratios suggest that the New Britain volcanics were derived from a residual mantle source with a light-REE depletion.

Table 1.2.7.8 Rb/Sr and Sr isotopic ratios for the New Britain Island Arc Basalt Reference Suite.

IA Number	Rb/Sr	$^{87}Sr/^{86}Sr$	σ
1	0.016	0.70323	±6
2	0.015	0.70305	±6
3	0.010	0.70351	±4
4	0.015	0.70340	±9
5	0.005	0.70344	±9
6	0.017	0.70328	±5
7	0.029	0.70332	±4
8	0.014	0.70336	±6
9	0.110	0.70301	±5
10	0.036	0.70330	±5
11	0.013	0.70331 } 0.70323 }	±8 ±9
12	0.009	0.70317	±6

Analyst: M. R. Perfit. Values normalized to NBS987 = 0.71022, Rb/Sr from XRF values. Measured value for NBS 987 = 0.71032. Measured value for E and A = 0.70813. σ = 1 standard deviation.

Table 1.2.7.9 Nd isotope ratios for the New Britain Island Arc Basalt Reference Suite.

Sample Number	$^{143}Nd/^{144}Nd$
IA-4	0.512209 ± 19
IA-11	0.512211 ± 26

Analyst: D. J. DePaolo. Measured BCR-1 (Cal Tech) = 0.511839 ± 22.

1.2.7.3 Review of island arc basalts

Introduction

Island arc basalts are the common mafic members (<53% SiO_2) of the island arc orogenic lava suites developed near convergent plate margins (Fig. 1.2.7.1). One of the most useful criteria used for classification of island arc suites is the SiO_2-K_2O relationship (Fig. 1.2.7.1). On this basis, basalts with less than 0.3–0.5% K_2O are classed as low-K tholeiites, which correspond approximately to the tholeiitic series of Kuno (1960), while those basalts with greater than about 0.3–0.5% K_2O correspond approximately to Kuno's high-Al basalt series. The pioneering work of Kuno (1960) and Sugimura (1960) was followed by Dickinson and Hatherton (1967) who emphasized the significance of chemical variation with depth, *h*, to the Benioff Zone. Jakeš and Gill (1970) subdivided the arc-trench-type rocks into an island arc tholeiite suite, a calc-alkaline suite and an alkali-rich shoshonitic suite, based primarily on K_2O–silica relationships, but incorporating trace element (e.g., REE) criteria into the classification. Donnelly and Rogers (pers. comm., 1978) distinguish a Primitive Island Arc series from the calc-alkaline and high-K series. All of these classifications attest to the importance of potassium (as the major element variable) and related trace elements in distinguishing among the various spatial and temporal relationships in island arcs. Most authors likewise note that they are subdividing a continuum. In the present context, these relationships are of less importance than the problem of distinguishing and contrasting island arc basalts from basalts formed in other tectonic environments.

The principal purpose of this section is to identify the distinguishing characteristics of the island arc basalts.

Distribution and abundance

Basaltic rocks have been described from most sequences of eruptive rocks formed at subducting plate boundaries. A comprehensive reference list would be too extensive for inclusion here. Instead, examples have been selected of suites which have been described well, and for which some combination of petrographic, major and trace element chemical, and isotopic data exist.

Aleutians	Kay (1977); Kay *et al.* (1978); Perfit (1979).
Caribbean	Brown *et al.* (1977); Donnelly and Rogers (pers. comm., 1978)
Chile	Lopez-Escobar *et al.* (1976, 1977)
Tonga	Ewart *et al.* (1973)
New Hebrides	Gorton (1977)
Papua-New Guinea	Johnson (1977); Johnson and Chappell (1979)
Indonesia	Whitford (1975); Whitford *et al.* (1979)
Scotia arc	Tarney *et al.* (1977); Hawkesworth *et al.* (1977)
Japan-Marianas-Kuriles	Kuno (1966); Philpotts *et al.* (1971); Masuda and Katsui (1975)

Although these rocks are common in some arc-trench systems, they are rare or lacking in others (e.g., N. Chile-Bolivia-S.Peru), so their volumetric importance is difficult to estimate. Thus Jakeš and Gill (1970) considered that the island arc tholeiitic series (which includes many rocks with >53% SiO_2; Fig. 1.2.7.1) makes up the bulk of the eruptive material in island arcs. This represents one extreme view. The average

Table 1.2.7.10 Compositions of average island arc volcanic rocks.†

	1	2	3	4
SiO_2	58	57.7	56.0	56.8
TiO_2	0.8	0.80	0.80	1.1
Al_2O_3	18	17.2	17.6	17.5
FeO	7.5	7.6	8.6	7.4
MgO	3.5	4.5	4.2	4.5
CaO	7.5	7.8	8.9	7.6
Na_2O	3.5	3.1	2.9	3.7
K_2O	1.5	1.3	1.0	1.4
SUM	100.3	100.0	100.0	100.0

†All values in wt.%.
1. Average island arc composition (Taylor, 1977, 1979).
2. Average of 195 "Continental" island arcs (Ewart, 1976).
3. Average of 153 "Intraoceanic" island arcs (Ewart, 1976).
4. Average of 145 Western American orogenic lavas (Ewart, 1976).

SiO_2 content of island arc volcanic rocks is much higher (56–58%) than that of the basaltic members. Estimates of the "average" composition of island arc eruptives by different authors are fairly uniform. Table 1.2.7.10 gives a representative set of recent averages from which it may be seen that intermediate compositions characterize the principal eruptive products near convergent plate margins. The most recent estimate of the relative frequency of different rock types in island arcs is given by Ewart (1976), who found that basalts comprise 23%, basaltic andesites 29%, pyroxene andesites 30% and hornblende andesites 18%, for compositions up to 63% SiO_2. When the amounts of more siliceous lavas (dacites

Table 1.2.7.11 Normative and modal mineralogy of orogenic basalts.†

	Island arc basalt	Continental margin basalt
Norm.		
Q	2.29	3.02
Or	4.96	6.14
Ab	22.24	27.06
An	33.11	30.26
Hy	16.14	15.73
Di	13.40	9.48
Mt	4.97	4.64
Il	1.67	2.17
Ap	0.45	0.62
n (sample number)	80	52
Mode		
Total crystal content	29	22
Modal plagioclase	20	15
Modal olivine	4	4
Modal augite	4.7	2.7
Modal magnetite	0.4	0.2
n (sample number)	59	27

†Ewart, 1976.

Table 1.2.7.12 Major elements in Island Arc Basalts.†

	1	2	3	4	5	6	7	8	9	10	11	12	16	17	18	Average	± σ	Average Reference suite*
wt.%																		
SiO_2	51.16	52.54	48.7	51.7	48.73	50.84	53.6	51.9	50.59	51.4	50.90	50.58	51.0	50.8	49.6	50.9	1.4	52.0
TiO_2	0.88	1.14	0.63	—	0.63	0.89	0.86	0.80	1.05	1.09	0.85	1.10	0.96	1.27	1.15	0.95	0.17	0.83
Al_2O_3	17.37	17.48	16.5	16.9	16.53	16.39	19.6	16.0	16.29	19.7	19.83	17.81	19.7	18.7	18.2	17.8	1.4	17.3
Fe_2O_3	3.43	3.20	3.4	—	3.37	1.76	—	—	3.66	—	2.51	3.55	—	—	—	—	—	—
FeO	6.16	5.44	8.4	10.4*	8.44	8.79	8.5*	9.56*	5.08	9.12*	6.97	6.55	9.4*	9.7*	9.8*	9.50*	1.1	7.53
MnO	0.17	0.14	0.29	—	0.29	0.21	0.16	0.17	0.17	0.19	0.19	0.17	0.18	0.12	0.20	0.19	0.05	0.19
MgO	6.14	5.85	8.2	6.5	8.24	5.94	4.1	6.77	8.96	4.74	3.74	5.48	5.1	5.0	5.1	6.0	1.7	7.39
CaO	10.27	8.82	12.2	11.0	12.25	10.72	9.0	11.8	9.50	9.13	11.10	10.51	9.9	9.5	10.2	10.4	1.2	11.7
Na_2O	2.63	3.20	1.2	3.1	1.21	3.03	3.4	2.42	2.89	3.29	2.50	2.62	3.1	3.3	3.0	2.73	0.7	2.26
K_2O	0.84	1.04	0.23	0.40	0.23	1.18	0.63	0.44	1.07	1.09	0.49	0.48	0.51	1.44	2.18	0.82	0.35	0.42
P_2O_5	0.19	0.26	0.10	—	0.10	0.26	0.18	0.11	0.21	0.26	0.12	0.20	0.16	0.32	0.40	0.23	0.06	0.13
H_2O^+	0.55	0.56	—	—	—	—	—	—	—	—	—	0.38	—	—	—	—	—	—
H_2O^-	0.23	0.15	—	—	—	—	—	—	0.81	—	—	0.20	—	—	—	—	—	—
	100.0	99.8	99.9	100.0	100.0	100.0	100.0	100.0	99.5	100.0	99.2	99.6	100.0	100.2	99.8	99.5	—	100.0
Mg/Mg + Fe	0.57	0.58	0.58	0.55	0.59	0.53	0.49	0.58	0.68	0.51	0.44	0.52	0.53	0.52	0.52	0.55	—	0.66

†Total iron expressed as FeO. *From Table 1.2.7.3.

1. Ewart (1976): "island-arc" basalt, Table 1. p.420. **2.** Ewart (1976): "W. American" basalt, Table 1. p. 420. **3.** Gill (1976): "arc tholeiite," Table 1. p. 1386. **4.** Trig K High-Al basalt, New Zealand (Taylor et al., 1969). **5.** Primary tholeiite, Izu-Hakone (Kuno, 1966, Table 1. p. 10). **6.** Aoba, New Hebrides (Gorton, 1977, Table 1, 51 6 high K) **7.** Sunda arc "tholeiite" (Whitford, 1975, Table 3.9 p. 63). **8.** Average island arc tholeiite (Jakěs and White, 1972, Table 2A). **9.** High Al basalt (Jakěs & White, 1972, Table 2A). **10.** High-Al Basalt Sundoro volcano, Java. (Whitford, 1975, 71-1026). **11.** Basalt, Dominica (Brown *et al.*, 1977, Table 6, Anal. 2). **12.** Average "calc-alkaline circumoceanic' basalt (Lewis, 1971, Table 3 p. 165). **13.** Hakone High-Al basalt (Philpotts *et al.*, 1971, Table 1, p. 91). **14.** Basalt SSW6.3, South Sandwich Islands (Hawkesworth *et al.*, 1977). **15.** Basalt SSR 1.3 South Sandwich Islands (Hawkesworth *et al.*, 1977). **16.** Lesser Antilles basalts (Lewis, 1971, Table 5). **17.** Average tholeiitic basalt, Sunda Arc (Whitford *et al.* , 1979). **18.** Average calc-alkaline basalt, Sunda Arc (Whitford *et al.*, 1979, Table 2). **19.** Average high K calc-alkaline basalt (Whitford *et al.*, 1979, Table 3).

Table 1.2.7.13 Large cations in island arc basalts.†

	3	4	6	7	8	9	10	11	13	14	15	16	17	18
Cs	—	—	—	—	—	—	0.66	—	—	—	—	0.5	1.5	4.1
Rb	—	9.6	20.2	15	5.0	10	26	16	2.43	14	1.75	11	32	61
K%	—	0.33	0.98	0.52	0.37	0.89	0.90	0.41	0.2137	0.46	0.09	0.42	1.20	1.81
Ba	—	115	430	150	75	115	310	105	117	132	55	120	340	480
Pb	2	1.7	—	5	—	—	5.1(id)	—	—	—	—	1.4	9.3	5.7
Sr	178	328	538	300	200	330	443	310	158	135	153	310	400	540
K/Rb	—	340	481	350	1000	340	340	240	879	329	514	392	377	297
Ba/Rb	12	21	10	15	11.5	12	6.6	48	9.4	31		10.9	10.6	7.87
Rb/Sr	—	0.029	0.038	0.050	0.025	0.030	0.059	0.052	0.015	0.10	0.01	0.035	0.080	0.11
Sr/Eu	—	252	517	—	—	—	430	—	206	119	300	360	308	415
K/Ba 1	29	23	35	49	77	29	13	13	18	34	36	34	35	38
$^{87}Sr/^{86}Sr$	0.7037	—	—	—	—	—	—	—	—	0.70399	0.70384	—	—	—

†All data in ppm except for K in wt.%.
Sources: See Table 1.2.7.12.

and rhyolites) are included, the volumetric proportion of basalt falls below 20%.

Mineralogy and petrography

CIPW norms and phenocryst abundances for island arc basalts (from Ewart, 1976) are given in Table 1.2.7.11, where the data are subdivided into island arc and continental margin regions. The island arc basalts are dominantly quartz-normative, in contrast with mid-ocean ridge basalts, which typically are olivine-normative. Total-phenocryst contents of about 25% (Ewart, 1976) are characteristic. The most abundant phenocryst phase is plagioclase. Clinopyroxene, olivine and magnetite are present in much lesser amounts (Table 1.2.7.11). Plagioclase is ubiquitous: An_{75} is the most frequent composition reported and the range is about An_{92} to An_{30} with a strongly skewed distribution. Many of the more sodic compositions are rims on phenocrysts. Olivine phenocryst compositions have a similarly skewed distribution, ranging from Fo_{55} to Fo_{90}, peaking at about Fo_{78}. Calcium-poor pyroxenes (typically En_{60-75}) appear rather uniformly as phenocrysts and augite is the common clinopyroxene phase. The dominant oxide mineral is a Ti-poor titanomagnetite. Ilmenite is usually absent.

Major element chemistry

A compilation of representative analyses and average compositions of island arc basaltic rocks is given in Table 1.2.7.12. The most distinctive features are an SiO_2 range from 48 to over 53%, generally high Al_2O_3 contents (16–20%), TiO_2 contents generally less than 1.2%, and K_2O contents usually less than 1.5%. CaO/Al_2O_3 ratios show a regular decrease with SiO_2 content, but Na_2O/K_2O ratios have an extremely variable range from <0.5 to >12. The overall average given in Table 1.2.7.12, with one-sigma limits, emphasizes the rather small variability in composition.

The major element showing the greatest dispersion is MgO. Computer-generated plots of published analyses show that there is an excellent correlation between Mg′ and MgO content, but no such correlation is observed with FeO (total). The Mg′-values given in Table 1.2.7.12 are generally low, averaging 0.55. The same values are observed for the low-Mg members of the reference suite, and only the high-Mg rocks of that suite have a Mg′-value of about 0.70. Most of the island arc basalts thus have Mg′-values which are too low for them to be in equilibrium with mantle olivine. This implies either that the basalts have come from a fractionated source, or that they underwent olivine and spinel fractionation en route to the surface.

Variation with depth h to the Benioff zone is usually less clearly evident when the basaltic rocks only are considered, but is usually clearer when the more silica-rich members of the suite are considered. Thus, in the Sunda arc of Indonesia (Whitford *et al.*, 1979), the basaltic andesites (52–57% SiO_2) show well-developed correlations for K_2O and P_2O_5. The other major elements show little significant correlation with depths h ranging from 130–240 km.

Table 1.2.7.14 Rare-earth elements in island arc basalts.

	3	4	6	7	8	9	10	13	14	15	16	17	18
La^{3+}	—	9.6	8.85	6.7	1.1	9.6	16.0	—	—	—	5.2	14.6	23.1
Ce^{3+}	3.8	19	—	—	2.6	19	33.8	10.0	12.8	3.09	14.6	32.1	53.6
Pr^{3+}	—	4.1	2.75	—	—	—	3.79	—	—	—	2.0	4.0	5.5
Nd^{3+}	—	20	12.82	—	—	—	15.6	7.57	10.5	3.18	9.2	17.6	21.3
Sm^{3+}	—	4.0	3.16	—	—	—	3.03	2.10	3.37	1.16	2.4	4.0	4.1
Eu^{3+}	—	1.3	1.04	—	—	—	1.03	0.765	1.13	0.51	0.86	1.3	1.3
Gd^{3+}	—	4.0	3.49	—	—	—	2.88	2.47	4.45	1.75	2.7	4.0	3.5
Tb^{3+}	—	0.8	0.60	—	—	—	0.45	—	—	—	0.48	0.68	0.54
Dy^{3+}	—	—	3.63	—	—	—	2.95	2.36	5.24	2.12	3.0	4.0	3.0
Ho^{3+}	—	1.2	0.77	—	—	—	0.65	—	—	—	0.69	0.89	0.67
Er^{3+}	—	4.2	2.23	—	—	—	1.82	1.36	3.39	1.41	2.1	2.5	1.77
Tm^{3+}	—	0.56	—	—	—	—	0.28	—	—	—	—	—	—
Yb^{3+}	—	2.7	2.18	—	1.4	2.7	1.71	1.24	3.29	1.40	2.0	2.4	1.5
Lu^{3+}	—	—	—	—	—	—	0.26	0.201	—	—	—	—	—
SUM	—	—	—	—	—	—	8.4	—	—	—	45	88	120
Y^{3+}	21	20	26.5	—	—	20	27	—	—	—	24	28	28
$^{143}Nd/^{144}Nd$	—	—	—	—	—	—	—	—	0.51304	—	—	—	—
La/Yb	—	3.6	4.1	—	—	3.6	9.4	—	—	—	2.5	6.1	15
Eu/Eu*	—	1.0	0.96	—	—	—	1.09	1.0	~0.9	1.11	1.0	1.0	1.0

Sources: See Table 1.2.7.12."

Eu* is the calcualted Eu abundance for zero Eu anomaly.

Trace element chemistry

A selection of trace element data from recent papers is given in Tables 1.2.7.13–16. In some cases, major element data are not available, and the coverage for the trace elements tends to be sporadic. The possibilities of interlaboratory bias, of analytical error, and of sampling problems need to be borne in mind. Nevertheless, some general patterns emerge from this compilation, particularly when these data are considered together with the values available for the reference suite. Table 1.2.7.13 provides data for the large cations. These data have a comparatively wide range, reflecting the sample variations in potassium contents. Ba is generally much higher than in MORB and K/Ba ratios average 34 compared with the reference suite average of 50. K/Rb ratios range from 240–1000, and are more variable than the K/Ba values. There is a general increase in K, Rb, Cs, Sr and Ba with increasing depth (*h*) to the Benioff or seismic zone. In the best documented case, the Sunda arc, the abundances of these elements increase by nearly an order of magnitude (Whitford *et al.*, 1979). The rare-earth element data given in Table 1.2.7.14 show a range similar to that in the reference suite, but few full REE data sets are available. Patterns range from light REE depleted, to flat, and light REE enriched, covering the range for MORB (including Project FAMOUS) values, as well as those for marginal basins (e.g., Tarney *et al.*, 1977).

An enrichment of Ba relative to REE appears to be a feature characteristic of island-arc basalts when compared with MORB and intraplate oceanic basalts (Kay, 1977; Sun *et al.*, 1979). This enrichment is exemplified by the ratio Ba_c/La_c. In Fig. 1.2.7.11 Ba_c/La_c is plotted against La_c/Sm_c, an index of light REE enrichment. MORB with depleted light-REE patterns have Ba_c/La_c values less than 1. With increasing light-REE fractionation (plume type to alkalic basalts), this ratio increases only slightly to less than 2. In contrast, the Ba_c/La_c ratios for island arc basalts with relatively flat REE patterns are generally >2 and may be as great as 10. As REE fractionation increases in island arc alkalic basalts, the Ba_c/La_c values overlap those in intraplate alkalic basalts. The Ba_c/La_c values in the *reference suite* show a decrease with increased distance from the trench. It is notable, however, that all reference suite samples with unusually low Ba_c/La_c are from the Witu Islands and plot near a number of basalts from marginal basins on Fig. 1.2.7.11.

The REE data for basalts from the Sunda arc (Whitford *et al.*, 1979) are all more fractionated than those of the reference suite (except IA-12; Fig. 1.2.7.10), and show a regular increase in La/Yb with distance from the trench. A regular feature of the REE patterns in island arc rocks in general is the absence of europium anomalies. Although plagioclase is a common phenocryst phase and readily incorporates Eu^{2+}, it is unusual to find any Eu enrichment or depletion in island arc rocks.

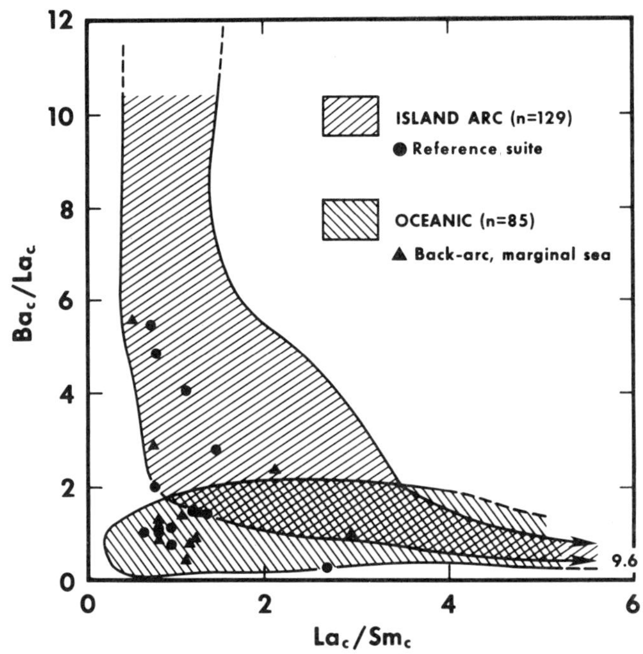

1.2.7.11 Chondrite-normalized ratios of Ba/La vs. La/Sm. Fields of island arc volcanic rocks and oceanic volcanic rocks (MORB and oceanic islands) are shown. N equals number of samples defining the fields. Basalts from marginal basins are plotted separately.

Table 1.2.7.15 provides data for the large high valency cations. The data from the Sunda arc (Whitford *et al.*, 1979) show an order of magnitude increase in Th and U with distance from the Benioff zone.

Although the island arc basalts tend to fall in an area distinct from MORB values on plots of Zr/Nb vs. K/Ba, they form a continuum with MORB data.

The data for the ferromagnesian elements in Table 1.2.7.16 show the distinctive features of island arc chemistry. Vanadium averages 290 ppm, compared with an average of 254 ppm for the reference suite. The overall Ni (average 17 ppm) and Cr values (average 45 ppm) are low and similar to values of 34 and 51 ppm respectively for the low-Mg members of the reference suite. The resulting high V/Ni and V/Cr ratios serve to distinguish many island arc rocks from other basaltic rock types, and appear to be characteristic of island arc rocks in general. The ferromagnesian trace elements do not show regular variations with depth, *h*, to the Benioff zone, but show much scatter about the average values, particularly for Cr and Ni.

Table 1.2.7.15 Large high-valence cations in island arc basalts.†

	3	4	6	7	8	9	10	11	17	18	19
Th^{4+}	0.13	1.1	1.18	1.3	0.5	1.1	3.2	—	0.79	3.6	6.5
U^{4+}	0.06	0.2	0.41	0.4	0.15	0.2	0.68	—	0.19	0.75	1.2
Zr^{4+}	44	100	63	80	70	100	88	44	74	129	113
Hf^{4+}	—	2.6	1.73	—	1.0	2.6	2.0	—	1.7	2.6	2.0
Nb^{5+}	1.0	4.0	2.1	—	—	—	4.3	—	2.7	5.8	11
Th/U	2.2	5.5	2.9	3.3	3.3	5.5	4.7	—	4.2	4.8	5.4
Zr/Hf	—	38	36	—	70	38	44	—	44	50	56
Zr/Nb	44	25	30	—	—	—	20	—	27	22	10

†Sources: See Table 1.2.7.12.

Isotope chemistry

$^{87}Sr/^{86}Sr$ values are generally greater than 0.703. Lewis (1971) summarized the older data for island arc basalts, which average 0.7036 ± 0.0006 (1σ). Data for the reference suite average 0.70328 ± 15, slightly lower than other values. Gill (1976a) gave an average of 0.7037 for island arc basalts, and Hawkesworth *et al.* (1977) considered that "the $^{87}Sr/^{86}Sr$ ratios of the parental magmas [of the South Sandwich Island basalts] were not far removed from 0.7038–0.7039." This tendency for the $^{87}Sr/^{86}Sr$ ratios to exceed 0.703 contrasts with "normal" MORB ratios which are generally lower than 0.703 (Sun *et al.*, 1979). However, Sun and others show that "plume" and "transitional" MORB have values generally greater than 0.703, corresponding to their more enriched trace element contents. Furthermore, Kay *et al.* (1978) and Perfit (1979) report $^{87}Sr/^{86}Sr$ values less than 0.703 for some individual Aleutian basalts and an average ratio of 0.7032, which is well within the range of all MORB and close to the ratio of the reference suite.

A limited number of measurements of Nd isotopes in island arc rocks have been made (DePaolo and Wasserburg, 1977; Hawkesworth *et al.*, 1977, 1979c; DePaolo and Johnson, 1979). In general, the results of Nd isotope analyses suggest that arc lavas have been derived from mantle sources in which Nd has been depleted relative to Sm, which results in $^{143}Nd/^{144}Nd$ values greater than those which would be expected in a present-day chondritic reservoir (CHUR, $^{143}Nd/^{144}Nd = 0.511836$). Consequently, nearly all of the island arc rocks measured thus far plot above the estimated bulk Earth line at $E_{Nd}^{CHUR} = 0$ on Fig. 1.2.7.12. The significant variations in $^{87}Sr/^{86}Sr$ in arc lavas cause the fields to diverge from the "mantle array," which is defined primarily by the linear trend of MORB and oceanic island basalts through the bulk Earth composition.

Calc-alkaline rocks from continental settings exhibit a much wider range of Sr and Nd isotope values (e.g., Chile, Patagonia), which extend toward the estimated fields of oceanic sediments and crustal rocks (Fig. 1.2.7.12). DePaolo and Johnson (1979) present Nd and Sr isotopic ratios for eight volcanic rocks from the New Britain arc, including two basalts from the reference suite (IA-4, IA-11, Table 1.2.7.9). The samples from New Britain have very uniform $^{143}Nd/^{144}Nd$ ratios which suggest a compositionally uniform source. The Sr-Nd characteristics are similar to those of oceanic island basalts. In other island arcs (e.g., Scotia, Mariana) the displacement to higher $^{87}Sr/^{86}Sr$ at rather constant $^{143}Nd/^{144}Nd$ could be attributed to the addition of seawater or of a slab-derived component from altered MORB. Basalts and ankaramites from Grenada Island

Table 1.2.7.16 Ferromagnesian trace elements in island arc basalts.†

	4	6	7	8	9	10	11	14	15	16	17	18	19	Average
Cr^{3+}	40	91	15	50	40	17	109	14	24	—	38	57	39	45
V^{3+}	255	460	—	270	210	—	—	—	270		280	320	280	290
Sc^{3+}	40	35	—	—	—	22	—	—	—	—	42	35	27	—
Ni^{2+}	25	32	8	30	25	9.6	36	15	27	13	15	23	19	17
Co^{2+}	40	—	—	—	—	24	—	—	—	24	30	31	32	—
Cu^{2+}	35	—	—	—	—	39	—	104	74	—	66	111	110	—
Li^{+}	10	—	—	—	—	—	—	—	—	—	—	—	—	—
Ga^{3+}	13	—	—	—	—	—	—	—	—	—	—	—	—	—
V/Ni	10.2	14.4	—	9.0	10.0	22	—	—	—	—	19	14	15	15
Cr/V	0.16	0.20	—	0.19	0.16	0.08	—	—	—	—	0.14	0.18	0.14	0.16

†Sources: See Table 1.2.7.12.

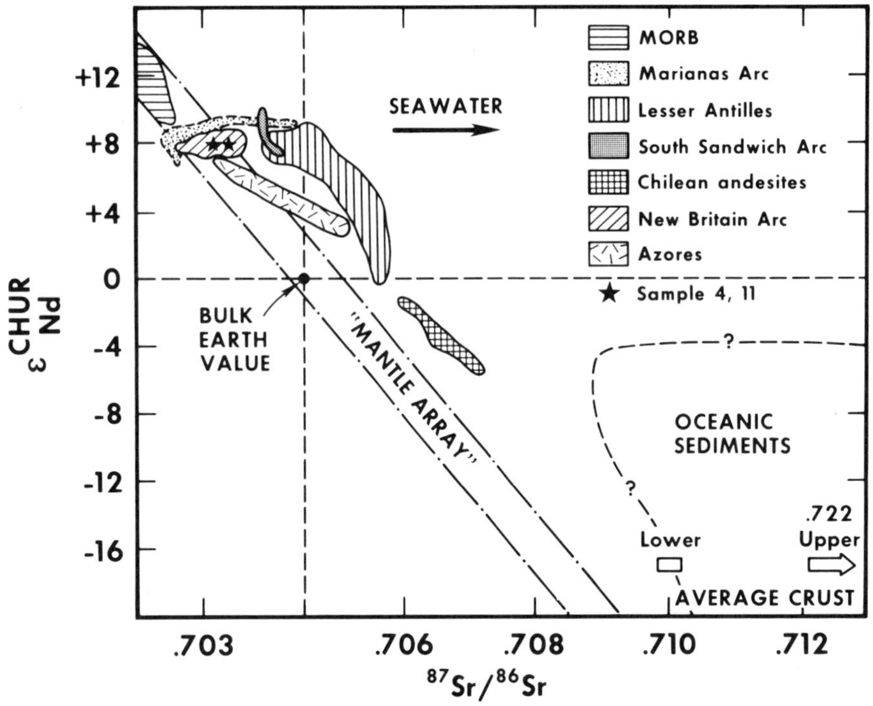

1.2.7.12 Nd and Sr isotopic compositions of island arc volcanic rocks. The "mantle array" is from DePaolo and Wasserburg (1979) and passes through the present-day bulk-Earth composition. The effects of seawater addition or alteration would drive ratios horizontally in the direction shown. The field of Chilean andesites (Hawkesworth *et al.*, 1979a) is shown for comparison. Approximate field of oceanic sediments is from DePaolo and Wasserburg (1979). Other fields are from the following sources: MORB; DePaolo and Wasserburg (1977); Marianas, DePaolo and Wasserburg (1977); New Britain, DePaolo and Johnson (1979); South Sandwich, Hawkesworth *et al.* (1977); Grenada, Lesser Antilles, Hawkesworth *et al.* (1979c); Azores, Hawkesworth *et al.* (1979b).

in the Lesser Antilles plot to the right but parallel with the "mantle array" due to their relatively high $^{87}Sr/^{86}Sr$. In this case there is little evidence to suggest extensive seawater or ocean floor basalt contamination, so the variability may be due to mixing of two (or more?) sources. The existence of enriched zones in the suboceanic mantle is indicated by the Sr and Nd isotopes of basalts from the Azores (Hawkesworth *et al.*, 1979b; see Fig. 1.2.7.12).

In New Britain, there is no evidence for the involvement of altered oceanic crust or ocean-floor sediments in the generation of magma. The possible addition of a small component of unaltered, slab-derived melt, which would account for the anomalously high Sr/Nd and undepleted REE patterns, cannot be discounted.

Comparisons with other related rocks

This topic is discussed more fully in section 1.3.1 of this chapter. Nevertheless, it is useful to emphasize the salient points here. The distinction between island arc basalts and other volcanic rocks formed near convergent plate margins is a matter of definition. Thus basaltic andesites have SiO_2 contents greater than 53%, although they may have values for K_2O, Ba, Ni, Cr, REE and Sr and Nd isotopes which are very similar to those found in island arc basalts.

The distinction from other basalts is somewhat more difficult to make. As discussed above in relation to the reference suite, the island arc basalt compositions grade into those of marginal basin basalts. Hart *et al.* (1972), Hawkins (1977), Tarney *et al.* (1977), and Hawkesworth *et al.* (1977) showed the difficulty of establishing absolute criteria to discriminate marginal-basin basalts from MORB. Hawkesworth *et al.* (1977) reported REE patterns for some back-arc basalts from the Scotia Sea which are close to those of MORB, so the extreme depletion in Large Ion Lithophile (LIL) elements may not be a unique characteristic of ridge basalts. The following generalizations may be used to distinguish island arc basalts from MORB, with the proviso that any one of these criteria may be mimicked in another environment. In comparison with MORB, many arc basalts have higher Al_2O_3, K_2O, Sr and Ba and lower TiO_2, K/Cs and K/Rb ratios. REE patterns and other trace element abundances show complete overlap and do not serve to discriminate arc basalts from MORB or marginal-basin basalts. Chondrite-normalized Ba in arc basalts is usually equal to or greater than chondrite-normalized LREE, whereas MORB commonly show the reverse trend. Commonly, but not universally, chondrite-normalized Ba is greater than La, so that a sharp upward spike appears (Fig. 1.2.7.9; Sun and Hanson, 1976; Kay, 1977).

Another observation is that arc basalts generally have lower Ni, Cr and Mg contents and lower Mg'-values than do MORB (Taylor *et al.*, 1969). The Mg'-

values typically are around 0.55 rather than the 0.70 characteristic of primitive MORB. Nevertheless, several members of the reference suite have high Mg′-values and Ni and Cr contents, indistinguishable from MORB. This may be a key observation for the genesis of island arc basalts.

The Sr isotopic constraints appear to be more rigorous discriminants. $^{86}Sr/^{86}Sr$ ratios of island arc basalts are greater than 0.7030 and average 0.7037, in contrast with an average value of 0.7026 for MORB. Neodymium isotopic ratios also distinguish these two groups fairly well (see above).

Petrogenesis

The genesis of the island arc basalts is intimately related to the origin of the wider range of eruptive rocks found near convergent plate margins. Controversy continues to surround this problem, and there is no agreed solution.

Although this section has dealt with the basaltic members of the suite, their origin is linked with that of the andesites, dacites and rhyolites, which display many of the same characteristics as the basalts. Similarities in chemistry and isotopic composition between the island arc basalts and the more siliceous rocks indicate that the controls generally responsible for the chemical characteristics of the volcanic rocks near convergent plate margins apply to the basalts as well as to the andesites and dacites. A variety of hypotheses have been advanced to explain the origin of these rocks. Most of these have addressed the origin of andesites rather than basalts, so they are not discussed further here.

Theories which derive andesites by fractional crystallization of basalt do not directly address the question of the ultimate origin of the distinctive island arc basalts. Those theories which invoke interaction with or contamination by continental crustal material are invalidated because sialic material is commonly missing in regions of island arc volcanism. Magma mixing models use basalt as one end-member without inquiring deeply into its ultimate origin.

One popular hypothesis for the origin of island arc basalts involves melting and dehydration of the subducted basaltic layer of the oceanic crust, with or without a sedimentary component (Green and Ringwood, 1968; Boettcher, 1977; Stern and Wyllie, 1978). The Pb, Sr and Nd isotope data rule out more than trivial amounts of sediment involvement, as does the absence of a sedimentary Eu depletion signature in the erupted lavas. It is widely believed that basalt transforms to eclogite at the depths at which subducted lithosphere occurs beneath island arc volcanoes (136 ± 41 km; 1σ; Gill, 1978). However, island arc rocks with their flat, heavy REE patterns bear no chemical evidence that garnet was more than a minor phase in the residue during partial melting. This evidence is particularly clear in the REE data. Gill (1978) notes similar problems with Sr, Ni and Sc, although the low Ni and Cr in many island arc basalts would be consistent with a basaltic, rather than a peridotitic source. A basic problem with this hypothesis is that nearly total melting of the subducted slab is required to produce the basaltic members of the island arc suite, and this unlikely event is not reconcilable with the Nd, Sr and Pb isotopic data.

If island arc basalts are produced by hydrous melting of the peridotite wedge overlying the subducted slab (e.g., Nicholls and Ringwood, 1973), their low Ni and Cr contents suggest that considerable fractionation of olivine and spinel must occur en route to the surface and that, therefore, the erupted melts are not primary.

Fundamental element-balance problems also exist for peridotite sources for the incompatible elements such as Ba, Pb, K and Cs. Indeed, the problem for hypotheses that propose melting of peridotite alone to account for the generation of island arc basalts is that they fail to explain the generally distinctive chemical characteristics of arc basalts compared with basalts, also presumed to have been generated from peridotite sources, found in other tectonic environments. Melting of peridotite mantle modified by introduction of water or melt from the subducted oceanic crust (e.g., Nicholls and Ringwood, 1973) provides some resolution for these problems, but the model has an elastic set of parameters which make it difficult to test. Major involvement of the downgoing slab is forbidden, and the role of LIL-enriched fluids derived from the hydrated oceanic crust or sediments is uncertain (see discussion above). If the water supplied to the mantle peridotite is primary, as may be deduced from $^{3}He/^{4}He$ ratios (Craig *et al.*, 1978) and $\delta^{18}O$ values (section 7.1.4), then the downgoing slab has no discernible effect on the chemistry of lavas erupted near convergent plate margins.

In conclusion, it is difficult to recognize the individual roles of the subducted slab (+ sediment) vs. overlying peridotite wedge in the genesis of arc basalts, and no simple, single-stage hypothesis is capable of explaining the entire spectrum of chemical data observed. Constraints on multistage hypotheses are presently loose, but this type of hypothesis offers the best chance of understanding the chemical differences observed in the comparison of island arc basalts with basalts from other tectonic environments.

1.2.7.4 References

Blake D. H. and Ewart A. (1974) Petrography and geochemistry of the Cape Haskins volcanoes, New Britain, Papua, New Guinea. *J. Geol. Soc. Aust.* **21**, 319–331.

Boettcher A. L. (1977) The role of amphiboles and water in circum-Pacific volcanism. In *High Pressure Research* (M. H. Managhnani, ed.), p. 107–126. Academic Press, N. Y.

Brooks C. K. (1976) The Fe_2O_3/FeO ratio of basalt analyses: an appeal for a standardised procedure. *Bull. Geol. Soc. Denmark* **25**, 117–120.

Brown G. M., Holland J. G., Sigurdsson H., Tomblin J. F. and Arculus R. J. (1977) Geochemistry of the Lesser Antilles volcanic island arc. *Geochim. Cosmochim. Acta* **41**, 785–801.

Craig H., Lupton J. E. and Horibe Y. (1978) A mantle helium component in circum-Pacific volcanic gases. In *Terrestrial Rare Gases* (E. C. Alexander and M. Ozima, eds.), p. 3–16. Japan Sci. Press, Tokyo.

Curtis J. W. (1973) Plate tectonics and the Papua New Guinea-Solomon Islands region. *J. Geol. Soc. Aust.* **20**, 21–26.

DePaolo D. J. and Johnson R. W. (1979) Magma genesis in the New Britain island arc: constraints from Nd and Sr isotopes and trace element patterns. *Contrib. Mineral. Petrol.* **70**, 367–369.

DePaolo D. J. and Wasserburg G. J. (1977) The sources of island arcs as indicated by Nd and Sr isotopic studies. *Geophys. Res. Lett.* **4**, 465–468.

DePaolo D. J. and Wasserburg G. J. (1979) Petrogenetic mixing models and Nd-Sr isotopic patterns. *Geochim. Cosmochim. Acta* **43**, 615–627.

Dickinson W. R. and Hatherton T. (1967) Andesitic volcanism and seismicity around the Pacific. *Science* **157**, 801–803.

Ewart A. (1976) Mineralogy and chemistry of modern orogenic lavas—some statistics and implications. *Earth Planet. Sci. Lett.* **31**, 417–432.

Ewart A., Bryan W. B. and Gill J. B. (1973) Mineralogy and geochemistry of the young volcanic islands of Tonga, S.W. Pacific. *J. Petrol.* **14**, 429–465.

Finlayson D. M., Drummond B. J., Collins C. D. M. and Connelly J. B. (1977) Crustal structures in the region of the Papuan Ultramafic Belt. *Phys. Earth Planet. Int.* **14**, 13–29.

Gill J. B. (1976a) Composition and age of Lau Basin and Ridge volcanic rocks: implications for evolution of an interarc basin and remnant arc. *Geol. Soc. Amer. Bull.* **87**, 1384–1395.

Gill J. B. (1976b) From island arc to oceanic islands: Fiji, South-western Pacific. *Geology* **4**, 123–126.

Gill J. B. (1978) Role of trace element partition coefficients in models of andesite genesis. *Geochim. Cosmochim. Acta* **42**, 709–724.

Gorton M. P. (1977) The geochemistry and origin of Quaternary volcanism in the New Hebrides. *Geochim. Cosmochim. Acta* **41**, 1257–1270.

Green T. H. and Ringwood A. E. (1968) Genesis of the calc-alkaline igneous rock suite. *Contrib. Mineral. Petrol.* **18**, 105.

Hart S. R., Glassley W. E. and Karig D. E. (1972) Basalts and seafloor spreading behind the Mariana island arc. *Earth Planet. Sci. Lett.* **15**, 12–18.

Hawkesworth C. J., Norry M. J., Roddick J. C., Baker P. E., Francis P. W. and Thorpe R. S. (1979a) $^{143}Nd/^{144}Nd$, $^{87}Sr/^{86}Sr$ and the incompatible element variations in the calc-alkaline andesites and plateau lavas from South America. *Earth Planet. Sci. Lett.* **42**, 45–57.

Hawkesworth C. J., Norry M. J., Roddick J. C. and Vollmer R. (1979b) $^{143}Nd/^{144}Nd$ and $^{87}Sr/^{86}Sr$ ratios from the Azores and their significance in LIL-element enriched mantle. *Nature* **280**, 28–31.

Hawkesworth C. J., O'Nions R. K. and Arculus R. J. (1979c) Nd- and Sr-isotope geochemistry of island arc volcanics, Grenada, Lesser Antilles. *Earth Planet. Sci. Lett.* **45**, 237–248.

Hawkesworth C. J., O'Nions R. K., Pankhurst R. J., Hamilton P. J. and Evensen N. M. (1977) A geochemical study of island-arc and back-arc tholeiites from the Scotia Sea. *Earth Planet. Sci. Lett.* **36**, 253–262.

Hawkins J., Jr. (1977) Petrologic and geochemical characteristics of marginal basin basalts. *Amer. Geophys. Union Monograph, Maurice Ewing Series* **1**, 355–365.

Heming R. F. (1974) Geology and petrology of Rabaul caldera, Papua, New Guinea. *Geol. Soc. Amer. Bull.* **85**, 1253–1264.

Jakeš P. and Gill J. (1970) Rare earth elements and the island arc tholeiitic series. *Earth Planet. Sci. Lett.* **9**, 17–28.

Jakeš P. and White A. J. R. (1972) Major and trace element abundances in volcanic rocks of orogenic areas. *Geol. Soc. Amer. Bull.* **83**, 29–40.

Johnson R. W. (1977) Distribution and major-element chemistry of late Cenozoic volcanoes at the southern margin of the Bismarck Sea, Papua, New Guinea. *Bur. Miner. Res. Aust. Rept.* **188**. 170 pp.

Johnson R. W. and Arculus R. J. (1978) Volcanic rocks of the Witu Islands, Papua, New Guinea. *Bull. Volcanol.* **41**, 609–655.

Johnson R. W. and Chappell B. W. (1979) Chemical analyses of rocks from the late Cenozoic volcanoes of north-central New Britain and the Witu Islands, Papua, New Guinea. *Bur. Miner. Res. Aust. Rept.* **209**, BMR Microform MF76.

Johnson T. and Molnar P. (1972) Focal mechanisms and plate tectonics of the southwest Pacific. *J. Geophys. Res.* **77**, 5000–5032.

Johnson R. W., Mackenzie D. W. and Smith I. E. M. (1978a) Volcanic rock associations at convergent plate boundaries: reappraisal of the concept using case histories from Papua, New Guinea. *Geol. Soc. Amer. Bull.* **89**, 96–106.

Johnson R. W., Mutter J. C. and Arculus R. J. (1979) Origin of the Willaumez-Manus Rise, Papua, New Guinea. *Earth Planet. Sci. Lett.* **44**, 247–260.

Kay R. W. (1977) Geochemical constraints on the origin of Aleutian magmas. In *Island Arcs, Deep Sea Trenches and Back-Arc Basins* (M. Talwani and W. C. Pitman III, eds.), p. 229–242. Maurice Ewing Series 1, Amer. Geophys. Union, Washington, D. C.

Kay R. W., Sun S. -S. and Lee-Hu C. -N. (1978) Pb and Sr isotopes in volcanic rocks from the Aleutian Islands and Pribilof Islands, Alaska. *Geochim. Cosmochim. Acta.* **42**, 263–273.

Krause D. C. (1973) Crustal plates of the Bismarck and Solomon Seas. In *Oceanography of the South Pacific 1972*, (R. Fraser, ed.), p. 217–280. N.Z. Nat. Comm. UNESCO, Wellington

Kuno H. (1960) High-alumina basalt. *J. Petrol.* **1**, 121–145.

Kuno H. (1966) Lateral variation of basalt magma type across continental margins and island arcs. *Bull. Volcanol.* **29**, 195–222.

Langmuir C. H., Bender J. F., Bence A. E., Hanson G. N. and Taylor S. R. (1977) Petrogenesis of basalts from the FAMOUS area, Mid-Atlantic Ridge. *Earth Planet. Sci. Lett.* **36**, 133–156.

Le Pichon X. (1970) Correction to paper by Xavier Le Pichon 'Sea-floor spreading and continental drift.' *J. Geophys. Res.* **75**, 2793.

Lewis J. F. (1971) Composition, origin and differentiation of basalt magma in the Lesser Antilles. *Geol. Soc. Amer. Mem.* **130**, 159–179.

Lopez-Escobar L., Frey F. A. and Vergara M. (1976) Andesites from central-south Chile: trace element abundances and petrogenesis. IAVCEI Symp. Int. Volc., 725–761.

Lopez-Escobar L., Frey F. A. and Vergara M. (1977) Andesites and high-alumina basalts from the central-south Chile high Andes: geochemical evidence bearing on their petrogenesis. *Contrib. Mineral. Petrol.* **63**, 199–228.

Lowder G. G. (1970) The volcanoes and caldera of Talasea, New Britain: mineralogy. *Contrib. Mineral. Petrol.* **26**, 324–340.

Lowder G. G. and Carmichael I. S. E. (1970) The volcanoes and caldera of Talasea, New Britain: geology and petrology. *Geol. Soc. Amer. Bull.* **81**, 17–38.

Masuda Y. and Katsui Y. (1975) Rare-earth and trace elements in the Quaternary volcanic rocks of Hokkaido, Japan. *Chem. Geol.* **15**, 251–271.

Nicholls I. A. and Ringwood A. E. (1973) Effect of water on olivine stability in tholeiites and production of silica-saturated magmas in the island arc environment. *J. Geol.* **81**, 285.

Osborne E. F. and Watson E. B. (1977) Studies of phase relations in sub-alkaline volcanic rock series. *Carnegie Inst. Wash., Yearb.* **76**, 472–477.

Page R. W. and Johnson R. W. (1974) Strontium isotope ratios of Quaternary volcanic rocks from Papua, New Guinea. *Lithos* **7**, 91–100.

Perfit M. (1979) The petrochemistry and strontium isotopic composition of mafic basalts from the Aleutian Islands. *Abstracts with Program,* Geol. Soc. Amer. **10**, 470.

Peterman Z. E., Lowder G. F. and Carmichael I. S. E. (1970) Sr^{87}/Sr^{86} ratios of the Talasea Series, New Britain, Territory of New Guinea. *Geol. Soc. Amer. Bull.* **81**, 39–40.

Philpotts J. A., Martin W. and Schnetzler C. C. (1971) Geochemical aspects of some Japanese lavas. *Earth Planet. Sci. Lett.* **12**, 89–96.

Roeder P. L. and Emslie R. F. (1970) Olivine-liquid equilibrium. *Contrib. Mineral. Petrol.* **29**, 275–289.

Saunders A. D. and Tarney J. (1979) The geochemistry of basalts from a back-arc spreading center in the East Scotia Sea. *Geochim. Cosmochim. Acta* **43**, 555–572.

Stern C. R. and Wyllie P. J. (1978) Phase compositions through crystallization intervals in basalt-andesite-H_2O at 30 kb with implications for subduction zone magmas. *Amer. Mineral.* **63**, 641–663.

Sugimura A. (1960) Zonal arrangement of some geophysical and petrological features in Japan and its environs. *J. Fac. Sci.*, Section II, **XII**, 133–153. Univ. Tokyo, Japan.

Sun S. -S. and Hanson G. N. (1976) Evolution of the mantle: geochemical evidence from alkali basalt: reply. *Geology* **4**, 626–631.

Sun S. -S., Nesbitt R. W. and Sharaskin A. Y. (1979) Geochemical characteristics of Mid-ocean Ridge basalts. *Earth Planet. Sci. Lett.* **44**, 119–138.

Tarney J., Saunders A. D. and Weaver S. D. (1977) Geochemistry of volcanic rocks from the island arcs and marginal basins of the Scotia arc region. *Maurice Ewing Series, Monograph 1*, p. 367–377. Amer. Geophys. Union, Washington, D.C.

Taylor B. (1979) Bismarck Sea: evolution of a back-arc basin. *Geology* **7**, 171–174.

Taylor S. R. (1977) Island arc models and the composition of the continental crust. *Maurice Ewing Series, Monograph 1*, p. 325–335. Amer. Geophys. Union, Washington, D.C.

Taylor S. R. (1979) Chemical composition and evolution of the continental crust: the rare-earth element evidence. In *The Earth: Its Origin, Structure and Evolution*, (M. W. McElhinny, ed.), p. 353–376. Academic Press, N.Y.

Taylor S. R., Kaye M., White A. J. R., Duncan A. R. and Ewart A. (1969) Genetic significance of Co, Cr, Ni, Sc and V content of andesites. *Geochim. Cosmochim. Acta* **33**, 275–286.

Tilley C. E. (1950) Some aspects of magmatic evolution. *Q. J. Geol. Soc. London* **106**, 37–61.

Weaver S. D., Saunders A. D., Pankhurst R. J. and Tarney J. (1979) A geochemical study of magmatism associated with the initial stages of back-arc spreading. *Contrib. Mineral. Petrol.* **68**, 151–169.

Whitford D. J. (1975) Geochemistry and petrology of volcanic rocks from the Sunda arc, Indonesia. Ph.D. Thesis, Australian Natl. Univ.

Whitford D. J., Nicholls I. A. and Taylor S. R. (1979) Spatial variations in the geochemistry of Quarternary lavas across the Sunda arc in Java and Bali. *Contrib. Mineral. Petrol.* **70**, 341–356.

1.2.8 THE BASALTIC METEORITES

1.2.8.1 Introduction

The known range of basaltic volcanism in the solar system is considerably extended in space and time by studies of basaltic material in meteorites. This material occurs in various types of meteorites: eucrites and shergottites consist entirely of basaltic material, whereas howardites and mesosiderites contain basaltic material variously mixed and metamorphosed with other types of materials. The eucrites, most of which are monomict breccias, are the most abundant of the basaltic meteorites. The eucrites crystallized 4.54 ± 0.02 b.y. ago (see Chapter 7) in magmatic events that occurred largely within 30 ± 20 m.y. after the earliest (dated) event in our solar system, which was the formation of the refractory particles now included in the Allende carbonaceous chondrite. Thus, the eucrites are the oldest known basalts and are basalts which may have formed by partial melting of primitive (i.e., little differentiated) condensed solar material. Investigations of the eucrites can lead to an understanding of very early magmatic processes and can be used to constrain hypotheses on the nature of their source material, and thereby the processes of condensation and accretion in the early solar system. Only small samples of a limited number of eucrites are available to us on Earth. We lack any direct information on their geologic setting, except that we can infer that most of them crystallized close to a planetary surface. The parent body or bodies from which the eucrites are derived have not been identified, although a strong argument can be made in favor of their derivation from asteroids. A specific asteroid, 4 Vesta, has been cited as a possible source for the eucrites (Consolmagno and Drake, 1977), based upon infrared spectrophotometric observations of its surface (McCord *et al.*, 1970; Larson and Fink, 1975).

Most eucrites are pyroxene-plagioclase rocks with the textures of rapidly cooled melts. They occur as unbrecciated basalts or, more commonly, as monomict breccias. A few eucrites have coarser, cumulate-like textures. Basalt fragments occur also as clasts in howardites and in mesosiderites. Howardites contain a variety of clast types from fine-grained basalts to coarse, monomineralic fragments of pyroxene and plagioclase with rare exotic fragments, including carbonaceous debris (Wilkening, 1976). Howardites have been interpreted as polymict regolith breccias, of which eucrite and diogenite fragments are prominent components. Diogenites are orthopyroxenites, most of which are monomict breccias, that may have formed initially as cumulates or as residua from partial melting events. Mesosiderites are stony-iron meteorites, half metallic iron and half silicate. The silicate portion is complex, but commonly contains a high proportion of eucritic material. Most investigators believe that eucrites, howardites, diogenites and the silicate portions of the mesosiderites are genetically related, but opinions differ as to the nature of the relationship.

The meteorites Shergotty and Zagami constitute a compositionally-distinct group of basaltic meteorites, the shergottites. These are unbrecciated, pyroxene-feldspar rocks, which differ from the eucrites in bulk composition, in mineral composition, and in having crystallized under more oxidizing (Earth-like) conditions. The shergottites are believed to be derived from a parent body which is distinct from that of the eucrites, and which has produced basaltic melts with a crystallization age of 0.65 b.y. (Nyquist *et al.*, 1979b), which would be younger than any other extraterrestrial basalt yet sampled. A possibly related meteorite, ALHA 77005, contains substantial amounts of olivine, but otherwise is mineralogically similar to Shergotty and apparently has a young crystallization age (Nyquist *et al.*, 1979a).

Other achondrites, mostly unique occurrences such as the nakhlites, angrites, and chassignites (Mason, 1962), apparently have been formed by igneous processes. These are ultrabasic materials which have not been shown to have close genetic relationships to either eucrites or shergottites, although Stolper *et al.* (1979)

Note: Eu = eucrite; Ho = howardite; Di = diogenite; Sh = shergottite. [a] Classification according to Mason *et al.* (1979) or from description for meteorites discovered since 1962. [b] Bp = polymict breccia; Bm = monomict breccia; NB = not brecciated. [c] E = equilibrated; U = unequilibrated; C = cumulate; R = recrystallized. [d] Reference to descriptive classification only (petrography or chemistry); (a) Mason *et al.* (1979); (b) Lomena *et al.* (1976); (c) Takeda *et al.* (1978a); (d) Olsen *et al.* (1978); (e) Antarctic Meteorite Newsletter V. 2 No. 3, 1979. (g) LaCroix (1926); (h) Duke and Silver (1967); (i) Zavaritskii and Kvasha (1952); (j) Mason (1967); (k) Wahl (1952); (l) Mason and Wiik (1966); (m) Steele and Smith (1976); (n) Bunch *et al.* (1976); (o) Fredriksson and Keil (1963); (p) Stolper (1977); (q) Reid (1974); (r) Jérome and Christopher Michel-Levy (1971); (s) Hutchison *et al.* (1977); (t) Lovering (1975); (u) Frost (1971); (v) Hess and Henderson (1949); (w) Binns (1967); (x) Lange and Keil (1976); (y) Foshag (1938); (z) Stolper and McSween (1979); (aa) Smith and Hering (1979); (bb) Engelhardt (1963); (cc) Takeda *et al.* (1978b); (dd) Takeda (1979).

Table 1.2.8.1 Basaltic material in meteorites.

Meteorite	Class[a]	Structure[b]	Eucrite[c] Type	Reference[d]
Adalia	Eu	NB	E	(a)
Aioun El Atrouss	Eu/Di	Bp	E	(b) (c)
Allan Hills-76005	Eu	Bm	U	(d)
Allan Hills-77302	Eu	Bm	U	(e)
Allan Hills-77005	Sh	NB	C	(e)
Allan Hills-78006	Ho	Bp	—	(f)
Allan Hills-78132	Eu	Bm	E	(f)
Allan Hills-78158	Eu	Bm	E	(f)
Allan Hills-78165	Eu	Bm	E	(f)
Béréba	Eu	Bm	E	(g)
Bholghati	Ho	Bp	—	(a)
Bialystok	Ho	Bp	—	(a)
Binda	Eu	Bm	C	(h)
Brient	Ho	Bp	—	(i)
Bununu	Ho	Bp	—	(j)
Cachari	Eu	Bm	E	(k)
Chaves	Ho	Bp	—	(h) (a)
Chervony Kut	Eu	NB	E	(k)
Emmaville	Eu	Bm	E	(a)
Frankfort	Ho	Bp	—	(l)
Haraiya	Eu	Bm	E	(h)
Ibitira	Eu	NB	R	(m)
Jodzie	Ho	Bp	—	(n)
Jonzac	Eu	Bm	E	(k)
Juvinas	Eu	Bm	E	(k)
Kapoeta	Ho	Bp	—	(o)
Kirbyville	Eu	Bm	E	(a)
Lakangaon	Eu	Bm	E	(p)
LeTeilleul	Ho	Bp	—	(k)
Luotolax	Ho	Bp	—	(k)
Macibini	Ho	Bp	—	(q)
Malvern	Ho	Bp	—	(r)
Mässing	Ho	Bp	—	(a)
Medanitos	Eu	NB?	C?	(s)
Millbillillie	Eu	Bm	E	(a)
Moama	Eu	NB	C	(t)
Molteno	Ho	Bp	—	(u)
Moore County	Eu	NB	C	(v)
Nagaria	Eu	Bm	E	this work
Nobleborough	Ho	Bp	—	(a)
Nuevo Laredo	Eu	Bm	E	(h)
Padvarninkai	Eu	Bm	E	(w)
Palo Blanco Creek	Eu	Bm	E	(x)
Pasamonte	Eu	Bm	U	(y)
Pavlovka	Ho	Bp	—	(h)
Peramiho	Eu	Bm	—	(k)
Petersburg	Ho	Bp	—	(h)
Pomozdino	Eu	Bm	?	(a)
Serra de Magé	Eu	NB	C	(l)
Shergotty	Sh	NB	C	(z) (aa)
Sioux County	Eu	Bm	E	(h)
Stannern	Eu	Bm	E	(bb)
Washougal	Ho	Bp	—	(h)
Yamato-7307(e)	Ho	Bp	—	(a)
Yamato-74159	Eu	Bp	U	(cc)
Yamato-74356	Eu	Nb	E	(dd)
Yamato-74450	Eu	Bm	U	(cc)
Yamato-75011	Eu	Bp	U	(cc)
Yamato-75015	Eu	Bp	U	(cc)
Yurtuk	Ho	Bp	—	(a)
Zagami	Sh	NB	C	(w)
Zmenj	Ho	Bp	—	(a)

discussed possible genetic relationships among the hypothetical parental materials for this range of meteorites.

1.2.8.2 Reference Suite

A current list of eucrites, howardites and shergottites (Table 1.2.8.1) contains 62 meteorites.* The sample suite assembled for this study has been chosen to exhibit the range of basaltic compositional and textural types among eucrites and shergottites. Selection was also constrained by sample availability; therefore, the suite is biased towards the larger falls, which are also those for which the most data already exist. Howardites have been excluded because of the small size of most of their clasts, and their polymict character. Six basaltic meteorites were selected: Juvinas, Pasamonte, Stannern, Nuevo Laredo, Aioun el Atrouss and Shergotty. Detailed descriptions and new analytical data for these meteorites are given in Appendix A-8.

In any plot of bulk eucrite compositions, there is a clustering of points around an apparent preferred composition (e.g., see Fig. 1.2.8.7). Eucrites with compositions close to this preferred composition are referred to here as **main group** eucrites. Juvinas is an example of the most common eucrite composition. The bulk composition of Pasamonte is close to the main group, though with a slightly lower Mg/(Mg + Fe) ratio. Most main group eucrites contain an equilibrated pyroxene assemblage consisting of pigeonite and augite with near constant Mg/(Mg + Fe) ratios. In contrast, Pasamonte has strongly-zoned, unequilibrated pyroxenes. Stannern is one of the most studied eucrites and is of particular interest because of its relatively high content of incompatible trace elements. Compared with the main group eucrites, Stannern has a similar Mg/(Mg + Fe) ratio but higher contents of large-ion-lithophile trace elements and of alkalies. For example, Stannern has higher REE contents than Juvinas and has a negative europium anomaly. Nuevo Laredo was selected because it also has higher incompatible element concentrations than the main group eucrites, but this occurs in conjunction with a substantially lower Mg/(Mg + Fe) ratio. Nuevo Laredo contains quench textures, and may be the most rapidly cooled of the eucrite melts. The inclusion of Aioun el Atrouss in the sample suite was governed by availability and by the virtual lack of any data on this unique meteorite. Aioun el Atrouss is a mixture dominantly of diogenite with distinct clasts of eucrite, which are evidence for a complex history following primary crystallization. Data are presented here only for the eucrite portion of Aioun el Atrouss. The Shergotty meteorite was included as a representative of the shergottite class of meteoritic basalts and as a sample apparently derived from a different parent body.

For Juvinas, Pasamonte, Stannern and Nuevo Laredo, studies were made of individual clasts in the breccias in order to assess the amount of interclast variability. Where possible, large clasts with igneous textures were identified and were separated to provide subsamples of unequivocal melt fragments with no adhering matrix. These subsamples were analyzed by instrumental neutron activation. Portions were cut from the clasts and polished thin sections were made for petrographic and mineral chemical studies by electron microprobe. About 90 mg to 1 g of material remains from each of the analyzed clasts and is available for further study.

1.2.8.3 Petrography of meteoritic basalts

The meteoritic basalts contain calcic plagioclase and clinopyroxene with minor amounts of accessory phases that include orthopyroxene, silica (cristobalite, tridymite, quartz), phosphate, oxides (chromite, ilmenite, magnetite), sulfide (troilite), and metallic iron. Fayalitic olivine occurs in Ibitira and Shergotty. Forsteritic olivine is present in ALHA 77005 and occurs as mineral fragments in some howardites, where it may represent a related cumulus or residual mineral. The eucrites and shergottites represent distinct basalt types and will be described separately below.

Eucrite basalts

The dominant minerals are calcic plagioclase (~35%), pigeonite and augite (~60%), with about 5% of accessory phases (cristobalite, tridymite, quartz, chromite, ilmenite, troilite, metal). Hydrous phases have not been observed. Grain size ranges from amorphous or cryptocrystalline to coarse, and textures include a variety of subophitic, variolitic (fan-spherulitic) and intersertal textures. Related rocks of possible cumulate origin have medium to coarse-grained equigranular textures. For the most part, the following descriptions are based on individual rock fragments from brecciated meteorites. Some rock fragments are modified by shock processes; no modification of textures by weathering is observed.

Vitric texture. Rare fragments of dark amorphous "glass" occur in the Nuevo Laredo meteorite, in which the fragments generally exhibit the finest-grained igneous textures found in the eucrites. Duke and Silver

**Mesosiderites and diogenites are excluded from Table 1.2.8.1.*

(1967) show an example of such a glass containing a phenocryst of clinopyroxene. Other examples of this vitric material in eucrites contain small cavities, possibly vesicles (Fig. 1.2.8.1a). This material is apparently structureless and is distinct from other vitric material apparently formed by shock processes, such as the irregular glass pods and veins in Stannern (Duke and Silver, 1967), Cachari (Fredriksson and Kraut, 1967), and Bereba, which also contain highly strained mineral fragments.

Intersertal texture. This texture occurs in fine-grained fragments, principally in Nuevo Laredo, where grain sizes of 0.02–0.2 mm are typical (Fig. 1.2.8.1a). Pyroxene grains typically lie between plagioclase laths in this texture, rather than being penetrated by plagioclase. Vesicles are not present.

Subophitic texture. Varieties of subophitic texture are the most abundant texture, and are found in all brecciated eucrites. Subophitic textures range from fine-grained to medium-grained varieties (Fig. 1.2.8.1b,c). In these textures, plagioclase occurs primarily as laths or plates, growing into and partially surrounded by clinopyroxene. Accessory phases are concentrated in interstitial areas, where they commonly form a fine-grained mesostasis-like intergrowth with plagioclase. The textures are nonporphyritic.

Two varieties of pyroxene crystallization are observed. In Pasamonte, pyroxenes have pigeonite cores, which are zoned outward toward either pigeonite or augite rims. Augite is texturally late. The pyroxenes do not contain exsolution lamellae which are visible optically, but are exsolved on a submicroscopic scale (Takeda *et al.*, 1976). In Juvinas, Nuevo Laredo, and Stannern, pigeonite is the predominant clinopyroxene (Nuevo Laredo has a small amount of primary augite) and does not show significant zonation. In these textures, pyroxenes show exsolution lamellae ranging in size from barely visible in the microscope to several microns in width.

"Vesicles" have been reported in the Juvinas meteorite (LaCroix, 1926) but have not been observed by the authors in fragments with subophitic texture, which are dominant in that meteorite.

Variolitic texture. A wide range of variolitic textures (also called fan-spherulitic, see Table 1.3.3.1) are present. Pasamonte, Y-74159, Y-74450 and Stannern show particularly good examples. In the finest-grained variolitic textures, sheaves of pyroxene and plagioclase are closely intergrown into a felted mass. Patches of this sheaf-like texture are intergrown with coarser-grained material in Y-74450. Coarser varieties occur with radiating plagioclase and pyroxene (Fig. 1.2.8.1d); unfortunately, these have not been seen in fragments large enough to observe the intervariole texture. The variolitic textures are not reported to be vesicular, except that the recrystallized meteorite Ibitira contains both cavities and a relict variolitic texture. Vesicles have been described in Ibitira by Wilkening and Anders (1975) and by Steele and Smith (1976).

Porphyritic texture. Only one eucrite, Y-74450, contains basalt with distinctly porphyritic textures (Takeda *et al.*, 1978a). In this texture, clinopyroxene cores, commonly equant in shape, are surrounded by a finer-grained, somewhat variolitic texture (Fig. 1.2.8.1e). Although the fragments are small and modal statistics are poor, the porphyritic basalt appears to be richer in pyroxene than basalts with subophitic textures.

Other textures. Rocks which appear to be cumulates have medium- to coarse-grained equigranular textures (Fig. 1.2.8.1f). Evidence for preferred orientation of pyroxene grains has been reported by Hess and Henderson (1949) and by Duke and Silver (1967). Exsolution lamellae are prominent in the pyroxenes, and the original pigeonite has partially or completely inverted to orthopyroxene (Hess and Henderson, 1949; Duke and Silver, 1967). The Sioux County meteorite contains some very coarse-grained rock fragments (grain size up to 1 cm). These have not been studied petrographically, but appear to fall between the subophitic and equigranular cumulate-like textures macroscopically.

In the Aioun el Atrouss meteorite, irregular patches of granoblastic material are closely intergrown with subophitic basalt (Fig. 1.2.8.1g). This texture has been interpreted to represent recrystallized basalt breccia, in which the granoblastic material corresponds with recrystallized breccia matrix (Duke, 1978). Small rock fragments with granoblastic textures, which occur commonly in howardites, may represent similar recrystallized material.

Shergottites

The textures of Shergotty and Zagami, the two examples of the shergottite class, consist of 75% prismatic clinopyroxene grains with strong planar orientation (Duke, 1968) and 25% interstitial tabular plagioclase and accessory phases (Fig. 1.2.8.1h). Both Ca-rich and Ca-poor pyroxenes are strongly zoned (Duke, 1968; Stolper and McSween, 1979). The accessory phases are silica, titano-magnetite, whitlockite, ilmenite and pyrrhotite (Smith and Hervig, 1979; Stolper and McSween, 1979). These meteorites are not brecciated but have

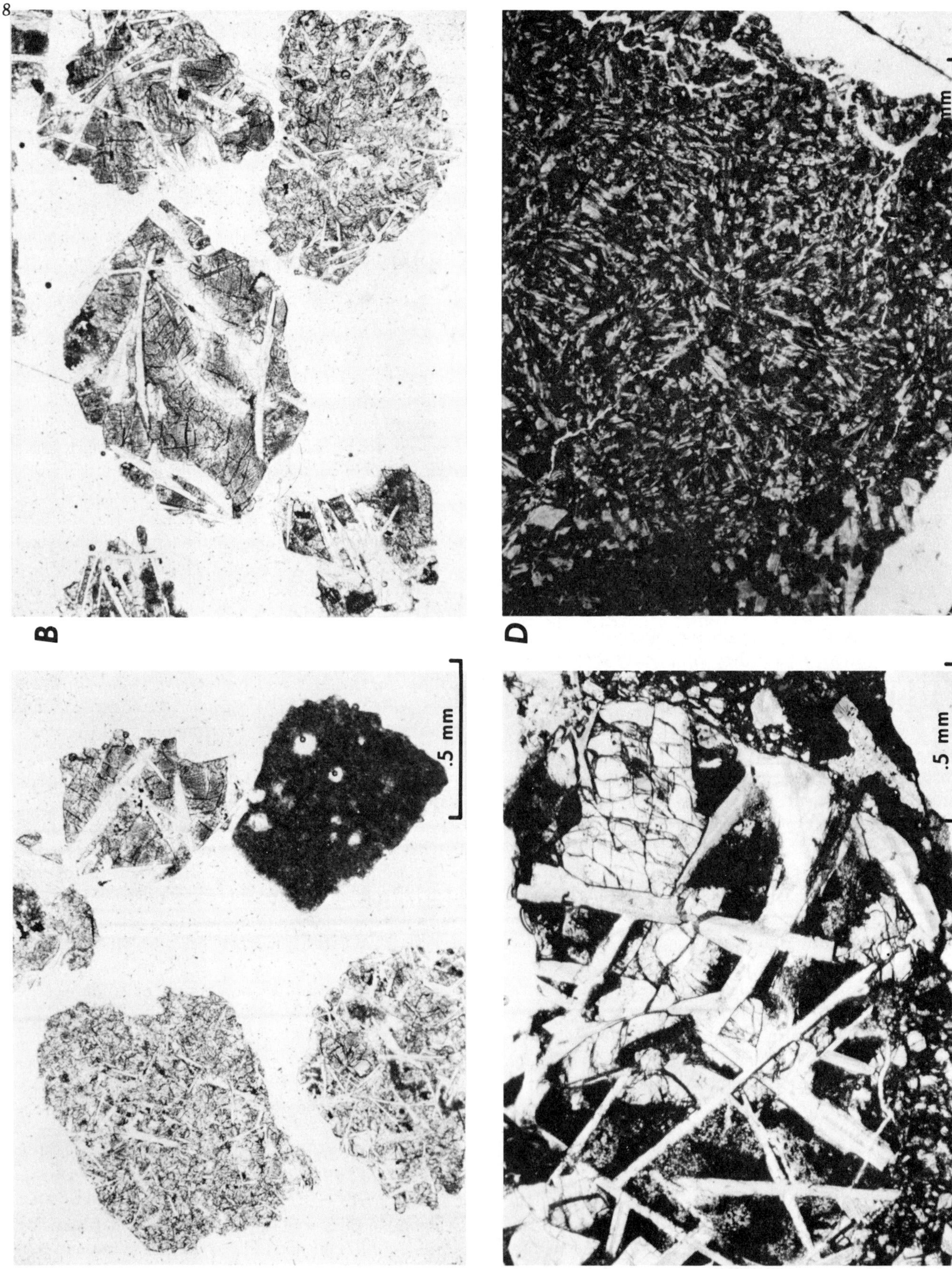
A
B
C
D
.5 mm
.5 mm
mm

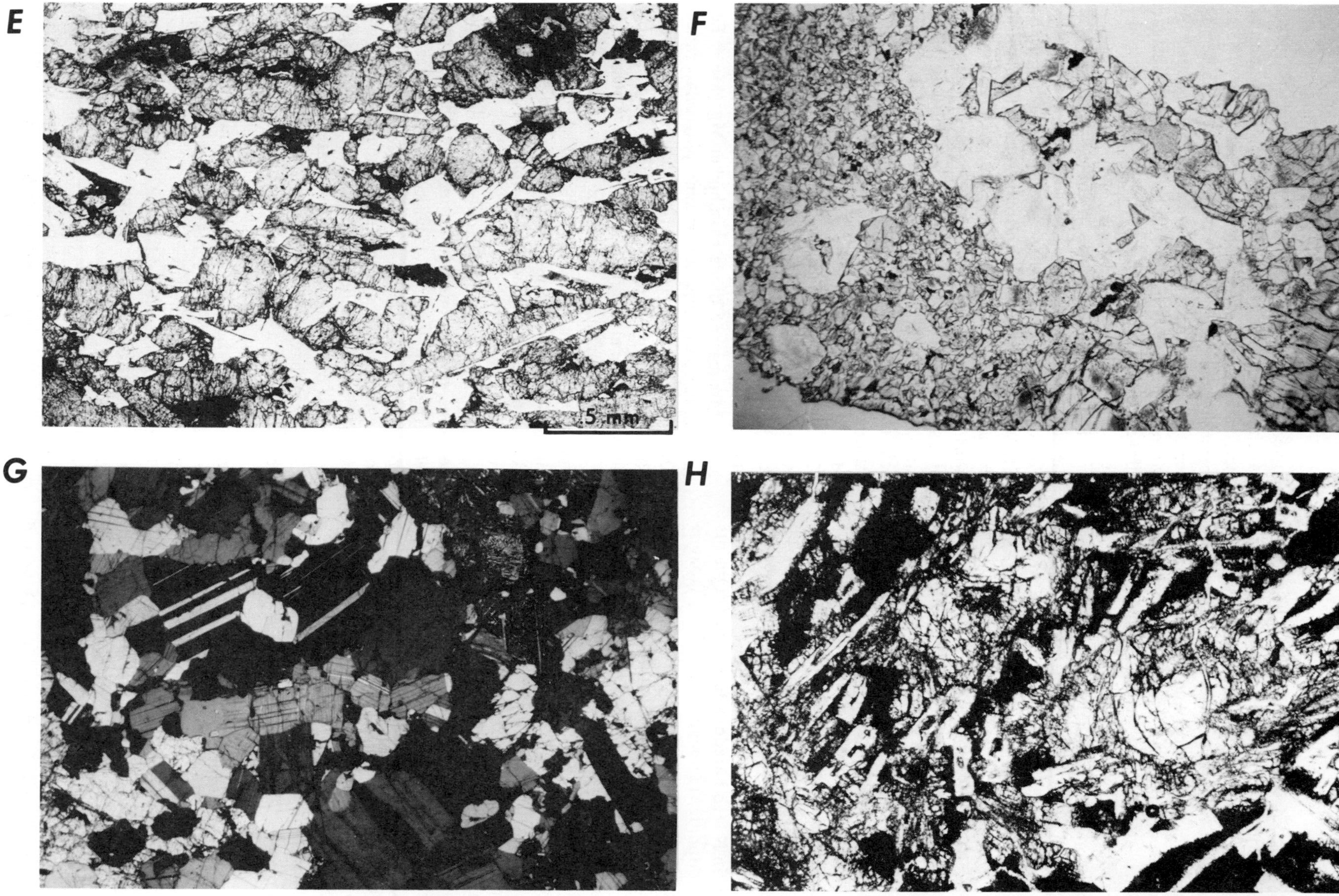

1.2.8.1 Photomicrographs of basaltic meteorite textures: **a**. Nuevo Laredo; vitric, vesicular breccia fragment on right, intersertal fine-grained fragment on left. **b**. Nuevo Laredo, varieties of subophitic texture. **c**. Pasamonte, subophitic texture. **d**. Stannern, variolitic texture. **e**. Yamato 74450, pigeonite phenocrysts. Cores are uniform magnesian pigeonite, grading to more iron-rich at rims. Tabular plagioclase apparently nucleated on pyroxene phenocrysts. **f**. Moore County, cumulate texture. Note complex exsolution in pyroxenes. **g**. Aioun el Atrouss; fine recrystallized breccia in contact with recrystallized subophitic basalt. **h**. Shergotty; preferred orientation of clinopyroxene and maskelynite (width of field is 8 mm.).

been strongly shocked, with all of the plagioclase transformed to a vitreous glass (maskelynite) and the silica transformed to cryptocrystalline cristobalite, without modification of the original texture. The origin of the texture is uncertain, but Stolper and McSween (1979) believe it to be cumulate. ALHA 77005 consists of olivine, Ca-rich and Ca-poor pyroxene, maskelynite and chromite with minor ilmenite, troilite and whitlockite in a hypidiomorphic granular to medium-grained texture (McSween *et al.*,1979b). This meteorite may represent a more basic differentiate related to the shergottites.

Range of textures in individual eucrites

Most brecciated eucrites show a prominent textural type among their clasts, in most cases a subophitic texture. Characteristic combinations of igneous texture, subsolidus inversion of pyroxenes, shock effects and brecciation tend to render each meteorite distinctive and recognizable to an experienced observer. However, there are minor variations of textural type observed in the clasts of each meteorite, which probably would be extended if larger samples were to be studied. For example, in Nuevo Laredo, a few vitric fragments and a few subophitic fragments containing pyroxene grains as large as 1 mm are found mixed with more common intersertal and subophitic fragments containing 0.1–0.3 mm pyroxene grains. The predominant fragments in Pasamonte are subophitic; however, a few variolitic clasts occur. In Y-74450, most fragments appear to be variolitic/porphyritic, with the exception of some that are more typically subophitic. Such variations are consistent with the environment of mechanical mixing that must have existed during breccia formation. However, to our knowledge, the varieties observed in any brecciated eucrite are consistent with derivations from closely associated igneous bodies. Mixing has been limited and can be interpreted as having been confined to a single lava flow or intrusion, or to very closely related bodies. This conclusion is supported by mineralogical data and by crystallographic study of the pyroxenes (Takeda, 1979).

1.2.8.4 Chemical composition

Major elements

Selected major element analyses for meteoritic basalts are given in Table 1.2.8.2. The quality of available analytical data on eucrites is variable and several older analyses that give extreme values should be repeated using modern techniques. The majority of modern analyses derive from the papers by Duke and Silver (1967) and by McCarthy *et al.* (1972, 1973, 1974). Much of our understanding of the bulk composition of meteoritic basalts and much of the discussion of their formation depend upon the data base from those papers. This data base has recently been expanded by Mason *et al.* (1979) and should be extended to include all meteoritic basalts where adequate material is available.

The analyses and norms in Table 1.2.8.2 represent the compositional range of eucrites and shergottites. For a eucrite, there is a close relationship between the CIPW norm derived from the major element composition and the modal analysis. The major element variations are principally related to two parameters: the pyroxene/plagioclase ratio and the Mg/(Mg + Fe) ratio in pyroxenes. Most eucrite compositions cluster near the compositions of Juvinas, Sioux County and Pasamonte. Nuevo Laredo is distinguished by a lower Mg/(Mg + Fe) ratio and by a slightly lower pyroxene/plagioclase ratio than the main group. The Yamato meteorites Y-74450 and Y-74159 have more modal pyroxene, with a higher Mg/(Mg + Fe) ratio than the main group. The meteorites Moore County, Serra de Magé, Moama and Binda have higher Mg/(Mg + Fe) ratios and show a range of pyroxene/plagioclase ratios. The alkali contents of eucrites generally are low, although Nuevo Laredo and Stannern are noticeably richer than the main group. The normative plagioclase contents of cumulate eucrites are slightly more calcic than those of the basaltic members. The eucrites, like lunar basalts, are free of ferric iron and water. A FeO/MnO ratio of about 33 (by weight) is characteristic of the eucrites.

The shergottites also have iron-rich basaltic compositions, but are much lower in Al_2O_3 than eucrites, reflecting their lower feldspar contents. On textural grounds, the shergottites are believed to be cumulates, so their compositions cannot be considered to represent basalt liquid compositions. The Ca/Al ratios are higher than those of the eucrites due to a greater abundance of Ca-rich pyroxene, which is approximately equal in abundance to Ca-poor pyroxene. The shergottites also are distinct from the eucrites in their higher alkali contents ($Na_2O \sim 1.3$, $K_2O \sim 0.2$ wt.%, i.e., more than twice eucritic abundances), in containing significant Fe^{3+}, and in their high phosphorus contents ($P_2O_5 \sim 0.5$ wt.%). The unique meteorite, ALHA 77005, differs in bulk composition from the shergottites, reflecting its substantial content of olivine, but bears a chemical signature that ties it closely to the shergottites, particularly with respect to trace element abundances, volatile/involatile element ratios and oxidation state.

Table 1.2.8.2 Chemical compositions of eucrites.

	Sioux County *(2)*	Pasamonte *(2)*	Juvinas *(2)*	Béréba *(2)*	Cachari *(2)*	Haraiya *(2)*	Macibini *(2)*	Nuevo Laredo *(1)*
SiO_2	49.03	48.07	49.34	49.13	48.26	48.69	49.32	49.46
TiO_2	0.62	0.72	0.64	0.70	0.63	0.56	0.72	0.95
Al_2O_3	12.84	12.25	13.00	12.75	12.85	12.49	12.12	11.78
Cr_2O_3	0.35	0.31	0.34	0.32	0.32	0.32	0.42	0.29
FeO	18.58	19.10	18.82	19.83	19.10	19.25	18.31	20.10
MnO	0.56	0.56	0.56	0.55	0.59	0.59	0.54	0.56
MgO	7.11	6.53	7.27	6.80	7.14	7.11	8.37	5.46
CaO	10.35	10.24	10.38	10.48	10.25	10.26	9.97	10.40
Na_2O	0.45	0.51	0.47	0.47	0.51	0.50	0.48	0.57
K_2O	0.04	0.04	0.04	0.04	0.05	0.03	0.05	0.05
P_2O_5	0.09	0.11	0.09	0.10	0.08	0.10	0.11	0.11
TOTAL	100.02	98.44	100.95	101.95	99.78	99.90	100.41	99.73
Qz	2.17	1.90	1.83	1.58	0.66	1.30	1.61	4.00
Or	0.24	0.24	0.24	0.24	0.30	0.18	0.30	0.30
Ab	3.81	4.32	3.98	3.98	4.32	4.23	4.06	4.82
An	32.90	31.02	33.25	32.56	32.63	31.75	30.77	29.44
Di	15.15	16.24	14.98	15.99	15.03	15.72	15.17	18.38
Hy	43.87	42.68	44.77	44.82	45.02	44.98	46.29	40.33
Ol	—	—	—	—	—	—	—	—
Cm	0.52	0.46	0.50	0.47	0.47	0.47	0.62	0.43
Ilm	1.18	1.37	1.22	1.33	1.20	1.06	1.37	1.80
Ap	0.20	0.24	0.20	0.22	0.17	0.22	0.24	0.24
Mg′	0.44	0.40	0.43	0.40	0.43	0.42	0.48	0.35

Table 1.2.8.2 *(Continued)*

	Stannern *(1)*	Yamato 74159	Yamato 74450 *(3)*	Moore County *(2)*	Serra de Magé *(2)*	Moama *(4)*	Binda *(2)*	Shergotty *(5)*
SiO_2	49.33	49.04	49.36	48.32	48.42	48.58	50.41	50.1
TiO_2	0.96	1.09	1.04	0.44	0.17	0.22	0.23	0.92
Al_2O_3	12.34	10.35	10.82	12.65	12.69	13.74	6.94	6.68
Cr_2O_3	0.28	0.44	0.33	0.44	0.54	0.61	0.75	0.18
FeO	17.92	19.23	18.26	17.24	16.18	14.85	16.83	18.66
MnO	0.50	0.53	0.51	0.52	0.55	0.50	0.54	0.50
MgO	6.36	8.29	8.06	9.41	11.35	11.89	17.75	9.40
CaO	10.58	9.48	9.52	9.37	9.08	9.47	5.82	10.03
Na_2O	0.60	0.58	0.51	n.d.	n.d.	0.22	0.20	1.28
K_2O	0.08	0.07	0.06	0.02	0.007	0.01	<.01	0.16
P_2O_5	0.13	0.07	0.10	0.04	0.06	—	0.04	0.71
TOTAL	99.08	99.17	98.57	98.45	99.05	100.09	99.52	100.11 †
Qz	3.67	1.94	3.44	2.43	0.68	—	—	0.89
Or	0.47	0.41	0.35	0.12	0.06	0.06	0.06	0.95
Ab	5.08	4.91	4.32	—	—	1.86	1.69	10.83
An	30.74	25.43	27.06	34.46	34.60	36.48	18.04	12.01
Di	17.77	17.83	16.47	9.85	8.37	8.67	8.67	28.09
Hy	38.83	45.78	44.26	50.02	54.10	49.16	65.69	41.63
Ol	—	—	—	—	—	2.55	3.76	2.16
Cm	0.41	0.65	0.49	0.65	0.80	1.27	1.10	0.27
Ilm	1.82	2.07	1.98	0.84	0.32	0.90	0.44	1.75
Ap	0.28	0.15	0.22	0.09	0.13	0.42	0.09	1.55
Mg′	0.41	0.46	0.47	0.52	0.58	0.61	0.68	0.48

References: *(1)* Duke and Silver (1967); *(2)* McCarthy *et al.* (1973); *(3)* Takeda *et al.* (1978a); *(4)* Lovering (1975); *(5)* Duke (1968).

†1.49% Fe_2O_3 included in total.

Trace elements

Trace element data for selected eucrites are given in Table 1.2.8.3. Ahrens and Danchin (1971) summarized the available eucrite data, but a substantial number of additional analyses are now available (e.g., Schmitt *et al.*, 1972; Laul *et al.*, 1972; McCarthy *et al.*, 1972, 1973; Wänke *et al.*, 1977; Morgan *et al.*, 1978). Figures 1.2.8.2a and b show trace element data for eucrites, two of which are cumulate eucrites, normalized to type 1 carbonaceous chondrite abundances. Refractory lithophile trace elements are enriched over C1 chondrite values (~ 13 times C1 chondrites) but are essentially unfractionated except for Sc, which is depleted relative to elements of similar volatility by about a factor of 2. The main group eucrites have REE patterns which are unfractionated relative to chondritic abundances at levels about 6–10 times chondritic (Fig. 1.2.8.3). Stannern and Nuevo Laredo have higher REE abundances and negative europium anomalies. The more magnesian (cumulate) eucrites have lower absolute REE abundances (1–5 times chondritic) and Moore County, Serra de Magé and Moama have positive europium anomalies (see summary in Consolmagno and Drake, 1977). In contrast, the more volatile lithophile trace elements are depleted, with Rb, Cs, Br and Bi present at approximately 6×10^{-2} times C1 chondrites.

Siderophile trace elements are strongly depleted in eucrites relative to chondritic abundances (Fig. 1.2.8.2b). There is no simple correlation between relative abundance and condensation temperature. Morgan *et al.* (1978) have suggested that siderophile abundances may be controlled by metal/silicate partitioning. These data suggest derivation of the eucritic liquids from sources that retained a metallic iron phase in the residuum or from sources that had lost metallic iron and the associated siderophile elements prior to formation of the eucrite basalts. Siderophile trace element abundances may also have been affected by meteoritic contamination subsequent to basalt crystallization, particularly if some eucrites are regolith breccias, so the observed values may represent only upper limits on original values. Figure 1.2.8.2b also shows relative abundances of chalcophile trace elements, which are depleted relative to C1 chondrites.

The shergottites are not characterized by the volatile element depletion that is so evident in the eucrite basalts. Laul *et al.* (1972), Stolper (1979) and Stolper and McSween (1979) have stressed similarities between trace element abundances in shergottites and in terrestrial basalts. This similarity extends to the volatile, refractory and siderophile trace elements as well, and contrasts strongly with the trace element abundance patterns in eucrites. The chondrite-normalized REE pattern of Shergotty is slightly heavy REE-enriched at about 6–8 times chondritic levels with no europium anomaly.

1.2.8.5 Mineral chemistry

Mineral assemblages in meteoritic basalts are simple, consisting of clinopyroxene (dominantly pigeonite with augite subordinate except in shergottites), and calcic plagioclase with minor free silica, ilmenite, chromite, troilite and metallic iron. Orthopyroxene and augite are present as products of the exsolution and inversion of pigeonite. Magnesian olivine is not present in rocks with basaltic textures, but ultramafic, apparently cumulate, material found in howardites contains minor olivine, and ALHA 77005, a cumulate which seems to be related to the shergottites, contains magnesian olivine as a major phase. Fayalite occurs in Ibitira and Sher-

Table 1.2.8.3 Selected trace element abundances in eucrites and Shergotty, *ppm.*

	Stannern	Nuevo Laredo	Pasamonte	Juvinas	Aioun el Atrouss	Moore County	Binda	Shergotty
K	550*(1)*	400*(4)*	355*(1)*	340*(1)*	308*(14)*	180*(1)*	85*(1)*	—
Rb	0.827*(3)*	0.380*(3)*	0.242*(3)*	0.202*(3)*	—	0.061*(3)*	—	6.1*(13)*
Ba	50*(1)*	44*(8)*	33*(1)*	31*(1)*	—	19*(1)*	11*(1)*	90 *(8)*
Sr	92*(1)*	84*(9)*	80*(1)*	78*(1)*	80*(14)*	64*(1)*	33*(1)*	65 *(8)*
Sm	3.17*(6)*	2.19*(5)*	1.90*(6)*	1.70*(6)*	2.6*(14)*	0.94*(7)*	—	1.25*(8)*
Eu	0.83*(6)*	0.75*(5)*	0.68*(6)*	0.62*(6)*	0.59*(14)*	0.59*(7)*	—	0.49*(8)*
Y	26*(1)*	—	17*(1)*	16*(1)*	—	—	4.4*(1)*	—
Sc	32*(9)*	38*(10)*	26*(10)*	29*(10)*	33.5*(14)*	24*(10)*	—	52.4*(8)*
Th	0.50*(11)*	0.47*(11)*	0.52*(2)*	0.60*(11)*	—	0.06*(12)*	0.06*(12)*	0.3*(8)*
Zr	87*(1)*	67*(6)*	53*(1)*	46*(1)*	—	—	15*(1)*	—

References: *(1)* McCarthy *et al.* (1973); *(2)* Morgan *et al.* (1978); *(3)* Papanastassiou and Wasserburg (1969); *(4)* Duke and Silver (1967); *(5)* Schmitt *et al.* (1963); *(6)* Schmitt *et al.* (1964); *(7)* Schnetzler and Philpotts (1969); *(8)* Gast (1965); *(9)* Gast (1962); *(10)* Schmitt *et. al.* (1972); *(11)* Rowe *et al.* (1963); *(12)* Morgan and Lovering (1964); *(13)* Laul *et al.* (1972); *(14)* D. Blanchard, analyst.

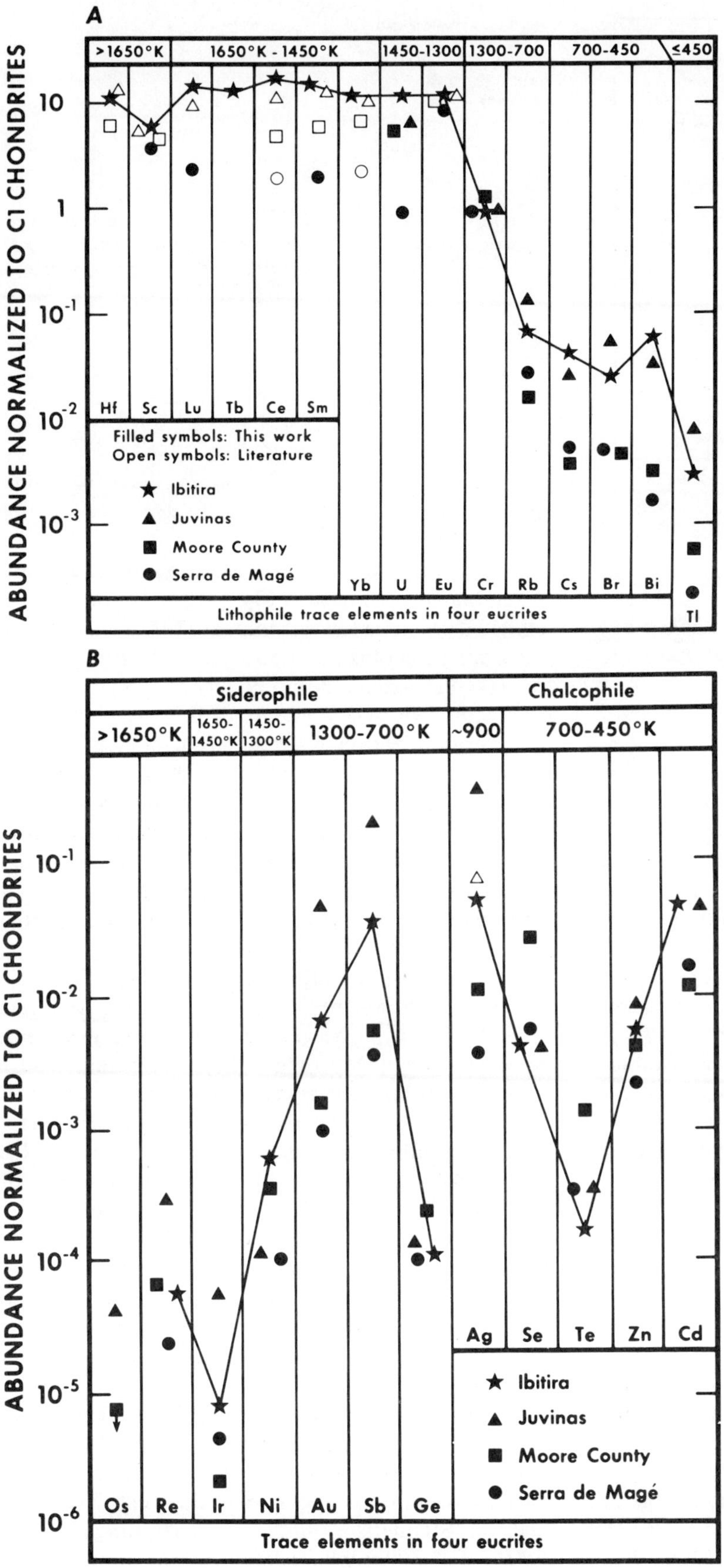

1.2.8.2 Trace elements in eucrites. **a**. Abundances relative to C1 chondrites of large-ion-lithophile trace elements, arranged in approximate order of condensation temperature, in four eucrites. **b**. Similar figure for siderophile and chalcophile trace elements in four eucrites. After Morgan *et al.* (1978).

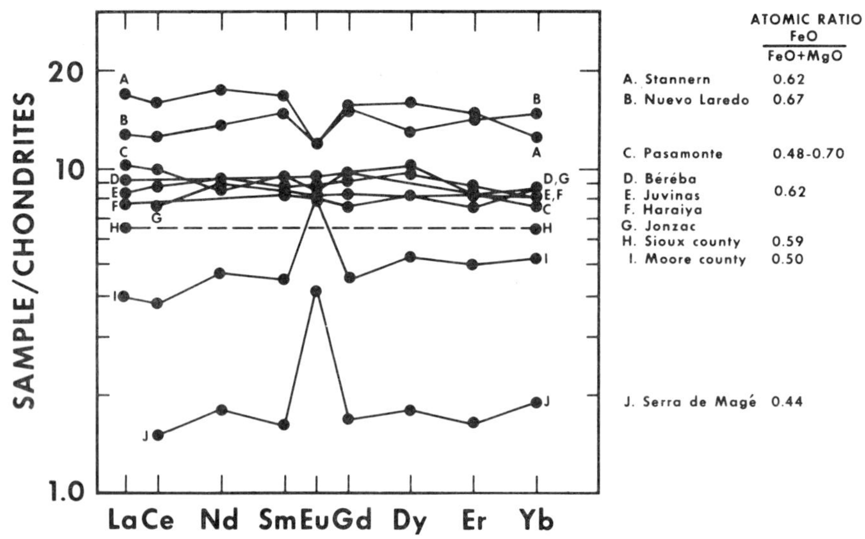

1.2.8.3 Chondrite-normalized rare-earth element abundances of eucrites. After Consolmagno and Drake (1977).

gotty. A complex, fine-grained mesostasis is present in some samples but has not been studied in detail. The shergottites also contain magnetite and whitlockite but lack metallic iron. Mineral compositional data are reported in Appendix A-8.

Pyroxene

The dominant primary pyroxene in all eucrites is pigeonite. Two specific types of pigeonite occurrences have been recognized. *Unequilibrated* eucrites are characterized by the presence of pigeonites with a wide range of Mg/(Mg + Fe) values expressed largely as normal zoning. These pyroxenes do not exhibit microscopic exsolution. *Equilibrated* eucrites have pigeonite with only a narrow range of Mg/(Mg + Fe) values; however, individual grains show a range of Ca/(Ca + Fe) values, due to the presence of microscopic exsolution lamellae. Table 1.2.8.1 classifies eucrites according to this characteristic. The classification has been made petrographically for meteorites which have not been studied with the electron microprobe, based upon the presence or absence of exsolution lamellae and visible compositional zoning.

The pyroxenes of the Pasamonte meteorite are typical of the unequilibrated eucrites (Fig. 1.2.8.4a). The full range of compositions is shown in individual subcentimeter-size basalt clasts. Cores of pigeonites are magnesium-rich, with normal zoning toward Ca-rich or Fe-rich rims and small interstitial grains. Takeda *et al.* (1978a) found Ca-enrichment toward contacts with mesostasis areas and Fe-enrichment toward plagioclase crystal contacts. Some of the scatter in values for calcium contents of the Fe-rich pyroxenes may be due to incipient exsolution, which can be observed in X-ray patterns but not optically (Takeda *et al.*, 1976). The bulk meteorite Mg/(Mg + Fe) ratio is 0.38. The most magnesian pigeonite has a Mg/(Mg + Fe) ratio of 0.69, which is consistent with Stolper's (1977) experimental determination of the composition of the first pyroxene to crystallize from a liquid of eucrite composition.

Minor element contents of the pyroxenes reflect their crystallization history. Al_2O_3 content is about 0.13% and shows little variation with fractionation. This is consistent with petrographic and experimental evidence that pigeonite and plagioclase crystallized simultaneously from the beginning of crystallization. Titanium contents in the high magnesium cores are low and increase with decreasing Mg/(Mg + Fe), particularly in the latest-formed Ca-rich pyroxenes, which is consistent with the late and minor role of ilmenite. Chromium diminishes in pyroxenes with decreasing Mg/(Mg + Fe), with a greater decrease in pigeonite less magnesian than Mg/(Mg + Fe) = 0.65 that may be related to the late crystallization of chromite, which also is a minor interstitial phase (Bunch and Keil, 1971).

Unequilibrated eucrites are uncommon. Y-74450 and Y-74159 are the only other unequilibrated eucrites that have been studied in any detail (Takeda *et al.*, 1978a, b). In Y-74450, a magnesian pigeonite [Mg/(Mg + Fe) = 0.72] crystallized first, followed by pigeonites, which were richer in iron and calcium. The early pigeonites of Y-74450 are Al-rich, consistent with some crystallization of pyroxene prior to initiation of plagioclase crystallization.

Stannern is an example of an equilibrated eucrite (Fig. 1.2.8.4b). The bulk composition of Stannern is similar to that of Pasamonte, but the pyroxene compositional variation is distinctively different. The pyroxenes in Stannern exhibit a narrow range of Mg/(Mg + Fe) values which define a trend approximately

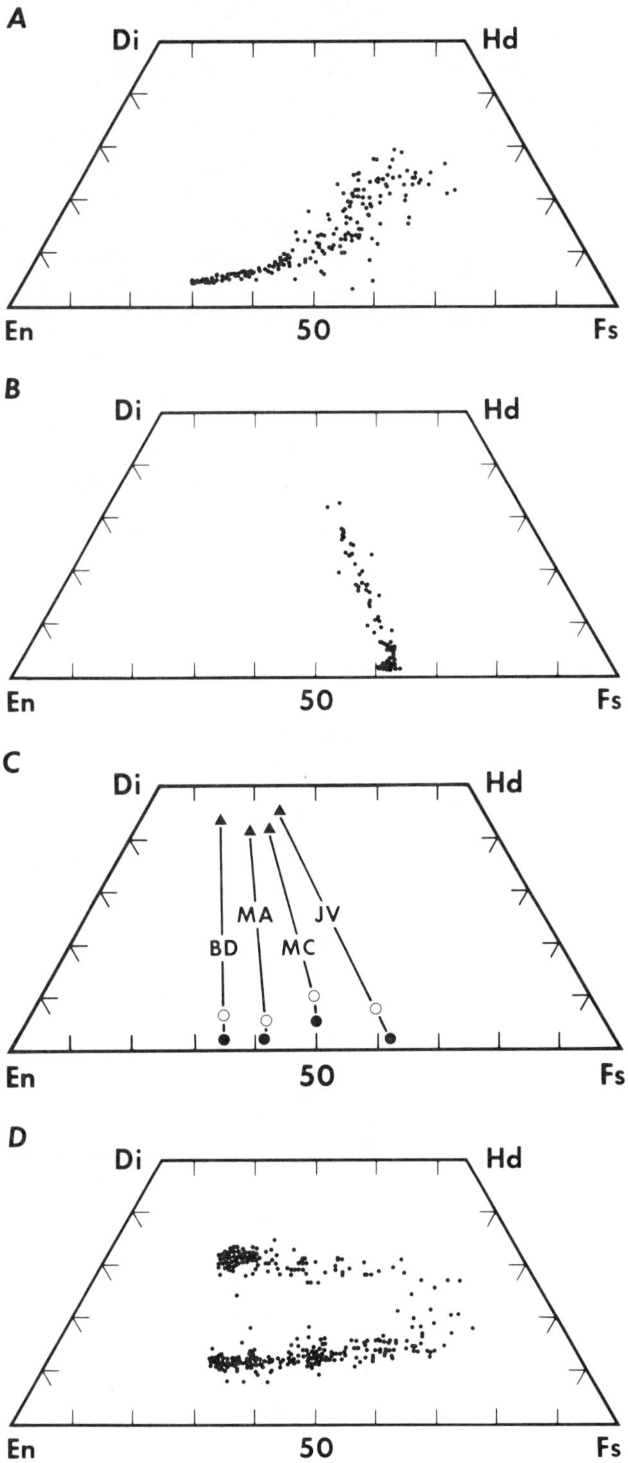

1.2.8.4 Pyroxene compositions in basaltic meteorites. **a**. Pasamonte, an unequilibrated eucrite. **b**. Stannern, an equilibrated eucrite. **c**. Three cumulate eucrites (BD-Binda, MA-Moama, MC-Moore County) plus one main group eucrite (JV-Juvinas). Open circles represent bulk compositions of pyroxenes; filled circles represent exsolution lamellae. **d**. Shergotty.

parallel to the Hd-Fs boundary of the quadrilateral. This trend in part or in whole reflects the presence of numerous thin lamellae of ferroaugite, which cannot be completely resolved by the microprobe, within the low Ca-pyroxene host. Within most equilibrated pyroxenes, a concentration of analyses at low-Ca values ($\sim Wo_2$) may indicate the composition of the host phase (clinohypersthene in Juvinas; Takeda *et al.*, 1976). The narrow ferroaugite lamellae mostly are too narrow for standard electron microprobe analysis; however, some analyses are as calcic as Wo_{30}. Average compositions are difficult to determine for pyroxenes, due to the complexity of exsolution and inversion features. Despite the obvious heterogeneity, the pyroxene analyses all plot along the same narrow trend. The constraint on $Mg/(Mg + Fe)$ values suggests that the general subsolidus equilibration between high and low-Ca phases occurred within a limited temperature range.

Stannern pyroxenes have low TiO_2 and Al_2O_3 contents (average TiO_2 ~0.25 wt.%). Ti and Al do not vary sympathetically with Fe or Ca content but there is a weak positive correlation between Ti and Al. Chromium is high and uniform (average 0.73 wt.%). Similar pyroxene relationships are found in the other equilibrated eucrites, Juvinas, Nuevo Laredo and Aioun el Atrouss. In Nuevo Laredo, texturally primary ferroaugite and pigeonite coexist in basaltic fragments (Takeda *et al.*, 1978a).

The fine-grained textures of the eucrites, such as Juvinas and Pasamonte, indicate that they crystallized from the magma rapidly. Under such crystallization conditions in the laboratory, pyroxenes grow with strong Fe/Mg zonation. Furthermore, exsolution of pigeonite does not occur in such rapid crystallization environments. These two characteristics, lack of exsolution and strongly-zoned pyroxenes are characteristic of the unequilibrated eucrites, such as Pasamonte. The equilibration observed in meteorites such as Juvinas implies that subsequent recrystallization has occurred either as a transformation during the initial crystallization and cooling episode, or through subsequent reheating and metamorphism.

The occurrence of two primary clinopyroxenes, augite and pigeonite, in Nuevo Laredo suggests that two pyroxenes would be stable near the solidus in the main group eucrites. However, in those meteorites, the primary pyroxene phase was pigeonite alone. Recrystallization to two discrete pyroxenes would be expected if the pyroxene zonation was homogenized much below the solidus. Exsolution in the main group eucrites has produced lamellae of relatively uniform dimensions and composition throughout the host pigeonite. This implies that exsolution occurred subsequent to homogenization. Since exsolution appears to have developed in homogeneous pigeonite grains, the homogenization must have occurred at near-solidus temperatures. Takeda *et al.* (1976) have concluded that the exsolution in Juvinas occurred at high temperature (~900°C). Liquidus temperatures are approximately 1160°C for Stannern and 1180°C for Juvinas and pyroxene, plagioclase and silica would all have begun to crystallize by 1090°C (Stolper, 1977). If homogenization took place slightly above solidus temperatures, the presence of residual liquid could have facilitated the process.

The pyroxenes of the cumulate eucrites, Moore County, Serra de Magé, Moama and Binda, show similarities to those of equilibrated eucrites (Fig. 1.2.8.4c). They typically have coarser exsolution lamellae and partial to complete inversion of primary pigeonite to orthopyroxene, as was shown for Moore County by Hess and Henderson (1949), who suggested its origin in a buried intrusive body. In howardites, such as Kapoeta, a wide range of pigeonite compositions is present (Fredriksson and Keil, 1963; Dymek *et al.*, 1976). It is probable that the brecciated pyroxene fragments include representatives of unequilibrated, equilibrated and cumulate eucrites; however, detailed studies of the total range of pyroxene types in howardites have yet to be made.

The shergottites contain two primary pyroxenes, pigeonite and augite (Stolper and McSween, 1979; Smith and Hervig, 1979), both of which are highly zoned (Fig. 1.2.8.4d). Our data show that the most magnesian pigeonite is $Wo_{12}En_{61}Fs_{27}$ and the most magnesian augite is $Wo_{33}En_{49}Fs_{18}$. Both have relatively homogeneous cores with highly-zoned outer margins. The two pyroxene trends are subparallel except for high Fe compositions where the two trends merge at approximately $Wo_{22}En_{16}Fs_{62}$. There is a natural break in compositions between high-Ca and low-Ca pyroxenes at approximately Wo_{24}, but many pigeonites have ferroaugite rims with Wo > 24. Exsolution lamellae with contrasting Ca contents, which are present in both types of pyroxene, record a lower temperature subsolidus history. The lamellae are very narrow (< 2 microns) and their presence contributes to the scatter of electron microprobe analytical values seen in Fig. 1.2.8.4d.

The shergottites apparently are cumulates. A well-defined planar orientation of prismatic pyroxene grains is present (Duke, 1968). The initiation of plagioclase crystallization probably was responsible for an abrupt decrease in aluminum content of the pyroxenes for compositions less magnesian than (augite) $Wo_{32}En_{41}Fs_{28}$ and (pigeonite) $Wo_{13}En_{48}Fs_{39}$.

Plagioclase

Eucrites contain plagioclase with high-Ca contents, generally in the range An_{80} to An_{90} with some interstitial grains as sodic as An_{65} (Duke and Silver, 1967). In unequilibrated eucrites such as Pasamonte, the range is from $Or_{0.06}Ab_5An_{95}$ to $Or_{1.2}Ab_{19}An_{80}$ and there are small differences, though with considerable overlap, from clast to clast. The feldspars have low but detectable contents of Ti, Fe and Mg. Pasamonte feldspars show apparent departures from stoichiometry (Si-Na-K averages 2.03, Al-Fe-Ca averages 0.93) and in this respect resemble the feldspars in lunar mare basalts.

Feldspars in equilibrated eucrites, such as Juvinas, show a comparable range in compositions from $Or_{0.5}Ab_6An_{93}$ to $Or_{1.7}Ab_{18}An_{80}$. Stannern, also equilibrated, has more sodic plagioclases, ranging from $Or_{0.4}Ab_{11}An_{89}$ to $Or_2Ab_{26}An_{72}$. The departures from stoichiometry in the equilibrated eucrites are less than in the unequilibrated. Stannern feldspars analyzed in this study average Si-Na-K of 2.004 and Al-Fe-Ca of 0.96. This close approach to stoichiometry is consistent with the occurrence of a thermal annealing event subsequent to primary crystallization. Relatively little work has been done on optical properties of eucrite plagioclase. Game (1957) reported varied optical properties in Juvinas plagioclase and Duke (1963) reported high but varied structural states from X-ray studies of the same meteorite. The persistence of evidence for disequilibrium in plagioclases contrasts with the equilibrated characters of the pyroxenes.

Plagioclase feldspar in the Shergotty meteorite has been converted by high shock pressures (≳250 kb) to maskelynite (Schaal and Hörz, 1977). The maskelynite preserves the original feldspar crystal forms and apparently preserves original compositions as well, including normal zoning. The range of compositions in Shergotty is from $Or_1Ab_{42}An_{56}$ to $Or_{4.5}Ab_{60}An_{39}$, reflecting the higher alkali contents of shergottites. The maskelynites contain around 0.6 wt.% FeO and low but detectable amounts of Ti and Mg. Departures from stoichiometry, assuming that maskelynites retain feldspar compositions, are greater than in the unequilibrated eucrites (Si-Na-K averages 2.04, Al-Fe-Ca averages 0.91).* Stolper and McSween (1979) have presented evidence that a potassium-rich phase present in the mesostasis is glass.

---

* *Errors in electron microprobe analysis make it very difficult to assess whether the apparent departures from strict stoichiometry in meteorite plagioclases are real. We believe that it is significant that the plagioclases in "unequilibrated" eucrites consistently show larger departures from stoichiometry than the plagioclases in "equilibrated" eucrites.*

Other phases

Free silica is present in all of the basalts. A summary of occurrences is found in Duke and Silver (1967). In cases where cristobalite or tridymite is the primary phase, some quartz typically is present, apparently as a late-stage crystallization product in interstices, rather than as an inversion product from the higher temperature polymorph. In Shergotty, the silica is a glassy form, devitrified to cristobalite (Duke, 1968).

Ilmenite and chromite are principal oxide phases, typically making up less than 0.5% by weight of the basalts. Their compositions in eucrites have been studied by Bunch and Keil (1971), who found that chromites contain 9–12% Al_2O_3, 3–4% TiO_2, and smaller amounts of V_2O_3, MgO, MnO and ZnO.

Ilmenite is rather pure $FeTiO_3$, containing less than 1% of Cr_2O_3, MnO, MgO, CaO and Al_2O_3. Titanomagnetite ($Mt_{37}Usp_{63}$) coexists with ilmenite ($Ilm_{95}Hm_5$) in Shergotty, defining an equilibration temperature of 860° C and $\log f_{O_2}$ of −4 (Stolper and McSween, 1979).

Sulfide (troilite) is found primarily in interstitial areas where it occurs with silica, plagioclase and minor metallic iron. Some rock fragments in the breccias are highly fractured, possibly by impact processes. Some of the fractures show localized grains of sulfide and appear bleached, apparently with loss of iron to the sulfide phase and, in some cases, crystallization of silica. In Shergotty, the sulfide is pyrrhotite, consistent with a more oxidized state for that meteorite (Smith and Hervig, 1979).

Metallic iron exists in the eucrite basalts in two forms (Duke, 1965): as a primary phase generally in the mesostasis, and as metal formed by reduction of pyroxene. The metal in the basalts is quite low in nickel (Lovering, 1964; Duke, 1965); the Co/Ni ratio is variable (Hewins and Klein, 1978).

Phosphate (whitlockite) occurs in abundances of 1–2% in Shergotty, but has not been observed in eucrites, except as grains on the interior of vesicles in Ibitira (R. Fruland, pers. comm., 1979).

Mineral proportions

Modal proportions of minerals are difficult to determine in most eucrites, due to the small size of breccia clasts and the difficulty of determining minerals in the breccia matrix. Because eucrite mineral compositions typically are simple, with little Al in pyroxenes, the modal and normative mineralogies are closely comparable. By weight, the proportions are typically 35–37% plagioclase, 59–60% pyroxene, 1–3% silica and 2% opaques. The modal silica content is less than that of the norm, but silica can be found petrographically in all of the eucrites. Nuevo Laredo is slightly poorer in plagioclase and richer in silica, which is consistent with the expected trend of cotectic compositions with decreasing Mg/(Mg + Fe) (see next section). Y-74159 and Y-74450 have slightly lower feldspar and higher pyroxene contents, which is consistent with the porphyritic nature of some of the lithic fragments. The equilibrated eucrites contain relatively small amounts of mesostasis. The cumulate eucrites, Moore County, Serra de Magé, Moama and Binda, show a range of feldspar-pyroxene proportions, with Binda being lowest in feldspar. Shergotty also is relatively depleted in feldspar, and apparently does not represent a basaltic liquid.

Variations within eucrite breccias

Most eucrites are breccias; therefore, before bulk analyses can be used to formulate and to test petrologic models, we must address the question of whether eucrite bulk analyses represent the compositions of individual rocks or are merely averages on a range of clast and matrix compositions. Most workers have treated the eucrites as monomict breccias because of the overall similarities observed among clasts in terms of petrography and mineral chemistry and because different samples of a eucrite have yielded the same bulk compositions. Other kinds of evidence that support this interpretation include the observation that unbrecciated eucrites (e.g., Ibitira; Wilkening and Anders, 1975) have compositions similar to the brecciated eucrites, and experimental studies (Stolper, 1977), which show that most eucrite compositions are multiply-saturated at one atmosphere liquidus temperatures.

Petrographic studies of eucrites reveal a range of textural types in clasts from a single meteorite, indicating a variety of thermal histories. Despite this range, the similarities in mineral assemblage and in mineral chemistry from clast to clast indicate that, even if the eucrites are polymict breccias, the individual fragments must be closely related and interclast differences must be minor.

We have studied several clasts from two eucrites, Stannern and Pasamonte, in order to determine whether clast-to-clast differences are significant (Reid and Barnard, 1979). The pyroxene compositions in Stannern are uniform within individual clasts (Fig. 1.2.8.5a), whereas strongly-zoned pyroxenes occur in most clasts in Pasamonte (Fig. 1.2.8.5b; see also the section on mineral chemistry and Appendix A-8).

The pyroxene and feldspar compositions in Stannern show little variation from clast to clast and INAA of three clasts show essentially identical REE patterns (Fig. 1.2.8.6a). The pyroxene and plagioclase compositions are more complex in Pasamonte clasts. Each clast shows a range of compositions with small differences from clast to clast. Within the clasts investigated,

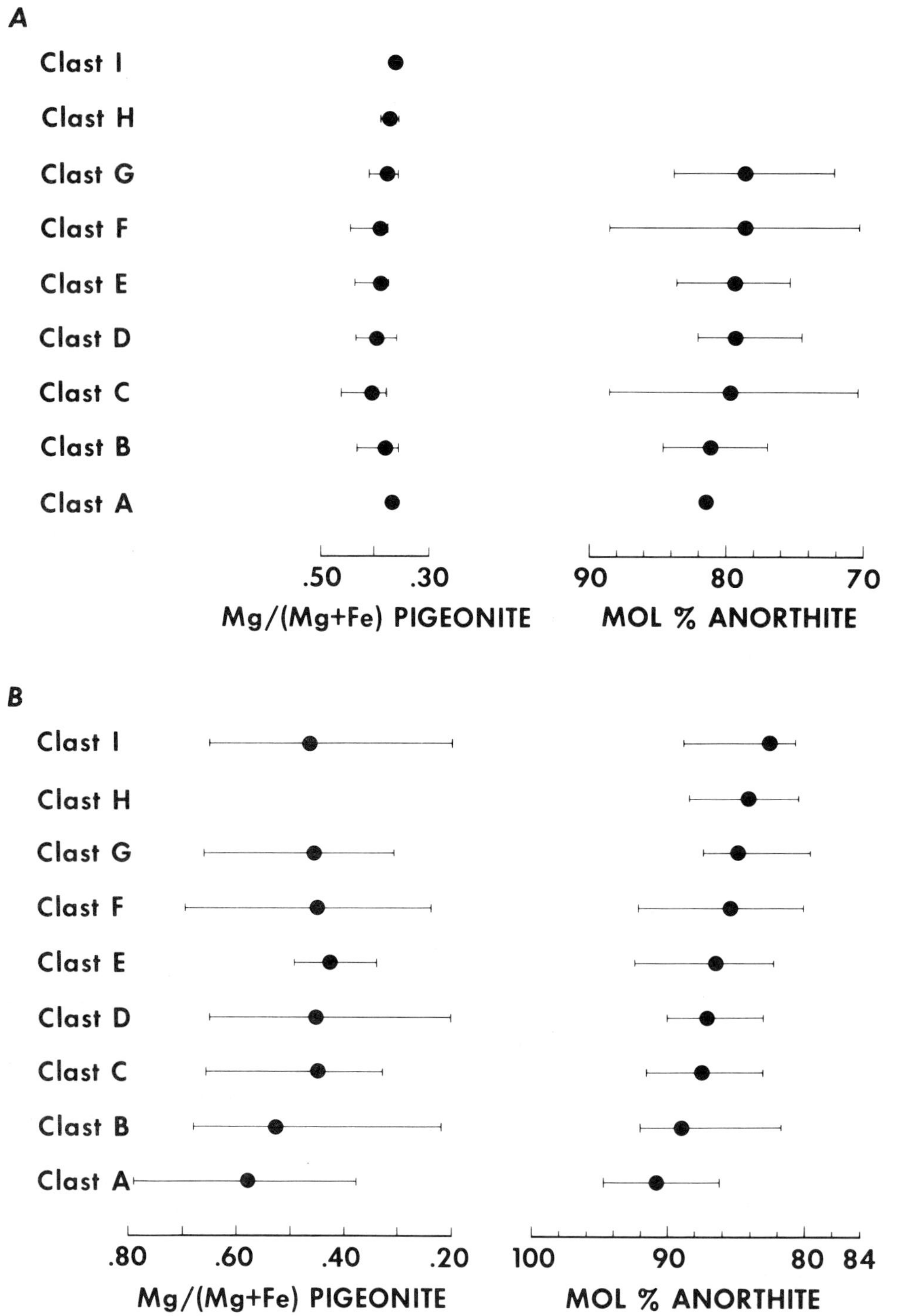

1.2.8.5 Ranges of compositions of pyroxene and plagioclase in clasts from: **a**. Stannern, an equilibrated eucrite; **b**. Pasamonte, an unequilibrated eucrite.

increases in albite content of plagioclases can be correlated with decreasing Mg/(Mg + Fe) ratios in pyroxene and with increases in the proportion of fine-grained, complex mesostasis. Large-ion-lithophile element abundances are also correlated with increasing amounts of mesostasis in the three analyzed clasts (Fig. 1.2.8.6b, also see Appendix A-8). Among three clasts analyzed, one had a small positive Eu anomaly, two had small negative anomalies. These variations observed in Pasamonte appear to have been due either to local inhomogeneities in the crystallizing magma or to sampling problems related to the small size of the clasts, rather than to the mixing of rocks of different composition.

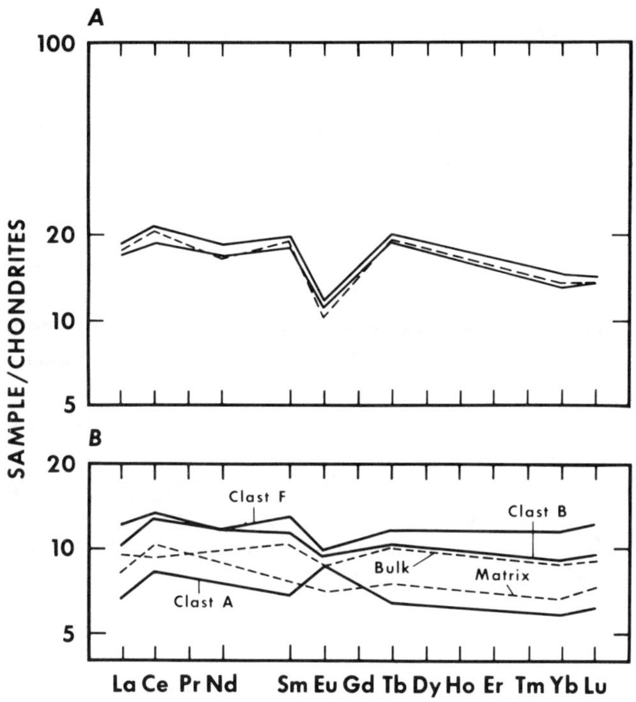

1.2.8.6 Chondrite-normalized rare-earth element abundances in separated clasts from: **a**. Stannern, an equilibrated eucrite; **b**. Pasamonte, an unequilibrated eucrite.

We have not encountered any "exotic" clasts in our sample suite study, i.e., clasts that are not eucrites or eucrite fragments. The possibility that such clasts exist cannot be excluded, but they certainly are not abundant.

We conclude that eucrites are monomict breccias wherefrom exotic clasts are rare or absent, and wherein interclast variations are minor and are consistent with limited crystal-liquid fractionation. Bulk eucrite analyses can thus be used to indicate original rock compositions despite the brecciation events.

1.2.8.6 Petrogenesis of meteoritic basalts

Eucrite compositional types

Figure 1.2.8.7 shows TiO_2 variation as a function of Mg/(Mg + Fe), which is one of several two-component diagrams that can be used to depict major element variations among the eucrites. The data, which come from a variety of sources, define several distinctive groups with some apparent anomalies:

(1) **Main group eucrites**: On binary plots of eucrite bulk compositions, there is a grouping of data points among the eucrites that have basaltic textures. This cluster of compositions defines what are here called the *main group* eucrites. Sioux County, Juvinas, Cachari and Haraiya are main group eucrites and Jonzac, Chervony Kut and Bereba have similar compositions. The bulk compositions of these main group eucrites represent the most common meteoritic basalt in our limited sample collection.

(2) **The Stannern trend**: A second group of eucrites defines a trend from the main group eucrites to Stannern and includes Chervony Kut, ALHA 76005 and Ibitira. Along this trend there are only minor differences in major element compositions and in Mg/(Mg + Fe) ratios. There are, however, changes in minor and trace element compositions from the main group eucrites to Stannern, which has substantially higher concentrations of alkalies, TiO_2 and large-ion-lithophile trace elements. The Stannern trend is extended further if we include a basaltic clast (KV-04) from the Nobleborough howardite, which was analyzed by Mittlefehldt (1979a).

(3) **The Nuevo Laredo trend**: The compositional trend from the main group eucrites to Nuevo Laredo includes Pasamonte, Lakangaon and Nuevo Laredo. Compositional changes along this trend encompass increases, relative to the main group eucrite values, in alkalies, TiO_2 and large-ion-lithophile trace elements.

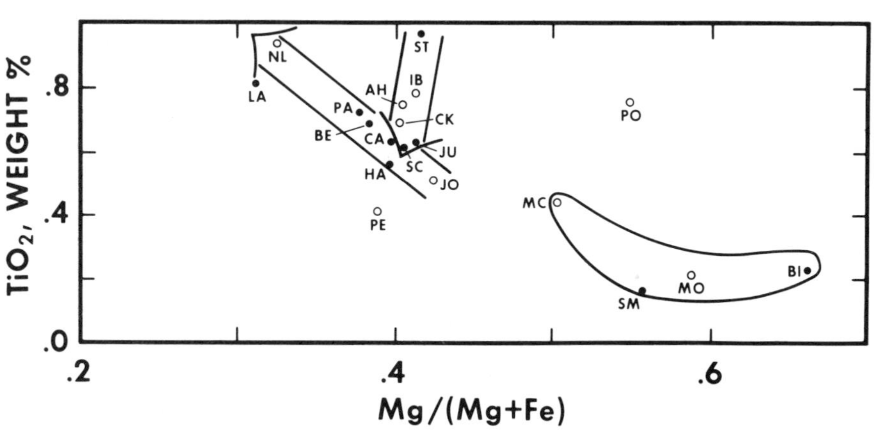

1.2.8.7 Bulk composition of eucrites expressed as a plot of wt.% TiO_2 vs. molar Mg/(Mg + Fe). AH-ALHA 76005; BE-Béréba; BI-Binda; CA-Cachari; CK-Chervony Kut; IB-Ibitira; JO-Jonzac; JU-Juvinas; LA-Lakangaon; MO-Moama; MC-Moore County; NL-Nuevo Laredo; PA-Pasamonte; PE-Peramiho; PO-Pomozdino; SC-Sioux County; SM-Serra de Magé; ST-Stannern. Filled circles are University of Cape Town analyses.

In contrast to the Stannern trend, the minor and trace element differences are coupled to systematic changes in major element composition and particularly to decreases in Mg/(Mg + Fe) ratios.

(4) **Cumulate eucrites**: The high-Mg eucrites plotted in Fig. 1.2.8.7 are those with Mg/(Mg + Fe) ratios substantially higher than the main group eucrites and with lower TiO_2 contents. They include Moore County, Serra de Magé, Moama and Binda. Unlike other eucrites they do not have basaltic textures, but are coarser-grained and show evidence of extensive subsolidus reequilibration of pyroxenes. Some members of this group have cumulate textures.

(5) **Anomalies**: Some eucrites, such as Pomozdino and Peramiho, do not fit into any of the above groups. These eucrites may represent products of separate source regions or of different fractionation processes, or of both. Further analytical work is needed on eucrites from this group, including comparative analyses from the laboratories responsible for the bulk of the eucrite data, so that their apparently anomalous positions can be evaluated properly.

Crystallization history models

The experimental petrologic data of Stolper (1977) have laid a framework for interpreting the history of these basalts. He demonstrated that the main group eucrites can be understood as near-peritectic compositions in the pseudoternary system Ol-SiO_2-An (see Fig. 3.5.1 and related discussion in Chapter 3). Stolper (1977) argues that the eucrite compositions are concentrated near this pseudoperitectic because the eucrites are partial melts of a source region which contained olivine (Fo_{65}), pigeonite (Wo_5En_{65}), plagioclase (An_{94}), Cr-rich spinel and metal. Magmas which were evolving by fractional crystallization would not be constrained to pause at the pseudoperitectic, so no concentration of compositions would be likely to result at that point. Additional modelling, directed at understanding the nature of the differences in eucrite compositions, has been performed largely within the framework defined by the experimental data.

Broadly, there are three processes which may contribute to compositional variations within the eucrites: (1) different degrees of partial melting of a common source material; (2) fractional crystallization of a common parent material; and (3) partial melting of slightly different source compositions. Models that relate all eucrites to processes (1) and (2), operating on a common source material, have been investigated by Schnetzler and Philpotts (1969), McCarthy *et al.* (1973), Consolmagno and Drake (1977), and Reid *et al.* (1979).

The abundances of the rare-earth elements are of particular value in attempts to refine the petrological models. Consolmagno and Drake (1977) accepted the general model proposed by Stolper and attempted to model the proposed fractionation process quantitatively to account for the rare-earth element abundances in eucrites. Fixing the mineral compositions of the source region according to Stolper's model, and assuming the derived liquid composition to be defined by the peritectic, Consolmagno and Drake were able to construct consistent models for: (1) derivation of the main group eucrites (Juvinas, Sioux County) by 10–15% equilibrium partial melting of a source with the relative and absolute REE abundances of chondrites, (2) derivation of the Stannern composition by about 4% partial melting of the same source, (3) derivation of the eucrites Moore County and Serra de Magé as cumulates from a Juvinas-like melt. They also suggested that Nuevo Laredo could be a residual liquid from such fractional crystallization.

Additional modelling of the type performed by Consolmagno and Drake, but also including the elements Sc, Ba, Sr, Zr, and Y, plus major element modelling to derive the Nuevo Laredo trend have been reported by Reid *et al.* (1979). Adding the other refractory trace elements and utilizing the Stolper model, the Juvinas composition can be generated by 12–13% equilibrium partial melting of a source with chondritic abundances. Sioux County can be modelled as a 14% equilibrium partial melt of the same source. The main group eucrites have flat chondrite-normalized REE patterns consistent with exhaustion of plagioclase from the source region, which occurs in these models at 12–13% melting. With smaller degrees of melting the liquid would have a negative Eu anomaly, as in Stannern, which can be derived as a 6% equilibrium partial melt of the same source. If trace elements are considered alone, this model suggests that Stannern, Ibitira, ALHA 76005, and main group eucrites (Juvinas, Sioux County) represent increasing degrees of partial melting of the same source region (the Stannern trend).

A possible problem with this model is that small but significant variations in Mg/(Mg + Fe) should accompany the different degrees of partial melting of the source. Thus, Stannern should be slightly less magnesian than the common eucrites. This difference, if present, is small.

Small differences in alkali contents of the source region can have significant effects on the compositions of partial melts. Such variations in source composition offer an alternative explanation for the differences between Stannern and Juvinas, but are difficult, if not impossible, to test. Stolper *et al.* (1979) have suggested a model relating eucrites and shergottites which is based on alkali variations in otherwise similar source regions.

The magnesium-rich eucrites, Moore County, Moama, Serra de Magé and Binda, have textures and compositions which are consistent with cumulate origins, being enriched to various degrees in pyroxene with respect to cotectic liquid compositions. Moore County and Serra de Magé cannot be cumulates directly from main group eucrite magmas, because the pyroxenes are too iron-rich to be in equilibrium with the eucrite melts (Stolper, 1977). However, fractional crystallization of main group eucrite melts produces liquids with lower Mg/Fe ratios (such as Nuevo Laredo) which could be parental to these cumulates. The Nuevo Laredo trend, including Bereba and Pasamonte, can be modelled by removal of pyroxene and plagioclase from a liquid of Sioux County composition. Cachari and Lakangaon lie close to the Nuevo Laredo trend but cannot be successfully modelled as derivatives from a main group eucrite liquid (Reid *et al.*, 1979).

Thermal history of eucrites

Walker *et al.* (1978) have investigated crystallization behavior in melts having the composition of the Stannern eucrite. They found that feldspar crystal size varies as a function of cooling rate approximately according to the relationship $\ln D = (-0.6)\ln R - 2.2$, where D is the size of feldspar grains in millimeters and R is the cooling rate in °C per hour. For Stannern, the observed textures are consistent with cooling rates of 0.1–100°C/hr. Extrapolating to Juvinas, which has some coarser-grained textures, cooling rates as slow as 0.01°C/hr can be inferred. For Sioux County, with plagioclase grains up to 1 cm in size, cooling rates on the order of 0.001°C/hr are inferred. It appears that all of the eucrite basalts cooled from liquidus to solidus (~ 100°C) in times which ranged from hours to a few tens of years. The coarser-grained magnesium-rich eucrites with cumulate textures may have cooled more slowly.

The unequilibrated eucrites have pyroxene compositional variations similar to those observed in the experiments of Walker *et al.* (1978). Early pyroxenes are magnesium-rich pigeonites, with a general iron-enrichment trend as crystallization proceeds. In unequilibrated eucrites, pigeonite grains show only submicroscopic exsolution lamellae, which is consistent with rapid cooling at subsolidus temperatures.

The equilibrated eucrites are characterized by a pigeonite-augite exsolution host-lamella relationship, with limited variability of pigeonite and augite compositions. In the eucrite basalts, exsolution lamellae range in thickness from less than 1 μm in some fragments in Nuevo Laredo to approximately 7 μm in some grains in Juvinas and Sioux County. Miyamoto and Takeda (1977) developed a model for the development of exsolution lamellae in several achondrites. They assumed a cooling rate defined by the cooling of a spherical asteroid. Under their assumptions, cooling rates of $4°/10^3$ years were required to form the exsolution lamellae of Juvinas at temperatures between 1200° and 850°C. These probably are relatively high temperature estimates, because exsolution of pigeonite must have begun at a temperature well below the Stannern liquidus at 1160°C (Stolper, 1977), and longer times (slower cooling rates) would be necessary if the lamellae actually formed over a smaller temperature range than assumed.

Although further work is required to improve the diffusional model of pyroxene exsolution, it is clear that the subsolidus cooling rates are much slower than the cooling rates through the crystallization interval. Pyroxene homogenization occurred prior to exsolution and at temperatures near the solidus, possibly above the solidus. This suggests that the onset of slower cooling rates may have occurred prior to complete solidification of the basalts. Various physical situations are consistent with a rapid slowing of cooling rates, such as extrusion of a basalt which then was buried in a layer of hot impact ejecta or was covered by additional flows. Alternatively, the eucrite basalts may have been emplaced in an environment whose ambient temperature was close to the eucrite solidus. The distinction among alternative models for eucrite thermal history will require much further detailed study of these meteorites, but can provide significant boundary conditions for the origins and evolution of their parent bodies.

The existence of brecciated eucrites, such as Juvinas, which show 4.5 b.y. internal isochron ages indicates either that brecciation occurred at an early stage of parent body formation or that brecciation was not accompanied by chemical fractionation. Most eucrite breccias, although they are somewhat indurated, do not have a crystalline matrix and cannot have been strongly heated after brecciation. In contrast, the Aioun el Atrouss meteorite contains evidence for thermal annealing of the eucrite breccia at high subsolidus temperatures. The recrystallization of Aioun el Atrouss indicates that, at least in some cases, substantial heating occurred after the brecciation event.

Younger disturbances, perhaps associated with brecciation events, are recorded in some meteorites. The howardite breccias, which contain eucrite basalt fragments, are more akin to regolith breccias. Some of the eucrite fragments in howardites give young ages, which are believed to represent disturbances rather than initial crystallization (Dymek *et al.*, 1976). These ages could have been produced by impact reheating, and imply that regolith formation continued at least through the period indicated by the youngest ages of 3.6 b.y.

The eucrite parent body

Eucrite basalts are partial melts from a source region that was chemically primitive in the sense that the relative abundances and probably the absolute abundances of the large-ion-lithophile trace elements were chondritic or near-chondritic. Abundances of the moderately volatile elements, such as K, Na and Rb, in the eucritic source are, in contrast, much lower than the chondritic values. Stolper (1977) has concluded that primary eucrite liquids could be generated by partial melting of a source composed of olivine ($\sim Fo_{65}$) + low-Ca pyroxene + plagioclase + Cr-spinel + metal; however, his model constrains only some aspects of the bulk composition of the source. Several authors (e.g., Dreibus *et al.*, 1976, 1977; Consolmagno and Drake, 1977; Morgan *et al.*, 1978) have made estimates of the bulk composition of the eucrite parent body based on the types of evidence outlined above. These estimates are presented and discussed in detail in Chapter 4 (section 4.5.6). The estimates are all to some extent model-dependent and dependent also on the authors' conceptions of the "primitive" composition of the solar system and of the differentiation processes that have contributed to the diversity of compositions within bodies in the solar system. The agreement among the various models is encouraging. All require a parent body which was chondritic or near-chondritic in terms of involatile trace element abundances but which was relatively depleted in alkalies and other volatile elements. Major element abundances are not well constrained but appear to be consistent with chondritic compositions, except for the volatile and the metallic components. The chemical characteristics of the source probably were established during accretion (rather than in a later planetary differentiation process) and therefore provide clues to the types of element fractionation processes which governed differences among solar bodies during or before the accretion event.

The source region for eucrites was one of low pressure (Stolper, 1977). The quartz-normative nature of the eucrite melts requires that source region pressures probably were less than ~2 kb; at higher pressures olivine is no longer in reaction relationship with liquid, and partial melts would be silica-undersaturated. If the parent body was similar in size to the Moon, then the source must have been within 7km of the surface. Eucrites must be derived from small bodies or from the very near-surface regions of larger bodies. This conclusion is consistent with other lines of evidence that meteorites in general are derived from small planetary (asteroidal) bodies. The asteroid 4 Vesta is the only candidate for a specific source for the eucrites which has been proposed seriously. Infrared spectrophometric observations by McCord *et al.* (1970) and Larson and Fink (1975) indicate that iron-rich pigeonite probably is a major mineral phase on the surface of Vesta. This consideration, plus the density estimate of 3.4 gm cm^{-3} by Morrison (1976), led Consolmagno and Drake (1977) to advocate 4 Vesta as the eucrite parent body. Pressures at the center of Vesta would not reach 2kb (see Fig. 3.5.2).

The relationship between eucrites and shergottites

Petrography, mineral chemistry and the experimental studies by Stolper and McSween (1979) indicate that Shergotty and Zagami do not represent basaltic liquid compositions. The bulk compositions are probably cumulus crystal-enriched, particularly with pyroxene, relative to the parent basaltic liquid. Because we can place only limited constraints on the parental basalt liquid composition, we find it even more difficult to constrain the shergottite source composition.

Differences in alkalies, volatile elements, oxygen fugacity and oxygen isotopes between shergottites and eucrites make it difficult to model them as derivatives from a single source. Stolper *et al.* (1979) have discussed the possibility that eucrites and shergottites are derived by partial melting (followed by some crystal-liquid fractionation) of source regions that are distinct but that could be related to one another rather simply. According to their arguments, eucrites, howardites, shergottites, nakhlites and chassignites all could be derived ultimately by partial melting of plagioclase peridotites that are compositionally distinct but which may form a simple compositional sequence. The putative source regions can be understood in terms of two-component mixtures generated by the increasing addition of a volatile-rich component to the volatile-poor peridotite source of the eucrites. Stolper *et al.* argue that this sequence of peridotites might be present on a single, heterogenous parent body, but more probably would require a series of related parent bodies.

The isotopic data of Nyquist *et al.* (1979b) indicate that Shergotty is the product of at least a two-stage process. The first stage, which occurred early in the history of the parent body, fractionated the rare-earth elements and rubidium. This may have been a planetary differentiation episode. Later, material was remelted to form the Shergotty parent liquid. This is in contrast with the eucrites, which apparently were essentially primary melts of primordial material, which melted shortly after the origin of the solar system. Thus, Shergotty is derived from a relatively evolved parent, the eucrites from a primitive parent(s). Stolper (1979), Stolper and McSween (1979) and McSween *et al.* (1979a) have emphasized the similarity between shergottites and terrestrial basalts.

The later history of the Shergotty parent body is currently under intense scrutiny. Nyquist *et al.* (1979b) have shown that the shergottites have very young crystallization ages for meteoritic material. Their best estimate of crystallization age for this complex disturbed system is 650 m.y., far younger than any other meteorite (the closest is Nakhla at 1.3 b.y.). The shergottites thus raise major problems in understanding how basaltic liquids can form on small planetary bodies 4 b.y. after formation of the solar system. This problem is compounded by the discovery of an Antarctic meteorite (ALHA 77005) which is an olivine-rich cumulate apparently related to the shergottites by igneous processes (McSween *et al.*, 1979a; 1979b).

The 650 m.y. melting of Shergotty could have been a magmatic event or could have been impact remelting of older basaltic material. If the shergottites and ALHA 77005 are related (and if shergottites and nakhlites are related), then it is difficult to envisage these different meteorites as having been generated by discrete impact events, rather than through normal igneous processes. On the other hand, it is very difficult to postulate a source of thermal energy sufficient to cause magmatism on a small planetary object so long after its initial accretion. If these were 600 m.y. internally-generated magmatic events, then either the thermal constraints on melting in smaller planetary bodies must be reevaluated or the possible parent bodies of shergottites must be reexamined (see Mittlefehldt, 1979b; Walker *et al.*, 1979).

Some conclusions

Meteoritic basalts fall very clearly into two groups, eucrites and shergottites, which derive from at least two different parent bodies. Eucrites, which formed in a small asteroidal-size body (or bodies), date from the first 50 m.y. of solar system history. The material from which they formed as partial melts was primitive in that it had chondritic or near-chondritic abundances for many elements. Alkalies and other volatile element abundances were depleted relative to chondrites. The chemical characteristics of the eucrite source probably were representative of the whole parent body, because basalt generation is likely to have been a one-stage process in an undifferentiated (hot?) body which had attained its compositional characteristics during accretion.

The shergottites are derived from a parent body that had more nearly chondritic alkali and volatile element abundances and which may have undergone an early planetary differentiation. These basalts have young ages compared with eucrites and the Moon. If they are not impact remelts then they represent extraterrestrial volcanism at a very late stage, when smaller bodies would lack obvious heat sources adequate to support magmatic events.

1.2.8.7 References

Ahrens L. H. and Danchin R. B. (1971) The chemical composition of the basaltic achondrites. *Phys. Chem. Earth* **8**, 267–303.

Antarctic Meteorite Newsletter (1978) vol. 1, no. 2. Johnson Space Center, Houston.

Antarctic Meteorite Newsletter (1979) vol. 2, no. 3. Johnson Space Center, Houston.

Binns R. A. (1967) Stony meteorites bearing maskelynite. *Nature* **213**, 1111–1112.

Bunch T. E. and Keil K. (1971) Chromite and ilmenite in non-chondritic meteorites. *Amer. Mineral.* **56**, 146–157.

Bunch T. E., Chang S., Neil J. M. and Burlingame A. (1976) Unique characteristics of the Jodzie howardite (abstract). *Meteoritics* **11**, 260–261.

Consolmagno G. J. and Drake M. J. (1977) Composition and evolution of the eucrite parent body. *Geochim. Cosmochim. Acta* **41**, 1271–1282.

Dreibus G., Kruse H., Spettel B. and Wänke H. (1977) The bulk composition of the Moon and the eucrite parent body. *Proc. Lunar Sci. Conf. 8th*, p. 211–227.

Dreibus G., Spettel B. and Wänke H. (1976) Lithium as a correlated element, its condensation behavior, and its use to estimate the bulk composition of the Moon and the eucrite parent body. *Proc. Lunar Sci. Conf. 7th*, p. 3383–3396.

Duke M. B. (1963) Petrology of basaltic achondrites. Unpublished Ph.D. thesis, Calif. Inst. of Technology, Pasadena, Calif.

Duke M. B. (1965) Metallic iron in basaltic achondrites. *J. Geophys. Res.* **70**, 1523–1527.

Duke M. B. (1968) The Shergotty meteorite: magmatic and shock metamorphic features. In *Shock Metamorphism of Natural Materials* (B. M. French and N. M. Short, eds.), p. 613–621. Mono Book Corp., Baltimore.

Duke M. B. (1978) Aioun el Atrouss: evidence for thermal recrystallization of a eucrite breccia. *Meteoritics* **13**, 443–448.

Duke M. B. and Silver L. T. (1967) Petrology of eucrites, howardites and mesosiderites. *Geochim. Cosmochim. Acta* **31**, p. 1637–1665.

Dymek F. R., Albee A. L., Chodos A. A. and Wasserburg G. J. (1976) Petrography of isotopically-dated clasts in the Kapoeta howardite and petrologic constraints on the evolution of its parent body. *Geochim. Cosmochim. Acta* **40**, 1115–1130.

Engelhardt W. von (1963) De eukrite von Stannern. *Beit. Mineral Petrog.* **9**, 65–94.

Foshag W. F. (1938) Petrology of the Pasamonte, New Mexico meteorite. *Amer. J. Sci.* **35**, 374–382.

Fredriksson K. and Keil K. (1963) The light-dark structure in the Pantar and Kapoeta stone meteorites. *Geochim. Cosmochim. Acta* **27**, 719–739.

Fredriksson K. and Kraut F. (1967) Impact glass in the Cachari eucrite. *Geochim. Cosmochim. Acta* **31**, 1701–1704.

Frost M. J. (1971) The Molteno meteorite. *Mineral. Mag.* **38**, 89–93.

Game P. (1957) Plagioclases from Juvinas meteorite and from allivalite from the Isle of Rhum. *Mineral. Mag.* **31**, 656–671.

Gast P. W. (1962) The isotopic composition of strontium and the age of stone meteorites. *Geochim. Cosmochim. Acta* **26**, 927–943.

Gast P. W. (1965) Terrestrial ratio of potassium to rubidium and the composition of the Earth's mantle. *Science* **147**, 858–860.

Hess H. H. and Henderson E. P. (1949) The Moore County meteorite: a further study with comment on its primordial environment. *Amer. Mineral.* **34**, 494–507.

Hewins R. H. and Klein L. C. (1978) Provenance of metal and melt rock textures in the Malvern howardite. *Proc. Lunar Planet. Sci. Conf. 9th*, p. 1137–1156.

Hutchison R., Bevan A. W. R. and Hall J. M. (1977) Appendix to the catalogue of meteorites with special reference to those represented in the collection of the British Museum (Natural History). British Museum (Natural History), London. 297 pp.

Jérome D. Y. and Christophe Michel-Lévy M. (1971) Quelques aspects remarquables de la météorite achondritique de Malvern (Afrique du Sud). *Bull. Soc. fr. Minéral. Cristallogr* **94**, 156–161.

LaCroix A. (1926) L'eucrite de Bereba (Haute-Volta) et les Meteorites feldspathiques en general. *Arch. Museum Hist. Nat. Paris Ser.* **6**, 1, 15–58.

Lange D. E. and Keil K. (1976) Meteorites of northeastern New Mexico. *New Mexico Geol. Soc. Guidebook, 27th Field Conference*, p. 293–299.

Larson H. P. and Fink U. (1975) Infrared spectral observations of asteroids 4 Vesta. *Icarus* **26**, 420–427.

Laul J. C., Keays R. R., Ganapathy R., Anders E. and Morgan J. W. (1972) Chemical fractionation in meteorites-V. Volatile and siderophile elements in achondrites and ocean ridge basalts. *Geochim. Cosmochim. Acta* **36**, 329–345.

Lomena I. S. M., Toure F., Gibson E. K., Jr., Clanton U. S. and Reid A. M. (1976) Aioun el Atrouss: a new hypersthene achondrite with eucritic inclusions. *Meteoritics* **11**, 51–57.

Lovering J. F. (1964) Electron microprobe analysis of the metallic phase in basic achondrites. *Nature* **203**, 70.

Lovering J. F. (1975) The Moama eucrite—a pyroxene-plagioclase adcumulate. *Meteoritics* **10**, 101–114.

Mason B. (1962) *Meteorites*. John Wiley and Sons, N.Y. 274 pp.

Mason B. (1967) The Bununu meteorite and a discussion of the pyroxene plagioclase achondrites. *Geochim. Cosmochim. Acta* **31**, 107–115.

Mason B. and Wiik H. B. (1966) The composition of the Bath, Frankfort, Kakangari, Rose City and Tadjera Meteorites. *Amer. Mus. Novitates No. 2272*.

Mason B., Jarosewich E. and Nelen J. A. (1979) The pyroxene-plagioclase achondrites. *Smithsonian Contrib. Earth Sci.* **22**, 27–45.

McCarthy T. S., Ahrens L. H. and Erlank A. J. (1972) Further evidence for the mixing model for howardite origin. *Earth Planet. Sci. Lett.* **15**, 86–93.

McCarthy T. S., Erlank A. J. and Willis J. P. (1973) On the origin of eucrites and diogenites. *Earth Planet. Sci. Lett.* **18**, 442–443.

McCarthy T. S., Erlank A. J., Willis J. P. and Ahrens L. H. (1974) New chemical analyses of six achondrites and one chondrite. *Meteoritics* **9**, 215–221.

McCord T. B., Adams J. B. and Johnson T. V. (1970) Asteroid Vesta: spectral reflectivity and compositional implications. *Science* **168**, 1445–1447.

McSween H. Y. Jr., Stolper E. M., Taylor L. A., Muntean R. A., O'Kelley G. D., Eldridge J. S., Biswas S., Ngo H. T. and Lipschutz M. E. (1979a) Petrogenetic relationship between Allan Hills 77005 and other achondrites. *Earth Planet. Sci. Lett.* **45**, 275–284.

McSween H. Y. Jr., Taylor L. A. and Stolper E. M. (1979b) Allan Hills 77005: a new meteorite type found in Antarctica. *Science* **204**, 1201–1202.

Mittlefehldt D. W. (1979a) Petrographic and chemical characterization of igneous lithic clasts from mesosiderites and howardites and comparison with eucrites and diogenites. *Geochim. Cosmochim. Acta* **43**, 1917–1935.

Mittlefehldt D. W. (1979b) The nature of asteroidal differentiation processes: implications for primordial heat sources. *Proc. Lunar Planet. Sci. Conf. 10th*, p. 1975–1993.

Miyamoto M. and Takeda H. (1977) Evaluation of a crust model of eucrites from the width of exsolution lamellae. *Geochem. J.* **11**, 161–169.

Morgan J. W. and Lovering J. F. (1964) Uranium and thorium abundances in stony meteorites II. The achondritic meteorites. *J. Geophys. Res.* **69**, 1989–1994.

Morgan J. W., Higuchi H., Takahashi H. and Hertogen J. (1978) A "chondritic" eucrite parent body: inference from trace elements. *Geochim. Cosmochim. Acta* **42**, 27–38.

Morrison D. (1976) The densities and bulk compositions of Ceres and Vesta. *Geophys. Res. Lett.* **3**, 701–704.

Nyquist L. E., Bogard D. D., Wooden J. L., Wiesmann H., Shih C. -Y., Bansal B. M. and McKay G. (1979a) Early differentiation, late magmatism and recent bombardment of the Shergottite parent planet (abstract). *Meteoritics* **14**, 502.

Nyquist L. E., Wooden J., Bansal B., Wiesmann H., McKay G. and Bogard D. D. (1979b) Rb-Sr age of the Shergotty achondrites and implications for metamorphic resetting of isochron ages. *Geochim. Cosmochim. Acta* **43**, 1057–1074.

Olsen E. J., Noonan A., Fredriksson K., Jarosewich E. and Moreland G. (1978) Eleven new meteorites from Antarctica, 1976–1977. *Meteoritics* **13**, 209–225.

Papanastassiou D. A. and Wasserburg G. J. (1969) Initial strontium isotopic abundances and the resolution of small time differences in the formation of planetary objects. *Earth Planet. Sci. Lett.* **5**, 361–376.

Reid A. M. (1974) The Macibini meteorite and some thoughts on the origin of basaltic achondrites (abstract). *Meteoritics* **9**, 398–399.

Reid A. M. and Barnard B. M. (1979) Unequilibrated and equilibrated eucrites (abstract). In *Lunar and Planetary Science X*, p. 1019–1021. Lunar and Planetary Institute, Houston.

Reid A. M., Duncan A. R. and LeRoex A. (1979) Petrogenetic models for eucrite genesis (abstract). In *Lunar and Planetary Science X*, p. 1022–1024. Lunar and Planetary Institute, Houston.

Rowe M. W., van Dilla M. A. and Anderson E. C. (1963) On the radioactivity of stone meteorites. *Geochim. Cosmochim. Acta* **27**, 983–1001.

Schaal R. B. and Hörz F. (1977) Shock metamorphism of lunar and terrestrial basalts. *Proc. Lunar Sci. Conf. 8th*, p. 1697–1729.

Schmitt R. A., Goles G. G., Smith R. H. and Osborn T. W. (1972) Elemental abundances in stone meteorites. *Meteoritics* **2**, 131–214.

Schmitt R. A., Smith R. H., Lasch J. E., Mosen A. W., Olehy D. A. and Vasilevskis J. (1963) Abundances of the fourteen rare-earth elements, scandium and yttrium in meteoritic and terrestrial matter. *Geochim. Cosmochim. Acta* **27**, 577–622.

Schmitt R. A., Smith R. H. and Olehy D. A. (1964) Rare earth, yttrium and scandium abundances in meteorite and terrestrial matter-II. *Geochim. Cosmochim. Acta* **28**, 67–86.

Schnetzler C. C. and Philpotts J. A. (1969) Genesis of the calcium-rich achondrites in the light of rare earth and barium concentrations. In *Meteorite Research* (P.M. Millman, ed.), p. 207–216. Springer-Verlag, N.Y.

Smith J. V. and Hervig R. L. (1979) Shergotty Meteorite: mineralogy, petrography and minor elements. *Meteoritics* **14**, 121–142.

Steele I. M. and Smith J. V. (1976) Ibitira eucrite: mineralogy and comparison with lunar samples and other eucrites (abstract). In *Lunar Science VII*, p. 833–835. The Lunar Science Institute, Houston.

Stolper E. (1977) Experimental petrology of eucritic meteorites. *Geochim. Cosmochim. Acta* **41**, 587–611.

Stolper E. (1979) Trace elements in shergottite meteorites: implications for the origins of planets. *Earth Planet. Sci. Lett.* **42**, 239–242.

Stolper E. and McSween H. Y. (1979) Petrology and origin of the shergottite meteorites. *Geochim. Cosmochim. Acta* **43**, 1475–1498.

Stolper E., McSween H. Y., Jr. and Hays J. F. (1979) A petrogenetic model of the relationships among achondritic meteorites. *Geochim. Cosmochim. Acta* **43**, 589–602.

Takeda H. (1979) A layered-crust model of a howardite parent body. *Icarus* **40**, 455–470.

Takeda H., Miyamoto M., Ishii T. and Reid A. M. (1976) Characterization of crust formation on a parent body of achondrites and the moon by pyroxene crystallography and chemistry. *Proc. Lunar Sci. Conf. 7th*, p. 3535–3548.

Takeda H., Miyamoto M., Duke M. B. and Ishii T. (1978a) Crystallization of pyroxenes in lunar KREEP basalt 15386 and meteoritic basalts. *Proc. Lunar Planet. Sci. Conf. 9th*, p. 1157–1171.

Takeda H., Miyamoto M., Yanai K. and Haramura H. (1978b) A preliminary mineralogical examination of the Yamato-74 achondrites. *Memoirs of the National Institute of Polar Research, Special Issue No. 8* (T. Nagata, ed.), p. 170–184. Nat'l. Inst. Polar Res., Tokyo.

Wahl W. (1952) The brecciated stony meteorites and meteorites containing foreign fragments. *Geochim. Cosmochim. Acta* **2**, 91–117.

Walker D., Powell M. A. and Hays J. F. (1978) Dynamic crystallization of a eucritic basalt. *Proc. Lunar Planet. Sci. Conf. 9th*, p. 1369–1391.

Walker D., Stolper E. M. and Hays J. F. (1979) Basaltic volcanism: the importance of planet size. *Proc. Lunar Planet. Sci. Conf. 10th*, p. 1995–2105.

Wänke H., Palme H., Braddenhausen H., Dreibus G., Kruse H. and Spettel B. (1977) Element correlations and the bulk compositions of the Moon. *Phil. Trans. Roy. Soc. Lond.* **A285**, 41–48.

Wilkening L. L. (1976) Foreign inclusions in stony meteorites-I. Chondritic inclusions in the Kapoeta howardite. *Geochim. Cosmochim. Acta* **37**, 1985–1989.

Wilkening L. L. and Anders E. (1975) Some studies of an unusual eucrite. *Geochim. Cosmochim. Acta* **39**, 1205–1210.

Zavaritskii A. N. and Kvasha L. G. (1952) *Meteorites of the U.S.S.R.* Academy of Sciences of the U.S.S.R., Moscow. 245 pp.

1.2.9 LUNAR MARE BASALTS

1.2.9.1 Introduction

In the period following the termination of the heavy lunar bombardments ~3.9 b.y. ago, many of the lunar basins and plains were flooded by volcanic flows. These mare basalts are mafic (modal plagioclase content ~25% to 50%) and have characteristically high FeO contents (~20 wt.%). Compared with rocks of the lunar highlands, they have high Ca/Al ratios which are manifested in a predominance of calcic pyroxenes in contrast to the calcium-poor pyroxenes of the highlands. Mare basalt bulk compositions are extremely varied in major (e.g., TiO_2) and trace (e.g., rare earth) element abundances and, like the bulk moon, are extremely depleted in volatiles.

Experimental studies of mare basalt compositions suggest that mare basalts were produced by partial melting of mafic sources at depths of about 150–500 km, within the lunar mantle. Experimental, isotopic, and trace element studies demonstrate that the sources are varied, and not primitive or bulk moon compositions. The consensus is that the sources are cumulate rocks produced during an early massive differentiation of the moon, but the processes of accumulation and remelting are by no means well-understood. Mare basalts provide important clues to the constitution of the lunar interior and to its thermal history.

According to Head (1976), mare basalts constitute only 1% of the lunar crust and cover 17% of the surface. Papike *et al.* (1976) have reviewed the crystal chemistry, mineralogy, and petrology of the returned samples, and Head (1976) has reviewed the broader aspects of lunar volcanism, based largely on remote-sensing (especially photographic) data. Since these reviews were written, several studies have added to our knowledge and understanding of the variation among mare basalts. Rhodes *et al.* (1977) and Dungan and Brown (1977) reexamined a large suite of Apollo 12 mare basalts and have given us some important new insights into these rocks. Vaniman and Papike (1977a) discovered an important new mare basalt type which is characterized by Very Low TiO_2 contents (VLT). This mare basalt type, which was discovered in the Apollo 17 core materials, is described in more detail by Taylor *et al.* (1977a) and by Vaniman and Papike (1977b). Simultaneously with the discovery of Apollo 17 VLT, the initial results of the Luna 24 mission were presented at the 8th Lunar and Planetary Science Conference. The Luna 24 mare basalts are also low in TiO_2 but have higher FeO contents and are more fractionated than the Apollo 17 VLT. Thus, to the broad two-fold division of mare basalts into high-Ti and low-Ti groups, we must now add a third, very low-Ti group.

The U. S. Geological Survey (Flagstaff) has developed a computerized system to synthesize and display on a common lunar grid such diverse data as: X-ray spectra, X-ray fluorescence spectra, geological maps, gravity maps, magnetometer surveys, electron reflection measurements, visible and near-infrared spectra, reflectance maps, maps of relative mare ages based on crater morphology, albedo maps, and laser altimetry profiles. This ambitious system has led to some remarkable new insights. Some aspects of these results are discussed below. Papers based on this synthesis appeared in the *Proceedings of the Eighth Lunar and Planetary Science Conference* and include Arnold *et al.* (1977), Bielefeld *et al.* (1977), Johnson *et al.* (1977a, b), Russell *et al.* (1977), Eliason and Soderblom (1977), Metzger *et al.* (1977) and Soderblom *et al.* (1977). With the addition of new ground data from Luna 24, it is clear that our ideas and models concerning mare basalt petrogenesis must be updated.

We begin by reviewing the areal and temporal distribution of mare basalt types. Then we discuss major and trace element chemistry, petrology and mineralogy of these lavas, with reference to data for a suite of 42 samples chosen to include representatives of all the mare basalt types that have been sampled. Finally, we discuss three different models which have been proposed for the petrogenesis of mare basalts.

1.2.9.2 Regional setting and temporal relations

Figure 1.2.9.1 summarizes the lunar sampling locations and the distribution of three spectrally-distinct mare lithologies. The map presents combined ultraviolet negatives and infrared positives of the full Moon. It is thought that the color differences represent, among other things, the different Ti contents of the mare units. Blue appears to represent high-TiO_2 basalts (8–14 wt.% TiO_2), orange the low-TiO_2 basalts (1.5–5 wt.% TiO_2), and red the very low-TiO_2 (VLT) basalts (< 1.5 wt.% TiO_2). This obviously is an over-simplification; nevertheless, this whole-moon map is a useful backdrop for discussions of regional setting.

The Apollo 11 landing site is located on the edge of Mare Tranquillitatis in the blue mare unit. It appears that the Apollo 11 samples are quite representative of

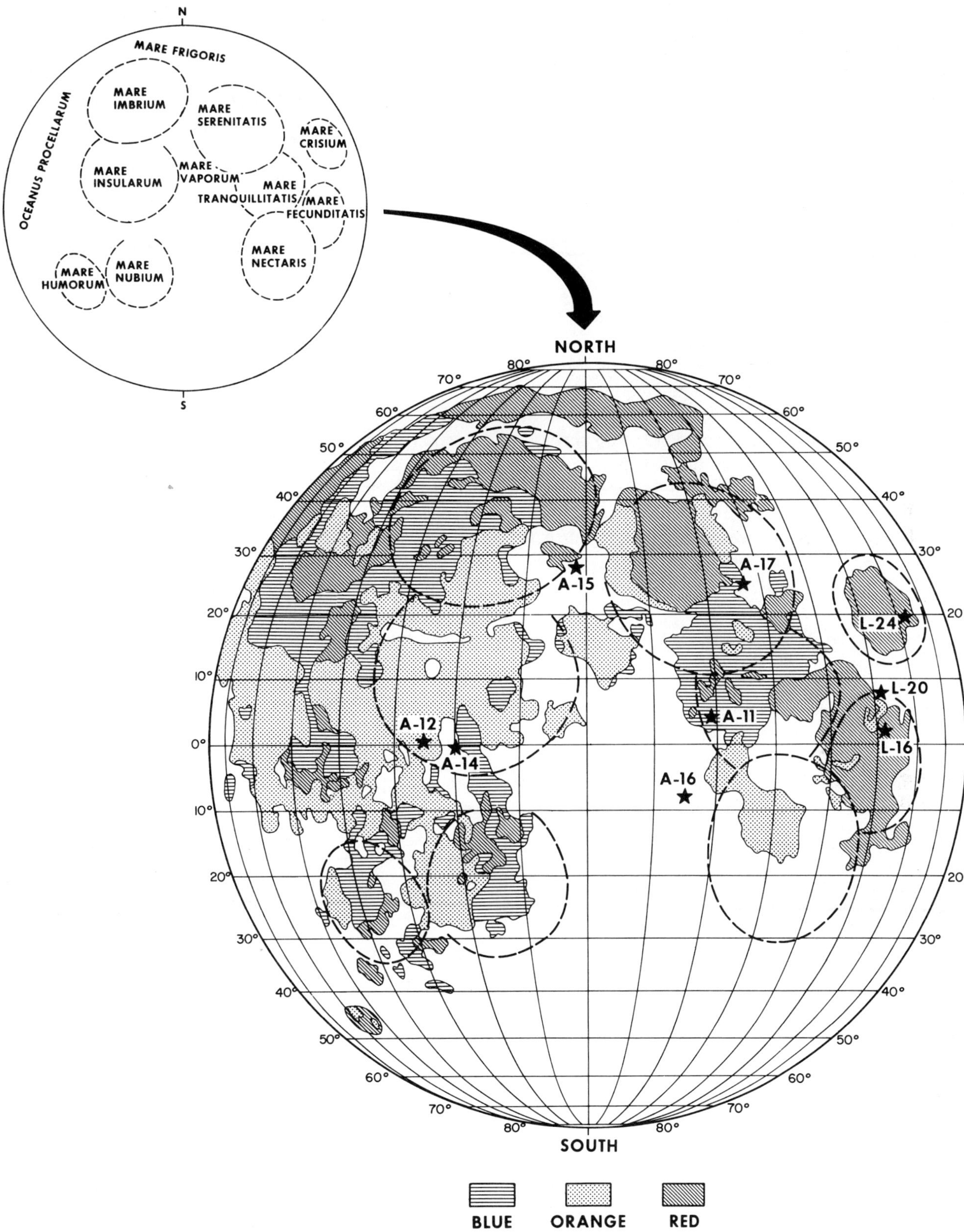

1.2.9.1 Lunar sample localities and spectral determinations of the color-groups of mare basalts: data from Head (1976), Strom (1972) and Whitaker (1972).

the surface flows which cover much of this basin. Apollo 12 is located in the orange unit which, although not yet mapped completely, probably extends as far as the Apollo 15 site.

Apollo 17 is located in the blue unit, and the Apollo 17 samples, which are high in TiO_2, are similar to those of Apollo 11. It is clear that the Apollo 17 basalts are not representative of the entire Serenitatis Basin; at least two additional units occur within this basin (Bryan and Adams, 1973; Boyce and Dial, 1973). It is possible that the orange annulus is composed of as yet unsampled mare basalts, while red represents the VLT basalts which were discovered as minor components of the Apollo 17 soil (Vaniman and Papike, 1977a). Luna 16 is located in a red region of Mare Fecunditatis, but the basalts are not VLT as would be predicted by our simplistic approach. However, VLT basalts are found in the Luna 16 collection (Green *et al.*, 1972), and perhaps the low-Ti and VLT basalts occur in close proximity to one another in this basin. Luna 24 is located in a red region of Mare Crisium and, as would be predicted, consists of VLT basalts. However, detailed, small-scale spectral mapping of this basin suggests that several other basalt types are present as well (Adams and Ralph, 1977; Andre *et al.*, 1977; Head *et al.*, 1977).

The discussion above presents a simple overview, but the true story is much more complex. We now realize that essentially the whole spectrum of basalt types occurs both on the eastern and on the western half of the Moon's nearside. Soderblom *et al.* (1977), however, point out that major regional differences exist between the eastern and the western maria. The eastern maria (Crisium, Tranquillitatis, Fecunditatis, Serenitatis) are older (≥3.4 b.y.), are more highly magnetized,

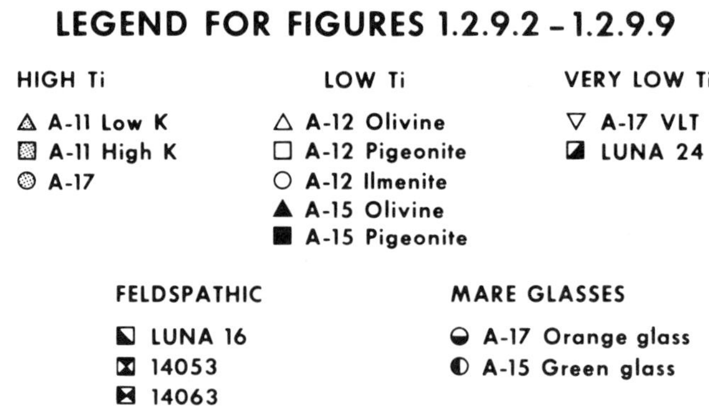

1.2.9.3 Legend for Figs. 1.2.9.2 through 1.2.9.9.

and are lower in radioactivity than are the western maria (Imbrium, Procellarum, Nubium). Nevertheless, there is a province in southwestern Oceanus Procellarum which is very much like the eastern maria. Fe and Ti contents of these basalts are not correlated with age, so models for mare basalt genesis in which the partial melting zone moves to greater depths with time are inadequate. Some of the lateral variations in chemistry could be due to large-scale planetary heterogeneities produced during the accretionary process. Johnson *et al.* (1977a, b) made spectral maps of the maria in the lunar northern hemisphere and show that the red units (= VLT basalts) have a broad areal distribution matching that of other spectrally-defined units.

Figure 1.2.9.2 summarizes the ages determined for lunar samples (see also Table 7.3.1). Although the samples that have been collected probably do not represent all major mare units, several important things are

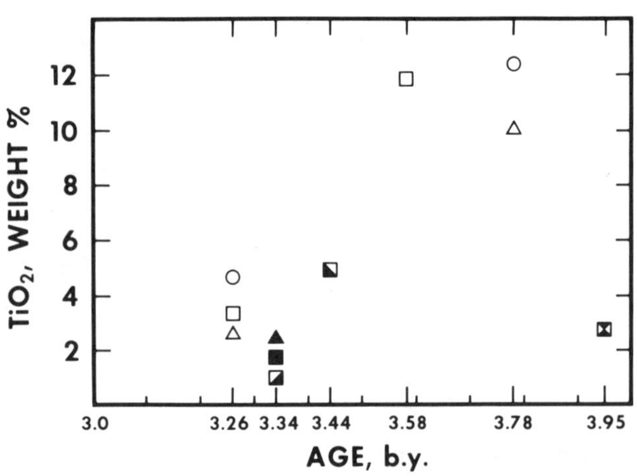

1.2.9.2 Radiometric ages of mare basalts compared with TiO_2 contents. Age data from Table 7.3.1.

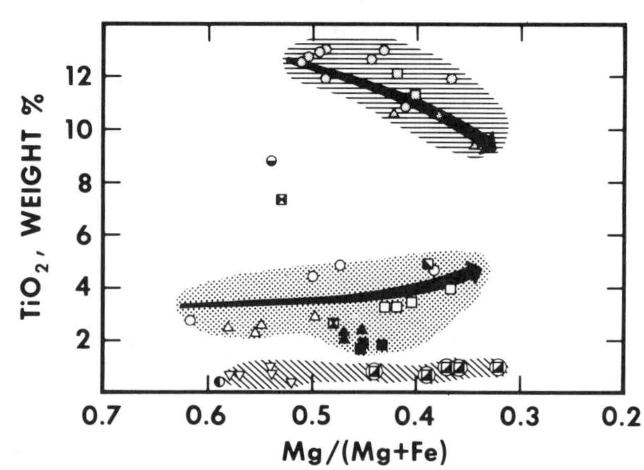

1.2.9.4 TiO_2 (wt.%) vs. Mg′ variation in mare basalts. Legend in Fig. 1.2.9.3.

clear: (1) the samples represent an interval of volcanism between 3.2 and 4.0 b.y. ago; (2) there is no correlation of TiO_2 content with age; (3) samples from eastern maria have both old and young ages, e.g., Apollo 11 and 17 compared with Luna 16 and 24; (4) a large time span of volcanism (~200 m.y.) is represented by samples from the Apollo 11 site; (5) very old mare samples exist in the Apollo 14 collection and may represent the earliest known period of extensive mare flooding. However, it is apparent that we have not sampled the entire temporal span of mare volcanism. The data of Boyce (1976), based on crater degradation studies, strongly indicate that basalts younger than 3.0 b.y. exist on the lunar surface. It is quite likely that such basalts also exist in the lunar soil samples but have not yet been recognized.

1.2.9.3 Major element chemistry

The major element chemistry of mare basalts was reviewed in detail by Papike *et al.* (1976). This report will repeat only the highlights of that discussion and will incorporate new results for the VLT basalts. Major

Table 1.2.9.1 Major element compositions of samples in the Mare Basalt Reference Suite (in wt. %).†

	Apollo 11 Low K			Apollo 11 High K		Apollo 12 Olivine			
	1	**2**	**3**	**4**	**5**	**6**	**7**	**8**	**9**
Sample No.	10003	10020	10047	10049	10072	12002	12009	12035	12040
SiO_2	39.76	39.79	42.16	41.00	40.26	43.56	45.03	43.17	43.88
TiO_2	10.50	10.62	9.43	11.30	12.07	2.6	2.90	2.28	2.45
Al_2O_3	10.43	9.93	9.89	9.5	7.93	7.87	8.59	8.03	7.27
FeO	19.80	19.21	19.11	18.7	19.58	21.66	21.03	22.20	21.09
MnO	0.30	0.26	0.28	0.25	0.23	0.28	0.28	0.29	0.27
MgO	6.69	7.93	5.67	7.03	7.8	14.88	11.55	15.49	16.45
CaO	11.13	11.32	12.15	11.0	10.41	8.26	9.42	8.08	8.01
Na_2O	0.40	0.39	0.45	0.51	0.53	0.23	0.23	0.21	0.17
K_2O	0.06	0.04	0.11	0.36	0.28	0.05	0.06	0.05	0.05
Cr_2O_3	0.25	0.39	0.18	0.32	0.36	0.96	0.55	0.49	0.63
SUM	99.32	99.88	99.43	99.97	99.45	100.35	99.64	100.29	100.27
Mg/Mg+Fe (atomic)	0.38	0.42	0.35	0.40	0.42	0.55	0.50	0.55	0.58
CaO/Al_2O_3 (wt. %)	1.07	1.14	1.29	1.16	1.31	1.05	1.10	1.01	1.10

	Apollo 12 Pigeonite				Apollo 12 Ilmenite				Apollo 15 Olivine		
	10	**11**	**12**	**13**	**14**	**15**	**16**	**17**	**18**	**19**	**20**
Sample No.	12011	12021	12052	12064	12005	12008	12022	12051	15016	15545	15555
SiO_2	46.63	46.68	46.40	46.30	41.56	42.75	42.77	45.31	44.08	45.21	44.57
TiO_2	3.29	3.53	3.28	3.99	2.76	4.45	4.85	4.68	2.28	2.41	2.10
Al_2O_3	9.77	10.78	10.16	10.73	5.30	7.98	9.08	9.95	8.38	8.59	8.69
FeO	19.53	19.31	20.15	19.89	22.27	21.94	21.75	20.22	22.74	22.15	22.53
MnO	0.29	0.26	0.27	0.27	0.30	0.30	0.25	0.28	0.32	0.30	0.29
MgO	8.26	7.39	8.22	6.49	19.97	12.33	11.01	7.01	11.30	10.28	11.36
CaO	10.63	11.38	10.80	11.77	6.31	8.97	9.47	11.39	9.27	9.82	9.40
Na_2O	0.25	0.31	0.27	0.28	0.16	0.25	0.38	0.29	0.27	0.31	0.27
K_2O	0.06	0.07	0.07	0.07	0.04	0.05	0.07	0.06	0.04	0.04	0.09
Cr_2O_3	0.59	0.40	0.52	0.37	0.75	0.61	0.56	0.31	0.85	0.68	0.61
SUM	99.30	100.11	100.14	100.16	99.42	99.63	100.19	99.50	99.53	99.79	99.91
Mg/Mg+Fe (atomic)	0.43	0.40	0.42	0.37	0.62	0.50	0.47	0.38	0.47	0.45	0.47
CaO/Al_2O_3 (wt. %)	1.09	1.06	1.06	1.10	1.19	1.12	1.04	1.14	1.11	1.14	1.08

†Data from Lofgren and Lofgren (1981).

Table 1.2.9.1 *(Continued)*

	Apollo 15 Pigeonite			Apollo 15 Green Glass	Apollo 17 Group A		Apollo 17 Group B		Apollo 17 Group C		
	21	**22**	**23**	**24**	**25**	**26**	**27**	**28**	**29**	**30**	**31**
Sample No.	15058	15499	15597		75055	76136	70215	70275	74245	74255	74275
SiO_2	48.14	47.81	47.98	44.14	40.60	38.6	37.79	39.37	38.59	38.18	38.43
TiO_2	1.69	1.78	1.80	0.37	10.79	12.64	12.97	11.90	11.92	12.47	12.70
Al_2O_3	8.89	9.11	9.44	7.81	9.67	8.65	8.85	10.23	8.72	8.69	8.72
FeO	19.86	20.19	20.23	21.05	18.01	19.12	19.66	18.61	18.06	18.05	18.14
MnO	0.27	0.28	0.30	—	0.29	0.28	0.27	0.28	0.27	0.28	0.26
MgO	9.28	9.33	8.74	16.72	7.05	8.61	8.44	6.09	9.65	10.61	10.36
CaO	10.27	10.34	10.43	8.41	12.35	10.53	10.74	11.65	10.59	10.27	10.32
Na_2O	0.28	0.32	0.32	0.13	0.43	0.38	0.36	0.38	0.36	0.37	0.35
K_2O	0.03	0.05	0.06	0.03	0.08	0.06	0.05	0.06	0.06	0.09	0.07
Cr_2O_3	0.66	0.57	0.48	0.33	0.27	0.44	0.41	0.26	0.54	0.60	0.65
SUM	99.37	99.78	99.78	98.99	99.54	99.31	99.54	98.83	98.76	99.61	100.00
Mg/Mg+Fe (atomic)	0.45	0.45	0.44	0.59	0.41	0.45	0.43	0.37	0.49	0.51	0.50
CaO/Al_2O_3 (wt. %)	1.16	1.14	1.10	1.08	1.28	1.22	1.21	1.14	1.21	1.18	1.18

	Apollo 17 Group U		A-17 Orange Glass	Luna 16	Luna 24		Apollo 17 VLT			Other Feldspathic	
	32	**33**	**34**	**35**	**36**	**37**	**38**	**39**	**40**	**41**	**42**
Sample No.	70017	70035	74220	B1	24182, 25-100	24109, 52-18	70008, 356-P	70007, 328-V	78526,1	14053	14063
SiO_2	38.54	37.84	38.57	43.80	46.82	45.2	48.1	48.8	46.7	46.4	45.9
TiO_2	12.99	12.97	8.81	4.90	1.00	0.89	0.36	0.69	0.92	2.64	7.3
Al_2O_3	8.65	8.85	6.32	13.65	12.58	13.8	11.2	10.0	10.0	13.6	14.4
FeO	18.25	18.46	22.04	19.35	20.46	20.5	18.2	17.9	18.6	16.8	10.7
MnO	0.25	0.28	—	0.20	0.24	0.27	0.26	0.30	0.24	0.26	—
MgO	9.98	9.89	14.44	7.05	6.65	6.35	11.0	11.8	12.2	8.48	6.7
CaO	10.28	10.07	7.68	10.40	12.49	12.7	10.2	9.4	10.0	11.2	11.3
Na_2O	0.39	0.35	0.36	0.33	0.10	0.24	0.15	0.06	0.12	—	1.0
K_2O	0.05	0.06	0.09	0.15	0.02	0.01	0.01	0.02	<0.01	0.10	0.1
Cr_2O_3	0.50	0.61	0.75	0.28	0.16	0.19	0.60	0.69	0.74	—	0.2
SUM	99.88	99.38	99.06	100.11	100.52	100.2	100.1	99.7	99.52	99.48	97.60
Mg/Mg+Fe (atomic)	0.49	0.49	0.54	0.39	0.37	0.36	0.52	0.54	0.54	0.48	0.53
CaO/Al_2O_3 (wt. %)	1.19	1.14	1.22	0.76	0.99	0.92	0.90	0.93	1.00	0.82	0.78

element compositions of basalts in the reference suite are given in Table 1.2.9.1 and the data are plotted in Figs. 1.2.9.3–9. More detailed variation diagrams may be found in Papike *et al.* (1976), where fractionation trends are discussed further. The most useful single chemical discriminant for separating the mare basalt groups is TiO_2 (Fig. 1.2.9.4). The high-Ti basalts include Apollo 11 low-K, Apollo 11 high-K and Apollo 17 basalts; low-Ti basalts include Apollo 12 olivine, Apollo 12 pigeonite, Apollo 12 ilmenite, Apollo 15 olivine and Apollo 15 pigeonite basalts; and the very low-Ti basalts include Apollo 17 VLT and Luna 24 VLT. In addition, there are feldspathic basalts (e.g., Luna 16 and 14053) whose TiO_2 contents are similar to those of the low-Ti group. Sample 14053 (251 grams) is the largest mare basalt fragment returned by the Apollo 14 mission. Sample 14063 deserves special mention. Basalt fragments in this Apollo 14 breccia were discussed by Ridley

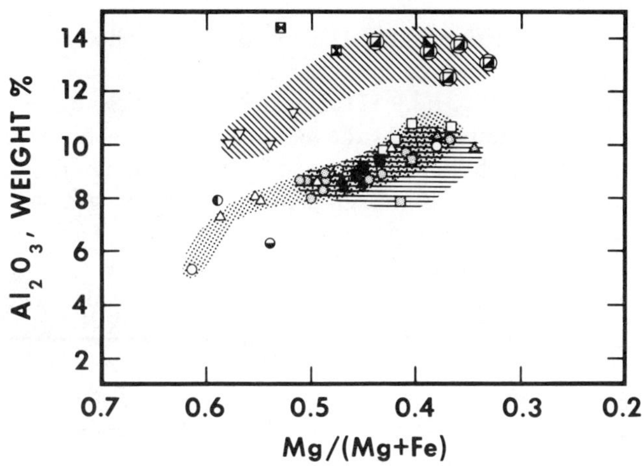

1.2.9.5 Al_2O_3 (wt.%) vs. Mg′ variation in mare basalts. Legend in Fig. 1.2.9.3.

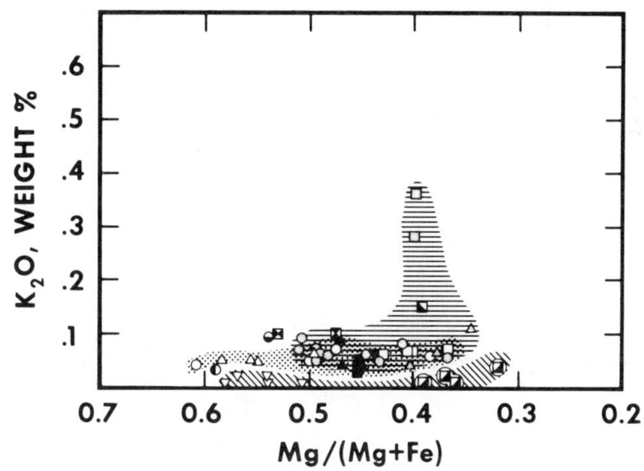

1.2.9.7 K_2O (wt.%) vs. Mg′ variation in mare basalts. Legend in Fig. 1.2.9.3.

(1975). One important feature of this basalt type is that it has a TiO_2 content which falls right in the middle of the "TiO_2 gap" between high- and low-Ti basalts. Data for two important mare glasses (the Apollo 15 emerald green glass and the Apollo 17 orange glass) are also plotted on these diagrams.

The high-Ti basalts show a trend in which TiO_2 content decreases with increasing degree of fractionation (Fig. 1.2.9.4). This reflects the subtraction from the melt of an Fe-Ti oxide mineral which occurs as a liquidus or near-liquidus phase. An important implication of this observation is that, if high-Ti lavas erupt from fissures and flow long distances, fractionation can produce significant Ti gradients along flow lines. High-resolution spectral mapping could interpret such a titanium difference in terms of more than one flow unit when, in truth, it may be a single, continuous flow unit which experienced flow fractionation. Combined TiO_2 and crater degradation maps might resolve this ambiguity. The low-Ti and VLT basalts behave in just the opposite way with respect to titanium. Fractionation of these magmas leads to an increase in TiO_2 content.

The VLT suite displays a range of Mg′-values, from the high values characteristic of Apollo 15 green glass to those of the fractionated Luna 24 ferrobasalts. Several authors have suggested this may represent a fractionation sequence (Taylor *et al.*, 1977a; Vaniman and Papike, 1977b, c). However, Grove and Vaniman

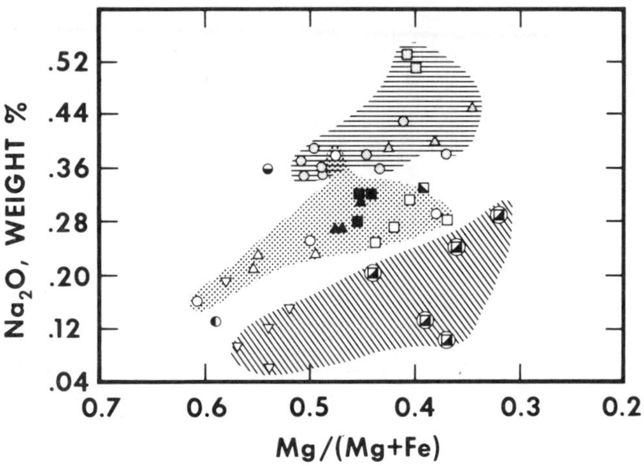

1.2.9.6 Na_2O (wt.%) vs. Mg′ variation in mare basalts. Legend in Fig. 1.2.9.3.

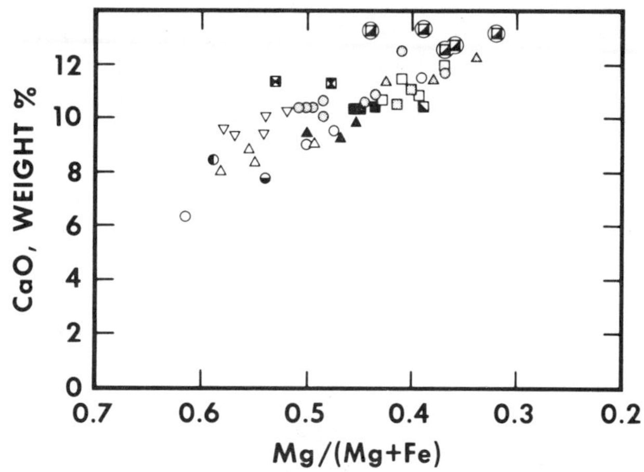

1.2.9.8 CaO (wt.%) vs. Mg′ variation in mare basalts. Legend in Fig. 1.2.9.3.

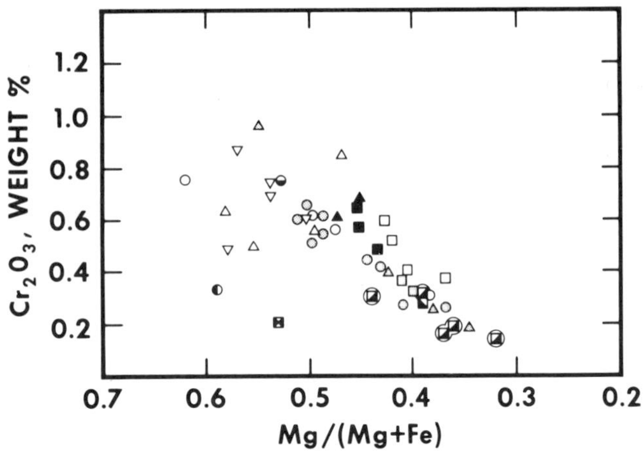

1.2.9.9 Cr_2O_3 (wt.%) vs. Mg′ variation in mare basalts. Legend in Fig. 1.2.9.3.

(1978) have shown that Luna 24 ferrobasalts cannot be obtained by low-pressure fractionation of an Apollo 15 green glass parent magma.

The Al_2O_3 vs. Mg′ diagram (Fig. 1.2.9.5) shows that the VLT basalts are distinct from the low- and high-Ti groups, but that Al_2O_3 contents of the latter two groups overlap strongly. We also see an Al_2O_3 increase with Mg′ decrease. Good group separation is observed on a Na_2O vs. Mg′ diagram (Fig. 1.2.9.6). That is, high-Ti basalts have higher Na_2O contents than low-Ti basalts. A Na_2O increase is also observed with increasing degree of fractionation.

The behavior of K_2O is more complicated than that of Na_2O (Fig. 1.2.9.7). For K_2O, the high-Ti and low-Ti groups, with the exception of the Apollo 11 high K basalts, overlap significantly. However, the K_2O contents of the VLT basalts are significantly lower than those of the other two groups. No clear group separation is apparent on the CaO vs. Mg′ diagram (Fig. 1.2.9.8). The main trend is one of increasing CaO content with decreasing Mg′. Likewise, the groups overlap in the Cr_2O_3 vs. Mg′ diagram (Fig. 1.2.9.9); however, Cr_2O_3 contents decrease with decreasing Mg′.

A histogram of Mg′-values is shown as Fig. 1.2.9.10. Papike *et al.* (1976) pointed out that the least-fractionated liquids of both the high-Ti and low-Ti suites have Mg′≈0.5; basalts from these groups with Mg′>0.5 are thought to be cumulates. In contrast, the Apollo 17 VLT basalts are characterized by Mg′>0.6. A VLT vitrophyre from Luna 24 has an Mg′≈0.6 (Taylor *et al.*, 1977b). These Mg′-values approach those of the Apollo 15 green glasses. The implication is that these VLT sources were more magnesian (and possibly deeper) or else experienced higher degrees of partial melting than sources of other mare basalts.

An additional useful chemical variation diagram is a three-axis plot of Mg′ vs. (CaO/Al_2O_3 wt. ratio) vs. TiO_2 wt.% (Fig. 1.2.9.11). This diagram serves not only to discriminate among the mare basalt groups but also to distinguish the mare basalts from the highland melt rocks. The six highland melt-rock groups are well separated from the mare basalts on the basis of CaO/Al_2O_3 and Mg′; however, TiO_2 contents overlap with those of the low-Ti and VLT mare basalts. The VLT basalts, Apollo 17 yellow/green glass and green glass are distinguished from the other mare basalts mainly by Mg′ and TiO_2 content. Luna 24 VLT basalts have the lowest Mg′ of any of the rock types plotted on the diagram.

1.2.9.4 Trace element chemistry

Trace element data provide important constraints on the natures of the source regions of mare basalts, on their fractionation histories, and on the evolutionary history and bulk composition of the Moon. Many of the first order observations, such as the depletion of the basalts in volatile and siderophile elements, and their enrichment in refractory elements, were made during the initial analysis of the Apollo 11 high-Ti basalts (LSPET, 1969). Trace element data for the mare basalt reference suite, which encompasses the many varieties, are reported in Tables 1.2.9.2–5.

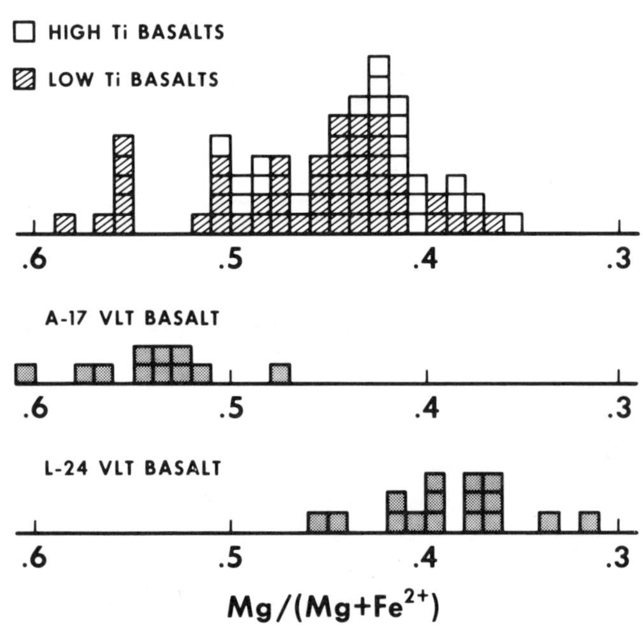

1.2.9.10 Comparative histograms of Mg′ in high-Ti, low-Ti and VLT mare basalts.

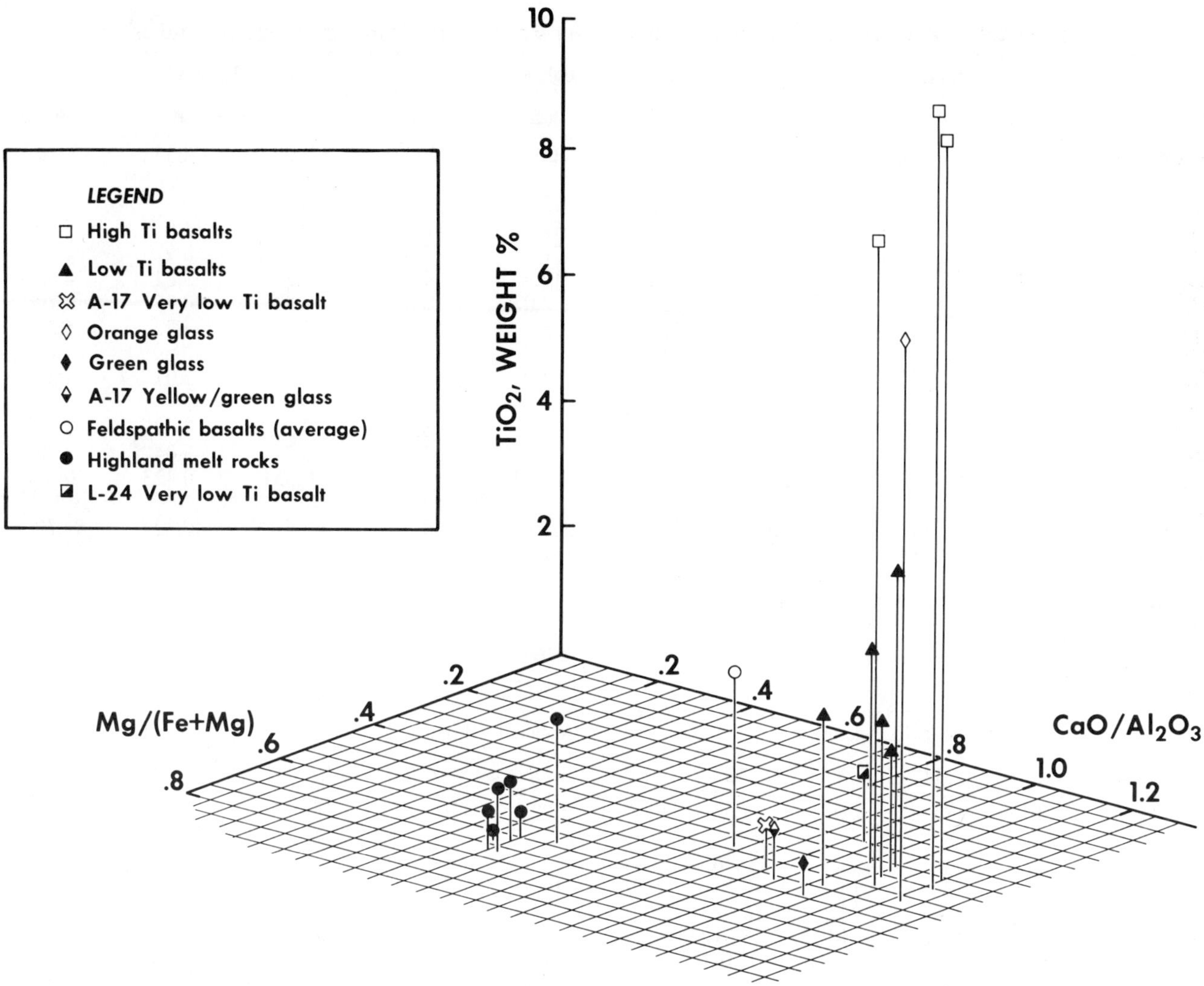

1.2.9.11 Mg′, weight % ratio CaO/Al_2O_3, and weight % TiO_2 in mare basalts and highland melt rocks. Averages of mare basalt types are compared with the averages of six highland melt rock groups plus the average feldspathic basalt. Data plotted include average compositions for the high-Ti and low-Ti basalt groups from Papike *et al.* (1976) and for Apollo 17 VLT and the Apollo 17 yellow/green glass cluster from Vaniman and Papike (1977d); the orange glass analysis of Rhodes *et al.* (1974); the average green glass analysis of Ridley *et al.* (1973); cluster groups for highland melt rocks and feldspathic mare basalts from Naney *et al.* (1977); and the Luna 24 ferrobasalt from Vaniman and Papike (1977c).

The rare-earth elements (REE) (Table 1.2.9.2) provide many important insights into the natures of the source regions of mare basalts. Typical REE patterns are shown in Figs. 1.2.9.12–13, in which nearly all classes are represented. With minor exceptions (Luna 24 and Apollo 17 VLT basalts), all mare basalts have negative Eu anomalies. Although many hypotheses have been advanced to account for this Eu depletion (see Taylor, 1975, pp. 154-159, for a review), it now is recognized as an intrinsic signature of the source regions, which indicates that prior removal of plagioclase had extracted Eu (and Sr) from the source regions of the mare basalts. These elements now reside principally in the highland crust (Fig. 1.2.9.14).

Isotopic systematics (Chapter 7) date this plagioclase removal as occurring prior to 4.4 b.y. ago, during crystallization of a global magma ocean. This primary evidence for the derivation of mare basalts from source regions which had already become differentiated is confirmed by a wide variety of other trace element evidence. REE patterns commonly show light REE depletion. The heavy REE patterns are commonly parallel to those of chondrites, but may be relatively depleted (high-Ti basalts) or enriched (Apollo 17 VLT

Table 1.2.9.2 Rare-earth element abundances (ppm) for samples in the Mare Basalt Reference Suite.

	Apollo 11 Low K			Apollo 11 High K		Apollo 12 Olivine			
	1	**2**	**3**	**4**	**5**	**6**	**7**	**8**	**9**
Sample No.	10003	10020	10047	10049	10072	12002	12009	12035	12040
La	15.5[b]	8.11	11.3	28.8	22.7	6.02	6.1	—	—
Ce	47.2[b]	25.8	46	82.8	69	17	16.8	11.5	15.3
Nd	40.0[b]	23.9	—	62.8	51	12.3	16	8.91	12
Sm	14.4[b]	9.47	18.9	22.3	17.9	4.24	4.53	3.22	4.03
Eu	1.81[b]	1.6	2.71	2.29	2.07	0.835	0.94	0.751	0.796
Gd	19.5[b]	12.8	—	29.3	26	5.65	5.2	4.32	5.6
Tb	—	—	4.1	—	4.3	—	1.11	—	—
Dy	12.9[b]	15.8	—	33.4	31.2	6.34	7.13	5.07	6.36
Ho	—	—	7.9	—	6.8	—	—	—	—
Er	13.6[b]	10.0	—	30.9	16	3.89	3.6	3.09	3.71
Yb	13.2[b]	9.87	18.2	20.2	16.6	3.78	3.74	3.04	3.38
Lu	1.0[b]	1.43	2.88	—	2.24	—	0.551	0.423	0.521
Σ REE	179.11	118.78	111.99	312.79	265.81	60.07	65.7	40.32	51.70
Y	112[d]	84	134	—	161	39	—	29	22
La/Yb	1.17	0.82	0.62	1.43	1.37	1.59	1.63	—	—

	Apollo 12 Pigeonite				Apollo 12 Ilmenite				Apollo 15 Olivine		
	10[f]	**11**	**12**	**13**	**14[f]**	**15[f]**	**16**	**17**	**18**	**19**	**20**
Sample No.	12011	12021	12052	12064	12005	12008	12022	12051	15016	15545	15555
La	—	—	—	6.76	—	—	—	6.53	5.58	4.93	—
Ce	19.9	19.8	18.8	17.5	10.2	16.9	17.4	19.2	15.6	13.9	8.06
Nd	—	14.4	14.7	16	—	—	14.4	15.4	11.4	9.82	6.26
Sm	5.0	4.84	4.91	5.51	2.99	5.35	5.38	5.68	4.05	3.29	2.09
Eu	0.95	1.116	1.04	1.161	0.62	1.06	1.26	1.23	0.97	0.895	0.688
Gd	—	6.59	6.87	7.2	—	—	7.71	7.89	5.4	4.48	2.9
Tb	1.06	—	—	1.27	0.77	1.39	—	—	0.9	—	—
Dy	—	7.86	7.74	9.03	—	—	9.37	9.05	5.74	4.68	3.27
Ho	—	—	—	1.72	—	—	—	—	1.1	—	—
Er	—	4.53	4.55	6	—	—	5.42	5.57	3.1	2.67	1.7
Yb	4.2	4.12	4.32	4.59	2.7	4.9	5.69	5.46	2.62	2.16	1.45
Lu	0.62	0.64	0.651	0.67	0.41	0.71	—	—	0.321	0.308	—
Σ REE	31.73	63.9	63.58	77.41	17.69	30.31	66.63	76.01	56.78	47.13	26.42
Y	39	50.5	—	41	28	45	68	48	21	33	23
La/Yb	—	—	—	1.47	—	—	—	1.20	2.13	2.28	—

basalts). The latter patterns provide evidence of derivation for this particular class of basalt from a source region dominated by olivine and orthopyroxene. Among other constraints, the REE evidence indicates an extreme diversity of sources for the mare basalts. The overall REE patterns for the Moon are probably parallel to those of chondritic meteorites, or at least there exists no definitive evidence to the contrary.

The large high-valency cations (Table 1.2.9.4) typically are enriched in lunar basalts relative to terrestrial basalts. Zr/Nb ratios in lunar basalts are constant at 12–15 when data from a single laboratory are employed (Duncan *et al.*, 1976). The enrichment of these refractory elements in the Moon, relative to the Earth and eucrites, appears to be well established. These elements are all "incompatible" as well as refractory. During

Table 1.2.9.5 *(Continued)*

	Apollo 15 Pigeonite			Apollo15 Green Glass	Apollo 17 Group A		Apollo 17 Group B		Apollo 17 Group C		
	21	**22**	**23**	**24[h]**	**25**	**26**	**27**	**28**	**29**	**30**	**31**
Sample No.	15058	15499	15597		75055	76136	70215	70275	74245	74255	74275
La	—	—	4.86	—	6.265	6.91	5.22	6.32	6.24	6.5	6.33
Ce	14.5	—	13	3.8	21.5	23.8	16.5	20.8	22.2	22.5	21.4
Nd	10.9	—	9.3	2.2	23.9	26.2	16.7	21.8	24.9	24.7	22.8
Sm	3.9	—	3.09	0.76	10.05	10.9	6.69	8.75	9.8	10.1	9.19
Eu	0.908	—	0.84	0.21	2.09	2.14	1.37	1.73	1.77	1.85	1.8
Gd	5.0	—	4.4	0.91	15.7	16.4	10.4	14.0	—	15.3	14.8
Tb	1.07	—	0.69	—	—	—	—	—	—	—	—
Dy	5.9	—	4.51	1.1	18.1	19.3	12.2	15.2	17.5	17.3	16.3
Ho	1.1	—	0.86	—	—	—	—	—	—	—	—
Er	3.2	—	1.9	0.8	10.72	11.4	7.4	9.14	9.68	10.0	9.66
Yb	2.54	—	2.13	0.93	9.79	10.2	7.04	8.3	9.13	8.93	8.47
Lu	0.388	—	0.301	0.14	—	1.42	1.03	—	1.25	—	—
Σ REE	57.04	—	45.88	10.85	118.12	128.67	84.55	106.04	102.47	117.18	110.75
Y	21.1	31.1	—	—	112	—	—	—	—	160.0	81.5
La/Yb	—	—	2.28	—	0.64	0.68	0.74	0.76	0.68	0.73	0.75

	Apollo 17 Group U		A-17 Orange Glass	Luna 16	Luna 24	Apollo 17 VLT	Other Feldspathic
	32	**33**	**34[e]**	**35**	**36[i]**	**37[g]**	**38**
Sample No.	70017	70035	74220	B-1	24174	78526	14053
La	—	4.79	6.25	—	2.87	1.2	13.0
Ce	10.7	16.4	19.0	—	8.6	—	34.5
Nd	12.1	18.2	17.8	—	7.0	—	21.9
Sm	5.13	7.63	6.53	—	2.10	1.0	6.56
Eu	1.62	1.82	1.80	—	0.83	0.30	1.21
Gd	—	11.0	8.52	—	—	—	8.59
Tb	—	—	—	—	—	—	—
Dy	10.2	14.1	9.40	—	2.9	2.0	10.5
Ho	—	—	—	—	—	—	—
Er	6.31	8.4	5.10	—	—	—	6.51
Yb	6.25	7.79	4.43	—	2.00	1.4	6
Lu	0.954	1.17	0.61	—	0.31	0.23	—
Σ REE	53.26	91.25	79.44	—	26.61	6.13	108.77
Y	71.2	75	—	—	—	—	54.7
La/Yb	—	0.61	1.41	—	1.44	0.86	2.17

[b]Gast *et al.* (1970) [d]Compston *et al.* (1970) [e]Rhodes *et al.* (1974) [f]Rhodes *et al.* (1977) [g]Laul and Schmitt (1975) [h]Taylor (1975) [i]Laul *et al.* (1978) Except as noted, all other data from Lofgren and Lofgren (1981).

melting and recrystallization of the magma ocean, they would have been excluded from the major mineral phases (olivine, pyroxenes, plagioclase). Increasing amounts would be trapped in interstitial liquids or in accessory minerals in the later stages of crystallization. Thus the high-Ti basalts typically contain greater concentrations than do low-Ti basalts. Extreme concentrations occur in KREEP basalts from the highland crust,

Table 1.2.9.3 Abundances of ferromagnesian elements (ppm, except where indicated) for samples in the Mare Basalt Reference Suite.

	Apollo 11 Low K			Apollo 11 High K		Apollo 12 Olivine			
	1	**2**	**3**	**4**	**5**	**6**	**7**	**8**	**9**
Sample No.	10003	10020	10047	10049	10072	12002	12009	12035	12040
Cr	1390[c]	—	—	—	—	5620	—	—	3826.67
V	63[c]	81	13	—	62	175	153	130	153
Sc	—	85	92	80.9	77	38.3	46	33.7	37.53
Ni	—	<2	<2	—	3	63.9	51.9	32.9	40
Cu	—	<2	16	—	2	4.6	41	2	13
Fe%	15.44	14.98	14.91	14.59	15.27	16.89	16.40	17.32	16.45
Mn	2310	2000	2160	1930	1770	2160	2160	2230	2080
Zn	7.4[a]	2	13	—	2	1.5	30	2	8
Mg%	4.01	4.76	3.40	4.22	4.68	8.93	6.93	9.29	9.87
Ga	3[a]	—	—	—	—	—	—	—	—
V/Ni	—	—	—	—	20.7	2.7	2.9	3.9	3.8
Cr/V	—	—	—	—	—	32.1	—	—	25.0

	Apollo 12 Pigeonite				Apollo 12 Ilmenite				Apollo 15 Olivine		
	10[f]	**11**	**12**	**13**	**14[f]**	**15[f]**	**16**	**17**	**18**	**19**	**20**
Sample No.	12011	12021	12052	12064	12005	12008	12022	12051	15016	15545	15555
Cr	4050	1870	3300	—	5200	4200	3300	—	6400	—	—
V	—	160	149	119	—	—	180	102	200	168	—
Sc	52.2	49.8	50.6	63.1	37.1	52.4	53.80	58	39.1	42.0	—
Ni	—	16	5.9	6.9	90	—	41.9	5.9	85.9	68.9	41.9
Cu	—	8.1	8	7	—	—	8	6	11	370	—
Fe%	15.23	15.06	15.72	15.51	17.37	17.11	16.97	15.77	17.74	17.28	17.57
Mn	2230	2000	2080	2080	2310	2310	1930	2160	2460	2310	2230
Zn	—	4.15	9	2	—	—	1.2	2	4	23	10
Mg%	4.96	4.43	4.93	3.89	11.98	7.40	6.61	4.21	6.78	6.71	6.82
Ga	—	—	—	—	—	—	—	—	—	—	—
V/Ni	—	10.0	25.2	17.2	—	—	4.3	17.3	2.3	2.4	—
Cr/V	—	11.7	22.1	—	57.8	—	18.3	—	32.0	—	—

which most probably were derived from the invasion of the highlands by residual melts produced during crystallization of the magma ocean (see section 1.2.10). Later partial melting within the source regions of the mare basalts would concentrate these incompatible elements in the melt, so that at least two stages of enrichment, over whole-moon abundances, have contributed to generating the mare basalts. Differentiation, caused by fractional crystallization, then would result in a further concentration of these elements in the basaltic melts.

The ferromagnesian trace elements (Table 1.2.9.3) provide additional evidence. Nickel contents of the high-Ti basalts are below detection limits (<2 ppm), but rise to values of 60 ppm in the low-Ti basalts. These values are low in comparison with those in terrestrial basalts, and probably reflect an overall depletion of the Moon in siderophile elements relative to the Earth. In contrast to Ni, Cr is enriched in mare basalts, relative to terrestrial basalts, by about an order of magnitude. A large proportion of Cr in the Moon is present as Cr^{2+}. Schreiber and Haskin (1976) have shown that this ion becomes concentrated in the melt in preference to entering olivine, so that pyroxenes would have become enriched with Cr during crystallization of the magma ocean. Thus, high concentrations of Cr could form in what later were to be the source regions of the mare basalts. Vanadium shows a distribution similar to that of Cr, while Sc parallels Ti. The ferromagnesian element distributions are entirely consistent with derivation of the mare basalts from source regions which had

Table 1.2.9.3 (Continued)

	Apollo 15 Pigeonite			Apollo 15 Green Glass	Apollo 17 Group A		Apollo 17 Group B		Apollo 17 Group C		
	21	**22**	**23**	**24**	**25**	**26**	**27**	**28**	**29**	**30**	**31**
Sample No.	15058	15499	15597		75055	76136	70215	70275	74245	74255	74275
Cr	2840	—	—	—	—	3790	3030	—	4630	—	—
V	—	189	—	—	—	—	50	—	—	65	79
Sc	42	—	—	—	82.7	82	85.9	85	77	74.2	75.1
Ni	31	19	30.0	—	2	—	3	—	—	17	3
Cu	—	3	—	—	—	—	6.39	—	—	36	3
Fe%	15.49	15.75	15.75	16.42	14.05	14.91	15.33	14.52	14.09	14.08	14.15
Mn	2080	2160	2310	—	2230	2160	2080	2160	2080	2160	2000
Zn	—	2	1.2	—	7	—	—	—	—	5.4	<2
Mg%	5.57	5.60	5.24	10.03	4.23	5.17	5.06	3.65	5.79	6.37	6.22
Ga	—	—	—	—	—	—	—	—	—	—	—
V/Ni	—	9.9	—	—	—	—	16.7	—	—	3.8	26.3
Cr/V	—	—	—	—	—	—	60.6	—	—	—	—

	Apollo 17 Group U		A-17 Orange Glass	Luna 16	Luna 24	Apollo 17 VLT	Other Feldspathic
	32	**33**	**34**	**35**	**36**	**37**	**38**
Sample No.	70017	70035	74220	B-1	24174[i]	78526	14053
Cr	3490	3890	4650	—	—	—	2860
V	146	—	—	—	—	—	—
Sc	80	82.5	—	—	—	—	55
Ni	<3	2	—	—	—	—	14
Cu	28	—	—	—	—	—	—
Fe%	14.24	14.40	17.19	15.09	17.24	14.51	13.1
Mn	1930	2160	—	1540	2160	1850	2000
Zn	<2	4	—	—	—	—	3.4
Mg%	5.99	5.93	8.66	4.23	3.60	7.32	5.09
Ga	—	—	—	—	—	—	—
V/Ni	—	—	—	—	—	—	—
Cr/V	23.9	—	—	—	—	—	—

[a]Ganapathy *et al.* (1970) [c]Goles *et al.* (1970) [e]Rhodes *et al.* (1974) [f]Rhodes *et al.* (1977) [i]Laul *et al.* (1978) Except as noted, data from Lofgren and Lofgren (1981).

undergone prior crystallization of olivine, pyroxene and plagioclase.

The volatile/involatile element ratios in mare basalts indicate that the Moon is depleted in volatile elements relative to the abundances in carbonaceous chondrites. K/U ratios are 2000–3000 compared with $>10^4$ for the Earth and 50,000–80,000 for meteorites (Fig. 1.2.9.15). The isotopic systematics confirm that this volatile depletion, which is greater for the highly volatile elements, occurred before the Moon accreted. Rb/Ba ratios (Fig. 1.2.9.16) provide a good example of the constancy of volatile/involatile element ratios. A striking characteristic is that the lunar highlands samples and the mare basalts show similar volatile/involatile element ratios (e.g., Duncan *et al.*, 1976). This similarity rules out models which call for heterogeneous accretion of the highland crust from a source enriched in refractory elements relative to the lunar interior, and implies that the mare basalts and the highlands share a common source. Such ratios (e.g., K/U, Rb/Ba) are useful discriminators between planetary bodies (Figs. 1.2.9.15–16).

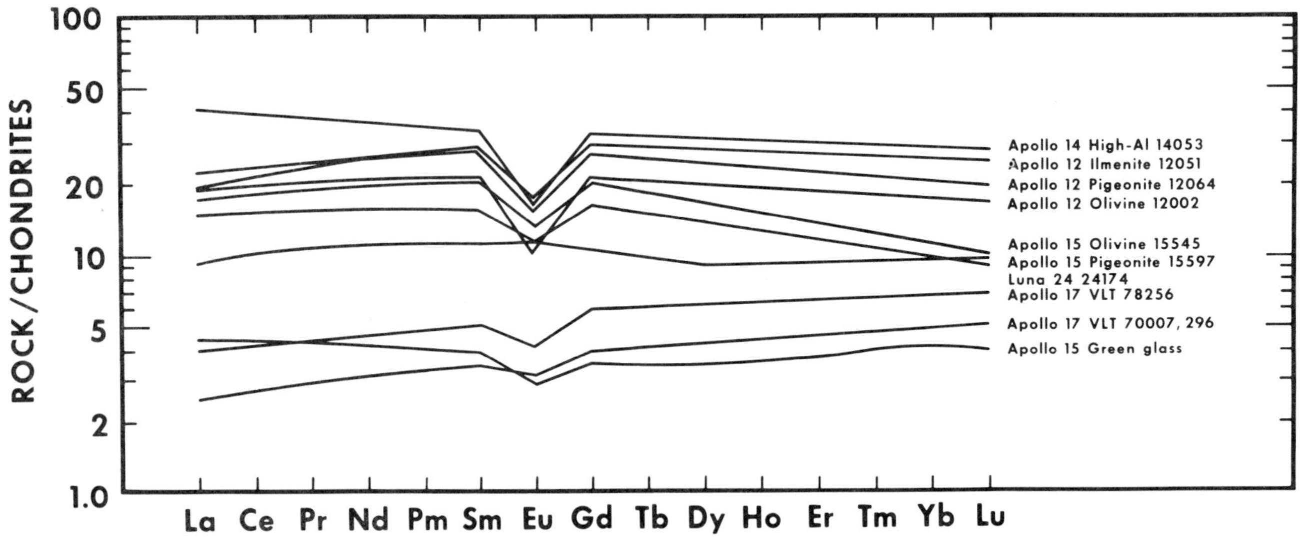

1.2.9.12 Chondrite-normalized REE patterns for selected low-Ti, VLT and feldspathic mare basalts. Note LREE depletion in the low-Ti basalts and the HREE enrichment in the VLT basalts. The Luna 24 high-Al basalt has no Eu anomaly, while the Apollo 14 high-Al basalt has LREE enrichment. The green glass, a very primitive basalt, has an Eu depletion, indicating derivation from an already fractionated source.

The trace element and isotopic data are consistent with the concept, which is referred to as the cumulate model (Fig. 1.2.9.14), that the mare basalts were derived from differentiated source regions which had crystallized from a magma ocean 4.4 b.y. ago. The hypothetical crystallization sequence is illustrated in Fig. 1.2.9.14, which shows a layered sequence ranging from olivine and orthopyroxene cumulates at the base to clinopyroxene and ilmenite cumulates at higher levels. Adcumulus growth (excluding trapped liquid) would predominate

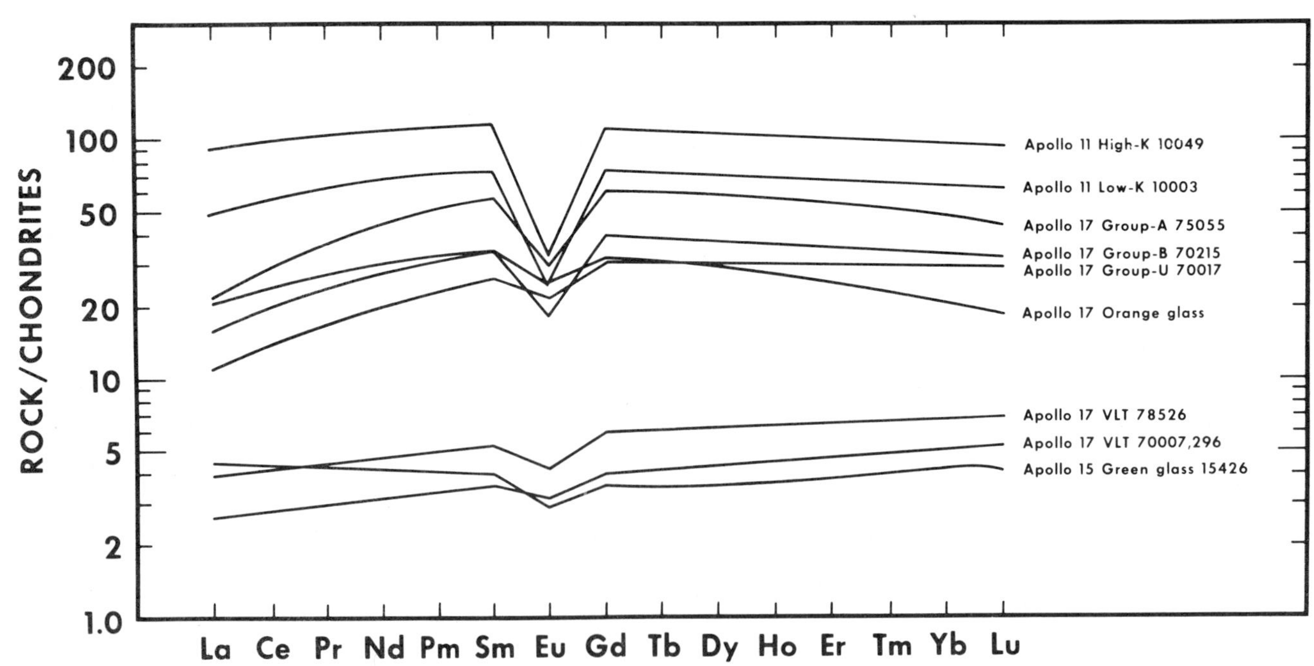

1.2.9.13 Chondrite-normalized REE patterns for high-Ti mare basalts, with those for Apollo 15 green glass and Apollo 17 VLT basalts shown for comparison. Note the strong Eu depletion in the high-Ti suite.

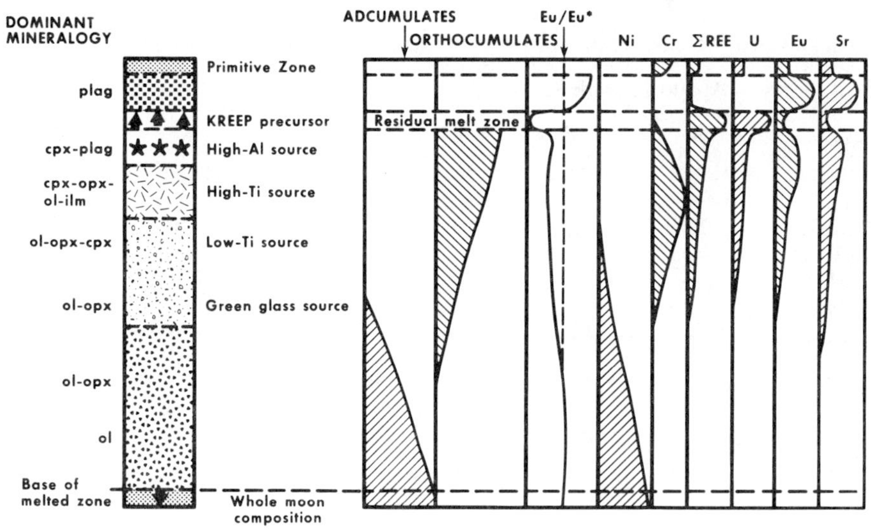

1.2.9.14 A schematic diagram of the crystallization of the lunar magma ocean 4.4 b.y. ago. This idealized section would be greatly modified by sinking of dense cumulates, convective overturn, and by planetesimal impacts, leading to a heterogeneous lunar interior source region for mare basalts. Idealized element distributions of Ni, Cr, ΣREE, Eu depletion or enrichment (Eu/Eu*, where Eu* is the abundance for no Eu anomaly), U and Sr are shown.

in the olivine-orthopyroxene layers, while orthocumulus growth (trapping of pore liquids) would occur in the later-crystallizing portions, by analogy with terrestrial layered intrusions (Wager and Brown, 1968). However, many factors may combine to upset this idealized situation. Sinking of dense cumulates, convective overturn and the effects of large planetesimal infalls can produce a heterogeneous interior. The model serves to illustrate the sorting out of the trace elements into differing zones which, several hundred million years later, became the source regions for the mare basalts. Eu and Sr (Fig. 1.2.9.14) were concentrated principally in the feldspar-rich highland crust. Incompatible elements (REE, Zr, Hf, Nb, Rb, Ba, Cs, Th, U) were concentrated in the residual melt which finally invaded and pervaded the highland crust to form the KREEP component. Ni became concentrated in the olivine and orthopyroxene layers, while Cr, even as Cr^{2+} (Schreiber and Haskin, 1976), was excluded from these layers and so became concentrated in the later-crystallizing minerals. These processes resulted in a lunar interior which was highly differentiated with respect to trace elements.

Several factors argue against models which imply that mare basalts were derived from a primitive lunar interior, and assimilated or became contaminated with differentiated material. There are two particularly decisive arguments: (1) the nickel abundances of high-Ti basalts are extremely low. No amount of contamination or assimilation could produce this effect. (2) There is a good inverse correlation between the abundances of incompatible elements in mare basalts and those in highland rocks (including KREEP). Those elements which are most depleted in mare basalts are most enriched in potential contaminating materials. This observation is the reverse of that which would be predicted by models requiring assimilation.

The chalcophile, volatile and siderophile trace element abundances are of special interest because of the evidence which they may provide for preaccretionary mass loss, core formation and interplanetary comparisons. The abundances of these elements (Os, Ir, Re, Ni, Pd, Au, Sb, Ge, Se, Te, Ag, Zn, In, Cd, Bi and Tl—listed in order of decreasing nebula condensation temperature) are not well established, except for Ni, due to a combination of analytical, sampling and contamination problems. Only a few laboratories have provided useful data, so the reliable data base is small. The spread

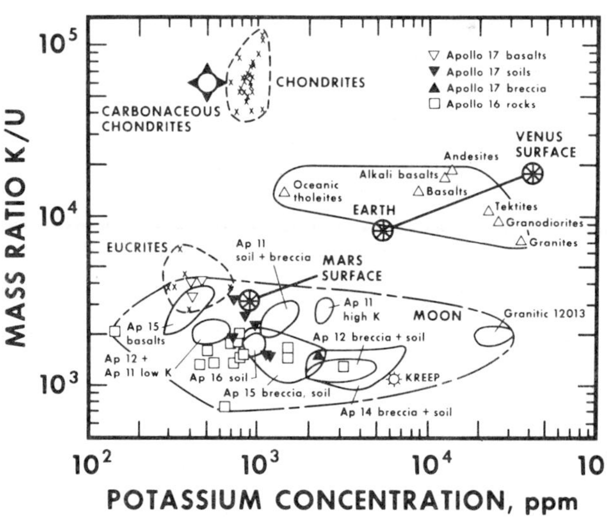

1.2.9.15 K/U ratios and K abundances showing the separation of lunar K/U ratios from those of the Earth and chondritic meteorites. Note that the eucrite and martian data overlap those for the Moon, while the surface data from Venus overlap those for the Earth. Adapted from Eldridge *et al.*, 1975; venusian and martian data from Surkov, 1977.

Table 1.2.9.4 Abundances of large high valency cations (ppm) for samples in the Mare Basalt Reference Suite.

	Apollo 11 Low K			Apollo 11 High K		Apollo 12 Olivine			
	1	**2**	**3**	**4**	**5**	**6**	**7**	**8**	**9**
Sample No.	10003	10020	10047	10049	10072	12002	12009	12035	12040
U	—	0.19	0.22	0.81	0.86	0.22	0.24	0.20	0.16
Th	1.1[d]	0.68	0.72	4.03	3.16	0.75	0.88	0.68	0.47
Zr	309[d]	224	334	—	504	106	107	81	57
Hf	—	—	13.2	17.3	12	2.49	4	1	2.39
Nb	—	17.6	23	—	27.8	8.5	6	4	2
Ti%	6.30	6.37	5.66	6.78	7.24	1.56	1.74	1.37	1.47
Th/U	—	3.53	3.23	4.95	3.68	3.41	3.63	3.43	0.29
Zr/Hf	—	—	25.3	—	42.0	42.6	26.7	81	23.8
Zr/Nb	—	12.7	14.5	—	18.1	12.5	17.8	20.2	28.5

	Apollo 12 Pigeonite				Apollo 12 Ilmenite				Apollo 15 Olivine		
	10[f]	**11**	**12**	**13**	**14[f]**	**15[f]**	**16**	**17**	**18**	**19**	**20**
Sample No.	12011	12021	12052	12064	12005	12008	12022	12051	15016	15545	15555
U	—	0.26	0.37	0.22	—	—	0.20	0.26	0.12	0.13	0.13
Th	—	0.93	1.28	0.84	—	—	0.71	1	0.50	0.43	0.46
Zr	128	123	121	114	66	117	180	128	86	190	76
Hf	3.7	4.09	3.8	3.9	2.4	3.8	3.4	3.1	2.6	2.2	—
Nb	7.4	14	7	7	4.3	5.9	6	7	10	—	4.3
Ti%	1.97	2.12	1.97	2.39	1.66	2.67	2.91	2.81	1.37	1.45	1.26
Th/U	—	3.57	3.51	3.81	—	—	3.59	3.85	4.17	3.31	3.64
Zr/Hf	34.6	30.1	31.8	29.2	27.5	30.8	52.9	41.3	33.1	86.4	—
Zr/Nb	17.3	8.8	17.3	16.3	15.3	19.8	30.0	18.3	8.6	—	17.7

	Apollo 15 Pigeonite			Apollo 15 Green Glass	Apollo 17 Group A		Apollo 17 Group B		Apollo 17 Group C		
	21	**22**	**23**	**24**	**25**	**26**	**27**	**28**	**29**	**30**	**31**
Sample No.	15058	15499	15597		75055	76136	70215	70275	74245	74255	74275
U	0.13	0.16	0.14	—	0.14	—	0.13	0.12	0.13	0.13	0.14
Th	0.52	0.59	0.53	—	0.45	—	0.34	0.43	0.4	0.45	0.46
Zr	98	109	—	—	272	—	192	—	—	238	248
Hf	2.16	—	—	—	7.2	9.4	6.33	—	8.7	—	8.55
Nb	4.9	5.4	—	—	25	—	20.8	—	—	<10	22.1
Ti%	1.01	1.07	1.08	0.22	6.58	7.58	7.78	7.14	7.15	7.48	7.62
Th/U	4.00	3.69	3.79	—	3.29	—	2.62	3.58	3.08	3.36	3.42
Zr/Hf	45.4	—	—	—	37.8	—	30.3	—	—	—	29.0
Zr/Nb	20.0	20.2	—	—	10.9	—	9.2	—	—	—	11.2

Table 1.2.9.4 (Continued)

	Apollo 17 Group U		A-17 Orange Glass	Luna 16	Luna 24	Apollo 17 VLT	Other Feldspathic
	32	**33**	**34**[e]	**35**[j]	**36**[i]	**37**	**38**
Sample No.	70017	70035	74220	B-1	24174	78526	14053
U	0.06	0.09	0.16	0.30	—	—	0.59
Th	0.20	—	—	—	0.2	—	2.1
Zr	218	217	185	—	50	—	215
Hf	6.4	—	—	—	—	—	9.8
Nb	18.5	20	—	—	—	—	15.7
Ti%	7.79	7.78	5.29	2.94	0.66	0.55	1.58
Th/U	3.30	—	—	—	—	—	3.55
Zr/Hf	34.1	—	—	—	—	—	21.9
Zr/Nb	11.8	—	—	—	—	—	13.7

References: [d]Compston *et al.* (1970). [e]Rhodes *et al.* (1974). [f]Rhodes *et al.* (1977). [i]Laul *et al.* (1978). [j]Albee *et al.* (1972). Except as noted, data from Lofgren and Lofgren (1981).

in abundances for these elements is large, so it is difficult to establish averages appropriate for use in interplanetary comparisons. Nevertheless, some consistent trends have emerged (Wolf *et al.*, 1979). The Moon appears to be systematically depleted in volatile elements relative to the Earth by relatively constant factors (Fig. 1.2.9.17; Wolf *et al.*, 1979). These trends are matched by depletions of the LIL volatile elements K, Rb, and Cs, with extent of depletion correlating with the order of relative volatilities given above. The depletion factors for chalcophile, volatile and siderophile elements in the Moon, relative to C1 chondrites, range from about 10^2

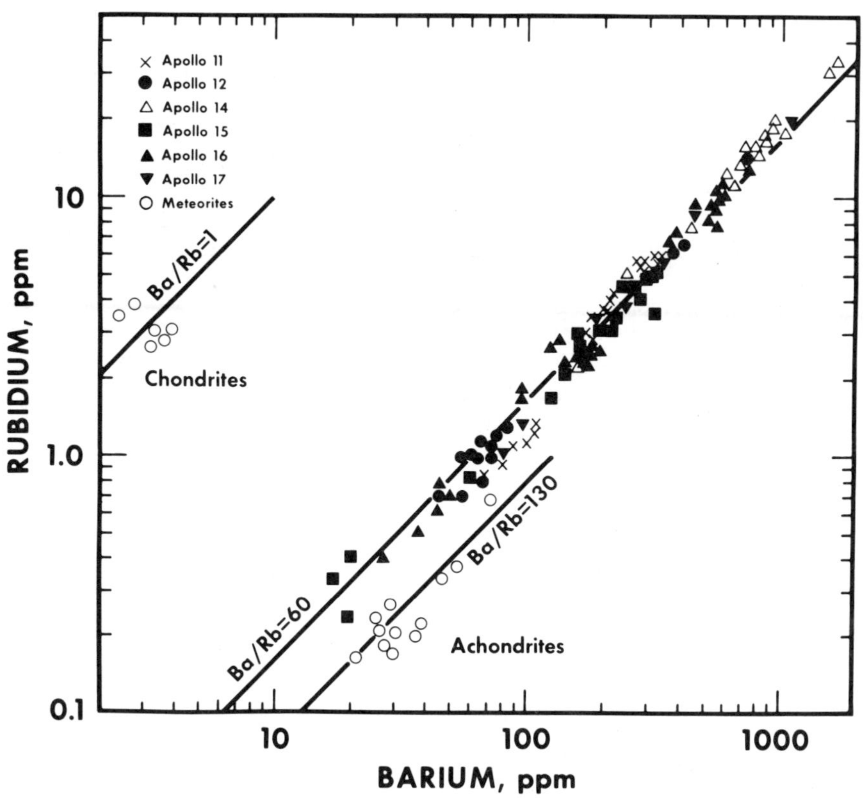

1.2.9.16 Correlation between the volatile element Rb and the refractory element Ba in mare basalts, highland samples and meteorites. The sources of the highlands and mare samples possess similar volatile/refractory element ratios. Thus these portions of the Moon (to a depth of several hundred kilometers) have accreted from homogeneous material, or were homogenized following accretion. Note that the meteorite samples, both differentiated (achondrites) and primitive (chondrite), have distinctly different ratios.

Table 1.2.9.5 Abundances of large cations (ppm, except where indicated) for samples in the Mare Basalt Reference Suite.

	Apollo 11 Low K			Apollo 11 High K		Apollo 12 Olivine			
	1	**2**	**3**	**4**	**5**	**6**	**7**	**8**	**9**
Sample No.	10003	10020	10047	10049	10072	12002	12009	12035	12040
Cs	0.022[a]	—	—	—	—	—	—	—	—
Rb	0.62[d]	0.63	0.93	6.2	5.7	1.04	1.05	0.69	1.0
K%	0.05	0.03	0.09	0.30	0.23	0.04	0.05	0.04	0.04
Ba	108[b]	77	88	330	294	67	60	47	57
Sr	161[d]	150	194	161	168	101	95.6	84.3	85.5
Ca%	7.90	8.04	8.63	7.81	7.39	5.86	6.69	5.74	5.69
Na%	0.30	0.28	0.33	0.38	0.39	0.17	0.17	0.16	0.13
K/Rb	806	476	968	484	404	385	476	580	400
Rb/Cs	28.2	—	—	—	—	—	—	—	—
Ba/Rb	174	122	94.6	53.2	51.6	64.6	57.1	68.4	57.2
Rb/Sr	0.004	0.004	0.005	0.04	0.03	0.01	0.01	0.01	0.01
Sr/Eu	88.9	93.6	71.6	70.2	81.3	118	102	112	107
K/Ba	4.6	3.9	10.2	9.1	7.8	5.9	8.3	8.5	7.0

	Apollo 12 Pigeonite				Apollo 12 Ilmenite				Apollo 15 Olivine		
	10[f]	**11**	**12**	**13**	**14[f]**	**15[f]**	**16**	**17**	**18**	**19**	**20**
Sample No.	12011	12021	12052	12064	12005	12008	12022	12051	15016	15545	15555
Cs	—	—	—	—	—	—	—	—	—	—	—
Rb	—	1.14	1.26	1.05	—	—	0.74	0.91	0.81	0.75	0.45
K%	0.05	0.06	0.06	0.06	0.03	0.04	0.06	0.05	0.03	0.03	0.07
Ba	71	71.1	75	70	35	51	55	73.6	61	46.7	32.2
Sr	113	129	116	135	83	130	143	148	91.4	104	84.4
Ca%	7.55	8.08	7.67	8.36	4.48	6.37	7.01	8.09	6.58	6.97	6.67
Na%	0.19	0.23	0.20	0.21	0.14	0.19	0.53	0.21	0.20	0.23	0.20
K/Rb	—	526	476	571	—	—	811	549	370	400	1570
Rb/Cs	—	—	—	—	—	—	—	—	—	—	—
Ba/Rb	—	62.4	59.5	66.7	—	—	74.3	80.9	75.3	62.3	72.4
Rb/Sr	—	0.01	0.01	0.01	—	—	0.01	0.01	0.01	0.01	0.01
Sr/Eu	119	115	111	116	134	123	113	120	94.2	116	123
K/Ba	7.0	8.4	8.0	8.6	8.6	7.8	10.9	6.8	4.9	6.4	21.7

	Apollo 15 Pigeonite			Apollo 15 Green Glass	Apollo 17 Group A		Apollo 17 Group B		Apollo 17 Group C		
	21	**22**	**23**	**24[h]**	**25**	**26**	**27**	**28**	**29**	**30**	**31**
Sample No.	15058	15499	15597		75055	76136	70215	70275	74245	74255	74275
Cs	—	—	—	—	—	—	—	—	—	—	—
Rb	2.0	0.99	1.13	0.58	0.58	0.67	0.36	0.45	1.17	1.22	1.2
K%	0.02	0.04	0.05	0.02	0.07	0.05	0.04	0.05	0.05	0.07	0.06
Ba	62	69	52	18	76.2	83.7	56.9	73.5	67.4	71.1	67.3
Sr	99.2	108	111	41	191	190	121	153	159	163	153
Ca%	7.29	7.34	7.41	5.97	8.77	7.48	7.63	8.27	7.52	7.29	7.33
Na%	0.21	0.24	0.24	0.10	0.32	0.28	0.27	0.28	0.27	0.27	0.26
K/Rb	100	402	442	345	1200	746	1120	1100	427	574	500
Rb/Cs	—	—	—	—	—	—	—	—	—	—	—
Ba/Rb	31.0	69.3	46.0	31.0	130	125	160	162	57.6	58.3	56.1
Rb/Sr	0.02	0.01	0.01	0.01	0.003	0.004	0.003	0.003	0.01	0.01	0.01
Sr/Eu	109	—	132	195	91.1	88.8	88.3	88.4	89.8	88.1	85.0
K/Ba	3.23	5.8	9.6	11.1	9.2	6.0	7.0	6.8	7.4	9.9	8.9

	Apollo 17 Group U		A-17 Orange Glass	Luna 16	Luna 24	Apollo 17 VLT	Other Feldspathic
	32	**33**	**34**[e]	**35**[j]	**36**[i]	**37**[g]	**38**
Sample No.	70017	70035	74220	B1	24174	78526	14053
Cs	—	—	—	0.054	—	—	—
Rb	0.28	0.46	1.11	1.58	—	—	2.19
K%	0.04	0.05	0.07	0.12	0.02	0.01	0.08
Ba	43	62.1	76.4	218	50	—	146
Sr	168	174	209	437	110	—	98
Ca%	7.30	7.15	5.45	7.38	8.24	7.10	7.95
Na%	0.29	0.47	0.27	0.24	0.19	0.09	—
K/Rb	1430	1080	630	760	—	—	365
Rb/Cs	—	—	—	—	—	—	—
Ba/Rb	157	135	68.8	138	—	—	66.7
Rb/Sr	0.002	0.003	0.01	0.004	—	—	0.02
Sr/Eu	104	95.6	116	—	132	—	81.0
K/Ba	9.3	8.1	9.2	5.5	4.0	—	5.5

[a]Ganapathy *et al.* (1970) [b]Gast *et al.* (1970) [d]Compston *et al.* (1970) [e]Rhodes *et al.* (1974) [f]Rhodes *et al.* (1977) [h]Taylor (1975) [i]Laul *et al.* (1978) [j]Albee *et al.* (1972) Except as noted, data from Lofgren and Lofgren (1981).

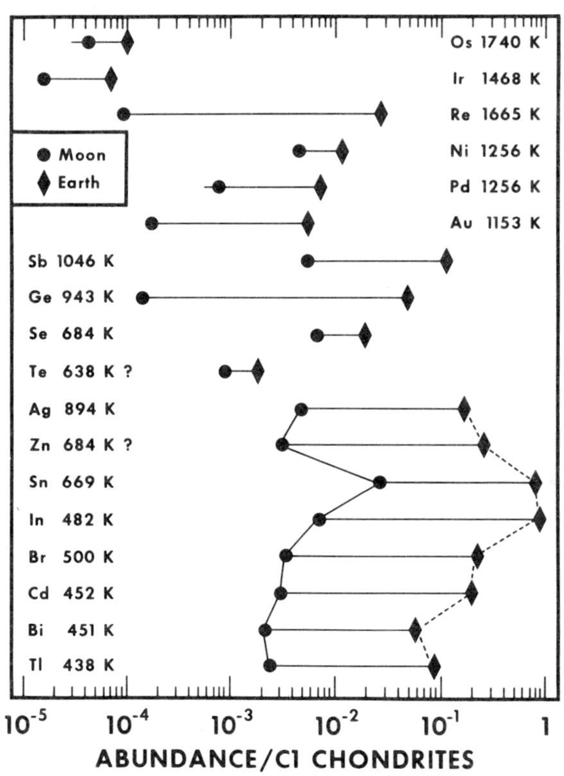

1.2.9.17 Comparison between abundances of siderophile and volatile elements in terrestrial and lunar basalts normalized to C1 chondrite abundances (adapted from Wolf *et al.*, 1979). Condensation temperatures for elements from a gas of solar composition at 10^{-5} atm are shown. Note the depletion of all these elements in lunar relative to terrestrial basalts.

to 10^5. Proposed lunar cores (Nakamura *et al.*, 1976) occupy only 1–2% of the total lunar volume. Accordingly, they are too small to be of much use as a geochemical sink for the siderophile or chalcophile trace elements, including Ni.

It is predicted that Ni would enter early-crystallizing olivines during magma ocean solidification, hence, it would be depleted in the source regions of most mare basalts. However, no such mechanism is available to explain siderophile trace element abundances. There is no indication in mare basalt chemistry that the source regions contained metallic phases. The Fe metal, which is observed in some basalts, crystallized late in the paragenetic sequence, so its presence is a near-surface effect (Sato, 1976). The conclusion emerges that the Moon as a whole is depleted in siderophile elements relative to C1 chondrites and to the Earth.

A general conclusion may be drawn with respect to abundances of the incompatible elements. During the crystallization of the magma ocean, and during subsequent processes of partial melting to generate basalts, these elements become concentrated preferentially in the residual liquids or enter early melts. Accordingly, there is a tendency for the ratios of these elements to remain relatively constant; therefore, ratios such as K/U, K/Rb, K/La, K/Cs, and ratios among the various chalcophile and siderophile elements may reflect whole-moon abundance ratios. Such aids to estimating the composition of the Moon must be treated with appropriate caution and with comprehension of the geochemical subtleties of varying distribution coefficients, and of the causative parameters of ionic radii, valency, and crystal field effects which may alter the simple picture.

1.2.9.5 Petrology

Figure 1.2.9.18 is a histogram relating the major element chemistry and modal petrology of the important basalt groups. A summary of the modal data is presented in Table 1.2.9.6. The diagram is ordered, left to right, in terms of decreasing TiO_2 content. A significant correlation is evident between TiO_2 content and modal opaques. It is interesting that basalts with the extremes in TiO_2 content are found at the Apollo 17 site, where the high-Ti basalts contain up to 13 wt.% TiO_2 and the Apollo 17 VLT ~0.6 wt.% TiO_2. It is apparent that, in the Serenitatis Basin, early flooding (~3.8 b.y.) involved basalts which were highly enriched in Ti, whereas the younger central basin fill is at the lower extreme in Ti content. The diagram also shows that Luna 24 ferrobasalts are enriched in Al_2O_3, CaO and FeO, but depleted in MgO relative to Apollo 17 VLT. These chemical differences result in higher modal feldspar in Luna 24 ferrobasalts than in Apollo 17 VLT. The higher modal olivine in Luna 24 basalts is less easily explained. The reason for there being more modal olivine in Luna 24 than in Apollo 17 VLT probably is in part the lower SiO_2 content of Luna 24 basalts and the different late-stage cooling histories of these two basalt groups. In Apollo 17 VLT basalts, pyroxenes crystallized metastably in the "forbidden region" (Lindsley and Munoz, 1969) and the stable assemblage hedenbergite-SiO_2-olivine did not form. Thus, some of the "olivine content" of Apollo 17 VLT basalts is still tied up in the pyroxenes. The higher Fe/(Fe + Mg) of Luna 24 ferrobasalt may explain this observation, since, with increasing Fe content, pyroxene and olivine no longer are in reaction relation and more olivine can crystallize.

The remaining discussion in this section concerns the petrology of the different basalt groups. However, since this topic has been reviewed extensively by Papike *et al.* (1976), the discussion here is brief except for the new results on Apollo 12 mare basalts, Apollo 17 VLT basalts, and Luna 24 VLT basalts.

High-Ti basalts

Apollo 11 basalts. Apollo 11 high-Ti basalts fall into two groups, low-K and high-K. However, Beaty and Albee (1978) have found that the low-K group may be represented by at least three flows. Apollo 11 low-K basalts are fine- to medium-grained, vesicular, and subophitic to ophitic. They show chemical and modal trends which indicate that they are products of limited near-surface fractionation. CaO, Na_2O, K_2O, P_2O_5 and Fe/(Fe + Mg) increase and FeO, MgO, TiO_2 and Cr_2O_3 decrease

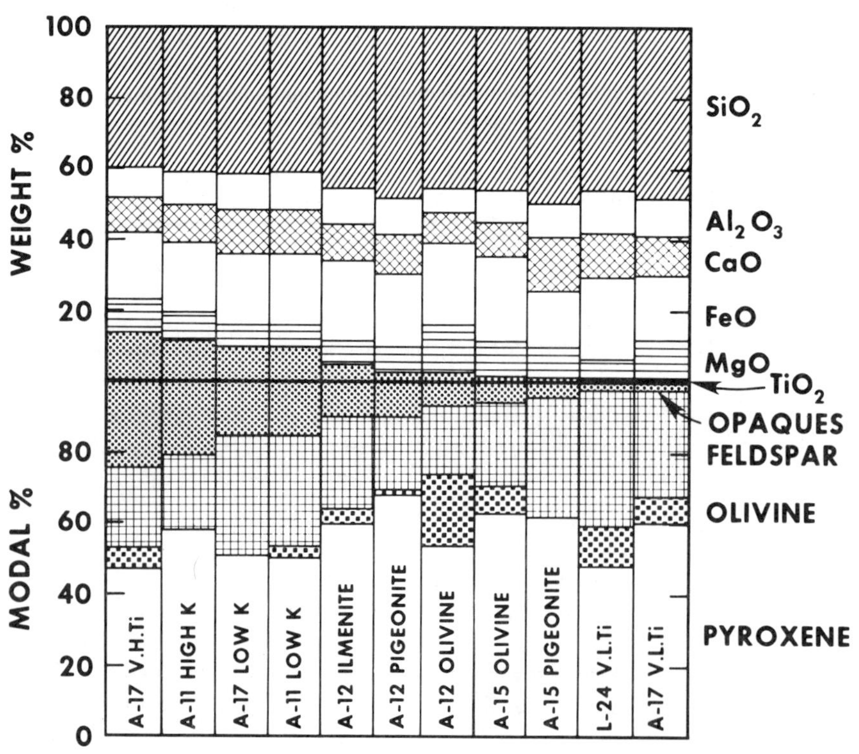

1.2.9.18 Diagram showing correlation between major element chemistry and modal mineralogy for mare basalts. Histogram is ordered from left to right by decreasing TiO_2 (or opaque oxide) content.

Table 1.2.9.6 Summary of modal data† for mare basalts.

	Opaques	Pyroxene	Feldspar	Olivine
A-17 very high Ti	24.4	47.7	23.4	4.6
A-11 high K	20.6	57.5	21.7	0.1
A-17 low K	15.1	51.6	33.3	—
A-11 low K	14.6	50.9	32.2	2.3
A-12 ilmenite	9.3	61.1	25.9	3.6
A-12 pigeonite	9.1	68.4	21.1	1.4
A-12 olivine	7.1	53.5	19.2	20.2
L-16	7.1	51.5	41.2	0.1
A-15 olivine	5.5	63.3	24.1	7.0
A-15 pigeonite	3.7	62.5	33.8	—
14053	3.2	53.8	43.0	—
L-24 ferrobasalt	1.8	48.6	39.1	10.4
L-24 recrystallized ferrobasalt	1.4	60.2	34.2	4.2
A-17 VLT	1.0	61.7	31.9	5.4

†Normalized to 100% for the four phases considered. Ordered from top to bottom in terms of decreasing modal content of opaques.

with increasing grain size. These trends indicate a progressive loss of olivine and ilmenite in the coarser-grained rocks. Thus the textural, modal and chemical data are consistent with a model in which the Apollo 11 low-K basalts are portions of lava flows that underwent minor near-surface fractionation of olivine and ilmenite.

Apollo 11 high-K basalts are fine-grained vesicular rocks characterized by highly-variable holocrystalline to intersertal textures. There is little chemical or modal variation within the high-K group and no systematic modal variation with grain size. The small amount of variation noted within this group, coupled with the limited area over which the samples were collected, indicates that the suite probably consists of samples of a single thin, rapidly-cooled flow or sill.

Apollo 17 basalts. The Apollo 17 high-Ti basalts are similar to the Apollo 11 high-Ti basalts (LSPET, 1973). They consist of two extreme types: a group of very high-Ti basalts which are unique to the Apollo 17 site, and a group of basalts that are texturally and chemically similar to the Apollo 11 low-K basalts. Chemical trends within the Apollo 17 very high-Ti basalts are consistent with low-pressure fractionation of olivine, armalcolite, and spinel; however, a simple correlation between major element chemistry and grain size, similar to that found in the Apollo 11 low-K basalts, does not appear to be present in the Apollo 17 samples. This most likely indicates that multiple cooling units were sampled. The chemical study of Rhodes *et al.* (1976) has confirmed this interpretation.

The Apollo 17 very high-Ti basalts are modally distinct from the Apollo 11 mare basalts in that they contain more modal olivine and opaque minerals. This is consistent with the higher normative olivine and TiO_2 contents of the very high-Ti basalts. A second set of samples returned from the Apollo 17 site (e.g., 75035 and 75055) is chemically, texturally, and modally similar to the Apollo 11 low-K basalts. Major element chemistry and experimental studies of Apollo 17 basalts (Kesson, 1975; Walker *et al.*, 1975) indicate that the low-K samples could be derived from the Apollo 17 very high-Ti basalts through low-pressure fractionation. An extensive chemical study by Rhodes *et al.* (1976) has demonstrated that the early two-fold classification of the Apollo 17 basalts was an oversimplification. In fact, there exists a broad compositional continuum between the most mafic olivine-normative basalts and the chemically- and mineralogically-fractionated quartz-normative type. Their study defined three self-consistent basalt types on the basis of fine-grained, rapidly-cooled samples. The chemical variation within each group is attributed to moderate amounts of near-surface crystal fractionation (from 4–22%) dominated by the removal of olivine, armalcolite/ilmenite, and chrome-spinel. Rhodes *et al.* (1976) suggest further that the three distinct groups of basalts and their respective parental magmas cannot be explained by near-surface crystal fractionation processes. The differences are attributed to partial melting of a heterogeneous source region.

Low-Ti basalts

Apollo 12 basalts. The mare basalts returned by the Apollo 12 mission consist of four chemically- and texturally-distinct rock groups: (1) olivine basalts, (2) pigeonite basalts, (3) ilmenite basalts, and (4) feldspathic basalts.

The Apollo 12 olivine basalts are a group of fine-to-medium-grained olivine-rich basalts ranging from olivine vitrophyre to microgabbro. The chemical, modal, and textural trends indicate that many of the coarser-grained samples are olivine cumulates. There is a good correlation between chemistry and grain size, with the coarser-grained rocks being distinctly more magnesian. Experimental studies (Green *et al.*, 1971; Newton *et al.*, 1971) indicate that liquids similar in composition to the coarse-grained samples should crystallize olivine considerably more magnesian than that found in the samples. Other studies of finer-grained samples, 12002 (Walker *et al.*, 1976) and 12009 (Green *et al.*, 1971; Donaldson *et al.*, 1975), indicate that they may be representative of liquids.

The Apollo 12 pigeonite basalts are characterized by large, zoned phenocrysts of pyroxene. Major element data indicate that the Apollo 12 pigeonite basalts could have been derived from the Apollo 12 olivine basalts through olivine fractionation.

The discussion of the Apollo 12 suite by Papike *et al.* (1976) was limited by the rather small number of samples that had been studied up to that time. Rhodes *et al.* (1977) obtained additional major and trace element chemical data which confirmed earlier classifications of the Apollo 12 basalts into the four compositional groups mentioned above. The new data also confirmed the fact that the olivine and pigeonite basalts are comagmatic and are related by olivine fractionation. Rhodes *et al.* (1977) demonstrated that the ilmenite basalts are a major basalt type which is indigenous to the Apollo 12 site and which has a compositional range equivalent to that of the olivine-pigeonite basalts. Rhodes *et al.* (1977) suggested further that the flows at the Apollo 12 site are thicker than the majority of lunar flows, and that the younger ilmenite basalts overlie the olivine-pigeonite basalts. Chemical variation within these basalt groups was attributed to extensive, olivine-dominated, near-surface crystal fractionation (about 30%) of parental magmas with compositions equivalent to the vitrophyres in each group. Dungan and Brown (1977) carried out a detailed petrologic study of the Apollo 12 ilmenite basalts. They pointed out that the chemical variations in the ilmenite suite can be reconciled with differentiation within a single cooling unit. The suite includes olivine porphyritic basalts, medium-grained basalts with evolved compositions and coarse-grained olivine cumulates. The chemical variation can be accounted for by fractionation of olivine equivalent to the composition of phenocrysts in vitrophyre 12008, which might represent the parental magma for the ilmenite basalt suite.

Apollo 15 basalts. Mare basalts returned by the Apollo 15 mission are similar to the olivine and pigeonite basalts returned by the Apollo 12 mission. Relatively little has been published on the petrology of these rocks, and a comprehensive study, including acquisition of high-quality major element chemical data, would be very valuable. The results of the Apollo 17 restudy (Rhodes *et al.*, 1976) and Apollo 12 restudy (Rhodes *et al.*, 1977; Dungan and Brown, 1977) point out the rich harvest that such a study might yield. Although much work remains to be done on the Apollo 15 mare basalts, the existing major and trace element data show quite conclusively that the olivine and pigeonite basalts represent distinct magma types and cannot be related by high- or low-pressure fractionation (see, for example, Papike *et al.*, 1976).

The Apollo 15 olivine basalts include a large group of vesicular to scoriaceous, fine- to medium-grained, granular to subophitic basaltic samples. Major element chemical variations, experimentally determined liquid-lines-of-descent (Kesson, 1975; Walker *et al.*, 1977), and textural evidence indicate that the Apollo 15 olivine basalts are products of a limited amount of olivine fractionation and are not olivine cumulates. The evidence for olivine fractionation, lack of cumulates, and vesicularity indicate that the Apollo 15 olivine basalts are largely samples from the upper portion of one or more cooling units.

The Apollo 15 pigeonite basalts are characterized by composite pyroxene phenocrysts with pigeonite cores and augite mantles. The major element variation is consistent with a limited degree of pigeonite fractionation. The controlled cooling rate studies of Lofgren *et al.* (1974, 1975) indicate that the porphyritic textures of the Apollo 15 pigeonite basalts can be produced *in situ* with a single cooling rate. Grove and Walker (1977) used controlled cooling rate studies of vitrophyre 15597, one of the Apollo 15 pigeonite basalts, to determine the cooling histories and physical properties of these rocks. The physical properties determined indicate that pyroxene phenocrysts do not settle appreciably during the cooling of a flow unit, an observation which is consistent with the small degree of chemical differentiation observed between members of the suite. They found that two kinds of cooling histories characterize these rocks: (1) the vitrophyres experienced initial slow cooling followed by rapid cooling and (2) the coarse-grained microgabbros experienced slow, nearly-linear cooling histories. Grove and Walker (1977) also pointed out that the textural evidence from the lunar vitrophyres is consistent with the crystallization of these samples near the top of a flow.

Very Low-Ti basalts

Apollo 17 VLT basalts. The Apollo 17 drill core contains a few phaneritic, vitrophyric and glass fragments of a very low-Ti (VLT) mare basalt (Vaniman and Papike, 1977a, b; Taylor *et al.*, 1977a). The basalt is characterized by TiO_2 ~0.5 wt.%, Mg' ~0.52, CaO/Al_2O_3 (wt. ratio) ~0.9, and low-alkali contents (Na_2O ~0.1 wt.%, K_2O ~0.0 wt.%). Phaneritic VLT basalts from the Apollo 17 drill core are medium-grained (0.2–0.4 mm) subophitic rocks. Subhedral pyroxene rims are intergrown with euhedral laths of later-crystallized plagioclase. The earliest phases to crystallize were chromian spinel and olivine. Euhedral spinels up to 0.05 mm across are characteristic of VLT basalts. Zoned and rounded olivine grains have resorbed boundaries against pyroxene. Intergrain (late-stage) areas contain glass and anhedral crystals of CaFe pyroxene, ulvöspinel, ilmenite, silica, Fe-sulfide and metallic iron. Vitrophyres contain chromian spinel and olivine ± pyroxene in a matrix of pale yellow/green to clear glass; olivine and pyroxene form skeletal crystals with grain sizes ranging from 0.001 mm–0.1 mm. Spinel is ubiquitous and occurs as small (0.005 mm) equant, octahedral or cruciform crystals in the vitrophyres. Yellow/green glass spheres (0.02–0.2 mm diameter) with VLT compositions occur in the Apollo 17 soil and form a tight compositional cluster (Vaniman and Papike, 1977d).

Olivine (Fo_{74-75}) and chromian spinel are liquidus or near-liquidus phases of these lavas; all vitrophyres contain at least these two minerals. Pigeonite exists with olivine in many vitrophyres, but in phaneritic rocks the olivine is zoned to Fo_{55} and the rims are rounded and resorbed against adjacent pyroxene. No plagioclase is observed in the vitrophyres. Phaneritic VLT basalts contain plagioclase laths which are intergrown only with the pyroxene rims. These observations suggest that the onset of plagioclase crystallization occurred late in the pyroxene crystallization history. Oxide crystallization had a two-stage history: (1) iron-enrichment of chromian spinel, with a slight increase in Ti, followed by cessation of spinel crystallization; and (2) a late-stage crystallization of ulvöspinel together with Fe-sulfide and Fe metal.

It has been suggested that Apollo 17 VLT basalts can be related to Apollo 15 green glass compositions (Taylor *et al.*, 1977a; Vaniman and Papike, 1977b). Grove and Vaniman (1978) conducted low-pressure equilibrium experiments in iron capsules on synthetic analogs of Apollo 15 green glass and Luna 24 brown glass to determine whether crystal fractionation of the Apollo 15 mafic green glasses could generate Apollo 17 VLT and/or Luna 24 VLT. Their results indicate that Apollo 17 VLTs can be derived by low-pressure crystal-liquid fractionation from a green glass type that is similar to, but not identical to, the Apollo 15 emerald green glass. Because of these relationships, Vaniman and Papike (1977b) speculated that the age of Apollo 17 VLTs might be similar to that of green glass and the low-Ti basalts (Apollo 12 and Apollo 15), and predicted an age of 3.3 b.y. However, no fragment of Apollo 17 VLT has yet been dated.

Luna 24 VLT basalts. Soil from Mare Crisium returned by the Soviet Luna 24 mission contains fragments of ferrobasalts and ferrogabbro. The ferrobasalt is very low in Ti content (~ 1.0 wt.% TiO_2), similar to Apollo 17 VLT. These rocks are also characterized by Mg' ~0.3–0.4, Al_2O_3 ~13%, CaO ~13% and Cr_2O_3 ~0.2–0.3%. Fragments of ferrogabbro are most abundant in the soil, with lesser amounts of ferrobasalt (Barsukov *et al.*, 1977). Ages of ~3.3 b.y. have been determined for these samples (Papanastassiou *et al.*, 1977; Schaeffer *et al.*, 1978).

Luna 24 VLT ferrobasalts are ophitic, containing pyroxene and olivine phenocrysts (up to 0.4 mm), small euhedral chrome spinels, and subhedral ulvöspinels up to 0.5 mm in diameter. Residual patches composed of ulvöspinel, ilmenite, silica, Fe-pyroxene and K-enriched glass (0.5% K_2O) occur in intergranular regions.

The first pyroxenes to crystallize range from Ca-rich to Ca-poor varieties. Zoning trends of Fe-enrichment pass through a low-Ca point, $Ca_{25}Mg_{25}Fe_{50}$, and then either rise in Ca toward hedenbergite or zone directly to a final Fe-pyroxene, $Wo_{20}En_5Fs_{75}$. This residual pyroxene composition is identical to that which occurs in the Apollo 17 VLT basalt. However, early pyroxenes in Apollo 17 VLT are more magnesian, consistent with the more magnesian nature of those rocks. Olivine is more abundant in Luna 24 ferrobasalt than in Apollo 17 VLT, and Luna 24 olivines are zoned to much more Fe-rich compositions (Fo_{58-5}) than are the olivines of Apollo 17 VLT basalt (Fo_{76-53}). Feldspar in Luna 24 ferrobasalts is ophitically intergrown with both olivine and pyroxene. It is zoned from $Ab_{90}Ab_4$ to $An_{86}Ab_{14}$, and the Or content is low but increases from ~0.0% in calcic plagioclase to ~1.0% in the more sodic varieties. Haggerty (1977) showed that the first opaque mineral to crystallize in Luna 24 ferrobasalt was chromian spinel. With subsequent crystallization, the spinel became enriched in Fe and Ti, and ulvöspinel + ilmenite appeared as residual phases. This spinel crystallization trend is approximately the same as that observed for Apollo 17 VLT (Vaniman and Papike, 1977b). Differences between these two basalts in Mg/Fe and Cr contents are reflected in minor differences between their spinel crystallization trends. The Luna 24 ferrobasalt spinels have initial compositions more Fe-rich than spinels from Apollo 17 VLT.

Based on results of the Apollo 17 VLT and Luna 24 VLT investigations, Vaniman and Papike (1977c) speculated on the vertical plumbing beneath the Serenitatis and Crisium Basins. They assumed that green glass-type liquids rose from mantle depths under both basins. It now is clear, based on the results of recent experimental investigations (Grove and Vaniman, 1978), that this model actually requires several different parental liquids similar to, but not identical to, Apollo 15 emerald green glass. These liquids, which could have been derived at ~200–400 km (Taylor and Jakeš, 1974), moved to the surface of the Serenitatis and Crisium Basins about 3.3 b.y. ago. Green glass erupted by fire fountaining (Chou *et al.*, 1975; Meyer *et al.*, 1975), while parental liquids of the Apollo 17 VLT and Luna 24 VLT were trapped in holding chambers beneath the Serenitatis and Crisium Basins, respectively.

Feldspathic basalts

Feldspathic basalts, containing up to ~15 wt.% Al_2O_3, have been found at several sites, and include samples described above, e.g., Luna 24 VLTs. All of the mare basalt fragments from the Apollo 14 site appear to be feldspathic, including 14072 (Longhi *et al.*, 1972), fragments in breccia 14321 (Grieve *et al.*, 1975) and samples described below. The Luna 16 regolith is dominated by feldspathic mare basalts.

Luna 16 basalts. Luna 16 sample B-1 is the largest fragment (62 mg) obtained in the sample exchange with the U.S.S.R. Albee *et al.* (1972) have described this fragment in detail. It is a fine-grained ophitic basalt composed predominantly of clinopyroxene (50%), plagioclase (40%), and ilmenite (7%) with minor amounts of Cr-ulvöspinel, olivine, troilite, mesostasis and other minor phases. Slender laths of plagioclase (~0.1 mm) are enclosed in pyroxene grains which range up to 0.7 mm in length. Ilmenite occurs in laths ~0.1 mm long and is predominantly interstitial to the pyroxene grains. The texture suggests a paragenetic sequence in which early olivine and pyroxene were followed by ilmenite and plagioclase. Mesostasis occurs in tiny patches interstitial to plagioclase laths. Kurat *et al.* (1976) have described many other feldspathic basalt fragments from the Luna 16 site.

Apollo 14 basalts. Sample 14053, one of the oldest known mare basalts (Fig. 1.2.9.2), is a coarse-grained microgabbro with strongly-zoned pyroxene grains up to 5 mm in length enclosing smaller subhedral laths of plagioclase. Small amounts of olivine, which occurs both as subrounded phenocrysts and in the late-stage residual assemblage, were reported by Kushiro *et al.* (1972). Other minerals reported include spinel, ilmenite, cristobalite, troilite, metallic Fe, a phosphate phase, and K-rich glass. The late-stage mineral assemblage in this sample indicates very reducing conditions, f_{O_2} ~10^{-16} at 1000° C (Haggerty, 1972).

A group of mare basalts found in Apollo 14 breccia 14063 (Ridley, 1975) are aluminous, but have higher TiO_2, Na_2O, Mg/Fe and lower FeO contents than other members of this group. The basalts have a uniform texture defined by lathy to acicular plagioclase, lathy and anhedral ilmenite, intergranular pyroxene, and rare, pale-brown mesostasis glass (Ridley, 1975). The texture suggests rapid cooling from a melt in which plagioclase and ilmenite nucleated prior to pyroxene. As mentioned above, these basalts are particularly interesting because they have TiO_2 contents of ~7%, in the middle of the compositional gap between the low-Ti and high-Ti basalts.

Mare glasses

Apollo 15 green glass. Ridley *et al.* (1973) have discussed the chemistry of the Apollo 15 green glasses. They pointed out that green glass spheres are a distinctive component of several Apollo 15 soils. They analyzed 28 spheres or fragments from spheres in the <1 mm soil fraction and found each grain to be devoid of inclusions and to be homogeneous. Chou *et al.* (1975), using leaching and etching experiments, demonstrated that Zn, Ge, Cd, In and Au are concentrated in surficial deposits on the Apollo 15 green glass spherules. They speculated that the surface deposits are condensates from the magmatic gas associated with the eruptive process by which the spheres were extruded from the lunar interior. Meyer *et al.* (1975) obtained ion microprobe mass analyses of the surface film on the green glass spheres, and showed that the film is composed of Zn, Ga, Pb, Cu, Tl, S, F, Cl and other elements. They also concluded that these elements condensed on the outside surface of these glasses in lava fountains.

Stolper (1974) and Wood and Ryder (1977) pointed out that, in fact, there is a range of green glass compositions with a systematic trend. Stolper *et al.* (1974) described a trend of olivine control that would require an olivine composition of Fo_{70}, a composition which is considerably at odds with the low-pressure liquidus olivine (~Fo_{82}). However, Grove and Lindsley (1978), based on a re-evaluation of compositional variations within the Apollo 15 green glass data, concluded that the chemical variations cannot be explained simply by olivine of Fo_{70} composition. They observed that the chemical variations involve removal of Al_2O_3, CaO, MgO and FeO and follow a trend of silica depletion or enrichment. They suggested that these trends resulted from variable degrees of batch partial melting of a two-pyroxene + olivine ± garnet source at depths >400 km.

Another study by Delano and Ringwood (1979) involved a detailed major element and Ni study of green glass beads from Apollo 15. The chemical systematics (e.g., inverse correlation of Mg/(Mg + Fe) with Ni content) appear to discount the batch partial melting model of Grove and Lindsley (1978). Delano and Ringwood concluded that green glasses were derived by deep partial melting but that some type of mixing had taken place before the liquids arrived at the lunar surface. Their data appear to be completely inconsistent with an impact melt origin for green glass.

Apollo 17 orange glass. A deposit of orange glass was discovered by the Apollo 17 astronauts at Shorty Crater. Initial speculation was that this might be a young fumarole deposit enriched in volatiles and perhaps even water. Detailed studies of this glass led to a different scenario. The soil consists almost entirely of glass spheres (0.1–0.2 mm) which are either orange in color or black because of devitrification resulting in crystallization of Fe-Ti oxides. The high TiO_2 content (~9 wt.%) imparts the orange color to the glass. The high-Ti content, close association with the high-Ti basalts at the Apollo 17 site, and the old age have led many investigators to consider the orange glass and Apollo 17 basalts to be members of one closely-related group. There is some uncertainty about the true age of Apollo 17 orange glass. At least three ages have been reported: 3.54 b.y. (Huneke *et al.*, 1973), 3.71 b.y. (Husain and Schaeffer, 1973), and 3.66 b.y. (Turner, 1977). In any case, the age is apparently somewhere in the interval between the 3.58 b.y. age of Apollo 11 high-K basalts and the 3.78 b.y. age of Apollo 11 low-K and Apollo 17 mare basalts (see Fig. 1.2.9.2).

Meyer *et al.* (1975) conducted an ion probe study of the surface coated volatiles and concluded that the glasses must have formed during a fire-fountaining event. Hubbard and Rhodes (1976) questioned the classification of Apollo 17 orange glass with Apollo 17 mare basalts. After examining the major and trace element characteristics of orange glass and mare basalts, they concluded that orange glass is most closely associated with the Apollo 12 ilmenite basalts. The chemical variation diagrams presented in Figs. 1.2.9.4–9 show that orange glass has a higher Mg′ (~0.54) than low- and high-Ti mare basalts (excluding the rocks thought to be cumulates) but lower than Apollo 15 emerald green glass (Mg′ ~0.58). The TiO_2 content of the orange glass is within the "TiO_2 gap" between high- and low-Ti basalts. It also is lower in Al_2O_3 than most mare basalts, but contains Na_2O and K_2O comparable to high-Ti basalts, and has CaO and "Cr_2O_3" contents falling on a trend common to all of the mare basalts. Green *et al.* (1975) conducted high-pressure experiments on orange glass compositions. They concluded that the liquids could have been derived by partial melting from olivine + orthopyroxene at depths ≥400 km.

An important point concerning both the Apollo 17 orange glass and Apollo 15 green glass is that neither can be related directly to any known lunar basalt type by any simple igneous process, despite their close associations with the mare basalts and despite their apparent volcanic origins.

1.2.9.6 Mineralogy

The crystal chemistry and mineralogy of the mare basalt mineral phases, pyroxene, feldspar, olivine, spinel, ilmenite and armalcolite, have been discussed extensively by Papike *et al.* (1976) and many high quality microprobe analyses are tabulated in that paper. Thus, the discussion presented here is brief and concentrates on major element variations.

Pyroxene

Pyroxene is the most abundant phase in mare basalts and the most faithful mineral recorder of the petrogenetic process. It usually crystallizes over a considerable span of the liquidus-solidus interval and accommodates most of the major elements present in the mare basalts. The pyroxene structure accepts Si, Al, Ti, Cr, Mg, Fe, Mn, Ca and Na. However, K is excluded because it is too large to be accommodated in any of the pyroxene crystallographic sites. Much of the pyroxene compositional variation can be displayed on a pyroxene quadrilateral plot which shows the relative proportions of Ca, Mg and Fe^{2+}. Pyroxene zoning trends for all of the major mare basalt groups are illustrated in Fig. 1.2.9.19.

Pyroxenes from Apollo 11 low-K mare basalts display strong Ca-Mg-Fe zonation with Ca depletion and Fe enrichment of the pyroxenoid, pyroxferroite ($Ca_{1/7}Fe_{6/7}SiO_3$). Pigeonite (low-calcium clinopyroxene) occurs in both Apollo 11 low-K and Apollo 11 high-K basalts but apparently is more abundant in the latter. Pyroxene grains in Apollo 12 mare basalts show significant chemical zoning with differences which reflect bulk chemical differences among the four major Apollo 12 rock types: Apollo 12 olivine, Apollo 12 pigeonite, Apollo 12 ilmenite and Apollo 12 feldspathic basalts. Although pigeonite is present in all of these groups, it is most abundant and best developed in the Apollo 12 pigeonite basalts, where it invariably is mantled by augite. The pyroxene chemical trends in both Apollo 12 olivine and Apollo 12 pigeonite basalts are very similar, an observation which is consistent with the conclusion discussed above that these basalts are comagmatic and are related by a fractionation sequence. Apollo 12 feldspathic basalts contain extremely Fe-rich pyroxene grains with a number of analyses reported near the hedenbergite (CaFe)-ferrosilite (FeFe) join.

Pyroxene grains from the Apollo 15 mare basalts show distinctive trends for the two major bulk chemical groups. These basalts have textures indicating a range of cooling histories: two different trends, "slower" and "faster", are illustrated in Fig. 1.2.9.19. Although both Apollo 15 olivine and Apollo 15 pigeonite basalts are characterized by pyroxene trends which commence at pigeonite compositions, pigeonite is better developed in the Apollo 15 pigeonite basalts where it is mantled by augite. In the more slowly-cooled basalts, the first appearance of feldspar is reflected by a distinct change in the zoning trends of the augitic pyroxene. This compositional break is not present in the quickly-cooled members of this series, because plagioclase nucleates very late or not at all and thus is not in competition with the augite for Ca. These crystal-liquid systematics have been discussed at length by Bence and Papike (1972) and have been quantified using controlled cooling rate experiments by Grove and Bence (1976).

Pyroxene grains from Apollo 17 basalts show at least three distinct chemical trends. The highly-fractionated Apollo 17 basalts that are virtually identical to Apollo 11 low-K basalts also have nearly identical pyroxene chemical trends. These pyroxenes show extreme Fe enrichment extending across the quadrilateral to pyroxferroite compositions. However, pyroxene grains in the Apollo 17 very high-Ti basalts do not show this extreme Fe zonation. This difference was attributed

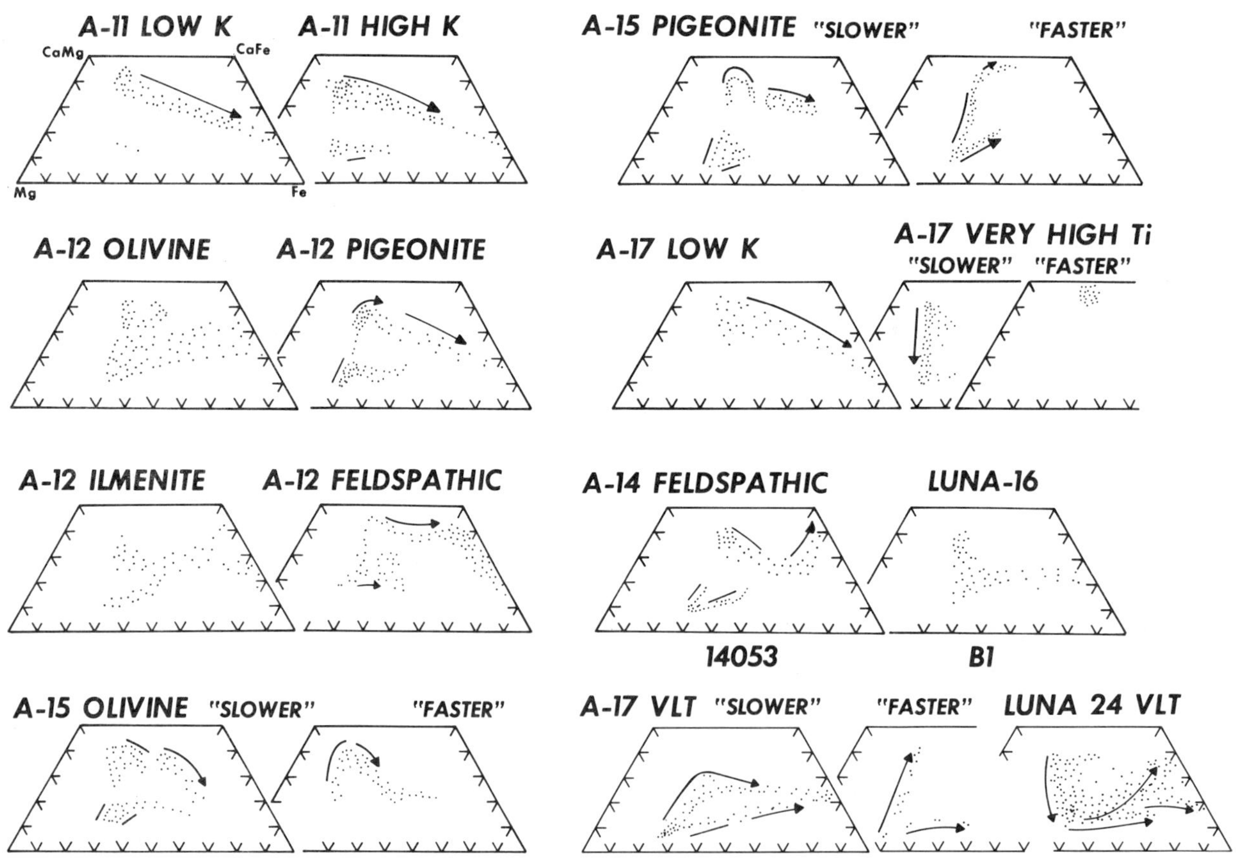

1.2.9.19 Ca-Fe-Mg variations in pyroxenes of mare basalts. Diagrams are schematic, with arrows indicating important crystallization trends.

by Papike *et al.* (1974) to the high activity of Ti, which led to the precipitation of abundant Fe-Ti oxides (armalcolite/ilmenite) which incorporated most of the available Fe. Pyroxene chemical trends in the Apollo 17 very high-Ti basalts differ markedly between the "faster" and "slower" cooled varieties. The more slowly-cooled basalts have pyroxene grains that first crystallized as augite and then zoned to pigeonite. This was ascribed to the resorption of early crystallizing olivine, which adds pigeonite components to the melt (Papike *et al.*, 1974). The more quickly-cooled Apollo 17 mare basalts, on the other hand, contain olivine that is mantled by pyroxene and was thus prevented from reacting with the melt. Pyroxene grains in these basalts exhibit a tight compositional cluster near the diopside-hedenbergite join.

Pyroxene grains from feldspathic mare basalt 14053 show distinctive chemical trends. Crystallization commenced with pigeonite followed by augite. Extreme Fe zonation is observed. The late-stage cooling of these rocks was slow enough so that the "forbidden region" of the pyroxene quadrilateral was not violated (Lindsley and Munoz, 1969). Pyroxene grains from Luna 16 feldspathic basalts start at more Fe-rich compositions than do pyroxene grains in most of the other groups. This is a reflection of the low Mg′-value of these highly-evolved basalts. Fe-enrichment trends in the Luna 16 pyroxene grains extend to the hedenbergite-ferrosilite join.

The Apollo 17 VLT pyroxene trends correspond to a "slower" cooling trend in the phaneritic fragments and a "faster" cooling trend in the vitrophyres. The data and discussion are from Vaniman and Papike (1977b). The first pyroxene to crystallize in both the phaneritic and vitrophyric Apollo 17 VLTs was a pigeonite, $Ca_6Mg_{68}Fe_{26}$. Pyroxenes in the ("faster") vitrophyres developed strong sectoral differences in Ca abundances, but these sectoral differences were suppressed in ("slower") phaneritic rocks. The Ca-enrichment trend in phaneritic VLT shows an inflection when plagioclase began to crystallize and to remove Ca from the melt. In contrast, the Ca-enrichment trend is not interrupted in VLT vitrophyres, where plagioclase nucleation was suppressed. The first pyroxenes which formed in Luna 24 ferrobasalts range from a dominant Ca-rich pyroxene at $Ca_{37}Mg_{41}Fe_{22}$ to a Ca-poor pyroxene at $Ca_{12}Mg_{52}Fe_{36}$. Zoning trends pass through a low-Ca point, $Ca_{25}Mg_{25}Fe_{50}$, and then either rise toward hedenbergite or move directly across the "forbidden" region to $\sim Ca_{20}Mg_5Fe_{75}$. The Luna 24 VLT pyroxene grains started crystallizing at more Fe-rich compositions than those in the Apollo 17 VLT, which is a reflection of the highly-fractionated nature of the Luna 24 ferrobasalt.

Olivine

The modal abundances of mare basalt olivines range from 0–36% (Papike *et al.*, 1976). However, when present, olivine is commonly an early-crystallizing phase that is a chief contributor to the fractionation process. The Fe-Mg variations of olivines from the major basalt groups are illustrated in Fig. 1.2.9.20. The most magnesian olivines reported are Fo_{75} to Fo_{80} and commonly are zoned to Fo_{30}. Relatively few analyses had been reported in the range Fo_{30-10} until recently, when it was found that Luna 24 basalts contain olivines which range in composition from Fo_{60-0}. The Fe-rich, early olivine in Luna 24 ferrobasalts illustrates their highly-evolved nature. For mare basalts in general the very Fe-rich olivines (Fo_{10-0}) are part of the stable

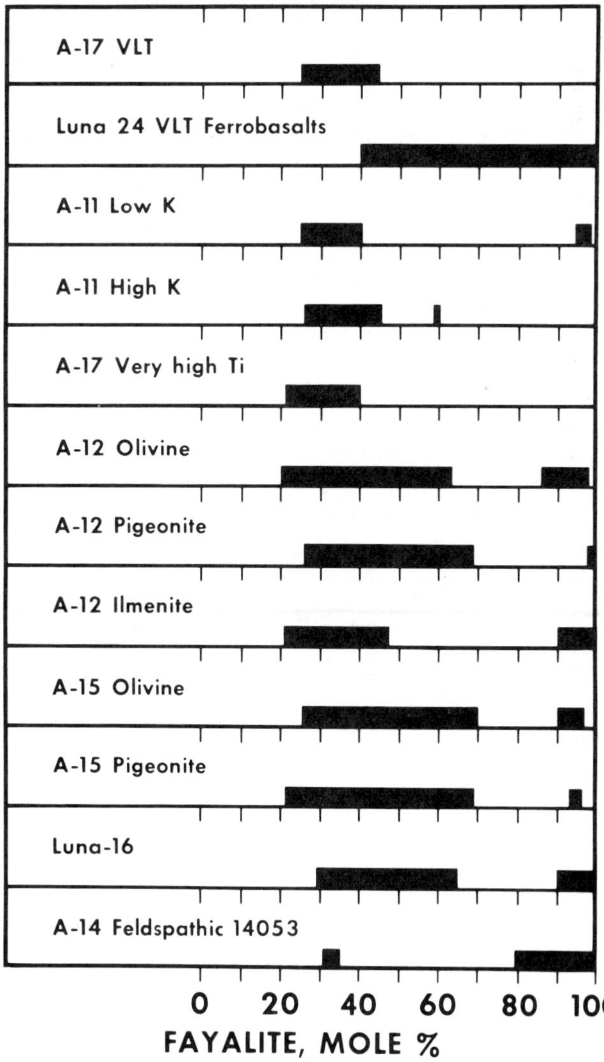

1.2.9.20 Fayalite contents of mare basalt olivines.

three-phase assemblage (hedenbergite-fayalite-silica) that crystallizes from relatively slowly-cooled, late-stage, iron-enriched portions of the basalt melts. Mare basalts which cooled rapidly during the late stages of crystallization contain Fe-rich pyroxene (in some cases zoning to pyroxferroite) which crystallized metastably relative to the stable three-phase assemblage. The most magnesian olivines in the mare basalt suite (Fo_{75-80}) should be in equilibrium with a melt with Mg′ ~0.47–0.57 (K_D estimates range from 0.26 for high Ti-basalts to 0.33 for low-Ti basalts; Longhi *et al.*, 1978). This is consistent with our estimates for the most primitive melts for the high- and low-Ti mare basalt suites, but is too low for some of the VLT vitrophyres (Mg′~0.6), where an olivine of ~Fo_{83} would be expected if the olivine grains were not too small to analyze by microprobe. The olivines plotted in Fig. 1.2.9.20 are from phaneritic lithic fragments that have lower Mg′-values.

Feldspar

Na-Ca (Ab-An) variations in feldspar grains from the major mare basalt groups are plotted in Fig. 1.2.9.21. A striking feature of the feldspar Ab-An variations (An_{73-98}) is the limited content of the albite ($NaAlSi_3O_8$) end-member, compared with terrestrial basalts (e.g., An_{40-80} for mid-ocean ridge basalts). Low Ab and Or($KAlSi_3O_8$) contents of the mare feldspar grains are logical consequences of the low-alkali contents of these basalts (e.g., Na_2O for mare basalts ≈0.21–0.55 wt.%; Na_2O for MORB ≈1.6–2.7 wt.%). Despite the low albite contents of mare basalt feldspars, there is some correlation between Ab content of feldspar and the Na_2O content of the host mare basalt (note Fig. 1.2.9.6). For example, the high-Ti basalts and Luna 16 basalts, which have relatively high Na_2O contents, contain the most Ab-enriched feldspar; the low-Ti basalts which have intermediate Na_2O contents contain feldspars with intermediate Ab contents, and the VLT basalts (especially Apollo 17 VLT) which have the lowest Na_2O contents contain the least Ab-enriched feldspars.

Spinels

A complete review of the crystal chemistry and mineralogy of the Fe-Ti oxide minerals in mare basalts is presented by Papike *et al.* (1976) and a comprehensive discussion of oxide minerals in lunar rocks in general is presented by El Goresy (1976). Therefore, our discussion here is brief and serves mainly to compare the spinel data for Apollo 17 VLTs and Luna 24 VLTs with data for the other basalt groups.

Typical spinels in mare basalts have compositions that can be represented, as first approximations, in the ternary system Fe_2TiO_4 (ulvöspinel)-$FeCr_2O_4$ (chromite)-$FeAl_2O_4$ (hercynite). The principal cationic substitutions are of the type, $Fe^{2+} + Ti^{4+} \rightleftarrows 2(Cr^{3+}, Al^{3+})$. In addition, substantial substitution of Mg for Fe occurs. These chemical variations can be conveniently portrayed on a spinel prism (Haggerty, 1972: note Fig. 1.2.9.22). The spinels in the Apollo 11 and Apollo 17 high-Ti basalt groups are very similar and show less compositional variation than spinels in the low-Ti or VLT mare basalts. The compositions of spinel grains from Apollo 12 olivine and Apollo 12 pigeonite basalts fall mainly into two clusters, chromite and chromian ulvöspinel. The trends for the spinels in both of these groups are quite similar, which is consistent with their comagmatic nature (discussed above). However, spinels in the Apollo 12 pigeonite basalts are zoned to more

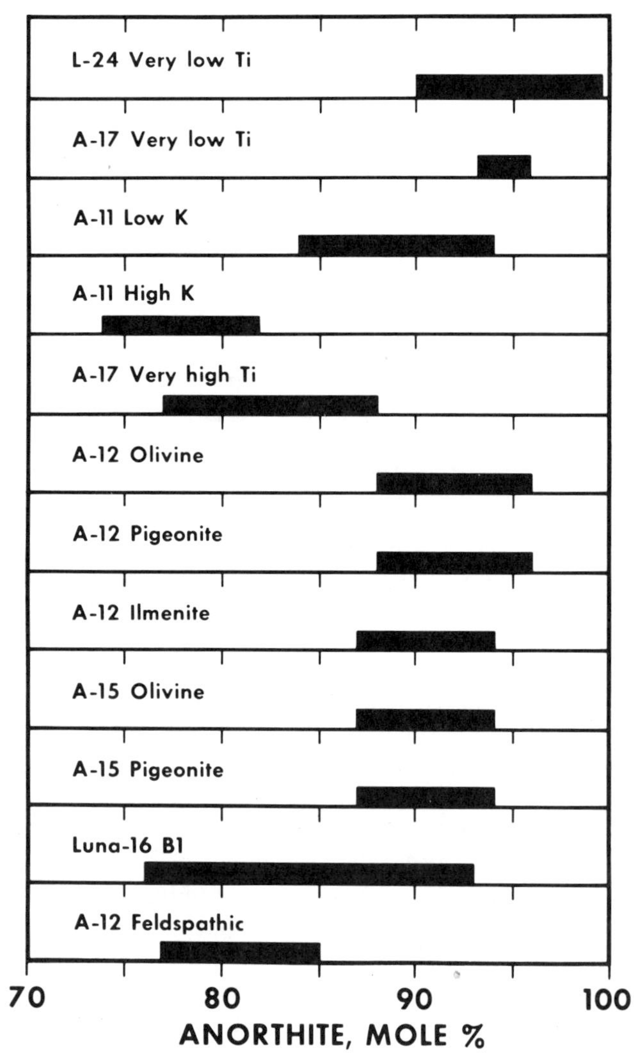

1.2.9.21 Anorthite contents of mare basalt feldspars.

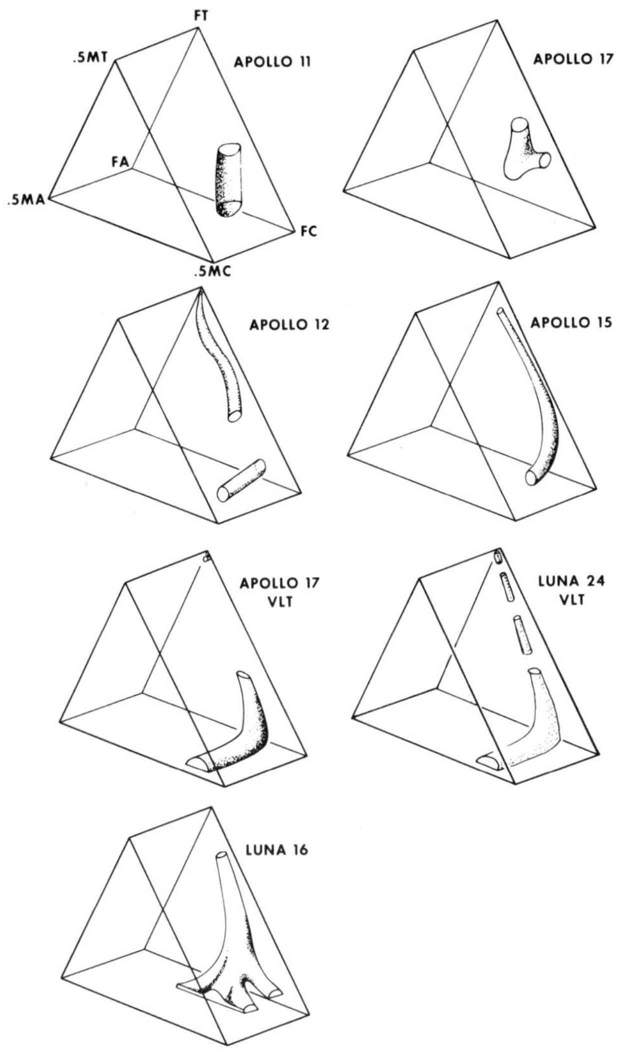

1.2.9.22 Variation diagrams for mare basalt spinel compositions. Modified from Haggerty, 1972.

Fe-rich compositions, which again is consistent with the Apollo 12 pigeonite basalts' being the more fractionated of the Apollo 12 olivine to pigeonite basalt series. In contrast, spinels from the Apollo 12 ilmenite group (not shown in Fig. 1.2.9.22) have compositions which cover the entire range from chromite to ulvöspinel. This observation is consistent with the Apollo 12 ilmenite basalts being a discrete mare group at the Apollo 12 site (as discussed above). The compositional trends of spinel grains in Apollo 15 olivine basalts are very similar to those of the Apollo 12 ilmenite basalts (i.e., forming a series from chromite to chromian ulvöspinel). The near-liquidus oxide phase of Apollo 17 VLT basalt (Vaniman and Papike, 1977b) is a chromian spinel with Fe/(Fe + Mg) ~0.65.

In phaneritic Apollo 17 VLTs, spinels are zoned to high-Fe contents with only slight Ti enrichment (Ti ~0.03 atoms per 4 oxygens). There is a distinct hiatus before discrete grains of late anhedral ulvöspinel grains appear. In contrast, Apollo 17 VLT vitrophyres contain spinels with the same initial Fe-enrichment trend but which are zoned to much higher Ti contents (Ti ~0.45 atoms per 4 oxygens). Zoning trends in spinels from Luna 24 VLT basalts were reported by Haggerty (1977). He showed that the first opaque mineral to crystallize was chromian spinel. During the subsequent crystallization, the spinel became Fe,Ti-enriched and zoned to ulvöspinel. The spinel zoning trend is strikingly similar to that of Apollo 17 VLTs. Differences between these two basalts in bulk Mg/Fe ratio and Cr contents are reflected in minor differences between their spinel crystallization trends. The Luna 24 spinels have initial compositions which are more Mg-poor than Apollo 17 VLT spinels [Mg/(Mg + Fe) = 0.33 vs. 0.40] and late-stage Ti-enriched spinels have lower Cr/Al ratios (<3) than spinel grains in Apollo 17 VLT (Cr/Al >3).

1.2.9.7 Petrogenesis

Integration of major and trace element data, isotopic systematics, experimental phase equilibria of both primitive mare basalt liquids and whole-moon compositions, and geophysical constraints is necessary to construct meaningful models of mare basalt petrogenesis.

Three basic types of models for mare basalt source regions have been presented: cumulate source models, primitive source models and assimilation models (that utilize aspects of both of the other models).

The cumulate model has become popular, because it explains the complementary geochemical natures of the lunar highlands and mare basalt source regions. In the model of Taylor and Jakeš (1974), which invokes extensive early melting and differentiation of the Moon to depths of 500 km (Taylor, 1978), the source regions of low-TiO_2 mare basalts (200–400 km depths) were composed of olivine-orthopyroxene cumulates. Late stage differentiates contained clinopyroxene and ilmenite and are the proposed sources of high-TiO_2 basalts. Prior extraction of plagioclase to form the feldspar-rich lunar highland crust had depleted the source regions of Eu, Sr and Al, which accounts for the low abundances of these elements in mare basalts. Recent experimental work on the depths of melting of mare basalts (Delano, 1980) suggests that most probably were derived from a complex zone of cumulates at about 400 km depth, which is consistent with sinking and convective overturn of the cumulate layers during crystallization of the magma ocean (Delano and Taylor, 1980).

The second model, the primitive source model, proposes that mare source regions were composed of primitive, unfractionated material of lunar bulk composition at the time of magma genesis (Ringwood, 1970; Ringwood and Essene, 1970; Green *et al.*, 1971). The range of basalt chemistries is attributed to different degrees of partial melting of this homogeneous source. Some problems confronting this model are:

(1) It cannot explain the geochemical characteristics and the wide variety of high- and low-Ti basalts (e.g., their REE patterns and TiO_2 contents).

(2) The presence of a negative Eu anomaly in mare basalts indicates prior separation of plagioclase from the source regions.

(3) It cannot account for the early differentiation event which is recorded in mare basalt Sm-Nd (Lugmair *et al.*, 1975), U-Pb (Tera and Wasserburg, 1975), and Rb-Sr isotopic systematics (Nyquist *et al.*, 1976).

(4) In general, the high-Ti basalts contain very little nickel (<2 ppm), but the low-Ti basalts contain up to 60 ppm and the Apollo 15 green glass contain 180 ppm. Thus the high-Ti basalts have no "memory" of a primitive lunar nickel content.

The third class of models involves hybridization or assimilation. Hubbard and Minear (1975) suggested that magmas from the primitive interior selectively assimilated parts of the 4.6–4.4 b.y. late-stage residuum, thereby giving rise to mare basalt hybrid liquids. Ringwood and Kesson (1976) pointed out that such a process should produce basalt liquids with extremely high Fe/Mg ratios, and that these liquids are not observed. The inverse correlation observed between trace element abundances in KREEP and those in mare basalts rules out any involvement of late-stage residual KREEP compositions, derived from crystallization of the magma ocean, in the genesis of mare basalts. Ringwood and Kesson (1976) suggested a "dynamic assimilation" model. They proposed that the outer few hundred kilometers of the Moon were involved in the 4.6–4.4 b.y. lunar differentiation event. The Fe, Ti, and incompatible element-rich, late-stage liquids thus formed were segregated in "pods" (having dimensions of 5–20 km) beneath the crust and above early cumulates and refractory residua. This gravitationally unstable situation resulted in the sinking of ilmenite-pyroxenite pods through the underlying cumulate and residual sequence. Meanwhile temperatures in the primitive undifferentiated interior were beginning to rise due to radiogenic heating, and the sinking pods began to melt and to react to form hybrid patches in the lunar mantle which *later* experienced partial melting to produce high-Ti basalts. The Fe and Ti-rich liquids thus produced reacted and reequilibrated with a large surrounding volume, giving rise to high-Ti mare basalts which retained the isotopic imprint of the ilmenite pyroxenites. Ringwood and Kesson (1976) further speculated that, by ~3.3 b.y. ago, continued radioactive heating had resulted in partial melting of a zone near the base of the differentiated sequence. Smaller and less-efficiently differentiated ilmenite-pyroxenite pods sank much more slowly through the differentiated sequence. The infall of these relatively cool pods into the partially molten zone allegedly produced low-Ti hybrid magmas. These low-Ti hybrid liquids would also carry the imprint of the 4.6–4.4 b.y. differentiation event and of the ilmenite pyroxenite cumulates. Ringwood and Kesson (1976) suggested that high-Al_2O_3 mare basalts were produced at the crust-mantle interface, when parental magmas from the deep interior assimilated limited amounts of plagioclase-rich crustal material. The high-Al_2O_3 basalts, however, have very low Ni contents (<2 ppm).

Derivation of mare basalts from a cumulate source region developed during the 4.4 b.y. differentiation event appears to be the most promising explanation. No mare basalts have been identified as having come from primitive unfractionated source regions. All bear the isotopic and trace element imprint of the early lunar differentiation event. As our understanding of the details of trace element behavior in the Moon increases (for example, lunar chromium which is mainly divalent, in contrast with its normally trivalent state in the Earth), so the fit of the basalt chemistry to the cumulate model improves. It is probable that most mare basalts have experienced some olivine removal en route to the surface (even such samples as 12009). It likewise is probable that most types of mare basalts were derived from depths of about 400 km (Delano, 1980). The simple layered sequence depicted in Fig. 1.2.9.14 would have been much disrupted by sinking, convective overturn, and the effects of giant planetesimal impacts.

1.2.9.8 Concluding statement

Major advances in our understanding of mare basalt petrogenesis are still being made. However, our efforts have been and will be compromised by the fact that we have been working with a very limited number of samples. This situation must be rectified by (1) continuing to quantify and correlate remote-sensing results; (2) continuing to look for and find small mare basalt fragments in regolith samples and breccias that are not represented in the large rock sample suite; (3) continuing to carry out unmanned sampling of strategic locations on the lunar surface (e.g., far lunar west side and backside).

1.2.9.9 References

Adams J. B. and Ralph R. L. (1977) Spectral reflectance of Luna 24 soils (abstract). In *Papers Presented to the Conference on Luna 24*, p. 1–4. The Lunar Science Institute, Houston.

Albee A. L., Chodos A. A., Gancarz A. J., Haines E. L., Papanastassiou D. A., Ray L., Tera F., Wasserburg G. J. and Wen T. (1972) Mineralogy, petrology, and chemistry of a Luna 16 basaltic fragment sample B-1. *Earth Planet. Sci. Lett.* **13**, 353–367.

Arnold J. R., Metzger A. E. and Reedy R. C. (1977) Computer-generated maps of lunar composition from gamma-ray data. *Proc. Lunar Sci. Conf. 8th*, p. 945–948.

Barsukov V. L., Tarasov L. S., Dmitriev L. V., Kolesov G. M., Shevaleevsky I. D. and Garanin A. V. (1977) The geochemical and petrochemical features of regolith and rocks from Mare Crisium (preliminary data). *Proc. Lunar Sci. Conf. 8th*, p. 3319–3332.

Beaty D. W. and Albee A. L. (1978) Comparative petrology and possible genetic relations among the Apollo 11 basalts. *Proc. Lunar Sci. Conf. 9th*, p. 359–463.

Bence A. E. and Papike J. J. (1972) Pyroxenes as recorders of lunar basalt petrogenesis: chemical trends due to crystal-liquid interaction. *Proc. Lunar Sci. Conf. 3rd*, 431–469.

Bielefeld M. J., Andre C. G., Eliason E. M., Clark P. E., Adler I. and Trombka J. I. (1977) Imaging of lunar surface chemistry from orbital X-ray data. *Proc. Lunar Sci. Conf. 8th*, p. 901–908.

Boyce J. M. (1976) Ages of flow units in the lunar nearside maria based on Lunar Orbiter IV photographs. *Proc. Lunar Sci. Conf. 7th*, p. 2717–2728.

Boyce J. M. and Dial A. L. (1973) Relative ages of some nearside mare units based on Apollo 17 metric photographs. *Apollo 17 Preliminary Science Report*, p. 29–26 to 29–28. NASA SP-330.

Bryan W. B. and Adams M. L. (1973) Some volcanic and structural features of Mare Serenitatis. *Apollo 17 Preliminary Science Report*, p. 30–9 to 30–12. NASA SP-330.

Chou C. L., Boynton W. V., Sundberg L. L. and Wasson J. T. (1975) Volatiles on the surface of Apollo 15 green glass and trace element distributions among Apollo 15 soils. *Proc. Lunar Sci. Conf. 6th*, p. 1701–1727.

Compston W., Chappell B. W., Arriens P. A. and Vernon M. J. (1970) The chemistry and age of Apollo 11 lunar material. *Proc. Apollo 11 Lunar Sci. Conf.*, p. 1007–1027.

Delano J. W. (1979) Apollo 15 green glass: chemistry and possible origin. *Proc. Lunar Planet. Sci. Conf. 10th*, p. 275–300.

Delano J. W. (1980) Apollo 15 red glass: chemistry and liquidus phase relations (abstract). In *Lunar and Planetary Science XI*, p. 210–212. Lunar and Planetary Institute, Houston.

Delano J. W. and Taylor S. R. (1980) Composition and structure of the deep lunar interior (abstract). In *Lunar and Planetary Science XI*, p. 225–227. Lunar and Planetary Institute, Houston.

Donaldson C. H., Usselman T. M., Williams R. J. and Lofgren G. E. (1975) Experimental modeling of the cooling history of Apollo 12 olivine basalts. *Proc. Lunar Sci. Conf. 6th*, p. 843–869.

Duncan A. R., Erlank A. J., Sher M. K., Abraham Y. C., Willis J. P. and Ahrens L. H. (1976) Some trace element constraints on mare basalt genesis. *Proc. Lunar Sci. Conf. 7th*, p. 1659–1671.

Dungan M. A. and Brown R. W. (1977) The petrology of the Apollo 12 ilmenite basalt suite. *Proc. Lunar Sci. Conf. 8th*, p. 1339–1381.

Eldridge J. S., O'Kelley G. D. and Northcutt K. J. (1975) Primordial and cosmogenic radionuclides in Descartes and Taurus-Littrow materials. *Proc. Lunar Sci. Conf. 6th*, p. 1407–1418.

El Goresy A. (1976) Oxide minerals in lunar rocks. In *Oxide Minerals* (D. Rumble, III, ed.) p. EG-1 to EG-46. Mineral. Soc. Amer. Short Course Notes. Mineral. Soc. Amer., Washington, D.C.

Eliason E. M. and Soderblom L. A. (1977) An array processing system for lunar geochemical and geophysical data. *Proc. Lunar Sci. Conf. 8th*, p. 1163–1170.

Ganapathy R., Keays R. R., Laul J. C. and Anders E. (1970) Trace elements in Apollo 11 lunar rocks: implications for meteorite influx and origin of moon. *Proc. Apollo 11 Lunar Sci. Conf.*, p. 1117–1142.

Gast P. W., Hubbard N. J. and Weismann H. (1970) Chemical composition and petrogenesis of basalts from Tranqillity base. *Proc. Apollo 11 Lunar Sci. Conf.*, p. 1143–1163.

Goles G. G., Randle K., Osawa M., Schmitt R. A., Wakita H., Ehmann W. D. and Morgan J. W. (1970) Elemental abundances by instrumental activation analyses in chips from 27 rocks. *Proc. Apollo 11 Lunar Sci. Conf.*, p. 1165–1176.

Green D. H., Ringwood A. E., Hibberson W. O. and Ware N. G. (1975) Experimental petrology of Apollo 17 mare basalts. *Proc. Lunar Sci. Conf. 6th*, p. 871–893.

Green D. H., Ware N. G., Hibberson W. O. and Major A. (1971) Experimental petrology of Apollo 12 basalts, I. sample 12009. *Earth Planet. Sci. Lett.* **13**, 85–96.

Green J. A., Kurat G., Keil K. and Prinz M. (1972) Electron microprobe analyses of lithic fragments, glasses, chondrites and minerals from Luna 16 fines. *Inst. Meteorites Spec. Publ. No. 5*, p. 1–16. Univ. New Mexico, Albuquerque.

Grieve R. A. F., McKay G. A., Smith H. D. and Weill D. F. (1975) Lunar polymict breccia 14321: a petrographic study. *Geochim. Cosmochim. Acta* **39**, 229–245.

Grove T. L. and Bence A. E. (1976) Experimental study of pyroxene-liquid interaction in quartz-normative basalt 15597. *Proc. Lunar Sci. Conf. 8th*, p. 1549–1579.

Grove T. L. and Lindsley D. H. (1978) Compositional variation and origin of lunar ultramafic green glasses (abstract). In *Lunar and Planetary Science IX*, p. 430–432. Lunar and Planetary Institute, Houston.

Grove T. L. and Vaniman D. T. (1978) Experimental petrology of very low Ti basalts and origin of Luna 24 ferrobasalt. In *Mare Crisium: The View From Luna 24* (R. B. Merrill and J. J. Papike, eds.), p. 445–477. Pergamon, N.Y.

Grove T. L. and Walker D. (1977) Cooling histories of Apollo 15 quartz normative basalts. *Proc. Lunar Sci. Conf. 8th*, p. 1501–1520.

Haggerty S. E. (1972) Luna 16; an opaque mineral study and a systematic examination of compositional variations of spinels from Mare Fecunditatis. *Earth Planet. Sci. Lett.* **13**, 328–352.

Haggerty S. E. (1977) Luna 24: opaque mineral chemistry of gabbroic and basaltic fragments from Mare Crisium. *Geophys. Res. Lett.* **4**, 489–492.

Head J. W. (1976) Lunar volcanism in space and time. *Rev. Geophys. Space Phys.* **14**, 265–300.

Head J. W., Zisk S., Adams J., McCord T. and Pieters C. (1977) Geologic setting and topography of the Crisium basin and the Luna 24 landing region (abstract). In *Papers Presented to the Conference on Luna 24*, p. 78–81. The Lunar Science Institute, Houston.

Hubbard N. J. and Minear J. W. (1975) A chemical and physical model for the genesis of lunar rocks. Part II: mare basalts (abstract). In *Lunar Science VI*, p. 405–407. The Lunar Science Institute, Houston.

Hubbard N. J. and Rhodes J. M. (1976) The chemical classification of Apollo 17 orange glass (abstract). In *Lunar Science VII*, p. 390–392. The Lunar Science Institute, Houston.

Huneke J. C., Jessberger E. K., Podosek F. A. and Wasserburg G. J. (1973) $^{40}Ar/^{39}Ar$ measurements in Apollo 16 and 17 samples and the chronology of metamorphic and volcanic activity in the Taurus-Littrow region. *Proc. Lunar Sci. Conf. 4th*, p. 1725–1756.

Husain L. and Schaeffer O. A. (1973) Lunar volcanism: age of glass in the Apollo 17 orange soil. *Science* **180**, 1358–1360.

Johnson T. V., Mosher J. A. and Matson D. L. (1977a) Lunar spectral units: a northern hemispheric mosaic. *Proc. Lunar Sci. Conf. 8th*, 1013–1028.

Johnson T. V., Saunders R. S., Matson D. L. and Mosher J. A. (1977b) A TiO_2 abundance map for the northern maria. *Proc. Lunar Sci. Conf. 8th*, p. 1029–1036.

Kesson S. E. (1975) Mare basalts: melting experiments and petrogenetic interpretations. *Proc. Lunar Sci. Conf. 6th*, p. 921–944.

Kurat G., Kracher A., Keil K., Warner R. and Prinz M. (1976) Composition and origin of Luna 16 aluminous mare basalts. *Proc. Lunar Sci. Conf. 7th*, p. 1301–1321.

Kushiro I., Ikeda Y. and Nakamura Y. (1972) Petrology of Apollo 14 high alumina basalt. *Proc. Lunar Sci. Conf. 3rd*, p. 115–129.

Laul J. C. and Schmitt R. A. (1975) Chemical composition of Apollo 17 samples: boulder breccias (2), rake breccias (8), and others (abstract). In *Lunar Science VI*, p. 489–491. The Lunar Science Institute, Houston.

Laul J. C., Vaniman D. T. and Papike J. J. (1978) Chemistry, mineralogy and petrology of seven >1 mm fragments from mare Crisium. In *Mare Crisium: The View From Luna 24* (R. B. Merrill and J. J. Papike, eds.), p. 537–568. Pergamon, N.Y.

Lindsley D. H. and Munoz J. L. (1969) Subsolidus relations along the join hedenbergite-ferrosilite. *Amer. J. Sci.* **267A**, 295–324.

Lofgren G. E. and Lofgren E. M. (1981) Catalog of lunar mare basalts greater than 40 grams. Part I. Major and trace element chemistry. *Lunar Planet. Inst. Contrib. 438*. Lunar and Planetary Institute, Houston.

Lofgren G., Donaldson C. H. and Usselman T. M. (1975) Geology, petrology, and crystallization of Apollo 15 quartz-normative basalts. *Proc. Lunar Sci. Conf. 6th*, p. 79–99.

Lofgren G., Donaldson C. H., Williams R. J. and Mullins O. (1974) Experimentally reproduced textures and mineral chemistry of A-15 quartz basalts (abstract). In *Lunar Science V*, p. 458–460. The Lunar Science Institute, Houston.

Longhi J., Walker D. and Hays J. F. (1972) Petrography and crystallization history of basalts 14310 and 14072. *Proc. Lunar Sci. Conf. 3rd*, p. 131–139.

Longhi J., Walker D. and Hays J. F. (1978) The distribution of Fe and Mg between olivine and lunar basaltic liquids. *Geochim. Cosmochim. Acta* **42**, 1545–1558.

LSPET (1969) Preliminary examination of lunar samples from Apollo 11. *Science* **165**, 1211–1227.

LSPET (1973) The Apollo 16 lunar samples: petrographic and chemical description. *Science* **179**, 23–34.

Lugmair G. W., Scheinin N. B. and Marti K. (1975) Sm-Nd age of Apollo 17 basalt 75075: two stage igneous processes in mare basalt genesis (abstract). In *Lunar Science VI*, p. 807–809. The Lunar Science Institute, Houston.

Metzger A. E., Lin R. P. and Russell C. T. (1977) On a correlation between surface remanent magnetism and chemistry for the lunar frontside and limbs. *Proc. Lunar Sci. Conf. 8th*, p. 1187–1190.

Meyer C., Jr., McKay D. S., Anderson D. H. and Bulter P. (1975) The source of sublimates on the Apollo 15 green and Apollo 17 orange glass samples. *Proc. Lunar Sci. Conf. 6th*, p. 1673–1699.

Nakamura Y., Latham G. V., Dorman H. J. and Duennebier F. K. (1976) Seismic structure of the Moon. *Proc. Lunar Sci. Conf. 7th*, p. 3113–3121.

Naney M. T., Papike J. J. and Vaniman D. T. (1977) Lunar highland melt rocks: major element chemistry and comparisons with lunar highland glass groups (abstract). In *Lunar Science VIII*, p. 717–719. The Lunar Science Institute, Houston.

Newton R. C., Anderson A. T. and Smith J. V. (1971) Accumulation of olivine in rock 12040 and other basaltic fragments in the light of analysis and synthesis. *Proc. Lunar Sci. Conf. 2nd*, p. 575–582.

Nyquist L. E., Bansal B. M. and Wiesman H. (1976) Sr isotopic constraints on the petrogenesis of Apollo 17 mare basalts. *Proc. Lunar Sci. Conf. 7th*, 1507–1528.

Papanastassiou D. A., McCulloch M. T., DePaolo D. J., Huneke J. C. and Wasserburg G. J. (1977) Ages and evolution of Luna 24 samples (abstract). In *Papers Presented to the Conference on Luna 24*, 3 pp. supplement. The Lunar Science Institute, Houston.

Papike J. J. and Vaniman D. T. (1978) Luna 24 ferrobasalts and the mare basalt suite: comparative chemistry, mineralogy, and petrology. In *Mare Crisium: The View From Luna 24* (R. B. Merrill and J. J. Papike, eds.), p. 371–401. Pergamon, N.Y.

Papike J. J., Bence A. E. and Lindsley D. H. (1974) Mare basalts from the Taurus-Littrow region of the Moon. *Proc. Lunar Sci. Conf 5th*, p. 471–504.

Papike J. J., Hodges F. N., Bence A. E., Cameron M. and Rhodes J. M. (1976) Mare basalts: crystal chemistry, mineralogy and petrology. *Rev. Geophys. Space Phys.* **14**, 475–540.

Rhodes J. M., Blanchard D. P., Dungan M. A., Brannon J. C. and Rodgers K. V. (1977) Chemistry of Apollo 12 mare basalts: magma types and fractionation. *Proc. Lunar Sci. Conf. 8th*, p. 1305–1338.

Rhodes J. M., Hubbard N. J., Wiesmann H., Rodgers K. V., Brannon J. C. and Bansal B. M. (1976) Chemistry, classification and petrogenesis of Apollo 17 mare basalts. *Proc. Lunar Sci. Conf. 7th*, p. 1467–1489.

Rhodes J. M., Rodgers K. V., Shih C., Bansal B. M., Nyquist L. E., Wiesmann H. and Hubbard N. J. (1974) The relationships between geology and soil chemistry of the Apollo 17 landing site. *Proc. Lunar Sci. Conf. 5th*, p. 1097–1117.

Ridley W. I. (1975) On high-alumina mare basalts. *Proc. Lunar Sci. Conf. 6th*, p. 131–145.

Ridley W. I., Reid A. M., Warner J. L. and Brown R. W. (1973) Apollo 15 green glass. *Phys. Earth Planet. Int.* **7**, 133–136.

Ringwood A. E. (1970) Petrogenesis of Apollo 11 basalts and implications for lunar origin. *J. Geophys. Res.* **75**, 6453–6479.

Ringwood A. E. and Essene E. (1970) Petrogenesis of Apollo 11 basalts: internal constitution of the Moon. *Proc. Apollo 11 Lunar Sci. Conf.*, p. 769–799.

Ringwood A. E. and Kesson S. E. (1976) A dynamic model for mare basalt petrogenesis. *Proc. Lunar Sci. Conf. 7th*, p. 1697–1722.

Russell C. T., Weiss H., Coleman P. J., Soderblom L. A., Stuart-Alexander D. E. and Wilhelms D. E. (1977) Geologic-magnetic correlations on the Moon: Apollo subsatellite results. *Proc. Lunar Sci. Conf. 8th*, p. 1171–1185.

Sato M. (1976) Oxygen fugacity and other thermochemical parameters of Apollo 17 high-Ti basalts and their implications on the reduction mechanism. *Proc. Lunar Sci. Conf. 7th*, p. 1323–1344.

Schaeffer O. A., Bence A. E., Eichorn G., Papike J. J. and Vaniman D. T. (1978) ^{39}Ar–^{40}Ar and petrologic study of Mare Crisium: age and petrology of Luna 24 samples 24077,13 and 24077,63 (abstract). In *Lunar and Planetary Science IX* p. 1007–1009. The Lunar Science Institute, Houston.

Schreiber H. D. and Haskin L. A. (1976) Chromium in basalts. *Proc. Lunar Sci. Conf. 7th*, p. 1221–1259.

Smith J. V., Anderson A. T., Newton R. C., Olson E. J., Wyllie P. J., Crewe A. V., Isaacson M. S. and Johnson D. (1970) Petrologic history of the Moon inferred from petrology, mineralogy, and petrogenesis of Apollo 11 rocks. *Proc. Apollo 11 Lunar Sci. Conf.*, p. 897–925.

Soderblom L. A., Arnold J. A., Boyce J. M. and Lin R. P. (1977) Regional variations in the lunar maria: age, remanent magnetism and chemistry. *Proc. Lunar Sci. Conf. 8th*, p. 1191–1199.

Stolper E. M. (1974) Lunar ultramafic glasses. B. A. Thesis. Harvard University, Cambridge, Mass. 178 pp.

Stolper E. M., Walker D., Longhi J. and Hays J. F. (1974) Compositional variation in lunar ultramafic glasses (abstract). In *Lunar Science V*, p. 749–751. The Lunar Science Institute, Houston.

Strom R. G. (1972) Lunar mare ridges, rings, and volcanic ring complexes. In *The Moon Symposium 47 of the International Astronomical Union*, (S. Runcorn and H. Urey, eds.), p. 187–215. Reidell, Dordrecht.

Surkov Yu. A. (1977) Gamma spectrometry in cosmic investigations. *Atomizdat*, Moscow. 238 pp.

Taylor G. J., Keil K., and Warner R. D. (1977a) Very low-Ti basalt. *Geophys. Res. Lett.* **4**, 207–210.

Taylor G. J., Warner R., Wentworth S., Keil K. and Sayeed U. (1977b) Petrology and chemistry of Luna 24 mare basalts and basaltic glasses (abstracts). In *Papers Presented to the Conference on Luna 24*, p. 189–192. The Lunar Science Institute, Houston.

Taylor S. R. (1975) *Lunar Science: A Post-Apollo View*. Pergamon, N.Y. 372 pp.

Taylor S. R. (1978) Geochemical constraints on melting and differentiation of the Moon. *Proc. Lunar Sci. Conf. 9th*, p. 15–23.

Taylor S. R. and Jakeš P. (1974) The geochemical evolution of the Moon. *Proc. Lunar Sci. Conf. 5th*, p. 1287–1305.

Tera F. and Wasserburg G. J. (1975) The evolution and history of mare basalts as inferred from U-Th-Pb systematics (abstract). In *Lunar Science VI*, p. 807–809. The Lunar Science Institute, Houston.

Turner G. (1977) Potassium-argon chronology of the Moon. *Phys. Chem. Earth* **10**, 145–195.

Vaniman D. T. and Papike J. J. (1977a) The Apollo 17 drill core: chemistry and stratigraphy of monomineralic fragments and the discovery of a new very low Ti (VLT) mare basalt (abstract). In *Lunar Science VIII*, p. 952–954. The Lunar Science Institute, Houston.

Vaniman D. T. and Papike J. J. (1977b) Very low Ti (VLT) basalts: a new mare rock type from the Apollo 17 drill core. *Proc. Lunar Sci. Conf. 8th*, p. 1443–1471.

Vaniman D. T. and Papike J. J. (1977c) Ferrobasalts from Mare Crisium: Luna 24. *Geophys. Res. Lett.* **4**, 497–500.

Vaniman D. T. and Papike J. J. (1977d) The Apollo 17 drill core: modal petrology and glass chemistry (sections 70007, 70008, 70009). *Proc. Lunar Sci. Conf. 8th*, p. 3161–3193.

Wager L. R. and Brown G. M. (1968) *Layered Igneous Rocks*. Oliver and Boyd, London. 588 pp.

Walker D., Kirkpatrick R. J., Longhi J. and Hays J. F. (1976) Crystallization history of lunar picritic basalt sample 12002: phase equilibria and cooling-rate studies. *Geol. Soc. Amer. Bull.* **87**, 646–656.

Walker D., Longhi J., Lasaga A. C., Stolper E. M., Grove T. L. and Hays J. F. (1977) Slowly cooled microgabbros 15555 and 15065. *Proc. Lunar Sci. Conf. 8th*, p. 1521–1547.

Walker D., Longhi J., Stolper E. M., Grove T. L. and Hays J. F. (1975) Origin of titaniferous lunar basalts. *Geochim. Cosmochim. Acta* **39**, 1219–1235.

Whitaker E. A. (1972) Mare Imbrium lava flows and the relationships to color boundaries. *Apollo 15 Preliminary Science Report*, p.25-83 to 25-84. NASA SP-289.

Wolf R., Woodrow A. and Anders E. (1979) Siderophile and volatile elements in the Earth and the Moon: similar or not? (abstract). In *Lunar and Planetary Science X*, p. 1361-1363. Lunar and Planetary Institute, Houston.

Wood J. A. and Ryder G. (1977) The Apollo 15 green clods and the green glass enigma (abstract). In *Lunar Science VIII*, p. 1026–1028. The Lunar Science Institute, Houston.

1.2.10 LUNAR HIGHLAND BASALTS

Before the Apollo landings, the lunar highlands were suspected to be largely volcanic (Head and Goetz, 1972; Milton and Hodges, 1972; Trask and McCauley, 1972). One Apollo mission, Apollo 16, was specifically directed to sample highland volcanic flows; however, most of the samples found are impact breccias. About 20–30% of the returned highlands samples have textures resembling those of extrusive basaltic lavas; however, the great majority of these are considered to be impact melt-rocks. Melt-textured highland rocks have been found at all lunar landing sites, but only a few samples at the Apollo 15 and 17 sites have been widely accepted as endogenic basalts. Conversely, many of the breccias and melt-rocks have basalt-like compositions and possibly represent pre-3.9 b.y. highland volcanics which have been modified by impact brecciation, mixing and melting. Since our definition of a basalt is designed to select those samples which give us some information about planetary interiors, a first-order problem is that of distinguishing endogenic highland volcanic rocks from exogenic highland impact melts.

Although endogenic highland basalts can be identified, there is a possibility that naming them as volcanic rocks taxes the common usage of the term "volcanic." The concept of volcanism implies that heat sources internal to a planetary body are used to generate magma. However, some models of lunar highland volcanism call on external energy sources (e.g., infalling planetesimals and solar luminosity; Hostetler and Drake, 1979) to melt the primordial outer Moon. A lunar magma ocean (Wood *et al.*, 1970; Smith *et al.*, 1970) would follow some temperature-time-crystallization history which is not well known but must in any event result in a crust overlying residual liquids which in turn overlie an even deeper mafic cumulate zone. The residual liquids, which are the end product of melting generated by a complex mix of external and internal energies, could be subsequently extruded onto the lunar surface. Here, a basalt is considered to be volcanic as long as it arises from magma which originated at depth as a part of some larger process affecting the mantle, the lower crust, or the primordial formation of these major planetary layers.

There is a compositional distinction between dark, mafic basalt flows of the lunar maria (high CaO/Al_2O_3, with $TiO_2 + FeO + MgO = 27$–42 wt.%) and the light-colored, feldspathic lunar highlands (low CaO/Al_2O_3, $TiO_2 + FeO + MgO < 22$ wt.%, except for rare dunites and troctolites). Such a distinction between mare and highland basalt types is useful for classification but in fact may be artificial. Ryder *et al.* (1977) describe some small boulder clasts at Apollo 17 that appear to be intermediate between low-Ti mare compositions and highland basalts. Perhaps it is not entirely fortuitous that these intermediate basalt clasts in a boulder at Serenitatis are dated at ~4.0 b.y. (Compston *et al.*, 1975), an age which coincides with the oldest mare basalt ages (basalt clasts in boulders dated at ~4.0 b.y.; Compston *et al.*, 1975) and overlaps with the youngest (~3.9 b.y.) evidence of highland volcanism (Head and McCord, 1978; Chapter 5; Carlson and Lugmair, 1979). However, the evidence for any form of volcanism on the Moon prior to 4.0 b.y. is largely circumstantial.

1.2.10.1 Source regions and time scales of highland volcanism

Melting of the outer Moon to depths of several hundred kilometers is inferred from analyses of the energetics of the accretion period, from petrologic evidence for the origin of mare and highland source regions by complementary fractionation from a single source composition, and from geochemical and geophysical data indicating that the products of this massive fractionation may be found at the appropriate depths for those source regions (Wood *et al.*, 1970; Mizutani *et al.*, 1972; Taylor and Jakeš, 1974; Ringwood and Kesson, 1976). The concept of a major melting event, leading to a magma ocean, and preceding cumulate formation and magmatic differentiation on a planetary scale, has been widely accepted but is difficult to model in detail (Walker, 1979). Raedeke and McCallum (1979a, b) describe features analogous to those proposed to be the result of magma ocean petrogenesis, but which occur on a petrologically manageable scale in the Stillwater layered intrusion. It is instructive to recall the complexities of large terrestrial layered intrusions and the still active disagreements over their internal processes (Wager and Brown, 1968). Thus it is not surprising that we are still a long way from a complete understanding of the crystal-melt processes in a magma shell hundreds of kilometers thick. Despite these difficulties, there is little doubt that such melting would lead to the formation of crust by cumulus feldspar flotation (Walker and Hays, 1977). This early crust would be initially transient, disrupted by meteoritic bombardment as fast as it formed. At some stage, by simple crustal thickening or by a lull in the meteorite bombardment, the crust should be stable enough to maintain standing magma conduits and to support volcanic features. The earliest *volcanism* (as defined here) could be the eruption of unsolidified remnants of the magma ocean onto this stabilized platform.

There is no direct evidence for such ancient volcanism. If intact samples from such events could be found, they would have distinctively ancient initial isotopic ratios. If some version of the cumulate remelting hypothesis is accepted for the origin of mare basalts (section 1.2.9.7), then the strong coincidence of Rb-Sr, Sm-Nd and U-Pb model ages indicates that the mare source regions were formed ~4.42 b.y. (Papanastassiou *et al.*, 1977), representing the age of isotopic closure in the deep cumulate zones (≃200–400 km depth; see section 1.2.9.7). The deep cumulate zones are depleted in a lighter residual component rich in potassium (K), the rare-earth elements (REE), phosphorus (P), and other incompatible elements (U, Th); these elements form the very distinctive lunar crustal component commonly referred to as "KREEP." It has been suggested that the Rb-Sr and Nd-Sm isotopic closure of the KREEP source region(s) is 4.3–4.4 b.y. (Lugmair and Carlson, 1978; Carlson and Lugmair, 1979), slightly younger than the 4.42 b.y. age of the mare source region(s).

This age difference, and the density contrast between KREEP residue and the earlier cumulus phases, indicate that the KREEP reservoirs are above the mare basalt source regions (Hess *et al.*, 1978). Dating from the initial establishment of source regions, mare-type volcanism could have commenced as early as 4.42 b.y., although partial melts from the mare source regions would have had great difficulty rising through the unsolidified dregs of the magma ocean without assimilating considerable amounts of this KREEP-rich liquid. Thus either hybrid mare/KREEP volcanism or the eruption of residual KREEP liquids may have been possible before 4.4 b.y., but highland basalts which bear the distinct isotopic character of the final KREEP reservoir could not have been formed prior to 4.4 b.y.

The La/Yb ratios in most highland rocks are constant at ~3–3.35, giving the chondrite-normalized REE plots a constant negative slope (Fig. 1.2.10.2). This constant negative slope is maintained over a wide range of concentration (from Sm ~3 to ~300 times chondritic abundances), suggesting that the KREEP contents of most highland rocks arise as mechanical mixtures of KREEP and KREEP-free upper-level cumulates (Taylor *et al.*, 1973; Hubbard *et al.*, 1973). The occurrence of a KREEP component in virtually all impacted highland samples argues both for the widespread distribution of KREEP-bearing igneous rocks and for thoroughness of impact mixing. However, Warner *et al.* (1977) describe a group of granulitic impactites from the Apollo 16 and 17 sites with ages >4.2 b.y. and without KREEP contamination. Further discoveries of such samples may help constrain a maximum age for the first abundant appearance of KREEP melts on the Moon's surface (volcanic) or near the Moon's surface (shallow plutonic).

1.2.10.2 Highland Reference Suite

Figure 1.2.10.1 shows two groups of lunar highland volcanic or impact-melt rocks plotted on the pseudoternary silica-olivine-anorthite projection of Walker *et al.* (1972). These two groups represent two of the six groups identified by Naney *et al.* (1977) from cluster analysis of all available analyses of highland volcanic or impact-melt samples. An independent classification was presented by Irving (1975). The two groups shown here comprise those samples for which there is a consensus favoring volcanic origin (Intermediate-K Fra Mauro or IKFM), and those samples which have definite evidence of being impact melts, but for which there

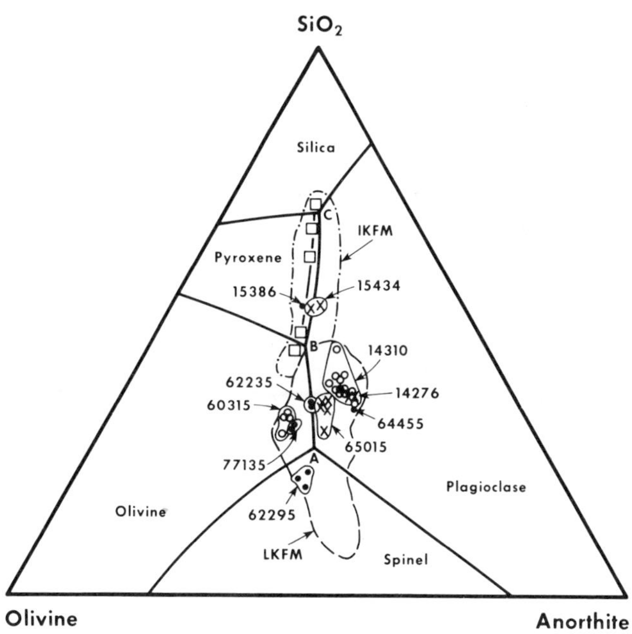

1.2.10.1 Low-K Fra Mauro (LKFM) and Intermediate-K Fra Mauro (IKFM) samples plotted in the anorthite-olivine-silica projection (projection method of Walker *et al.*, 1972). Fields labelled "LKFM" and "IKFM" represent results of highland melt-rock cluster analysis by Naney *et al.* (1977). Plotted symbols show the range of available analyses for each sample; differing symbols are used where different samples overlap. Connected boxes show the KREEP fractionation trend of Irving (1977).

is equivocal evidence that their compositions are representative of ancient volcanic rocks (Low-K Fra Mauro or LKFM). The names of both groups are based on the first recognition of KREEP compositions from the Fra Mauro Formation (Apollo 14). Ten representative samples (> 1 cm in size) have been selected from both of these groups to illustrate the petrologic and chemical characteristics of highland volcanic or impact-melt samples. Major and trace element data are summarized in Table 1.2.10.1, and rare-earth element patterns are shown in Fig. 1.2.10.2. Modal and age data are given in Tables 1.2.10.2–3.

1.2.10.3 Distinction of impact melt-rocks from volcanic rocks

The textures of highland impact-melt samples result from a complex mixture of shock melt with relatively cold, usually shock-generated, clastic debris. Simonds (1975) discusses the thermal history of an Apollo 17 impact melt rock and deduces a two-stage cooling history which has proved applicable to terrestrial as well as lunar impact melts. This two-stage cooling history is the result of mixing during impact of an impact-generated melt and clastic debris, resulting in emplacement of a clast/melt "melt sheet." The first stage of cooling is a rapid loss of heat from the superheated impact melt to the enclosed, cooler clasts, followed by slower second-stage cooling by conduction and radiation, as in a normal volcanic flow. This first stage is extremely rapid in all impact melt sheets (e.g., first-stage cooling rates may be on the order of tens of degrees per second; Onorato *et al.*, 1978). First-stage cooling rates do not depend on position within the melt sheet. The second-stage cooling, however, depends on the thickness of the melt sheet, depth of debris cover, position in the melt sheet, and other factors relating the melt sheet to its geologic environment.

In thicker and better insulated impact melts, grain sizes tend to be coarser and the clastic debris is more thoroughly digested, leaving fewer relic clasts. Textures may be intersertal, poikilitic, intergranular, or ophitic to subophitic. Fine-grained vitrophyric, radiate, or "fan spherulitic" variants may also occur.

In many cases an impact melt-rock can be identified by the presence of clasts, but the absence of textural evidence of impact does not itself allow the conclusion that a sample is endogenous. Among the chemical criteria proposed for the recognition of pristine highlands samples, including volcanics, is very low concentrations of siderophile elements (e.g., Au, Ir). The matrices of highlands breccias possess high and variable levels of these elements (e.g., Ir up to 40 ppb), usually taken to be a product of meteoritic contamination. Conversely, samples with extremely *low* siderophile abundances are taken to lack meteoritic contamination, and hence to be endogenous (Anders, 1978; Warren and Wasson, 1977, 1978; Wänke *et al.*, 1976). Warren and Wasson (1974) suggest that samples with less than 2×10^{-4} times chondritic abundances of any of the refractory siderophiles can be considered endogenous. Those IKFM samples which have been recognized as volcanic rocks (e.g., 15386 and 15434) are characterized by their sub-

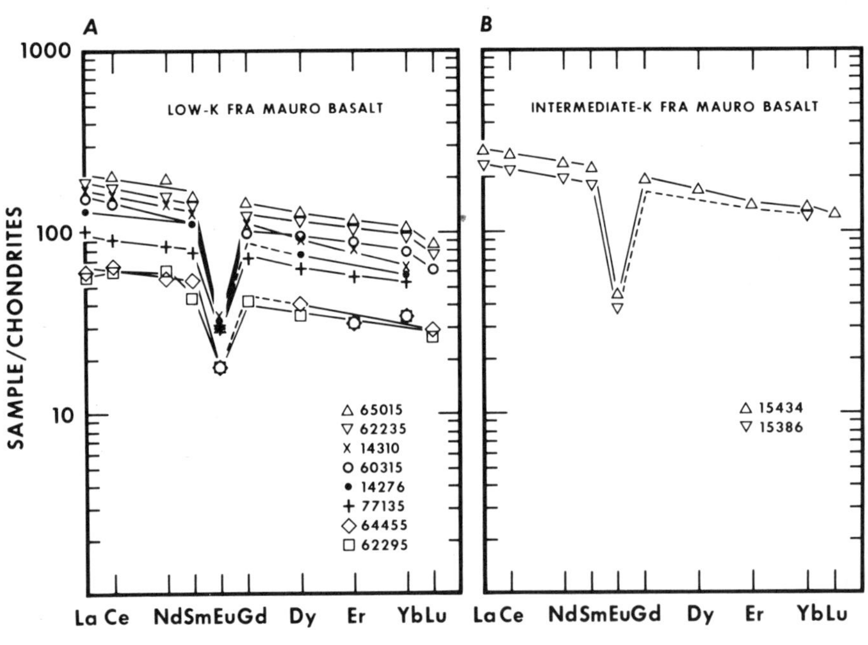

1.2.10.2 Chondrite-normalized rare-earth data for the ten LKFM and IKFM samples shown in Fig. 1.2.10.1.

Table 1.2.10.1 Chemical composition of samples in the Highland Reference Suite.

	Low-K Fra Mauro								Intermediate-K Fra Mauro	
wt. %	14276	14310	64455	62295	60315	62235	65015	77135	15386	15434
SiO_2	47.60	47.2	48.5	45.16	46.75	47.04	46.99	46.13	50.83	52.14
TiO_2	1.20	1.24	0.65	0.70	1.38	1.21	1.26	1.54	2.23	2.23
Al_2O_3	21.34	20.1	22.4	20.05	17.10	18.69	19.68	18.01	14.77	16.16
FeO	7.94	8.38	5.47	6.40	8.64	9.45	8.59	9.11	10.55	10.82
MnO	0.12	0.11	0.074	0.09	0.11	0.11	0.112	0.13	0.16	0.17
MgO	7.10	7.87	9.29	14.85	13.42	10.14	9.31	12.63	8.17	7.84
CaO	13.18	12.3	13.4	11.85	10.50	11.52	11.90	11.03	9.71	9.67
Na_2O	0.72	0.63	0.57	0.48	0.61	0.48	0.55	0.53	0.73	0.64
K_2O	0.48	0.49	0.245	0.11	0.49	0.34	0.36	0.30	0.67	0.61
Cr_2O_3	0.26	0.18	0.16	0.19	0.24	0.20	0.19	0.20	0.35	0.36
P_2O_5	0.40	0.34		0.15	0.48	0.41	0.41	0.28	0.70	
	100.34	98.8	100.8	100.03	99.72	99.59	99.35	99.89	98.87	100.64
Mg/(Mg+Fe)	0.61	0.63	0.75	0.81	0.73	0.66	0.66	0.71	0.58	0.56
CaO/Al_2O_3	0.62	0.61	0.60	0.59	0.61	0.62	0.60	0.61	0.66	0.60
ppm										
La	43	56.4	21.1	19.2	49.6	60.3	65.5	32.1	83.5	90.0
Ce		144.0	56.0	52.2	136	160	185	81.2	211	228
Pr		17		7.4	19.2	21.5	26.0			
Nd		87.0	34.0	33	88	94	118	51.6	131	150
Sm	20.0	24.0	9.8	8.10	20.6	25.2	26.7	14.6	37.5	40.4
Eu	1.9	2.15	1.23	1.21	1.95	1.93	1.97	1.99	2.72	3.00
Gd		28.1		10.4	26.0	31.7	36.0	18.5	45.4	48.0
Tb	4.8	5.1	1.97	1.76	4.67	5.56	6.28		7.9	7.9
Dy	23	32.7	12.2	11.0	28.5	35.0	38.1	19.1	46.3	51.7
Ho	7.3	6.5	2.7	2.4	6.0	6.8	8.3			11
Er		19.7	6.7	6.9	18.0	21.0	23.8	11.4	27.3	27
Tm		3.0			1.4					
Yb	11.7	18.4	6.6	6.29	15.8	19.4	21.1	10.5	24.4	27.2
Lu		2.5	0.96	0.905	2.10	2.59	2.84		3.4	4.12
Y	200	174		59	131	166	174	109		
La/Yb	3.68	3.07	3.20	3.05	3.14	3.11	3.10	3.06	3.42	3.31
Ir (ppb)		10.5	2.25	5.5	9.4	17	17		0.061	
Au (ppb)	0.3	4.31	1.56	7.0	16.7	17.6	10.9		0.22	
Re (ppb)		1.02	0.284	1.2	1.3	1.7	1.8			
Os (ppb)		10.5								

ophitic textures and lack of meteoritic (i.e., impact) contamination. The textures of two impact-melt rocks and the sample 15386 are compared in Fig. 1.2.10.3.

1.2.10.4 Petrography of selected LKFM and IKFM samples

Low-K Fra Mauro (LKFM)

The Low-K Fra Mauro melt rocks considered here can be divided into two subgroups: those with >20% Al_2O_3 and an intergranular to subophitic texture, and those with <20% Al_2O_3 and poikilitic texture. Although rocks with poikilitic textures tend to have lower Al_2O_3 contents, exceptions do occur (Simonds *et al.*, 1973).

Subophitic-intergranular (high Al_2O_3) LKFM. The three intergranular- to subophitic-textured LKFM samples 14276, 14310 and 64455 plot together on the An-SiO_2-Ol pseudoternary projection (Fig. 1.2.10.1); a fourth sample, the vitrophyre 62295, plots within the spinel field. Samples 14276 and 14310 are similar in composition, texture and mineralogy. Although 64455 plots with the Apollo 14 KREEP basalts in Fig. 1.2.10.1, it is significantly lower in Fe and alkalies, it lacks mineral zonation, and it contains olivine (which is absent from the Apollo 14 KREEP basalts).

Samples 14276 and 14310 have extensive feldspar and pyroxene zonation (Figs. 1.2.10.4–5). Pyroxenes vary from orthopyroxene through pigeonite to augite. Olivine is absent; experimental studies (Green *et al.*,

Table 1.2.10.1 *(Continued)*

	Low-K Fra Mauro								High-K Fra Mauro	
	14276	14310	64455	62295	60315	62235	65015	77135	15386	15434
Cr	~1800	1260	1100	1300	1600	1400	1280	1400	2400	2400
V	25	36		23	47	65				
Sc	18.2	20	7.8	10.3	15.1	16.5	17.1		23.6	38.7
Ni	113	271	540	330	878	830	730	86	12.5	
Cu	38	5		18.1	18.6	24.9	5.06			
Zn	1.7	1.8	4.0	22.1	2.37	11	1.46	4	3.5	
Ga	3.8	3.2	3.05	2.74	3.85	3.82	3.81			
V/Ni	0.22	0.13		0.07	0.05	0.08				
Cr/V	~72	35		~57	~34	~22				
U	2.5	3.10	1.430	0.76	2.19	2.08	2.22	1.42	2.8	
Th	8	10.42		2.70	6.84	7.83	8.92	5.5	10	
Zr	620	842		283	720	858	940	501	970	
Hf	16	21.0	7.8	6.55	17.1	20.7	21.6		31.6	40.0
Nb	33	52		18	31	43	48	33		
Th/U	3.20	3.35		3.55	3.12	3.76	4.02	3.87	3.57	
Zr/Hf	38.8	40.1		43.2	42.1	41.4	43.5		30.7	
Zr/Nb	18.8	16.2		15.7	23.2	19.9	19.6	15.2		
Cs	0.57	0.54	0.27	0.560	0.560	0.350	0.423			
Rb	14	12.8	6.0	5.58	9.54	9.32	10.2	7.32	18.46	15.77
K %	0.40	0.4075	0.203	0.09	0.41	0.28	0.30	0.25	0.56	0.51
Ba	540	630		197	479	530	570	337	837	
Pb	9.0	6.18		1.9	5.2		4.66			
Sr	170	250		132	159	160.7	146	172.2	187.4	140.7
K/Rb	~290	318	340	~160	~430	~300	~290	~340	~300	~320
Rb/Cs	24.6	23.7	22.2	10.0	17.0	26.6	24.1			
Ba/Rb	38.6	49.2		35.3	50.2	56.9	55.9	46	45.3	
Rb/Sr	0.08	0.0512		0.0423	0.0600	0.0580	0.0699	0.0425	0.0985	
Sr/Eu	90	116		110	82	83	74	87	69	47
K/Ba	~7	6.47		~4.6	~8.6	~5.3	~5.3	~7.4	~6.7	

14276: Brunfelt *et al.*, 1972; Rose *et al.*, 1972. **14310:** Hubbard *et al.*, 1972. **60315:** LSPET, 1973; Rose *et al.*, 1973; Wänke *et al.*, 1976. **62235:** LSPET, 1973; Wänke *et al.*, 1976. **62295:** Walker *et al.*, 1973; Wänke *et al.*, 1976; Rose *et al.*, 1973. **64455:** Haskin *et al.*, 1973. **65015:** Haskin *et al.*, 1973; Duncan *et al.*, 1973; Wänke *et al.*, 1973. **77135:** Rhodes *et al.*, 1974; LSPET, 1973a; Hubbard *et al.*, 1974. **15386:** Rhodes and Hubbard, 1973; Blanchard (unpublished); Warren and Wasson, 1978. **15434:** Helmke *et al.*, 1973; Helmke (unpublished).

1972; Walker *et al.*, 1973) show that olivine appears briefly at ~1200° C in the crystallization sequence, following feldspar and along with orthopyroxene, but the olivine is quickly resorbed. Common accessory phases are ilmenite, troilite, ulvöspinel, armalcolite, metal, schreibersite, apatite, whitlockite, K-Ba feldspar, and tranquillityite [$Fe_8(Y,Zr)_2Ti_3Si_3O_{24}$]. The Apollo 14 KREEP basalts have been studied intensively, and the interested reader may want to review the following articles: Kushiro *et al.* (1972), Longhi *et al.* (1972), Brown *et al.* (1972), Ridley *et al.* (1972), James (1973), Gancarz *et al.* (1972), and Crawford and Hollister (1974).

In contrast to 14276 and 14310, sample 64455 is strikingly restricted in mineral zonation (Figs. 1.2.10.4–5) and contains abundant olivine (Fo_{81-78}). This rock has been severely affected by shock; re-equilibration is suggested by the restricted mineral zonation. Impact and thermal annealing are also suggested by the glass coating and aerodynamically rounded shape of 64455: the glass coating is foreign to the crystalline sample and is of a more feldspathic composition (Grieve and Plant, 1973).

The LKFM group (as defined by Naney *et al.*, 1977) includes relatively Si-poor and Mg-rich troctolitic compositions such as sample 62295 (Mg′ = 0.81).

Table 1.2.10.2 Modal data for samples in the Highland Reference Suite.

	14276*	14310*	62295*	64455	60315	62235
MATRIX						
feldspar	64.6	59.0	35.5	59.9	43.2	34.1
pyroxene						
clinopyroxene	13.3	16.6	—	(32.7)	4.9	6.0
orthopyroxene	15.7	16.3	—		39.0	36.3
olivine	—	—	28.3	6.5	6.1	6.0
silica	—	—	—	—	—	—
ilmenite	1.2	1.8	—	—	2.1	2.8
ulvöspinel	<0.1	<0.1	—	—	—	—
Mg,Al spinel	—	—	6.5	—	—	—
metal	0.3	0.1	<0.1	0.2	0.7	1.2
troilite	<0.1	<0.1	—	0.3	—	0.1
phosphate	0.5	0.3	—	—	—	—
mesostasis	3.5	4.4	—	0.4	—	—
glass	—	—	30	—	—	—
accessories	schreibersite	schreibersite	schreibersite	schreibersite	K feldspar	
(<0.1%)	armalcolite	armalcolite		akaganeite		
	tranquillityite	tranquillityite				
	K-Ba feldspar					
CLASTS						
xenocrysts						
feldspar	?	?	0.1	—	3.0	9.0
olivine	—	—	<0.1	—	1.0	—
pyroxene	—	—	—	—	—	0.5
spinel	—	—	<0.1	—	—	—
xenoliths	—	—	—	—	—	ANT (3.4)[a]
						M.R. (0.6)[a]

	62235	65015*	77135	15434	15386
MATRIX					
feldspar	34.1	57.1	41.1	44.0	37.5
pyroxene					
clinopyroxene	6.0	6.4		18.9	18.4
orthopyroxene	36.3	28.9	(30.0)	16.0	20.0
olivine	6.0	1.1	15.1	—	—
silica	—	—	—	3.9	4.4
ilmenite	2.8	1.2	1.4	2.7	3.5
ulvospinel	—	—	—	0.2	—
Mg,Al spinel	—	—	—	—	—
metal	1.2	0.4	0.2	—	0.4
troilite	0.1	0.1	0.1	<0.1	0.3
phosphate	—	0.7	—	<0.1	—
mesostasis	—	3.6	—	—	15.3
glass	—	—	—	14.1	—
accessories					
(<0.1%)					
CLASTS					
xenocrysts	9.0	(<1)	6.2	—	—
feldspar	—	—	—	—	—
olivine	0.5	—	2.0	—	—
pyroxene	—	—	—	—	—
spinel	ANT (3.4)[a]	—	ANT (2.9)[a]	—	—
xenoliths	M.R. (0.6)[a]		M.R. (1.0)[a]		

* (14276 and 14310 from Gancarz *et al.*, 1972; 62295 from Walker *et al.*, 1973) (65015 from Albee *et al.*, 1973)

[a] The acronyms "ANT" and "M.R." refer to anorthosite/norite/ or troctolite coarse-crystalline rock fragments and melt rock fragments, respectively.

Table 1.2.10.3 Ages (b.y.) for samples in the Highland Reference Suite.

	$^{40}Ar/^{39}Ar$		Rb-Sr internal isochron	Quoted $^{87}Sr/^{86}Sr$ "T_{BABI}" model age	Nd-Sm isochron	$^{143}Nd/^{144}Nd$ "T_{ICE}" model age
	whole rock	feldspar				
Low-K Fra Mauro						
14276			3.88 ± 0.01 (*1*)			
14310	3.88 ± 0.06 (*2*) 3.88 ± 0.05 (*6*) 3.92 ± 0.01 (*8*)	3.89 ± 0.04 (*3*)	3.93 ± 0.04 (*4*) 3.93 ± 0.06 (*7*)	4.49 ± 0.13 (*5*)		
62295	3.89 ± 0.05 (*9*)		4.00 ± 0.06 (*10*)	4.28 ± 0.07 (*11*)		
60315	4.03 ± 0.03 (*12*) 3.907 ± 0.010 (*8*)			4.41 ± 0.06 (*11*)		
62235				4.48 ± 0.07 (*11*)		
65015	3.98 (*13*) 3.92 ± 0.04 (*12*)	4.47 (*13*)	3.93 (*14*)	4.45 ± 0.06 (*11*) 4.41 ± 0.08 (*16*)		4.32 ± 0.12 (*15*)
77135	3.90 ± 0.03 (*17*)					
Intermediate-K Fra Mauro						
15382	3.90 (*2*)			4.30 (*11*)		4.38 ± 0.14 (*15*)
15386			3.94 ± 0.04 (*18*)	4.25 ± 0.04 (*18*)	3.87 ± 0.12 (*19*)	
15434			3.91 ± 0.04 (*16*)	4.27 ± 0.05 (*16*)		

References: (*1*) Wasserburg and Papanastassiou (1971); (*2*) Stettler *et al.* (1973); (*3*) Turner *et al.* (1971); (*4*) Compston *et al.* (1972); (*5*) Nyquist *et al.* (1972); (*6*) Turner *et al.* (1972); (*7*) Murthy *et al.* (1972); (*8*) Schaeffer *et al.* (1976); (*9*) Turner *et al.* (1973); (*10*) Mark *et al.* (1974); (*11*) Nyquist *et al.* (1973); (*12*) Kirsten *et al.* (1973); (*13*) Jessberger *et al.* (1974); (*14*) Papanastassiou and Wasserburg (1972); (*15*) Lugmair and Carlson (1978); (*16*) Nyquist *et al.* (1974); (*17*) Stettler *et al.* (1975); (*18*) Nyquist *et al.* (1975); (*19*) Carlson and Lugmair (1979).

This rock is a fine-grained (<0.2 mm) olivine, feldspar, Mg-Al spinel vitrophyre with rounded xenocrysts of more magnesian olivine, spinel (subhedral and higher in Cr_2O_3 and FeO than the spinel inclusions in plagioclase), plagioclase, and metal + schreibersite. Walker *et al.* (1973) suggest that 62295 is a low-pressure spinel-plagioclase-olivine peritectic partial melt of a source similar in composition to pink-spinel olivine cumulate 67435.

Poikilitic (low Al_2O_3) LKFM. The samples 60315, 62235, 65015, and 77135 represent poikilitic-textured LKFM. All four of these samples plot near the feldspar-olivine cotectic (Fig. 1.2.10.1) and have similar mineral compositions (Figs. 1.2.10.4–5). The poikilitic texture is due to enhanced growth of pyroxene and ilmenite, enclosing smaller grains of feldspar and olivine. Olivine may also enclose feldspar laths. Simonds *et al.* (1973) suggest that this texture is a result of two-stage cooling: initial rapid cooling near the cotectic with nucleation of feldspar and olivine at many loci, followed by slower cooling and crystallization at the point where pyroxene saturation is reached, allowing the growth of large pyroxene grains encompassing the previous crystals. Ryder and Bower (1976) and Lofgren (1977) have suggested that nucleation effects (e.g., many nucleation sites) are important in the origin of this texture. Minor phases in the four poikilitic LKFM rocks include metal, troilite, ilmenite, pleonaste spinel, whitlockite and K-feldspar.

Intermediate-K Fra Mauro (IKFM)

Samples 15386 and 15434 are members of the Apollo 15 KREEP crystal-liquid equilibration series of

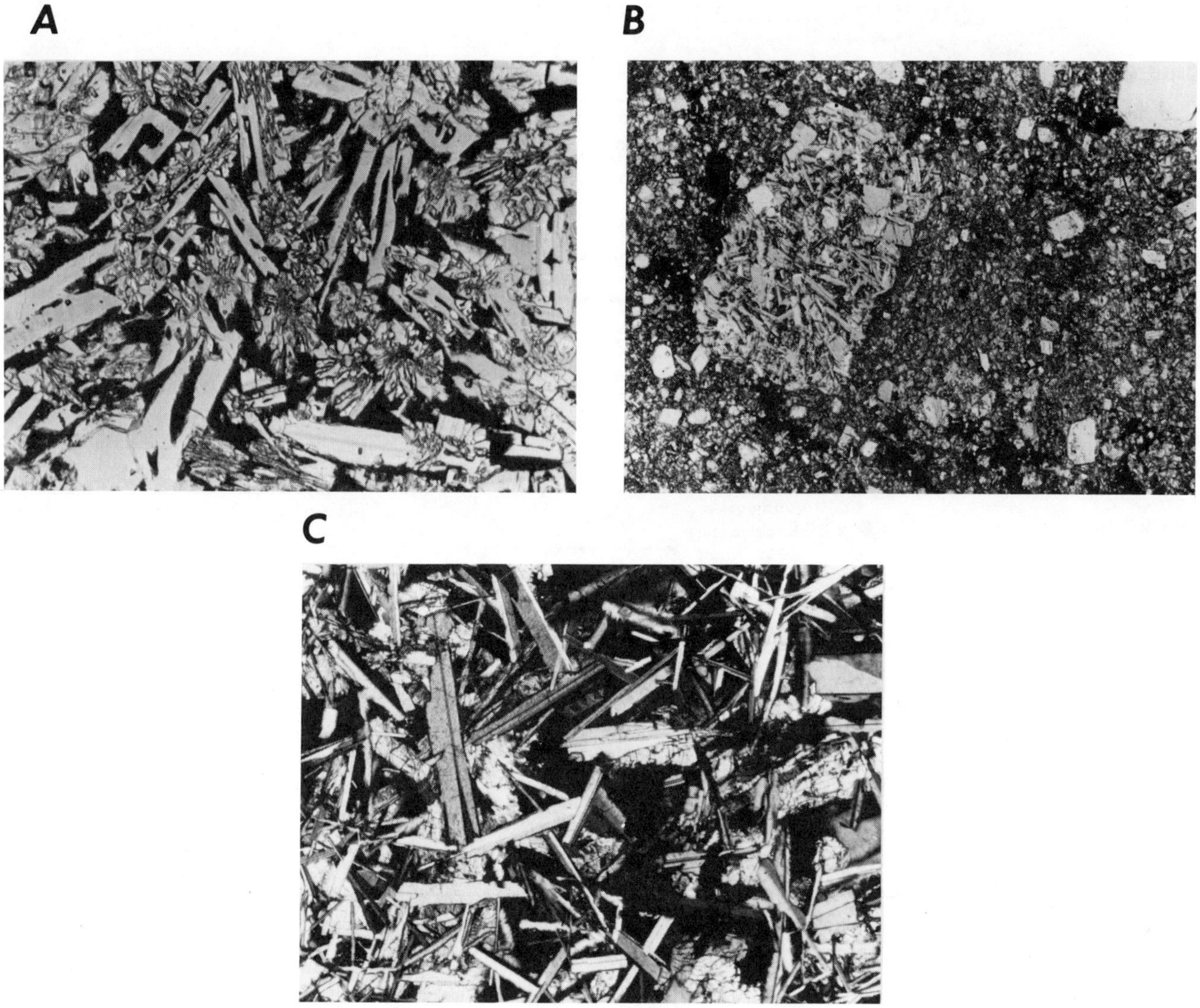

1.2.10.3 Representative textures: **a.** intersertal olivine-rich Low-K Fra Mauro, 62295,76; **b.** poikilitic Low-K Fra Mauro with an intergranular melt-rock inclusion, 62235,6; **c.** coarse intersertal Intermediate-K Fra Mauro basalt, 15386,3. Horizontal dimension in each photomicrograph is 1 mm.

Irving (1977). Irving has proposed that this series was produced by fractional crystallization of a parental partial melt from a troctolitic source deep in the lunar crust or upper mantle. The lack of xenolithic and xenocrystic debris and the low Ni content ($<$ 40 ppm) and low Ni/Co ratio ($\sim$2) in 15386 indicate a lack of meteoritic contamination. The low Ni content and Ni/Co ratio of intermediate-K Fra Mauro contrast sharply with high levels of meteoritic contamination in the LKFM samples (Ni = 60–1400 ppm, Ni/Co = 4.5–20). The metal compositions in 15382, a small $<$ 1 cm rock which is very similar to 15386 and 15434, indicate that IKFM samples are the only members of the suite which do not contain high-Ni metals of meteoritic rather than lunar composition (Irving, 1975). Wänke *et al.* (1976) and Warren and Wasson (1977, 1978) emphasize the significance of low trace siderophile contents (particularly Ir, Au, and Re) in distinguishing "pristine" lunar samples which have not been reworked by impact and therefore are not contaminated with siderophile-rich extralunar debris; petrographic criteria (plutonic texture, or basaltic texture without clastic debris) must also be considered. The IKFM sample 15386 is a particularly pristine volcanic rock by the criteria of Warren and Wasson (1978); both Ir and Au in this sample are below the ppb level (Table 1.2.10.1). This is consistent with the hypothesis of volcanic origin as defined above.

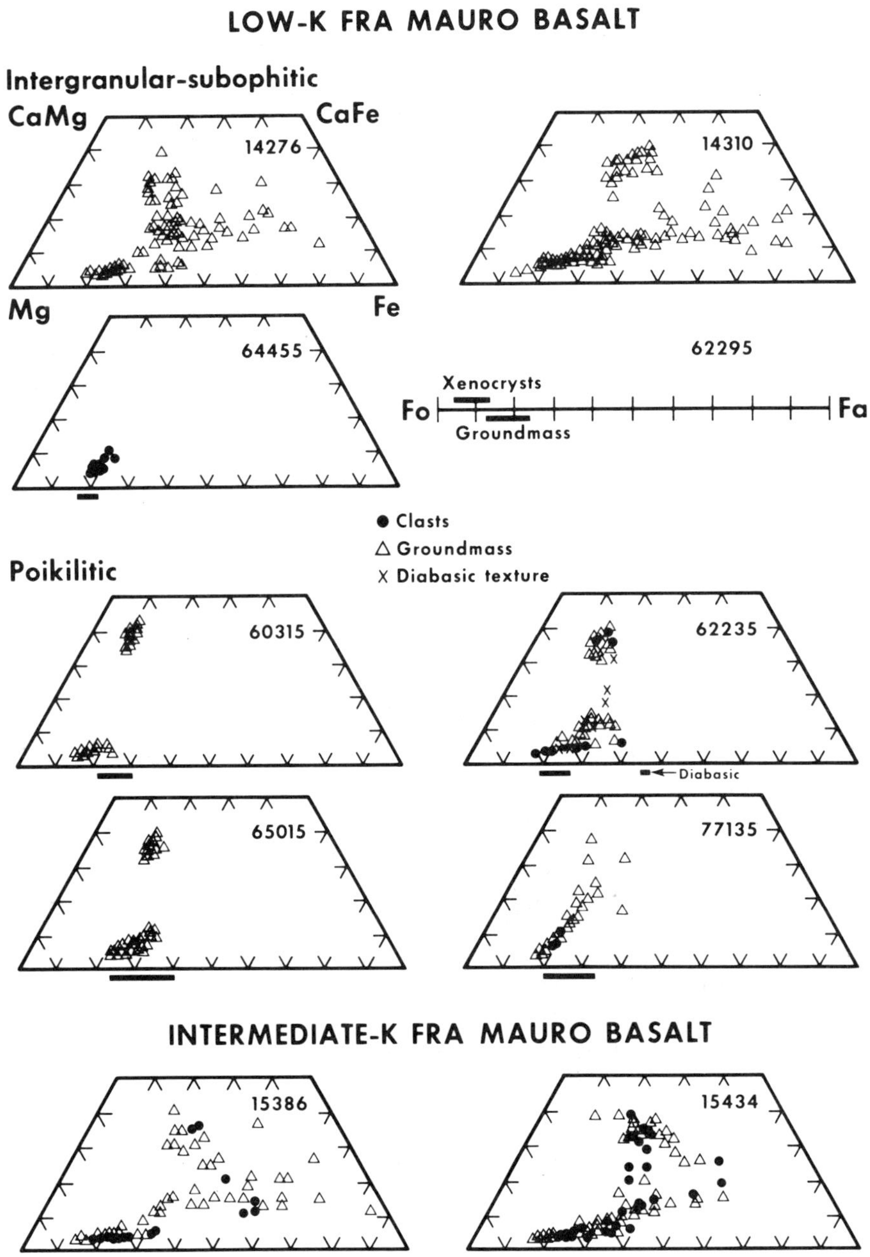

1.2.10.4 Ca-Fe-Mg composition of pyroxenes, and Fe-Mg composition of olivines in LKFM and IKFM samples. In addition to data collected for this study, data from the following studies are included: 14310: Brown *et al.* (1972), Crawford and Hollister (1974), Kushiro *et al.* (1972), and Ridley *et al.* (1972); 14276: Gancarz *et al.* (1972); 60315: Bence *et al.* (1973), Hodges and Kushiro (1973), and Walker *et al.* (1973); 60335: Walker *et al.* (1973); 62235: Crawford and Hollister (1974); 62295: Walker *et al.* (1973); 65015: Albee *et al.* (1973). Note that the use of various symbols denotes textural differences within poikilitic samples; among the IKFM samples, differing symbols are used for analyses from this study (circles) and analyses (triangles) from Steele *et al.* (1972).

Pyroxene and feldspar compositions of 15386 and 15434 are summarized in Figs. 1.2.10.4–5. Both samples are ophitic to intersertal; the grain size is ~0.5–1 mm. Experiments at 1–5 kb (Irving, 1977) indicate that subcalcic pyroxene and plagioclase are both near-liquidus phases in 15386. Similar results were obtained by Hess *et al.* (1977, 1978). Textural evidence and microprobe data indicate that both 15434 (thin section 116) and 15386 are nearly co-saturated with orthopyroxene ($En_{80}Wo_3Fs_{17}$) and plagioclase (An_{87}). The ilmenite in these rocks is particularly distinctive for it occurs as acicular grains (length/width >25) that contrast sharply with the stubby, anhedral ilmenite grains in impact melt rocks. Feldspars are zoned from An_{87} to $An_{70}Ab_{27}Or_3$.

Pyroxene zonation trends follow a pattern which is found in feldspathic impact melts (Vaniman and Papike, 1979); orthopyroxene cores become more Fe-rich toward their rims and are mantled by low-Ca pigeonite (Wo_7) and augite (as calcic as Wo_{35-40}). A broad range in Wo content is maintained as pyroxene becomes more Fe-rich. In minor-element composition the initial orthopyroxenes are Al-rich (~0.12 cations/6 oxygens) and become less aluminous as the pyroxenes become more Fe-rich. In the orthopyroxene crystallization series, Al/Ti ratios are high (>4). The first Ca-rich augites to form have an Al/Ti ratio of 2, indicating an increased content of Ti as $CaTiAl_2O_6$. With Fe-enrichment the Ti, Al and Cr contents of pyroxene decrease with the Al/Ti ratio maintained at 2. High initial Cr contents reflect the high Cr in the IKFM bulk composition (Table 1.2.10.1). The residual phases in 15386 include cristobalite, metal, troilite, and a more stubby ilmenite with higher Fe/Mg ratio (Mg'_{ilm} falls from ~0.03 in early-formed acicular grains to ~0.00 in late-formed grains). Small pyroxene grains isolated in the glassy mesostasis approach pyroxferroite composition.

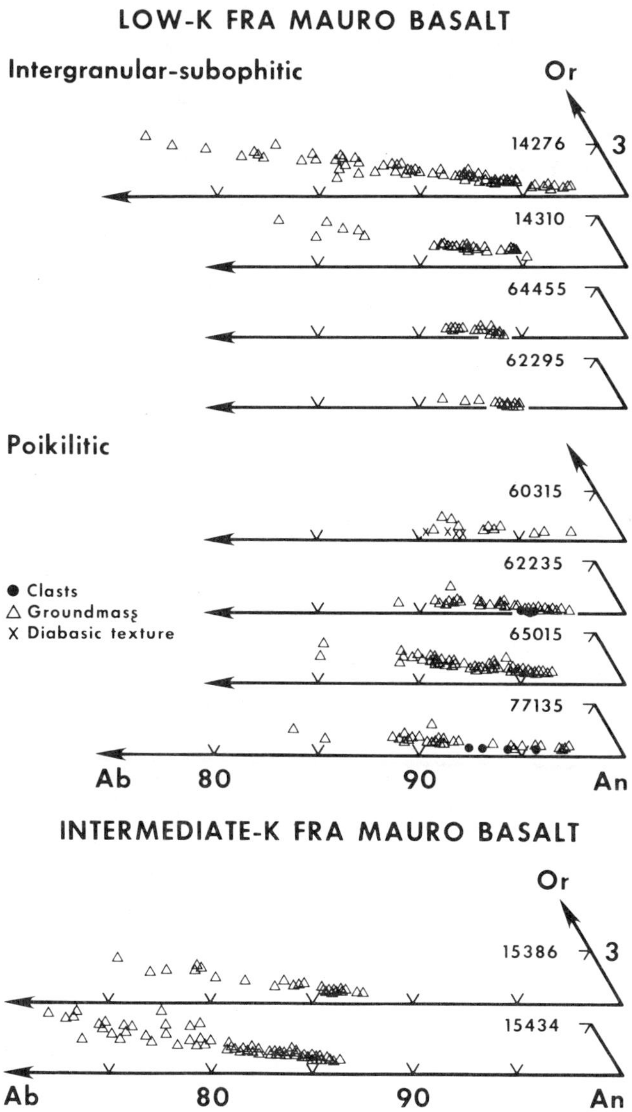

1.2.10.5 Ab-An-Or composition of feldspars in LKFM and IKFM samples. Data from this study, and from the sources cited for Fig. 1.2.10.4. Note that the use of various symbols denotes textural differences within poikilitic samples.

1.2.10.5 Petrogenesis of IKFM

Most workers agree that igneous-textured IKFM samples are volcanic in origin (Hubbard and Gast, 1971; Irving, 1975, 1977; Warren and Wasson, 1978; Dowty *et al.*, 1976; Meyer, 1977; Carlson and Lugmair, 1979; Hess *et al.*, 1977). The arguments for volcanic origin are based on several lines of evidence, including texture, lack of meteoritic contaminants, and apparent three-phase cotectic control of compositions parental to the IKFM fractionation series (point B, Fig. 1.2.10.1). However, there is still a question as to whether all such basalts are the products of partial melting events in the lower crust, or if some of these samples could be residual liquids of the magma ocean concentrated at 4.3–4.4 b.y.

Both partial melting and the residual melt hypotheses have received support in the literature. The Rb/Sr internal isochron ages of 3.94 ± 0.04 and 3.91 ± 0.04 b.y. for 15386 and 15434 and their initial $^{87}Sr/^{86}Sr$ values of 0.7004 and 0.7007 (Nyquist *et al.*, 1974, 1975) are compatible with an origin by partial melting, at ~3.9 b.y., of sources in which the Sr-isotopic system was last set at ~4.25 ± 0.04 to 4.27 ± 0.05 b.y. The Sm-Nd isotopic studies of Lugmair and Carlson (1978) show that IKFM breccias from Apollo 12 and 14 as well as IKFM basalt 15382 have remarkably constant Sm/Nd and model ages of 4.3–4.4 b.y. The first interpretations of the constant Sm/Nd, over a range of Sm and Nd concentrations in IKFM, leaned toward volcanism from a single homogeneous moon-wide reservoir, with the low Rb/Sr ages representing a much later thermal resetting following the initial volcanism. However, work by Carlson and Lugmair (1979) shows that the Nd/Sm internal isochron age of 15386 is 3.87 ± 0.12 b.y., completely in accord with the Rb/Sr internal isochron ages for IKFM basalts. Since the Nd/Sm isochron age is relatively resistant to thermal resetting and the Rb/Sr isochron is not, the congruent Rb/Sr and Nd/Sm age of ~3.9 b.y. indicates an IKFM highland volcanic event brought on by partial melting of a source region with Nd and Sr isotope systematics set at ~4.3 b.y.

Could IKFM volcanism have occurred prior to 3.9 b.y.? Although this seems to be a real possibility, the only evidence we have for IKFM volcanism is founded on radiometric ages of ~3.9 b.y. and the occurrence of volcanic domes of probable highland composition (not necessarily IKFM), also of Imbrian (~3.9 b.y.) age, in

Mare Procellarum (Head and McCord, 1978; Chapter 5). One can only speculate about earlier volcanism, though such volcanism does seem likely. It should not be surprising that pre-3.9 b.y. lunar volcanic samples have not been found, considering the continuation of cataclysmic meteorite bombardment(s) of the lunar surface from the time of formation of mare and highland source regions at 4.4–4.3 b.y. to the times recorded by the oldest preserved volcanic samples at 4.0–3.9 b.y. Speculation about pre-3.9 b.y. highland volcanism can also be extended to the LKFM composition.

1.2.10.6 Petrogenesis of LKFM

All samples of LKFM with "extrusive" textures have proved to be impact melt rocks. None of the LKFM samples described here or in the literature are free of meteoritic contamination. Moreover, most LKFM impact-melt samples have Rb/Sr model ages of ~4.41–4.49 b.y., indicating isotopic sources that were closed just before the ~4.3–4.4 b.y. isotopic equilibration in the IKFM source region (Table 1.2.10.3). Sample 62295 is an exception, with a model age of 4.28 ± 0.07 b.y., comparable to the Sr isotopic model age for IKFM. This sample is contaminated with meteoritic trace metals (Table 1.2.10.1), but it has a bulk composition which may be inherited from internal lunar igneous processes.

Walker *et al.* (1973) point out an apparent olivine-anorthite-liquid cotectic control in some low-K Fra Mauro (LKFM) compositions. They propose that some LKFM compositions which plot along the olivine-anorthite cotectic were derived by extensive partial melting of crustal cumulate rocks. Walker *et al.* (1973) also note that sample 62295 is displaced from the piercing point of olivine-anorthite-plagioclase-liquid cosaturation (Fig. 1.2.10.1), because it contains xenocrysts of spinel and olivine which are very similar to the spinel and olivine of "pink spinel troctolite" 67435. From this evidence they originally suggested that sample 62295 originated by partial melting of Mg-rich cumulate rocks (Mg' ~ 0.92) at shallow depths in the lunar crust. Other researchers (e.g., Reid *et al.*, 1977) have noted that there are no good basaltic-textured samples of LKFM that fall along the anorthite-olivine-liquid cotectic. The few melt rocks with anorthite-olivine cosaturation (i.e., the poikilitic-textured low-Al_2O_3 LKFM) contain evidence of meteoritic contamination, suggesting impact-melt origin. Yet, it is a suspicious coincidence that many highland poikilitic impact melts, glasses, breccias, and even the average Fra Mauro (Apollo 14) soil all plot at the low-pressure olivine-anorthite-liquid cotectic. The Fra Mauro soil is of particular interest in this respect, because no other lunar soils have this low-pressure cosaturated "volcanic" composition (Fig. 1.2.10.1).

The internal melting hypothesis for 62295 and other LKFM samples is difficult to reconcile with the evidence of impact modification. Is it possible to have impact-melting of an initially igneous rock *without* assimilating other lithic debris? This suggestion is not supported by studies of impact phenomena. Studies of Canadian impact craters (Manicouagan, West Clearwater, and Mistastin) and considerations of cratering mechanics indicate that, although a total melt is generated in a relatively small volume near the point of impact, this melt is charged with relatively cold debris and is turbulently mixed as it travels down and out along the walls of the transient impact cavity (Grieve and Floran, 1978; Simonds *et al.*, 1978; Grieve, 1975; Simonds, 1979). The intermixed clasts are melted as they draw heat from the surrounding superheated liquid; melting may be partial or nearly complete, but the result in either case is a rather homogeneous mixture of the rock types found throughout the area of the transient crater.

It is very unlikely that a thin flow of volcanic rock could be completely remelted without acquiring fragments of underlying rocks or overlying deposits. Thus sample 62295 and other LKFM basalts may be remelted, originally endogenic magmas only if they were (a) very thick (> 1 km) and (b) at or very near the lunar surface at the time of impact remelting. Hess *et al.* (1977) suggest that the LKFM compositions which are meteorite-contaminated, yet have apparent cotectic control, are the products of impact remelting of shallow plutons of LKFM composition that have fractionated internally to develop cumulate layering; impact remelting and remixing of these cumulate layers might reproduce the LKFM composition in impact melt rocks, breccias and ejecta blankets. Blanchard and Budahn (1979) contribute to this hypothesis by their description of mineral clasts, in breccias, which must originally have equilibrated with magmas of LKFM trace element composition. Blanchard and Budahn propose that shallow plutons with this composition were excavated by impacts.

A "cleaner" reproduction of the LKFM composition would occur if the protolith was a thick series of lava flows on the lunar surface, allowing the transient impact cavity to develop entirely within the LKFM composition. Large-scale cratering processes are still poorly understood, and the depths of excavation and melting relative to the diameters of very large craters are not known; therefore, it is difficult to estimate how thick and extensive such a pile of LKFM flows must be in order to contain a large impact crater (10–100 km diameter). However, it is possible that LKFM flows were brought to the lunar surface in significant volume, either by tapping residual magmas of the ancient crustal differentiation (~4.4 b.y.) or by partial melting at a later period. In the terrestrial environment, flood basalts derived by partial melting may accumulate thicknesses of > 2 km and cover > 10^5 km^2 (sections 1.2.2, 1.2.3), and there is no reason to reject the possibility of such occurrences on the Moon prior to 3.9 b.y. Schonfeld and Meyer (1973) suggest that such LKFM flows occurred in the Imbrium basin between 4.4–4.3 b.y. Because of the meteoritic contamination of all LKFM samples now in hand, such possibilities must remain purely speculative.

A Sm/Nd model age of 4.32 ± 0.12 b.y. has been determined for 65015 (Lugmair and Carlson, 1978), but it would be interesting to know the Sm-Nd isotopic systematics for more low-K Fra Mauro samples. If some of the LKFM rocks were formed originally by partial melting in cumulate crustal zones and were extruded onto the lunar surface before the isotopic closure of the IKFM source regions, then the Sm-Nd model ages of those LKFM melt-rocks may be significantly different from the <4.4 b.y. model ages of IKFM basalts. However, if all LKFM melt rocks are impact mixtures of cumulate rocks plus KREEP, they should have Sm-Nd systematics dominated by the REE-rich KREEP component and reflect the isotopic composition of the 4.3–4.4 b.y. KREEP reservoir.

1.2.10.7 Conclusions

Endogenic lunar highland basalts are represented by a small group of incompatible-element rich samples with ophitic to intersertal textures, without meteoritic contamination, and with volcanic ages of ~3.9 b.y. This age overlaps with the oldest stages of mare volcanism, and with the ages of the only known surface features (volcanic domes) that may be correlated with highland volcanism. Although evidence for volcanism prior to ~4.0 b.y. remains speculative, there is reason to believe that volcanic processes were operative during the extensive magmatic fractionation of the lunar crust. Despite pervasive meteoritic reworking and the lack of any pristine igneous samples, the very common but enigmatic lunar composition known as "Low-K Fra Mauro" remains a candidate for ancient highland-type volcanism because of its apparent relation to cotectic compositions, a relation which should not be expected in impact mixed crustal rocks.

1.2.10.8 References

Albee A. L., Gancarz A. J. and Chodos A. A. (1973) Metamorphism of Apollo 16 and 17 and Luna 20 metaclastic rocks at about 3.95 AE: samples 61156, 64423, 14-2, 65015, 67483, 15-2, 76055, 22006, and 22007. *Proc. Lunar Sci. Conf. 4th*, p. 569–595.

Anders E. (1978) Procrustean science: indigenous siderophiles in the lunar highlands, according to Delano and Ringwood. *Proc. Lunar Planet. Sci. Conf. 9th*, p. 161–184.

Bence A. E., Papike J. J., Sueno S. and Delano J. W. (1973) Pyroxene poikiloblastic rocks from the lunar highlands. *Proc. Lunar Sci. Conf. 4th*, p. 597–611.

Blanchard D. P. and Budahn J. R. (1979) Remnants from the ancient lunar crust: clasts from consortium breccia 73255. *Proc. Lunar Planet. Sci. Conf. 10th*, p. 803–816.

Brown G. M., Emeleus C. H., Holland J. G., Peckett A. and Phillips R. (1972) Mineral–chemical variations in Apollo 14 and Apollo 15 basalts and granitic fractions. *Proc. Lunar Sci. Conf. 3rd*, p. 141–157.

Brunfelt A. O., Heier K. S., Nilssen B. and Sundvoll B. (1972) Distribution of elements between different phases of Apollo 14 rocks and soils. *Proc. Lunar Sci. Conf. 3rd*, p. 1133–1147.

Carlson R. W. and Lugmair G. W. (1979) Sm–Nd study of pristine KREEP basalt 15386 (abstract). In *Lunar and Planetary Science X*, p. 178–180. Lunar and Planetary Institute, Houston.

Compston W., Foster J. J. and Gray C. M. (1975) Rb–Sr ages of clasts from within Boulder 1, Station 2, Apollo 17. *The Moon* **14**, 445–462.

Compston W., Vernon M. J., Berry H., Rudowski R., Gray C. M., Ware N., Chappell B. W. and Kaye M. (1972) Apollo 14 mineral ages and the thermal history of the Fra Mauro formation. *Proc. Lunar Sci. Conf. 3rd*, p. 1487–1501.

Crawford M. L. and Hollister L. S. (1974) KREEP basalt: a possible partial melt from the lunar interior. *Proc. Lunar Sci. Conf. 5th*, p. 399–419.

Dowty E., Keil K., Prinz M., Gros J. and Takahashi H. (1976) Meteorite-free Apollo 15 crystalline KREEP. *Proc. Lunar Sci. Conf. 7th*, p. 1833–1844.

Duncan A. R., Erlank A. J., Willis J. P. and Ahrens L. H. (1973) Composition and inter-relationships of some Apollo 16 samples. *Proc. Lunar Sci. Conf. 4th*, p. 1097–1113.

Gancarz A. J., Albee A. L. and Chodos A. A. (1972) Comparative petrology of Apollo 16 sample 68415 and Apollo 14 samples 14276 and 14310. *Earth Planet. Sci. Lett.* **16**, 307–330.

Green D. H., Ringwood A. E., Ware N. G. and Hibberson W. O. (1972) Experimental petrology and petrogenesis of Apollo 14 basalts. *Proc. Lunar Sci. Conf. 3rd*, p. 197–206.

Grieve R. A. F. (1975) Petrology of the impact melt at Mistastin Lake Crater, Labrador. *Geol. Soc. Amer. Bull.* **86**, 1617–1629.

Grieve R. A. F. and Floran R. J. (1978) Manicouagan impact melt, 2, chemical interrelations with basement and formational processes. *J. Geophys. Res.* **83**, 2761–2771.

Grieve R. A. F. and Plant A. G. (1973) Partial melting on the lunar surface, as observed in glass coated Apollo 16 samples. *Proc. Lunar Sci. Conf. 4th*, p. 667–679.

Haskin L. A., Helmke P. A., Blanchard D. P., Jacobs J. W. and Telander K. (1973) Major and trace element abundances in samples from the lunar highlands. *Proc. Lunar Sci. Conf. 4th*, p. 1275–1296.

Head J. W., III and Goetz A. F. H. (1972) Descartes region: evidence for Copernican-age volcanism. *J. Geophys. Res.* **27**, 1368–1374.

Head J. W., III and McCord T. B. (1978) Imbrian-age highland volcanism on the Moon: the Gruithuisen and Mairan domes. *Science* **199**, 1433–1436.

Helmke P. A., Blanchard D. P., Haskin L. A., Telander K., Weiss C. and Jacobs J. W. (1973) Major and trace elements in igneous rocks from Apollo 15. *The Moon* **8**, 129–148.

Hess P. C., Rutherford M. J. and Campbell H. W. (1977) Origin and evolution of LKFM basalt. *Proc. Lunar Sci. Conf. 8th*, p. 2357–2373.

Hess P. C., Rutherford M. J. and Campbell H. W. (1978) Ilmenite crystallization in nonmare basalt: genesis of KREEP and high-Ti mare basalt. *Proc. Lunar and Planet. Sci. Conf. 9th*, p. 705–724.

Hodges F. N. and Kushiro I. (1973) Petrology of Apollo 16 lunar highland rocks. *Proc. Lunar Sci. Conf. 4th*, p. 1033–1048.

Hostetler C. J. and Drake M. J. (1979) On the early global melting of the terrestrial planets (abstract). In *Papers Presented to the Conference on the Lunar Highlands Crust*, p. 78–80. Lunar and Planetary Institute, Houston.

Hubbard N. J. and Gast P. W. (1971) Chemical composition and origin of nonmare lunar basalts. *Proc. Lunar Sci. Conf. 2nd*, p. 999–1020.

Hubbard N. J., Gast P. W., Rhodes J. M., Bansal B. M., Wiesmann H. and Church S. E. (1972) Nonmare basalts: Part II. *Proc. Lunar Sci. Conf. 3rd*, p. 1161–1179.

Hubbard N. J., Rhodes J. M., Gast P. W., Bansal B. M., Shih C. -Y., Wiesmann H. and Nyquist L. E. (1973) Lunar rock types: the role of plagioclase in non-mare and highland rock types. *Proc. Lunar Sci. Conf. 4th*, p. 1297–1312.

Hubbard N. J., Rhodes J. M., Wiesmann H., Shih C. -Y and Bansal B. M. (1974) The chemical definition and interpretation of rock types returned from the non-mare regions of the Moon. *Proc. Lunar Sci. Conf. 5th*, p. 1227–1247.

Irving A. J. (1975) Chemical, mineralogical and textural systematics of non-mare melt rocks: implications for lunar impact and volcanic processes. *Proc. Lunar Sci. Conf. 6th*, p. 363–394.

Irving A. J. (1977) Chemical variation and fractionation of KREEP basalt magmas. *Proc. Lunar Sci. Conf. 8th*, p. 2433–2448.

James O. B. (1973) Crystallization history of lunar feldspathic basalt 14310. *U.S. Geol. Survey Prof. Paper 841*. 29 pp.

Jessberger E. K., Huneke J. C., Podosek F. A. and Wasserburg G. J. (1974) High resolution argon analysis of neutron-irradiated Apollo 16 rocks and separated minerals. *Proc. Lunar Sci. Conf. 5th*, p. 1419–1449.

Kirsten T., Horn P. and Kiko J. (1973) ^{39}Ar–^{40}Ar dating and rare gas analysis of Apollo 16 rocks and soils. *Proc. Lunar Sci. Conf. 4th*, p. 1757–1784.

Kushiro I., Ikeda Y. and Nakamura Y. (1972) Petrology of Apollo 14 high-alumina basalt. *Proc. Lunar Sci. Conf. 3rd*, p. 115–129.

Lofgren G. E. (1977) Dynamic crystallization experiments bearing on the origin of textures in impact-generated liquids. *Proc. Lunar Sci. Conf. 8th*, p. 2079–2095.

Longhi J., Walker D. and Hays J. F. (1972) Petrography and crystallization history of basalts 14310 and 14072. *Proc. Lunar Sci. Conf. 3rd*, p. 131–139.

LSPET (1973a) The Apollo 16 lunar samples: Petrographic and chemical description. *Science* **179**, 23–24.

LSPET (1973b) Apollo 17 lunar samples: chemical and petrographic description. *Science* **182**, 659–662.

Lugmair G. W. and Carlson R. W. (1978) The Sm-Nd history of KREEP. *Proc. Lunar Planet. Sci. Conf. 9th*, p. 689–704.

Mark R. K., Chin-Nan L. -H. and Wetherill G. W. (1974) Rb-Sr age of lunar igneous rocks 62295 and 14310. *Geochim. Cosmochim. Acta* **38**, 1643–1648.

Meyer C., Jr. (1977) Petrology, mineralogy and chemistry of KREEP basalt. *Phys. Chem. Earth* **10**, 239–260.

Milton D. J. and Hodges C. A. (1972) Geologic maps of the Descartes region of the Moon: Apollo 16 pre-mission map. *U.S.G.S. Misc. Geol. Inv. Map I-748*.

Mizutani H., Matsui T. and Takeuchi H. (1972) Accretion process of the Moon. *The Moon* **4**, 476–489.

Murthy V. R., Evensen N. M., Jahn B. -M. and Coscio M. R., Jr. (1972) Apollo 14 and 15 samples: Rb-Sr ages, trace elements, and lunar evolution. *Proc. Lunar Sci. Conf. 3rd*, p. 1503–1514.

Naney M. T., Papike J. J. and Vaniman D. T. (1977) Lunar highland melt rocks: major element chemistry and comparisons with lunar highland glass groups (abstract). In *Lunar Science VIII*, p. 717–719. The Lunar Science Institute, Houston.

Nunes P. D., Tatsumoto M., Knight R. J., Unruh D. M. and Doe B. R. (1973) U-Th-Pb systematics of some Apollo 16 lunar samples. *Proc. Lunar Sci. Conf. 4th*, p. 1797–1822.

Nyquist L. E., Hubbard N. J., Gast P. W., Church S. E., Bansal B. M. and Wiesmann H. (1972) Rb-Sr systematics for chemically defined Apollo 14 breccias. *Proc. Lunar Sci. Conf. 3rd*, p. 1515–1530.

Nyquist L. E., Hubbard N. J., Gast P. W., Bansal B. M., Wiesmann H. and Jahn B. -M. (1973) Rb-Sr systematics for chemically defined Apollo 15 and 16 materials. *Proc. Lunar Sci. Conf. 4th*, p. 1823–1846.

Nyquist L. E., Bansal B. M., Wiesmann H. and Jahn B. -M. (1974) Taurus-Littrow chronology: some constraints on early lunar crustal development. *Proc. Lunar Sci. Conf. 5th*, p. 1515–1539.

Nyquist L. E., Bansal B. M. and Wiesmann H. (1975) Rb-Sr ages and initial $^{87}Sr/^{86}Sr$ for Apollo 17 basalts and KREEP basalt 15386. *Proc. Lunar Sci. Conf. 6th*, p. 1445–1465.

Onorato P. I. K., Uhlmann D. R. and Simonds C. H. (1978) The thermal history of the Manicouagan impact melt sheet, Quebec. *J. Geophys. Res.* **83**, 2789–2798.

Papanastassiou D. A., DePaolo D. J., Tera F. and Wasserburg G. J. (1977) An isotopic triptych on mare basalts: Rb-Sr, Sm-Nd, U-Pb (abstract). In *Lunar Science VIII*, p. 750–752. The Lunar Science Institute, Houston.

Papanastassiou D. A. and Wasserburg G. J. (1972) Rb-Sr systematics of Luna 20 and Apollo 16 samples. *Earth Planet. Sci. Lett.* **17**, 52–63.

Raedeke L. D. and McCallum I. S. (1979a) The banded zone of the Stillwater Complex: stratigraphy, petrology, and a model for the generation of anorthosite zones. In *Ancient Crusts of the Terrestrial Planets*, p. 57–59. The Lunar and Planetary Institute, Houston.

Raedeke L. D. and McCallum I. S. (1979b) Fractionation in the Stillwater Complex: an analog for the lunar crust and upper mantle (abstract). In *Papers Presented to the Conference on the Lunar Highlands Crust*, p. 125–127. Lunar and Planetary Institute, Houston.

Reid A. M., Duncan A. R. and Richardson S. H. (1977) In search of LKFM. *Proc. Lunar Sci. Conf. 8th*, p. 2321–2338.

Rhodes J. M. and Hubbard N. J. (1973) Chemistry, classification, and petrogenesis of Apollo 15 mare basalts. *Proc. Lunar Sci. Conf. 4th*, p. 1127–1148.

Rhodes J. M., Rodgers K. V., Shih C., Bansal B. M., Nyquist L. E., Wiesmann H. and Hubbard N. J. (1974) The relationship between geology and soil chemistry at the Apollo 17 landing site. *Proc. Lunar Sci. Conf. 6th*, p. 1097–1117.

Ridley W. I., Brett R., Williams R. J., Takeda H. and Brown R. W. (1972) Petrology of Fra Mauro basalt 14310. *Proc. Lunar Sci. Conf. 3th*, p. 159–170.

Ringwood A. E. and Kesson S. E. (1976) A dynamic model for mare basalt petrogenesis. *Proc. Lunar Sci. Conf. 7th*, p. 1697–1722.

Rose H. J., Jr., Cuttitta F., Annell C. S., Carron M. K., Christian R. P., Dwornik E. J., Greenland L. P. and Ligon D. T., Jr. (1972) Compositional data for twenty-one Fra Mauro lunar materials. *Proc. Lunar Sci. Conf. 3th*, p. 1215–1229.

Rose H. J., Jr., Cuttitta F., Berman S., Carron M. K., Christian R. P., Dwornik E. J., Greenland L. P. and Ligon D. T., Jr. (1973) Compositional data for twenty-two Apollo 16 samples. *Proc. Lunar Sci. Conf. 4th*, p. 1149–1158.

Ryder G. and Bower J. F. (1976) Poikilitic impact melts in the Apollo 14 white rocks. *Proc. Lunar Sci. Conf. 7th*, p. 1925–1948.

Ryder G., Stoeser D. B. and Wood J. A. (1977) Apollo 17 KREEPy basalt: a rock type intermediate between mare and KREEP basalts. *Earth Planet. Sci. Lett.* **35**, 1–13.

Schaeffer O. A., Husain L. and Schaeffer G. A. (1976) Ages of highland rocks: the chronology of lunar basin formation revisited. *Proc. Lunar Sci. Conf. 7th*, p. 2067–2092.

Schonfeld E. and Meyer C., Jr. (1973) The old Imbrium hypothesis. *Proc. Lunar Sci. Conf. 4th*, p. 125–138.

Simonds C. H. (1975) Thermal regimes in impact melts and the petrology of the Apollo 17 Station 6 boulder. *Proc. Lunar Sci. Conf. 6th*, p. 641–672.

Simonds C. H. (1979) Low speed (8 km/sec) impact-accretional heating and early fractionation of the Moon (abstract). In *Papers Presented to the Conference on the Lunar Highlands Crust*, p. 148–150. Lunar and Planetary Institute, Houston.

Simonds C. H., Warner J. L. and Phinney W. C. (1973) Petrology of Apollo 16 poikilitic rocks. *Proc. Lunar Sci. Conf. 4th*, p. 613–632.

Simonds C. H., Floran R. J., McGee P. E., Phinney W. C. and Warner J. L. (1978) Petrogenesis of melt rocks, Manicouagan impact structure, Quebec. *J. Geophys. Res.* **83**, 2773–2788.

Smith J. V., Anderson A. T., Newton R. C., Olsen E. J., Wyllie P. J., Crewe A. V., Isaacson M. S. and Johnson D. (1970) Petrologic history of the Moon inferred from petrography, mineralogy, and petrogenesis of Apollo 11 rocks. *Proc. Apollo 11 Lunar Sci. Conf.*, p. 897–925.

Steele I. M., Smith J. V. and Grossman L. (1972) Mineralogy and petrology of Apollo 15 rake samples. In *The Apollo 15 Lunar Samples*, p. 158–160. The Lunar Science Institute, Houston.

Stettler A., Eberhardt P., Geiss J., Grögler N. and Guggisberg S. (1975) Age sequence in the Apollo 17 station 7 boulder (abstract). In *Lunar Science VI*, p. 771–773. The Lunar Science Institute, Houston.

Stettler A., Eberhardt P., Geiss J., Grögler N. and Maurer P. (1973) ^{39}Ar-^{40}Ar ages and ^{37}Ar-^{38}Ar exposure ages of lunar rocks. *Proc. Lunar Sci. Conf. 4th*, p. 1865–1888.

Taylor S. R. and Jakeš P. (1974) The geochemical evolution of the Moon. *Proc. Lunar Sci Conf. 5th*, p. 1287–1305.

Taylor S. R., Gorton M. P., Muir P., Nance W., Rudowski R. and Ware N. (1973) Lunar highland composition: Apennine Front. *Proc. Lunar Sci. Conf. 4th*, p. 1445–1459.

Trask N. J. and McCauley J. F. (1972) Differentiation and volcanism in the lunar highlands: photogeologic evidence and Apollo 16 implications. *Earth Planet. Sci. Lett.* **14**, 201–206.

Turner G., Cadogan P. H. and Yonge C. J. (1973) Argon seleochronology. *Proc. Lunar Sci. Conf. 4th*, p. 1889–1914.

Turner G., Huneke J. C., Podosek F. A. and Wasserburg G. J. (1971) ^{40}Ar-^{39}Ar ages and cosmic ray exposure ages of Apollo samples. *Earth Planet. Sci. Lett.* **12**, 19–35.

Turner G., Huneke J. C., Podosek F. A. and Wasserburg G. J. (1972) Ar^{40}-Ar^{39} systematics in rocks and separated minerals from Apollo 14. *Proc. Lunar Sci. Conf. 3rd*, p. 1589–1612.

Vaniman D. T. and Papike J. J. (1979) Silicate mineralogy of highland melt rocks (abstract). In *Papers Presented to the Conference on the Lunar Highlands Crust*, p. 178–180. Lunar and Planetary Institute, Houston.

Wager L. R. and Brown G. M. (1968) *Layered Igneous Rocks*. Oliver and Boyd, London. 588 pp.

Walker D. (1979) The lunar magma ocean revisited (abstract). In *Papers Presented to the Conference on the Lunar Highlands Crust*, p. 181–182. Lunar and Planetary Institute, Houston.

Walker D. and Hays J. F. (1977) Plagioclase flotation and lunar crust formation. *Geology* **5**, 425–428.

Walker D., Longhi J. and Hays J. F. (1972) Experimental petrology and the origin of Fra Mauro rocks and soil. *Proc. Lunar Sci. Conf. 3rd*, p. 797–817.

Walker D., Longhi J., Grove T. L., Stolper E. and Hays J. F. (1973) Experimental petrology and origin of rocks from the Descartes Highlands. *Proc. Lunar Sci. Conf. 4th*, p. 1013–1032.

Wänke H., Baddenhausen H., Dreibus G., Jagoutz E., Kruse H., Palme H., Spettel B. and Teschke F. (1973) Multielement analyses of Apollo 15, 16, and 17 samples and the bulk composition of the Moon. *Proc. Lunar Sci. Conf. 4th*, p. 1461–1481.

Wänke H., Palme H., Kruse H., Baddenhausen H., Cendales M., Dreibus G., Hofmeister H., Jagoutz E., Palme C., Spettel B. and Thacker R. (1976) Chemistry of lunar highland rocks. A refined evaluation of the composition of the primary matter. *Proc. Lunar Sci. Conf. 7th*, p. 3479–3499.

Warner J. L., Phinney W. C., Bickel C. E. and Simonds C. H. (1977) Feldspathic granulitic impactites and pre-final bombardment lunar evolution. *Proc. Lunar Sci. Conf. 8th*, p. 2051–2066.

Warren P. H. and Wasson J. T. (1977) Pristine nonmare rocks and the nature of the lunar crust. *Proc. Lunar Sci. Conf. 8th*, p. 2215–2235.

Warren P. H. and Wasson J. T. (1978) Compositional-petrographic investigation of pristine nonmare rocks. *Proc. Lunar Planet Sci. Conf. 9th*, p. 185–217.

Warren P. H., Afiattalab F. and Wasson J. T. (1978) Investigation of unusual KREEPy samples: pristine rock 15386, cone crater soil fragments 14143, and 12023, a *typical* Apollo 12 soil. *Proc. Lunar Planet. Sci. Conf. 9th*, p. 653–660.

Wasserburg G. J. and Papanastassiou D. A. (1971) Age of an Apollo 15 mare basalt: lunar crust and mantle evolution. *Earth Planet. Sci. Lett.* **13**, 97–104.

Wood J. A., Dickey J. S., Jr., Marvin U. B. and Powell B. N. (1970) Lunar anorthosites and a geophysical model of the Moon. *Proc. Apollo 11 Lunar Sci. Conf.*, p. 965–988.

1.2.11 ULTRAMAFIC XENOLITHS IN TERRESTRIAL VOLCANICS AND MANTLE MAGMATIC PROCESSES

1.2.11.1 Introduction

The concept that terrestrial basaltic magmas are generated by partial melting of mantle peridotite is reinforced by occurrences of ultramafic mantle rocks at the Earth's surface. The major types of occurrence are as tectonically-emplaced ultramafic massifs and ophiolites, and as xenoliths within basaltic and kimberlitic lavas. Examination of these samples has yielded direct information on mantle constitution in a broad sense, and their mineralogical and chemical variety has also provided important details on the nature of magmatic processes within the mantle. Although our knowledge is very limited, it has not yet been demonstrated that tectonic and volcanic processes on the other terrestrial planets result in surface outcrops of mantle rocks. For the Moon, impact mixing has allowed sampling of a large variety of rock types in spite of the small number of landing sites; however, no unequivocal mantle rocks have yet been identified, either as xenoliths in lavas or as clasts in breccias and soils (the rare lunar dunites and troctolites are probably crustal materials). Nevertheless, the mantles of the Moon and the other terrestrial planets are inferred indirectly from physical and chemical constraints to be broadly ultramafic silicate materials, which are the ultimate sources of basaltic volcanism on those bodies. Thus, direct observations of terrestrial mantle materials assume a wider importance, because insights into magma generation processes gained from them may be relevant for all the terrestrial planets.

The freshest terrestrial mantle samples, those least modified during or after their ascent to the surface, are found among ultramafic xenoliths. Even the freshest samples contain ubiquitous evidence of decompression reactions (disequilibrium partial melting and grain boundary metamorphic effects), and, particularly for garnet lherzolites in kimberlite, of late-stage deuteric introduction of serpentine, phlogopite and carbonates, probably from fluids associated with the host magma. Tectonically-emplaced peridotites generally show much more extensive metamorphic and hydrothermal effects than xenoliths, as a result of their much slower ascent rates. For xenoliths ascent times on the order of hours to days have been estimated from rheological constraints (Kushiro *et. al.*, 1976; Mercier, 1979).

The most common rock types found as xenoliths are spinel lherzolite and garnet lherzolite. Both occur predominantly in highly SiO_2-undersaturated host magmas, where their presence is most likely related to the high volatile contents and explosive eruptive styles of these liquids. Spinel lherzolite xenoliths are known from a number of oceanic islands as well as from many continental localities. Among analyzed basalts containing spinel lherzolite xenoliths from eruptive centers within the Tertiary-Quaternary eastern Australian provinces (Wass and Irving, 1976), the most common are basanite (23%), hawaiite (17%), olivine nephelinite (16%), alkali olivine basalt (13%), nepheline hawaiite (5%) and mugearite (5%). Other less common host magmas include olivine analcimite, olivine melilitite, nepheline mugearite, transitional basalt, nepheline benmoreite and phonolite. Lherzolite-bearing phonolites are also known from Nigeria, East Germany and New Zealand (Wright, 1969; Irving and Price, 1981). The global predominance of basanite as a host to spinel lherzolite xenoliths is evident from the literature.

Spinel lherzolite xenoliths and other high-pressure inclusions are virtually absent from tholeiitic basalts, possibly because of the difficulty of transporting such dense fragments in relatively fluid, volatile-poor tholeiitic magmas. However, the few known occurrences [e. g., at Andover, Tasmania (Sutherland, 1974) and in the 1950 Kaapuna flow from Mauna Loa (Finch and MacDonald, 1953)] are important confirmations of the high-pressure derivation of olivine tholeiite magmas. Further evidence for a high-pressure crystallization history is provided by the experimental duplication of aluminous pyroxene megacrysts from Saar Nahe, Germany and Nuanetsi, Rhodesia as near-liquidus phases of their host tholeiites at 6–12 kb (Nicholls and Lorenz, 1973; Cox and Jamieson, 1974).

Garnet peridotite xenoliths (dominantly garnet lherzolite with some garnet harzburgite) typically occur in kimberlite diatremes in stable continental regions (e. g., Nixon, 1973; Sobolev, 1977), but are found also in carbonatite at Lashaine, Tanzania (Dawson *et. al.*, 1970), in both oceanic and continental alkalic basalts (Jackson and Wright, 1970; Skewes and Stern, 1979), and in alnoite from the Solomon Islands (Nixon and Boyd, 1979). Garnet lherzolite and spinel lherzolite xenoliths occur together in some kimberlites and alkalic basalts (e. g., McCallum and Eggler, 1976; Skewes and Stern, 1979). Other peridotite xenoliths containing both garnet and spinel occur in basanitic to nephelinitic lavas or tuff-breccias at Ichinomegata, Japan (Kuno, 1967; Kuno and Aoki, 1970), Mt. Shadwell and Mt. Noorat, Victoria (Irving, 1974), Jugiong, New South Wales (Ferguson *et al.*, 1977), and at several localities in

France (Albert *et al.*, 1967; Berger and Brousse, 1976; Berger, 1977), on Oahu, Hawaii (Jackson, 1966; White, 1966; Jackson and Wright, 1970), and in southern Chile (Skewes and Stern, 1979).

Hydrous minerals, particularly pargasitic amphibole, are present in spinel lherzolite xenoliths from a small number of basalt localities (Varne, 1970; Wilshire and Trask, 1971; Aoki and Shiba, 1973; Frey and Green, 1974; Best, 1974a; Lloyd and Bailey, 1975; Francis, 1976a, b; Wilshire *et al.*, 1980; Kurat *et al.*, 1980; Stosch and Seck, 1980). Primary phlogopite occurs in some garnet peridotite xenoliths from kimberlite (e.g., Carswell, 1975; Delaney *et al.*, 1980), and a distinctive suite of hydrous and carbonate minerals (including tremolite-richterite amphibole, chlorite, titanoclinohumite, titanochondrodite and magnesite) has been recognized in spinel lherzolite xenoliths from kimberlitic diatremes on the Colorado Plateau (Smith, 1979). These occurrences have important implications for the role of volatile components in mantle processes.

1.2.11.2 Present study

A reference suite of 20 peridotite xenolith samples was assembled for detailed petrologic and chemical investigation. Half of the samples (from Kilbourne Hole and San Carlos) were collected specifically for this project and the remainder were donated. All samples are very fresh and most are very large (up to 30 cm in diameter). Two of the major objectives were (1) to obtain precise abundance data for siderophile and volatile trace elements, for which existing data were incomplete, imprecise or sparse; and (2) to augment the small amount of existing systematic data for rare-earth elements (REE) and other lithophile trace elements, with special emphasis on the ranges of abundances observed in samples from single localities (Kilbourne Hole and Matsoku). In addition, bulk major element data have been obtained on all samples, and the constituent phases of selected samples were analyzed for major elements and Sr, Nd and O isotopes. This report contains a brief evaluation of the new data and discussion in the light of recent relevant literature; more complete discussions by individual investigators appear elsewhere.

1.2.11.3 Geologic setting of samples

Samples in the reference suite are listed in Table 1.2.11.1. Fifteen of the 20 samples (including 8 samples from Kilbourne Hole) are spinel lherzolites from Quaternary alkalic basalt eruptive centers. The host basalts are scoriaceous basanites, except at Salt Lake Crater where the host is an olivine nephelinite tuff. These samples are typical of the Cr-diopside xenolith group (Wilshire and Shervais, 1975), which globally is by far the most abundant and widespread xenolith group in basalts. One spinel lherzolite sample from Nunivak Island, Alaska, contains hydrous mineral

Table 1.2.11.1 Samples in the Ultramafic Reference Suite.

Sample Number	Other Sample Number	Sample Type	Host Rock	Locality
UM-1	10050	paragasite spinel lherzolite	basanite	Nunivak Island, Alaska
UM-2	KH77-2	spinel lherzolite	basanite	Kilbourne Hole, New Mexico
UM-3	KH77-4	spinel lherzolite	basanite	Kilbourne Hole, New Mexico
UM-4	KH77-5	spinel lherzolite	basanite	Kilbourne Hole, New Mexico
UM-5	KH77-7	spinel lherzolite	basanite	Kilbourne Hole, New Mexico
UM-6	KH77-8	spinel lherzolite	basanite	Kilbourne Hole, New Mexico
UM-7	KH77-11	spinel lherzolite	basanite	Kilbourne Hole, New Mexico
UM-8	KH77-15	spinel lherzolite	basanite	Kilbourne Hole, New Mexico
UM-9	KH77-16	spinel lherzolite	basanite	Kilbourne Hole, New Mexico
UM-10	SC77-1	spinel lherzolite	basanite	San Carlos, Arizona
UM-11	SC77-2	spinel lherzolite	basanite	San Carlos, Arizona
UM-12	90-47	spinel lherzolite	basanite	Cochise Crater, Arizona
UM-13	P-1	spinel lherzolite	olivine nephelinite tuff	Salt Lake Crater, Hawaii
UM-14	SQ-7	spinel lherzolite	basanite	San Quintin, Mexico
UM-15	Q76-5	spinel lherzolite	basanite	Mount Quincan, Australia
UM-16	LBM 8	garnet lherzolite	kimberlite	Matsoku, Lesotho
UM-17	LBM 9	garnet lherzolite	kimberlite	Matsoku, Lesotho
UM-18	LBM 10	garnet lherzolite	kimberlite	Matsoku, Lesotho
UM-19	LBM 11	garnet lherzolite	kimberlite	Matsoku, Lesotho
UM-20	TZ-2-16	garnet-spinel harzburgite	ankaramite-carbonatite	Lashaine, Tanzania

phases (pargasitic amphibole, biotite and apatite). Most of the spinel lherzolites are from regions of prehistoric continental rifting (Kilbourne Hole, San Carlos, Cochise Crater, Mt. Quincan) or present rifting near continental margins (Nunivak Island, San Quintin). The single oceanic sample is associated with the rejuvenescent phase of volcanic activity at Salt Lake Crater, Hawaii located over a presumed melting anomaly beneath the Pacific Plate.

Four samples of garnet lherzolite from the Cretaceous Matsoku kimberlite pipe, Lesotho, were obtained for trace element analysis. The petrography and major element composition of these samples, which are unusually fresh among nodules in kimberlite, were previously discussed by Cox *et al.* (1973), Harte *et al.* (1975) and Gurney *et al.* (1975). One small sample of garnet-spinel harzburgite from the Quaternary Lashaine ankaramite-carbonatite tuff-ring, Tanzania was also studied.

1.2.11.4 Petrographic summary

Discussions of the textures and fabrics of ultramafic xenoliths, including samples from some of the present localities, have been given by Best (1974b), Mercier and Nicolas (1975), Pike and Schwarzman (1977), Basu (1977a), Francis (1978), Bouillier and Nicolas (1973, 1975) and Harte (1977). The textures of most of the present samples are coarse, allotriomorphic-granular to porphyroclastic (see Fig. 1.2.11.1 a, b, f), with average grain sizes ranging from 1–3mm. Five samples are texturally distinctive. Sample UM-1 from Nunivak Island is medium-grained and shows a preferred orientation of tabular grains (Fig. 1.2.11.1e). Samples UM-5, UM-7 and UM-8 from Kilbourne Hole are fine-grained (average grain size 0.5 mm) and also show a prominent preferred orientation of tabular grains (Fig. 1.2.11.1 c, d). Irving (1980) concluded that the two textural types of spinel lherzolite at Kilbourne Hole are genetically related, with the finer, tabular samples representing pieces of texturally and chemically modified wall rock (originally coarse and granular) adjacent to pyroxenite dikes, as found in composite xenoliths from that locality. The texture of UM-14 from San Quintin is porphyroclastic gneissose, the most common of the four textural types described by Basu (1977a), and has been ascribed to plastic flow during mantle diapiric upwelling associated with the generation of magmas like the host basanite.

The mineral modes of all the spinel lherzolites are generally similar. These rocks consist of 60–70% Fo-rich olivine, 10–30% aluminous enstatite, 8–12% aluminous chrome diopside, 3–13% Cr-Al spinel, and minor amounts of sulfides and glass ($\pm$ quench crystals). Trapped fluid phases and graphite occur in very minor amounts in some samples, especially those from Kilbourne Hole. The Nunivak Island sample (UM-1) contains 5% pargasitic amphibole, 1% biotite and 0.3% apatite, which tend to be associated with each other, but otherwise the rock is modally similar to the majority of the anhydrous spinel lherzolites; hydrous minerals in other xenoliths from this locality have been described by Francis (1976a, b). The samples from Kilbourne ~~Hole~~ and Cochise Crater are richest in sulfides (mainly pyrrhotite with minor chalcopyrite, pentlandite and other unidentified species); these make up as much as 0.5% modally (based on bulk rock sulfur analyses) and occur as composite spherical grains within silicates, as larger grains associated commonly with spinel, or along grain boundaries between other phases. The fine tabular samples from Kilbourne Hole are richer in spinel and sulfides than the associated coarse granular samples. MacRae (1979) has proposed that the interstitial sulfides in Victorian lherzolites are not remobilized primary mantle sulfides but represent precipitates from sulfur-bearing fluids associated with host magmas. Irving (1979, 1980) suggested that a similar process has affected the Kilbourne Hole lherzolites described here, but that it was operative at mantle pressures.

Every sample contains evidence of minor partial melting (probably caused by decompression during ascent) in the form of cuspate features ("fritting") on grain boundaries, thin glass veinlets within grains, vermicular ("spongy") zones of glass near grain margins, or fine-grained interstitial patches composed mainly of olivine, pyroxene, Fe-Ti oxides, vesicles and rare glass. Spinel and chrome diopside are most affected by this process. Bubbles of a fluid phase, which is probably CO_2-rich by analogy with observations on other spinel lherzolites by Roedder (1965), Green and Radcliffe (1975) and Murck *et al.* (1978), occur as a very minor component in glass inclusions within all phases of most samples, but particularly within olivine. Elemental carbon (believed to consist of a mixture of graphite, amorphous carbon, and possibly organic matter) has been confirmed in a Kilbourne Hole spinel lherzolite xenolith by Mathez and Delaney (1981). Optically identical material is present within cracks in grains and along grain boundaries in the samples from Kilbourne Hole and Cochise Crater studied here. Small amounts of fine-grained interstitial carbonate are present in the sample from Salt Lake Crater, but isotopic data suggest a deuteric origin (O'Neil *et al.*, 1970).

The Matsoku garnet lherzolites, UM-16, UM-17, UM-18, UM-19 (equivalent to LBM 8, LBM 9, LBM 10

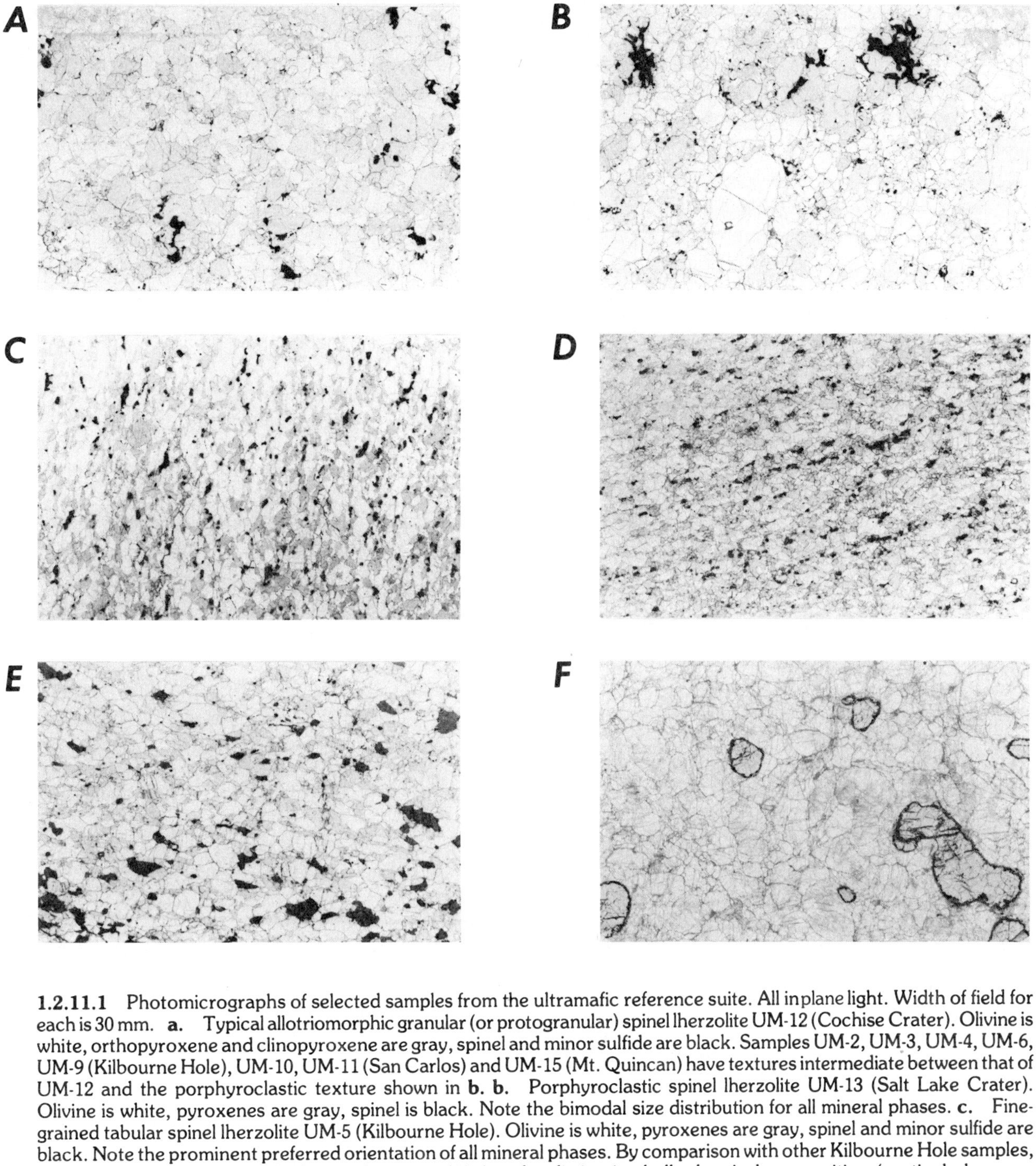

1.2.11.1 Photomicrographs of selected samples from the ultramafic reference suite. All inplane light. Width of field for each is 30 mm. **a.** Typical allotriomorphic granular (or protogranular) spinel lherzolite UM-12 (Cochise Crater). Olivine is white, orthopyroxene and clinopyroxene are gray, spinel and minor sulfide are black. Samples UM-2, UM-3, UM-4, UM-6, UM-9 (Kilbourne Hole), UM-10, UM-11 (San Carlos) and UM-15 (Mt. Quincan) have textures intermediate between that of UM-12 and the porphyroclastic texture shown in **b. b.** Porphyroclastic spinel lherzolite UM-13 (Salt Lake Crater). Olivine is white, pyroxenes are gray, spinel is black. Note the bimodal size distribution for all mineral phases. **c.** Fine-grained tabular spinel lherzolite UM-5 (Kilbourne Hole). Olivine is white, pyroxenes are gray, spinel and minor sulfide are black. Note the prominent preferred orientation of all mineral phases. By comparison with other Kilbourne Hole samples, this sample has a relatively high spinel content (12%) and a distinctive bulk chemical composition (particularly, more Fe-rich). **d.** Fine-grained tabular spinel lherzolite UM-7 (Kilbourne Hole). Olivine is light gray, pyroxenes are darker gray, spinel and minor sulfides are black. Sample UM-8 is texturally very similar to this sample. **e.** Medium-grained tabular amphibole-spinel lherzolite UM-1 (Nunivak Island). Olivine is white, pyroxenes are light gray, amphibole is dark gray, spinel is black. Minor amounts of biotite and apatite are also present, but are not clearly visible. Note the preferred orientation of all mineral phases, including amphibole. **f.** Typical granular garnet lherzolite UM-19 (Matsoku). Large garnet grains (darker gray, with black kelyphitic alteration) are irregularly distributed with respect to olivine and pyroxenes (lighter gray, and not clearly distinguishable from each other). Olivine and orthopyroxene have very minor serpentine alteration along fractures and grain boundaries.

and LBM 11 of Cox *et al.*, 1973; Mathias and Rickwood, 1969), consist of 48–67% Fo-rich olivine, 28–47% enstatite, trace to 2.7% chrome diopside, 2.2–5% pyrope garnet, and very minor (up to 1.3%) phlogopite. Minor serpentine occurs around olivine and enstatite (and within cracks in these phases). Garnets have thin (up to 0.1 mm wide) kelyphitic borders which grade outwards into zones containing phlogopite and rare perovskite. However, the near-pristine nature of these samples is remarkable.

Sample UM-20 from Lashaine is a small, inhomogeneous sample which possibly is not modally representative. It resembles some of the larger xenoliths from this locality (e.g., BD 797) described by Reid *et al.*

Table 1.2.11.2 Mineral chemistry of selected samples in the reference suite.

	UM-1[a]			UM-5[b]			
	Opx	Cpx	Amph	Ol	Opx	Cpx	Sp
SiO_2	56.1	53.9	43.6	38.7	53.7	50.8	0.08
TiO_2	0.05	0.17	0.95	—	0.19	0.78	0.31
Al_2O_3	3.10	3.79	14.2	0.03	3.9	6.0	53.5
Cr_2O_3	0.48	0.86	1.91	0.02	0.28	0.58	10.0
Σ FeO	6.1	2.92	3.90	14.4	8.9	4.5	16.3
MnO	0.12	0.08	0.06	0.22	0.20	0.07	0.12
MgO	34.0	16.6	17.8	46.4	31.4	15.2	13.6
NiO	—	—	—	0.13	0.02	0.02	0.30
CaO	0.67	20.3	10.9	0.07	0.73	21.2	—
Na_2O	0.13	1.51	3.0	—	0.07	0.89	—
K_2O	—	—	1.63	—	—	—	—
SUM	100.75	100.12	97.95	99.97	99.39	100.06	99.21

	UM-6[b]				UM-8[b]			
	Ol	Opx	Cpx	Sp	Ol	Opx	Cpx	Sp
SiO_2	40.1	54.3	51.9	0.06	39.9	55.0	51.0	0.06
TiO_2	—	0.17	0.50	0.17	—	0.13	0.79	0.15
Al_2O_3	0.05	4.5	7.5	58.1	0.05	4.1	6.8	58.5
Cr_2O_3	0.04	0.18	0.70	9.1	0.05	0.32	0.55	7.1
Σ FeO	10.4	6.5	3.0	11.1	11.3	7.4	3.5	12.3
MnO	0.21	0.09	0.05	0.04	0.17	0.14	0.09	0.16
MgO	48.6	32.7	15.6	20.9	48.5	32.1	15.5	20.6
NiO	0.34	0.16	0.04	0.49	0.13	0.07	0.05	0.42
CaO	0.07	0.79	19.3	—	0.05	0.69	20.5	—
Na_2O	—	0.10	1.7	—	—	0.04	1.2	—
SUM	99.81	99.49	100.29	99.96	100.24	99.99	99.98	99.29

	UM-9[b]				UM-17[c]			
	Ol	Opx	Cpx	Sp	Ol	Opx	Cpx	Ga
SiO_2	39.9	54.4	51.6	0.05	40.6	57.6	55.9	41.8
TiO_2	—	0.16	0.57	0.21	0.02	0.03	0.07	0.08
Al_2O_3	0.07	4.4	7.2	57.5	—	0.75	2.23	21.6
Cr_2O_3	0.04	0.21	0.55	9.2	0.03	0.28	1.72	4.0
Σ FeO	10.5	6.8	3.1	11.2	7.5	4.6	2.55	7.4
MnO	0.17	0.21	0.06	0.09	0.08	0.07	0.07	0.36
MgO	48.6	32.6	15.8	20.6	50.3	35.0	17.1	20.8
NiO	0.39	0.10	0.07	0.40	—	—	—	—
CaO	0.07	0.88	19.3	—	0.04	0.56	19.7	5.1
Na_2O	—	0.11	1.5	—	—	0.11	2.1	—
SUM	99.74	99.87	99.75	99.25	98.57	99.00	101.44	101.14

Table 1.2.11.2 *(Continued)*

	UM-19[c]				UM-20[d]			
	Ol	Opx	Cpx	Ga	Ol	Opx	Ga	Sp
SiO_2	40.3	57.5	55.7	41.9	41.0	56.9	42.9	—
TiO_2	0.02	0.03	0.09	0.13	—	0.05	0.08	—
Al_2O_3	—	0.77	2.48	19.7	—	1.5	21.5	15.4
Cr_2O_3	0.03	0.39	2.60	5.8	—	0.7	4.3	54.2
Σ FeO	7.5	4.6	2.43	7.3	6.6	3.8	5.5	14.5
MnO	0.07	0.07	0.09	0.35	0.1	0.1	0.3	0.3
MgO	50.3	34.9	16.3	20.5	51.4	36.7	25.5	15.3
NiO	—	—	—	—	—	—	—	—
CaO	0.08	0.48	18.7	5.2	0.07	0.2	1.5	—
Na_2O	—	0.14	2.5	—	—	0.2	0.03	—
SUM	98.3	98.88	100.89	100.88	99.17	100.15	101.61	99.7

All analyses by electron microprobe.
[a]Analyst D. M. Francis [b]Analyst G. Y. Bussod [c]From Cox, *et al.* (1973) [d]From Pike, *et al.* (1980)

(1975) and by Dawson *et al.* (1970). This garnet-spinel harzburgite (TZ-2-16) is one of a suite of samples described by Pike *et al.* (1980), and consists of 64% olivine, 21% enstatite, 5% pyrope garnet, 4% chromite, a trace of chrome diopside, and 6% secondary phases ("clay") associated with olivine and interstitial glassy patches.

1.2.11.5 Mineral chemistry and P-T estimates

Table 1.2.11.2 contains both new and previously published mineral chemical data for selected samples in the reference suite. Detailed discussions of geothermometry and geobarometry are beyond the scope of this section; however, some general inferences can be made from these data and from other published studies.

The recent applications of pyroxene thermometry to spinel lherzolite xenoliths from localities in France by Brown *et al.* (1980) have led to the conclusion that the highest temperature recorded is approximately 1100° C, with evidence for reequilibration down to at least 800° C. Equilibration temperatures estimated by two different methods for four Kilbourne Hole spinel lherzolites from the study suite range from 918–1080° C (Table 1.2.11.3). The two-pyroxene thermometer yields lower apparent temperatures for the two fine, tabular samples (UM-5, UM-8) than for the two coarse, granular samples (UM-6, UM-9), whereas the olivine-spinel thermometer gives the converse. The two thermometers agree within the estimated 50° C uncertainty of these methods only for the coarse, granular samples. The fine, tabular samples may have experienced a reheating event (possibly related to the formation of pyroxenite dikes), which is not recorded by the pyroxenes. However, much has still to be learned about element diffusion rates during thermal and rheologic adjustment in these mantle samples, and about the effects of bulk composition on calibration of the various geothermometers. The Nunivak Island amphibole-spinel lherzolite has an equilibration temperature of 988° C. Similar temperatures have been calculated from data for other samples from the same localities studied here (Kuno, 1969; Carter, 1970; Jackson and Wright, 1970; Basu, 1975, 1977b; Arculus *et al.*, 1977; Frey and Prinz, 1978; Reid and Woods, 1978; Evans and Nash, 1979; G. Y. Bussod, pers. comm., 1980). The majority of spinel lherzolite xenoliths in basalts show apparent equilibration temperatures in the range 900–1100° C.

Pressures of equilibration are more difficult to specify for spinel lherzolites. Limiting pressures are

Table 1.2.11.3 Estimated equilibration temperatures for selected xenoliths, Nunivak Island, Kilbourne Hole and Matsoku.

		UM-1	UM-5	UM-G	UM-8	UM-9	UM-17	UM-19
Temperature (°C)	A	988	918	1040	969	1007	984	980
	B	—	1080	1008	1081	997	—	—

Temperatures estimated by two methods:
A. diopside-enstatite miscibility gap (Wells, 1977).
B. Fe/Mg exchange between olivine and spinel (Fabriès, 1979).

provided by the experimental determinations of plagioclase and garnet stability in these bulk compositions (approximately 5–20 kb at 1100° C—see Chapter 3). However, most workers prefer values in the upper part of this range (see, for example, Brown *et al.* 1980).

Two of the garnet lherzolite xenoliths from Matsoku give equilibration temperatures of about 980° C (Table 1.2.11.3). Cox *et al.* (1973) suggested that these xenoliths last equilibrated at pressures of at least 30 kb. Reid *et al.* (1975) found that pyroxenes and garnets in the Lashaine peridotites are very similar to those in the Matsoku samples, and suggested equilibration conditions of 1050° C, 50 kb. Such high pressures are not inconsistent with the presence of a spinel phase in some of the Lashaine samples, because the spinel is a relatively Al-poor chromite, quite different from the

Table 1.2.11.4 Bulk compositions of samples in the reference suite: major element data in weight percent (X-ray fluorescence analyses by A. J. Irving).

	Nunivak Island	Kilbourne Hole								San Carlos
	UM-1	UM-2	UM-3	UM-4	UM-5	UM-6	UM-7	UM-8	UM-9	UM-10
SiO_2	45.06	44.20	44.45	44.70	42.16	46.12	44.14	44.82	45.21	43.43
TiO_2	0.08	0.10	0.13	0.11	0.15	0.14	0.11	0.12	0.11	0.05
Al_2O_3	2.16	3.36	2.86	3.57	3.07	4.07	3.20	3.37	3.19	1.37
Cr_2O_3[a]	0.48	0.46	0.42	0.45	0.47	0.48	0.49	0.49	0.43	0.54
Σ FeO	8.31	8.10	8.42	8.07	11.39	7.81	8.77	8.85	8.36	8.05
MnO	0.11	0.11	0.13	0.12	0.17	0.12	0.14	0.13	0.13	0.12
NiO[a]	0.30	0.28	0.28	0.26	0.23	0.25	0.27	0.25	0.26	0.30
MgO	41.52	39.80	40.85	38.89	38.70	37.67	38.89	39.19	39.91	44.14
CaO	2.02	2.91	2.75	3.04	3.96	3.45	3.12	3.19	2.70	1.48
Na_2O[a]	0.241	0.273	0.262	0.360	0.207	0.351	0.198	0.194	0.298	0.136
K_2O	0.14	0.03	0.01	0.05	0.02	0.03	0.03	0.03	0.03	0.02
P_2O_5	0.11	0.00	0.00	0.00	0.00	0.01	0.02	0.00	0.02	0.02
H_2O^+[b]	0.34	—	—	—	—	—	—	—	—	—
CO_2[c]	0.05	0.06	0.04	0.06	0.04	0.03	0.03	0.02	0.02	0.02
SUM	100.92	99.68	100.60	99.68	100.57	100.53	99.41	100.65	100.67	99.68
$\frac{100\ Mg}{Mg+\Sigma Fe}$	89.9	89.7	89.6	89.6	85.8	89.6	88.8	88.8	89.5	90.7

	San Carlos	Cochise Crater	Salt Lake Crater	San Quintin	Mt. Quincan	Matsoku				Lashaine
	UM-11	UM-12	UM-13	UM-14	UM-15	UM-16	UM-17[d]	UM-18[d]	UM-19[d]	UM-20
SiO_2	43.28	45.66	44.07	45.26	44.95	47.50	47.49	45.06	47.68	43.59
TiO_2	0.04	0.12	0.04	0.06	0.09	0.03	0.03	0.05	0.03	0.03
Al_2O_3	1.54	3.35	2.23	2.86	2.89	1.52	2.45	1.25	1.41	1.27
Cr_2O_3[a]	0.54	0.50	0.51	0.43	0.46	0.54	0.60	0.29	0.53	0.43
Σ FeO	8.38	8.10	7.60	7.90	8.21	5.63	6.14	6.57	6.17	5.71
MnO	0.13	0.12	0.11	0.12	0.13	0.09	0.12	0.10	0.12	0.07
NiO[a]	0.31	0.27	0.26	0.27	0.27	0.26	0.21	0.30	0.25	0.31
MgO	44.01	39.29	41.79	41.30	40.66	42.10	38.84	42.59	41.12	48.39
CaO	1.07	2.78	2.10	2.16	2.68	0.73	1.36	0.82	0.75	0.21
Na_2O[a]	0.118	0.199	0.295	0.207	0.221	0.123	0.165	0.161	0.279	0.059
K_2O	0.02	0.02	0.02	0.01	0.03	0.08	0.16	0.14	0.10	0.07
P_2O_5	0.00	0.01	0.00	0.01	0.01	0.02	0.03	0.03	0.03	0.00
H_2O^+[b]	—	—	0.63	—	—	1.81	1.65	1.82	0.99	—
CO_2[c]	0.03	0.01	0.18	0.02	0.02	0.08	0.08	0.11	0.09	0.07
SUM	99.47	100.43	99.84	100.61	100.62	100.51	99.63	99.29	99.55	100.21
$\frac{100\ Mg}{Mg+\Sigma Fe}$	90.3	89.6	90.7	90.3	89.8	93.0	91.9	92.0	92.2	93.8

Precision of new analyses 2%, except SiO_2, Al_2O_3 (3%) and NiO (4%). [a] Cr_2O_3, NiO and Na_2O by instrumental neutron activation analysis. [b] H_2O^+ by gravimetric analysis (analyst K. V. Rodgers). [c] CO_2 by Luft cell infrared spectroscopy (analyst E. K. Gibson, Jr.). [d] From Cox *et al.* (1973), with new values for Cr_2O_3, NiO, Na_2O and CO_2.

spinels typical of spinel lherzolite xenoliths in basalts. The garnet in UM-20 is richer in Mg and poorer in Ca than those in the samples examined by Reid *et al.* (1975), reflecting differences in bulk chemical composition.

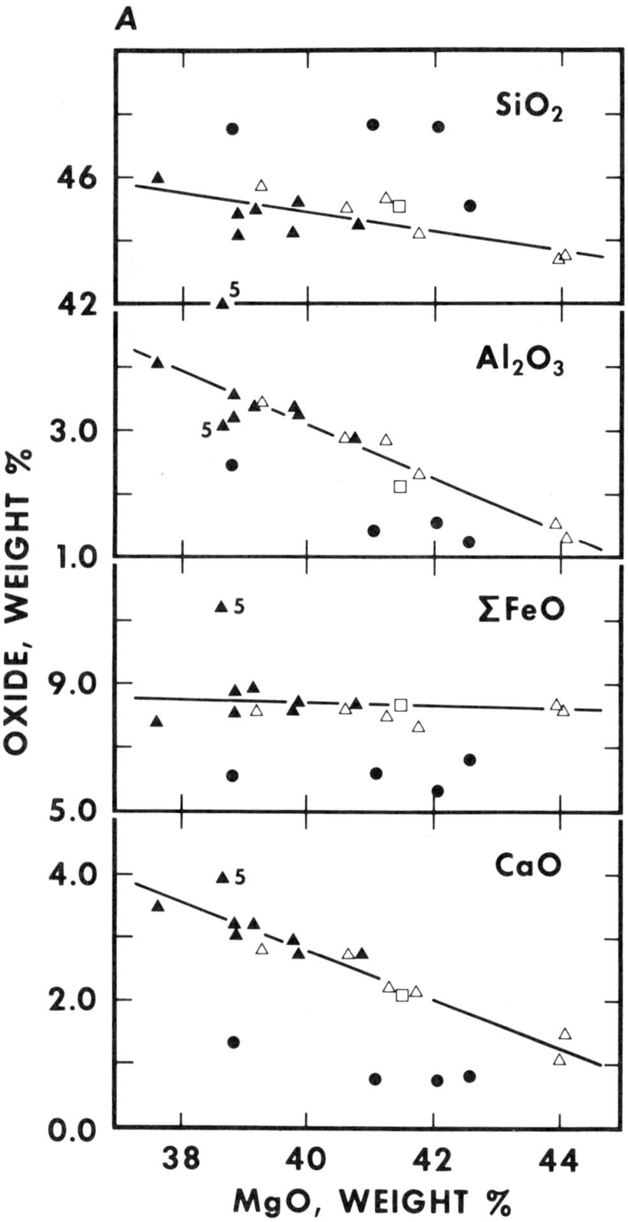

1.2.11.2 Variation diagram for bulk major and minor oxides as a function of bulk MgO content for samples from the reference suite. MnO, K_2O and P_2O_5 data have not been plotted. Data for Lashaine sample UM-20 (48.39% MgO) omitted for clarity. Regression lines are for spinel lherzolite data only, excluding data for Fe-rich sample UM-5 (labelled 5).

1.2.11.6 Major element and lithophile trace element chemistry

Bulk major and trace element data for all samples in the reference suite were obtained by X-ray fluorescence (XRF) and neutron activation (NAA) methods. Major element data for three of the Matsoku samples were given by Cox *et al.* (1973). Neutron activation analyses were made of pieces of their samples for this report.

Special attention was given to methods of sample preparation. Powders for XRF analysis were prepared in a tungsten carbide mill, whereas powders for NAA

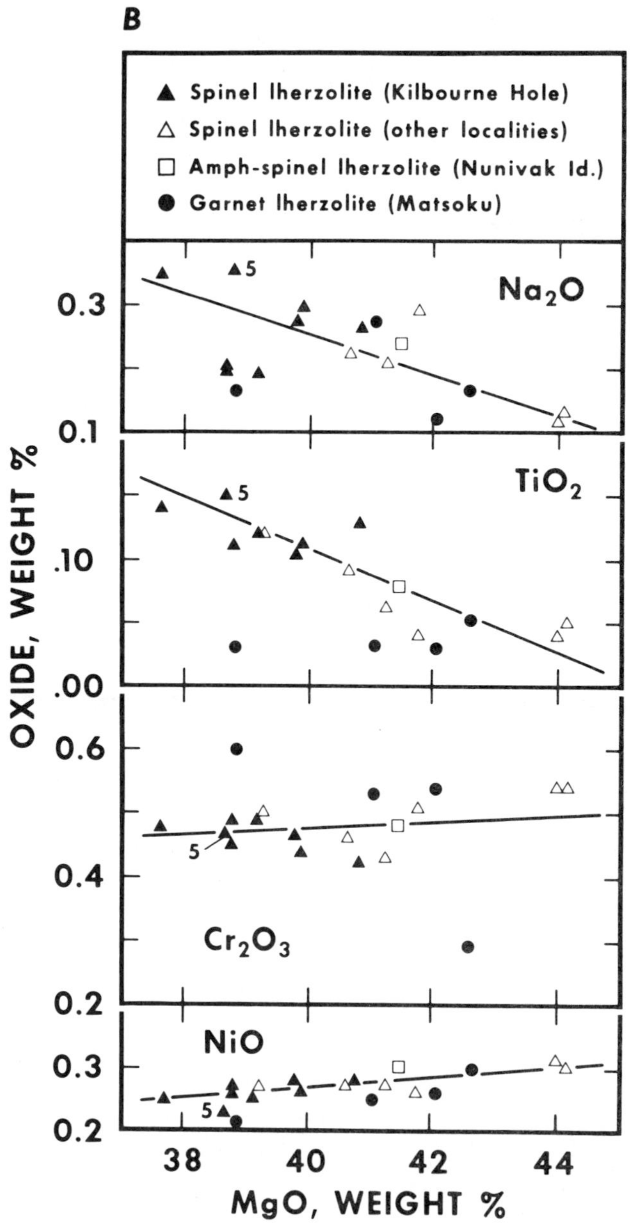

(both instrumental and radiochemical) were prepared in a high-purity alumina mill. Large (~20 cc), representative, intact pieces were sawn from the interior of each xenolith (with the exception of the small sample, UM-20). The friability of the samples probably insured negligible contamination from sawing. Adjacent rock pieces were used for XRF and NAA; the agreement between FeO data obtained by the two methods suggests that the

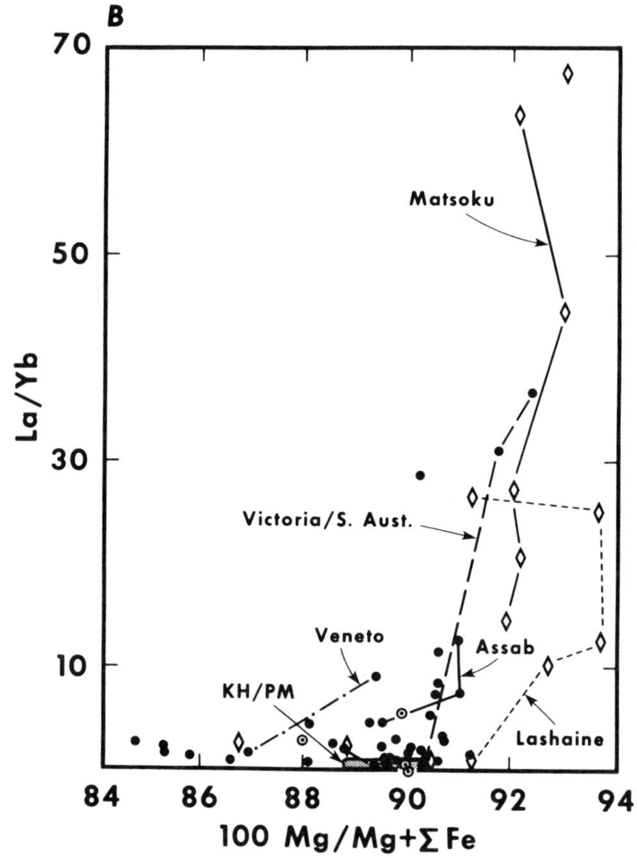

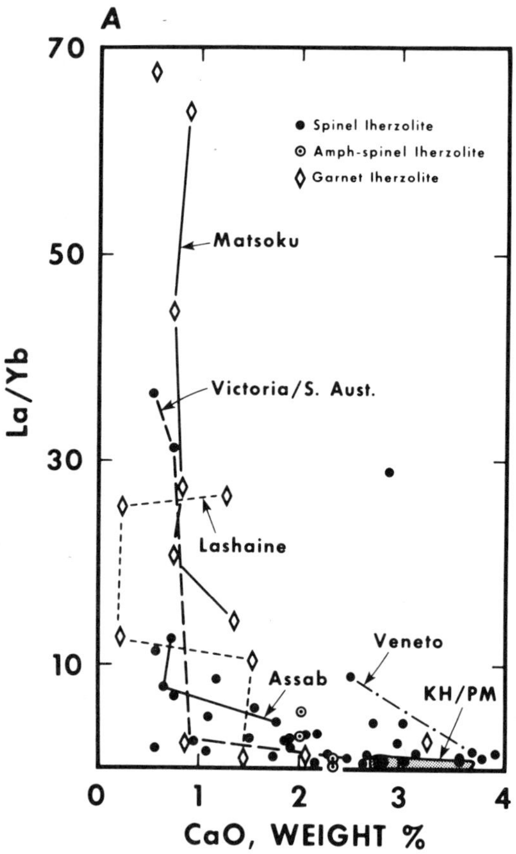

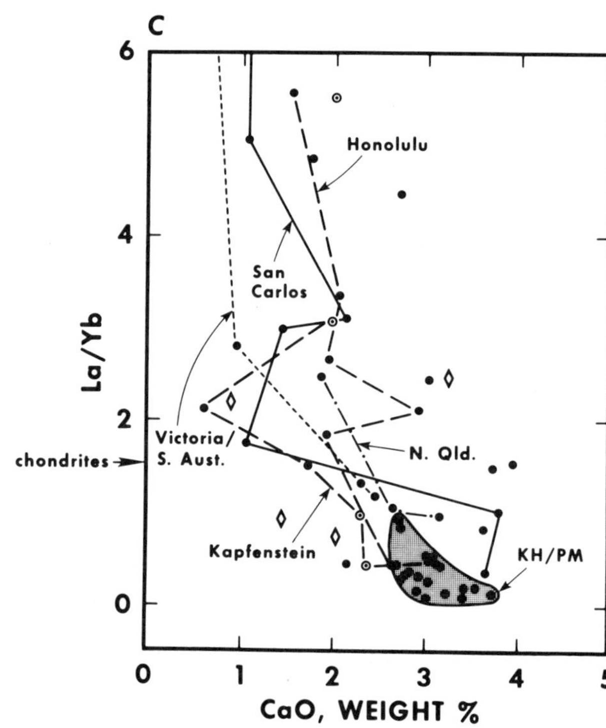

1.2.11.3 a. Plot of bulk La/Yb vs. bulk CaO content for spinel and garnet lherzolite xenoliths. Data from Tables 1.2.11.4, 1.2.11.5, Fig. 1.2.11.13, and as follows: Nagasawa *et al.* (1969), Kuno and Aoki (1970), Morten (1971, 1978), Nixon and Boyd (1973a, b), Frey and Green (1974), Ridley and Dawson (1975), Rhodes and Dawson (1975), Frey and Prinz (1978), Ottonello *et al.* (1978a, b), Jagoutz *et al.* (1979), Irving (1980, and unpubl. data), Kurat *et al.* (1980), F. A. Frey (unpubl. data). Lines connect data points for samples from single provinces or single localities. KH/PM = Kilbourne Hole (16 samples) + Potrillo Maar (1 sample). **b.** Plot of bulk La/Yb vs. bulk Mg/(Mg + Σ Fe) for spinel and garnet lherzolites. Data sources as in **a.c.** Expanded plot of the lower part of **a** to emphasize the inverse correlation of La/Yb with CaO for spinel lherzolite xenoliths.

samples are compositionally homogeneous on this scale.

Bulk major element compositions are reported in Table 1.2.11.4. The variations of major oxide contents as functions of MgO are shown in Fig. 1.2.11.2. The patterns of compositional variation for the spinel lherzolite samples taken together are very similar to patterns for other spinel lherzolite xenoliths compiled by Maaløe and Aoki (1977). Even at a single locality (Kilbourne Hole), the same variation patterns are evident in 7 of 8 samples analyzed (see Fig. 1.2.11.2); the exception is sample UM-5, which is notably Fe-rich and which also is texturally distinctive.

Data for the Matsoku garnet lherzolites are also plotted in Fig. 1.2.11.2; however, the compositional variations of garnet lherzolite xenoliths are better represented in the larger compilation of Maaløe and Aoki (1977). The patterns of variation are subparallel to, but displaced from, those observed in spinel lherzolites. Maaløe and Aoki suggest that the difference in

Table 1.2.11.5 Bulk compositions of samples in the reference suite: trace element data in ppm (instrumental neutron activation analyses by A. J. Irving).

	Nunivak Island	Kilbourne Hole								San Carlos
	UM-1	UM-2	UM-3	UM-4	UM-5	UM-6	UM-7	UM-8	UM-9	UM-10
La	0.77	0.045	0.095	0.19	0.52	0.070	0.159	0.134	0.157	0.25
Ce	3.5	0.27	—	—	—	—	—	0.51	—	0.57
Nd	3.4	—	—	—	—	—	—	—	—	—
Sm	0.75	0.193	0.245	0.225	0.479	0.260	0.182	0.220	0.240	0.120
Eu	0.226	0.083	0.110	0.094	0.175	0.108	0.080	0.097	0.103	0.046
Yb	0.14	0.31	0.33	0.36	0.34	0.40	0.31	0.31	0.36	0.084
Lu	0.021	0.054	0.058	0.061	0.052	0.064	0.056	0.050	0.060	0.014
Hf	0.66	0.12	0.16	0.13	0.20	0.12	—	0.17	0.17	—
Th	0.25	0.24	0.30	0.33	0.40	0.12	0.21	0.16	0.14	0.14
Sc	11.3	13.8	13.2	14.5	14.7	16.0	14.6	13.6	15.0	12.1
Co	115	107	111	106	127	101	109	110	103	116
Cr	3270	3130	2850	3060	3240	3280	3320	2910	2870	3690
Ni	2390	2210	2210	2060	1790	1980	2160	2000	2040	2330

	San Carlos	Cochise Crater	Salt Lake Crater	San Quintin	Mt. Quincan	Matsoku				Lashaine
	UM-11	UM-12	UM-13	UM-14	UM-15	UM-16	UM-17	UM-18	UM-19	UM-20
La	0.147	0.087	0.73	0.123	0.35	2.93	2.05	1.52	1.97	0.59
Ce	—	—	1.8	—	—	7.0	5.6	3.5	4.9	0.64
Nd	—	—	—	—	—	—	—	—	—	—
Sm	0.126	0.198	0.235	0.147	0.164	0.429	0.456	0.272	0.363	0.075
Eu	0.043	0.091	0.084	0.063	0.072	0.125	0.145	0.088	0.115	0.024
Yb	0.083	0.28	0.22	0.26	0.32	—	—	—	—	0.047
Lu	0.016	0.047	0.040	0.048	0.059	0.009	0.021	0.008	0.010	0.008
Hf	—	0.14	—	—	—	0.20	0.24	0.22	0.16	0.13
Th	0.15	0.13	—	—	0.14	0.41	0.30	0.31	0.34	0.17
Sc	9.4	13.2	12.9	13.0	14.1	7.8	11.7	4.56	7.5	6.4
Co	119	111	102	108	107	92	84	106	94	103
Cr	3700	3410	3520	2970	3120	3690	4100	1970	3600	2970
Ni	2440	2130	2050	2130	2090	2060	1680	2380	2000	2450

Analytical precision: Sc, Co, Cr (2%); Sm (2-3%); La (3-5%); Ni (4%); Yb, Lu (4-8%); Eu (6-8%); Hf, Th (20%).

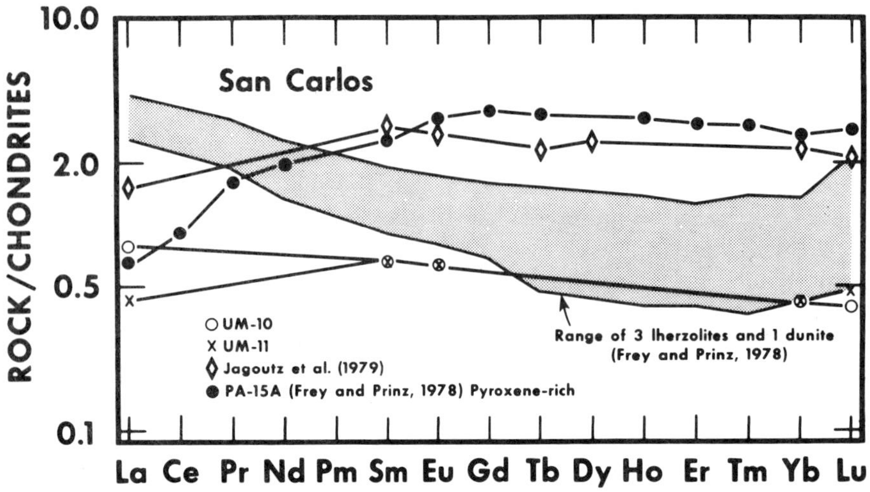

1.2.11.4 Chondrite-normalized rare-earth element data for spinel lherzolite xenoliths from San Carlos, Arizona.

bulk chemical composition between analyzed spinel and garnet lherzolites is the result of a sampling bias towards granular garnet lherzolite xenoliths from southern African kimberlites. They conclude that garnet lherzolite mantle beneath southern Africa is different from that elsewhere. At least some garnet lherzolites are compositionally similar to spinel lherzolite in both major and trace elements (see Fig. 1.2.11.3) and, overall, there is little evidence from the xenolith data to suggest a gross compositional change within the uppermost 100–200 km of the Earth's mantle. Although, undoubtedly, both lateral and vertical compositional heterogeneities exist within the mantle, ultramafic xenoliths may not be relevant for determining these heterogeneities on a small scale, because their compositions inevitably are affected by the very magmatic processes which are responsible for their occurrence at the Earth's surface (see Irving, 1976, for further discussion).

Despite these uncertainties concerning mantle composition deduced from xenoliths, important insights into mantle magmatic processes have been gained from studies of xenoliths, particularly from studies of correlations among lithophile trace elements and major elements. Frey and Green (1974) and Frey and Prinz (1978) found that, for two spinel lherzolite suites from Victoria and San Carlos, the most refractory samples (i.e., those having the highest Mg/(Mg + Fe) ratios, lowest CaO, Al_2O_3 and TiO_2 contents, and lowest modal abundances of clinopyroxene) possess the highest La/Yb ratios, highest K_2O contents and lowest Lu abundances. These correlations, based on small suites of samples, demonstrate that the lherzolites are not simply a sequence of residues after basaltic magma extraction by partial melting. In order to explain such features, these authors invoked a two-stage history involving varying degrees of metasomatic addition of a

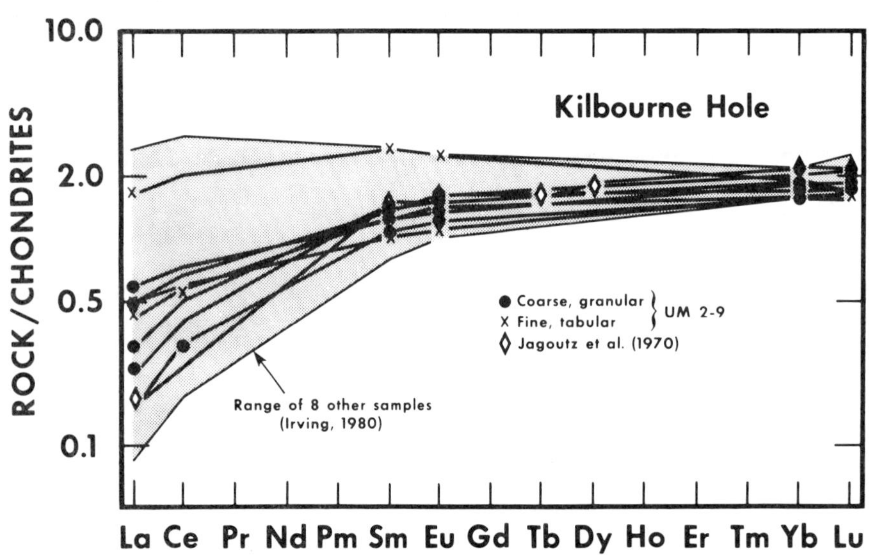

1.2.11.5 Chondrite-normalized rare-earth element data for spinel lherzolite xenoliths from Kilbourne Hole, New Mexico.

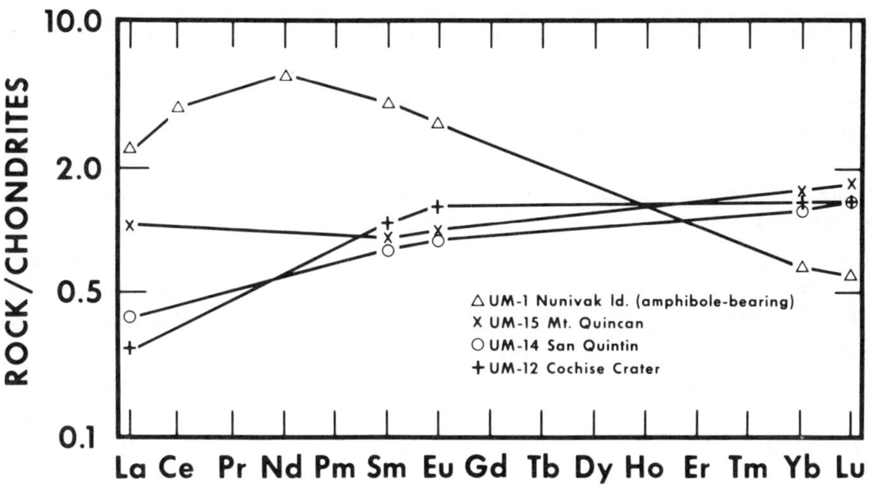

1.2.11.6 Chondrite-normalized rare-earth element data for other spinel lherzolite and amphibole-spinel lherzolite samples from the reference suite.

LIL (large ion lithophile) element-rich component (probably as a fluid) to partial melting residues. Similar compositional correlations are found in a larger data base containing analyses for the suite studied here and for other samples (see Fig. 1.2.11.3), which suggests the general applicability of multi-stage models.

Trace element data obtained by INAA for samples in the reference suite are given in Table 1.2.11.5. Chondrite-normalized REE abundances for these and other samples are shown in a series of diagrams (Figs. 1.2.11.4–1.2.11.13).

The REE data for the two San Carlos spinel lherzolites studied here extend the range of bulk light REE abundances found by Frey and Prinz (1978) to lower values (see Fig. 1.2.11.4). Light REE abundances for 17 Kilbourne Hole spinel lherzolites (including the 8 samples from the study suite) range over a factor of 50, whereas heavy REE abundances show a range of only a factor of 1.5 (see Fig. 1.2.11.5). Almost all of the Kilbourne Hole samples show bulk light REE depletion to varying degrees, as do clinopyroxenes separated from other samples from this locality (Reid and Frey, 1978). Two of the four analyzed fine, tabular xenoliths are enriched in light REE relative to the predominant coarse, granular xenoliths and have relatively flat chondrite-normalized REE distributions at 2 × chondrites; the other two fine, tabular samples have greater than average light REE abundances. For the large suite of Kilbourne Hole samples, the correlation between bulk La/Yb and bulk CaO (or modal clinopyroxene abundance) is imperfect, and can be interpreted in terms of multiple depletion (i.e., partial melting) and enrichment (i.e., metasomatism) events in the upper mantle beneath Kilbourne Hole (Irving, 1979, 1980).

Uranium concentrations in chrome diopsides from 6 of the Kilbourne Hole lherzolites, determined by

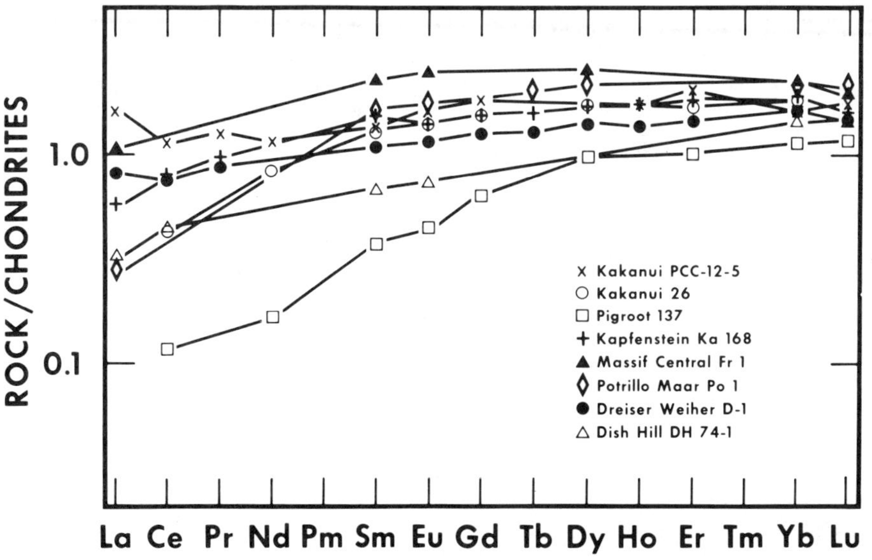

1.2.11.7 Chondrite-normalized rare-earth element data for light-REE depleted spinel lherzolite xenoliths from other localities. Data from Frey *et al.* (1971), Philpotts *et al.* (1972), Jagoutz *et al.* (1979) and A. J. Irving (unpubl. data).

S. Newman and J. D. Macdougall (pers. comm., 1979) using the induces fission track method, range from 0.5–9.7 ppb, and show a good correlation with bulk rock La abundances. Such a correlation is expected because most of the LIL elements in these rocks are contained in clinopyroxene rather than accessory phases (Philpotts *et al.*, 1972; Frey and Green, 1974). Further discussions of uranium distribution in spinel lherzolite xenoliths are provided by Kleeman *et al.* (1969), Fisher (1970), Nishimura (1972), Haines and Zartman (1973), Hamilton (1975) and Dostal and Capedri (1976).

Sample UM-12 from Cochise Crater (230 km west-southwest of Kilbourne Hole; locality 15 of Evans and Nash, 1979) also has a light REE depleted pattern (Fig. 1.2.11.6) similar to those shown by many Kilbourne Hole lherzolites, as does a sample from Potrillo Maar, 25 km south of Kilbourne Hole (Fig. 1.2.11.7; Jagoutz *et al.*, 1979). Light REE depleted patterns are also shown by sample UM-14 from San Quintin (Fig. 1.2.11.6) and by other spinel lherzolite xenoliths from New Zealand and Europe (Fig. 1.2.11.7). Sample UM-15 from Mt. Quincan has an approximately chondritic REE pattern but with some enrichment in heavy REE relative to Sm and Eu. Amphibole-bearing spinel lherzolite UM-1 has a distinctive REE pattern showing overall enrichment in light REE over heavy REE, but depletion in La relative to Nd and Sm (Fig. 1.2.11.6); this pattern is consistent with metasomatic introduction

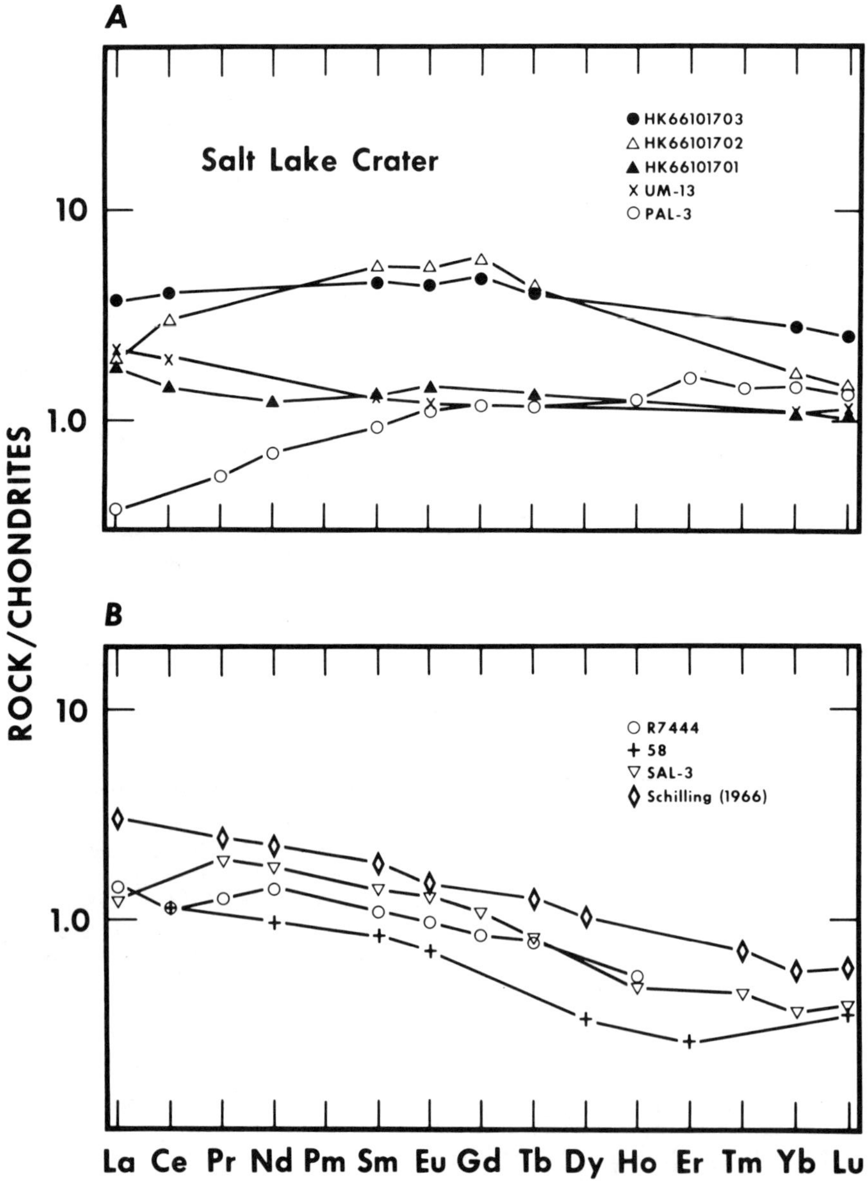

1.2.11.8 Chondrite-normalized rare-earth element data for spinel lherzolite xenoliths from Salt Lake Crater, Hawaii. **a.** Samples with relatively unfractionated light-REE depleted patterns. Note that sample UM-13 shows the greatest degree of light-REE enrichment in this group. **b.** Samples showing enrichment of light REE over heavy REE. Data from Table 1.2.11.5, Schilling (1966), Nagasawa *et al.* (1969), Philpotts *et al.* (1972) and F. A. Frey (unpubl. data).

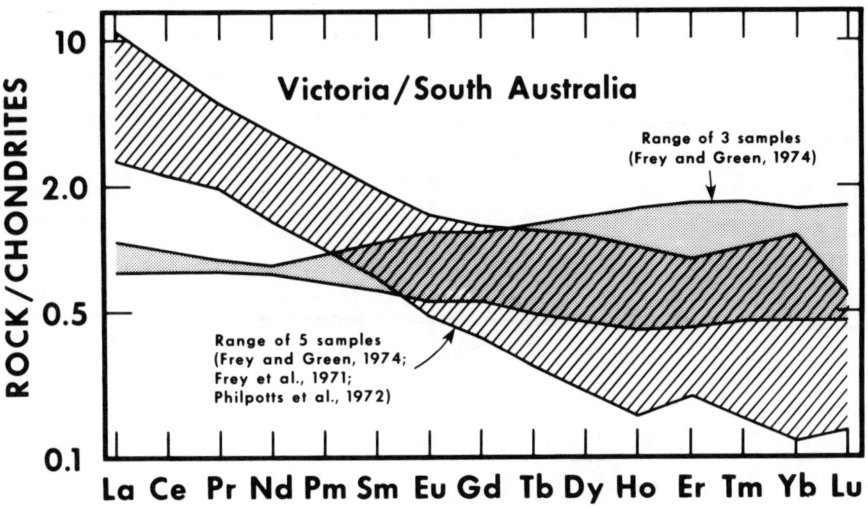

1.2.11.9 Chondrite-normalized rare-earth element data for spinel lherzolite xenoliths from the Newer Basalts province of Victoria and South Australia. Note the two distinct groups of patterns.

of amphibole into an anhydrous spinel lherzolite originally having a much flatter chondrite-normalized REE distribution (see Varne and Graham, 1971).

New REE data for sample UM-13 from Salt Lake Crater are compared with data for other samples from this locality in Fig. 1.2.11.8. There is some suggestion of two groups of samples at Salt Lake Crater—one with unfractionated to depleted light REE and heavy REE abundances at 1–3 × chondrites, and another showing light REE enrichment and heavy REE abundances at 0.3–0.6 × chondrites. Frey (1980) has noted that the latter group is more refractory (MgO >39 wt.%) than the former (MgO <39 wt.%), although sample UM-13 (MgO 41.8 wt.%) is intermediate between these two groups. Nevertheless, these observations suggest that metasomatic processes involving light REE enrichment and heavy REE depletion may also be operative in the oceanic mantle beneath Hawaii (compare data in Fig. 1.2.11.9 for the Newer Basalts province of Victoria and South Australia). A similar situation exists for the light

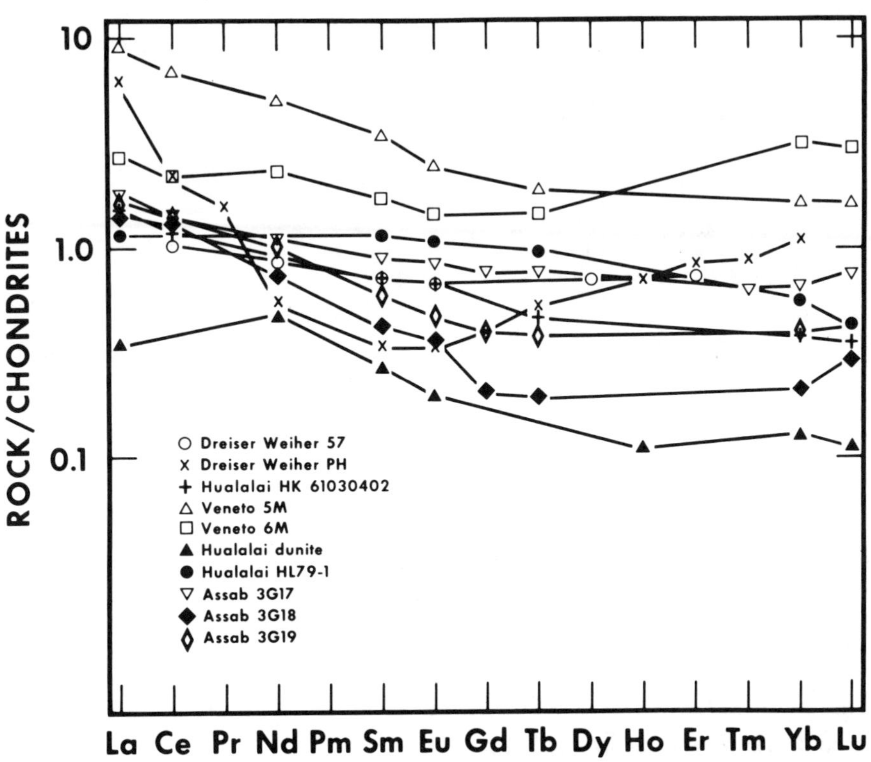

1.2.11.10 Chondrite-normalized rare-earth element data for other spinel lherzolite xenoliths which have light-REE enriched to relatively unfractionated patterns. Data from Schilling (1966), Nagasawa *et al.* (1969), Frey *et al.* (1971), Philpotts *et al.* (1972), Ottonello *et al.* (1978a, b), Morten (1978), and A. J. Irving (unpubl. data).

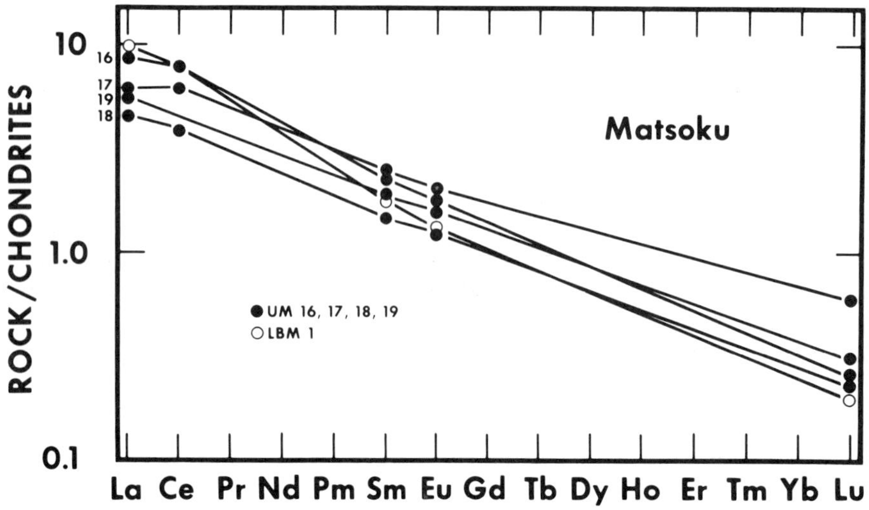

1.2.11.11 Chondrite-normalized rare-earth element data for garnet lherzolite xenoliths from the Matsoku kimberlite pipe. Data from Table 1.2.11.5 and A. J. Irving (unpubl. data)

REE enriched dunite and lherzolite xenoliths from Hualalai, which are compared with data for several light REE enriched European xenoliths in Fig. 1.2.11.10.

Five Matsoku garnet lherzolites have a relatively narrow range of light REE-enriched patterns (Fig. 1.2.11.11). Three samples from Lashaine have light REE-enriched patterns generally similar to these, but two other Lashaine samples (including UM-20) have flatter REE distributions (see Fig. 1.2.11.12). The Matsoku samples, which range in clinopyroxene content from trace to 4%, show some evidence for correlations among major and trace element parameters similar to those observed in spinel lherzolite suites. Ridley and Dawson (1975) found that the most refractory Lashaine sample has the highest La/Yb ratio.

More definitive evidence for mantle metasomatism comes from data for sheared and granular garnet lherzolites from Lesothan and South African kimberlites. Data compiled in Fig. 1.2.11.13 show that the granular samples have very steep REE patterns similar to those of the Matsoku xenoliths, whereas the sheared samples have approximately chondritic patterns (cf. Shimizu, 1975). Because the granular samples are consistently more refractory than the sheared samples (e.g., Nixon and Boyd, 1973a), it appears that metasomatic enrichment processes have affected the granular xenoliths.

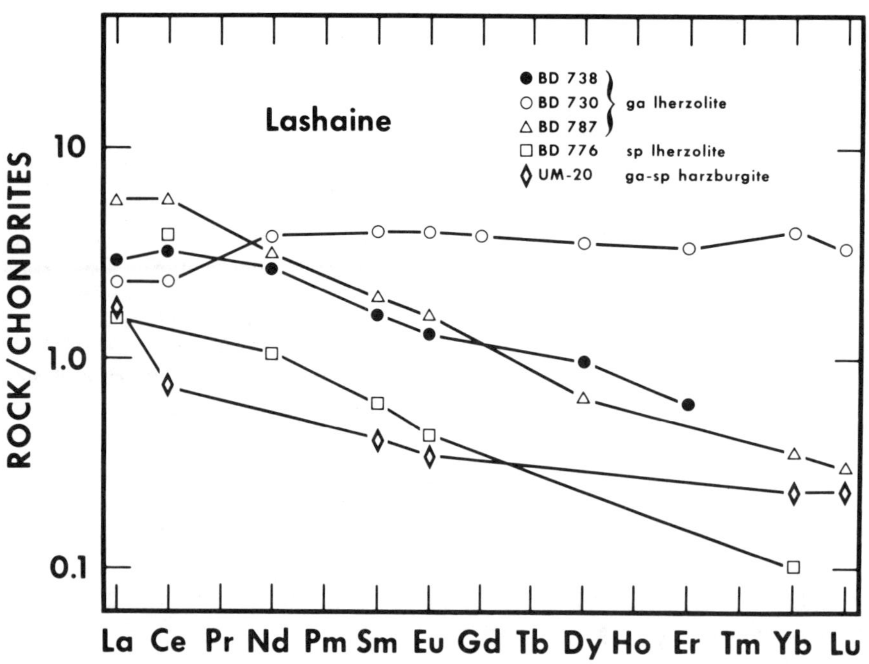

1.2.11.12 Chondrite-normalized rare-earth element data for xenoliths from Lashaine, Tanzania. Data from Table 1.2.11.5 and Ridley and Dawson (1975).

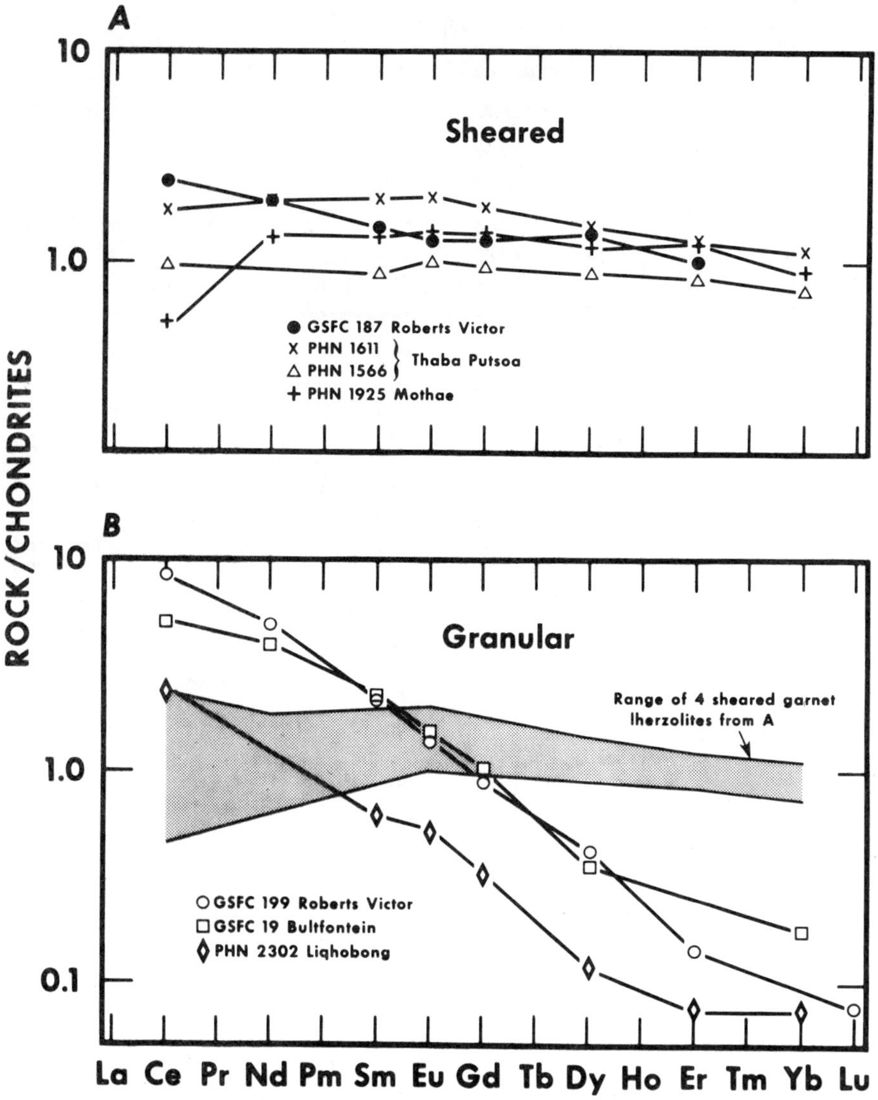

1.2.11.13 Comparison of chondrite-normalized rare-earth element patterns for sheared and granular garnet lherzolite xenoliths from other southern African kimberlites. Data for GSFC samples from Philpotts *et al.* (1972). Patterns for PHN samples calculated from data of Shimizu (1975), using revised modes computed from mineral and bulk major element data (Nixon and Boyd, 1973a, b) and clinopyroxene-orthopyroxene REE partition coefficients for GSFC 187 (Philpotts *et al.*, 1972).

Two such textural groups of garnet peridotite xenoliths occur also in minette at The Thumb and several other localities on the Colorado Plateau (Ehrenberg, 1978, 1979). Although the sheared samples consistently are enriched in Fe and Ti relative to the granular samples, bulk REE distributions calculated from mineral data given by Ehrenberg (1978) are relatively flat at 2 × chondrites for both textural types. Ehrenberg (1979) has suggested that, in contrast to the southern African xenoliths, the sheared xenoliths at The Thumb represent originally granular garnet peridotite mantle which has been deformed during metasomatism by an evolved Fe-Ti-rich liquid, possibly related to that commonly postulated (e.g., Wyatt, 1977) to have produced the "eutectoid" clinopyroxene-ilmenite intergrowths that are characteristic of many kimberlites and that are also present at The Thumb.

1.2.11.7 Siderophile and volatile element chemistry

For many years, the high abundance of Ni and other siderophile elements in the Earth's mantle has been an enigma which has been widely discussed by geochemists (e.g., Duke, 1965; Ringwood, 1966). Explanations have relied upon three mechanisms: (1) disequilibrium during core formation (Ringwood, 1966, 1971; Turekian and Clark, 1969); (2) increase in silicate/metal partition coefficients in response to pressure, temperature, complex formation, or high f_{O_2} (Brett, 1971; Kimura *et al.*, 1974; Ringwood, 1977; and (3) accretion of primordial (= chondritic) material after core formation (Turekian and Clark, 1969; Kimura *et al.*, 1974). The third mechanism may also provide the

Table 1.2.11.6 Bulk compositions of samples in the reference suite.†

Sample	Locality	Siderophile Elements									
		Ir	Os	Pd	Au	Re	Ge	Ni	Sb	Se	S*
UM-1	Nunivak Island	4.8	4.4	4.7	0.39	0.142	930	2260	—	10.7	55
UM-3	Kilbourne Hole	3.8	3.1	6.9	0.53	0.107	920	2210	—	38	107
UM-4	Kilbourne Hole	3.1	1.7	5.9	0.98	0.089	920	2020	—	28	86
UM-5	Kilbourne Hole	2.6	2.2	3.2	0.44	0.161	730	1780	1.6	51	152
UM-6	Kilbourne Hole	3.2	3.0	6.4	0.64	0.180	980	1910	—	47	134
UM-7	Kilbourne Hole	6.3	4.0	6.5	3.03	0.264	810	1950	4.9	61	210
UM-8	Kilbourne Hole	3.8	3.8	4.6	0.50	0.061	890	1970	—	25	93
UM-9	Kilbourne Hole	3.4	3.7	6.1	0.78	0.230	770	1890	1.3	61	200
UM-10	San Carlos	1.69	1.48	3.1	0.092	0.003	880	2270	—	1.3	50
UM-11	San Carlos	3.8	6.1	1.7	0.084	0.011	1100	2290	3.2	1.3	55
UM-12	Cochise Crater	3.9	4.9	5.3	0.96	0.014	730	2050	0.7	63	198
UM-13	Salt Lake Crater	4.4	2.9	6.9	0.60	0.071	920	2100	—	15.2	64
UM-14	San Quintin	3.5	3.1	1.30	0.158	0.055	840	2090	—	17.0	55
UM-15	Mt. Quincan	2.9	2.3	2.3	0.47	0.170	780	2120	—	7.1	52
UM-16	Matsoku	5.7	4.5	≤0.3	0.49	0.022	550	2020	9.1	8.2	68
UM-17	Matsoku	1.8	1.6	2.8	0.47	0.194	900	1640	—	7.6	85
UM-19	Matsoku	11.8	10.8	10.4	1.69	0.137	730	2070	—	9.8	72
UM-19	Matsoku	8.6	7.7	6.8	2.10	0.133	620	2040	2.8	8.4	—
UM-20	Lashaine	0.052	0.059	0.09	0.37	0.009	620	2420	—	2.2	49
D-10	Juan de Fuca Ridge (basalt)	0.023	—	0.77	0.57	1.42	1530	104	18	180	—
BCR-1	Columbia River (basalt)	~0.0006	~0.07	≤0.2	0.34	0.864	—	8.4	570	80	—
BCR-1	Columbia River (basalt)	0.0012	≤0.11	≤0.3	0.65	0.861	1500	18	610	78	—

Sample	Locality	Volatile Elements							
		Ag	Te	Zn	In	Cd	Bi	Tl	C*
UM-1	Nunivak Island	3.2	8	47	10	43	1.6	4.2	194
UM-3	Kilbourne Hole	4.8	11	43	10	—	1.5	1.0	138
UM-4	Kilbourne Hole	18.4	13	30	12	33	1.8	2.2	210
UM-5	Kolbourne Hole	9.6	10	67	15	57	2.0	1.1	139
UM-6	Kilbourne Hole	6.3	10	20	14	—	1.3	0.9	93
UM-7	Kilbourne Hole	8.1	9	60	7	48	2.9	1.3	114
UM-8	Kilbourne Hole	3.9	10	35	13	47	1.2	0.6	87
UM-9	Kilbourne Hole	6.3	18	53	13	61	1.0	0.8	75
UM-10	San Carlos	0.5	≤2	31	7	49	0.9	0.3	68
UM-11	San Carlos	0.5	8	54	5	30	1.0	1.4	122
UM-12	Cochise Crater	6.9	14	62	5	68	2.3	0.7	42
UM-13	Salt Lake Crater	9.6	8	42	9	49	2.2	2.5	670
UM-14	San Quintin	2.9	14	29	11	33	3.0	0.4	82
UM-15	Mt. Quincan	1.7	14	52	12	50	2.0	0.5	69
UM-16	Matsoku	3.6	—	34	1	67	3.4	4.4	292
UM-17	Matsoku	2.7	3	22	3	29	2.4	12	274
UM-19	Matsoku	3.4	5	30	2	—	5.0	22	320
UM-19	Matsoku	10.6	4	38	2	143	11	23	—
UM-20	Lashaine	1.3	≤2	30	2	9	1.0	4.2	269
D-10	Juan de Fuca Ridge (basalt)	104	—	99	100	152	9	50	—
BCR-1	Columbia River (basalt)	24	5	135	75	137	45	300	—
BCR-1	Columbia River (basalt)	23	6	134	95	134	45	280	—

†Siderophile and volatile elements in ppb, except for S, C, Ni and Zn in ppm (radiochemical neutron activation analyses by J. W. Morgan, G. A. Wandless and R. K. Petrie, except as noted.

* S and C analyses by Luft cell infrared spectroscopy (analyst E. K. Gibson, Jr.)

Earth's complement of highly volatile elements (Kokubu *et al.*, 1961; Ganapathy and Anders, 1974).

The distribution of siderophile elements in the Earth's upper mantle has recently become a topic of revived interest in deciphering the origin and evolution of the Earth-Moon system (Ringwood and Kesson, 1977; Chou, 1978). Although the distributions of such moderately-siderophile elements as Ni, Co, and Ge are reasonably well-known, the abundances of the more siderophile elements (e.g., Re, Au, and the Pt metals) are known only sketchily. Recent discussions of these elements have relied upon data from several laboratories for a variety of rock types, including komatiites, alpine peridotites and ultramafic xenoliths. Much of the scatter in the compiled data probably is due to secondary alteration of the samples, but other problems of analytical precision and interlaboratory accuracy have contributed to the confusion. Very few of the freshest ultramafic samples available, namely peridotite xenoliths, have been analyzed.

Morgan and Lovering (1967) determined Re and Os for one Australian spinel lherzolite, and Ehmann *et al.* (1970) measured the Au content of a Bultfontein garnet lherzolite. Paul *et al.* (1979) reported Pd, Ir and Au abundances for a number of garnet lherzolites and one phlogopite wehrlite from Lesotho and South Africa, but their data show very wide ranges in both absolute abundances and element ratios. Possibly the secondary alteration of their samples (evidenced by serpentine, phlogopite and magnetite) is in part responsible. Simultaneously with the present study, Jagoutz *et al.* (1979) analyzed six spinel lherzolites for many elements, including W, Ir, Au, Re, Os, and Pt (their data for the last three elements are mainly maximum values).

In this study, data for 16 siderophile and volatile elements (Ag, Au, Bi, Cd, Ge, In, Ir, Ni, Os, Pd, Re, Sb, Se, Te, Tl, Zn) were obtained on 18 samples by radiochemical neutron activation analysis. Abundances of S and C were obtained on the same alumina-ground powders by infrared spectroscopy. The data are reported in Table 1.2.11.6. Further discussion is given by Morgan *et al.* (1980, 1981).

The unusual freshness of the Matsoku garnet lherzolites may be very important in view of the data dispersion observed by Paul *et al.* (1979). The possibility of differential mobility and fractionation of siderophile elements during secondary alteration (or metasomatism) must be considered. Sulfides are known to be major sites for siderophile elements (e.g., Jagoutz *et al.*, 1979), but petrographic evidence from spinel lherzolite samples in the study suite suggests that they too are readily mobilized during secondary processes. Paul *et al.* (1979) also suggest that differential partitioning of these elements may occur during crystal/liquid fractionation processes. The average siderophile values from the new measurements of ultramafic nodules are

Table 1.2.11.7 Comparison of siderophile and volatile element abundances in ultramafic samples of the reference suite with previous compiled values and other recent analyses of ultramafic rocks (in ppb, except for Ge and Zn in ppm and Ni in percent).

	Ringwood and Kesson (1977)	Chou (1978)	Jagoutz *et al.* (1979)	Mitchell and Keays (1979)	Paul *et al.* (1979)	This Work		
Element	Compiled Values		Spinel lherzolite (6)	Spinel lherzolite (3)	Garnet lherzolite (5)	Spinel lherzolite (14)	Garnet lherzolite (4)	Garnet-spinel harzburgite (1)
Os	5	6	<5	—	—	3.3	5.1	0.06
Re	0.1	0.1	<0.3	—	—	0.11	0.19	0.01
Ir	2.4	2.4	3.4	4.0	13	3.7	5.3	0.05
Ni	0.2	0.16	0.21	—	—	0.21	0.19	0.24
Pd	13	8.2	—	6.3	3.8	4.5	4.0	0.09
Au	4.2	3.0	0.5	0.7	8.0	0.66	1.4	0.37
Ge	1	1.0	1.22	—	—	0.82	0.74	0.62
Sb	—	100	—	—	—	2	6	—
Ag	50	50	—	—	—	6	5	1.3
Zn	40	—	50	—	—	45	31	30
In	3	—	—	—	—	8	2	2
Cd	14	—	—	—	—	47	80	9
Bi	6	—	—	—	—	1.8	5	1.0
Tl	—	—	—	—	—	1.3	15	4.2
Se	50	—	40	—	—	47 [a], 9 [b]	9	2.2
Te	—	—	—	—	—	11	4	≤2

Number of samples averaged in parentheses. [a] Samples from Kilbourne Hole and Cochise Crater (8). [b] Samples from other localities (6).

compared in Table 1.2.11.7 with the abundances listed in the compilations of Ringwood and Kesson (1977) and Chou (1978) and with other recent analyses of fresh ultramafic nodules by Jagoutz *et al.* (1979) and by Mitchell and Keays (1979). Agreement is good for Ni and Ge, as might be expected for relatively abundant compatible elements, and is surprisingly good for Os, Re, Ir, and even for Pd. Gold is significantly lower in the newest sets of data, and our determinations of Ag and Sb are lower than previous estimates by factors of 10 and 30, respectively.

Cosmochemical vs. geochemical fractionation

Evidence from chondritic meteorites strongly indicates that cosmochemical fractionation is largely determined by volatility and is essentially independent of geochemical behavior (see review by Grossman and Larimer, 1974). In lunar breccias, therefore, the similarity in C1-normalized abundances of siderophiles of similar volatility is strong *prima facie* evidence for direct meteoritic contamination. A stepped pattern between groups of elements having different volatilities is apparently a reflection of cosmochemical fractionation. In lunar rocks, where elements of similar volatility but different geochemical character (for example, Re and Ir; or Au and Ni) have quite different C1-normalized abundances, there is a suspicion that the siderophiles are indigenous or represent meteoritic contamination that has been fractionated geochemically, perhaps in an impact melt sheet (see for example, the review by Morgan, 1977).

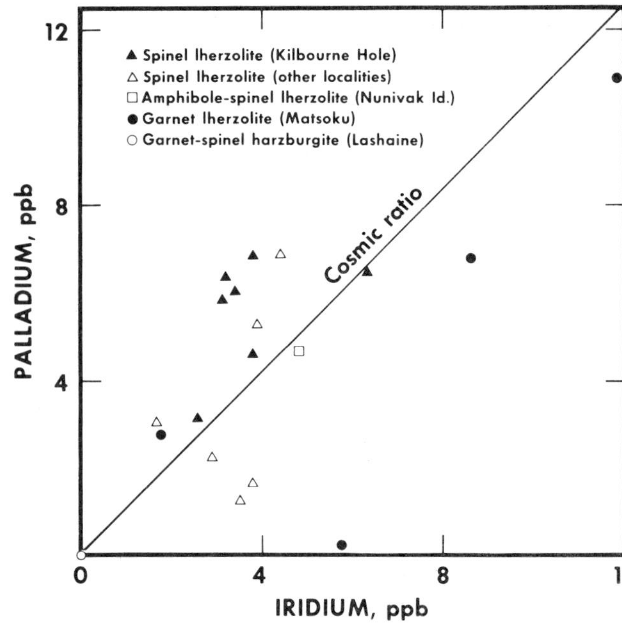

1.2.11.15 Palladium-iridium correlation for samples in the reference suite.

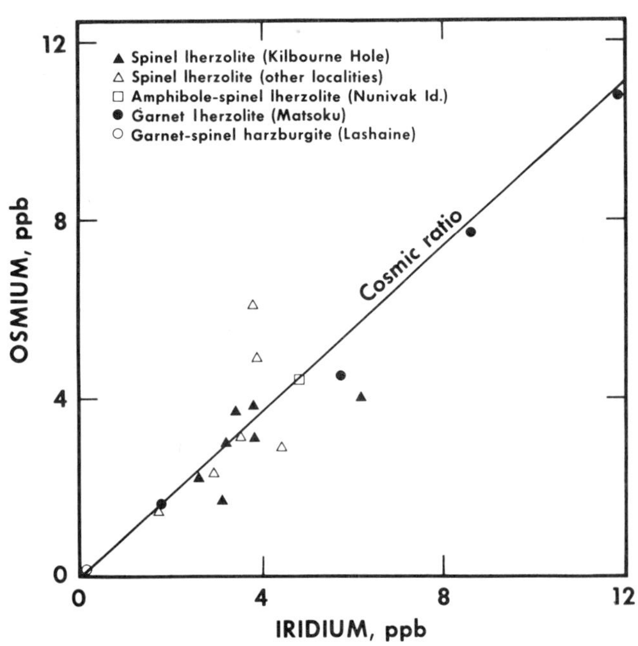

1.2.11.14 Osmium-iridium correlation for samples in the reference suite.

Siderophile and volatile element fractionation in ultramafic rocks

Iridium and osmium. Spinel lherzolites from all the localities studied are extremely uniform in Ir and Os contents. For 14 such rocks (including UM-1, the amphibole-spinel lherzolite), we find a mean Os content of 3.3 ± 1.3 ppb. For Ir, we can improve the global coverage by including the six values reported by Jagoutz *et al.* (1979). These include samples from Kapfenstein, Austria; Dreiser Weiher, Germany; the Massif Central, France; and the southwestern United States. The combined 20 samples have a mean Ir content 3.6 ± 1.0 ppb. The ratio of the two mean values is very close to the cosmic value: i.e., Os/Ir = 1.01 times C1. The garnet lherzolites from southern Africa are much more variable in Ir and Os, even if the very depleted Lashaine sample (UM-20) is excluded. Nevertheless, despite the very large variation, the garnet peridotites have C1-normalized Os/Ir ratios very close to unity.

Although there is a fairly large difference between the condensation temperatures of Os and Ir (1925° K and 1629° K, respectively, at 10^{-3} atm), both are considered to be refractory elements and are similarly enriched in Ca-Al inclusions in Allende (Grossman and Larimer,

1974). In addition, both elements appear to be highly siderophile, as shown by their depletion in eucrites (Morgan *et al.*, 1978), and tend to occur in the same minerals. It is not too surprising, therefore, that there is a strong Ir-Os correlation in the ultramafic nodule suite: r = 0.96; Fig. 1.2.11.14. The line drawn in Fig. 1.2.11.14 is not a least squares fit but represents the cosmic Os/Ir ratio. The largest deviation of an individual sample from this ratio is less than a factor of two, despite variations in absolute abundance of more than two orders of magnitude. Even though the garnet lherzolites have the greatest variation (and, of course, strongly influence the correlation), their Os/Ir ratios deviate insignificantly from the cosmic ratio.

Palladium. The less refractory, highly-siderophile elements, which condense along with Fe-Ni, may be represented by Pd (50% condensation from 10^{-5} atm solar gas at 1256° K for both Pd and Ni; Wolf *et al.*, 1979). The distribution of Pd in ultramafic rocks closely resembles that of Os and Ir, but with more scatter; the mean Pd content of the 14 spinel lherzolites is 4.6 ± 2.0 ppb. Individual rocks cluster about the cosmic Pd/Ir ratio, and there seems to be a tendency for the spinel lherzolites to have higher Pd/Ir ratios than the garnet lherzolites (Fig. 1.2.11.15). Except for garnet lherzolite UM-16, which is relatively depleted in Pd, no sample deviates from the cosmic Pd/Ir ratio by more than a factor of 3, and only 3 samples deviate by much more than a factor of 2. The mean Pd/Ir ratio (1.34 ± 0.14) is close to the chondritic ratio and, given all the uncertainties, is not significantly different. A regression line for 18 data points, forced through the origin, yields a Pd/Ir ratio of 1.21 C1.

Our results for garnet lherzolites are in marked contrast to those of Paul *et al.* (1979)—see Table 1.2.11.7. Although Pd abundances for their garnet lherzolites are similar to those for the Matsoku xenoliths, the Ir abundances in their samples are higher by up to a factor of 20, and in only 3 of their 11 samples is Pd/Ir within a factor of 3 of the chondritic ratio. These features may be attributable to secondary alteration in their samples.

Gold and rhenium. The Fe-Ni/silicate distribution coefficients, determined experimentally at 1700–1900° K and low pressure (Kimura *et al.*, 1974), are very large ($\geq 10^4$). The high abundance of gold and rhenium in terrestrial basalts, both in absolute terms and relative to the platinum metals, has been puzzling. (By contrast, the abundances in lunar basalts are more in keeping with the partition coefficients and with their having equilibrated with metallic Fe-Ni, particularly for Re.)

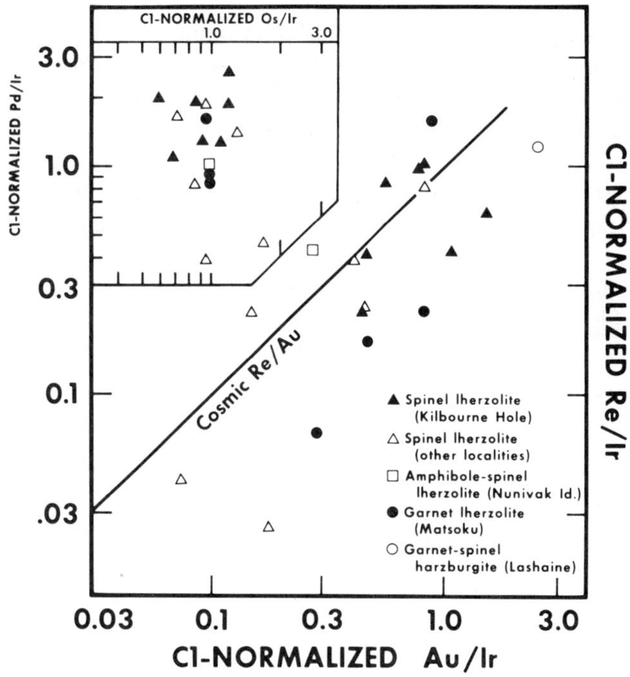

1.2.11.16 Distribution of osmium, palladium, gold and rhenium (normalized to C-1 chondrites and iridium) for samples in the reference suite.

In the study suite, Au and Re contents of the spinel lherzolites are highly variable, much more so than the Pt metals, Os, Ir and Pd. Figure 1.2.11.16 shows Os, Pd, Au and Re data normalized to C1 abundances and to Ir, and contrasts the relative constancy of the Pt metals with the variability of Re and Au. The highest normalized values of Re and Au cluster near the cosmic point (except for UM-20 in which Pt metals and Re are very depleted, but Au is not). It is tempting to suggest that fractionation effects may largely involve depletion of Re and Au relative to the Pt metals by preferential partitioning into the basaltic liquid formed on partial melting. The high abundances of Re and Au in terrestrial basalts may thus be a reflection of the general siderophile enrichment in the Earth's upper mantle, and not necessarily a consequence of some special anomaly in mantle-core partitioning.

The clustering of Re/Ir and Au/Ir ratios around the cosmic point need not be interpreted as indicating some type of primordial, unfractionated material. The four Kilbourne Hole samples in this cluster include all three of the fine-grained tabular spinel lherzolites. As discussed above, the fine-grained rocks have chondritic (UM-7) or nearly chondritic REE patterns, again suggesting that little or no fractionation has taken place. The field relationships and petrology imply otherwise, however. The fine-grained, tabular spinel lherzolites are similar in texture and chemical composition to the wall rocks which enclose pyroxenite dikes found in rare, composite nodules, and apparently have been subjected to deformation and to metasomatic enrichment in iron, large-ion-lithophile elements, sulfur, and Re and Au during dike injection. The chondritic ratios observed are apparently due to an approximate reconstitution of the composition by multiple events, rather than steadfast preservation.

Os/Re ratios for most samples are within a factor of 4 of the chondritic value (14.5), as is the Os/Re ratio for spinel lherzolite from Tumut-Eucumbene Tunnel, Australia (Morgan and Lovering, 1967). The garnet lherzolites show more dispersion about the chondritic value, but the largest deviations in this ratio (higher by a factor of 30) occur in the two San Carlos samples and the one sample from Cochise Crater. The latter values are higher by a factor of 4 than those for such ultramafic rocks as Twin Sisters dunite DTS-1 and Cazadero peridotite PCC-1 (Morgan and Lovering, 1967).

Cosmochemically, Re is a very refractory element, whereas Au is slightly more volatile than Ni and Pd (50% condensation temperatures from 10^{-5} atm solar gas are 1665° K for Re and 1153° K for Au). The

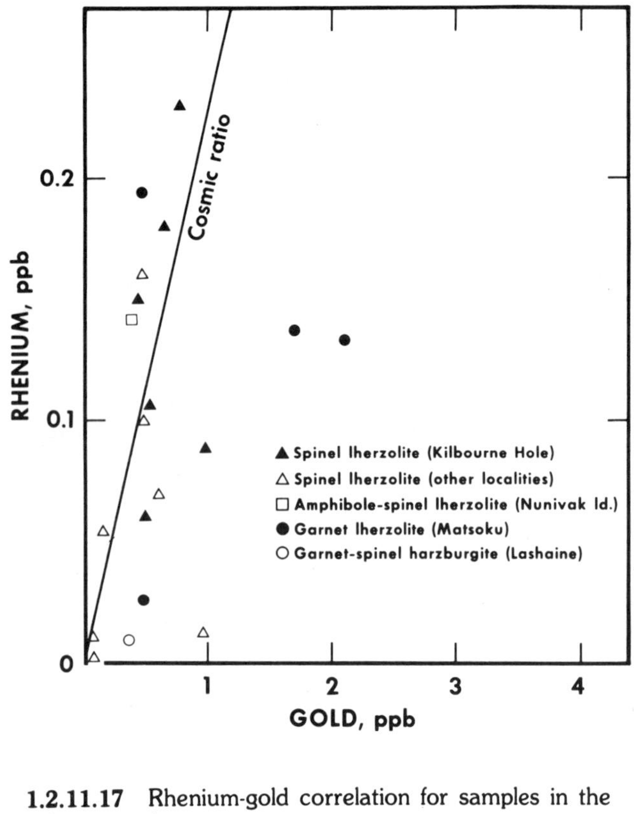

1.2.11.17 Rhenium-gold correlation for samples in the reference suite.

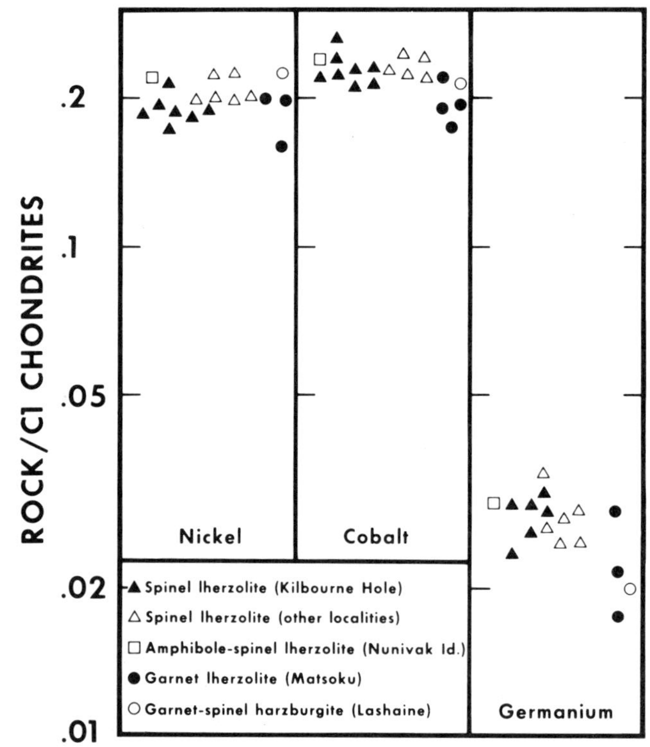

1.2.11.18 Chondrite-normalized abundances of moderately siderophile elements nickel, cobalt and germanium for samples in the reference suite. Ni and Ge data from Table 1.2.11.6, Co data from Table 1.2.11.5.

fractionation of Au and Re together from the other Pt metals, as illustrated in Fig. 1.2.11.16, tends to obscure the chondrite-like Au/Re ratio found in most samples. If the Au and Re abundances are plotted without normalization, as in Fig. 1.2.11.17, it becomes clear that the least-depleted samples generally lie rather close to the cosmic ratio. There are two exceptions, UM-7 and UM-19, but even these do not deviate by more than a factor of 2. It appears, therefore, that the five highly-siderophile elements determined in this suite of ultramafic rocks were present originally in chondritic proportions, and that these proportions have been remarkably well retained. In addition, the absolute amounts of siderophile elements appear to be extremely constant in spinel lherzolites on at least four continents (Europe, North America, Australia, Africa) and Hawaii.

Germanium, nickel and cobalt. Data for these moderately-siderophile elements are shown in Fig. 1.2.11.18. The dispersion and high abundance of Ge and particularly of Ni are well-known, and may be due to the lowering of metal/silicate distribution coefficients by a relatively high oxygen fugacity at the mantle-core boundary (Chou, 1978; Ringwood, 1977).

Antimony, selenium and sulfur. Two volatile siderophile elements, Sb and Se, are present in some spinel lherzolites in approximately C1 chondrite ratios relative to the platinum metals, but the similarity to cosmic ratios may be fortuitous. Antimony is an incompatible element that has been transferred from these rocks to the crust (levels tend to be higher in garnet lherzolites). Conversely, Se in spinel lherzolites may have been enhanced by the introduction of secondary sulfide. Sulfur and Se abundances are very well correlated (Fig. 1.2.11.19) and tend to be lower in the garnet lherzolites.

Volatile elements

Our results are compared in Table 1.2.11.7 with compiled values for six volatile elements. Indium and Tl show large differences between spinel and garnet lherzolites. Cadmium (whose abundance in ultramafics has previously been underestimated), Zn, Te and perhaps In (if the spinel lherzolite values are taken to be more representative of the depleted mantle) do not seem greatly fractionated from each other. Most of the Earth's complement of these elements may still reside in the mantle. Bismuth, Tl and Ag, on the other hand, evidently have migrated almost completely to the crust, as have Cl and Br (Jagoutz *et al.*, 1979).

Bulk carbon abundances show surprisingly little scatter considering the possibility of secondary mantle or crustal processes which could contaminate the nodules before, during and after ascent. The high C abundance in UM-13 from Salt Lake Crater is related largely to the observed deuteric carbonate in this sample (see also O'Neil *et al.*, 1970). More dispersion might have been expected in the samples from Arizona and New Mexico because of the prevalence of caliche in these desert environments (especially abundant at Kilbourne Hole). The garnet lherzolites are systematically richer in C than the spinel lherzolites; this could be partly due to mantle metasomatism or to deuteric alteration. Relatively high CO_2 contents are known to be features of many olivine-rich mantle xenoliths (e.g., Green and Radcliffe, 1975; Murck *et al.*, 1978; Kirby and Green, 1980). However, it is difficult to estimate indigenous upper mantle carbon abundances without detailed data for clean mineral separates.

Conclusions

The observed near-chondritic proportions of siderophile elements in these mantle-derived samples cannot be the result of metal-silicate fractionation processes, because it is very unlikely that the partition coefficients for all of these elements are similar. As Chou (1978) has concluded, the most plausible explanation is that the highly-siderophile elements were added to the upper mantle by late meteoritic influx after core formation. Chou estimates that 1% of C1 carbonaceous chondrite material would suffice. The relatively high mantle abundances of Ni, Co, Ge and other moderately-siderophile elements are interpreted by Chou as resulting from indigenous partitioning into silicates during core formation, followed by relatively minor enhancement by the meteoritic component. Jagoutz *et al.* (1979) have suggested addition of as much as 10% of a chondritic component in order to explain the factor of 10 depletion of the mantle in elements such as Cu, W, Fe,

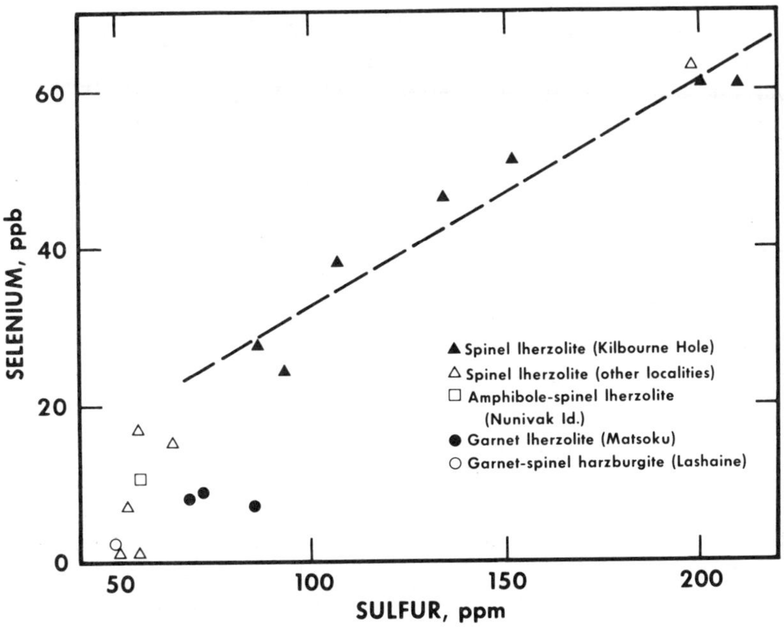

1.2.11.19 Selenium-sulfur correlation for samples in the reference suite. Regression line is for Kilbourne Hole samples.

Ga, F and Zn, but they must invoke subsequent metal segregation to remove the highly siderophile elements.

1.2.11.8 Sr and Nd isotope chemistry

Until recently, trace element and isotopic studies of mantle peridotites have been confined to anhydrous mantle assemblages (see Chapter 7). Despite the lherzolitic nature of these ultramafic rocks, the majority of the analyzed samples (e.g., Paul, 1971; Stueber and Ikramuddin, 1974; Dasch and Green, 1975; Burwell, 1975; Basu and Murthy, 1977; Menzies and Murthy, 1978) are very depleted in basaltic components. The LIL element content of these peridotites can, however, be increased during mantle metasomatism, which is a process that radically modifies the mineralogy of the source rock. Metasomatized mantle rocks commonly contain veins or patches of amphibole ± mica ± apatite. This petrographic observation and the importance of LIL elements in basalt genesis have led to an increased interest in enriched or metasomatized mantle (Lloyd and Bailey, 1975; Erlank, 1976; Francis, 1976a, b; Boettcher and O'Neil, 1980; Menzies and Murthy, 1980a, b).

A Sm-Nd and Rb-Sr study of metasomatized mantle peridotites from the southwestern United States (Wilshire *et al.*, 1980); 'Atāq, South Yemen (Varne and Graham, 1971); and Nunivak, Alaska (Francis, 1976a, b) was undertaken in an attempt to establish the nature of mantle metasomatism and its relationship to magma genesis. Amphibole-spinel lherzolite UM-1 from Nunivak Island was one of the samples studied.

Metasomatic fluids

The mineralogical, chemical and isotopic characteristics of crystallized mantle fluids can be assessed by analysis of metasomatic vein deposits. However, it first must be established that the introduction of these minerals occurred prior to entrainment of the nodule in the basanite or kimberlite. For example, in the case of the Nunivak sample UM-1, pargasite is seen to break down to a high temperature assemblage of diopside + spinel + glass (Francis 1976a, b) adjacent to the basanite. This clearly indicates that the pargasite is a mantle phase stable at temperatures lower than that of the basanite (~1200° C).

The mineralogy of crystallized mantle fluids is highly variable and commonly consists of amphibole (ranging in composition from potassic richterite to kaersutite and pargasite), clinopyroxene, apatite, phlogopite and sphene. Minor amounts of perovskite, calcite, plagioclase and garnet have also been reported (see Menzies and Murthy, 1980a for a summary). These minerals are found in veins that crosscut the mantle peridotite, as intergranular phases, or as replacement products of the original rock (Best, 1974a; Wilshire *et al.*, 1980; Lloyd and Bailey, 1975).

Careful separation and analysis of these minerals permits reconstruction of the chemistry of the original mantle fluid. Besides changing the major element chemistry of the peridotite, fluid migration tends to enhance the trace element content greatly. Fluids are frequently found to be enriched in Na, Al, P, K, Ca, Ti, Mn, Fe, Rb, Sr, Y, Zr, Ba, light REE, Ta, Th and U (Varne and Graham, 1971; Lloyd and Bailey, 1975; Erlank, 1976). Micas and amphiboles are found to contain F, Cl and O as well as OH (Boettcher and O'Neil, 1980). The approximate initial Nd and Sr isotopic composition of mantle fluids can be constrained by studying vein minerals with low Sm/Nd ratios (e.g., apatite and mica) and low Rb/Sr ratios (e.g., diopside). Information of this type is at present rather limited. Metasomatic mica and amphibole from Nunivak Island and the 'Atāq diatreme have $^{87}Sr/^{86}Sr = 0.703$–0.704 and $^{144}Nd/^{143}Nd = 0.51300$. However, preliminary data on micas with low Sm/Nd ratios indicates a considerable range in $^{144}Nd/^{143}Nd$ (0.5120–0.5130; M. A. Menzies, pers. comm., 1979). These fluids are somewhat similar to those invoked by Hawkesworth and Vollmer (1979) which had $^{87}Sr/^{86}Sr = 0.709$ and $^{144}Nd/^{143}Nd = 0.5122$.

Metasomatism and magma genesis

Mantle metasomatism, or the introduction of a KREEP-rich component, is believed to be a precursor to the genesis of alkaline magmas (Lloyd and Bailey, 1975; Boettcher and O'Neil, 1980; Menzies and Murthy, 1980a, b). Fluids carrying this LIL element-enriched component may also provide the volatile fluxes that trigger partial melting.

If metasomatism is indeed a precursor of alkalic magmatism, the isotopic composition of the modified mantle or of the minerals deposited by the metasomatic fluid should be identical to that of spatially-related alkalic basalts. Strontium and neodymium isotopic compositions for sample UM-1 and Nunivak host basalts are given in Table 1.2.11.8. Pargasitic amphiboles separated from the Nunivak amphibole lherzolites have a range in $^{87}Sr/^{86}Sr$ of 0.70270–0.70337, which is identical to that of the enclosing basalts ($^{87}Sr/^{86}Sr = 0.70251$–$0.70322$; Menzies and Murthy, 1980b). The Nd isotopic composition (normalized to $^{146}Nd/^{144}Nd = 0.7219$) of the Nunivak amphiboles and micas is also identical to that of the host alkalic basalts ($^{144}Nd/^{143}Nd = 0.51303$). Similarly, pargasites from the 'Atāq amphibole lherzolites have a range in $^{87}Sr/^{86}Sr$ of 0.70344–0.70408, indistinguishable from

Table 1.2.11.8 Strontium and neodymium isotopic compositions of rocks and minerals from Nunivak Island.

Sample	$^{87}Sr/^{86}Sr$	$^{143}Nd/^{144}Nd$
UM-1 whole rock	0.70312 ± 9	—
UM-1 amphibole	0.70337 ± 8	0.51302 ± 5
UM-1 clinopyroxene	0.70323 ± 8	0.51302 ± 2
Host basalts (5)	0.70251-0.70330	0.51289-0.51304

Analyst M. A. Menzies. Uncertainties are 2σ in the fifth place.

the range for associated alkalic basalts ($^{87}Sr/^{86}Sr$ = 0.70353–0.70426). Neodymium isotopic analyses indicate that the host alkaline magmas at Nunivak (and to a certain extent those at ‘Ataq) are derived from mantle source regions with a time-integrated light REE depletion. This contrasts markedly with the light REE enriched character of the lavas and necessitates either (1) 5–20% melting of recently enriched mantle (Menzies and Murthy, 1980b), or (2) less than 5% melting of a depleted source region. Option (1) is preferred since the included ultramafic nodules offer compelling evidence for a mantle enrichment event that introduced hydrous phases into the nodules prior to their incorporation in the basanite. The addition of metasomatic minerals by pneumatolytic fluids dramatically increases the light REE content of the source region, allowing alkalic basalts to be produced by higher degrees of melting than previously postulated (e.g., Gast, 1968). It should also be noted that the enhanced K content of the metasomatized mantle source region (K = 197–532 ppm, average 365 ppm) is compatible with several independent estimates of the K content of the mantle. Hofmann and Hart (1978) calculated a maximum value of 400 ppm K.

The timing of mantle metasomatic events can best be defined by isotopic analysis of the nodule suites or of the basalts. For example, potassic richterite-bearing peridotites from the Bultfontein kimberlite pipe, South Africa, define an apparent “age” of 148 ± 7 m. y. (Erlank and Shimizu, 1977) and an initial $^{87}Sr/^{86}Sr$ ratio of 0.7054. Since $^{87}Sr/^{86}Sr$ ratios of tholeiitic basalts and dolerites from the Karroo Basin are in the range 0.705–0.707, one can speculate that a genetic connection may exist between the mantle metasomatic event and Karroo volcanism. Alkalic basaltic volcanics along the south Arabian Coast have Rb-Sr ages 20–30 m.y. greater than their K-Ar ages (Dickinson *et al.*, 1969). These volcanic rocks have an initial $^{87}Sr/^{86}Sr$ ratio of 0.70368–0.70391 which compares favorably with that for metasomatized mantle nodules at ‘Atāq (average $^{87}Sr/^{86}Sr$ = 0.70383; Menzies and Murthy, 1980b). A mantle fractionation or enrichment event must have occurred prior to eruption of these basalts, perhaps associated with mantle degassing synchronous with continental rifting and fragmentation. Metasomatic episodes can also be dated by Sm-Nd methods because a considerable range of Sm/Nd ratios is commonly found in minerals such as amphibole, mica and apatite. Preliminary Nd isotopic analysis of mica, diopside and kaersutite from a nodule found at Dish Hill, California, gives an enrichment age of 200 m.y. (M. A. Menzies, pers. comm., 1979). Evidently, enrichment processes have been active recently in the subcontinental mantle producing a mixture of depleted and enriched mantle.

Mantle heterogeneity

If amphibole lherzolites are the source material for alkalic basalts, the survival of local isotopic heterogeneities during mantle metasomatism has profound implications with regard to disequilibrium melting models. Considerable controversy exists as to whether inter-mineral disequilibrium will persist through melt formation. Hofmann and Hart (1978) believe that isotopic disequilibrium will exist for long periods of time in a fluid-free mantle, but that the relatively rapid diffusion of Sr, REE, and other LIL elements through silicate melts will accelerate equilibration once the grain boundaries are coated with melt. However, Wood and Heming (1978) noted that melts in a non-static regime, such as that believed to exist in zones of magma generation, will approach trace element equilibrium much more slowly than melts in a static regime.

Neodymium and Sr isotopic data on the constituent minerals in amphibole lherzolites do not serve to distinguish between these conflicting viewpoints. Diopside and pargasite from a Nunivak amphibole lherzolite have identical $^{87}Sr/^{86}Sr$ = 0.70330 and $^{144}Nd/^{143}Nd$ = 0.51300 ratios. However, two amphibole lherzolites from ‘Atāq contain diopside and pargasite with identical $^{87}Sr/^{86}Sr$ ratios but different $^{144}Nd/^{143}Nd$ ratios. The possible survival of these Nd isotopic differences or heterogeneities during basalt genesis is reflected in the Nd isotopic composition of the host basalts. The basalts at ‘Atāq seem to tap different sources characterized by both light REE enrichment and depletion, in contrast to the uniform source of the Nunivak alkalic basalts (Menzies and Murthy, 1980a). Although this dual source characteristic of 'Atāq is shown by Nd isotopic data for both the host alkalic lavas and the lherzolite nodules, it is not clearly recognizable in the Sr isotopic data. Neodymium isotopic analyses of kaersutite veins in lherzolite nodules from Dish Hill, California, reveal that the vein amphibole and diopside in the surrounding lherzolite have either similar or very different $^{144}Nd/^{143}Nd$ ratios (M. A. Menzies, pers. comm., 1979) due to isotopic aging. Thus, the amount of isotopic (and chemical) exchange between the hydrous veins and the dry wall rock appears to be highly variable.

If these amphibole lherzolites are at all representative of the source material for alkalic basalts, then the trace element and isotopic composition of the melt fractions may be affected by

(1) the mineralogy and chemistry of the veins,
(2) the size and spacing of the veins,
(3) the degree of initial isotopic disequilibrium between the introduced metasomatic minerals and the dry mantle assemblage, and
(4) the age of the veining.

The sporadic distribution of hydrous vein or interstitial phases in mantle source regions, as at 'Atāq and Dish Hill, produces melts with diverse incompatible element ratios and Nd isotopic compositions, but with uniform $^{87}Sr/^{86}Sr$ ratios. Conversely, if the source region is similar to that at Nunivak, the melts will have both uniform Nd and Sr isotopic compositions, because it appears that the amphibole and diopside have fully equilibrated. These alternatives have important implications for basalt genesis, since previously proposed disequilibrium melting models (O'Nions and Pankhurst, 1974) could not adequately explain how chemically diverse magmas could be extracted from isotopically (Sr) homogeneous mantle. Further combined Sr and Nd isotopic studies are needed to reevaluate the importance of disequilibrium melting models.

1.2.11.9 Oxygen isotope chemistry

Oxygen isotope ratios are useful indicators of source materials for igneous rocks, and the fractionation of ^{18}O between coexisting phases as a function of temperature may be used as a geothermometer and genetic tracer (e.g., Taylor, 1968). The temperature dependence of oxygen isotope distributions has been applied to determine genetic relationships between basalts and xenoliths (Muehlenbachs and Kushiro, 1974; Kyser, 1979). These studies suggest that the variations in the oxygen isotope ratios of coexisting minerals from xenoliths may be used to distinguish cognate and accidental origins, as well as to estimate the internal equilibration temperature of mineral assemblages.

Three Kilbourne Hole spinel lherzolites and four Matsoku garnet lherzolites from the present suite were selected for oxygen isotope analysis. Data are presented in Table 1.2.11.9. The whole rock $\delta^{18}O$ values of the Kilbourne Hole samples are similar to those of spinel lherzolites from nearby San Carlos, Arizona, but are less ^{18}O-rich than spinel lherzolites from other areas such as Hawaii, San Quintin, the Grand Canyon and Europe (T. K. Kyser, pers. comm., 1980). The slight ^{18}O depletion in the nodules from Kilbourne Hole and San Carlos has been attributed to their interaction with ^{18}O-poor fluids that were generated at high temperatures (Boettcher and O'Neil, 1980). However, because the host basalts for these nodules have $\delta^{18}O$ values of 6.0, it is unlikely that the lherzolites are directly related

Table 1.2.11.9 Oxygen isotopic compositions ($\delta^{18}O$) of rocks and minerals from Kilbourne Hole and Matsoku, and estimated equilibration temperatures.

sample	olivine	orthopyroxene	clinopyroxene	spinel	whole rock	T1(°C)	T2 (°C)
UM-2	5.6	5.6	5.8	4.8	5.2	1151	1113
UM-5	4.9	5.6	5.4	4.6	5.5	997	1048
UM-9	5.1	6.1	5.6	4.0	5.3	910	1048
KH77-0†	—	—	—	—	5.8	—	—

sample	olivine	orthopyroxene	clinopyroxene	garnet	mica	whole rock	T1(°C)
UM-16	5.6	—	—	—	—	—	—
UM-17	5.5	5.1	5.2	5.1	—	5.2	1200
UM-18	5.6	4.8	4.9	—	6.2	5.3	1240
UM-19	5.8	4.9	—	4.9	—	5.4	1250

All $\delta^{18}O$ values in units of per mil. Precision and accuracy ± 0.1 per mil. Analyst T. K. Kyser. Whole rock isotope compositions of xenoliths calculated from estimated modes. †Host basanite. Temperatures estimated using oxygen isotope geothermometer of Kyser *et al.*, (1981): T1 for orthopyroxene-olivine pairs, T2 for clinopyroxene-olivine pairs.

to the alkalic volcanism that transported the nodules to the surface.

The oxygen isotope ratios of coexisting orthopyroxene and clinopyroxene may be used to evaluate whether or not the minerals in a nodule are in equilibrium because $\Delta^{18}O_{opx-cpx}$ should be zero at any temperature (Bottinga and Javoy, 1973; Kyser *et al.*, 1981). The data in Table 1.2.11.9 indicate that most of the samples satisfy this criterion (within $\pm$ 0.2 ‰) except for UM-9 (which has $\Delta^{18}O_{opx-cpx}$ of 0.5). Isotopic disequilibrium is also evident from the large difference between the isotopic ratio of spinel and other phases in UM-9. The reasons for this are not obvious, because UM-9 is very similar to UM-2 in texture and mineral chemistry.

Internal equilibration temperatures of the xenoliths calculated from the distribution of oxygen isotopes are also listed in Table 1.2.11.9. Because UM-9 is in isotopic disequilibrium, it is difficult to assess the significance of the temperatures calculated for this sample. Nevertheless, temperatures calculated for UM-5 and UM-9 using other chemical geothermometers (Table 1.2.11.3) are broadly similar to the oxygen isotope equilibration temperatures. However, for the Matsoku garnet lherzolites, the difference between equilibration temperatures calculated using the pyroxene solvus and using $\Delta^{18}O_{px-ol}$ is more than 200° C. The reasons for the discrepancy between results obtained using chemical and isotope geothermometers are not exactly known, but it should be noted that most chemical exchange geothermometers are dependent on pressure and bulk composition, whereas these factors have little effect on oxygen isotope distributions (Clayton *et al.*, 1975; Kyser *et al.*, 1981).

1.2.11.10 Concluding remarks

This multifaceted petrologic-chemical-isotopic study of mantle xenoliths has led to a number of new insights into the history of the terrestrial mantle and processes of basaltic magma generation. The extensive new data for siderophile elements have consolidated previous suggestions of near-chondritic mantle abundances of these elements, most likely resulting from addition of a chondritic meteorite component to the mantle after core formation. In contrast, the xenoliths do not in general possess chondritic abundances of lithophile elements, presumably because of extraction of basaltic magma from the mantle lithosphere. Nevertheless, the majority of ultramafic xenoliths from both alkalic basalts and kimberlites cannot be interpreted as simple residues after partial melting. Rather, the new data for the large suite of Kilbourne Hole lherzolites provide evidence for multiple depletion and enrichment of lithophile elements within the mantle by processes of partial melting and fluid metasomatism. Isotopic data for amphibole lherzolite suggest that fluid metasomatism is an essential precursor and trigger for the generation of SiO_2-undersaturated, alkali-rich liquids, such as basanites, olivine nephelinites, kimberlites and carbonatites in the terrestrial mantle. The general chemical analogy between terrestrial alkalic basalts and lunar KREEP basalts may mean that fluids (perhaps rich in F and Cl) play roles in magma generation in the deep lunar crust. Similar processes involving fluid metasomatism may be important also in generating basaltic or other lavas on the other terrestrial planets.

1.2.11.11 References

Albert D., Albert R. and Brousse R. (1967) Enclaves de péridotites à pyrope chromifère dans les pipes de la région de Bédarieux (Herault). *Comptes Rend. Acad. Sci. (Paris)* **D 265**, 657–659.

Aoki K. and Shiba I. (1973) Pargasite in lherzolite and websterite inclusions from Itinome-gata, Japan. *J. Jpn. Assoc. Mineral. Petrol. Econ. Geol.* **68**, 303–310.

Arculus R. J., Dungan M. A., Lofgren G. E. and Rhodes J. M. (1977) Lherzolite inclusions and megacrysts from the Geronimo volcanic field, San Bernadino Valley, southeastern Arizona (abstract). In *Extended Abstracts, Second Internat. Kimberlite Conf., Santa Fe.*

Basu A. R. (1975) Hot spots, mantle plumes and a model for the origin of ultramafic xenoliths in alkalic basalts. *Earth Planet. Sci. Lett.* **28**, 261–274.

Basu A. R. (1977a) Textures, microstructures and deformation of ultramafic xenoliths from San Quintin, Baja California. *Tectonophys.* **43**, 213–246.

Basu A. R. (1977b) Olivine-spinel equilibria in lherzolite xenoliths from San Quintin, Baja California. *Earth Planet. Sci. Lett.* **33**, 443–450.

Basu A. R. and Murthy V. R. (1977) Ancient lithospheric xenolith in alkali basalt from Baja California. *Earth Planet. Sci. Lett.* **34**, 246–253.

Berger E. (1977) Sur la présence d'une lherzolite à grenats en enclave dans le basalte alcalin de la Vestide du Pal (Ardèche): conditions d'équilibre, implications pétrogénétiques et géotectoniques. *Compt. Rend. Acad. Sci. (Paris)* **D284**, 709–712.

Berger E. and Brousse R. (1976) Une lherzolite à grenats du pipe d'Eglazines (Lozère, France) stabilisée à 1400° C et 30 kbars. *Compt. Rend. Acad. Sci. (Paris)* **D282**, 1477–1480.

Best M. G. (1974a) Mantle-derived amphibole within inclusions in alkalic-basaltic lavas. *J. Geophys. Res.* **79**, 2107–2113.

Best M. G. (1974b) Contrasting types of chromium-spinel peridotite xenoliths in basanitic lavas, Western Grand Canyon, Arizona. *Earth Planet. Sci. Lett.* **23**, 929–937.

Beswick A. E. and Carmichael I. S. E. (1978) Constraints on mantle source compositions imposed by phosphorus and the rare-earth elements. *Contrib. Mineral. Petrol.* **67**, 317–330.

Boettcher A. L. and O'Neil J. R. (1980) Stable isotope, chemical, and petrographic studies of high-pressures amphiboles and micas: evidence for metasomatism in the mantle source regions of alkali basalts and kimberlites. *Amer. J. Sci.*, Jackson vol. **280A**, 594–691.

Bottinga Y. and Javoy M. (1973) Comments on oxygen isotope geothermometry. *Earth Planet. Sci. Lett.* **20**, 250–265.

Bouillier A. M. and Nicolas A. (1973) Texture and fabric of peridotite nodules from kimberlite at Mothae, Thaba Putsoa and Kimberley. In *Lesotho Kimberlites* (P. H. Nixon, ed.), p. 56–66. Lesotho Nat. Devel. Corp.

Bouillier A. M. and Nicolas A. (1975) Classification of textures and fabrics of peridotite xenoliths from South African kimberlites. In *Physics and*

Chemistry of the Earth, v. 9, (L. H. Ahrens, J. B. Dawson, A. R. Duncan and A. J. Erlank, eds.), p. 467–475, Pergamon, N.Y.

Brett R. (1971) The Earth's core: speculations on its chemical equilibrium with the mantle. *Geochim. Cosmochim. Acta* **35**, 203–221.

Brown G. M., Pinsent R. H. and Coisy P. (1980) The petrology of spinel peridotite xenoliths from the Massif Central, France. *Amer. J. Sci.*, Jackson Vol. **280A**, 471–498.

Burwell A. D. M. (1975) Rb-Sr isotope geochemistry of lherzolites and their constituent minerals from Victoria, Australia. *Earth Planet. Sci. Lett.* **28**, 69–78.

Carswell D. A. (1975) Primary and secondary phlogopites and clinopyroxenes in garnet lherzolite xenoliths. In *Physics and Chemistry of the Earth*, v. 9, (L. H. Ahrens, J. B. Dawson, A. R. Duncan and A. J. Erlank, eds.), p. 417–429. Pergamon, N.Y.

Carter J. L. (1970) Mineralogy and chemistry of the Earth's upper mantle based on the partial fusion-partial crystallization model. *Geol. Soc. Amer. Bull.* **81**, 2021–2034.

Chou C. -L. (1978) Fractionation of siderophile elements in the Earth's upper mantle. *Proc. Lunar Planet. Sci. Conf. 9th*, p. 219–230.

Clayton R. N., Goldsmith J. R., Karel K. J., Mayeda T. K. and Newton R. C. (1975) Limits on the effect of pressure on isotopic fractionation. *Geochim. Cosmochim. Acta* **39**, 1197–1201.

Cox K. G. and Jamieson B. G. (1974) The olivine-rich lavas of Nuanetsi: a study of polybaric magmatic evolution. *J. Petrol.* **15**, 269–301.

Cox K. J., Gurney J. J. and Harte B. (1973) Xenoliths from the Matsoku Pipe. In *Lesotho Kimberlites* (P. H. Nixon, ed.), p. 76–91. Lesotho Nat. Devel. Corp.

Dasch E. J. and Green D. H. (1975) Strontium isotope geochemistry of lherzolite inclusions and host basaltic rocks, Victoria, Australia. *Amer. J. Sci.* **275**, 461–469.

Dawson J. B., Powell D. G. and Reid A. M. (1970) Ultrabasic xenoliths and lava from the Lashaine volcano, northern Tanzania. *J. Petrol.* **26**, 296–301.

Delaney J. S., Smith J. V., Carswell D. A. and Dawson J. B. (1980) Chemistry of micas from kimberlites and xenoliths - II. Primary- and secondary-textured micas from peridotite xenoliths. *Geochim. Cosmochim. Acta* **44**, 857–872.

Dickinson D. R., Dodson M. H., Gass I. G. and Rex D. C. (1969) Correlation of initial $^{87}Sr/^{86}Sr$ with Rb/Sr in some late Tertiary volcanic rocks of south Arabia. *Earth Planet. Sci. Lett.* **6**, 84–90.

Dostal J. and Capedri S. (1976) Uranium in spinel peridotite inclusions in basalts from Sardinia. *Contrib. Mineral. Petrol.* **54**, 245–254.

Duke M. B. (1965) Metallic iron in basaltic achondrites. *J. Geophys. Res.* **70**, 1523–1527.

Ehmann W. D., Baedecker P. A. and McKown D. M. (1970) Gold and iridium in meteorites and some selected rocks. *Geochim. Cosmochim. Acta* **34**, 493–507.

Ehrenberg S. N. (1978) Petrology of potassic volcanic rocks and ultramafic xenoliths from the Navajo volcanic field, New Mexico and Arizona. Ph.D. Thesis, Univ. of California, Los Angeles, 259 pp.

Ehrenberg S. N. (1979) Garnetiferous ultramafic inclusions in minette from the Navajo volcanic field. In *The Mantle Sample: Inclusions in Kimberlites and Other Volcanics* (F. R. Boyd and H. O. A. Meyer, eds.), p. 330–344. Amer. Geophys. Union, Washington, D.C.

Erlank A. J. (1976) Upper mantle metasomatism as revealed by potassic richterite bearing peridotite xenoliths from kimberlites (abstract). *EOS (Trans. Amer. Geophys. Union)* **57**, 597.

Erlank A. J. and Rickard R. S. (1977) Potassic richterite bearing peridotites from kimberlite and the evidence they provide for upper mantle metasomatism (abstract). In *Extended Abstracts, Second Internat. Kimberlite Conf., Santa Fe.*

Erlank A. and Shimizu N. (1977) Strontium and Sr isotope distributions in some kimberlite nodules and minerals (abstract). In *Extended Abstracts, Second Internat. Kimberlite Conf., Santa Fe.*

Evans S. H., Jr. and Nash W. P. (1979) Petrogenesis of xenolith-bearing basalts from southeastern Arizona. *Amer. Mineral.* **64**, 249–267.

Fabriès J. (1979) Spinel-olivine geothermometry in peridotites from ultramafic complexes. *Contrib. Mineral. Petrol.* **69**, 329–336.

Ferguson J., Ellis D. J. and England R. N. (1977) Unique spinel-garnet lherzolite inclusion in kimberlite from Australia. *Geology* **5**, 278–280.

Finch R. H. and MacDonald G. A. (1953) Hawaiian volcanoes during 1950. *U.S. Geol. Survey Bull. 996-B*, p. 72.

Fisher D. E. (1970) Homogenized fission track analysis of uranium in some ultramafic rocks of known potassium content. *Geochim. Cosmochim. Acta* **34**, 630–634.

Francis D. M. (1976a) The origin of amphibole in lherzolite xenoliths from Nunivak Island, Alaska. *J. Petrol.* **17**, 357–378.

Francis D. M. (1976b) Amphibole pyroxenite xenoliths: cumulate or replacement phenomena from the upper mantle, Nunivak Island, Alaska. *Contrib. Mineral. Petrol.* **58**, 51–61.

Francis D. M. (1978) The implications of the compositional dependence of texture in spinel lherzolite xenoliths. *J. Geol.* **86**, 473–485.

Frey F. A. (1980) The origin of pyroxenites and garnet pyroxenites from Salt Lake Crater, Oahu, Hawaii: trace element evidence. *Amer. J. Sci.*, Jackson vol. **280A**, 427–449.

Frey F. A. and Green D. H. (1974) The mineralogy, geochemistry, and origin of lherzolite inclusions in Victorian basanites. *Geochim. Cosmochim. Acta* **38**, 1023–1059.

Frey F. A. and Prinz M. (1978) Ultramafic inclusions from San Carlos, Arizona: petrologic and geochemical data bearing on their petrogenesis. *Earth Planet. Sci. Lett.* **38**, 129–176.

Frey F. A., Haskin L. A. and Haskin M. A. (1971) Rare-earth abundances in some ultramafic rocks. *J. Geophys. Res.* **76**, 2057–2070.

Frey F. A., Green D. H. and Roy S. D. (1978) Integrated models of basalt petrogenesis: a study of quartz tholeiites to olivine melilitites from south-eastern Australia utilizing geochemical and experimental petrological data. *J. Petrol.* **19**, 463–513.

Ganapathy R. and Anders E. (1974) Bulk compositions of the Moon and Earth, estimated from meteorites. *Proc. Lunar Sci. Conf. 5th*, p. 1181–1206.

Gast P. W. (1968) Trace element fractionation and the origin of tholeiitic and alkaline magma types. *Geochim. Cosmochim. Acta* **32**, 1057–1086.

Green H. W., II and Radcliffe S. V. (1975) Fluid precipitates in rocks from the Earth's mantle. *Geol. Soc. Amer. Bull.* **86**, 846–852.

Grossman L. and Larimer J. W. (1974) Early chemical history of the solar system. *Rev. Geophys. Space Phys.* **12**, 71–101.

Gurney J., Harte B. and Cox K. G. (1975) Mantle xenoliths in the Matsoku kimberlite pipe. In *Physics and Chemistry of the Earth*, v. 9, (L. H. Ahrens, J. B. Dawson, A. R. Duncan and A. J. Erlank, eds.), p. 507–523. Pergamon, N.Y.

Haines E. L. and Zartman R. E. (1973) Uranium concentration and distribution in six peridotite inclusions of probable mantle origin. *Earth Planet. Sci. Lett.* **20**, 45–53.

Hamilton E. I. (1975) The abundance and distribution of uranium in some oceanic, continental ultramafic inclusions and host basalts. *Chem. Geol.* **16**, 221–231.

Harte B. (1977) Rock nomenclature with particular reference to deformation and recrystallization textures in olivine-bearing xenoliths. *J. Geol.* **85**, 279–288.

Harte B., Cox K. G. and Gurney J. J. (1975) Petrography and geological history of upper mantle xenoliths from the Matsoku kimberlite pipe. In *Physics and Chemistry of the Earth*, v. 9 (L. H. Ahrens, J. B. Dawson, A. R. Duncan and A. J. Erlank, eds.), p. 477–506. Pergamon, N.Y.

Hawkesworth C. J. and Vollmer R. (1979) Crustal contamination versus enriched mantle: $^{143}Nd/^{144}Nd$ and $^{87}Sr/^{86}Sr$ evidence from the Italian volcanics. *Contrib. Mineral. Petrol.* **69**, 151–165.

Hofmann A. and Hart S. R. (1978) An assessment of local and regional isotopic equilibrium in the mantle. *Earth Planet. Sci. Lett.* **38**, 44–62.

Hofmann A. W., White W. M. and Whitford D. J. (1978) Geochemical constraints on mantle models: the case for a layered mantle. *Carnegie Inst. Wash., Yearb.* **77**, 548–562.

Irving A. J. (1974) Pyroxene-rich ultramafic xenoliths in the Newer Basalts of Victoria, Australia. *Neues Jahrb. Miner. Abh.* **120**, 147–167.

Irving A. J. (1976) On the validity of paleogeotherms determined from xenolith suites in basalts and kimberlites. *Amer. Mineral.* **61**, 638–642.

Irving A. J. (1979) Kilbourne Hole spinel lherzolites: samples of multiply depleted, enriched and deformed mantle (abstract). *EOS (Trans. Amer. Geophys. Union)* **60**, 418.

Irving A. J. (1980) Petrology and geochemistry of composite ultramafic xenoliths in alkalic basalts and implications for magmatic processes within the mantle. *Amer. J. Sci.*, Jackson vol. **280A**, 389–426.

Irving A. J. and Price R. C. (1981) Geochemistry and evolution of lherzolite-bearing phonolitic lavas from Nigeria, Australia, East Germany and New Zealand. *Geochim. Cosmochim. Acta.*.

Jackson E. D. (1966) "Eclogite" in Hawaiian basalts. *U.S. Geol. Surv. Paper 550-D*, 151–157.

Jackson E. D. and Wright T. L. (1970) Xenoliths in the Honolulu Volcanic Series, Hawaii. *J. Petrol.* **11**, 405–430.

Jagoutz E., Palme H., Baddenhausen H., Blum K., Cendales M., Dreibus G., Spettel B., Lorenz V. and Wänke H. (1979) The abundances of major, minor, and trace elements in the Earth's mantle as derived from primitive ultramafic nodules. *Proc. Lunar Planet. Sci. Conf. 10th*, p. 2031–2050.

Kimura K., Lewis R. S. and Anders E. (1974) Distribution of gold and rhenium between nickel-iron and silicate melts: implications for the abundances of siderophile elements on the Earth and Moon. *Geochim. Cosmochim. Acta* **38**, 683–701.

Kirby S. H. and Green H. W. II (1980) Dunite xenoliths from Hualalai Volcano: evidence for mantle diapiric flow beneath the island of Hawaii. *Amer. J. Sci.*, Jackson vol., **280A**, 550–575.

Kleeman J. D., Green D. H. and Lovering J. F. (1969) Uranium distribution in ultramafic inclusions from Victorian basalts. *Earth Planet. Sci. Lett.* **5**, 449–458.

Kokubu N., Mayeda T. and Urey H. C. (1961) Deuterium content of minerals, rocks, and liquid inclusions from rocks. *Geochim. Cosmochim. Acta* **21**, 247–256.

Kuno H. (1967) Mafic and ultramafic nodules from Itinome-gata, Japan. In *Ultramafic and Related Rocks*, (P. J. Wyllie, ed.), p. 337–342. John Wiley and Sons, N.Y.

Kuno H. (1969) Mafic and ultramafic nodules in basaltic rocks of Hawaii. *Geol. Soc. Amer. Mem. 115*, 189–233.

Kuno H. and Aoki K. (1970) Chemistry of ultramafic nodules and their bearing on the origin of basaltic magmas. *Phys. Earth Planet. Inter.* **3**, 273–301.

Kurat G., Palme H., Spettel B., Baddenhausen H., Hofmeister H., Palme C. and Wänke H. (1980) Geochemistry of ultramafic xenoliths from Kapfenstein, Austria: evidence for a variety of upper mantle processes. *Geochim. Cosmochim. Acta* **44**, 45–60.

Kyser T. K. (1979) The temperature dependence of oxygen isotope distributions and the origin of basalts and ultramafic nodules (abstract). *Abstr. with Programs* Geol. Soc. Amer. **11**, 462.

Kyser T. K., O'Neil J. R. and Carmichael I. S. E. (1981) Oxygen isotope thermometry of basic lavas and mantle nodules. *Contrib. Mineral. Petrol.*

Lloyd F. E. and Bailey D. K. (1975) Light element metasomatism of the continental mantle: the evidence and the consequences. In *Physics and Chemistry of the Earth*, v. 9 (L. H. Ahrens, J. B. Dawson, A. R. Duncan and A. J. Erlank, eds.), p. 389–416. Pergamon, N.Y.

Maaløe S. and Aoki K. (1977) The major element composition of the upper mantle estimated from the composition of lherzolites. *Contrib. Mineral. Petrol.* **63**, 161–173.

MacRae N. D. (1979) Silicate glass and sulfides in ultramafic xenoliths, Newer Basalts, Victoria, Australia. *Contrib. Mineral. Petrol.* **68**, 275–280.

Mathez E. A. and Delaney J. R. (1981) The nature and distribution of carbon in submarine basalts and peridotite nodules. *Earth Planet. Sci. Lett.*

Mathias M. and Rickwood P. C. (1969) Ultramafic xenoliths in the Matsoku kimberlite pipe, Lesotho. *Geol. Soc. South Africa Spec. Publ. 2*, 359–369.

McCallum M. E. and Eggler D. H. (1976) Diamonds in an upper mantle peridotite nodule from kimberlite in southern Wyoming. *Science* **192**, 253–256.

Menzies M. A. and Murthy V. R. (1978) Strontium isotope geochemistry of alpine tectonite lherzolites: data compatible with a mantle origin. *Earth Planet. Sci. Lett.* **38**, 346–354.

Menzies M. A. and Murthy V. R. (1980a) Nd and Sr isotope geochemistry of hydrous mantle nodules and their host alkali basalts: implications for local heterogeneities in metasomatically veined mantle. *Earth Planet. Sci. Lett.* **46**, 323–334.

Menzies M. and Murthy V. R. (1980b) Mantle metasomatism as a precursor to the genesis of alkaline magmas—isotopic evidence. *Amer. J. Sci.*, Jackson vol. **280A**, 622–638.

Mercier J-C. C. (1979) Peridotite xenoliths and the dynamics of kimberlite intrusion. In *The Mantle Sample: Inclusions in Kimberlites and Other Volcanics* (F. R. Boyd and H. O. A. Meyer, eds.), p. 197–212. Amer. Geophys. Union, Washington, D.C.

Mercier J-C. C. and Nicolas A. (1975) Textures and fabrics of upper-mantle peridotites as illustrated by xenoliths from basalts. *J. Petrol.* **16**, 454–487.

Mitchell R. H. and Keays R. R. (1979) Palladium, gold and iridium in mantle minerals: implications for models of magma genesis (abstract). In *Abstracts, Inter-Disciplinary Symposia, I.U.G.G. General Assembly, Canberra*, p. 259.

Morgan J. W. (1977) Chemical fractionation in the solar system. *J. Radioanal. Chem.* **37**, 79–118.

Morgan J. W. and Lovering J. J. (1967) Rhenium and osmium abundances in some igneous and metamorphic rocks. *Earth Planet. Sci. Lett.* **3**, 219–224.

Morgan J. W., Higuchi H., Takahashi H. and Hertogen J. (1978) A "chondritic" eucrite parent body: inference from trace elements. *Geochim. Cosmochim. Acta* **42**, 27–38.

Morgan J. W., Wandless G. A., Petrie R. K. and Irving A. J. (1980) Composition of the Earth's upper mantle—II. volatile trace elements in ultramafic xenoliths. *Proc. Lunar Planet. Sci. Conf. 11th*, p. 213–233.

Morgan J. W., Wandless G. A., Petrie R. K. and Irving A. J. (1981) Composition of the earth's upper mantle—I. siderophile trace elements in ultramafic nodules. *Tectonophys.* **75**, 47–67.

Morten L. (1971) Noduli ultrafemici nel basalto di S. Giovanni Ilarione (Valle dell'Alpone, M. Lessini, Verona). *Miner. Petrogr. Acta* **17**, 15–40.

Morten L. (1978) REE abundances in spinel-lherzolite nodules and host basalt from S. Giovanni Ilarione quarry, Veneto Region, North Italy. *Neues Jahrb. Miner. Mon.*, 159–165.

Muehlenbachs K. and Kushiro I. (1974) Oxygen isotope exchange and equilibrium of silicates with CO_2 or O_2. *Carnegie Inst. Wash., Yearb.* **73**, 223–236.

Murck B. W., Burruss R. C. and Hollister L. S. (1978) Phase equilibria in fluid inclusions in ultramafic xenoliths. *Amer. Mineral.* **63**, 40–46.

Nagasawa H., Wakita H., Higuchi H. and Onuma N. (1969) Rare earths in peridotite nodules: an explanation of the genetic relationship between basalt and peridotite nodules. *Earth Planet. Sci. Lett.* **5**, 337–381.

Nicholls I. A. and Lorenz V. (1973) Origin and crystallization history of Permian tholeiites from the Saar-Nahe Trough, SW Germany. *Contrib. Mineral. Petrol.* **40**, 327–344.

Nishimura S. (1972) Partition of uranium between peridotite nodules and host basalt. *Chem. Geol.* **10**, 211–221.

Nixon P. H., ed. (1973) *Lesotho Kimberlites*. Lesotho National Development Corporation, Maseru, 350 pp.

Nixon P. H. and Boyd F. R. (1973a) Petrogenesis of the granular and sheared ultrabasic nodule suite in kimberlites. In *Lesotho Kimberlites*, (P. H. Nixon, ed.), p. 48–56. Lesotho Nat. Devel. Corp.

Nixon P. H. and Boyd F. R. (1973b) The Liqhobong intrusions and kimberlitic olivine compositions. In *Lesotho Kimberlites*, (P. H. Nixon, ed.), p. 141–148. Lesotho Nat. Devel. Corp.

Nixon P. H. and Boyd F. R. (1979) Garnet bearing lherzolites and discrete nodule suites from the Malaita alnoite Solomon Islands, S. W. Pacific, and their bearing on ocean mantle composition and geotherm. In *The Mantle Sample: Inclusions in Kimberlites and Other Volcanics* (F. R. Boyd and H. O. A. Meyer, eds.), p. 400–423. Amer. Geophys. Union, Washington, D.C.

O'Neil J. R., Hedge C. E. and Jackson E. D. (1970) Isotopic investigations of xenoliths and host basalts from the Honolulu Volcanic Series. *Earth Planet. Sci. Lett.* **8**, 253–257.

O'Nions R. K. and Pankhurst R. J. (1974) Secular variations in the Sr-isotope composition of Icelandic volcanic rocks. *Earth Planet. Sci. Lett.* **21**, 12–21.

Ottonello G., Piccardo G. B., Mazzucotelli A. and Cimmino F. (1978a) Clinopyroxene-orthopyroxene major and rare earth elements partitioning in spinel peridotite xenoliths from Assab (Ethiopia). *Geochim. Cosmochim. Acta* **42**, 1817–1828.

Ottonello G., Piccardo G. B., Joron J. L. and Treuil M. (1978b) Evolution of the upper mantle under the Assab Region (Ethiopia): suggestions from petrology and geochemistry of tectonitic ultramafic xenoliths and host basaltic lavas. *Geol. Rundschu.* **67**, 547–575.

Paul D. K. (1971) Strontium isotope studies on ultramafic inclusions from Dreiser Weiher, Eifel, Germany. *Contrib. Mineral. Petrol.* **34**, 22–48.

Paul D. K., Crocket J. H. and Nixon P. H. (1979) Abundances of palladium, iridium and gold in kimberlites and associated nodules. In *Kimberlites, Diatremes and Diamonds: Their Geology, Petrology and Geochemistry*

(F. R. Boyd and H. O. A. Meyer, eds.), p. 272–279. Amer. Geophys. Union, Washington, D.C.

Philpotts J. A., Schnetzler C. C. and Thomas H. H. (1972) Petrogenetic implications of some new geochemical data on eclogitic and ultrabasic inclusions. *Geochim. Cosmochim. Acta* **36**, 1131–1166.

Pike J. E. N. and Schwarzman E. C. (1977) Classification of textures in ultramafic xenoliths. *J. Geol.* **85**, 49–61.

Pike J. E. N., Meyer C. E. and Wilshire H. G. (1980) Petrography and chemical composition of a suite of ultramafic xenoliths from Lashaine, Tanzania. *J. Geol.* **88**, 343–352.

Reid A. M., Donaldson C. H., Brown R. W., Ridley W. I. and Dawson J. B. (1975) Mineral chemistry of peridotite xenoliths from the Lashaine volcano, Tanzania. In *Physics and Chemistry of the Earth*, v. 9, (L. H. Ahrens, J. B. Dawson, A. R. Duncan and A. J. Erlank, eds.), p. 525–543. Pergamon, N.Y.

Reid J. B., Jr. and Frey F. A. (1971) Rare earth distribution in lherzolite and garnet pyroxenite xenoliths and the constitution of the upper mantle. *J. Geophys. Res.* **76**, 1184–1196.

Reid J. B., Jr. and Frey F. A. (1978) The nature of the upper mantle beneath the southern Rio Grande Rift: new rare earth evidence (abstract). *EOS (Trans. Am. Geophys. Union)* **59**, 1214.

Reid J. B. and Woods G. A. (1978) Oceanic mantle beneath the southern Rio Grande rift. *Earth Planet. Sci. Lett.* **41**, 303–316.

Rhodes J. M. and Dawson J. B. (1975) Major and trace element chemistry of peridotite inclusions from the Lashaine volcano, Tanzania. In *Physics and Chemistry of the Earth*, v. 9, (L. H. Ahrens, J. B. Dawson, A. R. Duncan and A. J. Erlank, eds.), p. 545–557. Pergamon, N.Y.

Ridley W. I. and Dawson J. B. (1975) Lithophile trace element data bearing on the origin of peridotite xenoliths, ankaramite and carbonatite from Lashaine volcano, N. Tanzania. In *Physics and Chemistry of the Earth*, v. 9, (L. H. Ahrens, J. B. Dawson, A. R. Duncan and A. J. Erlank, eds.), p. 559–569. Pergamon, N.Y.

Ringwood A. E. (1966) The chemical composition and origin of the earth. In *Advances in Earth Sciences* (P. M. Hurley, ed.), p. 287–356. The M.I.T. Press, Cambridge.

Ringwood A. E. (1971) Core-mantle equilibrium: comments on a paper by R. Brett. *Geochim. Cosmochim. Acta* **35**, 223–230.

Ringwood A. E. (1977) Composition of the core and implications for origin of the earth. *Geochem. J.* **11**, 111–135.

Ringwood A. E. and Kesson S. E. (1977) Basaltic magmatism and the bulk composition of the Moon II. siderophile and volatile elements in Moon, Earth, and chondrites: implications for lunar origin. *The Moon* **16**, 425–464.

Roedder E. (1965) Liquid CO_2 inclusions in olivine-bearing nodules and phenocrysts from basalts. *Amer. Mineral.* **50**, 1746–1782.

Schilling J. G. (1966) Rare-earth fractionation in Hawaiian volcanic rocks. Ph.D. Thesis, Mass. Inst. Technology.

Shimizu N. (1975) Rare-earth elements in garnets and clinopyroxenes from garnet lherzolite nodules in kimberlites. *Earth Planet. Sci. Lett.* **25**, 26–32.

Skewes M. A. and Stern C. R. (1979) Petrology and geochemistry of alkali basalts and ultramafic inclusions from the Palei-Aike volcanic field in southern Chile and the origin of the Patagonian plateau lavas. *J. Volcanol. Geotherm. Res.* **6**, 3–25.

Smith D. (1979) Hydrous minerals and carbonates in perodotite inclusions from the Green Knobs and Buell Park kimberlitic diatremes on the Colorado Plateau. In *The Mantle Sample: Inclusions in Kimberlites and Other Volcanics* (F. R. Boyd and H. O. A. Meyer, eds.), p. 345–356. Amer. Geophys. Union, Washington, D.C.

Sobolev N. V. (1977) *Deep-Seated Inclusions in Kimberlites and the Problem of the Composition of the Upper Mantle.* Amer. Geophys. Union, Washington, D.C. 279 pp.

Stosch H. -G. and Seck H. A. (1980) Geochemistry and mineralogy of two spinel lherzolite suites from Dreiser Weiher, West Germany. *Geochim. Cosmochim. Acta* **44**, 457–470.

Stueber A. M. and Ikramuddin M. (1974) Rubidum, strontium and the isotopic composition of strontium in ultramafic nodules, minerals and host basalts. *Geochim. Cosmochim. Acta* **38**, 207–216.

Sutherland F. L. (1974) High-pressure inclusions in tholeiitic basalt and the range of lherzolite-bearing magmas in the Tasmanian volcanic province. *Earth Planet. Sci. Lett.* **24**, 317–324.

Taylor H. P., Jr. (1968) The oxygen isotope geochemistry of igneous rocks. *Contrib. Mineral. Petrol.* **19**, 1–71.

Turekian K. K. and Clark S. P. (1969) Inhomogeneous accumulation of the earth from the primitive solar nebula. *Earth Planet. Sci. Lett.* **6**, 346–348.

Varne R. (1970) Hornblende lherzolite and the upper mantle. *Contrib. Mineral. Petrol.* **27**, 45–51.

Varne R. and Graham A. L. (1971) Rare earth abundances in hornblende and clinopyroxene lherzolite xenoliths: implications for upper mantle fractionation processes. *Earth Planet. Sci. Lett.* **13**, 11–18.

Wass S. Y. and Irving A. J. (1976) XENMEG: a catalogue of occurrences of xenoliths and megacrysts in basic volcanic rocks of eastern Australia. Australian Museum, Sydney, 455 pp.

Wells P. R. A. (1977) Pyroxene thermometry in simple and complex systems. *Contrib. Mineral. Petrol.* **62**, 129–139.

White R. W. (1966) Ultramafic inclusions in basaltic rocks from Hawaii. *Contrib. Mineral. Petrol.* **12**, 245–314.

Wilshire H. G. and Shervais J. W. (1975) Al-augite and Cr-diopside ultramafic xenoliths in basaltic rocks from western United States. In *Physics and Chemistry of the Earth*, v. 9, (L. H. Ahrens, J. B. Dawson, A. R. Duncan and A. J. Erlank, eds.), p. 257–272. Pergamon, N.Y.

Wilshire H. G. and Trask N. J. (1971) Structural and textural relationships of amphibole and phlogopite in peridotite inclusions, Dish Hill, California. *Amer. Mineral.* **56**, 240–255.

Wilshire H. G., Pike J. E. N., Meyer E. C. and Schwarzman E. C. (1980) Amphibole-rich veins in lherzolite xenoliths, Dish Hill and Deadman Lake, California. *Amer. J. Sci.*, Jackson vol. **280A**, 576–593.

Wolf R., Woodrow A. and Anders E. (1979) Lunar basalts and pristine highland rocks: comparison of siderophile and volatile elements. *Proc. Lunar Planet. Sci. Conf. 10th*, p. 2107–2130.

Wood B. J. and Heming R. F. (1978) Major and trace element disequilibrium during partial melting (abstract). In *U.S. Geol. Surv. Open-File Rept. 78-701*, p. 459–462.

Wright J. B. (1969) Olivine nodules and related inclusions in trachyte from the Jos Plateau, Nigeria. *Mineral. Mag.* **37**, 370–374.

Wyatt B. A. (1977) The melting and crystallisation behaviour of a natural clinopyroxene-ilmenite intergrowth. *Contrib. Mineral. Petrol.* **61**, 1–9.

1.3 PLANETARY COMPARISONS

1.3.1 BASALTS AS PROBES OF PLANETARY INTERIORS: CONSTRAINTS FROM MAJOR AND TRACE ELEMENT CHEMISTRY

1.3.1.1 Introduction

Basalts, as partial melts of planetary interiors, have compositions that are determined by many factors including pre-acretionary history of the planetary materials, the mass of the planetary body, the body's origin and early evolutionary history, the temporal evolution of the source, the nature of the partial melting and subsequent magmatic evolutionary processes, the residual mineralogy and phase chemistry of the source, and the intensive thermodynamic and kinetic parameters operating at the source and at the emplacement site. Chemical, isotopic, and experimental studies of basalt suites emplaced in different tectonic regimes can provide data that enable us to place constraints on these planetary formational and evolutionary processes and features.

We now have samples of basalts derived from at least three planetary objects: the Earth, the Moon, and the parent body of the basaltic achondrite meteorites (plagioclase-pyroxene achondrites, i.e., eucrites, howardites and shergottites). There is good evidence that these objects all experienced extensive early melting and differentiation events: the eucrite parent body (EPB) essentially at its time of formation ~4.5 b.y. ago (e.g., Huneke *et al.*, 1977); the Moon during the first 200–300 m.y. (e.g., Papanastassiou and Wasserburg, 1971, 1972; Nyquist *et al.*, 1974; Tera and Wasserburg, 1974, 1975; Lugmair, 1975; Lugmair *et al.*, 1975; Wasserburg *et al.*, 1977); and the Earth at the time of core separation within the first few hundred million years (Armstrong, 1968; Oversby and Ringwood, 1971; Murthy, 1975; Gancarz and Wasserburg, 1977; Vidal and Dosso, 1978).

The objective of this section is to compare and contrast the major and trace element chemistries of lunar mare basalts, of the eucrite group of the basaltic achondrites, and of selected terrestrial basalts from a variety of tectonic and temporal settings.

Temporal constraints which must be kept in mind are as follows:

(a) Mare basalts in the sampled suite erupted in the interval between 3.9–3.2 b.y. ago.
(b) Some eucrites have crystallization ages as great as ~4.5 b.y. (Huneke *et al.*, 1977).
(c) The terrestrial basalt suite includes Archean komatiites and basalts (3.8–2.6 b.y.), pre-Tertiary flood basalts, Tertiary flood basalts, island arc basalts ($\leq$ 150 m.y.), and mid-ocean ridge basalts (MORBs) ($\leq$ 180 m.y.).

Our discussion pertains to the development of the sources of these basalts *prior* to partial melting and to their chemical and mineralogical characteristics *at the time of* magma generation.

1.3.1.2 Recognition of primary magmas

Basaltic magmas should have chemistries characteristic of their parent body, provided that the effects of the processes of magma generation and subsequent modification through low- and high-pressure fractionation, mixing, assimilation, etc., can be recognized and isolated (see section 1.4.1). We can identify compositions that were generated internally as partial melts and that erupted and crystallized with minimal compositional modification, i.e., primary magmas (see section 1.4.2). This identification process requires careful assessment of the compositional ranges, the textures and mineralogy, and the phase relations at low and high pressures of each basalt group (see section 3.1).

Fortunately, in the case of terrestrial basalts, for which our sampled suites are vastly more extensive than for the Moon or for the meteorites, two important pieces of information enable us to constrain the compositions of the most primitive basalts generated in the upper mantle. *First*, experimental data indicate that the more primitive of such basalts (both alkaline and subalkaline) have olivine on the liquidus over a wide range of pressures. The consequences of olivine fractionation are easily predicted for many diagnostic major, minor, and trace elements. Furthermore, the most magnesian of these experimental olivines (Fo_{89-91}) (e.g., Green *et al.*, 1979) is comparable in composition to the most magnesian olivine phenocrysts observed in natural basalts. *Second*, from direct and indirect evidence, we have a fairly good idea of the mineralogy and mineral chemistry of terrestrial basalt mantle source regions: they consist of lherzolites dominated by olivine (Fo_{90}) and orthopyroxene with lesser clinopyroxene and a minor aluminous phase (spinel and/or garnet). Thus, from theoretical and experimental considerations, we are able to identify the most probable compositions for primary terrestrial basalts. Magmas derived by partial melting of mantle lherzolite with $Mg/(Mg + Fe^{2+}) \simeq 0.90$ should have a $Mg/(Mg + Fe^{2+})$ ratio $\approx$ 0.72–0.74

(e.g., Green, 1971; Green *et al.*, 1974; Ringwood, 1975; Hanson and Langmuir, 1978). This ratio (as well as the Ni content of the magmas) will be relatively insensitive to the percentage of partial melting involved in generating the magma, but will be highly sensitive to the amount of olivine fractionation which has taken place since.

No samples of the lunar mantle have been recognized, so we are forced to use geochemical arguments and indirect evidence to deduce the mineralogy of the lunar interior. The mare composition thought to be most primitive is the green glass in soil sample 15426 (Ridley *et al.*, 1973). However, this glass really consists of a spectrum of compositions (e.g., Delano 1979; Basu *et al.*, 1979). Low-pressure phase relationships of the glass have been explored, and have led to the conclusion that the compositional variations cannot be explained solely by low-pressure fractionation.

The low-pressure phase relationships are also known for several of the eucrites (Stolper, 1977). Their compositions appear to be saturated with olivine, plagioclase, and pyroxene at very low pressures. This may indicate that they all are products of extreme low-pressure fractional crystallization or that they are partial melts generated in a very small parent body. A fortuitous set of circumstances would have been required for the eucrite compositions to cluster as they do at a multiple saturation point, if crystal fractionation were the controlling process. Walker *et al.* (1979) argue that the eucrites are partial melts and that many of them are primary. If their arguments are valid, then the eucrites are capable of supplying information on the nature of their parent bodies.

1.3.1.3 Role of planet size

Planet size appears to play an important role in determining the chemistry of erupted basalts (Walker *et al.*, 1979). Of the several effects of planet size that influence the generation of basaltic magmas, we shall briefly discuss the effect of pressure. O'Hara (1968b), Kushiro (1972) and Presnall *et al.* (1979) have documented the effect of pressure on the composition of a partial melt derived from a lherzolitic source and Wyllie (1979) discusses the role of volatile components in terrestrial magmas. In a dry system, quartz-normative compositions are the first liquids produced by partial melting of lherzolite at low pressures. The effect of increasing pressure is to drive the cotectic melting reactions to increasingly olivine-normative compositions. Hence, a dry lherzolitic source region may yield a spectrum of melt compositions from quartz-normative to olivine-normative if partial melting occurs at different depths within the source. The presence of volatiles (CO_2, H_2O) will also affect the positions of cotectic reactions involved in the melting of a lherzolite source. Water has an effect opposite to that of increasing pressure, and allows quartz-normative magmas to be derived by small amounts of partial melting at high pressures (Kushiro, 1972). If CO_2 is the abundant volatile species, subsilicic magmas will be generated by partial melting at high pressures.

The preceding discussion is complicated by differences in source region chemistry of the planetary body because the bulk chemistry of the source region will also affect the compositions of magmas derived by partial melting at different depths. However, it is possible to apply the pressure control arguments to partial melts derived from the eucrite parent body and the Earth. The small planetary body that generated the eucrites (Stolper, 1977) had only a limited variation of pressure with depth and would have produced quartz-normative liquids throughout its interior, while a large planet (e.g., the Earth) is capable of producing a spectrum of liquid compositions from a single source, depending on the depth of partial melting.

1.3.1.4 Major element characteristics of planetary basalt

$Mg/(Mg + Fe^{2+})$ ratio

Planetary basalts form distinct major and minor element compositional groups which appear to be characteristic of their parent body. Variations within groups reflect, for the most part, fractional crystallization controls. The most magnesian, *noncumulate* compositions of the suite differ markedly between groups. Terrestrial basalts, with the exception of the Precambrian basaltic komatiites, have $Mg/(Mg + Fe^{2+})$ ratios (Mg') $\simeq 0.7$. Precambrian komatiites and basaltic komatiites have $Mg/(Mg + Fe^{2+})$ ratios approaching that of the mantle (~ 0.9) and appear to be the result of very high percentages of partial melting or of sequential melting.

Crystalline MORBs and oceanic intraplate (seamounts and islands) basalts having $Mg' \gtrsim 0.7$ are olivine cumulates. The least-fractionated MORB glasses have $Mg' = 0.70$–0.72 and are among the most likely candidates for primary basalt magmas.

The most magnesian noncumulate mare basalts [very low-titanium (VLT) basalt from Apollo 15 and 17, Vaniman and Papike (1976); and Apollo 15 green glass, Ridley *et al.* (1973)] have $Mg/(Mg + Fe^{2+})$ ratios $\simeq$ 0.60–0.64. Olivine compositions in these materials closely approximate equilibrium compositions calcu-

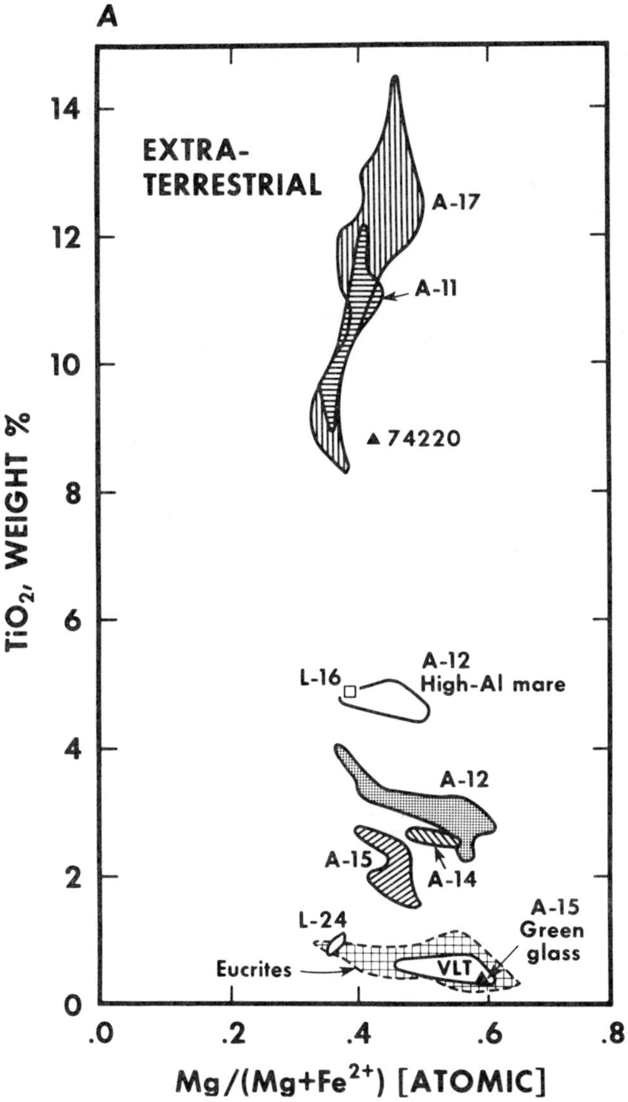

lated assuming a $K_D = 0.33$ (Longhi *et al.*, 1978). Furthermore, no olivine with Fo > 80 has been observed in mare basalts. Thus, there is no evidence that liquids more magnesian than $Mg/(Mg + Fe^{2+}) \simeq 0.60$–$0.64$ have existed at the lunar surface and, if olivine is the dominant phase in the source, its composition can be no more magnesian than $\sim Fo_{82-85}$.

The eucrites have $Mg/(Mg + Fe^{2+})$ ratios ranging from ~0.35 to ~0.65 with the majority in the range 0.38–0.45 (section 1.2.8). Only Binda (Mg′ = 0.65) has an $Mg/(Mg + Fe^{2+})$ ratio >0.6. This suggests that many of the eucrites are extensively evolved or that they are derivatives of source regions having $Mg/(Mg + Fe^{2+})$ as low as, and possibly lower than, that of the mare basalt source regions. If the eucrites are primitive partial melts, as suggested by Walker *et al.* (1979), then their chemistries reflect the chemistries and residual mineralogies of their sources. These sources thus appear to have had significantly lower $Mg/(Mg + Fe^{2+})$ ratios than most terrestrial basalt sources.

Titanium

Because Ti behaves largely as an incompatible element during partial melting of the Earth's mantle, its abundance in basalt magmas is a reflection of (a) its abundance in the source, (b) the percentage of partial melting, and (c) the extent and nature of crystal fractionation. The magnitude of the differences in TiO_2 abundances between and within the planetary basalt groups (Fig. 1.3.1.1) is striking.

The mare basalts fall into two major groups (high- and low-TiO_2 groups), each of which can be subdivided into smaller subgroups. The TiO_2 ranges of these groups

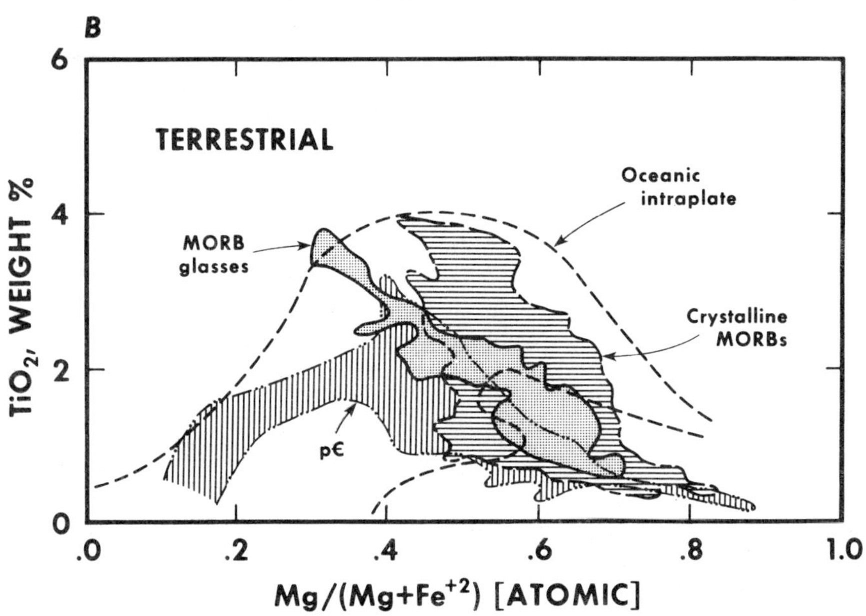

1.3.1.1 TiO_2 vs. $Mg/(Mg + Fe^{2+})$ fields of planetary basalt suites. **a.** Extraterrestrial basalts. Fields are shown for Apollos 11, 12, 15, 17 and Luna 16 and 24 mare basalts; Apollo 15 green glass; 74220, Apollo 17 orange glass; A-14, Apollo 14 mare-type basalts 14053 and 14072; VLT, very low-TiO_2 basalts from Vaniman and Papike (1976). Eucrite data from section 1.2.8. **b.** Terrestrial basalts. Fields are shown for oceanic intraplate basalts (Bence *et al.*, 1979a, b), crystalline MORBs and MORB glasses (Bence *et al.*, 1979a, b; see section 1.2.6), and earliest Archean (p€) basalts. For the terrestrial suites, $Mg' = Mg/(Mg + Fe^{2+})$ is calculated assuming $Fe^{3+}/\Sigma Fe = 0.1$ (atomic).

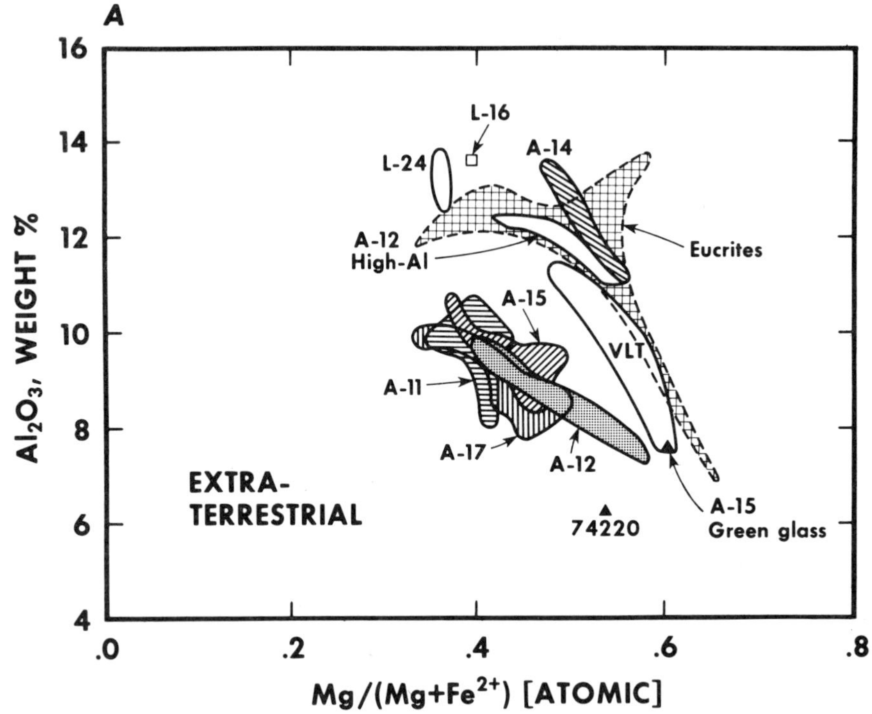

1.3.1.2 Al_2O_3 vs. Mg/(Mg + Fe^{2+}) fields of planetary basalt suites. **a.** Extraterrestrial basalts. **b.** Terrestrial basalts. Symbols as in Fig. 1.3.1.1.

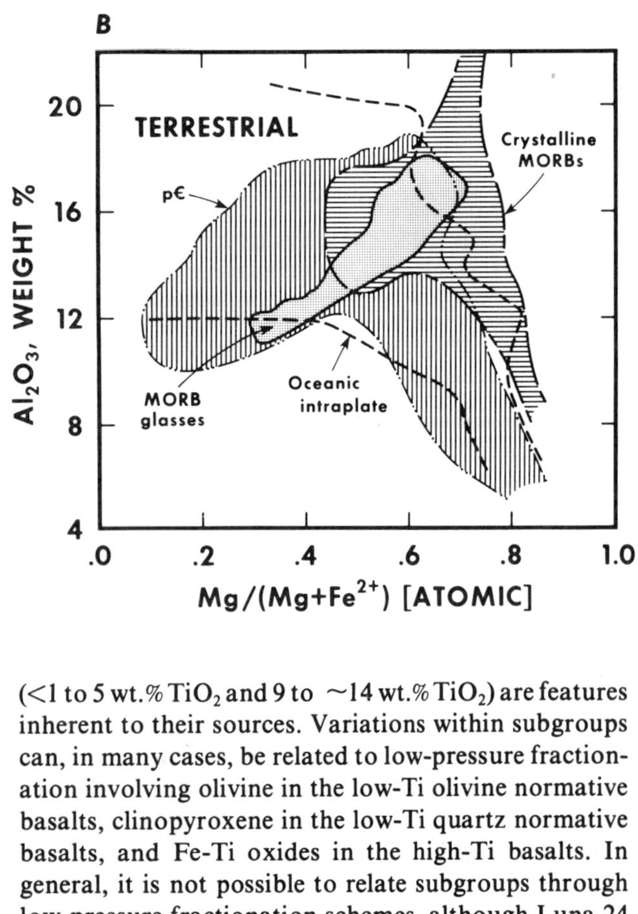

(<1 to 5 wt.% TiO_2 and 9 to ~14 wt.% TiO_2) are features inherent to their sources. Variations within subgroups can, in many cases, be related to low-pressure fractionation involving olivine in the low-Ti olivine normative basalts, clinopyroxene in the low-Ti quartz normative basalts, and Fe-Ti oxides in the high-Ti basalts. In general, it is not possible to relate subgroups through low-pressure fractionation schemes, although Luna 24 basalts may be related on major element grounds to the subgroup defined by the Apollo 17 VLT basalts and Apollo 15 green glasses. In contrast, however, rare-earth element data (Wentworth *et al.*, 1979) are interpreted to indicate different source materials for the Apollo 17 and Luna 24 VLT basalts.

The low-TiO_2 (0.5–1.0 wt.%) and variable, but generally low, Mg/(Mg + Fe^{2+}) ratios of the eucrites reflect the chemical characteristics of the source(s) of these basalts.

The most magnesian terrestrial Archean basalts and komatiites define an olivine control line, as do crystalline MORBs and basalts erupted in the oceanic intraplate regime. MORB glasses have a compositional field that reflects olivine-plagioclase cotectic control at high Mg'-values (Mg' $\gtrsim$ 0.6 and TiO_2 ~0.8–2 wt.%) and olivine-plagioclase-clinopyroxene control at lower Mg'-values (Mg' $\leq$0.55 and TiO_2 $\gtrsim$2 wt.%) (Bence *et al.*, 1979a). At ratios of Mg/(Mg + Fe^{2+}) ~0.70, MORB glasses and crystalline basalts and Archean basalts have TiO_2 concentrations in the range 0.5 to ~1 wt.%, which suggests sources with comparable TiO_2. If partial melting levels of 20–30% are assumed, these sources will have TiO_2 levels in the vicinity of 0.2 wt.%. Oceanic intraplate tholeiitic basalts have distinctly higher TiO_2 at the same Mg/(Mg + Fe^{2+}) ratio. For the oceanic intraplate basalts, this is a well-defined range (~1.5–2.5 wt.%), which has been interpreted (see section 1.2.6) to be a consequence of higher TiO_2 (by a factor of three) contents in the deeper sources of these basalts.

Earliest Precambrian (Archean) basalts with Mg′ ≃0.7 have TiO_2 concentrations similar to crystalline MORBs and MORB glasses. This is interpreted to be an indication of sources having comparable TiO_2 concentrations. Thus, with respect to TiO_2, early Precambrian basalt sources are "depleted" to the same extent as MORB sources.

Aluminum

The abundances of aluminum differ markedly among the planetary basalt groups. The most primitive (magnesian, noncumulate) mare basalts contain significantly less Al_2O_3 (6–9 wt.% Al_2O_3 at Mg′ = 0.55–0.60) than the most primitive MORB glasses (15–18 wt.% Al_2O_3 at Mg′ ≃0.7; Fig. 1.3.1.2). The eucrites and the aluminous mare basalts, defined by the Apollo 12 high-alumina basalts (e.g., Ridley, 1975), and the Apollo 14, Luna 16, Luna 24 and VLT basalt suites have similar Al_2O_3 abundances at similar Mg′-values. The Apollos 11, 15, 17 and a majority of Apollo 12 basalts have comparable Al_2O_3 contents. Variations within groups appear to be consequences of olivine or clinopyroxene fractionation. It is clear that there are two distinct mare basalt groups that can be recognized on the basis of their alumina contents. These groups cannot be related to each other through crystal fractionation. Instead, the differences must reflect either source region chemical and mineralogical differences or differences in the degree of partial melting.

Crystalline MORBs have a compositional spectrum that indicates the importance of both olivine and plagioclase accumulation. The latter effect appears to have been the more important in samples that have primitive chemistries (i.e., high Mg′). MORB glasses

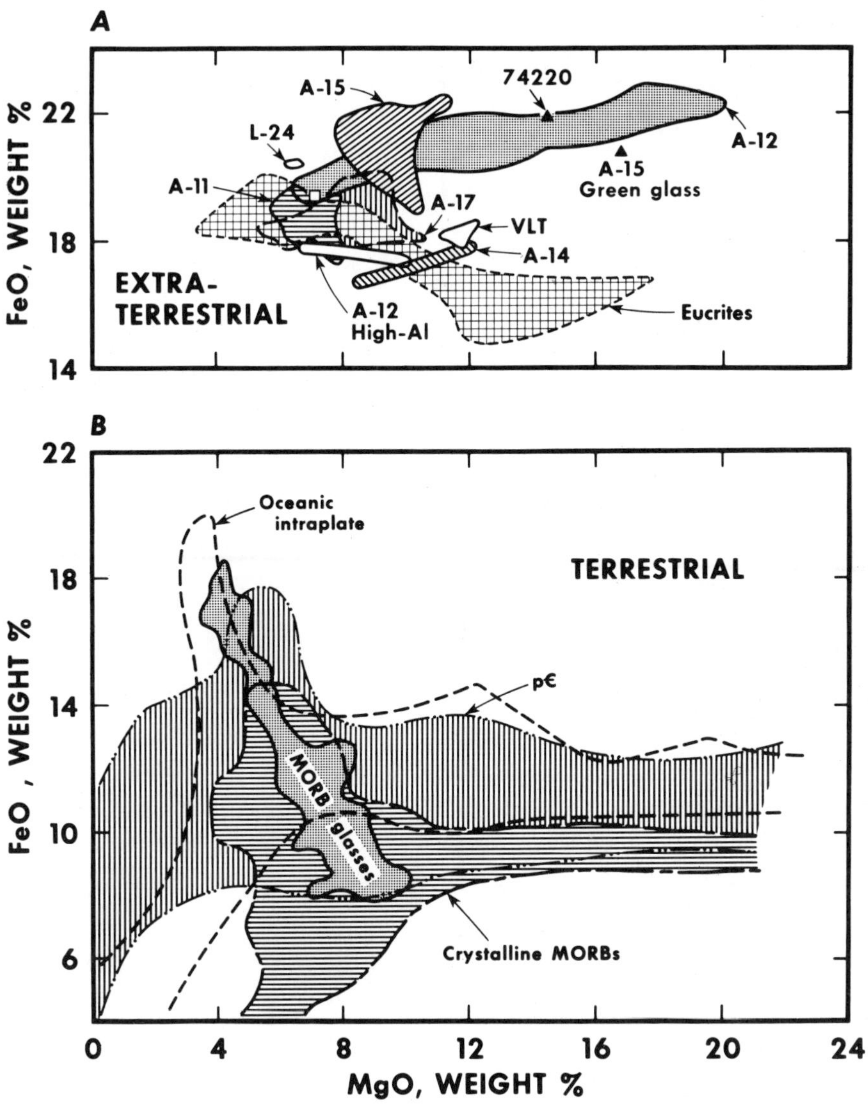

1.3.1.3 FeO vs. MgO fields for planetary basalt suites. FeO = total iron calculated as FeO. **a.** Extraterrestrial basalts. **b.** Terrestrial basalts. Symbols as in Fig. 1.3.1.1.

define a very tight field in which the most primitive [highest $Mg/(Mg + Fe^{2+})$] have the highest alumina (16–17.5 wt.% Al_2O_3). This field defines the low-pressure liquid line of descent which is followed by ocean ridge basalts that co-crystallize olivine + plagioclase but then become multiply-saturated with three crystalline phases (olivine + plagioclase + clinopyroxene) (Bence *et al.*, 1979a).

Ocean island (intraplate) tholeiitic basalts have systematically lower alumina contents than do MORBs with comparable values of $Mg/(Mg + Fe^{2+})$. This may be a consequence of source region chemical differences, or it may be a consequence of the retention of an aluminous phase (e.g., garnet) in the deeper sources for the oceanic intraplate basalts.

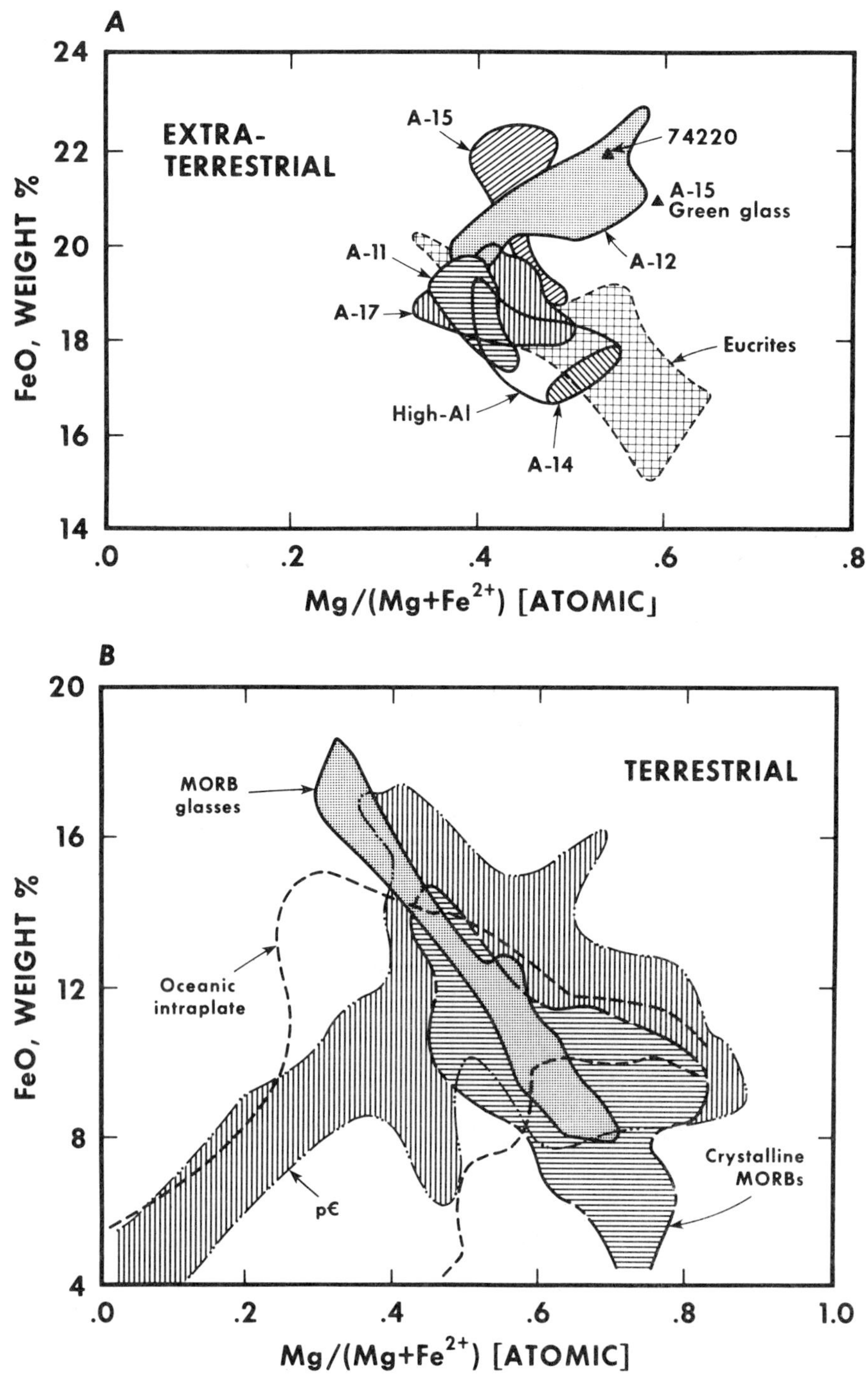

1.3.1.4 FeO vs. $Mg/(Mg + Fe^{2+})$ for planetary basalt suites. **a.** Extra-terrestrial basalts. **b.** Terrestrial basalts. Symbols as in Fig. 1.3.1.1.

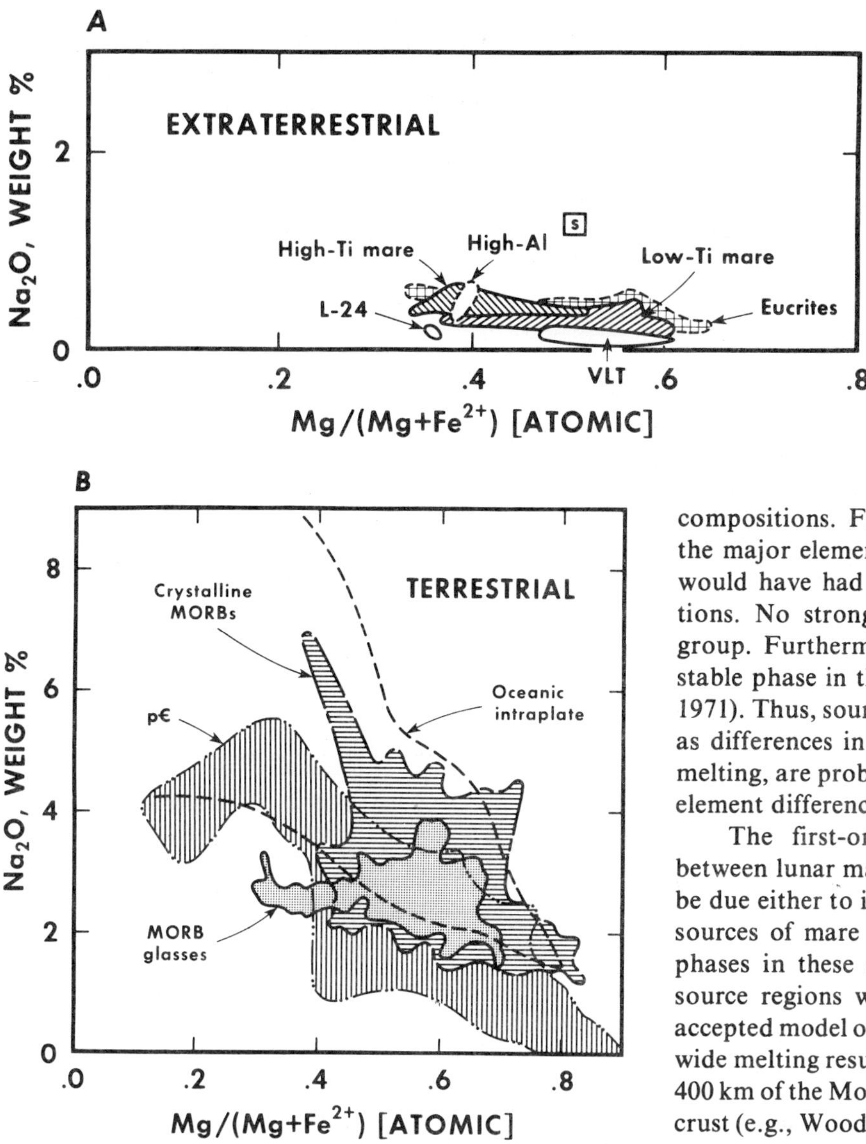

1.3.1.5 Na_2O vs. $Mg/(Mg + Fe^{2+})$ fields for planetary basalt suites. **a.** Extraterrestrial basalts, S = Shergotty. **b.** Terrestrial basalts. Symbols as in Fig. 1.3.1.1.

Earliest Precambrian (Archean) basalts and komatiites also have systematically lower Al_2O_3 contents than MORBs with comparable $Mg/(Mg + Fe^{2+})$ ratios. Variations within the group are readily explained by varying degrees of olivine, plagioclase or clinopyroxene fractionation. Noncumulate compositions appear to be the result of very high degrees of partial melting (e.g., Green *et al.*, 1975) or of sequential melting. The lower Al_2O_3 contents of the Archean basalts and komatiites relative to MORBs may result from source region chemical and mineralogical differences or differences in depth or in percentage of partial melting. Fractionation under any pressure cannot relate these two compositional groups. Olivine and clinopyroxene controls define compositional trends within the groups. Plagioclase is rarely a liquidus phase for the most primitive compositions. Fractionation of garnet could explain the major element chemistries of the two groups, but would have had a major effect on the REE distributions. No strong garnet effect is observed in either group. Furthermore, garnet does not appear to be a stable phase in the source regions of MORBs (Green, 1971). Thus, source region chemical differences, as well as differences in the depth and percentage of partial melting, are probably the causes of the observed major element differences.

The first-order differences in Al_2O_3 contents between lunar mare basalts and terrestrial basalts may be due either to inherently lower Al_2O_3 contents in the sources of mare basalts or to retention of aluminous phases in these sources. Lower alumina contents in source regions would be consistent with the widely-accepted model of lunar evolution wherein early Moon-wide melting resulted in the differentiation of the outer 400 km of the Moon and formation of a plagioclase-rich crust (e.g., Wood *et al.*, 1970; Taylor and Jakeš, 1974).

Iron

Major differences in total iron (calculated here as FeO) contents of the source regions are indicated by FeO variations in the derivative basalts. Mare basalts and eucrites have consistently higher FeO contents than any terrestrial basalts having comparable MgO abundances and $Mg/(Mg + Fe^{2+})$ ratios (Figs. 1.3.1.3–4). These must imply first-order planetary differences in the FeO contents of their source regions and cannot entirely be consequences of variable depths to the source or of variable degrees of partial melting.

There are several important observations to be made for the terrestrial basalts. First, oceanic intraplate tholeiitic basalts have consistently higher total iron contents than mid-ocean ridge basalts. This may be due to differences in the depths to the respective source regions; however, in view of the differences in TiO_2 (and other incompatible elements), it is likely that chemical differences play a significant role. Second, the earliest Precambrian basalts and komatiites have a much wider range of FeO abundances than crystalline MORBs. This reflects the variable iron contents which characterized basalt sources in the earliest Precambrian, and emphasizes the relative homogeneity of the modern oceanic upper mantle sources of MORBs.

Sodium

The abundance of Na_2O varies markedly in the basalts from Earth, on the one hand, and in the mare basalts and eucrites on the other. The lunar basalts and the eucrites, with the sole exception of Shergotty, are depleted in Na_2O relative to terrestrial basalts (Fig. 1.3.1.5). This is consistent with the overall depletion of the volatile elements and enrichment of refractory elements in the Moon. The overall depletion of volatiles in the Moon has been confirmed by studies of the highlands suite (e.g., Taylor, 1975). Consequently, volatile depletion must have occurred prior to lunar accretion. The similar depletion of Na_2O observed in the eucrites is consistent with a close genetic relationship between the eucrite parent body and the Moon.

1.3.1.5 Mantle source mineralogy

It is clear from the major element abundances in the various planetary basalt suites that there are compositional differences among them which reflect intrinsic chemical differences in their sources. What can these basalt suites tell us about their source region mineralogies? We must first identify the most primitive compositions—those that are the least-modified by post-segregation processes. If such liquids are not present, then we need to examine genetically-related basalt suites in order to identify initial magma compositions and to determine the liquid-line-of-descent. For example, a plot of the compositions of the Apollo 12 picrite basalt suite on the subprojection silica-olivine-anorthite (Fig. 1.3.1.6a) shows that they can be related by an olivine control line (Kushiro and Nakamura, 1970; Walker *et al.*, 1976). Such information, when combined with textural and phase chemical data, can be used to recognize cumulates and differentiates. In this way appropriate compositions may be selected for experimental studies.

Lunar mare basalt vitrophyre 12009 is a candidate near-primary liquid. Green *et al.* (1971) conducted melting experiments on a synthetic 12009 composition and determined that it was multiply-saturated with olivine and clinopyroxene near 7 kb (~150 km). The addition of 10% olivine (Fo_{75}) to the 12009 composition depresses the multiple saturation point to pressures >15 kb ($\gtrsim$300 km), where the crystallization sequence is olivine (Fo_{75-80}),

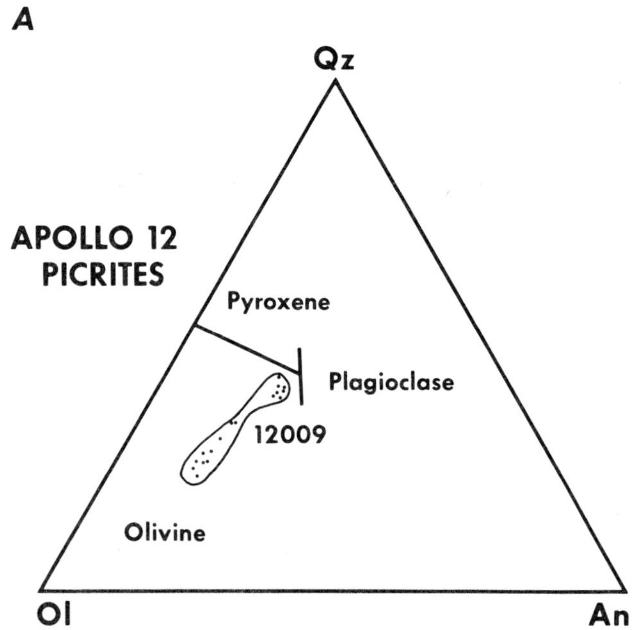

1.3.1.6a Apollo 12 mare basalt compositions on the subprojection olivine-quartz-anorthite. Sample 12009 is a candidate for a primary magma composition.

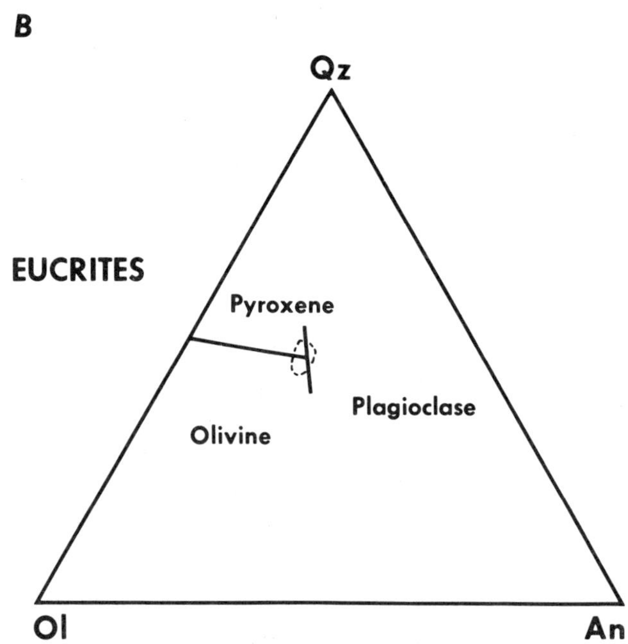

1.3.1.6b Eucrite compositions from Stolper (1977) on the subprojection olivine-quartz-anorthite.

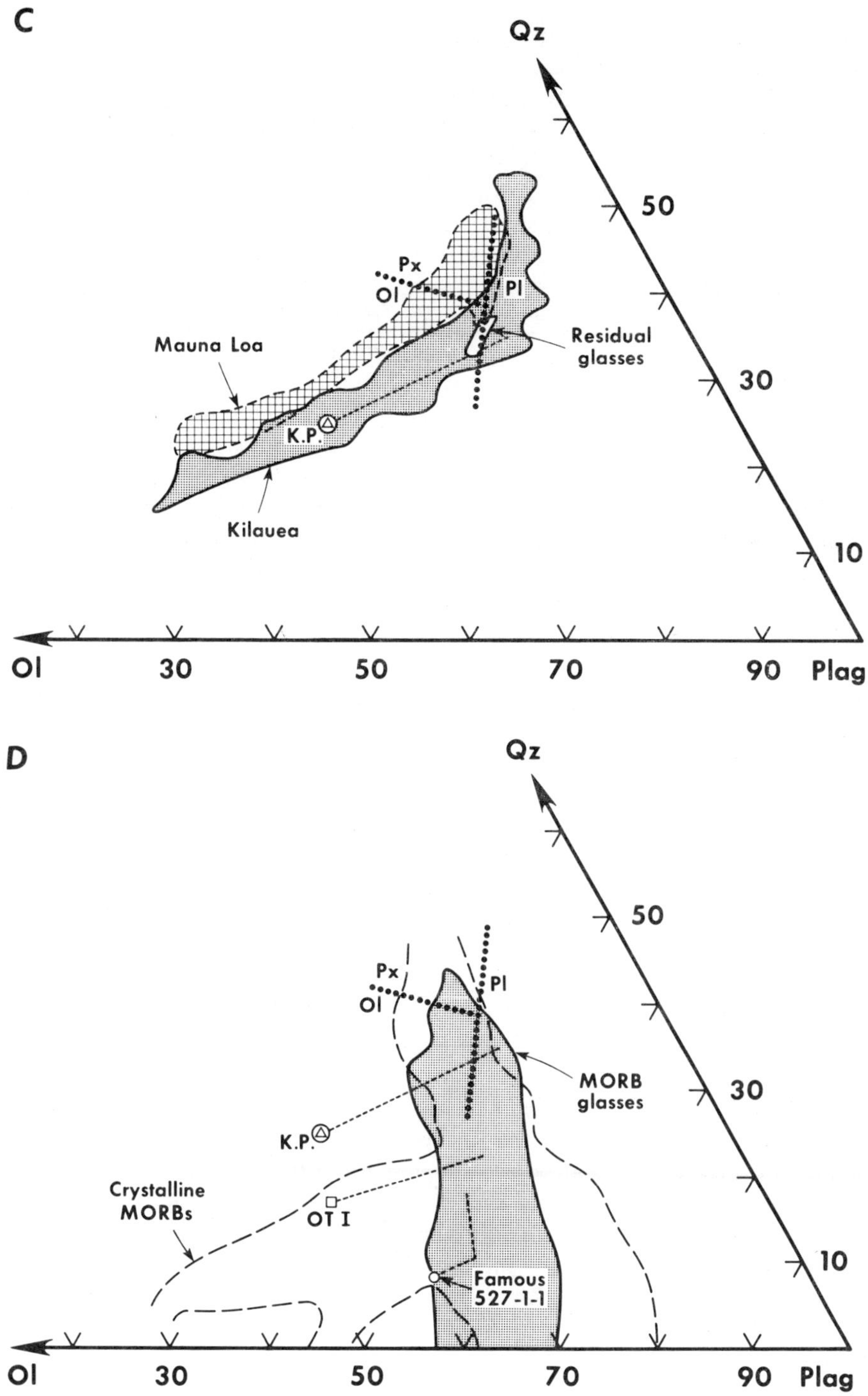

1.3.1.6c, d Terrestrial oceanic basalts on the subprojection olivine-quartz-plagioclase. MORB glass field from Bence *et al.* (1979a). Hawaiian Mauna Loa and Kilauea fields from section 1.2.6. Hypothetical Kilauea parent composition K. P. (section 1.2.6). O. T. I. is hypothetical Kilauea parent composition from Green and Ringwood (1967). The position of the multiple saturation point is schematic.

orthopyroxene [$Mg/(Mg + Fe^{2+})$ ~0.8], and clinopyroxene [$Mg/(Mg + Fe^{2+}) \simeq 0.7$–$0.8$]. Green *et al.* (1971) concluded that the parent liquid, from which 12009 was derived through olivine fractionation at low to moderate pressures, could have been produced by partial melting of a lunar mantle assemblage consisting of olivine, orthopyroxene, and clinopyroxene having $Mg/(Mg + Fe^{2+}) \simeq 0.75$–$0.80$ at depths $\gtrsim$200 km.

Another candidate for a lunar primary liquid composition, the Apollo 15 emerald green glass, is multiply-saturated at pressures equivalent to a depth of ~400 km (Stolper, 1974). This would be a minimum depth if olivine fractionation has occurred, but is a maximum if reduction and fractionation of metallic iron have taken place.

For a range of mare basalt compositions, there is a pressure at which each composition is multiply-saturated with olivine and pyroxene (e.g., Walker *et al.*, 1975). This pressure ranges from 5–20 kb (100–400 km). There is no consistent relationship between TiO_2 content and depth to the point of multiple saturation.

A number of eucrite compositions plot in the vicinity of the low-pressure olivine-pyroxene-plagioclase multiple saturation point in the pseudoquaternary subprojection olivine-plagioclase-silica (Fig. 1.3.1.6b). Stolper (1977) concluded that these compositions are saturated or very nearly saturated with these crystalline phases under low-pressure conditions. These results are consistent either with a petrogenesis involving partial melting in a small low-$Mg/(Mg + Fe^{2+})$ planetary object ($\leq$ several 100 km radius), or with an origin involving low-pressure crystal fractionation of more magnesian basaltic magma. If the eucrites are low-pressure partial melts, then the Eucrite Parent Body (EPB) sources contained olivine (Fo_{65}), low-Ca pyroxenes (Wo_5En_{65}) and plagioclase (An_{94}).

Terrestrial MORB glasses plotted on a similar projection (Fig. 1.3.1.6c, d) define a low-pressure liquid-line-of-descent on which there has been olivine-plagioclase control followed by olivine-plagioclase-clinopyroxene controls (O'Hara, 1968a; Bence *et al.*, 1979a; Walker *et al.*, 1979). The majority of these compositions appears to have equilibrated under low-pressure conditions. Crystalline MORBs also cluster in the vicinity of the olivine-plagioclase cotectic, but a large number of MORB compositions show the effects of olivine and plagioclase accumulation as well.

The most likely candidates for primary MORB compositions are magnesian vitrophyres and glasses (e.g., Bender *et al.*, 1978; Green *et al.*, 1979) which project into the olivine field to the left of the olivine plagioclase cotectic. However, experimental and trace element studies of primitive compositions identified by Green *et al.* (1979) and Bender *et al.* (1978) suggest that even these compositions have evolved through olivine fractionation. These compositions, corrected for olivine fractionation, contain 14–16 wt.% MgO [$Mg/(Mg + Fe^{2+}) \simeq 0.72$–$0.74$] and are multiply-saturated at pressures of 15–20 kb (50–65 km) with olivine ($\sim Fo_{90}$), clinopyroxene, and orthopyroxene.

Intraplate tholeiitic basalts tend to be uniformly more siliceous than MORBs; consequently, their liquid-line-of-descent differs from that of MORBs and can result in a different paragenesis. Kilauea lavas which contain uniformly less silica than Mauna Loa lavas (Fig. 1.2.6.10) have the paragenesis
ol → ol + plag → ol + plag + cpx. The Mauna Loa lava sequence is ol → ol + cpx → ol + cpx + plag. Chrome spinel precedes, or co-crystallizes with, olivine in both suites.

A candidate parent magma composition for Kilauea lavas (K.P. in Fig. 1.2.6.10) was calculated assuming that $Mg/(Mg + Fe^{2+}) = 0.70$–0.72 and using FeO, Al_2O_3, TiO_2, CaO, Na_2O, and K_2O restrictions imposed by the spectrum of Kilauea compositions. Silica contents obtained by difference must satisfy the restrictions imposed by projections in the olivine-plagioclase-clinopyroxene-silica psuedoquaternary. Note that the Kilauea Iki parent composition OT I, which was used by Green and Ringwood (1967) in their experimental studies, is not an appropriate composition due to too low SiO_2 and too high CaO and Al_2O_3 contents. However, the composition is sufficiently close to being correct that the anhydrous high-pressure melting experiments have some relevance. Olivine (Fo_{91}) is the liquidus phase to about 10 kb where it is replaced by orthopyroxene ($\sim Wo_4En_{88}Fs_8$) and it, in turn, is replaced by clinopyroxene at pressures of $\sim$20 kb. This information and REE data (e.g., Leeman *et al.*, 1977) suggest a mantle source consisting of olivine, orthopyroxene, clinopyroxene, and garnet, with tholeiitic liquids originating at depths on the order of 60–100 km.

How the chemistry of the Archean mantle differs from the modern oceanic mantle, and the conditions of magma genesis which prevailed in the Archean are subjects of considerable speculation. Basaltic and peridotitic komatiites erupted during the Archean may provide means of comparing the Archean mantle with the modern mantle. As has been pointed out, the Archean basalts with $Mg' \simeq 0.7$ have TiO_2 contents similar to those of MORBs, similar but much more variable contents of FeO, and lower Na_2O and Al_2O_3 contents. The variable FeO contents may imply that the upper mantle during the Archean was more heterogeneous than the modern oceanic upper mantle.

Low and high pressure melting studies on peridotitic komatiites indicate that these compositions erupted as high temperature ($\sim$1500–1650° C) liquids (Green *et al.*, 1975; Arndt, 1976; Bickle *et al.*, 1977). The mechanism by which these high-temperature liquids were generated is not clear. Green (1975) suggests that peridotitic komatiite (MgO > 30 wt.%) originated through diapirism of upper mantle material at depths $\geq$ 200 km. Subsequent fractionation of garnet could then generate the observed CaO/Al_2O_3 ratios, which differ from those of other Archean magmas and those of the model mantle compositions. Bickle *et al.* (1977) present a similar model, after recognizing that neither olivine cumulation nor equilibrium partial melting of a lherzolite source could generate the high-Mg liquids. They suggest that the important process is polybaric assimilation of solids into an initially high-Mg liquid, during its rapid ascent to the surface. Equilibrium partial melting of a lherzolite mantle can be discounted on the basis of CaO/Al_2O_3 relationships. Current models of mantle chemistry indicate that $CaO/Al_2O_3 \leq 1$, whereas these high-Mg peridotitic komatiites have $CaO/Al_2O_3 > 1$. This discrepancy could be a consequence of a variable ratio of clinopyroxene to garnet in the mantle or of garnet fractionation (Green, 1975).

1.3.1.6 Trace elements

Current analytical techniques allow the determination of about 35 trace elements in basaltic rocks with reasonable precision and accuracy. These elements may be conveniently divided into several groupings which show related geochemical behavior. These are: (1) large cations; (2) REE; (3) large high-valency cations; (4) ferromagnesian trace elements; (5) siderophile elements.

The large cations, the light REE (La-Sm) and the large high-valency cations comprise the LIL elements, or the "incompatible" (i.e., not occurring in 6-fold coordination sites) elements. The fourth group comprises the so-called "compatible" elements, which usually enter 6-fold coordination sites in common rock-forming minerals, although some may enter 8-fold sites in minerals such as garnet.

The radioactive elements and their daughter products, which provide information both on ages and on evolution with time, are discussed in Chapter 7, which deals with isotopic systematics. These elements include K, Rb, Sr, Sm, Nd, Th, U and Pb. Only passing reference to these particular LIL elements is made in this section.

Many subtle but important trends occur within the broad groupings noted above, so that there are widely varying degrees of substitution of trace elements for major elements in mineral structures. Thus, among the large cations, the size order ($Cs^+ > Rb^+ > K^+$) and valence order ($Ba^{2+} > K^+$) are reflected in variations in K/Rb, K/Cs and K/Ba ratios during partial melting or fractional crystallization episodes. Cs, Rb and Ba are concentrated in the liquid fraction during the first stages of partial melting, or during the last stages of fractional crystallization. Depletion in these elements relative to K is characteristic of basalt source regions which have been subjected to previous episodes of partial melting.

Among the 14 rare-earth elements, rather uniform fractionation trends occur based upon the monotonic decrease in ionic radius from La^{3+} to Lu^{3+}. The light REE (LREE = La–Sm) may be conveniently separated from the heavy REE (HREE = Gd–Lu), since the LREE commonly behave similarly to other LIL elements (Rb, Ba, Th, U) while the heavy REE are retained in some Ca^{2+} sites (but not in feldspar), because of their similarity to Ca in ionic radius. Y^{3+} behaves as a heavy REE. The occurrence of Eu as a divalent ion makes the behavior of Eu^{2+} equivalent to that of Sr^{2+} with consequent enrichment or depletion of Eu relative to the other REE. Although REE mobility is occasionally observed in low-grade metamorphosed volcanics, in most cases the REE appear to have been unaffected by metamorphism (Hamilton *et al.*, 1979; Whitford and Arndt, 1978).

Ionic size is an important parameter in controlling geochemical behavior among the large high-valency cations. Thus Th^{4+} and U^{4+} are usually truly incompatible elements, while the smaller Zr^{4+} and Hf^{4+} are more commonly accommodated in major mineral structures. Accordingly, mantle sources for basalts will become depleted in Th and U relative to Zr and Hf during successive melting episodes. Nb^{5+}, although smaller than Zr^{4+}, is more incompatible on account of its high valence, so that Zr/Nb ratios will rise in mantle areas depleted by extraction of basaltic melts.

The ferromagnesian trace elements show diverse behaviors based on size, valence and crystal-field effects. Cr^{3+} and Ni^{2+} enter the 6-fold Fe and Mg sites much more readily than V^{3+} or Co^{2+}, due to crystal field site preference energy, which leads to preferential retention of Ni in residual olivine during partial melting or in early cumulus olivine during fractional crystallization, whereas Cr may enter early-crystallizing spinel phases. In the Moon, Cr occurs mainly as Cr^{2+}. These effects lead to dramatic enrichment or depletion of Ni and Cr with consequent changes in Ni/Co and Cr/V ratios. The other two major groupings are the trace chalcophile elements which enter sulfide phases and the trace siderophile (Pt group) elements.

Rare-earth elements

Overall, rare-earth element absolute abundances in terrestrial and lunar basalts and in the eucrites are similar. However, the details of their relative abundances are diagnostic. Almost all of the lunar mare basalts have negative Eu anomalies whose magnitudes are roughly proportional to the overall REE abundances in the samples. Negative Eu anomalies may form in two ways: through the removal of plagioclase which preferentially incorporates Eu^{2+} relative to the trivalent REE, or through the accumulation of clinopyroxene which, to a limited extent, excludes Eu^{3+} relative to the other trivalent REE. The magnitude of the Eu anomaly in the majority of mare basalts (Fig. 1.3.1.7; section 1.2.9, Figs. 1.2.9.12–13) is too large to have been generated through clinopyroxene accumulation. The most

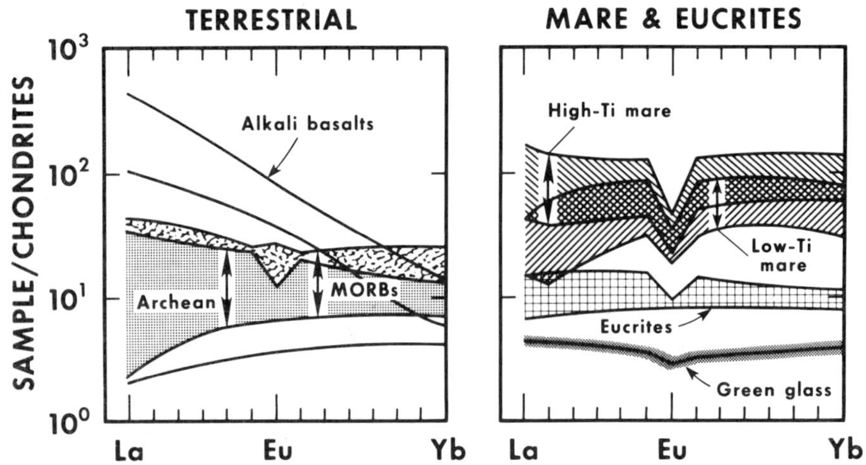

1.3.1.7 Chondrite-normalized rare-earth abundance relationships for the planetary basalt suites.

popular interpretation (e.g., Wood *et al.*, 1970; Taylor and Jakeš, 1974; Shih and Schonfeld, 1976; Drake and Consolmagno, 1976) is that the anomaly is related to plagioclase extraction. However, for many mare basalt compositions, plagioclase is not a liquidus or near-liquidus phase at any pressure (e.g., O'Hara *et al.*, 1970; Ringwood and Essene, 1970; Green *et al.*, 1971; Grove *et al.*, 1973; Walker *et al.*, 1975a, b). Exceptions include Apollo 12 aluminous basalt 12038 (Biggar *et al.*, 1971) and Luna 24 basalts (Grove and Vaniman, 1978) where plagioclase is a liquidus or near-liquidus phase at low pressure.

For those compositions in which plagioclase is not present at or near the liquidus, plagioclase fractionation during the process of magma genesis, emplacement, and crystallization is not a viable mechanism for the generation of the Eu anomaly in these basalts. Thus, the negative Eu anomalies must be intrinsic properties of the source. This is consistent with a lunar history involving an early (prior to 4.4 b.y.) Moon-wide melting and differentiation event which resulted in the formation of the mare basalt source regions through the accumulation of ferro-magnesian phases from the magma ocean (e.g., Walker *et al.*, 1975a, b; Wood, 1975; Longhi, 1977) and the formation of the plagioclase-rich lunar highlands through plagioclase flotation (e.g., Wood *et al.*, 1970; Taylor and Jakeš, 1974). Fractionation of Sm and Nd in the basalt sources took place at about 4.4 b.y. (Lugmair *et al.*, 1975), which places a lower limit on the time of formation of the mare basalt source regions. The U-Th-Pb (Nunes *et al.*, 1975; Tera and Wasserburg, 1975) and Rb-Sr (Papanastassiou and Wasserburg, 1972) systematics are consistent with this model.

Terrestrial basalts have no consistent negative Eu anomaly (e.g., Schnetzler and Philpotts, 1971), indicating that their source regions have not seen prior removal of plagioclase. Consequently, it may be concluded that, if the early differentiation of the Earth involved plagioclase fractionation, it cannot have been as important a process as it was on the Moon. Europium anomalies that are observed in some terrestrial basalts are the consequence of plagioclase fractionation (removal or accumulation), presumably under low to moderate

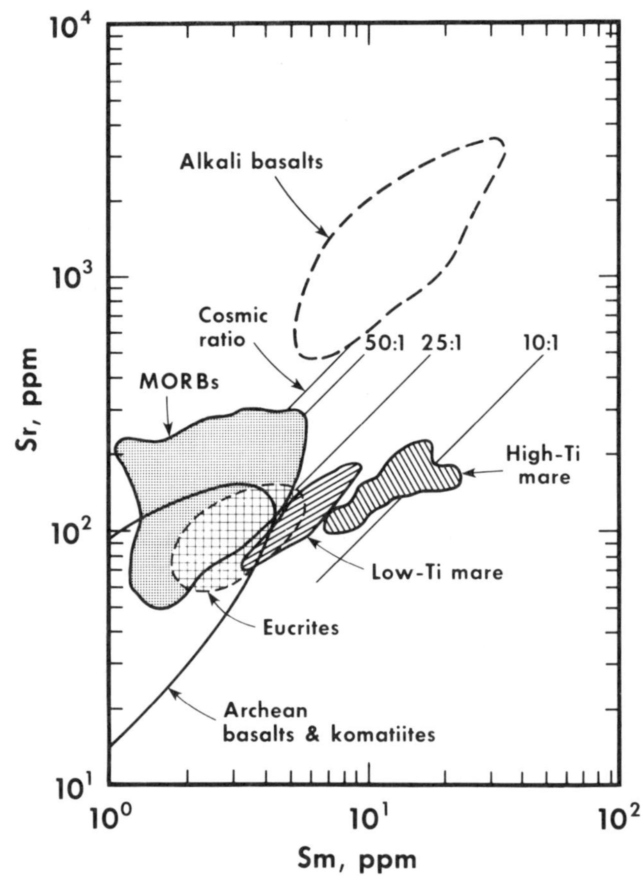

1.3.1.8 Sr vs. Sm fields for the planetary basalt suite.

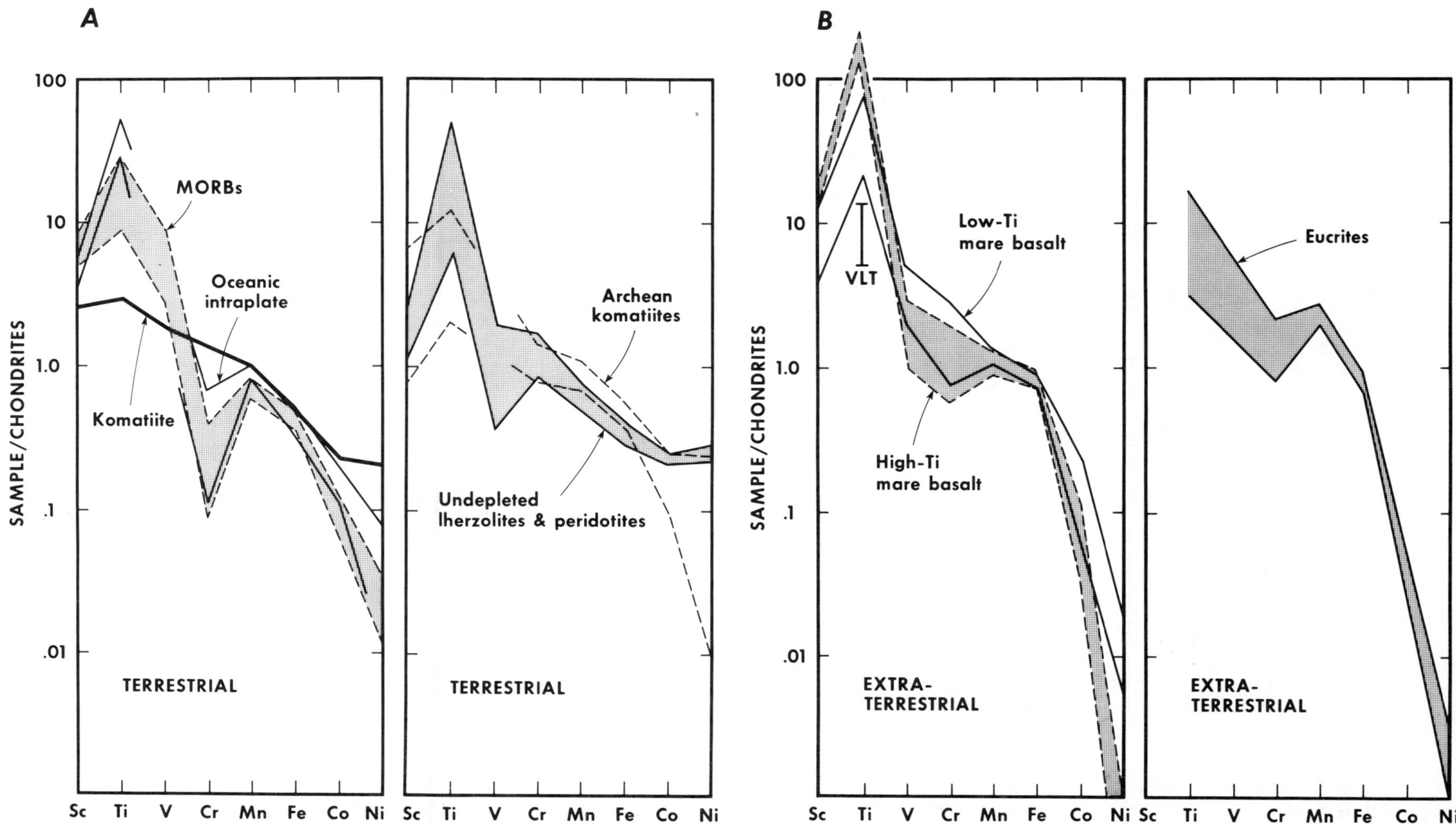

1.3.1.9 Chondrite-normalized transition metal abundances for planetary basalt suites. **a.** Terrestrial basalts. **b.** Extraterrestrial basalts.

pressure conditions, following the generation of the melt. Calcium-rich plagioclase is not stable in the Earth at depths exceeding 40 km.

The eucrites also have no consistent Eu anomaly. However, those that do have negative Eu anomalies also have the highest total abundances of rare-earth elements (e.g., Stolper, 1977). Gast and Hubbard (1970) and Ringwood (1970) suggest that the sequence Stannern-Nuevo Laredo-Pasamonte-Juvinas represents a series of compositions derived by increasing degrees of partial melting of a common source. The rare-earth data are then consistent with plagioclase saturation of the first-formed melts. Increased percentages of partial melting exhaust the plagioclase in the source, so subsequent liquids are no longer plagioclase-saturated and have lower total rare-earth abundances.

If the lunar mare basalt sources saw plagioclase extraction, this would have been recorded by elements other than Eu which are preferentially incorporated in plagioclase. One of these elements is Sr. Depletion of Sr relative to Sm (which is preferentially excluded from plagioclase) in the mare basalts (Fig. 1.3.1.8) is an intrinsic property of mare source regions and is consistent with separation of plagioclase-rich crust early in the Moon's history. The eucrites, however, are not depleted in Sr, which further confirms that plagioclase separation did not occur prior to the formation of the EPB or during the early differentiation of the EPB.

Ferromagnesian elements

Chondrite-normalized transition metal abundances of the planetary basalt suite reveal several prominent features (Fig. 1.3.1.9). There is a pronounced fractionation of these elements in all three suites, with Sc, Ti, and V contents ranging from 2 to nearly 100 times chondritic abundances, and with progressive depletion of Fe, Co, and Ni from slightly less than chondritic abundances (Fe) to a 1000-fold depletion (Ni). Fractionation which is similar but less extensive is observed in terrestrial lherzolites and komatiites. Thus, it appears that the source regions of all three planetary suites are fractionated with respect to these elements. The general sequence of transition metal *depletion* is: high-Ti mare basalts > eucrites > low-Ti mare basalts > terrestrial MORBs, intraplate basalts > terrestrial lherzolites.

Depletion of the siderophile elements in the Earth's mantle, relative to their chondritic abundances, has been interpreted (e.g., Ringwood, 1966) to be a consequence of their reduction and subsequent incorporation into a metallic core. The depletion of nickel in basalts derived from the Earth's mantle is also a consequence of (a) nickel retention in the residual source olivine, and (b) the scavenging of Ni from those liquids which have olivine as the low pressure liquidus phase and which experience olivine fractionation. However, as indicated by the lherzolite data, the Earth's mantle is fractionated with respect to the transition metals and the processes of basalt magma generation cannot be the sole source of this fractionation.

If separation of a metallic phase is responsible for the general trend of siderophile element depletion in the Earth, then a similar mechanism is indicated for the source region of mare basalts and for the eucrite parent body. In the lunar case, there is no compelling evidence for a core and if there is one, its mass cannot be significant (e.g., Nakamura *et al.*, 1974; Toksöz *et al.*, 1973; Dainty *et al.*, 1975). Furthermore, with the exception of the olivine normative Apollo 12 basalts, the mare basalts do not appear to be saturated with metallic iron. Thus, fractionation of a metallic phase, though a reasonable mechanism for fractionating and depleting the transition metals, cannot be responsible for their fractionated patterns in the mare basalts.

In the case of the eucrite parent body, the apparent size (up to several hundreds of kilometers) precludes formation of a significant metallic core. Experimental

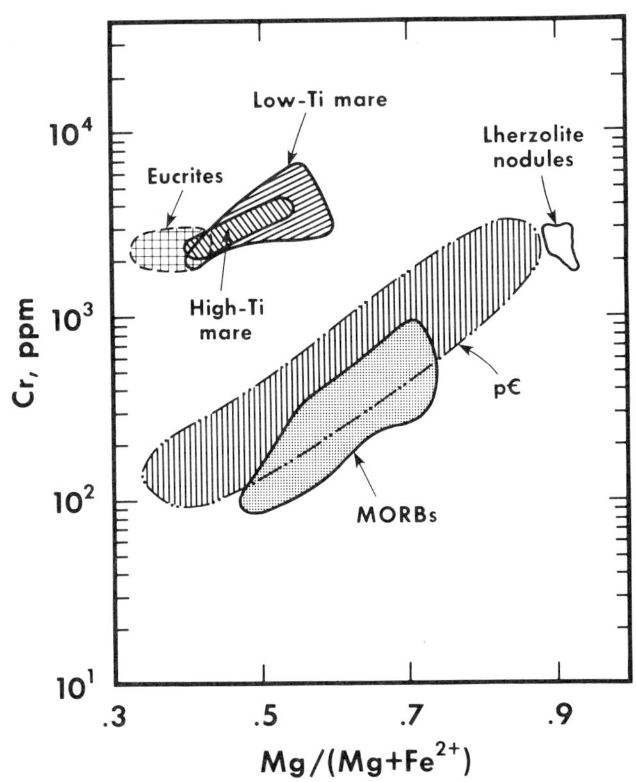

1.3.1.10 Chromium vs. Mg/(Mg + Fe^{2+}) fields of planetary basalt suites.

studies on selected eucrite compositions (Stolper, 1977) reveal that they are multiply-saturated with olivine, plagioclase and pyroxene at low pressures—consistent with their origin in a small object. Furthermore, metallic iron does not appear to be a stable phase in these compositions. We may conclude from these observations that fractionation of transition metals in the Moon and in the EPB must have occurred *prior to their accretion*.

A second major feature of the analytical data is the pronounced negative Cr anomaly in terrestrial basalts, which is not observed in mare basalts or in terrestrial komatiites and lherzolites. The apparent negative Cr anomaly in the eucrites is not real and is, instead, due to a positive Mn anomaly. Figure 1.3.1.10 shows that the Cr contents of the eucrites are similar to those of mare basalts and are quite different from those in terrestrial samples having comparable $Mg/(Mg + Fe^{2+})$ ratios. Three possible explanations of these relationships have been proposed (Papike and Bence, 1978):

(1) The Cr differences are intrinsic properties of the Earth, the Moon, and the eucrite parent body.

(2) Terrestrial samples have fractionated chrome spinel.

(3) The oxidation state of chromium, i.e., Cr^{2+}/Cr^{3+} ratio, differs in the three objects and reflects differences in the f_{O_2} of the sources (see also Sato, 1978).

There is no compelling reason to believe that the three objects have intrinsically different total Cr contents. To the contrary, the Cr contents of komatiites and lherzolites are comparable to those observed in the lunar mare basalts and the eucrites. Furthermore, no Cr anomaly is observed in the komatiites and lherzolites (compositions which more closely approximate that of the terrestrial upper mantle).

Chrome spinel is a liquidus or near-liquidus phase for some, but not all, of the most primitive terrestrial basalt compositions. Undoubtedly, some of those basalts which have experienced olivine fractionation have also experienced chrome spinel removal which would have contributed to the magnitudes of their Cr anomalies. The stability of spinel in terrestrial basalts is strongly f_{O_2}-dependent (e.g., Hill and Roeder, 1974; Fisk and Bence, 1979). Under conditions of very low f_{O_2} the reduced state of chromium (Cr^{2+}) may become a more important species than Cr^{3+}; thus the Cr content of a liquid generated by partial melting is strongly dependent upon f_{O_2}. Divalent Cr will be preferentially partitioned into the liquid during partial melting and will not be preferentially removed during spinel crystallization. Trivalent Cr will be retained in the source in clinopyroxene ($D_{Cr}^{cpx/liq} > 10$; Bender *et al.*, 1978; Bence and Bender, 1978) and in spinel ($D_{Cr}^{sp/liq} >> 10$) or garnet ($D_{Cr}^{gar/liq} \simeq 3$; Irving and Frey, 1978). Clinopyroxene and spinel are not important residual phases under conditions of very high percentages of partial melting as, for example, in the case of Archean komatiites, so virtually all of the Cr would enter the liquid.

We suggest that because komatiites are products either of high percentages of partial melting or of sequential melting, chrome spinel is not retained on their source regions. Furthermore, under low-pressure conditions, chrome spinel is not a liquidus or near-liquidus phase (e.g., Green *et al.*, 1975) so these liquids are not susceptible to spinel fractionation. The vast majority of terrestrial basalts have chrome anomalies. This anomaly originates in two ways. Under conditions of basalt genesis at shallow depths in the upper mantle $\leq$ 100 km, Cr-spinel is retained in the residuum. Second, liquids *en route* to the surface may fractionate Cr-spinel, thereby increasing the magnitude of the anomaly.

It has been suggested (Papike and Bence, 1978) that the variable oxidation state of Cr in lunar mare basalts and terrestrial MORBs accounts for the observed differences in total Cr through the above mechanism. If this is true, we may conclude that f_{O_2} conditions in the

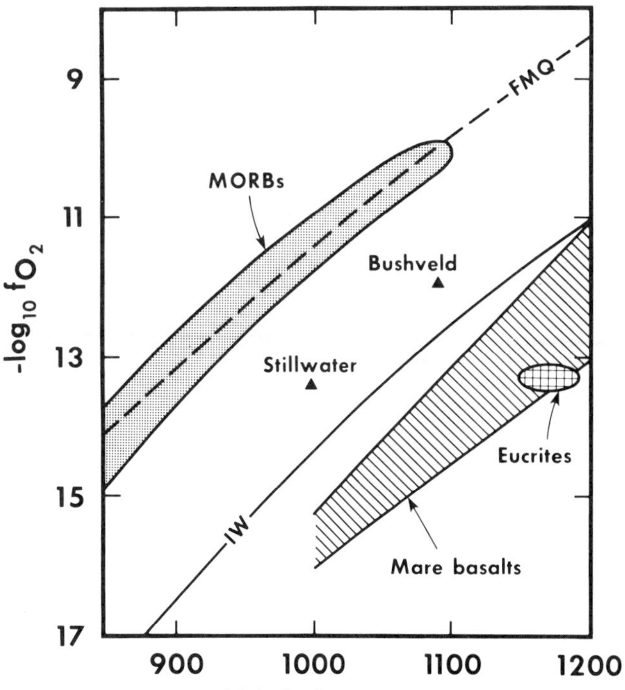

1.3.1.11 Log f_{O_2} vs. temperature fields for planetary basalts. Eucrites from Stolper (1977); mare basalts and terrestrial basalts from Haggerty (1976).

Moon and within the eucrite parent body were comparable. With the exception of T-f_{O_2} measurements obtained on ultramafic nodules (Sato, 1978), we have no direct observations on the T-f_{O_2} conditions prevailing within planetary basalt source regions.

Available data on f_{O_2} vs. temperature for the various basalt groups indicate lower f_{O_2} conditions for the crystallization of lunar mare and eucrite basalts than for terrestrial basalts (Fig. 1.3.1.11; section 1.3.5).

Nickel differences within the planetary basalt groups are related largely to olivine control lines (Fig. 1.3.1.12). However, differences between the groups are related to planetary and source region variations. The high-Ti mare basalts are highly-depleted in Ni relative to the low-Ti basalts at comparable Mg/(Mg + Fe^{2+}) ratios. This is consistent with their derivations from source regions with characteristically different nickel contents. Eucrite Ni abundances are comparable to those in the more evolved members of the low-Ti mare basalt suite. Terrestrial mantle lherzolites and other peridotites contain ~2–3 × 10^3 ppm Ni (e.g., Frey and Green, 1974). The most primitive (noncumulate) terrestrial basalts, excluding the basaltic komatiites, have Ni contents of 200–300 ppm. However, as was pointed out by Green *et al.* (1979) and Bender *et al.* (1978), the more primitive MORB compositions probably are olivine fractionates of more magnesian liquids. Such liquids would have Ni contents as high as 600–800 ppm (see also Stolper and Walker, 1980).

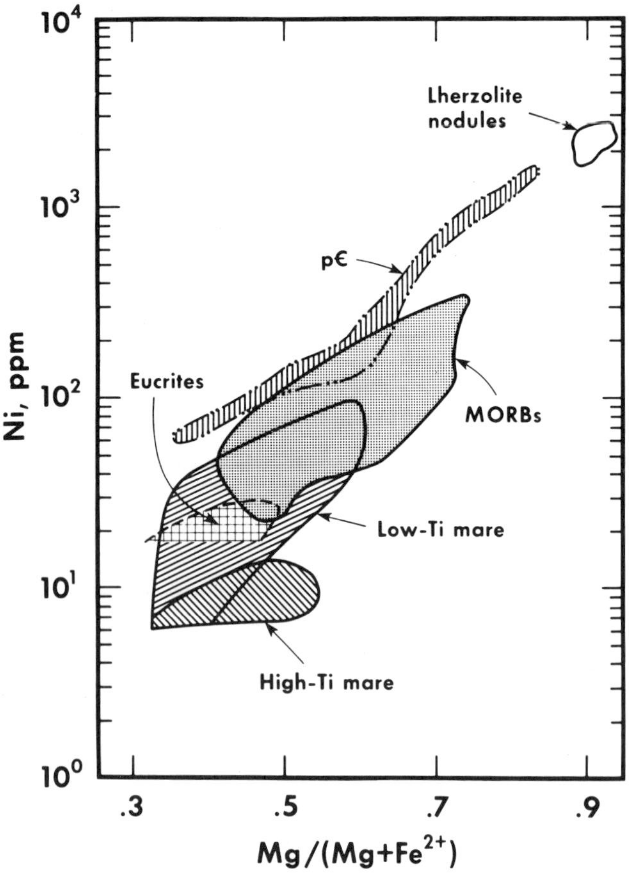

1.3.1.12 Ni vs. Mg/(Mg + Fe^{2+}) fields for planetary basalt suites.

Archean basalts and komatiites have consistently higher Ni contents than MORBs with comparable Mg/(Mg + Fe^{2+}) ratios. This difference may relate to depths and conditions of magma generation in the Archean and to the effects of pressure and temperature on the bulk distribution coefficient for Ni.

1.3.1.7 Implications of trace element abundances

The trace element data may be used to address several questions regarding the compositions and evolution of the planets. For example, we begin by considering the following questions relating to the Earth.

(1) Is the mantle at present homogeneous or heterogeneous?

(2) Has there been a change in mantle composition with time?

(3) What is the composition of the upper mantle and how does it relate to the composition of the whole mantle, and to that of the bulk Earth?

Once these questions have been discussed, we then can begin to use the terrestrial data to make valid interplanetary comparisons with data derived from the study of lunar rocks and meteorites (see also sections 7.4.2–3).

The Earth's mantle: homogeneous or heterogeneous at present?

This discussion will be limited to the composition of the upper mantle during the past 200 m.y., using primary evidence from mid-ocean ridge and oceanic island data. The major element data and compatible trace elements (Ni, Co) indicate a broadly homogeneous upper mantle (Ringwood, 1975). Minor element, LIL trace element and isotopic data (Chapter 7) combine to prove heterogeneity, with MORB and oceanic island basalts being derived from totally distinct mantle sources. There appears to be a major, first order difference between these mantle sources.

The best discriminator is the Sr isotopic information. MORBs generally have $^{87}Sr/^{86}Sr$ ratios < 0.7030, while the oceanic island basalts have values in excess of 0.7030 (Hofmann and Hart, 1978) and similar distinctions exist among the lead and neodymium isotope values (Figs. 7.4.8–10). The trace element data show many differences, although there are some cases of overlap. In general, the depletion of LIL elements in MORBs follows a systematic pattern. Large ions (e.g., Cs) are more depleted in MORBs than are smaller ions (e.g., K). The LREE are more depleted than are the heavy REE, and Nb is more depleted than Zr (Erlank and Kable, 1976).

The differences between MORBs and the oceanic island volcanics cannot be accounted for by variations in degree of partial melting of a common source, or by fractional crystallization (e.g., Kay and Hubbard, 1978), but demand that the source regions be distinctly different in composition. In general, the differences between these two sources are not explicable by secondary processes, such as interaction with or contamination by sediments, seawater or continental crust (e.g., O'Nions *et al.*, 1979a). These distinctions are most apparent between the oceanic island basalts and MORBs from ridge segments remote from islands (e.g., Iceland) which straddle the ridge. Those ridge sections which include islands (e.g., Iceland, Azores) are intermediate in character between these two extremes.

A very large debate has ensued, following proposals that the oceanic island basalts result from mantle plumes (Morgan, 1971; White and Schilling, 1978). The sources of the oceanic island basalts, which are smaller in volume than MORBs, are generally held to be deeper than those of the voluminous MORBs. A basic question is whether ridge-straddling islands such as Iceland are "accidentally" located on ridges. This would require the existence of only two types of mantle sources: a depleted, shallow source for MORBs and a deeper, more primitive source for the oceanic islands basalts, with the ridge-straddling islands representing an intermediate case, involving mixing between the two sources.

The differences between the two basalt types are not due to disequilibrium melting on a centimeter scale (Hofmann and Hart, 1978; see section 1.4.1), but must involve distinctive mantle regions on scales of tens (vertically) to hundreds of kilometers (laterally). However, variations in isotopic ratios, REE, K/Ba, Zr/Nb, etc., occur at closely-spaced sites in both MORBs and in oceanic island basalts. In addition, smooth trends in strontium isotopic compositions which correlate with distance from the East Pacific Rise (Hedge, 1978), may perhaps indicate some lateral mixing of basalts from two distinct sources (by advective transport? Richter and Ribe, 1979).

Such observations weaken the case for the "plume" model. A further problem with explanations which employ large scale mantle plumes as sources for the oceanic island volcanoes is that there are a very large number of seamounts, in addition to the islands, which apparently have very similar chemistries. Thus many "plumes" would be required.

The extent of MORB volcanism, which occurs on a much greater scale than oceanic island volcanism, indicates that large volumes of magma are homogenized during partial melting, segregation of the melt, and extrusion. (In detail this may be a very complex process such as Langmuir *et al.*, 1977, envisage in dynamic partial melting.) The sources of the lavas appear to be compositionally uniform on time scales of 10^8 years. The Emperor seamounts have major and trace element compositions identical to those of the present-day Hawaiian volcanoes, Mauna Loa and Kilauea (Bence *et al.*, 1979b), so the mantle source for the Hawaiian volcanic chain has not changed composition in 70 m.y. MORB sources in the Atlantic display similar uniformities over comparable periods of time (Carter *et al.*, 1979). If the mantle is heterogeneous on a large scale, will this situation persist on longer time scales? Would mantle recycling, remixing, convection or diffusion act to homogenize a heterogeneous mantle? The answer appears to be no. Diffusion rates (Hofmann and Hart, 1978) are too slow for these processes to proceed at geologically realistic rates, so that heterogeneities on a kilometer scale would persist throughout geological time. Thus large scale recycling of material through the mantle, as envisaged in some plate tectonic models, is unlikely to occur. The isotopic data suggest that these differences are maintained over periods of 10^9 years (see Chapter 7; O'Nions *et al.*, 1979a, b, c). This evidence appears to rule out the recycling of oceanic or continental crust through the mantle which has been suggested by many authors (e.g., Armstrong, 1968). The enrichment of LIL elements in the continental crust constitutes a one-way, irreversible depletion of the mantle in these elements.

Continental basaltic volcanic rocks frequently show LIL element enrichment which may indicate contamination from continental crustal material (O'Nions *et al.*, 1979c). The complexity of this magma-crust interaction is difficult to unscramble from trace element systematics, but the use of isotopic data is beginning to make headway in this field. Sm-Nd systematics are particularly useful (e.g., DePaolo and Wasserburg, 1976; Wasserburg and DePaolo, 1979), but the relative contributions of crust and mantle are unresolved at this stage. The possibility that the subcontinental mantle is chemically distinct from the suboceanic mantle (Jordan, 1978) thus remains an interesting suggestion. The indication from isotopic studies that the subcontinental lithosphere is "undepleted" (DePaolo and Wasserburg, 1976; Wasserburg and DePaolo, 1979) is compatible with the trace element data. However, it should be noted that basalts which erupted during the initial opening of the Atlantic appear to have come from sources identical to those which produce present-day MORBs (Carter *et al.*, 1979); therefore, LIL-depleted sources were available at the beginning of continental rifting.

The continental flood basalts have generally low Ni and Cr contents. Do these values represent the mantle source, or are they caused by fractionation *en route* to the surface, as appears to be the case for the island arc rocks? Plots of Ni vs. Mg or Mg/(Mg + Fe) are consistent with olivine removal from "primitive" magmas having Mg/(Mg + Fe) = 70 and Ni = 300 ppm, so there appears to be no need to postulate mantle regions which are inherently low in olivine or in Ni as source regions for the Tertiary or pre-Tertiary flood basalts (cf. Wright *et al.*, 1973; sections 1.2.2–3). Although it may appear surprising that such voluminous eruptives as flood basalts undergo such a degree of fractionation (see Stolper and Walker, 1980), a study of Ni vs. Mg/(Mg + Fe) shows an extremely wide dispersion of values, indicating that a continual fractionation process is depleting the magma of Mg and Ni. Since these lavas are also erupted through the continental crust, the possibility of LIL element contamination is present. Nevertheless, evidence for local mantle heterogeneity appears to be well established in the Karroo volcanic rocks (A. R. Duncan, pers. comm., 1978).

The island arc basalts present a different case. They are derived from the mantle, albeit mantle which is somewhat altered from the "traditional" peridotite or pyrolite. Their Sr isotopic values are distinct from those of MORBs, with $^{87}Sr/^{86}Sr$ values generally > 0.7030. The limited Nd isotope data indicate derivation from a source similar to that of oceanic island basalts. Most of the other trace element parameters overlap either with the oceanic island basalt or with MORB values, although for the LIL elements there is a closer resemblance to data from the oceanic islands (see section 1.3.7). A general difference is that Ni and Cr are notably lower in abundance in comparison with MORB or the oceanic island data, which indicates that the erupted lavas were not in equilibrium with mantle peridotite. The REE patterns show no Eu anomalies, indicating little or no plagioclase fractionation, nor do they provide any evidence for the presence of garnet in the source. These constraints rule out subducting oceanic crust as a *primary* source, so most petrogenetic models have become complex, belying the rather uniform nature of the erupted lavas.

The isotopic systematics of arc volcanics (Chapter 7) in some areas show the effects of contamination with Sr derived from oceanic sediments from sea water, or from continental crust in those areas where this is a possibility, but these effects appear to be accidental or localized, and not to be of fundamental genetic significance. Olivine and Cr-spinel removal appear to be responsible for the Ni and Cr depletion, which presumably happens *en route* to the surface (by analogy with the flood basalts) unless the source regions have themselves been depleted in these elements. Although higher values of the distribution coefficient for Ni between olivine and liquid (Davis and Hart, 1978; Leeman and Lindstrom, 1978; but see also Clarke and O'Hara, 1979; Elthon and Ridley, 1979) have been reported, particularly for hydrated sources, separation of *both* Cr spinel and olivine prior to eruption would be required. Rare examples of possible primary magmas with Mg/(Mg + Fe) = 70 occur (see section 1.2.7), but the great bulk of island arc rocks do not show these features. They are thus of less use than the oceanic samples as probes of primary mantle sources; however, they occupy a significant place in planetary evolution models, since they probably provide the mechanism for the supply of new material to the continental crust (Taylor, 1979a).

Does the mantle composition change with time?

It was noted above that sources of both MORBs and oceanic island basalts could be compositionally uniform on time scales of 10^8 years. The question of changes on longer time scales (10^9 years) is inherently difficult to study, since, if the mantle is heterogeneous at present, such heterogeneity is likely to mask variations in mantle composition with time. However, it has slowly been realized, mainly from isotopic studies, that large areas of the mantle, such as the source regions for MORB, have undergone Rb and LREE losses on long (> 1 b.y.) time scales (Chapter 7; McCulloch and Wasserburg, 1977). Thus the present distinction between the depleted source regions of MORBs and the sources of the oceanic island basalts cannot have arisen due to volcanic activity within the past 200 m.y. (Chapter 7; Sun and Nesbitt, 1977). However, extraction of LIL elements into the continental crust must be a prime cause of the observed heterogeneity of the upper mantle. The Archean *upper* crust is more primitive than the post-Archean *upper* crust (Taylor, 1979a), although the overall crustal compositions are probably similar. The differences between Archean and post-Archean *upper* crust probably are due to intracrustal processes (e.g., depletion of lower crust in LIL elements by extraction of granodiorite). These episodic intracrustal processes, which have been well-documented by McCulloch and Wasserburg (1977), are possibly related to mantle evolutionary processes, since the isotopic evidence forbids long crustal residence times prior to these events.

Secular trends among basalts derived from the mantle over geological time are also inherently difficult to disentangle from the effects of varying degrees of partial melting, fractional crystallization and other processes. Is there any evidence for such changes in basalt composition which might reflect a change in mantle composition? In this context, we can examine data for the Archean ultramafic and mafic rocks (section 1.2.1), and also can compare the pre-Tertiary and Tertiary plateau basalt data (sections 1.2.2–3).

For the Archean data, comparisons can be made only for those elements whose abundances are not affected by weathering and alteration processes. This excludes elements such as K, Rb, Cs, Ba and Pb. A total of seven distinctive ultramafic and mafic rocks are recognized (section 1.2.1). The komatiites have high Ni and Cr values and apparently are not represented among younger rocks. Does their absence represent a fundamental difference in source region chemistry, or merely variations in the degree of partial melting of the mantle? Magnesian basalts also are uncommon in younger periods, but are not unknown, and high magnesia, picritic liquids (> 18% MgO) have been proposed as the parent magmas for modern ocean floor basalts (Elthon, 1979; Clarke and O'Hara, 1979; Stolper and Walker, 1980). The magnesium-rich basalts and komatiites require higher degrees of partial melting for any reasonable model of mantle composition. Higher rates of heat flow during the Archean are usually postulated and again appear to be a reasonable explanation for the production of high MgO (> 20%) liquids in the Archean.

Elements such as Ti, Zr and Y have similar abundances in Archean basalts and in MORBs (Sun and Nesbitt, 1977). Although these authors suggested that Archean basalts were enriched in LIL elements relative to MORBs, the ranges of REE patterns for Archean basalts, for example, show complete overlap (e.g., Fig. 1.2.1.18, section 1.2.1) with those for rocks of younger periods. Whitford and Arndt (1978) concluded that portions of the Archean mantle were depleted in LIL elements to about the same extent as the sources of present-day MORBs. This indicates that depletion of these elements cannot be ascribed to a single episodic event 1.5–2.0 b.y. ago (Brooks *et al.*, 1976), but is a continuing process (Sun *et al.*, 1979; Whitford and Arndt, 1978).

As noted earlier, Archean basalts possess higher Ni and Cr contents than do present-day MORBs, when compared at the same Mg′-value (Fig. 12 in Gill, 1979). Whether this is due to intrinsically higher Ni and Cr contents in the Archean mantle (Gill, 1979) or to different depths of melting or to different thermal gradients is not yet clear. The difficulties of depleting the mantle in Ni and Cr make it unlikely that removal of basalt would produce such an effect. Alternative views include the possibility that the mantle was heterogeneous with respect to these elements, either laterally or vertically. This problem forms an interesting field for further research.

Archean calc-alkaline suites may resemble modern analogues very closely for all trace elements (e.g., Marda calc-alkaline suite; Taylor and Hallberg, 1977). This similarity indicates that some process analogous to present-day subduction was modifying mantle compositions in the Archean to allow the generation of magmas which mimic calc-alkaline lavas.

It is clear for the pre-Tertiary flood basalts that various combinations of partial melting, fractional crystallization and possibly crustal contamination mask any more subtle secular changes. The difficulty in identifying parental magmas is unusually acute for these rocks. There is considerable evidence from the wide variations in Mg′-value and Ni contents (see section 1.2.2) that widespread olivine depletion probably occurred prior to eruption. Accordingly, and surprisingly, these examples of voluminous eruptive episodes do not supply us with the expected samples of pristine mantle melts.

In conclusion, although mantle heterogeneity must develop through time as LIL elements are extracted into the continental crust, this process does not appear to be reflected in basaltic compositions as an identifiable secular change. Such heterogeneity is apparent in Archean basalts, resulting in the production of lavas indistinguishable from present-day MORB, except for higher Ni and Cr contents.

The primitive mantle and bulk earth compositions

A basic question is whether we can identify primitive mantle material in the Earth at present, or can reconstruct it from the trace element evidence. Both the isotopic and chemical data indicate that the Earth's mantle being sampled at present is both evolved and heterogeneous. The possibility that primitive mantle exists beneath the continents is obscured at present by the uncertainty of potential crustal interactions. However, evidence can be gleaned from the ratios of some refractory elements. These include the heavy REE (Gd-Lu), Zr and Hf, as well as Ti, Al and Ca among the major elements. These elements are not dramatically concentrated in the crust, and largely are contained in major mantle phases [unlike Th, U, Nb, Ba, Cs, K, Rb and the LREE (La, Ce, Pr, Nd)]. Accordingly, the mantle has not been greatly depleted in these elements. Calculations based on crustal abundances indicate that about 3% Y(= HREE), 2% of Zr, Ti and Al and 1% of Ca have been concentrated in the continental crust. In contrast, 38% of the Rb, 31% of the K, 30% of the U, 26% of the Ba and 14% of the La have been so concentrated (Taylor, 1979a).

Archean basalts and MORBs do not differ significantly in the abundances of these elements. Nor do the abundances differ much among MORBs, oceanic island basalts and island arc basalts. The source regions of the oceanic island basalts represent probably the nearest approach to "primitive" mantle which we can sample at present.

Various constraints on mantle compositions can be gained from the refractory elements discussed above. The heavy REE patterns are parallel to those of chondrites, and the Sm-Nd systematics are consistent with an overall chondritic pattern for the Earth (DePaolo and Wasserburg, 1976). The abundance levels for the REE are at least twice chondritic in the source regions of the oceanic island basalts. The Ca/Al ratios in the mantle are close to chondritic (Ringwood, 1975). All of these elements are highly refractory and condensed together from the solar nebula. Most estimates of the abundances of these elements in the bulk earth lie in the range 1–2 × chondritic (Ganapathy and Anders, 1974; Ringwood, 1975; Sun and Nesbitt, 1977; O'Nions *et al.*, 1979a).

We may thus reach the conclusion that the bulk earth *ratios* of the refractory lithophile elements are essentially similar to those for chondrites. Since U and Th are included among the refractory elements, the bulk earth U value will be in the range 10–20 ppb. From the K/U ratio of 10^4, this gives a K value of 100–200 ppm, which is close to estimates based on K-Ar systematics (Hurley, 1968; Ozima and Kudo, 1972; Hart and Hogan, 1978; Taylor, 1979b). These values imply that only about one-half of the present heat flow is due to steady state radioactive heat generation. The K value puts limits on possible mantle sources for K. For example, phlogopite contains 8% K, so 0.2% of this mineral will provide 160 ppm K. It seems unlikely, therefore, that phlogopite is more than a trace constituent of the mantle (Hofmann and Hart, 1978). The mantle Rb/Sr ratio is known to be 0.031 from the Sr-Nd isotopic systematics (O'Nions *et al.*, 1979a, b, c), and K/Rb = 300. The question of bulk mantle and Earth compo itions will be considered further in making compa isons with the lunar data.

Is there any evidence of original mantle heterogenc ties inherited from the accretion of the Earth? All of the above evidence is consistent with the operation of crystal-liquid equilibrium processes, which implies that presently-observable processes are responsible for the development of the heterogeneous mantle. There appears to be no apparent dependence of the trace element chemistry on volatility differences or on the operation of solid-gas equilibria. Thus we can make the important observation that the present heterogeneity of the mantle was not a primary feature of the Earth and it was not acquired due to differential accumulation of volatile-rich and refractory material during accretion of the Earth. Rather, it is a product of internal solid-liquid equilibria operating throughout geological time.

This conclusion is based on incompatible element data. A considerable body of data exists for trace siderophile elements in upper mantle samples (Chou, 1978; Morgan *et al.*, 1981; section 1.2.11) which indicates that these elements (Ru, Rh, Pd, Re, Os, Ir, Pt and Au) have nearly cosmic (C1) ratios. A popular interpretation is that they were added to the upper mantle during a late heavy meteoritic bombardment (after core formation). If so, then these elements apparently have remained in the upper mantle for over 4 b.y., which places considerable restrictions on the amount of whole mantle convection, recycling and mixing which can have occurred during that time.

Trace elements in the Moon

A brief overview of lunar structure and evolution is given by Taylor (1979c; see also Taylor, 1975). There are several striking differences between the trace element characteristics of lunar and terrestrial basalts. The alkali elements are depleted in lunar basalts, with Na in particular being lower (average 0.5%) than in terrestrial basalts. The REE patterns are distinctive. Most lunar basalts have negative Eu anomalies. The few samples of very low-titanium (VLT) basalts which have positive Eu anomalies (section 1.2.9, Figs. 1.2.9.12, 1.2.9.13) are so small (< 0.5 mg) that sampling problems (e.g., presence of plagioclase phenocrysts) cannot be excluded (Ma *et al.*, 1978). The patterns also show LREE depletion, particularly for the high-Ti basalts. Many show HREE depletion in addition, although the Apollo 17 VLT basalts, which are among the most primitive, show HREE enrichment (Wentworth *et al.*, 1979). There is no strong evidence that the overall lunar pattern is other than parallel to the chondritic pattern.

Barium is generally somewhat enriched in lunar basalts, as part of a general enrichment of refractory elements in the Moon compared to the Earth. Thus the lunar basalts have higher concentrations of refractory elements (e.g., Zr) than their terrestrial counterparts. The ferromagnesian elements show contrasting behaviors, with striking depletion in nickel and enrichment in Cr relative to terrestrial basalts.

The trace siderophile element abundances are very low in lunar basalts (section 1.2.9, Fig. 1.2.9.17). Potential lunar cores are not sufficiently large to act as a sink. Attempts at making lunar-terrestrial trace siderophile element comparisons (Wolf *et al.*, 1979) are exacerbated by the problem of determining the abundances in the terrestrial mantle (Chou, 1978; Morgan *et al.*, 1981). The overall spread in data for terrestrial ocean floor *basalts* covers two orders of magnitude for most siderophile elements. Crocket and Teruta (1977) draw attention to losses due to sea water leaching, in addition to the variation due to petrogenetic factors. They note that "sea water leaching of the platinoids from such readily altered sites as a glassy mesostasis, pillow rinds or sulfides, are all possible explanations for the wide dispersion of the Pt group elements." Thus some caution is needed in the use of terrestrial basalts in making interplanetary comparisons. The ultramafic nodule data (section 1.2.11) may be a more reliable indicator of upper mantle abundance patterns. The oceanic island data may also be a better indicator of primordial terrestrial mantle abundances, since it is inferred that these basalts come from less depleted mantle than do MORBs. However, since Ni and the siderophile elements are in general retained in the residual phases, rather than being partitioned into the melt, the MORB source may not be greatly depleted in these elements. Nevertheless, when all these caveats are borne in mind, the depletion of the Moon in siderophile elements relative to the Earth seems to be established (Wolf *et al.*, 1979; section 1.2.9, Fig. 1.2.9.17) and indicates a fundamental difference between the Earth and the Moon. Prior separation of a metal phase, as indicated by the low iron content (~11% FeO) of the Moon, must have occurred before accretion.

The lunar depletion, relative to the Earth, for such moderately volatile elements as K, Rb and Mn is not great (< 50%). Small differences in FeO/MnO ratios (Earth = 60; Moon = 80) can be ascribed to a small volatile loss consistent with the relative values for K, Rb, Na, etc., between the two bodies. The highly volatile elements (Ag, Zn, Sn, In, Br, Cd, Bi and Tl; section 1.2.9, Fig. 1.2.9.17) are rather uniformly depleted in the Moon relative to the Earth by factors of about 50.

The mare basalt trace element data (section 1.2.9) in general support the idea that these basalts are derived from a zoned lunar interior. This rather simple lunar evolutionary history contrasts with the much more complicated evolution of the terrestrial mantle. As noted earlier, the terrestrial mantle is heterogeneous vertically and probably laterally on a variety of scales. This has probably developed throughout geological time due mainly to the extraction of LIL elements into the continental crust. The most favored interpretation is that this heterogeneity has developed in a continuous, or episodic manner since at least 3.8 b.y. ago (e.g., McCulloch and Wasserburg, 1977). The lunar interior is different. The most probable scenario is that a zoned interior was established following early melting of the outer few hundred kilometers of the Moon (Taylor and Jakeš, 1974). Early crystallization of olivine and orthopyroxene was followed by plagioclase flotation to form the thick (60 km) highland crust. The isotopic systematics indicate that this stage was reached 4.5–4.3 b.y. ago. An idealized sketch of this situation is given in section 1.2.9, Fig. 1.2.9.14. Convective overturn, sinking of dense cumulates and the effects of the giant collisions would have produced a heterogeneous interior. Thus although the lunar scenario is much simpler than the terrestrial one, the lunar interior is likely to be equally or more heterogeneous. These lunar heterogeneities became frozen some 4.4 b.y. ago.

Subsequent partial melting produced the mare basalts. The depths at which this partial melting occurred, as deduced from experimental petrology, are still somewhat uncertain. Possibly all lavas come from about 400 km (J. W. Delano, pers. comm., 1980), in which case the variety of lavas would have been derived from a complex jumble of cumulate layers at that depth. In the cumulate model, the variety of lunar lavas erupted can be related to the various mineral zones which were produced during crystallization of the initial magma ocean. In this model, a thick zone of olivine and orthopyroxene, which contains very little Th, U, or K, formed first. This zone would have incorporated Ni. This makes it a little more difficult to answer the question of whether the Moon as a whole is depleted in nickel relative to the Earth. However, all lunar basalts have very high Fe/Ni ratios compared with terrestrial basalts, so that overall lunar depletion in Ni is reasonable. Cr^{3+} would have been excluded from early-crystallizing olivine and orthopyroxene. Cr^{2+} present in the highly reducing conditions of the Moon likewise would have been excluded from these early crystallizing minerals (Schreiber and Haskin, 1976) so that the effect of crystallization of the magma ocean was to concentrate Cr in the future source regions of the mare basalts.

The several types of mare basalts which have been identified recently (Beaty *et al.*, 1979; Ma and Schmitt, 1979) provide additional support for the cumulate model. The very primitive Apollo 17 VLT basalts (Wentworth *et al.*, 1979) possess, for example, in addition to the ubiquitous Eu depletion, an enrichment in HREE which is consistent with their derivation from an olivine-orthopyroxene source region which had already experienced plagioclase removal. Such patterns cannot be reconciled with melting of primitive lunar mantle, nor with contamination or assimilation of late stage lunar differentiates. Even such "primitive" lunar lavas as the Apollo 15 emerald green glass (15426) contain negative Eu anomalies, indicating prior removal of plagioclase from their source regions (see section 1.3.1.6). The net result is that no lunar basalts can be identified as having been derived from primitive unfractionated material of bulk lunar composition. As was noted above, we also cannot readily identify pristine unfractionated mantle in the Earth. The absence of identifiable "primary undifferentiated" mantle in the Earth and Moon raises some difficulties for interplanetary comparisons using trace elements. Direct matching of terrestrial and lunar basalt data is difficult, since they are derived from source regions with differing evolutionary histories. This means that interplanetary comparisons are best carried out at a more sophisticated level than simple matching of surface rock compositions.

Many incompatible elements became concentrated in the residual liquid phase during crystallization of the magma ocean. This phase, which is the precursor of the KREEP basalts, is pervasive throughout the highlands (Taylor, 1975). It is convenient to consider the highland melt rock suite at this stage. The anorthositic gabbro samples are melts generated by meteorite impact. Some of the Fra Mauro basalts have a similar origin. The derivation of melt rocks which appear to have primary volcanic origins is linked with the final stages of the crystallization of the magma ocean. The intrusion of the final residual liquid into the crust, with subsequent mixing and remelting by impacts during the final stages of accretion appears to account for the presence of these "volcanic" rocks. Evidence for the "Moon-wide" origin of this KREEP component, rather than a more localized origin, is provided by the uniformity of Sm-Nd isotopic systematics at the different Apollo sites (Lugmair and Carlson, 1978).

Comparisons of trace element contents between the Earth's mantle and the Moon

The basalt trace element data can further be used to provide some constraints on terrestrial mantle and lunar abundances. Note that values refer to a primitive mantle, or to mantle plus modern crust. Attention is focused on a few key elements, since these serve to constrain the abundances of many other correlated elements. In this context, U and K abundances are particularly useful because they are relevant for heat flow calculations and for dating. Uranium is refractory and K is volatile, so their abundances are related through K/U ratios, which appear to be planetary constants (Fig. 1.2.9.15).

Three separate methods are available to calculate the uranium content of the Earth:

(a) *Ratios of refractory elements (Ca/U, Al/U).* Current cosmochemical theory predicts that Ca, Al, U, Th and REE form part of an early refractory-rich condensate, with relatively constant inter-element ratios (Grossman and Larimer, 1974). Note that all of these elements will be present in the mantle, with negligible amounts in the core. Estimates for the Ca content of the Earth's mantle average about 2.1 wt.%, which for a Ca/Al ratio of 1.08 implies an Al value $\lesssim$ 2 wt.%. Bulk earth values are about 1.5–2 × chondritic (Taylor, 1979b), which is consistent with a major element composition for the Earth's mantle that does not seriously depart, for the refractory elements, from values typical of the primordial solar nebula (Smith, 1979; O'Nions *et al.*, 1979a, b, c). Unless we postulate fractionation of U relative to Ca and Al, then the Earth's mantle values for U will similarly be about 1.5–2 × chondritic (about 15–20 ppm U).

(b) *Heat flow estimates.* If a steady-state relationship between heat production from radioactive sources and heat flow exists, then the U content of the bulk earth is 30 ppb, or 40 ppb in the mantle, if no U entered the core. (In this case, K = 400 ppm from K/U = 10^4.)

However, if a substantial amount of modern heat flow is due to whole Earth cooling, fossil heat, etc., this relationship no longer holds. Daly and Richter (1978) and M. G. Langseth (pers. comm., 1978) have pointed out the consequences of a non-steady-state situation. On the basis of these models, at least half of the observed heat flow is not due to direct generation from K, U and Th. Thus the bulk U content of the Earth lies in the range 10–20 ppb (K = 100–200 ppm for K/U - 10^4), which is not significantly different from the chondritic average.

(c) *K/U ratios and K abundance in the Earth.* Based on ^{40}Ar systematics and K/Rb ratios, independent estimates of the K content of the Earth have been provided by Hurley (1968), Ozima and Kudo (1972), Hart and Hogan (1978) and Taylor (1979b). Most estimates fall within the range 100–200 ppm, well below the 400 ppm demanded by the steady-state heat production relationship. The terrestrial K/U ratio of 10^4 is well-established, contrasting sharply with chondritic ratios of 50,000–80,000 and the lunar value of

2500 (section 1.2.9, Fig. 1.2.9.15). For the K values derived above, terrestrial uranium lies in the range 10–20 ppb. It thus appears that the best estimate for the uranium content for the bulk earth is in the range 15–20 ppb, about 1.5–2 × chondritic (Smith, 1979).

Two methods exist for calculating the bulk uranium content of the Moon:

(a) *Heat flow data.* Keihm and Langseth (1977) assume a steady-state relationship between lunar radioactive heat production and heat flow, a view which is justified by the small size of the Moon (1/81 Earth mass), by the absence of volcanism for 3 b.y., by the lack of tectonic activity, and by the concentration of most of the radioactive elements close to the lunar surface. They estimate a U content of 35–46 ppb for the bulk moon. A non-steady-state solution (Schubert *et al.*, 1979) could reduce this estimate by about 20%.

(b) *Geochemical balance estimates for K and U.* Taylor (1976) has shown that 40 ppb U and 100 ppm K (K/U ~ 2500) are minimal bulk moon values required to account for the observed elemental abundances in near-surface lunar samples. Thus the best estimate for the lunar U abundance is about 40 ppb. Assuming a K/Rb ratio of 300 in both bodies gives 0.43 ppm Rb for the Earth and 0.30 ppm for the Moon. From the terrestrial Rb/Sr ratio of 0.031, this gives Sr = 14 ppm, close to the chondritic value of 11 ppm. Assuming Sr/U coherence yields a lunar Sr value of about 40 ppm, giving a lunar Rb/Sr ratio of 7.5×10^{-3}, which is consistent with the isotopic data.

Potassium/lanthanum ratios provide another example of a volatile/refractory element pair which is useful in planetary comparisons (Wänke *et al.*, 1973). Within individual suites, the absolute abundances of K and La may vary by up to two orders of magnitude; however, the ratio remains constant (Fig. 1.3.1.13). Ordinary chondrites have the highest K/La (wt.) ratio (~2300). There is a general progression of K depletion (relative to La) among terrestrial crustal rocks (~450), the eucrites (~140), and the Moon (mare basalts and highlands ~70) (Wänke *et al.*, 1973). The similarity of the K/La ratios in lunar mare basalts and highland rocks is consistent with the homogeneous accretion model (Wänke *et al.*, 1973; Taylor, 1973) and with the formation of the highlands during an early Moon-wide melting event.

Differences in the K/La ratio between members of the planetary suites are consistent with the relative depletion of other volatile elements: H_2O, C, N, Na_2O (as discussed above), Rb (Taylor *et al.*, 1973; Taylor, 1975), Tl and Cs (Krähenbühl *et al.*, 1973), etc., and have been interpreted by a number of workers (e.g., Anders *et al.*, 1971; Krähenbühl, 1973) to reflect differences in the temperatures at which dust in the solar nebula ceased to equilibrate with gas. If the relative depletions in the volatile elements accurately reflect these temperatures of equilibration, then the Moon accreted from material that equilibrated at the highest temperatures, the eucrite parent bodies at lower temperatures, and the Earth at the lowest temperatures.

Implications for lunar and terrestrial origin

It is reasonable to suppose that, for a planet of Earth's size, the elemental abundances might be close to the overall composition of the non-volatile elements in the solar nebula in the Earth's neighborhood. The general similarity of refractory element ratios and the fact that the whole-earth REE patterns are chondritic indicate that the Earth possesses close to the primordial contents of the refractory elements. Substantial loss of volatiles has occurred as attested by Rb-Sr systematics, so that the Earth contains less than 20% of the primitive abundances of K and Rb.

When we compare the Earth with the Moon, the K/U ratios clearly differ by a factor of 4. However, the K abundances in the Earth are less than twice those of the Moon. Accordingly, the Moon *must* be enriched in U relative to the Earth. This condition also applies to the amounts of other refractory elements, such as Ca and Al, assuming no selective loss or enrichment during planetary formation. On this basis, the concentration of refractory elements in the Moon is about 2–3 times that of the Earth, which is in agreement with much geochemical and petrological evidence and which provides *inter alia* appropriate amounts of Ca, Al, Eu and Sr for the thick anorthositic crust. The more volatile elements (e.g., Bi, Tl, etc., Fig. 1.2.9.17) are depleted in the Moon relative to the Earth by much larger factors (~50, Wolf *et al.*, 1979).

The following constraints are placed on Earth-Moon relationships:

(a) The Moon overall is enriched in refractory elements (Ca, Al, U, Th, Sr, REE) relative to the Earth by a factor between 2 and 3.

(b) The Moon has lower absolute abundance levels of the volatile elements K and Rb compared to the Earth, and has very much lower abundances of the very volatile elements.

(c) Hypotheses which derive the Moon from the mantle of the Earth must account for (i) volatile

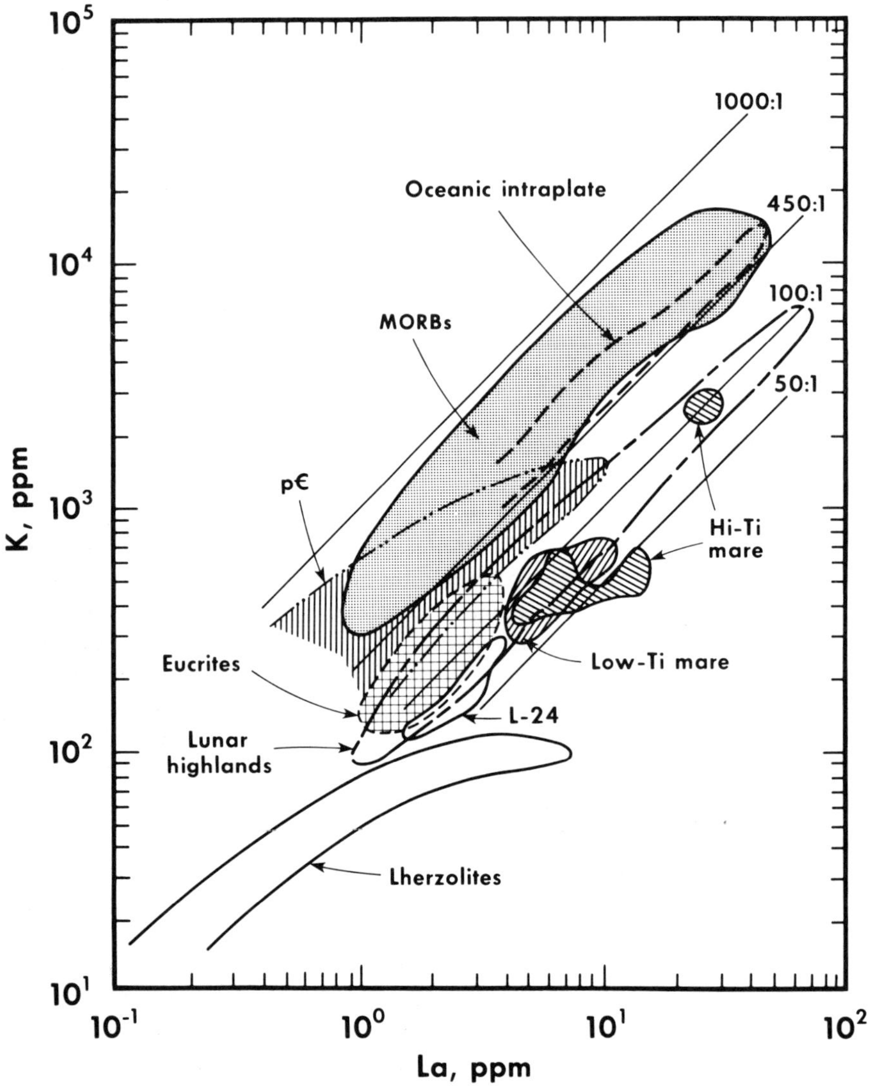

1.3.1.13 K vs. La fields for the planetary basalt suites.

loss and (ii) substantial refractory element enrichment during this process (Kaula, 1977).

Basaltic achondrites (eucrites and shergottites)

These form a third group of basaltic samples possessing many similarities in chemistry to lunar basalts. They have Cr, Ni and Na contents notably similar to those of low-Ti mare basalts, although they lack the evidence of depletion in light REE which is commonly present in lunar basalts. Europium depletions are known only for Stannern and Nuevo Laredo, while Serra de Magé and Moore County have positive Eu anomalies. Most of the others have REE patterns which parallel those of chondrites. The other related meteorites (diogenites and howardites) apparently are related simply, the diogenites being dominantly composed of pyroxene, whereas the howardites are mixtures of eucrites and diogenites.

The generally flat REE patterns of the eucrites are consistent with their derivation, by varying degrees of partial melting, from a chondritic source (Consolmagno and Drake, 1977). The experimental petrological data are consistent with their formation by varying degrees of partial melting of an olivine-pyroxene-plagioclase-spinel-metal source (Stolper, 1977), with occasional formation of cumulates and fractionation of plagioclase and pyroxene, all at low pressures. It is not clear whether the depletion in Ni and other siderophiles is due to pre-accretion loss of metal or to core formation in the parent body, although the former alternative appears more likely (Rammensee and Wänke, 1977). The asteroid Vesta has been suggested as the parent body of the eucrites by various workers (e.g., Consolmagno and Drake, 1977). A depletion in alkali elements similar to that of the Moon is consistent with the low alkali contents of the eucrites. There apparently is no need to postulate more than one parent body for the eucrites. The W/La and oxygen isotope ratios (Clayton and Mayeda, 1978) are close to those for the Moon and the Earth. All of these parameters point to a parent body with many characteristics similar to the Moon. However, the eucrites are derived apparently by single-stage melting of a volatile-depleted, and possibly siderophile-depleted but otherwise chondritic, parent body (see sections 1.2.8 and 3.5). The ancient radiometric ages (4.6×10^9 years) apparently preclude two-stage processes, whereas the trace element and petrological evidence is consistent with single-stage melting and crystal fractionation at low pressures. There is no inbuilt Eu anomaly, for example, in the source region. Thus, there are intriguing similarities and differences among the Earth, Moon and the eucrite parent body. All have suffered volatile element depletion relative to the chondrites. The Moon and possibly the eucrite parent body are depleted in siderophile elements (see sections 1.2.8 and 3.5). Internal partial melting in the eucrite parent body produced basaltic compositions with ages very close to the condensation age of the solar nebula.

The other group of meteoritic basalts considered in section 1.2.8 are the shergottites (Stolper *et al.*, 1979). In contrast to the eucrites, the isotopic data demand derivation by at least two stages, so they show analogies to lunar basalts in this respect. The latest melting episode, which generated the presently-observed meteorites, occurred at 650 m.y. (Nyquist *et al.*, 1979), which is a younger age than any hitherto observed for basaltic achondrites. Shergotty is of particular interest since it has very similar trace element abundances to terrestrial basalts. The relative abundances of the volatile and siderophile elements are close to those for terrestrial basalts. Stolper (1979) concludes that "in view of these similarities, Ringwood and Kesson's conclusion that the siderophile and volatile element pattern of the Earth's upper mantle was the result of complex uniquely terrestrial processes appears unwarranted. Their conclusion that the Moon was derived from the Earth's mantle after core formation or indeed at all, is also weakened." The recent evidence that trace siderophile abundances in the terrestrial upper mantle may be due to a pre-4 b.y. meteoritic bombardment (Chou, 1978; Morgan *et al.*, 1981; section 1.2.11 and discussion above) means that similar abundance patterns will be common throughout the inner solar system.

Venus and Mars

The extremely limited trace element data base for these planets indicates that the surface of Venus has terrestrial-like K/U ratios (Fig. 1.2.9.15; Surkov, 1977) while the martian surface compositions (Toulmin *et al.*, 1977; Clark and Baird, 1979) resemble those of low-Ti lunar mare basalts with similar K/U ratios (Fig. 1.2.9.16; Surkov, 1977).

1.3.1.8 Summary

Basaltic magmas, derived by the partial melting of planetary interiors, have compositions that reflect the preaccretionary history of the material from which the planet formed, the planet's subsequent evolutionary history, the chemistry and mineralogy of the source regions, and the intensive thermodynamic parameters existing at the source and emplacement sites. Studies of basalt suites from the Earth, the Moon, and the eucrite parent body reveal compositional differences intrinsic to their source regions which are, in turn, characteristic of each planet and of its formational and evolutionary history.

Major interplanetary differences are observed in iron, $Mg/(Mg + Fe^{2+})$, TiO_2, Al_2O_3, Na_2O, Cr, Ni, and in refractory and volatile element abundances. The most primitive lunar mare basalts have Mg′-values of about 0.6. On the Earth, MORBs have Mg′-values of 0.70–0.72, whereas Archean peridotitic komatiites have Mg′-values of up to 0.9. Eucrites have Mg -values approaching 0.5 (excepting Binda). These differences reflect inherent differences in $Mg/(Mg + Fe^{2+})$ values of their sources. Striking differences in the TiO_2 abundances of the planetary basalts reflect both inter- and intraplanetary variations in source chemistry. Primitive MORBs and primitive oceanic intraplate tholeiites have a factor of 2–3 difference in TiO_2 contents at comparable Mg′ (0.7–1.2 vs. 2–3 wt.% respectively). Three major TiO_2-content groups are recognized in the lunar mare suite: high TiO_2 (8–13 wt.%), low TiO_2 (2–5 wt.%) and very low TiO_2 (< 1 wt.%). The eucrites have TiO_2 contents less than 1 wt.%.

The mare basalts and eucrites have pronounced Na_2O and K_2O depletions relative to all terrestrial basalts. This is a consequence of the preaccretion loss of volatiles from the material that accreted to form the Moon and the eucrite parent bodies.

Mare basalts have consistently lower Al_2O_3 contents than the terrestrial basalts. This may be due either to an inherently lower content of Al_2O_3 in the mare basalt source regions or Al_2O_3 retention in a residual aluminous phase.

Overall rare-earth element abundances are comparable among all three objects. Mare basalts have pronounced negative Eu anomalies which are inherited from their source region and which record plagioclase removal from a crystallizing magma ocean early in lunar history (4.4–4.6 b.y.). Early separation of plagioclase appears to have been a relatively unimportant process on the Earth. The Archean basalt source regions did not experience plagioclase extraction, so one may conclude that this process was not important on the Earth.

The terrestrial mantle is heterogeneous at present, and this heterogeneity appears to have extended back into Archean time.

The transition metals are fractionated in all three planetary basalt suites. For terrestrial basalts this may reflect core separation; however, in the cases of the Moon and eucrite parent bodies, preaccretionary separation of metal and silicates is a more reasonable explanation. A pronounced Cr anomaly is observed in terrestrial MORBs but not in the mare basalts. The Cr anomaly in terrestrial basalts is a consequence of their higher f_{O_2} of crystallization, which stabilizes Cr-spinel and Cr^{3+}-bearing clinopyroxenes.

Bulk-earth ratios for refractory elements (e.g., REE) appear to parallel those in carbonaceous chondrites, and there is no definitive evidence to suggest that the Moon or the eucrite parent body had different ratios. The Earth is depleted in volatile elements relative to primitive solar nebula abundances. For the moderately volatile elements such as potassium, the Moon is depleted relative to the Earth by factors of 2–3. The highly volatile elements (e.g., Tl, Bi) are depleted in the Moon by factors of 50 relative to the Earth. In contrast the Moon is enriched relative to the Earth in refractory elements (Ca, Al, Ti, U, REE) by factors of 2–3. Potassium/uranium ratios in carbonaceous chondrites are 65,000. The Earth, the shergottite parent body, and (probably) Venus have K/U ratios of about 10,000. The Moon, eucrites, and (probably) Mars have K/U ratios of 2000–3000.

All three planetary objects underwent extensive, early melting and differentiation events. On the Moon, fractionation of a Moon-wide magma ocean (4.4–4.6 b.y.) saw the formation of a cumulate mare source and plagioclase differentiation through flotation to form an ancient crust. The mare basalts record the effects of this plagioclase extraction in their Eu depletion. The Eu anomaly is an intrinsic property of the sources which was acquired prior to basalt genesis.

Geochemical evidence suggests that the eucrite parent body probably had no crust and did not undergo a differentiation that involved plagioclase fractionation. The compositions of the eucrites are consistent with partial melting in small planetary objects ($\leq$ 200 km radius).

1.3.1.9 References

Anders E., Ganapathy R., Keays R. R., Laul J. C. and Morgan J. W. (1971) Volatile and siderophile elements in lunar rocks: comparison with terrestrial and meteoritic basalts. *Proc. Lunar Sci. Conf. 2nd*, p.1021–1036.

Armstrong R. L. (1968) A model for the evolution of strontium and lead isotopes in a dynamic earth. *Rev. Geophys. Space Phys.* **6**, 175–199.

Arndt N. T. (1976) Melting relations of ultramafic lavas (komatiites) at one atmosphere and high pressure. *Carnegie Inst. Wash., Yearb.* **75**, 555–561.

Basu A., McKay D. S., Moore C. H. and Shaffer N. R. (1979) A note on the Apollo 15 green glass vitrophyres. *Proc. Lunar Planet. Sci. Conf. 10th*, p. 301–310.

Beaty D. W., Hill S. M. R., Albee A. L., Ma M. -S. and Schmitt R. A. (1979) The petrology and chemistry of basaltic fragments from the Apollo 11 soil, part 1. *Proc. Lunar Planet. Sci. Conf. 10th*, p. 41–75.

Bence A. E. and Bender J. F. (1978) Clinopyroxenes in mid-ocean ridge basalts: experimental studies. *EOS (Trans. Amer. Geophys. Union)* **59**, 410.

Bence A. E., Baylis D., Bender J. F. and Grove T. L. (1979a) Controls on the major and minor element chemistry of mid-ocean ridge basalts and glasses. In *Deep Drilling Results in the Atlantic Ocean: Ocean Crust*, (M. Talwani, C. G. Harrison and D. E. Hayes, eds.), p. 331–341. Maurice Ewing Series 2. Amer. Geophys. Union, Washington, D.C.

Bence A. E., Taylor S. R. and Fisk M. (1979b) Major and trace element geochemistry of basalts from Ojin, Nintoku and Suiko seamounts of the Emperor Seamount Chain: DSDP-IPOD Leg 55. *Reports of the Deep Sea Drilling Project*, v. LV.

Bender J. F., Hodges F. N. and Bence A. E. (1978) Petrogenesis of basalts from the project FAMOUS area: experimental study from 0 to 15 kbars. *Earth Planet. Sci. Lett.* **41**, 277–302.

Bickle M. J., Ford C. E. and Nisbet E. G. (1977) The petrogenesis of peridotitic komatiites: evidence from high-pressure melting experiments. *Earth Planet. Sci. Lett.* **37**, 97–106.

Biggar G. M., O'Hara M. J., Peckett A. and Humphries J. (1971) Lunar lavas and the achondrites: petrogenesis of protohypersthene basalts in the Maria Lava lakes. *Proc. Lunar Sci. Conf. 2nd*, p. 617–643.

Brooks C., Hart S. R., Hoffmann A. and James D. E. (1976) Rb-Sr mantle isochrons from oceanic regions. *Earth Planet. Sci. Lett.* **32**, 51–61.

Carter S. R., Evensen N. M., Hamilton P. J. and O'Nions R. K. (1979) Basalt magma sources during the opening of the Atlantic. *Nature* **281**, 28–30.

Chou C. -L. (1978) Fractionation of siderophile elements in the earth's upper mantle. *Proc. Lunar Planet. Sci. Conf. 9th*, p. 219–230.

Clarke B. C. and Baird A. K. (1979) Chemical analyses of martian surface materials (abstract). In *Lunar and Planetary Science X*, p. 215–217. Lunar and Planetary Institute, Houston.

Clarke D. B. and O'Hara M. J. (1979) Nickel and the existence of high-MgO liquids in nature. *Earth Planet. Sci. Lett.* **44**, 153–158.

Clayton R. N. and Mayeda T. K. (1978) Multiple parent bodies of polymict brecciated chondrites. *Geochim. Cosmochim. Acta* **42**, 325–327.

Consolmagno C. J. and Drake M. J. (1977) Composition and evolution of the eucrite parent body. *Geochim. Cosmochim. Acta* **41**, 1271– 1282.

Crocket J. H. and Teruta Y. (1977) Pd, Ir and Au contents of mafic and ultramafic rocks drilled from the Mid-Atlantic Ridge, Leg 37 DSDP. *Can. J. Earth Sci.* **14**, 777–784.

Dainty A. M., Goins N. R. and Toksöz M. N. (1975) Natural lunar seismic events and the structure of the Moon. *Proc. Lunar Sci. Conf. 6th*, p. 2887–2897.

Daly S. F. and Richter F. M. (1978) Convection with decaying heat sources: a simple thermal evolution model (abstract). In *Lunar and Planetary Science IX*, p. 213–214. Lunar and Planetary Institute, Houston.

Davis K. E. and Hart S. R. (1978) Partitioning of Ni between olivine and silicate liquid. *Earth Planet. Sci. Lett.* **40**, 203–219.

Delano J. W. (1979) Apollo 15 green glass: chemistry and possible origin. *Proc. Lunar Planet. Sci. Conf. 10th*, p. 275–300.

DePaolo D. J. and Wasserburg G. J. (1976) Inferences about magma sources and mantle structure from variations in $^{143}Nd/^{144}Nd$. *Geophys. Res. Lett.* **3**, 743–746.

Drake M. J. and Consolmagno G. J. (1976) Critical review of models for the evolution of high-Ti mare basalts. *Proc. Lunar Sci. Conf. 7th*, p. 1633–1657.

Elthon D. (1979) High magnesia liquids as the parental magma for ocean floor basalts. *Nature* **278**, 514–518.

Elthon D. and Ridley W. I. (1979) Comments on "the partitioning of nickel between olivine and silicate melt." *Earth Planet. Sci. Lett.* **44**, 162–164.

Erlank A. J. and Kable E. J. D. (1976) The significance of incompatible elements in Mid-Atlantic Ridge basalts from 45° N with particular reference to Zr/Nb. *Contrib. Mineral. Petrol.* **54**, 281–291.

Fisk M. R. and Bence A. E. (1979) Experimental studies of spinel crystallization in FAMOUS basalt 527-1-1. *EOS (Trans. Amer. Geophys. Union)* **60**, 420.

Frey F. A. and Green D. H. (1974) The mineralogy, geochemistry, and origin of lherzolite inclusions in Victorian basanites. *Geochim. Cosmochim. Acta* **38**, 1023–1059.

Gancarz A. J. and Wasserburg G. J. (1977) Initial Pb of the Amitsoq gneiss, West Greenland, and implications for the age of the Earth. *Geochim. Cosmochim. Acta* **41**, 1283–1301.

Gast P. W. and Hubbard N. J. (1970) Rare-earth abundances in soil and rocks from the Ocean of Storms. *Earth Planet. Sci. Lett.* **10**, 94–101.

Gill R. C. O. (1979) Comparative petrogenesis of Archean and modern low-K tholeiites: a critical review of some geochemical aspects. *Phys. Chem. Earth* **11**, 431–447.

Green D. H. (1971) Composition of basaltic magmas as indicators of conditions of origin: application to oceanic volcanism. *Phil. Trans. Roy. Soc. Lond.* **A268**, 707–725.

Green D. H. and Ringwood A. E. (1967) The genesis of basaltic magmas. *Contrib. Mineral. Petrol.* **15**, 103–190.

Green D. H., Edgar A. D., Beasley P., Kiss E. and Ware N. G. (1974) Upper mantle source for some hawaiites, mugearites, and benmoreites. *Contrib. Mineral. Petrol.* **48**, 33–43.

Green D. H., Hibberson W. P. and Jacques A. L. (1979) Petrogenesis of mid-ocean ridge basalts. In *The Earth: Its Origin Structure and Evolution* (M. W. McElhinny, ed.) p.265–299. Academic Press, N.Y.

Green D. H., Nicholls I. A., Viljoen M. and Viljoen R. (1975) Experimental demonstration of the existence of peridotitic liquids in earliest Archean magmatism. *Geology* **3**, 11–14.

Green D. H., Ware N. G., Hibberson W. O. and Major A. (1971) Experimental petrology of Apollo 12 basalts: part 1, sample 12009. *Earth Planet. Sci. Lett.* **13**, 85–96.

Grossman L. and Larimer J. W. (1974) Early chemical history of the solar system. *Rev. Geophys. Space Phys.* **12**, 71–101.

Grove T. L. and Vaniman D. T. (1978) Experimental petrology of very low Ti (VLT) basalts. In *Mare Crisium: The View from Luna 24* (R. B. Merrill and J. J. Papike, eds.), p. 445–471. Pergamon Press, N.Y.

Grove T. L., Walker D., Longhi J., Stolper E. M. and Hays J. F. (1973) Petrology of rock 12002 from Oceanus Procellarum. *Proc. Lunar Sci. Conf. 4th*, p. 995–1011.

Haggerty S. E. (1976) Oxidation of opaque mineral oxides in basalts. In *Oxide Minerals* (D. Rumble, ed.), p. Hgl–Hgl00. Mineral Soc. Amer. Short Course Notes, v. 3.

Hamilton P. J., Evensen M. N., O'Nions R. K., Smith H. S. and Erlank A. J. (1979) Sm-Nd dating of Onverwacht Group volcanics, Southern Africa. *Nature* **279**, 298–300.

Hanson G. N. and Langmuir C. H. (1978) Modelling of major elements in mantle-melt systems using trace element approaches. *Geochim. Cosmochim. Acta* **42**, 725–742.

Hart R. and Hogan L. (1978) Earth degassing models and the heterogeneous versus homogeneous mantle. In *Terrestrial Rare Gases* (E. C. Alexander and M. Ozima, eds.), p. 193–208. Jap. Sci. Soc. Press.

Hedge C. E. (1978) Strontium isotopes in basalts from the Pacific Ocean basin. *Earth Planet. Sci. Lett.* **38**, 88–94.

Hill R. and Roeder P. (1974) The crystallization of spinel from basaltic liquid as a function of oxygen fugacity. *J. Geol.* **82**, 709–729.

Hofmann A. W. and Hart S. R. (1978) An assessment of local and regional isotopic equilibrium in the mantle. *Earth Planet. Sci. Lett.* **37**, 44–62.

Huneke J. C., Smith S. P., Rajan R. S., Papanastassiou D. A. and Wasserburg G. J. (1977) Comparison of the chronology of the Kapoeta parent planet and the Moon (abstract). In *Lunar Science VII*, p. 484–486. The Lunar Science Institute, Houston.

Hurley P. M. (1968) Absolute abundance and distribution of Rb, K and Sr in the Earth. *Geochim. Cosmochim. Acta* **32**, 273–283.

Irving A. J. and Frey F. A. (1978) Distribution of trace elements between garnet megacrysts and host volcanic liquids of kimberlitic to rhyolitic composition. *Geochim. Cosmochim. Acta* **42**, 771–787.

Jordan T. (1978) Composition and development of the continental tectosphere. *Nature* **274**, 544–548.

Kaula W. M. (1977) On the origin of the Moon, with emphasis on bulk composition. *Proc. Lunar Sci. Conf. 8th*, p. 321–331.

Kay R. W. and Hubbard N. J. (1978) Trace elements in ocean ridge basalts. *Earth Planet. Sci. Lett.* **38**, 95–116.

Keihm S. J. and Langseth M. G. (1977) Lunar thermal regime to 500 km (abstract). In *Lunar and Planetary Science IX*, p. 531–533. Lunar and Planetary Institute, Houston.

Krähenbühl U., Ganapathy R., Morgan J. W. and Anders E. (1973) Volatile elements in Apollo 16 samples: implications for highland volcanism and accretion history of the Moon. *Proc. Lunar Sci. Conf. 4th*, p. 1325–1348.

Kushiro I. (1972) The effect of water on the composition of magmas formed at high pressures. *J. Petrol.* **13**, 311–334.

Kushiro I. and Nakamura Y. (1970) Petrology of some lunar crystalline rocks. *Proc. Apollo 11 Lunar Sci. Conf.*, p. 607–626.

Langmuir C. H., Bender J. F., Bence A. E., Hanson G. N. and Taylor S. R. (1977) Petrogenesis of basalts from the FAMOUS area, Mid-Atlantic Ridge. *Earth Planet. Sci. Lett.* **36**, 133–156.

Leeman W. P. and Linstrom D. J. (1978) Partitioning of Ni^{2+} between basaltic and synthetic melts and olivines—an experimental study. *Geochim. Cosmochim. Acta* **42**, 801–816.

Longhi J. (1977) Magma oceanography 2: chemical evolution and crustal formation. *Proc. Lunar Sci. Conf. 8th*, p. 601–621.

Longhi J., Walker D. and Hays J. F. (1978) The distribution of Fe and Mg between olivine and lunar basaltic liquids. *Geochim. Cosmochim. Acta* **42**, p. 1545–1558.

Lugmair G. W. (1975) Sm-Nd systematics of some Apollo 17 basalts (abstract). In *Conference on Origins Of Mare Basalts and Their Implications for Lunar Evolution*, p. 107–110. The Lunar Science Institute, Houston.

Lugmair G. W. and Carlson R. W. (1978) The Sm-Nd history of KREEP. *Proc. Lunar Sci. Conf. 9th*, p. 689–704.

Lugmair G. W., Scheinin N. B. and Marti K. (1975) Sm-Nd age and history of Apollo 17 basalt 75075: evidence for early differentiation of the lunar interior. *Proc. Lunar Sci. Conf. 6th*, p. 1419–1429.

Ma M. -S. and Schmitt R. A. (1979) A chemical review of the various types of lunar mare low-K basalts (abstract). In *Lunar and Planetary Science X*, p. 753–755. Lunar and Planetary Science Institute, Houston.

Ma M.-S., Schmitt R. A., Taylor G. J., Warner R.D., Lange D. W. and Keil K. (1978) Chemistry and petrology of Luna 24 lithic fragments and soils: constraints on the origin of VLT basalts. In *Mare Crisium: The View From Luna 24*, (R. B. Merrill and J. J. Papike, eds.), p. 569–592. Pergamon, N.Y.

McCulloch M. T. and Wasserburg G. J. (1977) Sm-Nd and Rb-Sr chronology of continental crust formation. *Science* **200**, 1003–1011.

Morgan J. W., Wandless G. A., Petrie R. K. and Irving A. J. (1981) Composition of the Earth's upper mantle—I. siderophile trace elements in ultramafic nodules. *Tectonophys.* **75**, 47–67.

Morgan J. W. (1971) Convection plumes in the lower mantle. *Nature* **230**, 42–43.

Murthy V. R. (1975) Composition of the core and the early chemical history of the Earth. In *The Early History of the Earth*, (B. F. Windley, ed.), p. 21–31. Wiley, N.Y.

Nakamura Y., Latham G., Lammlein D., Ewing M., Duennebier F. and Dorman J. (1974) Deep lunar interior inferred from recent seismic data. *Geophys. Res. Lett.* **1**, 137–140.

Nunes P. D., Tatsumoto M. and Unruh D. M. (1975) U-Th-Pb systematics of anorthositic gabbros 78155 and 77017—implications for early lunar evolution. *Proc. Lunar Sci. Conf. 6th*, p. 1431–1444.

Nyquist L. E., Bansal B. M., Weismann H. and Jahn B. -M. (1974) Taurus-Littrow chronology: some constraints on Earth lunar crustal development. *Proc. Lunar Sci. Conf. 5th*, p. 1515–1539.

Nyquist L. E., Wooden J., Bansal B., Wiesmann H., McKay G. and Bogard D. D. (1979) Rb-Sr age of the Shergotty achondrite and implications for metamorphic resetting of isochron ages. *Geochim,. Cosmochim. Acta.* **43**, 1047–1074.

O'Hara M. J. (1968a) Are ocean flow basalts primary magma? *Nature* **220**, 683–686.

O'Hara M. J. (1968b) The bearing of phase equilibrium studies in synthetic and natural systems on the origin and evolution of basic and ultrabasic rocks. *Earth Sci. Rev.* **4**, 69–133.

O'Hara M. J., Biggar G. M., Richardson S. W., Ford C. E. and Jamieson B. G. (1970) The nature of seas, mascons and the lunar interior in the light of experimental studies. *Proc. Apollo 11 Lunar Sci. Conf.*, p. 695–710.

O'Nions R. K., Carter S. R., Evensen N. M. and Hamilton P. J. (1979a) Upper mantle geochemistry. In *The Sea*, **v. 7** (E. Emiliani, ed.).

O'Nions R. K., Carter S. R., Evensen N. M. and Hamilton P. J. (1979b) Isotope geochemical studies of North Atlantic ocean basalts and their implications for mantle evolution. In *Deep Drilling Results in the Atlantic Ocean: Ocean Crust* (M. Talwani, C. G. Harrison and D. E. Hayes, eds.), p. 342–351. Maurice Ewing Series 2. Amer. Geophys. Union, Washington, D.C.

O'Nions R. K., Carter S. R., Evensen N. M. and Hamilton P. J. (1979c) Geochemical and cosmochemical applications of Nd isotope analysis. *Ann. Rev. Earth Planet. Sci.* **7**, 11–38.

Oversby V. M. and Ringwood A. E. (1971) Time of formation of the Earth's core. *Nature* **234**, 463–465.

Ozima M. and Kudo K. (1972) Excess argon in submarine basalts and an Earth atmosphere evolution model. *Nat. Phys. Sci.* **239**, 23–24.

Papanastassiou D. A. and Wasserburg G. J. (1971) Lunar chronology and evolution from Rb-Sr studies of Apollo 11 and 12 samples. *Earth Planet. Sci. Lett.* **11**, 37–62.

Papanastassiou D. A. and Wasserburg G. J. (1972) Rb-Sr age of a Luna 16 basalt and the model age of lunar soils. *Earth Planet. Sci. Lett.* **13**, 368–374.

Papike J. J. and Bence A. E. (1978) Lunar mare versus terrestrial mid-ocean ridge basalts: planetary constraints on basaltic volcanism. *Geophys. Res. Lett.* **5**, 803–806.

Presnall D. C., Dixon J. R., O'Donnell T. H. and Dixon S. A. (1979) Generation of mid-ocean ridge tholeiites. *J. Petrol.* **20**, 3–35.

Rammensee W. and Wänke H. (1977) On the partition coefficient of tungsten between metal and silicate and its bearing on the origin of the Moon. *Proc. Lunar Sci. Conf. 8th*, p. 399–409.

Rhodes J. M., Dungan M. A., Blanchard D. P. and Long P. E. (1979) Magma mixing at mid-ocean ridges: evidence from basalts drilled near 22° N on the Mid-Atlantic Ridge. *Tectonophys.* **55**, 35–61.

Richter F. M. and Ribe N. M. (1979) On the importance of advection in determining the local isotopic composition of the mantle. *Earth Planet. Sci. Lett.* **43**, 212–222.

Ridley W. I. (1975) On high-alumina mare basalts. *Proc. Lunar Sci. Conf. 6th*, p. 131–145.

Ridley W. I., Reid A. M., Warner J. L. and Brown R. W. (1973) Apollo 15 green glass. *Phys. Earth Planet. Inter.* **7**, 133–136.

Ringwood A. E. (1966) The chemical composition and origin of the Earth. In *Advances in Earth Science* (P. M. Hurley, ed.), p. 287–356. MIT Press, Cambridge.

Ringwood A. E. (1970) Petrogenesis of Apollo 11 basalts and implications of lunar origin. *J. Geophys. Res.* **75**, 6453–6479.

Ringwood A. E. (1975) *Composition and Petrology of the Earth's Mantle.* McGraw-Hill, N.Y. 618 pp.

Ringwood A. E. and Essene E. (1970) Petrogenesis of Apollo 11 basalts, internal constitution and origin of the Moon. *Proc. Apollo 11 Lunar Sci. Conf.*, p. 769–799.

Sato M. (1978) Oxygen fugacity of basaltic magmas and the role of gas-forming elements. *Geophys. Res. Lett.* **5**, 447–449.

Schnetzler C. C. and Philpotts J. A. (1971) Alkali, alkaline earth and rare-earth element concentrations in some Apollo 12 soils, rocks, and separated phases. *Proc. Lunar Sci. Conf. 2nd*, p. 1101–1122.

Schreiber H. D. and Haskin L. A. (1976) Chromium in basalts: experimental determination of redox states and partitioning among synthetic silicate phases. *Proc. Lunar Sci. Conf. 7th*, p. 1221–1259.

Schubert G., Cassen P. and Young R. E. (1979) Subsolidus convective cooling histories of terrestrial planets. *Icarus* **38**, 192–211.

Shih C. and Schonfeld E. (1976) Mare basalt genesis: a cumulate remelting model. *Proc. Lunar Sci. Conf. 7th*, p. 1757–1792.

Smith J. V. (1979) Mineralogy of the planets: a voyage in space and time. *Mineral Mag.* **43**, 1–89.

Stolper E. M. (1974) *Lunar Ultramafic Glasses.* Master's Thesis, Harvard Univ. 178 pp.

Stolper E. (1977) Experimental petrology of eucritic meteorites. *Geochim. Cosmochim. Acta* **41**, 587–611.

Stolper E. (1979) Trace elements in shergottite meteorites: implications for the origins of planets. *Earth Planet. Sci. Lett.* **42**, 239–242.

Stolper E. and Walker D. (1980) Melt density and average composition of basalt. *Contrib. Mineral. Petrol.* **74**, 7–12.

Stolper E., McSween H. Y. and Hays J. F. (1979) A petrogenetic model of the relationships among achondritic meteorites. *Geochim. Cosmochim. Acta* **43**, 589–602.

Sun S. -S. and Nesbitt R. W. (1977) Chemical heterogeneity of the Archean mantle, composition of the Earth, and mantle evolution. *Earth Planet. Sci. Lett.* **35**, 429–448.

Sun S. -S., Nesbitt R. W. and Sharaskin A. Y. (1979) Geochemical characteristics of mid-ocean ridge basalts. *Earth Planet. Sci. Lett.* **44**, 119–138.

Surkov Y. A. (1977) Geochemical studies of Venus by Venera 9 and 10 automatic interplanetary stations. *Proc. Lunar Sci. Conf. 8th*, p.2665–2689.

Taylor S. R. (1973) Chemical evidence for lunar melting and differentiation. *Nature* **245**, 203–205.

Taylor S. R. (1975) *Lunar Science: A Post Apollo View.* Pergamon, N.Y. 372 pp.

Taylor S. R. (1976) Geochemical constraints on the composition of the Moon. *Proc. Lunar Sci. Conf. 7th*, p. 3461–3477.

Taylor S. R. (1979a) Chemical composition and evolution of the continental crust: the rare-earth element evidence. In *The Earth: Its Origin, History and Evolution*, (M. W. McElhinny, ed.), p. 353–376. Academic Press, N.Y.

Taylor S. R. (1979b) Lunar and terrestrial potassium and uranium abundances: implications for the fission hypothesis. *Proc. Lunar Planet. Sci. Conf. 10th*, p. 2017–2030.

Taylor S. R. (1979c) Structure and evolution of the Moon. *Nature* **281**, 105–110.

Taylor S. R. and Hallberg J. A. (1977) Rare-earth elements in the Marda calc-alkaline suite: an Archean geochemical analogue of Andean-type volcanism. *Geochim. Cosmochim. Acta* **41**, 1125–1129.

Taylor S. R. and Jakeš P. (1974) The geochemical evolution of the Moon. *Proc. Lunar Sci. Conf. 5th*, p. 1287–1305.

Taylor S. R., Gorton M. P., Muir P., Nance W. R., Rudowski R. and Ware N. (1973) Composition of the Descartes region, lunar highlands. *Geochim. Cosmochim. Acta* **37**, 2665–2683.

Tera F. and Wasserburg G. J. (1974) U-Th-Pb systematics on lunar rocks and inferences about lunar evolution and the age of the Moon. *Proc. Lunar Sci. Conf. 5th*, p. 1571–1599.

Tera F. and Wasserburg G. J. (1975) The evolution and history of mare basalts as inferred from the U-Th-Pb systematics (abstract). In *Lunar Science VI*, p. 807–809. Lunar Science Institute, Houston.

Toksöz M. N., Dainty A. M., Solomon S. C. and Anderson K. R. (1973) Velocity structure and evolution of the Moon. *Proc. Lunar Sci. Conf. 4th*, p. 2529–2547.

Toulmin P., Baird A. K., Clark B. C., Keil K., Rose H. J., Christian R. P., Evans P. H. and Kelliher W. C. (1977) Geochemical and mineralogical interpretation of the Viking inorganic chemical results. *J. Geophys. Res.* **82**, 4625–4634.

Vaniman D. T. and Papike J. J. (1976) Very low Ti (VLT) basalts: a new mare rock type from the Apollo 17 drill core. *Proc. Lunar Sci. Conf. 8th*, p. 1443–1471.

Vidal P. and Dosso L. (1978) Core formation: catastrophic or continuous? Sr and Pb isotope geochemistry constraints. *Geophys. Res. Lett.* **5**, 169–172.

Walker D., Longhi J. and Hays J. F. (1975a) Differentiation of a very thick magma body and implications for the source regions of mare basalts. *Proc. Lunar Sci. Conf. 6th*, p. 1103–1120.

Walker D., Longhi J., Kirpatrick R. J. and Hays J. F. (1976) Differentiation of an Apollo 12 picrite magma. *Proc. Lunar Sci. Conf. 7th*, p. 1365–1389.

Walker D., Longhi J., Stolper E. M., Grove T. L. and Hays J. F. (1975b) Origin of titaniferous lunar basalts. *Geochim. Cosmochim. Acta* **39**, 1219–1235.

Walker D., Stolper E. M. and Hays J. F. (1979) Basaltic volcanism: the importance of planet size. *Proc. Lunar Planet. Sci. Conf. 10th*, p. 1995–2015.

Wänke H., Baddenhausen H., Dreibus G., Jagoutz E., Kruse H., Palme H., Spettel B. and Teschke F. (1973) Multielement analyses of Apollo 15, 16 and 17 samples and the bulk composition of the Moon. *Proc. Lunar Sci. Conf. 4th*, p. 1461–1481.

Wasserburg G. J. and DePaolo D. J. (1979) Models of Earth structure inferred from Nd and Sr isotopic abundances. *Proc. Nat. Acad. Sci.* **76**, 3594–3598.

Wasserburg G. J., Papanastassiou D. A.,, Tera F. and Huneke J. C. (1977) Outline of a lunar chronology. In "The Moon—a new appraised from space missions and laboratory analyses." *Phil. Trans. Roy. Soc. Lond.*. **A285**, 7–22.

Wentworth S., Taylor G. J., Warner R. D., Keil K., Ma M. -S. and Schmitt R. A. (1979) The unique nature of Apollo 17 VLT mare basalts. *Proc. Lunar Planet. Sci. Conf. 10th*, p. 207–223.

White W. M. and Schilling J. G. (1978) The nature and origin of geochemical variation in Mid-Atlantic Ridge basalts from the central North Atlantic. *Geochim. Cosmochim. Acta* **42**, 1501–1516.

Whitford D. J. and Arndt N. T. (1978) REE abundances in a thick layered komatiite lava flow from Ontario, Canada. *Earth Planet. Sci. Lett.* **41**, 188–196.

Wolf R., Woodrow A. and Anders E. (1979) Siderophile and volatile elements in the Earth and the Moon: similar or not? (abstract). In *Lunar and Planetary Science X*, p. 1361–1363. Lunar and Planetary Science Institute, Houston.

Wood J. A. (1975) Lunar petrogenesis in a well-stirred magma ocean. *Proc. Lunar Sci. Conf. 6th*, p. 1087–1102.

Wood J. A., Dickey J. S., Marvin U. B. and Powell B. N. (1970) Lunar anorthosites and a geophysical model of the Moon. *Proc. Apollo 11 Lunar Sci. Conf.*, p. 965–988.

Wright T. L., Grolier M. J. and Swanson D. A. (1973) Chemical variation related to stratigraphy of the Columbia River basalt. *Geol. Soc. Amer. Bull.* **84**, 371–386.

Wyllie P. J. (1979) Magmas and volatile components. *Amer. Mineral.* **67**, 469–500.

1.3.2 SILICATE MINERALOGY OF PLANETARY BASALTS

1.3.2.1 Introduction

The mineralogy and mineral chemistry of a basalt are largely functions of its bulk chemical composition and, to some extent, the intensive variables (e.g., pressure, gas fugacities, cooling rate) operative during its crystallization. In section 1.3.1, we have emphasized the important inferences that can be made about planetary constitution and evolution from the bulk chemistry of basalts. Those inferences are, however, critically dependent on having representative rock samples in our possession. But what if our only sample from a planet were a small quantity of fine grained regolith soil? Or if our only information about a planetary surface were from remotely sensed spectra of a regolith? Could we still make reasonable inferences about the internal constitution of the planet?

Our experience with the lunar soil samples has resulted in ingenious techniques to obtain bulk chemical data from very small rock clasts. However, there is a limit to the clast size from which representative data can be obtained. In contrast, information on the major element chemistry of virtually every mineral grain in a soil sample is accessible by electron microprobe techniques. As discussed in Chapter 2, a number of remote-sensing techniques are capable of determining some bulk chemical characteristics of a planetary regolith, but, more importantly, spectral reflectance data can place constraints on mineralogy and mineral chemistry. If we are to be able to make the most of our information for other planetary bodies, it is essential that we have a good comparative mineral chemical data base for those planetary basalt samples that we have in hand.

With these goals in mind, this section is a synthesis of the chemistry of three major mineral groups for the ten basalt suites discussed in preceding sections: Archean, Keweenawan, Columbia Plateau, Rio Grande, ocean floor, Hawaiian, island arc, basaltic meteorite, lunar mare and lunar highland. Although this is a relatively small set of samples, it does serve to emphasize the variety among at least three different planetary bodies.

We emphasize that the mineral chemistry presented in this section is considered in overview fashion. More detailed discussions are included either in the individual chapter sections concerning the basalt suites or in the references cited. The reader is also referred to the review papers by Smith (1974, 1979) and Smith and Steele (1976), and the extensive references therein. Since this section is concerned with planetary comparisons, we will be especially interested in those characteristics of the mineral chemistry that reflect planetary constraints of the parent bodies on the host basalts.

The number of analyses of each silicate type (olivine, feldspar and pyroxene) for each suite is tabulated in Table 1.3.2.1. Many representative silicate analyses (by electron microprobe) are presented in the microfiche Appendix A-12 whereas certain averages for the groups are tabulated here. In addition, twelve of the pyroxene suites were used in a statistical analysis. In this analysis the cations ^{IV}Al, ^{VI}Al, Fe^{2+}, "Fe^{3+}", Mg, Mn, Ti, Cr, Ca and Na were variables treated with a factor analysis program (BMDP4M; Dixon, 1975). The cation-cation correlation coefficients obtained in this analysis give important insights concerning the substitutional couples in the pyroxene structure.

1.3.2.2 Olivines

We have fair coverage of the olivine assemblages in the basalt suites except for the basaltic meteorites, where olivine is absent, and for the Keweenawan basalts, where it is extensively altered. The complete data set referred to in this section is presented in Table 1 of Appendix A-12, whereas averages for the groups are presented in Table 1.3.2.2 and Fig. 1.3.2.1. Figure 1.3.2.1 presents a series of chemical variation diagrams that are all normalized to a common analysis listed as the grand average in Table 1.3.2.2. The plotted value for each oxide and for each group is obtained by the calculation x/(x + grand average), where x equals the value for each oxide listed in Table 1.3.2.2. This type of diagram has the advantage of portraying similarities and differences in concentrations for a large number of oxides on a single plot. Some interesting chemical varia-

Table 1.3.2.1 Numbers of silicate analyses from basalt suites used in synthesis.

Suite	Pyroxenes	Olivines	Feldspars
Archean	39	19	—
Columbia Plateau	18	24	27
Ocean Floor	140	38	85
Basaltic Meteorites	426	—	214
Hawaiian	60	45	34
Island Arc	165	74	106
Keweenawan	132	3	115
Lunar Highland	59	21	38
Lunar Mare	101	19	23
Rio Grande	123	26	66
	1263	269	708

Table 1.3.2.2 Mean compositions of olivines.

	Archean	Columbia Plateau	Ocean Floor	Hawaiian	Island Arc
SiO_2	40.64	35.35	38.16	38.82	38.37
Al_2O_3	0.14	0.11	0.02	0.09	0.07
FeO	8.94	37.72	22.02	18.86	21.51
MgO	49.63	26.34	39.51	42.00	39.91
MnO	0.06	0.56	0.38	0.28	0.23
Cr_2O_3	0.18	0.02	0.05	0.08	0.00
CaO	0.27	0.40	0.36	0.30	0.25
	99.86	100.50	100.50	100.43	100.34

	Keweenawan	Lunar Highland	Lunar Mare	Rio Grande	(Grand Average)
SiO_2	36.05	38.14	35.37	38.34	37.69
Al_2O_3	0.03	0.12	0.05	0.03	0.07
FeO	32.15	23.90	36.59	20.48	24.69
MgO	30.41	38.00	27.07	39.86	36.97
MnO	0.43	0.23	0.37	0.32	0.32
Cr_2O_3	0.00	0.10	0.18	0.00	0.07
CaO	0.15	0.23	0.35	0.18	0.28
	99.22	100.72	99.98	99.21	100.09

tions occur in the minor oxides Al_2O_3 and Cr_2O_3. Olivines from Archean basalts are relatively enriched in Al_2O_3 whereas olivines from ocean floor and Rio Grande basalts are depleted in Al_2O_3. Likewise, Cr_2O_3 enrichments occur in olivines from Archean, lunar highland and lunar mare basalts, and Cr_2O_3 depletions occur in olivines from Columbia Plateau, island arc, Keweenawan and Rio Grande basalts.

The forsterite (Mg_2SiO_4) - fayalite (Fe_2SiO_4) ranges for the suites are illustrated as a bar histogram in Fig. 1.3.2.2. The parameter $X_{Fe} = Fe/(Fe + Mg)$ atomic is used in this and subsequent variation diagrams to indicate Mg-Fe^{2+} substitutions. The most primitive olivines, i.e., those with the lowest X_{Fe} are found in Archean basalts where the most magnesian olivines approach $X_{Fe} = 0.05$. The komatiite liquids, which contain up to 33 wt.% MgO, appear to be unique to the terrestrial Archean. The olivines in komatiites are also significantly enriched in Cr ("Cr_2O_3" = 0.1–0.4 wt.%) relative to all other terrestrial basalts. These Cr contents are comparable to those found in olivines from lunar mare basalts (note Table 1.3.2.2, Fig. 1.3.2.9 and Fig. 1.3.2.1). The high Cr abundance in olivines in mare basalts has been explained in terms of reduced valence states of Cr from Cr^{3+} to Cr^{2+}. Since Cr^{2+} is more readily accepted into the olivine structure than Cr^{3+}, this would explain the high abundances (see Papike *et al.*, 1976, for a review). Could some of the Cr in Archean komatiites and basalts also be in the Cr^{2+} valence state? The olivines from Archean basalts are also enriched in Al relative to other terrestrial basalt olivines and have values (Ave = 0.14 wt.%) comparable to those in lunar highland melt rocks. The Mn contents (Fig. 1.3.2.1 and Fig. 1.3.2.10) of these olivines are low as would be expected from the low X_{Fe} values, since Mn is highly correlated with Fe^{2+} (Simkin and Smith, 1970).

The most magnesian olivines from ocean floor and Hawaiian basalts have $X_{Fe} = 0.10$, indicating that these olivines crystallized from liquids that underwent little or no modification after derivation from a presumed magnesian mantle source region. Olivines in lunar highland and island arc basalts are slightly less magnesian ($X_{Fe} = 0.12$).

Olivines from the Rio Grande basalts (lowest $X_{Fe} = 0.14$) and Columbia Plateau basalts (lowest $X_{Fe} = 0.25$) possibly reflect crystallization from more evolved liquids. The olivines from mare basalts with $X_{Fe} = 0.22$ are thought to have crystallized from some of the most primitive mare liquids (Papike *et al.*, 1976). The higher X_{Fe} of these mare olivines, and thus of the mare basalts from which they crystallized, are consistent with a lunar mantle source which is richer in Fe than the terrestrial mantle (Papike and Bence, 1978).

The Ca-X_{Fe} systematics for olivines from the basalt suites are presented in Fig. 1.3.2.11 and are consistent with rapid cooling histories (Simkin and Smith, 1970).

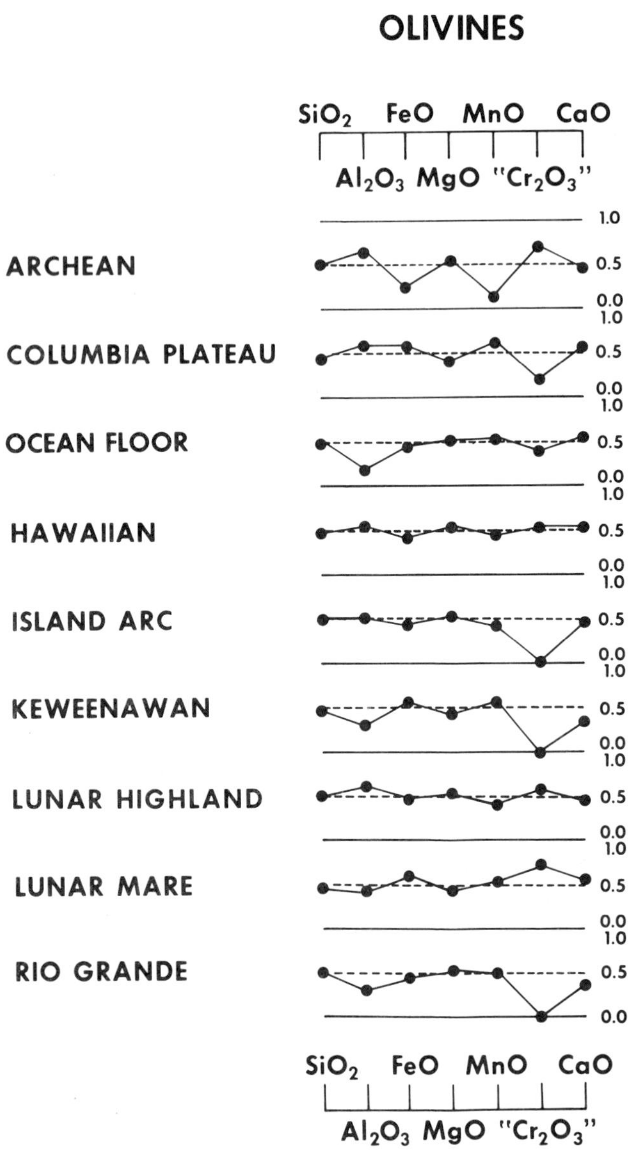

1.3.2.1 Variation diagrams for seven oxides for the olivine mineral suites. The actual numerical values plotted, which vary between zero and one, are derived from the ratio oxide/(oxide + average). "Oxide" represents the mean for each olivine suite (note Table 1.3.2.2) and "average" is the grand average for all of the olivine suites (note Table 1.3.2.2).

1.3.2.3 Feldspars

Feldspar chemical data are available for all the suites except the Archean, in which alteration has destroyed original feldspar compositions. The data are presented as follows: bar histogram of Ab($NaAlSi_3O_8$) -An($CaAl_2Si_2O_8$) variations, Fig. 1.3.2.3; Or-An-Ab variations for suites, Fig. 1.3.2.12a; Or-An-Ab variations for meteorites, Fig. 1.3.2.12b; complete set of feldspar analyses, Table 2 of Appendix A-12; average feldspar analyses for the suites, Table 1.3.2.3a; average feldspar analyses for six different meteorites, Table 1.3.2.3b.

Feldspars most enriched in the anorthite component or depleted in the albite component are found in lunar highland melt rocks, lunar mare basalts and eucrites, presumably reflecting the overall alkali depletion on the Moon and the eucrite parent body. However, feldspars from the Shergotty meteorite (see discussion of the basaltic meteorite suite, section 1.2.8) are considerably enriched in alkalies (An_{46-57}) relative to the eucrites (Fig. 1.3.2.12b).

The feldspars from the ocean floor, island arc and Hawaiian suites show maximum An contents that are similar and quite high (An $\sim$ 88%). The calcic composition of the feldspar and the high forsterite content of the olivines imply crystallization from relatively primitive liquids as discussed above. The lower maximum An content for feldspars from the Columbia Plateau suite correlates with more Fe-rich olivines in these rocks and implies equilibration with much more evolved liquids.

The Fe contents of the feldspars from the different suites are quite similar (Table 1.3.2.3a), except for the lower contents in feldspars from the lunar highland melt rock suite where the bulk FeO contents of the host rocks are also quite low (see section 1.2.10).

1.3.2.4 Pyroxenes

Of the major silicates in basaltic rocks, pyroxenes are the most complex chemically and are powerful recorders of the evolutionary history of these rocks. The pyroxenes contain most of the major elements of their host rocks. Ca, Na, Mn, Fe^{2+} and Mg are accommodated in the large distorted 8 to 6 coordinated M2 site; Mn, Fe^{2+}, Mg, Fe^{3+}, Cr^{3+}, Cr^{2+}, Ti^{4+}, Ti^{3+}, and Al may substitute in the octahedral M1 site; and Al and Si, and sometimes Fe^{3+}, can occupy the tetrahedral site. The only major element that cannot be accommodated in the pyroxene structure is K. In the basalts considered here, potassium resides either in feldspar or in the mesostasis.

The pyroxene mineral chemistry is presented as follows: listing of a large number of representative and high quality pyroxene analyses, Table 3 of Appendix A-12; average mineral chemistry for the suites, Table

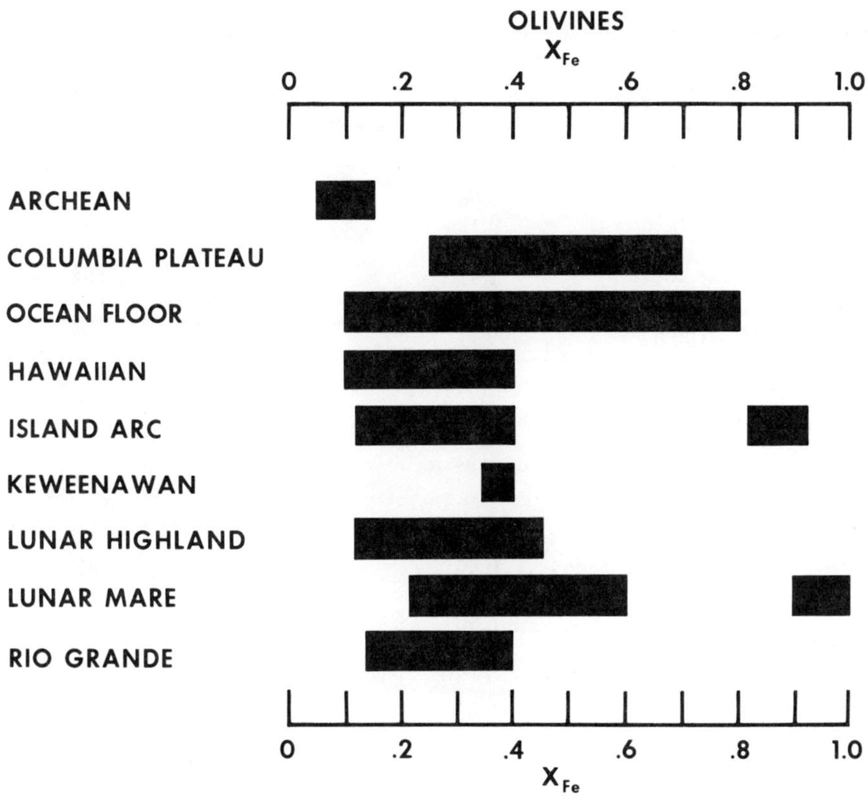

1.3.2.2 Olivine bar histogram showing Mg-Fe^{2+} variations. $X_{Fe} = Fe^{2+}/(Fe^{2+} + Mg)$ atomic.

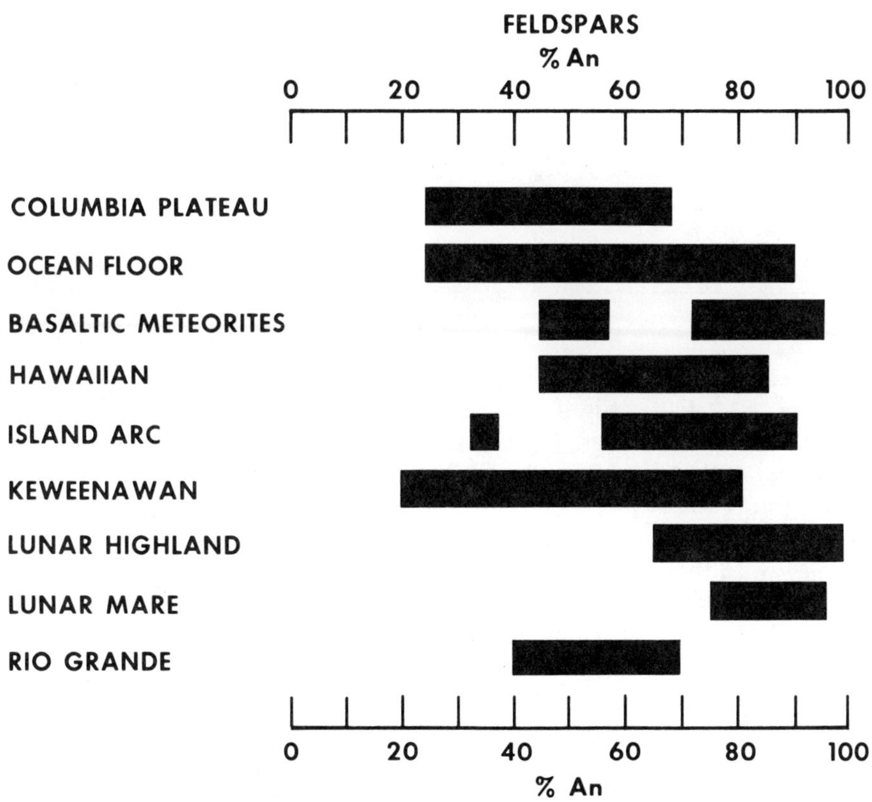

1.3.2.3 Feldspar bar histogram showing Na-Ca variations.

Table 1.3.2.3a Mean compositions of feldspars.

	Columbia Plateau	Ocean Floor	Basaltic Meteorites	Hawaiian	Island Arc
SiO_2	54.91	53.00	47.87	52.02	48.70
Al_2O_3	27.53	29.28	32.65	29.55	32.32
FeO	0.87	0.74	0.66	0.77	0.71
MgO	0.07	0.16	0.12	0.08	0.05
CaO	10.99	12.64	16.26	13.06	15.58
Na_2O	5.05	4.14	1.97	3.94	2.58
K_2O	0.40	0.06	0.18	0.21	0.19
	99.82	100.02	99.71	99.63	100.13

	Keweenawan	Lunar Highland	Lunar Mare	Rio Grande	(Grand Average)
SiO_2	54.93	46.92	48.09	53.18	51.07
Al_2O_3	28.30	33.68	32.38	29.28	30.55
FeO	0.71	0.28	0.69	0.72	0.68
MgO	0.13	0.14	0.20	0.10	0.12
CaO	10.19	17.48	16.90	11.58	13.85
Na_2O	5.15	1.29	1.45	4.52	3.34
K_2O	0.79	0.19	0.15	0.33	0.28
	100.20	99.98	99.86	99.71	99.89

Table 1.3.2.3b Mean compositions of feldspars in individual basaltic meteorites.

	Aioun el Atrouss	Juvinas	Nuevo Laredo	Stannern	Pasamonte	Shergotty	(Average)
SiO_2	45.13	45.31	46.98	47.82	46.89	55.81	47.99
Al_2O_3	34.67	34.60	32.72	32.61	33.41	27.44	32.58
FeO	0.19	0.61	1.44	0.40	0.68	0.68	0.67
MgO	0.03	0.14	0.32	0.08	0.10	0.09	0.13
CaO	17.62	17.73	15.92	16.25	17.50	10.90	15.99
Na_2O	1.17	0.96	1.66	2.06	1.40	5.17	2.07
K_2O	0.10	0.08	0.17	0.30	0.09	0.31	0.18
	98.91	99.43	99.21	99.52	100.07	100.40	99.61

1.3.2.4a; averages for basaltic meteorites, Table 1.3.2.4b; average chemical data variation diagram, Fig. 1.3.2.4; average chemical data variation diagram for basaltic meteorites, Fig. 1.3.2.5; composite quadrilateral diagrams, Fig. 1.3.2.6a, b; quadrilateral diagrams for individual suites, Fig. 1.3.2.13a; quadrilateral diagrams for individual basaltic meteorites, Fig. 1.3.2.13b; $(Na + {}^{IV}Al)$ vs. $({}^{VI}Al + 2Ti + Cr)$ diagrams for individual suites, Fig. 1.3.2.7a; $(Na + {}^{IV}Al)$ vs. $({}^{VI}Al + 2Ti + Cr)$ for basaltic meteorites, Fig. 1.3.2.7b; Ti-^{IV}Al diagrams for suites, Fig. 1.3.2.14a; Ti-^{IV}Al diagrams for basaltic meteorites, Fig. 1.3.2.14b; ^{IV}Al-X_{Fe} diagrams for suites, Fig. 1.3.2.15a; ^{IV}Al-X_{Fe} diagrams for basaltic meteorites, Fig. 1.3.2.15b; Cr-X_{Fe} composite diagrams, Figs. 1.3.2.8a, b; Cr-X_{Fe} diagrams for suites, Fig. 1.3.2.16a; Cr-X_{Fe} diagrams for basaltic meteorites, Fig. 1.3.2.16b; Mn-X_{Fe} diagrams for suites, Fig. 1.3.2.17a; Mn-X_{Fe} diagrams for basaltic meteorites, Fig. 1.3.2.17b; Na-X_{Fe} diagrams for suites, Fig. 1.3.2.18a; Na-X_{Fe} diagrams for basaltic meteorites, Fig. 1.3.2.18b; Ti-X_{Fe} diagram for suites, Fig. 1.3.2.19a; Ti-X_{Fe} diagrams for basaltic meteorites, Fig. 1.3.2.19b.

Figures 1.3.2.4 and 1.3.2.5 are derived in a similar manner to those for the olivines discussed above. Each oxide for each group is normalized against the grand average value given in Tables 1.3.2.4a and 1.3.2.4b. A considerable range of Fe_2O_3 values is observed for these pyroxenes (Fe_2O_3 contents were estimated by the method of Papike *et al.*,1974). Pyroxenes from basaltic meteorites, lunar highland melt rocks and mare basalts show essentially no Fe_2O_3, reflecting the low oxygen fugacities that existed during the extrusion of the parent

Table 1.3.2.4a Mean compositions of pyroxenes.

	Archean	Columbia Plateau	Ocean Floor	Basaltic Meteorites	Hawaiian	Island Arc
SiO_2	51.68	50.30	50.42	50.17	52.25	51.31
Al_2O_3	3.92	2.00	3.01	0.75	2.82	2.83
FeO	9.60	12.51	10.60	27.17	8.45	9.54
Fe_2O_3	0.70	0.62	1.61	0.00	1.02	1.79
MgO	16.75	13.05	14.89	13.36	21.05	16.02
MnO	0.24	0.27	0.28	0.79	0.14	0.19
TiO_2	0.46	1.27	0.90	0.24	0.85	0.47
Cr_2O_3	0.20	0.05	0.20	0.52	0.38	0.18
CaO	16.82	19.23	17.86	6.81	12.84	17.30
Na_2O	0.17	0.16	0.19	0.04	0.13	0.29
	100.54	99.46	99.96	99.85	99.93	99.92

	Keweenawan	Lunar Highland	Lunar Mare	Rio Grande	(Grand Average)
SiO_2	51.31	52.12	49.00	50.93	50.95
Al_2O_3	1.79	1.98	2.55	2.68	2.43
FeO	13.21	15.62	22.15	10.30	13.91
Fe_2O_3	1.02	0.00	—	1.43	1.17
MgO	15.37	21.18	13.06	15.73	16.05
MnO	0.34	0.24	0.36	0.29	0.31
TiO_2	0.88	0.92	1.44	0.95	0.65
Cr_2O_3	0.15	0.56	0.54	0.12	0.29
CaO	16.02	7.36	10.51	17.02	14.18
Na_2O	0.24	0.02	0.02	0.31	0.16
	100.33	100.00	99.63	99.76	100.10

Table 1.3.2.4b Mean compositions of pyroxenes in individual basaltic meteorites.

	Aioun el Atrouss	Juvinas	Nuevo Laredo	Stannern	Pasamonte	Shergotty	(Average)
SiO_2	49.99	49.36	49.00	49.22	50.11	51.87	49.93
Al_2O_3	0.26	0.42	0.33	0.30	1.32	0.98	0.60
FeO	31.57	30.63	32.63	32.85	25.90	19.02	28.75
Fe_2O_3	—	—	—	—	—	—	—
MgO	11.25	12.25	9.58	11.43	14.48	16.48	12.58
MnO	0.94	0.89	0.96	0.83	0.75	0.55	0.84
TiO_2	0.16	0.24	0.24	0.24	0.29	0.22	0.23
Cr_2O_3	0.12	0.46	0.51	0.61	0.65	0.44	0.47
CaO	5.70	5.26	6.80	4.44	6.32	10.11	6.44
Na_2O	0.00	0.00	0.00	0.01	0.02	0.14	0.03
	99.99	99.51	100.05	99.93	99.84	99.82	99.87

basalts. The presence of some Fe_2O_3 in pyroxenes from Shergotty is suggested by the calculation method of Papike *et al.* 1974. (However, in the comparative mineralogy discussed below, all Fe is considered as Fe^{2+} for Shergotty pyroxenes.) This possibility would be consistent with the estimates of higher oxygen fugacities for shergottites compared to eucrites (see discussion of basaltic meteorites in section 1.2.8). Relatively high values of Fe_2O_3 are estimated for pyroxenes from ocean floor (Schweitzer *et al.*, 1978, 1979), Hawaiian, island arc and Rio Grande basalts. It is interesting to note the overall similarity of the "patterns" for the lunar highland and mare basalt pyroxenes even though these rock suites have very different bulk chemistries. There appears to be a distinct lunar signature for these pyroxenes.

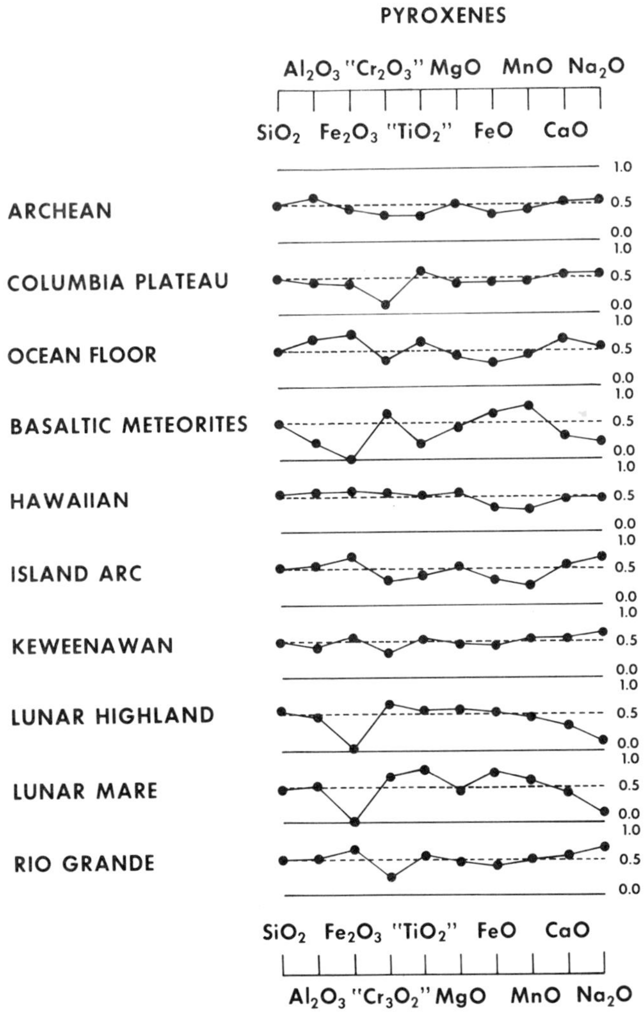

1.3.2.4 Variation diagrams for ten oxides for the pyroxene mineral suites. The actual numerical values plotted are derived from the ratio oxide/(oxide + average). "Oxide" represents the mean for each pyroxene suite (note Table 1.3.2.4a) and "average" is the grand average for all of the pyroxene suites (note Table 1.3.2.4a).

Figure 1.3.2.5 shows the intragroup comparisons for the basaltic meteorites. On this plot it is apparent that Juvinas, Nuevo Laredo and Stannern have very similar pyroxenes. Pyroxenes in Aioun el Atrouss are also similar, but appear to be relatively depleted in Al_2O_3, Cr_2O_3 and TiO_2 compared with the other equilibrated eucrites in the suite. Pyroxenes from the two unequilibrated meteorites in the basaltic meteorite suite, Pasamonte and Shergotty, have similarities in their major oxide patterns even though the host rock chemistries are quite different. This implies a cooling rate control, especially for Al, where pyroxenes from Pasamonte and Shergotty show significant Al enrichments relative to the average. An obvious difference in the major oxide patterns for Shergotty and Pasamonte pyroxenes is that Shergotty pyroxenes show Na enrichment.

Some of the major features of the "Quad" chemical systematics are illustrated in Figs. 1.3.2.6a, b and 1.3.2.13a, b. The two lunar suites (highland and mare) show the greatest chemical variability on these diagrams. The suites that have very magnesian compositions and most closely approach the CaMg (diopside) - MgMg (enstatite) sideline include the lunar highland, island arc, Archean, ocean floor and Hawaiian suites. Pyroxenes from the Columbia Plateau, Rio Grande and Keweenawan suites are more evolved (i.e., less magnesian); this is consistent with the olivine and feldspar data (presented earlier). The basaltic meteorite pyroxene assemblages plotted on the composite diagram (Fig. 1.3.2.6a) are only for the "equilibrated" groups (see section 1.2.8). The Pasamonte and Shergotty pyroxene chemical trends show significant zoning consistent with rapid cooling histories (Fig. 1.3.2.13b). The mare basalt pyroxenes are displaced towards the $CaFe^{2+}$ (hedenbergite)-$Fe^{2+}Fe^{2+}$ (ferrosilite) sideline.

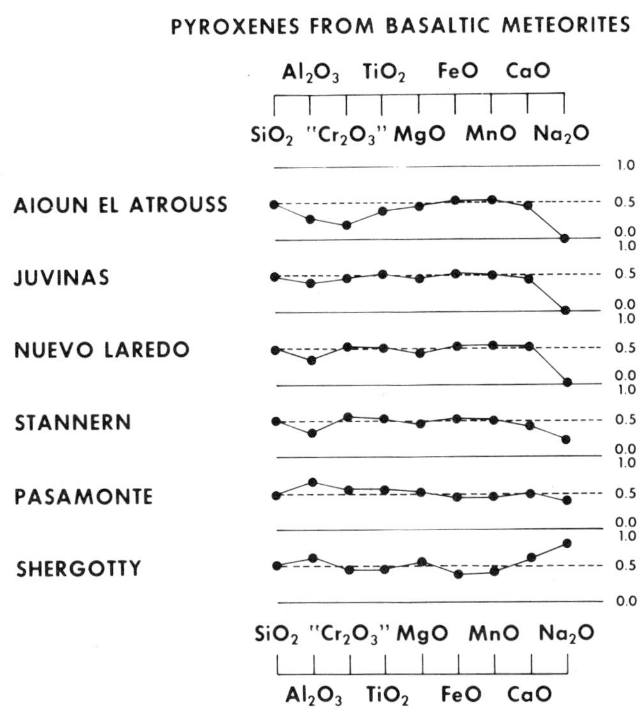

1.3.2.5 Variation diagrams for nine oxides for the pyroxenes from the six individual basaltic meteorites. The actual numerical values plotted are derived from the ratio oxide/(oxide + average). "Oxide" represents the mean for each meteorite suite (note Table 1.3.2.4b) and "average" is the grand average for all six meteorites (note Table 1.3.2.4b).

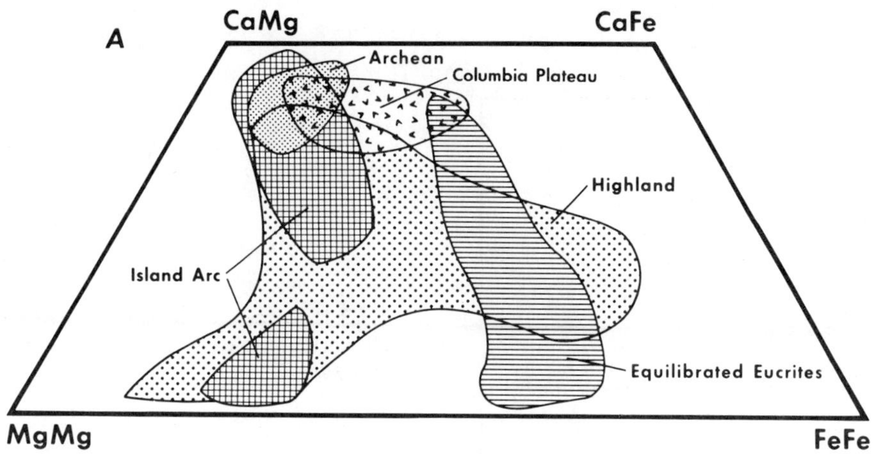

1.3.2.6a Pyroxene Wo ($CaSiO_3$) – En ($MgSiO_3$) – Fs ($Fe^{2+}SiO_3$) composite variation diagrams for five of the ten mineral suites. Note that the meteorite pyroxenes are only for the "equilibrated" eucrites.

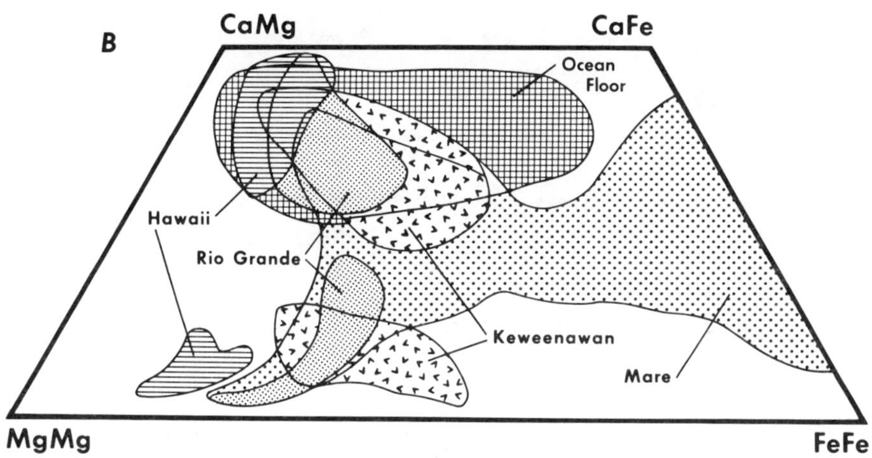

1.3.2.6b Pyroxene Wo ($CaSiO_3$) – En ($MgSiO_3$) – Fs ($Fe^{2+}SiO_3$). Composite variation diagrams for five of the ten mineral suites.

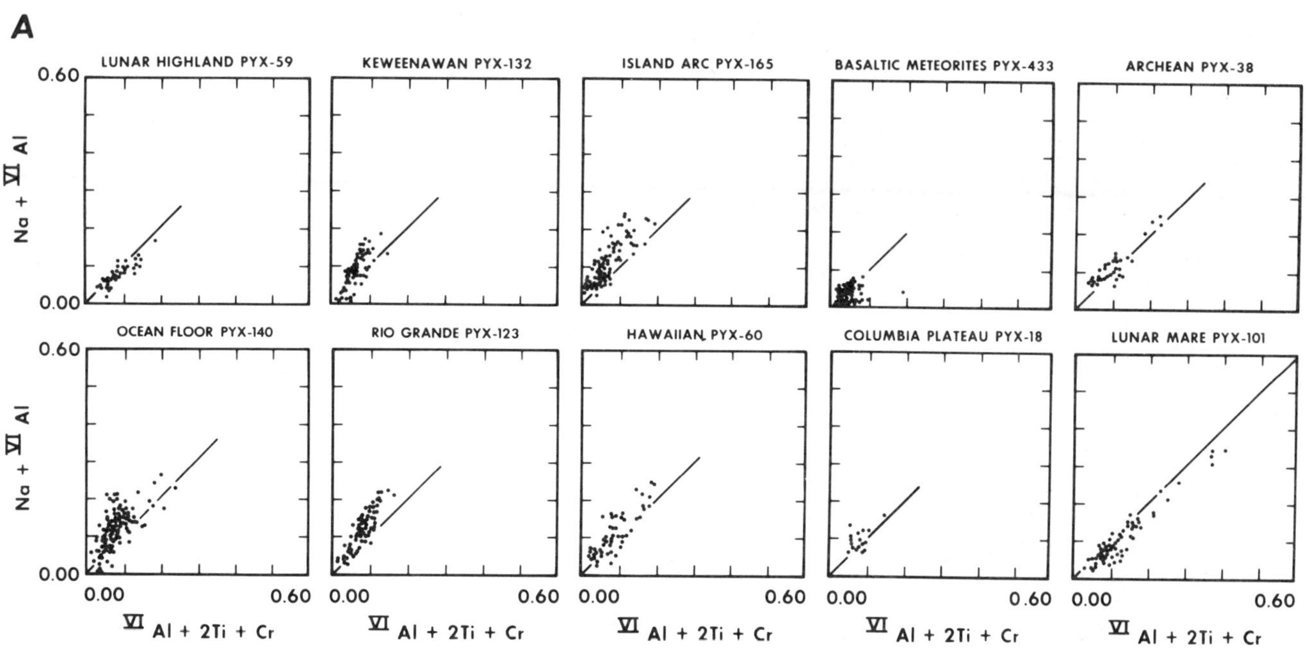

1.3.2.7a (Na + ^{IV}Al) vs. (^{VI}Al + 2Ti + Cr) variation diagrams for the ten pyroxene suites.

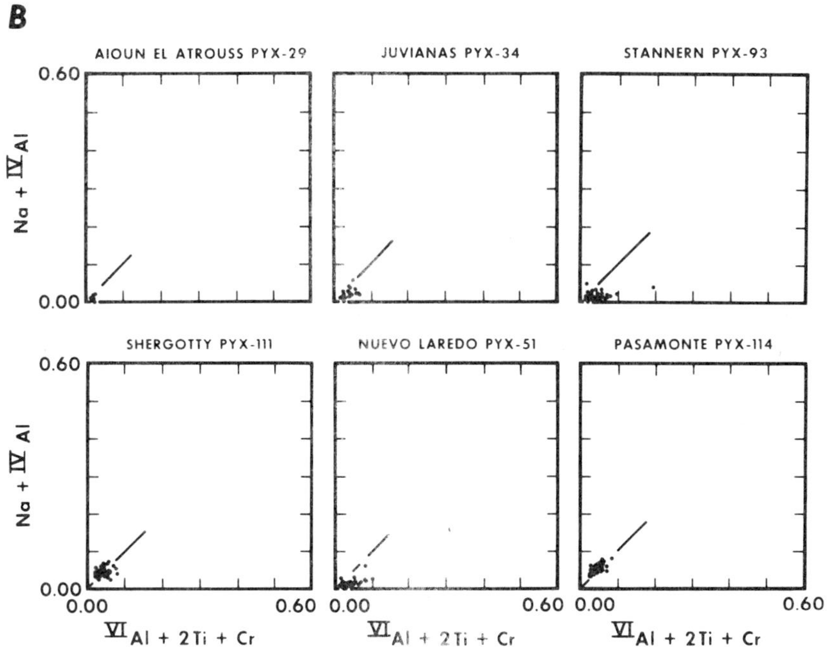

1.3.2.7b (Na + ^{IV}Al) vs. (^{VI}Al + 2Ti + Cr) variation diagrams for six meteorite pyroxene suites.

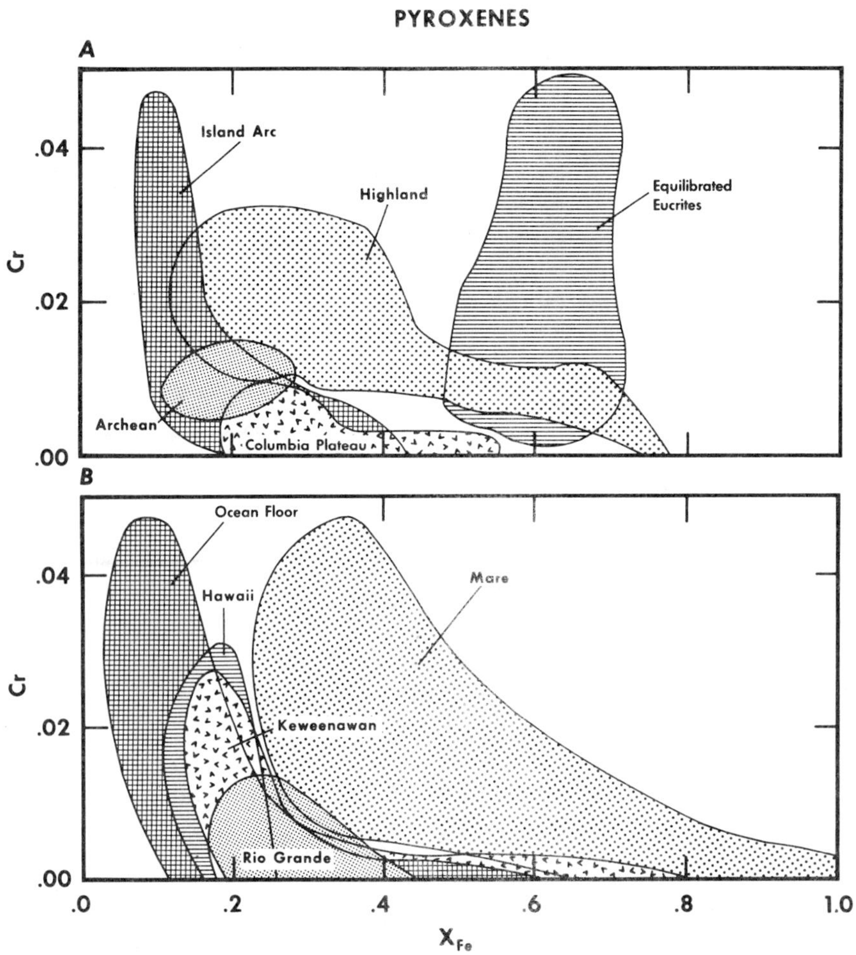

1.3.2.8a Pyroxene Cr-X_{Fe} composite variation diagrams for five of the ten mineral suites. Note that the meteorite pyroxenes are only for the "equilibrated" eucrites.

1.3.2.8b Pyroxene Cr-X_{Fe} composite variation diagrams for five of the ten mineral suites.

The plots of Na + ^{IV}Al against ^{VI}Al + 2Ti + Cr (Figs. 1.3.2.7 a, b) elucidate the oxidation state of these elements. If all the Fe in pyroxenes were Fe^{2+}, and Ti and Cr were in the 4+ and 3+ states, respectively, the data would plot on the 45° line on such diagrams. Chemical data for pyroxenes from the Archean, Columbia Plateau, ocean floor, Hawaiian, island arc, Keweenawan and Rio Grande suites plot towards the Na + ^{IV}Al axis indicating significant Fe^{3+} in these pyroxenes. The Fe^{3+} content reported in the tables for terrestrial assemblages was estimated by the method of Papike *et al.* (1974). In contrast, pyroxenes from lunar highland and lunar maria plot on the other side of the 45° line, towards ^{VI}Al + 2Ti + Cr which indicates the presence of Ti^{3+} and/or Cr^{2+} and reflects the lower oxygen fugacity that existed on the Moon relative to the Earth. These diagrams also give an indication of the amount of pyroxene components other than Wo ($CaSiO_3$) –En($MgSiO_3$) – Fs($Fe^{2+}SiO_3$). These so-called "Others" include Na, Al, Cr, Ti and Fe^{3+}. The highest upper limits for "Others" are found in lunar mare and Archean basalts, probably reflecting very rapid cooling histories. Intermediate concentrations are found in pyroxenes from Columbia Plateau, ocean floor, Hawaiian, island arc, Keweenawan, lunar highland and Rio Grande suites. The lowest concentrations of "Others" (< 5 mole %) are found in pyroxenes from the "equilibrated" eucrites, Juvinas, Stannern, Nuevo Laredo and Aioun el Atrouss (Fig. 1.3.2.7b). The "Others" content of Pasamonte pyroxenes probably results from a more rapid cooling history, whereas the "Others" content of Shergotty pyroxenes indicates both a more rapid cooling history and a different bulk host rock composition (e.g., higher Na).

The most important substitutional couples involved in pyroxene "Others" components are: $^{VI}R^{3+}$ – ^{IV}Al, $^{VI}R^{3+}$ – Na, ^{VI}Ti – ^{IV}Al, ^{VI}Ti – Na, where R^{3+} = Fe^{3+} + Al + Cr^{3+}. Plots of Ti against ^{IV}Al (Figs. 1.3.2.14a,b) give some insight into the relative importance of these couples; a more rigorous treatment of this subject is given below in connection with the factor analysis. If the data plots on the Ti/Al = 1/2 line, a major "Others" substitutional couple is $^{VI}Ti^{4+}$ – $2^{IV}Al^{3+}$; this couple is important, for example, in pyroxenes from mare basalts.

A useful equation, based on charge balance, describes the permissible cation substitutional couples other than those related to the pyroxene quadrilateral "Quad" ($CaMgSi_2O_6$ – $CaFe^{2+}Si_2O_6$ – $Fe_2^{2+}SiO_6$ – $Mg_2Si_2O_6$) (Papike and Cameron, 1976). It is

$$^{VI}Al + {}^{VI}Fe^{3+} + {}^{VI}Cr^{3+} + 2\,{}^{VI}Ti^{4+} = {}^{IV}Al\,{}^{M2}Na$$

When ^{VI}Al, $^{VI}Fe^{3+}$, or $^{VI}Cr^{3+}$ substitutes in the pyroxene M1 site, a charge excess of +1 results *relative* to the (Mg, Fe^{2+}, Mn^{2+}) occupancy of this site in "Quad" pyroxene components. By similar reasoning, a Ti^{4+} substitution in M1 causes a charge excess of +2 relative to "Quad." Thus, for charge balance to be maintained in the pyroxene structure, these site charge excesses must be compensated by site charge deficiencies. The substitution of Al for Si in the tetrahedral site causes a charge deficiency of –1. Similarly, the substitution of Na in the M2 site for (Ca, Fe^{2+}, Mg, Mn^{2+}) also results in a site charge deficiency of –1 relative to "Quad." Since the Na content of the pyroxenes in the reference suites is quite low (note Table 1.3.2.4a), most of the charge deficiencies in these pyroxenes are accounted for by ^{IV}Al. Thus, the most important "Others" components in these pyroxenes are: ^{VI}Al - ^{IV}Al, $^{VI}Fe^{3+}$ - ^{IV}Al, $^{VI}Cr^{3+}$ - ^{IV}Al and $^{VI}Ti^{4+}$ - $2^{IV}Al$. We will discuss the relative importance of these substitutional couples in more detail below.

The amount of ^{IV}Al also gives a good approximation of the percent "Others" in the pyroxenes or, in other words, the amount of compositional deviation from the pure pyroxene quadrilateral. Since the maximum amount of ^{IV}Al permissible in the pyroxene structure is 1 atom per formula unit, $^{IV}Al \times 100$ is a good estimate of percent "Others." Thus, Figs. 1.3.2.15a and b, which are variation diagrams of ^{IV}Al vs. X_{Fe}, give a good estimate of the relative amount of "Others" in the pyroxene suites. The maximum amount of "Others" are found in mare pyroxenes followed by Archean, deep sea, and Hawaiian pyroxenes. The lowest "Others" contents are found in the pyroxenes from basaltic meteorites. In fact, if we omit the relatively high ^{IV}Al values of Shergotty and Pasamonte pyroxenes, the pyroxenes from "equilibrated" eucrites are exceedingly low in "Others" (Fig. 1.3.2.15b). The significance of this observation is that the many studies of pyroxene phase equilibria for the pure "Quad" components are very applicable to pyroxenes from "equilibrated" eucrites.

Figures 1.3.2.8a, b and 1.3.2.16a, b show the Cr vs. X_{Fe} variations for pyroxenes from the basalt suites. Pyroxenes from ocean floor, island arc, Keweenawan and Hawaiian basalts show similar behavior in that they all show high Cr concentrations at low X_{Fe}. Pyroxenes from the "equilibrated" eucrites (Aioun el Atrouss, Juvinas, Nuevo Laredo and Stannern) show anomalous systematics in that they have both high and low Cr concentrations at high X_{Fe}. It is quite possible that a significant portion of the Cr in the eucrites is in the reduced valence state Cr^{2+} because of the low oxygen fugacities they experienced (see section 1.2.8).

The Cr contents of basaltic meteorite pyroxenes require more discussion. Reference to Fig. 1.3.2.16b shows a very unusual set of chemical systematics. Aioun el Atrouss and Shergotty show relatively well-behaved trends of Cr decreasing with increasing X_{Fe}. Pasamonte also shows a crude trend of Cr vs. X_{Fe}. However, the Cr systematics for Juvinas, Nuevo Laredo and Stannern are complex. The Cr does not appear to be correlated with Ca as one might expect (note Table 1.3.2.A-3 in Appendix A-12). We speculate that during the pyroxene equilibration, phase separation took place, not only involving two pyroxenes, but also a Cr-bearing phase (perhaps Cr-spinel). This possibility should be tested by close examination of the pyroxenes with TEM and electron diffraction techniques. The microprobe data indicate that the Cr is not evenly distributed in these pyroxenes.

The Mn vs. X_{Fe} systematics for the pyroxenes from the suites are illustrated in Figs. 1.3.2.17a and b. In most of the diagrams for the individual suites, Mn and X_{Fe} show a strong positive correlation (see discussion below); however, the slopes are different. The lowest Mn/X_{Fe} values are found in lunar pyroxenes, intermediate values in terrestrial basalt pyroxenes, and the highest values in meteoritic pyroxenes. The systematics have interesting complexities, however. For example, differences in Mn/X_{Fe} are found among the different suites for one planetary body, e.g., Keweenawan basalt pyroxenes show higher Mn/X_{Fe} than ocean floor basalt pyroxenes.

Even more interesting features are observed for pyroxenes from basaltic meteorites. Shergotty pyroxene systematics (Fig. 1.3.2.17b) show two trends of increasing Mn with X_{Fe}, one for high-Ca pyroxene and one for low-Ca pyroxenes. The trends are parallel to each other and show the expected positive correlation of Mn and X_{Fe}. However, the trend for low-Ca pyroxenes is displaced towards higher Mn values. The crystal chemical interpretation of this feature is that in the low-Ca pyroxenes the Mn^{2+} cations are strongly ordered into the pyroxene M2 site. Even more striking is the significantly different slope of the Mn-X_{Fe} trends for the equilibrated Aioun el Atrouss, Juvinas, Nuevo Laredo and Stannern pyroxenes compared with pyroxenes from Shergotty and Pasamonte. The Mn/X_{Fe} values for the pyroxenes from equilibrated meteorites are considerably higher than those for Shergotty and Pasamonte.

The reason for these differing systematics appears to be subsolidus exsolution of the pyroxenes in the equilibrated eucrites. As mentioned above in reference to the Shergotty meteorite, Mn preferentially enters the low-Ca pyroxenes. It is also observed that low-Ca pyroxenes are enriched in Fe^{2+} relative to the high-Ca pyroxenes. Thus, the diagrams of Mn vs. X_{Fe} are simply showing that low-Ca pyroxenes are enriched in both Mn and Fe^{2+} relative to the high-Ca pyroxenes. The crystal chemical explanation for this is straightforward: during the subsolidus re-equilibration, which causes exsolution in these pyroxenes, Ca is fractionated into the M2 sites of the augites, while Mn and Fe^{2+} are fractionated into the M2 site of pigeonite. The site preference for the pyroxene M2 site is Ca $> Mn^{2+} > Fe^{2+} >$ Mg. It is clear from the combined chemical systematics for pyroxenes from the equilibrated eucrite pyroxenes that the chemical trends on the pyroxene quadrilaterals (Fig. 1.3.2.13b) are simply "mixing" lines caused by lack of microprobe resolution of very fine exsolution lamellae of augite and pigeonite. This is consistent with the results of several X-ray diffraction studies (e.g., Takeda *et al.*, 1976).

Figures 1.3.2.18a and b illustrate Na-X_{Fe} systematics. The Na contents of the pyroxenes show considerable variation. Low values are found in pyroxenes from eucrites, lunar highland, lunar mare and Archean basalt pyroxenes. Intermediate values are found in pyroxenes from the Columbia Plateau, ocean floor and the Keweenawan basalts. Higher values are found in pyroxenes from Hawaiian, island arc and Rio Grande basalts. Although the Na contents of basaltic meteorite pyroxenes are low, variations are found among individual meteorites (Fig. 1.3.2.18b). The Na contents of pyroxenes from Aioun el Atrouss, Juvinas and Nuevo Laredo are negligible. Those from Stannern and Pasamonte are higher, but the highest Na contents are found in pyroxenes from Shergotty. The data for Shergotty also show that Na preferentially enters the M2 site of high-Ca pyroxenes relative to low-Ca pyroxenes.

The Ti-X_{Fe} systematics for the pyroxenes from the various suites are illustrated in Figs. 1.3.2.19a, b. The highest concentrations are found in mare basalts; intermediate concentrations in Columbia Plateau, ocean floor, Keweenawan, Hawaiian, lunar highlands and Rio Grande basalts; and lower values in island arc, Archean and basaltic meteorites. Within the basaltic meteorite suite, there are also variations (Fig. 1.3.2.19b). The equilibrated eucrites, Aioun el Atrouss, Juvinas, Nuevo Laredo, and Stannern have pyroxenes with extremely low-Ti contents. The Ti concentrations in pyroxenes from Shergotty and Pasamonte are higher. In these samples a positive correlation between Ti and X_{Fe} is observed.

1.3.2.5 Pyroxene substitution couples

The cation contents for twelve subgroups of the pyroxenes were treated statistically by a factor analysis program BMDP4M (Dixon, 1975). Results of the calculations are presented in Tables 1.3.2.5a–1.3.2.5d. Table 1.3.2.5a illustrates the relative importance of the "Others" substitutional couples.

The Archean pyroxenes have ^{VI}Al-^{VI}Al and $^{VI}Ti^{4+}$-^{IV}Al as the two most important "Others" couples. These couples are consistent with the high Al concentrations reported for these pyroxenes. In pyroxenes from the Columbia Plateau suite $^{VI}Fe^{3+}$-^{IV} Al and $^{VI}Ti^{4+}$-^{IV}Al are the two most important couples. Fe^{3+} is clearly important in these pyroxenes. In pyroxenes from the ocean floor basalts the same two couples are important, indicating similar petrogenetic scenarios for these two rock suites. In the equilibrated eucrites the significance of the correlation coefficients is equivocal because their values are low and the concentration of the "Others" components are also low. Nevertheless, $^{VI}Ti^{4+}$-^{IV}Al and ^{VI}Al-^{IV}Al substitution couples appear to be present. An equally high correlation coefficient is indicated for $^{VI}Cr^{3+}$-^{IV}Al, but there are problems with the reported Cr contents for these pyroxenes as discussed above.

In Pasamonte, which is virtually identical in bulk composition to the equilibrated eucrites, the most important substitutional couple is $^{VI}Ti^{4+}$-^{IV}Al. In Shergotty, which has a different bulk chemistry (e.g., higher Na), the two most important "Others" couples are $^{VI}Ti^{4+}$-^{IV}Al and Na-^{VI}Al (Jadeite). Pyroxenes from the Hawaiian suite show an extremely high correlation coefficient (0.932) for the ^{VI}Ti-^{IV}Al couple demonstrating the dominance of this couple in these pyroxenes. The Na containing couples Na-^{VI} Fe^{3+} and Na-$^{VI}Ti^{4+}$ are also important. In the island arc suite the dominant "Others" couples are ^{VI}Al-^{IV}Al, $^{VI}Fe^{3+}$-^{IV}Al and $^{VI}Ti^{4+}$-^{IV}Al. In pyroxenes from the Keweenawan suite the three most important couples are $^{VI}Fe^{3+}$-^{IV}Al, $^{VI}Ti^{4+}$-Al and Na-$^{VI}Ti^{4+}$. In the two lunar suites, highland and mare, $^{VI}Ti^{4+}$-^{IV}Al is the dominant "Others" couple. Pyroxenes from the Rio Grande basalts show $^{VI}Fe^{3+}$ -^{IV}Al as the dominant couple followed by ^{VI} Ti^{4+}-^{IV}Al. Table 1.3.2.5d summarizes the two most important "Others" couples for each of these groups.

Several things are apparent from this summary. First, the ^{VI}Ti-^{IV}Al couple is by far the most important "Others" couple ranking in the top two for all except the island arc suite. Second, Fe^{3+} is important in all of the terrestrial suites with the possible exception of the Archean; it is usually present as the $^{VI}Fe^{3+}$-^{IV}Al couple. This emphasizes the importance of f_{O_2} in pyroxene-liquid elemental partitioning. At the relatively high oxygen fugacities of terrestrial basalt petrogenesis, iron enters the pyroxene both as Fe^{3+} and Fe^{2+}, whereas in lunar and most basaltic meteorite pyroxenes, iron is only present as Fe^{2+}.

Tables 1.3.2.5b and 1.3.2.5c list the "Other"-"Quad" cation correlations. These correlation coefficients enable us to ascertain if there is any preferential linking of the "Others" couples with a specific quadrilateral component. We will simplify the discussion by considering only the two most important "Others" couples $^{VI}Ti^{4+}$–^{IV}Al and $^{VI}Fe^{3+}$–^{IV}Al. By thus observing the correlation coefficients in Tables 1.3.2.5b and c we can decide which of the following pyroxene components are most important:

$CaTiAl_2O_6$	$CaFe^{3+}SiAlO_6$	$Fe^{2+}TiAl_2O_6$
$MgTiAl_2O_6$	$MgFe^{3+}SiAlO_6$	$Fe^{2+}Fe^{3+}SiAlO_6$

Table 1.3.2.5a Pyroxene "Other" — "Other" cation correlations.†

Group	^{VI}Al-^{IV}Al	$^{VI}Fe^{3+}$-^{IV}Al	$^{VI}Cr^{3+}$-^{IV}AL	$^{VI}Ti^{4+}$-^{IV}Al	Na-^{VI}Al	Na-$^{VI}Fe^{3+}$	Na-$^{VI}Cr^{3+}$	Na-$^{VI}Ti^{4+}$
Archean	0.527	0.203	0.360	0.527	0.427	−0.319	−0.353	−0.282
Columbia Plateau	−0.436	0.568	0.072	0.566	−0.027	0.202	−0.121	0.395
Ocean Floor	0.232	0.584	0.068	0.590	−0.051	0.453	−0.137	0.349
Equilibrated Eucrites	0.270	—	0.342	0.328	0.021	—	−0.032	−0.078
Pasamonte	0.033	—	0.234	0.488	−0.059	—	−0.054	0.088
Shergotty	−0.100	—	0.129	0.260	0.234	—	0.173	0.160
Hawaiian	0.495	0.436	0.375	0.932	0.178	0.608	0.069	0.537
Island Arc	0.588	0.486	0.213	0.355	0.378	−0.223	−0.156	0.382
Keweenawan	0.091	0.794	0.360	0.635	−0.042	0.562	0.159	0.568
Lunar Highland	0.285	—	0.562	0.642	−0.136	—	0.050	0.611
Lunar Mare	0.675	—	0.300	0.833	0.158	—	0.068	0.598
Rio Grande	0.304	0.818	0.207	0.642	0.571	0.463	−0.210	0.001

†Correlation coefficients >0.6 are underlined.

Table 1.3.2.5b Pyroxene "Other" — "Quad" cation correlations.

Group	Ti-Ca	Ti-Mg	$Ti-Fe^{2+}$	$^{VI}Al-Ca$	$^{VI}Al-Mg$	$^{VI}Al-Fe^{2+}$	$Fe^{3+}-Ca$
Archean	0.519	−0.785	0.645	−0.180	−0.020	−0.198	0.266
Columbia Plateau	0.405	−0.541	0.357	−0.441	0.146	0.008	0.468
Ocean Floor	−0.014	−0.486	0.405	0.152	0.142	−0.306	0.342
Equilibrated Eucrites	0.159	−0.110	−0.147	0.083	−0.136	−0.107	—
Pasamonte	0.783	−0.841	0.808	−0.234	0.288	−0.370	—
Shergotty	0.323	−0.768	0.372	0.351	0.064	0.301	—
Hawaiian	0.737	−0.783	−0.550	0.494	−0.533	−0.290	0.278
Island Arc	0.101	−0.505	0.243	0.322	−0.687	0.008	0.009
Keweenawan	0.450	−0.529	−0.177	0.132	0.198	0.378	0.558
Lunar Highland	0.760	−0.493	−0.090	−0.357	0.392	−0.285	—
Lunar Mare	0.713	−0.213	−0.273	0.135	−0.187	−0.025	—
Rio Grande	0.372	−0.522	−0.105	0.123	−0.013	−0.411	0.637

Table 1.3.2.5c Pyroxene "Other" — "Quad" cation correlations.

Group	Na-Ca	Na-Mg	$Na-Fe^{2+}$	$^{IV}Al-Ca$	$^{IV}Al-Mg$	$^{IV}Al-Fe^{2+}$	Mn-Ca
Archean	−0.409	.0258	0.208	0.346	−0.498	0.131	0.272
Columbia Plateau	0.534	−0.555	0.225	0.726	−0.092	−0.331	0.041
Ocean Floor	0.335	−0.469	0.114	0.359	−0.227	−0.203	−0.345
Equilibrated Eucrites	0.000	−0.015	0.009	0.458	−0.154	−0.501	−0.664
Pasamonte	0.091	−0.135	0.163	0.445	−0.391	0.287	0.521
Shergotty	0.461	−0.149	−0.252	0.148	−0.044	−0.066	0.020
Hawaiian	0.720	−0.722	−0.682	0.733	−0.758	−0.691	−0.152
Island Arc	−0.231	−0.324	0.483	0.388	−0.611	−0.237	−0.852
Keweenawan	0.722	−0.613	−0.452	0.782	−0.493	−0.663	−0.403
Lunar Highlands	0.542	−0.327	−0.098	0.383	0.034	−0.526	0.294
Lunar Mare	0.434	−0.169	−0.124	0.513	−0.063	−0.357	−0.231
Rio Grande	0.618	−0.599	−0.584	0.538	−0.507	−0.593	−0.569

Table 1.3.2.5d Summary of relative importance of pyroxene "Others" cation couples.

Group	First Most Important	Second Most Important
Archean	Ti-Al	Al-Al
Columbia Plateau	Fe^{3+}-Al	Ti-Al
Ocean Floor	Fe^{3+}-Al	Ti-Al
Equilibrated Eucrites	(Cr^{3+}-Al)?	Ti-Al
Pasamonte	Ti-Al	Cr-Al
Shergotty	Ti-Al	Na-Al
Hawaiian	Ti-Al	Na-Fe^{3+}
Island Arc	Al-Al	Fe^{3+}-Al
Keweenawan	Fe^{3+}-Al	Ti-Al
Lunar Highland	Ti-Al	Na-Ti
Lunar Mare	Ti-Al	Al-Al
Rio Grande	Fe^{3+}-Al	Ti-Al

We can then see whether these couples are correlated with augitic pyroxenes (high correlation with Ca), whether they occur early in the crystallization sequence (high correlation with Mg) or late in the crystallization sequence (high correlation with Fe^{2+}). As a result of such an analysis we see that the Ti-Al couple is correlated with the calcium content of the pyroxene and with the degree of fractionation in Archean, Columbia Plateau, ocean floor, Pasamonte, and Shergotty basalts, while essentially it is correlated only with calcium in Hawaiian, island arc, Keweenawan, lunar and Rio Grande basalts. The Table 1.3.2.5b correlation coefficients show clearly that the $^{VI}Fe^{3+}$-^{IV}Al couple enters the pyroxene structure as $CaFe^{3+}SiAlO_6$ and thus is largely concentrated in the augitic pyroxenes.

Fe^{3+}-Mg	Fe^{3+}-Fe^{2+}	Cr-Ca	Cr-Mg	Cr-Fe^{2+}
−0.086	−0.106	0.035	0.150	−0.363
0.234	−0.546	−0.269	0.210	−0.116
−0.166	−0.196	0.284	0.382	−0.586
—	—	0.043	−0.086	−0.009
—	—	−0.224	0.218	−0.225
—	—	0.639	0.292	−0.792
−0.249	−0.522	0.194	−0.184	−0.358
0.064	−0.233	0.469	−0.153	−0.508
−0.382	−0.451	0.374	0.133	−0.621
—	—	0.201	0.233	−0.603
—	—	−0.033	0.645	−0.680
−0.574	−0.669	0.167	−0.052	−0.225

Mn-Mg	Mn-Fe^{2+}	^{IV}Al-XFe	Cr-XFe	Mn-XFe	Na-XFe	Ti-XFe
−0.373	0.431	0.304	−0.316	0.434	−0.249	0.773
−0.509	0.524	−0.209	−0.137	0.514	0.342	0.438
−0.675	0.938	−0.101	−0.572	0.919	0.256	0.446
0.175	0.729	−0.447	0.052	0.646	0.013	−0.099
−0.646	0.699	0.372	−0.239	0.660	0.136	0.844
−0.055	0.823	−0.019	−0.721	0.736	−0.145	0.557
0.098	0.382	0.108	−0.221	0.422	0.013	0.355
0.474	0.726	−0.016	−0.429	0.509	0.612	0.465
−0.176	0.681	−0.432	−0.686	0.657	−0.178	0.076
−0.656	0.698	−0.367	−0.506	0.711	0.056	0.132
−0.637	0.817	−0.178	−0.707	0.757	0.011	−0.090
0.337	0.709	−0.444	−0.193	0.542	−0.391	0.177

1.3.2.6 Summary and conclusions

The most important conclusions from this synthesis are: (1) The most "primitive" olivines, i.e., those with the highest $Mg/(Mg + Fe^{2+})$ are found in the Archean basalts. (2) The most "primitive" olivines from lunar mare basalts are more iron-rich than terrestrial olivines, indicating crystallization from more iron-rich liquids, which in turn originated from a mantle source enriched in iron relative to terrestrial mantle sources. (3) Feldspars from lunar mare, lunar highlands and meteoritic basalts are depleted in alkalis relative to those from terrestrial basalts, reflecting alkali depletion of their parent bodies. (4) Maximum anorthite contents of feldspars from the various suites correlate with maximum forsterite contents of olivines from the same suites. Thus, both are indices of relative "primitiveness" of the basalt liquids. (5) The most magnesian pyroxenes are found in the lunar highland, island arc, Archean, ocean floor and Hawaiian suites. This is a reflection of the fact that these suites include some of the most primitive basalts. (6) The highest concentrations of pyroxene "Others" components (e.g., Cr, Al, Ti, Fe^{3+}, Na) are found in lunar mare and Archean basalts implying, among other things, rapid cooling histories. (7) The Cr content of pyroxenes from "equilibrated" eucrites shows anomalous trends, indicating that the Cr now resides in finely dispersed phases formed by subsolidus annealing. (8) The most important "Others" substitutional couple for all of the planetary suites combined is $^{VI}Ti^{4+}$–^{IV}Al. It enters pyroxenes mainly as the component $CaTiAl_2O_6$. (9) A very important component in pyroxenes from terrestrial basalts is $CaFe^{3+}SiAlO_6$. This emphasizes the importance of f_{O_2} in pyroxene-liquid elemental partitioning. At the relatively high oxygen fugacities of terrestrial petrogenesis, iron enters the pyroxene structure both as Fe^{3+} and Fe^{2+}, whereas in lunar and most basaltic meteorite pyroxenes, iron is only present as Fe^{2+}.

In summary, inspection of a large number of high quality silicate analyses of olivines, feldspars and pyroxenes demonstrates conclusively that these phases carry a signature of the planetary body in which they evolved. Conversely, these phases are powerful planetary probes capable of constraining the temperature, pressure, f_{O_2} and compositional parameters that governed the petrogenesis of their host basaltic liquids.

1.3.2.7 References

Dixon W. J. (ed.) (1975) *BMDP Biomedical Computer Programs*. University of California Press, Berkeley.

Papike J. J. and Bence A. E. (1978) Lunar mare versus terrestrial mid-ocean ridge basalts: planetary constraints on basaltic volcanism. *Geophys. Res. Lett.* **5**, 803–806.

Papike J. J. and Cameron M. (1976) Crystal chemistry of silcate minerals of geophysical interest. *Rev. Geophys. Space Phys.* **14**, 37–80.

Papike J. J., Cameron K. L. and Baldwin K. (1974) Amphiboles and pyroxenes: characterization of *Other* than quadrilateral components and estimates of ferric iron from microprobe data (abstract). *Abstracts with Program,* Geol. Soc. Amer. **6**, 1053–1054.

Papike J. J., Hodges F. N., Bence A. E., Cameron M. and Rhodes J. M. (1976) Mare basalts: crystal chemistry, mineralogy, and petrology. *Rev. Geophys. Space Phys.* **14**, 475–540.

Schweitzer E. L., Papike J. J. and Bence A. E. (1978) Clinopyroxenes from deep sea basalts: a statistical analysis. *Geophys. Res. Lett.* **5**, 573–576.

Schweitzer E. L., Papike J. J. and Bence A. E. (1979) Statistical analysis of clinopyroxenes from deep-sea basalts. *Amer. Mineral.* **64**, 501–513.

Simkin T. and Smith J. V. (1970) Minor element distribution in olivine. *J. Geol.* **74**, 1–16.

Smith J. V. (1974) Lunar mineralogy: a heavenly detective story. Part I. *Amer. Mineral.* **59**, 231–243.

Smith J. V. (1979) Mineralogy of the planets: a voyage in space and time. *Mineral. Mag.* **43**, 1–89.

Smith J. V. and Steel I. M. (1976) Lunar mineralogy: a heavenly detective story. Part II. *Amer. Mineral.* **61**, 1059–1116.

Takeda H., Miyamoto M., Ishii T. and Reid A. M. (1976) Characterization of crust formation on a parent body of achondrites and the Moon by pyroxene crystallography and chemistry. *Proc. Lunar Sci. Conf. 7th*, p. 3535–3548.

1.3.2.8 Appendix

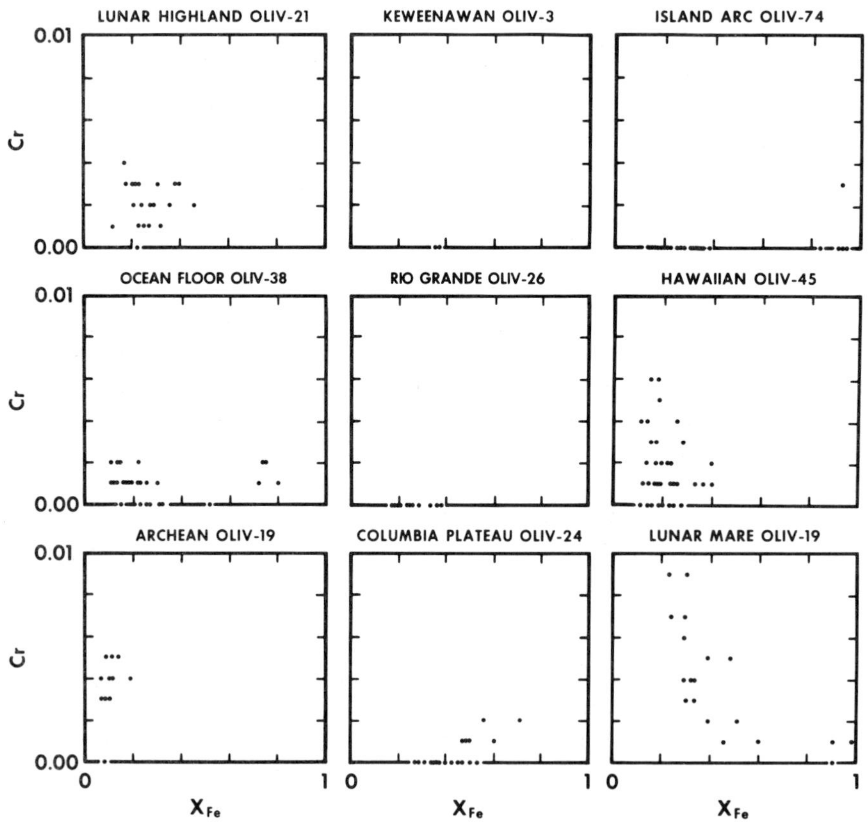

1.3.2.9 Olivine Cr-X_{Fe} variation diagrams.

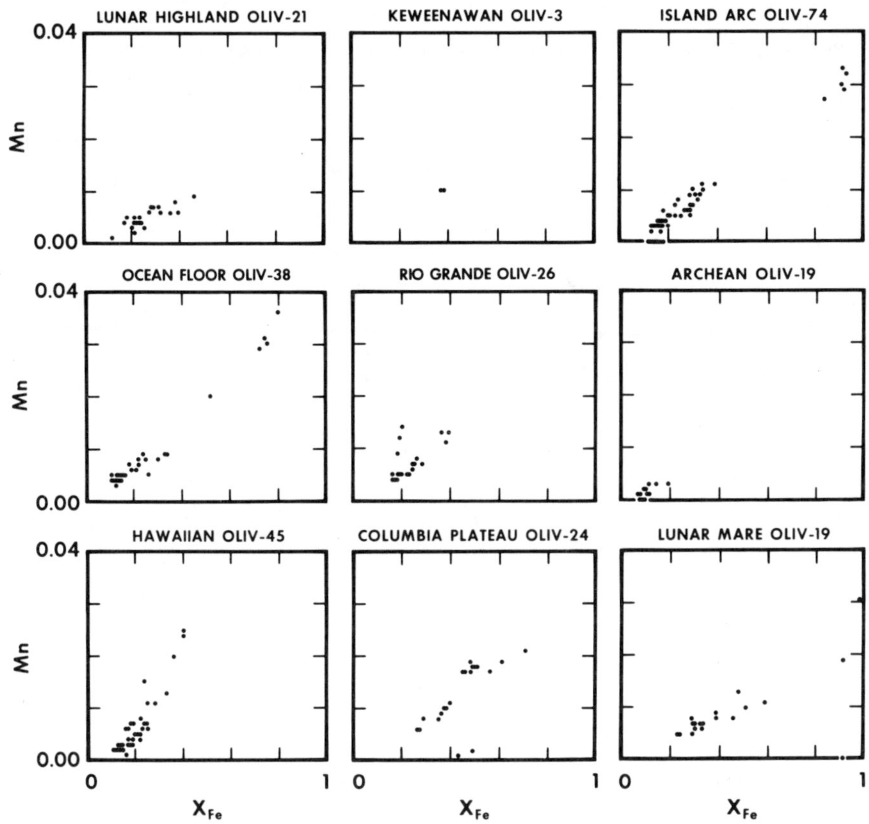

1.3.2.10 Olivine Mn-X_{Fe} variation diagrams

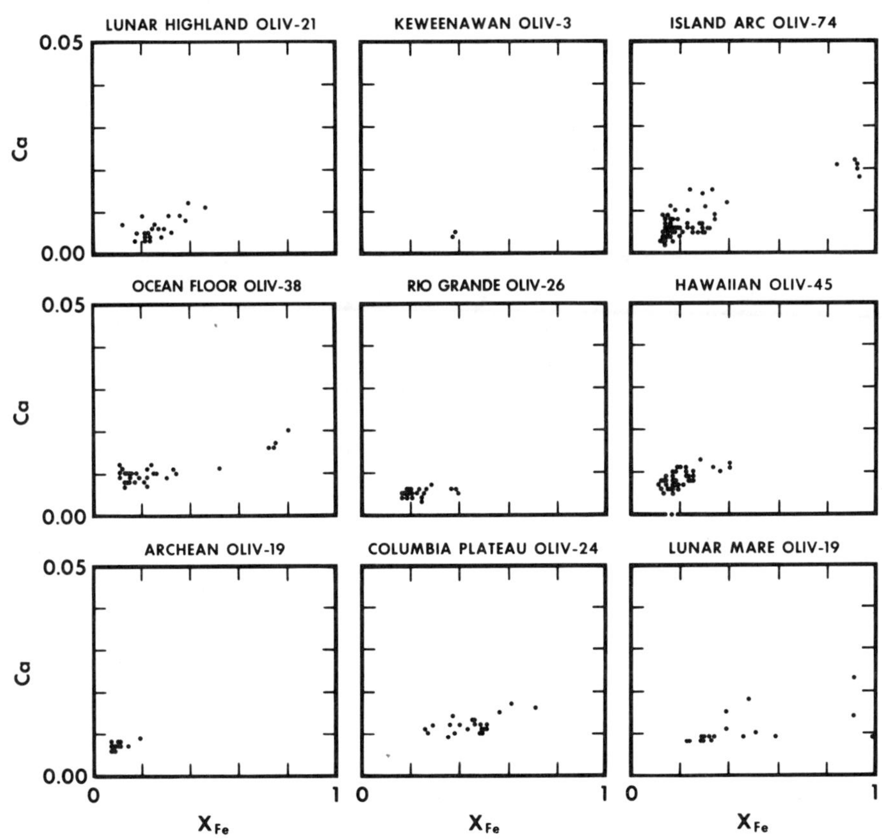

1.3.2.11 Olivine Ca-X_{Fe} variation diagrams.

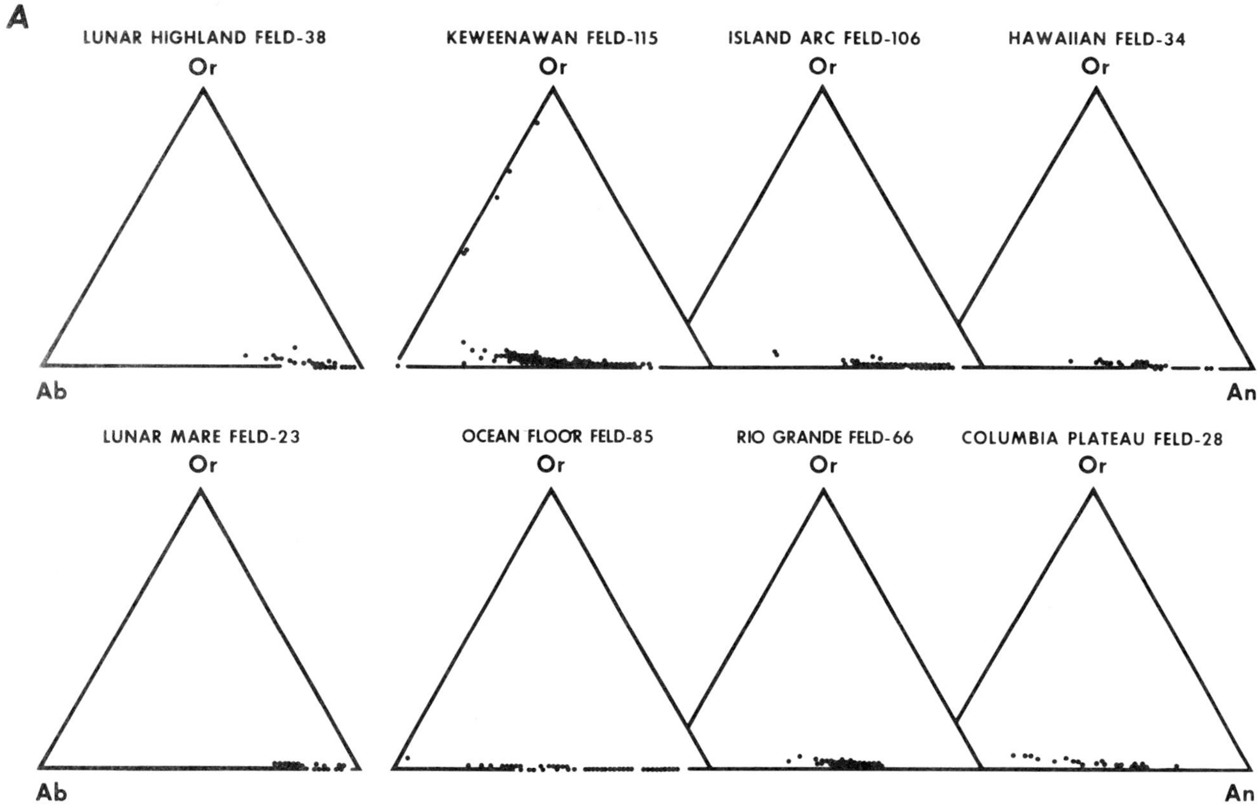

1.3.2.12a Feldspar Or($KAlSi_3O_8$) – An($CaAl_2Si_2O_8$) – Ab($NaAlSi_3O_8$) variation diagrams for eight feldspar suites.

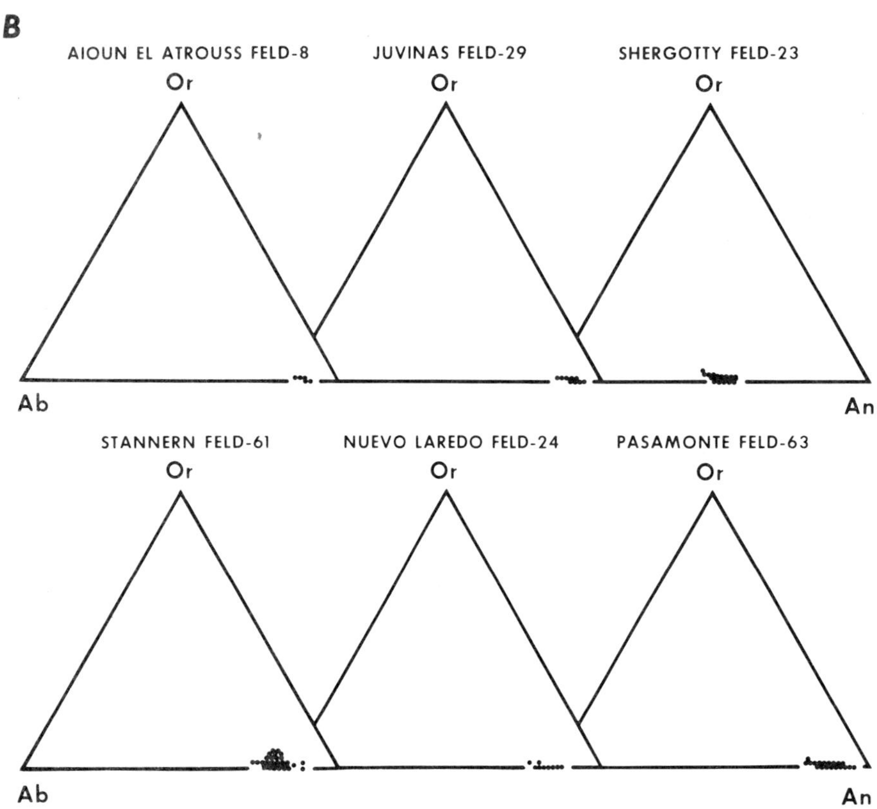

1.3.2.12b Feldspar Or($KAlSi_3O_8$) – An($CaAl_2Si_2O_8$) – Ab ($NaAlSi_3O_8$) variation diagrams for six meteorite suites.

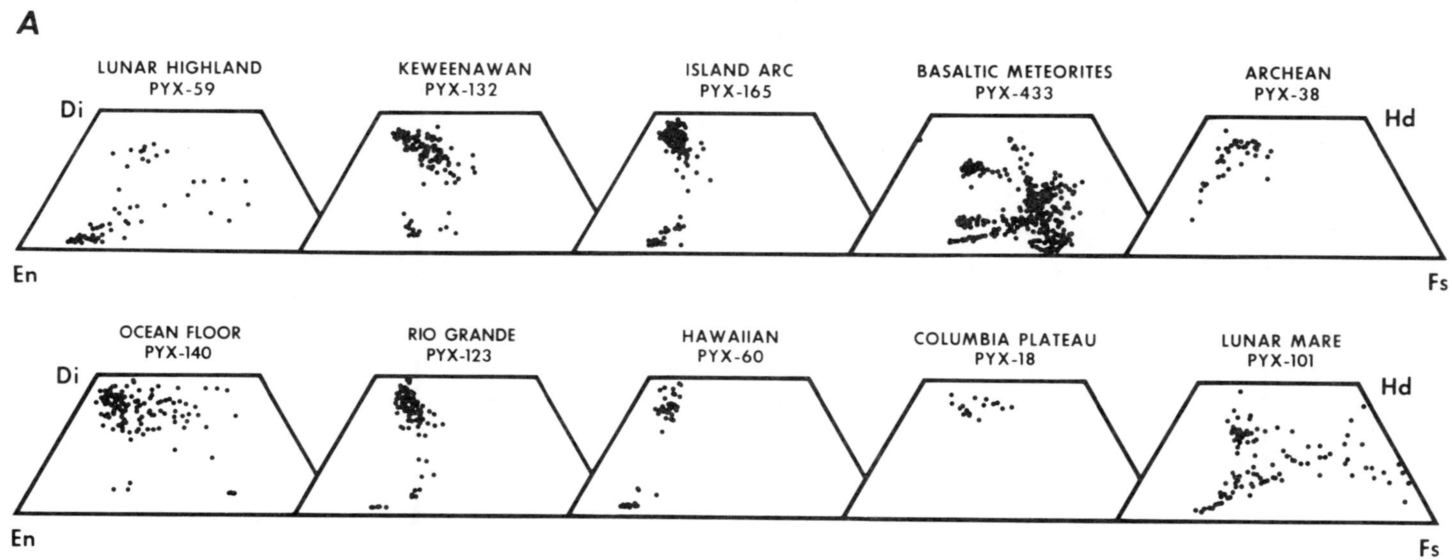

1.3.2.13a Pyroxene Wo ($CaSiO_3$) – En ($MgSiO_3$) – Fs ($Fe^{2+}SiO_3$) variation diagrams for the ten pyroxene suites.

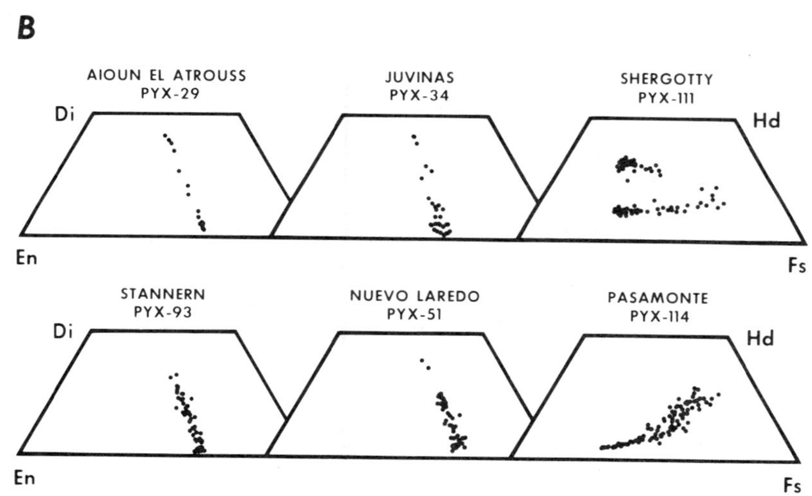

1.3.2.13b Pyroxene Wo ($CaSiO_3$) – En ($MgSiO_3$) – Fs ($Fe^{2+}SiO_3$) variation diagrams for the six meteorite pyroxene suites.

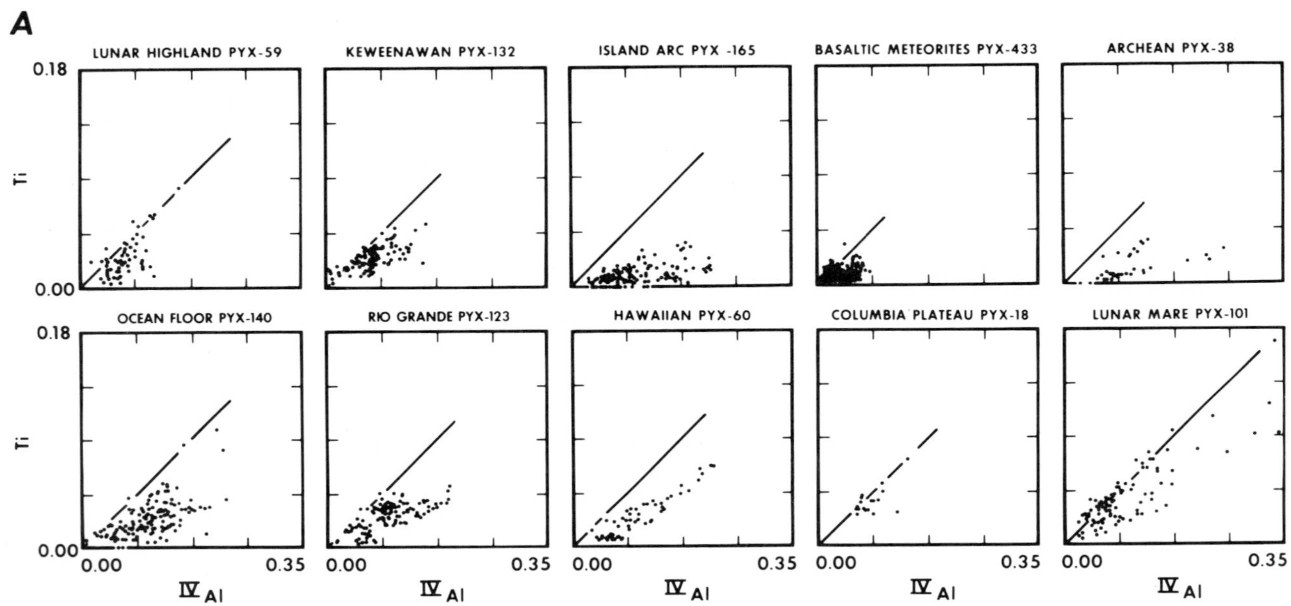

1.3.2.14a Ti-^{IV}Al variation diagrams for the ten pyroxene suites.

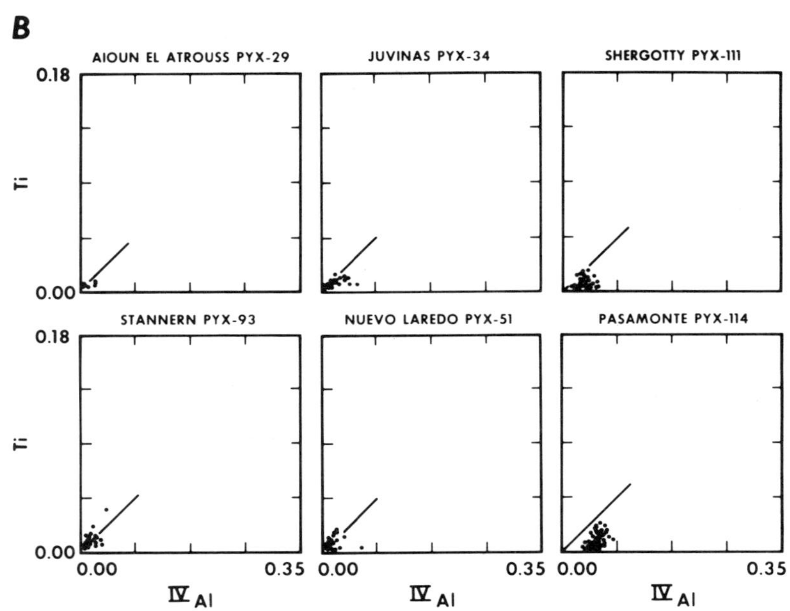

1.3.2.14b Ti-^{IV}Al variation diagrams for six meteorite pyroxene suites.

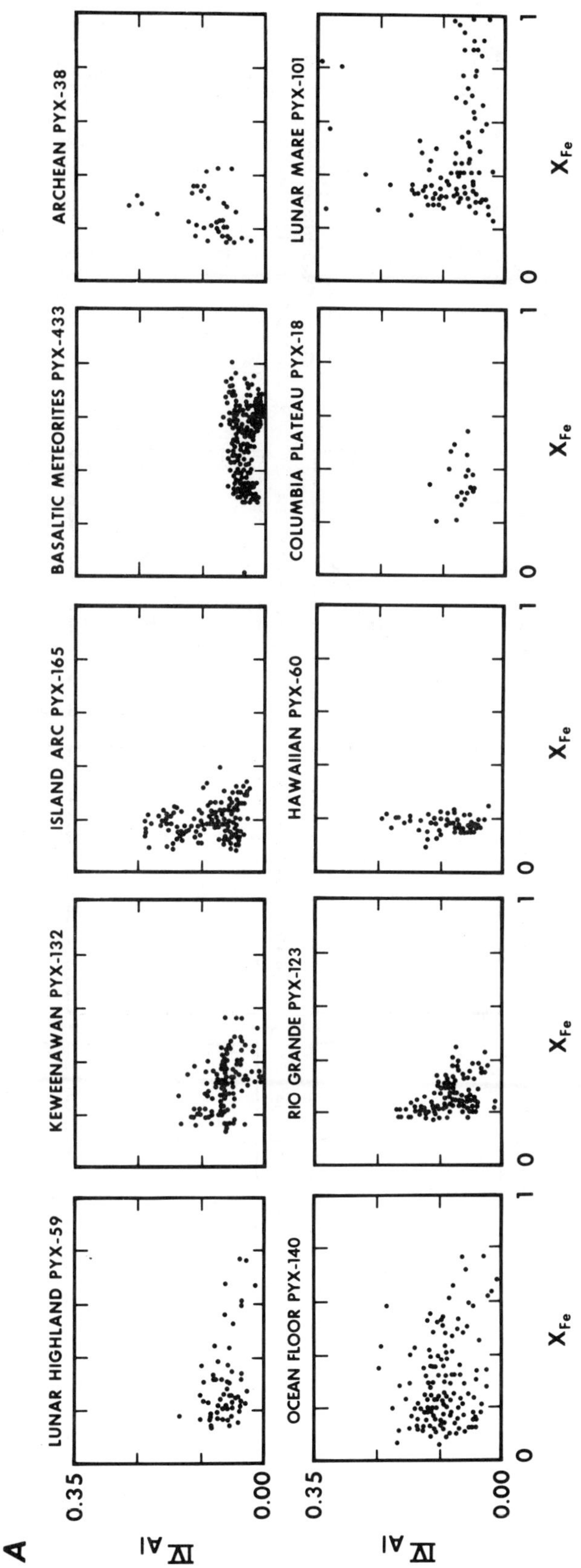

1.3.2.15a ^{IV}Al-X_{Fe} variation diagrams for the ten pyroxene suites.

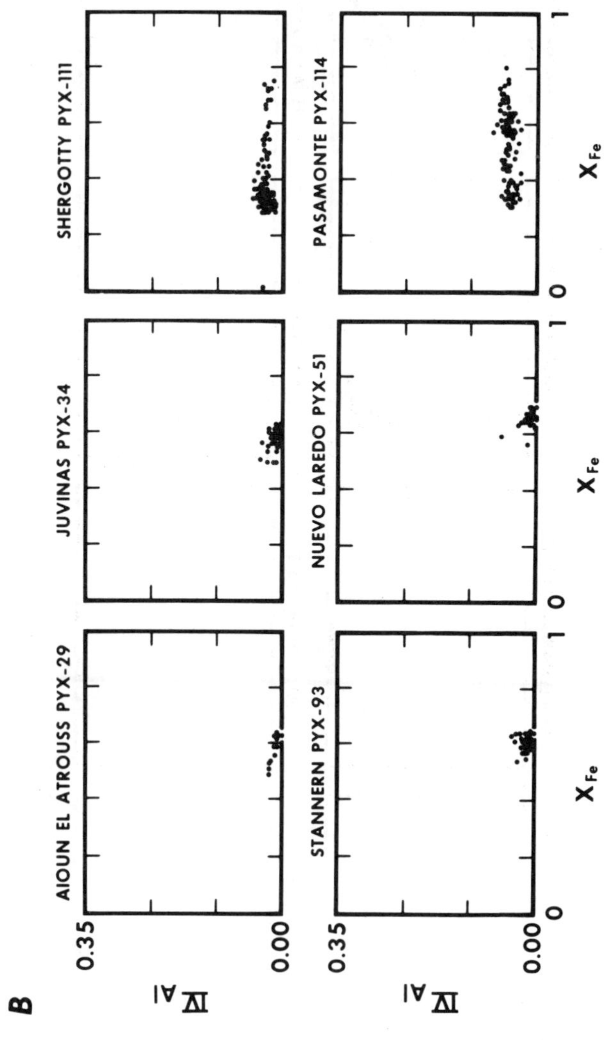

1.3.2.15b ^{IV}Al-X_{Fe} variation diagrams for six meteorite pyroxene suites.

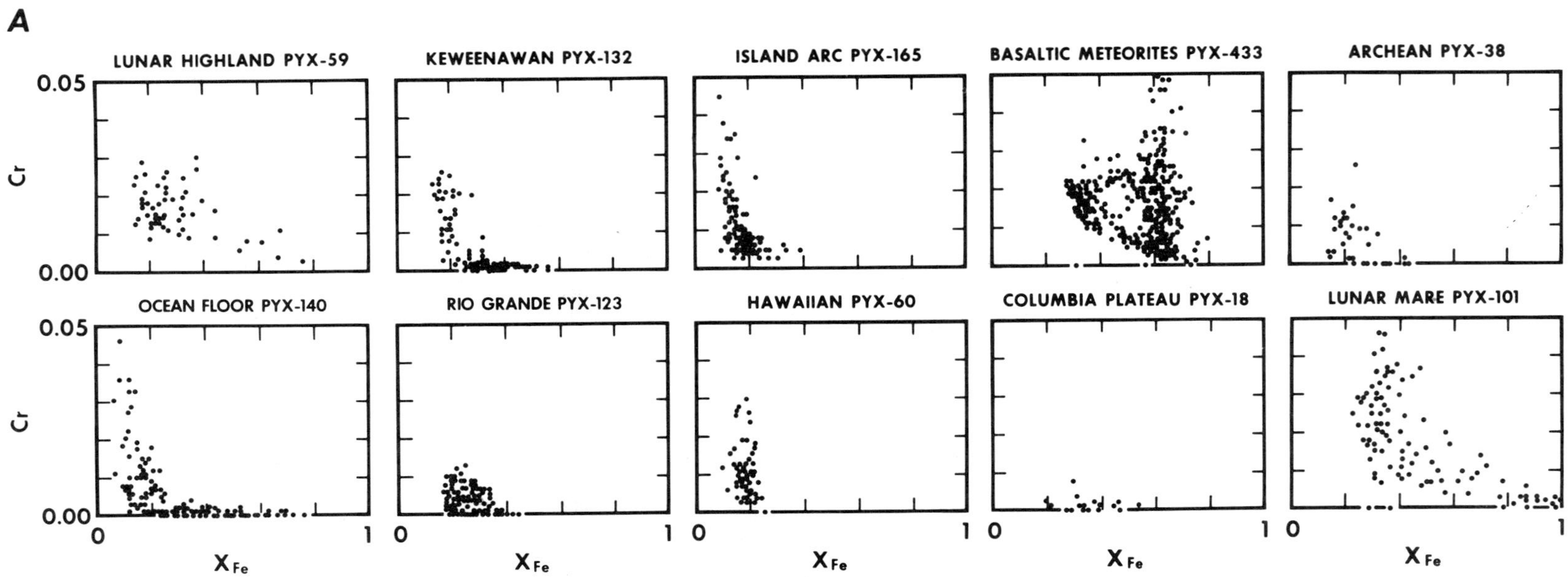

1.3.2.16a Cr-X_{Fe} variation diagrams for the ten pyroxene suites.

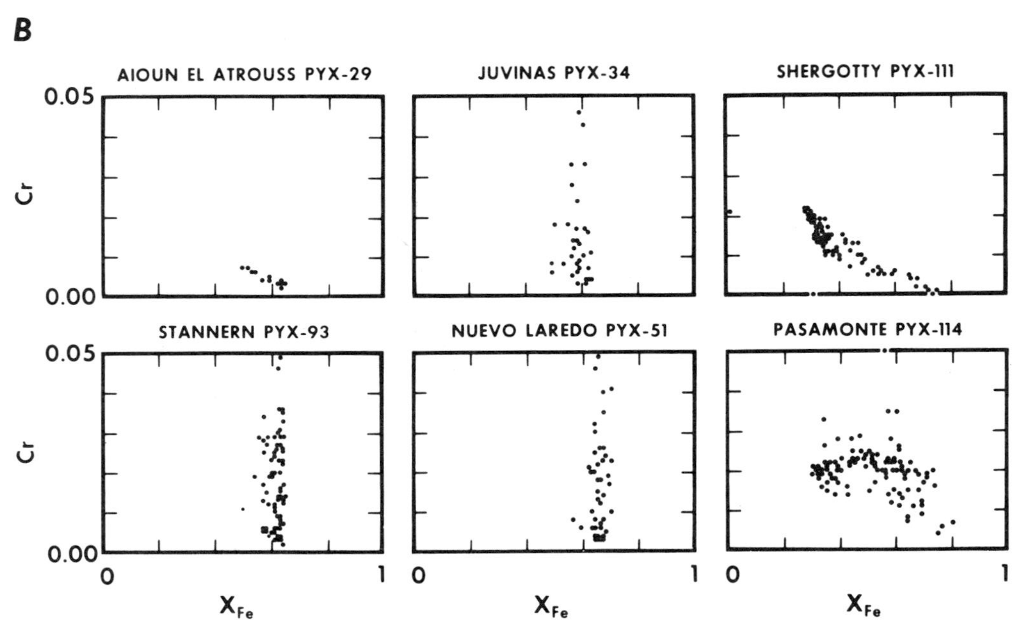

1.3.2.16b Cr-X_{Fe} variation diagrams for six meteorite pyroxene suites.

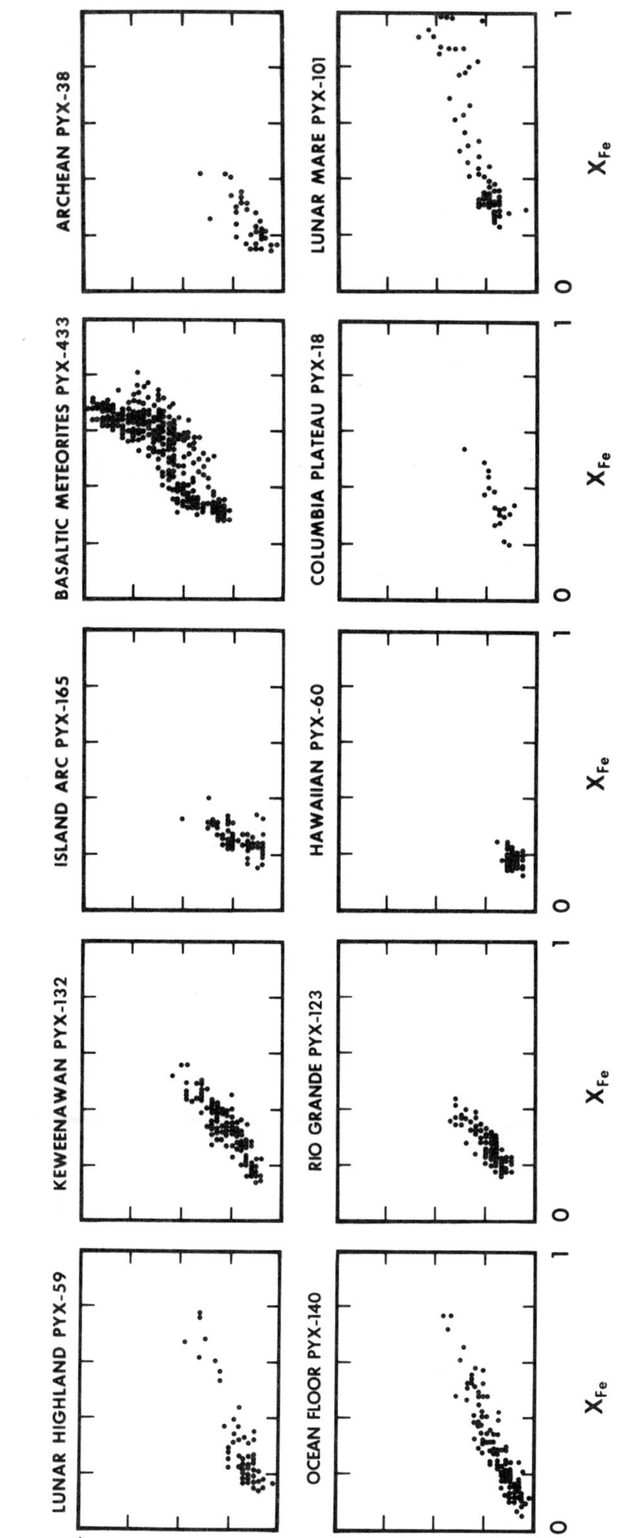

1.3.2.17a Mn-X_{Fe} variation diagrams for the ten pyroxene suites.

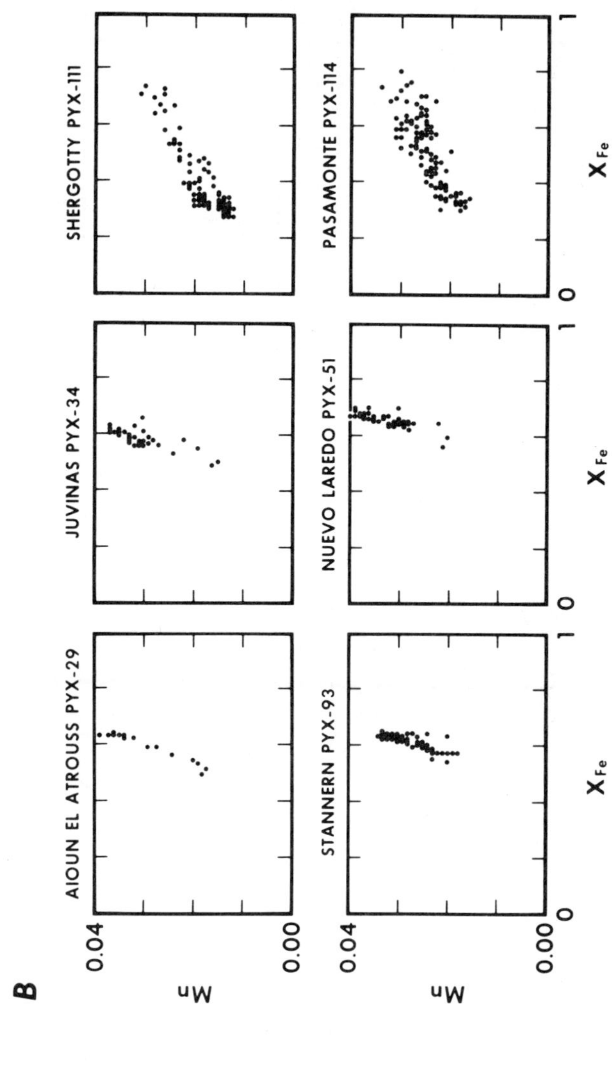

1.3.2.17b Mn-X_{Fe} variation diagrams for six meteorite pyroxene suites.

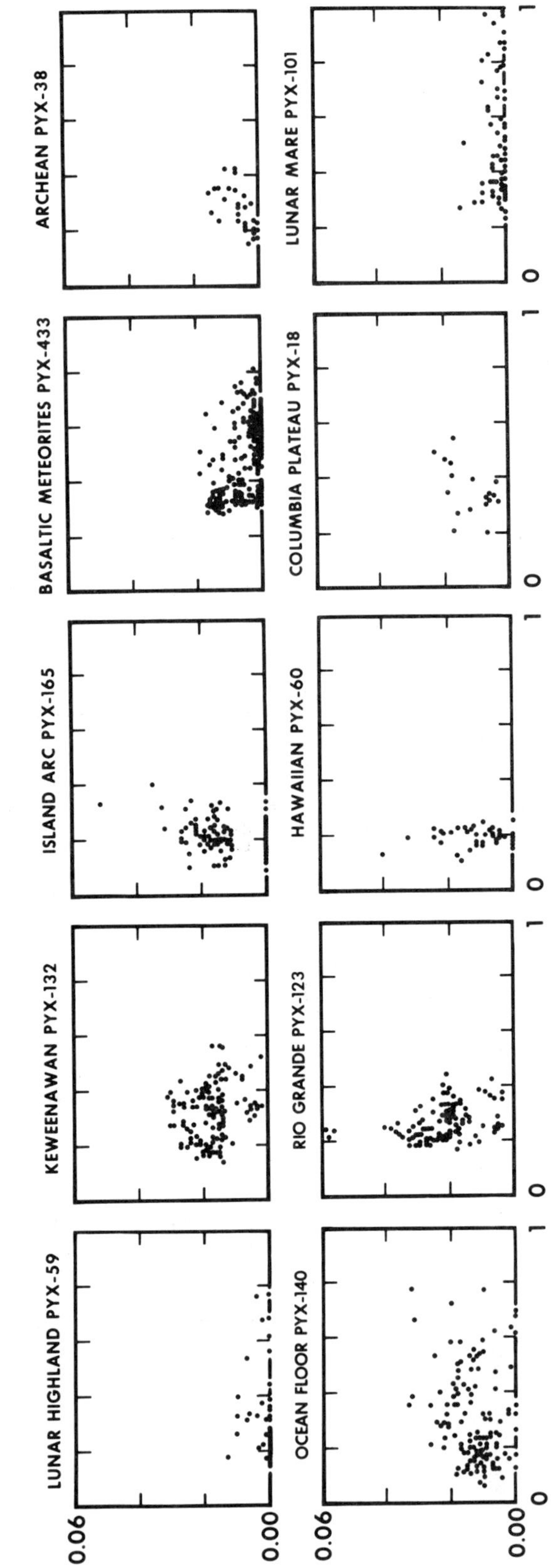

1.3.2.18a Na-X_{Fe} variation diagrams for the ten pyroxene suites.

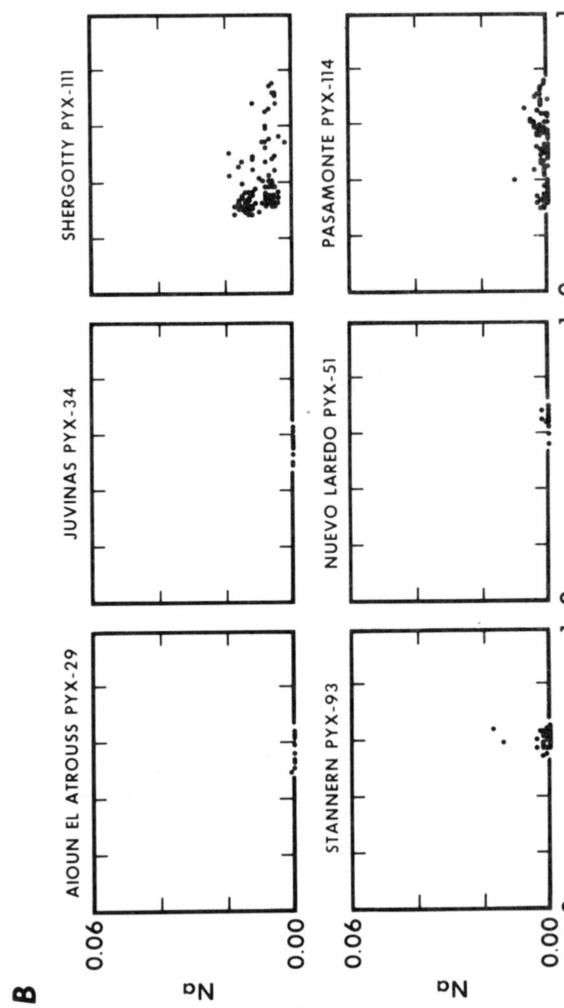

1.3.2.18b Na-X_{Fe} variation diagrams for six meteorite pyroxene suites.

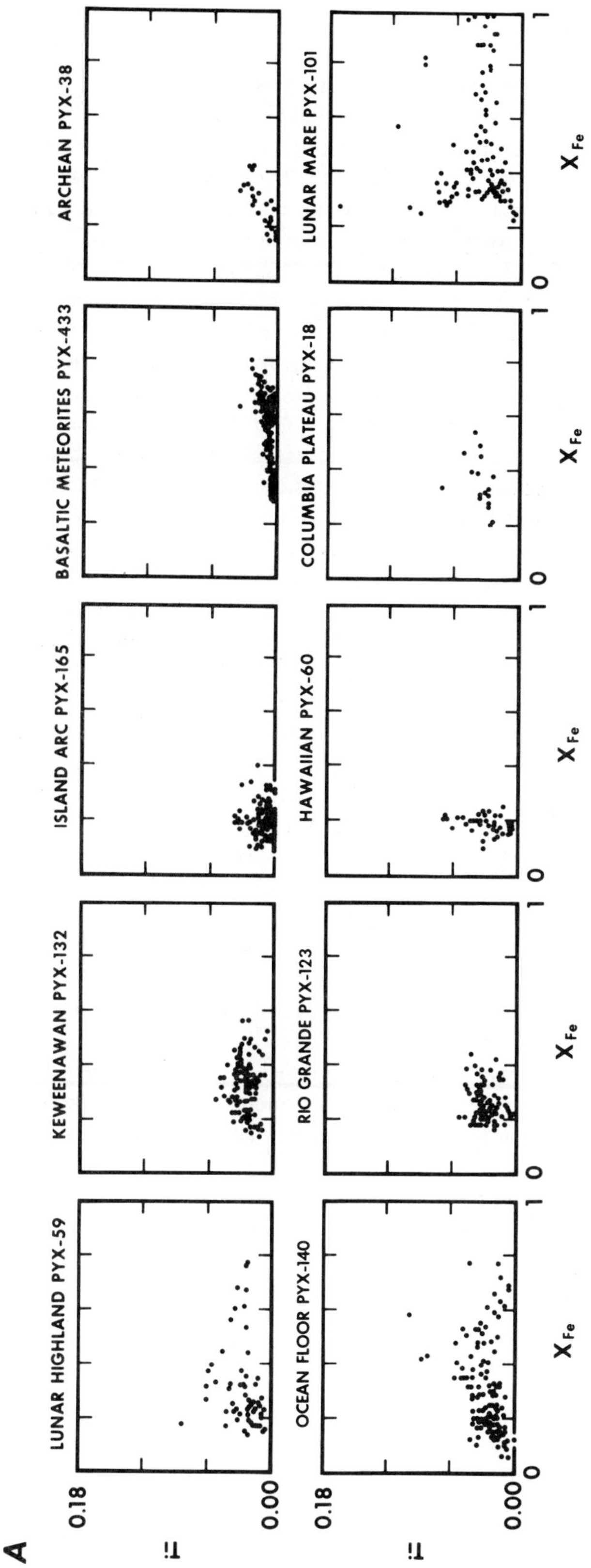

1.3.2.19a Ti-X_{Fe} variation diagrams for the ten pyroxene suites.

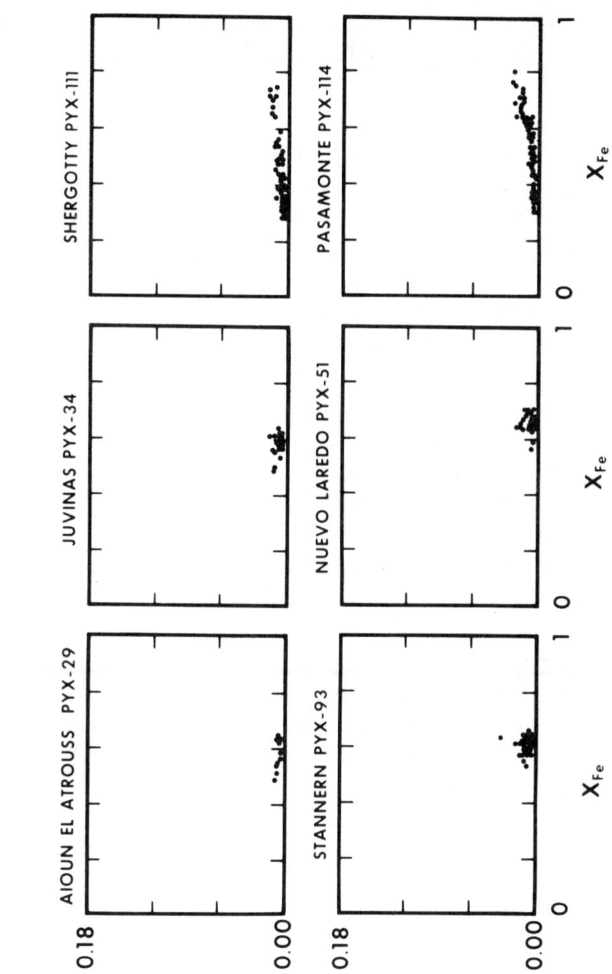

1.3.2.19b Ti-X_{Fe} variation diagrams for six meteorite pyroxene suites.

1.3.3 COMPARATIVE PETROGRAPHY AND COOLING RATES

The most detailed information on the cooling histories of basalts is contained in their textures, but the interpretation of these textures has been largely qualitative. Using a variety of methods, however, it is now possible to associate semiquantitative cooling rates with textures. For a given texture, it is now possible to place limits on the cooling rates of one group of basalts vs. another, and to determine minimum cooling rates for specific bulk-rock compositions. Before attempting to determine whether there are any systematic differences in the cooling histories or rates of basalts from the Earth, Moon and meteorites, however, it is necessary to briefly discuss recent experiments that have increased our understanding of factors which affect texture.

1.3.3.1 Controls on texture

The development of textures in basaltic rocks is controlled by the cooling rate, fluid flow, composition of the liquid, nucleation and growth rates, heterogeneous nucleation, and settling or floating of crystals. Recent dynamic crystallization experiments (i.e., controlled cooling-rate experiments) on basaltic materials from the Earth, Moon, and meteorites have provided an insight into the relative importance of some of these factors (Lofgren, 1980). In this type of experimentation an attempt is made to duplicate the cooling history of natural basalts; this involves the study of nucleation behavior, the variation of growth rate with cooling rate, and the settling of crystals.

Nucleation

Dynamic crystallization experiments have shown that the nature of the melt at the initiation of cooling, especially with regard to the presence of heterogeneous nuclei, can have a dramatic effect on the resulting texture (Donaldson *et al.*, 1975; Bianco and Taylor, 1977; Lofgren, 1977; Lofgren *et al.*, 1978). In general, if nucleation occurs more easily because, for example, nuclei are already present as a result of incomplete melting when cooling is initiated, the resulting product will be more completely crystallized. The grain size then will depend more on the number of nuclei per unit volume than on the growth rate. The effect of melt history is most dramatic at cooling rates less than 10°C/hr. This is particularly true for experiments on terrestrial basalts in which nucleation occurs less readily and the range of textural possibilities is greater than in lunar basalts (Lofgren, 1980).

Dynamic crystallization experiments on an olivine tholeiite from the Taos Plateau suite have shown that, for a cooling rate of 2°C/hr, the texture can vary from fine-grained dendritic or spherulitic to coarse-grained intersertal. This range of textures can be achieved simply by varying the temperature at which cooling is initiated and by varying the time spent at that temperature (Lofgren, 1979). In contrast, the textures of many lunar basalt compositions cooled at 2°C/hr are medium- to coarse-grained subophitic to intersertal and porphyritic and are independent of modest variations in starting temperatures.

This increased understanding of the effect of nucleation on texture has made it very difficult to relate the array of textures produced in experiments to the natural equivalents. It is now necessary to have some idea of the number of nuclei in the basaltic melt at the initiation of cooling. Ultimately, however, this knowledge of the precooling, nucleation condition of the melt will allow us to interpret not only the crystallization history but also the precooling melt history.

An important application of this concept of precooling melt history may be to interpret interflow grain-size variations. For example, the Dodge flow in the Columbia Plateau is significantly more coarse-grained than other flows of comparable composition from that area, which are much thicker and which must have cooled much more slowly. A thick flow could be fine-grained throughout, i.e., with small regard to cooling rate, if the number of nucleation centers were high. If, however, the nucleation density were much lower, the crystals would be larger even if the cooling rates were faster overall. Thus differences in grain size can indicate more about the number of nuclei and the nature of the melt than they do about the melt's cooling history.

Cooling rate

The influence of cooling rate generally is discussed in relation to nucleation rate (events per cm^3/sec) or nucleation density (number of crystals per cm^3). If the nucleation rate is low and growth rate is high, large crystals will result. If the nucleation rate is high, however, the same growth rate will produce much smaller crystals. Thus, while growth is a factor, it is not nearly as important an influence on the final texture as is nucleation.

The cooling rate dictates the time available for nucleation to occur and the chemical driving force for growth once nuclei are present. The influence of cooling rate on texture is greatest when the melt liquid is free of nuclei at the initiation of cooling because, if nucleation is delayed to high degrees of supercooling, the shapes and compositions of the resulting phases are more profoundly affected (Lofgren, 1980). Once nuclei are present, growth proceeds immediately, upon further cooling, at a rate which is dictated by the cooling rate. Since large degrees of supercooling are produced only at the most rapid cooling rates, there usually is a less pronounced effect of cooling rate on crystal shape and thus, on the final rock texture.

Composition

Composition of the melt is an important variable in comparing textures among basalts and in attempting to relate those textures to cooling rate. Composition affects texture in several ways. Viscosity of the liquid depends upon composition and nucleation generally becomes more difficult with increasing viscosity. Phase relations also depend on composition and affect the texture in a complex manner that is as yet not adequately understood. It is known that, if a composition is characterized by a significant temperature interval below the liquidus in which the liquidus phase crystallizes alone before the second phase appears, there is a greater tendency for a porphyritic texture to form during cooling (Lofgren *et al.*, 1974). The porphyritic texture will only form, however, if the nucleation density is low. Porphyritic textures do not form during cooling of basalts which are multiply-saturated or nearly so at the liquidus (Walker *et al.*, 1978; Lofgren, 1980), and would only occur in rocks produced from such melts if there had been inherited (intratelluric) phenocrysts. Intersertal and intergranular textures are more common in these basalts.

Flow and settling rates of crystals

The effects of a moving magma usually make themselves apparent in the aligned crystals typical of pilotaxitic textures. Flow can keep crystals in suspension that might otherwise settle. Flow may also cause heterogeneous nucleation. Crystal settling is an obvious factor in the development of textures in komatiitic flows, but should be considered in the interpretation of any cooling unit greater in thickness than approximately 10 m.

1.3.3.2 Textural comparisons

There are systematic differences in the textures of the basalts in the different suites under consideration. It is necessary, however, to distinguish differences that have resulted from different cooling histories from those that have resulted from the properties of the melts, such as the viscosity or the presence or absence of nuclei. Variations in viscosity, for example, affect the thickness of lava flows and thereby the cooling rates that the melt experiences. Mechanisms of eruption and local topography can also affect the thickness of flows. Imposed upon the natural variations is the choice of samples at the outcrop. To make textural comparisons most meaningful, all samples should come from the same relative position in the lava flow. The most informative relative position would be that of the most coarse-grained horizon. No such careful sampling program has been followed. A proper program would be very time-consuming. Several samples would be needed to document the grain size variations in each flow. Comparisons will be made among our reference suites, but the above reservations must be kept in mind.

Terrestrial basalts

Basalts in the ocean floor and Hawaiian suites generally contain more glass or fewer crystals than those of the other suites. This is typified by the porphyritic texture of OF-5 (Fig. 1.3.3.1a) which has a spherulitic matrix (see Table 1.3.3.1 for definitions of textural terms). This difference probably is a function of thinner flows and of sampling limited to near flow surfaces. The Taos Plateau lavas are porphyritic and occur in flows of thicknesses comparable to the ocean floor and Hawaiian flows. The matrices, however, are intersertal to intergranular (Fig. 1.3.3.1b) to subophitic or ophitic, reflecting the more nearly multiply-saturated nature of the Taos lavas. The Keweenawan and Columbia Plateau suites more commonly display intergranular, subophitic (Fig. 1.3.3.1c) and ophitic (Fig. 1.3.3.1d) textures and generally are more completely crystalline. The flows in these two suites commonly are thicker than those of the other suites, which implies slower cooling rates. It is not clear whether this difference in thickness is related to eruption rate or to viscosity or both.

The island arc basalts generally are well crystallized but are characteristically porphyritic. The matrix enclosing the phenocrysts most commonly is intergranular (Fig. 1.3.3.1e). The Archean suite contains numerous MgO-rich komatiitic basalts as well as proper komatiites; their spinifex textures (Fig. 1.3.3.1f) are unique among basalts, primarily because their compositions are unique. The tholeiites and other non-komatiitic rocks of the Archean suite display typical basaltic textures, which range from well-crystallized intergranular to porphyritic with fine-radiate to fan-spherulitic matrices.

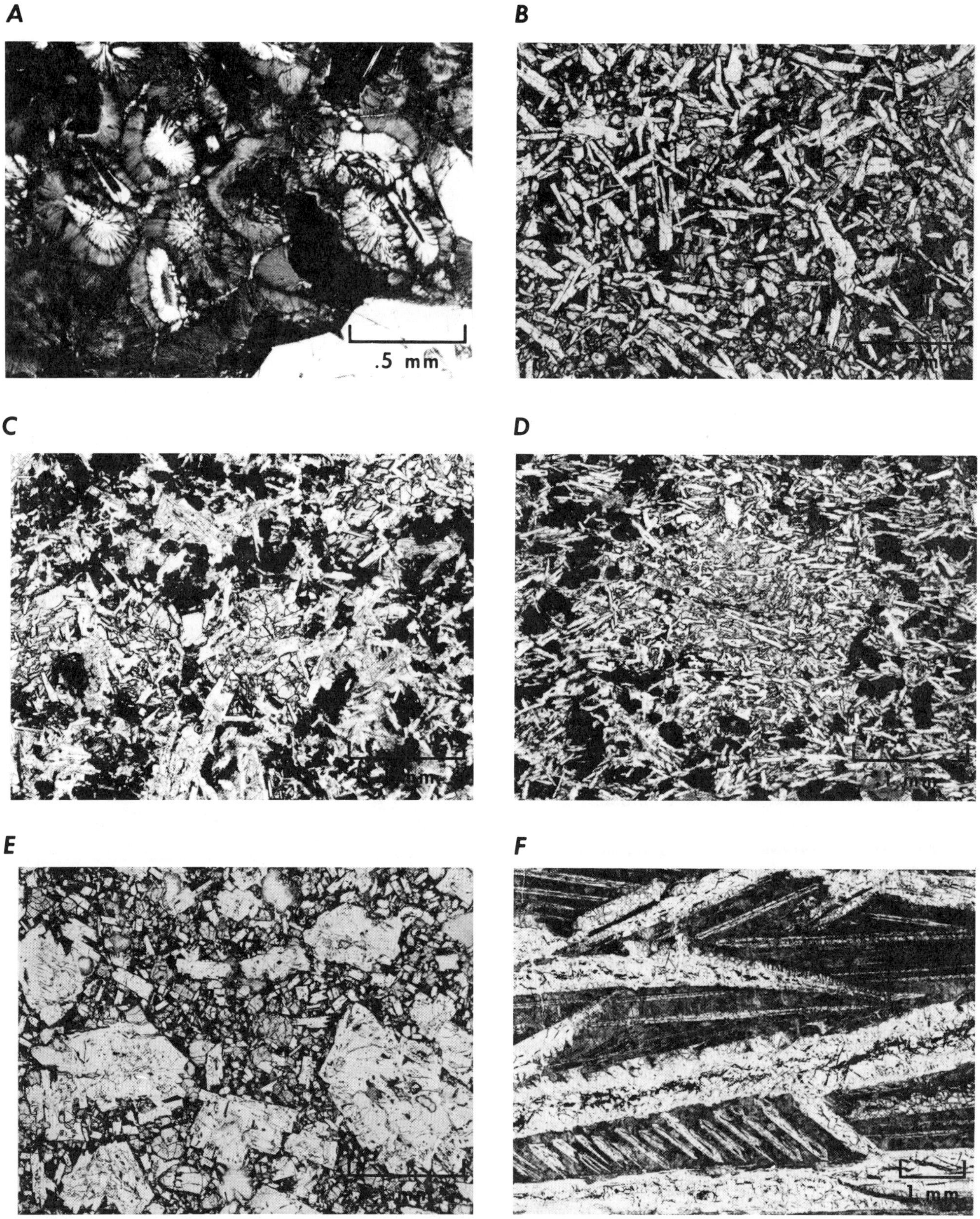

1.3.3.1 Photomicrographs of terrestrial basalts from the reference suites which show typical textural features. **a.** Microporphyritic with fan-spherulitic matrix, OF-7R. **b.** Intergranular with intersertal areas, TP-5. **c.** Subophitic with an irregular distribution of matrix and opaque minerals, KEW-12. **d.** Ophitic, large area in center of photomicrograph which is free of opaque minerals is the ophitic clinopyroxene, KEW-2. **e.** Porphyritic with largely intergranular matrix, IA-6. **f.** Spinifex texture, large dendritic olivine enclosed in a coarsely radiate clinopyroxene matrix, ACH-6.

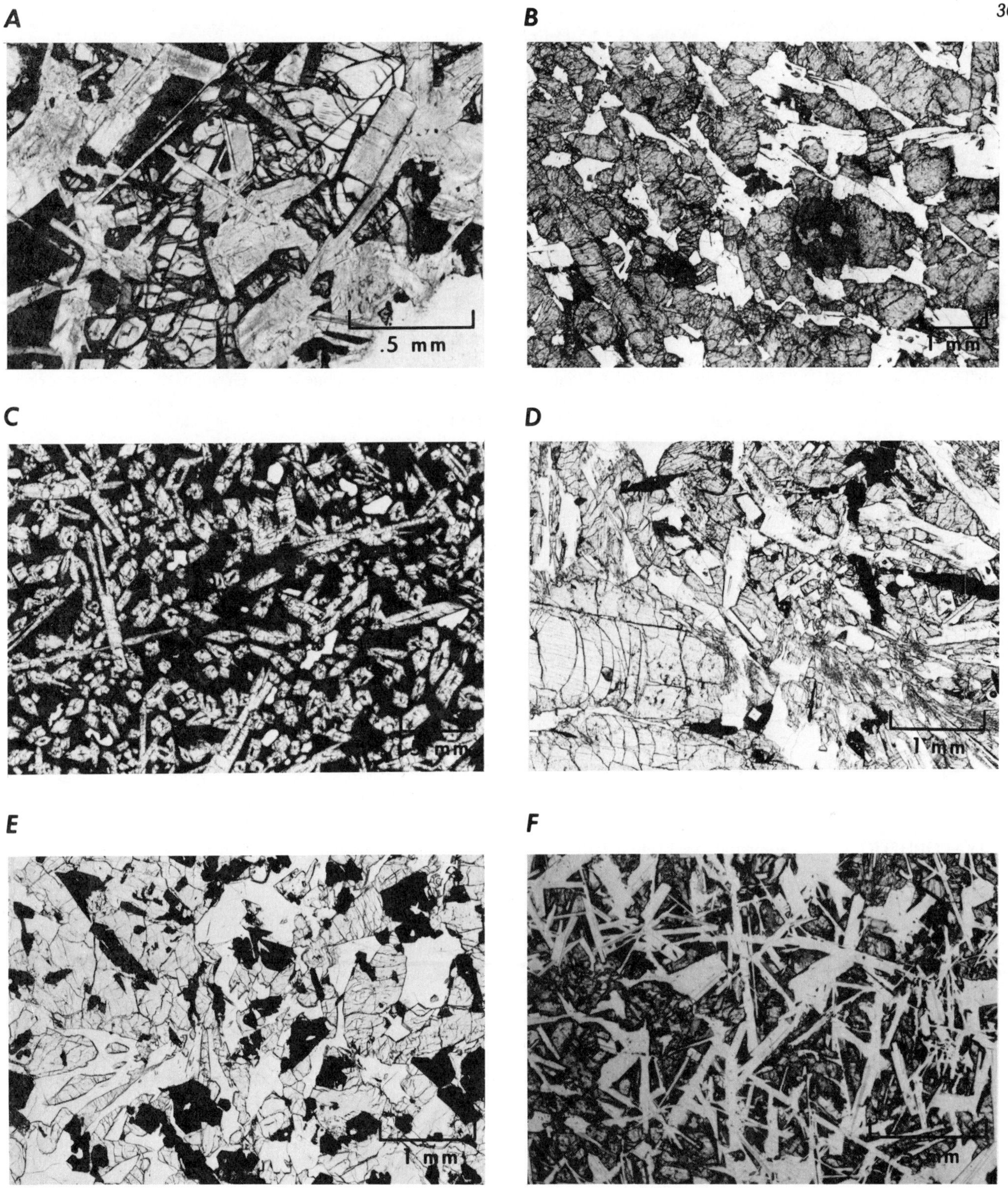

1.3.3.2 Photomicrographs of basaltic meteorites and lunar basalts. **a.** Pasamonte with a typical subophitic texture. **b.** Shergotty with its unusual apparently cumulate texture comprising clinopyroxene and plagioclase, which is now maskelynite. **c.** Clinopyroxene-phyric basalt, phenocrysts grew rapidly either on or near lunar surface, matrix is fan-spherulitic pyroxene, 15597. **d.** Coarse-grained example of porphyritic basalt with highly zoned somewhat skeletal clinopyroxene phenocrysts in a matrix of skeletal plagioclase and both granular and radiate to fan-spherulitic pyroxene, 12021. **e.** Ilmenite rich, anti-subophitic to anti-intergranular texture, 70017. **f.** Feldspathic basalt 14310, which has an intergranular to intersertal texture.

Table 1.3.3.1 Definitions of petrographic terms.

TEXTURE	PLAGIOCLASE	PYROXENE	MESOSTASIS/MATRIX
OPHITIC	Euhedral to subhedral, tabular to lath rare skeletal. Enclosed by pyroxene; crystal length does not exceed size of pyroxene.	Subhedral to anhedral, poikilitic, rarely euhedral. Larger than long dimension of and encloses plagioclase.	Completely crystalline, little or no glass remains. Olivine, opaque, or salic minerals may be present.
SUBOPHITIC	Euhedral to subhedral, lath to elongate; may be skeletal. Partially enclosed by pyroxene; crystal length commonly exceeds size of pyroxene.	Subhedral to anhedral, sometimes euhedral. Partially encloses plagioclase whose length exceeds size of pyroxene.	Completely or nearly crystalline; some glass, however, is often present in basalts. Olivine, opaque, or salic minerals may be present.
INTERGRANULAR	Euhedral to subhedral, tabular to lath or elongate; may be skeletal. Arranged in a fretwork which may be diverse, subradial, or subparallel.	Anhedral, sometimes subhedral to euhedral, granular to aggregate granular. Interstitial to plagioclase fretwork.	Granular olivine, opaque, or salic minerals may be present.
INTERSERTAL	Euhedral to subhedral, lath to acicular; commonly skeletal. Arranged in a fretwork which may be diverse, subradial, or subparallel.	Most pyroxene has not crystallized or occurs as fine spherulites (often fan-shaped). Granular crystals often present in small to modest amounts — sometimes are subhedral to euhedral.	Glass or palagonite or fine-grained alteration products thereof. Fan-spherulitic pyroxene or some granular pyroxene often present.
SPHERULITIC (FAN-)	Euhedral to subhedral, elongate to acicular; usually skeletal. Arranged in radial or subradial clusters; often fan-shaped and intimately interwoven with pyroxene fan spherulites.	Euhedral to subhedral, elongate to highly acicular; often skeletal, most often fan spherulitic, fan-shaped arrays of very fine-grained ($< 10\mu$) highly acicular crystals.	Glass or palagonite or fine-grained alteration products thereof. Skeletal or lattice olivine may be present.
HYALOPHITIC	Euhedral to subhedral, lath to elongate to acicular; usually skeletal. Arranged in diverse, subradial or subparalled orientations, usually not in contact; fretwork lacking.	Usually a small fraction crystallized as microlites, fan spherulites, or granular crystals.	Glass, palagonite or fine-grained alteration products thereof. Pyroxene usually considered part of matrix, expecially if fan spherulitic.
HYALOPILITIC	Euhedral to subhedral microlites; may be skeletal; sparse density, random orientation. When microlites are closely packed or interwoven texture is felted.	Usually a small fraction crystallized as microlites, spherulites, or granular crystals. Acicular prisms of sodic pyroxene in appropriate rock compositions.	Glass, palagonite or fine-grained alteration products thereof. Pyroxene usually considered part of matrix, especially if fan-spherulitic.
PILOTAXITIC	Euhedral to subhedral microlites; may be skeletal. Aligned subparallel to parallel, presumably by flow.	Small granular to microlitic.	Micro-, or cryptocrystalline or minute granular pyroxene, opaque minerals, apatite, and salic minerals.
TRACHYTIC	Euhedral to subhedral laths (includes sanidine). Aligned subparallel, presumably to flow.	Granular to microlitic.	Micro-, or cryptocrystalline material, devitrified glass, granular pyroxene or opaque minerals.

COMPARISONS	SYNONYM (generally not preferred usage)
Variety of poikilitic with included mineral being only plagioclase, which is usually well formed.	poikilophitic diabasic doleritic granitotrachytic
Distinguished from ophitic by smaller, often better-formed pyroxene, and commonly more skeletal plagioclase.	doleritic
Distinguished from intersertal by absence of mesostasis and from ophitic by granular, interstitial, nonpoikilitic pyroxene.	subdoleritic diabasic
Distinguished from intergranular by presence of interstitial glass or other fine-grained felsic material and from hyalophitic and hyalopilitic by presence of plagioclase fretwork.	vitrophyre porphyritic
Distinguished by dominance of fan-spherulitic pyroxene especially when fine grained. Plagioclase usually intergrown with pyroxene.	vitrophyre variolitic radiate fasicular
Distinguished from intersertal by increased fraction of matrix (usually glass) which totally encloses and is not interstitial to plagioclase.	vitrophyric porphyritic
Distinguished from hyalophitic by small microlitic plagioclase and increased fraction of matrix, or from intersertal by lack of plagioclase fretwork.	microporphyritic vitrophyric microlitic
Distinguished from hyalopilitic by much increased fraction of plagioclase and apparent lack of matrix or poorly crystallized material.	microlitic
Distinguished from pilotaxitic by larger feldspar and generally more pronounced alignment of crystals.	

Meteorites

The basaltic textures observed in meteorites range over the entire known spectrum of textures, but subophitic textures (Fig. 1.3.3.2a) are most common. Cumulate textures (Fig. 1.3.3.2b) are often found in close association (i.e., in the same meteorite). Because of the nature of the samples, their small sizes coupled with variable degrees of brecciation and alteration, it is difficult to make any general statements about the cooling histories of this group. Many of the basaltic fragments could have resulted from cooling of impact melts derived from a basaltic target. Porphyritic textures are rare in meteorites, probably reflecting their nearly cotectic compositions (Stolper, 1977; Walker *et al.*, 1978) and their (presumably) simple melt histories.

Lunar basalts

Lunar basalts display the entire range of basaltic textures from vitrophyric (Fig. 1.3.3.2c) to ophitic. In general, however, their lower viscosities have allowed more complete crystallization at higher cooling rates than is typical for terrestrial basalts. In addition, pyroxene is often modally dominant over plagioclase. This difference in modal mineralogy, which is due to differences in bulk composition, is reflected in the viscosities and therefore in the textures. Porphyritic textures are common in lunar basalts (Figs. 1.3.3.2c, d), but appear to be near-surface features due to rapid growth of phenocrysts followed by matrix in a single cooling event (Lofgren *et al.*, 1974; Walker *et al.*, 1976). The dominance of pyroxene over plagioclase often reverses the roles of plagioclase and pyroxene as textural elements from those in terrestrial basalts. Grove and Beaty (1980) have used the prefix anti- to indicate this reversal. In Fig. 1.3.3.2e the pyroxene forms the fretwork and plagioclase is interstitial to poikilitic, so the texture is anti-intergranular to anti-subophitic. However, classic intergranular and intersertal textures are found in the more feldspathic highland basalts (Fig. 1.3.3.2f).

1.3.3.3 Quantitative cooling rates

Recent dynamic crystallization experiments conducted on basaltic materials, principally on basalts from the Moon but also from meteorites and the Earth, have provided quantitative relationships between cooling rate and petrographic features for a variety of basalts. Several different petrographic features have been used to relate laboratory experiments to their natural counterparts. These include crystal shape, size, major and trace element zoning, and nucleation density. Most of these techniques are semi-quantitative or give order-of-magnitude estimates of the cooling rate. Most impressive are the direct comparisons of textures in different basalts cooled under identical conditions, which show that, for equivalent cooling rates, lunar basalts will be the coarsest-grained followed by meteoritic basalts and then terrestrial basalts.

Of the methods used to quantify cooling rates, the most successful is crystal size (Walker *et al.*, 1976, 1978; Grove and Walker, 1977). Crystal size, of course, depends somewhat on nucleation density, but if density is comparable then the crystal size is directly related to growth rate, which is usually related directly to cooling rate. According to this technique, some of the coarsest-grained low-Ti lunar basalts experienced cooling rates on the order of 0.1°C/hr. Walker *et al.* (1978) studied basaltic clasts in the Stannern meteorite and found a range of cooling rates from 100° to 0.1°C/hr. Their calculations indicate that this range of cooling rates could have been produced in a flow 3 m thick. Terrestrial basalts have not been studied extensively using the crystal size technique, but where results are compared directly at, for example, 2°C/hr, cooled from a complete liquid, terrestrial basalts do not crystallize nearly as readily as lunar basalts. Whereas a lunar basalt might show a medium- to coarse-grained porphyritic or subophitic texture at 2°C/hr, a terrestrial basalt is likely to have only dendritic olivine or pyroxene crystals (Lofgren, 1980). Such comparisons will vary with composition but the disparity is sufficiently large that the statement is generally true. The Stannern basalt crystallization behavior (Walker *et al.*, 1978) is intermediate between those of the lunar and terrestrial basalts.

1.3.3.4 Implications of textures

The differences in grain size between the lunar, meteoritic, and terrestrial basalts produced at equivalent cooling rates indicate that a higher percentage of lunar lava flows will be coarser-grained. This is manifestation of the different bulk compositions and viscosities of the basalts. Such considerations may be important to remote-sensing studies, such as spectral reflectance.

The presence and nature of phenocrysts in a basalt can have much to reveal about the early formational history of the basalt, which may have been influenced by processes such as fractionation, partial melting, or magma mixing. Terrestrial basalts display an infinite variety of phenocrysts from cognate to xenocrystic, intratelluric to late-growing, micro- to macro-. These phenocrysts provide strong evidence that all of the above-mentioned petrogenetic processes operate on terrestrial basalts. Lunar and meteoritic basalts do not contain such a complex array of phenocrysts. Most or all pyroxene phenocrysts present in lunar basalts have grown near or on the surface of the Moon (Lofgren, 1975). There is evidence that a few olivine phenocrysts in one sample have long, complex histories (Donaldson *et al.*, 1975), but their story is not clear. Olivines can grow rapidly and can react with almost equal rapidity with their host liquids, so often they are not good chemical recorders of growth histories. Plagioclase phenocrysts in terrestrial basalts are the best records of their histories. Unfortunately, lunar basalts do not often have plagioclase near the liquidus and, even where it is present, the plagioclase typically is too calcic to reveal much about its history. Thus, based on petrography, it is difficult to support arguments for any complex interplay of fractionation, partial melting, or magma mixing in lunar basalts or on the source bodies of meteoritic basalts.

1.3.3.5 References

Bianco A. S. and Taylor L. A. (1977) Applications of dynamic crystallization studies: lunar olivine normative basalts. *Proc. Lunar Sci. Conf. 8th*, p. 1593–1610.

Donaldson C. H., Usselman T. M., Williams R. J. and Lofgren G. E. (1975) Experimental modelling of the cooling history of Apollo 12 olivine basalts. *Proc. Lunar Sci. Conf. 6th*, p. 843–870.

Grove T. L. and Walker D. (1977) Cooling histories of Apollo 15 quartz normative basalts. *Proc. Lunar Sci. Conf. 8th*, p. 1501–1520.

Grove T. L. and Beaty D. W. (1980) Classification, experimental petrology and possible volcanic histories of the Apollo 11 high-K basalts. *Proc. Lunar Planet. Sci. Conf. 11th*, p. 149–177.

Lofgren G. E. (1977) Dynamic crystallization experiments bearing on the origin of textures in impact generated liquids. *Proc. Lunar Sci. Conf. 8th*, p. 2079–2095.

Lofgren G. E. (1979) The effect of nucleation on the basaltic textures (abstract). *Abstracts with Program*, Geol. Soc. Amer. **11**, 467–468.

Lofgren G. E. (1980) Experimental studies on the dynamic crystallization of silicate melts. In *Physics of Magmatic Processes*, (R. B. Hargraves, ed.), p. 487–551. Princeton Univ. Press.

Lofgren G. E., Donaldson C. H., Williams R. J., Mullins O., Jr. and Usselman T. M. (1974) Experimentally reproduced textures and mineral chemistry of Apollo 15 quartz normative basalts. *Proc. Lunar Sci. Conf. 5th*, p. 549–567.

Lofgren G. E., Smith D. P. and Brown R. W. (1978) Dynamic crystallization and kinetic melting of the lunar soil. *Proc. Lunar Planet. Sci. Conf. 9th*, p. 959–975.

Stolper E. M. (1977) Experimental petrology of eucritic meteorites. *Geochim. Cosmochim. Acta* **41**, 581–611.

Walker D., Kirkpatrick R. J., Longhi J. and Hays J. F. (1976) Crystallization history of lunar picritic basalt sample 12002: phase equilibria and cooling rate studies. *Geol. Soc. Amer. Bull.* **87**, 646–656.

Walker D., Powell M. A., Lofgren G. E. and Hays J. F. (1978) Dynamic crystallization of a eucrite basalt. *Proc. Lunar Planet. Sci. Conf. 9th*, p. 1369–1391.

1.3.4 TEMPERATURES AND GAS FUGACITIES OF PLANETARY BASALTS

1.3.4.1 Introduction

A major and as yet unresolved question is whether the oxidation states of magmas are buffered by their coexisting vapor phases or whether the vapor component is the overriding factor in controlling reduction and oxidation potentials of magmas. A question of equal significance is the effect of gas loss on extrusion in contrast to the importance that is attached to closed-system models of magmas cooling at depth. The relative solubilities of the dominant volatiles in basaltic liquids, coupled with volcanic gas analyses and with mineral equilibria data, provide a framework in which magma-gas equilibria may be assessed. These data also provide insights into the relative effects of redox potential in extrusive and intrusive environments, and into contrasts in the oxidation states of basalts that have evolved on other planets. The data bases on which such evaluations are dependent, however, vary considerably. A sufficiently large and internally consistent set of data is available for terrestrial basalts; the data for lunar basalts, although limited, define the conditions of crystallization adequately, whereas basaltic achondrites, with the least data, pose some tantalizing questions on the oxidation state of asteroidal parent bodies.

From the data available, we may conclude that the intensive parameters of temperature and fugacity are reasonably well defined for terrestrial basalts, are constrained for lunar basalts, and are within a wide range for the achondrites. Although the data are not equally distributed, the diversity of the suites available offers the opportunity to consider the types of redox conditions that may be present on other planets in the inner solar system.

1.3.4.2 Methods and analysis

The data presented in this section were obtained using a variety of techniques. Temperatures and oxygen fugacities are determined from the compositions of coequilibrated ulvöspinel-magnetite solid solutions and ilmenite-hematite solid solutions (Buddington and Lindsley, 1964). These data are widely employed in terrestrial basalts, are not applicable to lunar basalts, and are useful only in some achondrites. *In-situ* measurements of temperature and f_{O_2} have been obtained by drilling into crystallizing lava lakes and from gas vents, using thermocouples and the solid-state Zr-electrolyte cell technique (Sato and Wright, 1966; Sato, 1970, 1972). This same technique is employed in determining the intrinsic f_{O_2} of a basalt or of its component minerals in the laboratory and has been successfully applied to studies of terrestrial, lunar and achondritic basalts. Oxygen fugacities may also be determined by matching the $FeO:Fe_2O_3$ ratio of a rock experimentally, where temperature and the mixing ratios of furnace gases are adjusted to attain the required $Fe^{2+}:Fe^{3+}$ ratio (Fudali, 1965). Sulfur fugacities generally are estimated from phase equilibria studies, based on the limits of mineral stability, mineral composition, and sulfur solubilities in mafic melts (Naldrett, 1969; MacLean, 1969; Haughton *et al.*, 1974; Katsura and Nagashima, 1974; Shima and Naldrett, 1975). Phase equilibria studies and thermochemical calculations are also employed to bracket probable ranges of f_{S_2} and f_{O_2} during crystallization in natural basalt systems. Direct knowledge of sulfur fugacities in volcanic fumaroles is limited to that determined for Mt. Etna by Sato and Moore (1973).

Oxygen fugacity data reported in this discussion are taken from the literature as well from determinations obtained on samples in the reference suites. These data are illustrated in summary formats and the following buffer curves are used as internal references for the purposes of comparison:

MH, magnetite-hematite (Eugster and Wones, 1962);
MW, magnetite-wüstite (Eugster and Wones, 1962);
FMQ, fayalite-magnetite-quartz (Chou, 1978);
IW, iron-wüstite (Eugster and Wones, 1962); and
FIQ, fayalite-iron-quartz (Eugster and Wones, 1962).

1.3.4.3 Terrestrial basalts

Geothermometry and oxygen geobarometry from coexisting oxides in terrestrial basalt suites are summarized in Fig. 1.3.4.1. These data are separated into those for basalts in oceanic (ocean floor and oceanic island) environments and those for basalts in continental environments. This distinction has been made because of possible contrasts in primary volatile contents and compositions of the basalts, and because of the differences that may exist in the rates of volatile effusion in these settings. Oceanic island basalts, in common with continental basalts, are extruded under subaerial conditions, so some overlap in the data sets should be expected in the contexts of volatile loss and consequent effects on redox potentials.

All determinations fall within envelopes that broadly parallel and are constrained within limits defined by the

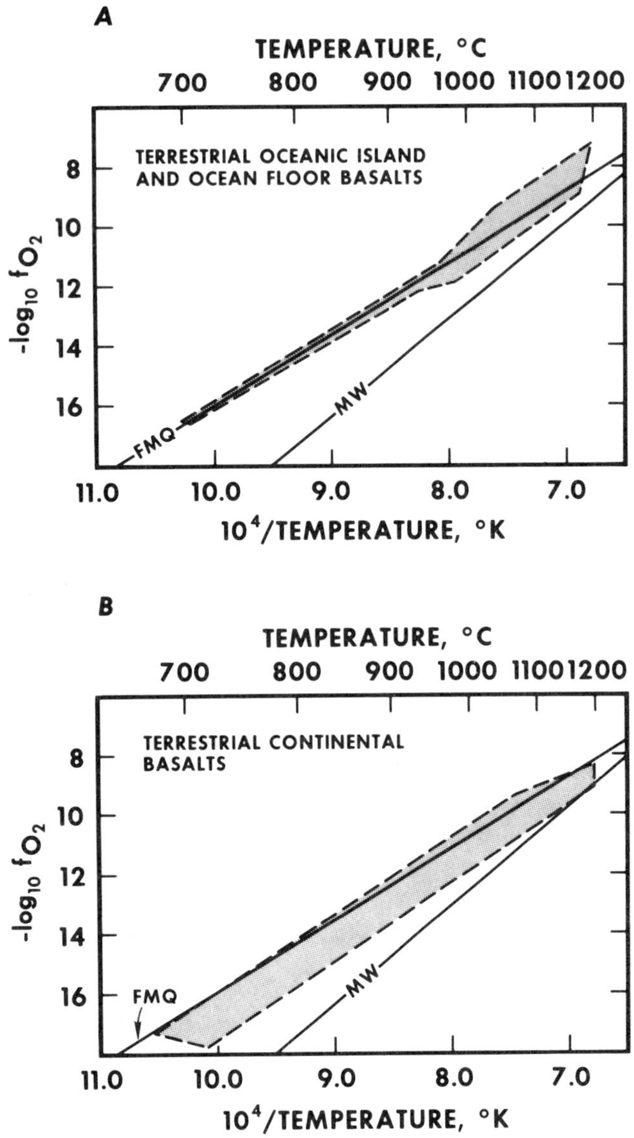

1.3.4.1 Temperature and oxygen fugacity data for terrestrial basalt suites, obtained by analyzing coexisting Fe-Ti oxides. **a.** Oceanic basalts. **b.** Continental basalts.

FMQ and MW buffer curves. Equilibration temperatures range between 650–1150° C, and the corresponding oxygen fugacities are 10^{-17} to 10^{-8} atms. At high temperatures (950–1200° C), oceanic basalts extend to f_{O_2} values that are greater than FMQ, whereas continental basalts in the same temperature range are more reducing and approach MW. On the average, ocean-floor basalts equilibrate at high temperatures (950–1150° C) and f_{O_2} between 10^{-12} and 10^{-8} atms; oceanic island basalts also equilibrate within this temperature and f_{O_2} range (e.g., basalts from Hawaii, Tristan Da Cunha, and Jan Mayen). Basalts from Iceland exhibit a large spread in temperature and f_{O_2}, which is typical of continental basalts. Continental basalts, however, are distinct from oceanic basalts in that their equilibration trends deviate rather substantially from the FMQ buffer curve, approaching in some instances levels of reduction that are one to two orders of magnitude f_{O_2} below the FMQ buffer. These distinctions may be attributed to the relatively rapid rates of cooling attendant on quenching in water in contrast to the moderately slow cooling which occurs under atmospheric conditions. Relative viscosities also play important roles in the differences that are present in the temperature and f_{O_2} trends, insofar as most oceanic island flows are thin and contrast clearly with the thick plateau basalts present on the continents; Iceland is perhaps the one exception to this general rule. Contrasts in the ranges of f_{O_2} between oceanic and continental basalts may of course be inherent properties of the volatile constituents of their respective source regions, but it may also be related to the effects of degassing which are discussed in section 1.3.5.

Experimental determinations of intrinsic f_{O_2}, data obtained from the crystallization of basalts under controlled temperature and f_{O_2} conditions, and values obtained from the Fe^{2+}:Fe^{3+} ratio matching technique are illustrated in Fig. 1.3.4.2 for suites of basalts from the Columbia River Plateau, Hawaii, and the Azores.

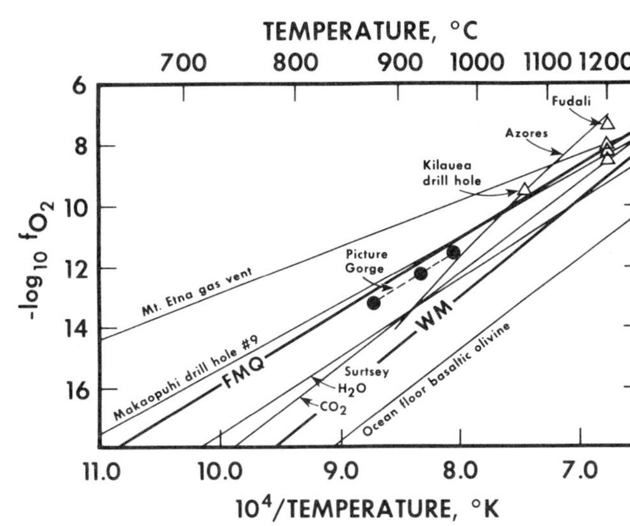

1.3.4.2 Temperature and oxygen fugacity data obtained by the Zr-electrolyte cell technique for Mt. Etna (Sato and Moore, 1973), Makaopuhi (Sato and Wright, 1966), Surtsey (Steinthorsson, 1972), Kilauea (Peck and Wright, 1966), and olivine (Sato, 1972). The 1200°C values are from Fudali (1965); those for the Azores are from Duke (1974); and those for Picture Gorge are from Helz (1973). The last sets of data are from laboratory determinations.

These values and their temperature and f_{O_2} profiles, with respect to the FMQ buffer curve, are similar to the mineral oxide determinations at high temperatures. Equilibration at lower temperatures (<800° C), however, has not been attempted, so we cannot determine whether the deviation from FMQ that is observed on cooling is the norm or the exception.

Temperatures and oxygen fugacities of volcanic gases from Surtsey and the Makaopuhi lava lake are also plotted in Fig. 1.3.4.2. The upper and lower limits defined by these analyses embrace the entire spectrum of temperature and f_{O_2} determinations obtained from mineral oxide data and from experimental data, which suggests that equilibrium between basalts and the coexisting gas phases was approached, if not in fact reached, in these measurements. Gas data for Kilauea (Gerlach, 1980a) and *in-situ* measurements (Peck and Wright, 1966) are within 0.2 log units of FMQ, a value that is very similar to that of gases emanating from the Erta' Ale lava lake, Afar, Ethiopia (Gerlach, 1980b). The Makaopuhi gases are slightly more reduced than the FMQ buffer at magmatic temperatures, but become more oxidizing at temperatures below ~1000° C. Gases from the Surtsey eruption are substantially more reduced

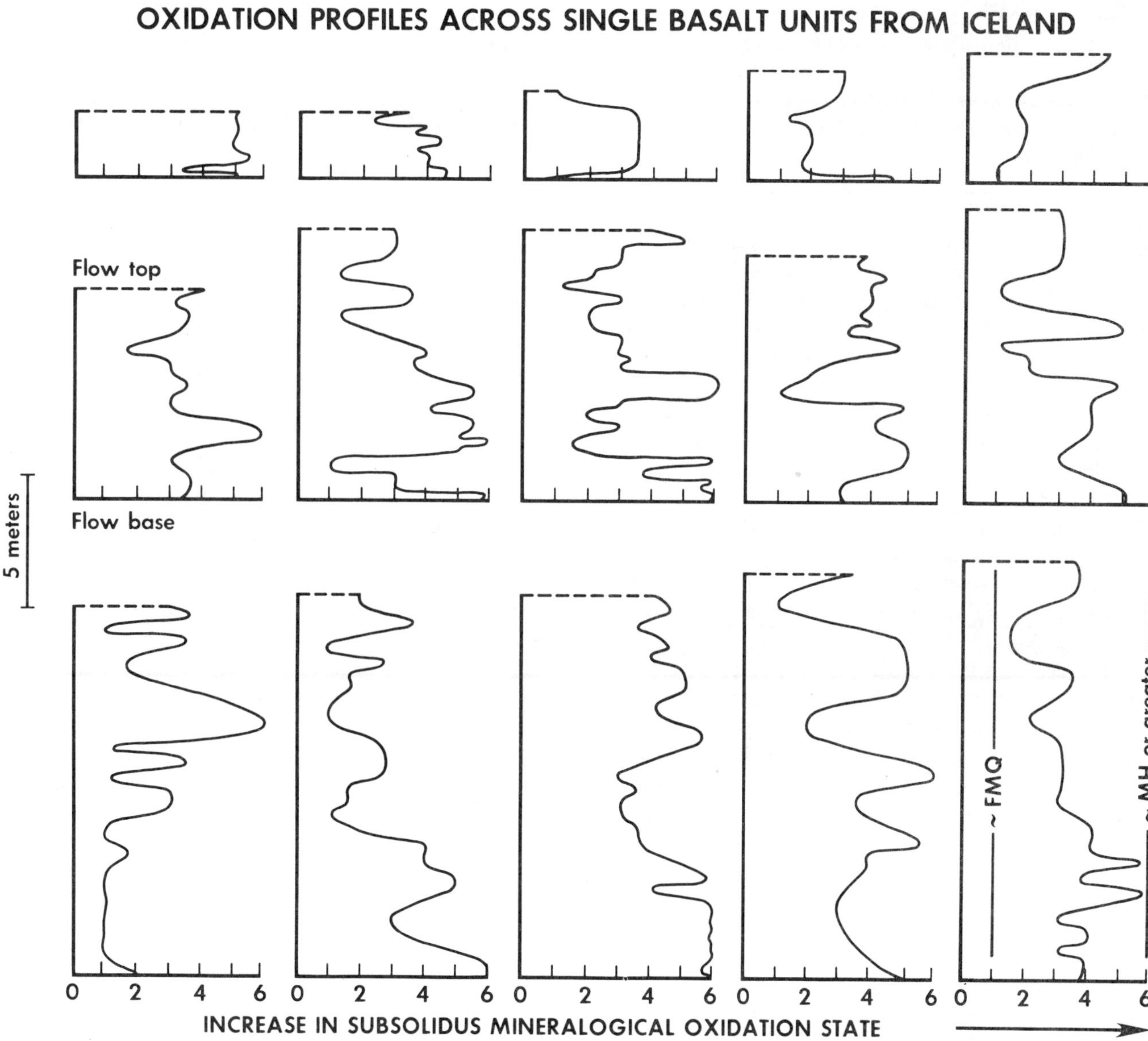

1.3.4.3 Relative oxidation variations for vertical profiles across a series of single basalt flows from Iceland. Data from Watkins and Haggerty (1967).

and are equivalent to MW at ~1200° C. Oxygen fugacity data for gases venting from Mt. Etna are approximately 0.2 log units more oxidizing than FMQ at 1200° C but become 2 orders of magnitude more oxidizing than FMQ at 800° C, for example. Based on the thermochemical treatment of these data by Gerlach and Nordlie (1975a, b, c), we may infer that the CO_2:H_2 ratios are greater in the Etna gases and lower in those from Surtsey, which is confirmed by gas analyses of sampled exhalatives when these have been suitably adjusted to account for the effects of contamination (Gerlach, 1979). A determination of the intrinsic f_{O_2} of cumulus olivine in an ocean-floor basalt (Fig. 1.3.4.2) lies close to the iron-wüstite buffer curve, which results from the fact that these olivines contain liquid + gaseous CO_2 inclusions (Roedder, 1965; Sato, 1970) and perhaps even elemental carbon (Sato, 1978).

The relative proportions of gases change with time and with the phase of volcanism. A significant property is the accumulation of water and the dissociation of H_2O into H_2 + ½ O_2 during basalt crystallization. The effects of oxidation that result from this process have been studied in vertical profiles across a series of olivine basalts and tholeiites (1 to 18 meters thick) from Iceland (Watkins and Haggerty, 1968; Wilson *et al.*, 1968; Haggerty, 1976a). The following is a summary of the most notable features:

(1) There are wide oxidation variations among the profiles, ranging from low oxidation states to intensely oxidized regions.
(2) There are differences which may be related to flow thickness, with thicker units exhibiting higher oxidation states than thinner flows.
(3) Maximum oxidation states in thick units (>3m) are preferentially concentrated towards flow interiors (Fig. 1.3.4.3).

Ponded lava lakes are expected to display oxidation distribution patterns similar to those of thick plateau basalts. In an example from the Makaopuhi lava lake (Sato and Wright, 1966; Haggerty, 1976a), where *in situ* temperature and f_{O_2} measurements were made and drill core samples were recovered, a zone of intense oxidation developed at T = 550–750° C (Fig. 1.3.4.4) which migrated as cooling and crystallization ensued.

The model proposed for these single unit variations depends on the rate of H_2O dissociation, on the relative ionic sizes of hydrogen and oxygen, and on a *semi-permeable membrane mechanism* which permits the effusive loss of H_2 but retention of oxygen in the lava interior. Computed profiles for the rate of cooling support the basic premise of this model, and temperature vs. time distribution patterns parallel those of oxidation (Watkins and Haggerty, 1968). Average oxidation pro-

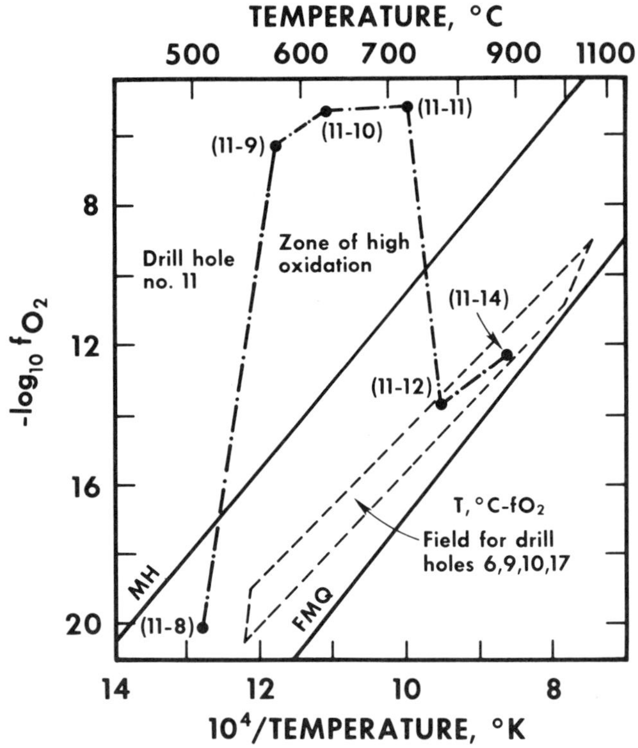

1.3.4.4 Temperature and oxygen fugacity relationships in holes drilled into the actively crystallizing Makaopuhi lava lake (Sato and Wright, 1966).

files are consistent, furthermore, with a phase compatibility treatment of Fe-Ti oxide assemblages at 750° C in a lava with a thickness of 10 meters, with an eruption temperature of 1000° C, a lower cooling face of basalt, and an upper cooling face of air. For this case, the lava interior temperature would be 650° C after 1.1 years and total cooling to 100° C for the entire unit occurs only after 4.0 years (Haggerty, 1976a). In thick terrestrial lavas, therefore, the effects of preferred volatile accumulation towards the interior of a flow are directly related to processes of deuteric cooling (i.e., are consequences of the initial cooling of a lava flow) and to the disproportionation of water into hydrogen and oxygen. The level of oxidation reached in the *interiors* of these cooling units is several orders of magnitude above the magnetite-hematite (MH) buffer (Fig. 1.3.4.4).

The only *in situ* measurements that have been obtained of sulfur fugacities are those by Sato and Moore (1973) for gases emanating from Mt. Etna. In that study f_{O_2} was $10^{-2.4}$ atms at 830° C. From restored gas analyses from Mt. Etna at 1075° C and an f_{O_2} of $10^{-9.47}$ atms an f_{O_2} range of $10^{+2.08}$ to $10^{+2.82}$ atms at 860° C has been calculated by Gerlach (1979). As outlined in section 1.3.5, the sulfur gas species vary as a function of

f_{O_2}. In parallel with the effects of hydrogen loss and consequent high oxidation zones in single cooling units, the loss of sulfur has a redox effect. This effect is the inverse of the hydrogen loss effect; sulfur degassing tends to reduce the magma. Immiscible sulfides or sulfide decorated vesicles are extremely rare in subaerially extruded basalts but are hallmarks of ocean-floor basalts. Sulfur loss in deep-sea basalts is minimal due to the pressure of overlying sea water. Because of the absence of sulfides from subaerially extruded flows, it must be assumed that sulfur loss has been extensive and, therefore, that partial reduction of the magma has been induced. Given the intense oxidation that occurs towards the interiors of lava flows, the effect of hydrogen loss (i.e., net oxidation) must be overwhelmingly greater than that of sulfur loss (i.e., net reduction) with prolonged cooling. However, considering the relative differences between oceanic basalts and continental basalts (Fig. 1.3.4.1) in the overall states of oxidation at T<900° C, and given the differences in sulfur contents between these suites, it is possible that sulfur devolatilization is largely responsible for those redox differences that are observed. This is also supported by the facts that gabbros and dolerites typically contain sulfides and that these rocks are also slightly more oxidized than are continental basalts relative to FMQ (Haggerty, 1976b). An alternative explanation is that the source regions of continental basalts are more reduced than those of oceanic basalts.

In summary, terrestrial basalts have average equilibrium f_{O_2} values that correspond approximately to the FMQ buffer. However, oceanic basalts extend to higher oxidation states than continental basalts at magmatic temperatures, approaching that of the MW univariant curve. On equilibration at lower temperatures (700–900° C), oceanic basalts span a narrow band around FMQ, whereas continental basalts are typically more reduced than FMQ. These data are in broad agreement with volcanic gas determinations at high temperatures, so it is concluded that the gases and the magmas are close to equilibrium. Gases from the Makaopuhi lava lake closely model oceanic basalt equilibration trends, whereas the effusive gases from Surtsey are comparable to levels of oxidation experienced by continental basalts. Sulfur loss and the consequent relative reduction that takes place in thin, subaerially extruded basalts contrasts with the extensive hydrogen diffusion that occurs in thick basalt flows and in ponded lava lakes. At magmatic temperatures, intrusive mafic suites have f_{O_2} profiles which are remarkably similar to those of extrusive suites, so that relative to one of the standard buffer curves (e.g., FMQ or MW) it is reasonable to assume, to a first approximation, that basalts are qualitatively representative of the f_{O_2} conditions at which they originated, provided of course that the effects of subsolidus oxidation or reduction are taken into account. Whether it is the gas that buffers the magma or the magma that buffers the gas remains unanswered. It would appear that the magma buffers the gas at high temperatures, in agreement with Gerlach and Nordlie (1957a, b, c), but that the reverse applies under slow cooling conditions and at lower temperatures.

1.3.4.4 Lunar basalts

The mineralogy of lunar basalts is characterized in all cases by the presence of metallic Fe and Ni-Fe alloys. To a first approximation, therefore, crystallization f_{O_2} conditions must be below the iron-wüstite (IW) buffer curve. In addition, mineral chemistry and spectral analyses indicate that Cr^{2+} and Ti^{3+} probably are present. Ferric iron is absent, so phases such as magnetite and hematite do not form. Lunar basalts are anhydrous but S, H_2, CO_2 and CO are released during heating experiments.

Intrinsic f_{O_2} determinations on a variety of basalts from the Apollo missions (Sato and Helz, 1971; Sato *et al.*, 1973) show that the ranges of temperature and f_{O_2} are relatively small, varying from 10^{-13} atms at 1200° C to 10^{-16} atms at 1000° C (Fig. 1.3.4.5). The range reported by Wellman (1970) at 1123° C and $f_{O_2} = 10^{-13.1}$

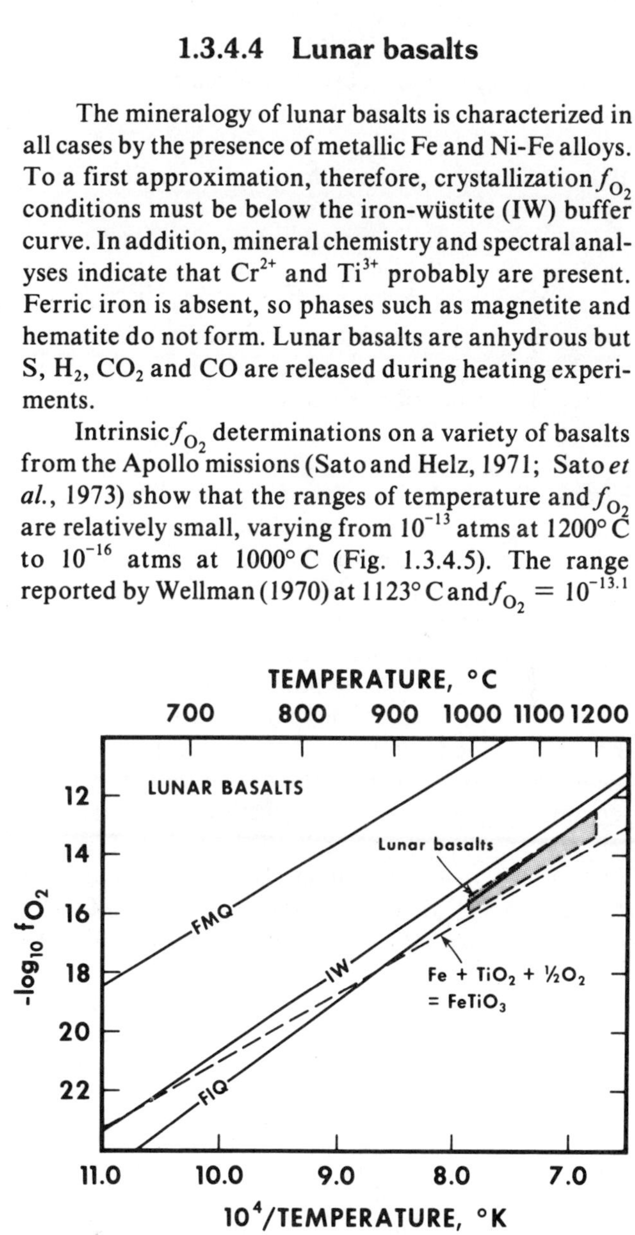

1.3.4.5 Temperature and oxygen fugacity data for lunar basalts obtained by the intrinsic f_{O_2} method (Sato *et al.*, 1973).

to $10^{-13.7}$, based on thermodynamic gas equilibria calculations, is in close agreement with the values determined experimentally by Sato and co-authors. Relative to standard buffer curves, these values are approximately one order of magnitude above fayalite-iron-quartz (FIQ), and one order of magnitude below IW (Fig. 1.3.4.5).

The lower thermal stability limits, as functions of f_{O_2}, for Fe^{2+}-bearing oxides and silicates (e.g., ulvöspinel, ilmenite, armalcolite and olivine) confirm that most lunar basalts crystallize within the interval of f_{O_2} defined by the IW and FIQ buffer curves. However, some lunar basalts, notably those from the Apollo 14 site, do show evidence of subsolidus reduction (El Goresy *et al.*, 1971; Haggerty, 1971, 1972) in which ulvöspinel decomposes to ilmenite + metallic Fe, and fayalite breaks down to cristobalite + iron.

These reactions take place within a limited range of f_{O_2}, that is below IW and below FIQ. Reduction cannot have taken place at an f_{O_2} much below that of the univariant curve that defines the stability of ilmenite which was determined experimentally by Taylor *et al.* (1972) and calculated by Haggerty (1971) and Nash and Hansel (1973). This constraint is imposed because ilmenite is a byproduct of ulvöspinel reduction and because discrete ilmenite is a stable, coexisting mineral phase. Of the variety of mechanisms proposed for reduction (meteoritic impact, volatilization, solar wind hydrogen implantation), the most plausible involve the presence of high concentrations of hydrogen, high H_2:CO(+ CO_2) ratios or of elemental carbon in association with iron carbonyl compounds (Sato, 1976, 1978a) in the initial magma. Although the nature of the reducing agent remains unknown, there is evidence from metal chemistry and from the distribution of troilite (commonly present in the reduction products) that reduction was active during the entire period of crystallization and deuteric cooling (Haggerty, 1977). The deuteric model yields reduction conditions of 700–1000° C and corresponding f_{O_2} values between 10^{-23} and 10^{-16} atms.

Lunar basalts crystallized at oxygen fugacities which were 4 to 5 orders of magnitude lower than those of terrestrial basalts at equivalent temperatures. However, upper and lower deuteric cooling limits are defined by high oxidation zones (equivalent to or greater than the MH buffer curve) in terrestrial basalts and lava lakes (Fig. 1.3.4.4), and by the ilmenite-iron-rutile curve in lunar basalts. The difference of 4 to 5 orders of magnitude f_{O_2} for *initial* crystallization is extended to a *subsolidus* difference that now spans 10 orders of magnitude in redox potential. The only known exceptions to this extreme difference are basalts and dolerite dikes from Disko Island, Greenland, where the mafic magmas incorporated carbonaceous shales. Metallic iron is present in the Disko rocks in association with immiscible sulfides and estimates at 1200° C yield f_{O_2} ~10^{-12} to 10^{-13} atms, and f_{S_2} ~10^{-5} atms (Pedersen, 1979).

Although no direct measurements of the redox state of the lunar crust are available, there is permissive evidence that the highly reduced state of lunar basalts is an intrinsic property of the Moon's interior. This is based on the observation that metallic iron is present in intrusive rocks (gabbros, troctolites, norites and dunites) that bear no indications of subsolidus reduction. Recall that the oxidation states of terrestrial basalts and their intrusive equivalents are, on the average, in close proximity to the FMQ buffer curve, and that the equilibrium redox condition is modified only slightly by surface extrusion or by crystallization at depth. If the same is true of lunar basalts and gabbros, then we must conclude that the differences in oxidation potential between lunar and terrestrial basalts are inherent properties of their respective magmas and of their coexisting volatile assemblages.

1.3.4.5 Basaltic achondrites

Stolper *et al.* (1979) have presented a petrogenetic model for achondritic meteorites in which the basaltic achondrites (eucrites, howardites, diogenites, and mesosiderites), the shergottites, the nakhlites, and the chassignites appear to define a coherent geochemical sequence. The sequence from basaltic achondrites to the chassignites is related in terms of an increase in alkali contents and in terms of a progressive increase in volatiles. They concluded that each of these meteorite types could have been derived by low-pressure partial melting of several plagioclase peridotites, each of which had a different composition.

Initial estimates of the f_{O_2} conditions under which these meteorites crystallized can be obtained from their respective mineralogies. Basaltic achondrites are characterized by metallic Fe, which places them in close proximity to the field of lunar basalts (Fig. 1.3.4.5) that is, below the IW buffer curve (Fig. 1.3.4.6). Eucritic meteorites are texturally, compositionally, and mineralogically similar to lunar basalts. The field of temperatures and oxygen fugacities shown in Fig. 1.3.4.6 is the estimated range based on these similarities.

The diogenites are cumulates (usually orthopyroxenites), but are also considered to be closely related to eucrites. These meteorites contain metallic iron coexisting with minute amounts of titaniferous magnetite and hercynite. An intrinsic f_{O_2} measurement on Angra dos

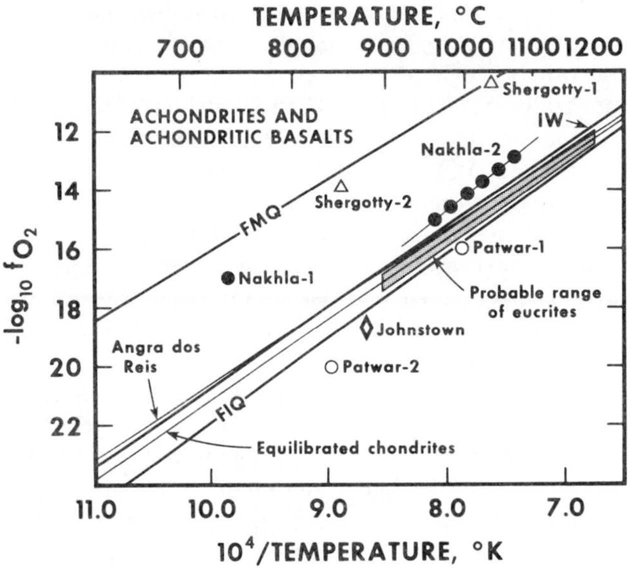

1.3.4.6 Temperature and oxygen fugacity data for achondrites. The probable field for achondritic basalts is based on the presence of metallic iron and the similarity of eucrites to lunar basalts. Angra dos Reis (Brett *et al.*, 1977); equilibrated chondrites (Williams, 1971b); Nakhla 1 (Reid and Bunch, 1975); Nakhla 2 (Delano and Arculus, 1980); Shergotty 1 (McSween and Stolper, 1978); Shergotty 2 (Smith and Hervig, 1979); Johnstown diogenite (Gooley and Moore, 1976); and Patwar 1 and 2 (Snellenburg *et al.*, 1979).

Reis (a clinopyroxenite) yields a T-f_{O_2} path that is approximately 0.2 log units below that of the IW buffer curve at high temperatures (T>800° C), and approximately 0.2 log units above IW at lower temperatures (Fig. 1.3.4.6). In their study of the Johnstown diogenite, which is characterized by metallic Fe + SiO_2 that has evidently resulted by subsolidus reduction of pyroxene, Gooley and Moore (1976) calculated that $f_{O_2} = 10^{-18.5}$ atm (Fig. 1.3.4.6) at a selected temperature of 850° C (based on data from Williams, 1971a). This value lies below that of Angra dos Reis and below that of the FIQ buffer, which is consistent with the meteorite's having experienced an episode of intense reduction.

Howardites and mesosiderites are complex breccias composed largely of eucrite and diogenite clasts, but the mesosiderites are dominated by metallic Fe (Stolper, 1977; Floran, 1979), which suggests that they also lie below IW. Using a model experimental system (olivine-pyroxene-silica-iron-liquid) and observing the presence of ilmenite + rutile, Snellenburg *et al.* (1979) estimated equilibration to have occurred at $f_{O_2} = 10^{-16}$ atms at ~1000° C for matrix pyroxene compositions (Patwar -1, Fig. 1.3.4.6), and 10^{-20} atms at 840° C for olivine corona assemblages (Patwar -2, Fig. 1.3.4.6) in Patwar, a mesosiderite. Both of these estimates are below both IW and FIQ.

The shergottites are similar to terrestrial dolerites, texturally and compositionally, but both of the meteorites in this class that have been recognized are intensely shocked. Their similarity to terrestrial basic rocks extends to equilibration temperatures and levels of f_{O_2} (Stolper and McSween, 1979). Using coexisting titanomagnetite and ilmenite solid solution members, two determinations have been made for Shergotty by McSween and Stolper (1978) and by Smith and Hervig (1979): 1042° C at $10^{-10.2}$ atms and ~850° C at 10^{-14} atms, respectively. The former of these determinations lies precisely on the FMQ buffer curve, whereas the latter is ~0.4 log f_{O_2} units below FMQ (Fig. 1.3.4.6).

Nakhlites are cumulates dominated by augite and olivine with subordinate feldspar and coexisting Fe-Ti oxides. Minor concentrations of apatite, pyrite, chalcopyrite and K-rich glass are also present (Bunch and Reid, 1975; Boctor *et al.*, 1976; Berkley *et al.*, 1979). Some of the olivine has been altered to iddingsite (smectite + hydrated Fe oxide) which may be indicative of extraterrestrial water. Because of the low temperature thermal stability of iddingsite, oxidation and hydration probably took place at T<500° C (Baker and Haggerty, 1967). In addition to the oxidation of olivine, there is also oxidation of titanomagnetite, which is manifested by the formation of ilmenite lamellae oriented along {111} planes in the spinel host. Such oxidation occurred in the subsolidus (Buddington and Lindsley, 1964), during deuteric cooling (i.e., T<1000° C) or during a post-crystallization oxidation event (Haggerty, 1976a). Reid and Bunch (1975) have estimated that Nakhla equilibrated at ~740° C and 10–17 atms, which is 1.2 orders of magnitude f_{O_2} below that of FMQ. This estimate, coupled with the presence of iddingsite and the formation of ilmenite in spinel, is in marked contrast to the intrinsic f_{O_2} determination reported for Nakhla by Delano and Arculus (1980). Their values are approximately 3 orders of magnitude more reducing than FMQ and 1 order of magnitude more oxidized than MW, in the temperature range 956–1061° C. Although these determinations are not consistent, the fact that Nakhla has been oxidized may make both the oxide geothermometer-oxygen geobarometer and the intrinsic f_{O_2} measurement open to question. It should, however, be noted that, if only the Reid and Bunch (1975) values are considered in association with Shergotty (Fig. 1.3.4.6), the deviation expected from FMQ on cooling is comparable to that of continental terrestrial basalts at low temperatures (i.e., at T<800° C) as illustrated in Fig. 1.3.4.1b.

The chassignites, in common with nakhlites, are cumulates dominated by olivine in association with augite, but are distinctive in that they contain kaersutite, a hydrous amphibole (Floran *et al.*, 1978). Plagioclase and alkali feldspars are also present. Spinel contains a large magnetite component. The large Fe^{3+} component and the presence of primary kaersutite may imply crystallization f_{O_2} values that are more oxidized than those of the FMQ buffer (Stolper *et al.*, 1979), but no direct determinations have yet been reported.

In summary, we conclude that there are major differences in the oxidation states of meteoritic achondrites. Eucrites have affinities that closely parallel those of lunar basalts, characterized by metallic iron with crystallization f_{O_2} below that of the IW buffer curve. The crystallization paths of some diogenites (e.g., Angra dos Reis) are equivalent to those calculated for equilibrated chondrites by Williams (1971b) and also lie below IW (Fig. 1.3.4.6). The Johnstown diogenite has undergone subsolidus reduction, with an f_{O_2} below the FIQ buffer curve. The Patwar mesosiderite has equilibration T and f_{O_2} that are approximately one order of magnitude below FIQ. Shergottites, nakhlites, and chassignites, however, have levels of oxidation similar to those of terrestrial basalts. Although petrological and geochemical continua exist among these meteorites, the general consensus is that the more reduced basaltic achondrites are derived from one parent body, and that the more oxidized suites have evolved from a different parent body.

1.3.4.6 The redox state of basalt source regions

The following questions are among the most important raised by the f_{O_2} data on planetary basalts:

(1) Why are planetary basalts at their respective f_{O_2} levels?

(2) Do these levels reflect the f_{O_2} conditions of their source regions?

(3) If the source regions differ from surface conditions, how are basalts modified during ascent to the surface?

The fundamental question is whether basalts are adequate probes of the f_{O_2} conditions at which they originate. Recall that terrestrial basalts have equilibrium crystallization f_{O_2} values (observed) that lie close to FMQ (e.g., 10^{-8} atms at 1200° C) or which lie between FMQ and MW at lower temperatures (e.g., 10^{-15} atms at 800° C). Yet these basalts are extruded into a highly oxidizing atmosphere under conditions where the f_{O_2} is $10^{-0.6}$ atms. In addition, it has been shown that mafic magmas, cooling under intrusive environments, have f_{O_2} values which are approximately equivalent to those of basalts. Although it has been demonstrated that volatile loss or retention is influenced by a variety of factors (section 1.3.5), the inevitable conclusion must be that loss or retention of volatiles has very little effect on the *primary* magmatic redox state of basalts. A clear endorsement of this supposition is the similarity in average magmatic f_{O_2} values between basalts which are extruded on oceanic islands and those which evolve on the ocean floors. Nevertheless, if we can independently demonstrate that source regions have f_{O_2} values comparable to those of the basalts, then similar relationships can be assumed in evaluating the mantles of other planets. Using the Earth as a model, it is indeed concluded that basalts retain their f_{O_2} signatures on extrusion, but that these signatures may be modified if cooling is insufficiently rapid.

Some insights may be gained into the levels of f_{O_2} that characterize the Earth's mantle by considering gas and fluid inclusions in diamonds, similar types of inclusions in ultramafic nodules, and f_{O_2} equilibria derived experimentally under mantle conditions.

Comprehensive studies of gases released from crushed diamonds have been undertaken by Giardini and Melton (1975) and Melton and Giardini (1975). They have shown that H_2O(33.3%), H_2(23.9%), CO_2(19.5%), N_2(8.2%), CH_4(7.6%), alcohols (0.6%), Ar(0.3%), other hydrocarbons (0.2%) and O_2(0.04%) typically are present. The relatively large concentrations of hydrogen and methane are indicative of highly reduced conditions but, because the proportions of H_2O and CO_2 are also relatively large, the overall state of oxidation is somewhat higher. Isotopic data on ^{13}C measured by Deines (1980) in carbon gas species liberated from diamonds show that a range of ~30 ‰ is present. Deines concluded that, among the various models (58 in total) tested to determine whether or not the vapor phase (i.e., H_2O, H_2, N_2, O_2, CH_4) participated in or was influenced by the growth of diamond, no unambiguous answer could be found. However, he concluded that the isotopic form of carbon does not play a major role and that the large variations in ^{13}C are related to carbon in the source. Although ~30 ‰ is an enormous range, it may only be representative of some diamonds; kimberlites and carbonatites typically contain $\delta 13C$ between −2 and −10‰. Trace amounts of ^{13}C in basalts are approximately −25‰, but values in the range of kimberlites have been reported for "popping" tholeiitic rocks (Hekinian *et al.*, 1973) from the Mid-Atlantic Ridge by Pineau *et al.* (1976). Concentrations of −7.6 ‰ are present in the latter, and the authors concluded that the $\delta^{13}C$ content is equivalent to that of the mantle from which these basalts originated.

The gaseous species of carbon (either CO_2 or CO) have long been recognized as major components of kimberlites and carbonatites. Support for the presence of such species in the mantle also is obtained from fluid inclusions in mantle-derived nodule suites. Green and Radcliffe (1975) and Murck *et al.* (1978) have studied the fluid precipitates in dunites, peridotites, and pyroxenite xenoliths from alkali basalts and kimberlites. They have shown, in support of the earlier work by Roedder (1965), that CO_2-rich fluid inclusions are ubiquitous and that most such inclusions are very nearly pure CO_2 but that minor concentrations of SO_2, H_2S, or CO may also be present. Another significant factor is that CO_2 is also present in olivine phenocrysts in basalts (Roedder, 1965). An intrinsic f_{O_2} determination was performed on olivine from a dunite in the 1801 Hualalai alkali basalt in Hawaii by Sato (1978b). The heating and cooling curves for Sato's determination lie, respectively, above and below the iron-wüstite (IW) buffer curve, a range which is well below those of terrestrial basalts or terrestrial volcanic gas determinations (Figs. 1.3.4.1–2). Sato (1978b) notes that CO_2 alone is insufficient to impose this reduced state of oxidation, and has suggested that elemental carbon must also be present. Credence to the proposition that carbon may

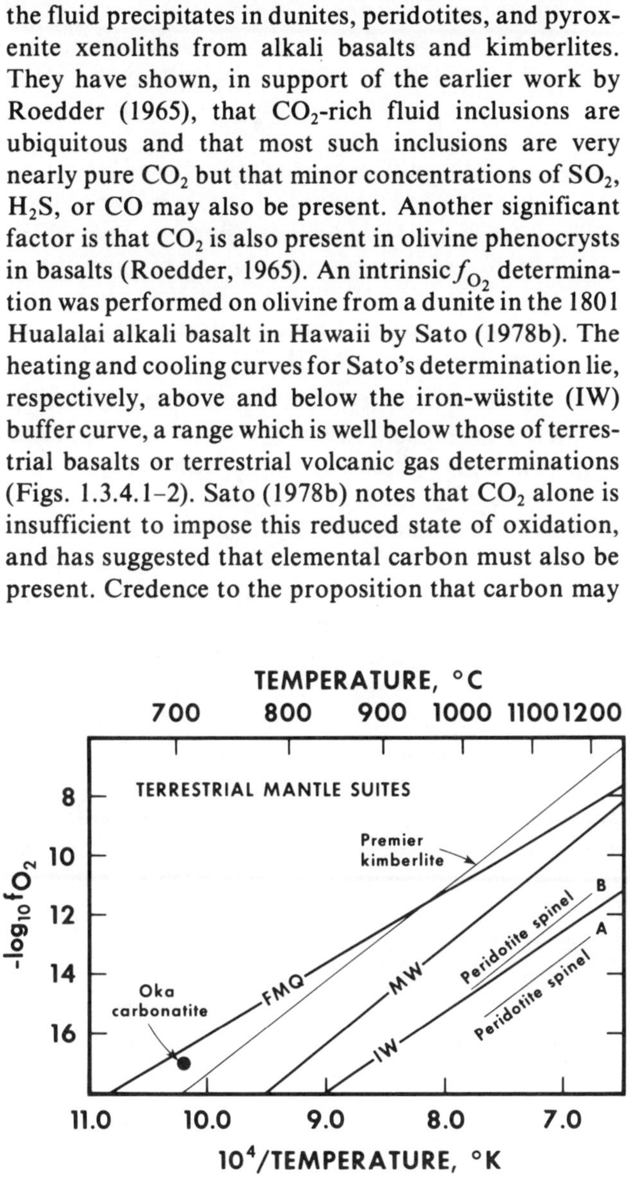

1.3.4.7 Temperature and oxygen fugacity data for terrestrial mantle suites. Data are from the following sources: Premier kimberlite (Ulmer *et al.*, 1976); Oka carbonatite (Friel and Ulmer, 1974); peridotite spinel (Arculus and Delano, 1980).

be a significant component of mantle olivine may be derived from the study by Freund *et al.* (1980) who have shown, first, that elemental carbon can indeed form in solid solution with forsterite and, second, that elemental carbon may also exsolve from olivine on cooling. Other intrinsic f_{O_2} determinations performed on spinels from peridotites (Arculus and Delano, 1980) lie either just above or just below the IW buffer (Fig. 1.3.4.7), thus supporting Sato's (1978b) data for olivine. Whether elemental carbon is also present in these spinels has not been determined. Apart from elemental carbon, it is significant to note that methane has been observed to be the predominant hydrocarbon in mantle-derived amphiboles, in association with ethane, ethylene, acetylene, propane, water and CO_2 (Garcia *et al.*, 1980). Intrinsic f_{O_2} determinations have also been made on a kimberlite (Ulmer *et al.*, 1976) and a carbonatite (Friel and Ulmer, 1974), which are plotted in Fig. 1.3.4.7. The kimberlite is more oxidized than FMQ, at high temperatures (>950 °C) but it is more reduced than FMQ at T<950°C. The carbonatite equilibrated at $f_{O_2} = 10^{-17.1}$ and T = 710°C.

The xenolith data from alkali basalts and kimberlites imply that the mantle is in a highly reduced state, approaching that of IW. However, the kimberlite and carbonatite data are remarkably similar to those of terrestrial basalts. The questions now are: What is the redox profile of the mantle, and what is the redox state in that region of the mantle in which basalts are generated?

An intrinsic property of any f_{O_2} buffer is that f_{O_2} increases with pressure in a manner rather similar to the way in which f_{O_2} increases with temperature. The effects of pressure, however, are far less significant than those of temperature. If we consider simple extrapolations of the low pressure buffer equations for FMQ and MW, selected because of the range of f_{O_2} values exhibited in terrestrial basalts and volcanic gases, we find that upper mantle f_{O_2} values, in the depth range of 125–175 km (40–45 kb) and at 1200° C, are $>10^{-6}$ atm for MW and approximately 10^{-6} atm for FMQ (Fig. 1.3.4.8). The f_{O_2} values for both of these buffers decrease systematically with decreasing pressure and temperature. Extrapolation to these high pressures is not entirely valid, because measurements of the buffers are limited to P<5kb and because other gas species have not been considered. Buffers involving the carbon gas species, in particular CO and CO_2 either in gaseous form or in solid carbonates (e.g., magnesite, $MgCO_3$) have been proposed to be viable and more practical alternatives to the more conventional 1 atm buffers when mantle environments are being evaluated (Woermann *et al.*, 1977; Sato, 1978b;

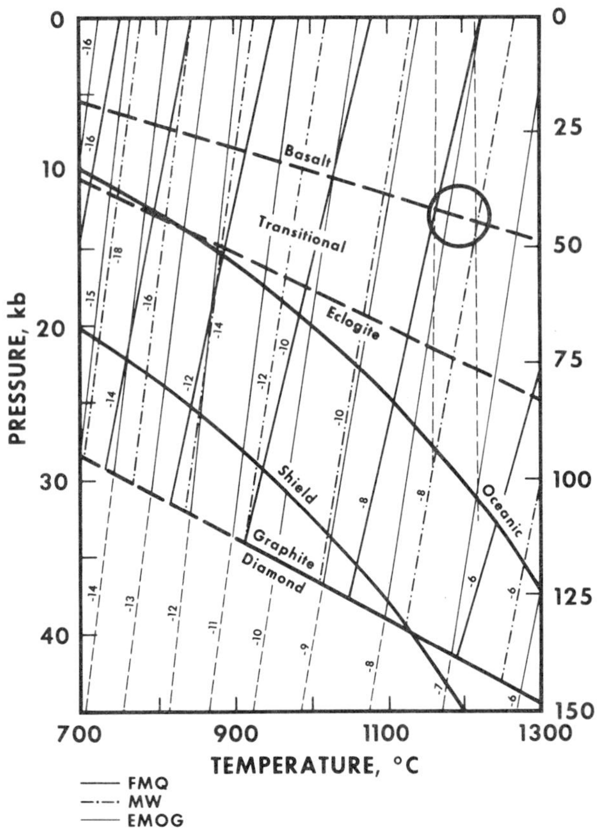

1.3.4.8 Oxygen buffer curves (indicated by negative values and expressed as $-\log_{10} f_{O_2}$) plotted as functions of temperature and pressure. Data are calculated from the following sources: FMQ (Wones and Gilbert, 1969); MW (Eugster and Wones, 1962); and EMOG (Eggler *et al.*, 1980). Values for EMOG are extended into the diamond stability field but do not represent EMOD; these extensions are merely for the purposes of clarity. A significant point to be noted on the diagram is that EMOG is more reducing than MW (e.g., 125 km and 1300°C), is intermediate in redox between MW and FMQ (e.g., 100 km and 1050°C), and is more oxidized than either FMQ or MW (e.g., 50 km and 780°C). The vertical dashed lines indicate the maximum temperatures obtained for terrestrial basalts, and the circle encompasses f_{O_2} isobars at 10^{-8} atms, a value close to that obtained for coequilibrated oxides in terrestrial basalts.

Eggler *et al.*, 1980). The data of Eggler *et al.* (1980) have been selected for inclusion in Fig. 1.3.4.8. In their study the f_{O_2} of the reaction

enstatite + magnesite ⇌ forserite + graphite + O_2

is defined by the P-T intersection of two reactions involving C-O fluids

enstatite + magnesite + graphite ⇌ forsterite + vapor

magnetite + graphite ⇌ wustite + vapor.

Entropy and molar volume data allow calculation of EMOG (En + Mag + O_2 + graphite) or EMOD (where D is diamond rather than graphite). For EMOG

$$f_{O_2} = 7.83 - \frac{23781}{T} + \frac{0.54(p-1)}{T}$$

Using this equation, f_{O_2} isobars have been calculated as functions of P and T (Fig. 1.3.4.8). Several noteworthy features are present:

(1) At high P(>40 kb) and T(>1200° C), EMOG is more reducing than either FMQ or MW.
(2) However, at intermediate P and T, EMOG lies between FMQ and MW.
(3) At low P and T, EMOG approaches FMQ and ultimately becomes more oxidizing than FMQ.
(4) At 1200° C, EMOG has an f_{O_2} equivalent to 10^{-7} atms.
(5) All of the buffers have relatively steep slopes, so that for a small range in T, an f_{O_2} isobar spans a large variation in total pressure.
(6) In spite of the apparently invalid extrapolations of FMQ and MW to high pressures, these data are *not* markedly different from those of the preferred values for EMOG.

Additional data pertinent to basalt genesis are also shown in Fig. 1.3.4.8, including the oceanic and shield geotherms and the basalt-eclogite transition. Assuming that basalts originate from depths on the order of 50–125 km, and assuming a temperature of 1300° C, basalts originating at greater depths will have f_{O_2} equivalents in which EMOG is more reduced or approaches that of MW. However, basalts generated at higher levels will have EMOG f_{O_2} equivalents that lie between FMQ and MW (Fig. 1.3.4.8). Referring back to the f_{O_2} determinations for terrestrial basalts (Figs. 1.3.4.1a, b) and for volcanic gas emanations (Fig. 1.3.4.2), we find that at 1200° C, for example, the corresponding range in f_{O_2} is between $10^{-7.5}$ and $10^{-9.0}$ atms. These absolute extremes lie just above FMQ and at MW. Considering the oceanic and the continental suites separately, average f_{O_2} levels are close to FMQ for the former, and between FMQ and MW for the latter. Taking the two extremes for the depths considered, magmas generated at shallower levels should be and are relatively more oxidized than those produced at greater depths (i.e., with respect to FMQ and MW) as illustrated in Fig. 1.3.4.8 and using EMOG as the reference. Thus it would appear that basalts do indeed reflect the f_{O_2} conditions under which they originate and, hence, because oceanic basaltic source regions are shallower than those giving rise to basalts beneath the continents, the latter basalts are expected to be more reduced. Admittedly, the correspondence in absolute f_{O_2} terms is close but not perfect; however, in a relative sense the match is excellent.

In an earlier discussion of deuteric oxidation in single lava flows and in the Makaopuhi lava lake it was noted that oxidation within these cooling units is best explained in terms of Sato and Wright's (1966) *semi-permeable membrane* model in which hydrogen loss plays a significant role. Sato (1978a) has argued that hydrogen loss probably is an important factor during basalt ascent. Although this would appear to be a reasonable assumption, the implied levels of f_{O_2} based on FMQ, MW, and EMOG (Fig. 1.3.4.8) clearly do not support it. Hydrogen loss undoubtedly takes place but it cannot be significant.

In summary, assuming that EMOG or its equivalent EMOD is a reasonable estimate of f_{O_2} conditions in the Earth's upper mantle, we may conclude that EMOG is more reduced than MW at T>1200° C and P>45kb. However, with decreasing P and T, EMOG lies between MW and FMQ; at low P and T, EMOG approaches FMQ and ultimately becomes more oxidized than FMQ. Within the region of basalt genesis (50–125 km), the level of f_{O_2} is remarkably similar to the measurements obtained on basalts and on gas emanations. The presence of elemental carbon and hydrocarbons in mantle-derived suites is consistent with a more reduced upper mantle, but a level much more reduced than MW appears to be unlikely. The intrinsic f_{O_2} values equivalent to IW measured for olivine and for spinel in peridotites are of considerable interest, but whether these values are anomalous or are artifacts of the method of analysis has yet to be resolved.

With respect to terrestrial basalts, the overall conclusion is that their source regions have f_{O_2} equivalents that lie between FMQ and MW and, therefore, that basalts which are extruded at the Earth's surface qualitatively reflect the conditions in their source regions. Lunar basalts, dunites, norites, troctolites and gabbros contain metallic iron. It is assumed, therefore, that the source region is highly reduced and that the level of f_{O_2} in the Moon's interior is below that of the IW buffer. It is concluded that, because basaltic achondrites contain metallic iron and because cumulate diogenites have high temperature f_{O_2} values that are below IW, the asteroidal mantle(s) that gave rise to these meteorites had a redox potential rather similar to that of the Moon. The more highly oxidized achondrites have f_{O_2} levels comparable to or greater than those of the Earth, so their source regions may also have been in an equivalent state of oxidation.

1.3.4.7 Inner solar system basalts

Based on the conclusions reached above and in order to arrive at some reasonable estimates of the redox states of basalts on Mercury, Venus and Mars, a variety of cosmogenic properties have been evaluated along the lines of reasoning suggested by Ringwood (1966), Lewis (1973), Grossman and Larimer (1974), and Anders and Owen (1977). These properties are:

(1) the primordial condensation sequences of minerals condensing in the early solar nebula;
(2) bulk planet densities, moments of inertia and planet sizes;
(3) thermal evolutionary models (Toksöz and Johnston, 1977);
(4) planetary interiors;
(5) the ratios of Fe^0/Fe^{2+} in planetary cores, and the proportions of Fe^0, Fe^{2+} and Fe^{3+} in model planetary crusts;
(6) planetary atmospheres.

The condensation of Mercury in close proximity to the proto-sun suggests that the planet is very likely to have been devoid of volatiles, and it certainly is true that it now has no atmosphere. Mercury has a magnetic field, a bulk density of 5.4 and an inferred Fe-Ni core which is 1800 km in diameter. The core is considered to constitute approximately 80% of the planet's mass. The ratio of refractory silicates to metal alloys is calculated to be 37:63. Venus is approximately the same size, mass and density as the Earth, and since the planet formed in the same region of the condensing solar nebula as the Earth, is assumed to have a similar composition and to have differentiated to form a core, a mantle, and a rocky crust. Venus has a density of 5.2, a silicate to metal alloy ratio of 74:26, and an inferred liquid core, but no magnetic field because of the planet's slow rotational rate. The atmosphere of Venus is largely CO_2 and, because the planet condensed closer to the proto-sun, it probably would have had a lower bulk volatile content than the Earth. Mars is now recognized as being the most highly oxidized of the terrestrial planets, with a bulk density of 3.93, a silicate to metal alloy ratio of 81:19, a (postulated) high pressure magnetite core (Ringwood and Clark, 1971), and a CO_2-rich atmosphere.

If the densities of the terrestrial planets are taken as indications of the relative proportions of silicates to metal alloys, with average densities of 3.3 and 7.9 g/cm^3 at 10 kb, respectively (Ringwood, 1966), and if the condensation sequence from the proto-sun (Mercury, Venus, Earth, Mars) is an accurate predictor of each planet's primordial volatile content, and of water content in particular, then the expected crustal oxygen fugacities of the planets should form an orderly sequence with increasing distance from the sun. The intrinsic redox state of each planet will influence the partitioning of the valency states of iron (Fe^0, Fe^{2+}, Fe^{3+}) among metals, silicates and oxides so that, at low f_{O_2} levels, we may expect large metal abundances and small concentrations of Fe^{2+}-bearing silicates, as is probably the case for the Earth's *lower* mantle. Similarly, at high oxygen fugacities, the amount of free metal alloys should be proportionately smaller and the proportion of Fe^{2+} in silicates and oxides proportionately larger. Core fractionation and volatile degassing, however, are influenced by planetary size and thermal history as well, and there is circumstantial evidence which suggests that the smaller the planet the greater is the incidence of volatile loss (Anders and Owen, 1977). Although these factors compromise the accuracy of estimates of basaltic redox states, an important result suggested by terrestrial and lunar mafic rocks is that the redox states of basalts are not depth dependent. Intrinsic levels of f_{O_2} of surface basalt samples are similar to those of the upper mantle source regions which produced those basalts.

Based on the six cosmogenic properties identified above, the following ranges of redox states appear to be reasonable for the terrestrial planets:

(1) Data for lunar basalts indicate that they crystallized at f_{O_2} levels below the IW buffer curve (Fig. 1.3.4.5). Basalts on Mercury probably have comparable or lower f_{O_2} values.

(2) Venusian basalts should be more oxidized than those on Mercury, but less oxidized than those on Earth. Crystallization of venusian basalts probably occurred in the wüstite stability field.

(3) Basalts on Earth crystallize along the FMQ buffer curve (Fig. 1.3.4.1), above the wüstite stability field and in the field of magnetite stability.

(4) Basalts on Mars should crystallize in the uppermost region of the magnetite stability field and well into the hematite field in terms of temperature and f_{O_2} relationships.

1.3.4.8 Summary

The redox states of planetary basalts are controlled by liquid-solid-gas equilibria. On the Earth, the oxidation state of basalts is approximately equivalent to that of the FMQ buffer at high temperatures and lies between FMQ and MW at lower temperatures. However, there are apparent differences in the overall states of oxidation at magmatic temperatures between basalts originating in oceanic environments (ocean floor and oceanic island) and those extruded on the continents. The latter basalts are more reduced and the former are more oxidized. It is concluded that volatile losses alone cannot account for these distinctions but, if the depths of origin are considered, basalts originating at greater depths (i.e., continental basalts) are likely to be more reduced, relative to standard buffers, than those that are generated at shallower levels in the upper mantle. On equilibration at lower temperatures, oceanic basalts are rigorously controlled to levels of f_{O_2} that correspond to FMQ, whereas continental basalts lie between an f_{O_2} range defined by FMQ and MW. On prolonged cooling of thick plateau basalts, or of ponded lava lakes, the dissociation of water produces high states of oxidation, equivalent to or greater than MH, in the interior portion of the magma body. It is considered that the magma buffers the gas at high temperatures, but that at lower temperatures, in the subsolidus region, the gas buffers the crystalline basalt.

Lunar basalts and basaltic achondrites have equilibrium crystallization f_{O_2} values equivalent to conditions below the IW buffer curve. Subsolidus reduction of some lunar basalts and diogenites has equilibrated these rocks at levels of f_{O_2} that are below IW and below FIQ.

By modelling the f_{O_2} conditions of basaltic source regions on the Earth, it is concluded that basalts qualitatively reflect the redox potentials of the environments in which they originate. Similarly, it is concluded that the interior of the Moon has an f_{O_2} level equivalent to conditions more reduced than IW, which is true also of the asteroidal mantle(s) that gave rise to the suites of meteoritic basaltic achondrites. However, the mantle(s) that generated the more oxidized shergottites, nakhlites, and chassignites have f_{O_2} levels that are equal to or greater than the mantle of the Earth.

If the current views of cosmological condensation models are accepted, then redox states of inner solar system basalts are considered to fall into the following stability fields in terms of temperature and oxygen fugacity: basalts on the Moon and Mercury are in the metallic iron field, while those on Venus are in the wüstite field. Basalts on Earth are in the magnetite field, and basalts on Mars are in the hematite field. Estimates for the Earth and the Moon have high confidence levels, those predicted for Mars are of a second-order level, and those for Mercury and Venus are speculative. A perplexing problem is that some basaltic achondrites have oxidation potentials which are similar to those of lunar basalts, while others have characteristically high f_{O_2} potentials equivalent to those of terrestrial basalts, yet these achondrites form an apparently coherent geochemical sequence.

1.3.4.9 References

Anders E. and Owen T. (1977) Mars and Earth: origin and abundance of volatiles. *Science* **198**, 453–465.

Arculus R. J. and Delano J. W. (1980) Intrinsic oxygen fugacities of the Earth's mantle: a new approach and results for spinel peridotites (abstract). In

Lunar and Planetary Science XI, p. 28–30. Lunar and Planetary Institute, Houston.

Baker I. and Haggerty S. E. (1967) The alteration of olivine in basaltic and associated lavas, part II: intermediate and low temperature alteration. *Contrib. Mineral. Petrol.* **16**, 258–273.

Berkley J. L., Keil K., Prinz M. and Gomes C. B. (1979) The Governador Valadares nakhlite and its relationship to other nakhlites (abstract). In *Lunar and Planetary Science X*, p. 101–103. Lunar and Planetary Institute, Houston.

Boctor N. Z., Meyer H. O. A. and Kullerud G. (1976) Lafayette meteorite: petrology and opaque mineralogy. *Earth Planet. Sci. Lett.* **32**, 69–76.

Brett R., Huebner J. S. and Sato M. (1977) Measured oxygen fugacities of the Angra Dos Reis achondrite as a function of temperature. *Earth Planet. Sci. Lett.* **35**, 363–368.

Buddington A. F. and Lindsley D. H. (1964) Iron titanium oxide minerals and synthetic equivalent. *J. Petrol.* **5**, 310–357.

Bunch T. E. and Reid A. M. (1975) The nakhlites part I: petrography and mineral chemistry. *Meteoritics* **10**, 303–315.

Chou I. -M. (1978) Calibration of oxygen buffers at elevated P and T using the hydrogen fugacity sensor. *Amer. Mineral.* **63**, 690–703.

Deines P. (1980) The carbon isotopic composition of diamonds. *EOS (Trans. Amer. Geophys. Union)* **61**, 394.

Delano J. W. and Arculus R. J. (1980) Nakhla: oxidation state and other constraints (abstract). In *Lunar and Planetary Science XI*, p. 219–221. Lunar and Planetary Institute, Houston.

Duke J. M. (1974) The effect of oxidation on the crystallization of an alkali basalt from the Azores. *J. Geol.* **82**, 524–528.

Eggler D. H., Baker D. R. and Wendlandt R. F. (1980) f_{O_2} of the assemblage graphite enstatite fosterite magnesite: experiment and application to mantle f_{O_2} and diamond formation (abstract). *Abstracts with Programs, Geol. Soc. Amer.* **12**, 420.

El Goresy A., Ramdohr P. and Taylor L. A. (1971) The geochemistry of the opaque minerals in Apollo 14 crystalline rocks. *Earth Planet. Sci. Lett.* **13**, 121–129.

Eugster H. P. and Wones D. R. (1962) Stability relations of the ferruginous biotite, annite. *J. Petrol.* **3**, 82–125.

Floran R. J. (1979) Silicate petrography, classification, and origin of the mesosiderites: review and new observations. *Proc. Lunar Planet. Sci. Conf. 9th*, p. 1053–1081.

Floran R. J., Prinz M., Hlava P. F., Keil K., Nehru C. E. and Hinthorne J. R. (1978) The Chassigny meteorite: a cumulate dunite with hydrous amphibole bearing melt inclusions. *Geochim. Cosmochim. Acta* **42**, 1213–1229.

Freund F., Kathrein H., Wengeler R. and Knobel R. (1980) Carbon in solid solution in forsterite—a key to the untractable nature of reduced carbon in terrestrial and cosmogenic rocks. *Geochim. Cosmochim. Acta* **44**, 1319–1333.

Friel J. J. and Ulmer G. C. (1974) Oxygen fugacity geothermometry of the Oka carbonatite. *Amer. Mineral.* **59**, 314–318.

Fudali R. F. (1965) Oxygen fugacities of basaltic and andesitic magmas. *Geochim. Cosmochim. Acta* **29**, 1063–1075.

Garcia M. O., Muenow D. W., Byers C. D. and Matson D. W. (1980) Hydrocarbons in mantle derived amphiboles, Grand Canyon area, Arizona (abstract). *Abstracts with Programs, Geol. Soc. Amer.* **12**, 430.

Gerlach T. M. (1979) Evaluation and restoration of the 1970 volcanic gas analyses from Mt. Etna, Sicily. *J. Volc. Geotherm. Res.* **6**, 165–178.

Gerlach T. M. (1980a) Evaluation of volcanic gas from Kilauea volcano. *J. Volc. Geotherm. Res.* **7**, 295–317.

Gerlach T. M. (1980b) Investigation of volcanic gas analyses and magma outgassing from Erta' Ale lava lake, Afar, Ethiopia. *J. Volc. Geotherm. Res.* **7**, 415–441.

Gerlach T. M. and Nordlie B. E. (1975a) The C-O-H-S gaseous system, part I: composition limits and trends in basaltic cases. *Amer. J. Sci.* **275**, 353–376.

Gerlach T. M. and Nordlie B. E. (1975b) The C-O-H-S gaseous system, part II: temperature, atomic composition, and molecular equilibria in volcanic gases. *Amer. J. Sci.* **275**, 377–394.

Gerlach T. M. and Nordlie B. E. (1975c) The C-O-H-S gaseous system, part III: magmatic gases compatible with oxides and sulfides in basaltic magmas. *Amer. J. Sci.* **275**, 395–410.

Giardini A. A. and Melton C. E. (1975) Gases released from natural and synthetic diamonds by crushing under high vacuum at 200°C, and their significance to diamond genesis. *Fortschr. Mineral.* **52**, 455–464.

Gooley R. and Moore C. B. (1976) Native metal in diogenite meteorites. *Amer. Mineral.* **61**, 373–378.

Green H. W., II and Radcliffe S. V. (1975) Fluid precipitates in rocks from the Earth's mantle. *Geol. Soc. Amer. Bull.* **86**, 846–852.

Grossman L. and Larimer J. W. (1974) Early chemical history of the solar system. *Rev. Geophys. Space Phys.* **12**, 71–101.

Haggerty S. E. (1971) Subsolidus reduction of lunar spinels. *Nat. Phys. Sci.* **234**, 113–117.

Haggerty S. E. (1972) Apollo 14: subsolidus reduction and compositional variations of spinels. *Proc. Lunar Sci. Conf. 3rd*, p. 305–332.

Haggerty S. E. (1976a) Oxidation of Opaque Mineral Oxides in Basalts. In *Oxide Minerals—Short Course Notes, Vol. 3* (D. Rumble, III, ed.), p. 1–100. Miner. Soc. Amer., Washington, D. C.

Haggerty S. E. (1976b) Opaque Mineral Oxides in Terrestrial Igneous Rocks. In *Oxide Minerals—Short Course Notes, Vol. 3* (D. Rumble, III, ed.), p. 101–301. Miner. Soc. Amer., Washington, D. C.

Haggerty S. E. (1977) Apollo 14: oxide, metal, and olivine mineral chemistries in 14072 with a bearing on the temporal relationships of subsolidus reduction. *Proc. Lunar Sci. Conf. 8th*, p. 1809–1829.

Haughton D. R., Roeder P. L. and Skinner B. J. (1974) Solubility of sulfur in mafic magmas. *Econ. Geol.* **69**, 451–467.

Hekinian R., Chaigneau M. and Cheminee J. L. (1973) Popping rocks and lava tubes from the Mid-Atlantic Rift Valley at 36° N. *Nature* **245**, 371–373.

Helz R. T. (1973) Phase relations of basalts in their melting range at PH_2O = 5kb as a function of oxygen fugacity. Part 1. mafic phases. *J. Petrol.* **14**, 249–302.

Katsura T. and Nagashima S. (1974) Solubility of sulfur in some magmas at 1 atmosphere. *Geochim. Cosmochim. Acta* **38**, 517–531.

Lewis J. S. (1973) Chemistry of the planets. *Ann. Rev. Phys. Chem.* **24**, 339–351.

MacLean W. H. (1969) Liquidus phase relations in the FeS-FeO-Fe_3O_4-SiO_2 system, and their application to geology. *Econ. Geol.* **64**, 865–884.

McSween H. Y. and Stolper E. M. (1978) Shergottite meteorites I: mineralogy and petrography (abstract). In *Lunar and Planetary Science IX*, p. 732–734. Lunar and Planetary Institute, Houston.

Melton C. E. and Giardini A. A. (1975) Experimental results and a theoretical interpretation of gaseous inclusions found in Arkansas natural diamonds. *Amer. Mineral.* **60**, 413–417.

Murck B. W., Burruss R. C. and Hollister L. S. (1978) Phase equilibria in fluid inclusions in ultramafic xenoliths. *Amer. Mineral.* **63**, 40–46.

Naldrett A. J. (1969) A portion of the system Fe-S-O between 900 and 1080° C and its application to sulfide ore magmas. *J. Petrol.* **10**, 171–201.

Nash W. P. and Hansel W. D. (1973) Partial pressures of oxygen, phosphorus and fluorine in some lunar lavas. *Earth Planet. Sci. Lett.* **20**, 13–27.

Peck D. L. and Wright T. L. (1966) Experimental studies of molten basalts *in-situ*: a summary of physical and chemical measurements on recent lava lakes of Kilauea volcano, Hawaii. *Geol. Soc. Amer., Annual Meeting 101*, 158.

Pedersen A. K. (1979) Basaltic glass with high-temperature equilibrated immiscible sulphide bodies with native iron from Disko, central west Greenland. *Contrib. Mineral. Petrol.* **69**, 397–407.

Pineau F., Javoy M. and Bottinga Y. (1976) $^{13}C^{12}C$ ratios of rocks and inclusions in popping rocks of the Mid-Atlantic Ridge and their bearing on the problem of isotopic composition of deep seated carbon. *Earth Planet. Sci. Lett.* **29**, 413–421.

Reid A. M. and Bunch T. E. (1975) The nakhlites, part II: where, when, and how. *Meteoritics* **10**, 317–324.

Ringwood A. E. (1966) Chemical evolution of the terrestrial planets. *Geochim. Cosmochim. Acta* **30**, 41–104.

Ringwood A. E. and Clark S. P. (1971) Internal constitution of Mars. *Nature* **234**, 89–92.

Roedder E. (1965) Liquid CO_2 inclusions in olivine-bearing nodules and phenocrysts from basalts. *Amer. Mineral.* **50**, 1746–1782.

Sato M. (1970) An electrochemical method of oxygen fugacity control of furnace atmostphere for mineral syntheses. *Amer. Mineral.* **55**, 1424–1437.

Sato M. (1972) Intrinsic oxygen fugacities of iron-bearing oxide and silicate minerals under low total pressure. *Mem., Geol. Soc. America* **135**, 289–307.

Sato M. (1976) Oxygen fugacity and other thermochemical parameters of Apollo 17 high-Ti basalts and their implications on the reduction mechanism. *Proc. Lunar Sci. Conf. 7th*, p. 1323–1344.

Sato M. (1978a) A possible role of carbon in characterizing the oxidation state of a planetary interior and originating a metallic core (abstract). In *Lunar and Planetary Science IX*, p. 990–992. Lunar and Planetary Institute, Houston.

Sato M. (1978b) Oxygen fugacity of basaltic magmas and the role of gas-forming elements. *Geophys. Res. Lett.* **5**, 447–449.

Sato M. and Helz R. T. (1971) Oxygen fugacity values of Apollo 12 basaltic rocks (abstract). In *Lunar Science II*, p. 144–145. The Lunar Science Institute, Houston.

Sato M. and Moore J. G. (1973) Oxygen and sulphur fugacities of magmatic gases directly measured in active vents of Mount Etna. *Phil. Trans. Roy. Soc. Lond.* **A274**, 137–146.

Sato M. and Wright T. C. (1966) Oxygen fugacities directly measured in magmatic gases. *Science* **153**, 1103–1105.

Sato M., Hickling N. and McLane J. E. (1973) Oxygen fugacity values of Apollo 12, 14 and 15 lunar samples and reduced state of lunar magmas. *Proc. Lunar Sci. Conf. 4th*, p. 1061–1079.

Shima H. and Naldrett A. J. (1975) Solubility of sulfur in an ultramafic melt and the relevance of the system Fe-S-O. *Econ. Geol.* **70**, 960–967.

Smith J. V. and Hervig R. L. (1979) Shergotty meteorite: mineralogy, petrography and minor elements. *Meteoritics* **14**, 121–142.

Snellenburg J. W., Nehru C. E., Culfield J. B. D., Zucker S. and Prinz M. (1979) Petrology temperature and oxygen fugacity indicating mineral assemblages in four grade mesosiderites (abstract). In *Lunar and Planetary Science X*, p. 1137–1139. Lunar and Planetary Institute, Houston.

Steinthorsson S. (1972) The opaque mineralogy of Surtsey. *Surtsey Prog. Rep.* **VI**, 152–157.

Stolper E. (1977) Experimental petrology of eucritic meteorites. *Geochim. Cosmochim. Acta* **41**, 587–611.

Stolper E. and McSween H. Y., Jr. (1979) Petrology and origin of the shergottite meteorites. *Geochim. Cosmochim. Acta* **43**, 1475–1498.

Stolper E., McSween H. Y., Jr. and Hays J. F. (1979) A petrogenetic model of the relationships among achondritic meteorites. *Geochim. Cosmochim. Acta* **43**, 589–602.

Taylor L. A., Williams R. J. and McCallister R. H. (1972) Stability relations of ilmenite and ulvöspinel in the Fe-Ti-O system and application of these data to lunar mineral assemblages. *Earth Planet. Sci. Lett.* **16**, 282–288.

Toksöz M. N. and Johnston D. H. (1977) The evolution of the moon and the terrestrial planets. In *The Soviet-American Conference on Cosmochemistry of the Moon and Planets*, pt. 1 (J. H. Pomeroy and N. J. Hubbard, eds.), p. 295–327. NASA SP-370.

Ulmer G. C., Rosenhauer M., Woermann E., Ginder J., Drory-Wolff A. and Wasilewski P. (1976) Applicability of electrochemical oxygen fugacity measurements to geothermometry. *Amer. Mineral.* **61**, 653–660.

Watkins N. D. and Haggerty S. E. (1968) Oxidation and magnetic polarity in single Icelandic lavas and dikes. *Geophys. J. Roy. Astron. Soc.* **15**, 305–315.

Williams R. J. (1971a) Reaction constants in the system Fe-MgO-SiO-O_2 at one atm, between 900° and 1300° C: experimental results. *Amer. J. Sci.* **270**, 334–347.

Williams R. J. (1971b) Equilibrium temperatures, pressures, and oxygen fugacities of the equilibrated chondrites. *Geochim. Cosmochim. Acta* **35**, 407–411.

Wilson R. L., Haggerty S. E. and Watkins N. D. (1968) Variation of palaeomagnetic stability and other parameters in a verticle traverse of a single Icelandic lava. *Geophys. J. Roy. Astron. Soc.* **16**, 79–96.

Woermann E., Knecht B., Rosenhauer M. and Ulmer G. C. (1977) The stability of graphite in the system C-O (abstract). In *Second International Kimberlite Conf., Santa Fe, New Mexico Abstracts*. American Geophysical Union, Washington, D.C.

Wones D. R. and Gilbert M. C. (1969) The fayalite-magnetite-quartz assemblage between 600° and 800°C. *Amer. J. Sci.* **267-A**, 480–488.

1.3.5 VOLATILES IN PLANETARY BASALTS: HYDROGEN, CARBON AND SULFUR GAS SPECIES

1.3.5.1 Introduction

The orderly progression of volatile-rich planets in the outer solar system and of the crust-core differentiated planets of the inner solar system is generally attributed to the temperature, pressure, and density gradients present across the nebular disc at the time of condensation. Extensive differentiation and degassing of the inner planets has produced atmospheres of varying composition; Mercury, which is closest to the sun, is devoid of an atmosphere.

Volatiles have played important roles in the evolution of the planets and exert a major influence on rock and mineral petrogenesis. The relative solubilities of H_2O and CO_2 in rocks of various composition are correlated with their estimated depths of origin. The quantity of each volatile species present influences the degree of partial melting of mantle material at a given temperature, and the composition of melt. Of comparable significance are the effects of volatiles on liquid-lines-of-descent and on processes that involve mass fractionation. The trends of crystallization will involve silica-enrichment or iron-enrichment, depending on whether the f_{O_2} is high or low, respectively. Magmatic gas compositions have overriding effects on the ionic states of elements, the distribution of elements between liquids and solids, and the partitioning of elements among crystalline species. The degree of subsolidus oxidation or reduction in a rock is also regulated by coexisting magmatic gases, and volatile retention or degassing may activate volcanism or influence the style of volcanism.

Apart from the effects of volatiles on igneous petrogenetic processes, the retention or outgassing of volatiles is a factor that may contribute to the formation and the distribution of certain classes of ore deposits. Massive sulfide deposits are commonly associated with basaltic subaqueous environments, whereas metallic deposits associated with volcanics (e.g., Cu) are in some cases related to subaerial conditions. Sulfur is also regarded by some to be the light element density-dilutant in the core, which is a component required to account for the bulk density of the Earth.

Volatiles are fundamental participants in many planetary processes; however, this review will be restricted in context, focusing only on the most abundant volatiles in known planetary basalts. Volatiles found in terrestrial volcanic emanations will be considered in detail, and volatiles found in basalts from the Earth, the Moon and basaltic achondrites will be discussed as well.

1.3.5.2 Volcanic gases

Considerable effort and attention have been applied to the collection of samples of the gases which escape from terrestrial basaltic volcanoes in hopes of determining the volatile content of the source region. There are major technical problems in sampling toxic gases at high temperatures, and there are problems of equal magnitude associated with detecting atmospheric contamination, ground water contamination, and the contribution made by the thermal activation of volatiles from surrounding country rocks. Sampling remains a hazardous process. Atmospheric contamination has been adequately addressed elsewhere, and it is generally agreed that, although the non-volcanic volatile component is difficult to assess, its contribution is probably negligible.

Anderson (1975) has provided the most recent and comprehensive review of gaseous emanations from basaltic volcanoes (Table 1.3.5.1). Based on classic studies by Shepard (1938), Jagger (1940), White and Waring (1963), Naughton *et al.* (1963), Finlayson *et al.* (1968), Sigvaldason and Elisson (1968), and Huntingdon (1973) it has been concluded that the most abundant volatile species is H_2O ($\sim$30 to $\sim$80 mole %), followed by CO_2 ($\sim$10 to $\sim$40 mole %) and SO_2 ($\sim$10 to 30 mole %). Hydrogen, H_2S, S, O_2, the halogens and the noble gases are all present in minor concentrations ($<$ 0.1 mole %). Gerlach and Nordlie (1975a), however, noted that many gas samples have high carbon to hydrogen and/or low carbon to sulfur ratios, and that most gas analyses lie along mixing lines defined either by (SO_2 + CO_2)-H_2O or by (SO_2 + CO_2)-H.

The ranges reported in Table 1.3.5.1 are extremely large. It is important to recognize that there is widespread evidence that the concentrations and the relative proportions of gases from one volcanic vent, or from a single volcanic province, may vary: (1) as functions of the evolution of the magma, which will depend on the initial degree of partial melting and on the extent of fractionation; (2) as functions of the amount of degassing of the magma during migration from source to surface; (3) depending upon the complexity of the plumbing system, whether it is closed or will readily

Table 1.3.5.1 Volcanic gas compositions (Mole %).

VOLCANO	H_2O	H_2	CO_2	CO	SO_2	S_2	H_2S	HCl	N_2	O_2	Ar
Kilauea, 1200°C (1918)	30	0.35	40	1.2	28	0.04	0.028	0.034	0		
	* 52.30	0.79	30.87	1.00	14.95	0.12	1.03	0.14			
Kileuea (Makaopuhi), 1000°C (1965)	67.1		8.5		24.3						
Kilauea, 1200°C 1 atm (calculated)	86.1	0.9	1.4	0.04	11.5	0.02	0.01	0.02			
Kilauea, 1200°C 490 atm (submarine)	21	0.1	10	0.2	59		9	0.1			
Surtsey, 1127°C (1964-1967)	86	4.7	5.7	0.38	2.67			0.4	0.07		
Erta'Ale, 1075°C (1971)	45.9	2.3	28.7	1.0	19.5		0.03		2.4	0.05	0.03
	* 69.56	1.57	17.80	0.78	8.79	0.51	0.95				
Erta'Ale, 1125-1135°C (1974)	* 78.18	1.60	11.11	0.50	7.04	0.26	0.87	0.42			
Nyiragongo, 1020°C	43.2	0.75	40.9	2.4	4.4				8.3	0.6	
Etna (1970)	† 18.27	29.59	22.19	2.44	23.35		0.24				
	* 49.14	0.53	23.41	0.49	25.94	0.25	0.21				

Note: Sets of analyses marked by an asterisk have been adjusted, using numerical procedures based on thermodynamic calculations to restore the probable composition of the erupted gases, taking into account the effects of contamination and the effects of disequilibrium. Data for Etna, Kilauea, and Erta'Ale are from Gerlach (1979, 1980a, 1980b), respectively. The unrestored (†) data for Etna are analyses without Ar and N_2 (Gerlach, 1979). After Anderson (1975).

allow volatile seepage; (4) in response to the pattern of volcanism with time; (5) with distance from the volcanic vent; and (6) as functions of cooling rate. In a broader context, if the magmatic source region is rich in volatiles, gas emanations on extrusion are voluminous, as is evidenced by the explosive volcanism in andesitic and rhyolitic vents. Gas compositions and their relative concentrations also vary as functions of bulk rock composition; for example, the high CO_2 (40 mole %) contents determined from the Nyiragongo alkaline complex (Table 1.3.5.1) contrast with the lower CO_2 abundances (10–20 mole %) which characterize most basaltic volcanoes.

Anderson (1975) has considered these modifications in the context of volatile solubilities in basaltic magmas, in terms of the isolation of gases and their losses during crystallization at constant pressure, and with respect to the behavior of volatiles during cooling and decompression. A few specific examples will serve to illustrate the underlying physical and chemical properties that are involved in magmatic gas evolution.

Experimental determinations and the thermodynamic treatment of H_2O and CO_2 solubilities in mafic melts are reviewed by Mysen (1977) and by Spera and Bergman (1980). Available data show that solubilities are dependent on melt compositions, that the solubility of H_2O is several times greater than that of CO_2, but that the solubility of H_2O is more strongly pressure and to a lesser extent temperature-dependent than that of CO_2. Comparable data for sulfur have been presented by Fincham and Richardson (1954), MacLean (1969), Haughton *et al.* (1974), Katsura and Nagashima (1974), Shima and Naldrett (1975), Buchanan and Nolan (1979), and Mysen and Popp (1980). In these studies, sulfur solubilities have been demonstrated to vary as functions of f_{O_2}, f_{S_2}, composition, temperature and pressure. A common and significant factor among H_2O, CO_2 and S is that the solubility of each of these species independently increases with increasing P and T. Although it is the net effect of mixed volatiles that ought to be considered, such data, except for $CO_2 + H_2O$, are not available. However, we may assume, at least in principle, that the least soluble gases will be released at early stages during crystallization with falling temperature, or during decompression and effervescence (Fig. 1.3.5.1).

The equilibrium compositions of gases that evolve on eruption vary as functions of temperature and pressure, and as functions of the partial pressures of oxygen and water (Fig. 1.3.5.2), as demonstrated by Heald *et al.* (1963), Nordlie (1971), and Gerlach and Nordlie (1975a,

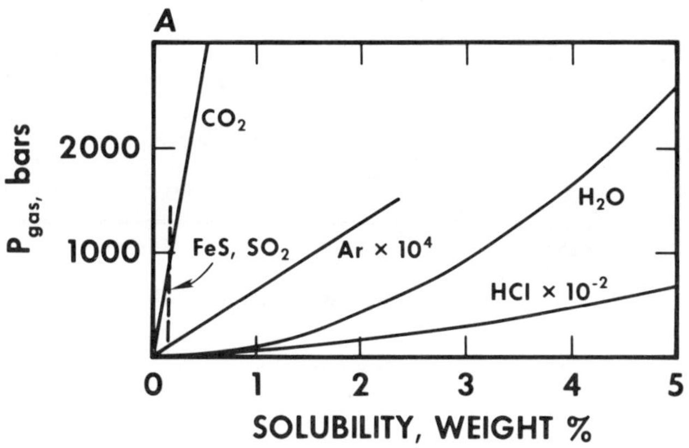

1.3.5.1a Solubilities of selected gases in basaltic melts as functions of gas pressure. Data from Anderson (1975).

b, c). From their thermodynamic calculations, relative concentrations of oxidized and reduced gas couples (e.g., H_2 and H_2O; H_2S and SO_2) can be determined. From these data, it is possible to evaluate the roles of various gas species, to determine the dominant gas species at any value of a selected intensive parameter, and to ascertain the changing nature of the gas as it evolves upon cooling and with time. For example, judging from the observed compositions of basaltic volcanic gases, H_2O and CO_2 are the dominant hydrogen-bearing and carbon-containing gas species. SO_2 is the dominant sulfur-bearing species at high temperatures (>800° C), whereas H_2S is present at lower temperatures (<500° C). Similarly, H_2S is the dominant sulfur-bearing species at low values of f_{O_2} ($<10^{-10}$ atms at ~1200° C), whereas SO_2 is the characteristic species at high f_{O_2} ($<10^{-8}$ atms at ~1200° C); high concentrations of SO_3 are generally interpreted as being due to atmospheric contamination (Nordlie, 1971).

Qualitatively, the data may be interpreted as implying that a gas sample containing high concentrations of SO_2 and H_2 is a high-temperature, reduced sample, whereas one with high SO_2 and low H_2 is a low-temperature, oxidized sample. Gerlach and Nordlie (1975a) have shown that basaltic gas compositions are constrained within an upper limit defined by the magnetite-hematite (MH) buffer and a lower limit

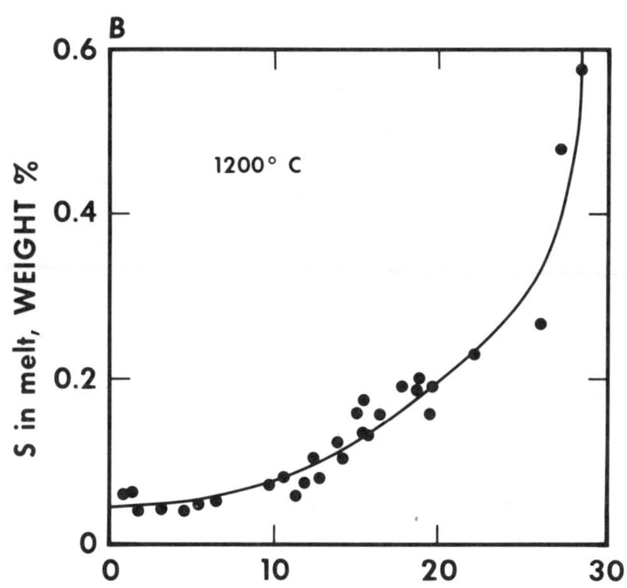

1.3.5.1b Sulfur and FeO contents of a basaltic melt in equilibrium with varying Fe-S-O liquids at 1200° C. Oxygen fugacities range from $10^{-8.9}$ to $10^{-12.04}$ atm, and sulfur fugacities range from $10^{-0.82}$ to $10^{-2.64}$ atm (Haughton *et al.*, 1974).

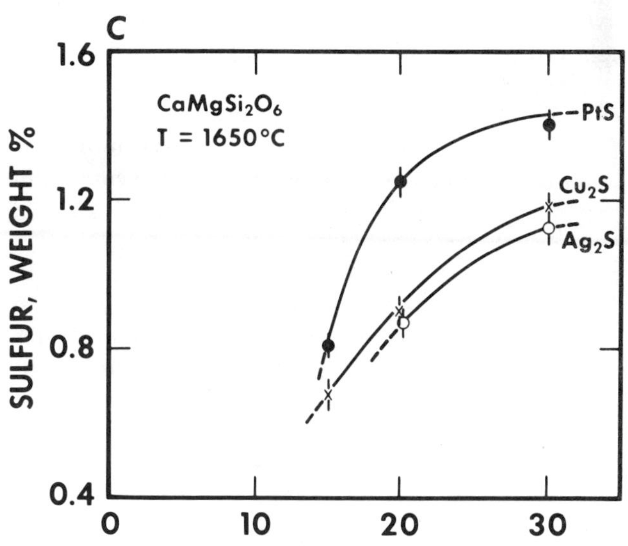

1.3.5.1c Solubility of sulfur in a diopside melt as a function of pressure, and as a function of sulfur buffers defined by PtS = Pt + S; Cu_2S = 2Cu + S; Ag_2S = 2Ag + S (Mysen and Popp, 1980).

defined by the fayalite-magnetite-quartz (FMQ) buffer. Because gas compositions are closely related to FMQ, they concluded that the gases are commonly buffered by the magma. If varying proportions of gas species are considered, those with extremely small carbon to hydrogen ratios (e.g., 1×10^{-6}) coincide with the FMQ buffer curve and become more reducing on cooling; this result follows regardless of the carbon to sulfur ratio. However, sulfur-rich and sulfur-poor gases both become more oxidizing upon cooling and, at lower temperatures (e.g., 650–700° C), these conditions are above the nickel-nickel oxide (NNO) buffer curve by a maximum of two orders of magnitude f_{O_2}.

The variation of basaltic gas emanations with time can be established from gas analyses taken during and after volcanic eruptions. Vents have been monitored in Hawaii (Heald *et al.*, 1963; Finlayson *et al.*, 1968) and in Iceland (Sigvaldason and Elisson, 1968). The results of observations of the Puu Puai cinder cone on Kilauea Iki and the Makaopuhi lava lake are illustrated in Fig. 1.3.5.3. In these examples, the partial pressure of water was found to be constantly high and to vary only slightly during the measurement period. Kilauea Iki showed gradual increases in CO_2 and SO_2 that were followed by rapid declines in concentrations after a period of two months; Makaopuhi, on the other hand, exhibited systematic decreases in the concentrations of these two gases. In the Surtsey eruption, hydrogen, water and carbon components were preferentially released during the initial degassing stage, whereas the sulfur species showed relative increases as degassing proceeded. These variations are best explained in terms of a gradual approach to equilibrium (Heald *et al.*, 1963), rather than solely on the basis of gas solubility or immiscibility in the magma, which is an interpretation that is consistent with gas samples obtained from lava lake drill holes at various depths and temperatures (Finlayson *et al.*, 1968). The increases in SO_2 contents in the Puu Puai cinder cone and in the Surtsey eruption probably resulted from lower magmatic temperatures and the concomitant increases in f_{O_2} conditions that prevailed as degassing continued.

Gas oxidation with magma cooling has been studied by Sato and Wright (1966) in the Makaopuhi lava lake, and by Naughton (1973) and Cruikshank *et al.* (1973) at Kilauea. Zones of high oxidation were detected on Makaopuhi in drill holes that were sunk into the actively crystallizing lava lake. These zones develop at temperatures between 550–750° C and migrate with cooling. Oxidation is attributed to the dissociation of H_2O into $H_2 + \frac{1}{2}O_2$, and because hydrogen has a smaller ionic radius and is less dense than oxygen, a *semi-permeable membrane* mechanism was proposed in

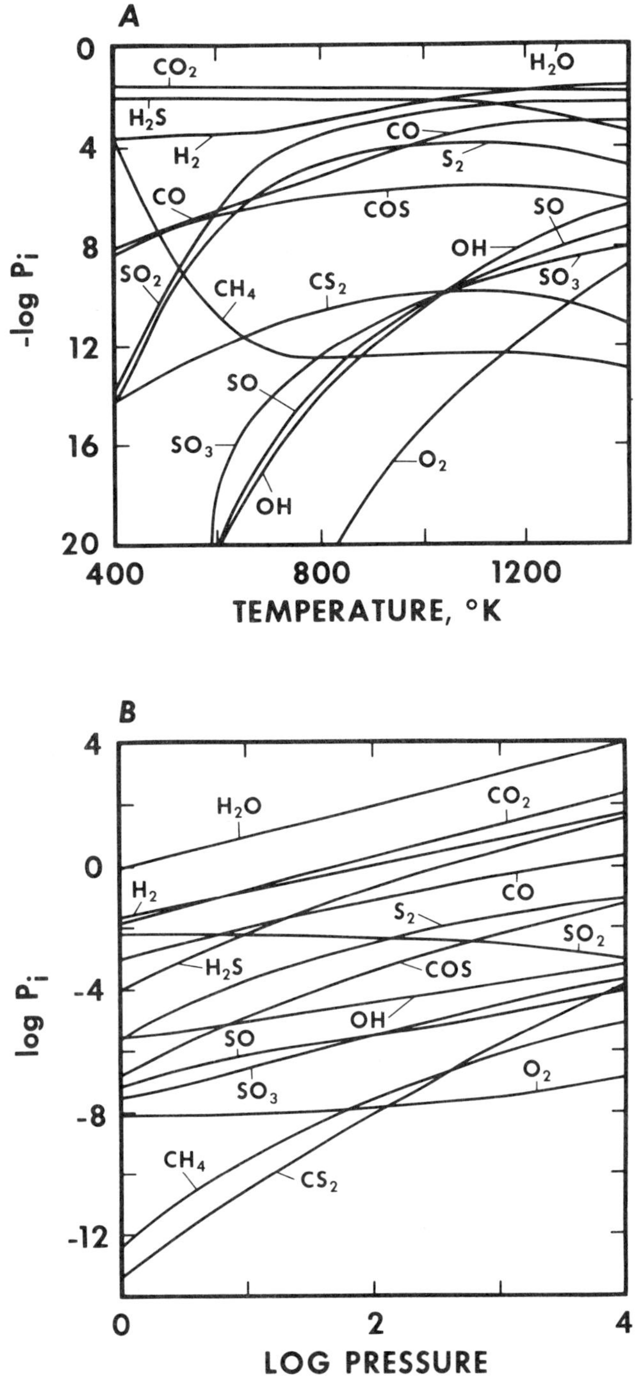

1.3.5.2 Variation in the equilibrium composition of volcanic gas phase as a function of various parameters. Based on equilibrium thermodynamic calculations for a variety of possible species in volcanic gases. The partial pressure of each gas is designated as $-\log P_i$. **a.** Partial pressures are illustrated as functions of temperature (1 atm pressure); **b.** as functions of total pressure; **c.** as functions of degree of oxidation at 900°K and 1 atm; **d.** at 1500°K and 1 atm; **e.** as functions of water content at 900°K and 1 atm. Data are from Heald *et al.* (1963).

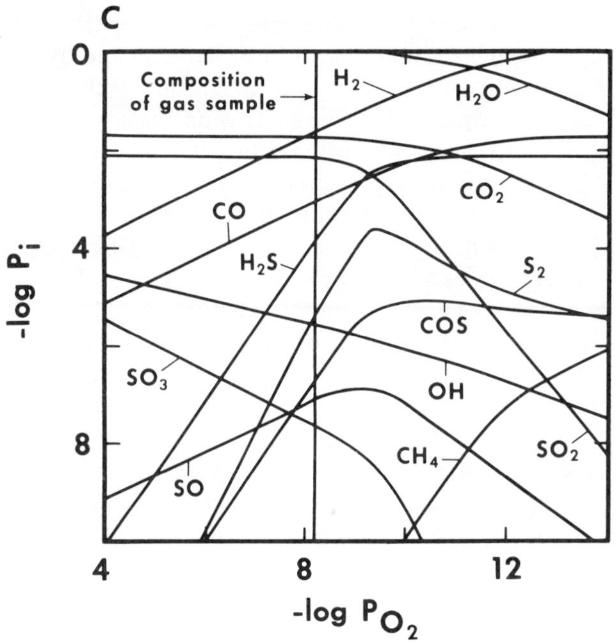

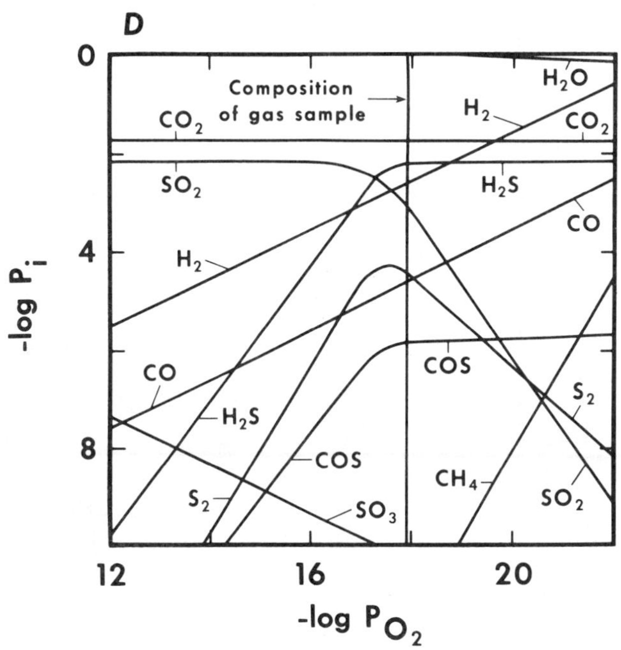

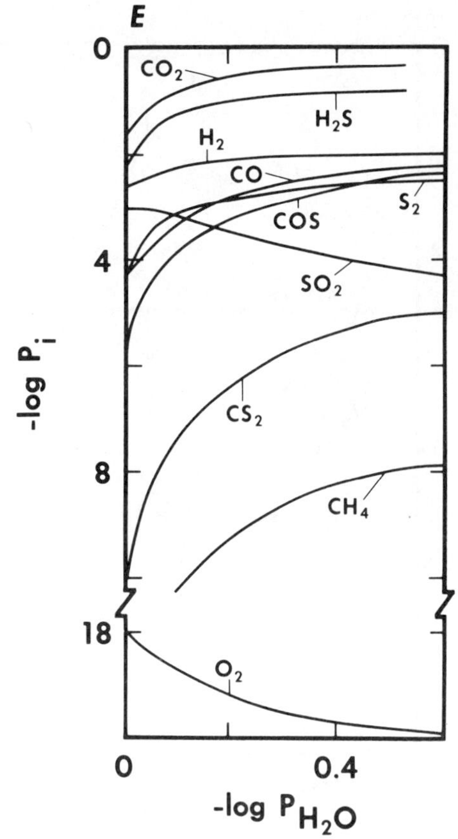

which H_2 is preferentially released to the atmosphere through the crystalline, membrane-interstices of the basalt, but oxygen is retained (Sato and Wright, 1966). This model is supported by the observation that volcanic flame burning takes place during the quiescent phases of eruptions on Kilauea, with gas ignition arising from the spontaneous combustion of vented hydrogen on contact with air at temperatures between 500–750° C (Naughton, 1973; Cruikshank *et al.*, 1973). Perched oxidation horizons and gas ignition are consistent with the presence of large quantities of water in the gas analyses. These data, therefore, provide evidence for the accumulation and decomposition of H_2O, the retention of oxygen, and the jet streaming of hydrogen during basalt crystallization.

The loss of volatiles during fountaining and the losses that are incurred with increasing flow distance from the vent have been determined by Swanson and Fabbi (1973) at Kilauea. Their results show that fountaining to heights of 450–540 m induces a 0.05 wt.% loss in H_2O and a 0.003 wt.% loss in S, from an initially saturated melt that contained 0.5 wt.% H_2O and 0.08 wt.% S. Flow over a distance of 12 km results in losses of 0.03–0.04 wt.% H_2O, and 0.007–0.008 wt.% S. The total amount of degassing that results from these two processes is therefore equivalent to a net loss of ~18% H_2O and ~14% S. Similar results were also obtained for desulfurization of the Mauna Ulu flow with a 33% loss in S over a distance of 7 km (Gibson *et al.*, 1975).

In summary, although the bulk gas contents of basaltic volcanoes are variable, and although the gas release profile patterns vary as a function of time and in response to the style of volcanism, it should be noted that these complexities may be entirely due to the

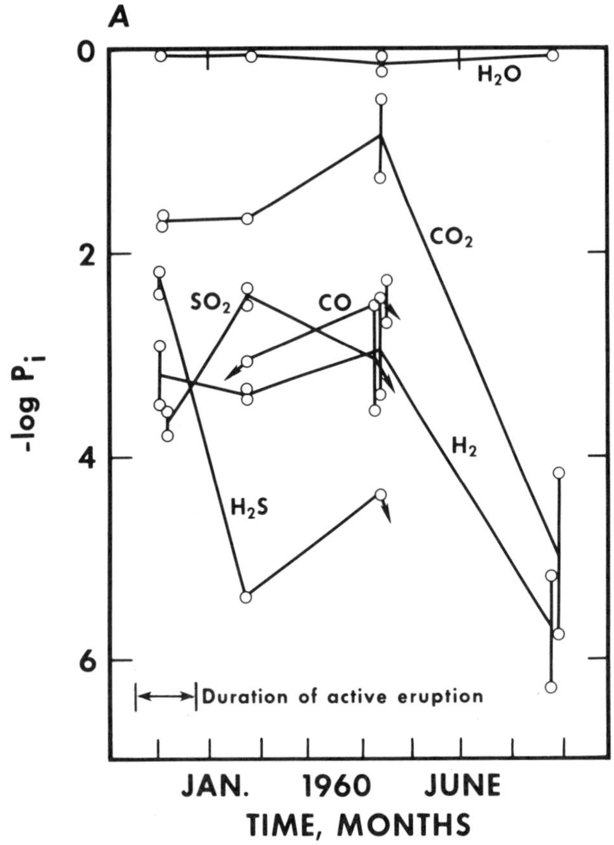

1.3.5.3a Changes in quantities (–log P_i = partial pressure) of the major gas species emanating from the Puu Puai cinder cone, Kilauea Iki, as functions of time (Heald *et al.*, 1963).

limited, but well-documented, examples that are cited. There is broad agreement between the results obtained from experimental gas solubility studies in mafic magmas (Fig. 1.3.5.1) and natural degassing trends, and there is also a close correspondence between gas equilibrium calculations and the species observed in volcanic emanations (Fig. 1.3.5.2). Basaltic gas oxidation states are equivalent to those of the FMQ buffer at magmatic temperatures, but in general become more oxidizing upon cooling.

1.3.5.3 Volatiles in basalts

The solubility relationships illustrated in Fig. 1.3.5.1 for the most abundant volatile species in basalts are consistent with the concentrations observed in natural samples. Important differences, however, exist between terrestrial basalts and extraterrestrial basalts. In terrestrial environments, water is the most abundant volatile and is also the most soluble, followed by the carbon and sulfur species. Lunar basalts are characterized by CO, CO_2 and hydrogen, with some proportion of these having been implanted by the solar wind. Basaltic achondrites, in common with lunar basalts, are anhydrous. These meteorites crystallized under moderately reducing conditions; free carbon may be present in association with sulfur gas species and dissolved CO + CO_2.

Estimates of the proportions of juvenile volatiles in basalts have been attempted in many studies. However, the values quoted in all of these determinations bear elements of uncertainty related to degassing. Volatile loss is an important factor in terrestrial volcanic processes and is an even greater consideration in lunar basalts that are extruded into a hard vacuum; meteoritic impact-volatilization is an added complexity. Water and the Earth's atmosphere are the main sources of secondary terrestrial contamination, whereas the solar wind, cometary components, assimilated or sublimated

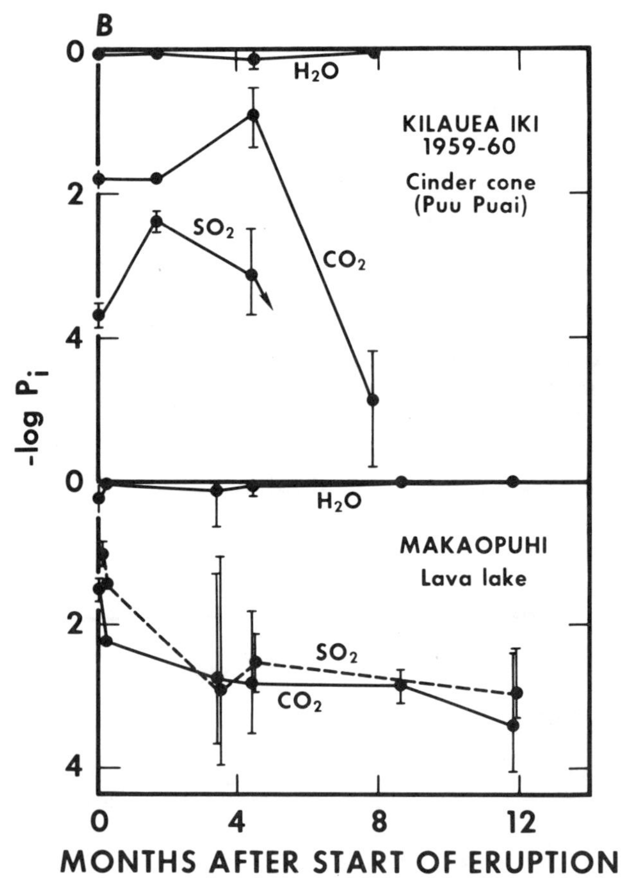

1.3.5.3b Changes in the content of water, carbon dioxide, and sulfur dioxide in gases from Kilauea Iki and the Makaopuhi lava lake as a function of time (Finlayson *et al.*, 1968).

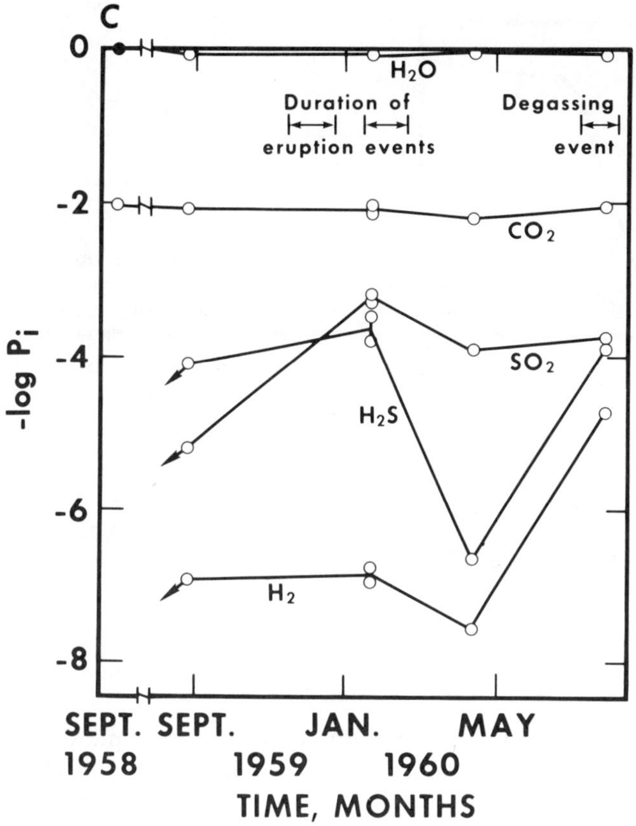

1.3.5.3c Changes in composition ($-\log P_i$ = partial pressure) of gases from Sulfur Bank, a solfataric fumarole at the rim of the Kilauea caldera, over the period 1958–1960 (Heald *et al.*, 1963).

meteoritic material, and low pressure condensation products are likely sources of contaminants in extraterrestrial basalts.

In most instances the extent of degassing and the degree of contamination are difficult to determine, but a number of studies have been undertaken to evaluate the rate and temperature of degassing, the effect of confining pressure on degassing, and the paragenesis of volatile entrapment. Examples of each of these classes of studies are presented to provide an overall perspective of some factors that have been demonstrated to influence the volatile contents of basalts; a few studies are quoted that permit the assessment of contamination, and other primary controlling factors are discussed, where appropriate, in the sections that follow. The objective is to determine the contents of volatiles in their source regions, recognizing that some basalts may be more suitable for this purpose than others.

(1) In the study by Swanson and Fabbi (1973), substantial decreases in volatile concentrations were shown to occur during various phases of lava fountaining activity. A much greater loss resulted from flowage. The most pronounced depletion was observed for sulfur, which decreased from 0.012 wt.% to 0.0075 wt.% between 0.4 and 12.0 km from the vent (the initial gas was assumed to be saturated with sulfur at a concentration of 0.08 wt.% S). A 33% loss in S was also determined by Gibson *et al.* (1975) for basalt flowage over a distance of 7 km. Volatile loss on flowage appears to be well established; however, arguments have also been presented to account for the volatile-rich coatings on lunar pyroclastic glasses, in which lava fountaining is invoked as a mechanism for deposition by sublimation (e.g., Heiken *et al.*, 1974).

Experimentally pyrolyzed samples of terrestrial and lunar basalts (Killingley and Muenow, 1975; Delaney *et al.*, 1978; Garcia *et al.*, 1979; Gibson and Moore, 1973) show that contaminated, loosely bonded volatiles are released below 500° C, with the dissolved, and hence primary, magmatic volatiles being released only above 800° C for H_2O, at ~900° C for CO_2, and at the softening point (~1100° C) for the sulfur species. Therefore, unless there is independent evidence that intense thermal metamorphism has occurred, the gas release profiles may be assumed to represent occlusion at magmatic crystallizing temperatures.

(2) Confining pressure is an important parameter that determines whether volatile gases will be retained in the melt or will be released. Brett (1976) has presented a series of arguments to account for the formation of free metallic Fe in lunar basalts that is related to sulfur volatilization under the intense vacuum conditions of extrusion. He notes that rapidly cooled rocks apparently lose S because they are close to the surface of a flow, whereas coarser-grained, more slowly cooled basalts lose comparable amounts of S because degassing takes place over an extended period of time.

Submarine basalts and basalts intruded at hypabyssal levels in the crust are readily distinguished from those that are extruded subaerially. The former are volatile-enriched and the latter are depleted. Moore (1965) showed that vesicle formation and gas release in ocean-floor basalts are directly related to the depth of water in which extrusion takes place (Fig. 1.3.5.4). Basalts extruded at 200 m are partially degassed, whereas those extruded at depths of >200 m appear to have retained their volatile components, almost to the point of saturation (Moore and Fabbi, 1971; Moore and Schilling, 1973). Sulfur retention has been confirmed by others (e.g., Mathez, 1976; Killingley and Muenow, 1975a; Delaney *et al.*, 1978; Naldrett *et al.*, 1978; Garcia *et al.*, 1979). It is generally concluded that basalts from >200 m contain ~0.08 wt.% S, basalts

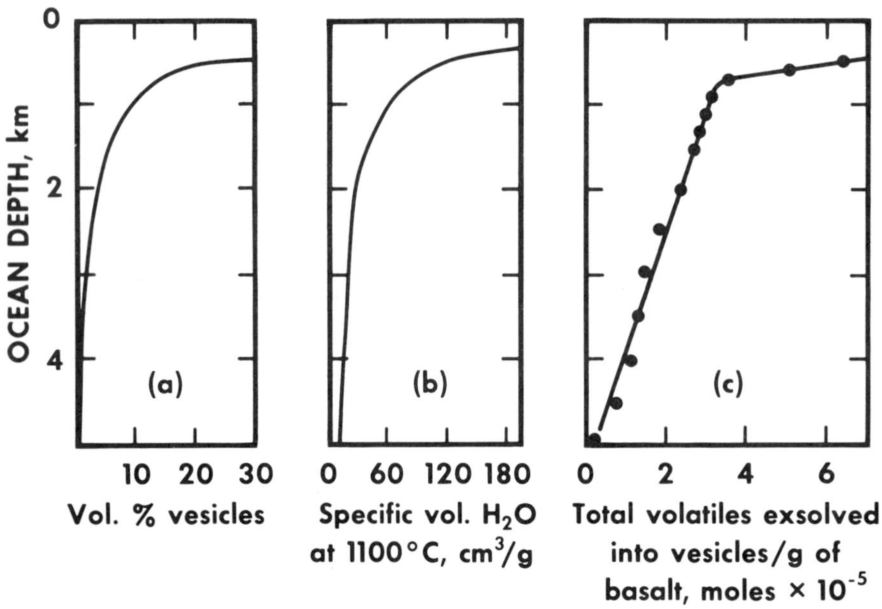

1.3.5.4 Variations in (**a**) volume of vesicles in basalt, (**b**) the specific volume of H_2O at 1100°C, and (**c**) the concentration of volatiles *exsolved* into vesicles in basalt as functions of ocean water depth. From these data it is inferred that basalts extruded at depths of <200 m will have undergone extensive devolatilization (Moore, 1965).

from <200 m contain ~0.017 wt.% S, and basalts subaerially extruded contain between 0.05–0.018 wt.% S. The degassing patterns for other volatiles are more complex, as discussed below.

(3) Information critical to the sampling of volatiles during basalt crystallization has been obtained by systematic dissection, in which the volatile compositions of glasses that are included in phenocrysts of silicates or oxides are distinguished from: (a) volatiles occluded in minerals; (b) volatiles that are present in the glassy matrix; or (c) volatiles that are present in late-stage residual mesostases. These anatomical studies have yielded some rather startling results. Anderson (1974) has shown that glasses trapped in subaerially extruded basalts contain 50–2800 ppm S, whereas vesiculated glasses contain 50–400 ppm S, a result that is confirmed by others (Killingley and Muenow, 1975a; Delaney *et al.*, 1978; Garcia *et al.*, 1979) on similar inclusions in ocean-floor basalts. Delaney and co-workers have also demonstrated that the most abundant volatile in phenocryst-enclosed glasses is CO_2; water is below the limit of detection (1×10^{-11}g) and was observed only in the glassy matrix of the pillow basalts studied. In the study by Garcia *et al.* (1979), the CO_2:H_2O ratio is 1:1 in glass-vapor inclusions in plagioclase from the quenched rims of pillow basalts. These studies imply that CO_2 is the dominant gas species in some magmas, that it may be as abundant as H_2O in others, and that S volatilization is significant. The high concentrations of CO_2 are, furthermore, consistent with Roedder's (1965) observation that olivine phenocrysts in basalts commonly contain liquid and gaseous CO_2 entrapped at relatively high P and T.

Perhaps the most extraordinary example of the effects of confining pressure on volatile retention are in "popping" rocks recovered from the Mid-Atlantic Ridge at depths of approximately 2700 m (Hekinian *et al.*, 1973). These rocks ". . . oval glassy pebbles 1–5 cms in diameter, were observed to jump and explode, one or two at a time, for 3 days after the dredge haul was emptied on to the deck of the ship." In comparison with the glassy margins of pillow basalts, the "popping" rocks were shown to have higher volume percentages of SO_2 (10.3 vs. 1.7) and CO (16.8 vs. 12.6), but lower concentrations of H_2 (26.7 vs. 34.7), and CO_2 (38.8 vs. 44.0). The authors concluded that it is largely the loss of CO_2 and H_2 that is responsible for the explosive behavior, and that gas release was induced by the sudden change in pressure of very recent lava that had not yet equilibrated on the ocean floor. This observed pressure phenomenon may be comparable to the bursting of CO_2 inclusions from ruptured olivines upon cooling from high temperatures (Killingley and Muenow, 1975b).

In summary, *extensive* volatile degassing takes place in lava fountaining, in lavas flowing from a volcanic vent, under vesiculating subaerial conditions, and at ocean depths of <200 m. Vacuum volatilization may be an important process in lunar and achondritic basalts but the extent to which this has been operative cannot be evaluated. Bearing the constraints of potential volatilization in mind, the abundances that are presented in the following section, with few exceptions, can only be regarded as *qualitative* estimates of the contents of volatiles in their source regions.

1.3.5.4 Water

Water is the principal volatile species in gaseous emanations from terrestrial volcanic vents and is the dominant volatile dissolved in terrestrial basalts; lunar basalts and basaltic achondrites are anhydrous, although H_2O may have played a role in their genesis.

The solubility of water in magmas of basaltic composition increases as a function of P_{H_2O} (Hamilton *et al.*, 1964; Khitarov and Kadik, 1973), and varies as a function of bulk composition, with the pre-eruption concentration of dissolved H_2O being greater in alkali basalts than in tholeiites and even greater in basalts related to convergent plate boundaries (Anderson, 1975). Moore (1970) showed, in a systematic study of subaerially extruded basalts on Hawaii, that water contents vary as functions of alkali concentrations: K-poor oceanic tholeiites contain ~0.25 wt.% H_2O; intermediate-K basalts contain ~0.5 wt.% H_2O; and alkali-rich basalts contain ~0.9 wt.% H_2O. Ocean-floor basalts also increase in H_2O content with increasing alkalinity and the range reported for a suite comparable to the island volcanics is 0.06–0.98 wt.% (Moore, 1970). Correlations between bulk K_2O contents and H_2O concentrations were also noted by Anderson (1974) and by Czamanske and Moore (1977). Concentrations of ~0.2 to ~1.0 wt.% H_2O appear to be typical of most ocean-floor basalts (Moore, 1965, 1970; Moore and Schilling, 1973; Killingley and Muenow, 1975a; Czamanske and Moore, 1977; Delaney *et al.*, 1978; Garcia *et al.*, 1979; Muenow *et al.*, 1980).

If the trapped glass inclusions in phenocrysts represent pre-eruption melt compositions, then the data of Delaney *et al.* (1978) suggest anhydrous conditions for ridge and interarc basalts. Concentrations of ~1 wt.% H_2O are present for arc and trough basalts (Garcia *et al.*, 1979), and 0.05 to ~4 wt. % H_2O exists in island and continental basalts (Anderson, 1974). The host, glassy matrices of these basalt suites contain ~0.2 to ~1.0 wt.% H_2O, and Delaney and co-workers have suggested that this enrichment resulted from an influx of H_2O into the magma, after inclusion formation but before crystallization on the ocean floor. Determinations of δD (relative to SMOW) yield −60 ‰ for pillow basalt glass (Moore, 1970) and a range of −60 ‰ to −80 ‰ in ^3He-rich glass rims (Craig and Lupton, 1976). These data provide convincing evidence that magmatic (Moore, 1970) or mantle-derived (Craig and Lupton, 1976) water is present. The water proposed to enter the magma (Delaney *et al.*, 1978) clearly is not seawater, and the apparent absence of H_2O from some encapsulated glasses (early entrapped liquids) may be due to diffusion losses, if the phenocrysts were suspended for prolonged periods of time at high temperatures in a magma chamber prior to eruption. An alternative explanation was suggested by Anderson (1975) whereby the vapor phase is enriched in the insoluble gases at high pressures, but at lower pressures the most soluble (and predominant) gases will be most abundant in the vapor. Because CO_2 is substantially less soluble than H_2O, the CO_2:H_2O ratio is higher in the vapor at higher pressures. However, in this model it is only the melt that is encapsulated and not the vapor. Nonetheless, the observed variations in the early H_2O contents of basalt magmas, as products of their tectonic settings, pose a number of interesting questions which have been outlined by Delaney *et al.* (1978), Garcia *et al.* (1979), and by Muenow *et al.* (1980) in the context of magma generation at ridge axes, in arc, interarc, and back-arc basin environments.

Degassing is an important process to be considered in making determinations of the H_2O contents of continental basalts (Swanson and Fabbi, 1973), but it may also be quite significant in ocean-floor basalts that are under relatively high confining pressures. At water depths of the order of ~5 km (Moore, 1970; Anderson, 1975), even glass-rich basalts show incipient vesiculation that amounts to ~0.1% by volume.

Whether water was ever a significant component of lunar and meteoritic basalts cannot now be assessed.

1.3.5.5 Carbon

Carbon may be present as finely dispersed elemental carbon or as carbide complexes in lunar basalts (Sato, 1978), and these phases may also exist in some basaltic achondrites (Brett *et al.*, 1977). The dissolved C species are dominated by CO_2 in terrestrial basalts (Delaney *et al.*, 1978) and by CO in extraterrestrial basalts (Gibson, 1977), with smaller concentrations of CO_2 also present. Although equilibrium-molar-species calculations demonstrate that CS_2, CH_4, and COS are possible compound species (Fig. 1.3.5.2) in basaltic gases (Heald *et al.*, 1963; Gerlach and Nordlie, 1975b), these compounds have not been detected in basalts.

Earth

Roedder (1965) noted that liquid and gas-rich CO_2 inclusions are widespread in phenocrystic olivines from terrestrial basalts, and estimated that CO_2 is trapped at 2.5 kb and ~1200° C. Bazarova *et al.* (1973) reported a range of CO_2 contents between 55% and 97% for trapped gases in a variety of phenocrysts enclosed in a suite of ankaramites, basanites, and melilitites from Cape Verde; H_2S, SO_2, NH_3, HCl, N_2 and the rare gases were also present in substantial concentrations. Large numbers of small CO_2-rich fluid inclusions were reported by Green and Radcliffe (1975) in xenoliths included in basalts from Hawaii and Nevada. These are of interest because the authors demonstrated that the CO_2 had been included prior to the xenoliths' incorporation into the basalt, basing this conclusion on the observation that the CO_2 bubbles are concentrated along dislocations and exsolution features within crystals in the xenoliths. Murck *et al.* (1978) analyzed a series of dunite, peridotite, and pyroxenite xenoliths associated with basalts, and showed that most inclusions contain nearly pure CO_2 but that in some cases minor concentrations of SO_2, H_2S or CO_2 are also present.

CO_2 contents of anhydrous glasses enclosed in olivine or plagioclase from ocean-floor basalts are estimated to be approximately 0.4 wt.% (Delaney *et al.*, 1978). Quantitative determinations of volatile contents in these trapped liquids (glasses) are difficult to obtain, but Garcia *et al.* (1979) reported CO_2:H_2O ratios of 1:2 to 2:1 for inclusions in Mariana trough basalts, and a ratio of 1:1 for inclusions in island arc basalts. The matrix glass on the rims of these pillow basalts contains from 0.14–0.43 wt.% CO_2, with average CO_2:H_2O ratios of 0.09; CO_2:H_2O averages in trough and arc basalts, however, are indistinguishable from one another (0.091 and 0.095, respectively). Back-arc basin basaltic glasses contain 0.129–0.378 wt.% CO_2 (Muenow *et al.*, 1980) and average values from other oceanic localities are as follows: Hawaii, 0.095 wt.% CO_2 (Garcia *et al.*, 1979), 0.2 wt.% CO_2 (Moore, 1965, 1970); Mid-Atlantic Ridge, 0.132 wt.% CO_2 (Delaney *et al.*, 1978); and Iceland, 0.05 wt.% CO_2 (Anderson, 1975; Björnsson, 1968).

Although CO_2 is the second most abundant gas in volcanic vent systems, the liquid CO_2 inclusions in olivine and the high proportions of CO_2 in trapped magmatic glasses indicate that CO_2 is the predominant component (rather than H_2O) in the environment of phenocryst crystallization. CO_2 is far less soluble in basaltic melts than most other gases (Fig. 1.3.5.1), and Anderson (1975) noted that carbon gas species should dominate the vapor phase at relatively high pressures (~4 kb), in accordance with the experimental data of Kadik *et al.* (1972).

CO_2 volatilization does not vary appreciably with fountaining activity or with distance from the vent (Swanson and Fabbi, 1973), and so it is assumed, because of the insoluble nature of CO_2, that rapid release takes place at the onset of volcanism, or during the effervescent stage that may precede violent volcanic activity.

Moon

The average carbon content of lunar basalts is ~30 ppm. The maximum average value of 67 ppm C is present in Apollo 11 basalts and the minimum of 15 ppm C is present in Apollo 15 basalts (Gibson, 1977). The two carbon gases that are released upon acid hydrolysis are CO and CO_2, which generally exhibit a good positive correlation, suggesting that the gases are not of terrestrial origin. CO:CO_2 ratios in the Apollo 15 basalts (Gibson *et al.*, 1975) are approximately 1:2 at low concentrations (2 ppm CO, 4 ppm CO_2) and 1:4 at high concentrations (12 ppm CO, 48 ppm CO_2). These authors noted that CO_2 abundances at high concentrations are contrary to those expected (CO:CO_2 = 1:2) for the gas phase that ought to be present during the crystallization of lunar basalts, assuming T = 1123° C and $f_{O_2} = 10^{-13.1}$ to $10^{-13.7}$ atm (Wellman, 1970). The explanation offered was that CO may react with metallic Fe to form Fe-carbides, thereby liberating excess oxygen for the formation of additional CO_2. This reaction would deplete the carbon species in CO and enrich the gas proportionately in CO_2, according to the reactions $CO + Fe = FeC + \frac{1}{2}O_2$, and $CO + \frac{1}{2}O_2 = CO_2$. Most lunar soils are enriched in C (average soil content of C is 115 ppm vs. 30 ppm for basalts). Concentrations of C in soils increase with soil maturity, which is attributed to increasing solar wind, meteoritic or cometary contamination. Although a percentage of the C in lunar basalts may stem from these sources, the largest proportions probably are indigenous to the Moon as primary volcanic gas components (Gibson *et al.*, 1975). The highly vesiculated nature of many lunar basalts is, in fact, generally ascribed to CO + CO_2 liberation (Wellman, 1970).

Meteorites

Refined analyses of carbon in achondrites are scarce. Those listed here are provided by C. B. Moore (pers. comm., 1980). Eucrites: Haraiya (4500–4900 ppm), Pasamonte (630–740 ppm), and Sioux County (550–680 ppm). Howardites: Kapoeta (700–740 ppm) and Yurtuk (820–1100 ppm). Although eucrites are similar to lunar basalts in many respects, meteoritic basalts are clearly distinctive in having substantially larger concentrations of carbon. Even the highly enriched and possibly contaminated lunar soils (115 ppm) are depleted in carbon relative to eucrites or howardites. For ordinary chondrites (H, L, and LL) Moore and Lewis (1967) reported an overall range of 1600–5700 ppm C, but noted that most chondrites have carbon contents of <2000 ppm and that a general trend exists in which the less metamorphosed chondrites have higher concentrations of carbon. Whether carbon in these meteorites is present exclusively as elemental carbon, as carbide, as carbonate, or in an organic form has not yet been determined.

1.3.5.6 Sulfur

The sulfur species observed in volcanic gases are H_2S, S_2, SO_2 and SO_3, with SO_2 dominating at magmatic temperatures (~1200° C) and with H_2S being the major component at lower temperatures (Fig. 1.3.5.2). Connolly and Haughton (1972) have shown that sulfur is present in basaltic glasses as the negative ion coordinated with positively charged ions, such as Fe^{2+}; this conclusion is substantiated in experimental studies at high temperatures (~1200° C) and low f_{O_2} conditions (10^{-8} to 10^{-10} atms) by MacLean (1969), Haughton *et al.* (1974), and Shima and Naldrett (1975). Between 1250–1350° C and f_{O_2} values $>10^{-8}$ atms, however, sulfur is present as the dissolved sulfate (Katsura and Nagashima, 1974). Terrestrial basalts, lunar basalts, and basaltic achondrites crystallize at f_{O_2} $<10^{-8}$ atms at 1000–1200° C, so we may assume that dissolved sulfur is present as the negative ion. Sulfur is also present in discrete sulfide minerals, such as pyrrhotite, pentlandite and chalcopyrite in terrestrial basalts, and troilite in lunar and achondritic basalts.

Sulfide liquid immiscibility

The solubility of sulfur in mafic melts varies as a function of FeO content (MacLean, 1969; Haughton *et al.*, 1974), increasing systematically from 500–2000 ppm as FeO increases from 5–20 wt.% (Fig. 1.3.5.1). The interdependency of sulfur and FeO are demonstrated in Archean basalts (Naldrett and Goodwin, 1977), in ocean-floor basalts (Mathez, 1976), and in lunar basalts (Gibson *et al.*, 1975), although these trends are not precisely identical to the experimental curve determined by Haughton *et al.* (1974). Sulfur solubility also varies as a function of f_{O_2} and f_{S_2} (MacLean, 1969; Haughton *et al.*, 1974; Katsura and Nagashima, 1974; Shima and Naldrett, 1975; Buchanan and Nolan, 1979), showing an increase with increasing f_{S_2} and decreasing f_{O_2}. Sulfur solubility passes through a minimum at f_{O_2} ~10^{-8} atms at 1250–1350° C, which results from a transformation of S^{-2} to the dissolved sulfate when conditions are modified from low to high oxygen fugacities.

A significant aspect of these experimental data and of the data of Naldrett (1969) for the system Fe-S-O is sulfide liquid immiscibility in basalts, a mechanism that is commonly invoked for igneous differentiates that yield large metallic sulfide ore deposits (e.g., Smith, 1962; Hawley, 1962; Naldrett *et al.*, 1967; Usselman *et al.*, 1979). Immiscible, spheroidal droplets of pyrrhotite, pentlandite and chalcopyrite are common constituents of ocean-floor basalts (Moore and Calk, 1971; Kanehira *et al.*, 1973; Mathez, 1976; Czamanske and Moore, 1977), usually found entrapped in phenocrysts or in the glassy matrix. The vesiculated glassy rinds of these basalts also are frequently decorated with sulfides along the vesicle walls (Moore and Calk, 1971; Propach, 1979), indicative either of liquid immiscibility or vapor deposition. Subaerially extruded basalts contain phenocryst-entrapped immiscible sulfide liquids (Desborough *et al.*, 1968; Anderson, 1974), which are also recognized in melt oozes from lava lakes (Skinner and Peck, 1969). These liquids contain magnetite as an associated phase, in contrast with the troilite + Ni-Fe assemblages identified in lunar basalts (e.g., Skinner, 1970; Roedder and Weiblen, 1978) and in achondrites (Stolper, 1977; Brett *et al.*, 1977).

In all instances, the textures and mineral assemblages bear striking resemblances to closely corresponding experimental analogs (Naldrett, 1969; MacLean, 1969). Sulfide liquid immiscibility may be induced (i.e., to separate or to *exsolve* from a mafic silicate melt) by a decrease in temperature, by an increase in f_{O_2}, or by a decrease in the activity of FeO brought about by the crystallization of Fe-bearing oxides or silicates. The fact that most sulfide droplets are either entrapped in phenocrysts or are present in deep-sea basalts from depths >200 m suggests that liquid separation takes place under confining pressures and at magmatic temperatures. The presence of such droplets indicates that extensive degassing has not taken place.

Earth

The sulfur contents of terrestrial ocean-floor basalts range between 60–800 ppm (Moore and Fabbi, 1971; Moore and Schilling, 1973; Killingley and Muenow, 1975; Delaney *et al.*, 1978; Garcia *et al.*, 1979), but the upper value is generally regarded as approaching the limit of saturation for lavas dredged at water depths that exceed 200 m (Moore and Calk, 1971; Mathez, 1976). A comparable range is reported by Naldrett *et al.* (1978) and by Mathez (1976). Recent subaerially extruded basalts contain 50–2000 ppm S, with the lower value indicative of degassing (Swanson and Fabbi, 1973), and the upper value being characteristic only of glass-entrapped inclusions in phenocrysts (Anderson, 1974). Naldrett and Goodwin (1977) and Naldrett *et al.* (1978) found that the sulfur contents of Archean basalts range between zero and 8000 ppm. There are few basalts at the upper limit, but a significant number occur in the range between 1000–2000 ppm, with a peak at 1000 ppm. Their interpretation of these S concentrations, which are extraordinarily high in contrast with those of recent basalts, is that the retention of sulfur, despite extensive redistribution by metamorphism, results from the rapid accumulation of Archean volcanic rocks and their limited interaction with seawater.

Moon

Sulfur contents of lunar basalts range from an average of 400 ppm in some Apollo 15 basalts to 2600 ppm in basalts from the Apollo 17 site (Gibson and Moore, 1973, 1974; Gibson *et al.*, 1975; Moore, 1973). The high FeO contents of many lunar basalts (16–20 wt.%) are within the experimental range of compositions in which S solubilities were determined by Haughton *et al.* (1974) at liquidus temperatures and oxygen fugacities of 10^{-8} to 10^{-12} atm. Sulfur saturation of these mafic melts varies between 1500–2000 ppm, indicating that the Apollo 11 and Apollo 17 mare basalts were saturated with respect to sulfur, whereas those from Apollo 12 and Apollo 15 were undersaturated (Brett, 1975; Gibson *et al.*, 1975). Brett (1976) has shown, based on the abundance of metallic iron, and on the negative correlation that is exhibited between Fe (metal) and S contents, that most lunar basalts have undergone some degree of reduction, with concomitant formation of metallic Fe, due to sulfur loss. Nash and Hausel (1973) have pointed out that the equilibrium sulfur fugacity for the reaction $Fe + \frac{1}{2}S_2 = FeS$ at 1200° C is eight orders of magnitude above the confining pressure of the lunar atmosphere, so S loss might be anticipated during eruption (Brett, 1975). In the study by Gibson *et al.* (1977) on a suite of Apollo 12 basalts (range in S content, 560–1520 ppm), no correlation was observed between sulfur and metallic iron contents. Sulfur abundances in these mare basalts increase with increasing degrees of fractionation, so cumulative S enrichment of Fe-FeS liquids is tentatively proposed. Titanium appears to play an important role in the fractionation model, an observation that is also made by Czamanske and Moore (1977) in ocean-floor basalts. The mechanism of fractionation and sulfur abundance correlations, however, is not fully understood.

Meteorites

Sulfur contents for ten achondrites have been provided by C. B. Moore (pers. comm., 1980). Shergotty has the lowest concentration with 360 ± 30 ppm (range = 320–420 ppm) and Stannern has the highest concentration with 9700 ± 100 ppm (range = 8240–11,400 ppm). Abundances in other eucrites are in the range 1000–1500 ppm. Howardites have somewhat higher concentrations: 1980 ppm S for Petersburg and 3270 ppm S for Kapoeta. The similarities in mineralogy, petrology and chemistry between eucrites and lunar basalts (400–2600 ppm S) extend also to the respective abundances of sulfur, with Stannern being the only known exception. Shergottites, on the other hand, are similar to terrestrial dolerites, have intrinsic oxygen fugacities that are comparable to those of terrestrial basalts, and have low sulfur contents that may be indicative of devolatilization, if these values are compared with ocean-floor basalts extruded at <200 m or with basalts extruded subaerially.

1.3.5.7 Summary

Water is the predominant volatile species in terrestrial volcanic gases, followed by CO_2 and SO_2. Relative abundances may change as functions of time and stage of volcanism. Thick extrusive flows and ponded lava lakes facilitate the accumulation of water and the dissociation of H_2O into hydrogen + oxygen. Hydrogen that is vented from this reaction may ignite in volcanic flames, and the retained oxygen may induce basalt oxidation during deuteric cooling. Volatile release from basaltic magmas, either as gases or as immiscible sulfide liquids, can be modelled in terms of solubility coefficients and such intensive parameters as pressure, temperature, f_{O_2}, f_{H_2O}, and f_{S_2}.

Terrestrial basalts are hydrous, contain CO_2 and S (either as dissolved sulfur or as immiscible sulfide liquids) and crystallize at f_{O_2} values close to those of the FMQ buffer (section 1.3.4). The volatile contents of basaltic glasses (i.e., of quenched liquids) are generally very low in contrast with those of their crystalline rock equivalents. Typical concentrations of H_2O are ~0.2 to ~1.0 wt.%, but values as high as 4 wt.% have been recorded in some basaltic glasses, whereas other glasses are evidently anhydrous. Concentrations of CO_2 and S are ~0.1 to ~0.4 wt.% and ~60 to ~800 ppm, respectively. However, basalts with a negligible S content and basalts containing as much as 8000 ppm S have also been reported.

Lunar basalts are anhydrous, are carbon- and sulfur-bearing, and crystallized within the field of metallic iron stability (section 1.3.4). The bulk-rock carbon and sulfur contents of lunar basalts are 15–67 ppm and 400–2600 ppm, respectively.

Meteoritic achondrites vary between those that are anhydrous (basaltic achondrites) to those that are hydrous (chassignites), with oxidation states that span an equivalently wide spectrum (section 1.3.4). Basaltic achondrites crystallized at f_{O_2} values below those of the iron-wüstite buffer, have carbon contents of 550–4900 ppm, and have sulfur contents between 1000–1500 ppm.

In conjunction with experimental data and from the concentrations of volatile species observed in terrestrial basalts, we may conclude that volatiles in the source regions of terrestrial basalts are probably dominated by CO_2 and have high CO_2:H_2O ratios or have CO_2:H_2O ratios that approach unity. There is evidence that these ratios may be influenced by tectonic environment, which is supported by the fact that volatile abundances show marked correlations with bulk composition, specifically with alkalinity. Lunar basalts have intrinsically low states of oxidation, low carbon contents and are on average more sulfur-enriched than typical terrestrial basalts. Basaltic achondrites, on the other hand, are both carbon and sulfur-enriched. Eucrites are comparable to lunar basalts mineralogically, compositionally, and in terms of the level of oxidation during crystallization. Marked differences, however, exist with respect to their volatile abundances. Shergotty is of specific interest because of its similarity to terrestrial dolerites, similarity in f_{O_2} to terrestrial basalts, similarity in content of sulfur to terrestrial basalts, and the distinctive contrast between this meteorite and eucrites.

1.3.5.8 References

Anderson A. T. (1974) Chlorine, sulfur, and water in magmas and oceans. *Geol. Soc. Amer. Bull.* **85**, 1485–1492.

Anderson A. T. (1975) Some basaltic and andesitic gases. *Rev. Geophys. Space Phys.* **13**, 37–55.

Bazarova T. Yu., Canilho E. M., Sobolev V. S. and Shugurova N. A. (1973) Temperatures and conditions of crystallization of alkaline effusives from the Majo Isle (Cap Vert). *Lithos* **6**, 403–411.

Björnsson S. (1968) Radon and water in volcanic gas at Surtsey, Iceland. *Geochim. Cosmochim. Acta* **32**, 815–821.

Brett R. (1975) Thicknesses of some lunar mare basalt flows and ejecta blankets based on chemical kinetic data. *Geochim. Cosmochim. Acta* **39**, 1135–1142.

Brett R. (1976) Reduction of mare basalts by sulfur loss. *Geochim. Cosmochim. Acta* **40**, 997–1004.

Brett R., Huebner J. S. and Sato M. (1977) Measured oxygen fugacities of the Angra Dos Reis achondrite as a function of temperature. *Earth Planet. Sci. Lett.* **35**, 363–368.

Buchanan D. L. and Nolan J. (1979) Solubility of sulfur and sulfide immiscibility in synthetic tholeiitic melts and their relevance to Bushveld-Complex rocks. *Can. Mineral.* **17**, 483–494.

Connolly J. W. D. and Haughton D. R. (1972) The valence of sulfur in glass of basaltic compositions formed under conditions of low oxidation potential. *Amer. Mineral.* **57**, 1515–1517.

Craig H. and Lupton J. E. (1976) Rare gases and volatiles in oceanic basalts (abstract). *EOS (Trans. Amer. Geophys. Union)* **57**, 408.

Cruikshank D. P., Morrison D. and Lennon K. (1973) Volcanic gases: hydrogen burning at Kilauea Volcano, Hawaii. *Science* **182**, 277–278.

Czamanske G. K. and Moore J. G. (1977) Composition and phase chemistry of sulfide globules in basalt from the Mid-Atlantic Ridge rift valley near 37° N lat. *Geol. Soc. Amer. Bull.* **88**, 587–599.

Delaney J. R., Muenow D. W. and Graham D. G. (1978) Abundance and distribution of water, carbon and sulfur in the glassy rims of submarine pillow basalts. *Geochim. Cosmochim. Acta* **42**, 581–594.

Desborough G. A., Anderson A. T. and Wright T. L. (1968) Mineralogy of sulfides from certain Hawaiian basalts. *Econ. Geol.* **63**, 636–644.

Fincham C. J. B. and Richardson F. D. (1954) The behaviour of sulphur in silicate and aluminate melts. *Phil. Trans. Roy. Soc. Lond.* **A223**, 40–62.

Finlayson J. B., Barnes I. L. and Naughton J. J. (1968) Developments in volcanic gas research in Hawaii. *Inter. Upper Mantle Project: Scientific Report No. 15*, p. 428–438. Amer. Geophys. Union, Washington, D.C.

Garcia M. O., Liu N. W. and Muenow D. W. (1979) Volatiles in submarine volcanic rocks from the Mariana Island arc and trough. *Geochim. Cosmochim. Acta* **43**, 305–312.

Gerlach T. M. (1979) Evaluation and restoration of the 1970 volcanic gas analyses from Mount Etna, Sicily. *J. Volc. Geotherm. Res.* **6**, 165–178.

Gerlach T. M. (1980a) Evaluation of volcanic gas analyses from Kilauea Volcano. *J. Volc. Geotherm. Res.* **7**, 295–317.

Gerlach T. M. (1980b) Investigation of volcanic gas analyses and magma outgassing from Erta' Ale lava lake, Afar, Ethopia. *J. Volc. Geotherm. Res.* **7**, 415–441.

Gerlach T. M. and Nordlie B. E. (1975a) The C-O-H-S gaseous system, part I: composition limits and trends in basaltic cases. *Amer. J. Sci.* **275**, 353–376.

Gerlach T. M. and Nordlie B. E. (1975b) The C-O-H-S gaseous system, part II: temperature, atomic composition, and molecular equilibria in volcanic gases. *Amer. J. Sci.* **275**, 377–394.

Gerlach T. M. and Nordlie B. E. (1975c) The C-O-H-S gaseous system, part III: magmatic gases compatible with oxides and sulfides in basaltic magmas. *Amer. J. Sci.* **275**, 395–410.

Gibson E. K., Jr. (1975) Distribution, movement, and evolution of the volatile elements in the lunar regolith. *The Moon* **13**, 321–326.

Gibson E. K., Jr. (1977) Volatile elements, carbon, nitrogen, sulfur, sodium, potassium and rubidium in the lunar regolith. *Phys. Chem. Earth.* **X**, 57–62.

Gibson E. K., Jr. and Moore G. W. (1973) Carbon and sulfur distributions and abundances in lunar fines. *Proc. Lunar Sci. Conf. 4th*, p. 1577–1586.

Gibson E. K., Jr. and Moore G. W. (1974) Sulfur abundances and distributions in the valley of Taurus-Littrow. *Proc. Lunar Sci. Conf. 5th*, p. 1823–1837.

Gibson E. K., Jr., Brett R. and Andrawes F. (1977) Sulfur in lunar mare basalts as a function of bulk composition. *Proc. Lunar Sci. Conf. 8th*, p. 1417–1428.

Gibson E. K., Jr., Chang S., Lennon K., Moore G. W. and Pearce G. W. (1975) Sulfur abundances and distributions in mare basalts and their source magmas. *Proc. Lunar Sci. Conf. 6th*, p. 1827–1301.

Green H. W., II and Radcliffe S. V. (1975) Fluid precipitates in rocks from the Earth's mantle. *Geol. Soc. Amer. Bull.* **86**, 846–852.

Hamilton D. L., Burnham W. C. and Osborn E. F. (1964) The solubility of water and effects of oxygen fugacity and water content on crystallization in mafic magmas. *J. Petrol.* **5**, 21–39.

Haughton D. R., Roeder P. L. and Skinner B. J. (1974) Solubility of sulfur in mafic magmas. *Econ. Geol.* **69**, 451–467.

Hawley J. E. (1962) The Sudbury ores: their mineralogy and origin. *J. Mineral. Assoc. Can.* vol. 7, part 1. 1–207.

Heald E. F., Naughton J. J. and Barnes I. L., Jr. (1963) The chemistry of volcanic gases 2. Use of equilibrium calculations in the interpretation of volcanic gas samples. *J. Geophys. Res.* **68**, 545–557.

Heiken G. H., McKay D. S. and Brown R. W. (1974) Lunar deposits of possible pyroclastic origin. *Geochim. Cosmochim. Acta* **38**, 1703–1718.

Hekinian R., Chaigneau M. and Cheminee J. L. (1973) Popping rocks and lava tubes from the Mid-Atlantic Rift Valley at 36° N. *Nature* **245**, 371–373.

Huntingdon A. T. (1973) The collection and analysis of volcanic gases from Mt. Etna. *Phil. Trans. Roy. Soc. Lond.* **A274**, 119–128.

Jagger T. A. (1940) Magmatic gases. *Amer. J. Sci.* **238**, 313–353.

Kadik A. A., Lukanin O. A., Lebedev Y. B. and Korovushkina E. Y. (1972) Solubility of H_2O and CO_2 in granite and basalt melts at high pressures. *Geochem. Internat.* **6**, 1041–1050.

Kanehira K. S., Sakai Y. H. and Sasaki A. (1973) Sulphide globules and sulphur isotope ratios in the abyssal tholeiite from the Mid-Atlantic Ridge near 30° N latitude. *Geochem. J.* **7**, 89–96.

Katsura T. and Nagashima S. (1974) Solubility of sulfur in some magmas at 1 atmosphere. *Geochim. Cosmochim. Acta* **38**, 517–531.

Khitarov N. I. and Kadik A. A. (1973) Water and carbon dioxide in magmatic melts and peculiarities of the melting process. *Contrib. Mineral. Petrol.* **41**, 205–215.

Killingley J. S. and Muenow D. W. (1975a) Volatiles from Hawaiian submarine basalts determined by dynamic high temperature mass spectrometry. *Geochim. Cosmochim. Acta* **39**, 1467–1473.

Killingley J. S. and Muenow D. W. (1975b) Thermal stress-induced release of CO_2 inclusions in olivine on cooling from high temperatures. *Amer. Mineral.* **60**, 148–151.

MacLean W. H. (1969) Liquidus phase relations in the FeS-FeO-Fe_3O_4-SiO_2 system, and their application in geology. *Econ. Geol.* **64**, 865–884.

Mathez E. A. (1976) Sulfur solubility and magmatic sulfides in submarine basalt glass. *J. Geophys. Res.* **81**, 4269–4276.

Moore C. B and Lewis C. F. (1967) Total carbon content of ordinary chondrites. *J. Geophys. Res.* **72**, 6289–6292.

Moore J. G. (1965) Petrology of deep-sea basalt near Hawaii. *Amer. J. Sci.* **263**, 40–52.

Moore J. G. (1970) Water content of basalt erupted on the ocean floor. *Contrib. Mineral. Petrol.* **28**, 272–279.

Moore J. G. and Calk L. (1971) Sulfide spherules in vesicles of dredged pillow basalt. *Amer. Mineral.* **56**, 476–488.

Moore J. G. and Fabbi B. P. (1971) An estimate of the juvenile sulfur content of basalt. *Contrib. Mineral. Petrol.* **33**, 118–127.

Moore J. G. and Schilling J. G. (1973) Vesicles, water, and sulfur in Reykjanes Ridge basalts. *Contrib. Mineral. Petrol.* **41**, 105–118.

Muenow D. W., Liu N. W. K., Garcia M. O. and Saunders A. D. (1980) Volatiles in submarine volcanic rocks from the spreading axis of the East Scotia Sea back-arc basin. *Earth Planet. Sci. Lett.* **47**, 272–278.

Murck B. W., Burruss R. C. and Hollister L. S. (1978) Phase equilibria in fluid inclusions in ultramafic xenoliths. *Amer. Mineral.* **63**, 40–46.

Mysen B. O. (1977) The solubility of H_2O and CO_2 under predicted magma genesis conditions and some petrological and geophysical implications. *Rev. Geophys. Space Phys.* **15**, 351–361.

Mysen B. O. and Popp R. K. (1980) Solubility of sulfur in $CaMgSi_2O_6$ and $NaAlSi_3O_8$ melts at high pressure and temperature with controlled f_{O_2} and f_{S_2}. *Amer. J. Sci.* **280**, 78–92.

Naldrett A. J. (1969) A portion of the system Fe-S-O between 900 and 1080° C and its application to sulfide ore magmas. *J. Petrol.* **10**, 171–201.

Naldrett A. J. and Goodwin A. M. (1977) Volcanic rocks of the Blake River Group, Abitibi Greenstone Belt, Ontario, and their sulfur content. *J. Earth Sci.* **14**, 539–550.

Naldrett A. J., Craig J. R. and Kullerud G. (1967) The central portion of the Fe-Ni-S system and its bearing on pentlandite exsolution in iron-nickel sulfide ores. *Econ. Geol.* **62**, 826–847.

Naldrett A. J., Goodwin A. M., Fisher T. L. and Ridler R. H. (1978) The sulfur content of Archean volcanic rocks and a comparison with ocean-floor basalts. *Can. J. Earth Sci.* **15**, 715–728.

Nash W. P. and Hausel W. D. (1973) Partial pressures of oxygen, phosphorus and fluorine in some lunar lavas. *Earth Planet. Sci. Lett.* **20**, 13–27.

Naughton J. J. (1973) Volcanic flame: source of fuel and relation to volcanic gas-lava equilibrium. *Geochim. Cosmochim. Acta* **37**, 1163–1169.

Naughton J. J., Heald E. F. and Barnes I. L., Jr. (1963) The chemistry of volcanic gases. 1. Collection and analysis of equilibrium mixtures by gas chromatography. *J. Geophys. Res.* **68**, 539–544.

Nordlie B. E. (1971) The composition of the magmatic gas of Kilauea and its behavior in the near surface environment. *Amer. J. Sci.* **271**, 417–463.

Propach G. (1979) Decorated vesicles in basaltic glasses from hole 396, drilled on leg 45, deep-sea drilling project. *Initial Reports of the Deep Sea Drilling Project* **XLV**, p. 551–556. U.S. Govt. Printing Office, Washington, D.C.

Roedder E. (1965) Liquid CO_2 inclusions in olivine bearing nodules and phenocrysts from basalts. *Amer. Mineral.* **50**, 1746–1782.

Roedder E. and Weiblen P. W. (1978) Melt inclusions in Luna 24 soil fragments. In *Mare Crisium: The View from Luna 24*. (R. B. Merrill and J. J. Papike, eds.), p. 495–522. Lunar and Planetary Science Institute. Pergamon Press, N.Y.

Sato M. (1978) Oxygen fugacity of basaltic magmas and the role of gas forming elements. *Geophys. Res. Lett.* **5**, 447–449.

Sato M. and Wright T. L. (1966) Oxygen fugacities directly measured in magmatic gases. *Science* **153**, 1103–1105.

Shepard E. S. (1938) The gases in rocks and some related problems. *Amer. J. Sci.* **35A**, 311–351.

Shima H. and Naldrett A. J. (1975) Solubility of sulfur in an ultramafic melt and the relevance of the system Fe-S-O. *Econ. Geol.* **70**, 960–967.

Sigvaldason G. E. and Elisson G. (1968) Collection and analysis of volcanic gases at Surtsey, Iceland. *Geochim. Cosmochim. Acta* **32**, 797–805.

Skinner B. J. (1970) High crystallization temperatures indicated for igneous rocks from Tranquillity Base. *Science* **167**, 652–654.

Skinner B. J. and Peck D. L. (1969) An immiscible sulfide melt from Hawaii. *Econ. Geol.* **4**, 310–322.

Smith F. G. (1962) Metallic sulphide melts as igneous differentiates. *Can. Mineral.* **6**, 663–669.

Spera F. J. and Bergman S. C. (1980) Carbon dioxide in igneous petrogenesis: I, Aspects of the dissolution of CO_2 in silicate liquids. *Contrib. Mineral. Petrol.* **74**, 55–66.

Stolper E. (1977) Experimental petrology of eucritic meteorites. *Geochim. Cosmochim. Acta* **41**, 587–611.

Swanson D. A. and Fabbi B. P. (1973) Loss of volatiles during fountaining and flowage of basaltic lava at Kilauea Volcano, Hawaii. *J. Res. U.S. Geol. Survey* **1**, 649–658.

Usselman T. M., Hodge D. S., Naldrett A. J. and Campbell I. H. (1979) Physical constraints on the characteristics of nickel sulfide ore in ultramafic lavas. *Can. Mineral.* **17**, 361–362.

Wellman T. R. (1970) Gaseous species in equilibrium with the Apollo 11 holocrystalline rocks during their crystallization. *Nature* **225**, 716–717.

White E. and Waring G. A. (1963) Volcanic emanations. *Data of Geochemistry, 6th edition, U.S. Geol. Survey Prof. Paper 440-K*, p. K1–K13.

1.4 PRIMARY MAGMAS AND MAGMA EVOLUTION

1.4.1 PROCESSES THAT AFFECT BASALTIC MAGMA COMPOSITION

1.4.1.1 Introduction

Many processes can alter the composition of magma formed within a planetary interior before it solidifies to produce basalt at the surface. Table 1.4.1.1 is a partial listing of important chemical processes. Each process requires an appropriate physical setting and time scale to be effective. Products of differentiation range in scale from a particular layer in a single zoned crystal to tens of cubic kilometers of lava. Because the mechanisms which result in melting in the magma source region can strongly affect magma composition, they are listed in Table 1.4.1.1.

This section is concerned principally with identification of the processes that affect magma compositions. Additional information pertinent to many of these processes is included in the sections of this chapter on primary magmas (1.4.2) and on volatiles (1.3.5), and in Chapters 3 (experimental petrology) and 7 (isotopes and trace elements).

A basaltic liquid produced directly from a mantle source and compositionally unmodified by any subsequent process is termed *primary*. Primary liquids are of special interest as compositional probes of their source regions. From them, one can infer whether planetary mantle source regions are homogeneous or inhomogeneous, and whether the sources themselves are differentiated relative to the primordial matter (taken to be represented by solar and chondritic relative elemental abundances) involved in planetary accretion (see section 3.1.2.).

How do we decide whether the composition of a basalt sample might represent that of a primary liquid? First it must be shown that the composition is that of a liquid, as opposed to that of a mixture of liquid and fractional crystallization products or that of a material affected by alteration. This requires textural studies and melting experiments (section 3.1.4). Then compositional properties must match those expected for primary liquids on the basis of current concepts of the nature of the planetary mantle and melting processes (section 1.4.2). As shown in section 1.4.2, very few basalts or even basaltic liquids possess these properties. Because opportunities abound for compositional alteration of primary liquids prior to emplacement of an actual basalt, this is perhaps not surprising.

In a series of basalts thought to be derived from a common source region, but modified individually to different extents by post-melting processes, those members that compositionally come closest to meeting the criteria for a primary basalt are termed the most "primitive" members of the series. Such basalts have the highest MgO concentrations, the highest values of Mg/(Mg + Fe), the highest compatible trace element concentrations, and the lowest incompatible trace element concentrations *among those members of the series whose compositions are those of liquids*. As shown in section 1.4.2, some characteristics of primitive basalts from different basalt suites converge toward common values which might be regarded as characteristic of primary basalts. In these, relative concentrations of incompatible trace elements deviate substantially from

Table 1.4.1.1 Some processes that can affect magma compositions.

Melting	Crystallization	Mixing	Fluid separation
Equilibrium	*Equilibrium*	*Equilibrium*	*Equilibrium*
Batch	Bulk equilibrium	Two or more magmas	Immiscible liquids
Incremental	*Kinetic control*	Wall-rock equilibration	Liquid-vapor
Each increment a flow	Surface equilibrium	Fluid addition	Solid-vapor
Increments pooled	Rate effects of	*Kinetic control*	*Kinetic control*
Inhomogeneous source	Nucleation	Two or more magmas	Immiscible liquids
Kinetic control	Quenching	Wall-rock assimilation or extraction	Liquid-vapor
Sudden local heating or pressure release & melt separation	Supercooling	Xenocrysts and xenoliths	Solid-vapor
	Crystal settling		Flow separation of crystals during transport
	Crystal flotation		
Uniform heating, rapid melt separation	Crystal plating		
	Diffusion		

chondritic values. Since these elements are thought to be extracted efficiently from the source regions during melting, it follows that the source regions themselves are differentiated from primordial matter. This conclusion is supported by studies on ultramafic nodules (e.g., Frey and Prinz, 1978). From these studies, Frey and co-workers have concluded that terrestrial mantle source regions consist of two independent components. The first is responsible for most major element trends, for the principal mineralogical composition, and for contents of compatible trace elements; it corresponds in composition to the residue from 5–25% partial melting of pyrolite. The other, more mobile component, is responsible for the remainder of the major elements (especially alkalies), the incompatible trace elements, and volatiles. Menzies and Murthy (1980) have proposed that an essential step toward formation of certain alkalic basalts is the recharging of refractory residues with the second component.

1.4.1.2 Crystal fractionation

The differentiation processes that affect the compositions of magmatic liquids on their way to the surface may occur simultaneously or sequentially. Their combined effects will depend on the extent of each and the order in which they occur. Crystal fractionation, i.e., partial or complete separation of crystallized products from residual parent liquid, is the process usually thought of as being mainly responsible for deviations of basalt compositions from that of the primary magma. Crystallization may serve only to change relative amounts of equilibrium crystals and liquid in a given portion of a magma. It may, however, lead to substantial changes in magma composition if extensive amounts of crystals are continuously isolated from the crystallizing parent liquid (fractional crystallization).

Crystal fractionation is expected to occur. A liquid cut off from its heat source will cool, producing crystals with densities and transport properties different from those of the liquid, so that gravitational settling or flow can lead to their separation. Compositional paths of magmas experiencing fractional crystallization (e.g., "liquid lines of descent") can be estimated by experimental petrology. Effects of fractional crystallization are manifest within single hand specimens of basalt (phenocrysts, variable groundmass, mesostasis) and on a much larger scale in layered intrusive bodies (e.g., the Skaergaard intrusion) and in thick sills (e.g., the Palisades).

Compositional paths for residual liquids are commonly represented on AFM diagrams or variation diagrams (e.g., Figs. 1.2.2.11, 1.2.3.8). The most common paths for mafic liquids lead toward decreasing Mg/(Mg + Fe), decreasing Ca/Na, and increasing SiO_2. Whenever appreciable olivine or pyroxene is crystallized and separated, concentrations of compatible elements such as Ni decrease drastically in the residual liquid. Concentrations of incompatible elements increase, approximately in inverse proportion to the fraction of the original material remaining liquid. As in the case of partial melting processes, trace element behavior during fractional crystallization can be described through the use of partition coefficients (e.g., Schilling and Winchester, 1967; Greenland, 1970; Paster *et al.*, 1974).

When compositions of individual basalts, or of basalts and associated, more silicic volcanic rocks, are examined, they commonly show the trends expected from fractional crystallization. Combining that observation with knowledge of layered intrusions has led to the concept of the magma chamber as the locus of activity for most differentiation. By magma chamber, we mean a reservoir in which magma resides temporarily, between the time of magma formation and eruption. During this interval of residence in a magma chamber, heat may be lost and the magma may partly crystallize. Geophysical studies (e.g., seismic and tiltmeter measurements) have been interpreted as indicating substantial amounts of magma beneath Hawaii, as would be expected in magma chambers (e.g., Koyanagi *et al.*, 1976). There seems to be a requirement for some sort of collecting chamber for magmas. For example, tholeiites are regarded as products of ~20% partial melting of their source regions (assuming a source similar in composition to pyrolite). It appears difficult, however, to retain as much as a percent or so of partial melt in contact with its source residue (Turcotte and Ahern, 1978a, b), so successive increments of melt apparently have to be pooled before they can erupt as tholeiite. Just how this happens remains unclear. The sizes of magma chambers which would have been required to store some of the Columbia River flows prior to their eruptions are impressive—typically 10–20 km^3, but as much as 700 km^3 per flow (see section 1.2.3).

What compositional relationships might be expected among members of a series of successive lava flows? In a very simple situation, such flows might represent successive residual liquids in the sense that, as magma in a chamber differentiates continuously, portions of the liquid remaining would occasionally be extruded. Among the characteristics which such a series of lava flows would display are the following: compositions that monotonically follow a differentiation trend as defined by experimental petrological studies; inclusion in the lavas of phenocrysts corresponding to the crystals being fractionated; quantitative agreement between observed liquid compositions and those predicted from mass-balance mixing models for major elements and from partitioning models for trace elements; isotopic homogeneity; and overall consistency with field and geophysical information. A search for all these types of evidence for a lava series requires extensive, detailed studies based on large numbers of well-documented samples and the combining of results from many disciplines. This, however, is a relatively recent style of investigation, so complete sets of evidence on single volcanic series are rare.

Even before such integrated studies were begun, it was evident from field and petrographic observations that successive lava flows in a given area did not support a model as simple as the quietly differentiating magma chamber suggested above. The more complicated patterns observed have not generally been taken as evidence against the simple magma chamber hypothesis, however. Results have been interpreted to imply multiple chambers, multiple conduits, or other features, so that the observed stratigraphic order is a product of the combined outputs of several chambers at various different stages of crystallization. The basic notion of crystal fractionation in magma chambers seems to enjoy firm acceptance as the major process for generation of a broad range of magma compositions.

As indicated by the examples below, it is important to define better the setting and mechanism of crystal fractionation in basaltic magmas and to be wary of interpretations based on a dozen or so samples collected to represent the widest variety of rock types observed in the field. The integrated studies now being carried out do not imply that simple crystal fractionation is the principal mechanism responsible for compositional variations among members of as many lava suites as might be expected. They indicate that crystal fractionation, whether in a magma chamber or in some other setting, is not the sole or principal cause of substantial compositional variation in some suites. Where crystal fractionation is most clearly implied by the evidence, compositional changes are not extensive. Where crystal fractionation remains the favored mechanism for extensive compositional variations, essential confirming evidence is lacking. Thus, the extent and the importance of this mechanism are poorly known.

1.4.1.3 Results from some selected studies

In this section we discuss a few studies selected from the literature that bear on processes proposed for producing or modifying basalt compositions. Definitive statements can be made about processes responsible for compositional variations in some suites of lavas. These studies indicate how heavily evidence from one discipline can weigh against mechanisms which might be accepted on the basis of evidence from other disciplines.

Cases in which crystal fractionation may predominate

A good example of the complexity that can be encountered is the study by Leeman *et al.* (1976) of 39 lavas from the Craters of the Moon National Monument. Lavas sampled ranged from alkali-rich ferrobasalts to ferrolatites in composition. On an AFM diagram, their compositions define a smooth enrichment of Fe and alkalies as would be expected for a series of successive liquids derived by fractional crystallization. A least-squares mixing model was used to demonstrate that the more evolved lavas could be related to one of the more primitive ones by loss of crystals with compositions like those of included microphenocrysts. Quantitative modelling showed that trace element compositions also could be accounted for by that model. The authors concluded only that, overall, the mineralogic, petrographic, and geochemical evidence *suggested* that the lavas were members of a genetically related suite of liquids derived from a common parent magma or from a common type of parent magma.

With so much evidence favoring an origin for the lavas by fractional crystallization from a common parent, why not make a stronger statement? The authors found no simple relationship between stratigraphic position and extent of compositional evolution. Phenocrysts observed were mainly microphenocrysts. Experiments based on anhydrous, low-pressure conditions did not produce the evolutionary path inferred from the petrography and compositions of the lavas. Rather high and somewhat variable $^{87}Sr/^{86}Sr$ ratios (0.708–0.712) were found, suggesting some incorporation of crustal material. Chemical and isotopic compositions did not support an origin for the most primitive ferrobasalt by direct melting of a mantle source, but neither did a crystal fractionation model support derivation of the ferrobasalt by partial crystallization of typical, primitive olivine tholeiite of the Snake River Plain. These problems could perhaps be overcome in a manner that leaves fractional crystallization as the principal process responsible for the trend of compositions among the sampled lavas (e.g., a hydrous rather than anhydrous differentiation path for the lavas sampled, a model for selective contamination with crustal radiogenic elements, etc.). However, this argument has not yet been made. The problem of the stratigraphy will remain and will require a mechanism more complex than simple crystallization of a single parent liquid in a magma chamber.

Ferrara and Trueil (1974) and Barberi *et al.* (1975) make a good case, based on quantitative modelling of incompatible trace elements and consistent petrographic information, that the suites of peralkaline silicic volcanics of Pantelleria and Boina constitute straightforward series derived by fractional crystallization of basalt [Pantelleria is a Mediterranean island; Boina is set on fissure basalt fields of the Central Western Afar rift]. Concentration ratios of incompatible elements are constant over an order of magnitude of change in concentration, as would be expected for pure liquid residues of fractional crystallization. These papers do not include detailed modelling of the behavior of major element or compatible trace element compositions. Low but slightly variable $^{87}Sr/^{86}Sr$ ratios were observed, requiring either selective contamination of the magma or storage of the magma prior to its eruption for long enough that Rb decay altered the initial isotopic composition (0.7046–0.7074, Boina; 0.7030–0.7078, Pantelleria). Details of stratigraphy are not reported, except that the more silicic lavas were among the last to erupt in these shield volcanoes.

Classical examples of lava suites which are believed to be related to common parents by fractional crystallization are those of the oceanic islands. These suites may range in composition from olivine basalt, nephelinite and basanite through a series of more silicic lavas all the way to rhyolite. In these settings, one cannot call upon contamination by assimilation of continental crust or by melting of siliceous crust to explain the origins of the silicic volcanics, because only mafic oceanic crust overlies the source regions. Zielinski (1975) studied 11 lavas of Reunion Island ranging in composition from olivine-rich basalt to trachyte. He concluded, on the basis of major and trace element modelling, that 8 of the rocks might be related to a common olivine basalt by fractional crystallization, but that 3 could not be so related to that olivine basalt or to any common source. After more extensive samplings, Ludden (1978) found two distinct sequences in one Reunion volcano (Piton de la Fournaise). Olivine was the principal phase fractionating throughout basalt evolution, followed by the cotectic crystallization of four phases wherever still more evolved lavas had formed. He concluded, on the basis of petrographic and major and trace element modelling of 9 samples, that fractional crystallization could account for all compositional variation observed among the basalts of that volcano. The earlier volcanics were the more silicic. Ludden suggested that, as the primary shield formed, magma input rates were relatively low and feeder systems poorly developed, so that magma had sufficient time to evolve into the cotectic crystallization mode prior to eruption. Later lavas rose more rapidly and through well-developed feeders, so their parent magmas were still crystallizing olivine alone at the time of eruption. Such a scheme would account for production of more differentiated lavas during buildup of the shield volcano, followed by olivine basalt during the later caldera stage. It is not clear, however, what physical setting would be required for this process to occur.

More complex behavior in magma chambers

O'Hara (1977) produced a mathematical model for trace element behavior in a single magma chamber which is periodically refilled with fresh magma. Successive lavas erupted from such a chamber would correspond to members of a series of liquids produced by an ever-increasing degree of monotonic fractional crystallization only until a new injection of magma entered the chamber. The new magma then would mix with the old, beginning a new fractionation sequence with a lava of less primitive compositional character than that just injected into the chamber, because the fresh magma has mixed with the previously-differentiated magma. Also, because the injection of new magma terminates the monotonic differentiation series of the previous batch, many sequences might not reach very silicic compositions before the next sequence begins. Yanagi and Ishizaka (1978) showed that the variations in Rb, Sr and K concentrations in 51 samples collected with stratigraphic control from 3 stratovolcanoes in Central Japan seemed to be consistent with O'Hara's model, with each volcano yielding several sequences. Data for major elements, other trace elements and isotopes are missing, however, and petrographic studies (Hayatsu, 1977) indicate abundant xenoliths whose presence was not accounted for by Yanagi and Ishizaka.

That such differentiation sequences have not been observed in more locations may simply reflect failure to collect and study sufficient samples in well-documented stratigraphic sequences. However, O'Hara's model does not seem to be applicable to continental flood or ocean floor basalts, judging from the evidence discussed earlier in this chapter (sections 1.2.3, 1.2.4 and 1.2.5). Nor do successive lavas of the North Shore Volcanic Group, or of the Siberian Traps (section 1.2.2), or of the several sequences discussed below follow such a pattern; rather, they trend toward less evolved basalt compositions with increasing stratigraphic height (younger age). Lavas of the Columbia River Plateau (section 1.2.3) do not yield any such sequence, and those of the Taos Plateau (section 1.2.4) hardly vary in composition with stratigraphic position. Nor can the more silicic volcanics of the Taos Plateau or the North Shore Volcanic Group be related to the basalts by simple fractional crystallization.

Numerous studies have been made of andesites and their associated volcanic rocks. Compositional and petrographic data have been used to support an origin of more evolved lavas by fractional crystallization of more primitive ones (e.g., Ewart *et al.*, 1973). Various schemes for fractional crystallization among andesites have been reviewed by Gill (1978), who has concluded that quantitative modelling of the processes does not yet match the observed compositions well enough to be convincing. However, sampling has been inadequate to provide a good test of whether fractional crystallization can reasonably account for the compositions or whether more complicated modes of origin are required.

Cases of "inverted" stratigraphy

Individual Galapagos Island volcanoes have yielded series of lavas ranging in composition from tholeiite to, in some cycles, siliceous rhyolite (Swanson *et al.*, 1974; Lindstrom, 1976). Compositions of the individual lavas within a series seem to be related to one another by fractional crystallization, on the basis of all chemical and petrographic information. Each sequence, however, begins with siliceous lava and trends toward the more primitive members. The physical setting in which such a process might occur has not been defined. It does not appear that a well-mixed magma chamber could differentiate so as to produce aphyric trachyte liquid at the top, while retaining fluid tholeiite near the base, then erupt in a manner which preserves that stratification in the lavas. Might such extensive differentiation occur within conduits, as suggested for the lavas from Piton de la Fourniase by Ludden (1978)? Irving (1980) has interpreted structures and compositions of composite mantle xenoliths in terms of a model in which magmatic differentiation occurs mainly through the plating of crystals onto walls of conduits during magma gathering and transport (see section 1.2.11). Could differentiation of earlier portions of a magma be extensive, while that for later portions be less so, in such a case?

That the first lavas emitted during an eruption cycle are relatively silicic apparently is not unusual. Thorarinsson and Sigvaldason (1972) point out that cycles of the Icelandic volcano Hekla begin that way. Furthermore, they show a monotonic relationship between the length of the period of quiescence between eruption cycles and the silica content of the initial tephra ejected when a new cycle begins. Silica content decreases with time as the eruption cycle continues. The authors suggest that the process is dominated by the assimilation of acid rocks during quiet periods, rather than fractional crystallization. Large phenocrysts are absent, and compositional variability observed among early tephras of the 1970 eruption suggests that there may have been two distinct liquid compositions involved, which were mixed incompletely.

Volumetrically, cases of "inverted" stratigraphy are important. The North Shore Volcanic Group and the Siberian Traps represent major rift plateaus. Inverted stratigraphy has been documented by Wood (1976, 1978, 1979) for flows from Iceland, the Faeroes, the Isle of Skye, and the Troodos Massif. He explains these results on the basis of dynamic melting (Langmuir *et al.*, 1977). Also, Batiza (1981) reports a weak but definite trend toward less differentiated lavas at higher elevations for MORB lavas from DSDP Leg 61.

Differentiation accompanying magma movement

Crystal fractionation during short-term storage in a dike-like conduit has been suggested by Rose *et al.* (1978) to account for the compositions and stratigraphic relationships among successive emissions of volcanic tephra of Fuego volcano, Guatemala. The magma which erupted earliest was rich in plagioclase and poor in mafic phenocrysts, the magmas that followed contained few phenocrysts of any kind, and late magmas were enriched in all phenocryst phases. This compositional sequence, in conjunction with theoretical and observed geophysical information, led Rose *et al.* to a model of a dike-shaped conduit that was filled with fresh magma which stayed in place for as much as a few tens of days before a series of eruptions. During that time, mafic crystals and some plagioclase crystals settled toward the lower regions of the conduit and some plagioclase, just too dense to float in the parent magma, was rafted upward by bubbles. The eruption sequence simply preserved the stratified compositions of the conduit.

Effects of crystal fractionation on compositions of Hawaiian lavas are well established. Wright and Fiske (1971) and Wright *et al.* (1975) made extensive use of a least-squares mixing model (Wright and Doherty, 1970), in conjunction with geophysical and stratigraphic information and knowledge of basalt crystallization trends, to provide a quantitative demonstration of the occurrence and extent of several composition-modifying processes involved in the production of Kilauea lavas. Very precisely determined major element compositions were used with the mixing model to establish what compositions of liquids plus phenocryst minerals (mainly olivine) or of liquids plus other liquids could explain the compositions of all lavas studied in terms of a few parental liquids. They were able to recognize batches of magma which were produced separately in the source region and to define the compositional differences among batches. They showed the effects of olivine loss or gain on portions of given magma batches and that mixing of different batches of magma had occurred to produce lavas in some of the eruptions. They showed that flow differentiation of olivine, augite, and plagioclase had occurred during transport in conduits of the rift zone and used this information to locate the rift zone system. They identified four separate processes which have affected the compositions of Kilauea basalts. Much of the variation observed corresponds to simple gain or loss of olivine without extensive compositional evolution of the residual lava due to fractional crystallization. No trace element studies were included.

Richardson (1979) used a similar mixing model with both major and trace element data to show that a sill of Karroo dolerite was stratified compositionally by flow differentiation that concentrated olivine, plagioclase, and augite phenocrysts toward the center of the sill.

Small-scale compositional heterogeneity

Single lava flows which are compositionally homogeneous, in the sense that they show no systematic effects of crystal fractionation by flow differentiation or gravity settling, can be heterogeneous on the small scale of a hand specimen. Such spatially-random compositional heterogeneity was observed in some 30 samples from a vertical section of an 11 m thick Iceland flow by Watkins *et al.* (1970). This type of heterogeneity appears not to be unusual and to result mainly from nonrandom distribution of the phenocryst and groundmass crystals and the products of crystallization of residual liquid, which apparently is still mobile even when the flow is ~90% crystallized (Lindstrom and Haskin, 1980). Such compositional heterogeneity has been observed chemically and petrographically in lunar basalts (Haskin and Korotev, 1977; Walker *et al.*, 1976a). Movement of relatively late-stage liquid is evident in segregation vesicles (Smith, 1967). Portions of Servilleta basalt (Taos Plateau) which are rich in segregation vesicle pipes have major and trace element compositions which are substantially different from those in more massive portions of the flows (Haskin *et al.*, 1978). Because of these short-range heterogeneities, the variability of compositions of samples of dense Servilleta basalt from a single flow approaches that observed among similar single samples from 15 separate flows. Some geochemical models for relating separate flows to each other as part of a fractionation sequence depend upon the accurate determination of small compositional differences. Short-range heterogeneities can produce compositional variations very similar to those which would be found in a series of flows related by limited crystal fractionation and could result in lavas from identical flows being interpreted as such a sequence.

The lunar mare basalts pose an especially interesting problem in this regard, because there is no field evidence to indicate how many different flows were sampled at each landing site. Petrologists, on the basis of petrography, mineral chemistry, and bulk compositions, suggest that relatively few flows were sampled (e.g., Rhodes *et al.*, 1977; Beaty and Albee, 1978). Some geochemists, on the other hand, have been unable to match the predictions of simple models for fractional crystallization and partial melting with the actual trace element compositions of the samples, so they suggested that many flows were sampled. Further, they inferred from the data that the partial melting events which produced those flows occurred in a slightly heterogeneous source region (e.g., Shih *et al.*, 1975). Recognition that the variability of trace element contents within the basalt suites defined by petrologists is similar to that seen in individual terrestrial flows (Haskin *et al.*, 1977), and that short-range heterogeneities were the cause and could be explained (Lindstrom and Haskin, 1980), removes the apparent requirement for a large number of flows. This does not negate the possibility that some lunar basalt suites consist of related samples derived by crystal fractionation, but emphasizes the need for supporting evidence beyond chemical compositions when the extent of differentiation is small. Walker *et al.* (1976b) on petrological grounds and Rhodes *et al.* (1977) on compositional grounds make good cases for relating the members of the suite of Apollo 12 olivine basalts to each other by crystal fractionation of olivine within a single flow.

1.4.1.4 Magma mixing

Dungan and Rhodes (1978) found that residual glasses from different basalt units in ocean floor basalts (DSDP Legs 45 and 46) could not be related to one another compositionally by simple fractional crystallization. Plagioclase phenocrysts in these lavas are partially resorbed, with cores more anorthitic in some cases and less anorthitic in others than would be expected to crystallize from their host liquids. The cores contain melt inclusions which have been interpreted as being related to the melts responsible for the partial resorption. Those crystals which are rimmed by plagioclase which is more anorthitic than the cores contain melt inclusions with higher Mg/(Mg + Fe) than the bulk rocks, but not as high as is observed in the most primitive magma of the series.

This situation can be explained if primitive magma has been mixed with more evolved magma which contained abundant phenocrysts. The specific model used (Rhodes *et al.*, 1979), based in part on that of O'Hara (1977), involved extensive fractional crystallization in a periodically refilled magma chamber, and was designed to account for a number of characteristics of ocean ridge basalts. These characteristics include the scarcity of primitive lavas, the apparent chemical requirement for some clinopyroxene crystallization (if lavas are to be related by crystal fractionation) despite the absence of abundant clinopyroxene phenocrysts, and the inability to account for incompatible trace element compositions of evolved lavas by simple fractional crystallization. Walker *et al.* (1979) found similar relationships in basalts from the Oceanographer Fracture Zone.

Magma mixing has also been proposed to account for compositional trends which extend over long distances along the Mid-Atlantic Ridge, e.g., the Reykjanes Ridge to Iceland (Schilling, 1973). To a first approximation, compositional trends there seem to grade from typical ocean ridge basalt at one extreme to incompatible-element-enriched island tholeiite at the other. A simple mixture of two magmas, however, falls far short of accounting for the details of the compositions of the individual suites of basalts collected. A review of recent work is included in the paper of Langmuir *et al.* (1978), which discusses trace element and isotope behavior during mixing processes. Zindler *et al.* (1979) argue, based upon compositional and isotopic studies of basalts from the Reykjanes Peninsula, that the mantle underneath must be vertically heterogeneous. They contend that the different lavas result from mixtures of separate magmas produced by different degrees of partial melting within the heterogeneous source region. Methods by which the effects of mixing can be recognized unambiguously and by which the relative importance of mixing processes can be determined are still in early stages of development.

Jacobsson *et al.* (1978) have shown that lavas of the Reykjanes Peninsula occur as swarms, beginning with picrite shield volcanoes, followed soon by tholeiite shields and then, after a time, by fissure tholeiites. They found evidence for low-pressure fractional crystallization, but concluded that the shield and fissure lava suites are not related to each other and that fractional crystallization cannot account in general for the volume-composition relationships among the lavas within each suite. They noted a positive correlation of the concentrations of incompatible elements and total iron with volume of flow, and suggested that vapor transport may be operative in producing the lavas. A model for trace element behavior in the presence of a vapor phase has been formulated (Shaw, 1978) but has not yet been applied to volcanic sources or systems.

Arguments for source region heterogeneity rest mainly on isotopic evidence and are discussed in Chapter 7 (see section 7.4.2.). Isotopic heterogeneity beneath the ocean crust is widespread and well-documented (e.g., Gast, 1968). Arguments for heterogeneity, as well as for mixing of magmas from different source regions of the mantle, are most easily made for volcanics derived from the oceanic environment, where conduits are not obliged to traverse continental crust. Isotopic heterogeneity among continental basalt flows can often be ascribed to assimilation of crustal material. Doe *et al.* (1969) found isotopic heterogeneity in Servilleta basalts and others from the Southern Rocky Mountains which correlated with the presence of alkali plagioclase xenocrysts (armored with calcic plagioclase) and quartz xenocrysts (armored with clinopyroxene). They ascribed this to assimilation of more silicic crustal rock.

1.4.1.5 Assimilation

Assimilation of crustal material becomes indistinct from magma mixing, according to the evidence compiled by Anderson (1976), Eichelberger (1975, 1978), and Eichelberger and Gooley (1977). Briefly, they suggest that the genetic relationships among basalts and spatially associated, more silicic magmas do not stem from fractional crystallization of primitive basalt magma. Rather, they suggest that the rise of basaltic magma from mantle source regions into the crust transfers sufficient heat to cause melting of crustal material, thus producing magmas more silicic than basalt. Eichelberger suggests that the partial melting usually produces rhyolitic magma, and that lavas intermediate between basalt and rhyolite in composition are mostly mixtures of the two end-member magma types. Anderson does not mention such an extreme composition as rhyolite for melts formed within the lower crust, but suggests that andesitic and silicic magmas are of crustal origin and that they mix extensively with each other and with basalt just prior to eruption. These conclusions are based mainly on extensive petrographic and field observations so far, and need to be supported by more comprehensive geochemical studies.

1.4.1.6 Partial melting

As shown most vividly by Gast (1968), the melting process exerts very strong control over basalt compositions and, especially for incompatible trace elements, can be the predominant factor responsible for the wide range of compositions observed in a suite. The comprehensive modelling study of Frey *et al.* (1978) demonstrates how this can work. The suite of basalts studied was from southeastern Australia and covered a compositional range from tholeiites through alkali olivine basalts and basanites to olivine melilitites. On the basis of criteria similar to those discussed earlier in this section, 6 of the 13 lavas studied were identified as possible primary liquids, including at least one of each lava type except tholeiite. Primary liquid compositions for the other individual lavas were estimated by adding sufficient olivine to each to give it a composition that would be in equilibrium with a source region with olivine of Fo_{88-90}. The observed range in concentrations for a typical incompatible element, La, was 16 to 103 ppm. Within the limits of knowledge of values for distribution coefficients, Frey *et al.* were able to account for the origins of all of the "primary" compositions by partial melting of a single, homogeneous source, with 4–6% being required to produce the olivine melilitites and 20–25% to produce the tholeiites. The source region, however, was inferred to be substantially different in composition from primordial matter with respect to trace elements, being relatively enriched in strongly incompatible elements (6–9 times chondritic abundances) as compared with moderately incompatible ones (2–3 times). It was proposed that a volatile-rich fluid might be responsible for such differentiation of the mantle and its consequent heterogeneity.

Partial or complete melting can occur at the surface of a planet when meteoroid impact occurs. This can result in large amounts of melt which, in an appropriate substrate, might have the composition of basaltic magma. (For a discussion of impact melting, see O'Keefe and Ahrens, 1977.) Such melting can be accompanied by extensive volatilization, which selectively depletes the magma of some elements. Housley (1979) has produced a model for vaporization during small scale impacts which indicates that minor to extensive loss of all but the most refractory constituents of the substrate would occur. On the Moon, it can be argued that some small glassy samples represent melts of crystalline rocks which have experienced substantial losses of volatiles by this mechanism (e.g., Ivanov and Florensky, 1975), although the extent to which this might be true for larger melt sheets is not clear. Floran *et al.* (1978), however, found that a principal characteristic of the Manicouagan impact melt sheet (Canada) was its extensive chemical homogeneity over distances of tens of kilometers and thicknesses up to 200 meters. No evidence for selective loss of any volatile elements (e.g., alkalies) relative to the substrate rocks was found by them or by Grieve and Floran (1978). Gibson and Hubbard (1972) have shown that, in the lunar environment (i.e., vacuum conditions), alkalies are lost at temperatures above 1050° C, but concluded that no significant fraction would be lost unless lava flows were stirred extensively.

1.4.1.7 A comment about use of models for composition-modifying processes

It is important to recognize that compositional modelling can provide useful constraints on the detailed nature of the processes by which basaltic magmas form and are modified. In order to use models most effectively, it is necessary to begin with well-defined concepts of the process of interest (e.g., crystal fractionation in a single magma chamber, in multiple chambers, in conduits, or by flow differentiation), then to tailor the mathematical model to fit that situation. The model of O'Hara (1977) for a periodically recharged magma chamber is such a model. Another is the dynamic melting model of Langmuir *et al.* (1977), which accounts for a multiplicity of processes that might cause the considerable compositional variations found in ocean floor basalts in the FAMOUS region. Paster *et al.* (1974) adapted the simple logarithmic model for fractional crystallization to the Skaergaard layered series, and Haskin and Korotev (1977) adapted it to describe crystallization of a basalt on a hand-specimen scale. Allègre *et al.* (1977) and Minster *et al.* (1977) have provided formulations to optimize agreement between models and data from rock systems. Shaw (1978) has produced a model to describe vapor transport. All of the models just mentioned are for trace elements. An effort to provide similar models for major elements has been begun by Hanson and Langmuir (1978).

1.4.1.8 Conclusions

Of the processes listed in Table 1.4.1.1 which may affect basalt magma compositions, several are particularly important, with varying degrees of certainty about our recognition and understanding of them. These include partial melting; fractional crystallization in magma chambers or conduits, with and without occasional recharging; crystal fractionation by gravity in conduits and sills, and by flow differentiation; magma mixing; assimilation; and possibly vapor transport. Similar products may result from different processes or from combinations of processes. The clear-cut identification of the dominant process responsible for formation of a single basalt or a suite of related basalts requires comprehensive observational, analytical, and experimental study.

1.4.1.9 References

Allégre C. J., Treuil M., Minster J. -F., Minster B. and Albarede F. (1977) Systematic use of trace element in igneous process. Part I: fractional crystallization processes in volcanic suites. *Contrib. Mineral. Petrol.* **60**, 57–75.

Anderson A. T. (1976) Magma mixing: petrological process and volcanological tool. *J. Volcan. Geotherm. Res.* **1**, 3–33.

Barberi F., Ferrara G., Santaccoce R., Treuil M. and Varet J. (1975) A transitional basalt-pantellerite sequence of fractional crystallization, the Boina Centre (Afar Rift, Ethiopia). *J. Petrol.* **16**, 22–56.

Batiza R. (1981) Trace-element characteristics of Leg 61 basalts. *Initial Reports of the Deep Sea Drilling Project*, **61**. U. S. Gov't. Printing Office.

Beaty D. W. and Albee A. L. (1978) Comparative petrology and possible genetic relations among the Apollo 11 basalts. *Proc. Lunar Planet. Sci. Conf. 9th*, p. 359–463.

Doe B. R., Lipman P. W. and Hedge C. E. (1969) Primitive and contaminated basalts from the Southern Rocky Mountains, U.S.A. *Contrib. Mineral. Petrol.* **21**, 142–156.

Dungan M. A. and Rhodes J. M. (1978) Residual glasses and melt inclusions in basalts from DSDP Legs 45 and 46: evidence for magma mixing. *Contrib. Mineral. Petrol.* **67**, 417–431.

Eichelberger J. C. (1975) Origin of andesite and dacite: evidence of mixing at Glass Mountain in California and at other circum-Pacific volcanoes. *Geol. Soc. Amer. Bull.* **86**, 1381–1391.

Eichelberger J. C. (1978) Andesitic volcanism and crustal evolution. *Nature* **275**, 21–27.

Eichelberger J. C. and Gooley R. (1977) Evolution of silicic magma chambers and their relationship to basaltic volcanism. *Amer. Geophys. Union Monograph 20*, p. 57–77.

Ewart A., Bryan W. B. and Gill J. B. (1973) Mineralogy and geochemistry of the younger volcanic islands of Tonga, S.W. Pacific. *J. Petrol.* **14**, 429–465.

Ferrara G. and Treuil M. (1974) Petrological implications of trace element and Sr isotope distributions in basalt-pantellerite series. *Bull. Volcanol.* **38**, 548–574.

Floran R. J., Grieve R. A. F., Phinney W. C., Warner J. L., Simonds C. H., Blanchard D. P. and Dence M. R. (1978) Manicouagan impact melt, Quebec, I, stratigraphy, petrology, and chemistry. *J. Geophys. Res.* **83**, 2737–2759.

Frey F. A. and Prinz M. (1978) Ultramafic inclusions from San Carlos, Arizona: petrologic and geochemical data bearing on their petrogenesis. *Earth Planet. Sci. Lett.* **38**, 129–176.

Frey F. A., Green D. H. and Roy S. D. (1978) Integrated models of basalt petrogenesis: a study of quartz tholeiites to olivine melilitites from South Eastern Australia utilizing geochemical and experimental petrological data. *J. Petrol.* **19**, 463–513.

Gast P. W. (1968a) Isotope geochemistry of volcanic rocks. In *Basalts*, v.1, p. 325–358. Interscience, N.Y.

Gast P. W. (1968b) Trace element fractionation and the origin of alkaline and tholeiitic magma types. *Geochim. Cosmochim. Acta* **32**, 1057–1086.

Gibson E. K. and Hubbard N. J. (1972) Thermal volatilization studies on lunar samples. *Proc. Lunar Sci. Conf. 3rd*, p. 2003–2014.

Gill J. B. (1978) Role of trace element partition coefficients in models of andesite genesis. *Geochim. Cosmochim. Acta* **42**, 709–724.

Greenland L. P. (1970) An equation for trace-element distribution during magmatic crystallization. *Amer. Mineral.* **55**, 455–465.

Grieve R. A. F. and Floran R. J. (1978) Manicouagan impact melt, Quebec, 2, chemical interactions with basement and formational processes. *J. Geophys. Res.* **83**, 2761–2771.

Hanson G. N. and Langmuir C. H. (1978) Modelling of major elements in mantle-melt systems using trace element approaches. *Geochim. Cosmochim. Acta* **42**, 725–741.

Haskin L. A., Jacobs J. W., Brannon J. C. and Haskin M. A. (1977) Compositional dispersions in lunar and terrestrial basalts. *Proc. Lunar Sci. Conf. 8th*, p. 1731–1750.

Haskin L. A. and Korotev R. L. (1977) Test of a model for trace element partitioning during closed system solidification of a silicate liquid. *Geochim. Cosmochim. Acta.* **41**, 921–939.

Haskin L. A., Lindstrom M. M. and Dungan M. A. (1978) Trace element fractionation during crystallization of thin tholeiite flows (abstract). In *Lunar and Planetary Science IX*, p. 468–470. Lunar and Planetary Institute, Houston.

Hayatsu K. (1977) Geologic study of the Myoko volcanoes, central Japan: Part 2, petrography. *Mem. Fac. Sci. Kyoto Univ., Series Geol. and Mineral., XLIII*, 1–48.

Housley R. M. (1979) A model for chemical and isotopic fractionation in the lunar regolith by impact vaporization. *Proc. Lunar Planet. Sci. Conf. 10th*, p. 1673–1683.

Irving A. J. (1980) Petrology and geochemistry of composite ultramafic xenoliths in alkalic basalts and implications for magmatic processes within the mantle. *Amer. J. Sci.* **280A**, 389–426.

Ivanov A. V. and Florensky K. P. (1975) The role of vaporization in lunar rock formation. *Proc. Lunar Sci. Conf. 6th*, p. 1341–1350.

Jacobsson S. P., Jansson J. and Shido F. (1978) Petrology of the Western Reykjanes Peninsula, Iceland. *J. Petrol.* **19**, 669–705.

Koyanagi R. Y., Eudo E. T. and Ward P. L. (1976) Seismic activity on the Island of Hawaii, 1970 to 1973. In *The Geophysics of the Pacific Ocean Basin and its Margin* (G. H. Sutton, M. H. Manghani and R. Moberly, eds.), p. 169–172. Geophys. Monograph 19, Amer. Geophys. Union, Washington, D.C.

Langmuir C. H., Bender J. F., Bence A. E., Hanson G. N. and Taylor S. R. (1977) Petrogenesis of basalts from the FAMOUS area: Mid-Atlantic Ridge. *Earth Planet. Sci. Lett.* **36**, 133–156.

Langmuir C. H., Vocke R. D., Jr., Hanson G. N. and Hart S. R. (1978) A general mixing equation with applications to Icelandic basalts. *Earth Planet. Sci. Lett.* **37**, 380–392.

Leeman W. P., Vitaliano C. J. and Prinz M. (1976) Evolved lavas from the Snake River Plain: Craters of the Moon National Monument, Idaho. *Contrib. Mineral. Petrol.* **56**, 35–60.

Lindstrom M. M. (1976) Geochemical studies of volcanic rocks from Pinzon and Santiago Islands, Galapagos Archipelago. Ph.D. dissertation, Univ. of Oregon. 174 pp.

Lindstrom M. M. and Haskin L. A. (1978) Causes of compositional variations within mare basalt suites. *Proc. Lunar Planet. Sci. Conf. 9th*, p. 465–486.

Lindstrom M. M. and Haskin L. A. (1980) Compositional inhomogeneities in a single Icelandic tholeiite flow. *Geochim. Cosmochim. Acta* **45**, 15–31.

Ludden J. N. (1978) Magmatic evolution of the basaltic shield volcanoes of Reunion Island. *J. Volcanol. Geotherm. Res.* **4**, 171–198.

Menzies M. and Murthy V. R. (1980) Mantle metasomatism as a precursor to the genesis of alkaline magmas-isotopic evidence *Amer. J. Sci.* **280A**, 622–638.

Minster J. -F., Minster J. B., Treuil M. and Allégre C. J. (1977) Systematic use of trace elements in igneous processes. Part II. fractional crystallization processes in volcanic suites. *Contrib. Mineral. Petrol.* **61**, 49–77.

O'Hara M. J. (1977) Geochemical evolution during fractional crystallization of a periodically refilled magma chamber. *Nature* **266**, 503–507.

O'Keefe J. D. and Ahrens T. J. (1977) Impact-induced energy partitioning, melting, and vaporization on terrestrial planets. *Proc. Lunar Sci. Conf. 8th*, p. 3357–3374.

Paster T. P., Schauwecker D. S. and Haskin L. A. (1974) The behavior of some trace elements during solidification of the Skaergaard layered series. *Geochim. Cosmochim. Acta* **38**, 1549–1577.

Rhodes J. M., Blanchard D. P., Dungan M. A., Brannon J. C. and Rodgers K. V. (1977) Chemistry of Apollo 12 mare basalts: magma types and fractionation processes. *Proc. Lunar Sci. Conf. 8th*, p. 1305–1338.

Rhodes J. M., Dungan M. A., Blanchard D. P. and Long P. E. (1979) Magma mixing at mid-ocean ridges: evidence from basalts drilled near 22° N on the Mid-Atlantic Ridge. *Tectonophys.* **55**, 35–61.

Richardson S. H. (1979) Chemical variation induced by flowage differentiation in an extensive Karroo dolerite sheet, Southern Namibia. *Geochim. Cosmochim. Acta* **43**, 1433–1441.

Rose W. I., Anderson A. T., Woodruff L. G. and Bonis S. B. (1978). The October 1974 basaltic tephra from Fuego Volcano: description and history of the magma body. *J. Volcanol. Geotherm. Res.* **4**, 3–53.

Schilling J. G. (1973) Iceland mantle plume; geochemical evidence along the Reykjanes Ridge. *Nature* **242**, 565.

Schilling J. G. and Winchester J. W. (1967) Rare-earth fractionation and magmatic processes. In *Mantles of the Earth and Terrestrial Planets,* (S. K. Runcorn, ed.), p. 267–283. Interscience, N.Y.

Shaw D. M. (1978) Trace element behavior during anatexis in the presence of a fluid phase. *Geochim. Cosmochim. Acta* **42**, 933–943.

Shih C. -Y., Haskin L. A., Wiesmann H., Bansal B. and Brannon J. C. (1975) On the origin of high-Ti mare basalts. *Proc. Lunar Sci. Conf. 6th*, p. 1255–1285.

Smith R. E. (1967) Segregation vesicles in basaltic lava. *Amer. J. Sci.* **265**, 696–713.

Swanson F. J., Baitis H. W., Lexa J. and Dymond J. (1974) Geology of Santiago, Rabida, and Pinzon Islands, Galapagos. *Geol. Soc. Amer. Bull.* **85**, 1803–1810.

Thorarinsson S. and Sigvaldason G. E. (1972) The Hekla eruption of 1970. *Bull. Volcanol.* **36**, 269–288.

Turcotte D. L. and Ahern J. L. (1978a) A porous flow model for magma migration in the asthenosphere. *J. Geophys. Res.* **83**, 767–772.

Turcotte D. L. and Ahern J. L. (1978b) Magma production and migration within the moon. *Proc. Lunar Planet. Sci. Conf. 9th*, p. 307–318.

Walker D., Longhi J. and Hays J. F. (1976a) Heterogeneity in titaniferous lunar basalts. *Earth Planet. Sci. Lett.* **30**, 27–36.

Walker D., Longhi J., Kirkpatrick R. J. and Hays J. F. (1976b) Differentiation of an Apollo 12 picrite magma. *Proc. Lunar Sci. Conf. 7th*, p. 1365–1389.

Walker D., Shibata T. and DeLong S. E. (1979) Abyssal tholeiites from the Oceanographer Fracture Zone. II. phase equilibria and mixing. *Contrib. Mineral. Petrol.* **70**, 111–125.

Watkins N. D., Gunn B. M. and Coy-yll R. (1970) Major and trace element variations during the initial cooling of an Icelandic lava. *Amer. J. Sci.* **268**, 24–49.

Wood D. A. (1976) Spatial and temporal variation in the trace element geochemistry of the Eastern Iceland flood basalt succession. *J. Geophys. Res.* **81**, 4353–4360.

Wood D. A. (1978) Major and trace element variations in the Tertiary lavas of eastern Iceland and their significance with respect to the Iceland geochemical anomaly. *J. Petrol.* **19**, 393–436.

Wood D. A. (1979) Dynamic partial melting: its application to the petrogenesis of basalt lava series from Iceland, the Faeroe Islands, the Isle of Skye (Scotland) and the Troodos Massif (Cyprus). *Geochim. Cosmochim. Acta* **43**, 1031–1046.

Wright T. L. and Doherty P. C. (1970) A linear programming and least squares method for solving petrologic mixing problems. *Geol. Soc. Amer. Bull.* **81**, 1995–2008.

Wright T. L. and Fiske R. S. (1971) Origin of differentiated and hybrid lavas of Kilauea Volcano, Hawaii. *J. Petrol.* **12**, 1–65.

Wright T. L., Swanson D. A. and Duffield W. A. (1975) Chemical compositions of Kilauea east-rift lava, 1968–1971. *J. Petrol.* **16**, 110–133.

Yanagi T. and Ishizaka K. (1978) Batch fractionation model for the evolution of volcanic rocks in an island arc: an example from Central Japan. *Earth Planet. Sci. Lett.* **40**, 252–262.

Zielinski R. A. (1975) Trace element evaluation of a suite of rocks from Reunion Island, Indian Ocean. *Geochim. Cosmochim. Acta* **39**, 713–734.

Zindler A., Hart S. R., Frey F. A. and Jakobsson S. P. (1979) Nd and Sr isotope ratios and rare earth element abundances in Reykjanes Peninsula basalts: evidence for mantle heterogeneity beneath Iceland. *Earth Planet. Sci. Lett.* **45**, 249–262.

1.4.2 CHARACTERISTICS OF PRIMARY BASALTIC MAGMAS

1.4.2.1 Introduction

A *primary* basaltic magma may be defined as a liquid of basaltic composition formed by melting within a planetary interior, whose composition has been unmodified by differentiation processes (such as crystal fractionation, contamination, magma mixing, liquid immiscibility and volatile loss) following *segregation* from the crystalline residue of the source and its eruption onto or emplacement near the planet's surface. Peridotite is widely recognized as the most probable source material. By definition, primary magmas are parental, giving rise through differentiation processes to more evolved basalts and to other magmatic derivatives. On the other hand, *parental* basaltic magmas may or may not be primary. They are liquid compositions from which other *associated* and cogenetic magmas can be derived by processes of crystal fractionation and accumulation, contamination, liquid immiscibility and magma mixing (see section 3.1.2). In the last case, more than one parental magma may be involved. As will be evident throughout this section, it is easier to demonstrate the parental nature of a magma than to ascertain whether or not it is primary. Parental magmas have higher liquidus temperatures than associated, related magmas, and crystallize phases with the highest melting points. Because in basaltic magmas these phases are Mg-rich olivines and pyroxenes and Ca-rich plagioclases, parental basaltic magmas have the highest Mg/(Mg + Fe) and Ca/(Ca + Na) ratios within the association, with the exception of rocks of accumulative origin. They are also characterized by high Ni and Cr contents and low incompatible element abundances.

The term *primitive* magma has been used loosely and has several connotations. It is often used synonymously with primary or parental magma, but it is also used to denote eruptions early in a magma sequence or copious eruptions of uniform magma compositions (e.g., MacDonald, 1949; Powers, 1955; Engel *et al.*, 1965). In isotopic and geochemical studies, the term "primitive" is frequently used in a relative sense to indicate compositional characteristics closest to those of chondrites (Chapter 7). Because of this diversity of usage, Carmichael *et al.* (1974) recommend avoiding the term altogether. The term *primitive* will be used here in a comparative sense to indicate that a particular basalt has a greater Mg/(Mg + Fe) ratio, Ca/(Ca + Na) ratio, etc. than some other basalt.

Following Bowen (1928), it is widely accepted that basaltic magmas are products of partial fusion of peridotite within planetary interiors or mantles (e.g., Ringwood, 1975; Taylor, 1975; Yoder, 1976; McGetchin and Smyth, 1978). Early studies of terrestrial basalts searched for a common primary basalt type from which other basalts and, ultimately, more evolved magmas could be derived (e.g., Bowen, 1928; Yoder and Tilley, 1962; Engel *et al.*, 1965; Green and Ringwood, 1967; MacDonald, 1968). It is now recognized that primary basaltic magmas may encompass a variety of types, including tholeiitic, alkaline and nephelinitic basalts (Yoder, 1976). Whether high-alumina basalts (Kuno, 1960) are primary or derivative is still not clear. Lunar primary basalts encompass a range of compositions, including high- and low-K varieties, a wide spectrum of TiO_2 concentrations, and some aluminous types.

In view of the compelling isotopic and trace element evidence for mantle heterogeneity (Hofmann and Hart, 1978), and the wide range of physical conditions under which basaltic magmas might potentially be generated (e.g., Green and Ringwood, 1967; Ringwood, 1975), a variety of primary basaltic magma compositions are to be expected.

Recognizing and characterizing primary magmas and understanding the processes responsible for their formation and evolution are fundamental problems in igneous petrology. Knowledge of their compositions and phase relationships under various physical conditions can provide constraints on the composition and mineralogy of their source regions, the planetary mantles, and on one of the principal mechanisms of planetary differentiation. However, despite their conceptual importance to petrologists, geochemists and planetary scientists, there are no generally accepted criteria that can be used to distinguish primary magmas from magmas whose compositions have been modified since separation from their source regions. Consequently, primary magmas have proven to be extremely elusive, and their identity and characterization are subjects of considerable and enduring contention. On the one hand is the view that most common terrestrial, lunar and achondritic basalts have eutectic-like compositions that are orderly consequences of polybaric crystallization (e.g., Yoder and Tilley, 1962; Mason, 1962; O'Hara, 1965, 1968a, b; O'Hara *et al.*, 1970, 1976; Yoder, 1976). The opposing view is that most basalts are close to primary compositions and have been erupted without *substantial* modification en route to the surface (e.g., Engel *et al.*, 1965; Green and Ringwood, 1967; Green, 1970; Ringwood and Essene, 1970; Green *et al.*, 1971a, b, 1975b; Carmichael *et al.*, 1974; Ringwood, 1975; Wilkinson and Binns, 1977; Presnall *et al.*, 1979).

For many years, Bowen's contention that most high-magnesium basalts were the consequence of crystal accumulation prevailed (Bowen, 1928). A notable exception to this viewpoint was the work of Drever and co-workers (e.g., Drever and Johnston, 1958, 1966), who presented persuasive evidence for the existence of picritic liquids. More recently, following the vigorous arguments of O'Hara (e.g., O'Hara, 1965, 1968a), opinion favors the view that most basalts are evolved and that they are derived from high-magnesium primary magmas (e.g., O'Hara, 1968a, b; Clarke, 1970; Yoder, 1976; Cox, 1978; Elthon, 1979; Green *et al.*, 1979). A major stumbling block to this argument is the apparent scarcity of high-magnesium basalts that are demonstrably close to liquid compositions. A number of explanations for this dearth have been offered: (1) Primary magmas are denser than evolved ones, and hence are less likely to attain the planetary surface (Sparks *et al.*, 1980; Stolper and Walker, 1980). (2) Basaltic magmas may fractionate at various "staging-posts" on the way to the surface, as suggested in Chapter 3. This model has been developed with greatest clarity for Kilauea (Eaton and Murata, 1960; Wright, 1971), and might be expected to pertain in regions of thick lithosphere and crust. Evidence supporting this suggestion comes from highly evolved continental flood basalts. Conversely, one might expect to find less evolved basalts in regions of lithospheric rifting, such as ocean-ridge basaltic volcanism. (3) Increasing evidence suggests that magma mixing is a fundamental volcanic process (e.g., Wright and Fiske, 1971; Wright, 1973; O'Hara, 1977; Rhodes *et al.*, 1979a), and that primary magmas may be mixed with more evolved magmas in shallow magma reservoirs.

On Earth, there are three major contenders for the role of primary basaltic magma: komatiites, picritic basalts and the most "primitive" of ordinary basalts. Consequently, the following possibilities must be evaluated:

1) All basalts are ultimately derived from komatiitic liquids by fractionation processes. This is an extreme statement of the O'Hara model, but one that must be considered since komatiites can differentiate towards normal basaltic compositions (e.g., O'Hara *et al.*, 1975; Nisbet *et al.*, 1977; Whitford and Arndt, 1978).

2) All basalts, with the exception of peridotitic komatiites, are derived from picritic magmas by fractionation processes (e.g., O'Hara, 1968a, b; Clarke, 1970; Wright, 1971; Elthon, 1979). Komatiites are a special case, involving high thermal gradients and substantial amounts of mantle melting, and are possibly restricted in time.

3) A spectrum of primary magmas exists, reflecting melting and remelting of compositionally different mantles under varying conditions of temperature, pressure, volatile content and oxygen fugacity.

Analogous problems exist for basalt genesis on the Moon and achondrite parent bodies. For example, are members of the main group of eucrites derivatives of a more "primitive" melt, the evidence for which is provided by the diogenites (Mason, 1962; Duke and Silver, 1967), or are they primary compositions produced by low-pressure partial melting of plagioclase peridotite (Stolper, 1977)? Similarly, are the most "primitive" lunar basalts identified to date primary magmas, or are they the derivatives of some unsampled, more magnesian magmas, perhaps analogous in composition to the Apollo 15 green glass (e.g., O'Hara *et al.*, 1970, 1976; Green *et al.*, 1971a, b, 1975b; Green and Ringwood, 1973; Walker *et al.*, 1975; Papike *et al.*, 1976; Rhodes *et al.*, 1976, 1977; Delano, 1979; Delano *et al.*, 1980)?

1.4.2.2 Characteristics of primary basalts

This section reviews the criteria that have been used to identify potential primary basalts. No single characteristic is diagnostic. Many are subjects of contention, and each has its advocates and detractors.

Liquid compositions

An essential requirement for any primary basalt candidate is that the bulk composition be equivalent to that of a liquid. Basaltic glass provides the only entirely conclusive evidence for liquid compositions. All other assessments rely on textural or experimental evidence and are subject to alternative interpretations.

Terrestrial basaltic glass is restricted in composition and, in contrast with associated phaneritic lavas, rarely contains more than 11% MgO (Thompson, 1973). A notable exception is the altered interstitial glass (Table 1.4.2.1, #7) found in some komatiite flows (Arndt *et al.*, 1977; Nisbet *et al.*, 1977). Examples of the most magnesian glasses found among ocean-floor basalts (#18–19) and Hawaiian lavas (#26) are given in Table 1.4.2.1. It was this conspicuous absence of high-Mg compositions among basalts manifestly close to liquid compositions that led Bowen (1928) to regard picritic or ultramafic lavas as products of phenocryst accumulation. This view prevailed for many years, and the absence of glassy or aphanitic basalts with more than 11% MgO is a major obstacle to the acceptance of high-Mg primary magmas.

Table 1.4.2.1 Proposed primary magma compositions.

	1	2	3	4	5	6	7	8	9	10
SiO_2	44.35	46.24	46.13	45.48	46.43	49.9	47.41	46.4	46.32	59.2
TiO_2	0.35	0.35	0.20	0.28	0.24	0.43	0.46	0.46	0.69	0.22
Al_2O_3	7.39	7.30	3.55	5.74	5.36	8.18	15.97	9.6	12.63	11.25
FeO^t	13.28	10.51	11.06	10.58	10.32	11.92	8.22	12.2	11.63	8.86
MnO	0.26	0.18	0.21	0.21	0.15	0.21	0.20	0.24	0.19	0.16
MgO	27.42	27.36	32.71	31.15	32.37	19.70	19.84	19.6	16.64	11.40
CaO	6.32	6.92	5.06	5.63	4.45	8.87	5.31	11.3	10.53	6.55
Na_2O	0.06	0.58	0.48	0.43	0.14	0.25	2.10	0.14	1.18	1.78
K_2O	0.07	0.12	0.19	0.02	0.08	0.15	0.49	0.00	0.02	0.43
P_2O_5	—	0.03	0.01	0.02	0.02	0.04	—	—	—	0.04
Cr_2O_3	0.48	0.40	0.42	0.44	0.43	0.40	—	—	0.17	0.17
Mg′	0.803	0.837	0.854	0.854	0.861	0.766	0.827	0.761	0.739	0.726
Fo (calc.)	93.2	94.5	95.1	95.1	95.4	91.6	94.1	91.4	90.4	89.8
Fo (max.)	93.8	93.4	93.6	—	92.2	—	—	90.0	91.6	92
T°C	1547	1538	1600	1581	1595	1443	1425	1442	1391	1285
Ni, ppm	1575	1354	1965	1540	1750	785	—	—	718	196

	11	12	13	14	15	16	17	18	19	20
SiO_2	46.3	45.8	49.96	46.41	47.47	49.4	47.07	49.7	49.1	50.06
TiO_2	0.78	1.22	2.89	0.41	2.03	2.67	0.81	0.72	0.62	0.82
Al_2O_3	11.1	10.5	9.24	13.75	9.48	8.86	15.46	16.4	16.5	15.84
FeO^t	10.5	11.2	11.04	8.43	10.93	12.2	10.70	7.89	8.78	7.53
MnO	0.18	0.17	0.15	0.15	0.17	0.15	0.17	0.12	0.15	0.15
MgO	20.2	19.2	14.93	18.23	15.57	15.2	12.86	10.1	10.3	10.14
CaO	9.4	10.0	7.42	11.04	11.35	7.64	11.48	13.0	12.4	13.09
Na_2O	1.07	1.41	2.09	1.33	1.79	1.63	1.36	1.98	1.92	1.97
K_2O	0.08	0.13	1.99	0.01	0.68	1.60	0.04	0.01	0.07	0.17
P_2O_5	0.09	0.14	0.52	0.02	0.29	0.42	0.04	—	0.06	0.08
Cr_2O_3	0.28	0.23	—	0.22	0.22	0.14	—	0.07	0.06	0.14
Mg′	0.792	0.772	0.728	0.810	0.738	0.712	0.704	0.717	0.699	0.727
Fo (calc.)	92.7	91.9	89.9	93.5	90.4	89.1	88.8	89.4	88.6	89.9
Fo (max.)	—	—	—	90	90	92	—	89.7	88.7	90
T°C	1443	1429	1359	1405	1371	1370	1320	1250	1258	1249
Ni, ppm	950	1030	—	—	665	799	718	200	232	249

	21	22	23	24	25	26	27	28	29	30
SiO_2	47.85	47.25	48.3	50.73	49.97	49.33	48.33	46.7	48.4	46.9
TiO_2	0.59	0.79	0.60	0.15	2.69	2.58	2.25	1.85	2.28	1.84
Al_2O_3	12.08	13.64	13.7	11.91	12.46	12.80	11.40	9.17	11.4	14.2
FeO^t	8.98	9.77	7.9	10.08	11.33	11.43	11.56	11.9	11.3	12.4
MnO	0.12	0.14	0.12	—	0.13	0.17	0.18	0.19	0.18	0.18
MgO	17.81	17.61	16.7	14.15	10.35	10.03	13.97	20.0	14.1	11.3
CaO	11.22	9.58	10.9	10.91	10.36	10.78	9.76	7.86	9.59	9.21
Na_2O	1.31	0.89	1.65	1.89	1.97	2.13	1.89	1.54	1.90	3.16
K_2O	0.03	0.06	0.01	0.03	0.45	0.51	0.44	0.40	0.49	0.58
P_2O_5	—	0.07	—	0.03	0.29	0.25	0.22	0.19	0.23	0.23
Cr_2O_3	—	0.20	0.06	0.11	—	—	—	0.19	0.14	—
Mg′	0.797	0.781	0.807	0.735	0.644	0.635	0.705	0.769	0.712	0.643
Fo (calc.)	92.9	92.3	93.3	90.3	85.8	85.2	88.9	91.7	89.2	85.7
Fo (max.)	—	—	89.7	—	—	88	88	88	88	84
T°C	1401	1401	1378	1342	1272	1264	1344	1444	1346	1292
Ni, ppm	—	—	(600)	270	253†	233†	479†	855†	487†	—

†Calculated from data of Gunn (1971).

Table 1.4.2.1 *(Continued)*

	31	32	33	34	35	36	37	38	39	40
SiO_2	45.4	44.2	45.5	43.44	43.6	48.07	45.13	38.31	45.2	42.9
TiO_2	2.13	3.69	2.4	2.40	2.9	2.14	2.78	2.72	2.91	4.46
Al_2O_3	12.4	12.1	13.8	12.15	10.6	13.93	12.73	9.59	8.61	8.00
FeO^t	11.1	10.87	12.5	12.73	12.3	10.95	11.69	12.87	21.1	22.0
MnO	0.16	0.19	—	0.20	0.2	0.16	0.16	0.24	0.28	0.30
MgO	13.3	13.1	10.1	12.59	16.2	11.41	12.59	16.42	11.6	12.4
CaO	11.5	10.1	10.0	11.29	10.8	8.36	8.77	13.20	9.45	8.99
Na_2O	2.70	3.56	3.5	3.28	2.0	3.23	3.48	3.77	0.23	0.25
K_2O	0.79	1.29	1.6	1.28	1.3	1.18	1.80	1.43	0.06	0.05
P_2O_5	0.41	0.83	—	0.63	—	0.51	0.81	1.35	0.07	0.08
Cr_2O_3	—	0.08	—	—	0.1	0.06	0.05	0.07	0.55	0.61
Mg′	0.703	0.705	0.627	0.663	0.723	0.674	0.681	0.717	0.495	0.501
Fo (calc.)	88.8	88.8	84.8	86.7	89.7	87.3	87.7	89.4	76.6	77.0
Fo (max.)	90	—	—	—	89	—	—	—	76	76
T°C	1328	1322	1276	1318	1384	1289	1314	1385	1348	1367
Ni, ppm	474	290	—	—	—	364	342	458	52	—

	41	42	43	44	45	46	47
SiO_2	38.4	38.7	35.6	42.9	45.5	49.31	44.4
TiO_2	12.7	9.46	13.8	3.6	0.39	0.57	—
Al_2O_3	8.72	5.98	7.14	8.6	7.78	12.80	11.5
FeO^t	18.1	22.60	21.9	22.0	19.73	18.40	14.4
MnO	0.26	0.30	0.25	0.26	0.23	0.55	—
MgO	10.4	14.20	12.1	12.9	17.15	7.23	17.7
CaO	10.3	7.73	7.88	8.7	8.64	10.24	9.7
Na_2O	0.35	0.35	0.49	0.46	—	0.43	1.8
K_2O	0.07	0.07	0.12	—	—	0.04	0.4
P_2O_5	0.06	—	—	—	—	0.09	—
Cr_2O_3	0.65	0.59	0.77	0.58	0.54	0.35	—
Mg′	0.506	0.528	0.496	0.511	0.608	0.412	0.687
Fo (calc.)	77.3	78.9	76.7	77.7	83.8	70.0	88.0
Fo (max.)	79.5	81	—	—	75	65	—
T°C	1316	1404	1365	1375	1436	1240	1418
Ni, ppm	—	—	<50	95	150	—	—

Glass spheres and fragments in the lunar soil may be of impact or volcanic origin. Those thought to be of volcanic origin are compositionally homogeneous, free of lithic or metallic inclusions, and have a surface-coating of volatile elements (e.g., Butler, 1978; Delano, 1979). Some, containing euhedral olivine phenocrysts in equilibrium with the enclosing glass, are most certainly of volcanic origin (Heiken *et al.*, 1974; Rhodes, 1978). With the exception of the Apollo 15 green glass (Table 1.4.2.1, #45), most contain from 12–14% MgO and, in contrast with terrestrial basaltic glasses, have Mg′-values (0.50–0.53) slightly higher than the least-fractionated of the mare basalts (Table 1.4.2.1, #39–44). Delano *et al.* (1980) suggest that these glasses are more likely candidates for primary magmas than their basaltic counterparts. The Apollo 15 green glass (Table 1.4.2.1, #45) has much higher Mg′-values, and is of equivocal origin (Green and Ringwood, 1973; Delano, 1979).

Basalts containing acicular, dendritic and skeletal phenocrysts, usually of olivine but also of pyroxene, are commonly thought to have cooled rapidly, and therefore to have retained the liquid composition. The existence of high-Mg picritic liquids has been championed by Drever and co-workers (e.g., Drever and Johnston, 1958, 1966). They have argued that the presence of skeletal crystals in intrusive bodies together with an absence of larger phenocrysts is evidence of rapid crystal growth, and that these non-porphyritic facies of the intrusives represent picritic liquids (Table 1.4.2.1, #17). Similar textures are found in many lavas containing more than 10% MgO, including boninites (Dietrich *et al.*, 1978; Cameron *et al.*, 1979), and, particularly, in komatiitic rocks (e.g., Viljoen and Viljoen, 1969; Nes-

Table 1.4.2.1 *(Concluded)*

All analyses are summed to 100 percent on a volatile-free basis. FeO^t is total Fe expressed as FeO. Mg′ = molar (Mg/(Mg + Fe) after adjusting Fe so that Fe^{2+}/total Fe as FeO = 0.9. Fo (calc.) is the olivine composition calculated from the Mg′-value assuming K_D = 0.30. Fo (calc.) = Mg′/(Mg′ + K_D (1 − Mg′)) Fo (max.) is the maximum forsterite content of olivine recorded in the rock or associated lavas. T°C is the calculated liquidus temperature based on olivine-melt K_D values for MgO and FeO in cation mole percent (Roeder and Emslie, 1970); Langmuir and Hanson, 1979). **1** Spinifex-textured peridotitic komatiite, (Arndt *et al.*, 1977, table 9, p. 118). Best estimates of liquid composition. **2** Average of ACH-1, ACH-2, ACH-4, ACH-5, ACH-6 (Section 1.2.1). Estimate of peridotic komatiite liquid composition. **3** Komatiite (49J) from Komati Formation, Barberton Mountain Land (Green *et al.*, 1975a, table 1). **4** Average of komatiites from Yilgarn Block, (Nesbitt and Sun, 1976, table 1, #1-7. **5** Komatiite with 30% MgO, Belingwe Greenstone Belt (Nisbet *et al.*, 1977, Table 4, D; estimated from linear regression). **6** Komatiite with 20% MgO, Belingwe Greenstone Belt (Nisbet *et al.*, 1977, Table 4, C; estimated from linear regression). **7** Interstitial "glass" in komatiite flow (Nisbet *et al.*, 1977, Table 2, SF 134). **8** Top of layered komatiite flow, Munro Township (Whitford and Arndt, 1977, table 38, pg-111). **9** Neogene komatiite, Gorgona Island (Gansser *et al.*, 1979). Olivine composition from Echeverria, 1980). **10** Boninite, Mariana trench (Dietrich *et al.*, 1978; Cameron *et al.*, 1979, #1403-44). **11** Average of Group II olivine basalts, Baffin Island (Clarke, 1970, table 1). **12** Average of Group II olivine basalts, Svartenhuk (Clarke, 1970, table 1). **13** Bezi dike, Nuanetsi province (Cox *et al.*, 1965, table 9, KC-37). **14** Picritic basalt, Reykjanes Peninsula (Jakobsson *et al.*, 1978; Maaløe and Jakobsson, 1980, table 3, RE-78). **15** Picritic basalt, Deccan Traps (Krishnamurthy and Cox, 1977, table 5, #252). Closest in composition to weighted average of Deccan Trap picrites. **16** Average of olivine-rich Karroo lavas, Nuanetsi province (Cox and Jamieson, 1974, table 8, A. Cr_2O_3 and Ni from table 9). **17** Average of non-porphyritic facies of picritic minor intrusions, Skye (Drever and Johnson, 1966, table 1, #2). **18** Most "primitive" ocean-floor basaltic glass, DSDP Leg 3 (Frey *et al.*, 1974, 3-18-7-1). **19** "Primitive" ocean-floor basaltic glass from FAMOUS area (Langmuir *et al.*, 1977, table 1, 527-1-1). **20** Aphyric ocean-floor basalt, N. Atlantic 45° (J.M. Rhodes, unpubl. data, Chain 43-1 — 4-23). **21** Primary ocean-floor basaltic magma based on estimate of ocean crust composition (Elthon, 1979, table 1). **22** High-magnesium dike from Tortuga ophiolite complex (Elthon, 1979, table 2, N-23). **23** Proposed primary ocean-ridge olivine tholeiite composition based on experimental studies on #18 (this table) + 17% olivine (Green *et al.*, 1979). **24** Aphyric basalt from lower lavas of Betts Cove ophiolite (Coish and Church, 1979). **25** Average of Kilauean lavas (MacDonald, 1949). **26** Most magnesium-rich glass, Kilauea Iki 1959 eruption (Murata and Richter, 1966). **27** Calculated composition of "primitive" Kilauean magma, based on weighted average composition of first eruptive phase of Kilauea Iki 1959 (Murata and Richter, 1966). **28** Proposed primary magma for Kilauean eruptions (Wright, 1971). **29** Estimated "primitive" liquid for 1959 Kilauean eruptions (Irvine, 1977). **30** Proposed primary alkali basalt, Skye (Thompson, 1974, #66018). **31** Microphyric basanite, intermediate lava series, Mauritius (Baxter, 1976 table 1, #B6). **32** "Primitive" basanite, Ross Island, Antarctica (Goldich *et al.*, 1975; Sun and Hanson, 1975 #15765). **33** Average of 34 basanites from the western Grand Canyon region (Best and Brimhall, 1974, table 3). **34** Nodule-bearing basanite, Anjouan, Comores Archipelago (Flower, 1973, table 5, AJ-218). **35** Nodule-bearing basanite, Tahiti (Tracy and Stolper, 1979). **36** Nodule-bearing alkali basalt, Mt. Frazier, Victoria (Frey *et al.*, 1978, table 2, 69-1036). **37** Nodule-bearing basanite, Mt. Shadwell, Victoria (Frey *et al.*, 1978, table 2, 2679. **38** Nodule-bearing olivine melilitite, Laughing Jack Marsh, Tasmania (Frey *et al.*, 1978, table 2, 2927). **39** Vitrophyric Apollo 12 olivine basalt 12009 (Green *et al.*, 1971b, table 1). **40** Vitrophyric Apollo 12 ilmenite basalt 12008 (Rhodes *et al.*, 1977, table 1). **41** Vitrophyric Apollo 17 high-titanium basalt 74275 (Rhodes *et al.*, 1976, table 1). **42** Apollo 17 orange glass (Heiken *et al.*, 1974). **43** Apollo 15 red glass, type A (Delano, 1980). **44** Apollo 15 yellow glass of volcanic origin (Delano, 1980). **45** Apollo 15 green glass, type A (Delano, 1979, table 1). **46** Sioux County eucrite (Section 1.2.8). **47** Calculated "primitive" martian melt (McGetchin and Smyth, 1978, table 8).

bitt, 1971; Pyke *et al.*, 1973; Gansser *et al.*, 1979). Typical analyses of these rocks are given in Table 1.4.2.1 (#1–10). Their textural similarity to quench textures in experimental runs has led to the interpretation of these "spinifex-textured" komatiites and some picrites as rapidly cooled lavas that are probably close to liquid compositions. Although persuasive, the petrographic evidence for these apparent quench crystals is insufficient on its own to prove that the bulk composition is necessarily that of a liquid, since such textures are not strictly products of rapid cooling, but of rapid crystallization in *slowly* cooling nuclei-free melts. Consequently, there may be opportunity for skeletal crystals to accumulate, as has been shown by Pyke *et al.* (1973), or to be redistributed relative to the melt by processes akin to filter-pressing.

Many aphyric basalts, or basalts containing minor phenocrysts and microphenocrysts, are probably close to liquid compositions. This certainly is true for many ocean-floor and Hawaiian basalts: fine-grained basalts are usually isochemical with glassy pillow margins or associated glassy spatter.

Lunar vitrophyres containing small olivine phenocrysts in a glassy-to-fine-grained matrix (Table 1.4.2.1, #39–41) have been widely accepted as the best available approximates of liquid mare basalt compositions, and the parental magmas from which other mare basalts have been derived (e.g., Green *et al.*, 1971b, 1975b; Rhodes *et al.*, 1976, 1977).

Cox (1978) argues that some fine-grained voluminous picritic lavas containing as much as 19–20% MgO, such as the Bezi dike of Nuanetsi (Table 1.4.2.1, #13), have liquid compositions, and that the minor olivine phenocrysts contribute little to the the total MgO content. Important examples of this type are the Tertiary picritic basalts from Baffin Bay (Table 1.4.2.1, #11–12; Clarke, 1970; O'Nions and Clarke, 1972).

Porphyritic lavas are most likely to have compositions modified by crystal accumulation or redistribution, and are therefore suspect as primary magmas.

However, Cox (1978) has proposed that some high-Mg porphyritic lavas, which he terms "primitive porphyritic magmas," may not have gained their composition through simple cumulus enrichment of more normal basalts, but may have retained the original liquid composition. He envisions that this may be accomplished by *compensated crystal settling* in conduits and pipes, where crystal loss downwards is roughly balanced by crystals gained from above, so that the bulk composition of the magma is not altered substantially as solidification proceeds (Cox and Bell, 1972). Examples come from Karroo basalts (Cox *et al.*, 1965; Cox and Jamieson, 1974) and Deccan Traps (Krishnamurthy and Cox, 1977), where large volumes of porphyritic picritic lavas (Table 1.4.2.1, #15–16) are overlain by lesser volumes of more evolved basalts. It is argued that there is insufficient evolved magma to have formed the volumetrically abundant picrites through crystal accumulation, and that the sequence is the reverse of what one would expect from the discharge of a stratified magma chamber. Such a view of magma systems is perhaps overly simplistic. Volcanological observations within the last decade (e.g., Wright and Fiske, 1971; Sigurdsson and Sparks, 1978; Björnsson *et al.*, 1979) indicate that magma reservoirs are not simple containers, but open systems, characterized by complex patterns of filling, fractionation, mixing, draining and eruption, and therefore lava sequences and volumes reflecting the orderly and quantitative evacuation of a magma chamber are unlikely.

In the absence of persuasive textural evidence, near-liquidus melting studies of natural rocks may provide evidence as to whether or not the bulk composition reflects that of a basaltic liquid (e.g., Green, 1970; Green *et al.*, 1971b, 1975a, b; Longhi *et al.*, 1974; Rhodes *et al.*, 1979b). If the experimentally determined liquidus phase (usually olivine) at low pressure and appropriate oxygen fugacity corresponds closely in composition with the natural phenocrysts, then it is a reasonable assumption that the basalt was once erupted as a liquid. Conversely, if there is a lack of correspondence, the basalts may have accumulated phenocrysts, picked up foreign crystals or xenocrysts, or cooled slowly and re-equilibrated while maintaining the liquid composition (see Fig. 3.1.4).

A similar test can be made by comparing observed olivine phenocryst compositions with those predicted from whole-rock compositions on the basis of Fe/Mg distribution between olivine and melt (Roeder and Emslie, 1970). This has been done in Table 1.4.2.1 wherever possible, comparing published values for the maximum forsterite content of olivine phenocrysts with calculated olivine compositions ($K_D = 0.30$). In these calculations (and others in this section) the total iron content, expressed as FeO*, has been adjusted so that $Fe^{3+}/(Fe^{3+} + Fe^{2+}) = 0.1$ (Ringwood, 1975). It is evident from Table 1.4.2.1 that not all examples proposed as primary magmas satisfy these requirements for liquid compositions, and they may well have accumulated some olivine. Examples include several picritic magmas (#14, 16, 23, 28), and some, but not all, komatiites and komatiite-like magmas (#3, 5, 8, 10). Komatiites with over 30% MgO, for which data are available, fall into this category. Some minor discrepancies may result from an inappropriate choice of K_D. Longhi *et al.* (1978) have shown that K_D for lunar basalts may range from 0.28 for the low-Ti variety to 0.33 for high-Ti basalts. Similarly, there is growing evidence that ocean-floor basalts have K_D values of about 0.28 (Bryan and Moore, 1977; Bender *et al.*, 1978; Rhodes *et al.*, 1979a) in contrast with the wider range of 0.3 ± 0.03 for basalts in general (Roeder and Emslie, 1970). Komatiites present a more serious problem. Green *et al.* (1975a) have compared liquidus olivine with olivine phenocrysts in a komatiite from the Barberton Mountain Land and conclude that the lava was extruded as a peridotitic liquid. However, both the experimentally produced olivine and the most Mg-rich olivine phenocrysts are only $Fo_{93.6}$ compared with olivine compositions of $Fo_{94.7-95.6}$ predicted on the basis of K_D values between 0.27 and 0.33 (Table 1.4.2.1, #3). This discrepancy corresponds to a K_D value of about 0.4, suggesting that higher K_D values may be more appropriate for ultramafic compositions, or that the discrepancies reflect failure to attain equilibrium in both the experiments and the flows as a consequence of super-cooling.

Volume and eruptive sequence

In contrast with the compositional diversity necessitated by crystal fractionation, partial fusion of a peridotitic source either at a eutectic-like invariant point or low-temperature cusp on the mantle solidus can produce large volumes of compositionally uniform magma (Presnall, 1969; Yoder, 1976; Presnall *et al.*, 1979). Consequently, vast volumes of relatively uniform basaltic magma, such as continental flood basalts and basalts of the ocean floor, have been interpreted as products of moderate amounts of fusion, perhaps modified by subordinate fractionation, rather than as the end products of extensive crystal fractionation (e.g., Engel *et al.*, 1965; Carmichael *et al.*, 1974; Wilkinson and Binns, 1977; Presnall *et al.*, 1979). The paradox is that although many of these basalts appear to be compositionally evolved, to derive them by extensive crystallization of a deep-seated picritic magma requires the improbable condition that successive magma batches must lose the *same* fraction of *identical* phases while successfully retracing an *identical* path of ascent (Carmichael *et al.*, 1974). On the other hand, there are also serious obstacles to the primary magma alternative. Assuming that the more magnesian of these lavas are primary, the mantle source for many of these basalts must be more iron-rich than is usually considered appropriate (e.g., Kuno and Aoki, 1970; Wright, 1971; Thompson, 1974, 1975; Wilkinson and Binns, 1977). In the specific case of the Columbia River basalts, Swanson *et al.* (1975) argue that the extrusion rate is too large to be commensurate with direct derivation by mantle melting. They suggest that the magma has gradually accumulated in vast storage reservoirs prior to rapid extrusion. Such a model must surely necessitate substantial crystal fractionation. Cox (1980) proposes that this extensive fractionation takes place in a sill complex at, or near, the base of the crust. Magma mixing has been proposed as an alternative mechanism to account for the relative uniformity of ocean-floor basalts, and the rarity of both "primitive" and highly differentiated basalts (O'Hara, 1977; Rhodes and Dungan, 1979; Walker *et al.*, 1979).

The average composition of a volcanic sequence has been taken by some authors as the best estimate of the primary or parental magma composition on the assumption that phenocryst-enriched lavas will be compensated for by more evolved variants. This was the approach adopted by MacDonald (1949) and Powers (1955) for Hawaiian lavas (Table 1.4.2.1, #25). More recently, Maaløe and Jakobsson (1980) selected a picritic basalt from Iceland as a primary magma (Table 1.4.2.1, #14) partly on the grounds that it closely approaches the average of the extrusive sequence (Jakobsson *et al.*, 1978). As noted earlier, growing evidence for the movement of magma within complex magma reservoirs (e.g., Wright and Fiske, 1971; Sigurdsson and Sparks, 1978; Björnsson *et al.*, 1979) suggests that it is unlikely that the products of differentiation will be erupted quantitatively. Perhaps the estimate of Murata and Richter (1966) for the parental magma of the 1959 Kilauea Iki eruption (Table 1.4.2.1, #27), which was based on combined geophysical measurements and careful documentation of the eruptive history, comes closest to a primary magma composition using this approach.

On a somewhat different scale, Elthon (1979) has suggested that the volumetrically weighted average of components in the oceanic crust (Table 1.4.2.1, #21) provides a better estimate of the composition of primary or parental ocean-floor basalt than any individual sample obtained to date. His estimate for the oceanic crust is based on the assumption that the weighted bulk composition of the pillow lavas, sheeted dikes and cumulate gabbroic and ultramafic rocks of ophiolite sections equal the composition of the parental magma which entered the sub-ridge magma chamber prior to an extended history of fractionation and mixing. However, not all ophiolites reflect segments of the oceanic crust. Some contain lavas with the compositional characteristics of those from island arcs and marginal basins, including lavas with calc-alkaline affinities and lavas similar to boninites, that are extremely depleted in magmaphile elements (e.g., Sun and Nesbitt, 1978; Cameron *et al.*, 1979; Coish and Church, 1979; Upadhyay and Neale, 1979; Stern, 1980). Consequently, so general a model should be treated with caution.

Nodule-bearing basalts

Ultramafic nodules in basalts, spinel peridotites, in particular, are thought to represent fragments of refractory mantle that have been carried upward by the ascending magma (see section 1.2.11). Such nodules are most common in alkali basalts; at least one occurrence has been described in a tholeiitic basalt from Tasmania (Sutherland, 1974). The presence of these inclusions has been widely accepted as evidence for rapid ascent velocities ranging between 1 and 50 cm/sec (e.g., Maaløe, 1973; Yoder, 1976; Spera, 1980). Such rapid ascent precludes olivine fractionation en route to the surface, and it is therefore unlikely that the magma composition would be modified during this period other than by the inclusion of mineral fragments resulting from the comminution of the entrained nodules. Consequently, there is a consensus, even among authors who favor the hypothesis that most basaltic magmas are modified during ascent (e.g., Yoder, 1976), that basalts (and kimberlites) containing mantle-derived nodules are close to primary compositions. This is an important point since, if correct, it implies that not all primary magmas are necessarily picritic or komatiitic. Several examples of possible nodule-bearing primary magmas have been described, with those from southeastern Australia receiving the most detailed attention (e.g., Flower, 1971, 1973; Kesson, 1973; Best and Brimhall, 1974; Sun and Hanson, 1975; Irving and Green, 1976; Frey *et al.*, 1978; Tracy and Stolper, 1978). Some potential nodule-bearing, primary basalt compositions are given in Table 1.4.2.1 (#33–38).

The presence of peridotite inclusions, however, is not by itself diagnostic of primary magmas. Sparks *et al.* (1977) suggest that transportation of nodules by basaltic magmas is a function of the magma's yield strength as well as of its ascent velocity, and that a high yield strength is a consequence of crystallization. Thus, even quite large inclusions (about 14 cm diameter) will tend not to sink in a magma that is about 35% crystalline

(Spera, 1980). Consequently, the possibility of fractionation in the mantle prior to the incorporation of the nodules cannot be ignored. O'Hara and co-workers (e.g., O'Hara, 1968a; O'Hara *et al.*, 1975) have argued for extensive mantle fractionation, on the basis of eclogite nodules in kimberlite, which they interpret as high-pressure cumulates derived from a fractionating magma. Other evidence for mantle fractionation comes from associated lavas with compositional trends that cannot be ascribed to low-pressure crystal fractionation (e.g., Wright, 1971; Strong, 1972; Cox and Jamieson, 1974; Thompson, 1974; Baxter, 1976; Tracy and Stolper, 1978) and from nodule-bearing hawaiites, mugearites, benmoreites and phonolites (Wilkinson and Binns, 1969; Price and Green, 1972; Green *et al.*, 1974; Irving and Green, 1976; Irving and Price, 1981).

Experimental constraints

The use of melting experiments to identify liquid compositions was described earlier. Here, the role of such experiments in identifying primary magmas and elucidating their conditions of formation is considered. Some of these arguments are dealt with more fully in Chapter 3 (section 3.1.4).

Experimental melting studies of both natural and synthetic basalt systems have provided some of the strongest arguments against the primary nature of common basaltic magma, and have led to the advocacy of other more "primitive" primary magmas despite their embarrassing rarity. One-atmosphere melting experiments tend to confirm petrographic interpretations that the majority of terrestrial, lunar and achondritic basaltic liquids are multiply saturated, their compositions apparently controlled by low-pressure cotectics or invariant points (e.g., Yoder and Tilley, 1962; O'Hara, 1968a, b; O'Hara *et al.*, 1970, 1975; Green *et al.*, 1971a; Thompson, 1972; Longhi *et al.*, 1974; Kesson, 1975; Stolper, 1977; Bender *et al.*, 1978; Walker *et al.*, 1979). Since there is a shift in multiple saturation points and cotectic surfaces with increasing pressure, principally as a consequence of the contraction of the olivine phase volume (e.g., O'Hara, 1968a; Presnall *et al.*, 1978), it is improbable that primary melts, formed at these pressures, will have compositions that are multiply saturated at low pressures. The obvious inference is that most basalts have arrived at their present compositions through low-pressure or polybaric crystal fractionation of more "primitive" liquids (e.g., O'Hara, 1968a, b; Yoder, 1976). Some achondrites are exceptions to this argument, having compositions that may reflect primary magmas produced by low-pressure melting of a plagioclase peridotite source on small planetary (asteroidal?) bodies (Stolper, 1977; Consolmagno and Drake, 1977; Stolper *et al.*, 1979).

A basic tenet of experimental petrology is that primary magmas produced by the partial fusion of peridotite are saturated with the residual phases at temperatures and pressures corresponding to the depth of segregation. High-pressure melting studies on natural peridotites or synthetic analogs (e.g., Mysen and Kushiro, 1977; Jaques and Green, 1980) suggest that three or more phases (ol + opx + cpx ± an Al-rich phase) will be on the liquidus up to about 20–25% melting, two or three phases (ol + opx ± cpx) between 25 and 40%, and one or two phases (ol ± opx) above about 40% melting. The Al-rich phase will be plagioclase, spinel or garnet, depending on pressure. It follows that all primary basalts derived from an olivine-rich mantle should have olivine on the liquidus together with other appropriate phases at pressures corresponding to their depths of origin. Thus, if a basalt is primary, the pressure at which olivine plus one or more phases are on the liquidus may provide an indication of the depth at which the magma has segregated from its source. (See section 3.1.4 for a more detailed discussion of this concept.)

This approach has worked well for the most "primitive" of the lunar mare basalts. These basalts are multiply saturated with olivine and various pyroxenes at pressures of 6–15 kb implying source regions consisting of olivine and pyroxenes for both low-Ti and high-Ti basalts at depths ranging from 120–300 km (Green *et al.*, 1971a, b, 1975b; Grove *et al.*, 1973; Longhi *et al.*, 1974; Kesson, 1975; Walker *et al.*, 1975, 1976). The more "primitive" orange and green glasses are multiply saturated at 20–25 kb, corresponding to depths of about 400 km (Green and Ringwood, 1973; Green *et al.*, 1975b; Walker *et al.*, 1975, 1976). Delano *et al.* (1980) have argued that these "primitive" glasses are the best candidates for primary lunar basalts, and that other basalts are probably derivative.

Considerable controversy has developed in the interpretation of high-pressure melting experiments on terrestrial basalts. Those compositions that are multiply saturated at high pressures commonly lack olivine on the liquidus at pressures above 10–15 kb (e.g., Green and Ringwood, 1967; Thompson, 1974; Kushiro, 1973; Bender *et al.*, 1978). Such low pressures are obviously unsatisfactory if a garnet peridotite source is assumed, and the implied depths of 30–40 km, although perhaps appropriate for ocean-ridge volcanism, are considered too shallow for intraplate and continental volcanism (e.g., O'Hara, 1968a, b; Yoder 1976).

Similar problems are encountered with orthopyroxene. It is widely accepted, on the basis of melting experiments and observation of mantle-derived peridotites (e.g., nodules, alpine peridotites and ophiolite sections), that harzburgites (ol + opx) are refractory residues of partial melting processes (e.g., Carter, 1970; O'Hara *et al.*, 1975). Consequently, it is to be anticipated that, with the exception of basalts produced by large amounts of melting which leave behind a refractory dunite residue, orthopyroxene should be on the liquidus of primary basalt magmas together with olivine at pressures corresponding to their depths of segregation. To the confusion of petrologists, few basalts studied have orthopyroxene on the liquidus at any pressure. Arguments such as these led O'Hara (1968a) to reject common basalts as primary magmas, and to propose that they are derived from picritic magmas by crystal fractionation. Recent experimental studies are ambivalent. Some (e.g., Green *et al.*, 1979; Stolper, 1980) show that picritic magmas containing 13–17% MgO can have both olivine and orthopyroxene on the liquidus at 10–20 kb. On the other hand, Bender *et al.* (1978) report orthopyroxene on the liquidus of a "primitive" ocean-floor basaltic glass at 10–12 kb, and Presnall *et al.* (1979) argue, on the basis of experiments in the system CaO-MgO-Al_2O_3-SiO_2, that the most "primitive" ocean-floor basalts will be saturated with both olivine and orthopyroxene at 9 kb and that therefore they may be primary. Maaløe and Jakobsson (1980) point out that because of possible reaction relationships among olivine, orthopyroxene and liquid, the absence of orthopyroxene from the liquidus may not be diagnostic.

Direct partial melting experiments on natural and synthetic peridotites at various pressures, although troubled by experimental difficulties, offer an alternative perspective on these problems (see section 3.3.2). Experiments by Mysen and Kushiro (1977) and Jaques and Green (1980) show that melt compositions resembling both alkalic and tholeiitic basalts can be produced by up to about 20–25% melting, leaving a residue of olivine, orthopyroxene and clinopyroxene (± spinel) at pressures between 10 and 20 kb. Garnet is an additional residual phase above 35 kb and plagioclase at pressures below 10 kb. Larger amounts of melting (about 25–40%) produce picritic melts leaving a residue of olivine and orthopyroxene up to 20 kb, plus clinopyroxene at higher pressures. More than 40% melting yields komatiite-like liquids at all pressures, with olivine, or olivine plus orthopyroxene, as the residue. Similar results are obtained for a "depleted" peridotite, except that the pyroxene components are consumed with lesser amounts of melting (Jaques and Green, 1980). Under hydrous conditions melts are produced that have many affinities with high-Mg andesites or boninites (Kushiro, 1972; Green, 1976). Such melts have been identified in island arcs and ophiolite sequences, and may be parental to andesites (e.g., Cameron *et al.*, 1979).

Three crucial points emerge from these studies. First, a wide spectrum of melt compositions are produced with varying amounts of melting, ranging in the anhydrous experiments from alkali basalt, through tholeiitic basalt and picrite, to komatiite. Second, in contrast with melting experiments on basaltic compositions, all these melts are apparently in equilibrium with orthopyroxene and/or olivine. Third, the refractory residue contains olivine that increases in forsterite content (Fo_{88}–Fo_{95}) as melting increases. It appears, therefore, that experimental studies provide conflicting evidence. Most, but not all, melting experiments on natural basalts and synthetic systems point to picritic primary melts, whereas the experimental melting of peridotites suggests that a plethora of primary magma compositions including high-Mg andesites or boninites, alkalic and tholeiitic basalts, picrites and komatiites may be produced.

Melt inclusion studies

Melt inclusions in phenocrysts provide an opportunity to examine liquid compositions during the evolution of a basaltic magma. Some inclusions may have been entrapped early in the magmatic history, providing an estimate of "primitive," possibly primary, compositions. Several studies of melt inclusions in ocean-floor basalts have been undertaken, using various approaches ranging from direct measurement (Donaldson and Brown, 1977) to extrapolations which attempt to compensate for entrapment at various stages of magmatic evolution and for reaction with the host minerals, including olivine, plagioclase and pyroxene (Watson, 1976; Dungan and Rhodes, 1978). Despite the different approaches the results are remarkably similar, suggesting a rather narrow compositional range for primary ocean-floor basaltic magmas. These compositions are very similar to the most mafic glass sampled from the ocean floor (e.g., Table 1.4.2.1, #18–19).

Melt inclusion studies on lunar basalts have focussed on the nature of the evolved liquids and the onset of liquid immiscibility (e.g., Roedder and Weiblen, 1971, 1977), and have not been concerned with the nature of primary magmas. Nevertheless, such studies offer promise for investigation of primary compositions in lunar basalts and returned samples from other planets.

Liquidus mineralogy

With the exception of eucrites, olivine (± spinel) is the first phase to crystallize from the majority of the basaltic melts that have been characterized as primary.

This is hardly surprising since, as noted earlier, at high pressures the primary phase volume of olivine contracts, largely at the expense of pyroxene (e.g., O'Hara, 1968a; Presnall *et al.*, 1978). Consequently, melting at these pressures of an olivine-bearing source at a eutectic-like invariant point or cotectic will produce a liquid that lies within the olivine phase volume at low pressures.

Some komatiites with MgO less than 16% have pyroxene spinifex textures. Petrographic evidence supported by variations in bulk chemistry points to the early crystallization of pyroxene rather than olivine in these lavas (Arndt, 1976; Arndt *et al.*, 1977). These observations contrast with experimental studies, which give olivine as the liquidus phase for these compositions. The status of such komatiites is not clear (see discussion in section 1.2.1): some appear to be derived by fractionation from more "primitive" melts, whereas others are thought to be primary magmas in their own right. The problem is complicated further by the possibility of chemical alteration (Arndt, 1976).

High-alumina basalts, identified by Kuno (1960) as a magma type distinct from alkaline and tholeiitic basalts, appear to have plagioclase as the liquidus phase. Shido *et al.* (1971) have classified ocean-floor basalts into olivine and plagioclase tholeiites on the basis of the first phase to crystallize. Implicit in this distinction is the assumption that plagioclase tholeiites are a primary magma type. Recent studies show that most basalts of this type are plagioclase phyric, and that the plagioclase phenocrysts are commonly too calcic to have crystallized from the host magma (e.g., Donaldson and Brown, 1977; Rhodes *et al.*, 1979). It appears that these basalts do not reflect liquid compositions, and that they gained their present compositions through phenocryst accumulation, associated with mixing of magmas (Rhodes and Dungan, 1979).

A few high-alumina basalts have plagioclase on the liquidus and are aphyric. These are unlikely to have attained their present composition through phenocryst accumulation. Examples include the Warner basalt from the Medicine Lake highlands (Yoder and Tilley, 1962), and a tholeiite from the Mid-Atlantic Ridge (Kushiro, 1973). Kushiro (1979) suggests that such magmas may have crystallized along the olivine-plagioclase cotectic at high P_{H_2O} prior to eruption. If correct, this excludes such basalts from consideration as primary magmas.

Magnesium-iron distribution between olivine and melt

Roeder and Emslie (1970) have shown that the distribution of iron and magnesium between olivine and melt (K_d) is temperature dependent, but the exchange reaction distribution (K_D) is relatively insensitive to temperature, composition and oxygen fugacity. Subsequent studies have confirmed these relationships and identified small changes in K_D as functions of pressure and composition (e.g., Bickle *et al.*, 1976; Bender *et al.*, 1978; Longhi *et al.*, 1978; Leeman, 1978). Since olivine is the low pressure liquidus phase for the majority of potential primary magmas, these relationships can be used to place constraints on the compositions of primary magmas. Thus, if the magma is primary the forsterite content of liquidus olivine will be the same as that of olivine in the residual source, and both can be predicted from the Mg′-value [mol. Mg/(Mg + Fe)] of the melt, and vice versa (e.g., Roeder and Emslie, 1970; Green, 1970; Langmuir and Hanson, 1980). The assumption that K_D is constant may not be strictly valid. If it increases with pressure as has been proposed (Longhi *et al.*, 1978), then olivine crystallizing from the melt at or near the surface will be slightly more forsteritic than the residual olivine in the mantle.

A number of authors have used assumptions concerning the composition of the Earth's mantle to predict, on the basis of a K_D of 0.3 ± 0.03, that the Mg′-value of primary terrestrial melts will range from about 0.68 to 0.75 (e.g., Green, 1970; O'Hara *et al.*, 1975; Irving and Green, 1976: Frey *et al.*, 1978). These estimates which are based largely on undepleted mantle compositions with Mg′-values of about 0.90, appear unduly restrictive because the Mg′-value of the melt will reflect the composition of the refractory residuum, and not the initial mantle composition. Samples of residual mantle material obtained from lherzolite and harzburgite nodules in kimberlites and basalts, and from harzburgites associated with ophiolites commonly contain olivine ranging in composition between Fo_{90}–Fo_{94}. (e.g., Ross *et al.*, 1954; O'Hara and Mercy, 1963; Meyer and Boyd, 1972; Frey and Green, 1974; Reid *et al.*, 1975; de Albuquerque *et al.*, 1977; Frey and Prinz, 1978; Malpas, 1978; Reid and Woods, 1978; Elthon, 1979; Pallister and Hopson, 1979). Consequently, primary melts in equilibrium with these olivines should exhibit a spectrum of Mg′-values, ranging from 0.70–0.83, assuming that K_D varies between 0.27 and 0.33. Olivine compositions in proposed primary basalts of between Fo_{88} and Fo_{94} (Table 1.4.2.1) lend support to this suggestion. So also do the peridotite melting experiments of Jaques and Green (1980), in which increased melting produced melts with Mg′-values between 0.69 and 0.86 associated with residual olivines ranging from Fo_{88} to Fo_{95}.

These relationships are explored in Fig. 1.4.2.1 where proposed primary magma compositions are compared with zones of constant Mg′-value. These

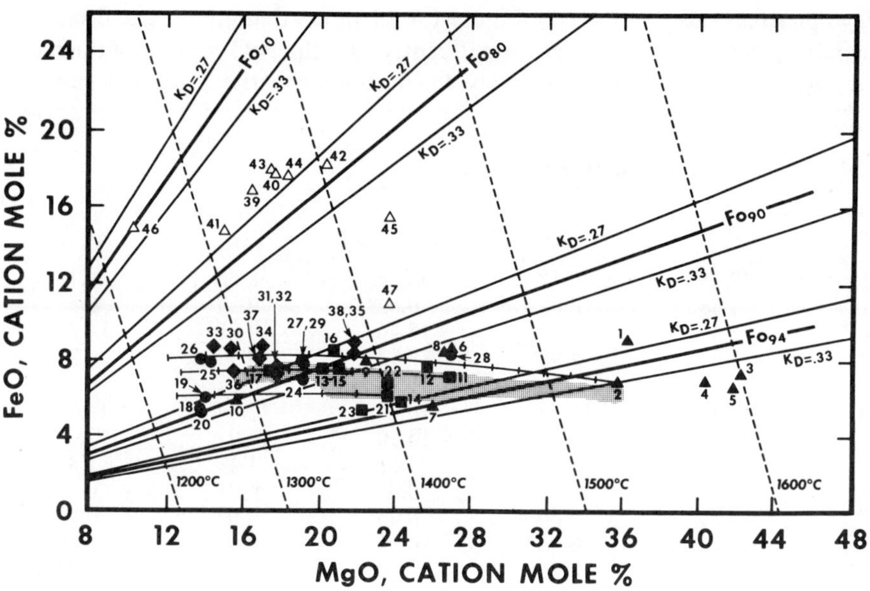

1.4.2.1 MgO-FeO relationships in cation mole % for possibly primary basaltic magma compositions (Table 1.4.2.1). Zones of constant olivine composition are calculated from Mg′-values [mol. Mg/(Mg + Fe) after adjusting terrestrial samples so that Fe^{3+}/(total Fe as Fe^{2+}) = 0.1] using $K_D = 0.30 \pm 0.03$. Olivine liquidus temperatures are shown as dashed lines, and are calculated from equations of Roeder and Emslie (1970) and Langmuir and Hanson (1980). Several olivine fractionation trends are shown by barred lines. The bars indicate 4% fractionation steps. Komatiites and komatiite-like rocks are shown as △; picrites as □; alkali and nodule-bearing basalts as ◇; tholeiites as ●; and extra-terrestrial samples as +. Numbers correspond to Table 1.4.2.1.

zones correspond to olivine of fixed composition, allowing for the uncertainty in the value of K_D. Thus, primary basalts should plot between Fo_{90} (K_D 0.27) and Fo_{94} (K_D 0.33), corresponding to Mg′-values of 0.71 and 0.83 respectively. Also shown on this diagram is a field of melt compositions resulting from the peridotite melting experiments of Jaques and Green (1980), and one-atmosphere liquidus temperatures which are based on K_d values for MgO and FeO in cation mole percent (Roeder and Emslie, 1970; Langmuir and Hanson, 1980).

Figure 1.4.2.1 shows that many proposed terrestrial primary basalts meet these requirements, forming a broad trend of increasing MgO, but roughly constant FeO, from "primitive" basalt to komatiite, which is coincident with the liquids produced by the progressive melting of peridotites (Jaques and Green, 1980). However, a number of basalts (e.g., #25, 26, 30, 33, 34, 36), including some that contain ultramafic nodules, have Mg′-values that are too low, indicating either that they are not primary, or that they are derived from an iron-rich mantle (Thompson, 1974; Wilkinson and Binns, 1977). Conversely, most komatiites, with MgO > 30%, have Mg′-values that are too high. Does this mean that despite spinifex textures these compositions result from olivine accumulation, or are the K_D values used inappropriate for these compositions? The experiments of Green *et al.* (1975a) point to a higher K_D of about 0.4, whereas Bickle *et al.* (1976) obtained values between 0.33–0.36. As noted earlier, komatiites with < 30% MgO contain olivines in agreement with the predicted forsterite content using K_D values of 0.30 ± 0.03 (Table 1.4.2.1, Fig. 1.4.2.1). This would seem to suggest that the K_D values used are not inappropriate, and that the more Mg-rich komatiites may perhaps have accumulated some olivine.

With the exception of the komatiites, most other proposed terrestrial primary compositions fall within the Fo_{90} (K_D 0.27–0.33) zone. That is, they have Mg′-values between 0.708 and 0.748. Included here are the most "primitive" of ocean-floor basalts (#18, 20) and nodule-bearing alkali basalts (#35, 38), estimated compositions for primary Kilauean lava (#27, 29), and many of the picritic magmas (#9, 13, 15, 16, 17). This is an important observation since picrites such as these are frequently cited as the best candidates for primary magmas from which other basalts evolve. Inspection of Fig. 1.4.2.1 reveals that although these picrites are high in MgO, they are also high in FeO and therefore of no greater standing in the "hierarchy" of potential primary magmas than other "primitive" basalts with Mg′-values also within this range. Note that "modern" komatiite-like rocks, such as those from Gorgona Island (Gansser *et al.*, 1979) and the boninites of island arcs (Cameron *et al.*, 1979), although texturally similar to komatiites, have compositional affinities with picrites having Mg′-values within the range 0.70–0.75.

Examples of picritic magmas, other than ancient komatiites, with Mg′-values above 0.708–0.748 are extremely rare. They include lavas from the Baffin Bay area (Clarke, 1970), dikes in Chilean ophiolites (Elthon, 1979), and a lava from the Reykjanes Peninsula (Jakobsson *et al.*, 1978). The last example is porphyritic, with a pronounced discrepancy between phenocryst and experimental or predicted olivine compositions (Table 1.4.2.1; Maaløe and Jakobsson, 1980), and may well be accumulative. Only these few picrites and komatiites are capable of producing "primitive" basalts through extensive (> 10%) olivine fractionation (Fig. 1.4.2.1). Figure 1.4.2.1 shows that olivine fractionation does not significantly alter the FeO content of the melt. Consequently, the Baffin Bay picrites, although suitable

Table 1.4.2.2. Calculated primary magma compositions

	1 DL	2 PL1	3 PL2	4 PL3	5 DL	6 PL1	7 PL2
SiO_2	49.70	49.61	49.18	48.35	48.88	48.56	48.18
TiO_2	0.72	0.71	0.68	0.61	0.62	0.59	0.57
Al_2O_3	16.40	16.24	15.44	13.82	15.46	14.85	14.12
FeO^t	7.89	7.92	7.99	7.98	8.93	8.99	9.01
MnO	0.12	0.12	0.11	0.10	0.16	0.15	0.15
MgO	10.10	10.49	12.41	16.44	10.74	12.24	14.06
CaO	13.0	12.87	12.24	10.96	13.50	12.97	12.33
Na_2O	1.98	1.96	1.86	1.67	1.55	1.49	1.41
K_2O	0.01	0.01	0.01	0.01	0.01	0.01	0.01
P_2O_5	0.00	0.00	0.00	0.00	0.03	0.03	0.03
Cr_2O_3	0.07	0.07	0.07	0.06	0.11	0.10	0.10
Mg′	0.717	0.724	0.753	0.800	0.704	0.728	0.754
Fo (calc.)	89.4	89.7	91.0	93.0	88.8	90.0	91.0
F%	0	1	6	17	0	4	9

	8 PL3	9 DL	10 PL1	11 PL2	12 PL3	13 DL	14 PL1
SiO_2	47.39	49.33	48.63	47.53	46.64	51.25	49.48
TiO_2	0.50	2.58	2.38	2.05	1.76	2.06	1.74
Al_2O_3	12.52	12.80	11.81	10.16	8.74	13.56	11.43
FeO^t	8.88	11.43	11.55	11.36	10.87	10.77	11.32
MnO	0.13	0.17	0.16	0.13	0.12	0.18	0.15
MgO	18.27	10.03	12.87	17.93	22.55	7.01	13.09
CaO	10.93	10.78	9.95	8.55	7.36	10.37	8.74
Na_2O	1.25	2.13	1.97	1.69	1.45	2.29	1.93
K_2O	0.01	0.51	0.47	0.40	0.35	0.36	0.31
P_2O_5	0.02	0.25	0.23	0.20	0.17	0.24	0.20
Cr_2O_3	0.09	—	—	—	—	—	—
Mg′	0.800	0.635	0.686	0.754	0.799	0.563	0.691
Fo (calc.)	93.0	85.3	88.0	91.0	93.0	81.1	88.2
F%	21	0	8	23	38	0	17

	15 PL2	16 PL3	17 DL=PL1	18 DL	19 PL1	20 DL	21 PL1
SiO_2	48.34	47.34	42.90	45.20	45.13	38.40	38.42
TiO_2	1.51	1.30	4.46	2.91	2.88	12.70	12.08
Al_2O_3	9.93	8.54	8.00	8.61	8.52	8.72	8.29
FeO^t	11.16	10.69	22.00	21.10	21.11	18.10	18.21
MnO	0.13	0.11	0.30	0.28	0.28	0.26	0.25
MgO	17.82	22.46	12.40	11.60	11.88	10.40	11.89
CaO	7.59	6.53	8.99	9.45	9.35	10.30	9.79
Na_2O	1.68	1.44	0.25	0.23	0.23	0.35	0.33
K_2O	0.27	0.23	0.05	0.06	0.06	0.07	0.07
P_2O_5	0.17	0.15	0.08	0.07	0.07	0.06	0.06
Cr_2O_3	—	—	0.61	0.55	0.54	0.65	0.62
Mg′	0.754	0.800	0.501	0.495	0.501	0.506	0.534
Fo (calc.)	91.0	93.0	77.0	76.6	77.0	77.3	79.5
F%	31	46	0	0	1	0	5

Table 1.4.2.2. Calculated primary magma compositions.

Starting with a derivative liquid composition (DL), successive primary liquids have been calculated to be in equilibrium with the most forsteritic olivine phenocryst recorded in the lavas (PL1), and with hypothetical mantle olivine compositions of Fo_{91} (PL2) and Fo_{93} (PL-3) (K_D = 0.30). F% indicates the amounts of olivine added to obtain the appropriate Mg′ value. **1-4** Ocean floor basalt. Derivative liquid (DL) is from Frey *et al.* (1974) 3-18-7-1. The most forsteritic olivine phenocryst composition recorded is $Fo_{89.7}$ **5-8** Picritic basalt from the Reykjanes Peninsula. The derivative liquid (DL) is sample RE46 (Jakobsson *et al*, 1978) which is described as a picrite containing only 1% olivine phenocrysts. The maximum forsterite content of olivine in these lavas is Fo_{90}. **9-12** Kilauea Iki 1959 lavas. The derivative liquid (DL) is the most magnesium-rich glass recorded by Murata and Richter (1966). The maximum forsterite content of olivine in these lavas is Fo_{88}. **13-16** Mauna Loa lavas. The derivative liquid (DL) is the average glass composition for historic lavas (J.M. Rhodes, unpubl. data). The maximum forsterite content of olivine in these lavas is Fo_{88}. **17** Apollo 12 ilmenite basalt 12008 (Rhodes *et al.*, 1977). The maximum forsterite content of olivine in these lavas is Fo_{77}, which is in accord with the calculated composition. Therefore, this is the best estimate available. **18-19** Apollo 12 olivine basalt 12009 (Green *et al.*, 1971b). The maximum forsterite content of olivine recorded in these lavas is Fo_{77}, in close agreement with the calculated composition. **20-21** Apollo 17 high-Ti basalt 74275 (Rhodes *et al.*, 1976). PL1 has been calculated on the basis of a maximum forsterite content of $Fo_{79.5}$ recorded in this sample (Green *et al.*, 1975).

progenitors for many intraplate basalts, are too iron-rich to be parental to the most "primitive" of the ocean-floor basalts. If these ocean-floor basalts are derivative and not primary, magmas similar to the Tortuga dikes or Elthon's calculated ocean crust (Elthon, 1979) have the most suitable parental compositions (Fig. 1.4.2.1).

Figure 1.4.2.1 shows that the iron content of the melt remains roughly constant not only during extensive olivine fractionation, but also throughout a wide melting interval (20–50%). Consequently, the difference in iron content between intraplate primary basalts and those erupted at plate margins (#10, 14, 18, 19, 20, 22) must reflect a fundamental difference in the source composition, rather than variation in melting or fractionation processes.

A common characteristic of these most "primitive" picrites and of many komatiites is that they have Mg′-values indicating that they would have been in equilibrium with residual mantle olivine between Fo_{92} and Fo_{94}. Although such olivines are found in garnet peridotite inclusions in kimberlite, and are typically between Fo_{92} and Fo_{94} with an average of Fo_{93} (e.g., O'Hara and Mercy, 1963; Meyer and Boyd, 1972; Reid *et al.*, 1975), they are uncommon in other mantle samples. For example, olivines from spinel peridotite nodules in alkali basalts vary between Fo_{90} and Fo_{92}, with an average of Fo_{91} (Ross *et al.*, 1954; Frey and Green, 1974; Frey and Prinz, 1978; de Albuquerque *et al.*, 1977; Reid and Woods, 1978). Similarly, those from tectonized harzburgites at the base of ophiolite sequences typically average Fo_{91} (Malpas, 1978; Elthon, 1979; Pallister and Hopson, 1979). Thus, it appears that there is a spectrum of refractory mantle types and complementary magmas. The most "primitive" picritic and komatiite magmas, if primary and not accumulative, may be complementary melts to residual mantle containing $Fo_{93\pm1}$ olivine. Such compositions are the likely consequence either of substantial melting, or of lesser amounts of melting of an already depleted mantle (Arndt, 1977a). The experiments of Jaques and Green (1980) suggest that 30–50% melting of a pyrolite-like mantle, or 20–40% of an already depleted source may be appropriate. On the other hand, the more common picrites and "primitive" basalts with Mg′-values betwen 0.70 and 0.75 may well be primary melts complementary to residual mantle containing $Fo_{91\pm1}$ olivine. These basalts may result from 20–30% melting of pyrolite-like mantle, or less than 20% melting of already depleted mantle.

Figure 1.4.2.1 shows that lunar mare basalts (#41–45) and those from other planets, such as eucrites (#46) or a hypothetical martian magma (#47), have higher iron contents than terrestrial basalts. Unfortunately, we lack the direct evidence afforded by mantle-derived samples for constraining the composition of these extraterrestrial primary magmas.

An alternative approach is to use the maximum forsterite content of olivine phenocrysts to provide information concerning the most "primitive" composition within a given suite of basalts. Such compositions may not be primary, but they are the closest approximation available.

Lunar mare basalts contain olivine phenocrysts with maximum forsterite contents of Fo_{77} for the low-Ti variety and $Fo_{79.5}$ for some high-Ti basalts (Table 1.4.2.1). These values correspond to Mg′-values of about 0.52, employing K_D values of 0.28 for high-Ti basalts and 0.33 for the low-Ti variety (Longhi *et al.*, 1978). These values are only slightly higher than the most "primitive" vitrophyres that have been identified as parental, possibly primary, magmas for several lunar mare basalt associations (e.g., Green *et al.*, 1971b; Rhodes *et al.*, 1976, 1977), and are within the range of a number of lunar glass compositions of purported volcanic origin (Heiken *et al.*, 1974; Delano, 1980a, b; Delano *et al.*, 1980). Thus, it appears that among mare basalts the best estimates of primary compositions are those with Mg′-values between 0.50–0.52. If correct, this raises the interesting question why primary magmas are relatively common on the Moon, and yet are so elusive on Earth.

Irvine (1977) has recently extended the use of olivine compositions for estimating the most discernibly "primitive" composition within a cogenetic basalt series. As discussed above, olivine phenocryst compositions provide an estimate of the highest Mg′-value in the series. A second composition, that of an evolved liquid, can be related to this Mg′-value through olivine fractionation to provide the most "primitive" composition that can be inferred from the combined chemical and petrographic data. Important, but not necessarily valid, assumptions latent in this approach are that crystal fractionation is the only operative process, that only olivine has been fractionated, and that the olivine is not exotic.

An analogous approach is used in Table 1.4.2.2, where hypothetical primary magma compositions are calculated from selected derivative melt compositions. In these calculations, equilibrium olivine compositions (K_D = 0.30) are added in small increments to the melt, until a prescribed olivine composition is attained. For terrestrial lavas, three potential primary liquids (PL) have been calculated from the derivative liquid (DL): (1) PL1 corresponds to the maximum forsterite content of olivine in the series. This is the most "primitive" liquid that can be inferred from the combined chemical and petrographic data, and corresponds to "primitive" liquid compositions calculated by Irvine (1977). (2) PL2 is a hypothetical primary melt complementary to the type of residual mantle containing $Fo_{91\pm1}$ olivine commonly found as inclusions in basalt and as tectonized harzburgite at the base of ophiolite sequences. (3) PL3 is a hypothetical primary melt complementary to residual mantle characterized by $Fo_{93\pm1}$ olivine, found almost exclusively as inclusions in kimberlite. In the case of lunar magmas only one primary liquid (PL1) is calculated, corresponding to the maximum recorded forsterite content in olivine. Included in Table 1.4.2.2 are examples of an ocean-floor basalt, an Icelandic picrite, Hawaiian basalts and both high- and low-Ti lunar mare basalts.

Several clarifying points emerge from these calculations. In the Icelandic and ocean-floor examples, if picritic magmas with MgO contents of about 16–18% are primary (PL3), they must be complementary to a highly refractory mantle residue containing $Fo_{93\pm1}$ olivine. If, however, a less refractory mantle comparable wth peridotite nodules in basalt or the basal tectonized harzburgite of ophiolites is considered more appropriate as the residue, then the primary magmas are unlikely to be picritic, but are "primitive" basalts containing 12–14% MgO (PL2). Both calculated (PL1) and commonly observed basaltic compositions can be derived from such primary magmas with only minor amounts (4–10%) of olivine fractionation. In the case of the Hawaiian lavas, although the compositions calculated for PL1 compare closely with the primary magma estimates of Murata and Richter (1966) and Irvine (1977), they differ markedly from the PL2 compositions. Thus, unless these magmas were derived from an iron-rich mantle, they are substantially evolved compared to a melt that could be complementary to typical ultramafic nodules in basalts (PL2). Note that as a consequence of the higher iron content, the PL2 composition is also much higher in magnesium (MgO ≃18%) than the comparable Icelandic or ocean-floor examples. The MgO content becomes even higher (≃22%) if a highly refractory residual mantle (PL3) is preferred. (This last point illustrates the danger of using only the MgO content of a basalt for diagnostic purposes.) For the lunar mare basalts, Table 1.4.2.2 shows that the most "primitive" samples available (DL) compare very closely with the calculated primary magmas (PL1), and that the two can be related by very minor (1–4%) olivine fractionation. The Mg′-values of the calculated compositions range between 0.50–0.54, and, with the exception of Apollo 15 and 16 green glasses, overlap the values found for all other "primitive" mare glasses (Delano *et al.*, 1980). Consequently, in the absence of other constraints, their compositions can be taken as the best estimates available of primary lunar mare basalts.

Nickel abundances

The partitioning of Ni between olivine and melt has been widely used to identify and delimit the compositions of primary mantle-derived melts (e.g., Kay *et al.*, 1970; Bougault and Hekinian, 1974; O'Hara *et al.*, 1975; Sato, 1977; Allègre *et al.*, 1977). Estimated primary Ni contents are relatively insensitive to the extent of melting, and range between 200–450 ppm for terrestrial magmas, depending on assumptions about mantle mineralogy and Ni content, and on the appropriate value for Ni distribution coefficients. On the basis of such calculations, many of the basalts listed in Table 1.4.2.1 appear to be too high in Ni for primary melts.

Similar calculations for lunar or achondritic magmas are complicated by their low oxygen fugacities and the potential presence of metallic iron both in the source and in subsequent fractionation processes. The low Ni content (<1–52 ppm) of "primitive" lunar vitrophyres (Table 1.4.2.1) indicates either extremely low Ni concentrations in the lunar mantle, or the presence of metallic iron.

Experimental studies have shown that the distribution of Ni between olivine and melt is not constant, but varies widely with composition and/or temperature (e.g., Irvine and Kushiro, 1976; Leeman and Scheidegger, 1977; Arndt, 1977b; Cawthorne and McIver, 1977; Hart and Davis, 1978; Leeman and Lindstrom, 1978; Irving, 1978).

The lack of agreement among these studies on appropriate distribution coefficient values is perplexing and requires further investigation. All show systematic variation with MgO content of the melt, but the values reported for synthetic systems and basaltic melts (Leeman and Scheidegger, 1977; Hart and Davis, 1978; Leeman and Lindstrom, 1978) are higher, for a given MgO content, than values determined for komatiitic lavas (Arndt, 1977b; Cawthorn and McIver, 1977). The problem is not resolved by turning to natural cogenetic lavas, and using either the log-log relationship between Ni and a highly incompatible element (Allègre *et al.*, 1977) or linear MgO-Ni trends in accumulative lavas to estimate olivine/liquid distribution coefficients. Calculations based on Kilauean lavas (Gunn, 1971) and ocean-floor basalts (J. M. Rhodes, pers. comm., 1980) give distribution coefficients which support the data obtained on basalts and synthetic systems, whereas distribution coefficients based on Baffin Bay lavas (Clarke, 1970) are commensurate with data obtained on komatiites.

These uncertainties in Ni partitioning data result in equivocal interpretations of Ni-MgO relationships for basalts. Thus, if the partitioning data obtained for basalts and experimental systems are used, one is led to the conclusion that basalts with high MgO and Ni contents are not suitable primary magmas, and that those with MgO $> 15\%$ are probably accumulative or derived from an unusually Ni-rich source (Hart and Davis, 1978). On the other hand, if partitioning data obtained on komatiites are considered more appropriate, then high-MgO basalts have suitable primary compositions and may produce common basalt types by olivine fractionation (e.g., Clarke and O'Hara, 1979; Hart and Davis, 1978; Elthon and Ridley, 1979).

This situation may possibly be clarified by use of an exchange partition coefficient K_D defined as

$$(Ni/MgO)_{olivine}/(Ni/MgO)_{melt}$$

with concentrations in weight percent and parts per million for convenience. This was the approach adopted by Sato (1977), who used an exchange partition coefficient of 2.3 ± 0.3 to estimate Ni abundances of 240–390 ppm, and an MgO content between 10–12% for primary magmas. On this basis some "primitive" ocean-floor, Hawaiian, and alkali basalts may represent primary magmas.

Recent studies have shown that, unlike the Mg-Fe exchange partition coefficient, the Mg-Ni exchange coefficient is not constant, as was assumed by Sato, but varies as a function of temperature (e.g., Arndt, 1977b; Hart and Davis, 1978; Leeman and Lindstrom, 1978). However, in contrast with standard partition coefficients, experimental data obtained for exchange partition coefficients in basalts, synthetic systems and komatiites are broadly in agreement, and mostly within the range of experimental error. Allowing for scatter in the data and for the linear relationships between ln K_D and inverse temperature (Hart and Davis, 1978; Leeman and Lindstrom, 1978), exchange partition coefficients (K_D) between 1.65 and 2.72 appear appropriate for melts between 1500°C and 1250°C. These two extremes can be used to place tentative upper and lower limits on the Ni and MgO content of primary magmas.

Although the uncertainty in the appropriate value for the exchange coefficient is large, useful inferences can still be made about the Ni-MgO relationships of terrestrial mantle-derived melts. As discussed earlier, nodules of residual peridotite in basalts and tectonized harzburgites in ophiolite sections typically contain Fo_{91} olivine. The Ni content of this olivine is remarkably constant, usually no more than 3,200 ppm (Forbes and Banno, 1966; Frey and Green, 1974; Sato, 1977). Consequently we can use this value, together with the MgO content of Fo_{91} (50.1%) and an appropriate value for K_D, to calculate the Ni/MgO ratio of a melt in equilibrium with residual olivine in mantle material of this type. These ratios are shown in Fig. 1.4.2.2 (lines A and B) for values of K_D of 2.72 and 1.65, respectively. Primary magmas in equilibrium with mantle of this type should fall within these limits. Note that these limitations are not necessarily diagnostic. There is considerable latitude for olivine fractionation between lines A and B, and evolved magmas derived from melts with higher Ni/MgO ratios will also plot in this field. The important point is that unless a magma has Ni/MgO ratios that are between 23–39, it is unlikely to be a primary melt complementary to this type of residual mantle. Magmas having lower ratios, such as typical basalts with MgO between 6–10% and Ni less than 200 ppm, are almost certainly evolved. Magmas with higher ratios, such as all of the picrites and komatiites plotted in Fig. 1.4.2.2, are unlikely to have been derived from this type of mantle, or are possibly accumulative.

Previous discussions of Mg-Fe relationships showed that mantle containing Fo_{93} olivine is an appropriate residual source for komatiites and the most primitive picrites. Mantle nodules of this type are commonly found in kimberlites, and contain olivine with higher Ni contents than nodules in basalt (e.g., O'Hara and

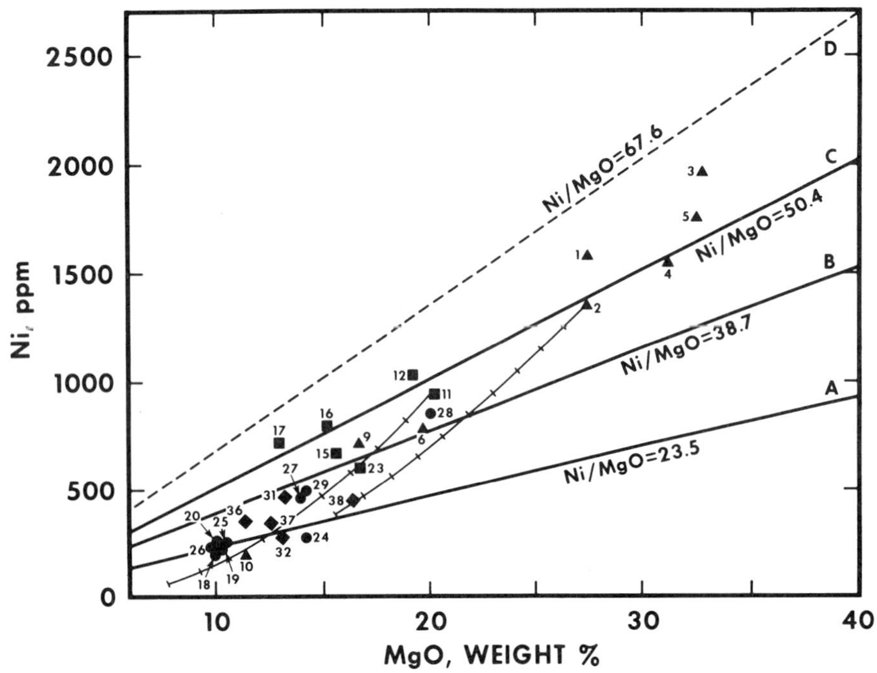

1.4.2.2 MgO-Ni relationships for potential terrestrial primary melts. Symbols are the same as in Fig. 1.4.2.1. Lines of constant Ni/MgO in the melt (A, B, C, D) are calculated from $(Ni/MgO)_{melt}$ = $(Ni/MgO)_{olivine}$ /K, where K is the Ni-Mg exchange partition coefficient. See text for further details. Olivine fractionation trends from a picritic melt (#11) and a komatiitic melt (#2) are shown by barred lines. The bars indicate 4% fractionation steps.

Mercy, 1963; O'Hara *et al.*, 1975; Reid *et al.*, 1975; Sato, 1977). The highest value recorded is 3,600 ppm, but olivines in some komatiites contain up to 4,300 ppm (Green *et al.*, 1975a; Arndt *et al.*, 1977). Using this maximum Ni content, Fo_{93} (MgO = 51.7%) and a value of 1.65 (1500° C) for K_D, the Ni/MgO ratio for a melt in equilibrium with residual olivine in mantle material of this type is 50.4, shown in Fig. 1.4.2.2 by line C. Most of the picrites and some of the komatiites fall below this line and are potential primary melts in equilibrium with residual mantle of this type. A few of the picrites plot above line C and may have accumulated a small amount of olivine. The same argument could be used for Ni-rich komatiites. However, it is probable that K_D should be as low as 1.2 for these rocks to allow for possible liquidus temperatures as high as 1700° C. Line D in Fig. 1.4.2.2 has been calculated on this assumption and shows that extremely high temperatures could account for these high Ni abundances.

Olivine fractionation trends are shown in Fig. 1.4.2.2 from a komatiite magma (#2) and from one of the Baffin Bay picrites (#11). These trends show that less "primitive" basaltic compositions could be derived from magmas such as these by olivine fractionation. This does not imply that these less "primitive" magmas are necessarily evolved from picrite or komatiite magmas. The previous calculations have shown that many could equally well be considered primary melts segregated from the kind of mantle that is preserved as nodules in basalt and at the base of ophiolite sections.

1.4.2.3 Discussion of primary terrestrial magmas

The characteristics and criteria developed in the preceding sections are used in the discussion that follows to examine selected examples of the three major terrestrial magma types—komatiites, picrites and basalts—that may potentially be primary. Central to this discussion is the question raised in the introduction. Is there a range of primary magma types, including komatiite, picrite and basalt, or are basalts the products of extensive differentiation of primary picritic or komatiitic magmas?

Komatiitic primary magmas

There can be little doubt that primary magmas are to be found among komatiites, and it is upon this assumption that inferences concerning the composition of the Archean mantle have been made (Bickle *et al.*, 1976; Sun and Nesbitt, 1977). Opinions differ on whether the full range of komatiite compositions, from peridotitic komatiites (MgO>20%) to basaltic komatiites (MgO 12–20%), are primary and result from variable amounts of partial melting (Green, 1975; Herrmann *et al.*, 1976; Nesbitt and Sun, 1976), or whether they are produced by extensive crystal fractionation (Viljoen and Viljoen, 1969), or by some combination of these two processes (Arndt *et al.*, 1977; Nisbet *et al.*, 1977). Detailed studies of single, thick lava flows demonstrate that a wide range of komatiitic compositions can differentiate from a single parental magma (Arndt *et al.*, 1977; Whitford and Arndt, 1978). However, this process alone cannot account for marked differences in trace element abundances and patterns observed in komatiites from different localities, nor for the covariance of light/heavy REE enrichment with decreasing MgO content in komatiites from a single region (Arth *et al.*, 1977; Sun and Nesbitt, 1977). Such variations, which require differences both in source composition and in the extent of melting, point to a spectrum of primary komatiite magma compositions.

Assuming that komatiite compositions have not been seriously modified by metamorphism, a major problem in understanding their petrogenesis is to account for the extremely high temperatures and extensive melting apparently required to produce them in the first place. Not all komatiites reflect liquid compositions; some are almost certainly accumulative. Arndt *et al.* (1977) suggest that lavas with MgO>30% probably contain accumulated olivine, and this appears to be borne out, as noted earlier, by K_D values for Mg/Fe and Mg/Ni distributions that differ from those of basalts and other komatiites. Nonetheless, liquid compositions are undoubtedly encompassed by spinifex-textured lavas with 25–30% MgO. It is widely held that such lavas have high liquidus temperatures (>1500° C) and probably formed as consequences of substantial (50–70%) mantle melting within unusually high temperature regimes (e.g., Green, 1975; Nesbitt and Sun, 1976; Nisbet *et al.*, 1977). An opposing viewpoint (O'Hara *et al.*, 1975), that these magmas were produced at great depth in the mantle by small amounts of invariant melting as a consequence of the contraction of the olivine stability field, is not supported by the low incompatible trace element abundances. Judging by the apparent absence of peridotitic komatiites from Phanerozoic and modern volcanic sequences, it seems that such high temperature regimes were confined to the Archean. Younger komatiites or komatiite-like rocks (e.g., Gale, 1973; Cameron *et al.*, 1979; Gansser *et al.*, 1979) are all basaltic komatiites, compositionally analogous to picrites, and have therefore probably formed under comparable thermal conditions.

A crucial dilemma for the concept of single, extensive melting event for the formation of peridotitic komatiites is to explain why the melt failed to segregate from the residue long before the required 50–70% melting was achieved (Arndt, 1977a). Estimates of the efficacy of melt segregation are varied, and the relative roles of permeability and shear are unclear (Yoder, 1976; Arndt, 1977a; Maaløe and Printzlau, 1979; Spera, 1980). However, accepting the guidelines of Spera (1980, table 6), that a 35% melt will segregate from the residue over 1000 times more effectively than 5% melt, and combining with this the observation that alkali basalts, widely held to result from small (2–10%) amounts of melting, are a common rock type, which attests to their ability to segregate effectively from the mantle, it seems likely that melt segregation will occur long before 50% melting is achieved. A more plausible explanation is that komatiites result from sequential melting processes in which the source material is the highly refractory residuum of a previous melting episode or episodes (Arndt, 1977a). The systematic change in light/heavy REE abundances with varying MgO content described by Arth *et al.* (1977) supports this hypothesis, as does the tendency for low-MgO differentiates of komatiites to have higher transition element abundances (Cr, Co, Ni) than basaltic rocks of comparable MgO content (Nesbitt and Sun, 1976). Melting experiments (Jaques and Green, 1980) show that, in contrast with the 40 to more than 50% melting required to produce peridotitic komatiites with MgO between 20–25% from "pyrolite-like" peridotite, similar liquids are produced from "depleted" peridotite following 30–40% melting. The residue in both cases contains Fo_{93-95} olivine, which reinforces the point made earlier that magmas which result from extensive melting or remelting will be complementary to highly refractory mantle containing $Fo_{93\pm1}$ olivine.

The sequential melting hypothesis for the origin of komatiites provides a satisfactory explanation for the common association and inter-layering of basaltic and komatiitic lavas (Arndt, 1977a; Arth *et al.*, 1977; Echeverria, 1980). If correct, this hypothesis has profound implications for the primary magma controversy, since primary basaltic and/or picritic magmas are *necessary precursors* to the formation of komatiites, and are not simply their derivatives.

Picritic primary magmas

Picritic magmas, typically with MgO contents between 12–20% have compositional affinities with basaltic komatiites. They are crucial to the primary magma problem, since it is widely held that it is from melts such as these that other basaltic magmas are derived through extensive crystal fractionation (e.g., O'Hara, 1968a, b; Wright, 1971; Yoder, 1976; Cox, 1978; Elthon, 1979; Green *et al.*, 1979; Stolper, 1980).

The major flaw in this concept of the universality of primary picritic magmas is the apparent rarity of picrites that can be shown to have erupted with liquid compositions. This is all the more surprising when one considers that lavas inferred to have liquid compositions are not uncommon among both Precambrian basaltic and peridotitic komatiites. It would appear, therefore, that the apparent absence of picritic liquid is a problem confined to Phanerozoic lavas. In part the problem is semantic, since there is a tendency for high-MgO lavas exhibiting quench-like textures to be called komatiites (e.g., Cameron *et al.*, 1979; Ganser *et al.*, 1979; Echeverria, 1980). Nonetheless, the vast majority of picrites are phenocryst-laden and far removed from liquid compositions. Explanations which may account for the paucity of picritic liquid include:

a) The melts fractionate within the mantle, either in an ascending magma column, or at various staging-posts en route to the surface (e.g., O'Hara, 1968a; O'Hara *et al.*, 1975).
b) Picritic melts mix with evolved consanguinous magmas in shallow crustal magma chambers to produce typical basaltic magmas (O'Hara, 1977; Rhodes *et al.*, 1979a; Sparks *et al.*, 1980).
c) Dense picritic liquids are trapped at the base of, or within, less dense crust, where they differentiate until the density differential between crust and magma is appropriate for the ascent and eruption of the differentiated lavas (Goles, 1976; Cox, 1980; Stolper and Walker, 1980).

According to these scenarios, picritic liquids will be erupted at the surface, only under exceptional circumstances, perhaps early or late in the magmatic sequence or under conditions of tectonic rifting (Cox, 1978). Why these processes should have been less prevalent for Precambrian lavas than for Phanerozoic and modern lavas is not clear.

The principal arguments favoring primary picritic magmas come from experimental studies, and in particular from melting experiments on basalts. However, as discussed earlier, melting studies on peridotites offer an alternative perspective, namely that picrites are only part of a spectrum of potential primary magma compositions. This view is enhanced when proposed primary picritic magma compositions are examined.

Many of the various picrites that have been proposed as primary magmas contain between 12–16% MgO and 11–12% total Fe as FeO. Although the MgO content is high, so is the FeO content, with the consequence that the Mg′-values are not significantly higher than those of "primitive" basalts. Thus, like the basalts, they could be derivative from more "primitive" picritic melts or they could be primary melts complementary to residual mantle containing $Fo_{91\pm1}$ olivine (Fig. 1.4.2.1). Unlike more "primitive" picrites and komatiites, these picrites tend to have low CaO/TiO_2, Al_2O_3/TiO_2 and CaO/Na_2O ratios similar to intraplate basalts. This would seem to exclude the possibility of fractionation involving only olivine from more "primitive" melts. They could be produced by substantial melting of an undepleted "pyrolite-like" mantle or by small amounts of melting of an already depleted mantle. Melting experiments on peridotites (Mysen and Kushiro, 1977; Jaques and Green, 1980) indicate that 20–30% melting is probably appropriate.

Since this type of picrite is transitional to "primitive" basalts, it follows that the latter could be derived from these picrites through small to moderate amounts (<10%) of olivine fractionation, or they could themselves be primary melts derived from a similar source region by lesser amounts of melting.

At the other end of the range are picrites with 16–20% MgO (Table 1.4.2.1, #11, 12, 14, 21, 22). Examples that may have been extruded as liquids are extremely rare, perhaps the most convincing being the Baffin Bay lavas (Clarke, 1970), and dikes in the Tortuga ophiolite complex (Elthon, 1979). The high Mg′-values and Ni concentrations are appropriate for melts complementary to a mantle residue containing $Fo_{93\pm1}$ olivine (Figs. 1.4.2.1–2), and produced either by extensive (30–40%) melting or by less extensive melting of an alredy depleted source. The second alternative appears to be the more likely, by analogy with komatiites, and because their CaO/Na_2O, CaO/TiO_2 and Al_2O_3/TiO_2 ratios are higher than those of less "primitive" picrites. The trace element data of O'Nions and Clarke (1972) also point to a depleted source for the Baffin Bay lavas. Similarly, lavas erupted in an island-arc or back-arc environment, of which the Tortuga ophiolite is thought to be an example (Stern, 1980), frequently have geochemical characteristics indicative of variable source depletion and multiple melting episodes (Sun and Nesbitt, 1978; Coish and Church, 1979; Duncan and Green, 1978).

Extensive differentiation (20–30%) of these "primitive" picritic magmas will produce basaltic liquids (Figs. 1.4.2.1–2). Evidence for this is provided by the Baffin Island and Svartenhuk lavas (Clarke, 1970) and by analogous komatiite sequences that have differentiated to low-MgO melts (e.g., Nisbet *et al.*, 1977; Cawthorn and McIver, 1977). An important question, therefore, is whether the *majority* of basalts result from such extensive differentiation. The use of Ni concentrations to resolve this issue, although potentially helpful, is hindered by uncertainties in appropriate distribution coefficients (e.g., Hart and Davis, 1978; Clarke and O'Hara, 1979). Figure 1.4.2.1 shows that picrites like the Baffin Bay examples might be parental to high-FeO intraplate basalts, whereas those like the Tortuga example could be parental to low-FeO ocean-floor basalts. However, in neither case do the majority of basalts exhibit the depleted characteristics appropriate to derivatives of these picrites. For example, most "primitive" high-FeO intraplate basalts have TiO_2, incompatible element abundances, and ratios of Na_2O/CaO, TiO_2/Al_2O_3 and TiO_2/CaO that are too high for them to be derivatives of lavas like the Baffin Bay picrites. Similarly, derivatives of the Tortuga lavas would have higher CaO/Na_2O ratios than are common in most "primitive" ocean-floor basalts. A few examples of high CaO/Na_2O ocean-floor basalts are known. They are characterized by high SiO_2 and CaO and low Na_2O and lithophile elements (Melson *et al.*, 1976; Autio and Rhodes, pers. comm., 1980), and might have evolved from Tortuga-like picrites. These lavas would crystallize highly anorthitic plagioclase (An_{84-88}) once they attain the olivine-plagioclase cotectic, providing a potential source for the highly calcic plagioclase phenocrysts found in mixed ocean-floor basalts (Donaldson and Brown, 1977; Rhodes *et al.*, 1979a). If this suggestion is valid, it points to the complex interplay of different primary ocean-floor magmas, possibly produced by successive melting episodes of a common source region (Duncan and Green, 1980).

Primary basaltic magmas

A central theme of the preceding discussion is that komatiitic and some picritic magmas are end-products of complex multiple-melting episodes, leaving behind highly depleted and refractory mantle residues. An important consequence of this hypothesis is that primary basaltic magmas are necessary precursors of the formation of these more "primitive" melts. This implies that basalts may be both the direct products of mantle melting and the differentiates of more "primitive" melts, and that a wide range of primary magma compositions is to be anticipated. There is corroboration for this suggestion from several sources. Melting experiments on peridotites (Mysen and Kushiro, 1977; Jaques and Green, 1980) show that "primitive" basalts, encompassing both alkaline and tholeiitic varieties and a continuum of compositions, may be produced by moderate amounts of melting (<20%). More extensive melting leads to the production of picritic and komatiitic melts. Samples of mantle, available world-wide as nodules in basalt, alpine peridotites and at the base of ophiolite sequences, typically contain $Fo_{91\pm1}$ olivine. If, as is widely supposed, these peridotites are residues left after previous magma extraction events, primary melts less "primitive" than picrite are required. Inspection of Fig. 1.4.2.1-2 indicates that there exist examples of both tholeiitic and alkaline basalts that have Mg′-values and Ni abundances appropriate for primary melts which would complement such residual mantle material.

The intense debate in recent years over the primary nature of ocean-floor basalts mirrors the wider arguments concerning primary basaltic magmas in general. On the one hand is the view that the most MgO-rich glasses identified to date ($MgO \simeq 10\%$) constitute the best estimates of primary magma compositions available (e.g., Frey *et al.*, 1974; Bender *et al.*, 1978; Bryan, 1979; Presnall *et al.*, 1979; Rhodes *et al.*, 1979a); and on the other, the view that these glasses represent differentiates of primary picritic melts which contained 17–20% MgO (O'Hara, 1968b; Elthon, 1979; Green *et al.*, 1979; Stolper, 1980).

Arguments favoring a picritic primary magma rest largely on the experimental evidence that orthopyroxene is not present on the liquidus of "primitive" ocean-floor basaltic liquids at any pressure, implying that these melts are not in equilibrium with a lherzolitic or harzburgitic mantle residue. Estimates for the bulk composition of the oceanic crust, based on crustal sections such as ophiolites, provide additional support for a picritic primary magma (Elthon, 1979). The major stumbling block to the primary picrite hypothesis, of course, is the lack of suitable picritic parental magmas within the oceanic environment. The Tortuga dikes described by Elthon (1979) are perhaps the most appropriate compositions but, as noted previously, differentiation of these magmas would lead to rare lavas with high Ca/Na ratios, not to typical ocean-floor basalts. A further difficulty with the primary picrite hypothesis is that the residue remaining after melting would be highly refractory, containing $Fo_{93\pm1}$ olivine. This is at variance with the widespread occurrence of Fo_{91} olivine in harzburgites at the base of many ophiolite sequences. These ultramafics have been widely interpreted as being residual material formed during melting and extraction of magma from the oceanic mantle (e.g., Malpas, 1978; Elthon, 1979). On the other hand, the most "primitive" ocean-floor basaltic glasses so far identified (Table 1.4.2.1, #18–20) have Mg′-values, Ni concentrations and olivine phenocryst compositions which are appropriate for melts that complement the harzburgite residue, or for derivation from such melts by minor olivine fractionation (<5%).

Basalts containing ultramafic inclusions are widespread, occurring in oceanic islands, tectonic rift zones and continental interiors; they include alkali basalts, basanites and nephelinites. The inclusions of spinel peridotite are evidence that the magma ascended rapidly from mantle depths greater than 30 km without opportunity for crystal fractionation en route to the surface, or for fractionation and mixing in mantle or crustal magma chambers (e.g., Maaløe, 1973; Yoder, 1976; Spera, 1980). Furthermore, the high ascent velocities required (1–50 cm/sec) preclude the possibility of density contrasts between crust and melt (Stolper and

Walker, 1980) acting as a filter against denser, more "primitive," melts. Consequently, these basalts are either primary magmas, or are results of differentiation within the mantle prior to the incorporation of the ultramafic inclusions.

The existence of nodule-bearing hawaiites, mugearites, benmoreites and even phonolites in addition to basalts demonstrates that extensive differentiation occurs within the mantle (Wilkinson and Binns, 1969; Price and Green, 1972; Green *et al.*, 1974; Irving and Green, 1976; Irving and Price, 1981). However, because of the rapid, uninterrupted ascent, the potential for selective eruption of evolved rather than "primitive" lavas, must be lower among nodule-bearing basalts than among other basalt types. Consequently, we might expect to find primary magma compositions among the most "primitive" nodule-bearing lavas. These are all basaltic, with Mg'-values of about 0.69–0.72 (e.g., Table 1.4.2.1, #31, 32, 35, 38). Figures 1.4.2.1 and 1.4.2.2 show that several of these nodule-bearing basalts have Mg'-values and Ni concentrations appropriate for primary magmas. More "primitive" compositions, such as oceanites and ankaramites, lack ultramafic inclusions and are widely interpreted as being accumulates.

Some of the most compelling evidence for the primary nature of nodule-bearing basalts comes from trace element studies. The strong enrichment in LREE relative to HREE characteristic of these basalts is usually considered to be a consequence of partial melting of a garnet-bearing mantle source, either through small amounts of melting (0.5–2%) of a source with relative chondritic abundances (Kay and Gast, 1973), or by more extensive melting (3–15%) of an LREE-enriched source (Sun and Hanson, 1975; Frey *et al.*, 1978). The model proposed by Frey *et al.* (1978), in which associated melilitites, basanites, alkali basalts and tholeiites are produced from a common LREE-enriched pyrolite-like source by 4–25% melting, is particularly convincing. The amount of melting and the residual mineralogy used in the trace element models are consistent with liquidus phases in melting experiments at appropriate pressures, with experimental and major element constraints on the extent of melting, and with refractory lherzolite and harzburgite mineral assemblages and chemical compositions.

It is fundamental to all of these models that, whereas the highly fractionated REE pattern may be readily produced by small amounts of partial melting, prohibitively large amounts of crystal fractionation involving garnet plus pyroxene or olivine (eclogite fractionation) would be required to achieve the same result. Consequently, it is improbable that the characteristic REE pattern could be produced from any known picritic composition by realistic amounts of crystal fractionation. The only way around this dilemma is to propose that the alkali basalts are derived by moderate amounts (10–20%) of crystal fractionation from some as yet unidentified LREE-enriched picrite composition. In turn, this would require an extremely unusual LREE-enriched mantle source for the picrite.

From these considerations, it seems appropriate to consider the "primitive" alkali basalts, basanites and nephelinites as potential primary magmas or as magmas that, at the most, have been modified only by small to moderate amounts (<10%) of crystal fractionation. An important consequence of this argument is that it offers additional evidence that similar basaltic magmas may be produced as primary melts as well as by differentiation from more "primitive" magmas.

1.4.2.4 Conclusions

The preceding sections have explored the characteristics of primary magmas and scrutinized some major rock types for these characteristics. Several useful generalizations can be made. A substantial body of evidence documents the existence of primary komatiitic and picritic magmas. Some are also accumulative, whereas others are evolved, and resolving these differences is not always an easy task. Undoubtedly some basalts are derivatives of these more "primitive" melts through extensive differentiation. However, if as appears to be the case, "primitive" primary melts such as komatiites and picrites are produced as consequences not of a single, extensive melting event, but of sequential melting episodes in which previously melted and depleted mantle is re-melted, it follows that primary basaltic magmas are necessary precursors of these more "primitive" melts. If correct, this argument points to a spectrum of primary magma compositions. Compositional data for basalts and melting experiments on peridotites substantiate this view. At one end of the range are "primitive" basalts produced from relatively undepleted mantle by small to moderate amounts of melting (<25%), leaving a complementary residue similar to ultramafic nodules in basalts, to alpine peridotites and to the basal layers of ophiolite complexes. At the other extreme are komatiites and picrites probably produced through similar amounts of melting of already-depleted peridotite, leaving behind an extremely refractory residue.

Thus, basalts may be both primary and derivative, and both types may erupt within a single magmatic province. In order to resolve these alternatives it will be necessary to remedy the discrepancies that currently exist in experimental melting studies, to arrive at a better understanding of the factors that control element distribution coefficients, and to obtain better quality comprehensive geochemical and isotopic data for basaltic terrains that are thoroughly documented, both geologically and petrologically.

1.4.2.5 References

Allègre C. J., Treuil M., Minster J. -F., Minster B. and Albarède F. (1977) Systematic use of trace element in igneous process, part I: fractional crystallization processes in volcanic suites. *Contrib. Mineral. Petrol.* **60**, 57–75.

Arndt N. T. (1976) Melting relations of ultramafic lavas (komatiites) at 1 atm. and high pressure. *Carnegie Inst. Wash., Yearb.* **75**, 555–562.

Arndt N. T. (1977a) Ultrabasic magmas and high-degree melting of the mantle. *Contrib. Mineral. Petrol.* **64**, 205–221.

Arndt N. T. (1977b) Partitioning of nickel between olivine and ultrabasic and basic komatiite liquids. *Carnegie Inst. Wash., Yearb.* **76**, 553–557.

Arndt N. T., Naldrett A. J. and Pyke D. R. (1977) Komatiitic and iron-rich tholeiitic lavas of Munro Township, northeast Ontario. *J. Petrol.* **18**, 319–369.

Arth J. G., Arndt N. T. and Naldrett A. J. (1977) Genesis of Archean komatiites from Munro Township, Ontario: trace-element evidence. *Geology* **5**, 590–594.

Baxter A. N. (1976) Geochemistry and petrogenesis of primitive alkali basalt from Mauritius, Indian Ocean. *Geol. Soc. Amer. Bull.* **87**, 1028–1034.

Bender J. F., Hodges F. N. and Bence A. E. (1978) Petrogenesis of basalts from the Project FAMOUS area: experimental study from 0 to 15 kbars. *Earth Planet. Sci. Lett.* **41**, 277–302.

Best M. G. and Brimhall W. H. (1974) Late Cenozoic alkalic basaltic magmas in the Western Colorado Plateaus and the Basin and Range Transition Zone, U.S.A., and their bearing on mantle dynamics. *Geol. Soc. Amer. Bull.* **85**, 1077–1090.

Bickle M. J., Hawkesworth C. J., Martin A., Nisbet E. G. and O'Nions R. K. (1976) Mantle composition derived from the chemistry of ultramafic lavas. *Nature* **263**, 577–580.

Björnsson A., Johnsen G., Sigurdsson S. and Thorbergsson G. (1979) Rifting of the plate boundary in North Iceland 1975–1978. *J. Geophys. Res.* **84**, 3029–3038.

Bougault H. and Hekinian R. (1974) Rift valley in the Atlantic Ocean near 36° 50′ N: petrology and geochemistry of basaltic rocks. *Earth Planet. Sci. Lett.* **24**, 249–261.

Bowen N. L. (1928) *The Evolution of the Igneous Rocks,* Princeton Univ. Press. 334 pp.

Bryan W. B. (1979) Regional variation and petrogenesis of basalt glasses from the FAMOUS area, Mid-Atlantic Ridge. *J. Petrol* **20**, 293–325.

Bryan W. B. and Moore J. G. (1977) Compositional variations of young basalts in the Mid-Atlantic ridge rift valley near 36° 49′ N. *Geol. Soc. Amer. Bull.* **88**, 556–570.

Butler P., Jr. (1978) Recognition of lunar glass droplets produced directly from endogenous liquids: the evidence from S-Zn coatings. *Proc. Lunar Sci. Conf. 9th*, p. 1561–1581.

Cameron W. E., Nisbet E. G. and Dietrich V. J. (1979) Boninites, komatiites and ophiolitic basalts. *Nature* **280**, 550–554.

Carmichael I. S. E., Turner F. J. and Verhoogen J. (1974) *Igneous Petrology.* McGraw-Hill, N.Y. 739 pp.

Carter J. L. (1970) Mineralogy and chemistry of the earth's upper mantle based on the partial fusion—partial crystallization model. *Geol. Soc. Amer. Bull.* **81**, 2021–2034.

Cawthorn R. G. and McIver J. R. (1977) Nickel in komatiites. *Nature* **266**, 716–718.

Clarke D. B. (1970) Tertiary basalts of Baffin Bay: possible primary magma from the mantle. *Contrib. Mineral. Petrol.* **25**, 203–224.

Clarke D. B. and O'Hara M. J. (1979) Nickel, and the existence of high-MgO liquids in nature. *Earth Planet. Sci. Lett.* **44**, 153–158.

Coish R. A. and Church W. R. (1979) Igneous geochemistry of mafic rocks in the Betts Cove ophiolite, Newfoundland. *Contrib. Mineral. Petrol.* **70**, 29–39.

Consolmagno G. and Drake M. J. (1977) Composition and evolution of the eucrite parent body: evidence from rare earth elements. *Geochim. Cosmochim. Acta* **41**, 1271–1282.

Cox K. G. (1978) Komatiites and other high-magnesia lavas: some problems. *Phil. Trans. Roy. Soc. Lond.* **A288**, 599–609.

Cox K. G. (1980) A model for flood basalt volcanism. *J. Petrol.* **21**, 629–650.

Cox K. G. and Bell J. D. (1972) A crystal fractionation model for the basaltic rocks of the New Georgia Group, British Solomon Islands. *Contrib. Mineral. Petrol.* **37**, 1–13.

Cox K. G. and Jamieson B. G. (1974) The olivine-rich lavas of Nuanetsi: a study of polybaric magmatic evolution. *J. Petrol.* **15**, 269–301.

Cox K. G., Johnson R. L., Monkman L. J., Stillman C. J., Vail J. R. and Wood D. N. (1965) The geology of the Nuanetsi igneous province. *Phil. Trans. Roy. Soc. Lond.* **A257**, 71–218.

De Albuquerque C. A. R., Capedri S. and Dostal J. (1977) Mineralogy of spinel peridotite inclusions of alkali basalt from Sardinia. *Geol. Soc. Amer. Bull.* **88**, 1493–1496.

Delano J. W. (1979) Apollo 15 green glass: chemistry and possible origin. *Proc. Lunar Sci. Conf. 10th*, p. 275–300.

Delano J. W. (1980a) Apollo 15 red glass: chemistry and liquidus phase relations (abstract). In *Lunar and Planetary Science XI*, p. 210–212. Lunar and Planetary Institute, Houston.

Delano J. W. (1980b) Apollo 15 yellow glasses: chemistry and possible origins (abstract). In *Lunar and Planetary Science XI*, p. 213–215. Lunar and Planetary Institute, Houston.

Delano J. W., Taylor S. R. and Ringwood A. E. (1980) Composition and structure of the deep lunar interior (abstract). In *Lunar and Planetary Science XI*, p. 225–227. Lunar and Planetary Institute, Houston.

Dietrich V., Emmerman R., Oberhansli R., and Puchelt H. (1978) Geochemistry of basaltic and gabbroic rocks from the West Mariana Basin and the Mariana Trench. *Earth Planet. Sci. Lett.* **39**, 127–144.

Donaldson C. H. and Brown R. W. (1977) Refractory megacrysts and magnesium-rich melt inclusions within spinel in oceanic tholeiites: indicators of magma mixing and parental magma composition. *Earth Planet. Sci. Lett.* **37**, 81–89.

Drever H. I. and Johnston R. (1958) The petrology of picritic rocks in minor intrusions—Hebridean group. *Trans. R. Soc. Edinburgh* **63**, 459–499.

Drever H. I. and Johnston R. (1966) A natural high-lime silicate liquid more basic than basalt. *J. Petrol.* **7**, 414–420.

Duke M. B. and Silver L. T. (1967) Petrology of eucrites, howardites and mesosiderites. *Geochim. Cosmochim. Acta* **31**, 1637–1665.

Duncan R. A. and Green D. H. (1980) The role of multi-stage melting in the formation of oceanic crust. *Geology* **8**, 22–28.

Dungan M. A. and Rhodes J. M. (1978) Residual glasses and melt inclusions in basalts from DSDP Legs 45 and 46: evidence for magma mixing. *Contrib. Mineral. Petrol.* **67**, 417–431.

Eaton J. P. and Murata K. J. (1960) How volcanoes grow. *Science* **132**, 925–938.

Echeverría L. M. (1980) Tertiary or Mesozoic komatiites from Gorgona Island, Columbia: field relations and geochemistry. *Contrib. Mineral. Petrol.* **73**, 253–266.

Elthon D. (1979) High magnesia liquids as the parental magma for ocean floor basalts. *Nature* **278**, 514–518.

Elthon D. and Ridley W. I. (1979) Comments on "The partitioning of nickel between olivine and silicate melt" by S. R. Hart and K. E. Davis. *Earth Planet. Sci. Lett.* **44**, 162–164.

Engel A. E. J., Engel C. J. and Havens R. G. (1965) Chemical characteristics of oceanic basalts and the upper mantle. *Geol. Soc. Amer. Bull.* **76**, 719–734.

Flower M. F. J. (1971) Rare earth element distribution in lavas and ultramafic xenoliths from the Comores Archipelago, western Indian Ocean. *Contrib. Mineral. Petrol.* **31**, 335–346.

Flower M. F. J. (1973) Evolution of basaltic and differentiated lavas from Anjouan, Comores Archipelago. *Contrib. Mineral. Petrol.* **38**, 237–260.

Forbes R. B. and Banno S. (1966) Nickel-iron content of peridotite inclusion and cognate olivine from an alkali-olivine basalt. *Amer. Min.* **51**, 130–140.

Frey F. A. and Green D. H. (1974) The mineralogy, geochemistry and origin of lherzolite inclusions in Victorian basanites. *Geochim. Cosmochim. Acta* **38**, 1023–1059.

Frey F. A. and Prinz M. (1978) Ultramafic inclusions from San Carlos, Arizona: petrological and geochemical data bearing on their petrogenesis. *Earth Planet. Sci. Lett.* **38**, 129–176.

Frey F. A., Bryan W. B. and Thompson G. (1974) Atlantic Ocean Floor: geochemistry and petrology of basalts from Legs 2 and 3 of the Deep Sea Drilling Project. *J. Geophys. Res.* **79**, 5507–5527.

Frey F. A., Green D. H. and Roy S. D. (1978) Integrated models of basalt petrogenesis: a study of quartz tholeiites to olivine melilitites from south eastern Australia utilizing geochemical and experimental petrological data. *J. Petrol.* **19**, 463–513.

Gale G. H. (1973) Palaeozoic komatiites and ocean-floor type basalts from northeastern Newfoundland. *Earth Planet. Sci. Lett.* **18**, 22–28.

Gansser A., Dietrich V. J. and Cameron W. E. (1979) Neogene komatiites from Gorgona Island. *Nature* **278**, 545–546.

Goldich S. S., Treves S. B., Suhr N. H. and Stuckless J. S. (1975) Geochemistry of the Cenozoic volcanic rocks of Ross Island and vicinity, Antarctica. *J. Geol.* **83**, 415–435.

Goles G. G. (1976) Some constraints on the origin of phonolites from the Gregory Rift, Kenya, and inferences concerning basaltic magmas in the rift system. *Lithos* **9**, 1–8.

Green D. H. (1970) The origin of basaltic and nephelenitic magmas. *Trans. Leicester Lit. Philos. Soc.* **64**, 28–54.

Green D. H. (1975) Genesis of Archean peridotitic magmas and constraints on Archean geothermal gradients and tectonics. *Geology* **3**, 15–18.

Green D. H. (1976) Experimental testing of "equilibrium" partial melting of peridotite under water-saturated, high pressure conditions. *Can. Mineral.* **14**, 255–268.

Green D. H. and Ringwood A. E. (1967) The genesis of basaltic magmas. *Contrib. Mineral. Petrol.* **15**, 103–190.

Green D. H. and Ringwood A. E. (1973) Significance of a primitive lunar basaltic composition present in Apollo 15 soils and breccias. *Earth Planet. Sci. Lett.* **19**, 1–8.

Green D. H., Ringwood A. E., Ware N. G., Hibberson W. O., Major A. and Kiss E. (1971a) Experimental petrology and petrogenosis of Apollo 12 basalts. *Proc. Lunar Sci. Conf. 2nd*, p. 601–615.

Green D. H., Ware N. G., Hibberson W. O. and Major A. (1971b) Experimental petrology of Apollo 12 basalts: part 1, sample 12009. *Earth Planet. Sci. Lett.* **13**, 85–96.

Green D. H., Edgar A. D., Beasley P., Kiss E. and Ware N. G. (1974) Upper mantle source for some hawaiites, mugearites and benmoreites. *Contrib. Mineral. Petrol.* **48**, 33–43.

Green D. H., Nichols I. A., Viljoen M. and Viljoen R. (1975a) Experimental demonstration of the existence of peridotitic liquids in earliest Archean magmatism. *Geology* **3**, 11–14.

Green D. H., Ringwood A. E., Hibberson W. O. and Ware N. G. (1975b) Experimental petrology of Apollo 17 mare basalts. *Proc. Lunar Sci. Conf. 6th*, p. 871–893.

Green D. H., Hibberson W. O. and Jaques A. L. (1979) Petrogenesis of mid-ocean basalts. In *The Earth: Its Origin, Structure and Evolution*. (W. H. McElhinny, ed.), p. 265–299. Academic Press, London.

Grove T. L., Walker D., Longhi J., Stolper E. and Hays J. F. (1973) Petrology of rock 12002 and origin of picritic basalts at Oceanus Procellarum. *Proc. Lunar Sci. Conf. 4th*, p. 995–1011.

Gunn B. M. (1971) Trace element partition during olivine fractionation of Hawaiian basalts. *Chem. Geol.* **8**, 1–13.

Hart S. R. and Davis K. E. (1978) Nickel partitioning between olivine and silicate melt. *Earth Planet. Sci. Lett.* **40**, 203–219.

Heiken G. H., McKay D. S. and Brown R. W. (1974) Lunar deposits of possible pyroclastic origin. *Geochim. Cosmochim. Acta* **38**, 1703–1718.

Herrmann A. G., Blanchard D. P., Haskin L. A., Jacobs J. W., Knake D., Korotev R. L. and Brannon J. C. (1976) Major, minor and trace element compositions of peridotitic and basaltic komatiites from the Precambrian crust of southern Africa. *Contrib. Mineral. Petrol.* **59**, 1–12.

Hofmann A. W. and Hart S. R. (1978) An assessment of local and regional isotopic equilibrium in the mantle. *Earth Planet. Sci. Lett.* **38**, 44–62.

Irvine T. N. (1977) Definition of primitive liquid compositions for basic magmas. *Carnegie Inst. Wash., Yearb.* **76**, 454–461.

Irvine T. N. and Kushiro I. (1976) Partitioning of Ni and Mg between olivine and silicate liquids. *Carnegie Inst. Wash., Yearb.* **75**, 668–675.

Irving A. J. (1978) A review of experimental studies of crystal/liquid trace element partitioning. *Geochim. Cosmochim. Acta* **42**, 743–770.

Irving A. J. and Green D. H. (1976) Geochemistry and petrogenesis of the Newer Basalts of Victoria and South Australia. *J. Geol. Soc. Aust.* **23**, 45–66.

Irving A. J. and Price R. C. (1981) Geochemistry and evolution of lherzolite-bearing phonolitic lavas from Nigeria, Australia, East Germany and New Zealand. *Geochim. Cosmochim. Acta.*

Jakobsson S. P., Jonsson J. and Shido F. (1978) Petrology of the western Reykjanes Peninsula, Iceland. *J. Petrol.* **19**, 669–705.

Jaques A. L. and Green D. H. (1980) Anhydrous melting of peridotite at 0–15 kb pressure and the genesis of tholeiitic basalts. *Contrib. Mineral. Petrol.* **73**, 287–310.

Kay R. W. and Gast P. W. (1973) The rare-earth content and origin of alkali-rich basalt. *J. Geol.* **81**, 653–682.

Kay R., Hubbard N. J. and Gast D. W. (1970) Chemical characteristics and origin of oceanic ridge volcanic rocks. *J. Geophys. Res.* **75**, 1585–1613.

Kesson S. E. (1973) The primary geochemistry of the Monaro alkaline volcanics, Southeastern Australia—evidence for upper mantle heterogeneity. *Contrib. Mineral. Petrol.* **42**, 93–108.

Kesson S. E. (1975) Mare basalts: melting experiments and petrogenetic interpretations. *Proc. Lunar Sci. Conf. 6th*, p. 921–944.

Krishnamurthy P. and Cox K. G. (1977) Picrite basalts and related lavas from the Deccan Traps of western India. *Contrib. Mineral. Petrol.* **62**, 53–75.

Kuno H. (1960) High-alumina basalt. *J. Petrol.* **1**, 121–145.

Kuno H. and Aoki K. (1970) Chemistry of ultramafic nodules and their bearing on the origin of ultramafic magmas. *Phys. Earth Planet. Inter.* **3**, 273–301.

Kushiro I. (1972) Effect of water on the composition of magmas formed at high pressures. *J. Petrol.* **13**, 311–334.

Kushiro I. (1973) Origin of some magmas in oceanic and circum-oceanic regions. *Tectonophys.* **16**, 211–222.

Kushiro I. (1979) Fractional crystallization of basaltic magma. In *The Evolution of the Igneous Rocks*, (H. S. Yoder, Jr., ed.), p. 171–203. Princeton Univ. Press.

Langmuir C. H., Bender J. F., Bence A. E. and Hanson G. N. (1977) Petrogenesis of basalts from the FAMOUS area: Mid-Atlantic Ridge. *Earth Planet. Sci. Lett.* **36**, 133–156.

Langmuir C. H. and Hanson G. N. (1980) An evaluation of major element heterogeneity in the mantle source of basalts. *Phil. Trans. Roy. Soc. Lond.* **A297**, 383–407.

Leeman W. P. (1978) Distribution of Mg^{2+} between olivine and silicate melt, and its implications regarding melt structure. *Geochim. Cosmochim. Acta* **42**, 789–800.

Leeman W. P. and Lindstrom D. J. (1978) Partitioning of Ni^{2+} between basaltic and synthetic melts and olivines—an experimental study. *Geochim. Cosmochim. Acta* **42**, 801–816.

Leeman W. P. and Scheidegger K. F. (1977) Olivine/liquid coefficients and a test for crystal-liquid equilibrium. *Earth Planet. Sci. Lett.* **35**, 247–257.

Longhi J., Walker D., Grove T. L., Stolper E. M. and Hays J. F. (1974) The petrology of Apollo 17 mare basalts. *Proc. Lunar Sci. Conf. 5th*, p. 447–469.

Longhi J., Walker D. and Hays J. F. (1978) The distribution of Fe and Mg between olivine and lunar basaltic liquid. *Geochim. Cosmochim. Acta* **42**, 1545–1558.

Maaløe S. (1973) Temperature and pressure relations of ascending primary magmas. *J. Geophys. Res.* **78**, 6877–6886.

Maaløe S. and Jakobsson S. P. (1980) The PT phase relations of a primary oceanite from the Reykjanes Peninsula, Iceland. *Lithos* **13**, 237–246.

Maaløe S. and Printzlau I. (1979) Natural partial melting of spinel lherzolite. *J. Petrol.* **20**, 727–741.

MacDonald G. A. (1949) Petrography of the island of Hawaii. *U.S. Geol. Survey Prof. Paper 214D*, p. 51–96.

MacDonald G. A. (1968) Composition and origin of Hawaiian lavas. In *Studies in Volcanology: a Memoir in Honor of Howel Williams*, (R. R. Coats, R. L. Hay and C. A. Anderson, eds.), p. 477–522. Mem., Geol. Soc. Amer. 116.

Malpas J. (1978) Magma generation in the upper mantle, field evidence from ophiolite suites, and application to the generation of oceanic lithosphere. *Phil. Trans. Roy. Soc. Lond.* **A288**, 527–546.

Mason B. (1962) *Meteorites*. Wiley, N.Y. 274pp.

McGetchin T. R. and Smyth J. R. (1978) The mantle of Mars: some possible geological implications of its high density. *Icarus* **34**, 512–536.

Melson W. G., Vallier T. L., Wright T. L., Byerly G. and Nelen J. (1976) Chemical diversity of abyssal volcanic glass erupted along Pacific, Atlantic, and Indian Ocean sea-floor spreading centers. In *The Geophysics of the Pacific Ocean Basin and its Margins*, p. 351–367. Amer. Geophys. Union, Washington, D.C.

Meyer H. O. A. and Boyd F. R. (1972) Composition and origin of crystalline inclusions in natural diamonds. *Geochim. Cosmochim. Acta* **36**, 1255–1274.

Murata K. J. and Richter D. H. (1966) Chemistry of the lavas of the 1959–60 eruption of Kilauea Volcano, Hawaii. *U.S. Geol. Survey Prof. Paper 537A*. 26 pp.

Mysen B. O. and Kushiro I. (1977) Compositional variations of coexisting phases with degree of melting of peridotite in the upper mantle. *Amer. Mineral.* **62**, 843–865.

Nesbitt R. W. (1971) Skeletal crystal forms in the ultramafic rocks of the Yilgarn block, western Australia: evidence for an Archean ultramafic liquid. *Spec. Publ. Geol. Soc. Aust.* **3**, 331–348.

Nesbitt R. W. and Sun S. -S. (1976) Geochemistry of Archean spinifex textured peridotites and magnesian and low magnesian tholeiites. *Earth Planet. Sci. Lett.* **31**, 433–453.

Nisbet E. G., Bickle M. J. and Martin A. (1977) The mafic and ultramafic lavas of the Belingwe Greenstone Belt, Rhodesia. *J. Petrol.* **18**, 521–566.

O'Hara M. J. (1965) Primary magmas and the origin of basalts. *Scottish J. Geol.* **1**, 19–40.

O'Hara M. J. (1968a) The bearing of phase equilibria studies in synthetic and natural systems on the origin and evolution of basic and ultrabasic rocks. *Earth Sci. Rev.* **4**, 69–133.

O'Hara M. J. (1968b) Are ocean floor basalts primary magmas? *Nature* **220**, 683–686.

O'Hara M. J. (1977) Geochemical evolution during fractional crystallisation of a periodically refilled magma chamber. *Nature* **266**, 503–507.

O'Hara M. J. and Mercy E. L.. P. (1963) Petrology and petrogenesis of some garnetiferous peridotites. *Trans. R. Soc. Edinburgh* **65**, 251–314.

O'Hara M. J., Biggar G. M., Richardson S. W., Ford C. E. and Jamieson B. G. (1970) The nature of the seas, mascons and the lunar interior in the light of experimental studies. *Proc. Apollo 11 Lunar Sci. Conf.*, p. 695–710.

O'Hara M. J., Humphries D. J. and Waterston S. (1976) Petrogenesis of mare basalts: implications for chemical, mineralogical, and thermal models for the moon. *Proc. Lunar Sci. Conf. 6th*, p. 1043–1055.

O'Hara M. J., Saunders M. J. and Mercy E. L. P. (1975) Garnet-peridotite, primary ultrabasic magma and eclogite; interpretation of upper mantle processes in kimberlite. *Phys. Chem. Earth* **9**, 571–604.

O'Nions R. K. and Clarke D. B. (1972) Comparative trace element geochemistry of Tertiary basalts from Baffin Bay. *Earth Planet. Sci. Lett.* **15**, 436–446.

Pallister J. S. and Hopson C. A. (1979) Samail ophiolite magma chamber: part II, evidence from cryptic variation and mineral chemistry. *Proc. Int. Ophiolite Sym.* Cyprus Geol. Survey Dept., Cyprus, Nicosia.

Papike J. J., Hodges F. N., Bence A. E., Cameron M. and Rhodes J. M. (1976) Mare basalts: crystal chemistry, mineralogy, and petrology. *Rev. Geophys. Space Phys.* **14**, 475–540.

Powers H. A. (1955) Composition and origin of basaltic magmas on the Hawaiian islands. *Geochim. Cosmochim. Acta* **7**, 77–107.

Presnall D. C. (1969) The geometrical analysis of partial fusion. *Amer. J. Sci.* **267**, 1178–1194.

Presnall D. C., Dixon S. A., Dixon J. R., O'Donnell T. H., Brenner N. L., Schrock R. L. and Dycus W. (1978) Liquidus phase relationships on the join diopside-forsterite-anorthite from 1 atm to 20 kbar: their bearing on the generation and crystallization of basaltic magma. *Contrib. Mineral. Petrol.* **66**, 203–220.

Presnall D. C., Dixon J. R., O'Donnell T. H. and Dixon S. A. (1979) Generation of mid-ocean ridge tholeiites. *J. Petrol.* **20**, 3–35.

Price R. C. and Green D. H. (1972) Lherzolite nodules in a "mafic phonolite" from northeast Otago, New Zealand. *Nature Phys. Sci.* **235**, 134–135.

Pyke D. R., Naldrett A. J. and Eckstrand A. R. (1973) Archean ultramafic flows in Munro Township, Ontario. *Geol. Soc. Amer. Bull.* **84**, 955–978.

Reid A. M., Donaldson C. H., Brown R. W., Ridley W. I. and Dawson J. B. (1975) Mineral chemistry of peridotite xenoliths from the Lashaine volcano, Tanzania. *Phys. Chem. Earth* **9**, 525–543.

Reid J. B., Jr. and Woods G. A. (1978) Oceanic mantle beneath the southern Rio Grande rift. *Earth Planet. Sci. Lett.* **41**, 303–316.

Rhodes J. M. (1978) Primary mare basalts and green glass (abstract). *Lunar and Planetary Science IX*, p. 958–960. Lunar and Planetary Institute, Houston.

Rhodes J. M. and Dungan M. A. (1979) The evolution of ocean-floor basaltic magmas. In *Deep Drilling Results in the Atlantic Ocean: Ocean Crust.* (M. Talwani, C. G. Harrison and D. E. Hayes, eds.), p. 262–272. Amer. Geophys. Union, Washington, D.C.

Rhodes J. M., Hubbard N. J., Wiesmann H., Rodgers K. V., Brannon J. C. and Bansall B. M. (1976) Chemistry, classification and petrogenesis of Apollo 17 mare basalts. *Proc. Lunar Sci. Conf. 7th*, p. 1467–1489.

Rhodes J. M., Blanchard D. P., Brannon J. C., Rodgers K. V. and Dungan M. A. (1977) Chemistry of Apollo 12 mare basalts: magma types and fractionation processes. *Proc. Lunar Sci. Conf. 8th*, p. 1305–1338.

Rhodes J. M., Dungan M. A., Blanchard D. P. and Long P. E. (1979a) Magma mixing at mid-ocean ridges: evidence from basalts drilled near 22° N on the Mid-Atlantic Ridge. *Tectonophys.* **55**, 35–62.

Rhodes J. M., Lofgren G. E. and Smith D. P. (1979b) One atmospheric melting experiments on ilmenite basalt 12008. *Proc. Lunar Sci. Conf. 10th*, p. 407–422.

Ringwood A. E. (1975) *Composition and Petrology of the Earth's Mantle.* McGraw Hill, N.Y. 618 pp.

Ringwood A. E. and Essene E. (1970) Petrogenesis of lunar basalts, internal constitution and origin of the moon. *Proc. Apollo 11 Lunar Sci. Conf.*, p. 769–799.

Roedder E. and Weiblen P. W. (1971) Petrology of silicate melt inclusions, Apollo 11 and Apollo 12 and terrestrial equivalents. *Proc. Lunar Sci. Conf. 2nd*, p. 507–528.

Roedder E. and Weiblen P. W. (1977) Compositional variation in late-stage differentiates in mare lavas, as indicated by silicate melt inclusions. *Proc. Lunar Sci. Conf. 8th*, p. 1767–1783.

Roeder P. L. and Emslie R. F. (1970) Olivine-liquid equilibrium. *Contrib. Mineral. Petrol.* **29**, 275–289.

Ross C. S., Foster M. D. and Myers A. T. (1959) Origin of dunites and olivine-rich inclusions in basaltic rocks. *Amer. Mineral.* **39**, 693–737.

Sato H. (1977) Nickel content of basaltic magmas: identification of primary magmas and a measure of the degree of olivine fractionation. *Lithos* **10**, 113–120.

Shido F. A., Miyashiro A. and Ewing M. (1971) Crystallization of abyssal tholeiites. *Contrib. Mineral. Petrol.* **31**, 251–266.

Sigurdsson H. and Sparks R. S. J. (1978) Lateral magma flow within rifted Icelandic crust. *Nature* **274**, 126–130.

Sparks R. S. J., Pinkerton H. and MacDonald R. (1977) The transport of xenoliths in magmas. *Earth Planet. Sci. Lett.* **35**, 234–238.

Sparks R. S. J., Meyer P. and Sigurdsson H. (1980) Density variation amongst mid-ocean ridge basalts: implications for magma mixing and the scarcity of primitive lavas. *Earth Planet. Sci. Lett.* **46**, 419–430.

Spera F. J. (1980) Aspects of magma transport. In *Physics of Magmatic Processes* (R. B. Hargreaves, ed.), p. 265–323. Princeton Univ. Press.

Stern C. R. (1980) Geochemistry of Chilean ophiolites: evidence for the compositional evolution of the mantle source of back-arc basin basalts. *J. Geophys. Res.* **85**, 955–966.

Stolper E. (1977) Experimental petrology of eucritic meteorites. *Geochim. Cosmochim. Acta* **41**, 587–612.

Stolper E. (1980) A phase diagram for mid-ocean ridge basalts: preliminary results and implications for petrogenesis. *Contrib. Mineral. Petrol.* **74**, 13–27.

Stolper E. and Walker D. (1980) Melt density and the average composition of basalt. *Contrib. Mineral. Petrol.* **74**, 7–12.

Stolper E., McSween H. Y. and Hays J. F. (1979) A petrogenetic model of the relationships among achondritic meteorites. *Geochim. Cosmochim. Acta* **43**, 589–602.

Strong D. F. (1972) The petrology of the lavas of Grande Comore. *J. Petrol.* **13**, 181–217.

Sun S. -S. and Hanson G. N. (1975) Origin of Ross Island basanitoids and limitations upon the heterogeneity of mantle sources for alkali basalts and nephelinites. *Contrib. Mineral. Petrol.* **52**, 77–106.

Sun S. -S. and Nesbitt R. W. (1977) Chemical heterogeneity of the Archean mantle, composition of the Earth and mantle evolution. *Earth Planet. Sci. Lett.* **35**, 429–448.

Sun S. -S. and Nesbitt R. W. (1978) Geochemical regularities and genetic significance of ophiolitic basalts. *Geology* **6**, 689–693.

Sutherland F. L. (1974) High-pressure inclusions in tholeiitic basalt and the range of lherzolite-bearing magmas in the Tasmanian volcanic province. *Earth Planet. Sci. Lett.* **24**, 317–324.

Swanson D. A., Wright T. L. and Helz R. T. (1975) Linear vent systems and estimated rates of magma production and eruption for the Yakima basalt on the Columbia Plateau. *Amer. J. Sci.* **275**, 877–905.

Taylor S. R. (1975) *Lunar Science: A Post-Apollo View*. Pergamon Press, N.Y. 372 pp.

Thompson R. N. (1972) The 1-atmosphere melting patterns of some basaltic volcanic series. *Amer. J. Sci.* **272**, 901–932.

Thompson R. N. (1973) One-atmosphere melting behavior and nomenclature of terrestrial lavas. *Contrib. Mineral. Petrol.* **41**, 197–204.

Thompson R. N. (1974) Primary basalts and magma genesis. *Contrib. Mineral. Petrol.* **45**, 317–341.

Thompson R. N. (1975) Primary basalts and magma genesis II. Snake River Plain, Idaho, U.S.A. *Contrib. Mineral. Petrol.* **52**, 213–232.

Tracy R. J. and Stolper E. M. (1978) Experimental petrology of a xenolith-bearing basanite from Tahiti—implications for a Tahitian parent magma (abstract). *Abstracts with Program*, Geol. Soc. Amer. **10**, 506.

Upadhyay H. D. and Neale E. R. W. (1979) On the tectonic regimes of ophiolite genesis. *Earth Planet. Sci. Lett.* **43**, 93–102.

Viljoen M. J. and Viljoen R. P. (1969) The geology and geochemistry of the lower ultramafic unit of the Onverwacht Group and a proposed new class of igneous rocks. Upper Mantle Project. Spec. Publ. 2, Geol. Soc. S. Africa, p. 87–112.

Walker D., Longhi J. and Hays J. F. (1976) Heterogeneity in titaniferous lunar basalts. *Earth Planet. Sci. Lett.* **30**, 27–36.

Walker D., Longhi J., Stolper E. M., Grove T. L. and Hays J. F. (1975) Origin of titaniferous lunar basalts. *Geochim. Cosmochim. Acta* **39**, 1219–1235.

Walker D., Shibata T. and DeLong S. E. (1979) Abyssal tholeiites from the Oceanographer Fracture Zone II: Phase equilibria and mixing. *Contrib. Mineral. Petrol.* **70**, 111–125.

Watson E. B. (1976) Glass inclusions as samples of early magmatic liquid: determinative method and application to a South Atlantic basalt. *J. Volcanol. Geotherm. Res.* **1**, 73–84.

Whitford D. J. and Arndt N. T. (1978) Rare earth element abundances in a thick, layered komatiite lava flow from Ontario, Canada. *Earth Planet. Sci. Lett.* **41**, 188–196.

Wilkinson J. F. G. and Binns R. A. (1969) Hawaiite of high pressure origin from north eastern New South Wales. *Nature* **222**, 553–555.

Wilkinson J. F. G. and Binns R. A. (1977) Relatively iron-rich lherzolite xenoliths of the Cr-diopside suite: a guide to the primary nature of anorogenic tholeiitic andesitic magmas. *Contrib. Mineral. Petrol.* **65**, 199–212.

Wright T. L. (1971) Chemistry of Kilauea and Mauna Loa lavas in space and time. *U.S. Geol. Survey Prof. Paper 735*. 40 pp.

Wright T. L. (1973) Magma mixing as illustrated by the 1959 eruption, Kilauea Volcano, Hawaii. *Geol. Soc. Amer. Bull.* **84**, 849–858.

Wright T. L. and Fiske R. J. (1971) Origin of the differentiated and hybrid lavas of Kilauea Volcano, Hawaii. *J. Petrol.* **12**, 1–65.

Yoder H. S. (1976) *Generation of Basaltic Magma*. National Academy of Sciences, Washington, D.C. 265 pp.

Yoder H. S. and Tilley C. E. (1962) Origin of basalt magmas: an experimental study of natural and synthetic rock systems. *J. Petrol.* **3**, 342–532.

1.5 CONCLUSIONS

Basaltic volcanism is a well-documented stage in planetary evolution. The volcanic rocks generated have the potential for providing information on planetary mantles that could not otherwise be sampled.

We have reviewed the existing data bases for planetary basalts and found that in many cases, especially for the terrestrial basalts, complete analyses of individual samples were unavailable. In contrast, the lunar sample data base, although regionally restricted, is more complete and of a higher quality. We have compiled representative reference suites in which the essential chemical and mineralogical parameters have been measured at high levels of accuracy and precision.

With this uniform data base, and a review of current thinking on basalt petrogenesis, we have addressed the following questions: How do planetary mantles differ, and why? How do planetary mantles evolve with time? How inhomogeneous are planetary mantles and on what scale do these inhomogeneities persist? What are the evolutionary processes that influence basalt compositions on the respective planets and how do we see through these processes to distinguish primary, mantle-derived liquids? How effectively do basalts record such important planetary parameters as size, physical properties, radioactive element content, accretionary history, and intensive thermodynamic parameters, e.g., P, T, f_{O_2}? The answers to these questions provide a framework against which we can evaluate the composition and evolution of the mantles of other solid bodies in the solar system.

1.5.1 HETEROGENEITIES IN INTER- AND INTRAPLANETARY BASALT SOURCE REGIONS

Major and trace element abundances and isotopic systematics of the planetary basalt suites, when considered within constraints imposed by experimental phase relationships, reveal chemical and mineralogical differences among their mantle sources. For example, primitive basalt magmas have bulk chemical compositions and liquidus phase compositions which indicate that major differences existed in the $Mg/(Mg + Fe^{2+})$ ratios of their sources. These sources, ranked in order of decreasing $Mg/(Mg + Fe^{2+})$, are those for the terrestrial, eucrite, and lunar mare basalts. Furthermore, the basalts from all three of these planetary sources are characterized by fractionated transition metal abundances. This effect, which results from metal-silicate fractionation, can occur prior to planetary accretion or during core formation, as it did evidently in the Earth. Volatile/involatile element ratios in terrestrial basalts differ from those in eucrites and lunar basalts. This difference has been widely interpreted in terms of preaccretional volatilization. Alumina differences in the basalts are related to bulk composition of the planet, early evolutionary history, lateral and vertical chemical variation, percentage of partial melting and residual mineralogy.

The abundances of the large cations and of the high-valency cations are highly variable both within and among planetary basalt suites, reflecting the presence of marked heterogeneities within the mantle sources. For example, mantle sources for mid-ocean ridge basalts are depleted in LIL trace elements, in contrast with the sources of oceanic island and continental basalts. These elements eventually become concentrated in the continental crust through the production of magmas in island arcs. Thus the Earth's mantle has evolved slowly in composition through geological time, and the heterogeneities persist on time scales of 10^9 years. The mantle reservoirs are large enough so that secular trends in basalt composition are not apparent. The mid-ocean ridge basalt sources have refractory element compositions which are 1–2 times chondritic abundances, whereas the source regions for the oceanic island and continental basalts have abundances which are 3–5 times chondritic.

In contrast to the Earth, the lunar mantle underwent an early massive differentiation more than 4.4 b.y. ago. Lunar mare basalts were derived by partial melting from cumulate source regions which had experienced prior extraction of plagioclase. Because of their different evolutionary histories, lunar and terrestrial basalts show many distinctive differences in trace element abundances. The overall composition of the Moon reflects these differences and shows that the Moon is enriched in refractory trace elements relative to the Earth.

1.5.2 SILICATE MINERALOGY OF PLANETARY BASALTS

The most magnesian olivines in the planetary basalt suite occur in terrestrial basalts. Mare and meteoritic basalt olivines are more enriched in iron, which indicates that they crystallized from more iron-rich liquids and implies a more magnesian mantle source for the Earth than for the Moon or the basaltic meteorite parent body. Feldspars from lunar mare, lunar highland and meteorite basalts have low Na_2O and K_2O contents relative to feldspars from terrestrial basalts. This reflects the low-alkali contents of their magmas, which in turn can be related both to depletion of volatiles in their parent body and to that body's early differentiation history. Apart from Wo, $CaSiO_3$; En, $MgSiO_3$; and Fs, $Fe^{2+}SiO_3$, the most important additional component in pyroxenes in the nonterrestrial suites is $CaTiAl_2O_6$. $CaFe^{3+}SiAlO_6$ is a very important component in pyroxenes of terrestrial basalts but not in those of lunar or meteoritic basalts. This emphasizes the importance of f_{O_2} in pyroxene-liquid elemental partitioning. At the relatively high oxygen fugacities of terrestrial petrogenesis, iron enters the pyroxene both as Fe^{3+} and Fe^{2+}, whereas iron is present only as Fe^{2+} in lunar and most basaltic meteorite pyroxenes. Thus, inspection of a large number of high quality analyses of olivine, feldspar and pyroxene demonstrates that these phases carry the signature of the planetary body in which they evolved.

1.5.3 TEMPERATURES, OXYGEN FUGACITY, VOLATILES

Temperatures and oxygen fugacities of basalt equilibration for terrestrial basalts are constrained to limits that approximate the fayalite-magnetite-quartz buffer. Lunar basalts and basaltic achondrites equilibrated under f_{O_2} condition which were defined by metallic iron stability. Careful determinations place the equilibrium f_{O_2} of these basalts one order of magnitude below that of the iron-wüstite buffer curve. Contrasts in f_{O_2} conditions between terrestrial basalts on the one hand, and lunar and achondritic basalts on the other, are adequately explained by the distinctly different volatile concentrations which prevailed in their respective source regions. The evidence for terrestrial basalts suggests that the H_2O:CO_2 concentration is high, whereas in extraterrestrial basalts we may infer, from the observed mineralogy, the presence of high H_2:CO ratios, or perhaps even the presence of elemental C or carbonyl compounds. Data from these three suites of basalts, and for related intrusives or cumulates, provide convincing evidence for the fact that redox paths of crystallization are not influenced by whether the system is open (surface extrusion) or closed (cooling under confined pressure), but are controlled instead by the intrinsic volatile contents that are present in the source regions of the planet or asteroidal body.

Volatile to involatile element ratios vary among terrestrial, lunar and eucritic basalts. These variations are interpreted in terms of differences in compositions during planetary accretion and in degrees of volatilization that followed planet formation. Contrasts in volatile abundances probably were reflected in the conditions which governed basalt generation in the mantle source regions of these bodies. Volatile element compositional differences are also expressed in the relative abundances of elements such as Na and K, which are large in terrestrial basalts but extremely low in eucrites and lunar basalts.

1.5.4 FACTORS IN BASALT VARIATION

Of the factors likely to bring about variation in the composition of a basaltic magma, the one for which we have best evidence in planetary basalts (meteoritic, lunar and terrestrial) is that of variation in the degree of partial melting in source regions. A second factor known to be important in the terrestrial and lunar samples is crystal/liquid fractionation within flows (or sills). When the terrestrial samples only are considered, there is evidence for a much broader range of phenomena, including fractionation in magma chambers, mixing of magmas, assimilation of solids by a magma, separation of liquid from crystals by flow mechanisms, and separation of vapor phases. The absence of evidence for all of these mechanisms in the lunar and basaltic achondrite samples may be due simply to inadequate sampling, but a few lines of evidence point to the rarity or absence of some of these processes in extraterrestrial samples.

Among the meteorites, the tight clustering of eucrite compositions, which are multiply saturated with the principal mineral phases, suggests a simple derivation by direct partial melting. Variations from the mean composition can be explained by small differences in degree of partial melting or small amounts of crystal/liquid fractionation. As there is no reason to believe that all eucrites were derived from a single magma chamber, their compositional similarity suggests a uniform source material, with lower Mg' and alkalies than mantle sources in the Earth or Moon. This uniformity of eucrite compositions suggests the absence of complex mechanisms of basalt variation (i.e., mechanisms other than partial melting or crystal/liquid fractionation).

Shergotty and its relatives were derived from a more evolved planet; crystal/liquid fractionation is indicated by the varying mineral proportions in the members of this group, pyroxene zoning and mineral preferred orientation. The degree of partial melting necessary to form the Shergotty liquid is uncertain; however, the source composition was richer in alkalies than that which produced the eucrites.

The phenocrysts in terrestrial basalts exhibit more diversity in mineralogy and growth-dissolution histories than those in meteoritic basalts. Demonstrably intratelluric phenocrysts are vanishingly rare in lunar and meteoritic basalts. These differences do in part reflect differences in major element compositions, particularly alumina and alkalies, and volatile compositions and contents, particularly H_2O; however, the greater complexity recorded in terrestrial phenocrysts is indicative of the greater array of processes which affect terrestrial basalts in transit from their source areas to the surface. Although the simplicity of phenocrysts in lunar and meteoritic basalts does not exclude the operation of differentiation mechanisms such as polybaric crystal fractionation, wallrock reaction and magma mixing, there certainly is a strong suggestion that they are not major factors.

Despite our brief and incomplete experience in planetary exploration there is compelling evidence to suggest that basalts vary between different planetary bodies or on a single planet primarily because of differences in mantle composition, and that basalts are further subject to variation because of differences in partial melting origins: e.g., the location of the melting zone, the type of melting process, and the mechanical controls of magma accumulation and eruption.

Authorship & Acknowledgments

Section 1.2.1 Authors: A. J. Naldrett, Department of Geology, University of Toronto; I. E. M. Smith, Department of Geology, University of Toronto. Major elements, Y, Sr, Rb, Zr, Nb, Pb, Ga by XRF: J. M. Rhodes and Joel Sparks, Department of Geology and Geography, Univ. of Massachusetts; K. A. Rodgers, LEMSCO, Houston. Trace element data for La, Ce, Sm, Eu, Tb, Yb, Lu, Sc, Hf, Cr, Th, Ba and Ni and values for Na_2O were obtained by neutron activation analysis by Marilyn M. Lindstrom, Jeffrey W. Jacobs and Larry A. Haskin, Department of Earth and Planetary Sciences and McDonnell Center for Space Sciences, Washington University, St. Louis. See appendix A-1 for information on procedures and estimates of analytical precision. Samples for the Archean suite provided by A. J. Naldrett and I. E. M. Smith.

Section 1.2.2 Authors: John C. Green, Department of Geology, University of Minnesota, Duluth; Larry A. Haskin, Department of Earth and Planetary Sciences and McDonnell Center for Space Sciences, Washington University, St. Louis. Major elements (some), Cs, Rb, Sr, Zr, Nb by XRF: Joyce C. Brannon, Department of Earth and Planetary Sciences and McDonnell Center for Space Sciences, Washington University, St. Louis. Trace element data for La, Ce, Sm, Eu, Tb, Yb, Lu, Sc, Hf, Cr, Th, Ba and Ni and values for Na_2O were obtained by neutron activation analysis by Joyce C. Brannon and Larry A. Haskin, Department of Earth and Planetary Sciences and McDonnell Center for Space Sciences, Washington University, St. Louis. See appendix A-2 for information on procedures and estimates of analytical precision. Samples for the Keweenawan suite provided by J. C. Green. J. C. Green acknowledges the National Research Council for providing a Senior Research Associateship under which much of the preparation of this section was carried out.

Section 1.2.3 Authors: S. E. Haggerty, Department of Geology and Geography, University of Massachusetts; A. J. Irving, Lunar and Planetary Institute, Houston, and Department of Geological Sciences, University of Washington, Seattle. Major elements, Y, Sr, Rb, Zr, Nb, Pb, Ga by XRF: J. M. Rhodes and Joel Sparks, Department of Geology and Geography, University of Massachusetts; K. A. Rodgers, LEMSCO, Houston. Trace element data for La, Ce, Sm, Eu, Tb, Yb, Lu, Sc, Hf, Cr, Th, Ba and Ni and values for Na_2O were obtained by neutron activation analysis by Marilyn M. Lindstrom and Larry A. Haskin, Department of Earth and Planetary Sciences and McDonnell Center for Space Sciences, Washington University, St. Louis. See appendix A-3 for information on procedures and estimates of analytical precision. Samples for the Columbia Plateau suite were donated by T. L. Wright, U. S. Geological Survey, Reston. S. E. Haggerty gratefully acknowledges the assistance of Richard Bence of the University of Massachusetts.

Section 1.2.4 Authors: M. A. Dungan, Department of Geological Sciences, Southern Methodist University; P. W. Lipman, U. S. Geological Survey, Denver; S. Williams, Department of Geology and Geophysics, University of Minnesota. Major elements, Y, Sr, Rb, Zr, Nb, Pb, Ga by

XRF: J. M. Rhodes and Joel Sparks, Department of Geology and Geography, University of Massachusetts; K. A. Rodgers, LEMSCO, Houston. Trace element data for La, Ce, Sm, Eu, Tb, Yb, Lu, Sc, Hf, Cr, Th, Ba and Ni and values for Na_2O were obtained by neutron activation analysis by Marilyn M. Lindstrom and Larry A. Haskin, Department of Earth and Planetary Sciences and McDonnell Center for Space Sciences, Washington University, St. Louis. Samples for the Taos Plateau suite were collected for this project and donated by P. W. Lipman and M. A. Dungan. M. A. Dungan was a member of the scientific staff of the Lunar and Planetary Institute during the early stages of this project. He gratefully acknowledges the Johnson Space Center for the use of the electron microprobe. Roy Brown and Tom Pronold are gratefully acknowledged for assistance. S. Williams gratefully acknowledges V. Rama Murthy for guidance.

Section 1.2.5 Authors: J. M. Rhodes, Department of Geology and Geography, University of Massachusetts; A. E. Bence, Department of Earth and Space Sciences, State University, New York, Stony Brook. Major elements, Y, Sr, Rb, Zr, Nb, Pb, Ga by XRF: J. M. Rhodes and Joel Sparks, Department of Geology and Geography, University of Massachusetts; K. A. Rodgers, LEMSCO, Houston. Trace element data for La, Ce, Sm, Eu, Tb, Yb, Lu, Sc, Hf, Th, Ba and Ni and values for Na_2O were obtained by neutron activation analysis by D. P. Blanchard, Johnson Space Center, Houston; J. Budahn, LEMSCO, Houston. Samples for the Ocean Floor suite were contributed by W. B. Bryan, Wood's Hole Oceanographic Institute; W. G. Melson, Smithsonian Institution; and S. R. Hart, Massachusetts Institute of Technology.

Section 1.2.6 Author: A. E. Bence, Department of Earth and Space Sciences, State University of New York, Stony Brook. Major elements, Y, Sr, Rb, Zr, Nb, Pb, Ga by XRF: J. M. Rhodes and Joel Sparks, Department of Geology and Geography, University of Massachusetts; K. A. Rodgers, LEMSCO, Houston. Trace element data for La, Ce, Sm, Eu, Tb, Yb, Lu, Sc, Hf, Cr, Th, Ba and Ni and values for Na_2O were obtained by neutron activation analysis by Marilyn Lindstrom and Larry A. Haskin, Department of Earth and Planetary Sciences and McDonnell Center for Space Sciences, Washington University, St. Louis. See appendix A-6 for information on procedures and estimates of analytical precision. Samples for the Hawaiian suite were donated by the Hawaiian Volcano Observatory, U. S. Geological Survey, with the help of P. W. Lipman. Authors wish to acknowledge the assistance of Fred Allen, Donna Baylis, Linda Bindman, Martin Fisk, Timothy Grove and Charles Langmuir, State University of New York, Stony Brook, in various aspects of this work.

Section 1.2.7 Authors: S. R. Taylor, R. Arculus, and M. R.Perfit, Research School of Earth Science, Australian National University, Canberra, Australia; R. W. Johnson, Bureau of Mineral Resources, Canberra, Australia. The authors are grateful to the following for work on the Island Arc Basalt Reference suite: B. W. Chappell, who contributed the XRF determinations; Ms. P. E. Oswald-Sealy, W. B. Nance and J. M. G. Shelley for the spark source trace element analyses; N. G. Ware for the electron probe analyses on samples IA-1–IA-5; E. Kiss for the potassium determinations; and D. Gust for the compilation of 1200 basalt analyses from numerous island arcs. The portions of this paper by R. W. Johnson are published by permission of the Director of the Bureau of Mineral Resources, Canberra. Samples for the Island Arc suite were donated by R. W. Johnson.

Section 1.2.8 Authors: A. M. Reid, Department of Geology, University of Capetown, South Africa; M. B. Duke, NASA Johnson Space Center, Houston. The authors would like to thank the following individuals for the donation of meteorite samples: Nuevo Laredo—L. T. Silver, California Institute of Technology; Juvinas—J. Fabies, Paris Museum; Stannern—G.Kurat, Vienna Museum; Pasamonte—C. Moore, Institute for Meteoritics, Tempe; Aioun al Atrouss—E. K. Gibson, Jr., NASA Johnson Space Center, Houston; Shergotty—B. Mason, Smithsonian Museum.

Section 1.2.9 Authors: J. J. Papike, Department of Earth and Space Sciences, State University of New York, Stony Brook; D. T. Vaniman, Department of Earth and Space Sciences, State University of New York, Stony Brook, and Los Alamos Scientific Laboratory; S. R. Taylor, Research School of Earth Science, Australian National University, Canberra, Australia.

Section 1.2.10 Authors: J. J. Papike, Department of Earth and Space Sciences, State University of New York, Stony Brook; D. T. Vaniman, Department of Earth and Space Sciences, State University of New York, Stony Brook, and Los Alamos Scientific Laboratory.

Section 1.2.11 Authors: Coordinator/Editor—A. J. Irving. Petrography and mineral chemistry: A. J. Irving, G. Y. Bussod; major elements and lithophile trace elements: A. J. Irving; siderophile and volatile trace elements: J. W. Morgan, G. A. Wandless, R. K. Petrie and E. K. Gibson, Jr.; Sr and Nd isotopes: M. A. Menzies; O isotopes: T. K. Kyser. Samples and data for the ultramafic suite were kindly provided by A. J. Irving, A. R. Basu, M. A. Dungan, D. M. Francis, F. A. Frey, B. Harte, J. E. N. Pike and D. C. Presnall. A. J. Irving acknowledges a Visiting Research Scientist appointment at the Lunar and Planetary Institute during the course of this work, and also thanks D. P. Blanchard and J. M. Rhodes for access to analytical facilities at NASA Johnson Space Center.

Section 1.3.1 Authors: A. E. Bence, T. L. Grove and J. J. Papike, Department of Earth and Planetary Sciences, State University of New York, Stony Brook; S. R.Taylor, Research School of Earth Science, Australian National University, Canberra, Australia. The assistance of Linda Bindman, State University of New York, Stony Brook, is gratefully acknowledged.

Section 1.3.2 Author: J. J. Papike, Department of Earth and Space Science, State University of New York, Stony Brook.

Section 1.3.3 Author: Gary E. Lofgren, NASA Johnson Space Center, Houston.

Section 1.3.4 Author: S. E. Haggerty, Department of Geology and Geography, University of Massachusetts.

Section 1.3.5 Author: S. E. Haggerty, Department of Geology and Geography, University of Massachusetts.

Section 1.4.1 Authors: Larry A. Haskin, Department of Earth and Planetary Sciences and McDonnell Center for Space Sciences, Washington University, St. Louis; Michael A. Dungan, Department of Geological Sciences, Southern Methodist University. The authors gratefully acknowledge assistance by Dr. Rodey Batiza.

Section 1.4.2 Author: J. M. Rhodes, Department of Geology and Geography, University of Massachusetts.

Review and scientific editing of sections were helpful and are gratefully acknowledged: N. T. Arndt, W. S. Baldridge, D. A. Clague, C. H. Donaldson, F. A. Frey, E. K. Gibson, Jr., J. C. Green, T. L. Grove, S. R. Hart, A. J. Irving, R. W. Kay, S. E. Kesson, R. J. Kirkpatrick, J. Longhi, I. S. McCallum, S. A. Morse, J. Nicholls, A. M. Reid, G. Ryder, M. Sato, C. H. Simonds, J. V. Smith, E. M. Stolper, G. J. Taylor, L. A. Taylor, D. Walker, P. W. Weiblen, T. L. Wright.

Individual authors wish to acknowledge the following grants which supported laboratories in which analyses were completed or individuals who contributed to the final product: A. E. Bence, NSF grant OCE-76-22193 and NASA grant NGL-33-015-130; S. E. Haggerty, NSF grant EAR-78-02539 and NASA grant NSG-7549; Larry A. Haskin, NSF grant EAR-79-08762 and NASA grant NSG-9055; J. J. Papike, NSG-9044; J. M. Rhodes, NSF grant EAR-78-19892, NSF grant OCE-78-26330 and NASA grant NSG-69069; J. W. Morgan, NASA contract T-4089F and NSR agreement EAR-79-11120; M. A. Menzies, NSF grant EAR-75-14806 (to V. Rama Murthy).

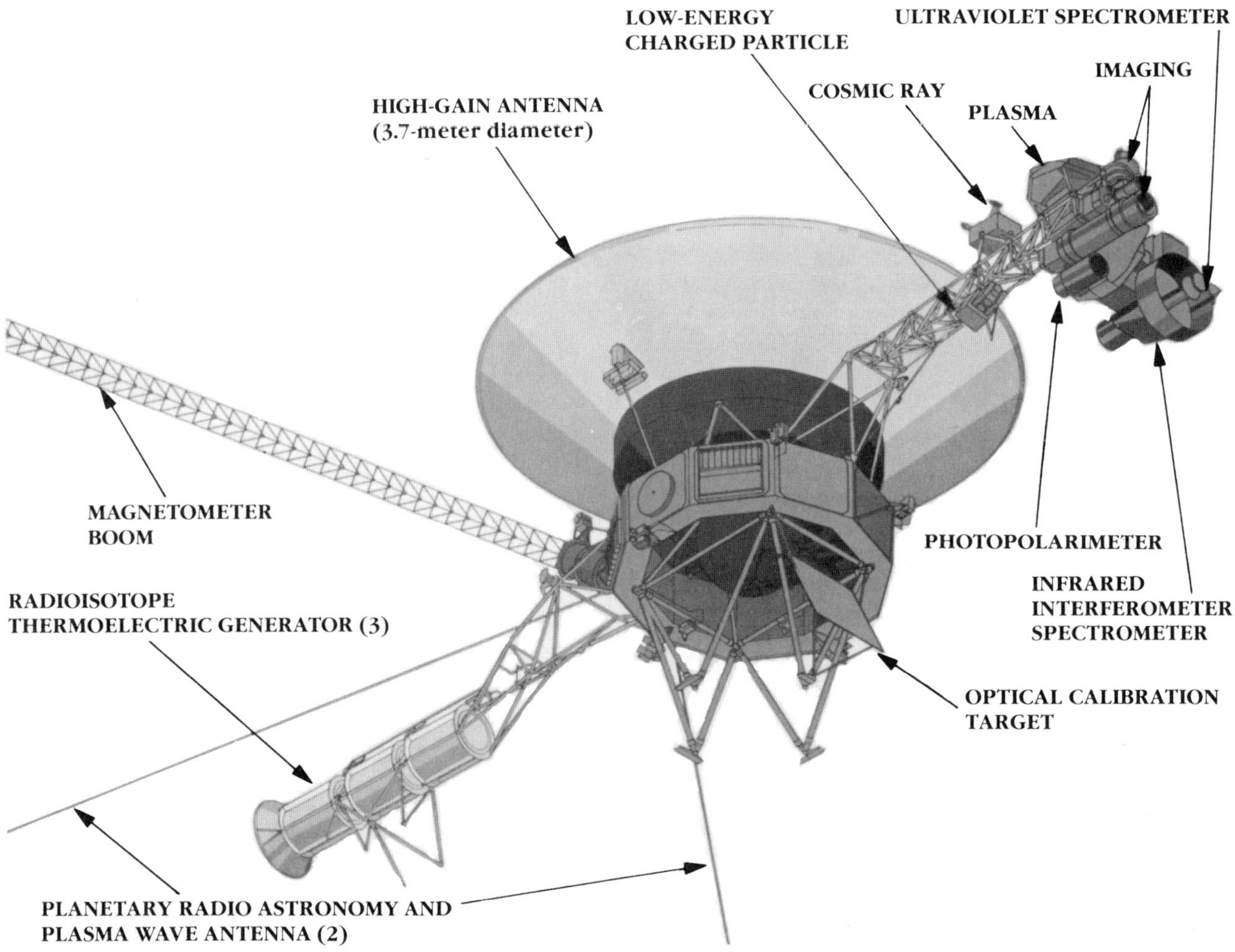

Voyager spacecraft and scientific instruments. Drawing courtesy of JPL, *Voyager 1 Encounters Saturn,* (1980)

2. Remote sensing of basalts in the solar system

John B. Adams, Team Leader
—*Department of Geological Sciences*
University of Washington

Carlé M. Pieters
—*Department of Geological Sciences*
Brown University

Albert E. Metzger
—*Jet Propulsion Laboratory*
Pasadena, California

Isidore Adler
—*Department of Chemistry*
University of Maryland

Thomas B. McCord
—*Institute of Geophysics*
University of Hawaii

Clark R. Chapman
—*Planetary Science Institute*
Tucson, Arizona

Torrence V. Johnson
—*Jet Propulsion Laboratory*
Pasadena, California

Michael J. Bielefeld
—*Computer Science Corporation*
Silver Springs, Maryland

2.1 INTRODUCTION

The study of basaltic volcanism as a stage in planetary evolution requires an understanding of where basalts occur in the solar system, and the geologic context in which basalts are found on a given body. For the Moon, which has been sampled directly but is still inaccessible for field work, remote sensing has been a means of establishing a geologic context for the samples. Observations based on detailed laboratory studies have been generalized to include regions beyond the Apollo and Luna landing sites. For other bodies that have not yet been sampled, remote sensing and surface landers have been the only ways of determining whether basalts are present. A wide variety of observations suggest that basalts occur on Mercury, Venus, Mars and Vesta. The evidence for the existence and occurrence of basalts in the solar system differs for each object, and an evaluation of this evidence is a principal goal of this chapter.

In discussing the spatial distribution of known basalts and the existence of probable basalts on bodies in the solar system, we concentrate on those remote-sensing measurements that yield compositional information (chemistry, mineralogy). For example γ-ray and X-ray orbital measurements of the Moon give information on the concentration of certain elements such as K, U, Th, Fe, Ti, Si, Al, Mg, and lunar spectral reflectance measurements yield data primarily on the surface mineralogy. Most other remote-sensing techniques, although important, do not yield *direct* compositional information. Gravitational, magnetic, electrical, thermal and optical polarization measurements are examples, and are not discussed in this chapter.

We further focus our discussion on compositional remote sensing of basalts. No attempt has been made to give a balanced view of remote sensing *per se*. If a technique has not been used for compositional analysis of basalts it is not central to this chapter. For example, a highly developed remote-sensing technique for terrestrial exploration is airborne γ-ray spectroscopy; however, the method is best suited for sensing high K, U and Th concentrations in granitic rocks and has been used only rarely for studying basalts.

Photogeologic studies of the morphology of planetary surfaces are discussed in Chapter 5. For completeness we also cover in this chapter the use of *in situ* analyses (e.g., the Viking Lander X-ray experiments); however, these are not remote-sensing measurements in the usual sense.

In this chapter we discuss a variety of compositional remote-sensing techniques that are used to define and characterize basalts on solid objects in the solar system. We use the terms "define" and "characterize" following Head *et al.* (1978b) where "defining" parameters can be distinguished in two dimensions, allowing a unit to be mapped, and "characterizing" parameters provide information about the composition of the surface material. With minimum information it may only be possible to define the presence of volcanic units by mapping some parameter that distinguishes them from other units. For example, landforms, albedo and color imply the presence of volcanic rocks on the Moon; however, they do not reveal the detailed nature of these rocks. With further information it is possible to characterize the units by determining their chemical, mineralogical and physical properties. For the Moon, telescopic and orbital remote-sensing data have been combined with data from lunar samples to define and characterize units. Some remote-sensing techniques provide indirect or supporting information on basalts, but when used alone they do not reveal the composition and areal extent of the materials.

Many of the defining, characterizing and supporting remote-sensing techniques discussed in this chapter have been developed within the past decade. The newness of the methods necessitates taking a critical look at how they have actually performed. The most important testing ground for compositional remote sensing of geologic targets has been the Moon: the Moon is readily measured by telescopic techniques from Earth; parts of it have been surveyed using orbital gamma-ray (γ-ray) and X-ray spectrometers as well as by other instruments; and samples taken from nine locations on the front side of the Moon have been studied in detail in terrestrial laboratories. Thus, not only is the Moon the object for which the most varied compositional remote-sensing data exist (including Earth), but the lunar exploration program is also largely the means by which γ-ray, X-ray, spectral reflectance and other techniques were developed.

In the following sections we have separated our discussions of results from those of techniques. The emphasis in section 2.2 is on what has been learned about the composition and distribution of basalts in the solar system by using remote-sensing techniques. In section 2.3 we have attempted to provide concise explanations of the principles behind the main remote-sensing techniques. An evaluation of their effectiveness and limitations is made and illustrated by using specific examples. This information is important as a background for understanding the conclusions reached about the occurrence of basalts on solar system bodies.

The reader is urged to keep in mind that no single remote-sensing method is able to satisfactorily define and fully characterize basalts or any other geologic units. Each method has strengths and weaknesses, and it is through the combination of data from several techniques that the best results are obtained. Furthermore, most of the geologic units measured do not occur in the form of clean outcrop surfaces having simple geometries. Instead, the available evidence indicates that the vast majority of non-icy solid surfaces in the solar system are covered with dust or soil. This cover may or may not be locally derived, and may or may not have been substantially altered in chemistry and in mineralogy from the parent material. The degree of surface alteration is itself important as a measure of the cumulative time of exposure to the alteration processes. It is essential, then, to understand the surface alteration process on a planet before interpreting remote compositional or mineralogical measurements. A detailed discussion of surface processes on each of the planets is beyond the scope of this chapter. In our assessment of the evidence for basalts, we have taken into account that the remote-sensing data typically were obtained on a surface soil layer. For further details the reader is referred to the papers cited in the text.

2.2 BASALTS ON SOLAR SYSTEM BODIES

2.2.1 Moon

Introduction

The Moon is the only planet, other than our own, from which samples have been collected for study. The Apollo and Luna samples left no doubt that the dark lunar maria are extensive fields of basalt. As discussed in detail in Chapter 1, there are, however, very distinct differences in chemistry, mineralogy, and age among the various lunar basalt types. A basic understanding of the evolution and thermal history of the Moon requires information on the chemistry and age of the major types of surface materials. A global understanding of lunar surface geochemistry also is essential for planning future utilization of lunar resources. Although basalts are known to exist on the Moon, it is now important to know how well the current lunar samples represent the major geochemical units on the Moon, and to find answers to such questions as: What is the geochemistry of unsampled regions? What is the global extent of various rock types? What is the temporal sequence of mare basalt types? Discussed in this section are answers derived from current remote-sensing techniques. We have limited the discussion to the composition and extent of lunar basalts. A complete presentation of lunar remote-sensing information is beyond the scope of this section.

There are three general types of remote-sensing data currently available that together allow lunar basaltic units to be defined and characterized:

1. Photographic images, discussed in detail in Chapter 5.
2. Orbital X-ray and γ-ray data from the spectrometers flown on Apollo 15 and 16.
3. Measurements of the reflectance properties of lunar soil, both multispectral images and spectral reflectance measurements.

The primary focus of this section is the definition and characterization of basaltic units on the lunar surface. The Apollo γ-ray and X-ray orbital data provide direct information on *elemental abundances* and are the most important data available for unit characterization. These data have been critical in establishing the basaltic nature of the lunar maria on a global scale. Although of low spatial resolution, the information derived from the γ-ray and X-ray experiments provides approximate boundaries for unit definition. Spectral reflectance data generally provide information on the *mineralogy* of surface material. Spectral reflectance and multispectral mapping have been utilized extensively for defining the boundaries of basalt units and for distinguishing among the various surface units. Compared to the X-ray and γ-ray data, multispectral imaging and spectral reflectance have a higher spatial resolution and a greater surface coverage. Because of this, they have been the primary techniques for defining particular basaltic units. Characterization is derived from all four techniques (X-ray, γ-ray, multispectral imaging, spectral reflectance).

Other remote-sensing techniques have been successfully utilized in studying the Moon (see Head *et al.*, 1978a). Although they have not provided direct information for defining and characterizing basalt units, they have been useful in understanding the overall geology and geophysics of specific regions. For example, the description of excess mass concentrations of mascons (Muller and Sjogren, 1968; Sjogren *et al.*, 1974) has put geophysical constraints on the evolution of the maria and on such parameters as thickness of mare basalts. The topography of the mare surfaces, derived from

Earth-based radar interferometry (Zisk, 1972) and orbital sounder information (Phillips *et al.*, 1973), has been used in understanding unit morphology and emplacement history. The orbital sounder data also have provided information on subsurface layers and on basalt stratigraphy (Head *et al.*, 1978a; Maxwell and Phillips, 1978). A useful collation of many Earth-based and orbital data for the Moon has been prepared by the La Jolla Consortium using the processing facilities of the U.S. Geological Survey at Flagstaff (Eliason and Soderblom, 1977). These data (see frontispieces of the *Proc. Lunar Sci. Conf 8th*, 1977) are all displayed in the same format so that they can be easily compared.

Ground data

Prior to the Apollo landings and the return of lunar samples there were strong indications that the lunar mare regions were basaltic in composition. The alpha-scattering experiments on the Surveyor landers in the 1960's (Turkevich *et al.*, 1969; see section 2.3.6) not only suggested that the maria were rich in iron and titanium, but also indicated the chemical distinction between the mafic maria and the more aluminous highlands. Early spectral reflectance measurements (Adams 1968; McCord and Johnson, 1969) indicated that various ferrosilicate minerals were likely candidates for mare surface material, and arguments based on the photometric function and albedo of the Moon implied a basaltic composition for the maria (Hapke, 1968). Only with the return of lunar samples, however, was the nature of lunar soils and rocks fully understood. The lunar soils are unique surface materials altered through time by an environment foreign to terrestrial laboratories. We now know from studying samples that lunar soils are fine-grained mixtures of mineral and lithic fragments, breccias, glass, and glass-welded aggregates generated by micrometeoroid bombardment and by exposure to solar wind. The samples provided "ground truth" information on soil chemistry and mineralogy, and put geochemical remote sensing on a firm foundation for exploration. Compositions of mature soils are presented in Table 2.2.1.

The major and minor element chemistry of returned lunar samples is discussed in Chapter 1. The distinction between basalt types in the returned samples is based largely on analyses of rock and mineral samples. The mineralogy, and to some extent the chemistry, of a particular sample is the result of its emplacement and excavation history. Geochemical relationships among rock samples must be determined from a given suite of samples in order to identify the primary rock types representative of a given region. The task of remote sensing is to extend these results to other unexplored lunar regions. Remote-sensing techniques typically do not provide direct information about lunar rocks or rock fragments; it is now understood that remote-sensing data pertains almost entirely to surface soils.

Because development of the lunar soil is a function of cumulative exposure time at the surface, it is convenient to refer to the state of soil maturation. From studies of lunar samples it is known that the lunar soils are not simply pulverized rock; they also contain glassy aggregates of particles known as agglutinates. Agglutinates are made when micrometeoritic impacts melt and destroy mineral and lithic fragments on the lunar surface. Fully mature soils appear to reach a near steady-

Table 2.2.1 Average compositions of lunar soil samples (<1 mm fraction).[†]

	Apollo 11 (%)	Apollo 12 (%)	Apollo 14 (%)	Apollo 15 (%)	Luna 16 (%)	Luna 20 (%)	Apollo 16 (%)
SiO_2	42.04	46.40	47.93	46.61	41.70	45.40	44.94
TiO_2	7.48	2.66	1.74	1.36	3.38	0.47	0.58
Al_2O_3	13.92	13.50	17.60	17.18	15.33	23.44	26.71
FeO	15.74	15.50	10.37	11.62	16.64	7.37	5.49
MgO	7.90	9.73	9.24	10.46	8.78	9.19	5.96
CaO	12.01	10.50	11.19	11.64	12.50	13.38	15.57
Na_2O	0.44	0.59	0.68	0.46	0.34	0.29	0.48
K_2O	0.14	0.32	0.55	0.20	0.10	0.07	0.13
P_2O_5	0.12	0.40	0.53	0.19	-	0.06	0.12
MnO	0.21	0.21	0.14	0.16	0.21	0.10	0.07
Cr_2O_3	0.30	0.40	0.25	0.25	0.28	0.14	0.12
Total	100.30	100.21	100.22	100.13	99.26	99.91	100.17

[†]From Taylor (1975), p. 62.

state equilibrium where further impacts mainly remelt the products of earlier impacts.

Two important aspects of the lunar alteration of rocks to soil should be noted before comparing basalt types identified by lunar rock samples (Chapter 1) and those identified as units with remote-sensing techniques: (1) All mature lunar soils contain up to 20% foreign or "exotic" material. For mare soils this component is generally terra material (although the source and nature of this material is still debated; e.g., see discussion in Hubbard and Woloszyn, 1977; Hörz, 1978). A mare basalt unit defined by soil properties would, therefore, have different chemical and modal analyses than a mare basalt type described by rock analyses. (2) Mature mare soils also have a different mineralogy than would result from a comminution of parent rock, because soils contain agglutinates and glass.

Although both of these soil alteration processes (addition of foreign components and development of agglutinates) are global in nature, neither the orbital nor the Earth-based techniques have detected changes of composition with distance from mare-highland boundaries. Boundaries are generally sharp within the resolution of measurement. Except for local variations caused by cratering, all surfaces appear to be affected equally by alteration. Thus, even though lunar soils have been transformed physically, and to some extent chemically, during the regolith formation process, the units that are defined and characterized by remote sensing closely mimic the underlying primary rock types.

Orbital chemistry

Two geochemical instruments were flown in lunar orbit on Apollo 15 and 16: a γ-ray spectrometer and X-ray spectrometer. A discussion of the details of these two remote-sensing techniques, γ-ray and X-ray spectroscopy, can be found in section 2.3. Some of the important information concerning lunar basalts resulting from these experiments is outlined below. A comparison of the field of view of the Apollo spectrometers and the effect on spatial resolution is shown in Fig. 2.2.1 (Bielefeld *et al.*, 1977). Both of the orbital experiments, as well as the Earth-based reflectance measurements discussed in the next section, measured the geochemistry of surface *soils*.

Gamma-ray spectroscopy. Gamma rays of characteristic energy are emitted from the surface of the Moon as a result of two processes. One is the decay of naturally radioactive isotopes, K, Th, U and their unstable daughter products; the other involves nuclear reactions induced by high-energy cosmic-ray bombardment. In both processes there are transitions between well-defined energy levels. The energy generated by such transitions is characteristic of the change in state, whereas the measured flux is proportional to the concentration of the element involved in the reaction.

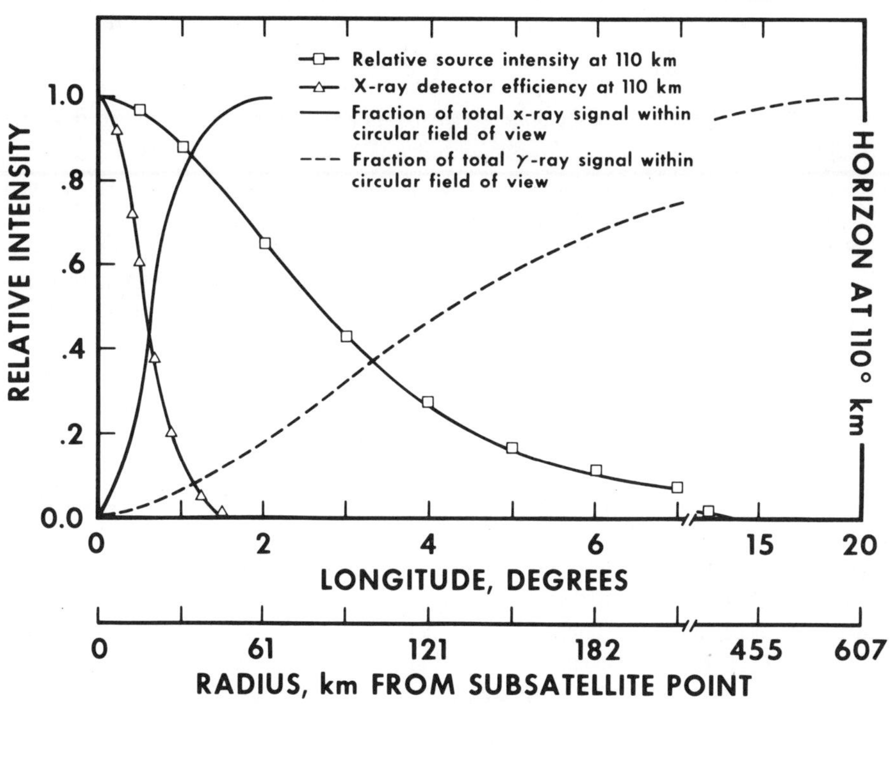

2.2.1 Circular field-of-view as function of longitude. Comparison of the spatial resolution of the Apollo X-ray and γ-ray spectrometers at an altitude of 110 km. Source term has assumed a uniformly emitting lunar surface.

Twenty percent of the lunar surface was mapped by the γ-ray spectrometers carried in orbit aboard Apollo 15 and 16. The sensitivity of the measurements depends on elemental concentration, nuclear emission characteristics, orbital altitude, duration of observation and detector sensitivity. Mapping resolution ranges from 2° × 2° (60 km × 60 km) to about 10° × 10° (300 km × 300 km). Preliminary results (Arnold *et al.*, 1972; Metzger *et al.*, 1973), have been substantially expanded and upgraded through improvements in data-reduction techniques. To date, relative and absolute values of total natural radioactivity for Th, K, Fe, Ti and Mg have been extracted from the data. Al and U may be added to the list in the future. Two other elements, Si and O, can also be measured, but these measurements have little significance because of the limited range of variation in lunar rocks.

Plate 2.1, a global map of combined Apollo γ-ray results (Arnold *et al.*, 1977), shows the surface coverage of the measurements. In this plate the data from the lower portion of the energy range (0.55–2.75 MeV) are grouped together to show the areal distribution of natural radioactivity across the lunar surface. In this energy range, regional differences in count rate are predominantly due to varying intensities in the lines of Th, U and K. Use of the integrated intensities increases the statistical precision and allows improved spatial resolution. The colors in Plate 2.1 represent about a factor of 30 range in concentration, and display the significant differences in areal distribution.

In general, most of the extensive terra regions are less radioactive than are the maria. A second general observation is that the western maria show a higher and greater range of radioactivity relative to the eastern maria. Three locations with particularly high concentrations of radioactivity, all within or adjacent to the western maria, are Aristarchus, Archimedes and Fra Mauro. The high radioactivity of the western maria may indicate the presence of KREEP, a highland-derived rock and soil component that is enriched in a number of minor and trace elements including potassium, uranium and thorium. To what extent the sources of these high concentrations are localized in the maria or contiguous highlands is difficult to determine directly from the data, due to the limits of spatial resolution. However, by removing the instrument response from the observed thorium data, Haines *et al.* (1978) were able to show that the high thorium concentrations in the Aristarchus region are more likely to be associated with the Aristarchus ejecta blanket and Plateau, than with the surrounding mare. A similar result has been found at Archimedes where the highest Th values are associated with the terra regions around Archimedes, and with the Apennine Bench and Apennine Mountains, rather than with Mare Imbrium or Palus Putredinus (Metzger *et al.*, 1979).

Using raw-data processing techniques, Metzger *et al.* (1977a) determined thorium concentrations for 144 regions on the lunar surface. Thorium concentrations in the western maria range from about 2.5–6 ppm. These are well above the values found in sampled igneous mare basalt rocks (Laul and Fruchter, 1976). Thorium concentrations in the eastern maria are lower (1.3–2 ppm). Orbital and sample measurements show very similar proportions of K to Th, the K/Th ratio being inversely proportional to the Th concentration (Schonfeld, 1977). The ratio of K to Th is significantly below the ratios found in terrestrial basalts and chondritic meteorites. Values of K found in the maria range from 800–2300 ppm, and are somewhat higher for the three western regions showing enhanced radioactivity. A consistent inverse correlation between Al and Th, as measured by the X-ray (see below) and γ-ray experiments respectively, was seen by Apollo 15 and 16 across the eastern nearside face of the Moon (Bielefeld, 1977; Haines *et al.*, 1978).

A wide γ-ray energy band (2.75–8.6 MeV) has been selected to display the combined variations of Fe and Ti (Frontispiece, 1977; Arnold *et al.*, 1977). An early estimation of Fe concentration was derived from the data and was based on a theoretical model relating the emitted lunar flux to the unit concentration (Bielefeld *et al.*, 1978). An improved estimation of Fe concentrations has been derived by correlating sampled soil averages with the energy band counting rate for the region that contains the sampled site (Davis, 1980). The band from 6.99 MeV to 8.89 MeV is representative of Fe and has been used to construct the recent map of Fe concentration presented in Plate 2.2. The highest Fe concentrations mapped by the γ-ray experiment are in mare basalt regions. This is in strong contrast with low-Fe terra regions in which the anorthosite rock suite predominates. The Fe concentrations (Davis, 1980) for mare regions range between 9–14.5% (12–18.5% FeO) as shown in Table 2.2.2.

A direct measurement of Ti concentrations from the current γ-ray data is more difficult due to interfering bands, primarily from Fe and O. Two techniques for separating Ti information from the interfering bands have been developed recently and have provided useful results. In one approach, a single band from 5.95–6.37 MeV has been used to obtain concentrations of Ti for about 140 regions having adequate counting statistics. These concentration values for Ti have been calibrated with ground site soil values (Metzger and Parker, 1980). The single band method requires a separate correction

Table 2.2.2 Regional iron and titanium concentrations from Apollo orbital γ-ray spectroscopy.†

Region	Name	FeO + σ (wt.%)	TiO_2 + σ (wt.%)
1	Mare Cognitum (16)	14.9 ± 1.8	0.8 ± 1.0
2	Mare Crisium (15)	18.5 ± 3.1	1.5 ± 1.5
3	Mare Fecunditatis		
	A. Total (15)	17.5 ± 1.9	3.8 ± 1.0
	B. Total (16)	13.9 ± 1.3	3.2 ± 0.7
	D. Limited (16)	13.5 ± 1.7	3.2 ± 1.0
4	Mare Imbrium (15)		
	A. Total	17.5 ± 1.3	3.3 ± 0.7
	B. West	18.0 ± 1.7	3.8 ± 0.8
	C. East	17.0 ± 1.7	2.8 ± 0.8
5	Mare Serenitatis (15)		
	A. Total	17.2 ± 1.5	3.8 ± 0.8
	B. Limited	17.8 ± 1.7	4.0 ± 1.0
	C. Inner	18.1 ± 1.8	3.8 ± 1.0
	E. West	-	5.2 ± 1.7
	F. East	17.4 ± 1.5	3.5 ± 0.8
6	Mare Smythii (16)		
	B. Total	10.8 ± 1.9	1.8 ± 1.0
	D. Flooded	13.1 ± 2.4	2.3 ± 1.3
	E. Unflooded	8.1 ± 2.8	1.3 ± 1.5
7	Mare Tranquillitatis (15)		
	A. Total	13.6 ± 1.4	4.8 ± 0.8
	B. Limited	12.9 ± 1.8	5.2 ± 1.0
	C. East	13.4 ± 1.7	4.8 ± 1.0
	E. South (High Ti)	12.0 ± 2.4	7.4 ± 1.5
9	Oceanus Procellarum		
	A. North (15)	17.2 ± 1.4	4.0 ± 0.8
	B. South (16)	13.4 ± 1.2	3.0 ± 0.7
	E. N. Mare Highlands Border (15)	14.9 ± 2.1	2.5 ± 1.0
	F. Northeast (15)	16.7 ± 1.8	3.3 ± 1.0
	G. Southeast (16)	13.9 ± 1.7	3.0 ± 0.8
	H. North-a (15)	17.5 ± 1.7	3.8 ± 0.8
	I. North-b (15)	16.8 ± 1.8	4.0 ± 1.0
	J. South-a (16)	14.4 ± 1.9	1.8 ± 1.0
	K. South-b (16)	14.2 ± 1.9	3.7 ± 0.8
20	Apennines (15)	11.6 ± 2.2	1.3 ± 1.2
21	Archimedes (15)	13.6 ± 2.2	0.8 ± 1.0
22	Aristarchus (15)		
	A. Total	13.6 ± 1.5	4.3 ± 0.8
	C. Plateau	11.3 ± 2.6	2.8 ± 1.5
	D. Plateau & mantling	10.9 ± 2.6	3.0 ± 1.3
	F. South of Plateau	16.5 ± 2.4	6.5 ± 1.5
24	Fra Mauro (16)		
	A. Total	9.9 ± 1.3	2.5 ± 0.7
	B. Lalande-Davy	9.3 ± 1.9	3.0 ± 1.0
	C. Limited	10.4 ± 1.9	2.7 ± 1.2
26	Grimaldi (16)	6.9 ± 2.3	3.8 ± 1.3
45	Taurus Littrow (15)	15.3 ± 2.3	4.5 ± 1.3

†Boundaries of each region are provided by Metzger *et al.* (1977a) and Metzger and Parker (1980). Apollo mission number in parentheses.

for interfering elements in the energy band of choice. A map of the Ti distribution based on the single 5.56–6.37 MeV band is shown in Plate 2.3. Regional values (from Metzger and Parker, 1980) are listed in Table 2.2.2. A second approach involves a multiband, multielement coefficient analysis in which interferences are removed directly (Davis, 1980). Results of the multiband approach are in good agreement with the single band analyses. The γ-ray Ti map of Plate 2.3 shows the predominance of low-Ti basalts over high-Ti basalts, and the presence of some very low-Ti basalts in the regions mapped.

A comparison between values for TiO_2 obtained from γ-ray data and those estimated from spectral reflectance measurements was made by Metzger *et al.* (1979). Neither technique is free from ambiguities. Nevertheless, for most regions for which data are available, the agreement is good, or within errors expected due to differences in spatial resolution and precision.

X-ray spectroscopy. Fluorescent X-rays, characteristic of some major rock-forming elements, are produced in measurable quantities at the lunar surface by the natural energy of solar radiation. X-ray fluorescence spectrometers were flown aboard the Apollo 15 and 16 spacecraft, and were used to measure that part of the secondary X-ray flux from the lunar surface that contains the characteristic X-rays of Si, Al and Mg. The intensity and characteristic energy distribution of these secondary X-rays are produced by the elemental concentration in the lunar surface as well as by the character and intensity of the exciting solar flux (see section 2.3 for more detail). The effects of variations of solar-flux and surface roughness were minimized by the use of Al/Si and Mg/Si ratios. These ratios, calculated from orbital data, are proportional to Al/Si and Mg/Si concentration ratios of lunar soils. Variations in these ratios are considered as variations in Al and Mg, because Si concentrations in returned soils are nearly constant. X-ray data were obtained from about 10% of the lunar surface, including coverage of many nearside maria and small portions of the western maria.

The early results (Adler *et al.*, 1972 and 1973) demonstrated that the terrae and the maria differed markedly on a global scale. The low Al/Si values of the maria (Plate 2.4), coupled with the high-Fe values from the γ-ray experiment, were consistent with the maria being basaltic. The clarity with which the Al/Si ratios and, to a lesser extent, Mg/Si ratios (Plate 2.5) delineated mare/terra boundaries indicates that the regolith has not been homogenized by horizontal transport (Adler *et al.*, 1973). Calculated concentration ratios were found to agree with values from returned lunar samples to within 20%. Additional good agreement was obtained by comparing the orbital concentration ratios with data from later missions such as Luna 24 (Andre *et al.*, 1978a).

Recent refinements in data processing have improved the overall quality of the data. These include methods of eliminating the effects of solar-flux variations (Hubbard and Keith, 1977; Clark and Adler, 1978), and reducing the random scatter of data (Andre *et al.*, 1977a; Bielefeld *et al.*, 1977; Hubbard and Keith, 1977). These techniques have also allowed spatial resolution to be improved. Under some conditions a resolution of 15–30 km can be obtained, depending on the chemical contrast between a particular feature and its surroundings (Andre and Adler, 1978; Fig. 2.2.2).

The chemical distinction between the highlands and the maria and variations within these major rock types can easily be seen when the Al/Si data are presented in image format. The most complete and current data (Andre *et al.*, 1977b) are shown in Plate 2.4. The data have been processed with a weighted filter that improves the signal-to-noise ratio without losing spatial resolution. Unwanted solar effects have been removed largely through empirical internal normalization. Mare basalts are clearly identified by their low Al/Si chemistry. There appears to be a slight increase in Al in the eastern maria.

The Mg/Si orbital data, like the returned-sample data, show minimal differences between maria and highlands. Major features of the Mg/Si data were obscured until recently by solar effects that produced variations in the Mg/Si ratio equal in size to the lunar chemical variations. Recent processing of this data (Hubbard *et al.*, 1978b; Clark *et al.*, 1978; Hubbard and Keith, 1978) allows a closer examination of intermare differences. The highest Mg/Si areas are largely restricted to maria as can be seen in Plate 2.5, an image representation of the data.

With the exception of central Mare Serenitatis, the highest Mg/Si ratios within the maria are concentrated in peripheral mare units and in the ejecta deposits of some mare craters. These craters, such as Piccard and Peirce in Mare Crisium and Messier A in Mare Fecunditatis, have excavated subsurface basalts with a higher magnesium composition (Andre *et al.*, 1979). At least in these areas basalts with a higher concentration of magnesium predate those which later flooded the maria.

Some areas of high Mg/Si correlate with dark mantle deposits such as those in the vicinity of the Apollo 17 landing site (Schonfeld and Bielefeld, 1978; Hubbard *et al.*, 1978b). Other observed differences probably are associated with different mare basalt units.

The Mg/Al ratio has several advantages over other elemental ratios for distinguishing chemical variations within mare basalt units. The Mg/Al ratio is less subject to effects from solar spectrum changes than is the Mg/Si ratio, and it has a greater dynamic range of values than either the Mg/Si or Al/Si ratios. However, there may be ambiguities in identifying mare basalt types because both numerator and denominator may vary independently. The Mg/Al ratio was particularly useful for locating areas thought to have a high proportion of green and brown glasses (low Al, high Mg) that

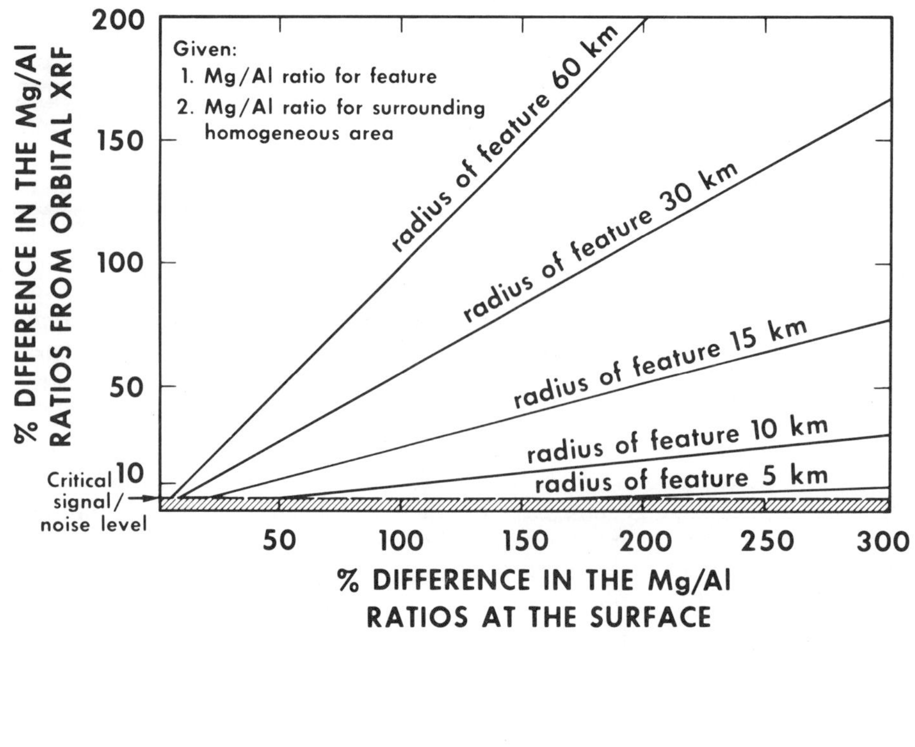

2.2.2 Orbital X-ray data. Discrimination of a chemical feature as a function of its size and chemical contrast to the surrounding soils. From this figure, given a feature size less than the X-ray field of view, the actual value can be reconstructed from the orbital value. For instance, a feature 15 km in radius with an orbital Mg/Si ratio of 1.25 vs. surrounding values of 1.0 (a 25% difference) would actually have a 100% difference. Therefore, the true Mg/Si value of the feature would be 2.0.

correlate with "dark mantle" deposits of low albedo (Schonfeld and Bielefeld, 1978). The ratio was also used to identify craters that might have penetrated through mare basalt layers (high Mg, low Al) thereby exposing submare crustal material (low Mg, high Al).

Earth-based reflectance spectroscopy

Reflectance spectroscopy involves the measurement of light reflected from a surface as a function of wavelength. Absorptions at visible and near-infrared wavelengths are controlled by the types and amounts of minerals present in surface materials (see section 2.3.3).

Spectral reflectance data are available for the nearside lunar surface from Earth-based telescopic measurements, and can be used to define and characterize geologic units. (Reflectance spectra have not yet been obtained from lunar orbit.) There are four principal types of data for the Moon:

1. Telescopic reflectance spectra of areas 10–20 km in diameter on the lunar surface (0.3–1.1 μm, 30 nm spectral resolution).
2. Telescopic infrared reflectance spectra of areas 20 km in diameter (0.65–2.5 μm with 10–20 nm spectral resolution).
3. Telescopic spectral images (spatial resolution 1–50 km; 2–5 band passes between 0.38 and 1.0 μm).
4. Laboratory reflectance spectra of returned lunar samples (0.35–2.5 μm, with 5 nm spectral resolution).

Laboratory reflectance spectra of lunar rocks typically show well-developed absorption bands that arise from electronic processes in constituent minerals and glasses. (The theory of reflectance spectroscopy is discussed in section 2.3.3, where examples of spectra of lunar minerals and rocks are shown). Telescopic spectra of certain bright craters and crater ejecta in the lunar maria show the same absorption features that occur in the laboratory spectra of mare basalts. For example, the telescopic spectrum of the crater Messier (Fig. 2.2.3) can be interpreted in terms of Ca-rich pyroxene (Adams, 1974)—a typical constituent of mare basalts. This interpretation is based on the position of prominent absorption features at 0.95 μm and near 2 μm. The crater Guericke C in the same figure has absorptions at 0.90 μm and near 1.8 μm, indicating the presence of Ca-poor orthopyroxene. Because orthopyroxene is a typical component of terra rocks it is probable that Guericke C exposes terra rocks from underneath the basalts in Mare Nubium. The strength of the absorption feature is much greater than that observed from most terra material, implying a large modal abundance of orthopyroxene in the Guericke C material.

It is difficult to obtain telescopic spectra of lunar rocks. Only a few of the bright craters in the lunar maria show well-developed absorption bands (Messier and Guericke C are notable examples but not the only

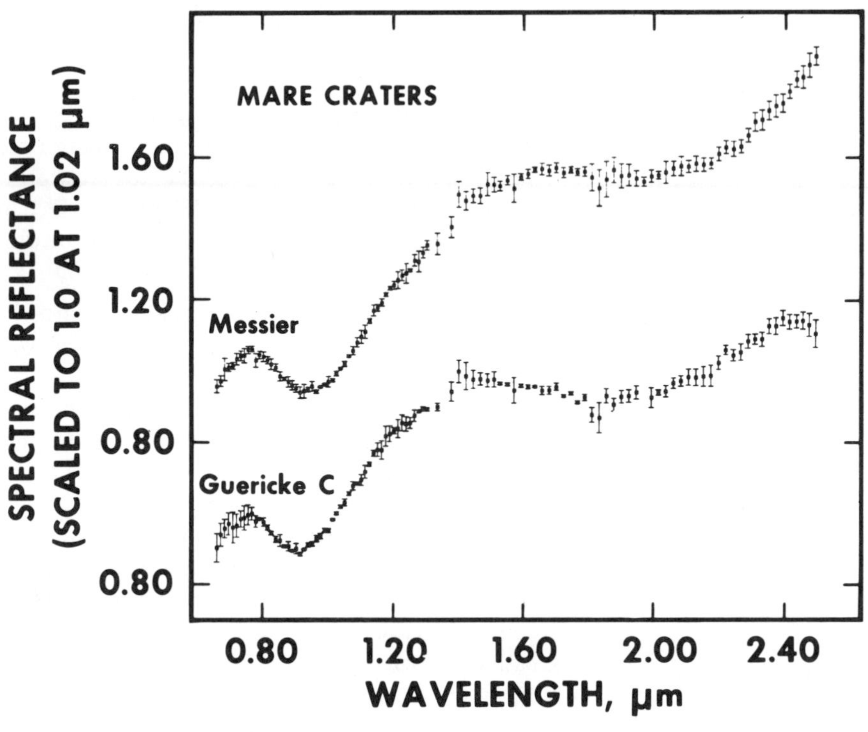

2.2.3 Telescopic reflectance spectra of two fresh craters in mare regions. The depth of the 1 and 2 μm absorption features indicate that mineral or lithic fragments are a major component of the surface material, and that the mineral pyroxene dominates the spectrum. The pyroxene composition for Guericke C (in Mare Nubium) is less calcium-rich than for Messier (in Mare Fecunditatis).

examples). This is not surprising as bright craters are few in number and most of them are smaller (<<10 km) than the spectral resolution of available telescopic measurements.

In contrast, there are vast areas of the lunar maria that have telescopic spectra that correspond to samples of lunar *soils*. Early in the Apollo program Adams and McCord (1970) showed that the telescopic spectra of

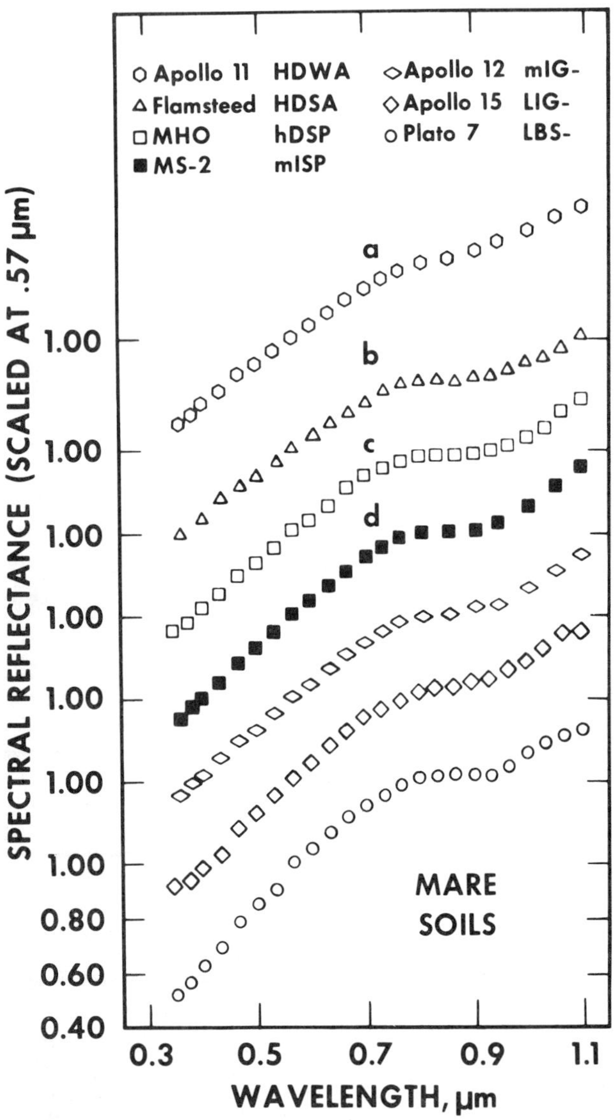

2.2.4 Telescopic reflectance spectra of representative mare soils from 0.3–1.1 μm. All spectra are scaled to unity at 0.57 μm. Distinctions between mare areas are made through measurements of the slope of the continuum (primarily UV/VIS ratio) and the strength and nature of the absorption feature near 1 μm. The four-letter designation to the right of the spot name is the unit designation discussed in the text.

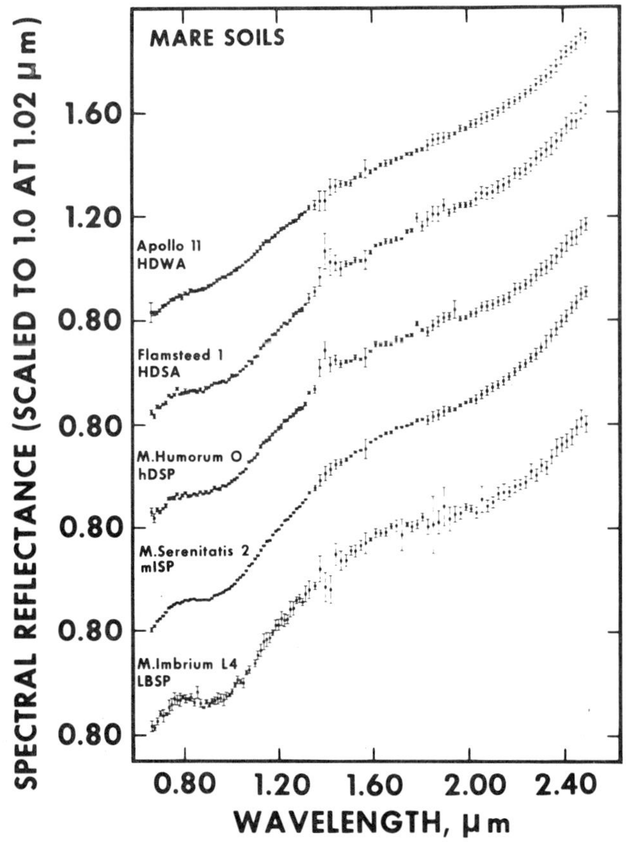

2.2.5 Telescopic reflectance spectra (0.7–2.5 μm) of representative mare soils measured on a scale of 10–20 km. All spectra are scaled to unity at 1.02 μm. Distinctions between mare areas are made by measuring the strength and nature of the absorption features near 1 and 2 μm.

10–20 km areas at the Apollo landing sites agreed with laboratory spectra of a few square millimeters of soil from the same sites. They concluded that the laboratory soil samples were representative of the lunar surface on a scale of at least tens of kilometers, and could be used as a guide to geologic mapping.

The absorption features in mature agglutinate-rich soils are more subdued. Agglutinates and otherwise altered soil particles are responsible for a variety of complex optical absorption and scattering processes. (See Adams and McCord, 1971, 1973; Nash and Conel, 1973; Adams and Charette, 1975; and Hapke *et al.*, 1975a, for discussions of the optical properties of soils having various maturities.) With maturation, soils darken, the continuum slope increases and electronic absorption bands of the original minerals become shallower. These changes can be understood in terms of the progressive destruction of the mineral grains and the build up of dark glassy particles and coatings.

Table 2.2.3 Effects of composition on measurable spectral parameters for mare soils.

UV/VIS	Albedo (0.62 μm)	1 μm band		2 μm band	
		Wavelength	Depth	Wavelength	Depth
In: glass UV absorption (In: $FeO \cdot TiO_2$)	D: % plag	glass	D: FeO in glass	px comp	D: % px
			D: % glass		
In: Fe°	In: glass absorption	px comp	D: % px		In: opaques (In: TiO_2, FeO)
In: albedo	In: opaques (In: TiO_2, FeO)		D: % ol		
	In: Fe°		In: opaques (In: TiO_2, FeO)		

In: inversely related px: pyroxene comp: composition
D: directly related ol: olivine

Telescopic spectra of mare soils are shown in Figs. 2.2.4–5. The broad absorption bands near 1 μm and 2 μm (that are caused by pyroxenes) are present in the soil spectra; however, these bands are much shallower than in the spectra of rocks and bright craters. In addition, the soil spectra show a regular increase in reflectance to longer wavelengths. It is the subtle variations in the continuum slopes and in the wavelength positions and depths of the shallow absorption bands that are used to define and characterize the soils that have formed on top of basalt flows on the Moon.

Analyses of telescopic spectra, such as those in Fig. 2.2.5, show that most of the lunar surface is covered with mature soil. This is an important conclusion because it follows that the differences among lunar spectra (other than those of bright craters) are due to real compositional differences in the soils and not to variations in maturity.

The spectral reflectance characteristics of lunar soils have been classified using the four measurable parameters discussed below. Each of these parameters is affected by the composition of the surface, a detailed discussion of which is beyond the scope of this section, but which is summarized in Table 2.2.3 (from Pieters, 1978).

UV/VIS spectral reflectance ratio. Because of their broad coverage and high spatial resolution, multispectral images have been used widely in the past ten years for mapping mare basalts. Shown in Fig. 2.2.6 is a photographic composite of images taken at 0.38 μm (negative) with those taken at 0.62 μm (positive). Soils with a high UV/VIS ratio (0.38/0.62 μm) appear dark. Although the data are qualitative, one of the major advantages of this type of image is that many units can be identified with relatively high spatial resolution (1–2 km). Digital multispectral images (McCord *et al.*, 1976; McCord *et al.*, 1979; Johnson *et al.*, 1977) allow measurements of the UV/VIS variations to be made for specific regions when high precision, high resolution

2.2.6 Full Moon color difference photograph. Eight sections have been mosaicked from photographic composites of images near 0.38 μm (negative) and 0.62 μm (positive). Bright indicates a low UV/VIS ratio, or low 0.38 μm relative to 0.62 μm. Photograph courtesy of E. Whitaker (1976).

data are required. The digital images are obtained with a silicon vidicon camera and have a broader spectral range (0.35–1.0 μm) than photographic film (see section 2.3.4).

Numerous spectra of small regions allow the variation in the UV/VIS ratio to be quantified and linked to specific geochemical properties, in particular to the TiO_2 content of lunar soils (Charette *et al.*, 1974). Calibrated spectra are required to allow two widely separated regions to be quantitatively compared. Using returned lunar soil samples as ground data for telescopic measurements, an empirical relation can be derived (Fig. 2.2.7) that allows the soils' content of TiO_2 to be estimated from 0.40/56 μm ratios. The reasons for this relationship are complex (Adams and Charette, 1975; Pieters, 1978), but largely are results of the effects on albedo and on spectral contrast of Ti- and Fe-bearing opaques and glasses. The stippled region of Fig. 2.2.7 indicates the estimated range of TiO_2 for a measured UV/VIS ratio. High-titanium basalts can be easily recognized by their relatively high UV/VIS ratio (and very low albedo); basalts with a very low UV/VIS ratio are likely to be very low in TiO_2. Note, however, the nonunique range of inferred TiO_2 for basalts with intermediate UV/VIS ratios. Although the UV/VIS ratio can now be measured precisely, the UV/VIS ratio should not be used alone to distinguish or characterize mare basalt types.

Albedo. Even to the unaided eye viewing from Earth, the Moon is seen to be composed of basically two kinds of material—dark and light. The earliest telescopes were used to describe in some detail the distribution of dark mare material within the brighter terra. More recently, several quantitative maps of albedo in the visible range have been made (Pohn and Wildey, 1970; Evsyukov and Barabashov, 1973a, b) as well as high quality photographic images. (Geometric, or normal albedo is defined as the reflectance of a surface at normal incidence compared to that of a Lambert, or perfectly diffusing surface.) Albedo contains compositional information for it is strongly influenced by the absorption coefficient of the surface material. The general correlation of albedo with Al/Si ratios measured from orbit during the Apollo missions (Adler *et al.*, 1972, 1973) demonstrates this relationship; the more feldspathic terrae are brighter and have higher Al/Si ratios than the basaltic maria. Albedo also is controlled by the physical nature of the surface material (grain size, packing, etc.), which affects both the efficiency of reflection and the photometric function (Adams and Filice, 1967). It is well-documented (e.g., Adams and McCord, 1973) that as fresh lunar material matures to well-developed soil, the albedo decreases to some equilibrium value. Albedo commonly is used in conjuction with spectral reflectance to resolve ambiguities in compositional interpretation, since materials of differing

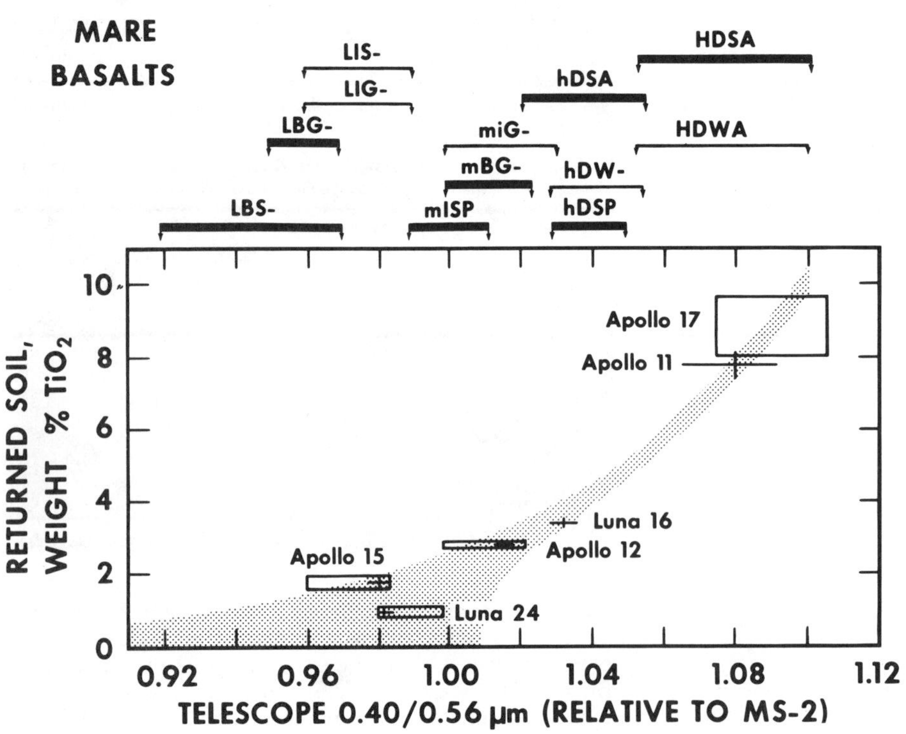

2.2.7 Relationship between wt. % TiO_2 in lunar mare soils and the 0.40–0.56 μm reflectance ratio for telescopic spectra relative to MS2 (after Charette *et al.*, 1974; and Pieters and McCord, 1976). The stippled area is the estimated range of TiO_2 that can be derived from a 0.40–0.56 μm ratio measurement of mature mare regions. Shown above the plot are the ranges of 0.40–0.56 μm ratio observed for each of the basalt types discussed in Pieters (1978). The heavy lines indicate unsampled basalt types.

absorption coefficients can have similar spectra. As noted in Table 2.2.3, the UV/VIS ratio for dark mare soils is inversely related to albedo, and it is generally true that the darkest maria have the highest UV/VIS ratio.

Absorption bands. Absorption bands near 1 μm and 2 μm in the spectra of mare soils can be used to distinguish among mare basalts. The weak 1 μm and 2 μm features of Fig. 2.2.5 are more easily examined after enhancement by removal of a general (straight line) continuum (Fig. 2.2.8). The residual absorption bands show differences in depth, in symmetry, in the wavelength center of the 1 μm band and in the depth of 2 μm band. Each of these parameters has been used by Pieters (1978) to assist in mapping basaltic soils on the lunar surface.

The 1 μm band for lunar mare soils is a composite band involving Fe^{+2}, primarily in pyroxene and to a lesser extent in glass, and olivine (Adams, 1975). As summarized in Table 2.2.3, the depth of this combined band depends on the composition (%FeO) and the proportion of iron-bearing glass in the soil. It also depends on the amount of pyroxene in the soil, the amount of olivine in the soil, and on the amount of opaques (which tend to mask absorption bands) in the soil. Only the relative depths of the 1 μm bands and 2 μm are used as parameters in this discussion. Examination of band symmetry and wavelength centers provides additional information on mineralogy. The spectra of mature soils in the lunar maria are presently being analyzed using computer band-fitting techniques (Adams and Ralph, 1978; McCord *et al.*, 1980; Farr *et al.*, 1980).

Mare basalt units

The orbital geochemical measurements and the four reflectance parameters discussed above have been assembled for nearside mare regions. Basalt types have been mapped and characterized using combinations of the available parameters. If all data were of equivalent quality and spatial resolution, a global compositional map would be easy to prepare and interpret. Given the variation of quality and resolution of the real data, however, the high spatial resolution information (i.e., reflectance data) has been used to *define* unit boundaries and establish a base map. Unit designation is therefore based on reflectance properties. All available remotely-sensed data have been used to *characterize* the basalt units. The use of the diverse data has been generally successful in describing basalt units, although broad geochemical trends, such as east-west difference in radioactivity and easterly increase in Al, must be mentioned separately. Disagreements between similar data for particular units are discussed earlier in this section.

The base map of mare basalt units, which is used in the following discussion of lunar basalt types, is presented in Plate 2.8 and Fig. 2.2.9. An itemization of the characteristics of each distinct basalt unit, measured with remote-sensing techniques, is presented in Table 2.2.4 and is cross-referenced to previous unit designations. Only extensive compositional units are discussed

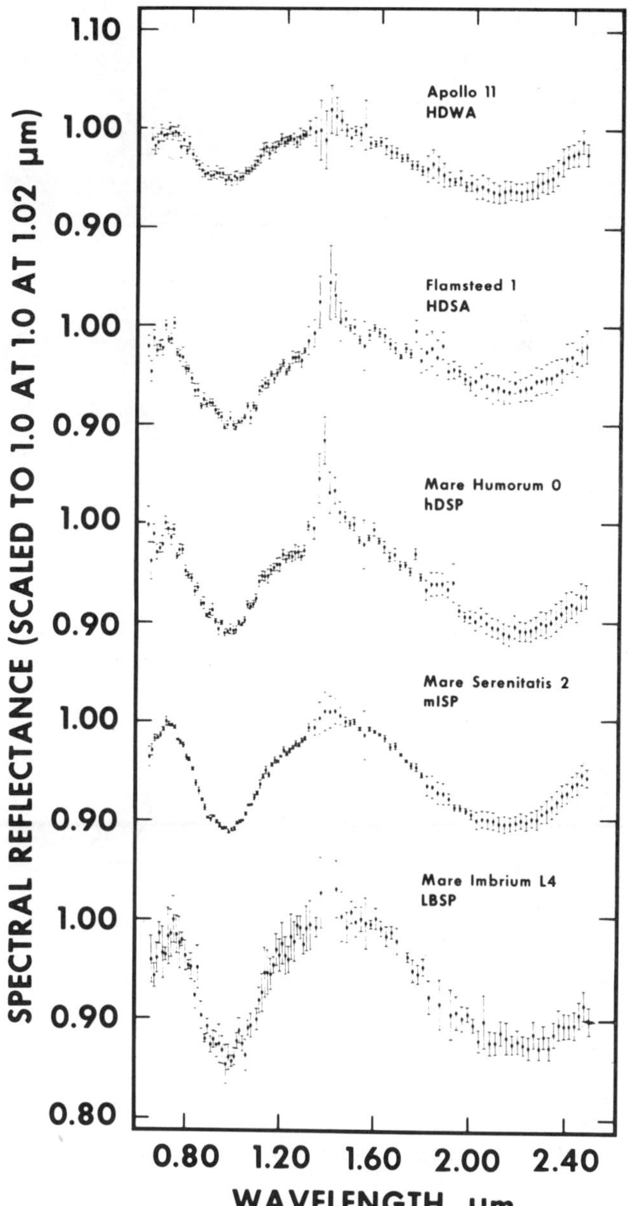

2.2.8 Residual absorption features for the telescopic spectra of mare soils shown in Fig. 2.2.5 after removal of a straight-line continuum (McCord *et al.*, 1980).

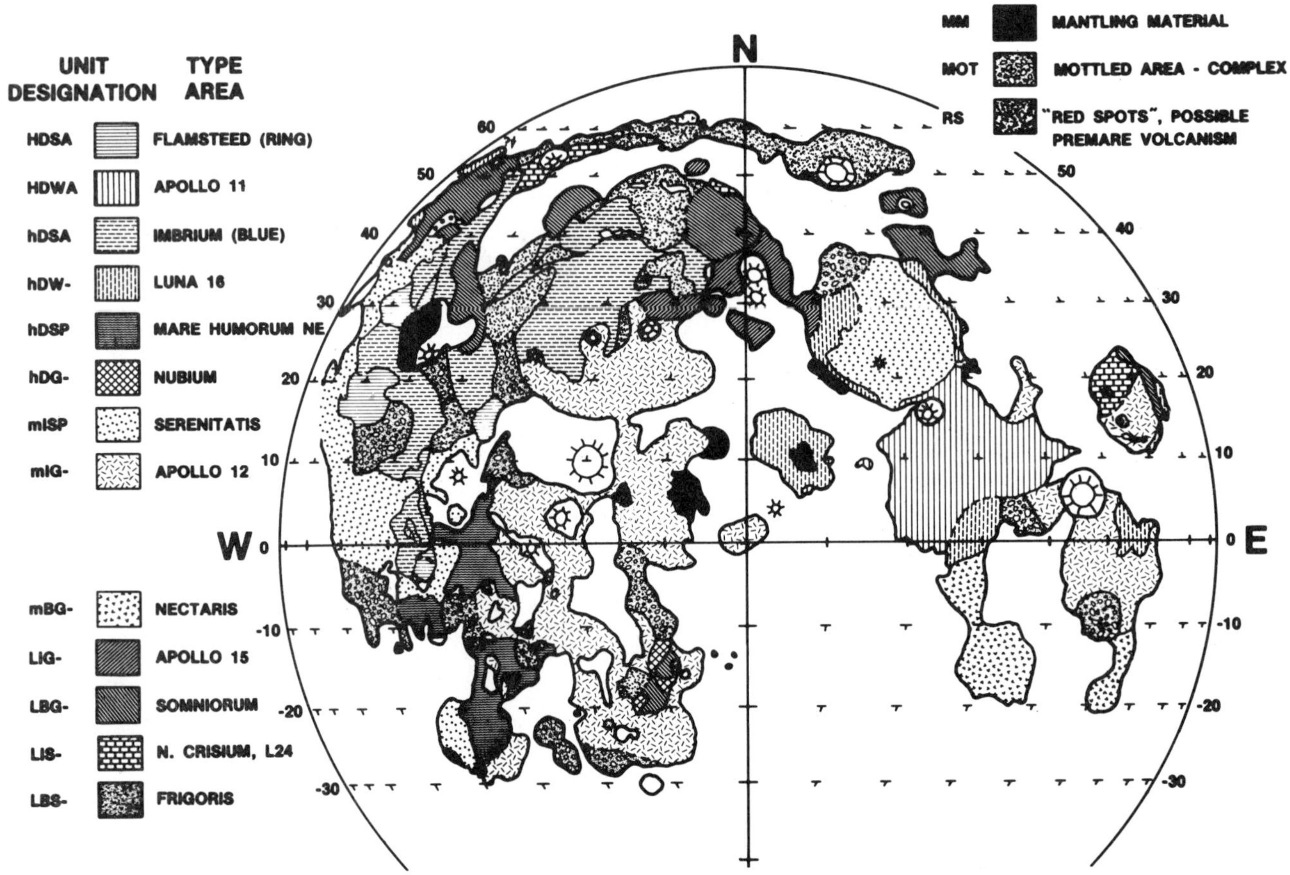

2.2.9 Major basalt types for the front side of the Moon as derived from the current spectral reflectance data. The unit designations represent values for four measurable parameters (UV/VIS ratio, albedo, strength of the 1μm band, and strength of the 2μm band). Since each of these parameters depends upon the composition of the surface, basalt units which are distinct from the sampled units can be identified and mapped on the basis of spectral properties. This same map appears in color as Plate 2.8.

here. Detailed mapping and interpretations at higher spatial resolution are referenced in the summary of synthesis studies.

In the following discussion, the name of the type area and a four-letter unit designation (described below) are used to reference each major mare basalt unit. Some of the mare basalt units have been formally named (e.g., Whitford-Stark and Head, 1980), and in a few regions the relationship between a given compositional unit and the geological evolution of its surroundings have been fully studied (e.g., Head *et al.*, 1978b; Pieters *et al.*, 1980).

The reflectance parameters that form the basis for the unit designations of mare basalt types are itemized in Table 2.2.4. As an example, the Apollo 11 site has a high UV/VIS ratio (H), is dark (D), has a weak 1 μm band (W), and exhibits an attenuated or weak 2 μm band (A). The area is thus mapped and labeled HDWA in Plate 2.8. Unit designations are derived in a similar manner for all nearside maria. The data are not of uniform quality for all regions (see discussion in Pieters, 1978); however, the mare basalt map of Plate 2.8 and Fig. 2.2.9 will not change significantly as more detailed information becomes available.

Sampled basalt units. Mare basalt units sampled by the Apollo and Luna missions are mapped in yellow in Plate 2.9. Units with properties similar to but not necessarily contiguous with sampled basalts are mapped in orange. A brief description of each of these sampled units follows.

HDWA Apollo 11. These are the early high-titanium basalts that were sampled by both Apollo 11 and Apollo 17. The only known surface exposure is on the

Table 2.2.4 Measurable parameters of lunar basalt types.

Unit Designation	Type Area	Areal Extent	UV/VIS	Albedo	1 μm Band	2μm Band	TiO_2†	γ-ray K, Th	γ-ray Fe	X-ray Al/Si	Relative Age	Remarks	Previous Designations
HDWA	Apollo 11	extensive east only	High	Dark	Weak	Attenuated	high	medium	medium	low	old	sampled; high-Ti basalt Apollo 11, Apollo 17	H_1[a]; T, Tr, Tb[c]
HDSA	Flamsteed	extensive west only	High	Dark	Strong	Attenuated	high-medium	high	high	-	young	unsampled; high-Ti basalt; likely to be more mafic than Apollo 11	H_2[a]
hDSA	Imbrium (blue)		medium high	Dark	Strong	Attenuated	medium	high	high	(low)	young	unsampled; grades into HDSA	h_2[a]
hDWA	Luna 16	significant mostly east	medium high	Dark	Weak	Attenuated	medium	medium	-	medium	middle	sampled; medium-Ti basalt Luna 16; portion possibly Mg-rich	h_1[a]; Sb, Tr[c], I[e]
hDSP	Humorum NE	significant southwest only	medium high	Dark	Strong	Prominent	medium	high	-	-	young	unsampled; medium high-Ti basalt; may have high pyroxene component	h_3[a]; H[b]; T[d]
hDG-	Nubium	minor south central	medium high	Dark	Gentle	-	medium	-	-	-	middle	unsampled; medium-Ti basalt	
mISP	Serenitatis	extensive east and west	medium	Intermediate	Strong	Prominent	low-medium	medium	high	low	middle	unsampled; low-Ti western basalts may be different from those in the east	m_4[a]; L_1[b]; S[c]; S[d]
mIG-	Apollo 12	extensive east and west	medium	Intermediate	Gentle	-	low-medium	medium-high	medium	low-medium	middle	sampled; low-Ti basalt Apollo 12; minor variations throughout unit	m_1[a]; L_2[b]; S, S_b[c]; S[d]; III[e]
mBG	Nectaris	significant east only	medium	Bright	Gentle	-	low	-	-	-	old	unsampled	m_1[a]
LISP	N. Crisium (Luna 24)	minor east and west	Low	Intermediate	Strong	Prominent	low	-	-	medium	middle	sampled; very low-Ti basalt Mare Crisium–Luna 24	m_1[a]; II_a[e]
LIG-	Apollo 15	significant east and west	Low	Intermediate	Gentle	-	low	high	-	medium	middle	sampled; low-Ti basalt Apollo 15	l_1, L_1[a]
LBG-	Somniorum	significant east and west	Low	Bright	Gentle	-	low	-	-	-	middle	unsampled; very low-Ti basalt; possibly has a high feldspar component	L_1[a]; II_b[e]
LBSP	Frigoris	extensive east and west	Low(-est)	Bright	Strong	Prominent	low-medium	high	-	-	middle	unsampled; very low-Ti basalt	L_4[a]; L_1[b]

- data not available

†If TiO_2 values from color and γ-ray differ, both are listed respectively.

References: [a]Pieters and McCord (1976); [b]Pieters *et al.* (1975); [c]Soderblom (1970); [d]Johnson *et al.* (1973); [e]Head *et al.* (1978a).

eastern nearside. Although all of Mare Tranquillitatis and most of the low albedo ring around Mare Serenitatis are similar (see Table 2.2.4), there are local variations due to differences in individual flows or to vertical mixing (see Plates 2.1–5 and Fig. 2.2.6). This basalt type is suspected of underlying some low-titanium basalts, such as those of the central nearside maria (Head, 1974) and of central Mare Serenitatis (May *et al.*, 1976; Solomon and Head, 1979).

hDWA Luna 16. This is a medium-titanium basalt. The sampled region in Mare Fecunditatis is surrounded by lower-titanium basalts. There are discrepancies in the FeO values for Mare Fecunditatis (Regions 3A, B, Table 2.2.2) perhaps due to mixing of units below the resolution limit of the γ-ray data. In Mare Crisium, basalts similar in spectral properties to these Fecunditatis basalts (Head *et al.*, 1978b) have been shown to be relatively high in magnesium (Andre *et al.*, 1978a, 1979). They underlie surface basalts of a different composition (mIG-and LBSP). The only known surface exposures of hDWA are located on the eastern nearside. Some small areas bordering northern Oceanus Procellarum might also be this basalt type (Whitford-Stark and Head, 1980).

mIG- Apollo 12. This low-titanium basalt is widespread. For the Mare Cognitum region a lower value of TiO_2 is obtained from the γ-ray data than from the reflectance data. The areas mapped as mIG- may include a variety of similar basalt types that are gathered into one general class but which are nevertheless distinct from other low-Ti basalts (see Table 2.2.4). The Ti content ranges from medium-low to low; western mIG- areas have a much higher radioactivity than do the eastern. High resolution multispectral images of the Apollo 12 landing site (Pieters and McCord, 1976) show a small medium-Ti flow to be within a few kilometers of the mIG- basalts of the landing site. This might be the source area for the Apollo 12 ilmenite basalts described in Chapter 1.

LIG- Apollo 15. This low-titanium basalt occurs only in the northern hemisphere. The unit is difficult to characterize for the basalts of the landing site are almost surrounded by terra within the resolution of the orbital techniques. In the spectral data these basalts are distinguished from most other low-Ti units, but they may form a gradational sequence with LBG- basalts.

LISP Mare Crisium, Luna 24. This unit comprises (very) low-titanium basalt. The γ-ray data indicate that the basalts of southern Mare Crisium have some of the lowest general values of Ti for a mare region and that they have relatively high values of iron. The stratigraphy of the basalts filling the basin is complex and involves a variety of basalt types (Head *et al.*, 1978a; Andre *et al.*, 1978a). Multispectral images indicate that the LISP basalts of the sampling site are surrounded by other low-Ti basalts (mIG-) and that the largest areal extent of LISP basalts occurs in northern Mare Crisium. There is good additional evidence that an early magnesium-rich basalt (hDWA) most likely flooded the basin (Andre *et al.*, 1979) prior to the emplacement of the low-titanium basalts.

Unsampled basalt units

Only a few of the basalt types mapped in Plate 2.8 and characterized in Table 2.2.4 were sampled by landed spacecraft. About two-thirds of the area of the lunar maria consist of units that do not correspond with rock types in the lunar sample collection returned by U.S. Apollo and Soviet Luna missions. The areal extent of unsampled surface basalts is shown in red in Plate 2.9. Discussed below are the specific unsampled basalt types that are regionally extensive, or are especially important.

HDSA and hDSA: Flamsteed and Imbrium basalts.

Extensive portions of Mare Imbrium and Oceanus Procellarum are covered with basalts classified as HDSA and hDSA. These two basalt types are Ti-rich (~8–4% TiO_2) and are grouped together since all characteristics are the same except for small differences in the value of UV/VIS ratio: HSDA has a 1–3% higher ratio than hDSA. The type area, Flamsteed 1, is in southern Oceanus Procellarum. These western HDSA basalts are similar to the eastern Apollo 11 HDWA basalts in Ti content, but their age, radioactivity, and near-infrared spectral character indicate they are a distinct and an unsampled basalt type. In particular, they are relatively young (Plate 2.6) and represent, perhaps, one of the last major episodes of basaltic volcanism on the Moon. They have a relatively high level of radioactivity (Plate 2.1), and their spectra exhibit a unique combination of a strong 1 μm feature with a relatively weak or attenuated 2 μm feature (Figs. 2.2.4 and 2.2.8). The Flamsteed region of Oceanus Procellarum has been studied intensively by Pieters *et al.* (1980) in order to better characterize these unique unsampled basalts. Spectral data for these HDSA Flamsteed basalts indicate either (1) the soils are enriched in Fe-bearing glass, (2) the glass in the soil is itself enriched in FeO (implying the bulk composition is Fe-rich), or (3) the olivine/pyroxene ratio for the soil is $\gg$1. The latter two interpretations are preferred by Pieters *et al.* (1980) and, if correct, would make the high-Ti HDSA basalts more mafic than any sampled. The γ-ray data for Fe content are inconclusive and do not clarify this interpretation;

however, the two regions which include the largest areal extent of HDSA and hDSA in Apollo 15 data (4B and 9H in Table 2.2.2) are slightly enriched in FeO relative to neighboring regions.

hDSP: Humorum basalts. A second type of young unsampled basalt found in the western maria occurs in northeastern Mare Humorum (see Johnson *et al.*, 1973; Pieters *et al.*, 1975; Greeley and Spudis, 1978). The hDSP unit is extensive in the lunar southwest, but it does not appear to occur elsewhere on the Moon. These basalts are relatively young (Plate 2.6), although some regions of hDSP appear to predate the hDSA/HDSA basalts discussed above. The hDSP Humorum basalts are moderately Ti-rich (3–5% TiO_2), and cannot be distinguished from the hDSA/HDSA Flamsteed basalts in multispectral maps (e.g., Fig. 2.2.6) or on the Fe and Ti orbital maps (Plates 2.2 and 2.3). The hDSP Humorum basalts are, however, sufficiently distinct in near-infrared characteristics (Figs. 2.2.4 and 2.2.8) to merit designation as a separate unsampled basalt type. The hDSP Humorum basalts not only have a strong 1 μm band, but their 2 μm (pyroxene) absorption band is also prominent. This implies that the hDSP Humorum soils are richer in pyroxene than the previously discussed hDSA/HDSA Flamsteed basalts that appear to be dominated by Fe-bearing glass or olivine. The HDSP Humorum basalts were covered by only a small part of the Apollo 16 orbital data (Plate 2.1) that show the southwestern maria to be less radioactive than the (older) eastern maria.

mISP: Serenitatis basalts. All remote-sensing data indicate that central Mare Serenitatis contains a distinct unsampled basalt type. Characterization of this mISP unit is ambiguous. Both the γ-ray and the spectral reflectance data indicate that mISP basalts of central Serenitatis have a distinctly lower Ti content than the high-Ti HDWA basalts of Tranquillitatis (Apollo 11 and 17). The upper limit for TiO_2 of mISP basalts from reflectance data (Fig. 2.2.7), however, barely overlaps the lower limit from the γ-ray data (2.8% TiO_2; Table 2.2.2, region 5C). Central Serenitatis mISP basalts appear from the orbital γ-ray measurements to be relatively Fe-rich (Table 2.2.2). A high-Fe content for these basalts is also a possible interpretation for the distinctions in near-infrared spectral characteristics of MS-2 (see Fig. 2.2.8), a standard area in Mare Serenitatis for reflectance data (McCord *et al.*, 1972). Serenitatis mISP basalts have a stronger 1 μm feature than do similar intermediate-to-low-Ti basalts elsewhere on the Moon (e.g., mIG- at Apollo 12). Since the 2 μm (pyroxene) band in the spectra of mISP basalts of Serenitatis is also prominent, an alternative interpretation of the 1 and 2 μm absorption characteristics is that mISP basalts are enriched in pyroxene.

The mISP basalts in western Oceanus Procellarum appear to have an even stronger 1 μm feature, but the orbital γ-ray data do not show them to be Fe-rich. Although the four parameters used for unit designation are the same for both Serenitatis and Procellarum mISP units, it appears likely that they are in fact two different low-Ti basalt types. Both the Procellarum and the Serenitatis mISP basalts show lower radioactivity than the unsampled Imbrium basalts.

Central Serenitatis mISP basalts have been suggested as a possible source material for the VLT (very low titanium) basalts discovered as minor components in the Apollo 17 drill core (Vaniman and Papike, 1977). The occurrence of minor VLT components at Apollo 17, for which the dominant basalt type is high titanium, implies either lateral transport from a distant surface unit or vertical mixing of a subsurface unit. The characterization of Serenitatis mISP outlined above, however, does not uniquely fit the description of these VLT basalts (see Chapter 1). Although mISP basalt is the most extensive surface basalt in central Serenitatis, there are other basalt types bordering Mare Serenitatis (some of which may be Mg-rich, Andre *et al.*, 1979). Two other well-defined possible source candidates for VLT basalts in the region, LBG- and LIG-, occur along the northern edge of the Serenitatis basin in Somniorum and along the eastern edge near Le Monnier, respectively, (see Plate 2.8) and could be present elsewhere as subsurface material. There is currently insufficient evidence to convincingly identify the VLT source area. A sample from central Serenitatis mISP would provide an ideal piece of ground data for orbital and telescopic remote-sensing studies.

LBSP and LBG-: Frigoris and Somniorum. These two extensive basalt types are the "reddest" basalts on the lunar surface, and generally are found near the edges of basins. Because most surface expressions of these basalts occur above 30° N (Plate 2.8), they were not measured by the Apollo orbital experiments. They are of medium age, possibly old in some areas (Plate 2.6), and are unlike any major basalts sampled. Their albedo is high, indicating low concentrations of opaques in the soil. The titanium content is probably low to very low, based on spectral reflectance measurements. The brightest, LBSP Frigoris, contains the strongest 1 μm and 2 μm absorptions for any mature mare soil. LBSP Frigoris thus probably contains more FeO (in soil glass) or more pyroxene than does LBG-Somniorum. LBG-Somniorum, which may be a subsurface basalt in the Serenitatis region, is a good candidate for the basalt type associated with the minor VLT basalts of Apollo 17.

Nonmare basalts and possible terra volcanism

Remote-sensing techniques and particularly the orbital γ-ray observations have contributed to the evidence for the existence of highland volcanism. Fragments of old basalts are found as minor components in terra samples, particularly Apollo 14, 15 and 17 (see Chapter 7). Irving (1977) presented evidence that the source of the basalts is a crustal endogenic melt rather than an impact melt. Remote observations suggest that terra basalts were likely to have been emplaced prior to and possibly contemporaneous with the emplacement of the maria.

Terra-associated KREEP material (see Chapter 1) has the highest thorium and uranium concentrations and the lowest K/Th ratios of any known lunar rock type. This allows γ-ray observations of Th and K to delineate the distribution of the KREEP basalt component over the surface (Metzger *et al.*, 1974). KREEP basalts appear to be concentrated in the western nearside quadrant. They are mixed with mare basalts in Mare Imbrium and Oceanus Procellarum and are also found in the adjoining terra regions east, west and south of the maria. The three regions of highest radioactivity (see Plate 2.1) are terra areas, one bordering and two within the western maria, which suggests that KREEP basalts were emplaced throughout the western nearside terra prior to mare basalt extrusion (Metzger *et al.*, 1973). The orbital γ-ray Fe values of these regions, and the spectral reflectance measurements discussed by Charette *et al.* (1977), suggest that the Archimedes plains and Fra Mauro regions may be relatively enriched in FeO. This places them within the range of Fra Mauro (terra) basalt compositions (FeO = 8.5–11.5%).

Many authors have concluded that KREEP is a residual late-phase liquid of the primary crustal differentiation process (e.g., Hubbard and Gast, 1971; Meyer, 1977). KREEP basalt liquids would be expected to rise to the surface because their density is less than that of the plagioclase terra crust (Solomon, 1975; Hubbard and Woloszyn, 1977). Modifying factors are the local elevation and the characteristics of the crustal flotation layer as a function of time (Minear and Fletcher, 1978). Hawke and Head (1978) have studied the history of KREEP basalts in the Fra Mauro and Imbrium basin regions by combining the remote-sensing chemical observations with data from the Apollo 14 and 15 sites and the lunar samples. They concluded that separate episodes of KREEP terra volcanism have occurred at both locations, and are responsible for emplacing a basaltic layer on top of the crust. This basaltic layer dominates the composition of the Apennine Bench formation. Heat-flow measurements (Langseth *et al.*,1976) dictate that this layer must be relatively thin.

By contrast, the lunar farside and eastern limb show little evidence of basaltic volcanism. Hubbard and Minear (1976) among others, have estimated that nonmare basaltic materials make up about 5% of the farside terra mapped by the orbital γ-ray experiment; orbital Fe and albedo values suggest similarly low amounts of mare material. Only in the topographically low region of Van de Graaff and Mare Ingenii is there substantial orbital evidence for farside terra volcanism in the form of Th, K and Fe values higher than those of the surrounding highlands. Eastern nearside and eastern limb terra regions appear to have much less KREEP than do the western nearside regions, despite the Th levels in the former areas being significantly above farside and western limb concentrations. Further deconvolution studies of γ-ray data have revealed Th-rich terra areas east of Mare Fecunditatis (Plate 2.1) that are associated with areas of relatively smooth plains. Haines *et al.* (1978) suggest the presence of localized regions having significant KREEP basaltic composition, implying that an episode of terra volcanism has occurred in this part of the Moon.

Of special interest are some small western nearside terra areas, commonly termed "red spots" (see Plate 2.8). They are below the resolution of the orbital measurements but are easily distinguished from other terra units in multispectral images because of their strong absorption in the ultraviolet (Malin, 1974; Head and McCord, 1978). The spectral character of these "red spots" is unique (T.B. McCord, pers. comm., 1980), and possible mineralogical interpretations are being investigated. The morphology of some of these features (see Chapter 5) strongly suggests viscous flows of volcanic origin.

Mantling material

Mantling material comprises a special category of lunar volcanic deposits. These deposits occur on both mare and highland surfaces, usually near the edge of a mare (see Plate 2.8). They are typically, but not always, very dark. Prior to Apollo 17, many mantled areas had been mapped (Wilhelms and McCauley, 1971) and studied with remote-sensing techniques (Pieters *et al.*, 1973), but only after the return of the soils from Shorty Crater at Apollo 17 did the nature of this unusual material become apparent.

The spectral characteristics of the darkest mantled areas (lunar "Black Spots") were shown to be dominated by material like the Apollo 17 black glass beads (Pieters *et al.*, 1974; Adams *et al.*, 1974). Some of these areas along the Apollo 15 ground track have noticeably high Mg/Al or Mg/Si ratios (Plate 2.5; Schonfeld and Bielefeld, 1978; Hubbard *et al.*, 1979). The black glass spheres, the related orange spheres, the homogeneous green spheres of Apollo 15, and now also a wealth of similar glass beads found as minor components of other lunar soils, all have characteristics that imply a fire-fountaining mode of emplacement (Heiken *et al.*, 1974). The orange, black and green glass spheres have unique spectral characteristics (Adams *et al.*, 1974). Five areas of black glass deposits have been identified telescopically using the spectral signature of samples. Unpublished spectral reflectance data for other mantled areas (e.g., Aristarchus, Dopellmayer, Alphonsus, Sulpicius Gallus) show them to be distinct from each other and from surrounding maria or terra. Recent remote-sensing studies of the Apollo 15 region (Hawke *et al.*, 1979) identified other possible mantled areas. Orange glass may exist at Sulpicius Gallus (Lucchitta and Schmitt, 1974) and perhaps at Aristarchus (Zisk *et al.*, 1977). As mentioned in Chapter 1, these glass beads may represent some of the most primitive lunar volcanic material sampled.

Relative ages of mare basalts

Crater counts are the only means available for estimating the age of an unsampled lunar surface. The scarcity of large lunar craters restricts the technique to areas of the Moon similar to, or larger than, whole mare basins. Because such large lunar provinces exhibit great heterogeneity of basalt units (Plate 2.8), meaningful studies of age with relation to compositional variations cannot be done easily. Some improvement has been obtained by restricting the examination of large-crater densities to a few units of appropriately large extent (defined in Plate 2.8), rather than the less meaningful topographical boundaries of mare basins (see Chapter 8).

Another useful approach has been to compare basalt units with ages inferred by the D_L crater degradation method (Soderblom and Lebofsky, 1972; Soderblom *et al.*, 1977; see Chapter 8). Although subject to ambiguities, this technique has the advantage of permitting measurement of relative ages for much smaller units. Models, describing the nature of small-scale lunar cratering and degradation processes, have been proposed that question the validity of the D_L technique. Comparing the D_L technique with other crater-density techniques indicates, nevertheless, that the ages inferred from D_L probably correlate well with true ages. The technique certainly can be used to define relatively young, medium, and old units, although finer distinctions may not be supportable. The degree of confidence with which absolute ages can be determined for units (e.g., many western maria) with much younger ages than the range of ages for which the technique has been calibrated is uncertain. A map of the distribution of D_L values and the inferred ages across the frontside lunar maria (Boyce, 1975; Boyce, 1976; Boyce and Johnson, 1978) is presented in Plate 2.6 and Plate 2.7. Correlation of these ages with each of the basalt units defined above is presented in Table 2.2.4 and utilized in the discussions of unsampled basalt types.

Depth of mare fill

The depth of basaltic material filling the basins and lowlands of the Moon is discussed in detail in Chapter 5. The technique used is largely photo-interpretation constrained by geophysical parameters. Basalt stratigraphy is inferred primarily from superposition relations supplemented by a small amount of Apollo 17 orbital radio-sounder data for a few maria.

The remote-sensing information discussed above defines and characterizes *surface* units, but generally does not reveal stratigraphic relationships. Because the lunar surface has been continuously altered by impacts of various sizes, a well-developed regolith (5–10 m thick) exists over most of the lunar surface with the uppermost part consisting of a fine-grained altered soil. Craters that excavate below this homogenized soil zone can be used as probes to examine the deposits of large craters and to determine the nature of excavated subsurface units. As discussed previously, when material excavated by a mare crater has the properties of terra material, it can be inferred that the crater has penetrated through the mare. Limits can then be set on the depth of mare fill from crater depth-diameter parameters.

The use of remote sensing for direct stratigraphic information requires the following: (1) spatial resolution sufficient to distinguish between a crater, its ejecta, and surrounding material; (2) available craters; and (3) precision and sensitivity great enough to enable the technique to distinguish compositional units. Current data have shown that the idea of using craters as stratigraphic probes is promising. One of the first applications concerned the crater Copernicus (McCord *et al.*, 1972). The spectral properties of Copernican ejecta are clearly those of mature terra material, implying not only that the mare is relatively thin in that area, but also that the soil formation process has reached equilibrium maturity since the event. More recently, analysis of data for two craters in Mare Crisium, Picard and Pierce, has clearly shown an excavated subsurface mare unit that is both magnesium-rich (from X-ray data; Andre *et al.*, 1978a) and relatively titanium-rich (from reflectance data; Head *et al.*, 1978a). Similarly, in Mare Fecunditatis the crater Messier has apparently excavated a Mg-rich basalt (Andre *et al.*, 1979); in Mare Imbrium the crater Timocharis is associated with Th-rich material (Metzger *et al.*, 1979); and in Mare Smythii a post-mare crater (~0.5 km deep) in the center of the largest basalt area has apparently penetrated the basalt to expose crustal material (from X-ray data; Andre and Adler, 1978). A few small craters (7–20 km diameter) on the edges (shelf topography) of Mare Humorum and Mare Serenitatis have also excavated crustal material (spectral reflectance data; Pieters, 1977b), implying a thin mare fill for those specific areas.

Synthesis studies

An important advance has been made in formatting lunar remote-sensing data to a common scale. This greatly facilitates correlations among data sets and aids geologic mapping. A computerized multiparameter approach has been developed which is discussed by Soderblom *et al.* (1977) and by Eliason and Soderblom (1977). Examples of these data were discussed earlier in this section.

For completeness we refer here to papers that use several types of remote-sensing data to study the distribution and geologic setting of lunar basalts.

Major regional studies emphasizing remote-sensing data interpreted in a geological context include (roughly from east to west): *Mare Smythii* (El-Baz and Wilhelms, 1975; Andre *et al.*, 1977c; Haines *et al.*, 1978); *Mare Crisium* (El-Baz and Wilhelms, 1975; Adams *et al.*, 1978; Head *et al.*, 1978b; Maxwell and Phillips, 1978); *Langrenus/Dorsa Geike/Mare Fecunditatis* (Andre *et al.*, 1977a; Hubbard and Andre, 1977); *Palus Somni* (Clark *et al.*, 1976); *Taurus Littrow and other dark mantling deposits* (Pieters *et al.*, 1973; Head, 1974; Adams *et al.*, 1974; Andre *et al.*, 1975; Schonfeld and Bielefeld, 1978); *Mare Serenitatis* (Thompson *et al.*, 1974; Johnson *et al.*, 1977); *Mare Imbrium* (Schaber, 1973; Head *et al.*, 1978a; Whitford-Stark, 1980); *Mare Humorum* (Johnson *et al.*, 1973; Pieters *et al.*, 1975; Greeley and Spudis, 1978); *Aristarchus/Harbinger* (Zisk *et al.*, 1977; Haines *et al.*, 1978); *Flamsteed region* of Oceanus Procellarum (Pieters *et al.*, 1980) and *Oceanus Procellarum* in general (Whitford-Stark and Head, 1977a, b, 1980).

Other studies have addressed specifically the distribution of mare basalt units in space and time (see review by Head, 1976). A more thorough understanding of the temporal sequence of basalt emplacement and evolution of current surface features in the context of tectonic and cratering events has been developed for some provinces (Pieters *et al.*, 1975; Zisk *et al.*, 1977; Head *et al.*, 1978b; Andre *et al.*, 1978a, b; Pieters *et al.*, 1980; and Whitford-Stark and Head, 1980). The compositions of these basalt units express the evolution of deep-seated magmatic processes. In addition to temporal changes in basalt composition (chemistry and mineralogy), there is appreciable evidence of regional compositional differences that may reflect original accretional lateral heterogeneities (Soderblom *et al.*, 1977; Metzger *et al.*, 1977b; Pieters, 1977a).

General summary

Remote sensing of lunar surface units has been remarkably successful within the constraints of the available data. A few general highlights concerning lunar basalts include:

1. All the dark maria of the Moon, which account for about 17% of the lunar surface, are basaltic in composition and volcanic in nature.
2. A wide variety of basalt types exists on the lunar surface, two-thirds of which have not been sampled. Some of this unsampled material is compositionally very different from anything studied to date in the laboratory.
3. The most voluminous basalt types are apparently low-Ti basalts of intermediate age. These are spatially and temporally intermingled with both high-Ti basalts and very low-Ti basalts.
4. Although extensive areas of a single basalt type exist on the present lunar surface, all mare basins studied show that the evolution of basaltic volcanism has produced a varied stratigraphy of mare basalt types.
5. From the current surface expressions of nearside basalts, it appears that at least three sections of the Moon have followed different evolutionary trends, with corresponding distinctions in basalt composition and age: (a) the eastern maria, (b) the southwestern maria, and (c) the northwestern maria.

2.2.2 Earth

Introduction

Because basalts and other rock types on Earth can be studied directly in the field and in the laboratory, the incentive for making remote compositional measurements of the Earth has been less than for the other planets. In addition, there are various difficulties involved in interpreting remote measurements of vegetated and weathered surfaces. These considerations have limited the application of remote sensing in the study of terrestrial basalts. Photogeologic methods

(Chapter 5) employing black and white, color and low sun angle images, taken at a variety of scales and altitudes, have been used extensively in studies of basaltic volcanism. Additional morphological information has been obtained by recent applications of side-looking airborne radar (e.g., Malin *et al.*, 1978; Evans, 1978). Most remote-sensing methods that yield direct compositional information about basalts, however, are still in the developmental stages.

Gamma-ray spectrometry

Airborne gamma-ray sensing has been used widely in prospecting for uranium and thorium. The method involves direct measurement of gamma-radiation from the naturally occurring isotopes ^{40}K, ^{238}U, ^{232}Th, as described in section 2.3.1. Gamma rays from natural sources are absorbed by about 30 cm of overlying soil; therefore, the technique is most effective over outcrops or over areas with a thin surficial cover. In addition, gamma rays are significantly attenuated by the Earth's atmosphere. Measurements typically are made from altitudes of 150 m or less. As an exploration technique, gamma-ray spectrometry was developed to locate rocks having high concentrations of ^{238}U and ^{232}Th; thus, less attention has been paid to rocks, such as basalts, that have low concentrations of these radioactive elements. Airborne gamma-ray surveys have been made for much of the United States and Canada, and for selected areas elsewhere, to search for nuclear fuels.

One known application of airborne gamma-ray spectrometry to basalts was discussed by Schmidt (1961) and Pitkin (1968), who were able to differentiate lava flows in the Columbia River Plateau where the loess cover was thin or absent. Variations in the radioactivity level of the Snake River Group (basalts) ranged from 300–500 counts per second (cps) for basalt southwest of Idaho Falls to 700–1100 cps over a serrate flow in Craters of the Moon National Monument, Idaho (Schmidt, 1961). Other basalts in the province had intermediate levels of radioactivity. Basalts formed from the magma in its last stages of differentiation had relatively high concentrations of ^{40}K. These were correlated with the higher gamma-ray counts, showing that the airborne measurements were able to distinguish compositional units and relative ages.

Spectral imaging

The LANDSAT multispectral scanners have been used for a variety of remote-sensing purposes including geologic studies. LANDSATs are sun-synchronous satellites in near-polar orbits at altitudes of ~920 km. They orbit fourteen times each day and repeat coverage of a given area every eighteen days at the same local time. The multispectral scanners image in each of four broad bands at 0.5–0.6 μm, 0.6–0.7 μm, 0.7–0.8 μm and 0.8–1.1 μm. LANDSAT III has an additional band from 10–12 μm.

Compositional information on rocks is not easily obtained from LANDSAT images or from other similar images. This is due to the broad and limited bandpasses used to obtain the images and to the problems of surface alteration, surface vegetation, atmospheric absorption and atmospheric scattering. In spite of these problems, discrimination of units and limited compositional information has been obtained. For example, Rowan *et al.* (1977) successfully mapped altered rocks of the Goldfield (Nevada) Mining District using LANDSAT data. Abrams *et al.* (1977) used aircraft multispectral data with a spectral range from (0.46–2.36 μm) to achieve improved identification of altered rocks in the Cuprite Mining District, Arizona.

As applied to basalts, LANDSAT and aircraft multispectral scanning systems have been useful primarily for studying the morphology of flows and for determining relative ages, based on the degree of surface weathering and the amount of vegetation. LeFebvre (1975) and LeFebvre and Abrams (1977) used LANDSAT images, aerial photographs and field observations of the Craters of the Moon area in Idaho to show that older flows have higher radiance values and smaller radiance differences between the blue (0.5–0.6 μm) and the red (0.8–1.1 μm) bands. Faye and Miller (1973) showed that fresh (~2000 years before present) pahoehoe flows at Craters of the Moon are dark and blue; weathering and masking by sediment and vegetation change the reflectance properties with time. A similar effect has been observed by several authors in LANDSAT images of basalt flows in the San Francisco Peaks, Arizona, area and for flows at Mauna Loa and Kilauea in Hawaii.

Infrared measurements

Emission spectra of basalts in the laboratory (Lyon, 1965; Vincent *et al.*, 1975) show Reststrahlen bands in the 8–12 μm region. Vincent and Thomson (1971) developed a method for displaying in image format the spatial distribution of materials having different band positions. Experiments at Pisgah Crater, California, by Vincent and Thomson (1972) and Lyon (1972) using airborne scanners showed that it was possible to distinguish basalt from other more siliceous rocks.

Infrared radiometric measurements were also made of Pisgah Crater from an aircraft. Kahle (1977) showed that thermal inertia images allowed discrimination between black basalt flows and black cinders, where the cinders had a lower thermal inertia. The investigators also were able to detect basalt under a thin layer of loose sand using the thermal inertia images. These techniques are still being developed, and so far have had limited application to studies of basalts.

2.2.3 Mercury

Mariner 10 images show smooth dark plains on Mercury that are reminiscent of some of the lunar maria. Strom *et al.* (1975), Trask and Guest (1975) and Murray *et al.* (1975) have presented morphological evidence that most of the smooth plains areas are volcanic in origin; some smooth plains may be the products of impact processes. Smooth plains on Mercury cover about 15% of the area photographed by Mariner 10. As primary volcanic landforms appear to be lacking in the Mariner 10 images (Strom *et al.*, 1975), the existence of volcanic rocks on Mercury depends on the origin of the smooth plains. The evidence for primary volcanic landforms on Mercury is discussed by Malin (1978). Further details on morphology are given in section 5.4.3.

The normal albedos of the smooth plains range from 0.09–0.21 and have an average value of about 0.14 (Hapke *et al.*, 1975b). The darkest smooth plains on Mercury thus have albedos that are higher than the average lunar maria, but which are comparable to the albedos of some of the brighter lunar mare units. The albedos and morphology are consistent with the presence of low-titanium basalts at the surface if the analogy with the Moon holds. Implicit in this comparison is the assumption that the soil-forming processes on the two planets are the same; therefore, mature soils on similar basalts darken to the same albedo in both cases.

There is evidence that the optical properties of the surface of Mercury are closely similar to those of the Moon. The presence of bright-rayed craters on both bodies indicates that the materials that were excavated from below the surface by impact events have higher albedos than the undisturbed surface. On both bodies rays are absent around craters that retain most of their original surface morphological features but are somewhat degraded compared to the fresh ray craters. The disappearance of rays implies that there is a process (or processes) that darkens the surface of both bodies with time. Hapke *et al.* (1975b), furthermore, have shown by means of Mariner 10 relative color images that the mercurian fresh bright craters and their ray systems are about 12% "bluer" than the background. Fresh craters and rays on the Moon are also bluer than are mature soils of the same composition. Although there has been controversy in the literature over the exact causes of the darkening of lunar soil, there is a convergence of evidence that the impact of small particles leads to the production of heterogeneous aggregates of glass and soil (agglutinates) where the glass is typically filled with inclusions of metallic iron and gas bubbles. These agglutinates and glass-coated particles lower the albedo and steepen the spectral curve, causing mature soil to appear redder than the fresh parent material.

Additional and independent evidence that a lunar-like soil exists on Mercury comes from telescopic spectral reflectance data (McCord and Adams, 1972a, b; Vilas and McCord, 1976; Hapke, 1977; McCord and Clark, 1979). Spectra of the integral illuminated portion of the planet are remarkably similar to spectra of the Moon and to laboratory spectra of mature lunar soils. No other solar system objects are known to have comparable optical properties. The telescopic measurements of Mercury include all terrain types, and do not resolve the smooth plains on Mercury. Adams and McCord (1977) concluded that the very weak absorption band near 1 μm in the spectra of Mercury implied a very low (<6%) FeO content, and that by analogy with the Moon, the observed surface was largely feldspathic or anorthositic. McCord and Clark (1979) have presented additional telescopic data, and have analyzed the available spectra using computerized band-fitting techniques. They reaffirm the presence of a broad shallow absorption band, and show that it is centered at 0.9 μm.

The significance of this band position is that it indicates Fe^{2+} in orthopyroxene (Adams, 1974). Possible ambiguity in assigning the 0.90 μm band to Fe^{3+} in a hydrated oxide such as goethite/limonite is easily eliminated. This is due to the absence of a steep Fe^{3+} charge-transfer absorption edge in the visible part of the spectrum (which would make Mercury appear red). A band at 0.90 μm from orthopyroxene on Mercury invites comparison with very similar spectra of lunar highland soils. Band fitting to the laboratory spectra of lunar highland samples (Farr *et al.*, 1980) shows a band centered at 0.90 μm. Agglutinitic glass in the lunar soil has the effect of straightening and steepening the continuum of the curve. It does not contribute a resolvable Fe^{2+} glass band near 1.0 μm or otherwise shift the position of the 0.90 μm pyroxene band. Farr *et al.* (1980) conclude that the details of the curve shape and of the 0.9 μm band structure in the telescopic data of Mercury (McCord and Clark, 1979) are reproduced in the spectra of lunar terra soil. This confirms earlier interpretations (Adams and McCord, 1977; Hapke, 1977) that the telescopic integral spectra of Mercury represent a mature impact soil developed on anorthositic gabbroic rocks. The planetary scale of this material implies an anorthositic gabbroic crust on both bodies.

The absence of telescopic spectral evidence for basalt on Mercury may be a consequence of the low spatial resolution. For example, the spectra of the full lunar disc is dominated by the more highly reflective terrae (McCord, 1968) even though basalts cover much of the near hemisphere. Even if all of the smooth plains on Mercury are basalts they amount to only 15% of the area photographed by Mariner 10, and their contribution to the spectral curve could be obscured. If basalts are present on Mercury, the evidence comes entirely from the albedo of the darker smooth plains observed by Mariner 10. The remarkable similarity of optical properties between the Moon and Mercury discussed above encourages comparisons of albedos on the two bodies as functions of composition. Thus, as Hapke *et al.* (1975b) have suggested, if basalts occur on Mercury they probably are of the low-Ti variety, perhaps similar to those sampled at Apollo 12 and 15. High-Ti basalts (low albedo) are unlikely to occur on the hemisphere observed by Mariner 10.

2.2.4 Mars

Introduction

Mars has been studied by a variety of flyby, orbiting and landing spacecraft and by Earth-based telescopes. Evidence for volcanism on Mars comes from Mariner 9 pictures and Viking Orbiter images. These show physiographic features such as volcanic mountains, flow fronts and lava plains (See section 2.3.4 and Chapter 5). The Viking Lander images, in which rocks appear vesicular, also suggest volcanism. Compositional information for the martian surface is derived from (1) Earth-based spectral reflectance measurements of the martian surface and of dust clouds; (2) Viking XRF analyses of martian soils; (3) broad-band spectrophotometric measurements of soils and boulders at the Viking landing sites; (4) heights of the lava flow fronts (Moore *et al.*, 1978); (5) Mariner 9 infrared observations of the dust clouds (Toon *et al.*, 1977); and (6) theoretical models of the chemistry and mineralogy of the mantle of Mars, based on density (see Chapter 4).

Earth-based reflectance spectroscopy

Earth-based telescopic measurements yielded the earliest information on the mineralogical composition of the martian surface. The red color of Mars has suggested to many that iron oxide is a constituent of the martian soil (e.g., Binder and Cruikshank, 1966; Wildt, 1934; Pollack and Sagan, 1969).

An early review of the martian reflectance spectra (McCord and Adams, 1969) included a 0.3–1.1 μm composite spectrum of the bright area Arabia and the dark area Syrtis Major. Adams and McCord (1969) modelled these spectra in the laboratory using various methods for oxidizing basalt samples. They deduced, because of the agreement between optical properties of their laboratory samples and the martian surface, that both regions were composed of combinations of ferric oxide and mafic silicate rock, probably basalt. Arabia was interpreted as having soils which were fine-grained, and more strongly oxidized than those of Syrtis Major.

Measurements of reflectance spectra of martian regions (200–1000 km) have been made almost regularly at each opposition since 1967 (McCord and Westphal, 1971; McCord *et al.*, 1977, 1978; T. B. McCord, pers. comm., 1978; Binder and Jones, 1972). Spectra show absorptions due to Fe^{3+} and Fe^{2+} in minerals, and there are also absorption bands attributed to water.

The spectra of bright areas and of dust clouds are quite similar and show no regional variations. This contrasts with the spectra of the dark areas which show substantial regional variation. The dark-area materials are also considerably different in optical properties (and mineralogy) from the bright material. As discussed by Adams and McCord (1969), the bright- and dark-area differences probably reflect differences in the extents to which these two units have been chemically weathered. The spectral variation among dark-area units indicates variety in the mafic mineralogy (McCord *et al.*, 1978; Singer *et al.*, 1979).

The spectra of mixed- and intermediate-albedo areas have the same absorption bands seen in both the bright- and the dark-area spectra. As those areas do not display any unique features, they probably consist of mixtures of high albedo dust and dark-area materials.

A more detailed analysis of the 1973 spectra (Huguenin *et al.*, 1977b) resulted in identification of some of the absorption features. The ferric oxide contents of the bright areas and dust clouds ranged from about 6% in the 1969 spectra to a few percent higher in the 1973 spectra; the airborne dust appeared to be more hydrated than the surface soil in 1973. The 1969 bright-area spectra indicated a multicomponent soil of a fine-grained dust that contained ferric oxide and possibly a more coarse-grained pyroxene-bearing soil. The 1973 bright-area spectra indicated no ferrosilicates, but these results are probably due to masking by dust. The dark-area spectra were reported to show a variety of ferrosilicate bands near 1.0 μm, suggesting the presence of several mafic minerals (pyroxenes, olivine). This is consistent with a basaltic composition.

Moderrazzo and Huguenin (1977) reported the composition of the bright material as 6–12% ferric oxide, 3–5% magnetite and some unidentified Fe^{2+}-poor material, possibly clay and minor salts. They pointed out that this is a composition consistent with weathered material from a mafic rock. Using the telescopic data as a basis, Huguenin (1974, 1976a, b) and Huguenin *et al.* (1977a) studied photo-oxidation as a method of producing the bright weathered material.

Recent whole-disk infrared spectra (0.6–2.5 μm) better defined the 1 μm region Fe^{2+} absorption and revealed several H_2O bands (McCord *et al.*, 1978). The H_2O bands were interpreted as indicating the presence of H_2O ice combined with a desiccated mineral-hydrate. Several earlier Earth-based observations discovered (Sinton, 1967) and better defined (Beer *et al.*, 1971; Houck *et al.*, 1973) a water of hydration band near 3 μm for Mars soil. The most recent observations have produced spectra of a number of martian areas from 0.6–2.5 μm (Singer *et al.*, 1979). Composite dark-area and bright-area spectra are shown in Fig. 2.2.10 from the spectral region 0.3–2.5 μm, the region where the most detailed information exists.

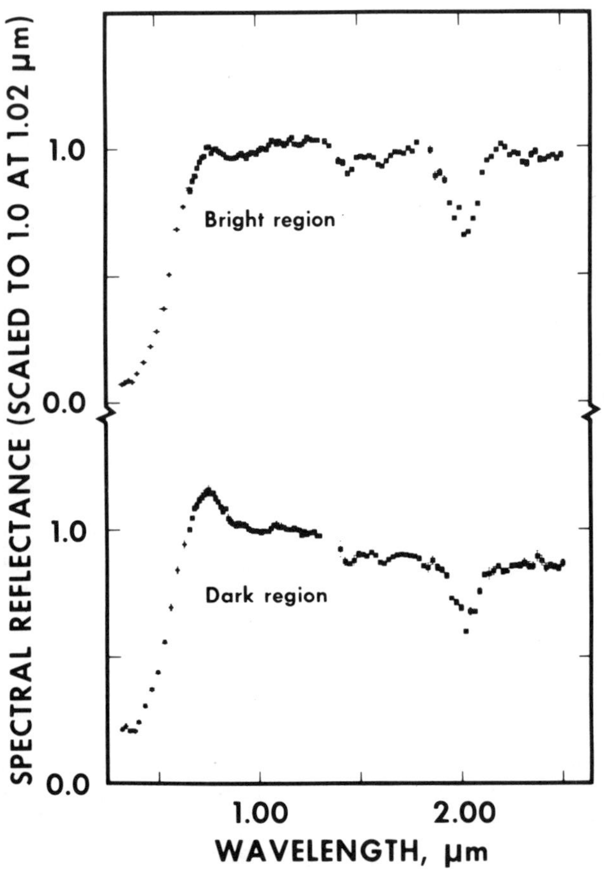

2.2.10 Representative bright and dark region reflectance spectra, scaled to unity at 1.02 μm. The bright region spectrum (top) is composed of an average of the brightest areas observed in 1973 (visible) and 1978 (infrared). The dark region spectrum (bottom) is a composite of data from two nearby locations in Iapygia: 69-6 (visible) and 78-10 (infrared). From Singer *et al.* (1979, p. 8419).

Thus, the evidence for basalts on Mars from Earth-based spectral reflectance measurements consists of the presence of mafic minerals as constituents of the dark area, and the oxidized bright material that is consistent with a weathering product of basalts.

Orbiter infrared spectroscopy

During the Mariner 9 mission, thermal emission spectra of Mars were measured by the infrared interferometer-spectrometer (IRIS). The IRIS system recorded spectra between 5 and 50 μm with a minimum spectral resolution of 2.4 cm^{-1} and a spatial resolution of about 125 km at periapsis (Hanel *et al.*, 1972). The spectra recorded during the first 100 orbits show strong absorptions which the experiment team reported as being characteristic of an SiO_2 content of 60 ± 10 wt.%. Hunt *et al.* (1973) and Logan *et al.* (1975) compared the Mariner 9 infrared spectrum with laboratory transmission spectra and suggested that the clay mineral, montmorillonite, could be the major component of the martian dust clouds. This suggestion was based on the position, shape and strength of the 9 μm Reststrahlen band in the Mars spectrum. Montmorillonite is a common weathering product of mafic minerals and its presence is consistent with basalts on Mars. They further reported that hematite and carbonates were excluded as major components of the martian dust cloud but that other salts, such as sulfates, could be present in large amounts. These interpretations agree with the subsequent Viking XRF results.

Aronson and Emslie (1975) also attempted to interpret the IRIS spectrum using modelling computations. They reported that the composition of the martian dust is biased toward silicate structures that involve a high degree of silicate-oxygen sharing. A feldspar-rich dust of about 1 μm particle size appeared to be the best candidate, although sheet silicates could not be ruled out.

Viking Lander reflectance spectroscopy

The Viking Lander cameras produced six-channel spectra for martian soils and rocks. The soil spectra were found to be in good agreement with bright-area spectra reported from Earth-based telescopic measurements and the compositional implications are the same (Huck *et al.*, 1977).

Arvidson and Bragg (1977) noted that a group of rocks approximately 20 m from the Viking 1 Lander are distinctly darker than most other rocks in the field of view. The albedos and the textures of the dark rocks suggest that they may be relatively unweathered mafic volcanics.

Viking Orbiter multispectral mapping

Images obtained by the Viking Orbiters have provided additional information about the regional extent of martian compositional units. The Viking Orbiter cameras obtained photographs at three different wavelengths: 0.45 ± 0.3 μm, 0.53 ± 0.05 μm and 0.59 ± 0.05 μm. Although the cameras have lower spectral coverage and resolution than do ground-based telescopes, their high spatial resolution allows examination of the local geologic setting for different spectral types (Soderblom *et al.*, 1978). With orbital spatial resolution, it has been possible to recognize a greater diversity of color units than have been observed from Earth.

Ratio images of 0.45/0.53 μm and 0.59/0.53 μm probably display compositional variations in surface materials (Soderblom *et al.*, 1978; McCord *et al.*, 1980). Several different color units appear in the dark areas, suggesting distinct compositional units, and supporting the earlier ground-based spectroscopic results (McCord *et al.*, 1977). According to Soderblom *et al.* (1978), the units within the dark regions range from the most red to the least red on Mars. The color within each of the units appears uniform and the boundaries are associated with changes in terrain type and relative age, which could indicate that the color variations are directly related to variations in bedrock lithology. In color maps of the dark equatorial zone between about 0° and 30° south, two general provinces have been identified:

1. An older unit (2.5–3.5 b.y.), which is the reddest on Mars. This unit is characterized by topographic highs; ridges, rugged plateaus, and rims of ancient large craters that are dissected by small dendritic channels.
2. An intermediate aged (2.0–2.3 b.y.), smooth intercrater plain, which is among the bluest lithologic units on Mars. Within this unit are numerous volcanic ridges similar to those found in the lunar maria.

X-ray fluorescence

Indirect evidence for the presence of an iron-rich basaltic bedrock comes from the Viking Lander X-ray fluorescence data (Toulmin *et al.*, 1977; Clark *et al.*, 1977). Each Viking Lander carried an energy-dispersive X-ray fluorescence spectrometer (XRFS) for elemental analysis of samples collected by the Viking surface sampler. Elemental analyses of fines in the martian regolith at Chryse Planitia and Utopia Planitia are very similar. This is interpreted to mean that chemically and mineralogically homogenous fines cover large areas of the martian surface, an interpretation which is in agreement with reflectance spectroscopy and multispectral mapping results.

According to Toulmin *et al.* (1977), silicon, iron, magnesium, sulfur, aluminum and calcium are major elements in the soil, with titanium and chlorine present in minor quantities (Table 2.2.5). Low amounts of the trace elements, the alkalies and aluminum suggest that the great preponderance of the material is of mafic to ultramafic derivation; highly differentiated sialic igneous rocks or their weathering products were found to be insignificant components of the samples. Toulmin *et al.* (1977) interpreted the elemental analyses to indicate iron-rich smectites (clays), carbonates, iron oxides and sulfates. They measured a SiO_2 content of $45 \pm 5\%$ by weight for the surface fines, which contrasts with the $60 \pm 10\%$ by weight SiO_2 derived from the Mariner 9 infrared spectrum for the dust particles (Hunt *et al.*, 1973). This disagreement in SiO_2 content remains unexplained.

Magnetic properties investigation

Each Viking Lander carried three arrays of magnets that were used to estimate the composition and abundance of magnetic minerals in the martian surface soil. Two magnet arrays (one weak and one strong) attached to the surface-sampler backhoe came in contact with the surface material during sample acquisition. The third magnet array was mounted in one of the

Table 2.2.5 Compositions of martian samples.†

	S1	S2	S3	U1	Estimated Absolute Error
SiO_2, wt.%	44.7	44.5	43.9	42.8	5.3
Al_2O_3, wt.%	5.7	n.y.a.	5.5	n.y.a.	1.7
Fe_2O_3, wt.%	18.2	18.0	18.7	20.3	2.9
MgO, wt.%	8.3	n.y.a.	8.6	n.y.a.	4.1
CaO, wt.%	5.6	5.3	5.6	5.0	1.1
K_2O, wt.%	<0.3	<0.3	<0.3	<0.3	-
TiO_2, wt.%	0.9	0.9	0.9	1.0	0.3
SO_3, wt.%	7.7	9.5	9.5	6.5	1.2
Cl, wt.%	0.7	0.8	0.9	0.6	0.3
Sum	91.8	n.y.a.	93.6	n.y.a.	-
Rb, ppm	≤30			≤30	
Sr, ppm	60 ± 30			100 ± 40	
Y, ppm	70 ± 30			50 ± 30	
Zr, ppm	≤30			30 ± 20	

†From Toulmin *et al.* (1977), p. 4629.
The abbreviation n.y.a means not yet available.

reference test charts atop each lander and attracted magnetic particles already in or raised into the martian atmosphere.

The results from the Viking 1 and Viking 2 backhoe magnet arrays were similar (Hargraves *et al.*, 1977). Both the weak and the strong magnets were covered with low albedo particles, most of sub-millimeter diameter. The large number of particles present on the weak magnet suggests that the martian surface material contains a relatively pure, highly magnetic mineral phase. Potential candidates for this phase are (1) metallic iron, Fe or NiFe; (2) magnetite, Fe_3O_4 or titaniferous magnetite; (3) pyrrhotite, approximately $Fe_{0.9}S$; and (4) maghemite, Fe_2O_3.

Summary

Although there is only indirect remote-sensing evidence for basalts on Mars, spectral reflectance measurements suggest the presence of pyroxene and olivine in the dark areas, and show that the bright areas contain Fe^{3+} oxides/hydroxides. Orbital infrared (IR) spectroscopy suggests that the bright material contains a clay similar to montmorillonite. Spectral reflectance and multispectral maps show that the bright material has uniform optical properties, and is therefore probably of uniform composition, but that the dark areas are heterogeneous. Viking 1 and 2 Lander X-ray fluorescence (XRF) data indicate that the bright material has essentially the same chemistry at both sites, and that the calculated modal minerals are consistent with a derivation by weathering of mafic igneous rocks.

2.2.5 Venus

Our perception of global-scale surface features on Venus is based on radar images obtained by terrestrial telescopes and on radar images from the orbiting Pioneer Venus spacecraft. These data and the evidence for possible volcanic landforms are discussed in section 5.6. Although radar imaging is an important remote-sensing technique, it is not normally possible to derive direct compositional information from radar returns. Radar images of the surface of Venus, therefore, do not reveal the presence or absence of basalts *per se*, except indirectly by expressions of surface morphology. The opaque atmosphere of Venus prevents our use of other remote-sensing techniques (e.g., γ-ray, X-ray, visible and infrared spectroscopy) that are sensitive to the chemistry and mineralogy of the surface materials. Thus, there are no telescopic, flyby or orbital remote-sensing measurements of the composition of the surface of Venus. Instead, the available direct compositional data comes from three very localized areas on the surface.

Three Soviet spacecraft have landed on the surface of Venus. Two of them, Venera 9 and 10, transmitted images of the venusian surface (Florensky *et al.*, 1977; Ronca *et al.*, 1978). The Venera 9 and 10 landing sites have rocks (possibly bedrock at the Venera 10 site) within view of the cameras; however, identification of the rock types is not possible from the images alone. It is not yet known what process or processes created the angular pieces of rock and rock slabs observed; a detailed discussion is given by Ronca *et al.* (1978).

All three Venera spacecraft carried γ-ray instruments that gave compositional information about the local surface at the landing sites. Venera 10 had a second γ-ray instrument on board that measured the bulk density of the surface materials. (For a discussion of these experiments, see section 2.3.1.) Most of what little we know about the composition of the venusian surface comes from these γ-ray measurements. Some additional information is derived from spectral photometric measurements of the surfaces at the Venera 9 and 10 sites.

The Venera 8 (1972) γ-ray experiment made the first direct compositional measurements of Venus (Vinogradov *et al.*, 1973). The data showed high abundances of K, U and Th (Table 2.2.6). The K value is comparable to granitic rock on Earth, and the U and Th values are similar to terrestrial alkali basalts. Vinogradov *et al.* (1973) concluded that such high radioactivity values could not be characteristic of the planet as a whole, and that they imply a differentiated crust. Measurements from the Venera 9 and 10 landings indicated sites with much lower γ-ray activity (Table 2.2.6), with values for K and U that, according to Surkov (1977b), are similar to those of terrestrial oceanic basalts.

As can be seen from Table 2.2.6, the concentrations of Th, U and K vary considerably among the sites. Uncertainties in these values are ascribed to the statistics of the net counts in the corresponding energy ranges. Also shown in Table 2.2.6 are K/U and Th/U

Table 2.2.6 Gamma-ray measurements by Venera spacecraft.

	Venera 8	Venera 9	Venera 10
K, %	4.0 ± 1.2	0.47 ± 0.08	0.30 ± 0.16
U, ppm	2.2 ± 0.7	0.60 ± 0.16	0.46 ± 0.26
Th, ppm	6.5 ± 0.2	3.65 ± 0.42	0.70 ± 0.34
K/U	20,000	8,000	7,000
Th/U	3.0	6.1	1.5

ratios derived from the data. The K/U ratio for terrestrial basalts averages 10^4 and that for lunar mare basalts is about 3×10^3. These values almost bracket the ratios derived from the Venera data which range from 7×10^3 to 2×10^4. The Venera Th/U ratios range from 1.5–6.1. By comparison, terrestrial and lunar values are close to 3.8. The range in Th/U between Venera 8, 9 and 10 indicates either a diversity of differentiation processes on Venus or uncertainties for the values in Table 2.2.6. (See section 2.3 for discussion of differences in the Venera instrumentation.)

The Venera 10 γ-ray density measurement gave a value of 2.88 ± 0.1 g/cm^3 (Surkov *et al.*, 1976). Massive terrestrial basalts have densities in the range 2.7–2.9 g/cm^3. Thus, evidence from the Venera 9 and 10 K and U values, and from the density measurements is at least consistent with the presence of basaltic rocks on Venus.

Spectral photometric measurements were made at the Venera 9 and 10 sites, and have been interpreted by Ekonomov *et al.* (1980) as evidence for a basaltic surface. Photometers measured light fluxes in five spectral ranges (from about 0.5–1.0 μm) at a variety of angles. Surface albedos as a function of wavelength were calculated, and the reconstructed spectral curves were compared with spectra of terrestrial samples. The curves for the Venera 9 and 10 sites are similar to one another in shape, with Venera 10 having a slightly higher overall reflectance. Both photometers measured spots on the ground about 10×30 meters in size.

The Venera 9 and 10 photometers each monitored temporary changes in light levels after touchdown. Ekonomov *et al.* (1980) presented evidence that surface dust was entrained in the atmosphere during the landings. The spectral photometric measurements of the surface, therefore, were interpreted in terms of dust between, and perhaps partially covering, the rocks that are observed in the images.

The low albedos (<0.05) measured at the Venera sites may be the result of the dark soil that is seen between the higher albedo rocks in the images. There is a fair agreement between the shapes of the Venera spectral photometry curves and curves for Apollo 17 black spheres (Adams, pers. comm.; Adams *et al.*, 1974) that appear to be mafic pyroclastic material. Further analyses of the spectral data are needed; however, the spectral photometric data are at least consistent with the possibility of basaltic material at the Venera 9 and 10 landing sites.

As is discussed in section 5.6, the Venera 9 and 10 landing sites are located at the border of a broad, low, conical radar feature, Beta, that has been interpreted by Malin and Saunders (1977) as a possible shield volcano. They also note that the radar reflectivity of Beta is reminiscent of terrestrial flows. The combined radar, surface γ-ray, and spectral reflectance evidence, therefore, suggest that basalts are likely to be present on the surface of Venus, and may even be widely distributed. It is important to keep in mind, however, that the direct compositional evidence for basalts on Venus is based primarily on measurements, of unknown accuracy, of K and U concentrations at two landing sites.

2.2.6 Asteroids and satellites

Asteroids

By the conventional wisdom of two decades ago, the asteroid belt would seem a poor place for studying basalts. Asteroids are all very small with the largest being only 1000 km in diameter. They are too small to be substantially heated by decay of long-lived radionuclides. To be sure, there are basaltic meteorites in museum collections, but two decades ago the potential origin of such meteorites in the asteroid belt was regarded as no more likely than origin on the Moon. Thus, it was a notable discovery when, in 1970, McCord, Adams and Johnson reported the results of the first application to an asteroid of the newly developed technique of spectrophotometry. The asteroid, 4 Vesta, the brightest and third largest asteroid, was found to have a reflection spectrum dominated by the signature of a particular pyroxene, pigeonite. Indeed, the spectrum matched very closely a laboratory spectrum of the meteorite Nuevo Laredo, a basaltic achondrite. McCord *et al.* (1970) concluded that Vesta has a surface composition "very similar to that of certain basaltic achondrites."

In addition, McCord *et al.* made a prediction: "Determination of Vesta's spectrum from 1.1 to 2.5 μm would help to confirm our conclusions; a band should exist centered near 2.0 μm." Within a few years, H. Larson and U. Fink had developed the capability of measuring the spectra of relatively faint objects in the 1–3 μm portion of the infrared. Their first application of the technique to an asteroid was again to Vesta. Larson and Fink (1975) reported "a broad absorption centered near 2 μm is the most conspicuous spectral feature." The latest detailed study of Vesta's spectrum is that of McFadden *et al.* (1977), who confirmed the McCord *et al.* interpretation. In particular, Vesta has abundant pigeonite with a calcium content, Ca/(Ca + Mg + Fe), of about 10–20%. There is little or no olivine. Such a mineral assemblage is not expected to result from direct condensation from the solar nebula and it is believed that such a mineral assemblage can be derived only by substantial differentiation of primordial material. The fact that the spectrum of Vesta so closely matches spectra of several basaltic achondritic meteorites (e.g., Kapoeta, Nuevo Laredo) suggests that Vesta is actually covered by basaltic materials. While such an interpretation is probably correct, it cannot be ruled out that Vesta has a nonbasaltic assemblage of minerals, necessarily including pigeonite, but perhaps with significant amounts of other relatively transparent minerals (such as plagioclase feldspars) that do not have strong spectral signatures. The conclusion that Vesta is covered with basalts is secured only by augmenting the interpretation of the spectral features with cosmochemical arguments about plausible minerals that can be associated with dominant pigeonite. Such arguments have been made by several authors and have been widely accepted.

Degewig (1977) has compared Vesta's photometric lightcurve with variations in polarization, which is sensitive to albedo, and has concluded that Vesta has albedo variations, or "spots." Vesta is also known to exhibit minor color variations as it rotates. Together, these results imply that Vesta is unusual among asteroids in having a surface that is not totally homogeneous. The spatial scales of these spots, and hence the magnitude of the compositional variations across the surface of Vesta, cannot be determined from Earth-based observations (as these are necessarily hemispheric averages). Closer examination of Vesta with spacecraft may reveal that Vesta has recognizable basaltic flows of differing chemistry, such as are found in the lunar maria.

Vesta is the only asteroid that has been found to exhibit a spectrum diagnostic of basalt. Certainly there exists no other asteroid larger than 100 km diameter, and probably none larger than 50 km diameter, with an albedo and spectrum similar to Vesta's (Chapman *et al.*,1975). Thus, Vesta is a prime candidate for being the parent body of the basaltic achondrites, in particular the eucrites (Consolmagno and Drake, 1977). Consolmagno and Drake (1977) demonstrated from an analysis of rare-earth elements, that eucrites could be produced by approximately 10% equilibrium partial melting of a source region composed of 85% olivine. This might occur during a single-stage melting and crystallization event involving the whole parent body. Since no known meteorites have the composition calculated for the residual solid mantle of the parent body, Consolmagno and Drake argue that the parent body must still be intact, and that it must be at least 200 km in diameter (Duke and Silver, 1967; Walker *et al.*, 1977). Vesta is the only large asteroid exhibiting a reflection spectrum consistent with eucritic composition, and therefore, Consolmagno and Drake conclude that Vesta is the eucrite parent body.

There are some difficulties and uncertainties associated with this argument. Of primary concern is the lack of a plausible dynamical mode for getting fragments from Vesta into Earth-crossing orbits. It is also difficult to explain how the presumably thin basaltic crust of Vesta can have been so well preserved for 4.55 b.y. (since the formation of the basaltic achondrites), given the high rate of asteroidal collision to which it has presumably been subjected. Despite such problems, the idea that Vesta is indeed the (sole) parent body for the basaltic achondrites is still the best hypothesis.

Since McCord *et al.*'s pioneering work on Vesta, reflection spectra have been obtained for approximately 150 additional asteroids (McCord and Chapman, 1975a, b). None of these is similar to that of Vesta. In their summary of a much larger body of asteroid data, Bowell *et al.* (1978) note that UBV colors of four small asteroids (149 Medusa, 1023 Thomana, 1326 Losaka, and 1566 Icarus) suggest similarity to Vesta, although no diagnostic reflection spectra are available.

One might ask whether any other asteroids have reflection spectra consistent with a basaltic composition even if these spectra differ significantly from that of Vesta. The mineralogical interpretation of asteroid spectra has been reviewed recently by Gaffey and McCord (1977). The most common asteroids (called C-type) show relatively flat spectra and have very low albedos; they are interpreted as being similar in composition to carbonaceous chondrites. Spectra of the second most abundant type of asteroids exhibit pyroxene absorption bands, but often in association with what is interpreted as the spectral signature of nickel-iron. No asteroid spectrum analyzed by Gaffey and McCord, except that of Vesta, appears to indicate basaltic mineralogy.

A possibility that has been occasionally discussed in the literature (e.g., Chapman *et al.*, 1975; Chapman, 1976; Gaffey and McCord, 1977) is that a subgroup of the C-type asteroids (called "C*" by Chapman, "F" by Gaffey and McCord) might have opaque-rich, basaltic compositions. There are difficulties in distinguishing a primitive carbonaceous assemblage from a magnetite-rich basaltic assemblage, because opaques (carbonaceous compounds or magnetite) tend to suppress the spectral features of the other minerals present. The recent discovery of a prominent 3 μm absorption feature due to water of hydration in the spectrum of Ceres, a typical C* (F-type) asteroid, leads Lebofsky (1977) to favor the conventional interpretation of these spectra as being of carbonaceous chondritic compositions. One cannot categorically exclude the possibility that some asteroids have abundant opaque-rich basalts on their surfaces and have undergone extensive basaltic volcanism.

A final point to be made about basaltic volcanism on asteroids is that, while only a single asteroid exists that directly exhibits a basaltic surface, there is good reason to believe that such processes once occurred on other asteroids. These other asteroids are believed to have been substantially broken up by asteroidal collisions. If the Gaffey-McCord interpretation of S-type asteroid mineral assemblages is correct, then a significant number of asteroids (though still a minority) may be stony-iron cores of differentiated proto-asteroids which have been stripped of their crusts and mantles by collisions. In addition, the M-type asteroids (see Bowell *et al.*, 1978, for definitions of the C, S, M, etc., classes) may be metallic cores with no residual silicates. The S- and M-type asteroids, which comprise about half of the population near the inner edge of the belt but have virtually no representatives at the outer edge, may be the collisional remnants of a population of bodies that once were volcanically active. This hypothesis has been discussed by Chapman (1974, 1976) and Chapman and Davis (1975). More recent work has cast some doubt on the quantitative result of Chapman and Davis, which estimated that there were about 50–150 proto-S asteroids. From what little is known about asteroid collision processes, it seems unlikely that there remains, in space, a significant reservoir of basalts from these now-fragmented parent bodies. Presumably, some insight into asteroidal volcanic processes could be gained from a more detailed study of remnant cores, if in fact that is what they are.

Phobos and Deimos

The satellites of Mars are difficult to study from Earth. However, measurements made by the Mariner and Viking spacecraft, combined with earlier Earth-based data, have allowed some constraints to be placed on the compositions of Phobos and Deimos.

Phobos' optical characteristics (a low albedo and a relatively flat reflection spectrum, which declines sharply in the ultraviolet) are similar to those of opaque-rich basalts, so a basaltic composition cannot be ruled out. However, two recent papers have concluded that the reflection spectrum of Phobos resembles that of carbonaceous chondrite more than it does that of a basalt.

Pang *et al.* (1978) demonstrated that the reflection spectrum of Phobos is similar to those of Ceres and Pallas, both asteroids of the "C*" type (Chapman, 1976; called "F" type by Gaffey and McCord, 1977). These have been interpreted most commonly as having some form of carbonaceous chondrite composition. Pollack *et al.* (1978) adopted a different approach when interpreting the reflection spectrum of Phobos. They compared the Phobos spectrum with laboratory spectra of three basalts and four carbonaceous chondrites. Since Phobos resembles three of the four carbonaceous meteorites better than it does any of the basalts, Pollack *et al.* concluded that the spectrum of Phobos is dominated by carbonaceous chondritic material. The problem with Pollack *et al.*'s approach is that the three basalts selected for measurement clearly were not representative of the opaque-rich basalts that are candidate analogs for Phobos; all had visible albedos two to four times that of Phobos. Hunt *et al.* (1974) published reflection spectra for eleven basalts of different types, four of which had 1 μm albedos of 6% or less and were much better matches for the Phobos spectrum than any of the Pollack *et al.* basalts. At least one basalt, a vesicular basalt from the U. S. Bureau of Mines (labeled as sample 57 by Hunt *et al.*), was as good a match for Phobos as the carbonaceous chondrites. Unfortunately, the Hunt *et al.* measurements did not extend to wavelengths shorter than 0.3 μm, so a comparison with the Mariner 9 UVS data of Pang *et al.* is not possible. Hunt *et al.* reported that their low albedo basalt spectra were dominated by the opaque mineral magnetite, which quenches the spectral features exhibited by the opaque-poor basalts measured by Pollack *et al.*

Mass and volume measurements from the Viking Orbiters have shown that the density of Phobos is 1.5–2.3 g/cm^3 and that of Deimos is 1.2–2.4 g/cm^3 (Duxbury *et al.* 1978). Such low densities are typical of the most primitive carbonaceous chondrites and are well below the values of 3 g/cm^3 which are typical of basalts.

From its spectral characteristics and density values, it seems reasonable to think that Phobos is similar in composition to carbonaceous chondrites. By extending this interpretation to Deimos, one could say that Phobos and Deimos probably have never been subjected to processes of basaltic volcanism.

Outer solar system satellites

In considering the problem of basaltic volcanism in the solar system, we need to identify the range of planetary conditions that result in basaltic surfaces. With the exception of Earth, the Moon is the most carefully studied object on which basaltic volcanism is known to have occurred. How these basaltic deposits are related to the Moon's early history, its bulk chemistry and its size is the focus of much lunar research. Among the terrestrial bodies, the Moon is a representative of a class of moderate-sized, mostly rocky objects. The occurrence or nonoccurrence of basalts on other planets in this class is thus an important question in generalizing our knowledge of lunar processes.

Beyond the asteroid belt, water ice is stable on unprotected planetary surfaces. It is not surprising, therefore, that we find abundant remote-sensing evidence for water ice and frost on the surfaces of many of the satellites of the outer planets. What perhaps is surprising and significant for a study of basaltic volcanism in the solar system is that among the dozen or so well studied satellites of the outer planets, there is no evidence for large exposures of basaltic material. This is true even on those bodies which are not apparently covered by water frost or other types of frost.

Satellites of Jupiter. Among the large satellites of Jupiter, Io and Europa are very similar to the Moon in mass and size. Ganymede and Callisto are comparable in size to Mercury but have bulk densities less than 2 g/cm^3. They are apparently composed of mixtures of rock and icy material. Figure 2.2.11 shows a plot of log radius vs. log density for a group of extraterrestrial bodies. It clearly shows the overall similarity of the Moon, Europa and Io. However, the surface properties of these bodies differ widely. The spectral albedos of Io and Europa are compared with that of the Moon in Fig. 2.2.12. Both have albedos in the visual region nearly an order of magnitude higher than the Moon's. In addition, Europa's spectrum shows strong infrared absorptions characteristic of water ice (Pilcher *et al.*, 1972; Fink *et al.*, 1973). The rocky surface of Europa is thus hidden by a blanket of snow and ice of undetermined thickness; therefore, the question of basaltic volcanism cannot be definitely answered by remote-sensing techniques.

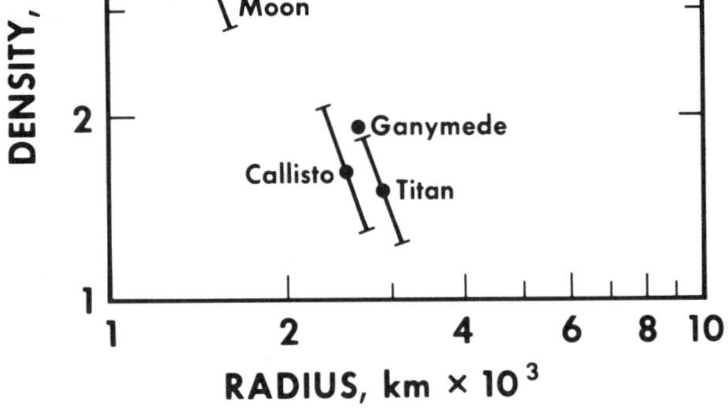

2.2.11 Log radius vs. log density for the inner planets, Galilean satellites and Titan. From Johnson (1978).

Although Voyager images of Europa show a highly complex striped and mottled surface, these images reveal no fundamental indications of large rock exposures which might be related to volcanism (Smith *et al.*, 1979c).

Io, on the other hand, shows no evidence of water frost or ice on its surface. Its highly unusual optical properties in the full disc telescopic spectra must result from some other material. Many compositional suggestions have been made, including evaporite salts and compounds containing sulfur (see Fanale *et al.*, 1974, 1977, for a review).

Data from Voyager 1's instruments have provided ample evidence for volcanism on Io (Smith *et al.*, 1979a), and also new information on surface composition. The magnetospheric heavy ion composition suggests that S and O are the major elements escaping from Io, along with smaller amounts of Na and K (Bridge *et al.*, 1979; Broadfoot *et al.*, 1979; Krimigis *et al.*, 1979; Vogt *et al.*, 1979). Discovery of SO_2 gas by the infrared instrument (Pearl *et al.*, 1979) and identification of SO_2 surface frost (Fanale *et al.*, 1979; Smythe *et al.*, 1979) confirm that sulfur-rich volcanic gases and compounds play a major role both in Io's volcanism (Smith *et al.*, 1979a, b) and in coloring Io's surface.

The question of what role basaltic volcanism has played in Io's history is more speculative. New models suggest that tidal heating caused by Io's resonance-forced eccentricity is the major energy source for Io's currently active state (Peale *et al.*, 1979). These models imply major melting of Io's interior, including the possibility that the current solid crust is only tens of kilometers thick. If this scenario is correct, then basaltic volcanism would certainly have occurred at some point in Io's history. The observed active volcanism, on the other hand, appears to be the result of explosive eruptions driven by SO_2 at relatively low temperatures (Smith *et al.*, 1979b). The absence of species other than S, O, Na and K in the magnetosphere of Io also suggests that the active volcanic plumes are not ejecting silicate or basaltic materials (as Si and Fe species would then be expected in the magnetosphere). There are, however, numerous calderas and flows which resemble terrestrial volcanic features except in having a sulfur related coloration. Both silicate and sulfur volcanism have been suggested to explain these features

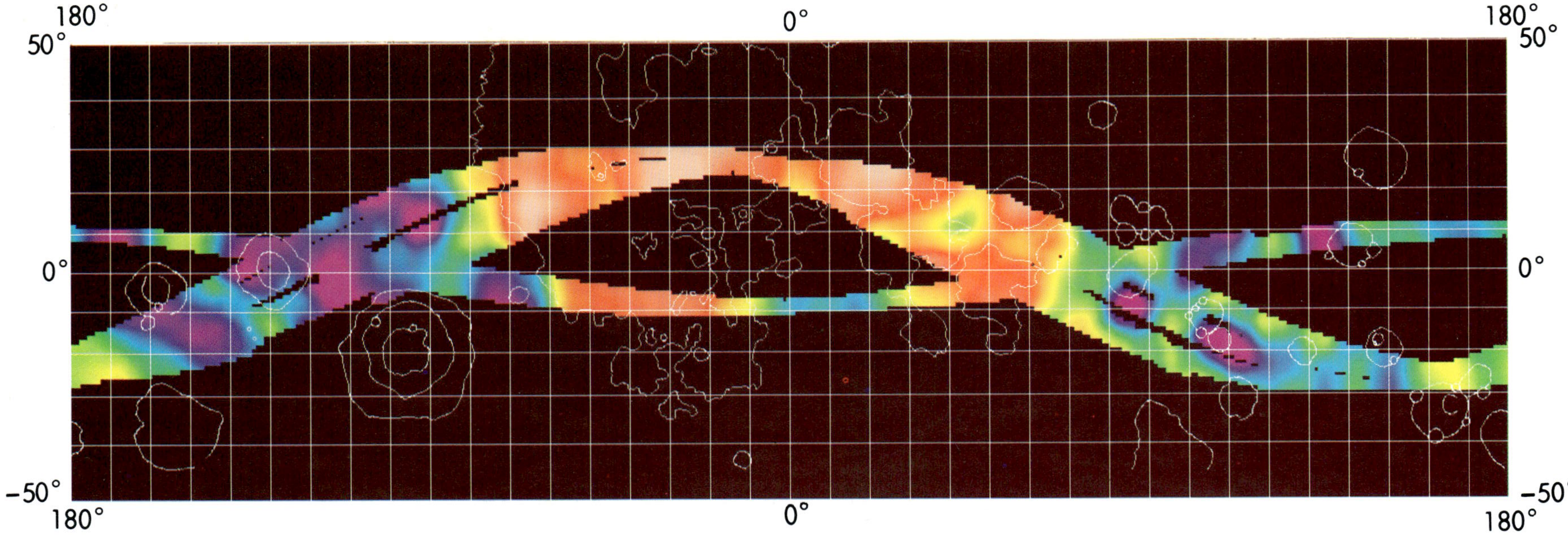

Plate 2.1 Apollo 15 and 16 γ-ray spectrometers mapped the γ-ray flux from 0.2–10 MeV in lunar orbit. The low energy channels, which cover the 0.55–2.75 MeV energy interval and are summed here measure the natural radioactive decay of Th, K and U. The spectrum of colors represents fluxes ranging from 61 c/s (pink) to 77 c/s (orange). (See Arnold *et al.*, 1977 and Metzger *et al.*, 1977a.) The pure colors correspond to Th concentrations as follows: pink = <0.9 ppm, purple = 0.9–1.9 ppm, blue = 1.9–2.8 ppm, green = 2.8–3.8 ppm, yellow = 3.8–4.7 ppm, orange = 4.7–5.7 ppm, and deep orange = >5.7 ppm.

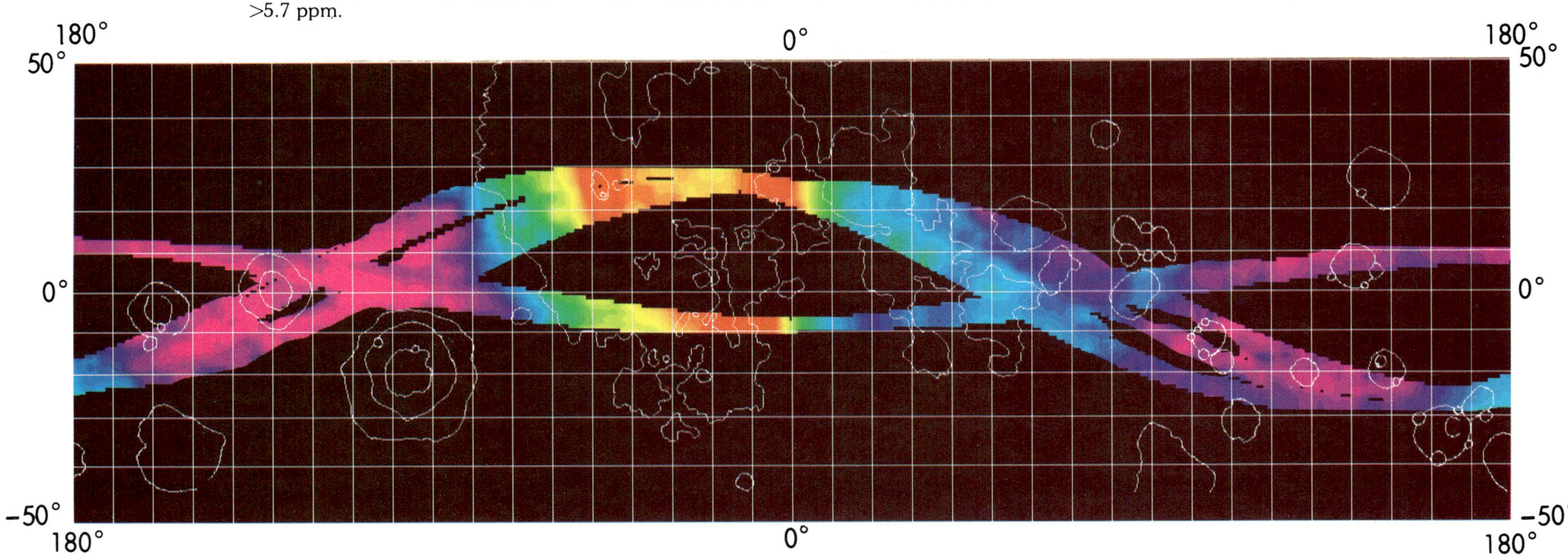

Plate 2.2 Variation in Fe concentration based on the 6.994–8.894 MeV γ-ray energy band. The color concentration correspondence is as follows: violet = <1.5%, blue = 3.0%, cyan = 4.4%, green = 5.6%, yellow = 7.4, red = 9.6 and bright red = >13.1% (Davis, 1980).

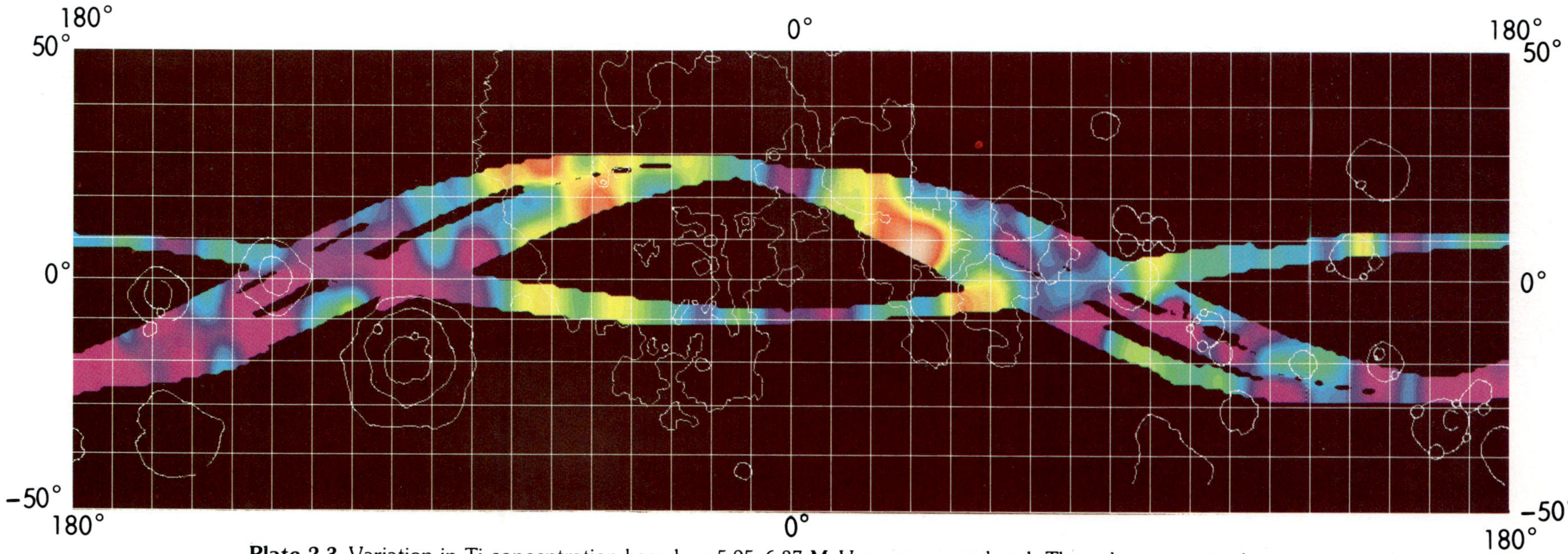

Plate 2.3 Variation in Ti concentration based on 5.95–6.37 MeV γ-ray energy band. The color concentration correspondence is as follows: pink = <0.36% Ti, violet = 0.36–0.78% Ti, blue = 0.78–1.20% Ti, green = 1.20–1.68% Ti, yellow = 1.68–2.16% Ti, orange = 2.16–2.70% Ti, and beige = >2.70% Ti (based on Metzger and Parker, 1979).

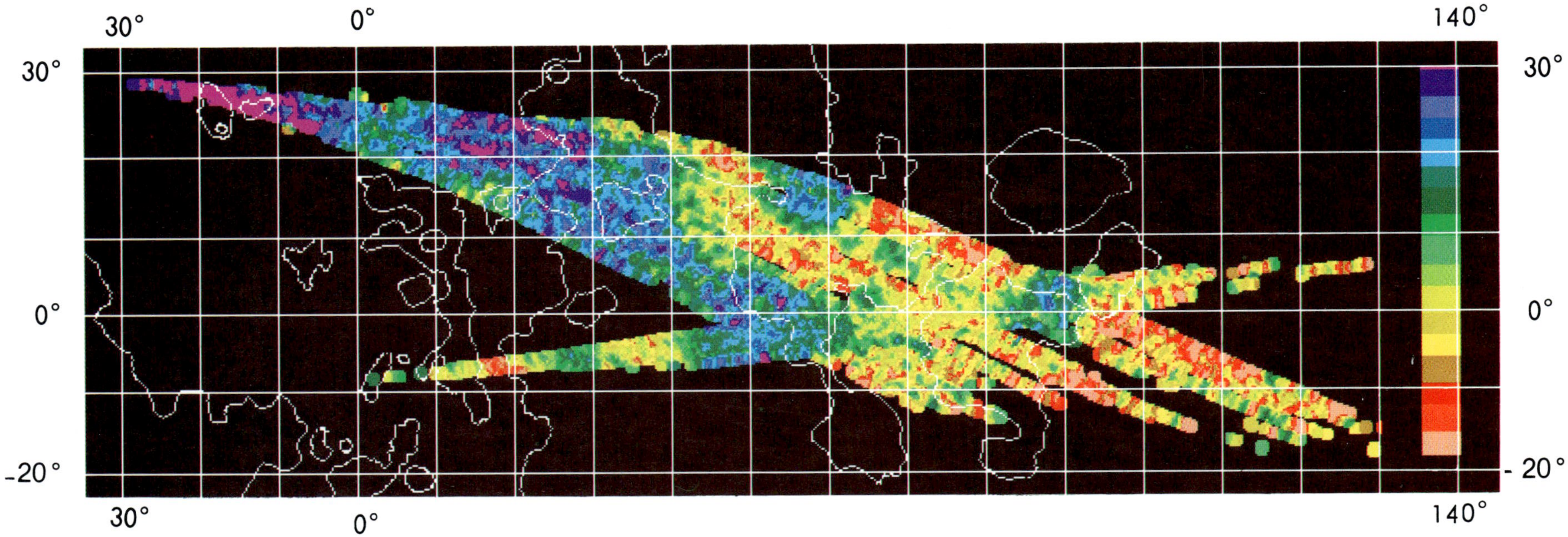

Plate 2.4 Concentration of Al_2O_3 on the lunar surface. Each color tone represents one weight percent of Al_2O_3 with the concentration increasing from violet (≤ 11 wt. % Al_2O_3) to light red (>26 wt. % Al_2O_3). Data is taken from the Apollo orbital fluorescent X-ray measurement of Al/Si intensity ratios (Adler *et al.*, 1973). Conversion to absolute Al_2O_3 values assumes a uniform SiO_2 concentration of 45% with ground data normalization (Bielefeld, 1977). In the more aluminous materials this assumption of uniform SiO_2 can cause errors of up to 2 wt. % Al_2O_3. Note three tones of yellow are not clearly distinguishable.

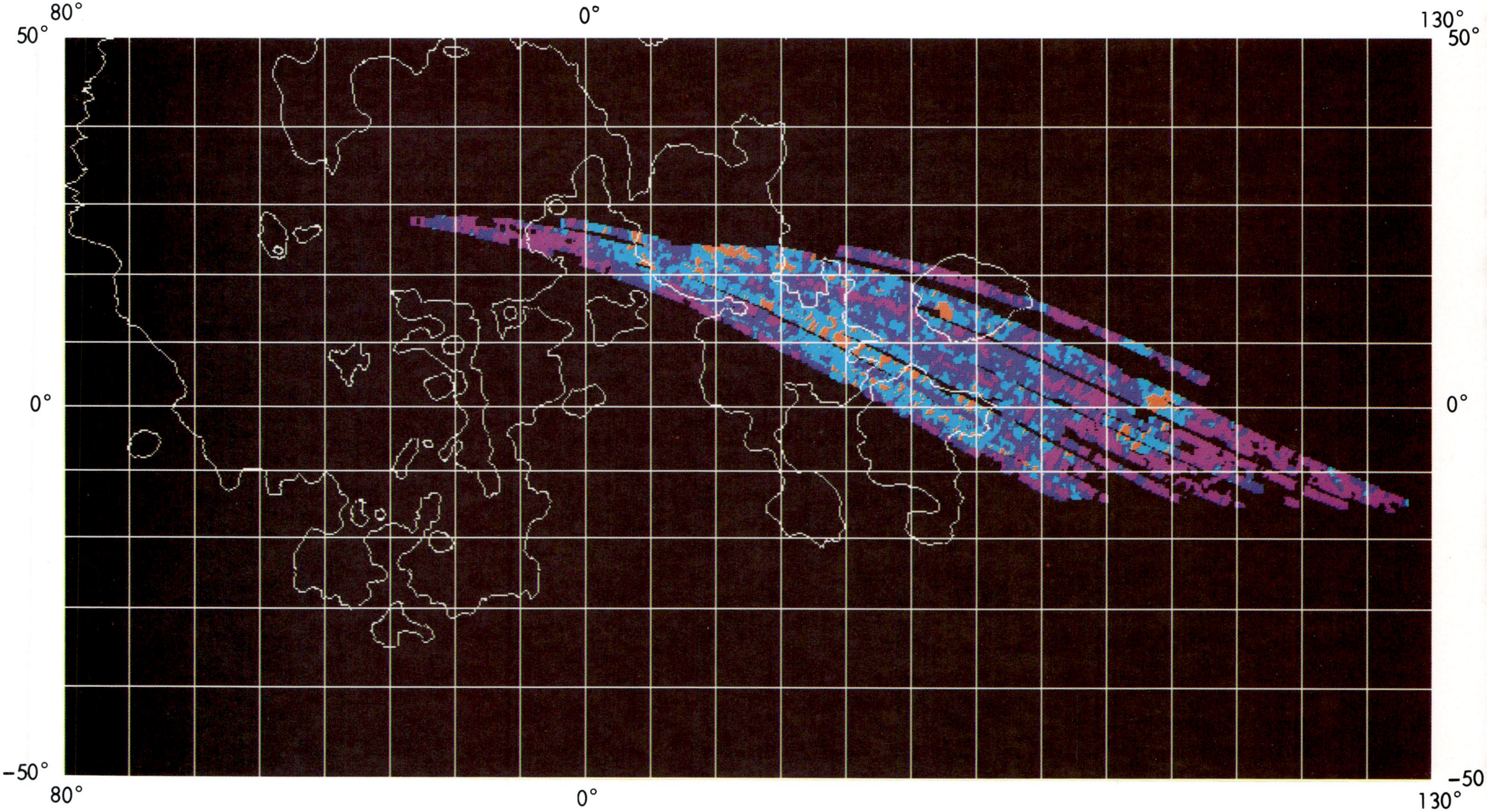

Plate 2.5 Variation in the Mg/Si intensity ratio for the lunar surface as measured by the Apollo 15 orbital X-ray fluorescence experiment. The Mg/Si intensity ratios are mapped using a color scheme ranging from pink (0.3–0.5 μm), purple (0.5–0.7 μm), blue (0.7–0.9 μm), to red (7.9μm). Conversion of these intensity ratios to MgO concentration ratios gives pink (3.3 wt.% MgO), purple (3.3–6.5 wt.% MgO), blue (6.5–9.6 wt.% MgO) and red (> 9.6 wt.% MgO). (Hubbard, 1979).

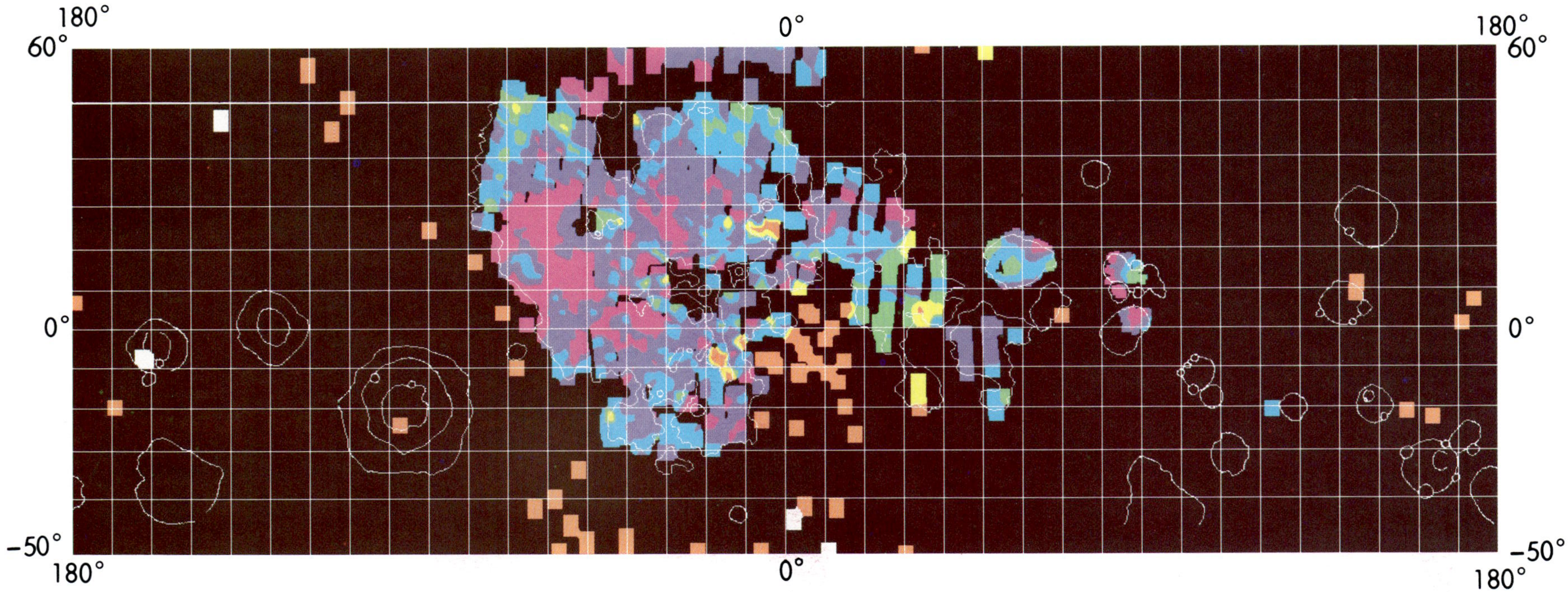

Plate 2.6 Relative ages of the maria and light plains. This map is the final product of a mapping effort that involved using both crater density and crater morphology techniques. The data for Mare Crisium, Marginis, and Smythii are based on crater density; all others are based on crater morphology. The two data sets have been correlated using the relationship of the crater density to D_L. The absolute age of each unit has been estimated based on radiometric ages of Apollo samples for absolute calibration. The ages are: 2.5 ± 0.5 b.y. (pink); 3.2 ± 0.2 b.y. (violet); 3.5 ± 0.1 b.y. (blue); 3.65 ± 0.05 b.y. (green); 3.75 ± 0.05 b.y. (yellow); 3.85 ± 0.05 b.y. (orange); 3.90 ± 0.5 b.y. (white). See Boyce and Johnson, 1978.

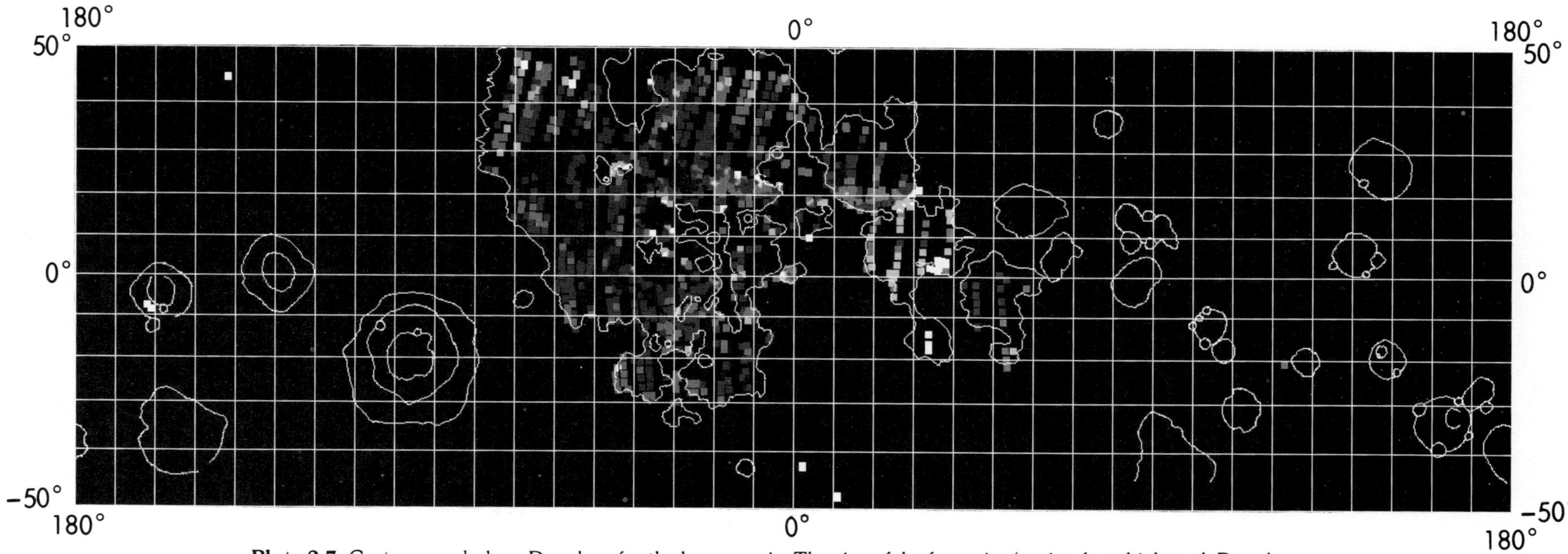

Plate 2.7 Crater morphology D_L values for the lunar maria. The size of the footprint (region for which each D_L value was measured) is presented in reduced size to allow data for areas of image overlap to be more easily discerned. Low D_L values are dark; high D_L values are bright.

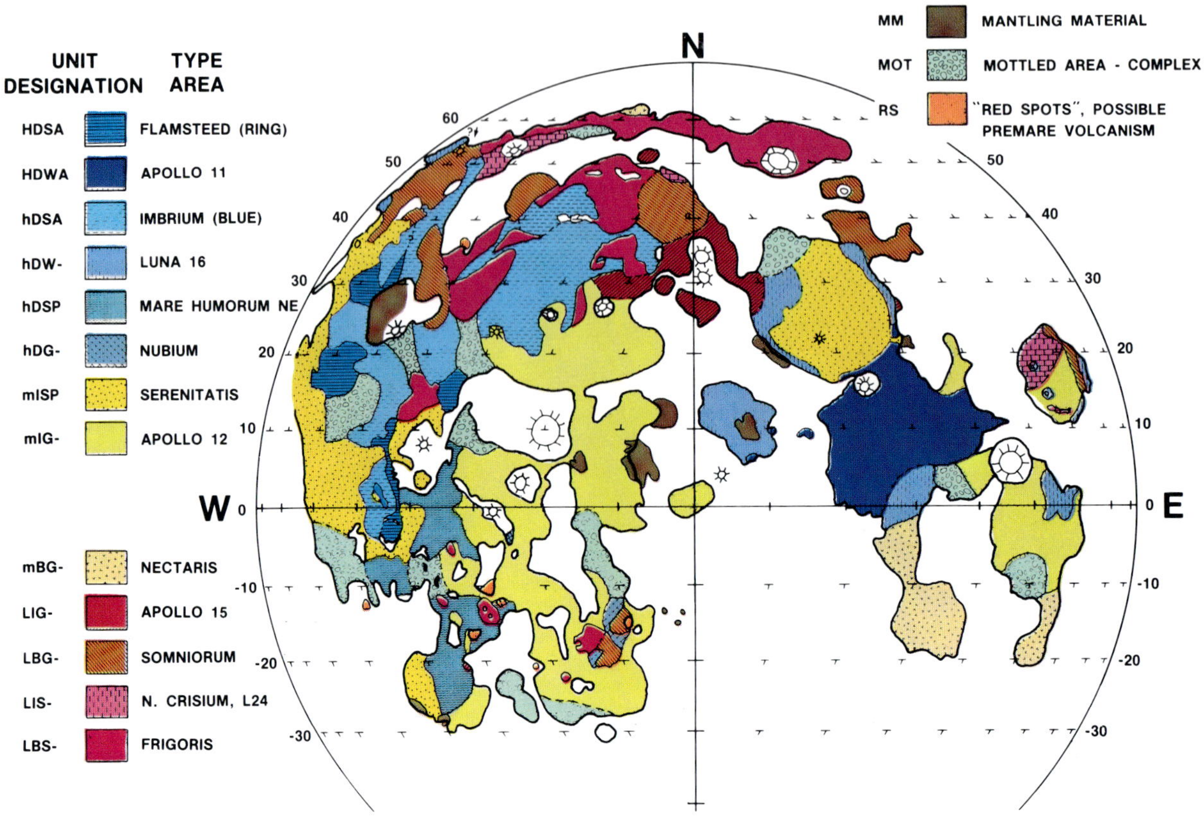

Plate 2.8 Major basalt types for the frontside of the Moon as derived from the current spectral reflectance data. The unit designations represent values for four measurable parameters (UV/VIS ratio, albedo, strength of the 1 μm band and strength of the 2 μm band). Since each of these parameters depends upon the composition of the surface, basalt units which are distinct from the sampled units can be identified and mapped on the basis of spectral properties. This same map appears in black and white as Fig. 2.2.9.

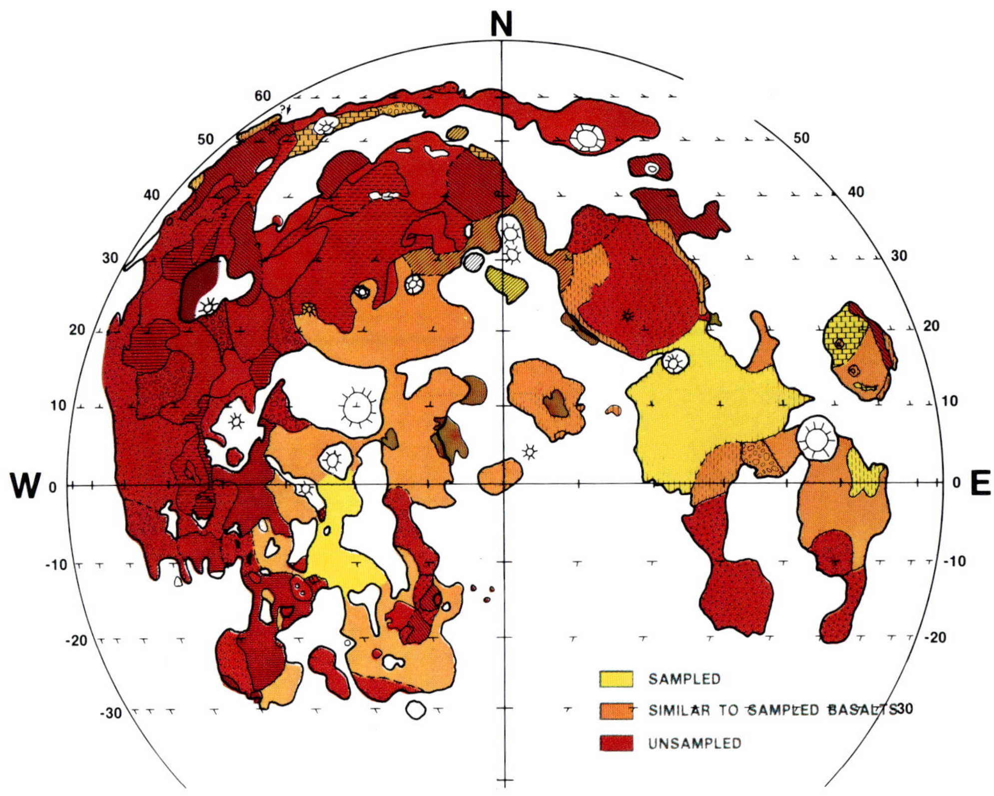

Plate 2.9 Sampled and unsampled basalt types on the frontside of the Moon. The base map shows the distribution of distinct mare basalt units as derived from all available spectral reflectance data and multispectral imagery (see Plate 2.8). Sampled units and units similar to sampled units are designated with yellow and orange overlays respectively. About two-thirds of the nearside mare surface remains unsampled. This unsampled mare includes up to eight distinct basalt types and is indicated with a red overlay.

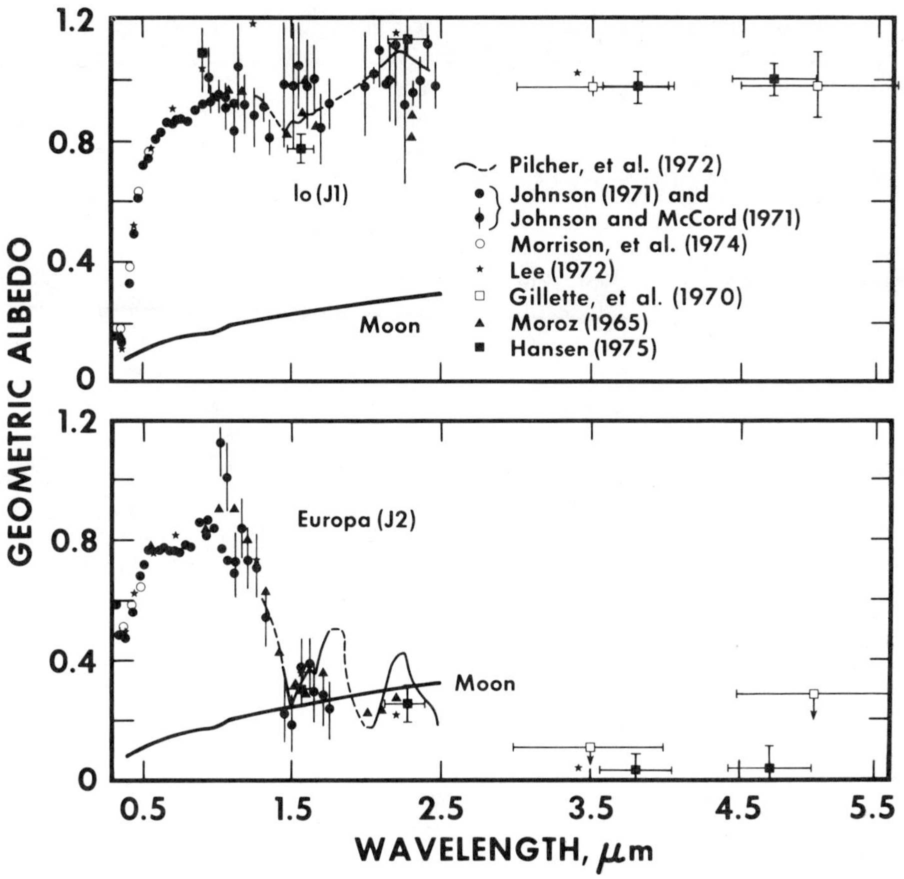

2.2.12 Geometric albedos as functions of wavelength for Io, Europa and the Moon. From Johnson and Pilcher (1977).

(Carr *et al.*, 1979; Sagan, 1979; Smith *et al.*, 1979b). The role of basaltic volcanism in the evolution of Io has yet to be discovered but will certainly be the subject of intense study in the coming years.

Callisto, while having a density indicative of significant quantities of water, does not show a bright, water frost covered surface. It is brighter than the Moon (albedo ~20%) but only half as bright as Ganymede. In addition, its infrared spectrum lacks the strong water absorption band near 1.4 and 1.8 μm found in the spectra of Europa and Ganymede. Its spectrum does have an absorption band near 3.0 μm, indicating some water on its surface. Thus Callisto can be characterized as having a "dirty" ice surface.

The nature of the non-icy material on Callisto (and the non-icy material forming dark spots and hemispheric markings on Europa and Ganymede) is not well determined. This material does not exhibit any strong electronic absorptions near 1.0 μm, a part of the spectrum where absorptions are related to mafic mineralogy (pyroxenes and/or olivines). Unaltered basalts thus appear to be ruled out. One of the current hypotheses is that this material is a hydrated, clay-type of silicate like that thought to compose many of the dark asteroids.

Satellites of Saturn. Densities of the saturnian moons are not as well known as those of the Galilean satellites (Morrison *et al.*, 1977). There are, however, no large, high-density satellites. The surfaces of most of the smaller satellites are apparently covered with water frost (Johnson *et al.*, 1975; Fink *et al.*, 1976). The surface of Titan cannot be seen due to its thick atmosphere, but the satellite has a low density. Iapetus (R ~900 km) is known to have one very dark hemisphere which is not covered by ice. This dark material, which is similar to the dark material on Callisto, does not appear to have a basaltic spectrum. It may resemble the matrix of carbonaceous chondrites.

Satellites of Uranus. Little is known of the Uranus satellites; however, recent infrared data suggest that they may be similar to the satellite Callisto (Cruikshank *et al.*, 1977).

Conclusion. Despite the fact that there are bodies of planetary size (lunar radii and larger) in the outer solar system, and that some of these bodies have silicate densities and surfaces not covered by ice, remote-sensing data indicate that none of these bodies has an exposed basaltic surface. Whether this is due to secondary processes having covered or altered previous basalt surfaces, or whether the initial conditions (composition, energy sources, etc.) in the outer solar system did not favor the production of basalts is an intriguing, but as yet unanswered question.

2.3 REMOTE-SENSING TECHNIQUES

2.3.1 Orbital gamma-ray spectroscopy

Introduction

To date, the technique of remotely sensing planetary geochemistry by using orbital γ-ray spectroscopy has been applied to investigations of the Moon and Mars by orbiting spacecraft, to Venus by surface landers, and to the Earth by low flying aircraft. The results from these experiments demonstrate that the technique is one of the most valuable tools for geochemical surveying. Several monographs and review articles cover the principles of orbital γ-ray spectroscopy (Adler and Trombka, 1970; Metzger and Arnold, 1970; Birks, 1964; Chupp, 1976; Haines *et al.*, 1976; Surkov, 1977a).

Electromagnetic radiation in the energy region 0.1–10 MeV contains information characteristic of nuclear transitions. Orbital γ-ray spectroscopy makes use of these characteristic photons to infer the elemental abundances of planetary surface materials. The emitted gamma rays come from the naturally radioactive elements, such as Th, U and K, and from elements activated by a variety of nuclear interactions. The most important of these interactions are neutron-capture reactions involving "slow" or thermal neutrons, and inelastic scattering reactions involving energetic neutrons. An example of the former is the capture of thermal neutrons by iron nuclei. This results in the emission of several gamma rays at energies characteristic of iron, e.g., 7.6 MeV. Alternatively, a high-energy neutron can inelastically scatter off an iron nucleus to produce a prompt γ-ray at 0.85 MeV. Other interactions leading to characteristic γ-rays are neutron-activation reactions producing longer-lived radioisotopes and charged-particle reactions. The latter are principally induced by cosmic-ray and solar protons.

Gamma-ray photons can interact with matter by complete absorption (emitting a photoelectron), by inelastic ("Compton") scattering from an electron, or by absorption followed by production of an electron-positron pair. Low-energy photons are readily removed by the photoelectric process; high-energy γ-rays are more penetrating and can interact by any of these three processes. Most γ-rays with energies below 1 MeV come from the upper 10 cm of the surface of a planet (and would be seriously attenuated by an atmosphere as thin as that of Mars). A photon with 10 MeV of energy has a total cross section of interaction that is about a quarter of that for a 0.5 MeV γ-ray; hence it can originate from greater depths within a planetary regolith (Reedy *et al.*, 1973). The γ-ray source depths are greater than the source depths for photons with wavelengths in the X-ray or visible regions.

The number and distribution of characteristic γ-ray lines vary from element to element, and depend on the elemental abundances, the capture and scattering probabilities and the energy distribution of the neutron flux. Of the major elements comprising the Moon, Fe, O, Si, Mg, Ti and Al were detected by Apollo NaI (Tl) scintillation detectors at orbital distances. Ca and Na, however, were not found in the spectra obtained by the scintillators. The naturally radioactive elements Th, K and U are all detectable; however, U is more difficult to detect than Th or K.

Although the photon flux from the naturally radioactive elements is in direct proportion to their elemental abundance, the flux from the induced radioactive elements depends both on the elemental abundance and on the flux and distribution of neutrons. Neutrons are the by-products of high-energy cosmic-ray bombardment of the surface material. A planetary surface may be protected from the isotropic cosmic-ray flux by a dense atmosphere or a strong magnetic field. The distribution of neutrons below the surface is a function of the material in the surface. This is also true of their energy distribution. For example, the presence of hydrogen can rapidly moderate the fast neutrons producing a preponderance of thermal neutrons. On the other hand, the presence of nuclei such as gadolinium, with very high cross sections for thermal neutron interactions, could cause a local depression of the slow neutron flux in the surface material.

Simulating the space environment for neutron production and the detection of the resultant γ-ray flux presents severe experimental problems. Measurements conducted at high energy particle accelerators are proving increasingly useful (Surkov *et al.*, 1977; Reedy, 1978), but most of the interpretation of planetary γ-ray spectra is founded on theoretical and semi-empirical modelling. Fortunately, the predictions for the case of the Moon have proven to be quite reliable, both in terms of neutron production and distribution (Reedy and Arnold, 1972) and in the resulting γ-ray fluxes (Reedy *et al.*, 1973). The major factors in the complex production of γ-radiation in nonhydrogenous planetary bodies that have neither an atmosphere nor a magnetic field are now understood.

Detectors

In the past the detector of choice for spaceborne γ-ray spectroscopy in the energy region of 0.1–10 MeV has been a single scintillating crystal of NaI or CsI doped with the impurity thallium. Gamma-ray photons interacting with such a detector will deposit some or all of their energy in the crystal. This produces scintillations that are detected by a photomultiplier tube optically coupled to the crystal. The intensities of the electronic pulses, which are proportional to the amount of energy deposited, are digitized by an analog-to-digital converter. They are either accumulated in a multichannel analyzer on board the spacecraft, or transmitted event-by-event to Earth to form a pulse height spectrum. It is important that the detector system be able to discriminate between γ-rays and charged particles (which also cause scintillations in the crystal). This can be accomplished by surrounding the main γ-ray detector with a second detector, such as a plastic scintillator, which is efficient in identifying charged particle events and is relatively transparent to γ-rays. An electronic coincidence circuit will then reject counts in the main detector which are registered within approximately 10 μ seconds of an event in the anti-coincidence detector. A detailed description of the Apollo lunar NaI (Tl) detector system is given by Harrington *et al.* (1974). Future γ-ray spectrometers will utilize high resolution semiconductor detectors, such as lithium-drifted germanium and high purity germanium crystals. The latter have recently been shown to be suitable for remote sensing (Nakano *et al.*, 1976; Pehl, 1977). Compared to an energy resolution of 8% attained with a NaI (Tl) crystal, germanium detectors can resolve the ^{137}Cs 0.66 MeV calibration line to approximately 0.2%.

Background signals

The line structure of a pulse height spectrum contains the geochemical information about a planetary surface. This characteristic line is masked by at least eight interfering background signals which have been isolated for the lunar Apollo γ-ray experiment. Table 2.3.1 gives the relative magnitude of each of these components. It is important to realize that the signal containing the geochemical information from the Moon comprises only 10–15% of the total γ-ray signal measured by the detector at orbital distances. About 20% of the total signal is generated by the spacecraft (by the material of the detector system, and by the NaI crystal), through many of the same processes that generate γ-rays in the lunar material. By far the largest background

Table 2.3.1 Percent composition of average Apollo 16 lunar pulse height spectrum.

	Pulse Height Energy (MeV)			
	0.4–0.8	0.8–2.0	2.0–5.0	5.0–9.0
Primary cosmic ray induced radioactivity in detector	1.8	4.5	4.6	0.8
Secondary neutron induced radioactivity in detector	7.1	12.3	1.5	0.0
Lines produced in local mass around detector	3.3	2.8	0.6	0.0
Electron bremsstrahlung from local mass	2.8	3.5	4.0	3.4
Cosmic gamma-rays	4.0	3.0	2.2	1.4
Induced and natural radioactivity in spacecraft	5.1	4.6	5.4	7.1
Multiple scatter from Moon due to variations in Th, U & K	4.8	1.8	0.1	0.0
Multiple scatter from Moon at lowest concentration of Th, U & K	56.1	58.0	71.3	73.0
Discrete lunar lines	15.0	9.5	10.2	14.1
Total (%)	100.0	100.0	100.0	100.0
Total (counts/second)	55.2	45.8	25.3	8.6

component (60–70%) is the continuum signal from multiple photon scatter on the lunar surface. The two most important factors in determining background levels are the shape of the continuum from the planetary surface and the intensity of the lines generated by the spacecraft and detector.

Data processing and analysis

Orbital γ-ray sensors have the same data storage problems that affect other planetary imaging experiments. The detection of transmission errors, the altitude and location of the spacecraft, and the operational mode and health of the instrument are all factors which have to be accounted for to ensure the integrity of data. The count rate due to γ-ray surface emission ordinarily is low, and decreases with increasing orbital height. Because the Apollo experiments measured a total of 70–90 counts per second between 0.8–9.0 MeV, spectra had to be accumulated over extended time intervals in order to ensure statistical significance. Given the precision of the spacecraft orbit, adequate observations over localized regions required adding spectra taken days and even months apart. Therefore, the pulse height spectra have to be corrected for gain shift and zero drift in the electronics over those periods. Changes in orbital altitude must also be compensated for before spectra can be added together.

The areal resolution of the detector system is determined by two factors: accumulation time and geometry. For Apollo, a multi-element analysis required some 2000 seconds of data to achieve statistical reliability. Over orbital node areas (e.g., eastern and western limbs of the Moon for Apollos 15 and 16) this requirement translated into a 3° × 3° surface element for the four-day coverage. Above nonnode areas the surface element containing 2000 seconds of data is larger, and therefore the resolution is much worse. Since the Apollo detector was uncollimated, it possessed a field of view stretching from horizon to horizon. However, geometrical analysis showed that approximately 50% of the measured γ-ray signal was produced within 120 km of the subsatellite point when the orbital height was 110 km.

The basic multi-element unfolding techniques used in the Apollo experiments were outlined by Reedy *et al.* (1973) and implemented in Metzger *et al.* (1974) and Bielefeld *et al.* (1976). The unfolding of the discrete lunar line spectrum (Fig. 2.3.1) required an understanding of how the photons deposit their energy in the detector. The Apollo analysis relied on theoretical response functions (Berger and Seltzer, 1972), which were generated by means of a Monte Carlo procedure. This took into account the particular size of detector and semi-empirical estimates of relative line intensities (Reedy *et al.*, 1973). The estimates for line production, from the neutron capture and inelastic scattering processes, were made from the best available cross section and neutron-flux data (Reedy, 1978). Estimates were also made for some of the less significant processes such as induced radioactivity and spallation-induced γ-rays. An alternative to the theoretical procedure is to experimentally measure the detector responses to a variety of induced radioactive materials. This has been done by using an accelerator to simulate the cosmic-ray bom-

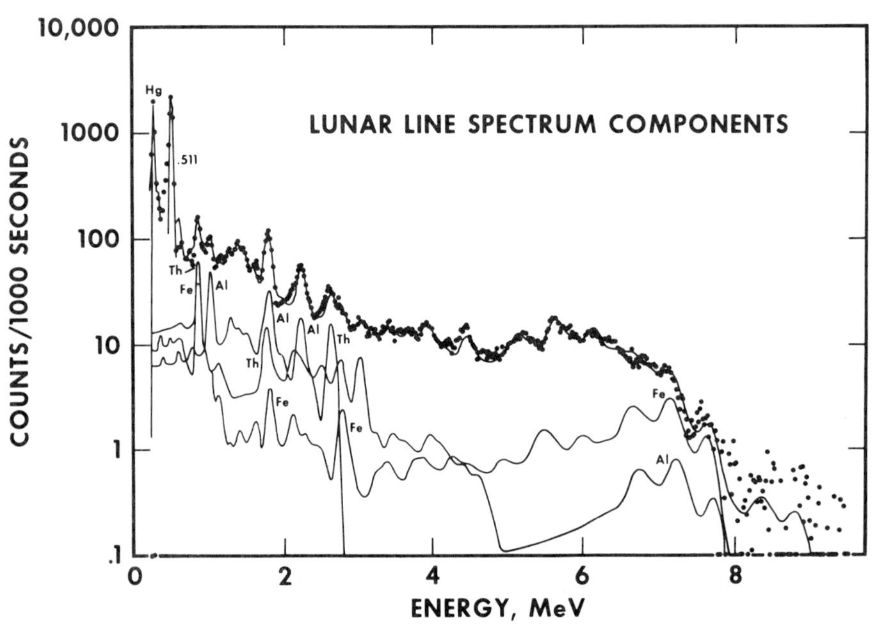

2.3.1 The net spectrum due to the lines in the lunar surface as determined by library function matrix inversion. The position line is at 0.511 MeV; the line at 0.279 MeV is due to a ^{203}Mg calibration source. The top curve represents the total synthesized spectrum. Shown below this curve are the individual element components for Al, Fe and Th out of a set of fourteen element functions used for the library. From Bielefeld *et al.* (1976), p. 2661.

bardment of planetary surfaces (Surkov *et al.*, 1977). Both methods have their own particular advantages and disadvantages.

Figure 2.3.1 graphically presents the synthesis of the measured lunar spectrum from a composite of twelve response functions characteristic of lunar elements combined by the least squares method to give a "best" fit. With regional pulse height spectra collected over the minimum time interval of 2000 seconds, the localized variation of the naturally radioactive elements Th, U and K and the induced radioactive elements Fe, Mg, Si, O, Ti and Al were estimated. Because these elements have complex line structures, their signals interfere with one another. For example, the signature for oxygen partially masks that of titanium. In order to interpret the data, certain physical restraints, derived from analyses of returned samples, had to be imposed on the unfolding procedure. For example, no regional variation was allowed for oxygen and silicon concentrations which were fixed at 44 and 21% by weight. The ratio of Th to U was also not allowed to vary from the 3.8 value seen in most returned lunar soil samples. For the nonradioactive elements, several "ground data" regions were used to normalize the theoretical concentrations to "known" surface soil concentrations. These "known" concentrations came from large uniform areas of the Moon characterized by returned lunar samples. The ground truth normalization factors were near unity for most of the elements, a fact that lends confidence to the unfolding technique and to the theoretical flux models. The geochemical results for some sixty regions are given by Bielefeld *et al.* (1977) using this procedure.

Metzger *et al.* (1977a) have developed a different procedure for the analysis of a single element that improves the areal resolution and sensitivity in certain cases by requiring less accumulation time. The method uses only those channels or energy bands where an element has a strong line. These intervals (energy bands) are narrow for most elements; however, the multiplicity of lines below 2.75 MeV from Th and U has allowed the relative abundance of natural radioactivity to be determined with as little as 25 seconds of observation. Compositions are obtained by plotting a regression curve of known concentration vs. the integral count rate (Fig.2.3.2). The known concentrations have come from a multi-element analysis or from lunar sample values. The method has been successfully applied to Th, Ti, Fe and K, and may also be applicable to O, Al and U. The methodology developed for single element analysis will be directly transferable to unfolding future high resolution spectra using germanium detectors.

The detectability of regional variations in elemental abundances is affected by the concentration of the element, by the degree of regional relief a particular element shows with respect to neighboring regions, by the photon flux per unit concentration, and by the interference of other signals. Although Th is a trace element, its enormous photon flux per unit concentration makes it an ideal candidate for orbital geochemistry. Oxygen and silicon comprise over 60% (by weight) on the Moon's surface, yet their relatively uniform distribution makes detection of variations difficult. The signal from titanium is partly hidden by oxygen and to some extent by iron. Measuring variations in the titanium concentration with uncertainties comparable to the more sensitive elements requires a high concentration or relatively long accumulation times. Calcium is a major element, but its γ-ray flux per unit concentration is low, and it has not been successfully isolated in the Apollo spectra.

The broad angular detection response that characterized the Apollo γ-ray spectrometer resulted in a loss of spatial resolution. This reduced the contrast among surface concentrations within lunar regions that are small compared to the field of view (Haines *et al.*, 1978). A computer deconvolution technique has been developed that removes much of this instrumental effect. This improves the spatial resolution but lowers the data's precision. Geometric models of regional concentration are convoluted through the response function of the instrument to yield a predicted distribution that is then compared with the observed data field for quality of fit. The method has been applied to the distribution of Th in areas such as Aristarchus, Mare Smythii and the Apennine Bench Formation.

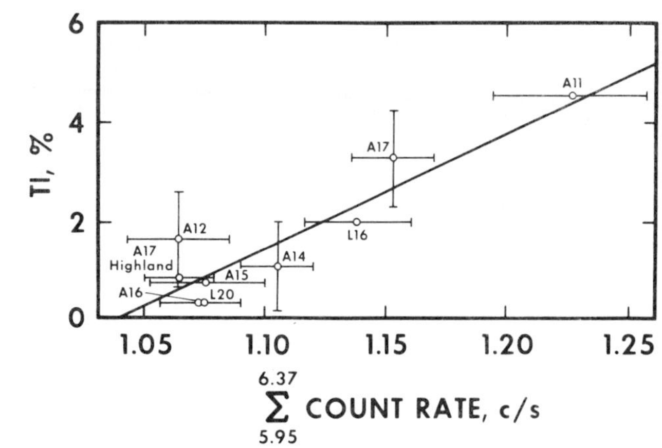

2.3.2 Regression curve for titanium. The concentration of Ti in lunar samples is plotted vs. the γ-ray integrated count rate for the region which contains the ground site. The count rates were made over a selected energy band, 5.95–6.37 MeV in this case.

Data presentation

An ideal format for presenting data acquired during a geochemical survey of a planetary surface is a two-dimensional array depicting elemental concentration. This forms a pseudo-image of the planetary surface much like a digitized TV image of the surface. However, unlike an ordinary image from a scanning instrument, the array of data is irregular in position and in statistical reliability. Nevertheless, several attempts have been made to image the Apollo γ-ray results following the single element analysis mentioned above (Frontispiece, 1973, 1976, 1977; Arnold *et al.*, 1977; Eberhardt, 1977). This type of presentation facilitates the correlation and interpretation of geochemical and geophysical information obtained by a variety of remote-sensing techniques. Variations in spatial resolution of the different detectors and the wide discrepancy in the statistical significance of each data set are difficulties encountered when comparing data in image format. These problems are minimized by using sliding filters of variable area for averaging and new methods of display (Bielefeld *et al.*, 1977).

2.3.2 Orbital X-ray

Introduction

X-ray techniques can be divided into passive-orbital measurements (which depend on some outside source of exciting radiation), and *in situ* measurements (which use an artificial excitation source). Both techniques have been applied to planetary studies. Orbital X-ray observations were made during the Apollo 15 and 16 flights (Adler *et al.*, 1972); a roving vehicle equipped with an X-ray spectrometer for *in situ* analysis of the surface was deployed by the U. S. S. R. (Adler *et al.*, 1972; Kocharov and Victorov, 1973); and the two Viking spacecraft carried X-ray fluorescence experiments to Mars (Clark *et al.*, 1977).

Orbital X-ray measurements rely on the fact that the solar system has two known excitation sources: X-rays emitted by the sun, and high energy particles trapped in Jupiter's magnetic field. Both sources are capable of producing characteristic X-rays from planetary surfaces. The methods employed in X-ray spectroscopy are tailored to the surface environment under study. Conditions which influence signal strengths, signal-to-noise ratios, and the problems of spectral unfolding must be considered in the experimental design.

Principles

X-rays result when inner shell electrons are ejected and the excited atom returns to its ground state. It does this by the emission of X-rays having wavelengths characteristic of the emitting element (line emission). The cross section for a given line emission decreases markedly with decreasing atomic number. This is due to the competing process of Auger electron emission, which becomes more dominant with decreasing atomic number.

Excitation leading to line emission may occur by X-ray irradiation (X-ray fluorescence), by bombardment with energetic particles (high energy electrons, protons or alpha particles) or by K capture and internal conversion. In the case of X-ray irradiation, the excitation process is photoelectric absorption and the lines produced are free of continuum. Heavy energetic particles such as protons or alpha particles also produce lines which are superposed in a minimal continuum. Energetic electrons on the other hand produce characteristic lines superposed on a very substantial bremsstrahlung continuum.

Solar excitation

The Sun-Moon system provides a good example of X-ray fluorescence. The Moon is constantly bombarded by high fluxes of soft (low energy) solar X-rays having approximately the energy distribution shown in Fig. 2.3.3. This is based on a simplified estimate of the solar X-ray flux under quiet sun conditions. The primary solar X-rays are capable of exciting K lines from those elements whose absorption edges lie on the low energy side of the spectral distribution. In the case of the Moon, secondary X-rays are produced which are characteristic of the lighter elements such as Mg, Al and Si. The energy distribution as seen in Fig. 2.3.3 rises almost exponentially with decreasing energy. This effect tends to compensate for the decrease in fluorescence yields with decreasing atomic number.

Because the source of excitation is the solar X-ray flux, special constraints are placed on the experiments designed to measure the secondary fluorescence X-rays. For example, one must constantly monitor the intensity as well as spectral distribution of the solar flux. Concomitant methods for unfolding the effects of the variations in the exciting flux had to be developed. In addition, the secondary fluorescence is likely to be low, and since short period measurements are desired (in order to provide better spatial resolution of the planetary surface), it is necessary to devise a highly efficient measuring apparatus. An example of such a device is the one flown on Apollo 15 and 16 (Adler *et al.*, 1972).

The lunar surface is bombarded by substantial fluxes of soft X-rays even under quiet sun conditions. Calculations show that a typical solar X-ray spectrum is energetically suitable for producing measurable amounts of characteristic X-rays from all the abundant

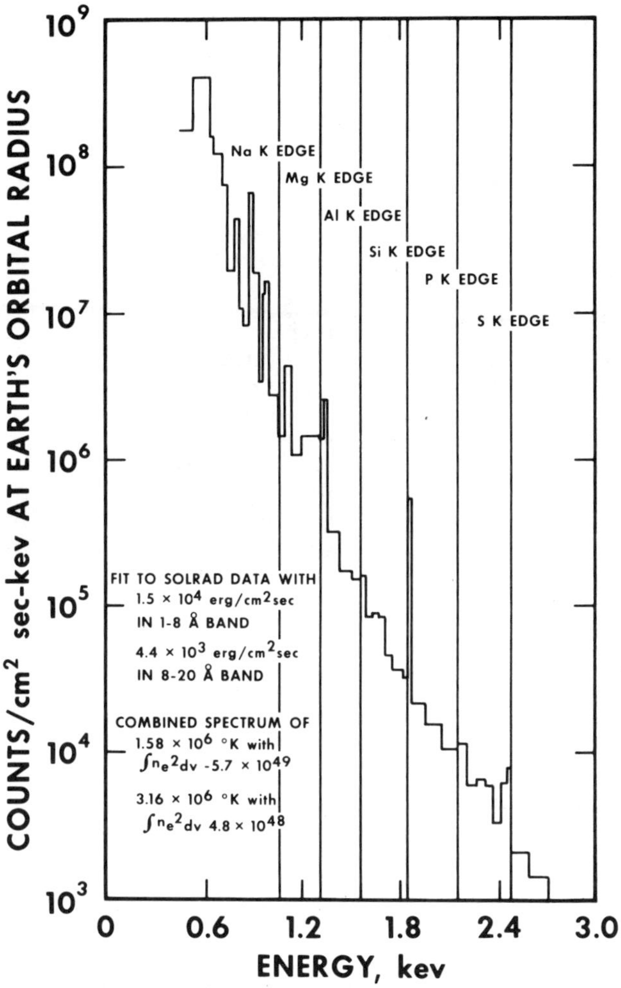

2.3.3 The Apollo 15 and 16 X-ray fluorescence experiments. An estimate of a typical quiet sun X-ray spectrum from Adler *et al.* (1972).

elements with atomic numbers less than or equal to 14 (Si). Observations during periods of enhanced solar activity (such as flares) show an increase in X-ray output. In addition, there is a hardened (more energetic) X-ray spectrum from which one can predict the production of characteristic X-rays from elements, such as K, Ca and Fe, of higher atomic number. The typical quiet sun spectrum (Fig. 2.3.4) has been attributed to the solar corona at temperatures between 10^6 and 10^7 °K. Changes in flux and spectral composition are due to changes in the coronal temperature. These variations mandate some form of solar monitoring. For example, a hardened solar spectrum would be expected to produce changes in the intensity ratios of the elements being measured. An example of the effect of the solar flux is shown in Fig. 2.3.4. As predicted, it is observed that a hardened flux enhances the silicon intensities relative to the aluminum.

Using the calculated X-ray solar fluxes and spectral distributions, and assuming some typical terrestrial rock composition such as basalt or granite, it is possible to calculate the expected yield of fluorescent X-rays. Such calculations were made by Gorenstein *et al.* (1969). The results, shown in Fig. 2.3.5, are the basis for predicting distinct responses from basalts relative to other rock types. The Apollo results confirm the value of these calculations and the model selected for predicting Al/Si and Mg/Si concentration ratios.

Complexity of X-ray measurements

Remote analysis involves the problems of particle size variations, matrix effects (absorption and enhancement), and scattering. All of these contribute to background noise, and can be largely suppressed by expressing elemental data as a ratio. The need for background correction is of paramount importance, particularly in light-element analysis where signal-to-noise

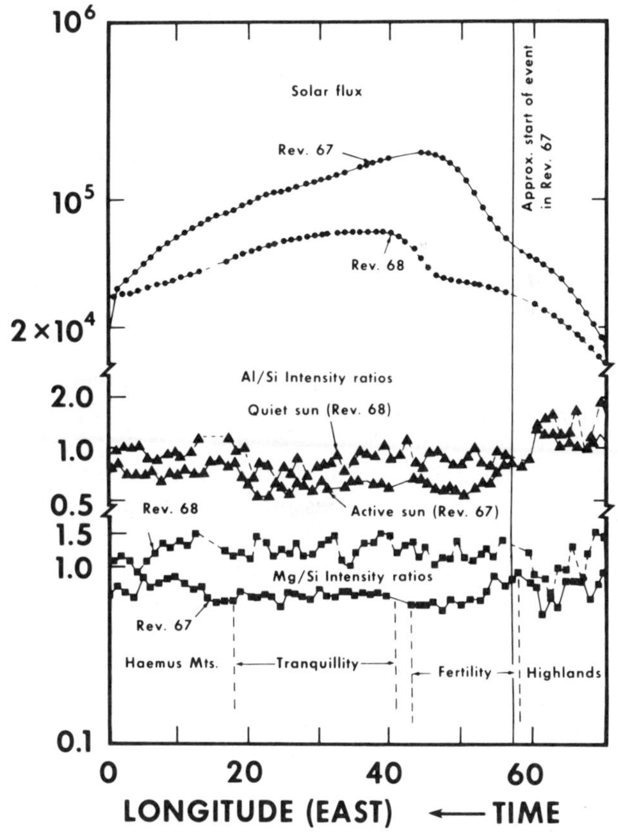

2.3.4 The Apollo 15 X-ray fluorescence experiment. Al/Si and Mg/Si intensity ratios for quiet and active sun conditions from Alder *et al.* (1972).

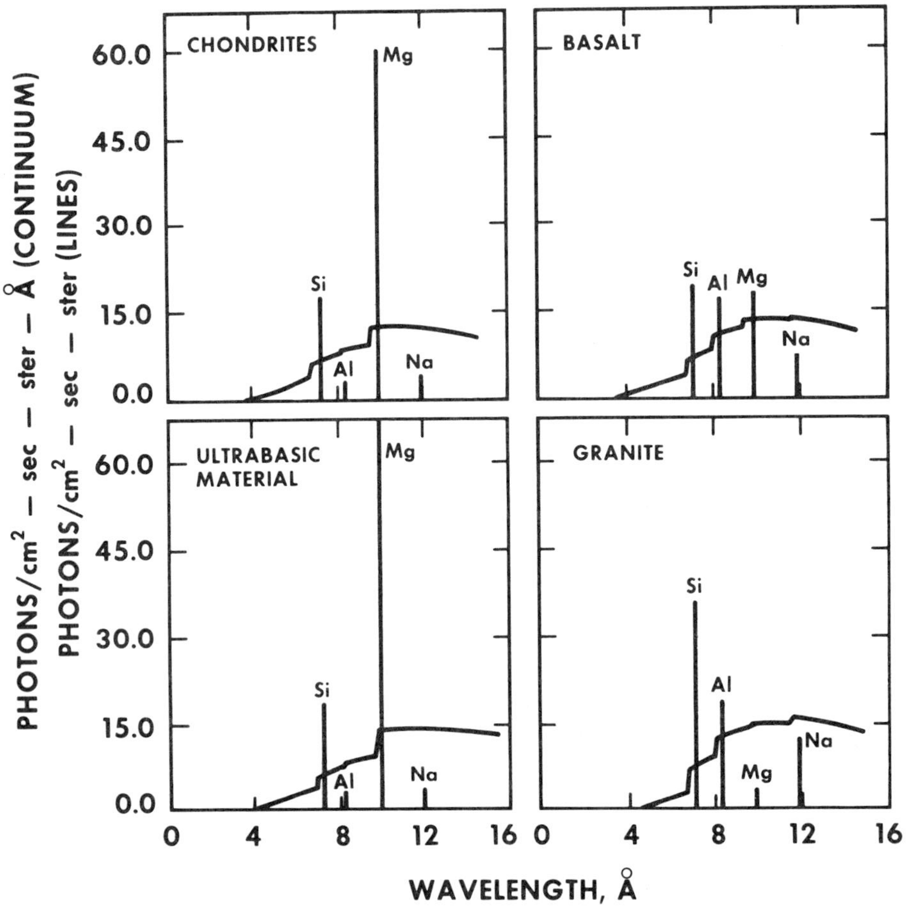

2.3.5 Calculated albedo X-ray spectra of lunar materials. The solar X-ray spectrum is a "free-free" continuum with $T = 4 \times 10^6$ °K. The solar intensity ($1 < \lambda < 8$ A) is normalized to the average daily intensity experienced in April 1967. Vertical lines denote K fluorescent radiation. The continuum is the scattered solar spectrum. From Gorenstein *et al.* (1969), p. 335.

ratios are inherently poor. The major source of background noise is X-ray scatter, particularly under solar excitation. In lieu of experimental observations of such backgrounds (which are very difficult to make with instrumentation of coarse energy resolution), one must rely on background estimates based on the principles of coherent scatter for the wavelength range being measured. While it is not possible to make a precise background correction, it is possible to accomplish some data refinement.

Some of the problems cited above are consequences of experimental design. The use of nondispersive systems (pulse height discriminators), which use proportional counters, make it very difficult to separate the characteristic energies emitted by the various elements from the background on which the spectral lines are superimposed. As an example, the Apollo 15 and 16 X-ray experiments used gas-filled proportional counters with an energy resolution of about 40% in the Al,K region. It is necessary to limit the Apollo experiments to eight discriminator channels covering 0.5–2.75 KeV in equal intervals. X-ray filters (Mg foil and Al foil) were placed in front of two of the three detectors to provide additional discrimination. The use of the filters in the data reduction is described below.

Data reduction

Data reduction in the Apollo experiments has been described in a number of publications (Adler *et al.*, 1972, 1973). Briefly, the data were treated in a simple fashion based on the energy discrimination afforded by the selected X-ray filters. The following assumptions were made:

1. All three proportional counters had identical characteristics.
2. The detectors were 100% efficient for the radiation transmitted through the detector windows.
3. Background corrections could be made using measured darkside fluxes. This assumption implies that the major background was due to low energy noise.
4. The effect of X-ray scattering on the intensity ratios could be ignored in first estimates.

Based on the above assumptions, three simultaneous expressions were written:

$$I_{Base} = T^{Be}_{Al} I_{Al} + T^{Be}_{Mg} I_{Mg} + T^{Be}_{Si} I_{Si}$$

$$I_{Al}\text{ filter} = T^{Be}_{Al}\left(I_{Al}\,T^{Al}_{Al}\right) + T^{Be}_{Mg}\left(I_{Mg}\,T^{Mg}_{Mg}\right) + T^{Be}_{Si}\left(I_{Si}\,T^{Al}_{Si}\right)$$

$$I_{Mg}\text{ filter} = T^{Be}_{Mg}\left(I_{Al}\,T^{Mg}_{Al}\right) + T^{Be}_{Mg}\left(I_{Mg}\,T^{Mg}_{Mg}\right) + T^{Be}_{Si}\left(I_{Si}\,T^{Mg}_{Si}\right)$$

The total intensities summed over all seven channels for the base detector, the detector with the Al filter, and the detector with the Mg filter are I_{Base}, I_{Al} filter and I_{Mg} filter, respectively. The transmission factors of the Al and Mg filters for the various energies are T^{Be}_{Al}, T^{Be}_{Mg}, T^{Be}_{Mg}, etc., respectively. These expressions have been solved by a least-squares method to yield intensity ratios for the photon fluxes in 1.25, 1.49 and 1.74 keV energy regions. It should be noted that the quotients of the fluxes are intensity ratios which are proportional, but not equal to the chemical ratios.

The above approach is at best approximate, because the only background correction made is for the ambient radiation determined during darkside passes. The failure to correct for scatter introduces errors. These can be minimized by the use of ratios and are small for Al/Si in any case.

Data quality

In order to determine the level of confidence in the Apollo X-ray data and the limits of application to geochemical studies, it is important to examine the quality of the data. It is now clear that the X-ray fluorescence data initially available were quite satisfactory in distinguishing first order effects. It was used successfully to measure chemical differences between maria and terrae and to define the general distribution of basaltic and anorthositic materials. These very significant results were published immediately following the Apollo 15 and 16 flights. Recently, the data have been extended to more detailed spatial analysis by considering second order effects. It has been possible to partially reclaim those data neglected in the earlier analyses. This is especially true of data where significant solar fluctuations were observed. Furthermore, as indicated in the text, some significant conclusions have been drawn about chemical features with a resolution on the order of 30 km.

The new ways in which the data are being handled take into account statistical errors, the variability of the sun as an exciting source, and the coherent X-ray scatter. Smoothing techniques have been used to minimize statistical problems. The instrumental approach of Apollo to the problem of solar variability had limitations. The solar monitor designed to measure the exciting flux (magnitude and spectral distribution) did not have sufficient spectral resolution to allow direct adjustments for the sun's variations. Thus it was necessary to carry out corrective procedures that were either empirical or based on independent solar observations made by other spacecraft such as Solrad 9 and 10. Unfortunately, even the Solrad data did not have adequate spectral or temporal resolution. Other approaches have involved a more theoretical treatment which predicts the effects on the intensity ratios as functions of the changing solar spectrum (Clark and Adler, 1978).

The interpretation of the X-ray data has facilitated correlating the processed data with other observations, such as those supplied by studies of photogeology, albedo, altimetry and γ-ray geochemistry. These efforts have been described by a number of investigators (Adler *et al.*, 1972, 1973; Andre *et al.*, 1978a, b; Schonfeld and Bielefeld, 1978; Hubbard *et al.*, 1978a, c; Bielefeld, 1977; Bielefeld *et al.*, 1978; Bielefeld, 1979a, b).

2.3.3 Visible and near-infrared reflectance spectroscopy

Introduction

Spectral reflectance is the fraction of incident radiation that is reflected from a surface as a function of wavelength. With sufficient spectral resolution and intensity precision, absorption bands in diffuse reflectance spectra can be resolved that are diagnostic of surface composition and mineralogy. For mineralogical interpretation the wavelength dependence of reflectance is of primary importance, while the albedo (or absolute amount of reflected light), the angular dependence, and the polarization of the reflected radiation are of secondary importance.

The wavelength range in which reflectance spectra can be measured is governed mainly by atmospheric opacity, the intensity of emitted thermal radiation, and limitations in radiation detectors. Laboratory and telescopic measurements discussed in this section fall between about 0.2 μm (50,000 cm^{-1}) and 2.5 μm (4000 cm^{-1}).

As an example of the relationship between reflected and emitted components of radiation, the spectral flux received at the Earth from a square kilometer of a lunar mare region is plotted in Fig. 2.3.6. The radiation consists of two parts: (1) reflected solar radiation, and (2) solar radiation that has been absorbed and reemitted as thermal radiation. The shapes of these two components of the spectral flux curve are similar for the surfaces of all solar system objects. The intersection of the two curves and their relative magnitudes vary with the semi-major axis of the planet's orbit: for Mercury, the crossover occurs near 1.8 μm (5550 cm^{-1}), for Mars it is near 4 μm (2500 cm^{-1}), and for Uranus it is near 20 μm (500 cm^{-1}).

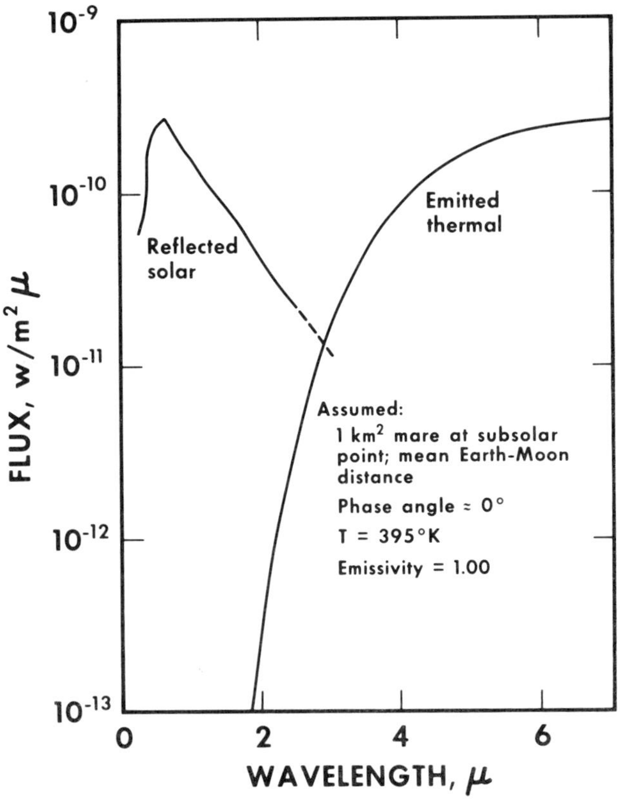

2.3.6 Radiation received from a one-kilometer-square mare area, albedo 0.06, at the mean Earth-Moon distance (Adams and McCord, 1970).

The surfaces of solar system objects typically are extensively covered with randomly oriented fragments of soil or dust. When observed in reflected solar light, the particulate material returns two components of radiation: (1) a specular component that consists of first surface reflections and is described by Fresnel's laws for absorbing dielectrics; and (2) a body component that is composed of light that has entered at least one grain and has been scattered back into space toward the observer. It is the body component that contains the most compositional information.

Interpretation of reflectance spectra

Reflectance spectra are interpreted using a combination of laboratory and theoretical data. Terrestrial, lunar, and meteoritic minerals and rocks have been chemically analyzed, and the laboratory spectra from these control samples have been studied as functions of particle size, phase angle, mixing ratios and mixing heterogeneity. Interpretations of absorption bands are based on ligand field theory (inter-electronic transitions), molecular orbital theory (charge transfer photoemission), and theories of molecular vibration. Relative modal abundances are determined by comparing the strengths of the bands contributed by the constituent minerals, and by making assumptions about relative grain sizes and homogeneity. Simple comparisons with catalogs of laboratory spectra may be misleading unless the absorption features are understood.

The absorption bands that appear in reflectance spectra are due primarily to (1) electronic transitions and charge transfers by d-shell electrons in transition metal ions (Fe^{2+}, Ti^{3+}, Cr^{3+}, etc.); (2) overtones of molecular vibrations; (3) photoelectric emission; and (4) photoconduction. The wavelength positions of the absorption band centers depend on the types of ions present and on the dimensions and symmetries of the sites in which the ions are situated. These two factors to a large extent also define the mineralogy of a sample. Note that reflectance spectroscopy is a method of remotely sensing *mineralogy*, in contrast with γ-ray spectroscopy and X-ray fluorescence, which remotely sense *elemental composition*.

Electronic transition absorption. When optical radiation is passed through a mineral or glass, certain regions of the spectrum are absorbed preferentially. As an example, the polarized-light transmission spectrum (Adams and McCord, 1972) of the mineral bronzite, an orthopyroxene, is shown in Fig. 2.3.7. Three different spectra are shown, corresponding to transmission along three crystallographic axes. The main spectral features are due to absorption of radiation by the d-shell electrons of iron ions (Fe^{2+}). These ions reside in octahedral coordination sites in the crystal lattice. Electronic transition and charge-transfer absorptions occur commonly for minerals containing transition element ions (Fe, Ti, Cr, etc.). Of particular interest to studies of planetary surfaces are the metal silicates and oxides, especially those of iron. The wavelengths (or energies) of these absorptions depend on the types of ions present and on the dimensions and symmetries of the sites in which the ions are located. The same factors vary with mineralogy; thus, a mineral often can be identified by its optical absorption spectrum. Further details on the principles of optical absorptions in minerals are given in Burns (1970).

Charge-transfer absorption. Another mechanism that produces optical absorption in the wavelength range of interest is charge-transfer. These absorptions arise from electronic transitions between orbital states of neighboring ions. Such transitions can occur between ligands and cations, cations and cations, and ligands and ligands. Crystal field theory does not adequately

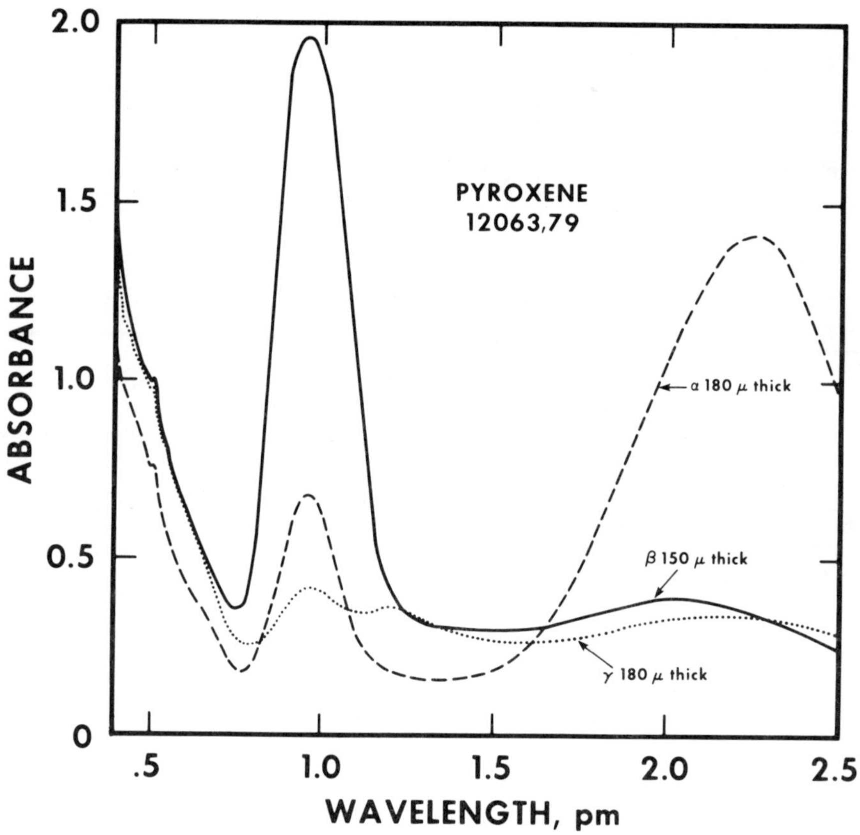

2.3.7 Polarized absorption spectrum of a single crystal of clinopyroxene from an Apollo 12 rock. The bulk of the crystal is pigeonite ($Wo_{20}\,En_{38}\,Fs_{42}$), but the rim is augite ($Wo_{40}\,En_{31}\,Fs_{29}$). Optical and microprobe analyses by Dr. Peter Bell, Geophysical Laboratory.

explain charge-transfer, so molecular-orbital theory is employed instead. For estimating ligand-cation charge transfers, a satisfactory approach is that used by Tossell *et al.* (1974).

In general, the centers of charge-transfer bands are located at blue and UV wavelengths, but their limbs may extend into the visible and near-infrared portions of the spectrum. As a result, electronic transition bands (discussed previously) in diffuse reflectance spectra of minerals often occur as subtle depressions on an intense charge-transfer limb.

The structure of the limb of a charge-transfer band involving Fe^{2+}, Ti^{4+} and Ti^{3+} ions in lunar glasses (Burns, 1970; Bell and Mao, 1972) was used by Charette *et al.* (1974) to determine the TiO_2 content of glasses in lunar soil. This was done from the telescope reflectance spectra and enabled a correct prediction of the TiO_2-rich soil at the Apollo 17 site.

Vibrational absorption. In the 1–4 μm wavelength region OH^-, H_2O and CO_2^{2-} ligands, as well as gaseous, liquid, and solid H_2O, and gaseous and solid CO_2 produce vibrational overtones and combination absorption bands. The wavelengths of the band centers and the relative intensities of the bands are sensitive indicators of the bonding configuration of the ligands and of the lattice vibrational frequencies of the host mineral. Ligands other than OH^-, H_2O, and CO_2^{2-} contribute little to the vibrational absorption features in the 1–4 μm wavelength region. The SiO_2 fundamentals occur near 10 μm (Si-O stretching modes) and 20 μm (O-Si-O bonding modes) and do not have appreciable overtones in the near IR.

Reflectance spectra of rocks and soils. The diffuse reflectance spectrum of a mineral in powdered form is an average of the spectra of the randomly oriented individual crystals. Absorption bands persist in the powder spectra, and they still arise from Fe^{2+} in the mineral crystal field. Transmission and reflectance spectra of a lunar pyroxene are compared in Figs. 2.3.7–8. Note that absorption maxima correspond to reflectance minima.

Figures 2.3.7 and 2.3.8 illustate that the spectral contrast in the absorption spectrum is stronger than in the reflectance spectrum. This is mainly because the reflectance spectrum includes a specular component. The proportions of specular and body components of radiation in the reflectance spectrum depend on the optical depth-to-grain-size ratio, i.e., how many scattering boundaries are traversed before the probability of photon absorption nears unity. The depth-to-grain-size

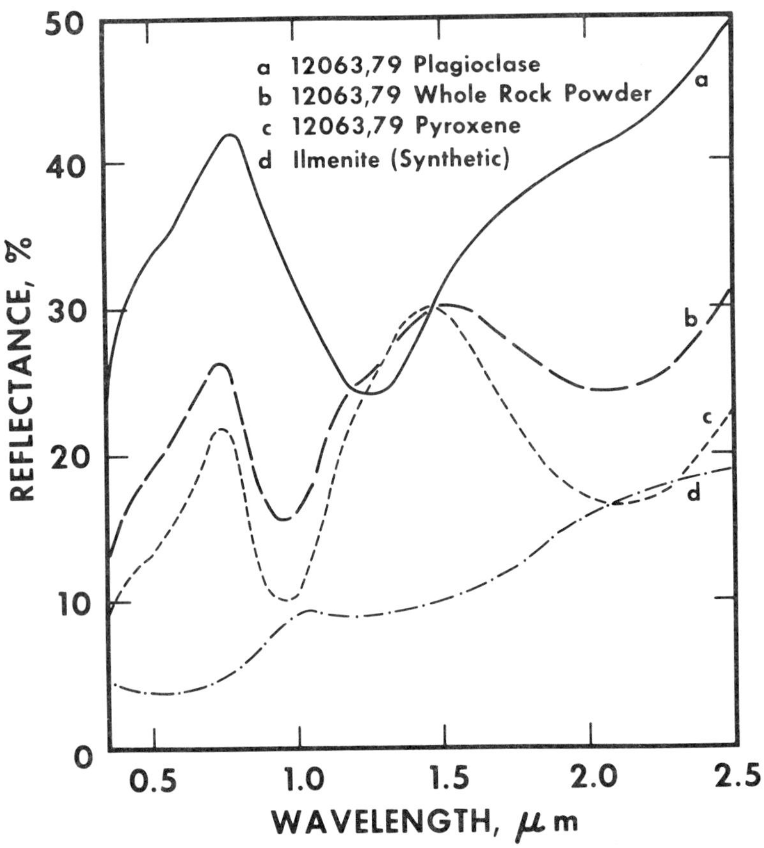

2.3.8 Spectral reflectivity of Apollo 12 basalt powder 12063 and of plagioclase and pyroxene separates from the same rock. Ilmenite is a synthetic sample (Adams and McCord, 1971).

ratio also varies slightly with the angles of illumination and viewing. The body component of reflected radiation has its maximum strength, and therefore deepest absorption bands, when grain size is about one optical depth. The wavelength positions of electronic transition absorption bands are not appreciably affected by grain size and opacity. The effect of grain size and opacity on absorption band strengths has been investigated by Adams and Filice (1967) and Pieters (1974).

Mixing of several mineral powders to produce a rock soil results in a reflection spectrum that is a weighted composite of the spectra of the individual soil constituents. Minerals with optical depth-to-grain-size ratios near unity will contribute absorptions to the reflection spectrum to a much greater extent than will constituents that are either transparent or opaque. An example of this behavior is illustrated in Fig. 2.3.8, where diffuse reflectance spectra for a powdered lunar rock and its major constituents are plotted. Pyroxene absorption dominates the spectrum over the relatively transparent plagioclase and the opaque ilmenite (Adams and McCord, 1971; Pieters, 1974; Gaffey, 1975).

Telescopic measurements

Spectral reflectance is measured at telescopes using a variety of spectrometers and imaging systems. The usual method of measurement (T. B. McCord, pers. comm., 1978) is as follows:

1. The flux from the object of interest is measured, $f_{OBJ}(\lambda)$.
2. At nearly the same time a nearby star is also observed, $f_S(\lambda)$.
3. The object/star flux ratio is formed $f_{OBJ}(\lambda)/f_S(\lambda)$, which eliminates instrument response and atmospheric effects.
4. In previous studies the spectral fluxes of series of standard stars have been determined relative to a master standard star α Lyrae, $[f_S(\lambda)/f_{\alpha L}(\lambda)]$ and the spectral flux of α Lyrae above the Earth's atmosphere, $F_{\alpha L}(\lambda)$, was determined (Oke, 1964; Oke and Schild, 1970; Hayes, 1970; Schild *et al.*, 1971).
5. Thus, the spectral flux received from the object is determined: $F_{OBJ}(\lambda) = [f_{OBJ}(\lambda)/f_S(\lambda)][f_X(\lambda)/f_{\alpha L}(\lambda)] F_{\alpha L}(\lambda)$.
6. Spectral reflectance is finally determined by using measurements and calculations of the spectral flux

from the solar disk above the Earth's atmosphere, $F_{sun}(\lambda)$ (Labs and Neckel, 1968):

$$R(\lambda) = F_{OBJ}(\lambda)/\ F_{sun}(\lambda).$$

Usually the spectral reflectance is displayed as a function scaled to unity at 0.56 μm in order to eliminate albedo effects and to concentrate on differential spectral reflectance. This technique has been used to obtain spectra of light reflected from small areas on Mars and the Moon, and of the integrated light from Mercury, many asteroids and the Galilean satellites.

The precision of the telescopic spectra has been tested by measurements of returned Apollo samples. Spectral reflectance curves of lunar soils measured in the laboratory have been compared with those measured by telescope of 10–20 km diameter areas containing the Apollo and Luna landing sites. The agreement at most sites is within a few percent (e.g., Adams and McCord, 1970). The telescopic data are accurate to about 1% and indicate that the telescopic measurements of 10–20 km diameter areas are mainly of lunar soil rather than of rocks or breccia.

It is significant for remote analysis that the lunar sample spectral curves show the same subtle, but important, absorption features as are seen in the telescopic curves. Studies of the lunar samples have made it possible to understand what chemical and mineralogical properties control the shape of the reflectance spectrum. This knowledge has been used to extend to the entire front surface of the Moon the compositional information obtained by direct sampling. Here spectra have provided information on the pyroxene composition of the soil, the TiO_2 concentration in lunar glasses, soil maturity, mare basalt types and the distribution of dark mantle deposits.

Reflectance spectroscopy using ground-based telescopes has also been successfully applied to other objects in the solar system. For example, information on the surface composition of Mars has been obtained; mineralogical compositions have been determined for over 100 asteroids (allowing asteroids to be correlated with certain types of meteorites); and on Mercury an upper limit has been placed on the FeO content of its lunar-like soil and a low-Ca pyroxene has been identified.

Spacecraft measurements

Spectral reflectance curves of the surfaces of Mars and Venus have been reconstructed from Viking Lander and Venera photometric data, respectively. Although the curves are approximate, due to the use of broad-band filters, the spectra have facilitated comparisons with terrestrial soil analogs.

2.3.4 Multispectral mapping

Instrumental techniques

Photography. The oldest technique for acquiring multispectral imaging is the use of filters and photographic emulsions sensitive to different wavelengths. Although these methods are inexpensive and relatively simple, photographic films have a number of problems including: (1) limited dynamic range; (2) nonlinear response and reciprocity failure; (3) difficulties in uniform development and calibration; and (4) limited spectral response. High-quality color difference photography has, however, achieved impressive results, particularly in the mapping of lunar mare basalts (Whitaker, 1966). Whitaker's relative color-difference images of the Moon (see Fig. 2.2.6) are in good agreement with photoelectric and vidicon data (over the limited photographic spectral range 0.35–0.8 μm). This is particularly true within mare regions where the technique is not strained by extreme albedo variations. These color difference maps have been successfully used in initial efforts to integrate spectral information with other lunar data.

Line-scan imaging. Line-scan imaging systems use a few detectors (usually with different spectral bandpasses) that are linked to a mechanical or optical scanning device. The principal advantages of such systems are: (1) ease of calibration (due to small number of detectors), (2) automatic registration of the images taken by the individual detectors, and (3) relatively simple optical design as good image-quality over a large area of the focal plane is not required. Difficulties include: (1) the need for mechanical stability and reliability in the scanning mechanism, (2) the difficulty in reconstructing a geometrically accurate image from the individual scan lines, and (3) lack of time synchronization among different parts of the image—e.g., if some intrinsic change in the target or in the target-observer geometry occurs while an image is being built up, spatial and temporal information becomes confused.

A variant of the line-scan principle has been developed by Soderblom (1970) for telescopic observations of the Moon. His system uses a single photomultiplier as a detector with multispectral capability being provided by a rapidly spinning filter wheel. The output of the detector is electronically "gated" into a separate counter for each filter. The scanning motion is provided by the natural motion of the Moon in the sky. Data at all wavelengths are not collected simultaneously as in an ideal line-scan system. However, in practice, adequate signal-to-noise for each filter is achieved because of the short time needed to make the measurement as compared with the time necessary for the Moon to "scan" one picture element (or pixel). The size of the pixel is defined by a focal-plane aperture in front of the filter wheel. This scheme results in effectively multiplexing in wavelength, and has the further advantage of requiring no interdetector calibration, as stability of the single detector during the imaging period is all that is required for highly accurate relative spectrophotometry. Soderblom's images of the Moon have high photometric accuracy, and Whitaker's images have high spatial resolution. The two methods are complementary for mapping mare basalts.

Vidicons. Scientific imaging with television-type vidicon tubes has been a major development of the unmanned lunar and planetary exploration program. The conventional type of vidicon used for these purposes has been the SeS vidicon (spectral range 0.3–0.7 μm). These devices have a number of features that have made them unsuitable for precise photometric studies, and they are, therefore, of limited use in multispectral imaging (Young, 1974). Greater understanding of the operation and calibration of the vidicons has recently allowed multispectral data to be obtained from Mariner 10 images (Hapke *et al.*, 1975b). This data will help determine whether or not basalts exist on Mercury.

Silicon vidicons are superior to photographic imaging systems and to older TV systems in photometric accuracy and in spectral range (0.35 μm–1.1 μm). A description of these devices and a discussion of how to calibrate them for astronomical imaging can be found in McCord and Westphal (1972) and in McCord and Frankston (1975). In a standard operation, multispectral data are acquired by taking a succession of images through narrow-band interference filters. If frames are taken rapidly (usually about every minute), atmospheric changes and changes in the telescope-target geometry will be small between the time the first and the last filter is used. Individual frames are then processed by computer to remove dark-current and to correct for the tube-optics response (McCord *et al.*, 1976).

Data handling

A basic set of calibrated, near monochromatic images can be used in several ways for multispectral mapping. Reconstruction to "natural" color is not common, as it is usually desirable instead to enhance very small color contrasts. Even with the expanded spectral range of a vidicon, considerable computer processing is required to display many of the subtle "color" variations revealed by spectrophotometry. An example applied to the lunar surface utilizes three spectral bands (0.38 μm, 0.56 μm, and 0.95 μm) and greatly enhances the dynamic range of hue and intensity to produce a three-color "enhanced color" image.

More commonly, the same ratioing techniques used in the spectrophotometric work are applied. The individual frames are geometrically registered and then ratioed, pixel by pixel, in the computer. These ratio frames can then be displayed in a number of ways for comparison with other data. Each ratio frame can be displayed as a black-and-white image with the gray level of each pixel being related to the digital value of the ratio. These ratio images are usually greatly contrast-enhanced to display the full range of the very small color variations. In typical enhancements a 10% variation in the spectral ratio spans the complete range from black to white. Such a ratio frame can be viewed as a map of a given spectral parameter (e.g., reflectance at 0.4 μm compared to reflectance at 0.56 μm). Two ratio images of this kind of the Moon are shown in Adams *et al.* (1974). The images display the 0.4–0.56 μm ratio and the 0.95–0.56 μm ratio data for a region centered on the Taurus Littrow Apollo 17 landing site. In mare soils the 0.4–0.56 μm ratio is related to TiO_2 content, while the 0.95–0.56 μm ratio is related to agglutinate content and soil maturity. These two images were used to investigate the extent of basalts and dark mantling material at the Apollo 17 site.

As the above example illustrates, it is often desirable to correlate the information from two or more ratio frames. Analysis of such correlations is made easier through the use of color composites and computer classification techniques. In a color-ratio composite, three black-and-white ratio frames are optically combined to produce a color image with each of the ratio frames controlling the intensity of one of the primary colors. Regions of a given color in the final image are areas having similar values of *all* the input ratios. Note that with this technique, information from up to six separate bandpasses can be displayed instead of the three for enhanced color or "false color" displays.

Classification maps can be made from an arbitrary number of input data sets. There are several ways for classification to proceed, depending on the nature of the problem: (1) if type localities are defined by the user, other areas within the type localities can be identified and mapped; (2) if particular values and ranges of a set of spectral parameters are specified, all areas within an image with the same spectral properties can be mapped; and (3) automatic classification using cluster analysis techniques can be performed by specifying the number of categories to be found in a given data set. Examples of color-ratio-composite of lunar basalts and a classification map showing Ti contents of basalts in the northern regions of the Moon are given by Johnson *et al.* (1977).

Summary

The strength of multispectral imaging lies in its ability to provide two-dimensional information. This allows the spatial extent of units to be studied in detail and facilitates correlation of spectral data with geological, morphological, chemical and other data sets. In ideal cases, multispectral images can display important data in a format that is easily interpreted.

One major problem with multispectral imaging results from the huge amount of information contained in a single image. A typical telescopic multispectral color ratio composite frame of a 300 km^2 region of the Moon contains on the order of 10^7 bits of information. It is impossible for a single display to show even a small fraction of this data in an interpretable manner. A poor choice of key parameters, such as spectral band positions or number of bands, can result in data that is difficult to understand. To avoid these difficulties, multispectral images need to be keyed to laboratory data and theoretical interpretations of spectral features.

2.3.5 *In situ* analyses

Gamma-ray measurements

As is discussed in section 2.2.5, *in situ* γ-ray measurements were obtained on the surface of Venus by Venera 8, 9 and 10. All three lander spacecraft contained a γ-ray spectrometer for determining the presence of the γ-ray emitting elements, K, Th and U. In addition, Venera 10 carried a γ-ray densitometer to measure the bulk density of the surface (Surkov *et al.*, 1976).

The γ-ray spectrometer was contained in the hermetically sealed spacecraft. The sensor consisted of a 6.3 cm × 6.3 cm NaI (Tl) scintillator. On Venera 8, a 60-channel ultrasonic delay-line pulse height analyzer acquired two 21-minute spectra from 0.3–3 MeV. An additional counter monitored the integral flux during descent and on the surface. No significant change of count rate was noted during descent showing that the short-lived cosmic-ray-induced component was not great. An induced component was present, however, as revealed by the presence of a flux above 2.62 MeV. In contrast with the 60 channels in the Venera 8 instrument, the Venera 9 and 10 spectrometers each had only 3 channels, one for each element. These covered a range which included the most resolvable lines of each element: K(1.46 MeV) range 1.15–1.55 MeV; U(1.76 MeV) range 1.55–2.3 MeV; Th(2.62 MeV) range 2.3–3.0 MeV. It was assumed that instrument gain would remain constant through the entire mission until being turned on just prior to landing. It was at this time that a background measurement was made.

The Venera 10 *density probe* operated on the principle that the probability of a γ-ray source's undergoing Compton scattering is proportional to the number of electrons in a unit of volume of matter and, thus, to its density, for a constant Z/A, where Z and A are respectively the atomic number and atomic weight. Because Z/A is essentially constant (~0.5) for rock-forming elements, the intensity of the scattered quanta is essentially independent of variations in the chemical composition of the rock.

The sensor unit was placed directly on the surface. It consisted of a strong ^{137}Cs source of 0.66 MeV γ-rays and three groups of sensors located at baseline separations of 100 mm, 180 mm and 250 mm from the source. Surkov *et al.* have demonstrated that multiple detectors and calibration curves are necessary because, in the case of the venusian surface, a single detector is not likely to yield a unique solution between counting rate and density.

The curves of count rate vs. density go through a maximum. This follows from the fact that there are two competing processes—scattering and absorption. Calibration curves for the three detectors were used to resolve the ambiguity. A radiation shield was located between the source and detectors to reduce the line-of-sight background.

X-ray

Radioactive sources are used to produce inner atomic shell ionizations that lead to emissions of characteristic X-rays. These are then measured for elemental analysis. Example of such experiments are the X-ray measurements made by the Russian Lunakhod Automatic Stations and those made by the Viking Landers.

Kocharov *et al.* (1971) have described their X-ray fluorescence experiment, designated as "RIFMA," which was mounted on the lunar-roving Lunakhods. The "RIFMA" experiment used radioactive sources such as $^3H/Zr$ and $^3H/Ti$ on an instrument deployable to the lunar surface. The device obtained hundreds of analyses of major and minor elements at the lunar surface, including mare basalts.

The Viking X-ray experiments on Mars involved collection and sieving of samples. It was also possible to select specific sites for sampling using the lander cameras. The samples delivered to the instrument were irradiated by two radioisotopic sources, ^{109}Cd and ^{55}Fe, and the resulting fluorescent radiation was analyzed in an energy-dispersive (pulse height) mode by four small proportional counters. The use of these two sources made available the following excitation energies: 5.9 keV from the ^{55}Fe, 22 keV and 87.7 keV from the ^{109}Cd. As a consequence, fluorescent emission from all the elements between Mg and U could be excited. The minimum energy giving useful sensitivity for the Viking configuration is the 1.25 keV K Mg line (Clark *et al.*, 1977).

Alpha scattering

The alpha back-scatter method was used successfully by Turkevich and co-workers on the three Surveyor lunar landings, two of which occurred on mare basalts (Mare Tranquillitatis and in Sinus Medii). Turkevich *et al.* (1969) were able to determine a large number of elements on the lunar surface by exploiting the Rutherford alpha back-scattering phenomenon as well as (α, p) reactions. The priniciples of the technique have been described by Patterson *et al.* (1965).

The energy spectra of back-scattered alpha particles are characteristic of the elements present in the scattering material. Rutherford *et al.* (1930) derived an expression for the maximum energy of the back-scattered alpha particles, which at high angles is given by

$$F(A, 180°) = \frac{T_{max}}{T_o} \left(\frac{A-4}{A+4} \right)$$

where T_o is the initial energy, T_{max} is the maximum energy and A is the atomic weight. The necessary relationships have been derived for obtaining quantitative data on a number of major elements (excluding hydrogen, helium and lithium). The method resolves the light elements ($Z < 15$) individually and the heavier elements by mass-number grouping.

The Surveyor instrument had six ^{242}Cm alpha sources distributed about the alpha scattering sensor head and collimated to direct the alpha particles through an opening to the lunar surface material. The (recoil) alpha particle detectors were of silicon surface barrier type. In addition there were four lithium-drifted, silicon detectors for the protons produced in the samples by (α, p) reactions. Alpha and proton spectra were determined by 120-channel analyzers.

2.4 SUMMARY

Remote sensing of basalts has been more effectively applied to the Moon than to other solar system bodies including Earth. The lunar maria have been mapped in terms of basalt units, and most of these units have been characterized according to some elemental and/or mineralogical properties. The result has been to delineate those basalts sampled by Apollo and Luna missions, and to recognize and map extensive regions of lunar maria that are covered with basalts that have not yet been sampled. Remote-sensing measurements have helped to work out specific geologic relationships (such as the contrasting compositions of basalts exposed by impact craters with those on the undisturbed surface), thereby contributing to an understanding of the geologic setting in which the basalts occur.

On Earth, compositional remote sensing has not been important for the study of basalts. Mars, Mercury and Venus are inferred to have surface basalts, based on a variety of remote-sensing evidence, and (for Mars and Venus) on measurements by surface landers. Vesta is notable among the asteroids in having the definitive optical properties of basaltic meteorites. The significance of this correlation lies not only in the availability of basaltic meteorites for laboratory study but also in the extension of basalt compositions to minor objects in the solar system. Other solar system bodies are not known to have basalts on their surfaces. A possible exception is Io; however, definitive remote-sensing measurements remain to be made.

ACKNOWLEDGMENTS—We acknowledge the help of J. Arnold, R. E. Arvidson, D. Evans, F. P. Fanale, B. R. Hawke, J. Head, N. Hubbard, R. Huguenin, M. Malin, J. Papike, R. C. Reedy, E. Schonfeld, R. B. Singer, L. Soderblom, and D. Wilhelms in discussing, assembling and reviewing the manuscript. S. Taylor-George and J. Heidanus bore the brunt of the editing and typing. R. Ridings, P. Jones, R. Merrill and the LPI staff exhibited remarkable patience through several last deadlines.

2.5 REFERENCES

Abrams M. J., Ashley R. P., Rowan L. C., Geotz A. F. H. and Kahle A. B. (1977) Mapping of hydrothermal alteration in the Cuprite mining district, Nevada, using aircraft scanner images for the spectral region 0.46 to 2.36 μm. *Geology* **5**, 713–718.

Adams J. B. (1968) Lunar and martian surfaces: petrologic significance of absorption bands in the near-infrared. *Science* **159**, 1453–1455.

Adams J. B. (1974) Visible and near-infrared diffuse reflectance spectra of pyroxenes as applied to remote sensing of solid objects in the solar system. *J. Geophys. Res.* **78**, 4829.

Adams J. B. (1975) Interpretation of visible and near-infrared diffuse reflectance spectra of pyroxenes and other rock forming minerals. In *Infrared and Raman Spectroscopy of Lunar and Terrestrial Minerals* (C. Karr, Jr., ed.), p. 91–116. Academic Press, N. Y.

Adams J. B. and Charette M. P. (1975) Effects of maturation on the reflectance of the lunar regolith: Apollo 16 a case study. *The Moon* **13**, 293–299.

Adams J. B. and Filice A. L. (1967) Spectral reflectance 0.4 to 2.0 microns of silicate rock powders. *J. Geophys. Res.* **72**, 5705–5715.

Adams J. B. and McCord T. B. (1969) Mars: interpretation of spectral reflectivity of light and dark regions. *J. Geophys. Res.*7, 4851–4856.

Adams J. B. and McCord T. B. (1970) Remote sensing of lunar surface mineralogy: implications from visible and near-infrared reflectivity of Apollo 11 samples. *Proc. Apollo 11 Lunar Sci. Conf.*, p. 1937–1945.

Adams J. B. and McCord T. B. (1971) Optical properties of mineral separates, glass and anorthositic fragments from Apollo mare samples. *Proc. Lunar Sci. Conf. 2nd.* p. 2183–2195.

Adams J. B. and McCord T. B. (1972) Electronic spectra of pyroxenes and interpretation of telescopic spectral reflectivity curves of the Moon. *Proc. Lunar Sci. Conf. 3rd*, p. 3021–3034.

Adams J. B. and McCord T. B. (1973) Vitrification darkening in the lunar highlands and identification of Descartes material at the Apollo 16 site. *Proc. Lunar Sci. Conf. 4th*, p. 163–177.

Adams J. B. and McCord T. B. (1977) Mercury: evidence for an anorthositic crust from the refletance spectra (abstract). In *Amer. Astron. Soc. Bull.* **9**, 457.

Adams J. B. and Ralph R. L. (1978) Diffuse reflectance spectra of Luna 24 soils. In *Mare Crisium: The View from Luna 24* (R. B. Merrill and J. J. Papike, eds.), p. 81–87. Pergamon, N.Y.

Adams J. B., Head J. W., McCord T. B., Pieters C. and Zisk S. (1978) Mare Crisium — regional stratigraphy and geologic history. *Geophys. Res. Lett.* **5**, 313–316.

Adams J. B., Pieters C. and McCord T. B. (1974) Orange glass: evidence for regional deposits of pyroclastic origin on the Moon. *Proc. Lunar Sci. Conf. 5th*, p. 171–186.

Adler I. and Trombka J. I. (1970) *Geochemical Exploration of the Moon and Planets.* Springer-Verlag, Berlin. 243pp.

Adler I., Gerard J., Trombka J., Schmadebeck R., Lowman P., Blodget H., Yin L., Eller E., Lamonthe R., Gorenstein P., Bjorkholm P., Harris B. and Gursky H. (1972) The apollo 15 X-ray fluorescence experiment. *Proc. Lunar Sci. Conf 3rd*, p. 2157–2178.

Adler I., Trombka J., Schmadebeck R., Lowman P., Blodget N., Yin L. and Eller E. (1973) Results of the Apollo 15 and 16 X-ray experiment. *Proc. Lunar Sci. Conf 4th* p. 2783–2791.

Andre C. G. and Adler I. (1978) Orbital XRF observations of the lunar subsurface through impact craters (abstract). In *Lunar and Planetary Science IX*, p. 20–21. Lunar and Planetary Institute, Houston.

Andre C. G., Adler I., Clark P. E., Weidner J. R. and Philpotts J. A. (1977a) A geochemical anomaly contiguous with the Dorsa Geike wrinkle ridge in Mare Fecunditatis. *Proc. Lunar Sci. Conf. 7th*, p. 2649–2660.

Andre C. G., Bielefeld M. J., Eliason E., Soderblom L. A., Adler I. and Philpotts J. A. (1977b) Lunar surface chemistry: a new imaging technique. *Science* **197**, 986–989.

Andre C. G., Hallam M. E., Weidner J. R., Podwysocki M. H., Philpotts J. A., Clark P. E. and Alder I. (1975) Correlation of Al/Si X-ray fluorescence data with other remote sensing data from the Taurus-Littrow area. *Proc. Lunar Sci. Conf. 6th*, p. 2739–2748.

Andre C. G., Wolfe R. W. and Adler I. (1978a) Evidence for a high-magnesium subsurface basalt in Mare Crisium from orbital X-ray fluorescence data. In *Mare Crisium: The View from Luna 24* (R. B. Merrill and J. J. Papike, eds.), p. 1–12. Pergamon, N. Y.

Andre C. G., Wolfe R. W. and Adler I. (1978b) Mare basalt depths from orbital X-ray data (abstract). In *Lunar & Planetary Science X*, p. 38–40. Lunar and Planetary Institute, Houston.

Andre C. G., Wolfe R. W. and Alder I. (1979) Are early magnesium-rich basalts widespread on the Moon? *Proc. Lunar Planet. Sci. Conf. 10th*, p. 1739–1751.

Andre C. G., Wolfe R., Adler I., Clark P., Weidner J. and Philpotts J. A. (1977c) Chemical character of the partially flooded Smythii basin based on Al/Si orbital X-ray data. *Proc. Lunar Sci. Conf. 8th*, p. 925–931.

Arnold J. R., Metzger A. E. and Reedy R. C. (1977) Computer-generated maps of lunar composition from gamma-ray data. *Proc. Lunar Sci. Conf. 8th*, p. 945-948.

Arnold J. R., Peterson L. E., Metzger A. E. and Trombka J. I. (1972) Gamma-ray spectrometer experiment. In *Apollo 15 Prelim. Sci. Report*, p. 16–1. NASA SP-289.

Aronson J. R. and Emslie A. G. (1975) Composition of the martian dust as derived by infrared spectroscopy from Mariner 9. *J. Geophys. Res.* **80**, 4925–4931.

Arvidson R. and Bragg S. (1977) Spectrophotometry of Mars by way of the Viking lander cameras. In *Reports of Planetary Geology Program 1977*, p. 204–206. NASA TM X-3511.

Beer R., Norton R. H. and Martonchik J. V. (1971) Astronomical infrared spectroscopy with a Connes-type interferometer: II. Mars, 2500 -3500 cm^{-1}. *Icarus* **15**, 1–10.

Bell P. M. and Mao H. K. (1972) Crystal-field effects of iron and titanium in selected grains of Apollo 12, 14 and 15 rocks, glasses and fine fractions. *Geochim. Cosmochim. Acta* **36**, 545.

Berger M. J. and Seltzer S. M. (1972) Response functions for NaI scintillation detectors. *Nucl. Instr. Methods* **104**, 317–333.

Bielefeld M. (1977) Lunar surface chemistry of regions common to orbital X-ray and gamma-ray experiments. *Proc. Lunar Sci. Conf. 8th*, p. 1131–1147.

Bielefeld M. (1979a) Remote sensing chemistry of geological units (abstract). In *Lunar and Planetary Science X*, p. 116–118. Lunar and Planetary Institute, Houston.

Bielefeld M. (1979b). Classification of the lunar surface using orbital altimetry, geochemistry and geology data (abstract). In *Lunar and Planetary Science X*, p. 113–115. Lunar and Planetary Institute, Houston.

Bielefeld M., Andre C., Eliason E., Clark P., Adler I. and Trombka J. (1977) Imaging of lunar surface chemistry from orbital X-ray data. *Proc. Lunar Sci. Conf. 8th*, p. 901–908.

Bielefeld M. J., Reedy R. C., Metzger A. E., Trombka J. I. and Arnold J. R. (1976) Surface chemistry of selected lunar regions. *Proc. Lunar Sci. Conf. 7th*, p. 2661–2676.

Bielefeld M., Wildey R. and Trombka J. (1978) Correlation of chemistry with normal albedo in the Crisium region. In *Mare Crisium: The View from Luna 24*, (R. B. Merrill and J. J. Papike, eds.), p. 123–132. Pergamon, N. Y.

Binder A. B. and Cruikshank D. P. (1966) Lithological and mineralogical investigation of the surface of Mars. *Icarus* **5**, 521–525.

Binder A. B. and Jones J. C. (1972) Spectrophotometric studies of the photometric function, composition and distribution of the surface materials of Mars. *J. Geophys. Res.* **77**, 3005–3020.

Birks J. B. (1964) *The Theory and Practice of Scintillation Counting.* Pergamon Press, Ltd., Oxford. 662 pp.

Bowell E., Chapman C. R., Gradie J. C., Morrison D. and Zellner B. (1978) Taxonomy of asteroids. *Icarus* **35**, 313–335.

Boyce J. M. (1975) Chronology of the major flow units in the western nearside maria (abstract). In *Origins of Mare Basalts and Their Implications for Lunar Evolution*, p. 11–14. The Lunar Science Institute, Houston.

Boyce J. M. (1976) Age of flow units in the lunar near-side maria based on Lunar Orbiter IV photographs. *Proc. Lunar Sci. Conf. 7th*, p. 2717–2728.

Boyce J. M. and Johnson D. A. (1978). Ages of flow units in the far eastern maria and implications for basin-filling history. *Proc. Lunar Planet. Sci. Conf. 9th*, p. 3275–3283.

Bridge H. S., Belcher J. W., Lazarus A. J., Sullivan J. D., McNutt R. L., Baginal F., Scudder J. D., Sittle E. C., Siscoe G. L., Vasyliunas V. M., Goertz C. K. and Yeates C. M. (1979) Plasma observations near Jupiter: initial results from Voyager 1. *Science* **204**, 987–991.

Broadfoot A. L., Belton M. J. S., Takacs P. Z., Sandel B. R., Shemansky D. E., Holberg J. B., Ajello J. M., Atreya S. K., Sonogue T. M., Moos H. W., Bextaux J. L., Blamont J. E., Strobel D. F., McConnell J. C., Dalgamo A., Goody R. and McElroy M. B. (1979) Extreme ultraviolet observations from Voyager 1 encounter with Jupiter. *Science* **204**, 979–982.

Burns R. G. (1970) *Mineralogical applications of crystal-field theory.* Cambridge Univ. Press. 224 pp.

Carr M. H., Masursky, H., Strom R. G. and Terrile R. J. (1979) Volcanic features of Io. *Nature* **280**, 729–733.

Chapman C. R. (1974) Asteroid size distribution: implications for the origin of stony-iron and iron meteorites. *Geophys. Res. Lett.* **1**, p. 341-344.

Chapman C. R. (1976) Asteroids as meteorite parent bodies: the astronomical perspective. *Geochim Cosmochim. Acta* **40**, 701–719.

Chapman C. R. and Davis D. R. (1975) Asteroid collisional evolution: evidence for a much larger early population. *Science* **190**, 553–556.

Chapman C. R., Morrison D. and Zellner B. (1975) Surface properties of asteroids: a synthesis of polarimetry, radiometry and spectrophotometry. *Icarus* **25**, 104–130.

Charette M. P., McCord T. B., Pieters C. and Adams J. B. (1974) Application of remote spectral reflectance measurements to lunar geology classification and determination of titanium content of lunar soils. *J. Geophys. Res.* **79**, 1065.

Charette M. P., Taylor S. R., Adams J. B. and McCord T. B. (1977) The detection of soils of Fra Mauro basalt and anorthositic gabbro composition in the lunar highlands by remote spectral reflectance techniques. *Proc. Lunar Sci. Conf. 8th*, p. 1049–1061.

Chupp E. L. (1976) *Gamma-Ray Astronomy*. D. Reidel, Boston. 317 pp.

Clark P. and Adler I. (1978) Utilization of independent solar flux measurements to eliminate nongeochemical variation in X-ray fluorescence data. *Proc. Lunar Sci. Conf. 9th*, p. 3029–3036.

Clark P. E., Andre C. G., Adler I., Podwysocki M. H. and Weidner J. (1976) Palus-Somni anomalies in correlation of Al-Si X-ray fluorescence intensity ratios and broad spectrum visible albedos. *Geophys. Res. Lett.* **3**, 421–424.

Clark P. E., Baird A. K., Rose H. J., Jr., Toulmin P., III, Christian R. D., Killiher W. C., Castro A. J., Rowe C. D., Keil K. and Huss G. R. (1977) The Viking X-ray fluorescence experiment: analytical methods and early results. *J. Geophys. Res.* **82**, 4577–4594.

Clark P. E., Eliason E., Andre C. G. and Alder I. (1978) A new color correlation method applied to XRF Al/Si ratios and other lunar remote sensing data. *Proc. Lunar Planet. Sci. Conf. 9th*, p. 3015–3027.

Consolmagno G. J. and Drake M. J. (1977) Composition and evolution of the eucrite parent body: evidence from rare earth elements. *Geochim. Cosmochim. Acta*, **41** 1271–1282.

Cruikshank D. P., Pilcher C. B. and Morrison D. (1977) Evidence for a new class of satellites in the outer solar system. *Astrophys. J.* **217**, 1006–1010.

Davis P. A., Jr. (1980) Iron and titanium distribution on the Moon from orbital gamma-ray spectrometry with implications for crustal evolutionary models. *J. Geophys. Res.* **85**, 3209–3224.

Degewij J. (1977) Lightcurve analyses for 170 small asteroids. *Proc. Lunar Sci. Conf. 8th*, p. 145–148.

Duke M. B. and Silver L. T. (1967) Petrology of eucrites, howardites and mesosiderites. *Geochim. Cosmochim. Acta* **31**, 1637–1665.

Duxbury T. C., Born G. H., Tolson R. H. and Veverka J. (1978) Phobos-Deimos flyby experiment results. *Trans. Amer. Geophys. Union* **59**, 245.

Eberhardt J. (1977) The whole Moon catalog. *Science News III*, p. 300–302.

Ekonomov A. P., Golovin Yu. M. and Moshkin B. E. (1980) Visible radiation observed near the surface of Venus: results and their interpretation. *Icarus* **41**, 65–75.

El-Baz F. and Wilhelms D. (1975) Photogeological, geophysical and geochemical data on the east side of the Moon. *Proc. Lunar Sci. Conf. 6th*, p. 2721–2738.

Eliason E. and Soderblom L. A. (1977) An array processing system for lunar geochemical and geophysical data. *Proc. Lunar Sci. Conf. 8th*, p. 1163–1170.

Evans D. E. (1978) Radar observations of a volcanic terrain: Askja Caldera, Iceland. *JPL Publications 78–81*, p. 36. Jet Propulsion Laboratory, Pasadena.

Evsyukov N. N. and Barabashov N. P. (1973a) Color map of the visible hemisphere of the Moon. Academy of Science, Kiev, U. S. S. R.

Evsyukov N. N. and Barabashov N. P. (1973b) Photometric map of the visible hemisphere of the Moon. Academy of Science, Kiev, U. S. S. R.

Fanale F. P., Brown R. H., Cruikshank D. P. and Clark R. N. (1979) Significance of absorption features in Io's infrared reflectance spectrum. *Nature* **280**, 761–763.

Fanale F. P., Johnson T. V. and Matson D. L. (1974) Io: a surface evaporite deposit? *Science* **186**, 922–924.

Fanale F. T., Johnson T. V. and Matson D. L. (1977) Io's surface and the histories of the Galilean satellites. In *Planetary Satellites* (J. A. Burns, ed), p. 379–405. Univ. of Arizona Press, Tucson.

Farr T. G., Bates B., Ralph R. L. and Adams J. B. (1980) Effects of overlapping optical absorption bands of pyroxenes and glass on the reflectance spectra of lunar soils (abstract). In *Lunar and Planetary Science XI*, p. 276–277. Lunar and Planetary Insitute, Houston.

Faye G. H. and Miller R. M. (1973 "Blue dragon" basalt from Craters of the Moon National Monument, Idaho: origin of color. *Amer. Mineral.* **58**, 1048–1051.

Fink U., Dekkers N. H. and Larson H. P. (1973) Infrared spectra of the Galilean satellites of Jupiter. *Astrophys. J.* **179**, L155–L159.

Fink U., Larson H. P., Gautier T. N., III and Treffers R. R. (1976) Infrared spectra of the satellites of Saturn: identification of water ice on Iapetus Rhaa, Dion and Tethys. *Astrophys. J.* **207**, L63–L67.

Florensky C. P., Basilevsky A. T., Burba G. A. Nikolaeva O. V., Pronin A. A. and Volkov V. P. (1977) First panoramas of the venusian surface. *Proc. Lunar Sci. Conf. 8th*, p. 2655–2664.

Frontispiece (1973) *Proc. Lunar Sci. Conf. 4th*, Vol. 1, Plate II.

Frontispiece (1976) *Proc. Lunar Sci. Conf. 7th*, Vol. 1, Plate III.

Frontispiece (1977) *Proc. Lunar Sci. Conf. 8th*, Vol. 1, Plate 24.

Gaffey M. (1975) Quantification of spectral reflectance characteristics for mineralogical and petrological information (abstract). In Sixth Annual Meeting of the AAS/DPS, p. 4.

Gaffey M. J. and McCord T. B. (1977) Asteroid surface materials: mineralogical characterizations and cosmological implications. *Proc. Lunar Sci. Conf. 8th*, p. 113–143.

Gillett F. C., Merrill K. M. and Stein W. A. (1970) Albedo and thermal emission of Jovian satellites I-IV. *Astrophys. Lett* **6**, 247–249.

Gorenstein P., Gursky H., Adler I. and Trombka J. (1969) System for non-dispersive analysis of lunar X-rays from Apollo. *Advances in X-ray Anal.* **13**, 330–341.

Greeley R. and Spudis P. D. (1978) Volcanism in cratered terrain hemisphere of Mars. *Geophys. Res. Lett.* **5**, 453–455.

Haines E. L., Arnold J. R. and Metzger A. E. (1976) Chemical mapping of planetary surfaces. *IEEE Transact. Geoscience Elect. Ge-14*, p. 141–153.

Haines E. L., Etchegaray-Ramirez M. I. and Metzger A. E. (1978) Thorium concentrations in the lunar surface. II: deconvolution modeling and its application to the regions of Aristarchus and Mare Smythii. *Proc. Lunar Planet. Sci. Conf. 9th*, p. 2985–3013.

Hanel R., Conrath B., Hovis W., Kunde V., Lowman P., Maguire W., Pearl J., Pirraglia J., Prabhakara C., Schlachman B., Levin G., Straat P. and Burke T. (1972) Investigation of the martian environment by infrared spectroscopy on Mariner 9. *Icarus* **17**, 423–442.

Hansen O. L. (1975) Infrared albedos and rotation curves of the Galilean satellites. *Icarus* **26**, 24–29.

Hapke B. (1968) Lunar surface: composition inferred from optical properties. *Science* **159**, 76–79.

Hapke B. (1977) Interpretation of optical observations of Mercury and the Moon. *Phys. Earth Planet. Int.* **15**, 264–274.

Hapke B., Cassidy W. and Wells E. (1975a) Effects of vapor deposition processes on the optical, chemical and magnetic properties of the lunar regolith. *The Moon* **13**, 339–353.

Hapke B., Danielson G. E., Klaasen K. and Wilson L. (1975b) Photometric observations of Mercury from Mariner 10. *J. Geophys. Res.* **80**, 2431–2443.

Hargraves R. B., Collinson D. W., Arvidson R. E. and Spitzer C. R. (1977) The Viking magnetic properties experiment: primary mission results. *J. Geophys. Res.* **82**, 4547–4558.

Harrington T. M., Marshall J. H., Arnold J. R., Peterson L. E., Trombka J. I. and Metzger A. E. (1974) The Apollo gamma-ray spectrometer. *Nucl. Instr. Methods* **118**, 401–411.

Hawke B. R. and Head J. W. (1978) Lunar KREEP volcanism: geologic evidence for history and mode of emplacement. *Proc. Lunar Planet. Sci. Conf. 9th*, p. 3285–3309.

Hawke B. R., MacLaskey D. and McCord T. B. (1979) Multispectral mapping of Apollo 15 — Apennine Region: the identification, distribution, and characterization of region pyroclastic deposits (abstract). In *Lunar and Planetary Science X*, p. 513–515. Lunar and Planetary Institute, Houston.

Hayes D. S. (1970) An absolute spectrophotogetric calibration of the energy distribution of twelve standard stars. *Astrophys. J.* **159**, 165.

Head J. W. (1974) Lunar dark-mantle deposits: possible clues to the distribution of early mare deposits. *Proc. Lunar Sci. Conf. 5th*, p. 207–222.

Head J. W. (1976) Lunar volcanism in space and time. *Rev. Geophys. Space Phys.* **14**, 265.

Head J. W. and McCord T. B. (1978) Imbrian-age highland volcanism on the Moon: the Gruithuisin and Mairan domes. *Science* **199**, 1433–1436.

Head J. W., Adams J. B., McCord T. B., Pieters C. and Zisk S. (1978a) Regional stratigraphy and geologic history of Mare Crisium. In *Mare Crisium: The View from Luna 24*, (R. B. Merrill, J. J. Papike, eds.), p. 43–47. Pergamon, N. Y.

Head J. W., Pieters C., McCord T., Adams J. and Zisk S. (1978b) Definition and detailed characterization of lunar surface units using remote observations. *Icarus* **33**, 145–172.

Heiken G. H., McKay D. S. and Brown R. W. (1974) Lunar deposits of possible pyroclastic origin. *Geochim. Cosmochim. Acta* **38**, 1703–1718.

Hörz F. (1978) How thick are lunar mare basalts? *Proc. Lunar Planet. Sci. Conf. 9th*, p. 3311–3331.

Houck J. R., Pollack J. B., Sagan C., Schaack D. and Decker J. A., Jr. (1973) High altitude infrared spectroscopic evidence for bound water on Mars. *Icarus* **18**, 470–480.

Hubbard N. J. (1979) Regional chemical variations in lunar basaltic lavas. *Proc. Lunar Plant. Sci. Conf. 10th*, p. 1753–1774.

Hubbard N. J. and Andre C. G. (1977) High resolution orbital X-ray results for the Mare Fecunditatis-Langrenus crater region (abstract). In *Lunar Science VIII*, p. 469–471. The Lunar Science Institute, Houston.

Hubbard N. J. and Gast P. W. (1971) Chemical composition and origin of non-mare lunar basalts. *Proc. Lunar Sci. Conf. 2nd*, p. 999–1020.

Hubbard N. J. and Keith J. E. (1977) Reevaluation of the Apollo orbital X-ray fluorescence data. *Proc. Lunar Sci. Conf. 8th*, p. 909–923.

Hubbard N. J. and Keith J. E. (1978) A declaration of independence for Mg/Si. *Proc. Lunar Planet. Sci. Conf. 9th*, p. 3049–3056.

Hubbard N. J. and Minear J. W. (1976) Petrogenesis in a modestly endowed Moon. *Proc. Lunar Sci. Conf. 7th*, p. 3421–3435.

Hubbard N. J. and Woloszyn D. (1977) Orbital gamma-ray data and large-scale lunar problems. *Phys. Earth and Planet. Inter.*, **15**, 287–302.

Hubbard N. J., Vilas F. and Keith J. E. (1978a) From Serenity to Langemak: a regional chemical setting for Mare Crisium. In *Mare Crisium: The View from Luna 24*, (R. B. Merrill and J. J. Papike, eds.), p. 13–32. Pergamon, N. Y.

Hubbard N. J., Vilas F. and Keith J. E. (1978b) Improvements in orbital X-ray fluorescence data reduction (abstract). In *Lunar and Planetary Science IX*, p. 555–557. Lunar and Planetary Institute, Houston.

Hubbard N. J., Vilas F. and Keith J. E. (1978c) Solar corrections for the Apollo orbital X-ray fluorescence data (abstract). In *Lunar and Planetary Science IX*, p. 558–560. Lunar and Planetary Institute, Houston.

Huck F. O., Jobson D. J., Park S. K., Wall S. D., Arvidson R. E., Patterson W. R. and Benton W. D. (1977). Spectrophotometric and color estimates of the Viking lander sites. *J. Geophys. Res.*, **82**, 4401–4411.

Huguenin R. L. (1974) The formation of goethite and hydrated clay minerals on Mars. *J. Geophys. Res.* **79**, 3895–3905.

Huguenin R. L. (1976a) Mars: chemical weathering as a massive volatile sink. *Icarus* **28**, 203–212.

Huguenin R. L. (1976b) Surface oxidation: a major sink for water on Mars. *Science* **192**, 138–139.

Huguenin R. L., Adams J. B. and McCord T. B. (1977a) Mars: surface mineralogy from reflectance spectra (abstract). In *Lunar Science VIII*, p. 478–480. The Lunar Science Institute, Houston.

Huguenin R. L., Prinn R. G. and Maderazzo M. (1977b). Mars: photodesorption from mineral surfaces and its effect on atmospheric stability. *Icarus* **32**, 270–298.

Hunt G. R., Logan L. M. and Salisbury J. W. (1973) Mars: components of infrared spectra and the composition of the dust cloud. *Icarus* **18**, 459–469.

Hunt G. R., Salisbury J. W. and Lenoff C. J. (1974) Visible and near IR spectra of minerals and rocks: IX. Basic and ultrabasic igneous rocks. *Mod. Geol.* **5**, 15–22.

Irving A. J. (1977) Chemical variation and fractionation of KREEP basalt magmas. *Proc. Lunar Sci. Conf. 8th*, p. 2433–2448.

Johnson T. V. (1971) Galilean satellites: narrowband photometry 0.30 to 1.10 microns. *Icarus* **14**, 94–111.

Johnson T. V. (1978) The Galilean satellites of Jupiter: four worlds. *Ann. Rev. Earth Planet. Sci.* **6**, 93–125.

Johnson T. V. and McCord T. B. (1971) Spectral geometric albedo of the Galilean satellites, 0.3 to 2.5 microns. *Astrophys. J.* **169**, 584–594.

Johnson T. V. and Pilcher C. B. (1977) Satellite spectrophotometry and surface composition. In *Planetary Satellites* (J. A. Burns, ed.), p. 232–268. Univ. of Arizona Press, Tucson.

Johnson T. V., Mosher J. A. and Matson D. L. (1977) Lunar spectral units: a northern hemispheric mosaic. *Proc. Lunar Sci. Conf. 8th*, p. 1013–1028.

Johnson T. V., Pieters C. and McCord T. B. (1973) Mare Humorum: an integrated study of spectral reflectivity. *Icarus* **19**, 224–229.

Johnson T. V., Veeder G. J. and Matson D. L. (1975) Evidence for frost on Rhea's surface. *Icarus* **24**, 428–432.

Kahle A. B. (1977) A simple thermal model of the Earth's surface for geologic mapping by remote sensing. *J. Geophys. Res.* **82**, 1673–1680.

Kocharov G. E. and Viktorov S. V. (1973) Lunar surface chemical composition in the region of activity of Lunkahod 2. *Doklady Akad. Nauk, SSR.*

Kocharov G. E., Viktorov S. V., Borodulin N. F., Voropayev O. M., Dzevanovskaya A. Yu., Ivanov O. G., Kir'yan G. V., Petrov V. V. and Sakul'skiy V. A. (1971) In the Symposium, *The Mobile Laboratory "Lunakhod 1" on the Moon*, Pt. 1. 89. NAUKA, in Russian.

Krimigis S. M., Armstrong T. P., Axford W. I., Bostrom C. O., Fan C. Y., Gloeckler G., Lanzerotti L. J., Keath E. P., Zwickl R. D., Carbary J. F. and Hamilton D. C. (1979) Low-energy charged particle environment at Jupiter: a first look. *Science* **204**, 978–1003.

Labs D. and Neckel H. (1968) The radiation of the solar photosphere from 2000 Å to 100 μ. *Z. Astrophys.* **69**, 1–73.

Langseth M. G., Keihm S. J. and Peters K. (1976) Revised lunar heat-flow values. *Proc. Lunar Sci. Conf. 7th*, p. 3143–3171.

Larson H. P. and Fink U. (1975) Infrared spectral observations of asteroid (4) Vesta. *Icarus* **26**, 420–427.

Laul J. C. and Fruchter J. S. (1976) Thorium and uranium variations in Apollo 17 basalts, and K-U systematics. *Proc. Lunar Sci. Conf. 7th*, p. 1545–1559.

Lebofsky L. A. (1978) Asteroid 1 Ceres: evidence for water of hydration. *Royal Astron. Soc. Monthly Notices* **182**, 17p–21p.

Lee T. (1971) Spectral albedos of the Galilean satellites. *Commun. Lunar and Planetary Lab., Univ. Arizona* **9**, 179–180.

Lefebvre R. H. (1975) Mapping in the Craters of the Moon Volcanic Field, Idaho, with Landsat imagery. *Proc. 10th Inter. Symposium on Remote Sensing of the Environment* **2**, 951–963. Environmental Research Institute of Michigan, Ann Arbor.

Lefebvre R. H. and Abrams M. J. (1977) Computer classification of pahoehoe flows in the Craters of the Moon Volcanic Field, Idaho, using Landsat data. In *Abstracts for the Planetary Geology Field Conference on the Snake River Plain, Idaho*, p. 436. NASA TM-78.

Logan L. M., Hunt G. R. and Salisbury J. W. (1975) The use of mid-infrared spectroscopy in remote sensing of space targets. In *Infrared and Raman Spectra of Lunar and Terrestrial Minerals* (C. Kaer, Jr., ed), p. 117–142. Academic Press, N. Y.

Lucchita B. K. and Schmitt H. H. (1974) Orange material in the Sulpicius Gallus formation at the southwestern edge of Mare Serenitatis. *Proc. Lunar Sci. Conf. 5th*, p. 223–234.

Lyon R. J. P. (1965) Analysis of rocks by spectral IR emission (8–25 μm). *Econ. Geol.* **60**, 715–736.

Lyon R. J. P. (1972) Infrared spectral emittance in geological mapping: airborne spectrometer data from Pisgah Crater, California. *Science* **175**, 983–985.

Malin M. C. (1974) Lunar red spots: possible pre-mare materials. *Earth Planet. Sci. Lett.* **21**, 331–341.

Malin M. C. (1978) Surfaces of Mercury and the Moon — effects of resolution and lighting conditions on discrimination of volcanic features. *Proc. Lunar Planet Sci. Conf. 9th*, p. 3395–3409.

Malin M. C. and Saunders R. S. (1977) Surface of Venus: evidence of diverse landforms from radar observations. *Science* **196**, 987–990.

Malin M. C., Evans D. and Elachi C. (1978) Imaging radar observations of Askja Caldera, Iceland. *Geophys. Res. Lett.* **5**, 931–934.

Maxwell T. A. and Phillips R. J. (1978) Stratigraphic correlations of the radar-detected subsurface interface in Mare Crisium. *Geophys. Res. Lett.* **9**, 811–814.

May T. W., Peeples W. J., Maxwell T., Sill W. R., Ward S. H., Phillips R. J., Jordan R. and Abott E. (1976) Subsurface layering in Mare Serenitatis and Crisium: Apollo lunar sounder results. *Proc. Lunar Sci. Conf. 7th*, p. 540–542.

McCord T. B. (1968). A double beam astronomical photometer. *Applied Optics* **7**, 475–478.

McCord T. B. and Adams J. B. (1969) Spectral reflectivity of Mars. *Science* **163**, 1058–1060.

McCord T. B. and Adams J. B. (1972a) Mercury: interpretation of optical observations. *Icarus* **17**, 585.

McCord T. B. and Adams J. B. (1972b) Mercury: surface composition from the reflection spectrum. *Science* **178**, 745.

McCord T. B. and Chapman C. R. (1975a) Asteroids: spectral reflectance and color characteristics I. *Astrophys. J.* **195**, 553–562.

McCord T. B. and Chapman C. R. (1975b) Asteroids: spectral reflectance and color characteristics II. *Astrophys. J.* **197**, 781–790.

McCord T. B. and Clark R. N. (1979) The Mercury soil: presence of Fe^{2+} *J. Geophys. Res.* **84**, 7664–7668.

McCord T. B. and Frankston M. J. (1975) Observational experience with silicon diode array vidicons at the telescope. *Applied Optics* **14**, 1437–1446.

McCord T. B. and Johnson T. V. (1969) Lunar spectral reflectivity (0.30–2.50 μm) and implications for remote mineralogical analysis. *Science* **169**, 855–858.

McCord T. B. and Westphal J. A. (1971) Mars: narrowband photometry 0.3 – 2.5 microns of surface regions during the 1969 apparition. *Astrophys. J.* **168**, 141.

McCord T. B. and Westphal J. A. (1972) Two-dimensional silicon vidicon astronomical photometer. *Applied Optics* **11**, 522–526.

McCord T. B., Adams J. B. and Johnson T. V. (1970) Asteroid Vesta: spectral reflectivity and compositional implications. *Science* **168**, 1445.

McCord T. B., Charette M., Johnson T. V., Lobofsky L. A., Pieters C. and Adams J. B. (1972) Lunar spectral types. *J. Geophys. Res.*77, 1349–1359.

McCord T. B., Clark R. and Huguenin R. L. (1978) Mars: near-infrared spectral reflectance and compositional implications. *J. Geophys. Res.* **83**, 5433–5441.

McCord T. B., Grabow M., Feierberg M. A., MacLaskey D., and Pieters C. (1979) Lunar Multispectral Maps: Part II of the Lunar Nearside. *Icarus* **37**, 1–28.

McCord T. B., Huguenin R. L., Mink D. and Pieters C. (1977) Spectral reflectance of martian areas during the 1973 opposition: photoelectric filter photometry 0.3 – 1.10 μ.m. *Icarus* **31**, 25–39.

McCord T. B., Pieters C. and Feierberg M. A. (1976) Multispectral mapping of the lunar surface using ground-based telescopes. *Icarus* **29**, 1–34.

McFadden L., McCord T. B. and Pieters C. (1977) Vesta: the first pyroxene band from new spectroscopic measurements. *Icarus* **31**, 439–446.

Metzger A. E. and Arnold J. R. (1970) Gamma-ray spectroscopic measurements of Mars. *Applied Optics* **9**, 1298–1303.

Metzger A. E. and Parker R. E. (1980) The distribution of titanium on the lunar surface. *Earth Planet. Sci. Lett.*, **45**, 155–171.

Metzger A. E., Haines E. L. and Etchegaray-Ramirez M. I. (1979) Thorium concentrations in the lunar surface: III deconvolution of the Apenninus region. *Proc. Lunar Planet. Sci. Conf. 10th*, p. 1701–1718.

Metzger A. E., Haines E. L., Parker R. E. and Radocinski R. G. (1977a) Thorium concentrations in the lunar surface: I. regional values and crustal content. *Proc. Lunar Sci. Conf. 8th*, p. 949–999.

Metzger A. E., Lin R. P. and Russel C. T. (1977b) On a correlation between surface remnant magnetism and chemistry for the lunar frontside and limbs. *Proc. Lunar Sci. Conf. 8th*, p. 1187–1190.

Metzger A. E., Trombka J. I., Peterson L. E., Reedy R. C. and Arnold J. R. (1973) Lunar surface radioactivity: preliminary results of the Apollo 15 and Apollo 16 gamma-ray spectrometer experiments. *Science* **179**, 800–803.

Metzger A. E., Trombka J. I., Reedy R. C. and Arnold J. R. (1974) Element concentrations from lunar orbital gamma-ray measurements. *Proc. Lunar Sci. Conf. 5th*, p. 1067–1078.

Meyer C. (1977) Petrology, mineralogy and chemistry of KREEP basalt. *Phys. Chem. Earth* **10**, 239–260.

Minear J. W. and Fletcher C. R. (1978) Crystallization of a lunar magma ocean. *Proc. Lunar Planet. Sci. Conf. 9th*, p. 263–282.

Moderazzo M. and Huguenin R. L. (1977) Petrologic implications of Viking XRF analysis based on reflection spectra and the photochemical weathering model (abstract). *Bull. Amer. Astron. Soc.* **9**, 527–528.

Moore H. J., Arthur D. W. G. and Schaber G. G. (1978) Yield strengths of flow on the Earth, Mars, and Moon. *Proc. Lunar Planet. Sci. Conf. 9th*, p. 3351–3378.

Moroz V. I. (1965) Infrared spectrophotometry of the Moon and the Galilean satellites of Jupiter. (1966) *Astron. Zh.* **42**, 1287–1295 (Translated in *Sov. Astron. AJ* **9**, 999–1066).

Morrison D., Cruikshank D. P. and Burns J. A. (1977) Introducing the satellites. In *Planetary Satellites*, (J. A. Burns, ed.), p. 3–17. Univ. of Arizona Press, Tucson.

Morrison D., Morrison N. D. and Lazarewicz A. R. (1974) Four color photometry of the Galilean satellites. *Icarus* **23**, 399–416.

Muller P. M. and Sjogren W. L. (1968) Mascons: lunar mass concentrations. *Science* **161**, 680–684.

Murray B. C., Strom R. G., Trask N. J. and Gault D. E. (1975) Surface history of Mercury — implications for terrestrial planets. *J. Geophys. Res.* **80**, 2431–2443.

Nakano G. H., Imhof W. L. and Reagan J. B. (1976) Cooled germanium spectrometers for X- and gamma-ray astronomy. *Space Sci. Instru.* **2**, 219–237.

Nash D. B. and Conel J. E. (1973) Vitrification darkening of rock powders: implications for optical properties of the lunar surface. *The Moon* **8**, 346–364.

Oke J. B. (1964) Photoelectric spectrometry of stars suitable for standards. *Astrophys. J.* **140**, 689.

Oke J. B. and Schild R. E. (1970) The absolute spectral energy distribution of alpha lyrae. *Astrophys. J.* **161**, 1015.

Pang K. D., Pollack J. B., Veverka J., Lane A. L. and Ajello J. M. (1978) The composition of Phobos: evidence carbonaceous chondrite surface from spectral analysis. *Science* **199**, 64–66.

Patterson J. H., Turkevich A. L. and Franzgrote E. J. (1965) Chemical analysis of surface using alpha particles. *J. Geophys. Res.* **70**, 1311–1327.

Peale S. J., Cassen P. and Reynolds R. T. (1979) Melting of Io by tidal dissipation. *Science* **203**, 892.

Pearl J., Hanel R., Kunde V., Maguire W., Fox K., Gupta S., Ponnamperuma C. and Roulin F. (1979) Identification of gaseous SO_2 and new upper limits for other gases on Io. *Nature* **280**, 755–758.

Pehl R. H. (1977) Germanium gamma-ray detectors. *Physics Today* **30**, 50–61.

Phillips R. J., Adams G. R., Brown W. E., Jr., Eggleton R. E., Jackson P., Jordan R., Linlor W. I., Peeples W. J., Procello L. J., Ryn J., Schaber G., Sill W. R., Thompson T. W., Ward S. H. and Zelenka J. S. (1973) Apollo lunar sounder experiment. *Apollo 17 Prelim. Sci. Rep.*, p. 22-1–22-26. NASA SP-330.

Pieters C. (1974) Polarization in a mineral absorption band. In *Planets, Stars and Nebulae* (T. Gehrels, ed.), p. 405–418. Univ. of Arizona Press, Tucson.

Pieters C. (1977a) Characterization and distribution of lunar mare basalt types using remote sensing techniques. Ph.D. thesis, Massachusetts Institute of Technology. 348pp.

Pieters C. (1977b) Characterization of lunar mare basalt types II: spectral classification of fresh mare craters. In *Proc. Lunar Sci. Conf. 8th*, p. 1037–1048.

Pieters C. (1978) Mare basalt types on the front side of the Moon: a summary of spectral reflectance data. *Proc. Lunar Planet. Sci. Conf. 9th*, p. 2825–2849.

Pieters C. and McCord T. B. (1976) Characterization of lunar mare basalt types: I. remote sensing study using reflection spectroscopy of surface soils. *Proc. Lunar Sci. Conf. 7th*, p. 2677–2690.

Pieters C., Head J. W., Adams J. B., McCord T. B., Zisk S. and Whitford-Stark J. (1980) Late high titanium basalts of the western maria: geology of the Flamsteed region of Oceanus Procellarum. *J. Geophys. Res.* **85**, 3913–3938.

Pieters C., McCord T. B., Charette M. P. and Adams J. B. (1974) Lunar Surface: identification of the dark mantling material in the Apollo 17 soil samples. *Science* **183**, 1191–1194.

Pieters C., McCord T. B., Charette M. P., Adams J. B. and Zisk S. (1975) Geochemical and geological units of Mare Humorum: definition listing remote sensing and lunar sample informatin. *Proc. Lunar Sci. Conf. 6th*, p. 2689–2710.

Pieters C., McCord T. B., Zisk S. and Adams J. B. (1973) Lunar black spots and the nature of the Apollo 17 landing area. *J. Geophys. Res.* **78**, 5867–5875.

Pilcher C. B., Ridgeway S. T. and McCord T. B. (1972) Galilean satellites: identification of water frost. *Science* **178**, 1087.

Pitkin J. A. (1968) Airborne measurements of terrestrial radioactivity as an aid to geologic mapping. *U. S. G. S. Professional Paper 516-F.*

Pohn H. and Wildey R. L. (1970) A photoelectric-photographic study of the normal albedo of the Moon. *U. S. G. S. Professional Paper (map) 559-E.*

Pollack J. B. and Sagan C. (1969) An analysis of martian photometry and polarimetry. *Space Sci. Rev.* **9**, 243–299.

Pollack J. B., Veverka J., Pang K. D., Colburn D., Lane A. L. and Ajello J. M. (1978) Multicolor observations of Phobos with the Viking lander cameras: evidence for a carbonaceous chondritic composition. *Science* **199**, 66–69.

Reedy R. C. (1978) Planetary gamma-ray spectroscopy. *Proc. Lunar Planet. Sci. Conf. 9th*, p. 2961–2985.

Reedy R. C. and Arnold J. R. (1972) Interaction of solar and galactic cosmic ray particles with the Moon. *J. Geophys. Res.* **77**, 537–555.

Reedy R. C., Arnold J. R. and Trombka J. I. (1973) Expected gamma-ray emission spectra from the lunar surface as a function of chemical composition. *J. Geophys. Res.* **78**, 5847–5866.

Ronca L. B. and Furlong R. B. (1977) Topographic changes on airless bodies: theoretical considerations. *The Moon* **17**, 233–237.

Rowan L. C., Goetz A. H. F. and Ashley R. P. (1977) Discrimination of hydrothermally altered and unaltered rocks in visible and near-infrared multispectral images. *Geophys.* **42**, 522–535.

Rutherford E., Chadwick J. and Ellis C. D. (1930) *Radiations from radioactive substances.* Cambridge Univ. Press. 244 pp.

Sagan C. (1979) Sulphur flows on Io. *Nature* **280**, 750–753.

Schaber G. G. (1973) Lava flows in Mare Imbrium: geologic evaluation from Apollo orbital photography (abstract). In *Lunar Science IV*, p. 653–655. The Lunar Science Institute, Houston.

Schild R. E., Peterson E. M. and Oke J. B. (1971) Effective temperatures of B-and A-type stars. *Astrophys. J.* **166**, 95.

Schmidt R. G. (1961) Aeroradioactivity of the Hanford area, Washington and Oregon. *U. S. G. S. Geophys. Inv. Map. GP-307.*

Schonfeld E. (1977) Comparison of orbital chemistry with crustal thickness and lunar sample chemistry. *Proc. Lunar Sci. Conf. 8th*, p. 1149–1162.

Schonfeld E. and Bielefeld M. J. (1978) Correlation of dark mantle deposits with high Mg/Al ratios. *Proc. Lunar Planet. Sci. Conf. 9th*, p. 3037–3048.

Singer R. B., McCord T. B., Clark R. N., Adams J. B. and Huguenin R. L. (1979). Mars surface composition from reflectance spectroscopy: a summary. *J. Geophys. Res.* **84**, 8415–8426.

Sinton W. M. (1967) On the composition of martian surface material. *Icarus* **6**, 222–228.

Sjogren W. L., Wimberly R. N. and Wollenhaupt W. R. (1974) Lunar gravity via the Apollo 15 and 16 subsatellites. *The Moon* **9**, 115–128.

Smith B. A., Shoemaker E. M., Kieffer S. W. and Cook A. F., II (1979a) The role of SO_2 in volcanism on Io. *Nature* **280**, 738–743.

Smith B. A., Soderblom L. A., Beebe R., Boyce J., Briggs G., Carr M., Collins S. A., Cook A. F., Danielson G. E., Davies M. E., Hunt G. E., Ingersoll A., Johnson T. V., Masursky H., McCauley J., Morrison D., Owen T., Sagan C., Shoemaker E. M., Strom R., Suomi V. F. and Veverka J. (1979c) The Galilean satellites and Jupiter: Voyager 2 imaging science results. *Science* **206**, 927–950.

Smith B. A., Soderblom L. A., Johnson T. V., Ingersoll A. P., Collins S. A., Shoemaker E. M., Hunt G. E., Masursky H., Carr M. H., Davies M. E., Cook A. F., II, Boyce J., Danielson G. E., Owen T., Sagan C., Beebe R. F., Veverka J., Strom R. G., McCauley J. F., Morrison D., Briggs G. A. and Suomi V. E. (1979b) The Jupiter system through the eyes of Voyager 1. *Science* **204**, 951–972.

Smythe W. D., Nelson R. M. and Nash P. B. (1979) Spectral evidence for SO_2 frost or adsorbate on Io's surface. *Nature* **280**, 766.

Soderblom L. A. (1970) The distribution of ages of regional units in the lunar maria. PhD thesis, California Institute of Technology, Pasadena.

Soderblom L. A. and Lebofsky L. A. (1972) Technique for rapid determination of relative ages of lunar areas from orbital photography. *J. Geophys. Res.* **77**, 279–296.

Soderblom L. A., Arnold J. R., Boyce J. M. and Lin R. P. (1977) Regional variations in the lunar maria: age, remnant magnetism, and chemistry. *Proc. Lunar Sci. Conf. 8th*, p. 1191–1199.

Soderblom L. A., Edwards K., Eliason E. M., Sanchez E. M. and Charette M. P. (1978) Global color variations on the martian surface. *Icarus* **34**, 446–464.

Solomon S. C. (1975) Mare volcanism and lunar crustal structure. *Proc. Lunar Sci. Conf. 6th*, p. 1021–1042.

Solomon S. C. and Head J. W. (1979) Vertical movement in mare basins: relation to mare implacement, basin tectonics and lunar thermal history. *J. Geophys. Res.* **84**, 1667–1682.

Strom R., Trask R. and Guest J. (1975) Tectonism and volcanism on Mercury. *J. Geophys. Res.* **80**, 2478–2507.

Surkov Yu. A. (1977a) Gamma-spektrometriya v kosmicheskikh issledovaniyakh. Atom. Press, Moscow.

Surkov Yu. A. (1977b) Geochemical studies of Venus by Venera 9 and 10 automatic interplanetary stations. *Proc. Lunar Sci. Conf. 8th*, p. 2665–2689.

Surkov Yu. A., Kirnozov F. F., Glazov V. N., Duchenko A. G. and Tatsy L. P. (1976) The content of natural radioactive elements in venusian rock as determined by Venera 9 and Venera 10. *Kosmich. Issled.* **14**, 704–709.

Surkov Yu. A., Moskaleva L. P., Kharyukova V. P., Manvelyan O. S. and Shcheglov O. P. (1977) Experimental simulation of the gamma radiation produced by nuclear interactions of high energy particles with the material of the Moon and planets. *Kosmich. Issled.* **15**, 144–151. (Translated in *Cosmic Research* **15**, 118–124).

Taylor S. R. (1975) *Lunar Science: A Post-Apollo View.* Pergamon, N. Y. 372 pp.

Thompson T. W., Shorthill R. W., Whitaker E. A. and Zisk S. H. (1974) Mare Serenitatis: a preliminary definition of surface units by remote observations. *The Moon* **9**, 89–96.

Toon O. B., Pollack J. B. and Sagan C. (1977) Physical properties of the particles composing the martian dust storm of 1971–1972. *Icarus* **30**, 663–696.

Tossell J. A., Vaughn D. J. and Johnson K. H. (1974) The electronic structure of rutile, wustile, and hematite from molecular orbital calculations. *Amer. Mineral.* **59**, 319.

Toulmin P., III, Baird A. K., Clark B. C., Keil K., Rose H. J., Jr., Christian R. P., Evans P. H. and Kelleher W. C. (1977) Geochemical and mineralogical interpretation of the Viking inorganic chemical results. *J. Geophys. Res.* **82**, 4625–4634.

Trask N. J. and Guest J. E. (1975) Preliminary geologic terrain map of Mercury. *J. Geophys. Res.* **80**, 2461–2477.

Turkevich A. L., Franzgrote E. J. and Patterson J. H. (1969) Chemical composition of the lunar surface in Mare Tranquillitatis. *Science* **165**, 277.

Vaniman D. T. and Papike J. J. (1977) Very low Ti (VLT) basalts: a new mare rock type from the Apollo 17 drill core. *Proc. Lunar Sci. Conf. 8th*, p. 1443–1471.

Vilas F. and McCord T. (1976) Mercury: spectral reflectance measurements (0.33 – 1.06 m) 1974–75. *Icarus* **29**, 593–599.

Vincent R. K. and Thomson F. J. (1971) Discrimination of basic silicate rocks by recognition maps processed from serial infrared data. *Inter. Symposium Proc. 7th on Remote Sensing of the Environment* **1**, 247–252. Willow Run Laboratory, Ann Arbor.

Vincent R. K. and Thomson F. J. (1972) Rock-type discrimination from ratioed infrared scanner images of Pisgah Crater, California. *Science* **175**, 986–988.

Vincent R. K., Rowan L. C., Gillespie R. E. and Knapp C. (1975) Thermal IR spectra and chemical analysis of 26 igneous rock samples. *Remote Sensing of Environment* **4**, 199–209.

Vinogradov A. P., Surkov Yu. A., Kirnozov F. F. and Glazov V. N. (1973) The content of natural radioactive elements in venusian rock according to the data obtained by the automatic probe "Venera 8". *Geochem. Int.* **10**, 1.

Vogt R. E., Cook W. R., Cummings A. C., Garrard T. L., Gehrels N., Stone E. C., Trainor J., II, Schardt A. W., Conlon T., Lal N. and McDonald F. B. (1979) Voyager 1: energetic ions and electrons in the Jovian magnetosphere. *Science* **204**, 1003–1007.

Walker D., Stolper E. M. and Hays J. F. (1977) Size of the eucrite parent body. *Meteoritics* **12**, 375.

Whitaker E. A. (1966) The surface of the Moon. In *The Nature of the Lunar Surface* (W. N. Hess, D. H. Menzel and J. A. O'Keefe, eds.), p. 78–98. John Hopkins Press, Baltimore.

Whitford-Stark J. L. (1980) The craters of Mare Imbrium (abstract). In *Lunar and Planetary Science XI*, p. 1242–1244. Lunar and Planetary Institute, Houston.

Whitford-Stark J. L. and Head J. W. (1977a) Oceanus Procellarum: volcanic and tectonic evolution (abstract). In *Lunar Science VIII.*, p. 1011–1014. The Lunar Science Institute, Houston.

Whitford-Stark J. L. and Head J. W. (1977b) The Procellarum volcanic complexes: contrasting styles of volcanism. *Proc Lunar Sci. Conf. 8th*, p. 2705–2724.

Whitford-Stark J. L. and Head J. W. (1981) Stratigraphy of Oceanus Procellarum sources and styles of emplacement. *J. Geophys. Res.*, in press.

Wildt R. (1934) Ozon and Sauerstoff in den Planeten Atmosphären. *Veroffentli Univ-Sternwaite Gottingen*, no. 38.

Wilhelms D. E. and McCauley J. (1971) Geologic map of the near side of the Moon. *U. S. G. S. Map I-703.*

Young A. T. (1974) Television Photometr: the Mariner 9 experience. *Icarus* **21**, 262–282.

Zisk S. H. (1972) Lunar topography: first radarinterferometer measurements of the Alphonsus-Ptolemaeus-Arzachel region. *Science* **178**, 977–980.

Zisk S. H., Hodges C. A., Moore H. J., Shorthill R. W., Thompson T. W., Whitaker E. A. and Wilhelms D. E. (1977) The Aristarchus-Harbinger region of the Moon: surface geology and history from recent remote-sensing observation. *The Moon* **17**, 59–99.

Overleaf:

Section through an experimental charge of basalt glass in a molybdenum capsule (width 5 mm) after an experimental run in gas media apparatus. Photo courtesy of A. J. Irving.

4

3. Experimental petrology of basalts and their source rocks

Peter J. Wyllie, Team Leader
—*Department of the Geophysical Sciences*
University of Chicago

Colin H. Donaldson
—*Department of Geology*
University of St. Andrews

Anthony J. Irving
—*Department of Geological Sciences*
University of Washington

Susan E. Kesson
—*Research School of Earth Sciences*
Australian National University

Russell B. Merrill
—*Lunar and Planetary Institute*
Houston, Texas

Dean C. Presnall
—*Institute for Geosciences*
University of Texas, Dallas

Edward M. Stolper
—*Division of Geological and Planetary Sciences*
California Institute of Technology

Thomas M. Usselman
—*Geophysics Research Board*
National Research Council

David Walker
—*Harvard Geology Museum*
Harvard University

3.1 INTRODUCTION

This volume is concerned with the basaltic effluent from planetary interiors, and this chapter is concerned with the extent to which experimental petrology of basalts can inform us about the composition and mineralogy of the source material within the interiors of planets. If the compositions of sources can be deduced using the experimental petrology of basalts, or in any other way (Chapter 4), then the relationship between source rock and basaltic magma can be investigated by complementary experiments using the proposed source material.

Basaltic volcanism appears to be a characteristic feature of most investigated planetary surfaces (Chapters 2 and 5), with basalts being produced either when internal temperatures become high enough to cause partial melting of the planetary rock, or when impacts by other bodies raise the temperature high enough to cause partial or complete melting of near-surface rocks. The main variables to be considered in these processes are the pressure and equivalent depth, the temperature, and the composition and mineralogy of the planetary interior and near-surface rocks.

Of these three variables, the value of pressure as a function of depth is the best known. A given pressure is achieved at depths that vary considerably from one planetary body to another as illustrated in Fig. 3.1.1.

Temperature is a variable that changes as a function of time, and as a function of process (Chapter 9). Convective movements within a planet, for example, influence both the rate of heat transport to the surface, and the development of regional variations in temperature distribution versus depth. The temperature produced by meteoritic or planetesimal bombardment depends upon the mass, velocity, and frequency of impacting bodies. Variations reviewed in Chapter 9 indicate the extent of our uncertainty about the temperatures of planetary interiors, even for our own planet.

Figure 3.1.2 shows three geotherms that have been proposed for a convecting mantle within the present Earth (Solomon, 1976). The geotherms drawn for the lithosphere (shallower than about 100 km) are passed through the zones plotted by Solomon (1976) for values estimated from study of peridotite nodules from the mantle; these geotherms are similar to those calculated for conduction models (e.g., Clark and Ringwood, 1964). Temperatures in a region of upwelling, beneath ocean ridges for example, are higher than temperatures beneath normal ocean plates. The temperatures shown beneath shield and ocean plates become identical at 200 km depth (contrast Jordan, 1975).

The third variable, the composition and mineralogy of basalt source regions, is what we seek to determine from experimental petrology of erupted basalts. For the Earth, however, we already have a fairly well-defined model for the structure and petrology of the mantle, based on cosmochemistry, geophysics, and petrology (Chapter 4). The composition corresponds to that of a peridotite dominated at low pressures by olivine and two pyroxene minerals. Figure 3.1.2 outlines the main phase relationships for this material up to pres-

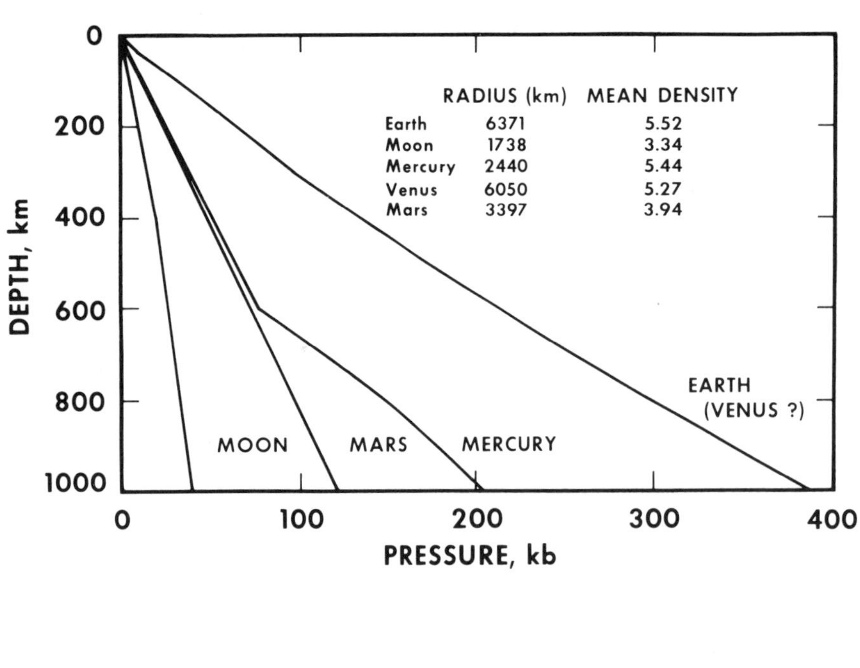

3.1.1 Pressure as a function of depth in the terrestrial planets. This function, details of which depend upon structural models of the planetary interiors, is best known for the Earth and least well-known for Venus. Internal structure of Venus is expected to resemble that of the Earth, to which the planet is similar in size and in mean density. Pressure at any given depth within Venus may be slightly less than at the same depth in Earth, because Venus is slightly smaller and less dense. Radius and mean uncompressed density are listed for each planet. Based on data from Jordan and Anderson (1974), Siegfried and Solomon (1974), Johnson and Toksöz (1977), and Masursky *et al.* (1977).

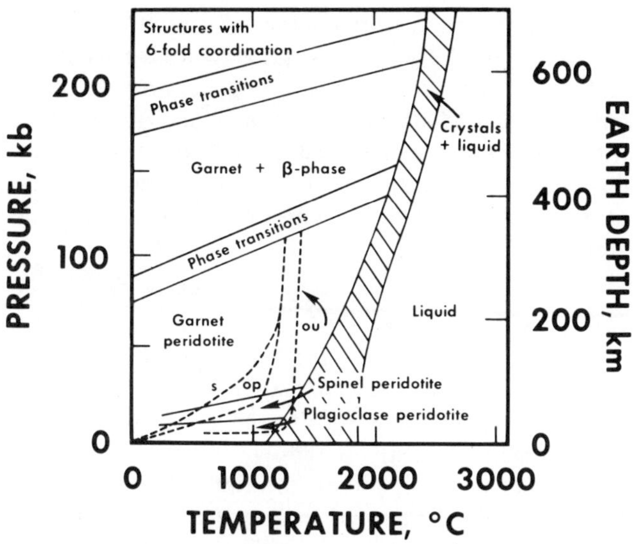

3.1.2 Schematic phase diagram for peridotite based on data from various sources (Wyllie, 1971a, Chapter 6; Wyllie, 1973, Fig. 2; Ringwood, 1975, Chapters 5 and 6; Yoder, 1976, Chapter 2). Depths are for Earth. The dashed lines are geotherms for Earth from Solomon (1976). Abbreviations: s = shield; op = ocean plate; ou = oceanic upwelling, as beneath ocean ridge (modified from Wyllie, 1979).

sures of 250 kb, corresponding to a depth of 700 km within the Earth. For other planetary bodies dominated by the components of olivine and pyroxenes, the phase relationships would be similar, but the specific depth scales would differ (Fig. 3.1.1).

Basaltic magmas are generated within the melting interval (crystals + liquid), when the temperature becomes high enough to cross the solidus. According to the terrestrial geotherms given in Fig. 3.1.2, melting does not occur at all beneath shield and ocean plates, and therefore some special circumstance is required to increase the temperature locally to account for the oceanic volcanoes and the continental flood basalts. It is generally assumed that upward convection of mantle material raises the temperature at relatively shallow levels, as indicated by the geotherm associated with mantle upwelling beneath ocean ridges in Fig. 3.1.2. In this tectonic environment, a geotherm crosses the solidus at depths of 70 ± 40 km (sections 3.3.7 and 3.3.8).

Each planetary body has characteristics that distinguish it from the others, even if it should turn out that most of them have similar major element compositions. The different pressure-depth relationships suggest that temperature distribution curves are likely to intersect solidus curves at different pressures on the different bodies. The phase relationships shown in Fig. 3.1.2 may be changed by variations in Fe, Mg, Ca, and Al proportions, and changed radically by the addition of volatile components (sections 3.3.2 and 3.3.3). For parts of the Earth's interior containing traces of CO_2 and H_2O, there is a wide interval of incipient melting below the solidus plotted in Fig. 3.1.2. For the Moon, in contrast, the concentration of volatile components appears to be negligible. The oxygen fugacity of lunar rocks is much lower than that of terrestrial rocks. Mars may be enriched in Fe and S compared with Earth and Moon (McGetchin and Smyth, 1978). The occurrence of partial melting and volcanism depends upon the maintenance of sufficiently high temperatures and thus upon the body's thermal history (Chapter 9). Parent bodies of basaltic achondrite meteorites are believed to have solidified at an early stage in the development of the solar system, and active volcanism on the Moon ceased relatively early in its history. Mars appears to have ceased evolving, and the Earth continues with active volcanism caused to a large extent by the mass movement associated with plate tectonics. Therefore, in considering the basaltic volcanism of each planetary body, we have different characteristics, and different ground rules for interpretation. The spectacular eruptions on Io were not anticipated according to preexisting ground rules.

The purpose of this chapter is to outline the methods used to determine whether a particular basalt is likely to contain direct information about its source region by application of the methods of experimental petrology. In addition, the experimental work on terrestrial, lunar, and meteoritic basalts, as well as likely source materials, will be reviewed to show the extent to which the link between the basalts and their source regions has been established, and to show the extent to which additional processes are responsible for the chemistry observed in the basalts. We will conclude that experimental petrology does provide an internally consistent framework for understanding basalts as the melting products of ultrabasic mantle assemblages, and that laboratory experiments can be used for the exercise of going back from basalt chemistry to source region constitution, provided that certain conditions are satisfied.

3.1.1 Historical developments

Few geologists would disagree with the statement that most basalts are derived from liquids produced by partial melting of peridotites in the Earth's upper mantle. This is the main premise of this chapter, indeed, of this book, and the primary ingredient of the foundation on which most current research on basaltic igneous rocks is built. Yet, in the first half of this century, this statement would not have been widely accepted, and few petrologists would have felt that proof or disproof of this statement would have had much impact on the nature or direction of their research. How then, did the concept of mantle peridotite sources of basaltic magmas evolve into a cornerstone of petrogenetic theory?

The idea of a peridotite mantle source for basalts is not new. It is usually attributed to Bowen (1928), and although he appears to have been the first to make a reasoned case for it on the basis of petrology, geophysics, and cosmochemistry, there were earlier statements of this hypothesis (e.g., Harper, 1915). The original source of the idea is not really important, however, because although it was well publicized in Bowen's authoritative book, it was not subsequently developed further—not even by Bowen himself. The principal concern of petrologists of that time, including Bowen, was not where magmas came from, but how one magma could be derived from another. Thus, the focus of petrological research was on understanding the processes by which one magma could evolve into another (e.g., fractional crystallization, liquid immiscibility, assimilation, vapor transfer, magma mixing, etc.), on detailed fractionation schemes relating known magma types, and on the identification of "primary magma" types.

The term "primary magma" meant something different to Bowen and his contemporaries than it does to us today. The change in the meaning of this term parallels the evolution of the focus of basalt petrology. Today, **primary magma** refers to a magma whose composition has not changed since it was generated by melting in its source region. Its importance lies in what it can tell us about its source region. This usage of the term appears to have been introduced by Turner and Verhoogen (1960). Prior to this, a "primary magma" referred to one which was widespread, uniform in composition, and for which no parent magma could be found. These magmas were important to Bowen and his contemporaries because these were the magmas from which they attempted to derive all other known magmas. Where they themselves came from—the mantle, the crust, or pockets or layers of melt of obscure origin—was of little immediate concern.

The most widely accepted theories from the 1920's through the mid-1940's stated that basaltic magmas originated in layers of basaltic composition. Views differed widely on the location, and extent of these layers. For example, Daly (1933) maintained that basalts originated in a glassy substratum of basaltic composition below the crust, while Kennedy and Anderson (1938) derived them by remelting of basaltic layers in the crust. In the intellectual framework of the time, the primary basalt compositions simply existed, prefabricated early in earth history and stored in the substratum or crustal layers. Any model of origin was regarded as speculative, and the main concern was with how the primary magmas fractionated after they were produced rather than with how they were produced. It was simpler to accept this sort of model than Bowen's, with its added complexities of partial fusion and degrees of freedom in peridotite chemistry.

Our current outlook began to take hold during the 1950's. In 1950, Tilley could still deliver a prescient Presidential Address to the Geological Society of London on "Some aspects of magmatic evolution," and not make a single reference to the source of primary basaltic magmas. Daly abandoned his glassy substratum model in 1946, but in the first edition of their *Igneous and Metamorphic Petrology*, Turner and Verhoogen (1951) gave equal billing to partial melting of peridotite and remelting of a crustal or subcrustal basaltic layer, and appeared to prefer a model based largely on the latter hypothesis. However, by the end of the 1950's, the notion of peridotite mantle which on partial melting gives rise to basaltic magmas began to appear regularly in reviews and specific petrogenetic models (e.g., Powers, 1955; Wager, 1958; Kuno, 1959; Turner and Verhoogen, 1960). By the early-1960's, this idea was nearly universally accepted as the starting point for research and thought on basalt petrogenesis.

A detailed analysis of what happened in the 1950's is beyond the scope of this chapter, but factors which may have been influential in the rapid evolution of attitude include the following: (1) Ross, Foster, and Myers (1954) forcefully brought to the attention of the geological community the significance of ultramafic nodules in basalts to models of mantle constitution and basalt genesis. (2) Renewed interest in cosmochemical models of earth composition, origin and evolution had been sparked by Urey and his co-workers in the early 1950's. (3) The petrological implications of geophysical constraints on mantle composition, mineralogy, and physical conditions were widely discussed (e.g., Birch, 1952; Verhoogen, 1956; Ringwood, 1958–62). Similar considerations had caused Daly (1946) to abandon his glassy substratum model. (4) The development of new equipment made possible the quantitative study of phase equilibria of basalts, peridotites, and their synthetic equivalents at high pressures and temperatures. Although relevant experimental data were not widely available until the 1960's (e.g., Yoder and Tilley, 1961, 1962), limited data available in the early 1950's (e.g., Yoder, 1952) opened up new vistas for speculation on petrogenetic schemes involving igneous processes at mantle depths.

The 1960's saw probably the greatest advances in our knowledge of the phase relationships of basaltic magmas at high pressures. Yoder and Tilley's ambitious pioneering synthesis of petrologic observations coupled with experiments from 1 atm to 40 kb, begun in the mid-1950's, was published in 1962. The development of the piston-cylinder apparatus by Boyd and England (1959, 1960) permitted routine experimentation at pressures relevant to the generation of basaltic magmas from mantle peridotite. Workers such as Boyd, Kushiro, O'Hara, Yoder and others at the Geophysical Laboratory studied several natural rock samples, and many simplified systems as analogs of natural basalts and peridotites.

Kuno was the first, in 1960, to propose that the compositions of different primary magmas were controlled by the depth of generation. The important effects of pressure were confirmed by Yoder and Tilley (1962), who demonstrated experimentally that thermal divides between different basaltic compositions shifted as a function of pressure. O'Hara (1965, 1968a) and O'Hara and Yoder (1967) pointed out that the compositions of liquids in equilibrium with peridotite mineral assemblages at high pressures do not correspond to the compositions of lavas commonly interpreted as primary magmas. This could be explained if magmas experienced fractional crystallization during part or all of their transit from source to surface. Yoder (1976) summarized the experimental data supporting his conclusion that the major commitment of basaltic types is made by melting processes at various depths, and that most extrusive basalts are residual liquids of advanced fractional crystallization.

Parallel studies by Green, Ringwood and coworkers at the Australian National University involved the crystallization of complex natural basalt and peridotite compositions, with analysis of phases by the electron microprobe (invented only in 1951). Green and Ringwood (1967a) determined the phase relationships of a series of synthetic basalt compositions, compared the near-liquidus minerals at various pressures with near-solidus minerals in peridotite, and presented a comprehensive scheme for the chemical variation among the major types of terrestrial basalts in terms of the physical conditions of partial melting of peridotite, and of subsequent fractional crystallization. Details were extended and modified in subsequent papers, and the effect of H_2O was included, as reviewed by Ringwood (1975, Chapter 4).

During the same period, Ito and Kennedy (1967, 1968) and Cohen *et al.* (1967) were working with natural peridotite, basalt and basanite, or glasses made from them, with the same objective, to determine the relationships among extrusive basalts and mantle peridotite source. They shared O'Hara's conclusion that common basalts are not primary, but derived from deep-seated picritic liquids which undergo extensive olivine fractionation during uprise.

The return of the first basalt samples from the Moon in 1969 provided an opportunity to apply the experimental approaches developed from the study of terrestrial basalts and peridotites to another planet, without having any samples of the source rocks. The approach of deducing planetary mantle constitution by experimental study of erupted basalts, although criticized by O'Hara and co-workers, was applied to basalts from succeeding Apollo and Luna missions through the 1970's particularly by Green, Kesson, Ringwood, Walker, and associates (section 3.4). The results provide the basis for present models of the lunar interior (Chapter 4). More recently, Stolper has applied this approach to the eucrite meteorites and developed a model for the constitution of their parent body (section 3.5).

In parallel with the studies of extraterrestrial basalts during the 1970's, there were important advances in our understanding of terrestrial basalts. The development of energy-dispersive and automated electron microprobes made possible the analysis of phases in experimental run products with precision that was inconceivable a few years earlier. Experiments on basalt and peridotite systems containing controlled amounts of volatile components, although previously carried out by a few workers, were begun in earnest in the late 1960's and were continued through the 1970's particularly by Boettcher, Eggler, Green, Kushiro, Wyllie, and their associates. Attention was first directed to the effects of H_2O, particularly because of the spatial association of andesites with subducted ocean floor, but also because no way had been found to generate highly SiO_2-undersaturated alkalic basalts and kimberlites from peridotite under volatile-free conditions. By the mid-1970's the dramatic effects of CO_2 and of H_2O-CO_2 mixtures were being documented in detail. At the beginning of the 1980's, we now have a reasonably self-consistent model for the evolution of most types of terrestrial basaltic magmas, although many important details remain undetermined or unresolved (section 3.3). In addition, our knowledge of the effects of H_2O and CO_2 on the behavior of silicate melts at high pressures, as well as the probable effects of other volatile species such as sulfur, should eventually make important contributions to our understanding of basalts on Venus and Mars.

Experimental approaches to the problem of the generation and evolution of basaltic magmas have involved determination of the phase relationships of erupted lavas, assumed primary magmas, and possible magma source rocks. A parallel approach involves the study of phase relationships in simple analog synthetic systems, representing the complex natural compositions. The two approaches can provide complementary insights to a given problem (e.g., Arculus, 1975). With the rock compositions, all components are present, but it is difficult to determine how the results for one composition apply quantitatively to an adjacent composition, it is difficult to evaluate schemes involving fractional fusion or crystallization, and it may be difficult to determine the composition of liquid present in small quantities. With the synthetic systems, the compositions of all phases can be represented geometrically, the chemical trends of all phases can be plotted for any scheme of fusion or crystallization with any arbitrary starting composition, and the phase relationships near the solidus for a particular composition can be studied in detail by determinations on other compositions near the appropriate eutectic or peritectic, but the results have to be extrapolated to systems containing additional components. Experiments using natural rock samples have been cited above. The complementary approach is represented by the study by Presnall *et al.* (1979) in the system CaO-MgO-Al_2O_3-SiO_2 with application to ocean floor tholeiites, and additional examples have been reviewed by Yoder (1976).

The 1970's also saw significant advances in other disciplines such as trace element and isotope chemistry, which provided important independent constraints on experimentally-based models for the generation of terrestrial, lunar, and eucritic basalts. Conversely, experimental studies of trace element partitioning in basaltic systems (summarized by Irving, 1978) resulted in more sophisticated trace element modelling of magmatic processes (Chapters 1 and 7).

3.1.2 Generation of basaltic magmas

The generation of basalts may be a complex multistage process. Figure 3.1.3 is a schematic representation of various stages in this process. An ascending diapir of partially-melted mantle material has been halted by the effective rigidity of the lower lithosphere. Melt has segregated into a pool at A, but it is still in equilibrium with the crystalline residuum from which it separated. This segregated melt is defined as **primary** liquid. It is to be distinguished from bulk suspensions of this liquid containing source region crystals which were parental before segregation was complete, and from any liquids to which it may give rise by any modification process. If partial crystallization or vesiculation should occur without change of bulk composition, the liquid becomes a primary magma. The segregated primary liquid may be erupted directly to the planet's surface as at A . Samples recovered from such an eruption, if they preserve the composition of the primary liquid, are the most informative for the experimental petrologist to investigate for insights into the source region.

If no adequate duct is available for escape of primary liquid from pool A, heat loss to the base of the lithosphere will induce crystallization. The primary magma may be in reaction relation with any of the phases of the crystalline residuum. Then, fractional crystallization of the primary magma will produce derivative liquid compositions that are no longer saturated with all of the residuum minerals. If the primary magma is not in reaction relation with the residuum (eutectoid situation), then fractional crystallization of primary magma at A produces liquids that remain in equilibrium with the residual minerals of the source rock, but the compositions of minerals with continuous solid solution would change. Loss of heat and crystallization at depth leads eventually to the crystallization of minerals from the derivative liquids that are incompatible with the source material.

Ducts may become available permitting primary (or fractionated) magma at pool A to ascend only part of the way to the surface. Suppose that the magma collects at level B in Fig. 3.1.3. While traversing cooler surroundings from A to B, there is opportunity for melt to crystallize partially and for crystals to settle or be plated on conduit walls, thereby further modifying the magma composition. In addition, some assimilation or contamination from the duct may occur. If the melt stagnates in pool B, heat loss will again induce crystallization. If A and B are at significantly different pressures, the proportions and even the types of minerals precipitated may be different in the two pools. Consequently, a sequence of fractionated magmas erupted from chamber A might exhibit chemical variation different from that of fractionated magmas erupted from chamber B, even though there was only one primary magma.

Magma from B may find access to a subvolcanic magma chamber such as C, just below the eventual site of eruption D. Again, there may be further differentiation in transit from B to C as well as magma contamination from the surroundings. The mineral compatibilities may have changed from B to C, and fractionation at C will produce magmas with yet a different chemical trend.

Further complications may be introduced during fractional crystallization if batches of magma from B find their way into chamber C. There is also a possibility

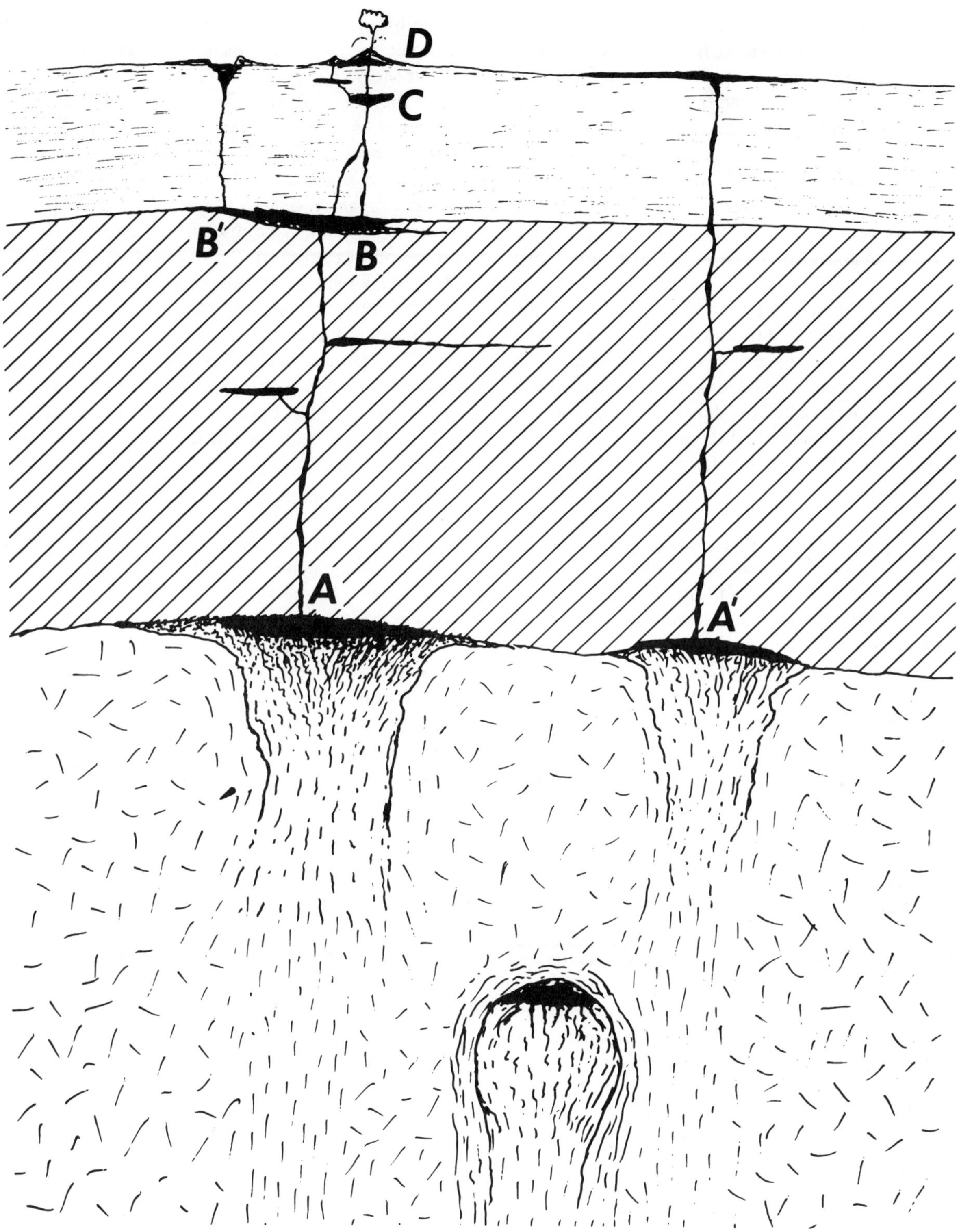

3.1.3 Model for magma generation. Primary magma produced either by *in situ* partial melting or diapiric uprise, occurs at A and A′. Fractionation of magma may occur both at A and B-B′, and further fractionation of B may occur at C, with final eruption at D. Of the samples available at the surface, A′ is the most useful to the experimental petrologist because it is less fractionated than B, C or D.

that magmas unrelated to the sequence A-B-C may invade the magma chambers at B or C. In subduction zones, for example, primary magmas of quite different compositions may be generated in the subducted ocean crust, in the overlying mantle, and in the crustal rocks of island arcs or continental margins. If successive magmas from A or B, or from some unrelated source rock, should mix with fractionating magmas in chamber C, it may become very difficult to unravel the sequence of events between source material at A and eruptive product at D.

Finally, magma from C may be erupted to the surface D where crystallization proceeds. During crystallization, two additional processes can cause differentiation. Gravitational crystal sinking can enrich basal portions of the magma body in dense, early-crystallized or intratelluric phenocrysts. If the magma continues to flow during its solidification, flowage differentiation preferentially moves crystals away from fixed boundary surfaces and into the flow stream. Through such processes, significant lateral and vertical heterogeneity may be introduced in a single magma cooling unit. During eruption and solidification, additional contamination by soil, surficial debris, and volcanic or atmospheric gases is also possible.

Clearly, the primary magma sampled at A′ may be very different from the basalt extruded at D. Petrologists must identify the possible modifications which could occur between A and D. Of necessity, this is done in the reverse order by checking the last possible modifications first.

3.1.3 Forward and inverse approach in basalt-source relationships

The experimental study of terrestrial basalts and analog systems has been concerned more with the causes of diversity among known basaltic magmas, than with elucidation of the petrology and constitution of the mantle. Preconceptions of the nature of the terrestrial source region have been used to constrain theories of the petrogenesis of basalts rather than the other way around.

This directionality in the approach to unravelling the relation between basalts and their source regions is so strong that if a basalt cannot be shown to equilibrate with plausible ultrabasic mineral assemblages under some set of physical conditions (e.g., any P, T, dry), the mineral assemblage is not abandoned as implausible. Instead, different physical conditions (i.e., wet, carbonated, oxidized, etc.) are investigated until the liquid composition can be related to a plausible ultrabasic assemblage. In this intellectual framework, it may be difficult to claim that basaltic rocks are informative about the petrology of their source regions in anything more than a trivial sense. However, the exercise is useful in that an internally consistent model of basalt generation is developed in which physical conditions of the generation process (P, T, activities of H_2O and CO_2, oxygen fugacity, degree of melting, etc.) are identified along with detailed compositional information on the phases of the residuum. The exercise, conducted experimentally, which shows how a known or assumed source material can give rise to a particular type of basalt, may be considered as a "forward" problem.

The related "inverse" problem is what one wishes to solve for other planets, and for the Moon or eucrite parent body. What does the observed basalt chemistry tell us about the unsampled source region on those planets? As with any inverse problem, the solution has an associated uncertainty or ambiguity even when the solution to the forward problem is known. In section 3.3 we examine both the forward and inverse approaches for terrestrial basalts to see how secure a foundation we may expect for a solution of the inverse problem elsewhere.

The ultimate achievement of experimental studies of natural rocks and simplified model systems would be to determine all phase relations in the complete multicomponent system relevant to igneous rocks as functions of pressure. Such a massive and complete data set would stand alone, independent of any philosophical approach to a petrological problem, either forward or inverse, and would provide a rigorous framework for constructing and testing petrogenetic models. Although progress toward this ultimate goal is being made (Ringwood, 1975; Yoder, 1976; Presnall *et al.*, 1979; this chapter), it is clear that the end result will not be achieved in the near future. In the meantime, important conclusions can be obtained from studies of specific, carefully-chosen compositions within the multicomponent system, as discussed in the next section.

The distinction between forward and inverse approaches may appear to be somewhat arbitrary. Geologists study the accessible rocks in order to infer the processes that occurred at inaccessible depths, in earlier times. In this respect, the inverse approach is the standard geological approach. Geologists use any additional information available to reconstruct the evolution of a rock assemblage, and this includes data that we have identified as related to the forward approach. Therefore, petrologists concerned with the origin of terrestrial basalts use both approaches more or less simultaneously. It is within the framework of experimental petrology that the forward approach tends to have been emphasized. On the other hand, geochemists attempting to interpret the distributions of rare-earth elements, trace elements, and isotopes have used their data to identify possible source rocks and to unravel possible processes between source and surface (Chapters 1 and 7), and this in an inverse approach. A combined approach, the geological approach, is obviously desirable, but when presented with a small set of basaltic samples from the surface of an unexplored planet, petrologists have no choice initially but to follow the inverse approach.

3.1.4 Experimental petrology in a flow diagram for the inverse approach

Starting with a set of basalt samples collected from the surface of a planet, as at D in Fig. 3.1.3, the strategy outlined in Fig. 3.1.4 illustrates the inverse approach for evaluation of the mineralogy and composition of the planet's interior at the level A in Fig. 3.1.3, where the magma was segregated from its crystalline source. The sequence of steps is designed to unravel the processes that might have occurred between A and D, and to reconstruct the path of the changing liquid composition back down to the primary source at A.

Step 1 is to identify the samples that represent erupted *liquids*, and step 2 is to determine if these are related by low-pressure processes (e.g., at D or C in Fig. 3.1.3). If they are so related, the parental liquid of the suite is identified (step 3), and experiments are conducted in step 5 to determine whether the parental liquid is characterized by a single mineral or multiple minerals on its low-pressure liquidus. Multiple saturation indicates that the effects of low-pressure processes obscure the earlier, deep-seated events that may have been involved. If the rock exhibits single-mineral saturation, or if the liquids are not related by low-pressure processes (step 2), the procedure follows through step 4 to the identification of primitive liquids, which could have formed at deep levels B or A in Fig. 3.1.3. This may require some high-pressure experiments (step 6). It is also common practice to identify primitive liquids on the basis of their geochemical characteristics even at stage 1 (Chapter 1, section 1.3.1).

The primitive liquids are then subjected to detailed high-pressure experimental studies to evaluate whether or not they may represent primary liquids formed at A in Fig. 3.1.3, and to see what constraints they may provide for the source region mineralogy. At this stage, the inverse approach is supplemented by whatever information may be gleaned from other sources about the mineralogy and composition of the planet's interior (Chapter 4), and models are developed for the mantle composition. Experimental testing of the mantle models involves the forward approach of determining the compositions of liquids produced from this material at various conditions. These liquid compositions are then related to the compositions of the primary and primitive liquids determined by the inverse approach. The procedure is deemed to be successful if an internally consistent model can be erected relating mantle composition and mineralogy to primary liquids and erupted lavas, through reasonable processes from source to surface.

Identification of liquid compositions (step 1)

The starting point in the flow chart is discrimination of basalts with compositions representing quenched liquids from phenocryst- or xenocryst-enriched samples, or cumulates. The traditional criteria for identification of quenched liquids have been largely petrographic. Glassy and aphyric rocks are usually assumed to represent quenched liquids. For a porphyritic rock, textural criteria alone cannot establish whether the phenocrysts were concentrated by gravity or by flowage differentiation (in which case the rock composition does not represent a liquid composition), or whether they crystallized in place (in which case the rock composition may represent a liquid composition). Figure 3.1.5 illustrates three different shapes for olivine in a single lunar basalt. Each morphology represents a different rate of heat loss during crystallization. The skeletal microphenocryst in Fig. 3.1.5a may have crystallized in place from liquid, but the equant olivines in Fig. 3.1.5b probably represent intratelluric phenocrysts. This basalt may have suffered some degree of olivine fractionation or enrichment. Another lunar basalt illustrated in Fig. 3.1.6 has groundmass dominated by dendritic pyroxene, but the presence of nonskeletal olivine phenocrysts indicates that it may not have erupted as a liquid.

The field relations may be useful if, for example, the sample was collected from a chilled margin or a porphyritic layer near the base of a lava flow. Unfortunately, for many important rock types such as dredged suboceanic basalts and lunar and meteoritic basalts, this type of information is usually unavailable.

Experimental petrology provides several tests which can aid in distinguishing quenched liquid compositions from phenocryst- and xenocryst-enriched compositions. In nonequilibrium experiments conducted at controlled cooling rates, a basalt is melted and then cooled at various rates. If the resultant textures, crystal morphologies and compositions are similar to those in the natural basalt, this is evidence of a liquid origin for the basalt. However, the large number of

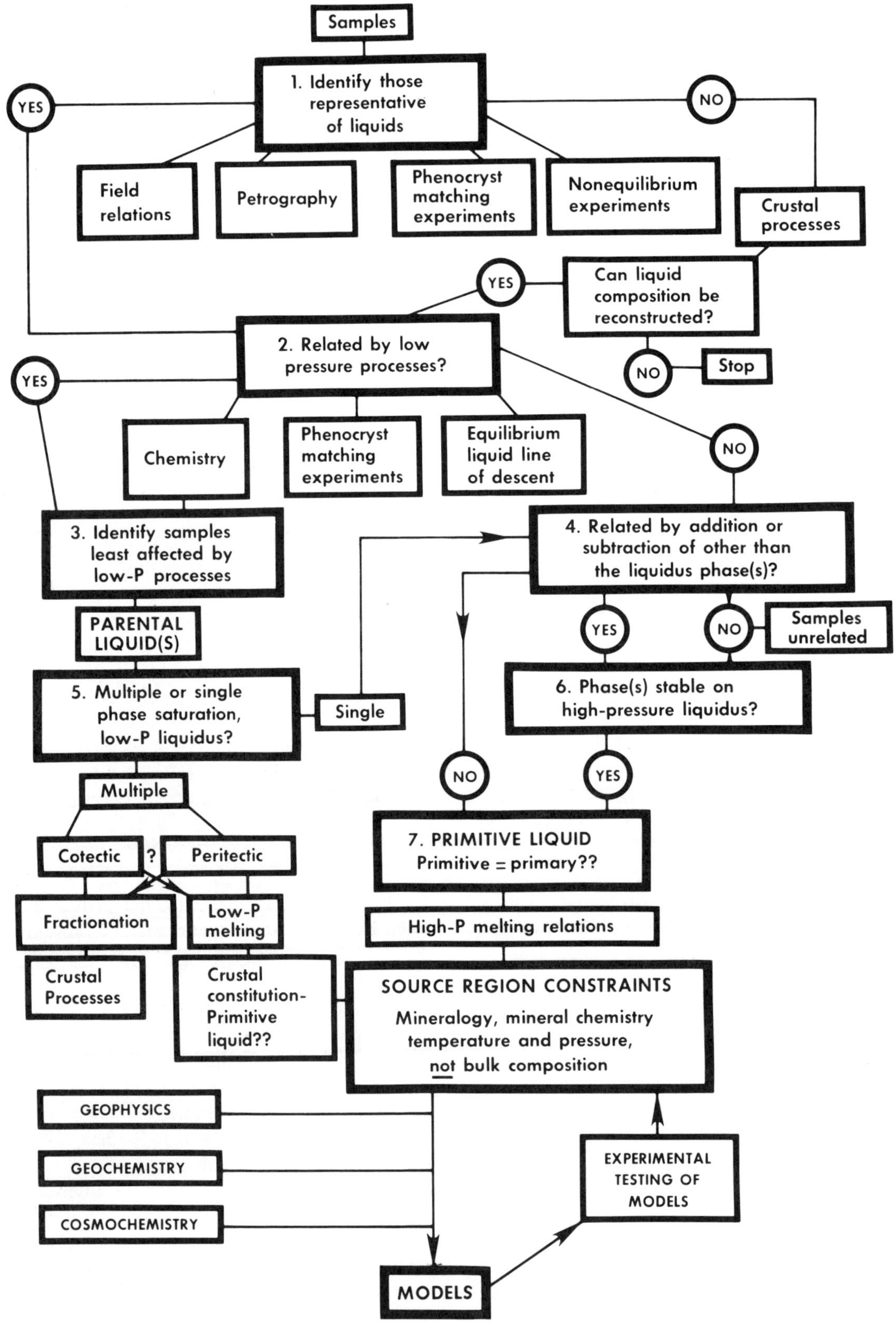

3.1.4 Flow chart for the interpretation of basaltic rocks by the inverse problem approach.

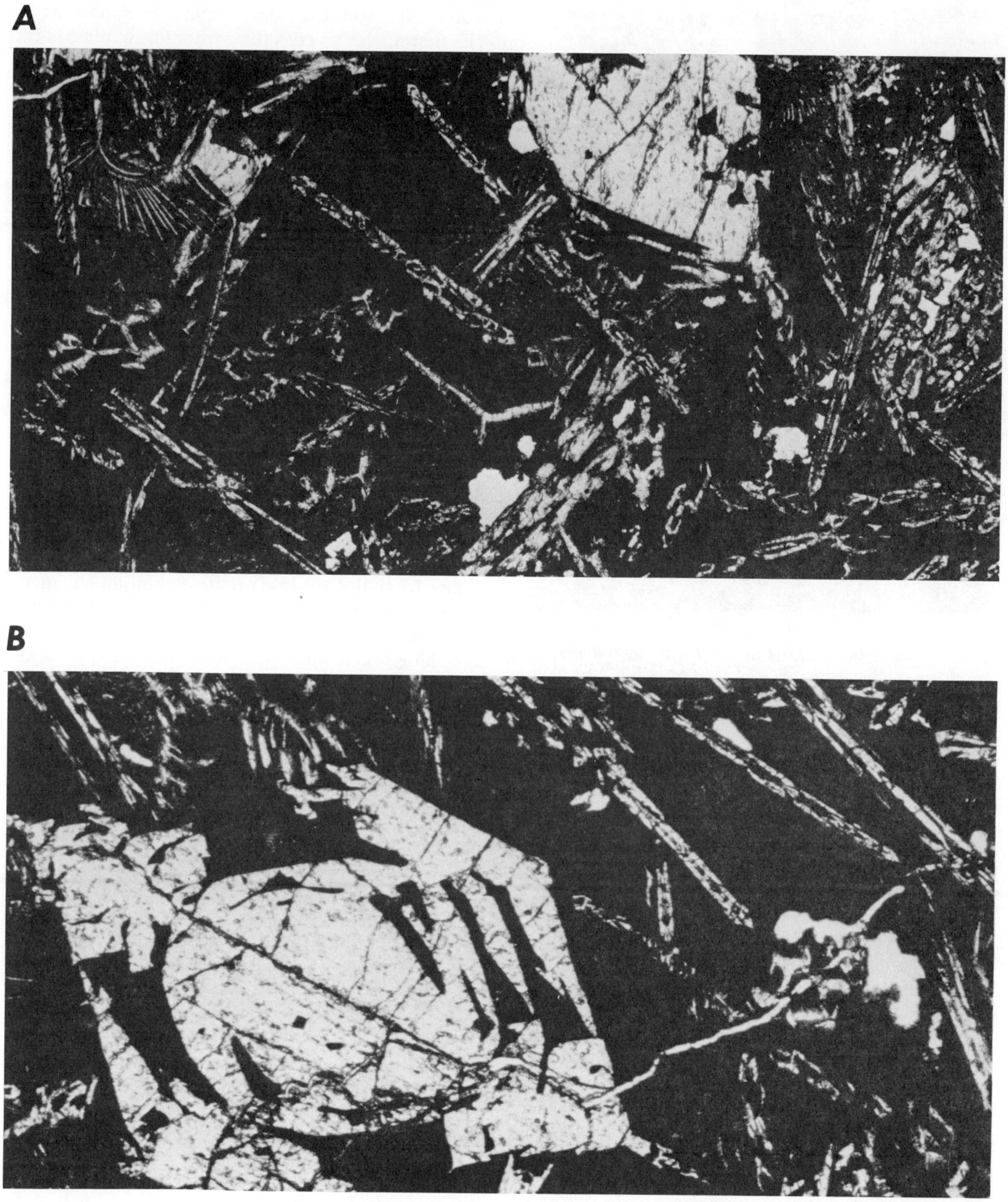

3.1.5 Three shapes of olivine in lunar basalt 12009 (from Donaldson *et al.*, 1975b). Skeletal microphenocrysts in (a) and equant microphenocrysts in (b) each occur with chain olivines. Each morphology represents different rates of heat loss with the equant olivine (b) possibly representing intratelluric phenocrysts. Thus this basalt may have suffered a certain degree of olivine fractionation or enrichment.

3.1.6 Olivine phenocrysts in a dendritic pyroxene-rich groundmass (lunar rock 74325 from Apollo 17). This rock may not have been completely liquid on eruption.

degrees of freedom in experimental and natural crystallization sequences limits the usefulness of this approach. Nevertheless, it has been used with some success in several studies of lunar basalts (Lofgren *et al.*, 1974; Donaldson *et al.*, 1975b; Usselman *et al.*, 1975; Walker *et al.*, 1976a and b; Usselman and Lofgren, 1976).

The phenocryst matching test in equilibrium experiments is another means of testing for phenocryst- or xenocryst-enrichment. The compositions of the earliest crystallized minerals in the rock itself are compared with the compositions of those present near the liquidus of the rock composition. If the experimentally produced liquidus minerals are enriched in the higher temperature component [e.g., olivine has higher Mg/(Mg + Fe), or plagioclase has higher Ca/(Ca + Na)] relative to those minerals in the natural sample which appear to have crystallized first, this indicates that the rock composition is phenocryst-enriched. Green *et al.* (1975b) and Walker *et al.* (1976b) have published examples of this test.

The principle is easily grasped with the aid of Fig. 3.1.7, which shows the Mg_2SiO_4-Fe_2SiO_4 crystallization interval at 1 atm. Consider a liquid of composition A. The first olivine to crystallize from this liquid has the composition C. If a rock were formed by crystallization of a liquid of composition A at a rate sufficiently rapid that the first-formed crystals were preserved in the cores of the crystals in the rock, then the most magnesian olivines in the rock do have the same composition as the olivine near the liquidus of the rock. Consider, on the other hand, a rock B, produced by the accumulation of olivines of composition C into the liquid A with which they were in equilibrium. If crystallized in the laboratory under equilibrium conditions, the first olivines to precipitate from the phenocryst-enriched composition B would be of composition D. The most magnesian olivines produced in the experiment on the rock would thus be more magnesian than the most magnesian olivines found in the rock (C). The phenocryst matching test can be applied to any mineral with a continuous solid solution (e.g., pyroxenes, feldspars), and a distribution coefficient with the liquid significantly different from 1.

Unfortunately, the phenocryst matching test is ambiguous. Suppose the liquid of composition B in Fig. 3.1.7 cooled so slowly that equilibrium was maintained, and then no olivines of composition D would be preserved in the rock. Application of the phenocryst matching test to this rock would suggest that it was produced by phenocryst-enrichment (e.g., liquid A plus crystals C). This failure of the test may result in the misclassification of a liquid composition as a phenocryst-enriched composition, but the test will rarely suggest that a rock which is in fact phenocryst-enriched had been a liquid. From the point of view of the inverse

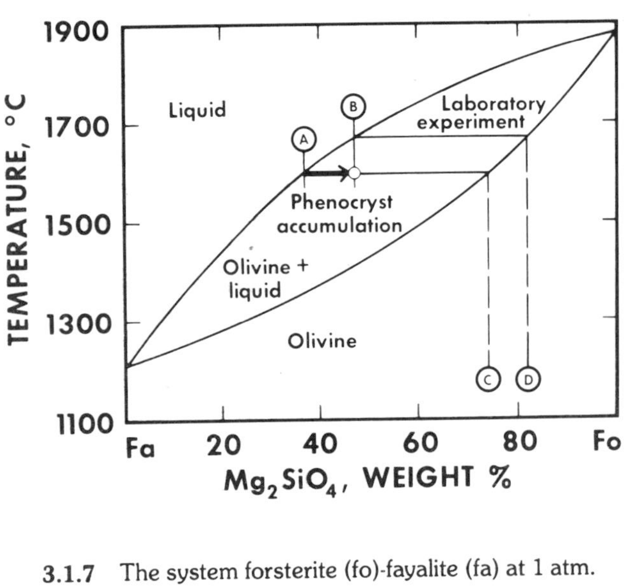

3.1.7 The system forsterite (fo)-fayalite (fa) at 1 atm.

problem approach under consideration, this is important because it means that non-liquid compositions will not generally be misidentified as liquids and then used to constrain the source region properties incorrectly.

There are, however, other potential difficulties with the phenocryst matching test. Several of these are experimental: The difficulties of maintaining constant the composition of the sample and of properly reproducing the oxygen fugacity relevant to natural crystallization in the experiments are the most critical. For example, loss of iron to the capsule in which the experiment is conducted or improper Fe^{2+}/Fe^{3+} balance in the experiments due to improper oxygen fugacity control can greatly affect the composition of near-liquidus olivines produced in experiments.

Another difficulty with the phenocryst matching test is associated with the possibility that equilibrium experiments may not faithfully reproduce the compositions of minerals crystallizing early in the natural crystallization sequence. In particular, the results of Lofgren *et al.* (1975) suggest that the first pyroxenes produced during rapid crystallization are more iron-rich than the liquidus pyroxenes produced in equilibrium experiments. This suggestion is not universally accepted, but if it is confirmed, it means that phenocryst matching arguments based on equilibrium experiments could incorrectly classify liquid compositions as phenocryst-enriched compositions.

In some rocks, the most magnesian natural olivine is more magnesian than the most magnesian olivines produced during experimental crystallization of the rock. For example, suppose a rock of composition A in Fig. 3.1.7 contains olivines of composition D. This rock then cannot be interpreted as a liquid composition, but as a liquid which contains xenocrystal material (Walker *et al.*, 1976c).

A variant of the phenocryst matching test which may be helpful in recognizing phenocryst-enriched compositions exploits the discontinuous reaction relationships between liquid and some minerals. Consider Fig. 3.1.8. Experimental crystallization of a rock of composition B would show olivine as the liquidus phase. However, if the petrography indicated that only pyroxene phenocrysts were present in the rock, the experiments could indicate that composition B was produced by enrichment of pyroxene phenocrysts (P) into a liquid such as A which had pyroxene on its liquidus. If, however, a liquid of composition B had experienced polybaric crystallization, the application of this reasoning could lead to an invalid conclusion, because phenocrysts could have grown by slow cooling of liquid B at high pressures, when the olivine-pyroxene liquidus boundary corresponded to the dashed line in Fig. 3.1.8.

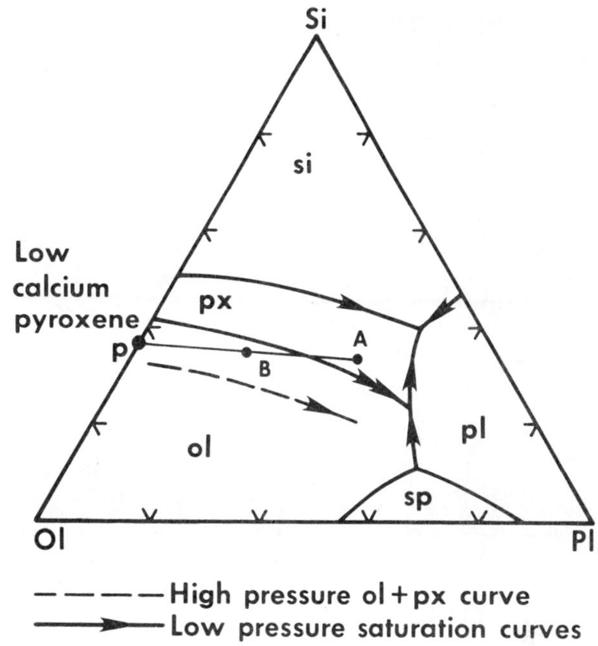

3.1.8 Liquidus surface for the system olivine (ol)-anorthite (pl) -silica (Si). The primary phase field of pyroxene expands at the expense of olivine, with increasing pressure. Abbreviations: px = low calcium pyroxene; sp = spinel; pl = plagioclase; si = silica.

If two or more minerals crystallize simultaneously at the liquidus of a rock composition at 1 atm, this may indicate that the composition represents a liquid composition. Consider the liquid at x in Fig. 3.1.9 that crystallizes minerals a and b simultaneously. Phenocryst-enrichment of these minerals in liquid x can produce bulk compositions plotting anywhere in the shaded region. It would thus be fortuitous for a phenocryst-enriched composition to be on the a-b cotectic curve. On the other hand, liquid compositions move inexorably toward these multiple saturation curves so that compositions on the curves are likely to represent liquids. One limitation for this line of reasoning is that if the cotectics are approximately straight as in Fig. 3.1.9, then phenocryst-enrichment in the proportions in which the minerals crystallize from the liquid can produce phenocryst-enriched compositions plotting on the cotectic. For example, the bulk solid separating from liquid x in Fig. 3.1.9 is y. If this is concentrated in the liquid with no differential movement of minerals a and b, then the phenocryst-enriched composition (e.g., z) would also plot on the a-b cotectic.

It is often supposed that a basalt containing dense xenoliths of high-pressure origin must represent a liquid

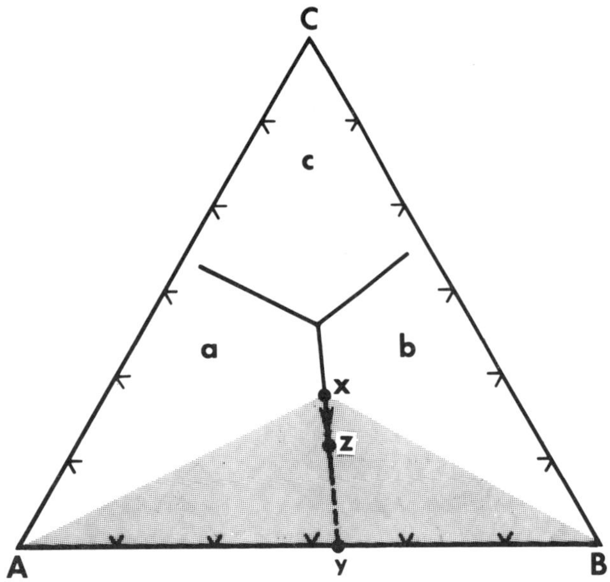

3.1.9 Model system A-B-C, illustrating compositional modifications which may result from phenocryst-enrichment processes.

composition. If the transport of the magma to the surface was sufficiently rapid that the large xenoliths (e.g., from A′ in Fig. 3.1.3) were not able to settle out of the ascending liquid, it is argued that the smaller phenocrysts which are often present in the host basalt could not have been added to or removed from the basalt.

This conclusion follows from calculations, using Stokes' Law, of the particle settling velocity which depends on the square of the particle size. If the magma ascent rate can counteract the settling velocity of larger nodules, smaller crystals with correspondingly lower settling velocity will certainly have no opportunity for separation. These conclusions are valid if the fluid supporting the nodules has a Newtonian rheology, an implicit feature of the Stokes' formulation. Magmas in the wholly liquid state do appear to be Newtonian (Bottinga and Weill, 1972). Partially crystalline magmas, however, appear to have a definite yield strength before viscous flow begins and hence are not Newtonian (Shaw *et al.*, 1968) but more nearly Bingham plastics.

Sparks *et al.* (1977) pointed out that the Bingham behavior of partially crystalline magmas will increase their nodule-carrying capacity because the yield strength must be overcome before the nodules can begin to settle, no matter what the ascent rate. They proposed that magmas carry nodules because of their rheology rather than their ascent rate. They suggested that the small volumes of (terrestrial) alkalic lavas (G. P. L. Walker, 1973) lose heat to their conduits and suffer partial crystallization. Consequently, the alkalic magmas acquire a rheology more conducive to nodule transport than the larger volumes of more rapidly erupted tholeiitic magma which do not exchange enough heat with their conduits to acquire a nodule-carrying rheology. If this interpretation is correct, then the magma which bears nodules has suffered partial crystallization since segregation from the source region. Furthermore, because the local liquid on the scale of small phenocrysts is Newtonian in rheology (as opposed to the macroscopic partially crystalline Bingham magma suspending the nodules), phenocryst separation or enrichment may occur even in the presence of suspended nodules. It is also possible that significant amounts of xenocrystal debris may be included in the host lava (thus inhibiting meaningful chemical analysis), or that xenocryst or phenocryst resorption may have occurred under conditions of pressure release during ascent. These considerations cast some doubt on the concept that the presence of nodules can be taken as a firm indication that the host lava is an unmodified liquid.

The preceding discussion demonstrates that the identification of quenched liquid compositions is not a trivial task. Nevertheless, a balanced application of textural criteria, petrographic analysis, experimental evidence, considerations of the chemical variations among related members of a suite of rocks, and field evidence (where available) can permit the distinction of liquid from nonliquid compositions.

If all compositions in a suite of basaltic rocks appear to be cumulates or phenocryst-enriched compositions, it may be possible to place constraints on the compositions of the liquids into which phenocryst-enrichment occurred by subtracting the phenocryst compositions from the bulk rock composition to obtain an estimate of the groundmass composition. The amount of a phenocrystal component to be subtracted may be determined by modal analysis, or the sample may be experimentally partially melted until the compositions of the residual phases in the experiment correspond to the phenocryst compositions. The experimentally produced liquid at this point should have the composition of the liquid into which the phenocrysts accumulated. The difficulty in distinguishing accumulated phenocryst material from material overgrown on phenocrysts limits the applicability of this type of analysis. Suites of rocks in which only phenocryst-enriched or cumulate compositions are known are common among meteorites: the shergottite, nakhlite, chassignite, and angrite meteorite classes contain only cumulate members. Thus it is often necessary to resort to this type of analysis to identify the compositions of liquid basalts on meteorite parent bodies.

Are the liquid compositions related by low pressure processes? (steps 2 and 3)

Once the liquid compositions in a suite of basaltic rocks have been identified, the next problem is to establish the relationships among them, if any exist. The starting point is to determine if the compositions can be related by igneous processes operative at relatively low pressures.

The primary inputs into this analysis are the bulk compositions of the liquids. Continuous trends on variation diagrams are usually taken as evidence that the liquid compositions were produced by crystallization differentiation of some parental composition. Experimental studies can be used to verify the plausibility of various differentiation schemes and to constrain the conditions under which the differentiation occurred.

Consider, for example, the hypothetical variation diagram shown in Fig. 3.1.10 for a suite of liquid compositions from a basaltic province. The variation diagram might be provisionally interpreted in terms of a sequence of liquids produced by crystallization differentiation of a liquid of composition A; olivine fractionated alone between A and B and an augite pyroxene joined the crystallization sequence at B. By melting composition A to different degrees at 1 atm and analyzing the glasses produced by quenching of the experimentally generated liquids, one can deduce the equilibrium liquid line of descent of composition A at 1 atm. If the experimental line coincides with the trend A-B-C in Fig. 3.1.10, and if the experiment in which clinopyroxene first appears in the crystallization sequence coincides with the experiment in which the liquid has the composition B, this can be interpreted as confirmation of the hypothesis that this suite of liquids was produced by crystallization differentiation of composition A at low pressure.

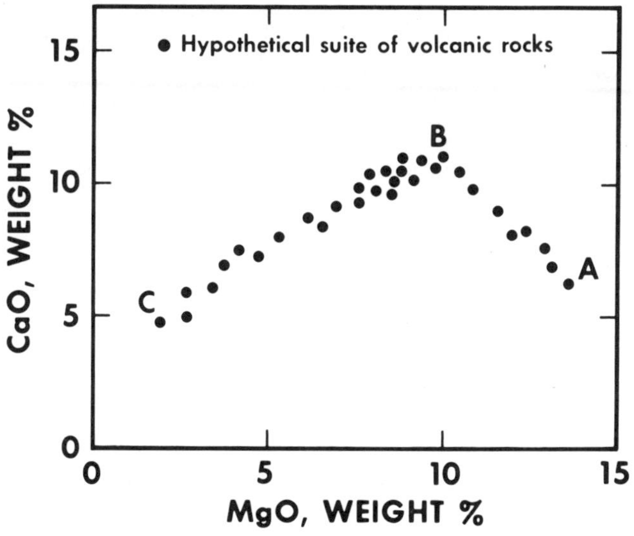

3.1.10 Hypothetical CaO vs. MgO variation diagram for a typical basaltic province.

These experiments yield the equilibrium liquid line of descent of composition A. The liquid line of descent produced by natural processes would probably correspond more closely to that for fractional crystallization than that for the equilibrium crystallization conducted in the laboratory. This may result in differences between the natural and experimental liquid lines of descent. Other causes of differences between natural and experimental liquid lines of descent include differences in oxygen fugacity between the experimental and natural crystallization conditions, differences in volatile components and progressive contamination in the natural sequence.

Are the liquid compositions related by high pressure processes? (steps 4 and 6)

The liquid trend shown in Fig. 3.1.10 is reproduced in Fig. 3.1.11. Suppose that rather than following the path A-B-C, the 1 atm liquid line of descent followed the path A-B-E-F. This experimental liquid line of descent does not substantiate the initial suggestion that the trend A-B-C was developed by the crystallization differentiation of a liquid of composition A at 1 atm by the fractionation of olivine alone from A→B and olivine and augite from B→C. Ideally, this might have been deduced from the petrography of rock B by noting that its natural crystallization sequence was olivine followed by olivine + augite rather than near-liquidus coprecipitation of olivine + clinopyroxene.

Suppose that equilibrium melting experiments were then carried out on composition A at a pressure of 10 kb, and that the experimental liquid line of descent at this pressure was close to the path A→B→C. (This reflects the fact that the pyroxene liquidus field expands relative to the olivine field with increasing pressure, as illustrated schematically for low calcium pyroxenes in Fig. 3.1.8.) The correspondence of the high-pressure liquid line of descent to the natural liquid line of descent would support the interpretation that the suite of liquids sampled was related by crystallization differentiation at high pressure. This might correspond to successive tappings of a deep magma chamber (e.g., chambers A or B in Fig. 3.1.3).

Another indication that high pressure fractionation has been important in the evolution of a suite of liquids is the presence of phenocrysts or megacrysts, which can crystallize from the lava compositions only at elevated pressures. An example is the occurrence of

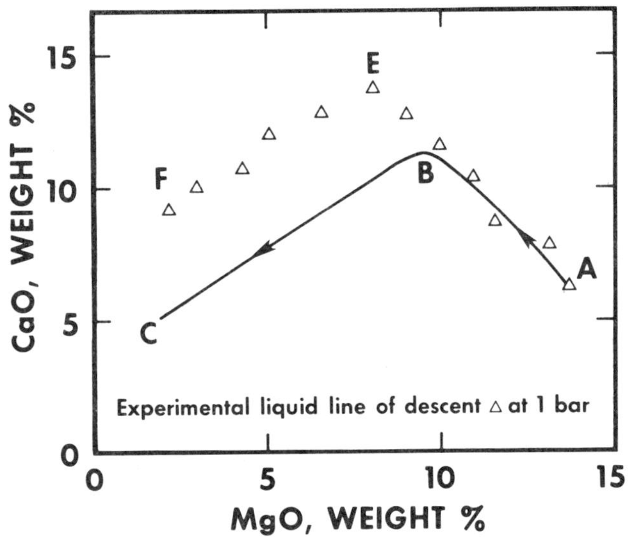

3.1.11 Hypothetical CaO vs. MgO diagram illustrating effects of high-pressure fractionation.

orthopyroxene megacrysts in alkalic basalts, which do not have orthopyroxene near the liquidus at low pressure. Experimental determination of the range of experimental conditions under which the high pressure mineral does crystallize near the liquidus can help to constrain the conditions under which fractionation occurred (step 6 in Fig. 3.1.4).

The preceding examples have emphasized simple, single-stage types of fractionation histories to explain the relationships between members of a basaltic suite, but polybaric fractionation may also occur. For example, it is possible for fractionated liquids to develop in successively shallower magma chambers or along the system of conduits that carry them to the surface. The more complex histories are probably more realistic than the simple isobaric differentiation histories discussed above.

Parental and primary liquids (steps 3, 4, 5 and 6)

In both Fig. 3.1.10 and 3.1.11 it appears that a liquid of composition A could yield all compositions from A→B→C by crystallization differentiation. Thus, liquid A can be regarded as a composition which is parental to the other members of the A→B→C suite. The next step is to consider the origin of the parental liquid A, and to decide whether it is primary, or to what extent it has differentiated at high or low pressures from a primary liquid. For terrestrial rocks we have fifty years of petrological wisdom and preconceptions to guide us, but for basalts from meteorites and other planetary bodies we cannot rely on preconceptions.

Two lines of reasoning of general significance in distinguishing primary from differentiated compositions can be useful under special circumstances. Consider a suite of liquids which are virtually identical in major element chemistry but vary significantly in contents of incompatible elements. This can be interpreted as indicating that the sequence of liquids was produced by increasing but small degrees of partial melting of a single type of source material. There are alternate interpretations of this same trend, however, and the relative merits of the competing interpretations must be assessed. For example, the trend could be produced by periodic tapping of a magma chamber in which successive pulses of fresh primary liquid mix with the differentiated liquids from early pulses (O'Hara, 1977).

The second line of reasoning is valid only for parental compositions which are multiply saturated under the physical conditions at which the other members of the suite developed from the parental liquid by crystallization differentiation. This line of reasoning states that (1) if there appears to be a clustering of liquid compositions at the parental liquid composition, (2) if the parental liquid composition is multiply saturated, and (3) if one or more of the minerals with which the liquid is saturated is in reaction relation with the liquid, then it is likely that the parental liquid composition is a primary liquid composition rather than a differentiated liquid composition. This type of argument was developed in detail by Walker *et al.* (1972) with respect to lunar KREEP basalts (section 3.4.1) and by Stolper (1977) in a discussion of the origin of eucritic meteorites (section 3.5.4).

For terrestrial rocks, the criteria for assessing the significance of a parental liquid and deciding if it may be a primary liquid composition are relatively straightforward. Our petrological wisdom tells us that basalts are generated by partial melting in the mantle, where the olivines may be as magnesian as Fo_{94}. Therefore, we expect a primary liquid to be capable of equilibrating with olivine-orthopyroxene bearing residues at high pressures, and the liquid to be magnesian enough to equilibrate with olivines as magnesian as Fo_{90}. The liquidus would not be multiply saturated at low pressures except by a fortuitous accident. A parental liquid with major and trace element contents appropriate for equilibration with our notion of mantle source rocks is described as a primitive magma. The more primitive the geochemistry of a parental basalt, the more closely we would expect it to approximate a primary liquid composition.

As an example, suppose that the parental liquid in a suite of continental basalts has olivine (Fo_{65}), pyroxene and plagioclase simultaneously on its liquidus at 1 atm pressure (step 5 in Fig. 3.1.4). Most petrologists would reject this parental basalt as a possible primary magma, because (1) a liquid in equilibrium with olivine of composition Fo_{65} could not be a primary liquid from our preconceived notion of the mantle source material, and (2) the experimental evidence for multiple phase saturation suggests that it was produced by shallow igneous processes.

What if this same parental basalt composition appeared as a meteorite? With no preconceptions to guide us, we cannot discard this as a possible primary meteoritic liquid composition. For all we know, meteoritic source regions could be at low pressures and have iron-rich olivines. The point of this example is to demonstrate that assessment of the significance of a parental liquid is of necessity subjective, and quite often based on circular reasoning. When we try to generalize the philosophy behind petrologic constraints on planetary interiors, we find that nearly all of our terrestrial ground rules concerning source region composition, mineralogy, and depth are inapplicable.

In general, the only course open, barring evidence to the contrary, is to assume that the parental liquid is primary and to examine the implications of this assumption for the constitution of the basalt source region. If a parental liquid is not primary, it may still be possible to infer the type of fractionation which produced the differentiated parental liquid and, given this, to use the phase relations of the parental liquid to set constraints on the phase relations of the actual primary liquids. For example, while most olivine tholeiites are not primary, it is considered by some that they differ from primary liquids only by virtue of extensive fractionation of olivine. If this is correct, and if the phase relations of olivine tholeiites are known, one can reasonably infer the phase relations of the primary compositions which differ from olivine tholeiites by being richer in the olivine component.

Primary basaltic liquids as probes into planetary interiors

By definition, a primary magma is a liquid that has not changed in composition since it was generated in its source region by melting. Assuming that the liquid was in equilibrium with the crystalline phases in the source region, it is possible, by examining the liquidus minerals of a primary liquid, to identify some or all of the minerals (and their compositions) left in the source region as residues, and to constrain the pressure and temperature in the source region. The procedure can be illustrated with Fig. 3.1.12, the liquidus diagram for a typical lunar primary liquid. As with most inverse problems, the solution is not unique. The liquidus diagram outlines the range of possible solutions: (1) the primary liquid was generated between 0-12 kb and between T_1 and T_2 leaving olivine in the residue, or (2) it was generated at 12 kb and T_2 leaving both olivine and clinopyroxene in the residue, or (3) it was generated between 12 and 25 kb and between T_2 and T_3 leaving clinopyroxene in the residue, or (4) it was generated at 25 kb and T_3 leaving both clinopyroxene and garnet in the residue, or (5) it was generated at pressures above 25 kb and a temperature above T_3 leaving garnet in the residue.

Again, prejudice and data from other sources are commonly used to place further constraints on the range of likely solutions. Suppose, for example, given certain assumptions about the initial rare earth pattern of the source, that the basalt rare earth pattern indicates garnet in the residue. This would suggest that the liquid was generated at $P \geq 25$ kb and $T \geq T_3$. Or suppose that geochemical arguments or geophysical limitations on the amount of liquid in a source region suggest that the liquid was generated by small degrees of partial melting. Because the source in general must have been polymineralic, it is usually assumed that more than one mineral would remain in the residue at small degrees of partial melting (section 3.3.3). This would suggest that the

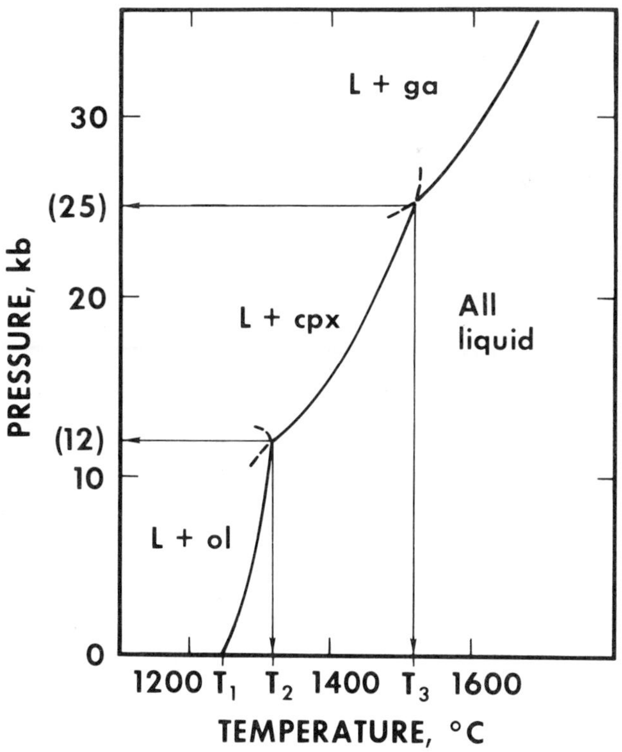

3.1.12 Model liquidus mineralogy for typical lunar or terrestrial basaltic compositions. Abbreviations: ol = olivine; cpx = clinopyroxene; ga = garnet; L = liquid.

liquid was generated at P = 12 kb, $T = T_2$ or at P = 25 kb, $T = T_3$.

In general, any randomly chosen bulk composition will have only one mineral on its 1 atm liquidus. By increasing pressure, the liquidus mineral will generally change at discrete T and P, as in Fig. 3.1.12. For a random composition, it is extremely unlikely that more than two phases will appear simultaneously on the liquidus. If, for a given primary liquid composition, a point is found on the liquidus at which three or more minerals coexist, it is likely that the P and T of this multiple saturation represent the conditions of the separation of the liquid, and that the three or more saturating phases were residual minerals left in the source region. Figure 3.1.12 is typical of many lunar and terrestrial basaltic compositions. For meteoritic basalts, where our preconceptions tell us that only relatively low pressures are applicable to the asteroidal source regions, it is the low pressure phase diagrams which are applicable.

The P-T diagram in Fig. 3.1.12 is for experiments conducted without H_2O or CO_2. This is applicable to lunar basalts, but it is not strictly applicable to most terrestrial basalts. Where volatile components are suspected to play a role in magma genesis, the inverse approach requires that the liquidus of the primary liquid composition must be investigated under conditions of controlled activities of volatile components. The interpretation can proceed much as outlined above for the dry liquids, except that the degrees of freedom allowed in the interpretation are substantially increased. Therefore, as the liquidus is investigated as a function of P, f_{O_2}, f_{H_2O}, f_{CO_2} etc., the multiple saturation points in Fig. 3.1.12 extend to lines and areas, and the condition of multiple saturation, where two or more minerals coexist with the liquid, loses the precise significance assigned to it in a system where P and T are the only intensive variables.

What if we knew that the parental liquid composition of a given suite of rocks was not primary? Could any constraints be placed on the mineralogy and composition of the source region and the conditions which prevailed there during generation of the primary liquid? The answer may be yes, if the type of fractionation which has produced the parental liquid from a primary liquid can be ascertained. Consider, for example, the composition whose phase diagram is shown schematically in Fig. 3.1.12. Suppose it is known that this basalt differs from a primary liquid only in that some unknown amount of olivine has crystallized from it. With addition of olivine, the point where the liquidus mineral changes from olivine to pyroxene moves to higher pressure (steps 4 and 6 in Fig. 3.1.4). If we assume that olivine and pyroxene were left as residual phases in the source region, then this point on the phase diagram of the differentiated basalt composition gives us a lower limit for the P-T conditions prevailing in the source region of the primary liquid. The effect of addition of olivine to lunar basaltic compositions is very well known (Walker *et al.*, 1976c), and if the amount of olivine which separated from the primary magma can be estimated, then the conditions of pressure and temperature of the cotectic of the deduced primary magma can be predicted with confidence.

The success of olivine-addition experiments depends critically on the assumption that olivine was the only mineral to crystallize from the primary liquid as it migrated from source region to surface. Assume that the liquid p in Fig. 3.1.13a is a primary liquid, produced at 20 kb from a source rock composed of forsterite, enstatite, and diopside. Equilibrium crystallization of olivine and diopside from this liquid at 1 atm pressure would yield fractionated liquids such as liquid f. Addition of forsterite to liquid f would not yield the original primary magma p.

It has been implicitly assumed that if a mineral is present in the source region after the generation of a primary liquid, then this mineral will crystallize at the liquidus of the primary liquid. However, this is not so if a mineral in the residue is in reaction relation with the liquid in the source region. In experiments conducted under conditions similar to those prevailing in the source region, the mineral in reaction relation will not appear on the experimental liquidus, even though the liquid is saturated with this phase under these conditions. In Fig. 3.1.13a, reaction relationships exist at 1 bar pressure between liquids a and forsterite, and b and forsterite, and at high pressures between liquids p and enstatite, and q and enstatite. The consequence is illustrated in Fig. 3.1.14, which shows the binary phase diagram from one side of Fig. 3.1.13a. The liquid a that coexists with forsterite and enstatite, if removed, would precipitate only enstatite. There would be no evidence that it could coexist with forsterite. Addition of a few percent of forsterite to a, however, would produce the assemblage forsterite, enstatite, and liquid at temperature T_1.

The mineral in reaction relationship may appear in experiments conducted at pressures slightly higher or lower than the actual pressure prevailing in the source region, but Fig. 3.1.13 demonstrates that this is not necessarily so. The primary magma p is in reaction relationship with enstatite in the source region at 20 kb. The phase diagram for the specific composition p is given in Fig. 3.1.13b. Although the liquid coexists with enstatite at 20 kb, enstatite does not occur at any pressure or temperature on the isopleth of Fig. 3.1.13b.

If a reaction relation exists, then one or more minerals present in the source region may not be recognized by the inverse method approach, even if a primary liquid has been identified. Studies of synthetic systems

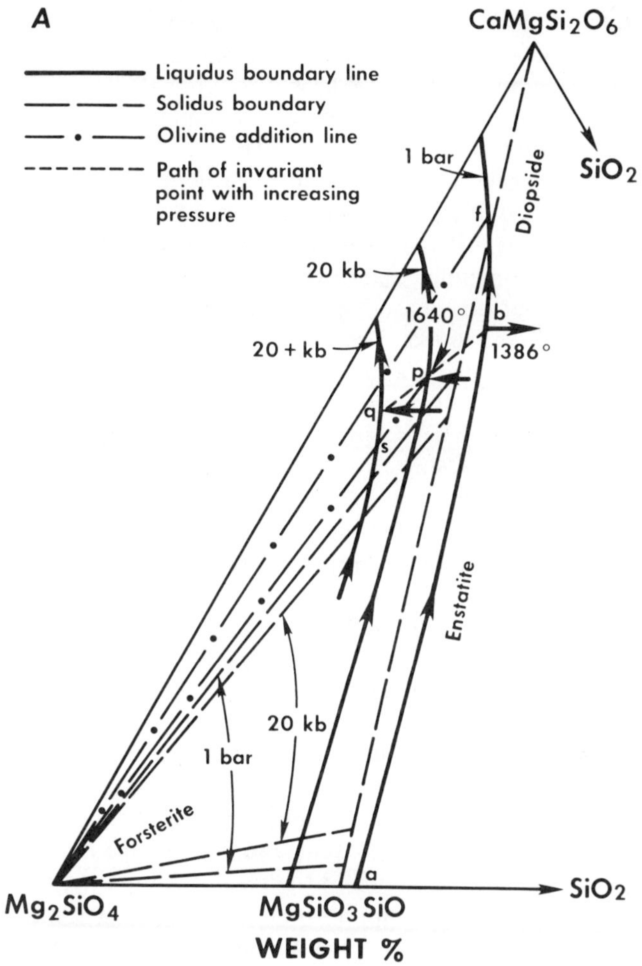

3.1.13 Phase relationships in the system Mg_2SiO_4-CaMg Si_2O_6-SiO_2 based on Bowen (1914), Schairer and Yoder (1962), Kushiro and Schairer (1963), Kushiro (1969), and Chen and Presnall (1975), but modified in order to illustrate graphically some possible effects of olivine—addition experiments and reaction relationships. **a.** Effect of pressure on the phase boundaries. Short-dashed line b-p-q shows the path followed by the isobaric invariant point for liquid coexisting with forsterite and two pyroxenes. Long-dashed lines connect forsterite to pyroxene compositions. Dash-dot lines show effect of adding forsterite to liquid compositions p and f. **b.** Phase relationships for the composition p (isopleth) in **a**, the isobaric invariant liquid at 20 kb. Note that although liquid p is in equilibrium with enstatite solid solution at 20 kb (**a**), enstatite does not appear anywhere in its equilibrium crystallization sequence at this pressure.

and reference to the liquidus diagrams of several basaltic compositions can aid the petrologist in recognizing situations in which this complication may be significant (section 3.3.8). In general, if a mineral is present at the liquidus of a primary basalt at the pressure of magma separation from the source, then that mineral must be present in the source. If a mineral is absent under the same conditions, then that mineral may or may not be present in the source.

Reaction relationships in systems more complex than binary systems (Fig. 3.1.14) may also confuse the rationale and interpretation of olivine-addition experiments. Assume that the liquid composition p, defined as primary at 20 kb in Fig. 3.1.13a, was sampled as a lava at the Earth's surface, and interpreted as a primitive and possibly primary composition. The phase diagram for this composition in Fig. 3.1.13b shows the liquid to be saturated with forsterite and diopside at 20 kb, but enstatite is absent. The inverse approach has been followed through the flow chart of Fig. 3.1.4 to this stage, but it is commonly abandoned when it comes to considerations of source region constraints and other applicable evidence. For example, petrologists expect a terrestrial primary basalt to be saturated with olivine and orthopyroxene at the source depth (section 3.3.1). Given the phase diagram in Fig. 3.1.13b, a petrologist

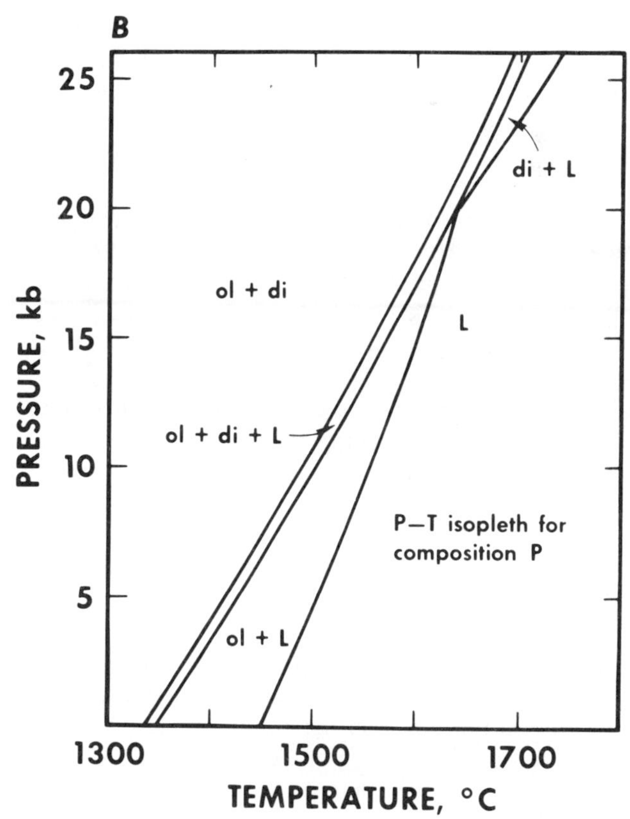

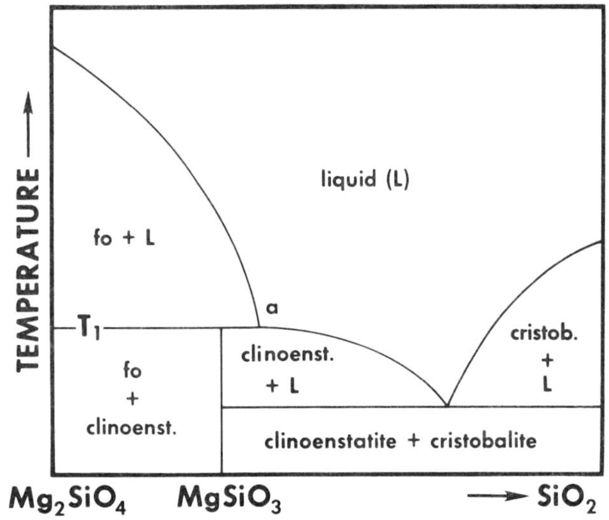

3.1.14 A portion of the system Mg_2SiO_4-SiO_2 at 1 bar (not to scale) illustrating the effect of a reaction relationship in confusing the application of the inverse approach. If liquid a, saturated with forsterite and clinoenstatite, is separated from its source, it will at no time crystallize forsterite during subsequent cooling and equilibrium crystallization.

might therefore conclude that this composition was a primitive magma which had precipitated some olivine en route from source to surface, and an experimentalist might conduct olivine-addition experiments to determine conditions under which olivine and enstatite would coexist at the liquidus. Figure 3.1.13a shows that with addition of forsterite to the liquid p, the field boundary between forsterite and enstatite would be crossed at higher pressure than 20 kb, e.g., at point s. The liquid s is picritic compared with the primary magma p, and it is saturated simultaneously with olivine and orthopyroxene. But it would be incorrect to infer that the picritic liquid s has the composition of a primary magma, generated at great depth, or that it was parental to liquid p. This is a hypothetical example where application of assumed source region constraints to the high-pressure melting relationships (Fig. 3.1.4) leads to incorrect conclusions, simply because enstatite is in reaction relationship with liquid at points along the polybaric line pq.

Enstatite-addition experiments may also be misleading. Based on the absence of enstatite at the liquidus of a postulated primary magma composition, one might undertake experiments on mixtures of this composition with varying amounts of enstatite in order to see how far the composition lies from the enstatite field (Green *et al.*, 1979). For the topology shown in Fig. 3.1.13, addition of enstatite to the composition p would show enstatite at the 20 kb liquidus even for mixtures containing the smallest amount of added enstatite, and one could therefore conclude that enstatite was present in the source. However, the topology of Fig. 3.1.13 could be modified so that no amount of added enstatite would produce enstatite at the 30 kb liquidus even though the composition p remained in equilibrium with enstatite. This would occur if enstatite melted incongruently at 20 kb and liquid p remained at its location shown in Fig. 3.1.13. Of course, enstatite does not melt incongruently at 20 kb, but the purpose here is to illustrate permissible lines of reasoning for complex compositions where the detailed topological relationships are not known. Thus, the generalization observed earlier holds also for enstatite-addition or other types of addition experiments. That is, if a phase is present at the liquidus, it can safely be concluded that this phase is also present in the source, but if a phase is not observed, it may still be present in the source. The only way to avoid this ambiguity in the inverse approach is to explore completely the liquidus phases in the composition space around the postulated primary magma composition. Partial explorations, such as enstatite-addition experiments, are a help, but they are not definitive.

Another major limitation on the inverse approach is that while it can identify the residual minerals in source regions, their compositions (obtained from the compositions of the liquidus minerals in the experiments), and the physical conditions prevailing during magma generation, it cannot set constraints on the proportions of the minerals left in the residue. For example, suppose we concluded that the basalt in Fig. 3.1.12 was generated at $P = 12$ kb and $T = T_2$ leaving olivine and clinopyroxene in the residue. The liquid could just as well have left 1% pyroxene and 99% olivine or 99% pyroxene and 1% olivine in the residue. Similarly, the liquid would be in equilibrium with the olivine and pyroxene if there were 99% liquid and 1% residue, or if there were 1% liquid and 99% residue, so that the experiments cannot set limits on the degree of partial melting. Constraints on the proportions of residual phases and on the percent of partial melting require the application of geochemical or geophysical arguments, or petrological prejudice concerning the nature of the source region.

In principle, the minerals at the liquidus of a primary liquid composition can be used to set limits on the mineralogy of the residue left in the source region after extraction of the primary liquid, and on the conditions in the source region during magma genesis. This is one of the most powerful of available probes into planetary interiors, despite the pitfalls and limitations emphasized in the preceding discussion. Petrologists have relied on this approach largely to provide support and detail for their already firm notions of the nature of basalt source regions on the Earth. However, in applications to extraterrestrial petrology where supplementary data in the form of field relations, ultramafic rocks and other petrological information are unavailable, the application of this approach is our main clue to the nature of basalt source regions.

3.2 EXPERIMENTAL PROCEDURES AND INTERPRETATIONS

The details of experimental procedures for the various types of apparatus used in experimental petrology have been adequately covered by Ulmer (1971), Edgar (1973) and Ernst (1976). The following discussion concentrates on the problems and limitations of experimental techniques, and their influence on the interpretation of results. It is important to understand these problems, because interpretations of the experimental results place constraints on the conditions and processes under which natural rocks form, and these should provide cornerstones for petrogenetic theory. Experimental problems have contributed significantly to bitter controversies about fundamental aspects of petrology, such as the effect of water on the composition of magma generated from peridotite (sections 3.2.2 and 3.3).

3.2.1 Documentation of experiments

It is a maxim that scientific reports should include comprehensive documentation of experimental techniques. Unfortunately, the technical descriptions accompanying many published experimental studies are inadequate to permit their re-evaluation in light of more recently-discovered experimental difficulties. This renders the data unreliable or obsolete. Table 3.2.1 is a list of items which we consider necessary for a thorough documentation of any petrological experiments.

The normal procedure for petrological experiments is to observe the effects of experimental run conditions upon a phase assemblage. Changes in the phase assemblage depend primarily on pressure and temperature, but may also depend on the starting material and the experimental technique.

Starting material

It is essential to document the bulk composition and the composition and structural state of each phase present in the starting material. The experimenter should determine the bulk composition of the starting material after it has been prepared, so that effects of the preparative steps (oxidation, reactions with its container, volatilization, etc.) are accounted for in the final analysis.

The starting materials are normally either crystalline rock powders, glasses or gels, crystallized glasses or gels, or chemical mixtures (principally for experiments involving volatile phases or components). For rock powders it is important to document (1) efforts to insure

Table 3.2.1 Items necessary to document an experimental study properly.

1. Aim of experiments.
2. Sample documentation:
   (a) Natural or synthetic.
   (b) Preparation (oxide mix or gel, if synthetic; grinding method; temperature and oxygen fugacity of preparation).
   (c) State (e.g., crystalline, glass, gel).
   (d) Mineralogy, if crystalline.
   (e) Grain size and size distribution.
   (f) Storage methods.
   (g) Composition before preparation, after preparation and after experiment.
3. Experimental apparatus.
4. Thermocouple type, calibration and error limits.
5. Pressure measurement method, calibration and error limits.
6. Methods of temperature and pressure control.
7. Oxygen fugacity (controlled or uncontrolled, method of measurement and error).
8. Method of addition of volatiles (if any) and error.
9. Sample container:
   (a) Material.
   (b) Shape.
   (c) Mass and/or ratio of surface area to that of the sample.
   (d) Open or sealed.
10. Details of change in P and T during ascent to run conditions, or during run.
11. Time at each pressure and temperature.
12. Quenching method and time to reach room temperature (or some arbitrary temperature between run temperature and room temperature).
13. Any special treatment during opening of charge (e.g., freezing a volatile phase).
14. Evaluation of attainment of equilibrium.

homogeneity of the powder, so that each experiment starts with the identical composition, and (2) the grain size of the powder. The grain size (or the surface area) influences reaction rates and thus aids in evaluating the duration of the experiment necessary to approach equilibrium. Gels have been used for their ease of homogeneous mixing and their reactivity, but they have problems which need to be evaluated for each investigation. Homogeneous glass can be prepared from rock powders, gels, or chemical mixtures. The use of glass insures an identical composition of each experiment and eliminates the possible metastable persistence of a crystalline phase. This advantage is offset by the fact that some minerals (e.g., feldspar) have lengthy incubation periods before nucleation from glass or melt, even when the conditions are well within their true stability fields. The nucleation and incubation period may be dependent on the amount of superheat (above the equilibrium liquidus temperature) used in preparing the glass. To avoid nucleation problems or for preparation of starting material for reversal experiments, the glass can be partially or wholly crystallized.

Techniques

It is essential to document carefully the techniques employed before, during, and after each run, because these are critical to interpretation of the experimental data. It is important to know what changes occur as the charge is approaching run conditions. Crystals which nucleate and grow within a starting material during the "run-up" and persist metastably throughout the experiment may compromise the run product assemblage sufficiently to affect its interpretation. The accuracy of the pressure and temperature conditions may substantially affect interpretation of the results. Also crucial to the data are composition control during the run, charge-capsule interaction, and monitoring and control of volatile components. Reactions which occur during the quench may vary according to quenching techniques and rates. The methods of determining run duration and evaluating the approach to equilibrium (not a factor in controlled cooling rate runs) must also be documented.

Run products

In order to permit evaluation of the quality of an experiment and to understand the reaction and crystallization processes which have occurred, run products ideally should be described in terms of the phases present, the proportions of phases, phase chemistry, crystal morphology, grain size and crystal abundance, and structural state.

The phases present may be identified by use of the petrographic microscope, X-ray diffractometer and electron microprobe. Additional information can be obtained from density measurements, thermogravimetric analyses, transmission electron microscope studies and infrared absorption spectra.

The proportions of phases may be estimated by inspection (which is inadequate), measured by microscopic modal analysis, or calculated by chemical mass balance. This information is used in checking for changes in bulk composition of the charge, in constructing quantitative models of fractional crystallization, and in testing for equilibrium by reproduction of the same proportions of phases in runs identical in all respects but duration.

Approximate estimates of mineral compositions may be obtained by optical methods, including refractive index measurements, for example, and from X-ray diffractometer measurements. Precise results are obtained by electron microprobe analysis of all run-product phases, including homogeneity or zoning within crystals, and the detection of minute inclusions. An indication of how well the phase assemblage that formed during experimentation was retained during quenching may be obtained by analysis of chemical gradients in glass adjacent to crystals—if the condition of an experiment is preserved during quenching there will be no composition gradients; however, these may be too small to observe. The trace element, water, and carbon contents of melts and crystals may be determined by the β-track mapping method (Mysen and Seitz, 1975). In experiments involving a vapor phase, a minimum estimate of the content of silicate material dissolved in the vapor may be obtained by microchemical analysis of the condensed vapor (Holloway, 1971).

Crystal shape is a sensitive index of growth conditions, growth rate and growth mechanism. Crystals grown under conditions of slight super-saturation are facetted and complete, those grown at substantial super-saturation generally grow more rapidly as skeletons or dendrites. The crystal shape can be useful in distinguishing crystals grown isothermally from those grown as run conditions changed, or during the quenching process. Note the two different morphologies of olivine in Fig. 3.2.1.

Grain size is an important factor in the quality of electron microprobe analyses, because interference and absorption effects may be severe for crystals less than 10 microns in size. In addition, grain size and crystal abundance contain information about the relative rates of nucleation and growth of crystals in glass or melt.

The stability and composition of certain minerals—notably pyroxenes and plagioclases—are sensitive to

3.2.1 Basaltic charge contained within a platinum capsule, which was run in the horizontal position shown; the height is 3 mm. The material was heated to melting, then cooled at a regular rate, and then quenched. Note the two different olivine morphologies (Donaldson, 1976).

the internal structure of the crystal. When comparing experimental results with natural rocks, the crystals being compared should be in the same structural state.

3.2.2 Experimental techniques

Equipment

Most experimental studies of basaltic and peridotitic materials have been conducted in the temperature range 600–1600°C at pressures from 1 atm to 50 kb. Many different types of 1 atm quench furnaces have been utilized including several in which f_{O_2}, f_{H_2} and f_{S_2} can be controlled by gas-mixing techniques (see Edgar, 1973, for design details). The more commonly used types of high-pressure apparatus are cold-seal gas vessels (0.1 to 6 kb; 200–800°C), internally-heated gas apparatus (0.1 to 10^{+} kb; 200–1400°C), solid-media piston-cylinder apparatus (7 to 60 kb; 200–1800°C), and other types of solid-media apparatus (belt, tetrahedral anvil, etc.) for pressures greater than 25 kb.

Control and calibration of temperature

Temperatures during experiments are normally measured by one or more thermocouples, and maintained within narrow limits by automatic controllers. Commonly used thermocouples (and their operational ranges) are composed of Pt-Rh alloys (up to 1600°C), W-Re alloys (1100–2200°C) and the Ni-rich alloys Chromel and Alumel (up to 1200°C). Temperatures recorded by thermocouples may be monitored during experiments by use of a chart recorder and measured using a calibrated potentiometer. All thermocouples are subject to poisoning (contamination) by extraneous elements, which results in erroneously measured temperatures. Such effects may be small or large depending on thermocouple type, the host material, temperature, and time, and they need to be evaluated for each set of experiments. In some types of apparatus, thermocouple contamination can be reduced by using sheaths of Inconel or other materials, or by using inert wafers between the thermocouple bead and sample container.

Temperatures within all types of apparatus are calibrated by determining one or more simple reactions within the temperature-pressure range of interest. Some of the more commonly used are the melting points of Au, Ag, Ni, AgCl, NaCl and diopside, and the eutectic melting reactions of Fe-FeS and anorthite-wollastonite-gehlenite. The temperature distribution within the furnace in the vicinity of the sample container is another important consideration, which can be determined by monitoring a calibration reaction with one or more thermocouple probes.

Thermocouple emf is also affected by pressure. This effect is commonly disregarded, although some calibrations have been made (e.g., Getting and Kennedy, 1970). A statement should be made whether the temperatures have been corrected and, if so, using what calibration. Merrill and Wyllie (1975) concluded from available studies that the maximum temperature uncertainty due to pressure effects for their melting experiments on basaltic compositions in piston-cylinder apparatus was ±13°C (at 1000°C, 30 kb). The temperature uncertainty from other sources in all types of modern apparatus is commonly ±5°C or better.

Control and calibration of pressure

In gas-media apparatus, the hydrostatic pressures are monitored with mechanical gauges attached directly to the pressure vessel. Gauges may be calibrated against a dead-weight piston.

In solid-media apparatus, pressures are non-hydrostatic and cannot be monitored directly. Hydraulic ram pressures are monitored instead. Each particular apparatus and furnace assembly must be individually calibrated for friction effects and "anvil" (or "pistoning") effects over the pressure-temperature range of interest. This is usually done by using reactions such as albite ⇌ jadeite + quartz (e.g., Johannes *et al.*, 1971) and the quartz ⇌ coesite transition, electrical resistivity transitions in Ba, Th, Cs or Bi, and some of the melting reactions mentioned above. The precision of pressure measurement in gas-media apparatus is commonly better than ±0.5%, but in solid-media apparatus it may reach ±5%.

Control of activities of volatile species

Because of the significant effects of volatile components such as H_2O, CO_2, H_2 and S_2 on reactions in basaltic and peridotitic systems, many experiments have been conducted in which the activities of volatile species are controlled or at least closely defined.

In 1 atm experiments, f_{H_2}, f_{O_2} and f_{S_2} may be controlled using mixtures of gases continually flowing past the open charge. Oxygen fugacity is normally controlled by CO-CO_2 or CO_2-H_2 gas mixtures. In the latter mixture there is a hydrogen fugacity as well as a water fugacity. Oxygen fugacity and sulfur fugacity can be controlled with CO_2-H_2S mixtures. The fugacities of the mixtures are calculated from thermodynamic data (Robie *et al.*, 1978). The oxygen fugacity can be measured independently using solid ceramic oxygen electrodes (Williams and Mullins, 1976).

At higher pressures, f_{O_2} may be fixed using solid buffer assemblages with a multiple capsule (see Ulmer, 1971 for further discussion of the solid buffer technique; in reality, it is f_{H_2} that is controlled). To date there are no techniques for control of sulfur fugacity at high pressures.

Activities of H_2O and CO_2 have been controlled or monitored in high-pressure experiments on basalts and peridotites. Most such experiments have involved vapor-saturated (vapor-excess) conditions, although the activities of H_2O and CO_2 in the vapor may be fixed at any value between 0 and 1. H_2O is usually added to experimental charges in liquid form, unless the starting material contains sufficient structurally-bound water that can be released under the experimental conditions. CO_2 may be added as oxalate (e.g., AgC_2O_4) or, especially in synthetic starting mixtures, as carbonate. Mixtures of water and oxalate can yield all activities of H_2O and CO_2, and an activity of 0.5 can also be obtained from crystalline oxalic acid dihydrate. With the use of a solid buffer assemblage in addition, the activities of H_2O, CO_2 and O_2 can be controlled independently (Mysen and Boettcher, 1975a; Allen and Boettcher, 1978).

Experiments also have been conducted under vapor-absent conditions, but with the volatile components dissolved in melts or structurally-bound in solids. Such experiments require the addition of smaller amounts of H_2O- or CO_2-generating compounds than those required to saturate the charge at experimental conditions. For H_2O contents less than about 5 wt.%, weighing errors can be significant (especially with the 5–15 mg charges used in piston-cylinder apparatus) and H_2O contents may not be reproducible from run to run. Unless the starting material is a natural rock containing structurally-bound H_2O, it is preferable to prepare from the dry starting material a larger volume of experimentally-hydrated subsolidus material for use in subsequent experiments.

Very dry conditions are readily attained in 1 atm furnaces and in gas-media high pressure apparatus. In solid-media apparatus, H_2O produced by dehydration of components of the furnace assembly (e.g., talc) can be excluded from the container environment fairly effectively by use of pyrex or BN sleeves. However, diffusion of H_2 through noble metal containers can cause the formation of H_2O in iron-bearing anhydrous mixtures through loss of iron to the capsule, leaving oxygen for combination with the hydrogen. Because of the effect of even small amounts of H_2O on melting reactions, diffusion of H_2 into charges should be monitored, and run durations selected accordingly.

Quenching

During the period of time while an experiment is cooling to room temperature, both equilibrium and nonequilibrium phases can grow in the experimental charge. The run products then examined would not correspond to the equilibrium assemblage produced during the experiment. Failure to identify modifications of this type can lead to major errors in the interpretation of experimental results. Normally, however, quench phases can be identified by their skeletal, dendritic or feathery morphologies. Moreover, their compositions may be inconsistent with equilibrium. For example, quench pyroxenes precipitated in melting experiments may have abnormally high FeO and alkali contents compared with equilibrium pyroxenes.

In melting experiments, the precipitation of quench-phases, either individually or as overgrowths on equilibrium crystals, modifies the liquid composition with the result that the composition of the glass analyzed differs from that of the liquid which was present under run conditions. In many experimental studies, determination of the equilibrium liquid composition is critical. An estimate can be obtained by combining the compositions of glass plus quench phases in appropriate proportions, but this estimate must be refined and confirmed by a complementary series of experiments. The compositions of minerals coexisting with the liquid under run conditions can be determined. The complementary experiment involves determining the liquidus minerals for the postulated equilibrium liquid composition under identical experimental conditions. If the compositions of these liquidus minerals match those of the same minerals in the original experiment, then the identity of the equilibrium liquid is confirmed. Alternatively, if large clear areas of glass are found to be homogeneous even in the presence of quench crystals around the margins of such areas, the glass compositions may be considered to be valid. Any effect of the quench crystals on the glass far removed from the quench crystals would be revealed by composition gradients in the glass.

An important example of this type of problem is provided by the determination of the first liquids produced during the melting of peridotite in the presence of water (Green, 1976a). The problems of quenching and testing equilibrium liquid compositions have also been well-documented by Cawthorn *et al.* (1973a) and Nicholls (1974).

It is commonly desirable to confirm the presence or absence of a free vapor phase. Quench problems may interfere with this interpretation. For example, in many systems which are undersaturated under run conditions, vapor bubbles may appear in the charge during quenching.

It is not possible to preserve the exact equilibrium conditions of some experiments during the quench. High temperature magnetic states may not be preserved. Structural modifications may occur, with high-pressure polymorphs reverting to low-pressure forms, as observed in studies of $MgSiO_3$ pyroxenes. In very high pressure experiments, the high-pressure mineralogy may revert to a glass or to the low-pressure mineralogy upon quenching. This reversion has been confirmed by studies using *in situ* X-ray diffraction of minerals produced in diamond anvil cells.

3.2.3 Composition control during experimentation: the container problem

Differences in the interpretation of experimental results often arise from lack of control of the bulk composition of the charge during the experiment. The major problems are interactions of the sample with its container, resulting in the gain or loss of iron, and changes in the activity of FeO due to lack of oxygen fugacity control. The container problems have not yet been adequately resolved. A discussion follows of the various materials which have been used.

Platinum group metals

Platinum, palladium, rhodium, and their various alloys have classically been used for high-temperature experimentation. They have relatively high melting points (circa 1600°–1800° C) and are easily sealed by welding techniques. The metals are malleable, and thin-walled metal capsules collapse, effectively transmitting the pressure to the sample.

Containers of these metals are notorious for alloying with iron in iron-bearing systems (Merrill and Wyllie, 1973; Nehru and Wyllie, 1975). The alloying of Fe (from FeO in the charge) with these metals affects the a_{FeO} during the experiment. These metals are also very permeable to hydrogen. In high-pressure experiments where H_2O is present as a component, or where H_2O may be generated from the pressure media (e.g., by dehydration of talc or pyrophyllite), H_2 from dissociated H_2O may diffuse into or out of the capsule. This will not only directly affect the f_{O_2} and the complementary a_{FeO}, but may also change the observed phase relations due to the presence of H_2O in experiments designed to be anhydrous, or to a reduction in the amount of H_2O in hydrous experiments. The problem of hydrogen diffusion can be minimized by careful experimental design as mentioned in section (3.2.2).

Two methods of minimizing iron loss to these metals merit discussion. One is to presaturate the platinum or alloy with the proper amount of iron so that no iron loss or gain occurs at the temperature and pressure of the experiment (Huebner, 1973). This technique requires knowledge of the a_{FeO} of the charge at the experimental P and T and saturation of the metal with the proper amount of iron so that no interchange occurs at the given P and T. This is an almost ideal sample container, but it requires detailed thermochemical data for the sample and tedious capsule preparation procedures. More recent methods involving the use of soluble iron silicates (Ford, 1978) and electroplating of iron on capsule walls (Johannes and Bode, 1978) have given promising results.

The other method of minimizing iron loss to these metals has been to suspend the melt from wire segments or loops in controlled atmosphere experiments at 1 atm (Riboud and Muan, 1962; Presnall and Brenner, 1974; Donaldson *et al.*, 1975a). The iron loss to the metal is minimized due to the large mass of the sample compared to the mass of the wire loop (commonly from 200:1 to 50:1). An experimental charge is illustrated in Fig. 3.2.2. Relative iron losses of up to 10% have been recorded in experiments of more than a few hours duration at temperatures of 1100°–1200° C (Usselman and Lofgren, 1976; Irving *et al.*, 1978; Corrigan and Gibb, 1979; Donaldson, 1979). Huebner's (1973) technique of presaturating platinum with iron can substantially reduce the iron loss. The loss of volatile elements such as sodium is a disadvantage of the wire loop technique at high temperatures.

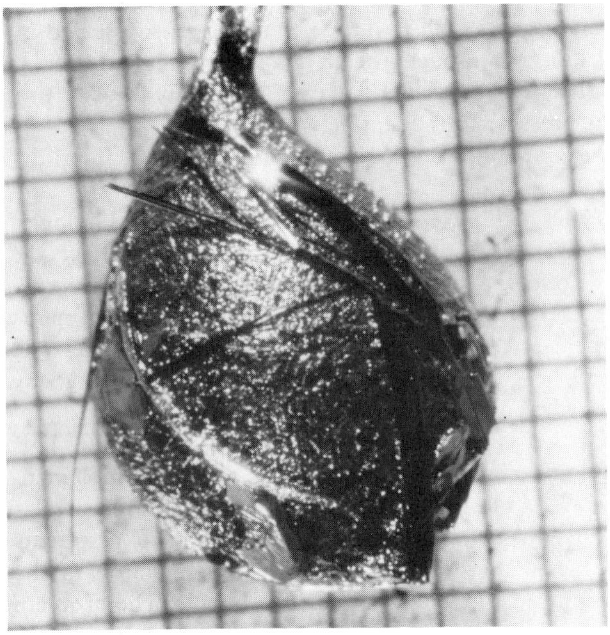

3.2.2 Quenched experimental charge suspended on a Pt-wire loop. A pellet of high-titanium lunar basalt (74275) was placed on the loop and melted. The molten sphere, adhering to the wire by surface tension, was cooled from the liquid field to 1030°C at a rate of 7°C/hr. The bladed crystals are armalcolite and ilmenite. The grid spacing is 0.5 mm.

The compositional problems encountered with these platinum group metals are not limited to the overwhelming iron loss, but extend to the loss of other elements such as Mn, Co, Ni, and S. The effects of these problems on the interpretation of experimental results are normally insignificant, but they can be significant in some investigations.

For long runs at very high temperatures (> 1500° C), Chen and Presnall (1975) found that silicon diffuses from the charge into a Pt capsule, forming a Pt-Si alloy, and causing failure of the capsule by melting far below the normal melting temperature of Pt.

Silver and silver-palladium alloys

These capsule materials behave similarly to platinum in malleability and sealability, but their useful temperature ranges are restricted: silver melts at 961° C, whereas some Ag-Pd alloys can be used up to about 1200° C. Pure silver does not alloy readily with iron from iron-bearing samples, but as the palladium content of the alloy increases, the magnitude of iron losses approaches that encountered with platinum. The iron loss is not as rapid as to platinum (Nehru and Wyllie, 1975). In addition, these alloys are also permeable to hydrogen and possess the same problems as platinum in this respect.

Gold

Gold capsules exhibit superior qualities. Gold is malleable, is easy to seal, does not readily alloy with iron, is almost impermeable to hydrogen, and is unreactive with sulfur. It melts at 1063° C, which makes it suitable for only a limited range of hydrous basalt experiments. The alloying of small amounts of gold with platinum only slightly decreases the iron loss problems of platinum.

Iron and various iron alloys

Closed capsules of these materials have been used (primarily in studies of lunar and meteoritic basalts) in attempts to buffer the oxygen fugacity in closed capsules, and thus the a_{FeO} of basaltic charges. Capsules are normally used with a tight fitting cap, because it is extremely difficult to seal these metals. For 1 atm experiments, the capsules are placed in an evacuated sealed silica glass tube. Alkali loss and volatile loss to the evacuated volume of the silica glass tube is often recorded (Walker *et al.*, 1976a). Devitrification of the silica glass tube, in long term experiments, allows oxygen to diffuse through with a potential oxidation of the sample.

The a_{FeO} of the charge equilibrates with the capsule material as the oxygen fugacity is buffered by the capsule and the phase assemblage of the charge. Depending on the initial a_{FeO} of the sample, the resultant iron content can increase, decrease, or remain the same (O'Hara and Humphries, 1977). The buffering capacity of the capsule can be changed from iron saturation (pure iron capsules) to nickel saturation (about three orders of magnitude more oxidizing); Fe-Ni alloys produce intermediate levels of oxidation. The actual oxygen fugacity imposed on the samples is not known for any of the iron or alloy containers.

Open iron capsules have been used in controlled atmosphere experiments for limited ranges of oxygen fugacity. They can only be used for systems which are iron saturated; at oxygen fugacities above that of iron saturation there are significant increases in the FeO content of the charge.

Molybdenum

Molybdenum capsules have been used in experiments where the oxygen fugacity desired is near the iron-wüstite buffer (MoO_2-Mo buffer is reportedly near Fe-FeO; Visser and Koster Van Groos, 1979). Molybdenum behaves mechanically much like iron, but it is more brittle. Molybdenum will alloy with iron from the charge if the sample is iron saturated. It can readily substitute in some phases, especially oxide minerals.

Graphite

Graphite capsules have been used in some high-pressure experimentation. At about 150 bars, the graphite buffer crosses the quartz-fayalite-iron buffer. At higher pressures the graphite buffer probably lies between the Fe_3O_4-FeO and Fe-FeO buffers. It has been suggested by Sato (1978) that terrestrial basalts and possibly lunar basalts originated under conditions of graphite saturation. Graphite containers should be further calibrated before their utility in high pressure experiments can be adequately assessed. Graphite capsules are not acceptable below 150 bars due to their very strong reducing capacity. At high pressures, C dissolves in some liquids producing carbonate ions.

Boron nitride

Boron nitride behaves similarly to graphite, but there are no indications that its buffer curve crosses over into a useable basaltic range of oxygen fugacity at high-pressures. It is, however, a useful capsule material for high-pressure sulfide experimentation (Usselman, 1975), although in some systems boron from the BN capsule may alloy with metallic iron, with a significant effect on the determined phase relations.

Capsule summary

For 1 atm experimentation, the preferred technique of oxygen fugacity control is the use of gas-mixing systems (e.g., Williams and Mullins, 1976). This, however, does not overcome the container interactions, and it introduces another constraint to the experimental design. One must have a good estimate of what the intrinsic f_{O_2} of the basalt was at formation; this can be approximately measured by various techniques, or inferred from mineral assemblages and thermodynamic considerations. Other minor, but potentially important compositional changes may occur with gas-mixing techniques. Loss of sodium and other volatile elements may affect the determinations of phase boundaries. Sodium loss will increase the temperature of plagioclase appearance. This error can be assessed experimentally if identical runs are made at critical conditions as a function of time.

The problems of constant composition and f_{O_2} control become more acute in high-pressure experiments, because the sample containers that can be used have unacceptable properties, and because f_{O_2} control is possible only by using solid buffer techniques (which require water to buffer the charge through hydrogen equilibration). Furthermore, we have inadequate knowledge of oxygen fugacity at the pressures and temperatures of basaltic evolution.

In an attempt to avoid the iron loss problems, Coons *et al.* (1976) investigated an analogue basalt composition where the ferrous iron was replaced by divalent cobalt. Initial results offer some promise, but further calibration and work on the effect of Co^{2+} (instead of Fe^{2+}) is required before the utility of such analogues can be confirmed.

This review of sample containers and experimental design may appear somewhat pessimistic. However, documented information on the magnitude of compositional changes during an experiment allow constraints to be placed on the interpretation of the results. Determination of the phase relations over a range of oxygen fugacities allows interpretations which are independent of preconceptions of oxygen fugacity at the time of evolution or fractionation.

3.2.4 Equilibrium and its recognition

A closed system at constant pressure and temperature is in an **equilibrium state** if the condition of the system does not change spontaneously with increasing time. Distinctions between different sorts of equilibrium states depend on how the system responds to externally imposed disturbances such as fluctuations in temperature, pressure, or substitution of different phase assemblages of equivalent bulk composition. Systems which do not return to their original state following such disturbances are in **unstable equilibrium.**

Examples of equilibrium states are illustrated by the positions of the spheres on the slope in Fig. 3.2.3. The horizontal axis represents some reaction variable, which is associated with a change in energy for the system. The sphere in position A is in equilibrium until a slight change causes it to roll down to position B. It is in unstable equilibrium because it has an inherent capacity to react. Equilibria which are stable against small temperature and pressure fluctuations or phase substitutions may not return to their original state for larger fluctuations or other phase substitutions. The activation energy required to displace the sphere from equilibrium position B is much less than that required to displace the sphere from equilibrium position C. All equilibrium states in Fig. 3.2.3. are **metastable** with respect to the position D, the state with the lowest free energy for the stipulated conditions. Metastable states may persist indefinitely. If the magnitude of the activation energy involved in a transition is large (as for position C in Fig. 3.2.3.), or if all possible equilibrium states are not anticipated, it may become impractical to

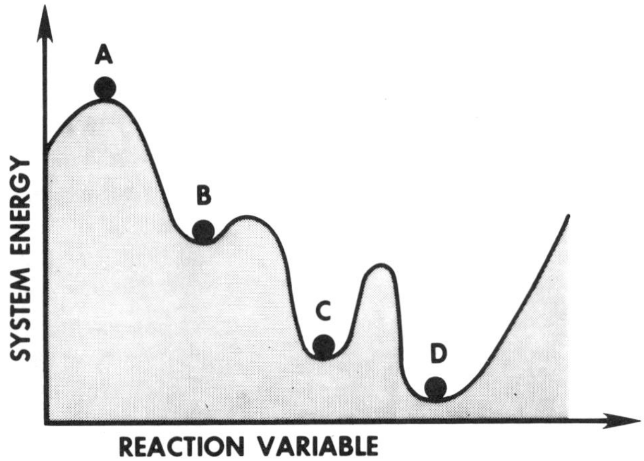

3.2.3 Diagram illustrating positions of unstable (A), metastable (B and C), and stable (D) equilibrium. It requires more energy to change from metastable condition C to stable condition D than from condition B to C, because of the higher threshold (energy barrier) between C and D than between B and C.

determine experimentally whether a state of lower energy exists for the system.

Departures from an equilibrium state in natural mineral systems are introduced by changes of physical conditions such as pressure and temperature or changes in the system's energy (by heat flow). If the changes are introduced slowly enough so that the system continues to maintain an equilibrium state throughout the process, the process is **reversible**. Such a transformation is a succession of equilibrium states which are each characterized by a particular set of physical conditions. To the extent that a natural process is reversible, laboratory observation of the equilibrium state of the system under each set of physical conditions will provide an adequate understanding of the possible system behavior. Many natural and experimental systems, however, are not the result of equilibrium states or of reversible processes. These departures from equilibrium may be of interest in themselves, but they are most instructive when compared relative to the equilibrium state. It is of critical importance, therefore, to demonstrate whether the results in experimental studies correspond to an equilibrium state, using the methods outlined in the following sections.

Variation of experiment length

A system at equilibrium does not change with increasing time of observation. If the run products in short runs are different from those run at a variety of longer times under the same physical conditions, the short-term run products are not equilibrium products. Although products of the longer runs are more likely to be the equilibrium products, this is not assured. Long runs may promote interaction of charge with container as discussed in section 3.2.3., and place an operational limit on the useful length of an experiment. Furthermore there is no guarantee that long runs give stable rather than metastable products. The persistence of metastable glass and non-growth of crystalline phases below their melting temperature is legendary for albite (Schairer, 1959) and common for plagioclase from basaltic compositions (Biggar *et al.*, 1972). Increasing run time without changing run products is not a rigorous demonstration of equilibrium.

Synthesis and use of different starting materials

A common experimental technique employed in reconnaissance studies aimed at outlining possible system configurations is the use of reactive starting materials such as glass or gels. Products of such experiments may show changes in phase appearance and changes in solid solution compositions with variations of experimental conditions. The synthesis boundaries determined in this way may not correspond with equilibrium boundaries, because all that has been demonstrated is that the product is more stable than the starting materials. The stability of the products with respect to alternate states is not addressed. Although synthesis products may correspond with equilibrium results, counter examples are numerous. For instance, Green (1976b) has shown that intermediate-calcium pyroxenes synthesized near the liquidi of many lunar mare basalt compositions at elevated pressures are metastable with respect to mixtures of high- and low-calcium pyroxenes.

Phase rule

Chemical thermodynamics (Gibbs' phase rule) places a limit on the number of phases which may be present in a system of a given complexity at equilibrium. If this number is exceeded, equilibrium has not been established. Typical basaltic systems have enough minor components which can be accommodated by solid solutions in the major mineral phases that an improbable number of phases must be present to see a clear violation of the phase rule. Therefore, this criterion of disequilibrium is most readily applicable to simple laboratory systems with limited compositional variability. Furthermore, even if the product obeys the phase rule with respect to numbers, there is no assurance that the phases present are stable so we can only use the phase rule as a criterion of disequilibrium.

Textural features

The equilibrium state, when considered as an energy minimum, has textural features which are a result of the minimization of the surface energy of the phases. Although the energy involved in this consideration may be relatively small, the effects are readily observable and well known. Coarseness of grains correlates with equilibration. Also, skeletal crystal morphologies with large surface areas may be interpreted as disequilibrium products as in the case of "quench" crystals. Once again, however, we have an essentially negative criterion for equilibrium, since it is perfectly possible to grow coarse, faceted crystals which do not correspond to the equilibrium composition or structural state. Absence of disequilibrium characteristics is a necessary but not a sufficient condition to demonstrate equilibrium.

Reversals and convergences

Equilibria of interest in basalt petrogenesis commonly involve the appearance or disappearance of a phase, or change in composition of a phase as physical conditions are changed. Consider the high-pressure reaction represented by the boundary between phase assemblages B and C in Fig. 3.2.4a. The starting material M, of suitable composition to determine this reaction, may be composed of glass, gel, oxide mixture, or crushed rock. If a series of one-stage runs, 1a, 1b, and 1c, produce the phase assemblages A, B, and C at high pressures, the synthesis boundary is located between the runs producing B and C.

To test for stability, it is common practice to introduce initial perturbations of the system by substitution of different starting materials. If the system reacts to the same final states from the different starting materials, the equilibrium is at least demonstrated to be stable against perturbations as large as the difference in starting materials. For sluggish reactions, seeds of the desired products in B or C may be added to starting material M in order to overcome nucleation problems. If the phase assemblage A includes high pressure minerals absent from the starting mixtures, M, it may be useful to repeat with two-stage runs, 2ab and 2ac, in which starting material M is first converted to A by routine 1a, and then increased in temperature. If phase assemblages B and C are produced at the same conditions from D as they were from M in the one-stage runs, confidence in the synthesis boundary is increased.

For demonstration of reversible equilibrium, it is necessary to perform two-stage runs 2bc and 2cb, where starting mixture M is first converted to B (via first stage 1b) or to C (via first stage 1c), and then converted by change of temperature in the second stage from B to C,

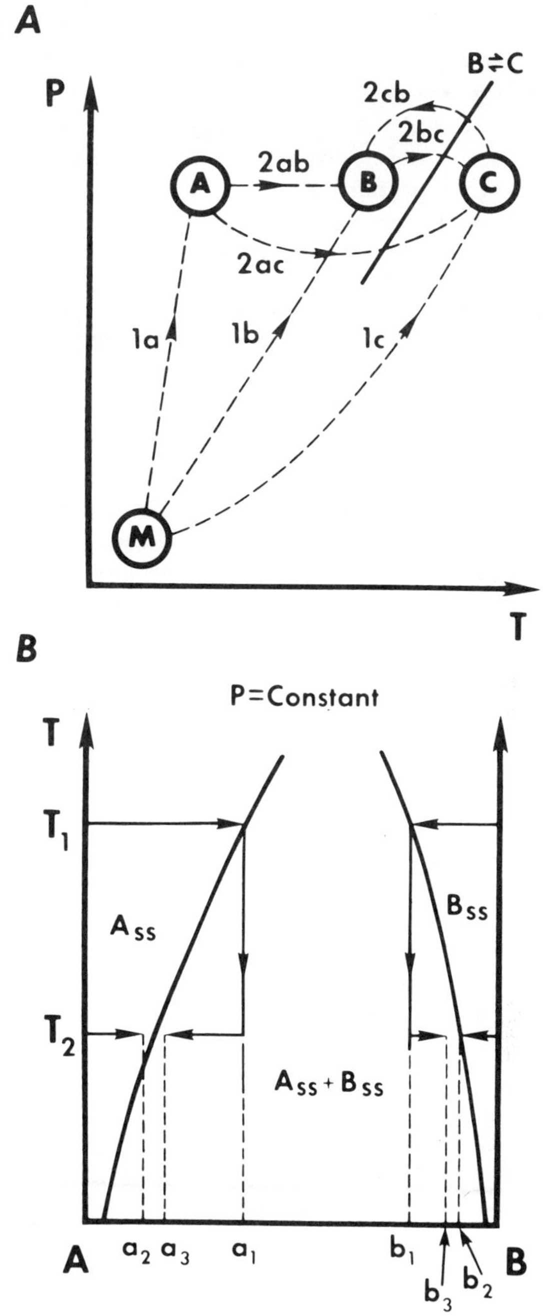

3.2.4 Experimental routines for approaching and testing a state of equilibrium: reversals and convergence. **a.** Starting mixture M is converted to assemblage A, B or C by synthesis runs 1a, 1b or 1c. Reversal is demonstrated by converting B to C, and C to B, via two-stage runs 2bc and 2cb (see text, and Stern and Wyllie, 1975). **b.** Mixtures of minerals A and B can be reacted to coexisting solid solutions a_1-b_1 and a_2-b_2 at temperatures T_1 and T_2, respectively. These compositions may define a solvus, but for confirmation, the high-temperature assemblage a_1-b_1 is lowered to T_2 where it is converted to a_3-b_3. Convergence of a_3-b_3 with a_2-b_2 is evidence for equilibrium.

and from C to B, respectively. In simple systems, complete reaction in both directions may be achieved but, commonly, the demonstration that B is partially converted to C and that C is partially converted to B is accepted as proof of reversible equilibrium. Similarly, the change in proportions of phases in a mixture of B and C is used to indicate direction of reaction. In complex silicate systems where the phase assemblages B and C include minerals of variable composition, it may be impossible to achieve complete reaction to phases of identical proportions and compositions on either side of the equilibrium boundary.

Equilibria involving phases of variable composition may be investigated to establish the stable composition by the method of convergence of composition from two different directions. Consider, for example, the solvus between minerals A and B in Fig. 3.2.4b. At temperatures T_1 and T_2, the mineral pairs a_1-b_1 and a_2-b_2 were synthesized from a mixture of minerals A and B. The mixture a_1-b_1 when run at temperature T_2 reacted to produce the minerals a_3-b_3. The compositions of the minerals at temperature T_2 converged towards the same composition, and the equilibrium phase boundary is thus located between a_2-a_3 and b_2-b_3.

The solvus may be defined by the synthesis boundaries represented by minerals a_1-a_2, and b_1-b_2 in a reactive system, but convergence tests are required to prove that the synthesis products have reached the equilibrium boundaries. For example, the reaction boundaries obtained by Nehru and Wyllie (1974) for the solvus between enstatite and diopside in the presence of hydrous liquid at 30 kb remained suspect until Lindsley and Dixon (1976) confirmed them by the convergence method.

Finally, it should be pointed out that metastable phases may show compositional convergence in the same way that metastable phases may participate in a reversal demonstration. Experimental distinction between the metastable and stable valleys in Fig. 3.2.3 is very difficult.

Liquidus reversals

The temperature of first appearance of crystalline phases in the melting interval of basaltic compositions at constant pressure may be shown to be the equilibrium temperature by the following procedure. In Fig. 3.2.4a let C be liquid, and B the assemblage liquid plus primary crystals. Crystalline starting material (M) is run at successively higher temperatures (runs 1a, 1b, 1c) and examined until the product of the experiment is all liquid (or glass if examination follows a quench). A charge is then run at this temperature (run 1c) for a time previously shown to give a liquid product (now a liquid starting material) and then the temperature is lowered until crystals are observed to have grown (run 2cb). The equilibrium liquidus temperature may be between the temperature at which crystalline material melts and liquid materials begin to crystallize.

This statement is necessarily qualified for several reasons. If the crystalline assemblage melted (M) is not the same as the one grown (in assemblage B), the demonstration is incomplete. This may occur if the starting material is a slightly quartz-normative crystalline composition without olivine, which has olivine as the liquidus phase. A more subtle problem can arise if the liquidus phase entertains solid solution so that the crystalline starting material may have the liquidus species present but of the wrong composition. These problems may be overcome by taking a charge up to the temperature shown to grow crystals from a liquid starting material (run 1b) and determining that the crystals in the two situations correspond. The demonstration is only completed when a charge processed in this way is shown to melt completely when raised to the higher temperature above the liquidus (run 2bc).

Bulk vs. surface equilibrium

Reversal of phase appearance temperatures within the crystallization interval suffers additional complications, largely as a result of possible solid solutions and the difficulty of establishing homogeneous equilibrium within the crystalline phases. Zoning may be produced and preserved in phases which grow as temperature is lowered from above to below the appearance temperature of the phase of interest. Zoning present in initially crystalline starting material may also be preserved. Besides indicating a type of disequilibrium, the presence of zoning will affect the composition of the liquid phase in the experiment. This has the undesirable consequence that there will be difficulty in establishing the correspondence of products and starting materials on either side of a phase appearance temperature, which is necessary for a demonstration of reversal.

The recognition of the greatly different rates of homogenization in liquid and crystals has led to the concept of surface equilibrium, in which homogeneous liquid is in equilibrium with the surface of a crystal without regard as to whether the crystal is in homogeneous equilibrium. This condition is not really an equilibrium state, because the system would change as the crystal homogenized with increasing length of time. Surface equilibrium is a useful concept, however, because it can provide a framework for understanding the zoning in crystals so often observed in basaltic systems. The effect on the temperature, and even the order, of phase appearance within a basaltic crystallization interval may be very large, and the difference between liquid compositions in a bulk equilibrium vs. a surface equilibrium situation may be substantial (compare Walker *et al.*, 1976a and b). Demonstrations of phase appearance temperature during controlled cooling experiments, where surface equilibrium conditions obtain, may be of interest because rapidly crystallizing basaltic melts are not governed by the bulk equilibrium condition.

Summary and caution

Once obvious indications of disequilibrium are lacking, demonstrations of reversal or convergence constitute the most nearly reliable evidence of equilibrium. Accordingly, Kesson and Lindsley (1976) placed high importance on these criteria in judging whether equilibrium had been achieved in experiments on lunar mare basalts. However, it must be remembered that such demonstrations are only as thorough as the range of initial perturbations explored. There always remains the prospect of metastability relative to some untried state or with respect to some longer period of observation, especially when one considers the disparity between the time scale of the laboratory experiment and the time available for geological processes. Experimental results must be reproducible both by the original investigators and by others. Failure to meet this requirement indicates that at least some of the results are not consistent with equilibrium. It is unfortunately common for different investigators to report different results for similar experiments, and potential users of the results should pay attention to the tests made for stability, reversibility, convergence, and reproducibility. It should be clear that a completely adequate demonstration of the stable equilibrium state can rarely if ever be made.

3.3 TERRESTRIAL ROCKS

While it is a relatively simple matter to arrive at a working definition of a basalt, definition of limits to what is meant by "basaltic volcanism" is less straightforward. Indeed, most terrestrial igneous rocks have field associations with basalts and an understanding of their origins would presumably clarify their relationships to basalts and thereby aid our understanding of basaltic volcanism. For example, while neither komatiites nor rhyolites are basalts, both are often closely associated with basalts and are often clearly related to them. Most petrologists would agree that the experimental petrology of komatiites can yield valuable information relevant to the origins of basalts and the nature of the mantle. Some petrologists have suggested that rhyolites and basalts are partial melts of the same source regions, and others have suggested that rhyolites are often the residual liquids produced by extensive fractionation of basaltic melts. Either of these suggestions of rhyolite origin can be tested by an experimental approach and either, if shown to be valid, would yield valuable constraints on models of basalt genesis and on our understanding of the Earth's interior. However, it is clearly beyond the scope of this Chapter to review the experimental petrology of all igneous rocks which are associated with or related to basalts. If andesites are generated in a hydrous mantle above subducted oceanic plates, then the experimental petrology of andesites should yield additional constraints on the mantle mineralogy. Where do we draw the line? Should we include andesites, trachytes, ultrapotassic lavas and anorthosites? The list could become very long indeed. A somewhat arbitrary choice of topics must be made, and we will focus on rocks which can yield the most direct insights into the nature of the Earth's mantle. These include basalts, komatiites, highly SiO_2-undersaturated lavas, and andesites. Detailed discussions and definitions of the rock types and their petrographic associations are given in Chapter 1, section 1.2.

Basalts are the most abundant extrusive igneous rocks on Earth. According to Daly's (1933) estimates from continental outcrops, basalt has at least five times the total volume of all the other extrusive rocks combined; basalt and pyroxene andesite together have at least fifty times the volume of all other extrusive rocks combined. In addition, it is now established that the ocean basin is floored by a layer of basalt (Chapter 1, section 1.2.5). The prospects for interpretation of the experimental petrology of basalts in terms of planetary interiors must be far more favorable for Earth than for other planets, given this large volume of terrestrial basalts available for study, with a wide range of compositions (section 3.3.1).

3.3.1 Source rocks and magmas

The historical development of ideas related to basalt petrogenesis were outlined in section 3.1.1. It was only in the late 1950's that petrologists began to pay much attention to mantle peridotite as a source for basaltic magmas.

In the discussion of the flow chart in Fig. 3.1.4, we maintained that through an integrated application of the results of petrographical, geochemical and experimental petrological studies, constraints could be placed on the mineralogy and composition of basalt source regions and the physical conditions of basalt genesis in these source regions. While we hailed the virtues of this approach as applied to lunar and meteoritic basalt studies, we emphasized that it has seen little detailed application to terrestrial basalt petrology because knowledge of the petrology and chemistry of ultramafic rocks and of the geophysical properties of the crust and upper mantle had historically provided us with preconceptions of the nature of basalt source regions. In effect, terrestrial petrologists already knew too much about the Earth when the techniques needed to apply the inverse approach became available. Evidence from other sources had already constrained the nature of the upper mantle, so that it has never been necessary to use the basalts for any more than filling in the details of mantle mineralogy, the building of detailed models within the framework of already firm notions of source region properties, and as a check on internal consistency. In this section, therefore, we begin at the end of Fig. 3.1.4 and present the source region constraints provided by model mantle compositions, and then consider the compositions of the samples in step 1 of Fig. 3.1.4.

Peridotites

Five estimates of the composition of the Earth's mantle plus crust are listed in Chapter 4, Table 4.3.2. These were derived from considerations of the chemistry of the solar system, physical and geochemical models for the origin of the terrestrial planets, and the gross geophysical properties of the Earth. The crust constitutes a small fraction of the total, and the estimated compositions correspond effectively to that of the mantle. More than 90% by weight of these compositions is represented by the components MgO-FeO-SiO_2, and another 6.9–9.7% by the components Na_2O-CaO-Al_2O_3. More than 99% of the mantle is composed of these six components. Material of this composition at upper mantle pressures would crystallize to form a peridotite dominated by olivine. These compositions are represented in terms of normative mineralogy at various pressures in Figs. 3.6.1–3.

Earlier estimates for the composition of mantle peridotite using extraterrestrial chemistry and geochemical models for the origin of the solar system, and other approaches involving the study of peridotites derived from the mantle, or the computation of hypothetical peridotite compositions (pyrolite = pyroxene + olivine rock) were reviewed by Wyllie (1971a, Chapter 6), Ringwood (1975, Chapters 3, 5, 6 and 16), and Yoder (1976, Chapter 2). Whatever approach is used, the result is consistent with the compositions cited above, although there are large uncertainties about the concentrations and distributions of trace elements and volatile components; e.g., estimated K_2O contents vary by a factor of ten (about 0.015–0.15%).

The **pyrolite** model for the chemistry of primitive mantle was formulated by Ringwood (1962a, and subsequent versions) by recombining basalt with a refractory peridotite such as harzburgite or dunite, presumed to be a residue after basalt extraction. The specific basalt and refractory material chosen, and the proportions in which they ought to be recombined, have been subjects of debate. Ringwood (1975, Chapter 5) has reviewed the substance and history of the pyrolite model.

The ultramafic nodules brought up in alkalic basalts and kimberlites include candidates for the peridotite source rocks of basaltic magmas (Chapter 1, section 1.2.11). The observed range of compositions of peridotite nodules is wide (e.g., Maaløe and Aoki, 1977), and much attention has been devoted to finding the Holy Grail among them, i.e., primitive, undepleted, or fertile mantle. Behind this search lie the assumptions that the mantle at one time was comprised wholly of this material, and that this was the most readily fusible peridotite in the mantle.

The peridotite nodules demonstrate that the upper mantle is chemically heterogeneous, although mineralogically fairly uniform. Samples could range from original undifferentiated mantle, to depleted modifications produced by the partial or complete removal of the more fusible fractions, or to enriched modifications produced by the infiltration of melts or metasomatic solutions. Trace element and isotopic studies have generally supported the view that the source material of basalts in space and time can be neither unique nor uniform (Chapters 1 and 7).

Many of the peridotite nodules have mineral assemblages which indicate that they equilibrated at considerable depth, and it is commonly assumed they may have come from the neighborhood of the source region of the host magma. The relevant chemical, petrological and textural data for these natural mantle samples have been presented in exhaustive detail in several recent volumes (Nixon, 1973; Ahrens *et al.*, 1975; Boettcher, 1976; Sobolev, 1977; Boyd and Meyer, 1979), and only a brief outline is needed here.

Spinel lherzolite with Cr-diopside is by far the most common type of nodule in alkalic basalts (Wass and

Irving, 1976), and there is abundant chemical and isotopic evidence that these nodules are accidentally-incorporated fragments of chemically-modified mantle residues after partial melting (Frey and Prinz, 1978). Kimberlites contain a different nodule suite dominated by garnet lherzolite, which also appears to be of accidental origin (e.g., Barrett, 1975). The pyrolite compositions derived from recombining basalt and refractory residue crystallize as a garnet-lherzolite assemblage at pressures greater than about 25 kb, or to spinel- or plagioclase-lherzolite at successively lower pressures, as depicted in Fig. 3.1.2. The mineralogy of the peridotite nodules is thus consistent with the pyrolite model, although the nodules encompass a wider range of compositions.

The forward approach to terrestrial basaltic volcanism, that of determining the compositions of liquids derived by partial melting of the source peridotite under various conditions, thus involves the measurement of liquid compositions generated from assemblages containing olivine, orthopyroxene, clinopyroxene, and garnet or spinel or plagioclase. We should emphasize at this point that any knowledge of the source region of terrestrial basalts is indirect, even though various lines of evidence are convergent. The foundation of the forward approach is no more secure than our inferences. Moreover, the forward approach need not have a unique foundation. There may be various combinations of minerals in variable proportions which will yield basalt on partial fusion.

Given the known range of compositions for mantle peridotite nodules, models of magma genesis based on primitive pyrolite represent only one of many possible schemes, since the composition of the "blood" obtained from a "stone" depends on the composition of the stone itself (Mysen and Boettcher, 1975b). The phase relationships of a range of peridotite compositions from primitive or fertile to depleted are relevant, particularly as subduction processes may recycle depleted peridotites into the potential source regions of basaltic magmas. The compositions of peridotites used in experimental studies (Table 3.3.1) are listed in Table 3.3.2, together with an estimate of primitive mantle composition based on peridotite nodules. These analyses merit comparison with those in Table 4.3.2, Chapter 4.

Basaltic magma types

Solution of the inverse problem illustrated in Fig. 3.1.4 involves an experimental demonstration of the conditions under which basaltic liquids can equilibrate with plausible mineral assemblages of the mantle. This demonstration is more useful as a test for internal consistency in theories of basalt petrogenesis than as an avenue for fresh insight into terrestrial mantle petrology, because of our well-developed ideas about what constitutes a plausible mantle assemblage. Nevertheless, further compositional and mineralogical details of the mantle residuum assemblage can be provided by such experiments. For the procedures to provide real insight into mantle constitution, we must pay particular attention to step 1 of the flow chart in Fig. 3.1.4. Ideally, the basalts studied must be demonstrated to have erupted in the liquid state rather than as crystal-liquid mixtures, and they must be reasonably inferred to have come directly from the mantle without crystal fractionation or contamination from the time of segregation of the liquid from its residuum until the experimental work is performed.

The unequivocal identification of a primary magma in any given tectonic environment or geographic region on the Earth is notoriously difficult. This state of affairs persists despite the availability of large numbers of samples, not only of lavas, but also of the residual material left after magma generation (ultramafic nodules). Under the optimum conditions in which data are available for a large number of samples, it is sometimes possible to make a reasonable estimate of a primary magma composition (Chapter 1, section 1.4.2). The results of crystallization experiments on this composition would not generally lead to a unique characterization of the source material, but would provide a set of constraints that could be used alongside other geochemical data to narrow the possible alternatives.

How many terrestrial samples warrant experimental investigation for information about the mantle? To answer this question, we must have some idea of the range in composition of natural basaltic liquids, and the distribution of this range of compositions relative to atmospheric pressure thermal divides and liquidus saturation curves. This information can only be gained by experimental studies (steps 1–5 in Fig. 3.1.4). If basalts are found to be distributed on opposite sides of a thermal divide, they cannot have been produced from the same parent in a fractionation sequence near the Earth's surface, since such sequences do not cross low pressure thermal divides. For terrestrial basalts, there are at least two important low pressure thermal divides.

The concept of the basalt tetrahedron was developed by Yoder and Tilley (1962), and reviewed by Yoder (1976, Chapter 7).

The thermal divides are illustrated in Fig. 3.3.1, the tetrahedron with apices normative olivine, diopside, quartz and nepheline. These normative constituents also provide the basis for nomenclature of basaltic rocks (see Chapter 1). The *olivine basalt divide* is the plane including olivine, diopside and plagioclase solid

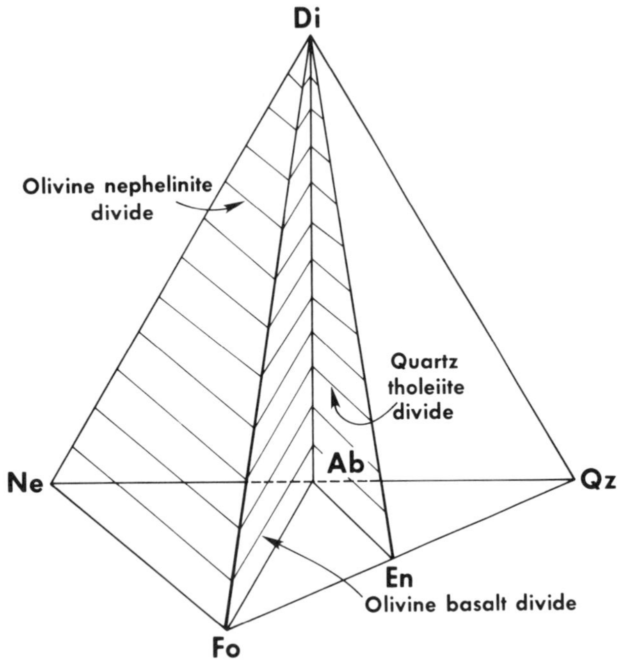

3.3.1 Schematic representation of the system diopside-forsterite-nepheline-quartz showing the plane of silica saturation, Di-En-Ab, and the critical plane of silica undersaturation, Di-Fo-Ab. This iron-free simple basalt system includes the principal components of the major phases of basalt.

solutions, and the *olivine nephelinite divide* is the plane including olivine, diopside, and nepheline solid solutions. The compositional range of terrestrial basaltic rocks extends across both divides, as shown in Fig. 3.3.2. Hypersthene-normative tholeiitic basalts occupy the compositional range to the high-silica side of the *olivine basalt divide*. Nepheline-normative alkali olivine basalts and basanites without hypersthene or larnite in their norms occupy the range between the two divides. Olivine nephelinites and olivine melilitites with nepheline and either very minor albite or larnite in the norm populate the highly alkaline, subsilicic side of the *olivine nephelinite divide*.

Alkali olivine basalts are here defined as nepheline-normative magmas with no more than 5% normative nepheline. They are compositionally separate from more SiO_2-undersaturated magmas such as basanites. The distinction can also be drawn from trace element data. Basanites have steep REE patterns enriched in light REE, suggesting that garnet is a residual phase in their generation (e.g., Kay and Gast, 1973; Frey *et al.*, 1978), whereas the most primitive alkali olivine basalts show lower light REE abundances and probably have never equilibrated with garnet (e.g., Frey *et al.*, 1978).

Yoder (1976) traced the flow sheet of univariant liquid paths connecting invariant points at 1 atm for a tetrahedron expanded to include larnite-normative compositions, using experimental data from many synthetic systems including relevant minerals. He concluded (p. 128) that three parental magmas are required to derive the magma sequences observed at 1 atm: olivine tholeiite, basanite, and olivine-melilite nephelinite. These three magma types are separated by the two thermal divides, and they cannot be interrelated by crystal fractionation at low pressures. Therefore, at least these three compositions need to be experimentally investigated at high pressures (step 6 in Fig. 3.1.4).

Figure 3.3.2 shows something of the range and distribution of the compositions of basic terrestrial igneous rocks. The data base for this plot is the PETROS compendium of igneous rock analyses prepared by Mutschler *et al.* (1976). Included are the 13600 rock analyses which have SiO_2 less than 56% and are not too oxidized ($Fe_2O_3/FeO < 1$), hydrated ($\Sigma\ H_2O < 4\%$), or otherwise too obviously degraded to be of petrologic interest. The analyses have been recast into a modified version of the CAMS system proposed by O'Hara (1968a). Molar oxides are combined as follows:

$$C = (CaO - P_2O_5/3 + K_2O + 2Na_2O) \times 56.08$$
$$A = (Al_2O_3 + Cr_2O_3 + Fe_2O_3 + Na_2O) \times 101.96$$
$$M = (MgO + FeO + MnO) \times 40.30$$
$$S = (SiO_2 - 2Na_2O - 4K_2O) \times 60.08$$

This modification has the property that all feldspars still plot together at CAS_2, but that leucite and nepheline now plot together at CA. TiO_2 has been ignored to suppress the undesirable effects of TiO_2 concentration during fractional crystallization which are incurred in the original CAMS recalculation. Figure 3.3.2 is a subprojection in the CAMS system which uses S, M, and CA as the image plane and CS-M as the projection point (CS-M is equivalent to $CaMg_{-1}SiO_2$). This unconventional choice of components was made so that both low pressure thermal divides would appear as lines in projection. The choice of CS-M as a projection point makes diopside plot on olivine at M_2S so that the *olivine basalt divide* is the line connecting M_2S and CAS_2, the point where feldspar compositions plot. The recombination procedure used for K_2O which puts leucite with nepheline at CA, makes the line connecting CA with M_2S equivalent to the *olivine nephelinite divide*. Akermanite plots at M and other parts of the melilitite solid solution plot below the M to CA line.

Of the 13600 basic igneous rock compositions plotted in Fig. 3.3.2, 62% fall above the *olivine basalt divide* and, hence, are tholeiitic. Alkali olivine basalts to

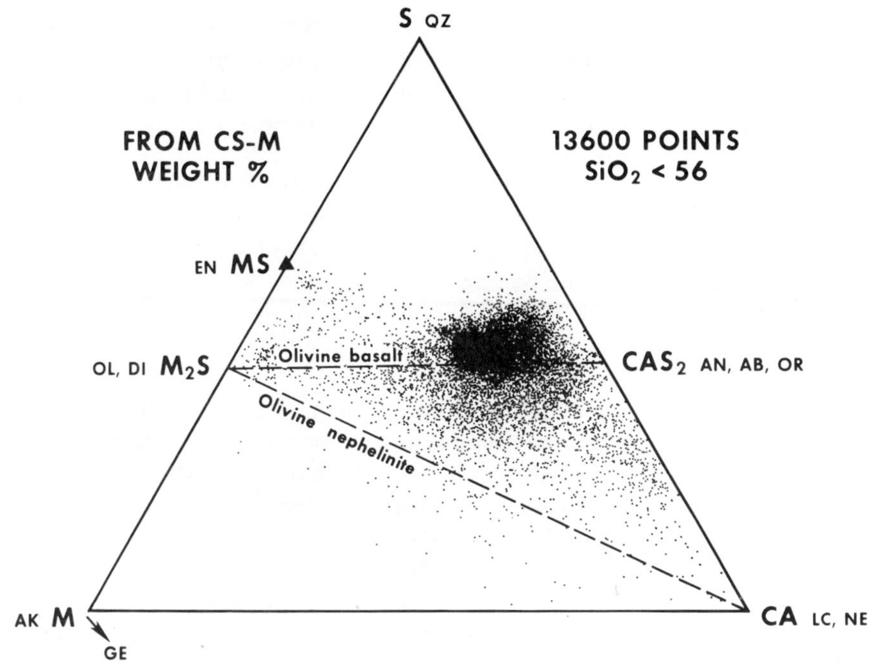

3.3.2 The range of compositions of terrestrial basaltic rocks, projected into the system CaO-Al_2O_3-MgO-SiO_2 by a modified version of the procedure proposed by O'Hara (1968a), as described in the text. Two of the thermal divides shown in Fig. 3.3.1 appear as lines in this projection. The heaviest concentration corresponds to tholeiites.

basanites falling between the *olivine basalt divide* and the *olivine nephelinite divide* comprise 37% of the population, whereas olivine nephelinites and olivine melilitites falling below the *olivine nephelinite divide* are slightly less than 1% of the population. This breakdown of the population may not be an accurate reflection of the natural abundances of these rock types because of sampling procedures, a point made with some force by Chayes (1972, 1977). It is easy to imagine, for example, that the melilitites are collected and analyzed out of proportion to their natural abundance as a direct consequence of their novelty. However, the central point here is that, of basalts which come to the attention of science through collection and analysis, the distribution is heavily weighted towards tholeiites.

An important feature of Fig. 3.3.2 is that there is a very dense concentration of compositions rather than a uniform spread over the available composition space. This cluster includes a substantial part of the tholeiites but partly straddles the *olivine basalt divide* to include some alkali olivine basalts and basanites. In contrast, the more alkalic rocks (with the exception of the alkali olivine basalt subsection of the big cluster) show a considerably more diffuse distribution over composition space.

Parental compositions within the three crystallization cells in Fig. 3.3.1 and 3.3.2 should be chosen to avoid the low-pressure liquidus saturation curves. Determination of the field boundaries representing multiple saturation at the liquidus requires many experiments, involving steps 2, 3, and 5 of the inverse flow chart of Fig. 3.1.4. It turns out that the densest part of the cluster in Fig. 3.3.2, near the *olivine basalt divide*, coincides with the low-pressure liquidus field boundaries of multiple saturation with plagioclase, olivine, and pyroxene (section 3.3.8). This implies that the tholeiite compositions include liquids, because crystal-liquid mixtures would only have these special compositions by chance, but the low-pressure control on their compositions implies that they are derivative liquids, and therefore not candidates for primitive or primary magmas. This conclusion applies to the basalts produced at submarine ridges (section 3.3.8). Considerable care must be exercised in the selection of samples to avoid the low-pressure derivative material which will not be relevant to exploration of basalt generation at higher pressures.

It should be recognized that uniformity of composition and overall abundances are not necessarily indicative of primary character. In their pioneering experimental studies, Yoder and Tilley (1962) concluded that the compositions of most magmas generated in the mantle had apparently adjusted successively on the way to the surface, or equilibrated near the surface in magma reservoirs; the common basalts appear to be the orderly end-products of polybaric fractional crystallization (Fig. 3.1.3). This theme was developed by O'Hara (1965, 1968a), O'Hara and Yoder (1967), and Yoder (1976).

In the tholeiitic crystallization cell, on the silica side of the *olivine basalt divide*, many liquids are found to be significantly displaced from low-pressure multiple saturation curves, in the direction of oversaturation both with plagioclase and with olivine. The komatiite lavas are examples of liquids that are extremely oversaturated with olivine at low pressures.

The absence of clustering among the olivine nephelinites and olivine melilitites can probably be taken as an indication that no single control, such as low-pressure fractionation, is responsible for their present range of compositions.

One criterion which is used in the search for primary magmas is that their Fe^{2+}/Mg ratios must be within the range expected for liquids equilibrated with mantle olivines. The distribution coefficient between olivine and basaltic liquids

$$K_D^{Fe^{2+}/Mg} = (Fe^{2+}/Mg)_{olivine}/(Fe^{2+}/Mg)_{liquid}$$

is virtually independent of temperature over the melting interval of basaltic compositions and typically has a value between 0.30 and 0.33 (Roeder and Emslie, 1970; Roeder, 1974a; Longhi *et al.*, 1978). The olivine-liquid distribution coefficient exhibits some dependence on pressure (Longhi *et al.*, 1978; Bender *et al.*, 1978) and liquid composition (Longhi *et al.*, 1978). Although K_D is largely independent of oxygen fugacity, the composition of the olivine in equilibrium with a particular liquid will vary as a function of oxygen fugacity since Fe^{2+}/Mg of the liquid varies with oxygen fugacity as $Fe^{2+} \rightleftharpoons Fe^{3+}$ equilibrium changes. The composition of olivine in equilibrium with a particular liquid can vary significantly over the range of oxygen fugacities which may be relevant to basalt genesis (Biggar *et al.*, 1974; Biggar, 1974; Hill and Roeder, 1974; Mysen, 1975).

Assuming $K_D = 0.33$, liquids in equilibrium with mantle olivines of composition Fo_{86}-Fo_{90} would have Mg-values between 66 and 75. Erupted liquids with Mg-values less than 66 can therefore be inferred to be derivative liquids produced by fractionation of more primitive liquids which could have equilibrated with mantle olivines. Application of this line of reasoning to determine whether a specific lava composition could be primary requires: (1) demonstration that the lava composition represents that of a liquid; (2) the assumption that mantle olivines are always more forsteritic than Fo_{86}; (3) demonstration that the value of the K_D employed is relevant for the liquid composition and the likely pressure of equilibration with mantle olivines; and (4) demonstration that Fe^{3+}/Fe^{2+} of the lava used in calculating its Mg-value is consistent with the likely oxidation state of the source regions. The Mg-value only includes ferrous iron, but since many analyses only report total iron and in others Fe^{3+}/Fe^{2+} has been affected by weathering, hydrogen loss or other processes which tend to oxidize the lava, the Mg-values actually used to compare with probable primary Mg-values often depend on an assumed Fe^{3+}/Fe^{2+} ratio for the lava.

A further criterion is available for the identification of primitive basic alkalic rocks. Unlike all but a few tholeiites (e.g., Sutherland, 1974), these rocks commonly contain inclusions of dense mantle-derived material (dominantly spinel lherzolite nodules). Assuming a Newtonian flow model, it may be argued (e.g., Maaløe, 1973; Kushiro *et al.*, 1976) that, if the ascent of liquid had been slow enough to allow fractionation by crystal settling en route, these nodules should have been removed more efficiently than the crystals because of their greater size and bulk density (see Section 3.1.4). However, Sparks *et al.* (1977) have shown that for a Bingham plastic flow model, the presence of nodules could imply slow ascent rates which allowed the liquid to acquire a finite yield strength by partial crystallization (not necessarily accompanied by crystal fractionation). Another factor which may be even more important in governing xenolith transport is the volatile content of the liquid, because xenoliths are observed to be most abundant in explosively-erupted kimberlite and highly alkalic basalt magmas, which are known to have high contents of dissolved H_2O and CO_2.

The presence of peridotite xenoliths has been generally accepted as a criterion for identifying the more primitive liquids of a particular province which have not suffered low-pressure fractionation by crystal settling, despite the questions raised by Sparks *et al.* (1977). The chemical systematics of xenolith-bearing magmas themselves are certainly consistent with this interpretation (e.g., Irving and Green, 1976).

Picrite and komatiite magmas

Picrite and komatiite magmas are of considerable interest because they span the composition gap between basalt and its assumed source, peridotite. Picrites occur in both the tholeiitic and alkaline associations whereas komatiites appear to belong to a distinct magma series whose existence has been recognized only in recent years. Geological studies of Archean greenstone terranes have confirmed the identity of the komatiite magma series, which ranges from peridotitic through basaltic compositions (Chapter 1, section 1.2.1). With only a very few exceptions (e.g., in the Proterozoic of Newfoundland—Gale, 1973), these rocks are Archaean in age. Field relations (e.g., pillow structures, brecciated flow tops, chert horizons) indicate that the lavas were erupted under submarine conditions. The ultramafic members of this rock series, the komatiites, represent a unique type of ultrabasic liquid, in which varying degrees of crystallization and accumulation of olivine phenocrysts have occurred. These rocks characteristically exhibit **spinifex texture** whereby acicular bladed and skeletal olivine (also pyroxene) crystals form sheaf-like aggregates. These textures are analogous to experimentally-produced quench textures and provide key supporting evidence for the eruption of ultrabasic liquids (Green *et al.*, 1975a; Donaldson, 1976), although considerable loss or gain of olivine by crystal settling nevertheless has occurred in many hand-specimens.

The chemical characteristics of komatiitic peridotites have been discussed in some detail (e.g., Arndt *et al.*, 1977), and they are reviewed in Chapter 1, section 1.2.1. MgO contents typically range from 20–30%, 100 $Mg/(Mg + Fe^{2+})$ (Mg-value) from 60 to 80. At any particular Al_2O_3 content, the Mg-value of a member of the komatiite magma series is higher than that of a corresponding member of the tholeiitic magma series. Table 3.3.2 lists analyses of picrites and komatiites used in experimental studies. Most komatiitic peridotites have cosmic CaO/Al_2O_3 ratios ($\leqq 1$). The high values (> 1) reported for the Barberton Mountain Land komatiites—if indeed this is a primary igneous feature and does not reflect secondary alteration—may be due to minor garnet fractionation at high pressures. Green *et al.* (1975a) pointed out that this explanation seems plausible in examples where high CaO/Al_2O_3 is accompanied by heavy REE depletions. Nesbitt and Sun (1976) and Sun and Nesbitt (1978) have shown that the abundances of incompatible elements, particularly those little affected by secondary alteration (e.g., Ti and heavy REE), are generally 2–2.5 times chondritic. Many komatiites have REE patterns characterized by marked depletions in the light REE.

Initial studies of komatiites were directed to demonstrating that some were liquids (step 1 in Fig. 3.1.4). This has now been established for the tops of flows from petrography (Viljoen and Viljoen, 1969; Nesbitt, 1971), from experimental duplication of the distinctive crystal shapes and textures of komatiites (Green *et al.*, 1975a; Donaldson, 1976; Arndt, 1977), and for sample 49J from Barberton Mountain Land (Table 3.3.2, No. 13) by demonstration that the most magnesian olivine in the sample matched that of the experimental liquidus olivine (Green *et al.*, 1975a). Although there still remains some doubt as to the amount of olivine which has settled from these liquids before, during and following emplacement (Lewis and Williams, 1973; Arndt *et al.*, 1977; Walker *et al.*, 1977a), they probably are depleted in olivine relative to the time of magma segregation from the source.

Some picrites are undoubtedly olivine-enriched, whereas others containing skeletal olivines (phenocrysts and some groundmass), represent liquids (Drever and Johnston, 1966; Donaldson, 1976). The same is probably true of komatiitic basalts (sometimes called "high-magnesium basalts"). To our knowledge, no phenocryst-matching tests have ever been performed on a picrite.

Highly SiO_2-undersaturated basaltic rocks

Highly SiO_2-undersaturated alkalic basalts, such as basanites, olivine nephelinites, olivine melilitites and certain highly potassic lavas, are volumetrically very minor by comparison with tholeiites (Fig. 3.3.2). However, many of these lavas contain xenoliths and megacrysts of high-pressure origin which has ensured their widespread study (see Chapter 1 section 1.2.1). Green and Ringwood (1967a) conjectured that, just as alkali olivine basalts could be derived from olivine tholeiites by high pressure fractional crystallization processes dominated by aluminous orthopyroxene, so basanites and other more SiO_2-undersaturated liquids might be produced from alkali olivine basalts by continued fractional crystallization. In order to account for the very high abundances of incompatible elements in these rocks they invoked a mechanism of "wall rock reaction." However, basanites, olivine nephelinites, etc., do not have systematically lower Mg-values than other lavas as would be expected from orthopyroxene removal: in fact the opposite is true. Furthermore, the earliest high-pressure experiments on such compositions under dry conditions (Ito and Kennedy, 1968; Bultitude and Green, 1971) showed that orthopyroxene was conspicuously absent as a liquidus phase at any pressure, and this has been confirmed by Arculus (1975). Thus, the concept of a high-pressure fractionation series from tholeiite through alkali olivine basalt to basanite was abandoned as a general model in favor of the concept of a spectrum of primary liquids (Green, 1970a). Experimental studies (sections 3.3.2 and 3.3.6) have since confirmed the influence of H_2O and CO_2 in the generation of low-SiO_2 liquids from mantle peridotite.

Some kaersutite megacrysts from basanites have planetary-type rare gas isotope patterns and primitive Sr and O isotope ratios (Saito *et al.*, 1978; Basu, 1978; Boettcher and O'Neil, 1980) consistent with mantle derivation, and they contain small quantities of H_2O (0.1–0.8 wt.%) as well as F and Cl (Boettcher and O'Neil, 1980).

Rare-earth element data for natural basanites were interpreted by Frey *et al.* (1978) to indicate a garnet-bearing residuum and degrees of melting of 5–7%.

However, in order to reconcile the low Sm/Nd ratio (< 1) measured for basanite lavas (e.g., Frey *et al.*, 1978) with the high time-averaged Sm/Nd ratio (> 1) of their mantle sources inferred from Nd isotopes (e.g., DePaolo, 1978), it is necessary to postulate a recent enrichment of the source peridotite in light rare-earth and other lithophile elements (Carter *et al.*, 1978; see also Chapter 7). Petrologic, trace element and isotopic data for xenoliths and megacrysts from basanites (e.g., Frey and Prinz, 1978; Boettcher and O'Neil, 1980; Menzies and Murthy, 1980) support the idea that metasomatism by migrating fluids is a common precursor to the generation of highly SiO_2-undersaturated magmas (Chapter 7). Degrees of melting greater than 5–7% (perhaps as high as 10–15%) would then be required to produce basanites, with the exact value for a given lava dependent on the extent of local pre-enrichment of its source.

Trace element data for olivine nephelinites and olivine melilitites, as for basanites, strongly indicate garnet as a residual phase in the source regions of these magmas (e.g., Frey *et al.*, 1978).

There are other alkalic lavas which contain high-pressure inclusions. These include nepheline hawaiites, nepheline mugearites, nepheline benmoreites and phonolites, which are especially well-represented in eastern Australia (e.g., Irving and Green, 1976; Irving and Price, 1980). Most examples of these volcanic rock types, as well as trachytes, tristanites, alkali rhyolites and other evolved lavas, have been interpreted classically as the products of low-pressure fractional crystallization of basanites, alkali olivine basalts or transitional basalts (e.g., Bowen, 1928; Mukherjee, 1967). The lack of high-pressure inclusions in well-studied evolved volcanic suites from oceanic islands (e.g., Reunion, Tristan da Cunha, Comoro Islands, Hawaii) and continental provinces (e.g., Kenya Rift, Hebridean, Rhine Graben), along with chemical evidence for feldspar removal (or addition) in these suites (e.g., Zielinski, 1975; Baker *et al.*, 1977), are consistent with the classic interpretation.

In contrast, the less common inclusion-bearing examples show no evidence that feldspar was involved in their evolution (e.g., Irving and Price, 1980), yet they have Mg-values (45–64) that are too low for primary liquids (unless unusually iron-rich sources are invoked). This leaves the possibility that the inclusion-bearing lavas evolved at higher pressures before they acquired their mantle xenoliths.

Certain alkalic lavas enriched in K_2O (5–7%), TiO_2 (4–5%) and other large ion lithophile elements, best known from Bufumbira, East Africa, and Leucite Hills, Wyoming, were thought by Carmichael (1967) to be primary magmas. Rare-earth element data (Kay and Gast, 1973) indicate equilibration with garnet and even smaller degrees of melting (or alternatively greater degrees of source pre-enrichment in light REE) than inferred for olivine nephelinites and olivine melilitites (Frey *et al.*, 1978). Olivine leucitites are the most typical lavas, but a range of compositions (and names) has been included in this potassic kindred. The minette from Shiprock, New Mexico, is unusual because it contains xenoliths of garnet lherzolite (Ehrenberg, 1977).

Subduction zone source rocks and magmas

The subduction of cool oceanic lithosphere through the seismic low-velocity zone to depths of at least 700 km introduces variety to the possible magmatic source materials, with the prospect of magmatic processes that differ significantly from those occurring in other tectonic environments. An essential prerequisite for comprehension of the processes associated with subduction is a knowledge of the temperature distribution. This is discussed in Chapter 9. Although there is a reasonable consensus for the thermal structure of the descending lithosphere plate, uncertainties remain with respect to the actual temperatures at specific locations in the subducted ocean crust, and with respect to the thermal structure associated with asthenospheric convection and back-arc spreading. These uncertainties give great flexibility to the selection of source rocks for the generation of primary magmas. At various times all possible sources have been claimed by investigators.

Magmatic products associated with subduction are dominated by the calc-alkaline rock series, basalt-andesite-dacite-rhyolite of volcanic arcs, and gabbro-tonalite-granodiorite-granite of batholiths in continental margin arcs. Possible sources for the magmas in these associations are the subducted oceanic crust, the overlying mantle, and, for the batholiths and some acidic lavas, the base of the continental crust. Progressive dehydration during subduction of hydrothermally altered oceanic crust provides water for participation in magma generation.

Many of the processes that have been invoked to explain the origin of the volcanic rocks are summarized in schematic form in Fig. 3.3.3. Magma could be generated from the subducted oceanic crust by partial melting of sediments, or by partial melting of amphibolite or quartz/coesite eclogite, either dry or with hydrothermal fluids. Magma could be generated in overlying mantle if the hydrothermal fluids cross the peridotite-H_2O solidus, if hydrous magmas from the subducted crust react with mantle producing pyroxenite or hybrid magmas, or if the hydrous fluids produce metasomatic phlogopite or other hydrous minerals in the peridotite that are subsequently transported by convecting mantle down to melting temperatures. Rising magmas could experience fractionation through the crystallization of minerals that varied as a function of depth.

There is still no general agreement about which lavas, if any, represent primary magmas. Kuno (1950) and Tilley (1950) were the first to recognize the existence of tholeiitic basalts in island arcs. Subsequently, Kuno (1959) showed that in Japan the volcanic rocks could be grouped into three series, the tholeiite (pigeonitic) series nearest the trench, the calc-alkali (hypersthenic) series at an intermediate position, and the alkali series farthest from the trench. Each series shows a wide range of rock types, from tholeiitic basalt through andesite and dacite to rhyolite for the tholeiite and calc-alkali series and from alkali olivine basalt through mugearite to trachyte for the alkali series. The tholeiite series typically shows hypersthene grading into pigeonite in intermediate members of the series whereas the

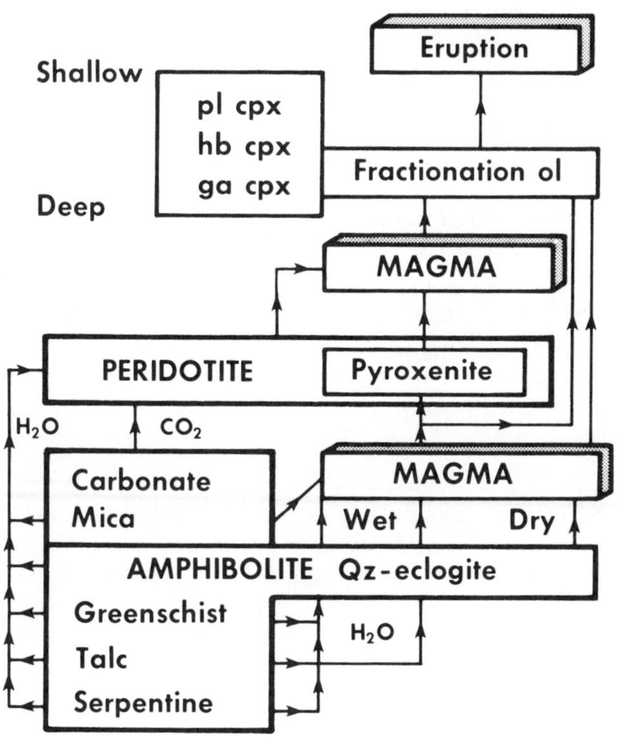

3.3.3 Processes that have been proposed for subduction zones involve the release and migration of volatile components, and the generation of magmas in ocean crust, overlying mantle peridotite, or in hybrid rocks generated from magma reacting with mantle. The diagram is a schematic representation of paths followed by H_2O and CO_2, of proposed sites of magma generation, and the main types of fractional crystallization that may modify magma compositions during uprise (after Wyllie, 1979a, b).

calc-alkali series shows only hypersthene. Also, the tholeiite series shows more iron enrichment and lower $Na_2O + K_2O$ than the calc-alkali series (see Chapter 1, section 1.2.7).

Jakeš and Gill (1970) and Jakeš and White (1971, 1972) argued in favor of a temporal sequence in which the tholeiite series appears first and is followed sequentially by the calc-alkali and alkali series. Jakeš and White (1971) also concluded that basalt is the most abundant member of the tholeiite and alkali series but andesite is the most abundant member of the calc-alkalic series. Arculus and Johnson (1978) accepted the existence of the three rock series but presented evidence that generalizations about their temporal and geographic distribution do not always apply. For example, in the Lesser Antilles the rocks change from tholeiitic to calc-alkalic to alkalic from north to south parallel to the arc rather than perpendicular to it, and this change is not accompanied by an increase in depth to the Benioff zone. Moreover, the three rock types have formed more or less contemporaneously in post-Miocene time.

If, as in the Lesser Antilles, each of the three rock series is capable of being produced in geographic isolation from the others, it may be argued that at least three primary magmas are required. Also, because basalts are associated with each of the three series, it appears that some sort of basaltic primary magma must be associated with each series, although it is not a necessary consequence that more felsic members of the same series must be derived from this primary basalt by fractional crystallization. In this regard, it should be noted that rocks with less than 56% SiO_2 are rare or absent in at least some continental arcs (Jakeš and White, 1972, p. 36), and many have argued the case for primary andesites derived either from mantle peridotite, or from subducted ocean crust (see review by Boettcher, 1973).

It is clear that the minimum requirement for the construction of more satisfactory petrogenetic models for the production of island arc volcanic rocks is more detailed information on the spatial and temporal distribution of rock types. In a review of the island arc volcanic rock series, Miyashiro (1975) wrote: "the existing data are so complicated that they are not properly understood and used in the existing discussions, and that incorrect generalizations are widespread."

Summary

In this section, we have considered the range of basaltic magma compositions with reference to their candidacy as primary liquids from the mantle, but without specific reference to their tectonic environment, except for subduction zone magmas. It should be evident that experimental petrology has become an important adjunct to geochemistry in the identification of primary liquids (see also section 3.3.8). The significance of tectonic environment will be discussed in section 3.3.4, and the petrogenesis of some basaltic rocks in selected tectonic environments will be reviewed briefly in section 3.3.7, using the results of experimental petrology (sections 3.3.2, 3.3.3, and 3.3.4) and of geochemistry.

3.3.2 Phase relationships of peridotite

Peridotite has assumed a critical role in discussions of the origin of terrestrial basalts, and many experimental studies have been directed towards the subsolidus mineralogy, the determination of solidus curves, and the amount and composition of the liquids produced as a function of temperature above the solidus at various pressures. Unfortunately, the results of the various studies have not always appeared to be compatible and this fact, combined with the acrimony of some of the debate about the incompatibility of the data, has done little to improve the confidence of geochemists and petrologists in the experimentalist. The result has been to foster suspicion about the quality of the work conducted, whereas the real culprit is probably inadequate appreciation by the experimentalists themselves of the differences between their laboratories in such matters as capsule materials, open versus closed capsules, f_{O_2}, run durations, iron loss, and the content and composition of volatiles. These and related experimental topics were reviewed in detail in section 3.2.

Lavas contain dissolved gases, some of which are derived from the source regions, so it is necessary to determine the effect of these volatile components on the phase relationships. In a detailed review of gases from volcanoes, from hot springs, and from mineral inclusions, Anderson (1975) concluded that the major components of volcanic gases are H_2O, CO_2, SO_2, and H_2, commonly in that order. This composition differs from the inferred composition of the magmatic gas at Kilauea at 1200° C and 500 bars, corresponding to seafloor pressure: 60% SO_2, 20% H_2O, 10% H_2S, and 10% CO_2. The relevant system for study is peridotite-C-H-S-O. Data are available for peridotite-CO_2-H_2O.

The evidence that minerals and magmas derived from the upper mantle contain H_2O and CO_2 has been reviewed by Irving and Wyllie (1975), Wyllie (1977, 1979) and Eggler (1978). The existence of diamonds and graphite in peridotite nodules demonstrates the existence of carbon at great depths, and indicates that the oxygen fugacity is not high enough to oxidize all carbon at these depths. According to M. Sato (pers. comm., 1977), there is probably a depth interval within which carbon and carbonate can coexist in the mantle, a view shared by Rosenhauer *et al.* (1977) and Woermann *et al.* (1977). It appears that at some level, the deep-seated components C-H-O are sufficiently oxidized that H_2O and CO_2 become dominant. The variation of temperature and oxygen fugacity as a function of depth in different tectonic environments is inadequately understood.

We will consider first the phase diagrams for different peridotite compositions studied in different laboratories, and the changes produced by adding H_2O, CO_2, or H_2O-CO_2 mixtures. Given this experimental framework, we then review the compositions of liquids produced as a function of peridotite composition, pressure, temperature, and the amounts of volatile components.

Table 3.3.1 summarizes the significant experimental details for published phase equilibrium studies of peridotite compositions. The compositions of the rocks or synthetic equivalents used are given in Table 3.3.2.

Synthetic and natural peridotite

The phase relationships of peridotite have been determined by the two complementary approaches of studying reactions in synthetic systems simple enough that the phase rule can be applied with confidence, and of studying the phase assemblages produced in complex peridotite compositions. Figure 3.1.2 is a generalized phase diagram using data from both approaches. The results presented in this section illustrate in more detail the phase transitions and phase assemblages in the low-pressure part of Fig. 3.1.2.

Presnall *et al.* (1979) studied a simplified lherzolite composition in the system CaO-MgO-Al_2O_3-SiO_2. The composition crystallizes to an assemblage of olivine and two pyroxenes, together with an aluminous mineral such as anorthite, spinel, or garnet, occupying the phase fields shown in Fig. 3.3.4. The three subsolidus phase assemblages correspond to the fields for plagioclase-, spinel-, and garnet-peridotite in Fig. 3.1.2. Two cusps are produced by the changes in slope of the solidus curve at each triple-point in Fig. 3.3.4. Similar changes in slope must exist for the complex peridotite compositions, although the precision of experiments has not yet been adequate to define them (Fig. 3.3.5).

In Fig. 3.3.5, the phase relationships determined for a garnet lherzolite nodule by Ito and Kennedy (1967) are compared with results determined for pyrolite composition by Green and Ringwood (1967b). Both investigations reproduced the garnet-peridotite and

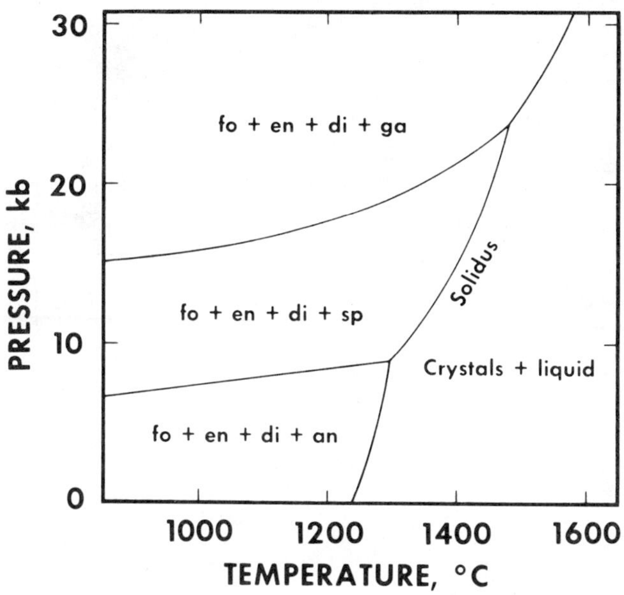

3.3.4 Solidus curve and stability limits for simplified plagioclase lherzolite (fo + en + di + an), spinel lherzolite (fo + en + di + sp), and garnet lherzolite (fo + en + di + ga) in the system CaO-MgO-Al_2O_3-SiO_2, after MacGregor (1965), Kushiro and Yoder (1966), O'Hara *et al.* (1971), Presnall (1976), Presnall *et al.* (1979), and Jenkins and Newton (1979). fo = forsterite, en = enstatite, di = diopside, an = anorthite, ga = garnet, sp = spinel.

spinel-peridotite facies shown in Figs. 3.1.2 and 3.3.4, but neither located the boundary with the plagioclase-peridotite field. Ito and Kennedy (1967) reported three runs in a divariant field between spinel- and garnet-peridotite. Green and Ringwood explored the effect of changing bulk composition. The pyrolite-III composition has a near-solidus field free of both spinel and garnet (KLF in Fig. 3.3.5b). Increasing the Al_2O_3 content produced spinel and garnet in addition to the aluminous pyroxenes in this area, and the phase relationships then corresponded to those for the model peridotite of Fig. 3.3.4. Kushiro *et al.* (1968) reported similar subsolidus relationships for a spinel lherzolite nodule, with no garnet present below 20 kb, but with some spinel persisting with garnet to pressures of at least 40 kb. The five-phase assemblages reported by Ito and Kennedy (1967) and Kushiro *et al.* (1968) could be caused by failure to reach equilibrium. Smaller Mg-values and higher Al, Fe^{3+}, Ca and Na contents stabilize garnet to lower pressures, whereas higher Cr^{3+}contents stabilize spinel to higher pressures.

The solidus curve for peridotite depends upon the bulk composition, as shown for four materials in Fig. 3.3.6. The solidus for the garnet lherzolite from Fig. 3.3.5a is somewhat lower than that for pyrolite from Fig. 3.3.5b, whereas the solidus obtained by Kushiro *et al.* (1968) for spinel lherzolite is somewhat higher. In view of the lower Mg-value for the spinel lherzolite, its higher solidus temperature may reflect iron loss during the runs. Nishikawa *et al.* (1970) studied a spinel lherzolite nodule, and found a solidus at temperatures above the liquidus for basalts. They concluded that this sample represented residual mantle.

The phase fields above the solidus have not yet been adequately defined for any peridotite. For garnet lherzolite, Ito and Kennedy (1967) found complete melting of clinopyroxene at the solidus, and substantial stability fields for spinel and garnet with increasing temperature (Fig. 3.3.5a), whereas Green and Ringwood (1967b) observed the disappearance of spinel and garnet at the solidus in pyrolite (Fig. 3.3.5b). These

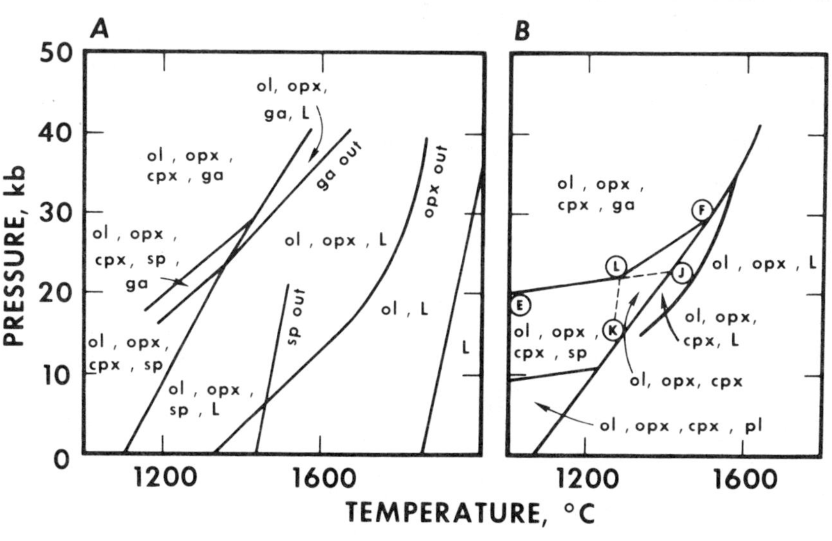

3.3.5 **a**. Anhydrous melting of peridotite (Ito and Kennedy, 1967). The liquidus boundary was not actually determined. **b.** Anhydrous phase relations of pyrolite III (Green and Ringwood, 1967b). ol = olivine, sp = spinel, opx = orthopyroxene, cpx = clinopyroxene, ga = garnet, pl = plagioclase, L = liquid.

Table 3.3.1 Experimental melting studies of complex peridotite, peridotitic komatiite and picrite compositions.[†]

Sample	Starting Material	Apparatus	Container	Run Pressures	Volatile Activities	Oxygen Fugacity	Temperature Extent of Runs	Reference
Synthetic lherzolite	sintered oxide mix	nitrogen atmosphere furnace	open Pt capsules	1 atm	dry	?	3.9–28.6% molten	*(27)*
Mixtures of minerals from garnet lherzolite A.3/10596, Wesselton Mine	powders	piston-cylinder	open Pt capsules	30 kb	dry	?	near-solidus	*(26)*
Garnet lherzolite KA 64–16, Dutoitspan Mine	rock powder dried at 1000°C	piston-cylinder	graphite or Fe capsules	1 atm–40 kb	dry	?	subsolidus to near liquidus	*(15)*
Pyrolite III minus 40% olivine ($Fo_{91.6}$)	crystallized oxide mix	piston-cylinder	graphite, sealed Pt, $Ag_{75}Pd_{25}$ or $Ag_{50}Pd_{50}$ capsules	5–40.5 kb	dry H_2O-deficient (0.3 wt.%) H_2O-excess (10 wt.%)	?	subsolidus to 200°C above solidus	*(8)*, *(10)*, *(11)*
Pyrolites I and II minus 50% olivine ($Fo_{91.5}$)	crystallized oxide mix	piston-cylinder	graphite or sealed Pt capsules	18–40.5 kb	dry	?	subsolidus to solidus	*(10)*, *(11)*
Spinel lherzolite, HK66101703, Salt Lake Crater	rock powder dried at 650°C	tetrahedral-anvil piston-cylinder gas mixing furnace	open Mo or $Ag_{60}Pd_{40}$ and sealed Pt capsules	1 atm–50 kb	dry H_2O-deficient (uncontrolled) H_2O-excess	controlled only at 1 atm	near-solidus	*(17)*, *(20)*, *(21)*
Spinel lherzolite M6-128, Beni Bouchera	rock powder	piston-cylinder	sealed Pt capsules	15–25 kb	dry	?	near-solidus	*(16)*
Spinel-amphibole lherzolite HK 63111602a, Ichinomegata, and two mineral mixtures	powders dried at 130°C	tetrahedral-anvil piston-cylinder gas mixing furnace	open Mo and Fe-saturated Pt capsules	1 atm–50 kb	dry	controlled only at 1 atm	near-solidus	*(25)*
Mixture of minerals from Dutoitspan Mine garnet lherzolite + 2% fluorphlogopite	powder	tetrahedral-anvil	open Mo capsules	20–30 kb	dry	?	near-solidus to ~ 10% molten	*(21)*

[†]See Table 3.3.2.

Table 3.3.1 *(Continued)*

Sample	Starting Material	Apparatus	Container	Run Pressures	Volatile Activities	Oxygen Fugacity	Temperature Extent of Runs	Reference
Mixture of minerals from Bultfontein garnet lherzolite JJG-352	powder heated at 1000°C	piston-cylinder	graphite capsules	10–25 kb	dry	?	near-solidus	*(18), (19)*
Spinel lherzolite from Salt Lake Crater + 1% anhydrous phlogopite	powder heated at 1000°C	piston-cylinder	graphite capsules	10–20 kb	dry	?	near-solidus	*(18)*
Paragasite-spinel peridotite 18–900, St. Paul's Rocks	rock powder dried at 120°C	piston-cylinder	sealed Pt capsules	10–30 kb	H_2O deficient (5.7 wt.%)	?	solidus to liquidus	*(22), (31)*
Iron-rich garnet lherzolite 1032, Matsoku	rock powder	piston-cylinder	sealed Pt capsules	35–45 kb	dry	?	solidus to liquidus	*(13)*
Garnet lherzolite ga-p(1), Wesselton Mine	rock powder	piston-cylinder	sealed $Ag_{50}Pd_{50}$ capsules	7.5–30 kb	H_2O-excess H_2O+CO_2-excess (X_{H_2O} = 0.25–0.75)	uncontrolled or MH	solidus to ~ 60% molten	*(23)*
Spinel lherzolite 618–138 b.1 Salt Lake Crater	rock powder	piston-cylinder	sealed $Ag_{50}Pd_{50}$ capsules	7.5–30 kb	H_2O-excess H_2O+CO_2-excess (X_{H_2O} = 0.25–0.75)	uncontrolled MH, IW, MW or NNO	solidus to ~ 60% molten	*(23)*
Spinel lherzolite 66 PAL-3 Pali dike ± 2.7–10.7% phlogopite	rock powders	piston-cylinder	sealed $Ag_{50}Pd_{50}$ capsules	7.5–30 kb	H_2O-excess H_2O+CO_2-excess (X_{H_2O} = 0.25–0.75)	uncontrolled or MH	solidus to liquidus	*(23)*
Garnet-olivine websterite 66SAL-1, Salt Lake Crater	rock powder	piston-cylinder	sealed $Ag_{50}Pd_{50}$ or $Pt_{95}Au_5$ capsules	7.5–30 kb	dry H_2O-excess H_2O+CO_2-excess (X_{H_2O} = 0.25–0.75)	uncontrolled or MH	solidus to ~ 60% molten	*(23), (24)*
Garnet lherzolite PHN 1611, Thaba Putsoa	rock powder dried at 600°C in air or fired at 1150°C (QFM)	piston-cylinder	graphite or sealed $Pt_{95}Au_5$ capsules	10–35 kb	dry H_2O-deficient (1.9 wt.%) CO_2-deficient (5 wt.%) CO_2-excess	?	near-solidus to ~ 70% molten	*(19), (24), (30)*

Table 3.3.1 *(Concluded)*

Sample	Starting Material	Apparatus	Container	Run Pressures	Volatile Activities	Oxygen Fugacity	Temperature Extent of Runs	Reference
Spinel lherzolite 618–138 b.1 + 10% phlogopite	powder dried at 110°C	piston-cylinder	sealed $Ag_{50}Pd_{50}$ or $Pt_{95}Au_5$ capsules	10–30 kb	H_2O-deficient (0.56 wt.%) H_2O-excess (10 wt.%)	?	near-solidus to > 200°C above solidus	*(29)*
Peridotitic komatiite 49J, Barberton Mountain Land	rock powder	piston-cylinder	graphite or sealed Pt capsules	2–25 kb	dry H_2O-deficient (5 wt.%, 10 wt.%) H_2O-excess (30 wt.%)	?	near-liquidus	*(12)*
Peridotitic komatiite SA 3091, Munro Township	rock powder dried at 900°C	piston-cylinder gas mixing furnace	graphite within sealed Pt capsules, Pt wire loops	1 atm–42 kb	dry	1 atm runs controlled at 10^{-9} – 10^{-12} atm	solidus to liquidus	*(1)*
Peridotitic komatiite (5), Belingwe	rock powders dried at 1075°C (IW)	piston-cylinder gas mixing furnace	graphite capsules, Pt wire loops	1 atm–40 kb	dry	1 atm runs controlled at CCO or IW	near-liquidus	*(3), (4)*
Synthetic picrite	glass	piston-cylinder argon atmosphere furnace	sealed Pt capsules	1 atm–27 kb	dry	?	solidus to liquidus	*(9)*
Picrite dikes, Skye	rock powders	gas mixing furnace	open Mo, $Ag_{40}Pd_{60}$, $Ag_{60}Pd_{40}$ or Pt capsules	1 atm	dry	$10^{-7.2}$ – $10^{-11.8}$ atm	near-solidus to near-liquidus	*(5), (7)*
Parental picrites, Rhum	gels	gas mixing furnace	Pt wire loops	1 atm	dry	$10^{-6.7}$ – $10^{-9.1}$ atm	near-solidus to near-liquidus	*(6)*
Picrite basalt, Kilauea Iki	rock powder	air atmosphere furnace	sealed Pt capsules	1 atm	dry	air	liquidus to near-solidus	*(28)*
Oceanite series lavas, Reunion	rock powders	gas mixing furnace	open Mo capsules	1 atm	dry	IW	near-liquidus to solidus	*(14)*
Oceanite A219M, Mauritius	rock powder dried at 1000°C	piston-cylinder	sealed Pt capsules	5–14 kb	dry	?	near-liquidus	*(2)*

References:
(1) Arndt (1976); *(2)* Baxter (1972); *(3)* Bickle *et al.* (1977); *(4)* Bickle (1978); *(5)* Biggar and Gibb (1976); *(6)* Forster (1978); *(7)* Gibb (1971); *(8)* Green (1973a); *(9)* Green and Ringwood (1967a); *(10)* Green and Ringwood (1967b); *(11)* Green and Ringwood (1970); *(12)* Green *et al.* (1975a); *(13)* Howells *et al.* (1975); *(14)* Humphries (1972); *(15)* Ito and Kennedy (1967); *(16)* Kornprobst (1970); *(17)* Kushiro (1970); *(18)* Kushiro (1972); *(19)* Kushiro (1973b); *(20)* Kushiro *et al.* (1968); *(21)* Kushiro *et al.* (1972a); *(22)* Millhollen *et al.* (1974); *(23)* Mysen and Boettcher (1975a, b); *(24)* Mysen and Kushiro (1977); *(25)* Nishikawa *et al.* (1970); *(26)* O'Hara and Yoder (1967); *(27)* Reay and Harris (1964); *(28)* Tilley *et al.* (1965); *(29)* Wendlandt and Eggler (1980); *(30)* Wendlandt and Mysen (1978); *(31)* Nehru and Wyllie (1975).

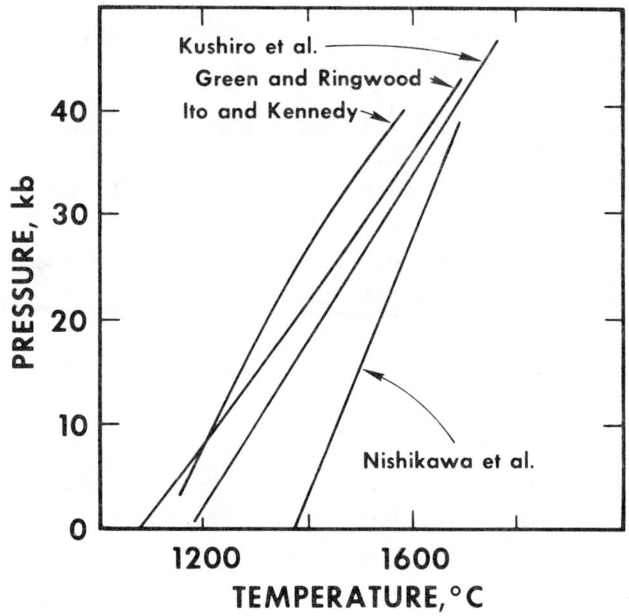

3.3.6 Comparison of anhydrous solidi for peridotite compositions 3,4,5 and 7 in Table 3.3.2.

different results can probably be attributed to the effects of different bulk compositions, because the peridotite is quite magnesian (Mg-value 94), with CaO and Al_2O_3 contents roughly half those of the pyrolite compositions (see Table 3.3.2). At pressures greater than about 20 kb, there appears to be agreement that clinopyroxene, spinel and garnet are completely melted within 50° C of the solidus, producing a wide field for the coexistence of olivine and orthopyroxene with liquid. Mysen and Kushiro (1977) obtained a similar result at 20 and 35 kb for a garnet peridotite nodule (see Fig. 3.3.20).

Peridotite-H_2O

Water reacts with peridotite to produce hydrous minerals, including serpentine, amphibole, phlogopite, humite-minerals, and various hydroxylated silicates stable only at high pressures. If more H_2O is added than that required to generate the maximum amount of hydrous minerals appropriate for the specific bulk composition, pressure, and temperature, the excess H_2O exists as vapor. The presence of vapor causes a marked lowering of the solidus temperature, due to the high solubility of H_2O in silicate liquids. The liquid at the solidus is H_2O-saturated, but unless there is sufficient H_2O present to saturate the whole rock, the liquid becomes H_2O-undersaturated at some temperature above the solidus. If all H_2O is stored in hydrous minerals, the solidus is lowered to the temperature where the hydrous mineral reacts with the peridotite assemblage, breaking down to release H_2O for solution in H_2O-undersaturated liquid.

The first determination of the solidus for peridotite with excess H_2O was by Kushiro *et al.* (1968; see also Kushiro, 1970, for run points at 15 and 20 kb). Other H_2O-excess solidus curves were reported by Green (1973a) for pyrolite-III, by Millhollen *et al.* (1974) and Nehru and Wyllie (1975) for a mantle-derived amphibole-peridotite, and by Mysen and Boettcher (1975a) for four nodule specimens (selected to give a broad range of composition). These solidus curves are compared in Fig. 3.3.7. The differences among the first three determinations are no greater than might reasonably be expected for the different bulk compositions, and the different experimental techniques employed in three different laboratories, but the results obtained by Mysen and Boettcher (1975a) are lower than the earlier determinations by 50–200° C. Mysen and Boettcher (1975a) explained the difference of 120–200° C between their highest and lowest solidus determinations in terms of variations in the chemistry of the four samples, and

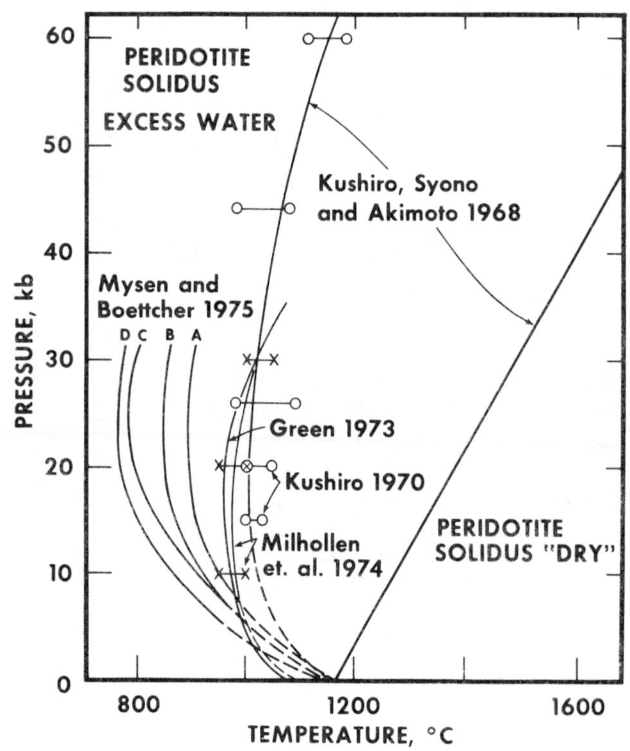

3.3.7 Comparison of results of several studies on the solidus of peridotite in the presence of excess water. Brackets denoted by circles are from Kushiro *et al.* (1968) and Kushiro (1970), and those denoted by crosses are from Millhollen *et al.* (1974). The four curves A-D are from Mysen and Boettcher (1975a).

Table 3.3.2 Compositions and norms of complex peridotites, peridotitic komatiites and picrite used in experimental studies, and comparisons with three estimated primitive mantle compositions (columns 18, 19, 20).

	1	2	3	4	5	6	7	8	9	10	11	12	13	14
SiO_2	44.6	47.38	43.86	47.8	48.02	45.36	44.02	48.06	49.46	42.22	45.52	45.7	43.7	45.10
TiO_2	0.4	0.13	0.39	1.2	0.22	-	-	0.05	-	0.30	0.15	0.05	0.20	0.13
Al_2O_3	4.3	9.70	1.96	5.9	4.88	4.39	1.03	4.93	6.83	4.42	4.83	1.6	4.0	3.92
Cr_2O_3	-	1.27	0.46	0.7	0.25	0.25	0.42	2.16	1.70	0.50	0.58	0.41	0.40	0.31
Fe_2O_3	2.4	1.05	0.69	0.8	1.94	-	0.83	-	-	2.86	1.61	0.77	0.89	1.00
FeO	9.5	5.69	6.21	8.2	8.15	6.59†	7.06	5.73†	5.10†	4.45	10.40	5.21	8.09	7.29
MnO	0.15	0.24	0.14	0.1	0.14	-	0.11	0.18	0.13	0.13	0.18	0.09	0.12	0.14
NiO	-	0.12	0.25	0.2	-	-	-	0.14	-	0.27	0.22	0.26	0.24	0.25
MgO	33.4	28.62	39.82	28.7	32.35	38.73	46.99	32.90	28.62	34.61	30.34	42.8	37.4	38.81
CaO	4.2	6.02	1.68	5.1	2.97	2.24	0.04	4.99	7.70	3.92	4.26	0.70	3.50	2.66
Na_2O	0.76	0.38	0.14	0.95	0.66	0.26	0.05	0.62	0.46	0.43	0.34	0.09	0.38	0.27
K_2O	0.13	0.01	0.08	0.22	0.07	0.02	-	0.25	-	0.11	0.04	0.04	0.01	0.02
P_2O_5	-	0.03	0.03	-	0.07	0.18	-	-	-	0.05	0.02	0.01	0.00	0.01
SUM	99.84	100.64	95.71	99.87	99.72	98.02	100.55	100.01	100.00	94.27	98.49	97.73	98.93	99.91
$\frac{100\,Mg}{Mg + \Sigma Fe}$	83.6	88.5	91.2	85.2	85.3	91.3	91.5	91.1	90.9	89.8	82.0	92.8	88.2	89.4
						*C. I. P. W. Weight Percent Norms**								
or	0.77	0.06	0.47	1.30	0.41	0.12	-	1.48	-	0.65	0.24	0.24	0.06	0.12
ab	6.43	3.22	1.18	8.04	5.58	2.20	0.42	5.25	3.89	3.64	2.88	0.76	3.22	2.28
an	7.94	24.73	4.48	11.19	10.15	10.05	0.20	9.93	16.57	9.81	11.54	3.41	9.18	9.43
cor	-	-	-	-	-	0.26	0.88	-	-	-	-	0.16	-	-
di	10.22	3.91	2.89	11.17	3.30	-	-	11.65	17.01	7.32	7.54	-	6.46	2.93
hy	13.01	32.55	26.14	27.37	38.38	28.06	20.06	27.82	33.14	17.20	30.70	32.95	13.28	22.15
ol	57.23	32.46	57.80	36.13	38.15	55.76	77.17	39.76	26.26	49.82	41.87	58.11	64.23	60.58
mt	3.48	1.52	1.00	1.16	2.81	0.87	1.20	0.75	0.68	4.15	2.33	1.12	1.29	1.45
chr	-	1.87	0.68	1.03	0.37	0.37	0.62	3.18	2.50	0.74	0.85	0.60	0.59	0.46
ilm	0.76	0.25	0.74	2.28	0.42	-	-	0.09	-	0.57	0.28	0.09	0.38	0.25
ap	-	0.07	0.07	-	0.15	0.39	-	-	-	0.11	0.04	0.02	-	0.02

Notes:
†Total Fe as FeO.
*Norms for compositions 6, 8, 9, 20, 22–27 calculated for $Fe_2O_3/Fe) = 0.1$.
1. Synthetic lherzolite modeled on Nockolds' average peridotite (Reay and Harris, 1964). **2.** Mixture of minerals ($Cpx_{20}\ Opx\text{-}_{20}Ol_{20}Ga_{40}$) from Wesselton Mine garnet lherzolite A.3/10596 (O'Hara and Yoder, 1967). **3.** Garnet lherzolite KA 64-16, Dutoitspan Mine (Ito and Kennedy, 1967). **4.** Pyrolite III minus 40% olivine (Green and Ringwood, 1967b, 1970; Green, 1973a). **5.** Spinel lherzolite HK 66101703, Salt Lake Crater (Kushiro *et al.*, 1968, 1972a; Kushiro, 1970). **6.** Spinel lherzolite M6-128, Beni Bouchera (Kornprobst, 1970). **7.** Spinel-amphibole lherzolite HK 63111602a, Ichinomegata (Nishikawa *et al.*, 1970). **8.** Mixture of minerals ($Ol_{30}Opx_{28}Cpx_{20}Ga_{20}$) from Dutoitspan Mine garnet lherzolite + 2% fluorphlogopite (Kushiro *et al.*, 1972a). **9.** Mixture of minerals ($Ol_{20}Opx_{20}Cpx_{30}Ga_{30}$) from Bultfontein garnet lherzolite JJG-352 (Kushiro, 1972b, 1973b). **10.** Pargasite-spinel peridotite 18–900, St. Paul's Rocks (Millhollen *et al.*, 1974). **11.** Iron-rich garnet lherzolite 1032, Matsoku (Howells *et al.*, 1975). **12.** Garnet lherzolite ga-p(1), Wesselton Mine (Mysen and Boettcher, 1975a, b). **13.** Spinel lherzolite 618–138b.1, Salt Lake Crater (Mysen and Boettcher, 1975a, b). **14.** Spinel lherzolite 66PAL-3, Pali dike (Mysen and Boettcher, 1975a, b). **15.** Garnet-olivine websterite 66SAL-1, Salt Lake Crater (Mysen and Boettcher, 1975a, b; Mysen and Kushiro, 1977). **16.** Garnet lherzolite PHN1611, Thaba Putsoa (Mysen and Kushiro, 1977; Kushiro, 1973b; Wendlandt and Mysen, 1978). **17.** Spinel lherzolite 618–138b.1 + 10% phlogopite (Wendlandt and Eggler, 1980). **18.** Pyrolite I (Green and Ringwood, 1967b). **19.** Pyrolite III (Green and Ringwood, 1967b). **20.** Proposed primitive mantle composition (Maaløe and Aoki, 1977). **21.** Peridotitic komatiite 49J, Barberton Mountain Land, South Africa (Green *et al.*, 1975a). **22.** Peridotitic komatiite SA3091 Munro Township, Ontario (Arndt, 1976). **23–27.** Peridotitic komatiites (NG157, NG152, NG7638, NG7621, NG7628, respectively), Belingwe Greenstone Belt, Rhodesia (Bickle *et al.*, 1977; Bickle, 1978). **28.** Synthetic picrite (Green and Ringwood, 1967a).

Table 3.3.2 *(Concluded)*

	15	16	17	18	19	20	21	22	23	24	25	26	27	28
SiO_2	44.82	43.70	43.7	43.20	45.20	44.71	46.46	45.4	47.59	46.90	47.96	48.76	51.83	45.51
TiO_2	0.52	0.25	0.18	0.58	0.71	0.16	0.19	0.39	0.29	0.22	0.36	0.42	0.37	1.93
Al_2O_3	8.21	2.75	4.84	4.01	3.54	2.46	3.58	7.5	6.34	5.08	7.44	8.44	6.65	12.44
Cr_2O_3	0.20	0.28	0.36	0.42	0.43	0.42	0.43	0.40	0.42	0.44	0.41	0.43	0.45	-
Fe_2O_3	2.07	1.38	0.80	0.35	0.48	-	1.00	-	-	-	-	-	-	0.92
FeO	7.91	8.81	7.28	7.88	8.04	8.15†	9.38	12.9†	10.79†	10.62†	11.39†	12.09†	10.71†	8.67
MnO	0.19	0.13	0.11	0.13	0.14	0.18	0.21	0.28	0.17	0.14	0.20	0.19	0.17	0.15
NiO	0.20	-	0.22	0.39	0.20	0.26	0.27	-	0.18	0.21	0.13	0.13	0.13	-
MgO	26.53	37.22	36.6	39.54	37.48	41.00	32.97	25.1	28.18	31.81	24.35	20.45	22.47	18.79
CaO	8.12	3.26	3.15	2.67	3.08	2.42	5.10	7.4	5.91	5.05	7.46	8.58	7.24	9.67
Na_2O	0.89	0.33	0.36	0.61	0.57	0.29	0.49	0.73	0.16	0.02	0.64	0.39	0.26	1.64
K_2O	0.03	0.14	1.02	0.22	0.13	0.09	0.18	-	0.44	0.03	0.06	0.50	0.19	0.08
P_2O_5	0.04	-	-	-	-	0.06	0.01	-	0.02	0.01	0.03	0.04	0.00	0.20
SUM	99.73	98.25	98.62	100.00	100.00	100.20	100.27	100.10	100.49	100.53	100.43	100.42	100.47	100.00
$\frac{100\ Mg}{Mg + \Sigma Fe}$	82.9	86.8	89.1	89.6	88.7	89.9	85.1	77.6	82.3	84.2	79.2	75.1	78.9	77.9
						*C. I. P. W. Weight Percent Norms**								
or	0.18	0.83	6.03	1.30	0.77	0.53	1.06	-	2.60	0.18	0.35	2.95	1.12	0.47
ab	7.53	2.79	3.05	5.16	4.82	2.45	4.15	6.81	1.35	0.17	5.42	3.30	2.20	13.88
an	18.32	5.61	8.58	7.55	6.72	5.15	7.04	17.19	15.28	13.68	17.25	19.80	16.42	26.35
di	17.22	8.35	5.56	4.49	6.76	5.13	14.44	15.63	11.08	8.98	15.65	18.10	15.59	16.31
hy	7.09	14.97	5.64	4.32	16.54	16.74	19.99	14.56	31.65	33.51	28.35	32.93	55.38	3.04
ol	44.83	62.81	67.52	74.56	61.52	67.89	50.86	43.56	35.80	41.39	30.52	20.19	6.95	34.52
mt	3.00	2.00	1.16	0.51	0.70	1.07	1.45	1.77	1.44	1.42	1.51	1.61	1.42	1.33
chr	0.29	0.41	0.53	0.62	0.63	0.62	0.63	0.59	0.62	0.65	0.60	0.63	0.66	-
ilm	0.99	0.47	0.34	1.10	1.35	0.30	0.36	0.74	0.55	0.42	0.68	0.80	0.70	3.67
ap	0.09	-	-	-	-	0.13	0.02	-	0.04	0.02	0.07	0.09	-	0.44

they gave reasons for suspecting the validity of the results of Kushiro *et al.* (1968) and of Green (1973a). The results of Millhollen *et al.* (1974), independently determined, were consistent with the latter results. Nehru and Wyllie (1975) documented the problems of iron loss and discussed other experimental problems in peridotite-H_2O systems, and Green (1976a) presented a detailed review of experimental problems, including specific reference to the difference between solidus curves in Fig. 3.3.7. Green stated that he had foreseen and minimized or surmounted the experimental problems, and he concluded that Mysen and Boettcher (1975a) had misinterpreted glassy deposits from the vapor phase as glass quenched from liquid (see also Brey and Green, 1976b). We remain uncertain of the reason for the different results in Fig. 3.3.7.

Phase relationships through at least part of the melting interval of peridotite with excess H_2O were presented by Kushiro *et al.* (1968), Green (1973a), Millhollen *et al.* (1974), and Mysen and Boettcher (1975a). Results for pyrolite-III and a spinel lherzolite nodule are compared in Fig. 3.3.8. Mysen and Boettcher (1975a) found a similar pattern of phase relationships in all four of their samples, with the positions of the phase boundaries sensitive to the bulk composition. The amphibole breakdown curves in each of their samples had a shape similar to that in Fig. 3.3.8b. This is compared in Fig. 3.3.9 with amphibole breakdown

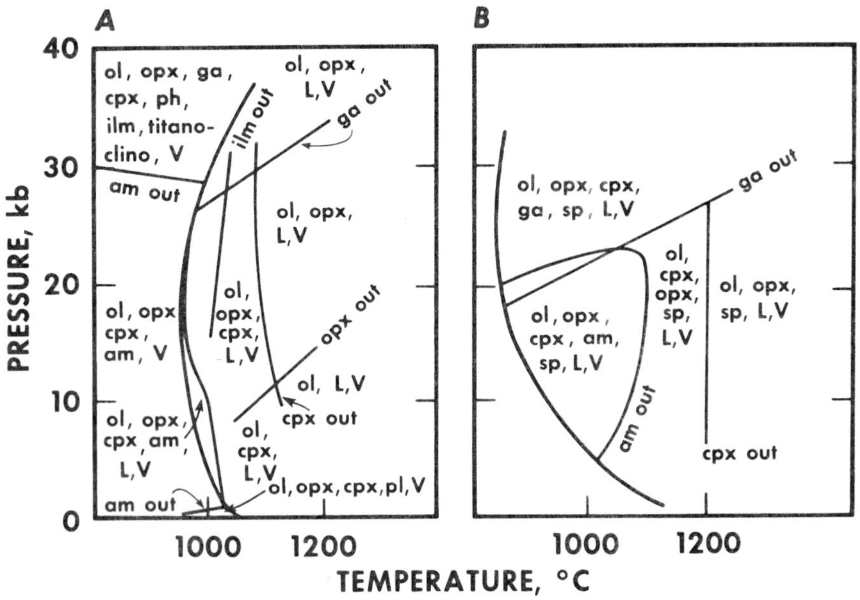

3.3.8 a. Water-saturated phase relations of pyrolite III (Green, 1973a). **b.** Phase relationships for peridotite with excess H_2O. (Mysen and Boettcher, 1975a). Abbreviations as in Fig. 3.3.5.

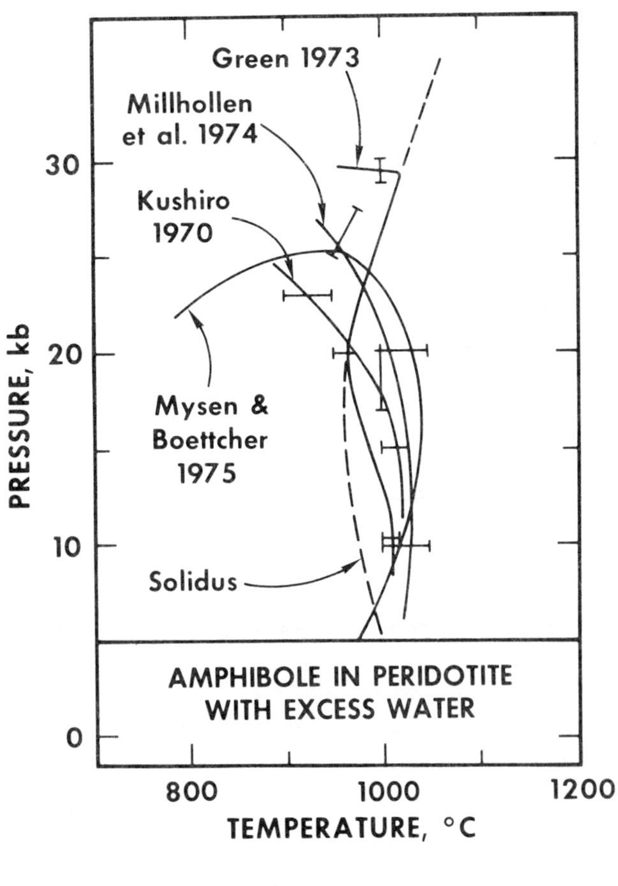

3.3.9 Comparison of results of several studies on the stability of amphibole in peridotite in the presence of excess H_2O. Amphibole is stable on the low temperature/low pressure side of each curve. The solidus curve shown is after Green (1973a), as given in Fig. 3.3.7.

curves determined in three other studies of peridotite with excess H_2O.

The differences between the shapes of amphibole phase boundaries are due in part to differences in chemistry, but some of the experimental problems discussed above and in section 3.2 probably contribute. Amphibole breaks down in a pressure-sensitive reaction producing garnet and other minerals (Lambert and Wyllie, 1968), and one might therefore expect a change in slope of the amphibole boundary like that reported by Mysen and Boettcher. Attempts to establish the subsolidus reaction interval have not been satisfactory, on account of the complexity and sluggishness of both the reaction to garnet, and the synthesis of amphibole of the correct composition. On the basis of the differences shown in Figs. 3.3.8 and 3.3.9, it appears that similar problems may persist under conditions with liquid present.

For more realistic mantle conditions, it is necessary to know the phase relationships in the presence of small amounts of H_2O. Kushiro *et al.* (1968) determined the solidus for peridotite with low H_2O contents, using open capsules, but the uncertainty about the conditions renders the results inapplicable. Green (1973a) used a pyrolite composition, minus the components of 40% olivine, crystallized to amphibole-peridotite containing 0.3% H_2O, to produce the phase diagram shown in Fig. 3.3.10a; this corresponds to pyrolite with 0.2% H_2O. The solidus curve is reproduced in Fig. 3.3.10b, and compared with other experimentally based and deduced solidus curves for peridotite containing less than about 0.4% H_2O, the amount required to produce the maximum amphibole (about 30%) in fertile peridotite.

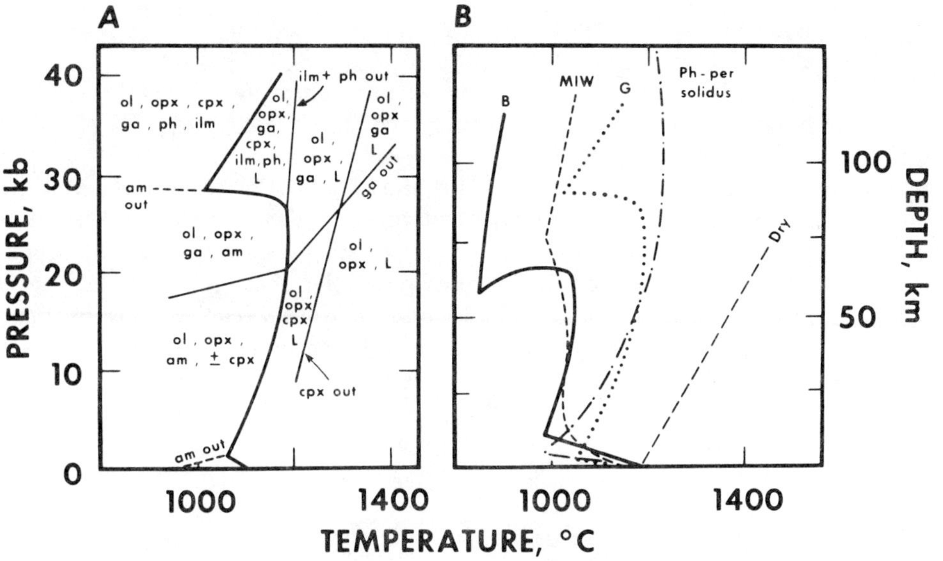

3.3.10 **a.** Solidus of pyrolite III with 0.2% H_2O (Green, 1973a). ilm = ilmenite, am = amphibole, ph = phlogopite. **b.** Experimentally based and deduced curves for the solidus of amphibole-peridotite, with <0.4 wt.% H_2O, and for phlogopite-peridotite with <0.02 wt.% H_2O. Depths are for the Earth. MIW = Millhollen *et al.* (1974); B = Boettcher (Mysen and Boettcher, 1975a); G = Green (1973a). See Wyllie (1978a).

The solidus curves in Fig. 3.3.10b correspond to the excess-H_2O curves of Fig. 3.3.7, except where amphibole is stable. In this interval, all H_2O is stored in amphibole, and melting begins only where amphibole reacts. The solidus curves in Fig. 3.3.10b are therefore related to the amphibole curves in Fig. 3.3.9. The solidus curve presented by Boettcher (1973) was based on studies in peridotite-H_2O-CO_2 for low H_2O activities (Mysen and Boettcher, 1975a). The amphibole-peridotite solidus of Millhollen *et al.* (1974) was based on the curve determined with excess H_2O (Fig. 3.3.9), with temperature increased according to the demonstration of Holloway (1973) that the melting temperature of pargasite at 10 kb increases from 1065° C with

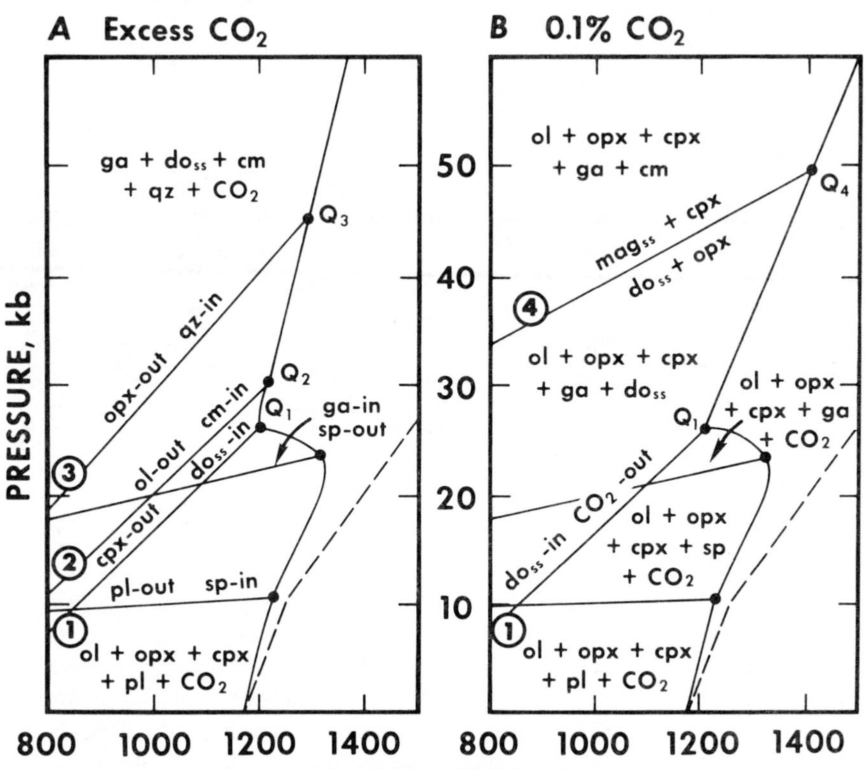

3.3.11 Partly-schematic phase diagrams for peridotite-CO_2 with **a.** excess CO_2 and **b.** 0.1% CO_2. See Eggler (1976, 1978) and Wyllie (1977, 1978a) for data sources. Abbreviations: ol = olivine, opx = orthopyroxene, ga = garnet, sp = spinel, pl = plagioclase, qz = SiO_2 polymorphs, do = dolomite solid solution, cm = magnesite solid solution, V = vapor, dashed line = dry peridotite solidus.

excess H_2O to a maximum of 1110°C with decreasing H_2O content. The experimentally determined solidus for vapor-absent amphibole pyrolite is similarly at higher temperature than the amphibole boundary in pyrolite with excess H_2O, the difference being about 80°C at 10 kb, and 150–160°C at 20 kb. This increase in solidus temperature is higher than that anticipated by others, which may indicate that the basis for anticipation was incorrect, or that amphibole persisted above its stability limit in the vapor-absent experiments of Green (1973a).

Phlogopite is another hydrous mineral which may play an important role in magma generation. A peridotite containing 0.1% K_2O would require about 0.02% H_2O to convert all of the K_2O into about 0.5% phlogopite. Figure 3.3.10b shows an estimate of the solidus curve for vapor-absent phlogopite-peridotite, based on the results of Modreski and Boettcher (1973) and Holloway and Eggler (1976).

Peridotite-CO_2

On the basis of studies in the model mantle system CaO-MgO-SiO_2-CO_2 by Wyllie and Huang (1975, 1976) and Eggler (1974, 1976, 1978), the approximate phase relationships for peridotite-CO_2 have been determined within limits that are fairly well constrained (Wyllie, 1977, 1978a; Eggler, 1978). Figure 3.3.11 shows the reactions caused by CO_2 superimposed on the subsolidus phase boundaries separating the fields for plagioclase-, spinel-, and garnet-peridotite facies (see Figs. 3.1.2 and 3.3.4). The solidus for peridotite alone is marked by the dashed line.

Carbon dioxide reacts with peridotite at high pressures to produce calcic dolomite (Wyllie and Huang, 1975, 1976; Eggler, 1976) by the reaction:

$$\text{olivine} + \text{clinopyroxene} + CO_2 = \text{orthopyroxene} + \text{dolomite}_{ss}. \quad (1)$$

This reaction, plotted in Fig. 3.3.11, divides the subsolidus region into two parts. For peridotite with less than about 5% CO_2, free CO_2 vapor exists only on the low-pressure, high-temperature side of this reaction. On the high-pressure side, as shown in Fig. 3.3.11b, all of the CO_2 reacts with the peridotite to produce calcic dolomite. If there is more than 5% CO_2 present, all of the clinopyroxene reacts away and, with increasing pressure, two additional carbonation reactions occur as shown in Fig. 3.3.11a (Wyllie and Huang, 1975, 1976).

$$\text{olivine} + \text{dolomite}_{ss} + CO_2 = \text{orthopyroxene} + \text{magnesite}_{ss} \quad (2)$$

and:

$$\text{orthopyroxene} + \text{dolomite}_{ss} + CO_2 = \text{quartz/coesite} + \text{magnesite}_{ss}\,. \quad (3)$$

It requires approximately 23% CO_2 to convert peridotite completely to coesite-garnet-marble via the three reactions. For peridotite with less than 5% CO_2, the vapor-absent dolomite-peridotite shown in Fig. 3.3.11b is stable to high pressures until it reaches the subsolidus exchange reaction between pyroxenes and carbonates (Kushiro *et al.*, 1975):

$$\text{orthopyroxene} + \text{dolomite}_{ss} = \text{clinopyroxene} + \text{magnesite}_{ss}. \quad (4)$$

According to G. Brey and D. H. Green (pers. comm., 1977), this reaction has a steeper slope than that depicted, which brings the magnesite-peridotite stability field down to lower pressures at moderate temperatures.

The very low solubility of CO_2 in silicate liquids at low pressures causes only minor depression of the solidus for peridotite-CO_2 compared with that for peridotite, as shown in Fig. 3.3.11. Mysen *et al.* (1976) determined that the solubility of CO_2 in the model mantle system increased to about 5% at 20 kb, with moderate depression of the solidus, as shown in Fig. 3.3.11 and by Eggler (1974). Wyllie and Huang (1975, 1976) outlined evidence supporting their conclusion that the liquid at the invariant point Q_1 and along the solidus at higher pressures in Fig. 3.3.11 was essentially a carbonate liquid, with only 10–15% dissolved silicates. The marked increase in CO_2 solubility, which is associated with the stabilization of dolomite at the solidus, is responsible for the sharp curvature and decrease in temperature of the solidus within the small pressure interval below Q_1. At pressures above Q_1, the solidus in Fig. 3.3.11b for vapor-absent dolomite-peridotite remains unchanged for CO_2 contents up to 5%, and the solidus in Fig. 3.3.11a for carbonated-peridotite in the presence of CO_2 occurs at lower temperatures (the temperature difference has not been determined, even in the model system).

Peridotite-H_2O-CO_2

Mysen and Boettcher (1975a, 1975b) published the only comprehensive experimental investigation of the effects of H_2O-CO_2 mixtures (with controlled oxygen fugacity) on the melting relationships of natural peridotites. Eggler (1975) studied their effects on the melting of synthetic forsterite + enstatite. These investigations established the effects of H_2O + CO_2 at pressures up to about 20 kb, but they did not take into account the effect of carbonation reactions. This effect, discovered by Wyllie and Huang (1975) and Eggler (1976), produces striking changes in near-solidus phase relationships between about 20 and 30 kb, as shown for the CO_2 system in Fig. 3.3.11.

Figure 3.3.12 shows the divariant solidus surface for peridotite-H_2O-CO_2 with contours for constant vapor-phase composition, based at pressures up to 20 kb on experimental data of Mysen and Boettcher (1975a), and fitted between the solidus curves for excess H_2O (Fig. 3.3.7, solidus of Millhollen *et al.*, 1974) and for excess CO_2 (Fig. 3.3.11a). There is a similar contoured solidus surface at pressures above Q_1, extending from the dolomite-peridotite-CO_2 solidus (Fig. 3.3.11a) to lower temperatures as H_2O is added. Whereas this surface is convex upward with respect to temperature, the low-pressure solidus surface is concave upward. The univariant carbonation reaction (1) terminating at Q_1 in Fig. 3.3.11 becomes a divariant surface if H_2O is added, and the shape of this surface is shown in Fig. 3.3.12 by the contours for constant vapor-phase composition (Eggler *et al.*, 1976; Wyllie, 1977). Similar divariant surfaces for the carbonation reactions extending from Q_2 and Q_3 (Fig. 3.3.11a) are not illustrated, because they can be attained only for CO_2 contents greater than 5%.

The carbonation surface meets the solidus surface along the line Q_1N, which gives the univariant beginning of melting of partly carbonated peridotite in the presence of H_2O-CO_2 vapors of buffered compositions. The significance of this vapor-buffer line was discussed by Eggler (1978) and Wyllie (1977), who both extended the contour lines on the low-pressure solidus surface directly to the corresponding compositions on the buffer line. Wyllie (1978a), after consideration of the distribution of H_2O and CO_2 between liquid and vapor, concluded that this geometry was incorrect, and that the contours on the solidus surface at low pressures had to be passed over a temperature maximum (the light line mn in Fig. 3.3.12) and then down to the buffer line Q_1N. The temperature maximum, a ridge on the solidus and liquidus surfaces, is a significant addition to the previous interpretations. The limiting contour for CO_2 has a shape consistent with this family of solidus contours. The shapes of the contours can be more easily visualized in Fig. 3.3.13.

The temperature maximum, mn, divides the solidus surface for peridotite-vapor into two parts. For the area extending from low pressures to the line mn, H_2O is preferentially dissolved in the liquid, leaving the vapor enriched in CO_2 (Mysen and Boettcher, 1975a; Eggler, 1975). For the area between the temperature maximum and the dolomite-peridotite buffer line, $mnNQ_1$, CO_2 is preferentially dissolved in the liquid, leaving the vapor relatively enriched in H_2O. The same condition holds for the carbonated-peridotite-vapor solidus surface at

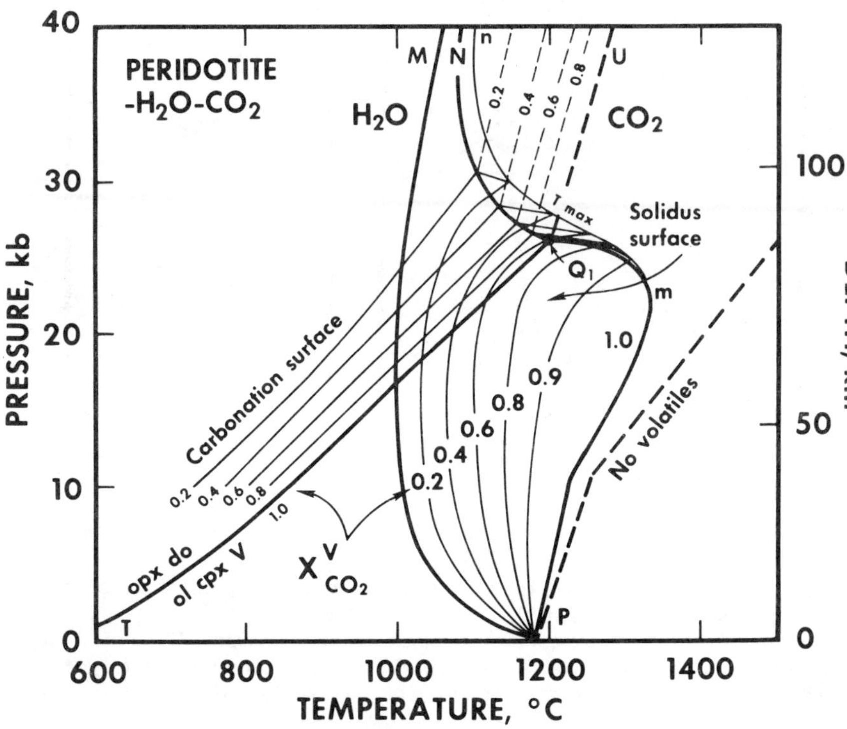

3.3.12 Peridotite-H_2O-CO_2 with excess vapor, omitting amphibole and phlogopite, pieced together and averaged from various data sources (see Wyllie, 1977, 1978a; Eggler *et al.*, 1976). The divariant carbonation surface divides the solidus surface into (1) the area PQNM passing through a temperature maximum along mn, for the melting of peridotite-vapor, and (2) the area QNU for the melting of carbonated peridotite-vapor. Two other carbonation reactions (not shown) intersect the latter area (see Fig. 3.3.11). The temperature maximum mn is consistent with phase relationships deduced in the system MgO-SiO_2-CO_2-H_2O (Ellis and Wyllie, 1979), but its magnitude is not known (except for PmQ, pure CO_2). Depths are for the Earth. $X^V_{CO_2}$ = mole fraction of CO_2 in vapor phase.

higher pressures. Along the line mn, CO_2/H_2O is the same in liquid and vapor.

The phase diagrams for peridotite-H_2O-CO_2 given here do not include the transitions from plagioclase- to spinel- to garnet-peridotite. These phase boundaries (Fig. 3.3.5) can be superimposed as they were for peridotite-CO_2 in Fig. 3.3.11. The aluminous minerals appear to remain unaffected by carbonation reactions through the PT range under consideration.

The occurrence of amphibole was omitted from Fig. 3.3.12 for simplicity, in order to illustrate the buffering effect of dolomite in peridotite-H_2O-CO_2. The

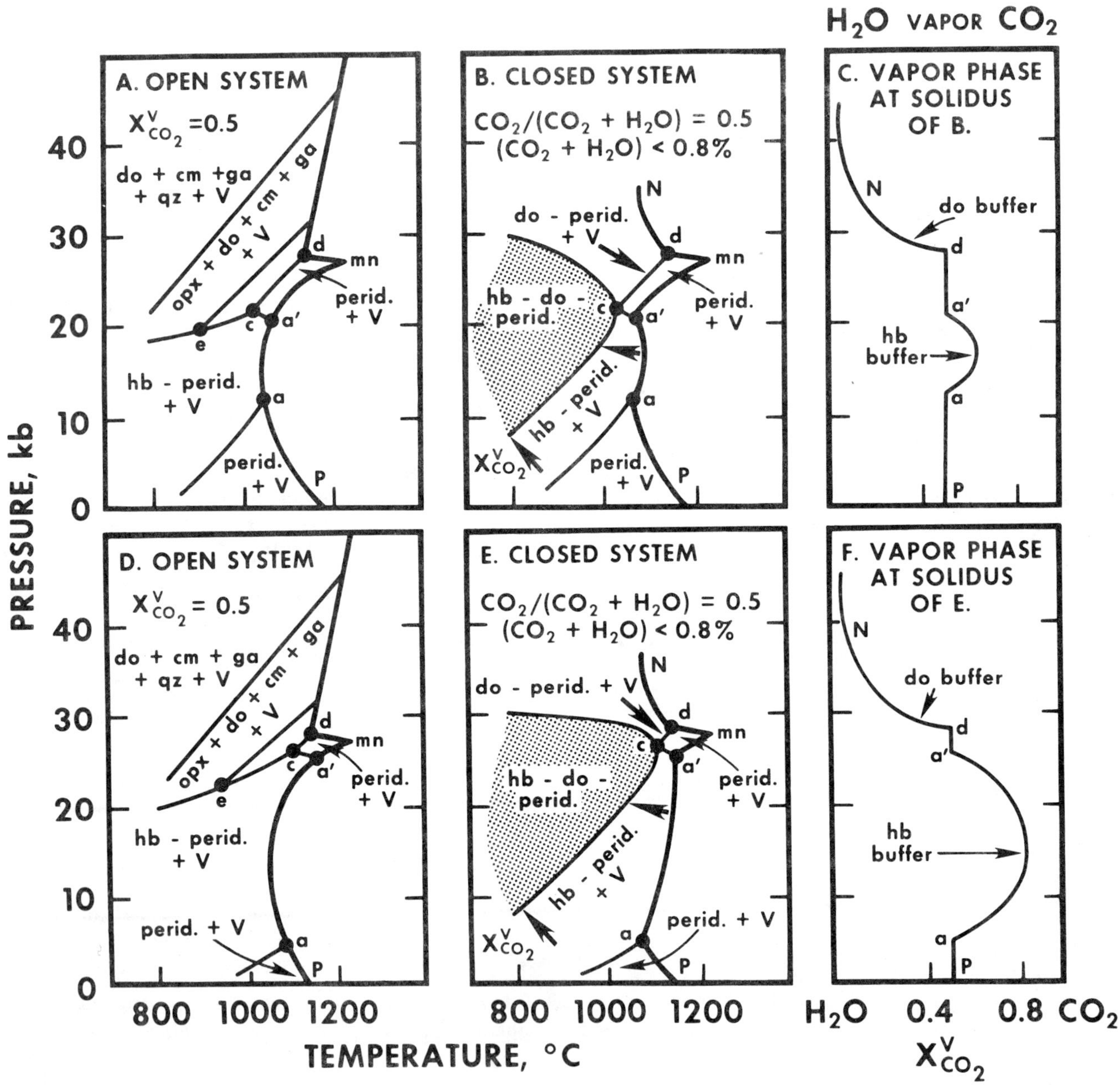

3.3.13 Schematic phase relationships in the system peridotite-CO_2-H_2O using two different experimental results for amphibole (from Wyllie, 1979). A, B and C after Millhollen *et al.* (1974); D, E and F after Green (1973a). The heavy arrows show the directions of increasing $X^V_{CO_2}$ on divariant vapor-buffered reaction surfaces. No vapor exists within the shaded areas. Points a, c and d are located on univariant curves; d, for example, is on Q_1N in Fig. 3.3.12. Point mn is on the T_{max} line mn of Fig. 3.3.12. Abbreviations: hb = amphibole, do = dolomite, opx = orthopyroxene, ga = garnet, cm = magnesite solid solution, qz = SiO_2 polymorphs.

subsolidus divariant surface for the dehydration of amphibole in the presence of H_2O-CO_2 vapors intersects the solidus surface along an amphibole vapor-buffer curve, and there is also a subsolidus vapor-buffer curve where the decarbonation and dehydration surfaces meet (Wyllie, 1978a).

The main features of the system peridotite-H_2O-CO_2, assuming a composition without phlogopite, are illustrated in Fig. 3.3.13 (Wyllie, 1979). Figure 3.3.13a shows the phase fields for an open system, where H_2O or CO_2 is added to or subtracted from the system as required to maintain a constant vapor phase composition. Each phase boundary is thus a contour from a divariant surface, like those projected in Fig. 3.3.12. The temperature maximum on the solidus surface is clearly displayed by the point mn. Figure 3.3.13b shows the phase fields for a closed system, with less than 0.4% H_2O. With the formation of dolomite, the vapor-phase composition is buffered, and the solidus dN for dolomite-peridotite-vapor is part of the line Q_1N in Fig. 3.3.12. With the formation of amphibole, the vapor is buffered to the solidus for amphibole-peridotite-vapor, along the curve aa′. The composition of the vapor phase as a function of pressure along the solidus in Fig. 3.3.13b is given in Fig. 3.3.13c. In the shaded area of Fig. 3.3.13b, all H_2O and CO_2 are stored in vapor-absent amphibole-dolomite peridotite. This area is bounded by the subsolidus vapor-buffer line.

The top row of diagrams in Fig. 3.3.13 uses the amphibole data of Millhollen *et al.* (1974), and the bottom row shows equivalent diagrams using the amphibole data of Green (1973a). Although the shapes and positions of phase boundaries differ, the basic geometry remains the same. Green's larger field for amphibole (Fig. 3.3.10b) buffers the vapor along aa to higher CO_2/H_2O. An important feature of Fig. 3.3. 13 is the demonstration that at pressures greater than about 30 kb, vapor coexisting with peridotite must be H_2O-rich. It requires only a low ratio of CO_2/H_2O to produce dolomite, and the formation of dolomite absorbs CO_2 and forces the vapor to very high H_2O/CO_2.

For a wide range of total CO_2/H_2O in the system, and for systems with $H_2O < 0.4\%$ and $CO_2 < 5\%$, the phase relationships are similar to those shown in Fig. 3.3.13 (Wyllie, 1979). The solidus buffer lines are fixed in position, but other phase fields' boundaries change position, following the vapor contours on divariant surfaces (Fig. 3.3.12) for the specific values of total CO_2/H_2O. In the limiting cases, the amphibole field disappears as in Fig. 3.3.11b, and the dolomite field disappears as in Fig. 3.3.10b.

The phase relationships become more complex if there is sufficient K_2O present for the growth of phlogopite. Holloway and Eggler (1976) determined the solidus temperatures for vapor-absent phlogopite-peridotite and phlogopite-dolomite-peridotite at 30 kb. Other details were reviewed by Eggler (1978) and Wyllie (1978a). According to Wyllie (1978a), the occurrence of phlogopite introduces a phlogopite-dolomite-vapor sub-solidus reaction which intersects the dolomite-peridotite vapor-buffer line Q_1N at an invariant point very close to Q_1 (Fig. 3.3.12). Extending from this point to higher pressures there are two additional solidus curves: the phlogopite-dolomite-peridotite vapor-buffer curve situated at somewhat lower temperatures than Q_1N, and the vapor-absent phlogopite-dolomite-peridotite solidus situated at somewhat higher temperatures than Q_1N. The buffering capacity of dolomite is far greater than that of phlogopite, and the buffered vapor composition with phlogopite-dolomite-peridotite is very similar to that for dolomite-peridotite, illustrated by dN in Fig. 3.3.13c. The vapor has high H_2O/CO_2 above 30 kb.

3.3.3 Compositions of liquids from partial melting of peridotite

Some of the volatile components dissolved in magmas (section 3.3.2) are derived from mantle sources. The compositions of initial liquids produced from mantle peridotite may be strongly influenced by the amount and composition of the volatile components. This has been established for H_2O and CO_2 (Kushiro, 1972b; Eggler, 1973, 1974; Mysen and Boettcher, 1975b; Wyllie and Huang, 1975, 1976).

Small quantities of volatile components can have a large effect on solidus temperatures (Fig. 3.3.10 and 3.3.11b), but they have little effect on liquidus temperatures and relationships. Recent data by Delaney *et al.* (1978) and Muenow *et al.* (1979) on the volatile content of glass-vapor inclusions contained within olivine and plagioclase phenocrysts from Hawaiian and mid-ocean ridge tholeiites bear on this question. They found 0.2–0.4 wt. % CO_2 and no detectable H_2O, with a detectability limit for H_2O of 0.0004 wt. %. In agreement with Hart and Nalwalk (1970) and Moore (1970), they found 0.1–0.7 wt. % H_2O in the matrix glass surrounding the phenocrysts. Data are not yet available on the volatile contents of inclusions from continental basalts, island arc basalts, and alkalic basalts from oceanic islands. Moore (1970) found that matrix glasses of alkalic basalts from Hawaii contain about 1 wt.% H_2O, and Garcia *et al.* (1979) have found 1 wt.% H_2O, and 0.2 wt.% CO_2 in the matrix glass of andesites and basalts from the Marianas Islands. If the same relative difference between matrix and inclusion glass in oceanic tholeiites holds also for basalts from these other environments, 1 wt.% would be an extreme upper limit for the H_2O content of the magmas at depth.

The upper limits of the volatile contents of both tholeiitic and alkali olivine basalt, if uncontaminated, are thus about 1% H_2O and 0.4% CO_2, and actual contents are commonly much lower. These concentrations of volatile components would cause a negligible change in composition of basalts derived from peridotite, compared with basalts derived under similar conditions but without the volatile components. This was pointed out by Presnall *et al.* (1978a) in their discussion of melting in the model mantle system Mg_2SiO_4-$CaMgSi_2O_6$-SiO_2-H_2O at 20 kb. It has been confirmed experimentally that addition of a few tenths of a percent H_2O causes only minor changes in the liquidus temperature and mineralogy, as shown in Fig. 3.3.29. The volatile-free melting relationships of peridotite thus appear to be an adequate approximation of actual melting relations involving the production of the abundant basalts.

Some magmas contain higher proportions of dissolved volatile components. Anderson's (1975) analyses of glasses trapped in large crystals in pumice indicate that 3–5% H_2O is commonly dissolved in basaltic and andesite magmas below volcanoes of continental margins associated with compressive plate boundaries. Water may therefore be influential in modifying the compositions of magmas generated from peridotite in subduction zones (Fig. 3.3.3) (Boettcher, 1973; Mysen and Boettcher, 1975b; Green, 1976a). Kimberlite magmas contain high proportions of CO_2 and H_2O, and CO_2 appears to be important in controlling the composition of kimberlite and other subsilicic alkaline magmas (Irving and Wyllie, 1975; Eggler, 1974, 1976; Wyllie and Huang, 1975, 1976; Brey and Green, 1975, 1976a, b, 1977).

The following discussion will be divided into two sections, the first discussing melting relations of peridotite without volatile components, and the second discussing melting relationships with H_2O and CO_2. Determination of the composition of the experimentally-produced liquids is typically beset with uncertainties related to quenching the liquid to a homogeneous glass (sections 3.2.2 and 3.3.3). Some experimental results are listed in Table 3.3.3. For each study it is desirable to carry out a complementary set of experiments to confirm that the compositions of the liquidus minerals in the glass match those of the peridotite residuum after partial melting, but this confirmatory step has been taken in only a few of the experiments conducted to date (Green, 1976a).

Models of magma genesis will not be developed here. Instead, the experimental data will be treated as one set of constraints that must be taken into account when developing such models, as illustrated briefly for selected examples in sections 3.3.5–7.

Types of fusion processes

Before we consider the specific compositions of liquids that can be derived from partial melting of peridotites under various conditions, we should consider the possible types of melting processes that can take place, and understand how these processes can affect the compositions of the magmas that are separated from the source rocks. This is most important for evaluation of a suite of basalts from a given region in an attempt to determine the primary magma composition or compositions.

Because of the emphasis for many years on crystallization differentiation (section 3.1.1), petrologists are familiar with the limiting cases of equilibrium and fractional crystallization of liquids. It was only in the late 1950's and 1960's that petrologists became concerned with the processes of partial fusion, and it was Presnall (1969) who systematized the types of processes that could be discerned from consideration of simple phase diagrams. Processes in nature would not exactly duplicate those in simplified systems, but they would be analogous.

Presnall (1969) defined **partial fusion** as fusion of some portion less than the whole. Two types of partial fusion were distinguished. With **equilibrium fusion**, the liquid produced during heating remains in equilibrium with the crystalline residue until it is separated from the source. With **fractional fusion**, each infinitesimal amount of melt is removed from the source as soon as it is formed and is thereby prevented from further reaction with the crystalline residue. A process intermediate between these two extremes can be thought of as a series of equilibrium fusion events from the same source, and as the number of equilibrium fusion events increases to infinity, the process approaches the extreme case of perfect fractional fusion. Methods for deducing crystal and liquid paths followed during equilibrium and fractional fusion in simplified systems have been developed by Presnall (1969) and Roeder (1974b), and some further quantitative aspects of these processes have been discussed by Morse (1976).

Presnall (1969) pointed out that during fractional fusion in simplified systems the liquid path is frequently discontinuous. That is, large volumes of a series of liquids of constant composition are produced without intermediate compositions. Also, he emphasized the point previously made by Yoder and Tilley (1962) that if fusion takes place at a liquidus invariant point, wide variations in the composition of the source can exist

Table 3.3.3 Selected compositions and norms of partial melts obtained from peridotites in the presence of excess H_2O and/or CO_2.

	1	2	3	4	5	6	7	8	9
Pressure (kb)	26	15	15	15	10	20	15	15	15
Temperature (°C)	1190	950	1150	1150	1100	1100	1415	1425	1600
$(X^V_{H_2O})^\dagger$	1.0	1.0	0.75	0.4	1.0	1.0	0.0	0.0	0.0
SiO_2	58.9	62.2	58.0	52.0	56.1	48.5	51.5	50.1	45.7
TiO_2	0.5	0.8	0.4	0.7	2.5	2.7	0.75	0.79	0.38
Al_2O_3	14.0	20.1	16.1	14.1	12.4	12.2	10.5	10.8	4.3
Cr_2O_3	-	-	-	-	0.1	1.1	0.48	0.49	0.38
ΣFeO	1.0	1.7	5.2	9.4	7.3	9.6	9.9	10.8	13.9
MnO	0.1	0.2	0.1	0.2	0.1	0.2	0.24	0.24	0.15
MgO	0.5	2.1	4.7	6.7	9.3	11.3	9.1	9.1	29.3
CaO	8.9	10.4	12.0	14.3	9.5	10.7	9.3	9.1	5.0
Na_2O	2.7	2.1	3.8	2.5	2.0	2.1	0.53	0.67	0.52
K_2O	0.3	0.2	0.5	0.1	0.5	0.5	0.37	0.42	0.21
Sum	86.9	99.8	100.00	100.00	99.8	98.9	92.67	92.51	99.84
$\frac{100\,Mg}{Mg + \Sigma Fe}$	47.1	69.0	62.0	56.0	69.5	72.0	62.1	60.0	79.0
C. I. P. W. weight percent norms (for $Fe_2O_3/FeO = 0.1$)									
qz	25.88	24.40	5.04	-	9.95	-	11.33	8.52	-
or	1.77	1.18	2.95	0.59	2.95	2.95	2.19	2.48	1.24
ab	22.85	17.77	32.16	21.16	16.92	17.77	4.48	5.67	4.40
an	25.20	44.83	25.40	26.96	23.38	22.39	25.18	25.22	8.78
di	4.52	5.43	27.88	36.21	19.03	24.66	17.10	16.30	12.81
hy	-	4.52	5.96	10.98	21.77	14.18	29.03	30.75	21.63
ol	-	-	-	1.61	-	9.00	-	-	47.98
mt	0.13	0.16	0.70	1.25	0.97	1.29	1.32	1.45	1.86
chr	-	-	-	-	0.15	1.62	0.71	0.72	0.56
ilm	0.95	1.52	0.76	1.33	4.75	5.13	1.42	1.50	0.72

$^\dagger X^V_{H_2O} = 1.0 - X^V_{CO_2}$.
1. Kushiro *et al.* (1972a, Table 4). Starting material: Column 5, Table 3.3.2.
2. Mysen and Boettcher (1975b, Table 9). Starting material: Column 13, Table 3.3.2.
3,4. Mysen and Boettcher (1975b, Table 11). Starting material: Column 13, Table 3.3.2, f_{H_2} = magnetite – hematite buffer.
5,6. Green (1973a, Tables 3 and 8). Melt compositions computed from bulk composition and crystal compositions. Starting material: Column 4, Table 3.3.2.
7,8,9. Wendlandt and Mysen (1978, Table 20). Starting material: Column 16, Table 3.3.2.

without changing the composition of the initial melt. In order to maintain the same initial melt composition, it is necessary only to retain the same crystalline phases in the source; variations in the proportions of these phases do not affect the initial melt composition.

These features can be understood more easily by reference to Fig. 3.3.14. Suppose the simplified mantle composition m consisting of forsterite, enstatite (b) and diopside (a) is fractionally fused at 20 kb. The initial liquid is produced at the peritectic p and the composition of this initial liquid would not change regardless of the proportions of forsterite, enstatite, and diopside. That is, the initial liquid would lie at p for any starting composition within the triangle Mg_2SiO_4-a-b. For

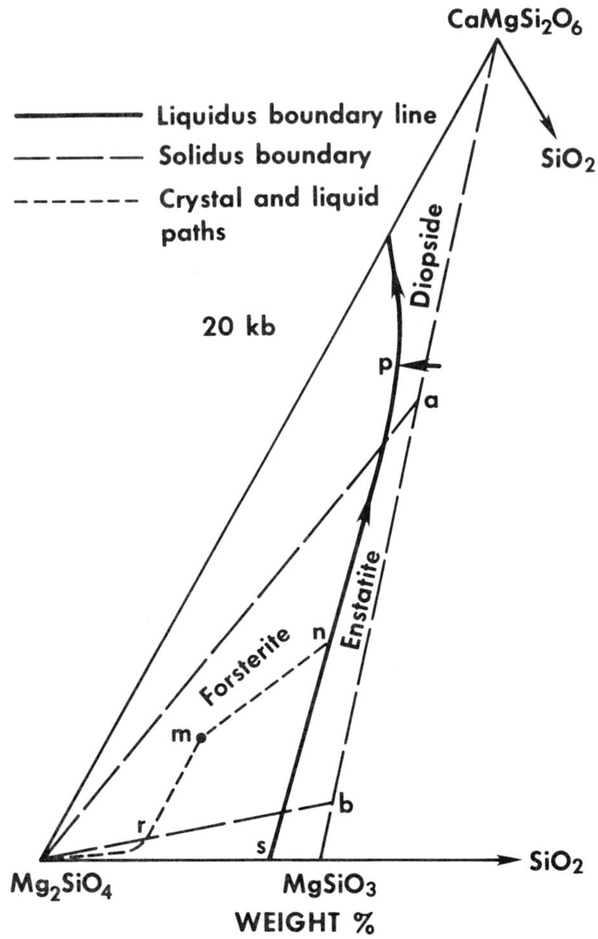

3.3.14 Liquidus (heavy boundary lines with arrows indicating directions of decreasing temperature) and solidus (long-dashed lines) phase relationships in the system Mg_2SiO_4-$CaMgSi_2O_6$-SiO_2 at 20 kb, modified after Kushiro (1969) and Chen and Presnall (1975).

the starting composition m, liquid at p would continue to be extracted until 30% of the source was melted. The crystal path would move across the solidus along the line m-r. Then, liquids along the univariant line p-s would be produced as the crystal path moves along a solidus fractionation line (not shown) close to but not exactly coincident with the dashed curved line r-Mg_2SiO_4. Just as about 62% of the source is fused (see Morse, 1976, for use of the lever rule during fractional fusion), the liquid path reaches s and the crystal path reaches pure forsterite. Further addition of heat results in an increase in temperature without the production of any further liquids until the temperature reaches the melting temperature of pure forsterite. Finally, the remaining 38% of the starting composition m is produced as pure forsterite liquid. Thus, large amounts of liquid of constant composition are produced at two points, the peritectic p and pure forsterite, 32% of the source is fused to produce liquids of variable composition along p-s, and no liquids occur in the gap between s and forsterite.

The spectrum of primary magmas produced would depend on the amount of thermal energy available and the nature of the fusion process. For example, if there is only enough thermal energy to fuse up to 30% of the source, only one primary magma composition would be produced regardless of whether the magma is extracted in one step or many steps. If there is enough thermal energy to fuse up to 62% of the source and if the magma is extracted in one step, the composition of the magma could lie anywhere along the boundary line p-s. If the magma is extracted in a large number of small steps, a range of compositions from p to a point slightly toward p from s would be produced, with those at p making up about half of the total. This illustrates the point that in some areas, not one but many different primary magma compositions could be produced from the same source. Therefore, selecting the "most primitive" lava with the highest value for $Mg/(Mg+Fe^{2+})$ might lead to an incorrect petrogenetic model.

In the mantle, melting relationships would not conform to the simplified situation just described. There are many components in the mantle and only a few phases. Thus, fusion would not take place at an invariant point, and the compositions of minerals would change during melting (e.g., Mg/Fe). Also, the extraction of finite amounts of melt from the source implies that real fusion processes would at best only approach perfect fractional fusion. Finally, it is likely that part of the liquid produced at each fusion step would be left behind in the source region. It might be supposed that these complications would cause important differences between the above simplified discussion and real fusion processes in the mantle. Presnall (1979) has argued, however, that even though these complications would cause some changes, the general types of effects deduced from simplified phase diagrams would be retained, and in particular, large volumes of primary magma showing only small variations in composition would still be expected.

As discussed above, it is possible for the composition of the primary melt to remain fairly constant even though the proportions of phases in the source vary widely. Thus, if one has samples only of basalts produced from this source, as on the Moon, it is possible from experimental studies to place restrictions on the phases that must exist in the source, but it is important to recognize that the proportions of these phases are difficult to determine from constraints of experimental petrology alone. Thus, very large uncertainties in the bulk composition of the source can exist. On the Earth, samples of the residual material remaining after fusion (ultramafic nodules) are also available, and with this added information, the proportions of phases in the unfused mantle can be estimated with greater certainty.

Peridotite without volatile components

In general, the composition of a primary magma depends on (1) the bulk composition of the source material, (2) the amount of fusion, and (3) the depth of separation of magma from the source. In the following discussion, the source will be considered to consist essentially of olivine, orthopyroxene, clinopyroxene, and one of the aluminous phases plagioclase, spinel, or garnet (see Figs. 3.1.2, 3.3.4 and 3.3.5). By assuming a four-phase peridotite source, the first factor will be held relatively constant, although small changes in derived liquids would be expected depending on the assumed compositions and proportions of phases in the peridotite. It should be kept in mind, however, that detailed knowledge about variations in the composition of the mantle source for basalts does not presently exist.

Consider first the dependence of the primary magma composition on the depth at which it is generated. Numerous studies of both simplified systems and natural compositions have shown that as pressure increases in the approximate range 0–24 kb, near-solidus liquids from peridotite become increasingly depleted in SiO_2 and eventually become nepheline-normative (O'Hara, 1965, 1968a; Green and Ringwood, 1967a; Kushiro, 1968, 1972b, 1973a; Presnall *et al.*, 1978a, 1979).

In studies of simplified model systems, Presnall *et al.* (1979) have achieved the closest approach to the compositions of real liquids generated from partial fusion of lherzolite. In the system CaO-MgO-Al_2O_3-SiO_2, they determined the compositions of initial liquids in equilibrium with simplified plagioclase lherzolite (forsterite + enstatite + diopside + anorthite) and spinel lherzolite (forsterite + enstatite + diopside + spinel) from 1 atm to 20 kb. They plotted their results in the simplified tholeiitic basalt tetrahedron forsterite-diopside-anorthite-quartz (compare Fig. 3.3.1) and projected the compositions onto three faces of this tetrahedron (Fig. 3.3.15–17). These compositions correspond to liquids along the univariant solidus curve for simplified lherzolite given in Fig. 3.3.4. As pressure increases, the liquids change from quartz-normative compositions below about 9 kb to olivine-normative compositions above 9 kb. The sharp change in direction at 9 kb occurs at the invariant point in P-T space at which forsterite, enstatite, diopside, anorthite, spinel,

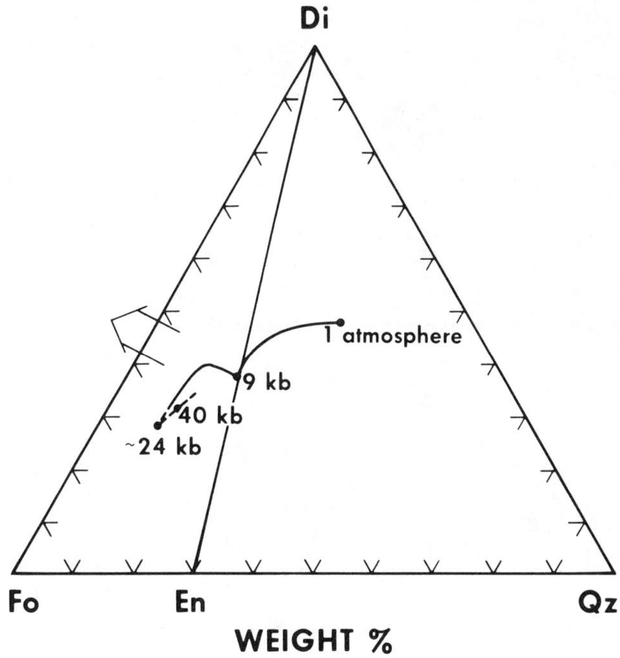

3.3.16 Projection from an. See caption to Fig. 3.3.15 for explanation.

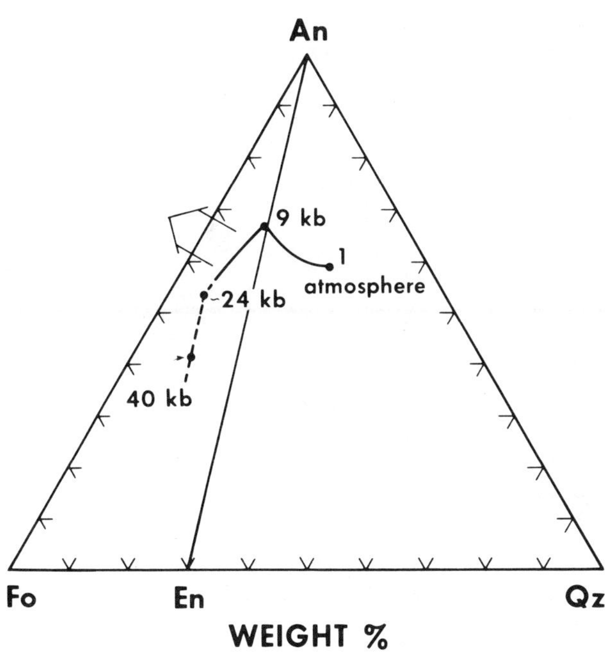

3.3.15 Projection from di showing compositions of first liquids (dashed where inferred) along the solidus curve of Fig. 3.3.4. Di = $CaMgSi_2O_6$, Fo = Mg_2SiO_4, En = $MgSiO_3$, Qz = SiO_2, An = $CaAl_2Si_2O_8$. After Presnall *et al.* (1979) and Davis and Schairer (1965).

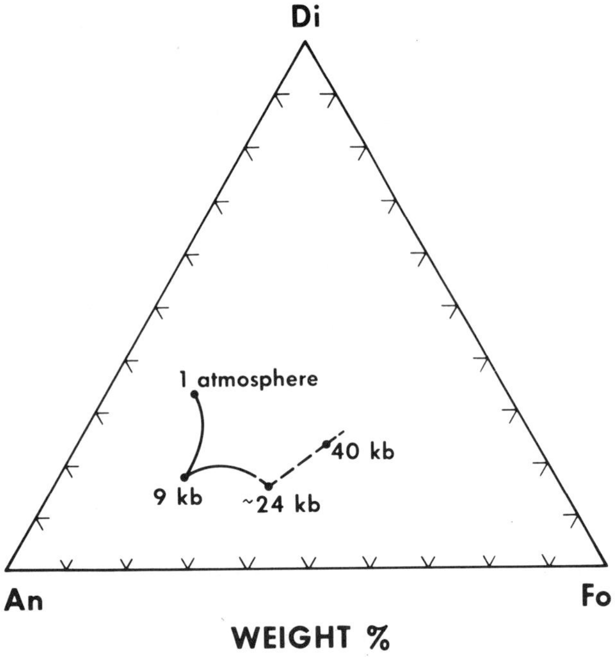

3.3.17 Projection from qz. See caption to Fig. 3.3.15 for explanation.

and liquid are all in equilibrium. This invariant point marks the intersection of the P-T univariant solidus curve with the univariant subsolidus transition between plagioclase and spinel lherzolite.

Data on first liquid compositions are not yet available above 20 kb in the system CaO-MgO-Al_2O_3-SiO_2 except for one point estimated by Davis and Schairer (1965), who reported a hypersthene-normative picritic liquid in equilibrium with forsterite, enstatite, diopside, and garnet at 40 kb. This composition is plotted in Figs. 3.3.15–17. O'Hara (1965, 1968a) also discussed the production of picritic liquid as an initial melt from lherzolite at high pressures. In the system CaO-MgO-Al_2O_3-SiO_2, it is probable that first liquids produced from simplified lherzolite will never lose normative hypersthene. Liquids with the lowest content of normative hypersthene will probably occur at approximately 24 kb (Figs. 3.3.15–16), where the spinel lherzolite to garnet lherzolite transition intersects the solidus to form an invariant point at which forsterite, enstatite, diopside, spinel, garnet, and liquid are in equilibrium (Fig. 3.3.4).

As discussed by O'Hara (1965, 1968a), Green and Ringwood (1967a), and Kushiro (1968), the reason for the shift to liquids with lower SiO_2 content as pressure increases in the range 0–24 kb is the expansion of the orthopyroxene primary phase field at the expense of the olivine field, as indicated in Fig. 3.1.13 (Boyd *et al.*, 1964; Kushiro, 1969; Chen and Presnall, 1975). Kushiro (1969) and Presnall *et al.* (1978a) have shown that the diopside primary phase field also expands with pressure at the expense of the olivine field (Fig. 3.3.18). At higher pressures, the production of hypersthene-normative picritic liquids is caused by expansion of the garnet field.

Consider now the effect of additional components on the trends shown in Figs. 3.3.15–17. Although detailed information is not available, several studies of the melting behavior of compositions containing other components give an indication of the changes to be expected (Kushiro, 1968, 1972b, 1973a, 1973b; Green and Ringwood, 1967a, b; Mysen and Kushiro, 1977). Based on these studies, Presnall *et al.* (1978a, 1979) concluded that for the real mantle, the entire trend would be shifted away from the SiO_2 apex, as shown in Figs. 3.3.15–16 by the large arrow, and that the minimum pressure at which initial liquids could be alkalic is approximately 12 kb. Thus, the data available at present suggest that if all nonvolatile components are present, initial liquids will be quartz-normative at the lowest pressures and will become increasingly depleted in SiO_2 as pressure increases up to approximately 24 kb. At intermediate pressures up to about 12 kb, the liquids

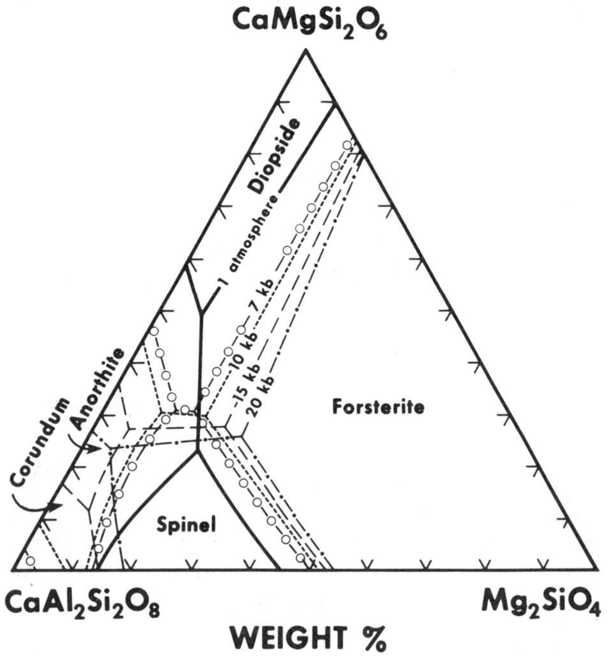

3.3.18 Composite diagram of the join Mg_2SiO_4-$CaAl_2Si_2O_8$-$CaMgSi_2O_6$ showing the effect of pressure on liquidus boundary lines. Labels of primary phase fields refer to the 15 kb diagram. After Presnall *et al.* (1978a).

would retain normative hypersthene but would contain normative nepheline at pressures above about 12 kb. With increasing pressure above the spinel lherzolite to garnet lherzolite transition at about 24 kb, initial liquids would become increasingly picritic and would move back toward hypersthene-normative compositions. The pressure at which the change would occur from nepheline-normative back to hypersthene-normative compositions is not known, but is probably in excess of 35 kb (Mysen and Kushiro, 1977). The overall trend from quartz-normative to olivine- and hypersthene-normative to nepheline-normative and back to hypersthene-normative picritic compositions is qualitatively similar to the trend proposed by O'Hara (1965, 1968a).

The discussion so far has been concerned for the most part only with primitive or fertile peridotites, and with the compositions of initial liquids, those produced near the solidus. Liquid compositions also change significantly as a function of temperature above the solidus. A depleted peridotite from which basaltic liquid had previously been extracted would produce liquid of a different composition if it was remelted, and the solidus temperatures would be higher (Fig. 3.3.6).

The information needed for the forward approach to basaltic volcanism could be plotted in a three-dimensional diagram, with axes for pressure, temperature, and percent melting. The result would be a surface on which changes in liquid composition could be mapped. There would be a different surface for different source peridotites, varying from fertile to depleted.

The changes in percentage and composition of liquid produced through the melting interval at constant pressure can be illustrated in detail in the system Mg_2SiO_4-$CaMgSi_2O_6$-SiO_2. Figure 3.3.14 gives the phase relationships at 20 kb. Assume that the undepleted mantle source consists of 45% forsterite, 28% enstatite (b), and 27% diopside (a), with bulk composition given by the point m. The pyroxene compositions a and b are those in equilibrium at 20 kb with the peritectic liquid p. These proportions are similar to those listed by Carter (1970, Table 2A) for undepleted mantle beneath Kilbourne Hole, New Mexico. During equilibrium fusion, the liquid path is the line p-n-m, and the crystal path (Presnall, 1969) is the dashed line m-r-Mg_2SiO_4.

Melting begins at 1640°C with the formation of liquid p, and the same liquid is produced at constant temperature until 30% of the mixture has melted. Separation of liquid at this stage would leave a crystalline residue consisting of forsterite, enstatite, and diopside. When all of the diopside a has melted, the temperature increases as the liquid changes composition along the field boundary p-n, with composition becoming enriched in Mg/Ca. During this stage of fusion, which continues up to 62%, the crystalline residue consists of forsterite and enstatite, with the proportion of enstatite decreasing as fusion proceeds. For very large amounts of fusion in excess of 62%, the liquid path would move from n toward m with increase in temperature and dissolved forsterite component. During this stage, the crystalline residue would be pure forsterite.

The curve for percent melting as a function of temperature has changes in slope associated with the disappearance of each mineral in the crystalline residue. The curve is isothermal for 30% melting, followed by a change in slope as the liquid composition leaves point p along p-n, with another change in slope at n. For a natural peridotite, the effects of solid solution in the

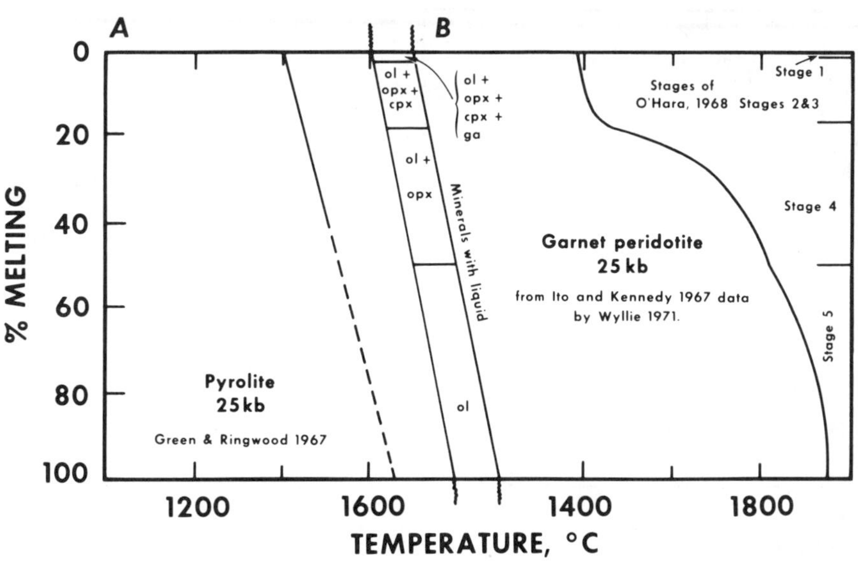

3.3.19 Estimated fusion curves for dry peridotite, showing the percentage liquid produced at successive temperatures above the solidus, and the minerals coexisting with the liquid in different parts of the melting interval (after Wyllie, 1971a). **a.** Dry curve from Green and Ringwood (1967b). **b.** Dry curve estimated by Wyllie (1971a, p. 204) from partial experimental data by Ito and Kennedy (1967), see Fig. 3.3.5a, with stages according to O'Hara (1968a).

minerals would be to change the 30% fusion corresponding to point p in Fig. 3.3.14 from constant temperature through a temperature interval, whose extent is a subject of debate.

Two comprehensive, experimentally based petrogenetic schemes relating temperature and percentage melting of peridotite to liquid composition were compared by Wyllie (1971a, p. 196–208). These schemes were developed by O'Hara (1965, 1968a) and Green and Ringwood (1967a, b; see also Ringwood, 1975, Chapter 4). Green and Ringwood (1967a, b) developed a model for the fusion of pyrolite estimating the percentage of melting as a linear function of temperature above the solidus and assigning compositions to various percentages of liquid produced. An example of their scheme is shown for 25 kb in Fig. 3.3.19a. Presnall (1969) has emphasized, however, that in simplified systems such as Fig. 3.3.14 described above, the percentage of melting is not a linear function of temperature, and would not be expected to be a linear function in more complex natural compositions. Wyllie (1971a) estimated the stepped curve shown in Fig. 3.3.19b, based on the experimental data of Ito and Kennedy (1967) (see Fig 3.3.5b), and on consideration of the shapes of liquidus surfaces and field boundaries in simpler systems. The steps correspond also to what O'Hara (1968a) referred to as stages of melting. Subsequently, Green (1970b, 1971) agreed that a curve with changes in slope is more likely.

Mysen and Kushiro (1977) conducted melting studies of two natural garnet-peridotite nodules using ^{14}C tracer to determine accurately the proportions of liquid. The results for one specimen at 20 and 35 kb are reproduced in Fig. 3.3.20. They were interpreted in terms of stepped curves (although smooth curves with changes in slope could be drawn through the error boxes). Mysen and Kushiro noted that the initial melting step closely resembled the isobaric invariant melting of a synthetic composition (similar to the melting described above for composition m in Fig. 3.3.14).

Jaques and Green (D. H. Green, pers. comm., 1979) have completed experiments on the melting of anhydrous pyrolite and the Tinaquillo lherzolite composition, "eliminating problems of starting material and coping with the Fe-loss problems." They found that the solid solution character dominates over the cotectic character at least for melting greater than 10%. "The stepped character of melting is so inconspicuous as to be undetectable," as shown by their results for pyrolite at 15 kb in Fig. 3.3.20.

Figure 3.3.21 is a sketch diagram showing some of the results obtained by Jaques and Green for part of the P-T-% melting surface for pyrolite. Contours are given for the percentage melting up to 50%, with other contours showing the amount of normative olivine in the liquid. The compositions of liquids given in this diagram may be compared with earlier results and conclusions reviewed above for initial liquids, and below for liquids within the melting interval.

At intermediate pressures (about 12–35 kb), studies of both simplified and natural compositions have indicated that initial liquids are alkalic. With increasing amounts of equilibrium fusion, the liquids become tholeiitic (Green and Ringwood, 1967a, b Kushiro, 1968, 1972b, 1973a, b; Mysen and Kushiro, 1977). In addition, Green and Ringwood (1967a, b) noted a shift

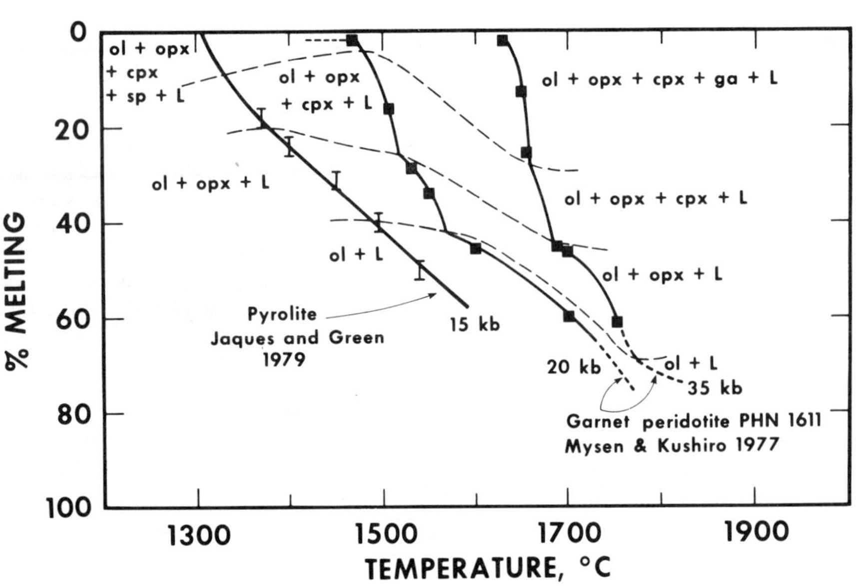

3.3.20 Fusion curves for anhydrous peridotite at 20 and 35 kb (Mysen and Kushiro, 1977, Fig. 3), compared with curve for pyrolite at 15 kb (Jaques and Green, manuscript in preparation, from D. H. Green, pers. comm., 1979; see legend to Fig. 3.3.21).

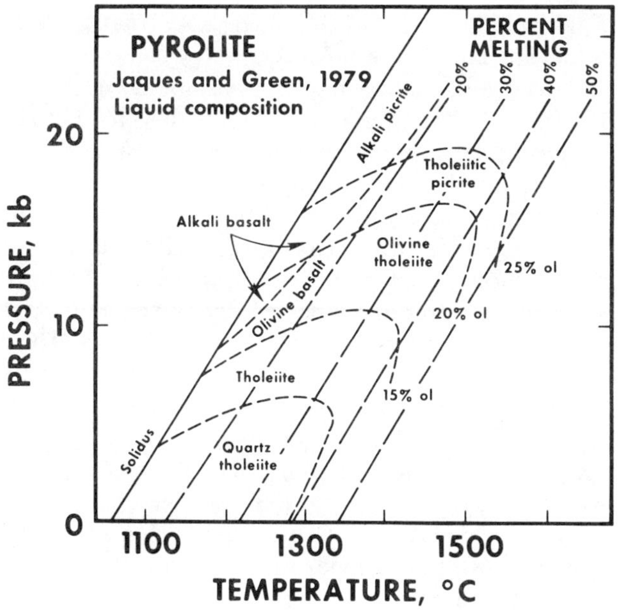

3.3.21 Anhydrous melting of pyrolite composition, showing percent melting contours, and the change in composition of liquids as a function of pressure and temperature; the normative olivine content of liquids is shown by contours. This is based on a sketch diagram kindly provided by D. H. Green from a manuscript by A. L. Jaques and D. H. Green, "Anhydrous melting of peridotite compositions at 0–15 kb and the genesis of tholeiitic basalts" (D. H. Green, pers. comm., 1979).

produced by 40% melting at 27 kb toward more olivine-rich (picritic) liquids. Similarly, Mysen and Kushiro (1977) found that at 20 kb, nepheline normative liquids are produced by less than 2% fusion, but with increasing amounts of fusion, the liquid compositions change from olivine tholeiite through picrite to komatiite. Microprobe analysis of liquids produced at 35 kb was hampered by the presence of quench crystals, but Mysen and Kushiro (1977) suggested that initial liquids (up to 25%, in equilibrium with garnet) resemble alkali picrite and that liquids produced by greater degrees of fusion range from olivine tholeiite through picrite to komatiite, as observed at 20 kb. Phase relations in the system CaO-MgO-Al_2O_3-SiO_2 (Figs. 3.3.14–18) considered together with probable shifts in the phase boundaries produced by the addition of other components (Presnall *et al.*, 1978a, 1979), also indicate that larger amounts of equilibrium fusion would produce liquids that not only shift from alkalic to tholeiitic compositions but become richer in olivine.

With increasing pressure, the composition of the initial liquid would vary, as described above, from tholeiite to alkalic basalt to hypersthene normative picrite. Very large amounts of fusion, however, would tend to suppress differences among primary magmas produced at different pressures. That is, the liquid will become richer in olivine and approach the composition of the source regardless of the pressure at which fusion occurs.

Peridotite with H_2O and CO_2

In a volatile-free peridotite mantle, the only liquids expected as primary magmas are tholeiitic basalts, alkali olivine basalts, picrites and komatiites, as discussed in the previous section. However, if volatile components are present in the source, other types of primary magmas may be produced. If H_2O is present, phase boundaries are shifted toward more silica-rich compositions, far enough to allow the production of andesite as a primary liquid. If CO_2 is present, phase boundaries are shifted in the opposite direction away from silica, reaching compositions equivalent to the silica-undersaturated liquids such as melilitite, kimberlite, and carbonatite. The main thrust of this book concerns the abundant basalt types, but the possibility that andesites and highly silica-undersaturated magmas may be produced directly from a volatile-enriched peridotite mantle justifies some discussion.

The compositions of liquids generated at the solidus surface for peridotite-H_2O-CO_2 have been reviewed by Mysen and Boettcher (1975b), Eggler (1978), and Wyllie (1978a, 1979). No definitive scheme emerges from the available data, and Fig. 3.3.22 is Wyllie's (1979a) attempted compromise which illustrates the major trends. It will serve as a basis for comparison of the various experimental results discussed. Figure 3.3.22c is a projection of the divariant solidus surface from Fig. 3.3.12, with vapor-phase contours omitted, and with the chemical variation of near-solidus liquids shown by lines representing changes in the normative composition. Figures 3.3.22a and b show the P-X^V projection of the solidus surface, giving the compositions of the liquids as a function of pressure and vapor-phase composition.

As discussed in section 3.3.2, the solidus surface is traversed by a series of vapor-buffer lines involving dolomite, amphibole, and phlogopite. With excess of $H_2O + CO_2$, i.e., with enough to complete hydration and carbonation of the peridotite (Figs. 3.3.7, 3.3.8, and 3.3.11a), the vapor phase is not buffered, and melting may begin anywhere on the surface. The dolomite-peridotite vapor-buffer line, Q_1N, cannot be crossed if the CO_2 content is less than 5%, as shown by the shaded area in Fig. 3.3.22 (see discussion of Figs. 3.3.11–12). Similarly, if the H_2O content is less than 0.4%, the amphibole-peridotite vapor-buffer line AaB in Fig. 3.3.22 becomes effective, and no vapors can exist in the shaded area within AaB in Fig. 3.3.22 (compare the buffer line aa′ in Figs. 3.3.13c and f).

The compositions of liquids near the solidus of peridotite without volatile components are given in Fig. 3.3.21. Figure 3.3.22 shows how the addition of CO_2, or H_2O, or mixtures of CO_2+H_2O lowers the solidus temperature and changes the liquid compositions.

Consider first the effect of CO_2. At pressures to more than 20 kb, the solubility of CO_2 is low, and the CO_2-saturated peridotite solidus is depressed only about 100°C below the peridotite solidus (Fig. 3.3.11), compared with more than 400°C for the H_2O-saturated solidus. Eggler (1974, 1978) studied the synthetic peridotite assemblage forsterite + enstatite + diopside in the system Na_2O-CaO-Al_2O_3-MgO-SiO_2, and established that with increasing pressure in the presence of CO_2, the near-solidus liquid becomes more alkalic and depleted in SiO_2, in comparison with the results given for peridotite in Fig. 3.3.21. Dramatic changes occur above about 20 kb. The forsterite-enstatite-normative liquid becomes larnite-normative as the liquid migrates towards the lower-temperature silicate-carbonate liquids (Wyllie and Huang, 1975, 1976). The CO_2-saturated solidus drops through about 300°C, and the solubility of CO_2 increases to high values as the liquid approaches the invariant point Q_1. Estimates of CO_2 contents of the liquid at Q_1 are 20% (Eggler, 1976), 27% (Eggler, 1978), or 35–40% (Wyllie and Huang, 1975, 1976). The liquid is carbonatitic.

According to these results, as shown in Fig. 3.3.22, the effect of CO_2 on near-solidus liquids is to shift the compositional boundaries down to somewhat lower pressures, with minor changes in composition until the pressure exceeds about 20 kb, where the composition changes within a narrow pressure interval to that of a carbonatitic liquid (coexisting with dolomite). According to results reported by Wendlandt and Mysen (1978) on a garnet peridotite nodule with CO_2, using ^{185}W as a radioactive tracer to determine the degree of melting, carbonatitic liquids are produced at 30 kb up to at least 20% melting, but at 15 kb the analyzed compositions of quenched liquids are "approximately tholeiitic at temperatures near the solidus." The weight of evidence from the other experiments outlined above suggests that at 15 kb, the tholeiitic compositions are developed not at the solidus, but at temperatures somewhat higher than the generation of nepheline-normative liquids at or near the solidus (e.g., Fig. 3.3.21).

The effect of H_2O on liquid compositions is less dramatic, but more complicated. The solubility of H_2O reaches high levels at moderate pressures, in contrast to CO_2, and the largest effect occurs at these pressures, where the tholeiitic liquids of Fig. 3.3.21 become ande-

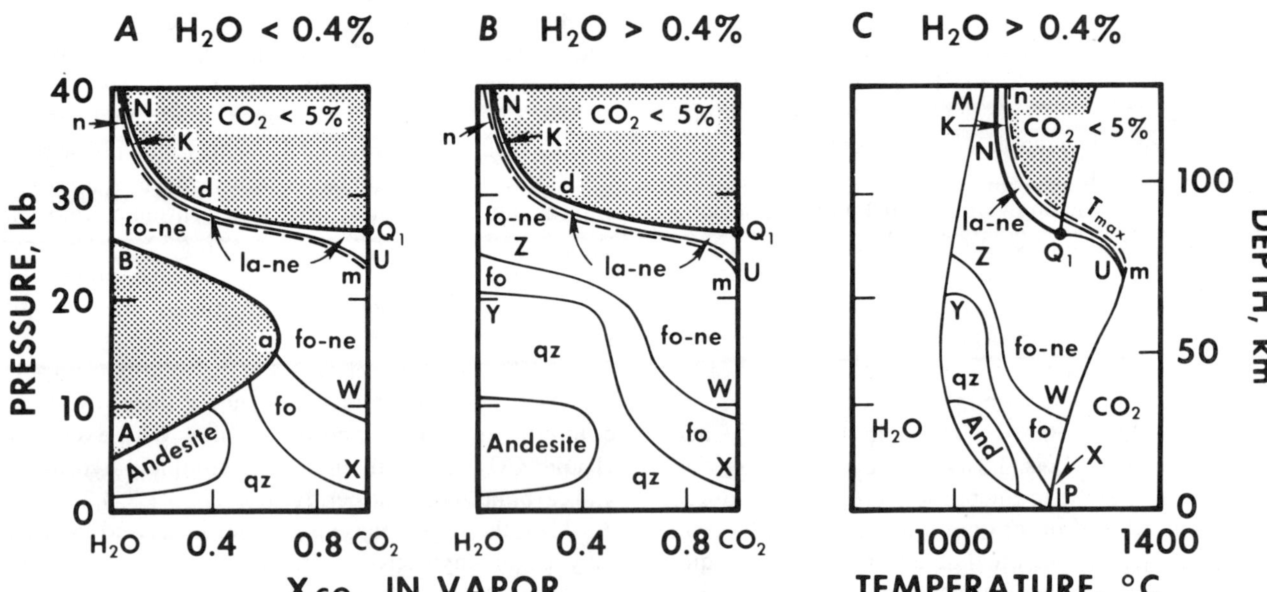

3.3.22 Projections of the solidus surface for peridotite-CO_2-H_2O (compare Figs. 3.3.12, 3.3.13c and f) with tentative boundaries showing the changes in normative composition of the near-solidus, vapor-saturated liquids. Conditions in the shaded areas cannot be achieved for the stipulated conditions. Abbreviations: fo = forsterite; ne = nepheline; la = larnite; qz = quartz; and = andesitic; fo-ne = forsterite-nepheline-normative liquid; la-ne = larnite-nepheline-normative liquid; (after Wyllie, 1979).

sitic, as shown in Fig. 3.2.22. With increasing pressure, the relative SiO_2-enrichment of the liquid decreases, and the quartz-normative liquid becomes forsterite-normative at some pressure which remains a topic for debate (Nicholls and Ringwood, 1973; Mysen and Boettcher, 1975b; Nehru and Wyllie, 1975; Green, 1976a).

O'Hara (1965) appears to have been the first to use experimental data (very limited) to support the suggestion that andesites might be generated as partial melts from a hydrous peridotite mantle, but this was omitted from his subsequent petrogenetic model (O'Hara, 1968a). Kushiro (1972a) studied several simplified systems under H_2O-saturated conditions, and concluded that andesitic or dacitic liquids could be produced by partial fusion of peridotite at depths down to about 80 km (25 kb). In further support of this conclusion, Kushiro *et al.* (1972a) gave an analysis of a glass produced by less than 20% melting of a spinel lherzolite at 26 kb, 1190° C, which they argued was broadly andesitic (Table 3.3.3), although it was enriched in CaO and depleted in MgO and iron in comparison to andesites. Mysen and Boettcher (1975b) melted peridotite in the presence of excess H_2O, and their analyses of the quenched glasses corresponded to aluminous, silica-rich liquids between 7.5 and 17 kb, and at temperatures up to 200° C above the solidus. They concluded that these liquids, which contained 10–16% dissolved H_2O, resembled andesites and dacites, and they suggested that andesites could be produced from peridotite in the presence of H_2O at pressures up to at least 25 kb.

In contrast, Nicholls and Ringwood (1972, 1973), Nicholls (1974), Green (1976a), and Ringwood (1977a) argued that, although the presence of water increases the silica content of partial melts, the effect is not as great as that indicated by Kushiro, Mysen, and Boettcher, and that andesitic liquids cannot be generated as primary liquids from wet peridotite at pressures greater than about 10 kb. Further details concerning the various arguments are contained in a discussion paper by Mysen *et al.* (1974), and reviews of the experimental problems which may have contributed to the controversy were given by Nehru and Wyllie (1975) and Green (1976a).

Green (1973a) published the compositions of all phases obtained in H_2O-saturated melting experiments on the material pyrolite III less 40% olivine. Because of changes in Fe content of the charge during experimental runs and crystallization during quenching, Green concluded that glass compositions were not reliable estimates of equilibrium liquid compositions, and he argued that liquid compositions should be calculated from a knowledge of the amounts and compositions of crystalline phases. He concluded that the SiO_2 content of liquid obtained from pyrolite is never likely to exceed 56 wt. %, and thus will never be dacitic (contrast Kushiro *et al.*, 1972a). With additional experimental data in hand, Green (1976a) concluded that quartz tholeiites and more siliceous liquids cannot be equilibrium partial melts from pyrolite under any conditions at pressures greater than 20 kb. At lower pressures, H_2O-saturated melting of peridotite could produce magnesian quartz tholeiites.

Nicholls (1974; Nicholls and Ringwood, 1972, 1973) approached the problem from two directions (H_2O-saturated): melting the pyrolite less 40% olivine composition, and determining the liquidus relationships of andesite and basaltic andesite compositions olivine components added. He demonstrated convincingly that at pressures above 15 kb, andesitic liquid (about 60% SiO_2) cannot be produced by direct partial melting of peridotite. The liquids produced under these conditions have 5–10% normative olivine, and a maximum of 56% SiO_2. At pressures below 10 kb, Nicholls (1974) found that more siliceous liquids could be generated, but that they differed from most natural andesites in having much higher MgO and lower alkalies.

The most recent study of the effects of H_2O on melting relationships of peridotite is that of Mysen and Kushiro (1977), who compared anhydrous and H_2O-undersaturated melting relationships at 20 kb. Inspection of their Table 3 in conjunction with their Figure 2 reveals that hydrous liquids in equilibrium with olivine and enstatite are considerably richer in normative olivine than corresponding anhydrous liquids, a result in conflict with the findings of almost all other studies (e.g., Kushiro *et al.*, 1968), which indicate that water causes the olivine-enstatite phase boundary to shift away from olivine toward silica. Thus, a serious problem with their data appears to exist independent of the comments and correction by Presnall *et al.* (1978b). It seems that the last word on the effect of H_2O on phase boundaries has yet to be written.

The major changes in liquid composition at pressures up to Q_1 (Fig. 3.3.22) are caused by changes in the relative sizes of the liquidus fields for olivine and orthopyroxene. Kushiro (1972a) showed that whereas pressure alone causes the stability field of olivine to shrink relative to that of orthopyroxene, H_2O under pressure causes expansion of the olivine field. Eggler (1974) demonstrated that the effect of CO_2 under pressure is to expand the primary phase field of orthopyroxene relative to olivine. Therefore, in the system peridotite-CO_2-H_2O, the compositions of liquids coexisting with olivine and orthopyroxene vary systematically as a function of pressure and CO_2/H_2O, across the divariant solidus surface.

There are only two relevant experimental studies on the compositions of glasses quenched from liquids with $CO_2 + H_2O$. Eggler (1975) studied forsterite-enstatite at 20 kb, and Mysen and Boettcher (1975a,b) studied four peridotite compositions from 10–20 kb. These results indicate that near-solidus liquids between 10 and 20 kb are quartz-normative for vapors up to about 50 mole fraction CO_2. In the presence of vapors with higher CO_2/H_2O, Mysen and Boettcher reported glasses that were nepheline-normative, and even larnite-normative at 10 and 15 kb. The liquid compositions represented in terms of their normative character on the solidus surface in Fig. 3.3.22 take into account these results, and those reviewed above for the effects of H_2O, and of CO_2, but they are not consistent with all results, because many of these remain controversial.

The line XY marks the fundamental change from quartz-normative to olivine-normative (Fig. 3.3.22). The small area for andesitic liquids is based on the results of Nicholls (1974) with H_2O, but Mysen and Boettcher (1975b) would expand the andesite field upwards in pressure at least to the line YX. The line ZW for normative nepheline (Fig. 3.3.22) is quite arbitrary. The results of Mysen and Boettcher (1975b) suggest that ZW is close to XY between 10 and 20 kb. Kushiro *et al.* (1972a) and Mysen and Boettcher (1975b) would probably prefer to see point Z located at higher pressures. Wyllie (1978a) suggested that the change to larnite-normative liquids (Eggler, 1974) was associated with the change in distribution of H_2O and CO_2 between liquid and vapor, and therefore with the temperature maximum on the solidus surface. The boundary proposed is UK, located near the line mn. This is not consistent with the larnite-normative glasses at 10 kb reported by Mysen and Boettcher (1975b).

For H_2O and CO_2 contents less than 0.4% and 5%, respectively, the area of divariant melting is restricted by the vapor-buffer lines, as shown in Fig. 3.3.22a. The amphibole buffer AaB eliminates a large area for the development of quartz-normative liquids. At pressures higher than 25 kb, melting is controlled by the dolomite buffer, Q_1N. The near-solidus liquid along Q_1N is dominated by dolomite, and even in the presence of the buffered H_2O-rich vapor it is probably carbonatitic, with $Mg/Ca < 1$ (Wyllie, 1977, 1978). With phlogopite present in addition, Mg/Ca would probably be increased in the first liquid developed at the phlogopite-dolomite buffer line near Q_1N. Holloway and Eggler (1976) studied a garnet lherzolite with phlogopite and dolomite added, and estimated that the liquid developed in the vapor-absent assemblage at 30 kb was melilititic in composition. This could coexist with vapor buffered to high H_2O/CO_2, probably with little change in composition.

The data available on the compositions of liquids along Q_1N and at higher temperatures are as yet preliminary. For increasing amounts of fusion above Q_1N, with low contents of H_2O and CO_2 corresponding to upper mantle conditions, there is a very wide temperature interval through which liquids present in very small quantities can follow diverse paths through a complex system, with decreasing CO_2, H_2O, and alkalies, increasing SiO_2, and variable Ca/Mg. The liquids range from carbonatitic, through various high-alkali, low-SiO_2 compositions including kimberlites and melilitites, and eventually approach the composition of an alkali picrite at temperatures near the solidus in Fig. 3.3.21.

3.3.4 Phase relationships of abundant basalts

The influence of experimental studies on the historical development of petrogenetic theory concerning the origin of basalts was outlined in section 3.1.1. The study of haplobasaltic compositions in synthetic systems has provided important insights, as shown by accounts in the book by Yoder (1976), and in the extensive study by Presnall *et al.* (1978a, 1979) in the system CaO-MgO-Al_2O_3-SiO_2 (see section 3.3.2 and Figs. 3.3.14–18). In this section, we concentrate on experimental studies of natural basalt compositions.

In Appendix 3.1 we have attempted to document all the high-pressure studies (mainly at pressures up to 40 kb) of terrestrial basaltic compositions published through 1979. Selected 1-atm studies are also listed. Details of the experimental methods that should be known in order to evaluate experimental results are tabulated (see also section 3.2, and Table 3.2.1). The chemical compositions of the basalts that have been used in phase equilibrium studies are listed in Table 3.3.4.

Not all of the experimental melting studies on terrestrial basalt compositions that have been carried out during the past twenty years have direct relevance to problems of generation of primary magmas from the Earth's mantle. Most of the 1-atm studies were completed without reference to the philosophical framework given by the flow diagram of Fig. 3.1.4. One-atmosphere melting studies by Tilley and co-workers (reviewed by Thompson, 1973) were interpreted in terms of liquidus temperature and Mg/Fe to provide an estimate of the parental magma composition in a given suite of basalts. These and other 1-atm results have served to identify the liquidus phases and to establish the probable crystallization sequences, and to determine the effect of oxygen fugacity on crystallization sequences, as illustrated in Fig. 3.3.23. Dynamic crystallization studies at 1 atm have not yet been widely applied to the identification of natural liquid compositions, as advocated in the flow diagram (Fig. 3.1.4).

One-atmosphere studies could provide much of the basis for selection of samples to be used in high-pressure studies for insights into magma genesis, as discussed in connection with the association of basaltic compositions with boundaries of multiple saturation in Fig. 3.3.2. However, with the exception of the experiments of Bender *et al.* (1978) and Green *et al.* (1979) which are described in section 3.3.8 as examples of the inverse approach, detailed 1-atm studies have not been conducted as a prelude to high-pressure studies of specific terrestrial basaltic compositions. Instead, the selection of compositions for high-pressure study has been based on other criteria (particularly Mg-value), without rigorous experimental confirmation that the compositions represent erupted liquids. Several investigated compositions, such as the olivine tholeiite used by Green and Ringwood (1967a) and the basanite used by Green (1973b), are not naturally-occurring magmas but inferred or model compositions beyond the extremes of variation of natural magmas in a particular province, and evaluation of such studies must take into account the necessarily subjective means of composition selection.

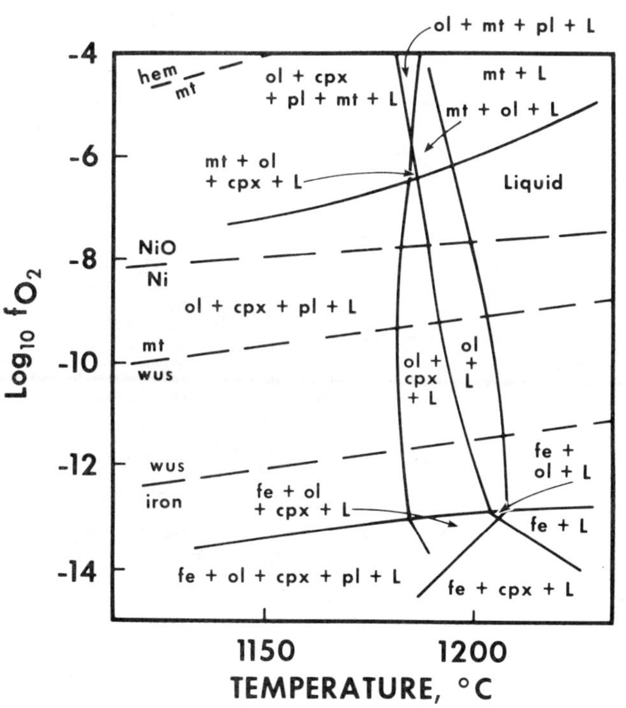

3.3.23 Generalized diagram showing the crystallization sequence of a typical basaltic composition at 1 atm as a function of oxygen fugacity (Biggar, 1974). Dashed lines are oxygen buffers. Similar diagrams for specific basalt compositions are given in Biggar (1974), Hill and Roeder (1974) and Thompson (1975).

Olivine tholeiites

Olivine tholeiites are by far the most voluminous basaltic magmas erupted on the Earth. Three major suites can be distinguished on the basis of tectonic setting: seafloor olivine tholeiites, ocean island olivine tholeiites and continental flood (plateau) olivine tholeiites. The more primitive examples in these suites are reasonably similar in major element chemistry, so that experimental studies would be expected to yield broadly similar conclusions regarding the physical conditions of their origin. The three suites differ, however, in major element variation (see Chapter 1), suggesting that the detailed source chemistry, source mineralogy and the processes of evolution of primitive mantle partial melts in these three environments may be different.

We have previously noted the low contents of volatile components dissolved in the tholeiites (section 3.3.3). This fact establishes the relevance of melting experiments with dry basalts.

Oceanic tholeiites are discussed in considerable detail in the case study of the inverse approach to basalt petrogenesis in section 3.3.8. Experiments by Bender *et al.* (1978) on a primitive FAMOUS basalt and by Kushiro (1973a) and Fujii and Kushiro (1977) on less primitive olivine tholeiites (Mg-values 57–64) are interpreted in terms of a close approximation to multiple saturation of the liquidus with olivine, calcic pyroxene, sub-calcic pyroxene, and plagioclase, at 8–10 kb, as illustrated in Fig. 3.3.24. However, following their own work on similar primitive oceanic tholeiite, Green *et al.* (1979) expressed doubts that orthopyroxene saturation had really been demonstrated, and concluded that these basalts had lost 5–20% olivine since their segregation from the source region. When the olivine is recombined with the primitive ocean tholeiite composition, a multiple-saturation point on the liquidus involving olivine, orthopyroxene, and clinopyroxene is discovered at about 20 kb. Green *et al.* (1979) thus endorsed the model of O'Hara (1968b) in which tholeiitic picrite segregates from the mantle at depths greater than 60 km. The petrogenetic discussion associated with these experimental results will be considered in sections 3.3.7 and 3.3.8.

Table 3.3.4 Compositions and norms of selected complex terrestrial basaltic and kimberlitic compositions used in experimental studies.

	1	2	3	4	5	6	7	8	9	10	11	12
SiO_2	49.11	46.95	48.2	49.7	50.0	50.3	52.9	48.21	49.93	45.91	46.53	45.39
TiO_2	2.51	2.02	0.73	0.72	1.64	1.7	1.5	2.99	1.34	0.94	2.28	2.52
Al_2O_3	12.74	13.10	16.3	16.4	15.0	17.0	16.9	14.38	16.75	17.19	14.31	14.69
Cr_2O_3	-	-	0.05	0.07	-	-	-	-	-	-	0.06	-
Fe_2O_3	3.23	1.02	-	-	-	1.5	0.3	-	-	2.33	3.16	1.87
FeO	8.40	10.07	8.92†	7.9†	11.2†	7.6	7.9	13.55†	11.40†	7.67	9.81	12.42
MnO	0.17	0.15	0.25	0.12	0.22	0.1	0.2	0.21	0.18	0.22	0.18	0.18
MgO	10.31	14.55	10.7	10.1	8.40	7.8	7.0	7.18	7.59	7.48	9.54	10.37
CaO	10.73	10.16	12.0	13.1	10.6	11.4	10.0	9.15	9.33	13.54	10.32.	9.14
Na_2O	1.97	1.73	1.95	2.0	2.70	2.8	2.7	2.70	2.92	1.63	2.85	2.62
K_2O	0.49	0.08	0.09	0.01	0.13	0.2	0.6	0.73	0.37	0.14	0.84	0.78
P_2O_5	0.27	0.21	0.06	-	-	-	-	0.56	0.19	0.04	0.28	0.02
H_2O^+	0.07	-	-	-	-	-	-	-	-	1.78	0.08	-
CO_2	-	-	-	-	-	-	-	-	-	-	-	-
SUM	100.00	100.04	99.25	100.12	99.89	100.4	100.0	99.66	100.00	98.87	100.24	100.00
$\frac{100\,Mg}{Mg + \Sigma Fe}$	61.9	70.2	68.1	70.9	57.2	64.7	60.4	48.6	54.3	57.7	57.3	56.7
C. I. P. W. weight percent norms§												
qz	-	-	-	-	-	-	1.26	-	-	-	-	-
or	2.90	0.47	0.53	0.06	0.77	1.18	3.55	4.31	2.19	0.83	4.96	4.61
lc	-	-	-	-	-	-	-	-	-	-	-	-
ab	16.67	14.64	16.50	16.92	22.85	23.69	22.85	22.85	24.71	13.79	20.39	17.82
an	24.47	27.74	35.46	35.74	28.43	33.23	32.22	24.96	31.51	39.18	23.77	26.02
ne	-	-	-	-	-	-	-	-	-	-	2.02	2.36
ac	-	-	-	-	-	-	-	-	-	-	-	-
ns	-	-	-	-	-	-	-	-	-	-	-	-
ks	-	-	-	-	-	-	-	-	-	-	-	-
di	21.80	17.25	19.22	23.63	19.81	18.91	14.22	14.16	11.25	22.59	20.87	15.69
hy	23.81	12.52	8.90	10.68	14.28	11.74	22.63	15.51	15.39	6.94	-	-
ol	0.25	21.64	15.94	10.56	9.29	6.25	-	9.29	10.57	8.51	18.53	25.97
ln	-	-	-	-	-	-	-	-	-	-	-	-
mt	4.68	1.48	1.19	1.13	1.45	2.17	0.43	1.80	1.54	3.38	4.58	2.71
chr	-	-	0.07	0.10	-	-	-	-	-	-	0.09	-
ilm	4.77	3.84	1.39	1.37	3.11	3.23	2.85	5.68	2.54	1.79	4.33	4.79
ap	0.59	0.46	0.13	-	-	-	-	1.22	0.42	0.09	0.61	0.04
cc	-	-	-	-	-	-	-	-	-	-	-	-

Notes:
† Total Fe as FeO.
* CoO as analogue for total Fe.
§ Norms for compositions 3, 4, 5, 8, 9, 22, 24 calculated using $Fe_2O_3/FeO = 0.1$.

1. 1921 Kilauea olivine tholeiite (Fudali, 1965; Hill and Boettcher, 1970; Holloway and Burnham, 1972; Helz, 1973, 1976; Allen *et al.*, 1975; Allen and Boettcher, (1978). Similar compositions were used by Yoder and Tilley (1962) and Hill and Roeder (1974). **2.** Synthetic olivine tholeiite glass (Green and Ringwood, 1967a; Green, 1970a, b). **3.** FAMOUS olivine tholeiite glass 527-1-1 (Bender *et al.*, 1978). **4.** Synthetic olivine tholeiite glass modelled on DSDP3-18-17-1 (Green *et al.*, 1979). **5.** DSDP Leg 45 olivine tholeiite glass 395A-8-1-9 (Fujii and Kushiro, 1977). **6.** Synthetic abyssal olivine tholeiite glass (Green and Ringwood, 1968; Green *et al.*, 1967). **7.** Synthetic quartz tholeiite glass (Green *et al.*, 1967; T. H. Green and Ringwood, 1967; Green and Ringwood, 1968). **8.** Snake River Plain olivine tholeiite 59-P-13 (Thompson, 1975; Tilley and Thompson, 1970). **9.** High-alumina olivine tholeiite glass prepared from Buffalo Butte basalt NM5 (Cohen *et al.*, 1967; Yoder and Tilley, 1962). **10.** Soay high-alumina olivine tholeiite (Lambert and Wyllie, 1972; Stern and Wyllie, 1973, 1978; Stern *et al.*, 1975). **11.** Prehistoric Hualalai alkali olivine basalt (Yoder and Tilley, 1964; Nesbitt and Hamilton, 1970). A similar composition was used by Haygood *et al.* (1971), Allen *et al.* (1975), Allen

Table 3.3.4 *(Concluded)*

	13	14	15	16	17	18	19	20	21	22	23	24
SiO_2	46.12	44.7	44.19	41.20	49.1	40.37	38.57	39.4	42.6	55.14	43.98	37.68
TiO_2	1.81	2.9	2.75	3.20	2.17	4.38	2.79	2.9	5.3	2.58	1.85	2.16
Al_2O_3	13.94	11.7	13.26	16.35	15.4	14.90	11.71	9.8	8.3	10.35	9.91	5.11
Cr_2O_3	-	-	0.03	-	0.04	-	0.06	-	0.2	0.04	0.16	0.23
Fe_2O_3	1.95	1.8	2.82	2.87	1.11	3.30	5.21	1.2	0.2	3.27	3.06	-
FeO	10.46	10.5	8.79	11.40	9.0	7.95	7.78	10.3	7.8	0.62	8.10	(11.27)*
MnO	0.18	0.15	0.18	0.20	0.18	0.09	0.11	0.4	0.2	0.06	0.17	-
MgO	11.08	13.9	10.30	10.93	7.1	12.80	13.08	16.3	15.8	6.41	20.30	25.85
CaO	9.05	7.7	8.77	9.61	6.3	10.30	12.84	13.4	10.7	3.45	8.75	10.62
Na_2O	3.11	3.65	3.71	2.33	5.6	2.60	4.22	3.7	0.8	1.27	1.55	0.22
K_2O	0.57	2.0	2.02	0.86	2.9	2.05	1.20	1.4	7.1	11.77	1.74	0.91
P_2O_5	0.23	1.0	0.97	0.14	0.68	-	1.11	1.3	0.4	1.40	0.37	0.74
H_2O^+	1.49	-	0.86	0.38	-	0.90	0.59	-	-	1.23	0.47	-
CO_2	-	-	0.19	-	-	-	0.27	-	-	0.20	-	5.22
SUM	99.99	100.00	98.84	99.47	99.58	99.64	99.58	100.1	99.4	97.79	100.41	100.01
$\frac{100\ Mg}{Mg + \Sigma Fe}$	61.8	67.1	61.8	58.2	55.9	67.6	65.2	71.9	77.9	76.2	76.9	(80.3)*

C. I. P. W. weight percent norms§

	13	14	15	16	17	18	19	20	21	22	23	24
qz	-	-	-	-	-	-	-	-	-	3.51	-	-
or	3.37	11.82	11.94	5.08	17.14	2.01	-	-	-	56.51	10.28	5.38
lc	-	-	-	-	-	7.92	5.56	6.49	32.9	-	-	-
ab	21.22	11.07	13.28	7.73	20.33	-	-	-	-	-	4.49	1.86
an	22.39	9.63	13.56	31.62	8.32	22.93	9.47	6.00	-	-	14.94	10.27
ne	2.76	10.73	9.81	6.49	14.66	11.92	19.34	16.96	1.71	-	4.67	-
ac	-	-	-	-	-	-	-	-	0.58	0.93	-	-
ns	-	-	-	-	-	-	-	-	0.69	2.26	-	-
ks	-	-	-	-	-	-	-	-	-	3.62	-	-
di	17.16	18.24	18.51	12.34	15.49	22.26	26.61	18.75	16.65	6.00	20.84	4.03
hy	-	-	-	-	-	-	-	-	-	14.97	-	15.04
ol	24.83	28.20	18.97	25.29	16.38	18.60	18.15	32.70	26.46	-	35.72	44.11
ln	-	-	-	-	-	-	3.83	9.12	9.18	-	-	-
mt	2.83	2.61	4.09	4.16	1.61	4.78	7.55	1.74	-	-	4.44	1.49
chr	-	-	0.04	-	0.06	-	0.09	-	0.29	0.06	0.24	0.34
ilm	3.44	5.51	5.22	6.08	4.12	8.32	5.30	5.51	10.07	4.90	3.51	4.10
ap	0.50	2.19	2.12	0.31	1.49	-	2.43	2.84	0.87	3.06	0.81	1.62
cc	-	-	0.43	-	-	-	0.61	-	-	0.45	-	11.87

Notes:
and Boettcher (1978), and another similar composition (1801 Hualalai flow) was used by Helz (1973, 1976) and Tilley *et al.* (1965). **12.** Synthetic alkali olivine basalt glass (Green and Ringwood, 1967a; Essene *et al.*, 1970). **13.** Skye alkali olvine basalt 66018 (Thompson, 1974). **14.** Basanite glass prepared from Mt. Leura basanite 2650 + 10% Fo_{90} olivine (Green, 1973). **15.** Mt. Shadwell basanite 2164 (Arculus, 1975). **16.** Kakanui kaersutite eclogite K-14 (Merrill and Wyllie, 1975). **17.** Nepheline mugearite glass prepared from The Anakies (east) nepheline mugearite 2102 (Irving and Green, 1980). **18.** Kakanui kaersutite megacryst K-1 (Merrill and Wyllie, 1975). **19.** Honolulu olivine-melilite nephelinite (Allen *et* al., 1975; Allen and Boettcher, 1978). **20.** Synthetic olivine melilitite glass modelled on Laughing Jack Marsh olivine melilitite 2927 (Brey and Green, 1975, 1976, 1977). **21.** Synthetic biotite mafurite glass modelled on natural lava from Bufumbira, Uganda (Edgar *et al.*, 1976). A similar composition was used by Ryabachikov and Green (1978). **22.** Leucite Hills orendite A1770 (Barton and Hamilton, 1978, 1979b; Carmichael, 1967). **23.** Olivine ugandite lava from Bufumbira, Uganda (Edgar and Condliffe, 1978; Edgar *et al.*, 1980). **24.** Synthetic kimberlite analogue composition (oxide-silicate-carbonate-phosphate mix) modelled on average Lesotho kimberlite with all Fe replaced by Co (Eggler and Wendlandt, 1979).

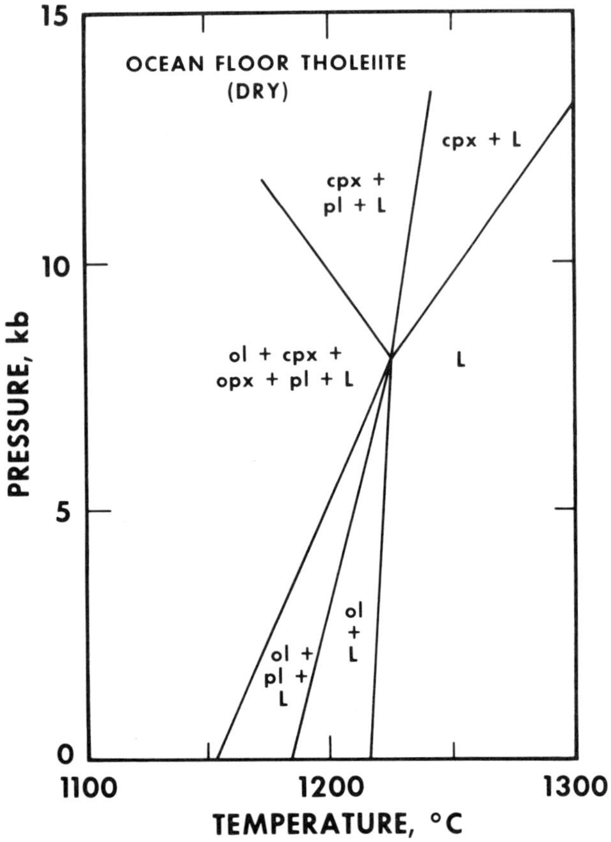

3.3.24 Anhydrous liquidus phase relationships for ocean floor basalt from Fujii and Kushiro (1977). The liquidus is simultaneously saturated with olivine, orthopyroxene, clinopyroxene and plagioclase at about 8 kb and 1250°C.

Chemical and petrographic work by Macdonald and Katsura (1961) and Murata and Richter (1966) identified a primitive Hawaiian olivine tholeiite composition (containing 22% normative olivine) inferred to be parental to the 1959–60 Kilauea Iki lavas. Green and Ringwood (1967a) determined the phase relationships for this model composition, with results given in Fig. 3.3.25. The liquidus is cosaturated with olivine and aluminous orthopyroxene at 11.5 kb under dry conditions, indicating that it could have been derived as a primary liquid by partial melting of peridotite at about 35 km depth leaving a residuum of olivine + orthopyroxene. Although the measured relative iron losses to platinum capsules reached about 30% (even in near-liquidus runs of 40–60 minutes duration), later experiments on the same composition using graphite capsules (Green, 1970a) confirmed olivine-orthopyroxene cosaturation between 9 and 13.5 kb. An illustration of how critically the inferred depth of origin depends upon bulk composition is provided by Green's (1970a) results for a slightly more magnesian olivine tholeiite composition, prepared by adding 5% olivine (Fo_{90}) to the original composition. This second composition also shows cosaturation with olivine and orthopyroxene, but at 16 kb (equivalent to about 50 km depth).

Little systematic experimental work has been done on continental flood tholeiites. Thompson (1975) studied an evolved tholeiite (Mg-value 49.1) from the Snake River Plain and concluded that it may be a product of fractional crystallization at about 8 kb from a more Mg-rich tholeiite parental liquid. The most primitive Snake River Plain (and Columbia Plateau) tholeiites have Mg-values of 65 and, if primary, could be partial melts from a mantle containing Fo_{85} olivine (slightly more Fe-rich than the Fo_{87}-Fo_{89} inferred for other olivine tholeiites and for nodule-bearing basanites). By analogy with dry melting relations for a Skye alkali olivine basalt (Fig. 3.3.26), Thompson (1975) suggested that the parental Snake River Plain magma might be derived from lherzolite at about 50 km depth (by larger degrees of melting than required to yield the Skye basalt). Alternatively, more primitive flood tholeiites may exist unsampled which could be primary melts

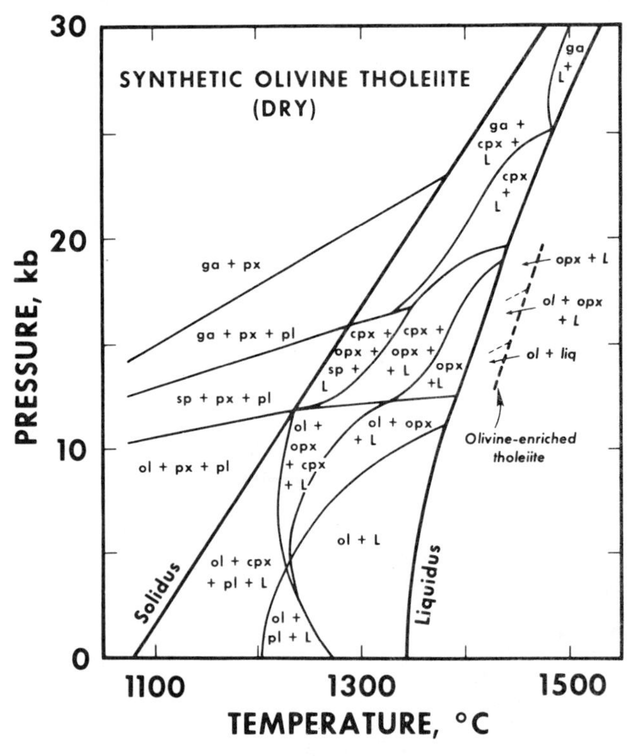

3.3.25 Anhydrous phase relations for Hawaiian olivine tholeiite (Green and Ringwood, 1967a; Green, 1970a). Liquidus mineralogy changes from olivine to orthopyroxene to garnet with increasing pressure.

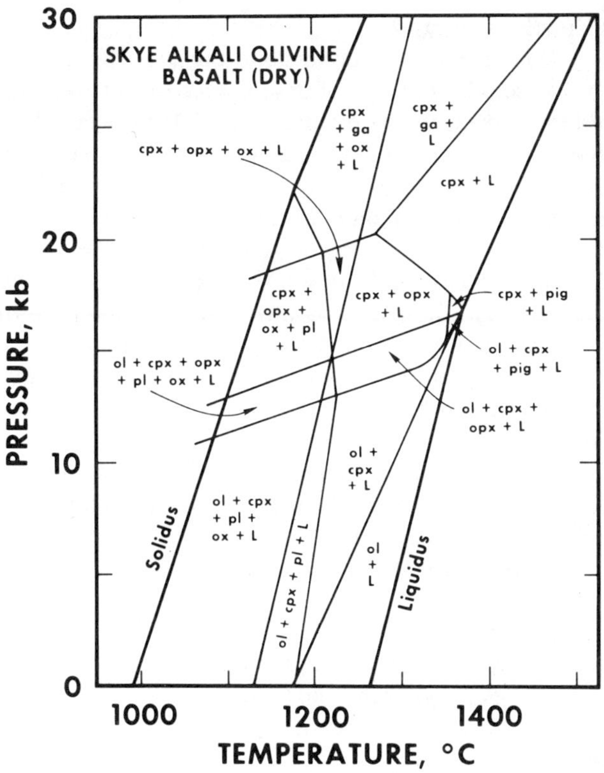

3.3.26 Anhydrous phase relations for alkali olivine basalt (Thompson, 1974). Liquidus is multiply saturated with olivine and high Ca-pyroxene (± pigeonite) at about 16 kb and 1380°C.

from a source similar in major element chemistry to that inferred for the FAMOUS sample.

Unusual olivine-rich lavas of the Karroo Province at Nuanetsi have been studied by Cox and Jamieson (1974). Although these tholeiitic lavas are not typical of the Karroo flood basalts (especially because of their very high MgO and K_2O contents), they are of interest because they contain megacrysts of aluminous orthopyroxene and olivine of probable high-pressure origin. Melting relations of several lavas were determined at 7 kb and 10 kb. Most samples (including one of the most magnesian, with Mg-value 79) show near-liquidus olivine followed closely by orthopyroxene at 7 kb and near-liquidus orthopyroxene (± calcic pyroxene ± olivine) at 10 kb. Cox and Jamieson (1974) suggested that the Nuanetsi suite may be derived from a parental K-rich magma (originally in equilibrium with olivine and orthopyroxene at about 12 kb, i.e., about 36 km depth), which underwent either polybaric fractional crystallization of olivine and orthopyroxene or polybaric reaction with harzburgite wall-rock during ascent to depths of 18 km (i.e., pressures of about 6 kb).

The conclusion that emerges is that the parental magmas of the three most voluminous basalt suites on Earth originated as primary liquids which segregated from the upper mantle at depths of at least 35–50 km. Near-liquidus experimental studies provide little information on the extent of partial melting involved, except that the residual olivine + subcalcic pyroxene mineralogy suggests that the degree of melting was sufficiently large to eliminate calcic clinopyroxene and any Al-rich phase presumed to have been present in the original source peridotite. Partial melting experiments on peridotite (section 3.3.3) and trace element data for tholeiites (e.g., Frey *et al.*, 1978) show that generation of primary olivine tholeiite liquids requires degrees of melting of 20–25%, with the exact composition of a particular batch of segregated liquid being a function of source bulk composition and residual mineralogy at the particular depth of segregation, as well as the exact degree of melting involved. Such relatively high degrees of melting ensure very low volatile contents in the partial melts, even if minor amounts of volatile-bearing phases were present in the original source. The subsequent evolution of a primary olivine tholeiite liquid is likely to be complex, involving such processes as polybaric crystal fractionation (e.g., O'Hara, 1965, 1968a) and magma mixing (e.g., Rhodes *et al.*, 1979) during ascent. The extent to which these modifying processes operate, and hence the extent of compositional diversity among erupted lavas, will be controlled in large part by tectonic factors, both in the plasticly-deforming asthenosphere and the relatively rigid lithosphere.

Alkali olivine basalts and transitional basalts

Green and Ringwood (1967a) conjectured that if the olivine basalt thermal divide of Fig. 3.3.1 was breached at pressures greater than about 12 kb (Yoder and Tilley, 1962), then liquids resembling alkali olivine basalts could be produced by fractional crystallization of orthopyroxene + olivine from olivine tholeiite at depths of about 40 km. Transitional basalts (just saturated with normative hypersthene or nepheline) were envisaged by Green and Ringwood as intermediate residual liquids of this process, and the occurrence of aluminous orthopyroxene megacrysts in some of them (e.g., Green and Hibberson, 1970; Merrill and Irving, 1977) is supporting evidence for high-pressure orthopyroxene crystallization from such liquids. The high-pressure fractional crystallization hypothesis is adequate for some alkali olivine basalts, but, because of the relatively rapid iron enrichment involved, this process cannot explain those examples with Mg-values as great as that of the assumed tholeiite parental liquid. In view of this problem, Green (1970a, p. 227) favored the idea that the more primitive alkali olivine basalts are primary liquids.

The most primitive sample studied experimentally is an aphyric alkali olivine basalt (Mg-value 65) from Skye, which Thompson (1974) found to be multiply saturated with olivine, calcic clinopyroxene and pigeonite at about 17 kb, as shown in Fig. 3.3.26. Thompson interpreted this lava as a primary partial melt of relatively Fe-rich (spinel) lherzolite at depths of about 50 km. Alternatively, if it suffered minor olivine loss during ascent, a more magnesian source lherzolite at slightly greater depths would be appropriate.

Spinel lherzolite xenoliths do not occur in the Hebridean alkali olivine basalts, but they do occur in more primitive examples, such as in the Newer Basalts of Victoria, Australia (Irving and Green, 1976). Frey *et al.* (1978) found that trace element data for the latter basalts and xenoliths are consistent with degrees of melting of 11–15% from a garnet-free lherzolite source. Such degrees of melting are sufficiently great that the volatile content of alkali olivine basalt liquids remains low enough for dry melting relations to be applicable, even though the content of volatile components is greater than in olivine tholeiites (e.g., Moore, 1970; section 3.3.3).

Eclogite

Yoder and Tilley (1962) experimented with several basaltic compositions at high pressures and demonstrated that they could all be converted into eclogite, consisting essentially of garnet and clinopyroxene, with minor amounts of hypersthene, quartz, or kyanite, depending upon the bulk composition of the basalt. Subduction of basaltic ocean crust introduces gabbro (or amphibolite) or eclogite as possible source rocks for magma generation at depth below the andesitic volcanoes of volcanic arcs. The basalts and andesites in this tectonic environment may therefore need special consideration in connection with the flow chart of Fig. 3.1.4, and source region constraints.

Experimental studies of the transition from basalt to eclogite and of related simple mineral reactions were reviewed by Wyllie (1971a, Chapter 5) and Ringwood (1975, Chapter 1). Figure 3.3.27 is a generalized phase diagram, based on the reviewed results. Attempts to determine the precise mineralogical changes occurring within the transition interval, and the slope of the transition at low temperatures, have been hampered by the sluggish reactions in the complex compositions. It is significant that compositions corresponding to many ocean tholeiites, without too much normative olivine, can be transformed into quartz eclogite (Green and Ringwood, 1967c; Ito and Kennedy, 1971).

The effect of H_2O

It is necessary to understand the effect of H_2O on the phase relationships of basalt, because petrogenetic models appeal to the involvement of H_2O released by dehydration of subducted ocean crust in the generation of magmas from the overlying mantle (section 3.3.7), and because Anderson (1975) has reported 3–5% H_2O dissolved in some basaltic magmas associated with subduction. Yoder and Tilley (1962) determined the effect of excess H_2O on several basaltic compositions up to 10 kb, including the olivine tholeiite and alkali olivine basalt shown in Fig. 3.3.28. These results have been extended to 30 kb, using data from the other sources cited.

The main effects of H_2O are to lower the melting temperatures, increase the melting interval, and introduce a hydrous mineral, amphibole. The amphibole masks the gabbro-eclogite phase transition (compare Figs. 3.3.27 and 3.3.28), but it breaks down in a complex pressure-sensitive reaction involving the formation of garnet, and the production of eclogite plus vapor. Phase boundaries of pyroxene and plagioclase are not given in Fig. 3.3.28, but the effect of H_2O is to lower the phase boundary for plagioclase to a greater extent than for other minerals; the boundary reaches the solidus at about 15 kb, where it becomes unstable by reaction to

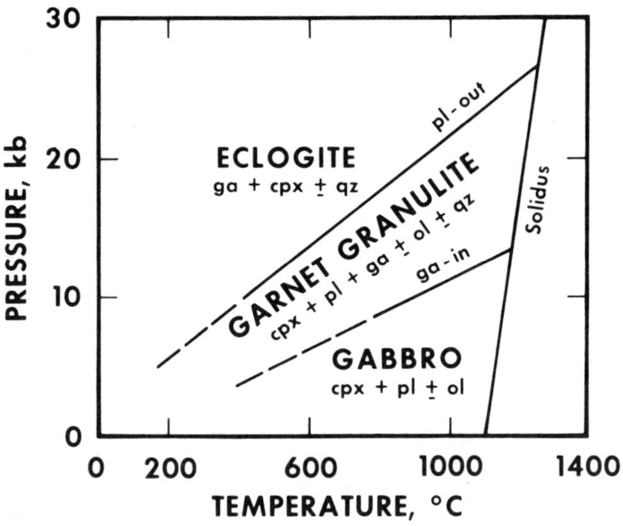

3.3.27 Generalized diagram illustrating the transformation of gabbro to eclogite. The width of the reaction interval and mineralogical details vary with bulk composition. There are changes in slope of the solidus at each phase change. The basic transition involves reaction of plagioclase and augite to garnet and omphacite, but a gabbro containing a small percentage of olivine can be converted into an eclogite containing quartz.

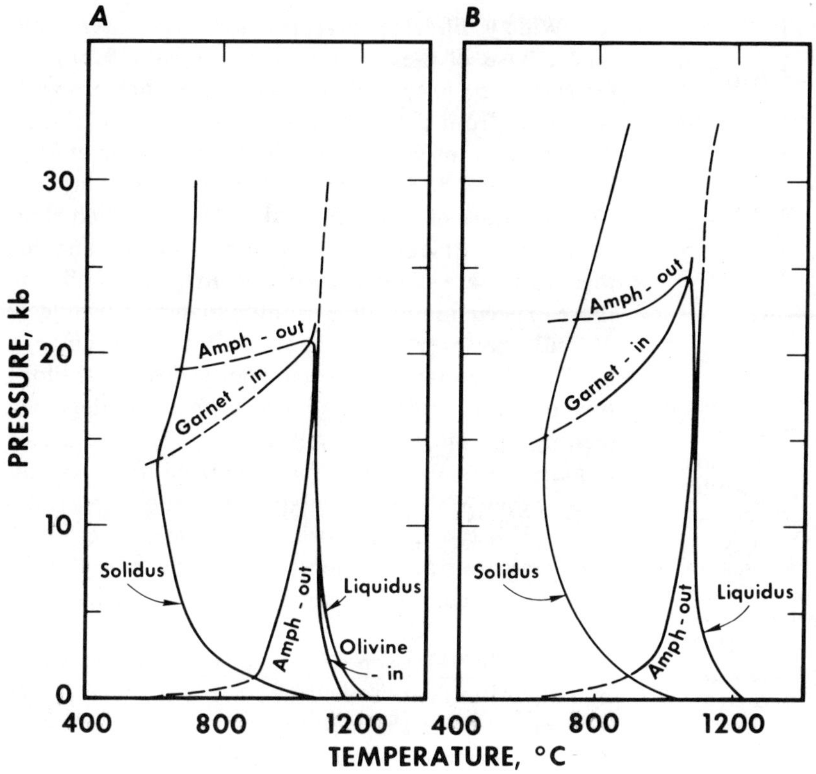

3.3.28 Selected phase boundaries for basalt with excess H_2O. **a.** Kilauea 1921 olivine tholeiite. **b.** Prehistoric Hualalai alkali olivine basalt. Compiled from Allen *et al.* (1975), Haygood *et al.* (1971), Helz (1973), Hill and Boettcher (1970), Holloway and Burnham (1972), Nesbitt and Hamilton (1970), and Yoder and Tilley (1962).

eclogitic minerals. Allen *et al.* (1975) determined the effect of oxygen fugacity on phase relationships between about 10 and 20 kb, and reviewed previous work at lower pressures.

There are only two conditions under which the excess-H_2O results are likely to be relevant to magmas. In subducted ocean crust at high pressures, melting of amphibolite or eclogite may begin near the solidus in the presence of hydrothermal fluids, but the magmas would not be basaltic in composition. In subvolcanic chambers, crystallizing basaltic magmas with initial H_2O contents higher than normal, because of the subduction environment, may reach H_2O-saturation.

More generally relevant are the phase relationships in the presence of small amounts of H_2O. Allen and Boettcher (1978) determined the crystallization sequences of the olivine tholeiite in Fig. 3.3.28a at lower and more realistic values of H_2O fugacity by using H_2O-CO_2 vapors, with oxygen fugacity buffered. They were especially concerned with amphibole stability in the range 10–20 kb.

Figure 3.3.29 shows the subsolidus phase fields and melting interval for H_2O-saturated gabbro (a high-alumina olivine tholeiite), and the H_2O-undersaturated liquidus surface connecting the H_2O-saturated liquidus with the dry basalt liquidus (compare Figs. 3.1.12, and 3.3.16–24). This is Wyllie's (1979) modification of a diagram published by Stern *et al.* (1975), based on data obtained at 10 kb and above by Lambert and Wyllie (1972), Stern and Wyllie (1973, 1978), and Huang (reported in Stern *et al.*, 1975), and estimated at lower pressures on the basis of published data on similar compositions. The contours on the surface give the liquidus temperatures for specific H_2O contents, and show the increase in H_2O solubility for saturated liquids as a function of pressure. The liquidus or near-liquidus minerals are shown. The addition of H_2O increases the maximum pressure at which olivine is a liquidus phase in this composition (Nicholls and Ringwood, 1973), but there is no area on this surface where the liquid exhibits multiple saturation with mantle minerals olivine + orthopyroxene + clinopyroxene.

3.3.5 Phase relationships and petrogenesis of picrites and komatiites

Tables 3.3.1 and 3.3.2 list the compositions and experimental details for one picrite and six komatiites for which full or complete phase relationships have been published. Phase diagrams are given in Fig. 3.3.30. High-pressure melting experiments on picrites and komatiites show an extensive field of olivine crystallization, extending to both higher temperatures and higher

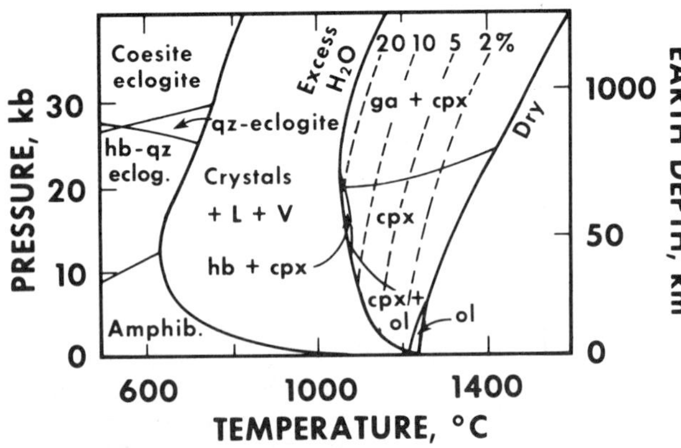

3.3.29 Basalt-H_2O phase relationships for a high-alumina olivine tholeiite (Lambert and Wyllie, 1972) compiled by Stern *et al.* (1975) and Wyllie (1979) from several experimental studies. The presence of excess H_2O produces amphibole (see Fig. 3.3.28) which masks the gabbro-eclogite transition (compare Fig. 3.3.27). The melting interval with excess H_2O corresponds to those in Fig. 3.3.28. The purpose of the diagram is to illustrate the H_2O-undersaturated liquidus surface, connecting the excess-H_2O liquidus (compare Fig. 3.3.28) and the dry liquidus (see Figs. 3.3.24–26). Dashed contour lines show the liquidus for specific H_2O contents, and they give the H_2O solubility as a function of pressure along the excess H_2O liquidus. The minerals stable at or near the liquidus surface are shown. Abbreviations: Amphib. = amphibolite, hb = amphibole, qz = quartz, eclog. = eclogite, cpx = clinopyroxene, ol = olivine, ga = garnet.

pressures than the olivine field in basaltic compositions (compare Fig. 3.3.30 with Figs. 3.3.24–26). Moreover, the pressure of multiple-phase saturation at the liquidus increases from basalt to picrite to komatiite. This is consistent with the experimental determination that the content of normative olivine in liquids derived by partial melting of peridotite increases as a function of pressure (Fig. 3.3.21). Primary liquids are picritic if segregated at pressures greater than 15–20 kb.

Figure 3.3.30a indicates that typical ultrabasic komatiites have high liquidus temperatures at the Earth's surface. Arndt (1976) deduced experimentally that a sample with 25% MgO and Mg-value of 78 has a dry liquidus temperature of about 1500°C at 1 atm. Green *et al.* (1975a) determined the liquidus for a komatiite both dry, and H_2O-saturated. Green (1975) deduced the sequence of crystallization at pressures greater than 40 kb to be olivine, then orthopyroxene, then garnet (compare Fig. 3.3.30a). Green *et al.* (1975a) and Green (1975) showed that extrusion temperatures would be as high as 1650°C for a sample with 33% MgO and Mg-value of 85. Green *et al.* (1975a) concluded that the natural magmas contained less than 0.2% H_2O.

Widely differing interpretations have been made on the basis of these results. Green *et al.* (1975a) and Green (1975) proposed that komatiites are probably generated from the same source as basalts at similar depths, and thus very extensive partial melting of up to 70% is required. Supporting their arguments are the chemical data of Sun and Nesbitt (1978), which show that incompatible element abundances are low, and that their ratios are commonly chondritic, i.e., similar to those expected in fertile peridotite mantle. It would be difficult, however, to retain such a high percentage of liquid in deep, partially melted mantle through the long time interval required to achieve such a high degree of melting. In contrast, Howells *et al.* (1975) interpreted komatiites as minimum melts derived from depths greater than 120 km, and suggested that ultrabasic liquids of this type are the primary magmas from which high-pressure fractionation produces the parental liquids for the common eruptive basaltic magma series (e.g., olivine tholeiites).

Arndt (1977) proposed an appealing model whereby komatiites are obtained by small degrees of partial melting at fairly shallow depths (<120 km) from a depleted peridotite, a source material which has already been subjected to a previous partial melting episode. This model is consistent with the phase equilibria and it also accounts for field relationships. Moreover, it appears to be consistent with geochemical constraints, because remelting of depleted peridotite should produce liquids with light REE depletions, as observed in many komatiites.

Bickle *et al.* (1977) also suggested that a depleted peridotite is the source for komatiites, but proposed a somewhat more complex petrogenesis. They suggested that during adiabatic uprise, the melt assimilated and digested approximately its own mass of garnet peridotite. This appears to be very unlikely from consideration of the constraints imposed by the thermal budget.

3.3.6 Phase relationships and petrogenesis of highly SiO_2-undersaturated magmas

The SiO_2-undersaturated basaltic rocks were discussed in section 3.3.1. It is evident that they cannot be primary melts from volatile-free mantle peridotite (section 3.3.3, Fig. 3.3.21), but the review in section 3.3.3 showed that some of these magmas could possibly represent primary liquids from mantle containing H_2O and CO_2. A few experiments have been completed on possible primary liquids to determine the conditions under which they might have equilibrated with mantle peridotite.

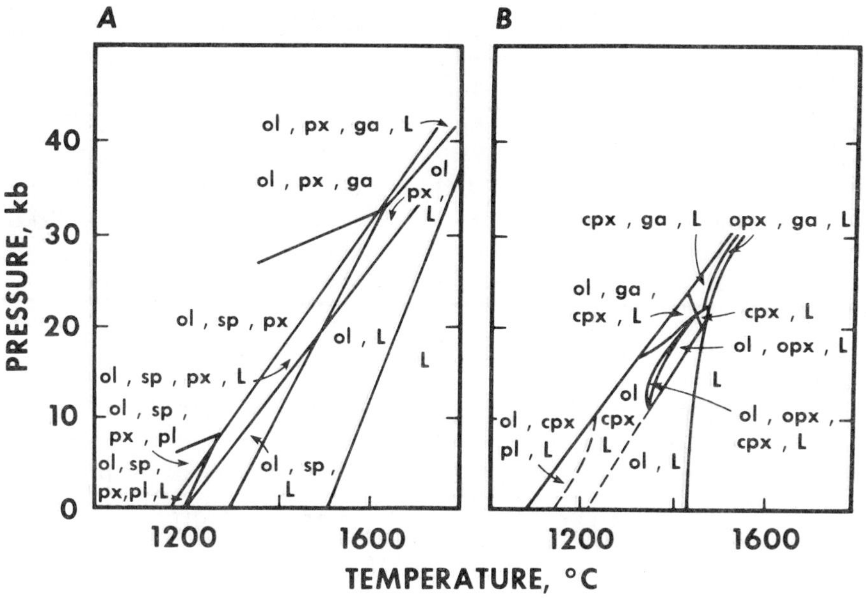

3.3.30 **a.** Phase relationships of komatiite, sample SA 3091 of Arndt (1976). **b.** Phase relationships for anhydrous picrite (Green and Ringwood, 1967a). Phase fields very approximate. Abbreviations as for Fig. 3.3.29, plus sp = spinel, pl = plagioclase, opx = orthopyroxene, px = pyroxene.

Bultitude and Green (1968, 1971) reasoned that H_2O was probably involved in the generation of basanites. This would be consistent with the explosive eruptive style of these magmas, and their common content of megacrysts of kaersutite and biotite, generally thought to be high-pressure crystallization products (e.g., Irving, 1974, 1977a). Initial experiments by Bultitude and Green (1968) were conducted in open capsules with low but uncontrolled H_2O activities. In his search for a basanite possessing liquidus minerals consistent with an origin by partial fusion of the mantle, Green (1973b) postulated that such magmas, formed at 25–30 kb, should contain about 25% normative olivine. Therefore, in constructing an experimental starting material, he added 10% olivine components to a natural basanite. His experiments in sealed Pt and Ag-Pd capsules with a range of H_2O contents yielded the H_2O-undersaturated liquidus surface shown in Fig. 3.3.31 (compare Fig. 3.3.29 for a basalt). The liquidus is saturated with the mantle minerals olivine, orthopyroxene, clinopyroxene and garnet within a limited area between 25–30 kb, 1200–1300° C, and 2–7% H_2O. Green developed from this the specific model for the generation of basanite by 6% partial melting at about 100 km depth of a garnet lherzolite containing 0.2–0.4% H_2O. The liquid, containing 2–7% H_2O, "could have dropped olivine prior to picking up mantle xenoliths at 10–20 kb," and thus could have attained the composition of the basanite erupted at the surface.

Merrill and Wyllie (1975) determined the H_2O-undersaturated liquidus surface, corresponding to Fig. 3.3.31, for a kaersutite eclogite nodule from Kakanui, with bulk composition corresponding to that of

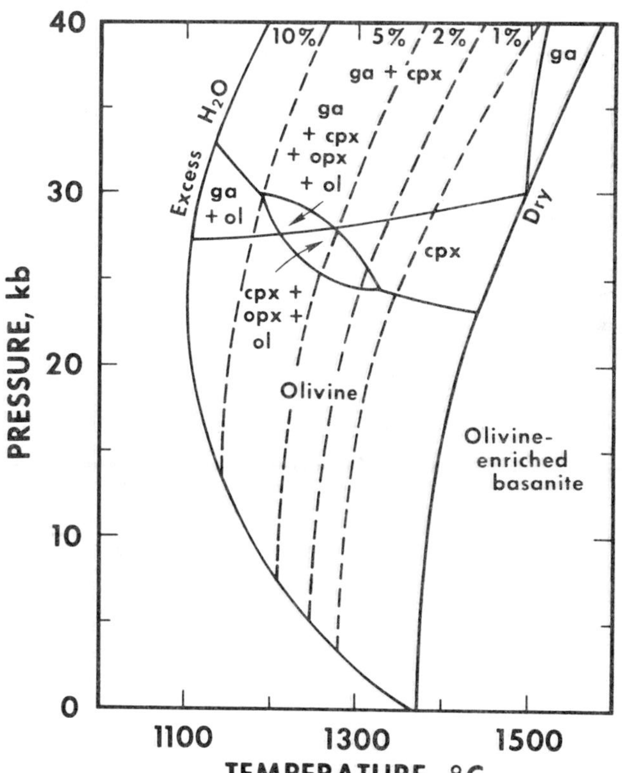

3.3.31 Liquidus surface as a function of H_2O content. H_2O-undersaturated liquidus surface for basanite (compare Fig. 3.3.29 for basalt). Olivine, orthopyroxene, clinopyroxene and garnet occur simultaneously on the liquidus at about 1250° at 28 kb within a limited range of H_2O contents (2–10%).

olivine basanite. They located a more extensive area for multiple saturation with orthopyroxene, clinopyroxene and garnet, extending from about 20 kb at the H_2O-excess solidus, and opening up to lower H_2O contents with pressure increasing to greater than 30 kb. The olivine field on the liquidus did not quite reach this area, but with addition of normative olivine to the composition to allow for olivine fractionation (Green, 1973b), the area of four-phase saturation could be considerably more extensive than that defined for the composition in Fig. 3.3.31.

Merrill and Wyllie (1975) determined a field for kaersutite on the H_2O-undersaturated liquidus surface for their basanitic composition, between about 12 and 21 kb, extending to almost 1200°C. The crystallization of kaersutite on or near the liquidus is restricted to liquids containing at least 2–3% H_2O, and the occurrence of kaersutite megacrysts in basanites, implying near-liquidus crystallization, indicate that basanite magmas at depth contain at least 2–3% H_2O.

Chemical variations among basanites and nephelinites are irregular (Bultitude and Green, 1971), as indicated by the scatter of points in Fig. 3.3.2. The experimental results suggest that many of these lavas are produced by deep-seated fractionation of olivine from hydrous low-SiO_2 picritic liquids. There is evidence that other minerals may be involved in deep-seated fractionation. High-pressure experiments on a lherzolite-bearing nepheline mugearite as a function of H_2O content (Irving and Green, 1972 and pers. comm., 1980) showed that kaersutitic amphibole (similar in composition to the kaersutite megacrysts typically found in such evolved lavas) is a near-liquidus phase at special conditions, with about 3.5% H_2O at about 14 kb. This observation reinforces arguments from major and trace element data (Irving and Price, 1981) that the lineage basanite to nepheline hawaiite to nepheline mugearite may be produced at moderate pressure (depths of 40–45 km) by fractionation of predominantly olivine + kaersutite, and that further fractionation dominated by aluminous clinopyroxene + biotite will yield nepheline benmoreites and phonolites. Irving (1980) suggested that the dominant mechanism by which fractionation takes place at mantle pressures is flow crystallization within narrow dikes rather than gravity crystal settling.

Olivine nephelinites and olivine melilitites are also candidates for primary magmas, but in experiments with H_2O, conditions for multiple saturation of the liquidus with an appropriate assemblage of mantle minerals have not been found. Clinopyroxene and not orthopyroxene appeared on the liquidus, as shown for a synthetic olivine nephelinite by Bultitude and Green (1968), an olivine nephelinite from Honolulu by Allen *et al.* (1975), an olivine nephelinite composition (natural kaersutite) by Merrill and Wyllie (1975), and an olivine melilitite by Brey and Green (1975, 1977).

While Eggler (1974, 1978), Huang and Wyllie (1974) and Wyllie and Huang (1975, 1976) investigated the effect of CO_2 in producing low-SiO_2 liquids from model mantle compositions, Brey and Green (1975, 1976a, 1977) followed the complementary approach for a specific olivine melilitite composition, determining the liquidus mineralogy as a function of pressure, and contents of H_2O and CO_2. At 30 kb, they determined a small range of $CO_2/(CO_2 + H_2O)$ for which the liquidus was close to multiple saturation with garnet lherzolite minerals. They concluded that this specific composition could be a 5% melt from garnet lherzolite at about 27 kb, between 1150°C and 1200°C, with 7–8% H_2O and 6–7% CO_2 dissolved in the liquid. Eggler and Mysen (1976) argued that the mole fraction of CO_2 required to produce olivine melilitite must be much higher (see also reply by Brey and Green, 1976b).

Eggler and Wendlandt (1979) similarly determined the phase relationships for a kimberlite composition with variable amounts of H_2O and CO_2 at 30 and 55 kb, in order to locate conditions where kimberlite could be a primary liquid from the mantle. In order to overcome the problem of Fe-absorption by Pt capsules, they substituted CoO for FeO (Coons *et al.*, 1976). Their kimberlite composition crystallized subsolidus to garnet lherzolite minerals with H_2O-rich buffered vapor, plus phlogopite and dolomite (30 kb) or magnesite solid solution (55 kb) (see Fig. 3.3.11b and the dolomite vapor-buffer Q_1N in Figs. 3.3.12 and 3.3.22). They found that kimberlite containing 5% H_2O and 5% CO_2 is likely to be a primary magma at 55 kb, but not at 30 kb. At 55 kb, liquids resembling those of the kimberlite composition coexist with the four lherzolite minerals through at least 200°C. The conclusion that kimberlite could be a primary liquid from the mantle is consistent with REE patterns (Mitchell and Brunfelt, 1975; Frey *et al.*, 1977).

Several potassic lavas have been studied experimentally at high pressures. Cundari and O'Hara (1976) found clinopyroxene as the sole liquidus phase of an olivine leucitite under dry conditions from 20–40 kb, and Thompson (1977) obtained similar results for a clinopyroxene leucitite. If these magmas were dry, they must be derivative liquids, but general chemical analogies with olivine nephelinites, olivine melilitites and kimberlites suggest that H_2O and CO_2 are involved in their genesis.

Three lavas from the Leucite Hills were investigated by Barton (1976) and Barton and Hamilton (1978, 1979). Under vapor-absent conditions (with small quantities of both H_2O and CO_2 present from the natural rocks), clinopyroxene was the sole liquidus phase of a madupite to 30 kb, but an orendite showed a near-liquidus assemblage of olivine + orthopyroxene + clinopyroxene + garnet above 26 kb. Barton and Hamilton concluded that the orendite was a primary magma and that other K-rich magmas might be produced by high-pressure fractionation of olivine + pyroxenes or lower-pressure fractionation of mica + clinopyroxene (possibly correlated with xenoliths of biotite pyroxenite characteristically associated with K-rich lavas). Experimental data for a biotite mafurite composition with varying amounts of H_2O and CO_2 (Edgar *et al.*, 1976; Ryabchikov and Green, 1978) suggest that the generation of primary potassic magmas requires the presence of both CO_2 and H_2O and small degrees of melting. It has been suggested that phlogopite may be a residual phase in the source region of some potassic magmas; however, the experimental evidence to support this suggestion is inconclusive.

The conclusion from the foregoing discussion is that highly SiO_2-undersaturated magmas including basanites, olivine nephelinites, kimberlites, and various potassic magmas can be generated in the mantle as primary magmas only if small proportions of CO_2 and H_2O are present. There is growing evidence that the generation of primary SiO_2-undersaturated magmas involves the local influx into the upper mantle of metasomatizing fluids rich in H_2O, CO_2, and lithophile elements such as light REE, K, Rb, etc. The evidence for mantle metasomatism raises some intriguing questions. It is not clear to what extent these components are derived from deeper mantle sources, or through recycling of the hydrosphere via subduction. If they come from deeper sources, they must be transported to the regions of metasomatism by magmas.

If interstitial magma is associated with the seismic low-velocity zone (Lambert and Wyllie, 1968), it belongs to this family of highly SiO_2-undersaturated liquids (Green, 1973b; Eggler, 1976). Whether or not the liquid can become concentrated and escape in an eruptive event depends upon the tectonic conditions. For low degrees of melting, Figs. 3.3.11b and 3.3.13b show that magmas migrating upward under adiabatic conditions would reach the solidus at a depth of 80–100 km. The solidus barrier in the system peridotite-CO_2-H_2O is related to the temperature maximum mn in Fig. 3.3.12 (point mn in Fig. 3.3.13b). Under equilibrium conditions, the magma would crystallize at that level, releasing dissolved volatile components, which could represent one source for metasomatizing solutions at shallower levels. In a region of tectonic tension, however, the gases could enhance the prospect of crack propogation, and explosive eruption of magma and gas. This feature could certainly contribute to the eruptive mechanism of kimberlites and related magmas. If tectonic conditions were not suitable, that is, without tension, the magmas would not be erupted. For higher degrees of melting, or higher values of H_2O/CO_2 in the system (Wyllie, 1979a, Fig. 7), the barrier mn becomes less effective. Thus, the compositions of erupted magmas depend not only on the physical chemistry at the source, but on the physical conditions of the tectonic environment.

3.3.7 Petrogenesis of basalts and andesites

This is an open-ended subject, and the geochemical data interpreted in Chapters 1 and 7 provide an essential prerequisite for its discussion. It is not our intent to review all aspects of petrogenesis; we intend only to demonstrate that petrologists do not yet agree on the compositions of primary magmas.

The petrogenesis of komatiites, picrites, and the highly SiO_2-undersaturated magmas was discussed briefly along with the presentation of phase relationships in sections 3.3.5–6. In this section, with similar brevity, we combine the phase relationships reviewed for basalts and peridotite from sections 3.3.2–4 with other constraints, mainly geochemical (see Chapters 1 and 7 for details), and consider the compositions of likely primary magmas in each of three different tectonic environments: mid-ocean ridges, oceanic islands, and subduction zones. Continental basalts will not be reviewed, because there are insufficient experimental data to justify additional discussion. Basalts in regions of compressive plate boundaries cannot be discussed without reference to andesites and their sources (section 3.3.1).

Primary magmas produced at mid-ocean ridges

The discussion of this topic is not completed in this section, which constitutes the background for a more searching inquiry into methods given in section 3.3.8, where the question is asked: can we take oceanic basalts, follow the inverse approach, and reach definitive conclusions about the nature of the mantle below?

Engel and Engel (1964) and Engel *et al.* (1965) were the first to present extensive chemical data on mid-ocean ridge tholeiites. They commented on the uniform compositions of these basalts and proposed that they represent primary melts from the mantle. Green *et al.* (1967) also argued that these basalts are relatively unfractionated and suggested that they represent direct partial melts separated from the mantle source at about 30 km depth (9 kb). In contrast, O'Hara (1968b) proposed that the primary magma is picritic and is generated at 25–30 kb pressure. According to this model, the compositions of mid-ocean ridge tholeiites as observed at the surface are the result of extensive fractionation of olivine.

More recently, many new data on mid-ocean ridge tholeiites have appeared including many electron microprobe analyses of glasses from chilled pillow margins (Melson *et al.*, 1976a, b; Bryan and Moore, 1977; O'Donnell and Presnall, 1980; Bryan, 1979). It is now a well-documented observation that mid-ocean ridge tholeiites show evidence of crystal fractionation (for example, see Shido *et al.*, 1971; Clague and Bunch, 1976; Bryan and Moore, 1977; O'Donnell and Presnall, 1980) and attention has recently been focused on the least fractionated of these basalts as likely candidates for primary magmas. Frey *et al.* (1974) reported analyses of three basalts from DSDP Legs 2 and 3 that they considered to be good candidates for primary magmas because of their high Mg-values [$Mg/(Mg + Fe^{2+})$] of 70–71. Bender *et al.* (1978) studied a basalt from the FAMOUS area with an Mg-value of 68 and concluded that this basalt was itself not primary but the result of crystallization of about 10% olivine from a more primitive parent. Rhodes and Dungan (1979) argued that primary abyssal tholeiites are characterized by Mg-values of 70–73, high CaO (12.0–13.5%), low TiO_2 (< 0.9%), low Na_2O (< 2.1%), and Ni between 250 and 300 ppm. Although Donaldson and Brown (1977) are in general agreement with these characteristics, they have proposed a more magnesian composition with Mg-values of 75–79. See Chapter 1, section 1.2.5.

Presnall *et al.* (1979) also argued that the least fractionated abyssal tholeiites with Mg-values of 70–72 have been modified very little during their passage to the surface and are close to the compositions of primary magmas. They presented a model to explain the production of large volumes of these primary magmas at about 30 km depth based on high pressure experimental data in the system $CaO\text{-}MgO\text{-}Al_2O_3\text{-}SiO_2$. They determined the solidus for simplified plagioclase lherzolite and spinel lherzolite, found a low temperature cusp at 9 kb, 1300° C (Fig. 3.3.4), and noted that the composition of the initial melt at the 9 kb cusp corresponds closely to the composition field of the least fractionated mid-ocean ridge tholeiites. They postulated continuous production of primary mid-ocean ridge tholeiites at the cusp from mantle material rising along the ascending limb of a mantle convection cell, with the geotherm directly beneath ridges being controlled by the production and upward transport of magma at the cusp.

However, the nature of primary magmas produced at oceanic ridges remains controversial. At about the same time that Presnall *et al.* (1979) presented strong arguments favoring a shallow depth of origin of 30 km, as proposed earlier by Green and Ringwood (1967a), Green *et al.* (1979) abandoned this earlier model and proposed that a picritic primary magma produced at depths of about 70 km is parental to mid-ocean ridge tholeiites, a model very similar to that of O'Hara (1968b). Bender *et al.* (1978) reached a similar conclusion. The arguments are reviewed in detail below, in section 3.3.8.

One of the most important constraints on the type of fusion process occurring at mid-ocean ridges comes not from experimental petrology but from rare-earth element (REE) data. Mid-ocean ridge tholeiites typically show nearly flat or slightly light-REE-depleted patterns. Given reasonable assumptions about REE abundances in the source, such patterns require that the magmas separated from the source are the result of a large amount of fusion, about 10–30% (Gast, 1968; Kay *et al.*, 1970; Schilling, 1971, 1975). That is, the liquid must remain in equilibrium with the source until a large amount of this liquid is formed, and models involving a close approach to fractional fusion are eliminated (section 3.3.3). This conclusion applies regardless of the disagreement discussed above concerning the nature of the primary magma, because those postulating a picritic primary magma would obtain the primary magma of Presnall *et al.* (1979) and Rhodes and Dungan (1979) by crystallization of olivine. This would change the REE concentrations, but would not alter the slope of the REE distribution pattern.

Primary magmas produced at oceanic islands

One of the most important petrologic conclusions from isotopic studies of oceanic basalts is the finding, first reported by Tatsumoto (1966), that the source regions for mid-ocean ridge basalts and oceanic island basalts are different and have existed separately for a time on the order of 1–2 billion years (see Chapter 1). If it is assumed that the entire oceanic lithosphere is created at mid-ocean ridges, the isotopic data require that the source region for oceanic island basalts must lie, at least in part, beneath the lithosphere. Thus, inferences about the nature of the mantle based on studies of mid-ocean ridge basalts will not necessarily apply to the source region for oceanic island basalts.

The locations of volcanically active oceanic islands show no special correlation with the locations of mid-ocean ridges. Iceland lies astride the Mid-Atlantic

Ridge, several others such as Ascension, Tristan da Cunha, and the Galapagos Islands are near active ridges, and others such as the Hawaiian Islands are far removed from ridges. The compositions of the erupted lavas also vary. For example, at many islands, only alkalic lavas are found, but at Hawaii, the most abundant basalt is tholeiite. Hawaiian tholeiites show very pronounced shallow-level crystal fractionation controlled in the initial stages by olivine (Powers, 1955; Murata and Richter, 1966), a feature that is absent from the Galapagos tholeiites (McBirney and Williams, 1969). Thus, it appears that a single petrogenetic model for oceanic islands cannot be universally applied. Only the Hawaiian Islands will be discussed here, mainly because far more data exist for this island group than any other. It is not to be implied, however, that the Hawaiian Island chain is typical of oceanic islands in general. See Chapter 1, section 1.2.6.

Dalrymple *et al.* (1973) presented an overview of the geology of the Hawaiian Archipelago-Emperor Seamount chain. Very briefly, the Hawaiian Archipelago consists of a broadly linear chain of islands extending northwestward for about 3500 km across the central Pacific Ocean. Near Yuryaku Seamount, the chain abruptly changes direction and becomes the Emperor Seamount chain, which continues northward for another 2500 km. The ages of the volcanoes decrease progressively southeastward to the presently active island of Hawaii. The entire chain is generally believed to be the result of movement of the Pacific plate across a "hot-spot" (Wilson, 1963; Morgan, 1972) or "melting anomaly" (Shaw and Jackson, 1973). In the model of Morgan, the source material for the volcanoes starts in the deep mantle and rises along a thermal plume, whereas in the model of Shaw and Jackson, melting takes place in the asthenosphere with formation of a residue that is denser than either the surrounding undepleted asthenosphere or the mantle beneath the asthenosphere. The residue thus sinks into the mantle and forms what they called a "gravitational anchor." However, O'Hara (1975) and Boyd and McCallister (1976) concluded that the residue of partial fusion would be less dense, not more dense than the unfused parental material, and the residue would not sink. It is possible, however, that the source region might be in the asthenosphere even though the residue does not sink, although if Schilling (1975) is correct that the source of mid-ocean ridge basalts is in the asthenosphere, the isotopic constraints mentioned at the beginning of this section would appear to favor a deeper source.

On a given volcano, the earliest observable lavas are invariably tholeiitic, and these lavas are erupted throughout the main shield-building stage. As the volcano matures, the tholeiitic flows are typically followed by alkalic lavas that form a thin cap on the tholeiitic shield. After the alkalic stage, the eruption rate decreases enough to allow dissection of the volcano and the formation of deep canyons. Finally, in the terminal stages of the eruptive cycle, small amounts of nephelinitic lavas are produced from vents scattered over the shield.

Green and Ringwood (1967a) argued that in Hawaii, primary olivine tholeiite magmas are generated by a high degree of partial fusion at 35–70 km depth. They accepted the arguments of previous investigators that fractionation at shallow depths, initially of olivine and later of hypersthene, augite, and plagioclase, is primarily responsible for the trend toward quartz-normative tholeiites. They proposed that the alkali olivine basalts are produced from primary olivine tholeiites by fractionation of aluminous enstatite also in the depth range of 35–70 km (see section 3.3.4). However, O'Hara (1968a), Kushiro (1969), and Presnall *et al.* (1978a) pointed out that derivation of alkali olivine basalt from an olivine tholeiite parent must be accompanied by the crystallization of olivine as well as enstatite. This is because the olivine field expands as pressure drops. A magma generated at any given pressure must therefore crystallize olivine at any lower pressure. All the available evidence indicates that derivation of alkali olivine basalts from tholeiites by fractional crystallization at shallow depths within the crust is not possible (Powers, 1955; Yoder and Tilley, 1962; O'Hara, 1965, 1968a; Presnall *et al.*, 1978a). For the nephelinites, Green and Ringwood (1967a) preferred a separate source. At least for the Koolau shield on Oahu, a separate source for the nephelinites and tholeiites appears to be required by the systematically lower $^{87}Sr/^{86}Sr$ ratios of the nephelinites (Powell and De Long, 1966; Lanphere and Dalyrmple, 1980).

Tholeiites, alkali olivine basalts, and nephelinites from the same volcano can be characterized by heterogeneous $^{87}Sr/^{86}Sr$ ratios (Powell and De Long, 1966; O'Nions *et al.*, 1977), which suggests that a single, well mixed source is not appropriate even for successive eruptions of a single basalt type from the same volcano. Data available at present indicate that the range of $^{87}Sr/^{86}Sr$ values in tholeiites and alkali olivine basalts from the same volcano overlap. Thus, even though multiple source regions are indicated for the same volcano, it may still be possible for some Fe-rich alkali olivine basalts to be derived by high pressure fractional crystallizaton of primary olivine tholeiites.

Wright (1971) carried out a detailed study of Kilauea and Mauna Loa lavas and suggested that the primary Kilauea magma is a picritic olivine tholeiite (oceanite) with about 25% MgO. He based this on two considerations. First, Murata and Richter (1966) had found that the lava from the highest temperature part of the 1959 Kilauea eruption contains slightly more than 15% MgO. Wright considered this olivine-phyric lava to be representative of the magma supplied to the 2–4 km deep reservoir beneath Kilauea, and that the primary magma generated at depth in the mantle must therefore contain at least this amount of MgO. Second, he assumed with Jackson and Wright (1970) that dunite is the residue produced by the removal of tholeiite from the mantle. The undepleted mantle would then lie on a line between dunite and tholeiite. He then calculated the composition of the undepleted mantle by assuming it contained 35% MgO, and from reasonable assumptions about the type and compositions of minerals present, he calculated a mode. He arrived at a minimum MgO content of 15–20% for the primary magma by assuming all minerals except olivine were incorporated into the primary melt. Because some olivine would also be melted, he increased the MgO content of the primary magma to 25%.

In summary, for oceanic islands, as for mid-ocean ridges, many basalts appear to be derived from a primary, magnesian olivine tholeiite, the precise composition of which is still being defined. It is evident that there is no universal petrogenetic model for oceanic islands, and that there are multiple source regions even for the same volcano. The nephelinites are derived from an independent magmatic episode involving a primary liquid different from the picritic tholeiite (section 3.3.6).

Primary magmas produced at subduction zones

This discussion begins with section 3.3.1, where it was concluded that there may be three independent primary liquids associated with the three magmatic series: tholeiitic, calc-alkalic, and alkalic, which exhibit spatial and perhaps temporal relationships with the geometry and history of subducted oceanic lithosphere (see Chapter 1, section 1.2.7.) It was also pointed out that more detailed geological information is required before satisfactory petrogenetic models can be constructed. The wide variety of possible source rocks and magmatic processes that have been proposed is summarized in Fig. 3.3.3. A distinctive feature of this environment, compared with the oceanic environments, is the availability at depth of H_2O carried down with the hydrated ocean crust. The geochemistry of rare-earth elements, other trace elements and isotopes offers promise for discrimination among the possible sources (Chapters 1 and 7). In order to unravel the petrogenesis, we must know, in addition to the geology, geochemistry, and geophysics, the phase diagrams for possible source rocks and observed magmatic products through the range of pressure and temperature from deep source to surface, including the effects of H_2O. We are still far from the attainment of this goal.

We have nothing to add to previous discussions about primary liquids and primary magmas associated with the tholeiitic and alkalic rocks. Most of the recent petrogenetic models for island arcs have focused on mechanisms for producing the andesitic members of the calc-alkalic series. Some of these invoke partial melting of amphibolite or hydrous quartz eclogite from the subducted oceanic crust (Green and Ringwood, 1968; Green, 1972; Marsh and Carmichael, 1974), possibly with an intermediate stage of hybridization of the liquid with mantle peridotite (Nicholls and Ringwood, 1973; Nicholls, 1974). Others invoke the direct production of primary andesitic liquids by the wet partial fusion of the peridotite wedge above the Benioff zone at depths as great as 80 km (Kushiro, 1972a; Kushiro *et al.*, 1972a; Mysen and Boettcher, 1975b; Anderson *et al.*, 1978). Reviews of these proposals based on a framework of experimental petrology have been given by Boettcher (1973), Ringwood (1975, Chapters 7 and 8), Stern and Wyllie (1978), and Wyllie (1979b).

Allen *et al.* (1975, 1978) studied the phase relationships of amphiboles in basalts and andesites under a variety of conditions (section 3.3.4), and concluded that amphiboles are a major carrier of H_2O in subducted oceanic slabs, and that crystal-liquid equilibria involving amphiboles are prominent in the genesis of calc-alkaline magmas. The process begins by partial melting of the oceanic crust. The compositions of liquids were determined in experiments, but were not specified for the natural processes. However, it appears that amphibole fractionation ensures that the primary liquids would not reach the surface as primary magmas.

Marsh and Carmichael (1974) tested earlier models of Green and Ringwood (1968) and, by ingenious calculations, they related primary andesites to a source of quartz eclogite. Stern and Wyllie (1978) combined their new experimental data on the phase relationships of basalt-andesite-H_2O at 30 kb with the results of Green and Ringwood (1968) and Green (1972), and estimated the effects of bulk composition, pressure, and H_2O on the compositions of liquids through the melting intervals of the rocks, and of garnets and clinopyroxenes coexisting with the liquids. They concluded that partial melting of subducted oceanic crust at about 100 km depth does yield liquids with a range of intermediate SiO_2 contents, but that these must be modified by fractionation at shallower depths if they are to reach the surface with chemistry corresponding to the average calc-alkalic lavas.

The data of Stern and Wyllie (1978) can be used to test the primary character of andesite with respect to the mineralogy of possible residual basaltic source rock, following step 6 and source region constraints in the flow chart of Fig. 3.1.4. Figure 3.3.32 shows for basalt the effect of H_2O content on the compositions of liquidus garnet and clinopyroxene and, in addition, the paths followed by these two minerals during the crystallization of dry basalt, and of basalt with 5% H_2O. These crystallization paths are compared with the liquidus mineralogy of andesite as a function of H_2O content. The phase relationships suggest that neither dry nor hydrous andesite liquid can coexist with residual minerals during the partial fusion of this basaltic composition (quartz eclogite) at a depth of about 100 km, for the following reasons.

During the crystallization of dry basalt (or equilibrium fusion of quartz eclogite, section 3.3.3), the garnet composition passes very close to the liquidus garnet for

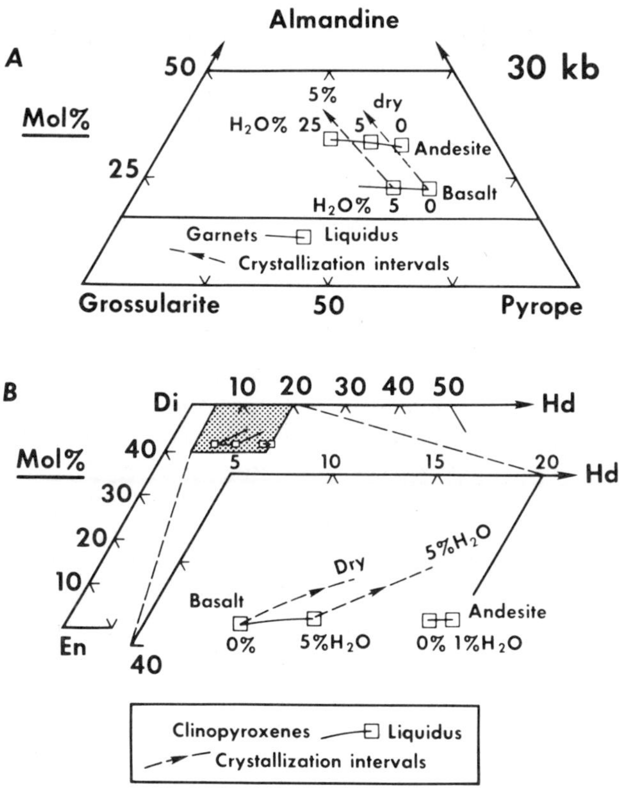

3.3.32 The compositional variation of liquidus garnets and clinopyroxenes for basalt and andesite as a function of H_2O content, compared with garnet and clinopyroxene compositions through the equilibrium crystallization intervals of basalt dry, and with 5% H_2O (after Wyllie, 1978b). Based on data of Green and Ringwood (1968, 1972) and Stern and Wyllie (1978).

dry andesite, indicating a possible match. However, the clinopyroxene in the dry basalt has a composition quite different from the liquidus clinopyroxene of the andesite. More experiments are required for basalts and andesites of different compositions.

During the crystallization of basalt with 5% H_2O (or equilibrium fusion of hydrous quartz eclogite), the composition path of the garnet crosses the garnet trend for the andesite liquidus at a point corresponding to H_2O content within the range 10–15%. At the point of intersection, the H_2O content of the liquid in the basaltic assemblage would be between 8% and 13%, according to the results of Stern and Wyllie (1978), which is a reasonable match for the andesite liquidus, considering the uncertainties in these estimates. Furthermore, the temperature on the basalt path is between 1000° C and 1100° C, which is in the correct range for the andesite liquidus with this H_2O content. A match is denied, however, by the fact that for andesite liquids with more than about 1% H_2O, clinopyroxene does not crystallize at the liquidus.

If the dehydration of subducted oceanic crust triggers melting of overlying peridotite, with the generation of primary andesites, then these andesites should be capable of providing information about the mantle source via the flow chart of Fig. 3.1.4. The processes of dehydration and melting in and above a subducted slab are commonly described in qualitative terms, without too much attention to specific phase diagrams, temperatures, and pressures. Experimental data are now available to provide some constraints for the conditions under which andesite might be produced as a primary magma from mantle peridotite.

The progressive thermal metamorphism of a descending slab of ocean crust produces a series of dehydration fronts, which can be mapped according to whatever thermal structure is adopted (see Chapter 9; Wyllie, 1973; Delany and Helgeson, 1978; Anderson *et al.*, 1978). Anderson *et al.* (1978) took into account the cooling effect of endothermic dehydration reactions and, using a finite difference numerical program, they computed isotherms at various stages during the history of a descending slab. They concluded that, because of the heat sink created by dehydration reactions within the depth interval 75–125 km, the oceanic crust does not reach a high enough temperature to be melted before complete desiccation. They concluded that arc volcanism must be produced by melting induced in the overlying peridotite by aqueous solutions rising from the subducted slab.

Experimental data needed for evaluation of this proposal include the solidus for peridotite-H_2O (a variety of results is shown in Fig. 3.3.7), and the range of pressures and temperatures above the solidus for peridotite-H_2O-CO_2 through which andesitic liquids can be generated (results for the solidus surface are summarized in Fig. 3.3.22). For many published thermal structures, hydrothermal fluids released from the subducted oceanic crust could not cause melting in the overlying peridotite because the overlying mantle would be too cool for temperatures to exceed the peridotite-H_2O solidus. For the model of Anderson *et al.* (1978) to be effective, appeal must be made to vigorous asthenospheric convection in the mantle wedge, raising temperatures to high levels at shallow depths, and producing high temperature gradients above the oceanic crust without overcoming the cooling effect of the dehydrating slab (Wyllie, 1978b). Even if these conditions are achieved, Fig. 3.3.22 indicates that the production of primary andesites is limited to depths shallower than about 40 km. The H_2O-saturated liquids that would be generated from mantle peridotite at depths of 80–100 km are not even quartz-normative according to Fig. 3.3.22; they may be forsterite-normative, or even nepheline-normative. It is difficult to devise a reasonable physical process to generate andesitic liquids from mantle peridotite flushed by H_2O from 100 km depth.

Nicholls and Ringwood (1973) proposed that subducted oceanic crust dehydrates in the depth interval of 70–100 km and releases H_2O into the mantle wedge above. They argued that this would reduce the viscosity and increase the mobility of the overlying mantle, which would cause diapirs of wet peridotite to rise. Partial fusion of these rising diapirs would then produce primary SiO_2-saturated tholeiites that would fractionate to produce the SiO_2-oversaturated basalts and basaltic andesites of the tholeiite series.

At greater depths of 100–150 km along the subducted plate, Nicholls and Ringwood (1973) proposed the generation of highly siliceous melts from wet oceanic crust. These melts then rise, react with the overlying peridotite and produce wet garnet pyroxenites. Finally, these wet garnet pyroxenites rise diapirically and partially melt to produce calc-alkalic magmas, mainly andesite.

The approximate H_2O content of H_2O-saturated magmas as a function of pressure is given by the contours intersecting the excess-H_2O liquidus for basalt in Fig. 3.3.29. Values range from about 15% to nearly 30% between about 40 km and 100 km depth. Garcia *et al.* (1979) have recently found that the H_2O content of the glassy rims of andesitic and basaltic pillow lavas from the Marianas Island Arc and trough is only about 1%. If this is a measure of the H_2O content at the site of magma generation, then the compositions of the primary liquids would not be significantly altered from those expected from anhydrous melting, and a serious problem exists for all the models that invoke wet partial fusion of peridotite, regardless of whether basalt or andesite is considered to be a product of this fusion. Andesite would be immediately eliminated as a partial fusion product from peridotite at any depth and the primary SiO_2-saturated tholeiite of Nicholls and Ringwood would have to be separated from the rising mantle diapir at less than about 15 km depth (see Figs. 3.3.22 and 3.3.29). However, seismic evidence indicates that the crust directly beneath island arcs, although highly variable from arc to arc, averages about 22 km (Condie, 1976, p. 57) and a rising peridotite diapir would be incapable of reaching this level.

The question of the compositions of primary magmas in subduction zone regions thus remains unanswered. The processes are complex. Even if a primary magma composition can be identified, the possibility that hydrothermal fluids released from the subducted oceanic crust have migrated into the overlying peridotite may limit severely its usefulness as a guide to normal mantle material.

It is well known that aqueous fluids at mantle pressures are concentrated solutions containing tens of percent of dissolved solid components (Holloway, 1971; Nakamura and Kushiro, 1974; Wyllie, 1979, Figure 18). Such fluids would be expected to metasomatize peridotite. Mysen (1978) determined the rare-earth element partition coefficients for the minerals of eclogite and peridotite in equilibrium with H_2O at upper mantle pressures and temperatures. From these results, he argued that solutions released from subducted eclogite would equilibrate with peridotite and cause considerable light-REE enrichment. He related this to metasomatized garnet peridotite nodules (section 3.3.6). He also argued that partial melting of this type of hydrated peridotite would produce liquids of andesitic bulk composition with REE patterns closely resembling those found in andesites of island arcs. We have reviewed the problem of deriving andesites as primary magmas from mantle peridotite. These results show that the inverse approach applied to some primary magmas may yield chemical information about a mantle source that has been modified by hydrothermal alteration.

3.3.8 Does the inverse approach work on Earth?

The procedure outlined in the flow chart of Fig. 3.1.4 is designed to lead from studies of surface basalts to information about the mineralogy and constitution of the internal source material, using the results of experimental petrology together with whatever other information could be used to apply constraints. For terrestrial basalts, we commonly apply a forward approach in conjunction with the inverse approach, as discussed in some detail in section 3.1.4.

Would the inverse approach work on the Earth? If Martians landed on Earth, scooped up some basalt and took it back to their laboratories and applied the flow-chart, would they come up with what we perceive as the right answer? This is, after all, what we claim to have done on the Moon (see section 3.4) and what we propose to do for the other planets (section 3.6.4). Let us examine how the inverse approach fares when applied to the Earth as it has been applied to extraterrestrial basalt petrology.

The inverse approach: a case study

For our case study we will deal with oceanic tholeiites (sections 3.3.4 and 3.3.7). These basalts are particularly suitable for our purposes, because they are surely the most abundant surface basalts on Earth, and therefore have some analogy with what would be returned by a grab sampling of lavas from this planet if the oceans were absent. Their chemistry has been dealt with in detail in Chapter 1, section 1.2.5, and there are recent experimental data available.

The flow chart first directs us to identify liquid compositions and to reject phenocryst-enriched or cumulate compositions (step 1, Fig. 3.1.4). For oceanic tholeiites, this is a relatively simple task, because most contain less than 5% microphenocrysts and many glassy samples are available.

From those samples that are wholly glass or that contain few (e.g., $< 5\%$) phenocrysts, the flow chart then directs us to establish whether these liquid compositions are related by low pressure fractionation (step 2, Fig. 3.1.4). Figure 3.3.33 shows the variations of $Fe/(Fe + Mg)$ and TiO_2 concentrations for basalt glasses from oceanic tholeiites. Both of these parameters increase in residual liquids during fractionation of olivine or pyroxene. Because olivine phenocrysts are present in most phenocryst-bearing samples, olivine was probably involved in any differentiation which might have affected these samples. Therefore, the liquid sample with the lowest TiO_2 and $Fe/(Fe + Mg)$ is a reasonable choice for the most primitive liquid composition. Several other lines of evidence point to a liquid with molar $Fe/(Fe + Mg)$ about 0.28 and TiO_2 about 0.7% as the most primitive sample available. These are discussed in detail in Chapter 1.

Experiments at low pressure on vitreous samples near the low $Fe/(Fe + Mg)$ and low TiO_2 end of the spectrum have been performed by Bender *et al.* (1978) and Usselman and Hodge (1978). These samples had the same composition and are plotted on Fig. 3.3.33 and listed in Table 3.3.4. A similar oceanic basalt composition was studied by Green *et al.* (1979) and is also shown in Fig. 3.3.33 and Table 3.3.4. These liquids are primitive compared to the rest of the spectrum.

A summary of experimental phase relations for oceanic tholeiites at 1 bar from Walker *et al.* (1979) is

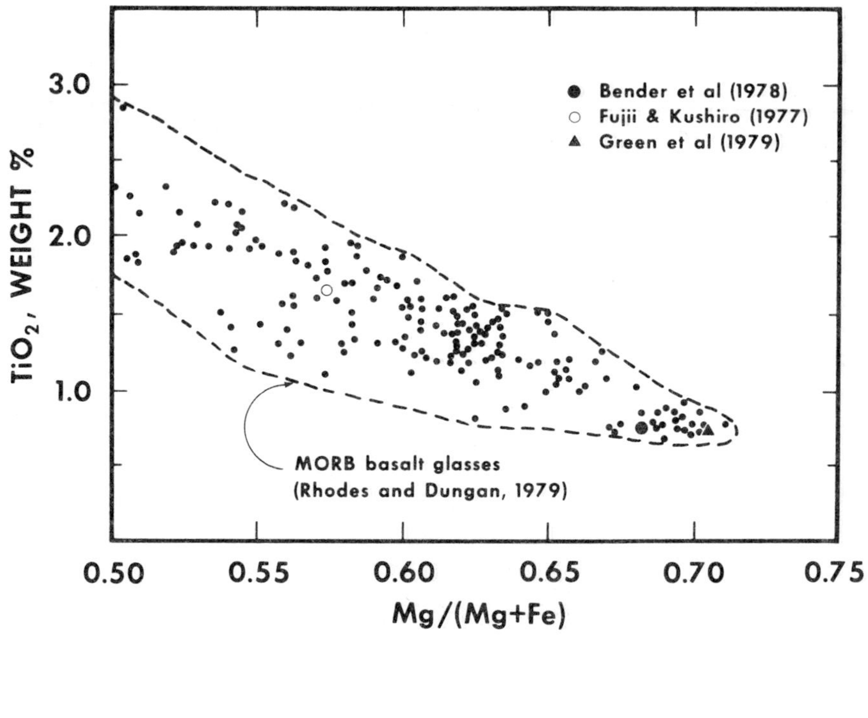

3.3.33 Compositions of abyssal volcanic glass (after Rhodes and Dungan, 1979), with bulk compositions of basalts studied experimentally and discussed in the text.

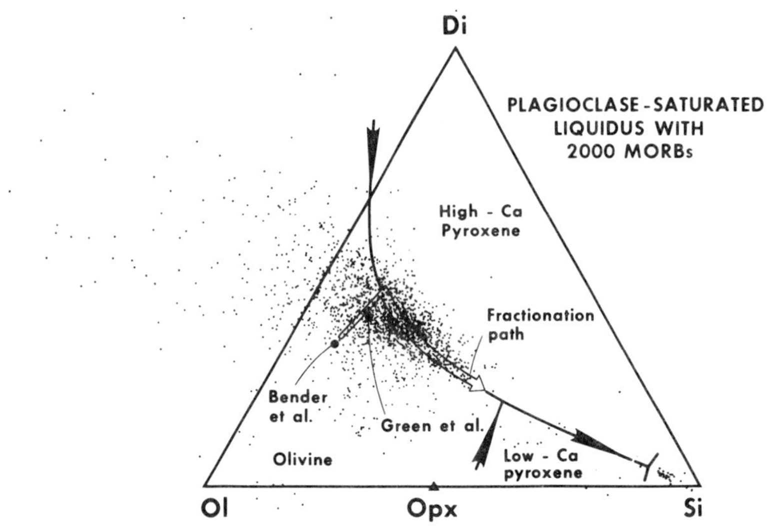

3.3.34 The compositions of oceanic tholeiites compared with experimental phase relationships for liquids saturated with plagioclase feldspar (after Walker *et al.*, 1979). The two primitive liquids plotted in Fig. 3.3.33 precipitate first olivine at low pressures, followed by plagioclase and then clinopyroxene.

shown in Fig. 3.3.34 along with a comprehensive compilation of oceanic basalt compositions and the compositions of the primitive basalts studied by Bender *et al.* (1978) and Green *et al.* (1979). Figure 3.3.34 is a simplified liquidus projection for liquids saturated with plagioclase feldspar, and may be used to evaluate the fractionation of oceanic basalts at low pressure. Arrows indicate the direction of falling temperature on boundary curves between phase volumes. Low pressure crystallization of both primitive samples begins with olivine and is followed by plagioclase before clinopyroxene. Fractionation of the primitive samples on Fig. 3.3.34 produces residual liquids which move from the primitive compositions away from olivine and towards the boundary curve olivine and clinopyroxene (+ plagioclase) and then down the boundary curve (steps 3 and 4 of Fig. 3.1.4).

One point to emerge from Fig. 3.3.34 is that the largest cluster of oceanic basalt compositions is on the olivine + clinopyroxene (+ plagioclase) boundary curve and appears to be the result of crystal-liquid fractionation at low pressure (see discussion of Fig. 3.3.2). Furthermore, this dense cluster of oceanic basalt compositions is on the liquid line of descent from the primitive compositions and, therefore, these primitive compositions could be parental to a large number of oceanic basalts (step 5, Fig. 3.1.4). Another point to emerge from Fig. 3.3.34 is the noticeable scatter about the densest cluster. Some of this is undoubtedly due to analytical and sampling problems but may also include effects introduced by admixture of various phenocryst proportions. Further complications may result from fractionation processes at high pressures, or from mixing of magma batches at various stages of evolution in a magma chamber feeding the ocean ridge volcanism. Evidence of the existence of such complex processes has been presented by Cann (1974), Dewey and Kidd (1977), Bryan and Moore (1977), Donaldson and Brown (1977), O'Hara (1977), Dungan and Rhodes (1978), O'Donnell and Presnall (1980), Rhodes *et al.* (1979) and Walker *et al.* (1979).

No matter what the nature and complexity of processes that supplement low pressure fractionation of oceanic basalts, the primitive compositions selected by Bender *et al.* (1978) and Green *et al.* (1979) are reasonable points of departure in the search for primary oceanic basalts. Both compositions are primitive in terms of TiO_2 and Fe/(Fe + Mg) criteria and can fractionate to produce many other oceanic basalts. Bender's composition is more primitive in terms of the abundance of normative olivine shown in Fig. 3.3.34, whereas Green's composition is more primitive in terms of Fe/(Fe + Mg) shown in Fig. 3.3.33. This discrepancy, if significant, may indicate that primary compositions can differ slightly in Fe/(Fe + Mg) ratio in different regions of the ocean. It is also possible that the operation of complex fractionation or mixing processes, or both, have influenced even these rather primitive oceanic basalts. Nonetheless, the prescription of the flow chart (Fig. 3.1.4) is that we proceed with the most primitive compositions available.

We therefore make the interim assumption that these primitive compositions are primary. Next, we examine the consequences of this assumption with the aid of Fig. 3.3.35, which shows the results of high pressure experiments on these compositions. Both compositions have similar near-liquidus phase relations. The results of Green *et al.* are shown in detail, with only the liquidus of Bender *et al.* given for comparison. Bender's composition (ALV-527-1-1) has more normative oliv-

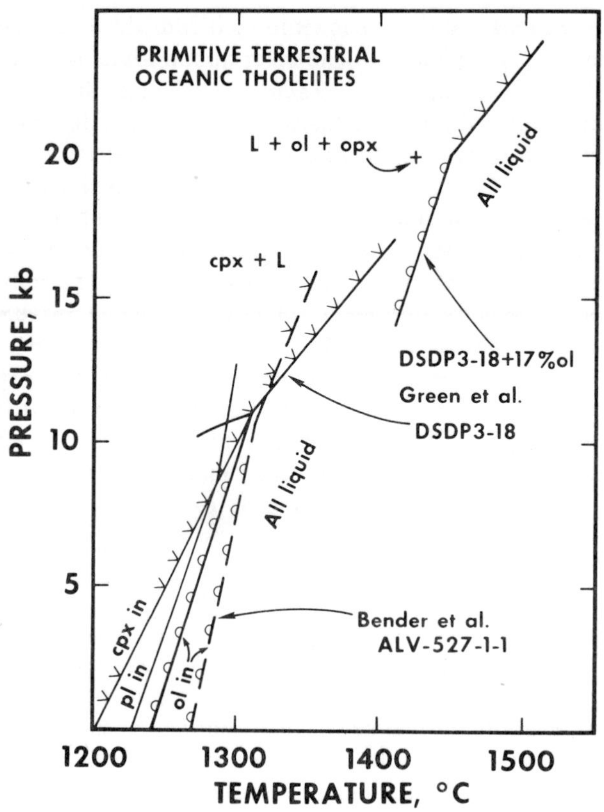

3.3.35 The liquidus and near-liquidus minerals for olivine tholeiite DSDP3-18-7-1, and an olivine-enriched composition, after Green *et al.* (1979). This is the primitive composition plotted in Figs. 3.3.33–34. The liquidus results for the primitive composition of Bender *et al.* (1978) are given by the dashed line (ALV-527-1-1).

ine and, hence, a higher liquidus temperature than Green's (DSDP3-18-7-1). These results may be compared to those of Fujii and Kushiro (1977) on a mid-ocean ridge basalt that is less primitive, based on its slightly higher Fe/(Fe + Mg) ratio and TiO_2 content (Fig. 3.3.24).

All three compositions show a major change of slope in the liquidus at 8–12 kb and 1250–1300° C where the primary liquidus phase changes from olivine to clinopyroxene with increasing pressure (Fig. 3.1.12). At this change in slope, Fujii and Kushiro (1977) found that their basalt was multiply saturated at the liquidus at 8 kb with olivine, clinopyroxene, plagioclase and orthopyroxene. In contrast, Green *et al.* (1979), found approximate multiple saturation with olivine, clinopyroxene, and plagioclase, while Bender *et al.* (1978) found only olivine and clinopyroxene.

Because of the presence of orthopyroxene at the multiply saturated inflection on the liquidus, Fujii and Kushiro argued that their basalt could be in equilibrium with plagioclase lherzolite in the mantle at about 25 km depth. If an alien petrologist following the discussion in section 3.1.4 on the flow chart had only a limited number of samples available to him and had chosen this basalt as the most primitive, he would conclude that it is primary and was derived at about 25 km depth at 1225° C from a mantle containing olivine, orthopyroxene, clinopyroxene and plagioclase. The alien petrologist might also equate the compositions of liquidus crystals at the inflection with those in the residual mantle. Indeed, a line of reasoning similar to this was followed by Fujii and Kushiro.

However, a more extensive set of samples is available to us and we have chosen the basalts of Bender *et al.* (1978) and Green *et al.* (1979) as provisional primary magmas. An alien petrologist might conclude that the basalt studied by Green *et al.* was primary and derived at about 35 km depth at 1300° C from a mantle constituted of residual olivine, clinopyroxene and plagioclase; or that the basalt of Bender *et al.* was derived at a similar depth and temperature from a mantle consisting only of residual olivine and clinopyroxene.

It must be concluded, as we might have expected, that the inverse approach, when used with only a limited type of data such as basaltic lava compositions, may lead to a solution that is not definitive or is even incorrect. Much of what is known about the Earth's mantle has been learned from studies of ultramafic nodules, but these nodules would have a low probability of being sampled by a brief series of missions to Earth by alien petrologists. Largely because of studies on these nodules, it is generally believed that orthopyroxene must be present in the mantle. This additional constraint caused Green *et al.* (1979) to disregard the liquidus inflection in Fig. 3.3.35, and to abandon the idea that their basalt was primary, because orthopyroxene was not present at the liquidus at any pressure. To correct this deficiency, they added olivine to DSDP3-18-7-1 until orthopyroxene became a saturating phase near the liquidus inflection. This condition was encountered after a 17% olivine addition at 20 kb, 1430° C, as shown by the additional experiments in Fig. 3.3.35. Green *et al.* concluded from this result that picritic basalt produced by adding about 17% olivine was the primary liquid and that it was separated from its source at about 70 km. Bender *et al.* (1978) also concluded that their basalt was not quite primary, and that it was produced by fractionation of 5–10% olivine from a more picritic primary magma. Some of the problems associated with the interpretation of olivine-addition experiments are discussed in section 3.1.4.

Before proceeding, it should be emphasized that phase relationships involving reaction relations are possible in which a primary magma could be in equilibrium with a mineral in the mantle yet never show that mineral at the liquidus at any pressure. This situation is reviewed in section 3.1.4. Thus, the absence of orthopyroxene from the basalts of Green *et al.* and Bender *et al.* is not *a priori* evidence that these basalts were never in equilibrium with orthopyroxene in the mantle. Nevertheless, the simplified phase diagrams for the tholeiitic basalt tetrahedron at various pressures given by Presnall *et al.* (1979, Fig. 7–10) suggest that, for this example, orthopyroxene would be observed at the liquidus if the basalt had been in equilibrium with orthopyroxene in the mantle.

Green *et al.* (1979) assumed that their basalt DSDP-3-18-7-1 was produced by fractional crystallization only of olivine from a picritic composition, and their experiments verified that this explanation is possible. This is consistent with O'Hara's (1968a) earlier advocacy of primary picrite magmas. However, it is also possible that basalt DSDP-3-18-7-1 was derived by fractional crystallization of other phases, possibly clinopyroxene and plagioclase, in addition to olivine, and that the shortest compositional route to the orthopyroxene field lies in another direction from that taken by Green *et al.* The discussion of Fig. 3.1.13 is relevant. If minerals in addition to olivine were involved, then the primary magma would not necessarily be picritic, and could be close to the composition of DSDP-3-18-7-1.

Presnall *et al.* (1979) have argued that basalts like DSDP-3-18-7-1 are either primary magmas derived at about 9 kb or are the result of only a small amount of fractionation in this pressure range. They argued against extensive fractionation alone of olivine from a picritic parent because olivine-controlled fractionation trends do not exist among reported analyses of mid-ocean ridge basalt glasses. They presented evidence that the enstatite field lies close to the composition of primitive basalts such as DSDP-3-18-7-1 at about 9 kb.

Thus, there are presently two views of the depth of origin and composition of primary magmas produced at mid-ocean ridges. It appears that considerably more work will be required to resolve the controversy. The results of Fujii and Kushiro (1977) remain tantalizing: it appears to be possible that their (seemingly more fractionated) basalt may actually be a primary basalt derived at about 8 kb from a fertile mantle somewhat richer in Fe/(Fe + Mg) than has usually been assumed.

We have only discussed the minerals in the mantle residue, not their relative proportions. There are several ways in which these proportions and the bulk composition of the source could be constrained within the framework of the interpretation of the experimental data. One of these might be to assume a rare-earth distribution pattern in the source. Using the rare-earth analysis of the basalt, known crystal/liquid partition coefficients for basalt at 20 kb, 1425° C, and a specific partial melting model, one could solve for the degree of partial melting represented by the primitive liquid and the proportions of phases in the residue after adjustment is made for some olivine fractionation. The compositions of the phases in the residue and the basaltic liquid composition could then be mixed in the proportions determined by the rare-earth analysis to obtain an estimate of the initial source composition. Although this exercise has not been carried through, it illustrates how the inverse approach could be used to provide a detailed estimate of the source region composition.

Appraisal of the inverse approach

It is clear from the discussion in section 3.3.7 of primary magmas on Earth that petrologists are a long way from agreement on the compositions of primary magmas in various tectonic environments. Successful application of the inverse approach to infer the constitution of the basaltic source material on another planet requires that primary magmas be identified, yet the data on which these identifications will be based will, in the foreseeable future, be much more limited than the data base for Earth. Gaps in understanding are certain to exist for a long time, yet it is a remarkable advance that we are even able to address problems of magma genesis on other planets.

Despite the remaining uncertainty about the primary magma composition for the ocean tholeiite case study, the application of the inverse approach prescribed in Fig. 3.1.4 was fairly straightforward through step 6. Liquids can be identified and the most primitive ones can be chosen. They can be related to their differentiates, they can be assumed to be primary, and their high pressure melting relations can be determined. At this point a critical logical crossroad is reached. The terrestrial petrologist, realizing the importance of orthopyroxene as a residual mantle phase, may abandon the connection between the primitive oceanic tholeiite and a primary magma since no liquidus saturation point with olivine and orthopyroxene is discovered. A primary liquid is then reconstructed in the simplest way possible—adding olivine to the primitive liquid—and an internally consistent solution is demonstrated. But this is not a unique solution, as discussed in the preceding section (see section 3.1.4).

An alien petrologist, without the benefit of experience with terrestrial peridotites, might well take a different turn at the critical crossroad. He might assign petrologic significance to the multiply-saturated inflection on the liquidus of the primitive samples at 11–12 kb and interpret the liquidus phases as components of the mantle residual to primitive oceanic tholeiite generation. The primitive basalts would be accepted as primary, just as they were not accepted by the terrestrial petrologist. Alternately, the alien petrologist may be influenced by experience on his home planet and conclude that the primitive samples are not primary, but for reasons other than the absence of orthopyroxene at the inflection, which caught the attention of the terrestrial petrologist. Furthermore, the alien petrologist might well choose a different expedient for reconstructing the primary magma from the constraints imposed by high pressure melting studies of the primitive samples. He could add any combination of the high pressure liquidus minerals (or minerals in reaction equilibrium not seen on the liquidus) to reconstruct his preferred primary magma and not do violence to the experimental data. Needless to say, the alternate choices may well lead to what we would consider unrealistic models of the Earth's mantle and basalt genesis.

Most of our choices involved assuming the simplest model, and an understanding of what we knew would lead us to our perception of a "correct" answer. A Martian, using martian experience, might well have made different assumptions or choices. The lesson to be learned from this is, of course, that regardless of the objectivity with which we may claim to apply our flow chart to other basalts and other planets, we always do so within the framework of what we already know about the Earth. We must emphasize the importance of the cycle at the end of the flow chart which requires testing the models derived from basalt studies against other constraints on the source regions. With basalt studies, internally consistent solutions can always be produced. With intelligent choices at branches in the flow chart and sufficient additional constraints there is a reasonable presumption of arriving at the correct solution, but no guarantee. We will thus start our tour of "experimental petrology on the other planets" with confidence in our vehicle but with some circumspection as to the sufficiency of road signs at critical turns.

3.4 LUNAR BASALTS

The return of lunar basalt samples provided the first opportunity to test the scheme that we have plotted, in retrospect, in Fig. 3.1.4. The basalts themselves are our only petrologic clues to the natures of igneous processes and source regions within the Moon. None of the rocks returned from the Moon includes fragments analogous to the ultramafic xenoliths in terrestrial basalts and kimberlites, so we possess no direct evidence of mineral assemblages in the lunar mantle. Such fragments occur at relatively few terrestrial localities, so even if they were erupted on the lunar surface, it is not surprising that we have failed to find them in our limited sampling.

The lower oxygen fugacity, very low alkalis and volatile components, and the virtual absence of hydrous minerals are the most obvious features distinguishing lunar from terrestrial basalts. The differences in oxygen fugacity are reflected in lower Cr^{3+}/Cr^{2+} and Fe^{3+}/Fe^{2+} of lunar basalts. In fact lunar basalts appear to crystallize Fe(Ni) metallic alloy phases. These features indicate that the oxygen fugacity and temperature at the beginning of melting differ considerably in the interiors of the two planets. According to Sato (1978), the lower oxidation state of lunar basalts is a consequence of the absence of H_2O from the lunar mantle. In both lunar and terrestrial mantles, the oxygen fugacity appears to be buffered by the graphite-CO_2 system. Lunar basalts continue to be buffered in this way throughout ascent, but in terrestrial basalts, reaction between graphite and H_2O produces CO_2 and H_2, H_2 escapes due to its high mobility, and exhaustion of the graphite eliminates the buffer; subsequent thermal dissociation of H_2O, with continued loss of H_2, leads to increased oxygen fugacity.

Basaltic samples returned from the Moon's two major physiographic provinces—highlands and maria—are distinctly different in composition and in age. The highly-shocked and brecciated, plagioclase-rich rocks of the lunar highlands have crystallization ages of 3.9–4.1 b.y. Samples of the more mafic mare basalts range in crystallization age between 3.1–3.9 b.y., although photogeologic studies indicate that unsampled basalt flows as young as 2.5 b.y. exist in some mare regions (Chapter 8). Detailed discussions of lunar rock types, including their chemical and mineral constituents and ages, are given in Chapters 1, 7 and 8; relevant points will be reiterated where necessary in this discussion. Details for experimental studies of compositions related to lunar mare volcanism and highlands volcanism are listed in Tables 3.4.1 and 3.4.2, respectively.

Table 3.4.1 Experimental studies of lunar mare basalt compositions.
a. Melting experiments at 0–1 bar. [a]

Sample Description	Starting Material	Container	Technique	Reference	Criteria Met[b]
Apollo 17					
74275	natural	Pt wire loop	CO-CO_2 atmosphere, controlled f_{O_2} range	*(14)*	2, 3, 4
Luna 24 low-Ti ferrobasalts					
Luna 24 average	synthetic	high purity Fe	evacuated SiO_2-glass tubes, controlled cooling rates	*(4)*	2, 3, 4
Apollo 12 olivine-normative					
12009 analog	synthetic glass	Pt wire loop	controlled cooling rates, CO-CO_2 atmosphere	*(2)*	2, 3, 4
Apollo 15 olivine-normative					
15555 analog	synthetic glass	Pt wire loop	controlled cooling rates, CO-CO_2 atmosphere	*(2)*	2, 3, 4
15555	natural	high purity Fe	evacuated SiO_2-glass tubes, controlled cooling rates	*(15)*	1, 2, 3, 4
Apollo 15 quartz-normative					
15597	synthetic	high purity Fe	evacuated SiO_2-glass tubes, controlled cooling rates	*(6)* *(5)* *(7)*	2, 3, 4
15065	natural	high purity Fe	evacuated SiO_2 tubes	*(15)*	1, 2, 3, 4

b. Miscellaneous experimental studies relevant to mare basalt petrogenesis. [c]

Description of Experiments	Reference
Oxidation states of Ti, Eu and Cr in synthetic lunar basalts a function of f_{O_2}.	*(12)*
Cr partitioning among olivine, spinel and liquid.	*(1), (8)*
Cr partitioning between crystals and liquid.	*(13)*
Intrinsic f_{O_2} measurements on Apollo 16 and 17 rocks.	*(11)*
Model hybrid origin for mare basalts.	*(10)*
Armalcolite stability relations.	*(3)*
Sm, Tm, Sc, Cr and Mn partitioning among olivine, ilmenite, armalcolite and liquid.	*(9)*

c. Melting experiments at high pressure.[d]

Sample Description	Pressure Range, kb	Temperature Range, °C	Starting Material	Container	Technique	Duration (hours)	Reference	Criteria Met
Apollo 15 quartz-normative								
15065	5–30	1130–1375	natural	Mo	instroke-8%	2–42	*(15)*	2, 3, 4
Apollo 15 olivine-normative								
15555	5–30	1125–1450	natural	Mo, Fe, C	instroke-8%	2–42	*(15)*	2, 3, 4

[a] Continuation of Table 1 from Kesson and Lindsley (1976). See also comments by O'Hara and Biggar (1977).
[b] The criteria are as follows: (1) demonstration of equilibrium (not relevant for rate studies), entry of phases in parentheses being bracketed; (2) temperature control; (3) oxygen fugacity controlled or monitored at levels appropriate to lunar petrogenesis; and (4) bulk composition maintained.
[c] Continuation of Table 2 from Kesson and Lindsley (1976).
[d] Continuation of Table 3 from Kesson and Lindsley (1976).

References:
(1) Akella *et al.* (1976); *(2)* Bianco and Taylor (1977); *(3)* Friel *et al.* (1977); *(4)* Grove (1978); *(5)* Grove and Bence (1977); *(6)* Grove and Raudsepp (1978); *(7)* Grove and Walker (1977); *(8)* Huebner *et al.* (1976); *(9)* Irving *et al.* (1978); *(10)* Ringwood and Kesson (1976); *(11)* Sato (1976); *(12)* Schreiber (1977); *(13)* Schreiber and Haskin (1976); *(14)* Usselman and Lofgren (1976); *(15)* Walker *et al.* (1977b).

Table 3.4.2 Experimental studies of compositions related to lunar highlands volcanism.
a. Experiments at 0–1 bar.

Sample	Starting Material	Container	Technique	Duration (hours)	References	Criteria Met[†]
			Apollo 14			
14053	rock powder and glass	Mo	H_2-CO_2 atm at Fe/FeO	5–12	*(2)*	1, 2, 3, 4
14072	rock powder	Mo	evacuated SiO_2	17–148	*(16)*	1 (sp, px, plag), 2
14162	soil powder	Mo	H_2-CO_2 atm at Fe/FeO	5–72	*(2)*	2, 3, 4
14259	soil powder	Fe	evacuated SiO_2	16–64	*(13)*	2, 3, 4
14259	soil powder	Mo	evacuated SiO_2	1–10	*(16)*	2
14310	rock powder and glass	Mo	H_2-CO_2 atm at Fe/Feo	5–72	*(2), (4)*	1 (plag, sp,ol,px), 2, 3, 4
14310	rock powder $+1.5\% Na_2O+0.5\% K_2O$	Mo	H_2-CO_2 atm at FeO	5	*(2)*	2, 3, 4
14310	glass and devitrified glass	Fe	evacuated SiO_2	1?	*(5)*	1 (plag), 2, 3?
14310	rock powder	Mo	evacuated SiO_2	21–73	*(6), (8)*	2, 3, 4
14310	synthetic glass	Pt loops	CO-CO_2 atm below Fe/FeO, controlled cooling		*(10)*	2, 3
14310	rock powder?	Fe	evacuated SiO_2	16	*(13)*	2, 3, 4
14310	rock powder	Mo, Fe	evacuated SiO_2	4–111	*(16)*	1 (plag, px, silica),2
14321	breccia powder	Mo, Fe	Fe/FeO and H_2 90% – CO_2 10% $f_{O_2} \approx 10^{-14}$	5–11	*(2)*	2, 3, 4
			Apollo 15			
15405 KREEP	reduced gel	Mo	evacuated SiO_2	20–40	*(14)*	2, 3, 4
15382	reduced gel	Mo	evacuated SiO_2	2–188	*(8)*	2, 3, 4
			Apollo 16			
60025	rock powder	Mo	H_2-CO_2 atm at Fe/FeO	?	*(3)*	2, 3
60315	rock powder	Mo	H_2-CO_2 atm at Fe/FeO	?	*(3)*	2, 3
60335	rock powder	Mo	H_2-CO_2 atm at Fe/FeO	?	*(3)*	2, 3
62295	rock powder	Mo	H_2-CO_2 atm at Fe/FeO	?	*(3)*	2, 3
62295	rock powder	Fe	H_2-CO_2 atm below Fe/FeO	1–4	*(17)*	2, 3, 4
63545	synthetic glass ± crystals	high purity Fe	evacuated SiO_2	1.5–51	*(1)*	1, 2, 3, 4
68415	rock powder	Mo	H_2-CO_2 atm at Fe/FeO	3	*(3), (4)*	2, 3
68415	rock powder	Mo	evacuated SiO_2	52–163	*(17)*	2

[†]See Table 3.4.1.

References for **Table 3.4.2a**:

(1) Delano (1977); *(2)* Ford *et al.* (1972); *(3)* Ford *et al.* (1974); *(4)* Ford *et al.* (1977); *(5)* Green *et al.* (1972); *(6)* Hess *et al.* (1975); *(7)* Hess *et al.* (1977); *(8)* Hess *et al.* (1978); *(9)* Lipin (1978); *(10)* Lofgren (1977); *(11)* McKay and Weill (1977).

Table 3.4.2a *(Continued)*

Sample	Starting Material	Container	Technique	Duration (hours)	References	Criteria Met†
			Apollo 17			
76015	synthetic glass ± crystals	high purity Fe	evacuated SiO_2	5–21	*(1)*	2, 3, 4
76055	synthetic glass ± crystals	high purity Fe	evacuated SiO_2	1–64	*(1)*	1, 2, 3, 4
77135	rock powder, glass and devitrified glass	Mo	H_2-CO_2 atm at Fe/FeO	?	*(15)*	1, 2, 3
			Related Compositions			
"Monzonite"	reduced gel ± crystals	Mo	evacuated SiO_2	3–89	*(14)*	1 (px), 2, 3, 4
Apollo Soil Survey (1971) Glass C	glass ± (Na_2O, K_2O)	Mo	H_2-CO_2 atm at Fe/FeO	5–72	*(2)*	2, 3, 4
Apollo Soil Survey (1971) Glass B	glass ± (Na_2O, K_2O)	Mo	H_2-CO_2 atm at Fe/FeO	5–72	*(2)*	2, 3, 4
Model KREEP	glass	Fe	evacuated SiO_2	1?	*(5)*	1 (ol), 2, 3?
Apollo Soil Survey (1971) Glass E	glass ± (Na_2O, K_2O)	Mo	H_2-CO_2 atm at Fe/FeO	5	*(2)*	2, 3, 4
Low-K Fra Mauro	synthetic glass	pre-soaked Pt-Rh loops	H_2-CO_2 atm	24–48	*(11)*	2, 3, 4
Low-K Fra Mauro	reduced gel	Mo	evacuated SiO_2	24–144	*(7)*	2, 3, 4
Evolved Low-K Fra Mauro	reduced gel	Mo	evacuated SiO_2	48	*(8)*	2, 3, 4
Average Highlands	glass?	Pt loops	H_2-CO_2 atm	24–48	*(18)*	2, 3, 4
"Anorthositic Gabbro"	glass	Mo	evacuated SiO_2	3–115	*(17)*	2
$CaAl_2Si_2O_8$-$MgSiO_3$-$FeSiO_3$	synthetic glass + gel	Pt loops	CO-CO_2 atm at Fe/FeO	17–163	*(12)*	2, 3, 4
Mg_2SiO_4-Fe_2SiO_4-$CaAl_2Si_2O_8$-SiO_2	synthetic crystalline mixture	high purity Fe	evacuated SiO_2	17–161	*(9)*	1, 2, 3, 4

(12) Merrill and Williams (1975); *(13)* Muan *et al.* (1972); *(14)* Rutherford *et al.* (1976); *(15)* Storey *et al.* (1974); *(16)* Walker *et al.* (1972); *(17)* Walker *et al.* (1973a); *(18)* Weill and McKay (1975).

Table 3.4.2b *(Continued)*

b. 1 bar related experiments.

Sample	Starting Material	Experiment	References	Criteria Met[†]
14259	synthetic glass	crystal growth rate measurements	*(14)*	2
14259	synthetic glass + liquid	viscosity measurements	*(1)*	2
14310	synthetic glass + liquid	viscosity measurements	*(1)*	2
14310	synthetic glass	crystal growth rate measurements	*(14)*	2
15286	synthetic glass + liquid	viscosity and crystal growth measurements	*(15)*	2, 3
15418	synthetic glass + liquid	viscosity and crystal growth measurements	*(16)*	2
15498	synthetic glass + liquid	viscosity and crystal growth measurements	*(15)*	2, 3
60095	synthetic glass + liquid	viscosity and crystal growth measurements	*(16)*	2
60095	synthetic glass + liquid	viscoisty and crystal growth measurements	*(16)*	2
60095	natural glass	crystal growth measurements	*(10)*	2
65016	synthetic glass + liquid	viscosity and crystal growth measurements	*(16)*	2
68502	synthetic glass + liquid	viscosity and crystal growth measurements	*(2)*	2

HIGH PRESSURE EXPERIMENTS

Sample	Starting Material	Pressure range,bk	Temp. range °C	Container	Technique	Duration (hours)	References	Criteria Met[†]
				Apollo 14				
14053	rock powder ± H_2O	2–5	1200–950	$Ag_{70}Pd_{30}$ Pt sealed	internally heated gas vessel	0.17–17	*(5)*	2, 3 (FeO/Fe_3O_4)
14072	rock powder	5–20	1150–1440	Mo	piston in -8%	3–15	*(17)*	2
14259	powdered soil	5–20	1150–1425	Mo	piston in -8%	1.5–24	*(17)*	2
14310	rock powder	2–10	1336–1175	Fe sealed	internally heated gas vessel	0.5–22	*(5)*	2
14310	rock powder ± H_2O ± Na_2O	2–5	1200–1000	Pt, Fe $Ag_{70}Pd_{30}$	internally heated gas vessel sealed	0.5–22	*(5), (6)*	2, 3 (FeO/Fe_3O_4)
14310	rock powder ± Na_2O ± K_2O	10–25	1450–1175	Pt Fe	piston out	1–4	*(5)*	2
14310	glass + seeds	5–25	1100–1350	Fe	piston in -10%	1?	*(7)*	2,3,4
14310	rock powder and glass	5–25	1470–1100	graphite	piston out	0.5–4	*(11)*	2, 4
14310	rock powder + glass + 3.2% Na_2O + ~1% K_2O	2.5–10	1300–1200	graphite	piston out	1–3.5	*(11)*	2, 4
14310	rock powder	5–20	1150–1405	Mo	piston in -8%	2–24	*(17)*	2
14321	rock powder	2–10	1336–1175	Fe sealed	internally heated gas vessel	2–88	*(5)*	2
14321	breccia powder	10–25	1440–1150	Pt Fe	piston out	0.17–3	*(5)*	2
14321	rock powder ±H_2O±Na_2O±K_2O	2–5	1200–950	$Ag_{30}Pd_{30}$ Pt sealed	internally heated gas vessel	0.17–22	*(5)*	2, 3 (FeO/Fe_3O_4)

[†]See Table 3.4.1, footnote b.

References for **Table 3.4.2b**:
(1) Cukierman *et al.*; (1972) *(2)* Cukierman *et al.* (1973); *(3)* Delano (1977); *(4)* Ford (1976); *(5)* Ford *et al.* (1972); *(6)* Ford *et al.* (1977).

Table 3.4.2b *(Concluded)*

Sample	Starting Material	Pressure range, bk	Temp. range °C	Container	Technique	Duration (hours)	References	Criteria Met†
				Apollo 15				
15386	synthetic glass	1–5	1200–1125	Mo	internally heated gas apparatus	19–20	*(9)*	2, 3, 4
				Apollo 16				
62295	rock powder	5–30	1525–1250	graphite Mo	piston in -8%	2–24	*(18)*	2
62295	rock powder	5–12.5	1375 1200	graphite	piston out	1–4	*(8)*	2, 4
63545	synthetic glass ± crystals	5–30	1650–1230	high purity Fe, Mo	overpump 1 kb adjust to nominal P	0.7–3.3	*(3)*	1, 2, 3, 4
68415	rock powder	5–30	1500–1225	Mo	piston in -8%	1–24	*(18)*	2
68416	rock powder	5	1400–1225	graphite	piston out	1–3	*(8)*	2, 4
				Apollo 17				
76015	synthetic glass ± crystals	5–15	1395–1295	high purity Fe	overpump 1 kb, adjust to nominal P	0.7–1.8	*(3)*	1, 2, 3, 4
76055	synthetic glass ± crystals	7.5–25	1560–1380	high purity Fe, Mo	overpump 1 kb adjust to nominal P	0.7–1.8	*(3)*	1, 2, 3, 4
				Related Compositions				
Model KREEP	synthetic glass	5–25	1100–1350	Fe	piston in -10%	1?	*(7)*	2, 3, 4
Apollo 17 KREEP	synthetic glass glass	1–5	1200–1125	Mo	internally heated gas apparatus	19–20	*(9)*	2, 3, 4
"Anorthositic Gabbro"	glass	5–20	1500–1275	graphite Mo	piston in -8%	1–19	*(18)*	2
"Highland Basalt"	synthetic glass ± crystals		1450–1000	Fe	piston in -10%	.2≃100	*(13)*	1, 2, 3
$CaAl_2Si_2O_8-FeSiO_3$	synthetic glass + gels	2	1300–1150	Mo	internally heated gas apparatus	1–7	*(12)*	2, 3, 4
				Related Experiments				
68415	rock melt + H_2O	0–1	1250	Pt	depressurization-devolatilization quench		*(4)*	
60335	rock melt + H_2O	0–1	1250		depressurization-devolatilization quench		*(4)*	
60315	rock melt + H_2O	0–1	1250		depressurization-devolatilization quench		*(4)*	
62295	rock melt + H_2O	0–1	1250		depressurization-devolatilization quench		*(4)*	
77135	synthetic glass + H_2O	0–.25	1250–1180	Pt	depressurization devolatilization quench		*(4)*	

(7) Green *et al.* (1972); *(8)* Hodges and Kushiro (1973); *(9)* Irving (1977b); *(10)* Klein and Uhlmann (1976); *(11)* Kushiro *et al.* (1972b); *(12)* Merrill and Williams (1975); *(13)* Råheim and Green (1974); *(14)* Scherer *et al.* (1972); *(15)*Uhlmann and Klein (1976); *(16)* Uhlmann *et al.* (1974); *(17)* Walker *et al.* (1972); *(18)* Walker *et al.* (1973a).

Table 3.4.1 updates the bibliography for the previous tabulation and evaluation of mare basalt experiments by Kesson and Lindsley (1976).

Estimates of the abundance of Fra Mauro basalts are somewhat variable. They range from a very conservative estimate that basalts comprise a few percent of the crust—a delicate lacing of partial melt in a predominantly anorthositic norite crust—to estimates that basalts comprise a major portion of the lower crust (Ryder and Wood, 1977). Whichever estimate one chooses, it seems inescapable that these rocks are volumetrically more important on the Moon than are the mare basalts, which make up less then 1% of the crust (Head, 1976; DeHon and Waskom, 1976; Hörz, 1978).

3.4.1 Highland basalts

The lunar highland crust appears to be a 60–100 km thick layer of plagioclase-rich basalts, anorthositic norites, anorthosites, and noritic anorthosites which have been almost completely reconstituted by many impact events subsequent to their initial crystallization. Primary igneous textures cannot be distinguished unambiguously from textures produced by the crystallization of impact melts. Interpretations of the ultimate origins of these rocks have been based primarily upon their compositions and upon equilibrium phase relations, rather than textures.

Plagioclase-rich basalts were found as important components of the highland samples at Apollo 14, 15, 16, 17, and Luna 20 sites and were sampled as soil fragments at other mare landing sites, presumably as the result of ballistic transport from neighboring highlands. Their first abundant return was from the Apollo 14 mission to the Fra Mauro Hills; hence, this suite became known as the Fra Mauro basalt suite. The nomenclature and compositional variations of these rocks are reviewed in Chapter 1. Certain variants among the Apollo 15, 16 and 17 sample returns were not recognized until later. The relatively high abundances of K, REE and P and of other incompatible elements in these samples were incorporated in the term KREEP basalts. The extent of enrichment of this collection of elements is highly variable.

Age dating of members of this suite has indicated that most samples crystallized from 3.9–3.95 b.y. ago, with some Apollo 17 samples having ages near 4.1 b.y. ago (Shih, 1977). Recent Nd/Sm work has suggested that the last REE fractionation to affect these samples took place about 4.3 b.y. ago (Lugmair and Carlson, 1978). The picture which emerges is that the chemical fractionation which established the element abundance patterns in Fra Mauro basalts took place 0.2–0.4 b.y. before the eruption and solidification of the magmas.

The Fra Mauro rocks are characterized by a predominance of normative low-Ca pyroxene over high-Ca pyroxene. Chemical variations in the Fra Mauro basalt system can be described approximately with the components $(Mg,Fe)_2SiO_4$-$CaAl_2Si_2O_8$-SiO_2. One-atmosphere liquidus diagrams for this compositional system have been constructed on the basis of experimental work on the Fra Mauro samples themselves (Walker *et al.*, 1972, 1973a, b; Kushiro *et al.*, 1972b) and on synthetic analog systems (McKay and Weill, 1976; Lipin, 1976, 1978). Additional experimental work on subsystems and on specific complex compositions approximately within the system has been performed by Ford *et al.* (1972), Green *et al.* (1972), Hodges and Kushiro (1973), Råheim and Green (1974), Merrill and Williams (1975), Irving (1977b) and Hess *et al.* (1977).

Results of these experiments are summarized in Fig. 3.4.1, a pseudoternary liquidus diagram. The liquidus saturation curves are for 1 bar pressure, dry, and low oxygen fugacity. The projections of the Fra Mauro basalt compositions on this diagram are concentrated

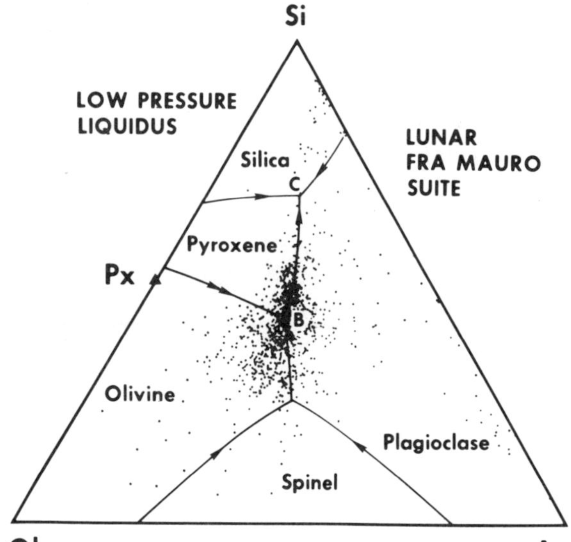

3.4.1 Fra Mauro basaltic compositions plotted on the low-pressure liquidus surface of the system olivine-anorthite-silica (Walker *et al.*, 1973a). Data base includes rocks, lithic fragments, and glass compositions compiled by Wood (1977). Fra Mauro suite rocks are separated from ANT suite rocks on the basis of their higher Mg/(Mg + Fe) and lower normative plagioclase. These two non-mare populations are separated by a line on plots of Mg/(Mg + Fe) vs. an/(an + ol + qz) with slope 1.

along the low-temperature, plagioclase-saturated trough, which suggests that the plagioclase-saturated equilibria are responsible for their compositional distribution. These phase relations probably do not change dramatically for pressures less than about 5 kb (Walker *et al.*, 1972), so the events leading to the observed compositional distribution could have occurred anywhere in the outer 100 km of the Moon (Fig. 3.1.1). It has been common practice to assume that these events took place in the outer 60 km of plagioclase-rich lunar crust. These experiments provide a positive answer for step 2 of the flow diagram of Fig. 3.1.4. We have concluded that the compositions of Fra Mauro basalts are related by low-pressure processes.

Walker *et al.* (1972, 1973b) proposed that partial melting of highland crust composed of the mineral assemblage olivine, plagioclase, low-Ca pyroxene(s) was responsible for the production of parental Fra Mauro basalts. Their reason for favoring partial melting as the generating mechanism, instead of fractional crystallization within crustal-level magma chambers, was the observation that a large proportion of the sample population has compositions near point B in Fig. 3.4.1, a liquid which is saturated with plagioclase, olivine and low-Ca pyroxene. Because this is a reaction-relation equilibrium in which olivine disappears with falling temperature, it was argued that fractional crystallization could not produce a preferred liquid of this composition because fractional crystallization would not allow olivine to remain present on the liquidus with falling temperature.

Whereas most of the samples plotted in Fig. 3.4.1 do not have igneous textures and could possibly have some admixture of anorthositic material, at least fifty Apollo 15 KREEP basalt samples (including 15405 quartz monzodiorite) have been identified which appear to be igneous liquids on the basis of texture, metal compositions and siderophile trace element abundances (Irving, 1975, 1977c). The compositions of these are plotted in Fig. 3.4.2 (a portion of Fig. 3.4.1). The distribution of liquid compositions along the low-Ca pyroxene-plagioclase cotectic from B to C is most simply interpreted as a result of low pressure fractional crystallization of a parental liquid at B (Walker *et al.*, 1973b), a hypothesis consistent also with trace element variation of these samples (Irving, 1977c).

An embarrassment for the partial melting model for the parental Fra Mauro or KREEP magmas is the difficulty of finding a suitable heat source to remelt a portion of the lunar crust once it has crystallized initially to form the required three-phase assemblage. For this and other reasons, Lipin (1976) resurrected the model of Smith *et al.* (1970) and Wood *et al.* (1970) in which the plagioclase-rich basalts were the end products of fractional crystallization of large volumes of magma. Lipin (1976) proposed that the KREEP (or Fra Mauro) basalts were residual liquids from crystallization of a magma ocean. This alternate interpretation appears to be inconsistent with the relatively high Mg/Fe (about 50–70) and relatively high Cr contents of the samples, because these values should be decreased by a fractional crystallization process.

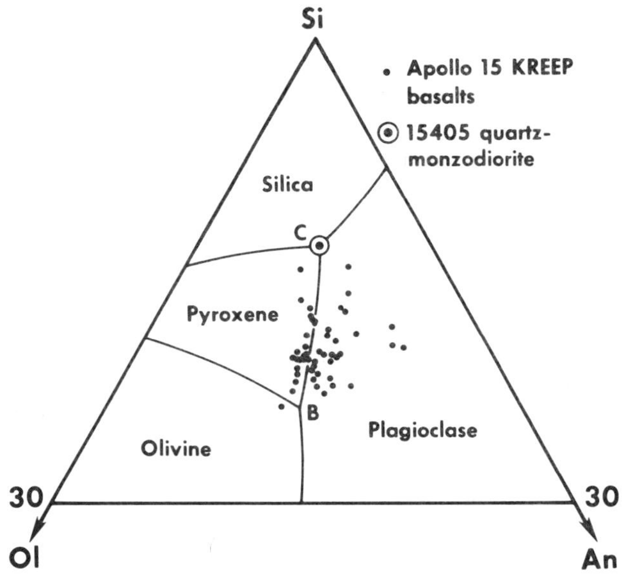

3.4.2 Part of the pseudo-ternary diagram of Fig. 3.4.1 after Irving (1977c). Data from melt-textured Apollo 15 KREEP basalts and 15405 quartz monzodiorite. The effects of low-pressure crystal fractionation have produced a distribution of compositions from melts near B to residual liquids at C.

Recently, more complex models have been proposed in which either crustal assimilation (Hubbard and Minear, 1975; Warren and Wasson, 1979) or magma mixing (O'Hara, 1977) are involved. In these models, hot liquid generated in the lunar interior intrudes the crust. The thermally more massive crust imposes upon this foreign magma the major element chemistry appropriate to saturation with crustal phases (plagioclase, magnesian olivine and low-Ca pyroxene). However, assimilation of the crustal material, which is poor in incompatible elements, has relatively little effect upon the incompatible element abundances of the hot foreign liquid. Additional complications may result if the hot liquid intrudes pools of stagnant liquid which have already accommodated their chemistries to the crustal saturating phases (O'Hara, 1977). In these models, the hot foreign magma is itself the heat source for generating crystal-liquid equilibration relations within the crust. It remains to be shown, however, whether the major and trace elements can be so effectively decoupled.

Whatever the detailed mechanism of Fra Mauro basalt production, it is clear that the major element compositions of the magmas are controlled by processes operating at rather low pressures, and that we do not have samples in this suite whose major element compositions were established by partial melting events at depths greater than 60–100 km in the lunar mantle. In terms of the flow diagram (Fig. 3.1.4), these basalts lead us via step 5 (multiply-saturated liquid at low pressure) to conclude that the suite was generated by processes occurring within the lunar crust.

3.4.2 Mare basalts

The mare basalt samples are generally less shocked and brecciated than the highlands rocks, so step 1 in the flow chart of Fig. 3.1.4 is facilitated, because textural criteria can be applied in the search for parental magmas.

Search for parental magmas

Fine-grained aphyric or vitrophyric basalts, products of rapid cooling and solidification, are the mare basalt samples most likely to have preserved original liquid compositions. Such basalts are relatively rare in the lunar collection, and much emphasis has been placed upon the study of these few. None of the basalts is entirely convincing as a likely liquid composition on the basis of petrographic examination, because all contain microphenocrysts or significant textural inhomogeneities or both. These features comprise the interpretive link between rock and liquid chemistries because the magma could have been enriched in microphenocrysts prior to solidification and because the textural domains, often characterized by compositional differences, may introduce sampling problems.

O'Hara *et al.* (1975, and earlier references) have rejected the mare basalt hand specimens themselves as valid samples of magmas erupted at the various landing sites, contending that the local soil compositions, corrected for obviously extraneous highlands debris, are a better statistical sample of local magma compositions. The results of their exercise, with the exception of soils at the Apollo 15 landing site, indicate that the average lava erupted should have had the composition of a plagioclase-saturated cotectic liquid. Hand specimens corresponding to such cotectic liquids do exist among the samples (e.g., 12038, 10047, 75035), but they do not include the fine-grained, quench-textured samples. According to their argument, hand specimens which do not correspond to the average cotectic compositions must have crystallized from magma which had become enriched with microphenocrysts. If this were true, the microphenocrysts would have compositions which were in equilibrium with the cotectic liquid upon eruption. However, experiments consistently have shown that the equilibrium compositions of olivines and pyroxenes grown in equilibrium with the cotectic liquid compositions are more Fe-rich than microphenocrysts actually present in the hand specimens. Similar experiments employing liquids having the compositions of the hand specimens themselves produce ferromagnesian liquidus phases which do match microphenocrysts in the hand specimens (Green *et al.*, 1971a, b; Grove *et al.*, 1973; Kesson, 1975).

In addition, it has been shown that armalcolite, which is found as microphenocrysts in high-Ti basalts, is not stable at the plagioclase-saturated cotectic (Longhi *et al.*, 1974), so the quench-textured, armalcolite-bearing samples could not have been derived by simple enrichment of such a plagioclase-saturated cotectic liquid with cognate armalcolite phenocrysts.

These results are comprehensible only if the hand specimen compositions represent liquid compositions which have not been modified significantly by admixture of crystals. The experimental arguments have been sufficiently convincing that most petrologists have reached the point of trying to identify the most primitive liquid compositions among the fine-grained, quench-textured mare basalt hand specimens. For relatively small compositional differences, phenocryst-matching tests do distinguish liquids from crystal-liquid mixtures reliably (step 1 in Fig. 3.1.4). Although the tests discussed above demonstrate that the hand specimens are not cotectic liquids enriched in crystals, they do not rule out the possibility that the hand specimens represent a microphenocryst-enriched parental liquid which was not cotectic.

Low-titanium mare basalts

Low-titanium mare basalts are representative of the Apollo 12 and 15 sites. The Apollo 12 suite consists of olivine-normative and quartz-normative (pigeonite) basalts which appear to be related by olivine fractionation (Rhodes *et al.*, 1977). Other distinctive Apollo 12 basalt suites are the ilmenite basalts and feldspathic basalts (see Chapter 1). The Apollo 15 basalts also include olivine- and quartz-normative suites, but these two suites (each of which shows internal chemical fractionation trends) are unrelated to one another by low-pressure fractionation processes (e.g., Papike *et al.*, 1976).

The low-pressure equilibrium experiments (Green *et al.*, 1971a, b; Kushiro and Haramura, 1971; Grove *et al.*, 1973; Kesson, 1975; Walker *et al.*, 1976a, b) substantiate that the various low-Ti basalt suites are characterized by low-pressure fractionation trends and that many of the samples can be considered to represent liquids (steps 1 and 2 in Fig. 3.1.4). The least-fractionated Apollo 12 and 15 olivine-normative basalts have olivine as their sole liquidus phase and can be treated as the parental liquids of their respective suites (step 3 in Fig. 3.1.4).

The phenocryst-matching test has proven sufficiently sensitive for the Apollo 12 olivine basalt suite, for example, to show that the most olivine-normative samples probably are crystal-liquid admixtures and do not represent unmodified liquid compositions. This conclusion rests upon the observation that liquidus olivines grown experimentally in those samples are significantly more magnesian than the olivine phenocrysts actually in the samples. The probable compositions of liquids parental to this suite lie between the most olivine-normative basalts and the cotectic liquid.

Green *et al.* (1971b) proposed that one vitrophyre, 12009, has an appropriate composition for such a parental liquid. Walker *et al.* (1976b) have shown that a quantitative and internally consistent explanation of the chemical differentiation within the olivine basalt suite can be based upon a parental 12009-liquid, in terms of kinetic and physical properties measured experimentally as functions of cooling rate.

Donaldson *et al.* (1975b) have shown that a few percent of the olivine microphenocrysts in 12009 were present before eruption and may be accidental inclusions, as suggested by their heterogeneous distribution from one section to the next. The calculations by Walker *et al.* (1976b), however, suggest that the amount of admixed olivine in 12009 must be less than about 3% on the basis of the effect a greater amount would have had upon magma differentiation. The conclusion, then is that sample 12009 probably represents the most primitive liquid in the Apollo 12 collection, and therefore is the best available candidate for a parental magma.

Arguments have not been made in similar detail for the analogous Apollo 15 suite, but liquidus phase compositions suggest that samples 15016 and 15555 are suitable choices for parental magmas (Kesson, 1975).

These parental magmas have primitive geochemical characteristics (Chapter 1), and this leads us through steps 4, 5 and 6 in the flow diagram of Fig. 3.1.4 to determination of the high-pressure melting relationships of the primitive liquid compositions. Figure 3.4.3 represents the phase relationships of Apollo 12 picrite liquid, based on results for 12009 (Green *et al.*, 1971b) and 12002 (Walker *et al.*, 1976). Note the arrowed point near 15 kb where the picrite liquid equilibrates with olivine and two pyroxenes. If the liquid is a primary

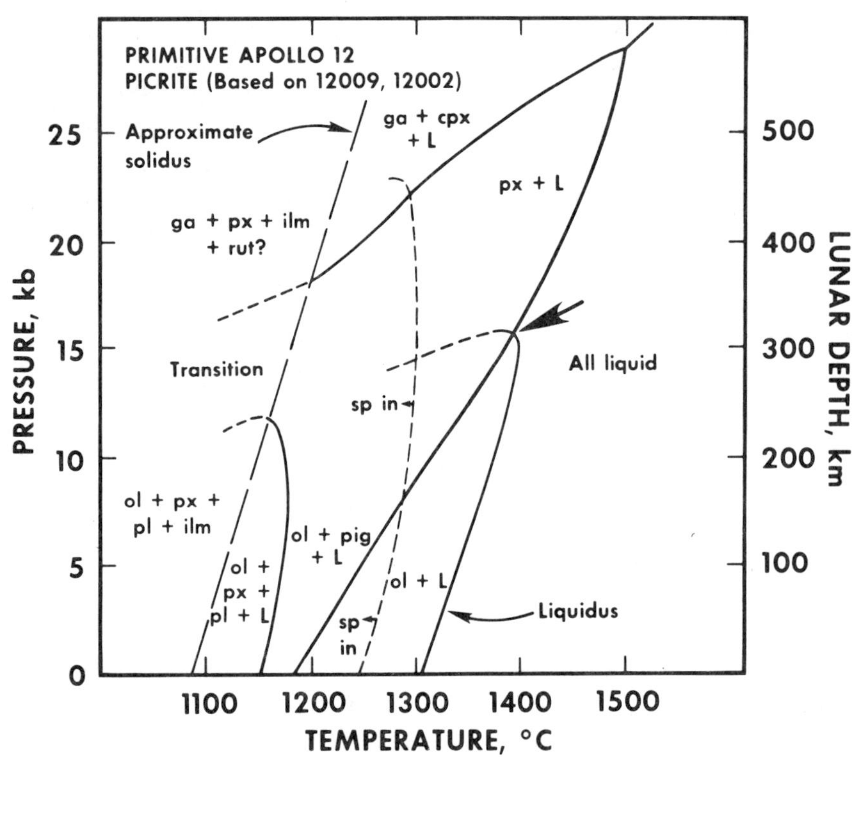

3.4.3 Phase relations of a primitive Apollo 12 picrite liquid as a function of pressure and temperature. Based on 12009 (Green *et al.*, 1971b) and 12002 (Walker *et al.*, 1976a). Note liquidus region at about 15 kb, where picritic liquid equilibrates with olivine and two pyroxenes. This has been interpreted in terms of the segregation conditions and residual mineralogy for primary lunar picritic magma.

magma generated from an olivine-pyroxenite source region (Chapter 4), then this point corresponds to the depth where the magma was segregated from the source. The cosaturation pressure for olivine and pyroxene(s) is commonly used as an index of depth of origin from primary lunar magmas.

Green *et al.* (1971b), Chappell and Green (1973) and Green and Ringwood (1973) have coupled geochemical data with their high-pressure experimental studies, and interpreted the olivine-normative low-Ti basalts as primary magmas, partial melts of an olivine-pyroxenite source at depths of 200–400 km for Apollo 12 basalts and 300–400 km for Apollo 15 basalts. These interpretations concur with those of Walker *et al.* (1976a) for 12002. Kesson (1975) concluded that the minimum source depth for a primitive Apollo 15 olivine-normative basalt was 240 km.

The range in inferred source depths could be real, reflecting a spectrum of partial melts generated over a wide range of depths within the lunar interior. However, the position of the arrowed point in Fig. 3.4.3 is very sensitive to the abundance of normative olivine in the sample. Small errors introduced by including cumulus olivine in the bulk chemical sample translate into large depth uncertainties.

The effect of normative olivine abundance on the pressure for the coexistence of olivine and pyroxenes on the liquidus of lunar basalt is illustrated in Fig. 3.4.4. The two lines correspond to the variation of olivine abundance (horizontal axis) in high-titanium and low-titanium basalts. The error bars for 70215 and 74275 show the range of analyses reported for these two fine-grained samples. The sampling and analytical range alone for 70215 introduces an uncertainty of about 75 km in the depth of magma segregation. Much of the apparent difference in depth estimates may simply reflect minor modifications in magma compositions by crystal settling processes (Walker *et al.*, 1976c). Similarly, a primitive liquid magma which has lost olivine by fractionation during uprise would give a depth corresponding to the arrow in Fig. 3.4.3 which would represent only the minimum depth of magma separation.

Relatively little attention has been given to the search for primary magmas among the quartz-normative pigeonite basalts from Apollo 12 and 15, despite their vitrophyric textures. This neglect results from the experimental observation that these compositions are multiply saturated with both olivine and pyroxene at pressures below 5 kb, which suggests that they last equilibrated at rather shallow depths, perhaps even within the lunar crust (step 5 in Fig. 3.1.4). Hence, they are not useful probes of the deep lunar interior. Chappell and Green (1973) concluded that the Apollo

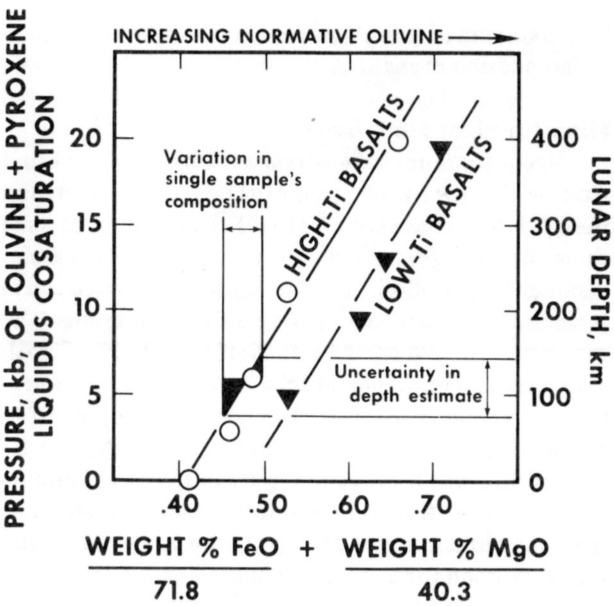

3.4.4 The pressure at which olivine and pyroxene(s) coexist on the liquidus of lunar basalts vs. (FeO/71.8) + (MgO/40.3) which is an index of normative olivine abundance. Error bars (on 70215) show the range of analyses reported for this fine-grained sample (after Walker *et al.*, 1976c). Small additions or subtractions of olivine phenocrysts in liquid compositions translate into significant shifts of the estimated depth of derivation.

15 quartz-normative basalts experienced shallow-level (200 km) partial melting or magma separation.

Although there is general agreement as to the residual mineralogy and source depths for low-Ti partial melts, divergent models have been proposed for the petrogenesis of the source regions. Green *et al.* (1971a, b) proposed that the source region constituted the primordial lunar interior, an interpretation which, although consistent with experimental data, left many geochemical questions unanswered. Moreover, they proposed that the differences between the high-Ti and low-Ti basalt suites reflected either widely different degrees of partial melting, or inhomogeneous source regions. The latter explanation was preferred by Green *et al.* (1975b). Walker *et al.* (1975a) suggested that the source regions for the low-Ti suite consisted of early ferromagnesian cumulates formed during the 4.6–4.4 b.y. differentiation events, an interpretation which is compatible with the trace element and isotopic characteristics of the basalts. Ringwood and Kesson (1976) have subsequently taken an intermediate position. They proposed that low-Ti basalts constitute partial melts of the primordial lunar interior, but suggested that these source regions had experienced minor contamination by the sinking of cumulates produced in the 4.6–4.4 b.y. differentiation event.

High-titanium mare basalts

High-titanium basalts collected by Apollo 11 and Apollo 17 show a broad range in chemical composition (Chapter 1). Rhodes *et al.* (1976) showed that at least three chemically-distinct groups of high-Ti basalts occur at the Apollo 17 site. Each group has been interpreted to represent various stages of differentiation by subtraction of low pressure phases (step 2 in Fig. 3.1.4). The most differentiated samples in the Apollo 17 collection are chemically similar to all of the samples from Apollo 11. This implies that none of the Apollo 11 basalts is as primitive as the most primitive Apollo 17 samples, so the Apollo 11 basalts must have crystallized from fractionated liquids. Figure 3.4.5 is a schematic phase diagram for a primitive high-Ti basalt.

Plagioclase-saturated cotectic liquid compositions are present among the titaniferous basalts of Apollo 11 and 17 as hand specimens (e.g., 10047 and 75035), but such rocks are among the coarsest-grained members of their suites. They record the slowest cooling histories with the greatest opportunities for differentiation, and contain the least-convincing petrographic evidence for well-preserved liquid compositions. It is the compositions themselves which provide the strongest indications that these samples represent simple liquids, because it would require an unlikely coincidence for crystal-liquid mixtures to have precisely the composition of the special multiply-saturated liquid.

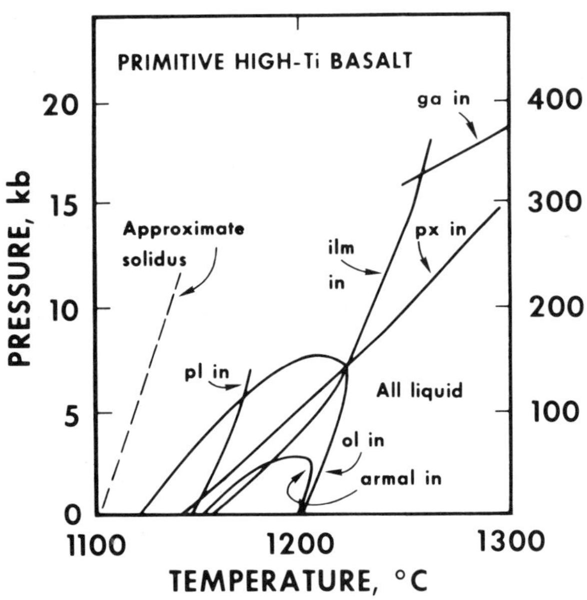

3.4.5 Schematic phase diagram for a primitive high-Ti basalt composition (corresponding to sample 70215) based on results in Longhi *et al.* (1974), Kesson (1975), Green *et al.* (1975b), and Walker *et al.* (1976c). Solidus approximates from results of O'Hara *et al.* (1970) on sample 10017. Spinel may be present near the liquidus at low to intermediate pressures. Phase boundaries are labeled by showing the side of the boundary on which the phase is stable.

The more primitive samples [higher Mg/(Mg + Fe) and more normative olivine] are not plagioclase saturated (Fig. 3.4.5) and are sometimes very fine-grained. It has been difficult to select a possible parental magma composition from these primitive samples for two reasons. First, the most primitive sample of all (74275), using the Mg/(Mg + Fe) ratio as criterion, is contaminated with fragments of polycrystalline olivine amounting to as much as 4.7% of the sample (Walker *et al.*, 1976c). Second, these rocks (including 74275) exhibit textural domains with different bulk chemistries, which are of large size compared with a standard thin section or the typical sample allocated for analytical or experimental work. Walker *et al.* (1976c) suggested that these domains may be the source of the analytical and experimental inconsistencies which have plagued the study of high-titanium basalts. If they are correct, then it may be that none of the samples studied so far strictly represents a liquid composition.

Sample 70215 seems to be the most primitive titaniferous basalt without obvious admixed material; hence it would seem to be the best candidate for a parental liquid composition. However, Kesson (1975) reported that the average 70215 composition is multiply saturated with olivine, armalcolite and spinel (but not with plagioclase) at 1 bar (see Fig. 3.4.5), suggesting that though this sample may be one of the most primitive available to us in the titaniferous basalt suite, it may have evolved at low pressures by fractionation of still more primitive liquids.

There are two further considerations in choosing possible parental liquids from the high-Ti basalt suite. Usselman and Lofgren (1976) found that they were unable to reproduce the natural titaniferous basalt textures in controlled cooling rate experiments unless their charges were partially crystalline at the commencement of cooling, but not so crystalline that the residual liquid was saturated with plagioclase. The implication is that the magmas from which the natural rocks crystallized contained crystals before eruption.

The second consideration is the disparity between compositions of the soils and of hand specimens at high-Ti basalt localities. A similar disparity at the low-Ti Apollo 12 locality was resolved by noting that the inferred parental magma composition could have differentiated to form a large volume of cotectic residual liquid, plus a smaller volume of olivine-rich cumulate rocks. The residual compositions from flow tops became degraded and entered the soils, and the cumulates were sampled as hand specimens. No similar solution to the problem can be postulated at high-Ti basalt localities because no examples have been found of the cumulus-crystal-enriched rocks needed to complement the liquids which would be represented in the soils (O'Hara *et al.*, 1975). Haphazard sampling techniques are unlikely to account for their absence, because at least three separate flows were sampled at Apollo 17, and two at Apollo 11. These difficulties cloud any straight-forward identification of the sample most useful for study as a primitive parental magma in the high-Ti basalt suite.

Longhi *et al.* (1974) and Walker *et al.* (1975a) have determined experimentally the liquid line-of-descent for the chemically least-fractionated high-Ti basalts [highest Mg/(Mg + Fe), highest Cr_2O_3, and lowest incompatible element content]. This low-pressure liquid line-of-descent corresponds to the low-pressure fractionation sequence interpreted on geochemical grounds by Rhodes *et al.* (1976). Green *et al.* (1975b) observed that the experimental liquidus olivine compositions matched the compositions of the cores of olivine phenocrysts in the least-fractionated Apollo 17 samples, and concluded that these samples may represent liquids. Proceeding on the joint assumptions that some of the Apollo 17 basalts do represent liquid compositions, and that the least-fractionated samples can be identified, we can equate these samples to three parental liquids (one for each chemical group observed by Rhodes *et al.*, 1976). Of these samples, only 70215 (Group B of Rhodes *et al.*) and 74275 (Group C) have been studied experimentally. The least-differentiated samples of Group A, 71175 and 76136, have not been studied experimentally to date. A schematic liquidus diagram for a typical primitive high-Ti basalt is shown in Fig. 3.4.5. References to specific studies on high-Ti basalts are listed in Table 3.4.1.

O'Hara *et al.* (1970) interpreted the high-Ti suite as products of advanced near-surface fractional crystallization of lava lakes filling mare basins. They concluded that high-pressure experiments on such extremely fractionated samples would be meaningless and could not establish the nature of the source regions of the parental magmas.

In contrast, Ringwood and Essene (1970) concluded that near-surface fractionation processes did not play an important role, and that high-pressure studies of Apollo 11 basalts, which they interpreted as primitive partial melts, could reveal substantial information on the nature of the source region. They proposed that high-Ti basalts were produced by very small degrees of partial melting of the primordial lunar interior at depths of 200–400 km. This single-stage partial melting model has since been abandoned (e.g., Green *et al.*, 1975b) in view of the demonstrably derivative status of the composition studied, and the mounting geochemical and isotopic evidence for a more complex history.

It has been inferred from trace element and isotope geochemistry that Ti-rich cumulates were formed during the early melting and differentiation of the Moon at about 4.6–4.4. b.y., and then later were involved in high-Ti basalt genesis. Longhi *et al.* (1974) and Walker *et al.* (1975a) found that a primitive high-Ti basalt, 70215, was multiply saturated with olivine, pyroxene and ilmenite at pressures corresponding to a depth of 120 km, as shown in Fig. 3.4.5. They attached considerable importance to the presence of ilmenite on the high-pressure liquidus and interpreted their results in terms of the then popular cumulate-remelting hypothesis (e.g., Schnetzler and Philpotts, 1971; Taylor and Jakeš, 1974), suggesting that the high-Ti suite was produced by the remelting of an ilmenite-bearing residuum immediately beneath the crust. However, Green *et al.* (1975b) claimed that one of the most primitive Apollo 17 basalts, 74275, was nowhere saturated with ilmenite on its high-pressure liquidus, and Kesson (1975) obtained similar results for a pimitive high-Ti glass composition. Kesson (1975) confirmed that the 70215 composition is simultaneously saturated with olivine, pyroxene and ilmenite at high pressures (see Fig. 3.4.5). However, she also showed that it is saturated with olivine, spinel and armalcolite at 1 atm, and it may therefore be a differentiated liquid, whose high-pressure equilibria are simply not relevant. Thus, both Green *et al.* (1975b) and Kesson (1975) concluded that high-Ti basalt source regions did not contain ilmenite and, moreover, that those source regions lay at considerable depths (below 200 km) within the lunar interior.

Hubbard and Minear (1975) suggested that the high-Ti suite might be of hybrid origin (step 4 in Fig. 3.1.4). Kesson and Ringwood (1976) and Ringwood and Kesson (1976) developed this idea further, proposing that large, dense ilmenite-pyroxenite cumulate bodies, produced during the final stages of the 4.6–4.4 b.y. differentiation event, sank through the differentiated lithosphere and into the primordial interior. Hybrid regions were thus produced. Subsequent partial melting of these regions at about 3.8 b.y. produced the high-Ti basalts. Ringwood and Kesson (1976) demonstrated experimentally that the assimilative processes proposed for the hybrid regions are plausible.

Throughout the lunar program, experimental petrologists have agreed that pyroxene was present in the residuum following the genesis of the high-Ti basalt. All experimental results (e.g., Kesson, 1975; Walker *et al.*, 1975a; Green *et al.*, 1975b) have indicated that this pyroxene was pigeonitic. However, Green (1976b) has demonstrated that these experimental pigeonites are in fact metastable and that the equilibrium mineralogy of the residuum probably included both orthopyroxene and calcic clinopyroxene.

Biggar *et al.* (1972) extended the arguments of the Edinburgh laboratory concerning the fractionation of high-Ti basalts to cover the low-Ti basalts as well. They maintained that near-surface fractionation phenomena have so modified magma chemistry on the hand specimen scale that high-pressure experimental studies are futile for petrogenetic interpretations. It is otherwise generally accepted that selected low-Ti basalts and, in particular the more olivine-normative low-Ti basalts, may usefully serve as probes of the deep lunar interior. An extreme view is held by some (e.g., Ringwood and Kesson, 1976), who maintain that only the more primitive low-Ti basalts provide adequate geochemical information on the nature of the primordial lunar interior, and that the high-Ti suite is irrelevant in this context.

Ultramafic glasses

The ultramafic green and orange glasses have also been considered as samples of primitive, possibly primary, liquids. There is little doubt that these materials once were liquids. However, very few samples have yet been identified which can be considered possible derivatives of these compositions. Grove and Vaniman (1978) suggest that the very low titanium (VLT) basalt fragments which were returned from Mare Crisium by Luna 24 may be derived by near-surface fractionation from the ultramafic glasses found associated with them, but no similar relationships have been established at other sites. High-pressure studies of the 74220 orange glass (Green *et al.*, 1974; Walker *et al.*, 1975a) found the liquidus to be multiply saturated with olivine and orthopyroxene at around 20 kb. This implies that the orange glass could represent a partial melt of an olivine-pyroxenite source at depths of 400 km or more or, alternatively, it might represent a partial melt produced at shallower depths, where partial melting was so extensive that olivine alone remained in the residuum (see Fig. 3.4.3 and 3.4.5).

Many of the ultramafic lunar glass suites are characterized by internal compositional variations (Stolper *et al.*, 1974) which are inconsistent with equilibrium or fractional melting of a planetary interior. So, even if the glasses should prove to be samples of primary ultramafic liquids, experimental studies of these variable compositions may provide little in the way of information directly applicable to deep source regions.

3.4.3 Experimental constraints on the bulk composition of the Moon

The methods and techniques of experimental petrology can be used to evaluate lunar bulk composition models. It can often be a simple matter to prove a model unsatisfactory. For example, following the Apollo 11 mission, Ringwood and Essene (1970) and O'Hara *et al.* (1970) demonstrated that Apollo 11 basalt with its very dense high-pressure mineralogy could not represent the bulk composition of the Moon because neither the moment of inertia nor mean density constraints would be satisfied. Similarly, following their high-pressure studies, Hodges and Kushiro (1973) concluded that bulk compositions resembling the (Ca, Al)-rich aggregates in the Allende meteorite were unsuitable. Råheim and Green (1974) reached similar conclusions and showed in addition that lunar highland basalt was an unsatisfactory model for the Moon's bulk composition.

The various estimates of the bulk composition of the Moon obtained using geophysical and geochemical constraints differ principally in the proportions of involatile elements (Ca, Al, Ti. . . .) relative to elements of intermediate volatility (Fe, Mg, Si. . . .). Low abundances of volatile elements (Na, K. . . .) are common to all model compositions (see Chapter 4). Any acceptable lunar bulk composition must satisfy the constraint that it be capable of generating mare basalt magmas. Because this criterion is model-dependent, all plausible petrogenetic schemes must be considered. These include the formation of mare basalts by the following mechanisms: (a) by direct melting of the primordial lunar interior, e.g., Ringwood and Essene (1970); (b) by assimilation at shallow levels, where parental partial melts from the primordial interior assimilate differentiated material beneath the crust, producing hybrid mare basalt magmas, e.g., Hubbard and Minear (1975); (c) by assimilation at depth, where dense cumulates formed during the early lunar differentiation sank down into the primordial interior, causing local contaminaton and ultimately giving rise to mare basalts, e.g., Kesson and Ringwood (1976), Ringwood and Kesson (1976); and (d) by remelting of cumulates formed during an early differentiation event (about 4.6 b.y. ago), e.g., Schnetzler and Philpotts (1971), Taylor and Jakeš (1974), Walker *et al.* (1975a).

The probable behavior of any assumed bulk composition if subjected to melting and differentiaton can be tested by means of melting experiments at high pressures, provided that one assumes differentiation to be a straightforward process analogous to that interpreted for terrestrial stratiform intrusions (an alternative viewpoint was expressed by Walker *et al.*, 1975b). From the high-pressure equilibria, Ringwood (1976) deduced the mineralogical zonation resulting either from total lunar melting or from melting only of the outer regions, followed by differentiation. He concluded by emphasizing the unsuitability of lunar bulk compositions with comparatively high abundances of involatile elements, such as those of Anderson (1973), Ganapathy and Anders (1974), Wänke *et al.* (1974), and Taylor and Jakeš (1974). Irrespective of the particular petrogenetic model being considered, these (Ca,Al)-rich compositions would give rise to high-Al liquids quite unlike primitive mare basalts. Similar conclusions were reached by Råheim and Green (1974). Even the model compositions suggested by Taylor and Bence (1975) with lower CaO and Al_2O_3 contents than any of those discussed above appear to be unsatisfactory for this reason (Kesson and Ringwood, 1977). Moreover, in all of these model compositions, the pyroxenes that co-precipitate with plagioclase (which represent those which were co-precipitating during formation of the lunar crust) are Ca-rich and unlike the normal pyroxenes of the lunar highlands.

These experimental evaluations of model compositions place petrologic limits on the bulk composition of the Moon. In this manner, Ringwood (1976) and Kesson and Ringwood (1977) have argued that the bulk composition of the Moon is very similar to that of the Earth's mantle, having similar abundances of the involatile elements Ca, Al, REE, U, etc. Their conclusions are in accordance with the most recent lunar heat flow values of Langseth *et al.* (1976). Chapters 4 and 9 give more details of lunar composition and thermal history.

3.5 METEORITIC BASALTS

Meteorites with compositions, mineralogies, and textures similar to terrestrial basalts have been known and studied since the middle of the nineteenth century. Although these meteoritic basalts have not been studied as extensively as terrestrial or lunar basalts, sufficient information is available to begin to characterize igneous processes on the parent bodies of these meteorites, and in so doing, to extend our understanding of basaltic volcanism to (1) bodies outside of the Earth-Moon system, to (2) bodies smaller than the Earth and the Moon, and to (3) the period immediately following planetary accretion.

Five types of differentiated meteorites believed to have been formed by igneous processes similar to those involved in the formation of terrestrial and lunar basalts are currently recognized: the basaltic achondrites, the shergottites, the nakhlites, the chassignites, and Angra dos Reis (see Chapter 1, section 1.2.8). The relationships between these igneous meteorites and cosmochemical models for planetary compositions will be reviewed in section 3.6.3 (see Fig. 3.6.4). In this section, further discussion will be limited to the basaltic achondrites, including the eucrites, howardites, and diogenites and the silicate portions of mesosiderites. These are the most abundant and extensively studied of the differentiated meteorites. The discussion will illustrate how the methods of experimental petrology could be used to elucidate the origins of other meteorite groups.

3.5.1 Terminology and brief description of basaltic achondrites

Wasson (1974) reported that 71 basaltic achondrite meteorites are known, of which 52 were falls. These amount to 7% of all falls. Descriptions of the petrography, structure, and chemistry of the basaltic achondrites were given by Mason (1963), Duke and Silver (1967) and Powell (1969, 1971). Briefly, eucrites are composed dominantly of low-calcium pyroxene (initially pigeonite) and a calcic plagioclase (An_{80-96}), with minor amounts of silica polymorphs, ferroaugite, chromian spinel, ilmenite, and metallic iron. Igneous textures ranging from vitrophyric to gabbroic are often present, but are usually somewhat obscured by brecciation and thermal metamorphism. The term *eucrite* is restricted to unbrecciated and monomict brecciated achondritic meteorites fitting the above description. Diogenites are composed dominantly of orthopyroxene (En_{72-76}) with accessory olivine, plagioclase, spinel and metal. They are generally coarse-grained and usually brecciated. Ideally, the term *diogenite* should be restricted to unbrecciated and monomict brecciated meteorites, but most diogenites contain some fragments of eucritic material and are, in fact, polymict. *Howardites* are polymict breccias containing a spectrum of eucritic and diogenitic mineral and lithic fragments, many of which are unknown from monomict brecciated or unbrecciated meteorites. Howardites are similar in many respects to lunar soil breccias. *Mesosiderites* are metal-rich meteorites whose silicate portions are similar to howardites. Whole-rock and mineral isochron dating of eucrites and fragments in howardites and the presence in these meteorites of ^{129}Xe from decay of ^{129}I indicate that the igneous processes involved in the formation of these meteorites were restricted to a narrow time interval near 4.6 b.y. (see Chapter 7).

3.5.2 Experimental studies

Sample selection

One-atmosphere melting experiments on fourteen eucritic meteorites were described by Stolper (1975, 1977). Samples were chosen largely on the basis of availability and to span the range of known eucritic compositions. Experimentation was limited to eucrites because these unbrecciated and monomict brecciated meteorites are most likely to have bulk compositions controlled by igneous processes alone. The polymict characters of howardites and mesosiderites indicate that their bulk compositions were controlled mainly by mechanical mixing processes. Despite their monomict characters, however, the bulk compositions of brecciated eucrites could be influenced by impact processes, such as contamination from the impacting projectiles or impact differentiation. Additional potential difficulties with the starting materials used in these studies were: (1) the metamorphism which affected most eucrites may have modified their original compositions; and (2) the compositions of the starting materials used in experiments may not have been representative of the bulk compositions of the meteorites, due to the small sample size available for study (generally less than 2 grams).

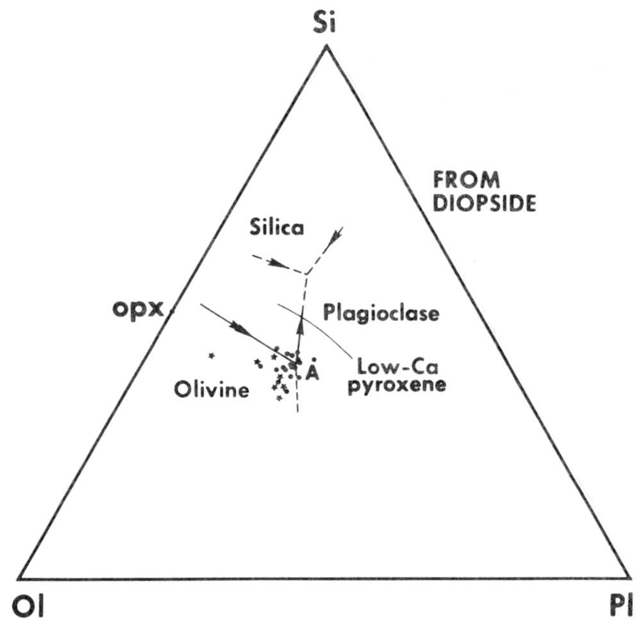

3.5.1 Metallic iron and chromian spinel saturated plagioclase-silica-olivine "pseudo-liquidus" diagram for eucritic compositions. Phase boundaries for molar Mg/(Mg + Fe) = 0.60 (Bowen and Schairer, 1935) based on multiply-saturate liquid compositions from experiments on the Sioux County and Juvinas eucrites (Stolper, 1977). Filled circles are projected compositions of eucrites taken from the literature. Stars are projected compositions of "anomalous" eucrites, some or all of which may be cumulates. Projection as in Stolper (1977). Note that the field boundaries for Mg/(Mg + Fe) = 0.60 do not define the paths followed by residual liquids produced by fractionation of liquid compositions with Mg/(Mg + Fe) = 0.60.

Experimental results

Eucrites appear to have crystallized at low oxygen fugacities similar to those of lunar basalts (Duke, 1965). Experiments at 1 atm have been conducted at oxygen fugacities between the iron-wüstite and iron-ilmenite-ferropseudobrookite buffers. The major features of the results of these experiments are summarized in Fig. 3.5.1. The clustering of nearly all eucritic compositions near the 1-atm peritectic A, olivine (Fo_{65})-pigeonite (Wo_5En_{65})-plagioclase (An_{94})-chromian spinel-metal, demonstrates that these meteorites have the compositions of liquids controlled by equilibration with these crystalline phases at low pressures (steps 1 and 5, Fig. 3.1.4).

The clustering of the meteorite compositions near this peritectic, despite the variable extent of brecciation and degree of metamorphism, demonstrates that these two processes did not significantly modify the initial compositions produced by igneous processes. All of the eucrites whose compositions plot near the peritectic A have a small degree of crystallinity ($<5\%$) at multiple saturation, which demonstrates that the restriction to small sample size introduced no significant problems. Therefore, heterogeneities similar to those observed by Walker *et al.* (1976c) in 70215 are not important in the eucrites. If they were developed at some stage, brecciation has homogenized them. Before the experimental data were available, Mason (1962) and Duke and Silver (1967) had concluded that most eucrites have the compositions of liquids on the basis of petrographic studies of the original igneous textures of eucrites (step 1, Fig. 3.1.4). Duke and Silver (1967) also predicted the 1-atm multiple saturation of eucrite liquids (step 5, Fig. 3.1.4) on the basis of interpolation between the anorthite-enstatite (Anderson, 1915) and anorthite-ferrosilite

(Schairer, 1942) joins, and they discussed its significance. The experimental studies of Stolper (1975, 1977) confirm these predictions. The textures of the eucrites whose compositions cluster at peritectic A in Fig. 3.5.1 vary from vitrophyric to gabbroic, and if textural criteria alone were used to distinguish liquid compositions from cumulate compositions, some of these eucrites would be suspected of containing accumulated material. This observation emphasizes the difficulty, so well exemplified in the lunar literature, of distinguishing quenched liquids from cumulates by textural criteria alone.

3.5.3 Cumulate eucrites

The eucrites shown as stars in Fig. 3.5.1 have compositions which deviate from the olivine-pyroxene-plagioclase peritectic at 1 atm. These eucrites are more magnesian than those near the peritectic. Three of them have unbrecciated poikilitic or ophitic textures and are identified as adcumulates: Moama, Moore County and Serra de Magé. The Binda eucrite has a composition displaced from the 1-atm peritectic. Two interpretations for this have been offered. McCarthy *et al.* (1973) suggested that the composition of Binda may represent that of a magnesian parental liquid from which the eucrites at peritectic A developed by differentiation. Stolper (1975, 1977) suggested that Binda may represent a cumulate, on the basis of its coarse-grain size and the fact that its most magnesian pyroxenes are similar in composition to the pyroxenes near the liquidi of the eucrites at peritectic A. Three other eucrites, Brient, Medanitos and Pomozdino could be subject to similar interpretations, but no detailed petrographic descriptions are available for these meteorites.

These two interpretations of the Binda eucrite demonstrate that the petrology of meteoritic basalts has not entirely escaped the ambiguities of distinguishing liquid from cumulate compositions. In fact, the difficulties of interpreting the textures of brecciated and metamorphosed eucrites can render the problem even more intractable than for lunar rocks.

3.5.4 Primary or differentiated liquids?

The major problem once the compositions of liquids have been established is to determine to what extent these liquid compositions were controlled by melting processes, and to what extent they have been modified by crystal fractionation and other post-melting processes. This distinction has been controversial, just as it has for lunar and terrestrial basalts. Some petrologists interpret those eucrites with compositions near the 1-atm peritectic (A in Fig. 3.5.1) as residual liquids produced by extensive fractionation of abundant orthopyroxene plus minor olivine, spinel and metal from magnesian parental magmas, perhaps similar in composition to chondrites or to Binda (Mason, 1962, 1967; Moore, 1962; Duke and Silver, 1967; Schnetzler and Philpotts, 1969; McCarthy *et al.*, 1973; Reid, 1974; Shimizu and Allègre, 1975; Bunch, 1975a, b; Dymek *et al.*, 1976). Others have suggested that the liquids with compositions near the peritectic were produced by low-pressure partial melting of source regions containing olivine (Fo_{65}) + low-calcium pyroxene (Wo_5En_{65}) + plagioclase (An_{94}) + Cr-rich spinel + metal, followed by only minor fractionation of pigeonite and plagioclase (Stolper, 1975, 1977; Consolmagno and Drake, 1977; Ma *et al.*, 1977).

Geophysical and geochemical constraints

Experimental study of the compositions identified as liquids is generally insufficient to establish whether a given liquid composition is primary or differentiated (Fig. 3.1.4). This distinction, for terrestrial and lunar basalts, is largely based on preconceptions concerning the depth of generation of primary magmas, the nature of the source region(s), and the relationships of basaltic magmas to certain ultramafic rocks (sections 3.3–4). Likewise, there are geophysical and geochemical constraints on the generation of meteoritic basalts.

Low-pressure multiple saturation in lunar and terrestrial basalts would generally be regarded as evidence against their interpretation as primary magmas. This means of identifying differentiated liquids is not valid for meteoritic basalts, if the belief that these meteorites originate in the asteroid belt is correct. Because all igneous processes on asteroids must occur at low pressures, low-pressure multiple saturation is equally consistent with a partial melting origin as with a residual liquid origin. Figure 3.5.2 gives the general relation between central pressure and planet size and mean density. The central pressure in the largest asteroid, Ceres ($R = 510$ km, $\rho \cong 2.4$ g/cc) is approximately 2 kb, while the central pressure in Vesta ($R = 275$ km, $\rho \cong 3.5$ g/cc), often regarded as similar to the parent body of the basaltic achondrites, is approximately 1 kb. The heat source which caused melting is unknown, and the primary magmas for the basaltic achondrite suite may have been generated at depth in the parent body, or near the surface.

Foremost among the constraints affecting the identification of primary meteoritic magmas is the assumption that the eucrites, howardites, and probably the

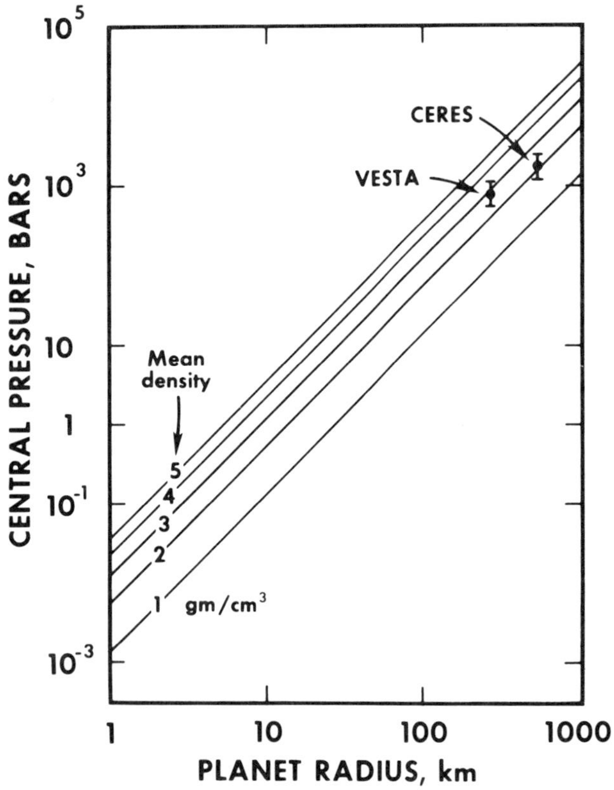

3.5.2 Relationships between the pressure at the center of a small planet, and the size and density of the planet.

diogenites were produced by the same or similar igneous processes on the same or similar parent bodies. The juxtaposition of eucritic and diogenitic fragments in howardite polymict breccias bears out this assumption. This has been the most influential constraint placed on the origin of the basaltic achondrites and, in particular, on whether eucrites represent primary or differentiated magmas. This is because the orthopyroxenes in diogenites (En_{72-76}; Dymek *et al.*, 1976) and the most magnesian orthopyroxenes in howardites (En_{85}) are significantly more magnesian than the pigeonites present at the liquidi of the eucrites with compositions near peritectic A (Wo_5En_{65}). This implies that liquids more magnesian than these eucrites must have existed and that these magnesian liquids were related to the eucrites.

Model 1: eucrites as residual liquids

Most petrologists have concluded from the evidence outlined above that the implied liquids more magnesian than the eucrites were probably parental to the eucrites, and hence that the eucrites are probably far removed from primary liquid compositions. This petrogenetic model for the origin of basaltic achondrites is portrayed in Fig. 3.5.3a.

According to this interpretation, the continuous trend of pyroxene compositions observed in howardites (Dymek *et al.*, 1976) reflects the trend of cryptic variation in cumulus pyroxenes accumulated during fractionation of the parental liquids in plutonic environments. It is doubtful that any known meteorites represent quenched examples of the hypothesized magnesian parental liquids. Binda is a possible example, but the sparse petrographic data available appear to be more consistent with a cumulate origin for this meteorite. Brient and Pomozdino are other possibilities, but their compositional similarities to the Serra de Magé cumulate strongly suggest that they too are cumulates. A cumulate origin for Medanitos is probable as well, judging from its compositional similarity to the Moama cumulate.

This petrogenetic model has the advantage of producing all of the basaltic achondrites in a single fractionation sequence, sometimes identified with the differentiation of the initially molten planet, but there are several difficulties which have been discussed in detail by Stolper (1975, 1977). Briefly, these are: (1) residual liquids similar in composition to the eucrites are unlikely to form by fractionation of the proposed parental magmas (section 3.5.4); (2) residual liquids are unlikely to be produced preferentially at a peritectic by fractional crystallization; (3) the eucrites have constant major element chemistry but variable incompatible element chemistry, which is not consistent with a series of residual liquids produced by fractionation.

Model 2: eucrites as primary magmas

Figure 3.5.3b illustrates schematically a contrasted model, in which the eucrites are regarded as primary magmas or as magmas modified only slightly by crystal fractionation. According to the details given by Stolper (1975, 1977) and Consolmagno and Drake (1977), and summarized below, this appears to be more consistent with most of the known features of basaltic achondrites than does the fractionation model 1.

Low-pressure partial melting (5–20%) of a source comprising olivine + low-Ca pyroxene + plagioclase + Cr-spinel + metal, produced eucritic liquids with compositions at peritectic A in Fig. 3.5.1, until plagioclase was exhausted from the residue. Minor (<20%) fractionation of pigeonite and plagioclase from these primary magmas produced the compositions of most of the eucrites with compositions near peritectic A. Two known eucrites appear to require 30–35% fractionation of the primary magmas to account for their lower Mg/(Mg + Fe) ratios and incompatible element concentrations. Further melting after the exhaustion of

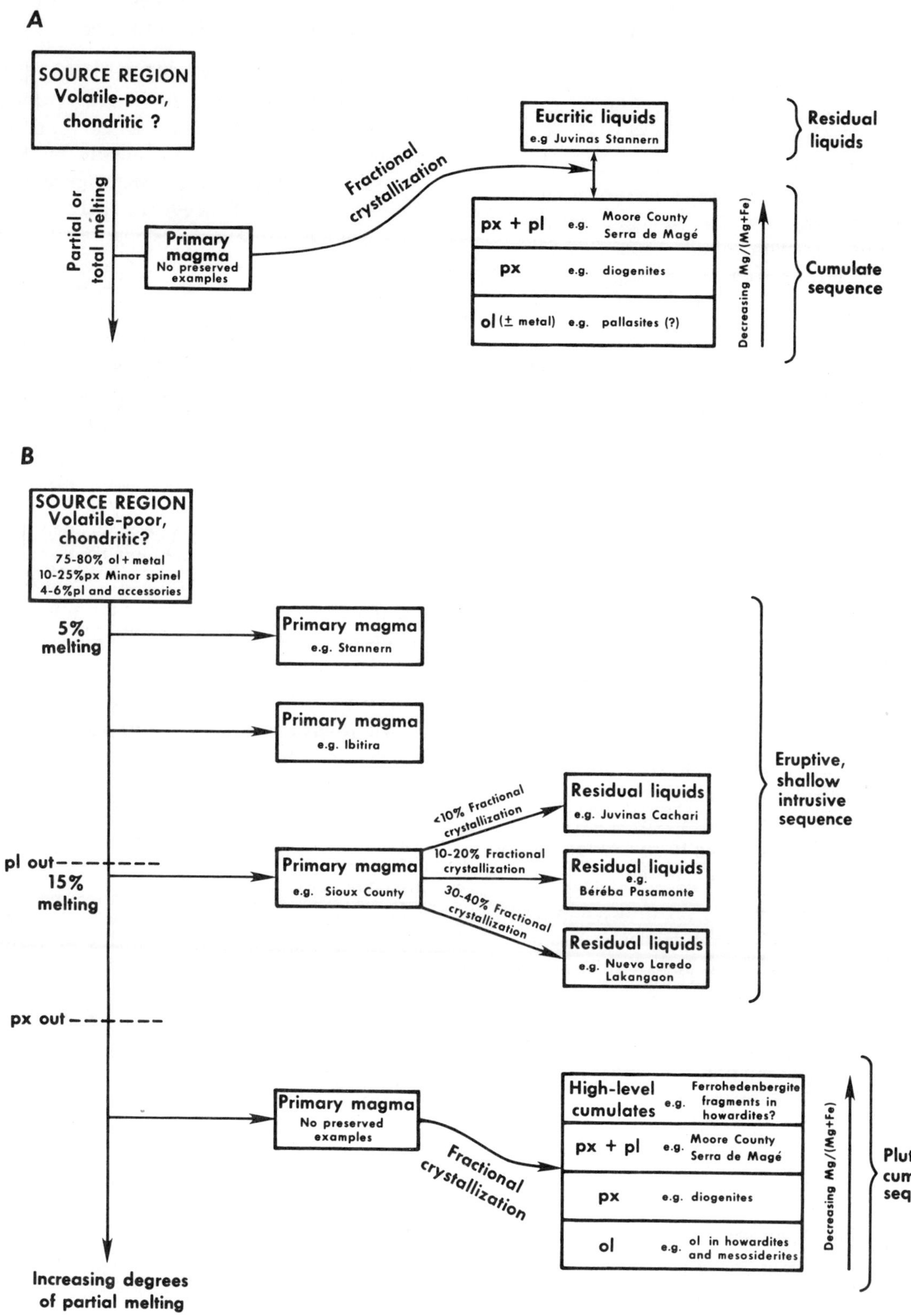

3.5.3 Schematic flow diagrams illustrating proposed models of basaltic achondrite genesis. **a**. After Mason (1962, 1967) and Bunch (1975b). **b**. After Stolper (1977).

plagioclase from the source regions produced liquids becoming progressively more magnesian, and moving away from peritectic A on the olivine-pyroxene reaction surface. These liquids became magnesian enough to crystallize pyroxenes as magnesian as the most magnesian pyroxenes in howardites and diogenites. Eventually, pyroxene became exhausted from the source regions and liquids produced by larger degrees of partial melting moved away from the olivine-pyroxene reaction surface into the olivine volume. Crystallization of these magmas first produced olivine, perhaps accounting for the rare olivines observed in howardites and diogenites.

The magnesian melts produced by more extensive partial melting would have resembled the parental magmas of the fractionation model of eucrite genesis. Crystallization of these magmas in plutonic environments would have produced olivine cumulates, overlain by orthopyroxene cumulates, and finally overlain by pyroxene-plagioclase cumulates. Ultimately, silica and ferrohedenbergite might have crystallized, as in the high levels of large terrestrial intrusions. The composition of the liquid in such an intrusion would have started in the olivine field; moved away from olivine until it reached pyroxene saturation; then, because of the fractional crystallization, it would have moved across the pyroxene volume until it reached plagioclase saturation; and then moved along the pyroxene + plagioclase surface until it became saturated with a silica polymorph. At no time would the residual liquid have been similar in composition to known eucrites. Diogenites and the cumulate eucrites would, according to this model, be samples of the cumulates from such plutons.

The scarcity of known meteorites with compositions which could represent the more magnesian liquids could indicate that these advanced partial melts were restricted to crystallization in plutonic environments. One possible explanation of this could be that the eucrites generally produced near-surface flows, and underwent only minor fractionation. The earlier eucrite flows may have formed an impermeable crust, thus limiting the later-formed magnesian melts to development in plutons beneath the crust. The rarity of meteorites with compositions similar to the residual liquids which would have been produced during the fractionation of these postulated plutons could suggest that the plutons crystallized as closed systems. One would expect that high-level cumulates containing iron-rich pyroxenes, such as ferrohedenbergite, would be the least abundant of all basaltic achondrites, and, in fact, they are only known as rare fragments in howardites (Dymek *et al.*, 1976).

3.5.5 The bulk composition of the source regions of eucritic magmas

Consolmagno and Drake (1977) calculated a bulk composition of the source region, using the major-element compositions of the minerals near the liquidi of the eucrites investigated experimentally by Stolper (1977), assuming that the eucrites were approximately primary magmas as outlined above (model 2), and using rare-earth element patterns of the eucrites. The calculated bulk composition of the source region resembles ordinary chondrites and Cl chondrites in all but the most volatile elements (e.g., H, C, N), except for a significant depletion in moderately volatile elements such as Na and K. It also resembles the bulk composition of the eucrite parent planet calculated by Morgan *et al.* (1978) and Hertogen *et al.* (1977) on independent bases. When coupled with the 4.6 b.y. age of the eucrites, this similarity suggests that the eucrites may represent the quenched products of melting of the bulk composition of a planet soon after homogeneous accretion, and that the basaltic achondrites as a whole represent the products of the initial differentiation of a small planet. The estimates of the composition of the eucrite parent body and further discussion are given in Chapter 4, section 4.

3.5.6 Experimental petrology as a guide to the origins of meteoritic basalts

The available evidence appears to favor model 2, the interpretation that most of the eucrites are quenched primary or nearly primary magmas produced by 5–20% low-pressure partial melting of a source region assemblage: olivine + low-calcium pyroxene + plagioclase + spinel + metal. As shown in Fig. 3.5.3b, most of the other basaltic achondrites could then be produced by physical mixing of these eucrites and the cumulates formed during fractional crystallization of more advanced partial melts of the same source regions in plutonic environments.

The details of the considerations which led to this model and the model itself are of interest for their own sakes, because if this preferred model of igneous evolution of the basaltic achondrite parent body is valid, it gives us an idea of (1) the nature of planetary differentiation on small bodies, (2) the bulk composition of a differentiated meteorite parent body, and (3) the nature of basalt development on bodies outside of the Earth-Moon system.

The details are also of interest in that they demonstrate the applicability to other bodies of the petrologic principles developed for the Earth and the Moon, and the possibilities and hazards involved in extending this approach to meteorite groups and other planets where important data (e.g., field relationships) can only be guessed or at best inferred. These considerations also demonstrate some of the specific problems associated with application of petrologic reasoning to meteorites and other planets. Some of these problems are the smallness of the sample, the possibly unrepresentative nature of the sample, and the necessity of assuming that different meteorites come from the same or similar planets.

Although the specific conclusions reached above will undoubtedly be modified or discarded in the future, either due to new insights, new data, or perhaps due to newly discovered meteorites which will change our views of processes or relationships between basaltic achondrites, the approach serves as an example of the type of approach which will be needed in the study of other meteorites, and the exploration of other planets where the samples will be few and not random.

3.6 OTHER PLANETS

Experimental petrology will probably play a central role in future efforts to understand the evolution of the terrestrial planets and the asteroids. In connection with basaltic samples from the Earth, Moon and meteorites, we have reviewed the application of an inverse approach from basalt to source rock, following procedures outlined in the flow chart of Fig. 3.1.4. For terrestrial basalts, it is clear that the interpretations and conclusions are actually influenced very strongly by the forward approach, according to our knowledge of the composition and mineralogy of the mantle, and of liquids produced by partial melting of this material. We are not yet in a position to attempt the inverse approach for basaltic samples from Mercury, Venus and Mars, but there is sufficient information available for cosmochemists, geophysicists and geochemists to estimate the compositions of the mantles of these planets. Therefore, we can start at the end of the flow chart in Fig. 3.1.4, and follow the forward approach involving experimental testing of models of mantle source rocks. This will allow predictions about basaltic volcanism on planets which we have not yet sampled.

3.6.1 The composition and petrology of planetary mantles

Model compositions for the terrestrial planets are presented in Table 4.3.2. These compositions are speculative, as indicated by the range of estimates listed for each specific planet, including those that have been sampled. The compositions given are for mantle + crust, which limits the applicability of the following discussion, because we assume that basalts are generally produced by partial melting of planetary mantles. However, estimates of the composition of the Earth's mantle are similar to those of mantle + crust, and this limitation therefore may not affect our conclusions. For the Moon, in contrast, the crust constitutes a considerably larger proportion of the mantle + crust system, and the mantle source regions of basalts may bear little compositional or petrological resemblance to the primordial mantle. We will nevertheless treat the mantle + crust compositions as if they do provide adequate approximations of the compositions of the mantle sources of basaltic magmas.

Chemical compositions as they are presented in Table 4.3.2, provide only limited petrological insights. Comparisons are made more usefully if the chemical analyses are recast from oxide components into the norm, which represents the chemistry in terms of a series of standard mineral components which could make up a rock of the given composition. For the chemical compositions given in Table 4.3.2, graphical equivalents of the norm calculation are schematically displayed in Figs. 3.6.1–2 for pressures between 1 atm and about 65 kb. The recalculated compositions from Table 4.3.2 are plotted in these diagrams, and the positions of each point shows the subsolidus mineral assemblage for each composition. The figures do not take into account the effects of variations in Mg/(Mg + Fe), and probably do not fully account for the effects of the abundances of minor components, such as Na_2O, Cr_2O_3, and TiO_2, on the subsolidus mineral assemblages. However, they do provide a good expression of the compositional and mineralogical differences between hypothetical planetary mantles, and the effects of pressure on the mantle mineral assemblages.

Quantitative calculation of the norms shows that 49–90% of these compositions, by weight, would be composed of olivine. Figure 3.6.1 shows that all of the subsolidus assemblages are olivine-rich. Variations in the amounts of normative constituents other than olivine are shown in Fig. 3.6.2. On this diagram, all of the estimates of planetary composition vary approximately along a line emanating from the MS corner. This reflects the fact that all of these compositions have Ca/Al near the cosmic ratio of 1.10.

The phase diagrams from Fig. 3.6.1 have been reproduced in Fig. 3.6.3, but only the planetary compositions based on the equilibrium condensation model and its feeding zone modifications are shown (see Chapter 4 and Table 4.3.2). This permits the trends in planetary compositions to be compared within the framework of a single model.

Earth

The mineralogical changes with depth in the Earth are reasonably well known (section 3.3.2), and the subsolidus phase changes are illustrated in Fig. 3.1.2. The phase assemblages repesented in Fig. 3.1.2 correspond closely with the normative mineral assemblages calculated from the estimated mantle composition, near its solidus, as shown in Fig. 3.6.3: (a) 0–10 kb (0–33 km depth)—olivine + diopside + enstatite + plagioclase (plagioclase lherzolite); (b) 10–25 kb (33–80 km depth)—olivine + diopside + enstatite + spinel (spinel lherzolite); (c) 25–130 kb (80–400 km depth)—olivine + diopside + enstatite + garnet (garnet lherzolite). The pressure (depth) limits of these assemblages, as well as those discussed below, are only approximate because: (1) the transformations do not take place at a discrete

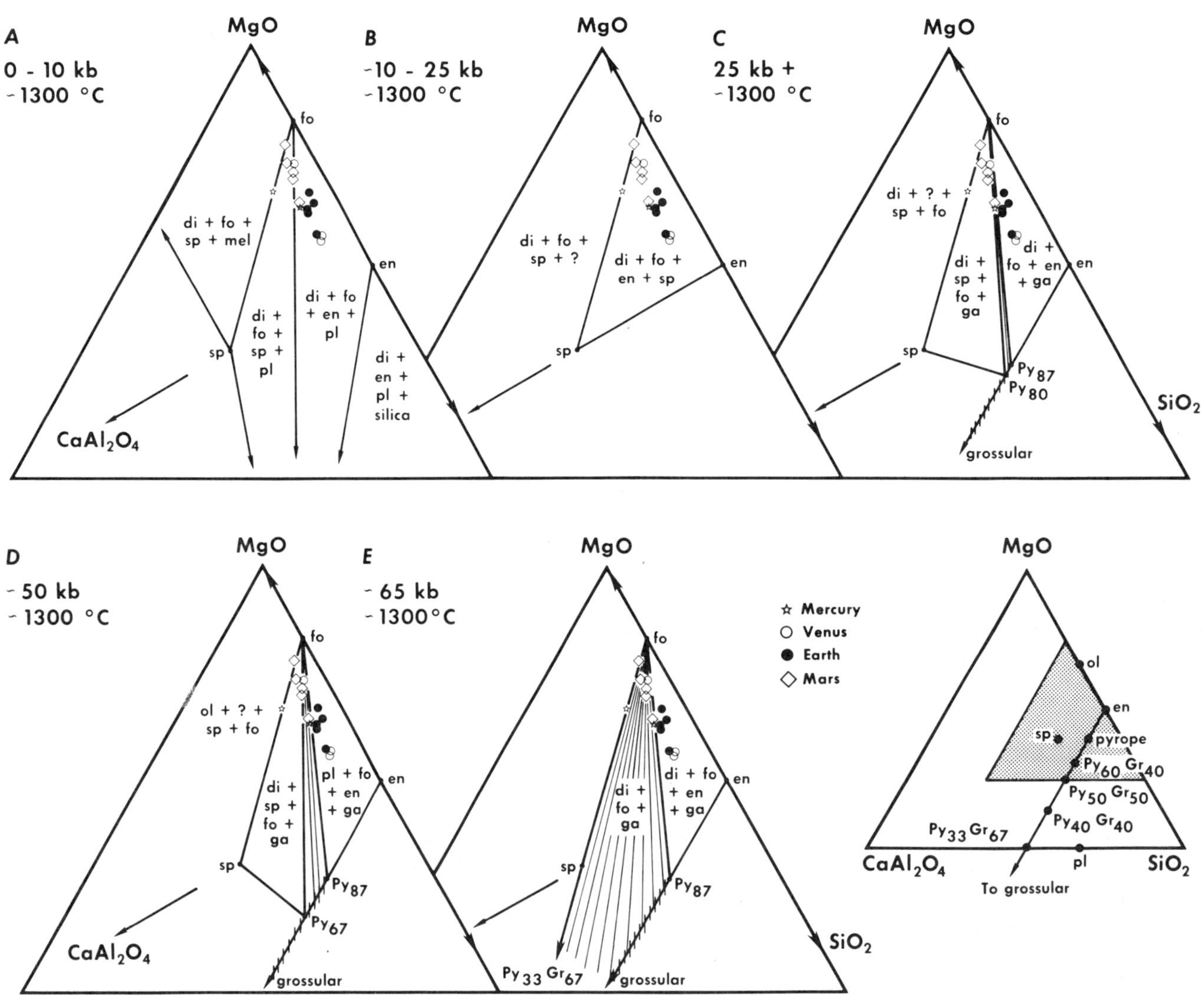

3.6.1 Near-solidus phase diagrams (largely schematic) between 1 atm and 65 kb for estimates of planetary bulk composition (Table 4.3.2). Molar CMAS coordinates (O'Hara, 1968a) recast into molar coordinates M-S-CA-CMS_2 then projected from diopside (CMS_2). All assemblages are diopside-saturated. Inset shows portion of CA-M-S plane plotted in a-e. Garnet compositions in di-sp-fo-ga and di-fo-en-ga assemblages are based on data in MacGregor (1965). Mel = melilite.

pressure, but rather, through a pressure interval; and (2) the locations of the reactions in P-T space have not been unambiguously determined. If the mantles are not at the solidus, the sequence of mineral assemblages will generally be the same as the solidus sequence, but the exact pressures of the transformations will depend on the geotherm (Fig. 3.1.2).

Venus

The venusian mantle composition is qualitatively similar to the terrestrial mantle and has the same sequence of assemblages as the function of pressure (Fig. 3.6.3). The depth ranges of these assemblages would also be approximately the same on the two planets (see Fig. 3.1.1). For the equilibrium condensation model, Venus' mantle composition is richer in orthopyroxene relative to olivine and has a lower Mg/(Mg + Fe) ratio than the terrestrial mantle. However, when the full range of venusian and terrestrial mantle compositional estimates are compared (Figs. 3.6.1–2; Table 4.3.2), these differences are diminished.

Mars

From 0–10 kb (0–80 km depth), the estimate of the martian mantle composition from the equilibrium condensation model would be close to the olivine + plagioclase + diopside join, (Fig. 3.6.3a), but most of the other estimates of the martian mantle composition would also include spinel (Fig. 3.6.1a). From 10–25 kb (80–210 km depth) the martian mantle composition would have the same spinel lherzolite assemblage as at intermediate depths in the Earth and Venus: spinel + olivine + diopside + enstatite (Fig. 3.6.3b). At pressures greater than 25 kb (210 km depth), the martian mantle would consist of olivine + spinel + garnet + diopside (Fig. 3.6.3c). This would eventually be replaced by an olivine + garnet ± (diopside or spinel) assemblage at high pressure (by about 50 kb, 400 km

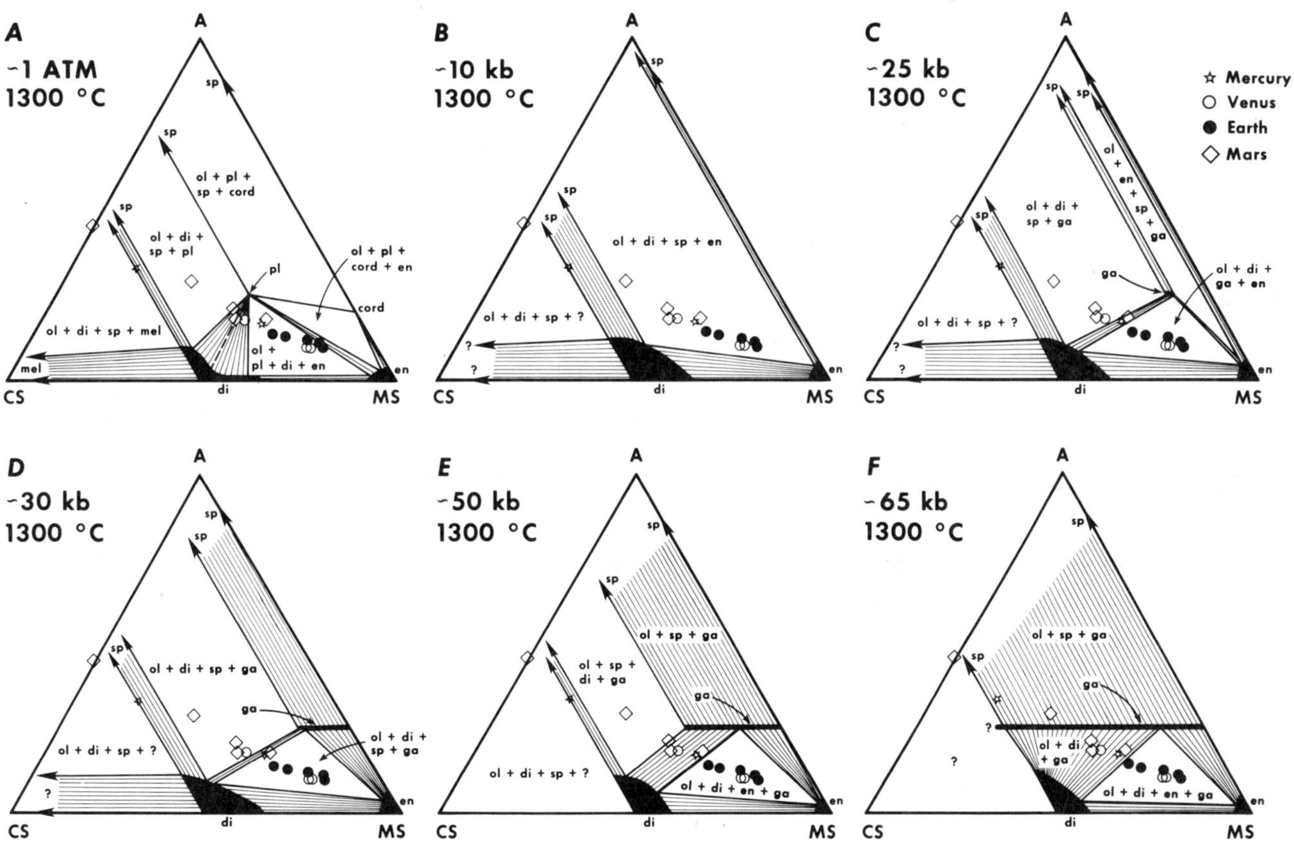

3.6.2 Near-solidus phase diagrams (largely schematic) between 1 atm and 65 kb for estimates of planetary bulk composition (Table 4.3.2). Molar CMAS coordinates (O'Hara, 1968a) recast into molar coordinates CS-MS-A-M_2S then projected from olivine (M_2S). All assemblages are olivine-saturated. Diopside compositions based loosely on data from Ferguson *et al.* (1977), Herzberg (1978), and Presnall *et al.* (1978a). Garnet compositions and phase relations at pressures above 25 kb based on 1300°C data and extrapolations in MacGregor (1965). Cord = cordierite. Mel = melilite.

depth). These high pressure assemblages (>25 kb), containing spinel and lacking orthopyroxene, would differ from those of the venusian and terrestrial mantles at equivalent pressures. McGetchin and Smyth (1978) also concluded that the martian mantle would lack orthopyroxene, but suggested that an (Mg, Fe)O phase, rather than spinel, would be stable at high pressures. There is, however, experimental evidence that the spinel + olivine + garnet + diopside assemblage is stable to at least 40 kb and that the olivine-spinel join is not broken by an (Mg, Fe)O-garnet tie line (MacGregor, 1965). The martian mantle also differs from the terrestrial mantle in that it has a significantly higher Fe/(Mg + Fe) ratio according to all available estimates (0.22–0.34 for Mars vs. 0.11–0.15 for Earth).

Mercury

Although less magnesian [Mg/(Mg + Fe)<0.01] and virtually devoid of alkalis and other volatile elements, Mercury's mantle would be most similar in petrological terms to that of Mars, if it has a composition intermediate between Me1 and Me2 (Table 4.3.2, Chapter 4). Its sequence of mineral assemblages, and the depths at which they occur, are the same as those given above for the martian mantle (Fig. 3.6.3). The Me1 composition actually falls slightly to the MgO-rich side of the diopside + olivine + spinel join, and could conceivably have normative periclase, monticellite, or melilite rather than plagioclase, orthopyroxene, or garnet in addition to olivine + spinel + diopside. The Me2 composition would have normative enstatite

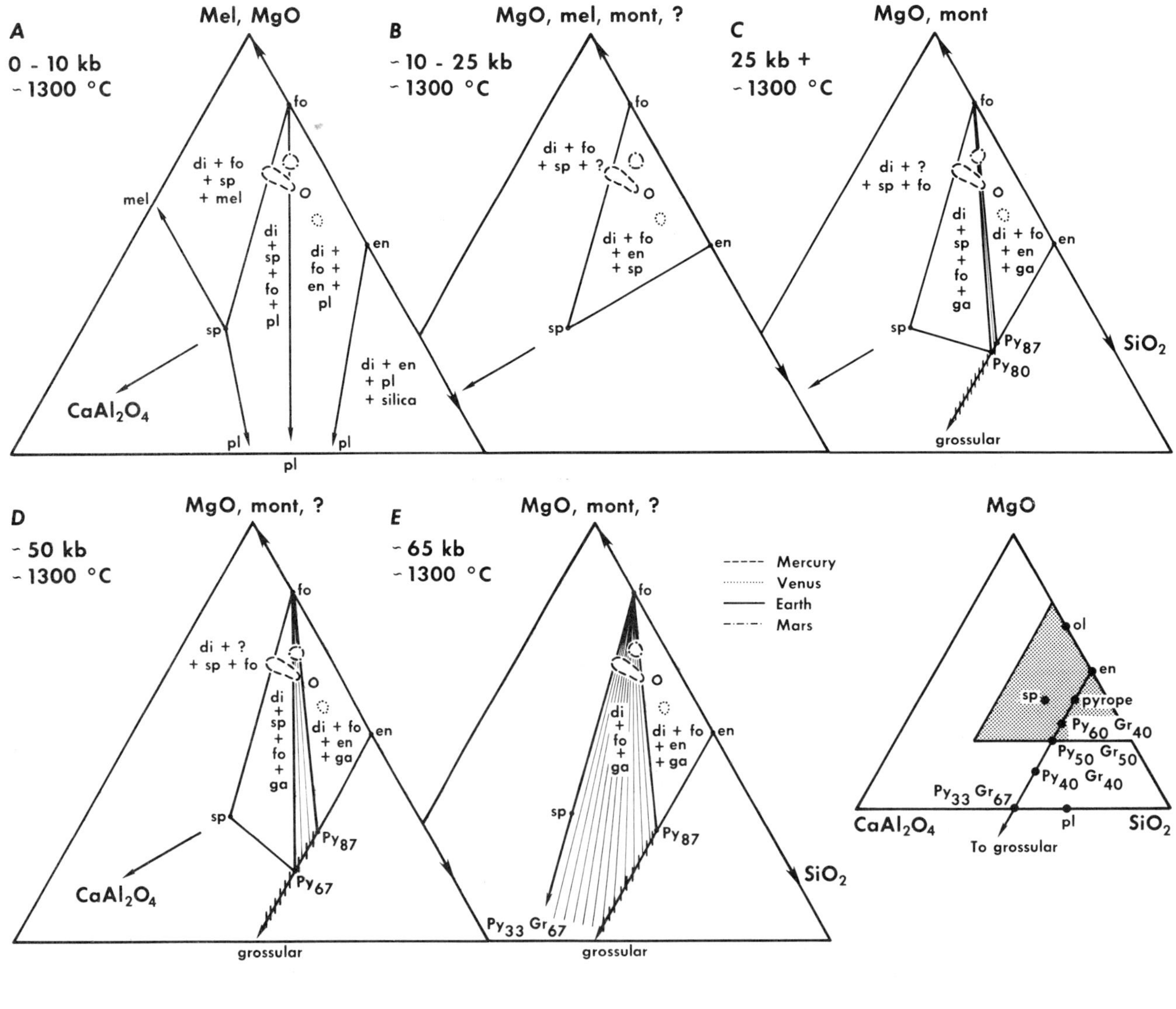

3.6.3 As in Fig. 3.6.1, except only planetary compositions from equilibrium condensation model and its "feeding zone" modifications are shown (see Chapter 4). Mel = melilite. Mont = monticellite.

in the low pressure (<10 kb) region. However, most of the compositions falling between the Me1 and Me2 compositions would have the Mars-like mineralogy described above.

Comparative mantle mineralogies

Despite the differences in the estimates of the compositions of planets given in Table 4.3.2, we may draw the following tentative conclusions: at low pressures (0–10 kb), the venusian and terrestrial mantles would be plagioclase lherzolites (ol + di + en + plag) while the martian and mercurian mantles would be spinel-plagioclase wehrlites (ol + di + sp + plag). At intermediate pressures (10–25 kb), mantles on all four planets would have the same mineral assemblage—spinel lherzolite (ol + di + en + sp). At high pressures (>25 kb), the planetary mantles would probably again divide into two groups based on their subsolidus mineralogies: Earth and Venus would be garnet lherzolites (ol + di + en + garnet) while Mercury and Mars would be spinel-garnet wehrlites (ol + di + sp + garnet).

3.6.2 Testing of model mantle compositions

Examination of the subsolidus and melting interval phase relations of these model mantle compositions and of the liquids produced by partial melting could yield insights into the modes of differentiation of other planets, and into the igneous activity expected on them. McGetchin and Smyth (1978) have made a start in the application of this approach by using the results of experiments on simple systems to make inferences about the mineralogy and melting relations of a model martian interior.

Stolper *et al.* (1979) took a less specific approach to the testing of model mantle or bulk planet compositions. Most current cosmochemical models of planetary composition involve mixtures of several nebular components (Chapter 4). Some of these are relatively simple two-component models in which planets are composed of varying proportions of a volatile-poor and a volatile-rich component (e.g., Wood, 1962; Larimer and Anders, 1967; Dreibus *et al.*, 1977; Ringwood, 1977b).

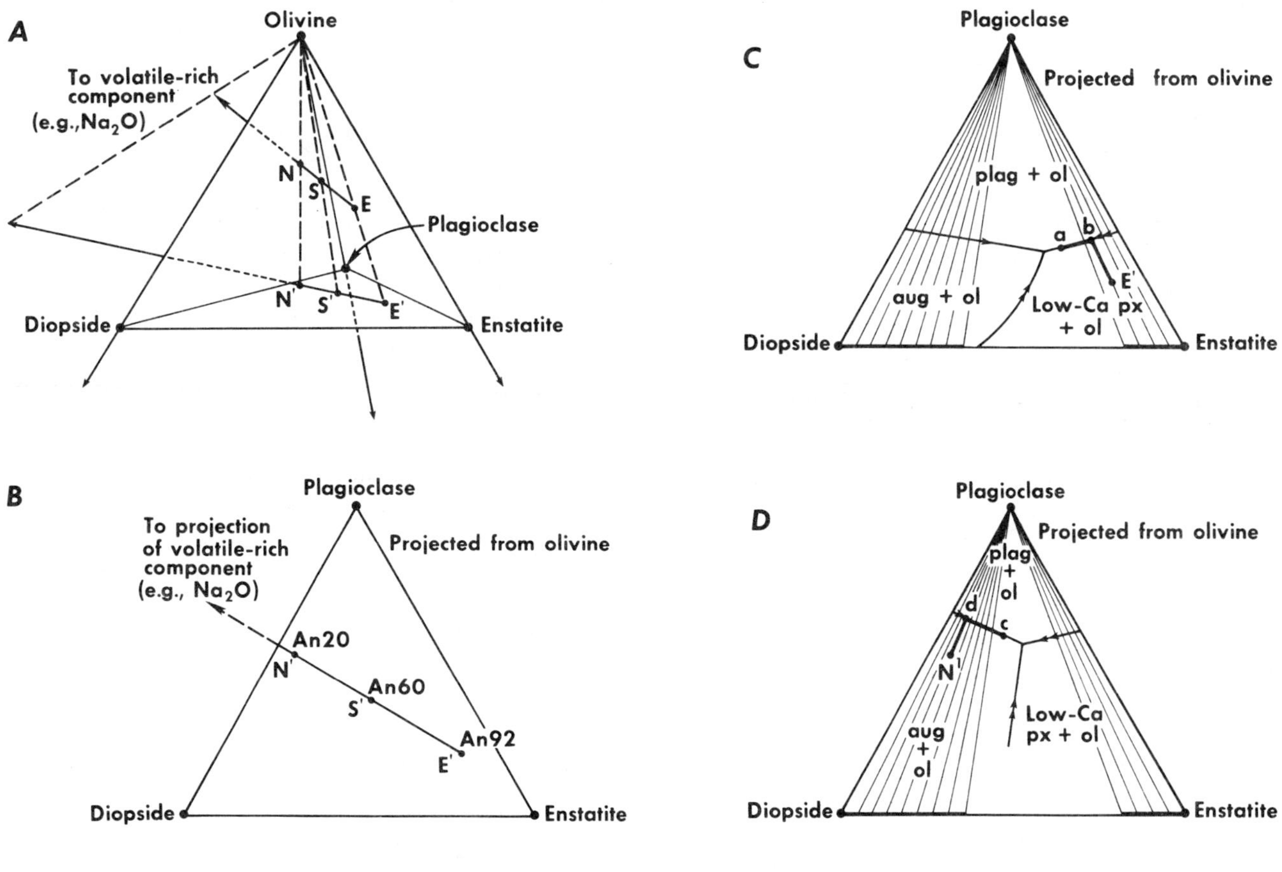

3.6.4 Phase relations in the system olivine-diopside-enstatite-plagioclase, projected from olivine, representing two-component models for mantle or bulk planet compositions.

Stolper *et al.* (1979) examined the petrological consequences of such two-component models in order to elucidate the relationships between four groups of igneous meteorites—the basaltic achondrites, the shergottites, the nakhlites, and the chassignites. We will discuss these petrological consequences in detail in order to show briefly how they can be used to constrain our thoughts on the igneous petrology of other planets, and to provide a petrological framework for Chapter 4.

The volatile-poor component in these models at low pressures is essentially a feldspathic harzburgite; that is, largely olivine with lesser amounts of low-calcium pyroxene and minor calcic plagioclase. The bulk composition of the eucrite parent body (section 3.5.5) is an example of this component. The volatile-rich components of these models often have normative feldspathoids or alkali silicates. The effect of addition of the volatile-rich components to the volatile-poor peridotite component is to increase the albite content of the plagioclase and to increase the high-calcium pyroxene/low-calcium pyroxene ratio in the peridotite. This effect is summarized in the following normative reaction:

$$\underset{\text{(alkali-rich component)}}{Na_2O} + \underset{\text{(anorthite)}}{CaAl_2Si_2O_8} + \underset{\text{(low-Ca pyroxene)}}{11\ MgSiO_3} \rightleftharpoons$$

$$\underset{\text{(albite)}}{2\ NaAlSi_3O_8} + \underset{\text{(high-Ca pyroxene)}}{CaMgSi_2O_6} + \underset{\text{(olivine)}}{5Mg_2SiO_4}$$

Figure 3.6.4a illustrates the effect of adding a pure Na_2O component to the bulk composition of the eucrite parent body. The effect is similar if a carbonaceous chondrite composition is used as the volatile-rich component. It is clear from Figs. 3.6.4a and b that a series of planetary compositions produced by these two-component models will vary continuously from peridotites such as the eucrite parent body (E) with calcic plagioclase and dominantly low-calcium pyroxene, to peridotites such as N with sodic plagioclase and dominantly high-calcium pyroxene. If the volatile-rich component is also oxidized and hydrous, as it is in some models, peridotites such as N would also be more oxidized and hydrous than E.

The compositions of the liquids produced on melting of peridotites with compositions ranging from E to N (Fig. 3.6.4a) will differ. The low pressure phase boundaries for calcic compositions such as E and for sodic compositions such as N are projected in Figs. 3.6.4c and d, respectively.

Melting of E will begin with liquid a and will follow the path a-b-E at low pressure (Fig. 3.6.4c). The products of fractional crystallization of liquids between b and E will be low-Ca pyroxene, low-Ca pyroxene + plagioclase, joined late in the crystallization sequence by augite and a silica polymorph. The crystallization sequences of the liquids produced by melting of E correspond well to basaltic achondrite meteorites.

Melting of N begins with liquid c and follows the path c-d-N (Fig. 3.6.4d). The early portions of the fractional crystallization sequences of liquids between c and N would be olivine + augite, and olivine + augite + sodic plagioclase. The crystallization sequences from the liquids produced by melting of N correspond to those inferred for the nakhlite and chassignite meteorites. Similarly, the shergottite meteorites correspond in terms of their compositions and mineralogy to crystal fractions from liquids produced by the melting of a source peridotite such as S, intermediate between E and N.

In the sequence basaltic achondrites to shergottites to nakhlites-chassignites, the plagioclases become more sodic, the oxidation state increases, the K/U ratio increases, and hydrous phases begin to appear. These facts, together with the crystallization sequences described above (Figs. 3.6.4c and d), suggest that a link can be made between the sequence of igneous meteorites, and the cosmochemical models which predict a sequence of planetary compositions between E and N (Fig. 3.6.4).

This example has dealt with only a small range of meteorites and peridotites at low pressure, but the integration of (1) experimentally determined phase diagrams; (2) the petrographic, chemical and experimental petrological studies of the meteorites; and (3) cosmochemical models is illustrative of the type of study which must become important in future attempts to synthesize models of the terrestrial planets. Extension of this same approach to other models and higher pressures may be the next step. For the model mantle compositions discussed in this section and in Chapter 4, detailed phase diagrams corresponding to Fig. 3.1.2 should be determined in order to confirm that the subsolidus modal mineralogy is consistent with the deduced normative mineralogy illustrated in Figs. 3.6.1–3. The compositions of liquids derived by partial melting of these model mantles should also be determined (see section 3.3.3 for terrestrial examples).

3.6.3 Implications of model mantles for basalt genesis

Although the analysis of mantle mineral assemblages in section 3.6.1 was based on the equilibrium condensation model of planetary compositions, reference to Figs. 3.6.1–2 shows that the trends established by these compositions generally hold for the other estimates of planetary composition from Table 4.3.2 in Chapter 4. We will now apply the analyses of near-solidus mineralogy on these different planets to a discussion of the compositions of those primary magmas that would be produced by degrees of partial melting at which no minerals are exhausted from the residues. The conclusions depend to a considerable extent on phase relationships reviewed in sections 3.3.2–3.

The first variation to be anticipated is in the Mg-values of mantle liquids. We should expect a progression from liquids with Mg-values near 100 on Mercury, to liquids with Mg-values near 70 on the Earth, to iron-rich liquids with Mg-values near 40 on Mars [based on an olivine-liquid K_D(Fe/Mg) of about 0.3]. There is no general agreement on the relative Mg-value of the venusian mantle. This simple sequence could be disturbed if the Mg-values of basalt source regions on Mars, Venus, or Mercury had been altered during previous episodes of differentiation.

At low pressures (<10 kb), incipient melts of the venusian and terrestrial mantles would have compositions controlled by equilibration with olivine + diopside + plagioclase + enstatite; such liquids would be similar to each other, and to common terrestrial tholeiites, though they might differ from typical terrestrial tholeiites in their Mg-values. Compositions of liquids produced by melting of the mercurian and martian mantles would be controlled by equilibration with olivine + spinel + plagioclase + diopside, and would therefore be petrologically distinct from terrestrial and venusian low pressure melts. At pressures below about 5 kb (Presnall *et al.*, 1978a), these liquids would be undersaturated—nepheline-normative on Mars, and larnite-normative on Mercury. At pressures higher than 5 kb, liquids produced by melting of these source regions would be tholeiitic according to Presnall and his co-workers.

At intermediate pressures (10–25 kb), all of the mantle mineralogies are the same (olivine + spinel + diopside + enstatite), and consequently the liquids will be petrologically similar. At pressures below about 12–15 kb, the liquids from these mantles will be tholeiitic. At higher pressures within this interval, liquids on Venus, Earth and Mars would be alkali olivine basalts, i.e., nepheline-normative. On Mercury, with its low alkali contents, it is unlikely that liquid compositions would cross the olivine-plagioclase-diopside join into undersaturated compositions even at these higher pressures (Presnall et al., 1978a).

At higher pressures (25 kb +), the mantle assemblages again divide into two groups: Earth and Venus, olivine + diopside + enstatite + garnet, and Mars and Mercury, olivine + diopside + garnet + (spinel). Liquids from these pressures on Venus and Earth will thus be similar, and bear the stamp of equilibration with the garnet lherzolite assemblage. Martian and mercurian basalts will differ from them, however, since they would be produced by melting of a different assemblage. Just how different these liquids would be is difficult to determine, because the locations in composition space of liquids saturated with olivine + diopside + garnet + (spinel) are unknown. However, the liquids are probably more aluminous and undersaturated than orthopyroxene-saturated liquids produced on Earth and Venus.

In summary, the most notable differences between primary magmas produced on these planets would be in Mg/(Mg + Fe) ratio: mercurian magmas would have very low Mg/(Mg + Fe) ratios, terrestrial (and venusian?) basalts would have intermediate ratios, and martian basalts would have very high ratios (but no higher than lunar mare basalts and achondritic meteorites and their parent magmas). Mercurian basalts would also be nearly devoid of alkalis and more volatile elements. The phase equilibria of primary liquids on Earth and Venus would differ from those of primary liquids on Mars and Mercury if the liquids were produced by melting at pressures less than 10 kb or greater than 25 kb. Primary liquids from these pressure ranges on Mars and Mercury would probably be more aluminous and undersaturated than terrestrial and venusian liquids from similar pressures.

We emphasize that in this analysis we have tried to characterize only the compositions of the most abundant types of basalts on a planet, analogous to mid-ocean ridge basalts or continental tholeiites on Earth. Just as there are carbonatites and lamprophyres on the Earth, undoubtedly there will be less abundant magma types on other planets which do not conform to the patterns outlined above. The diversity of magma types may be related to planetary size or richness in volatiles (sections 3.6.4 and 3.7), so that mercurian magmas may have less diversity than those from Venus, the Earth, and Mars.

3.6.4 Solar system exploration and the inverse approach

The approach of analyzing hypothetical planetary compositions in terms of the growing body of phase equilibrium data, and the completion of subsolidus and melting experiments on the hypothetical compositions themselves, will not only yield more detailed insights, but will also provide a framework for mission planning and for the interpretation of sample returns, remotely sensed surface compositions, and geological features observed via satellites and surface-based cameras.

Experimentation on returned samples, on compositions obtained by automated analysis (e.g., the Martian XRF experiment), and on compositions obtained by remote sensing techniques (Chapter 2) is inevitable once these samples and compositions become available. The utility of this approach will depend upon whether igneous compositions are returned, or whether surface compositions are extensively weathered or altered, as they may be on the sampled regions of Mars. Assuming the availability of igneous compositions, experimental techniques can be employed to place constraints on the possible relationships between different samples, on the physical conditions and processes involved in their genesis, and on the mineralogy and composition of the source regions in the planetary interiors from which the igneous rocks were ultimately generated by partial or complete melting.

The flow chart introduced in Fig. 3.1.4 and discussed in connection with terrestrial, lunar and meteoritic basalt studies summarizes the experimental approach which we anticipate will be applied to the study of igneous rocks from other planets. The ground rules may, of course, differ from planet to planet. The geophysical, geochemical and petrological constraints which provide the framework for the application of the flow chart will almost certainly vary. Planetary density and heliocentric distance, or the discovery of rock types other than basalt (e.g., ultramafic rocks or granites) on a planet will probably influence the types of experiments conducted and their interpretations. We have referred to Fig. 3.1.4 as the inverse approach, but the procedures followed actually represent an integrated approach, as exemplified in sections 3.6.1–3 by the formulation and testing of hypothetical planetary mantle compositions before basaltic samples have been obtained.

We have emphasized in this chapter that primary liquid compositions are required before meaningful inferences concerning the mineralogy and conditions (e.g., P, T, activities of volatile components) of the source regions can be made from experimental petrology. Is it likely that primary liquid compositions will be returned by automated "grab" sampling of planetary surface? Are average surface compositions obtained by remote sensing likely to yield primary liquid compositions? Is field judgment required for the return of these petrologically significant primary liquid compositions? Questions of this sort are vital to mission planners as well as to petrologists. Can petrologists provide insight into the answers to these questions by application of their experience on the Earth, the Moon, and the eucrite parent body?

Walker and Hays (1977) suggested that the answer to this last question may be yes. They noted that the most abundant basalt types on the Earth, the Moon, and the eucrite parent body—oceanic tholeiites, Fra Mauro basalts, and eucrites, respectively—appear to have compositions which were controlled by low pressure equilibration with olivine, pyroxene and plagioclase. They then inferred, by analogy, that "the most abundant basalts on Mercury and Mars will not be primary magmas from great depth." If valid, this inference has important implications for planetary exploration. However, a direct comparison between the Earth, the Moon, and the eucrite parent body may not be strictly valid. Even though the most abundant basalts are multiply saturated at low pressure, they may be so for three different reasons: the Earth because of fractionation between the source and the surface; the Moon because melting occurred in the lunar crust; and the eucrite parent body because of its small size and low maximum internal pressures.

Several factors control whether primary magmas are likely to fractionate en route to the surface or to erupt unfractionated: the depth of melting, the size of the body and $\bar{g}$, the layering of the planet (discontinuities in density may provide sites for formation of magma chambers), the widths and spatial density of conduits, the dynamics of melting, tectonics, and the thermal structure of the planet. On the Earth, the large value of $\bar{g}$, the layering of the outer regions of the mantle and crust, and the generally deep level of melting make fractionation of melts en route to the surface the rule rather than the exception. These factors probably play important roles in other large planets, so that pessimism regarding the likelihood of obtaining primary liquid compositions in a random, small sample may be realistic. The change from a tensional to a compressive stress regime on Mars and Mercury at the close of their volcanic histories (Solomon and Chaiken, 1976) would also favor the eruption of fractionated compositions, suggesting perhaps that the latest volcanic rocks on both of these planets—those which would be sampled selectively on the surface—will be heavily weighted towards differentiated rather than primary compositions.

Our experience has been that the abundant basalts on each planet sampled are multiply saturated at low pressure, and we can think of reasons why this might be so, but it would be unfortunate to conclude that sampling of basalts is a useless exercise. On small planets, for example those parental to basaltic meteorites, low pressure fractionation control does not rule out the sampling of primary magmas since all processes, including primary melting, must occur at low pressure (Fig. 3.5.2). Furthermore, the lunar experience was profitable despite the fact that the sampling incidence of primary material was quite low, as it is on the Earth. Given these precedents, we may expect to find useful primary material, but we should not be surprised if it is low in abundance.

Let us consider the incidence with which primary material has been sampled. For the Earth, oceanic basalts provide a useful analog for planetary sampling in that they are of widespread occurrence, and have been sampled by procedures such as dredging and drilling which afford relatively little geological control. Reference to Fig. 3.3.34 shows that the incidence of primitive magmas close to compositions which might be primary melts is extremely low. These likely candidates comprise a fraction of a percent of the total population of oceanic basalts, which generally show the effect of low pressure fractionation. In lunar experience the fraction of basalts which were eventually decided to be nearly primary was also low, but the proportion of the total population was measured in percents. For the eucritic meteorites, the unmodified melts certainly did not comprise the majority of samples either. However, the proportion of samples of primary character reached 25–33% of the population.

Thus, it appears that the proportion of primary samples in a population decreases dramatically with increasing size of the planet on which the population occurs. This observation is consistent with the notion that larger planets with stronger gravity fields have more opportunity to impose differentiation on their primary melts. Consequently it might be concluded that smaller objects require smaller sampling programs to extract the required information contained in primary melts. This conclusion is dangerous for two reasons. It is very difficult, on the basis of present experience, to know how to scale a sampling program to planet size—the correlation noted can only be a qualitative guide at best. Second, the proportion of primary material also is inversely correlated with the total size of the populations collected. The fewer samples we have, the more primary material seems to be present on a planet-by-planet survey (of only three planets!). If the anti-correlation of the proportion of primary material with planet size is more than an illusion of population size, then it has significant implications for differentiation dynamics. These implications also correlate with intuition. However, it would be unfortunate to design sampling programs as if the apparent correlation were universal, thereby forfeiting an opportunity to test the premise.

It is difficult to anticipate the limitations which will be placed on our ability to use experimental petrology to place constraints on planetary interiors by the availability of only a small number of samples from a given planet, none of which represent primary liquids. Stolper and McSween (1979) were presented with this problem, in their study of the shergottites, a two member group of basaltic meteorites. Both of the shergottites are cumulates, and experimental melting studies demonstrated that the liquids from which the cumulates formed were not olivine-saturated. Assuming that the liquids parental to the shergottite suite were produced by low pressure partial melting of a peridotite, the lack of olivine-saturation in the liquids from which the cumulates formed indicated that these liquids were not primary. Nevertheless, it was possible to set broad constraints on the mineralogy and bulk composition of the source regions of primary shergottite liquids, given the assumption of a low pressure peridotite source, and the composition and liquidus phase relations of the differentiated liquid from which the shergottite cumulates formed, and the rare-earth pattern of this liquid. Therefore, even without the availability of primary liquid compositions from other planets, meaningful constraints on the mineralogy and composition of magma source regions will be attainable from the inverse approach outlined in the flow chart of Fig. 3.1.4, if this is integrated with the forward approach using whatever other information and deductions that may be available in the normal, geological way.

3.7 OVERVIEW SUMMARY

Some of the lessons we have learned from this review of the experimental petrology of basalts and their source regions can now be summarized, and some inferences can be drawn about the basalts we are likely to sample on other planets.

The rationale for assembling the present chapter is that the experimental study of crystal-liquid equilibria is relevant to an understanding of basalt petrogenesis, and the review has confirmed that this is true. For example, among the abundant basalts sampled on each planet, a sense of the importance of crystal-liquid saturation equilibria can be appreciated by a glance at Figs. 3.3.34, 3.4.1 and 3.5.1. It is seen that these basalts have compositions which generally conform to crystal-liquid saturation boundaries determined by the methods of experimental petrology. Therefore, we pursue laboratory investigations with some confidence that the relationships discovered have applications to the problems we try to solve. Furthermore, we have every reason to believe that the petrogenesis of basalts on planets not yet sampled may be addressed by the same methods.

In our review of the melting relations of terrestrial peridotites we learned that a variety of liquid compositions could be generated at different conditions of temperature, pressure, and activity of volatile species, and indeed that some of these liquids corresponded to basalts that have been observed to erupt. Furthermore it was found that certain primitive basalts could be shown to equilibrate with the mineralogy of peridotite under conditions relevant to the Earth's interior and that an internally consistent scheme of basalt petrogenesis by melting of peridotite could be constructed. This exercise produces information on the pressure, temperature, and residual mineralogy of the magma segregation event.

The results of experimental work on primary magmas from planets sampled so far have converged in that the planetary mantles reconstructed as source regions bear a generic resemblance to one another. Olivine and pyroxene(s) emerge as the principle constituents along with some aluminous phase. This result is not particularly surprising in terms of cosmochemical and bulk density constraints. However, it may not be free of a component of circular reasoning. In any event, the convergent results are internally consistent and investigations of basalts from other planets will probably use this framework as a point of departure. Nonetheless, we should not discount alternate possibilities. In fact, a case can be made for expecting that oxides may be more important than pyroxenes on Mars on the basis of its mean density and moment of inertia. Furthermore, the enstatite achondrites may point to a planet where pyroxene is more important than olivine.

We emphasize that models of planetary mantles based on the primary basalts which issue therefrom have not converged to the point where they are indistinguishable—far from it. Each mantle appears to have distinct characteristics, and the proportions of phases and details of mineral chemistry in each mantle may be quite different. The abundance of volatile components demonstrably shows enormous variation from planet to planet. Indeed, it may prove possible to relate many of the important interplanetary differences to the proportion of volatile components such as alkalis incorporated during planet formation.

In contrast to the important role volatile components may play in establishing the constitution of planetary mantle source regions, we discovered that the composition of the most abundant basalts erupted on all the planets sampled so far is influenced to a negligible extent by H_2O or CO_2. For Earth, where these volatile components are present (as they are not on the Moon or eucrite planet), this is because abundant basalts are generated by a high degree of melting of the source region. When a source region produces a large amount of basalt, the volatiles are either bled off in the early-generated melts, or diluted to such an extent that their influence is small. This is not to say that H_2O and CO_2 have no important role in terrestrial petrogenesis. They clearly have a profound influence on the first melting of peridotite, and the composition and physical behavior of some magmas. But, because of their low abundance, the volume of liquids produced under hydrated or carbonated conditions is trivial compared with the essentially anhydrous, voluminous lavas of ocean ridges.

By analogy with Earth, we may expect volatile components to play an important role on Venus and Mars in determining such things as the first melting of the mantle, the presence or absence of a low-velocity zone, the compositions of rarer basalts, and eruptive mechanisms. We may also expect that the compositions of the most voluminous eruptions are little influenced unless the abundance of volatile components on these planets exceeds that on Earth by orders of magnitude.

We find that some results can be correlated with planetary size for the planets sampled so far, in order of decreasing size: the Earth, the Moon, and the sources of basaltic meteorites (presumed to be small). A large planet has a stronger gravitational field than a small planet and has more energy/unit mass available through processes such as core formation and pressure release. Because of the increase of solidus temperatures with increasing pressure, melts produced at depth in a large planet with a large pressure gradient must be hotter than melts produced at similar depths and degrees of melting in a small planet. Following this reasoning, we might expect bigger planets to be able to produce hotter magmas. This seems to be valid, for the Earth has generated komatiites (~1600°C), the Moon has produced ultramafic glasses (~1400°C), and the eucrites erupted as relatively cool liquids (~1180°C). Of course, energetic heating mechanisms (unspecified) could raise temperatures on small planets as well, so deviations from this general expectation would not be surprising.

Larger planets have better insulation in terms of their surface/volume ratio. They may be expected to remain sufficiently hot and active enough to produce basalts longer than small planets. Once again, we find our expectations crudely confirmed. The Earth is still volcanically active, the Moon seems to have been dead in this respect for at least 2.5 b.y., and the eucrite planet burned out shortly after its formation about 4.5 b.y. ago. Again, this correlation between size and endurance may be upset by any number of additional factors. Indeed, if the shergottite, nakhlite, and chassignite meteorites should prove to come from a small planet or planets, their ages of about 1.3–0.6 b.y. would be exceptions. However, the crater chronologies reviewed in Chapter 8 are consistent with the expectation, based on size, that Mercury and Mars have had a volcanic endurance intermediate between the Earth and the Moon.

Larger planets have a larger range of pressure in their interior than small planets. Also, once they reach a certain size, they are able to contain effervescent volatile components when hot. Furthermore, the longer-lived activity of larger planets gives more opportunity for multiple and continuous differentiation events to produce a spectrum of source regions. These factors might lead us to expect a greater complexity of volcanic rock types on a large planet, because melt composition is influenced by volatile content, pressure of segregation, and previous depletion of the source region. Clearly, the Earth has a variety of volcanic rocks reflecting these influences. The lunar collection is quite pedestrian by comparison and the eucrites are barren of variety. We should hasten to point out that the variety we find correlates strongly with the number of samples we have from these bodies, a correlation which may be more than coincidental. Hence, it is not safe to conclude that smaller planets need smaller sampling programs.

We emphasized the importance of identifying the actual primary liquid segregated if the exercise of working back from the basalt to the source region constitution is to be meaningful, and we suggested a strategy for accomplishing this goal in Fig. 3.1.4. In our case study of how well the approach works (section 3.3.8), we pointed out the importance of having as many additional constraints on the source region mineralogy as possible, because the process of working backward from the basalt to the planetary interior could give ambiguous or misleading results, as with any inverse problem.

In our survey of the planets sampled we found that the majority of abundant basalts had experienced crystal-liquid fractionation at low pressures, as is graphically apparent in Figs. 3.3.34, 3.4.1 and 3.5.1. Even for the Moon, where the primary melting occurred at low pressure in the lunar crust, and for the eucrite parent planet, which presumably was too small to have the primary melting occur at other than low pressure, the majority of compositions had experienced some low-pressure fractionation. Low-pressure fractionation is also ubiquitous on the Earth, where primary melting does occur at elevated pressures. As a result, the primary magmas we require in order to work back to basalt source regions are decidedly in the minority in sample populations. It would be surprising if this were not also true on the planets remaining to be sampled.

This expected low incidence of primary magmas is of great importance to mission planners, and two points must be emphasized. The first is that our experience shows that this incidence is not zero and important results are derived from the few that are found. The second is that sampling must be designed to encounter as much variability as possible. In the search for primary magmas, the population is as important as the individual—it serves as a background against which the individual acquires more definite significance. Consequently, missions which produce a single average or grab sample are of decidedly less use to the petrologist and planetologist than the mission which produces a suite of samples.

ACKNOWLEDGMENTS—We thank M. J. O'Hara and D. H. Green, associate members of this team. M. J. O'Hara was initially a team member, participating in discussions at the first workshop in 1976. D. H. Green provided some background information early in the program, and provided unpublished material including data for Figs. 3.3.20–21.

Most of the material for this chapter was gathered by December 1978, although some references were updated during subsequent revisions.

Team members wish to acknowledge general support from other sources as follows: D. C. Presnall, NSF Grants EAR 74-22571 A01 and EAR 78-22766; E. M. Stolper and D. Walker, NASA Grant NGL-22-007-247, NSF Grants EAR 79-06321 and EAR 79-23977 (J. F. Hays, Principal Investigator); D. Walker, NSF Grant OCE 79-09699; P. J. Wyllie, NSF Grants EAR 76-20410 and EAR 76-20413; R. Merrill, NASA Contracts NSR 09-051-001 and NASW 3389.

3.8 APPENDIX

Appendix 3.1 Experimental studies of complex terrestrial basaltic compositions.[†]

Composition	100 Mg / Mg+Fe	Normative Features (*pl* = *an*/ *ab* + *an*)	Nature of Starting Material	f_{O_2} of 1 Atm. Runs	1 Atm. Sample Container	High Pressure Apparatus	High Pressure Sample Container	Run Pressures	Volatile Activities	High Pressure f_{O_2}	Temperature Extent of Runs	References
1. *Ocean Island Olivine Tholeiites*												
1921 Kilauea	65.6	*ol* 19.14, *hy* 15.35, *pl* 55.1	rock powder	air N_2	sealed Pt open Fe	cold seal internally-heated Ar	sealed Pt	1–10 kb	H_2O-excess	? ~ HM	liquidus to subsolidus	Yoder & Tilley (1962)
1921 Kilauea	68.6	*ol* 0.25, *hy* 23.8, *pl* 59.5	rock powder	-	-	piston-cylinder	sealed $Ag_{50}Pd_{50}$	10–30 kb	H_2O-excess $H_2O + CO_2$–excess ($X_{H_2O} = 0.5$)	~ NNO	solidus to near liquidus	Hill & Boettcher (1970)
1921 Kilauea	68.6	*ol* 0.25, *hy* 23.8, *pl* 59.5	rock powder	-	-	internally-heated Ar cold seal	sealed $Ag_{40}Pd_{60}$	2–8 kb	$H_2O + CO_2$–excess ($X_{H_2O} = 0.59$) H_2O-excess	NNO QFM	solidus to near liquidus	Holloway & Burnham (1972)
1921 Kilauea	68.6	*ol* 0.25, *hy* 23.8, *pl* 59.5	rock powder	$10^{-8.5}$–10^{-8}	$Ag_{60}Pd_{40}$	piston-cylinder	sealed $Ag_{50}Pd_{50}$	10–21 kb	H_2O-excess	~ NNO HM MW	liquidus and near-liquidus	Allen *et al.* (1975); Allen & Boettcher (1978); Fudali (1965)
1921 Kilauea	63.6	*ol* 17.30, *hy* 19.77, *pl* 59.5	rock powder	-	-	internally-heated Ar	sealed $Ag_{40}Pd_{60}$	5 kb	H_2O-excess	QFM HM	solidus to ~50% molten	Helz (1973, 1976)
1921 Kilauea	56.6	*ol* 1.6, *hy* 18.7, *pl* 57.2	glass	$10^{-0.68}$–10^{-14}	open Pt, $Ag_{60}Pd_{40}$, $Ag_{55}Pd_{45}$ or Al_2O_3	-	-	-	-	-	liquidus to ~50% crystalline	Hill & Roeder (1974)
1959–60 Kilauea	various	various	rock powders	air	sealed Pt	-	-	-	-	-	liquidus to ~50% crystalline	Thompson & Tilley (1969)
Synthetic Kilauea Iki 1959 "Parent Magma"	72.0	*ol* 21.9, *hy* 12.3, *pl* 65.2	glass	?	sealed Pt	piston-cylinder	sealed Pt	5–27 kb	dry	? ~ NNO	subsolidus to liquids	Green & Ringwood (1967a); Green (1970)
Anjouan	70.2–63.6	*ol* 15.88–11.76, *hy* 7.23–6 .62, *pl* 51.1–59.6	rock powders	Ar	open $Ag_{70}Pd_{30}$ or Pt	-	-	-	-	-	liquidus to ~60% crystalline	Thompson & Flower (1971)
Reunion	various	various	rock powders	Ar	open $Ag_{70}Pd_{30}$ or Pt	-	-	-	-	-	liquidus to ~50% crystalline	Tilley *et al.* (1971)
Galapagos	various	various	rock powders	NNO	sealed Pt, $Ag_{40}Pd_{60}$ or Ag	-	-	-	-	-	solidus to liquidus	Bow (1976)
Iceland	various	various	rock powders	air QFM	sealed Pt	-	-	-	-	-	liquidus	Tilley *et al.* (1967); Fisk *et al.* (1976)

[†]See Table 3.3.4.

Appendix 3.1 *(Continued)*[†]

Composition	100 Mg / Mg+Fe	Normative Features (*pl* = *an*/*ab* + *an*)	Nature of Starting Material	f_{O_2} of 1 Atm. Runs	1 Atm. Sample Container	High Pressure Apparatus	High Pressure Sample Container	Run Pressures	Volatile Activities	High Pressure f_{O_2}	Temperature Extent of Runs	References
2. *Abyssal Tholeiites*												
Mid-Atlantic Ridge 24°–30° N (T87, T89)	66.5 63.1	*ol* 11.0, *hy* 10.9, *pl* 54.4. *ol* 7.7, *hy* 10.3, *pl* 55.2	rock powders	Ar	Pt	piston-cylinder	graphite sealed Pt	5–12 kb	dry – ~ 2% H_2O	>IW ~HM	near-liquidus	Kushiro & Thompson (1972); Tilley *et al.* (1972)
Mid-Atlantic Ridge 36° N FAMOUS (572-1-1)	68.1	*ol* 15.9, *hy* 8.9, *pl* 68 .2	natural glass	MMO	Mo (sealed in SiO_2)	piston-cylinder	graphite	5–15 kb	dry	> IW	near-liquidus	Bender *et al.* (1978)
Mid-Atlantic Ridge DSDP3-18-17-1 (synthetic)	70.9	*ol* 10.6, *hy* 10.7, *pl* 67 .9	devitrified glass glass (± olivine)	?	Fe (sealed in SiO_2)	piston-cylinder	Fe graphite	5–25 kb	dry	?	near-liquidus	Green *et al.* (1979)
Mid-Atlantic Ridge FAMOUS (6 samples)	?	?	rock powders	10^{-5}–10^{-14}	open AgPd or Mo	-	-	-	-	-	liquidus to near solidus	Benhamou & Biggar (1978a)
Mid-Atlantic Ridge DSDP45-395A-8-1-9	57.2	*ol* 9.3, *hy* 14.3, *pl* 55.4	rock powder		Mo (sealed in SiO_2)	piston-cylinder	graphite	5–12 kb	dry	> IW	near-liquidus	Fujii & Kushiro (1977)
Mid-Atlantic Ridge	67.9	*ol* 120.18, *hy* 4.25, *pl* 63.5	rock powder	air	sealed Pt	-	-	-	-	-	near-liquius	Tilley *et al.* (1965)
Mid-Atlantic Ridge DSDP45-395A	63.9 70.4	*ol* 7.3, *hy* 11.5, *pl* 51.2; *ol* 5.5, *hy* 10.6, *pl* 60.7	rock powders	0.5 log unit below QFM	Pt wire loops	-	-	-	-	-	liquidus to near-solidus	Dungan *et al.* (1979)
Reykjanes Ridge	various	various	rock powders	QFM	-	-	-	-	-	-	near-liquidus	Fisk *et al.* (1976)
Synthetic oceanic tholeiite	60.7	*ol* 6.2, *hy* 11.9, *pl* 58.4	glass	?	Pt	piston-cylinder	sealed Pt unsealed Pt	4.5–36 kb	dry H_2O–deficient (uncon-trolled)	? ~NNO	liquidus to solidus	T. H. Green & Ringwood (1968); T.H. Green *et al.* (1967)
3. *Quartz Tholeiites*												
Picture Gorge	42.0	*qz* 0.73, *hy* 20.9, *pl* 46.3	rock powder	-	-	internally-heated Ar	sealed $Ag_{40}Pd_{60}$	5 kb	H_2O-excess	QFM HM	solidus to ~ 50% molten	Helz (1973, 1976)

[†]See Table 3.3.4.

Appendix 3.1 *(Continued)*[†]

Composition	100 Mg / Mg+Fe	Normative Features ($pl = an/ab + an$)	Nature of Starting Material	f_{O_2} of 1 Atm. Runs	1 Atm. Sample Container	High Pressure Apparatus	High Pressure Sample Container	Run Pressures	Volatile Activities	High Pressure f_{O_2}	Temperature Extent of Runs	References
Picture Gorge	47.9	*qz* 3.82, *hy* 14.39, *pl* 46.3	rock powder	10^{-8}–$10^{-5.8}$	$Ag_{60}Pd_{40}$	piston-cylinder	sealed $Ag_{50}Pd_{50}$	10–21 kb	H_2O-excess	~ NNO HM MW	near-liquidus	Allen *et al.* (1975); Allen & Boettcher (1978); Fudali (1965)
Picture Gorge	47.9	*qz* 3.82, *hy* 14.39, *pl* 46.3	rock powder	air	sealed Pt	internally-heated Ar	sealed Pt	1 kb	H_2O-excess	QFM HM MW	near-liquidus	Hamilton *et al.* (1964); Tilley *et al.* (1964)
Hirschberg	79.3	*qz* 2.1, *hy* 27.5, *pl* 50.1	glass	-	-	piston-cylinder	sealed Pt or $Ag_{75}Pd_{25}$	2–15 kb	2% H_2O, 4% H_2O	~ NNO	near-liquidus	Nicholls & Lorenz (1973)
Glenelg eclogite	51.3	*qz* 1.80, *hy* 18.71, *pl* 54.5	rock powder	air	sealed Pt	internally-heated Ar piston-cylinder	sealed Pt	10–31.4 kb	dry H_2O-excess	? ~ HM	liquidus to subsolidus	Yoder & Tilley (1962)
Synthetic quartz tholeiite	60.4	*qz* 1.3, *hy* 22.6, *pl* 58.5	glass	?	Pt	piston-cylinder	sealed Pt unsealed Pt	4.5–36 kb	dry H_2O-deficient (uncontrolled)	? ~ NNO	liquidus to subsolidus	T. H. Green *et al.* (1967); T.H. Green & Ringwood (1967, 1968)
Synthetic albite-rich quartz tholeiite	61.0	*qz* 2.5, *hy* 20.6, *pl* 53.6	glass	-	-	piston-cylinder	sealed Pt or $Ag_{75}Pd_{25}$	13.5–29 kb	H_2O-excess	~ NNO	near-solidus and subsolidus	Essene *et al.* (1970)
Synthetic SiO_2-saturated tholeiite	65.0	*qz* 0.2, *hy* 27.8, *pl* 53.1	olivine-seeded glass, glass	? ~ IW	Fe	piston-cylinder	sealed $Ag_{50}Pd_{50}$ or $Ag_{75}Pd_{25}$	5–22.5 kb	H_2O-deficient H_2O-excess	?	near liquidus	Nicholls & Ringwood (1973)
Kingston diabase RHB	45.3	*qz* 1.9, *hy* 26.2, *pl* 63.8	glass	$10^{-0.68}$–10^{-14}	open Pt, - $Ag_{60}Pd_{40}$, $Ag_{55}Pd_{45}$ or Al_2O_3	-	-	-	-	-	liquidus to ~ 50% crystalline	Hill & Roeder (1974)
Skaergaard chilled margin	51.2	*qz* 1.2, *hy* 21.7, *pl* 52.8	rock powder	NNO, QFM, MW	Pt wire loops	-	-	-	-	-	liquidus to ~ 50% crystalline	Hoover (1978)
4. *Continental Olivine Tholeiites*												
Snake River Plain	various 49.1	various *ol* 8.37, *hy* 15.49, *pl* 55.4	rock powders	Ar air	open $Ag_{70}Pd_{30}$ or Pt (sealed in air)	piston-cylinder	graphite	5–35.5 kb	dry	> IW	liquidus to solidus	Tilley & Thompson (1970); Thompson (1975)
Nuanetsi	various	various	rock powders	- heated gas	- 10 kb	internally- piston-cylinder	unsealed Pt	7 kb, ~ 60%	dry (1974)	? ~ NNO	liquidus to crystalline	Cox & Jamieson
Deccan Traps	various	various	rock powders	NNO IW	$Ag_{40}Pd_{60}$ MO	-	-	-	-	-	liquidus to ~60% crystalline	Krishnamurthy & Cox (1977)

[†]See Table 3.3.4.

Appendix 3.1 *(Continued)*[†]

Composition	100 Mg/(Mg+Fe)	Normative Features ($pl = an/ab + an$)	Nature of Starting Material	f_{O_2} of 1 Atm. Runs	1 Atm. Sample Container	High Pressure Apparatus	High Pressure Sample Container	Run Pressures	Volatile Activities	High Pressure f_{O_2}	Temperature Extent of Runs	References
5. *High-Alumina Tholeiites*												
Yakuno	?	?	rock powder	air	open Pt	piston-cylinder	sealed Pt open Pt	9.7–33 kb	dry–H_2O-excess	?	liquidus to solidus	Shimada (1966a,b)
Buffalo Butte	54.3	*ol* 14.06, *hy* 13.20, *pl* 56.0	glass, seeded glass	? ~ IW	Fe sealed Pt	piston-cylinder	Fe unsealed Pt	6.3–49.8 kb	dry	? ~ IW	liquidus to subsolidus (1962)	Cohen *et al.* (1967); Yoder & Tilley
Synthetic	69.8	*ol* 14.5, *hy* 9.7, *pl* 58.6	glass	-	-	piston-cylinder	sealed Pt	4.5–9 kb	dry	? ~ NNO	liquidus to near-solidus	T. H. Green *et al.* (1967)
Warner	65.8	*ol* 20.55, *hy* 0.0, *pl* 60.5	rock powder	air	sealed Pt	cold seal internally-heated Ar	sealed Pt	2–10 kb	H_2O-excess	? ~ HM	liquidus to subsolidus	Yoder & Tilley (1962)
Soay	63.5	*ol* 8.6, *hy* 7.0, *pl* 73.9	rock powder	-	-	piston-cylinder	sealed Pt, $Ag_{30}Pd_{70}$ or $Ag_{75}Pd_{25}$	10–35 kb	H_2O-excess H_2O-deficient	<NNO	liquidus to subsolidus	Lambert & Wyllie (1972); Stern & Wyllie (1973, 1978); Stern *et al.* (1975)
6. *"Pseudotholeiites"*												
Roberts Victor eclogite	67.7	*ol* 33.40, *hy* 10.17, *pl* 69.3	mineral mix; powder	$CO+CO_2$ (1:1)	Fe-bearing Pt	piston-cylinder	graphite Fe	5.5–30.5 kb	dry	? ≥ IW	solidus to liquidus	Ito & Kennedy (1974)
Salt Lake Crater garnet websterite	83.1	*ol* 20.0, *hy* 8.6, *pl* 74.6	rock powder	-	-	piston-cylinder	sealed $Ag_{50}Pd_{50}$	7.5–30 kb	H_2O-excess	~ NNO	liquidus to solidus	Mysen & Boettcher (1976)
Synthetic "olivine tholeiite"	65.0	*ol* 22.7, *hy* 9.7, *pl* 84.0	glass (fused mineral mix)	-	-	piston-cylinder	sealed Pt, $Ag_{50}Pd_{50}$ or $Ag_{75}Pd_{25}$; graphite	5–30 kb	dry, H_2O-deficient, H_2O-excess	~ NNO	near-liquidus	Nicholls & Ringwood (1973)
Lock Duich eclogite	43.2	*ol* 10.62, *hy* 15.83, *pl* 75.1	rock powder	air	sealed Pt	internally-heated Ar piston-cylinder	sealed Pt	10–30 kb	dry H_2O-excess	? ~ NNO	liquidus to subsolidus	Yoder & Tilley (1962)
Silberbach eclogite	84.9	*ol* 2.61, *hy* 16.95, *pl* 73.7	rock powder	air	sealed Pt	internally-heated Ar	sealed Pt	10 kb	dry–H_2O-excess	? ~ NNO	liquidus to subsolidus	Yoder & Tilley (1962)
Matsoku olivine-garnet websterite, 2 eclogites	various	various	rock powders	-	-	piston-cylinder	sealed Pt	20–40 kb	dry	?	solidus to liquidus	Howells *et al.* (1975)

[†] See Table 3.3.4.

Appendix 3.1 *(Continued)*[†]

Composition	100 Mg / Mg+Fe	Normative Features ($pl = an/ab + an$)	Nature of Starting Material	f_{O_2} of 1 Atm. Runs	1 Atm. Sample Container	High Pressure Apparatus	High Pressure Sample Container	Run Pressures	Volatile Activities	High Pressure f_{O_2}	Temperature Extent of Runs	References
South African and Tanzanian eclogites (3)	various	various	rock powders	-	-	piston-cylinder	open Pt	30 kb	dry	?	solidus to liquidus	O'Hara & Yoder (1967)
Beni Bouchera garnet pyroxenites (4)	various	various	rock powders	-	-	piston-cylinder	Pt	15–30 kb	dry	?	solidus to liquidus	Kornprobst (1970)
7. *Transitional Alkalic Basalts*												
Synthetic olivine basalt	68.1	*hy* 1.3, *ol* 21.9, *pl* 59.1	glass	-	-	piston-cylinder	sealed Pt	9–27 kb	dry	~NNO	near-liquidus	Green & Ringwood (1967a)
Auckland Islands	66.2	*hy* 2.7, *ol* 18.8, *pl* 43.2	glass	-	-	piston-cylinder	open or sealed Pt, graphite	9–18 kb	dry H_2O-deficient	~NNO	near-liquidus	Green & Hibberson (1970)
Skye (SK971)	58.2	*ne* 0.19, *pl* 55.2	rock powder	Ar	open $Ag_{70}Pd_{30}$ or Pt	piston-cylinder	graphite	5–26 kb	dry	>IW	liquidus to near-solidus	Thompson (1974); Thompson *et al.* (1972)
Mt. Baldy	43.1	*hy* 0.25, *ol* 13.81, *pl* 35.3	glass	-	-	piston-cylinder	sealed $Ag_{70}Pd_{30}$ sealed Pt	5–15 kb	dry to 5% H_2O	~NNO	near-liquidus	Knutson & Green (1975)
Alpine	50.6	*ne* 0.11, *pl* 45.7	glass	QFM IW	Pt wire loops	internally-heated Ar piston-cylinder	open Mo graphite	1–9 kb	dry	~MMO ? ~CCO	near-liquidus	Merrill & Irving (1977)
8. *Transitional "Pseudobasalt"*												
Ronda plagioclase-garnet clino-pyroxenite	69.2	*ne* 0.12, *pl* 69.3	rock powder glass seeded glass	-	-	piston-cylinder	graphite; sealed $Pt_{95}Au_5$	5–30 kb	dry	? ~HM	liquidus to subsolidus	Obata & Dickey (1976)
9. *Alkali Olivine Basalts*												
Prehistoric Hualalai	63.4	*ne* 2.20, *pl* 54.1	rock powder	air	sealed Pt	cold seal internally-heated Ar	sealed Pt	2–10 kb	H_2O-excess	? ~HM QFM	liquidus to solidus	Yoder & Tilley (1962); Nesbitt & Hamilton (1970)
Prehistoric Hualalai	66.6	*ne* 0.20, *pl* 52.8	rock powder	-	-	piston-cylinder	sealed $Ag_{50}Pd_{50}$	10–33 kb	H_2O-excess	~NNO MW HM	liquidus to solidus	Allen *et al.* (1975); Allen & Boettcher (1978); Haygood *et al.* (1971)

[†] See Table 3.3.4.

Appendix 3.1 *(Continued)*[†]

Composition	$\frac{100\ Mg}{Mg+Fe}$	Normative Features ($pl = an/ab + an$)	Nature of Starting Material	f_{O_2} of 1 Atm. Runs	1 Atm. Sample Container	High Pressure Apparatus	High Pressure Sample Container	Run Pressures	Volatile Activities	High Pressure f_{O_2}	Temperature Extent of Runs	References
1801 Hualalai	57.7	*ne* 3.02, *pl* 55.4	rock powder	air	sealed Pt	internally-heated Ar	sealed $Ag_{40}Pd_{60}$	5 kb	H_2O-excess	QFM HM	solidus to ~50% molten	Helz (1973, 1976); Tilley *et al.* (1965)
Synthetic	59.8	*ne* 2.2, *pl* 59.3	glass crystallized glass	? ~ IW	sealed Pt	piston-cylinder	sealed Pt, $Ag_{75}Pd_{25}$ or Au	9–32.5 kb	dry H_2O-excess	~NNO	liquidus to subsolidus	Green & Ringwood (1967a); Essene *et al.* (1970)
Skye (66018)	65.4	*ne* 2.84, *pl* 51.6	rock powder	-	-	piston-cylinder	graphite	8–31 kb	dry	>IW	liquidus to near-solidus	Thompson (1974)
Kilsyth Hills ZP422	?	?	rock powder	-	-	internally-heated Ar	sealed Pt, sealed or open Fe-sat. Pt	1–10 kb	dry	~ NNO	near-liquidus	Ford & Macdonald (1978)
Grenada (286)	66.6	*ne* 2.38, *pl* 67.4	rock powder	-	-	internally-heated Ar	sealed $Ag_{70}Pd_{30}$	5 kb	H_2O-excess	NNO	liquidus to ~ 50% crystalline	Cawthorn *et al.* (1973b)
Galapagos	various	various	rock powders	NNO	sealed Pt, $Ag_{40}Pd_{60}$ or $Ag_{70}Pd_{30}$	-	-	-	-	-	solidus to liquidus	Bow (1976)
Horta, Azores	57.3	*ne* 1.63, *pl* 47.9	glass	CO_2 $CO+CO_2$ (1:2) $CO+CO_2$ (1:10)	$Ag_{60}Pd_{40}$	-	-	-	-	-	liquidus to ~ 50% crystalline	Duke (1974)
1964 Surtsey	60.8	*ne* 4.2, *pl* 58.4	rock powder	air	sealed Pt	-	-	-	-	-	near-liquidus	Tilley *et al.* (1966)
Guyra	59.6	*ne* 2.3, *pl* 51.2	rock powder	air	sealed Pt	-	-	-	-	-	near-liquidus	Tilley *et al.* (1966)
10. *Basanites*												
Synthetic	73.7	*ne* 9.4, *pl* 74.3	glass	-	-	piston-cylinder	graphite	13.5 – 31.5 kb	dry	< QFM	liquidus to ~ 30% crystalline	Bultitude & Green (1971)
Mt. Leura + 10% olivine (Fo_{90})	70.2	*ne* 10.5, *pl* 45.5	glass	? ~ IW	Fe	piston-cylinder	sealed Pt, $Ag_{75}Pd_{25}$ or $Ag_{50}Pd_{50}$	5–36 kb	dry to H_2O-excess	~ NNO	liquidus to near-solidus	Green (1973b)
Dish Hill	?	?	rock powder	? ~ IW	Fe	piston-cylinder	Fe	10.8 – 40.8 kb	dry	? ~ IW	liquidus only	Ito & Kennedy (1968)
Mt. Shadwell	67.6	*ne* 11.92, *pl* 49.8	rock powder	-	-	piston-cylinder	graphite	10–30 kb	dry	> IW	liquidus to near-solidus	Arculus (1975)
Grenada (531)	73.7	*ne* 7.84, *pl* 78.4	rock powder	-	-	piston-cylinder	graphite	10–35 kb	dry	> IW	liquidus to near-solidus	Arculus (1975)

[†]See Table 3.3.4.

Appendix 3.1 *(Continued)*[†]

Composition	100 Mg/Mg+Fe	Normative Features ($pl = an/ab + an$)	Nature of Starting Material	f_{O_2} of 1 Atm. Runs	1 Atm. Sample Container	High Pressure Apparatus	High Pressure Sample Container	Run Pressures	Volatile Activities	High Pressure f_{O_2}	Temperature Extent of Runs	References
Grenada (209, 184)	73.4, 68.4	*ne* 10.84, *pl* 86.3 *ne* 8.78, *pl* 82.8	rock powders	-	-	internally-heated Ar	sealed $Ag_{70}Pd_{30}$	5 kb	H_2O-excess	NNO	liquidus to ~ 50% crystalline	Cawthorn *et al.* (1973b)
Mt. Kenya and Kilimanjaro	various	various	rock powders	-	-	internally-heated Ar piston-cylinder	open AgPd or Pt; sealed AgPd or Pt	2–15 kb	dry H_2O-excess CO_2-excess	MH	liquidus to near-solidus	Rock (1978)
Kakanui kaersutite eclogite	63.0	*ne* 6.4, *pl* 79.9	rock powder	-	-	piston-cylinder	sealed Pt sealed $Pd_{70}Ag_{30}$	10–30 kb	H_2O-excess H_2O deficient (0.4%)	<NNO	liquidus to subsolidus	Merrill & Wyllie (1975)
Inverell	62.7	*ne* 10.7, *pl* 67.3	rock powder	air	sealed Pt	-	-	-	-	-	near-liquidus	Tilley *et al.* (1966)
11. *Evolved Alkalic Basalts*												
The Anakies (east) nepheline mugearite (2102)	55.9	*ne* 4.7, *pl* 29.0	glass; seeded glass; crystal-lized glass	? ~ IW	Fe	piston-cylinder	sealed Pt sealted $Ag_{75}Pd_{25}$ sealed $Ag_{50}Pd_{50}$	2.2–31.5 kb	dry to H_2O-excess	~ NNO (NNO) (MH) (QFM)	liquidus to near-solidus	Irving & Green (1972, 1980)
Skye hawaiite (SK 931)	44.2	*ne* 2.29, *pl* 34.4	rock powder	Ar	open $Ag_{70}Pd_{30}$ or Pt	piston-cylinder	graphite	10–22 kb	dry	> IW	liquidus to ~ 50% crystalline	Thompson (1974); Thompson *et al.* (1972)
Mauna Kea hawaiite	49.9	*hy* 6.43, *pl* 45.3	rock powder	air	sealed Pt	cold seal internally-heated Ar	sealed Pt	2–5 kb	H_2O-excess		liquidus to subsolidus	Yoder & Tilley (1962); Tilley *et al.* (1965)
Anjouan nepheline hawaiites, neph. mugearites, neph. benmoreite	53.7–29.4	*ne* 4.48–14.06, *pl* 42.5–20.4	rock powders	Ar	open $Ag_{70}Pd_{30}$ or Pt	-	-	-	-	-	liquidus	Thompson & Flower (1971)
Aden trachybasalts and trachytes	various	various	rock powders	IW	Mo	-	-	-	-	-	liquidus to near-solidus	Humphries & Cox (1972)
Hebridean hawaiites, mugearites, benmoreites, trachyte	various	various	rock powders	Ar air	open $Ag_{70}Pd_{30}$ or Pt; sealed Pt	-	-	-	-	-	near-liquidus	Tilley *et al.* (1965, 1966); Thompson *et al.* (1972)
Hawaiian hawaiite, mugearite, benmoreite	various	various	rock powders	air	sealed Pt	-	-	-	-	-	near-liquidus	Tilley *et al.* (1965)
Reunion	various	various	rock powders	IW	Mo	-	-	-	-	-	liquidus to solidus	Humphries (1972)

[†]See Table 3.3.4.

Appendix 3.1 *(Continued)*[+]

Composition	100 Mg/Mg+Fe	Normative Features ($pl = an/ab + an$)	Nature of Starting Material	f_{O_2} of 1 Atm. Runs	1 Atm. Sample Container	High Pressure Apparatus	High Pressure Sample Container	Run Pressures	Volatile Activities	High Pressure f_{O_2}	Temperature Extent of Runs	References
Gough Island trachybasalt, tristanite, trachyte	various	various	rock powders	air	sealed Pt	-	-	-	-	-	near-liquidus	Tilley *et al.* (1966)
Tristan da Cunha trachybasalt, tristanite, trachyte	various	various	rock powders	air	sealed Pt	-	-	-	-	-	near-liquidus	Tilley *et al.* (1965, 1966)
Afar ferrobasalts, trachytes, alkali rhyolites, pantellerites	various	various	rock powders	NNO CO_2+0.2% H_2	open $Ag_{70}Pd_{30}$ or Pt	-	-	-	-	-	liquidus to near-solidus	Benhamou & Biggar (1978b)
12. *Olivine Nephelinites*												
Synthetic	71.0	*ab* 2.0, *ol* 25.9	glass	?	sealed Pt	piston-cylinder	sealed Pt graphite	4.5–36 kb	dry H_2O-deficient (uncontrolled)	~ NNO	near-liquidus	Bultitude & Green (1968, 1971)
Synthetic (picritic)	73.8	*ab* 3.4, *ol* 35.7	glass	-	-	piston-cylinder	graphite sealed Pt	9–36 kb	dry H_2O-deficient (uncontrolled)	~ NNO	near-liquidus	Bultitude & Green (1968, 1971)
St. Paul's Rocks brown hornblende mylonite	56.5	*ab* 2.14, *ol* 11.30	rock powder	-	-	piston-cylinder	sealed Pt	10–32 kb	H_2O-excess H_2O-deficient	< NNO	liquidus to solidus	Millhollen & Wyllie (1974)
Honolulu	75.0	*lc* 5.6, *ol* 18.2	rock powder	-	-	piston-cylinder	sealed $Ag_{50}Pd_{50}$	10–36 kb	H_2O-excess	~ NNO MW HM	near-liquidus	Allen *et al.* (1975); Allen & Boettcher (1978)
Monchique camptonite CDM43	61.3	*lc* 3.5, *ne* 26.6	rock powder	-	-	internally-heated Ar piston-cylinder	open AgPd or Pt sealed AgPd or Pt	2–15 kb	dry H_2O-excess CO_2-excess	MH	near-liquidus to near-solidus	Rock (1978)
Kakanui kaersutite	74.2	*lc* 6.2, *ol* 17.3	mineral powder	-	-	piston-cylinder	sealed P or $Ag_{30}Pd_{70}$	10–30 kb	H_2O-excess H_2O-deficient	< NNO	liquidus to solidus	Merrill & Wyllie (1975)
Kakanui kaersutite	72.8	*lc* 5.2, *ol* 16.8	mineral powder	air	? Pt	piston-cylinder tetrahedral anvil	sealed Pt	5–40 kb	H_2O-deficient	?	near-solidus and subsolidus	Yagi *et al.* (1975)
Dish Hill kaersutite	?	?	mineral powder	-	-	piston-cylinder	sealed AgPd or Pt	10–22 kb	H_2O-excess H_2O-deficient	?	liquidus to near-solidus	Stewart *et al.* (1979)

[+]See Table 3.3.4.

Appendix 3.1 *(Continued)*[+]

Composition	100 Mg / Mg+Fe	Normative Features ($pl = an/ab + an$)	Nature of Starting Material	f_{O_2} of 1 Atm. Runs	1 Atm. Sample Container	High Pressure Apparatus	High Pressure Sample Container	Run Pressures	Volatile Activities	High Pressure f_{O_2}	Temperature Extent of Runs	References
Pali	74.1	*kl* 4.1, *cs* 3.2, *ol* 17.2	rock powder	air	sealed Pt	-	-	-	-	-	liquidus to near-solidus	Yoder & Tilley (1962); Tilley *et al.* (1965)
13. *Olivine Melilitites*												
Laughing Jack Marsh	73.8	*lc* 6.5, *cs* 10.9, *ol* 36.3	glass	-	-	piston-cylinder	sealed $Ag_{77}Pd_{23}$, $Ag_{50}Pd_{50}$ or Pt	5–40 kb	dry, H_2O-excess to CO_2-excess, varying H_2O/CO_2	~ NNO HM	near-liquidus	Brey & Green (1975, 1976a,b, 1977)
Ca-Mg rich olivine melilitite (synthetic)	74.8	*kl* 4.4, *cs* 19.8, *ol* 44.4	glass and oxides or carbonates	-	-	piston-cylinder	sealed $Ag_{77}Pd_{23}$ or Pt	30–35 kb	H_2O-excess to CO_2-excess, varying H_2O/CO_2	HM	near-liquidus	Brey (1978)
Moiliili	66.0	*kl* 6.0, *cs* 5.1, *ol* 16.9	rock powder	air	sealed Pt	-	-	-	-	-	liquidus to near-solidus	Tilley *et al.* (1965)
Uvalde Co.	72.8	*kl* 4.1, *cs* 6.5 *ol* 18.2	rock powder	air	sealed Pt	-	-	-	-	-	liquidus to near-solidus	Tilley *et al.* (1965)
14. *Potassic Basalts*												
Synthetic biotite mafurites	77.6 76.1	*lc* 33.1 *ne* 1.4; *lc* 33.6, *ne* 0.45	glasses	-	-	piston-cylinder	Fe, graphite; sealed Pt, $Ag_{50}Pd_{50}$ or $Ag_{75}Pd_{25}$	10–30 kb	dry to H_2O-excess CO_2-excess CO_2+H_2O-excess	~ NNO ≤ HM	near-liquidus	Edgar *et al.* (1976); Ryabchikov & Green (1978)
Begargo Hill mafic leucitite	79.1	*ab* 0.93, *ne* 9.08	rock powder	NNO	Pt	piston-cylinder	sealed Pt	18–40 kb	dry	?	liquidus to near-solidus	Cundari & O'Hara (1976)
Leucite Hills madupite (LH16)	77.0	*lc* 33.3 *ne* 0.19	rock powder	?	sealed Pt, Au	cold seal internally-heated Ar piston-cylinder	sealed Pt or Au	0.5–30 kb	H_2O-excess H_2O-deficient	?	liquidus to solidus	Barton & Hamilton (1978, 1979a); Carmichael (1967)
Leucite Hills orendite (LH12)	76.2	*qz* 3.5, *ks* 3.6	rock powder; glass	?	sealed Pt, Au	cold seal internally-heated Ar piston-cylinder	sealed Pt or Au	0.5–30 kb	H_2O-excess H_2O-deficient	?	liquidus to solidus	Barton & Hamilton (1978, 1979b); Carmichael (1967)
Leucite Hills wyomingite (LH2)	75.3	*lc* 9.0, *ne* 1.0	rock powder; glass	?	sealed Pt, Au	cold seal internally-heated Ar piston-cylinder	sealed Pt or Au	0.5–5 kb	H_2O-excess H_2O-deficient	?	liquidus to solidus	Barton & Hamilton (1978, 1979b); Carmichael (1967)

[+]See Table 3.3.4.

Appendix 3.1 *(Concluded)*[†]

Composition	$\frac{100\ Mg}{Mg+Fe}$	Normative Features ($pl = an/ab + an$)	Nature of Starting Material	f_{O_2} of 1 Atm. Runs	1 Atm. Sample Container	High Pressure Apparatus	High Pressure Sample Container	Run Pressures	Volatile Activities	High Pressure f_{O_2}	Temperature Extent of Runs	References
Olivine ugandite, Bufumbira	81.7	*ne* 4.7, *pl* 76.9	rock powder; glass	-	-	piston-cylinder	graphite, sealed Pt, $Ag_{75}Pd_{25}$ or $Ag_{50}Pd_{50}$	10–40 kb	dry H_2O-deficient H_2O-excess H_2O+CO_2 excess	?	near-liquidus	Edgar & Condliffe (1978); Edgar *et al.* (1980)
Leucite Hills wyomingite (1101)	73.7	*lc* 38.8, *ne* 1.42	rock powder	air	Pt	-	-	-	-	-	near-liquidus	Sobolev *et al.* (1975)
Alban Hills leucitite (EU13)	63.4	*lc* 32.5, *ne* 13.6	rock powder	Ar	Pt	piston-cylinder	graphite	5–45 kb	dry	>IW	liquidus to near-solidus	Tilley & Thompson (1972); Thompson (1977)
Latera leucitite	73.8	*lc* 26.7, *ne* 8.6	rock powder	NNO	Pt	-	-	-	-	-	liquidus to near-solidus	Trigila (1969)
15. *Ankaramites*												
Anjouan	69.3 67.4	*ol* 17.39, *hy* 1.62, *pl* 51.0 *ol* 0.43, *hy* 24.44, *pl* 53.5	rock powders	Ar	open $Ag_{70}Pd_{30}$ or Pt	-	-	-	-	-	liquidus to ~60% crystalline	Thompson & Flower (1971)
Deccan (R-9)	?	?	rock powder	NNO	$Ag_{40}Pd_{60}$ or Pt	-	-	-	-	-	near-liquidus	Krishnamurthy (1976)

[†]See Table 3.3.4.

3.9 REFERENCES

Akella J., Williams R. J. and Mullins O. (1976) Solubility of Cr, Ti and Al in coexisting olivine, spinel and liquid at 1 atm. *Proc. Lunar Sci. Conf. 7th*, p. 1179–1194.

Arculus R. J. (1975) Melting behavior of two basanites in the range 10–35 kbar and the effect of TiO_2 on the olivine-diopside reactions at high pressures. *Carnegie Inst. Wash. Yearb.* **74**, 512–515.

Arculus R. J. and Johnson R. W. (1978) Criticism of generalized models for the magmatic evolution of arc-trench systems. *Earth Planet. Sci. Lett.* **39**, 118–126.

Arndt N. T. (1976) Melting relations of ultramafic lavas (komatiites) at 1 atm and high pressure. *Carnegie Inst. Wash. Yearb.* **75**, 555–562.

Arndt N. T. (1977) Ultrabasic magmas and high-degree melting of the mantle. *Contrib. Min. Petrol.* **64**, 205–221.

Arndt N. T., Naldrett A. J. and Pyke D. R. (1977) Komatiitic and iron-rich tholeiitic lavas from Munro Township, north-east Ontario. *J. Petrol.* **18**, 319–369.

Ahrens L. H., Dawson J. B., Duncan A. R. and Erlank A. J., editors (1975). *Physics and Chemistry of the Earth* **v. 9**, Papers presented at the First International Conference on Kimberlites. Pergamon, N.Y. 940 pp.

Allen J. C. and Boettcher A. L. (1978) Amphiboles in andesite and basalt: II. Stability as a function of P-T-f_{H_2O}-f_{O_2}. *Amer. Mineral.* **63**, 1074–1087.

Allen J. C., Boettcher A. L. and Marland G. (1975) Amphiboles in andesite and basalt: I. Stability as a function of P-T-f_{O_2}. *Amer. Mineral.* **60**, 1069–1085.

Anderson A. T. (1975) Some basaltic and andesitic gases. *Rev. Geophys. Space Phys.* **13**, 37–55.

Anderson D. (1973) The composition and origin of the Moon. *Earth Planet. Sci. Lett.* **18**, 301–316.

Anderson O. (1915) The system anorthite-forsterite-silica. *Amer. J. Sci.* **39**, 407–454.

Anderson R. N., DeLong S. E. and Schwarz W. M. (1978) Thermal model for subduction with dehydration in the downgoing slab. *J. Geol.* **86**, 731–739.

Apollo Soil Survey (1971) Apollo 14; nature and origin of rock types in soil from the Fra Mauro formation. *Earth Planet. Sci. Lett.* **12**, 49–54.

Baker B. H., Goles G. G., Leeman W. P. and Lindstrom M. M. (1977) Geochemistry and petrogenesis of a basalt-benmoreite-trachyte suite from the southern part of the Gregory Rift, Kenya. *Contrib. Mineral. Petrol.* **64**, 303–332.

Barrett D. R. (1975) The genesis of kimberlites and associated rocks: strontium isotopic evidence. In *Physics and Chemistry of the Earth* **v. 9**, (L. H. Ahrens *et al.*, eds.), p. 637–653. Pergamon, N.Y.

Barton M. (1976) Melting relations of some ultrapotassic volcanic rocks. In *Progress in Experimental Petrology*, (G. M. Biggar, ed.), p. 91–94. The Nat. Environ. Res. Council, London.

Barton M. and Hamilton D. L. (1978) Water-saturated melting relations to 5 kilobars of three Leucite Hills lavas. *Contrib. Mineral. Petrol.* **66**, 41–49.

Barton M. and Hamilton D. L. (1979) The melting relationships of a madupite from the Leucite Hills, Wyoming, to 30 kb. *Contrib. Mineral. Petrol.* **69**, 133–142.

Basu A. R. (1978) Trace elements and Sr-isotopes in some mantle-derived hydrous minerals and their significance. *Geochim. Cosmochim. Acta* **42**, 659–668.

Baxter A. (1972) Depth of derivation of oceanite lavas in Mauritius. In *Progress in Experimental Petrology*, (G. M. Biggar, ed.), p. 117. Nat. Environ. Res. Council.

Bender J. F., Hodges F. N. and Bence A. E. (1978) Petrogenesis of basalts from the Project FAMOUS area: Experimental study from 0 to 15 kbars. *Earth Planet. Sci. Lett.* **41**, 277–302.

Benhamou G. and Biggar G. M. (1978a) Melting experiments on Mid-Atlantic Ridge basalts. In *Progress in Experimental Petrology*, (G. M. Biggar, ed.), p. 202–204, Nat. Environ. Res. Council.

Benhamou G. and Biggar G. M. (1978b) Atmospheric pressure melting studies of some Ethiopian volcanic rocks. In *Progress in Experimental Petrology*, Natural Environment Research Council Rept. 4, p. 197–200.

Bianco A. S. and Taylor L. A. (1977) Applications of dynamic crystallization studies: lunar olivine-normative basalts. *Proc. Lunar Sci. Conf. 8th*, p. 1593–1610.

Bickle M. J. (1978) Melting experiments on peridotitic komatiites. In *Progress in Experimental Petrology*, Natural Environment Research Council Rept. 4, p. 187–195.

Bickle M. J., Ford C. E. and Nisbet E. G. (1977) The petrogenesis of peridotitic komatiites: evidence from high-pressure melting experiments. *Earth Planet. Sci. Lett.* **37**, 97–106.

Biggar G. M. (1974) Phase equilibrium studies of the chilled margins of some layered intrusions. *Contrib. Mineral. Petrol.* **46**, 159–167.

Biggar G. M. and Gibb F. G. F. (1976) Experimental recrystallization of an ultrabasic dyke rock from Skye. In *Progress in Experimental Petrology*, Natural Environment Research Council Rept. 3, p. 216–217.

Biggar G. M., O'Hara M. J. and Humphries D. J. (1972) Maria lavas, mascons, layered complexes, achondrites and the lunar mantle. In *The Moon*, (S. K. Runcorn and H. C. Urey, eds.), p. 124–164. Reidel, Dordrecht.

Biggar G. M., Humphries D. J. and O'Hara M. J. (1974) Experimental crystallization and geochemical interpretation of low titanium basalts (abstract). In *Lunar Science V*, p. 60–62. The Lunar Science Institute, Houston.

Birch F. (1952) Elasticity and constitution of the Earth's mantle. *J. Geophys. Res.* **57**, 227–286.

Boettcher A. L. (1973) Volcanism and orogenic belts—the origin of andesites. *Tectonophys.* **17**, 204–223.

Boettcher A. L., ed. (1976) Geothermometry-Geobarmetry. *Amer. Mineral.* **61**, Nos. 7 and 8, 816 pp.

Boettcher A. L. and O'Neil J. R. (1980) Stable isotope, chemical, and petrographic studies of high-pressures amphiboles and micas: evidence for metasomatism in the mantle source regions of alkali basalts and kimberlites. *Amer. J. Sci., Jackson v.* **280A**, 594–621.

Bottinga Y. and Weill D. F. (1972) The viscosity of magmatic silicate liquids: a model for calculation. *Amer. J. Sci.* **272**, 438–475.

Bow C. S. (1976) Melting relations of Galapagos basalts. In *Progress in Experimental Petrology*, Natural Environment Research Council Rept. 3, p. 221–223.

Bowen N. L. (1914) The ternary system diopside-forsterite-silica. *Amer. J. Sci.* **38**, 207–264.

Bowen N. L. (1928) *The Evolution of the Igneous Rocks*. Princeton Univ. Press, 332 pp.

Bowen N. L. and Schairer J. F. (1935) The system MgO-FeO-SiO_2. *Amer. J. Sci.* **24**, 151–217.

Boyd F. R. and England J. L. (1959) Experimentation at high pressures and temperatures. *Carnegie Inst. Wash. Yearb.* **58**, 82–89.

Boyd F. R. and England J. L. (1960) Apparatus for phase equilibrium measurements at pressures up to 50 kilobars and temperatures up to 1750° C. *J. Geophys. Res.* **65**, 741–748.

Boyd F. R., England J. L. and Davis B. T. C. (1964) Effects of pressure on the melting and polymorphism of enstatite, $MgSiO_3$. *J. Geophys. Res.* **69**, 2101–2109.

Boyd F. R. and McCallister R. H. (1976) Densities of fertile and sterile garnet peridotites. *Geophys. Res. Lett.* **3**, 509–512.

Boyd F. R. and Meyer H. O. A., eds. (1979) *The Mantle Sample: Inclusions in Kimberlites and Other Volcanics*. Amer. Geophys. Union, Washington, D.C. 424 pp.

Brey G. (1978) Origin of olivine melilitites—chemical and experimental constraints. *J. Volcanol. Geotherm. Res.* **3**, 61–88.

Brey G. and Green D. H. (1975) The role of CO_2 in the genesis of olivine melilitite. *Contrib. Mineral. Petrol.* **49**, 93–103.

Brey G. P. and Green D. H. (1976a) Solubility of CO_2 in olivine melilitite at high pressures and role of CO_2 in the Earth's upper mantle. *Contrib. Mineral. Petrol.* **55**, 217–230.

Brey G. P. and Green D. H. (1976b) Reply to Eggler and Mysen. *Contrib. Mineral. Petrol.* **55**, 237–239.

Brey G. and Green D. H. (1977) Systematic study of liquidus phase relations in olivine melilitite + H_2O + CO_2 at high pressures and petrogenesis of an olivine melilitite magma. *Contrib. Mineral. Petrol.* **61**, 141–162.

Bryan W. B. (1979) Regional variation and petrogenesis of basalt glasses from the FAMOUS area, Mid-Atlantic Ridge. *J. Petrol.* **20**, 293–325.

Bryan W. B. and Moore J. G. (1977) Compositional variations of young basalts in the Mid-Atlantic Ridge rift valley near lat. 36°49 N. *Geol. Soc. Amer. Bull.* **88**, 556–570.

Bultitude R. J. and Green D. H. (1968) Experimental study at high pressures on the origin of olivine nephelinite and olivine melilite nephelinite magmas. *Earth Planet. Sci. Lett.* **3**, 325–337.

Bultitude R. J. and Green D. H. (1971) Experimental study of crystal-liquid relationships at high pressures in olivine nephelinite and basanite compositions. *J. Petrol.* **12**, 121–147.

Bunch T. E. (1975a) Howardites: surface breccias of parent bodies. *Meteoritics* **10**, 91–92.

Bunch T. E. (1975b) Petrography and petrology of basaltic achondrite polymict breccias (howardites). *Proc. Lunar Sci. Conf. 6th*, p. 469–492.

Cann J. R. (1974) A model for oceanic crustal structure developed. *Geophys. J. Roy. Astron. Soc.* **39**, 169–187.

Carmichael I. S. E. (1967) The mineralogy and petrology of the volcanic rocks from the Leucite Hills, Wyoming. *Contrib. Mineral. Petrol.* **15**, 24–66.

Carter J. L. (1970) Mineralogy and chemistry of the Earth's upper mantle based on the partial fusion–partial crystallization model. *Geol. Soc. Amer. Bull.* **81**, 2021–2034.

Carter S. R., Evensen N. M., Hamilton P. J. and O'Nions R. K. (1978) Continental volcanics derived from enriched and depleted source regions: Nd- and Sr-isotope evidence. *Earth Planet. Sci. Lett.* **37**, 401–408.

Cawthorn R. G., Ford C. E., Biggar G. M., Bravo M. S. and Clarke D. B. (1973a) Determination of the liquid composition in experimental samples: discrepancies between microprobe analysis and other methods. *Earth Planet. Sci. Lett.* **21**, 1–5.

Cawthorn R. G., Curran E. B. and Arculus R. J. (1973b) A petrogenetic model for the origin of the calc-alkaline suite of Grenada, Lesser Antilles. *J. Petrol.* **14**, 327–337.

Chappell B. W. and Green D. H. (1973) Chemical composition and petrogenetic relationships in Apollo 15 mare basalts. *Earth Planet. Sci. Lett.* **18**, 237–246.

Chayes F. (1972) Silica saturation in Cenozoic basalt. *Phil. Trans. Roy. Soc. Lond.* **A271**, 285–296.

Chayes F. (1977) The oceanic basalt-trachyte relation in general and in the Canary Islands. *Amer. Mineral.* **62**, 666–671.

Chen C.-H. and Presnall D. C. (1975) The system Mg_2SiO_4-SiO_2 at pressures up to 25 kbar. *Amer. Mineral.* **60**, 398–406.

Clague D. A. and Bunch T. E. (1976) Formation of ferro-basalt at East Pacific midocean spreading centers. *J. Geophys. Res.* **81**, 4247–4256.

Clark S. P., Jr. and Ringwood A. E. (1964) Density distribution and constitution of the mantle. *Rev. Geophys. and Space Phys.* **2**, 35–88.

Cohen L. H., Ito K. and Kennedy G. C. (1967) Melting and phase relations in an anhydrous basalt to 40 kilobars. *Amer. J. Sci.* **265**, 475–518.

Condie K. C. (1976) *Plate Tectonics and Crustal Evolution.* Pergamon, N.Y., 288 pp.

Consolmagno G. and Drake M. J. (1977) Composition and evolution of the eucrite parent body: evidence from rare earth elements. *Geochim. Cosmochim. Acta* **41**, 1271–1282.

Coons W. E., Holloway J. R. and Navrotsky A. (1976) Co^{2+} as a chemical analogue for Fe^{2+} in high-temperature experiments in basaltic systems. *Earth Planet. Sci. Lett.* **30**, 303–308.

Corrigan G. and Gibb F. G. F. (1979) The loss of Fe and Na from a basaltic melt during experiments using the wire-loop technique. *Mineral. Mag.* **43**, 121–126.

Cox K. G. and Jamieson B. G. (1974) The olivine-rich lavas of Nuanetsi: a study of polybaric magmatic evolution. *J. Petrol.* **15**, 269–301.

Cukierman M., Tutts P. M. and Uhlmann D. R. (1972) Viscous flow behavior of lunar compositions 14259 and 14310. *Proc. Lunar Sci. Conf. 3rd*, p. 2619–2625.

Cukierman M., Klein L., Scherer G., Hopper R. W. and Uhlmann D. R. (1973) Viscous flow and crystallization behavior of selected lunar compositions. *Proc. Lunar Sci. Conf. 4th*, p. 2685–2696.

Cundari A. and O'Hara M. J. (1976) Experimental study at atmospheric and high pressures of a mafic leucitite from New South Wales, Australia. In *Progress in Experimental Petrology*, Natural Environment Research Council Rept. 3, p. 260–261.

Dalrymple G. B., Silver E. A. and Jackson E. D. (1973) Origin of the Hawaiian Islands. *Amer. Sci.* **61**, 294–308.

Daly R. A. (1933) *Igneous Rocks and the Depths of the Earth.* McGraw-Hill, N.Y. 598 pp.

Daly R. A. (1946) Nature of the asthenosphere. *Bull. Geol. Soc. Amer.* **57**, 707–726.

Davis B. T. C. and Schairer J. F. (1965) Melting relations in the join diopside-forsterite-pyrope at 40 kilobars and at one atmosphere. *Carnegie Inst. Wash. Yearb.* **64**, 123–126.

DeHon R. A. and Waskom J. D. (1976) Geologic structure of the eastern mare basins. *Proc. Lunar Sci. Conf. 7th*, p. 2729–2746.

Delaney J. R., Muenow D. W. and Graham D. G. (1978) Abundance and distribution of water, carbon, and sulfur in the glassy rims of submarine pillow basalts. *Geochim. Cosmochim. Acta* **42**, 581–594.

Delano J. W. (1977) Experimental melting relations of 63545, 76015, and 76055. *Proc. Lunar Sci. Conf. 8th*, p. 2097–2123.

Delany J. M. and Helgeson H. C. (1978) Calculation of the thermodynamic consequences of dehydration in subducting oceanic crust to 100 kb and >800° C. *Amer. J. Sci.* **278**, 638–686.

DePaolo D. J. (1978) Nd and Sr isotope systematics of young continental igneous rocks. *U. S. Geol. Survey Open-File Rept. 78-701*, 91–93.

Dewey J. F. and Kidd W. S. F. (1977) Geometry of plate accretion. *Geol. Soc. Amer. Bull* **88**, 960–968.

Donaldson C. H. (1976) An experimental investigation of olivine morphology. *Contrib. Mineral. Petrol.* **57**, 187–213.

Donaldson C. H. (1979) Composition changes in a basalt melt contained in a wire loop of $Pt_{80}Rh_{20}$: effects of temperature, time, and oxygen fugacity. *Mineral. Mag.* **43**, 115–119.

Donaldson C. H. and Brown R. W. (1977) Refractory megacrysts and magnesium-rich melt inclusions within spinel in oceanic tholeiites: indicators of magma mixing and parental magma compositions. *Earth Planet. Sci. Lett.* **37**, 81–89.

Donaldson C. H., Williams R. J. and Lofgren G. E. (1975a) A sample holding technique for study of crystal growth in silicate melts. *Amer. Mineral.* **60**, 324–326.

Donaldson C. H., Usselman T. M., Williams R. J. and Lofgren G. E. (1975b) Experimental modeling of the cooling history of Apollo 12 olivine basalts. *Proc. Lunar Sci. Conf. 6th*, p. 843–869.

Dreibus G., Kruse H., Spettel B. and Wänke H. (1977) The bulk composition of the Moon and eucrite parent body. *Proc. Lunar Sci. Conf. 8th*, p. 211–227.

Drever H. I. and Johnston R. (1966) A natural high-lime liquid more basic than basalt. *J. Petrol.* **7**, 414–420.

Duke J. M. (1974) The effect of oxidation on the crystallization of an alkali basalt from the Azores. *J. Geol.* **82**, 524–528.

Duke M. B. (1965) Metallic iron in basaltic achondrites. *J. Geophys. Res.* **70**, 1523–1527.

Duke M. B. and Silver L. T. (1967) Petrology of eucrites, howardites, and mesosiderites. *Geochim. Cosmochim. Acta* **31**, 1637–1665.

Dungan M. A. and Rhodes J. M. (1978) Residual glasses and melt inclusions in basalts from DSDP Legs 45 and 46: evidence for magma mixing. *Contrib. Mineral. Petrol.* **67**, 417–431.

Dungan M. A., Long P. E. and Rhodes J. M. (1978) The petrography, mineral chemistry and one-atmosphere phase relations of basalts from Site 395–Leg 45: DSDP. *Initial Rep. D.S.D.P.* **45**, 461–477.

Dymek R. F., Albee A. L., Chodos A. A. and Wasserburg G. J. (1976) Petrography of isotopically-dated clasts in the Kapoeta howardite and petrologic constraints on the evolution of its parent planet. *Geochim. Cosmochim. Acta* **40**, 1115–1130.

Edgar A. D. (1973) *Experimental Petrology: Basic Principles and Techniques.* Clarendon Press, Oxford. 217 pp.

Edgar A. D. and Condliffe E. (1978) Derivation of K-rich ultramafic magmas from a peridotitic mantle source. *Nature* **275**, 639–640.

Edgar A. D., Condliffe E., Barnett R. L. and Shirran R. J. (1980) An experimental study of an olivine ugandite magma and mechanisms for the formation of its K-enriched derivatives. *J. Petrol.* (in press).

Edgar A. D., Green D. H. and Hibberson W. O. (1976) Experimental petrology of a highly potassic magma. *J. Petrol.* **17**, 339–356.

Eggler D. H. (1973) Role of CO_2 in melting processes in the mantle. *Carnegie Inst. Wash. Yearb.* **72**, 457–467.

Eggler D. H. (1974) Effects of CO_2 on the melting of peridotite. *Carnegie Inst. Wash. Yearb.* **73**, 215–224.

Eggler D. H. (1975) CO_2 as a volatile component of the mantle: the system Mg_2SiO_4-SiO_2-H_2O-CO_2. In *Physics and Chemistry of the Earth* v. **9**, (L. H. Ahrens, J. B. Dawson, A. R. Duncan and A. J. Erlank, eds.), p. 869–881. Pergamon, N.Y.

Eggler D. H. (1976) Does CO_2 cause partial melting in the low-velocity layer of the mantle? *Geology* **4**, 69–72.

Eggler D. H. (1978) The effect of CO_2 upon partial melting of peridotite in the system Na_2O-CaO-Al_2O_3-MgO-SiO_2-CO_2 to 35 kb, with an analysis of melting in a peridotite-H_2O-CO_2 system. *Amer. J. Sci.* **278**, 305–343.

Eggler D. H. and Mysen B. O. (1976) The role of CO_2 in the genesis of olivine melilitite: discussion. *Contrib. Mineral. Petrol.* **55**, 231–236.

Eggler D. H. and Wendlandt R. F. (1979) Experimental studies of the relationship between kimberlite magmas and partial melting of peridotite. In *Kimberlites, Diatremes and Diamonds: Their Geology, Petrology and Geochemistry*, (F. R. Boyd and H. O. A. Meyer, eds.) Amer. Geophys. Union, Washington, D. C., p. 330–338.

Eggler D. H., Kushiro, I. and Holloway J. R. (1976) Stability of carbonate minerals in a hydrous mantle. *Carnegie Inst. Wash. Yearb.* **75**, 631–636.

Ehrenberg S. N. (1977) Garnet peridotite xenoliths in minette from the Navajo Volcanic Field (abstract). *Extended Abstracts, 2nd Internat. Kimberlite Conf., Santa Fe.*

Ellis D. E. and Wyllie P. J. (1979) A model of phase relations in the system MgO-SiO_2-H_2O-CO_2 and prediction of the compositions of liquids coexisting with forsterite and enstatite. In *Kimberlites, Diatremes and Diamonds: Their Geology, Petrology and Geochemistry*, (F. R. Boyd and H. O. A. Meyer, eds.) Amer. Geophys. Union, Washington, D. C., p. 313–318.

Engel A. E. J. and Engel C. G. (1964) Composition of basalts from the Mid-Atlantic Ridge. *Science* **144**, 1–4.

Engel A. E. J., Engel C. G. and Havens R. G. (1965) Chemical characteristics of oceanic basalts and the upper mantle. *Bull. Geol. Soc. Amer.* **76**, 719–734.

Ernst W. G. (1976) *Petrologic Phase Equilibria*. W. H. Freeman, San Francisco. 333 pp.

Essene E. J., Hensen B. J. and Green D. H. (1970) Experimental study of amphibolite and eclogite stability. *Phys. Earth Planet. Inter.* **3**, 378–384.

Ferguson J., Ellis D. J. and England R. N. (1977) Unique spinel-garnet lherzolite inclusion in kimberlite from Australia. *Geology* **5**, 278–280.

Fisk M. R., Schilling J.-G. and Sigurdsson H. (1976) Melting relations of tholeiites from the Reykjanes Ridge and Iceland (abstract). *EOS* (*Trans. Amer. Geophys. Union*) **57**, 343.

Ford C. E. (1976) Effects of explosive depressurization on lunar anorthositic gabbro melts. In *Progress in Experimental Petrology*, Natural Environment Research Council Rept. 3, p. 176–178.

Ford C. E. (1978) Platinum-iron alloy sample containers for melting experiments on iron-bearing rocks, minerals, and related systems. *Mineral. Mag.* **42**, 271–275.

Ford C. E. and Macdonald R. (1978) Melting relations of a Midland Valley alkali basalt. In *Progress in Experimental Petrology*, Natural Environment Research Council Rept. 4, p. 196–197.

Ford C. E., Biggar G. M., Humphries D. J., Wilson G., Dixon D. and O'Hara M. J. (1972) Role of water in the evolution of the lunar crust; an experimental study of sample 14310; an indication of lunar calc-alkaline volcanism. *Proc. Lunar Sci. Conf. 3rd*, p. 207–229.

Ford C. E., Biggar G. M., O'Hara M. J., Humphries D. J. and Spencer P. N. (1974) Origin of the lunar highlands (abstract). In *Lunar Science V*, p. 239–241. The Lunar Science Institute, Houston.

Ford C. E., O'Hara M. J. and Spencer P. N. (1977) The origin of lunar feldspathic liquids. *Phil. Trans. Roy. Soc. Lond. A285*, 193–197.

Forster R. M. (1978) Preliminary atmospheric pressure melting relationships of some proposed Rhum parent magmas. In *Progress in Experimental Petrology*, Natural Environment Research Council Rept. 4, p. 200–201.

Frey F. A. and Prinz M. (1978) Ultramafic inclusions from San Carlos, Arizona: petrologic and geochemical data bearing on their petrogenesis. *Earth Planet. Sci. Lett.* **38**, 129–176.

Frey F. A., Bryan W. B. and Thompson G. (1974) Atlantic Ocean floor: geochemistry and petrology of basalts from Legs 2 and 3 of the Deep-Sea Drilling Project. *J. Geophys. Res.* **79**, 5507–5527.

Frey F. A., Ferguson J. and Chappell B. W. (1977) Petrogenesis of South African and Australian kimberlite suites (abstract). In *Extended Abstracts, Second Internat. Kimberlite Conf., Santa Fe.*

Frey F. A., Green D. H. and Roy S. D. (1978) Integrated models of basalt petrogenesis: a study of quartz tholeiites to olivine melilitites from south-eastern Australia utilizing geochemical and experimental petrological data. *J. Petrol.* **19**, 463–513.

Friel J. J., Harker I. and Ulmer G. C. (1977) Armalcolite stability as a function of pressure and oxygen fugacity. *Geochim. Cosmochim. Acta* **41**, 403–410.

Fudali R. F. (1965) Oxygen fugacitites of basaltic and andesitic magma. *Geochim. Cosmochim. Acta* **29**, 1063–1075.

Fujii T. and Kushiro I. (1977) Melting relations and viscosity of an abyssal tholeiite. *Carnegie Inst. Wash. Yearb.* **76**, 461–465.

Gale G. H. (1973) Paleozoic basaltic komatiite and ocean-floor type basalts from Northeastern Newfoundland. *Earth Planet. Sci. Lett.* **18**, 22–30.

Ganapathy R. and Anders E. (1974) Bulk compositions of the Moon and Earth, estimated from meteorites. *Proc. Lunar Sci. Conf. 5th*, p. 1181–1206.

Garcia M. O., Liu N. W. K. and Muenow D. W. (1979) Volatiles in submarine volcanic rocks from the Mariana Island arc and trough. *Geochim. Cosmochim. Acta* **43**, 305–312.

Gast P. W. (1968) Trace element fractionation and the origin of tholeiitic and alkaline magma types. *Geochim. Cosmochim. Acta* **32**, 1057–1086.

Getting I. C. and Kennedy G. C. (1970) The effect of pressure on the emf of chromel-alumel and Pt/Pt 10% Rh thermocouples. *J. Appl. Phys.* **41**, 4552–4556.

Gibb F. G. F. (1971) Crystal-liquid relationships in some ultrabasic dykes and their petrological significance. *Contrib. Mineral. Petrol.* **30**, 103–118.

Green D. H. (1970a) A review of experimental evidence on the origin of basaltic and nephelinitic magmas. *Phys. Earth Planet. Inter.* **3**, 221–235.

Green D. H. (1970b) The origin of basaltic and nephelinitic magmas. *Trans. Leicester Lit. Phil. Soc.* **44**, 26–54.

Green D. H. (1972) Magmatic activity as the major process in the chemical evolution of the Earth's crust and mantle. *Tectonophys.* **13**, 47–71.

Green D. H. (1973a) Experimental melting studies on a model upper mantle composition at high pressure under water-saturated and water-undersaturated conditions. *Earth Planet. Sci. Lett.* **19**, 37–53.

Green D. H. (1973b) Conditions of melting of basanite magma from garnet peridotite. *Earth Planet. Sci. Lett.* **17**, 456–465.

Green D. H. (1975) Genesis of Archean peridotitic magmas and constraints on Archean geothermal gradients and tectonics. *Geology* **3**, 15–18.

Green D. H. (1976a) Experimental testing of "equilibrium" partial melting of peridotite under water-saturated, high-pressure conditions. *Canad. Mineral.* **14**, 255–268.

Green D. H. (1976b) Orthopyroxene in the lunar interior and constraints on early lunar differentiation (abstract). In *Lunar Science VII*, p. 336–338. The Lunar Science Institute, Houston.

Green D. H. and Hibberson W. O. (1970) Experimental duplication of conditions of precipitation of high-pressure phenocrysts in a basaltic magma. *Phys. Earth Planet. Inter.* **3**, 247–254.

Green D. H. and Ringwood A. E. (1967a) The genesis of basaltic magmas. *Contrib. Mineral. Petrol.* **15**, 103–190.

Green D. H. and Ringwood A. E. (1967b) The stability fields of aluminous pyroxene peridotite and garnet peridotite and their relevance in upper mantle structure. *Earth Planet. Sci. Lett.* **3**, 151–160.

Green D. H. and Ringwood A. E. (1967c) An experimental investigation of the gabbro to eclogite transformation and its petrological applications. *Geochim. Cosmochim. Acta* **31**, 767–833.

Green D. H. and Ringwood A. E. (1970) Mineralogy of peridotitic compositions under upper mantle conditions. *Phys. Earth Planet. Inter.* **3**, 359–371.

Green D. H. and Ringwood A. E. (1973) Significance of a primitive lunar basaltic composition present in Apollo 15 soils and breccias. *Earth Planet. Sci. Lett.* **13**, 85–96.

Green D. H., Hibberson W. O. and Jaques A. L. (1979) Petrogenesis of mid-ocean ridge basalts. In *The Earth: Its Origin, Structure, and Evolution* (M. W. McElhinny, ed.), p. 265–299. Academic Press, N.Y.

Green D. H., Nicholls I. A., Viljoen M. and Viljoen R. (1975a) Experimental demonstration of the existence of peridotitic liquids in earliest Archean magmatism. *Geology* **3**, 11–15.

Green D. H., Ringwood A. E., Hibberson W. O. and Ware N. G. (1975b) Experimental petrology of Apollo 17 mare basalts. *Proc. Lunar Sci. Conf. 6th*, p. 871–893.

Green D. H., Ringwood A. E., Ware N. G. and Hibberson W. O. (1972) Experimental petrology and petrogenesis of Apollo 14 basalts. *Proc. Lunar Sci. Conf. 3rd*, p. 197–206.

Green D. H., Ringwood A. E., Ware N. G. and Hibberson W. O. (1974) Petrology and petrogenesis of Apollo 17 basalts and Apollo 17 orange glass (abstract). In *Lunar Science V*, p. 287–289. The Lunar Science Institute, Houston.

Green D. H., Ringwood A. E., Ware N. G., Hibberson W. O., Major A. and Kiss E. (1971a) Experimental petrology and petrogenesis of Apollo 12 basalts. *Proc. Lunar Sci. Conf. 2nd*, p. 601–615.

Green D. H., Ware N. G., Hibberson W. O. and Major A. (1971b) Experimental petrology of Apollo 12 basalts. I. Sample 12009. *Earth Planet. Sci. Lett.* **13**, 85–96.

Green T. H. (1972) Crystallization of calc-alkaline andesite under controlled high-pressure hydrous conditions. *Contrib. Mineral. Petrol.* **34**, 150–166.

Green T. H. and Ringwood A. E. (1967) Crystallization of basalt and andesite under high pressure hydrous conditions. *Earth Planet. Sci. Lett.* **3**, 481–489.

Green T. H. and Ringwood A. E. (1968) Genesis of the calc-alkaline-igneous rock suite. *Contrib. Mineral. Petrol.* **18**, 105–162.

Green T. H. and Ringwood A. E. (1972) Crystallization of garnet-bearing rhyodacite under high-pressure hydrous conditions. *J. Geol. Soc. Aust.* **19**, 203–212.

Green T. H., Green D. H. and Ringwood A. E. (1967) The origin of high-alumina basalts and their relationships to quartz tholeiites and alkali basalts. *Earth Planet. Sci. Lett.* **2**, 41–51.

Grove T. L. (1978) Cooling histories of Luna 24 very low-Ti (VLT) ferrobasalts: an experimental study. *Proc. Lunar Planet. Sci. Conf. 9th*, p. 565–584.

Grove T. L. and Bence A. E. (1977) Experimental study of pyroxene-liquid interaction in quartz-normative basalt 15597. *Proc. Lunar Sci. Conf. 8th*, p. 1549–1580.

Grove T. L. and Raudsepp M. (1978) Effects of kinetics on the crystallization of quartz normative basalt 15597: an experimental study. *Proc. Lunar Planet. Sci. Conf. 9th*, p. 585–599.

Grove T. L. and Vaniman D. (1978) Experimental petrology of very low Ti (VLT) basalts. In *Mare Crisium: The View from Luna 24*, (R. B. Merrill and J. J. Papike, eds.), p. 445–471. Pergamon, N.Y.

Grove T. L. and Walker D. (1977) Cooling histories of Apollo 15 quartz-normative basalts. *Proc. Lunar Sci. Conf. 8th*, p. 1501–1520.

Grove T. L., Walker D., Longhi J., Stolper E. and Hays J. F. (1973) Petrology of rock 12002 and origin of picritic basalts at Oceanus Procellarum. *Proc. Lunar Sci. Conf. 4th*, p. 995–1011.

Hamilton D. L., Burnham C. W. and Osborn E. F. (1964) The solubility of water and effects of oxygen fugacity and water content on crystallization in mafic magmas. *J. Petrol.* **5**, 21–39.

Harper L. F. (1915) Geology and mineral resources of the southern coal-field. Part I. — The south coastal portion. *Geol. Survey New South Wales Memoir No. 7*, p. 280.

Hart S. R. and Nalwalk A. J. (1970) K, Rb, Cs and Sr relationships in submarine basalts from the Puerto Rico Trench. *Geochim. Cosmochim. Acta* **34**, 145–155.

Haygood C., Allen J. C. and Boettcher A. L. (1971) Phase relationships and the stability of amphiboles in the Hualalai alkaline basalt at high pressure (abstract). *Abstracts with Programs, Geol. Soc. Amer. Mtgs.* **3**, 594–595.

Head J. W. (1976) Lunar volcanism in space and time. *Rev. Geophys. Space Phys.* **14**, 265–300.

Helz R. T. (1973) Phase relations of basalts in their melting range at $P_{H_2O} = 5$ kb as a function of oxygen fugacity. Part I. Mafic phases. *J. Petrol.* **14**, 249–302.

Helz R. T. (1976) Phase relations of basalts in their melting ranges at $P_{H_2O} = 5$ kb. Part II. Melt compositions. *J. Petrol.* **17**, 139–193.

Hertogen J., Janssens M.-J., Takahashi H., Palme H. and Anders E. (1977) Lunar basins and craters: evidence for systematic compositional changes of bombarding population. *Proc. Lunar Sci. Conf. 8th*, p. 17–45.

Herzberg C. T. (1978) Pyroxene geothermometry and geobarometry: experimental and thermodynamic evaluation of some subsolidus phase relations involving pyroxenes in the system CaO-MgO-Al_2O_3-SiO_2. *Geochim. Cosmochim. Acta* **42**, 945–957.

Hess P. C., Rutherford M. J. and Campbell H. W. (1977) Origin and evolution of LKFM basalt. *Proc. Lunar Sci. Conf. 8th*, p. 2357–2373.

Hess P. C., Rutherford M. J. and Campbell H. W. (1978) Ilmenite crystallization in nonmare basalt: genesis of KREEP and high-Ti mare basalt. *Proc. Lunar Planet. Sci. Conf. 9th*, p. 705–724.

Hess P. C., Rutherford M. J., Guillemette R. N., Ryerson F. J. and Tuchfeld H. A. (1975) Residual products of fractional crystallization of lunar magmas: an experimental study. *Proc. Lunar Sci. Conf. 6th*, p. 895–909.

Hill R. E. T. and Boettcher A. L. (1970) Water in the Earth's mantle: melting curves of basalt-water and basalt-water-carbon dioxide. *Science* **167**, 980–982.

Hill R. E. T. and Roeder P. L. (1974) The crystallization of spinel from basaltic liquid as a function of oxygen fugacity. *J. Geol.* **82**, 709–729.

Hodges F. N. and Kushiro I. (1973) Petrology of Apollo 16 lunar highland rocks. *Proc. Lunar Sci. Conf. 4th*, p. 1033–1048.

Holloway J. R. (1971) Composition of fluid phase solutes in a basalt H_2O-CO_2 system. *Geol. Soc. Amer. Bull.* **82**, 233–238.

Holloway J. R. (1973) The system pargasite-H_2O-CO_2: a model for melting of a hydrous mineral with a mixed-volatile fluid. I. Experimental results to 8 kbar. *Geochim. Cosmochim. Acta* **37**, 651–666.

Holloway J. R. and Burnham C. W. (1972) Melting relations of basalt with equilibrium pressure less than total pressure. *J. Petrol.* **13**, 1–29.

Holloway J. R. and Eggler D. H. (1976) Fluid-absent melting of peridotite containing phlogopite and dolomite. *Carnegie Inst. Wash. Yearb.* **75**, 636–639.

Hoover J. D. (1978) Melting relations of a new chilled margin sample from the Skaergaard intrusion. *Carnegie Inst. Wash. Yearb.* **77**, 739–743.

Hörz F. (1978) How thick are lunar mare basalts? *Proc. Lunar Planet. Sci. Conf. 9th*, p. 3311–3331.

Howells S., Begg C. and O'Hara M. J. (1975) Crystallization of some natural eclogites and garnetiferous ultrabasic rocks at high pressures and temperatures. In *Physics and Chemistry of the Earth, Vol. 9*, (L. H. Ahrens, J. B. Dawson, A. R. Duncan and A. J. Erlank, eds.), p. 895–904. Pergamon, N.Y.

Huang W.-L. and Wyllie P. J. (1974) Eutectic between wollastonite II and calcite contrasted with thermal barrier in MgO-SiO_2CO_2 at 30 kilobars, with applications to kimberlite-carbonatite petrogenesis. *Earth Planet. Sci. Lett.* **24**, 305–310.

Hubbard N. J. and Minear J. W. (1975) A physical and chemical model of early lunar history. *Proc. Lunar Sci. Conf. 6th*, p. 1057–1085.

Huebner J. S. (1973) Experimental control of wüstite activity and mole fraction (abstract). *Geol. Soc. Amer. Abstracts with Programs* **5**, 676–677.

Huebner J. S., Lipin B. R. and Wiggins L. B. (1976) Partitioning of chromium between silicate crystals and melt. *Proc. Lunar Sci. Conf. 7th*, p. 1195–1220.

Humphries D. J. (1972) Melting behaviour of Reunion igneous rocks. In *Progress in Experimental Petrology*, Natural Environment Research Council Rept. 2, p. 113–115.

Humphries D. J. and Cox K. G. (1972) Melting data for volcanic rocks from Aden, South Arabia. In *Progress in Experimental Petrology*, Natural Environment Research Council Rept. 2, p. 116.

Irving A. J. (1974) Megacrysts from the Newer Basalts and other basaltic rocks of southeastern Australia. *Geol. Soc. Amer. Bull.*. **85**, 1503–1514.

Irving A. J. (1975) Chemical, mineralogical and textural systematics of non-mare melt rocks: implications for lunar impact and volcanic processes. *Proc. Lunar Sci. Conf. 6th*, p. 363–394.

Irving A. J. (1977a) Origin of megacryst suites in basaltic dikes near 96 Ranch, west Texas and Hoover Dam, Arizona (abstract). *EOS* (*Trans. Amer. Geophys. Union*) **58**, 526.

Irving A. J. (1977b) Chemical and experimental constraints on the genesis of Apollo 15 and Apollo 17 KREEP basalts (abstract). In *Lunar Science VIII*, p. 493–495. The Lunar Science Institute, Houston.

Irving A. J. (1977c) Chemical variation and fractionation of KREEP basalt magmas. *Proc. Lunar Sci. Conf. 8th*, p. 2433–2448.

Irving A. J. (1978) A review of experimental studies of crystal/liquid trace element partitioning. *Geochim. Cosmochim. Acta* **42**, 743–770.

Irving A. J. (1980) Geochemistry of composite ultramafic xenoliths in alkalic basalts and implications for magmatic processes in the mantle. *Amer. J. Sci., Jackson v.* **280A**, 389–426.

Irving A. J. and Green D. H. (1972) Experimental study of phase relationships in a high-pressure mugearitic basalt as a function of water content (abstract). *Abstracts with Programs, Geol. Soc. Amer. Mtgs.* **4**, 550–551.

Irving A. J. and Green D. H. (1976) Geochemistry and petrogenesis of the Newer Basalts of Victoria and South Australia. *J. Geol. Soc. Australia* **23**, 45–66.

Irving A. J. and Price R. C. (1981) Geochemistry and evolution of high-pressure phonolitic rocks from Nigeria, Australia, Eastern Germany, and New Zealand. *Geochim. Cosmochim. Acta* **45**, in press.

Irving A. J. and Wyllie P. J. (1975) Subsolidus and melting relationships for calcite, magnesite, and the join $CaCO_3$-$MgCO_3$ to 36 kilobars. *Geochim. Cosmochim. Acta* **39**, 35–53.

Irving A. J., Merrill R. B. and Singleton D. E. (1978) Experimental partitioning of rare earth elements and scandium among armalcolite, ilmenite, olivine, and mare basalt liquid. *Proc. Lunar Planet. Sci. Conf. 9th*, p. 601–612.

Ito K. and Kennedy G. C. (1967) Melting and phase relations in a natural peridotite to 40 kilobars. *Amer. J. Sci.* **265**, 519–538.

Ito K. and Kennedy G. C. (1968) Melting and phase relations in the plane tholeiite-lherzolite-nepheline basanite to 40 kilobars with geological implications. *Contrib. Mineral. Petrol.* **19**, 177–211.

Ito K. and Kennedy G. C. (1971) An experimental study of the basalt-garnet granulite-eclogite transition. In *The Structure and Physical Properties of the Earth's Crust*, (J. G. Heacock, ed.), *Amer. Geophys. Union Mon.* **14**, p. 303–314.

Ito K. and Kennedy G. C. (1974) The composition of liquids formed by partial melting of eclogites at high temperatures and pressures. *J. Geol.* **82**, 383–392.

Jackson E. D. and Wright T. L. (1970) Xenoliths in the Honolulu volcanic series. *J. Petrol.* **11**, 405–430.

Jakes P. and Gill J. (1970) Rare earth elements and the island arc tholeiitic series. *Earth Planet. Sci. Lett.* **9**, 17–28.

Jakes P. and White A. J. R. (1971) Composition of island arcs and continental growth. *Earth Planet. Sci. Lett.* **12**, 224–230.

Jakes P. and White A. J. R. (1972) Major and trace element abundances in volcanic rocks of orogenic areas. *Geol. Soc. Amer. Bull.* **83**, 29–40.

Jenkins D. M. and Newton R. C (1979) Experimental determination of the spinel peridotite to garnet peridotite inversion at 900° C and 1000° C in the system CaO-MgO-Al_2O_3-SiO_2, and at 900°C with natural garnet and olivine. *Contrib. Mineral. Petrol.* **68**, 407–419.

Johannes W. and Bode B. (1978) Loss of iron to the Pt-container in melting experiments with basalts and a method to reduce it. *Contrib. Mineral. Petrol.* **67**, 221–225.

Johannes W., Bell P. M., Mao H. K., Boettcher A. L., Chipman D. W., Hays J. F., Newton R. C. and Siefert F. (1971) An interlaboratory comparison of piston-cylinder pressure calibration using the albite breakdown reaction. *Contrib. Mineral. Petrol.* **32**, 24–38.

Johnston D. H. and Toksöz M. N. (1977) Internal structure and properties of Mars. *Icarus* **32**, 73–84.

Jordan T. H. (1975) The continental tectosphere. *Rev. Geophys. Space Phys.* **13**, 1–12.

Jordan T. H. and Anderson D. L. (1974) Earth structure from free oscillations and travel times. *Geophys. J. Roy. Astr. Soc.* **36**, 411–459.

Kay R. W. and Gast P. W. (1973) The rare earth content and origin of alkali-rich basalts. *J. Geol.* **81**, 653–682.

Kay R., Hubbard N. J. and Gast P. W. (1970) Chemical characteristics and origin of oceanic ridge volcanic rocks. *J. Geophys. Res.* **75**, 1585–1613.

Kennedy W. Q. and Anderson E. M. (1938) Crustal layers and the origin of magmas. *Bull. Volc.* **2–3**, 23–82.

Kesson S. E. (1975) Mare basalts: melting experiments and petrogenetic interpretations. *Proc. Lunar Sci. Conf. 6th*, p. 921–924.

Kesson S. E. and Lindsley D. H. (1976) Mare basalt petrogenesis—a review of experimental studies. *Rev. Geophys. Space Phys.* **14**, 361–373.

Kesson S. E. and Ringwood A. E. (1976) Mare basalt petrogenesis in a dynamic Moon. *Earth Planet. Sci. Lett.* **30**, 155–163.

Kesson S. E. and Ringwood A. E. (1977) Further limits on the bulk composition of the Moon. *Proc. Lunar Sci. Conf. 8th*, p. 411–431.

Klein L. C. and Uhlmann D. R. (1976) The kinetics of lunar glass formation, revisited. *Proc. Lunar Sci. Conf. 7th*, p. 1113–1121.

Knutson J. and Green T. H. (1975) Experimental duplication of a high-pressure megacryst/cumulate assemblage in a near-saturated hawaiite. *Contrib. Mineral. Petrol.* **52**, 121–132.

Kornprobst J. (1970) Les péridotites et les pyroxénolites du massif ultrabasique des Beni Bouchera: une étude expérimentale entre 1100 et 1550° C, sous 15 à 30 kilobars de pression sèche. *Contrib. Mineral. Petrol.* **29**, 290–309.

Krishnamurthy P. (1976) Atmospheric pressure melting studies of some Deccan Trap lavas from Western India. In *Progress in Experimental Petrology*, Natural Environment Research Council Rept. 3, 218–221.

Krishnamurthy P. and Cox K. G. (1977) Picrite basalts and related lavas from the Deccan Traps of western India. *Contrib. Mineral. Petrol.* **62**, 53–75.

Kuno H. (1950) Petrology of Hakone volcano and the adjacent areas, Japan. *Geol. Soc. Amer. Bull.* **61**, 957–72.

Kuno H. (1959) Origin of Cenozoic petrographic provinces of Japan and surrounding areas. *Bull. Volc.* **2–20**, 37–76.

Kuno H. (1960) High-alumina basalt *J. Petrol.* **1**, 121–145.

Kushiro I. (1968) Compositions of magmas formed by partial zone melting of the Earth's upper mantle. *J. Geophys. Res.* **73**, 619–634.

Kushiro I. (1969) The system forsterite-diopside-silica with and without water at high pressures. *Amer. J. Sci.* **267A**, 269–294.

Kushiro I. (1970) Stability of amphibole and phlogopite in the upper mantle. *Carnegie Inst. Wash. Yearb.* **68**, 245–247.

Kushiro I. (1972a) Effect of water on the composition of magmas formed at high pressures. *J. Petrol.* **13**, 311–334.

Kushiro I. (1972b) Partial melting of synthetic and natural peridotites at high pressures. *Carnegie Inst. Wash. Yearb.* **71**, 357–362.

Kushiro I. (1973a) Origin of some magmas in oceanic and circum-oceanic regions. *Tectonophys.* **17**, 211–222.

Kushiro I. (1973b) Partial melting of garnet lherzolites from kimberlites at high pressures. In *Lesotho Kimberlites*, (P. H. Nixon, ed.), p. 294–299. Lesotho National Development Corporation, Maseru.

Kushiro I. and Haramura H. (1971) Major element variation and possible source materials of Apollo 12 crystalline rocks. *Science* **171**, 1235–1237.

Kushiro I. and Schairer J. F. (1963) New data on the system diopside-forsterite-silica. *Carnegie Inst. Wash. Yearb.* **62**, 95–103.

Kushiro I. and Thompson R. N. (1972) Origin of some abyssal tholeiites from the Mid-Atlantic Ridge. *Carnegie Inst. Wash. Yearb.* **71**, 403–406.

Kushiro I. and Yoder H. S., Jr. (1966) Anorthite-forsterite and anorthite-enstatite reactions and their bearing on the basalt—ecologite transformation. *J. Petrol* **7**, 337–362.

Kushiro I., Syono Y. and Akimoto S. (1968) Melting of a peridotite nodule at high pressures and high water pressures. *J. Geophys. Res.* **73**, 6023–6029.

Kushiro I., Shimizu N. and Nakamura Y. (1972a) Compositions of coexisting liquid and solid phases formed upon melting of natural garnet and spinel lherzolites at high pressures: a preliminary report. *Earth Planet. Sci. Lett.* **14**, 19–25.

Kushiro I., Ikeda Y. and Nakamura Y. (1972b) Petrology of Apollo 14 high-alumina basalt. *Proc. Lunar Sci. Conf. 3rd*, p. 115–129.

Kushiro I., Satake H. and Akimoto S. (1975) Carbonate-silicate reactions at high pressures and possible presence of dolomite and magnesite in the upper mantle. *Earth Planet. Sci. Lett.* **28**, 116–120.

Kushiro I., Yoder H. S., Jr. and Mysen B. O. (1976) Viscosities of basalt and andesite melts at high pressures. *J. Geophys. Res.* **81**, 6351–6356.

Lambert I. B. and Wyllie P. J. (1968) Stability of hornblende and a model for the low velocity zone. *Nature* **219**, 1240–1241.

Lambert I. B. and Wyllie P. J. (1972) Melting of gabbro (quartz eclogite) with excess water to 35 kilobars, with geological applications. *J. Geol.* **80**, 693–708.

Langseth M. G., Keihm S. J. and Peters K. (1976) The revised lunar heat flow values (abstract). In *Lunar Science VII*, p. 474–476. The Lunar Science Institute, Houston.

Lanphere M. A. and Dalrymple G. B. (1980) Age and strontium isotopic composition of the Honolulu Volcanic Series, Oahu, Hawaii. *Amer. J. Sci., Jackson v.* **280A**, 736–751.

Larimer J. W. and Anders E. (1967) Chemical fractionations in meteorites—II. Abundance patterns and their interpretation. *Geochim. Cosmochim. Acta* **31**, 1239–1270.

Lewis J. D. and Williams I. R. (1973) The petrology of an ultramafic lava near Murphy Well, Eastern Goldfields, Western Australia. *Geol. Surv. West. Aust. Ann. Rept.*, 60–68.

Lindsley D. H. and Dixon S. A. (1976) Diopside-enstatite equilibria at 850° to 1400°C, 5 to 35 kb. *Amer. J. Sci.* **276**, 1285–1301.

Lipin B. R. (1976) The origin of Fra Mauro basalts (abstract). In *Lunar Science VII*, p. 495–497. The Lunar Science Institute, Houston.

Lipin B. R. (1978) The system Mg_2SiO_4-Fe_2SiO_4-$CaAl_2Si_2O_6$-SiO_2 and the origin of Fra Mauro basalts. *Amer. Mineral.* **63**, 350–364.

Lofgren G. E. (1977) Dynamic crystallization experiments bearing on the origin of textures in impact-generated liquids. *Proc. Lunar Sci. Conf. 8th*, p. 2079–2095.

Lofgren G. E., Donaldson C. H. and Usselman T. M. (1975) Geology, petrology, and crystallization of Apollo 15 quartz-normative basalts. *Proc. Lunar Sci. Conf. 6th*, p. 79–99.

Lofgren G., Donaldson C. H., Williams R. J., Mullins O., Jr. and Usselman T. M. (1974) Experimentally reproduced textures and mineral chemistry of Apollo 15 quartz normative basalts. *Proc. Lunar Sci. Conf. 5th*, p. 549–567.

Longhi J., Walker D. and Hays J. F. (1978) The distribution of Fe and Mg between olivine and lunar basaltic liquids. *Geochim. Cosmochim. Acta* **42**, 1545–1558.

Longhi J., Walker D., Grove T. L., Stolper E. M. and Hays J. F. (1974) The petrology of the Apollo 17 mare basalts. *Proc. Lunar Sci. Conf. 5th*, p. 447–469.

Lugmair G. W. and Carlson R. W. (1978) The Sm-Nd history of KREEP. *Proc. Lunar Planet. Sci. Conf. 9th*, p. 689–704.

MacDonald G. A. and Katsura T. (1961) Variations in the lava of the 1959 eruption of Kilauea Iki. *Pacific Sci.* **15**, 358–369.

MacGregor I. D. (1965) Stability fields of spinel and garnet peridotites in the system MgO-CaO-Al_2O_3-SiO_2. *Carnegie Inst. Wash. Yearb.* **64**, 126–134.

McBirney A. R. and Williams H. (1969) Geology and petrology of the Galapagos Islands. *Geol. Soc. Amer. Mem.* **118**, 197.

McCarthy T. S., Erlank A. J. and Willis J. P. (1973) On the origin of eucrites and diogenites. *Earth Planet. Sci. Lett.* **18**, 433–442.

McGetchin T. R. and Smyth J. R. (1978) The mantle of Mars: some possible geological implications of its high density. *Icarus* **34**, 512–536.

McKay G. A. and Weill D. F. (1976) Petrogenesis of KREEP. *Proc. Lunar Sci. Conf. 7th*, p. 2427–2447.

McKay G. A. and Weill D. F. (1977) KREEP petrogenesis revisited. *Proc. Lunar Sci. Conf. 8th*, p. 2339–2355.

Ma M.-S., Murali A. V. and Schmitt R. A. (1977) Genesis of the Angra dos Reis and other achondritic meteorites. *Earth Planet. Sci. Lett.* **35**, 331–346.

Maaløe S. (1973) Temperature and pressure relations of ascending primary magmas. *J. Geophys. Res.* **78**, 6877–6886.

Maaløe S. and Aoki K. (1977) The major element composition of the upper mantle estimated from the composition of lherzolites. *Contrib. Mineral. Petrol.* **63**, 161–173.

Marsh B. D. and Carmichael I. S. E. (1974) Benioff zone magmatism. *J. Geophys. Res.* **79**, 1196–1206.

Mason B. (1962) *Meteorites.* John Wiley and Sons, Inc., N.Y. 274 pp.

Mason B. (1963) The hypersthene achondrites. *American Museum Novitates, No. 2155.*

Mason B. (1967) The Bununu meteorite, and a discussion of the pyroxene-plagioclase achondrites. *Geochim. Cosmochim. Acta* **31**, 107–115.

Masursky H., Kaula W. M., McGill G. E., Pettengill G. H., Phillips R. J., Russell C. T., Schubert G. and Shapiro I. I. (1977) The surface and interior of Venus. *Space Sci. Rev.* **20**, 431–449.

Melson W. G., Byerly G. R., Nelen J. A., O'Hearn T., Wright T. L. and Vallier T. (1976a) A catalog of the major element chemistry of abyssal volcanic glasses. In *Mineral Sciences Investigations, 1974*–1975, (B. Mason, ed.), *Smithsonian Contrib. Earth Sci.* **19**, 31–60.

Melson W. G., Vallier T. L., Wright T. L., Byerly G. and Nelen J. (1976b) Chemical diversity of abyssal volcanic glass erupted along Pacific, Atlantic, and Indian Ocean sea-floor spreading centers. In *The Geophysics of the Pacific Ocean Basin and Its Margin, Geoph. Mon. 19*, (G. H. Sutton, M. H. Manghnani and R. Moberly, eds.), Amer. Geophys. Union, p. 351–367.

Menzies M. and Murthy V. R. (1980) Mantle metasomatism as a precursor to the genesis of alkaline magmas—isotopic evidence. *Amer. J. Sci., Jackson v.* **280A**, 622–638.

Merrill R. B. and Irving A. J. (1977) Chemistry and phase relations of an orthopyroxene-bearing transitional alkalic basalt (abstract). *EOS (Trans. Amer. Geophys. Union)* **58**, 526.

Merrill R. B. and Williams R. J. (1975) The system anorthite-forsterite-fayalite-silica to 2 kbar with lunar petrologic applications. *Proc. Lunar Sci. Conf. 6th*, p. 959–971.

Merrill R. B. and Wyllie P. J. (1973) Absorption of iron by platinum capsules in high pressure rock melting experiments. *Amer. Mineral.* **58**, 16–20.

Merrill R. B. and Wyllie P. J. (1975) Kaersutite and kaersutite eclogite from Kakanui, New Zealand—water-excess and water-deficient melting relations to 30 kilobars. *Geol. Soc. Amer. Bull.* **86**, 555–570.

Millhollen G. L. and Wyllie P. J. (1974) Melting relations of brown-hornblende mylonite from St. Paul's Rocks under water-saturated and water-undersaturated conditions to 30 kilobars. *J. Geol.* **82**, 589–606.

Millhollen G. L., Irving A. J. and Wyllie P. J. (1974) Melting interval of peridotite with 5.7 percent water to 30 kilobars. *J. Geol.* **82**, 575–587.

Mitchell R. H. and Brunfelt A. O. (1975) Rare earth element geochemistry of kimberlite. In *Physics and Chemistry of the Earth*, **v. 9**, (L. H. Ahrens, J. B. Dawson, A. R. Duncan and A. J. Erlank, eds.), p. 671–686. Pergamon, N.Y.

Miyashiro A. (1975) Island arc volcanic series: a critical review. *Petrologie* **1**, 177–187.

Modreski P. J. and Boettcher A. L. (1973) Phase relationships of phlogopite in the system K_2O-MgO-CaO-Al_2O_3-SiO_2H_2O to 35 kilobars: a better model for micas in the interior of the Earth. *Amer. J. Sci.* **273**, 385–414.

Moore C. B. (1962) The petrochemistry of the achondrites. In *Researches on Meteorites*, (C. B. Moore, ed.), p. 164–178. Wiley, N.Y.

Moore J. G. (1970) Water content of basalt erupted on the ocean floor. *Contrib. Mineral. Petrol.* **28**, 272–279.

Morgan J. W., Higuchi H., Takahashi H. and Hertogen J. (1978) A "chondritic" eucrite parent body: inference from trace elements. *Geochim. Cosmochim. Acta* **42**, 27–38.

Morgan W. J. (1972) Deep mantle convection plumes and plate motions. *Amer. Assoc. Pet. Geol. Bull.* **56**, 203–213.

Morse S. A. (1976) The lever rule with fractional crystallization and fusion. *Amer. J. Sci.* **276**, 330–346.

Muan A., Hauck J. and Löfall T. (1972) Equilibrium studies with a bearing on lunar rocks. *Proc. Lunar Sci. Conf. 3rd*, p. 185–196.

Muenow D. W., Graham D. G., Liu N. W. K. and Delaney J. R. (1979) Water-poor glass-vapor inclusions in phenocrysts from Hawaiian tholeiitic pillow-basalts. *Earth Planet. Sci. Lett.* **42**, 71–76.

Mukherjee A. (1967) Role of fractional crystallization in the descent, basalt →trachyte. *Contrib. Mineral. Petrol.* **16**, 139–148.

Murata K. J. and Richter D. H. (1966) Chemistry of the lavas of the 1959–60 eruption of Kilauea volcano, Hawaii. *U. S. Geol. Surv. Prof. Paper* **537-A**, 1–26.

Mutschler F. E., Rougon D. J. and Lavin O. P. (1976) PETROS—a data bank of major-element chemical analyses of igneous rocks for research and teaching. *Comput. Geosci.* **2**, 51–57.

Mysen B. (1975) Partitioning of iron and magnesium between crystals and partial melts in peridotite upper mantle. *Contrib. Mineral. Petrol.* **52**, 69–76.

Mysen B. O. (1978) Experimental determination of crystal-vapor partition coefficients for rare earth elements to 30 kbar pressure. *Carnegie Inst. Wash. Yearb.* **77**, 689–695.

Mysen B. O. and Boettcher A. L. (1975a) Melting of a hydrous mantle: phase relations of a natural peridotite at high pressures and temperatures with controlled activities of water, carbon dioxide, and hydrogen. *J. Petrol.* **16**, 520–548.

Mysen B. O. and Boettcher A. L. (1975b) Melting of a hydrous mantle: II. Geochemistry of crystals and liquids formed by anatexis of mantle peridotite at high pressures and high temperatures as a function of controlled activities of water, hydrogen and carbon dioxide. *J. Petrol.* **16**, 549–593.

Mysen B. O. and Boettcher A. L. (1976) Melting of a hydrous mantle: III. Phase relations of garnet websterite + H_2O at high pressures and temperatures. *J. Petrol.* **17**, 1–14.

Mysen B. O. and Kushiro I. (1977) Compositional variations of coexisting phases with degree of melting of peridotite in the upper mantle. *Amer. Mineral.* **62**, 843–865.

Mysen B. O. and Seitz M. G. (1975) Trace element partitioning determined by beta track mapping: an experimental study using carbon and samarium as examples. *J. Geophys. Res.* **80**, 2627–2635.

Mysen B. O., Eggler D. H., Seitz M. G. and Holloway J. R. (1976) Carbon dioxide in silicate melts and crystals. I. Solubility measurements. *Amer. J. Sci.* **276**, 455–479.

Mysen B. O., Kushiro I., Nicholls I. A. and Ringwood A. E. (1974) A possible mantle origin for andesitic magmas: discussion of a paper by Nicholls and Ringwood. *Earth Planet. Sci. Lett.* **21**, 221–229.

Nakamura Y. and Kushiro I. (1974) Composition of the gas phase in Mg_2SiO_4-SiO_2-H_2O, at 15 kbar. *Carnegie Inst. Wash. Yearb.* **73**, 255–258.

Nehru C. E. and Wyllie P. J. (1974) Electron microprobe measurement of pyroxenes coexisting with H_2O-under-saturated liquid in the join $CaMgSi_2O_6$-$Mg_2Si_2O_6$-H_2O at 30 kilobars, with applications to geothermometry. *Contrib. Mineral. Petrol.* **48**, 221–228.

Nehru C. E. and Wyllie P. J. (1975) Compositions of glasses from St. Paul's peridotite partially melted at 20 kilobars. *J. Geol.* **83**, 455–471.

Nesbitt R. W. (1971) Skeletal crystal forms in the ultramafic rocks of Yilgarn Block, Western Australia: evidence for an Archean ultramafic liquid. *Geol. Soc. Aust. Spec. Pub.* **3**, 331–347.

Nesbitt R. W. and Hamilton D. L. (1970) Crystallization of an alkali-olivine basalt under controlled P_{O_2}, P_{H_2O} conditions. *Phys. Earth Planet. Inter.* **3**, 309–315.

Nesbitt R. W. and Sun S.-S. (1976) Geochemistry of Archean spinifex-textured peridotites and magnesian and low-magnesian tholeiites. *Earth Planet. Sci. Lett.* **31**, 433-453.

Nicholls I. A. (1974) Liquids in equilibrium with peridotitic mineral assemblages at high water pressures. *Contrib. Mineral. Petrol.* **45**, 289–316.

Nicholls I. A. and Lorenz V. (1973) Origin and crystallization history of Permian tholeiites from the Saar-Nahe Trough, SW Germany. *Contrib. Mineral. Petrol.* **40**, 327–344.

Nicholls I. A. and Ringwood A. E. (1972) Production of silica-saturated tholeiitic magmas in island arcs. *Earth Planet. Sci. Lett.* **17**, 243–246.

Nicholls I. A. and Ringwood A. E. (1973) Effect of water on olivine stability in tholeiites and the production of silica-saturated magmas in the island-arc environment. *J. Geol.* **81**, 285–300.

Nishikawa M., Kono S. and Aramaki S. (1970) Melting of lherzolite from Ichinomegata at high pressures. *Phys. Earth Planet. Inter.* **4**, 138–144.

Nixon P. H., ed. (1973) *Lesotho Kimberlites.* Lesotho National Development Corporation, Maseru. 350 pp.

Obata M. and Dickey J. S., Jr. (1976) Phase relations of mafic layers in the Ronda Peridotite. *Carnegie Inst. Wash. Yearb.* **75**, 562–566.

O'Donnell T. H. and Presnall D. C. (1980) Chemical variations of the glass and mineral phases in basalts dredged from 25°–30° N. along the Mid-Atlantic Ridge. *Amer. J. Sci., Jackson v.* **280A**, 845–868.

O'Hara M. J. (1965) Primary magmas and the origin of basalts. *Scott. J. Geol.* **1**, 19–40.

O'Hara M. J. (1968a) The bearing of phase equilibria studies in synthetic and natural systems on the origin and evolution of basic and ultrabasic rocks. *Earth Sci. Rev.* **4**, 69–133.

O'Hara M. J. (1968b) Are ocean floor basalts primary magma? *Nature* **220**, 683–686.

O'Hara M. J. (1975) Is there an Icelandic mantle plume? *Nature* **253**, 708–710.

O'Hara M. J. (1977) Geochemical evolution during fractional crystallization of a periodically refilled magma chamber. *Nature* **266**, 503–507.

O'Hara M. J. and Biggar G. M. (1977) Comments on a paper by S. E. Kesson and D. H. Lindsley. *Rev. Geophys. Space Phys.* **15**, 253–255.

O'Hara M. J. and Humphries D. J. (1977) Problems of iron gain and loss during experimentation on natural rocks: the experimental crystallization of five lunar basalts at low pressures. *Phil. Trans. Roy. Soc. Lond.* **A286**, 313–330.

O'Hara M. J. and Yoder H. S., Jr. (1967) Formation and fractionation of basic magmas at high pressures. *Scott. J. Geol.* **3**, 67–117.

O'Hara M. J., Richardson S. W. and Wilson G. (1971) Garnet-peridotite stability and occurrence in crust and mantle. *Contrib. Mineral. Petrol.* **32**, 48–68.

O'Hara M. J., Biggar G. M., Richardson S. W., Ford C. E. and Jamieson B. G. (1970) The nature of seas, mascons and the lunar interior in the light of experimental studies. *Proc. Apollo 11 Lunar Sci. Conf.*, p. 695–710.

O'Hara M. J., Humphries D. J. and Waterston S. (1975) Petrogenesis of mare basalts: implications for chemical, mineralogical, and thermal models for the Moon. *Proc. Lunar Sci. Conf. 6th*, p. 1043–1055.

O'Nions R. K., Hamilton P. J. and Evensen N. M. (1977) Variations in $^{143}Nd/^{144}Nd$ and $^{87}Sr/^{86}Sr$ ratios in oceanic basalts. *Earth Planet. Sci. Lett.* **34**, 13–22.

Papike J. J., Hodges F. N., Bence A. E., Cameron M. and Rhodes J. M. (1976) Mare basalts: crystal chemistry, mineralogy and petrology. *Rev. Geophys. Space Phys.* **14**, 475–540.

Powell B. N. (1969) Petrology and chemistry of mesosiderites—I. Textures and composition of nickel-iron. *Geochim. Cosmochim. Acta* **33**, 789–810.

Powell B. N. (1971) Petrology and chemistry of mesosiderites—II. Silicate textures and compositions and metal-silicate relationships. *Geochim. Cosmochim. Acta* **35**, 5–34.

Powell J. L. and DeLong S. E. (1966) Isotopic composition of strontium in volcanic rocks from Oahu. *Science* **153**, 1239–1242.

Powers H. A. (1955) Composition and origin of basaltic magmas of the Hawaiian Islands. *Geochim. Cosmochim. Acta* **7**, 77–107.

Presnall D. C. (1969) The geometrical analysis of partial fusion. *Amer. J. Sci.* **267**, 1178–1194.

Presnall D. C. (1976) Alumina content of enstatite as a geobarometer for plagioclase and spinel lherzolites. *Amer. Mineral.* **61**, 582–588.

Presnall D. C. (1979) Fractional crystallization and partial fusion. In *The Evolution of the Igneous Rocks: Fiftieth Anniversary Perspectives*, (H. S. Yoder, Jr., ed.), p. 59–75. Princeton Univ. Press, Princeton.

Presnall D. C. and Brenner N. L. (1974) A method for studying iron-silicate liquids under reducing conditions with negligible iron loss. *Geochim. Cosmochim. Acta* **38**, 1785–1788.

Presnall D. C., Dixon S. A., Dixon J. R., O'Donnell T. H., Brenner N. L., Schrock R. L. and Dycus D. W. (1978a) Liquidus phase relations on the join diopside-forsterite-anorthite from 1 atm. to 20 kbar: their bearing on the generation and crystallization of basaltic magma. *Contrib. Mineral. Petrol.* **66**, 203–220.

Presnall D. C., Mysen B. O. and Kushiro I. (1978b) Correction to the paper "Compositional variations of coexisting phases with degree of melting of peridotite in the upper mantle," by B. O. Mysen and I. Kushiro. *Amer. Mineral.* **63**, 597.

Presnall D. C., Dixon J. R., O'Donnell T. H. and Dixon S. A. (1979) Generation of mid-ocean ridge tholeiites. *J. Petrol.* **20**, 3–35.

Råheim A. and Green D. H. (1974) Experimental petrology of lunar highland basalt composition and applications to models for the lunar interior. *J. Geol.* **82**, 607–622.

Reay A. and Harris P. G. (1964) The partial fusion of peridotite. *Bull. Volc.* **27**, 115–127.

Reid A. M. (1974) The Macibini meteorite and some thoughts on the origin of basaltic achondrites. *Meteoritics* **9**, 398–399.

Rhodes J. M. and Dungan M. A. (1979) The evolution ocean-floor basaltic magmas. In *Deep Drilling Results in the Atlantic Ocean: Ocean crust, M. Ewing Ser. 2*, p. 262–272. Amer. Geophys. Union.

Rhodes J. M., Dungan M. A., Blanchard D. P. and Long P. E. (1979) Magma mixing at mid-ocean ridges: evidence from basalts drilled at 22° N on the Mid-Atlantic Ridge. *Tectonophys.* **55**, 35–61.

Rhodes J. M., Blanchard D. P., Dungan M. A., Brannon J. C. and Rodgers K. V. (1977) Chemistry of Apollo 12 mare basalts: magma types and fractionation processes. *Proc. Lunar Sci. Conf. 8th*, p. 1305–1338.

Rhodes J. M., Hubbard N. J., Wiesmann H., Rodgers K. V., Brannon J. C. and Bansal B. M. (1976) Chemistry, classification and petrogenesis of Apollo 17 mare basalts. *Proc. Lunar Sci. Conf. 7th*, p. 1467–1489.

Riboud P. V. and Muan A. (1962) Phase equilibria in a part of the system "FeO"-MnO-SiO_2. *Trans. Amer. Inst. Mining Met. Eng.* **224**, 28–33.

Ringwood A. E. (1958a) The constitution of the mantle—I. Thermodynamics of the olivine-spinel transition. *Geochim. Cosmochim. Acta* **13**, 303–321.

Ringwood A. E. (1958b) The constitution of the mantle—II. Further data on the olivine-spinel transition. *Geochim. Cosmochim. Acta* **15**, 19–28.

Ringwood A. E. (1958c) The constitution of the mantle—III. Consequences of the olivine-spinel transition. *Geochim. Cosmochim. Acta* **15**, 195–212.

Ringwood A. E. (1962a) A model for the upper mantle. *J. Geophys. Res.* **67**, 857–867.

Ringwood A. E. (1962b) A model for the upper mantle, 2. *J. Geophys. Res.* **67**, 4473–4477.

Ringwood A. E. (1975) *Composition and Petrology of the Earth's Mantle.* McGraw-Hill, N.Y. 618 pp.

Ringwood A. E. (1976) Limits on the bulk composition of the Moon. *Icarus* **28**, 325–349.

Ringwood A. E. (1977a) Petrogenesis in island arc systems. In *Island Arcs, Deep Sea Trenches and Back-Arc Basins, M. Ewing Ser. 1,* p. 311–324. Amer. Geophys. Union.

Ringwood A. E. (1977b) Composition of the core and implications for origin of the Earth. *Geochem. J. (Japan)* **11**, 111–136.

Ringwood A. E. and Essene E. (1970) Petrogenesis of Apollo 11 basalts, internal constitution and origin of the Moon. *Proc. Apollo 11 Lunar Sci. Conf.,* 769–799.

Ringwood A. E. and Kesson S. E. (1976) A dynamic model for mare basalt petrogenesis. *Proc. Lunar Sci. Conf. 7th,* p. 1697–1772.

Robie R. A., Hemingway B. S. and Fisher J. R. (1978) Thermodynamic properties of minerals and related substances at 298.15 K and 1 bar (10^5 pascals) pressure and at higher temperatures. *U. S. Geol. Surv. Bull.* **1452**, 456.

Rock N. M. S. (1978) The effect of CO_2 on the liquidus relations of alkali basalts and its relevance to the nephelinite problem: a preliminary investigation. In *Progress in Experimental Petrology*, Natural Environment Research Council Rept. 4, p. 164–170.

Roeder P. L. (1974a) Activity of iron and olivine solubility in basaltic liquids. *Earth Planet. Sci. Lett.* **23**, 397–410.

Roeder P. L. (1974b) Paths of crystallization and fusion in systems showing ternary solid solution *Amer. J. Sci.* **274**, 48–60.

Roeder P. L. and Emslie R. F. (1970) Olivine-liquid equilibrium. *Contrib. Mineral. Petrol.* **29**, 275–289.

Rosenhauer M., Woermann E., Knecht B. and Ulmer G. C. (1977) The stability of graphite and diamond as a function of the oxygen fugacity in the mantle (abstract). In *Extended Abstracts, 2nd Internat. Kimberlite Conf., Santa Fe.*

Ross C. S., Foster M. D. and Myers A. T. (1954) Origin of dunites and of olivine-rich inclusions in basaltic rocks. *Amer. Mineral.* **39**, 693–736.

Rutherford M. J., Hess P. C., Ryerson F. J., Campbell H. W. and Dick P. A. (1976) The chemistry, origin and petrogenetic implications of lunar granite and monzonite. *Proc. Lunar Sci. Conf. 7th,* p. 1723–1740.

Ryabchikov I. D. and Green D. H. (1978) The role of carbon dioxide in the petrogenesis of highly potassic magmas. In *Problems of Petrology of the Earth's Crust and Upper Mantle,* (N. L. Dobretsov *et al.*, eds.), *Akad. Nauk SSSR, Sib. Otd., Inst. Geol. Geofiz, Trudy,* No. 403, p. 49–64.

Ryder G. and Wood J. (1977) Serenitatis and Imbrium impact melts: implications for large-scale layering in the lunar crust. *Proc. Lunar Sci. Conf. 8th,* p. 655–668.

Saito K., Basu A. R. and Alexander E. C., Jr. (1978) Planetary type noble gases in an upper mantle-derived amphibole. *Earth Planet. Sci. Lett.* **39**, 274–280.

Sato M. (1976) Oxygen fugacity and other thermochemical parameters of Apollo 17 high-Ti basalts and their implications on the reduction mechanism. *Proc. Lunar Sci. Conf. 7th,* p. 1323–1344.

Sato M. (1978) Oxygen fugacity of basaltic magmas and the role of gas-forming elements. *Geophys. Res. Lett.* **5**, 447–449.

Schairer J. F. (1942) The system CaO-FeO-Al_2O_3-SiO_2. I. Results of quenching experiments on five joins. *J. Amer. Ceram. Soc.* **25**, 241–274.

Schairer J. F. (1959) Phase equilibria, with particular reference to silicate systems. In *Physicochemical Measurements at High Temperatures,* (J. O. Bockris, J. L. White and J. D. Mackenzie, eds.), p. 117–134. Butterworths, London.

Schairer J. F. and Yoder H. S., Jr. (1962) The system diopside-enstatite-silica. *Carnegie Inst. Wash. Yearb.* **61**, 75–82.

Scherer G., Hopper R. W. and Uhlmann D. R. (1972) Crystallization behavior and glass formation of selected lunar compositions. *Proc. Lunar Sci. Conf. 3rd,* p. 2627–2637.

Schilling J.-G. (1971) Sea-floor evolution: rare-earth evidence. *Phil. Trans. R. Soc. Lond.* **A286**, 663–706.

Schilling J.-G. (1975) Rare-earth variations across 'normal segments' of the Reykjanes Ridge, 60°–53° N, Mid-Atlantic Ridge, 29° C, and East Pacific Rise, 2°–19° S, and evidence on the composition of the underlying low-velocity layer. *J. Geophys. Res.* **80**, 1459–1473.

Schnetzler C. C. and Philpotts J. A. (1969) Genesis of the calcium-rich achondrites in light of rare-earth and barium concentrations. In *Meteorite Research* (Peter M. Millman, ed.), p. 206–216. Reidel, Dortrecht-Holland.

Schnetzler C. C. and Philpotts J. A. (1971) Alkali, alkaline earth, and rare-earth element concentrations in some Apollo 12 soils, rocks, and separated phases. *Proc. Lunar Sci. Conf. 2nd,* p. 1101–1122.

Schreiber H. D. (1977) Redox states of Ti, Zr, Hf, Cr and Eu in basaltic magmas: an experimental study. *Proc. Lunar Sci. Conf. 8th,* p. 1785–1807.

Schreiber H. D. and Haskin L. A. (1976) Chromium in basalts: experimental determination of redox states and partitioning among synthetic silicate phases. *Proc. Lunar Sci. Conf. 7th,* p. 1221–1260.

Shaw H. R. and Jackson E. D. (1973) Linear island chains in the Pacific: result of thermal plumes or gravitational anchors? *J. Geophys. Res.* **78**, 8634–8652.

Shaw H. R., Peck D. L., Wright T. L. and Okamura R. (1968) The viscosity of basaltic magma: an analysis of field measurements in Makaopuhi lava lake, Hawaii. *Amer. J. Sci.* **266**, 225–264.

Shido F., Miyashiro A. and Ewing M. (1971) Crystallization of abyssal tholeiites. *Contrib. Mineral. Petrol.* **31**, 251–266.

Shih C. Y. (1977) Origins of KREEP basalts. *Proc. Lunar Sci. Conf. 8th,* p. 2375–2401.

Shimada M. (1966a) Melting of basalt at high temperatures. *J. Seismolog. Soc. Japan, Ser. II* **19**, 167–175.

Shimada M. (1966b) Effects of pressure and water on the melting of basalt. *Spec. Contrib. Kyoto Univ. Geophys. Inst.* no. 6, 303–311.

Shimizu N. and Allègre C. J. (1975) The genetic relationships among diogenites, howardites and eucrites. *Meteoritics* **10**, 488.

Siegfried R. W. and Solomon S. C. (1974) Mercury: internal structure and thermal evolution. *Icarus* **23**, 192–205.

Smith J. V., Anderson A. T., Newton R. C., Olsen E. J., Wyllie P. J., Crewe A. V., Isaacson M. S. and Johnson D. (1970) Petrologic history of the Moon inferred from petrography, mineralogy, and petrogenesis of Apollo 11 rocks. *Proc. Apollo 11 Lunar Sci. Conf.,* p. 897–925.

Sobolev N. V. (1977) *Deep-Seated Inclusions in Kimberlites and the Problem of the Composition of the Upper Mantle.* Amer. Geophys. Union, Washington. 279 pp.

Sobolev V. S., Bazarova T. J. and Yagi K. (1975) Crystallization temperature of wyomingite from Leucite Hills. *Contrib. Mineral. Petrol.* **49**, 301–308.

Solomon S. C. (1976) Geophysical constraints on radial and lateral temperature variations in the upper mantle. *Amer. Mineral.* **61**, 788–803.

Solomon S. and Chaiken J. (1976) Thermal expansion and thermal stress in the Moon and terrestrial planets: clues to early thermal history. *Proc. Lunar Sci. Conf. 7th,* p. 3229–3243.

Sparks R. S. J., Pinkerton H. and MacDonald R. (1977) The transport of xenoliths in magmas. *Earth Planet. Sci. Lett.* **35**, 234–238.

Stern C. and Wyllie P. J. (1973) Melting relations of basalt-andesite-rhyolite-H_2O and a pelagic red clay to 30 kbar. *Contrib. Mineral. Petrol.* **42**, 313–323.

Stern C. R. and Wyllie P. J. (1975) Effect of iron adsorption by noble-metal capsules on phase boundaries in rock-melting experiments at 30 kilobars. *Amer. Mineral.* **60**, 681–689.

Stern C. R. and Wyllie P. J. (1978) Phase compositions through crystallization intervals in basalt-andesite-H_2O at 30 kbar with implications for subduction zone magmas. *Amer. Mineral.* **63**, 641–663.

Stern C. R., Huang W.-L. and Wyllie P. J. (1975) Basalt-andesite-rhyolite-H_2O: crystallization intervals with excess H_2O and H_2O-undersaturated liquidus surfaces to 35 kilobars, with implications for magma genesis. *Earth Planet. Sci. Lett.* **28**, 189–196.

Stewart D. C., Boettcher A. L. and Eggler D. H. (1979) Phase relations of kaersutite at upper mantle conditions: implications for its subsolidus origin in ultramafic nodules (abstract). *EOS* (*Trans. Amer. Geophys. Union*) **60**, 418.

Stolper E. (1975) Petrogenesis of eucrite, howardite and diogenite meteorites. *Nature* **258**, 220–222.

Stolper E. M. (1977) Experimental petrology of eucritic meteorites. *Geochim. Cosmochim. Acta* **41**, 587–611.

Stolper E. M. and McSween H. Y., Jr. (1979) Petrology and origin of shergottite meteorites. *Geochim. Cosmochim. Acta* **43** (in press).

Stolper E. M., McSween H. Y., Jr. and Hays J. F. (1979) A petrogenetic model of the relationships among achondritic meteorites. *Geochim. Cosmochim. Acta* **43**, 589–602.

Stolper E., Walker D., Longhi J. and Hays J. F. (1974) Compositional variation in lunar ultramafic glasses (abstract). In *Lunar Science V*, p. 749–751.

Storey W. C., Humphries D. J. and O'Hara M. J. (1974) Experimental petrology of sample 77135. *Earth Planet. Sci. Lett.* **23**, 435–438.

Sun S.-S. and Nesbitt R. W. (1978) Petrogenesis of Archean ultrabasic and basic volcanics: evidence from rare earth elements. *Contrib. Mineral. Petrol.* **65**, 301–325.

Sutherland F. L. (1974) High-pressure inclusions in tholeiitic basalt and the range of lherzolite-bearing magmas in the Tasmanian volcanic province. *Earth Planet. Sci. Lett.* **24**, 317–324.

Tatsumoto M. (1966) Genetic relations of oceanic basalts as indicated by lead isotopes. *Science* **153**, 1094–1101.

Taylor S. R. and Bence A. E. (1975) Evolution of the lunar highland crust. *Proc. Lunar Sci. Conf. 6th*, p. 1121–1141.

Taylor S. R. and Jakes P. (1974) The geochemical evolution of the Moon. *Proc. Lunar Sci. Conf. 5th*, p. 1287–1305.

Thompson R. N. (1973) One-atmosphere melting behavior and nomenclature of terrestrial lavas. *Contrib. Mineral. Petrol.* **41**, 197–204.

Thompson R. N. (1974) Primary basalts and magma genesis. I. Skye, North-West Scotland. *Contrib. Mineral. Petrol.* **45**, 317–341.

Thompson R. N. (1975) Primary basalts and magma genesis. II. Snake River Plain, Idaho, U.S.A. *Contrib. Mineral. Petrol.* **52**, 213–232.

Thompson R. N. (1977) Primary basalts and magma genesis. III. Alban Hills, Roman Comagmatic Province, Central Italy. *Contrib. Mineral. Petrol.* **60**, 91–108.

Thompson R. N. and Flower M. F. J. (1971) One-atmosphere melting and crystallization relations of lavas from Anjouan, Comores Archipelago, western Indian Ocean. *Earth Planet. Sci. Lett.* **12**, 97–107.

Thompson R. N. and Tilley C. E. (1969) Melting and crystallization relations of Kilauean basalts of Hawaii. The lavas of the 1959–60 Kilauea eruption. *Earth Planet. Sci. Lett.* **5**, 469–477.

Thompson R. N., Esson J. and Dunham A. C. (1972) Major element chemical variation in the Eocene lavas of the Isle of Skye, Scotland. *J. Petrol.* **13**, 219–253.

Tilley C. E. (1950) Some aspects of magmatic evolution. *J. Geol. Soc. Lond.* **106**, 37–61.

Tilley C. E. and Thompson R. N. (1970) Melting and crystallization relations of the Snake River Basalts of Southern Idaho. *Earth Planet. Sci. Lett.* **8**, 79–92.

Tilley C. E. and Thompson R. N. (1972) Melting relations of some ultra alkalic volcanics. *Geol. J.* **8**, 65–70.

Tilley C. E., Yoder H. S., Jr. and Schairer J. F. (1964) New relations on melting of basalts. *Carnegie Inst. Wash. Yearb.* **63**, 92–97.

Tilley C. E., Yoder H. S., Jr. and Schairer J. F. (1965) Melting relations of volcanic tholeiite and alkali rock series. *Carnegie Inst. Wash. Yearb.* **64**, 69–82.

Tilley C. E., Yoder H. S., Jr. and Schairer J. F. (1966) Melting relations of volcanic rock series. *Carnegie Inst. Wash. Yearb.* **65**, 260–269.

Tilley C. E., Yoder H. S., Jr. and Schairer J. F. (1967) Melting relations of igneous rock series. *Carnegie Inst. Wash. Yearb.* **66**, 450–457.

Tilley C. E., Thompson R. N., Wadsworth W. J. and Upton B. G. J. (1971) Melting and crystallization relations of some lavas of Reunion Island, Indian Ocean. *Mineral. Mag.* **38**, 334–353.

Tilley C. E., Thompson R. N. and Lovenbury P. A. (1972) Melting relations of some oceanic basalts. *Geol. J.* **8**, 59–64.

Trigila R. C. (1969) Sulla genesi dei magmi a carattere mediterraneo. *Periodico Mineral.* **38**, 625–660.

Turner F. J. and Verhoogen J. (1951) *Igneous and Metamorphic Petrology*. 1st Edition. McGraw-Hill, N.Y.

Turner F. J. and Verhoogen J. (1960) *Igneous and Metamorphic Petrology*. 2nd Edition. McGraw-Hill, N.Y. 694 pp.

Uhlmann D. R. and Klein L. C. (1976) Crystallization kinetics, viscous flow, and thermal histories of lunar breccias 15286 and 15498. *Proc. Lunar Sci. Conf. 7th*, p. 2529–2541.

Uhlmann D. R., Klein L., Kritchevsky G. and Hopper R. W. (1974) The formation of lunar glasses. *Proc. Lunar Sci. Conf. 5th*, p. 2317–2331.

Ulmer G. C. (1971) *Research Techniques for High Pressure and High Temperature*. (G. C. Ulmer, ed.). Springer Verlag, N.Y. 367 pp.

Usselman T. M. (1975) Experimental approach to the state of the core: Part I. The liquidus relations of the Fe-rich portion of the Fe-Ni-S system from 30 to 100 kb. *Amer. J. Sci.* **275**, 278–290.

Usselman T. M. and Hodge D. S. (1978) Thermal control of low-pressure fractionation processes. *J. Volcanol. Geotherm. Res.* **4**, 265–281.

Usselman T. M. and Lofgren G. E. (1976) The phase relations, textures, and mineral chemistries of high-titanium mare basalts as a function of oxygen fugacity and cooling rate. *Proc. Lunar Sci. Conf. 7th*, p. 1345–1363.

Usselman T. M., Lofgren G. E., Donaldson C. H. and Williams R. J. (1975) Experimentally reproduced textures and mineral chemistries of high-titanium mare basalts. *Proc. Lunar Sci. Conf. 6th*, p. 997–1020.

Verhoogen J. (1956) Temperatures within the Earth. In *Physics and Chemistry of the Earth, Vol. 1* (L. H. Ahrens, F. Press, K. Rankama and S. K. Runcorn, eds.), p. 17–43. Pergamon, London.

Viljoen M. J. and Viljoen R. P. (1969) Evidence for the existence of a mobile extrusive peridotitic magma from the Komati Formation of the Onverwacht Group. *Geol. Soc. S. Africa, Spec. Publ.* **2**, 89–112.

Visser W. and Koster Van Groos A. F. (1979) Phase relations in the system K_2O-FeO-Al_2O_3-SiO_2 at 1 atmosphere with special emphasis on low temperature liquid immiscibility. *Amer. J. Sci.* **279**, 70–91.

Wager L. R. (1958) Beneath the Earth's crust. *Advancement Sci.* **15**, 31–45.

Walker D. and Hays J. F. (1977) Observations on basaltic volcanism (abstract). *Geol. Soc. Amer. Abstracts with Programs* **9**, 1215.

Walker D., Longhi J. and Hays J. F. (1972) Experimental petrology and origin of Fra Mauro rocks and soil. *Proc. Lunar Sci. Conf. 3rd*, p. 797–817.

Walker D., Longhi J., Grove T. L., Stolper E. M. and Hays J. F. (1973a) Experimental petrology and origin of rocks from the Descartes Highlands. *Proc. Lunar Sci. Conf. 4th*, 1013–1032.

Walker D., Grove T. L., Longhi J., Stolper E. M. and Hays J. F. (1973b) Origin of lunar feldspathic rocks. *Earth Planet. Sci. Lett.* **20**, 325–336.

Walker D., Longhi J., Stolper E. M., Grove T. L. and Hays J. F. (1975a) Origin of titaniferous lunar basalts. *Geochim. Cosmochim. Acta* **39**, 1219–1235.

Walker D., Longhi J. and Hays J. F. (1975b) Differentiation of a very thick magma body and implications for the source regions of mare basalts. *Proc. Lunar Sci. Conf. 6th* p. 1103–1120.

Walker D., Kirkpatrick R. J., Longhi J. and Hays J. F. (1976a) Crystallization history of lunar picritic basalt sample 12002: phase-equilibria and cooling-rate studies. *Geol. Soc. Amer. Bull.* **87**, 646–656.

Walker D., Longhi J., Kirkpatrick R. J. and Hays J. F. (1976b) Differentiation of an Apollo 12 picrite magma. *Proc. Lunar Sci. Conf. 7th*, p. 1365–1389.

Walker D., Longhi J. and Hays J. F. (1976c) Heterogeneity in titaniferous lunar basalts. *Earth Planet. Sci. Lett.* **30**, 27–36.

Walker D., Kirkpatrick R. J. and Hays J. F. (1977a) Differentiation of a komatiite lava (abstract). *EOS (Trans. Amer. Geophys. Union)* **58**, 527.

Walker D., Longhi J., Lasaga A. C., Stolper E. M., Grove T. L. and Hays J. F. (1977b) Slowly cooled microgabbros 15555 and 15016. *Proc. Lunar Sci. Conf. 8th*, p. 1521–1548.

Walker D., Shibata T. and DeLong S. E. (1979) Abyssal tholeiites from the Oceanographic Fracture Zone II. Phase equilibria and mixing. *Contrib. Mineral. Petrol.* **70**, 111–125.

Walker G. P. L. (1973) Lengths of lava flows. *Phil. Trans. Roy. Soc. Lond.* **A274**, 107–118.

Wänke H., Palme H., Baddenhausen H., Dreibus G., Jagoutz, E., Kruse H., Spettel B., Teschke F. and Thacker R. (1974) Chemistry of Apollo 16 and 17 samples: bulk composition, late stage accumulation and early differentiation of the Moon. *Proc. Lunar Sci. Conf. 5th*, p. 1307–1335.

Warren P. H. and Wasson J. T. (1979) The origin of KREEP. *Rev. Geophys. Space Phys.* **17**, 73–88.

Wass S. Y. and Irving A. J. (1976) *XENMEG: a catalogue of occurrences of xenoliths and megacrysts in basic volcanic rocks of eastern Australia*. Australian Museum, Sydney. 455 pp.

Wasson J. T. (1974) *Meteorites*. Springer-Verlag, N.Y. 316 pp.

Weill D. F. and McKay G. A. (1975) Partitioning of Mg, Fe, Sr, Ce, Sm, Eu and Yb in lunar igneous systems and a possible origin of KREEP by equilibrium partial melting. *Proc. Lunar Sci. Conf. 6th*, p. 1143–1158.

Wendlandt R. F. and Eggler D. H. (1980) The origins of potassic magmas: II. Stability of phlogopite in natural spinel lherzolite and in the system $KAlSiO_4MgO$-SiO_2-H_2O-CO_2 at high pressures and high temperatures. *Amer. J. Sci.* **280**, 421–458.

Wendlandt R. F. and Mysen B. O. (1978) Melting phase relations of natural peridotite + CO_2 as a function of degree of partial melting at 15 and 30 kbar. *Carnegie Inst. Wash. Yearb.* **77**, 756–761.

Williams R. J. and Mullins O. (1976) A system using solid ceramic oxygen electrolyte cells to measure oxygen fugacities in gas-mixing systems. NASA TM X-58167.

Wilson J. T. (1963) A possible origin of the Hawaiian Islands. *Canad. J. Phys.* **41**, 863–870.

Woermann E., Knecht B., Rosenhauer M. and Ulmer G. C. (1977) The stability of graphite in the system C–O (abstract). In *Extended Abstracts, 2nd Internat. Kimberlite Conf., Santa Fe.*

Wood J. A. (1962) Chondrules and the origin of the terrestrial planets. *Nature* **194**, 127–130.

Wood J. A. (1977) A survey of lunar rock types and comparison of the crusts of Earth and Moon. In *The Soviet-American Conference on the Cosmochemistry of the Moon and Planets*, (J. H. Pomeroy and N. J. Hubbard, eds.), NASA SP-370.

Wood J. A., Dickey J. S., Jr., Marvin U. B. and Powell B. N. (1970) Lunar anorthosites and a geophysical model of the Moon. *Proc. Apollo 11 Lunar Sci. Conf.*, p. 965–988.

Wright T. L. (1971) Chemistry of Kilauea and Mauna Loa lava in space and time. *U. S. Geol. Survey Prof. Paper* **735**, 40 p.

Wyllie P. J. (1971a) *The Dynamic Earth: A Textbook in Geosciences.* John Wiley and Sons, N.Y. 416 pp.

Wyllie P. J. (1971b) Role of water in magma generation and initiation of diapiric uprise in the mantle. *J. Geophys. Res.* **76**, 1328–1338.

Wyllie P. J. (1973) Experimental petrology and plate tectonics—a preview. *Tectonophys.* **17**, 189–209.

Wyllie P. J. (1977) Mantle fluid compositions buffered by carbonates in peridotite—CO_2-H_2O. *J. Geol.* **85**, 187–207.

Wyllie P. J. (1978a) Mantle fluid compositions buffered in peridotite-CO_2-H_2O by carbonates, amphibole, and phlogopite. *J. Geol.* **86**, 687–713.

Wyllie P. J. (1978b) Water and magma generation at subduction zones. *Bull. Volc.*, **41**, 360–377.

Wyllie P. J. (1979) Magmas and volatile components. *Amer. Mineral.* **64**, 469–500.

Wyllie P. J. and Huang W.-L. (1975) Peridotite, kimberlite, and carbonatite explained in the system CaO-MgO-SiO_2-CO_2. *Geology* **3**, 621–624.

Wyllie P. J. and Huang W.-L. (1976) Carbonation and melting reactions in the system CaO-MgO-SiO_2-CO_2 at mantle pressures with geophysical and petrological applications. *Contrib. Mineral. Petrol.* **54**, 79–107.

Yagi K., Hariya Y., Onuma K. and Fukushima N. (1975) Stability relation of kaersutite. *J. Fac. Sci. Hokkaido Univ., Ser. IV* **16**, 331–342.

Yoder H. S., Jr. (1952) Change of melting point of diopside with pressure. *J. Geol.* **60**, 364–374.

Yoder H. S., Jr. (1976) *Generation of Basaltic Magma.* National Academy of Sciences, Washington. 265 pp.

Yoder H. S., Jr. and Tilley C. E. (1961) Simple basalt systems. *Carnegie Inst. Wash. Yearb.* **60**, 106–113.

Yoder H. S., Jr. and Tilley C. E. (1962) Origin of basalt magmas: experimental study of natural and synthetic rock systems. *J. Petrol.* **3**, 342–532.

Zielinski R. A. (1975) Trace element evaluation of a suite of rocks from Reunion Island, Indian Ocean. *Geochim. Cosmochim. Acta* **39**, 713–734.

Overleaf:

Ultramafic xenoliths, probably derived from the upper mantle, in a sample of basanite from San Carlos.

4. Geophysical and cosmochemical constraints on properties of mantles of the terrestrial planets

John A. Wood, Team Leader
—Harvard-Smithsonian Center for Astrophysics

W. Roger Buck
—Department of Earth and Planetary Sciences Massachusetts Institute of Technology

Edward Anders
—Enrico Fermi Institute University of Chicago

John W. Morgan
—U. S. Geological Survey Reston, Virginia

Edward Stolper
—Division of Geological and Planetary Sciences California Institute of Technology

Don L. Anderson
—Seismological Laboratory California Institute of Technology

William M. Kaula
—Department of Earth and Planetary Sciences University of California, Los Angeles

Guy J. Consolmagno
—Harvard-Smithsonian Center for Astrophysics

A. E. Ringwood
—Australian National University Canberra, Australia

Heinrich Wänke
—Max-Planck-Institut für Chemie Mainz, Germany

4.1 INTRODUCTION

Basalts are understood to be derived by partial melting from the mantles of their planets, and properties of the basalt samples we have been able to obtain (from Earth, the Moon, and the parent planets of the basaltic achondrite meteorites) have been an invaluable source of information about the composition and nature of the mantles of these bodies (Chapter 3). There is little doubt that basaltic volcanism has also been active on the planets that we have not been able to sample; in several cases, the surface morphologies and photometric properties of these planets strongly indicate the action of basaltic volcanism (Chapters 5 and 2). The purpose of this chapter is to infer as much as we can, by means other than sampling, about the compositions of planetary mantles. Our estimates of mantle compositions provide a basis for experimental petrologists to project the possible compositions of basalts that would be erupted on unsampled planets (Chapter 3).

The clues we have to mantle compositions, apart from directly sampling them and the basalts that issue from them, are (1) the gross geophysical properties of the planets in question and (2) our estimates of their bulk compositions. Fragmentary chemical information comes to us from sources such as spectral studies of planetary atmospheres and surface regoliths, but most of our chemical constraints are based on broader considerations of the overall composition of the solar system and the early processes that appear to have caused the terrestrial planets to incorporate varying proportions of the elements.

The geophysical constraints are reviewed in section 4.2, below; the admittedly speculative question of chemical processes that affected the origin and bulk compositions of the planets (cosmochemistry) is treated in section 4.3. Section 4.4 describes the process of planetary modelling by which the most probable mantle compositions, or compositional profiles, are derived from these inputs. Finally, section 4.5 carries out this modelling exercise for the terrestrial planets and Earth's moon, discussing the most likely mantle properties for each body.

4.2 GEOPHYSICAL CONSTRAINTS

In this section the types of geophysical data available and the general constraints they impose on planetary models are summarized. For a more complete review of planetary physics, the reader is referred to Kaula (1968).

4.2.1 Geodetic constraints

Mean density

The determination of the mean density of a planet involves three quite different types of measurement: (1) the product GM, gravitational constant times mass of the planet, which is derived from its effect on some device that measures acceleration: typically, the deflection of trajectory of a passing spacecraft is measured by radio Doppler tracking; (2) the mean radius R, which is most precisely measured by radar, either Earth- or spacecraft-based; and (3) the gravitational constant G, which has been determined by laboratory measurements of the mutual gravitational attraction of two masses. For large terrestrial bodies, the limitation on accuracy of the mean density is usually the uncertainty in the gravitational constant G; for small bodies (asteroids, etc.), it may be in either GM or R.

Planetary densities ($\bar{\rho}$) and other parameters are summarized in Table 4.2.1.

The GM of the Earth is determined most accurately from spacecraft tracking: the mean rate-of-motion n and semimajor axis a of a spacecraft's orbit are measured, and Kepler's law, $GM = n^2a^3$, is applied.

The radius of the Earth comes from geometrical analyses of geodetic triangulation and spacecraft triangulation. It is conventionally given as the equatorial radius a_e of the "mean Earth ellipsoid," which is the ellipsoid that best fits the geoid. Hence the flattening f of the ellipsoid is needed to get the mean radius R

$$R = a_e(1-f)^{1/3}$$

The flattening is best determined dynamically, from the precessions of artificial satellite orbits.

Spacecraft have been placed in orbit around the Moon, Mars, and Venus. This enables accurate determinations to be made of both the GM (from the periods and semimajor axes of spacecraft orbits) and the mean radius R (preferably from altimetry, or, secondarily, from photogrammetric analysis). A refined estimate for

Table 4.2.1 Geodetic parameters* of planets and satellites.

Body	GM (10^{18}cm^3/sec^2)	R (km)	$\bar{\rho}$ (gm/cm^3)	I/MR^2	$\bar{D}$ (km)	References
Earth	398.6005 ±0.0002	6371.003 ±0.005	5.514 ±0.005	0.3315 ±0.0000	14	Moritz (1976).
Moon	4.9028 ±0.0001	1737.5 ±0.1	3.344 ±0.003	0.391 ±0.002	75	Bills and Ferrari (1977); Blackshear and Gapcynski (1977).
Mars	42.8283 ±0.0001	3390 ±1	3.934 ±0.004	0.365 ±0.005	>28	Anderson (1975); Reasenberg (1977); Bills and Ferrari (1978); Kaula (1979b).
Venus	324.858 ±0.001	6051.4 ±0.1	5.245 ±0.005	?	≳21	Anderson (1975); Pettengill *et al.* (1979).
Mercury	22.04 ±0.12	2439 ±2	5.435 ±0.008	?	≳7	Anderson (1975); Pettengill *et al.* (1979).
Vesta	0.016 ±0.003	269 ±10	2.9 ±0.5	?	?	Morrison (1977); Chapman *et al.* (1978).
Io	5.934 ±0.028	1816 ±5	3.55 ±0.03	?	?	Null (1976); Johnson (1978); Davies *et al.* (1979).
Europa	3.196 ±0.032	1536 ±10	2.99 ±0.06	?	?	Null (1976); Johnson (1978); Davies *et al.* (1979).
Ganymede	9.885 ±0.037	2638 ±10	1.93 ±0.02	?	?	Null (1976); Johnson (1978); Davies *et al.* (1979).
Callisto	7.172 ±0.024	2410 ±10	1.83 ±0.02	?	?	Null (1976); Johnson (1978); Davies *et al.* (1979).

*GM, product of gravitational constant and planetary mass; R, mean radius; $\bar{\rho}$, mean density; I/MR^2, moment-of-inertia factor; $\bar{D}$, mean crustal thickness, assuming a density contrast between crust and mantle of 0.4 g/cm^3.

the GM of Venus from the Pioneer Venus orbit has not been made at the time of writing.

Values of GM and R for other terrestrial bodies (Mercury, Vesta, Io, Europa) are less well known, since there have been only transitory flybys or measurements made at a distance. The GM's for Venus and Mercury are estimated from Mariner 10 trajectories, while those for the Galilean satellites are estimated from Pioneer 10 and 11 and Voyager 1 and 2 trajectories. The masses of the asteroids are very poorly known, being inferred for Ceres, Pallas, and Vesta by mutual perturbations. An attempt is underway to infer asteroidal masses from perturbations of the Viking orbiter.

The radius of Mercury is determined from Earth-based radar and from occultation of the Mariner 10 spacecraft signal. The radii of the Galilean satellites are best determined from analysis of Voyager photography. The radius of Vesta comes from radiometry and polarization measurements.

The gravitational constant, G, is the only component whose determination is the same for all planets. The current adopted value is 6.672 (±0.007) × 10^{-8} dyne cm^2/sec (Cohen and Taylor, 1973), thus leading to an uncertainty of a part in a thousand in the absolute density of any planet. Relative densities can be known much more accurately, however.

Moment of inertia

The moment of inertia (I) is the primary measure of the degree of radial differentiation in a planet

$$I = \frac{A + B + C}{3} = \frac{8\pi}{3} \int_0^R \rho(r) r^4 dr$$

where A and B are moments about equatorial axes, C is the moment about the rotation axis, and ρ is the density. The quantities which can be inferred from observation are the oblateness, $J_2 = [C-(A+B)/2]/MR^2$, and the precessional constant, $H = [C-(A+B)/2]/C$, whence

$$\frac{I}{MR^2} = J_2 \left(\frac{1}{H} - \frac{2}{3}\right)$$

where I/MR^2 is the moment-of-inertia factor (Table 4.2.1). The quantity J_2 can be determined from precessions of satellite orbits or perturbations of flyby trajectories. However, for the precessional constant H to be perceptible requires that the planet be torqued by another solar system body. For all bodies except satellites and the Earth, the sun is the major source of such torque, and the resultant precessions are slow enough to require observations from the planet's surface. Hence measurements of H have been made only for the Earth and Moon. However, if a planet is distended enough (R) and rotating (ω) fast enough so that $\omega^2R^2\rho$ is greater than the long-term yield stress of the planet's material, then I/MR^2 can be estimated from the functional relationship between I/MR^2 and J_2 for a fluid body. However, among the terrestrial bodies, only the Earth and Mars satisfy this criterion. The I/MR^2 given in Table 4.2.1 for Mars is calculated from the hydrostatic relationship using a J_2 from which an estimated nonhydrostatic component has been removed. This estimate is based on the remarkable coincidence of the equatorial ellipticity of Mars's gravity field with the topographic ellipticity, which is dominated by the Tharsis bulge. If it is assumed that the second-degree components of the nonhydrostatic mass distribution are symmetric about the Tharsis axis, then the correction to J_2 can be calculated from J_{22} (Reasenberg, 1977; Kaula, 1979b).

4.2.2 Seismic data

Determination of Earth composition from seismology

Seismic methods determine the velocities of compressional and shear waves, and density, as functions of depth in a planet. These in turn can be used to calculate such elastic parameters as bulk modulus, shear modulus, and Poisson's ratio. All three parameters depend on composition, the crystal structure of minerals, temperature, and pressure. If it can be assumed that seismic discontinuities in the mantle are not due to changes in chemical composition but to phase boundaries, the locations of these phase changes can be used to estimate temperatures at the depths where they occur, compositions, and crystal structures. This requires, of course, laboratory determination of the phase relations in mantle minerals.

Laboratory measurements are also necessary to infer compositions from seismic velocities. Ultrasonic measurements on crustal and mantle minerals, including the effects of temperature and pressure, provide the necessary data for inferring compositions of the crust and upper mantle. Since "elastic" velocities are frequency dependent, some care must be taken in comparing high-frequency laboratory data with low-frequency seismic data. The velocity difference can amount to 8% for shear waves.

Below the upper mantle the usual technique for inferring mantle mineralogy is to use velocity-density systematics, first applied by Birch (1952) and given a theoretical foundation by Anderson (1967). In general, seismic velocities increase with density for a given mean atomic weight, but decrease with the addition of FeO. The bulk modulus is relatively insensitive to Fe-Mg substitution, but is a fair indicator of crystal structure (Anderson, 1977). The simple relationships between velocity, density, and mean atomic weight break down when a change of crystallographic coordination is involved. This occurs, for example, in the quartz-stishovite transformation or the pyroxene-perovskite transformation. Coordination change is only important below some 670 km in the Earth's mantle, and is not pertinent in bodies of Mars's size or smaller. A relatively minor component such as Al_2O_3 can have a large influence on density and seismic velocities, since it can control the stability fields and abundances of phases such as garnet.

For the lower mantle, one must rely principally on data from shock wave experiments and static pressure data using the diamond-anvil pressure cell. In these techniques density as a function of pressure is well determined, but the bulk modulus and pressure derivative of the bulk modulus are obtained by differentiation of the experimental data, and hence are less accurate. Applications of diamond-anvil cell data are limited by the effects of nonhydrostatic pressures in the cell and by inaccuracy of temperature measurements. Nevertheless, diamond-anvil cell experiments have provided useful information about phase transformations, bulk modulus, and the pressure derivative of bulk modulus (especially when combined with ultrasonic data), electrical conductivity, valence states of Fe (via Mössbauer spectroscopy), and bond characteristics (via visible and infrared spectroscopy). High-pressure single-crystal elastic constants have been determined by Brillouin scattering. At lower-mantle pressures, however, the data are still very incomplete with respect to the range of possible compositions and temperatures. This incompleteness of the data and the fact that large corrections or extrapolations must be made in order to compare laboratory and seismic data result in uncertainties which are sufficiently large at present to preclude unambiguous resolution between the competing hypotheses for the composition of the lower mantle. These hypotheses (section 4.5.1) are that the terrestrial lower mantle is the same as the upper mantle, i.e., pyrolite (Ringwood, 1975); or that it is closer to the chondritic composition (Anderson, 1977). The former model can be accommodated if the mantle has denser phase assemblages than are inferred from shock-wave data, and if temperatures are high. The latter model is consistent with the shock-wave data as it stands, and requires a cooler lower mantle. The temperature differences are not great, and both models can probably be accommodated by current uncertainties in the actual temperature of the lower mantle.

Other information must be utilized to test the viability of the alternative hypotheses. The seismic data can determine the radial homogeneity of the various regions of the mantle. The mantle has a uniform velocity gradient between about 220 and 400 km, indicating chemical homogeneity and the absence of phase changes. Between 400 and 650 km the seismic gradients are steep. This region has been interpreted in terms of phase changes in the major mantle minerals: olivine, pyroxene, and garnet. The discontinuity at 650 km is primarily due to a change in crystal structure with depth, but it is not yet clear whether a change in bulk chemistry is involved as well. A phase change would not necessarily prohibit mantle convection or penetration by sinking slabs. An increase in intrinsic density due to a change in composition could inhibit convection and slab penetration (Richter and Parsons, 1975). A compositional boundary at this depth could explain the termination of earthquakes at 700 km.

The seismic data for the lower mantle indicate that it is not radially homogeneous in mineralogy. Seismic travel times, apparent velocities and amplitudes show variations with distance, although these are not nearly as pronounced as in the upper mantle or the lowermost 200 km of the mantle. These inhomogeneities may be due to minor phase changes, in which case they would provide no barrier to convection. They could also be compositional changes, in which case they would separate the mantle into convectively separate regimes (Elsasser *et al.*, 1979). The question of layered vs. whole-mantle convection is discussed in section 9.5.1.

4.2.3 Gravity data

Because of the limited strength of planetary material, geodetic data can be used to estimate the minimum extent to which there has been crustal differentiation. In general, isostasy prevails: topography variations are compensated by density variations at a depth that is small compared to the planet radius. The only way in which such density variations can have been maintained in the long run ($\gtrsim 10^9$ yrs) is compositionally: either by the differentiation of silicic from ultramafic rocks (the main subject of this volume), or by lateral variations in the mantle Fe/Mg ratio.

Given a topographic minimum of depth $-\Delta h_x$ with respect to the mean elevation of the surface of a planet, where isostatic compensation is by variations in the thickness of a crust of constant density, the minimum mean crustal thickness must be $\bar{D}_{min} \approx \Delta h_x \rho_c / \Delta\rho$. Here ρ_c is the crustal density, and $\Delta\rho$ the difference between mantle and crustal densities. For typical values of ρ_c and $\Delta\rho$, $\bar{D}_{min} \approx 7\Delta h_x$. In case the planet is not isostatically compensated, either or both of two devices can be applied to allow for elastic support: (1) Δh_x can be the maximum for some smoothed representation of topography, and (2) the free-air gravity anomaly Δg_x at a given point can be used to calculate the excess load (or deficiency) with respect to isostatic equilibrium, assuming the load to be at the surface.

If the gravity anomaly and topography are known in a region where the crustal thickness has been measured by seismology, then gravimetric and topographic measurements over other parts of the planet can be used to extrapolate a global mean crustal thickness ($\bar{D}$). These estimates of crustal thickness are given in Table 4.2.1.

4.2.4 Other geophysical inputs

Magnetic fields, surface heat flow, electrical conductivity as a function of depth, the chronology of igneous activity, and the nature of surface tectonic features are planetary properties that further constrain interior compositions. They are primarily applicable to estimations of the thermal evolution and present internal thermal profiles of planets, which are discussed in Chapter 9.

Thermal profiles in turn bear on the content of this chapter, since the equations of state relating the interior compositions of model planets to mean densities and other observable properties involve temperature as a variable (sections 4.4–5). Pressure-temperature curves can be expected to be similar in planets of all sizes, since both planetary mass (hence g, and internal pressures) and surface heat flow (hence the temperature gradient with depth) increase with the planets' ratios of volume to surface area.

4.3 COSMOCHEMICAL CONSTRAINTS

4.3.1 The overall composition of the solar system

For the last two decades it has been widely believed that the sun and planets formed contemporaneously from a common mass of interstellar gas and dust that had been drawn together by self-gravitational forces (section 10.2). It is believed that rotation of the contracting mass caused it to take on a spun-out disk-like form. In the center of the disk or **solar nebula**, the infant sun took form; from the whirling nebula, the planets precipitated. As evidence of a common source material for the sun and planets, one can point to the close similarity of abundances of condensable elements in carbonaceous chondritic meteorites (which are surviving samples of primitive, undifferentiated planetary material) and in the atmosphere of the present sun (Fig. 4.3.1). To a first approximation, it is not unreasonable to suppose that all the planets incorporated the condensable elements in the proportions displayed by the sun and chondrites. An up-to-date assessment of the solar system element abundances, based on both meteoritic and solar spectral data, appears in Table 4.3.1.

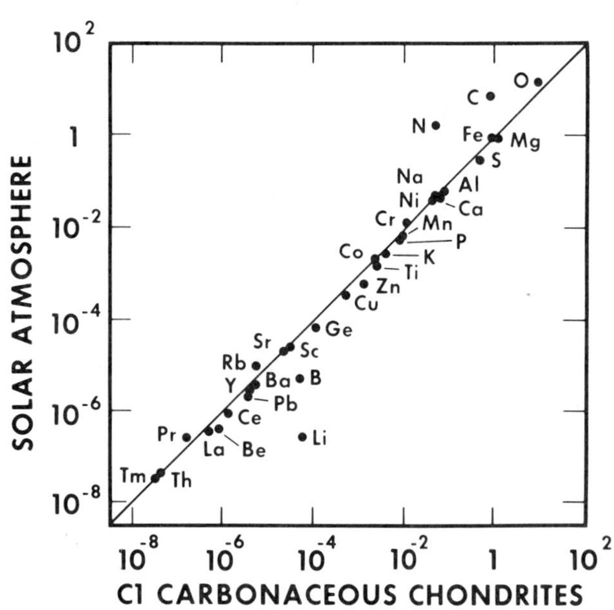

4.3.1 Comparison of abundances of condensable elements in C1 carbonaceous chondrites with abundances in the sun (Holweger, 1977). In both cases abundances are in terms of atoms per Si atom.

Table 4.3.1 Estimated solar system abundances of the elements, atoms per 10^6 Si atoms.[a]

Element		Abundance	Element		Abundance
1	H	2.66×10^{10}	44	Ru	1.9
2	He	1.9×10^9	45	Rh	0.40
3	Li	60	46	Pd	1.3
4	Be	1.2	47	Ag	0.46
5	B	45	48	Cd	1.55
6	C	1.11×10^7	49	In	0.19
7	N	2.31×10^6	50	Sn	3.7
8	O	1.84×10^7	51	Sb	0.31
9	F	780	52	Te	6.5
10	Ne	2.4×10^6	53	I	1.27
11	Na	6.0×10^4	54	Xe	5.84
12	Mg	1.06×10^6	55	Cs	0.39
13	Al	8.5×10^4	56	Ba	4.8
14	Si	1.00×10^6	57	La	0.37
15	P	6500	58	Ce	1.2
16	S	5.0×10^5	59	Pr	0.18
17	Cl	4740	60	Nd	0.79
18	Ar	1.06×10^5	62	Sm	0.24
19	K	3500	63	Eu	0.094
20	Ca	6.25×10^4	64	Gd	0.42
21	Sc	31	65	Tb	0.076
22	Ti	2400	66	Dy	0.37
23	V	254	67	Ho	0.092
24	Cr	1.27×10^4	68	Er	0.23
25	Mn	9300	69	Tm	0.035
26	Fe	9.0×10^5	70	Yb	0.20
27	Co	2200	71	Lu	0.035
28	Ni	4.78×10^4	72	Hf	0.17
29	Cu	540	73	Ta	0.020
30	Zn	1260	74	W	0.30
31	Ga	38	75	Re	0.051
32	Ge	117	76	Os	0.69
33	As	6.2	77	Ir	0.72
34	Se	67	78	Pt	1.41
35	Br	9.2	79	Au	0.21
36	Kr	41.3	80	Hg	0.14
37	Rb	6.1	81	Tl	0.19
38	Sr	22.9	82	Pb	2.6
39	Y	4.8	83	Bi	0.14
40	Zr	12	90	Th[b]	0.036
41	Nb	0.9	92	U[b]	0.010
42	Mo	4.0			

[a] Cameron, 1981.
[b] Present-day abundances of Th and U are given.

4.3.2 Compositional differences among the planets

On the other hand, it is clear from the mean densities of the planets that they cannot be composed of materials having exactly the same proportions of elements. The density contrast between terrestrial (3.9–5.5

g/cm^3) and Jovian (0.7–1.7 g/cm^3) planets is especially striking; it is clear that Jupiter, Saturn, Uranus and Neptune must contain much larger proportions of low-atomic weight elements than Mercury, Venus, Earth and Mars. Most workers ascribe this primarily to a temperature effect in the solar nebula at the time of planetary formation. The formative process, whatever the details of this may have been, was critically influenced by higher temperatures in the inner part of the nebula than in the outer, so that icy compounds of low atomic weight elements were able to join the Jovian planets but not the terrestrial planets. An obvious possible reason for the temperature difference was proximity of the zones of planet-formation to the protosun.

Among the subset of terrestrial planets there are also density differences (Table 4.2.1); not as dramatic as those of terrestrial vs. Jovian planets, but substantial enough to demonstrate unequivocally that there are important differences in bulk chemical composition. Knowledge of a planet's density does not, of course, specify its chemical composition; even when other geophysical parameters are known, multiple compositional models are admissible. Cosmic abundance considerations do suggest, however, that large differences in density among the terrestrial bodies are probably due to variations in their contents of Fe. Smaller density differences can be attributed to, among other things, the oxidation state of the Fe.

To make intelligent guesses as to the compositions of the planets, we open the question of how the planets formed, and what there was about the process that might have left the various planets with other than the solar system abundances of the elements. (The problem is further discussed in section 10.2.) Currently there are several competing views of the chemical and physical processes that led to compositional differences among the terrestrial planets. Factors that might have contributed to differing compositions of the terrestrial planets include, in addition to the temperature distribution in the nebula (already mentioned), the density distribution in the nebula; changes of temperature and density with time; the greater tendency of larger planets to retain volatile elements and compounds; the interplay between time scales of mineral condensation in the

Table 4.3.2a Model compositions (major and heat-producing elements) of Mercury.

	Me1	Me2	Me3	Me4	Me5
Mantle + Crust					
SiO_2	40.8	45.0	47.2	43.5	43.6
TiO_2	0.49	0.37	0.33	-	-
Al_2O_3	9.6	7.2	6 4	4.7	0
Cr_2O_3	-	-	3.3	-	-
MgO	40.5	40.8	33.7	47.7	54.6
FeO	0.05	0.04	3.7	0	0
MnO	-	-	0.06	-	-
CaO	8.6	6.6	5.2	4.1	1.8
Na_2O	0	0	0.08	0	0
H_2O	0	0	0.016	-	-
K(ppm)	0	0	69	0	0
U(ppm)	0.053	0.040	0.034	0.026	0
Th(ppm)	0.19	0.14	0.122	0.12	0
Core					
Fe	94.1	92.4	93.5†	94.5	94.5
Ni	5.9	7.6	5.4	5.5	5.5
S	0	0	0.35	0	0
O	-	-	-	0	0
Relative Masses					
Mantle + crust	61.6	60.9	32.0	35.2	35.2
Core	38.4	39.1	68.0	64.8	64.8

†Also 0.25% Co, 0.57% P.

Me1: Equilibrium Condensation. **Me2**: Equilibrium Condensation modified by use of feeding zones of Weidenschilling. **Me3**: Chondrite Model (Morgan and Anders, 1980). **Me4**: Extreme "dynamically mixed" model: cosmic Al/Mg ratio; sufficient SiO_2 for anorthite, diopside, and forsterite phases in mantle; core/(mantle + crust) mass ratio satisfying mean density (section 4.5.3). **Me5**: Extreme "collisionally differentiated" model, with core/(mantle + crust) mass ratio satisfying mean density (section 4.5.3).

Table 4.3.2b Model compositions (major and heat-producing elements) of Venus.

	V1	V2	V3	V4	V5
Mantle + Crust					
SiO_2	52.9	53.9	49.8	40.4	45.9
TiO_2	0.20	0.20	0.21	0.24	0.3
Al_2O_3	3.8	3.9	4.1	3.4	3.9
Cr_2O_3	-	-	0.87	0.3	0.3
MgO	37.6	38.3	35.5	33.3	38.0
FeO	0.24	2.1	5.4	18.7	8.1
MnO	-	-	0.09	0.2	0.1
CaO	3.6	3.6	3.3	3.4	3.2
Na_2O	1.6	1.5	0.28	0.15	0.1
H_2O	0	0	0.22	-	-
K(ppm)	1442	1318	221	185	129
U(ppm)	0.021	0.021	0.022	0.019	0.021
Th(ppm)	0.073	0.075	0.079	0.066	0.075
Core					
Fe	94.4	84.7	88.6†	78.7	86.2
Ni	5.6	5.2	5.5	6.6	4.8
S	0	10	5.1	4.9	1.0
O	-	-	-	9.8	8.0
Relative Masses					
Mantle + crust	69.8	69.1	68.0	76.4	71.8
Core	30.2	30.9	32.0	23.6	28.2

† Also 0.26% Co, 0.58% P.

V1: Equilibrium Condensation. **V2**: Equilibrium Condensation modified by use of feeding zones of Weidenschilling. **V3**: Chondrite Model (Morgan and Anders, 1980). **V4**: Model of Ringwood, discussed in section 4.3.2. **V5**: Iron-deficient model: same as model E5 for the Earth, with core/(mantle + crust) mass ratio satisfying lower mean density (section 4.5.4).

nebular gases, accretion of the mineral grains into planetesimals, and dissipation of the nebular gases (presumably by early vigorous coronal activity of the protosun); and also dynamic and other mechanical effects.

The major currently-advocated models of planet formation are reviewed below. It is difficult to compare these systematically, because they differ greatly in emphasis and completeness. Some focus on thermodynamics; others on dynamics; others on the demonstrated chemical variability of available planetary samples. The reader is warned that these models are presented uncritically. There exists a voluminous and passionate literature written by their proponents, each critical of the others, which we have not attempted to summarize. For this, the reader is referred to the literature citations that appear below.

Bulk chemical compositions of the terrestrial planets, as predicted by the models to be discussed, are collected in Table 4.3.2. (In all cases U and Th abundances are derived from Al abundances and the U/Al and Th/Al ratios of Table 4.3.1. These refractory elements are assumed to have condensed completely, and to have been incorporated in the planets in their cosmic proportions.) The models of Table 4.3.2 are tested against geophysical constraints in section 4.5. Selected trace element levels in the Earth and Moon, according to one of the models to be discussed, appear in Table 4.3.3.

Equilibrium Condensation Model

It is appropriate to begin with the Equilibrium Condensation Model (Lewis, 1972, 1973, 1974a, 1974b; Barshay and Lewis, 1975, 1976; Goettel and Barshay, 1978), since this is simply an extension of the principle held earlier to account for the compositional difference between the terrestrial and Jovian planets. The equilibrium assemblage of solid compounds that would exist in a system of solar composition (Table 4.3.1) depends only upon the system's temperature and pressure (Lord, 1965; Larimer, 1967; Grossman, 1972). At a pressure of 10^{-3} atm, for example, the system would be completely vaporized at temperatures greater than ~1900°K. Between 1750° and 1900°, the only stable solids would be the minuscule amounts of refractory metals such as W, Re and Os in the system. Between 1500° and 1750°, a series of refractory oxides and silicates of Al, Ca, Mg and Ti would also be stable. The equilibrium assemblage continues to change as regimes of lower temperature are considered, as summarized in Fig. 4.3.2 and Table 4.3.4. Beneath ~1000° K, Na and K are condensed as alkali feldspar, and a portion of the Fe in the system is stable as fayalite and ferrosilite in solution with Mg silicates; the lower the temperature, the greater the equilibrium value of Fe/(Mg + Fe) in olivine and pyroxene. Beneath ~700° K, FeS (troilite) is stable. If equilibrium prevails, reaction with nebular S and O to form troilite and Fe silicates will completely eliminate metallic Fe. Beneath ~300° K, hydrated Mg phyllosilicates are stable in lieu of olivine and pyroxene. Beneath ~200° K, the icy compounds that contribute to the low-density characters of the Jovian planets are stable.

This series of mineral assemblages is commonly referred to as the **condensation sequence**, since it names the succession of equilibrium assemblages that condensation and reaction would produce in a cooling gas of solar composition, but it is important to keep in mind that the concept also applies in situations where condensation played no role. For example, an initially cold mixture of gas and particulate solids having an overall composition equivalent to that of the solar atmosphere, if raised to a particular temperature in the condensation sequence range and allowed to equilibrate, would transform to the mineral assemblage indicated by the condensation sequence after vaporization of some of the solid material.

The Equilibrium Condensation Model as applied to planetary formation postulates that temperatures in the primordial solar nebula were highest in the vicinity of the protosun, and progressively lower at greater distances outward in the nebula. Everywhere in the nebula those solid compounds were present whose stable existence is predicted by the condensation sequence. Thus, at a particular time, the nebula may have been completely gaseous close to the protosun, it may have contained particles of refractory metals somewhat farther

Table 4.3.2c Model compositions (major and heat-producing elements) of Earth.

	E1	E2	E3	E4	E5	E6	E7
Mantle + Crust							
SiO_2	48.5	47.2	48.0	47.9	45.8	45.9	45.7
TiO_2	0.18	0.17	0.27	0.20	0.22	0.3	0.1
Al_2O_3	3.5	3.4	5.2	3.9	4.1	3.9	3.4
Cr_2O_3	-	-	1.1	0.9	0.5	0.3	0.4
MgO	34.5	33.6	34.3	34.1	37.3	38.0	38.4
FeO	8.3	10.6	7.9	8.9	7.6	8.1	8.0
MnO	-	-	0.12	0.14	0.13	0.1	0.1
CaO	3.3	3.2	4.2	3.2	3.7	3.2	3.1
Na_2O	1.5	1.5	0.33	0.25	0.37	0.1	0.4
H_2O	0.13	0.026	0.11	0.21	-	-	0.3
K(ppm)	1324	1289	262	200	260	129	387
U(ppm)	0.019	0.019	0.029	0.021	0.026	0.021	0.026
Th(ppm)	0.067	0.066	0.100	0.076	0.094	0.075	0.108
Core							
Fe	68.9	71.8	89.2	84.5†	*	86.2	77.7
Ni	5.0	5.5	5.7	5.6	-	4.8	6.0
S	26.1	22.7	5.1	9.0	-	1.0	16.4
O	-	-	-	-	-	8.0	-
Relative Masses							
Mantle + crust	69.2	71.9	63.9	67.6	-	68.9	68.3
Core	30.8	28.1	36.1	32.4	-	31.2	31.7

†Also 0.26% Co, 0.59% P.
*Composition and relative mass of core unspecified.
E1: Equilibrium Condensation. **E2**: Equilibrium Condensation modified by use of feeding zones of Weidenschilling. **E3**: Ganapathy and Anders (1974b). **E4**: Morgan and Anders (1980). **E5**: Jagoutz *et al.* (1979), Palme *et al.* (1978). **E6**: Model of Ringwood, 15% component A + 85% component B (section 4.3.2). **E7**: Empirical model: mantle and core from Ringwood (1977a), crust from Smith (1977).

Table 4.3.2d Model compositions (major and heat-producing elements) of Mars.

	Ma1	Ma2	Ma3	Ma4	Ma5	Ma6
Mantle + Crust						
SiO_2	43.6	43.9	41.6	36.8	40.0	39.4
TiO_2	0.16	0.16	0.3	0.2	0.1	0.6
Al_2O_3	3.1	3.2	6.4	3.1	3.1	3.1
Cr_2O_3	-	-	0.6	0.4	0.6	-
MgO	31.0	31.2	29.8	29.9	27.4	32.7
FeO	17.2	16.7	15.8	26.8	24.3	20.8†
MnO	-	-	0.15	0.1	0.2	-
CaO	2.9	3.0	5.2	2.4	2.5	2.7
Na_2O	1.4	1.4	0.1	0.2	0.8	0.5
H_2O	0.47	0.44	0.001	-	0.9	-
K(ppm)	1190	1199	59	218	573	1100
U(ppm)	0.017	0.018	0.033	0.017	0.017	0.017
Th(ppm)	0.060	0.062	0.113	0.060	0.077	0.060
Core						
Fe	59.8	60.4	88.1*	63.7	72.0	-
Ni	5.9	5.8	8.0	8.2	9.3	-
S	34.3	33.8	3.5	9.3	18.6	-
O	-	-	-	18.7	-	-
Relative Masses						
Mantle + crust	74.7	74.3	81.0	81.8	88.1	85.0
Core	25.3	25.7	19.0	18.2	11.9	15.0

†Including 0.8% Fe_2O_3.
*Also 0.4% Co.
Ma1: Equilibrium Condensation. **Ma2**: Equilibrium Condensation modified by use of feeding zones of Weidenschilling. **Ma3**: Morgan and Anders (1979). **Ma4**: Model of Ringwood, 30% component A + 70% component B (section 4.3.2). **Ma5**: Anderson (1972a). **Ma6**: McGetchin and Smyth (1978), pyrolite III + 12.8% FeO.

out from the center of the system, and so forth. When planetary accretion occurred, the solid material incorporated by each planet consisted of those minerals stable within the range of nebular temperatures that corresponded to that planet's position in the solar system. Since temperature varied with distance from the protosun in the nebula, mineral compositions and ultimately planetary densities also would have varied. Note that it is not necessary to postulate a constant temperature and equilibrium mineral assemblage for a particular zone of the nebula during the entire (probably protracted) time period required for complete accretion of a planet in that zone; all that is required is for mineral particles to accrete, at some characteristic temperature, into aggregations large enough to remove most of the mineral grains they incorporate from effective reactive contact with the nebular gases surrounding the aggregations. The compositions of these aggregations would then remain constant during their assembly into planets.

A simple model to examine would be one in which each terrestrial planet accreted from minerals that aggregated within a particular annular "feeding zone" in the nebula. Lewis (1972) assumes these annuli to be concentric (Fig. 4.3.3, top) in his discussion of the Equilibrium Condensation Model. He considers the temperature distribution in the nebula at the time of aggregation to be a free parameter, and adopts an adiabatic P,T gradient (Fig. 4.3.4) that would produce mineral assemblages at the positions of the planets whose densities correspond to the observed densities of the planets. Bulk compositions of Mercury, Venus, Earth and Mars, according to this model, appear in Tables 4.3.2a–d (Me1, V1, E1, Ma1).

These model compositions were calculated by J. A. Wood; they are similar to compositions derived by Goettel (1977, 1978). Grossman's (1972) diagrams showing percentage condensation of the elements as a function of temperature were used in computing the composition of stable solid material at various temperatures. The feeding zone of each planet occupied the range of mean distances and temperatures indicated in Figs. 4.3.3–4. The temperature/mineralogy relationships of Grossman were not followed strictly, since these were calculated at a constant pressure (10^{-3}atm), whereas the adiabat defined by Lewis (1972) in Fig. 4.3.4 spans a range of pressures. The temperature at which a mineral or mineral group begins to be stable (with declining temperature) was taken from Fig. 4.3.4; Grossman's diagrams were then used to establish the degree of condensation at temperatures beneath the beginning of condensation. The composition of each terrestrial planet was taken to be a simple average, over equal increments of mean distance, of the range of compositions of solid material stable within its feeding zone.

Inspection of models Me1, V1, E1, and Ma1 in Tables 4.3.2a–d shows that differences in the abundance and form of chemical combination of condensed Fe would be chiefly responsible for variability in the mass density of the solid material incorporated in the terrestrial planets in the Equilibrium Condensation Model. In Mercury, virtually all Fe is present as metal (the maximum density state). Magnesium silicates, the chief density-lowering component of planets, are only incompletely condensed; much Mg and Si remains in the vapor state. At Venus's position in the nebula, Fe is still essentially all in the metallic state, but Mg silicates are completely condensed, leading to a lower aggregate density of the condensate. At Earth's position the temperature is low enough to allow the reaction of nebular S with condensed Fe, transforming much of it FeS. Addition of a major amount of S (the atomic weight of which is somewhat higher than the mean atomic weight of other condensed elements) to the condensate leads, in this model, to a mean condensate density slightly greater than that of the material from which Venus formed. Finally, the still lower temperature at Mars's position in the nebula permits oxidation of all remaining Fe to Fe silicates. In addition, the Mg silicates are partially hydrated to lower-density compounds in this temperature range.

The computational procedure followed led to abundances of Fe core material in Mercury much smaller than those needed to account for that planet's mean density (Me1 vs. Me4, Me5 in Table 4.3.2a). However, the net Fe metal/Mg silicate ratio is extremely sensitive to the relative amounts of condensate that

Table 4.3.2e Model compositions (major and heat-producing elements) of Earth's Moon (Mo) and the eucrite parent asteroid (Eu).

	Mo1	Mo2	Mo3	Eu1	Eu2	Eu3	Eu4
Mantle + Crust							
SiO_2	43.3	45.6	44.6	39.8	41.4	46.0	39.0
TiO_2	0.4	0.2	0.3	0.1	0.12	0.15	-
Al_2O_3	7.6	4.6	3.7	2.5	2.4	3.1	1.8
Cr_2O_3	0.3	0.4	0.4	-	0.5	0.8	-
MgO	29.1	32.4	33.4	28.5	29.5	32.5	29.7
FeO	13.0	13.0	13.9	26.6	24.2	14.4	28.3
MnO	0.15	0.18	-	-	0.46	0.4	-
CaO	6.1	3.8	3.4	2.1	2.0	2.5	1.2
Na_2O	0.15	0.06	0.05	0.05	0.05	0.07	0.04
H_2O	-	-	-	-	-	-	-
K(ppm)	78	60	100	40	39	62	-
U(ppm)	0.041	0.027	0.028	-	0.013	0.018	-
Th(ppm)	0.145	-	-	-	0.047	0.07	-
Core							
Fe	61.5*	†	†	86.1	81.7	†	†
Ni	32.5	-	-	13.1	12.2	-	-
S	4.0	-	-	-	0.9	-	-
O	-	-	-	-	-	-	-
Relative Masses							
Mantle + crust	98	-	-	87	91.8	-	-
Core	2	-	-	13	8.2	-	-

†Core may be present, but composition and relative mass are unspecified.
*Also 1.5% Co.
Mo1: Morgan *et al.* (1978a), Model 4b. **Mo2**: Wänke *et al.* (1977). **Mo3**: Ringwood (1977b), Ringwood and Kesson (1977a, b). **Eu1**: Morgan *et al.* (1978b). **Eu2**: Hertogen *et al.* (1977). **Eu3**: Dreibus *et al.* (1977), Dreibus and Wänke (1979). **Eu4**: Consolmagno and Drake (1977).

Table 4.3.3 Abundances of selected trace elements in Earth and Moon, per the Chondrite Model.[(a)]

Model	E3	E4	Mo1	Mo[(b)]
C	350	446	11	9.9
N	9.1	4.1	0.10	0.26
F	53	13.5	24	30
P	2150	1920	690	538
Cl	25	20	0.47	0.70
Ga	5.5	3.1	0.61	0.66
Ge	13.8	7.6	1.53	1.66
Br(ppb)	134	106	2.6	3.8
Rb	0.58	0.46	0.26	0.33
Sr	18.2	14.5	41	60
Zr	6.8	7.2	15	22
Ba	5.1	4.0	11.4	16.8
La	0.48	0.38	1.06	1.57
W	0.25	0.18	0.51	0.75
Tl(ppb)	4.9	3.9	0.092	0.136
^{204}Pb(ppb)	1.97	1.58	0.037	0.055

(a) Measured in ppm, unless otherwise indicated.
(b) Model 3a, Ganapathy and Anders (1974b).

equilibrated in various parts of the Mercury feeding zone; patterns of sources other than a simple average of the condensates stable over the mean distance range 0.3–0.45 AU can be postulated which would lead to an aggregate density equal to that of Mercury.

The Equilibrium Condensation Model and other simple condensation models do not rationalize the composition of Earth's Moon. To account for the major difference in uncompressed densities between Earth and Moon (~4.4 vs. 3.3 g/cm^3) presents a stimulating challenge to all theorists considering planetary origin.

Proponents of the Equilibrium Condensation Model recognize its oversimplifications (e.g., Goettel and Barshay, 1978), and submit the model as an end-member in a range of more complex possible histories of planet formation. Major limitations of the model are:

(1) Uncertainties in our knowledge of the solar abundances (perhaps ± 10–20%) have a negligible effect on calculated condensation temperatures, but have a drastic effect on calculated planetary compositions and mineralogies (e.g., Fe/Si, Mg/Si, etc.).

(2) The idea that a planet has a single formation temperature that characterizes the condensates from which the planet accretes is obviously unrealistic. Mixing of materials with different formation temperatures would have been unavoidable. More realistic models would say that mixing occurred, yet perhaps not so much mixing as to obscure the major differences in planetary compositions predicted by the simple model. One exception to this is the volatile (atmophile) elements. Since N, C, the noble gases, etc., constitute such a negligible mass fraction of the terrestrial planets, their abundances are subject to drastic modification by the addition of small amounts of volatile-rich material (e.g., Lewis, 1974c).

(3) The assumption that gas/solid equilibrium prevailed becomes increasingly unrealistic at lower temperatures. The temperature regime at which kinetic constraints would seriously thwart attainment of equilibrium is unknown, but it could well be where the formation of FeS, Fe silicates, or hydrated silicates is supposedly occurring.

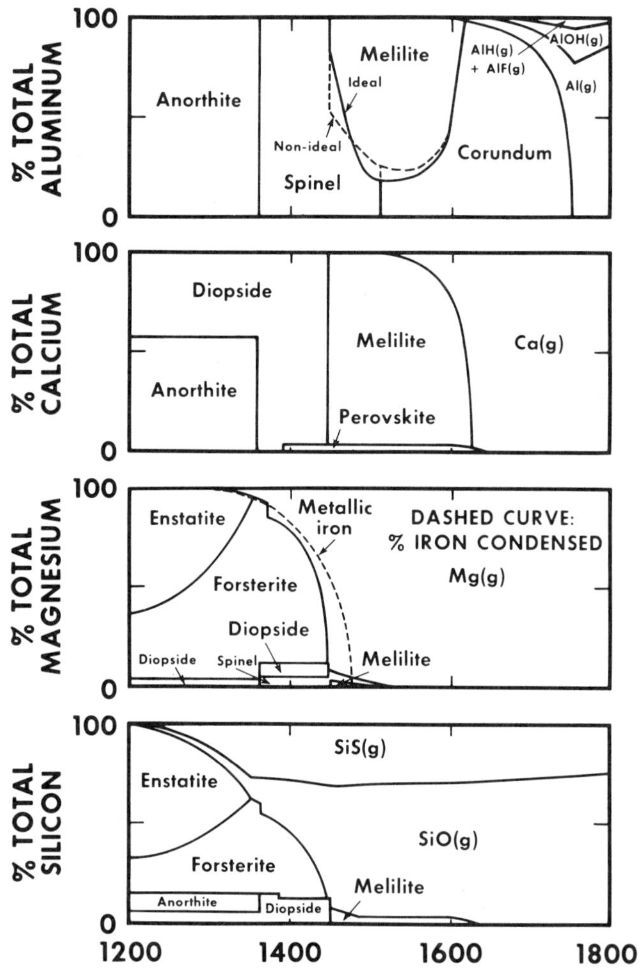

4.3.2 The equilibrium distribution of Al, Ca, Mg, Fe and Si between crystalline and vapor (g) phases, at 10^{-3} atm total system pressure (Grossman, 1972). Mineral formulas are given in Table 4.3.4.

(4) The feeding zones pictured by the Equilibrium Condensation Model (Fig. 4.3.3) would not have remained simple concentric annuli as the planets accreted; the growing masses of the planets, during the later stages of accretion, would tend to perturb remaining unaccreted condensate particles into increasingly eccentric and inclined orbits. Effectively, the annular feeding zones would increase in width and come to overlap. Condensed matter that equilibrated over a wider range of temperatures than in the previous model would contribute to each planet. Weidenschilling (1976) has considered simple dynamical accretion models, and finds that perturbations during accretion would expand the feeding zones to dimensions such as those shown in Fig. 4.3.3 (bottom). If we average the composition of equilibrium condensates from these ranges of mean distances and temperatures, instead of the narrower ranges assumed for the previous model, the planetary compositions Me2, V2, E2 and Ma2 of Tables 4.3.2a–d are obtained. (Actually this procedure exaggerates the differences that expansion of feeding zones might have made, since in the early stages of accretion—during accumulation of the first 50% of the mass of each planet, perhaps—the planets would not have been massive enough to perturb the orbits of condensate particles importantly, and feeding zones would have remained narrow.) This consideration appears to make little difference except in the case of Venus, where substantially more FeO and FeS are introduced from the broadened feeding zones..

The Heterogeneous Accretion Model

There is another way in which interactions between gaseous nebula and solid condensate particles might have controlled the composition of the planets. Equilibrium Condensation postulates that all or most of the material joining a planet equilibrated in a particular,

Table 4.3.4 Sequence of condensation of major minerals from a gas of solar system composition, at 10^{-3} atm total pressure. [(a)]

		Appearance [(b)]	Disappearance [(c)]
Hibonite	$CaAl_{12}O_{19}$	~1770° K	~1500° K
Perovskite	$CaTiO_3$	1647	1393
Melilite	$Ca_2Al_2SiO_7$-$Ca_2Mg_2Si_2O_7$	1625	1450
Spinel	$MgAl_2O_4$	1513	1362
Metallic iron	(Fe,Ni)	1473	
Diopside	$CaMgSi_2O_6$	1450	
Forsterite	Mg_2SiO_4	1444	
Anorthite	$CaAl_2Si_2O_8$	1362	
Enstatite	$MgSiO_3$	1349	
Alkali feldspar	$(Na,K)AlSi_3O_8$	~1000	
Olivines, pyroxenes of intermediate composition	$(Mg,Fe)_2SiO_4$, $(Mg,Fe)SiO_3$	500–1000	
Troilite, pentlandite	FeS, (Fe,Ni)S	700	
Magnetite	Fe_3O_4	405	
Phyllosilicate minerals	e.g., $Mg_3Si_2O_7 \cdot 2H_2O$	420	
Calcite	$CaCO_3$	<<400	
Ices	H_2O,NH_3,CH_4, etc.	<200	

(a) Grossman, 1972; Fuchs *et al.*, 1973.
(b) Temperature at which condensation begins, in a system of declining temperature.
(c) Temperature at which a mineral is consumed by continued reaction with cooling nebular gases, forming phases lower in the condensation sequence. Most of the minerals for which disappearance temperatures are not given are gradually consumed, or diluted by other miscible minerals, over a range of temperatures.

relatively narrow, range of temperatures peculiar to that planet. Suppose instead all the inner solar nebula was at the outset very hot, and completely vaporized. As it cooled, one mineral after another condensed as required to maintain thermodynamic equilibrium in the nebula, and was immediately accreted: i. e., removed from further interaction with the gas. At some point the process was interrupted, presumably by dissipation of the nebula. Then differences in the mean compositions of the planets would result from the place each planet had reached in the condensation sequence when the nebula, with any remaining uncondensed elements it contained, was removed. If temperatures declined with distance outward in the nebula at the time when condensation ended, it could account qualitatively for the density differences in the planets. Mercury might have gotten only through the Ca, Al-silicate and metallic Fe stages of condensation, and partway through the condensation of Mg silicates, when condensation and accretion were terminated; planets farther out in the solar system might have condensed and accreted increasingly large proportions of low-density silicates.

This hypothetical process, termed Heterogeneous Accretion (Turekian and Clark, 1969; Clark *et al.*, 1972; Turekian *et al.*, 1973; Anderson, 1972a, b), might seem only trivially different from the Equilibrium Condensation Model, but in fact there are two important differences. First, since it is pictured that minerals accrete to the growing Earth and other terrestrial planets as soon as they condense, the planets formed would be layered from the outset, having the highest-temperature condensates at their centers, and successively lower-temperature condensates closer to their surfaces. Thus planets would contain metallic cores from the outset as a direct result of accretion, not because of subsequent interior melting and differentiation. The name "Heterogeneous Accretion" refers to this property. Second, the condensation sequence applicable to this model differs from the scheme presented in Fig. 4.3.2. The latter might more appropriately be called the condensation/reaction sequence, since both processes are involved. For example, metallic Fe condenses in the temperature range 1470–1300° K according to the condensation/reaction sequence. If the Fe grains remain dispersed in the nebula and in reactive contact with it during subsequent cooling, they should react in

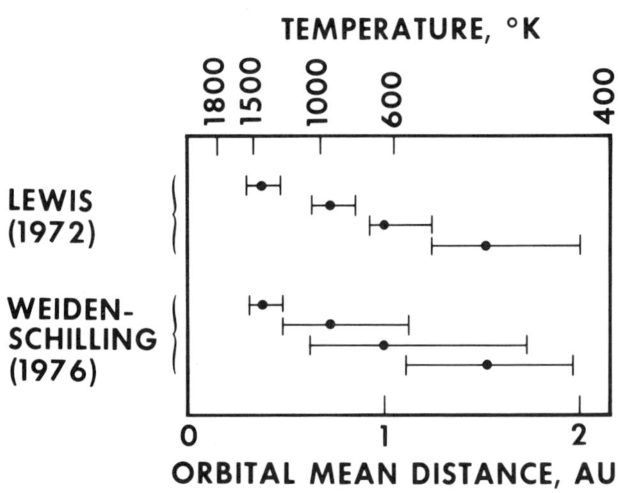

4.3.3 Widths of annular feeding zones that contributed solid material to (left to right) Mercury, Venus, Earth and Mars; according to the models of Lewis (1972) and Weidenschilling (1976). Nebular temperatures at which the solids equilibrate in the model of Lewis are shown.

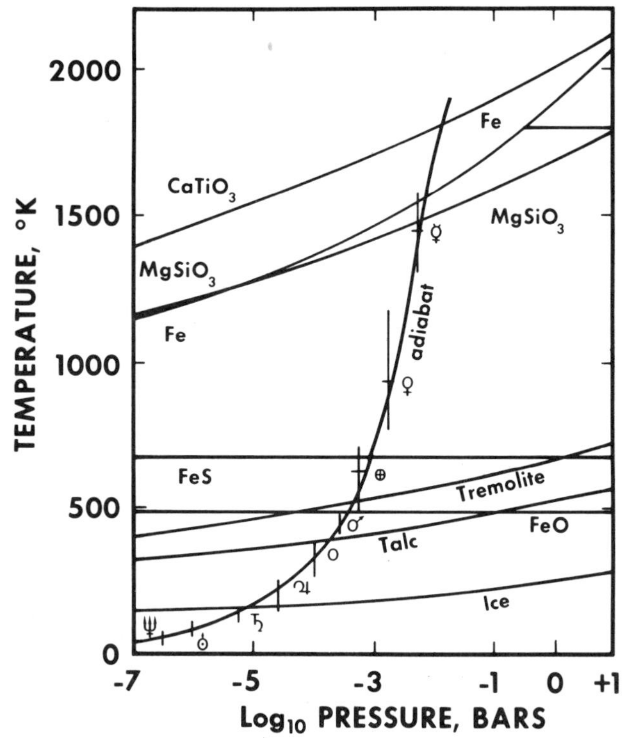

4.3.4 A plot showing P, T conditions under which various representative minerals begin to condense ($CaTiO_3$, Fe, $MgSiO_3$, ice) or begin to be produced by reaction of a preexisting mineral with nebular gases (FeS, tremolite, talc) in a cooling gaseous system of cosmic composition (Lewis, 1972). The diagram is a generalization of Fig. 4.3.2: it shows a range of pressures, but an incomplete set of minerals. An adiabat is shown; if equilibrium condensation and accretion occurred in the P, T ranges accompanying the planetary symbols (top to bottom: Mercury, Venus, Earth, Mars, asteroids, Jupiter, Saturn, Uranus, Neptune), differences in planetary densities would be largely accounted for.

the temperature range 700–500° K to form FeS, until all the S in the nebula is exhausted. The condensation/reaction sequence involves several such transactions. These are prohibited in the Heterogeneous Accretion Model, strictly interpreted. The Fe grains mentioned above would be removed from reactive contact with the nebula immediately upon condensation, and there would be no way for FeS to form.

Strict application of the Heterogeneous Accretion principle leads to extreme and impossible results, such as no FeS or FeO in the Earth's interior since all Fe would have been buried in the core of the Earth, surrounded by a mantle of pure Mg silicates, before the nebula cooled to the temperatures at which it could have reacted with Fe to form FeS or Fe silicates. The proponents of this model do not advocate undeviating Heterogeneous Accretion, however, only a tendency in this direction; thus, for example, part of the metallic Fe condensed is held to have remained unaccreted when the cores of the planets came together, so it was eligible for further reactions in the nebula. Since the proportions of components that followed these differing paths before joining each planet are undefined, there is no way to uniquely derive planetary compositions according to the principle of Heterogeneous Accretion, and none are given in Table 4.3.2.

Meteorites as clues to planet formation: the Chondrite Model

Another way to deal with the concept of condensation and/or equilibration of solids in the solar nebula is to accept the proposition that actual samples of these solids are available to us for inspection, making it unnecessary to rely on abstract conceptions of the nature of the solid condensates that accreted to form the planets. The chondritic meteorites are these surviving samples of primitive planetary material. When examined in detail, many of them are found to consist not of solids that equilibrated at a particular temperature or in a narrow range of temperatures, as the strictly applied Equilibrium Condensation Model anticipates, but of two structural components—fine-grained matrix and coarse-grained chondrules and metal grains—that have experienced very different temperature histories (Wood, 1962). These components are mingled on a scale of <1 mm. The matrix appears to have equilibrated with the nebula and condensed volatiles down to low temperatures, aided by its fine grain size ($\sim 10^{-4}$–10^{-6}cm) and slow cooling rate. Chondrules and metal particles, having been quenched rapidly (seconds to minutes) after their formation at >1500° K, and being of coarser size (0.01–0.1 cm), are generally volatile-poor.

Originally, the Chondrite Model tried to explain the planets, too, as analogous two-component mixtures of volatile-rich and volatile-poor materials (Wood, 1962; Anders, 1964; Larimer and Anders, 1967), similar to the present Canberra Model (described below), but allowing for some compositional variation of the two components with heliocentric distance. Subsequent studies of the bulk composition of the several chondrite subclasses (Larimer and Anders, 1967, 1970) and the condensation sequence of the elements from a solar gas (Larimer, 1967; Grossman, 1972) have led Anders and co-workers to a more elaborate scheme (Larimer and Anders, 1970; Anders, 1971, 1977; Krähenbühl *et al.*, 1973; Ganapathy and Anders, 1974a, b; Morgan *et al.*, 1978a, b; Morgan and Anders, 1979, 1980). It involves sequential condensation of three main condensates—refractories, metal, and magnesium silicates—which are modified by reaction with nebular volatiles on cooling, and are fractionated from each other by three processes before accretion:

1. Gain or loss of early condensate, by settling to the median plane of the nebula.

Table 4.3.5 The seven chemical components in planets, according to the Chondrite Model.

Component	Condensation Temperature (° K)	Elements
Early condensate	1800–1400	Ca, Al, Ti, and 34 other refractories
Remelted silicate	1400–1300	Mg, Si, Li, (Fe)
Unremelted silicate	1400–600	Mg, Si, Li, (Fe) + alkalis, Mn, F
Remelted metal	1400–1300	Fe, Co, Ni, Pd, Au, Cr, P, As
Unremelted metal	1400–600	Fe, Co, Ni, Pd, Au, Cr, P, As + Cu, Ag, Ga, Ge, Sn, Sb
Troilite	700–600	Fe, S, Se, Te, Zn
Volatile-rich component	<600	H, C, N, and 13 other volatiles

2. Gain or loss of metal relative to silicate, by ferromagnetic attraction and faster settling.
3. Chondrule formation, by remelting of the micron-sized, primary condensate (by collisions or electric discharges).

Proponents of the Chondrite Model of planet formation suggest that the same fractionation processes and the same physical components that rationalize differences in the chondrite subclasses are also responsible for differences in the compositions and mass densities of the terrestrial planets. They define seven chemical components (Table 4.3.5), most of them identifiable with visible physical components, which combined in various proportions account for compositions of the chondrite subclasses (Anders, 1971). Each component corresponds to a range of temperatures in the condensation sequence, and embraces a group of elements that are believed to remain relatively coherent in cosmochemical fractionations. Because some of these components are interdependent, the abundances of only four index elements (U, Fe, K and Tl) are needed to determine the proportions of the seven components.

Given the overall abundance of one of these index elements in a planet (or chondrite subclass), proponents of the Chondrite Model derive the amount of that particular component in the planet, and the abundances of all the other elements it embraces: e.g., if the abundance of U is estimated, the abundances of 36 other refractory elements can be derived. This approach appears to work for most of the 83 elements, but is an oversimplification for about 6 elements at the fringes of the volatility groups (Li, Na, Cs, Mn, Zn and Te), which are depleted by more variable factors.

The index elements are estimated from geophysical or geochemical data: Fe from the bulk density of a planet; U from the heat flow or, preferably, thermal history calculations that include the effects of initial heat; and K and Tl from K/U and Tl/U ratios in surface rocks (Ganapathy and Anders, 1974b; Morgan *et al.*, 1978a; Morgan and Anders, 1979).

Compositions derived by this means for the bulk Earth and Moon are shown in Fig. 4.3.5, and appear as models E3 and E4 in Table 4.3.2c, and Mo1 in Table 4.3.2e. Compositions have also been derived for Mercury, Venus, Mars and the eucrite parent body. Only the major elements are listed here; complete tables for all elements are given in the original papers. Abundances of a selection of trace elements in the Earth and Moon models appear in Table 4.3.3. Chemical fractionation factors derived for several planets and meteorite classes are summarized in Fig. 4.3.6.

The Earth tends to fall within the chondritic range in this figure, missing it only for Fe content (and Fe/Si). This is consistent with previous arguments for a "chondritic earth." However, according to Fig. 4.3.6, H-chondrites are not a very good match to the Earth, and neither are other known chondrite subclasses. One can call the Earth "chondritic" in a general way, but it cannot be matched up with a specific chondrite class.

The Moon and eucrite parent body also show fractionations like those displayed by chondrites, but of a more extreme degree: 8 of the 14 points plotted in Fig. 4.3.6 fall outside the chondritic range. Perhaps this reflects limited sampling of the asteroid belt, which causes us to underestimate the chondritic range. But perhaps the difference is fundamental, and tells us that small magmatically differentiated bodies tend to be more strongly fractionated (by nebular processes, before accretion) than are small undifferentiated bodies.

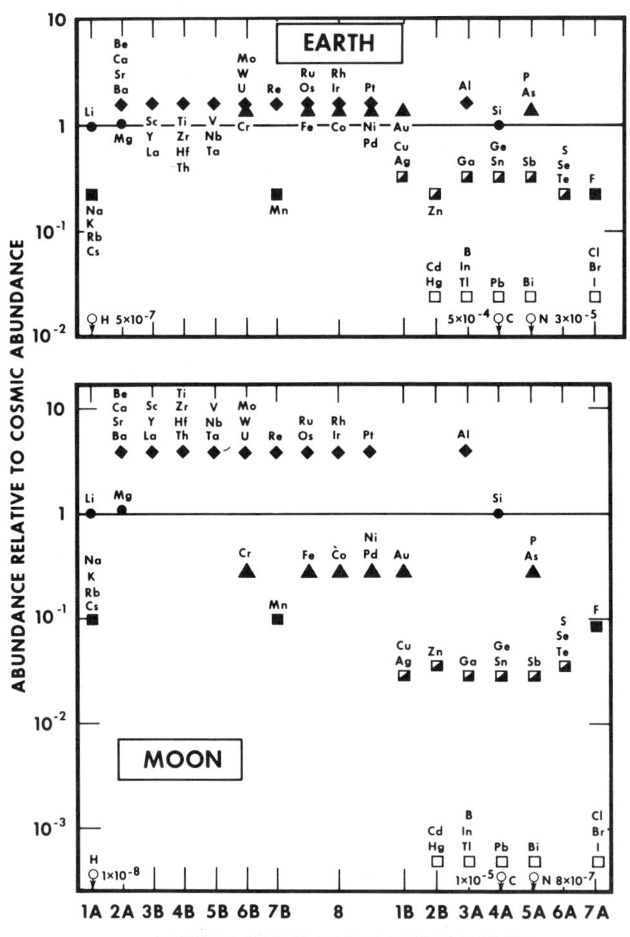

4.3.5 Model compositions of the Earth and Moon (Anders, 1977), based on the assumption that each chondritic component contains its cosmic complement of trace elements. Chemical components are defined in Table 4.3.5.

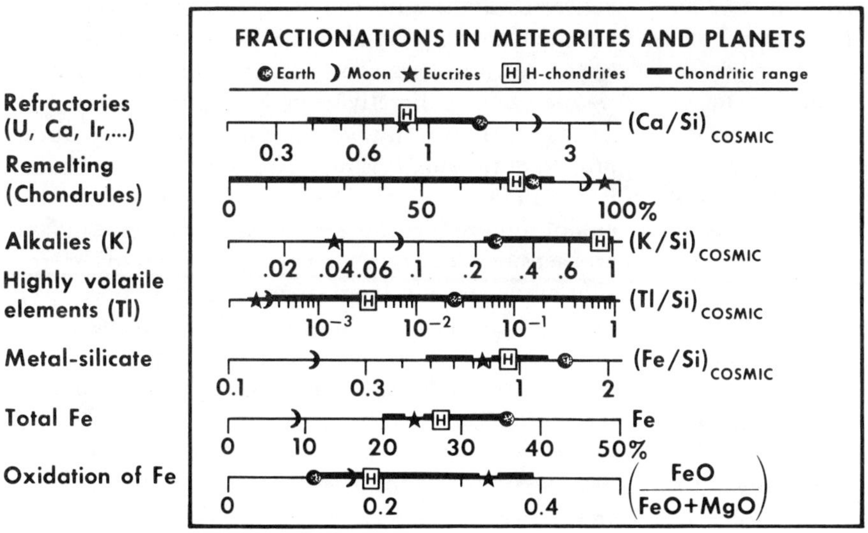

4.3.6 Degrees of fractionation of chemical components recognized by Anders and co-workers; in the Earth, Moon, eucrite parent body, and H-group chondrites.

The Chondrite Model assumes that highly volatile elements were added toward the end of accretion as a layer of volatile-rich material (some mix of C-chondrites). Though this involves the dubious tacit assumption that the previously accreted material was wholly devoid of highly volatile elements, the model nonetheless appears to have been successful in predicting the abundances of such elements on the Moon and Earth, using Tl as the index element.

When Tl is not known (e. g., for Mars), ^{36}Ar has been used as a stand-in on the grounds that the $^{36}Ar/Tl$ ratio in chondrites is constant within a factor of 10 (Anders and Owen, 1977; Morgan and Anders, 1979). However, it has become clear that no single volatile-rich material will fit all inner planets; Venus, with its high ^{36}Ar content, needs material of very high noble-gas content and high $^{36}Ar/Tl$ ratio. Among known materials, the C2 inclusion from the Krymka LL3 chondrite seems representative of the best source of volatiles for Venus (Lewis *et al.*, 1979; Morgan and Anders, 1979). As little as 0.1 $M_♀$ of such material would account for the ^{36}Ar on Venus. However, ^{36}Ar is not a perfect stand-in for Tl, as the 10-fold variation in $^{36}Ar/Tl$ ratio introduces a corresponding uncertainty in the abundances of 12 other volatiles that scale with Tl. Indeed, the 16 highly volatile elements actually comprise 3 distinct subgroups of markedly different condensation chemistry: trace metals (Tl, Bi, Hg, etc.), major elements (H, C, N), and noble gases. Thus an index element from one subgroup will give less accurate predictions for the other two subgroups (Morgan and Anders, 1979).

An interesting trend that seems to emerge from this work (Anders and Owen, 1977; Morgan and Anders, 1979) is that small, magmatically differentiated planets (Moon, Mars, eucrite parent body, but not the shergottite parent body) tend to be much poorer in highly volatile elements than are large differentiated planets (Earth, Venus). This is apparent from Fig. 4.3.7, which shows the abundances of the four index elements in five planets (Morgan and Anders, 1979).

The Chondrite Model unfortunately cannot predict the abundances of some major rock-forming elements (Mg, Ca, Al) to the accuracy needed by petrologists (i.e., ~30% vs. ~10% accuracy). Magnesium is virtually unconstrained: in the past, proponents of the Chondrite Model have simply assumed the cosmic

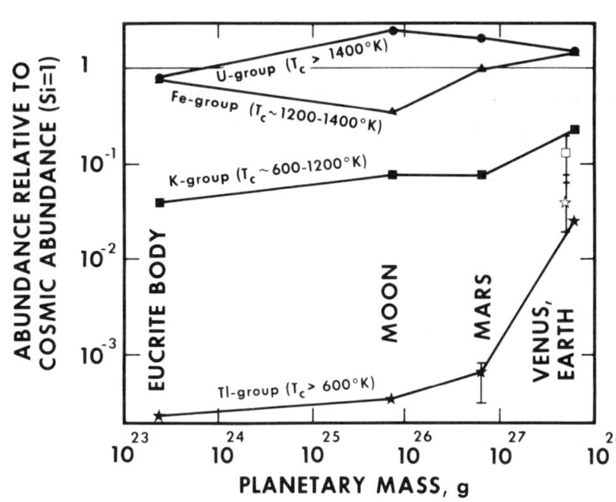

4.3.7 Abundances of the four index elements recognized by Anders and co-workers, in five inner solar system planets. From Morgan and Anders (1979).

Mg/Si ratio, but could not exclude ratios up to 30% lower, such as are found in ordinary and enstatite chondrites. Calcium, Al and Ti are coupled to each other, as well as to REE and the index element U. But since the U values of even the Earth and Moon are still rather uncertain, the abundances of Ca, Al and Ti are correspondingly uncertain. They can be constrained by experimental petrology, taking advantage of the fact that Ca, Al, Ti are coupled and using suitable constraints for Mg/Si, MgO/(MgO + FeO), etc. Studies by Walker *et al.* (1975), Longhi (1977), and Weill and McKay (1975) show that the Moon Model 2a of Ganapathy and Anders (1974a) gives reasonable results, after some fine-tuning. The most recent estimate of the lunar heat-flow value, and consequently the inferred abundance of U in the Moon (section 9.5.2), has effectively re-instated this model, with its low content of refractories (Morgan *et al.*, 1978a).

The oxidation state of a planet, as expressed by MgO/(MgO + FeO), is also not well constrained by the model. Early attempts to constrain it by the MnO/FeO ratio had to be abandoned, because the volatilization of MnO in chondrule formation is incomplete and variable. If FeO were controlled by the temperature-sensitive equilibrium (Larimer, 1968; Grossman, 1972):

$$2MgSiO_3 + 2Fe + 2H_2O \rightleftharpoons Fe_2SiO_4 + Mg_2SiO_4 + 2H_2$$

in the nebula, then the FeO content of a planet could be predicted from its heliocentric distance *a*, if the relation

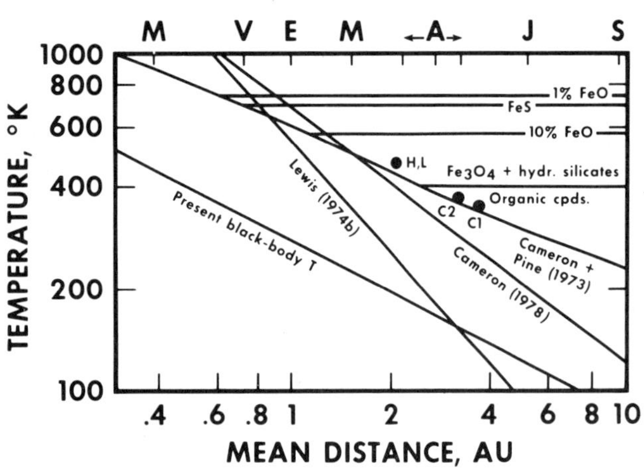

4.3.8 Temperature distribution in the solar nebula at the time of planet formation, according to several models (sloping curves). Horizontal lines show the oxidation state of Fe as a function of temperature (Grossman, 1972), and hence of heliocentric distance, for each model. 1%, 10% FeO refer to molar FeO content of mafic silicates.

between T and *a* were known. However, this relation is not well known. Figure 4.3.8 illustrates several suggested relationships. The Cameron-Pine (1973) curve was determined directly from the authors' graphs, with allowances for the inward movement of material during growth of the sun. It is much less steep ($T \propto a^{-0.5}$) than Lewis's (1974b) curve, based on his empirical fit of equilibrium condensate assemblages to observed planetary densities. (Lewis mistakenly states that the Cameron-Pine curve, like his, has a slope of −1.1.) A more recent model of Cameron (1978) shows an intermediate slope of −0.75.

A reasonable requirement of any such curve is that the 400° K isotherm, corresponding to the transition from Ol-Px-metal to carbonaceous chondrite mineralogy, pass through the asteroid belt, since the two main classes of asteroids do appear from spectral photometry studies to have such mineralogies. On this basis, the Cameron and Pine (1973) curve fits best. It also allows the principal chondrite types (formation temperatures, 350–560° K; Grossman and Larimer, 1974) to come from the asteroid belt. In terms of the simpler Equilibrium Condensation Model, this curve predicts somewhat less than 10 mol % FeO at Earth, 2–3% at Venus, marginal formation of FeS on Venus but not Mercury, and some 15–20% FeO but no hydrated silicates at Mars. Actually, FeO contents in the inner solar system seem even less variable than this (Anders, 1971; Morgan and Anders, 1980), which implies either a less steep temperature gradient, incomplete equilibration with the gas, or considerable mixing of Fe-rich materials.

Of course the late-accreted volatile-rich layer, being highly oxidized, would alter the redox state of the planets. For the Earth, this layer may have been quite massive—if it is to account for all the Earth's ^{36}Ar and H, it would have to comprise 0.063 $M_{\oplus}$ of C3V material (Anders and Owen, 1977). A similar amount, 0.1 $M_{♀}$ of Krymka C2 xenolith material, is required for Venus, judging from the abundances of ^{36}Ar, C and N on the two planets. Since FeO/(FeO + MgO) is more than 0.3 in C3V and C2 material, this could significantly raise the redox state of the planet.

The model can also be applied to unsampled or incompletely sampled planets. Iron is generally known from the density, U and K can be estimated from the K/U ratio, with the help of the trends in Fig. 4.3.7 or thermal history calculations, Tl can be replaced (at some risk) by C, N, or ^{36}Ar, and FeO/(FeO + MgO) can be estimated from some relation between T and *a*, as discussed above (Anders and Owen, 1977; Morgan and Anders, 1979, 1980). Tentative Mercury and Venus models (Me3, Table 4.3.2a; V3, Table 4.3.2b) have been calculated in this way (Morgan and Anders, 1980).

Planetary mixing models involving smaller numbers of components: the Mainz Model

Several cosmochemists accept the proposition that the bulk compositions of planets can be estimated from the abundance ratios of certain critical correlated elements, measured in admittedly differentiated crustal samples from the planets; but they disagree with the approach of Anders and co-workers (above) on the number of discrete components or element groups, the cleanness with which they can be discriminated, the element pairs most useful in tracing them, and/or on equating the element groups that shaped the compositions of planets with those that are demonstrably variable in the chondrite subclasses. (*Correlated elements* have very similar patterns of geochemical behavior, even though they may be strongly fractionated during nebular condensation. Thus abundance patterns established during condensation tend not to be disturbed by subsequent planetary melting and igneous fractionation.)

From the point of view of H. Wänke and co-workers, there is strong evidence that the most refractory elements condensed from the solar nebula as a group, unfractionated from one another, at temperatures above the condensation temperature of the Mg-silicates (Larimer and Anders, 1970; Grossman, 1972, 1973; Wänke *et al.*, 1974a). Hence, the lithophile refractory elements can be treated as one discrete component in each planet. The lithophile refractories are Al, Ca, Ti, Be, Sc, V, Sr, Y, Zr, Nb, (Mo), Ba, REE, Hf, Ta, (W), Th and U. From the observed abundance ratios of correlated elements in samples from the Moon, Earth, and achondrites, there is excellent proof that these elements are present in the same ratios as in C1 chondrites. The abundance of the refractory elements in a given planet can be calculated from the inferred abundance of their heat producing members U and Th, provided reliable data on the global heat flux are available (Ganapathy and Anders, 1974b). (This assumes, of course, that surface heat flow represents the current rate of heat production; see section 9.4.1 for discussion.)

Other element groups are less perfectly coherent in chondrites as well as planets, as also noted by the Chicago group. For example, of the six moderately volatile lithophiles in Table 4.3.5, five (Na, K, Rb, Cs, Mn) show appreciable variations, with Mn being less depleted than Na and K (Larimer and Anders, 1967; Laul *et al.*, 1973; Ganapathy and Anders, 1974b; Dreibus *et al.*, 1977).

As an example of the variability of abundances within a group as well as between groups, the refractory elements are depleted in both ordinary (H, L) and enstatite (E) chondrites relative to C1 chondrites (Larimer and Anders, 1970). Normalized to Si, the depletion factors relative to C1 chondrites for the refractory elements in ordinary and enstatite chondrites are 0.81 and 0.52, respectively. Element ratios in these four chondrite subclasses are shown in Table 4.3.6. The group behavior of the refractory elements is evident from the near-constancy of the Ca/Al ratio (Ahrens and Von Michaelis, 1969). The depletion of refractories in ordinary and enstatite chondrites can be seen in the decrease of Al/Si in these chondrites. However, the Mg/Si ratio varies in the same direction as the Al/Si ratio, but to a smaller extent, indicating that Mg is more refractory than Si but less than Al, as is to be expected from the calculated condensation temperature ranges (Larimer and Anders, 1970; Grossman, 1972; see Fig. 4.3.2).

Since the Mg/Si ratio (together with Fe^0/FeO) determines the ratio of normative olivine to pyroxene in the planets, it plays a crucial role in establishing planetary mantle compositions. If the chondritic pattern of element abundance variations cited above can be extrapolated to the terrestrial planets, Mg/Si ratios higher than the C1 value (which have been postulated for the Earth by a number of authors, and which are necessary for the presence of large amounts of normative olivine) should be accompanied by increases in Al/Si, and in refractory element abundances generally, by factors about twice as great. Due to the fact that the solid-liquid distribution coefficient of the FeO/MnO ratio is slightly above 1 for olivine and slightly below 1 for pyroxene, this ratio stays nearly constant during magmatic fractionations (Laul *et al.*, 1972a; Wänke *et al.*, 1973). If we neglect the possibility of major Mn depletion during planet formation, the FeO/MnO ratio in samples from a planet can be used to determine the planet's Fe^0/FeO ratio if the total Fe abundance is known or can be assumed to be chondritic. The FeO/MnO ratio is ~100 in the fully oxidized C1 chondrites, but only ~35 in eucrites, howardites and diogenites. This appears to require that less than half of the total Fe in the eucrite parent body be present in the oxidized form (as is also the case for Earth).

Table 4.3.6 Element ratios (by weight) in four subtypes of chondritic meteorites.

	C1	H	L	E6
Al/Si	0.080	0.063	0.063	0.044
Mg/Si	0.91	0.80	0.79	0.71
Ca/Al	1.10	1.11	1.08	1.06
Cr/Mg	0.025	0.025	0.026	0.024

The correlated elements W and La are of great importance, as they provide a direct link between metal and silicate abundances. When metal and silicates coexist (as during the hypothetical early stages of planet formation) W enters the metallic phase preferentially, and thus follows Fe^0 and Ni during core formation. The oxidized portion of W behaves as a strongly incompatible element and accompanies La during magmatic fractionations (Wänke *et al.*, 1974b). Since W and La both belong to the group of refractory elements that does not fractionate during condensation and accretion, the observed La/W ratio together with the metal-silicate partition coefficient of W can be used to calculate the amount of metallic iron which was once in equilibrium with the silicates (Rammensee and Wänke, 1977).

Dreibus *et al.* (1976) found evidence that Li condenses with the Mg silicates. However, a comparison of Li with refractory incompatible elements to estimate the abundance of refractory elements relative to Mg silicates leads to an overestimation, due to the fact that Li, unlike the truly incompatible elements, enters olivine and orthopyroxene to some extent. The refractory-nonrefractory pair V-Cr may be more useful for this purpose (Dreibus *et al.*, 1978).

The pair K-U traditionally has been used for the estimation of K relative to the refractory elements. Since it has been proven that K follows La more closely than U, however, it is preferable to use the K/La ratio for this purpose (Wänke *et al.*, 1974b).

For the estimation of bulk planetary compositions, Wänke and co-workers argue that only the refractory lithophile elements are present in all planetary objects in the same relative abundances as in C1 chondrites. For the other elements they have used observed element correlations, and attached great importance to the analysis of samples that are as nearly primitive as possible, i. e., which have been minimally affected by igneous differentiation (Palme *et al.*, 1978). The Al/Sc ratio has been used as a criterion to recognize the most primitive planetary samples, since the ratio of these two refractory elements should be constant in the bulk compositions of all planetary objects. Separation of olivine will change this ratio only slightly, but it is affected strongly by partial melting or late-stage crystal fractionation.

Such an approach does not permit the prediction of compositions of unsampled planets, however, as it is based on the compositions of a variety of samples from each planet under consideration. The estimated compositions of the eucrite parent body (EPB), the Moon, and the Earth obtained by the Mainz group are listed in the columns Eu3, Mo2 and E5 of Tables 4.3.2e and 4.3.2c. The particular approach used for each of these three objects was as follows.

Jérome and Goles (1971) and McCarthy *et al.* (1972) proposed a mixing hypothesis for the howardites (achondritic meteorites) according to which the howardites are mechanical mixtures of eucrites and diogenites. Dreibus *et al.* (1977) confirmed the validity of this mixing hypothesis for the eucrite parent body through mixing computations. Dreibus *et al.* were then able to use eucrites (feldspar and clinopyroxene), diogenites (orthopyroxene), and the olivine phase of pallasites as the three building blocks of the silicate part of the eucrite parent body (EPB). The proportions of the three end members were fixed in such a way that the resulting

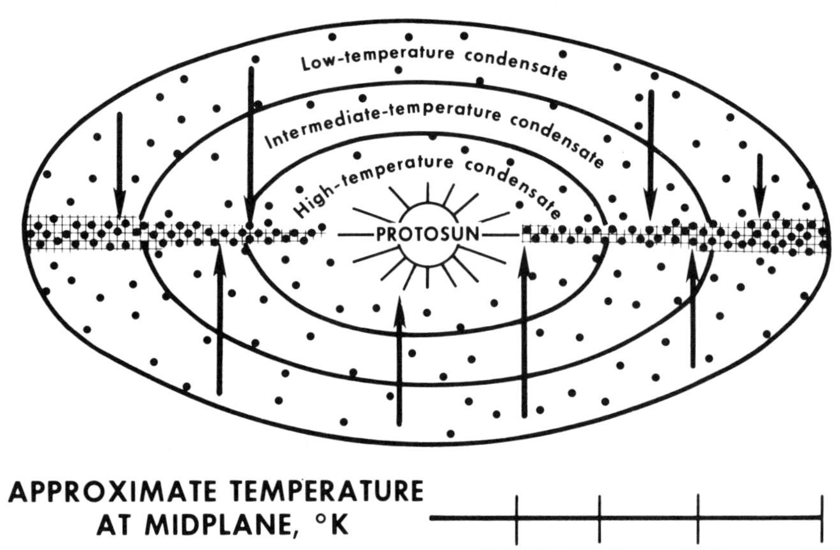

4.3.9 An impression of the protosun with its "cocoon nebula" prior to accretion and sedimentation of planetesimals into the ecliptic plane (Ringwood, 1977a).

bulk composition has chondritic Al/Sc and Mg/Si ratios. As their modelled EPB has almost chondritic abundances of all except the more volatile elements, this approach seems to be well justified.

To estimate the composition of the Moon, Wänke *et al.* (1977) relied on compositions of the siderophile-rich highland breccias, which they believe contain almost unfractionated primary material, along with other components.

In the case of the Earth, the Mainz group (Palme *et al.*, 1978; Jagoutz *et al.*, 1979) used the compositions of least-fractionated ultramafic xenoliths in basalts, the assumption of chondritic abundance ratios for all refractory elements, and measurements of certain correlated element pairs (K/La, K/U, etc.) to model the composition of the Earth's mantle.

The Canberra Model of planet formation

A. E. Ringwood's model for planet formation, which also involves mixtures of materials that equilibrated with the nebula in different temperature ranges, operates in an astrophysical framework similar to those proposed by Cameron (1963), Safronov (1972), Goldreich and Ward (1973), and others. According to these, the primitive solar nebula resembled an infrared protostar or **cocoon star** in which the protosun was surrounded by a shell of dust and gas with temperature decreasing radially outwards. Ringwood (1977a) proposes that dust particles began to coalesce to form small (1–100 cm) planetesimals within the shells pictured (Fig. 4.3.9). The planetesimals were henceforth chemically isolated from the gas phase. Owing to gas-solid dissipative processes combined with self-gravitation perpendicular to the ecliptic plane, the planetesimals sank rapidly (100–1000 yr) to the ecliptic plane. Figure 4.3.9 suggests that the planetesimals reaching the ecliptic plane would have reached equilibrium with surrounding gas over a wide range of temperatures. The condensates would therefore display a high degree of chemical heterogeneity, varying from highly oxidized, volatile-rich planetesimals, resembling C1 chondrites and formed in cool regions of the nebula, far away from the ecliptic plane, to highly-reduced, devolatilized, silicate-iron planetesimals formed at higher temperatures closer to the ecliptic plane and to the sun. The overall ratio of low-temperature to high-temperature condensates tends to increase with distance from the sun, and this may be reflected in the ultimate compositions of planets.

Accretion of planets in the ecliptic plane may have followed the Safronov (1972) model (section 10.2.3). Once sizeable planetary nuclei formed, gravitational perturbations caused strong mixing of planetesimals originally derived from different radial distances in the nebula. The material accreting in any single planet was therefore highly heterogeneous on a small scale, having been subject to widely varying degrees of reduction and devolatilization, but was well-mixed on a large scale. Moreover, the energetic processes of accretion would have caused further intimate mixing, so it is likely that the bulk of the planet accreted homogeneously.

Ringwood (1977a) considers that this homogeneous accretion model is strongly supported by detailed studies of the bulk composition of the Earth. He argues that the distribution of crystal-chemically compatible elements such as (Mg^{2+}, Fe^{2+}, Ni^{2+}, Zn^{2+}), (Al^{3+}, Ga^{3+}), and (Si^{4+},Ge^{4+}) appears to be extremely uniform throughout the mantle and thus apparently represents an original feature of the Earth immediately following accretion and core-segregation. Yet these elements would have displayed vastly different cosmochemical behaviors in the nebula, because of their greatly contrasting lithophile, siderophile and volatile characteristics. Thus it appears to Ringwood that the Earth accreted homogeneously from a mixture of siderophile, lithophile, volatile and involatile components.

The abundance ratios of major elements in the Earth, as estimated by Ringwood, are shown in Table 4.3.7. The relative abundances of the major elements (Mg, Fe, Si, Ca, Al) are very similar to their relative abundances in C1 chondrites and the sun. This suggests that the formation of the Earth was not accompanied by any major fractionation of metallic iron from silicates, or of high-temperature condensates (Ca, Al, Ti, etc.) from intermediate-temperature condensates. Such fractionation is also not necessary to explain the compositions of Mars and Venus. The high density of Mercury, however, implies a substantial degree of enrichment of Fe. Apparently, this planet represents a special case of

Table 4.3.7 Comparison of major element ratios (atomic) in Earth with corresponding ratios for C1 carbonaceous chondrites.[a]

	C1 Chondrites	Earth [b]
Si/Mg	0.95	0.80
Fe/Mg [c]	0.90	0.92
Al/Mg	0.08	0.07
Ca/Mg	0.07	0.06

(a) Average of Orgueil and Ivuna (Wiik, 1956). From Ringwood, 1977a.
(b) Core assumed to contain 10 wt.% of oxygen.
(c) Ni has been included with Fe.

metal/silicate fractionation arising from differences in condensation temperature of these components and Mercury's proximity to the sun, as suggested by a number of authors. The Moon, on the other hand, presents a different sort of anomaly. Whatever its mode of origin, it clearly has been formed from material which originated dominantly in the general region between Mars, Earth and Venus. Yet the Moon is drastically fractionated, both in siderophile and volatile elements, relative to the Earth. The extent of this difference suggests a fundamentally different origin for the Moon from that of the terrestrial planets (section 10.2.3).

Ringwood (1977a) suggests that the bulk composition of the Earth and its physical properties can be explained if it accreted from a mixture of about 10–15% low-temperature condensates similar to C1 chondrites and 85–90% of high-temperature, reduced, metal-rich, devolatilized condensates. Compositions of these components are given in Table 4.3.8. The constitution of Mars can be explained by a mixture of ~30% low-temperature condensate and ~70% high-temperature condensate. In consequence, it is more oxidized than Earth and probably does not possess a metallic iron core (Ringwood, 1959). Venus may represent an intermediate stage of oxidation, having collected about 20% of the low-temperature condensates.

In the simplest model we would expect the proportion of low-temperature condensate in a planet to increase with increasing distance from the sun. This expectation is fulfilled in the present model in the cases of Mercury, Earth and Mars. Venus appears to be an exception, showing that the model is oversimplified. Another factor which influences the proportion of high- and low-temperature condensates is the temporal sequence of accretion for given planets. It is possible that accretion of Venus was appreciably delayed relative to Earth, and that its feeding zone captured a larger component from the asteroid region than it would have otherwise, as a result of gravitational perturbations arising from the growth of Jupiter.

During the early stages of accretion of Earth, the accretion energy was low (Fig. 4.3.10). This permitted the trapping of a significant proportion of highly volatile substances introduced by component A (Table 4.3.8) at this stage. After the mass of the growing nucleus had increased to about one-tenth of the present Earth mass, planetesimals would have arrived at the Earth's surface with sufficiently high velocities to cause strong transient heating and evaporation of most of the volatiles to form an atmosphere. A relatively small proportion of incoming planetesimals of component A may have escaped this devolatilization process. Beyond this

Table 4.3.8 Compositions (wt.%) of Orgueil C1 chondrite and of reduced, devolatilized component condensing at elevated temperatures (~1000° C) in solar nebula.†

	Orgueil A	High Temperature Component B
Metallic nickel-iron (~5% Ni)	-	34.1
SiO_2	21.7	32.8
TiO_2	0.1	0.2
Al_2O_3	1.6	2.8
Cr_2O_3	0.35	0.2
MnO	0.2	0.1
FeO	22.9	-
NiO	1.2	-
MgO	15.2	27.7
CaO	1.2	2.3
Na_2O	0.7	-
K_2O	0.07	-
P_2O_5	0.3	-
H_2O	19.2	-
Organic compounds	9.7	-
S	5.7	-

†Ringwood, 1977a.

stage, the situation was such that a great excess (85%) of component B, containing 29% metallic Ni-Fe, was falling at high velocities into an atmosphere mainly composed of steam derived from the devolatilization of component A. Fifteen percent of this component A would provide 3% of water vapor in the atmosphere, so the principal chemical interaction of infalling planetesimals with the primitive atmosphere was between metallic iron and water vapor. The equilibrium constants for the relevant reaction are given in Table 4.3.9.

Ringwood expects that most of the water vapor in the primitive atmosphere would have been reduced by infalling excess metallic Fe to form FeO, which entered ferromagnesian silicates in the accreting Earth, and H_2. The H_2 would have been continually blown off from the Earth during accretion, carrying with it other volatiles of higher molecular weight. [The applicable mechanism of blowoff, which differs from traditional Jeans-Spitzer thermal escape, is discussed by Öpik (1963).]

Simple mass-balance considerations show that most of the FeO finally incorporated in the Earth would have been produced by reaction of metallic Fe in component B with water vapor from component A. The

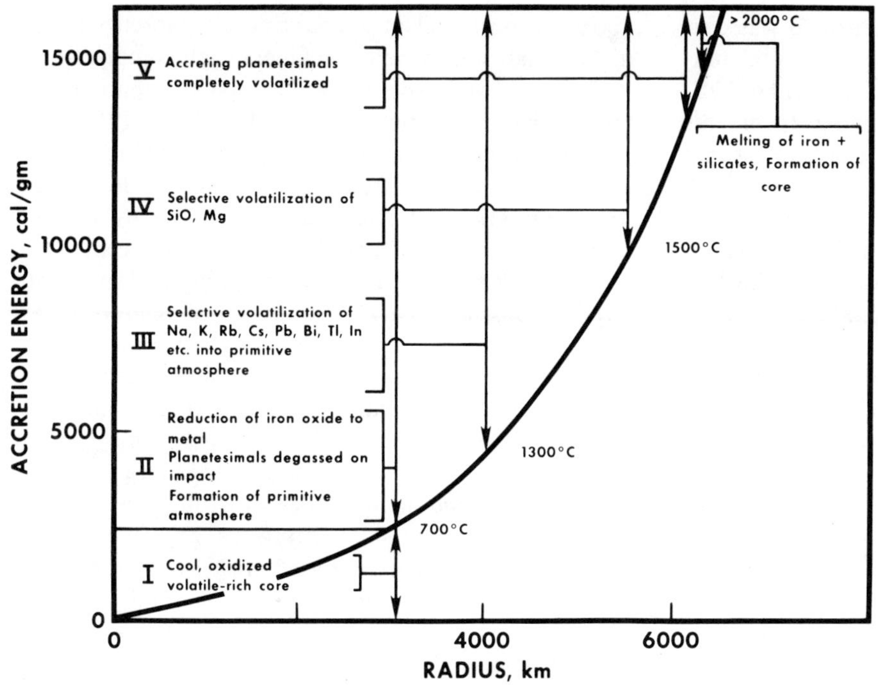

4.3.10 Relationship between energy of accretion and radius of the growing Earth (Ringwood, 1970). See also section 9.4.1.

proportion of FeO introduced *directly* from component A would have been much smaller.

A relatively small proportion of component A would have reached the Earth's surface in a cool state, retaining its volatiles, and thereby became incorporated in the growing Earth. However, most of the water in this component eventually would have been reduced by excess metallic Fe as temperatures and pressures rose

$$Fe + H_2O \rightleftharpoons FeO + H_2(\text{dissolved in Fe}) \quad (\Delta V \text{ negative})$$

Hydrogen would have formed interstitial solutions with metallic Fe, and ultimately was carried into the Earth's core.

At very high pressures and temperatures, deep in the Earth's interior, Fe would begin to melt and dissolve extensive quantities of FeO, which would substantially reduce its melting temperature (Ringwood, 1977a). This permitted the molten Fe-FeO alloy to segregate into large bodies which commenced to sink into the core, driving strong convection throughout the mantle. This process would have been accompanied by two important redox equilibria.

First, when O dissolved in the molten Fe, it would have reacted with H and C present in solid solution in the Fe to form H_2O and CO_2. The fugacities of these components would have been greatly increased, so that they migrated from the metal phase to the silicates of the mantle where they became fixed as OH^{-1} and CO_3^{-2}.

Second, the entry of O into solution in the Fe would have caused a marked increase in the O fugacity of the mantle, which would have a large effect on partition coefficients of siderophile elements (e.g., Ni, Co, Cu, Au, Pt, Re, Ir) between Fe and silicate phases. In Ringwood's view, the high abundances of siderophile elements in the Earth's mantle and its relatively high O fugacity are to be attributed to this effect.

Ringwood proposes that the evolution of Venus was similar of that of Earth, except that a larger amount of component A was accreted, leading to a higher oxidation state and a smaller core. In the case of Mars, a still larger proportion of component A was accreted, so that

Table 4.3.9 Equilibrium constants for the simplified reaction $MgSiO_2 + Fe + H_2O \rightleftharpoons 1/2 (Fe_2SiO_4 + Mg_2SiO_4) + H_2$, corresponding to olivine and pyroxene compositions with $FeO/(FeO + MgO) = 0.22$.†

Temperature (°C)	"K" = P_{H_2}/P_{H_2O}
400	132
600	35
800	17
1000	11
1200	7.6
1400	6

†Based on Mueller, 1964.

all, or nearly all, metallic Fe in component B was oxidized. Mars, therefore, probably possesses a core composed mainly of iron oxide, with a small amount of FeS.

A terrestrial composition stemming from this model appears as E6 in Table 4.3.2c. Here we assume 10.1% of Fe^0 from component B is oxidized to 13.0% FeO, of which 2.1% joins the mantle and 10.9% joins the core. The amount of oxidation was adjusted to match Ringwood's (1977a) empirical estimate of the Earth's composition, which appears as E7. A Mars model in which all Fe and Ni are oxidized, and the FeO is arbitrarily partitioned 50% into the mantle and 50% into the core, appears in Table 4.3.2d as Ma4. The Venus model, V4 in Table 4.3.2b, is oxidized to the extent necessary to provide a mantle ratio $Mg/(Mg + Fe) = 0.76$, in the spirit of Ringwood and Anderson (1977), and a mass of (S + O) in the core equivalent to that favored by these authors.

The fractionation of preplanetary solids by mechanical processes

Two other processes or effects, essentially mechanical in character, have been identified as potentially important if not crucial in establishing the compositions of the terrestrial planets. Unfortunately, these are not constrained closely enough to offer more than qualitative predictions of planet compositions.

The first is differential orbital decay (Laplace, 1805; See, 1910). The solar nebula would have been rotating at somewhat less than the Keplerian (dynamic orbital) velocity, since each element of the nebula was sustained at a particular mean distance from the center of mass of the solar system by a balance among gravitational force (inward), centrifugal force due to its orbital motion (outward), and the pressure gradient in the nebula (also outward). Condensate particles or clusters of particles entrained in the nebula, on the other hand, would not have felt a force due to the gas pressure gradient; in their case the gravitational force had to be offset entirely by centrifugal force, so the planetesimals had to revolve in orbit somewhat faster than the nebula turned. Consequently, they experienced gas drag in the nebula, were continually decelerated by it, so tended to spiral inward toward the protosun. For planetesimals $\gtrsim$1 m in dimension, the greater the surface/mass ratio of the planetesimal the more rapidly it would have spiralled in. Weidenschilling (1978) has argued that the inner part of the solar system would have become depleted in orbiting solids earliest by this mechanism, and that low-density silicate-rich planetesimals would be removed more rapidly than high-density metal-rich ones, unless the metal-rich objects were systematically smaller than the silicate planetesimals. These tendencies have the potential of accounting for Mercury's relatively small size as well as its metal-rich composition.

The second process is collisional differentiation, a mechanism championed early by Urey (1963) and Whipple (1964). Internal melting and differentiation occurred very soon after the accretion of at least some first-generation planetesimals, as shown by the old radiometric ages—essentially as old as the accepted age of the solar system—of many differentiated meteorite types (irons, achondrites, stony-irons; section 7.2.4). It is expected from dynamical considerations that the aggregation of a population of many small first-generation planetesimals into the present small number of much larger planets would have involved destruction as well as construction—i.e., many of the first-generation planetesimals would have been disrupted by mutual collisions, their debris then being available for re-accretion. Masses of differentiated nickel-iron are much tougher than silicate masses, so the metallic debris is likely to have consisted of systematically larger and denser objects than the silicate debris. Metal debris would therefore be less rapidly removed from the zone of accretion of terrestrial planets than finely-broken silicate debris, whether the removal process was orbital decay (above), radiation pressure, or the Poynting-Robertson effect (the latter two operative after dissipation of the solar nebula). Since collision velocities and frequencies would be greatest at small heliocentric distances, the effect would be strongest close to the sun, and would enhance the metal/silicate ratio most at Mercury's position among the terrestrial planets.

Collisional differentiation has been invoked as a means of explaining the anomalous composition of the Moon (Orowan, 1969; Ruskol, 1972; Kaula, 1977a, b). None of the models of planetary formation discussed earlier predict the composition of the Moon when extrapolated to an object that small or at that mean distance from the sun [though some mechanisms discussed elsewhere (section 10.2.3), such as rotational or collisional fission from Earth's mantle, probably do]. Collisional differentiation rationalizes the metal-poor character of the Moon as being due to the greater likelihood of capture of small, low-density silicate debris particles than large, dense metal fragments from heliocentric orbits into a geocentric debris swarm that was to become the Moon.

Effect of gravitational instability of the nebula

For completeness, one other model of planetary formation should be noted. The preceding discussions have assumed accretion of the planets from mineral grains that circulated at large in an integral nebula. Another possibility is that the nebula was gravitationally unstable, and separated into a number of discrete gaseous protoplanets, each orbiting the protosun (Kuiper, 1951; Cameron, 1978; section 10.2.2). For this to occur would require the coexistence of a relatively massive nebula with a small, incompletely accreted protosun. Solid particles would have tended to settle to the center of each protoplanet and there form a dense mass. After dissipation of the gas envelopes of the protoplanets, their solid cores would be left as planets (or would further assemble to form planets). Events within the protoplanets offer several opportunities for chemical fractionation. The model, which is discussed further in section 10.2, has not been developed to the point where it predicts planetary compositions.

Bulk planetary compositions vs. mantle compositions

The ultimate goal of this chapter is to estimate possible mantle compositions. To derive these from cosmochemical models of bulk compositions, it is necessary to allow for the differentiation of the silicate portion of each model into mantle and crust. If we can generalize from the Earth, however, the volume of the crust is so small (~0.5%) and the abundances of *compatible* elements are sufficiently similar in crust and mantle [see, e.g., the mean crustal and mantle compositions estimated by Smith (1977) and Ringwood (1977a), respectively] that the mantle compositions can be taken to be the same as the bulk silicate compositions, to well within the errors of estimation of the bulk composition. The comparison just mentioned indicates, however, that incompatible elements (i.e., elements whose charges and ionic radii prevent them from being readily incorporated in the abundant minerals of the mantle: examples are Cs, Rb, K, Ba, Pb, La, Ce, U, Th, Ti, Nb, B) are significantly fractionated into the Earth's crust. Roughly 30% of the Earth's inventory of these elements is believed to be concentrated in its crust, which occupies only 0.5% of its volume. According to Smith's and Ringwood's assessments, therefore, the mantles of the model planets under discussion would contain only about two-thirds of the abundances of incompatible elements that appear in the hypothetical bulk compositions of Table 4.3.2. [Anderson (1979), however, has recently proposed a terrestrial mantle composition model based on kimberlite xenoliths and a thick eclogite layer in the transition region of the upper mantle, in which the incompatible trace elements in the crust represent only 3–13% of the total crust-plus-mantle inventory of these elements.]

Models Me4, Me5 and V5 will be discussed in section 4.5. Other models in Table 4.3.2 that have not previously been referenced (E7, Ma5, Ma6, Eu4) are empirical in character, based on geophysical and geochemical parameters rather than on a model of planet formation that aspires to general applicability.

4.4 PLANETARY MODELLING

4.4.1 General considerations

To derive the density profile for a hypothetical planet, it is necessary to estimate the effects temperature and pressure would have on rock of a postulated composition if it were placed at various depths in the planet. Two distinct classes of effects concern us. First, the rock would simply be squeezed by pressure and expanded by temperature by certain amounts, changing its volume and thus its density. *Equations of state* for rocky materials, which are based on laboratory measurements and theoretical considerations, describe (within limits) the relationship between temperature, pressure and density. Second, however, there are regimes of pressure and temperature where rock of a given composition exists as a different set of stable mineral phases than would be the case at the surface of the Earth. Rock undergoes a major and discontinuous change in density when it experiences *phase changes* upon being translated into a new stability field, which transcends the effects predicted by an equation of state.

One of the ways in which we construct models of the other terrestrial planets is to use Earth, the planet we know most about, as a guide, and consider deviations from its properties. However, certain problems should be kept in mind, i. e., (1) the Earth may be enriched in Fe and possibly other refractory elements compared to other inner solar system objects, with the exception of Mercury; and (2) the upper mantle of the Earth may not be representative of the average composition of the mantle. As already noted, mantle materials at depth are subject to phase changes, which affect the mean density of the planet. These phase changes depend on composition. Iron-rich silicates, in general, transform at lower pressures than the Fe-poor silicates (Fig. 4.4.1). Also, the phase changes in pyroxene occur at different pressures and temperatures than those in olivine. The effects of composition and temperature on the locations of

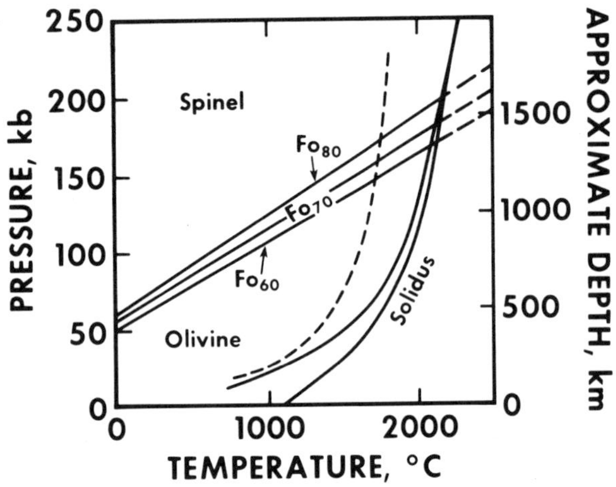

4.4.1 Pressure-temperature relationship (straight lines) for the transition of olivine to a spinel-structure mineral, for several different Mg contents of olivine; shown with two possible temperature profiles for Mars and the depths of the given pressures in Mars. The dashed curve is the profile assuming convective heat transfer, and the solid curve represents conductive heat transfer. From Johnston and Toksöz (1977).

phase boundaries are more important in assessing density profiles than are refinements in the equations of state. Densities for minerals which should be major components of planetary mantles are given in Table 4.4.1 and in Fig. 4.4.2. The variation of density with the mole ratio Mg/(Mg + Fe) is shown in Fig. 4.4.2 for olivine, garnet and the pyroxenes.

In small planets such as the Moon and Mars, temperature is as important as pressure in controlling the density; however, it is much more uncertain than pressure. The pressure vs. depth profile is almost an invariant in planetary models that have the same radius, mass and moment of inertia (see Fig. 3.1.1). It can be computed from the assumption of hydrostaticity. Even if the planet is grossly out of hydrostatic equilibrium, the effective pressure, which controls the density, can be estimated quite accurately. The actual density at a given depth is another matter, however. There is a trade-off between temperature and composition. When we have only two observational parameters, such as the mean density and moment of inertia, we can only determine two parameters; these can be, for example, two of the following three: mean density of the mantle, mean density of the core, and radius of the core. If information about the crust is available from gravity or seismology, this can be stripped off.

A simple equation of state that is adequate for small planets is the Murnaghan equation

$$P = \frac{K_0}{n}\left[\left(\frac{\rho}{\rho_0}\right)^n - 1\right]$$

where K_0 is the zero-pressure bulk modulus, ρ is the density, ρ_0 is the zero-pressure density, and n is the pressure derivative of the bulk modulus, $n = (\partial K/\partial P)_0$, which is generally between 3.0 and 4.0. To a first approximation, the effect of temperature can be included simply by writing

$$P = \frac{K_0(T)}{n}\left[\left(\frac{\rho}{\rho_0(T)}\right)^n - 1\right]$$

A higher-order equation of state is the Birch-Murnaghan equation, which is a member of a class of equations that can be written in the general form

$$P = \frac{3K_0}{m-n}\left[\left(\frac{\rho}{\rho_0}\right)^{\frac{1}{3}(m+3)} - \left(\frac{\rho}{\rho_0}\right)^{\frac{1}{3}(n+3)}\right]$$

where n is generally 1 or 2, and m is 2 or 4, with n<m.

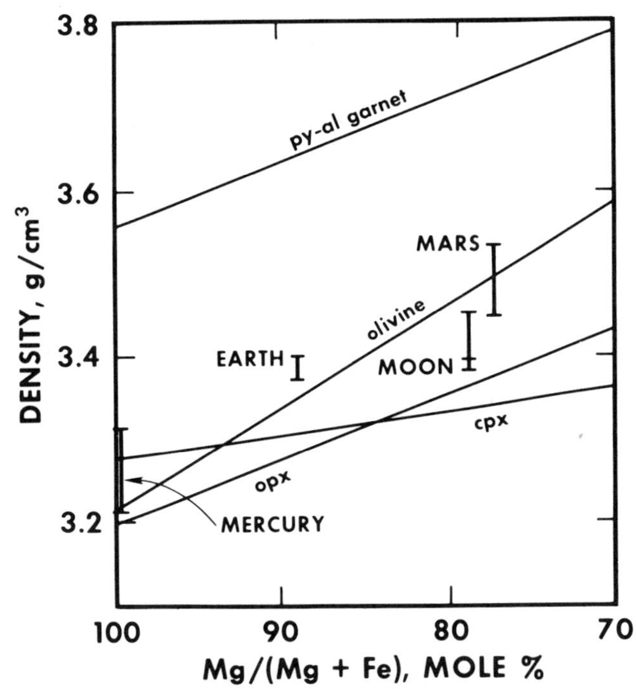

4.4.2 Zero-pressure densities of pyrope-almandine garnet, olivine, diopside-hedenbergite (cpx), and orthopyroxene (opx) as functions of Fe^{2+} content. Also shown are mantle density ranges for the acceptable models (this work) of Earth, Mars, Mercury and the Moon. The range shown for the Moon is divided in two segments, corresponding to the lunar upper mantle (below) and the lower mantle (above). Venus models vary widely, and are not shown.

Table 4.4.1 Formulas and mass densities of minerals used to calculate net densities of the mineral assemblages of Tables 4.5.3, 4.5.7, 4.5.10, 4.5.11 and 4.5.13.

Mineral	Formula	Density, g/cm^3
	P = 1 bar	
Jadeite	$NaAlSi_2O_6$	3.315
Tschermak's molecule (Ca)	$CaAl_2SiO_6$	3.440[a]
Tschermak's molecule (Mg)	$MgAl_2SiO_6$	3.390[b]
Diopside-hedenbergite	$Ca(Mg,Fe)Si_2O_6$	3.277–3.550
Orthopyroxene	$(Mg,Fe)SiO_3$	3.198–3.960
Olivine	$(Mg,Fe)_2SiO_4$	3.224–4.393
Pyrope-almandine garnet	$(Mg,Fe)_3Al_2Si_3O_{12}$	3.559–4.318
Grossular garnet	$Ca_3Al_2Si_3O_{12}$	3.595
Spinel	$(Mg,Fe^{2+})(Al,Fe^{3+},Cr)_2O_4$	3.582–4.280[c]
Plagioclase (anorthite)	$CaAl_2Si_2O_8$	2.76
Ilmenite	$FeTiO_3$	4.786
Chromite	$FeCr_2O_4$	4.500[d]
Magnetite	Fe_3O_4	5.820[e]
Pyrrhotite	$Fe_{0.9}S$	5.40[f]
Iron	Fe	7.874
Nickel	Ni	8.911
	P = 1.2 Mb[g]	
Olivine, Fo_{91}	$Mg_{1.82}Fe_{0.18}SiO_4$	5.26
Pyroxene, En_{85}	$Mg_{0.85}Fe_{0.15}SiO_3$	5.40
Wollastonite	$CaSiO_3$	5.30
Corundum	Al_2O_3	5.16

Unless marked, values come from Robie *et al.* (1966).
[a] From Okamura *et al.* (1974). [b] From Skinner and Boyd (1964). [c] From Wang and Simmons (1972). [d] From Deer *et al.* (1963). [e] From Ringwood and Clark (1971). [f] From King and Ahrens (1973). [g] Shock-wave data of Ahrens (1980).

Again, to first order, the effect of temperature can be included in K_0 and ρ_0. Equivalent expressions can be written for the seismic velocities (Sammis *et al.*, 1970).

The pressure at depth z in a planet is

$$P(z) = \int_{R-z}^{R} \rho(r)\, g(r)\, dr$$

where

$$g = \frac{GM(r)}{r^2}$$

and r is radius within the planet, M(r) is the mass inside radius r, R is total planetary radius, and g is the local value of gravity.

Much more sophisticated equations have been applied to the Earth, but they depend on a detailed knowledge of the seismic velocities, which are unknown on the other planets.

For the Moon, the effect of pressure is small, and a direct comparison of observed and laboratory seismic velocities can be made with only a small correction for pressure. Uncertainties in the internal temperatures, and in temperature coefficients of the physical parameters, are the main sources of uncertainty in constructing and interpreting models of Mars, the Moon, Mercury, and the asteroid parent bodies. Thermal history modelling and the likely present thermal states of the terrestrial planets are discussed in Chapter 9.

4.4.2 Planetary models for the compositions of section 4.3.2

Our goal in section 4.5 will be to translate the chemical models of section 4.3.2 into mineralogical models of the planets; to calculate density profiles for these models, allowing for local compression of materials by the weight of overlying layers, thermal expansion and phase changes; and from these profiles, to calculate mean densities and moment-of-inertia factors for comparison with the observed mean densities and (where known) moment-of-inertia factors of the planets. More strictly, this is the procedure that will be followed for the small planets Mercury, Mars and the Moon. Properties of Venus can be estimated more reliably by investigating the consequences of projected deviations from the Earth model.

The first step in our modelling studies was to convert the compositions of models in Table 4.3.2 to mineral assemblages that would be stable in the relevant ranges of pressures and temperature. If a substantial portion of the silicate fraction of the planet was assumed (from surface features, or seismic data, or cosmochemical considerations) to be segregated into a crust, then a composition and weight fraction of crustal material had to be subtracted from the model composition. The remaining mantle composition was then converted to mineral assemblages by use of a norm calculation. This algorithm determines the stable mineral assemblage for a particular rock composition by modification of its CAMS coordinates (O'Hara, 1968; discussed in section 3.3.1). A complication in the procedure comes in estimating the degree of substitution of Al_2O_3 (i.e., the Tschermak molecule) in the pyroxene structure. The data of Akella (1976) are used in making the estimate. There is considerable uncertainty in the disposition of Na_2O at high pressures (>100 kb), where it may fit into the garnet structure and not appear as jadeite (see Ringwood, 1975, p. 456).

The S and O in planetary cores were assumed to go into the pyrrhotite high-pressure structure of $Fe_{0.9}S$ (King and Ahrens, 1973) and the high-pressure phase of magnetite, Fe_3O_4 (Ringwood and Clark, 1971), respectively. The average densities, and bulk properties such as bulk modulus and thermal expansion, were then calculated for the mineral assemblages.

The densities then had to be extrapolated to higher temperatures and pressures. The second order Birch-Murnaghan equation with $n = 2$, $m = 4$ (see section 4.4.1) was used to make the pressure extrapolation, except in the cases of Earth and Venus. The thermal expansion α, its temperature derivative, and the first and second temperature derivatives of the bulk modulus (K) were used to make temperature corrections.

A two-stage computational procedure of first going up in temperature and then in pressure (see Fig. 4.4.3) was used in adjusting the densities of interior volume elements. Parameters used in the procedure are given in Table 4.4.2, and the zero-pressure densities of planetary materials are given for each model in tables in the subsections of 4.5. The zero-pressure bulk modulus, K, was taken to be an average of the values for the

Table 4.4.2 Values of physical parameters used in modelling internal densities of the planets.

	K_o (Mb)	$\frac{dK}{dT}$ $\left(\frac{Mb}{°C} \times 10^4\right)$	$\frac{d^2K}{dT^2}$ $\left(\frac{Mb}{°C^2} \times 10^8\right)$	α $(°C^{-1} \times 10^6)$	$\frac{d\alpha}{dT}$ $(°C^{-2} \times 10^8)$
Olivine	1.31	-1.31	-1.08	26	1.40
Pyroxene	1.04	-1.31	-1.08	24	1.40
Iron	1.70	-1.31	0	38	2.00
$Fe_{0.9}S$	1.20	-1.31	0	38	2.00
Earth, Venus Mantles	2.35	-1.15	-1.90	50	0
Spinel Phase	2.00	-1.31	-1.08	39	0.30

Bulk modulus values (K) are from: Simmons and Wang (1971) for olivine, pyroxene and iron; King and Ahrens (1973) for $Fe_{0.9}S$; and Sato (1977) for the spinel phase. The first and second temperature derivatives of bulk modulus were calculated from data on olivine in Simmons and Wang (1971) and assumed to hold for the other mineral phases, except in the Earth and Venus lower mantles, which were derived by using a bulk modulus value appropriate to mixed oxides (Ahrens *et al.*, 1970; Simmons and Wang, 1971) and by adjusting the temperature derivatives to match the density-with-depth profile of the Anderson and Hart Earth model (1976), using a temperature profile from Stacey (1977). All data on thermal expansion are from Skinner (1966).

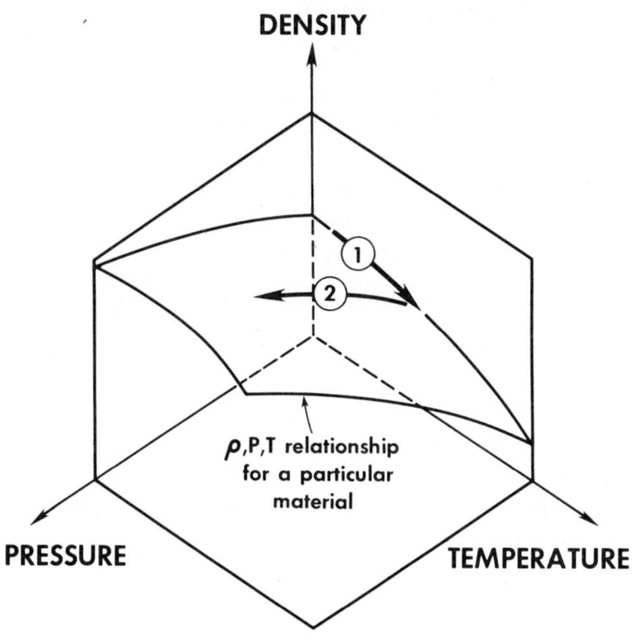

4.4.3 Illustration of the procedure used to estimate the densities of materials at high temperatures and pressures. First the effect of temperature was calculated, using thermal expansion coefficients, then the isothermal Birch-Murnaghan equation was used to make the correction to high pressure.

mineral phases, weighted according to their abundances in the norm. The total mass and moment-of-inertia factor for each model was then computed and compared with measured values. This procedure works well enough in the pressure ranges of the smaller planets, but the models of Venus, Earth, and to some extent Mars, suffer from the effects of high-pressure phase changes, and pressures too great for reliable extrapolations using the Birch-Murnaghan equation.

Temperature distributions assumed for the planets are based on published thermal models. Uncertainties in these temperature profiles combine with uncertainties in the temperature derivatives of Table 4.4.2 to contribute more than any other factor to error in the density modelling. The temperature profiles used (Fig. 4.4.4) are those of Toksöz *et al.* (1978), with the exception of that of Earth. The terrestrial thermal profile is that of Stacey (1977), and is the hottest of recent geotherms (Jeanloz and Richter, 1979; O. L. Anderson, 1980; Brown and Shankland, 1981). A detailed discussion of planetary thermal models appears in section 9.5.

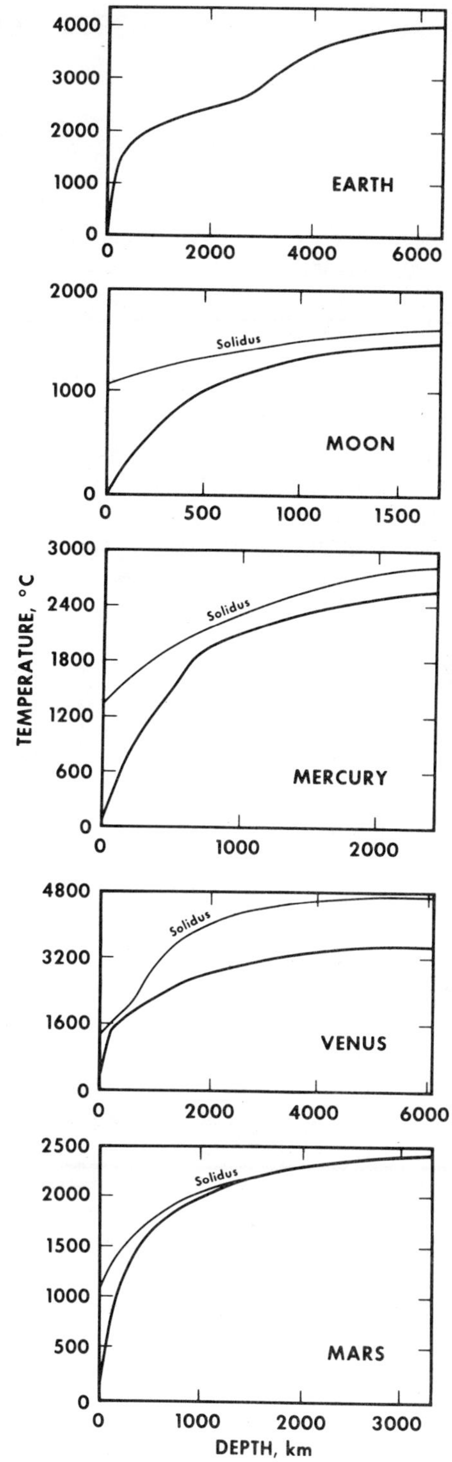

4.4.4 Temperature profiles used in density-modelling of Earth (Stacey, 1977), Moon, Mercury, Venus and Mars (Toksöz *et al.*, 1978).

4.5 MODELS OF THE TERRESTRIAL PLANETS

4.5.1 Earth

The Earth is the best-studied planet. It is logical to test cosmological models against the structure and composition of Earth as deduced from petrology and seismology, and also interesting to work in the opposite direction.

Inferences from geophysics

Earth is a differentiated body: the crust, mantle, outer core and inner core are the principal apparent subdivisions. Petrological models of the uppermost mantle agree well, generally, with the seismological data (Ringwood, 1975; Anderson, 1977; Anderson *et al.*, 1971). Below about 220 km the seismic data are not definitive in distinguishing among the various petrological possibilities.

The oceanic crust averages 6 km in thickness, and is underlain by a layer of undetermined thickness, which has seismic properties appropriate to dunite or harzburgite. This layer is probably 20–50 km thick. The suboceanic mantle, below the dunite layer, has properties consistent with garnet peridotite or pyrolite down to about 200–250 km. The continental crust averages about 30 km in thickness. The subcontinental mantle appears to be peridotite depleted in basaltic components, and cooler than the suboceanic mantle. Between about 200 and 400 km depth, mantle properties under both oceans and continents are consistent with either pyrolite (Ringwood, 1975; Anderson, 1977), or eclogite (Anderson, 1979). There appears to be a discontinuity in mantle properties at about 200 km depth, which could represent either the bottom of a partially molten layer or a chemical discontinuity (e.g., the boundary between basalt-depleted material and fertile or primitive mantle material; or between peridotite and eclogite). The seismic data cannot discriminate between pyrolite and eclogite mineralogies in the depth range 220–670 km. In the pyrolite model the increase of velocity in this region, over that due to self-compression and thermal expansion, would be due to collapse of olivine to β-spinel and γ-spinel and the interaction of pyroxene and garnet to form a complex garnet solid solution (Ringwood, 1975). In the eclogite model, the garnet solid solution reaction would be responsible for the 400-km discontinuity. Whether the 670-km discontinuity is due to phase changes in olivine or garnet solid solution, or whether it represents a chemical discontinuity is a controversial subject (succeeding subsection). This interface is sharp, and is near the maximum depths of earthquakes.

Below 670 km the garnet component of the mantle transforms to minerals having ilmenite and perovskite structures, while the olivine component transforms to (MgO + stishovite) and (perovskite analogue + MgO). Aluminous compounds may transform to minerals with the calcium ferrite structure at some depth (Ringwood, 1975).

The lower mantle contains a number of small seismic discontinuities. Some of these may be due to the phase changes mentioned above, but no P,T estimates are available for their locations. Discontinuities in bulk chemical composition of the lower mantle cannot be ruled out.

The lowermost 200 km of the mantle is inhomogeneous both radially and laterally. This may be due to a thermal boundary layer at the core-mantle interface (Elsasser *et al.*, 1979); but a layer having variable chemistry is also a possibility, since a thermal boundary layer would probably be less than 100 km thick.

In summary, the elastic properties and densities of various regions of the mantle can be satisfied by a range of petrological models. The strongest constraints

Table 4.5.1 Stable mineral assemblages in a mantle of pyrolite/garnet peridotite composition.†

Approximate Range of Depth	Mineralogy
100–150 km	olivine + orthopyroxene + Ca-rich clinopyroxene + pyrope-rich garnet
150–380 km	olivine + orthopyroxene + clinopyroxene + Ca-rich clinopyroxene + garnet solid solution
380–400 km	modified spinel (β-spinel) + olivine + Ca-poor clinopyroxene + Ca-rich clinopyroxene
400–500 km	β-spinel + garnet solid solution + Ca-poor clinopyroxene + Ca-rich clinopyroxene
500–600 km	spinel + β-spinel + garnet solid solution
600–650 km	spinel + garnet solid solution + stishovite
Lower mantle	dense phases in the ilmenite, perovskite, calcium ferrite and oxide structures

†Ringwood, 1975; Akaogi and Akimoto, 1977; Anderson, 1976.

on mineralogy are likely to be the locations and details of the upper mantle discontinuities. If phase changes can be found in plausible mineral assemblages corresponding to one particular bulk chemistry, at pressures appropriate for the 200- and 670-km discontinuities, then the homogeneity of the upper mantle will be established and the positions of the discontinuities can be used to infer thermal gradients. If no such phase changes are forthcoming, then a stronger case can be made for chemical discontinuities in the mantle. The 400-km discontinuity is conventionally interpreted as being due to the olivine-spinel phase change. Details of the alternative possibility, a change from eclogite to garnetite chemistry, have not been worked out. The overall composition of the mantle depends critically on the SiO_2 content of the lower mantle. If this region is pyroxene-rich, then the overall mantle can have a chondritic MgO/SiO_2 ratio. Otherwise, it seems likely that the Earth is deficient in SiO_2, as proposed by Ringwood (1975).

The question of chemical homogeneity of the mantle

Table 4.5.1 summarizes estimates of the stable mineral assemblages in the Earth's mantle, on the assumption that it is chemically uniform and has a composition corresponding to that of garnet peridotite or pyrolite. In this type of model the seismic discontinuities in the mantle are due to phase changes. In an alternative model (Anderson, 1979), the discontinuities at 220 and 670 km represent the boundaries of a buried eclogite layer which transforms to garnetite at 400 km.

As noted above, the lowermost 200 km of the mantle is physically inhomogeneous. Ringwood (1975) has suggested that this region consists of Fe-rich plates subducted from the lithosphere and formed from chondritic material that accreted on the Earth subsequent to core-formation. On the other hand, Anderson (1972b) suggests its composition is enriched in refractory elements, a relic of the formation of Earth by heterogeneous accretion.

Ringwood (1975) and Dickinson and Luth (1971) conclude that a large part of the mantle (~30–70%) has had basalt removed from it, possibly resulting in chemical stratification. They believe that plates subducted from the oceanic lithosphere sink into the lower mantle. Oxburgh and Parmentier (1977) point out that depleted peridotite would probably remain in the upper mantle, and only the basalt/eclogite portion of the plates would sink through undepleted peridotite. Anderson (1979) concluded that eclogite would stop at 670 km or above. Ringwood (1975) had previously assumed that phase transitions in the slab would occur at shallower depths than in normal mantle material. This would permit the slab to penetrate into the lower mantle. An eclogite layer may be the result of crystal fractionation in a magma ocean. In a lunar magma ocean, the Al_2O_3 of the melt is removed primarily by crystallization of plagioclase. At the higher pressures in the Earth's mantle an eclogite cumulate may form at depth (O'Hara, 1968).

Table 4.5.2 Estimates (wt. %) of terrestrial mantle composition.

	(1)	(2)	(3)	(4)	(5)	(6)	(7)	(8)
SiO_2	44.2	45.1	44.5	44.7	45.7	47.3	45.0	48.0
TiO_2	0.09	0.5	0.15	0.19	0.09	0.2	0.3	0.3
Al_2O_3	2.8	4.1	2.6	2.8	3.4	4.1	5.0	5.2
Cr_2O_3	0.3	0.3	-	0.5	0.4	0.2	0.2	1.1
Fe_2O_3	1.2	2.0	1.5	-	-	-	2.0	-
FeO	7.3	7.9	7.3	7.9	8.0	6.8	8.5	7.9
MnO	0.14	0.2	0.14	0.12	0.14	0.2	0.15	0.12
NiO	0.27	0.2	0.2	0.31	0.27	0.2	-	-
MgO	41.0	36.7	41.7	41.0	38.4	37.9	33.1	34.3
CaO	2.5	2.3	2.3	2.6	3.1	2.8	4.7	4.2
Na_2O	0.22	0.6	0.25	0.30	0.4	0.5	0.6	0.3
K_2O	0.04	0.02	0.02	0.12	-	0.2	0.2	0.2

(1) Mean of five high-Ca, high-Al nodules, Harris *et al.* (1967). (2) Estimate of volatile-free upper mantle, Nicholls (1967). (3) From frequency histograms of 168 ultramafic rocks, White (1967). (4) Average of 20 published analyses of garnet peridotites, Carswell (1968). (5) Pyrolite; Ringwood (1975). (6) 20% eclogite, 80% garnet lherzolite (Anderson, 1980). (7) 20% eclogite, 80% garnet peridotite (sheared) nodules (Anderson, 1980). (8) Cosmochemical model (Ganapathy and Anders, 1974b).

In either the Ringwood or Anderson model, plate subduction would displace more primitive mantle material upwards. An important question is whether this mantle material would be the residuum of a previous partial-melting event, or truly virgin mantle. As discussed in section 7.4.2, the isotopic evidence indicates that continental and ocean-rise tholeiites come from distinct reservoirs which cannot have been significantly mixed in the last ~2 b.y. (Wasserburg and DePaolo, 1979; O'Nions *et al.*, 1979). There may, of course, be unsampled regions of the mantle that have escaped differentiation processes. Plate tectonic, geophysical and geochemical considerations all point to a chemically stratified mantle, or at least a chemically inhomogeneous mantle. This complicates the process of estimating the composition of the primitive mantle, or the average composition of the present mantle. A useful review of irreversible differentiation of the Earth is given in Ringwood (1975).

Chemical composition of the bulk mantle

A variety of nodules and xenoliths are entrained in basalts and kimberlites. These are obviously samples from the mantle, and they have been used by a number of authors to estimate the composition of the upper mantle (section 1.2.11). Some of these estimates are given in Table 4.5.2. Most of the models based on xenoliths have a higher MgO/SiO_2 ratio than the cosmochemical models. This may indicate that the upper mantle is enriched in olivine compared to the lower mantle. Olivine is a residual phase of partial melting processes, and is lighter than undifferentiated mantle material below ~45 km. It may therefore concentrate in the upper mantle. The composition of the lower mantle is highly uncertain, but Anderson (1977), Sawamoto (1977) and Liu (1977) have proposed that the lower mantle is enriched in pyroxene relative to the upper mantle. Ringwood (1975), on the other hand, argues that pyrolite is appropriate for the lower as well as the upper mantle. The petrological (xenolith) models are also generally lower in Al_2O_3 and CaO than the cosmochemical model of Ganapathy and Anders (1974b). This may indicate that the latter overestimates the refractory content of the Earth, or it could mean there is a buried eclogite layer that is unaccounted for in all petrological models except those in columns (6) and (7) of Table 4.5.2. Such a layer would raise the SiO_2, Al_2O_3, and CaO content of the mantle over those estimates which are based only on garnet peridotite or its presumed differentiation products.

Cosmochemical (meteorite) and condensation models, all of which assume no fractionation between Mg and Si, give MgO/SiO_2 ratios of the order of 0.69–0.72 for the terrestrial planets (excluding Mercury) and the eucrite parent body (Table 4.3.2). Ringwood's models in this table have higher MgO/SiO_2 ratios (0.81–0.84) for the same bodies (except for the Moon, where it is 0.75). The first six petrological entries in Table 4.5.2 also have high ratios, 0.80–0.94. The lowest ratio (0.74) is given by a mantle (column 7) which contains a substantial amount of eclogite plus fertile garnet peridotite. This model is based on eclogite and garnet peridotite xenoliths from kimberlites. The isotopic evidence indicates that there are at least two distinct source regions in the mantle, although these data do not reveal the major element compositions of the regions (see section 7.4.2). If these reservoirs have remained separate for 1–2 b.y., then it seems probable that they differ in intrinsic density. This suggests that the reservoirs differ in their garnet content, and therefore in their bulk chemistry.

Modelling the interior of the Earth

In attempting to accommodate the terrestrial chemical models of Table 4.3.2c to geophysical models, the Earth's crust was taken to be 20 km thick and to have an STP density of 2.75 g/cm^3 (Ronov and Yaroshevsky, 1969). The crust was not considered significant enough volumetrically to warrant the subtrac-

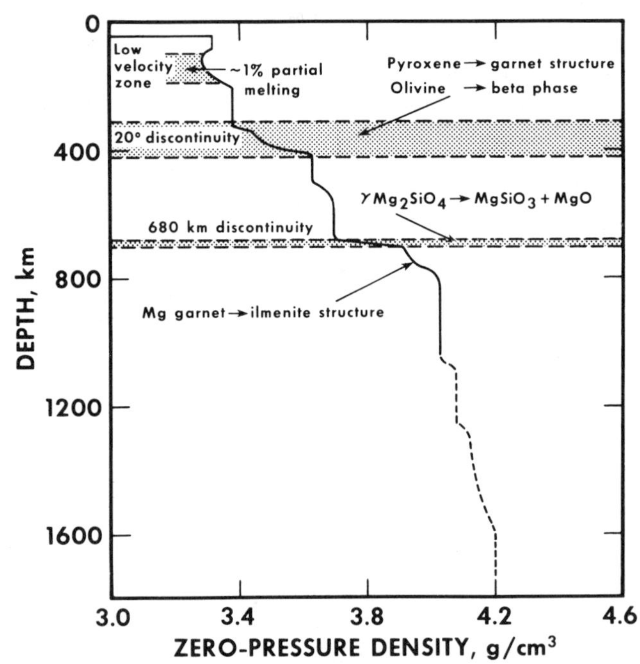

4.5.1 Density (corrected to zero pressure; approximate) as a function of depth, in Earth's mantle; showing the depth ranges in which major density discontinuities occur, and phase changes that may be responsible for them. After Ringwood (1977a).

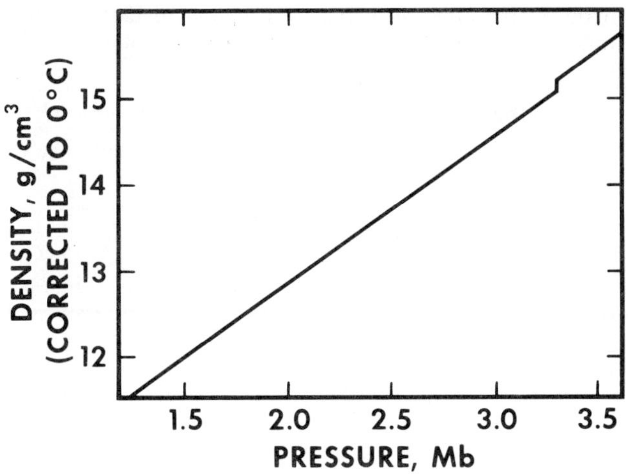

4.5.2 Relationship between density and pressure in the core of the Earth, corrected to 0°C (see section 4.5.1).

tion of its composition from that of the bulk mantle. The mantle was assumed to be uniform in composition: we did not attempt to model chemically stratified mantles.

As already noted, seismic data for Earth indicate a transition region between roughly 400 and 700 km (Anderson, 1977; Fig. 4.5.1), where velocities and densities increase. In this region the densities were calculated for olivine transforming to a spinel structure, and for pyroxene going to an ilmenite structure (Ringwood, 1975). The average density increase over the lower-pressure mineralogies was 11%.

To approximate the densities in Earth's lower mantle, the Birch-Murnaghan equation of state was used, but with some adjustment of the zero-pressure parameters to match the Anderson and Hart (1976) Earth model (a density model based on seismic data, and relatively independent of preconceptions about the composition of the Earth). Lower-mantle densities (Table 4.4.1) were calculated using the shock wave data of Ahrens (1980).

The major use made of analogy in calculating these models was in assuming that the pressure/density relationship in the core of the Earth is better known than the Birch-Murnaghan equation-of-state parameters for Fe + FeS (or Fe + FeO). Combining the Anderson and Hart (1976) model of the Earth with Stacey's (1977) thermal model (Fig. 4.4.4), and assuming a constant thermal expansion for the material in the core, a plot of the relationship of zero-temperature density to pressure was obtained. The relationship between density and pressure was found to be approximately linear (Fig. 4.5.2), so a simple relationship could be used to relate density at pressure to zero-pressure density and pressure

$$\rho = \rho_0 \, (a + bP)$$

Following Anderson's (1977) assumption of 59% Fe, 31% $Fe{\cdot}S^x$, and 10% Ni in the core, a value of 6.97 g/cm³ was obtained for ρ_0, giving

$$a = 1.348$$

$$b = 0.2488 \; (Mb^{-1})$$

Parameters of the geophysical models calculated for E1–E7 appear in Table 4.5.3, and density profiles for the models are shown in Fig. 4.5.3. The table shows that E1, E2, and E7 deviate most from the real Earth in aggregate mass. While E1, the earth formed by equilibrium condensation, is significantly overweight, the modification to equilibrium condensation made by Weidenshilling (E2) improves the match and is closest to the Anderson and Hart (1976) model in lower mantle density. Models E1 and E2 are less than 1.5% off in lower mantle density yet are about 10% too low in core density. Models E3, E4 and E7 are also approximately 2% low in lower mantle density, but 10% high in core density. The most realistic of these models are E4, E5 and E6, which are about 2% off in lower mantle density, and less than 1% off in core density. (The S content of the core of model E4 has been adjusted to match the observed mass of the Earth.)

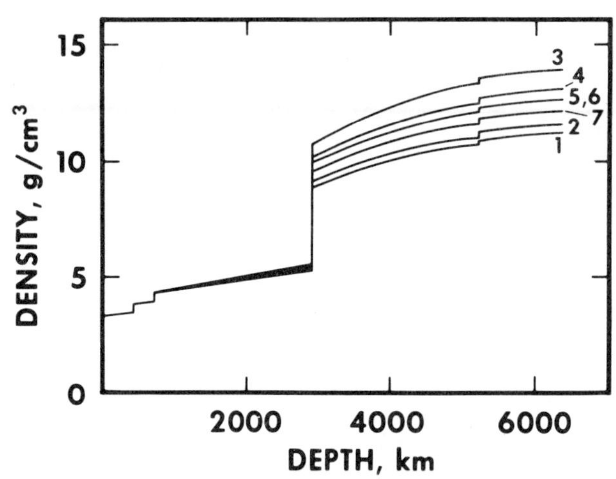

4.5.3 Density profiles for models having compositions E1-E7, in Table 4.3.2c. The profiles for E5 and E6 closely match the geophysical Earth model of Anderson and Hart (1976).

Table 4.5.3 Abundances of normative minerals (wt. %) in upper mantles, and other properties of geophysical Earth models constructed from chemical models of Table 4.3.2c. Numbers in parentheses refer to components of total mineral type.

	E1	E2	E3	E4	E5	E6	E7
Silica	0.0	0.0	0.0	0.0	0.0	0.0	0.0
K-Feldspar	0.0	0.0	0.0	0.0	0.0	0.0	0.0
Jadeite	9.79	9.79	2.15	1.63	2.41	0.65	2.61
Ca-Ts	(0.13)	(0.11)	(1.16)	(0.92)	(1.04)	(0.87)	(0.89)
Diopside	(12.59)	(12.25)	(13.46)	(10.71)	(12.03)	(10.14)	(10.30)
Total Ca-rich Pyroxene	12.72	12.36	14.62	11.63	13.07	11.01	11.19
Mg-Ts	(1.28)	(1.05)	(2.50)	(3.19)	(1.63)	(1.93)	(1.68)
Hypersthene	(27.06)	(22.37)	(31.21)	(39.54)	(20.07)	(23.70)	(20.74)
Total Ca-poor Pyroxene	28.34	23.42	33.71	42.73	21.70	25.63	22.42
Olivine	47.17	52.26	36.50	35.21	51.82	51.42	56.01
Py-Alm	(1.37)	(1.47)	(10.34)	(5.29)	(7.94)	(8.58)	(5.38)
Grossular	(0.16)	(0.17)	(1.25)	(0.64)	(0.96)	(1.04)	(0.65)
Total Garnet	1.53	1.64	11.59	5.93	8.90	9.62	6.03
Ilmenite	0.18	0.17	0.51	0.38	0.42	0.57	0.19
Chromite	0.0	0.0	1.62	1.35	0.74	0.44	0.59
Metal	0.0	0.0	0.0	0.0	0.0	0.0	0.0
Total	99.73	99.64	100.70	98.86	99.06	99.34	99.04
Density (STP) of upper mantle, g/cm^3	3.314	3.371	3.373	3.363	3.358	3.362	3.351
Density (1200 kb) of lower mantle, g/cm^3	5.341	5.380	5.305	5.330	5.300	5.308	5.308
Mg/(Mg + Fe) in mantle	0.881	0.850	0.886	0.880	0.900	0.893	0.895
Density (STP) of outer core, g/cm^3	6.06	6.24	7.74	7.18	7.00	7.00	7.63
Density (STP) of inner core, g/cm^3	6.22	6.40	7.91	7.38	7.16	7.16	6.89
Wt. % of Earth in core	32.2	31.6	32.1	29.1	32.4	32.5	32.5
Mass of Earth model ($\times 10^{27}$g)	5.59	5.75	6.09	5.95	5.88	5.89	5.79
Percent deviation from true mass	-6.36	-3.68	+2.01	-0.12	-1.57	-1.42	-3.10
I/MR^2 of Earth model	0.3383	0.3372	0.3247	0.3291	0.3306	0.3306	0.3335
Percent deviation from true I/MR^2	+2.05	+1.71	-2.05	-0.72	-0.27	-0.27	+0.60

4.5.2 The Moon

Geophysical data and other observations bearing on the internal structure of the Moon

More is known about the Moon than any other planetary body save Earth. The lunar data base, although much smaller than the Earth's, has been intensly studied. Many important differences between Moon and Earth have emerged. The Moon's very low mean density, 3.34 g/cm^3, implies that it has a much lower content of total Fe than does Earth. However, the mantle of the Moon has a higher Fe^{2+} content than the upper mantle of Earth [Fe/(Fe + Mg) = 0.16–0.22 vs. ~0.11]. The crust of the Moon appears to be much thicker than that of Earth [~75 vs. ~20 km, average values; based on the interpretation of the observed isostatic compensation of lunar topography as due to variable thickness of a ~2.95 g/cm^3 crust (Kaula *et al.*, 1974); see Table 4.5.5]. This latter property probably reflects the difference in pressure regimes in the Earth and Moon. On the Moon, feldspar is stable to a depth of ~180 km, whereas in the Earth's mantle it is stable only down to ~30 km (Fig. 4.5.7). Thus the Moon could have had a much thicker magma source region than Earth, from which feldspar might have floated to join the crust. The Moon is now cooler than Earth, and basalts apparently have not been extruded on its surface for the past 2.5 b.y. or longer (sections 2.2.3, 7.3.4, 8.6–7).

Returned lunar samples fall into two broad categories: mare basalts and lithologies from the lunar highlands, which are markedly different in composition from the mare materials. The average highland composition has been estimated by Taylor and Bence (1975) to be anorthositic gabbro, a material much richer in aluminous phases (i.e., feldspars) than the basaltic mare material. Radiometric dating has shown that most highland materials are substantially older than the mare basalts (section 7.3.5). An important difference between these two categories of lunar rock is in their patterns of rare-earth element abundances (REE). Not only do different lunar rock types contain different overall abundances of REE, but mare basalts are depleted to differing degrees in Eu relative to the other REE (section 1.3.2), while highland rocks are enriched in this element. Several hypotheses have been advanced to explain this, but the fact that Eu in the reduced (divalent) state is readily accommodated in the feldspar lattice has led to the prevalent view that early large-scale differentiation of the crust from what were to become the source regions of the mare basalts depleted these zones in Eu (Philpotts and Schnetzler, 1970).

The different types of lunar basalt and what we can infer from them are discussed in Chapters 1 and 3. Most pertinent to the geophysical modelling of the lunar mantle are estimates of the depths, bulk compositions, and trace element abundances of their source regions (Walker *et al.*, 1975; Kesson, 1975). Since petrologic studies indicate that basalts are not derived by simple partial melting of mantle material but are also affected by fractional crystallization and contamination *en route* to the surface, inferences made from them about mantle source regions are path dependent. Minimum depths for the source regions extend down to 400 km (Ringwood, 1977b). Although the compositions of the source regions are not well resolved, there is general agreement that Mg/(Mg + Fe) in these regions is between 0.75 and 0.80 (Ringwood and Essene, 1970; Morgan *et al.*, 1978a). The trace element patterns combined with depth estimates place a strong constraint on models of mantle evolution. The regions from which large amounts of plagioclase have been extracted

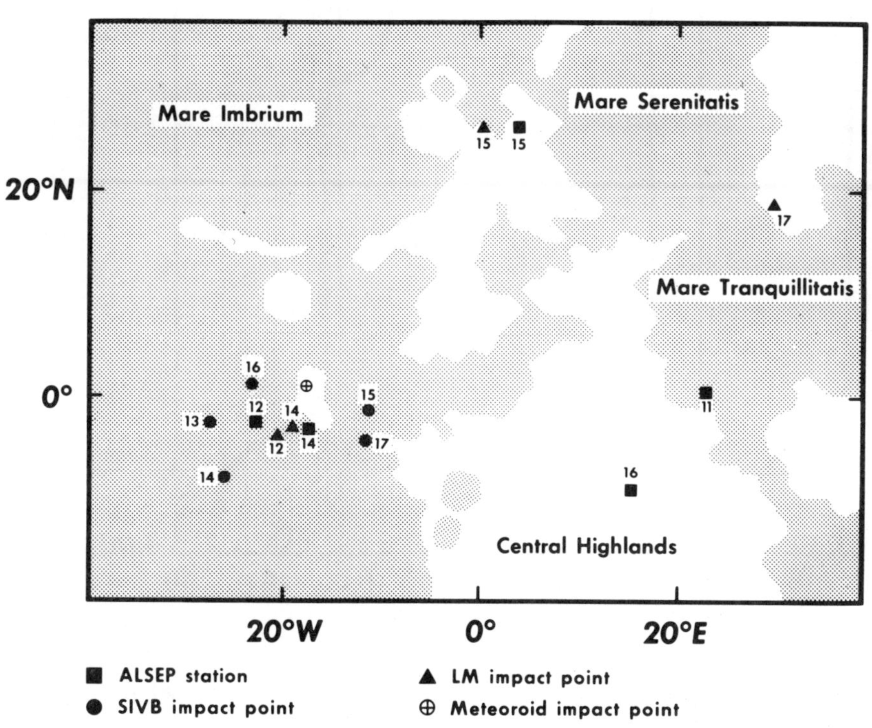

4.5.4 Position of the lunar seismic array (A11-A16 ALSEP Stations) on the nearside of the Moon, and the locations of several impact events. Numbers refer to the six Apollo missions.

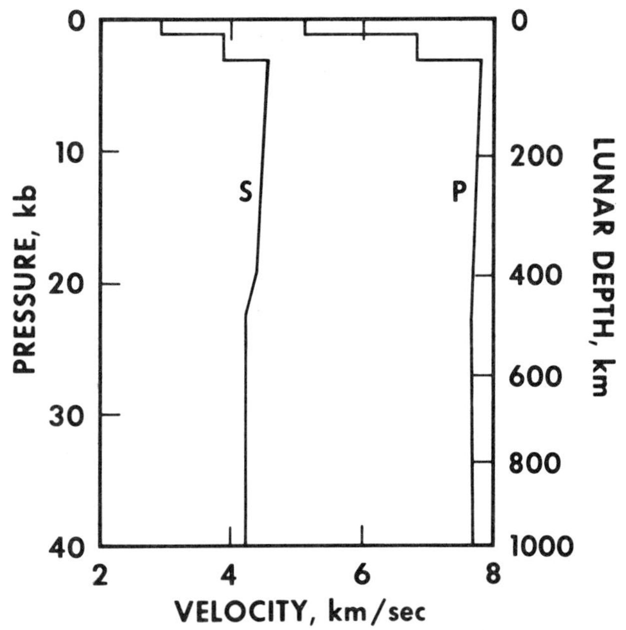

4.5.5 Seismic velocity profiles for the interior of the Moon for shear and compressional waves, according to the model of Goins *et al.* (1978).

should display negative Eu anomalies, and the residual liquids after crystal fractionation should be enriched in rare earths (Philpotts and Schnetzler, 1970; Drake, 1976).

Geophysical data taken from orbit and from the seismometers left on the Moon's surface provide constraints on the density distribution, internal structure, and composition of the Moon. The mean density of 3.34 g/cm^3 is the best known of the lunar geophysical parameters. The Moon is the only terrestrial planet whose uncompressed (STP) density is probably greater than its mean density, being about 3.38 g/cm^3 for the models in this chapter. This follows from the relatively low pressures in most of the interior of the Moon, which means the effects of thermal expansion due to probable high interior temperatures more than offset the compression due to pressure. Since the Moon's low density cannot be attributed to a high content of volatiles, this requires the Moon to have the lowest total Fe content of all the terrestrial planets. The models considered in this section, when adjusted to match the observed mean density, moment-of-inertia factor, and seismic veloci-

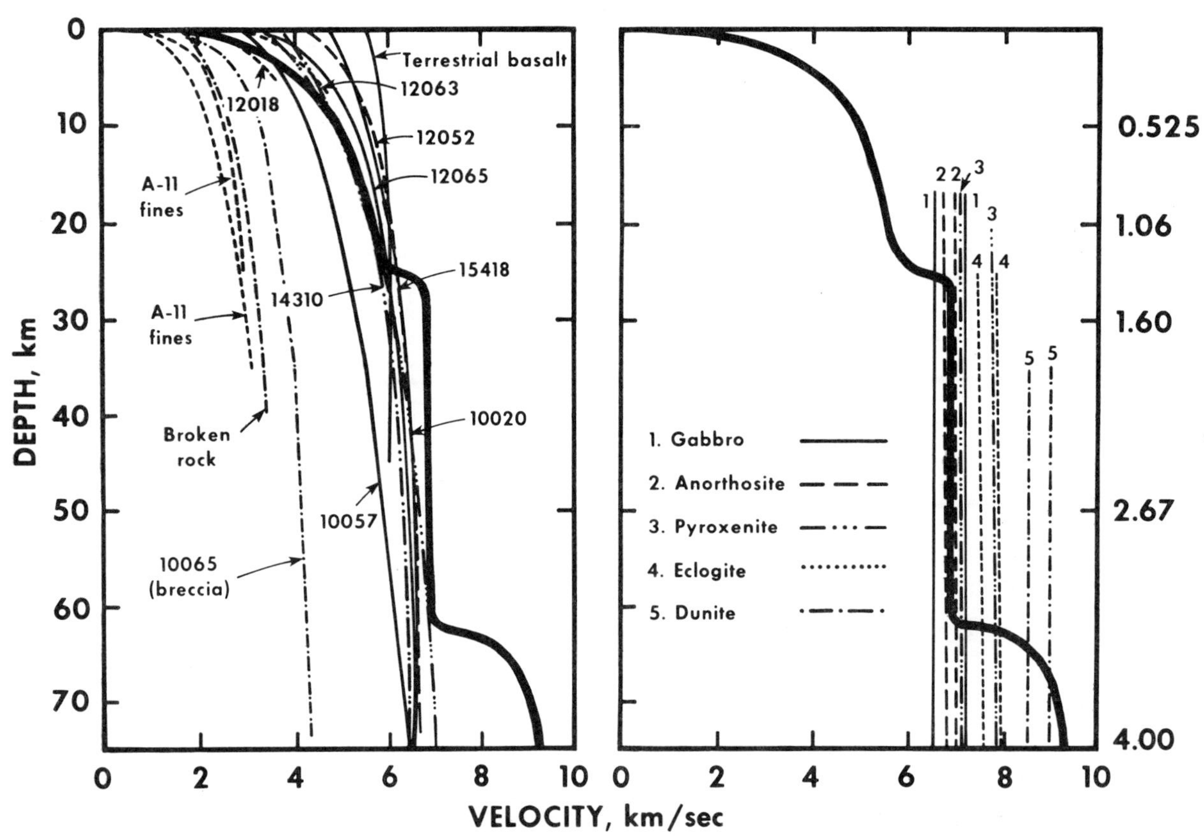

4.5.6 Observed seismic compressional wave velocity profile in the lunar crust (heavy curve), compared with measured velocities in lunar samples (left) and terrestrial rock types (right). From Toksöz *et al.* 1972.

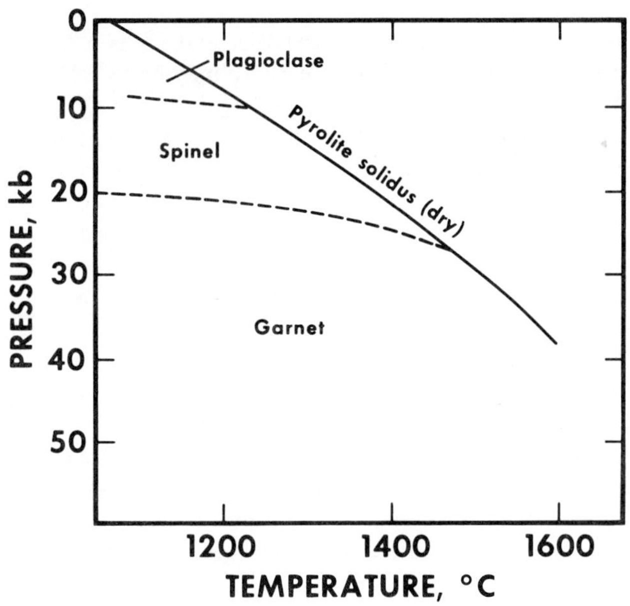

4.5.7 Stability fields of aluminous phases in mineral assemblages similar to those of the Earth's upper mantle (after Ringwood, 1975).

ties of the Moon, have Fe contents between 13 and 15 wt.%; the Earth contains ~35 wt.% Fe.

The currently accepted value of the lunar moment-of-inertia factor I/MR^2 is 0.391 (Table 4.2.1). This requires some kind of density increase with depth. Either layering of the silicate portion of the lower mantle, or the presence of a small core, is implied.

Seismic data from the Apollo Passive Seismic Network (Fig. 4.5.4 shows the station positions) give information on the velocity structure of the nearside crust, both under the maria and in the highlands, as well as the velocities of compressional and shear waves in the mantle. The major discontinuity indicated by the seismograms (Fig. 4.5.5) is taken to mark the interface between the crust and mantle. It lies between 45 and 60 km beneath Oceanus Procellarum (Toksöz *et al.*, 1974; Koyama and Nakamura, 1979; Goins *et al.*, 1978), while it may be as much as 75 km deep under the highland station of Apollo 16 (Goins, 1978). There is also a smaller discontinuity in the crust at about 20 km depth under both the mare and highlands. The cause of this break in velocities is disputed. An early hypothesis (Toksöz *et al.*, 1972) that this marked the base of the mare basalts was shown to be implausible by the continuation of the discontinuity under the highlands and by photogeologic estimates of much smaller basalt thickness (Head, 1979; section 5.4.2). Two views are now: (1) the discontinuity marks the depth of crack closure in the rocks (Simmons *et al.*, 1972), or (2) there is a change in the crustal composition at ~20km depth (Todd *et al.*, 1973). The second alternative is now the prevalent view (Simmons *et al.*, 1980). Seismic velocities in the crust are consistent with compositions of gabbro or anorthosite or a mixture of these and other rocks, as shown in Fig. 4.5.6 (Toksöz *et al.*, 1972; Lieberman and Ringwood, 1976). The density of the crust depends upon the assumed composition, ranging from ~2.75 for anorthosite to ~3.1 for gabbro.

The lunar seismic data indicate one discontinuity in the mantle (Nakamura *et al.*, 1976a; Dainty *et al.*, 1976; Goins *et al.*, 1978). Different workers have found various depths for the discontinuity, and slightly different velocity profiles. We use the most recently published velocity profile (Goins *et al.*, 1978), shown in Fig. 4.5.5. In this model the S-wave velocities drop sharply and the P-wave velocities decrease less sharply between 400 and 500 km depth. This decrease in velocities could be due to a compositional change (Dainty *et al.*, 1976), a phase change, or the onset of partial melting (Goins, 1978). There is a phase change from spinel- to garnet-bearing assemblages which would take place in this region (Fig. 4.5.7), but it is not likely that such a change would result in a decrease in velocities (Simmons and Wang, 1971; MacGregor, 1970; Ryzhova *et al.*, 1966). There seems to be too small a change in S-wave attenuation between the upper and lower mantle for the decrease in the velocity to be due to partial melting (Nakamura *et al.*, 1976b). The most likely possibility is that there is a

Table 4.5.4 Seismic velocities and temperature derivatives of velocity used to calculate seismic velocities of olivine and pyroxene at depth in the Moon.

	Seismic Velocities† (km/sec)(d,e)			
Mg/(Mg + Fe)	*Olivine*		*Pyroxene*	
mole %	V_P	V_S	V_P	V_S
90	8.30	4.80	7.96	4.82
85	8.20	4.72	7.86	4.76
80	8.10	4.64	7.77	4.70
75	7.99	4.55	7.67	4.65

	Velocity Derivatives			
	$\partial V_P/\partial P$(a)	$\partial V_P/\partial T$(b)	$\partial V_S/\partial P$(a)	$\partial V_S/\partial T$(b)
Olivine	10.8(c)	4.84(d)	3.60(f)	3.40(d)
Pyroxene	13.0(c)	4.15(g)	4.33(g)	3.39(h)

†Shown in Fig. 4.5.8

(a) $\times 10^{-3}$ km/sec-kb. (b) $\times 10^{-4}$ km/sec-°C. (c) Christensen (1974). (d) Simmons and Wang (1971). (e) Kumazawa (1969). (f) Kumazawa and Anderson (1959). (g) Birch (1969). (h) Birch (1943).

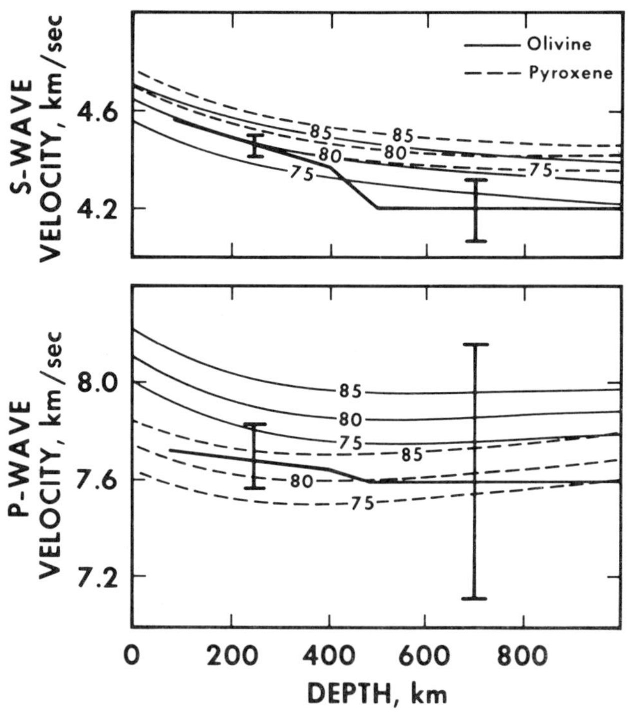

4.5.8 Variation of V_p and V_s in the Moon for olivines and pyroxenes of differing Mg/(Mg + Fe), shown with the Goins *et al.* (1978) velocity profiles (heavy lines). Associated error bars express allowable local excursions from the velocity profiles shown; velocities vertically averaged within the upper and lower mantles, respectively, are more closely constrained than this.

compositional change at this level. Two suggestions have been made: (1) a decrease in Mg/(Mg + Fe) of the silicates (Dainty *et al.*, 1976); or (2) the presence of a small amount of Fe-FeS trapped in the undifferentiated lower mantle (Buck and Toksöz, 1980).

Seismic velocities in olivines and pyroxenes having a range of Mg/(Mg + Fe) ratios were extrapolated to the pressures and putative temperatures of the lunar interior, using the velocity derivatives shown in Table 4.5.4. The extrapolated velocities are shown in Fig. 4.5.8, along with Goins *et al.*'s velocity profiles. These indicate that a mixture of olivine and pyroxene of Mg/(Mg + Fe) ~0.80 is required in the upper mantle (Nakamura *et al.*, 1974), and that an increase in Fe content with depth would explain the velocity discontinuity. The estimate of Fe content in the upper mantle is in agreement with the value derived from basalt petrogenesis, cited above. It is worth noting that the Fe content of the mantle of Earth is lower than that of the lunar mantle; the Mg/(Mg + Fe) ratio in the upper mantle of Earth is ~0.89 (Ringwood, 1975). This does not necessarily imply a bulk compositional difference between the Moon and Earth (Buck and Toksöz, 1980), since the lower mantle of the Earth may be enriched in FeO (section 4.5.1). The greater drop in S- than P-wave velocities at the mantle discontinuities (Fig. 4.5.5) seems to be related to the onset of the garnet stability field, combined with increased Fe content. It should be noted that the P-wave profile is not as well constrained as the S-wave profile.

There do not appear to be any additional discontinuities in the lower mantle, but S-waves are attenuated below ~1000 km (Fig. 4.5.9). This could indicate a small amount of partial melting in this zone (Nakamura *et al.*, 1974), though high attenuation does not require partial melting (Goetze, 1977). The seismic data indicate the presence of a small core, <360 km in radius (Nakamura *et al.*, 1974).

It is important to infer the average crustal thickness of the Moon, since the seismically-determined thickness is applicable only for a small area of the nearside. The center-of-mass of the Moon is offset toward Earth by 2.4 km from the center-of-figure (Kaula *et al.*, 1974). This is probably due to a greater thickness of low-density crust on the farside, and helps explain the lack of basin-filling by mare basalts there (Wood, 1973). Assume that the crustal depth at the point closest to Earth is either 60 km (Toksöz *et al.*, 1974) or 45 km (Koyama and Nakamura, 1979), and that the part of the Moon below the crust is a sphere displaced relative to

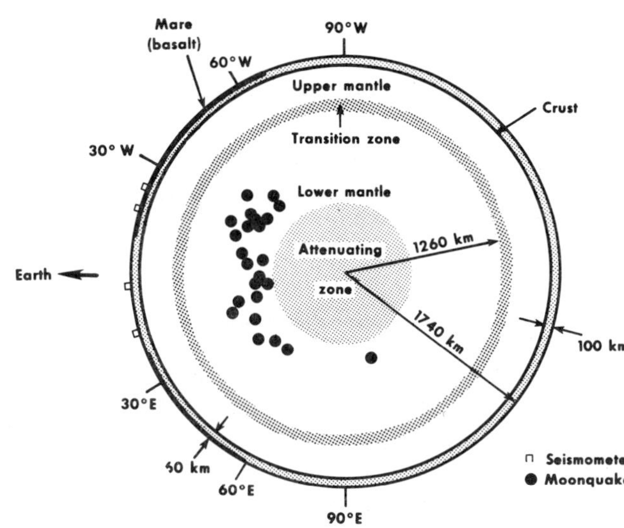

4.5.9 Interior structure of the Moon (equatorial section), showing the transition zone of seismic velocities at a depth of 400–500 km (radius of 1260 km), and the attenuating zone in the deep interior of the Moon. Heavy dots show epicenters of observed lower-mantle moonquakes. After Goins (1978).

Table 4.5.5 Crustal thicknesses, densities, and Al_2O_3 contents consistent with the lunar density, moment-of-inertia factor, and center-of-mass/center-of-figure offset.

Model[(a)]	1	2	3	4
Assumed crustal density, ρ_1(g/cm^3)	2.90	2.95	3.00	3.05
Mean subcrustal density, ρ_2 (g/cm^3)	3.40	3.39	3.38	3.37
Thickness difference (km) between near- and farside crusts, ΔT	38	43	51	61
Mean crustal thickness (km), T, assuming:				
Nearside crust thickness 60 km	76	79	81	86
Nearside crust thickness 45 km	61	63	67	72
Al_2O_3 content of crust (wt. %)	29.0	26.5	22.7	20.0
Al_2O_3 content of crust, expressed as wt. % of total Moon[(b)]	2.84	2.76	2.50	2.38

(a) Assumed crustal composition: 1 = material more aluminous and less dense than anorthositic gabbro; 2 = anorthositic gabbro; 3 = troctolite; 4 = material less aluminous and more dense than anorthositic gabbro.

(b) Mean crustal thickness assumed to be intermediate between the cases cited for two nearside thicknesses.

the center-of-figure. The amount of offset of this interior sphere required to account for the center-of-mass offset will depend on the density assumed for the crust, since this is not known. A simple equation relates the farside and average crustal thickness to the crustal density, ρ_1

$$\Delta T = \frac{2\Delta C\,[r_2^3(\rho_2-\rho_1) + r_1^3\rho_1]}{r_2^3(\rho_2-\rho_1)}$$

where ΔT is the thickness difference between nearside and farside, ΔC is the center-of-mass to center-of-figure offset (2.5 km), ρ_2 is the mean density of subcrustal material, r_1 is the radius of the whole Moon (1738 km), and r_2 is the radius of the subcrustal sphere, equal to r_1 - (T_n + ΔT/2), T_n being the nearside crustal thickness determined by seismology. Table 4.5.5 gives the values of T and ΔT obtained for several assumed crustal densities.

The assumptions used in making this simple calculation are somewhat questionable. In the first place, the two different estimates of crustal thickness in the area of the seismic array vary by 15 km. Second, possible lateral variations in crustal composition, and thus density, would affect the calculation of mean crustal thickness (Thurber and Solomon, 1978). Third, there could be lateral variations in the upper mantle composition, especially Fe/Mg. Finally, the density of the deep crust is not known and could be as high as 3.12 g/cm^3 (gabbro) and still be consistent with the seismic velocities. Mean crustal thickness can also be calculated from detailed gravity and topographic data as outlined in section 4.2.3 (Kaula *et al.*, 1974; Solomon, 1977; Bills and Ferrari, 1977), but this approach rests on the same assumptions as does the offset calculation.

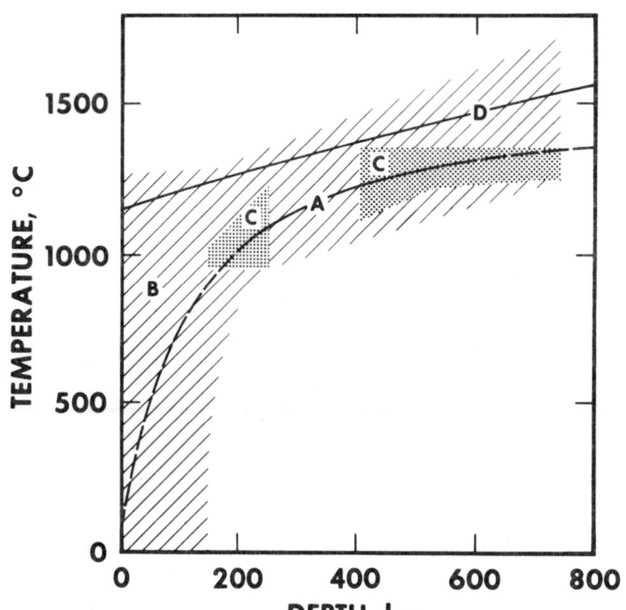

4.5.10 Ranges of temperature allowed for the interior of the Moon, based on measurements of electrical conductivity (from Duba *et al.*, 1976). Curve A: temperature profile estimated by Duba *et al.* from conductivity/depth data, assuming a mantle of orthopyroxene. Hatched area (B): range of temperature allowed by uncertainty of the conductivity/depth data of Dyal and Parkin (1973). Dotted areas (C): range of temperature allowed by uncertainty in the conductivity/depth data of Sonett *et al.* (1972). Curve D: solidus for the lunar pyroxenite of Ringwood and Essene (1970).

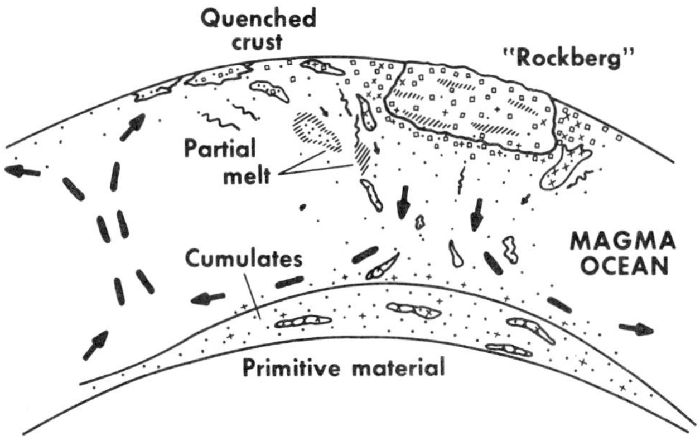

4.5.11 Cartoon of early crustal formation in the lunar magma ocean. Not to scale. Dots, olivine; squares, plagioclase; crosses, orthopyroxene; x's, pigeonite and augite. From Longhi (1978).

In spite of their uncertainty, the estimates of crustal thickness imply important things about the bulk composition of the Moon. A crust 60–80 km thick of anorthositic to gabbroic composition contains as much Al_2O_3 and CaO as a whole Moon of chondritic composition (see Table 4.5.5). Since we consider the crust to have been derived from only the outer part of the Moon, it appears that the latter must be enriched in Al_2O_3 and CaO relative to chondrites. Mineralogy of the lunar interior will be affected by the large quantities of these oxides in the unmelted part of the system, if such a zone survives. [A dissenting opinion on this point is held by A. E. Ringwood. He argues that the Moon appears enriched in CaO and Al_2O_3 only because it is depleted in Fe relative to chondrites, and that Ca/(Mg + Si) and Al/(Mg + Si) are not demonstrably enhanced in the Moon relative to Cl chondrites.]

The heat flow measurements made on the Moon (Langseth *et al.*, 1976) place limits on the lunar content of radioactive isotopes, and thus constrain thermal history models (see section 9.5.2). Electrical conductivity work also provides broad constraints on the present-day temperature profile, but the inference of temperature from conductivity depends upon assumptions of composition (Duba and Ringwood, 1973). Differing conductivity profiles have been reported (Wiskerchen and Sonett, 1977), and this range combined with the assumption about composition makes the temperature profiles rather uncertain. A common feature of the profiles derived from conductivity work is a steep increase in temperature to about 200 km and then a leveling off of the profile in the deep interior (e.g., Duba *et al.*, 1976; Fig. 4.5.10). Inferences from magnetic data about the size of a lunar core are also dependent on assumptions, but have been used to put an upper limit of ~580 km on the radius of the core (Goldstein *et al.*, 1976).

The deep interior of the Moon remains a region of mystery. Below ~1000 km we only know that S-waves are attenuated (Nakamura *et al.*, 1973). There are no reliable seismic velocities for this region. Lunar density models having the cosmochemical bulk compositions Mo1-Mo3 (Table 4.3.2e), calculated in the following subsection, have moment-of-inertia factors higher than

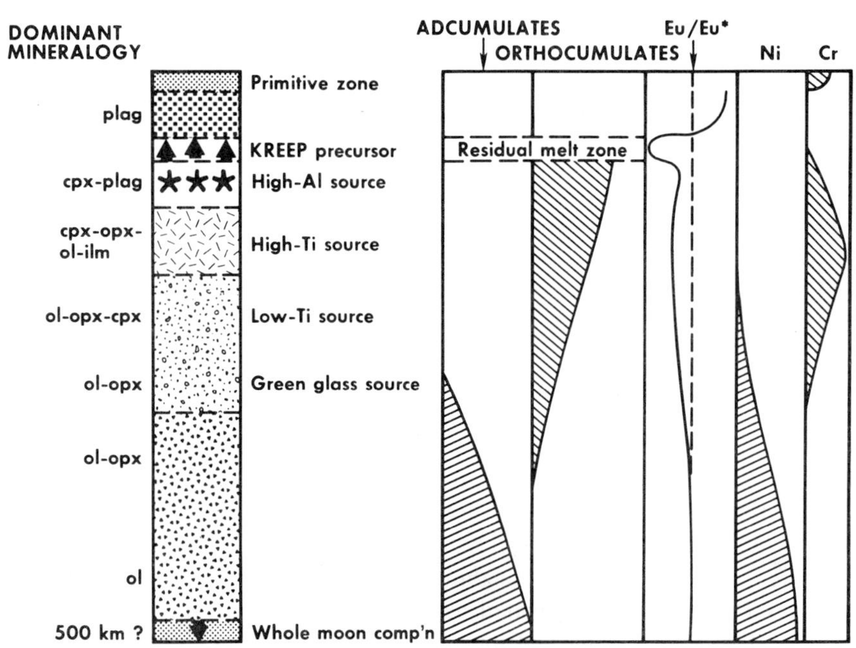

4.5.12 Left, mineral layering that might result from differentiation of a 500-km-deep magma ocean of the composition proposed by Taylor and Jakeš (1977). Also shown (right) are relative proportions of adcumulates and orthocumulates, and the distributions of Eu/Eu* (Eu* = the average europium content), Ni and Cr. From Taylor (1978).

that observed for the Moon. It appears necessary to postulate the presence of a small (~400 km radius) core, consistent with the seismic evidence for one (Nakamura *et al.*, 1974) and not inconsistent with electrical conductivity models (Wiskerchen and Sonett, 1977). How the core segregated is unclear. Total melting of the Moon seems inconsistent with considerations of thermal and tectonic history (Solomon and Chaiken, 1976) (see section 9.5.2). Inhomogeneous accretion or segregation due to partial melting are possibilities. Assuming the Moon formed by homogeneous accretion, there are two mechanisms that could account for the formation of a core. These are related to the possible reasons for lower seismic velocities in the lower mantle. First, an increase in Fe^{2+} in mantle silicates with depth (Dainty *et al.*, 1976) would be consistent with the formation of a core if reduction had occurred in the magma ocean during or after accretion. The free molten iron produced would eventually work its way to the center of the Moon to form a core. Second, it is possible that the Moon accreted homogeneously from material containing several wt.% of free Fe-FeS, and that this was only segregated from the upper mantle region (the former magma ocean). This material would also sink through the plastic lower mantle to form a core.

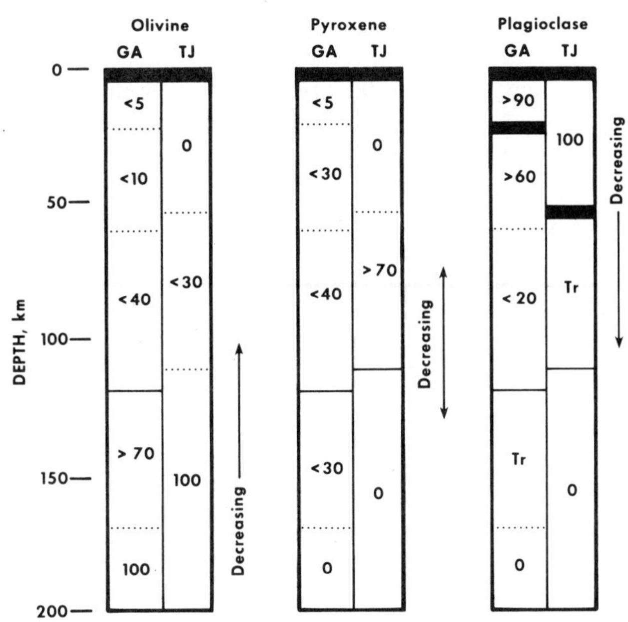

4.5.13 Distribution of olivine, pyroxene and plagioclase estimated for differentiation of a 200-km-thick magma ocean of the compositions proposed by Taylor and Jakeš (1974), TJ; and Ganapathy and Anders (1974b), GA. Numbers refer to modal percentages of that phase. From Minear and Fletcher (1978).

A framework in which to fit the knowledge we have about the lunar mantle and how it evolved is that of the "magma ocean" first proposed by Wood *et al.* (1970) and Smith *et al.* (1970) (see section 9.5.2). This refers to a thick molten zone which apparently occupied a large portion of the outer part of the Moon soon after it formed, and from which a plagioclase-rich crust separated (e.g., Longhi, 1978). Figure 4.5.11 is a cartoon suggesting some of the processes that operated during formation of the crust from the magma ocean. This model explains the thick lunar crust, but it is difficult to understand how so extensive a molten region could have formed without melting the whole Moon. The action of short-lived radioisotopes such as ^{26}Al has been proposed (Runcorn, 1978), but this requires improbably rapid accretion of the Moon, and (since Al is refractory, hence an early condensate) would be at least as effective in melting the Moon at depth as it would be near the surface. Heating by electrical induction has been put forward as a mechanism (Sonett *et al.*, 1975), but it is not clear how effective this would be. Accretional heating would have affected the outer portion of the Moon, since late-accreting material would have impacted with greater energy. However, recent studies show this may have been insufficient to produce the energy needed to cause melting (Wetherill, 1975; Kaula, 1979a).

Despite the problem of heat, the magma ocean model does explain many aspects of crust formation and basalt petrogenesis. It is of great interest to estimate the initial depth of melting of the magma ocean; Solomon (1977) has put the maximum depth of initial melting at ~400 km. His conclusion is based on the paucity of compressive features on the Moon's surface, and an estimate of the thermal contraction that would have resulted from the cooling of the near-solidus cumulates of the magma ocean. He argues that if a larger volume had been involved in the magma ocean, more compressional tectonic features should be seen at the lunar surface (section 9.5.2).

In a cooling magma ocean, the first mineral to crystallize would be Mg-rich olivine, which would have been denser than the magma. The olivine crystals would have sunk to form a layer at the bottom of the magma ocean. Taylor (1978) has pointed out that the decrease in seismic velocity now seen between 400 and 500 km is just what would be expected if this were the base of the olivine cumulate layer. He also points out that the crustal and whole-moon estimates of Th, U and K correspond well to crustal differentiation from a system of initial depth about 500 km. This depth also fits well with the estimated depths of basalt origin mentioned previously. Ringwood (1975) points out that the sunken olivine cumulate should have been rich in trapped liquid, and Minear and Fletcher (1978) note that the primitive undifferentiated interior would have a lower melting point than the olivine cumulate; mare basalts could have come from both sources.

The mineral assemblages left at different depths as a result of fractional crystallization of magma oceans having different bulk compositions have been calculated by a number of workers, who treated the petrologic and thermal complications of such a system with differing degrees of sophistication. Results of several of these studies are shown in Figs. 4.5.12–13. One success of these models is their prediction that the incompatible elements would be concentrated in the residual liquid of the fractionation process, which would have been left at shallow depths below the crust (e.g., Minear and Fletcher, 1978). Such a liquid may have played a role in the origin of KREEP basalts (section 3.4.1). The cooling times for magma oceans of different sizes have been estimated by Herbert *et al.* (1978), who find solidification times of 1–3 $\times$ 10^8 years (section 9.5.2).

Effects of liquid- and solid-state convection in the magma ocean are not completely understood. Fractional crystallization processes would have been complicated by convection (Longhi, 1978). Solid-state convection might have acted to homogenize the lunar mantle, but the preservation of the 400- to 500-km seismic discontinuity through time seems to show that convection either has not operated in this region or has been limited by a density contrast there; or, garnet's greater affinity for Fe than that of olivine and pyroxene (Ahrens, 1972) could have produced a difference in Fe content even if convection operated throughout.

Modelling the interior of the Moon

The Moon is the only extraterrestrial body for which we have information about seismic velocity as a function of depth; however, there are not enough other data (i.e., free oscillation periods) to constrain the density distribution well. Therefore, apparent velocity profiles, such as the one given by Goins (1978), are an important constraint on the composition of the mantle of the Moon. The pressures and temperatures in the Moon are also low enough that one can have some confidence in extrapolating seismic velocities to the center.

Extrapolation of the seismic velocities (V_p and V_s) of the major mineral phases likely to be present in the mantle of the Moon to values they should have at the depths shown in Fig. 4.5.8 are important in constraining possible mantle compositions. The seismic profiles of the Moon developed by Goins *et al.* (1978) are also plotted in Fig. 4.5.8. Extrapolation of the zero-pressure velocities of olivines and pyroxenes of several Fe contents to conditions at depth in the Moon was accomplished using the velocity derivatives ($\partial V/\partial P$ and $\partial V/\partial T$) of Table 4.5.4, the temperature profile of Toksöz *et al.* (1978) shown in Fig. 4.4.4, and pressures calculated for different depths. The greatest uncertainties in this procedure are in the determination of zero-pressure velocities for the Fe-rich minerals. This is the case because olivines and pyroxenes naturally occurring on Earth are of lower Fe content than those on the Moon. Consequently, mostly Fe-poor minerals have been studied; there are velocity measurements for only a few synthetic samples of higher Fe content. The error in the estimated velocities for minerals in the 0.75–0.80 Mg/(Mg + Fe) mole-ratio range is about $\pm$0.1 km/sec. Bars in Fig. 4.5.8 indicate maximum local velocity variations from the average velocities for the upper or lower mantle. It is unlikely that there are such great excursions from the average velocities shown, however, since we should see reflections from layers with very different velocities. The velocity derivatives and assumed temperature profile also contribute smaller uncertainties.

In testing the geophysical properties of the Moon models of Table 4.3.2e, the composition of the lunar crust was taken to depend on the Al_2O_3 content of the model composition. This was done since bulk compositions with less than 4.8 wt.% Al_2O_3 (Mo2 and Mo3) do not permit removal of a crust with anorthositic gabbro composition from the outer 450 km of the Moon. The mean composition of the lunar crust is not known, but there is considerable evidence that the average composition of the highlands surface is close to that of anorthositic gabbro (Taylor and Bence, 1975). Crusts which have lower Al_2O_3 contents than anorthositic gabbro

Table 4.5.6 Composition (wt.%) of the lunar upper mantle after removal of a crust from the outer 450 km of the Moon. The thickness and composition of the crust vary with the Al_2O_3 content of the overall model composition.

	Mo1	Mo2	Mo3
SiO_2	43.0	45.86	45.28
TiO_2	0.4	0.15	0.3
Al_2O_3	3.46	0.0	0.0
Cr_2O_3	0.35	0.5	0.5
MgO	33.8	37.41	35.72
FeO	14.6	14.7	16.52
MnO	0.15	0.18	0.0
CaO	4.22	1.4	1.36
Na_2O_3	0.12	0.0	0.0

Table 4.5.7 Abundances of normative minerals (wt. %) in mantle layers, and other properties of geophysical Moon models constructed from chemical models of Table 4.3.2e.

	Mo1			Mo2		Mo3	
	Plagioclase	Spinel	Lower	Upper	Lower	Upper	Lower
Silica	0.0	0.0	0.0	0.0	0.0	0.0	0.0
K-Feldspar	0.0	0.0	0.0	0.0	0.0	0.0	0.0
Plagioclase	9.91	0.0	0.0	0.0	0.0	0.0	0.0
Jadeite	0.0	0.76	0.98	0.02	0.39	0.0	0.33
Ca-Ts	(0.0)	(1.31)	(1.60)	(0.0)	(1.01)	(0.0)	(0.83)
Diopside	(9.62)	(15.42)	(18.74)	(5.42)	(11.88)	(5.42)	(9.70)
Total Ca-rich Pyroxene	9.62	16.73	20.34	5.42	12.89	5.42	10.53
Mg-Ts	(0.0)	(1.41)	(0.35)	(0.0)	(2.10)	(0.0)	(2.15)
Hypersthene	(12.3)	(17.81)	(4.39)	(34.4)	(26.46)	(35.43)	(27.10)
Total Ca-poor Pyroxene	12.3	19.22	4.74	34.4	28.56	35.43	29.25
Olivine	66.87	58.48	45.18	59.16	44.32	57.46	49.34
Py-Alm	(0.0)	(0.0)	(23.93)	(0.0)	(11.00)	(0.0)	(7.98)
Grossular	(0.0)	(0.0)	(2.84)	(0.0)	(1.31)	(0.0)	(0.95)
Total Garnet	0.0	0.0	26.77	0.0	12.31	0.0	8.93
Spinel	0.0	2.83	0.0	0.0	0.0	0.0	0.0
Ilmenite	0.76	0.76	0.76	0.29	0.38	0.51	0.57
Metal	0.0	0.0	0.0	0.0	0.0	0.0	0.0
Chromite	0.52	0.52	0.44	0.71	0.59	0.74	0.59
Total	99.99	99.31	99.20	100.00	99.44	99.55	99.53
Density (STP) of mantle layer	3.35	3.43	3.50	3.40	3.43	3.42	3.44
Mg/(Mg + Fe)	0.80	0.80	0.80	0.82	0.82	0.82	0.81
Mass of Moon model ($\times 10^{25}$ g)		7.36		7.32		7.39	
Percent deviation from true mass		+0.23		-0.37		+0.53	
I/MR^2 of Moon model		0.3943		0.3960		0.3977	
Percent deviation from true I/MR^2		+0.84		+1.28		+1.71	
Mg/Si, weight ratio		0.87		0.92		0.97	
Upper mantle V_s (km/sec)		4.45		4.52		4.48	
Lower mantle V_s (km/sec)		4.40		4.42		4.40	
Changes made to correct mass, I/MR^2, and seismic velocities							
Mg/Si wt. ratio		0.67		0.70		†	
Fe-FeS wt. % added to lower mantle		3.2		3.3			
Wt. % core added		0.26		1.0			
Core radius (km)		180		280			

† Mo3 could not be corrected without becoming quartz-normative, as discussed in text.

have to be correspondingly thicker (unless there is large-scale lateral compositional and density heterogeneity of the mantle). This is because the farside thickness has to be greater in the case of higher density (lower Al_2O_3) crusts in order to match the offset of center-of-figure to center-of-mass, as shown in Table 4.5.5. The total Al_2O_3 contents relative to the mass of the Moon are given there for different assumed crustal compositions. Model Mo1 with 7.6 wt.% Al_2O_3 overall was taken to have a crust of anorthositic gabbroic composition, while the other models were given crusts of lower Al_2O_3 content and therefore higher density and thickness. The assumed crustal composition was removed from the outer 450 km of the Moon, which is a slightly thicker zone than the 400 km limit on the depth of the magma ocean set by Solomon (1977), as discussed above. The base of the melted zone is assumed to lie in the middle of the depth range of the seismic velocity transition determined by Goins (1978). The lower mantle is taken to be undifferentiated.

The densities in the model mantles depend on the phases stable in the various depth zones of the mantle. Only in the case of the Moon, where 50% of the mantle lies above the garnet stability field (<20 kb), do we include the calculation of low-pressure norms (section 4.4.2). The reactions assumed for pressure-sensitive phase changes are shown in Fig. 4.5.7, and are discussed for the pyrolite composition by Green and Ringwood (1967) and other mineral systems by MacGregor (1974) (see section 3.3.2 of this volume). The plagioclase-to-spinel transition was taken to occur at 180 km (8 kb) in the Moon, and spinel-to-garnet at 400 km (20 kb). The low-pressure norm scheme was used to calculate mineral assemblages above 180 km, and between 180 and 400 km depth these assemblages were amended according to the reaction (from Kushiro and Yoder, 1966)

$$\underset{\text{plagioclase}}{CaAl_2Si_2O_8} + \underset{\text{olivine}}{2(MgO)_2SiO_2} \rightarrow \underset{\text{Ca-pyroxene}}{(CaO)(MgO)(SiO_2)_2} + \underset{\text{orthopyroxene}}{2(MgO)SiO_2} + \underset{\text{spinel}}{MgOAl_2O_3}$$

The pressure-induced reaction which should occur at ~400 km in the Moon (Fig. 4.5.7) can be written in simple form as

$$\underset{\text{pyroxene}}{4(MgO)SiO_2} + \underset{\text{spinel}}{MgOAl_2O_3} \rightarrow \underset{\text{olivine}}{(MgO)_2SiO_2} + \underset{\text{garnet}}{(MgO)_3(Al_2O_3)(SiO_2)_3}$$

(from MacGregor, 1970). The reaction is, of course, more complicated than this, involving Fe as well as Mg silicates and solid solutions of aluminous pyroxenes (Akella, 1976), which are taken into account in the high-pressure as well as the low-pressure norm schemes. In the above reaction a small quantity of spinel reacts with pyroxene to give olivine and a large quantity of garnet relative to the spinel. The resulting density increase depends on the amount of Al_2O_3 present. Since the Al-rich crust was assumed to be subtracted only from the uppermost 450 km of the Moon (the spinel and plagioclase stability fields), assemblages occurring in the garnet field are enriched in aluminous phases. The results of subtracting a crust of anorthositic gabbro from a 450-km-thick outer layer of the Moon are given in Table 4.5.6.

In our modelling scheme, Mo2 and Mo3 were taken to have differentiated all their Al into the crust. The upper-mantle mineralogies for Mo2 and Mo3 were taken to be the C.I.P.W. assemblages, and mineralogies for Mo1 were calculated for the plagioclase and spinel stability fields. All the assemblages used are shown in Table 4.5.7. We have not attempted to model a Moon in which the center-of-figure/center-of-mass offset is due to a more magnesian farside mantle (Wasson and Warren, 1980) instead of a thicker farside crust.

Results of the density modelling, also shown in Table 4.5.7, are that Mo1 is 0.2% too dense, Mo2 is 0.4% underdense, and Mo3 is ~0.5% too dense. All the models have too high a moment of inertia, with Mo1 being the best at $I/MR^2 = 0.394$. All models have seismic velocity profiles close to the observed upper-mantle values, but the model velocities for the lower mantle are much too high. This is due to the abundance of garnet, which has high seismic velocities, in the normative mineralogies for the lower mantle. The upper mantle is depleted in Al relative to the lower mantle since the

Table 4.5.8 Bulk and upper-mantle compositions of lunar models corrected (see text) to agree with observed mean density and moment of inertia of the Moon.

	Bulk Moon		Upper mantle†	
	Mo1	Mo2	Mo1	Mo2
SiO_2	48.7	44.5	49.7	52.8
TiO_2	0.4	0.2	0.4	0.15
Al_2O_3	7.6	4.6	3.46	0.0
Cr_2O_3	0.3	0.4	0.35	0.48
MgO	25.3	27.8	29.2	31.7
FeO	11.3	12.0	12.6	13.5
MnO	0.15	0.18	0.15	0.18
CaO	6.1	3.8	4.22	1.37
Na_2O	0.15	0.06	0.12	0.0

†After removal of crustal constituents, as discussed in text, from outer 450 km of the Moon.

Table 4.5.9 Normative mineralogies of layers in mantles having the corrected compositions of Table 4.5.8.

	Mo1			Mo2	
	Plagioclase†	Spinel	Lower	Upper	Lower
Jadeite	0.0	0.76	0.98	0.02	0.39
Plagioclase	9.91	0.0	0.0	0.0	0.0
Ca-Ts	0.0	1.31	1.67	0.0	1.09
Diopside	9.62	15.42	19.57	5.43	12.77
Mg-Ts	0.0	4.25	3.02	0.0	4.92
Hypersthene	51.21	53.72	38.27	75.97	62.25
Olivine	27.98	21.74	13.37	17.59	10.66
Py-Alm	0.0	0.0	18.77	0.0	5.60
Grossular	0.0	0.0	2.23	0.0	0.67
Spinel	0.0	0.75	0.0	0.0	0.0
Ilmenite	0.76	0.76	0.76	0.29	0.38
Chromite	0.52	0.52	0.44	0.71	0.59
Total	99.99	99.23	99.09	100.01	99.32
Density (STP) of mantle layer	3.31	3.39	3.44	3.37	3.39
Mg/(Mg + Fe)	0.812	0.812	0.807	0.812	0.810

†Upper and lower refer to the lunar mantle above and below 450 km. Plagioclase and spinel refer to the upper-mantle mineral stability zones discussed in the text.

crust has been derived from it. For this reason it does not contain normative spinel, which would raise its seismic velocities also.

The models have been adjusted to match four constraints to within one standard deviation of the accepted value: (1) the upper mantle seismic velocity; (2) lower mantle seismic velocity; (3) the moment-of-inertia factor; and (4) the mean density of the Moon. To do this, four parameters were varied: (1) the Fe content of the lower mantle was adjusted to correct seismic velocities there; (2) a core was added to correct I/MR^2; (3) Mg/Si (overall) was lowered to correct the mean density; and (4) Mg/(Mg + Fe) was adjusted to correct upper-mantle velocities. Inclusions of Fe-FeS were added to decrease the lower-mantle velocity, because this brings about a smaller density increase than would

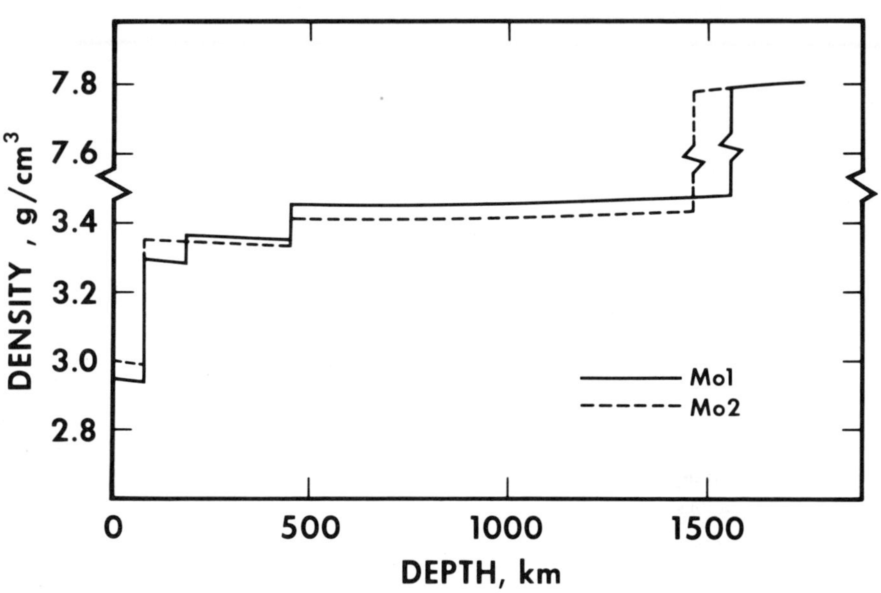

4.5.14 Density profiles for corrected models of the Moon, Mo1 and Mo2. Mo3 could not be corrected realistically.

addition of Fe^{2+} to the silicates. Also, we consider that a simpler evolutionary history attaches to the retention of dispersed Fe-FeS in the lower mantle than to heterogeneity of silicate composition in the mantle.

The results for the bulk composition of the mantle are somewhat surprising. The Mg/Si ratio must be much less than that of C1 chondrites (<0.91, by weight). Because the dense crust assumed for Mo3 increased the moment-of-inertia factor of the model, it turned out not to be possible to match the model to observed geophysical constraints without lowering its Mg/Si ratio to the point that free silica appeared in the upper mantle norms. This was considered unacceptable, since the source regions of mare basalts must have been olivine normative.

Corrected bulk- and upper-mantle compositions and mantle mineralogies for the models are shown in Tables 4.5.8 and 4.5.9, respectively. The corrected density profiles appear in Fig. 4.5.14. Note the large increase in normative pyroxenes, relative to olivine, in the corrected as opposed to uncorrected mantles. Figure 4.4.2 shows olivine, pyroxene and garnet densities as a function of Fe contents, and the ranges of acceptable upper- and lower-mantle densities found in this study. The origin of the core which is required by all these models is problematical, but its small size (as little as 0.26% of the mass of the Moon) means that only a small amount of free Fe need have been part of the bulk composition.

4.5.3 Mercury

Consequences of the density constraint on Mercury's composition, and other considerations

The planet Mercury has a mass of 3.30 (±0.02) × 10^{26} grams and a mean radius of 2.439 (±0.002) × 10^8 centimeters (Table 4.2.1), and thus a mean density of 5.43 grams/cm^3. Any plausible bulk composition satisfying this density must include well over 50% Fe. It is generally agreed that this Fe must be largely differentiated into a core, because (1) Mercury has a perceptible magnetic field, appreciably more than that of either Venus or Mars; and (2) Mercury's surface has the appearance of being predominantly silicate, its albedo being too high for Fe and its morphology suggestive of high viscosities. Mercurian spectrophotometry has detected a weak absorption band indicative of Fe^{2+} ions, consistent with a surface FeO content of about 5% (McCord and Clark, 1979; section 2.2.3). A further inference from Mercury's surface is that the separation of the Fe core probably took place rather early in its history. Such a separation would have resulted in a significant expansion of Mercury (from the fact that silicates are more compressible than Fe, and also as a result of heating due to the sinking of core materials: see section 9.5.4); yet surface tectonic features (arcuate thrust faults) point to contraction since early times (section 6.5.3).

The high likelihood of its Fe being separated into a core, together with Mercury's small size, simplifies the correction of its density to that which would obtain at 10 kb pressure (the customary standard for measuring mechanical properties): ~5.3 g/cm^3. Hence the problem of inferring the mercurian mantle composition is largely a problem of inferring its bulk composition, since almost certainly Fe and the siderophiles have gone into its core, while any free silica or calc-aluminous silicates would have differentiated into a crust very early. The latter inference is based primarily on lunar experience. The optical properties of the crusts of the Moon and Mercury are characteristic of silicates of low TiO_2 and FeO content (sections 2.2.1–3), and the morphology of their surfaces is largely due to impact cratering. Early differentiation of a crust seems characteristic of a small planet composed of refractory material.

The primary factor affecting the bulk composition of Mercury is, of course, the probable high temperature that existed in its zone of the solar nebula very close to the sun, so that it is formed of predominantly high-temperature condensates, as discussed in section 4.3.2. If the temperature was held around 1400° K until most of the uncondensed nebular material was blown away, then a composition satisfying Mercury's mean density can be rationalized, since most of the Fe but only a minor part of the magnesian silicates would have been condensed (Ringwood, 1966; Lewis, 1972). However, the band of temperatures at which this condition prevails is quite narrow, as indicated by the closeness of the forsterite and metallic Fe curves in Fig. 4.3.2 (Grossman and Larimer, 1974). Hence other factors must be considered in addition to condensation temperature. Two other principal factors affecting the mercurian bulk composition have been emphasized in recent years: (1) dynamical interaction among the terrestrial planet zones, leading to compositional mixing and nonsystematic planetary growth patterns, and (2) collisional differentiation (section 4.3.2).

The basis for consideration of (1), dynamical mixing, is primarily observational: the more data that become available, the more it appears that compositions from Venus to the asteroid belt depart from simple gradients of increasing volatile content and oxidation level. The apparent depletion in Earth of *all* elements condensing at temperatures below ~1200° K makes it unlikely that the Earth has a sulfur content approaching cosmic.

Hence the lower density of Venus compared to the Earth may be more plausibly accounted for by a higher oxidation level or lower Fe content, as discussed in section 4.5.4. On the other hand, the Viking data on gaseous isotope ratios in Mars suggest that Mars is depleted in volatiles compared to the Earth (Anders and Owen, 1977). While the chondritic meteorites must have resided in the asteroid belt during most of solar system history (Wetherill, 1974), their wide ranges of volatile contents and degrees of Fe oxidation suggest a wide range of formation conditions, perhaps spanning the entire range of terrestrial planet zones.

Factor (2), collisional differentiation, has been invoked as a means of separating differentiated meteorite types from one another, and as an explanation for the anomalous composition of the Moon (Orowan, 1969; Ruskol, 1972; Kaula, 1977b). If a planetesimal is differentiated into metallic and oxidized parts, and subsequently subjected to breakup by collision, then the mechanically weaker oxides or silicates will tend to be associated with the smaller fragments when the planetesimal is disrupted by a collision. Any subsequent dynamical process that is dependent on the size of bodies (such as gas drag or capture by a satellite swarm) will then act as a compositional differentiation process. These effects will be appreciably enhanced if the planetesimal has been gravitationally layered into a core and mantle, or a mantle and crust prior to disruption. If this effect is at all perceptible, then it would have operated most strongly in Mercury's zone, where collisional velocities resulting from interzone scattering were the highest, and where high orbital velocities relative to a gaseous nebula would make it most effective as a means of discriminating among planetesimals on the basis of area/mass ratio, as suggested by Weidenschilling (1978). Weidenschilling applied this drag mechanism to the problem of depletion in Mercury of all oxides with respect to nickel-iron. However, if the hypothesis is applied to fragments from the breakup of planetesimals which have differentiated a crust from a mantle, then it also would act (though less strongly) to deplete aluminous silicates with respect to magnesian silicates as well, because in collisional breakup the smaller fragments tend to be derived from the outer layers. If drag by nebula gas were appreciable, then the small fragments of high area/mass ratio would lose energy and spiral into the sun. This could happen rapidly enough not to require an implausibly long-lived nebula.

The necessarily-high Fe content makes it unlikely that Mercury's mean density is perceptibly affected by any element that has a condensation temperature below 1000° K (at 10^{-4} atm, the presumed nebular gas pressure). Mercury is also unlikely to contain a significant amount of SiO_2 beyond that necessary to combine with the MgO, Al_2O_3 and CaO to make forsterite, enstatite, diopside and anorthite. It may contain appreciably less SiO_2, if most of the Al_2O_3 it accreted was incorporated in spinel, as suggested by Fig. 4.3.2 for higher temperatures. Any model of Mercury must have a ratio of Fe + Ni to Al_2O_3 + CaO + MgO + SiO_2 higher than that calculated from the solar or cosmic abundances to satisfy its mean density. (The principal neglect here is Cr, 24% as abundant as Ni.) The range of possible compositions of Mercury can be characterized in terms of their Al/Mg ratios. The extreme condensation model (complete condensation of refractories + metallic Fe, partial condensation of forsterite) would have an Al/Mg ratio somewhat higher than solar, and a solar Al/Fe ratio. An extreme dynamically-mixed model would have a solar Al/Mg ratio. The consequences of an extreme collisionally-differentiated model would depend on the composition of the planetesimals on which it operated, and thus on the condensation temperatures of their materials. Referring to the uppermost part of Fig. 4.3.2, if these temperatures had all been above ~1360° K, the Al/Fe ratio would be solar, since all the Al would be retained in the denser spinel phase; while if temperatures had been somewhat lower, sufficient SiO_2 could have condensed to put all the Al into the lighter mineral anorthite. This mineral would have moved to the surface layer during planetesimal differentiation, where it would have been exposed to fragmentation into especially-small debris and selectively removed from the Mercury zone of the nebula by the dynamical effects discussed above. The result would have been an Al/Mg ratio approaching zero in solid material remaining in the Mercury zone.

If we assume: (1) all Fe-Ni is in metallic form (density 7.9); (2) Al is incorporated either in spinel, $MgAl_2O_4$ (density 3.55) or anorthite, $CaAl_2Si_2O_8$ (density 2.7); (3) the Ca/Al ratio is always solar; (4) the Ca in excess of that in the anorthite is incorporated in diopside, $CaMgSi_2O_6$ (density 3.3); and (5) the rest of Mercury is forsterite, Mg_2SiO_4 (density 3.3), then we can write two equations in four unknowns applicable to all models

$$1 = A + I + P + S$$

$$\frac{1}{5.3} = \frac{A}{2.7} + \frac{I}{7.9} + \frac{P}{3.3} + \frac{S}{3.55}$$

where A is the mass fraction of anorthite, I the mass fraction of Fe-Ni, P the mass fraction of diopside plus forsterite, and S the mass fraction of spinel. [The principal neglect in this model is enstatite, $MgSiO_3$, which becomes a significant condensate phase below ~1300° K

(Grossman and Larimer, 1974), but whose density is very close to that of forsterite.]

The additional equations to make the system solvable must come from model assumptions as to A/S and Al/Mg or Al/(Fe + Ni) ratios. Using the cosmic ratios of Cameron (1981; see Table 4.3.1), the pure condensation (to 1360° K) model then requires

$$\frac{S}{I} = 0.114, \qquad \frac{A}{I} = 0.000$$

whence S = 0.07, I = 0.64, P = 0.29. This value of P represents condensation of only 12% of the forsterite that the system's Mg is capable of making. The extreme dynamical mixing model requires, from the cosmic Al/Mg and Ca/Mg ratios

$$\frac{A}{P} = 0.148$$

whence A = 0.04, I = 0.65, P = 0.31. The corresponding composition is given as Me4 in Table 4.3.2a. (The differences from the more elaborate dynamical mixing model Me2 arise from the latter's not being forced to satisfy the mean density equation.) An extreme collisional differentiation model gives the same solution as the pure condensation model above 1360° K. For temperatures below 1360° K, if the model is taken to imply negligible Al content, then

$$A = 0.00, \qquad S = 0.00$$

whence I = 0.65, P = 0.35. The resulting composition is given as Me5 in Table 4.3.2a.

Thus, there appears to be little flexibility about the metallic portion of Mercury; the main question is the spinel/anorthite/(diopside + forsterite) ratio, on which the dynamical mixing model is an interpolation point 70% of the way from pure condensation to the extreme <1360° K collisional differentiation process. It is difficult to imagine the dynamical mixing or collisional differentiation processes being very efficient, however. The samples that seem nearest to being relict material from Mercury's primordial zone are type I enstatite chondrites. However, features of their composition, such as their 7–15% troilite content (Mason, 1962, p. 80), make them of minor relevance to Mercury's composition.

If the Al/Fe and Ca/Fe ratios were solar and all the Al was incorporated in spinel, then Mg/Fe would be about 23% of solar.

An additional consideration not taken into account in the foregoing discussion is the anorthite ⇄ garnet + kyanite + quartz phase transition, which would occur about 150 km deep in Mercury, assuming $dT/dZ \approx 6°/km$ (and the MgO in enstatite). However, if we adopt the popular (but unproven) assumption that the energy pulse from Mercury's core differentiation would have led to rather effective crustal differentiation, then this phase transition is of minor significance.

In summary, Mercury probably has an Al/Mg ratio appreciably higher than cosmic but an Al/Fe ratio somewhat lower. A significant uncertainty is the SiO_2 content, which depends in turn on the condensation temperatures of the material. If these temperatures were predominantly below 1360° K and if Mercury had an efficiency in upwardly differentiating crustal materials similar to the Moon's, on the order of 90% of this Al would be differentiated (in the form of anorthite) into a crust, which thus would be 75–150 km thick. On the other hand, condensation temperatures predominantly above 1360° K would lead to most of the Al's being buried as spinel, and a mafic surface. In any case, the low oxidation level leads to a mantle Mg/(Mg + Fe) molar ratio well over 90% and a mantle about 500 km thick. The free silica content of any crust plus mantle would be negligible, and probably only minor traces of more volatile elements would occur. The chemistry relevant to basaltic volcanism would probably be more similar to the Moon's than to that of any other terrestrial body, because of Mercury's paucity of volatiles and its low oxidation level. However, there still must be major differences: a higher Mg/Fe ratio, a higher spinel/anorthite ratio, and even greater depletions in both siderophiles and chalcophiles in Mercury than the Moon, because of the much greater metal content.

The radioactive heat sources U and Th probably have ratios to Al close to cosmic: about 10^{-6} for U/Al and 4×10^{-6} for Th/Al. These ratios have been used in Table 4.3.2a.

Modelling the interior of Mercury

Depth/density profiles for Mercury models Me1–Me4 (Table 4.3.2a) were derived in a manner analogous to that described for the Moon (section 4.5.2). In the case of Mercury, the assumed composition and thickness of crust influence the mantle composition importantly. We have assumed the crustal thickness to be 50 km, comparable with Mars, and since we have extremely meager data on its composition, we have tested two possible crust compositions: anorthositic gabbro and olivine tholeiite. Given the probable very large core size of Mercury and this conservative estimate of the crustal thickness, the crust is ~11% of the mass of the crust-plus-mantle. The crust was taken to have been derived from the whole mantle, and accordingly each crustal composition was subtracted from the crust-plus-mantle compositions. The remaining mantle oxide abundances were converted to mineralogies using the high-pressure norm scheme of section 4.4, and average densities were calculated for the assemblage (see Table 4.5.10). The crustal STP density was taken to be 3.0 g/cm^3.

Table 4.5.10 Abundances of normative minerals (wt.%) in mantles, and other properties of geophysical Mercury models constructed from chemical models of Table 4.3.2(a).

	Me1		Me2		Me3		Me4		Me5
	Olivine† Tholeiite	Anorthositic† Gabbro	Olivine Tholeiite	Anorthositic Gabbro	Olivine Tholeiite	Anorthositic Gabbro	Olivine Tholeiite	Anorthositic Gabbro	
Silica	0.0	0.0	0.0	0.0	0.0	0.0	0.0	0.0	0.0
K-Feldspar	0.0	0.0	0.0	0.0	0.0	0.0	0.0	0.0	0.0
Jadeite	0.0	0.0	0.0	0.0	0.0	0.39	0.0	0.0	0.0
Ca-Ts	(2.05)	1.6	(1.74)	(1.60)	(1.21)	(1.20)	(1.02)	(0.13)	(0.0)
Diopside	(28.85)	(27.76)	(19.84)	(18.23)	(13.92)	(13.77)	(11.64)	(10.52)	(0.0)
Total Ca-rich Pyroxene	30.9	29.36	21.61	19.83	15.13	14.97	12.66	10.65	0.0
Mg-Ts	(0.0)	(0.0)	(0.0)	(0.0)	(2.56)	(2.99)	(0.0)	(0.0)	(0.0)
Hypersthene	(0.0)	(0.0)	(0.0)	(0.0)	(30.76)	(35.82)	(0.0)	(0.0)	(1.7)
Total Ca-poor Pyroxene	0.0	0.0	0.0	0.0	33.32	38.41	0.0	0.0	1.7
Olivine	53.21	57.03	67.33	67.43	30.93	20.81	82.3	86.53	98.3
Py-Alm	(0.0)	(0.0)	(6.83)	(10.67)	(12.65)	(8.19)	(1.24)	(1.80)	(0.0)
Grossular	(0.0)	(0.0)	(0.85)	(1.35)	(1.56)	(1.01)	(0.16)	(0.22)	(0.0)
Total Garnet	0.0	0.0	7.68	12.02	14.21	9.20	1.40	2.02	0.0
Spinel	10.66	8.77	4.22	0.31	0.0	0.0	2.88	1.00	0.0
Ilmenite	0.37	0.51	0.23	0.37	0.42	0.61	0.0	0.0	0.0
Metal	4.04	3.2	0.0	0.0	5.30	5.30	0.0	0.0	0.0
Total	99.19	98.88	101.07	100.81	99.31	99.19	99.24	100.2	100.0
Density (STP) of mantle	3.309	3.298	3.278	3.279	3.346	3.334	3.246	3.239	3.224
Mg/(Mg + Fe) in mantle	1.00	1.00	1.00	1.00	0.980	0.977	1.00	1.00	1.00
Density (STP) of core	7.903		7.944		7.903		7.925		7.925
Wt.% of Mercury in core	38.4		39.1		68.0		67.0		65.0
Mass of Mercury model (× 10^{26} g)	2.53		2.55		3.30		3.150		3.245
Percent deviation from true mass	-23.3		-22.8		0.0		- 4.57		-1.70
I/MR^2 of Mercury model	0.341		0.340		0.335		0.331		0.334
Change made to correct mass:									
New wt.% core	68.4		67.9		68.0		69.1		69.4
I/MR^2	0.3350		0.3346		0.335		0.334		0.334

†Assumed compositions of crusts subtracted from bulk compositions to yield mantle compositions.

Testing of models Me1 and Me2 against the observed density of Mercury shows, of course, that both models are underdense by about 23%. The core masses had to be nearly doubled to bring the models into correspondence with the observed density, as shown in Table 4.5.10. The corrected density profiles appear in Fig. 4.5.15.

Chemical models Me3–Me5 were constrained to match the mercurian mean density at the outset. The only reason there are any discrepancies in total mass is that the norms used in deriving the chemical models are simpler than those used in the geophysical modelling. Model Me3 matches the total mass of Mercury exactly.

4.5.4 Venus

The relatively small differences in mass, mean density (see Table 4.2.1), and heliocentric distance of Venus from those of Earth (together with the paucity of compositional data) encourage us to discuss the bulk constitution of Venus in terms of differences from the Earth.

If Venus had a bulk composition and structure identical to that of Earth, then its mean density would be 5.34 g/cm^3, about 1.9% greater than the observed value (Table 4.2.1). "Identical structure" is defined here as: (1) most of the Fe is in the core, leaving an Mg/(Mg + Fe) molar ratio of 0.88 in the mantle; (2) essentially all the S is in the core; (3) the crust comprises about 0.6% of the total mass; and (4) at more than about 100 km depth, the temperature gradient is maintained by solid state convection. The density differential cited is rather accurately known: probably within 0.01 g/cm^3 (Ringwood and Anderson, 1977). Although there are appreciable uncertainties in absolute densities inferred from equations-of-state, as discussed in section 4.4, these equations should be quite adequate to assess effects of the relatively small pressure differences between Venus and Earth. Similarly, although there are uncertainties about just what temperature gradient is maintained by mantle convection, as discussed in section 9.6, the same physics applies to both bodies. The measured high surface temperature of Venus, 740° K, would have minor influence on the general pattern of convection: see section 9.5.5. (It may, of course, have major influence on crustal evolution through its effect on basaltic differentiation.)

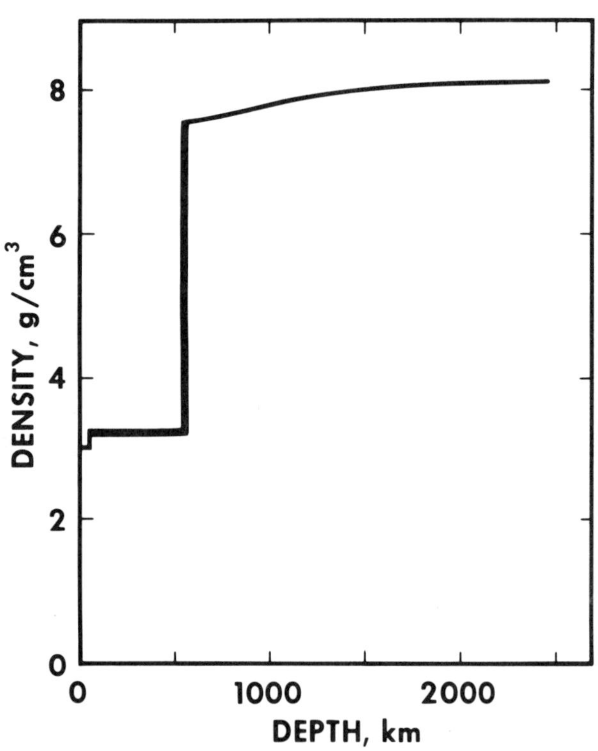

4.5.15 Density profiles for corrected Mercury models Me1-Me5. These are identical to within the width of the line shown.

Compositional considerations

The main debate about Venus thus centers on its primordial bulk composition, which must explain a mean density 1.9% lower than that of a Venus made of the same material as Earth. Hypotheses which have been suggested are: (1) a lower content of Fe in Venus; (2) less S; (3) a higher content of oxygen (O); and (4) the effect of Venus's high near-surface temperatures on its basalt/eclogite ratio. Debate exists because the evidence in favor of each of these hypotheses is empirical and indirect.

The principal rationale for lower Fe in Venus (model V5 in Table 4.3.2b) is probably the fact that it is necessary to invoke Fe contents for the Moon, Mercury and Mars that are different from Earth's. Also weakly in favor of less Fe in Venus is the absence of a magnetic field (Russell *et al.*, 1979), but this has alternate explanations. However, none of the hypothetical nebula processes discussed in section 4.3 has ever been used to rationalize a reduced Fe content for Venus, which is bigger than Mercury and closer to the sun than the Earth.

The principal factor favoring lower S in Venus than in Earth (model V1 in Table 4.3.2b) is that it was formed closer to the sun than the Earth was. The temperature gradient in the nebula before volatiles were blown away (Fig. 4.3.8) could plausibly have been such that the temperature in the proto-Venus zone was higher than the condensation temperature of FeS,

~700° K (Grossman and Larimer, 1974), but lower than this in the proto-Earth zone (Ringwood, 1966; Lewis, 1974b). Earth's core is too low in density to be pure Fe, and, previously, S has been considered the most likely light element in the core. However, if condensation temperature effects led to lower S in Venus, then the latter should also be appreciably more reduced than Earth. Ringwood and Anderson (1977) estimate that the density differential attainable by this effect would be only 0.4%.

The principal evidence in favor of a higher oxidation level in Venus than in Earth (model V4 in Table 4.3.2b) is probably its massive CO_2 atmosphere. Venus's atmosphere was probably outgassed like the Earth's. Despite having a primordial Ar content much higher than Earth's (Hoffman *et al.*, 1979), Venus is still orders-of-magnitude lower than the sun in inert gas abundances, and hence probably acquired its volatiles incorporated in planetesimals. To outgas an oxidized atmosphere requires the presence in the planet of substantial Fe^{2+} and Fe^{3+} (Ringwood and Anderson, 1977). As with the Fe-depletion hypothesis, a mechanism that would enrich Venus in O relative to Earth is difficult to find. Ringwood and Anderson (1977) suggest that Venus is sufficiently smaller than Earth to have been appreciably less heated by accretional energy, and possibly less reduced in the process. The absence of a moon around Venus also favors a higher oxidation level. If the Earth's moon was formed by binary accretion (a process entailing many small events), then Venus is also likely to have had a high-silicate moon, which eventually spun into Venus from tidal friction effects, effectively decreasing Venus's metal/silicate ratio. If the Earth's moon was formed by a major impact (one big event), on the other hand, then Venus could plausibly have escaped this devolatilizing catastrophe.

A concomitant of CO_2 outgassing is H_2O outgassing, implying a venusian mantle which once was as "wet" as the Earth's (Ringwood and Anderson, 1977). The H_2O history of the venusian atmosphere is still uncertain. Most models assume H_2O circulates to the upper atmosphere, where it is photodissociated; subsequently, H escapes from the exosphere, and O is incorporated in rocks (Walker, 1977).

Finally, tectonic considerations can also act to diminish Venus's mean density relative to the Earth's. Oxburgh and Parmentier (1978) discuss the conditions for instability of the Earth's lithosphere and calculate the age at which it became denser than the upper mantle. At present, lithosphere being subducted by the Earth contains less than 10% of crustal material. If Venus's surface was always as hot as it is now, it is likely that the lithosphere never became unstable, i.e., it has always been less dense than the upper mantle. D. L. Anderson (1981) proposed that in the absence of subduction, much of the basalt generated in the mantle of Venus has remained near the surface. The high surface temperature implies that it has a thinner thermal boundary layer than the Earth. The lower pressures and higher temperatures in Venus's upper mantle place it outside the stability field of eclogite. The resulting higher basalt/eclogite ratio in Venus might lead to as much as a 3% lower density for Venus than for Earth. If a more earth-like geotherm is assumed for Venus, the density difference is diminished.

The principal evidence that Venus has differentiated an appreciable crust consists of (1) gamma-ray measurements at Venera landing sites, indicating significant lithophile enrichment (Vinogradov *et al.*, 1973; Surkov, 1977; section 2.2.5); (2) elevation differences too large to maintain by mechanisms other than isostasy (Masursky *et al.*, 1977; Pettengill *et al.*, 1979); and (3) a surface morphology indicative of appreciable internally-generated thermotectonic activity (Pettengill *et al.*, 1979). Venus's size, its probable U and Th content, and manifestations of crustal differentiation indicate that its reduced Fe was differentiated into a core fairly early in its evolution, as discussed in section 9.5.5. Given a core, the composition of the mantle depends on the bulk planetary composition. An Fe- or S-depleted Venus would have an MgO/(MgO + FeO) ratio like the Earth's, ~0.88 (since the amount of O is unchanged). An O-enriched Venus would have an appreciably lower ratio: about 0.76 (Ringwood and Anderson, 1977). A consequence of all these hypotheses is a core proportionally smaller in Venus than in the Earth. Also indicative thereof (again, weakly) is the essential absence of a magnetic field (Russell *et al.*, 1979), though this is more likely due to the probable lack of inner core solidification as a dynamo energy source (Stevenson *et al.*, 1980).

Modelling the interior of Venus

A variant of the model of Earth's interior (section 4.5.1) was used to represent Venus. The same crust thickness and density were assumed, and the crust was considered to negligibly perturb the bulk composition of the planet. The transition zone occurring at 400–700 km depth in Earth is shifted 50 km deeper in Venus because of lower pressures in that slightly smaller planet. The Birch-Murnaghan equation-of-state was used to approximate densities in the lower mantle of Venus, as it was for the Earth model.

Density profiles were computed for the chemical models of Table 4.3.2b. These are summarized in Table 4.5.11. All five models are somewhat low in total mass. Possibly this is because the thermal profile used in geophysically modelling Venus is significantly hotter than Earth's. Model V3 came closest to the observed Venus mass, being 0.15% too low. (However, the density and size of the core in this model were adjusted to match the observed mass of Venus.)To correct the total masses of the models, the core radii were adjusted. Alternatively, it would have been possible to change the mantle or core densities to match this value. Density/depth

Table 4.5.11 Abundances of normative minerals (wt.%) in mantles, and other properties of geophysical Venus models constructed from chemical models of Table 4.3.2b.

	V1	V2	V3	V4	V5
Silica	0.0	0.0	0.0	0.0	0.0
K-Feldspar	0.0	0.0	0.0	0.0	0.0
Jadeite	10.44	9.79	1.83	0.98	0.65
Ca-Ts	(0.16)	(0.19)	(0.96)	(0.93)	(0.87)
Diopside	(13.74)	(13.51)	(11.08)	(10.94)	(10.14)
Total Ca-rich Pyroxene	13.90	13.70	12.04	11.87	11.01
Mg-Ts	(2.02)	(2.01)	(3.88)	(0.0)	(1.93)
Hypersthene	(41.68)	(41.78)	(47.30)	(0.0)	(23.70)
Total Ca-poor Pyroxene	43.70	43.79	51.18	0.0	25.63
Olivine	31.21	34.45	28.03	74.87	51.42
Py-Alm	(0.44)	(1.34)	(4.20)	(9.66)	(8.58)
Grossular	(0.0)	(0.17)	(0.51)	(1.14)	(1.04)
Total Garnet	0.44	1.51	4.71	10.08	9.62
Spinel	0.0	0.0	0.0	0.22	0.0
Ilmenite	0.20	0.20	0.38	0.46	0.57
Chromite	0.0	0.0	1.18	0.44	0.44
Metal	0.0	0.0	0.0	0.0	0.0
Total	99.93	103.44	99.28	99.62	99.34
Density (STP) of upper mantle	3.237	3.263	3.312	3.515	3.362
Density (STP) of lower (mixed oxides) mantle	5.223	5.247	5.277	5.513	5.308
Mg/(Mg + Fe) in mantle	0.99	0.97	0.93	0.76	0.90
Density (STP) of core	7.93	7.09	7.47	6.70	7.11
Wt. % of Venus in core	30.2	30.9	32.0	23.6	28.2
Mass of Venus model ($\times 10^{27}$ g)	4.74	4.71	4.86	4.74	4.71
Percent deviation from true mass	-2.55	-3.29	-0.15	-2.46	-3.29
I/MR^2 of Venus model	0.3283	0.3318	0.3290	0.3440	0.3360
Change made to correct mass:					
New wt. % core	33.22	35.14	32.63	27.53	32.52
I/MR^2	0.3245	0.3273	0.3286	0.3399	0.3311

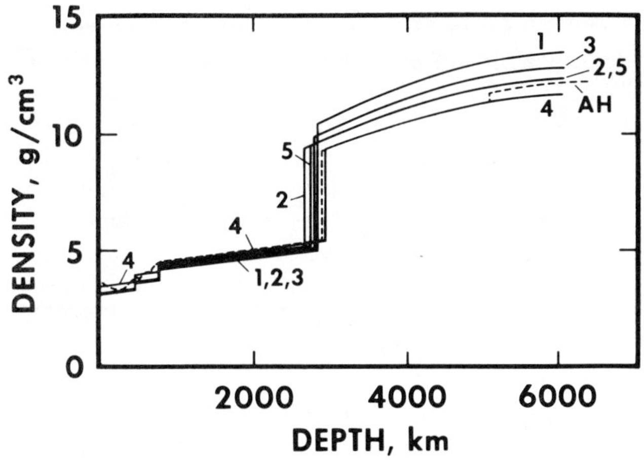

4.5.16 Density profiles for corrected Venus models V1-V5 compared with the Anderson and Hart (1976) model for Earth (AH, dashed line).

plots for these corrected models are shown in Fig. 4.5.16, along with the Anderson and Hart (1976) Earth model for comparison.

The mantles of models V1 and V2, both based on Equilibrium Condensation, are very similar. They are characterized by large pyroxene/olivine ratios. Both have very low zero-pressure densities and high Mg/(Mg + Fe) ratios (0.99 and 0.97 molar, respectively).

Model V4 is quite different from the others. It has a much higher zero-pressure mantle density, 3.51, and most of the mantle consists of olivine. This model also has much more Fe in the mantle, as seen in a mole ratio of 0.76 for Mg/(Mg + Fe). Here the mantle can have a higher density because the core is presumed to be enriched in O, and hence lower in density than cores of the other models.

4.5.5 Mars

Geophysical constraints and a review of modelling studies

Any compositional model of Mars has to account for its low mean density and high moment-of-inertia factor compared with Earth. The small size of Mars must reflect some process that occurred during its accretion, which acted to reduce the amount of material going into the planet. The uncompressed density is a more relevant indication of planetary composition, but this is necessarily uncertain due to lack of knowledge of temperatures in the planet and material properties at high pressure. The ranges of estimated uncompressed densities are 3.8–3.9 g/cm^3 for Mars and 4.4–4.5 g/cm^3 for Earth. The moment-of-inertia factor measured for Earth is 0.3315 ± 0.0000, while that for Mars is 0.365 ± 0.005 (Table 4.2.1). The uncertainty in the martian value arises because it depends on the assumed degree of symmetry of nonhydrostatic density about the Tharsis axis (section 4.2.1). The value obtained without considering the effect of the Tharsis Bulge is 0.377. The moment-of-inertia factor indicates less concentration of mass toward the center of Mars than of Earth.

There are two ways to achieve the lower mean density of Mars compared with Earth. Mars could have a lower Fe content; or it could be enriched in a light element, such as oxygen. The additional O could be in the core or it could be in the mantle, associated with Fe^{2+}, which would otherwise occur as Fe^{0} in the core. A smaller or lighter core alone would not account completely for the higher moment-of-inertia factor of Mars: it is also necessary that Mars's average mantle density be greater than that of Earth's mantle. No unique density can be specified by the geophysical constraints, however, since this value depends on assumptions of

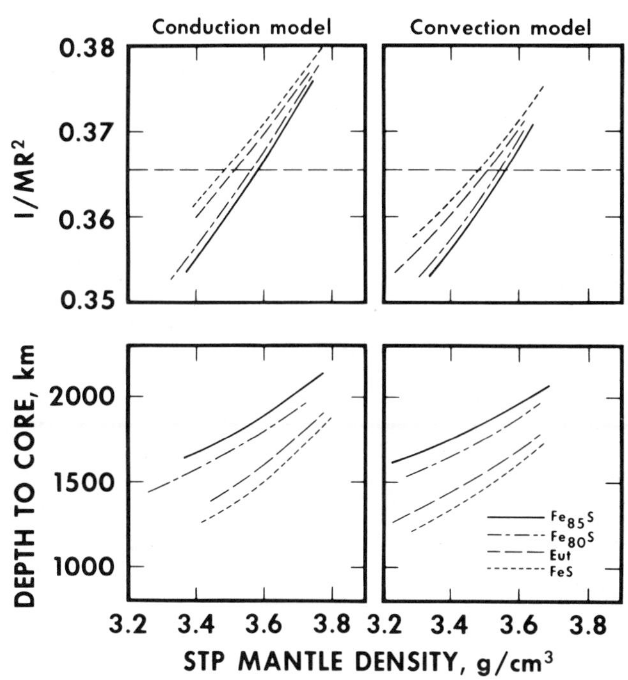

4.5.17 Relationship between uncompressed density of the Mars mantle and (above) the moment-of-inertia factor, (below) depth to the core for a model planet having the correct mean density. Horizontal dashed line in the upper plots represents the observed moment-of-inertia factor. Two different martian thermal models (conduction, convection) and four possible core compositions (key at lower right) are considered. Eut = Fe-FeS eutectic composition. From Johnston and Toksöz (1977).

crustal thickness and density and a temperature profile for the planet. Figure 4.5.17 shows ranges of permissible zero-pressure mantle densities for two temperature profiles. The upper and lower bounds are about 3.45 and 3.55 g/cm^3. The uncompressed density of material likely to be in Earth's upper mantle is about 3.34 g/cm^3 (Ringwood, 1975).

The small size of Mars, which reverses the trend of increasing size with heliocentric distance that holds for the other terrestrial planets, is most plausibly explained as arising from dynamic disturbances during accretion, generated by Jupiter. Jupiter scattered planetesimals to the zone of Mars, thus knocking materials out of that region. In addition, major mass changes in the Jupiter zone may have lead to a sweeping of resonance bands (similar to those believed to affect asteroid belt structure) through the Mars zone. It is plausible that the Mars zone would have originally contained several earth-masses of solids.

Given that Mars must contain less Fe and/or more of a light element than the planets closer to the sun, there are several ways of accomplishing this within the cosmochemical frameworks reviewed in section 4.3.2. The Equilibrium Condensation Model and its modification by Weidenschilling assume that the temperatures and pressures in the zone of accretion of Mars were low enough to promote condensation of a very different suite of minerals there than in the zones of the other planets. Low temperature would allow more complete reaction of S with Fe to form FeS, affecting core density; and more Fe would react to form ferrous silicates, resulting in an increased mantle density. Another interesting consequence of the Equilibrium Condensation Model is that silicates in the mantle would be partly hydrated; this, along with the Fe^{2+} content of its minerals, would lower the melting temperature of the mantle.

A method of inferring the bulk composition of Mars from estimates of present-day abundances of a few elements is illustrated by the model of Morgan and Anders (1979) (see section 4.3.2). These authors estimate a K/U ratio for Mars from Soviet Mars 5 orbital spectrometry data, and adjust the absolute abundances of K and U to match the amounts of heat-producing elements used in the thermal history work of Toksöz and Hsui (1978). They estimate Fe content from the mean density of the planet (thus this is not a purely cosmochemical model). Finally, the abundances of volatile elements are based on amounts of ^{36}Ar and ^{40}Ar measured in the martian atmosphere by Viking. Morgan and Anders assume that the inner planets received their volatile elements in the form of a veneer of carbonaceous chondrite material (Anders, 1968), and use the $^{36}Ar/Tl$ ratio in C3V chondrites to establish volatile element abundances. Since the only measurement of ^{36}Ar is for the atmosphere, the global content of this nuclide has to be estimated. To do this, the authors assume the same proportion of Mars's total ^{36}Ar is in the atmosphere as is the case for ^{40}Ar; they calculate the total amount of ^{40}Ar in Mars from the estimate of K abundance mentioned above.

The assumptions involved in this modelling exercise do not constrain two important parameters: the Mg/Si ratio, which bears on the mantle density; and the S content, which would affect core density. Morgan and Anders (1979) assume that Mg/Si and S/K equal the cosmic ratio, but as pointed out by Morgan *et al.* (1978a), both ratios actually have considerable latitude. However, their analysis unequivocally predicts a substantial depletion in volatiles, relative to Earth.

Two methods of estimating the composition of Mars that are less constrained by cosmochemistry are the approaches of Anderson (1972a) and McGetchin and Smyth (1978). Anderson combines amounts of different classes of chondritic meteorites to arrive at a material which, differentiated, would yield the correct mean density and moment-of-inertia factor. Figure

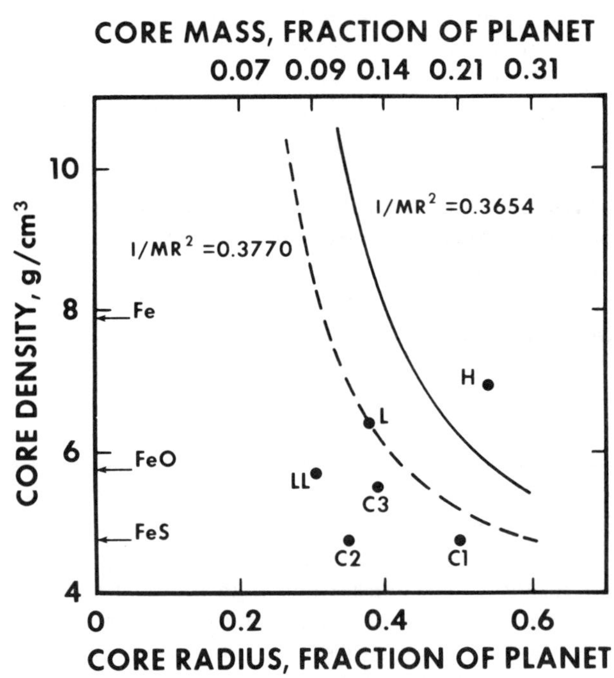

4.5.18 Relationship between core density, mass and radius for Mars model having the observed martian mean density (Okal and Anderson, 1978). Curves corresponding to two moment-of-inertia factors are shown. Points corresponding to Mars models having bulk compositions of six chondrite subclasses are shown, based on the relative fraction of core elements (Fe, Ni, S).

Table 4.5.12 Mantle mineral assemblages consistent with the mean density of Mars.†

Mineral Assemblage	Assumed Behavior of Major Chemical Components or Major Phases
1. Ol-cpx-garnet-periclase	FeO is principal dense component; CaO in diopside, not garnet; Fe^{2+}/Fe^{3+} partitioned between pyroxene, olivine and garnet.
2. Monticellite-ol-cpx-garnet	Clinopyroxene not stable in presence of periclase-wüstite, reacts to form monticellite plus forsterite.
3. Garnet-spinel-cpx-ol	If garnet is unstable in the presence of periclase, it may react to form olivine plus spinel.
4. Magnetite-garnet-cpx-opx-(forsterite)	If f_{O_2} is very high, olivine (fo) may be unstable, breaking down to orthopyroxene and oxide.
5. Garnet (andradite-rich)-opx-cpx (possibly acmitic)-ol	High f_{O_2} and abundant Fe^{3+}.
6. Pyrite (or pyrrhotite)-garnet-cpx-opx-ol	Fe present and abundant, as FeS.

† McGetchin and Smyth, 1978.

4.5.18 shows the effect on core densities and core sizes of using different chondrite classes to make models of Mars. The core sizes and densities required by geophysical constraints are also shown. A range of core sizes and densities is clearly possible, but Anderson has pointed out that the total Fe content of Mars is essentially independent of core size.

The McGetchin and Smyth method of modelling is to add FeO to a pyrolite composition corresponding to the most likely material of Earth's upper mantle. Partial melting of this mantle material yields an ultrabasic picritic-alkali-basalt melt of low viscosity, which seems to match observed martian basalt flow features and reflection spectrometry. McGetchin and Smyth also consider other compositions which would satisfy the constraint of high mantle density. Table 4.5.12 illustrates the wide range of geophysically permissible mantle assemblages.

Finally, it is possible to explain the low mean density and high moment of inertia of Mars through a higher oxidation state than that of Earth. Ringwood and Clark (1971) assume that Mars, Earth and Venus have the same abundances of the common metals Fe, Mg, Si and Al, but differing amounts of O. This method can account for the mean density of Mars, but the planet's moment-of-inertia factor depends upon a specification of the internal distribution of elements.

These different approaches show the variety of ways in which existing constraints for Mars can be matched. However, we can be sure of a few things about Mars: it has to have a core, though this can have a wide range of densities. Its mantle density cannot be uniquely specified, but must be relatively high.

Modelling the interior of Mars

In our modelling of this planet the crust was assumed to be 50 km thick, following Phillips *et al.* (1973). This is roughly 3% of the mass of the crust-plus-mantle, so no effort was made to subtract it from the model compositions of Table 4.3.2d. The crustal density was again taken to be 3.0 g/cm^3. Because of the high pressures in the mantle of Mars, the phase transitions olivine-to-spinel-structure and pyroxene-to-ilmenite-structure (Ringwood, 1975) had to be considered. Figure 4.4.1 shows the pressure of the olivine-to-spinel transition, as a function of temperature. This transition is thought to occur in the Earth at about 400 km (Anderson, 1977; see Fig. 4.5.1), corresponding to pressures of ~135 kb; however, since pressure increases with depth less rapidly in Mars than Earth, the transition is taken to occur in the martian interior at greater depth. This pressure is also affected by the higher Fe contents in the model compositions of the mantle of Mars (see Fig. 4.4.1).

The depth of this transition in the martian mantle was taken to be between 1250 and 1375 km, depending on Fe content. A density increase of ~10% is involved. Central pressures in Mars are only about 400 kb, so extrapolations of the Birch-Murnaghan equation-of-state were used.

The Mars model compositions fall into two groups: four that come close to matching the present estimate of

Table 4.5.13 Abundances of normative minerals (wt.%) in mantles, and other properties of geophysical Mars models constructed from chemical models of Table 4.3.2d.

	Ma1	Ma2	Ma3	Ma4	Ma5	Ma6
Silica	0.0	0.0	0.0	0.0	0.0	0.0
K-Feldspar	0.0	0.0	0.0	0.0	0.0	0.0
Jadeite	9.14	9.14	0.65	1.31	5.29	3.35
Ca-Ts	(0.08)	(0.1)	(1.53)	(0.24)	(0.2)	(1.74)
Diopside	(11.15)	(11.48)	(17.99)	(9.46)	(8.50)	(10.22)
Total Ca-rich Pyroxene	11.23	11.58	19.52	11.01	8.70	10.96
Mg-Ts	(0.57)	(0.6)	(0.0)	(0.0)	(0.09)	(0.0)
Hypersthene	(10.13)	(10.62)	(0.0)	(0.0)	(2.91)	(0.0)
Total Ca-poor Pyroxene	10.70	11.22	0.0	0.0	3.0	0.0
Olivine	65.93	64.95	54.78	83.37	72.86	74.75
Py-Alm	(1.92)	(2.22)	(20.03)	(0.0)	(8.05)	(7.57)
Grossular	(0.23)	(0.26)	(2.34)	(0.0)	(0.95)	(0.5)
Total Garnet	2.15	2.48	22.37	0.0	9.00	8.07
Spinel	0.0	0.0	0.25	4.57	0.0	0.4
Ilmenite	0.17	0.17	0.58	0.21	0.1	1.14
Metal	0.0	0.0	0.0	0.47	0.0	0.0
Chromite	0.0	0.0	0.89	0.0	0.0	0.0
Total	99.32	99.54	99.04	99.67	98.95	99.67
Density (STP) of mantle	3.463	3.454	3.539	3.607	3.590	3.528
Mg/(Mg+Fe) in mantle	0.76	0.77	0.77	0.66	0.67	0.74
Density (STP) of core	5.53	5.53	7.51	5.82	6.51	7.40
Wt. % of Mars in core	25.3	25.7	19.0	18.2	11.9	15.0
Mass of Mars model ($\times 10^{26}$ g)	6.32	6.31	6.52	6.43	6.31	6.34
Percent deviation from true mass	-1.45	-1.73	+1.68	+0.25	-1.69	-1.12
I/MR^2 of Mars model	0.3684	0.3682	0.3628	0.3726	0.3749	0.3682
Percent deviation from true I/MR^2	+0.93	+0.88	-0.71	+1.97	+2.60	+0.88
Changes made to correct mass and I/MR^2:						
Density (STP) of core	5.86	5.59	7.90	†	†	7.84
Wt. % of Mars in core	26.0	29.5	15.2			16.3

†Mantle material too dense to permit a match to observed mass and moment of inertia by adjusting core parameters.

Mars's moment-of-inertia factor (0.365), and two (Ma4 and Ma5) that are closer to the older value (0.377; Table 4.5.13). None of the models err by as much as 2% in total mass. The match would be worse if a conductive instead of a convective temperature profile (Fig. 4.5.17) were assumed; in this case, however, the models that are too dense would have more nearly the correct moment-of-inertia factor.

The corrections to these models were accomplished by varying the core size and density until the total mass and moment of inertia of the model matched the accepted quantities for Mars. The two models (Ma4 and Ma5) that were closer to the old value of the moment of inertia have mantles that are too dense to match the new moment and also the mean density. The corrected models are plotted in Fig. 4.5.19, and the weight percentages and densities of the corrected cores are shown in Table 4.5.13.

There is a wide range of mantles among these models, but all have a lower Mg/(Mg + Fe) molar ratio

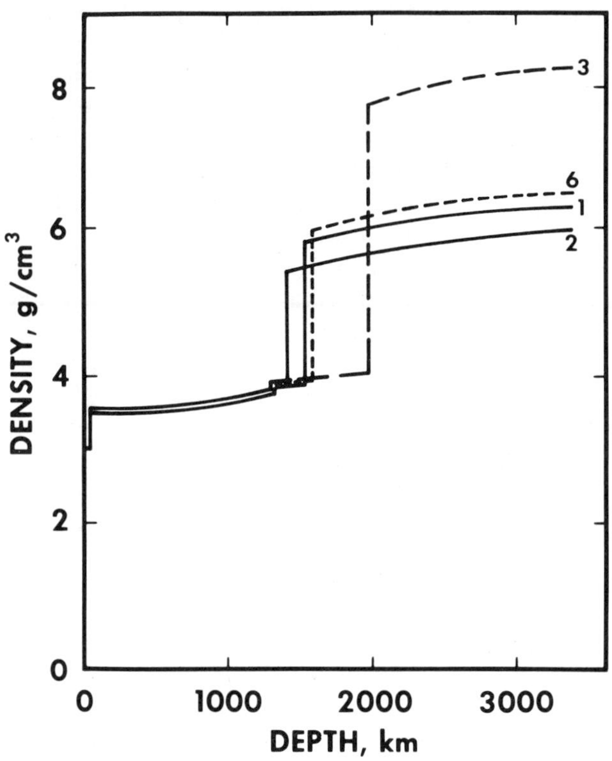

4.5.19 Density profiles of four of the Mars models of Table 4.3.2d, after adjustment of Mg/(Mg + Fe) and core density and radius to bring the models into correspondence with the observed mean density and moment-of-inertia factor of Mars. Ma4 and Ma5 have mantles too dense to allow correction by this means.

than the mantle of Earth, which should be reflected in the low viscosity of lavas derived from them (McGetchin and Smyth, 1978).

It should be stressed that the total mass and the moment-of-inertia factor can only put bounds on the mantle density of Mars. Even these bounds are dependent upon assumed crustal composition and thickness, and the temperature profile used. With the assumptions used in the present work, the range of permissible zero-pressure densities for the mantle is between 3.426 and 3.539 g/cm^3. Ranges for other assumed temperature profiles are shown in Fig. 4.5.17.

4.5.6 The eucrite parent body

Eucrites are the most abundant and thoroughly studied group of basaltic meteorites. Aspects of their chemistry, petrology, and ages are reviewed elsewhere in this volume (sections 1.2.8, 3.5, 7.2). Here we briefly review estimates of the bulk composition of the eucrite parent body (EPB), their petrological and cosmochemical implications, and current ideas about the size and identity of the EPB.

Composition and petrology

Estimates of the composition of the EPB are given in Table 4.3.2e. Their norms are graphically displayed in Figs. 4.5.20–21. Most of these estimates are based on the assumption that the compositions of the source regions of eucritic melts are the same as the bulk composition of the EPB. This assumption would not be valid if the source regions had undergone differentiation prior to the melting events which produced the eucrites; however, such a two-stage evolution of the source regions of eucritic magmas is unlikely in view of the great age of the eucrites (4.5–4.6 b.y., section 7.2). All of the estimates of the composition of the EPB are rich in olivine (Fig. 4.5.20); they contain 50–85 wt.% olivine in their low-pressure norms. Their pyroxene is dominantly, or entirely, low-Ca pyroxene at near-solidus temperatures (Fig. 4.5.21). These estimates of the composition of the EPB are poor in alkalis and other volatile elements. Their level of depletion relative to cosmic abundances is similar to that of model lunar bulk compositions. The estimates of bulk composition all have relatively high Fe/(Fe + Mg) ratios (Fig. 4.5.22). Low abundances of siderophile elements in eucrites suggest that free metal was present in their source regions. Estimates of the amount of metal in the EPB are uncertain but are about 10 wt.% (Morgan *et al.*, 1978b). An FeS abundance of 0.22% has been estimated (Hertogen *et al.*, 1977) on the assumption that S correlates with K, but the true amount could have been 2–3 times larger, as discussed by Morgan and Anders (1979) for Mars.

The bulk composition of the silicate portion of the EPB is in many respects similar to the compositions of ordinary chondrites and the Earth's mantle. Figures 4.5.20 and 4.5.21 illustrate that all have approximately the same high proportion of olivine relative to other phases, and have low-Ca pyroxene dominant over high-Ca pyroxene. This is not the case for all estimates of planetary composition; Mars and Mercury (according to most of the models in Tables 4.3.2a and 4.3.2d) would be rich in olivine but would have little, if any, low-Ca pyroxene at low pressures (section 3.6.1). The Earth's mantle has a lower Fe/(Fe + Mg) ratio than the EPB, while L and LL chondrites overlap with the EPB (Fig. 4.5.22). The Earth's mantle and ordinary chondrites are significantly richer in alkalis and other volatiles than the EPB.

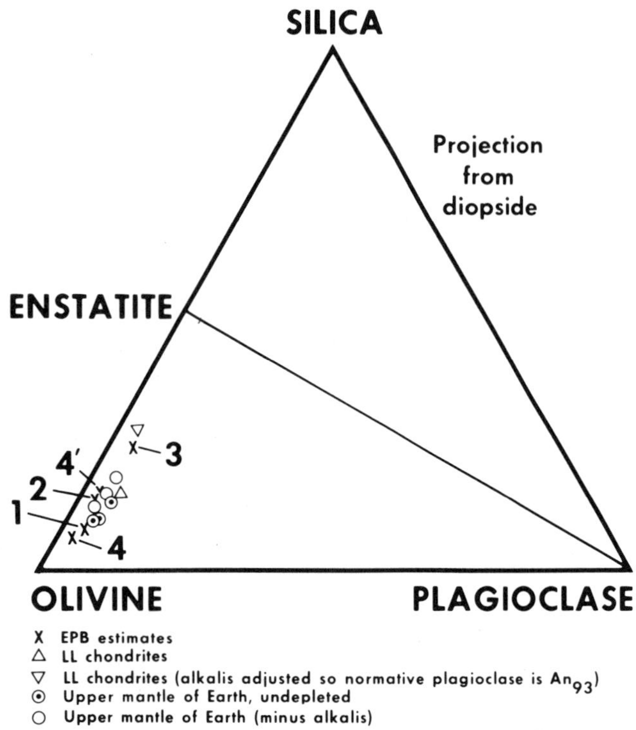

4.5.20 Estimates of the composition of the eucrite parent body (from Table 4.3.2e), compared with the Cherokee Springs LL-chondrite (with and without alkalies; Jarosewich and Mason, 1969) and estimates of the composition of the Earth's upper mantle (with and without alkalies; Ringwood, 1977a; Smith, 1977; Harris *et al.*, 1967). Composition 4 is a second EPB model from Consolmagno and Drake (1977). Molar coordinates. SiO_2, Al_2O_3, FeO, MgO, CaO, Na_2O, and K_2O are recast into the coordinates $(MgMnFe)_2SiO_4$, SiO_2, $CaAl_2Si_2O_8$, $NaAlSi_3O_8$, $KAlSi_3O_8$, and $Ca(MgMnFe)Si_2O_6$. $(MgMnFe)_2SiO_4$, SiO_2, and $[CaAl_2Si_2O_8 + (NaK)AlSi_3O_8]$ components are then recalculated to 100%.

Size and identity of the EPB

Prior to the Apollo program, it was widely believed that the Moon might be the EPB (e.g., Duke and Silver, 1967). Currently, an asteroidal EPB is favored, since eucrites are distinct from known lunar rock types and the dynamics of deriving them from a major terrestrial planet are not favorable (Wetherill, 1974). Spectral reflectance studies (e.g., McCord *et al.*, 1970; section 2.2.6) and petrological arguments (Stolper, 1977; Consolmagno and Drake, 1977) have led to the suggestion that the asteroid 4 Vesta may be the EPB. Although there are dynamical difficulties with deriving the eucrites from a body such as 4 Vesta (a main belt asteroid far from a resonance with Jupiter), Hostetler and Drake (1978) have developed a scenario, in which eucrites are derived by secondary ejection from large fragments ejected from Vesta, that could overcome these difficulties. Wetherill (1977) and Wetherill and Williams (1979) have suggested that the eucrites and other differentiated meteorites may come from the S-type asteroids (see section 2.2.6) near the inner rim of the asteroid belt. They have further suggested that these bodies may be related to the planetesimals from which the Earth and Venus accreted.

The size of the EPB is difficult to constrain. The asteroid 4 Vesta has a radius of about 270 km (Table 4.2.1). The asteroid sources preferred by Wetherill and Williams have radii up to 100 km. The maximum pressures in these bodies are probably less than 2 kb (Fig. 3.5.2). Walker *et al.* (1978) have estimated, based on dynamics of melt segregation on the EPB, that the EPB must have had a radius at least on the order of 10–100 km.

The probable size of the EPB and the fact that it appears to have differentiated immediately after or perhaps even during accretion make it unlikely that the decay of long-lived radionuclides (^{40}K, U, Th) was responsible for partial melting on the EPB. Short-lived radioactivity (e.g., ^{26}Al), electromagnetic induction heating (section 9.4.1), or impact heating might all be adequate as heat sources. Currently, heating due to

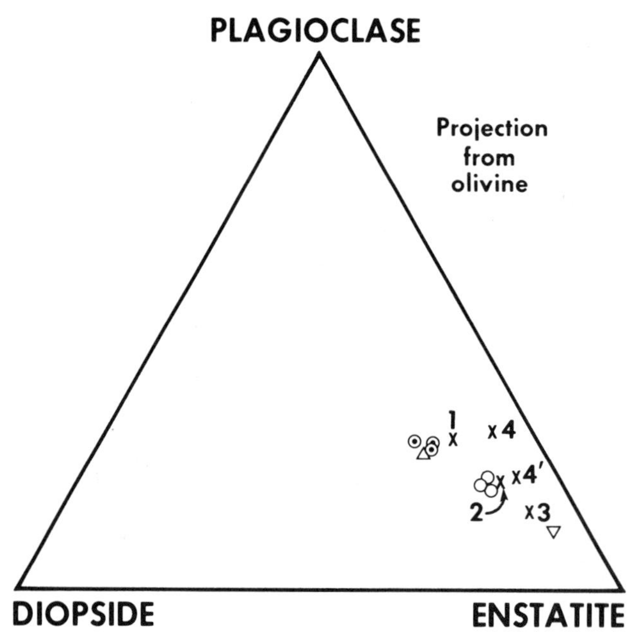

4.5.21 Same as Fig. 4.5.20, except new coordinates are $(MgMnFe)_2SiO_4$, $Ca(MgMnFe)Si_2O_6$, $(MgMnFe)_2Si_2O_8$, $NaAlSi_3O_8$, and $KAlSi_3O_8$. $(MgMnFe)_2Si_2O_6$, $Ca(MgMnFe)Si_2O_6$, and $[CaAl_2Si_2O_8 + (NaK)AlSi_3O_8]$ components are then recalculated to 100% to make this plot. Symbols as in Fig. 4.5.20.

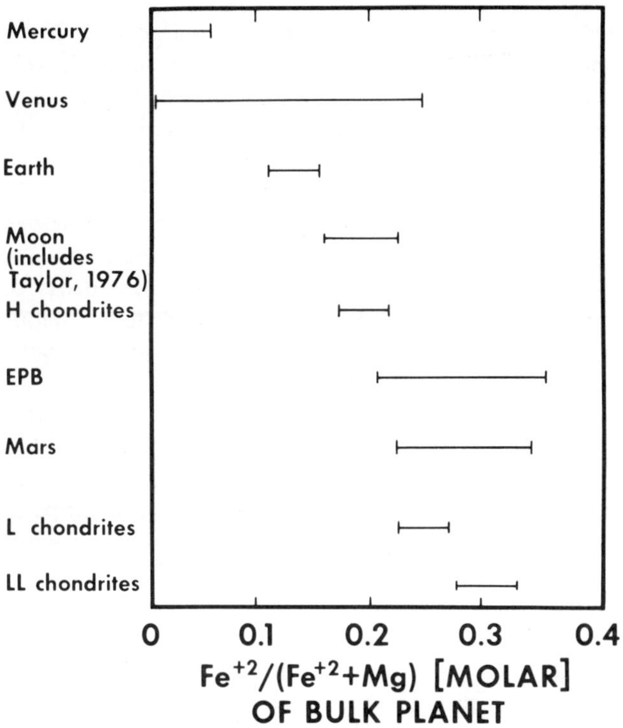

4.5.22 Fe/(Fe + Mg) ratios of silicates in estimates of planetary composition and ordinary chondrites. Data from Tables 4.3.2, Taylor (1976), and Hertogen *et al.*, 1977; analysis given in Morgan *et al.*, 1978b. Ordinary chondrite values are for olivines in LL6, L6, and H6 chondrites (compiled in Wasson, 1974).

decay of ^{26}Al is regarded by some cosmochemists as the most favorable of these mechanisms, since the existence of ^{26}Al in primitive solar system objects has now been verified (e.g., Lee *et al.*, 1976).

Significance of the EPB compositions

The essentially chondritic compositions of the EPB and the Earth's mantle reinforce the notion that materials of approximately chondritic composition were important building blocks of planetary bodies. The general similarities in bulk composition and oxygen isotopic characteristics (Clayton, 1977) between the EPB and the Earth also support the suggestion of Wetherill and Williams that the planetesimals from which the Earth accreted might have been related to the EPB. The three principal differences between the Earth and the EPB are the higher Fe/(Fe + Mg) ratio, the lower oxidation state, and the lower volatile content of the EPB. The combination of high FeO content and low volatile content estimated for the EPB is not compatible with simple equilibrium condensation models of planetary compositions, since by the time temperatures are reached at which extensive quantities of FeO would have formed, large amounts of alkalies and more volatile elements would also have condensed. Heating of FeO-and volatile-rich materials could produce the characteristics of the EPB, since alkalis and other volatiles can be lost without reduction of FeO. The Chondrite Model (section 4.3.2) is also capable of accounting for the combination of high FeO and low volatile content, as these two parameters are decoupled in this model: high FeO implies low nebular equilibration temperature, whereas low volatiles imply extensive remelting (= chondrule formation).

The fact that the Moon, Mars, and EPB are extensively depleted in volatiles relative to the Earth has led to the suggestion that among differentiated bodies, planetary size is a major factor in volatile depletion (Anders and Owen, 1977; Morgan and Anders, 1979; section 4.3.2). However, these authors point out that the shergottites, though undoubtedly derived from a small body, have an Earth-like abundance of volatiles (Laul *et al.*, 1972b; Stolper, 1979), so that the size alone cannot be the governing factor.

The characteristics of the EPB and its bulk composition which were most important in controlling the compositions of its basaltic magmas and in distinguishing them from basalts on other planets are:

(1) Its small size. All igneous processes on the EPB took place at low pressures. Thus, while primary magmas on the Earth are generally produced by melting of spinel- or garnet-bearing source regions at pressures greater than 9 kb, eucritic magmas evolved by equilibration with plagioclase-bearing sources at low pressures. This pressure effect accounts for many of the petrological differences between eucrites and most other basalts known or expected from other planets.

(2) Its low content of alkalies and more volatile elements. This is responsible, along with the low pressures of the EPB, for much of the petrological and geochemical distinctiveness of eucritic magmas. (See section 3.6.2 for a general discussion of the influence of alkalies on the petrology of source regions and their partial melts at low pressures.) This volatile-depletion may not be universal in small, differentiated bodies.

(3) The presence of metallic iron in the source regions. This resulted in highly reduced basalts, considerably different from terrestrial and probably venusian and martian basalts. It also produced a distinctive depletion in siderophile elements not observed in other basalts.

(4) The high Fe/(Fe + Mg) ratios for eucrite source regions. This results in magmas with Fe/(Fe + Mg)

ratios and FeO contents much higher than primary magmas on the Earth. Martian primary magmas may be similar to eucrites in this respect.

4.5.7 Galilean satellites

The Galilean satellites are primarily thought of as collections of hydrous silicates and water ice; considering their low densities and the fact that water ice dominates the spectra of two of them, it is not surprising that most modelling has considered primarily the interaction of water in its various physical states with some vaguely defined "silicate" material. But it is clear even from these models that the silicate material, regardless of its original composition, will almost inevitably have been raised to temperatures where basaltic volcanism can take place. It is worth remembering that even Europa, the smallest of these bodies, is comparable in size and density to the Moon, and that Ganymede has a larger diameter than Mercury. Indeed, the silicate core predicted by Consolmagno and Lewis (1976) for Ganymede's interior is itself larger than the Moon—as much as 29% larger, depending on the parameters of the model.

These satellites are in four very different places, and it should not be surprising if we discover that they evolved in four very distinct ways. Furthermore, the kind of basalt (if any) each of these objects produced is likely to differ not only from satellite to satellite, but also to differ significantly from that produced by any other planet in the solar system.

Masses, radii and densities of the Galilean satellites are given in Table 4.2.1. Although each satellite is unique, all are likely to share a similar origin. All four are in nearly circular, coplanar orbits around Jupiter, and a popular hypothesis has been that they coalesced out of a protojovian nebula not too different from the solar nebula but on a smaller scale. Kuiper (1952) suggested that the heat from the collapsing proto-Jupiter raised the temperature of the material forming the satellites so that those closer to Jupiter incorporated lesser amounts of the more volatile materials (primarily water ice); this would explain the regular decrease in mean densities going outwards from Jupiter. Graboske *et al.* (1975) modelled the luminosity of such a proto-Jupiter as a function of time, and Pollack and Reynolds (1974) estimated the pressure-temperature conditions in the mini-nebula where the satellites formed. From this work, and considering the mean densities of the bodies and the observed presence or absence of water ice absorption bands in the spectra of each, Fanale *et al.* (1977) predicted that Io was made primarily of low-temperature, hydrous silicates. Europa possibly had a small water-ice component, Ganymede considerably more water ice, and Callisto may have had its full cosmic component of water ice, in addition to hydrous silicates. Given such compositions, one can model how each satellite may have evolved.

Fanale *et al.* (1977) attempted to match the observed spectrum and high albedo of Io with evaporite salts, postulating that as Io heated up (due to the decay of long-lived radionuclides), various hydrous minerals would be altered, with the developed water carrying dissolved salts to the surface. In connection with this work, they made a simple model of Io to demonstrate that such heating would occur. Their model showed that, even assuming no initial thermal gradient, all of Io below a depth of 300 km would be raised to a temperature greater than a typical basalt solidus within 1.4 b.y. after formation. As they mentioned, this could lead to "conceivably, even the generation of surface igneous magmas," although they were only looking for a thermal gradient to drive water out of the interior.

More recent work (e.g., Peale *et al.*, 1979; Drobyshevski, 1979) has suggested that other heat sources may be even more important than radioactive decay. Because of its resonances with the other satellites and its proximity to Jupiter, tidal stresses are set up in Io which may continuously add five to ten times as much energy to its interior as radioactive sources alone could provide. The dissipation of electromagnetic energy inside Io as it traverses Jupiter's magnetic fields may add a similar amount of energy. Based on these calculations, Peale *et al.* (1979) predicted widescale and intensive volcanism on present-day Io. This prediction was dramatically confirmed by Voyager 1 (Morabito *et al.*, 1979).

Consolmagno (1979) and others have pointed out, on theoretical grounds, that Io's volcanoes must be essentially anhydrous; if such activity prevailed throughout the age of Io, no primordial gases could remain. Consolmagno suggested that the volatile substance driving the volcanoes is elemental sulfur. Hapke (1979) made a similar suggestion, also noting the possibility of SO_2 and S_2O gases. Acting on such suggestions, Smythe *et al.* (1979) looked for the observed spectral evidence of SO_2 frozen on the surface of Io.

Propelled by the excitement of Voyager, models for ionian volcanism have been developing rapidly, and it would be premature to attempt any definitive statements concerning them. However, a brief discussion of the types of models being considered may be of interest.

All current models start with the observation that sulfur dominates the surface of Io. This S is constantly being emplaced on the surface by volcanic processes. At least two types of volcanism are present: flows of what appear to be liquid S, and explosive plumes of S or SO_2-bearing gases and other possible constituents. The heat source for this volcanism is presumably tidal and electromagnetic heating; such heating should be sufficient to allow at least partial melting and widescale convection of rock throughout the interior of Io. Since solid S is a poor conductor of heat, the sulfurous surface may act as an insulating layer, allowing steep thermal gradients and hence high temperatures quite close to the surface.

For this situation, at least three models of ionian volcanism have been proposed. These can be labelled the "magma trigger" model, the "chemical equilibrium" model, and the "escape valve" model.

The "magma trigger" model was proposed by Consolmagno (1979). Given a layer of S near the surface, any rising column of hot basaltic magma from the interior will eventually come in contact with this S. Upon being heated to temperatures in excess of 1000° C, the S vaporizes. Its vapor pressure being higher than the ambient lithostatic pressure, an explosion results, giving rise to the explosive plumes seen on the surface. Following such an explosion, liquid S can flow from these vents.

The "chemical equilibrium" model has been developed by Lewis (Lewis and Consolmagno, 1979), who studied the thermodynamics of Fe-bearing minerals under the high f_{O_2} and f_{S_2} conditions predicted for Io. His studies showed that liquid S and high SO_2 pressures can occur at moderate temperatures in such a system, resulting in two distinct types of volcanism—SO_2-rich explosive volcanism, and liquid S- and FeS-rich volcanic flows.

The "escape valve" model (Smith *et al.*, 1979a), envisions a relatively thick S layer on Io, thick enough to allow molten S to exist at its base. This S, being less dense than the solid form, is gravitationally unstable and tends to flow out onto the surface through any cracks in the solid crust.

It should be obvious that none of these models are exclusive; nor do they exhaust the possibilities for ionian volcanic activity. With an abundance of both S and energy, Io should keep volcanologists busy for some time to come.

Both Fanale *et al.* (1977) and Consolmagno and Lewis (1976) have emphasized that the surface layer of ice on Europa cannot much exceed 75 km in thickness. From the satellite's spectrum, it is clear that ice makes up the bulk of its surface; but from its density, this ice cannot be a major constituent of the planet. The thermal models of Consolmagno and Lewis (1976) indicate that temperatures in the rocky core would exceed 1000° K within 1 b.y. Above this core would be a layer of water, either liquid with a thin icy crust at the surface, or frozen, but efficiently transporting heat from the core via liquid or solid-state convection. In this case, lava eruptions from the silicate core could result in unusual formations when they come in contact with the ice. Depending on the thickness of the local ice sheet (<75 km), such features as table mountains, maars and lahars might be visible on the surface (Allen, 1977).

The densities of Ganymede and Callisto are so low that a large proportion of water must be present. Even if these bodies have large rocky cores, it is unlikely that direct evidence of the magmatic activity, which must certainly occur in such cores, would be seen on their surfaces. However, indirect evidence of such processes might be seen in accurate moment-of-inertia data (which could reveal differentiation within any such cores), or by examining trace materials in the surface ice which might have been leached from the core/liquid mantle boundary. Unfortunately, a surface-sample-return mission from Ganymede is not likely to occur for some time yet; and the moment-of-inertia data from the Jupiter orbiter mission will, at best, determine whether the cores themselves exist (Hubbard and Anderson, 1978; Dermott, 1979).

There is one other way in which the icy Galilean satellites are relevant to the study of basaltic volcanism. These satellites are made of a low density, easily-melted material, and thus have interiors where melting, differentiation, or solid-state convection will occur at temperatures and pressures low enough to be easily simulated in the laboratory (we admit to some difficulty in reproducing the relevant length and time scales for such experiments). Evidence for all these processes has been seen in Voyager images (Smith *et al.*, 1979b). By substituting the properties of water ice for the properties of basalts in theoretical models, one should be able to use the icy satellites as tests of those models.

4.6 CONCLUSIONS

In order to assess the likely properties of mantle source regions of basalts, we have summarized proposed compositions of the Moon and terrestrial planets, modelled the density profile for each composition, and compared the geophysical properties of the density model with what is known of the planet it attempts to represent. It is important to keep in mind what an uncertain venture this is. The compositional estimates are, in some cases, very distant extrapolations of the knowledge we have of the terrestrial and lunar mantles. Not enough is known about the geophysical properties of most planets to constrain compositional models closely. And the density modelling procedure is necessarily highly simplified; for example, only in the case of the Moon was the possibility of a chemically stratified mantle tested.

Our best knowledge of the source region of basalts on Earth—the upper mantle—comes from direct observation of basalt and mantle xenolith compositions and from seismic studies, not from modelling studies of the Earth as a whole. The pyrolite composition of Ringwood (Table 4.5.2) is probably a good approximation to this source material, under ocean basins and continents. The pyrolite under ocean basins tends to be depleted in incompatible elements. Pyrolite contains less than the cosmic complement of Si (Si/Mg~0.84× the cosmic ratio). A major question remains: does the entire mantle have this composition, or is it chemically stratified, containing more Si beneath 200–250 km so that its bulk composition resembles that of chondritic silicates? Seismic studies of the lower mantle are unable to discriminate between these possibilities. Both models have vocal advocates. In our density-modelling study, Earth models having mantles of uniform pyrolite composition (E5, E6) more closely reproduced the observed density and moment-of-inertia factor of Earth than did models containing chondritic proportions of the elements (E1–E3). However, the latter compositions were not fairly tested because, lacking the ability to model the compositional stratification of the mantle with any confidence, we also assumed homogeneous mantle compositions in calculating the E1–E3 density models.

The composition of the Moon's upper mantle is, again, best read from the basalts that come from it. Estimates of the overall mantle composition are strongly constrained by the lunar seismic profile and density distribution. The predictions of cosmochemical models are of limited assistance here, because no straightforward interpolation of the processes held responsible for the run of properties of the terrestrial planets comes close to accounting for the Moon. However, the lunar composition is well enough constrained by geophysical properties and sample studies to eliminate major uncertainties. The three compositional models proposed (Mo1–Mo3) are similar except for their content of refractory elements, which ranges from ~1× to ~2× cosmic (relative to Si). Unless the assumed conventional model of the lunar crust is wrong, it appears these compositional models must be adjusted somewhat, principally by a ~25% decrease of Mg/Si, for their corresponding density models to best reproduce the geophysical properties of the Moon.

We have no geophysical constraints for Mercury and Venus except their mean densities, so cosmochemical modelling looms much larger in our estimates of their mantle compositions, and the latter are correspondingly more uncertain. Mercury's high mean density (5.435 g/cm^3) shows that it must contain more than the cosmic complement of Fe. Spectral studies of the mercurian surface (section 2.2.3) indicate a low abundance of Fe^{2+}, hence a highly reduced system. These are properties of the minerals that condense at high temperatures in the condensation sequence. There is a strong tendency among planetologists to extrapolate from this a Mercury composed largely or entirely of refractory condensates. This is reflected in the high values for Al_2O_3, CaO, U, and Th in models Me1–Me2. It means, for the mercurian mantle, a large value of Mg/(Mg + Fe); and major uncertainty about the mantle mineralogy, depending upon whether condensation temperatures were high enough to incorporate Al in spinel, or not quite so high, making anorthite the stable Al condensate. Anorthite would have tended to differentiate to the surface during planetary melting, but spinel would have remained in the mantle.

Venus is similar enough to Earth in mass and density for us to use the Earth as a starting point, and postulate departures from the composition of the terrestrial mantle. Venus is 1.9% less dense than it would be if it were made from the same material as the Earth. Several explanations for this have been proposed. Models V1–V5 are similar in major element abundances; only their contents of Fe^{2+} and volatile elements vary substantially. Density models of V1–V5 all came out lighter than Venus's observed mean density, by a few percent (except for V3, which was trimmed to match the Venus density).This may, however, be due to the relatively hot thermal profile assumed in modelling the planet's density.

Because we know the moment-of-inertia factor for Mars as well as its mean density, the properties of its mantle are better constrained than those of Mercury or Venus; but less well constrained than those of the Moon, where the seismic velocity profiles are also known. Mars must have a core, though not as substantial a core as Earth's; its mantle must be denser than Earth's. All the chemical models tested contain more mantle FeO than does the terrestrial mantle (16–27% vs. 8–11%). They vary somewhat in their content of refractory and volatile elements. With minor adjustments, all can be reconciled with the known geophysical parameters of Mars.

The planetary mantle from which eucritic (meteoritic) basalts were derived, presumably the mantle of an asteroid, appears to have contained more FeO and less refractory elements than the lunar or terrestrial mantles. This is inferred from sample studies. Model studies were not carried out for the eucrite parent body, since the geophysical parameters of the latter are not known and comparisons cannot be drawn.

ACKNOWLEDGMENTS—The Team Leader gratefully acknowledges the patience and support of Team members through six iterations of the chapter, and also the help of the LPI Staff, particularly Rosanna Ridings. D. H. Johnston, M. N. Toksöz and S. J. Weidenschilling contributed to early planning of the chapter. M. J. Drake, K. A. Goettel, G. V. Latham, E. Schonfeld, N. H. Sleep, S. C. Solomon and D. Walker provided thoughtful reviews of the product. F. P. Fanale, J. S. Lewis and K. K. Turekian participated in an LPI workshop on cosmochemical models early in the project. M. J. Drake and G. McLaughlin contributed the frontispiece photograph. The contributions of Team members were supported in part by the following research grants: NASA NGL 09-015-150 (JAW), NGL 14-001-167 (EA), NGL 05-002-069 (DLA), NSG-7081 (WRB), NGL 22-007-269 (GC), NGL 05-007-002 (WMK), and NASA contract T4089 (JM). H. Wänke gratefully acknowledges the financial support of the Deutsche Forschungsgemeinschaft.

4.7 REFERENCES

Ahrens L. H. and von Michaelis H. (1969) The composition of stony meteorites III. Some inter-element relationships. *Earth Planet. Sci. Lett.* **5**, 395–400.

Ahrens T. J. (1972) The mineralogic distribution of iron in the upper mantle. *Phys. Earth Planet. Inter.* **5**, 267–281.

Ahrens T. J. (1980) Dynamic compression of Earth materials. *Science* **207**, 1035–1041.

Ahrens T. J., Takahashi T. and Davies G. F. (1970) A proposed equation of state for stishovite. *J. Geophys. Res.* **75**, 310–316.

Akaogi M. and Akimoto S. (1977) Pyroxene-garnet solid solution. *Phys. Earth Planet. Inter.* **15**, 90–106.

Akella J. (1976) Garnet pyroxene equilibria in the system $CaSiO_3$-$MgSiO_3$-Al_2O_3 and in a natural mineral mixture. *Amer. Mineral.* **61**, 589–598.

Allen C. C. (1977) Interactions of volcanism and ice—Earth, Mars, and icy satellites. *Bull. Amer. Astron. Soc.* **9**, 541.

Anders E. (1964) Origin, age and composition of meteorites. *Space Sci. Rev.* **3**, 583–714.

Anders E. (1968) Chemical processes in the early solar system, as inferred from meteorites. *Acc. Chem. Res.* **1**, 289–298.

Anders E. (1971) Meteorites and the early solar system. *Ann. Rev. Astron. Astrophys.* **9**, 1–34.

Anders E. (1977) Chemical compositions of the Moon, Earth, and eucrite parent body. *Phil. Trans. Roy. Soc. Lond.* **A 285**, 23–40.

Anders E. and Owen T. (1977) Mars and Earth: origin and abundance of volatiles. *Science* **198**, 453–465.

Anderson D. L. (1967) A seismic equation of state. *Geophys. J. Roy. Astron. Soc.* **13**, 9–30.

Anderson D. L. (1972a) The internal composition of Mars. *J. Geophys. Res.* **77**, 789–795.

Anderson D. L. (1972b) Implications of the inhomogeneous planetary accretion hypothesis. *Comments on Earth Sciences: Geophysics* **2**, 93–98.

Anderson D.L. (1976) The 650 km mantle discontinuity. *Geophys. Res. Lett.* **3**, 347–349.

Anderson D. L. (1977) Composition of the mantle and core. *Ann. Rev. Earth Planet.* **5**, 179–202.

Anderson D. L. (1979) The upper mantle transition region: eclogite? *Geophys. Res. Lett.* **6**, 433.

Anderson D. L. (1981) A global geochemical model for the evolution of the mantle. In *Geophysical Geochemical Evolution of the Earth*, in press. Amer. Geophys. Union, spec. mono. Washington, D.C.

Anderson D. L. and Hart R. (1976) An Earth model based on free oscillations and body waves. *J. Geophys. Res.* **81**, 1461–1475.

Anderson D. L., Sammis C. G. and Jordan T. H. (1971) Composition and evolution of the mantle and core. *Science* **171**, 1103–1112.

Anderson J. D. (1975) Planetary geodesy. *Rev. Geophys. Space Phys.* **13**, 275–276.

Anderson O. L. (1980) The temperature profile of the upper mantle. *J. Geophys. Res.* **85**, 7003–7010.

Barshay S. S. and Lewis J. S. (1975) Chemistry of solar material. In *The Dusty Universe* (G. B. Field and A. G. W. Cameron, eds.), p. 34–36. Neale Watson Academic, N.Y.

Barshay S. S. and Lewis J. S. (1976) Chemistry of primitive solar material. *Ann. Rev. Astron. Astrophys.* **14**, 81–94.

Bills B. G. and Ferrari A. J. (1977) A harmonic analysis of lunar topography. *Icarus* **31**, 244–259.

Bills B. G. and Ferrari A. J. (1978) Mars topography harmonics and geophysical implications. *J. Geophys. Res.* **83**, 3497–3507.

Birch F. (1943) Elasticity of igneous rocks at high temperatures and pressures. *Geol. Soc. Amer. Bull.* **54**, 263–286.

Birch F. (1952) Elasticity and constitution of the Earth's interior. *J. Geophys. Res.* **57**, 227–286.

Birch F. (1969) Density and composition of the upper mantle; first approximation as an olivine layer. In *The Earth's Crust and Upper Mantle*, Amer. Geophys. Union Monogram No. 13, 18–36.

Blackshear W. T. and Gapcynski J. P. (1977) An improved value of the lunar moment of inertia. *J. Geophys. Res.* **82**, 1699–1701.

Brown J. M. and Shankland T. J. (1981) Thermodynamic parameters in the Earth as determined from seismic profiles. *Geophys. J. Roy. Astron. Soc.*, in press.

Buck W. R. and Toksöz M. N. (1980) The bulk composition of the Moon based on geophysical constraints. *Proc. Lunar Sci. Conf. 11th*, in press.

Cameron A. G. W. (1963) Formation of the solar nebula. *Icarus* **1**, 339–342.

Cameron A. G. W. (1978) The primitive solar accretion disk and the formation of the planets. In *The Origin of the Solar System* (S. F. Dermott, ed.), p. 49–73. John Wiley and Sons, N.Y.

Cameron A. G. W. (1981) Elementary and nuclidic abundances in the solar system. In *Essays in Nuclear Astrophysics* (C. A. Barnes, D. D. Clayton and D. N. Schramm, eds.). Cambridge Univ. Press, N.Y.

Cameron A. G. W. and Pine M. R. (1973) Numerical models of the primitive solar nebula. *Icarus* **18**, 377–406.

Carswell D. A. (1968) Possible primary upper mantle peridotite in Norwegian basalt gneiss. *Lithos* **1**, 322–355.

Chapman C. R., Williams J. G. and Hartmann W. K. (1978) The asteroids. *Ann. Rev. Astron. Astrophys.* **16**, 33–75.

Christensen W. I. (1974) Compressional wave velocities in possible mantle rocks to 30 kilobars. *J. Geophys. Res.* **79**, 407.

Clark S. P., Turekian K. K. and Grossman L. (1972) Model for the early history of the Earth. In *The Nature of the Solid Earth* (E. C. Robertson, ed.), p. 3–18. McGraw-Hill, N.Y.

Clayton R. N. (1977) Genetic relations among meteorites and planets. In *Comets, Asteroids, Meteorites: Interrelations, Evolution and Origins* (A. H. Delsemme, ed.), p. 545–550. Univ. of Toledo Press, Toledo.

Cohen E. R. and Taylor B. N. (1973) The 1973 least-squares adjustment of fundamental constants. *J. Phys. Chem. Ref. Data* **2**, 663–734.

Consolmagno G. J. (1979) Sulfur volcanoes on Io. *Science* **205**, 397–398.

Consolmagno G. J. and Drake M. J. (1977) Composition and evolution of the eucrite parent body: evidence from rare earth elements. *Geochim. Cosmochim. Acta* **41**, 1271–1282.

Consolmagno G. J. and Lewis J. S. (1976) Structural and thermal models of icy Galilean satellites. In *Jupiter* (T. Gehrels, ed.), p. 1035–1051. Univ. of Arizona Press, Tucson.

Dainty A. M., Toksöz M. N. and Stein S. (1976) Seismic investigation of the lunar interior. *Proc. Lunar Sci. Conf. 7th*, p. 3057–3075.

Davies M. E., Hauge T. A., Katayama F. Y. and Roth J. A. (1979) Control networks for the Galilean satellites. *Rand Corp. Rep. R-2532JPL/NASA*, 61 pp.

Deer W. A., Howie R. A. and Zussman J. (1963) *The Rock Forming Minerals*, **V. 2**, Longmans, Green and Co., London, 379 pp.

Dermott S. F. (1979) Shapes and gravitational moments of satellites and asteroids. *Icarus* **37**, 575–586.

Dickinson W. and Luth W. C. (1971) A model for plate tectonic evolution of mantle layers. *Science* **174**, 400–404.

Drake M. J. (1976) Evolution of major mineral compositions and trace element abundances during fractional crystallization of a model lunar composition. *Geochim. Cosmochim. Acta* **40**, 401–411.

Dreibus G. and Wänke H. (1979) On the chemical composition of the Moon and the eucrite parent body and a comparison with the composition of the Earth, the case of Mn, Cr, and V (abstract). In *Lunar and Planetary Science X*, p. 315–317. Lunar and Planetary Institute, Houston.

Dreibus G., Blum K., Spettel B. and Wänke H. (1978) Element correlations and their significance for the determination of the bulk composition of planetary objects (abstract). In *Lunar and Planetary Science IX*, p. 267–269. Lunar and Planetary Institute, Houston.

Dreibus G., Kruse H., Spettel B. and Wänke H. (1977) The bulk composition of the Moon and the eucrite parent body. *Proc. Lunar Sci. Conf. 8th*, p. 211–227.

Dreibus G., Spettel B. and Wänke H. (1976) Lithium as a correlated element, its condensation behavior, and its use to estimate the bulk composition of the Moon and the eucrite parent body. *Proc. Lunar Sci. Conf. 7th*, p. 3383–3396.

Drobyshevski E. M. (1979) Magnetic field of Jupiter and the volcanism and rotation of the Galilean satellites. *Nature* **282**, 811–813.

Duba A. and Ringwood A. E. (1973) Electrical conductivity, internal temperatures and thermal evolution of the Moon. *The Moon* **7**, 356–376.

Duba A., Heard H. C. and Schock R. N. (1976) Electrical conductivity of orthopyroxene to 1400°C and the resulting selenotherm. *Proc. Lunar Sci. Conf. 7th*, p. 3173–3181.

Duke M. B. and Silver L. T. (1967) Petrology of eucrites, howardites and mesosiderites. *Geochim. Cosmochim. Acta* **31**, 1637–1665.

Dyal P. and Parkin C. W. (1973) Global electromagnetic induction in the Moon and planets. *Phys. Earth and Planet. Inter.* **7**, 251–256.

Elsasser W. M. (1963) Early history of the Earth. *Earth Science and Meteorites* (J. Geiss and E. D. Goldberg, eds.), p. 1–30. Interscience Publishers, N.Y.

Elsasser W. M., Olson P. and March B. (1979) The depth of mantle convection. *J. Geophys. Res.* **84**, 147–155.

Fanale F. P., Johnson T. V. and Matson D. L. (1977) Io's surface and the histories of the Galilean satellites. In *Planetary Satellites* (J. Burns, ed.), p. 379–405. Unv. of Arizona Press, Tucson.

Frey F. A. and Prinz M. (1978) Ultramafic inclusions from San Carlos, Arizona: petrologic and geochemical data bearing on their petrogenesis. *Earth Planet. Sci. Lett.* **38**, 129–176.

Fuchs L. H., Olsen E. and Jensen K. J. (1973) Mineralogy, mineral-chemistry, and composition of the Murchison (C2) meteorite. *Smithsonian Contrib. Earth Sci.* **No. 10**, 39 pp.

Ganapathy R. and Anders E. (1974a) Bulk compositions of the Moon and Earth, estimated from meteorites (abstract). In *Lunar Science V*, p. 254–256. The Lunar Science Institute, Houston.

Ganapathy R. and Anders E. (1974b) Bulk compositions of the Moon and Earth estimated from meteorites. *Proc. Lunar Sci. Conf. 5th*, p. 1181–1206.

Goettel K. A. (1977) Composition of the terrestrial planets. In *Reports of Planetary Geology Program, 1976-1977*, p. 7–10. NASA TM X-3511.

Goettel K. A. (1978) The composition of Mars. In *Reports of Planetary Geology Program, 1977-1978*, p. 116–117b. NASA TM 79729.

Goettel K. A. and Barshay S. S. (1978) The chemical equilibrium model for condensation in the solar nebula: assumptions, implications, and limitations. In *The Origin of the Solar System* (S. F. Dermott, ed.), p. 611–627. John Wiley and Sons, N.Y.,

Goetze C. (1977) A brief summary of our present day understanding of the effect of volatiles and partial melt on the mechanical models of the upper mantle. In *High Pressure Research Applications in Geophysics* (M. H. Manghnani and S. Akimoto, eds.), p. 3–23. Academic Press, N.Y.

Goins N. R. (1978) Lunar seismology: the internal structure of the Moon. Ph.D. Thesis, M.I.T., 666 pp.

Goins N. R., Toksöz M. N. and Dainty A. M. (1978) Seismic structure of the lunar mantle: an overview. *Proc. Lunar Planet. Sci. Conf. 9th*, p. 3575–3588.

Goldreich P. and Ward W. R. (1973) The formation of planetesimals. *Astrophys. J.* **183**, 1051–1061.

Goldstein B. E., Phillips R. J. and Russell C. T. (1976) Magnetic evidence concerning a lunar core. *Proc. Lunar Sci. Conf. 7th*, p. 3321–3341.

Graboske H. C., Pollack J. B., Grossman A. S. and Olness R. J. (1975) The structure and evolution of Jupiter: the fluid contraction stage. *Astrophys. J.* **199**, 265–281.

Green D. H. and Ringwood A. E. (1967) The stability field of aluminous pyroxene peridotite and garnet peridotite and their relevance in upper mantle structure. *Earth Planet. Sci. Lett.* **3**, 151–160.

Grossman L. (1972) Condensation in the primitive solar nebula. *Geochim. Cosmochim. Acta* **36**, 597–619.

Grossman L. (1973) Refractory trace elements in Ca-Al-rich inclusions in the Allende meteorite. *Geochim. Cosmochim. Acta* **37**, 1119–1140.

Grossman L. and Larimer J. W. (1974) Early chemical history of the solar system. *Rev. Geophys. Space Phys.* **12**, 71–101.

Hapke B. (1979) Io's surface and environs: a magmatic-volatile model. *Geophys. Res. Lett.* **6**, 799–802.

Harris P. G., Reay A. and White I. G. (1967) Chemical composition of the upper mantle. *J. Geophys. Res.* **72**, 6359–6369.

Head J. W. (1979) Lava flooding of earth planetary crusts: geometry, thickness, and volumes of flooded impact basins (abstract). In *Lunar and Planetary Science X*, p. 516–518. Lunar and Planetary Institute, Houston.

Herbert F., Drake M. J. and Sonett C. P. (1978) Geophysical and geochemical evolution of the lunar magma ocean. *Proc. Lunar Planet. Sci. Conf. 9th*, p. 249–262.

Hertogen J., Vizgirda J. and Anders E. (1977) Composition of the parent body of eucrite meteorites. *Bull. Amer. Astron. Soc.* **9**, 458–459.

Hoffman J. H., Hodges R. R., McElroy M. B., Donahue T. M. and Kolpin M. A. (1979) Composition and structure of the Venus atmosphere: results from Pioneer Venus. *Science* **205**, 49–52.

Holweger H. (1977) The solar Na/Ca and S/Ca ratios: a close comparison with carbonaceous chondrites. *Earth Planet Sci. Lett.* **34**, 152–154.

Hostetler C. J. and Drake M. J. (1978) Quench temperatures of Moore County and other eucrites: residence time on eucrite parent body. *Geochim. Cosmochim. Acta* **42**, 517–522.

Hubbard W. B. and Anderson J. D. (1978) Possible fly-by measurements of Galilean satellite interior structure. *Icarus* **33**, 336–341.

Jagoutz E., Palme H., Baddenhausen H., Blum K., Cendales M., Dreibus G., Spettel B., Lorenz V. and Wänke H. (1979) The abundances of major, minor and trace elements in the Earth's mantle as derived from primitive ultramafic nodules. *Proc. Lunar Planet. Sci. Conf. 10th*, p. 2031–2050.

Jarosewich E. and Mason B. (1969) Chemical analyses with notes on one mesosiderite and seven chondrites. *Geochim. Cosmochim. Acta* **33**, 411–417.

Jeanloz R. and Richter F. M. (1979) Convection, composition, and the thermal state of the lower mantle. *J. Geophys. Res.* **84**, 5497–5504.

Jérome D. J. and Goles G. G. (1971) A re-examination of relationships among pyroxene-plagioclase achondrites. In *Activation Analys. in Geochem. Cosmochem.* (A. O. Brunfelt and E. Steinnes, eds.), p. 261–266. Universitetsforlarget, Oslo.

Johnson T. V. (1978) The Galilean satellites of Jupiter: four worlds. *Ann. Rev. Earth Planet. Sci.* **6**, 93–125.

Johnston D. H. and Toksöz M. N. (1977) Internal structure and properties of Mars. *Icarus* **32**, 73–84.

Kaula W. M. (1968) *An Introduction to Planetary Physics: The Terrestrial Planets.* Wiley, N.Y. 490 pp.

Kaula W. M. (1977a) On the origin of the Moon, with emphasis on bulk composition. *Proc. Lunar Sci. Conf. 8th,* p. 321–332.

Kaula W. M. (1977b) Mechanical processes affecting differentiation of protolunar material. In *Proc. Soviet-American Conf. on Cosmochem. Moon and Planets* (J. H. Pomeroy and N. J. Hubbard, eds.), p. 805–813. NASA, U.S. Government Printing Office No. 033-000-00633-1.

Kaula W. M. (1979a) Thermal evolution of Earth and Moon growing by planetesimal impacts. *J. Geophys. Res.* **84**, 999–1008.

Kaula W. M. (1979b) The moment of inertia of Mars. *Geophys. Res. Lett.* **6**, 194–196.

Kaula W. M., Schubert G., Lingenfelter R. E., Sjogren W. L. and Wollenhaupt W. R. (1974) Apollo laser altimetry and inferences as to lunar structure. *Proc. Lunar Sci. Conf. 5th,* p. 3049–3058.

Kesson S. E. (1975) Mare basalts: melting experiments and petrogenetic interpretations. *Proc. Lunar Sci. Conf. 6th,* p. 921–944.

King D. A. and Ahrens T. J. (1973) Shock compression of FeS. *Nature Phys. Sci.* **243**, 82–84.

Koyama J. and Nakamura Y. (1979) Re-examination of the lunar seismic velocity structure based on the complete data set (abstract). In *Lunar and Planetary Science X,* p. 685–687. Lunar and Planetary Institute, Houston.

Krähenbühl U., Ganapathy R., Morgan J. W. and Anders E. (1973) Volatile elements in Apollo 16 samples: implications for highland volcanism and accretion history of the Moon. *Proc. Lunar Sci. Conf. 4th,* p. 1325–1348.

Kuiper G. P. (1951) On the origin of the solar system. *Proc. Nat. Acad. Sci.* 37, 1–14.

Kuiper G. P. (1952) Planetary atmospheres and their origin. In *Surfaces and Interiors of Planets and Satellites* (A. Dollfus, ed.), p. 306–405. Univ. of Chicago Press, Chicago.

Kumazawa M. (1969) The elastic constants of single-crystal orthopyroxene. *J. Geophys. Res.* **74**, 5973–5980.

Kumazawa M. and Anderson O. L. (1969) Elastic moduli, pressure derivatives, and temperature derivatives of single-crystal olivine and single-crystal forsterite. *J. Geophys. Res.* **74**, 5961–5972.

Kushiro I. and Yoder H. S., Jr. (1966) Anorthite-forsterite and anorthite-enstatite reactions and their bearing on the basalt-eclogite transformation. *J. Petrol.* **7**, 337–362.

Langseth M. G., Keihm S. J. and Peters K. (1976) Revised lunar heat flow values. *Proc. Lunar Sci. Conf. 7th,* p. 3143–3171.

Laplace P. S. (1805) *Traité de Mécanique Céleste,* liv. X., Sur Differens Points Relatifs au Systêm du Monde, chap. VII, p. 9118.

Larimer J. W. (1967) Chemical fractionations in meteorites—I. Condensation of the elements. *Geochim. Cosmochim. Acta* **31**, 1215–1238.

Larimer J. W. (1968) Experimental studies on the system Fe-MgO-SiO_2-O_2 and their bearing on the petrology of chondritic meteorites. *Geochim. Cosmochim. Acta* **32**, 1187–1201.

Larimer J. W. and Anders E. (1967) Chemical fractionations in meteorites—II. Abundance patterns and their interpretation. *Geochim. Cosmochim. Acta* **31**, 1239–1270.

Larimer J. W. and Anders E. (1970) Chemical fractionations in meteorites—III. Major element fractionations in chondrites. *Geochim. Cosmochim. Acta* **34**, 367–387.

Laul J. C., Ganapathy R., Anders E. and Morgan J. W. (1973) Chemical fractionations in meteorites—VI. Accretion temperatures of H-, LL-, and E-chondrites, from abundance of volatile trace elements. *Geochim. Cosmochim. Acta* **35**, 337–363.

Laul J. C., Keays R. R., Ganapathy R., Anders E. and Morgan J. W. (1972b) Chemical fractionations in meteorites—V. Volatile and siderophile elements in achondrites and ocean ridge basalts. *Geochim. Cosmochim. Acta* **36**, 329–345.

Laul J. C., Wakita B., Showalter D. L., Boynton W. V. and Schmitt R. A. (1972a) Bulk, rare earth, and other trace elements in Apollo 14 and 15 and Luna 16 samples. *Proc. Lunar Sci. Conf. 3rd.,* p. 1181–2000.

Lee T., Papanastassiou D. and Wasserburg G. J. (1976) Demonstration of ^{26}Mg excess in Allende and evidence for ^{26}Al. *Geophys. Res. Lett.* **3**, 109–112.

Lewis J. S. (1972) Metal/silicate fractionation in the solar system. *Earth Planet. Sci. Lett.* **15**, 286–290.

Lewis J. S. (1973) Chemistry of the planets. *Ann. Rev. Phys. Chem.* **24**, 339–351.

Lewis J. S. (1974a) The chemistry of the solar system. *Sci. Amer.* **230, No. 3**, 50–65.

Lewis J. S. (1974b) The temperature gradient in the solar nebula. *Science* **186**, 440–443.

Lewis J. S. (1974c) Volatile element influx on Venus from cometary impacts. *Earth Planet. Sci. Lett.* **22**, 239–244.

Lewis J. S. and Consolmagno G. J. (1979) Io: geochemistry and geophysics of sulfur. *Bull. Amer. Astron. Soc.* **11**, 599.

Lewis R. S., Alaerts L., Hertogen J., Janssens M. -J., Palme H. and Anders E. (1979) A carbonaceous inclusion from the Krymka LL-chondrite: noble gases and trace elements. *Geochim. Cosmochim. Acta* **43**, 897–903.

Lieberman R. C. and Ringwood A. E. (1976) Elastic properties of anorthite and the nature of the lunar crust. *Earth Planet. Sci. Lett.* **31**, 69–74.

Liu L. -G. (1977) The system enstatite-pyrope at high pressures and temperatures and the mineralogy of the Earth's mantle. *Earth Planet. Sci. Lett.* **36**, 237–243.

Longhi J. (1977) Magma oceanography 2: chemical evolution and crustal formation. *Proc. Lunar Sci. Conf. 8th,* p. 601–621.

Longhi J. (1978) Pyroxene stability and the composition of the lunar magma ocean. *Proc. Lunar Planet. Sci. Conf. 9th,* p. 285–306.

Lord H. C., III (1965) Molecular equilibria and condensation in a solar nebula and cool stellar atmospheres. *Icarus* **4**, 279–288.

MacGregor I. D. (1970) The effect of CaO, Cr_2O_3, Fe_2O_3 and Al_2O_3 on the stability of spinel and garnet peridotites. *Phys. Earth Planet. Inter.* **3**, 372–377.

MacGregor I. D. (1974) The system MgO-Al_2O_3-SiO_2: solubility of Al_2O_3 in enstatite for spinel and garnet peridotite compositions. *Amer. Mineral.* **59**, 110–119.

Mason B. (1962) *Meteorites.* John Wiley and Sons, N.Y. 274 pp.

Masursky H., Kaula W. M., McGill G. E., Pettengill G. H., Phillips R. J., Russell C. T., Schubert G. and Shapiro I. I. (1977) The surface and interior of Venus. *Space Sci. Rev.* **20**, 431–449.

McCarthy T. S., Ahrens L. H. and Erlank A. J. (1972) Further evidence for the mixing model for howardite origin. *Earth Planet. Sci. Lett.* **15**, 86–93.

McCord T. B. and Clark R. N. (1979) The Mercury soil: presence of Fe^{2+} (abstract). In *Lunar and Planetary Science IX,* p. 789–791. Lunar and Planetary Institute, Houston.

McCord T. B., Adams J. B. and Johnson T. V. (1970) Asteroid Vesta: spectral reflectivity and compositional implications. *Science* **168**, 1445–1447.

McGetchin T. and Smyth J. R. (1978) The mantle of Mars: some possible geological implications of its high density. *Icarus* **34**, 512–536.

Minear J. W. and Fletcher C. R. (1978) Crystallization of a lunar magma ocean. *Proc. Lunar Planet. Sci. Conf. 9th,* p. 263–283.

Morabito L. A., Synnott S. P., Kupferman P. N. and Collins S. A. (1979) Discovery of currently active extraterrestrial volcanism. *Science* **204**, 892–894.

Morgan J. W. and Anders E. (1979) Chemical composition of Mars. *Geochim. Cosmochim. Acta* **43**, 1601–1610.

Morgan J. W. and Anders E. (1980) Chemical composition of the Earth, Venus, and Mercury. *Proc. Nat. Acad. Sci.* **77**, 6973–6977.

Morgan J. W., Hertogen J. and Anders E. (1978a) The Moon: composition determined by nebular processes. *Moon and Planets* **18**, 465–478.

Morgan J. W., Higuchi H., Takahashi H. and Hertogen J. (1978b) A "chondritic" eucrite parent body: inference from trace elements. *Geochim. Cosmochim. Acta* **42**, 27–38.

Moritz H. (1976) Fundamental geodetic constants. *Trav. Assoc. Int. Géodésie* **25**, 411–418.

Morrison D. (1977) Asteroid sizes and albedos. *Icarus* **31**, 185–220.

Mueller R. F. (1964) A comparison of oxidative equilibria in meteorites and terrestrial rocks. *Geochim. Cosmochim. Acta* **27**, 273–278.

Nakamura Y., Duennebier F., Latham G. and Dorman H. J. (1976a) Structure of the lunar mantle. *J. Geophys. Res.* **81**, 4818–4824.

Nakamura Y., Lammlein D., Latham G., Ewing M., Dorman H., Press F. and Toksöz N. M. (1973) New seismic data on the state of the deep lunar interior. *Science* **181**, 49–50.

Nakamura Y., Latham G. V., Dorman H. J. and Duennebier F. K. (1976b) Seismic structure of the Moon; a summary of current status. *Proc. Lunar Sci. Conf. 7th*, p. 3113–3121.

Nakamura Y., Latham G. V., Lammlein D., Ewing M., Duennebier F. and Dorman H. J. (1974) Deep lunar interior inferred from recent seismic data. *Geophys. Res. Lett.* **1**, 137–140.

Nicholls G. D. (1967) Geochemical studies in the ocean as evidence for the composition of the mantle. In *Mantles of the Earth and Terrestrial Planets* (S. K. Runcorn, ed.), p. 285–304. Wiley-Interscience, N.Y.

Null G. W. (1976) Gravity field of Jupiter and its satellites from Pioneer 10 and Pioneer 11 tracking data. *Astron. J.* **81**, 1153–1161.

O'Hara M. J. (1968) The bearing of phase equilibria studies in synthetic and natural systems on the origin and evolution of basic and ultrabasic rocks. *Earth Sci. Rev.* **4**, 69–133.

Okal E. A. and Anderson D. L. (1978) Theoretical models for Mars and their seismic properties. *Icarus* **33**, 514–528.

Okamura F. P., Ghose S. and Ohashi H. (1974) Structure and crystal chemistry of calcium, Tschermak's pyroxene, $CaAlAlSiO_6$. *Amer. Mineral.* **59**, 549–557.

O'Nions R. K., Evensen N. M. and Hamilton P. J. (1979) Geochemical modelling of mantle differentiation and crustal growth. *J. Geophys. Res.* **84**, 6091–6101.

Opik E. J. (1963) Selective escape of gases. *Geophys. J. Roy. Astron. Soc.* **7**, 490–509.

Orowan E. (1969) Density of the Moon and nucleation of planets. *Nature* **222**, 867.

Oxburgh E. R. and Parmentier E. M. (1977) Compositional and density stratification in oceanic lithosphere: causes and consequences. *J. Geol. Soc. Lond.* **133**, 343–355.

Oxburgh E. R. and Parmentier E. M. (1978) Thermal processes in the formation of continental lithosphere. *Phil. Trans. Roy. Soc. Lond.* **A288**, 415–430.

Palme H., Baddenhausen H., Blum K., Cendales M., Dreibus G., Hofmeister H., Kruse H., Palme C., Spettel B., Vilcsek E., Wänke H. and Kurat G. (1978) New data on lunar samples and achondrites and a comparison of the least fractionated samples from the Earth, the Moon and the eucrite parent body. *Proc. Lunar Planet. Sci. Conf. 9th*, p. 25–57.

Peale S. J., Cassen P. and Reynolds R. T. (1979) Melting of Io by tidal dissipation. *Science* **203**, 892–894.

Pettengill G. H., Ford P. G., Brown W. E., Kaula W. H., Keller C. H., Masursky H. and McGill G. H. (1979) The Pioneer Venus radar mapper experiment. *Science* **203**, 892–894.

Phillips R. J., Saunders R. S. and Conel J. E. (1973) Mars: crustal structure inferred from Bouguer gravity anomalies. *J. Geophys. Res.* **78**, 4815–4820.

Philpotts J. A. and Schnetzler C. C. (1970) Apollo 11 lunar samples: K, Rb, Sr, Ba and rare earth concentrations in some rocks and separated phases. *Proc. Apollo 11 Lunar Sci. Conf.*, p. 1471–1486.

Pollack J. B. and Reynolds T. R. (1974) Implications of Jupiter's early contraction history for the composition of the Galilean satellites. *Icarus* **21**, 243–253.

Rammensee W. and Wänke H. (1977) On the partition coefficient of tungsten between metal and silicate and its bearing on the origin of the Moon. *Proc. Lunar Sci. Conf. 8th*, p. 399–409.

Reasenberg R. D. (1977) The moment of inertia and isostasy of Mars. *J. Geophys. Res.* **82**, 369–388.

Richter F. and Parsons B. (1975) On the interaction of two scales of convection in the mantle. *J. Geophys. Res.* **80**, 2529–2541.

Ringwood A. E. (1959) On the chemical evolution and densities of the planets. *Geochim. Cosmochim. Acta* **15**, 257–283.

Ringwood A. E. (1966) Chemical evolution of the terrestrial planets. *Geochim. Cosmochim. Acta* **30**, 41–104.

Ringwood A. E (1970) Phase transformations and the constitution of the mantle. *Phys. Earth Planet. Inter.* **3**, 109–155.

Ringwood A. E. (1975) *Composition and Petrology of the Earth's Mantle.* McGraw-Hill, N.Y. 618 pp.

Ringwood A. E. (1977a) *Composition and Origin of the Earth.* Publication No. 1299, Res. School of Earth Sciences, Australian National Univ., Canberra. 65 pp.

Ringwood A. E. (1977b) Basaltic magmatism and the composition of the Moon I. Major and heat-producing elements. *The Moon* **16**, 389–423.

Ringwood A. E. and Anderson D. L. (1977) Earth and Venus: a comparative study. *Icarus* **30**, 243–253.

Ringwood A. E. and Clark S. P. (1971) Internal constitution of Mars. *Nature* **234**, 89–92.

Ringwood A. E. and Essene E. (1970) Petrogenesis of Apollo 11 basalts, internal constitution and origin of the Moon. *Proc. Apollo 11 Lunar Sci. Conf.*, p. 769–799.

Ringwood A. E. and Kesson S. E. (1977a) Composition and origin of the Moon. *Proc. Lunar Sci. Conf. 8th*, p. 371–389.

Ringwood A. E. and Kesson S. E. (1977b) Basaltic magmatism and the composition of the Moon II. Siderophile and volatile elements in Moon, Earth, and chondrites: implications for lunar origin. *The Moon* **16**, 425–464.

Robie R. A., Bethke P. M., Toulmin M. S. and Edwards J. L. (1966) X-ray crystallographic data, densities and molar volumes of minerals. In *Handbook of Physical Constants* (S. P. Clark, Jr., ed.), p. 27–73. Geol. Soc. Amer. Mem. 97.

Ronov A. B. and Yaroshevsky A. A. (1969) Chemical composition of the Earth's crust. In *The Earth's Crust and Upper Mantle* (P. J. Hart, ed.), p. 37–57. Amer. Geophys. Union Monograph 13.

Runcorn S. K. (1978) Superheavy elements in the primeval Moon? (abstract). In *Lunar and Planetary Science IX*, p. 979–981. Lunar and Planetary Institute, Houston.

Ruskol E. L. (1972) The origin of the Moon, 3. Some aspects of the dynamics of the circumterrestrial swarm. *Sov. Astron. A. J.* **15**, 646–654.

Russell T., Coleman P. J., Kennel C. F., McPherson R. L. and Siscoe G. L. (1979) The Pioneer Venus magnetometer experiment. *Science* **203**, 745–748.

Ryzhova T. V., Reshchikova L. M. and Alexandrov K. S. (1966) Elastic properties of rock forming minerals: 6—garnets. *Izv. Acad. Sci., U.S.S.R., Phys. Solid Earth* **7**, 447–450.

Safronov V. S. (1972) Accumulation of the planets. In *On the Origin of the Solar System* (H. Reeves, ed.), p. 89–113. Centre Nationale de Reserche Scientifique, Paris.

Sammis D. G., Anderson D. L. and Jordan T. H. (1970) Application of isotropic limit strain theory to ultrasonic and seismological data. *J. Geophys. Res.* **75**, 4478–4480.

Sato Y. (1977) Equation of state of mantle minerals determined through high pressure X-ray study in high pressure research applications. In *Geophysics* (M. H. Manghani and S. Akimoto, eds.), p. 307–323. Harcourt, Brace, Jovanovich, N.Y.

Sawamoto H. (1977) Orthorhombic perovskite (Mg,Fe) SiO_3 and constitution of the lower mantle. In *High-Pressure Research* (M. Manghnani and S. Akimoto, eds.), p. 219–244. Academic Press, N.Y.

Schonfeld E. (1975) A model for the lunar anorthositic gabbro. *Proc. Lunar Sci. Conf. 6th*, p. 1375–1384.

See T. J. J. (1910) *Researches on the Evolution of the Stellar System*, Vol. II, Thos. P. Nichols and Sons, Lynn, Mass.

Simmons G. and Wang H. (1971) *Single Crystal Elastic Constants and Calculated Aggregate Properties: a Handbook*, 2nd edition. M.I.T. Press, Cambridge. 370 pp.

Simmons G., Batzle M. L. and Harlow A. L. (1980) Thermal modification of microcracks in lunar rocks and revised estimates for the elastic properties of the shallow Moon (abstract). In *Lunar and Planetary Science XI*, p. 1030–1032. Lunar and Planetary Institute, Houston.

Simmons G., Todd T. and Wang H. (1972) The 25 km discontinuity: implications for lunar history. *Science* **182**, 158.

Skinner B. J. (1966) Thermal expansion. In *Handbook of Physical Constants* (S. P. Clark, ed.), p. 75–96. Geol. Soc. Amer. Mem. 97.

Skinner B. J. and Boyd F. R. (1964) Aluminous enstatites. *Carnegie Inst. Wash. Yearb.* **63**, 163–165.

Smith B. A., Shoemaker E. M., Kieffer S. W. and Cook A. F., II (1979a) The role of SO_2 in volcanism on Io. *Nature* **280**, 738–743.

Smith B. A., Soderblom L. A., Johnson T. V., Ingersoll A. P., Collins S. A., Shoemaker E. M., Hunt G. E., Masursky H., Carr M. H., Davies M. E., Cook A. F., III, Boyce J., Danielson G. E., Owen T., Sagan C., Beebe R. F., Veverka J., Strom R. G., McCauley J. F., Morrison D., Briggs G. A. and Suomi V. E. (1979b) The Jupiter system through the eyes of Voyager 1. *Science* **204**, 951–972.

Smith J. V. (1977) Possible controls on the bulk composition of the Earth: implications for the origin of the Earth and Moon. *Proc. Lunar Sci. Conf. 8th*, p. 333–369.

Smith J. V., Anderson A. T., Newton R. C., Olsen E. J., Wyllie P. J., Crewe A. V., Isaacson M. S. and Johnson D. (1970) Petrologic history of the Moon inferred from petrography, mineralogy, and petrogenesis of Apollo 11 rocks. *Proc. Apollo 11 Lunar Sci. Conf.*, p. 897–925.

Smythe W. D., Nelson R. M. and Nash D. B. (1979) Spectral evidence for SO_2 frost or adsorbate on Io's surface. *Nature* **280**, 766.

Solomon S. C. (1977) The relationship between crustal tectonics and internal evolution in the Moon and Mercury. *Phys. Earth Planet. Inter.* **15**, 135–145.

Solomon S. C. and Chaiken J. (1976) Thermal expansion and thermal stress in the Moon and terrestrial planets: clues to early thermal history. *Proc. Lunar Sci. Conf. 7th*, p. 3229–3243.

Sonett C. P., Colburn D. S. and Schwartz K. (1975) Formation of the lunar crust: an electrical source of heating. *Icarus* **24**, 231–235.

Sonett C. P., Smith B. F., Colburn D. S., Schubert G. and Schwartz K. (1972) The induced magnetic field of the Moon: conductivity profiles and inferred temperatures. *Proc. Lunar Sci. Conf. 3rd*, p. 2309–2335.

Stacey F. D. (1977) A thermal model of the Earth. *Phys. Earth and Planet. Inter.* **15**, 341–348.

Stevenson D. J., Schubert G., Cassen P. and Reynolds R. T. (1980) Core evolution and magnetism of the terrestrial planets (abstract). In *Lunar and Planetary Science XI*, p. 1091–1093. Lunar and Planetary Institute, Houston.

Stolper E. (1977) Experimental petrology of eucritic meteorites. *Geochim. Cosmochim. Acta* **41**, 587–611.

Stolper E. (1979) Trace elements in shergottite meteorites: implications for the origins of planets. *Earth Planet. Sci. Lett.* **42**, 239–242.

Surkov Yu. A. (1977) Geochemical studies of Venus by Venera 9 and 10 automatic interplanetary stations. *Proc. Lunar Sci. Conf. 8th*, p. 2665–2689.

Taylor S. R. (1976) Geochemical constraints on the composition of the Moon. *Proc. Lunar Sci. Conf. 7th*, p. 3461–3477.

Taylor S. R. (1978) Geochemical constraints on melting and differentiation of the Moon. *Proc. Lunar and Planet. Sci. Conf. 9th*, p. 15–23.

Taylor S. R. and Bence A. E. (1975) Evolution of the lunar highland crust. *Proc. Lunar Sci. Conf. 6th*, p. 1121–1141.

Taylor S. R. and Jakeš P. (1974) The geochemical evolution of the Moon. *Proc. Lunar Sci. Conf. 5th*, p. 1287–1305.

Taylor S. R. and Jakeš P. (1977) Geochemical evolution of the Moon revisited. *Proc. Lunar Sci. Conf. 8th*, p. 433–446.

Thurber C. H. and Solomon S. C. (1978) An assessment of crustal thickness variations on the lunar nearside: models, uncertainties, and implications for crustal differentiation. *Proc. Lunar and Planet. Sci. Conf. 9th*, p. 3481–3497.

Todd T., Richter D. A., Simmons G. and Wang H. (1973) Unique characterization of lunar samples by physical properties. *Proc. Lunar Sci. Conf. 4th*, p. 2639–2662.

Toksöz M. N. and Hsui A. T. (1978) Thermal history and evolution of Mars. *Icarus* **34**, 537–547.

Toksöz M. N., Dainty A. M., Solomon S. C. and Anderson K. (1974) Structure of the Moon. *Rev. Geophys. and Space Phys.* **12**, 539–567.

Toksöz M. N., Hsui A. T. and Johnston D. H. (1978) Thermal evolutions of the terrestrial planets. *Moon and Planets* **18**, 281–320.

Toksöz M. N., Press F., Anderson K., Latham G., Ewing M., Dorman J., Lammlein D., Nakamura Y., Sutton G. and Duennebier F. (1972) Velocity structure and properties of the lunar crust. *The Moon* **4**, 490–504.

Turekian K. K. and Clark S. (1969) Inhomogeneous accretion of the Earth from the primitive solar nebula. *Earth Planet. Sci. Lett.* **6**, 346–348.

Turekian K. K., Clark S., Davis A. and Grossman L. (1973) Heterogeneous accretion revisited (abstract). *Meteoritics* **8**, 77.

Urey H. C. (1963) The origin and evolution of the solar system. In *Space Science* (D. P. LeGalley, ed.), p.123. John Wiley and Sons, N.Y.

Vinogradov A. P., Surkov Yu. A. and Kirnozov F. F. (1973) The contents of uranium, thorium and potassium in the rocks of Venus as measured by Venera 8. *Icarus* **20**, 253–259.

Walker D., Longhi J. and Hays J. F. (1975) Differentiation of a very thick magma body and implications for the source regions of mare basalts. *Proc. Lunar Sci. Conf. 6th*, p. 1103–1120.

Walker D., Stolper E. and Hays J. F. (1978) A numerical treatment of melt/solid segregation: size of the eucrite parent body and stability of the terrestrial low-velocity zone. *J. Geophys. Res.* **83**, 6005–6013.

Walker J. C. G. (1977) *Evolution of the Atmosphere*. Macmillan, N.Y. 318 pp.

Wang H. and Simmons G. (1972) Elasticity of some mantle crystal structures, 1. Pleonaste and hercynite spinel. *J. Geophys. Res.* **77**, 4379–4392.

Wänke H., Baddenhausen H., Blum K., Cendales M., Dreibus G., Hofmeister H., Kruse H., Jagoutz E., Palme C., Spettel B., Thacker R. and Vilcsek E. (1977) On the chemistry of lunar samples and achondrites. Primary matter in the lunar highlands: a re-evaluation. *Proc. Lunar Sci. Conf. 8th*, p. 2191–2213.

Wänke H., Baddenhausen H., Dreibus G., Jagoutz E., Kruse H., Palme H., Spettel B. and Teschke F. (1973) Multielement analyses of Apollo 15, 16, and 17 samples and the bulk composition of the Moon. *Proc. Lunar Sci. Conf. 4th*, p. 1461–1481.

Wänke H., Baddenhausen H., Palme H. and Spettel B. (1974a) On the chemistry of the Allende inclusions and their origin as high temperature condensates. *Earth Planet. Sci. Lett.* **23**, 1–7.

Wänke H., Palme H., Baddenhausen H., Dreibus G., Jagoutz E., Kruse H., Spettel B., Teschke F. and Thacker R. (1974b) Chemistry of Apollo 16 and 17 samples: bulk composition, late stage accumulation and early differentiation of the Moon. *Proc. Lunar Sci. Conf. 5th*, p. 1307–1355.

Wasserburg G. T. and DePaolo D. J. (1979) Models of Earth structure inferred from neodymium and strontium isotopic abundances. *Proc. Nat. Acad. Sci.* **76**, 3594–3598.

Wasson J. T. (1974) *Meteorites*. Springer-Verlag, N.Y. 316 pp.

Wasson J. T. and Warren P. H. (1980) Contribution of the mantle to the lunar asymmetry (abstract). In *Lunar and Planetary Science XI*, p. 1220–1221. Lunar and Planetary Institute, Houston.

Weidenschilling S. J. (1976) Accretion of the terrestrial planets. II. *Icarus* **27**, 161–170.

Weidenschilling S. J. (1978) Iron/silicate fractionation and the origin of Mercury. *Icarus* **35**, 99–111.

Weill D. F. and McKay G. A. (1975) The partitioning of Mg, Fe, Sr, Ca, Sm, Eu, and Yb in lunar igneous systems and a possible origin of KREEP by equilibrium partial melting. *Proc. Lunar Sci. Conf. 6th*, p. 1143–1158.

Wetherill G. W. (1974) Solar system sources of meteorites and large meteoroids. *Ann. Rev. Earth Planet. Sci.* **2**, 303–331.

Wetherill G. W. (1975) Possible slow accretion of the Moon and its thermal and petrologic consequences (abstract). In *Origins of Mare Basalts and their Implications for Lunar Evolution*, p. 184–188. The Lunar Science Institute, Houston.

Wetherill G. W. (1977) Evolution of the Earth's planetesimal swarm subsequent to the formation of the Earth and Moon. *Proc. Lunar Sci. Conf. 8th*, p. 1–16.

Wetherill G. W. and Williams J. G. (1979) Origin of differentiated meteorites. In *Origin and Distribution of the Elements, Proc. of the Second Symposium, Vol. II* (A. H. Ahrens, ed.), p. 19–30. Pergamon Press, Oxford.

Whipple F. L. (1964) The history of the solar system. *Proc. Nat. Acad. Sci.* **52**, 565–594.

White I. (1967) Ultrabasic rocks and the composition of the upper mantle. *Earth Planet. Sci. Lett.* **3**, 11–18.

Wiik H. B. (1956) The chemical composition of some stony meteorites. *Geochim. Cosmochim. Acta* **2**, 91–117.

Wiskerchen M. J. and Sonett C.P. (1977) A lunar metal core? *Proc. Lunar Sci. Conf. 8th*, p. 515–535.

Wood J. A. (1962) Chondrules and the origin of the terrestrial planets. *Nature* **194**, 127–130.

Wood J. A. (1973) Bombardment as a cause of the lunar asymmetry. *The Moon* **8**, 73–103.

Wood J. A., Dickey J. S., Jr., Marvin U. B. and Powell B. N. (1970) Lunar anorthosites and a geophysical model of the Moon. *Proc. Apollo 11 Lunar Sci. Conf.*, p. 965–988.

Descending full moon over shadow of Mauna Kea Volcano from summit of Mauna Kea volcano (elevation 15,800 ft.) at dawn looking west towards Hualalai and Kosia. Photo by Carl A. Wilmington.

5. Distribution and morphology of basalt deposits on planets

James W. Head, III, Team Leader
—Department of Geological Sciences
Brown University

Wilfred B. Bryan
—Woods Hole Oceanographic Institute
Woods Hole, Massachusetts

Ronald Greeley
—Department of Geology
Arizona State University

John E. Guest
—University of London Observatory
London, England

Peter H. Schultz
—Lunar and Planetary Institute
Houston, Texas

R. S. J. Sparks
—Department of Mineralogy and Petrology
University of Cambridge

G. P. L. Walker
—MacDonald Chair of Volcanology
University of Hawaii

James L. Whitford-Stark
—Department of Geology
University of Missouri, Columbia

Charles A. Wood
—Johnson Space Center
Houston, Texas

Michael H. Carr
—Branch of Astrogeologic Studies
U. S. Geological Survey, Menlo Park

5.1 INTRODUCTION

The major objective of this chapter is to describe and assess the distribution and morphology of basalt deposits on the planets. Basalt distribution and morphology provide the geologic framework for understanding basalt petrogenesis and its relation to planetary tectonics and crustal evolution. The geometric relationships, age relationships, and areal distribution of basalts will be combined in this and other chapters in order to document the significance of basaltic volcanism in planetary evolution. This record of basaltic volcanism offers clues to the location of sources of magma in the interior, to the modification of the outer layers of a planet by the redistribution of magma, and to the planet's thermal history.

The morphology and geometry of basalt deposits are useful factors in determining deposit volumes, modes of emplacement, and tectonic environment. Extrusive morphology at all scales can also provide important information on composition, volume, extrusion rates, fluidity, and volatile content.

Terrestrial basaltic volcanism provides a key to interpretation of the deposits of other planets because of the wide variety of basalt environments on Earth and because of our access to and history of investigation of these environments. We begin by analyzing the factors governing eruption mechanisms and morphology on Earth and then examine the nature of the major types of volcanic features and the environment in which they occur. We then turn to the Moon and planets to document the distribution and morphology of known and suspected basaltic deposits there.

5.2 FACTORS GOVERNING ERUPTION MECHANISMS AND BASALTIC VOLCANO MORPHOLOGY

The interpretation of the morphology of volcanic landforms requires study of the physical and dynamic volcanic processes which influence the behavior of magma ascending and reaching the surface of a planet. Some of these are particular to a single planet, for example the value of gravity and the pressure and composition of the atmosphere. Other parameters, such as magma rheology, magma volatile content, magma discharge rate and ground slope may vary considerably even for one volcano. It is not surprising, therefore, that lava flows, volcanic craters and pyroclastic deposits show great diversity in their shape, size and surface expression features. For example, basaltic lava flows occur in a wide variety of surface textures, including (1) **pahoehoe**, with smooth billowing, and ropy surfaces; (2) **aa** lavas that are characterized by jagged, clinkery surfaces; and (3) **block** lavas that consist of massive blocks of basalt ~20 cm in diameter and larger. Each of these three types of lava, particularly pahoehoe and aa, can be further subdivided, and each can be gradational to another, as described by Sato (1950), Wentworth and Macdonald (1953), Swanson (1973), and others. Despite these variations, there are several morphological features, for example lava levees, lava tubes, and shield volcanoes, which are common on the terrestrial planets. The similarities of these first-order features indicate that their formation is controlled principally by factors common to all the planets.

5.2.1 Magma ascent and magma rheology

Geological and physical observations and constraints have been applied to the development of a model of the ascent and emplacement of basaltic magma on the Earth and Moon (Wilson and Head, 1980a). Relatively simple mathematical models of the motion of gas/liquid mixtures (McGetchin and Ullrich, 1973; Pai *et al.*, 1978) are adequate to treat basaltic eruptions provided that allowance is made for the coalescence of gas bubbles and that realistic geological and petrochemical constraints are applied to the numerical values of the variables involved. Gas exsolution from magmas on the Earth and Moon (and Mars) only occurs at relatively shallow depths (commonly less than 2 km for terrestrial and lunar basalts). As a result it is generally convenient to consider separately the rise of bubble-free magmatic liquid at depth in a planetary crust and the more complex motions which occur near the surface as gas is exsolved. A lower limit is set to the width of the dike or conduit up which a magma can rise from a given depth, either by the mechanical properties of the magma, or by the need to avoid excessive cooling of the magma during its ascent. The latter constraint is controlled mainly by the magma viscosity. For a wide range of magma types it is found that if the yield strength at depth is greater than a few hundred N/m^2, then yield strength rather than viscosity will limit the ability of the magma to erupt. Bubble coalescence, leading to intermittent explosive activity of strombolian style, will occur in any magma near the surface if the rise speed is sufficiently slow or the viscosity (and yield strength) are sufficiently small. For the commonly occurring basalts on both the Earth and Moon, the rise speed in the crust must be greater than 0.5–1 m/sec if relatively steady fire fountaining is to take place.

For terrestrial basalts commonly having viscosities not less than 10^3 poises and negligible yield strengths, dike or conduit widths in the range 0.2–0.6 m are needed to allow eruptions from depths between 500 m and 20 km. Corresponding minimum mass eruption rates lie in the range 10–300 kg/sec for isolated central conduits and in the range 30–500 kg/sec per meter length of fissure for elongate vents. To accommodate the output rates of 3×10^4 kg $s^{-1}m^{-1}$ for the Columbia River flood basalt eruptions, fissures up to about 4 m wide would be needed. Fire fountain height is dictated by the vertical velocity of the magma/gas dispersion emerging through the vent, and both quantities increase with increasing magma gas content, and with increasing mass eruption rate, and decrease with increasing magma viscosity. Fire fountains up to 500 m high imply the release of up to about 0.4 wt.% of water from the magma, corresponding to initial water contents up to about 0.6 wt.% (Wilson and Head, 1980a).

The final dimensions and shape of a lava flow depend on the rheology of the lava and changes in the rheology as the lava cools. Magma rheology also influences discharge rate, lava thickness, and surface structure. Similarly, the ease with which magma loses volatiles, and hence its explosivity, also depends on the magma rheology. Despite its importance in virtually all igneous processes, however, there is relatively little experimental data on magma rheology.

Shaw (1972) and Bottinga and Weill (1972) have summarized all the available data on the rheology of both natural and artificial silicate melts. They have demonstrated that *above the liquidus,* silicate melts behave as Newtonian fluids and that their viscosity can be predicted as a function of temperature and chemical composition. Within silicate melts molecular chains are formed by covalent bonding between silicon and oxygen atoms. Viscosity increases as the degree of polymerization of a melt increases. Alkalies and alkaline earth elements exist as cations within silicate melts and serve to break up silicate chains, forming weaker ionic bonds and hence reducing the viscosity of the melt. The viscosity of a melt thus depends on the proportions of network-forming species. In broad terms, low silica melts have low viscosity (Fig. 5.2.1). However, some elements, notably aluminum, have an ambiguous role in that they may act as both network formers or breakers depending on their concentration. Shaw (1963) has demonstrated that viscosity is lowered in silica-rich melts as the quantity of dissolved water is increased, although Scarfe (1973) has shown that the effect of water in basaltic melts is of less importance. Kushiro *et al.* (1976) have shown that pressure has little effect on viscosity in olivine tholeiite and andesite melts up to 30 kb at constant temperature. However, experiments by Kushiro (1976) on $NaAlSi_2O_6$ compositions indicate that viscosity can be reduced by the change of coordination of Al as pressure is increased.

Volcanologists have also attempted to estimate the viscosity of basaltic lavas, assuming Newtonian behavior, by a variety of methods summarized by Macdonald

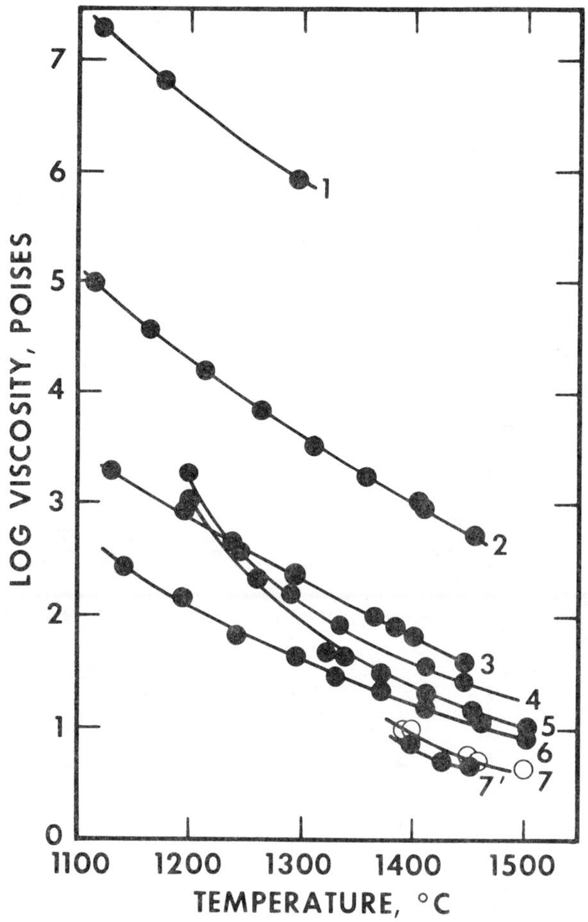

5.2.1 The viscosity of some common igneous rock melts as a function of temperature (Murase and McBirney, 1973): (1) rhyolite; (2) andesite (Mount Hood); (3) Columbia River basalts; (4) Skaergaard liquid at 35% crystallization; (5) Skaergaard liquid at 76% crystallization; (6) olivine basalt (Galapagos); (7) synthetic lunar basalt in air; (7′) synthetic lunar basalt in argon.

(1972). One of the most widely used techniques is to measure the velocity, dimensions and slope of lava in a channel and to apply the equations of open channel flow to estimate viscosity. Apparent lava viscosities estimated in this manner range from slightly less than 10^3 to over 10^5 poises for terrestrial basaltic flows. Experimental and theoretical data (Fig. 5.2.1) indicate that lunar basalt flows are somewhat more fluid at comparable temperatures: 10–100 poises at temperatures exceeding 1400° C (Murase and McBirney, 1973).

The problem with both the experimental studies and field techniques described above is that very few lavas are extruded above their liquidus. Furthermore, there is increasing evidence that many magmas beneath their liquidus do not behave as Newtonian fluids. Robson (1967) first suggested that Etna lava flows behaved as Bingham fluids as they ceased to flow. A Bingham fluid is one which requires the application of shear stress before flow commences. The value of the shear stress required for motion is called the yield strength — a quantity that varies with different fluids. Shaw *et al.* (1968) conducted rheological experiments on Makaopuhi lava lake in Hawaii, and demonstrated that the lava behaved as a Bingham fluid for strain rates less than 1 per sec (Fig. 5.2.2 a). They reported yield strengths of 70 and 120 newton/m^2 with plastic viscosities of 8000 and 6000 poises for a lava at 1135° C and phenocryst content of 35%. Shaw (1969), Murase and McBirney (1973), and McBirney and Noyes (1979) have reported non-Newtonian behavior in some of their experiments at subliquidus temperatures. Pinkerton and Sparks (1978) present measurements of the rheology of Etna lava flows (Fig. 5.2.2b), which have yield strengths ranging from 250–400 newton/m^2 and plastic viscosities of 10^5 poises at temperatures of 1070–1090° C.

One major reason that many lava flows exhibit non-Newtonian properties is the presence of dispersed crystals and gas bubbles. Even relatively minor proportions of dispersant can substantially change the rheological properties of a liquid (Rutgers, 1962). The effect depends on the size, shape, concentration and mechanical properties of the dispersant. There has not yet been a systematic attempt to relate many empirical industrial data on dispersions to the question of lava rheology. Murase and McBirney (1973) present some experimental data which shows the important effect of crystallization on magma rheology (Fig. 5.2.3). Another possible cause of non-Newtonian behavior of magma at subliquidus temperatures is a structuring of the melt

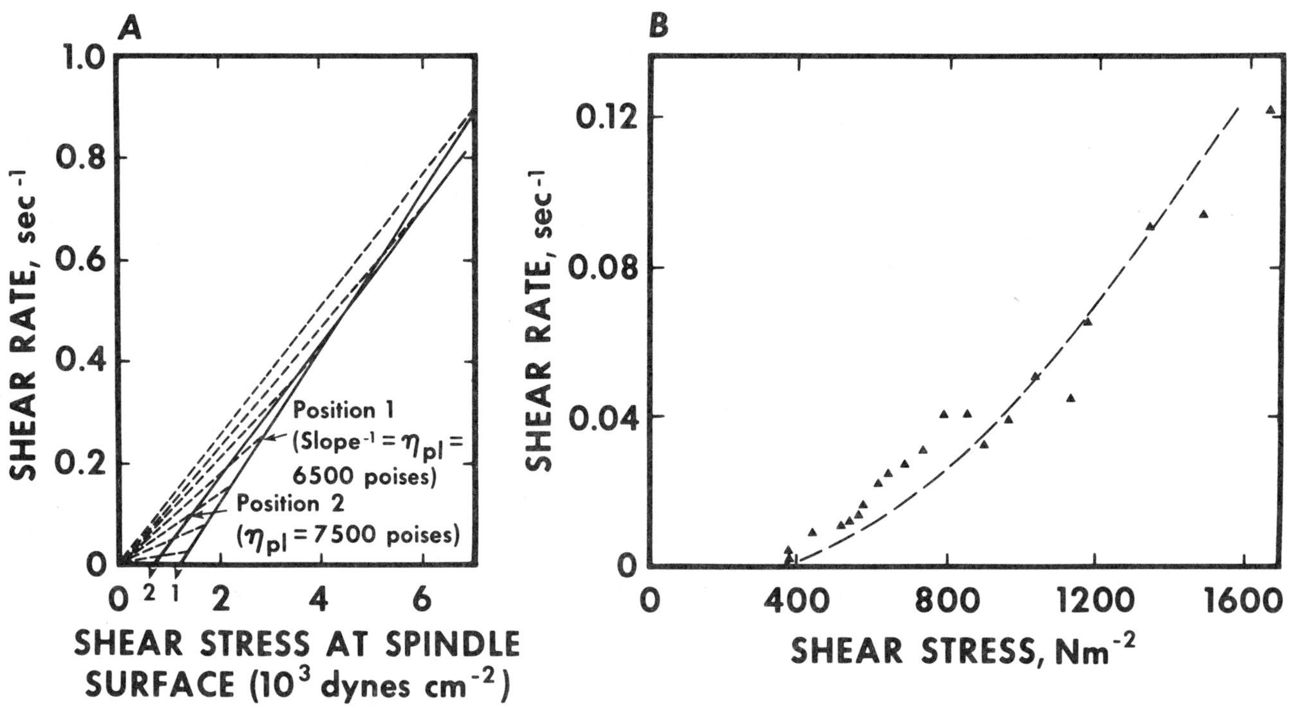

5.2.2 a. Flow curves for Hawaiian tholeiite at 1030–1035° C in Makaopuhi lava lake: shear stress vs. shear rate at steady state (solid lines). The slope of each dashed line gives the apparent viscosity of the lava at a given shear rate (Shaw *et al.*, 1968). **b.** Flow curve for Etna lava flow at 1086° C (Pinkerton and Sparks, 1978).

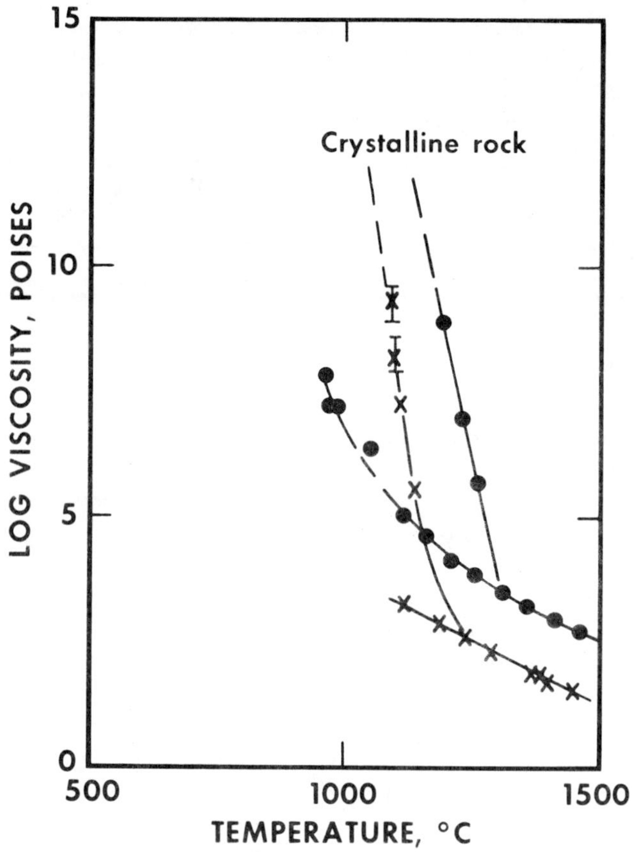

5.2.3 The relationships between viscosity and temperature are shown for glassy and crystalline rocks: (●) Columbia River basalt; (X) andesite (Mount Hood) (Murase and McBirney, 1973).

itself (McBirney and Noyes, 1979). Rheology at subliquidus temperatures is identified as a major area of future research if we are to understand and successfully interpret volcanic morphology.

Non-Newtonian behavior of magma has many implications. For example, many lavas containing acicular or tabular microlites or microphenocrysts will probably behave in a pseudoplastic or even thixotropic manner. Not only will such lava probably have a yield strength, but its apparent viscosity may decrease at higher rates of shear as dispersed crystals and silicate chains in the melt become aligned. In such a pseudoplastic fluid the relationship between shear stress and strain rate is nonlinear and the lava would appear much more fluid at high than low flow rates. This is illustrated in Fig. 5.2.2b, where the apparent viscosity decreases with increasing shear rate. The viscosity estimates derived from a slow moving flow, using the open-channel method, would be much higher than those estimated from a fast lava stream, though it is entirely possible that there is a little difference in the rheological relationships describing the flow behavior of the lava. Lava may also be strongly thixotropic, gaining a high yield strength if allowed to stand for a long time, and developing a low yield strength as it is sheared and the internal structure is broken down.

Several observations on flow shape also led Hulme (1974 a) to argue that lavas behave as non-Newtonian fluids. If lava were a true Newtonian fluid, there would be no limit on the flow length because the lava would continue to flow until it ponded in a depression or until surface tension forces became dominant. In contrast, lavas can come to rest on considerable slopes. As a consequence of yield strength, there is a critical lava thickness for a given slope beneath which flow will not take place. Lavas on Etna show a relationship between thickness and slope (Walker, 1967) which matches that predicted for a Bingham fluid (Fig. 5.2.4). Furthermore, lavas do not spread laterally as would be expected in Newtonian fluid, but are confined to well-defined channels bounded by levees (Fig. 5.2.5). Finally, the Bingham fluid concept predicted that the aspect ratio (width/thickness) of a flow would be independent of slope (Wadge, 1978). Hulme demonstrated theoretically that atmospheric cooling of the lateral margins of a flow was not the cause of lava becoming confined to a channel of approximately fixed width. Hulme argued that these and other features could be adequately explained by

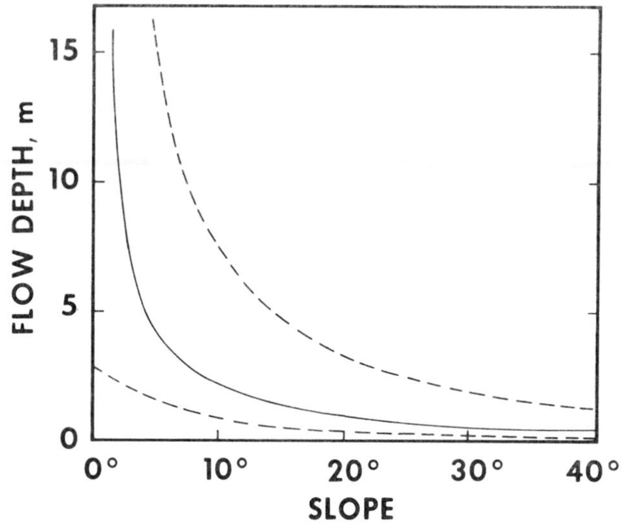

5.2.4 Comparison of observed and theoretical relationships between flow thickness and slope on Etna (envelope of measurements from Walker, 1967; Hulme, 1974a). Theoretical curve (solid) is for a yield strength of 1000 dynes/cm^2.

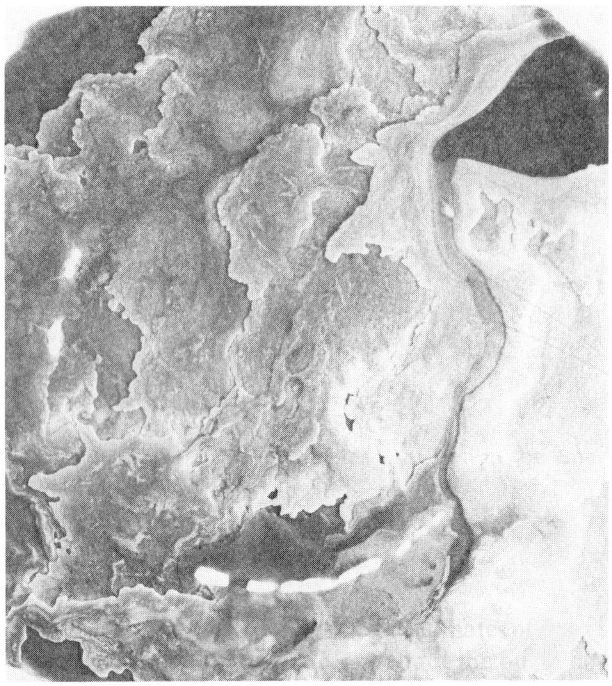

5.2.5 View of laboratory simulations of fluid lavas using Carbowax, showing a channel with levees and multiple flow units. (Photo by R. Greeley.)

assuming that lava, in fact, behaves as a non-Newtonian fluid with a well-defined yield strength.

Hulme has theoretically modelled the unconfined flow down an inclined plane of fluid which approximates Bingham behavior, and has conducted experiments using kaolin/water mixtures to test the theory. His models have reproduced some major features of lavas, notably levees and channels as well as surface crevasses. He has found that stationary bodies of fluid form as the fluid attempts to spread out laterally. The formation and dimensions of these stationary bodies (levees) depend on the yield strength of the liquid and they act to confine the flow to a fixed channel. Since Hulme's theoretical study, Sparks *et al.* (1976) have observed similar initial levees produced in this manner on three flows on Mt. Etna and have confirmed that, at least on Etna, lava levees form because the lava has a yield strength. Although levees of this type are almost always modified by subsequent processes in the lava channel (Fig. 5.2.6), for example overflows (Peterson and Swanson, 1974; Sparks *et al.*, 1976), these studies indicate that lava flow morphology is strongly linked to rheology. Levees and channel morphology are one of the most characteristic features of volcanic terrains on the Earth, Moon, and Mars (Fig. 5.2.7).

The widths of lava tubes, formed within fixed open channels (Greeley, 1970) are consequently also linked to magma rheology. Hulme (1974 a), Moore and Schaber (1975), Hulme and Fielder (1977) and Carr *et al.* (1977) have applied theoretical treatments of Bingham fluid flow to interpret lava-flow morphology on the Moon and Mars. Yield strengths from 1.4×10^3 to 2.0×10^5 dynes/cm^2 have been estimated for both the Moon and Mars, suggesting a considerable range in lava compositions or effusion temperatures. For example, surface flows have much higher yield strengths on Olympus Mons than on Ascraeus Mons (see Fig. 5.2.7). We are not, however, at a stage in our understanding to interpret these differences in terms of detailed lava compositions.

5.2.2 Gas loss

A significant control on magma rheology in volcanic eruptions is gas loss. Nearly all magmas originally contain dissolved volatiles either inherited from the mantle or assimilated during ascent to the surface. As confining pressure is reduced these gasses exsolve and, if present in sufficient quantity, result in explosive activity at the surface. Gas expansion is probably the major driving mechanism in generating hawaiian-style fire fountains and strombolian explosions (Fig. 5.2.8) which are so common in terrestrial basaltic eruptions.

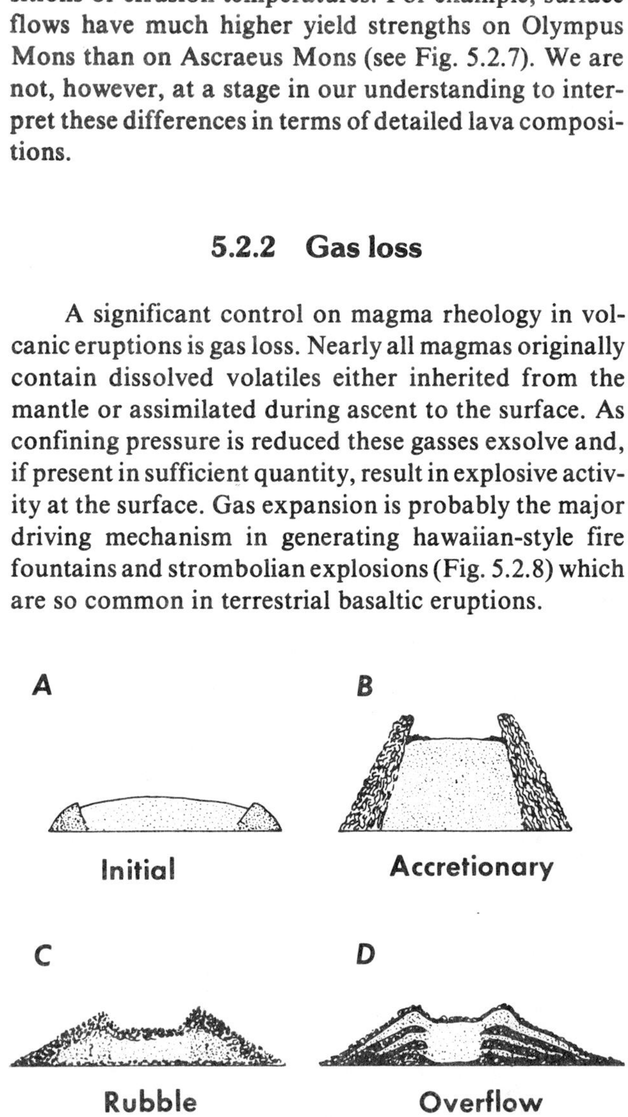

5.2.6 Cross-sections of four types of lava levee observed on Mount Etna (Sparks *et al.*, 1976): **a.** Initial levees formed due to yield strength of lava; **b.** Accretionary levees formed by continuous smearing and accretion of lava onto the tops and inside walls of levees; **c.** Rubble levees formed by talus of aa rubble at foot of flow; **d.** Overflow levees formed by repeated overflows of lava draping lava sheets over levee walls.

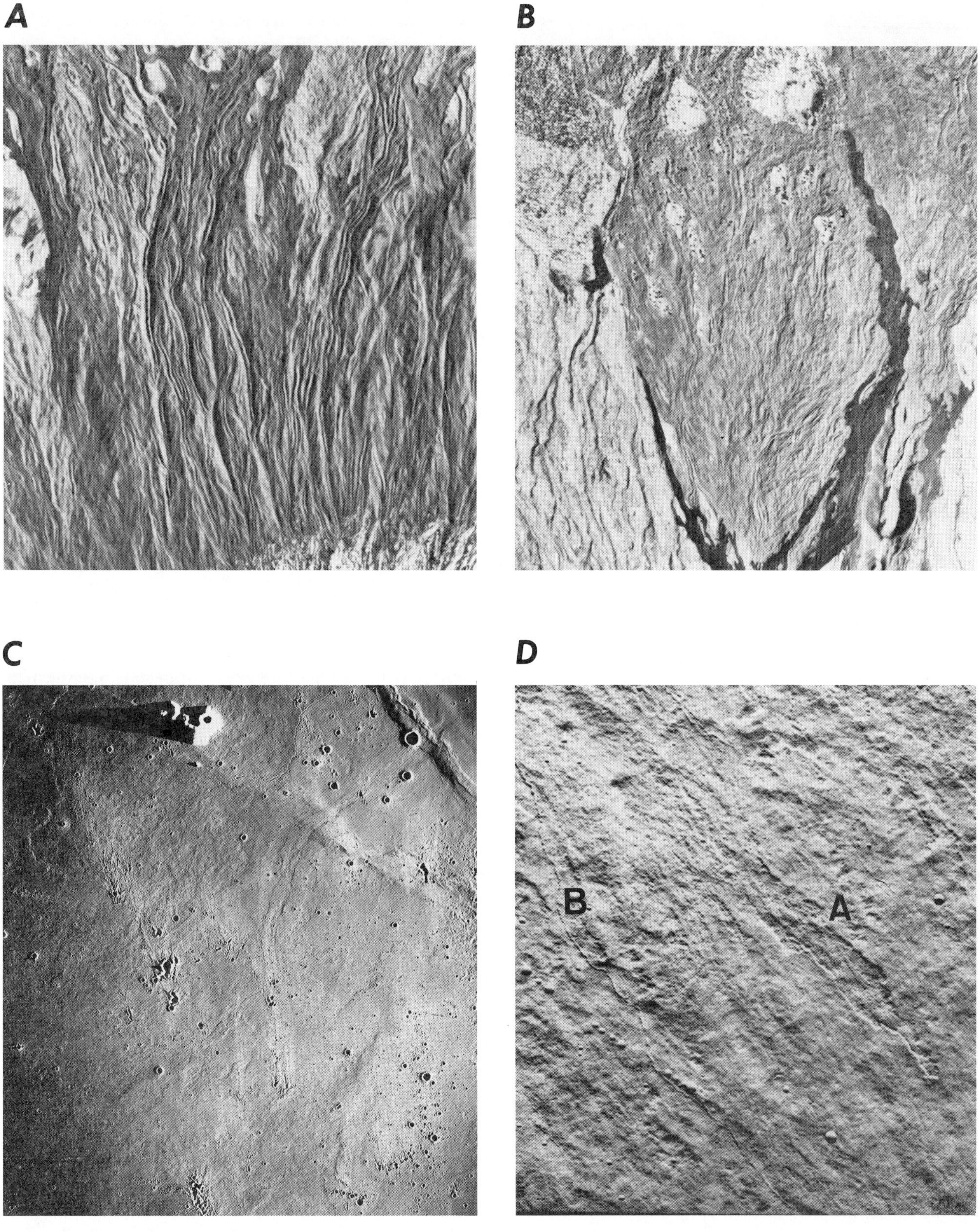

5.2.7 a. and **b.** Photographs of levee and channel morphology on Earth *et al.*, **c.** Photograph of levee and channel morphology on Moon (NASA). **d.** Photograph of levee and channel morphology on Mars (Viking Orbiter 401B12).

5.2.8 Photograph of strombolian eruption of Tolbachik cinder cones, Kamchatka. Tallest cone has basal diameter of ~1.6 km and height of ~0.3 km. (Photo by G. Steinberg.)

Loss of volatiles has several important effects. Dissolved volatiles, such as H_2O, bond in silicate melts in such a manner that they depolymerize the melt and decrease viscosity (Shaw, 1963; Burnham, 1975). Exsolution of water, therefore, increases viscosity. Gas exsolution also leads to adiabatic cooling of the total gas/magma system and an increase in magma viscosity. Exsolution of CO_2 has the opposite effect as it polymerizes the melt on solution (Mysen *et al.*, 1975), and therefore, a decrease in viscosity could be expected to accompany CO_2 loss. The adiabatic cooling amounts to approximately 20° C per 1% dissolved water. Dissolved volatiles, water in particular, generally lower the liquidus temperature of magma. Thus when gas is lost, the liquidus temperature of the magma is increased, resulting in effective undercooling of the melt. In addition, the heat of exsolution may be substantial (Boyd, 1961), further adding to undercooling.

Relatively small degrees of undercooling have a profound effect on the nucleation and growth rates of crystals (Kirkpatrick, 1976, 1977). A basalt magma with (originally) 1% water can be anticipated to be undercooled by at least 30° C below its dry liquidus temperature. In fact, the greater the initial water content and its subsequent loss, the larger the undercooling of the lava flow. Undercooling may be even greater because additional heat loss occurs during fire fountaining and further gas loss occurs during lava flowage (Swanson and Fabbi, 1973). On Hawaii, the degree of gas loss increases with greater fountain height, and fountain height increases with greater original gas contents (Swanson and Fabbi, 1973).

Undercooling leads to rapid formation of crystallites throughout the lava, whereas atmospheric cooling only affects the margins of a flow. If the data of Kirkpatrick (1976, 1977) are used, plagioclase crystals in basalt at 20–50° undercooling should grow at rates of the order of a few microns/minute. Thus crystals can grow to significant sizes (10^{-2} – 10^{-1} cm) on typical time scales of lava flow emplacement (hours, days or even weeks). Sparks and Pinkerton (1978) consider this to be a major cause of non-Newtonian properties developing in lava flows, which take on yield strength and higher viscosity as a result of an all-pervasive growth of quench crystals. There is thus a fundamental change in the rheology of magma accompanying degassing and undercooling of the melt.

The morphology and surface structures of lava flows and the explosivity of basaltic eruptions could be influenced by undercooling. In magmas with little gas, the undercooling will not change rheology substantially. Such lavas may often behave in a Newtonian manner if close to the liquidus. Some large flood tholeiites from Iceland, found tens of kilometers from sources associated with monogenetic shield volcanoes, are notable for the predominance of pahoehoe surfaces and may be examples of gas-poor basalts. On the other hand, a basalt magma with a high water content will be strongly undercooled. The increase in viscosity and yield strength resulting from rapid quench crystal growth is postulated to explain the strong explosivity and high apparent viscosity of lavas in eruptions of calc-alkaline basalts. The eruptions of Paricutin, Mexico, and Mt. Ulawan, Papua New Guinea (Johnson *et al.*, 1972), produced blocky aa lavas and intense strombolian activity, although in terms of silica content the magmas involved were little different from the Icelandic flood tholeiites.

Some aphyric tholeiitic and alkali basaltic lavas in Iceland produce predominantly aa flows. These eruptions, for example the 1961 eruption of Askja (Thorarinsson, 1968), also produce vigorous explosive activity and usually form substantial cinder cones. Although some of these lavas are characterized by lower SiO_2 content than flood tholeiites and are aphyric and therefore close to the liquidus, their flow morphology (well-developed levees and channels) indicates highly non-Newtonian properties. The vigorous explosive activity also indicates high initial gas contents. Gas loss accompanying the extrusion of these flows profoundly influences the style of volcanic activity and the morphology of basaltic landforms. The degree of degassing is also important in controlling the small-scale structures and textures of lava flows (Swanson, 1973; Swanson and Fabbi, 1973; Sparks and Pinkerton, 1978). The formation of aa surfaces has been attributed to the inability of surface tension forces to overcome a high yield strength (Hulme, 1974a). Thus, lava flows change from pahoehoe to aa more rapidly in flows with high yield strength. As suggested by Sparks and Pinkerton (1978), many aa flows may have suffered considerable gas loss resulting in a high yield strength.

5.2.3 Discharge rate

Discharge rate of magma is a major factor in determining lava flow dimensions. Walker (1973a) showed that lavas with orders of magnitude differences in viscosity had a considerable overlap in their flow lengths (Fig. 5.2.9a). Additionally, Walker showed that there was a close relationship between flow length and lava discharge rate (Fig. 5.2.9b). There is no doubt that viscosity has some control on length, but it is of second-order importance whereas effusion rate is of first-order importance. It is, however, clear that tube-fed pahoehoe flows can reach lengths much greater than aa flows of equivalent effusion rate (Swanson, 1973; Wadge, 1978). These observations have led workers to propose that some of the mare basalts on the Moon, which are several hundred kilometers in length, had extremely high effusion rates up to 2×20^5 m^3/sec (Hulme and Fielder, 1977). Carr *et al.* (1977) have also estimated high discharge rates for Olympus Mons lavas. Swanson *et al.* (1975) calculated volume rates of eruption for Columbia River basalt flows of up to 1 km^3/day/km of vent.

The observations of Walker can be explained on a theoretical basis (Hulme and Fielder, 1977; Wadge, 1978). Consider a hot viscous fluid flowing in a laminar fashion down a cold pipe. As the fluid moves down the pipe, heat is progressively lost to its surroundings, isotherms migrate into the center, and the fluid increases in viscosity. The amount of cooling in this situation can be conveniently expressed as a dimensionless number, often referred to as the Graetz number (Gz)

$$Gz = \frac{ud}{Kx}$$

where u is the flow velocity, d is the pipe diameter, x is the distance from the source down the pipe, and K is the thermal diffusivity of the fluid. Fluids with the same Graetz number have reached the same stage of cooling. This theory can be applied to lavas by making x equal to the final length of the lava, d the hydraulic diameter at the source vent, and u the effusion velocity. On Mt. Etna, where lava lengths range from less than 100 m to over 10 km, and effusion rates range from less than 0.001 to over 100 m^3/sec, all the lavas have Graetz numbers in the limited range 100–3000. The high effusion rate lavas, however, tend to have higher Graetz numbers, implying that their flow fronts have to reach a more advanced stage of cooling before movement is inhibited.

Shaw and Swanson (1974) have presented detailed fluid dynamic models of lava flowage applicable to large flood basalts and demonstrated that lavas with high discharge rates can flow hundreds of kilometers, especially if the flow is turbulent. There is some critical state of cooling at a lava flow front which must be

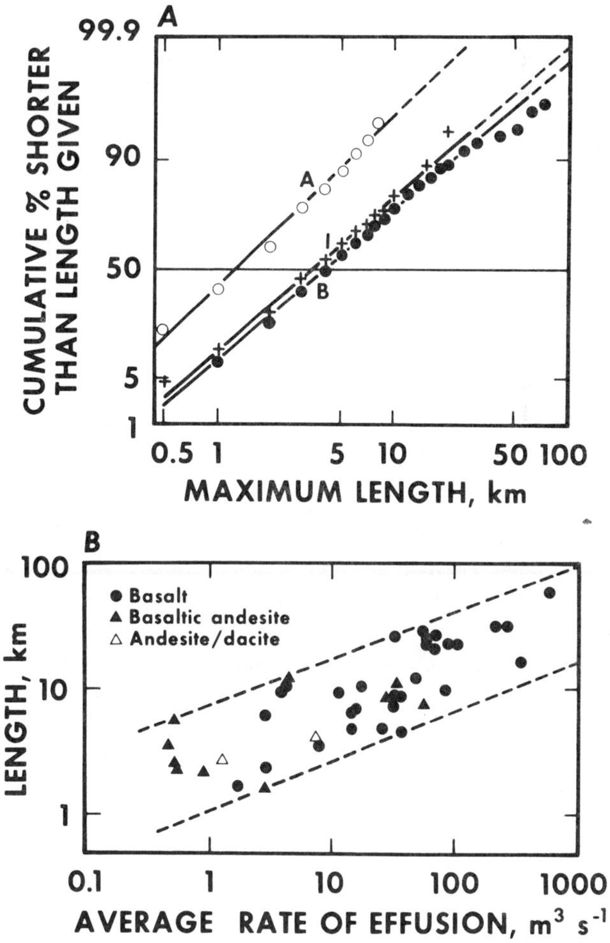

5.2.9 a. Logarithmic probability plot of the lengths of 479 group B (mostly basalt), 135 group I (phonolite and basaltic andesite) and 417 group A (trachyte, andesite/dacite and rhyolite) lava flows. **b.** Maximum lava flow length plotted against average discharge rate (from Walker, 1973).

reached to inhibit further flow. The higher the discharge rate the greater the distance that the lava will travel before the flow front reaches this critical state. The importance of the flow front acting as a barrier has been demonstrated in small Etna lava flows (Pinkerton and Sparks, 1976), where temperature measurements indicate that the interior of flows which had ceased movement had cooled less than 10°C. The flow fronts, however, had cooled sufficiently that their strength inhibited further flow.

Wadge (1978) has explored further the relationship between discharge rate and flow length in application to Mt. Etna lavas. In an alternative treatment to the Graetz tube flow, he has shown theoretically that flow distance is proportional to discharge rate for the simple case of a Bingham fluid in a confined channel. Wadge, however, points out that in order for a flow to achieve its maximum potential length, flow must be maintained for a certain length of time, which he estimates as ten days for Etna. On the basis of this model, Wadge has developed a method for estimating the variation of discharge rate with time from the shape of a lava flow field.

Although magma rheology does not have a direct influence in the simple model described, the discharge rate is certainly influenced by magma rheology, which controls the rate at which magma flows through fractures and ascends through the crust. Magma discharge rates are also influenced by pressure and density gradients, but that topic is beyond the scope of this discussion. Although there is a considerable overlap in the distribution of flow lengths of silicic and basic lavas, the longest lavas known are basaltic, partly because this type of magma (provided that its density is not too great) reaches a planet's surface with the greatest ease.

Discharge rate also determines the structure of a lava flow. When the discharge rate is high, extensive simple lava sheets are formed, whereas compound lava flows are formed with low discharge rates (Walker, 1970). Lava tubes are observed to form readily in the restricted range of 10–100 m^3/sec flow rates (Walker, 1970). Discharge rate is also one of the factors determining whether a pahoehoe or aa lava flow forms. Pinkerton and Sparks (1976) have observed Etna lava flows of similar effusion temperatures and rheologies that form aa flows when discharge rates exceed $2 \times 10^{-3} m^3$/sec, whereas pahoehoe flows form at lower discharge rates. Finally, it is clear that over extended time periods discharge rate has a profound effect on the morphology of polygenetic volcanoes. At one extreme, volcanoes with very high discharge rates will form no positive relief as these lavas would be extensive sheets infilling all available depressions, as in the Columbia River basalt province. Such a region is not generally regarded as "volcano" in the conventional sense. Moderately high discharge rates produce large shields with low slopes as in Hawaii. At the other extreme, volcanoes which consistently produce lavas at low effusion rates often construct high cones with very steep upper slopes, for example the basaltic volcanoes of Pico in the Azores, and Mayon in the Phillipines.

5.2.4 The lava tube mechanism

The formation of lava tubes is a major process which cannot be related solely to rheology. These features are common on the Earth and collapsed tubes may account for some sinuous rilles on the Moon and Mars (Greeley, 1971a; Cruikshank and Wood, 1972). A lava tube system provides a highly efficient mechanism of transporting lava over long distances with little heat loss. Tubes can form in well-developed, open channels by roofing-over processes (Plate 5.1) during periods of many hours or more (Greeley, 1970; Macdonald, 1972; Peterson and Swanson, 1974). Several variants of the formation of a roof by crusting over of the flowing lava surface have been documented on Hawaii. Another major process of tube formation, which can be observed on Hawaii, occurs when pahoehoe toes inflate like balloons, then break open and produce a new lava toe at the front (Wentworth and Macdonald, 1953; Peterson and Swanson, 1974). This process occurs repeatedly and a tube is formed linking the toes and feeding the front of the lava (Fig. 5.2.10a).

In a prolonged eruption a master channel is usually established (Figs. 5.2.10b, c). In the 1969–71 activity of Mauna Ulu on Kilauea, a master tube fed lava almost continuously for over 10 months (Peterson and Swanson, 1974). Other established tubes carried lava from Alae lava lake 12 km to the coast at rates up to $10^6 m^3$ per day. The tubes proved to be remarkably well-insulated with temperature drops of only 10–20°C being recorded over the 12-km journey. Hulme and Fielder (1977) have argued on the basis of theoretical calculations that thermal erosion by melting of a tube floor should occur in long-lived terrestrial tubes where flow is laminar. Peterson and Swanson (1974) observed erosion rates in Hawaiian lava tubes of 1 meter/month, which compares favorably with Hulme's theoretical modelling. Hulme (1973) showed that thermal erosion would have been even more effective on the Moon where many lavas may have had greatly increased heat transfer rates resulting from turbulent flow. Some deeply incised lunar sinuous rilles could have formed by lava erosion of a preexisting surface.

A

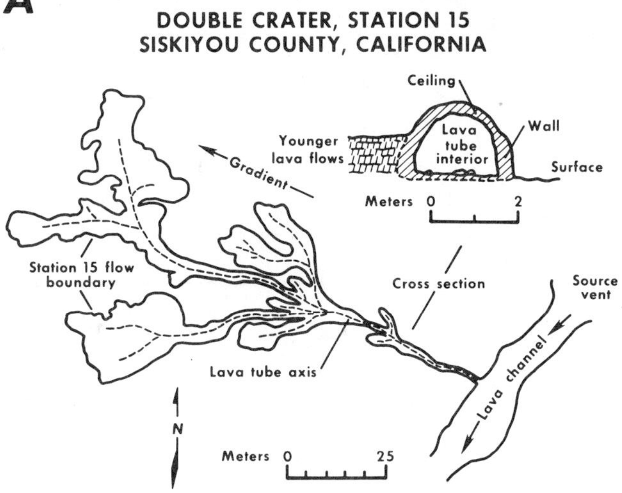

5.2.10 a. Diagram of small lava tube system in northern California showing relationship of lava tubes (dashed line) to the lava flow, and one cross section (from Greeley, 1971c).

5.2.5 Topography

Another important control on morphology is preexisting topography. Lava flows are channeled by valleys and are ponded in depressions. Many of the large-volume flood lavas of the Columbia River basalts are believed to be ponded and are characteristically thick, colonnaded and flat-topped (Shaw and Swanson, 1974), resulting in the featureless morphology observed from spacecraft. On shield volcanoes craters often trap flows to form lava lakes (Fig. 5.2.11), which act as complex supply reservoirs during prolonged eruptions (Wright *et al.*, 1968; Swanson *et al.*, 1972).

Ground slope affects lava flow rates and lava thickness (Walker, 1967). However, when unconfined, some lavas can flow tens of kilometers on slopes as small as ¼°, whereas other flows may move only a few meters on steep slopes. These facts suggest that slope is not of great importance in determining flow length. On very steep slopes, aa lava often breaks up at the flow front, increasing the proportion of clinker or rubble. In extreme cases a coherent flow cannot advance but continually breaks up, producing a scree slope composed of clinker.

5.2.6 Effect of water

Contact of lava with water has a profound influence on the style of basaltic eruptions and on the morphology of lava structures. On land, magma comes into contact with water in several situations: in shallow lakes or seas, in rivers, in a glacier or ice cap and also where ground waters are at a high level and the sediment

B

5.2.10 b. Low oblique aerial view of a lava channel developed in flows erupted from Mauna Ulu, Hawaii, showing the topographic "arch" along the axis of the channel that was constructed by overflow of lava from the channel (from Greeley, 1974).

C

5.2.10 c. Oblique aerial photograph of active lava channel in flows from Mauna Ulu, Hawaii, 1970, showing developing "cut-off" branch to the main channel (from Greeley, 1970).

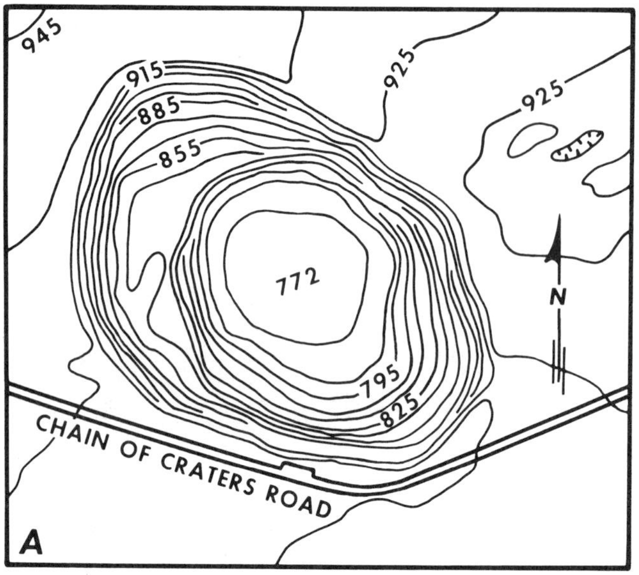

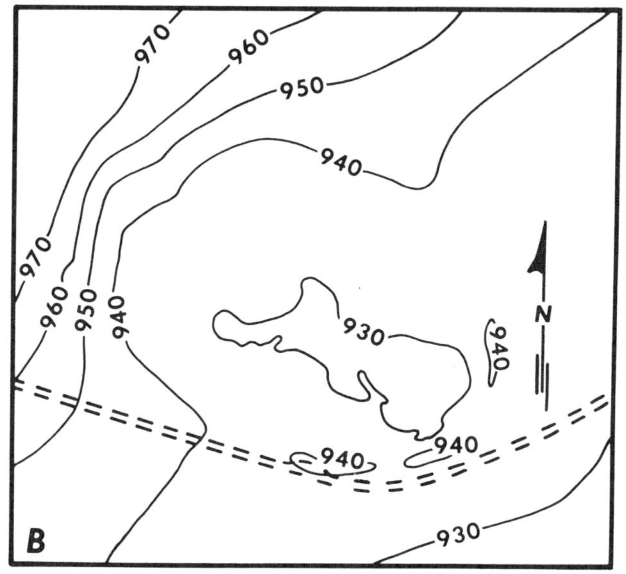

5.2.11 Topographic map of Alae crater, Hawaii: **a.** Before 1963; **b.** November 1970. The crater was filled in eruptions between 1963 and 1971 (Swanson *et al.*, 1972).

is soft with high porosity. In explosive activity both the intensity of explosions and the degree of magma fragmentation increase dramatically if water is available in copious quantities (Fig. 5.2.12). In phreatomagmatic eruptions such as Surtsey in 1963–64 (Thorarinsson, 1967b) a large proportion of the magma is erupted as fragmental material, and a tuff-ring rather than a cinder cone forms. Access of water to a vent in fact inhibits the formation of lava flows. During fire fountaining a considerable proportion of the ejecta is often recycled back to a lava lake and both the lava lake and the fountain itself feed lava flows. However, when the ejecta contacts water, further fragmentation and chilling result in much reduced recycling of molten lava and a higher proportion of ejecta entering the pyroclastic deposit. In many cases, such as Surtsey, no lava could be formed while abundant water had access to the vent.

Several morphological effects can be produced by lava flowing into shallow water. Littoral cones, or rootless vents, can form by explosions generated by entrapment and heating of water pockets beneath the lava flow, or by breaking of waves over the surface of flowing lava complexes (Thorarinsson, 1961; Moore *et al.*, 1973; Peterson, 1976). Degassed basalts flowing into the sea or a lake usually form a lava delta. As the lava front advances, it breaks up on contact with the sea to

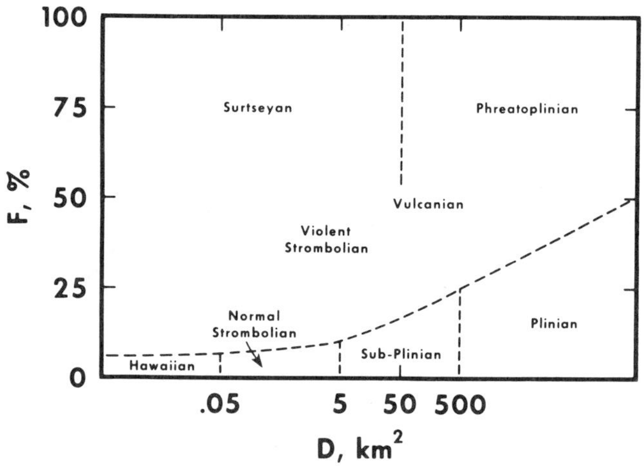

5.2.12 Plot of fragmentation index (F) against dispersal index (D) for various types of eruptions. F is the percentage of pyroclastics finer than 1 mm at the point where the 0.01 T_{max} isopach crosses the axis of dispersal, where T_{max} is the maximum thickness of the deposit at the source. D is the area in km^2 enclosed by the 0.01 T_{max} isopach. Note that a change in eruption style from strombolian to surtseyan (i.e., due to the presence of abundant water) results in a dramatic decrease in fragment size. Based on Walker (1973b) and Self and Sparks (1978).

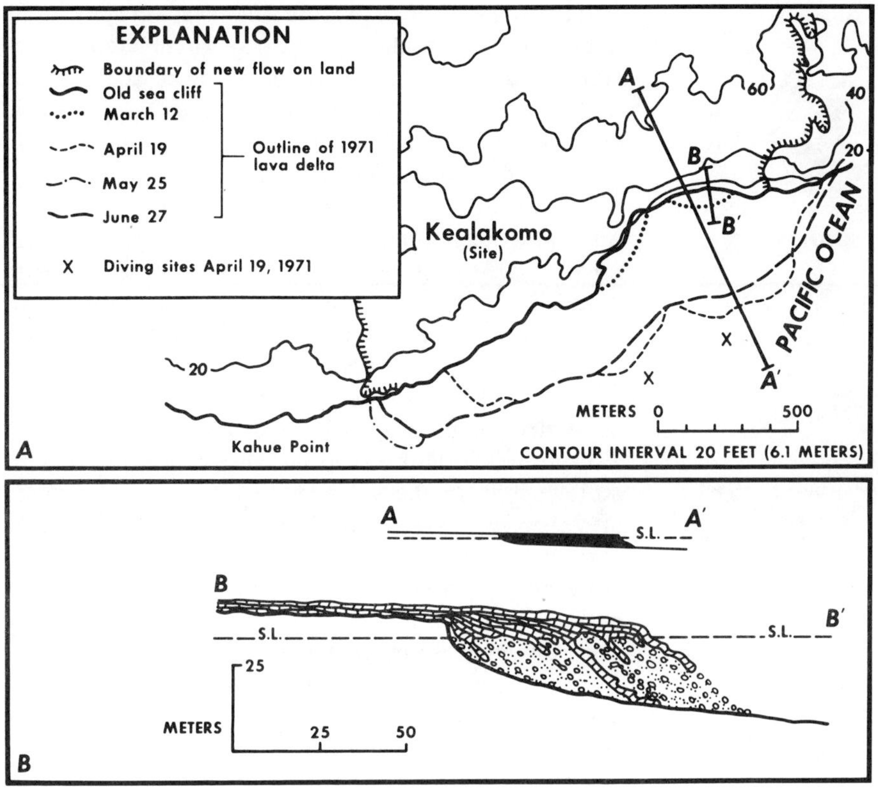

5.2.13 **a.** Map showing growth of the Kealakomo lava delta, Hawaii, in April, 1971. **b.** Cross-section of lava delta showing internal structure (Moore *et al.*, 1973). Cross-section A–A′ at same scale as in **a.**

form hyaloclastite, which accumulates as foreset beds over which the lava slowly advances (Fig. 5.2.13). Descriptions of the geology and formation of lava deltas are given by Moore *et al.* (1973) and Furnes and Fridleifsson (1974).

Walker (1975a, b) proposed that ground water provides a significant barrier to magma. He maintains that in Iceland, the upper 500 m of water-saturated crust, composed of high porosity tuff and lava, often prevents magma reaching the surface. Many Icelandic eruptions are preceded by several days or weeks of phreatic activity, presumably the result of magma attempting to penetrate this barrier. A good example of this is the 1961 eruption of Askja (Sigvaldason, 1964). In some circumstances, however, ground water interaction with shallow level magma could trigger a volcanic eruption (Schmincke, 1977).

5.2.7 Tectonic control

Tectonic control of volcanic superstructures is of fundamental importance in understanding eruption mechanisms and morphology of basaltic landforms. The volcanic edifice itself creates a localized structural environment superimposed on regional stresses (Fig. 5.2.14). Stresses within this edifice influence the orientation of fractures (Nakamura, 1977). For example, on Mt. Etna and Piton de la Fournaise, Reunion, (Wadge, 1976; Ludden, 1978) many flank fissures are radial to a central conduit system. On Kilauea and Mauna Loa, the influence of neighboring volcanoes and gravitational slumping of the flanks have combined to exert a strong influence on the location of eruptive fissures (Fig. 5.2.15) and well-defined rift zones have

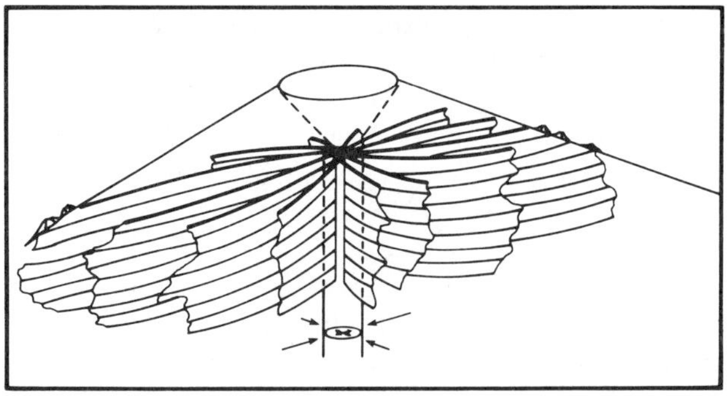

5.2.14 The radial dike and fissure pattern of a polygenetic volcano under a differential horizontal stress (Nakamura, 1977).

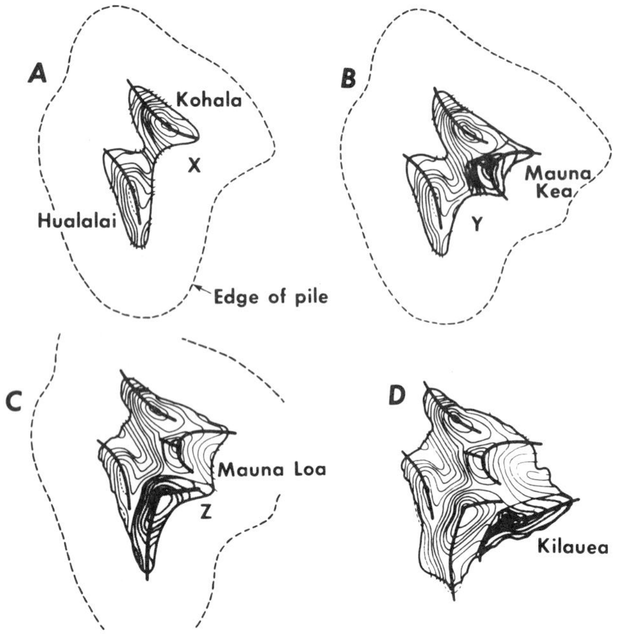

5.2.15 Schematic growth of Hawaii, showing the effect of neighboring volcanoes on the orientation and location of the main rift zones. Heavy lines represent rifts (Fiske and Jackson, 1972).

developed (Fiske and Jackson, 1972). On Mt. Etna, basement tectonic directions have influenced the orientation of rift zones and nonradial fissure systems (Wadge, 1976). In Icelandic central volcanoes, fissure rift zones are strongly aligned in the direction of the regional tectonic fabric (Saemundsson, 1974).

It is essential to understand the subsurface plumbing system of a volcano to interpret volcano morphology; however, rudimentary data of this type are only available for a few basaltic volcanoes. Good geophysical monitoring, an extended record of historical observations, and detailed geological control are required. Kilauea, Krafla (Iceland) and Mt. Etna have been studied most in terms of this crucial problem. Both Kilauea and Krafla (during the 1975–80 period of activity) show broadly analogous behavior (Macdonald, 1972; Björnsson *et al.*, 1977, 1979). In each case a shallow level magma reservoir is slowly replenished by the influx of new magma, causing inflation of the volcano. Björnsson *et al.* (1979) document a steady influx of 5 m^3/sec over the last few years in Krafla. The maximum inflation is centered a few kilometers beneath the summit region and continues over a period of weeks or months. Inflation is usually followed by lateral injection of magma into the rift zones. The magma propogates a fracture (Pollard, 1976; Aki *et al.*, 1977) which may intersect the surface in the rift zone to cause a flank lava eruption and probably forms a dike (Fig. 5.2.16). These lateral injections release pressure, cause deflation of the volcano and can cause caldera collapse (Wright *et al.*, 1968; Fiske and Kinoshita, 1969; Decker and Kinoshita, 1971; Walsh and Decker, 1971).

Wadge (1976, 1977) suggests that Etna behaves in an analogous manner except that a cylindrical central conduit is the postulated magma reservoir. He presents evidence that magma is continuously fed into the Etna reservoir and that a flank eruption is caused when magma pressure is sufficient to propagate a fracture out to the flanks. However, Murray and Guest (1980) show that there is no gross cumulative inflation of Mt. Etna prior to flank eruptions, implying little magma storage high in the volcano. Sigurdsson and Sparks (1978) present evidence that many Icelandic central volcanoes, including Krafla, Askja, Grimsvötn and Katla, store magma in large shallow-level reservoirs. During periods of tectonic quiescence pressure builds up in these chambers as magma is fed into them, and then is relieved by small to medium volume eruptions in the central volcano. However, in major crustal rifting episodes, substantial volumes of magma are injected into the rift zones to form fissure eruptions, often with extensive lava flows tens of kilometers from the central volcano. Sigurdsson and Sparks further propose that the 1783 Laki lava eruption, the largest historic lava flow (12 km^3), occurred in this way by draining of a magma chamber beneath Grimsvötn (Fig. 5.2.17). They suggest that many Icelandic calderas form subsequent to such eruptions.

The mechanisms of magma storage, pressure buildup, sudden release by fracture propagation, and the subsequent eruption as documented from Kilauea, Etna, and Icelandic central volcanoes, show that the internal plumbing of a basaltic volcano is a basic control on discharge rate, eruption style, and eruption frequency. In the case of Etna, Wadge (1978) demonstrates that flow morphology can be used to deduce both the variation of discharge rate during one eruption and also the range of discharge rates in a series of eruptions. The estimated discharge rates can be linked to the internal pressure variations in a volcano, the volcano's strength, and the volcano's role as an escape valve for magma (Wadge, 1977). The plumbing also partly controls magma composition, rheology, and gas content, as these are affected by long periods of storage.

5.2.8 The replenished reservoir model

The concept of a repeatedly replenished magma reservoir has recently emerged as an idea of fundamen-

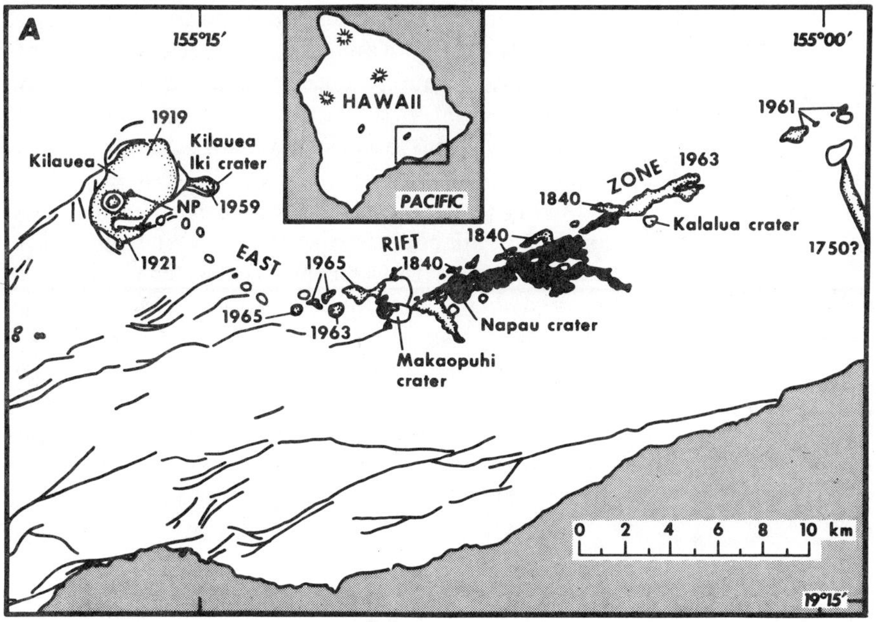

5.2.16 a. Map of Kilauea summit caldera and east rift zone. The black areas show the March 1965 lava which was associated with subsidence at the caldera (Wright *et al.*, 1968). **b.** Map of Krafla central volcano, Iceland, showing caldera and linear fissure swarms. In the activity from 1975 to 1978, magma was injected laterally into the fissures to the north and south, causing subsidence of the caldera (Björnsson *et al.*, 1977).

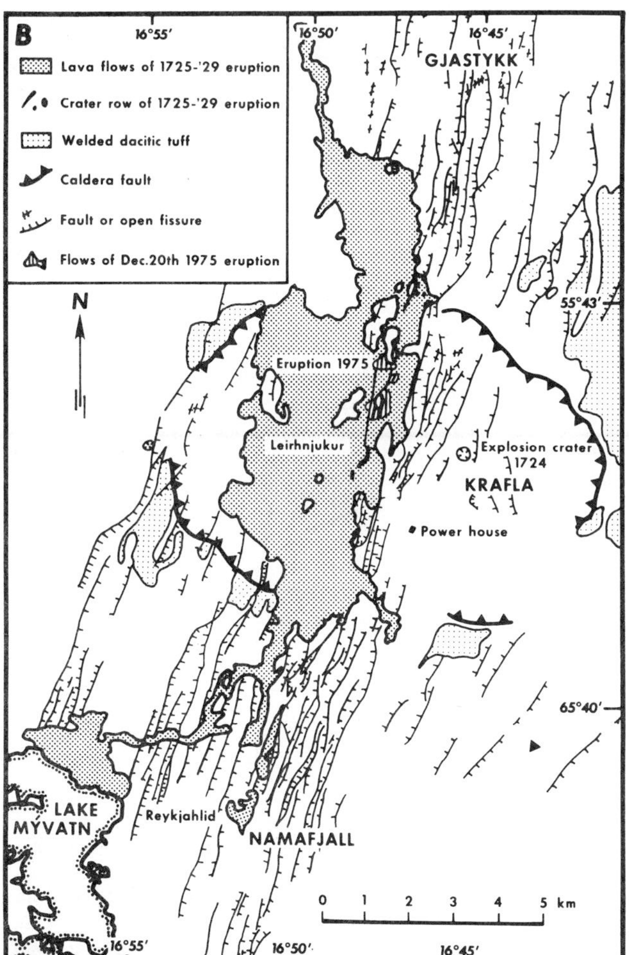

tal importance for understanding the surface manifestations of volcanism. Evidence for such a model (Fig. 5.2.18) has emerged from petrological, geochemical, geophysical and theoretical studies. It is envisaged that many basaltic volcanoes, including those on the mid-ocean ridges, are fed by a high-level magma chamber which is periodically replenished by new, more primitive magma from depth. The chamber need not be the stylized conventional shape of a sphere, but could be a plexus of interconnecting sills and dikes. These high-level reservoirs are periodically replenished by a pulse of new melt. In many cases the ascent of the new pulse is associated with tectonic activity, as in the case of Krafla, where a major rifting event in the Icelandic crust triggered inflation of the volcano reservoir.

Probably a great many volcanic eruptions are the direct result of influx of new magma. Thus the basalts observed to erupt are sampling magma chambers during periods of instability. Recent discussions of the evidence for this concept and its significance can be found in O'Hara (1977), Smith (1979) and Sparks *et al.* (1980). These and other studies suggest that the density contrast between the reservoir magma and the incoming magma can have profound effects on both the petrological characteristics of the erupted basalts and on morphological features evident at the surface. Two possible scenarios are shown in Fig. 5.2.18. In one case (Fig. 5.2.18a) the incoming magma is less dense than the reservoir magma and rises convectively to the top. At the same time magma is pushed out at the surface. Sparks *et al.* (1980) have shown that, among common

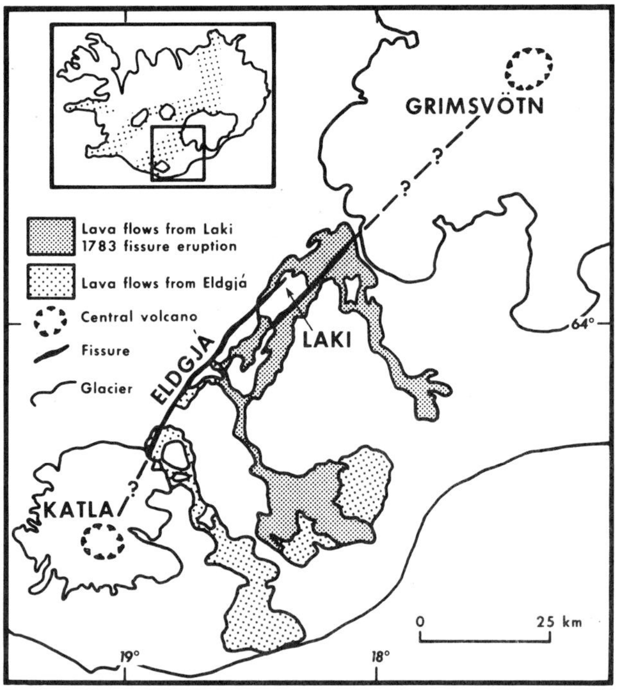

5.2.17 Map of the southern part of eastern volcanic zones of Iceland illustrating two examples of lateral magma flow. The 1783 Laki basalt is postulated to have formed by lateral flow from a high-level reservoir beneath Grimsvötn. Similarly, the Eldgjá fissure alkali basalt is believed to have formed by withdrawal of magma from beneath Katla. The large calderas of the two volcanoes were formed by this magma withdrawal (Sigurdsson and Sparks, 1978).

types of basalts, the buoyant primitive basalt will often mix turbulently with the reservoir magma. Such behavior may characterize many terrestrial basalt volcanoes, where crystal fractionation of plagioclase and pyroxene have led to iron enrichment and increasing magma density. New relatively primitive basalt will be lighter than the reservoir magma. When turbulent mixing is combined with probable lateral magma flow there is great scope for heterogeneity in the products of basaltic volcanoes.

Another possible mode of behavior is shown in Fig. 5.2.18b, where the incoming magma is denser than the reservoir magma. In this case the fractionated reservoir magma is pushed out the top (or alternatively intruded sideways) and the new magma stratifies at the base of the chamber. In many terrestrial magmas such a scenario is probable, where the course of crystal fractionation leads to a density decrease. Important examples include the members of the calc-alkaline magma series and picritic basalts, where olivine is the only fractionating phase. Clearly, high-level magma chambers act as very efficient magma filters in such cases.

The concept of replenished magma reservoir is identified as a subject area where a great deal of research is warranted. The morphology of many basaltic volcanic landforms is fundamentally a reflection of the internal dynamic processes within such reservoirs, which are controlled by factors such as the size, shape and depth of the reservoir, the properties of the reservoir and available primitive magmas, and by the tectonic setting of the volcano. Using such models as depicted in Fig. 5.2.18, the information on volcano morphology can ultimately be used to interpret the internal behavior of volcano supply systems. The recent advances in understanding terrestrial volcanoes should herald much improved understanding of extraterrestrial volcanoes.

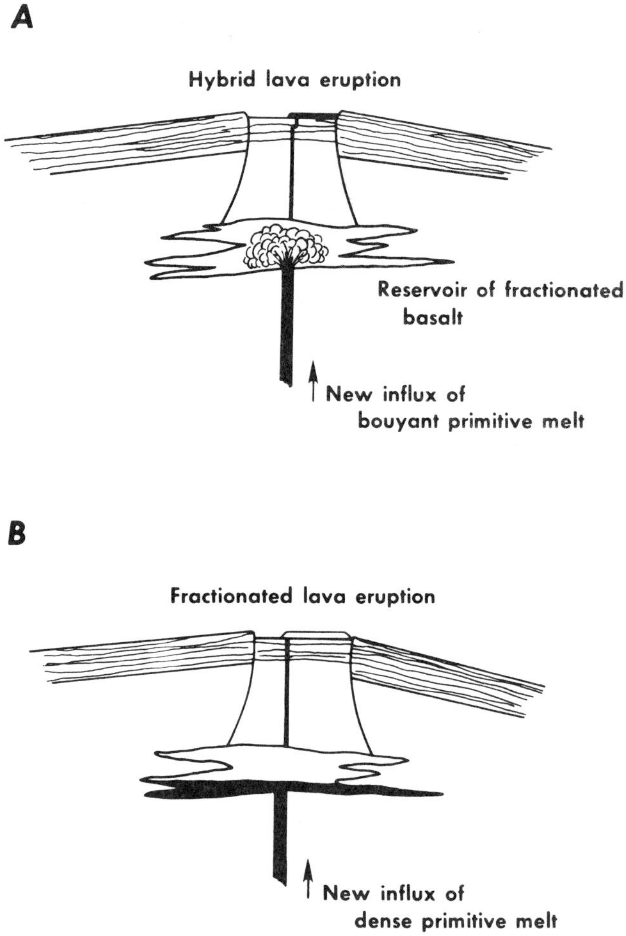

5.2.18 Diagram illustrating a high-level magma reservoir replenished by primitive magma. In **a** the intruding magma is lower in density and rises as a plume or thermal producing a mixed lava at the surface. In **b** the intruding magma is denser than the reservoir magma and stratifies at the base pushing out fractionated magma at the surface.

5.3 TERRESTRIAL BASALTIC VOLCANOES

Terrestrial volcanoes are commonly classified either by their morphologies or by the eruption style which led to their formation. Of the other terra-like solar system bodies, only Io has been observed to exhibit active volcanism. In this section, therefore, we have classified basaltic volcanoes on the basis of their morphologies but describe their known or inferred eruption styles. A further distinction is drawn between volcanoes erupted subaerially and those associated with an aqueous environment. It must be pointed out that these morphological classes are gradational with one another, and many provinces on Earth may display more than one class. For example, in the Pacific Northwest, large shield volcanoes (Medicine Lake Highlands, California) occur near basaltic plains (Modoc Plateau, California, Oregon) and flood basalt plateaus (Columbia Plateau). Moreover, plains-style flows occur within the flood basalts of the Columbia Plateau. This leads to the conclusion that the identification of these classes is useful for interpreting the styles of eruption for the latest (exposed) phase of volcanic activity but that caution must be exercised in applying the interpretation to the early history of any given region.

5.3.1 Subaerial basaltic volcanism

Plateau flood basalt complexes

The term **flood basalt** follows the usage of Greeley (1977), which limits the term to one particular style of volcanism and terrain, and which separates basaltic lava plains (see below). Flood basalt plateaus are characterized by thick flows with low surface relief. Flood-type eruptions result from extremely high rates of effusion. The resulting flows often become ponded in topographic depressions and may require tens of years to cool. Consequently, the flows typically display colonnade-entablature jointing patterns (Fig. 5.3.1). Flow structures such as lava tubes and lava channels are seldom present, either because they did not form during eruptions, or because they were destroyed during the emplacement of the ponded lavas.

More than a dozen flood basalt provinces have been described (Table 5.3.1; see also section 1.2). Typically, the tholeiitic lavas that dominate flood basalt regions cover areas of $0.05\text{–}1.5 \times 10^6$ km^2 and their volumes range from about $0.2\text{–}1.0 \times 10^6$ km^3. Fragmentary flood basalt provinces are known from the Precambrian, but most are younger than 200 m.y., with the interval from 200–1100 m.y. free of major deposits (Ahmad, 1972). Nearly all flood basalt provinces are at the edges of continents and appear to be associated with the break-up of continental plates.

Most flood basalt flows are between 5–10 m thick (Macdonald, 1972), although thicknesses up to 45 m are known from the Karroo of South Africa (Tyrrell, 1937). Soil horizons between flows indicate that often tens to thousands of years passed between eruptions, although in general the period of eruptive activity was strongly concentrated (50% of dated samples of the Serra Geral lavas cluster between 115–125 m.y.; Herz, 1966).

Many of the lava flows in the Columbia Plateau of the Pacific Northwest are examples of flood basalt flows. One of the flows that has been studied in detail, the Roza Member of the Yakima Basalt (Swanson *et al.*, 1975), originally covered an area of at least 40,000 km^2 and had a volume greater than 1500 km^3, or about 0.5% of the total Columbia River basalt volume. For the most part, it is made up of one or two thick flow units and is considered to be a simple cooling unit, although

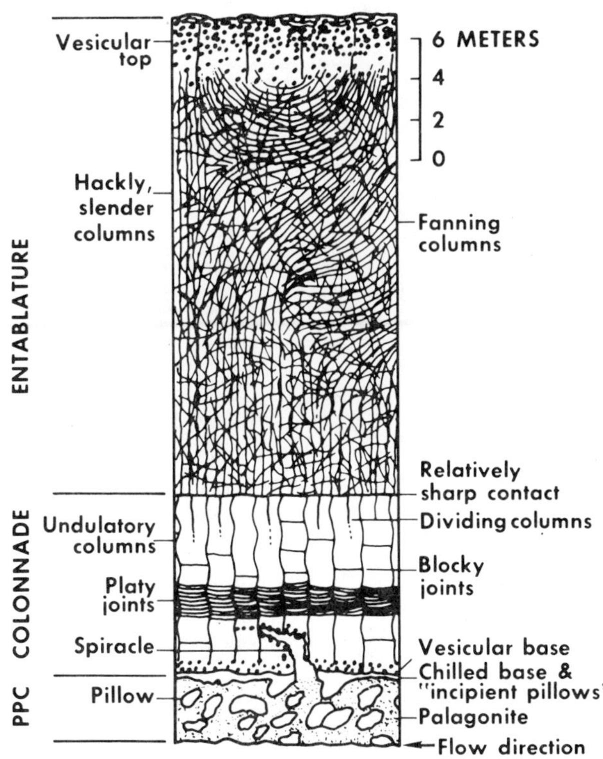

5.3.1 An idealized cross-section of a typical flow in the Yakima Basalt Subgroup (Swanson and Wright, 1978).

in some places it may be a compound unit. The distinction between a compound and a simple flow is that the former is divisible into separate flow units believed to have been emplaced at low effusion rates while simple flows are not divisible and are thought to form at high extrusion rates (Walker, 1970). The Roza Member was erupted during a very short period of time—a few hundred years for the entire unit. The vent system appears to be a fissure zone less than 15 km wide and at least 120 km long.

From the volume, areal extent, possible eruption duration, and vent dimensions, Swanson *et al.* (1975) have estimated the rate of eruption for the Roza Member to be 1 km^3/day/km of the fissure vent, using the rheological model of Shaw and Swanson (1974). Even with all the assumptions involved in such an estimate, this figure is on the conservative side, yet it is three to four orders of magnitude higher than average rates of eruption in Hawaii.

The vents for flood basalts are thought to be fissures that are often hidden by their own eruptive products; erosion, however, often exposes dike swarms. Enormous concentrations of dikes are exposed in the British and Greenland portions of the Thulean basalt province, as is true for the Karroo and Keweenawan basalts. Dikes range up to 60 m in width (Deccan field; Macdonald, 1972), compared to a range of 0.6–6 m for dikes cutting shield volcanoes (Macdonald, 1972). Massive sills and intrusive sheets are also commonly associated with flood basalts (e.g., Karroo, Keweenawan, and Serra Geral; Table 5.3.2), indicating emplacement of an even greater volume of magma than suggested by surface lava deposits. Commonly the centers of flood basalt provinces subside through time; Swanson *et al.*

Table 5.3.1 Major terrestrial flood and plains basalt provinces.

Name	Location	Age (m.y.)	Area ($10^6 km^2$)	Volume ($10^5 km^3$)	Reference
Siberia	U.S.S.R.	250–280	1.5	2.5	*(10)*
Serra Geral/Paraná	Brazil	115–135	1.2	3.5	*(7)*
Caribbean Crust	Caribbean	70–93	1.0	10.0 (?)	*(4)*
Patagonia	Argentina	190–210 (?)	0.75	-	*(14)*
Deccan	India	65	0.52	7.0	*(5)*
Antrim Plateau	Australia	580 (?)	0.15	-	*(2)*
Columbia River	U.S.A.	6–16.5	0.2	2.0	*(15)*
Modoc Plateau	U.S.A.	-	0.08	-	*(8)*
Snake River	U.S.A.	0.002–15	0.03	<0.7	*(6)*
Karroo	South Africa	110–200	1.4	4.2	*(3)*
Traps	Ethiopia	20–40	-	1.5	*(9)*
Keweenawan	U.S.A.	900–1135	0.2	1.0	*(16)*
Bushveld	South Africa	2000	0.06	1.0	*(12)*
West Greenland	Greenland	50–65	0.5	1.5 (?)	*(11)*
East Greenland	Thulean Province (traditional)	45–60	0.17	5.25	*(1)*
Faeroes	Thulean Province (traditional)	45–60	0.001	0.04	*(11)*
Britain (Present)	Thulean Province (traditional)	45–60	0.005	0.05	*(13)*
Iceland	Thulean Province (traditional)	7–20 (?)	0.04	1.8 (?)	*(11)*

References:
(1) Brooks *et al.* (1976); *(2)* Bultitude (1976); *(3)* Cox (1972); *(4)* Donelly *et al.* (1971); *(5)* Ghose (1972); *(6)* Greeley and King (1977); *(7)* Herz (1966); *(8)* Macdonald (1972); *(9)* Mohr (1972); *(10)* Nalivkin (1973); *(11)* Noe Nygaard (1974); *(12)* Rhodes (1975); *(13)* Richey *et al.* (1961); *(14)* Rittman (1962); *(15)* Swanson and Wright (1978); *(16)* Windley (1977).

Table 5.3.2 Chronology of some dike swarms and basalt plateau and greenstone belts.[†]

	Basalt Plateaus and Greenstones	Dike Swarms
Quaternary	South Vietnam Hsing An (Manchuria)	
Miocene	Canadian Cordillera, Snake River, Columbia, Iceland, East Africa, Antarctic (James Ross)	Iceland
Eocene	Ethiopia	Ethiopia
Cretaceous	Verkoyan'e kolym massif (U.S.S.R.), Deccan, Eastern Greenland, Hebridean, Antrim (Scotland), Spitzbergen, Red Sea	Eastern Greenland, Hebridean
	Raj Mahal (India), Perth and Tasmania, Antarctic, Paraná (Brazil), Kaoko (Southwest Africa)	Troodos
Jurassic	Karroo (South Africa)	Atlantic seaboard U.S.A., West Africa, northern South America
Triassic	Watchung (New Jersey), Tungus (Siberia), Yunnan (China), Tauden (West Africa)	
Permian		
Carboniferous	Clyde (Scotland)	West Scotland
Cambrian	Antrim (Australia)	600 Volyn (Ukraine) Maisk (U.S.S.R.) 700–800 Gwalier Bijawar N. Chhattisgarh (India), Kamennoborsk and Salminsk (U.S.S.R.), Krestetsk and Polesje (U.S.S.R.), Kaltasinsk (U.S.S.R.) 950–1250 Bukoban (East Africa), Baffin Island, Arizona, Death Valley 1100 Cuddapah (India) 1100–1350 Jotnian (Baltic)
950–1300*	Keweenawan (North America), Bukoban (East Africa)	1200–1400 Gardar 1350 Kimberly (Australia) 1450–60 Pilansberg (South Africa), northwestern Sudbury 1500 Roraima (Guyana) 1600 Granite Mts. (Wyoming), Soriname (South Africa) 1650–1800 Jatulia-Suisaria, Vespia (Baltic) Teisk (U.S.S.R.)
1800–2100	Birrimian (Africa)	1740–1950 Ivory Coast, Waterberg (South Africa), Mashonoland (Rhodesia), Umkondo (Rhodesia), Hart (Australia), India, Makunsk Billyakhsk Soolooiisk (Siberia) 2093 Northeastern Slave 2147 Abitibi

[†]After Read and Watson (1975), Windley (1977), and Makarenko (1977).
*Units on numbers in first column are 10^6 years.

Table 5.3.2 *(Concluded)*

Basalt Plateaus and Greenstones		Dike Swarms
		2155 Nipissing (Canada) 2150-2174 Northwestern Slave (Canada) 2200 Scourie, Bighorn (Wyoming)
2345	Dharwar (India)	2370 Chopan (India) 2400 Yilgarn (Australia)
2600–2700	Bulaweyan (Rhodesia), Superior and Slave (North Africa), Yilgarn (Kalgoorlie, Australia)	2690 Matachewan (Superior) 2692 ENE set Slave (North America) 2826 Bighorn (Wyoming)
3360±100	Sebakwian (Rhodesia)	
3500±200	Onverwatch (Transvaal)	3500 Ameralik (Greenland)
3760	Isua? (Greenland)	

(1975) have suggested that such subsidence of the Columbia River Plateau is due to magma withdrawal, isostatic adjustment of the thickening basalt pile, and regional tectonism.

With the exception of a few thin flow units, spatter, and other vent-related features near the vent system, surface features (other than flow fronts) large enough to be seen on spacecraft pictures are absent from flood basalts. This implies that flood-type flows may be difficult to identify on other planets on the basis of photogeology alone.

Basaltic lava plains

Basaltic plains regions (Greeley, 1976, 1977) are composed of a complex sequence of overlapping, coalescing low shields, with intervening tube-fed flows, and fissure flows (Fig. 5.3.2). The low shields with flank slopes of 1° or less typically are made up of compound lava flows erupted from a central vent. Summit regions of the low shields steepen to several degrees, evidently reflecting lower rates of effusion and/or differences in the rheological properties of the lavas. The surfaces of basaltic lava plains show moderate relief (~30–40 m) with surface features including pit craters on the vents (typically aligned along rift zones), fissures and sets of fractures, and collapsed lava tubes. These features allow basaltic lava plains to be distinguished from flood basalt plateaus.

The plains style of basaltic volcanism is contained within the category of "multiple-vent" volcanism, which Tyrrell (1937, p. 92) defined as basalt accumulations "arising from the confluence of lava flows from a large number of small and closely-spaced volcanoes." Tyrrell included in this category accumulations from concentrations of small shields (e.g., the Carboniferous basalts centered around Glasgow), and from pyroclastic and lava vents (e.g., the Auvergne region of France). In essence, the basaltic plains are the hawaiian-style equivalent of the platform cinder cone field. This interpretation is suggested by the length of the individual flows and the presence of complex lava tube-channel systems. Eruption rates in basaltic lava plains regions also appear to be less than those of flood basalts, and perhaps comparable to typical Hawaiian rates.

Basaltic plains combine aspects of both flood lavas and shield volcanoes. Like flood eruptions, the vent systems in the lava plains regions are controlled by rift zones; many of the vents are fissures or elongated central vents. Using the Snake River Plain as a type example, the flows that comprise basaltic lava plains are seldom more than about 10 m thick and are typically of the compound type—that is, they are multiple-flow units that behaved as individual cooling units.

Shield volcanoes

Shield volcanoes are broad, gently sloping cones, commonly convex-upward, with summit craters or calderas. They are circular to elongate in outline, and are composed predominantly of basaltic lavas (Fig. 5.3.3). Shields composed of andesitic lavas (Hayli Gub), alkaline/carbonatitic lavas (some African rift shields), and ignimbrites (Vulsini) are less common (Wood, 1977). On Earth, shields range in size from low shields (Greeley, 1977) less than 10 km across and having 1°

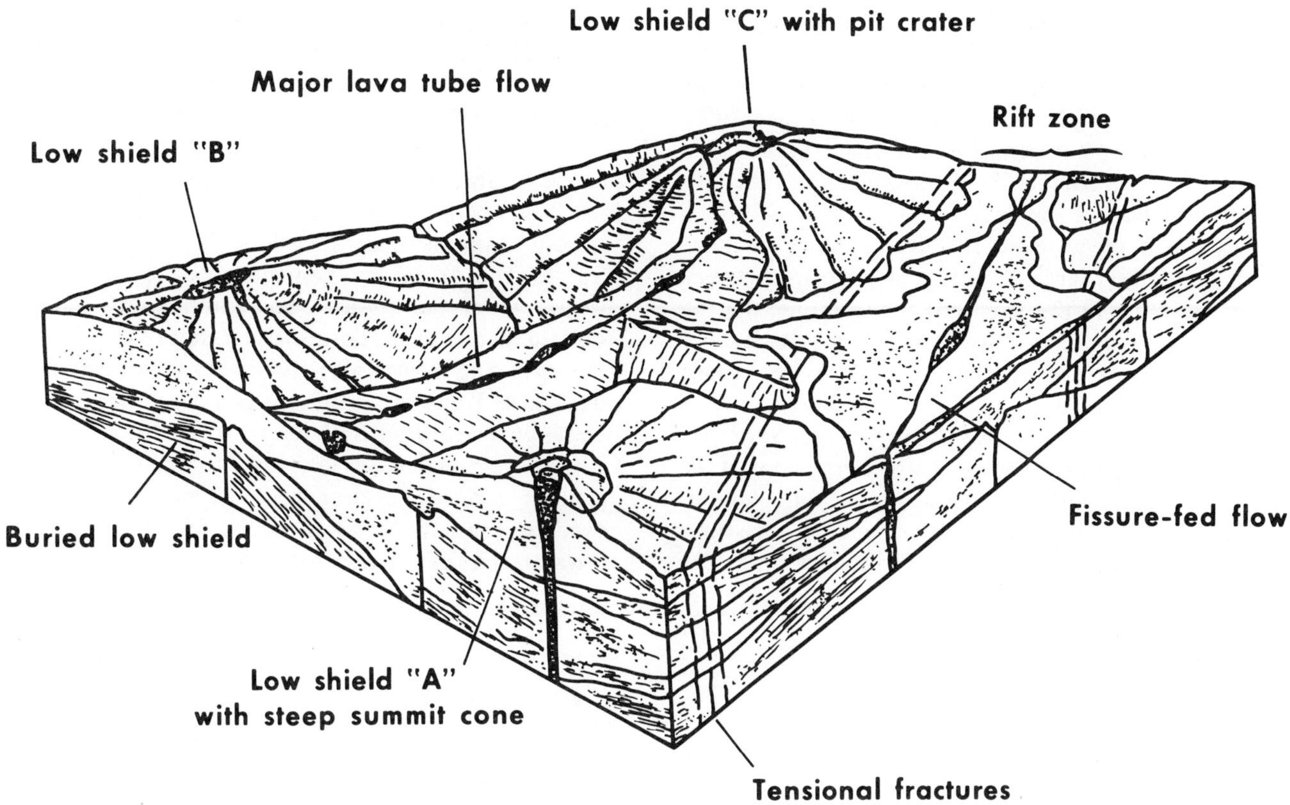

5.3.2 Diagram showing the features of "plains" volcanism (Greeley, 1977).

slopes, to the steeper (up to 35°) and larger Galapagos volcanoes. The larger shields are often surmounted by cinder cones, collapse pits, and smaller shields; rifting and fissuring are common features (Figs. 5.3.3–4). The fissures and rift zones may be concentric to the summit (Fernandina), radial (Haleakala), or both radial and concentric.

Using the Hawaiian volcanoes as typical examples, the flows that comprise large shield volcanoes are relatively thin, averaging about 4 m. They are erupted from both fissures and central vents and are commonly fed through lava tube systems and channels. For example, as much as 80% of the flows exposed on the present surface of the Mauna Loa shield were emplaced at least

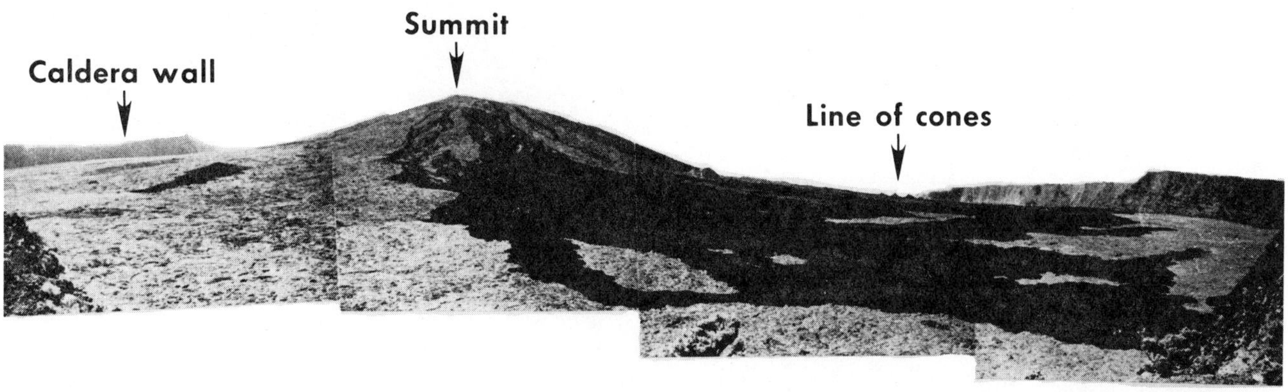

5.3.3 Photomosaic of Piton de la Fournaise, Reunion, showing old fissure eruption near summit parasitic cones and caldera wall. Width of caldera approximately 5 km. (Photo by P. Mouginis-Mark.)

partly by lava tubes or lava channels (Greeley *et al.*, 1976).

Average eruption rates for the lavas of the Hawaiian shields are typically of the order of 4–300 m^3/sec with a mean of about 60 m^3/sec. For the small Mauna Ulu shield, formed during 1969–73, the velocity of the lava flow fronts was of the order of 0.5–1 m/sec, though values of 5 m/sec were recorded on steep slopes. Some shields have very large flows that were formed during the early stages of the eruption; for example, the 150 km long flow produced by Trölladyngja in Iceland implies eruption rates in excess of 1000 m^3/sec (based on Walker, 1973a).

Shield volcanoes are usually the product of hawaiian-style volcanism typified by the historic eruptions of the Hawaiian shield volcanoes (Macdonald and Abbott, 1970). The eruptions commonly commence at fissures and are usually accompanied by fire fountaining. The deposits from these fountains build up banks on either side of the fissure and are known as spatter ramparts. As the eruption proceeds, the fissure commonly closes to a point source vent and a larger structure is constructed over this vent.

Fire fountains can attain heights of 500 m and temperatures of 1165°C (Swanson, 1973). At low eruption rates the gas-rich fountains produce glass droplets, Pele's hair, and reticulite; all are characterized by few to no phenocrysts or microlites, being essentially 100% sideromelane (Heiken, 1972). The low density and small size of these particles result in a strongly wind-dependent dispersal. Spatter from the higher fountains collects into lava exhibiting pahoehoe morphology near the vent but changing to aa downslope. These fountain-fed lavas tend to be denser than shelly pahoehoe erupted directly from the fissure (and confined to several hundred meters of the fissure) as a result of the gas loss accompanying fountaining. A further lava form associated with Hawaiian eruptions is a dense hummocky pahoehoe fed through an intricate system of lava tubes from the vent. All these flows are extremely fluid, with eruption viscosities of the order of 10^3 poises, and often cover large areas (50 km^2). Many of these lavas on Hawaii reach the sea and form lava deltas (e.g., Moore *et al.*, 1973; Peterson, 1976).

The shape of the volcano resulting from Hawaiian eruptions appears to be strongly dependent upon the eruption rate. With high discharge rates, extensive simple lava sheets of flood basalt type form with slopes as low as 1° (Fig. 5.3.5a). With moderate or low discharge rates, compound lava shields with slopes up to 10° are

5.3.4 High oblique aerial view of Kilauea caldera, Hawaii, showing typical scalloped outline that results from slumping of the rim; Halemaumau pit crater is within the caldera and is about 1 km across; Mauna Kea volcano is seen on the distant horizon (U. S. Navy photograph 0010, November, 1954).

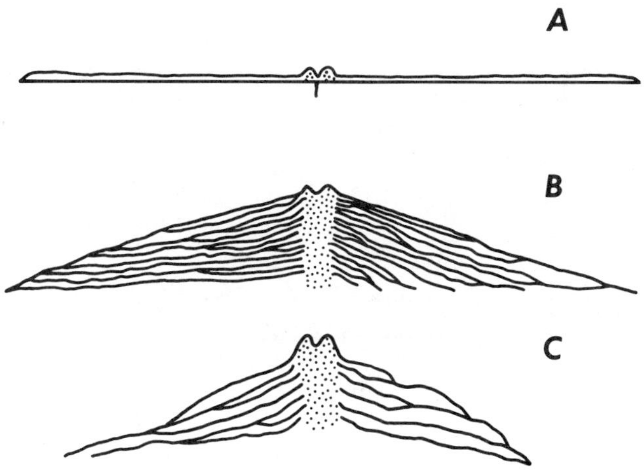

5.3.5 Variation in volcano morphology resulting from differences in discharge rate. **a.** High discharge hawaiian. **b.** Low to moderate discharge hawaiian. **c.** Low discharge strombolian.

formed (Fig. 5.3.5b). These shields frequently, but not inevitably, have summit lava lakes (Swanson *et al.*, 1972).

The scenario for shield formation appears to be as follows: (1) opening of a fissure and the eruption of lava; (2) concentration of the eruption around a point source; (3) repeated overflows of lava from the summit crater and from fissures cutting the crater floor and volcano flanks; (4) steepening of the shield toward the summit. This would mark the end of a monogenetic eruption, but for a polygenetic shield, additional stages *might* include: (5) opening of further vents on the earlier edifice with additional lavas and shields being superimposed; (6) change in the rheology of the magma and the formation of parasitic cinder cones; (7) caldera collapse; (8) almost complete masking of the original shield by a stratovolcano; and (9) erosion and burial. This change in volcanic style could reflect changes in volatile content, changes in composition (e.g., Hawaiian lavas become more alkalic with time), or decrease in eruption rate.

The majority of low shields are found in extensional environments (i.e., Iceland, Snake River Plain) usually where two plates are separating or have separated. Small to medium-sized shields are abundant in basalt plateaus (Deccan) and basaltic plains (Snake River). Small shields appear to be the product of single eruption episodes; that which produced Mauna Ulu lasted approximately five years.

It is also noteworthy that the localities where large shields are most abundant, such as the Hawaiian and Galapagos Islands, are also hypothesized to be the sites of mantle thermal plumes. Large shields are typically found on oceanic islands (Fig. 5.3.6) though more silicic shields also occur in mid-continent, such as the Tibesti complex in north-central Africa.

Because shield volcanoes are usually recognized by their external form, such structures are difficult to identify when eroded or buried. Many Tertiary volcanic centers, such as those of western Scotland, were probably the sites of large shield volcanoes, and early plateau basalts, such as the Jurassic Paraná deposits of South America, may also have been produced in part by shield eruptions.

Basaltic cones

Cinder cones are one of the most common volcanic landforms on Earth and are usually the product of single (monogenetic) eruptions (Wood, 1979d). Scoria cones typically have heights of the order of 20–350 m, with an average of about 150 m. Porter (1972) and Wood (1978) have gathered statistical information on cinder cone and composite cone volcanoes. From these data some linear relationships have been obtained. For example, Porter obtains the relationship $W_{cr} = 0.40 W_{co}$, where W_{cr} is the crater width and W_{co} is the cone width of cinder cones.

McGetchin *et al.* (1974) have modelled the growth of cinder cones based on short observations of Northeast Crater, Etna. The cinder cone is typically the product of strombolian-style eruptions, characterized by discrete explosions at intervals of less than a second to several hours. The magma tends to be more viscous (10^4–10^7 poises; Fedotov, 1978) than hawaiian magmas and has a greater gas content, with a gas:liquid magma ratio of the order of 2.5:1 to 16:1 (Chouet *et al.*, 1974). Typical gas velocities are of the order of 50–200 m/sec (Chouet *et al.*, 1974; Blackburn *et al.*, 1976; Fedotov, 1978), while average particle velocities were calculated at 2.5–72 m/sec on Stromboli, and exit velocities for 0.5–1 m diameter particles were estimated at 80–100 m/sec (Self *et al.*, 1974) on the more explosive Heimaey. At Heimaey, 1.0–50 mm diameter particles were carried to heights of 1 km above the rim (Self *et al.*, 1974), while at Tolbachik the eruption column grew to a height of 8 km (Fedotov, 1978).

The tephra from strombolian eruptions is largely scoria but also contains spatter and a small admixture of ash. The average diameter of tephra ejected at Stromboli was estimated at 2.5 cm (Chouet *et al.*, 1974), while less than 5.2% of a fall from Heimaey was less than 1 mm in diameter (Self *et al.*, 1974). Unlike the ash of hawaiian eruptions, strombolian ash contains phenocrysts and consists mainly of broken glass droplets (Heiken, 1972). Spindle-shaped bombs are commonly

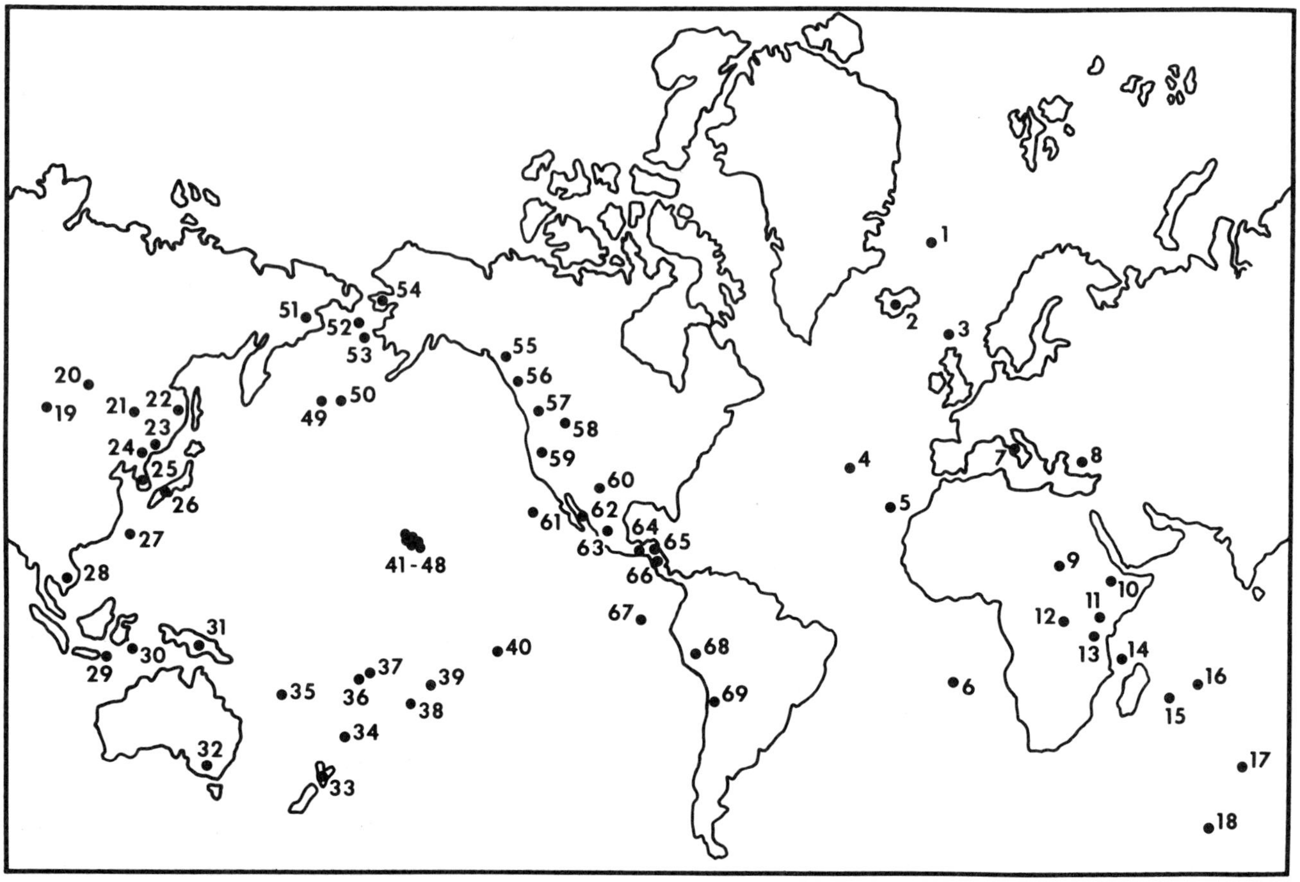

5.3.6 Distribution of some terrestrial shield volcanoes; *indicates nonbasaltic. **1.** Jan Mayen Is.: Beerenberg. **2.** Iceland: Almannagja, Arnasetur, Asbyrgi, Bjarnarfell, Bjornsfell, Blafjoll, Blagnippa, Borgarholar, Burfell, Dyngjuháls, Eftstadalsfjall, Eldborg, Geitahlith, Haleyjarbunga, Hestvatn, Kerlingardyngja, Ketildyngja, Kjalhraun, Lagafell, Lambahraun, Langhóll, Leggjabrjotúr, Lulli-hratur, Lyngdalsheith, Mithdalsfjall, Ok, Rauthamelskulur, Sandfell, Selvogsheithi, Skalapnes, Skersli, Skjaldbreithur, Skoflunger, Skrithan, Skuggadyngja, Solkath, Trolladulur, Trölladyngja, Vatnasheith. **3.** Faeroe Is. **4.** Azores. **5.** Canary Is. **6.** St. Helena. **7.** Italy: Alban Hills*, Sabatini*, Vico*, Vulsini*. **8.** Turkey: Tendürük Dágt. **9.** Sudan: Jebel Marra. **10.** Ethiopia: Alayta, Erta Ale, Hayli 'Gub, Mat'ala. **11.** Kenya: Kafkandal*, Kanatim*, Kaptagni*, Kilombe*, Mt. Kulal*, Nasakan*, Oliyamur*, Ribko*, Suswa*. **12.** Congo: Nyamuragira. **13.** Tanzania: Kiejo. **14.** Comores: Anjouan, Karthala, La Grille. **15.** Reunion. **16.** Mauritius. **17.** St. Paul Is., Amsterdam Is. **18.** Kerguelen: Cook, De Courbet, Jeanne d'Arc, L'Oranchet. **19.** U.S.S.R.: Derbi-Tayga. **20.** Chepe. **21.** China: ChiaTs'ung, Erh-ko-shan, Wu-ta-lien-chi. **22.** U.S.S.R.: Uysky. **23.** Merkushovsky. **24.** China: Chang-pai-shan, Chien-shan, Chi-hsing-shan, Ta-ping-shan. **25.** Korea: Halla. **26.** Japan: Abu, Aira*, Amagi, Aogoshima, Asama, Aso*,Ayame-Daira, Azuma, Duruma, Hakue-Zima, Hakusan, Ibusiki*, Izari, Kannabe, Kikai*, Kiriga-Mine, Kirisima, Kokuzo, Midagahara, Moriyosi, Muine, Numazawa, Oki-Dogo, Omuro-Yama, Oshima, Ozima-Zima, Sapporo, Siga, Soranuma, Takenotuzi, Takura, Teine, Ueno, Wasiba-Kumontaira, Zyaisi. **27.** Taiwan: Peng-hu, Tsao-ling-shan. **28.** Vietnam: Bu Jerman. **29.** Bali: Batur. **30.** Emperor of China. **31.** New Guinea: Mt. Giluwe. **32.** Australia: Mt. Porndon, Tweed. **33.** New Zealand: Akaroa, Karioi, Lyttleton, Pirogia, Pukekawa, Pukekohe Hill, Rangitoto, Whatitiri. **34.** Kermedec: Raoul Is. **35.** New Hebrides: Ambrym, Aoba. **36.** Niuafo'ou. **37.** Samoan Group: Matavanu. **38.** Raratonga. **39.** Society Is. **40.** Marquesas: Eiao, Fatu Hiva, Hiva Oa, Nuka Hiva, Tahuata, Ua Huka, Ua Pou. **41.** Hawaii: Hualalai, Kilauea, Kohala, Mauna Kea, Mauna Loa, Mauna Ulu. **42.** Oahu: Koolau, Waianea. **43.** Maui: East Maui, West Maui. **44.** Molakai: East Molakai, Kalaupapa, West Molakai. **45.** Kahoolawe. **46.** Lanai. **47.** Kauai. **48.** Niihau. **49.** Aleutians: Cereberus. **50.** Moffett, Agadak. **51.** U.S.S.R.: Kyrgannivaan, Marakveen, Shchitovaya, Terpukhoitskii. **52.** St. Lawrence Is.: Kookooligit Mts. **53.** Nunivak Is. **54.** Alaska: Imuruk Lake. **55.** Canada: Heart Peaks, Level Mt., Mt. Edziza. **56.** Itcha range, Ilgachuz range, Rainbow range. **57.** Oregon: Brown Mt., Mt. Newberry, Three Sisters, Timber crater. **58.** Idaho: Snake River Plains. **59.** California: Badger Mt., Black Mt., Crater Lake Mt., Harkness Peaks, Medicine Lake, Prospect Peak, Raker, Red Mountain, Simcoe Complex, Table Mt., Whaleback. **60.** New Mexico: Aden Crater. **61.** Guadalupe Is. **62.** Tortuga Is. **63.** Mexico: San Martin, Tancitaro. **64.** Guatemala: Cerro las Piedras, Volcan Jumaytepeque, Volcan las Viboras. **65.** Honduras: Cerro de Hule, Cerro Gavilantepeque, Cerro Santa Elena, Laguna el Pedregal, Montana de Izopo. **66.** Nicaragua: Coseguina, Masaya. **67.** Galapagos Is.: Abingdon, Alcedo, Bindloe, Cape Berkeley, Cerro Azul, Charles, Culpepper and Wenmann, Darwin, Indefatigable, James, Narborough, Sierra Negra, Tower, Wolfe. **68.** Peru: Suquihua. **69.** Chile: El Negrilla.

produced in strombolian eruptions and generally a lava flow forms part of the erupted basaltic material. This flow is most commonly of aa type and frequently rafts large sections of cinder cone wall.

Like hawaiian eruptions, strombolian eruptions frequently commence as fissure eruptions and close to a point source(s) (e.g., Fedotov, 1978). These eruptions are frequently of short duration (Wood, 1979a), although they can extend for many years. With a low discharge rate a strombolian eruption can build a moderate to high angle (>10°) lava shield (Fig. 5.3.5c). On a flat surface the scoria cone forms a central boss; the scoria cone extends down to the base of the shield. With a less common high discharge rate the lava forms an extensive lava sheet.

Sometimes a magma which is distinctly more viscous than normal strombolian, or possesses a substantial yield strength, gives rise to violent strombolian eruptions. These eruptions generally produce basaltic andesites though some are basaltic. Pyroclastics tend to occur in amounts comparable to lava flows, but sometimes pyroclastics are the sole eruption products. The main difference between this and the normal strombolian type is the greater proportion and great degree of fragmentation of the pyroclastics. The eruption columns are black due to the high ash content and often extend 5–10 km above the vent (Wilcox, 1954; Cole and Nairn, 1975). The 1886 eruption of Tarawera, New Zealand, appears to have emitted ash, lapilli and mud at a rate of 10^5 m^3/sec (Cole and Nairn, 1975). An increase in explosive activity at Paricutin from strombolian to violent strombolian appears to have been associated with a 0.2 wt.% increase in the total H_2O content of the magma (Wilcox, 1954).

Lavas produced during violent strombolian eruptions tend to be aa and form steep-angle (>10°) shields. On high cones violent strombolian activity can generate glowing avalanches which, as in the case of Fuego, can travel some 7 km from the vent at an average speed of 60 km/hr (Davies *et al.*, 1978).

Composite cones are second only to oceanic shields in terms of height (Wood, 1978). Although composite volcanoes (stratovolcanoes) are commonly andesitic landforms, some stratovolcanoes (Fuji, Fuego, Izalco) are composed entirely of basaltic rocks. Composite volcanoes are most commonly associated with convergent plate margins, although many in Iceland (Hekla) and the East African Rift (Ayelu) are at diverging margins.

Composite cones are composed of interbedded layers of pyroclastics and lavas, representing alternating styles of eruptive activity. Small cones are straight-sided (slopes ~30°) but larger cones have wide bases that feather into the surrounding terrain, having logarithmically (Milne, 1878) or hyperbolically curving slopes that gently steepen from 10–15° to 35–40° near the summit. Steinberg and Solov'yev (1976) have shown that as cones grow, their increased weight causes stresses within the base of the cone. When the height exceeds 2–2.5 km, distortions of the lower portion of the cone accentuate the concave upward profile of large composite volcanoes.

Basal diameters of relatively fresh composite volcanoes range from about 2–25 km, but older, more degraded cones have diameters up to 50–60 km. A tentative relationship between cone age (A, in years) and basal diameter (W_{co}, in km) derived by Wood (1978) on the basis of 9 composite cones, is

$$A = 13.15W_{co}^{5.56}$$

This relation appears valid for cones up to about 10 km across, but larger cones must have grown at a faster rate. The total eruptive lifetimes of composite volcanoes are probably of the order of 10^5–10^6 years.

Relations between cone heights, basal diameters, and crater diameters are listed in Table 5.3.3. The normal eruptive vent for composite volcanoes is a small crater, but collapse commonly produces a broad caldera when cones reach heights of about 3 km (Francis and Abbott, 1973)—the critical height for collapse possibly being governed by lithosphere thickness (Vogt, 1974; Mohr and Wood, 1976). The same stresses that affect the shapes of large cones result in the formation of basal radial fissures that become conduits for flank eruptions. Fuji and Etna are good examples of high composite cones with radial fissures (Nakamura, 1977; Wadge, 1977).

The eruptions that contribute most to the volumes of composite volcanoes are vulcanian. Like violent strombolian eruptions, vulcanian eruptions involve a relatively viscous magma; however, since the discharge rate is low, the activity is intermittent. In consequence, the vent becomes plugged with solid or near-solid lava and the explosive disruption of this material generates pyroclastics rich in nonvesiculated material and bread-crust bombs.

The lavas associated with vulcanian eruptions tend to form domes and again nuée ardentes are commonly generated on high cones. Although vulcanian eruptions are typically andesitic, some basaltic volcanoes display vulcanian activity, e.g., Merapi (Java), Manam (New Guinea), and Lopevi (New Hebrides). Nuée ardentes are common on basaltic cones such as Ulawan in New Britain (Melson *et al.*, 1972), while lahars (mudflows) often strip recent ash from the upper flanks of composite cones and deposit it far downslope (Macdonald,

Table 5.3.3 Average morphometry of terrestrial basaltic volcanoes.

Volcano type	Volume (km^3)	Wcr	Wco	Hco	Number	Example
Shield/caldera	1300	3.3	48.1	2.3	18	Mauna Loa, Hawaii
Composite (strato-)[a]	22	0.3	8.0	1.4	22	Fuji, Japan
Shield/ Icelandic	9	0.5	8.6	0.5	15	Skjaldbreidur, Iceland
Shield/low	0.4	0.4	4.8	0.07	49	Mauna Iki, Hawaii
Table moutain	0.2	0.4	1.7	0.3	10	Herdhubreidh, Iceland
Tuff ring	0.06	0.8	1.6	0.1	34	Diamond Head, Hawaii
Shield/steep	0.05	0.2	1.6	0.07	10	Mauna Ulu, Hawaii
Cinder cone[b]	0.02	0.3	0.8	0.14	83	Paricutin, Mexico
Maar	0.01	0.8	1.4	0.03	70	Elegante, Mexico
Spatter cone	0.001	0.1	0.4	0.03	14	Ducrot, Reunion
Pseudocrater	0.000	0.06	0.1	0.01	10	Haey, Hawaii
Pit crater[c]	0	0.5	0.5	0	16	Hiiaka, Hawaii

Notes:
All data from Pike (1978) except for: [a] Wood (1978); [b] Wood (1979a); [c] Wood, (unpublished data). Dimensions are averages, not values for given examples. Wcr = crater diameter; Wco = basal diameter of cone; Hco = cone height.

1972). Particularly violent eruptions have also been interpreted to have taken place on the presently docile Hawaiian volcanoes. Swanson and Christiansen (1973) have interpreted the deposits of a 1790 eruption of Kilauea to have been the product of a base surge similar to those produced during nuclear explosions.

Volcanoes do not always erupt material of the same composition throughout their lives. This is particularly true of those that have a long history, and thus have sufficient time for the magma reservoir to differentiate, or for conditions in the mantle source area to change. Often in basaltic volcanoes there is a regular progression from tholeiite to alkali basalt with time and this may cause a volcano to change its form because of the lesser eruption rates and higher viscosities of the more differentiated materials.

On Jan Mayen Island, the volcano Beerenberg (Fig. 5.3.7) consists of a large basal shield of ankaramitic ashes and lavas on which is built a lava cone of plagioclase basalt (Fitch, 1964). The island of Tenerife (Fig. 5.3.7) likewise is composed of a large basal ankaramitic shield on which has been constructed a phonolite cone within the summit caldera, and which is draped by extensive phonolitic ignimbrites (e.g., Ridley, 1972; Booth, 1973). The Ch'ang-pai Shan volcano (Fig. 5.3.7) on the Chinese-Korean border comprises a basal plateau of augite-olivine basalt 120–150 km in diameter, surmounted by a 30–40 km diameter alkaline trachyte dome, itself surmounted by a 10 km diameter pantellerite dome (Ogura, 1969). Less dramatic changes are represented by the Hawaiian volcanoes, where the greatest bulk of the shields are tholeiitic basalts overlain by small-volume, late cinder cones which tend to be composed of alkalic basalts, mugearites, and hawaiites (Macdonald, 1972).

Interpretation of the chemistry and petrological characteristics of a volcano based on its gross morphology can thus be misleading. Ash flows can produce broad shields that are, in general form, similar to basaltic shields (Wood, 1977); lavas of more evolved compositions may thinly mantle a basalt shield or stratovolcano and thus mimic its form; or a voluminous eruption of highly evolved lava can tend to obscure the original form of the underlying basaltic volcano.

The overall shape of polygenetic volcanoes is controlled both by the relative properties of material that is emitted from the summit vent or flank vents, and the location of the flank vents. For example, if a substantial proportion of low viscosity lavas comes from flank vents, the overall form of the volcano is with upwardly increasing slopes, as exemplified by Mauna Loa. Alternatively, if a substantial proportion comes from the central vent, or vents close to the central vent, the volcano is domelike, as exemplified by some Galapagos volcanoes. The steepness of the volcano is a function of (1) the proportion of magma coming out the central vent, (2) the volume of magma (per eruption) coming out from the central vent, (3) the discharge rate during eruptions, and (4) the explosivity. Thus a steep cone is favored by a large proportion from the central vent, a small volume per eruption from the central vent, a low discharge rate, and a large percentage of pyroclastic materials. Steep polygenetic basaltic cones include Pico (Azores) and Fuji.

Plate 5.1 Low oblique aerial-photograph of active lava "river" at Mauna Ulu, Hawaii, in 1970, showing crust being rafted downflow (toward top of photograph) and developing roof over the channel to form a lava tube (from Greeley, 1970).

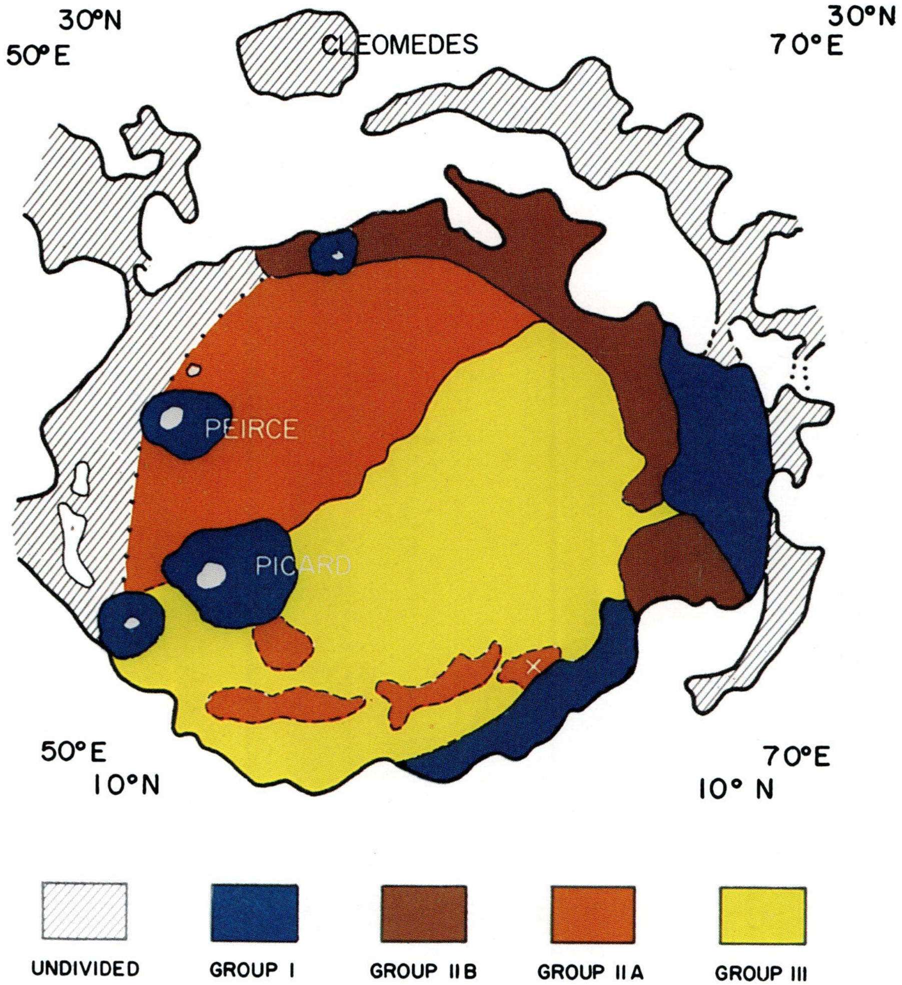

Plate 5.2 The distribution and stratigraphy of basalts in Mare Crisium based on photography and remote-sensing data (from Head *et al.*, 1978a).

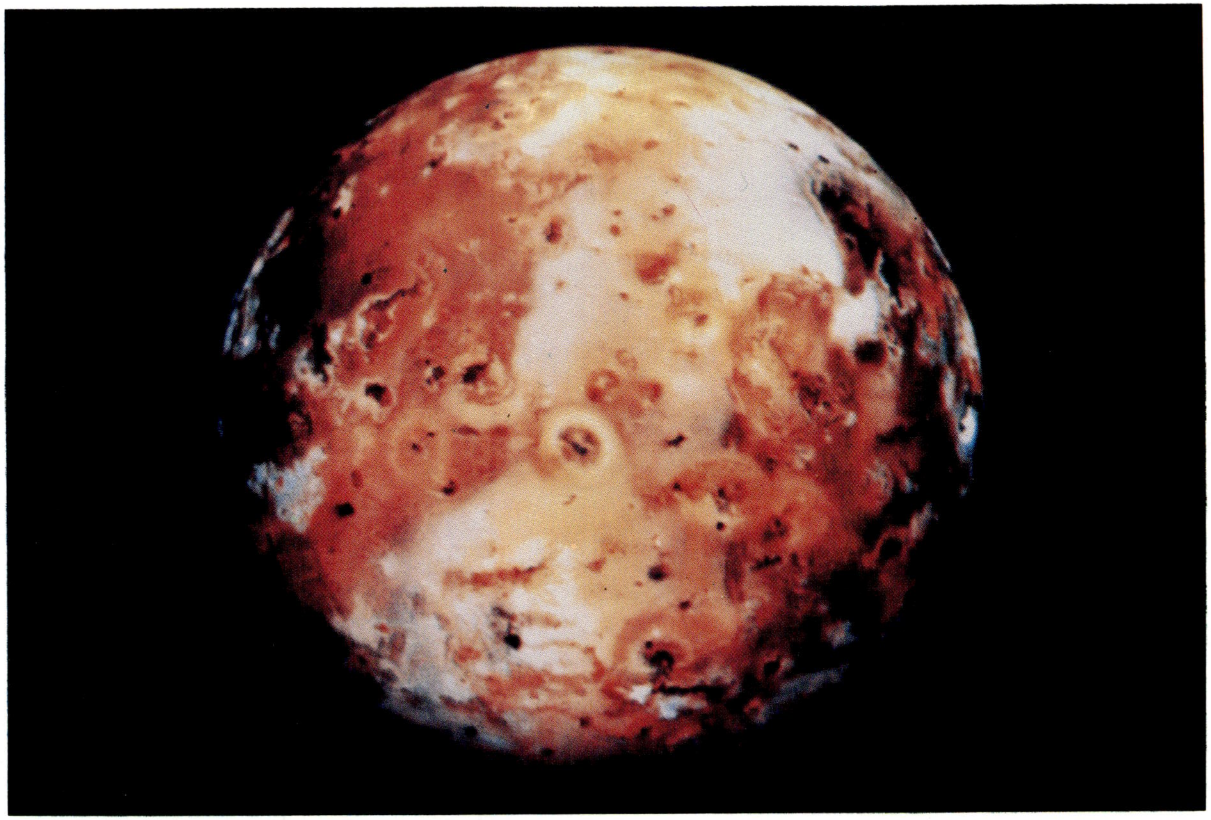

Plate 5.3 Color image of Io. Dark spots are calderas; diffuse circular features are eruption columns and ash deposits.

Monogenetic cones often occur in "fields" covering substantial areas. Any attempt to analyze the characteristics of these fields must consider both the stage of evolution that the field has reached and the degree of erosion that has taken place (Gutmann, 1979). For example, if the field were still actively forming, then a calculation of the volcano density might differ next year from this, while erosion and burial of early volcanoes within a field would result in significant differences in morphometric properties between these and younger volcanoes in the same field (Bloomfield, 1975).

Settle (1979) has separated cinder cone fields into two groups: (1) volcano cone fields, and (2) platform cone fields. The volcano cone fields are built on the flanks of polygenetic volcanoes, while the platform cone fields are relatively planar raised surfaces. In a study of six areas, Settle (1979) established that cones of volcano cone fields tended to be smaller (0.3–0.4 km basal diameter as against 0.9–1 km) and less widely spaced (0.6–0.8 km as against 1.0–1.2 km) than cones of platform fields. Table 5.3.4 lists some of the major features of these particular fields. In both types of fields the distribution of cones is often related to fractures (e.g., Breed, 1964; Porter, 1972). A study by Swanson *et al.* (1974) found that explosive activity of a more violent kind is associated with Galapagos volcanoes having a lower overall cone density; they recorded cone concentrations of $1/km^2$ on the upper slopes of Santiago, $0.5/km^2$ on Rábida, and $0.1/km^2$ on Pinzón.

Like cinder cones, composite cones can form fields. The result is clusters of closely-spaced, steep-sided cones. Examples are the Telica-El Liston-Santa Clara and Las Pilas-Cerro Negro-Cerro Grande clusters in Nicaragua (Mooser *et al.*, 1958).

5.3.2 Basaltic volcanism in aqueous environments

Section 5.3.1 described volcanic landforms produced by the eruption of magma directly onto the

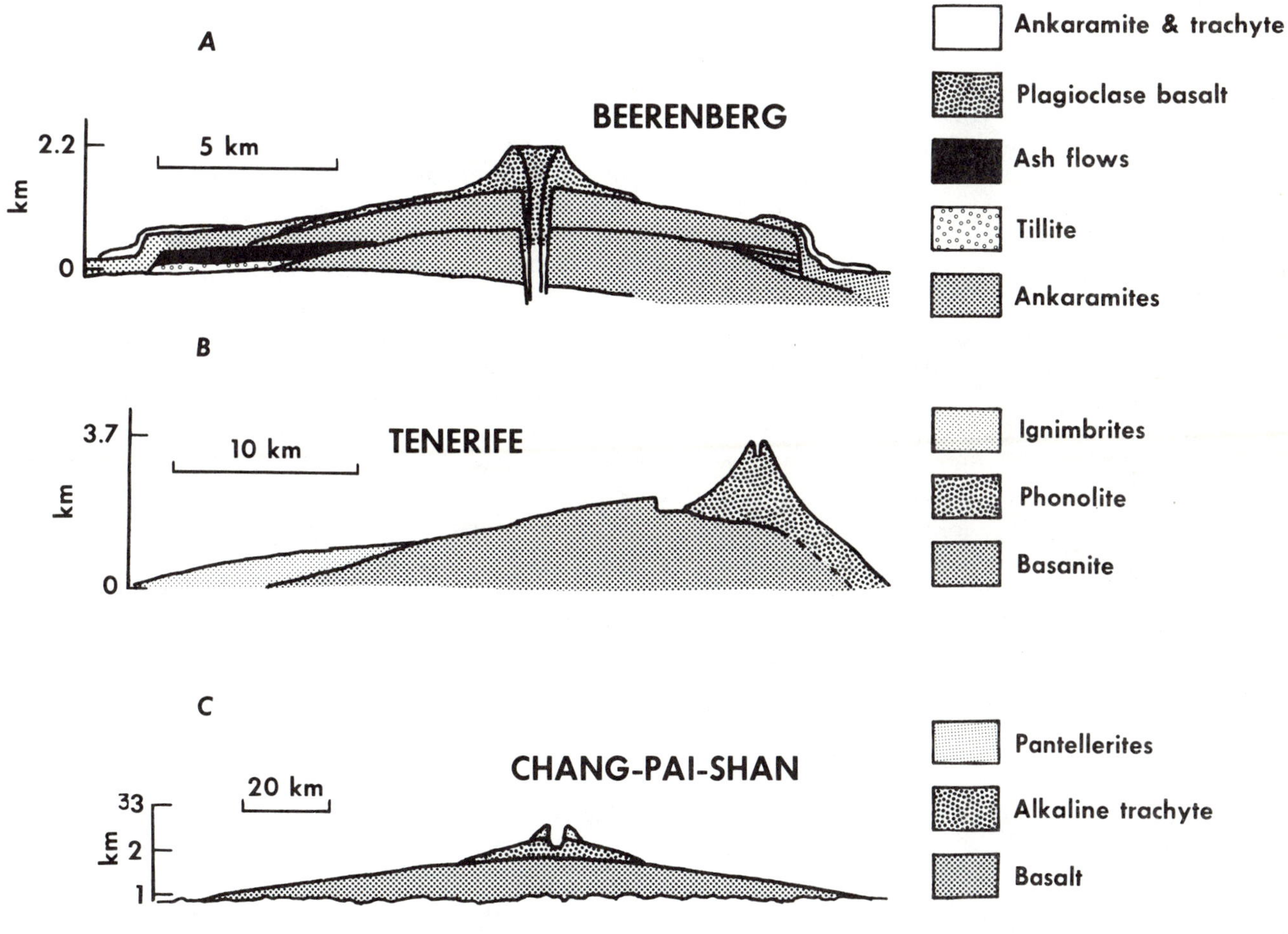

5.3.7 Diagrammatic sections of composite volcanoes (after Fitch, 1964; Ridley, 1972; and Ogura, 1969).

Table 5.3.4 A comparison of volcano cone and platform cone fields.†

	Age	Number of Cones	Area (km^2)	Average Areal Density	Maximum Density
Volcano Cone Fields					
Mauna Kea	majority < 0.3 m.y.	168	1500	11.1/100 km^2	4.7/10 km^2
Mt. Etna	all < 0.3 m.y.	87	800	11.3/100 km^2	3.9/10 km^2
Kilimanjaro	majority < 0.5 m.y.	205	5400	3.8/100 km^2	4.0/10 km^2
Platform Cone Fields					
San Francisco Mt.	majority < 1.0 m.y.	376	6100	6.3/100 km^2	3.2/10 km^2
Paricutin region	majority < 0.1 m.y.	170	2300	7.5/100 km^2	1.9/10 km^2
Nunivak Island	majority < 0.3 m.y.	83	3000	2.8/100 km^2	1.8/10 km^2

†After Settle, 1979.

Earth's surface at standard temperatures and pressures. This section examines the different landforms resulting from the interaction between magma and aqueous phases (groundwater, freshwater, ice, and seawater), sometimes under conditions which do not correspond to standard temperatures and pressures.

Phreatomagmatic (Surtseyan) volcanism

The volcanic constructs built by phreatomagmatic eruptions are known as tuff cones (Fig. 5.3.8), tuff rings, and maars (Fig. 5.3.9). While all three of these features are built of ash, they differ in morphology. Uneroded maars generally have deep craters cut into the surrounding terrain, suggesting that collapse played an important role in their formation (Wood, 1974), whereas the floors of tuff cones/rings are higher than their surroundings. Tuff volcanoes generally have relatively steep rims that dip both inwards and outwards with nearly equal slopes, but maars generally lack inward dipping deposits (Wood, 1979b). The main differences between tuff cones and tuff rings is that the latter have the larger craters and lesser heights. The Menan Buttes, Idaho, are tuff cones (Hamilton and Myer, 1963), and Diamond Head, Hawaii, is a typical tuff ring. Lorenz (1973) believes that the main factor leading to maar formation, rather than a tuff cone/ring, is the depth to the magma chamber. Near-surface contact of magma and water leads to wide dispersal of ejecta and formation of tuff rings/cones (Heiken, 1971). For deeper magma chambers and sufficiently long periods of activity, the vent is enlarged by spalling and slumping, and ultimately subsidence of wall materials and vent deposits along ring fractures produces a maar (Lorenz, 1975).

Walker and Croasdale (1972) point out that the ashes produced by phreatomagmatic eruptions (which they named surtseyan deposits) are considerably different from strombolian tephra. Surtseyan deposits are finer grained (2 to < 1/8 mm) and less well sorted than strombolian scoria deposits (>1 mm to <1 m, with 95% > 4 mm). Surtseyan ashes are roughly equant with fracture boundaries, whereas strombolian ashes are broken droplets or sprays of magma. Individual beds of surtseyan material are often very thin (to 1 mm) and continuous, whereas strombolian ash layers are often > 5 cm thick. Clasts and bomb sags are common in surtseyan deposits but rare in strombolian units; spatter and lava occur only in strombolian activity. The mechanism of magma/water interaction that transforms strombolian eruptions into strongly explosive eruptions has been discussed by numerous authors, including McBirney (1963), Sigvaldason (1968), Bennett (1971), Colgate and Sigurgeirsson (1973), and Peckover *et al.* (1973). The latter authors consider eruptions resulting from the thermal interaction of magma and water (phreatomagmatic eruptions) to be examples of a wider class of events called fuel-coolant interac-

5.3.8 Cerro Colorado tuff cone (crater diameter: 0.85–1.1 km; maximum depth: ~0.1 km). Pinacate Volcanic Field, Mexico.

tions. Rapid heat transfer from hot magma to cool water results in the vaporization and explosive expansion of the water. Pressure generated by the expanding water bubbles leads to further mixing of magma and water and thus more explosions. Cyclic explosions continue as long as both the fuel (magma) and coolant (water) are available. The 1963 eruption of Surtsey illustrates this mechanism. During the first six months of activity rhythmic eruptions of jets of fine grained tephra (Fig. 5.2.12) built a broad rimmed cone rising above the north Atlantic Ocean. However, after the first lava flows armored the cone and prevented access of seawater to the vent, the explosive eruptions stopped and much less energetic strombolian activity became dominant (Thorarinsson, 1967a).

Within 3 days Surtsey built a cone 60 m high and 600 m long; the average eruption rate throughout the first 4 months was about 50 m^3/sec (Thorarinsson, 1967b). These figures are similar to eruptions of strombolian style (Paricutin grew 135 m in its first week; the average eruption rate over the 12 days of activity of Cerro Negro, Nicaragua, during 1971 was 43 m^3/sec), the subaerial equivalent of surtseyan activity. Indeed, the surtseyan activity at Surtsey, Capelinhos, and Falcon Island changed to strombolian activity when seawater was finally excluded from their vents.

Although Ollier (1967) believes that maars and tuff cones form only from eruptions of basaltic magma, Lorenz (1973) points out that the chemistry of the magma is unimportant; the thermal contact of magma and water is alone responsible for the unique characteristics of maars and tuff volcanoes. Phreatomagmatic eruptions are most commonly basaltic because that is the most abundant magma type, but there are maars

5.3.9 Elegante Maar (crater diameter 1.5 km, depth 0.24 km), Pinacate Volcanic Field, Mexico. (Photo by S. Larson.)

and tuff cones with tholeiitic, alkalic basaltic, carbonatitic, andesitic, and dacitic compositions (Lorenz, 1973). Thus, maars are not characteristic landforms for a particular magma type, but rather are the morphometric signature of magma-volatile interaction.

Maars do not usually form on polygenetic volcanoes; more commonly they form in platform fields, in structurally controlled zones, or apparently at random. A strong northeast-southwest structural alignment with a spacing of 3–4 km is exhibited by the seven maars of the Lower Chindwin area of Burma (Chibber, 1934). A series of four maars exhibit a northeast-southwest alignment in central Anatolia (Keller, 1974). In western Victoria, Australia, a series of seventeen maars is arranged along the synclinal axis of Tertiary sediments (Ollier, 1972). Apart from being formed in stream valleys, the maars of the Eiffel region appear to be randomly distributed. In all, about 20 maars occupy a 1100 km^2 area of the Eiffel (Lorenz, 1973). Dense concentrations of young tuff rings (+60) are found within a 12 km radius of Auckland, New Zealand (Searle, 1964), and in some areas of southwest Uganda, scores of contiguous maars occur, in places cross-cutting one another.

Subglacial volcanism

Volcanism in areas with current extensive ice cover, such as Iceland (Nielsen, 1936), or past ice cover, such as British Columbia (Mathews, 1947), results in the production of unique landforms. Basaltic eruptions beneath thick glacial ice sheets create their own meltwater lake, into which they are extruded, and within which they are confined in lateral extent. The resulting table mountain or moberg hill may ultimately project above the ice cover, or the lake may be drained, so that

5.3.10 View of Herdhubreidh, Iceland, looking southwest. The slopes below the subaerial lava capping are predominantly hyaloclastic with some interbedded lava. The lower ridge extending to the left is Herdhubreidhartögl. (Drawing by W.B. Bryan.)

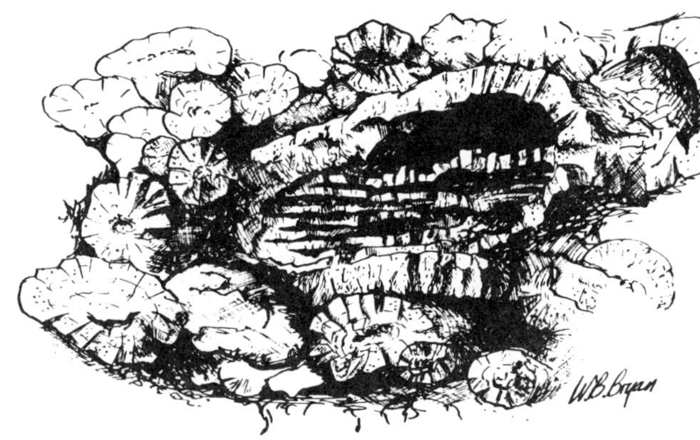

5.3.11 An exposure of truncated subglacial lava pillows, with a small lava tube with septa showing successive stands of falling lava levels. Compare to Fig. 5.3.27. Longest diameter of lava tube is about 1 m. (Drawing by W. B. Bryan.)

the latest eruptions are subaerial (Van Bemmelen and Rutten, 1955). After volcanic activity ceases, the hill may be over-ridden by glacial ice. Melting of the ice removes support from the flanks of the hill, and the slopes are further modified by meltwater erosion and by gravity slumping and sliding of the volcanic sequence.

In Iceland, subglacial eruptions have taken place along rifts extending for distances of 10–20 km, producing elongate, hummocky ridges composed of palagonite tuff-breccia called moberg (Walker, 1965). In gross morphology and dimensions these subglacial ridges are not unlike some submarine volcanic ridges.

Where a subglacial eruption is concentrated at a central vent, a table mountain (tuya) is built. The classic example of a table mountain is Herdhubreidh, southeast of Mývatn in northern Iceland (Fig. 5.3.10). The lower slopes are made up of hyaloclastic breccia and interbedded pillow lava. The mountain, which rises 1 km above the surrounding plain, is capped by a subaerial basalt sequence forming a 2 km diameter plateau surmounted by a cinder cone. A linear, hummocky ridge (Herdhubreidhartögl) extends some 10 km south of Herdhubreidh and parallels the general direction of fissuring. It is made up entirely of subglacial products and probably is part of the same rift system on which Herdhubreidh was erupted.

Subglacial eruptions are not confined to ice sheets; volcanic activity beneath thick valley glaciers would result in grossly similar morphological products. Walker and Blake (1966) have recorded an instance in Iceland where a lava flow travelled some 35 km beneath a valley glacier.

Subglacial pillow lavas are strikingly similar to submarine pillow lavas in dimensions and most mor-

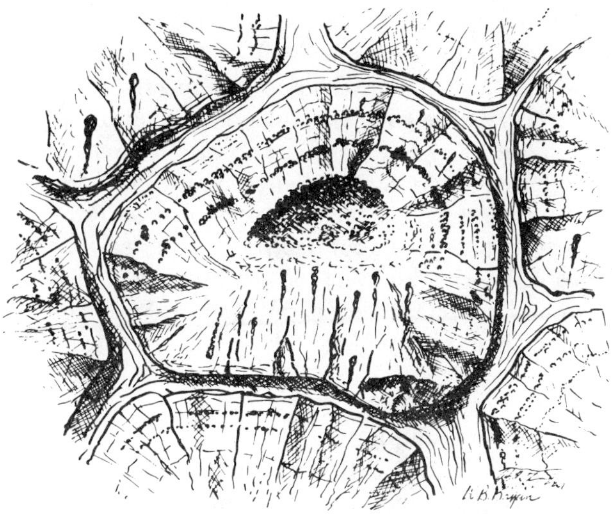

5.3.12 Detail of a vesicular subglacial basalt pillow exposed in a quarry near Keflavik, Iceland. Note matrix of hyaloclastite between pillows, elongated pipe vesicles in lower part of pillow, scoriaceous pillow core, and concentric zones of spherical vesicles in upper part of pillow. (Drawing by W. B. Bryan.)

phologic details (Fig. 5.3.11). The subglacial pillows do differ in the much greater amount of associated hyaloclastite, which commonly forms an inter-pillow matrix. In many exposures the amount of hyaloclastite exceeds the amount of pillowed lava, and the pillows appear as isolated tubular fragments dispersed in hyaloclastite. The subglacial pillows also differ in their higher degree of vesicularity (Jones, 1969), and the tendency for vesicles to occur as concentric zones of spherical and pipe vesicles just inside the pillow margin (Fig. 5.3.12).

Glacial transport of basaltic blocks and wind redistribution of the readily erodible hyaloclastite and basaltic ashes have led to Icelandic terrains extremely similar to those observed at the Viking landing sites on Mars (Fig. 5.3.13).

Submarine volcanism

The nature and extent of volcanic activity in the deep ocean environment have been recognized only recently. This deep-sea volcanism is a necessary consequence of seafloor spreading, which requires the generation of new seafloor along the locus of plate separation at mid-ocean ridges. Probably a great deal of additional volcanic activity occurs on the seafloor in areas remote from spreading mid-ocean ridges. Certain areas of seafloor, such as the eastern and northeastern Pacific, seem to contain especially large concentrations of seamounts and guyots, presumed to be volcanic cones.

Photographic surveys and direct observation of volcanic features from manned submersibles have been made in only a few small areas on the Mid-Atlantic Ridge, the Galapagos Rise and the East Pacific Rise. The deepest submarine volcanic activity yet documented in this way is in the Cayman Trough, a fracture zone separating the North American and Caribbean plates (Ballard *et al.*, 1979).

Acoustic imaging is an additional important technique for obtaining morphologic information on the seafloor. Wide-beam echo sounding is the most widely used method; its limitation arises from the fact that sound is reflected from a large area of seafloor, and the

5.3.13 Surface of a glacially stripped shield volcano in Iceland. Note effects of wind action (pitting, dunes) and similarity to surface of Mars at the Viking Lander sites.

first return is not necessarily that from directly beneath the ship if the seafloor is very rough, or slopes steeply. Better resolution can be obtained with narrow-beam systems; the most sophisticated methods use multiple-beam sonar arrays giving real-time, computer-reduced output as contoured strip charts of the seafloor. Contoured charts prepared from these strips may have a form line interval of 5–10 meters, and approach the resolution of topographic maps prepared from aerial stereo photography. Side-looking sonar also has been employed from deep-towed instruments. Potentially, sonar images can give "shadow graphs" with textural detail indicative of surface roughness, much like airborne side-looking radar images but, in most existing systems, with lower resolution.

In spite of limited sampling of volcanic features present on the seafloor, there is good reason to suppose that these submarine features are typical. This is suggested by gross similarities between forms observed in widely separated locations, and by their similarity to many subaerial volcanic features. The principal effects of the deep-sea environment are the almost complete elimination of explosive volcanism at depths greater than a few hundred meters, and somewhat more rapid cooling rate due to the greater thermal conductivity of water compared to air. Thus, virtually all submarine volcanic features are lava cones, domes, flows, or lakes, and flows tend to be shorter and thicker than their subaerial counterparts. Overall, submarine lava surfaces bear a strong resemblance to subaerial pahoehoe lava. To date, no good examples of a submarine equivalent of blocky or aa lava are known. This may simply reflect the rather limited range of seafloor geologic environments which have been studied so far. Such flows may prove to be common among submarine eruptions of basaltic andesite in island arcs, but there are presently no detailed observations of submarine morphology in any island arc.

Seamounts and guyots

The pronounced linear patterns of the principal Pacific island-seamount chains are indicated in Fig. 5.3.14. Islands are invariably located at the southeastern ends of the chains, and may include presently active volcanos. The Hawaiian-Emperor chain is perhaps the best-studied example, extending from Kilauea northwestward through successively older and more dissected volcanic islands to reefs and atolls capping drowned volcanic shields, and continuing in the line of guyots and seamounts making up the Emperor chain. The Cook-Austral chain includes the volcanically active Macdonald Seamount at its southeastern end. Furthermore, Pitcairn Island near the southeastern end of the

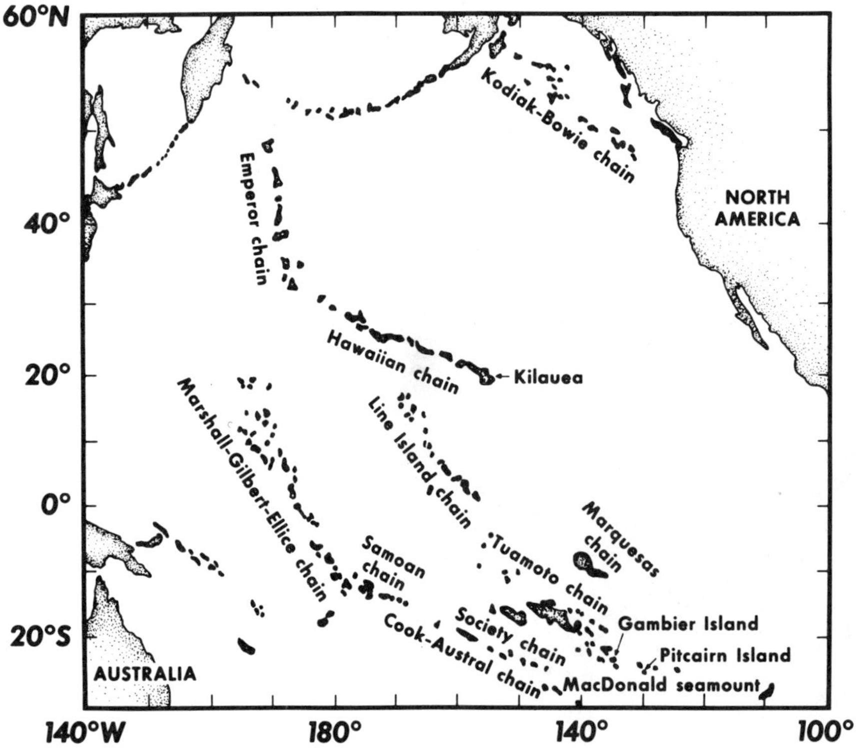

5.3.14 Seamount and island chains in the northwestern Pacific Ocean basin. Active or recently active volcanoes are located at Kilauea, Pitcairn, and Macdonald Seamount (after Jackson, 1976).

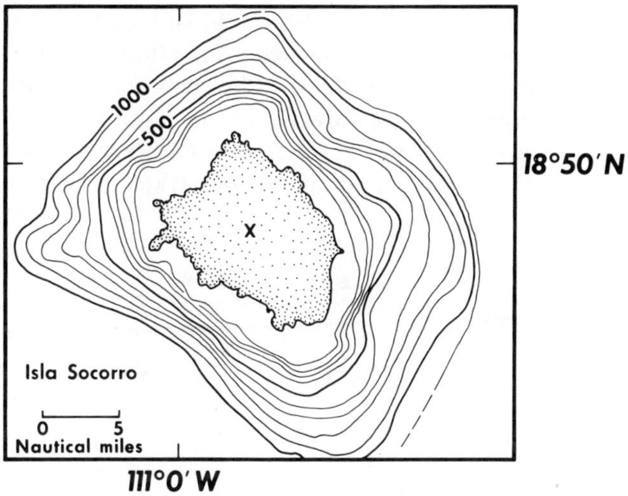

5.3.15 Bathymetry of Isla Socorro in the eastern Pacific (Bryan, 1964). Elongate ridges at north, west and southeast coincide with trends of fissures and vent alignments mapped on the island.

Tuamoto-Gambier line has been isotopically dated as less than a million years old (Jackson, 1976).

The essential morphologic continuity between islands, guyots, and seamounts is well portrayed by three of the islands in the Revillagigedo group, near the southeast side of the Baja California seamount province. The irregularities in the gross form of these islands are directly related to fault trends which control the distribution of volcanic vents. Isla Socorro (Fig. 5.3.15) is essentially a distorted cone, with extensions to the north, west and southeast directly reflecting the trends of fissure systems extending from the summit area (Bryan, 1964). Isla San Benedicto (Fig. 5.3.16) represents the tip of a seamount elongated in a north-south direction, with recognizable vents on the island oriented on an approximately north-south line (Richards, 1966). Isla Roca Partida (Fig. 5.3.17) illustrates the last stages in the formation of a guyot (Richards, 1964), the rock representing the remaining fragment of a much larger volcanic island.

Menard (1964) figures examples of seamounts (Fig. 5.3.18) and guyots (Fig. 5.3.19) with a variety of outlines, probably reflecting their varied underlying fracture patterns. The guyots are conventionally supposed to represent wave-truncated volcanic cones, by analogy with islands such as Roca Partida; however, on some guyots a suggestion of a slight depression possibly indicates a summit crater or caldera.

Hollister *et al.* (1978) have provided the first definitive evidence of deep-sea cratered cones in the northwest Pacific from high-resolution multibeam sonar contouring. Examples are shown in Figs. 5.3.20–21. These cones are at sufficient depth that they probably never were exposed to wave planation; the flattened summits are very likely surfaces of lava ponds in summit calderas (Fig. 5.3.20). Distinct summit craters are well defined on other cones (Fig. 5.3.21) which range in size from about 180–1800 m in height and from 0.5–36 km^3 in volume. Most of these individual cones are in small groups aligned in a north-northwest direction, essentially parallel to the spreading direction. This is consistent with the likelihood that they are small deep-sea volcanoes representing extrusive activity along fissures. They are assumed to be basaltic, though

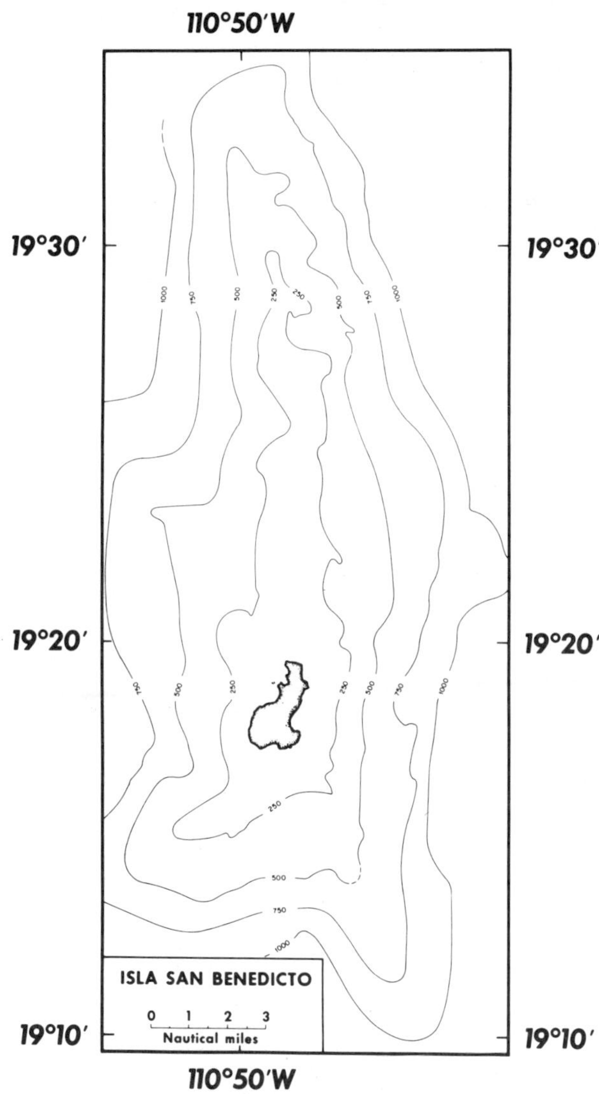

5.3.16 Bathymetry of Isla San Benedicto (Richards, 1966) showing strong north-south elongation which coincides with vent alignment on the island.

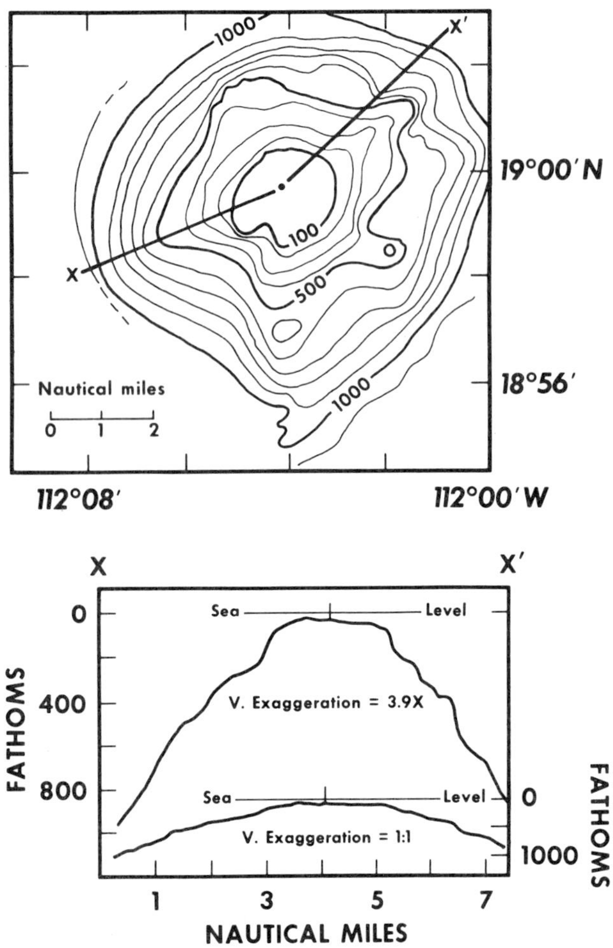

5.3.17 Bathymetry and profile of Isla Roca Partida (Richards, 1964), showing the last stage in the formation of a guyot from a volcanic island. Contour interval 100 fathoms.

Houghton *et al.* (1977) have described the morphologies of several seamounts in the New England chain which exhibit a diversity of forms and, as in the examples shown, may have flattened tops with wave-rounded pebbles and dead shallow-water fauna. These seamounts must have projected close to, or above, the sea surface, and are transitional to guyots. The most surprising feature documented in the submersible dives is the stepped profile observed on the upper slopes of the seamounts. On Rehoboth Seamount, these steps show slickensiding and other evidence of tectonism, and are clearly fault scarps (Fig. 5.3.22), although some flows have been extruded onto the benches from the base of the scarps. This is interpreted to mean that volcanic activity was synchronous with tectonic movements. Faulting appears to be more intensely developed where it parallels the trend of the seamount chain.

Balanus Seamount (Fig. 5.3.23) shows radial downslope ridges and a stepped profile that appear to be purely constructional, with the steps representing overlapping extrusive flow fronts. It is inferred that these flows were fed by coalescing lava tubes radial to a central vent. Such a structure, somewhat analogous to a

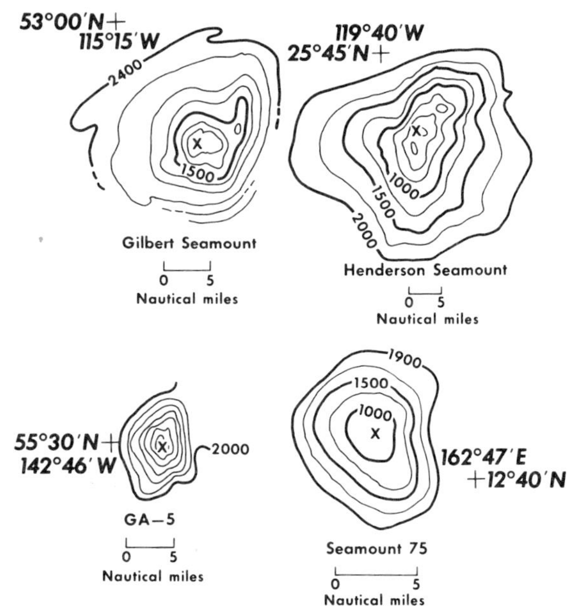

5.3.18 Bathymetry of four seamounts from the Pacific, showing variations in morphology (after Menard, 1964). Contour interval 250 fathoms.

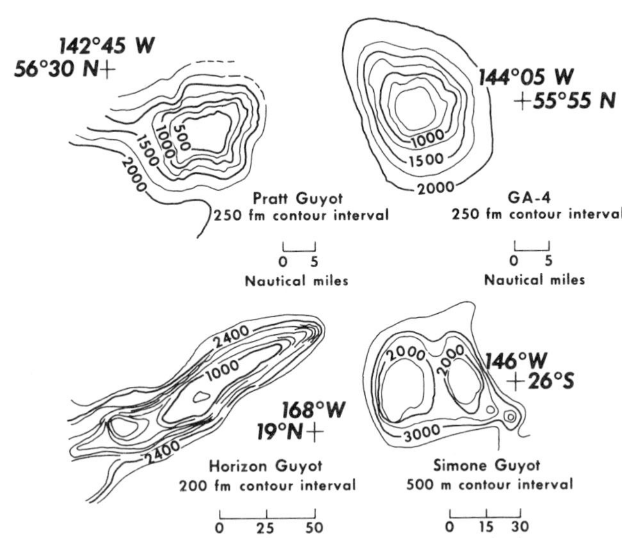

5.3.19 Bathymetry of four guyots from the Pacific, showing variations in morphology (after Menard, 1964).

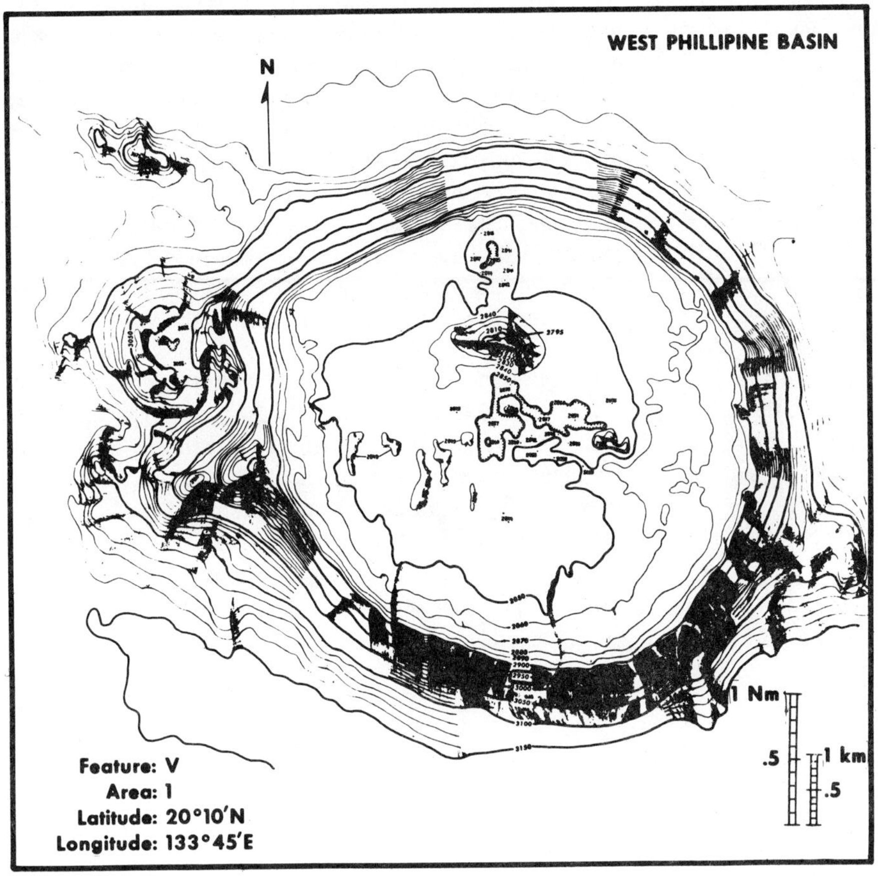

5.3.20 High resolution sonar-ray bathymetry of a seamount in the West Philippine Basin. Note apparent satellite cone on west flank, and nearly flat summit, possibly representing a filled caldera with a remnant of a cinder cone (Hollister *et al.*, 1978). Reprinted courtesy of Elsevier Scientific Publishing Company.

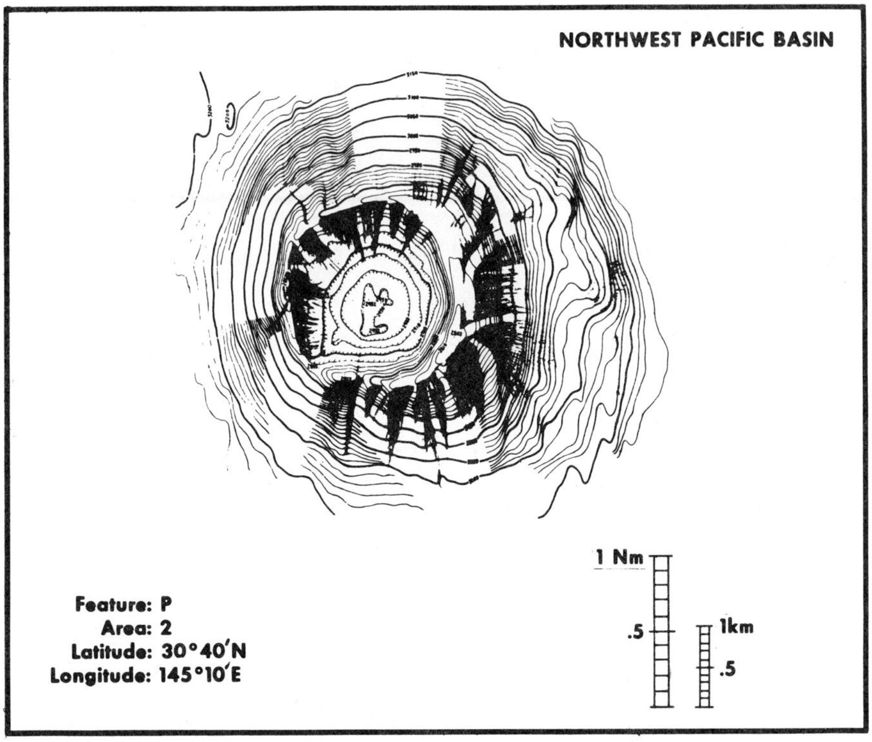

5.3.21 High-resolution sonar-ray bathymetry of a cratered seamount in the northwest Pacific (Hollister *et al.*, 1978). Reprinted courtesy of Elsevier Scientific Publishing Company.

lava "Christmas tree," would yield the radially arranged, outward-sloping stubby flows which seem characteristic of Balanus.

Slow-spreading ridges

Detailed observation of volcanic morphology on a slow-spreading ridge is so far limited to the FAMOUS area on the Mid-Atlantic Ridge between 36–37° north latitude (Fig. 5.3.24). Contours are based on multi-narrow-beam bathymetric surveys (Phillips and Fleming, 1978). The charts show a 25–30 km wide valley bounded by rift mountain crests rising to depths less than 800 fathoms. Within this broader valley, there is a well-defined narrow inner valley 3–9 km in width. Present volcanic activity appears to be confined to the floor and inner walls of this inner valley.

Direct observation of volcanic structures in the FAMOUS area is limited to the northern valley (Fig. 5.3.25). The central volcanic hills, Mt. Pluto and Mt. Venus, define the main axis of present-day volcanic activity. The active zone extends out to about 2 km either side of this axis, with the most recent activity confined to a narrower zone within 1 km of the central axis. The central volcanic hills appear to be almost entirely constructional volcanic forms. In contrast, the morphology of flank forms is strongly influenced by tensional and normal faults oriented parallel to the valley walls. This fault topography is veneered with thin flows, which originate on the valley walls and inner floor, and which flow inward toward the valley axis.

Ballard and Van Andel (1977) have outlined a model in which they show that the central volcanic hills are generally not split centrally as spreading continues, but are carried as distinct units to one side of the valley. In these flank positions, the hills are identifiable as marginal highs, and begin to be dismembered by faulting. Tensional rifts through these hills are the source of secondary flank eruptions which mantle the older volcanic hills with secondary lava flows. These flows over-

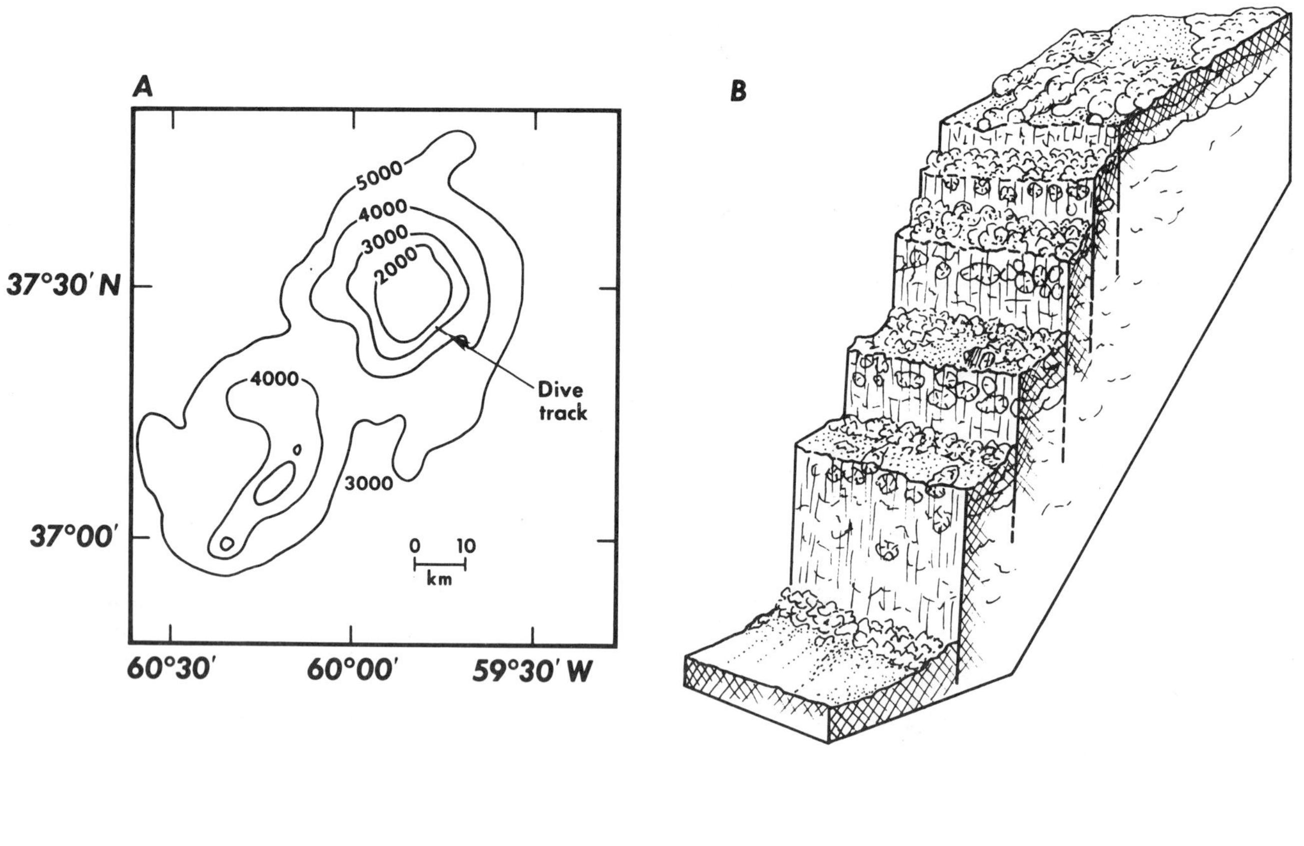

Rehoboth Seamount

5.3.22 Rehoboth Seamount in the New England seamount chain. **a.** Wide-beam bathymetry, showing location of R. V. *Alvin* dive 540. **b.** Conceptual drawing of step-like faulted profile observed on dive 540, showing truncated lava pillows and flows issuing from base of fault scarps. (Drawing by W. B. Bryan.)

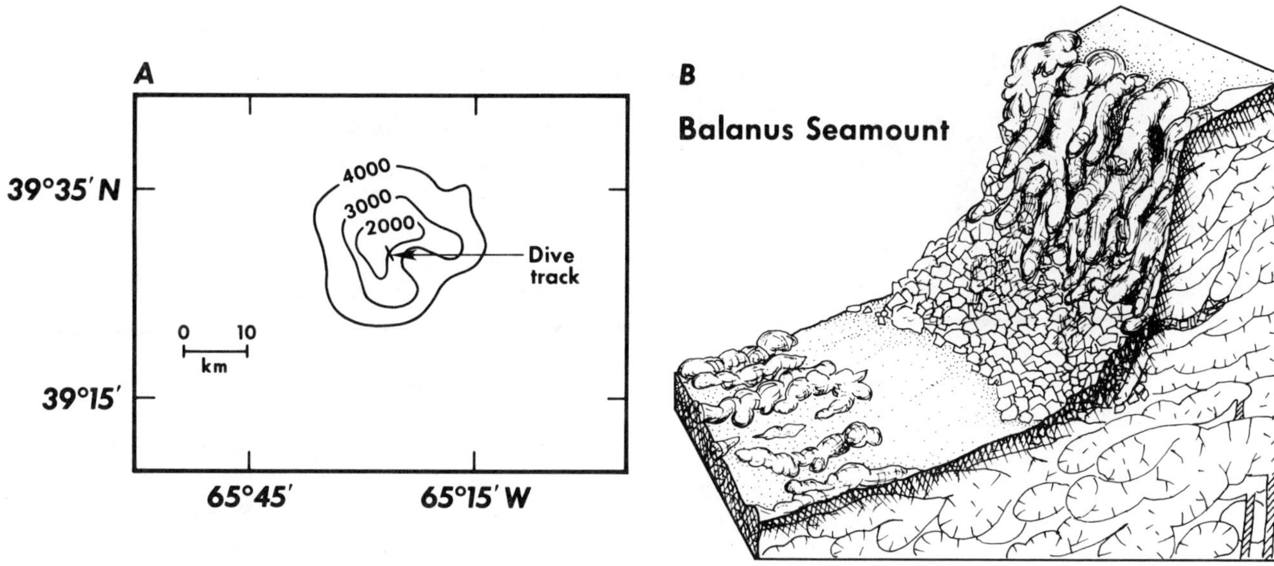

5.3.23 Balanus Seamount in the New England Seamount chain. **a.** Wide-beam bathymetry, showing location of R. V. *Alvin* dive 542. **b.** Conceptual view of stepped profile formed of overlapping, unfaulted flow fronts, as observed on dive 542. (Drawing by W. B. Bryan.)

lap in age with much more voluminous flows which build new central volcanic highs along the main spreading axis. The fact that these volcanic hills are physically distinct features shows that volcanic activity in the valley is neither continuous along the length of the valley, nor continuous in time. The maximum episodes of extrusive activity appear to be separated by intervals of about 100,000 years (Ballard and Van Andel, 1977; Ramberg and Van Andel, 1977).

Since the volcanic hills remain largely intact with transport, there are no deep exposures and their internal structure remains somewhat conjectural. Submersible traverses up the flanks of the central hills show that they have a stepped profile. The impression is that the lowest steps are produced by the most voluminous first flows, and are followed by lesser flows as the hill increases in height. The individual flows tend to be thick and bulbous, and it is their crudely shingle-like overlap which creates the step-like profile. The summit area of Mt. Pluto is capped by several conical peaks, referred to as "haystacks," which in some respects resemble subaerial spatter cones (Fig. 5.3.26). These haystack forms were also observed on the western flanks of the valley, and appear to mark the vents from which both central and flank lavas issue.

All the lava flows are pillowed with a variety of forms (Ballard and Moore, 1977). The nature of the pillows is closely related to their position on the flow, and to the slope on which the flow is erupted. Further subtle morphologic controls are probably introduced

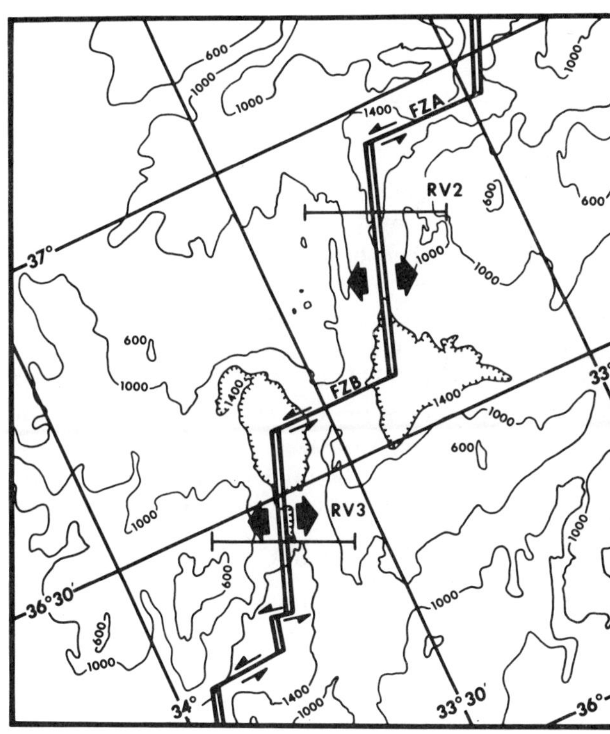

5.3.24 Location map of FAMOUS area on the Mid-Atlantic Ridge. Northern valley lies between fracture zone A (FZA) and fracture zone B (FZB); southern valley extends south from FZB. Double lines with large arrows indicate spreading centers; single lines with small arrows are present-day locations of transform faults (after Ballard and Moore, 1977).

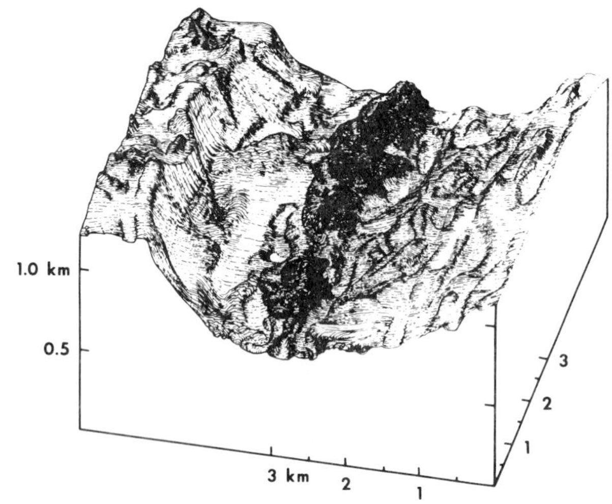

5.3.25 High-oblique perspective view of floor of inner valley, FAMOUS area, showing central extrusive hills, Mt. Pluto and Mt. Venus, on main volcanic axis. Flank flows descend slopes on either side of central hills, and are offset by normal faults and partly buried in sediment. (Drawing by Tau Rho Alpha, courtesy of Dr. J. G. Moore, U. S. Geological Survey.)

by differences in composition and temperature, but these relationships are not yet documented. Flow fronts tend to be steep (60–80°) and are typically 10–20 m in height. Flow front pillows are highly elongated, tubular forms that are elongated directly downslope. Tubes bud and branch and, where broken open, often show hollow cores due to lava drain-away; these cores may be bridged by horizontal lava septa showing different stillstands of the falling lava level within them (Fig. 5.3.27). These very elongated pillow forms also are characteristic of thin flank flows erupted over steep fault scarps on the valley walls (Fig. 5.3.28).

Upper surfaces of the thick central flows, and of flank flows on gentle slopes, are characterized by rounded, bulbous pillow forms. Many of these also are hollow due to lava drain-away, and appear as blister-like remnants, often with part of the roof and side caved in (Fig. 5.3.29).

The boundary between the central volcanic hills and the flank lavas is delineated by the gross morphology of the central hills with their relatively thick, voluminous flows, in contrast to the generally low relief of the thin flank flows. Pillow elongations show that flows from the central hills are outward-pointing while the flank flows are characterized by inward-oriented pillows. The relatively good morphologic preservation of most flank flows and their conformance to present-day topography attests to their very young age, as is also indicated by thicknesses of manganese crusts and palagonite rims on glass (Bryan and Moore, 1977).

Fast-spreading ridges

Detailed photographic surveys have been made of the East Pacific Rise at 21° N (Normark, 1976), and of the Galapagos Rift (Lonsdale, 1977). Direct observations from a manned submersible in the Galapagos Rift are summarized by Ballard *et al.* (1979). Spreading rates in both areas are about 6–7 cm/year; this is generally considered a "medium" rate. Volcanic forms at ridges with maximum rates of spreading (10–12 cm/year) have not yet been documented.

Bathymetric profiles from the East Pacific Rise and Galapagos Rift are shown in Fig. 5.3.30. Compared to the FAMOUS area, the major difference in gross morphology is the lack of a well-defined central inner

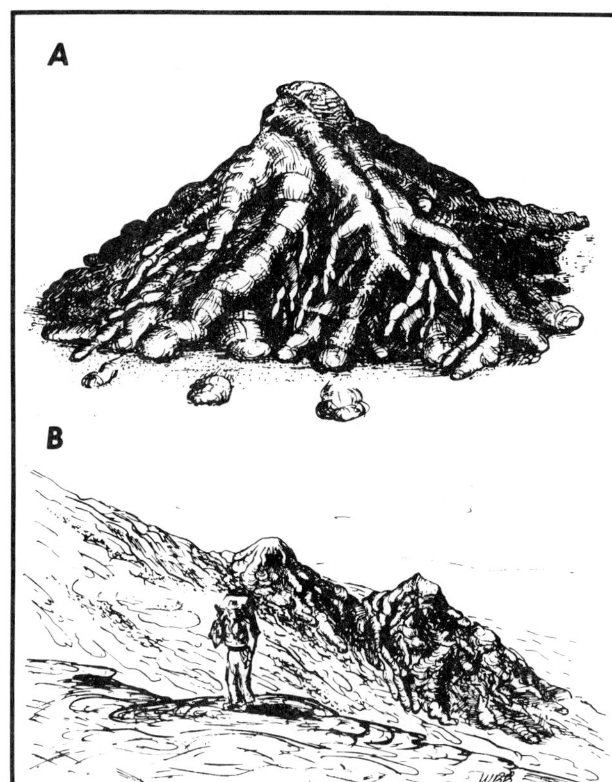

5.3.26 a. An artist's conception of a pillowed cone, or haystack, in the FAMOUS area. These cones mark extrusive vent locations and may be from 5–20 m in height; flank slopes are typically about 45 degrees. A line of these cones marks the top of Mt. Pluto and appears to define the position of an underlying fissure vent. **b.** Driblet cones, northwest slope of Mauna Loa, Hawaii. (Drawing by W. B. Bryan.)

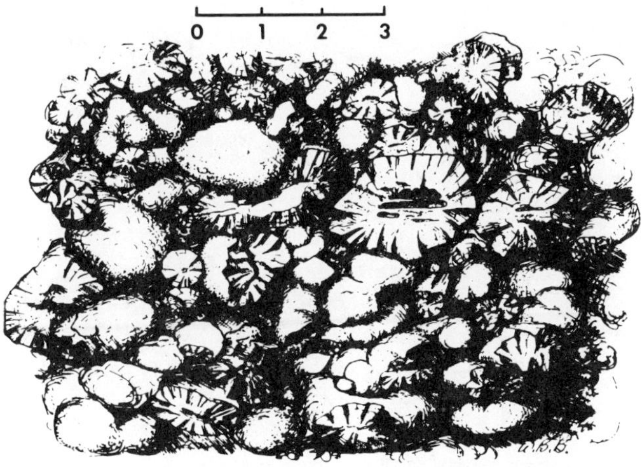

5.3.27 Typical view in cross-section of lava pillows, as observed on fault scarps in the FAMOUS area. Drawing is a composite of forms recorded in four different photographs. Note bulbous terminal ends of tubular pillows which happen to be tangent to the fault plane, and typical radial joint pattern in truncated pillows. The large layered lava tube records several stands of falling lava levels. Scale in meters. (Drawing by W. B. Bryan.)

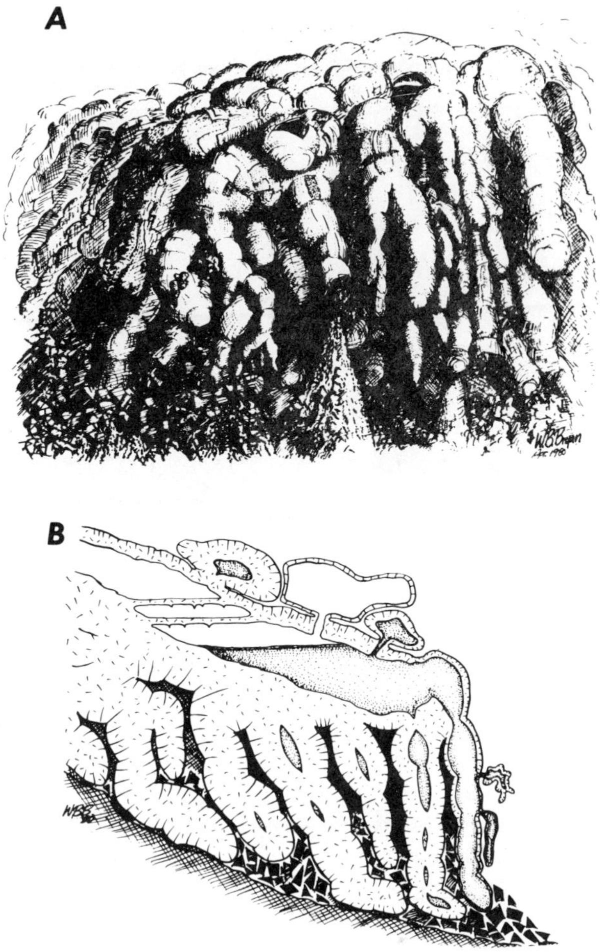

5.3.28 Elongated tubular lava pillows on a steep flow front. **a.** Conceptual view of development of foreset pillows on flow front as flow thickens and over-rides earlier tubular lava forms. **b.** As flow front grows in height, fragments of tubes break away and accumulate as angular talus at the foot of the flow. (Drawing by W. B. Bryan.)

valley. In other respects, there is a remarkable similarity in the dimensions and sequence of extrusive forms and the manner in which the morphology evolves from constructional-volcanic in the extrusive axis, to destructional-tectonic along the margins of the spreading axis. In both areas, the central, active extrusion zone is marked by a discontinuous, low ridge of knobby hills showing fresh, almost sediment-free pillow lava. These hills are 50–100 m high and 0.5–1.0 km wide, similar in dimensions to Mt. Venus and Mt. Pluto in the FAMOUS area. Flanking these central extrusive hills, plains of low relief are made up of ponded lava with relatively smooth surfaces. These ponded lavas tend to have wrinkled, ropy surfaces (Lonsdale, 1977) and apparently correspond to flank lava flows in the FAMOUS area, but are very unlike them in morphology. This extrusive zone, embracing both central hills and ponded flows, is about 2.5–3.0 km in width. Beyond this zone, fissures and normal faults begin to dominate, and horst-like marginal highs appear, flanked by inward-facing normal faults. These faults have much less displacement than the major faults bounding the inner valley in the FAMOUS area, but are of sufficient height to confine the flows within the central extrusive zone.

Pillowed lava flows resemble those observed in the FAMOUS area. The smooth-surfaced lava plains constitute the principal difference in volcanic morphology

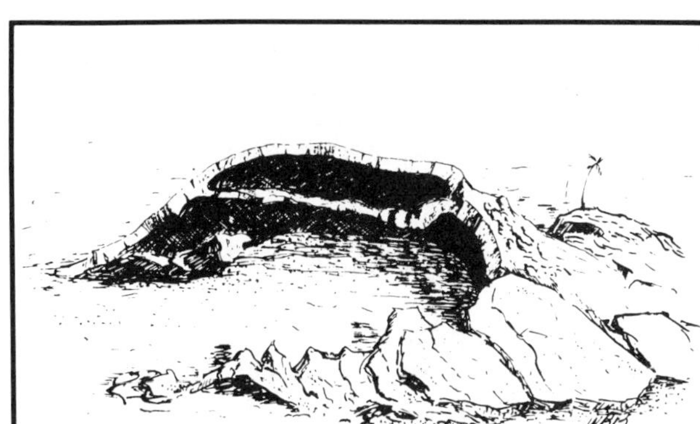

5.3.29 Collapsed blister pillow, about 1.5 m in diameter, FAMOUS area. (Drawing by W. B. Bryan.)

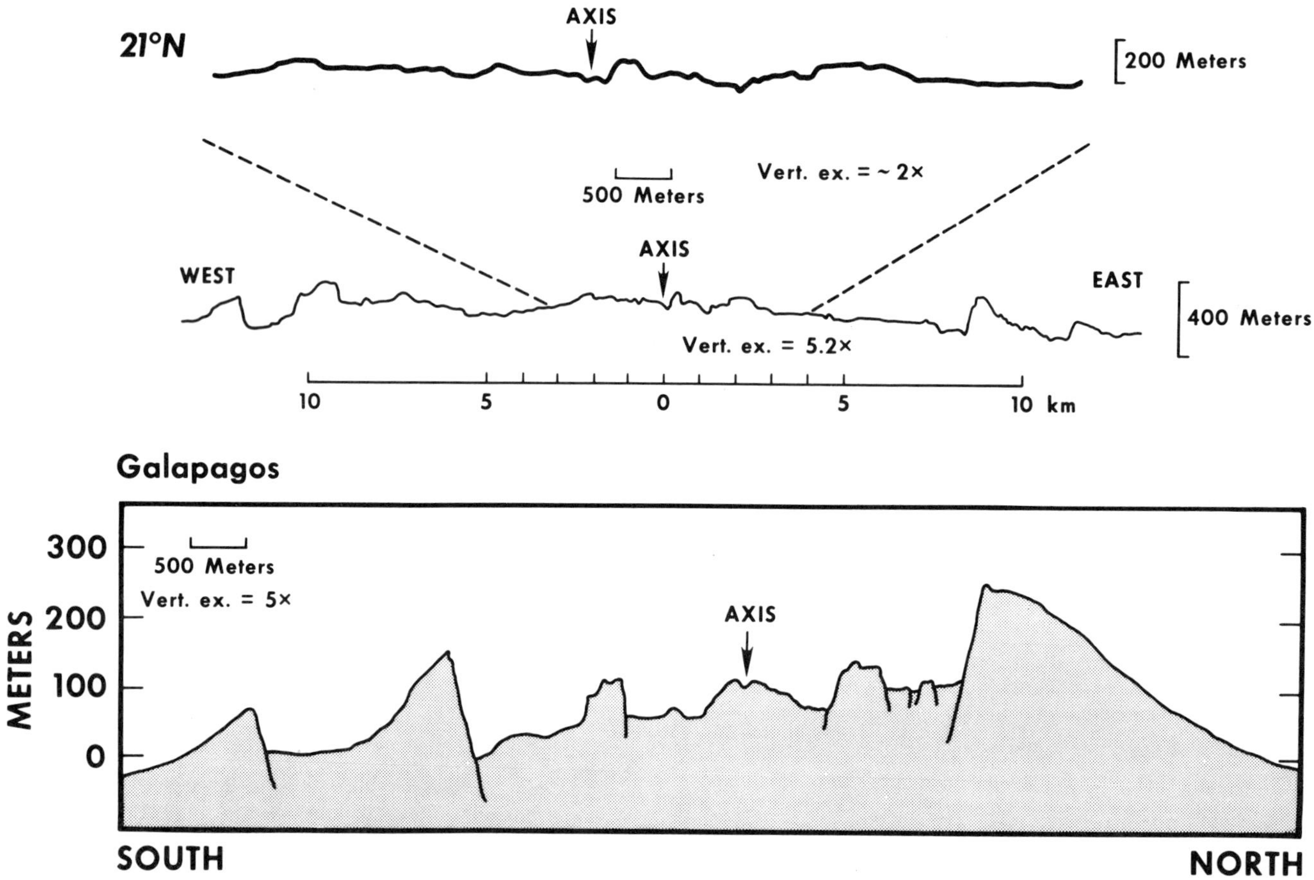

5.3.30 Profiles form East Pacific Rise at 21°N (upper) and Galapagos Rift (lower).

between these faster-spreading ridges and a slow-spreading ridge. These lava plains may be 0.5 km or more in width and may extend 2–3 km along the length of the valley. These plains have been provisionally called "sheet flows" by Ballard *et al.* (1979). Surfaces of some flows commonly consist of flattened pillows, giving a low, hummocky relief; other surfaces are flat, featureless, and divided into polygonal plates 2–10 m in diameter. This flat skin may be deformed and wrinkled to various degrees. This may produce intense crumpling and fracturing of the surface, or may result in whorls and ropy coils (Fig. 5.3.31). The resulting flow-forms strongly resemble the various forms observed on smooth-surfaced subaerial pahoehoe lava and lava lakes (Ballard *et al.*, 1979). The various flow-forms do not necessarily delineate distinct flows, but may grade one into another on the surface of a single flow, and evidently result from different degrees of deformation of an initially smooth skin during progress of the flow. The formation of these sheet flows is attributed to relatively high rates of extrusion compared to that required to form pillowed flows (Ballard *et al.*, 1979).

Collapse pits are common in many of the sheet flows and expose some details of their internal structure (Fig. 5.3.32). Unlike pillow flows, which show the typical "stacked log" appearance in exposed vertical cross section (see Fig. 5.3.27), the sheet flows display a layered structure where exposed on the walls of these pits. The pillow exposures are produced by faulting, whereas these layers are lava "bathtub rings" left as the lava level falls due to withdrawal from the pit. The pits vary in size from a few to hundreds of meters in diameter and 2–25 m in depth. Floors of the pits are partly covered by platy rubble from the collapsed roof. Pillars resembling subaerial tree molds are common along the margins of the pits. These are typically about 1 m in diameter and extend from the floor to the roof of the pit, and remnants of roof lava may form a bridge between the pillar and the adjacent wall.

The pits are reasonably explained as either the result of "drain-away" when a thick, ponded flow breaks through a marginal dam, or as a result of drain-back into a vent. The pillars are tentatively explained as large pipe vesicles or spiracles which formed as trapped water escaped through the marginal, more stagnant portions of the flow (Ballard *et al.*, 1979). These pits and pillars are remarkably similar to the Dimmurborger near Mývatn, Iceland (Barth, 1950), which appears to be a large subaerial basaltic lava blister that collapsed when a channel broke out on the southwestern side of the structure. Columns within the resulting pit show most of the features described by Ballard *et al.* (1979). Examination of several of the Dimmurborger columns by W. B. Bryan in 1977 showed that they have hollow cores and apparently are the source of lava tunnels feeding the basaltic shield prior to its collapse. Drainback of lava into the vents evidently exposed the ponded lava to air, quenching shortly before the collapse event. Water quenching of lava during vent drainback could create similar features on the seafloor.

A series of five active hydrothermal fields were discovered during the 1977 investigation of the Galapagos Rift. These fields vary in area from 400–4000 m^2 and are located along a 2.8 km long section of the axis of the central rift (Crane and Ballard, 1980). All but one of the vents were found in the pillow terrain, the exception occurring at the boundary between pillow and sheet terrain. A subsequent investigation of the East Pacific Rise (Spiess *et al.*, 1980) revealed 25 vents located within a 100–200 m wide zone along about 6 km of the Rise axis. These vents, much hotter (380 ± 30° C) than those along the Galapagos Rift (up to 20° C above

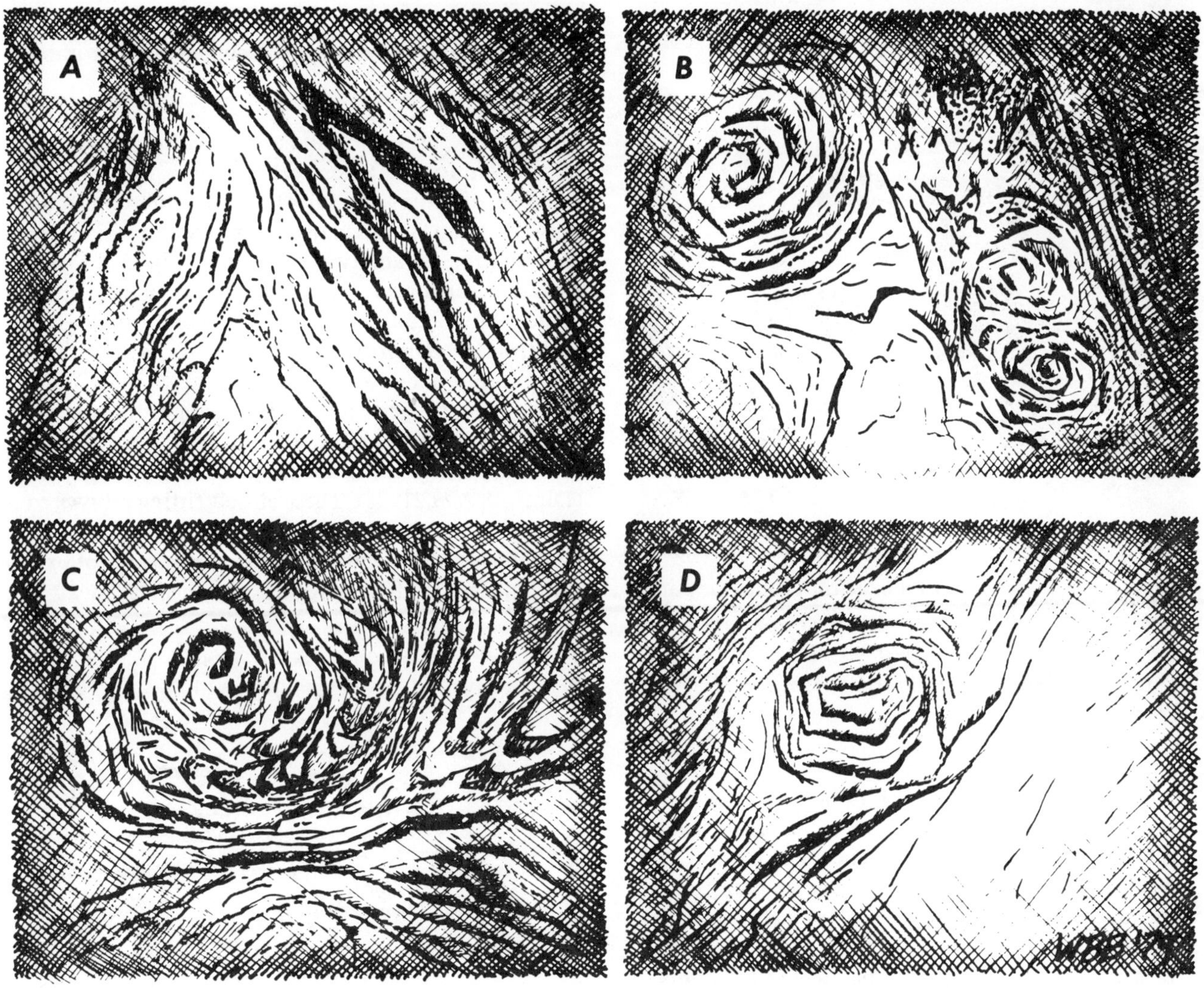

5.3.31 Examples of sheet-flow pahoehoe, Galapagos Rift . **a.** Crumpled surface. **b.** Right-handed (left) and left-handed (right) lava coils. **c.** Large right-handed lava coil. **d.** Boundary between smooth-surfaced pahoehoe (right) and coiled pahoehoe (left; after Lonsdale, 1977).

5.3.32 a. Large collapse pit in lava plain at Galapagos Rift. Note lava pillars and overhanging surface of sheet flow at upper right. Center of pit contains crumpled sheet lava (Fig. 5.3.31a) surrounded by coiled pahoehoe (Fig. 5.3.31b,c). There is a sharp boundary (Fig. 5.3.31d) between coiled and smooth-surfaced lava at the margin of the depression. **b.** Detail of lava pillar and bridging roof at the edge of a lava pit, showing platy internal structure of sheet flow and "bathtub rings" on walls of pit and column)(from Ballard *et al.*, 1979).

ambient), are arranged in *en echelon* groups oriented 15° to the axis. Chimneys, 30 cm in diameter, rise 1–5 m above mounds up to 15–30 m wide and 2 m high, and have associated massive copper, iron and zinc sulphides (Spiess *et al.*, 1980).

Archean pillow lavas

Exposures of extensive pillow lava sections in the Archean are of particular interest to planetary geologists because of the possibility that conditions on Earth at that time were similar to those observed today on the Moon and some other terrestrial planets. Glickson (1976) has suggested that Archean lavas may have formed in an environment similar to the lunar mare lavas. However, the many analogies between Archean lavas and modern seafloor basalts suggest that even at this early time (2.5–3.5 b.y. b.p.), many of these flows must have been erupted into shallow seas. Possibly, these seas may initially have occupied circular impact-induced basins.

Archean lava pillows appear to have formed by budding and branching and are basically cross-sections of tubular forms. Ghost selvages disappear into the interior of pillows, indicating the interconnections between them. Lenticular openings and bridging septa representing "drain-away" features (Fig. 5.3.33), and dynamic sorting of crystals into concentric zones in some pillows imply significant flowage through a tubular structure (Dimroth *et al.*, 1978; Wells *et al.*, 1979). The exceptionally close fit between the Archean pillows (Fig. 5.3.34) suggests that they were highly plastic when emplaced. Locally, pillows with fused, ghost selvages grade laterally into massive basalt along strike (Dimroth *et al.*, 1978). These closely fitting pillows contrast with the generally loose fit observed in modern lava pillows and suggest higher temperatures of extrusion, possibly of somewhat super-heated lava. The

5.3.33 Lenticular gas cavities in glaciated exposure of an Archean lava pillow, Abijevis Hills, Quebec (from Wells *et al.*, 1979). Reprinted courtesy of the University of Chicago Press.

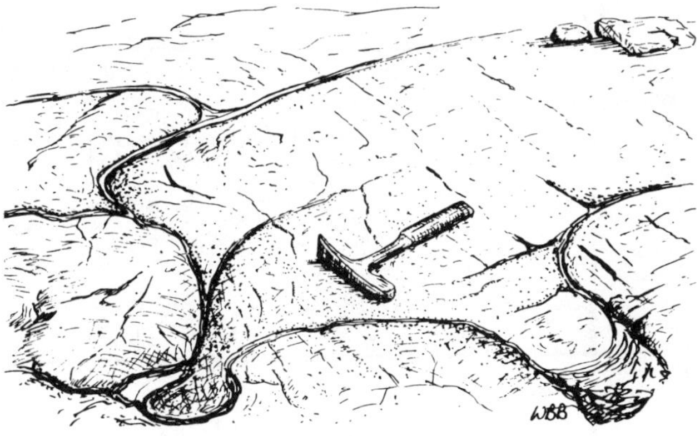

5.3.34 Glaciated exposure of Archean lava pillow, showing very close intricate fit of pillow margins and internal ghost selvage (above hammer) indicative of budding(from Wells *et al.*, 1979). Reprinted courtesy of the University of Chicago Press.

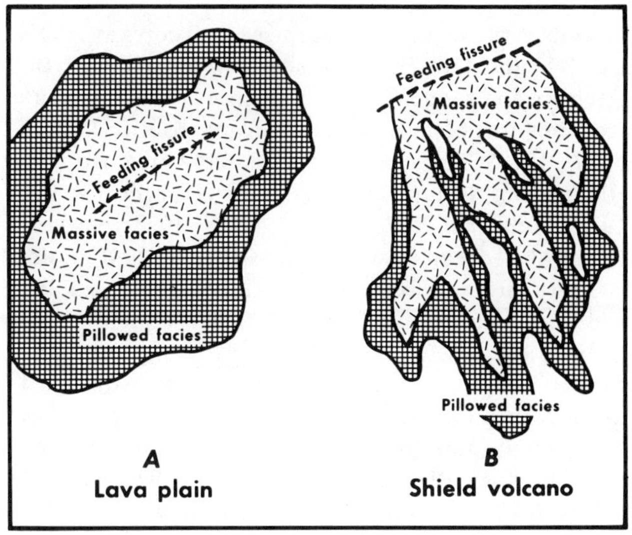

5.3.36 Conceptual plan view of Archean lava flow. **a.** Lava plain resulting from ponding above central fissure. **b.** Elongate flows from fissure on flank of submarine shield volcano (after Dimroth *et al.*, 1978).

Archean pillows also differ from typical modern deep-sea pillows in their generally greater vesicularity, and in their tendency to show concentric zones of vesicles and pipe vesicles near their outer margins (Fig. 5.3.35). In this respect they more closely resemble subglacial pillow lavas. This vesicularity may indicate a generally shallower depth of Archean seas.

Archean pillow lavas are interbedded with massive flows which do not, however, imply subaerial eruption. Wells *et al.* (1979) show that some massive flows have pillowed tops, and show also that the thickness and sequence of interbedded pillowed and massive flows

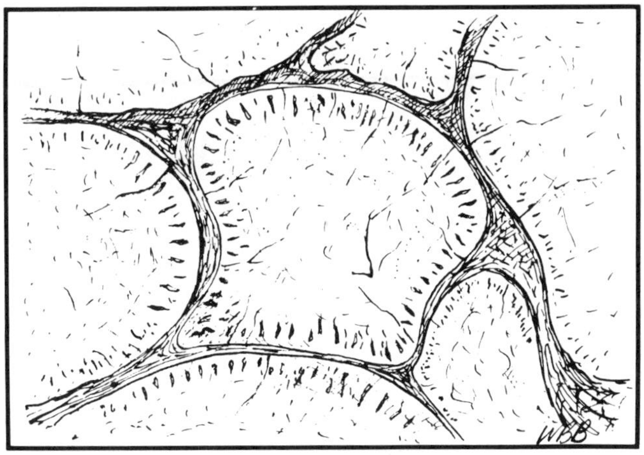

5.3.35 Vesicular Archean lava pillow, showing concentric marginal arrangement of pipe vesicles (from Wells *et al.*, 1979). Reprinted courtesy of the University of Chicago Press.

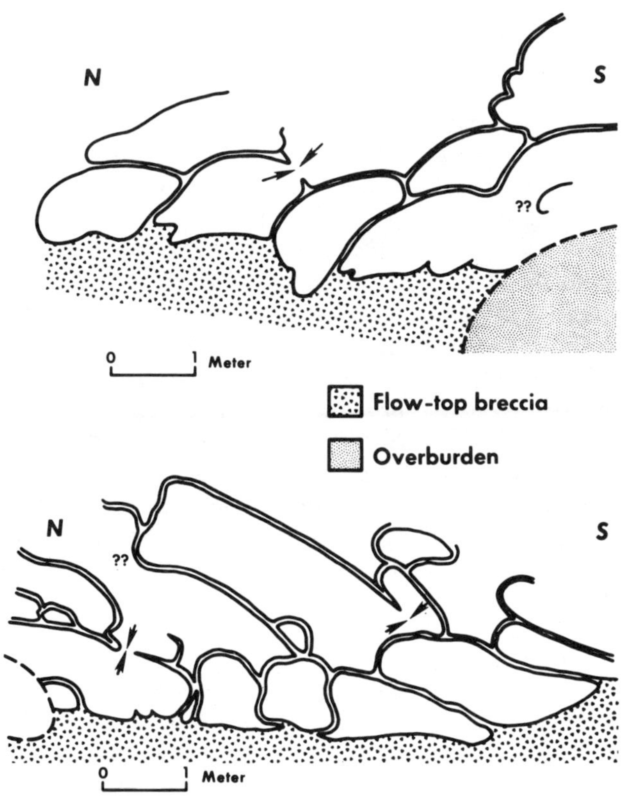

5.3.37 Cross-sectional exposures of foreset Archean lava pillows (after Dimroth *et al.*, 1978). The flow-top breccia is interpreted by Wells *et al.* (1979) as a flow-front breccia which has been over-ridden by the flow.

resembles the sequence in deep-sea drill cores at IPOD sites 417 and 418. These massive flow units in both the Archean and modern basalt sequences probably result from ponding of lava, as on the Galapagos Rift (Ballard *et al.*, 1979). Thickness of these units varies from 2–100 m, but is typically 10–20 m (Dimroth *et al.*, 1978; Wells *et al.*, 1979). Dimroth *et al.* (1978) infer that a typical Archean flow has a massive core and grades laterally into a pillowed marginal facies (Fig. 5.3.36), and they deduce a maximum planar dimension of 10 × 10 km. These pillowed marginal facies appear to consist of foreset pillow tubes similar to those observed in modern pillowed flows (Fig. 5.3.37). Inter-pillow breccia units, interpreted as flow-top breccia (Dimroth *et al.*, 1978) are regarded by Wells *et al.* (1979) as overridden flow-front breccias, by analogy with modern pillowed flows.

5.3.3 Volcanic craters

Craters are common features of terrestrial volcanoes. Volcanism is, however, not the only mechanism which results in the formation of craters, and not all volcanoes have craters. Morphologically similar nonvolcanic craters can be produced by meteorite impact, by collapse in limestone terrains, or by erosional processes such as glaciation. The mode of formation of terrestrial craters can, however, be established by analysis of the associated deposits and rarely by direct observation of the crater-forming event. This has led to a preference for genetic classification schemes for terrestrial craters and their geometric properties have largely been ignored. Nonterrestrial craters, with the exception of those on Io, have not been observed in the process of formation; neither have their associated deposits been subjected to *in situ* analysis. These nonterrestrial craters have therefore been classified on the basis of size, shape, age, and geologic associations. Clearly, it would be extremely profitable if correlations could be made between genetic and geometric classification schemes, since the mode of origin of nonterrestrial craters could then be inferred more confidently.

Figure 5.3.38a is a log/log plot of volcano height vs. crater diameter, showing an unusually large degree of scatter. Does this mean that there is no correlation between these geometric properties, or that trends are obscured because craters with different origins are compared, or are there post-formational modifications of the structures induced by nonvolcanic processes?

Erosion obviously influences both crater diameter and volcano height. Continued erosion will remove material from the crater rim, lowering its height, and increasing (initially) the crater's diameter. Therefore the position of a particular volcano on the height/diameter plot will move diagonally down to the right as the volcano ages, inducing a scatter when fresh and old craters are compared together. Similarly, on the crater depth/crater diameter plot (Fig. 5.3.39a), a volcano's position will move diagonally down to the right as the

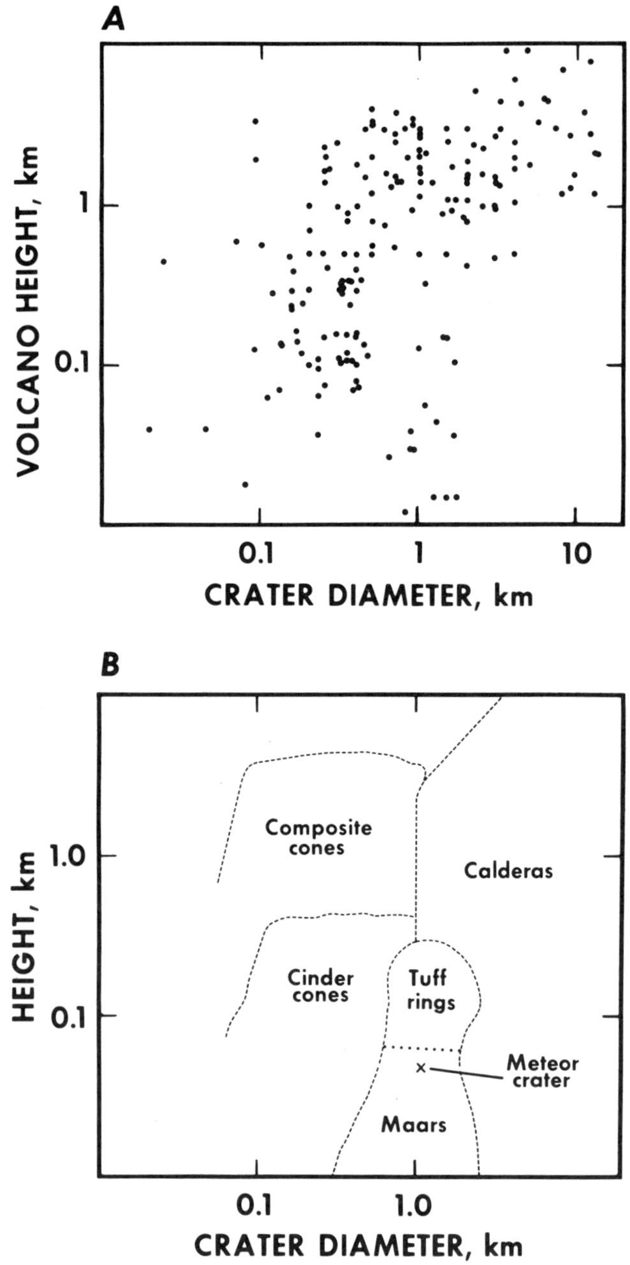

5.3.38 a. Crater diameter/volcano height plot for terrestrial volcanoes. Note the wide range of dimensions. **b.** The data from Fig. 5.3.38a divided into approximate fields associated with different volcano types.

crater enlarges and the floor infills. Exceptions are found when a crater is breached and material is removed from the crater interior (producing a wider, deeper crater with time) and where repeated eruptions infill the floor. The floor could be infilled by a lava lake, the construction of a somma or nested cone, or through resurgence (Smith and Bailey, 1968) of the crater.

Figure 5.3.38b subdivides the crater diameter/height data into fields of different volcano types as classified in the literature. A distinction is often made between craters, calderas and volcano-tectonic depressions, although definitions tend to be somewhat arbitrary. Calderas were defined as being craters greater than 1 mile in diameter (e.g., Macdonald, 1972) though now they tend to be defined as being greater than 1 km in diameter (Smith, 1979). Volcano-tectonic depressions are generally larger than 25 km in diameter. Table 5.3.5 summarizes characteristic features and modes of origin of various volcanic craters.

Calderas in basaltic shield volcanoes have been observed to form in only a few instances. In the case of Kilauea, withdrawal of magma from a shallow-level reservoir caused subsidence at the volcano summit. The magma itself was intruded into the rift zone to form

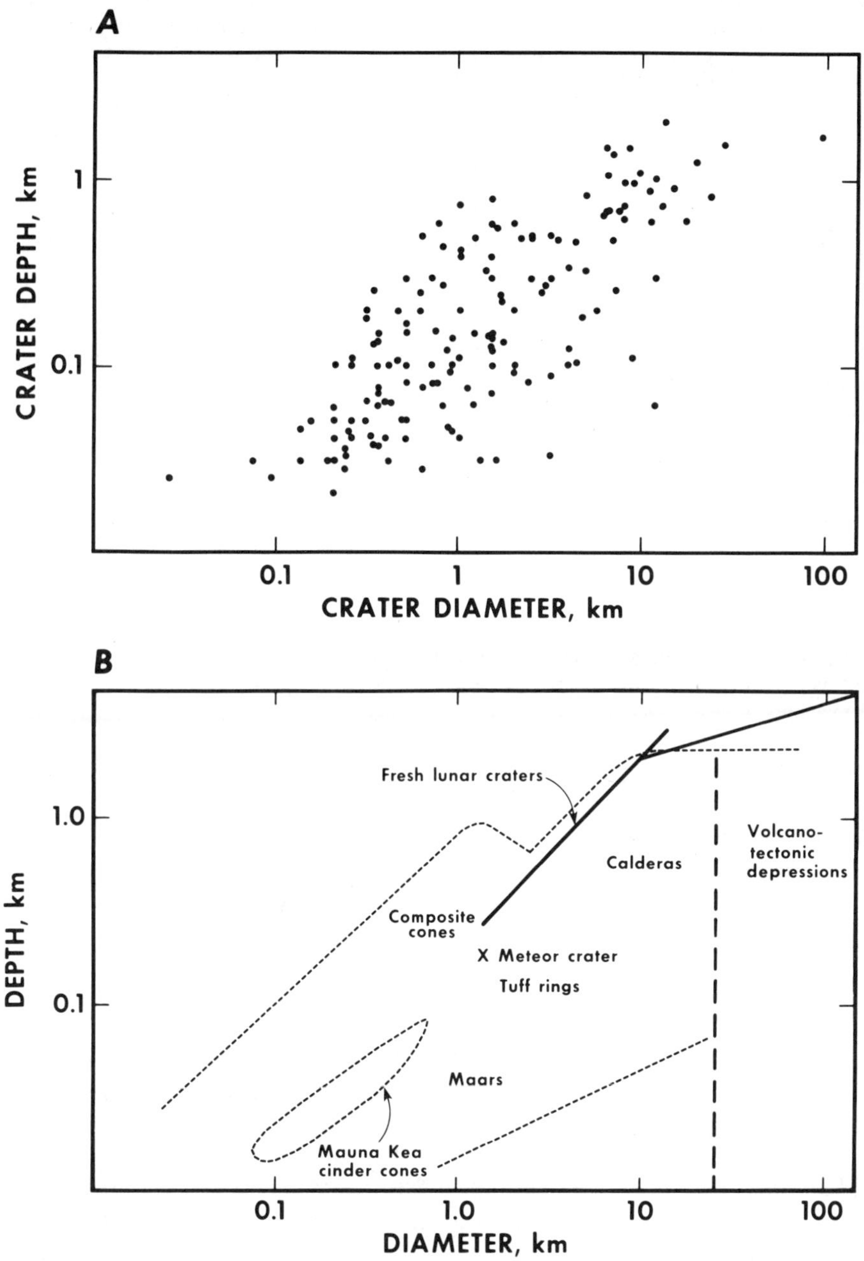

5.3.39 a. Crater depth/crater diameter plot for terrestrial volcanoes. These are not necessarily the same volcanoes as in Fig. 5.3.38a. **b.** Fields of different volcano types superimposed on the crater depth/diameter plot. The curve for fresh lunar craters is from Pike (1974) and the field of Mauna Kea cinder cones is from Porter (1972).

Table 5.3.5 Summary table of characteristics of volcanic craters and calderas.

Feature	Description	Origin
Craters	Normally less than 1 km diameter (maars can be up to 2.5 km): subcircular in plan view if formed from single vent, but elliptical or clustered craters if more than one vent. Depths variable; maar craters have floors below surrounding terrain; most others are at top of volcanic construct such as cinder cones.	Excavated by volcanic explosions, typically strombolian but also plinean and phreatomagmatic. Enlargement of crater by collapse and rim-slumping both during and after eruption.
Pits	Normally a few tens to a few hundred meters diameter. Subcircular to elliptical in plan view; shape usually reflects tectonic trends. Depths variable, exceptionally up to 1 km. Usually no raised rim. Often occur on rift zones (e.g., Hawaii) but also at volcano summit (e.g., Etna).	Formed entirely by collapse, presumably as result of magma withdrawal. Sometimes the site of repeated lava effusion long after initial collapse. Raised rims may form subsequent to collapse by accumulation of spatter or ash from fire fountains and strombolian explosion in pit.
Calderas	Usually greater than 1 km and less than 25 km diameter for basaltic volcanoes. Elliptical to irregular in plan view; rarely subcircular. May form single features or complex of calderas. Mostly occur over central conduit of volcano.	Major collapse, presumably as result of magma drainage from central conduit, often related to large flank eruption. Repeated collapse commonly occurs in the same caldera, or clusters of calderas can form.
Flank Depressions	Usually between 5–10 km diameter, occurring on the flanks of steeper volcanic edifices (slopes > 20°). Depression open on downslope often with "keyhole" shape in plan view (e.g., Valle del Bove, Etna; Orotava Valley, Tenerife).	Considered to be formed by combination of collapse and gravity sliding of volcano flank.
Sector Graben	Linear fault-bounded depressions radial to the volcanic edifice; often cut caldera rim.	Normal faulting giving graben-like troughs presumably as result of withdrawal of magma in a rift zone or tensional movement associated with volcanic activity.
Pseudocraters	Crater-like structures found in volcanic terrains which lack vents. Typically less than 80 m wide and 15 m high.	Caused by explosive reaction between lava and surface water or groundwater.

dikes and emerged lower down the volcano flanks from the rift zone to generate lava flows. In the 1968 collapse of Fernandina caldera (Galapagos) little new basalt was erupted, but substantial collapse (volume 1–2 km^3) occurred as a series of short drops over a period of 12 days (Simkin and Howard, 1970). The collapse was attributed to magma injection into sills and effusive eruptions from circumferential fractures. The large basaltic calderas of the Iceland central volcanoes have also been interpreted as being due to lateral magma injection along the rift zones and collapse of the calderas along ring fractures (Sigurdsson and Sparks, 1978).

Many small craters associated with basaltic eruptions are the result of explosive activity generated where magma contacts water. Depending on the extent of interaction, craters from 50 m to over 1 km in diameter and with depths from a few to over 100 m can be excavated. (Such craters are often termed maars.) Kienle *et al.* (1980) have recently documented the formation of two explosion craters of maar-type during the eruption of alkali basalt in Alaska in 1977.

The problems of employing purely geometric criteria to differentiate between volcanic and nonvolcanic craters are illustrated by Fig. 5.3.39b. From this figure it can be seen that both fresh lunar craters of impact origin and the impact-formed Meteor Crater, Arizona, fall within the same field as volcanic calderas and tuff rings. A multivariate approach to this geometric analysis of craters, such as that of Pike (1974), is required to separate craters of varied origins. Table 5.3.6 contrasts the properties of volcanic craters and those formed by meteorite impact.

Little information is available on either the growth or erosion of volcanic craters. The most comprehensive compilation of crater and volcano dimensions to date is that of Pike (1978).

To summarize, volcanic craters are the products of an interplay between a large number of forces including the eruption velocity, eruption rate, ejecta composition and character, wind strength and erosional processes.

5.3.4 Subvolcanic processes and their surface expression

Terrestrial erosion often exposes the feeder systems of subaerial volcanoes. The surface expression of these feeders can be one of positive or negative relief depending upon the ease of erosion in comparison to that of surrounding country rocks. It must be remembered, however, that not all subsurface magma bodies feed surface volcanics (e.g., Walker, 1975b). Because erosion appears to have been most important on the Earth (compared to the other terrestrial planets), only a brief description of subvolcanic features is given here. The mechanisms of formation of these features are presented in section 6.6.

The subsurface bodies can be conveniently subdivided into linear-tabular and circular-globular groups. The linear-tabular group includes dikes and sills—both most commonly associated with tensional environments such as divergent plate margins. In Ethiopia and Iceland, dikes have a mean thickness of 3.5 m, lengths varying from meters to many kilometers, and can achieve intensities of 120 m of dike to 1 km of country rock (Walker, 1960; Mohr, 1971). In the Troodos massif, Cyprus, thought to represent uplifted suboceanic crust, dikes comprise 90–100% of the sheeted complex (Gass and Smewing, 1973). By way of contrast, the ultrabasic Great Dike of Rhodesia has a length in excess of 500 km and a width of about 6 km (e.g., Hughes, 1970). Dikes frequently occur in large clusters or zones and often exhibit *en echelon* patterns and curving, pinched extremities (P. Delaney, pers. comm., 1980). Recent activity at Krafla, Iceland, indicates that magma is episodically being injected along the rift system for distances of the order of 30 km (e.g., Einarsson, 1979), the area above the dikes being marked by fissures. Table 5.3.2 lists some of the major dike swarms and basalt plateaus and greenstones of comparable age. Sills are intruded conformable to bedding planes in sedimentary sequences, commonly giving rise to a stepped topography. The Whin Sill of northern England has an average thickness of 25–30 m and an estimated area of 5000 km^2 (Taylor *et al.*, 1971). Frequently, both dikes and sills exhibit evidence for multiple episodes of magma passage.

Circular and globular subvolcanic bodies include ring dikes and cone sheets (Fig. 5.3.40 Reynolds, 1956; Elston and Erb, 1981), laccoliths (Johnson and Pollard, 1973), lopoliths (Wager and Brown, 1968), diatremes (Lorenz *et al.*, 1970) and plugs and necks (P. Delaney, pers. comm., 1980). Diatremes, plugs and necks apparently represent the infilled vents of extinct volcanoes. Ring dikes and cone sheets appear to be related to the ring faults associated with caldera formation (Clough *et al.*,1909). Lopoliths, laccoliths, and other layered mafic intrusions may represent subvolcanic magma storage areas. These structures vary in size from diameters of a few meters to about 65,000 km^2 in the case of the Bushveld (Wager and Brown, 1968).

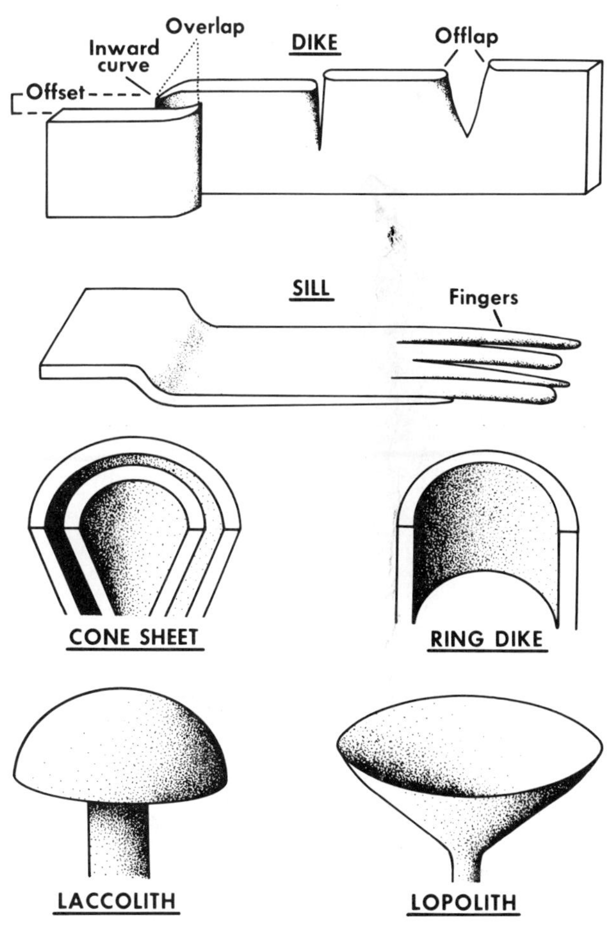

5.3.40 Idealized 3-D forms of intrusive bodies.

Table 5.3.6 Contrasting properties of impact and volcanic craters/calderas.

Property	Volcanic	Impact
I. Morphometry		
A. Plan	• Ranges from circular to elliptical and irregular; strongly dependent on source geometry and regional structure.†	• Generally circular although at small sizes (D < 1 km) noncircular plans commonly occur and reflect projectile properties (low velocity, multiple body, oblique impact) or target properties (complex stratigraphy, topography); at large sizes (D > 15 km), slumping controlled by topography and structure produces scalloped plans.
	• Breached rims result from rim collapse, peripheral eruptions, and lava rafting.	• Breached rims result from low-velocity impact and multiple impacts, and flooding by externally derived lava for small craters (D < 1 km).
	• "Nested" craters reflect multiphased eruption.	• Concentric plans reflect layering in target; at large sizes (D > 100 km), concentric plans reflect broad-scale crustal adjustment, crustal layering, or shock-induced phenomena.
B. Profile	• Floor lies from below (maars) to above (cones) surrounding terrain.†	• Floor lies below surrounding terrain roughly in proportion to crater size up to approximately 15 km in diameter, above which crater floor is typically 1–3 km below surrounding terrain. Unusually shallow floors can result from low-velocity impacts, unusual targets, or infilling by impact debris and lava.
	• Wall slope may be vertical (collapse) or equal to angle of repose.†	• Wall slopes generally approach angle of repose for craters with D < 15 km with lower mean slopes for larger craters with slumping. Secondary craters commonly exhibit more subdued profiles.
	• Rim relief ranges from essentially zero to edifice relief.†	• Rim relief approximately scales with crater size although contrasting layered targets can produce notable variation in small size craters (D < 1 km).
	• Central or acentral constructs may exceed rim crest height.	• Central positive relief much less than height of rim crest for most craters of all sizes.
II. Morphology		
	• Wide variety of floor from lava-filled plains to "bottomless" pits; commonly littered with talus or fallback.	• Floors typically littered with blocky debris (D < 1 km); ruffled and slumped debris (1 km ≲ D ≲ 30 km); flat floors with impact-melt plains units (20 km ≲ D) and lava.
	• May exhibit subsequent vent constructs commonly along floor margin.	• In large craters (D > 20 km), pseudo-vents and mantling deposits may result from cooling melt-capped floors.
	• Walls may exhibit vertical cliffs, ash-draped relief.	• Block strewn walls (D < 10 km); slumped or landslide/melt-dissected walls (D ≳ 15 km).

†Useful diagnostic criteria.

Table 5.3.6 *(Concluded)*

Property	Volcanic	Impact
	• Ballistic ejecta deposits are typically fine-size unshocked ash or spatter and limited in ballistic range, reflecting launch velocities <0.5 km/sec. Wind-borne material may be carried large distances on planets with atmospheres (typically asymmetric distribution).†	• Variable shocked ejecta exhibit wide range in size (partly dependent on target) and extend to enormous ranges reflecting velocities > 1 km/sec; large craters (D > 10 km) typically exhibit secondary craters with maximum size ~5% parent crater diameter (typically symmetric distribution).
	• Flank eruptions produce peripheral constructs and flows that are topographically controlled; lava volumes may be large.†	• Impact melt occurs as small ponds or leveed channels exhibiting topographic control; not associated with well-defined vents and commonly exhibit evidence of considerable horizontal momentum. Although most noticeably overlying ejecta deposits, complex stratigraphic relations indicate emplacement contemporary with ejecta.
III. Occurrence		
	• Typically located in regionally or locally active volcanic areas.†	• Independent of regional setting.
	• Typically occur as part of a long-term sequence of volcanic events.†	• Independent of geologic history.
	• Commonly occur as a member of a group of volcanic features, including other craters; may be tectonically controlled.	• Generally independent events with the important exceptions of meteoriod swarms and secondary craters; independent of tectonic control.

†Useful diagnostic criteria.

The relative volumes of intruded and extruded magma produced during eruptions are difficult to estimate since it is not possible to observe both at the same time. Walker (1975b), however, has estimated that more than half of the uprising magma in Iceland has been diverted to form dike swarms. This has important consequences for determining the volumes of material produced during eruptions on other planets where, usually, only the extrusive products are observed.

5.4 BASALTIC VOLCANISM ON THE MOON

5.4.1 Introduction

Information concerning basaltic volcanism on the Moon comes from several sources. Compositional and age information is available for several parts of the Moon from analysis of samples returned by Apollo and Luna Missions (see Chapter 1). Additional compositional information is obtained from experiments carried in spacecraft orbiting the Moon and from Earth-based telescopic observations (see Chapter 2). Data on the physical properties of lunar surface materials have been derived from surface exploration, orbital experiments, and Earth-based observations. Images of the Moon at visible wavelengths have been extremely useful in defining the morphology of volcanic deposits and their stratigraphic relationships. Reviews of the application of these data to the geologic mapping of the Moon can be found in Wilhelms (1970) and Head *et al.* (1978b). Emphasized here are those aspects of basaltic volcanism that can be understood through the study of morphology and geologic relationships. Other

Table 5.4.1 Techniques employed in the characterization of the properties of the lunar surface.

Type of Data	Technique	Wavelength	Information Obtained	Maximum Resolution	Area	Inception Date
Chemistry	Alpha particle (orbital)	4.7–8.8 MeV	^{222}Rn, ^{210}Po	5°	Apollo 15, 16 groundtrack	1971
	gamma-ray (orbital)	0.5–8.0 MeV	K, Th, U, Fe, O, Si, Mg, Ti, Al	60 × 60 km	Apollo 15, 16 groundtrack	1966
	X-ray fluorescence (orbital)	0.5–2.75 KeV	atomic numbers < 14	13 km	Apollo 15, 16 groundtrack	1971
	reflection spectroscopy	0.13–2.5 μm	mineralogy, glass content, TiO_2,FeO	4–20 km	nearside	1909
	color photography	0.35–0.82 μm	TiO_2	2–3 km	nearside	1910
	line scan maps	0.5–0.8 μm	TiO_2	30 km	nearside	1970
	digital vidicon	0.35–1.05 μm	TiO_2	2 km	nearside	1972
	radar (earthbase)	3.8 cm 70 cm	TiO_2?	500 m	nearside	1975
	polarimetry	0.3–1 μm	basic-acid	25 km	nearside	1927
	returned samples	-	everything	-	Apollo & Luna sites	1969
Age & Maturity	crater counting	-	density of craters	1 m	whole Moon	1907
	D_L	-	crater degradation	1 m	whole Moon	1972
	photometry	0.5–0.6 μm	albedo/age variation	1–5 km	nearside	1847
	returned samples	-	isotopic dating	-	Apollo & Luna sites	1969
	telescope	0.4–0.7 μm	areal dimensions & altitude	500 m	nearside	1610
	photometry	-	altitude?	20 km	nearside	1951
Morphological	Earth-based photography	0.4–0.8 μm	areal dimensions & altitude	500 m	nearside	1840
	orbital photography	0.4–0.8 μm	areal dimensions & altitude	1 m	whole Moon	1959
	orbital laser altimeter	0.69 μm	altitude	±10 m	Apollo 15, 16, 17 groundtrack	1971
	Lunar Sounder	2 m, 20 m, 60 m	altitude & subsurface structure		Apollo 17 groundtrack	1972
	radar interferometry	3.8 cm	altitude	1 km	nearside	1971
Physical & Textural Properties	U.V. spectroscopy	1180–1680Å	refractive index, surface structure	10 Å	Apollo 17 groundtrack	1972
	polarimetry	0.3–1 μm	refractive index, porosity, particle size, shadows	25 km	nearside	1811
	photometry	0.5–0.6 μm	porosity, particle size, slope distribution, albedo	1–5 km	nearside	1847
	infrared radiometry	3–1000 μm	refractive index, surface structure	20 km	nearside	1930
	scanning infrared radiometric (orbital)	1.2–70 μm	surface structure	2–2.6 km	Apollo 17 groundtrack	1972
	Earth-based radar	3.5 cm, 23 cm, 70 cm, 75 cm, 185 cm, 7.5 m	dielectric constant, thermal conductivity, porosity & slope distribution	500 m	nearside	1946
	orbital bistatic radar	75 cm, 185 cm	dielectric constant, thermal conductivity, porosity & slope distribution	5–10 km	Apollo-Explorer-Orbiter I groundtrack	1968 1966
	returned samples	-	grain size, porosity	-	Apollo & Luna sites	1969

data sets do, however, form an indispensable framework for the interpretation of basaltic volcanism on the Moon (Table 5.4.1).

The lunar maria, because of their distinctive albedo and morphology, are the most striking volcanic deposits and are treated first. Additional nonmare deposits may be basaltic in composition, but much of their morphology has been obscured or destroyed by the early cratering history of the Moon. Morphologic features associated with each type of volcanism and the geometry of the deposits are then outlined. These characteristics and relationships are used to determine the nature of eruption conditions, and then a framework for the volcanic history of the Moon is developed.

5.4.2 The lunar maria

Identification as basalts

The return of samples from the lunar maria established the basaltic composition of mare deposits at the sample sites. In addition, spectral reflectance and other remote-sensing data have shown that the lunar maria elsewhere are basaltic in nature, although there are compositional variations (Papike *et al.*, 1976; Pieters, 1978; see also Chapter 2).

Distribution and geometry of mare deposits

Lunar mare deposits, identified on the basis of albedo and spectral reflectance, are exposed over an area of about 6.3×10^6 km^2, or approximately 17% of the total lunar surface area (Head, 1975b; Fig. 5.4.1). There is a fundamental asymmetry in the distribution of post-Imbrium mare deposits with the vast majority occurring on the Earth-facing hemisphere.

Mare materials are primarily located in three types of regions (Fig. 5.4.2): (1) as fillings in the centers of the large circular multiringed basins (mascon basins such as Imbrium and Serenitatis) or in troughs concentric to the basin center (such as Lacus Veris and Autumni in the Orientale basin); (2) as flooding of large irregular depressions (such as Tranquillitatis or Oceanus Procellarum); and (3) as small patches in the highlands or on crater floors (such as in Archimedes or Plato).

Impact basins are topographic depressions formed early in lunar history (Howard *et al.*, 1974). The mode of occurrence of mare deposits (Fig. 5.4.2) reflects this basin-dominated topography (Fig. 5.4.3). The filling of individual mare basins was largely determined by basin geometry. Old degraded basins (Tranquillitatis, Fecunditatis) exhibit shallow mare fill of generally similar composition. More detailed analysis invariably

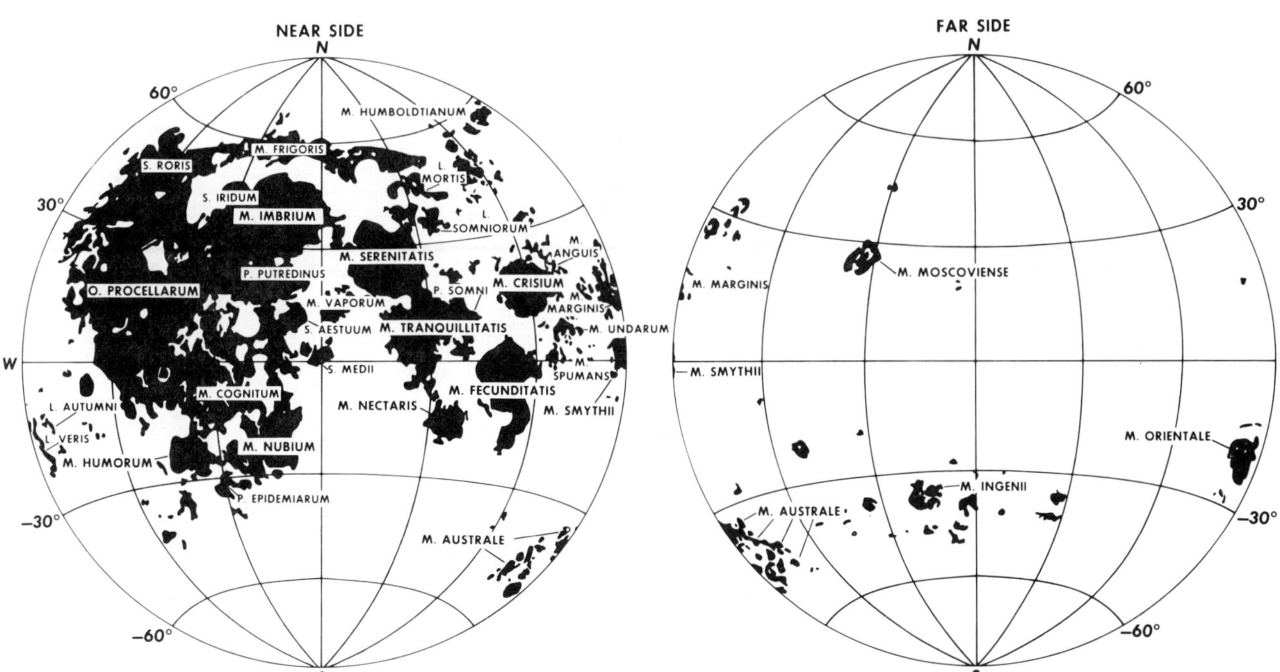

5.4.1 Distribution of mare deposits on the lunar near- and farside. M denotes mare; O, oceanus; S, sinus; P, palus; and L, lacus (Head, 1976).

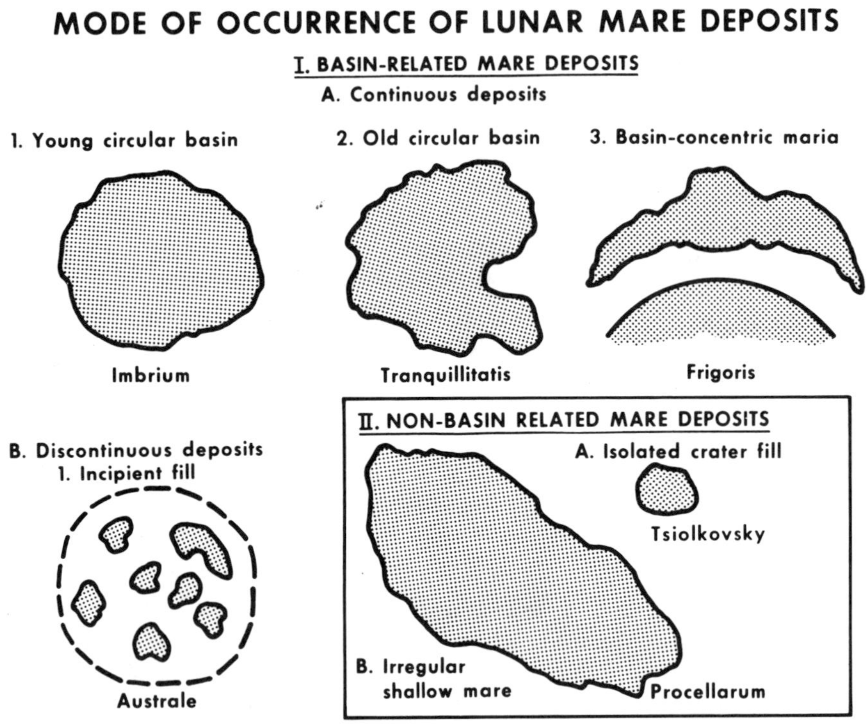

5.4.2 Classification of mode of occurrence of lunar mare deposits according to their location, shape, and areal distribution. Names indicate well-developed examples of mare types (Head, 1975b).

shows a complex array of subunits illustrating a complex filling process. Younger basins (Imbrium, Humorum, Serenitatis) exhibit this style of initial fill, but also display extensive younger surface flows and deposits that often originate along basin margins and tend to flood the basin interiors to varying degrees. The process of filling of mare basins took place over extensive periods of time (~0.5–1.5 b.y.) and produced complex deposits. The response of the lunar lithosphere to the early mare lava loads (downwarping) often caused topographic changes resulting in later lava accumulations in the basin centers (Solomon and Head, 1979, 1980).

Determination of the thickness and geometry of various mare units and sequences is of interest because of volume implications for thermal history, indications of depth of major basins, the loading history of the lithosphere, and the interpretation of seismic and gravity data. Detailed work in numerous areas provides data on gross geometry of mare deposits with estimates of 1–2 km thickness for irregular maria such as Tranquillitatis (DeHon, 1974), and over 4 km for central basin fill in areas such as Mare Imbrium (Hartmann and Wood, 1971; DeHon and Waskom, 1976; Hörz, 1978; Head, 1979a). Furthermore, reduction of data from the Lunar Sounder experiment flown on Apollo 17 has provided evidence for subsurface layering extending to depths of several kilometers in the maria (Phillips *et al.*, 1973; see also Table 8.9.1).

Studies of the nature of volcanic filling of cratered terrain and basins have confirmed that there are large local variations in lava thickness on the Moon (Head, 1979a, b). Although there is some disagreement concerning the actual thickness in specific parts of the Moon, no volume estimates exceed $\sim 10 \times 10^6$ km^3 (Head, 1975b), a value that is less than 1% of the total volume of the lunar crust. There remain two further, but volumetrically uncertain, contributions that could modify this statistic. First, possible ancient basalt surfaces now buried by impact-derived debris (Schultz and Spudis, 1978, 1979) could raise the total volume of extruded basalts. A second contribution is represented by shallow mafic reservoirs within the lunar crust. Such reservoirs are inferred to occur beneath numerous circum-mare craters and basins, with surface basalts representing a relatively thin veneer (Whitford-Stark, 1974; Schultz, 1976a). Consequently, although the relative volume of exposed surface basalts to total crustal volume is only 1%, the relative volume of exposed, buried, and intrusive basaltic materials may be significantly higher.

Age relations

The span of mare basalt ages determined from returned samples is 3.16–3.96 b.y. and is discussed in Chapter 7. Dating techniques using craters strongly suggest that volcanism extended past this time in Mare Imbrium and Oceanus Procellarum into the period

Overleaf:

5.4.4 Volcanic and tectonic map of part of the nearside of the Moon mapped from Orbiter IV high resolution photos. The domes in the Marius Hills have been omitted for clarity. (From Guest and Murray, 1976).

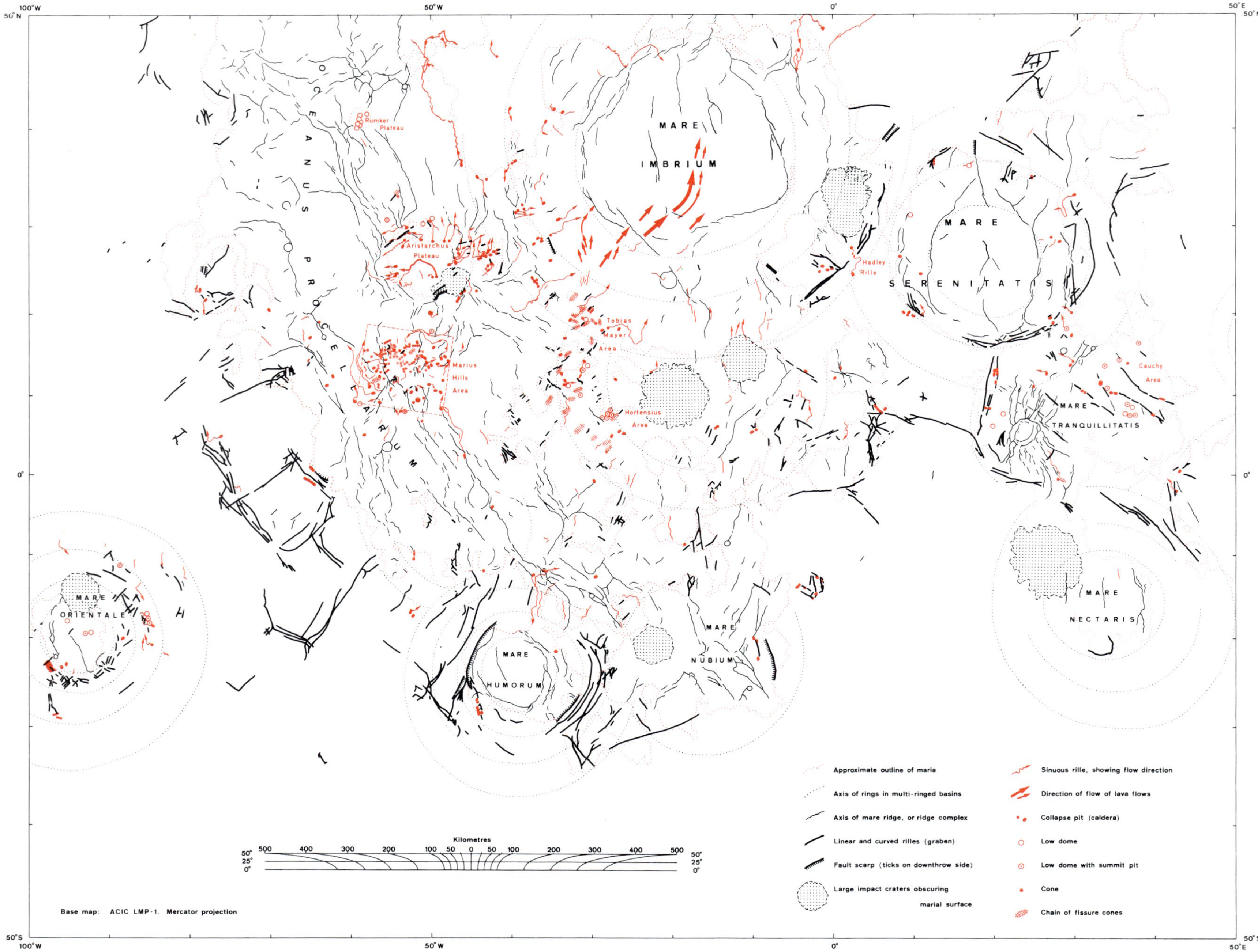
100°W
50°W
0°
50°E
50°N
0°
50°S
MARE IMBRIUM
OCEANUS PROCELLARUM
MARE SERENITATIS
MARE TRANQUILLITATIS
MARE NECTARIS
MARE NUBIUM
MARE HUMORUM
MARE ORIENTALE
Rumker Plateau
Aristarchus Plateau
Marius Hills Area
Tobias Mayer Area
Hortensius Area
Hadley Rille
Cauchy Area
Approximate outline of maria
Axis of rings in multi-ringed basins
Axis of mare ridge, or ridge complex
Linear and curved rilles (graben)
Fault scarp (ticks on downthrow side)
Large impact craters obscuring marial surface
Sinuous rille, showing flow direction
Direction of flow of lava flows
Collapse pit (caldera)
Low dome
Low dome with summit pit
Cone
Chain of fissure cones
Kilometres
500 400 300 200 100 50 0 50 100 200 300 400 500
50° 25° 0°
Base map: ACIC LMP-1. Mercator projection

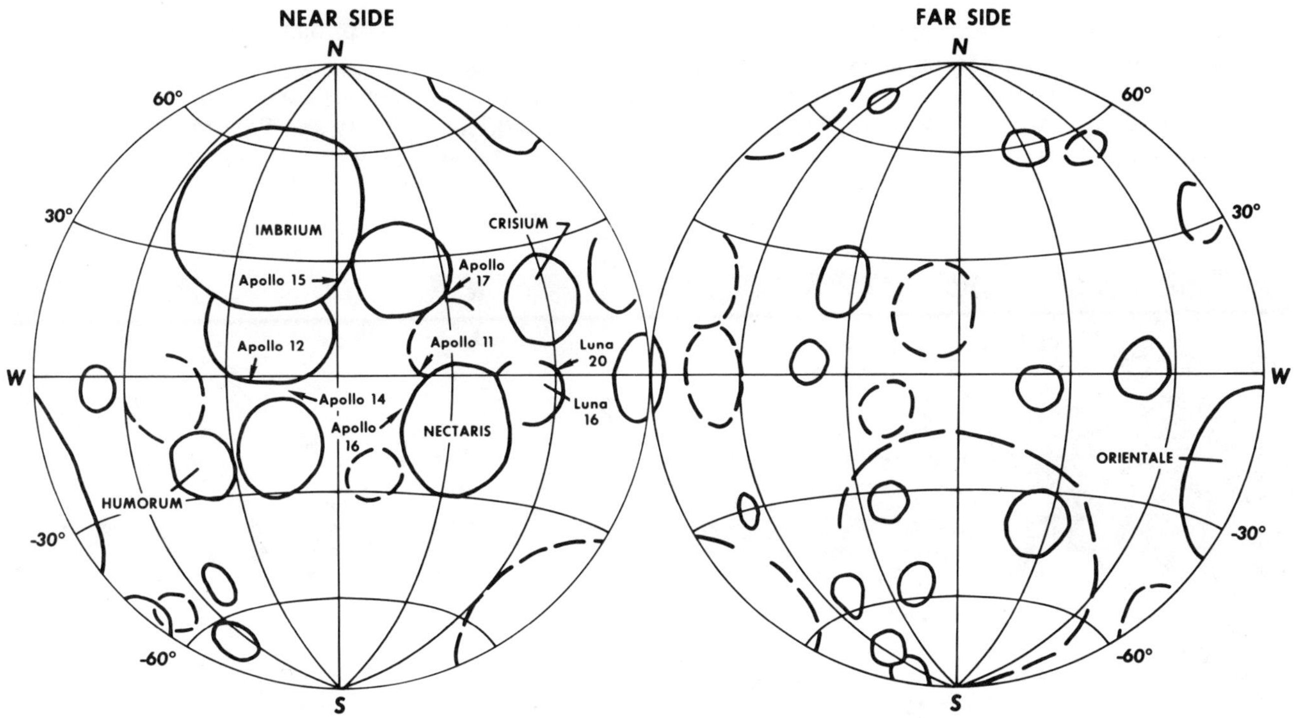

5.4.3 Distribution of basins on the lunar near- and farside. Enclosing line delineates approximate position of major topographic prominence that defines the basin positions (after Stuart-Alexander and Howard, 1970; Hartmann and Wood, 1971; and Howard *et al.*, 1974).

between 2–3 b.y. ago (see Chapter 8). The onset of mare volcanism is obscured by the impact record, but mare basalt clasts in breccias produced by basin-forming events provide evidence for the beginning of mare volcanic activity prior to the formation of the Imbrium basin (Ryder and Taylor, 1976). In addition, recent studies of dark-halo impact craters, which have excavated materials from beneath basin ejecta (Schultz and Spudis, 1979), provide evidence for mare deposits emplaced prior to the end of the heavy bombardment.

5.4.3 Mare volcanic structures

The morphology of features associated with the lunar maria provides evidence for the eruption styles and emplacement mechanisms of the lavas. This section describes the major morphologic features of the lunar maria and their distribution (Fig. 5.4.4. See Head, 1976, and Schultz, 1976b, for more extensive description and discussion).

Flow fronts

One of the most striking features associated with the maria are long, sinuous scarps which have been interpreted as lava flow fronts. Those in Mare Imbrium were sufficiently prominent to be seen during Earth-based telescopic observation (Strom, 1965; Wilhelms, 1970; Schaber, 1969), and subsequently, high-resolution photography of them was obtained by Lunar Orbiter and Apollo missions (Fig. 5.4.5). Photogrammetric measurements of the youngest flow scarps in Mare Imbrium indicate average scarp heights of 30–35 m with a range from 10–63 m (Schaber, 1973a, b). These flow scarps are the boundaries of a series of relatively young lava flows, which appear to have a common source in southwest Mare Imbrium near the crater Euler (Schaber, 1973a, b).

5.4.4 Volcanic and tectonic map of part of the nearside of the Moon mapped from Orbiter IV high resolution photos. The domes in the Marius Hills have been omitted for clarity (from Guest and Murray, 1976).

The vast majority of lunar mare deposits are older than the youngest Mare Imbrium flows and, although they often exhibit color boundaries, only occasionally exhibit flow scarps. Very low sun (near-terminator) photography obtained on Apollo missions has shown occasional flow fronts in older deposits (Gifford and El-Baz, 1978).

Using morphometric data from the Imbrium flows and considering viscosity and flow rates of lunar and terrestrial basalts, Schaber (1973a), has shown that:

A

B

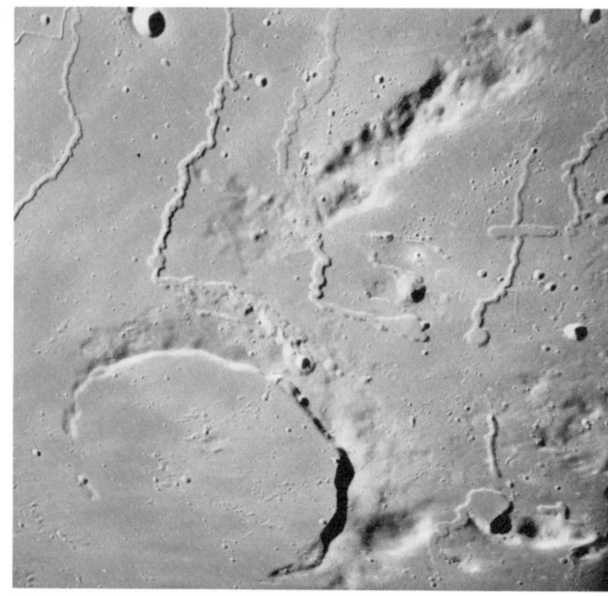

C

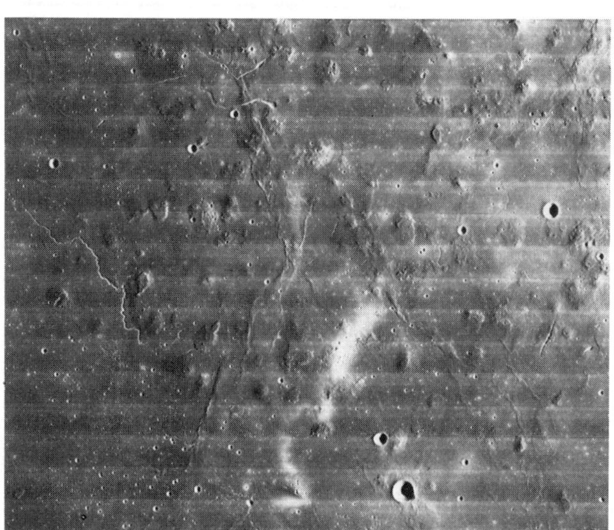

D

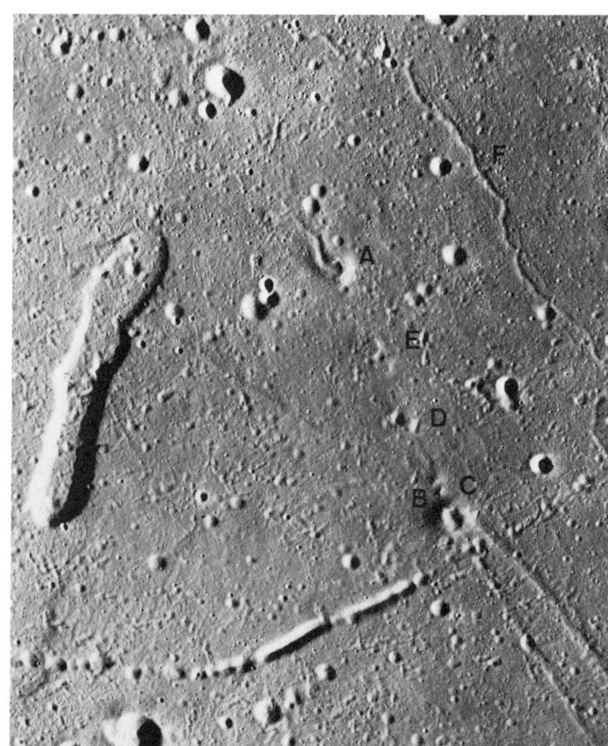

E

5.4.5 a. Lava flow lobes and flow fronts in Mare Imbrium. This oblique view of Imbrium looks north and the flow direction was from lower left to upper right. The flow lobes are 10–20 km in width and the two large craters on the horizon are about 35 km apart (NASA photo AS-15-1556). **b.** Sinuous rilles in the Harbinger Mountain area. Note the circular or elongate depressions at the head of each rille. These particular rilles are among the largest on the Moon, averaging 1–2 km width and 100–300 m in depth (Strain and El-Baz, 1977; NASA photo AS 15-12608). **c.** Lunar Orbiter IV photograph of part of the southwest Marius Hills showing steep domes, low-profile domes, cones, and sinuous rilles. Width of photograph approximately 120 km. **d.** Cinder cones (A and B), volcanic mounds (C, D, and E) and a rille (F) in Mare Serenitatis (from Scott, 1973; NASA photo AS 17-2317). **e.** Mare ridges and arches in southern Oceanus Procellarum. A major series of ridges extends from upper left to lower right. Most ridges are associated with broad arches. Width of view is about 190 km. Highland area in the lower left is the rim of Letronne (NASA photo AS 16-2839).

(1) about 1.7 times greater flow thickness on the Moon than on Earth would be required to produce a given flow length because of the lower lunar gravity; (2) a mathematical model employing lunar gravity, a slope of 1:1000, and a temperature of 1250°C as parameters result in calculated flow heights of 30.6 m for 200 km long flows and 35 m for 300 km flows, agreeing with the average flow heights for the youngest Imbrium lavas; (3) flow lengths on the Earth and the Moon can be increased by six times on extremely small slopes with only a doubling of flow thickness. Schaber suggests that of the two factors influencing the formation of far-reaching lunar lava flows (low melt viscosity and high extrusion rates), the more significant factor appears to be large driving pressures at the vents associated with very high rates of extrusion (the result being increased flow heights).

Although lunar flow front heights should be 1.7 times higher than analogous terrestrial flow fronts (Schaber, 1973a), flow front scarps are uncommon. Obliteration by regolith formation can only account for the lack of small scarps, probably less than 5–15 m in height (Gifford and El-Baz, 1978). Indeed, numerous areas of Oceanus Procellarum exhibit plateau/depression morphologies resembling the thin (10 m) repeated flows of the Snake River Plain (Greeley and Schultz, 1978).

Brett (1975) has estimated the thicknesses of lunar mare cooling units by combining data on the kinetics of some chemical reactions with data on thermal conductivity and specific heats of lunar samples. Estimates for Apollo 11, Apollo 12, and Apollo 15 mare basalts indicated that they were derived from cooling units no thicker than 10 m, in apparent agreement with the photogeologic evidence. Therefore, it appears that the vast majority of mare lava flows were thinner than the 30–35 m average for the relatively young Imbrium flows. Possible evidence for thicker flows is provided by lava terraces and subsided lava lakes with 10–50 m relief. These thicker cooling units may simply have lacked well-defined termini (Schultz, 1976b).

Sinuous rilles

Meandering channels known as sinuous rilles (Fig. 5.4.5b) are strikingly prominent features associated with maria. These generally range from tens of meters up to 3 km in width, and from a few kilometers up to 300 km in length (Schubert *et al.*, 1970). Sinuous rilles are distinguished from other troughs of similar size on the Moon by their curvilinear and commonly meandering patterns. Such rilles exhibit a variety of planimetric shapes ranging from meander wavelengths of tens of kilometers to cutoff meanders. Quantitative analysis of planimetric characteristics (Oberbeck *et al.*, 1972) holds promise for the detailed comparison with terrestrial features and in the detection of controls on shapes between and within rilles on the lunar surface. Not all sinuous rilles are continuous; many break up into small segments or crater chains. Some sinuous rilles originate in elongate depressions, others in craters and near or on mare ridges. Recent studies of the large elongate and circular depressions at the head of sinuous rilles (Wilson and Head, 1980b; Head and Wilson, 1980) suggest that if they are erosional in origin, they represent direct evidence for very high effusion rates (in excess of

0.3×10^7 kg/sec). The cross-sectional shape is also varied, including V-shaped, U-shaped, flat-floored, and nested (one rille along the floor of another).

Sinuous rilles occur predominantly in the maria, but often originate in adjacent highlands, and continue in the highlands for considerable distances. They sometimes terminate abruptly in the maria, merge into linear rilles or, most commonly, become shallower until they are no longer detectable. The vast majority of lunar sinuous rilles are located on the Earth-facing side of the Moon along the edges of the major mare-filled basins, or are concentrated in centers in Oceanus Procellarum (Schubert *et al.*, 1970; Murray, 1971). A number of hypotheses have been proposed for rille origin (summarized in Head, 1976), but return of Apollo samples and exploration of Hadley Rille have established their relation to basaltic lava extrusion and formation as lava channels or tubes.

The many similarities in environment and morphology of lunar sinuous rilles and terrestrial lava tubes and channels have been outlined by Oberbeck *et al.*, 1969; Greeley, 1971a, b, c, 1972; Greeley and Hyde, 1972; Murray, 1971; and Cruikshank and Wood, 1972. Major differences are as follows: (1) many lunar rilles are larger than terrestrial channels by an order of magnitude, (2) lunar rilles can be much more sinuous than terrestrial channels/tubes, and (3) some sinuous rilles appear to originate in and extend through nonbasaltic terrain before they enter the mare. In part, these differences are attributed to the lower viscosities of lunar lavas, lower densities at liquidus temperatures (the implication being that flow velocities are much higher), and lower thermal conductivity for lunar basalts (Murase and McBirney, 1970). These factors, combined with a lack of atmospheric convection, suggest slow cooling rates. Low lunar regional slopes may tend to favor meandering (Cruikshank and Wood, 1972).

Erosion by lava also appears to have been an important element in the formation of some sinuous rilles. Hulme (1973) has argued that many lunar flows were turbulent in nature. Since turbulent flow transfers heat to underlying material more efficiently than laminar flow, subflow melting and erosion may have been significant. Carr (1974) has examined the possibility of lunar lava erosion by thermal incision and found erosion rates comparable to those seen terrestrially. He has proposed that thermal incision operated to produce the sinuous channels leading from the uplands to the maria. Initial channeling was probably enhanced by the more incoherent nature of the regolith underlying lunar sinuous rilles compared with that associated with terrestrial tubes and channels, but appears to be at least partly due to the thermal erosion mechanism (Carr, 1974; Hulme, 1973). Thus the majority of sinuous rilles are interpreted to be related in origin to lava channels, collapsed lava tubes, and in some cases to collapse along sinuous fissure vents. Thermal erosion has been an important factor in producing upland channels and in enlarging many rilles (Carr, 1974).

Sinuous rilles thus interpreted can provide important insights into the distribution and characteristics of volcanic processes. Although lava channels and tubes can be reused by later flows, the structures may provide a distance estimate for length of travel of flows associated with them. This is of considerable interest in defining both the distribution and characteristics of flows. Over 75% of the rilles measured by Schubert *et al.* (1970) were 10–100 km in length and 16% were 100–340 km in length. These data illustrate that lunar sinuous rilles are longer on average than terrestrial lava channels and tubes (as pointed out by Oberbeck *et al.*, 1969; Cruikshank and Wood, 1972; and others). This fact may be related to the higher lunar extrusion rates (Schaber, 1973a). Tracing rilles back to their sources shows that many lava flows originate at basin edges from radial and concentric fractures, from circular depressions (often located near uplifted plateaus and associated with other volcanic activity), and from mare ridges (Guest and Murray, 1976). For the most part these vent locations suggest that lava flows associated with sinuous rilles emerge along major structural fractures, and that the magma probably utilizes these to travel to the surface.

Mare domes

Dome-like features occur in both the highlands and maria (Smith, 1973). A classification of over 200 mare domes (Head and Gifford, 1980) shows two major modes of occurrence: (1) low, flat, generally circular structures with convex shapes, having slopes less than about 5°, and displaying summit craters; and (2) irregular structures often adjacent to highland regions and rarely containing summit craters. On the basis of morphologic and morphometric similarities, the first dome type appears to be analogous to small terrestrial shield volcanoes, and to represent primary volcanic constructs, while the second class of domes appears to result from secondary volcanic effects (flooding of highland material to produce kipukas and draping of lavas to produce irregular dome-like topography).

Domes comparable to small shield volcanoes generally range from 3–17 km in diameter (Fig. 5.4.5c), are up to several hundred meters high, and occur predominantly in groups in the lunar equatorial region. Domes in the Marius Hills generally lack summit craters and have a rough surface texture formed in part by superposed cones and steep-sided flows. Elsewhere, volcanic domes are smooth-surfaced and commonly contain a summit crater. These domes are similar in general morphology to small terrestrial lava shields. They are generally intermediate in volume, slope, and height between small shields of terrestrial basaltic plains (such as the Snake River Plains) and larger Icelandic shields. Summit craters on lunar domes are considerably larger than craters on terrestrial shields with comparable diameters. Apparently, this difference is due to a combination of factors including vent enlargement during extrusion, possibly higher lunar extrusion rates, different amounts of collapse, impact erosion, and gravity (Head and Gifford, 1980; Wood, 1979a).

Most vent-related domes appear to be associated with and are thus approximately the same age as surrounding lava plains, although relationships in specific areas have not yet been established. Extrusion rates were apparently relatively low compared to the very high values characteristic of flows associated with major lunar sinuous rilles and terrestrial flood basalts, but may have been relatively high compared to similar terrestrial shields. Large shield volcanoes equivalent to the terrestrial Hawaiian-type or to the martian edifices such as Olympus Mons, do not occur on the Moon. Lack of these features may be due to the low viscosities and high effusion rates typical of many lunar eruptions and the lack of continuous eruptions from single sources.

Cones

A distinctive complex of domes and cones in Oceanus Procellarum known as the Marius Hills has been described by McCauley (1967). Several structures in this complex are reminiscent of terrestrial cinder cones and have been described as such (McCauley, 1969; Karlstrom *et al.*, 1968; Greeley, 1971a). In addition, Scott (1973), Bryan and Adams (1973), and Young *et al.* (1973) describe cones at the southeastern margin of Mare Serenitatis. Cones have also been reported from several other areas of the Moon including Mare Nubium and Cognitum (Mattingly *et al.*, 1972), Mare Fecunditatis (Schultz, 1976b) and Oceanus Procellarum (Murray, 1971; Guest, 1971; Fig. 5.4.5d). Cones can be subdivided into two groups (Head, 1975a): small cones (less than 100 m in height, less than 3 km in diameter) and large cones (larger than 3 km, up to 300–500 m in height). Cones are typically elongated along fissures, and sinuous channels originate from several examples. The cones in Marius Hills and Serenitatis have been interpreted as cinder cones (McCauley, 1967; Scott, 1973; Bryan and Adams, 1973) that originated by the explosive eruption of pyroclastic material.

Terrestrial cinder cones usually result from the accumulation of ejecta from strombolian-style eruptions. Wood (1979d) has shown that there are basic differences in the geometry of lunar and terrestrial cinder cones. Lunar cinder cones are likely to represent strombolian-style activity modulated by variations in volatile content, lunar gravity, and lack of an atmosphere.

Dark-halo craters

A number of craters have been mapped on the Moon that are surrounded by haloes of low albedo, rather than typical bright haloes. The central craters are circular to irregular, and the associated haloes are up to several times the crater diameter. Dark-halo craters can be of either impact or volcanic origin and are associated with both the lunar maria and nearby highland regions (Salisbury *et al.*, 1968). We concentrate here on those of volcanic origin (for a discussion of criteria for distinguishing origins see Head, 1976; and Head and Wilson, 1979).

The Alphonsus dark-halo craters represent volcanic structures and are characterized by: (1) asymmetric craters of about 1 km radius; (2) dark haloes extending to ranges of 3–4 km; (3) substantial amounts of non-juvenile material within the deposits; and (4) lack of associated lava flows or constructional features. An eruption process analogous to terrestrial vulcanian explosive activity provides a plausible explanation for the characteristics of the deposits (Head and Wilson, 1979). It is envisioned that the dark-halo craters formed contemporaneously with the emplacement of lavas in adjacent Mare Nubium. Magma intruded the breccia lens underlying Alphonsus and rose to approximately the same level as lavas in Mare Nubium in the manner described by Schultz (1976a) for volcanically modified impact craters. In the slightly lower Nubium region, lavas were extruded onto the surface to form mare deposits, while in the topographically higher Alphonsus crater they rose to the level of the breccia lens and crater floor where they cooled. Localized volatile accumulations exceeded the tensile strength of the cooled lava cap, causing explosive eruptions, which emplaced a deposit of fine-grained juvenile and non-juvenile material around the crater. The absence of evidence for near-surface volatiles such as water in the regolith argues against an eruption mechanism such as that usually envisioned for terrestrial maar-type structures.

The deposits of Alphonsus-type dark-halo craters have been compared to regional dark-mantle deposits. The differences in characteristics between the regional mantles and these craters, however, argue against a direct relationship between the eruption conditions responsible for the two. Dark-mantle deposits are more likely related to conditions characterized by strombolian or continuous eruption activity. It is unlikely that extensive dark-mantle deposits are the result of coalescing deposits of Alphonsus-type dark-halo craters.

Dark-mantle deposits

Extensive and thick deposits of a low-albedo mantling material are located in the highlands adjacent to maria in several areas (Head, 1974a). For example, those of Sinus Aestuum cover an area of 40,000 km^2. These deposits were thought to be of pyroclastic origin prior to Apollo sample return (Wilhelms, 1970). This interpretation was confirmed through exploration at Apollo 17, a dark-mantle area (Heiken *et al.*, 1974; Head, 1974a; Pieters *et al.*, 1974). Similar characteristics elsewhere (Pieters *et al.*, 1973; Lucchitta and Schmitt, 1974; Head, 1974a; Zisk *et al.*,1977) suggest a pyroclastic origin for most dark-mantle deposits.

Examination of the spectral and stratigraphic characteristics of the dark-mantle materials suggests that they have a compositional variety perhaps comparable to the mare basalts and are associated with mare basalts erupted at different periods of lunar history (Whitford-Stark and Head, 1980).

Mare ridges

Mare ridges, or wrinkle ridges, are prominent in essentially all lunar maria. They typically form concentric patterns in circular maria and occasionally extend into the adjacent lunar highlands (Fig. 5.4.4). In general, mare ridges are distinguished from broader arches on which they often occur (Fig. 5.4.5e; Strom, 1972; Colton *et al.*, 1972). Strom (1972) has described the broad arching associated with many ridges and the lobate plan of many mare ridges. Noting that mare ridge orientation parallels lineaments in the highlands, Strom has concluded that the arches are dikes, sills, or laccolithic-type near-surface intrusions that have emerged along major fractures. The lobate mare ridges often seen atop the arches have been interpreted by Strom as areas where lava has broken through to the surface. Mare ridges, however, appear to have complex origins. It can be demonstrated that many arches and ridges were formed during the period of mare volcanism, and thus the sinuous lobate ridges may in places represent squeeze-ups of lava liberated during settling, arching, and buckling of a mare lava crust (Hodges, 1973). In fact, Schaber (1973a) and Bryan (1973) have shown that mare ridges grew between successive flows in Mare Imbrium to such an extent that they diverted succeeding flows. These same ridges continued to grow after the termination of flow production. Bryan (1973) has summarized morphological relationships in Imbrium and Serenitatis and has concluded that mare wrinkle ridges were formed by localized compression of a relatively thin crust essentially uncoupled from underlying structure and topography. The extension of mare ridges into the highlands appears to support tectonic rather than volcanic origin (Bryan, 1973; Howard and Muehlberger, 1973; Head, 1974b). Colton *et al.*, (1972) cite additional areas where mare ridges appear to be caused by thrusting of mare deposits. As Colton *et al.* point out, a complex origin for arches and ridges is probable, including such processes as differential subsidence, faulting, intrusion, and extrusion; however, the fundamental processes of ridge formation appear to be tectonic in nature (Lucchitta, 1976, 1977).

Other features

Several other features associated with the lunar maria are discussed in detail in Schultz (1976b) and Head (1976). Although such features may be less obvious than the previous examples, they nevertheless provide important clues for the styles of lunar eruptions. These include the following:

Large craters of possible volcanic origin: craters characterized by smooth rims, lack of pronounced terraces, and absence of satellitic craters and marked radial rim deposits. These craters range from 20–40 km in diameter, are generally circular in form, and for the most part appear to be of Imbrian age. Large craters cited to have been formed by volcanic processes include Kopff, Ritter and Sabine (McCauley, 1968; DeHon, 1971).

Mare ring structures and ghost craters: most of these features result from partial lava flooding of cratered topography and draping of lava deposits over buried craters.

Collapse craters: rimless depressions scattered throughout the maria.

Kipukas: islands of preflow material including earlier lava deposits.

Lava terraces: small terraces of lava located at the highland-mare boundary and resulting from subsidence of the lava crust.

Ring moat structures: small ($<$200 m diameter) features defined by a narrow ($<$50 m wide) moat encircling a low-relief dome. They occur in clusters (commonly exceeding 200) within late-stage basalt units and have been interpreted as volcanic tumuli and cones subsequently surrounded by thin lava units (Schultz, 1976b).

Irregular plateaus: low-relief features with highly textured surfaces, common in the western maria (Schultz, 1976b).

Volcanic complexes

Attention has often been drawn to "volcanic complexes" or areas where morphologic features associated with volcanic sources are concentrated (Guest, 1971). The most spectacular examples are in Oceanus Procellarum, including the Marius Hills (McCauley, 1967; Karlstrom *et al.*, 1968), the Aristarchus Plateau (Moore, 1965, 1967; Zisk *et al.*, 1977), Harbinger Mountain region (Fig. 5.4.5b; Strain and El-Baz, 1977), and the Rümker Hills (Scott and Eggleton, 1973; Smith 1974). These complexes display a wide variety of morphologic features such as low domes, steep domes, collapse craters, scarps, mare ridges, sinuous rilles, and cones. The concentration of these features in these major areas strongly suggests that they were major centers for volcanic activity in the Procellarum region. Several of the plateaus (Rümker and Aristarchus in particular) appear to be highland blocks on which a thin volcanic veneer has been deposited. Although the complexes bear a general similarity, they show considerable variations in detail (Whitford-Stark and Head, 1977).

5.4.4 Relationship of mare volcanic features to eruption conditions

Observations of lunar surface deposits and landforms are consistent with expected variations in eruption styles due to variations in exsolved gas weight fractions and mass eruption rates (Fig. 5.4.6; Housley, 1978; Head, 1979a, b; Head and Wilson, 1980).

Strombolian explosions result from the rise of bubbles or groups of bubbles through a magma to disrupt its surface (Blackburn *et al.*, 1976). Small bubbles of exsolving gas can coalesce into larger bubbles if the magma rise rate is sufficiently low—less than 10–20 m/sec for a lunar basalt with viscosity of 30 poises (Wilson and Head, 1980a)—and the slower the magma rise speed, the larger the bubbles that can form. Calculations (Wilson and Head, 1980b) show that the effect of expansion of released magmatic gas, from preburst bubble pressures of order 0.1 bar to the negligible ambient pressure above the lunar surface, is to accelerate $\leqslant 1$ mm size fragments or clots of magma to high speeds. Small particles could be expelled to ranges of at least 300 km to form an extremely sparse mantle. Clots larger than a few millimeters, however, suffer minimal acceleration and are projected from the site of the bubble burst with speeds determined largely by the rise velocity of the bubble. As a bubble reaches the surface, the magma cannot respond to deformation sufficiently fast to avoid its surface being raised at a maximum rate equal to the sum of the rise rates of the bubble and the magma itself. Since the rise velocity of even a 20 m diameter bubble is unlikely to be more than 6 m/sec (Wilson and Head, 1979b) and that of the magma more than a few m/sec, if such a large bubble can form at all, it follows that coarse clasts are unlikely to be ejected to more than a hundred meters. Thus, strombolian-style explosions in lunar basalts are likely to produce a two-zone deposit: clasts larger than a few millimeters will be confined to the region within a hundred meters of the vent while clasts smaller than 1 mm will be transported to ranges of at least many tens of kilometers. Strombolian-style eruptions seem to explain the nature of many cone-like lunar deposits.

The steady eruption of a lunar basaltic liquid can lead to production of a fire fountain with an associated pyroclastic deposit and a lava flow, but there are important differences in style between such events on the Moon and the Earth (Wilson and Head, 1979b). If a

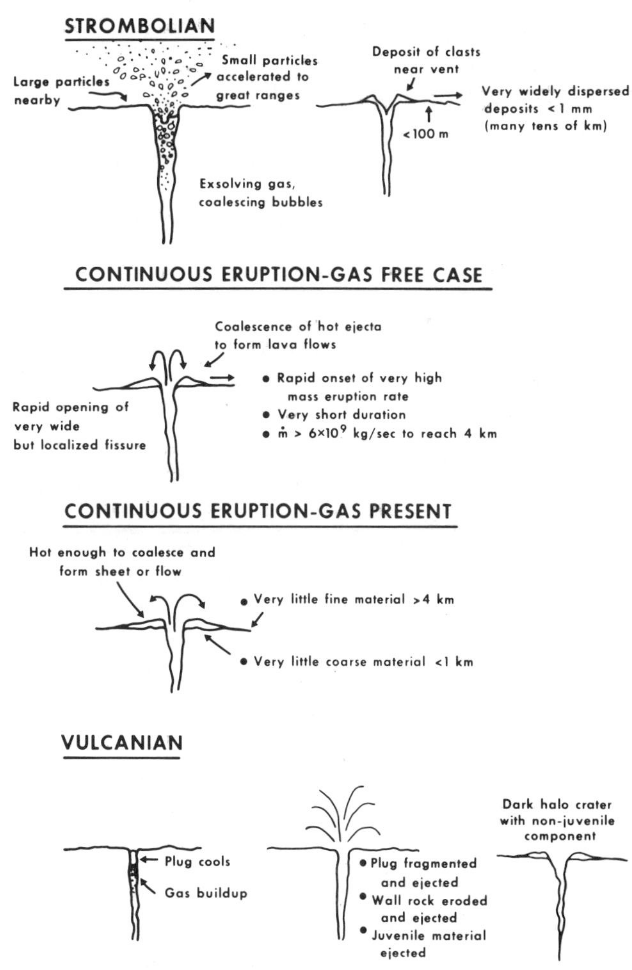

5.4.6 Diagram showing the relationships between volcanic ejecta and eruption style (from Wilson and Head, 1980a).

lunar basalt is gas-free, its exit velocity in the vent can be related easily to its viscosity, to the effective density difference driving the eruption at depth, and to the smallest dimension of the fissure or conduit up which it flows. Calculations (Wilson and Head, 1979b) show that the rate of accumulation of ejected magma is usually so great that coalescence of the deposit would form a continuous fluid sheet which would begin to move downslope as lava flows. Evidence from observed lunar flows suggests that mass eruption rates often must have been very high (Schaber, 1973a).

The presence of released gas in the eruption products will initiate the sorting mechanism described for strombolian explosions. A pyroclastic blanket of a given radius and associated lava flows can be produced, in principle, by a wide range of combinations of mass eruption rate and magma gas content (McGetchin and Ullrich, 1973; Wilson and Head, 1980a). Lunar volcanic cones have diameters less than 4 km and occur in groups, as at the Marius Hills, or in association with fissures or small vents. Morphologically distinct volcanic cones are not commonly associated with regional dark-mantle deposits. Regional dark-mantle deposits are usually characterized by a few relatively large volcanic vents (sinuous rilles, irregular depressions) and deposit diameters less than 200 km. The presence of pyroclasts provides strong evidence for the presence of released gas in at least some eruptions, particularly those associated with the dark-mantle deposits. The range of lunar morphologic features from flood basalts to small cones and associated pyroclastic deposits suggests a wide range of lunar mass eruption rates and gas contents (Wilson and Head, 1980a).

Interestingly, the expected effusion rates through fissures or conduits of a given size are not, however, grossly different on the Earth and Moon despite given differences in magma rheologies and crustal environments. The recognition of high effusion rate features such as long lava flows and sinuous rilles on the Moon implies only that the tectonic and other forces associated with the onset of some lunar eruptions were such as to allow wide fissures or conduits to form. It is possible that the common occurrence of high effusion rate eruptions could explain the apparent absence of large lunar basaltic shield volcanoes. Recognizable individual shields can only be built up when the mean distance flowed by lavas from one source area is substantially less than the mean spacing between sources. The surface widths of elongate fissure vents need be no wider than 10 m to permit the occurrence of mass eruption rates up to ten times larger than those proposed for terrestrial flood basalt eruptions; effusion rates one hundred times larger could be accommodated by 25 m wide vents. As a result, elongate fissure-like structures with widths of many tens to hundreds of meters commonly identified as lunar vents rarely are likely to represent the true vent sizes directly. These structures may be the results of collapse around a vent after eruption ceases, or may be produced by accumulation of pyroclastic debris in some circumstances.

5.4.5 The filling of the mare basins

The emplacement of mare basalts overlapped basin formation (Ryder and Taylor, 1976; Schultz and Spudis, 1979) but the early high impact flux has obscured the timing, nature, and volumetric significance of mare lavas prior to the Imbrium event. Stratigraphic studies of mare deposits postdating the Imbrium event have shown that the earliest exposed deposits (~3.5–3.8 b.y.) are the most significant volumetrically (summarized in Solomon and Head, 1980) followed in volume by those emplaced between about 3.0 and 3.5 b.y. Although young deposits (~2.5–3.0 b.y.) cover large areas of Imbrium and Procellarum, their volume is small and they obscure large volumes of earlier lavas (Howard *et al.*, 1974; Head, 1974c; Pieters *et al.*, 1975; Head *et al.*, 1978a; Pieters *et al.*, 1980; Whitford-Stark and Head, 1980).

Mare deposits are concentrated on the lunar nearside and the vast majority of lavas are located within the major ringed basins (Figs. 5.4.1 and 5.4.3). In general, extensive early flooding filled the inner part of each basin (Head, 1979c). The distribution of later deposits depended on the initial degradation state of the basin and the response of the basin to the early lava load. Detailed studies of mare surfaces invariably suggest a complex picture for detailed surface unit distributions (Johnson *et al.*, 1975) and ages (Boyce, 1976).

The Serenitatis basin provides an excellent example of the nature of mare basin fill. The present surface distribution of mare units in Serenitatis has been mapped by Thompson *et al.* (1973) and Howard *et al.* (1973). The generalized stratigraphy of Howard *et al.* (1973) supplemented by Solomon and Head (1979) is presented in Fig. 5.4.7. Earliest basalt deposits of Serenitatis (Fig. 5.4.7, unit I) occur today in a belt around the southern rim, continuous with Mare Tranquillitatis and including the Apollo 17 site. Ages of basalt samples from the site fall within the range 3.65–3.85 b.y. (Wasserburg *et al.*, 1977; Geiss *et al.*, 1977), and the dark-mantle soil component appears to be contemporaneous (~3.7 b.y.; Husain and Schaeffer, 1973). Subsequent to their emplacement, these basalts were downwarped toward the basin center, and a series of

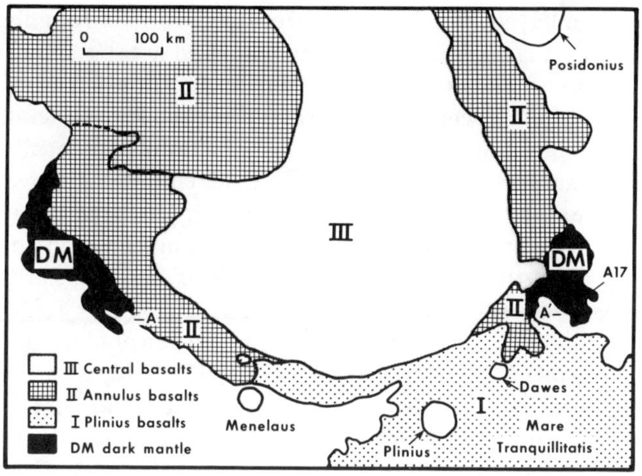

5.4.7 The distribution and stratigraphy of basalts in Mare Serenitatis based on photography and remote-sensing data (from Solomon and Head, 1979).

basin-concentric grabens formed in these units and adjacent highlands along the edges of Serenitatis. A second stage of filling (unit II) is represented by basalts presently exposed as part of an annulus along the eastern and southwestern parts of the basin. These lavas flood grabens developed in the earlier units and are mapped as Eratosthenian-Imbrian in age although, locally, patches may be younger. The unit II lavas have been substantially downwarped toward the basin center, but contain only minor fractures and grabens. The central downwarped part of Serenitatis was filled with a third stage of lavas (unit III) which postdate virtually all graben formation. Mare ridges are developed extensively in this unit, and appear to be compressional features (Howard *et al.*, 1973; Howard and Muehlberger, 1973; Muehlberger, 1974; Head, 1974b; Maxwell *et al.*, 1975; Lucchitta, 1977) associated with downwarping. The deformed surface indicates continued downwarping well past emplacement of the youngest lavas in Serenitatis. The central part of Serenitatis is now generally low relative to the rest of the mare, but also contains a slight topographic high at present. Mare ridges in unit III show as much as 300- to 450-m elevation relative to surrounding parts of unit III, suggesting that at least this much relative deformation occurred during and subsequent to the emplacement of unit III.

The Crisium basin shows a generally analogous style of basin fill although there is no evidence for associated grabens. Three major basalt groups are recognized (Head *et al.*, 1978a) (Plate 5.2): Group I—a soil derived from Fe- and Mg-rich titaniferous basalt similar to Luna 16 samples; Group II—soils derived from very low-titanium ferrobasalts similar to Luna 24 samples; and Group III—a soil derived from a low-titanium ferrobasalt similar to samples from Apollo 12. Group I basalts were apparently emplaced over most of the basin, but are presently exposed as a topographic shelf in the southwest part of the basin. The second group of basalts were emplaced subsequently in two stages and now appear as part of the outer shelf and topographic annulus. Subsidence of the basin interior preceded and continued after the emplacement of the third basalt unit. Accounts of the filling history of other mascon basins can be found in Solomon and Head (1980).

The inundation sequence of the irregular maria (e.g., Oceanus Procellarum) and old circular basins (e.g., Mare Smythii, Mare Australe) is closely tied to the volcano-tectonic modification of preexisting impact craters (Schultz, 1976a; Whitford-Stark, 1979). Volcanic modification is most clearly demonstrated by mare flooding of crater interiors. Additionally, many

5.4.8 A cluster of craters, modified by volcanic and tectonic processes, near the northwest border of Oceanus Procellarum. Note that the fractures and basalt-infilling are limited to the crater floors. The largest crater has a diameter of 70 km. (Lunar Orbiter IV 189-H2.)

craters exhibit extensive tectonic alteration as a consequence of mare emplacement or other volcanic activity. Such craters, termed "floor-fractured" or "internally-rilled" craters (Fig. 5.4.8), typically occur within the highlands bordering the irregular maria (Whitford-Stark, 1974; Schultz, 1976a). Observational and topographic data (Schultz, 1976a; Hall *et al.*, 1979) suggest that the floors of these craters were uplifted, probably as a result of subfloor intrusions and viscous relaxation. Complete flooding of these craters and the intercrater areas in topographically low locations may have been the mode of formation of some irregular maria. The irregular maria show a gradation between small patches with little evidence of deformation or complex filling, to large regions such as Procellarum, where multiple units and many tectonic structures are seen (Whitford-Stark and Head, 1980).

5.4.6 Nonmare volcanism

Three major classes of features have been cited historically as representing possible sites of lunar nonmare volcanic activity (Head, 1976): certain craters and crater chains (such as Davy), smooth and pitted plains (such as the Cayley Formation) and pre-Imbrian pitted plains (Wilhelms and McCauley, 1971), and domes and cones (such as the Descartes Formation; Milton and Hodges, 1972). Prior to Apollo 16, there was a strong feeling in the photogeologic community that many areas of smooth plains, domes, and cones were of extrusive volcanic origin. Indeed, the Apollo 16 mission was targeted to the Descartes region to sample areas of such plains and domes (Hinners, 1972). The samples from these two terrain types proved to be breccias of impact origin (LSPET, 1972), casting doubt on the interpretation of *any* such morphologic features on the Moon as possibly representing extrusive volcanism. Several subsequent studies have combined morphologic *and* remote geochemical observations to suggest that there are indeed some examples of lunar nonmare volcanic features (Malin, 1974; Wood and Head, 1975; Head and McCord, 1978). The geologic setting and spectral reflectance characteristics of three examples of such features in the Oceanus Procellarum region of the Moon show that these features have distinctive morphologies, are of nonmare composition (yet distinct from surrounding highlands), are relatively young parts of the highlands crust, and may represent a type of feature more common in earlier crustal history, but obscured by the early impact-crater modification of the surface:

(1) *Gruithuisen* γ (319° 20', 36° 30'N) is superposed on the edge of an upland plateau and is partly embayed by adjacent mare deposits. It is roughly circular (18 × 24 km at its base) and is capped by a broad, flat crest about 1200 m in elevation, superimposed on which are several domes, the largest being over a kilometer in diameter. The flanks are locally characterized by a distinctive ridged and furrowed texture oriented downslope, which has been interpreted to be flow units emanating from the vicinity of a northwest-trending linear (Head and McCord, 1978). The nearby *Gruithuisen* δ (320° 30', 36° 10'N) is similar to Gruithuisen γ in surface morphology but is rectangular in plan view (33 × 13.5 km), with the long axis oriented northwest, parallel to the summit lineament of Gruithuisen γ. Estimates of the dome height range from 1470–1660 m.

(2) The Mairan domes are located about 225 km northwest of the Gruithuisen domes. *Mairan T* (311° 50', 41° 50'N) is surrounded and apparently embayed by mare, is about 6.7 km at its base, and about 930 m in height. It differs from the other domes in that its summit is characterized by coalescing craters together forming a summit *depression* about 3–4 km in diameter. The concentration, location, and coalescence of the craters suggest an internal, nonimpact origin possibly as a volcanic cone or dome whose summit has collapsed.

(3) *Hansteen* α is located in southwest Oceanus Procellarum (309°, 12°S) just north of the crater Billy. It is triangular in plan view (informally called "Arrowhead"), each side being about 28 km. It is surrounded and apparently embayed by mare material and does not appear to be covered by ejecta from the nearby, Imbrian-aged Hansteen and Billy craters. This suggests an Imbrian age for the dome.

The Gruithuisen domes are prominent in color-difference images obtained at 0.40/0.56 μm but not at 0.95/0.56 μm. The relative reflectance spectrum for the domes and Hansteen α reveals that strong ultraviolet absorption is responsible for the red color. Other regions of the lunar surface with high albedo may have similar spectral reflectance properties and differ from the usual mature upland regions in this spectral range only in the strong ultraviolet absorption. No known mare area spectra are similar to the red spot spectra.

The morphology and surface texture suggest a volcanic origin for the domes, distinct from mare basalt eruption styles, and comparable to styles typical of more acidic eruptions. The age suggests synchronous eruption of these materials and mare basalts. Although these features could comprise crustal material melted by the ascending mare basalts, they may also represent the only preserved vestige of an earlier stage of highland volcanism not obliterated by the high impact flux characteristic of the first 600 m.y. of lunar history. The spectra indicate that the dome areas are covered with mature (agglutinate-rich) soils with a lower content of iron and titanium than mare soils, and perhaps lower than most highland soils. However, the specific compositional property causing the ultraviolet absorption in the red spot spectra is unknown. Correlations of red spots and KREEP basalts have been suggested (Malin, 1974).

The specific identification of other early basaltic units (e.g., the KREEP basalts) indicated by Apollo samples but not possessing distinctive morphologies must rely on remote-sensing data or more indirect evidence. Spudis (1978) and Hawke and Head (1978) have argued that spectrally and geochemically distinct plains units on the Apennine Bench in Mare Imbrium represent exposed patches of KREEP basalts. Furthermore, the high thorium contents associated with the ejecta blankets of impact craters in southeastern Mare Imbrium (Metzger *et al.*, 1979) suggest that KREEP basalts underlie the mare basalts in that area. Spudis (1979) has proposed that KREEP volcanism was a moonwide phenomenon early in lunar history. He has based this on his identification of orbital radioactive highs with areas of thin farside crust (<60 km thick) and topographic lows. The widespread distribution of impact-produced, dark-halo craters suggests that areally-extensive, prebasinal basaltic materials may underlie basin ejecta deposits (Schultz and Spudis, 1979), these basalts themselves infilling now-destroyed craters and basins (see also section 8.9).

5.4.7 Summary of lunar basaltic volcanism

Morphologic characteristics of mare features, petrologic characteristics, scale of deposits, and age information (see Chapter 8) suggest that lunar mare volcanism was characterized by massive outpouring of fluid, volatile-poor lava analogous to terrestrial flood basalt eruptions. This took place over a period of about 1.3 b.y. from 3.83 to about 2.5 b.y. ago. Photogeologic, geochemical, and petrologic evidence suggests that significant volumes of basalts were erupted at least back to 4.1 b.y., prior to the formation of the last major basins. Extrusive volcanism was essentially completed by about 2.5 b.y. ago, but the absolute age of the youngest flows is not exactly known. Many of the identifiable sources for mare lavas are major crustal fractures, mostly associated with the major multiringed basins, and smaller impact craters contained within these basins. The size of these fractures may help explain the high mass eruption rates typical of many lunar flows. Pyroclastic eruptions occasionally accompanied lava extrusion, indicating that volatiles were locally significant. Association with major crustal fractures and high rates and volumes of extrusion suggest that mare lavas originated at subcrustal depths and worked their way to the lunar surface through an essentially passive, although fractured, lunar crust (Fig. 5.4.9). Isolated shallow crustal magma reservoirs probably occurred below numerous impact craters but little evidence exists for the massive shallow reservoirs present on the Earth and, perhaps, Mars.

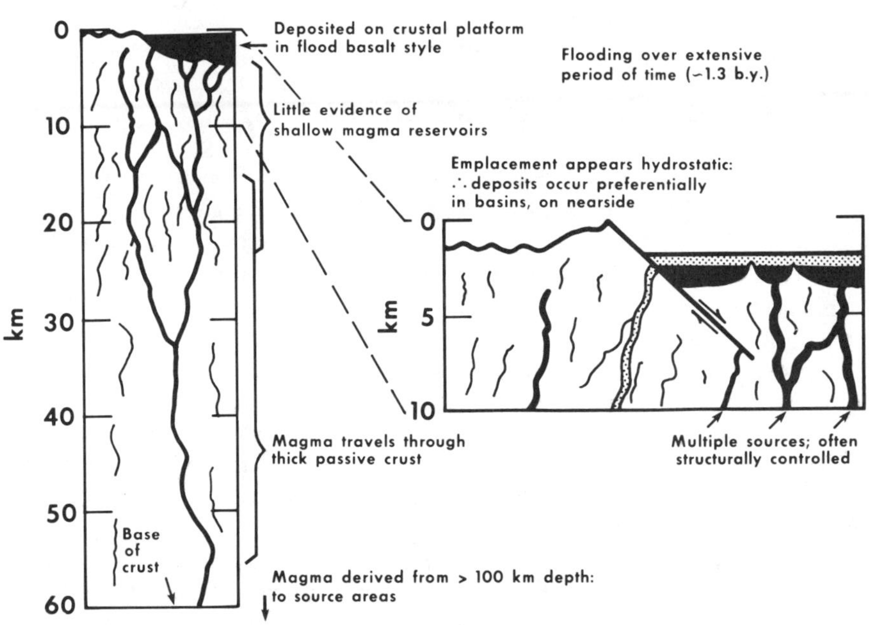

5.4.9 Summary of lunar mare volcanism (from Head, 1976). Total volume of surface mare deposits is very small compared to total crustal volume.

Lavas appear to have erupted at high rates, to have been very fluid, and to have spread out into thin flows. As with terrestrial flood basalts, the crust essentially acted as a passive platform on which the lavas were deposited. Lavas preferentially infilled the existing low regions on the Moon, thus dominating the interiors of the major basins. Crustal readjustment subsequent to mare deposition was concentrated in the mascon basins. The preferential occurrence of mare deposits in large lunar impact basins appears to be genetically unrelated to basin formation, since a considerable timespan seems to have separated basin formation and basalt eruption in Australe, but a short timespan separated them in Imbrium. Hulme (1974b), however, has argued that recovery of a crater from a transient cavity gives rise to a great increase in temperature beneath the crater and probably causes melting, possibly in narrow zones along shear planes. Such melted material may, in part, contribute to the basin fill, but in most cases it would appear that the low topography associated with the basin and the presence of fractures are the predominant factors determining basalt distributions.

Individual mare basins have a history of filling largely determined by basin geometry. Old degraded basins (Tranquillitatis, Fecunditatis) exhibit shallow mare fill of generally similar composition. More detailed analysis invariably shows a complex array of subunits illustrating a complex process of filling, in many instances controlled by old impact craters. Younger basins (Imbrium, Humorum, Serenitatis) exhibit this style of initial fill but also display extensive younger surface flows and deposits that often originate along basin margins and tend to flood basin interiors to varying degrees. These later deposits were emplaced after the initial load downwarped the mare basin. The process of filling of mare basins took place over extensive periods of time and produced complex deposits. There is some evidence (Greeley, 1976) that early mare deposits tended to be of flood basalt-type while later deposits appeared more related to hawaiian-type eruptions. Although there is no evidence that a mare lava lake on the scale of a major mare basin ever existed on the Moon, the numerous lava terraces within and surrounding the maria may indicate broad-scale cooling units of at least 50 m thickness (Schultz, 1976b). Small deposits of young lava-like material associated with Copernican age craters appear to be impact melt (Howard and Wilshire, 1975) rather than internally derived magma (Strom and Fielder, 1970).

Lunar mare deposits are generally thin, the major expanses being less than 2 km in thickness and approaching 6- to 8-km thickness locally in the center of the young mare basins (Imbrium, Serenitatis). On the basis of volume estimates, that portion of lunar mare deposits emplaced on the surface of the Moon represents only a very small percentage of the total volume of the lunar crust (much less than 1%).

There is geochemical and petrologic evidence for nonmare basaltic volcanism, but there is little unequivocal morphologic evidence. It is believed that the early intense bombardment of the highlands surface destroyed any morphologic evidence of nonmare volcanism that might have existed in early lunar history. The basaltic nature of morphologically distinct, nonmare domes has not yet been established. KREEP volcanism apparently played a significant role in the evolution of the lunar crust over a period extending back to at least 4.1 b.y. ago.

5.5 BASALTIC VOLCANISM ON MERCURY

5.5.1 Introduction

Prior to the Mariner 10 encounter in 1974, knowledge of the surface of Mercury was limited to Earth-based telescopic views having at best a resolution of 300 km and showing only faint albedo markings (Murray *et al.*, 1972). Mariner 10 made three flybys of Mercury with each encounter at the same lighting geometry, so the same hemisphere was in darkness on each occasion; thus, our knowledge of Mercury is based only on observations of less than one half of the planet.

The second encounter resulted in coverage of parts of the southern hemisphere previously photographed during the first encounter. Because the spacecraft was on a different trajectory for each encounter, pictures of the same area can be used as stereoscopic pairs. In addition, differences in color, although slight, can be detected in photographs of adjacent areas with different color filters; for example, bright cratered plains appeared redder than dark smooth plains. With an assumed photometric function to correct for different sun angles, Hapke *et al.* (1975) derived albedo differences for such areas and found that albedos of mercurian features are systematically higher than those of their lunar counterparts.

The varying resolution and illumination over different parts of Mercury present formidable problems for interpreting its surface record. Conditions of low sun-angle illumination permit identification of small surface features such as volcanic landforms while high-illumination conditions mask such detail but can reveal significant information on the contacts of different geologic units having strong albedo contrasts. Under relatively low angle illuminations (≲25° from the terminator), the normally quoted resolution limit is approximately 2.2 pixels (Saunders *et al.*, 1975). However, comparison of craters visible in Mariner 10 and

5.5.1 Incoming mosaicked view of Mercury.

5.5.2 Outgoing mosaicked view of Mercury. The large crescent-shaped mountain range represents the probable rim of the 1300-km-diameter Caloris impact basin.

Lunar Orbiter images of the Moon (calibration images produced as Mariner 10 left the Earth-Moon system) revealed that at this resolution 50% of the lunar craters 2.4 pixels in diameter were lost in smooth areas and 90% were lost in hilly areas (Schultz, 1977a). Consequently, reliable image resolution approaches 4 pixels, or 4–5 km, and deteriorates at sun angles ≳25°. The best resolution (2.2 times the picture element width) for contiguous pictures for the whole of the visible disc is just over 1 km. As a result, mosaics such as those shown in Figs. 5.5.1–2 are comparable in resolution to Earth-based telescopic views of the Moon. Other pictures at higher resolution (100 m or better in some cases) were taken, but such high-resolution coverage corresponds to a very small fraction of the imaged surface of Mercury.

5.5.2 Plains units as volcanic units

Smooth plains within and surrounding basins

Much of the imaged surface of Mercury closely resembles the lunar highlands, having numerous impact craters ranging in size from the resolution limit up to that of the 1300-km diameter Caloris Basin. Extensive plains areas with relatively few superposed large craters surround the Caloris Basin and occur in the northern hemisphere (Fig. 5.5.3). Although such plains were mapped by Trask and Guest (1975) as smooth plains (Fig. 5.5.4), the smoothness refers to relief and dimensions no less than 100 m due to the available resolutions of Mariner 10 photographs. One interpretation of the smooth plains is that they are analogous to the lunar maria, thereby implying extensive flooding by lava flows. However, they differ from the lunar maria in several respects, the most obvious of which is the general absence of a strong albedo contrast with the surrounding highlands.

No unambiguous volcanic landforms, comparable to those of the lunar maria, have been observed on the mercurian smooth plains although many parts of the lunar maria are also devoid of conspicuous volcanic features. The absence of obvious volcanic landforms on the smooth plains led Wilhelms (1976) to suggest that a better analogy could be drawn with the light Cayley plains of the lunar highlands. Prior to the Apollo 16 Mission, these plains were interpreted as the result of premare volcanism. Samples since returned from Cayley plains in the Descartes region consist of breccias interpreted to be of impact origin, implying that the Cayley plains (or at least the surface units) may be impact debris produced during impact basin-forming events. Lunar-like mare ridges on the mercurian smooth plains are, however, rare on lunar Cayley deposits.

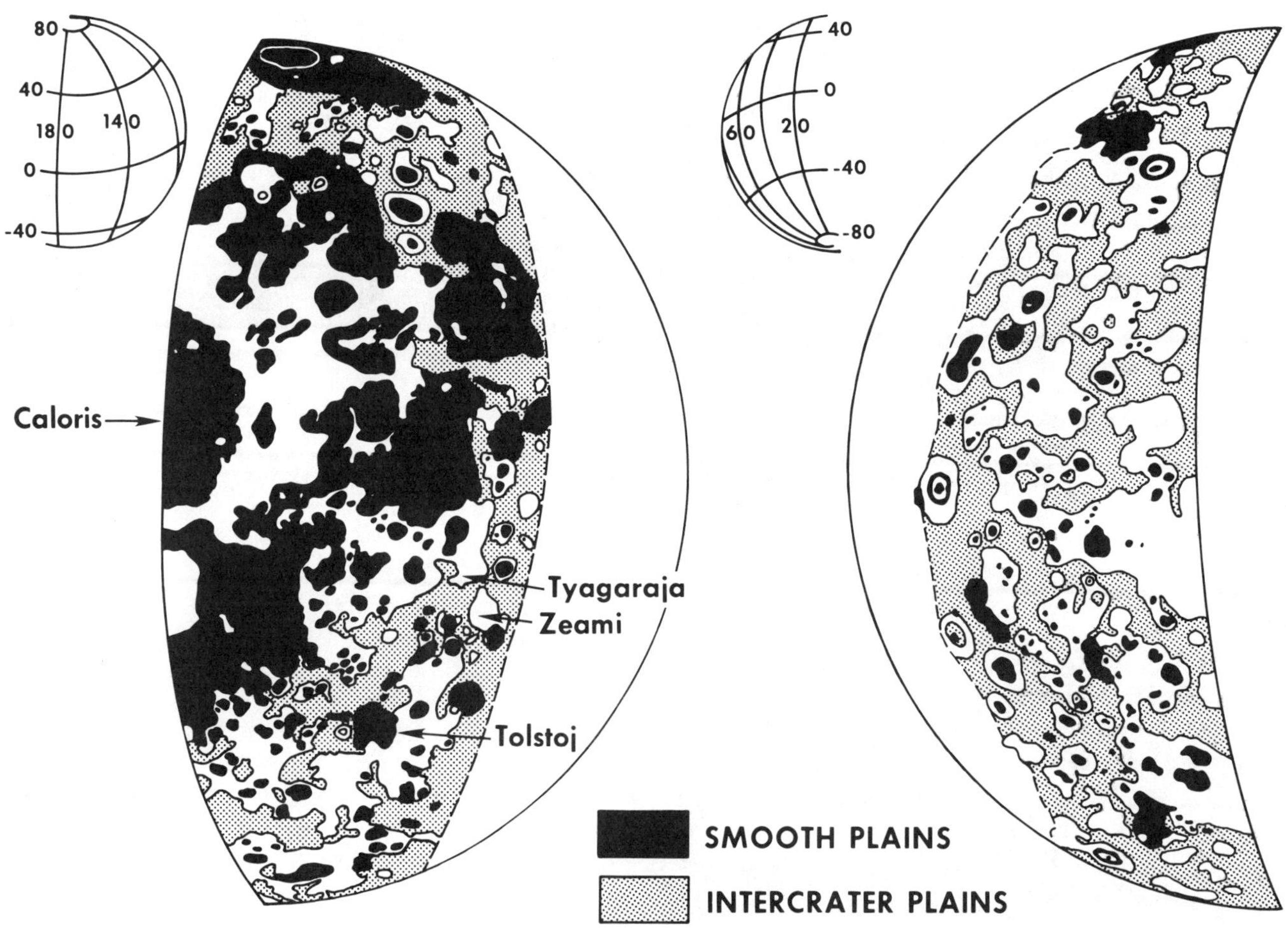

5.5.3 Map showing the distribution of the major plains units on Mercury (after Trask and Guest, 1975).

5.5.4 Oblique view of the mercurian smooth plains, (120°, + 75°) whose origin is under debate. Do such plains represent lava plains comparable to the lunar maria or merely impact-produced debris? Mariner 10 FDS 0152.

The distribution of the majority of plains observed on Mercury in a large swath around the Caloris Basin further suggests to Wilhelms that the mercurian plains are not volcanic but consist of breccias produced by basin-forming impacts. It should be noted, however, that Oceanus Procellarum and Mare Frigoris on the Moon form a swath around the Imbrium Basin, so the Caloris plains could be analogous to these lunar mare regions. This analogy is pointed out by DeHon (1979), who further calculates that the maximum thickness of the plains concentric to Caloris is of the order of 1.8 km, with an average thickness of 600 m. Wood *et al.* (1977) argue that if Caloris formed at the end of the period of intense mercurian bombardment, like Imbrium on the Moon, then a hiatus of unknown but significant duration existed between the time of basin formation and smooth plains deposition. This would imply that these plains are not impact related but may be volcanic.

Trask and Strom (1976) point out that other mercurian basins such as Bach (200 km diameter) have smooth plains inside the basin but very few outside (Fig. 5.5.5). Bach exhibits a well-preserved inner ring and rim as well as a narrow ejecta blanket and a dense field of secondary impact craters. Trask and Strom compare this with the North Polar Basin, another basin of similar size, that has smooth plains both inside and outside the basin. Apparently these plains have almost entirely buried the inner ring and covered the ejecta blanket. Trask and Strom therefore question the likelihood of two similar-sized basins having such different terrains, if the smooth plains were a result of the impact rather than of subsequent volcanic deposits. Strom *et al.* (1975a) also note that the floor of the Tolstoj Basin contains post-basin craters flooded by smooth plains, and that the crater density exterior to the basin is much greater than on the plains occupying the basin. They then argue that the plains postdate the basin and cannot be impact melt formed by the Tolstoj impact event. An external source for these smooth plains is deemed unlikely since no younger basins are close by.

The recognition of large, dark-haloed impact craters lacking bright rays located near or within large plains-filled basins has led to the suggestion (Schultz, 1977a) that the inner ejecta deposits of those basins indicate crustal heterogeneities associated with basin inundation. Color ratio data (Hapke *et al.*, 1975) show the haloes to be spectrally distinct, being redder than the surrounding terrain. Similar spectral contrasts on the Moon characterize craters that have excavated compositionally distinct units associated with mare flooding or early lunar volcanism. On Mercury, they may indicate similar stratigraphic histories or perhaps penetration of relatively thin crust. High-illumination views also reveal (Schultz, 1977a) sharp contacts between certain dark basin fill and surrounding rough, cratered terrains (Fig. 5.5.6). On the Moon, such a contact, although exhibiting greater contrast, is generally acknowledged as a useful indicator of mare surfaces.

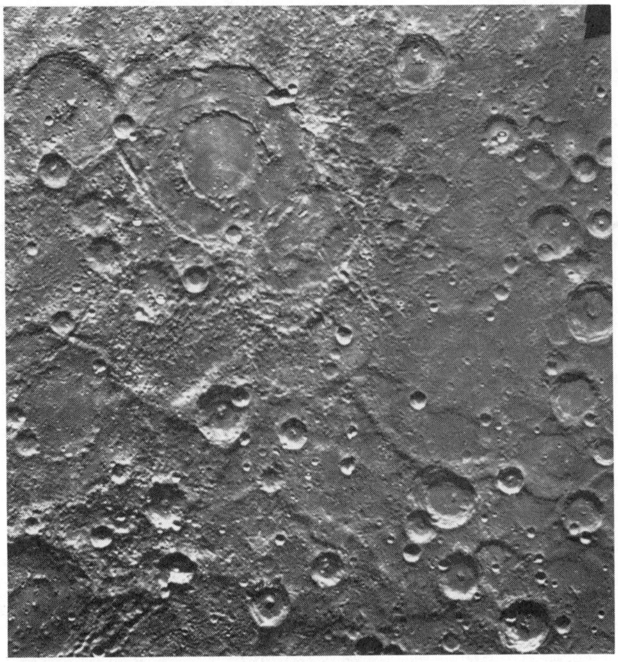

5.5.5 The mercurian crater Bach (103°, 68°S) which has a diameter of 225 km and two pronounced rings.

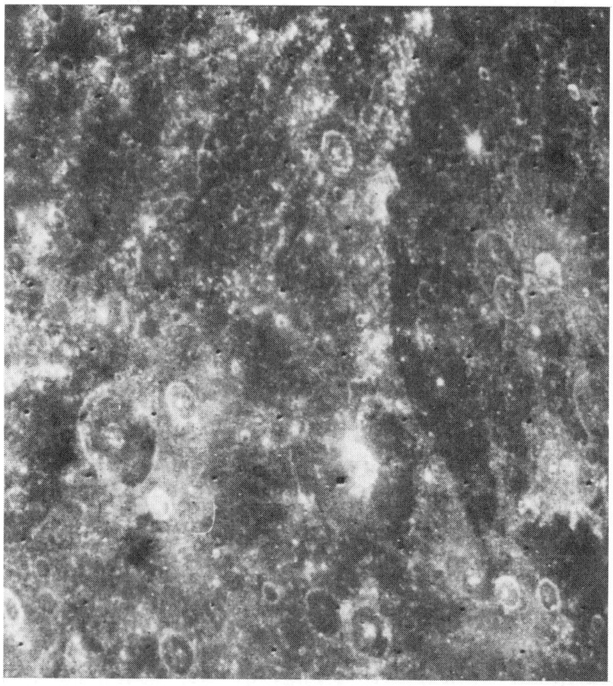

5.5.6 Sharp contrast between relatively dark plains material and relatively light hummocky surroundings (107°, + 15°). This high-illumination view permits recognition of albedo contrasts resembling lunar-maria/highlands contacts; however, the degree of contrast is less than the postulated lunar analogs. (Mariner 10 FDS 166785.)

Cintala *et al.* (1977), employing crater morphology statistics, have found significant differences in the occurrence of central peaks and wall failure features for lunar mare craters compared to highland craters. This has been interpreted as reflecting different physical characteristics of the two substrates. Similar differences have been found for mercurian craters; the response of the mercurian smooth plains is very similar to that of the lunar maria, indicating a competent rock substrate perhaps of volcanic origin.

Analogy with the Moon does not satisfactorily resolve whether mercurian smooth plains units are lavas similar to those of the lunar maria. If the majority of the plains were formed by volcanic processes, however, then they were clearly formed from low viscosity flows, either of lava or ash, emplaced as widespread flow units. If they are lava, terrestrial experience would imply a basaltic or ultramafic composition. On the basis of color and albedo relations, the plains resemble the low-Ti, low-Fe lunar maria in parts of Oceanus Procellarum and Mare Imbrium (Hapke *et al.*, 1975). The presence of an iron-bearing pyroxene detected in the mercurian visible reflectance spectrum, however, is suggestive of basaltic surface volcanism (McCord and Clark, 1979). Nonetheless, the possibility that at least some of the plains units were derived by impact-related processes must remain an equally important working alternative.

Intercrater plains

Within the cratered highlands of Mercury are relatively flat areas between large craters (Trask and Guest, 1975; Murray *et al.*, 1975). Regions mapped (Trask and Guest, 1975) as intercrater plains include all areas between large craters except those characterized by closely spaced hills and lineated terrain. These latter units are interpreted as terrains modified by basin ejecta. Intercrater plains vary from smooth plains (similar to those described in the previous section but with a higher crater density, indicative of an older age) to cratered plains that appear to consist of a thin deposit overlying a cratered surface. Although some intercrater plains may represent old surfaces that predate most of the large craters in the mercurian highlands, clearly some intercrater plains units overlie cratered terrain, indicating that cratering episodes pre- and postdated the formation of intercrater plains (Guest and O'Donnell, 1976; Malin, 1976; Strom, 1977). Leake (1979) further suggests that if the resurfacing potential of existing mercurian craters is smaller than that of comparable lunar craters, more basins than are visible are required to create the amount of visible intercrater plains through ejecta deposition, again, suggestive of a volcanic source for some of the plains. Cintala *et al.* (1977) additionally have found that morphologies of craters in the mercurian intercrater areas are similar to those of the smooth plains and lunar maria but distinctly different from those of the lunar highlands, again supporting a volcanic origin for intercrater plains. The albedos of the smooth and intercrater plains are also similar (Hapke *et al.*, 1975), suggesting that they have similar compositions, perhaps basaltic.

5.5.3 Possible volcanic landforms

Large shield volcanoes such as those on the Earth and Mars would have been visible on the imaged portions of Mercury, but have not been recognized. This absence might prompt comparisons with lunar eruptive styles where low viscosity lavas resulted in vast, relatively featureless lava plains. At present, only a few indicators of mercurian volcanism have been recognized. As seen from Table 5.5.1, this paucity of examples could be due to the relatively poor image resolution over large areas.

Table 5.5.1 Identification of volcanic features on Mercury.

Feature	Typical Dimension (km) Plan	Typical Dimension (km) Relief	Required[a] Resolution (km/pxl)	Mariner 10 Coverage (% total surface)
Earth				
Shield/caldera	50[b]	2[b]	10	40
Composite	8	1.4	5	30
Shield/low	5	0.1	0.1[c]	0
Cinder cone	0.8	0.1	0.1[c]	0
Lava channel/tube[d]	<0.1	0.01	0.05	0
Lava flow	10–500	<0.03	0.03[c]	0
Spatter cone	0.4	0.03	0.01[c]	0
Moon				
Dark-haloed vent[e]	<20(5)	<0.5	2	15
Crater-floor fracture[d]	<3	<0.5	2	15
Lava channel/tube[d]	<3	<0.1	1	5
Lava dome	<20	1.2	1	5
Cinder cone	<3	<0.5	0.5	<.01
Mare dome	<17	<0.3	0.3[c]	<.002
Subsided lava lake[d]	<20	<0.05	0.05[c]	0
Lava flow	300	<0.06	0.05[c]	0

Notes:
[a]Resolution for identification is typically greater than the nominal 2.2 km/pxl resolution for detection and depends not only on dimensions of the feature as listed, but also on relief, disinctive morphology (e.g., summit features, linearity, etc.), and illumination. [b]Shield/calderas on Mars are significantly larger. [c]Estimated for illumination of about 30°; lower illumination angles decrease limiting resolution of detection about a factor of 5. [d]Plan represents width; relief indicates depth. [e]Diameter of deposit (diameter of vent) and vent depth given.

Examples of irregular rimless pits on smooth plains units occur within basins and craters (Strom *et al.*, 1975b; Schultz, 1977a). Figure 5.5.7 illustrates one of several examples, in this case adjacent to a large wrinkle ridge. Whether or not such structures correspond to volcanic vents is debatable. On the Moon, similar rimless structures are formed by collapse and/or by complete surrounding of a crater by lava (Cruikshank *et al.*, 1973). Both origins suggest surface expressions of internal and, by inference, volcanic processes, either to create the subsurface voids or to produce the encompassing flows.

Two highly unusual crater structures on the margins of the intercrater plains northwest of Beethoven Basin contain large central mounds which have very distinctive red spectra (Fig. 5.5.8; Schultz, 1977a). Both craters have low rims, elongate forms in plan view, and almost featureless ejecta blankets. The long axis of one feature extends along an elongate scarp. This unusual morphology coupled with associated diffuse red haloes suggests an origin that may be analogous to red lunar domes interpreted as relatively viscous nonmare lavas (Malin, 1974; Head and McCord, 1978). Two mercurian domical features also have been compared with the red lunar domes (Malin, 1978).

Strom *et al.* (1975b) have interpreted certain mercurian lobate scarps and wrinkle ridges as being suggestive of lava flooding by analogy to mare wrinkle ridges associated with mare lavas on the Moon. This has been challenged by Wilhelms (1976), who interprets most wrinkle ridges as tectonic rather than volcanic in origin. A detailed analysis of the mercurian scarps and ridges by Dzurisin (1978) has led him to the conclusion that they are of tectonic origin, being planimetrically arcuate thrust or high-angle reverse faults, but that some of these faults have provided surface access for volcanic material.

On the Moon, low albedo patches commonly occur along fractures and around craters, such as those within Alphonsus, and are interpreted as volcanic features (Head, 1974a; Lucchitta and Schmitt, 1974). On Mercury, similar low albedo patches occur along the margins of large smooth-floored craters (Fig. 5.5.9) and along the larger mountain ring of Wang Meng Basin (Fig. 5.5.10). This similarity has prompted the suggestion (Schultz, 1977a) that the mercurian low albedo

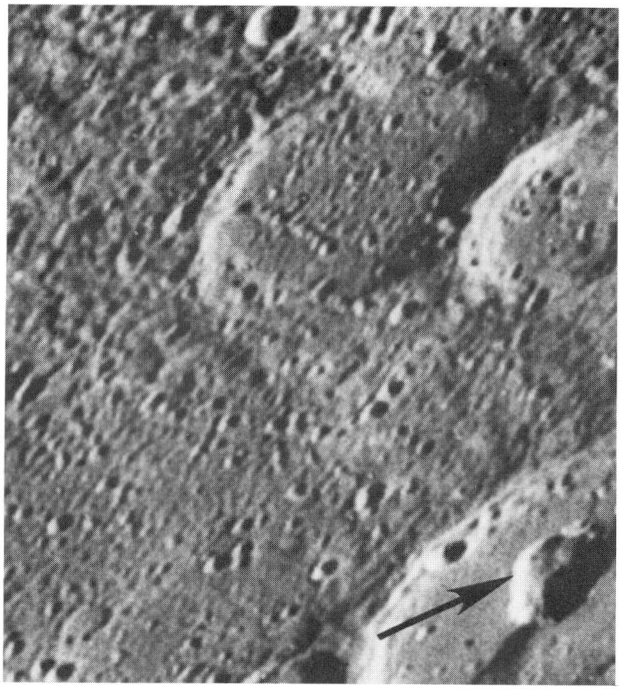

5.5.7 Irregular rimless pit (20 km) at the base of Antoniadi Ridge (29°, + 26°; FDS 528818). The well-defined wall scarp, yet absence of a raised rim, suggests that this feature (arrowed) was produced by collapse or flows encompassing an old crater to the rim. Both processes indicate endogenism that may be related to volcanic activity. This is not a unique example.

deposits could have been formed by fire fountaining and pyroclastic eruptions.

The floors of some lunar craters, like the basins, were once the sites of volcanic vents as illustrated by both unbreached mare-flooded craters, such as Archimedes and Plato, and floor-fractured craters, such as Gassendi and Encke. The albedo and spectral contrasts between the walls and floors of lunar mare-filled craters are also exhibited by several craters on Mercury (Schultz, 1977a). In particular, the crater Tyagaraja (Fig. 5.5.11) possesses a dark floor with bright central peak, wall, and floor deposits. In addition, the dark floor unit corresponds to spectrally red material that surrounds the bright blue patches of the central peak, wall, and floor materials (Hapke *et al.*, 1975). The absence of a bright ray system around this crater suggests that the crater is spectrally mature; i.e., it is sufficiently old to have produced a high percentage of agglutinates, or the excavated material was dark. Consequently, the observed spectral contrast is considered to correspond to compositionally distinct units unless sloping surfaces are rejuvenated by some process.

A

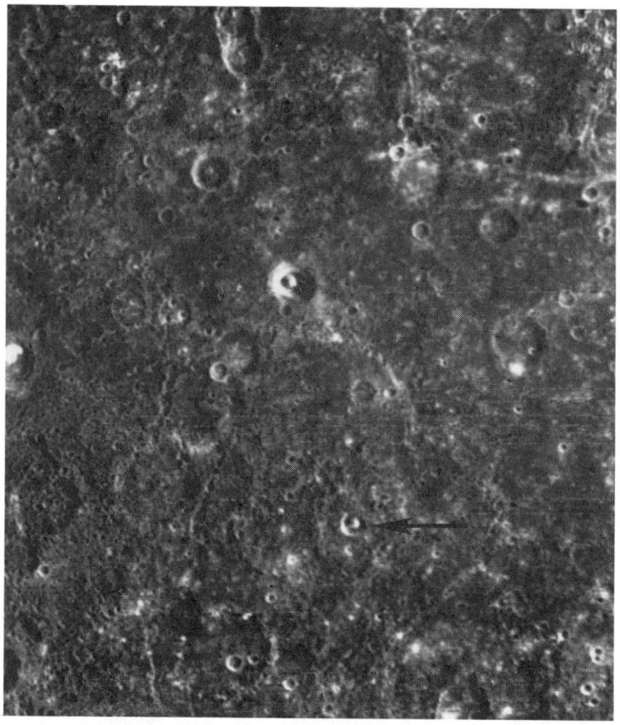

B

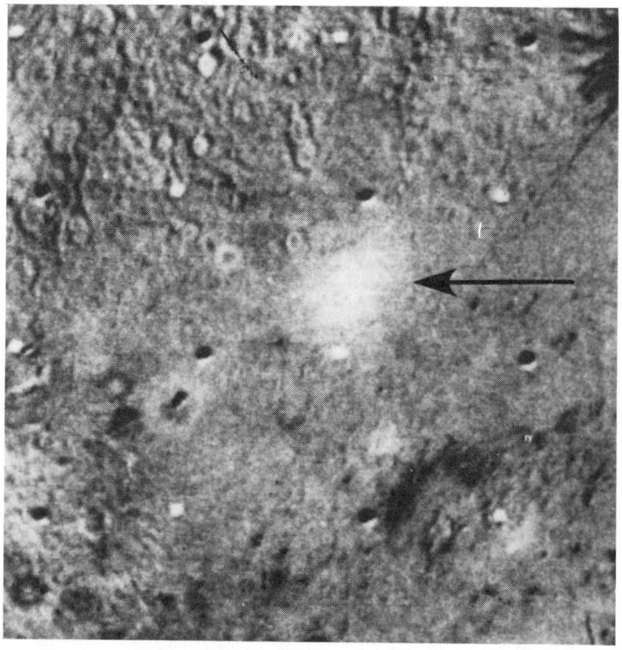

5.5.8 a. Unusual low-rimmed and elongate 45 km-diameter crater with large central mound. **b.** Major axis of this elongate crater corresponds to a northwest-trending ridge. The ejecta facies is very poorly defined in the visible yet exhibits a pronounced red coloration in the color-ratio images produced by Hapke *et al.* (1975).

Possible analogs to lunar floor-fractured craters also exist (Schultz, 1977a). The crater Zeami (Fig. 5.5.12), in particular, exhibits some of the same structural elements as the lunar crater Posidonius: shallow

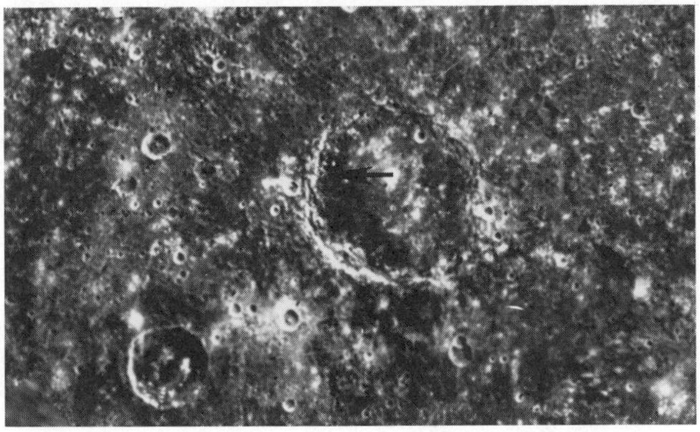

5.5.9 Dark patches on the floor margin of the 130-km-diameter crater Bello (120°, -18°; FDS 166815). Dark patches may correspond to pyroclastic deposits analogous to dark-haloed craters within Alphonsus and other volcanically modified craters on the Moon.

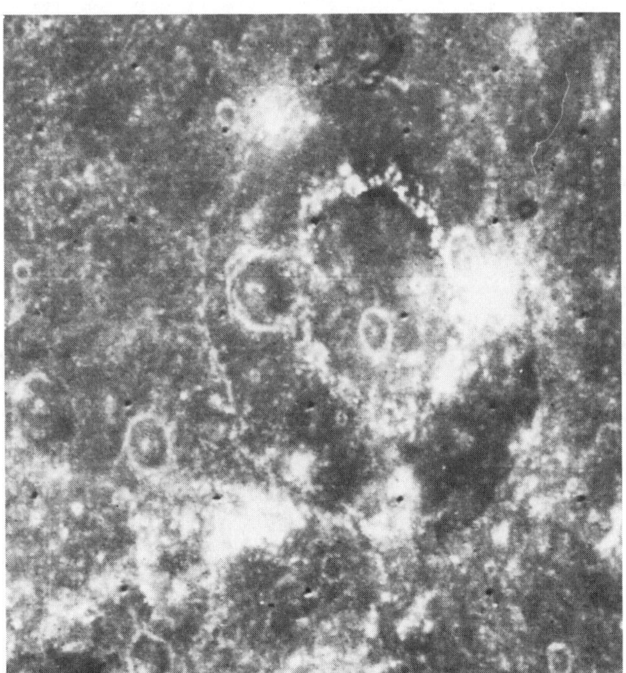

5.5.10 Dark material associated with inner ring of the 170-km-diameter multiringed basin Wang Meng (104°, +9°). On the Moon such dark materials correspond to mantling by dark volcanic glasses associated with fire-fountain-style eruptions (FDS 166783).

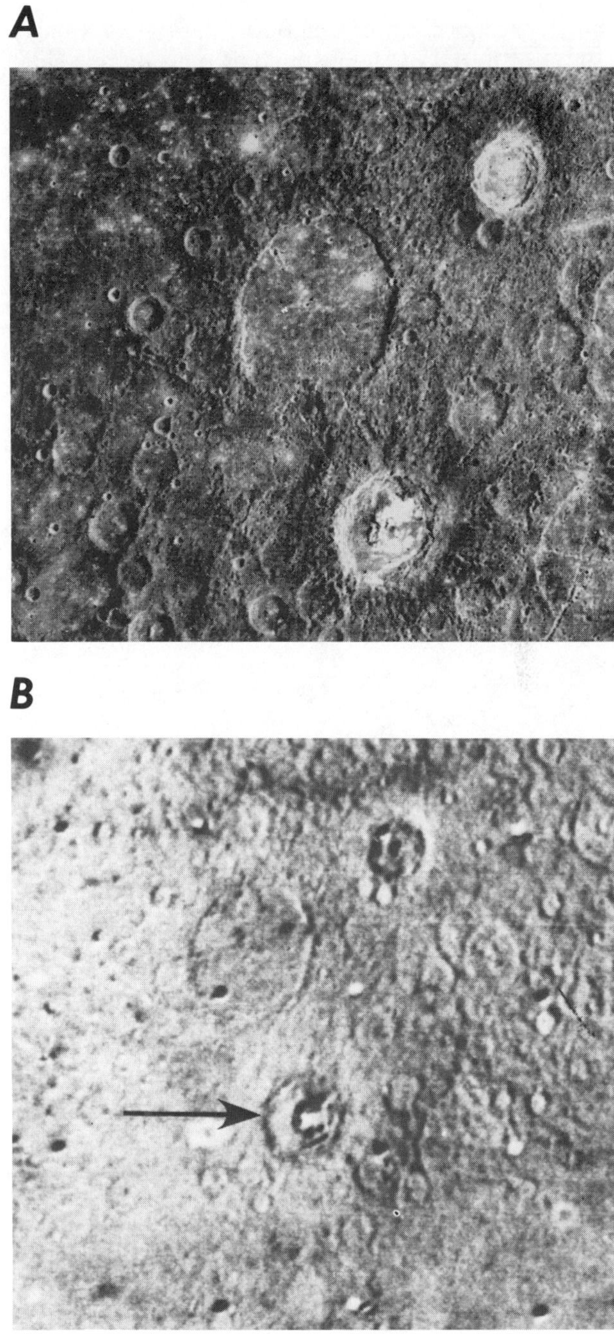

5.5.11 a. The 100-km-diameter crater Tyagaraja (148°, +4°; FDS 214) exhibiting a pronounced albedo contrast between the bright central peaks and floor patches and the dark, smooth floor unit. **b.** Color-ratio images by Hapke *et al.* (1975) reveal that the bright material is blue and the dark floor unit is red. This image also reveals that the dark halo corresponding to the ejecta facies is red. Dark, red-haloed craters are common on Mercury, particularly within smooth-floored basins. At the very least, they indicate a stratified crust and in several instances this stratification may represent sequences of different volcanic units perhaps comparable to the multistaged flooding history of Mare Crisium.

floor, concentric pattern of floor, dark floor (mare) units, and burial of slumps by dark unit on one side.

5.5.12 The 120-km-diameter crater Zeami (148°, -3°; FDS 215) exhibiting not only the same albedo and spectral contrast of Tyagaraja (Fig. 5.5.11), but also a very shallow floor and a concentric floor pattern that resembles the pattern found in floor-fractured craters on the Moon.

Well-defined fractures of crater floors on Mercury, however, have not yet been identified, although several examples exhibit annular arrangements of bright floor patches, shallow floors, wide moats surrounding a central hummocky floor, and relatively dark annuli.

5.5.4 Summary

Volcanism as an important process in shaping the surface of Mercury is suggested by a number of features including the following: rimless pits, unusual mounded crater floors, contrasting floor/rim morphologies of basins, dark plains units, age differences between some smooth plains and the basins they occupy, impact crater morphology, dark-haloed craters within large craters, shallow dark-haloed craters within basins, and spectrally distinct units and ejecta facies.

The confirmation of this interpretation of mercurian volcanic activity must await high resolution images of the entire planet. A possible similar age for the Caloris Basin fill and the lunar maria (see Chapter 8) suggests that mercurian volcanism extended over a significant time period and the youngest episodes may in part be of similar age or younger than the lunar maria. Although subtle indications of volcanism may seem unimportant in comparison to the complexity of Mars, an understanding of this activity on Mercury is fundamental to linking the relatively simple history on low-mass planetary bodies (the Moon) to the complex history on larger-mass bodies (Mars and the Earth).

5.6 BASALTIC VOLCANISM ON MARS

5.6.1 Introduction

Mars has a rich and diverse array of volcanic landforms that range in size from several hundred kilometers across down to the resolution limit of available images. They are distributed unevenly across the planet's surface (Fig. 5.6.1a) and through time (Fig. 5.6.1b). The most prominent are numerous large shield volcanoes and extensive areas of sparsely cratered volcanic plains which occur in the northern hemisphere. In contrast, most of the southern hemisphere is densely cratered, with crater abundances comparable to the lunar highlands. It is generally believed that most of the densely cratered terrain dates from very early in the planet's history. Although volcanic features are present, discrete volcanic edifices are rare, and burial of the cratered surface by younger sequences has not been sufficiently extensive to erase the early cratering record.

Martian volcanoes were first unambiguously identified on Mariner 9 photographs (Masursky, 1973). Three previous missions to the planet had failed to give any indication of the complex volcanic history of the planet, partly because of the limited resolution of the images, and partly because the areas covered were in the southern hemisphere, which has fewer volcanic features. Extensive imagery of the planet has since been obtained by the Mariner and Viking orbiters. Image quality may, however, still be biasing our impression of the volcanic history of the planet because the resolution of the Mariner and Viking images varies according to location. Mariner 9 images cover almost the entire planet at resolutions that nominally range from approximately 1 km close to the equator to 4 km at the poles. In addition, 2–3 % of the planet has been imaged at nominal resolutions of 100–400 meters. In practice these nominal resolutions have been rarely achieved because of clouds and the general atmospheric opacity. Coverage from the Viking orbiters is uneven both in terms of resolution and area covered. During the early part of the Viking mission much of the high resolution (< 100 m) coverage was concentrated in the Chryse Basin and the area immediately to the southwest, although images were acquired of other areas. In addition, much of the southern hemisphere was imaged at approximately 500 meters resolution and large fractions of the volcanic region of Tharsis and the northern plains between 40° and 50° N were imaged at approximately 250 meters resolution. In contrast to Mariner 9, the nominal resolution of the Viking images was usually achieved because the atmosphere was extremely clear. At this writing, approximately 30% of the planet has been imaged at 50-m resolution, 100% at 200-m resolution. The distribution and characteristics of large (>1 km) volcanic features are thus reasonably well known, but considerable uncertainty remains concerning the smaller features.

Martian volcanic features may differ from their terrestrial counterparts, first, because of possible fundamental differences in the style of martian volcanism, and second, because of different local conditions such as gravitational acceleration (371 cm/sec^2), atmospheric pressure (6–11 mb, according to season and location), wind (gusts up to 35 m/sec) and violent, annual, planet-wide storms that redistribute fine material over wide

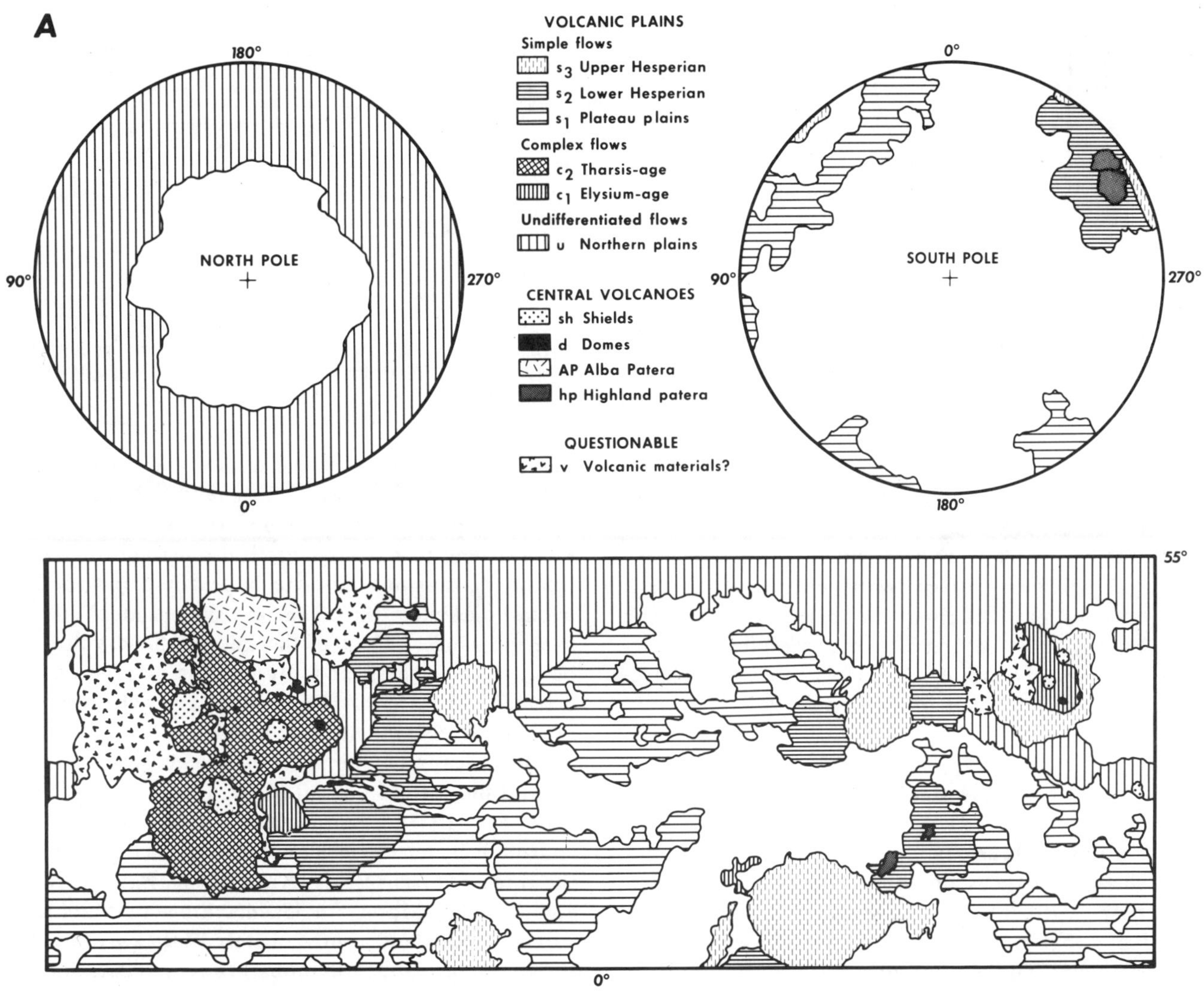

5.6.1 a. Map showing principal volcanic regions on Mars by volcanic type (from Greeley and Spudis, 1981).

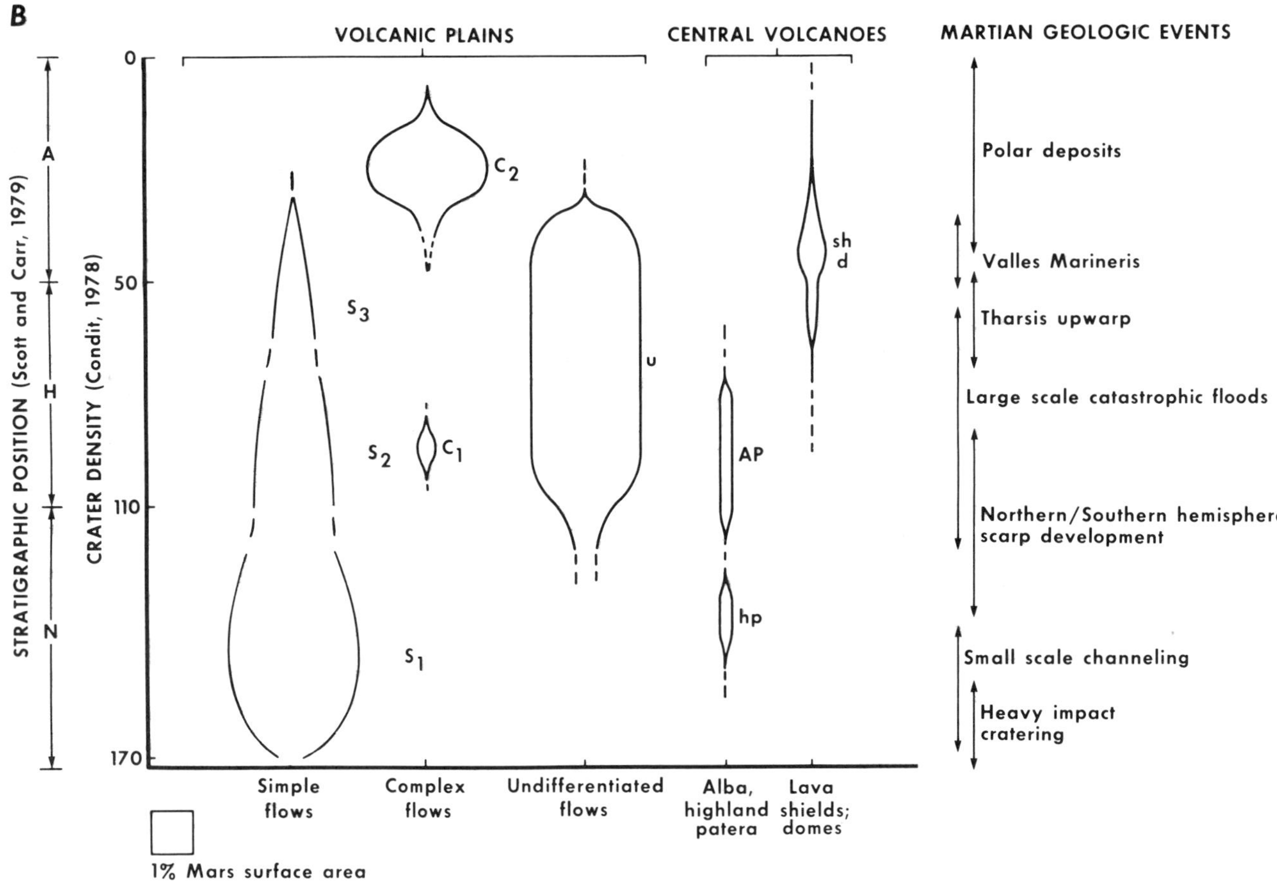

5.6.1 b. Sequential diagram showing the evolution of volcanic types and regions as a function of time; size of each event is proportional to the area covered on Mars, scaled to the box at the bottom of the diagram (from Greeley and Spudis, 1981).

areas. In the polar regions, volcanic features may be modified by the CO_2 ice caps that advance and recede with the seasons, and by the residual H_2O cap left at midsummer. There is also considerable indirect evidence of thick, near-surface, permanent ice over the rest of the planet, particularly in the ancient, cratered terrain. Finally, most features have been modified to some extent by impact cratering. All these factors could have influenced the final form of martian volcanoes.

5.6.2 Central constructs

Large shield volcanoes

Shield volcanoes on Mars are roughly circular mountains with central calderas and gently ($<10°$) sloping flanks that are inferred to be built largely of low viscosity lavas. The large shield volcanoes are discussed separately from domes (tholii) and paterae. The distinction is somewhat arbitrary and largely an artifact of nomenclature; many of the domes and paterae may also be shield volcanoes. Domes tend to have convex-upward flanks with steeper slopes than the large shields. Paterae tend to have less vertical relief and they have central depressions that are large compared to their overall diameter; thus, they are shaped much like an inverted saucer. Most of the known shield volcanoes occur in the Tharsis and Elysium regions. Those that occur elsewhere tend to be older and less well preserved.

The three large shield volcanoes of Tharsis and the nearby volcano Olympus Mons resemble each other in general but differ in detail (Carr, 1975; Blasius, 1976; Carr *et al.*, 1977; Crumpler and Aubele, 1978). All are enormous by terrestrial standards. The main edifices of the three Tharsis shields, Arsia Mons, Pavonis Mons and Ascraeus Mons, are 350–450 km across and have summits that stand approximately 17–19 km above the surrounding plains. Olympus Mons is 700 km across and reaches elevations as high as 20 km above the surrounding plains. All four shields have complex

5.6.2 Photomosaic of Olympus Mons. The volcano is about 700 km across and rises 27 km above the mean Mars datum; it is surrounded by a scarp that is up to 4 km high. Viking images show that lava flows drape across the scarp in many places and extend hundreds of kilometers onto the surrounding plains (Mosaic 211–5360).

summit calderas and flanks with average slopes of 4–6°. The slopes are not uniform; they are shallow in the summit region and steeper farther out. In the case of Olympus Mons, the slopes toward the outer edge of the edifice are also shallow and give the volcano a sinusoidal profile. As will be discussed below, the outer flanks of the three Tharsis shields appear to be buried by younger flows.

Olympus Mons

The main edifice of Olympus Mons (Fig. 5.6.2) can be divided into three parts: a summit caldera, terraced upper flanks and unterraced lower flanks (Fig. 5.6.3). The summit caldera consists of several intersecting depressions (Fig. 5.6.4), suggesting a complex history of subsidence about different centers. Several ridges on the caldera floor resemble those on the lunar maria, while old parts of the floor are cut by numerous concentric fractures. The upper flanks of the volcano exhibit a roughly concentric pattern of terraces. These terraces are typically 15–50 km across, have a slightly convex-upward surface, and are separated by sharp breaks in slope. Farther down the flanks, as the slopes become shallower, the terraces tend to disappear, although on the east flanks, terraces are visible out to the escarpment marking the edge of the main edifice. The flanks everywhere have a faint radial texture, which in many places can be resolved into individual flows (Fig. 5.6.5). These flows are generally long and narrow, sometimes with central channels which may have levees. Individual flows can be traced for several tens of kilometers but are generally less than 1 km wide. A roughly circular, outward-facing, discontinuous escarpment 550 km in

5.6.3 Oblique view of the summit of Olympus Mons showing the multiple craters and slump-blocks typical of calderas associated with shield volcanoes; the Olympus Mons caldera is more than 80 km across (Viking Orbiter Frame 41A36).

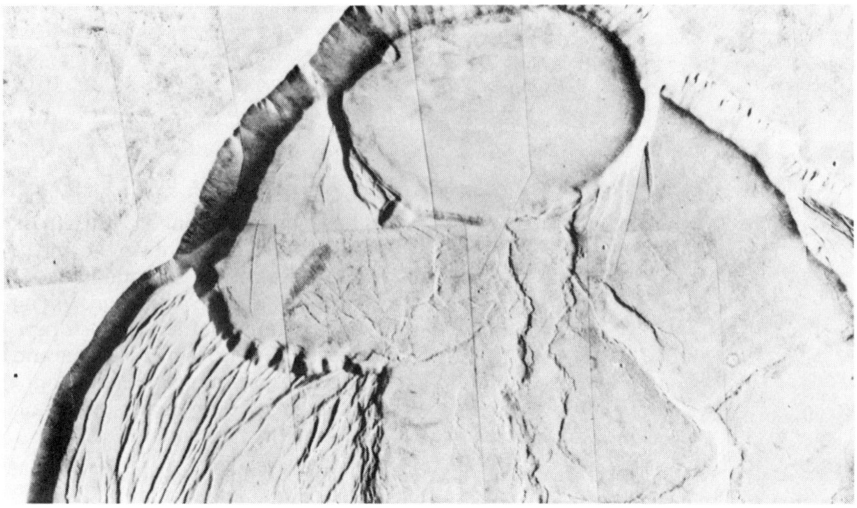

5.6.4 Mosaic of high resolution (18 m) Viking Orbiter images showing details in the summit caldera of Olympus Mons. The youngest feature seen here is the crater at the top of the picture; a substantial caldera in its own right, it measures more than 23 km across and is nearly 3 km deep. The oldest part of the caldera appears to be the area to the lower left consisting of foundered segments of the floor. Intermediate in age is the part of the caldera floor having numerous "wrinkle" ridges, similar in form to mare ridges found on lunar basalt flows (Mosaic P–19381B).

diameter surrounds the main part of the shield. The escarpment was originally thought to mark the edge of the Olympus Mons shield, but recent Viking data show that in many places lava flows drape the escarpment and extend many tens of kilometers beyond (Schaber *et al.*, 1978). The mean diameter of the shield is therefore approximately 700 km rather than 550 km as formerly thought. Numerous landslides occur along the basal scarp, particularly where it is not covered by flows. These generally have a marginal zone of conformal ridges and troughs and a hummocky interior zone.

Most of the features of the main Olympus Mons edifice are similar to those of terrestrial shield volcanoes and are consistent with the slow accumulation of low viscosity, low yield strength lavas (Hulme, 1976; Moore *et al.*, 1978), and repeated eruptions both from summit and from roughly concentric fractures on the flanks. Many eruptions were probably accompanied by swelling and collapse of the summit region, resulting in the formation of the central caldera.

One feature of Olympus Mons that has no obvious terrestrial counterpart is its aureole. This aureole completely surrounds the main edifice and extends in places to more than 1000 km from its center. It comprises blocks of very distinctively textured terrain, the texture on each block being dominated by closely spaced paral-

5.6.5 View of part of the Olympus Mons scarp-mantling lava flows; note that some parts of the scarp are not covered with lava. Although the mechanism of scarp formation is not clearly understood, the process appears to involve both faulting and mass wasting; geometric relations indicate that scarp formation and lava flow emplacement were concurrent. Image area is about 80 km by 80 km (Viking Orbiter frame 222 A 64).

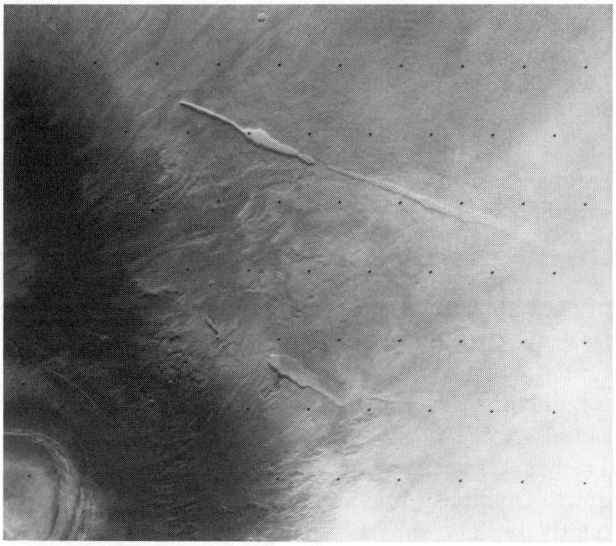

5.6.6 Views of lava flows extending down sides of Arsia Mons. Note graben-like features on flank have been partially buried by subsequent flows. Image width is 480 km. (Viking Frame 39A04.)

lel ridges. The outer boundary of some of the blocks transects the preexisting terrain, but in general, the blocks are embayed in the surrounding plains. The general impression is of blocks of intensely fractured terrain, tilted slightly inward toward Olympus Mons. The aureole is clearly related in origin to Olympus Mons and has probably been modified by wind erosion, but its origin remains a mystery. Suggestions include remnants of preexisting volcanoes, eroded ash-flow tuffs, gigantic landslides, subglacial pillow lavas and tuffs, and vast thrust sheets (Carr, 1973; King and Riehle, 1974; Harris, 1977; Hodges and Moore, 1979; Lopes *et al.*, 1980).

Arsia Mons

The southernmost Tharsis shield, Arsia Mons, has a simple 110-km-diameter summit depression surrounded by concentric grabens. Some of the numerous grabens on the volcano flanks grade into simple breaks in slope or strings of rimless depressions. A line of low mounds extends northeast-southwest across the generally featureless floor of the central caldera; occasional impact craters and some small (< 100 m across) elongate depressions are also present. Several flows into the caldera originate at the walls. On the northeast and southwest flanks of the main shield are many rimless depressions, some which coalesce to form large embayments in the edifice. These embayments are the source of numerous flows which fan out over the adjacent lava plain to the northeast and southwest and extend as far as several hundred kilometers from their sources. Those adjacent to the main shield cut across its radial and concentric fabric. It appears, therefore, that after the main Arsia Mons edifice was built, subsidiary vents developed on the northeast and southwest flanks and these were the sources of large volumes of lava which extended 1600 km across the adjacent plains (Schaber *et al.*, 1978) and buried part of the preexisting shield (Fig. 5.6.6). The shield also appears to have been modified by mass-wasting processes. A tongue-like feature extends 500 km to the northwest of the shield transecting the flows on the adjacent plains. Like the landslides at the base of the Olympus Mons scarp, the feature consists of a striated marginal zone which encloses a hummocky interior zone. It has been tentatively interpreted as an enormous landslide (Carr *et al.*, 1977).

Crater counts on Arsia Mons give relative ages that are consistent with the relations described above (Blasius, 1976; Crumpler and Aubele, 1978). The main shield gives the oldest ages with, for craters larger than 1 km diameter, counts of approximately 400 craters/

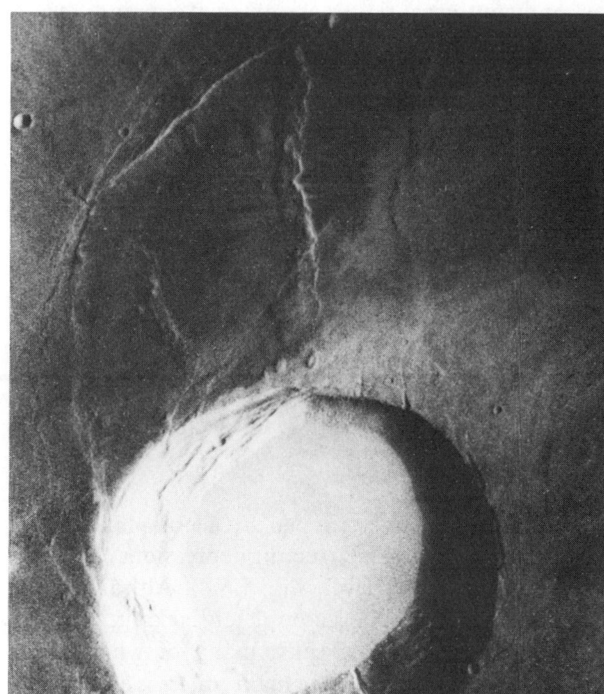

5.6.7 Medium resolution image of the summit of Pavonis Mons, middle member of the three Tharsis shield volcanoes. The prominent caldera at the bottom of the picture is about 38 km across and appears to transect an older summit structure outlined by the circular and linear ridges (Viking Orbiter frame 210 A 34).

10^6 km^2. The central caldera gives significantly lower counts of approximately 150 craters/10^6 km^2. Flows surrounding Arsia Mons give a wide range of counts. Those close to Arsia and derived from the satellitic sources on the northeast and southwest flanks give counts of less than 200 craters/10^6 km^2. Counts increase with increasing distance from Arsia Mons until values as high as 2000/10^6 km^2 are reached 1000 km to the south of the volcano. Schaber *et al.* (1978) recognize at least six major eruptive events on the south flank of the volcano.

Pavonis Mons

Pavonis Mons has a simple 45-km-diameter circular caldera (Fig. 5.6.7) with relatively little terracing. Outside the central caldera and displaced to the north is a 100-km-diameter shallow depression outlined by a low ridge resembling a lunar mare ridge. Other mare ridge-like features occur within the depression, roughly radial to the central caldera. On the flanks of the volcano, generally 120 km beyond the summit, are numerous grabens arranged concentric to the 100-km-diameter summit depression. Many of the grabens grade into lines of depressions. As in the case of Arsia Mons, the northeast and southwest flanks are extensively modified. A relatively simple embayment on the northeast flank of the shield appears to have erupted large volumes of lava which fan out over the adjacent plains to the north. On the southwest flanks are numerous rille-like features, each with a depression at its upper end. These depressions commonly are continuous with the concentric grabens. Many rilles have coalesced and appear to have been sources for flows which cover the plains to the southwest of the main shield. Fractures, and possible rilles with roughly radial and concentric orientations, occur within the flow field to the south and southwest of the shield. Prominent concentric fractures also occur to the north just beyond the edge of the shield.

Ascraeus Mons

Ascraeus Mons is topped by a complex caldera with several mutually intersecting depressions each with a floor at a different level (Fig. 5.6.8). Although concentric grabens are almost completely lacking, there are numerous pits on the flanks, many of which form roughly concentric lines. The main edifice is extensively modified to the northeast and southwest by intersecting rilles, fractures, pits and spatulate depressions. Most of the fractures and rilles are radially aligned but some are concentric and merge with faint lineaments that transect the dominant radial fabric of the shield.

Each of the three Tharsis shields appears to have followed a similar evolutionary trend. The first recognizable event was building of the main shield, probably accompanied by repeated subsidence at the summit to form the central caldera. In the late stages of shield building, eruption was confined mainly to the northeast and southwest flanks where the shields became considerably modified as parasitic vents developed and merged together. These peripheral vents supplied vast quantities of lava which fanned out over the surrounding plains and buried the lower flanks of the shields. The main part of each shield was also modified either by concentric fractures or by formation of numerous pit craters. In the case of Olympus Mons a northeast-southwest asymmetry did not develop. The volcano retains a roughly radial symmetry with no prominent peripheral vents. The basal escarpment and surrounding aureole are unique to Olympus Mons and both appear to have formed relatively late in its history by processes as yet undefined.

Elysium Mons

The second most prominent volcanic region on Mars is Elysium, within which Elysium Mons is the largest shield. It is 170 km in diameter, 14 km high and has a single, central caldera 12 km across (Malin, 1977). The shield is somewhat asymmetric having two distinct ridges which extend to the west and southeast of the central caldera. The flanks have a vague hummocky texture on a scale of about 10 km and a faint radial

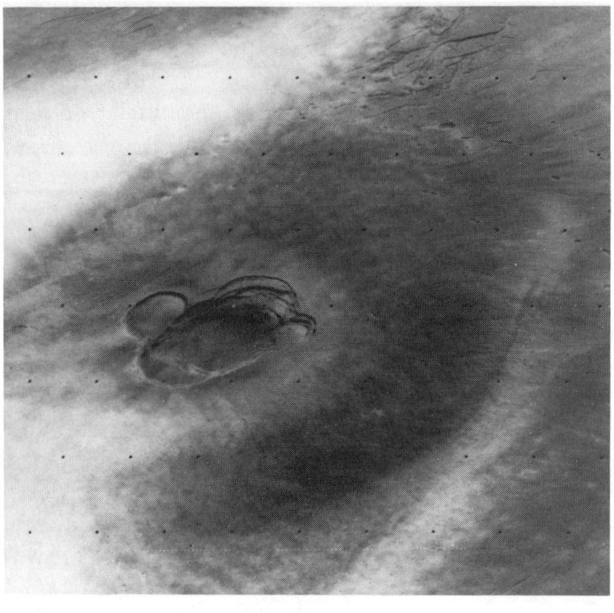

5.6.8 High-altitude oblique aerial view of Ascraeus Mons, northernmost member of the three large Tharsis shield volcanoes (Viking Orbiter frame 41 A77).

pattern caused partly by lines of craters and partly by very thin flows, just at the resolution limit of available images. The flank slopes (10–12°) are significantly steeper than those of Olympus Mons and the Tharsis shields. The outer boundary of the volcano is indistinct and merges with the surrounding plains. Malin (1977) has proposed that the steeper slopes and paucity of individual flow features is indicative of more silicic volcanism than at the Tharsis shields. Recent Viking data, however, reveal numerous flows with many of the characteristics of Tharsis shield lavas. The difference in shield shape may thus be caused by different eruption rates rather than by lavas of different composition.

Paterae and tholii

The terms patera and tholus have been used to denote positive-relief features considered to be of volcanic origin. The paterae are generally low profile, inverted, saucer-shaped constructs, while the tholii are steeper-sided, domical constructs. The terminology is not entirely consistent and both types of features will be discussed together, except for the unique case of Alba Patera. Alba Patera is by far the largest volcano on the planet in terms of its area, being 1500–2000 km across.

Alba Patera

Although it covers a large area, Alba Patera is only 3 km high at its summit. Therefore, it has an extremely low profile, with flanks sloping at only fractions of a degree. At the center is a caldera complex about 150 km in diameter. Numerous north-south grabens cut the edifice and wrap around the summit to form a fracture ring approximately 250 km in diameter. On the flanks are numerous flow features. The most conspicuous of the flow types are referred to as *sheet flows*. These are tabular flows with well-developed lobate flow fronts. They tend to have fairly level surfaces and generally lack surface features such as channels or unroofed lava tubes. Sheet flows are large, usually being several hundred kilometers long and tens of kilometers wide. Vents are not readily identifiable, except for a few flows near the summit that appear to originate from radial, fissure-like vents. Many of the large flows appear, however, to have sources near the ring of fractures.

5.6.9 Lava flows associated with Alba Patera; numerous tube and channel fed flows comprise the main mass of the volcano. Some of the dendritic-like channel patterns seen here may be fluvial; however, most appear to be volcanic in origin (Mosaic 211–5065).

The *tube-fed flows* of Carr *et al.* (1977) consist of long radial ridges on the volcano flanks (Fig. 5.6.9). These, again, may be hundreds of kilometers long—the longest measured at 350 km. The ridges are generally 5–10 km across, although the outer margins are often obscured by later flows channeled between the ridges. The ridges are identified as flows because of axial channels and partly collapsed tube systems over most of their lengths. Some of the ridges have dome-like features at the distal end, interpreted as lava built up over psuedo-vents fed from the axial tube system. The *tube/channel* flows of Carr *et al.* (1977) consist of extensive flows with complex channel-tube systems forming anastomosing patterns. There are local vents on the flow surfaces considered to be pseudo-vents, again fed from tube systems. Although individual flows cannot be readily identified, on many parts of the volcano there are recognizable lava flow features. These areas probably represent older lava surfaces that have become degraded and partly buried by younger flows.

At the summit of the volcano is a caldera complex unlike those of the previously described shield volcanoes. There appear to be several generations of calderas. The rim of the largest is only partly preserved, having been buried around much of its circumference by younger lava flows. The well-defined ring fractures mentioned above, at a radius of about 250 km from the summit, appear to post-date many of the flows on the volcano's flanks. However, young flows from the summit region may well post-date the fracture systems.

From the present preliminary studies of Alba Patera, it may be concluded that the lavas have a rheology similar to lavas of basaltic composition. This conclusion is based on observations of lava forms and from the presence of extensive lava tube systems which, on Earth, occur mainly in basaltic rocks and only occasionally in lavas of other compositions, such as phonolites.

Walker's (1973) demonstration that the rate of effusion is the single most important parameter in determining the length of flow leads to the suggestion that the very long sheet flows are formed as a result of the rapid eruption of large volumes of lava. The tube-fed ridge flows also involve large volumes of lava that may have been erupted at a somewhat slower rate. The tube/channel flows are more characteristic of prolonged eruption of lavas which built up thick piles of overlapping flow units. Flows of this type on Mt. Etna, for example, result from what is known as persistent activity, where lavas may be continuously erupted at a relatively slow rate over periods of several years.

One puzzling feature about Alba Patera is the enormous size of the lava flows. Many of the morphologic features of the flows are comparable to lava features on Earth, but they are as much as ten times larger. Almost certainly the answer to this problem lies in understanding the relations between effusion rate, viscosity, yield strength of the lavas, and surface conditions on Mars, but as yet, no theoretical model has been constructed to satisfy the martian condition.

Miscellaneous paterae and tholii

Paterae and tholii described in this section are central volcanoes with well-defined summit calderas

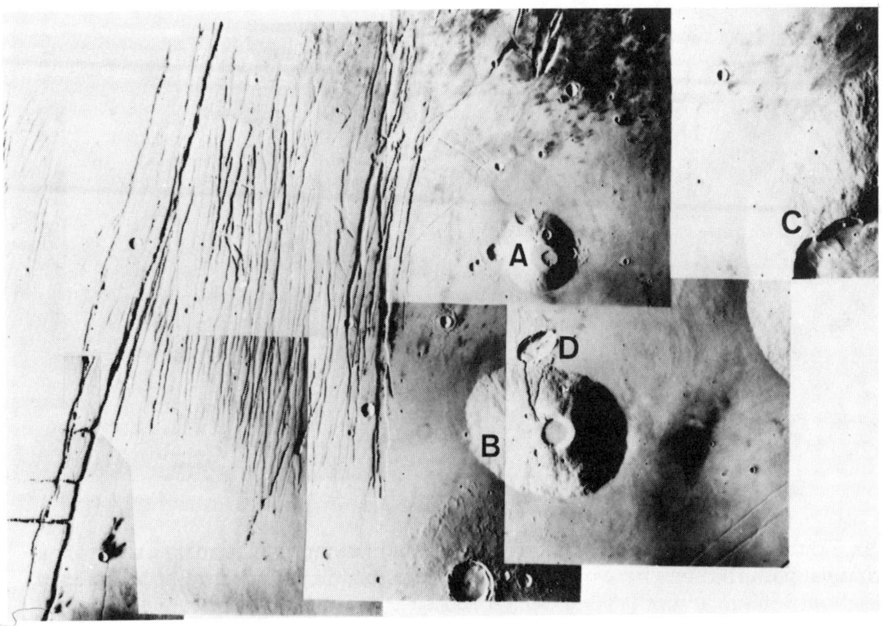

5.6.10 View of the northern Tharsis area showing Uranius Tholus ("A"), Ceraunius Tholus ("B"), part of Uranius Patera ("C"), and an impact crater at "D". Fractures to the west (left) are part of the tectonic pattern associated with Alba Patera, north of this area. Ceraunius Tholus is about 110 km by 90 km (Viking Orbiter Mosaic 211–5593).

approximately 50 km across and basal diameters of 100–200 km. They are found mostly in the Tharsis and Elysium regions. Prominent examples are the Biblis and Ulysses Paterae and the Uranius, Ceraunius and Tharsis Tholii, all in Tharsis (Fig. 5.6.10), and the Albor and Hecates Tholii (Fig. 5.6.11) in Elysium. The surfaces of these structures resemble those of the large shields and have a comparable range of textures. Some, such as Ceraunius Tholus, have numerous radial channels; others, such as Biblis and Ulysses Paterae, have a faint radial texture caused by long thin flows; still others, Hecates Tholus, for example, have radial lines of pits. Most have concentric grabens like the larger shield volcanoes. Where transection relations can be determined, the flanks are transected and buried by younger lavas of the surrounding plains. The original size of the volcanoes is thus unclear. Most evidence suggests that the styles of eruptions which formed these features were very similar to those which formed the large shields. They may not have achieved the dimensions of the major shields because of the smaller volumes of lava involved.

A feature deserving special mention is Apollinaris Patera (Fig. 5.6.12), located at 8° S, 186° W, outside the Elysium and Tharsis regions and within the ancient cratered terrain. The central caldera is approximately 70 km across, whereas the shield itself is approximately 200 km across and surrounded by an outward-facing cliff similar to Olympus Mons. Most of the patera has a distinct radial texture, but to the east a fan-like array of flows appears to emanate from an east-west fissure and extends almost 250 km from the center of the volcano, much like flows to the northeast and southwest of the Tharsis shields. The frequency of superposed impact craters suggests a relatively old age compared to most features so far described.

Ancient paterae

Some of the oldest recognizable volcanic constructs on Mars are seen to the northeast and south of the Hellas Basin. The most prominent are Tyrrhena Patera (Fig. 5.6.13), Hadriaca Patera and Amphitrites Patera (Peterson, 1978). They are shield-like forms with a low profile, each having numerous radial depressions on its flanks. Hadriaca Patera has a fairly well-defined caldera at its summit. The extent of the shield itself is indicated by a very subdued radial texture. The shield is relatively young compared to surrounding impact craters, but the degree of degradation and density of impact craters superimposed on its flanks

5.6.11 Mosaic showing part of Hecates Tholus, a volcano about 400 km northeast of Elysium Mons, showing a complex summit caldera, numerous radial lava flows, collapsed lava tubes, lava channels, and younger lava flows that have partly buried the volcano. Hecates Tholus is about 180 km across (Viking Orbiter Mosaic 211–5787).

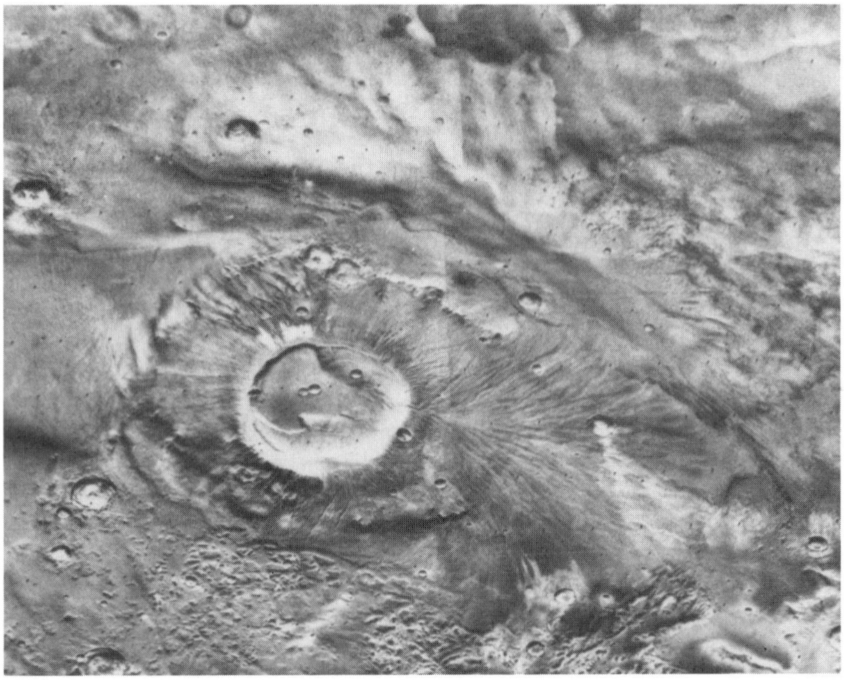

5.6.12 The volcano Apollinaris Patera at 8° S, 186° W. The volcano is probably relatively old as suggested by the number of superposed craters, some of which are probably of impact origin (Mosaic P–18143).

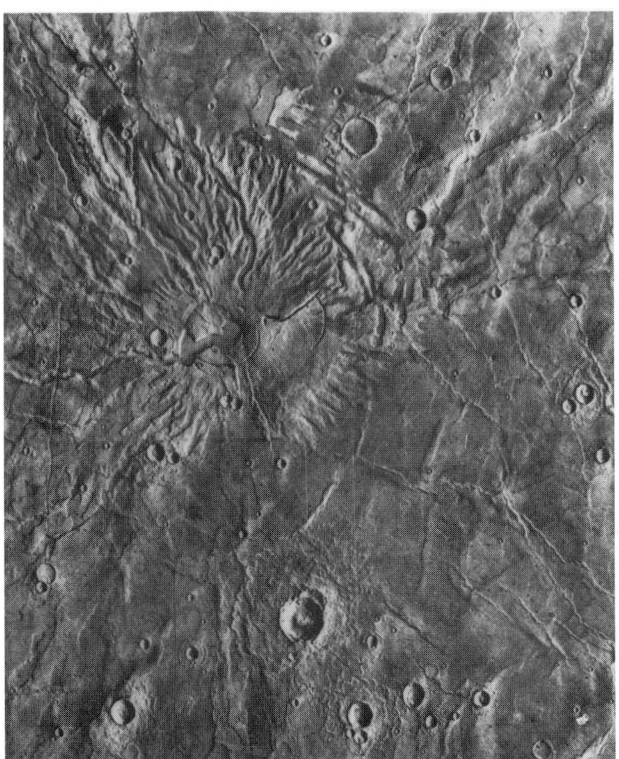

5.6.13 Mosaic of the region around Tyrrhena Patera at 22°S, 253°W made of frames taken from Viking Orbiter 1 on revolution 87. The flanks of the volcano appear to be partly covered by the surrounding lava plains (section of Mosaic 211–5213; from Carr *et al.*, 1977).

suggest that it is older than Tharsis volcanoes. Tyrrhena Patera is modified by radial channels and the summit region is marked by an irregular collapse feature which extends into a valley running down the outer flank. There are also concentric graben fractures around the summit. This volcano, like Hadriacum Patera, does not have well-defined surface features to indicate the type of volcanic flows that formed it. However, at the outer flanks, the units making up the volcano appear to have eroded into mesa-like forms. On Earth, this type of erosion often occurs in ash-flow deposits and it is possible that these ancient paterae are ash-flow shields rather than being built entirely of lava.

Cinder and spatter cones

Although many of the lavas observed on Mars do not have conspicuous vents marked by pyroclastic cones, there are certain areas where such features may be seen (e. g., West, 1974; Wood, 1979d). For example, in the summit caldera area of Arsia Mons, there are fissure vents with ramparts on either side indicative of local spatter build-up around the vent. Elsewhere on Mars, for example in the southern part of Chryse, there are small conical features with summit craters that might be interpreted as cinder cones. In some areas of this region the impact craters appear to have excavated an underlying dark layer (Greeley *et al.*, 1977) that may be a local deposit of volcanic ash.

Because pyroclastic cones are generally small, those on Mars may well be close to the Viking resolution limit and many may be smaller. If this is the case, there may well have been more pyroclastic activity on Mars than we have been able to detect with the present photographic coverage. However, with present resolution, it is clear that the major volcanic constructs do not have large numbers of parasitic cones of the type typical of many basaltic volcanoes on Earth, such as on Mauna Kea or Mt. Etna. This may imply that either the lavas were erupted in such a fluid state that gases escaped without violent explosive activity, or that martian lavas tend to be depleted in volatiles. Certainly, many of the flows observed so far are similar to the flood basalts on the Earth and Moon where, although gases are evolved rapidly at the eruption site, the fluidity of the lava is such that pyroclastic constructs are not formed.

5.6.3 Volcanic plains

Plains units of all types constitute more than 60% of the surface of Mars. Some of these units are clearly volcanic and some are clearly of other origins; a great many fall into a category of uncertain origin. In addition, some plains appear to be volcanic units that have a thin mantling of sediments, perhaps derived through aeolian processes. Part of the problem of recognition of volcanic plains stems from insufficient image resolution for large areas of Mars. However, even if high resolution images were available, it is unlikely that the origins of many plains units could be resolved simply due to a lack of diagnostic surface features.

One class of surface feature that is frequently used to identify volcanic units is wrinkle ridges of the type that commonly occur on lunar maria. However, a substantial body of literature exists supporting origins by both volcanic and tectonic processes, and although there are lines of evidence to distinguish tectonic ridges from volcanic ridges in some cases, most ridges cannot be classified definitively. Thus, unless specific wrinkle ridges can be identified as being volcanic in any given area, they should not be used as the sole means for interpreting a volcanic origin for the plains unit on which they occur. These ridges do, however, only appear to form in competent rock units.

Possible volcanic plains on Mars can be classified into five groups, based on geographic location and/or morphological characteristics: (1) plains units in the cratered plateau region (intercrater plains areas); (2) plains associated with impact basin floors and other topographic depressions; (3) sparsely cratered upland plains regions; (4) plains associated with central volcanoes; and (5) plains of the northern latitudes.

Plains of the cratered plateau

Physiographically, Mars can be divided into two hemispheres (Wilhelms, 1974)— a northern half that is sparsely cratered, and a southern half that is moderately to heavily cratered. The boundary between these two provinces, however, is not the equator, but corresponds to a great circle inclined to the equator at an angle of about 28° (Mutch *et al.*, 1976). Cratered plateau plains occupy much of the cratered southern hemisphere, as mapped by Scott and Carr (1978) from Mariner 9 images. Cratered plateau plains also are found east of Tempe Fossae in the northern hemisphere. The total areal extent is about 2.9×10^7 km^2 (Spudis and Greeley, 1977). The surface morphology of the cratered plateau plains ranges from relatively smooth and featureless to units having abundant wrinkle ridges; flow scarps are found in a few areas, but typically are absent. Based on the frequency of superposed impact craters, cratered plateau plains are among the oldest plains surfaces on Mars (Scott and Condit, 1977). Because of the lack of clearly volcanic surface features, these plains are also among the most questionable of the martian volcanic units. It should be noted, however, that there is a close correlation between the distribution of cratered plateau plains and floor-fractured (Fig. 5.6.14) craters (Greeley and Spudis, 1978) interpreted by Schultz (1977b) as being volcanically modified impact craters.

Plains units in basins

The floors of the large impact basins Hellas, Argyre, and Isidis, and the plains in topographic

5.6.14 Floor-fractured crater (90 km diameter) on Mars on the margin of Isidis Planitia (266°, 5°N) that exhibits many of the characteristics of the lunar floor-fractured craters: elevated central peak, moat at floor/wall contact (Viking Orbiter frame 067B65).

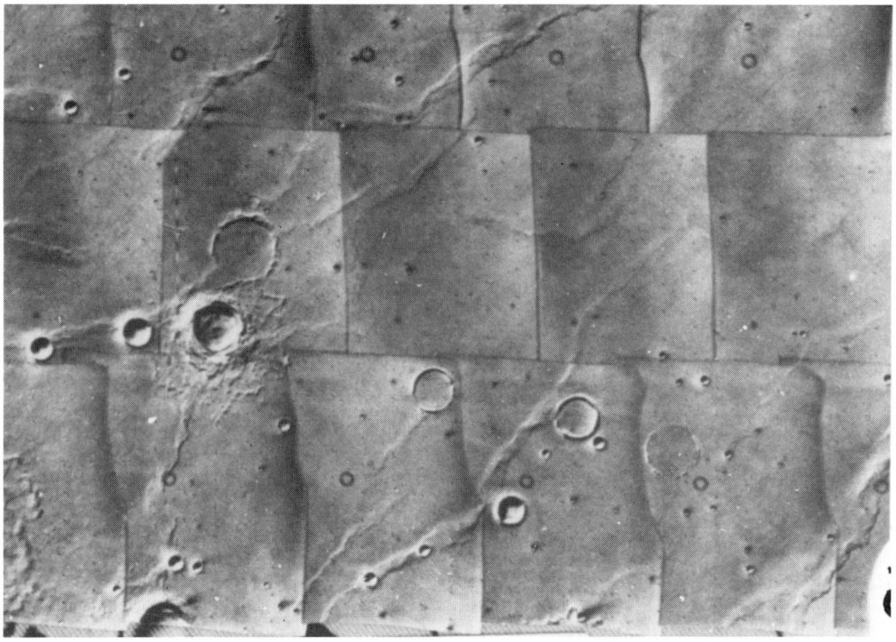

5.6.15 Photomosaic of an 84 × 105 km area of southwest Chryse Planitia (21°N, 47°W) showing mare ridges and impact craters in various stages of flooding and degradation (Viking Orbiter frames 20A 84–92).

depressions, such as Chryse Planitia and Syrtis Major, all show wrinkle ridges and possible flow fronts. Orbital images of Chryse Planitia show many wrinkle ridges that are interpreted to be volcanic (Fig. 5.6.15; Greeley *et al.*, 1977). Ring type ridges on the margins of the plains appear to result from settling of flood-type lava flows over preflow craters. Images from the Viking lander showing the surface of Chryse Planitia illustrate rocks interpreted as vesicular basalts (Mutch *et al.*, 1976). A geological map of Syrtis Major (Meyer and

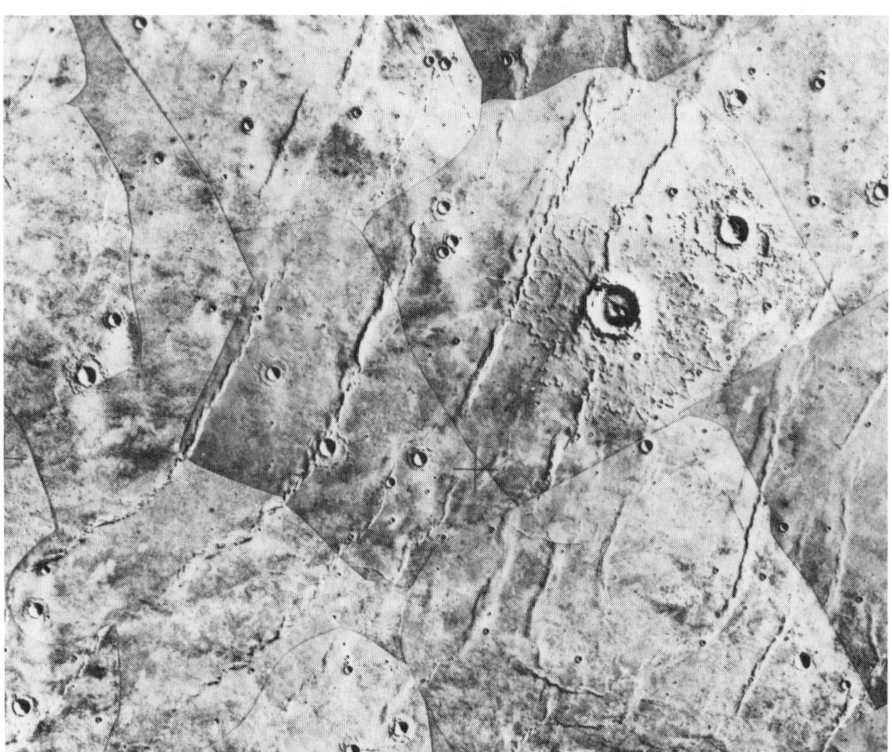

5.6.16 Photomosaic of part of Solis Planum showng mare ridges, flow fronts, and part of the older cratered terrain (U.S.G.S. Map I-1183).

Grolier, 1977) shows numerous features interpreted as volcanic, including a large caldera-like feature in the middle of the plain.

Sparsely cratered upland plains

Lunae Planum, Syria Planum, Sinai Planum, and Hesperia Planum all display features suggestive of volcanic plains, including flow fronts and wrinkle ridges (Fig. 5.6.16). These surfaces resemble those of the lunar mare, and like the lunar analogs, lack recognizable vent regions. In some areas such as Lunae Planum and Syria Planum layered units exposed in the walls of canyons and channels may represent multiple lava flows of the type exhibited in plateau basalt regions. Based on the geological time scale proposed by Scott and Carr (1978), most of these plains units are intermediate in age (Hesperian) and some, such as the units in Sinai Planum, are relatively young (Amazonian).

Two upland plains regions display numerous small (~5 to 10 km) surface features that enable a more positive volcanic identification than the plains described above. Numerous cones, shields, and wrinkle ridges—many in the form of ring structures—are identified in the Amphitrites Patera region and appear to have involved eruptions built on early flood-type flows. A similar area occurs east of Hellas where well-preserved cones, shields and some flow scarps suggest relatively young volcanic activity (Greeley and Spudis, 1978).

Plains associated with central volcanoes

Viking Orbiter images clearly reveal that some volcanic plains are derived from central volcanoes. Most notable are some of the plains units northeast of Olympus Mons that can be traced to the flanks of the shield. Although it is not clear that these units originated from the summit vent or from vents on the flank, they flowed over the shield scarp and onto the older volcanic plain for a distance of more than 400 km. Plains flow units south of Arsia Mons (Fig. 5.6.6) can be traced more than 600 km back to their apparent source at the base of the shield volcano (Carr *et al.*, 1977). Similarly, some of the plains flows around Elysium appear to be related to the central volcanoes. Many of these flows appear to have been emplaced through numerous lava tubes and lava channels, suggesting a prolonged volcanic history. Crater counts on some of the flows associated with Arsia Mons (Carr *et al.*, 1977) show a wide range of ages, supporting the interpretation of prolonged eruptive activity.

Plains in the northern latitudes

Lowland plains units in a latitudinal band south of the north polar cap and north of the southern cratered hemisphere (Fig. 5.6.17) are among the most enigmatic of the possible volcanic units on Mars. Orbital coverage does not shed much light on the characteristics and origin of these plains, and despite the fact that Viking

5.6.17 Photomosaic of possible volcanic plains materials in Northern Hemisphere of Mars (Mosaic 211–5818).

Lander 2 landed on these units, they remain poorly known.

Most investigators consider the northern plains to be lava flows that were at least partly buried by sediments, with subsequent stripping of part of the sedimentary mantle. Evidence for both the mantle and stripping is seen in the Cydonia region (Guest *et al.*, 1977). The sequence is probably more complex involving perhaps repeated emplacement of lavas, sedimentary mantling, and erosion, as discussed for the Chryse Planitia region (Greeley *et al.*, 1977), site of Viking Lander 1.

Some parts of the northern plain commonly ascribed to stripping, however, may represent primary lava flow surfaces. The irregular hummocky surfaces, presence of possible collapse depressions, and scarps similar to irregular flow fronts could well be primary basalt flow features. These features are particularly evident in the southern part of Utopia Planitia, southwest of Viking Lander 2. Unfortunately, definitive lines of evidence for volcanic origins—either for primary flow surfaces or for eroded lava flows—are lacking, and the problem remains open until further study can be devoted to the recently acquired data. An alternative suggestion (Allen, 1979) is that the features of the northern plains represent the remnents of past subglacial volcanism. The possible volcanic origin of the plains is therefore based on the presence of vesicular-appearing rocks observed in lander images and the presence of wrinkle ridges, flow scarps, and possible table mountains observed from orbit in some areas.

In summary, despite the fact that plains units on Mars constitute a substantial portion of the surface area of Mars, for the most part their origin is ambiguous. With the exception of a few regions where obvious volcanic features are observed, most of the evidence for volcanic origins is only weakly founded. The lack of recognizable vent areas and surface features—either because they were never present, or because they have been degraded beyond recognition—make the interpretation of the style of volcanism difficult, if not impossible. On Earth, the two types of volcanic regions that form featureless plains are flood basalts and extensive ash flows, either of which could be analogs to many plains on Mars. The former interpretation is most consistent with the style of landforms, and is also indirectly supported by the lack of sources (large calderas on Earth) for major ash flows.

5.6.4 Features of uncertain origin

With the increased resolution and coverage obtained by the Viking Orbiters, many landforms have been observed that are of uncertain origin and which had not been previously seen. Their morphology, or their occurrence in volcanic regions, indicates that some of these features may be of volcanic origin. Although confirmation of the features as volcanic would help in better defining the extent and time span of volcanism on Mars, at present there are no definitive means for their identification.

During the search for a safe site for Viking Lander 1, a field of about 60 small, low-profile structures was discovered in southern Chryse Planitia (Fig. 5.6.18). These features (Greeley *et al.*, 1977) average 2.5 km across; many have central craters, central knobs, or both craters and knobs. The features are comparatively young, being superimposed on both the plains and erosional channels. Although other origins cannot be ruled out, the features may be "low shields" of the type described in Idaho (Fig. 5.3.2), or they may be analogous to "pseudo-craters" generated by flow of basaltic lavas over marshy, water-saturated terrain (Frey *et al.*, 1979). The martian features are located in an ancient channel where water may have been localized in the regolith, tending to support this latter hypothesis.

Several lines of evidence point to the possibility of widespread water on Mars in the past—water that may have been on the surface and/or in the regolith. The interaction of magma, both at vents and as flowing lava, may have generated numerous phreatomagmatic features, including those described above. Additional features could include maar craters, table mountains, and vast deposits of palagonite, perhaps in the form of plains (Allen, 1979). Hodges and Moore (1978a) describe several features on Mars that may be similar to table mountains found in Iceland. It has been suggested that Olympus Mons may be a gigantic table mountain (Hodges and Moore, 1978b).

Many large ($\gtrsim$ 20 km) impact craters on Mars appear to have been modified by volcanic processes (Schultz, 1978) in a manner similar to some lunar craters. Termed "floor-fractured craters" (Schultz, 1978), these features appear to have served as the loci for rising magma. As the magma entered the brecciated zone of the impact craters' floors, the entire mass was elevated, generating fractures and fissures in the floor deposits. In some cases, magma erupted to emplace lava flows on the crater floors. The floor-fractured craters on Mars are distributed primarily at the boundary zone between the plains and crater highlands, similar to the distribution seen on the Moon (Whitford-Stark, 1974; Schultz, 1976a), as well as near the chaotic terrain east of Valles Marineris. Contrasting morphologies of lunar and martian floor-fractured craters have been ascribed to the probable presence of trapped water ice within the martian crust (Schultz and Glicken, 1979).

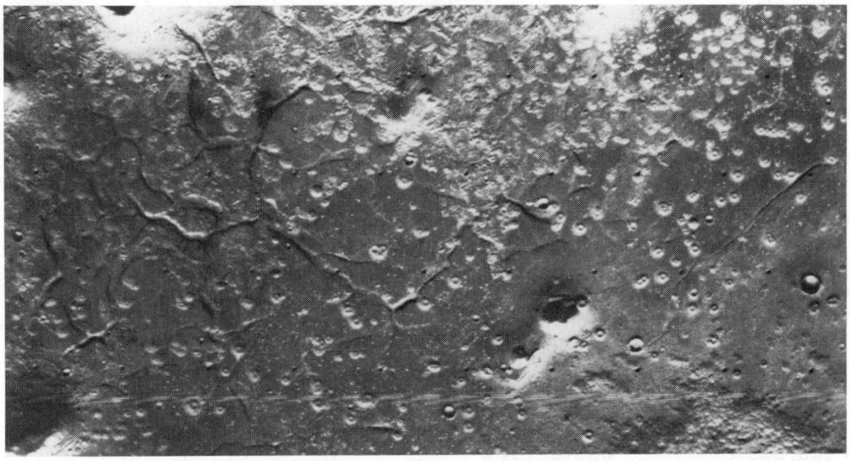

5.6.18 Numerous small domes, possible pseudo-craters are evident in this image of the northern plains of Mars. Image width is 50 km. (Viking Orbiter 72A02).

5.6.5 Summary

The volcanic features of Mars include several central constructs and extensive plains deposits. The most prominent of the central constructs are the Tharsis and Elysium shields, and the lower profile paterae and domical tholii. These structures are 100–2000 km in diameter and exhibit varying degrees of impact cratering, modification by caldera and rift formation, burial by younger materials, and erosion. The large volumes and surface morphologies of the lavas associated with these volcanoes suggest that they had rheologies similar to those of basalts. The three maria plains units include the old intercrater plains found predominantly in the southern hemisphere and characterized by a smooth surface morphology or abundant wrinkle ridges, the intermediate age plains with flow fronts and wrinkle ridges found within the large basins (e.g., Chryse Planitia), and the young volcanic plains associated with the shields or located nearby, but not necessarily erupted from the shields (e.g., Memnonia).

The earliest record of volcanic activity on Mars is possibly provided by the intercrater plains of the southern hemisphere. These plains appear to have been produced largely by flood-type eruptions, but some plains-type activity is suggested by possible lava channels and cinder cones. However, the cones cannot be reliably dated and may be younger than the plains. The old age of these intercrater plains is indicated by a high crater frequency and the presence of dendritic erosion channels formed early in the planet's history (Carr, 1979). Volcanism then focused on a central style to produce the early stage paterae (Tyrrhena Patera, Apollinaris Patera). This may have resulted from a decrease in eruption rate or a change in magma composition. A renewal of flood-style volcanism led to the emplacement of the plains of the large basins, Lunae Planum, and possibly the early flows of the Tharsis area. The identification of these plains as basaltic, or even volcanic, is tenuous, based on the presence of lunar mare-type ridges and some ill-defined flow fronts, but is supported by remote-sensing data and the information obtained from the Viking Lander 1 site. A return to shield-building volcanism led to the construction of the Elysium and Tharsis shields. The youngest obvious volcanic features are smooth units in the summit region of Arsia Mons and flows that drape the northeastern part of the Olympus Mons scarp.

An attempt to portray the change in martian volcanic style with time is made in Fig. 5.6.1b, from which it should be noted that the previously described sequence could be overlapping and not a discrete series of events; it is difficult to distinguish with certainty between the two. Much of this age sequence is based on crater counts (see Chapter 8) which indicate that volcanism has been an active though declining process throughout the entire "visible" history of the planet since the period of heavy cratering.

Many problems remain to be solved concerning martian volcanism. Prominent among these might be:

(1) What is the total time span of martian volcanism and is Mars a volcanically active planet today?

(2) Is the more than 50% of Mars surface that is occupied by plains all, partly, or not of volcanic origin, and if volcanic, what is the total volume?

(3) Does Alba Patera represent a unique style of volcanism; are there other volcanic constructs which are not being recognized because of our unfamiliarity with them; and why are martian volcanoes so much larger than terrestrial volcanoes?

(4) What has been the role of subsurface and surface ice in the modification and formation of volcanic landforms?

(5) Are the northern plains truly "stripped" (eroded) flows as generally considered, or are they primary lava flows of a type not known on other planets? If they are stripped surfaces, then the relative crater dating is not of the emplacement of the lavas, but rather of the timing of the erosion.

(6) Are the features which have morphologies and rheological characteristics of basalts really of basaltic composition?

Many of these questions may be answered by analysis of the now abundant orbital Viking photography and the return of martian samples at a future date.

5.7 BASALTIC VOLCANISM ON VENUS (?)

5.7.1 Introduction

Although the clouds of Venus have long prevented study of the planet's surface, the similarity in diameter, mass and density for Venus and the Earth has led to speculations that the two planets may have generally similar surfaces. The measurement of Th, U, K concentrations comparable to terrestrial basaltic and possibly granitic rocks for rocks on Venus (summarized in Florensky *et al.*, 1977; Surkov, 1977) suggests that, as on the Earth, planetary differentiation into crust, mantle, core, has occurred. The existence of an appreciable crust on Venus is also indicated by an offset of the center of figure from the center of mass (Masursky *et al.*, 1977).

Earth-based radar data, with a resolution varying between 10–100 km has allowed penetration of the cloud cover and has indicated that Venus has a diverse topography including heavily cratered surfaces, a major linear feature thought to be a canyon (Goldstein *et al.*, 1976), a rift valley (Wood and Head, 1977), or a young volcanic crater chain (Schaber, 1975), and positive relief features interpreted to be volcanoes (Saunders and Malin, 1977). The heavily cratered terrain and possible large tectonic structures suggest that Venus has a more rigid and stable lithosphere than that of the Earth, perhaps comparable to that of Mars.

The recent acquisition by Pioneer Venus (Pettengill *et al.*, 1979a, b) of orbital radar altimetry and imagery with a vertical resolution of 100 m and a possible horizontal resolution of 20–60 km has permitted more accurate characterization of venusian surface features. At present, 93% of the surface of Venus has been topographically mapped at 100 km horizontal resolution by the Pioneer Venus orbiter. A result of this mapping is that Venus has been found to have a unique unimodal topography distribution—60% of the mapped area lying within 500 m of the modal altitude (Pettengill *et al.*, 1980), which in turn is only 300 m below mean altitude (6051.4 ± 0.1 km).

The mapped area of Venus has been subdivided (Masursky *et al.*, 1980) into 20% lowland plains, 70% rolling uplands, and 10% highlands. The lowland plains vary in altitude from 0 to −2.9 km (from the mean), contain few rocks or radar absorbing materials, and may be basaltic lava flows. The uplands vary in altitude from 0 to 3 km, contain shallow craters and wispy linear features, and may be granitic in composition. The highlands, rising to 11.2 km, comprise two large "continents"—Ishtar Terra and Aphrodite Terra—and a group of "islands" of small areal extent. The Australia-sized Ishtar Terra in the northern hemisphere comprises a 5-km high plateau (Lakshmi Planum) with a 2-km high boundary scarp and a number of superposed mountain ranges and mountains, which may be basaltic volcanoes. Aphrodite Terra, about half the size of Africa, rises 6 km above the mean altitude and has a radar-rough surface. A plateau to the east of Aphrodite rises an average of 2 km above the mean and contains a complex of curvilinear ridges and trenches.

5.7.2 Environmental controls on venusian volcanism

The high surface temperature (about 730° K) and atmospheric pressure (about 90 atmospheres) (Florensky *et al.*, 1977) of Venus would profoundly affect volcanic landforms. A terrestrial analog to the high pressure at the surface of Venus occurs at ocean floor depths of 1 km. Eruptions at these depths are purely effusive, and theoretical studies show that explosive eruptions should not occur at depths greater than 500 m (McBirney, 1963) to 700 m (Peckover *et al.*, 1973). Thus, pyroclastic eruptions, and the resulting landforms (cinder cones and stratovolcanoes), should be uncommon on Venus, assuming similar volatile contents as terrestrial magmas.

If pyroclastic material were erupted into the venusian atmosphere it would be strongly affected by atmospheric drag and buoyancy. The presumed density difference between volcanic gases and the atmosphere would be much greater than on Earth, and vigorous convective thrusting (Sparks and Wilson, 1976) could carry ejecta to great heights. This buoyancy effect could distribute ejecta over much of the planet and prevent the buildup of a volcanic edifice at the vent; however, for ballistically transported material, the strong air drag would promote concentrations of material at the vent (Wood, 1979c).

It has been suggested (Wood, 1979c) that if an effusive eruption of basaltic lava occurred on Venus, the high atmospheric temperature would prolong cooling, and the high pressure would inhibit degassing (a major mechanism of cooling). The effective viscosity would thus be greatly reduced, resulting in the formation of longer flows than similar flows erupted on Earth.

Low measured surface wind speeds of 1.8–3.6 km/hr suggest that wind erosion is not an important landform modifier on Venus (Florensky *et al.*, 1977), although present data do not rule out the possibility of local or seasonal increases in wind velocity.

5.7.3 Volcanic constructs

From Earth-based radar data, Saunders and Malin (1977) interpreted the feature Beta to be a large volcanic construct similar to terrestrial and martian shield volcanoes. The structure was interpreted to have a diameter of 700 km, a height of 10 km relative to its surrounds and a 60 × 90 km summit depression. The basaltic composition of this structure was supported by analyses at the Venera 9 and 10 sites, both located on its borders (Florensky *et al.*, 1977). Subsequent analysis of orbital radar data (Masursky *et al.*, 1980) showed the Beta Regio to be dominated topographically by the two mountains Theia Mons and Rhea Mons, rising 5.4 km above the mean level. The presence of streaks of high radar scattering efficiency radiating away from the central region, and the summit depression on Theia Mons, support a volcanic origin for the mountains. Orbital data (Pettengill *et al.*, 1979b) further indicate that both Theia Mons (Beta) and Rhea Mons are significantly less than 10 km in height (closer to 5 km; Fig. 5.7.1). If these structures are volcanic, it would therefore appear that they are similar to very low shields or lava plateaus. Saunders and Malin (1977) suggest that the high radar reflectivity (produced by rough surfaces) of Theia Mons probably results from an aa surface texture. Another domal structure (10° S, 40° E) with a height of 1 km and 60–70 km summit "crater" has also been interpreted as a volcano with a summit caldera (Saunders and Malin, 1976).

A 1500-km-diameter semicircular feature located at 35° S, 140° E comprises two ridges with an intervening trough. This feature may have a volcano-tectonic origin (Masursky *et al.*, 1980).

The presence of topographically smooth areas on Venus (Pettengill *et al.*, 1979a) indicates plains-like deposits, perhaps comparable to those on the Moon, Mars and Mercury. The identification of small volcanic constructs will have to await a future orbital probe such as Venus Orbiting Imaging Radar (VOIR) with a nominal ground resolution of about 100 m.

5.7.4 Summary

Geochemical and geophysical data suggest that Venus is a differentiated planet. Linear topographic

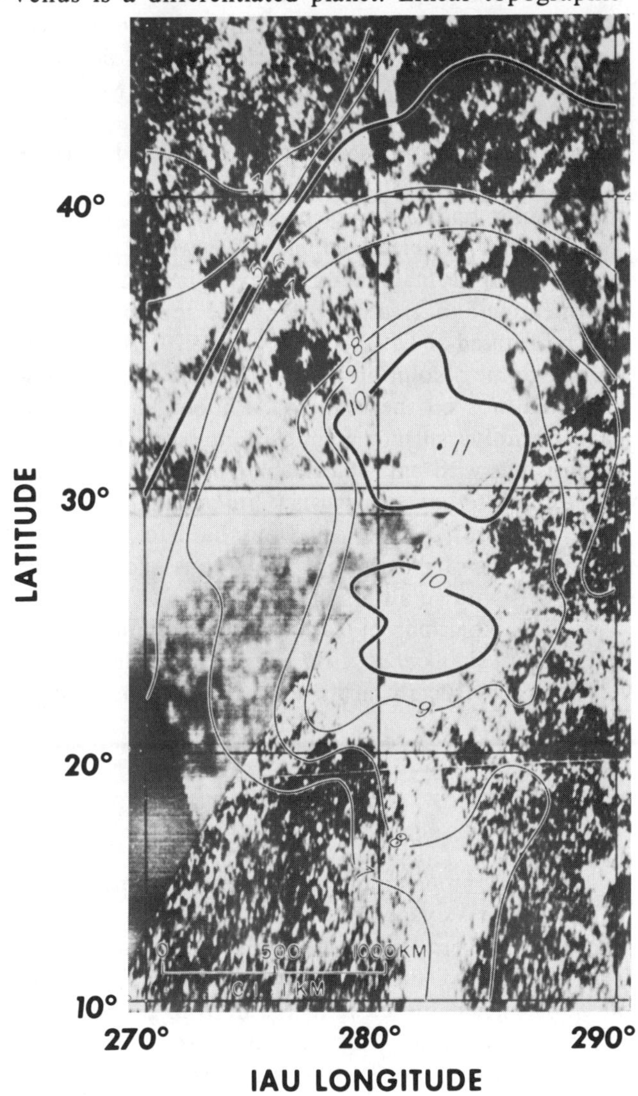

5.7.1 Contour map of the Beta Complex, Venus, based on Earth-based radar images and topography derived from orbital altimetry data (from Pettengill *et al.*, 1979b). Reprinted courtesy of *Science*; copyright 1979, by the American Association for the Advancement of Science.

elements and a large plateau suggest that tectonism has played an important role in shaping the surface of the planet, while geochemical and topographic data suggest that basaltic volcanism may have constructed large, low shields or plains-like volcanoes. Higher resolution radar images are required to identify the geological processes responsible for the observed topography.

5.8 SATELLITES AND ASTEROIDS

5.8.1 Introduction

Volcanism as a major geomorphic process in shaping surface structure is not restricted to the major planets of the solar system; the Earth's moon (see section 5.4) exhibits abundant evidence of past volcanism, the Galilean satellite Io is volcanically active (Morabito *et al.*, 1979), while the asteroid Vesta has a basaltic surface composition (see section 2.2.6) and various meteorites are igneous rocks. The erupted products on these minor bodies do not necessarily have compositions typical of terrestrial magmas; water has been suggested as a magma on Ganymede and Callisto (Parmentier and Head, 1979), while molten sulphur may be erupted on Io (Consolmagno, 1979). At present, photographic resolution is insufficient to identify volcanic features on the asteroids. Phobos and Deimos have primitive surface compositions (see Chapter 2) inconsistent with past volcanism. The saturnian satellites appear to have icy crusts (Cruikshank, 1979) like those of Europa, Ganymede and Callisto, while the compositions of the satellites of Uranus and Neptune are poorly known although they appear to be of low density like their parent planets (Newburn and Gulkis, 1975; Cruikshank *et al.*, 1977). Only Io has identifiable volcanic constructs which may be, in part, of basaltic composition.

5.8.2 Io

A total of nine eruption columns were detected on Io during the Voyager 1 flyby (Strom *et al.*, 1979). Of these, seven were seen to be still erupting during the Voyager 2 flyby; the most active had ceased, and one was not sufficiently close to the limb to be discriminated. These plumes ranged in height from 70–280 km and in diameter from 75–1000 km (Strom *et al.*, 1979). Four possible eruption sites lacking plumes were also identified on the Voyager 1 photographs (Strom *et al.*, 1979) and many more inactive volcanoes have been cataloged (Carr *et al.*, 1979). Io is therefore the most volcanically active body so far found in the solar system. The following descriptions of the volcanic features are taken largely from Carr *et al.* (1979) and Smith *et al.* (1979).

Calderas

There are five calderas larger than 20 km diameter for every 10^6 km ^{2} of the surface of Io and the largest has a diameter approaching 200 km (Plate 5.3). The calderas are separable into those with and those without associated flows; all are apparently rimless, have steep inner walls and flat floors (Fig. 5.8.1). Shapes range from circular with scalloped walls through rectilinear to elongate. The floors of the calderas commonly have low albedos (though some have higher albedos than their surrounds) and, in some, the floors have different elevations like the martian calderas. The relief of the structures on which the calderas are built is extremely low, virtually undetectable even where the structure is silhouetted at the limb, but suggested by the radial pattern of the associated flows.

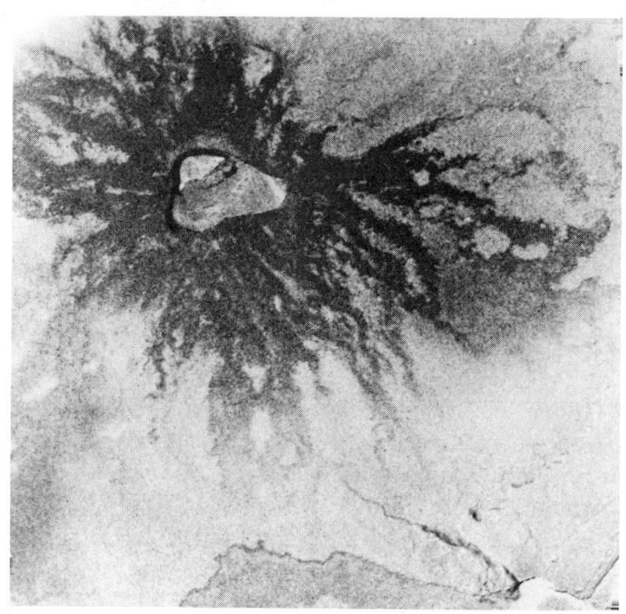

5.8.1 Pear-shaped caldera (–53°, 330°) on Io with dark radiating flows. Photo approximately 225 km in width (Voyager 1 0199J1).

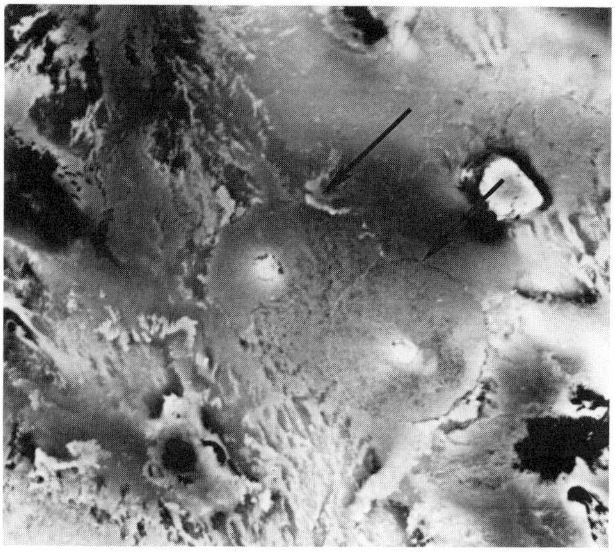

5.8.2 Two pancake-like features with central depressions at 345°W, 25°S on Io. Both features are surrounded by a low scarp, 150 km in diameter (Voyager 1 0071J1).

Positive relief features

"Pancakes" are structures with low outer scarps approximately 150 km in diameter and central, bright-floored, small craters (Fig. 5.8.2). They appear to cut, and thus be younger than, flows from nearby calderas.

"Butterflies" are comprised of calderas surrounded by wings of broad sheets up to 300 km across, each surrounded by a low escarpment. The sheets overlap one another, have no distinctive albedo pattern, and have a similar albedo to surrounding plains.

Faulted domes (Fig. 5.8.3) with diameters of about 200 km appear to have formed by uplift of an originally smooth, flat surface. The centers are usually occupied by irregular depressions connected to radial grabens.

Mountains, located predominantly in the polar regions, have a relief approaching 10 km and are characterized by parallel lineations (Fig. 5.8.4). Although not distinctively volcanic in appearance, one mountain has a caldera on its limb and is surrounded by a halo of high albedo material or sheets like those observed around the "butterflies."

Flows

Flows are observed around about half of the calderas and have been separated into dark and light groups. The dark flows (Fig. 5.8.5) have lengths ranging up to several hundred kilometers and widths up to tens of kilometers. They are further separable into long, narrow, lobate, radial and broad groups. Commonly, the dark flows are surrounded by a bright halo. The dark flows around a number of calderas have fairly regular lengths and commence right at the caldera wall or a dark apron surrounding the caldera. These features suggest that the flows may have arisen from overflow from lava lakes rather than being discrete eruptions. Light flows are also lobate and radial to calderas, extending, in places, 250 km from the caldera.

Smooth plains

Large areas of Io which do not contain the previously described features are featureless smooth plains. Some plains are cut by scarps which suggests that the plains are multiple layers of uniform thickness. The complete absence of impact craters larger than 1 km diameter on Io (Johnson *et al.*, 1979) argues against the plains being of impact origin. The most obvious candidates for the plains are low viscosity lavas or pyroclastic debris.

Sulphur or silicate volcanism?

The strong concentrations of sulphur in the atmosphere of Io have led to the proposition that many of the surface volcanic landforms on the planet may be the result of the eruption of liquid sulphur. Carr *et al.* (1979) suggest that the light flows could be sulphur but that the dark flows may be basaltic or at least silicate. Sulphur

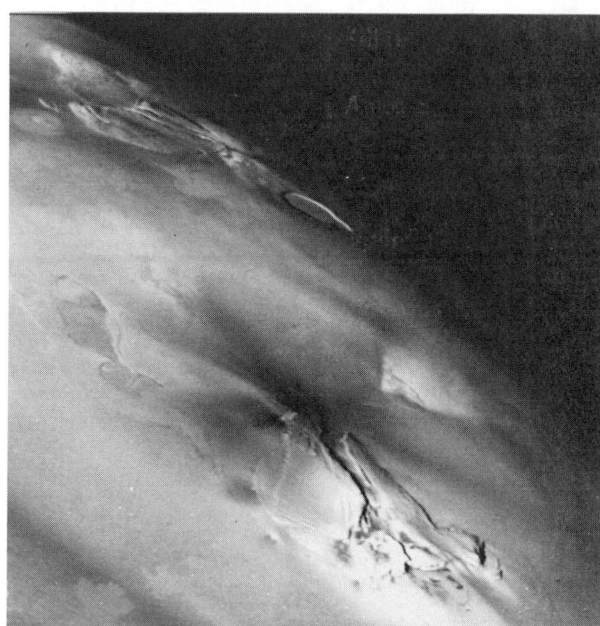

5.8.3 A large-domed, fractured feature of positive relief about 200 km across at about 250°W, 30°S. The dark central region appears to be the vent of one of the plumes (Voyager 1 0127J2).

5.8.4 Mountain in the south polar region of Io (39.6°W, 75.25° S). The mountain is approximately 200 km in width, has a pronounced low albedo halo, strong surface lineations and an apparent associated caldera (Voyager 1 0157J1).

lava flows are not unknown on Earth; Siretoko-Iwo-Zan extruded 200,000 tons of molten sulphur in 1936, while Kirisima has repeatedly erupted molten sulphur (Kuno, 1962), most recently in the winter of '78–'79 (SEAN bulletin). Furthermore, a possible sulphur lake occurs on Volcán Poás (Costa Rica; Bennett and Raccichini, 1978). Terrestrial sulphur-rich eruptions are, however, more typical of andesitic than basaltic volcanoes, although sulphide-rich layers are found in some ultrabasic intrusions (Wager and Brown, 1968) and small flows of sulphur have been reported on Hawaiian and Galapagos volcanoes (Colony and Nordlie, 1973; Skinner,1970). Andesitic eruptions, on the other hand, do not typically produce low profile volcanoes similar to those of Io although some pyroclastic shield volcanoes are of this form (Wood, 1977). The morphologic evidence would therefore indicate that the volcanic constructs on Io could be solid sulphur, or basalt, or solid sulphur and salic magma. Flows of sulphur share all the characteristic surface morphologies of basaltic flows (e.g., aa, pahoehoe, channels), differing only in color, which varies through yellow, red, brown, and black depending on temperature and cooling rate (Watanabe, 1940; Sagan, 1979). The common morphologies of erupted products of different compositions highlight the difficulties in identifying basaltic magmas from these criteria alone (Whitford-Stark, 1980).

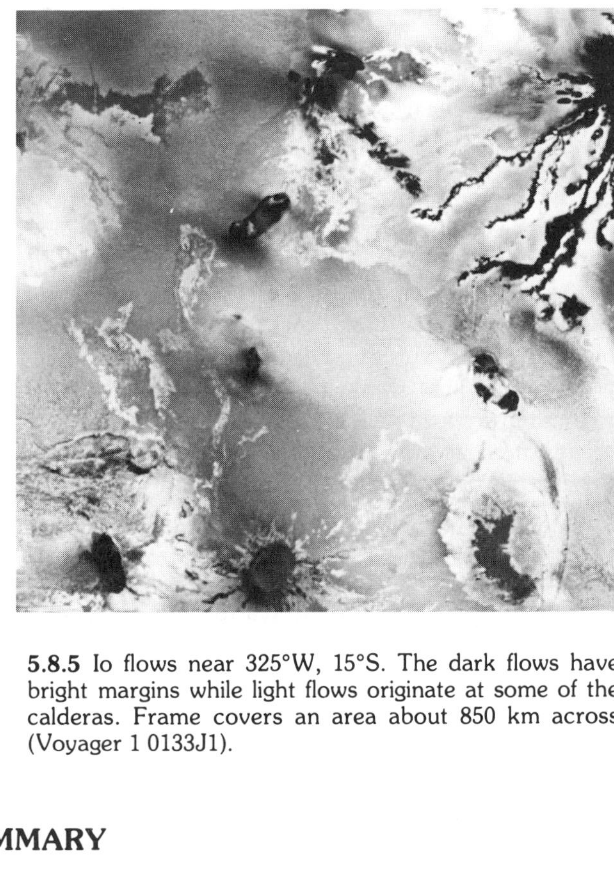

5.8.5 Io flows near 325°W, 15°S. The dark flows have bright margins while light flows originate at some of the calderas. Frame covers an area about 850 km across (Voyager 1 0133J1).

5.9 SUMMARY

This chapter has considered only the morphologic products of basaltic volcanism, but those products were shown to be extremely varied, depending upon the properties of the erupting material and the environment into which they were erupted. Magma rheology, gas content, and eruption rate were shown to be important controllers of the primary morphologies of basaltic volcanoes, while the local environment acted to produce a variety of specific landforms. Environmental factors included topography, the ambient pressure, the presence of water or ice, and the tectonic regime at the eruption site.

Terrestrial basaltic volcanoes vary from flat sheets to high cones and are predominantly concentratred on the ocean floor and at plate margins. Although rocks as old as 3.7 b.y. have been identified, most pristine volcanic landforms have been produced since the last ice age (<1 m.y.). Ice and water have greatly influenced both the construction and erosion of these landforms.

The Moon is dominated by the flat-lying mare basalts which infill impact basins and craters. These mare basalts were apparently produced at high eruption rates over a period extending from about 4.1–2.5 b.y. ago. Small constructs such as domes and cones have also been identified. Nonmare, KREEP, or highland basalts appear to have been widespread, though little morphologic evidence for this volcanism remains (scattered deposits of plains and isolated, steep-sided domes.) Extensive plains units in the lunar highlands may have been of volcanic origin, although their formation as impact ejecta is an alternative possibility.

Mercury is dominated by plains units which apparently formed over a time period similar to that evidenced by lunar volcanism. A volcanic origin for the mercurian plains has not been firmly established. Like the Moon, many of the mercurian plains appear to be basin-associated.

Mars has several large shield volcanoes but no large steep-sided cones like those of the Earth. As on the Earth, water and ice may have played important roles in both the formation and modification of volcanic constructs on Mars. Extensive plains units, recognized as volcanic by the presence of flow lobes, wrinkle ridges and cones, occupy large areas of the martian surface. The inception of shield-building volcanism (which commenced after the deposition of the early plains units) is believed to have taken place between 2.9 and 3.7 b.y. ago (Plescia and Saunders, 1979). Activity of Olympus Mons may continue to the present.

The surface of Venus is still poorly known, but the uncratered lowlands which comprise 20% of the mapped area of the planet may be young basaltic plains, while possible basaltic shield volcanoes are apparently associated with the continental areas. Ambiguous, linear ridges and trenches may be of tectonic origin and related to volcanism.

Io is the most volcanically active body so far observed in the solar system. There is not a single feature observed on its surface that can be ascribed a nonvolcanic origin. There is some debate as to whether the low relief volcanic constructs and their associated lavas are of sulphur or silicate composition or a mixture of both.

Many questions remain unanswered concerning the production of volcanic landforms on the terrestrial planets. Research into the processes involved in terrestrial volcanic eruptions is of prime importance to the understanding of volcanic activity both on this and other bodies. Accurate estimates of the total volumes of material erupted and the times when those eruptions took place would provide important constraints on the thermal and geochemical evolution of the planets. Such information is not available because of the difficulties in correctly identifying volcanic landforms and the lack of accurately defined chronologies. The return of samples from these other planets would help to rectify this situation.

ACKNOWLEDGMENTS—We would like to thank Peter Mouginis-Mark, Sally Bosworth and Doreen Hrabcsak for their assistance in manuscript preparation; I. B. Fridleiffson for his assistance and hospitality; H. Sigurdsson for expert field guidance; R. J. Strom, P. D. Spudis and P. W. Lipman for their meticulous reviews of an early manuscript draft. Photographs were provided by NSSDC.

A part of this work was performed under NASA Grant NGR 40-002-116, from the Planetary Geology Program (JWH).

5.10 REFERENCES

Ahmad M. U. (1972) Flood traps through space and time and their bearing on some problems of geotectonics. *Bull. Volc.* **35**, 539–563.

Aki K., Fehler M. and Das S. (1977) Source mechanism of volcanic tremor; fluid-driven crack models and their application to the 1963 Kilauea eruption. *J. Volcanol. Geotherm. Res.* **2**, 259–287.

Allen C. C. (1979) Volcano-ice interaction on Mars. *J. Geophys. Res.* **84**, 8048–8059.

Ballard R. D. and Moore J. G. (1977) *Photographic Atlas of the Mid-Atlantic Ridge*. Springer-Verlag, N.Y. 114 pp.

Ballard R. D. and van Andel T. H. (1977) Morphology and tectonics of the inner rift valley at latitude 36°50′N on the Mid-Atlantic Ridge. *Geol. Soc. Amer. Bull.* **88**, 507–530.

Ballard R. D., Holcomb R. T. and van Andel T. H. (1979) The Galapagos rift at 86°W 3. Sheet flows, collapse pits and lava lakes of the rift valley. *J. Geophys. Res.* **84**, 5407–5422.

Barth T. F. W. (1950) *Volcanic geology, hot springs and geysers of Iceland*. Publ. 587, Carnegie Inst. of Washington. 174 pp.

Bennett F. D. (1971) Vaporization waves in explosive volcanism. *Nature* **234**, 538–539.

Bennett F. D. and Raccichini S. M. (1978) Subaqueous sulphur lake in Volcan Poas. *Nature* **271**, 342–344.

Björnsson A., Johnsen G., Sigurdsson O., Thorbergsson G. and Tryggvason E. (1979) Rifting of the plate boundary in northern Iceland. *J. Geophys. Res.* **84**, 3029–3038.

Björnsson A., Saemundsson K., Einarsson P., Tryggvason E. and Grönvold K. (1977) Current rifting episode in North Iceland. *Nature* **266**, 318–323.

Blackburn E. A., Wilson and Sparks R. S. J. (1976) Mechanisms and dynamics of strombolian activity. *J. Geol. Soc. Lond.* **132**, 429–440.

Blasius K. R. (1976) The record of impact cratering on the great volcanic shields of the Tharsis region of Mars. *Icarus* **29**, 343–361.

Bloomfield K. (1975) A late-Quaternary monogenetic volcano field in Central Mexico. *Geol. Rundsch.* **64**, 476–497.

Booth B. (1973) The Granadilla pumice deposit of southern Tenerife, Canary Islands. *Proc. Geol. Assoc.* **84**, 353–370.

Bottinga Y. and Weill D. F. (1972) The viscosity of magmatic silicate liquids: a model for calculation. *Amer. J. Sci.* **272**, 438–475.

Boyce J. M. (1976) Ages of flow units in lunar nearside maria based on Lunar Orbiter 4 photographs. *Proc. Lunar Sci. Conf. 7th*, p. 2717–2728.

Boyd F. R. (1961) Welded tuffs and flows in the rhyolite Plateau of Yellowstone Park, Wyoming. *Geol. Soc. Amer. Bull.* **72**, 387–426.

Breed W. J. (1964) Morphology and lineation of cinder cones in the San Francisco volcanic field. *Mus. N. Ariz. Bull.* **40**, 65–71.

Brett R. (1975) Thickness of some lunar mare basalt flows and ejecta blankets based on chemical kinetic data. *Geochim. Cosmochim. Acta* **39**, 1135–1141.

Brooks C. K., Nielsen T. F. D. and Petersen T. S. (1976) The Blosseville Coast basalts of East Greenland: their occurrence, composition, and temporal variations. *Contrib. Mineral. Petrol.* **58**, 279–292.

Bryan W. B. (1964) Relative abundance of intermediate members of the oceanic basalt-trachyte association: evidence from Clarion and Socorro Islands, Revillagigedo Islands, Mexico. *J. Geophys. Res.* **69**, 3047–3049.

Bryan W. B. (1973) Wrinkle ridges as deformed surface crust on ponded mare lava. *Proc. Lunar Sci. Conf. 4th*, p. 93–106.

Bryan W. B. and Adams M. L. (1973) Some volcanic and structural features of Mare Serenitatis. *Apollo 17 Prelim. Sci. Rept.*, p. 3009–3012. NASA SP-330.

Bryan W. B. and Moore J. G. (1977) Compositional variations of young basalts in the Mid-Atlantic rift valley near latitude 36°49′N. *Geol. Soc. Amer. Bull.* **88**, 556–570.

Bultitude R. J. (1976) Flood basalts of probable early Cambrian age in northern Australia. In *Volcanism in Australasia* (R. W. Johnson, ed.), p. 1–20. Elsevier, Amsterdam.

Burnham C. W. (1975) Water and magmas; a mixing model. *Geochim. Cosmochim. Acta* **39**, 1077–1084.

Carr M. H. (1973) Volcanism on Mars. *J. Geophys. Res.* **78**, 4049–4062.

Carr M. H. (1974) The role of lava erosion in the formation of lunar rilles and martian channels. *Icarus* **22**, 1–23.

Carr M. H. (1975) The volcanoes of Mars. *Sci. Amer.* **234**, 32–43.

Carr M. H. (1979) Formation of martian flood features by release of water from confined aquifers. *J. Geophys. Res.* **84**, 2995–3007.

Carr M. H., Greeley R., Blasius K. R., Guest J. E. and Murray J. B. (1977) Some martian volcanic features as viewed from the Viking orbiters. *J. Geophys. Res.* **82**, 3985–4015.

Carr M. H., Masursky H., Strom R. and Terrile R. (1979) The volcanic features of Io. *Nature* **280**, 729–733.

Chibber H. L. (1934) *Geology of Burma*. Macmillian, London. 538 pp.

Chouet B., Hamisevicz N. and McGetchin T. R. (1974) Photoballistics of volcanic jet activity at Stromboli, Italy. *J. Geophys. Res.* **79**, 4961–4976.

Cintala M. J., Wood C. A. and Head J. W. (1977) The effects of target characteristics on fresh crater morphology: preliminary results from the Moon and Mercury. *Proc. Lunar Sci. Conf. 8th*, p. 3409–3426.

Clough C. T., Maufe H. B. and Bailey E. B. (1909) The cauldron subsidence of Glen Coe and the associated igneous phenomena. *Quart. J. Geol. Soc. Lond.* **65**, 611–676.

Cole J. W. and Nairn I. A. (1975) *Catalogue of the Active Volcanoes of the World. Part XXII, New Zealand.* International Association of Volcanology & Chemistry of the Earth's Interior, Tokyo. 156 pp.

Colgate S. A. and Sigurgeirsson T. (1973) Dynamic mixing of water and lava. *Nature* **244**, 552–555.

Colony W. E. and Nordlie B. E. (1973) Liquid sulphur at Volcan Azufre, Galapagos Islands. *Econ. Geol.* **68**, 371–380.

Colton G. W., Howard K. A. and Moore H. J. (1972) Mare ridges and arches in southern Oceanus Procellarum. *Apollo 16 Prelim. Sci. Rept.*, p. 29.90–29.93. NASA SP-315.

Consolmagno G. J. (1979) Sulphur volcanoes on Io. *Science* **205**, 397–398.

Cox K. G. (1972) The Karroo volcanic cycle. *J. Geol. Soc. Lond.* **128**, 311–336.

Crane K. and Ballard R. D. (1980) The Galapagos Rift at 86°W, 4, Structure and morphology of hydrothermal fields and their relationship to the volcanic and tectonic processes of the Rift Valley. *J. Geophys. Res.* **85**, 1443–1454.

Cruikshank D. P. (1979) The surfaces and interiors of Saturn's satellites. *Rev. Geophys. Space Phys.* **17**, 165–176.

Cruikshank D. P. and Wood C. A. (1972) Lunar rilles and Hawaiian volcanic features: possible analogues. *Moon* **3**, 412–447.

Cruikshank D. P., Hartmann W. K. and Wood C. A. (1973) Moon: "ghost" craters formed during mare filling. *The Moon* **7**, 440–452.

Cruikshank D. P., Pilcher C. B. and Morrison D. (1977) Identification of a new class of satellites in the outer solar system. *Astrophys. J.* **217**, 1006–1010.

Crumpler L. S. and Aubele J. C. (1978) Structural evolution of Arsia Mons, Pavonis Mons, and Ascraeus Mons: Tharsis region of Mars. *Icarus* **34**, 496–511.

Davies D. K., Quearry M. W. and Bonis S. B. (1978) Glowing avalanches from the 1974 eruption of the volcano Fuego, Guatemala. *Geol. Soc. Amer. Bull.* **89**, 369–384.

Decker R. W. and Kinoshita W. T. (1971) Geodetic measurements. In *The Surveillance and Prediction of Volcanic Activity: A Review of Methods and Techniques.* UNESCO, Paris, **8**, 44–72.

DeHon R. A. (1971) Cauldron subsidence in lunar craters, Ritter and Sabine. *J. Geophys. Res.* **76**, 5712–5718.

DeHon R. A. (1974) Thickness of mare material in the Tranquillitatis and Nectaris basins. *Proc. Lunar Sci. Conf. 5th*, p. 53–59.

DeHon R. A. (1979) Preliminary thickness study of plains-forming materials in the Caloris Basin Region of Mercury. In *Reports of Planetary Geology Program, 1978–1979*, p. 65–57. NASA TM-80339.

DeHon R. A. and Waskom J. D. (1976) Geologic structure of the eastern mare basins. *Proc. Lunar Sci. Conf. 7th*, p. 2729–2746.

Dimroth E., Cousineau P., Leduc M. and Sanschagrin Y. (1978) Structure and organization of Archean subaqueous basalt flows, Rouyn-Noranda area, Quebec, Canada. *Can. J. Earth Sci.* **15**, 902–918.

Donnelly T. W., Rogers J. J. W., Pushkar P. and Armstrong R. L. (1971) Chemical evolution of the igneous rocks of the eastern West Indies. *Geol. Soc. Amer. Mem.* **130**, 181–224.

Dzurisin D. (1978) The tectonic and volcanic history of Mercury as inferred from studies of scarps, ridges, troughs, and other lineaments. *J. Geophys. Res.* **83**, 4883–4906.

Einarsson P. (1979) Seismological evidence for horizontal propogation of dykes along the Krafla fault swarm during deflation events of the Krafla volcano in NE-Iceland. *Hawaiian Symposium on Intraplate Volcanism and Submarine Volcanism*, p. 39.

Elston W. and Erb E. E. (1981) Ring complexes. *Encyclopedia of Volcanology*.

Fedotov S. A. (1978) Ascent of basic magmas in the crust and the mechanism of basaltic fissure eruptions. *Int. Geol. Rev.* **20**, 33–48.

Fiske R. S. and Jackson E. D. (1972) Orientation and growth of Hawaiian volcanic rifts: the effect of regional structure and gravitational stresses. *Proc. Roy. Soc. Lond., Ser. A.* **329**, 299–326.

Fiske R. S. and Kinoshita W. T. (1969) Inflation of Kilauea volcano prior to its 1967–1968 eruption. *Science* **165**, 341–349.

Fitch F. J. (1964) The development of the Beerenberg Volcano, Jan Mayen. *Proc. Geol. Assoc.* **75**, 133–165.

Florensky C. P., Ronca L. B., Basilevsky A. T., Burba G. A., Nikolaeva O. V., Pronin A. A., Trakhtman A. M., Volkov V. P. and Zazetsky V. V. (1977) The surface of Venus as revealed by Soviet Venera 9 and 10. *Geol. Soc. Amer. Bull.* **88**, 1537–1545.

Francis P. W. and Abbott B. M. (1973) Sizes of conical volcanoes. *Nature* **244**, 22–23.

Frey H., Lowry B. L. and Chase S. A. (1979) Pseudocraters on Mars. *J. Geophys. Res.* **84**, 8075–8086.

Furnes H. and Fridleifsson I. B. (1974) Tidal effects on the formation of pillow lava/hyaloclastite deltas. *Geology* **5**, 381–384.

Gass I. G. and Smewing J. D. (1973) Intrusion, extrusion and metamorphism at constructive margins: evidence from the Troodos Massif, Cyprus. *Nature* **242**, 26–29.

Geiss J., Eberhardt P., Grögler N., Guggisberg S., Maurer P. and Stettler A. (1977) Absolute time scale of lunar mare formation and filling. *Phil. Trans. Roy. Soc. Lond.* **A285**, 151–158.

Ghose N. C. (1972) Chemical characteristics of some basaltic rock of India. *Bull. Volcanol.* **35**, 1022–1036.

Gifford A. W. and El-Baz F. (1978) Thickness of mare flow fronts (abstract). In *Lunar and Planetary Science IX*, p. 382–384. Lunar and Planetary Institute, Houston.

Glickson A. Y. (1976) Earliest Precambrian ultramafic-mafic volcanic rocks: ancient oceanic crust or relic terrestrial maria? *Geology* **4**, 201–205.

Goldstein R. M., Green R. R. and Rumsey H. C. (1976) Venus radar images. *J. Geophys. Res.* **81**, 4807–4817.

Greeley R. (1970) Observations of actively forming lava tubes and associated structures, Hawaii. *Mod. Geol.* **2**, 207–223.

Greeley R. (1971a) Lava tubes and channels in the lunar Marius Hills. *Moon* **3**, 289–314.

Greeley R. (1971b) Lunar Hadley Rille: considerations of its origin. *Science* **172**, 722–725.

Greeley R. (1971c) Geology of selected lava tubes in the Bend Area. *Bull. 71. State of Oregon Dept. Geol. and Miner. Indust.*, Portland, Oregon. 47 pp.

Greeley R. (1972) Additional observations of actively forming lava tubes and associated structures, Hawaii. *Mod. Geol.* **3**, 157–160.

Greeley R. (1974) Geologic guide to the Island of Hawaii. *A field guide for comparative planetary geology*, p. 257. NASA CR-152416.

Greeley R. (1976) Modes of emplacement of basalt terrains and an analysis of mare volcanism in the Orientale basin. *Proc. Lunar Sci. Conf. 7th*, p. 2747–2759.

Greeley R. (1977) Basaltic "plains" volcanism. In *Volcanism of the Eastern Snake River Plain, Idaho* (R. Greeley and J. S. King, eds.), p. 23–44. NASA CR-154621.

Greeley R. and Hyde J. H. (1972) Lava tubes of the Cave basalts, Mount St. Helens, Washington. *Geol. Soc. Amer. Bull.* **83**, 2397–2418.

Greeley R. and King J. S. (1977) *Volcanism of the Eastern Snake River Plain, Idaho*, 308 pp. NASA CR-154621.

Greeley R. and Schultz P. H. (1978) Possible planetary analogs to Snake River Plain basalt features. In *Volcanism of the Eastern Snake River Plain, Idaho* (R. Greeley and J. S. King, eds.), p. 233–251. NASA CR-154621.

Greeley R. and Spudis P. D. (1978) Volcanism in the cratered terrain hemisphere of Mars. *Geophys. Res. Lett.* **5**, 453–455.

Greeley R. and Spudis P. (1981) Volcanism on Mars. *Rev. Geophys. Space Phys.*, in press.

Greeley R., Storm D. and Wilbur C. (1976) Frequency distribution of lava tubes and channels on Mauna Loa Volcano, Hawaii. *Geol. Soc. Amer. Abs.* **v. 8**, 892.

Greeley R., Theilig E., Guest J. E., Carr M. H., Masursky H. and Cutts J. A. (1977) Geology of Chryse Planitia. *J. Geophys. Res.* **82**, 4093–4109.

Guest J. E. (1971) Centers of igneous activity in the maria. In *Geology and Geophysics of the Moon* (G. Fielder, ed.), p. 41–53. Elsevier, N. Y.

Guest J. E. and Murray J. B. (1976) Volcanic features of the nearside equatorial lunar maria. *J. Geol. Soc. Lond.* **132**, 251–258.

Guest J. E. and O'Donnell W. P. (1976) Surface history of Mercury: a review. *Vistas in Astron.* **20**, 273–300.

Guest J. E., Butterworth P. S. and Greeley R. (1977) Geological observations in the Cydonia Region of Mars from Viking. *J. Geophys. Res.* **82**, 4111–4120.

Gutmann J. T. (1979) Structure and eruptive cycle of cinder cones in the Pinacate volcanic field and the controls of strombolian activity. *J. Geol.* **87**, 448–454.

Hall J. L., Head J. W. and Solomon S. C. (1979) Lunar floor-fractured craters: quantitative tests of hypotheses for their origin. In *Reports of Planetary Geology Program, 1978–1979*, p. 129–131. NASA TM-80339.

Hamilton W. and Myer W. B. (1963) Menan Buttes cones of glassy tuff in the Snake River Plain, Idaho. *U. S. Geol. Surv. Prof. Paper 450 E-2*, E114–118.

Hapke B., Danielson G. E., Jr., Klausen K. and Wilson L. (1975) Photometric observations of Mercury from Mariner 10. *J. Geophys. Res.* **80**, 2431–2443.

Harris S. A. (1977) The aureole of Olympus Mons, Mars. *J. Geophys. Res.* **82**, 3099–3107.

Hartmann W. K. and Wood C. A. (1971) Moon: origin and evolution of multiring basins. *The Moon* **3**, 3–78.

Hawke B. R. and Head J. W. (1978) Lunar KREEP volcanism: geologic evidence for history and mode of emplacement. *Proc. Lunar Planet. Sci. Conf. 9th*, p. 3285–3309.

Head J. W. (1974a) Lunar dark-mantle deposits: possible clues to the distribution of early mare deposits. *Proc. Lunar Sci. Conf. 5th*, p. 207–222.

Head J. W. (1974b) Morphology and structure of the Taurus-Littrow highlands (Apollo 17): evidence for their origin and evolution. *The Moon* **9**, 355–395.

Head J. W. (1974c) Orientale multi-ring basin interior and implications for the petrogenesis of lunar highland samples. *The Moon* **11**, 327–356.

Head J. W. (1975a) Morphology of pyroclastic lunar volcanic deposits: implications for eruption conditions and localized sources of volatiles (abstract). In *Lunar Science VI*, p. 349–351. The Lunar Science Institute, Houston.

Head J. W. (1975b) Modes of occurrence and style of emplacement of lunar mare deposits (abstract). In *Origins of Mare Basalts and Their Implications for Lunar Evolution*, p. 61–65. The Lunar Science Institute, Houston.

Head J. W. (1976) Lunar volcanism in space and time. *Rev. Geophys. Space Phys.* **14**, 265–300.

Head J. W. (1979a) Lava flooding of early planetary crusts: geometry, thickness, and volumes of flooded impact basins (abstract). In *Lunar and Planetary Science X*, p. 516–518. Lunar and Planetary Institute, Houston.

Head J. W. (1979b) Lava flooding of early planetary crusts: geometry, thickness, and volumes of flooded lunar highland terrain (abstract). In *Lunar and Planetary Science X*, p. 519–521. Lunar and Planetary Institute, Houston.

Head J. W. (1979c) Serenitatis multi-ringed basin: regional geology and basin ring interpretation. *The Moon and Planets* **21**, 439–462.

Head J. W. and Gifford A. (1980) Lunar mare domes: classification and mode of origin. *The Moon and Planets* **22**, 235–258.

Head J. W. and McCord T. B. (1978) Imbrian-age highland volcanism on the Moon: the Gruithuisen and Mairan domes. *Science* **199**, 1433–1436.

Head J. W. and Wilson L. (1979) Alphonsus-type dark-halo craters: morphology, morphometry and eruption conditions. *Proc. Lunar Planet. Sci. Conf. 10th*, p. 2861–2897.

Head J. W. and Wilson L. (1980) The formation of eroded depressions around the sources of lunar sinuous rilles: observations (abstract). In *Lunar and Planetary Science XI*, p. 426–428. Lunar and Planetary Institute, Houston.

Head J. W., Adams J. B., McCord T. B., Pieters C. and Zisk S. (1978a) Regional stratigraphy and geologic history of Mare Crisium. In *Mare Crisium: The view from Luna 24* (R. B. Merrill and J. J. Papike, eds.), p. 43–74. Pergamon, N. Y.

Head J. W., Adams J. B., McCord T. B., Pieters C. and Zisk S. (1978b) Definition and detailed characterization of lunar surface units using remote observations. *Icarus* **33**, 145–172.

Head J. W., Settle M. and Wood C. A. (1976) Origin of Olympus Mons escarpment by erosion of pre-volcano substrate. *Nature* **263**, 667–668.

Heiken G. H. (1971) Tuff rings: examples from the Fort Rock-Christmas Lake Valley Basin, southcentral Oregon. *J. Geophys. Res.* **76**, 5615–5626.

Heiken G. H. (1972) Morphology and petrography of volcanic ashes. *Geol. Soc. Amer. Bull.* **83**, 1961–1988.

Heiken G. H., McKay D. S. and Brown R. W. (1974) Lunar deposits of possible pyroclastic origin. *Geochim. Cosmochim. Acta* **38**, 1703–1718.

Herz N. (1966) Tholeiitic and alkaline volcanism in southern Brazil. In *Guidebook Int. Field Inst. Brazil*, Chapter 5. p. V1–V6 Tab., Map A Washington.

Hinners N. W. (1972) Apollo 16 site selection. *Apollo 16 Prelim. Sci. Rept.*, p. 1.1–1.3. NASA SP-315.

Hodges C. A. (1973) Mare ridges and lava lakes. *Apollo 17 Prelim. Sci. Rept.*, p. 31.12–31.21. NASA SP-330.

Hodges C. A. and Moore H. J. (1978a) Table mountains of Mars (abstract). In *Lunar and Planetary Science IX*, p. 523–525. Lunar and Planetary Institute, Houston.

Hodges C. A. and Moore H. J. (1978b) The subglacial birth of Olympus Mons. *Abs. with Program, Geol. Soc. Amer.* **10**, 422.

Hodges C. A. and Moore H. J. (1979) The subglacial birth of Olympus Mons and its aureoles. *J. Geophys. Res.* **84**, 8061–8074.

Hollister C. D., Glenn M. F. and Lonsdale P. F. (1978) Morphology of seamounts in the West Pacific and Philippine Basin from multi-beam sonar data. *Earth Planet. Sci. Lett.* **41**, 405–418.

Hörz F. (1978) How thick are lunar mare basalts? *Proc. Lunar Planet. Sci. Conf. 9th*, p. 3311–3332.

Houghton R. L., Heirtzler J. R., Ballard R. D. and Taylor P. T. (1977) Submersible observations of the New England seamounts. *Naturwissenschaften* **64**, 348–355.

Housley R. M. (1978) Modeling lunar volcanic eruptions. *Proc. Lunar Planet. Sci. Conf. 9th*, p. 1473–1484.

Howard K. A. and Muehlberger W. R. (1973) Lunar thrust faults in the Taurus-Littrow region. *Apollo 17 Prelim. Sci. Rept.*, p. 31.22–31.25. NASA SP-330.

Howard K. A. and Wilshire H. G. (1975) Flows of impact melt at lunar craters. *J. Res. U. S. Geol. Surv.* **3**, 237–251.

Howard K. A., Carr M. H. and Muehlberger W. R. (1973) Basalt stratigraphy of southern Mare Serenitatis. *Apollo 17 Prelim. Sci. Rept.*, p. 29.10. NASA SP-330.

Howard K. A., Wilhelms D. E. and Scott D. H. (1974) Lunar basin formation and highland stratigraphy. *Rev. Geophys. Space Phys.* **12**, 309–327.

Hughes C. J. (1970) Major rhythmic layering in ultramafic rocks of the Great Dyke of Rhodesia with particular reference to the Sebakwe Area. In *Symposium on the Bushveld Igneous Complex and other layered intrusions.* Geol. Soc. S. Africa Spec. Pub. 1, 594–609.

Hulme G. (1973) Turbulent lava flow and the formation of lunar sinuous rilles. *Mod. Geol.* **4**, 107–117.

Hulme G. (1974a) The interpretation of lava flow morphology. *Geophys. J. Roy. Astron. Soc.* **39**, 361–383.

Hulme G. (1974b) Generation of magma at lunar impact crater sites. *Nature* **252**, 556–558.

Hulme G. (1976) The determination of the rheological properties and effusion rate of an Olympus Mons Lava. *Icarus* **27**, 207–213.

Hulme G. and Fielder G. (1977) Effusion rates and rheology of lunar lavas. *Phil. Trans. Roy. Soc. Lond.* **A285**, 227–234.

Husain L. and Schaeffer O. A. (1973) Lunar volcanism: age of the glass in the Apollo 17 orange soil. *Science* **180**, 1358–1360.

Jackson E. D. (1976) Linear volcanic chains on the Pacific plate. In *The geophysics of the Pacific Ocean basin and its margins* (G. H. Sutton, M. H. Manghnani and R. Moberly, eds.), p. 319–335. Geophysical monograph 19. *Amer. Geophy. Union,* Washington, D. C.

Johnson A. M. and Pollard D. D. (1973) Mechanics of growth of some laccolithic intrusions in the Henry Mountains, Utah I. Field observations, Gilbert's model, physical properties and flow of the magma. *Tectonophys.* **18**, 261–309.

Johnson R. W., Davies R. A. and White A. J. A. (1972) Ulawun Volcano, New Britain. *Bur. Min. Res. Geol. & Geophys. Australia Bull.* **142**, 42.

Johnson T. V., Cook A. F., Sagan C. and Soderblom L. A. (1979) Volcanic resurfacing rates and implications for volatiles on Io. *Nature* **280**, 746–750.

Johnson T. V., Matson D. L., Phillips R. J. and Saunders R. S. (1975) Vidicon spectral images: a geophysical study of the Lamont feature (abstract). In *Lunar Science VI*, p. 444. The Lunar Science Institute, Houston.

Jones J. G. (1969) Pillow lavas as depth indicators. *Amer. J. Sci.* **267**, 181–195.

Karlstrom T. N. V., McCauley J. F. and Swann G. A. (1968) Preliminary lunar exploration plan of the Marius Hills region of the Moon. *U. S. Geol. Surv. Interagency Report: Astrogeology* **5**, 42 pp.

Keller J. (1974) Quaternary maar volcanism near Karapinar in central Anatolia. *Bull. Volcanol.* **38**, 378–396.

Kienle J., Kyle P. R., Self S., Motyka R. J. and Lorenz V. (1980) Ukinrek maars, Alaska 1. April 1977 eruption sequence, petrology and tectonic setting. *J. Volcanol. Geotherm. Res.* **7**, 11–38.

King J. S. and Riehle J. R. (1974) A proposed origin of the Olympus Mons escarpment. *Icarus* **33**, 300–317.

Kirkpatrick J. R. (1976) Towards a kinetic model for the crystallization of magma bodies. *J. Geophys. Res.* **81**, 2565–2571.

Kirkpatrick J. R. (1977) Nucleation and growth of plagioclase, Makaopuhi and Alae lava lakes, Kilauea Volcano, Hawaii. *Geol. Soc. Amer. Bull.* **88**, 78–84.

Kuno H. (1962) *Catalogue of the active volcanoes of the world. Part XI: Japan, Taiwan, and Marianas.* International Association of Volcanology & Chemistry of the Earth's Interior, Tokyo. 332 pp.

Kushiro I. (1976) Changes in viscosity and structure of melt of $NaAlSi_2O_6$ composition at high pressure. *J. Geophys. Res.* **81**, 6247–6350.

Kushiro I., Yoder H. S., Jr. and Mysen B. O. (1976) Viscosities of basalt and andesite melts at high pressure. *J. Geophys. Res.* **81**, 6351–6356.

Leake M. A. (1979) The intercrater plains of Mercury. In *Reports of the Planetary Geology Program, 1978-1979*, p. 154–155. NASA TM-80339.

Lonsdale P. (1977) Abyssal pahoehoe with lava coils at the Galapagos Rift. *Geology* **5**, 147–152.

Lopes R., Guest J. E. and Wilson L. (1980) Origin of the Olympus Mons aureole and perimeter scarp. *Moon and Planets* **22**, 221–234.

Lorenz V. (1973) On the formation of maars. *Bull. Volcanol.* **37**, 183–204.

Lorenz V. (1975) Formation of phreatomagmatic maar-diatreme volcanoes and its relevance to kimberlite diatremes. *Phys. Chem. Earth* **9**, 17–27.

Lorenz V., McBirney A. R. and Williams H. (1970) *An investigation of volcanic depressions, Pt. III. Maars, tuff-rings, tuff-cones, and diatremes.* 198 pp. NASA CR-115236.

LSPET (1972) A petrographic and chemical description of samples from the lunar highlands. *Apollo 16 Prelim. Sci. Rept.*, p. 7.1–7.24. NASA SP-315.

Lucchitta B. K. (1976) Mare ridges and related highland scarps—result of vertical tectonism? *Proc. Lunar Sci. Conf. 7th*, p. 2761–2782.

Lucchitta B. K. (1977) Topography, structure, and mare ridges in southern Mare Imbrium and northern Oceanus Procellarum. *Proc. Lunar Sci. Conf. 8th*, p. 2691–2703.

Lucchitta B. K. and Schmitt H. H. (1974) Orange material in the Sulpicius Gallus Formation at the southwestern edge of Mare Serenitatis. *Proc. Lunar Sci. Conf. 5th*, p. 223–224.

Ludden J. N. (1978) Eruptive patterns for the volcano Piton de la Fournaise, Reunion Island. *J. Volcanol. Geotherm. Res.* **2**, 385–395.

Macdonald G. A. (1972) *Volcanoes.* Prentice-Hall, N. J. 510 pp.

Macdonald G. A. and Abbott A. T. (1970) *Volcanoes in the Sea.* Univ. of Hawaii Press, Honolulu. 440 pp.

Makarenko G. F. (1974) Trap areas of continents and basalt areas of oceans—areal comparison. *Moscow Univ. Geol. Bull.* **29**, 1–19.

Malin M. C. (1974) Lunar red spots; possible pre-mare materials. *Earth Planet. Sci. Lett.* **21**, 331–341.

Malin M. C. (1976) Observations of intercrater plains on Mercury. *Geophys. Res. Lett.* **3**, 581–584.

Malin M. C. (1977) Comparison of volcanic features of Elysium (Mars) and Tibesti (Earth). *Geol. Soc. Amer. Bull.* **88**, 908–919.

Malin M. C. (1978) Surfaces of Mercury and the Moon: effects of resolution and lighting conditions on the discrimination of volcanic features. *Proc. Lunar Planet. Sci. Conf. 9th*, p. 3395–3409.

Masursky H. (1973) An overview of geologic results from Mariner 9. *J. Geophys. Res.* **78**, 4000–4031.

Masursky H., Eliason E., Ford P. G., McGill G. E., Pettengill G. H., Schaber G. G. and Schubert G. (1980) Pioneer Venus radar results: geomorphology from imaging and altimetry (abstract). In *Lunar and Planetary Science XI*, p. 683–685. Lunar and Planetary Institute, Houston.

Masursky H., Kaula W. M., McGill G. E., Pettengill G. H., Phillips R. J., Russell C. T., Schubert G. and Shapiro I. I. (1977) The surface and interior of Venus. *Space Sci. Rev.* **20**, 431–449.

Mathews W. H. (1947) "Tuyas", flat-topped volcanoes in northern British Columbia. *Amer. J. Sci.* **245**, 560–570.

Mattingly T. F., El-Baz F. and Laidley R. A. (1972) Observations and impressions from lunar orbit. *Apollo 16 Prelim. Sci. Rept.* p. 28.1–28.16. NASA SP-315.

Maxwell T. A., El-Baz F. and Ward S. H. (1975) Distribution, morphology, and origin of ridges and arches in Mare Serenitatis. *Geol. Soc. Amer. Bull.* **86**, 1273–1278.

McBirney A. R. (1963) Factors governing the nature of submarine volcanism. *Bull. Volcanol.* **26**, 455–469.

McBirney A. R. and Noyes R. M. (1979) Crystallization and layering of the Skaergaard Intrusion. *J. Petrol.* **20**, 487–554.

McCauley J. F. (1967) Geologic map of the Hevelius region of the Moon. *U. S. Geol. Surv. Misc. Inv.*, Map I-491.

McCauley J. F. (1968) Geologic results from lunar precursor probes. *Amer. Inst. Aeronautics & Astronautics J.* **6**, 1991–1996.

McCauley J. F. (1969) The domes and cones in the Marius Hills region: evidence for lunar differentiation? *EOS (Trans. Amer. Geophys. Union)* **50**, 229.

McCord T. B. and Clark R. N. (1979) The Mercury soil: presence of Fe^{2+} (abstract). In *Lunar and Planetary Science X*, p. 789–791. Lunar and Planetary Institute, Houston.

McGetchin T. R. and Ullrich G. W. (1973) Xenoliths in maars and diatremes with inferences for the Moon, Mars and Venus. *J. Geophys. Res.* **78**, 1833–1853.

McGetchin T. R., Settle M. and Chouet B. A. (1974) Cinder cone growth modeled after Northeast Crater, Mount Etna, Sicily. *J. Geophys. Res.* **79**, 3257–3272.

Melson W. G., Jarosewich M. E., Switzer G. and Thompson G. (1972) Basaltic nuée ardentes of the 1970 eruption of Ulawun Volcano, New Britain. *Smithsonian Contrib. Earth Sci.* **9**, 15–32.

Menard H. W. (1964) *Marine geology of the Pacific.* McGraw-Hill, N. Y. 271 pp.

Metzger A. E., Haines E. L., Etchegaray-Ramirez M. I. and Hawke B. R. (1979) Thorium concentrations in the lunar surface: III. Deconvolution of the Apenninus region. *Proc. Lunar Planet. Sci. Conf. 10th*, p. 1701–1718.

Meyer J. D. and Grolier M. J. (1977) Geologic map of the Syrtis Major Quadrangle of Mars. *U. S. Geol. Surv. Misc. Geol. Inv.*, Map 1-995.

Milne J. (1878) On the form of volcanoes. *Geol. Mag. Dec. II* **5**, 337–345.

Milton D. J. and Hodges C. A. (1972) Geologic map of the Descartes region of the Moon. Apollo 16 Pre-mission Maps. *U. S. Geol. Surv. Misc. Inv. Ser.*, Map I-748.

Mohr P. A. (1971) Ethiopian tertiary dike swarms. *Smithsonian Astrophys. Obs. Spec. Rep.*, 339.

Mohr P. A. (1972) Regional significance of volcanic geochemistry in the Afar triple junction. *Geol. Soc. Amer. Bull.* **83**, 213–221.

Mohr P. A. and Wood C. A. (1976) Volcano spacings and lithospheric attenuation in the Eastern Rift of Africa. *Earth Planet. Sci. Lett.* **33**, 126–144.

Moore H. J. (1965) Geologic map of the Aristarchus region of the Moon. *U. S. Geol. Surv. Misc. Inv. Ser.*, Map I-465.

Moore H. J. (1967) Geologic map of the Seleucus Quadrangle of the Moon. *U. S. Geol. Surv. Misc. Inv. Ser.*, Map I-527.

Moore H. J. and Schaber G. G. (1975) An estimate of the yield strength of the Imbrium flows. *Proc. Lunar Sci. Conf. 6th*, p.101–118.

Moore H. J., Arthur D. W. G. and Schaber G. G. (1978) Yield strengths of flows on the Earth, Mars and Moon. *Proc. Lunar Planet. Sci. Conf. 9th*, p. 3351–3378.

Moore J. G., Phillips R. L., Grigg R. W., Peterson D. W. and Swanson D. A. (1973) Flow of lava into the sea, 1969–1971, Kilauea Volcano, Hawaii. *Geol. Soc. Amer. Bull.* **84**, 537–546.

Mooser F., Meyer-Abich H. and McBirney A. R. (1958) *Catalogue of the active volcanoes of the world. Part VI: Central America.* International Association of Volcanology & Chemistry of the Earth's Interior, Tokyo. 146 pp.

Morabito L. A., Synnott S. P., Kupfernan P. N. and Collins S. A. (1979) Discovery of currently active extraterrestrial volcanism. *Science* **204**, 972.

Muehlberger W. R. (1974) Structural history of southeastern Mare Serenitatis and adjacent highlands. *Proc. Lunar Sci. Conf. 5th*, p. 101–110.

Murase T. and McBirney A. R. (1970) Viscosity of lunar lavas. *Science* **167**, 1491–1493.

Murase T. and McBirney A. R. (1973) Properties of some common igneous rocks and their melts at high temperature. *Geol. Soc. Amer. Bull.* **84**, 3563–3592.

Murray B. C., Strom R. G., Trask N. J. and Gault D. E. (1975) Surface history of Mercury: implications for terrestrial plains. *J. Geophys. Res.* **80**, 2508–2514.

Murray J. B. (1971) Sinuous rilles. In *Geology and Physics of the Moon* (G. Fielder, ed.), p. 27–39. Elsevier, N. Y.

Murray J. B. and Guest J. E. (1980) Vertical ground deformation by precise levelling techniques, U. K. research on Mount Etna 1977–79. *Phil. Trans. Roy. Soc. Lond.*, 18–19.

Murray J. B., Dollfus A. and Smith B. (1972) Cartography of the surface markings of Mercury. *Icarus* **17**, 576–584.

Mutch T. A., Arvidson R. E., Head J. W., Jones K. L. and Saunders R. S. (1976) *The Geology of Mars.* Princeton Univ. Press, Princeton, N. J. 400 pp.

Mysen B. O., Arculus R. J. and Eggler D. (1975) Solubility of CO_2 in melts of andesite, tholeiite, and olivine nephelinite composition to 30 kb pressure. *Contrib. Mineral. Petrol.* **53**, 227–239.

Nakamura K. (1977) Volcanoes as possible indicators of tectonic stress orientation-principal and proposal. *J. Volc. & Geotherm. Res.* **2**, 1–16.

Nalivkin D. U. (1973) *Geology of the U. S. S. R.* (Trans. by N. Rast) Oliver and Boyd, Edinburgh. 855 pp.

Newburn R. L. and Gulkis S. (1975) Planets and satellites of the outer solar system, asteroids, and comets. In *Foundations of space biology and medicine* **v. 1** (M. Calvin and O. G. Gazenko, eds.), p. 168–197. NASA SP-374.

Nielsen N. (1936) A volcano under an ice cap, Vatnajokull, Iceland. *Geol. J.* **40**, 6–23.

Noe-Nygaard A. (1974) Cenozoic to recent volcanism in and around the North Atlantic basin. In *The Ocean Basins and Margins,* **v. 2** (A. E. Nairn and F. G. Stehli, eds.), p. 391–443. Plenum Pub. Co., N. Y.

Normark W. R. (1976) Delineation of the main extrusion zone of the East Pacific Rise at latitude 21° N. *Geology* **4**, 681–685.

Oberbeck V. R., Aoyagi M., Greeley R. and Lovas M. (1972) Planimetric shapes of lunar rilles. *Apollo 16 Prelim. Sci. Rept.*, p. 29.80–29.88. NASA SP-315.

Oberbeck V. R., Quaide W. L. and Greeley R. (1969) On the origin of sinuous rilles. *Mod. Geol.* **1**, 75–80.

Ogura T. (1969) Volcanoes in Manchuria. In *Geology and Mineralogy of the Far East* (T. Ogura, ed.), p. 373–413. Univ. of Tokyo Press., Tokyo.

O'Hara M. J. (1977) Geochemical evolution during fractional crystallization of a periodically refilled magma chamber. *Nature,* **266**, 503–507.

Ollier C. D. (1967) Maars: their characteristics, varieties and definition. *Bull. Volc.* **31**, 45–73.

Ollier C. D. (1972) *Volcanoes.* The M. I. T. Press, Cambridge. 177 pp.

Pai S. I., Hsu Y. and O'Keefe J. A. (1978) Similar explosive eruptions on lunar and terrestrial volcanoes. *Proc. Lunar Planet. Sci. Conf. 9th,* p. 1485–1508.

Papike J. J., Hodges E. N., Bence A. E., Cameron M. and Rhodes J. M. (1976) Mare basalts; crystal chemistry, mineralogy and petrology. *Rev. Geophys. Space Phys.* **14**, 475–540.

Parmentier E. M. and Head J. W. (1979) Internal processes affecting surfaces of low-density satellites: Ganymede and Callisto. *J. Geophys. Res.* **84**, 6263–6276.

Peckover R. S., Buchanan D. J. and Ashby D. E. T. F. (1973) Fuel-coolant interactions in submarine volcanism. *Nature* **247**, 272–273.

Peterson D. W. (1976) Processes of volcanic island growth, Kilauea Volcano, Hawaii, 1969–73. *Proc. Symp. on Andean and Antarctic Volcanology Problems Santiago, Chile,* p. 172–189.

Peterson D. W. and Swanson D. A. (1974) Observed formation of lava tubes. *Speleology* **2**, 209.

Peterson J. E. (1978) Volcanism in the Noachis — Hellas region of Mars, 2. *Proc. Lunar Planet. Sci. Conf. 9th,* p. 3411–3432.

Pettengill G. H., Ford P. G., Brown W. E., Kaula W. M., Keller C. H., Masursky H. and McGill G. E. (1979a) Pioneer Venus radar mapper experiment. *Science* **203**, 806–808.

Pettengill G. H., Ford P. G., Brown W. E., Kaula W. M., Masursky H., Eliason E. and McGill G. E. (1979b) Venus: preliminary topographic and surface imaging results from the Pioneer orbiter. *Science* **205**, 90–93.

Pettengill G. H., Ford P. G., Loriot G. B., Masursky H., Eliason E. and McGill G. E. (1980) Venus topography and meter-scale surface roughness as seen by the Pioneer orbiter radar (abstract). In *Lunar and Planetary Science XI,* p. 876. Lunar and Planetary Institute, Houston.

Phillips J. D. and Fleming H. S. (1978) Multi-beam sonar study of the Mid-Atlantic Ridge rift valley 36°–37°. *Map. Serv. MC-19.* Geol. Soc. Amer., Boulder, Co.

Phillips R. J., Adams G. F., Brown W. E., Jr., Eggleton R. E., Jackson P., Jordan R., Linlor W. I., Peeples W. J., Porcello L. J., Ryer J., Schaber G., Sill W. R., Thompson T. W., Ward S. W. and Zelenka J. S. (1973) Apollo Lunar Sounder experiments. *Apollo 17 Prelim. Sci. Rept.*, p. 22.1–22.26. NASA SP-330.

Pieters C. M. (1978) Mare basalt types on the front side of the Moon: a summary of spectral reflectance data. *Proc. Lunar Planet. Sci. Conf. 9th,* p. 2825–2849.

Pieters C. M., Head J. W., Adams J. B., McCord T. B., Zisk S. and Whitford-Stark J. L. (1980) Late high titanium basalts of the western maria: geology of the Flamsteed region of Oceanus Procellarum. *J. Geophys. Res.* **85**, 3913–3938.

Pieters C. M., Head J. W., McCord T. B., Adams J. B. and Zisk S. (1975) Geochemical and geological units of the Mare Humorum: definition using remote sensing and lunar sample information. *Proc. Lunar Sci. Conf. 6th,* p. 2689–2710.

Pieters C. M., McCord T. B., Charette M. P. and Adams J. B. (1974) Lunar surface identification of the dark mantling material in the Apollo 17 soil samples. *Science* **183**, 1191–1193.

Pieters C. M., McCord T. B., Zisk S. and Adams J. B. (1973) Lunar black spots, and nature of Apollo 17 landing area. *J. Geophys. Res.* **78**, 5867–5875.

Pike R. J. (1974) Craters on Earth, Moon and Mars: multivariate classification and mode of origin. *Earth Planet. Sci. Lett.* **22**, 245–255.

Pike R. J. (1978) Volcanoes on the inner planets: some preliminary comparisons of gross topography. *Proc. Lunar Planet. Sci. Conf. 9th*, p. 3239–3273.

Pinkerton H. and Sparks R. S. J. (1976) The 1975 sub-terminal lavas, Mount Etna: a case history of the formation of a compound lava field. *J. Volcanol. & Geothermal Res.* **1**, 167–182.

Pinkerton H. and Sparks R. S. J. (1978) Field measurements of the rheology of lava. *Nature* **276**, 383–385.

Plescia J. B. and Saunders R. S. (1979) The chronology of martian volcanoes. *Proc. Lunar Planet. Sci. Conf. 10th*, p. 2841–2859.

Pollard D. D. (1976) On the form and stability of open hydraulic fractures in the Earth's crust. *Geophys. Res. Lett.* **3**, 513–516.

Porter S. C. (1972) Distribution, morphology and size frequency of cinder cones on Mauna Kea Volcano, Hawaii. *Geol. Soc. Amer. Bull.* **83**, 3607–3612.

Ramberg I. B. and Van Andel T. H. (1977) Morphology and tectonic evolution of the rift valley at latitude 36°37 N Mid-Atlantic Ridge. *Geol. Soc. Amer. Bull.* **88**, 577–586.

Read H. H. and Watson J. (1975) *Introduction to Geology Vol. 2. Earth History, Part I Early Stages of Earth History.* Halsted, N. Y. 221 pp.

Reynolds D. L. (1956) Calderas and ring complexes. *Koninkl. Ned. Geol. Mijnbouwk. Genoot. Verhandel* **16**, 355–379.

Rhodes R. C. (1975) New evidence for impact origin of the Bushveld Complex, South Africa. *Geology* **3**, 549–554.

Richards A. F. (1964) Geology of the Islas Revillagigedo, Mexico 4. Geology and petrography of Isla Roca Partida. *Geol. Soc. Amer. Bull.* **75**, 1157–1164.

Richards A. F. (1966) Geology of the Islas Revillagigedo, Mexico 2. Geology and petrography of Isla San Benedicto. *Calif. Acad. Sci. Proc.* **33**, 361–414.

Richey J. E., MacGregor A. G. and Anderson F. W. (1961) *Scotland: The Tertiary Volcanic Districts.* Her Majesty's Stationary Office Press, London. 119 pp.

Ridley W. I. (1972) The field relations of the Las Cañadas Volcanoes, Tenerife, Canary Islands. *Bull. Volc.* **35**, 318–334.

Rittmann A. (1962) *Volcanoes and Their Activity.* Wiley-Interscience, N. Y. 305 pp.

Robson G. R. (1967) Thickness of Etnean lavas. *Nature* **216**, 251–252.

Rutgers R. (1962) Relative viscosity of suspensions of rigid spheres in Newtonian liquids. *Rheol. Acta* **2**, 202–210.

Ryder G. and Taylor G. J. (1976) Did mare-type volcanism commence early in lunar history? *Proc. Lunar Sci. Conf. 7th*, p. 1741–1755.

Saemundsson K. (1974) Evolution of the axial rifting zone in Northern Iceland and the Tjornes Fracture Zone. *Geol. Soc. Amer. Bull.* **85**, 495–504.

Sagan C. (1979) Sulphur flows on Io. *Nature* **280**, 750–753.

Salisbury J. W., Adler J. E. and Smalley V. G. (1968) Dark-haloed craters on the Moon. *Roy. Astron. Soc. Mon. Not.* **138**, 245–249.

Sato M. (1950) Geomorphological classification of lava flows. *Bull. Geog. Inst.*, Tokyo **No. 1**, 114–132 (in Japanese with English summary).

Saunders R. S. and Malin M. C. (1976) Venus: geologic analysis of radar images. *Geologica Romana* **15**, 507–515.

Saunders R. S. and Malin M. C. (1977) New observations of the surface of Venus. *Geophys. Res. Lett.* **4**, 547–550.

Saunders R. S., Mutch T. A. and Jones K. L. (1975) Guide to the use of Mariner images. *JPL Tech. Mem.*, 33–723.

Scarfe C. M. (1973) Viscosity of basic melts at varying pressure. *Nature* **241**, 101–102.

Schaber G. G. (1969) Geologic map of the Sinus Iridum quadrangle of the Moon. *U. S. Geol. Surv. Misc. Inv. Ser.*, Map I-602.

Schaber G. G. (1973a) Eratosthenian volcanism in Mare Imbrium: source of the youngest lava. *Apollo 17 Prelim. Sci. Rept.*, p. 30.17–30.21. NASA SP-330.

Schaber G. G. (1973b) Lava flows in Mare Imbrium: geologic evidence from Apollo orbital photography. *Proc. Lunar Sci. Conf. 4th*, p. 73–92.

Schaber G. G. (1975) Geologic application of radar data to planetary exploration. *Internat. Coll. Plan. Geol.*, expanded abstracts (Italian Cons. for Planet. Studies, ed.), p. 363–365.

Schaber G. G., Horstman K. C. and Dial A. L. (1978) Lava flow materials in the Tharsis region of Mars. *Proc. Lunar Planet. Sci. Conf. 9th*, p. 3433–3458.

Schmincke H. U. (1977) Phreatomagmatische Phasen in quataren Vulkanen der Osteifel. *Geol. Jb.* **39**, 3–45.

Schubert G., Lingenfelter R. E. and Peale S. J. (1970) The morphology, distribution, and origin of Lunar sinuous rilles. *Rev. Geophys. Space Phys.* **8**, 199–224.

Schultz P. H. (1976a) Floor-fractured lunar craters. *The Moon* **15**, 241–273.

Schultz P. H. (1976b) *Moon Morphology.* University of Texas Press, Austin. 626 pp.

Schultz P. H. (1977a) Endogenic modification of impact craters on Mercury. *Phys. Earth Planet. Int.* **15**, 202–219.

Schultz P. H. (1977b) Lunar and martian floor-fractured craters (abstract). In *Basaltic Volcanism Second Inter-team Meeting.*, p. 53–55. The Lunar and Planetary Institute, Houston.

Schultz P. H. (1978) Martian intrusions: possible sites and implications. *Geophys. Res. Lett.* **5**, 457–460.

Schultz P. H. and Glicken H. (1979) Impact crater and basin control of igneous processes on Mars. *J. Geophys. Res.* **84**, 8033–8047.

Schultz P. H. and Spudis P. (1978) The dark ring of Orientale: implications for pre-basin mare volcanism and a clue to the identification of the transient cavity rim. *Proc. Lunar Planet. Sci. Conf. 9th*, p. 1033–1035.

Schultz P. H. and Spudis P. D. (1979) Evidence for ancient mare volcanism. *Proc. Lunar Planet. Sci. Conf. 10th*, p. 2899–2918.

Schultz P. H., Greeley R. and Gault D. E. (1976) Degradation of small mare surface features. *Proc. Lunar Sci. Conf. 7th*, p. 985–1003.

Scott D. H. (1973) Mare Serenitatis cinder cones and terrestrial analogs. *Apollo 17 Prelim. Sci. Rept.*, p. 30.7–30.8. NASA SP-330.

Scott D. H. and Carr M. H. (1978) Geologic map of Mars. *U. S. Geol. Surv. Misc. Geol. Inv.*, Map. I–1083.

Scott D. H. and Condit C. D. (1977) Correlations: martian stratigraphy and crater density., p. 56–68. NASA TM X-3511.

Scott D. H. and Eggleton R. E. (1973) Geologic map of the Rümker quadrangle of the Moon. *U. S. Geol. Surv. Misc. Inv. Ser.*, Map I–805.

Searle E. J. (1964) Volcanic risk in the Auckland metropolitan district: New Zealand. *New Zealand J. Geol. and Geophys.* **7**, 94–100.

Self S. and Sparks R. S. J. (1978) Characteristics of widespread pyroclastic deposits formed by the interaction of silicic magma and water. *Bull. Volcanol.* **41**, 196–212.

Self S., Sparks R. S. J., Booth B. and Walker G. P. L. (1974) The 1973 Heimaey strombolian scoria deposit, Iceland. *Geol. Mag.* **III**, 539–548.

Settle M. (1979) The structure and emplacement of cinder cone fields. *Amer. J. Sci.* **279**, 1089–1107.

Shaw H. R. (1963) Obsidian-H_2O viscosities at 1,000 and 2,000 bars temperature range 700° to 900°C. *J. Geophys. Res.* **68**, 6337–6343.

Shaw H. R. (1969) Rheology of basalt in the melting range. *J. Petrol.* **10**, 510–535.

Shaw H. R. (1972) Viscosities of magmatic silicate liquids: an empirical method of prediction. *Amer. J. Sci.* **272**, 870–893.

Shaw H. R. and Swanson D. A. (1974) Eruption flow rates of flood basalts. Proceed. 2nd. *Columbia River Basalt Symposium*, 271–299. East Washington State College Press.

Shaw H. R., Wright T. L., Peck D. L. and Okamura R. (1968) The viscosity of basaltic magma: an analysis of field measurements in Makaopuhi Lava Lake, Hawaii. *Amer. J. Sci.* **266**, 225–264.

Sigurdsson H. and Sparks R. S. J. (1978) Lateral flow of magma in rifted Icelandic crust. *Nature* **274**, 126–130.

Sigvaldason G. E. (1964) Some geochemical and hydrothermal aspects of the 1961 Askja eruption. *Beitrage zur Min. Petrog.* **10**, 263–274.

Sigvaldason G. E. (1968) Structure and products of subaquatic volcanoes in Iceland. *Contrib. Mineral. Petrol.* **18**, 1–16.

Simkin T. and Howard K. A. (1970) Caldera collapse in the Galapagos Islands, 1968. *Science* **169**, 429–437.

Skinner B. J. (1970) A sulphur lava flow on Mauna Loa. *Pacific Sci.* **24**, 144–145.

Smith B. A., Soderblom L. A., Johnson T. V., Ingersoll A. P., Collins S. A., Shoemaker E. M., Hunt G. E., Masursky H., Carr M. H., Davies M. E., Cook A. F., II, Boyce J., Danielson G. E., Owen T., Sagan C., Beebe R. F., Veverka J., Strom R. G., McCauley J. F., Morrison D., Briggs G. A. and Suomi V. E. (1979) The Jupiter system through the eyes of Voyager 1. *Science* **204**, 951–972.

Smith E. I. (1973) Identification, distribution and significance of Lunar volcanic domes. *The Moon* **6**, 3–31.

Smith E. I. (1974) Rümker Hills: a lunar volcanic dome complex. *The Moon* **10**, 175–181.

Smith R. L. (1979) Ash-flow magmatism. *Geol. Soc. Amer. Spec. Paper* **180**, 5–25.

Smith R. L. and Bailey R. A. (1968) Resurgent cauldrons. *Geol. Soc. Amer. Mem.* **116**, 613–662.

Solomon S. C. and Head J. W. (1979) Vertical movement in Mare basins: relation to mare emplacement, basin tectonics, and lunar thermal history. *J. Geophys. Res.* **84**, 1667–1682.

Solomon S. C. and Head J. W. (1980) Lunar mascon basins: lava flooding, tectonics, and evolution of the lithosphere. *Rev. Geophys. Space Phys.* **18**, 107–141.

Sparks R. S. J. and Pinkerton H. (1978) The effect of degassing on the rheology of basaltic lava. *Nature* **276**, 385–386.

Sparks R. S. J. and Wilson L. (1976) A model for the formation of ignimbrite by gravitational column collapse. *J. Geol. Soc. Lond.* **132**, 441–451.

Sparks R. S. J., Meyer P. and Sigurdsson H. (1980) Density variation amongst mid-ocean ridge basalts: implications for magma mixing and the scarcity of primitive lavas. *Earth Planet. Sci. Lett.* **46**, 419–430.

Sparks R. S. J., Pinkerton H. and Hulme G. (1976) Classification and formation of lava levees on Mount Etna, Sicily. *Geology* **4**, 269–271.

Spiess F. N., Macdonald K. C., Atwater T., Ballard R., Carranza A., Cordoba D., Cox C., Diaz Garcia V. M., Francheteau J., Guerrero J., Hawkins J., Maymon R., Hessler R., Juteau T., Kastner M., Larson R., Luyendyk B., Macdougall J. D., Miller S., Normark W., Orcutt J. and Rangin C. (1980) East Pacific Rise: hot springs and geophysical experiments. *Science* **207**, 1421–1433.

Spudis P. D. (1978) Composition and origin of the Apennine Bench Formation. *Proc. Lunar Planet. Sci. Conf. 9th*, p. 3379–3394.

Spudis P. D. (1979) The extent and duration of lunar KREEP volcanism (abstract). In *Conference on the Lunar highlands crust*, p. 157–159. Lunar and Planetary Institute, Houston.

Spudis P. and Greeley R. (1977) Volcanism in the cratered uplands of Mars: a preliminary Viking View. *EOS (Trans Amer. Geophys. Union)* **58**, 1182 pp.

Steinberg G. S. and Solov'yev Ts. V. (1976) The shape of volcanoes and the position of subordinate vents. *Izv. Earth Physics*, 334–336.

Strain P. and El-Baz F. (1977) Topography of sinuous rilles in the Harbinger Mountain region of the Moon. *The Moon* **16**, 221–229.

Strom R. G. (1965) In, *Ranger VII, Part II Experimenters Analyses and Interpretations. Chapter III.* JPL Tech. Rept. 32–700. Small-scale lineaments on the Ranger VII photographs, p. 62–69.

Strom R. G. (1972) Lunar mare ridges, rings, and volcanic ring complexes. In *The Moon, Symposium 47 of the I. A. U.* (S. Runcorn and H. Urey, eds.), p. 187–215. D. Reidel, Dordrecht, Netherlands.

Strom R. G. (1977) Origin and relative age of lunar mercurian intercrater plains. *Phys. Earth Planet. Inter.* **15**, 156–172.

Strom R. G. and Fielder G. (1970) Multiphase eruptions associated with Tycho and Aristarchus. *Comm. Lunar Planet. Lab. Tuscon Ariz.*, p. 235–288.

Strom R. G., Murray B. C., Belton M. J. S., Danielson G. E., Davies M. E., Gault D. E., Hapke B., O'Leary B., Trask N., Guest J. E., Anderson J. and Klaasen K. (1975a) Preliminary imaging results from the second Mercury encounter. *J. Geophys. Res.* **80**, 2345–2356.

Strom R. G., Terrile R. J., Masursky H. and Hansen C. (1979) Volcanic eruption plumes on Io- Voyager 1 results. *Nature* **280**, 733–736.

Strom R. G., Trask N. J. and Guest J. E. (1975b) Tectonism and volcanism on Mercury. *J. Geophys. Res.* **80**, 2478–2507.

Stuart-Alexander D. E. and Howard K. A. (1970) Lunar maria and circular basins—a review. *Icarus* **12**, 440–456.

Surkov Yu. A. (1977) Geochemical studies of Venus by Venera 9 and 10 automatic interplanetary stations. *Proc. Lunar Sci. Conf. 8th*, p. 2665–2689.

Swanson D. A. (1973) Pahoehoe flows from the 1969–1971 Mauna Ulu eruption Kilauea Volcano, Hawaii. *Geol. Soc. Amer. Bull.* **84**, 615–626.

Swanson D. A. and Christiansen R. L. (1973) Tragic base surge in 1790 at Kilauea Volcano. *Geology* **1**, 83–86.

Swanson D. A. and Fabbi B. P. (1973) Loss of volatiles during fountaining and flowage of basaltic lava at Kilauea Volcano, Hawaii. *J. Res. U. S. Geol. Surv.* **1**, 649–658.

Swanson D. A. and Wright T. L. (1978) Bedrock geology of the Northern Columbia Plateau and adjacent areas. In *The Channeled Scabland* (V. R. Baker and D. Nummedal, eds.). NASA, p. 37–57.

Swanson D. A., Duffield W. A., Jackson D. B. and Peterson D. W. (1972) The complex filling of Alae Crater, Kilauea Volcano, Hawaii. *Bull. Volc.* **36**, 105–126.

Swanson D. A., Wright T. L. and Helz R. T. (1975) Linear vent systems and estimated rates of magma production and eruption for the Yakima basalt of the Columbia Plateau. *Amer. J. Sci.* **275**, 877–905.

Swanson F. J., Baitis H. W., Lexa J. and Dymond J. (1974) Geology of Santiago, Rábida and Pinzón Islands, Galapagos. *Geol. Soc. Amer. Bull.* **85**, 1803–1810.

Taylor B. J., Burgess I. C., Land D. H., Mills D. A. C., Smith D. B. and Warren P. T. (1971) *British Regional Geology: Northern England.* Her Majesty's Stationary Office Press, London. 121 pp.

Thompson T. W., Howard K. A., Shorthill R. W., Tyler G. L., Zisk S. H., Whitaker E. A., Schaber G. G. and Moore H. J. (1973) Remote sensing of Mare Serenitatis. *Apollo 17 Prelim. Sci. Rept.*, p. 33.3–33.10. NASA SP-330.

Thorarinsson S. (1961) The post-glacial volcanism: on the geology and geophysics of Iceland. *21st Intern. Geol. Congr. Norway. Guide to Exc. A-2.*, 33–45.

Thorarinsson S. (1967a) Some problems of volcanism in Iceland. *Geol. Rundsch.* **57**, 705–718.

Thorarinsson S. (1967b) *Surtsey, the New Island in the North Atlantic.* Viking Press, N. Y. 47 pp.

Thorarinsson S. (1968) On the rate of lava and tephra production and the upward migration of magma in four Icelandic eruptions *Geol. Rundsch.* **57**, 705–718.

Trask N. J. and Guest J. E. (1975) Preliminary geologic terrain map of Mercury. *J. Geophys. Res.* **80**, 2461–2477.

Trask N. J. and Strom R. G. (1976) Additional evidence of mercurian volcanism. *Icarus* **28**, 559–563.

Tyrrell G. W. (1937) Flood basalts and fissure eruption. *Bull. Volc.* Ser. 2. **1**, 89–111.

Van Bemmelen R. W. and Rutten M. G. (1955) *Tablemountains of Northern Iceland.* Brill, Leiden. 217 pp.

Vogt P. R. (1974) Volcano spacing fractures and thickness of the lithosphere. *Earth Planet. Sci. Lett.* **21**, 235–252.

Wadge G. (1976) Deformation of Mount Etna, 1971–1974. *J. Volcanol. and Geothermal Res.* **1**, 237–263.

Wadge G. (1977) The storage and release of magma on Mount Etna. *J. Volcanol. and Geothermal Res.*, **2**, 361–384.

Wadge G. (1978) Effusion rate and the shape of aa lava flow-fields on Mount Etna. *Geology* **6**, 503–506.

Wager L. R. and Brown G. H. (1968) *Layered Igneous Intrusions.* Oliver and Boyd, Edinburgh and London. 588 pp.

Walker G. P. L. (1960) Zeolite zones and dike distribution in relation to the structure of the basalts of eastern Iceland. *J. Geol.* **68**, 515–528.

Walker G. P. L. (1965) Some aspects of Quaternary volcanism in Iceland. *Leicester Lit. Phil. Soc. Trans.* **59**, 25–40.

Walker G. P. L. (1967) Thickness and viscosity of Etnean Lavas. *Nature* **213**, 484–485.

Walker G. P. L. (1970) Compound and simple lava flows and flood basalts. *Bull. Volcanol.* **35**, 579–590.

Walker G. P. L. (1973) Length of lava flows. *Phil. Trans. Roy. Soc. Lond.* **A274**, 107–118.

Walker G. P. L. (1975a) Eruption mechanisms in Iceland. In *Geodynamics of Iceland and North Atlantic Area* (L. Kristjansson, ed.), p. 189–201. NATO Advanced Studies Institute Series, Boston.

Walker G. P. L. (1975b) Intrusive sheet swarms and the identity of crustal layer 3 in Iceland. *J. Geol. Soc. Lond.* **131**, 143–162.

Walker G. P. L. and Blake D. H. (1966) The formation of a palagonite breccia mass beneath a valley glacier in Iceland *Quart. J. Geol. Soc.* **122**, 45–61.

Walker G. P. L. and Croasdale R. (1972) Characteristics of some basaltic pyroclastics. *Bull. Volc.* **35**, 303–317.

Walsh J. B. and Decker R. W. (1971) Surface deformation associated with volcanoes. *J. Geophys. Res.* **76**, 3291–3302.

Wasserburg G. J., Papanastassiou D. A., Tera F. and Huneke J. C. (1977) Outline of a lunar chronology. *Phil. Trans. Roy. Soc. Lond.* **A285**, 7–22.

Watanabe T. (1940) Eruptions of molten sulphur from the Siretoko-Iosan Volcano Hokkaido, Japan. *Jap. J. Geol. Geog.* **17**, 289–310.

Wells G., Pierce T. and Bryan W. B. (1979) A comparison of ancient and modern pillow lavas. *J. Geol.* **87**, 427–440.

Wentworth C. K. and Macdonald G. A. (1953) Structures and forms of basaltic rocks in Hawaii. *Bull. U. S. Geol. Surv.* **994**, 1–90.

West M. (1974) Martian volcanism: additional observations and evidence for pyroclastic activity. *Icarus* **21**, 1–11.

Whitford-Stark J. L. (1974) Internal origin for lunar rilled craters and the maria? *Nature* **248**, 573–574.

Whitford-Stark J. L. (1979) Charting the Southern Seas: the evolution of the lunar Mare Australe. *Proc. Lunar Planet. Sci. Conf. 10th*, p. 2975–2994.

Whitford-Stark J. L. (1980) Problems associated with the identification of magma composition from volcanic landforms. In *Reports of Planetary Geology Program, 1979–1980*. p. 195–197. NASA TM-81776.

Whitford-Stark J. L. and Head J. W. (1977) The Procellarum volcanic complexes: contrasting styles of volcanism. *Proc. Lunar Sci. Conf. 8th*, p. 2705–2724.

Whitford-Stark J. L. and Head J. W. (1980) Stratigraphy of Oceanus Procellarum basalts: sources and styles of emplacement. Submitted to *J. Geophys. Res.* **85**, 6579–6609.

Wilcox R. E. (1954) Petrology of Paricutin volcano, Mexico. *U. S. Geol. Surv. Bull.* **965-C**, 281–353.

Wilhelms D. E. (1970) Summary of lunar stratigraphy - telescope observations. U. S. Geol. Surv. Prof. Paper, 599-F.

Wilhelms D. E. (1974) Comparison of martian and lunar geologic provinces. *J. Geophys. Res.* **79**, 3933–3941.

Wilhelms D. E. (1976) Mercurian volcanism questioned. *Icarus* **28**, 551–558.

Wilhelms D. E. and McCauley J. F. (1971) Geologic map of the near side of the Moon. *U. S. Geol. Surv. Misc. Inv. Ser.*, Map I-703.

Wilson L. and Head J. W. (1979a) Ascent of magma with volatiles on the Earth and Moon. *Proc. Lunar Planet. Sci. Conf. 10th*, p. 1350–1352.

Wilson L. and Head J. W. (1979b) Lunar volcanic cones and dark mantling deposits: consequences of patterns of volatile release. *Proc. Lunar Planet. Sci. Conf. 10th*, p. 1353–1355.

Wilson L. and Head J. W. (1980a) Ascent and eruption of basaltic magma on the Earth and Moon. *J. Geophys. Res.*, in press.

Wilson L. and Head J. W. (1980b) The formation of eroded depressions around the sources of lunar sinuous rilles: theory (abstract). In *Lunar and Planetary Science XI*, p. 1260–1262. Lunar and Planetary Institute, Houston.

Windley B. F. (1977) *The Evolving Continents*. John Wiley and Sons, N. Y. 385 pp.

Wood C. A. (1974) Reconnaissance geophysics and geology of the Pinacate Craters, Sonora, Mexico. *Bull. Volcanol.* **38**, 149–172.

Wood C. A. (1977) Non-basaltic shield volcanoes. In *Abstracts for the Planetary Geology Field Conference on the Snake River Plains, Idaho* (R. Greeley and D. Black, eds.), p. 436. NASA TM-78.

Wood C. A. (1978) Morphometric evolution of composite volcanoes. *Geophys. Res. Lett.* **5**, 437–439.

Wood C. A. (1979a) Morphometric studies of planetary landforms: impact craters and volcanoes. Unpublished PhD thesis, Brown Univ., Providence.

Wood C. A. (1979b) Maars. *Encyclopedia of Volcanology*. (In press).

Wood C. A. (1979c) Venusian volcanism: environmental effects on style and landforms. *Reports of the Planetary Geology Program 1978, 1979* p. 244–246. NASA TM-80339.

Wood C. A. (1979d) Monogenetic volcanoes of the terrestrial planets. *Proc. Lunar Planet. Sci. Conf. 10th.*, p. 2815–2840.

Wood C. A. and Head J. W. (1975) Geologic setting and provenance of spectrally distinct pre-mare material of possible volcanic origin (abstract). In *Origins of mare basalts and their implications for lunar evolution*, p. 189–193. The Lunar Science Institute, Houston.

Wood C. A. and Head J. W. (1977) Rift valleys on Earth, Mars and Venus. In *Tectonics and Geophysics of Continental Rifts* (B. Ramberg and E. -R. Neumann, eds.). p. 401–408. D. Reidel, Netherlands.

Wood C. A., Head J. W. and Cintala M. J. (1977) Crater degradation on Mercury and the Moon: clues to surface evolution. *Proc. Lunar Sci. Conf. 8th*, p. 3503–3520.

Wright T. L., Kinoshita W. T. and Peck D. L. (1968) March 1965 eruption of Kilauea Volcano and the formation of Makaopuhi lava lake. *J. Geophys. Res.* **73**, 3181–3205.

Young R. A., Brennan W. J., Wolfe R. W. and Nichols D. J. (1973) Volcanism in the lunar maria. In *Apollo 17 Prelim. Sci. Rept.*, p. 31.1–31.11. NASA SP-330.

Zisk S. H., Hodges C. A., Moore H. J., Shorthill R. W., Thompson T. W., Whitaker E. A. and Wilhelms D. E. (1977) The Aristarchus-Harbinger region of the Moon: surface geology and history from recent remote-sensing observations. *The Moon* **17**, 59–99.

Overleaf:

Two of the four tectonic settings of basaltic volcanism on Earth are represented in this image of S. W. Iceland: accreting plate boundaries (spreading ridges), and hot spots. Most of the volume of basalt being erupted on Earth is emplaced at spreading ridge axes. Much lesser amounts are emplaced near convergent plate boundaries in volcanic arcs, and in zones of continental collision.

LANDSAT image number 81392-12191-5 printed as a *negative* to show features that are obscure in positive print. Black areas are ice caps—Myrdalsjökull (right) and Langjökull (upper center). Light areas are water. Reykanes peninsula in SW connects with Mid-Atlantic ridge. The zone of abundant active volcanism, fissuring and normal faulting in Iceland is typically about 30-km wide and extends from this peninsula to Langjökull (S. W. Iceland rift zone), and from W. of Myrdalsjökull to beyond the N.E. corner of the image (central Iceland rift zone). Hekla volcano is the darker area north of Myrdalsjökull, Surtsey is the small island near the lower edge of the print. Scale bar represents 20 km.

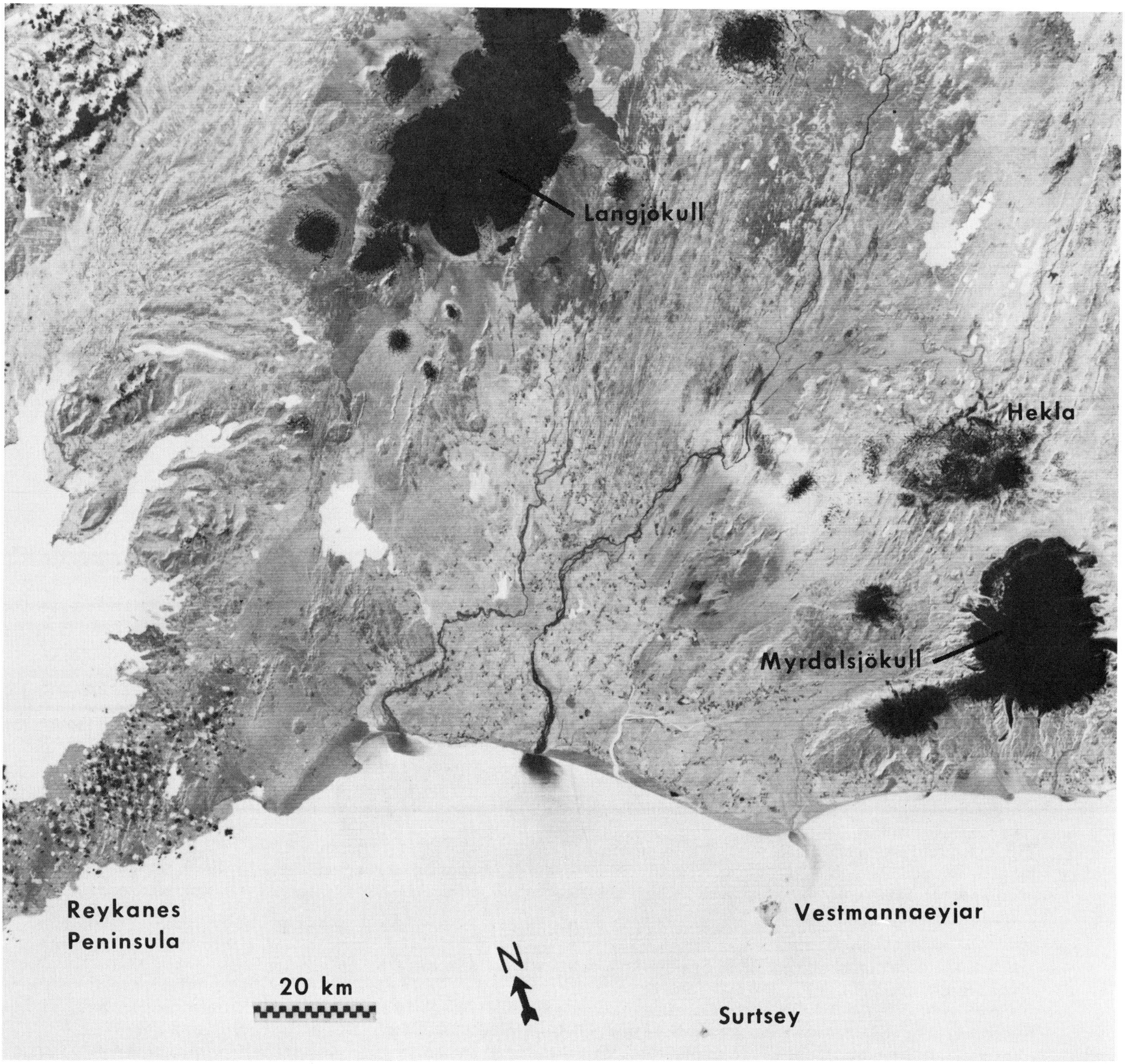
Langjökull
Hekla
Myrdalsjökull
Reykanes
Peninsula
Vestmannaeyjar
20 km
N
Surtsey

6. Tectonics of basaltic volcanism

Kevin C. Burke, Team Leader
—Department of Geological Sciences
State University of New York at Albany

William S. F. Kidd
—Department of Geological Sciences
State University of New York at Albany

Donald L. Turcotte
—Department of Geological Sciences
Cornell University

John F. Dewey
—Department of Geological Sciences
State University of New York at Albany

Peter J. Mouginis-Mark
—Department of Geological Sciences
Brown University

E. Marc Parmentier
—Department of Geological Sciences
State University of New York at Albany

A. M. Celâl Şengör
—Department of Geological Sciences
State University of New York at Albany

Paul E. Tapponnier
—Université de Montpellier
Montpellier, France

6.1 INTRODUCTION

Tectonics is the name given to the large-scale structural evolution of planetary lithospheres. On the Moon, Mercury, and to some extent Mars, nearly all of this structural evolution took place during a time when surface development was dominated by impacts. On Venus, the temporal relationship between lithospheric evolution and the impact history is enigmatic. For this reason, we have no unambiguous record of the tectonics of these planetary objects and emphasis in this chapter is necessarily very strongly on the Earth. A discussion of martian tectonics, especially that of the Tharsis region, is included here and comparisons are made with some terrestrial environments. Shorter reviews of the tectonics of the Moon, Mercury, and Venus are also presented.

Planetary tectonics, although essentially lithospheric, is the result of interaction between convecting mantle and overlying lithosphere, and very close links therefore exist between tectonics and the thermal histories of the planets. This is nowhere better seen than on the Earth where the plate tectonic process (Wilson, 1965a) plays an essential part in removing heat from the surface of the planet (Sclater and Francheteau, 1970; Turcotte and Burke, 1978).

In discussing the relationship of basaltic volcanism to tectonics, classical geological procedure is followed by considering first the presently active systems and then extending to the past the lessons learned from these. Known active volcanism in the solar system is restricted to the Earth and to Io; our discussion must start with the Earth because little is known yet about the volcanism on Io.

6.2 TECTONICS OF ACTIVE BASALTIC VOLCANISM

A revolution in understanding the lithospheric structure of the Earth was achieved when Wilson (1965a) pointed out that the lithosphere is broken into rigid plates that move toward, away from, or past each other. (The boundaries of these plates are shown in Fig. 6.2.14.) Tectonic and igneous activities are strongly concentrated at these boundaries. For students of basaltic volcanism it is convenient to consider basaltic activity at the three types of plate boundaries separately, and then to go on to consider the several varieties of non-plate-margin volcanism that together account for only a very small proportion of total terrestrial igneous activity. Active volcanism, possibly basaltic, is also occurring on Io, so its tectonics are briefly considered in this section. For a discussion of the possible mechanisms by which the plates are driven (all traceable to the decay of radioactive nuclides in the mantle) see section 9.5.1.

6.2.1 Tectonics of basaltic eruption at plate margins

Tectonics of basaltic eruption at accreting plate margins

Introduction. The majority of the basalt that is presently generated on Earth is emplaced along accreting plate boundaries (oceanic spreading centers). Menard (1967), Deffeyes (1970), and Dickinson and Luth (1971) have estimated the rates of production of basalt at spreading centers from the volumes produced since the late Mesozoic. The rates obtained are about 5–15 km^3/yr, which is only about one or two orders of magnitude greater than the present rate of magma production in the single most active non-plate-margin volcano on Earth, Hawaii (Lipman, pers. comm., 1978). The generally constant thickness of the (basaltic) oceanic crust shows that magma production rates at spreading ridges generally are remarkably steady, in contrast to intraplate-type volcanic areas. The latter, as shown by Hawaii and the Hawaiian-Emperor chain, may produce basaltic magma at a relatively high rate for the small area involved, but only for short times (on the order of 1 m.y. or less). Flood basalt areas, which are clearly associated with the largest hotspots and particularly with continental rifting (see section 6.3.2), probably have the highest magma production rate known, but only for the very short overall lifetime of these events (Baksi and Watkins, 1973). Thus, averaged over tens of millions of years, the volume of basalt produced at spreading ridges greatly exceeds that in other environments.

The recognition of the extent and continuity of the oceanic ridge system (Ewing and Heezen, 1956; Heezen, 1960), and the explanation of symmetrical magnetic anomalies by the seafloor spreading process (Vine and Matthews, 1963; Morley and Larochelle, pers. comm., 1962) are results of extensive oceanographic surveys. Heezen (1960) recognized that much of the ridge system in the Atlantic and Indian Oceans has an axial rift valley. The ridge in the Pacific Ocean was shown mostly not to have such a rift (Menard, 1964). The ridges, viewed on the broadest scale, are about 2000 km across; the axial part is generally about 2.5 km below sea level and the flanks subside to 5 km depth at a distance of about 1000 km from the axis. The subsidence of the ocean floor is clearly related to the cooling and thickening of the lithosphere away from the axis, and follows a quantitatively predictable path proportional to the square root of time, at least out to a crustal age of about 80 m.y. (Parker and Oldenberg, 1973; see Chapter 9).

The more detailed topography (see Chapter 5) is characterized by elongate ridges and valleys parallel to the axis of the ridge. This topography is more pronounced on ridges that possess an axial rift valley. It becomes progressively blanketed and obscured by sedimentation away from the ridge axis. The axial region of spreading ridges coincides with a narrow zone of shallow seismicity (Ewing and Heezen, 1956; Barazangi and Dorman, 1969; Sykes, 1967). Basaltic volcanism is also restricted to a zone a few kilometers or less in width along the ridge axis, and is almost wholly restricted to the inner floor of the rift valley when one is present (ARCYANA, 1975; Ballard *et al.*, 1975; Macdonald, 1977). The interaction of tectonics and volcanism in this axial zone is the key to understanding the formation of the oceanic crust and upper mantle and hence to understanding the occurrence of most of the basalt erupted today in the solar system.

Tectonics of spreading ridge axes. The presence or absence of an axial rift valley can be correlated to a large degree with spreading rate; above about 3 cm/yr half-spreading rate, an axial rift is not usually observed. Exceptions do occur: the Galapagos spreading ridge possesses a shallow rift where it is spreading at about 3.5 cm/yr half-rate (Hey, 1977), and the well-known Reykjanes ridge, spreading at about 1.5 cm/yr half-rate, does not have a rift. Both of these ridges are near sites of anomalous basaltic volcanism (Galapagos, Iceland) that have been termed "hot spots" (see section 6.2.4). Although there are likely to be differences between oceanic crust made at rifted and non-rifted (and hence slow- and fast-spreading) ridges, the tectonic processes involved are exclusively extensional and there must be more similarity than difference between crust formed in the two situations. Much detailed investigation of rifted spreading centers has been accomplished (see section 5.2.8), whereas there is much less information on nonrifted, fast-spreading ridges. Most of the detailed information on slow-spreading ridges has been obtained by deep-towed instrument packages and manned submersibles (Atwater and Mudie, 1968, 1973; ARCYANA, 1975; Ballard and Van Andel, 1977; Bryan and Moore, 1977; Macdonald and Luyendyk, 1977; CAYTROUGH, 1979). In addition, in two places (Iceland and the Afar region) the axis of the spreading ridge is emergent. Although, for that very reason, these two places are not typical, much can be learned from them about spreading ridge tectonics (Walker, 1975; Palmason, 1973; Palmason and Saemondsson, 1974; Needham *et al.*, 1976).

The rift valley, which typically is found in axial regions of slow spreading ridges, is a 30- to 35-km-wide, 1- to 2-km-deep trough, the inward-facing walls of which consist of a series of steps. The steps are comprised of a series of steeply dipping inward-facing scarps having relief of tens of meters to a few hundred meters. Some of the steps show clear outward tilt. The deepest part of the rift valley is commonly characterized by a relatively flat region 2–10 km wide called the inner floor (Needham and Francheteau, 1974). Most of the volcanic activity occurs in the inner floor, and lava flows build lenticular piles that are up to 200 m high in the FAMOUS area (Bellaiche *et al.*, 1974; ARCYANA, 1975; Ballard *et al.*, 1975). In Iceland, the zone of surface magmatism is largely between 30 and 40 km wide; most eruptions occur through fissures 2–10 km long and a few meters wide that are not concentrated in any one part of the rift. Dikes therefore feed subaerial flows that travel laterally for large distances (>5 km); this contrasts markedly with the situation in the areas (FAMOUS, Galapagos ridge) that have been observed by submersibles, although it is possible that some flows travel larger distances laterally in the submarine environment than have been observed. The observation of chemical zonation in basalts (all of very young age) in the FAMOUS area (Ballard *et al.*, 1975; Bryan and Moore, 1977) implies that the flows seen at the present surface apparently have not moved more than a few hundred meters from their points of extrusion; this observation also implies a zoned magma chamber below. The FAMOUS study demonstrated that few young extrusive rocks occur beyond the first major rift wall fault scarps. In Afar (Ghoubbat-Assal rift), extrusion also is essentially limited to the rift inner floor (4–10 km wide). Flows tend to be dammed by the rift walls and/or to flow along strike (Needham *et al.*, 1976). The locus of active volcanism and thus of dike intrusion appears to migrate within a 2-km-wide axial region of the Galapagos spreading ridge on a time scale of thousands of years (Van Andel and Ballard, 1979).

The lateral extent of flows, and therefore the width and constancy of position of the zone of extrusion, must affect the magnetic signature of the oceanic crust. Most of the magnetic signal comes from the upper 500 m or so of the crust (e.g., Lowrie, 1979; Schouten and Denham, 1979; Hall and Robinson, 1979), which appears to be largely composed of extrusive lavas. Atwater and Mudie (1973) inferred from the magnetic data that the zone of extrusion on rifted ridges is not more than a few kilometers wide; Larson and Spiess (1969) inferred by the same method that the zone of extrusion on a segment of the fast-spreading East Pacific Rise crest was no more than 300 meters across; this result has yet to be confirmed by extensive direct observations, but CYAMEX (1978) report a 0.4–1.0 km wide extrusion zone at 21°N.

The rift valley is usually grossly symmetrical with respect to the inner floor, but in some places (e.g., FAMOUS area, Fig. 6.2.1.a) there is moderate asymmetry. The whole structure is best explained as a series of not-necessarily-centered, nested grabens, the steep scarps of the walls being normal faults. Normal faulting takes place over a width of 10–15 km (Needham *et al.*, 1976; Lépine and Ruegg, 1976; Klein *et al.*, 1973). In most profiles, it is inferred that the outward tilts of the steps increase away from the axis of the valley together with a decrease in the dip of the scarps (Fig. 6.2.1). The hummocky terrain, with the peaks and troughs that constitute the rift mountains province, stands at a remarkably uniform elevation above the inner floor (Needham and Francheteau, 1974; Fig. 6.2.1). The floor of the rift valley deepens along strike toward the rift valley-transform fault intersection (Schroeder, 1977); for example, a 500-m deepening near fracture zone A has been observed in the FAMOUS area (Renard *et al.*, 1975).

The two mechanisms by which crustal extension occurs are fissuring and conjugate normal faulting (e.g., White and Stroup, 1979). In the FAMOUS area, Iceland, and Afar, the two processes are equally active in the inner floor of the rift, but normal faulting becomes predominant in the outer part of the rift floor and in the rift walls. One of the conjugate sets of faults becomes predominant in the rift walls but neither set is dominant in the floor of the rift (White and Stroup, 1979).

Thicknesses and layering of the oceanic crust are obtained from seismic refraction studies. These studies (Raitt, 1963; Drake and Nafe, 1968; Ludwig *et al.*, 1970; Shor *et al.*, 1970) indicate that a two-layer crust having a total and rather constant thickness of around 6.5 km or slightly more is commonly present. These seismic studies provide ambiguous information on the lithologies composing the crust; dredging, deep-tow camera observations and comparison with ophiolites better constrain the possible constituents of the oceanic crust. The upper layer, usually termed Layer 2 (about 2-km thick), appears to consist of basaltic lavas, mostly pillowed, on the present surface of the oceanic crust. The deeper parts of Layer 2 are likely to consist of fine-grained basaltic rocks, which may be either intrusive or extrusive. The lower crustal layer, termed Layer 3 (about 4.5 km thick), is generally considered to be mostly gabbros (e.g., Fox *et al.*, 1973); this overlies mantle with velocities appropriate for fresh peridotite. Layer 2 can be divided near the ridge axis into two portions (2A and 2B; Houtz and Ewing, 1976). Fox and DeLong (1976) suggest that this division, which dies out farther from the ridge axis, is due to highly fractured and rubbly basalts in the upper part near the ridge axis, that become consolidated farther away by cementation in the cracks. In some places in the Pacific, Layer 3 has been found to have a lower division, termed Layer 3B (Maynard, 1970; Sutton *et al.*, 1971; Snydsman *et al.*, 1973). This may consist of slightly serpentinized peridotite (Bottinga and Allègre, 1973) or of pyroxene-rich cumulate gabbroic rocks (Fox *et al.*, 1973). Two studies have detected this higher velocity, lower part of Layer 3 in the Atlantic, in crust formed at a slow-spreading ridge (Fowler, 1976; Steinmetz *et al.*, 1977). The extent of its development is not known in either the Atlantic or Pacific oceans, and earlier suggestions that it characterizes crust generated at fast-spreading ridges are not well-established.

Refraction studies have been conducted near the axes of rifted ridges at several locations. On the Mid-Atlantic Ridge at 45°N, Keen and Tramontini (1970) have results only for crust of age 2–6 m.y. (western rift mountains). The Reykjanes Ridge (Talwani *et al.*, 1971) is anomalous in depth and morphology. In the FAMOUS area, the refraction data point to the presence of Layer 2 at the axis as well as a Layer 3 (Whitmarsh, 1973, 1975; Poehls, 1974; Fowler, 1976). The depth of the mantle is on the order of 5–10 km. A point of dispute is the velocity of the mantle, either normal (Whitmarsh, 1973; Poehls, 1974) or reduced (Fowler and Matthews, 1974; Fowler, 1976). The basement layer (Layer 2) has a constant thickness away from the axis as shown by Houtz and Ewing (1976). Depth of the mantle is thus approximately 5 km below the ocean floor at least as close as 10 km away from the axis. Similar results have been obtained from precise refrac-

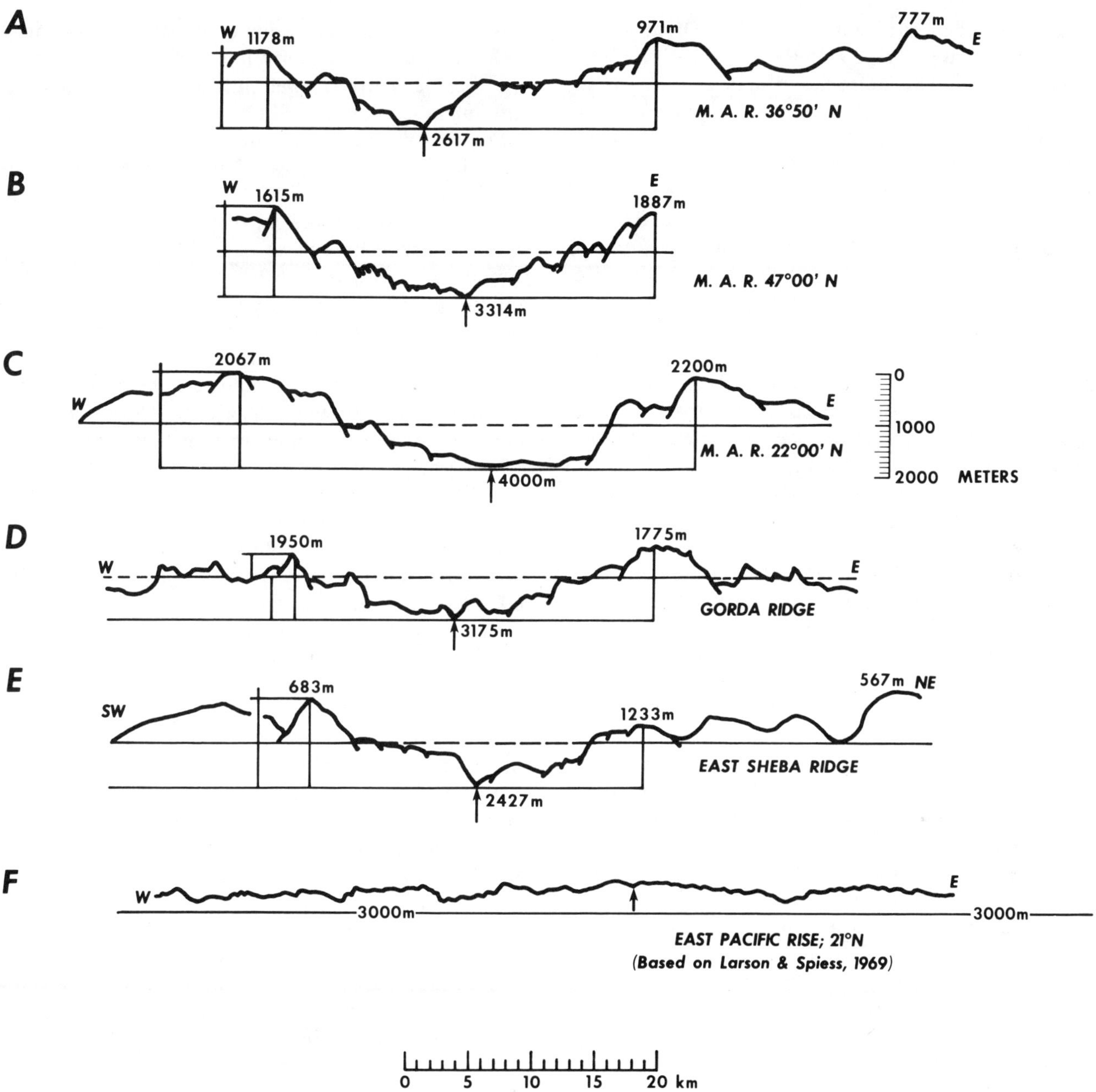

6.2.1 Profiles of slow-spreading, rifted (**a**, **b**, **c**, **d**, **e**) and fast-spreading, nonrifted (**f**) ridges. Arrows indicate axes; M. A. R. = Mid-Atlantic Ridge.

tion surveys in the axial region of the Tadjourah-Ghoubbet-Asal Rift in Afar (Lépine *et al.*, 1972; Ruegg, 1975).

The work of Francis and Porter (1973) at 45° N in the Atlantic and of Whitmarsh (1975) in the FAMOUS area suggests the presence of a magma chamber at a depth of about 3–5 km. Seismic refraction and multichannel seismic reflection investigations of the East Pacific Rise crest indicate the presence of a low velocity zone, interpreted to represent the top of a shallow level magma chamber, at a depth of 2–3 km (Orcutt *et al.*, 1976; Rosendahl *et al.*, 1976; Reid *et al.*, 1977; Herron *et al.*, 1978). This is also the estimate of J. C. Ruegg (pers. comm., 1976) for the region of Tadjourah. There is evidence for strong attenuation of P and S waves below the axis (Molnar and Oliver, 1969; Solomon, 1973; Solomon and Julian, 1974; Ruegg, 1974; Pontoise *et al.*, 1976). Steinmetz *et al.* (1976) and Sapin (1975) show that at 40°N in the Atlantic, the zone of strong P wave attenuation is about 50 km wide, with sharp lateral boundaries and an upper surface deeper than 6 km below the seafloor. Thus, energy is transmitted across the ridge in the uppermost lithosphere but not deeper down. This is also true for the FAMOUS area, where shear waves propagate across the axis (Fowler, 1976). This effect was not seen with the approach of Molnar and Oliver (1969).

Weidner and Aki (1973) have shown that earthquakes occur at the axis at shallow depths (2 or 3 km). This estimate may be inaccurate, but J. C. Ruegg (pers. comm., 1976) has found, with a local seismometer network in Afar, that most seismic foci are in the crust (i.e., are at depths less than 6 km). This reflects the thickness of the layer in which deformation is brittle in the axial region.

Earthquakes in the rift valleys have low magnitudes (≤ 5.6; Francis, 1974). Earthquake swarms are common (Sykes, 1970; Tadjourah and Gulf of California), and high b values have been related to low stress (Thatcher and Brune, 1971; Francis, 1974). Teleseismic data for large earthquakes ($M \geq 5.5$) yield normal fault mechanisms (Sykes, 1967; and for Iceland, Klein *et al.*, 1973), which is in agreement with field observations where no evidence for compressive tectonics has ever been found. The accuracy of epicenter locations determined from teleseismic data is not sufficient to provide constraints on the width of the active seismic zone. The width of this zone can be estimated from local observations in Tadjourah (Ruegg, 1975) and at the Mid-Atlantic Ridge at 45°N (Francis and Porter, 1973), where seismicity is confined to the median valley (perhaps in the central 15 km of this valley), which is in agreement with tectonic observations.

Evidence from ophiolite complexes. Models for the geometry of plate accretion at oceanic ridges have been developed by several authors who have adopted the view that ophiolite complexes (Penrose field conference participants, 1972) represent detached slices of oceanic crust and lithosphere (Dewey and Kidd, 1977; Kidd, 1977; Church and Riccio, 1974; Dewey *et al.*, 1973; Greenbaum, 1972; Moores and Vine, 1971). Numerous workers have pointed out that the thickness and petrology of the internal units that compose well-developed ophiolite sequences match our present understanding of the oceanic crust very well in a general way and that, although the thickness of individual ophiolite complex units may vary considerably, the range of thickness variation is comparable with layer thickness variations recorded in the oceans (Coleman, 1971; Dewey and Bird, 1971; Moores and Jackson, 1974).

The general sequence of rock types in ophiolite complexes consists (Fig. 6.2.2), from base to top, of the following units: harzburgite-minor dunite-rare lherzolite, with metamorphic textures, and a complex and pervasive high-temperature deformation history; cumulate dunite below a cumulate transition sequence of dunite, feldspathic dunite, wehrlite, feldspathic wehrlite, clinopyroxenite, and gabbro-containing zones affected by penetrative high-temperature deformation; layered cumulate gabbro, clinopyroxenite, and anorthositic gabbro; more isotropic and homogeneous, somewhat leucocratic gabbro with irregular pods and veins of trondhjemite near the top and with diabase dikes increasing in abundance toward the top; a sheeted diabase dike complex with minor trondhjemite bodies near the base; and a pillow lava complex with an increasing percentage of dikes, and sometimes sills, toward the base. The following constraints imposed on the geometry of plate accretion are drawn mainly from observations on the Bay of Islands, Mings Bight and Betts Cove Ophiolite Complexes in western Newfoundland (Dewey and Kidd, 1977), but include those on other ophiolite complexes, particularly Troodos (Gass and Smewing, 1973; Greenbaum, 1972; Kidd and Cann, 1974; Moores and Vine, 1971) and Vourinos (Jackson *et al.*, 1975; Moores, 1969).

In most ophiolite complexes, from 500 m to 1 km of pillow lavas are developed, within which the percentage of dike increases downward until the base is defined as the level at which the lowermost screen of pillow lava is found, indicating that the lateral extension strain from fissuring increases downward. This shows that the complete pillow complex cannot be accumulated in batches of full thickness, but must grow to its full thickness some distance from the axis. Sills become common toward the base of the pillow complex in the Betts Cove Ophiolite, and in the Point Sal ophiolite, a 100-m-thick sill zone separates the homogeneous gabbro and diorite zone from the pillow complex (Hopson and Frano, 1977), indicating that sill injection may account for a part of the growth in thickness of the lower part of the pillow complex.

Sheeted-complex dike widths are commonly in the range of 20 cm to 1 m, which, for a half-spreading rate in the range of 1–10 cm/yr, indicates an average of one dike emplaced every few years to a few hundred years. Sheeted-complex dikes commonly preserve only one chilled margin with a preferred chilling direction, indicating that dikes are split to allow further dike injection and that most injection takes place over a very narrow zone away from which dikes are injected at a rapidly decreasing rate (Kidd, 1977). These dikes are notably and consistently aphyric. The fissures in which the dikes form must either be tapping magma without phenocrysts or tapping fluid segregating from a mush of crystals and interstitial fluid. It is possible that the crystal mush, where present, may ascend some way up from the bottom of any dike-filled fissure; this may partly account for the absence of sharply defined bases of individual dikes in ophiolite gabbros.

There is a strong contrast in grain size between dikes and gabbro, indicating a rapid cooling of gabbro before the emplacement of a particular dike. This evidence supports the idea (Anderson, 1972; Lister, 1972) of hydrothermal convective (rather than conductive)

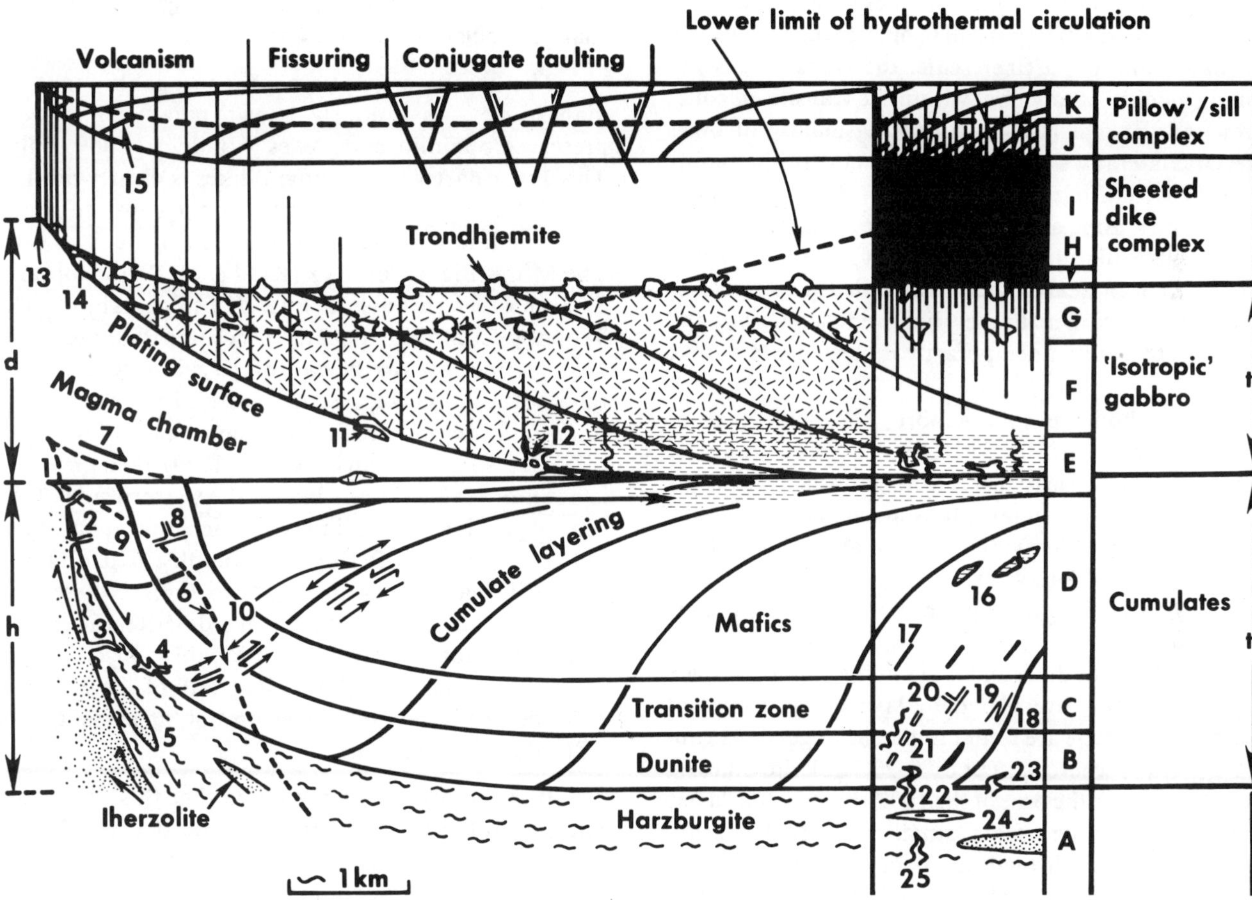

6.2.2 Steady-state model for development of oceanic crust and upper mantle at an accreting plate margin. Top of pillow lavas taken as level datum. 1—magma supply axis to magma chamber. 2, 3, 4, 8—dikes intruded in residual harzburgite and lower cumulates (dunite, orthopyroxenites, websterites, clinopyroxenites, gabbros). 2, 8—undeformed. 3—folded. 4—strongly folded and disrupted. 5—lherzolite lens trapped in residual harzburgite. 6—zone undergoing penetrative ductile deformation due to shear between subsiding cumulates and rising melt plus residuum. 7—possible slumping due to non-steady-state rise of axial welt. 9—lens of chromitite from deformation of cumulate layers. 10—contrasting bending deformation due to subsidence of cumulates. 11—block stoped from magma chamber roof. 12—local intrusive gabbro. 13—axis of spreading (axial singularity); top of magma chamber. 14—trondhjemite bodies forming near axis. 15—sills injected into lower volcanics. 16–25—features seen in fully formed ophiolite complex due to spreading process; column at right side of diagram gives vertical section of fully assembled oceanic crust. 16—xenoliths from roof (11). 17—small-scale cumulate layering attitude near top of layered cumulates (oblique to large-scale layering). 18—folded layers in deformed cumulates. 19—undeformed dikes (gabbro, pyroxenite) in cumulates. 20—buckle folded layers. 21—boudinaged layers. 22, 23, 25—folded dikes (pyroxenite, dunite) in cumulates and harzburgite. 24—gabbro lenses in residual harzburgite (dikes or bodies of trapped magma). See Fig. 6.2.4 for terminology.

cooling of the upper part of the oceanic crust as it travels away from the axial singularity. The depth of the transition from sheeted-complex to gabbro and, therefore, the thickness of the 100% sheeted dike layer are most probably controlled by the lower limit of hydrothermal circulation at the axial singularity. The numerous anastomosing breccia zones observed within the sheeted-complex in some ophiolite complexes also suggest pervasive hydrothermal circulation (Williams and Malpas, 1972). In western Newfoundland ophiolite complexes, the transition from 100% sheeted dike complex to gabbro with less than 5% dikes takes place over a vertical thickness of about 250 m or less when viewed on a broad scale. On a finer scale, the transition is not homogeneous; at any level within the transition, zones a few tens of meters wide consisting mainly of dikes alternate laterally with similar zones consisting mainly of gabbro. In Mings Bight (Kidd *et al.*, 1978) one example of such a zone about 300 m wide consists, near its base, of 95% dikes penetrating about 200 m lower into gabbro than in nearby sections. Each zone of greater dike concentration up to a few hundred meters wide and deep was therefore a temporarily preferred site of dike injection.

In the homogeneous gabbro, the percentage of dikes decreases with depth. Dikes do not gradually increase in grain size to merge with gabbro. This means that the amount of lateral extension decreases from the base of the sheeted-complex to zero at the base of the homogeneous gabbro where the last dike dies out. Individual dikes cut solidified gabbro and are then underplated by gabbro as they move away from the axial singularity, just as dikes and flows in the pillow complex are progressively overplated by lava flows. Rocks of trondhjemitic type are found in the homogeneous gabbro, generally where it is cut by 5–20% dolerite dikes. A few occur within the base of the 100% sheeted dike layer. These bodies range from small individual tension gashes and veins through areas of net-vein breccia a few meters to tens of meters across and, exceptionally (Troodos), bodies as much as 1 km in original horizontal width. They are clearly associated with the sheeted-complex–gabbro transition and are part of the ophiolitic magma sequence, because many examples cut some dolerite dikes but are cut by others. One example in Troodos is so greatly dissected by dikes that it must have formed very near the axis of injection. Many of these bodies show clear intrusive relationships (net-vein breccias); their mineralogy, texture and relationships to adjacent rocks, particularly where they crosscut dolerite or contain dolerite or gabbro xenoliths, show that they are not hydrothermally altered gabbro, even though such alteration may occur locally.

The upper part of the overall gabbro layer, including that within the transition into the dike complex, is generally very homogeneous but locally shows diffuse layering a few centimeters or less thick. This homogeneous gabbro is somewhat richer in plagioclase when compared to gabbros of, for example, isolated thick tholeiitic sills and dikes. Plagioclase flotation in a magma chamber probably is not necessary to explain this slight enrichment in plagioclase, particularly because the clinopyroxene content is still large and grains are not interstitial. In addition, the presence of plagioclase-rich layers in the underlying cumulate gabbros suggests that plagioclase flotation is not important.

The thickness of homogeneous gabbro cut by diabase dikes is a poorly known parameter in well-preserved ophiolite complexes with thick gabbro units. The upper part with abundant dikes is clearly not less than 300 m thick, but firm data on the depth to which less abundant dolerite dikes occur are difficult to find. The Mings Bight Complex (Kidd *et al.*, 1978) contains diabase dikes through at least 750 m of gabbro. The Troodos Complex appears to contain diabase dikes nearly as far down as the base of the gabbroic layer; the thickness that this represents depends on the structural interpretation preferred for the central part of the complex. If it is a gentle dome (Wilson, 1959), then the depth is not more than about 1 km; if the western side of the central part is the steep limb of a monocline, it may be as much as 3 km. The Betts Cove Complex (Upadhyay *et al.*, 1971) has a very thin gabbro layer; where it is clearly cumulate, it contains no dolerite dikes, but where it is homogeneous and probably not cumulate, dikes are found throughout almost all of the 250 m of gabbro. The other complexes do not, at present, provide additional data.

It is important to emphasize that this discussion refers to dolerite dikes that do not cut the cumulate or the noncumulate ultramafic rocks. The fact that the dikes of the sheeted-complex are "rootless" means that they were fed from a magma chamber and that the gabbro they cut was plated onto the roof of that chamber. Later events which affect the already-consolidated oceanic crust, are local and may include hot-spot-type magmatism, arc volcano construction, and effects where a crustal block passes a ridge-segment end across a transform-fault–fracture-zone transition. In all cases, these dikes would be expected to cut cumulate gabbros and ultramafic rocks and the noncumulate ultramafic rocks. This can be seen in the Mings Bight Ophiolite Complex (Kidd *et al.*, 1978) where large, distinctively porphyritic, dolerite dikes cut all members of the ophiolite complex and feed large sills within the volcaniclastic sediment/pillow lava sequence that lies on the ophiolite complex proper.

Because of the difficulty of distinguishing plated roof gabbro from rapidly deposited, homogeneous, cumulate gabbro, the maximum thickness of cumulate gabbro in ophiolite complexes is poorly known. Not more than 500 m of layered cumulate gabbro has been proved to exist in well-preserved ophiolite complexes, although 2 km or more may exist in the Bay of Islands, Papua and perhaps the Oman; Betts Cove and Mings Bight contain not more than about 300 m.

The most reliably established section of cumulate ultramafic rocks comes from Vourinos (Jackson *et al.*, 1975); even though it is composite and crossed by faults, a minimum thickness of 1 km of ultramafic cumulates is present. This section shows cycles of deposition, which may indicate a batch or an intermittent supply of magma. Interpretation of the structure of the western margin of the Troodos massif as a steep monocline limb leads to a maximum thickness of 1.5 km for the ultramafic cumulates, including the lower 300 m of dunite. The Bay of Islands Complex probably contains a similar thickness, with the lower part also being cumulate dunite and feldspathic dunite. Although cumulate gabbros and transition-zone rocks and particularly cumulate dunite contain zones affected by penetrative high-temperature deformation, regular undisturbed cumulate layering in the gabbros and ultramafic cumulates demonstrates that ophiolite magma chambers have generally flat floors. Yet, in the Bay of Islands Complex, the orientation of cumulate layering relative to gross unit layering varies from roughly parallel at the base and top of the cumulate sequence to as much as 90° in the middle of the cumulate sequence. Although detailed cumulate layering cannot be traced laterally for long distances as it can in many layered mafic plutons, there is no evidence that crosscutting individual plutons can be defined within the gabbro and ultramafic cumulate layers. Therefore, a steady-state mechanism must be inferred for the development of cumulates and homogeneous gabbro.

The boundary between the cumulate and deformed noncumulate ultramafic rocks has been shown to be sharp and traceable for 1 km in the Vourinos complex; it is possible that this contact actually is between undeformed and deformed cumulates rather than between cumulates and noncumulates. In the Bay of Islands Complex, this boundary is somewhat obscured because the lower ultramafic cumulates are penetratively deformed, a situation that also seems to be the case in Troodos (George, 1978). The basal cumulates must therefore have been deposited on a noncumulate residuum which was sufficiently consolidated to preserve the gross cumulate layering but plastic enough to allow their deformation. As the spreading process pulls the lithosphere apart, the top of this harzburgite "basement" has to be formed at the magma supply axis immediately prior to deposition of cumulate material on it; it cannot be much older basement as some have previously suggested.

The lack of intrusive, crosscutting, mafic material in the residual harzburgite and overlying ultramafic cumulates of well-preserved ophiolite complexes shows that the extraction of partial melt at the supply axis is extremely efficient and must occur over a very narrow zone closer to the supply axis than the position at which the first cumulates are deposited. The occurrence of residual harzburgite immediately below cumulates and the existence of an at least 7-km thickness of harzburgite in part of the Bay of Islands Complex show that efficient melt extraction and residuum plating occur from the level of cumulate deposition and that they must continue progressively to at least 7 km below that level.

Models for plate accretion. Models dealing with plate accretion have addressed two fundamental classes of problems. One is that of the large-scale surface topography and tectonics; in particular, explaining the existence of a steady-state axial rift valley on slow-spreading ridges. The other concerns the generation of the various lithological units of the oceanic crust and upper mantle.

There are two broad sets of models that purport to explain axial rift valleys or the general topography of spreading axes. Kinematic models have been proposed by Deffeyes (1970) and by Anderson and Noltimier (1973), who suggested that the existence of a rift valley is a function of a budget between strain in the axial region and lava accumulation. The model requires that formation of the oceanic crust takes place in a region which is wider than the tectonic zone. This is not supported by the seismic results (Whitmarsh, 1973, 1975; Poehls, 1974) or by the field observations (ARCYANA, 1975; Ballard *et al.*, 1975; Ballard and Van Andel, 1977; Bryan and Moore, 1977) in the FAMOUS area.

Palmason's (1973) model provides an interesting kinematic analysis for Iceland. However, his boundary condition of a steady-state horizontal topography makes his approach not useful to the problem of rift valley formation. Stressing the importance of isostatic compensation, Daignieres *et al.* (1975) have proposed a similar numerical model based on Iceland data, where both tectonics and volcanics are considered as successions of discrete events.

The following mechanical models have been proposed: Piper and Gibson (1972) have suggested that the presence of rift valleys is a function of the width over which stress is applied through hydrostatic pressure in a magma chamber below the axis. The analysis yields a zone of horizontal compressive stress very close to the axis. For example, a magma chamber with a 2-km radius at a depth of 6 km produces compression over a region extending about 6 km away from the axis, beyond which active normal faulting takes place. Other authors (Lliboutry, 1976; Koide and Bhattacharji, 1975) have suggested a similar active role for the magma intrusion in the axial region.

Sleep (1969) has suggested a viscodynamic explanation for the origin of rift valleys in slow-spreading ridges. The central depression is explained by viscous head loss of the rising material in a vertical-walled cleft below the axial region. Recovery lifts up the crust and lithosphere in both walls of the rift valley. This idea has been the starting point for more elaborate models such as those of Osmaston (1971), Lachenbruch (1973), and Sleep and Rosendahl (1979).

Osmaston (1971) and Harrison (1974) have predicted reverse faulting on rift valley walls, which is in contradiction with observations (ARCYANA, 1975; Needham *et al.*, 1976). The viscodynamic model of Sleep (1969) and Sleep and Biehler (1970) raises some basic difficulties. Sleep (1969) proposed his model because he rejected the idea that rift valleys could be the result of graben formation on the basis of Vening Meinesz's (1950) analysis for grabens within continental crust. The latter is based upon a set of questionable boundary conditions: normal faults are proposed to cut through the whole rigid lithosphere, and no friction and zero bending movements are assumed on the fault plane. Also, Sleep (1969) attempted to model a rift valley as one single graben and found actual valleys to be too deep to represent a "normal" isostatic graben. In reality, a rift valley is a series of nested grabens.

Lachenbruch (1973) has made a quantitative treatment of Sleep's viscodynamic mechanism with the aim of explaining second-order structural features at the axes of fast- and slow-spreading ridges. In fact, only the gross features are modelled: valley for slow-spreading ridges and rise for fast-spreading ridges; the top crustal layer is ignored; the graben nature of the valley is not considered. The large number of independent parameters and the restrictive conditions which Lachenbruch (1973) set for the geometry of the conduit and the behavior (Newtonian viscous) of the rising material make his model difficult to test. Lachenbruch (1976) extended the model to conduits of triangular shape. Little attention was given to fitting observations. Sleep and Rosendahl (1979) have constructed steady-state viscous flow models of the topography of fast-spreading, slow-spreading, and 3 cm/yr ridges, using rheology inferred from detailed thermal models. If the material in the crustal magma chamber was assumed to have low density and low viscosity, reasonable fits to the observed topography were obtained. At 3 cm/yr spreading rate, the presence of a rift or axial high was sensitive to the crustal thickness. A greater crustal thickness resulted in a larger crustal magma chamber and thus an axial high as at fast-spreading ridges. A lower crustal thickness resulted in a small magma chamber as at slow-spreading ridges. Slow-spreading ridges with very thick crusts, such as the Reykjanes Ridge, are supposed to have axial highs for the same reason.

Francis (1974) has proposed a model in which a rift valley is a zone of constant caldera collapse. The inner floor is viewed as the top of a lid floating on a magma chamber and is, at the time of collapse, decoupled from the adjacent rift valley walls. Caldera collapse may occur locally in the inner floor and is observed in Ardoukoba (Needham *et al.*, 1976), but it cannot be a dominant process, since active normal faults are observed at any scale over a broad width (10–15 km) and magma chamber widths are (supposedly) small (< 1–2 km) for slow-spreading ridges (Kusznir and Bott, 1976; Sleep and Wolery, 1975).

Tapponnier and Francheteau (1978) have proposed that because an oceanic rift valley is a steady-state feature, necking must be maintained in a steady-state situation. In a typical extension or squeezing experiment, the sample that is pulled shrinks in the region of necking down to zero thickness or to fracture. In the case of the rift valley, this situation is never reached, since a constant thickness of the lithosphere (4–5 km) is preserved near the axis of maximum necking because of a steady-state temperature gradient. The stretched lithosphere is healed from below through cooling and from above through lava accumulation. Cracks that provide access for the magma to the surface occur in the neck but heal rapidly.

Steady state is achieved only statistically, since the accretion process is a series of discrete successive events (normal faulting, dike intrusion, and lava accumulation; see, for illustrative sketches, Bayer *et al.*, 1973; and Daignieres *et al.*, 1975). The downward-diverging flow lines near the axis of Fig. 6.2.3 illustrate the cumulative effect of combined normal faulting, fissuring and lava piling (Palmason, 1973; Daignieres *et al.*, 1975). Also shown in Fig. 6.2.3 are the flow lines for plastic horizontal extension (Backhofen, 1972) in which vertical thinning balances horizontal stretching. The plastic flow field provides a consistent link between the downward diverging flow in the crust and the upward diverging flow in the mantle.

A consequence of the necking of the lithosphere is a redistribution of mass in the axial region. When the rift inner floor is affected by normal faulting as a result of necking, a small graben is created that results in local lowering of the floor relative to the walls, whereas at a greater depth, plastic deformation takes place. The effect of the plastic deformation is to transfer mass sideways so as to conserve volume (Backhofen, 1972; Fig. 6.2.3, dotted area). Since the density contrast at the bottom of the 5-km layer is negligible, the net result is a mass deficit (water replacing basalt) at the surface in the down-faulted region of the new graben. A net upward force due to buoyancy will affect the layer on top of the asthenosphere in the region immediately below the graben. Since the top layer is pictured as being continuous and with strength, the isostatic balance will not be restored locally but will possess a degree of regionality: the top layer will "bend" upward, and the wavelength of the response will be wider than the graben itself. The topography of nonrifted ridges is not addressed by these models except by that of Lachenbruch (1973).

Models that address the lithological assembly of part or all of the oceanic crust and upper mantle include those by Cann (1974), Greenbaum (1972), R. Kidd (1977) and Dewey and Kidd (1977). The treatments of R. Kidd (1977) and Dewey and Kidd (1977) basically agree on the large-scale features of the accumulation of the upper part of the crust (lavas, sills and dikes of basalt and diabase), although which of the various possible detailed geometries of accretion is dominant in this part of the crust cannot be decided from present data. All proposed models depend on the presence of a permanent magma chamber beneath the spreading axis. They differ significantly on the geometry of such a chamber. Dewey and Kidd (1977) also require the lower parts of the cumulate rocks produced from the chamber

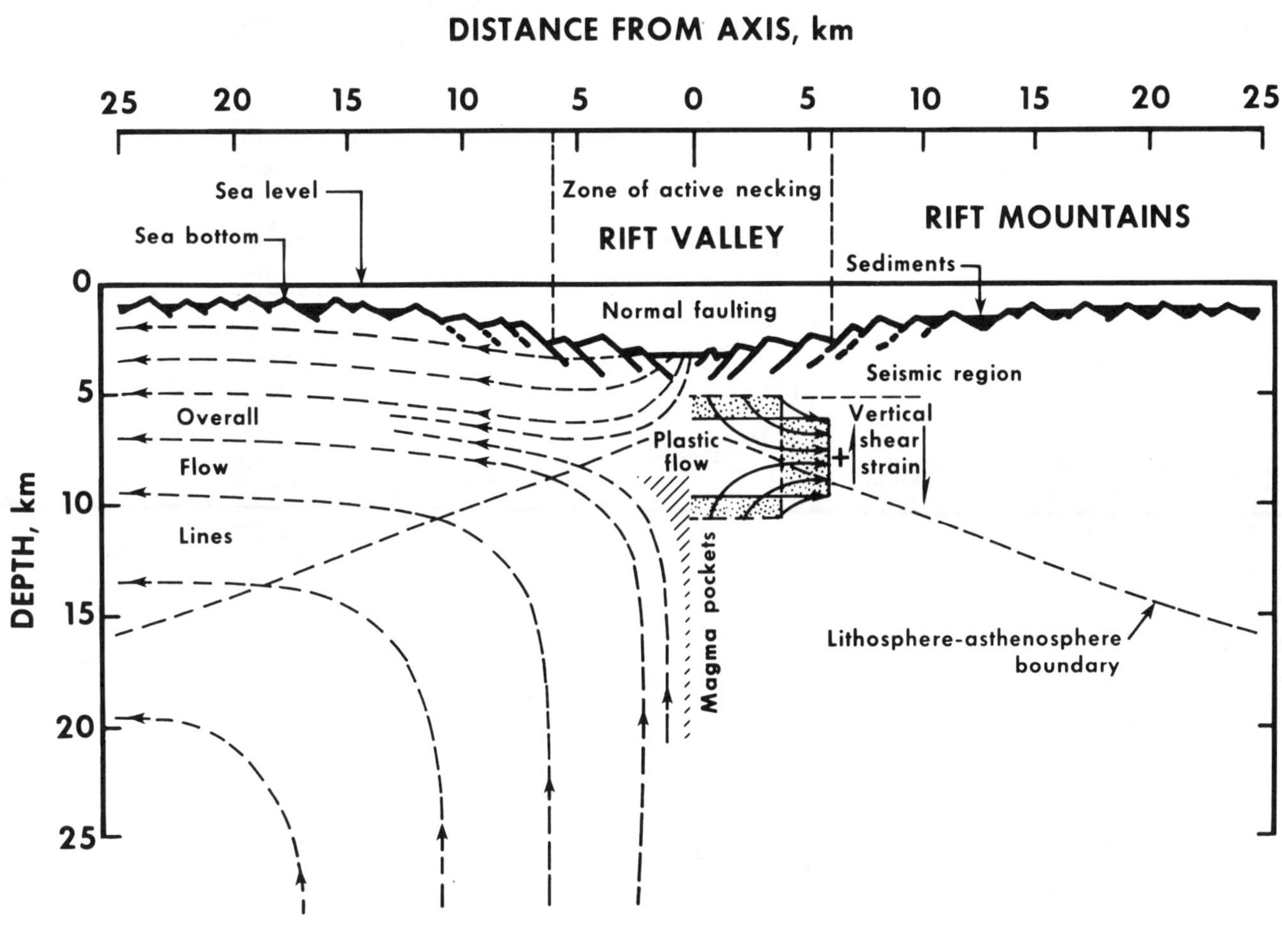

6.2.3 Major features of the mechanical behavior of the lithosphere in the region of plate accretion at plate boundaries with slow spreading rates. Steady-state necking of the lithosphere is suggested to occur. After Tapponnier and Francheteau (1978).

to undergo plastic deformation at high temperature; evidence seen in ophiolite complexes supports the latter contention. The main components of the model suggested by Dewey and Kidd (1977) are summarized in the following paragraphs (see Figs. 6.2.2–6.2.5).

Pillow lavas accumulate progressively, in any one section, from essentially zero thickness at the axis to full thickness 1 or 2 km from the axis. The thickness of 100% sheeted dikes (typically 1 km) is controlled by the depth of hydrothermal circulation at the axial singularity. At least 300 m of gabbro is plated, starting at the axial singularity, onto the roof of sheeted dikes in a complementary, progressive manner to the lavas above. As this gabbro is plated, it must be rapidly cooled by the

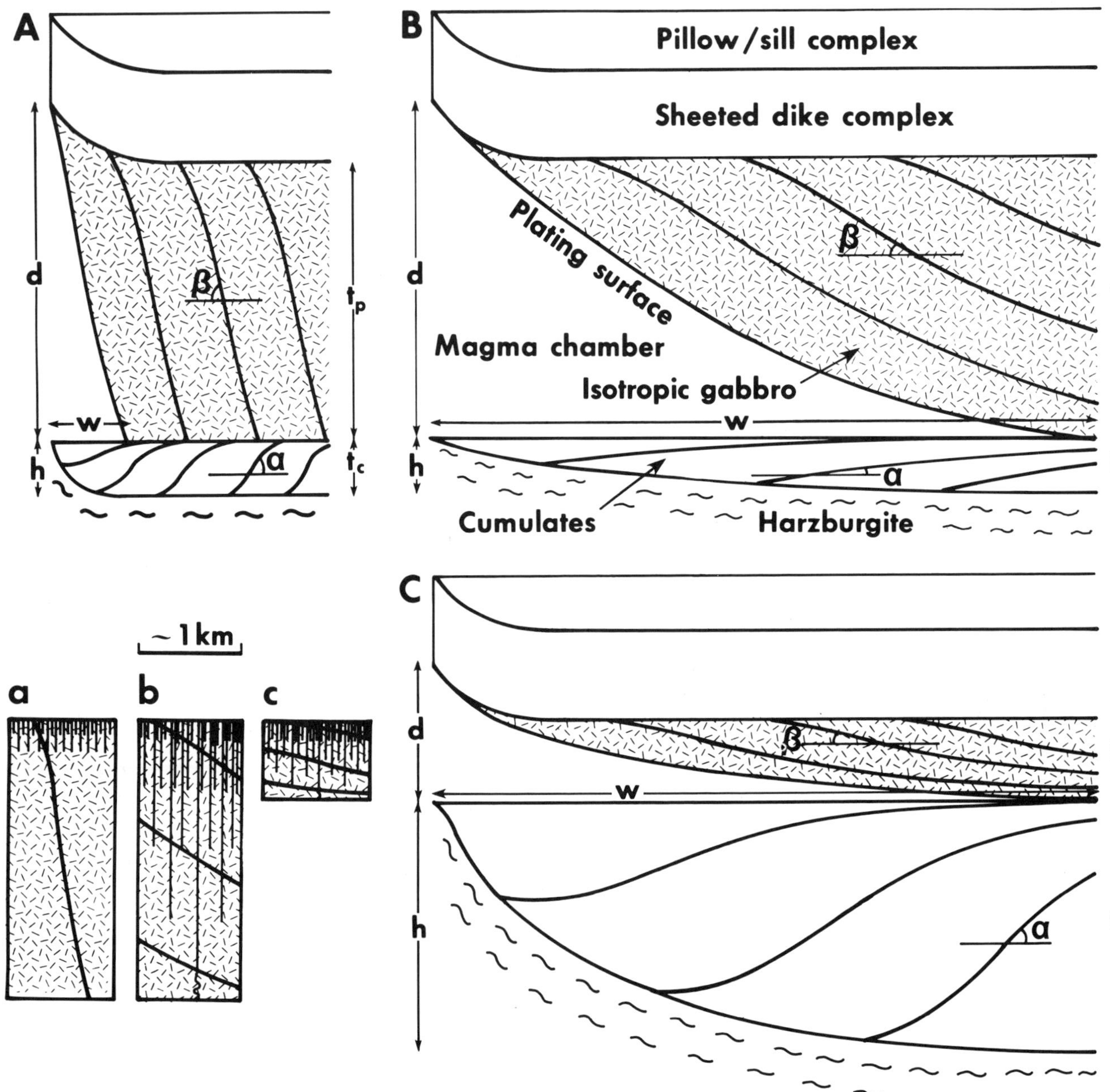

6.2.4 Diagrams showing how magma chamber width (w) and depth (d), and the height of the axial welt (h) control the thicknesses of roof-plated gabbro (t_p) and of cumulates (t_c), and the angles (α, β) between small-scale layering and the horizontal. **a**, **b**, and **c** show suggested distribution of dikes in roof-plated gabbro of **A**, **B**, and **C**, respectively. After Dewey and Kidd (1977).

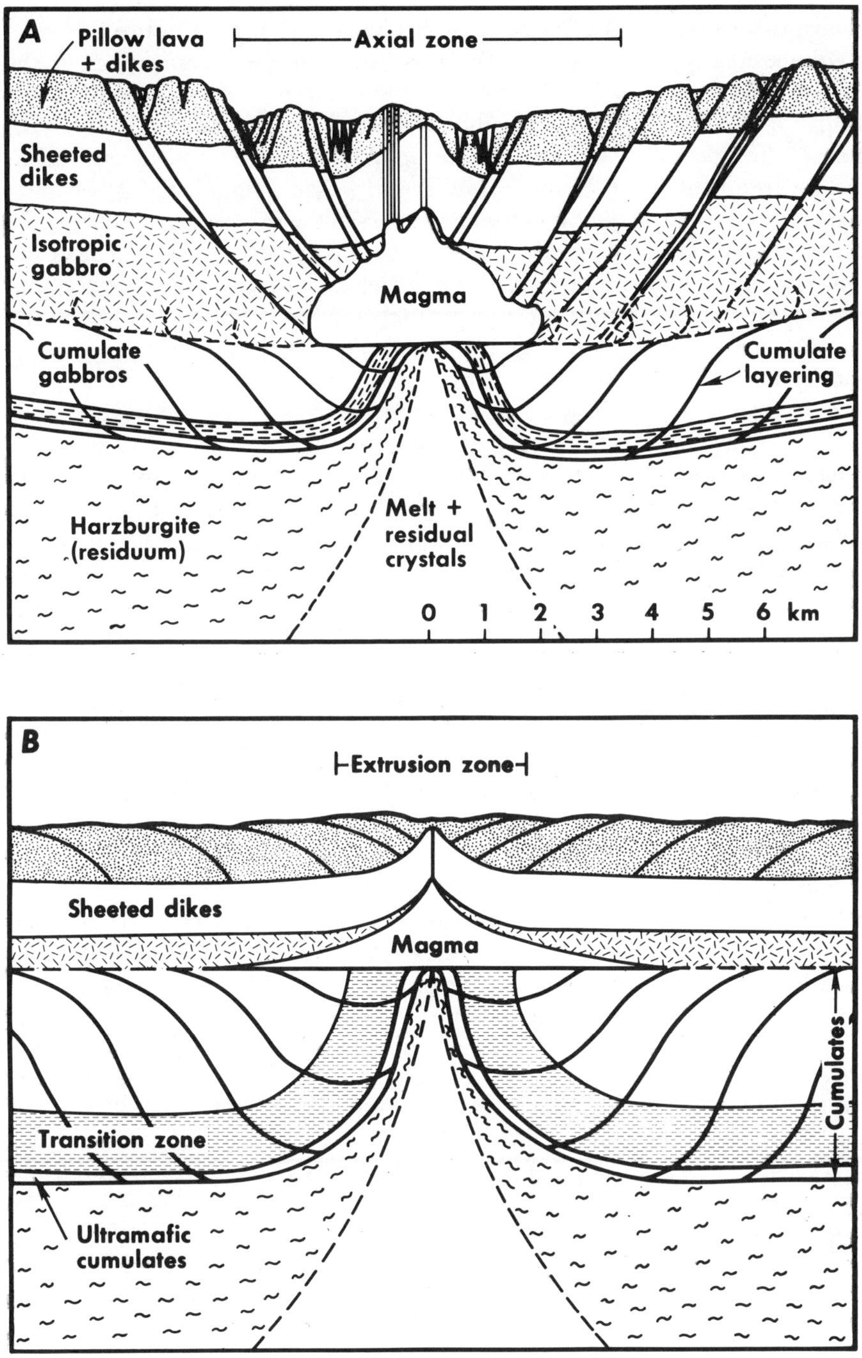

6.2.5 Sketched cross sections of plate accretion model drawn on (**a**) rifted, slow-spreading ridge profile, and (**b**) nonrifted, fast-spreading ridge profile. Profile in **a** from North Atlantic, 36°50′ N (ARCYANA, 1975). Profile in **b** redrawn from Larson and Spiess (1969); East Pacific Rise, 21°N. After Burke and Kidd (1980). Both diagrams at the same scale; no vertical exaggeration.

hydrothermal circulation and, at least near the axial singularity, cut by tensional cracks that are filled by dikes.

The magma chamber is zoned, with dominantly olivine cumulates being deposited near the magma supply axis, grading out to dominantly clinopyroxene-rich cumulates that in turn pass into clinopyroxene-plagioclase cumulates, which continue to the edges of the magma chamber. Magma chamber zonation is strongly suggested by the spatial variation of the composition of erupted magmas (Ballard *et al.*, 1975; Bryan and Moore, 1977). The fact that all well-preserved ophiolite complexes consistently show the vertical sequence of gross layering required by this zonation indicates that the thermally controlled system and the environment producing it are more or less steady-state, and that rapid convective overturn and total mixing are in some way inhibited. If the supply of new magma to the chamber is roughly balanced, on a short time scale (100 yr), by eruptive removal and by the volume created by spreading, then new hot magma should rise directly to the apex of the chamber and should not grossly disturb the somewhat cooler magma nearer the sides. The fact that the zonation exists is a strong argument for a fairly continuous supply of new magma; this is the only reasonable mechanism to keep the central zone in the magma chamber hotter than the sides when the whole roof is being cooled (mainly by hydrothermal circulation). D. Walker points out (pers. comm., 1980) that picritic magma, though hot, may not rise because it is too dense and thus enhances chamber zonation.

The modal compositions and relative volumes of the plated and cumulate rocks show that the phases involved in crystal fractionation are dominantly clinopyroxene and plagioclase; olivine is dominant near the axis but rapidly becomes negligible away from the axis and is, overall, a subordinate phase. Orthopyroxene is always a minor phase; opaque minerals are conspicuously absent from all the rocks, except for chromite in the dunite. Because fresh magma is being supplied fairly continuously to the chamber, fractionation models must allow for some mixing between this and fractionating magma. The magma chamber, above the section in which cumulates are forming, must be filled with magma (liquid + crystals), not with mush (framework of crystals with interstitial liquid). Mush may exist only below the cumulate deposition surface and above or outside the roof gabbro-plating-surface.

A flat-floored magma chamber is required, according to Dewey and Kidd (1977), because of the large thickness of cumulate rocks and of the need to form them within a reasonable distance of the axis. To do this with an essentially steady-state process requires that there be a very narrow axial welt consisting of a mush of residual crystals and melt at its center, and of plated residual harzburgite thickening downward and sideways at its sides. The central part of the welt will be about 70:30 in relative abundances of residuum:melt at the depth where harzburgite plating starts and will progressively change upward to near 100% melt at the magma chamber floor. The geometry requires rapid subsidence of the lower cumulates and implies the presence of a narrow zone of high shear strain between rising mush and subsiding cumulates. The precise shape of the axial welt, defined by the cumulate ultramafic-plated (noncumulate) harzburgite boundary, is at present unconstrained. The shape chosen gives a maximum amount of room to accommodate the cumulate pile and the zonation required in the magma chamber.

The penetrative deformation, which is seen to affect the harzburgite and zones in the cumulate ultramafics and lower gabbro, is suggested not to be due to horizontal shear by lithosphere plate movement on the asthenosphere, and particularly not to horizontal shear between asthenosphere diverging sideways under the ridge axis more quickly than the lithosphere, as has been implied in several previous treatments. It is evident that material transport in the mantle at shallow depths near the axis under the magma chamber must be dominantly vertical in order to compensate for the gap that would otherwise develop between the two diverging plate edges. Therefore, it is suggested that the penetrative deformation is due mainly to vertical shear stress between the rising mush of residual crystals with basalt melt and the subsiding plated harzburgite and lower cumulates. Some deformation in the cumulates may result from slump or flow processes caused by a rise or fall in the level of the top of the axial welt and to bending stresses in the differentially subsiding cumulate pile. Although the shear stress is vertical, the resulting structures end up nearly horizontal because of the geometry of the subsiding welt and cumulate pile, and nearly parallel with cumulate and other layering because even a small shear strain will rotate these markers into near parallelism with foliation and lineation. Episodic rise or fall in the level of the axial welt requires episodically reduced or accelerated shear strain below; this mechanism may be responsible for zones of more strongly and more weakly deformed harzburgites.

There are some problems that cannot be resolved from data available at present. These are as follows: the relative thicknesses of plated gabbro, cumulate gabbro, and cumulate ultramafics; changes in these as a function of magma chamber width; the origin of large trondhjemite bodies; and correlations between spreading rate, magma chamber width and thicknesses of the various units. The amount and composition of magma supplied for a given increment of spreading on normal ridge segments appear to be fairly constant, because the total thickness of crust in normal oceanic situations is fairly consistent at 6–7 km, and because compositional variations of erupted magmas are small along the spreading ridge and insignificant compared with the variation in composition at any one site.

This model, as illustrated in Fig. 6.2.2, has been deliberately drawn with approximately equal thicknesses of plated and cumulate gabbro. Their combined thickness is shown in this figure in the same proportion to the combined thickness of dikes and pillow lava as oceanic seismic Layer 3 (broadly speaking, gabbro) to oceanic seismic Layer 2 (broadly speaking, dikes and pillows). This has been done because the thicknesses and any differences in thickness of plated and cumulate gabbro in different ophiolite complexes are so poorly known. The model is easily modified (as illustrated in Fig. 6.2.4) to accommodate any necessary changes in the relative thicknesses of plated and cumulate gabbro when these are defined from study of thick, well-preserved ophiolite complexes.

Large-scale thermal models for the cooling of the oceanic lithosphere (Sleep, 1975; Lister, 1977) do not have the resolution necessary to provide the details of crustal accretion even if hydrothermal circulation is included in the analysis. These thermal models do, however, give information on whether it is possible to sustain a steady-state magma chamber for a given spreading rate. The faster the spreading rate, the more likely it is that such a magma chamber exists, but it is not yet clear from the thermal model calculations whether spreading rates typical of rifted ridges (1–2 cm/yr half-rate) allow one to exist. If it turns out that a magma chamber is not a steady-state feature at low spreading rates, then the well-preserved ophiolite complexes are likely to be a skewed sample, representing crust formed only on fast-spreading ridges. However, the lack of extreme fractionation in lavas from slow-spreading ridges is strong evidence for the existence of a steady-state magma chamber along slow-spreading ridges (Shibata *et al.*, 1979).

Tectonics of basaltic volcanism at convergent plate margins

The most readily observed active volcanism on Earth is that in Andean and island arcs. Activity is concentrated in a relatively small number of large volcanoes, many of which are subaerial and hence accessible for study, but it is not clear exactly how much of this activity is basaltic. Although andesitic volcanism dominates in arc-related systems, Ewart (1976) has estimated that about 20% of all the igneous rock produced in island arcs is basaltic, and our own estimate is higher, so that the convergent boundary environment apparently rates second (after divergent boundaries) in the volume of basalt currently being produced.

Discussion in this section is divided into considerations of:

(1) the distribution of arc volcanism in relation to subducted lithosphere;
(2) intra-arc tectonic features;
(3) contrasts between island arcs and Andean volcanic arcs;
(4) effects of ridge subduction on arc volcanism;
(5) the origin and evolution of marginal basins in relation to arcs; and
(6) volcanism at arc-continent and continent-continent collisions.

Distribution of arc volcanism in relation to subducted lithosphere. Wadati (1935) recognized that volcanoes in arcs overlie places where the seismic plane, now identified with the descending lithospheric slab, lies at depths greater than 100 km. This relationship has since been observed in many arcs and is now one of the best established associations in plate tectonics. Where the narrow zone of active volcanoes is separated by only a short distance (l) from the trench, the average dip of the seismic zone (defined by the angle θ from the horizontal) is steep and θ is large; conversely, where l is large

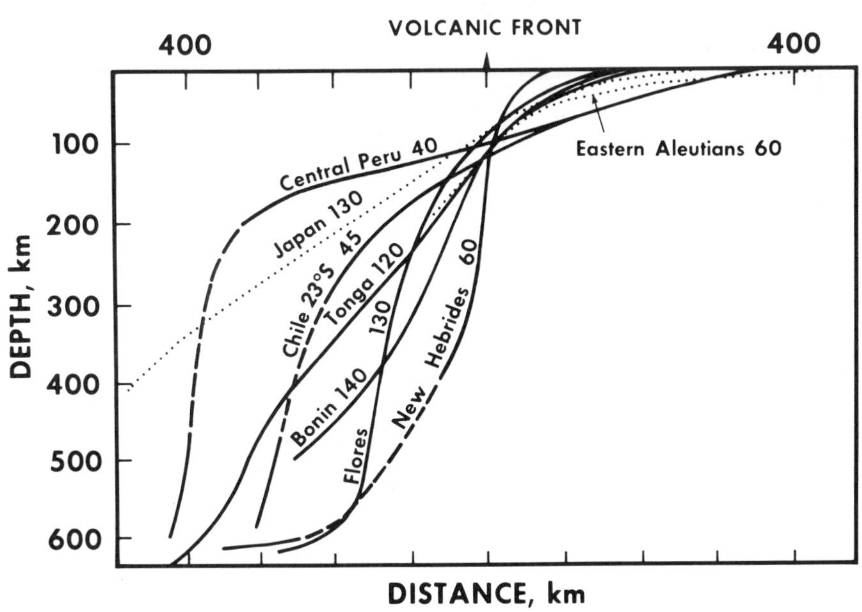

6.2.6 Volcanic arcs generally are located 100–150 km above their controlling subducted slabs. Numbers on the eight slab traces indicate the age (m.y.) of ocean floor being subducted at the respective trenches. Based on a compilation by J. F. Dewey.

(up to a maximum of 500 km in the Makran), θ is small. Estimates of l and θ from numerous arcs indicate that the volcanic zones generally occur 100–150 km above the seismic plane (Fig. 6.2.6), and it is therefore inferred that lithosphere descending at these depths is involved in production of arc magmas, although the question of exactly how and where the magmas are produced is as yet unresolved. Although the simple relationship works well, changes in the dip of the subduction zone can also be accommodated (Karig *et al.*, 1976).

A second unresolved question is that of the amount of magma produced along or above the descending slab. More precisely, what is the rate of production of new crustal material at convergent boundaries? The reason why these questions are unresolved is that most of the obvious ways of addressing them are subject to error. Examinations of volcanic rocks erupted in an arc through time have yielded some general proportions and quantities (Sugimura, 1968; Markhinin, 1968), but sampling errors are hard to assess and preservation is very incomplete and is biased against explosive and clastic material. Inferences about eruption rates based upon the apparent volume of the arc are difficult to make because it is not possible to separate that part of the volume which is maintained dynamically by hot, low-density mantle underlying the arc from the part representing newly produced crust. One possible way to obtain a maximum value for magma production at convergent boundaries would be to assume that magma production is related solely to slab dehydration (Anderson *et al.*, 1976), to assume the slab is uniformly (say, 1%) hydrated for the top 3 km of its 100 km thickness, and to calculate how much energy dehydration could release for partial melting of overlying pyrolite. From this calculation, a rate of crustal production (as basalt or basalt equivalent in calc-alkaline material) might be obtained. It seems likely that this value will be a great deal lower than the rate of crustal production at oceanic ridges.

Variation in composition within arcs has been related to a range of tectonic controls (see discussion below). Miyashiro (1972) has suggested that tholeiitic rocks are more abundant where plate convergence is fast and that alkaline rocks are more abundant where convergence is slow. Correlation of variations in alkali contents of volcanic rocks with depth to the underlying slab has been elaborated by Dickinson and his collaborators (Dickinson and Hatherton, 1967; Hatherton and Dickinson, 1968, 1969; Dickinson, 1970; following earlier work by Rittman, 1953; Kuno, 1966; Sugimura, 1968). Although each area should be considered singly because of local complexities, a general increase in alkalies in igneous products with increasing slab depth seems established.

Tectonics within arcs. Arc igneous activity is closely linked with local extensional phenomena. Compressional tectonics (operating in the arc-trench gap between the volcanoes and the site of subduction in the trench) have metamorphic but few igneous phenomena associated with them. Nakamura (1978) has suggested that dike orientation in various arcs can be interpreted as indicating the direction of maximum extensional stress within the arc and that the elliptical shapes of many calderas can be used in a similar way.

Very large calderas are prominent features of Andean arcs, especially in Sumatra and the Peruvian Andes. These structures imply extreme thinness of the lithosphere along the line of volcanoes; the localized lithospheric weakness of this narrow zone is important in the development of marginal basins.

Along-strike tectonic variations in arcs have been recognized by several workers and have been related to igneous activity. A fairly regular 75-km spacing of active volcanoes has been alleged by Marsh (1974) in the Aleutians and Lesser Antilles which he associates with a characteristic thickness of the magma source layer. Segmentation in seismicity and faulting have been linked by Stoiber and Carr (1973) to variations in volcanic behavior, but doubts have been cast on this correlation (e.g., by Isacks and Barazangi, 1977). A number of within-arc variations can be related to corresponding variations in the properties of the underlying slab, such as the subduction of shallow, buoyant ocean floor (Vogt *et al.*, 1976) and the subduction of lithosphere modified by the occurrence of prominent transform fault zones (Ravenne *et al.*, 1977). Both appear to generate within-arc tectonic variations (such as variations in faulting style and elevation).

Contrasts between island and Andean arcs. Island arcs are constructed on oceanic lithosphere and Andean arcs on continental lithosphere. Where marginal basins form, Andean arcs may become sandwiched by oceanic lithosphere and may evolve into island arcs. The character of the continental lithosphere on which Andean arcs form is itself variable. In many places (e.g., the eastern Aleutians) presently active arcs lie on top of Mesozoic arc material accreted to the continent within the last 100 m.y., but in other places (e.g., the Andes of Peru) arcs lie on rocks that have been continental for a large fraction of geological time. Thus, highly evolved continental crust may lie beneath such an area. There are all gradations of arc foundations, from mature, thick, continental lithosphere to oceanic lithosphere. An arc that has been the site of convergence for tens of millions of years apparently develops a substrate similar in age, thickness and related properties to that of a relatively young addition to a continent.

The variety in arc features becomes more understandable when consideration is given to the range of complexity implied by the variety of processes that lead to arc formation. Some features and changes in tectonic style can be satisfactorily related to changes in such properties as the dip of the subducting slab (e.g., Isacks and Barazangi, 1977) and to the age of the subducted lithosphere, but there is still no clear explanation of many prominent arc features. For example, evidence that the Altiplano of Peru was uplifted within the last 6 m.y. by 3 km or more is convincing (Audebaud *et al.*, 1973), but the mechanism by which it was uplifted is not clear; nor do we know yet why parts of the Andean arc are volcanically active today while others are not. Papers presented at the first Ewing symposium (Pitman and Talwani, 1977) give a good idea of general strengths and weaknesses of our understanding of arc activity. Against this framework, it is not surprising that our appreciation of variations in basaltic and related igneous activity in arcs is still problematical.

The most prominent contrast in igneous activity between island and Andean arcs is that silicic (~ 70% SiO_2) volcanism, often in the form of large-volume ignimbrites with high contents of sodium and potassium, is concentrated in—if not confined to—arcs underlain by continental lithosphere (Andes, New Zealand, Japan and Java-Sumatra). Some involvement of the underlying continental crust in the production of these volcanics appears very likely.

The effects of ridge subduction on arc volcanism. DeLong *et al.* (1978) and DeLong and Fox (1977) pointed out that ridge subduction causes a temporary decrease or cessation of arc magmatism while inducing an episode of low-grade regional metamorphism and wholesale elevation of the arc. The effects of ridge subduction have been observed in the Late Eocene to Miocene geology of the Aleutian arc, where the Kula ridge was subducted about 30–35 m.y. ago (DeLong *et al.*, 1978). Kidd *et al.* (1977) reported on a similar event from the medial Ordovician of Central Newfoundland. Theoretical analyses showing why subduction of a hot ridge can be expected to lead to a temporary shut-off in arc volcanism have been presented by DeLong *et al.* (1978). Such topics as the time of initiation of volcanism relative to the beginning of subduction; correlations between rates of extension and rates of subduction; migration of the volcanic arc with time; and changes in chemical characteristics of volcanics with time are perhaps best addressed in the context of a full kinematic analysis of the motions in the arc system (Dewey, 1980).

Origin and evolution of marginal basins. **Marginal basins** are defined as areas with oceanic water depths and crustal structure lying between a continent and a nearby island arc. Karig (1971a) first explicitly recognized that some of these basins are at anomalously shallow depths for their predicted or known ages, that they have steep, faulted margins and contain little sediment. Some high heat flow values have been obtained in marginal basins. Karig drew the conclusion that active or recently active seafloor spreading was responsible for generating these particular marginal basins. Packham and Falvey (1971) came independently to the same conclusion. It is important to distinguish this environment of seafloor spreading from that of the main ocean ridges, because in marginal basins it is a consequence of the existence of the island arc and the underlying subduction zone. The spreading is not a primary part of the global plate-boundary system. McKenzie and Weiss (1975) have suggested that main ocean basins were generated by propagation of rifts from marginal basin sites. This is clearly not applicable to, for example, the early Jurassic opening of the central North Atlantic, nor to the early Tertiary opening of the Labrador Sea and Baffin Bay; thus, it is doubtful that this proposal has any general validity.

Until recently, studies of marginal basins thought to contain active or recently active spreading centers did not report well-defined symmetrical sets of magnetic anomalies such as those that characterize the major oceanic spreading ridges. Also, the topography in these basins is much rougher than that near normal oceanic spreading ridges that have rates of spreading similar to those inferred for the basins. Except for the Andaman Sea (Rodolfo, 1969; Curray and Moore, 1974), no well-defined rifts such as those that characterize most slow-spreading ridge axes have been found. Since the supposed spreading axis or axes cannot be located accurately and because the topographic profiles are very "rough," it is hard to demonstrate the existence of a systematic subsidence pattern for marginal basin crust like that so well-established in the major oceans. Recently, however, fragmentary magnetic anomaly patterns due to seafloor spreading have been detected in several active or recently active marginal basins (Barker, 1972; Weissel, 1977; Watts and Weissel, 1975; Kobayashi and Isezaki, 1976). These patterns seem to indicate that the spreading process in marginal basins is prone to jumping of the ridge axis, particularly in a direction towards the island arc, so that extensive, simple, symmetrical patterns of anomalies should not be expected. Although the rate of spreading has not been determined very precisely in most active marginal basins, Weissel (1977) has given data indicating that the

Lau basin has opened at an average rate of about 5 cm/yr (total rate) near the north end of the basin. Kroenke and Scott (1978) have given data that yield average total opening rates of about 5 cm/yr for the Parece Vela basin and 2 cm/yr for the Mariana Trough. Although there is a crudely defined axial high in the Mariana Trough (Karig, 1971b; Anderson, 1975), except very locally, the well-defined axial rift that might be expected at the inferred spreading rate does not exist. Anderson (1975) has shown that heat flow measurements support the idea of active spreading and an active hydrothermal circulation system on the axial high of the Mariana Trough.

One possible explanation for the rough topography and ill-defined nature of the spreading magnetic anomalies in active marginal basins was given by Kidd (1977), who suggested that widespread tholeiitic volcanism was likely to occur in marginal basins away from the places where crust was accreted. Klein *et al.* (1978) have subsequently found evidence for such activity in the Shikoku basin.

A limited amount of data are available on the geochemistry of basalts from active marginal basins, and more will shortly be forthcoming in studies following Legs 58–60 of the Deep Sea Drilling Project in the Shikoku Basin and Phillipine Sea. Dredged samples from the Marianas Trough have been described by Hart *et al.* (1972), and Hawkins (1976, 1977) and Gill (1976) have documented the chemistry of rocks from the Lau-Havre Trough. Reay *et al.* (1974) have described the rocks from an active volcano (Niuafo'ou) in the northern end of the Lau Basin. Basalts from the actively spreading marginal basin in the Scotia Sea have been described by Tarney *et al.* (1977). All of these investigators have commented on the close similarity of the young basalts in these active marginal basins to the tholeiitic basalts of the better-known oceanic spreading ridges. However, it should be noted that none of these studies has been made with the idea that there may be widespread "off-axis" or subsequent volcanism in marginal basins, which might conceal much of the crust formed at the axis or axes of accretion. It has been suggested (Kidd, 1977) that basaltic and perhaps ultramafic komatiites might be found forming part of the crust in some marginal basins, particularly narrow ones such as are well-developed in the New Hebrides arc (Karig and Mammerickx, 1972). These suggestions follow from the occurrence of basaltic komatiites (Kidd, 1977) and perhaps ultramafic komatiites (Upadhyay, 1978) in sequences of the northwestern Newfoundland Appalachians, which have been interpreted as oceanic crustal floor and as sedimentary and volcanic fill of small marginal basins of Ordovician age. Detailed geochemical information has been obtained from a fossil Mesozoic marginal basin in the southern Andes (Saunders *et al.*, 1979). Basalts there are said to closely resemble oceanic spreading ridge basalts. A relationship of particular importance for students of marginal basin volcanism is the passage of the Lau-Havre Trough southward onto land in the Taupo volcanic rift zone of North Island, New Zealand. This relation, first clearly pointed out by Karig (1970), shows the largest along-strike variations in presently active marginal basin volcanism, with abundant dominantly ignimbrite volcanism of the Taupo zone passing laterally to the wholly basaltic spreading in the Lau basin. It seems likely that the variation in amount of extension, increasing northward along the zone, is connected with the northward increasing subduction rate. At present, it is not clear if the silicic magmas of the Taupo zone are dominantly of crustal or of mantle origin, and hence it is unclear to what extent the continental-type crust of New Zealand modifies the volcanic expression of the marginal basin.

It is important to note that marginal basins formed by the entrapment of oceanic lithosphere (generated in a major ocean by part of the global plate boundary network) between a new island arc and a continent are not discussed in this section. Examples of these "trapped" marginal basins, which often show better-developed magnetic anomalies at high angles to the nearest piece of bounding island arc, are the Bering Sea (Cooper *et al.*, 1977), possibly the West Philippine Basin (Karig *et al.*, 1973) and, perhaps, the South Fiji Basin (Watts *et al.*, 1977).

Volcanism at arc-continent and continent-continent collisions. Active arc-continent collisions are occurring at Timor, eastern New Guinea and Taiwan but are poorly known. Neogene collision of an arc with Australia in New Guinea has been followed by development of subduction in a different direction. This may be a general consequence of arc-continent collision. Continent-continent convergence, however, is persisting after collision from the eastern syntaxis of the Himalaya to Turkey (with the exception of the Makran gap). A recent comparative discussion of the tectonics of Turkey, Iran and Tibet by Şengör and Kidd (1979) illustrates the volcanic complexity that can occur in this environment.

The Turkish-Iranian Plateau is a high region with an average elevation of about 1.5 km. During the late Miocene, the last piece of oceanic lithosphere between the Eurasian and Arabian continents was eliminated at the Bitlis/Zagros suture zone (Fig. 6.2.7). Continued convergence across the collision site resulted in the shortening of the plateau across strike by thickening and along strike by sideways motion. Predominantly calc-alkaline volcanism is present on the highest portions of the area, despite the absence of a descending slab of lithosphere. Surface geology and volcanism of the Turkish-Iranian Plateau greatly resemble those of the Tibetan Plateau.

The post-Miocene tectonic picture of the Turkish-Iranian Plateau is one of active shortening across the plateau. In Turkey a considerable portion of this convergence is taken up by wedging out the Anatolian Plate along the North and East Anatolian transform faults with throws of 85 ± 5 km and 22 ± 5 km, respectively (Seymen, 1975; Arpat and Şaroglu, 1972; Seymen and Aydin, 1972). Active shortening has resulted in some strike-slip faulting to the east of where these faults meet (Toksöz *et al.*, 1977), but thickening is predominant on numerous thrusts. In Iran, thrusts and folds are abundant and probably greatly predominate over sideways motion (see summary in Şengör and Kidd, 1979). This may be because (1) Iran has less space to "escape" into than does Turkey and (2) its lower elevation requires less work for the present uplift. Negative Bouguer anomalies (−150 mgal) over a large part of the Turkish-Iranian Plateau (Özelçi, 1973a, b) are most easily, but not uniquely, interpreted in terms of crustal thickening, especially in the light of the geological data.

The Turkish-Iranian Plateau, especially in its western and central most elevated sections, is also the locus of intense late Tertiary-Quaternary volcanism (Fig. 6.2.8) and contains volcanoes that have erupted during historic times (e.g., Mt. Nemrud eruption, 1441 A.D.; Erinç, 1953). Although parts of the region were characterized by Cretaceous to Miocene calc-alkaline volcanism prior to the Pliocene to Recent volcanic phase (Ketin, 1961; Altinli, 1966), this was probably related to a subduction zone consuming Bitlis-Zagros ocean floor and dipping north-northeast under the future Turkish-Iranian Plateau as shown by the progressive increase in K_2O/SiO_2 ratios in the associated igneous rocks from south to north across the plateau (Adamia *et al.*, 1977). This phase of volcanic activity came to an end in Turkey about 20 m.y. ago (Innocenti *et al.*, 1976) and in Iran sometime during the late Miocene (Jung *et al.*, 1976). Volcanic activity recommenced during the Pliocene and is still active (Ketin, 1961; Gansser, 1966). There is no evidence of a descending lithospheric slab beneath the plateau that can be connected with the active volcanism; the age of collision (approximately

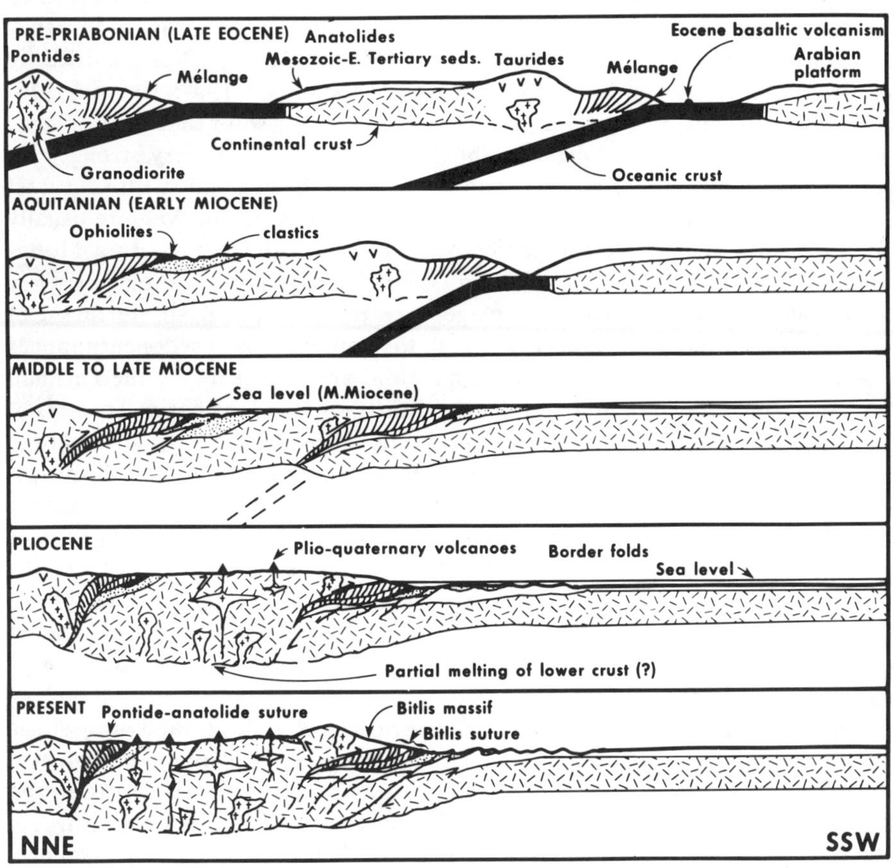

6.2.7 Schematic sections illustrating the evolution of the Turkish-Iranian plateau where active basaltic volcanism is related to continental collision. After Şengör and Kidd (1979).

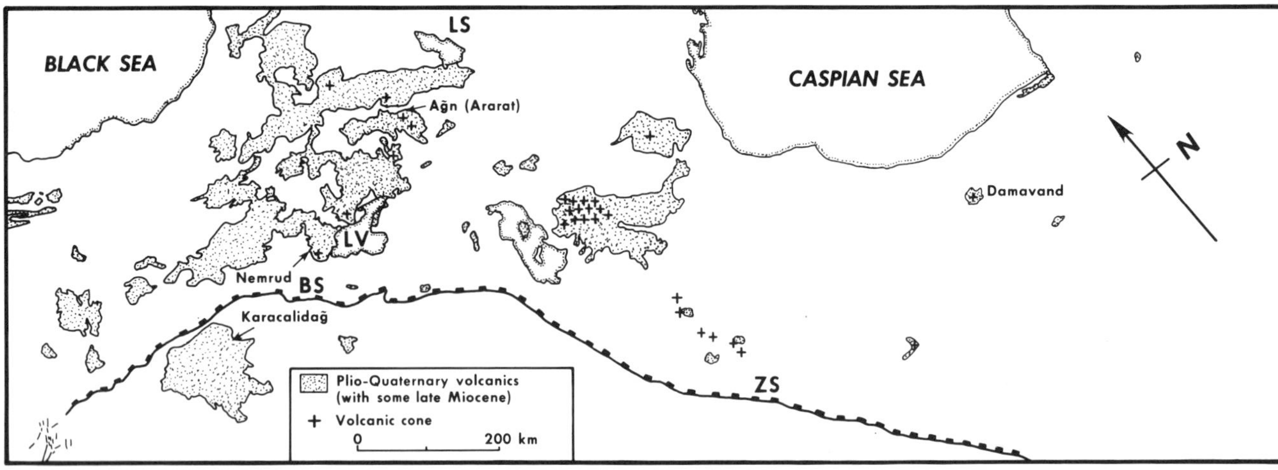

6.2.8 Distribution of Pliocene to Quaternary volcanic rocks on the Turkish-Iranian plateau. This volcanic activity is of great compositional range and is associated with collision between the Arabian continental mass (at the bottom of the map) and the complex arc and micro-continental terrains of Turkey and Iran across the Bitlis Suture (BS) and the Zagros Suture (ZS; LV is Lake Van; LS is Lake Sevan). Modified from Şengör and Kidd (1979).

15 m.y. ago) and the average convergence rate over this time period (about 4.5 cm/yr, McKenzie, 1972) indicate that the slab must long be past the 100–150 km depth where the majority of the calc-alkaline melts are generated. Pliocene to Recent volcanism is very extensive in the highest parts of the plateau (Fig. 6.2.8) and includes both calc-alkaline and alkaline associations, although the former greatly predominate over the latter. The calc-alkaline association is represented by andesites, dacites and rhyolites with some ignimbrites, whereas basalts and very limited phonolites and trachytes represent the alkaline association. Some volcanoes (e.g., Nemrud; Özpeker, 1973) appear to have erupted both alkaline and calc-alkaline rocks. Lambert *et al.* (1974) have reported a $^{87}Sr/^{86}Sr$ ratio of 0.7050 ± 0.0005 from the calc-alkaline lavas of Mt. Agri. If the Devonian and Permian rocks that outcrop very near the volcano represent the total basement beneath it, then this ratio precludes the continental crust as the source material for the melt. But in a terrane as complex as eastern Turkey, which contains a huge amount of accretionary mélange material, this may not be the case. The observations that the Plio-Quaternary volcanics are almost entirely confined to the highest parts of the plateau and that there is a zone of seismic attenuation beneath the plateau (Toksöz and Bird, 1977) support the view that a great portion of the calc-alkaline volcanics here may be products of the partial melting of lower levels of the thickened continental crust.

Over large areas of the Tibetan Plateau, as far as 92° E, the deformed pre-Mesozoic rocks are unconformably overlain by a sequence of Jurassic and Cretaceous sedimentary rocks. Clastic rocks, mainly red sandstones at the base, pass up into rudist-bearing limestones. In the northern part of the plateau the rocks are mainly red sandstones and locally contain gypsum. These Mesozoic sediments are found over the whole plateau south to near the Indus Suture, including near Lhasa, where they are strongly folded (Hayden, 1907). Dips recorded by Hedin (Backstrom and Johannsen, 1907; Hennig, 1915) and analysis of Landsat images show that the Mesozoic sequence is buckle-folded, in places very strongly so, over the whole width of the plateau, with east-west trending axial surfaces. Folds of this type are usually accompanied by thrust faults (e.g., in the Jura Mountains), although it is difficult to identify these positively from the imagery. In the Tsaidam Basin northeast of Tibet (Fig. 6.2.9), the youngest playa sediments appear to be folding now and are overthrust by the Permian rocks on the northeast border of the basin. The axis of maximum shortening, as judged from the strike of the fold axial planes, is northeast as opposed to the north-south shortening axis given by the east-west axial planes of Tibet. This may be due to a recent reorganization of the active deformation as it spreads away from the relatively stable Tarim Block. In terms of style of folding, overall morphology and tectonic setting, the Tsaidam Basin is very similar to the Dasht-i Kavir Depression of Iran.

Young volcanic rocks of andesitic and more silicic calc-alkaline compositions are widely developed over the Tibetan Plateau as far as 92° E (Backstrom and Johanssen, 1907; Hennig, 1915; Norin, 1946; Burke *et al.*, 1974; Kidd, 1975). Those in the northern and central parts of the plateau clearly postdate the folding of the Mesozoic strata in many places. They are known to consist of both alkaline and calc-alkaline potassic lavas (Deng, 1978). In the southern part of the plateau, a 200-km-wide belt of volcanics adjoins the northern side of the Indus Suture, and stretches from the Indus in the west to near Lhasa in the east. It contains a few volcanic features which are obviously young, judging from the LANDSAT imagery, but the youth of most of these abundant volcanics is not morphologically obvious. Specimens collected by Hedin from this terrane are mostly ignimbrites and subordinate related igneous rocks. While this belt could be the remains of an Andean-type magmatic arc, the reports of abundant hot and boiling springs in this terrane indicate the widespread presence of magma at shallow depth. For this reason, and because draping relations to folded Cretaceous sediments can be seen on the Landsat images in the northern part of this terrane, it has been suggested that the bulk of these volcanics are young (Neogene and younger), although a small portion could be of early Tertiary or late Cretaceous age (Burke *et al.*, 1974). The uplift of Tibet, although poorly dated, clearly predated the ongoing uplift of the Himalayas; this is shown by the antecedent Indus and Brahmaputra rivers.

The great resemblance between the Turkish-Iranian Plateau and Tibet, with respect to overall morphology and tectonics, was first emphasized by Von Zahn

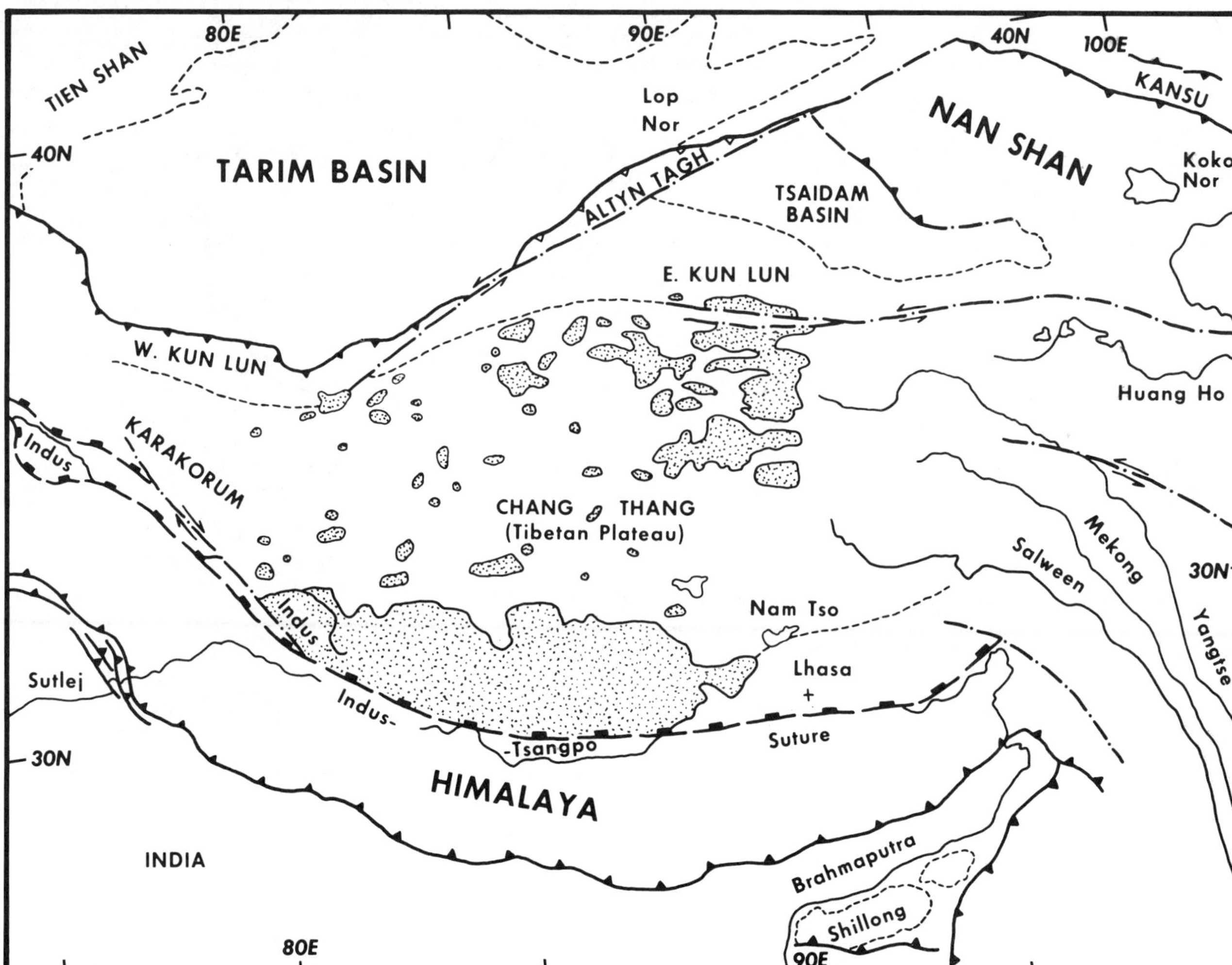

6.2.9 Stippled areas represent Neocene and more recent (including active) volcanism of great compositional range over nearly 10^6 km^2 associated with the collision of India with Asia. Lines with black triangles are active thrust faults; lines with open triangles are inactive thrust faults; lines with black rectangles are suture zones; and dash-dot lines are strike-slip faults. Modified from Şengör and Kidd (1979).

(1906). The foregoing descriptions show that the geological resemblances between the Tibetan and the Turkish-Iranian Plateau are readily apparent. Both plateau areas adjoin a suture where continental apposition has occurred and collision is in progress. The earlier onset of the collision, perhaps 30–40 m.y. ago (Dewey and Burke, 1973; Molnar and Tapponnier, 1975), is perhaps one reason why Tibet is higher and more extensive than the Turkish-Iranian Plateau. In both areas the plateaus were uplifted before the suture zone formed. We take this and the available seismic evidence to indicate that large-scale continental underthrusting, as suggested by Argand (1924), is not the cause of uplift. Shortening and resultant thickening, as proposed by Dewey and Burke (1973; Fig. 6.2.10) and favored by Le Fort (1975) for the origin of Tibet, appear to be the model most consistent with the currently available data from the Turkish-Iranian Plateau, as well as those from Tibet. Therefore, the lower elevation of the Turkish-Iranian Plateau, especially of the Iranian segment, may be also due to the existence of cratonic nuclei within the Central Iranian Plateau (Stöcklin, 1974) that resist deformation by shortening better than the relatively weaker accretionary prism material that appears to make up a large portion of the Tibetan basement.

Both plateau areas exhibit folding of covering sedimentary rocks in at least part of their areas. In the Turkish-Iranian Plateau, this folding, accompanied by

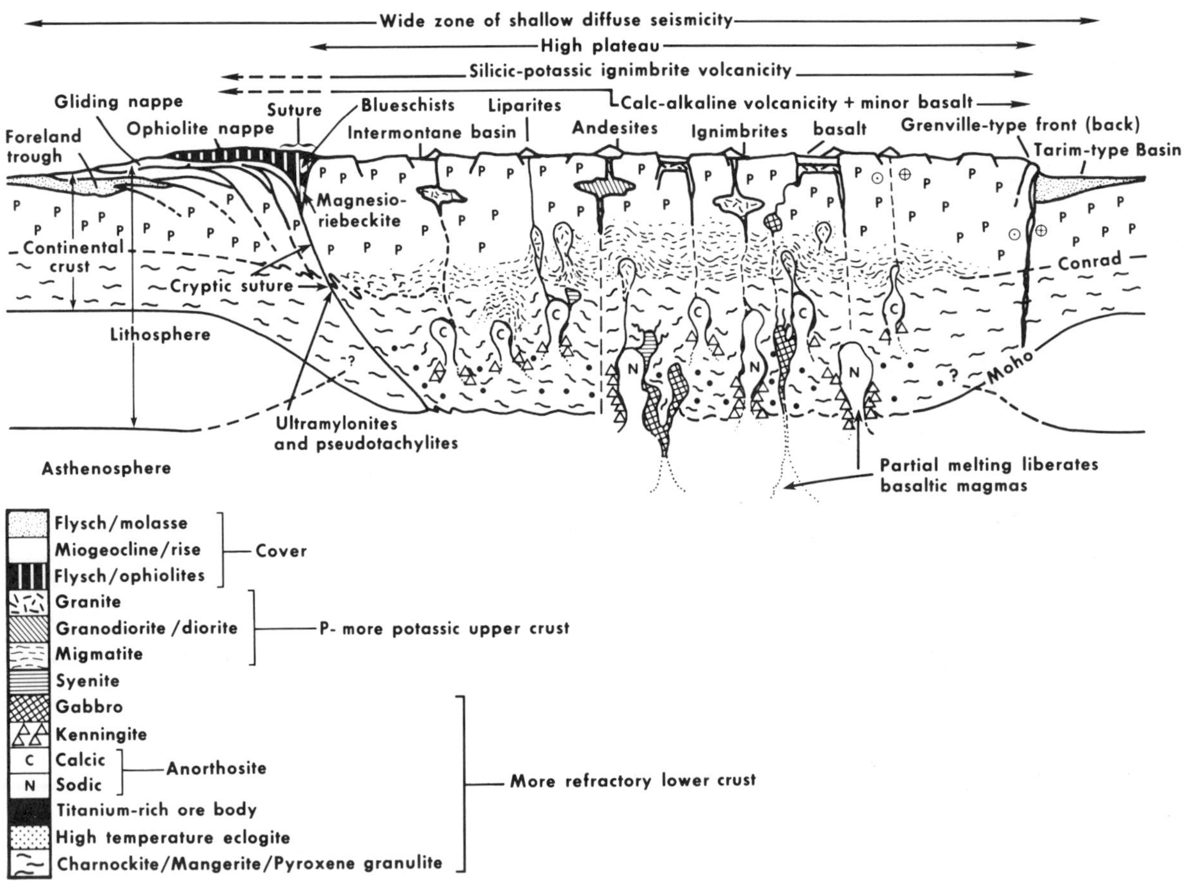

6.2.10 Basaltic and other volcanism produced at continental collision. Collision of two continents commonly results in the thickening of one, producing a Tibet-like plateau. This thickening is produced by horizontal shortening including thrusting and folding at shallow levels not shown on figure. The thickened continental crust is raised in temperature mainly by being depressed into hotter regions. The complex igneous and metamorphic results that follow are depicted in this figure. Partial melting of the continental crust yields highly-potassic volcanic rocks and shallow intrusives, leaving refractory residue at depth. Basaltic volcanism develops where the mantle below the thickened continent (>70 km depths) becomes sufficiently hot. Note that the lithosphere-asthenosphere boundary is poorly defined under the elevated plateau and may be no more than a few kilometers down. This style of modification represents an important stage in continental evolution that has happened widely but not universally. Areas that have escaped collision preserve Andean and island arc characters. Modified from Dewey and Burke (1973).

extensive thrusting, started nearly synchronously with the collision. The time of the folding in Tibet is less well-constrained, but the huge area affected by folding, particularly across strike, is remarkable and suggests that it is unlikely to have happened before the collision, considering the active folding in the Tsaidam Basin. Thrusting in Tibet does not seem to be as widespread as it is in the Turkish-Iranian Plateau, but we believe this to be an artifact of recognizing thrusts on Landsat imagery and not the result of the actual absence of the process. Both plateaus show minor normal faulting at high angles and strike-slip faulting at low angles to the suture.

Volcanism on the Tibetan Plateau greatly resembles that found in eastern Turkey and western Iran. It is similar in its composition, its wide extent, its occurrence on high ground, and, at least for a large proportion of the Tibetan volcanics, in its post-collisional age. Alkaline volcanics are present on both plateaus, although these are subordinate in amount to the very potassic calc-alkaline volcanics. From the point of view of basaltic volcanism, the remarkable thing about geologically young, collisional, orogenic volcanism is the great rarity of basalts relative to more silicic rocks. The only other comparable volcanic environment is the altiplano of the Andes, where similar volcanism occurs on an elevated plateau, but without accompanying continental collision.

Tectonics of basaltic volcanism at transform plate margins

Summary of transform fault tectonics. **Transform faults** were recognized and defined by Wilson (1965a) as plate boundaries across which lithosphere is undergoing relative displacement but is neither being created by seafloor spreading nor destroyed by subduction. This hypothesis was quickly and elegantly confirmed by the studies of earthquake first motions made by Sykes (1967). By implication, because extension and compression are not primary features of transform faults, Wilson's hypothesis suggested that volcanism would not be found to be a major component of their tectonics. Secondary extension or compression may occur across irregularities in the trends of transform faults. These local minor departures from pure strike-slip tectonics, when they lead to extension, may give rise to volcanism in some places along transform faults. Reasons for such secondary tectonics on transform faults have been discussed by Menard and Atwater (1968, 1969), Van Andel *et al.* (1971) and Dewey (1975). Another minor departure from pure strike-slip tectonics on oceanic transforms is the production of large, elongate ridges flanking one or, in places, both margins of the transform trough (Bonatti, 1971, 1973; Thompson and Melson, 1972; Bonatti *et al.*, 1979). These consist mostly of peridotite and serpentinized derivatives which may originally be intrusive, but which almost certainly have been uplifted by tectonic means, probably aided by, or perhaps largely due to, density reduction through serpentinization. Although some authors have proposed that sections of normal oceanic crust and upper mantle may be exposed on the slopes of these fault-bounded ridges and on the walls of transform and fracture zone troughs, Francheteau *et al.* (1976) convincingly demonstrated that, in general, this is unlikely to be true. A complex distribution of lithologies along transform faults and their fracture zone extensions is suggested by data from the Oceanographer Fracture Zone (Fox *et al.*, 1976). The complexity presumably results from the protracted history of tectonic slicing and differential horizontal and vertical displacements that rocks may suffer within the transform domain. Furthermore, it has been suggested (Fox, 1978; Gallo and Fox, 1979) that large-scale thermal effects are responsible for significant changes in oceanic crust near transform faults, due to the abutment of older, colder lithosphere against spreading ridge ends. The most prominent change suggested—a general marked thinning of both of the basaltic layers (and hence of the overall crust)—has been confirmed in at least one study (Detrick and Purdy, 1978). Therefore, even if essentially intact sections of oceanic crust are exposed locally in transform or fracture zone walls, the thickness and perhaps the lithologies and their sequence will not be representative of oceanic crust away from transform faults and fracture zones.

Volcanism on oceanic transform fault/fracture zones. Apart from volcanism at the ends of spreading ridges where they join transform faults, the evidence for volcanism along oceanic transform faults and their fracture zone extensions is not abundant. The young alkalic basalts and gabbros dredged along the St. Paul and Romanche fracture zones (Melson *et al.*, 1967; Honnorez and Bonatti, 1970; Bonatti *et al.*, 1971) are the only rocks now known that are likely to have been erupted on transform/fracture zones away from the places where spreading ridge axes abut such zones. Even in these two cases, the tectonic details of their occurrence and the local abundance of the alkaline volcanics (or lack of it) are poorly known. It is also hard to prove that these volcanics are not related to areas of "hot spot" volcanism, rather than to general transform/fracture zone tectonics. This region could be a hot spot with relatively little associated volcanism, like the Jos Plateau and Air regions of the African plate. The latter are clearly identifiable since they are not submarine, lie within a plate, and are obviously associated with a structurally and topographically defined uplift. Perhaps the fact that there is an uplift at St. Paul's Rocks favors the idea that these volcanics are hot-spot-related rather than related to secondary extension across the transform/fracture zone, since the presence of an anomalously uplifted block on a transform fault might be taken to indicate that the zone is under compression, a condition not favorable for volcanism.

ARCYANA (1975) and Choukroune *et al.* (1978) observed a few wall-like features trending parallel to the transform in the outer, less tectonically active part of the transform valley at the northern end of the FAMOUS area (Fig. 6.2.11). These were tentatively identified as dikes that had just reached the surface but failed to produce flows. If true, this indicates that some very minor magmatism occurs at some places along the transform zone. However, as these dikes (?) have thick manganese coatings, they could have formed close to the ridge axis. It also seems possible that they are lithified breccia zones rather than dikes, so their significance is uncertain.

Other evidence for young volcanism on oceanic transform fracture zones is indirect, consisting of seamount massifs that occur on the Oceanographer and Kane Fracture Zones. In the case of the Oceanographer Fracture Zone, the seamounts are located at the place where there is a significant change in the orientation of the fracture zone (P. J. Fox, pers. comm., 1978). This change is such that there is likely to have been an extensional zone developed during the change in spreading direction recorded by the different orientations of the fracture zone. Such an extensional zone, i.e., one that results in volcanism but not in the establishment of a new spreading ridge segment, has been termed a "leaky" transform; however, there are no well-documented examples of currently active "leaky" transform faults in the ocean, and there are few examples on land. The term should probably not be used for places where a short, spreading ridge segment joins two long transform faults, such as the Cayman Trough. It seems that even small extensional zones such as this develop a moderately coherent spreading center geometry, although there is probably some critical size range below which the intrusion and extrusion of basalt are less regular; such pull-apart zones without coherent spreading centers may legitimately be called "leaky transforms." It is unlikely that these leaky transform zones will persist for very long. They will disappear either because a further slight change in relative plate motion removes the component of extension across the transform, or because new ridge-transform segments develop which are more concordant with the new spreading direction. The ephemeral character and probable small scale of "leaky" transforms may be the reasons why current examples are hard to find in the oceans.

Most basalts dredged from oceanic ridge-ridge transform faults and fracture zones most probably erupted near one or the other of the two places where spreading ridge segments abut the fracture zone-transform fault intersections. Such basalts are more properly regarded as products of the spreading ridge

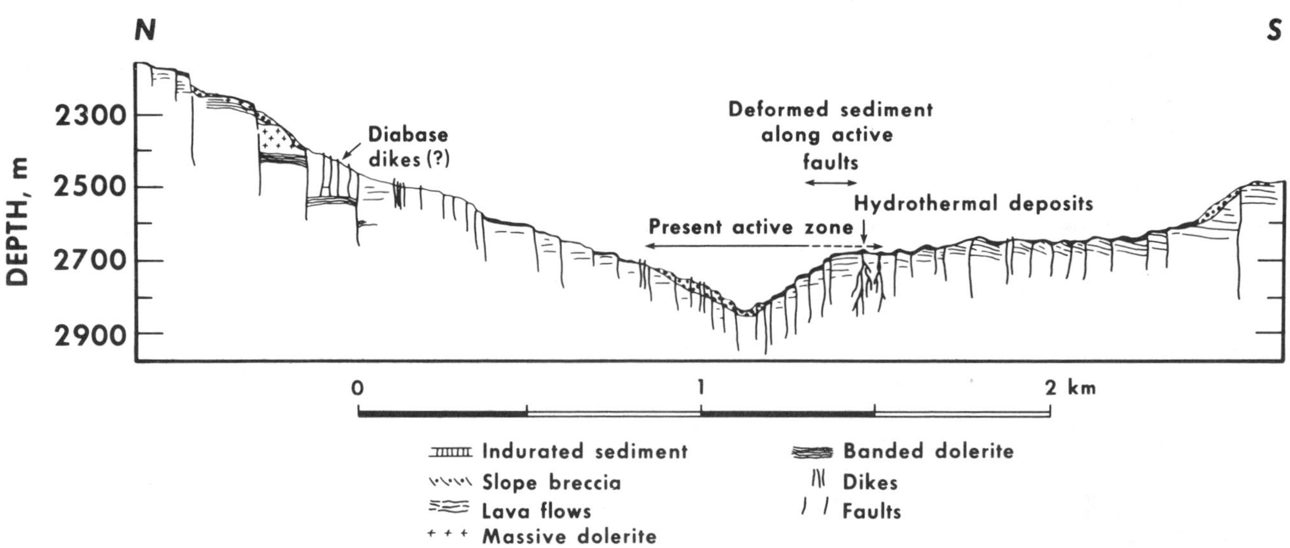

6.2.11 Diabase dikes (?) observed in the FAMOUS area are associated with a transform fault. See text for details. Slightly modified from ARCYANA (1975).

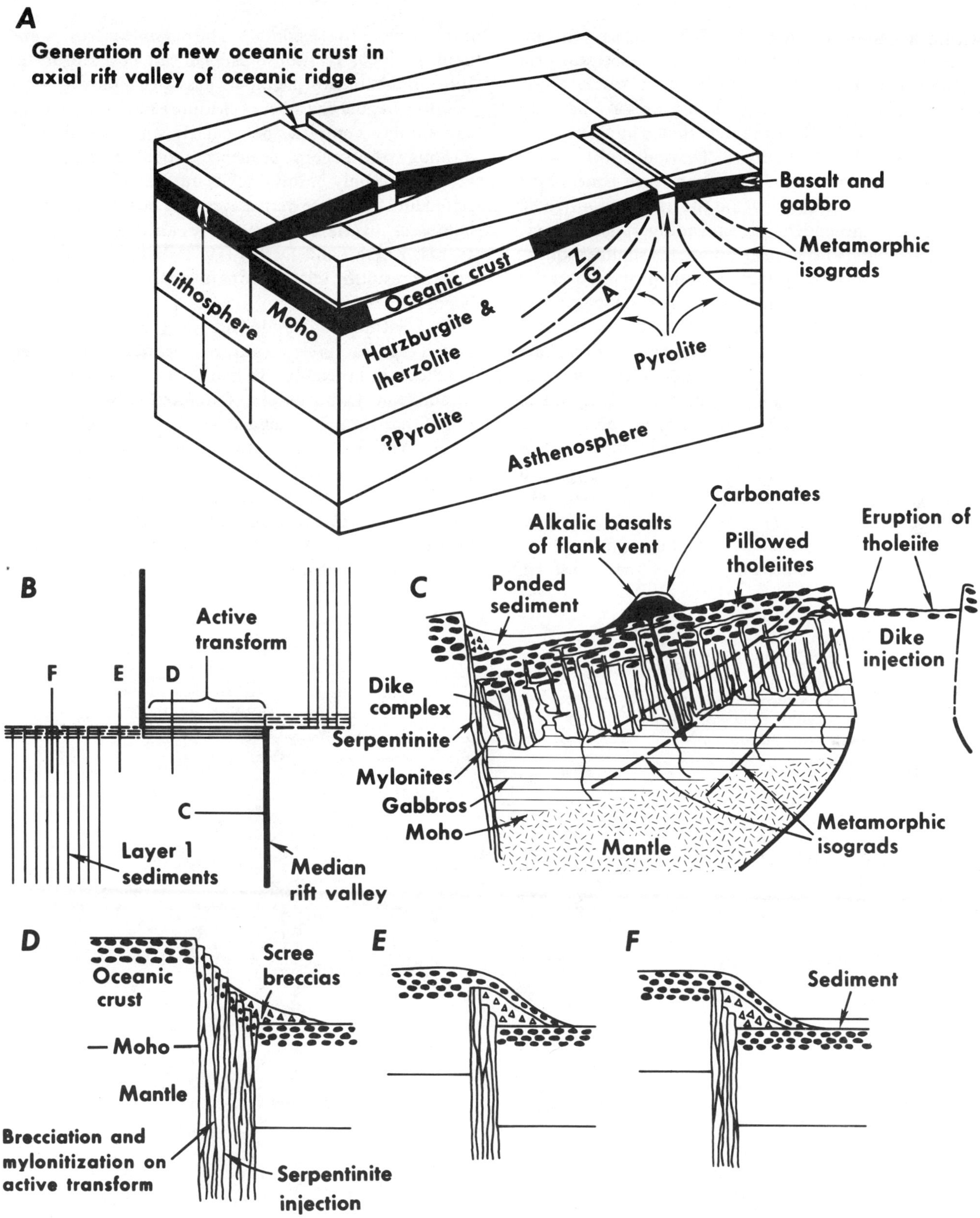

6.2.12 Block diagram (**a**), and map (**b**) showing ridge-transform-ridge intersection. **c**. Detail of accreted oceanic crust. **d**–**f**. Proposed structures and stratigraphic relationships developed along a transform fault and fracture zone; position of sections given in **b**. After Dewey and Bird (1971).

segments; however, interesting and complex relationships between basalts and the products of transform tectonics are likely to be developed in this system (Karson and Dewey, 1978). Two relationships are likely to be common, the first being that predicted by Dewey and Bird (1971; Fig. 6.2.12): this is the production of scree breccias on fault scarps in the transform zone which then become overlain by lavas as the transform-affected crust passes the end of the spreading ridge. Pelagic sediments might also be incorporated in similar sequences, and interbedded pelagic sediment and volcanics were observed at a ridge-fracture zone-transform intersection by ARCYANA (1975). The second tectonic relationship is required by the abutment of lithosphere that has suffered deformation along the transform fault zone against the spreading ridge axis (Karson and Dewey, 1978; Fig. 6.2.13). The predicted sequence, in the most general terms, consists of the juxtaposition of highly tectonized and faulted rocks, which have passed through the transform domain, with untectonized rocks found at the ridge axis. The contact is an intrusive one, and high temperature post-kinematic metamorphism should affect the deformed rocks of the transform domain close to the contact. The possible intrusive and deformational relationships that can develop as the rocks of the transform domain pass across the end of the basaltic magma chamber of the spreading ridge axis are exceedingly complex, particularly on a small scale, although the general sequence should be that given above. Not only basaltic rocks but also their differentiated and cumulate derivatives may be involved in the complex, overlapping intrusive and deformational events. Karson and Dewey (1978) discuss the process and the resulting relationships in much more detail than is possible here.

Collette (1974) and Turcotte (1974b) have suggested that fracture zones are subject to extension perpendicular to their lengths due to thermal contraction; in addition, Oxburgh and Turcotte (1974) have proposed that suitably oriented cracks, including fracture zones, would be subject to extension because of the change in Earth's radius which would occur if the plate in which they exist moved toward the equator. The amounts and rates of extension produced by either process seem insufficient to provoke significant magmatism or volcanism; no very convincing evidence of

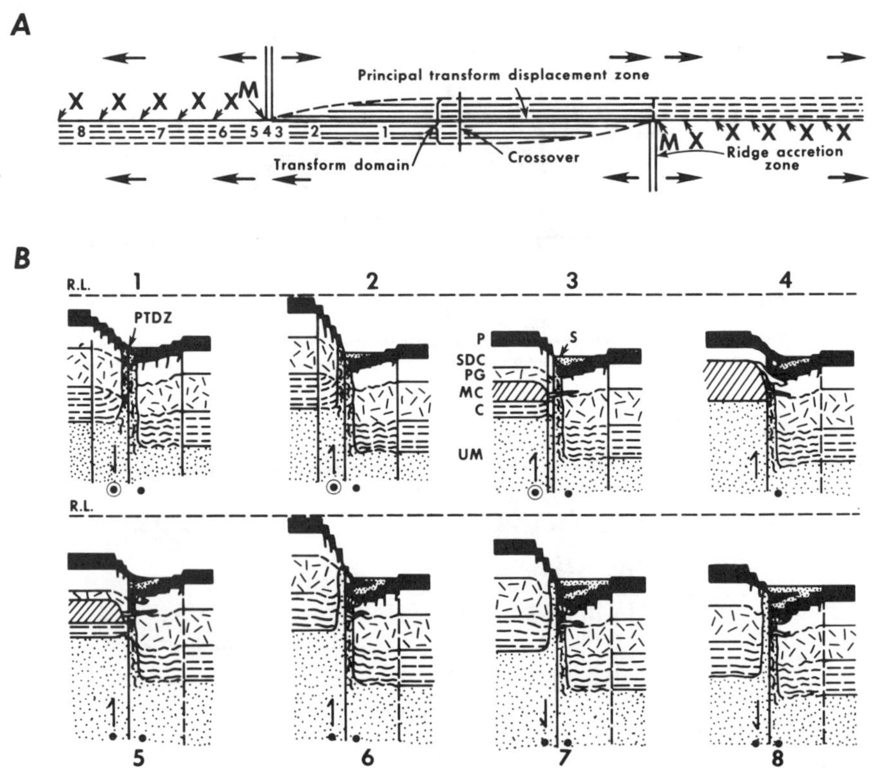

6.2.13 Proposed geologic and geometric development of ridge-ridge transform fault and its fracture zone extensions. (After Karson and Dewey, 1978). **a**. Transform displacement occurs within transform domain (horizontal lines), concentrated for short periods at a principal transform displacement zone (PTDZ). Fracture zone continuations consist of crust with a prior transform history (dashed horizontal lines) against which younger, underformed oceanic crust and mantle are accreted at point M. The contact between these two kinds of crust (X's) is basically intrusive. Numbers 1–8 show position of sections in **b**. **b**. Sketch sections across transform or fracture zone continuation, illustrating geological relationships built up by interaction of horizontal and vertical displacement and magmatism. Half arrows give sense of vertical motion. Positions of sections given on **a**; all views to the right. R. L.—reference level; thick vertical lines—PTDZ; thin vertical lines—limits of transform domain; dashed thin vertical lines—limits of lithosphere with prior transform history; fine stipple—sediments (s); black (p)—lavas; white (SDC) sheeted complex; random dashes (PG)—roof gabbros; oblique lines (MC)—magma chamber; horizontal dashed lines (C)—cumulate gabbros and ultramafics; stipple (UM)—mantle peridotites; vertical wavy lines—peridotite and serpentinite diapirs.

volcanism due to the latter process has been demonstrated.

Pronounced, closed topographic basins at the intersections of spreading ridge axes and transform faults have been ascribed to a magmatic process: viscous head loss (Sleep and Biehler, 1970). These authors have suggested that viscous dissipation due to magma moving upward in a restricted conduit system results in the magma's emplacement at a level below normal isostatic equilibrium. This effect, also proposed to be responsible for the rift along slow-spreading ridge axes, is suggested to be more pronounced at the fracture zone intersections because of the proximity of the older and colder plate edge across the transform-fracture zone.

Volcanism along subaerial transform faults. Where pull-apart basins occur along large strike-slip faults, volcanism may be expected. Examination of such situations—for example, the Salton Sea, the Dead Sea rift and small basins along the North Anatolian fault—reveals that such pull-aparts are not common, that they are usually small relative to the extent of the transform system along which they occur, and that the volume of volcanics directly associated with the pull-apart is also rather minor. It may be that larger volumes of magma occur at depth and that the surface expression is not representative of the amount of magmatism in these pull-aparts; the geothermal activity in the Salton Sea may be evidence of this.

It has been proposed that other volcanic rocks occurring in the general area of some parts of some large continental transform fault systems (for example, the young Arabian basalts, the Hsing-An basalts of China, and some of the young volcanics in the vicinity of the San Andreas Fault) are related to transform-zone tectonics in a less direct way. The evidence that there is a strong connection is less than convincing at the moment; such volcanics are better treated under intraplate volcanism.

6.2.2 Tectonics of non-plate-margin basalt

The relatively small volumes of basaltic igneous material produced in environments other than plate margin environments can be considered from various perspectives. We here make a two-fold distinction between hot-spot basalts and intracontinental rift basalts. No classification of basaltic volcanism is altogether coherent (see, for comparison, section 1.2). Thus there are a number of hot-spot-type volcanic objects on divergent plate boundaries of which the Azores and Iceland are most familiar. Perhaps it is more appropriate to consider the plate boundary as lying on or across these objects than to think of them as representing an unusual kind of plate boundary volcanism.

Hot spot volcanism

A large fraction of the Earth's volcanism occurs at plate boundaries. Continuous volcanism occurs along the world-wide oceanic ridge system where the surface plates are formed. Volcanism also occurs behind much of the world-wide trench system where the plates are subducted into the mantle. However, significant volcanism also occurs within plate interiors. Since there is no evidence for plate tectonics on Mars, Mercury or the Moon, this intraplate volcanism is likely to resemble the volcanism on those bodies.

Spectacular examples of intraplate volcanism are the Hawaiian Islands. These active volcanoes occur at the end of a chain of islands and seamounts that extends to the Aleutian Islands. Wilson (1963) proposed that the Hawaiian Islands are the result of a melting anomaly in the upper mantle; the motion of the lithosphere over the mantle caused the island chain to propagate. A straightforward explanation for a mantle melting anomaly is that mantle convection is occurring beneath the lithospheric plates without the fracturing that would create new plate boundaries. Ascending convection currents lead to pressure-release melting and the resultant magmas penetrate the plates, leading to volcanism. This is known as the mantle plume hypothesis.

Mantle plumes. Morgan (1971, 1972a,b), following Wilson (1963), hypothesized that deep mantle plumes were responsible for areas of anomalous volcanism, both intraplate and on ridges. In his original paper, Morgan (1971) considered sixteen plumes and suggested that they were stationary with respect to each other. However, it was soon recognized that the number of areas of intraplate and ridge volcanism that would qualify as plumes is much larger than sixteen. These areas of anomalous volcanism are often referred to as "hot spots." The world-wide distribution of hot spots proposed by Burke and Kidd (1975) is shown in Fig. 6.2.14.

From the figure it is obvious that hot spots are not uniformly distributed over the Earth's surface. There are numerous hot spots in Africa, none in South America and relatively few in North America, Europe and Asia. A classification of many of the hot spots has been given by Wilson (1973). Many, but not all, of the hot spots shown in Fig. 6.2.14 are also associated with local crustal upwarping.

One of the assumptions of the deep mantle plume hypothesis is that hot spots move very little with respect

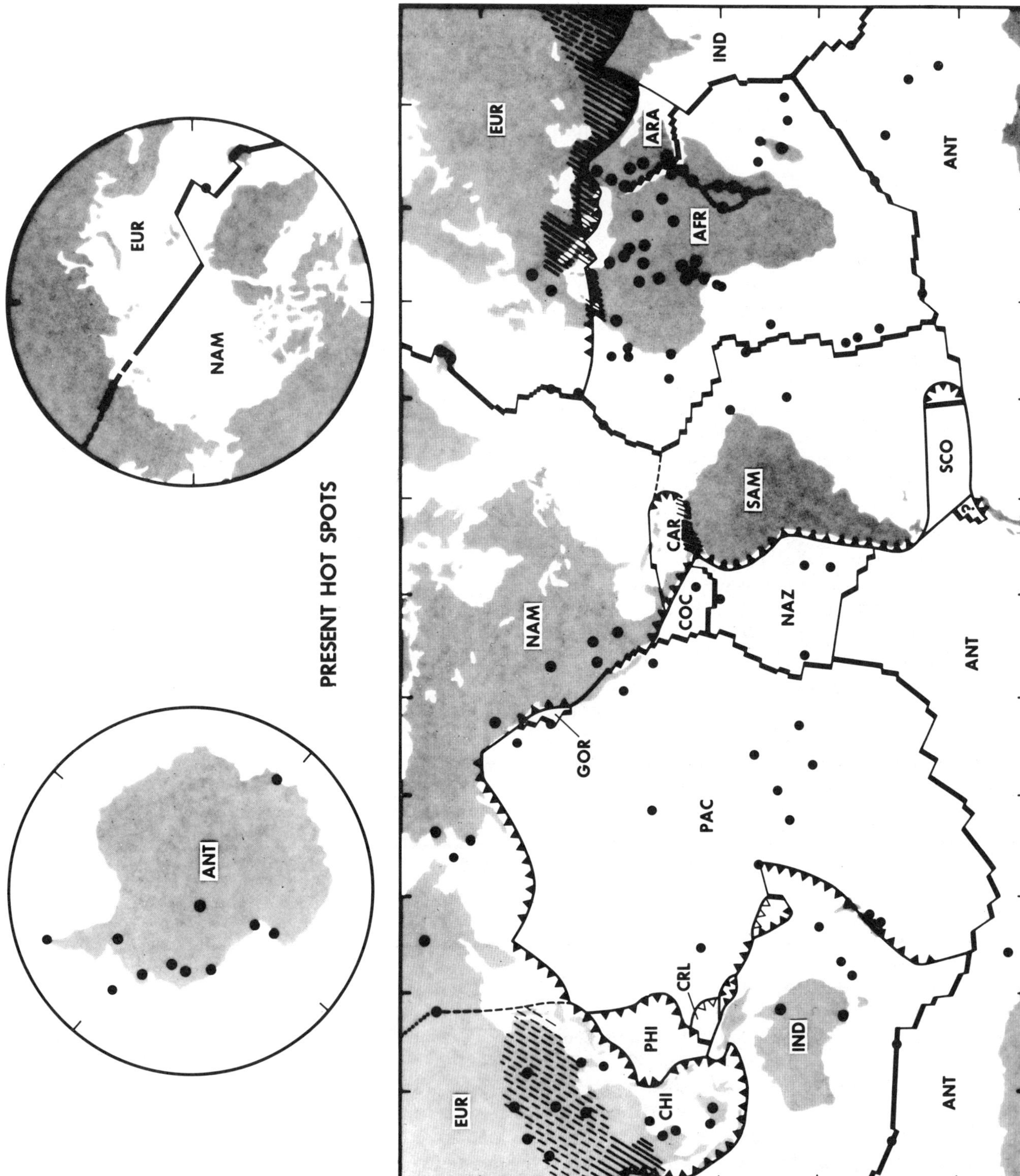

6.2.14 Hot spots (defined as non-plate-margin volcanism) are abundant but not uniformly distributed over the Earth. From Burke and Kidd (1975).

to each other. A number of studies have been carried out to determine whether this is in fact the case. Clague and Jarrard (1973), Molnar and Atwater (1973), Winterer (1973) and Jackson (1976) concluded that the trends of several Pacific island chains can be explained if the plumes migrate with respect to each other at a few millimeters per year. Studies of relative hot spot motions in the Atlantic (Burke *et al.*, 1973; Molnar and Atwater, 1973; Molnar and Francheteau, 1975) also show relative velocities of several centimeters per year. However, since the plate velocities in the Atlantic are also on the order of several centimeters per year, the relative migration of the hot spots is large. In particular, the motion of the Iceland hot spot does not appear to be consistent with the motion of the hot spots in the South Atlantic.

The trends of the Pacific volcanic chains and age progression of volcanism provide the strongest evidence favoring the fixed mantle plume hypothesis. Selected Pacific volcanic chains are illustrated in Fig. 6.2.15. Probably the most comprehensive study of plate motions on a world-wide basis was carried out by Minster *et al.* (1974). These authors inverted the trends of twenty linear volcanic chains and aseismic ridges, utilizing the fixed hot spot hypothesis, and minimized hot spot migration. [Other slowly moving reference frames yield similar results. See, for example, Kaula (1975).] The pole they obtained for the rotation (with respect to a fixed mantle) of the Pacific plate is at 67.3° N and 59.4° W. The angular velocity of the plate is 0.83° / m.y. The predicted directions of propagation of the Pacific Island chains are shown in Fig. 6.2.15. Although the general trends of the volcanic chains are in the predicted directions, significant deviations are seen only if the chains are drawn as lines, rather than as swaths of finite width (Fig. 6.2.16).

In addition to predicting a direction of propagation for volcanic islands, this model also predicts a rate of propagation that may be tested by dating volcanic rocks from islands and seamounts. The data have been summarized by Jackson (1976); ages for the island chains shown in Fig. 6.2.15 are given in Fig. 6.2.17. The age of the volcanic rocks on each island or seamount is plotted against the distance of that island or seamount from the currently active volcanic island at the end of the chain. Also included in Fig. 6.2.17 is the speed of propagation of the chain predicted by the rotation pole and angular velocity given by Minster *et al.* (1974).

The propagation of the Hawaiian-Emperor chain shown in Fig. 6.2.17 is in good agreement with the predicted rate of 8.9 cm / yr. This is even true beyond the change in the trend of the chain about 40 m.y.b.p. Several other chains exhibit similar age progressions, but there are striking exceptions such as the Cook-Austral chain. The mantle plume hypothesis is an attractive means of explaining these Pacific volcanic chains, although without extension and / or modification there are features that remain unexplained, particularly synchronous volcanism along extended lengths of a chain. Most early reports of examples of the latter phenomenon have proven to be erroneous, but some have survived.

The original hypothesis of deep mantle plumes proposed by Morgan (1971) envisioned narrowly confined (r $\simeq$ 100 km), cylindrical, thermally convected flows. It was pointed out by Tozer (1973) that the concept of narrowly confined flows is not applicable to high Prandtl-number fluids. Thermal convection at high Prandtl-numbers may result in narrowly confined thermal boundary layers or thermal plumes because of the slow diffusion of heat, but high viscosity leads to a

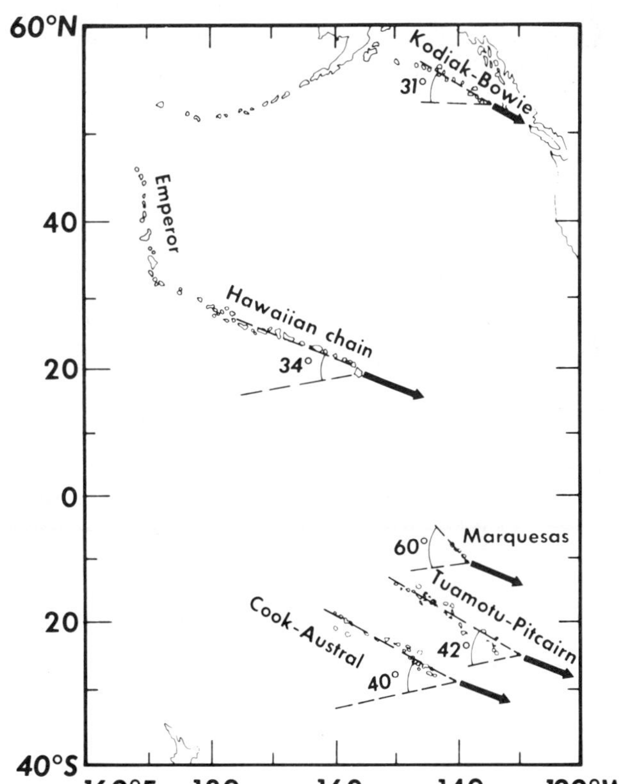

6.2.15 Some linear volcanic chains of the Pacific with arrows showing directions of propagation about a rotational pole (that of Minster *et al.*, 1974). Dashed lines mark the direction of seafloor spreading that formed the seafloor at the locations of some active hot spot volcanoes and the angles recorded are estimated as separating the trend of the volcanic chains from the local directions of seafloor spreading and fracture zones.

rapid diffusion of vorticity and a diffuse velocity structure.

With an exponentially temperature-dependent viscosity it might be expected that the high temperature region would also be a high velocity region because of the reduction in viscosity. Two-dimensional plumes of this type have been studied by Yuen and Schubert (1977). They find that the flow is primarily within the thermal plume if the temperature difference is sufficiently large. Numerical calculations for the structure of cylindrical mantle plumes have been carried out by Parmentier *et al.* (1975) using an exponentially temperature- and depth-dependent viscosity. These authors found that relatively narrow mantle plumes could be obtained if the convective flow were confined to the upper mantle and if the heating were primarily from below (from the lower mantle). An illustration of this type of shallow mantle plume obtained from the numerical calculations is given in Fig. 6.2.18.

Deffeyes (1972) suggested that deep mantle plumes could be driven by phase changes. A phase change is a local driving force which might drive a local flow; however, it would not induce a narrowly confined, deep mantle plume.

An alternative to thermal plumes is chemical (or compositional) plumes. Anderson (1975) has proposed that surface hot spots result from compositional plumes which rise from the lower mantle. Certainly the diapiric upwelling of lighter material from the lower mantle can result in relatively narrow plumes impinging upon the lithosphere. However, the instability mechanisms that lead to compositional plumes have not been clearly delineated. Also, there is a general depth constraint on all types of mantle convection. If material rises adiabatically from great depths, the degree of partial melting will greatly exceed that which is observed. Anderson's hypothesis implies that the chemical plumes (refractory, radial rods) have been in position since early in Earth's history.

Shaw and Jackson (1973) have suggested that dissipative heating and the consequent partial melting in the asthenosphere produce dense residues that sink in the mantle. The sinking residues create a "gravitational anchor" that fixes the melting anomaly to the mantle.

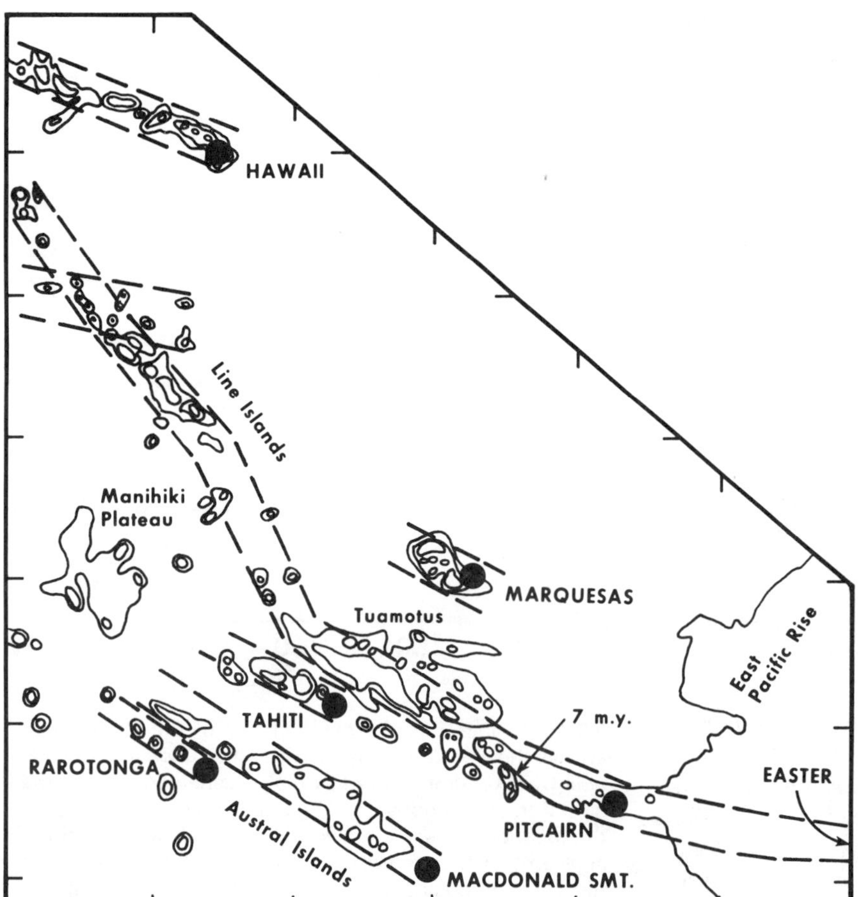

6.2.16 Another map of Pacific island chains, for comparison with Fig. 6.2.15, illustrating that choice of trends is more realistic when the finite size of hot spots is allowed for, and that two hot spots, at Rarotonga and Macdonald, describe the Cook-Austral track better than one.

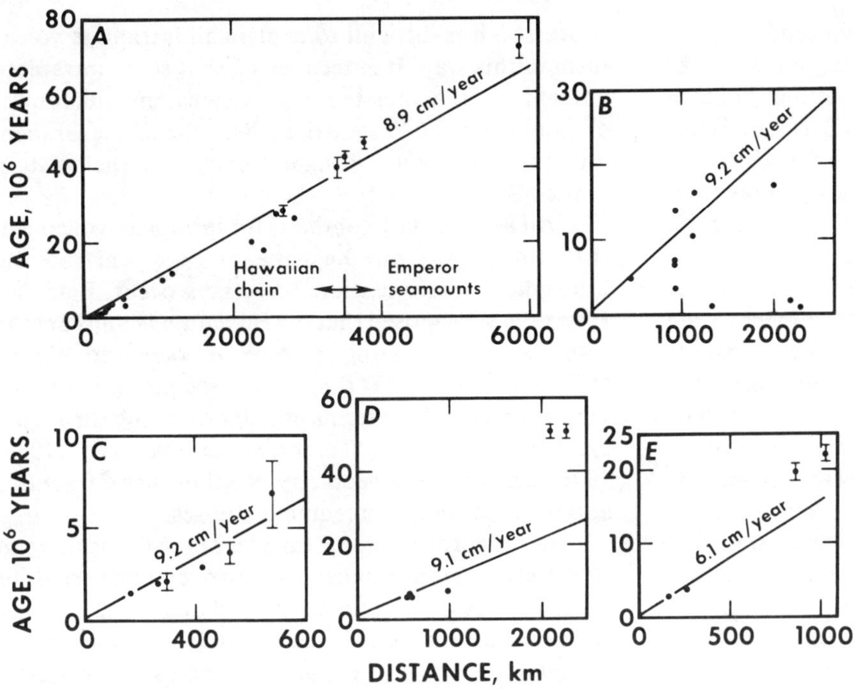

6.2.17 Pacific linear volcanic chains with active volcanoes at one end are shown as aging with distance along the chains. The fit with the rate of propagation for the chains predicted by Minster *et al.* (1974) is good and later work has improved it. The Cook-Austral islands are not readily attributable to a single hot spot.
a. Hawaii-Emperor Chain.
b. Cook-Austral Chain.
c. Marquesas Chain.
d. Pitcairn-Tuamotu Chain.
e. Kodiak-Bowie Chain.

The descending residuum induces further flow and shear heating and melting. This suggestion has yet to be reconciled with the commonly recognized idea that melt and residue are less dense than pristine upper mantle.

The Hawaiian swell is a prominent feature of the topography of the central Pacific. It is a near-circular elevation of the seafloor centered on the Hawaiian Islands. The amplitude of the swell is nearly 1.5 km and its radius is approximately 500 km. There is a long-wavelength positive free-air gravity anomaly of about 15 mgal. The magnitudes of both the topographic swell and of the gravity anomaly are such that they could be associated with a mantle plume beneath Hawaii.

In addition, however, to the many island and seamount chains, some of which we have discussed, there also exists a large number of isolated volcanic islands and seamounts. An example that has been studied in some detail is Bermuda (Pirsson, 1914; Reynolds and

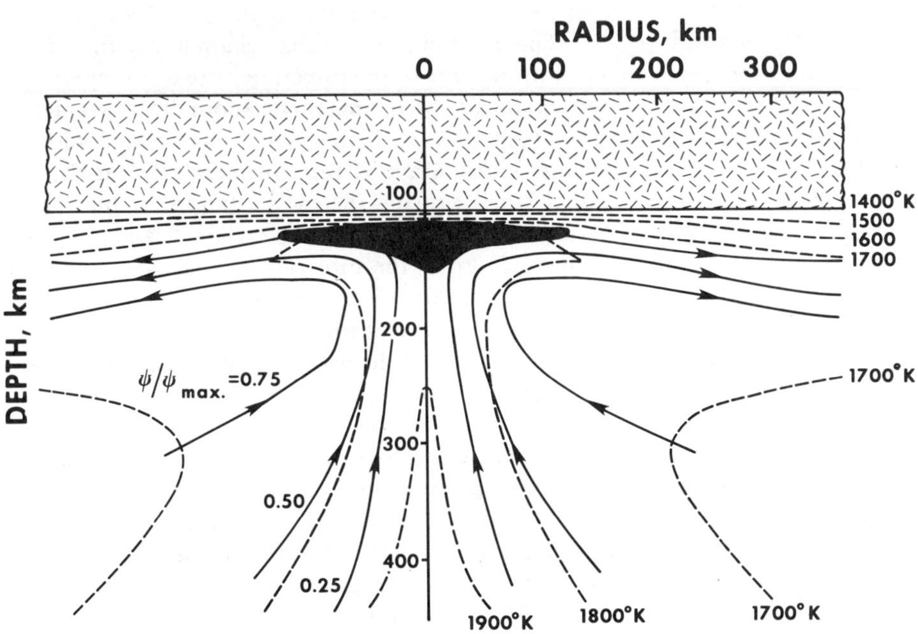

6.2.18 A lithospheric plate 100 km thick (dashed ornament) is underlain by a rising shallow mantle plume. Continuous lines are flow lines (normalized values) and dashed lines are isotherms. Black is zone of partial melting (Parmentier *et al.*, 1975).

Aumento, 1974). A suite of younger volcanics has an Early Oligocene age of about 38 m.y. and an older group an age of 60–110 m.y.; the seafloor beneath Bermuda has an age of about 110 m.y. Bermuda is not part of a volcanic chain and is difficult to explain by the mantle plume hypothesis. It is interesting to note that, like Hawaii, Bermuda is associated with a lithospheric swell. The amplitude of the swell is smaller (approximately 750 m) than the Hawaiian, but the radius is similar. Since Bermuda is moving with respect to the mantle, a mantle plume which had produced volcanism at Bermuda 38 m.y. ago would not be centered at Bermuda today. One preferred interpretation is that, if Bermuda's elevation is due to a mantle inhomogeneity, this inhomogeneity is not that which was responsible for the 38-m.y.-old volcanism.

Important sources of evidence for the existence of deep or shallow mantle plumes are seismic studies. Seismic studies can be used to search for velocity anomalies and/or regions of seismic attenuation in the mantle. An anomalous zone of seismic velocities and attenuation has been reported in the vicinity of the core-mantle boundary beneath Hawaii by Kanasewich *et al.* (1972, 1973) and by Kanasewich and Gutowski (1975). A number of deep mantle inhomogeneities, including one beneath Hawaii and several others not associated with surface hot spots, was reported by Julian and Sengupta (1973). However, the interpretation of these data has been questioned by Wright (1975) and Green (1975). These authors suggest that the observed anomalies are due to upper mantle inhomogeneities beneath the receiving seismic arrays in North America. At present, the seismic evidence for deep mantle plumes beneath hot spots is ambiguous at best. Further verification of the velocity anomaly beneath Hawaii is required. Deep anomalies beneath other hot spots must be sought and some are being found: for example, those in Yellowstone (Hadley *et al.*, 1976) and at Trinidad (Okal and Anderson, 1975).

Ellsworth and Koyanagi (1977) have reported on travel delays associated with teleseismic arrivals at 21 stations on Kilauea Volcano in Hawaii. They found no velocity anomalies at depths greater than 30 km. If any type of mantle convection were occurring beneath surface hot spots, a zone of anomalous mantle would be expected, such as that which occurs beneath ocean ridges and beneath back-arc spreading centers. The failure to find anomalous mantle beneath intraplate hot spots such as Hawaii suggests that these hot spots are not associated with special mantle features.

Lithospheric fractures. Although some fraction of intraplate volcanism may be associated with convective processes, it is difficult to explain all intraplate volcanism in this way. It is recognized that some intraplate volcanism is associated with extensional tectonics; examples are the East African Rift, the Rhine Graben, and the Basin and Range Province in the western United States.

An alternative hypothesis for intraplate volcanism is that magmas from the asthenosphere penetrate the lithosphere where tensional fractures occur. Betz and Hess (1942) proposed that the Hawaiian Islands are the result of a propagating fracture. Jackson and Wright (1970) and Green (1971) discussed the petrology of the Hawaiian Islands in terms of magmas rising through a cracked lithosphere. Turcotte and Oxburgh (1973, 1976) discussed the geometry of lithospheric fractures and the sources of the required stresses.

In order for tensional failures of the lithosphere to occur, elastic stresses must build up over long periods of time. The presence of an outer swell due to the bending of the lithosphere under the load of the Emperor seamounts (Watts and Cochran, 1974) indicates that elastic stresses are not relaxed for at least 40 m.y. The failure of the lithosphere under extension is not likely to resemble the tensional failure of a rock in the laboratory. Turcotte and Oxburgh (1973) have suggested that the lithosphere will fail plastically according to the failure criteria given by Bijlaard (1935). A thin, malleable plate under tension will fail plastically at an angle

$$\theta = \tfrac{1}{2}(\cos^{-1}\tfrac{1}{3}) = 35°16'$$

where θ is measured from a line perpendicular to the direction of the tensional stress. However, other workers have thought that it is inappropriate to apply membrane theory to terrestrial stresses.

The trend of the Hawaiian chain lies at an angle of 34° with respect to the direction of seafloor spreading defined locally by the magnetic anomalies and fracture zones (Fig. 6.2.15). If this angle is associated with the plastic failure angle θ, the conclusion is that the tensional stress causing the plastic failure of the Pacific plate in the vicinity of the Hawaiian Islands is nearly perpendicular to the direction of seafloor spreading. Stresses with this orientation that are large enough to cause failure of the lithosphere could result from thermal contraction of the oceanic plate (Turcotte, 1974a).

The rate of propagation of the tensional fractures is not predicted directly by the hypothesis. If thermal stresses cause the tensional failures, then it might be expected that the failure would propagate in the direction of seafloor spreading at about the seafloor spreading velocity. Taking the seafloor spreading velocity to be 6 cm/yr, the propagation velocity would be 6/cos 35° = 7.3 cm/yr.

Seismic studies are also relevant to the propagating fracture hypothesis. Focal mechanism studies of major earthquakes can give information on the state of stress which led to the seismic rupture. On June 27, 1962, a magnitude 6.1 earthquake occurred near the Kaoiki fault system on the island of Hawaii. This earthquake has been studied by Koyanagi *et al.* (1966). They have concluded that the earthquake was due to either right-lateral strike-slip movement on a fault trending N25° E or left-lateral strike-slip movement on a fault trending N65° W. The latter direction follows the trend of the island chain, the former direction is roughly parallel to the Kaoiki fault trace. If this earthquake were due to a simple shear failure of the brittle lithosphere under tension, the direction of the tension would be N15° W. This is within two degrees of the direction of tension (N13° W) given by the plastic failure hypothesis.

On April 26, 1973, a magnitude 6.2 subcrustal earthquake occurred beneath the northeast coast of the island of Hawaii. First motion studies reported by Koyanagi *et al.* (1976) suggest that the earthquake was due either to right-lateral strike-slip movement on a fault trending N30° E or to left-lateral strike-slip movement on a fault trending N70° W. If this earthquake were due to a simple shear failure of the brittle lithosphere under tension, the direction of the tension would be N20° W. Despite being at different depths and different locations, the focal mechanisms and the direction of the inferred tensional stress for the two earthquakes are in remarkably good agreement. They are also in good agreement with the predictions of the plastic failure hypothesis.

An essential part of the lithospheric fracture hypothesis is that the fracture allows an essentially artesian flow of magma from the asthenosphere to the surface (see section 6.6). Normal lithosphere behaves as an impermeable cap rock, but under tensional failure the hydrostatic head drives the magma to the surface. The physical processes by which magmas are transported through the lithosphere are poorly understood. However, Weertman (1971) has argued that magma in the asthenosphere can nucleate a crack in the lithosphere, provided that the lithosphere is in tension. The magma then follows the propagating fracture to the surface. Anderson and Grew (1977) have suggested that magma migrates through a lithospheric plate in tension due to "stress corrosion." Solomon and Sleep (1974) have attributed seamount chains to the propagation of the plate relative to the intraplate stress field maintained at the boundaries of the plate. Fujita and Sleep (1978) have extended the hypothesis to small seamounts associated with ridge-transform intersections. At present, these hypotheses are difficult to test because the intraplate stress is poorly determined. They also suffer from the inability to account for all, or even many, hot spots.

Africa. As shown in Fig. 6.2.19, Africa is the site of extensive intraplate volcanism. Burke and Wilson (1972) argued that Africa is stationary with respect to the present mantle convection pattern; therefore, mantle plumes could produce fixed volcanic centers on the continent. However, the world-wide studies of Minster *et al.* (1974) concluded that Africa is moving towards the northeast at about 2 cm/yr. This is within the limits of error predicted by the Burke and Wilson hypothesis.

The intraplate volcanism of Africa is associated with several different tectonic settings as shown in Fig. 6.2.19. Much of the volcanism is related to the East African Rift system. Burke and Wilson (1976) have associated the East African Rift with seven hot spots. These hot spots are defined by crustal doming, which is an important aspect of the rift system. Oxburgh and Turcotte (1974) have suggested that the East African Rift is formed by tensional failure of the African plate due to membrane stresses, but supporting evidence is not persuasive. They have argued that the doming and volcanism are secondary features resulting from the tensional failure.

Studies of the focal mechanisms of earthquakes in southern Africa have been carried out by Maasha and Molnar (1972) and by Scholz *et al.* (1976). These authors conclude that an extensional failure of the lithosphere is occurring in southern Africa and that the East African Rift system is propagating southward along a preexisting zone of weakness.

Other areas of extensive recent volcanism in Africa are the Tibesti area in northeast Chad and in Ahaggar to the west (Fig. 6.2.20). In addition to the volcanism, these areas are domed. North of Tibesti, the Haruj volcanics are associated with apparent doming of the basement. Adjacent to the continental margin in the Gulf of Guinea is the Cameroon Line. This is a nearly linear series of active and recently active volcanic centers, which passes from oceanic onto continental crust.

Farther to the north on the continental margin of Africa there is the recent volcanism of the Canary Islands. This is a group of volcanic islands similar to the Cape Verde Islands farther to the south. The association of the Canary Islands with a propagating fracture has been discussed by Anguita and Hernan (1975), but the evidence for this association is not persuasive. Such volcanic centers, which lie near the continental margins of the South Atlantic, may be associated with large stresses caused by changes in crustal thickness (Bott and Dean, 1972); however, there is not a great concentration of volcanic centers in this tectonic position, particularly on a global scale.

Western United States. The western United States is an area of extensive, young volcanism. The distribution of Quaternary (< 1.5 m.y.b.p.) and Pliocene (1.5–7 m.y.b.p.) volcanics is given in Fig. 6.2.21. Since the San Andreas fault is recognized as a major plate boundary between the Pacific and North American plates, the volcanism of this area may be classified as plate margin volcanism. However, it extends over a width of 1500 km. The spreading direction as defined by the Blanco Fracture Zone is not parallel with the trend of the San Andreas fault. This difference has led Atwater (1970) to conclude that active subduction is occurring along the Oregon-Washington coast. This subduction explains the line of active andesitic volcanoes extending from Mt. Baker in northern Washington to Mt. Lassen in northern California, although the lack of seismic activity associated with the hypothesized subduction zone must be noted.

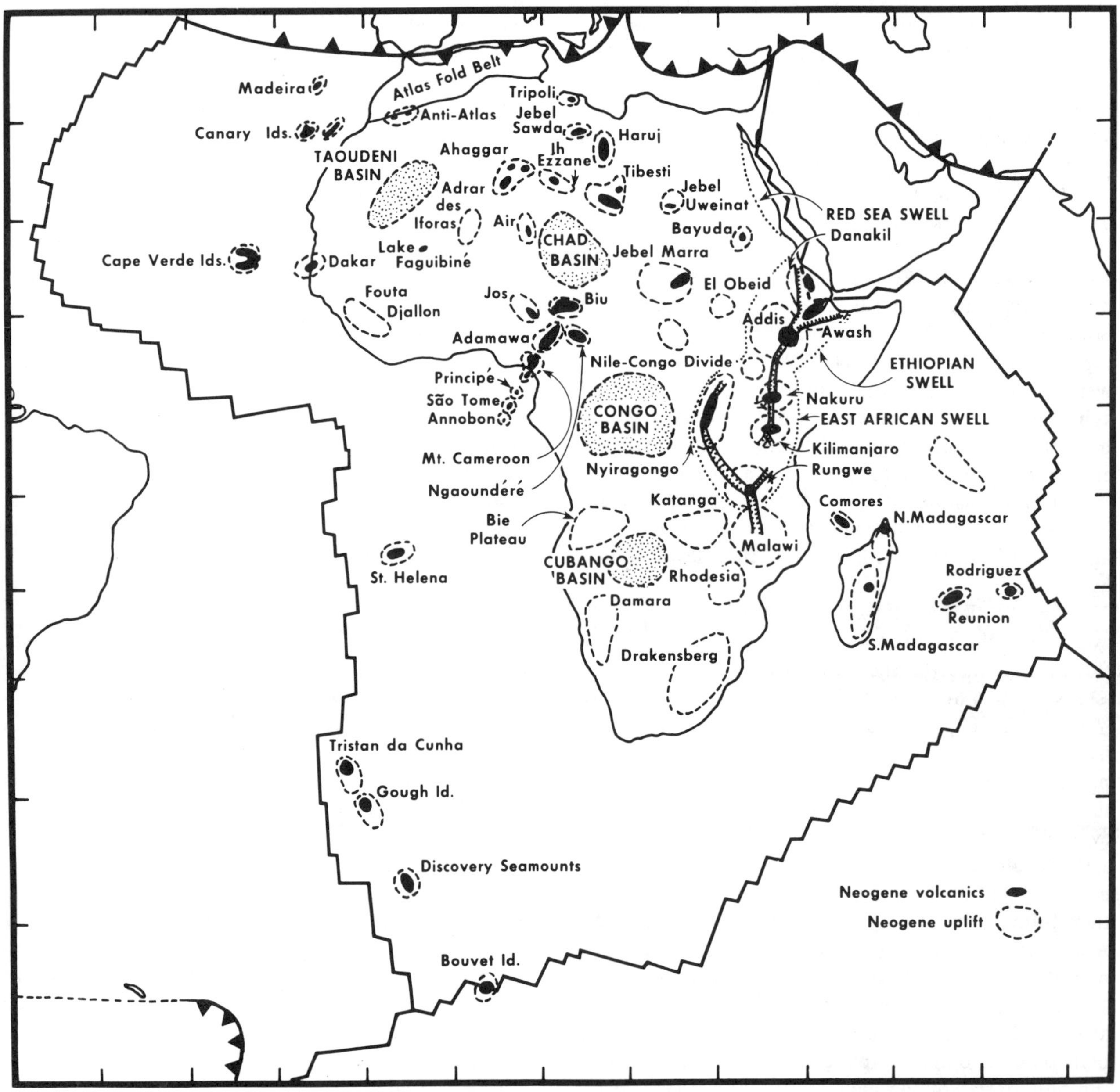

6.2.19 Hot spots on the African plate are associated with Neogene uplifts and with the Neogene and active East African rift system. Note the numerous uplifts without volcanism and the intracontinental basins lying between the uplifts.

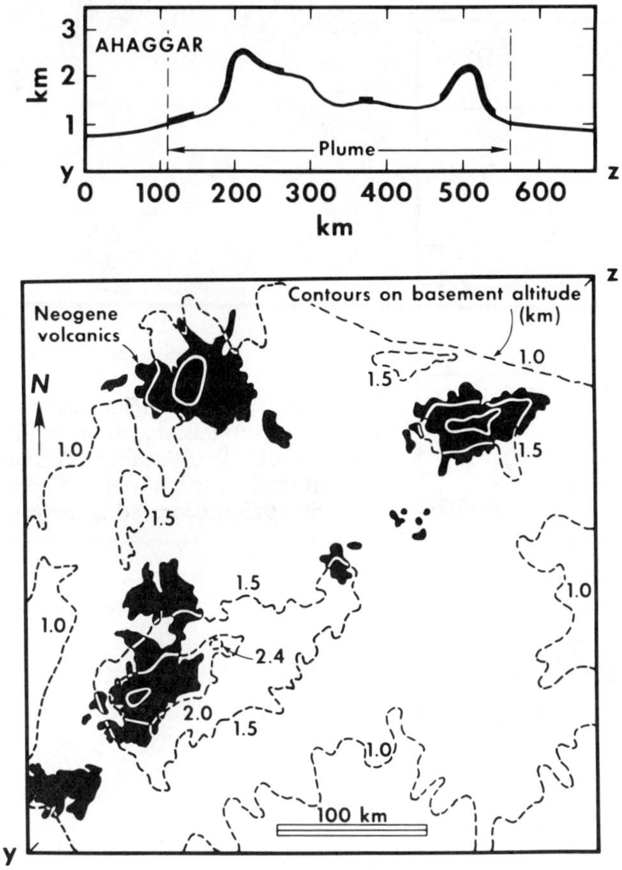

6.2.20 Map and cross section (after Black and Girod, 1970) illustrating that uplift is the dominant property of the Ahaggar hot spot and that Neogene volcanics, mainly basalt flows, cap the most elevated areas.

The distribution of seismicity in the western United States is shown in Fig. 6.2.22. The main area of active volcanism, seismicity and tectonics is bounded on the east by the Rio Grande Graben, and on the north by the Snake River Plain, although there is considerable seismicity in the northern Rockies. The Basin and Range Province is an extensive zone of horst and graben tectonics, high heat flow, hot springs and recent volcanism. Focal mechanism studies (Smith and Sbar, 1974) indicate the entire area is in tension with the directions indicated by the arrows in Fig. 6.2.22.

Atwater (1970) and Christiansen and Lipman (1972) have attributed the tensional tectonics and the volcanism of the western United States to forces exerted at the San Andreas plate boundary, with tensional failures of the lithosphere permitting the flow of magma to the surface. Scholz *et al.* (1971) have related the volcanism and tectonics of the western United States to behind-arc spreading.

Morgan (1971) has associated one of his original 16 plumes with the extensive recent volcanics of the Snake River Plain. If such a plume or hot spot is postulated in the vicinity of Yellowstone caldera at the eastern end of the plain, the trend of the eastern section of the plain is in good agreement with the direction of propagation predicted by Minster *et al.* (1974). A large number of volcanic rocks in the Snake River Plain have been dated by Armstrong *et al.* (1975). The ages of these (and other) rocks as a function of the distance from the Yellowstone caldera are given in Fig. 6.2.23. Also included is the rate of propagation of the volcanics predicted by Minster *et al.* (1974) on the basis of the fixed hot spot hypothesis. Although the age of the oldest volcanics tends to increase with distance from Yellowstone, the agreement with the predicted rate is rather poor. It is also evident from Fig. 6.2.23 (as well as Fig. 6.2.21) that there are recent volcanics along much of the length of the Snake River Plain. It is difficult to associate extensive basaltic volcanism over a distance of 500 km directly with a hot spot beneath Yellowstone. Smith *et al.* (1974, 1977) have suggested that the Snake River Plain volcanics have followed a preexisting zone of weakness and that a wedge-shaped fracture is propagating eastward in response to intraplate deformation.

Suppe *et al.* (1975) have attributed much of the volcanism of the western United States to two hot spot tracks. In addition to the Yellowstone hot spot, they have proposed that a hot spot is located near Raton, New Mexico. This is shown in Fig. 6.2.21. They would extend this hot spot track to the White Mountains of eastern Arizona and possibly as far as the Gulf of California. These authors have conceded, however, that these hot spots cannot explain all the intraplate volcanism of the western United States. Burke and Wilson (1976) have located four hot spots in the western United States (see Fig. 6.2.14). Even this number of hot spots cannot explain all the volcanics of the area.

It was shown by Gruntfest (1963) that the heat produced by viscous dissipation in a fluid with a strongly temperature-dependent viscosity could lead to an instability. Anderson and Perkins (1974) have suggested that this type of thermal runaway could be responsible for magma generation in the asthenosphere and the type of volcanism observed in the western United States.

Eurasia. Much of the active volcanism over this huge continental area can be related to extensional features developed in association with the Alpine-Himalayan collision (see section 6.2.1; Molnar and Tapponnier, 1975). The identification of volcanic areas related to mantle heterogeneity in a continental area where much

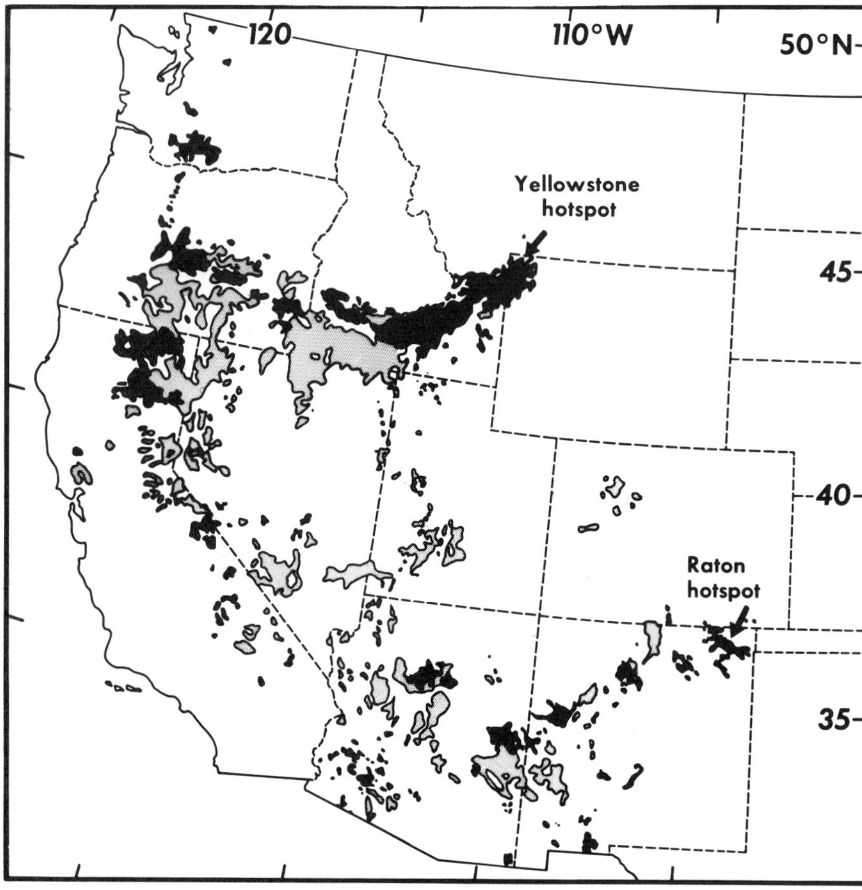

6.2.21 Quaternary (filled) and Pliocene (open) volcanic areas in the western United States. The Yellowstone and Raton hot spots of Suppe *et al.*, 1975, are located by arrows.

of the volcanism is associated with continental collision remains a major problem. Distinction between the two kinds of basaltic activity may be extremely difficult; it may prove necessary to resort to geochemical indices.

Volcanism in continental rifts

The word "rift" is used in geology in a variety of ways, most of which have nothing to do with major structures. Gregory (1921) was the first to use and later to popularize the word rift for a large-scale structure when he described the East African Rift. He defined a rift valley as a long depression between parallel normal faults. Here a modified version of Gregory's definition is used and **rifts** are defined as "elongate depressions overlying places where the entire thickness of the lithosphere has ruptured in extension." This omits reference to parallel faults because many familiar rifts (for example, the Connecticut Rift) have a major fault on only one side. In some rifts it is hard to be sure whether a boundary is a fault or a steep monoclinal flexure. By referring to "the entire thickness of the lithosphere," emphasis is given to rifts as large-scale structures and small depressions are excluded. The reference to extension is to distinguish rifts from the other major fractures, such as transforms and thrusts at convergent plate boundaries, that penetrate the whole lithosphere.

Rifts are the most common major lithospheric fractures. Since rupture of the lithosphere is involved in rift formation, they are most readily developed where the lithosphere is thinnest—that is, in the oceans. This section is solely concerned with continental rifts, which are now—and probably always have been—much less abundant than oceanic rifts (see section 6.2.1).

Two observations have been gleaned from continental rift studies made over the last ten years: first, intracontinental rifts are very numerous; second, rifts have developed within continents in a range of structural environments. These are all environments with dominantly extensional tectonics. In this section, active and ancient rift occurrences associated with a number of different extensional plate-tectonic regimes are described. It is necessary for us to define the Wilson cycle, which is the concept that permits application of plate tectonic principles to the record of historical geology within the continents.

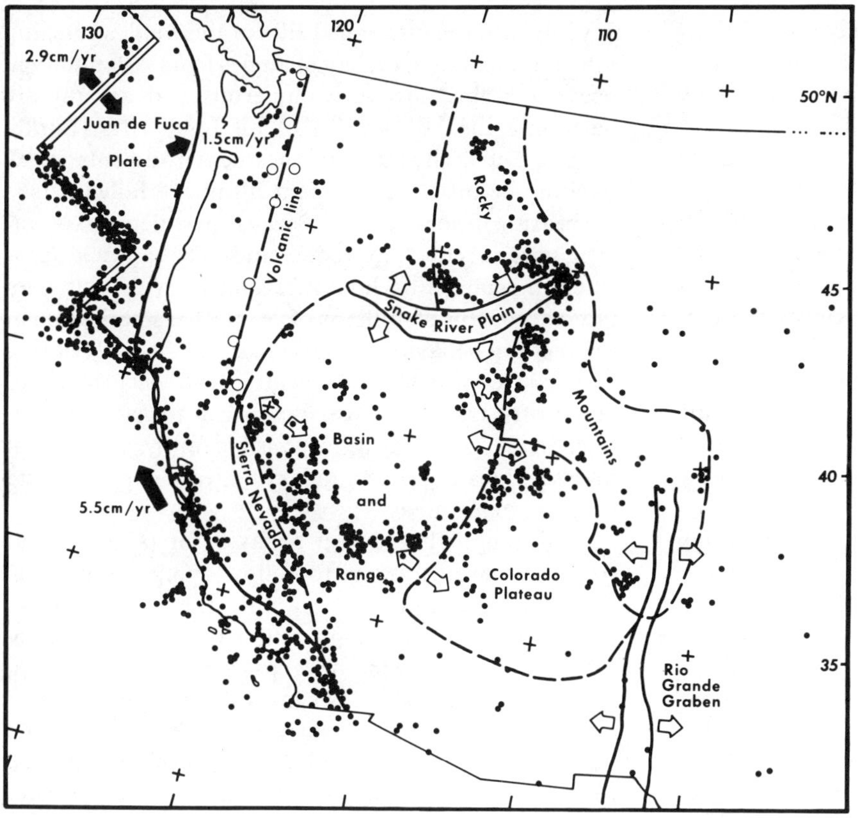

6.2.22 Earthquakes of the western United States (shown as filled circles) superimposed on a map showing geological provinces. Open circles are the Cascade province volcanoes. Solid arrows are directions of motion across the Juan de Fuca ridge and along the San Andreas fault. Open arrows are directions of motion inferred from seismic focal mechanisms. (Smith and Sbar, 1974).

Rifts in the Wilson cycle. Our discussion of basaltic volcanism in rifts comes twelve years after Tuzo Wilson first drew attention to the best way in which to apply the tenets of plate tectonics to studies of continental geology. Wilson (1968a) argued, before the American Philosophical Society, that plate tectonics showed that oceans are opening in some places on the surface of the Earth and closing in others. He suggested that the history of the Earth's surface, therefore, can be considered as a record of the opening and closing of oceans. Since ocean floor disappears by subduction or by erosion during or after obduction, the record of these cycles has to be sought within the continents. Looking for records of these cycles is the most powerful way of interpreting continental geology. Wilson analyzed the cycles in terms of oceanic evolution from youth (continental rupture) to old age (continental collision). Dewey and Burke (1974) later suggested that complex interwoven cycles of ocean opening and closing of the kind recorded within the continents be called Wilson cycles.

Rifts at continental breakup. Rifts occur at all stages of the Wilson cycle because extensional tectonics can develop locally at all stages; however, those associated with continental rupture are both particularly well-developed and particularly accessible to study. The East

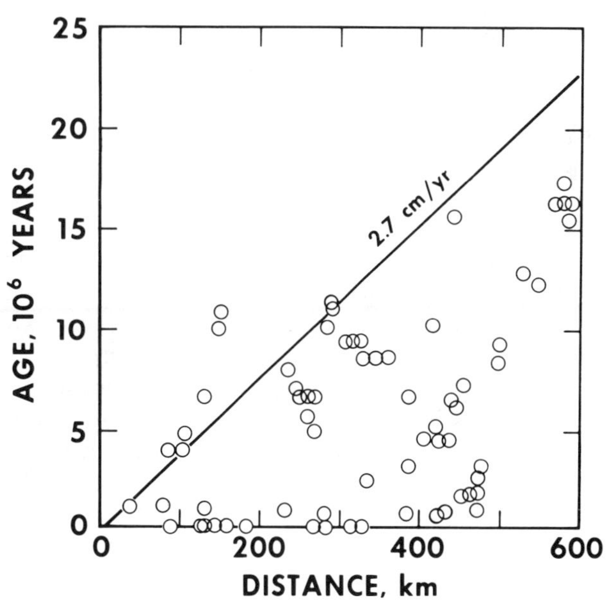

6.2.23 Ages of volcanic rocks on the Snake River Plain as a function of distance from the Yellowstone Caldera (Armstrong *et al.*, 1975). The solid line is a rate of propagation for a Yellowstone hot spot as predicted by Minster *et al.*, (1974). One simple interpretation of this figure is that volcanism started as the hot spot passed a locality, and continued in relation to subsequent rifting.

African Rift system is the best known of continental rift systems. Progress in its understanding has accelerated with appreciation that its present activity dates only from the beginning of the Neogene, when the African plate appears to have come to rest with respect to the underlying mantle convection pattern (Burke and Wilson, 1972). Distinction of the current episode of rifting from a very similar episode in Africa 100–200 m.y. ago that was associated with the break-up of Gondwana has proven to be important. Rift studies that define the timing of events accurately have become of particular importance in recent years, and seem likely to remain important in the future. It will only be possible to establish how rifting relates to other tectonic events if the relative timing of the phenomena is accurately known.

Two other rift properties that have been firmly established in the East African Rift system are the following: rift faults commonly follow and reactivate old structures (McConnell, 1974); rift igneous rocks are almost wholly mantle-derived. Crustal material seems to be involved only to a very small extent in igneous rocks of the East African and nearly all other rifts. Detailed studies of igneous rocks associated with the East African Rift system have emphasized the widespread occurrence of alkaline rocks, most of which appear to have last equilibrated with the mantle at depths of 60–100 km, and many of which appear to have suffered complex subsequent histories. Comparable alkaline rocks occur in many rift systems, but so do tholeiitic rocks, suggesting equilibration at much higher mantle levels.

Teleseismic studies have been among the most fruitful of geophysical investigations in the East African Rift system. These studies have shown the peculiar nature of the mantle below the rift system as well as the extensional character of most rift earthquakes. Although local seismic networks have yielded important results, the very limited number of modern seismic stations on the African continent has proven to be something of a barrier to research. Gravity studies are well advanced in East Africa and have proven most susceptible to analysis where information from them has been coupled with other data. Seismic refraction studies, for example, have indicated the presence of high velocity, dike-like objects in some rifts, which has helped control gravity interpretations.

If the East African system continues to propagate and eventually becomes a system extending across the African continent, then a new ocean may develop and the African plate may split in two. If this happens, many presently active rifts will be left in one or another of the two new continents. Such rifts will have failed to develop into oceans. Failed rifts of this kind, stretching into continents at a variety of angles from Atlantic-type ocean margins, are very numerous and are rapidly becoming the best-known of all fossil rifts (Burke, 1976a). Our knowledge of these environments is expanding because the sediment fill of the failed rifts is oil-bearing in some places. Because petroleum exploration, seismic reflection studies, and wells can yield accurate information on how rifts subside with time over intervals of tens of millions of years, characterization of the thermal behavior of individual rifts or parts of rifts may prove possible. The widespread occurrence of failed rifts, trending at various angles to Atlantic continental margins, may be responsible for the difficulty of precisely locating the boundary between continent and ocean in many areas.

Although the general behavior of failed rifts at Atlantic margins seems to follow a regular pattern, there are exceptions. For example, the South Atlantic opened about 120 m.y. ago, and many failed rifts associated with this event have passed through normal sequences of events and became filled with marine sediments in the Aptian. However, the Rio Salado and Colorado rifts (Urrien and Zambrano, 1973) continued to accumulate nonmarine sediments until much later (about 80 m.y. ago), and the Benue Trough in Nigeria behaved like a small ocean that appears to have opened and shut over the interval between 125 and 80 m.y. ago (Burke and Dewey, 1974).

Convergent boundary rifts. Rift developments at Andean continental margins and in island arcs are most prominent at places where the lithosphere is thinnest, that is, along the line of the volcanoes themselves (see section 6.2.2). These rifts, which are known in such active Andean arcs as New Zealand and Sumatra-Java, achieve their greatest tectonic significance only when the volcanic arc splits along them to form a marginal basin as happened, for example, in the Japan Sea at the beginning of the Miocene (Sillitoe, 1977). Rifts of Basin and Range type appear to be associated with imperfect transform motion along continental boundaries, but few fossil examples have been recognized as yet.

Collisional rifts. Perhaps the greatest variety of rift development is associated with continental collisions. Failed-rift systems formed at rupture become reactivated as rifts striking into the collisional mountain belts, forming the population of objects that Shatskiy (1947) called **aulacogens**. Also, a whole new set of rifts, of which the Lake Baikal and Rhine rifts seem the best active examples, are set up as a result of intracontinental strains established during the collision. These objects have been called **impactogens**, and they can be readily distinguished from aulacogens where their geological history is well-established.

In addition to these two major classes of collision-related rifts, final closing of an ocean is accompanied by very complex rift production (Tapponnier, 1977). This can be seen very well in the Mediterranean, where formation of the rifts of Corsica and Sardinia 20 m.y. ago accompanied rotation of these islands away from France (Dewey *et al.*, 1973), and where formation of the Aegean rifts within the last 20 m.y. has been related to the westward motion of Turkey with respect to the Black Sea and the Mediterranean (Dewey and Şengör, 1979). Rift development during the closing of the Mediterranean has been considered by Tapponnier (1977), who concluded that all of the rifts represent secondary extensional phenomena. If the closure of the Mediterranean continues, rifts of the Aegean and Corsican-Sardinian types are likely to be compressed out of all recognition. Such structures are probably preserved only as obscure areas within mountain belts that show signs of late-stage extension before final compression. For this reason, rifts of this kind are of less general interest in continental geological history than the four major rift classes (continental rupture rifts, failed rifts at Atlantic margins, aulacogens, and impactogens). A generalization of this type always has exceptions and the Devonian rift basins of Norway (Horn, Solund and Hornelen) form just such an exception (Steel, 1976). They appear to have formed within the Caledonides at continental collision and closely resemble the Thakkola rift of the high Himalaya today.

(a) *Aulacogens*. Although the aulacogen concept is over thirty years old, widespread use of the term among scientists in the United States dates only from attempts to interpret the aulacogen concept in plate tectonic terms (Burke, 1977). Shatskiy (1947) used the word only for rifts striking into fold belts, and it seems unwise to use it, as has sometimes happened, for all ancient rifts that have failed to develop into oceans. Shatskiy also emphasized that the depositional and structural history of the aulacogen begins at the same time as the depositional history of the fold belt into which it leads, thus eliminating the structures now being called impactogens. An unfortunate fact in the history of the use of the word "aulacogen" is that one of Shatskiy's type examples, the Dnieper-Donets structure, is a failed rift that strikes not into a fold belt but into the North Caspian depression. The North Caspian depression is an area which is probably underlain by ocean floor formed in the Devonian that has been covered with a great thickness (~14 km) of sediments but has, so far, escaped both subduction and obduction (Kidd, pers. comm., 1975).

Although aulacogen histories can be very diverse, it is often possible to make a two-fold division: an earlier continental rupture and ocean opening phase with development of typical rift features, including alkaline igneous rocks, basal coarse clastics and higher evaporites; and a distinct, later post-collisional phase with development of coarse clastic sediments derived from the collisional mountain belt, and of strike-slip faulting. The outstanding structural features of many aulacogens include the enormous thicknesses of sediments and volcanics that they contain. Shatskiy (1947) pointed out that the upper member of an aulacogen sediment pair commonly extends over an area wider than that of the initial rift. This seems a particular case of the more general observation that numerous major intracontinental basins, such as the Chad, Paris and Michigan basins, overlie older rifts (Burke, 1976b).

(b) *Impactogens*. In map view an impactogen resembles an aulacogen because it also is a rift structure striking into a fold belt. The distinction between the two is made initially by ascertaining whether rift history goes back as far as the opening of the ocean whose closure is represented in the fold belt. If it does not, and if rifting dates only from the time of ocean closure in the fold belt, then the rift is an impactogen. For example, both the Rhine graben and the Polish trough strike from northern Europe into the Alpine fold belt. The start of the geological history of the former is contemporaneous with the mid-Eocene Meso-Alpine collisional event (Şengör *et al.*, 1978), while that of the latter started in the Triassic. Thus, the Rhine graben is an impactogen while the Polish trough is an aulacogen.

Although the Rhine graben reaches to the edge of the Alpine System in the Jura, impactogenal rifts associated with the Himalayan collision (the Baikal and Shansi grabens) develop far from the main collision zone. Molnar and Tapponnier (1975) have suggested that these rifts formed as consequences of the eastward escape of China from crushing between the vise-like jaws of India and Asia. The Baikal and Shansi rifts are linked to the main collisional mountains in Tibet and the Himalaya through a system of strike-slip faults and folds. Timing of the Himalayan collision is not yet as well-established as that of the main Alpine collision, so it is hard to know how closely the start of faulting in Shansi and Baikal coincides with the collision marked by the Indus Suture.

Impactogens seem particularly well-developed in association with the Hercynian fold belt of Europe. The Hercynian ocean appears to have opened in two stages: one in medial Devonian times along a rift extending from western Britain to the Urals, and the other in Early Carboniferous times along a rift with a very different strike in Iberia. The timing of the Devonian event indicates that the rift may have been formed as an impactogen related to collision in the Caledonides. The second rift could have been an Acadian impactogen (Burke and Sawkins, 1977). Although evidence of the formation of the Hercynian ocean is very incomplete, evidence of its closure at the beginning of the Permian is much stronger. Rifts developed over much of northwestern Europe at the beginning of the Permian (the Oslo graben being the best known such rift). These rifts are strong candidates for interpretation as impactogens. The rifts of northwest Europe established in this Permian event have been repeatedly reactivated in Triassic, Jurassic, Cretaceous and Cenozoic times (Whiteman *et al.*, 1975). This is most obvious in the North Sea where geological development of the huge oil province has been dominated by rift reactivation (Woodland, 1975). A possible explanation of renewed subsidence in a long inactive rift, so typical of the North Sea, is that a thermal pulse in the mantle induces transition from a basaltic to an eclogitic mineralogy in a sub-rift axial dike system.

Relative timing of rifting and volcanism. Basaltic volcanism is a feature common to most rifts. In continental rifts, with which we are concerned here, these basalts are mainly alkaline. The relative timing of basaltic volcanism with respect to the associated rifting shows considerable variation. Currently, there are two main hypotheses to explain the origin of basaltic magmas beneath rifts. One favors the preliminary cracking of the lithosphere due to differential stresses resulting from two-dimensional plate evolution e.g., membrane stresses: Turcotte and Oxburgh, 1973; and Turcotte, 1974a; stresses due to the collision of continents: Molnar and Tapponnier, 1975; and Şengör, 1976; stresses resulting from the propagation of existing accreting plate margins into continents: McKenzie and Weiss, 1975 . This cracking upsets the temperature/pressure balance of the underlying mantle, resulting in its partial melting. The partially melted area experiences a volume increase that may induce a post-rifting uplift of the overlying lithosphere, such as appears to have happened in the Upper Rhine Graben (Şengör *et al.*, 1978). The other hypothesis suggests that a complicated convection pattern in the mantle is responsible for doming and cracking the lithosphere, thereby giving rise to extensional fractures and eventually to rifts (Burke and Whiteman, 1973; Burke and Dewey, 1973b). In the former view, the expected sequence of events is rifting-doming-volcanism, whereas in the latter it is doming-volcanism-rifting (Fig. 6.2.24). It is our contention that both hypotheses are compatible with the available data, but are applicable to fundamentally different tectonic environments. The first hypothesis is applicable to regions where rifting is the result of horizontal movements of plates and their interaction, and the mantle plays a passive role. The second hypothesis seems to be valid where (small-scale?) convection in the mantle (McKenzie and Weiss, 1975) directly affects the overlying lithosphere and induces rifting as a combined result of primary vertical tectonics (uplift) and possibly later horizontal motion (initial spreading). It appears that this latter mode of rifting occurs mainly on plates that are fixed with respect to the underlying mantle, as Africa has been suggested to have been since 25 m.y. ago (Burke and Wilson, 1972). At present, the former mode of rifting is by far the more widespread of the two.

Passive mantle hypothesis. The horizontal movement of plates and their interaction give rise to differential stresses within the lithosphere that may result in rifting. Turcotte and Oxburgh (1973) and Turcotte (1974a) argued that membrane stresses can be generated within a plate moving longitudinally on the surface of the Earth, due to the ellipticity of the Earth, and this may result in rifting. Burke and Dewey (1974), however, indicated that the elastic stresses thus generated cannot be stored for long periods of time and therefore may not be as important in generating rifts as Turcotte and Oxburgh (1973) suggested. Moreover, the supporting evidence that Oxburgh and Turcotte (1974) used to apply this idea to the East African rifts consisted of incorrect K/Ar data and misinterpretation of paleomagnetic results.

Molnar and Tapponnier (1975) and Şengör (1976) pointed out that continental collision, especially of irregular margins, also generates extension within continents, resulting in rifting more or less perpendicular to the direction of convergence. If collision-induced strike-slip faulting also causes rift formation, as is the case for the Baikal rift (Sherman, 1978), such rifts are not likely to have any systematic orientation with respect to the convergence direction. In the foreland of the Alps in Europe, for example, rifting events are well-correlated with the Mesoalpine collision (Trümpy, 1973; see Şengör, 1976, Table 1). Here rifting appears to have predated major basaltic volcanism. In the Upper Rhine Graben, rifting started during medial Eocene with some minor volcanism of 42 m.y. age along the master faults (Şengör *et al.*, 1978). Major basaltic volcanic activity began at the northern end of the graben about 25 m.y. ago (Lotze, 1974). This volcanic center, the Vogelsberg, appears to have continued its activity until very recently. In the south, well within the major graben trough, the Kaiserstuhl volcano began its activity about 18 m.y. ago; this, however, unlike the Vogelsberg, was a short-lived event. In the Lower Rhine Graben there are few volcanics, but the maars of Eifel appear to be related to Plio-Pleistocene northwest-trending faults (Greiner and Illies, 1977). In Bohemia, both along the Thuringian disturbance and in the graben between the Bohemian and Thuringian blocks (*Schollen*), volcanism began during the Oligocene and lasted through the Pliocene; rifting along these lines had begun in the earliest Tertiary and accelerated during late Eocene (Lotze, 1974). It appears that in Europe it is this kind of inter-Scholle activity that caused basaltic volcanism which is largely of olivine-nepheline-bearing alkaline type.

Western Turkey and the Basin and Range Province of the western United States are areas of extensive rift development, characterized by numerous subparallel grabens in "rift clusters." Both areas are related to large, imperfect strike-slip plate boundary zones (western Turkey: Dewey and Şengör, 1979; western U.S.: Atwater, 1970). In western Turkey, rifting began during the late Miocene and is currently active. Most rifts here

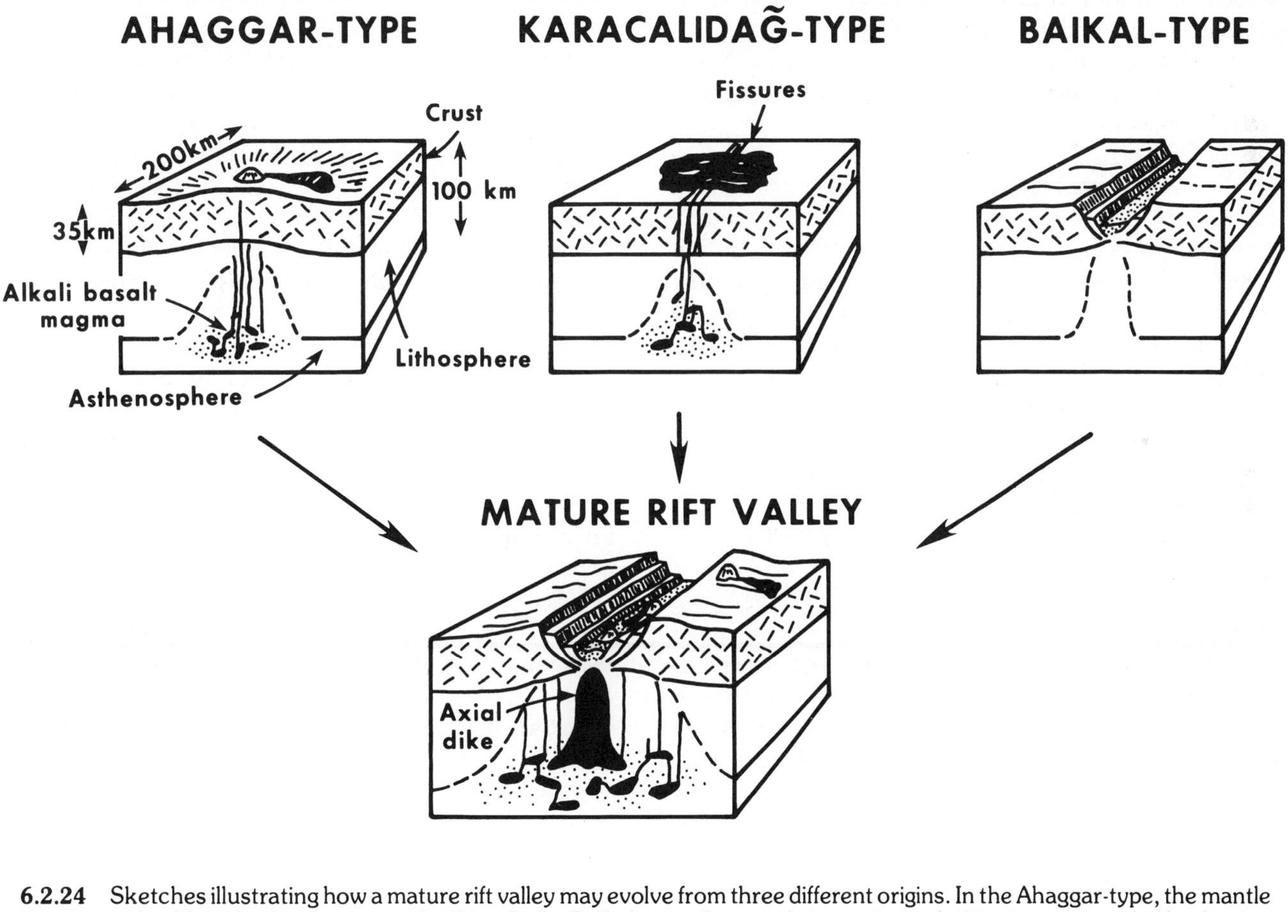

6.2.24 Sketches illustrating how a mature rift valley may evolve from three different origins. In the Ahaggar-type, the mantle is active and uplift and volcanism precede rifting. In the Baikal-type, the mantle is passive and rifting is not preceded by doming or volcanism. The Karacalidağ-type illustrates a complexity: horizontal extension forms fissures through which magma wells up prior to major down-faulting and rift formation. The sequence of events documented in the geological record may make a Karacalidağ-type rift look very like an Ahaggar-type rift, although the mode of origin more closely resembles that of a Baikal-type rift. However rifts originate, their structures develop in such a way that all mature rift valleys look rather similar as illustrated in the bottom figure. Because of this similarity, working out rift origins depends on a good record of geological history.

are devoid of volcanics with the exception of the Gediz and Simav grabens (Dewey and Şengör, 1979). On the northern shoulder of the Gediz graben, an area 50 km long and 20 km wide is covered with Pliocene to recent alkaline basalts that are nepheline-, leucite- and hornblende-bearing (i.e., kulaite: Washington, 1894; Erinc, 1970). These basalts emanated from fissures that can be shown to be controlled by faults related to the Gediz graben. Zeschke (1954) documented a similar history for the Simav basalts.

In the Basin and Range Province major rifting related to the present regime began about 18 m.y. ago (Noble, 1972) and was closely followed by basaltic/rhyolitic bimodal volcanism (McKee *et al.*, 1970). The geographic extent and the possible close temporal association with a strike-slip regime (Hamilton, 1970) of the western Siberian rift system of Triassic age (Logatchev, 1977) suggest that it might be a fossil analog of the Basin- and Range-type rift regimes, where generally basaltic volcanism postdates rifting.

Active mantle hypothesis. Rifts active on the African Plate can be interpreted as products of relatively simple mantle-lithosphere interaction. The control on rifting appears to be the lack of relative motion between the African Plate and underlying convective circulation over the last 25 m.y. (Burke and Wilson, 1972; Thiessen *et al.*, 1979). The sub-African Plate mantle-lithosphere interaction is manifested in several phenomena: the distinctive basin and swell structure; the occurrence of intraplate volcanism (largely on swell crests); the development of rifts and the evolution (in the Red Sea and in the Gulf of Aden) of rifts into oceans (see Burke, 1977, for a review).

Throughout southern Africa, swell crests carry no volcanoes and rifts are mainly not on swells but are reactivated old structures. This indicates that a sequential development from volcanoes on swells to rifts such as Burke and Whiteman (1973) distinguished is not universally recognizable. Further evidence that rift development is complex can be seen in the active rifts of East Africa. Volcanism there is very unevenly distributed along the length of the rifts. In some areas (for example, near Addis Ababa) volcanism is dominantly tholeiitic, whereas in other areas it is dominantly alkaline (Baker *et al.*, 1972). The implication of this observation is that the petrology of the igneous rocks is a very poor indicator of rift style compared with more direct structural/stratigraphic features such as topography, faulting and sediment fill. The general absence of signs of interaction between magma and continental crust in the rift igneous rocks is perhaps the most significant feature and is to be expected in a regime where axial dikes and extension dominate.

Older episodes of rifting induced by similar mantle interactions are generally hard to identify, but the lack of motion between Africa and the spin axis during the break-up of Pangaea has been taken as evidence that African rifting during that time was induced by this interaction (Burke and Dewey, 1973b). The Pangaean-rupturing rifts on either side of the Atlantic, which formed just before that ocean opened, provide further examples of the diversity of styles of volcanism. Some Pangaean rifts appear to be without volcanic material; others are associated with extensive pre-rift igneous activity; others show igneous activity only when the rifts are well-developed. As is evident in East Africa at present, compositional diversity is the rule. In some areas carbonatites and alkaline syenites abound (e.g., Los, Bagbe, Songo), whereas in others tholeiitic basalts are the only igneous rocks. It seems that terrestrial rifts may develop in a variety of ways, and only in exceptional circumstances can the record within individual rifts be deciphered in sufficient detail to permit clear understanding of all the processes that have operated in their development.

Flood basalts. Young volcanism not related to plate boundaries is widespread (Burke and Kidd, 1975, 1980); although in some ways there is great variety in its expression, there is sufficient similarity between the occurrences of non-plate-margin volcanism that it seems appropriate to treat them in a single category. Many areas of active intraplate volcanism are well-exposed in Africa and have been well-studied; the association of volcanics with structurally and topographically high areas is well-established (Burke and Whiteman, 1973; Thiessen *et al.*, 1979). The volcanics of Dakar, which are located near sea level, actually are not exceptions to this observation since subsurface data (Spengler and Delteil, 1966) reveal an underlying youthful structural elevation. There is a complete spectrum of expression of intraplate volcanic areas in Africa, from uplifts without any volcanism on them (e.g., Fouta Djallon) through minor volcanism (e.g., Aır) of alkaline type to more abundant alkaline volcanism (e.g., Tibesti), to areas of voluminous tholeiitic flood basalts together with comparatively minor alkaline volcanics (e.g., Ethiopia). In all of these cases there is a structural uplift associated with the volcanic area; the size of this uplift is somewhat variable although most tend to be 100–200 km across (Burke and Whiteman, 1973). There is perhaps a tendency for the areas with larger volumes of volcanics to have larger diameter uplifts, although it is possible to resolve subsidiary uplifts within the larger ones (e.g., Ahaggar; Fig. 6.2.20). Thus it is not wholly clear if the larger structures are merely groups of smaller ones, or whether the smaller diameter uplifts in them are secondary to the larger (but generally lower amplitude) uplifts.

This spectrum of intraplate volcanism can be seen, although less well-displayed, on other continents. More significantly, it is also developed in the oceans and the alkaline and tholeiitic volcanics in oceanic hot spots are essentially indistinguishable from those in continental hot spots. Once again the African plate provides many of the best examples. Oceanic intraplate volcanic areas also show a variation in volume from regions like Hawaii, with abundant tholeiite and minor alkaline volcanics, to relatively small edifices like Ascension, that are mainly alkaline volcanics. Very small volume intraplate volcanic areas, like some of those on the African continent, are harder to detect in the oceans, and uplifts without accompanying volcanism are even more difficult to detect without making direct attempts to find them (Menard, 1973).

Because alkaline magmas are erupted in places in present island arcs and collisional orogens, it is not yet possible to detect "intraplate type" or "hot spot type" activity within such zones of convergence even though it may well occur there. However, it is possible to detect such volcanism along divergent plate boundaries because of the alkaline character of some of the magmas, excess volumes of magma, and the associated structural uplift. The occurrence of hot-spot-type volcanism at discrete and long-lived sites along divergent plate boundaries is one of the most significant properties of their distribution and helps considerably to winnow the many hypotheses put forward for their origin. In particular, since the lithosphere is thin to virtually nonexistent at spreading ridge crests and is being created progressively away from them, it is unlikely that the crack propagation hypotheses of Turcotte and Oxburgh (1973) and Oxburgh and Turcotte (1974) and the dense anchor-asperity hypothesis of Shaw and Jackson (1973) are correct. An "active mantle" hypothesis for hot spot origin, such as that proposed by Wilson (1963), is more satisfactory.

A similar spectrum of size and volume of magmas can be seen in those hot spots sited on divergent plate boundaries. The central and north Atlantic contain the best examples. These range from the anomalously-elevated area on the ridge at 45° N where there seems to be relatively little excess volcanism, through the Azores with mostly alkaline volcanism constructing small islands above sea level, to the extreme of Iceland with its voluminous excess tholeiitic volcanism and, in relative terms, minor alkaline magmatism. Anomalous (with respect to spreading ridge basalt) geochemical signatures (see section 7.4.2), particularly of $^{87}Sr/^{86}Sr$ values (White *et al.*, 1976), correlate exactly with discrete anomalously elevated ridge crest areas (in Iceland, 45° N, Azores, and near 34° N at the Colorado Seamount). The distinctive geochemistry of the "excess" magmas argues strongly for a separate (deeper) source for them (Schilling, 1973) compared with normal spreading ridge basalts. It is because the latter are produced at most places in fairly constant amounts through a wide range of spreading rates that the "excess" character of the hot spot magmatism along ridge axes can easily be detected.

The connection between flood basalts and the largest hot spots is well shown by Iceland (Burke and Kidd, 1980). The island itself consists of flood basalts. The hot spot tracks that lead away from Iceland go to large areas of flood basalts in East Greenland and the northern British Isles (Fig. 6.2.25) that were erupted during the initial rifting that led to successful seafloor spreading in the northern Atlantic beginning about 60 m.y. ago. The relics of associated central alkaline volcanoes of that age are well-known from Scotland. A younger but similar situation is seen in the Afar (Fig. 6.2.25) with the Ethiopian Traps, which erupted before and up to the opening of the Red Sea and the Gulf of Aden. If the Pacific plate were not moving so rapidly with respect to the source of the Hawaiian magmas, the accumulation of igneous material at present rates of production would clearly rival Iceland. Although the Hawaiian-Emperor hot-spot trace (Fig. 6.2.25) does not lead back to a site of rifting and a large flood basalt pile, it is appropriately grouped with other flood basalt-producing objects. Several other large hot spots have tracks leading back to rifting sites and to accumulations of flood basalts (Fig. 6.2.25). In particular, Tristan/Gough leads to the Kaokoveld and Paraná basalts, erupted just before and up to the opening of the South Atlantic 120 m.y. ago, and Reunion leads back to the Deccan Traps erupted about 65 m.y. ago just before rifting in the Gulf of Khambhat and removal of the Seychelles from India (McKenzie and Sclater, 1971). The Galapagos hot spot, although its tracks do not lead back to preserved flood

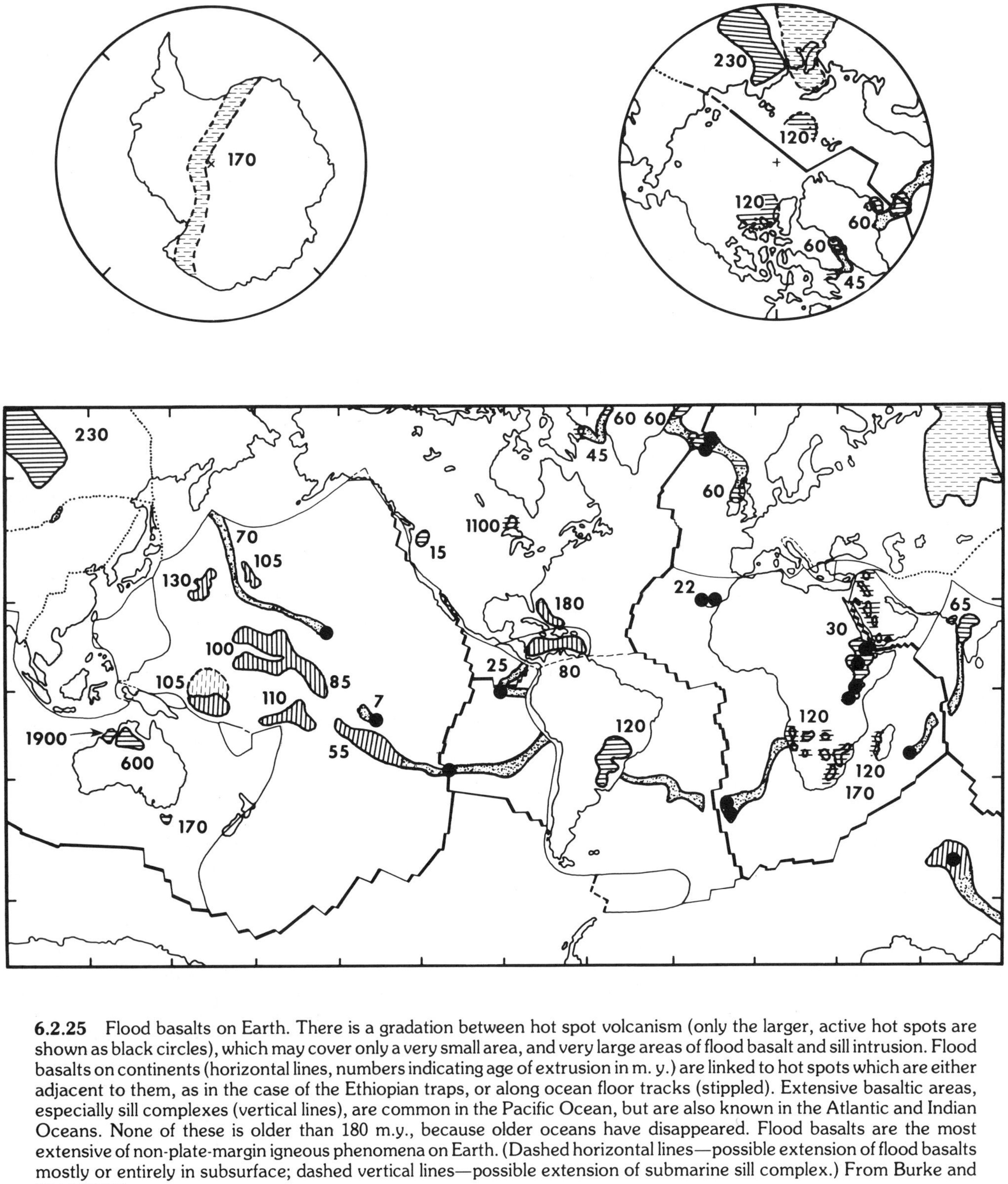

6.2.25 Flood basalts on Earth. There is a gradation between hot spot volcanism (only the larger, active hot spots are shown as black circles), which may cover only a very small area, and very large areas of flood basalt and sill intrusion. Flood basalts on continents (horizontal lines, numbers indicating age of extrusion in m. y.) are linked to hot spots which are either adjacent to them, as in the case of the Ethiopian traps, or along ocean floor tracks (stippled). Extensive basaltic areas, especially sill complexes (vertical lines), are common in the Pacific Ocean, but are also known in the Atlantic and Indian Oceans. None of these is older than 180 m.y., because older oceans have disappeared. Flood basalts are the most extensive of non-plate-margin igneous phenomena on Earth. (Dashed horizontal lines—possible extension of flood basalts mostly or entirely in subsurface; dashed vertical lines—possible extension of submarine sill complex.) From Burke and Kidd (1980).

basalts, is sited on a spreading ridge that started about 25 m.y. ago by rifting across older oceanic crust generated at the East Pacific Rise (Hey, 1977). It may not be a coincidence that the Galapagos hot spot, as shown by its area and the volume of magmatic products in its tracks, is a relatively large one, like others that have been listed as associated with flood basalt production and initial ocean opening. Ocean opening in the Labrador Sea-Baffin Bay was accompanied by flood basalt volcanism about 60 m.y. ago which now is seen on Disko Island and Cape Dyer of Baffin Island. The hot spot that produced this volcanism then produced the shallow ocean floor of Davis Strait as a track, but this hot spot is now obviously extinct. Perhaps it is also not coincidence that spreading ceased in the Labrador Sea-Baffin Bay at the same time or shortly after the hot spot died. This case illustrates the point emphasized by Vogt (1972), that the volumes of magma generated in hot spot sites vary with time, although we do not think there is strong evidence for synchroneity in this variation among many hot spots. Thus, the amount of magma produced by the Hawaiian hot spot, judged by the volume of the track, was very small at the time of and shortly after the bend formed in the Hawaiian-Emperor chain. It may have been no more volumetrically impressive at that time than one of the smaller present ocean island hot spots, such as St. Helena. The points we wish to make here are that voluminous flood basalt-producing hot spots are the same kind of object as the volumetrically smaller ones, and that one can change into the other and back again with time. They may also die out, in terms of their volcanic expression, temporarily or permanently. They vary a great deal in the length of time during which large quantities of tholeiitic magma have been produced, from the short burst of the Columbia River basalts (Baksi and Watkins, 1973) and Deccan Traps (Wellman and McElhinny, 1970) to the more extended histories of Iceland or Hawaii.

Large flood tholeiite events in the oceans seem, in some instances, to produce huge sill complexes, probably because of the limited abilities of even mafic lava to travel underwater before chilling. Such sill complexes (Fig. 6.2.25) underlie large portions of the Caribbean (Burke *et al.*, 1978) and the submarine plateaus of the western Pacific (Winterer, 1976). The Mid-Pacific Mountains, one of the latter, may perhaps be traced to the hot spots of either Easter Island and/or Pitcairn Island through the Tuamotu-Line Islands track. Sill complexes are, of course, also important components of many continental flood basalt events, particularly the early Jurassic Karoo and Ferrar dolerites of South Africa and Antarctica, and the Mid-Cretaceous Isachsen diabase of the Canadian Arctic (Fig. 6.2.25).

It is of note that all major flood basalt events, where at least some of the extensive lavas and/or sills are still preserved, were connected with extensive rifting and, in most cases, with successful opening of an ocean. Two large and well-known ones not associated with successful ocean opening are the Columbia River basalts and the Siberian Traps.

6.2.3 Tectonics of active volcanism elsewhere in the solar system

Tectonics of active volcanism on Io

Elsewhere in the solar system, active volcanoes have been recognized only on Io, the innermost of the Galilean satellites of Jupiter (Morabito *et al.*, 1979). The density of Io is such that these volcanoes could be erupting basalt as well as the more volatile sodium and sulfur-rich products that have been recorded.

The occurrence of what appear to be calderas on at least one volcano on Io indicates the existence, at least locally, of a rigid lithosphere, since calderas on Earth are formed by fault collapse. The surface of Io appears to be highly mobile although no coherent large-scale plate structure has emerged. In this respect the tectonic style of Io may be similar to a style that existed on other objects of planetary dimensions early in the history of the solar system and that probably existed only during the period of the high impact flux. The reason for the intense tectonic activity on Io appears to be that jovian tides generate very large amounts of energy within Io which is dissipated as heat (Peale *et al.*, 1979). Tidally induced tectonics do not appear to be significant elsewhere in the solar system although tidal influences triggering volcanic eruptions have been recognized on Earth (e.g., Mauk and Johnston, 1973).

An apparent peculiarity of the tectonics of Io is that, although the tidally induced volcanic activity seems likely to have been in progress for a large proportion of the history of the satellite, today's activity still involves large amounts of volatile material. There does not appear to have been on Io, as on the Earth and the Moon, irreversible fractionation of low density material to make outer parts like the lunar highlands and the terrestrial continents, oceans and atmosphere.

6.3 TECTONICS OF BASALTIC VOLCANISM ON EARTH IN THE LATTER HALF OF EARTH HISTORY

6.3.1 Tectonics of basaltic eruption at plate margins in the past

Tectonics of basaltic volcanism at past accreting plate margins: ophiolites

The idea that ophiolites represent small samples of old oceanic crust and upper mantle emplaced tectonically into orogenic belts was first clearly put forward by Dietz (1963a), Hess (1964), Gass (1968), Peyve (1969) and Laubscher (1969). An elaborate justification of this idea is not appropriate since it has been made by, among others, Thayer (1969), Coleman (1971) and Moores and Vine (1971). These and other workers have pointed out that the thickness and petrology of the internal units that compose well-developed ophiolite sequences match our present understanding of the oceanic crust very well in a general way. Although the thickness of individual ophiolite complex units may vary considerably, the range of thickness variation is comparable with layer thickness variation recorded in the ocean floors (Coleman, 1971; Dewey and Bird, 1971; Moores and Jackson, 1974).

The general sequence of rock types in ophiolite complexes has been presented above (section 6.2.1). Sediments overlying the igneous rocks are not regarded as being part of the ophiolite suite. They may consist of pelagic sediments like those covering much of the present ocean floor, or they may be clastics derived from relatively nearby continents or volcanic arcs. All but a very small proportion of the oceanic crust has been subducted. Of the crust that for various reasons has escaped this fate, most is badly shredded and dismembered; this is particularly true of the material preserved in steeply inclined orogenic zones and/or at deeper structural levels, even in Mesozoic and Tertiary orogenic belts. Thus, the general observation that recognizable ophiolite complexes become scarcer in older orogenic belts is probably explicable because more uplift and erosion have generally occurred in most parts of older belts compared with younger ones. Other reasons for relative scarcity may include the fact that earlier workers who studied older orogenic belts were unfamiliar with the utility of the ophiolite concept, the possibility that major continental collision and suturing was generally more effective and intense further back in time, and the fact that there is proportionately a smaller length of older orogenic belts exposed or preserved compared with younger ones. Because the process of ophiolite obduction is so important, a review of how it operated in a single mountain-belt is included below.

Well-preserved ophiolite complexes are not common, even in Mesozoic and younger orogenic belts. For example, the subduction of the ocean between India and Eurasia has been highly efficient, in conjunction with uplift and erosion, in removing all but a few tiny and tattered remnants of the vast ocean floor that once lay between them. Well-preserved Mesozoic and younger ophiolite complexes that have been more than cursorily described include Troodos (Wilson, 1959; Moores and Vine, 1971; Kidd and Cann, 1974; George, 1978); Vourinos (Moores, 1969; Jackson *et al.*, 1975); Canyon Mountain (Thayer, 1963; Ave'Lallement, 1976); Oman (Reinhardt, 1969; Glennie *et al.*, 1974); Antalya (Juteau *et al.*, 1977); and Papua (Davies, 1971).

Useful compendia of papers and/or reference sources on ophiolites may be found in Coleman (1977), Coleman and Irwin (1977) and the Proceedings of the International Ophiolite Symposium (Panyiotou, 1980). Well-described ophiolite complexes of Paleozoic age are known from the Appalachians and the Caledonides (Bird *et al.*, 1971; Kidd *et al.*, 1978; Dewey and Bird, 1971; Church and Stevens, 1971; Upadhyay *et al.*, 1971; Casey, 1979; Rosencrantz, 1979; Karson, 1977; Karson and Dewey, 1978; Sturt *et al.*, 1979) and less detailed descriptions of some of the widespread Paleozoic complexes in the Urals and central Asian orogenic belts are available (for example, Abdulin *et al.*, 1974; Makarychev and Shtreys, 1973).

Definite examples of pre-Paleozoic ophiolite complexes have been identified only in Pan-African orogenic belts in Morocco (LeBlanc, 1976), and Saudi Arabia (reported in Brown, 1978) and in the older Baikal orogenic belt in the U. S. S. R. (Klitin and Pavlova, 1974). Many other dismembered pieces surely remain to be identified in Proterozoic orogenic belts, since the overall geology and tectonics of both the Labrador-Cape Smith-Nelson (Wilson, 1968b) and the Coronation orogenic belts (Hoffman, 1973, 1980) so clearly involve rifting and later collision (Wilson cycles), starting in both cases about 2.15 b.y. ago. The occurrence of an extensive greenstone belt terrain of similar age to the Coronation orogen (the Birrimian of West Africa) makes it likely that ophiolites will be found eventually in greenstone belts. Extensive areas of mafic volcanics of tholeiitic compositions that are similar to compositions of oceanic crust exist in the Archean greenstone belts; what is generally lacking are equally widespread plutonic equivalents. Moores (1973) has suggested that older Precambrian oceanic crust was characterized by more abundant development of anorthosite than is younger material; this has yet to be verified. It is possible that the extensive tholeiitic submarine lavas of the greenstone belts include, besides ocean floor, and island arc-generated lavas, large thicknesses generated in the ocean by intraplate flood-basalt type magmatism.

The ophiolite obduction process. In recent years there has been a great revival of interest in Alpine-type mafic/ultramafic complexes mainly as a consequence of the hypothesis that ophiolite complexes represent slices of oceanic crust and mantle generated by seafloor spreading which subsequently were emplaced tectonically into various portions of orogenic belts (Gass, 1968; Temple and Zimmerman, 1969; Moores, 1970; Moores and Vine, 1971; Church and Stevens, 1971; Dewey and Bird, 1971). This chapter takes the generally accepted view that ophiolite complexes originated as oceanic crust and mantle either at the accreting plate margins of oceanic ridges (Moores and Vine, 1971) or in marginal basins behind or between island arcs (Dewey and Bird, 1971), and were subsequently emplaced at convergent plate margins (subduction zones).

There are many problems related to the origin and emplacement of ophiolite complexes, but four are particularly difficult to resolve. First, sedimentary sequences lying upon the igneous rocks of ophiolite sequences vary greatly in character, from cherts and maganiferous lutites to volcaniclastics, and bear a variety of strikingly different relationships to the igneous rocks of the ophiolite suite, from conformable to strongly unconformable. These variable relationships are of great importance in understanding the evolution of ophiolite sequences and will be discussed below. Second, the time between igneous origin (minimum age determined by the age of sediments lying upon the pillow lavas) and tectonic emplacement may vary considerably but is often very short. This means that ophiolite generation by some form of plate accretion often is rapidly followed by either the development of a subduction zone or lateral transport to a subduction zone. Possibly related to a short time interval between ophiolite origin and emplacement are the high temperature garnet-amphibolite "aureoles" beneath the ultramafics in several transported ophiolite nappes. If portions of oceanic lithosphere were tectonically emplaced shortly after generation, they would retain much of the heat developed at their generation site and would form a thermal blanket as the ophiolite nappes moved over subjacent rocks. Several of the obduction mechanisms discussed below may provide partial solutions to this problem. The third problem is that, in many orogenic belts, ophiolites of a particularly narrow time span are preserved. For example, in the Appalachian/Caledonian System, most ophiolite complexes are of Early Ordovician age; in the western part of the Alpine System they are of Late Jurassic/Early Cretaceous age, and in the eastern part of the Alpine System in the Middle East they are of Late Cretaceous age. In both these orogenic belts it is clear from stratigraphic considerations that a much longer oceanic history was involved. Thus, although ophiolite complexex provide key insights to the oceanic origins of many orogenic belts, they do not uniquely dterminne the life spans of the ocenas involved (Dewey *et al.*, 1973). Fourth, there are great variations in the internal sequence of igneous and metamorphic rock types, although the following general sequence, from base to top, applies to all the major well-developed occurrences: dunite, harzburgite, and minor lherzolite, with strong tectonic and metamorphic recrystallization fabrics (these rocks have tectonic relationships with the country rocks); cumulate ultramafic rocks; partly cumulate, mafic, plutonic rocks; sheeted diabase; pillow lava capped by sediments. The complex variations on this basic sequence are probably a function of variation in rates of plate accretion (Dewey and Kidd, 1977) and fracture zone influence (Karson and Dewey, 1978) at oceanic ridges and in marginal basins.

Ophiolite complexes are superbly developed and exposed in the Newfoundland Appalachians. These complexes may serve as a broad but not exclusive template for other, less well-developed, Appalachian/Caledonian ophiolite sequences. The general sequence of units developed in these complexes has been presented above (section 6.2.1). There are particular problems related to the mechanisms by which ophiolite complexes are emplaced in orogenic belts and how these mechanisms may be linked to processes at convergent plate boundaries. It is implicit, in the plate tectonics model, that oceanic crust and mantle are mainly subducted in oceanic trenches. It is, therefore, of great interest to understand the mechanisms by which large slabs of oceanic crust and mantle are detached from the evolving oceanic lithosphere and tectonically incorporated into orogenic belts. Coleman (1971) has coined the term **obduction** to describe this process and to make the distinction with the usual fate (subduction) of oceanic crust and mantle.

Ophiolite complexes vary greatly in size, occur in a wide variety of tectonic positions within orogenic belts, and bear highly variable relationships to sequences against and onto which they are emplaced. They occur commonly as blocks in blueschist mélanges, olistostromes and wildflysch deposits, as highly deformed slices in steep zones of high strain, as well-preserved sequences in faulted synclines, as autochthonous or parautochthonous basement to such sequences as the Great Valley Jurassic assemblage of California, and as giant nappes transported as thin sheets over platform or exogeosynclinal sequences. The last type of occurrence has been studied in great detail in the Bay of Islands Complex, Newfoundland (Stevens, 1970; Church and Stevens, 1971; Williams, 1971), in the Semail Complex in the Oman (Reinhardt, 1969; Glennie *et al.*, 1973) and in the Papuan ophiolites (Davies, 1971). The Bay of Islands Complex and the Semail Complex bear a striking resemblance to one another in tectonic emplacement styles and tectonic/stratigraphic relationships (Figs. 6.3.1a,b). A possibly important difference between the Bay of Islands ophiolite slice and the Semail ophiolite slice is that the former faces in the direction in which the sheet was transported, so that the basal thrust contact cuts across progressively higher units of the ophiolite sequence in that direction. The reverse is true of the Semail ophiolite slice. This may reflect the initial dip direction of the thrusts along which the ophiolite slices became progressively detached from the oceanic lithosphere. Nevertheless, in both western Newfoundland and in the Oman, obducted ophiolite sheets bear very similar tectonic/stratigraphic relationships to underlying rocks and in both cases a probable continental shelf, continental rise, ocean basin paleogeography was progressively telescoped into an identical stacking order.

If ophiolite complexes originate as oceanic crust and mantle generated by seafloor spreading at oceanic ridges or in marginal basins, the tectonic emplacement (obduction) of ophiolite sheets and slices must involve some form of decoupling of oceanic lithosphere prior to emplacement and the expulsion of relatively dense oceanic rocks onto less dense continental rocks. The major problems are the mechanism by which this decoupling takes place, the extent to which the decoupling fractures penetrate the entire lithosphere, and the mechanism and geometry of the tectonic emplacement process (that is, the extent to which compressional vs. gravity-sliding mechanisms predominate). Several writers (Coleman, 1971; Stevens, 1970; Church and Stevens, 1971; Temple

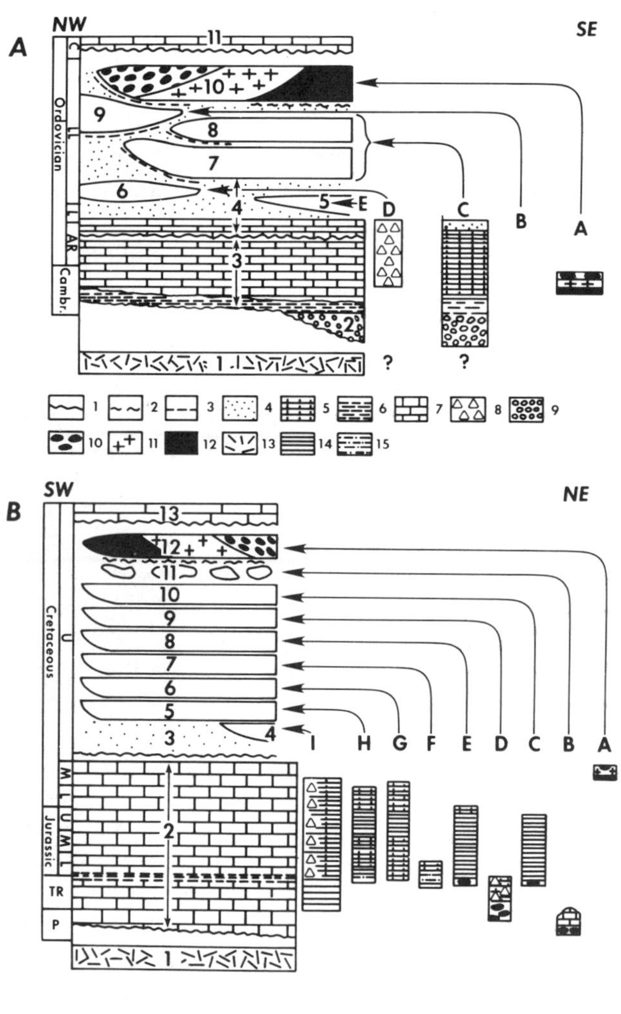

6.3.1 **a**. Schematic illustration of the stratigraphic/structural relationships of the Bay of Islands ophiolite sheet, western Newfoundland (data from Stevens, 1970; Williams, 1971). Key to symbols: large numerals refer to stratigraphic/tectonic units described in the text; 1—Precambrian (Grenville) crystalline basement; 2—Late Precambrian/Early Cambrian arkosic clastics and basalts; 3—Cambrian/Ordovician clastics and carbonates; 4—Ordovician carbonates and ophiolite-bearing flysch; 5—Penguin Cove allochthon; 6—Cow Head allochthon; 7 and 8—Humber Arm allochthon; 9—Coastal Complex allochthon; 10—Bay of Island allochthon; 11—Ordovician carbonates. Large letters refer to provenance and order of movement of the allochthons. Small letters in columns at left refer to stratigraphic age; AR—Arenigian; L—Llanvirnian; LL—Llandeilian; C—Caradocian. Small numerals in legend: 1—unconformity; 2—garnet-amphibolite; 3—ophiolite-bearing mélange; 4—ophiolite bearing flysch; 5—calcareous flysch; 6—shallow marine clastics; 7—shallow marine carbonate; 8—carbonate conglomerate and breccia; 9—red and green arkose, greywacke, shale and basalt; 10—pillow lava; 11—gabbro; 12—ultramafic rocks; 13—crystalline silicic basement rocks; 14—radiolarian chert; 15—deep-water quartz-rich clastics. **b**. Schematic illustration of the stratigraphic/structural relationships of the Semail ophiolite sheet, Oman. Data from Glennie *et al.* (1973). Key to symbols: large numerals refer to stratigraphic/tectonic units described in the text; 1—Precambrian crystalline basement; 2—Permian to Cretaceous carbonates; 3—Cretacrous flysch; 4, 5, 6, 7, 8, 9, 10 and 11—Hawasina allochthons; 12—Semail allochthon; 13—Upper Cretaceous carbonates. Large letters refer to provenance and movement order of the allochthons. Small letters in columns at left refer to stratigraphic age; P—Permian; TR—Triassic.

and Zimmerman, 1969; Dewey and Bird, 1970, 1971; Williams, 1971) have discussed these obduction problems and have offered various kinds of solutions. These solutions, among others, are discussed below.

Upwedging in subduction zones. Upwedging is a process that may occur beneath and behind the inner walls of oceanic trenches (subduction zones), whereby slices of oceanic crust and mantle are ripped from the upper part of the descending plate and then wedged and packed in high-pressure mélanges (glaucophane, lawsonite, jadeite assemblages) against the leading edge of the other plate (Fig. 6.3.2). Seismic first motion studies

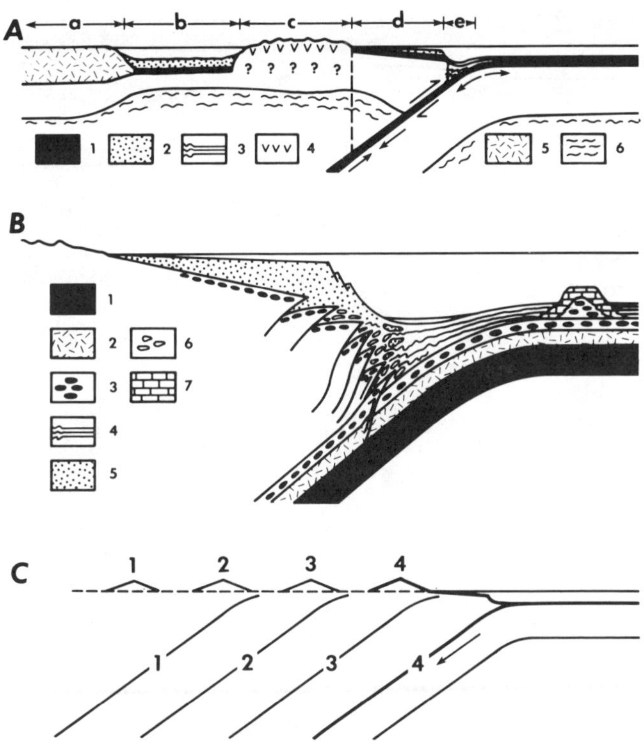

6.3.2 **a**. Schematic section from a continental margin (a), across a rear-arc oceanic basin (b), volcanic arc (c), arc-trench gap (d), trench (subduction zone, e) to an ocean. 1—oceanic crust; 2—volcanogenic sediments; 3—oceanic sediments; 4—silicic/intermediate volcanics; 5—continental crust; 6—low-velocity zone. **b**. Schematic illustration of some possible relationships at a subduction zone. 1—ultramafic rocks (mantle); 2—gabbro (oceanic layer 3); 3—pillow lavas and diabase (oceanic layer 2); 4—argillite and chert (oceanic layer 1); 5—volcanogenic sediments of the arc-trench gap; 6—blueschist and ophiolite mélanges; 7—carbonate fringing a submerged seamount. **c**. Diagrammatic illustration of continental accretion by the successive oceanward movement of the site of subduction. 1, 2, 3—former positions of volcanic arc and Benioff Zones; 4—present position of volcanic arc and Benioff Zone.

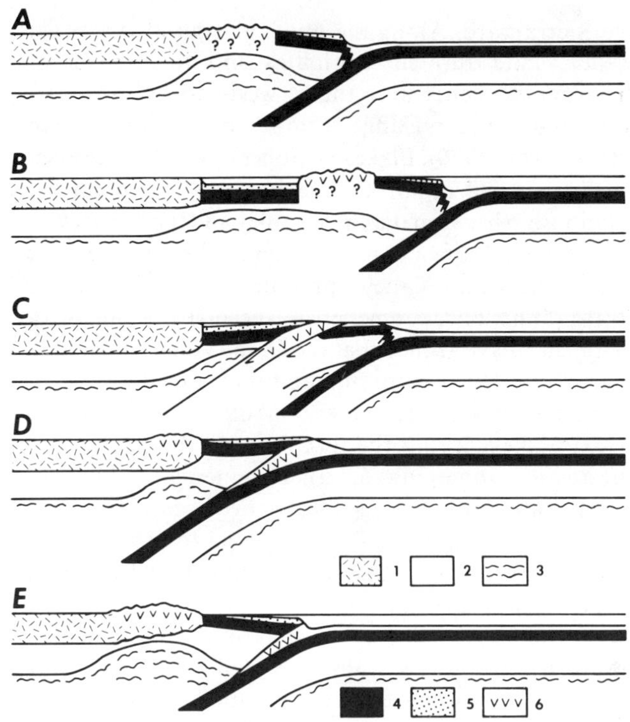

6.3.3 Schematic illustration of Blake and Jones (1973) hypothesis of the obduction of rear-arc basin ophiolites by the subduction of the associated volcanic arc. 1—continental crust; 2—mantle; 3—low-velocity zone; 4—oceanic crust; 5—volcanogenic sediments; 6—silicic/intermediate volcanics.

(Isacks and Molnar, 1969) have shown that the upper part of the descending plate is in tension where it bends to descend, and that thrust fault solutions characterize the region below the inner walls of trenches (Fig. 6.3.2). Weakening and cracking of oceanic crust and upper mantle is likely to occur in the tensional regime, following which it becomes subjected to strong shearing forces in which slabs of ophiolite could become detached and incorporated into mélange sequences. The Mesozoic Franciscan mélange assemblage contains ophiolite slabs of varying sizes believed by some authors (Hamilton, 1969; Dewey and Bird, 1970) to have been obducted in this way. At low structural levels, such ophiolite slices are likely to be encased in tectonic blueschist mélanges; at high levels they may be tectonically overlain by olistostrome and flysch sequences (Fig. 6.3.2).

It is now supposed that many arc-trench gap terranes (Dickinson, 1971) have an ophiolitic basement (Blake and Jones, 1973; Dewey and Bird, 1971). This is clearly the case for at least the outer margin of the Great Valley of California. During the Cretaceous, this terrane lay between the Franciscan subduction zone and the Sierran arc. Along the outer margin of the Great Valley, Late Jurassic ophiolites and their cover of Jurassic sediments were thrust westwards across various flysch and mélange terranes of the Franciscan (Bailey *et al.*, 1970). Blake and Jones (1973) have argued for a new and important mechanism to explain the relationship between the Great Valley and the Franciscan (Fig. 6.3.3). From a consideration of the Late Jurassic age of the Great Valley ophiolite basement, and the stratigraphy and geometry of successive Franciscan belts, they have argued that the Great Valley originated by rear-arc basin spreading and that, during the Cretaceous, the arc was largely subducted beneath the Great Valley during the inception of the Franciscan subduction zone (Fig. 6.3.3). Fragments of this supposed Late Jurassic volcanic arc are preserved in the Franciscan. This hypothesis explains the short time-span between origin and emplacement of the Great Valley ophiolite basement.

Progressive packing of ophiolite slices and arc fragments against the leading edge of a continent may continue over a long period of time and lead to a form of continental accretion (Fig. 6.3.2), perhaps exemplified by the Klamath Mountains of northern California and southern Oregon. Here, a series of ophiolitic thrust plates are stacked against one another and the plates become successively younger westwards. They are intruded by arc-type igneous rocks, which are probably the result of the progressive or intermittent westward-migration of volcanic arcs as the subduction zones migrated westwards, so that new volcanic arcs developed on older obducted ophiolite foundations.

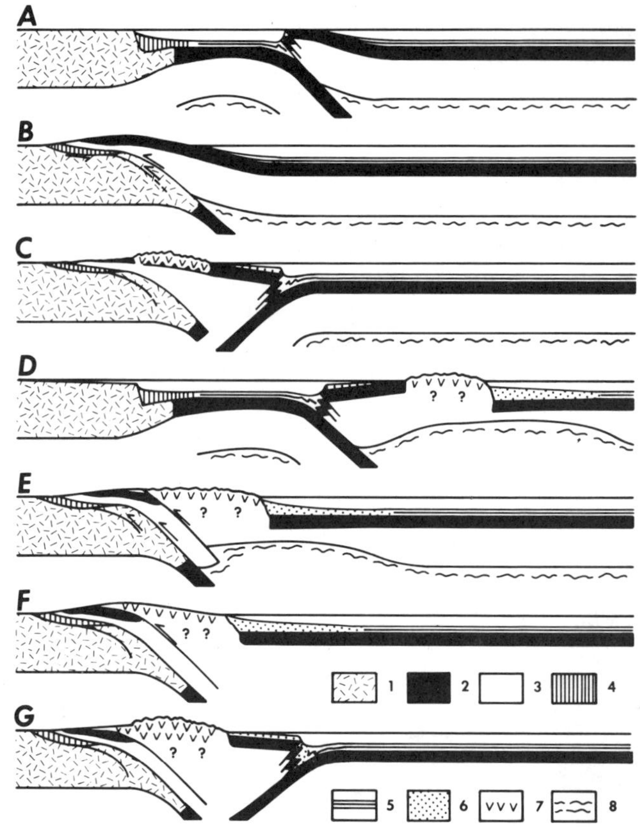

6.3.4 a,b,c. Obduction of an ophiolite sheet by the partial subduction of an Atlantic-type continental margin. **d,e,f,g.** Obduction of an arc-trench gap ophiolite sheet by the collision of an Atlantic-type continental margin with a volcanic arc. 1—continental crust; 2—oceanic crust; 3—mantle; 4—continental rise sediments; 5—oceanic sediments; 6—volcanogenic sediments; 7—silicic/intermediate volcanics; 8—low-velocity zone.

Obduction by compressional telescoping onto Atlantic-type continental margins. The simplest form of this type of obduction may follow from the development of a subduction zone near the continental margin with the geometry shown in Fig. 6.3.4. Above and behind the subduction zone, a welt of oceanic crust and mantle rides up over the descending plate. The ocean, intervening between the continental margin and the subduction zone, is progressively swallowed until the continental margin arrives at the subduction zone and a giant wedge or slice (nappe) of oceanic crust and mantle is pushed across the continental margin (Fig. 6.3.4). McKenzie (1969) has argued that buoyancy of the relatively light continental crust is likely to prohibit its extensive subduction; therefore, a flip in subduction polarity will occur so that the continental margin with its obducted ophiolite sheet will now lie above a descending plate (Fig. 6.3.2). This mechanism should yield a distinctive arrangement of tectonic and stratigraphic units. The stacking order should be similar to that observed in western Newfoundland, with a giant ophiolite sheet overlying oceanic, continental rise, and continental margin facies allochthons. The movement and assembly order should be the reverse of the stacking order. This model also yields the corollary that blueschist, ophiolite-bearing, subduction-mélanges should intervene, at least in the initial stages of nappe assembly, between the high-level ophiolite slice and subjacent continental rise slices. As movement progresses, these blueschist mélanges could be progressively attenuated into thin metamorphic "aureoles" beneath the ophiolite nappe and should indicate a protracted polyphase history of subduction and obduction tectonics. Complex relationships are likely to occur between different kinds of flysch, olistostrome, wildflysch, and tectonic mélanges. Early subduction mechanisms, involving low-level tectonic mélanges and high-level ophiolite-bearing olistostromes in the trench, may progressively feed flysch and olistostrome wedges advancing up the continental rise. These will subsequently be transported as new allochthonous sheets are successively detached from their basement in advance of the moving ophiolite sheet. Finally, the allochthon moves into foreland flysch basins on top of the former continental shelf.

The extent to which the high-level ophiolite sheet carries a cover of pre-obduction arc volcanics will depend upon the history of pre-obduction subduction. If only a short time intervenes between initiation of subduction and final obduction, arc volcanics may be scarce or absent—for example, Papua (Davies, 1971). If, however, a large tract of ocean intervenes between the continental margin and the subduction zone, a fully developed trench, arc-trench gap, arc, and rear-arc basin (Fig. 6.3.4) may eventually arrive at and collide with the continental margin. In this case, an arc-trench gap ophiolite basement, with a cover of arc-trench gap sediments, will be obducted onto the continental margin with a subjacent sequence of blueschist mélanges and continental rise allochthons (Fig. 6.3.4). Further convergence may lead to overthrusting of the volcanic arc assemblage, as a high-level nappe (Fig. 6.3.4). This may be followed by flipping of subduction polarity so that the obducted sequence is now underthrust from the oceanward side (Fig. 6.3.4). This may yield a situation where a new assemblage of arc volcanics and intrusives develops on and invades the obducted complex, and a new arc-trench gap, with a young rear-arc ophiolite basement, develops on the oceanward side.

This general mechanism, although providing an effective way of obducting ophiolite sheets, can yield an exceedingly complex series of relationships between continental rise allochthons, flysch and mélange sequences, volcanic arcs and ophiolites obducted by different mechanisms and originating in different ways. In particular, this mechanism does not yield an *a priori* time relationship between generation of oceanic crust and mantle by seafloor spreading and its subsequent obduction. If an arc were partially subducted beneath the rear-arc basin just prior to, or during, obduction onto the continental margin, the young obducted ophiolite sheet with its rear-arc basin sediment cover (andesitic flysch?) would be separated from the continental rise allochthons by a smeared-out volcanic arc. Such a smeared-out arc could have an older ophiolite basement which has been complexly deformed, metamorphosed, and injected by arc intrusives.

Obduction by gravity sliding onto Atlantic-type continental margins. This is a concept advocated by Reinhardt (1969) for the emplacement of the Semail Complex and argued by Church (1972) and Church and Stevens (1971) for the emplacement of the Bay of Islands sheet in western Newfoundland. It involves the progressive uplift of an actively spreading oceanic ridge, the detachment of slices from the *upper part* of the lithosphere and the subsequent gravity sliding of these slices onto the continental margin (Fig. 6.3.5). This mechanism could explain basal contact aureoles, since the young upper lithosphere would still be hot during emplacement. It would also account for a short time interval between the origin by seafloor spreading and tectonic emplacement. It could further explain why, in the Semail Nappe, the ophiolite sequence faces in the opposite direction (Fig. 6.3.1b) from the nappe transport direction. It seems to the writers, however, that this mechanism does not explain the progressive detachment of underlying nappes. It is also difficult to accept gravity

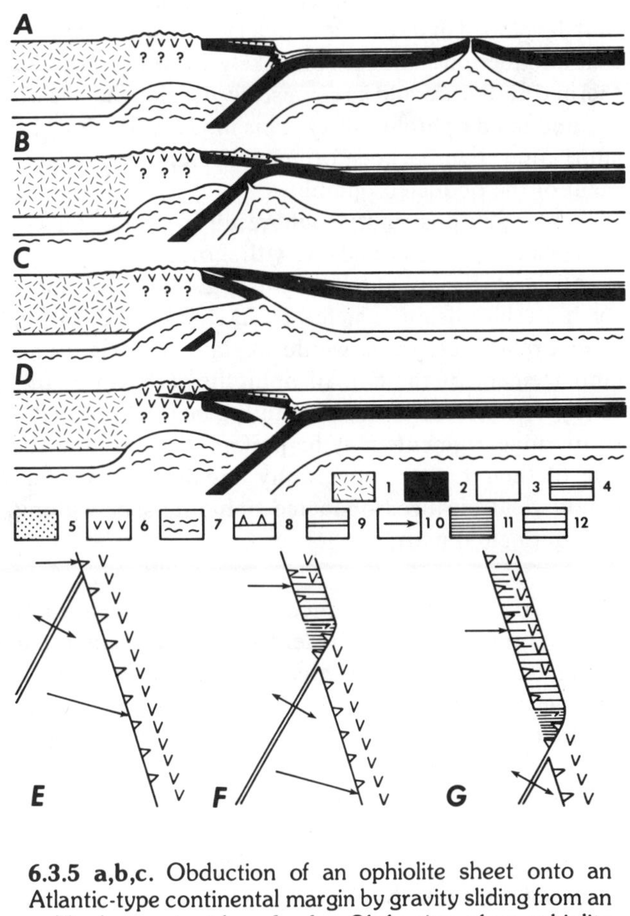

6.3.5 a,b,c. Obduction of an ophiolite sheet onto an Atlantic-type continental margin by gravity sliding from an uplifted oceanic ridge. **d,e,f,g.** Obduction of an ophiolite sheet onto an Atlantic-type continental margin by the transformation of a spreading oceanic ridge to a subduction zone. 1—continental crust; 2—oceanic crust; 3—mantle; 4—continental rise sediments; 5—oceanic sediments; 6—volcanogenic sediments; 7—silicic/intermediate volcanics; 8—low-velocity zone; 9—garnet-amphibolite.

sliding for a distance of 1200 km as claimed by Glennie *et al.* (1973). Last, it is difficult to visualize or account for the pre-gravity-sliding topographic inversion.

Obduction by transformation of a spreading ridge to a subduction zone adjacent to an Atlantic-type continental margin. This model is illustrated in Fig. 6.3.5. The change from a spreading plate boundary to a subduction plate boundary may result from rapid rearrangements of relative plate motion. However, it is more likely that it results from a progressive change in position of the plate boundary (with respect to the instantaneous rotation pole describing the relative motion of the two plates), mandated by the evolution of rigid plates on a sphere (Dewey, 1975). Whatever the causes of the relative change in motion, slices of young, hot, oceanic crust and mantle may be obducted onto the continental shelf. In many respects, a tectonic/stratigraphic arrangement of stacked allochthons similar to that developed in the second model would result, except that subduction mélanges are likely to be absent, and thermal aureoles could be extensively developed beneath the obducted ophiolite sheet. This model would yield a short time span between origin and tectonic emplacement of the obducted ophiolites.

A similar possibility could result from the conversion of a transform boundary, with short ridge segments parallel to the continental margin, to a subduction zone. Such a relationship, which may have occurred during Late Cretaceous times, would explain the origin and emplacement of the Semail ophiolites (Dewey *et al.*, 1973). The oceanic crust and mantle generated along the short ridge segments may be preferentially obducted, due to their higher topography and heat flow. This would yield localized obducted ophiolite sheets along the continental margin.

Obduction by the interference of a spreading ridge and a subduction zone. Consider the situation illustrated in Fig. 6.3.6a–d, where a spreading ridge approaches a subduction zone. Eventually, the ridge impinges on the subduction zone, at which time there will develop a complex interaction of subduction-related tectonic-sedimentary, and spreading-related tectonic-igneous activity. At this stage, one of several things may happen. Following the swallowing of the intervening plate, the decaying ridge may be subducted (Fig. 6.3.6). Alternatively, it is conceivable that a hot ridge flank rides upward, across the trench, to be obducted onto the arc-trench gap and arc terranes as a hot ophiolite slice. This would yield a very short interval between origin and emplacement of the ophiolite sheet, but this model cannot be used to explain obduction onto Atlantic-type continental margins. It seems unlikely that the ridge and trench intersections with the continental margin would exactly parallel the margin. If the intersections were oblique, ophiolite sheets with successively younger sea-floor spreading ages could be progressively emplaced across the arc diachronously (Fig. 6.3.6e–g).

Obduction from rear-arc oceanic basins. It has been suggested (Dewey and Bird, 1971) that a common form of ophiolite obduction is related to the closure of rear-arc marginal basins, and that during such closure, by subduction, slices of oceanic crust and mantle may be expelled onto adjacent continental forelands and emplaced as ophiolite sheets. Packham and Falvey (1971) have suggested that, following the generation of a rear-arc basin by the migration of a volcanic arc away

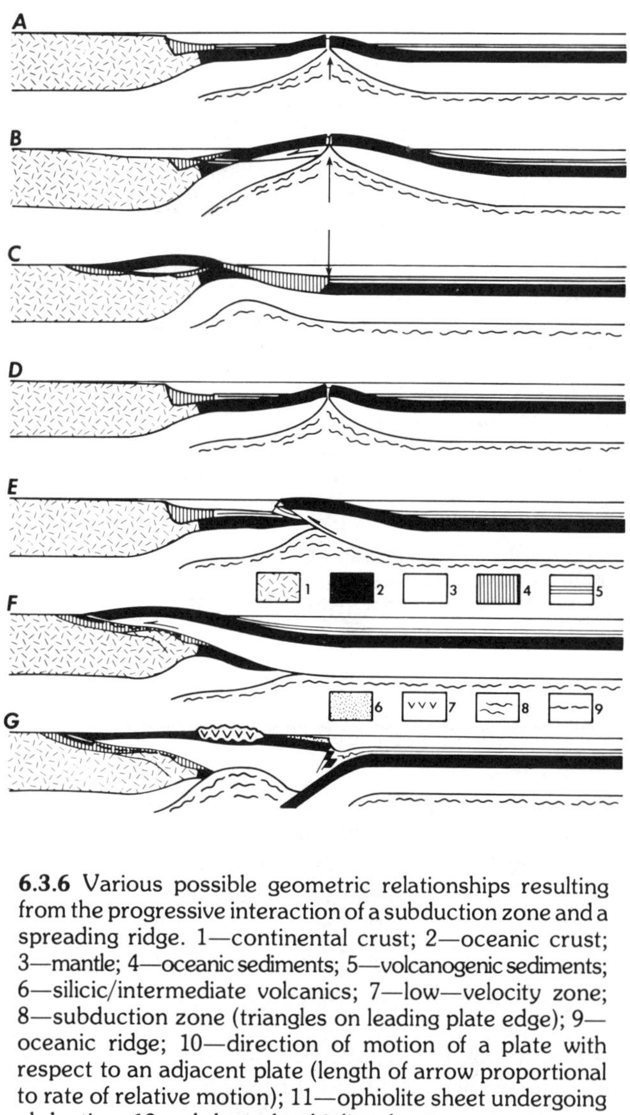

6.3.6 Various possible geometric relationships resulting from the progressive interaction of a subduction zone and a spreading ridge. 1—continental crust; 2—oceanic crust; 3—mantle; 4—oceanic sediments; 5—volcanogenic sediments; 6—silicic/intermediate volcanics; 7—low—velocity zone; 8—subduction zone (triangles on leading plate edge); 9—oceanic ridge; 10—direction of motion of a plate with respect to an adjacent plate (length of arrow proportional to rate of relative motion); 11—ophiolite sheet undergoing obduction; 12—obducted ophiolite sheet.

from its adjacent continental margin (Karig, 1971a), it becomes progressively filled with thick clastic sequences derived from both the continent and the arc. As the sediment pile thickens, anatectic fusion is supposed to occur as a result of thermal blanketing. The marginal basin is supposed to collapse and telescope with remobilized mantle peridotites flowing up as mobile screens and extrusion nappes moving over the continental margin. This model, therefore, involves the shortening, deformation and metamorphism of an oceanic tract without subduction. It is difficult to visualize how, in this model, intact ophiolite nappes become obducted, although the model could explain many serpentinite screens in high grade metamorphic terranes. Dewey and Bird (1971) have suggested that ophiolite sheets may originate from a rising welt in advance of a subduction zone, similar to that of the Mediterranean Ridge (see discussion below).

Another mechanism could account for the obduction of thin, hot sheets of oceanic lithosphere from marginal basins, which would explain most of the more problematical features of the emplacement of, for example, the Bay of Islands ophiolite complex. In the high heat flow region of a volcanic arc and rear-arc basin, the lithosphere is particularly thin (Fig. 6.3.2). The thin oceanic marginal basin lithosphere may preferentially fail along gently dipping thrust surfaces if a compressional stress field is applied to the region. Under these circumstances a thin sheet of marginal basin lithosphere may become detached and begin to ride over adjacent lithosphere, finally to become emplaced as a thin ophiolite sheet on the adjacent continental foreland (Fig. 6.3.7). This is basically a form of plate convergence whereby a thin, initially hot, but progressively cooling sheet of oceanic lithosphere is obducted over progressively cooler and thicker lithosphere.

Initially, the sheet would be attached to its hot mobile infrastructure (low-velocity zone), but the zone of dislocation would gradually become localized along a progressively cooler and thinner zone of high shearing strain. During the early stages of detachment, primitive mantle materials of the low-velocity zone (lherzolite and ariégite?) could be dragged along, and could partially lubricate the movement zone. As they cooled, these rocks would move into cataclastic fields of strain and could be represented by the lherzolites and ariégites at the base of the Bay of Islands ophiolite slice. During the initial phases of movement, the hot sheet could peel off and dynamothermally metamorphose thin slices of the mafic oceanic crust and its sediment cover. As the cooling sheet moves out across the inner margin of the basin onto the continental foreland, exotic metamorphosed (garnet amphibolite?) mafic slivers would be carried along at the base of the sheet and become sandwiched between the ophiolite sheet and continental-rise nappes successively detached from their roots. Because heat for the metamorphic reactions is derived from the overlying hot sheet, inverted prograde metamorphic gradients would characterize these transported mafic slices. During the early prograde metamorphic phase, fluids driven from wet sediments below the moving sheet would lubricate the shearing zone and reduce frictional resistance to movement. Progressive movement and cooling of the ophiolite sheet would lead to progressively semibrittle then to brittle conditions so that, during the late and final stages of emplacement, retrograde metamorphism and cataclastic textures would become superimposed upon earlier prograde ductile textures. Serpentinite

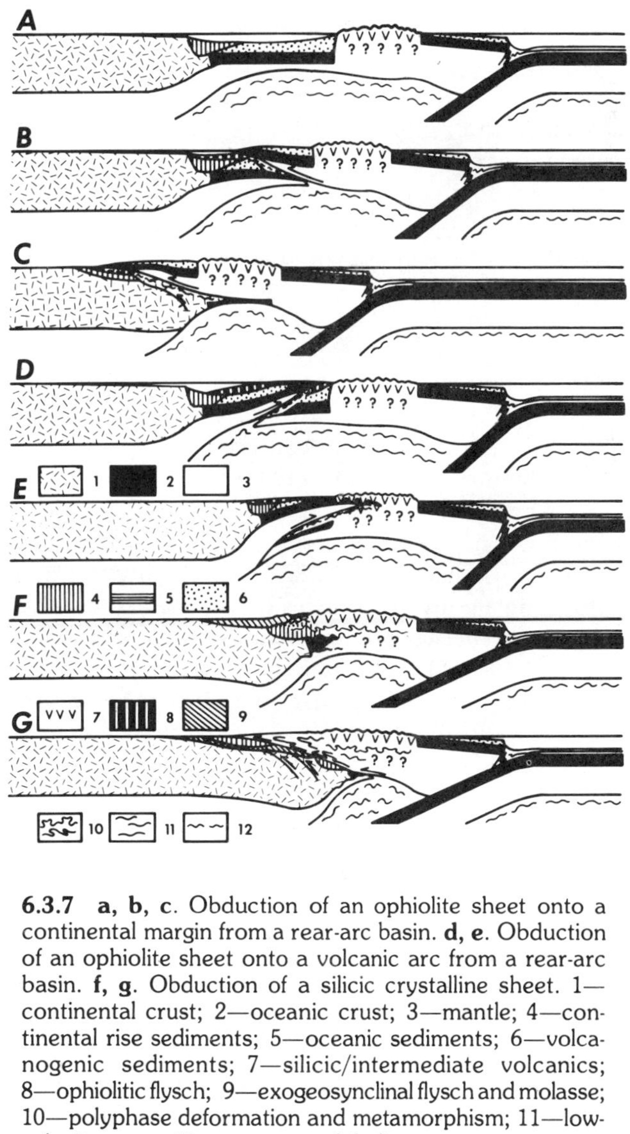

6.3.7 a, b, c. Obduction of an ophiolite sheet onto a continental margin from a rear-arc basin. **d, e**. Obduction of an ophiolite sheet onto a volcanic arc from a rear-arc basin. **f, g**. Obduction of a silicic crystalline sheet. 1—continental crust; 2—oceanic crust; 3—mantle; 4—continental rise sediments; 5—oceanic sediments; 6—volcanogenic sediments; 7—silicic/intermediate volcanics; 8—ophiolitic flysch; 9—exogeosynclinal flysch and molasse; 10—polyphase deformation and metamorphism; 11—low-velocity zone; 12—garnet amphibolite.

mylonites would be formed along the base of the ophiolite sheet and rodingite mineral assemblages would be developed along the contact zone in the underlying metamorphic mafic slivers.

There could be a whole range of transitions from rear-arc basin ophiolite obduction to the backthrusting of continental rocks over continental lithosphere depending upon the width, or absence, of rear-arc basins. Where rear-arc basins are absent, sheets of silicic crystalline rock could be thrust back across the foreland (Fig. 6.3.7). In addition, there is no *a priori* reason why ophiolite sheets should not be driven from the marginal basin onto the associated volcanic arc (Fig. 6.3.7). This model has the particular advantage of accounting for a short time interval between the generation of oceanic crust and mantle by spreading and its emplacement by obduction.

Obduction by continental collision. The various ocean basins of the Mediterranean region are believed by some authors (Smith, 1971; Dewey *et al.*, 1973) to be remnants of oceans generated at different times by seafloor spreading. The Alpine Orogenic Belt, in addition to these remnant oceans, is believed to register a complex history of plate interactions during the general convergence of the Eurasian and African Plates. This process of convergence is continuing at the present day. Several features of the Mediterranean may provide clues to the processes leading to ophiolite obduction during the final stages of ocean closure that eventually lead to continent/continent collision.

Rabinowitz and Ryan (1970) have argued, from gravity and seismic-profiling data, that the Mediterranean Ridge has a core of upthrust basement wedges of oceanic crust and mantle, which are wedging into and deforming the overlying oceanic sediment pile. This ridge lies between the Ionian-Cretan trench, where subduction of the Levantine Ocean is occurring, and the advancing stable north African continental margin. Dewey and Bird (1971) have suggested that continued convergence in this tectonic situation could lead to the oceanic basement wedges being driven (obducted) onto the continental margin. A possible sequence of events illustrating this obduction model is shown in Fig. 6.3.8. As the rising wedges of oceanic crust and mantle rise and are caught in the jaws of the continent/continent vise, they detach and begin to move up the advancing continental rise, detaching and transporting underlying allochthonous sheets (Fig. 6.3.8). Such an ophiolite sheet is likely to be detached, transported and emplaced as a relatively cold nappe so that there is less potential for such a sheet to develop subjacent thermal metamorphism. Continued convergence may lead to the overthrusting of the arc-trench gap (perhaps as a higher ophiolite sheet) and, eventually, to overthrusting of the metamorphic plutonic and volcanic rocks of the volcanic arc (Fig. 6.3.8). This sequence would generate the following stacking order from base to top: continental rise allochthons lying on exogeosynclinal flysch and wildflysch mélanges; a lower ophiolite sheet; attenuated blueschist mélanges; a higher ophiolite sheet; and a high-level crystalline sheet of arc metamorphics, plutonics, and volcanics.

Following total subduction of oceanic tracts, and perhaps a minor amount of continental subduction, continuing convergence may lead to a further sequence of intracontinental mechanisms of crustal shortening. Initially, crystalline basement nappes may break loose in a progressive sequence away from the main suture zone (Figs. 6.3.8–9). During the early parts of this phase the ophiolites are expelled, in a downward-narrowing

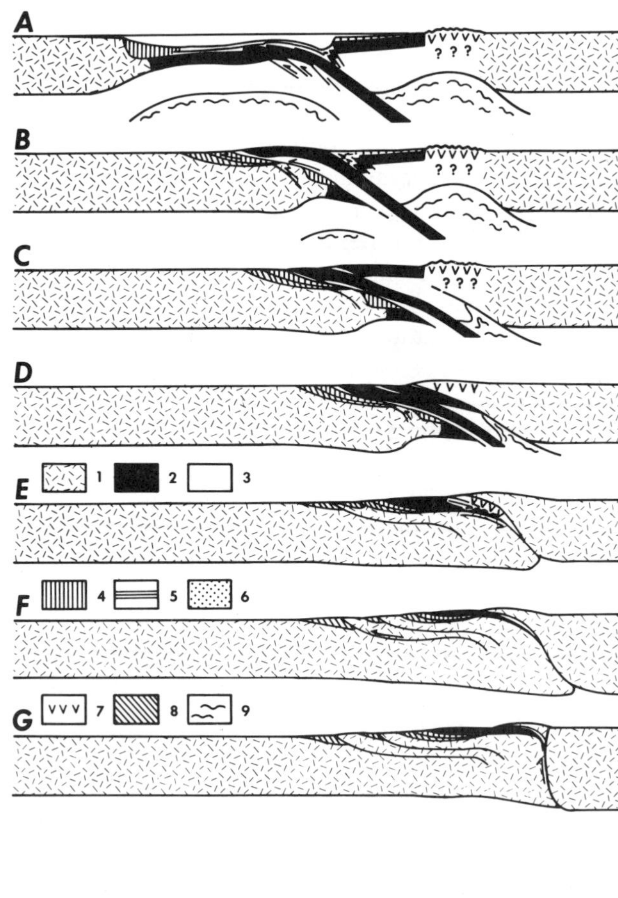

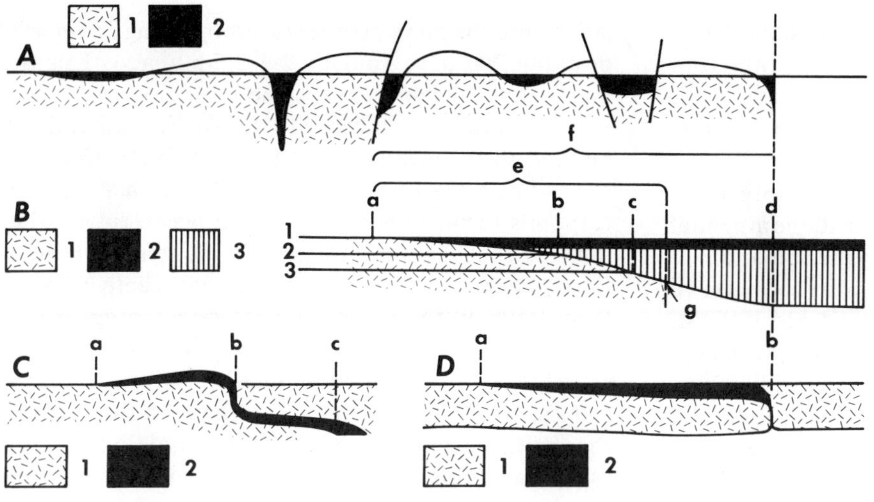

6.3.9 **a**. Preservation of ophiolite occurrences, which once formed parts of a continuous obducted sheet, by post-obduction folding and faulting. 1—continental crust; 2—ophiolites. **b**. Schematic illustration of the rooting of a continuous obducted ophiolite sheet. 1—continental crust; 2—pillow lava, diabase and gabbro; 3—ultramafic rocks. **c**. Large scale monocline in obducted ophiolite sheet. 1—continental crust; 2—ophiolites. **d**. Obducted ophiolite sheet. 1—continental crust; 2—ophiolites.

envelope, from the main suture zone. At deep levels, the suture may be steeply dipping, cryptic and reduced to a narrow zone of high strain (Dewey and Burke, 1973). At high levels, away from the suture, the ophiolites may be the highest unit of the stacking sequence but will be overlain by a crystalline nappe unit where the suture zone flattens above its cryptic root. Continued continental convergence beyond this phase may lead to a stage where the whole continental crust begins to thicken so that potassic/silicic partial melts originate in the lower crust (Dewey and Burke, 1973). These potassic/silicic liquids rise to form late-stage potash-granite intrusions and dacitic/rhyolitic ignimbrite calderas. Finally, isostatic readjustment may lead to high-angle back-thrusting (*retrocharriage*) of the nappe units (Fig. 6.3.9).

No single model is likely to account for the emplacement of all obducted ophiolite sheets. Ophiolite obduction is probably a process involving several—if not many—mechanisms. These may include some of the models discussed above. Which, if any, of these models was responsible for the emplacement of particular ophiolite occurrences can only be answered by detailed field studies of their structural and stratigraphic relationships. In any particular orogenic belt, great complexities may arise from different basic obduction templates being assembled in a variety of temporal sequences. Great geometric complexities may arise as a result of irregularities in the colliding margins of continents, and the diachronous evolution of plate boundaries and triple junctions (Dewey, 1975). Projections in approaching continental margins will collide first and result in intense suturing; salients may never collide and may preserve oceanic remnants buried beneath thick sedimentary sequences (Dewey and Burke, 1973). If orogenic belts result from "cycles" of ocean opening and closing, large parts of the geologic record will be destroyed. This includes not only the obvious subduction of oceanic lithosphere but also more subtle possibilities such as the subduction of volcanic arcs, as suggested by Blake and Jones (1973).

We are, therefore, faced with the problem of reconstructing complex events from an incomplete record. Many of these lost relationships may be critical to a full understanding of orogenic evolution. Although plate tectonics has in many ways increased our understanding of orogenic evolution, and thereby invigorated studies of orogenic belts, it has given us fresh insights into the kinds of questions to which we may never know the answers, and made us aware of the bewildering array of complexities inherent in orogenic evolution. For example, there has been discussion (Cady, 1972) on whether the Appalachian/Caledonian orogenic belt evolved dominantly on a sialic or an oceanic basement. The present distribution of sialic basement rocks, according to the plate tectonics model, is a poor guide to the nature of the basement during the total evolution of the orogen, because ocean closing must involve subduction and, therefore, the destruction of oceanic basement and the preservation of continental basement. Furthermore, the age of origin and emplacement of ophiolite complexes may be the worst possible guide to the total oceanic history involved in the evolution of an orogenic belt. It cannot be supposed that the age of the oldest ophiolite complex in an orogenic belt designates the oldest possible age of oceanic tracts involved in the history of that belt, and that the age of ophiolite obduction always marks the total destruction of those oceanic tracts. The possibility of rear-arc and interarc oceanic basins must be considered. Ophiolites generated in such basins will be considerably younger than the oldest oceans involved in the evolution of the orogen, and the time between origin and emplacement may be very short. The oldest oceanic tracts may be completely destroyed and their existence may be argued only from other facies (Bird and Dewey, 1970; Dewey and Bird, 1971) and geometric considerations (Smith, 1971; Dewey *et al.*, 1973).

The writers suggest that rear-arc basin spreading, with the volcanic arc moving away from the continental margin, followed shortly thereafter by the obduction of marginal basin ophiolites, appears to offer solutions to the following three particular ophiolite problems. First, as mentioned above, ophiolite origin and emplacement are commonly separated by only a short time span. Second, as Sclater *et al.* (1971) have shown, there is a clear relationship between age (and heat flow) and elevation of the ocean floor. Obduction of young, high, hot, thin, oceanic lithosphere would be mechanically easier than the detachment and overthrusting of older, low, cool, thick, oceanic crust and mantle. Third, the common direct association of ophiolites with rocks having volcanic arc affinities (andesitic flysch, andesites, tonalites, granodiorites) favors a rear-arc basin site for their origin.

Changes in basalt production through time

One important question is whether (and if so by how much) the amount of volcanism on Earth has fluctuated with time (see section 7.4.1). This question is fraught with difficulty. Generally, smaller proportions of older rocks are preserved so that the occurrence of volcanic rocks per unit time could be expected to be less; however, long ago, heat generation was greater so that volcanism may have been greater and these effects may partly cancel. Oceanic rocks are destroyed by subduction and obduction so that the record is necessarily incomplete and, generally, only arc and continental rocks are preserved. In spite of these difficulties, some workers have discerned episodicity in igneous activity, especially in the Precambrian. The crude technique of plotting histograms of numbers of isotopic ages, which was popular twenty years ago, has fallen into disfavor with the recognition of the complexity of the record (see Chapter 7). A particularly significant observation is that the continental areas representative of a particular time interval become comparatively much smaller for older rocks. For example, the Superior Province, which is the largest single area of Archean rocks on Earth (about half the world's total area of Archean not involved in orogenic reworking), amounts to only 1% of the present continental area. Several authors have taken a more sophisticated approach to episodicity by considering the isotopic compositions of Nd, Sr and Pb (e.g., McCulloch and Wasserberg, 1978). They have inferred that many Precambrian continental rocks became isolated from the main mantle reservoir during an interval of about 200 m.y., roughly 2.7–2.5 b.y. ago. This is a most interesting and unexpected result.

Turcotte and Burke (1978) applied an indirect method to the estimation of volcanic fluctuation with time during the Phanerozoic. Realizing that sea level responds to the volume of the mid-ocean ridges, they inferred that the times when the continents were most flooded were times when ridges were most active. Calculating, crudely, the proportions of heat escaping through conduction and ocean floor aging, they were able to estimate that nearly twice as much heat was escaping the Earth through the cooling of aging ocean floor during the late Cretaceous episode of continental flooding as is escaping by this means at present. In order to keep the Earth at roughly the same volume, these authors inferred that plate consumption was also at a peak during the late Cretaceous—a result consistent with the familiar high concentration of circum-Pacific batholithic emplacements during the Late Cretaceous and with the Late Cretaceous concentration of emplacements of Tethyan ophiolites.

Tectonics of basaltic eruptions at converging margins in the past

The igneous record of old convergent boundaries lies mainly in the intrusive bodies revealed by erosion in the cores of ancient mountain belts. These bodies, often batholithic in scale, have predominantly calc-alkaline compositional trends. For this reason, as well as reasons based upon structural considerations, they are commonly interpreted as representing the roots of volcanic arcs. Basalts, commonly as basaltic dikes, occur in small quantities in association with the calc-alkaline rocks of mountain-belt cores, but because of intense tectonism it is rarely possible to assign an individual occurrence to an unambiguous structural environment. Thus, basalts associated with extension (for example, in a pull-apart basin) in an overall convergent regime are hard to separate from basalts erupted from arc-volcanoes where both have been involved in subsequent deformation. Because of this kind of complexity, basaltic rocks representative of this environment have been studied relatively little, as yet.

Plate convergence eventually leads to continental collisions. These, in recent examples (Fig. 6.2.10), have been shown to give rise to dominantly calc-alkaline volcanics very similar to those found in Andean arcs (see section 6.2.1; Burke *et al.*, 1974; Kidd, 1975; Şengör and Kidd, 1979). Older examples of such calc-alkaline, K-rich volcanics and/or their subjacent post-kinematic granites are well known in collisional orogens up to about 2 b.y. old. Examples include the granites and other plutons of the Appalachian-Caledonian belt (Dewey and Kidd, 1974), the post-kinematic granites and local rhyolites of the Pan-African, the rhyolites of the St. Francois Mountains (Bickford and Mose, 1975) related to the collision-induced Elsonian orogeny, and the volcanics of the Bear Province (Hoffman, 1980) related to collision about 1.8 b.y. ago at the end of the Wilson cycle on the site of the Coronation orogen. Pieces of continental crust older than 2 b.y. are either too deeply uplifted and eroded, or are too small, or are too poorly known to allow us to distinguish collisional from arc-related magmatic products with any confidence; however, we see no reason to suppose that they are absent. As with the more recent examples (described in section 6.2.1), these magmas are conspicuously characterized by the dominance of intermediate and silicic rocks. Basalts are very subordinate and may even be absent in many areas.

Tectonics of basaltic eruptions at transform margins in the past

Oceans older than 200 m.y. are not preserved, so volcanic activity associated with old oceanic transforms is unknown except as fragmentary pieces in obducted ophiolite complexes (Karson and Dewey, 1978). Evidence of past activity has to be sought within the continents. Perhaps because of the small volume of transform-related volcanics and their susceptibility to later tectonic disruption, no well-documented examples are known from older, inactive transform zones. For example, well-studied pull-apart basins formed on an extensive, large-displacement Carboniferous strike-slip fault zone (a probable old transform system) through the Canadian Maritime Province (Belt, 1969) contain only very minor examples of contemporaneous volcanics. The exposed portion of the possibly correlative (Wilson, 1962) Great Glen Fault system does not show any obvious pull-apart structures. A much older example (~1.8 b.y.) of a preserved piece of a large-displacement strike-slip fault, the McDonald Fault (near the Great Slave Lake), also has no known volcanics associated with its movement, despite having very thick accumulations of coarse clastic sediments (P. F. Hoffman, pers. comm., 1977) that probably were preserved by pull-apart tectonics.

6.3.2 Tectonics of non-plate margin basalt in the past

Tectonics of basaltic eruption at hot spots in the past

Ophiolites, representing old ocean floor, locally contain alkali basalts, some of which may represent hot spot volcanic rocks carried into convergent plate boundary zones and later involved in obduction. Thrust faults are widespread in obducted terrains so that the alkali-basaltic assemblages in ophiolites are generally tectonically isolated. Middle Mountain, among the Franciscan mélanges of California (Berkland, 1972), constitutes an obvious example of what may be an alkali-basalt seamount. In general, the main tectonic significance of alkali-basalt occurrences in ophiolites is that they indicate that the ancient ocean floor, like the modern ocean floor, was studded with hot spot volcanoes. However, no alkaline and tholeiitic rock association that is large enough to represent an ancient Hawaii or an ancient Iceland has yet been recognized in mountain belts.

Just as there are active hot spot volcanoes within continents today, as best seen in Africa (e.g., Thiessen *et al.*, 1979), so there have been similar rocks in continents throughout the latter half of Earth history. Some ancient hot spots, like those of today, were associated with rift systems but others were isolated. It is difficult to assign exact tectonic significance to ancient isolated continental hot spot occurrences. In some cases, an ancient hot spot may have been contemporaneous with rifting. For example, North America was the site of active rift formation about 1200 m.y. ago and numerous isolated intracontinental igneous rocks of about the same age may be analogues of the isolated hot spots (e.g., Cameroun, Ahaggar) of Africa today. Burke and Wilson (1972) have associated the unusual concentration of hot spots in Africa with its being at rest over the mantle convective circulation. It is intriguing to speculate as to whether other outbreaks of intracontinental igneous activity remote from contemporaneous plate margins may also have been associated with a continental standstill.

The most complete records of intracontinental hot spot igneous activity are found in the Canadian Shield and in Africa, where large areas of old Precambrian rock are exposed. These areas display a record of hot spot igneous activity over a longer time interval than do continents with more extensive Phanerozoic cover. Two main kinds of intracontinental hot spot igneous activity can be distinguished in Africa and the Canadian Shield: those in which mantle-derived products such as kimberlites, basalts and carbonatites occur alone, and those in which mantle-derived products are associated with products of partial melting of the overlying continental crust. In some cases the distinction is clear; for example, the Jurassic granites of Nigeria (Jacobson *et al.*, 1958) are an excellent example of the latter type of igneous activity's being dominated by silicic material, which from its isotopic and trace element composition can clearly be seen to have been derived from continental melting. In other cases, such as the anorthosite terrain of Labrador, there is no predominance of continental crust-derived material, but it is not clear whether all of the hot spot igneous rocks in this province are mantle-derived.

Rifts in the past

In section 6.2.2 on active rifting, examples of ancient rifts were given to illustrate the main argument of the section: "rifts (elongate depressions overlying places where the lithosphere has ruptured in extension) occur in a great many different tectonic environments." Characterization of past tectonic environments is obviously more difficult than characterization of active regimes, and becomes progressively more difficult farther back in the past because the geologic record is less complete. A second characteristic feature of rifts emphasized in section 6.2.2 is that, whatever the environments in which they originate, rifts evolve to show very similar structures, stratigraphy and igneous activity. The continents contain very numerous ancient rifts (Burke *et al.*, 1978), but it is not easy to determine in all of these localities the type of tectonic regime in which they originated. For example, it is difficult, because of lack of accuracy of age determinations, to tell whether the Keweenawan system and its associated rifts formed in a continental rupture event of African type, as Sawkins (1976) suggested, or in the later collisional event that is commonly called the Grenville Orogeny. Somewhat similar rifts of lesser age are associated with either environment. One exciting possibility for future determinations is that features of rift-related basaltic volcanism may help in distinguishing the particular tectonic environment in which the rift formed. At present this has not been achieved. Basaltic and other volcanism in ancient rifts is varied in quantity, style and composition, but no correlations between volcanism and locality have been established. As in modern rifts, intense volcanism occurs in localized areas within the rift, rather than throughout the rift.

Flood basalts in the latter half of Earth history

Well-preserved remnants of older flood basalt and/or sill complexes are seen back to 2.15 b.y. ago. Most remnants are found within old rifts and aulacogens, because the accidents of erosion through geological time have claimed any more extensive basalts lying outside them. The most prominent flood basalt remnants of Paleozoic or older age lying outside old rifts and of any significant extent are the 600-m.y.-old Antrim basalts of Northwest Australia, the 1100-m.y.-old Keweenawan lavas and sills north of Lake Superior, and the 1900-m.y.-old Kimberly Plateau basalts and sills, also in Northwest Australia. Examples of flood basalts are more common within old rifts. Prominent ones in North America include the rifts of Keweenawan age (about 1100 m.y.), which include the Coppermine River (Northwest Territories), Seal Lake (Labrador), Gardar (Southern Greenland), Keweenawan and its extension in the Mid-Continent rift and gravity high, and various aulacogens along the Cordilleran margin of the North American craton, including one exposed in part in the Grand Canyon. Another prominent rifting episode is that shown by the extensive tholeiitic sill complex that invades the initial rifting facies clastics and volcanics (arkoses, etc., of the Seward and part of the Attikamagen Formations) preserved in the fold and thrust belt of the Labrador Trough (Baragar, 1967), whose age is about 2.15 b.y.

Associated with this 2.15 b.y. (Fahrig and Wanless, 1963) rifting and ocean-opening event is an extensive dike swarm (Stevenson, 1968) that runs (in Archean basement) subparallel with the northern end of the Labrador Trough and then turns to run obliquely along the Cape Smith belt, which is the continuation of the Labrador Trough in northern Ungava. This dike swarm is the last relic of a flood basalt event; a very similar, although smaller, example is seen close to and subparallel with the western margin of the Appalachians in northern Newfoundland (Williams, 1967), where it (and nearby small relics of flood basalts) was associated (Bird and Dewey, 1970) with the early Cambrian or slightly older opening of the ocean which Tuzo Wilson called the Proto-Atlantic (1966), now termed the Appalachian ocean or Iapetus.

Dike swarms as relics of flood basalt and rifting events are common. The MacKenzie dikes, which extend a huge distance (2500 km) north-northwest across the western part of the Canadian shield are of the same age as the Keweenawan rifting and flood basalt event. There are no well-preserved segments of similar continental rifted margins which are dated prior to 2.15 b.y. ago—the opening age of the Labrador Trough and Coronation oceans. However, there are examples of extensive tholeiitic dike swarms and we infer from these that rifting events, like those recorded in younger rocks, took place. The Ameralik dikes, which are found in West Greenland (McGregor, 1973) and are at least 3.0 b.y. and perhaps as much as 3.6 b.y. old (Pankhurst *et al.*, 1973), are evidence of the oldest rifting event directly recorded on Earth.

6.4 TECTONICS OF BASALTIC VOLCANISM ON EARTH IN THE FIRST HALF OF EARTH HISTORY

6.4.1 Tectonics of Archean basaltic volcanism

Jakeš (1973), among others, has shown that the average composition of continental crust resembles that of island arcs. He suggested that continents have been made by a two-stage fractionation of mantle material, the first stage producing ocean floor igneous rocks at ridges, and the second stage producing island arc igneous rocks above sinking, inclined slabs of oceanic lithosphere. Continents form from arcs by collision following horizontal motion. In the first stage, partial melting of mantle material to produce basalt and leave depleted pyrolite involves only vertical movement; however, once these rocks have been differentiated and emplaced, horizontal motion comes into operation as the new lithosphere is carried away from the spreading ridge. Cooling and thickening of aging lithosphere in this process is important in the convective dissipation of heat generated within the Earth. Because the thickness and elevation of cooling lithosphere change as the square root of the lithosphere's age (Parker and Oldenburg, 1973), more heat is dissipated by fast-spreading ridges than by slow-spreading ridges.

Figure 6.4.1 shows that much more heat was being generated by radioactive decay in the Earth 2.5 b.y. ago than now (see section 9.5.1). Many authors (for example, Goodwin, 1973) have emphasized the close resemblance of the rocks formed in the Archean and preserved in greenstone belts to those formed at ridges and above sinking slabs today. It is evident (Jakeš, 1973) from this resemblance that the two-stage fractionation of mantle material at ridges and above descending slabs was operating then much as today. Because two or three times as much heat was being generated, and because there is no evidence of the operation of any heat-dissipating process peculiar to ancient times, we infer that ridge processes were operating two or three times as fast as today. This greater intensity of operation could have been achieved either by greater average spreading rates or by greater total ridge length, or by both. Assuming that the volume of the Earth was about the same as now, there would have been either a greater length of subduction zone to consume the extra ocean floor or subduction would have worked more rapidly. Since there is some evidence that there is a general average for subduction rates at about 6 cm/yr (with respect to one of the slowly moving reference frames), a substantially greater length of subduction zone seems more likely.

The picture that emerges is that of an ocean in Archean time with a great length of fast-spreading ridge and a large number of long trenches, the two processes producing, at a great rate, ocean floor and arc rocks very like those of today. There seems little likelihood that oceanic thermal gradients were much greater in those times. Partial melting of pyrolite to make basalt presumably took place at temperatures and pressures similar to those of today. The greater amount of heat coming from within the Earth was dissipated mainly by the faster convective removal of basalt and depleted pyrolite at the faster-spreading ridges. Ultramafic lavas appear to have been rather more common in the Archean than in later rocks and their occurrence perhaps indicates local melting of the ancient pyrolite at higher temperatures. Since even in the Archean the proportion of these ultramafic lavas within basaltic piles was low, it seems unlikely that thermal gradients at ridges were much steeper than now. We suggest, as did Moores (1973), that Archean oceanic crust is likely to have been somewhat thicker than present oceanic crust, both because the mantle was then likely to have been a little less depleted, and because a slightly higher percentage of partial melting occurred due to slightly higher thermal gradients. In contrast, the Archean oceanic lithosphere (and continental lithosphere) is likely to have been (on average) somewhat, but not very significantly, thinner than at present because of somewhat higher conductive thermal gradients in the lithosphere.

Because all oceans soon "self-destruct," it is impossible to make direct observation of extensive areas of old oceanic material and inferences about ancient ocean can only be made indirectly, or from the dismembered slivers of oceanic and arc rocks to be found in greenstone belts. In contrast, continents are "non-geodegradable" and once made, they persist, although they are subject to episodic internal fractionation (Dewey and Burke, 1973) due to continent-continent collision.

Hargraves (1976) presented a non-uniformitarian model of an Archean Earth with a continuous sialic cover about 8 km thick overlain by 2 km of water. Mantle beneath the sial was suggested to convect and partially melt to yield basalt that broke the sial and erupted to form greenstone belts. This model is quite unrealistic. Greenstone belts are not piles of undeformed deep-water basalt; the rocks they contain and the structures that affect them so closely resemble those formed later in Earth history that there is no reason for invoking tectonic environments peculiar to early times.

We also reject the idea that greenstone belts record early impact history as Dietz (1965) suggested and as others have repeated. This is because the high impact flux in the solar system declined before the oldest preserved terrestrial rocks were formed and because the dynamic processes at the Earth's surface have been so intense as to destroy any possible record of the high impact flux. The condition of the Canadian shield as a recorder of late impacts shows how trivial these effects are when compared with terrestrial tectonic processes.

The 1850 m.y.-old Sudbury structure (lopolith) is the only well-documented terrestrial example of infilling of a large meteorite crater by basaltic magma where, from the coincidence in age, the impact probably triggered internally-fed magmatism (Dietz, 1964; French, 1970). The Bushveld Complex (about 2000 m.y. old) may be another, perhaps due to simultaneous impact by several large bodies (Dietz, 1963b; Rhodes, 1975). No other comparable examples are known of these rather distinctive, layered noritic rocks in clear association with indications of impact and impact melts. Most well-documented sites of large impacts, for example the 214 m.y.-old Manicouagan structure (Phinney *et al.*, 1978), show no evidence of associated basaltic magmatism but only of directly impact-generated non-basaltic melts and breccias. The geologic record thus suggests that impact-triggered basaltic magmatism has been a minor process on the Earth since heavy impact bombardment ceased about 3.9 b.y. ago.

Archean continental geothermal gradients

Archean geothermal gradients have often been supposed to have been much steeper than those of today (Fyfe, 1973; Baer, 1977; Strong and Stevens, 1974; Windley, 1977; Lubimova, 1969). Because heat generation from the decay of radioactive nuclides in the Earth was then much greater (Lambert, 1976; Lee, 1967; see Chapter 9), extra heat, if it escaped by conduction, must have been carried along steeper thermal gradients. Measurements of terrestrial conductive thermal gradients are very difficult to obtain, and estimates of thermal gradients in the past are even harder to make. Estimates made from igneous and metamorphic rock occurrences are of limited value, both because igneous rocks occur at levels in the Earth shallower by unknown amounts than those at which they form, and because nearly all igneous and metamorphic rocks are formed at plate boundaries where thermal gradients are much steeper than they are within plates. In plate margin areas, advective and convective processes dominate, so the normal conductive gradient of the lithosphere can only be measured or estimated in areas away from plate boundaries. The rocks of the Superior Province (formed roughly in the interval 3100–2500 m.y. ago) were not subjected to steep regional geothermal gradients after their assembly into a continent by lateral accretion of island arcs (Burke *et al.*, 1976), and hence the extra Archean heat escaped the Earth in some way other than by conduction through the continent. It has been suggested that this escape was mainly by cooling of the ocean floor as a boundary conduction layer (Burke and Kidd, 1978).

The Superior Province (Goodwin, 1972) consists of greenstone belts, the rocks of which have been compared to rocks formed in various island arc and marginal basin environments, and of intervening gneiss belts that have been compared to more deeply eroded arcs or microcontinents (Burke *et al.*, 1976). Although the gneiss belts show great structural complexity and may include some high-grade metasediments, granodioritic and tonalitic compositions greatly predominate. The gneissic terrains can be regarded as having a granodiorite-tonalite average composition (Goodwin, 1972). Wise (1974a), by considering the control on continental thickness exercised by the depth of water in the ocean, argued that continents 2500 m.y. ago were about as thick as they are now. Seismic refraction measurements (Hall, 1971) in the Superior Province indicate a depth to the base of the crust of about 35 km. Rocks of granodioritic composition subjected to pressures appropriate to a depth of 35 km (about 10 kb) begin to melt at a temperature close to 700° C, leaving a dry and refractory residuum.

The products of melting, which would tend to rise buoyantly in the crust and thus become exposed by erosion at the surface, typically are granites of minimum-melting composition (roughly equal amounts of quartz, albite and potash feldspar). Such granites are abundant in many crystalline terrains (for example, the Grenville, Pan-African and Churchill provinces) and have been compared to the acid volcanics of Tibet produced by partial melting of thickened continent (Dewey and Burke, 1973). Minimum-melting granites are extremely rare in the Superior Province. We interpret this scarcity as an indication that the temperature 35 km down at the base of Superior Province crust in Archean time did not generally exceed 800° C. We reject the idea that minimum-melting granites may have occurred at higher levels now eroded from above the Superior Province, because the preservation of low-grade assemblages in the greenstone belts of the province precludes deep burial and because of the limited occurrence of granulitic rocks in the province. As the temperature at the base of the continental crust today is commonly estimated (Clark and Ringwood, 1964) to be about 500° C, we conclude that in Archean times a typical continental average geothermal gradient was less than 23° C/km compared with 17° C/km today (Fyfe, 1973). Because heat generation rates (see Chapter 9) in the Earth between 2500 m.y. and 3100 m.y. ago may have been as much as three times greater than at present (Lee, 1967), the possible difference of 30% in the slope of the continental average thermal gradient would have been insufficient to remove all the extra heat then being generated.

How might the extra heat have left the Earth? Heat leaves the Earth's surface today in three ways, each of which accounts for roughly one-third of the total: by conduction through the continents, by conduction through the ocean floor, and by making and aging oceanic lithosphere as a boundary conduction layer to mantle convection (see section 9.5.1). An accurate assessment of the proportions of heat removed by these three processes is not possible, but it is suggested (see Sclater *et al.*, 1980) that one-third may be too low an estimate for making and aging oceanic lithosphere. During the Archean, the two conduction processes could have accounted for more heat, as both the mantle and the overlying continental and oceanic crust would have contained more heat-generating nuclides, but the evidence from the Superior Province (most other Archean terranes are similar) shows that increased conduction of mantle-generated heat through the continents was not significant. Although conduction through the largely-unpreserved Archean ocean floor may have been greater than at present, another way of removing very much more heat from the Earth is by making and cooling extra ocean floor. This could have been done by having a greater total length of ridge in the Archean ocean, or by spreading and thus making ocean floor faster at ridges, or by both processes. Since the rate at which ocean floor cools decreases as a function of age, more heat is removed from the Earth if the average age at which the ocean floor is subducted is reduced.

Ultramafic komatiites (Brooks and Hart, 1974), rocks that may represent a much larger proportion of partial melting of the mantle than do normal basalts, are apparently almost restricted to the Archean and have sometimes been considered as evidence for very steep Archean thermal gradients (Strong and Stevens, 1974). These rocks occur as a small proportion of piles basalts very similar to mid-ocean ridge basalts. We follow the suggestion (McKenzie and Weiss, 1975) that ultramafic komatiites are melts from great depths that have risen diapirically to the surface and have escaped the near-surface equilibration that is represented by basalt. We associate this process with the more intense convective circulation that we believe characterized the Archean rather than with a steep conductive oceanic thermal gradient. Arndt (1977) showed that komatiite chemistry was consistent with partial melting of upwelling material which had already fractionated basalt. Only a percent or two additional melting is necessary. See section 1.2.2.

Archean tectonics: the style of the Superior Province

A basaltic dike swarm not younger than 3 b.y. and perhaps as old as 3.7 b.y. cuts the oldest rocks yet recognized on Earth (McGregor, 1973). This fact clearly shows that the upper lithosphere of that time behaved in a brittle manner under stresses similar to those responsible for younger dike swarms of which most in the Proterozoic and Phanerozoic are directly connected to rifting and the opening of oceans.

A general and sustained surface temperature of less than about 100° C since 3.4 b.y. ago is implied by the presence of thick sections of that age and of younger ages containing pillow lavas, although higher temperatures could have been sustained below deep oceanic waters and/or under a high atmospheric pressure. This observation, combined with reasonable estimates of heat generation (Fig. 6.4.1) and a likely distribution of the heat-producing elements, also requires that there was a rigid lithosphere of significant thickness (~20–40 km) by 3.4 b.y. ago. Because mantle heat production was at least twice the present rate at the end of the Archean, strain rates due to first-order lithospheric tectonics are unlikely to have been less than at present. Therefore, such tectonics must have operated within the brittle regime from the top of the lithosphere to a depth perhaps slightly less than, but essentially similar to, that of today. We therefore infer:

(1) that a rigid lithosphere has been present since at least 3.7 b.y. ago;
(2) that lithospheric relative movement has been by ocean-floor spreading, transform faulting, and subduction at least since this time, even though there probably was a significantly greater length of plate boundary, faster average relative plate motion, and many small "plates" prior to 2.5 b.y. ago; and
(3) that continental fragments behaved in a manner similar to those of later times, and that the compressive deformation seen within them is due to orogenic events localized in time and space due to subduction and/or collision-related processes, which may have involved the local and temporary occurrence of greatly elevated thermal gradients and large volumes of silicic magmas.

Sequences composed dominantly of mafic rocks over 10 km thick, all in low greenschist metamorphic facies

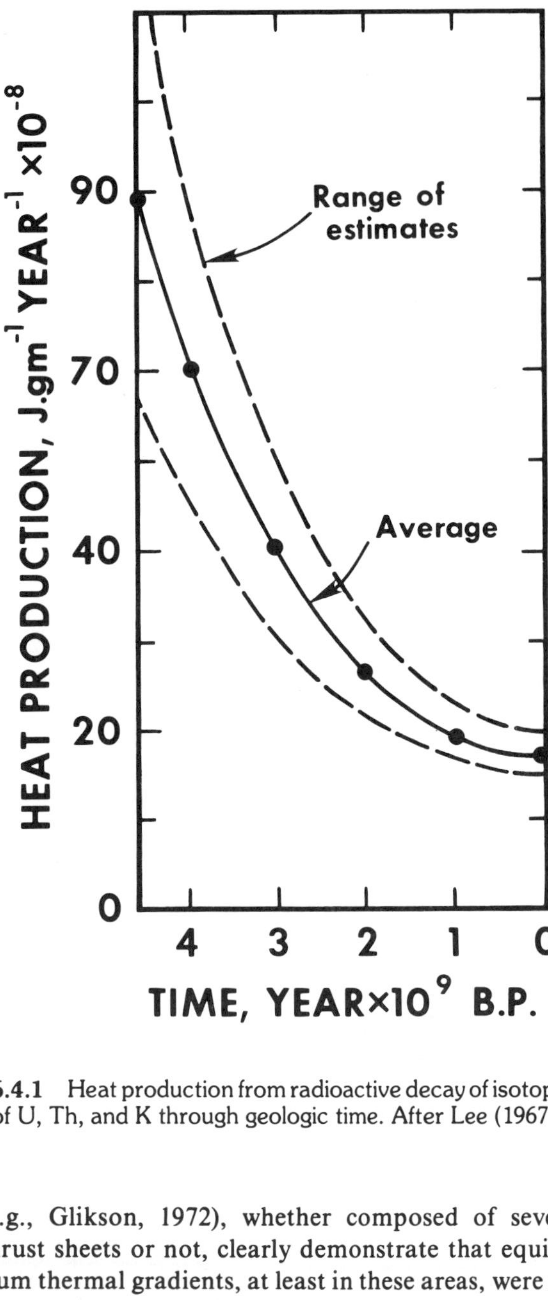

6.4.1 Heat production from radioactive decay of isotopes of U, Th, and K through geologic time. After Lee (1967).

(e.g., Glikson, 1972), whether composed of several thrust sheets or not, clearly demonstrate that equilibrium thermal gradients, at least in these areas, were not extreme in Archean times (not above 25° C/km).

Alkaline volcanic and/or plutonic rocks seem to be absent from rocks formed prior to 2.5 b.y. ago (with one exception noted below). This absence of alkaline magmas perhaps indicates that in environments now occupied by alkalic magmas (hot spots with or without associated rifting or spreading, in both continents and oceans), the percentage of mantle partial melt was relatively high (>10%), that perhaps this melting took place at shallower depths (<40 km), and may suggest that a somewhat thinner lithosphere was present prior to 2.5 b.y. ago.

No failed rift arms have been recognized prior to about 2.2 b.y. ago (2.4 b.y. ago if the southernmost part of the Huronian is in an aulacogen). We suggest that the faster pace of plate motion and the thinner lithosphere prior to 2.5 b.y. ago resulted in very few failures of rifts to evolve into oceans and, therefore, that there are no Archean aulacogens preserved in the small sample of essentially unmodified Archean crust remaining for inspection. Two types of Archean crustal terrain seem to be present. One is a "granitic," "gneissic" or "continental" terrain, which has been most thoroughly studied in West Greenland (Bridgwater *et al.*, 1974). The other kind is the greenstone/granodiorite or "oceanic" terrain. We consider the Superior Province the best example of such a terrain because its area greatly exceeds that of other greenstone/granodiorite terrains and because it has been so thoroughly studied. It is evident that the tectonic environment responsible for greenstone/granodiorite terrains such as the Superior must be compatible with plate tectonics because the Birrimian greenstone/granodiorite terrain of West Africa (~2.1–1.8 b.y.) is the same age and younger than the age of opening (~2.15 b.y.) of the ocean now represented by the Labrador Trough–Circum-Ungava–Nelson Front suture, which gives clear evidence of the operation of plate tectonics from this time onward (Burke and Dewey, 1972, 1973a).

Island arc complex assemblages and tectonics compared with greenstone/granodiorite terrains

Several authors (e.g., Engel, 1968; Folinsbee *et al.*, 1968; Anhaeusser *et al.*, 1968; Jakeš and Gill, 1970; Burke and Dewey, 1972) have put forward arguments suggesting that the granodiorite/greenstone terrains represent island arc rocks. Present-day island arcs and related surroundings contain five significantly different tectonic environments, all of which might be represented somewhere in the granodiorite/greenstone terrains if, as is likely, the analogy is valid. These environments are (Fig. 6.3.2): (a) remnant arc; (b) marginal basin, including oceanic crust formed by back-arc spreading (Karig, 1971a); (c) island arc; (d) arc/trench gap; and (e) trench mélange. Environments (a), (c) and (d) include oceanic crust on which the arc was built (Fig. 6.3.2), and (e) includes pieces of oceanic crust and overlying pelagic sediment in thrust slices and as tectonic blocks in mélange. Upon arc-arc or arc-microcontinent collision, the arc/trench gap and trench mélange are extremely likely, judging by their rarity in Phanerozoic orogenic belts, to be destroyed by being telescoped, sutured out, overthrust and eroded, perhaps because they are founded on essentially unmodified oceanic crust, which is readily subducted and/or obducted. Arcs involved in collisions tend to be uplifted and the calc-alkaline volcanic and volcaniclastic pile eroded off, which exposes the granodioritic plutonic terrain underneath. An example of this behavior includes the Ladakh "granite" (mostly granodiorite) in the Himalayas, which represents the roots of a volcanic arc that collided with the Asian continent in the Cretaceous. It is surrounded on both sides by narrow belts of ophiolitic rocks including, on the southern side, "flysch" derived by erosion from the arc (Gansser, 1964). Another example of this behavior is the Mesozoic batholithic belt of southern Chile, where the only evidence of the previously overlying volcanic rocks is now found in volcaniclastic sediments overlying the ophiolite complex floor of a Cretaceous marginal basin to the east (Dalziel *et al.*, 1974). This marginal basin opened by spreading behind the arc and closed by collision of the arc with South America, deforming the oceanic crust and overlying sediments in upright folds and steep tectonic slide zones (Dalziel *et al.*, 1974), and forming a greenstone belt (Tarney *et al.*, 1976). Even without collisions, volcanic arcs tend to end up being represented by their plutonic roots; for example, in the Sierra Nevada batholith complex of California. These examples illustrate that it is more likely that the volcanics and volcaniclastics preserved in geenstone belts are, where they are narrow anastomosing belts separated from one another by large areas of granodioritic material (such as in the Kenora area of the Superior Province), more likely to represent marginal basins, and possibly arc/trench gaps and "main" ocean sutures than they are to represent the island arc itself. However, where most of the outcrop area is greenstone with little granodiorite (as in the Abitibi area of the Superior Province), the arc volcanics, as might be predicted, seem to be well-preserved (Goodwin and Ridler, 1970). It must be emphasized that significant quantities of marginal basin and arc/trench gap rocks are also likely to be present in this and similar areas.

Although the bulk of lava flows in any one marginal basin—arc—arc/trench gap complex may be largely confined to the arc and to oceanic crust of various ages, they are not wholly restricted to the arc. Arc/trench gaps at any one time are defined by the trench and by the volcanic front which is controlled by the depth to the descending slab (Fig. 6.2.6). If the dip of the descending slab increases or decreases, volcanism will occur in what formerly was the arc/trench gap or marginal basin, respectively. Therefore, in an area representing an arc complex built over a significant time, the distinction between arc and arc/trench gap and between arc and marginal basin may not be clear-cut. Significant amounts of basaltic pillow lava are known within the mafic volcaniclastic sediment sequences overlying the oceanic crusts of two originally very narrow (< 50 km) fossil marginal basins in western Newfoundland (Upadhyay *et al.*, 1971; Kidd, 1977). These may represent a temporary shallowing of the dip of a subducted slab, or perhaps some other process peculiar to marginal basins. It is important that the ophiolite-complex pillow lavas (oceanic crust) of these two "fossil" marginal basins—but not the lavas in the sediments above—appear to be basaltic komatiites (Kidd, 1977). If, in addition to these processes and the opening and closing of marginal basins, arc complexes are constructed with continuous or episodic oceanward migration of the subduction zone (such as has been suggested for the Klamath Mountains in California; Fig. 6.3.2), great complexities of tectonic and magmatic chronology are likely, quite apart from additional complexities superimposed during arc-arc or arc-microcontinent collisions.

As well as granodioritic plutonic complexes representing the roots of arcs, granodioritic to granitic plutonism during and immediately following collision is probable, due to partial melting, caused by over thickening during collision, of arc or microcontinental crust, or of the underlying mantle forming the rest of the lithosphere (Dewey and Kidd, 1974). This process may have contributed a large proportion of the plutons in granodiorite/greenstone terrains, including most or all of the widely reported plutons that are synkinematic to some of the deformation sequences.

Two examples of alkaline volcanics and sills are reported from greenstone terrains. Both are in the Abitibi belt of the Superior Province: one near Kirkland Lake (Ridler, 1970), the other near Timmins (Goodwin, 1974). These may be island arc rocks, as suggested by Cooke and Moorhouse (1969) or, if there are more tectonic boundaries present than have presently been recognized, Archean seamounts tectonically incorporated into a trench complex (as is shown to be imminent in Fig. 6.3.2). Alternatively, it is possible that these two occurrences might be younger than the surrounding Archean rocks. Perhaps, judging by the largest presently observed, most of the hot spots were large in the Archean and erupted large quantities of flood basalt-type tholeiite relative to alkalic lavas. Lavas and sills generated in this way, and preserved from subduction, might explain the origin of at least some of the extensive basalts preserved in greenstone belts. Alternatively, efficient subduction of Archean oceanic lithosphere may have removed essentially all of the "hot spot" generated material and have preserved island arc edifices preferentially, which even today contain only rare occurrences of alkalic volcanics. Besides the small area of preserved Archean rocks, the small chance of preservation of alkalic volcanics may be a reason why they are so rare in such terranes.

Relatively small blocks, composed principally of quartzo-feldspathic rocks, occur within the dominantly greenstone/granodiorite terrain of the Superior Province, with the English River Gneiss Belt and the Quetico block being the best known examples (Goodwin, 1972). It has been reported that the rocks are la gely "paragneiss," but further study seems to be supporting the view that they consist mostly of extremely deformed plutonic rocks, such as reported for the "gneissic" terrain of West Greenland by Bridgwater *et al.* (1974). The boundaries of the two blocks mentioned are reported as zones of intense faulting and mylonitization (Goodwin, 1972). In some areas, particularly on the northern side of the English River Belt in Manitoba (Wilson *et al.*, 1972), and to a lesser extent on the northern side of the Quetico block, tectonic slivers of serpentinite, and serpentinized ultramafic rock are clearly associated with these zones, giving them a similar character to oceanic suture zones of younger orogenic belts. Burke *et al.* (1976) suggested that these two blocks and others like them are microcontinents in terms of their present relationships to the granodiorite/greenstone terrains. Remnant arcs and inactive arc complexes can also be regarded for tectonic purposes as microcontinents with respect to active convergent plate boundaries and their active arcs. Such microcontinents may in some cases be identifiable in that they may contain a mixture of plutonic rocks of two greatly different ages, one reflecting its arc history, and a younger one reflecting its collision with an active arc that will only give ages near and including the younger age range. Such a relationship was suggested by Lawson (1913) to be common in the Superior greenstone/granodiorite terrane on the basis of clasts of granitic rock in greenstone belt conglomerates.

Lithologic associations very similar to those in greenstone/granodiorite terrains and deformed in a similar style can be found locally within Phanerozoic orogenic belts. One example is the Round Pond area of south-central Newfoundland (Fig. 6.4.2). This area, which includes a well-developed ophiolite complex, was deformed and intruded by silicic magmas during the Acadian continental collision, which was relatively weak in the salient of central Newfoundland (Dewey and Kidd, 1974). The "triangular syncline" morphology, the scale, and the lithological assemblage in the Round Pond area are very similar to those in Archean greenstone/granodiorite terrains, illustrated for comparison in Fig. 6.4.2 by the Bulawayan greenstone-belt (Amm, 1940). The steep metamorphic gradients shown by the narrow dynamothermal aureoles of the silicic plutons in the Round Pond area are a particular point of resemblance to the Archean terrains. An even better example is found in the northwestern Sierra Nevada of California (Fig. 6.4.3); the resemblance of this area to Archean greenstone/granodiorite terrains is remarkable. The ultramafic and gabbroic rocks in this area are almost all or perhaps entirely variably dismembered ophiolite complexes, including (just to the south of the area shown) one with well-preserved sheeted dikes (R. G. W. Kidd, pers. comm., 1975; Moores, 1975).

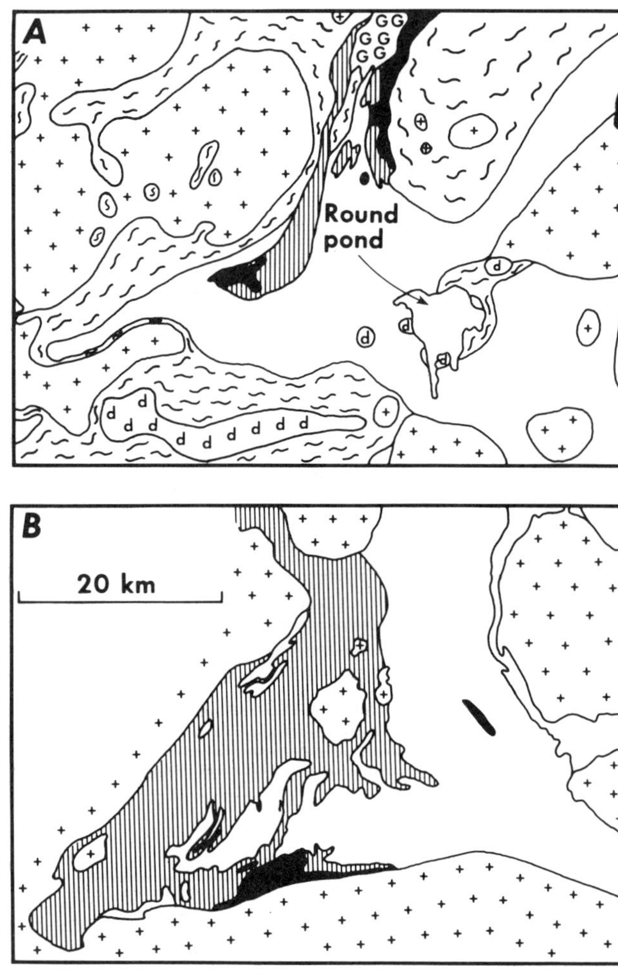

6.4.2 Geologic maps illustrating similar rock types, structural relationships and gross morphology in part of a Paleozoic orogenic belt and part of an Archean greenstone belt. **a**. Round Pond area, south-central Newfoundland (Williams, 1970). **b**. Bulawayan greenstone belt, Rhodesia (Amm, 1940). d—diorite; crosses—"granites"; wavy broken lines—high-grade metamorphics (probably dynamothermal aureoles around "granites"; black—ultramafic rocks; G—gabbros; vertical lines—mafic volcanics; white—sediments, including volcaniclastics. After Burke *et al.* (1976).

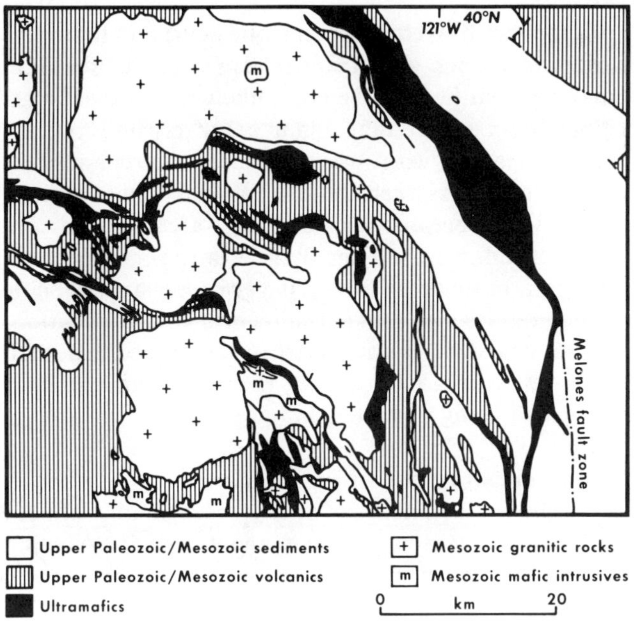

6.4.3 Map of pre-Cenozoic geology of part of northwestern Sierra Nevada, California, showing great resemblance in morphology and scale to Archean greenstone belts. Modified from Burnett and Jenkins (1962).

Tectonic boundaries and ductile high-strain zones are abundant (E. Nisbet, pers. comm., 1972). The ophiolite complexes and the mafic volcanics and volcaniclastics are probably at least partly of marginal basin origin.

It is important to recognize that the apparently swirly and plastic deformation and intrusion features seen in the map patterns of the Phanerozoic analogues cited for the Archean greenstone/granodiorite terrains are the sum of a complex series of events. In each case, these occurred during a significant time period in an area where the instantaneous situation at any time was either relative plate motion at sharply defined boundaries, or, during short-lived orogenic episodes, significant horizontal shortening with consequent upward motion of material to compensate. These episodes may have been due either to collision across a marginal basin or a "main" ocean, or to other (at present) poorly understood compressive orogenic episodes within and behind Andean and island arc complexes. There seems no compelling reason to suppose anything different for the Archean terrains.

An extensive gneissic terrain containing granulite-facies rocks forms the part of the Superior Province in the Ungava Peninsula. The radiometric ages (2.5–2.4 b.y.) and the nature of the rocks (Stevenson, 1968) show this to be an area of basement reactivation, which Burke *et al.* (1976) interpret as being due to convergence after a continental collision about 2.6 b.y. ago. This type of orogeny, where internal continental reactivation takes place (Burke and Dewey, 1973a), seems to have been relatively uncommon in Archean time, as shown by the large areas of granodiorite/greenstone terrain that have survived. Burke *et al.* (1976) suggested that it was the general incompleteness or weakly developed nature of collisions in the Archean that has led to widespread preservation of the greenstone belt material which is only locally preserved in younger orogenic belts.

The supposedly "vertical" tectonic style reported from greenstone belts seems to be based on the alleged control of deformation by the silicic plutons, the allegedly simple synclinal structures of the belts and, perhaps, on the subvertical elongation lineation commonly observed. The latter is a necessary consequence of horizontal shortening across a steep cleavage, because vertical displacement of plastically deforming material is usually easier than lateral displacement. The supposedly simple synclinal structure of greenstone belts is a myth (Ramsay, 1963). The steep inclination of cleavage and the overall steep structure of greenstone belts are quite compatible with collisional tectonics, being observed in Phanerozoic orogenic belts where incomplete collision has occurred (Dewey and Kidd, 1974). They also are found in very complete collisions in tightly sutured zones such as the Indus suture (Gansser, 1964). Although the Indus suture and others like it represent oceans driven out by low to moderately dipping subduction and thrusting, the suture zone itself becomes oversteepened in the later stages of the collisional process. This effect is also demonstrable in the Baie Verte Lineament (Kidd, 1977) on a scale more compatible with an individual greenstone belt. Ramsay (1963) has shown that the silicic diapirs do not significantly control the deformation in the Barberton greenstone belt. It is difficult to see how a strong regional cleavage and associated horizontal shortening could be impressed on relatively hard, cold volcanics and sediments by soft, ductile granitoid diapirs. An externally applied, horizontal compressive stress, such as is available during arc/arc or arc/microcontinent collision, is mechanically more reasonable, and is more compatible with the behavior of a planet on which occurred the major horizontal shortening clearly documented by Bridgwater *et al.* (1974) for the Archean of West Greenland.

Ophiolite complexes and Archaean oceanic crust and mantle

No examples of oceanic crust and mantle similar to fully developed ophiolite complexes have as yet been reported from greenstone belts. The suggestion (Moores,

1973) that Archean oceanic crust was thicker than present oceanic crust leads to the idea that possible examples of tectonic slices of Archean oceanic crust often may not include the depleted harzburgite-dunite that is a diagnostic and characteristic feature of their Phanerozoic counterparts. The somewhat higher heatflow and perhaps less-depleted mantle in Archean times may have resulted in a significantly greater thickness of essentially ultramafic cumulates and a proportionately lesser thickness of gabbro than is seen in Phanerozoic ophiolite complexes. As Moores (1973) pointed out, the stratiform anorthosites of the Early Precambrian (e.g., Fiskenaesset and the Limpopo) may also represent the upper part of the cumulate gabbro of the thick Archean oceanic crust. Many potential examples of Archean oceanic crust (regarding ultramafic cumulates as crust for brevity) are stated to be sills. A large number of Phanerozoic ophiolite complexes, now known to be variably dismembered and tectonically bounded slices of oceanic crust and mantle, were originally described as intrusive sills, despite the essential absence of significant metamorphic aureoles and the presence, usually overlooked, of narrow tectonic boundaries with large displacements. One example includes the Bay of Islands Complex (Smith, 1958). More dismembered examples, the Mings Bight and related Baie Verte Lineament Ophiolite Complexes (Bird *et al.*, 1971; Dewey and Bird, 1971; Kidd, 1977; Kidd *et al.*, 1978) were also described originally as concordant lenticular ultramafic and gabbro sills (Watson, 1943). Many examples of supposed sills which are very similar to the cumulate ultramafic and gabbroic parts of ophiolite complexes are known in Archean greenstone belts, and it has been suggested (Burke *et al.*, 1976) that many of them are good candidates for samples of Archean oceanic crust, perhaps including, in some instances, depleted noncumulate upper mantle as well.

These possible ocean floor fragments in greenstone belts with associated mafic volcanics and volcaniclastics closely resemble those dismembered ophiolites interpreted as being of marginal basin origin in Phanerozoic mountain belts (e.g., Bird *et al.*, 1971; Dalziel *et al.*, 1974). Burke *et al.* (1976) suggested that it is very likely that the contents of marginal basins compose a large proportion of Archean greenstone belts, and that there are two reasons why any Archean oceanic crust that may be identified will be more likely to represent marginal basin rather than main ocean crust. First, the Archean ocean with many scattered microcontinents and arcs appears likely to have had many more areas in which young ocean floor was close to arc and microcontinental sediment sources than in later times. Second, marginal basin ocean floor has a greater chance of being young (and hence thin and hot) near to a subduction-resistant arc or remnant arc (microcontinent) than does main ocean floor, and therefore has a greater chance of being obducted and preserved.

Summary of Archean tectonics

The characteristics of ancient ocean and continent outlined in this section have been inferred by assuming that rocks similar to those forming today were made by similar processes, and by allowing for the effects of faster heat generation in the past. Because much of the heat generated in the Earth today is dissipated in making oceanic lithosphere at spreading ridges, in aging it on ocean floors, in partly melting descending slabs of oceanic lithosphere below island arcs, and in emplacing the igneous products of this melting, Burke *et al.* (1976) have inferred that these processes operated more effectively during the Archean prior to 2.5 b.y. ago. The picture outlined by considering these intensive properties is of an Earth at the end of the Archean covered much as today with about one-third continent and two-thirds ocean, and the volume of water produced from the mantle concurrently with lithosphere being also similar to the present (Wise, 1974a). There are no minerals that will hold the water in the mantle at depths of much more than 100 km, so a hydrosphere containing much of the Earth's water must have existed from early times. Burke *et al.* (1976) have suggested that the length of plate boundary was, however, much greater and that this explains many of the differences. The extensive characters of exposed Archean terrains are consistent with this picture and permit its refinement. It is inferred that the environments of basaltic volcanism represented in Archean rocks were not very different from those of today.

6.5 TECTONICS OF BASALTIC VOLCANISM ON OTHER TERRESTRIAL PLANETS

Mars, Mercury and the Earth's moon share the property of having at present a stable global lithosphere. This is in contrast to Io, where lithospheric mobility is so great that all lithosphere is probably recycled frequently (on a time scale perhaps less than 10^7 years), and to the less extreme, plate-structured Earth. On Earth, oceanic lithosphere is fully recycled on a time scale of about 10^8 years, but most material that accreted to continents remains, forming part of the record of Earth history. Mars, the Moon and Mercury show no sign of comparable lithospheric mobility after the end of the period of heavy bombardment because, at least on part of their surfaces, they preserve a record of the great impacts. Other, presumably younger, surfaces show no morphologic evidence of lithospheric recycling. Data from Venus are sparse, but the evidence points towards a lack of present-day plate tectonics.

Solomon (1978) has reviewed the tectonic style of planets with immobile lithospheres ("one-plate planets") and has suggested that, in general, after they become one-plate planets, they may record on their surfaces an early stage of expansion and a later stage of contraction (Fig. 6.5.1). In the former stage extensional features, especially normal faults, and igneous activity are general, while in the latter stage compressional structures and an absence of igneous activity (eruption being inhibited by compressive stresses) are characteristic. The Moon and Mercury have passed through the first stage and are now in the second, while Mars was in the earlier stage until relatively recently, and it is not clear whether it has yet entered the second (Fig. 6.5.2).

Volume change alone results in an isotropic distribution of stress in the lithosphere and should produce randomly distributed tectonic features (folds, faults, fracture sets). Superimposed on this secular and spherically symmetric stress state are regional stress systems associated with thermal inhomogeneities in the mantles of these one-plate planets. Closely related are stresses due to thermally induced variations in lithospheric thickness. The mantles of these planets generate basaltic magmas at various stages of their history. When and where these magmas reach the surface are functions of the lithospheric stress history, specifically of the environments for extensional tectonics, which in turn are dependent on the superposition of regional and global stress systems. Additionally, basaltic magmas reach the surface through fractures created by large-scale impact events. However, a favorable stress history is required to keep these fractures open between the time of impact and the time of magma generation.

In addition to mantle thermal inhomogeneities, other phenomena concentrate tectonic stresses. These include lithospheric stresses due to loading by volcanic material, polar wander of the lithosphere, tidal despinning and, for the Moon, orbital recession. The relative importance of these various mechanisms varies among the planets and also during the evolution of an individual planet.

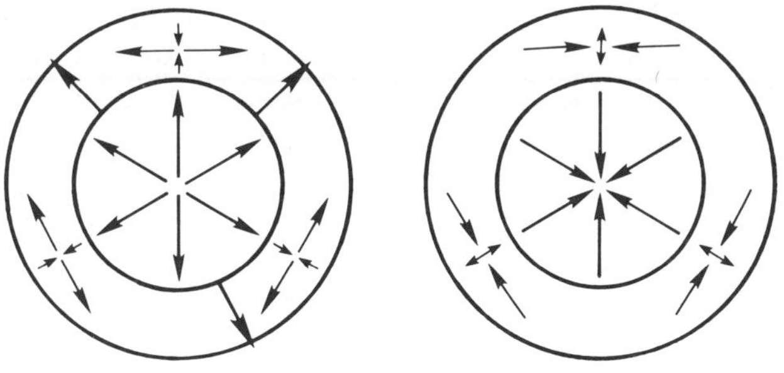

6.5.1 Sketch illustrating the behavior of a planet with a continuous immobile lithosphere. An earlier warming phase is associated with expansion, surface igneous activity and normal faulting of the lithosphere (annular region not shown to scale). Later cooling is associated with shrinking and absence of igneous extrusion. If the lithosphere fractures in this phase it does so in compression. Based on Solomon (1978).

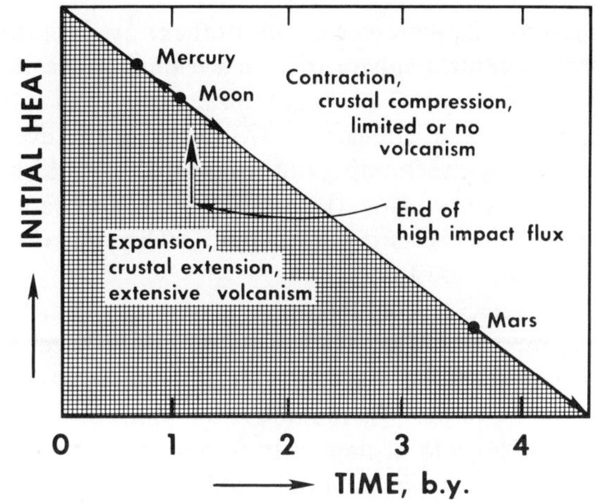

6.5.2 One-plate planets go through a secular evolution from interior expansion, crustal extension, and widespread volcanism to interior contraction, crustal compression, and cessation of volcanism. The time of transition from expansion to contraction is a function of the magnitude of initial extensive heating, shown here with an arbitrary scale. The transition, shown by the diagonal line, marks the approximate peak of planetary volume. The peak volumes (with uncertainties) are shown for Mercury, Moon, and Mars. Note that only Mars went through the transition after the end of the high impact flux. Modified after Solomon (1978).

6.5.1 Tectonics of basaltic volcanism on the Moon

Introduction

On the Moon, tectonics (large-scale lithosphere evolution) has been overwhelmingly dominated by impact-related phenomena. Almost the entire evolution of the lunar lithosphere took place during the high impact flux prior to 3.9 b.y. ago (for reviews, see Taylor, 1975; Guest and Greeley, 1977). Subsequent events involved only the flooding of the maria 3.9 to 3.65 b.y. ago (Lucchitta and Watkins, 1978) and later, small-volume eruptions lasting until about 3.0 b.y. ago (Wasserburg *et al.*, 1977).

The contrast between the Moon and Earth is almost complete. Tectonics on Earth is readily related to the long-term thermal evolution of the interior of the planet (see Chapter 9), and active tectonics persists to this day. The role of impact in terrestrial tectonics is generally regarded as minimal (but see, for contrasting views, Seyfert and Murtagh, 1977; and Weiblen and Schulz, 1977). On the Moon, impact dominance is so great that unrelated tectonism cannot be distinguished (Muehlberger, 1977). What we have learned about lunar thermal behavior and its variation with time has come from indirect inferences from the secular variation in structural response to both impact and subsequent volcanic emplacement (Solomon and Head, 1980a). In particular, the gravity ("load") tectonics associated with basalt infilling of the large impact basins is decipherable and places constraints on the global stress system associated with lunar expansion and contraction. However, there is no unequivocal direct evidence for global stresses large enough to have caused widespread lithospheric failure comparable to the global distributions of scarps on Mercury (see below).

Lunar highlands

The lunar highlands are inferred, from petrological and geochemical evidence, to represent material formed by global melting early in lunar history (see section 1.2.10). The highlands have been so severely modified by impact that no structural or tectonic evidence that could help to show how this process operated has yet been distinguished. The general elevation of the lunar highlands above the level of the basins (Kaula *et al.*, 1973) is evidence of crustal buoyancy, but the obvious analogy between buoyant highlands above lunar basins and buoyant terrestrial continents above ocean basins cannot be pursued too far. Highlands once covered much, if not all, of the lunar surface, whereas it is unlikely that continents on Earth ever were more extensive than they are now. Although ocean basins are also younger than the neighboring continents, the processes by which ocean basins form are quite unlike those that created the younger mare basins within the lunar highlands.

Early Earth-based observations (Fielder, 1961, 1965; Baldwin, 1963) identified an extensive, supposedly systematic, global distribution of lineaments that has been collectively referred to as the lunar grid. Similar lineament systems have been recognized by some on Mercury and Mars and in the past have been suggested for the Earth as well. The lack of distinction, or ignorance, of the relative ages of development of included features (such as scarps, ridges, crater walls) makes it difficult to interpret or assess the tectonic importance of these supposed systems. The inevitable inclusion in these systems of a large proportion of lineaments generated in various ways by impacts leads us to consider the tectonic significance of such "grids" as minimal.

No direct correlation can be seen between ancient lunar structural features and highland volcanism. Highland volcanism on the nearside shows evidence for structural control related to the Imbrium Basin ring system (Head and McCord, 1978; Hawke *et al.*, 1979) or adjacent mare infilling (Head and Wilson, 1979). Structural and volcanic features become increasingly rare toward the central farside (Scott *et al.*, 1977), with mare deposits concentrated in topographically low areas (Schultz, 1976; Stuart-Alexander, 1978).

Whitford-Stark (1974) and Bryan *et al.* (1975) drew attention to the pronounced fractures and evidence for volcanic processes that have modified large lunar impact craters. Many such craters are located within the lunar highlands close to the mare-highlands boundary, although a significant number also occur within highland regions far from such boundaries (Schultz, 1976). Floor-fractured craters may mark the locations of igneous intrusions that have uplifted and fractured the crater floor. Not all mare-highland boundaries show examples of fractured crater floors, but the mechanism for their formation is interpreted to be the intrusion of magmas within the fractured crustal material close to major impact basins. A similar mechanism has also been proposed by Head and Wilson (1979) to explain the occurrence of dark-halo craters close to basin margins.

Mare basins

The mare basins are large, basalt-flooded, impact-generated structures up to 2000 km across that have modified the lunar highlands. The distribution of basins across the lunar surface is not uniform because they are the products of only a few tens of random impacts. Consequently, the concentration of basins on the lunar nearside is not attributable to any large-scale structural heterogeneity within the Moon. Linear rilles and ridges (see Chapter 5; Fig. 6.5.3) are structures associated with mare basins that have been considered tectonic in origin, but "most rilles and ridges can be ascribed to subsidence of the mare basin region in response to the

6.5.3 A view of the western edge of Mare Serenitatis showing both linear rilles (lower center) and mare ridges (center and right). Note that linear rilles are flooded by later mare basalts. Apollo 17 metric photo 0953.

load produced by the basalt fill" (Solomon and Head, 1980a), and they are thus the result of gravity tectonics rather than internally generated tectonic phenomena.

The role of the high-density basalt fill (mascons) in the circular maria as a means of loading the lunar lithosphere has long been considered important (e.g., Baldwin, 1968; Phillips *et al.*, 1972) and structural features within maria (Fig. 6.5.3) often have been attributed to basin loading and subsidence (e.g., Baldwin, 1968; Muehlberger, 1974; Runcorn, 1974; Scott *et al.*, 1978; Maxwell *et al.*, 1975; Maxwell, 1978).

Solomon and Head (1979) made an ingenious tectonic inference from the distribution and age of these features. They pointed out that both linear rilles and ridges were formed before 3.6 b.y. ago (Lucchitta and Watkins, 1978), while only ridges continued to form after that time, persisting until at least 3.0 b.y. ago. Such a chronology was interpreted to be due to a global stress system that was superimposed on local mare basin stresses. This global stress field was extensional prior to 3.6. by ago and compressional after that time, inhibiting the formation of the linear rilles during the final 600 m.y. of lunar igneous activity. Thermally induced stresses in the lunar lithosphere, too small to have caused rupture, can be expected to have changed from extensional, when heat generation rates were relatively high, to compressional as heat generation rates declined. Solomon and Head suggest that the cessation of rille formation records the transition from global extension to global contraction about 3.6 b.y. ago (see Fig. 9.5.16).

The tectonic interpretation of structural features within the maria is not without controversy. As many as four different sets of ridges and arches within a lunar mascon basin have been recognized (Maxwell, 1978): (1) a circular inner-ring system; (2) ridges oriented radially to the basin; (3) north-trending ridges; (4) local systems associated with post-mare craters. Muehlberger (1974) advocated the idea that thrust faulting due to global contraction produced the ridges in southeastern Serenitatis. This is consistent with a hypothesis presented by Dvorak and Phillips (1979) suggesting that the topographic trough occupying western Mare Tranquillitatis is synclinal, resulting from global contraction of the lithosphere. Although Bryan (1973) had suggested that mare ridges formed by local sagging of the central portion of the Serenitatis Basin, Muehlberger (1977) found that only a fraction of the required crustal shortening could be accounted for by basin subsidence alone. Other evidence, however, also indicates that basin subsidence did occur during the emplacement of mare basalts. Based on the flow directions of sinuous rilles in the maria, Scott *et al.* (1978) inferred large topographic variations in the mare surfaces to represent multiple periods of tectonic deformation. Such an interpretation was corroborated by the lunar radar sounder experiment, which detected subsurface reflections in Serenitatis. The apparent subsidence of these reflectors suggests that deformation occurred prior to the last episode of basin infilling (Maxwell, 1978). Thus, for Serenitatis at least, the ridge system appears to be the result of both global stresses and local loading of the lithosphere.

Arcuate rilles peripheral to the circular lunar mare have been interpreted as grabens formed in response to subsidence of the filled lunar impact basins (e.g., Baldwin, 1963; Ronca, 1965; Quaide, 1965; Fig. 6.5.3). The existence of lunar mascon gravity anomalies (Muller and Sjogren, 1968) is consistent with the interpretation that the load of volcanic fill is wholly supported by the strength of the lithosphere (Phillips and Lambeck, 1980), but leads to local near-surface extensional failure. Several quantitative interpretations of arcuate rilles and mare ridges, based on flexure of an elastic lithosphere shell, have been reported (Melosh, 1978; Solomon and Head, 1979, 1980a). As discussed above, Solomon and Head (1979, 1980a) present an interpretation consistent with mascon loading and secular evolution of global stresses.

Melosh (1978) analyzed the tectonics of mascon loading and predicted an inner zone of radial thrust faults surrounded by an annulus of strike-slip faults encircled by concentric normal faults. This analysis, which yields a different result from that of Solomon and Head, did not consider modifications of the stress field by faulting, pre-existing zones of weaknesses in the lithosphere, residual stresses, regional stresses or global stresses. Further, Melosh's analysis, based on the principal stress-faulting theory of Anderson (1951), fails to accommodate the possibility of extensional failure. Photogeology could perhaps be applied in the future in an attempt to discriminate between the models of Solomon and Head and of Melosh. It is worth noting that as yet only Tjia (1970) has suggested that mare ridges near the edge of mare basins involve strike-slip motion, although the difference between central mare-ridge trends and peripheral ridge trends reported by Worral *et al.* (1978) may be tectonically significant.

The thickness of the lunar lithosphere modelled as an elastic layer over a hydrostatic interior has been estimated using evidence from the mascons. Variations in lithospheric thickness as functions of both age and location have been inferred. Solomon and Head (1980a) provide the most comprehensive analysis to date. They treated the volcanic and tectonic histories of eight major lunar mascons by considering that these histories are sensitive both to the thickness of the lunar lithosphere and to the presence of any superposed regional or global stress field. Estimates were obtained for the thickness of the elastic lithosphere in each basin region during the period of rille formation and during the later period when only mare ridges were formed. Substantial spatial and temporal variations in lunar lithospheric thickness emerged. Elastic lithospheric thickness during the interval 3.6–3.8 b.y., the time at which linear rilles formed, ranged from 25 km (at Grimaldi) through about 50 km (at Orientale, Serenitatis, Humorum and Imbrium) to 75 km or more (at Nectaris, Crisium and Smythii). Younger lithospheric thickness estimates (3.0–3.2 b.y. ago), with the exception of Grimaldi, were more than 100 km.

Factors influencing the derived variation in the earlier lithospheric thickness include age and size of basins, rate of basin filling, age of basin filling and temperature variations independent of the history of basin formation and fill. The first four factors cannot among them account for all the calculated variations in lithospheric thickness, and it is necessary to infer regional heterogeneity. The western nearside appears to have had a regionally thinner elastic lithosphere at the time of mare basin formation and filling than the eastern nearside (with Nectaris, Crisium and Smythii). Perhaps significantly, this observation correlates with distributions of both volcanism and radioactivity (Whitford-Stark and Head, 1980; Malin, 1974; Wood and Head, 1976; Trombka *et al.*, 1977).

The lunar lithosphere apparently increased in thickness over about 700 m.y. by a factor of about 2 or 3 and this indicates a general cooling of the outer part of the Moon, and of the mascon mare basins in particular. Variability in thickness appears to have decreased with time, perhaps indicating that shallow sources were responsible for much of the earlier heterogeneity.

The relative lack of mare flooding in some young multi-ringed impact basins, particularly on the farside, perhaps can be understood in the light of the heterogeneity recorded by Solomon and Head (1980a) in the thickness of the lunar lithosphere at this time. At the time of formation of these young basins, the lithosphere was probably colder and more depleted in basalt material than earlier in lunar history, and it was these depleted regions which could no longer yield basalt to flood the basin interiors. The distribution of the depleted regions was presumably established early in lunar history during a time for which the tectonic record has subsequently been obscured by the effects of the high impact flux. Alternatively, the lack of farside volcanism could be attributed to the offset, towards the farside, of the center of figure from the center of mass. If the heights reached by basaltic magmas on the lunar lithosphere were controlled hydrostatically, then the ratio of extruded to intruded basalt would be less on the farside, and extruded basalts would only reach the topographically lowest regions, as is observed.

The result of all of these tectonic studies does not totally clarify the relationships between impact-induced fractures, lithospheric stress state, and basaltic extrusion. An important result that emerges from Solomon and Head's studies is that most of the volume of the mascons is formed by rapid and voluminous basalt flooding within 0.1 to 0.2 b.y. after impact. Mare basalts are not, therefore, like basalts on Earth, recording the long-term evolution of a planetary mantle, but are predominantly erupted in response to impact and the attendant stress relief associated with the creation of a topographic low. As such they do not record a comparable history of tectonic development. The fact that these impact-induced fractures remained open to allow basalt migration to the surface attests to the global extensional regime of the lunar lithosphere. Once a fraction of the load was emplaced, the circumferential grabens may have formed and served to localize the regions of extrusion. Although the Solomon-Head model predicts a transition to a lithospheric compressional stress regime about 3.6 b.y. ago, the final 900 m of basalt fill in Serenitatis was emplaced approximately 3.2 b.y. ago (Peeples *et al.*, 1978). Thus, the mechanism of extrusion late in lunar volcanic history, during a period of global compression, requires clarification.

6.5.2 Tectonics of basaltic volcanism on Mars

Introduction

Although the occurrence of volcanoes and volcanic rocks on Mars is abundantly clear from the images returned to Earth, there is no certainty that the rocks forming these volcanoes are basaltic in composition (see, for example, McGetchin and Smyth, 1978). For the purposes of this section, the assumption is made that they are sufficiently like basalts to be comprehended within the terms of this survey.

The intermediate place of Mars in planetary volcanism has long been clear (see, for example, Mutch *et al.*, 1976, p. 319). Lunar volcanism persisted only until 3 b.y. ago and was then mostly focused on the sites of the last great impacts. Terrestrial volcanism and that on Io persist today, on the latter with continued ubiquitous vigor and on the former, concentrated at plate boundaries. Mars records volcanic activity like that of the early Moon on part of its surface, but also shows evidence of activity persisting until relatively recently, long after major impacts ceased to occur.

The surface of Mars falls into two morphologically distinct hemispheres: plains in the north and cratered terrains in the south, separated approximately along a great circle inclined at 35° to the equator. This hemispheric asymmetry of terrain types on Mars may be a consequence of an early martian tectonic evolution that is distinct from that of the Moon and Mercury. An isostatically compensated elevation difference of about 3 km between the northern plains and southern cratered highlands leads to the hypothesis of different crustal thicknesses and/or densities in the two regions (Phillips and Saunders, 1975; Mutch *et al.*, 1976; Phillips and Ivins, 1979; Wise *et al.*, 1979a).

Isolated regions of what may be highlands terrain occur within the northern plains. Fretted terrain at the boundary between highlands and plains has been interpreted to be a result of erosional retreat of the highlands, although stratigraphic relationships do not necessarily support this view (Scott, 1978). The northern plains are generally interpreted as lava flows with varying amounts of eolian cover. Certain morphologic features may be tectonic in origin, but their interpretation is highly ambiguous (Scott, 1979).

Although it has been suggested that erosional processes produced the northern plains, it has also been argued that the elevation difference cannot be explained simply by removing crustal material from the plains and redepositing it in the highlands (Phillips and Ivins, 1979; Wise *et al.*, 1979b). If martian crust and mantle densities were comparable to those of continental crust and mantle of the Earth, at least a 20 km thickness of material would have to be transferred from the northern plains to the highlands to account for the present elevation difference and maintain isostasy. Therefore, if Mars had an early crust of nearly uniform thickness and composition as is thought to occur on the Moon, an internal process must be strongly coupled to the formation of the asymmetry.

Alternatively, because Mars is a planet which has escaped both large-scale horizontal lithospheric motion and subduction (Phillips and Ivins, 1979), at least since the high impact influx ended (before that time there is no record), the present morphological division is one that might have persisted since early martian history (~3.8 b.y. ago). That is, thick, light, more differentiated crust apparently could have been made early over an area of about half the planet in contrast with both the Moon, which was probably entirely covered by highlands, and the Earth which is even now only one-third covered by continents. Because of the Earth's lithospheric mobility, its continental third is usually in several pieces; however, these fragments were nearly all assembled into one landmass during the short life of Pangea (~70 m.y.), ending 180 m.y. ago, and may also have been on previous occasions. On Mars, the concentration of light crust in one hemisphere may have happened stochastically. Mobility was perhaps lost before the concentration could break up.

Ancient volcanism

The heavily cratered terrain of the southern highlands is thought to be the oldest widely exposed surface material on Mars (Scott and Carr, 1978) and is the closest analog to the lunar highlands. Despite significant modification by the early martian cratering flux, extensive plains units believed to be volcanic in origin have been identified (Scott and Carr, 1978; Greeley and Spudis, 1978). Complex patterns of mare-like ridges and flow scarps, together with numerous volcanic cones are identifiable in the plateau plains material described by Greeley and Spudis (1978). The ridges in these plains lack any recognizable pattern in their distribution, but may perhaps indicate that regional (or global?) compressional tectonism was prevalent during early modification of the martian highlands. Numerous examples of floor-fractured craters have been identified by Schultz and Glicken (1979), and their distribution correlates well with the location of the highland-lowland boundary. As on the Moon, floor-fractured craters are interpreted to be examples of volcanically modified impact craters and, from their varied degradational states, indicate that highland volcanism was a protracted process on Mars.

Martian basin volcanism

In addition to the plains-forming units within the southern highlands, several old volcanic constructs (patera and tholii) have been identified. Australis Tholus (unit 47 in Chapter 8) and Tyrrhena Patera (unit 45) are both surrounded and embayed by floor volcanics that display complex distributions of wrinkle ridges. Peterson (1978), in discussing the Noachis-Hellas volcanoes, suggested that the occurrence of five volcanic centers in that region may be localized on the second and third basin rings of the Hellas structure where the rings are intersected by a northeast-trending deep fracture system of which there is some evidence in the gravity field. According to Peterson, localization of activity by the rings might imply that volcanism started not long after basin formation, but relatively youthful-looking features indicate either persistent or recurrent volcanism around Hellas. Radial grabens may also exist around the basin Schiaparelli (Fig. 6.5.4) and are genetically similar to features around Mare Imbrium on the Moon.

More recognizable as basin-related tectonic features are the concentric ridge systems within Schiaparelli and Huygens basins. These features may result from subsidence during basin loading similar to the mode of formation of lunar mare ridge systems (Solomon and Head, 1980a). Circumferential grabens, similar to those around Mare Serenitatis, are also present around Isidis basin on Mars, which contains a lunar-like mascon (Sjogren, 1979). The gravity field of Isidis has been

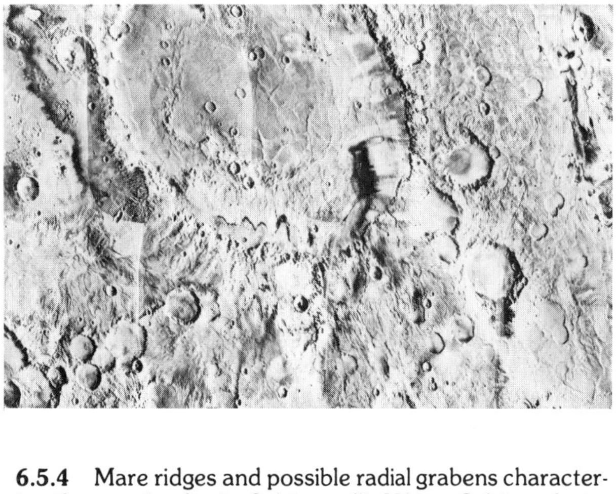

6.5.4 Mare ridges and possible radial grabens characterize the martian basin Schiaparelli. Viking Orbiter photomosaic.

analyzed as a flexural load and indicates an elastic lithosphere thickness formation of about 100 km (Solomon *et al.*, 1979). Although the volcanic origin of martian basin fill remains unproven, by analogy with lunar tectonic features, it appears that the martian basins may also mark the location of voluminous magma eruptions at an unspecified time after basin formation.

Tharsis province: volcanism

Later martian volcanism is generally confined to two broad regions: Tharsis and Elysium. The volcanic structures of the Elysium area are generally less degraded than those of Hellas but not as fresh as those of the Tharsis province. The Tharsis province, because of its size, constitutes one of the most tectonically remarkable volcanic structures in the solar system.

Elysium and Tharsis both form large-scale topographic (but not necessarily structural) domes that rise 6 and 10 km above the surrounding plains and have diameters of 40° and 90° of the surface arc, respectively. The Tharsis Dome (about 4000 km across) is flanked by depressions in Chryse and Amazonis Planitiae. Both domes have associated positive free air gravity anomalies and the flanking depressions of Tharsis have negative free air anomalies. The Tharsis anomaly dominates the second and third degree harmonics in the global gravity field (Phillips and Saunders, 1975; Lambeck, 1979; Sjogren, 1979), thus strongly emphasizing its global importance.

Both the Tharsis and Elysium provinces are capped by several individual volcanoes that rise in height an additional 5-15 km to a maximum elevation of approximately 25-27 km above the mean Mars datum. Multiple periods of caldera collapse, producing summit craters as much as 100 km in diameter, are evident on these volcanoes. Many individual lava flows on the flanks are in excess of 500 km in length (for an extended review of martian volcano morphology, see Chapter 5).

Tharsis province: interior structure

Evidence of the relative ages of late martian volcanism comes from cratering intensity studies as well as from structural and volcanic stratigraphic relationships. Although we have no reliable way of tying martian crater chronology to that of the isotopic age determinations on the Moon, we can probably assume that the high impact flux died away at about the same time for both objects. From this it appears that the later structural and volcanic events of Mars extended over a lengthy period (see Chapter 8). This implies that the total stress transmitted to the lithosphere at any one time was probably readily withstood without shear failure.

As there are no plate boundaries on Mars and the lithosphere is in one piece, later martian volcanoes are most like large intraplate volcanoes on Earth, such as those of Hawaii and Northern Africa. The Hawaiian and African volcanoes may reflect underlying mantle heterogeneities or may be responses to lithospheric fracturing; both hypotheses have their supporters.

Geophysical data, mainly gravity and topography, and surface structural features provide important constraints on the genesis and present interior structure of Tharsis. As reviewed by Mutch *et al.* (1976), Tharsis has been explained as an updomed region of the surface. This model is supported by the high elevation of old, densely cratered surfaces.

The radial pattern of Tharsis faulting (see below) has also been cited as evidence for doming. Wise *et al.* (1979a) and Plescia and Saunders (1979b), however, suggested that these features represent superimposed linear trends rather than a contemporaneous radial pattern. Alternatively, the fracture system may be a response to loading.

The relative importance of updoming in the Tharsis region has recently been questioned by Solomon and Head (1980b). They pointed out that the traditional view that the elevated, densely cratered terrain of Tharsis implies updoming of an earlier surface formed at lower elevations is not required by photogeologic observations. It is proposed that Tharsis and Elysium may be equally well explained as thick constructional volcanic piles partially supported by an elastic lithosphere.

Early studies of gravity and topography indicated that the topography could be only partially supported by shallow compensation due to variations in crustal thickness or density (Phillips and Saunders, 1975). Possible dynamic support of the Tharsis region by mantle convection is discussed by Phillips and Ivins (1979). An axially symmetric, first harmonic mode of thermal convection in the sub-lithospheric mantle with upwelling beneath Tharsis would be suggested. The absence of an antipodal, basin-like feature with a negative gravity anomaly might be explained by the diffuse nature of the downgoing convective flow that results from the temperature dependence of viscosity. Another mechanism for localizing convection beneath Tharsis is a concentration of heat release due to core formation. Wise *et al.* (1979a) speculated that a mechanism involving mantle convection and core formation resulted in subcrustal erosion beneath the northern plains and the emplacement of this crustal material beneath Tharsis.

Sleep and Phillips (1979) have shown that an isostatic model will satisfy the gravity data if the Tharsis province can be described in terms of a thin crust and a low-density upper mantle. Typical values, with a lithospheric thickness greater than 300 km, are a martian crust about 100 km thick, in general, thinning to about 50 km in the center of Tharsis, and an upper mantle (above about 300 km) beneath Tharsis contrasting in density with normal martian upper mantle by about 0.2 gm/cm^3. Sleep and Phillips (1979) have suggested that there is a resemblance between Tharsis volcanism, elevation and gravity field and those of the Basin and Range province. As also pointed out by Sleep and Phillips, a much closer resemblance exists to the great uplifts of Northern Africa described by Burke and Whiteman (1973) and, following them, Fairhead (1979), who suggested that these uplifts were precursors of tectonics of the African Rift valley type.

The terrestrial analogues to Tharsis are sites of active volcanism. The uplifts are dynamically maintained because of the underlying partial melting, and the lithosphere is thin under both the Basin and Range province and the African uplifts. Fairhead and Reeves (1977) inferred this for the African areas by correlating P delays for Kazakhstan nuclear tests with Bouguer anomalies; Burke and Whiteman inferred it simply because the erupted basalt comes from depths of about 50 km. If, in the future, volcanic activity ceases on the African uplifts, the elevations will be rapidly removed by erosion, the underlying mantle will revert to being part of the lithosphere (consequently becoming denser) and the associated gravity anomaly will not persist. In the Sleep and Phillips (1979) model, martian lithosphere under Tharsis is regarded as rather thick to maintain lateral density inhomogeneities, and the terrestrial analogues cannot be applied. However, dynamically maintained, low-density upper mantle is also consistent with the model, but if the terrestrial analogues are to be pressed very far, evidence of currently active volcanism at Tharsis seems necessary.

Tharsis province: lithospheric evolution

It seems probable that the thickness of the martian lithosphere in the Tharsis area has varied with time. By analogy with Africa and the Basin and Range, the lithosphere in Tharsis thinned during the time of volcanic activity. Locally thinned lithosphere causes stress concentration and localizes rupture, but it is not usually possible to tell whether the rupture happens first and volcanism is concentrated at the fractures, or whether magmatism first thins the lithosphere. In Tharsis there is some evidence of faulting earlier than much of the volcanism. Whatever the sequence of events, a locally developed, thin lithosphere would have characterized Tharsis during active volcanism and the huge load would have been maintained by the underlying, mass deficient, partly melted mantle. When volcanic activity ceased at Tharsis, the lithosphere probably reverted to the contemporary thickness of the lithosphere over the rest of the planet. Consequently, if the elastic lithosphere below Tharsis is supporting the vast load without continuing extensive failure (and this seems to be the case), then it may prove to be some hundreds of kilometers thick.

Other estimates for martian lithospheric thickness may be obtained by assuming that the magma source region for surface volcanism was located at the bottom of the lithosphere. The hydrostatic rise of magma to the summit of Olympus Mons suggests a magma source region at a depth of only a few hundred kilometers (Vogt, 1974a; Phillips and Ivins, 1979). Several studies (Carr, 1974; Vogt, 1974b; Plescia and Saunders, 1979a) have indicated increasing height and spacing of Tharsis volcanoes with decreasing inferred age. Plescia and Saunders (1979a) in particular have noted that older shield volcanoes were small low features while younger shields are significantly broader and taller. Following the earlier suggestion of Vogt (1974a) for terrestrial oceanic islands, it is interesting to speculate that increasing volcano heights with decreasing relative age are a result of a lithosphere that thickens with time.

Tharsis province: tectonic deformation

Four kinds of evidence are relevant in considering tectonic character of the later volcanism on Mars: the associated structural style, the gravity field, the distribution of the volcanism and its duration.

The surface of Mars is highly fractured (Mutch *et al.*, 1976, p. 226) and very many of these fractures are associated with extension and with graben formation. This structural pattern is best known in the Tharsis plateau where radial swarms of grabens extend south, southwest, north, and northeast of the uplift for as much as 2500 km (Fig. 6.5.5). Such grabens are 2-5 km in width, and in places (e.g., Tempe Fossae; Fig. 6.5.6) form a broad belt of fractures about 500 km wide. There appears to have been a long history of deformation in the plateau region where several sets of grabens show different orientations. Wise (1974b) documented a sequence of events in the Alba and Tempe regions on the northeastern fringe of the Tharsis uplift that involved five geologic episodes including two volcanic episodes separating and succeeded by fracturing and faulting. Complexes of interlocking grabens are particularly well-developed in the Noctis Labyrinthus area. The most prominent fracture system radiating from Tharsis is the huge Valles Marineris whose straight walls attest to a structural origin; it has been compared to the great rifts of Africa (Hartmann, 1973). In addition, an extensive series of mare-like wrinkle ridges is located to the east of the Tharsis dome in Lunae Planum (Fig. 6.5.7). Those ridges have a pronounced north-south orientation, are typically 300–400 km in length, and have a regular spacing of approximately 400 km.

Several recent studies examine the development of the Tharsis tectonic features. Wise *et al.* (1979a) identified two main centers of faulting: (1) an early center in the Thaumasia area southeast of the Tharsis dome with four distinct fault systems with linear trends and (2) a younger, northwest-southeast, elongated, radial set with a center near the crest of the present Tharsis dome. Lunae Planum wrinkle ridges are reported to form an elliptical pattern about a different center. Based primarily on features in the vicinity of Valles Marineris, Frey (1979a) also identified a dome in Thaumasia as an early center of faulting and uplift. He described the faulting as radial about the Thaumasia and Tharsis domes, and suggested that the Lunae Planum wrinkle ridges are radial to the Thaumasia center. Plescia and Saunders (1979a) also believed that they could identify two centers of fracturing, one near Pavonis Mons and the second within Syria Planum.

Despite the lack of general agreement on the history of faulting, it is clear that several periods of faulting generally radial to Tharsis took place. Apparently, the largest of these faulting episodes produced the grabens in Memnonia and Tempe Fossae with the possible initiation of the Valles Marineris rifting (Wise *et al.*, 1979a; Frey, 1979a). In addition, the three Tharsis summit volcanoes (Arsia, Pavonis and Ascraeus Montes) all lie on the same great circle, which also passes through the heavily faulted Tempe Fossae region over 2000 km to the northeast (Fig. 6.5.8). This suggests that fractures may have provided an important control on surface volcanism.

The tectonic development of Tharsis has been compared to the East African Rift System and to the tectonic province which comprised portions of North America, South America and Africa during the initial

6.5.5 Tectonic map of Mars. Fractures/grabens (central circles) and faults (circles on downthrown side) are shown in green. Mare-type ridges, which may be of compressional origin, are shown in red. Adapted from Scott and Carr (1978).

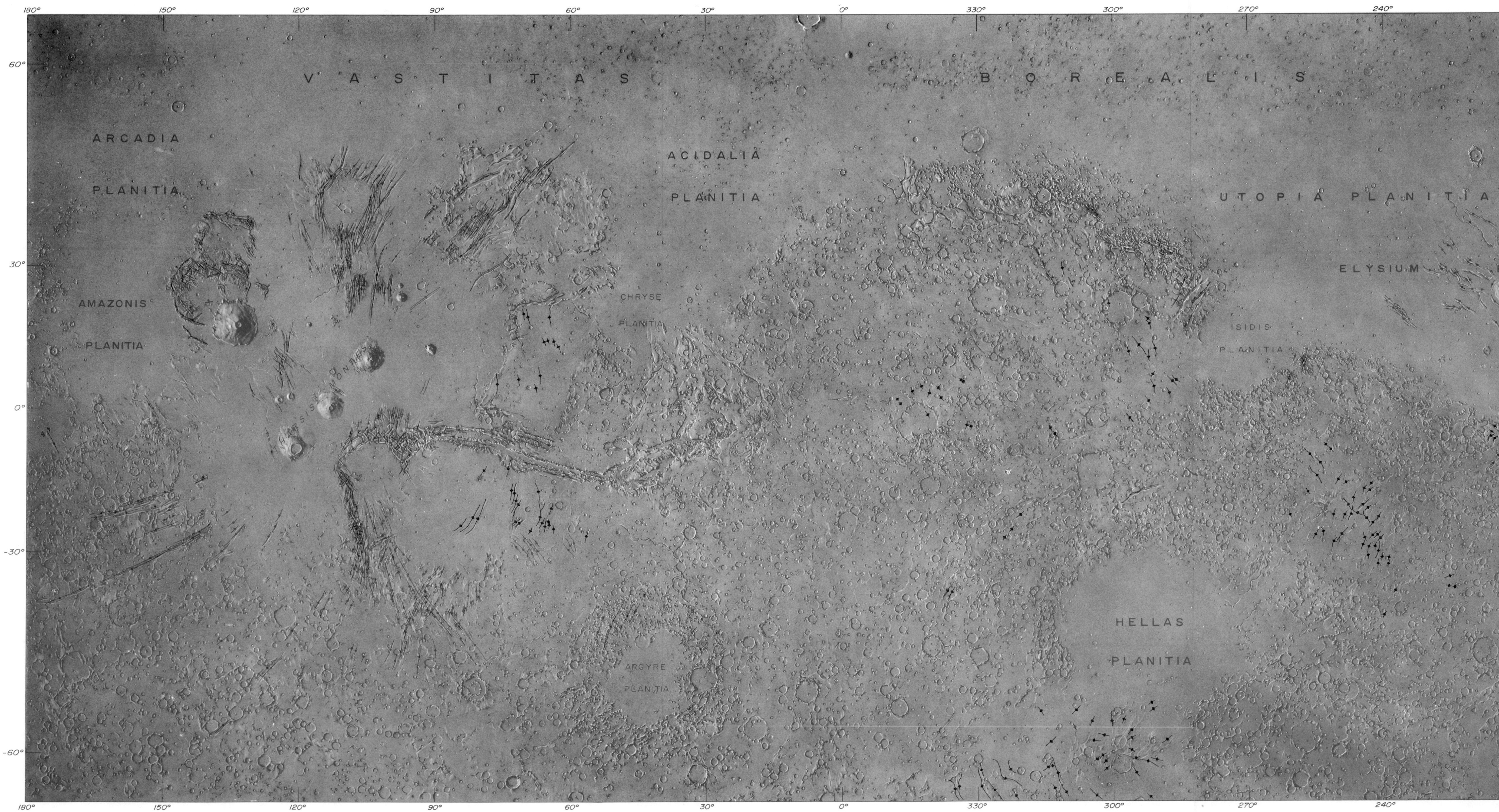
180°
150°
120°
90°
60°
30°
0°
330°
300°
270°
240°
60°
30°
0°
-30°
-60°
VASTITAS BOREALIS
ARCADIA
PLANITIA
ACIDALIA
PLANITIA
UTOPIA PLANITIA
ELYSIUM
AMAZONIS
PLANITIA
CHRYSE
PLANITIA
ISIDIS
PLANITIA
HELLAS
PLANITIA
ARGYRE
PLANITIA

Overleaf:

6.5.5 Tectonic map of Mars. Fractures/grabens (central circles) and faults (circles on downthrown side) are shown in green. Mare-type ridges, which may be of compressional origin, are shown in red. Adapted from Scott and Carr (1978).

opening of the Atlantic Ocean (see Mutch *et al.*, 1976). Possible similarities have been identified between the Valles Marineris rifting episode and features observed in the East African Rift (Frey, 1979b). Several attempts have been made to correlate the creation of the Valles Marineris (Coprates) canyon system with incipient plate tectonics on Mars (Courtillot *et al.*, 1975; Frey, 1979a; Schonfeld, 1979). This idea of incipient lithospheric spreading has, however, been dismissed by Sharp (1973), Phillips and Ivins (1979), Carr (1980) and Frey (1979a).

Nearly all structures recognized around Tharsis are extensional. Şengör and Jones (1975) and a very few other workers have drawn attention to sigmoidal structures which are strongly suggestive of horizontal shear in the Tharsis region and they, almost alone, have also seen evidence of convergent structures. The wrinkle ridges, seen in Lunae Planum, are probably compressional in origin, but are minor features when contrasted with the extensive graben systems. Operation of planet-wide tectonics with a predominance of extensional

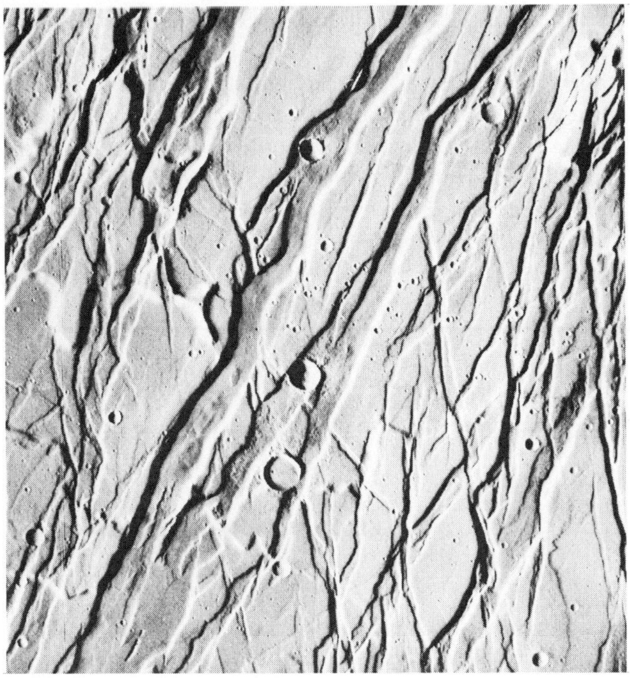

6.5.6 Multiple grabens 2000 km northeast of Tharsis are typical of the Tempe Fossae region on Mars. Viking frame number 627A36.

6.5.7 Wrinkle ridges morphologically similar to lunar mare ridges are common features in the Lunae Planum region of Mars. Viking frame number 610A22.

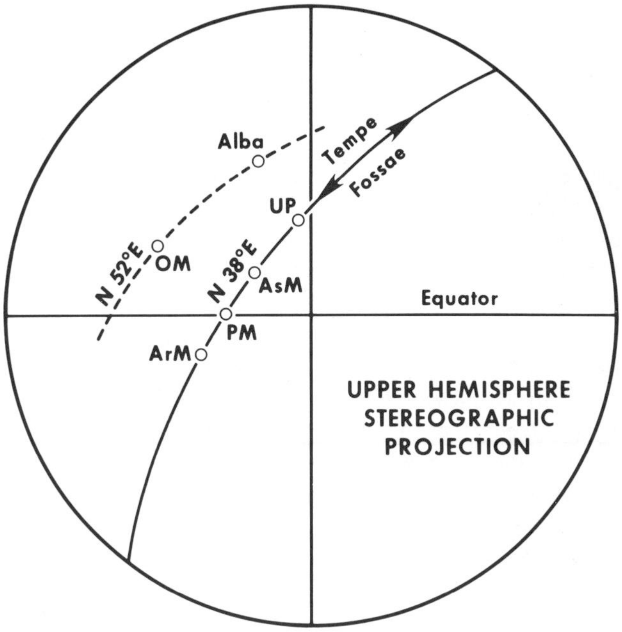

6.5.8 Alignment of the Tharsis Montes line of volcanoes. Locations of major volcanoes have been plotted on the upper hemisphere of a stereographic projection with 0° longitude at the right and 180° longitude at the left. The four volcanoes lie within a half-degree of a great circle oriented N38E. Tempe Fossae lies on the northeast projection of the same great circle. Olympus Mons and Alba Patera lie on a different great circle. Abbreviations: OM, Olympus Mons; UP, Uranius Patera; AsM, Ascraeus Mons; PM, Pavonis Mons; ArM, Arsia Mons. From Wise *et al.* (1979a).

features is possible because rocks are stronger in compression than in extension.

A variety of alternative interpretations based on Tharsis tectonic features has been proposed. Solomon and Chaiken (1976) emphasized the possible relationship between extensional tectonics and volume change due to planetary thermal evolution. Mutch *et al.* (1976) considered that expansion might be due to phase transitions in the mantle. McAdoo and Burns (1975) suggested that polar wandering of the lithosphere might also explain tensional faulting in the Valles Marineris complex. Melosh (1980) has also recently discussed tectonic patterns that might have resulted from such a reorientation of the lithosphere relative to the rotation axis. The movement of the excess mass of Tharsis toward the rotational equator of the planet, as a reorientation of the principal moment axes, is a possible driving mechanism. These studies, which use a faulting criterion different from that of McAdoo and Burns (1975), suggest that extensional structures radial to Tharsis are not related to polar wandering.

Gravity and topography only partially constrain the possible range of models which might explain Tharsis tectonism. From these constraints alone, it remains unclear if the Tharsis topography and the gravity anomaly are supported isostatically by density variations within the lithosphere, are supported by the strength of the lithosphere, or are supported by dynamic processes in the deeper viscous interior. Some selection from among these models is possible, however, with lithospheric stress modelling based on the gravity and topographic loading (Phillips *et al.*, 1981a). Using the general formulation for a self-gravitating elastic lithospheric shell with a hydrostatic interior, both an isostatic model patterned after Sleep and Phillips (1979a) and a model in which Tharsis is supported by the finite elastic strength of the lithosphere have been tested. The second model is consistent with the notion of Solomon and Head (1980b) that Tharsis, instead of being dominantly a structural dome, is for the most part a vast volcanic pile acting as a flexural load on the lithosphere. The resulting global surface patterns of principal deviatoric stresses show moderate to good agreement in the orthogonality of tensile stresses with inferred tensional surface features (e.g., grabens) and of compressive stresses with inferred compressional features (e.g., wrinkle ridges in Lunae Planum). However, the flexural model is clearly superior in comparison to the isostatic model in its agreement with the geologic features, indicating the likelihood of at least partial support by the finite strength of the lithosphere.

Whichever model (flexural or isostatic) is correct, however, a key result is that the tectonic features associated with Tharsis, including Valles Marineris, are clearly related to the incumbent loading of the lithosphere, and are not the result of structural updoming, which would have produced the conjugate set of structures. The lack of angular isotropy and contemporaneity of the fractures most likely reflects lithospheric heterogeneity (and evolution thereof). While the preferred (flexural) model is suggestive of volcanic constructional loading, significant internal loading (intrusives, mantle thickening, etc.) is not ruled out. Independent of its origin, Tharsis topography, once created, would tend to sustain access of magmas through the lithosphere because of the tensile stresses associated with the mere presence of the relief itself.

6.5.3 Tectonics of basaltic volcanism on Mercury

Mercury is little known in comparison with Earth, the Moon and Mars, so consideration of tectonics in relation to volcanism is extremely speculative. Mariner 10 images provide the sole basis for studies of the mercurian surface. For example: Mercury's smooth plains may be formed either of volcanic material (Trask and Strom, 1976) or of ejecta (Wilhelms, 1976; Oberbeck *et al.*, 1977), and it is clearly very difficult to know whether these deposits are basaltic, even if they are volcanic (Malin, 1978).

Strom *et al.* (1975a) suggested that irregular scarps restricted to crater floors may be lava flow fronts. If so, their heights of 100–500 m show that they were formed by very viscous lava compared to that of the lunar maria. A strong argument in favor of volcanic activity at some time in Mercury's early history is that the planet's magnetic field requires at least a partially molten metallic core. Formation of such a core might dissipate enough energy to ensure volcanic activity.

The absence of recognizable surface features which can be related to plate tectonics has led to the general consensus that Mercury, like the Moon, has had a mechanically continuous lithospheric shell dating at least from the end of the period of heavy bombardment. Tectonic features on Mercury show some similarities but also important differences from those on the Moon. Unlike those of the Moon, the major tectonic features are thought to have developed in response to tidal deformation and planetary volume change as well as basin loading (Burns, 1976; Melosh, 1977; Pechmann and Melosh, 1979).

If there is volcanic activity recorded on the surface of Mercury, the passage for the magma to the surface is probably, like that on the Moon, overwhelmingly linked to fracturing associated with impact cratering. There are also resemblances in morphology and distribution between the smooth plains of Mercury and the lunar maria (Strom *et al.*, 1975a). One very large crater, 1300 km in diameter, the Caloris Basin, was surveyed by Mariner 10. Wrinkle ridges similar in morphology to those on the lunar maria are developed in smooth plains units within this basin (Fig. 6.5.9). These ridges have orientations radial and concentric to the basin and are interpreted as compressional features (Strom *et al.*, 1975a). Ridges range in width from 1 to 12 km, attain heights of 500 m, and are continuous along their strike for as much as 300 km. Within the Caloris Basin, the ridges are cut by trough-like features forming a polygonal pattern with radial and concentric trends that have been interpreted as grabens or tensional fractures (Strom *et al.*, 1975a; Fig. 6.5.9). The fractures become progressively wider and deeper toward the basin center. Similar features have not been observed on the lunar maria. The wrinkle ridges and troughs in smooth plains within and around the Caloris Basin have been interpreted as structural features developed in response to basin tectonics (Maxwell and Gifford, 1980). Wrinkle ridges are hypothesized to be compressional features formed due to subsidence of the basin resulting from the loading of smooth plains-forming volcanics. Fractures or grabens, with trough-like morphologies, are related to uplift of the central portion of the basin. These fractures may have provided the only conduits for basaltic magmas to reach the surface.

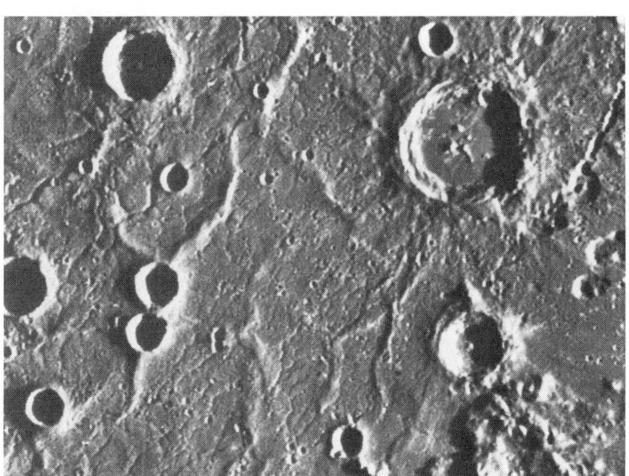

6.5.9 Part of the floor of the Caloris Basin showing the ridges and fractures. The length and width of the fractures increase toward the center of the basin (lower right to upper left) and the fractures transect the ridges at various angles (photo FDS 126). After Strom *et al.* (1975a).

6.5.10 Examples of scarps on Mercury extend off to the horizon in this Mariner 10 image (photo FDS 27328).

The 30% of Mercury imaged by Mariner 10 does, unlike the Moon, show evidence of a regional, perhaps even global, tectonic development (Fig. 6.5.10). This evidence lies in the lobate scarps, recognized by Strom *et al.* (1975a), that vary in length from 20 to 500 km and cross several different types of terrain (Fig. 6.5.11). Because they are cross-cutting, generally linear and lengthy, and because of a horizontal offset of a crater-rim (the wall of crater Guido Arezzo is displaced 10 km by the Vostok scarp, Fig. 6.5.12; Strom, 1979), the scarps are considered to be faults. Strom *et al.* (1975a), noting a contrast between scarp morphology and that of normal faults on the Moon and Mars, suggested that the scarps on Mercury are thrust faults. Because some of these scarps are disrupted by fresh craters larger than 40 km, it is believed that scarp formation was occurring near the end of heavy bombardment, and continued beyond the emplacement of smooth plains.

Melosh and Dzurisin (1978) indicated that arcuate scarps formed in smooth plains surrounding the Caloris Basin are smaller and fewer in number than scarps in older terrains. Because the lobate scarps are relatively uniformly distributed over the part of the surface of Mercury viewed by Mariner 10, it seems possible that an overall decrease in the surface area of the planet may be recorded in this thrusting. Strom *et al.* (1975a) estimated a decrease equivalent to a 1–2 km reduction in the radius of Mercury. Solomon (1976) calculated a comparable decrease by considering the thermal history of the planet (see Chapter 9). It is interesting to consider that a process which was widely considered during the 19th century to have been responsible for mountain building on Earth, but which was abandoned after the

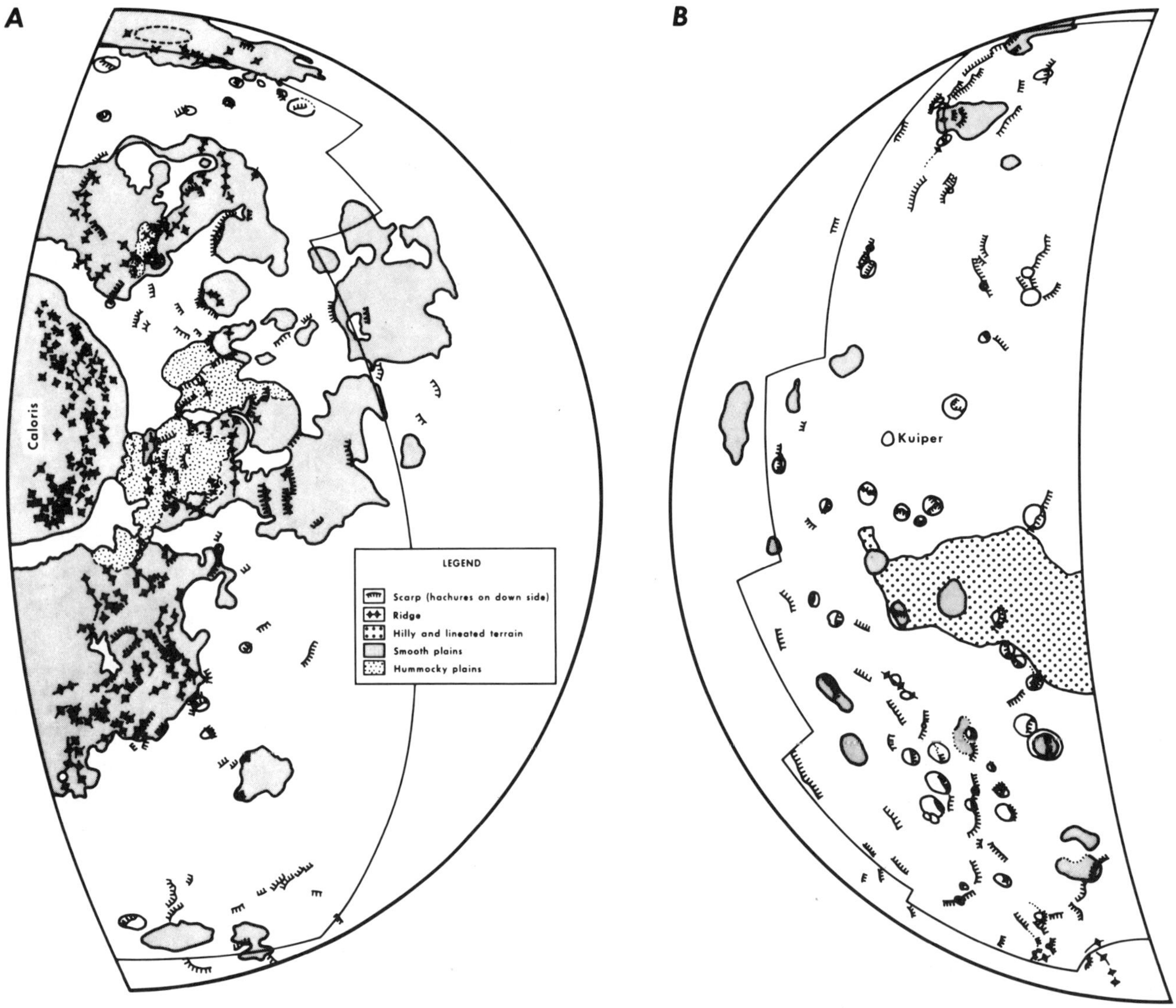

6.5.11 A tectonic map of Mercury showing lobate scarps and ridges on the incoming (**b**) and outgoing (**a**) side of Mercury. Quasi-circular features are craters. Modified from Strom *et al.* (1975a).

discovery of radioactivity, may have been dominant in tectonic evolution on Mercury.

Burns (1976) first examined the stresses and patterns of deformation that might result from tidal despinning. The combined effects of planetary volume change and tidal despinning were treated by Melosh (1977) and by Pechmann and Melosh (1979). For the tidal despinning of a planet with a thin elastic lithosphere, an equatorial zone of strike-slip faults and a polar region of east-west striking normal faults are predicted. Simultaneous contraction suppresses the region of normal faulting and introduces an equatorial zone of north-south striking thrust faults. In the absence of stresses due to tidal despinning, contraction should result in isotropic compressional stresses within the lithosphere and in a random orientation of thrust faults on the surface.

The relative importance and timing of tidal despinning and contraction to the development of surface structural patterns remains a subject of discussion. Cordell and Strom (1977) considered that the orientation of lobate scarps is random and not the north-south orientation expected for tidal despinning. Melosh and Dzurisin (1978) suggested that the more restricted class of arcuate scarps has a preference for north-south orientation and that the northeast and northwest trending

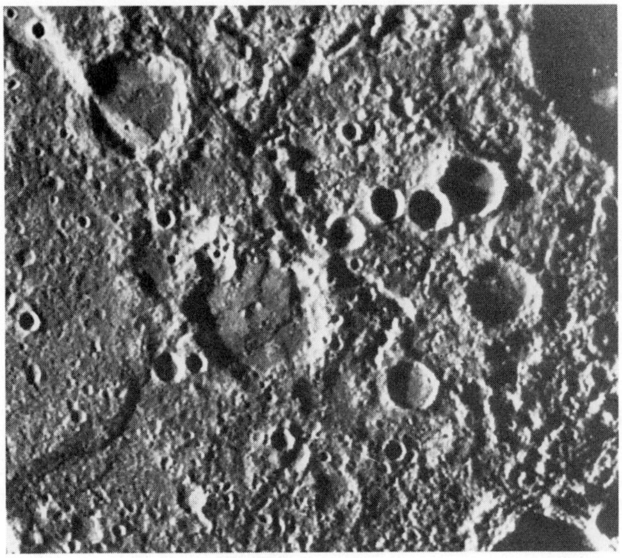

6.5.12 Two craters (unnamed and Guido d'Arezzo) are cut by the Vostok scarp (about 130 km long). The northwestern rim of Guido d'Arezzo, 65 km in diameter, has been offset about 10 km by the scarp. This offset appears to have been caused by shortening of the crater due to thrusting of the eastern part of the crater over the western part (photo FDS 27380).

structural lineaments were produced by strike-slip faulting. They proposed that continued contraction after tidal despinning could result in the observed range of arcuate scarp orientations. Normal faulting in polar regions appears to be a critical test of the importance of tidal despinning in controlling planetary tectonic patterns. Such features have not yet been observed. Strom *et al.* (1975b) indicated that numerous lobate scarps occur in south polar regions, but the ages of these scarps relative to tectonic features elsewhere have not been established.

The fundamental lesson to be learned from photogeologic examination of the surface of Mercury is that the lack of clearly identifiable volcanic flows on the surface of the planet is probably related to a lithosphere which has been in compression over much of the history of the planet (Solomon, 1978). Since scarp formation was occurring near the end of the period of heavy bombardment (~4 b.y. ago?), the lithosphere was probably in compression at that time or earlier and conditions were not favorable for magma migration to the surface. Very early surface volcanism, during a period of lithosphere extension, would be obscured by the high impact flux. Subsequent volcanism, if it occurred at all, must have been closely related in space and time to major impact events.

6.5.4 Tectonics of basaltic volcanism on Venus

Venusian tectonics remain less well understood than the tectonics of the other terrestrial planets due to the lack of detailed surface observations. Both Earth-based radar (Malin and Saunders, 1977) and the Pioneer Venus orbiter (Pettengill *et al.*, 1979, 1980; Masursky *et al.*, 1980a, b), however, provide low resolution topographic data. Precision tracking of the Pioneer spacecraft has also been used to determine the global gravity field (Phillips *et al.*, 1979b; Sjogren *et al.*, 1980; Ananda *et al.*, 1980).

In early Earth-based radar observations (Saunders and Malin, 1975), circular features with the gross morphologies of impact craters, one large trough-like depression comparable to Valles Marineris or to the East African Rift, and several mountainous regions of high relief were identified. Confirmation of these earlier observations and more complete coverage of surface topography has been obtained using the orbiting Pioneer Venus spacecraft (Pettengill *et al.*, 1979, 1980). From these data, Masursky *et al.* (1980) reported that about 27% of the surface of the planet consists of lowlands with elevations (relative to the mean radius of the planet) less than 0 km, 65% consists of upland plains with elevations of 0–2 km, and the remaining 8% is highlands with regional elevations in excess of 2 km. The highlands are dominated by two continent-sized regions, Aphrodite and Ishtar, with mountainous features reaching elevations of slightly more than 11 km. The uplands terrain is characterized by shallow circular features which may be impact craters and by radar-bright linear features which may be fault zones. The distribution of relief is reported to be unimodal with large areas of the surface at a nearly uniform elevation. Pettengill *et al.* (1980) indicate that the center of mass-center of figure offset is less than 400 m.

Although smaller than Aphrodite or Ishtar, the elevated Beta Regio appears to be associated with basaltic volcanism. On the basis of radar interpretation, the elevated southwestern crest of Beta may be a shield volcano (Saunders and Malin, 1977). The Venera 9 and 10 spacecraft landers determined the surface rocks on the flanks at Beta to have K, U, and Th contents similar to those of terrestrial basalt (Surkov *et al.*, 1976). Whether other highland regions are basaltic or not is problematical. Aphrodite occurs along a linear equatorial zone of elevated topography that includes Beta Regio and by inference might be associated with basaltic volcanism.

A depression to the east of the Beta "shield" has been mapped by Pioneer Venus and appears to be part of a chain of depressions defining a trough extending southward from 35° N the full north-south length of Beta Regio (McGill *et al.*, 1981). A similar trough south of Beta is seen in Earth-based radar images (Goldstein *et al.*, 1976), so that there exists a moderately well-defined north-south elevated trend with a median trough to 20° S, and possibly as far as 45° S. When compared with the East African Rift system, the morphological similarity is striking (Fig. 6.5.13). However, the latter tectonic system is not characterized by basaltic shield volcanism. Other "ridge and trench" and lineament systems occur on Venus (Masursky *et al.*, 1980). In total, there is ample evidence for horizontal stress in the lithosphere of Venus, but there is no way to clearly decide whether the origin of this stress is due to lithospheric plate motion or to regional loading of the lithosphere.

Gravity anomalies on Venus are relatively small compared to those for the Moon or Mars and have magnitudes more comparable to those of the Earth

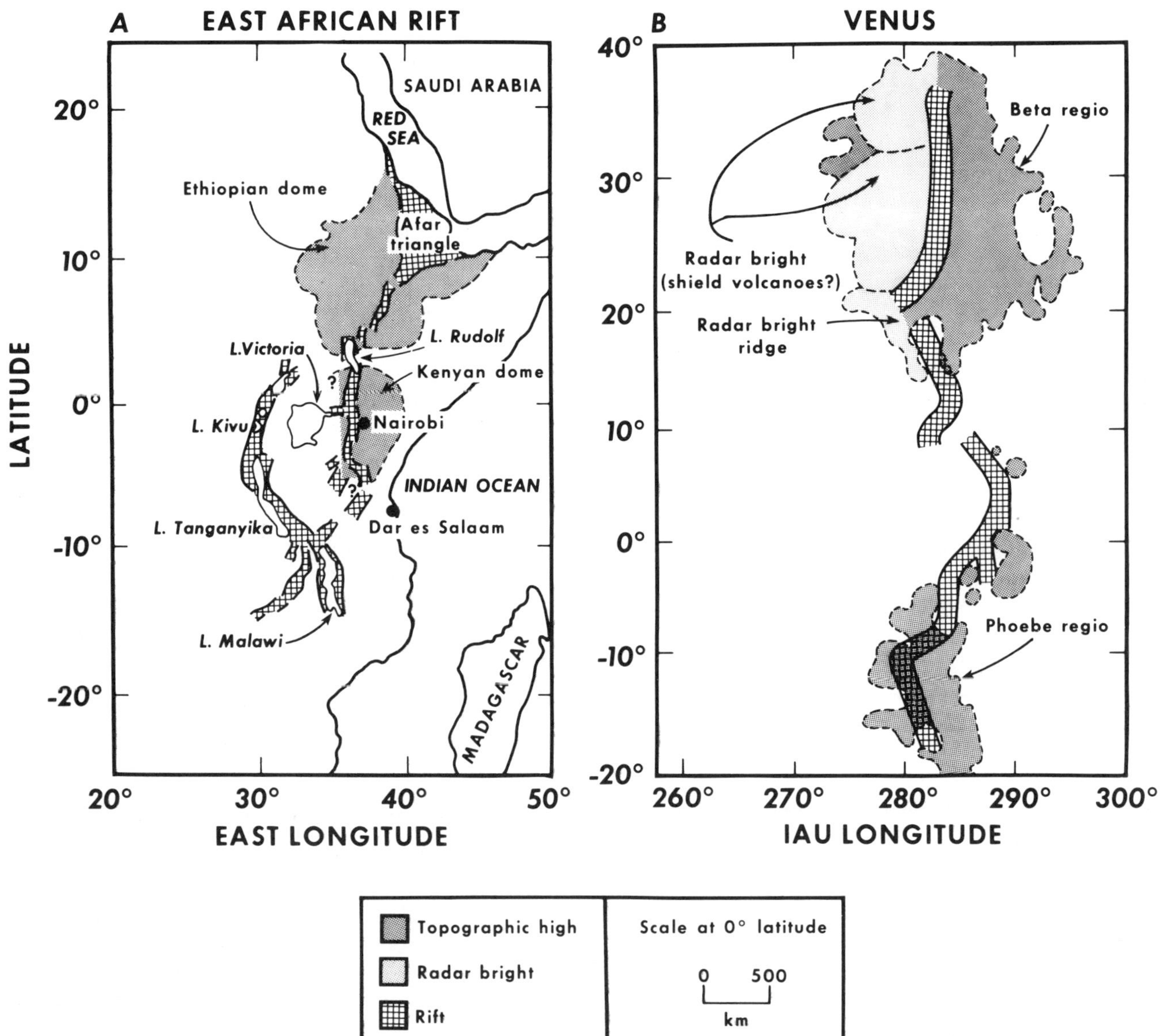

6.5.13 Comparison of East African Rift System with proposed rift system associated with Beta Regio and Phoebe Regio regions of Venus. Topographically high areas that are bright in Earth-based radar images are noted; they have been proposed as shield volcanoes. (Figure from George McGill, pers. comm., 1981.)

(Phillips *et al.*, 1979; Sjogren *et al.*, 1980). In contrast to the Earth, however, there appears to be a good correlation between gravity and topography at long wavelengths. Gravity anomalies are a factor of 4 or 5 smaller than would be predicted for uncompensated topography with a crustal density of 3 gm/cm^3 and are most consistent with compensation at a depth of about 100 km (Phillips *et al.*, 1981b). If Venus contains heat sources similar to the Earth, then the high Venus surface temperature implies an elastic lithosphere thickness no greater than 50 km, for even the most creep-resistant (as measured in the laboratory) geologic material. Thus the compensation of 100 km depth suggests that the topography is dynamically maintained (e.g., by convection) or is very young geologically.

If the impact crater interpretation of the upland circular features is correct, then vast areas of Venus are geologically old, and it is unlikely that present-day crustal generation by plate tectonics similar to that on Earth is taking place. An apparent absence of the characteristic concave cross-sectional topographic profiles associated on Earth with cooling diverging oceanic plate boundaries is further evidence against a present-day global system of tectonic plate motions on Venus. The linear zones of elevated topography found on Venus may be in essence the "pile-up" of new (basaltic?) crust in a lithospheric regime which does not allow spreading. An inability of the Venus lithosphere to subduct crustal materials might be attributed to the present surface temperature, which is about 450° higher than that of the Earth (Phillips *et al.*, 1981b). The lithosphere on Venus might be of relatively lower density and thus more buoyant and difficult to subduct because of a deeper basalt-eclogite transition (Anderson, 1980) and because of, if all other factors are equal, a higher Mg/Fe ratio in basaltic magmas.

The lack of contemporary plate tectonics does not rule out the possibility of significant horizontal stress in the venusian lithosphere. The great elevations of the highlands regions, particularly at Ishtar, impart stresses to the interior of at least several kilobars, even if they are in isostatic equilibrium. Such great elevations might require support both from density compensation and horizontal lithospheric stress, as has been suggested for the Tibetan Plateau on Earth (Molnar and Tapponnier, 1978). Phillips *et al.* (1981b) suggested that Ishtar might be the only true terrestrial-style continent on Venus and that it might be supported by the stresses associated with basaltic loading of the lithosphere in the equatorial highlands region, made up in part by Beta Regio and Aphrodite. Thus, a variety of important factors which may govern the tectonics on Venus has been proposed. It is clear, however, that better resolution of surface features is required to provide further constraints on venusian tectonics and its relationship to basaltic volcanism.

6.6 TECTONICS OF BASALTIC MAGMA MIGRATION

The origin of the magma that flows from most volcanoes on Earth is now reasonably well understood. A large fraction of the magma is produced by partial melting in the mantle, usually in the asthenosphere. (See section 1.4.1 for processes that influence basaltic magma composition. Also see section 9.4.3 for rheology.) The volcanism at ocean ridges is associated with pressure release melting during ascending mantle convection. Ascending convection of mantle rock is required to provide the material that makes up the spreading lithospheric plates. As the mantle rock ascends adiabatically and crosses its solidus, partial melting occurs. The magma then ascends through the mantle rock to form the oceanic crust.

The volcanism that forms the island arcs behind ocean trenches must be related to the descending lithospheric plate. Frictional heating occurs on the slip zone between the descending lithosphere and the overlying mantle. This heating may directly melt the descending oceanic crust including entrained sediments or it may result in the dehydration of hydrated minerals in the downgoing slab of crust. The resulting water may flux and cause partial melting of overlying mantle rocks. If these processes occur beneath continental crust, secondary melting may occur when the ascending magmas reach the lower crust. Intraplate volcanism is less well understood. Some fraction of this volcanism may be the result of pressure-release melting in mantle plumes. An alternative explanation is that the magma reaches the surface from the asthenosphere due to lithospheric fractures under tensional stresses.

Whatever the cause of the partial melting of the mantle rock, there is strong observational evidence that the initial melt fraction will form along the intersection of grain boundaries (Waff and Bulau, 1977, 1979). This is illustrated in Fig. 6.6.1a. When the degree of partial melting has reached a few percent, the magma which has collected along grain boundaries will form interconnected channels. The magma will then be free to migrate along these channels.

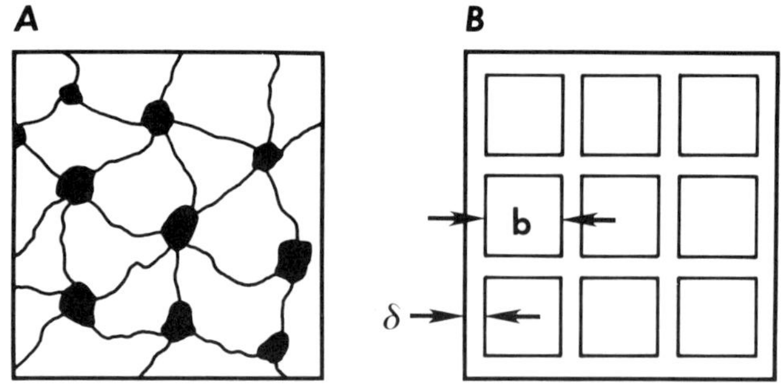

6.6.1 Magma migration: **a**. Initial melt forms along the intersection of grain boundaries in a rock. **b**. A model is used to allow permeability to be related to porosity. The rock is modelled as a cubical matrix of tubes of circular section. The matrix has dimensions b and the tubes have diameters δ.

In order to model the flow of magma along grain boundaries, the magma may be considered to be a fluid in a porous matrix, the solid crystals making up the matrix. This type of model has been proposed by Frank (1968) and by Turcotte and Ahern (1978). The volume fraction of magma present defines the porosity, X, of the porous medium. The velocity of the magma through the crystalline matrix v_l, is given by Darcy's law (Bear, 1972)

$$Xv_l = -\frac{k}{\eta_l}\frac{dp}{dz} = \frac{k}{\eta_l}(\rho_s - \rho_l)\,g \qquad (6.6.1)$$

where k is the permeability, η_l the viscosity of the magma, and dp/dz the vertical pressure gradient on the fluid. The vertical pressure gradient is assumed to be equal to the differential buoyancy of the magma and crystalline matrix, i.e., to the product of the density difference (ρ_l magma density, ρ_s crystalline matrix density) and the acceleration of gravity g. Implicit in this assumption is the negligence of any pressure difference between the fluid and matrix. At the high temperatures associated with partial melting, and on the relevant time scales, the crystalline matrix does not have sufficient strength to maintain a significant pressure difference.

In order to relate the permeability, k, to the porosity, X, a reasonable model is a cubical matrix of circular tubes; the matrix has dimensions, b, and the tubes have diameters, δ. This is illustrated in Fig. 6.6.1b. The resulting porosity is given by

$$X = \frac{3}{4}\pi\frac{\delta^2}{b^2} \qquad (6.6.2)$$

And for slow (laminar) flow in the channels the permeability is (Bear, 1972)

$$k = \frac{X\delta^2}{9b} \qquad (6.6.3)$$

Elimination of δ from Eqs. (6.6.2–3) gives

$$k = \frac{X^2b^2}{72\pi} \qquad (6.6.4)$$

The velocity of magma migration is obtained by substituting Eq. (6.6.4) into Eq. (6.6.1) with the result

$$v_l = \frac{Xb^2(\rho_s-\rho_l)g}{72\pi\eta_l} \qquad (6.6.5)$$

It is implicit in this model for magma migration that the crystalline matrix can deform to accommodate the loss of the liquid fraction as it percolates upward due to its differential buoyancy. Solid-state deformation processes, either ionic or dislocation, are sufficient to provide the deformation.

It is of interest to evaluate the magma migration velocity for typical asthenospheric conditions. The matrix dimension is taken to be b = 0.2 cm. This is a typical grain size for mantle rocks. The density difference is taken to be $\rho_s - \rho_l = 0.6$ gm/cm^3 and g = 10^3 cm/sec^2. The migration velocity is given in Fig. 6.6.2 as a function of the volume fraction of liquid for magma viscosities of 10, 100 and 1000 poises. A viscosity of 10 poises should be a reasonable value for a basaltic magma in the asthenosphere and andesitic magmas should have viscosities near 10^3 poises (Kushiro *et al.*, 1976; see also section 9.4.3). Typical mantle flow velocities are of the order of 1–10 cm/year. It is seen from Fig. 6.6.2 that the magma migration velocities are much larger than this when the liquid fraction is between 1–10%. Our conclusion is that magma will percolate upwards through the asthenosphere as rapidly as it is produced, although some recent studies (e.g., Waff, 1980; Walker *et al.*, 1979) suggest that relatively small magma bodies may not escape. The liquid fraction present will be close to that required to produce interconnected channels (that is, a few percent). At small percentages of partial melt the rate of magma flow is strongly dependent on the fraction of melt present, according to Sleep (1974).

An important question to address is whether the magma, as it percolates through the asthenosphere, will coalesce to form larger bodies of magma. In order to move through the crystalline matrix, these larger bodies would have to deform the matrix. It is appropriate to consider the motion of a spherical body of magma of diameter d. The crystalline matrix may be modelled as a Newtonian fluid with a solid-state viscosity η_s. Since this solid-state viscosity is much larger than the magma viscosity η_l the velocity at which the magma body will rise, v_l, is given by Lamb, 1945 as

$$v_l = \frac{d^2 g(\rho_s - \rho_l)}{12\eta_s} \qquad (6.6.6)$$

This velocity is given as a function of the diameter of the body in Fig. 6.6.3 for several values of the solid-state viscosity and other parameters as previously given. From studies of postglacial rebound, a reasonable value for the solid-state viscosity of the asthenosphere is $\eta_s = 4 \times 10^{20}$ poises. Because of the resistance of the crystalline matrix to deformation, the buoyancy-driven velocity of small magma bodies is infinitesimal. The percolation along the intersections of grain boundaries is a much more efficient mechanism for magma migration and it is likely to dominate in the asthenosphere.

Under the influence of gravity, magma is expected to rise vertically through the asthenosphere until it reaches the lithosphere. At oceanic ridges the lithosphere is thin and this flow is likely to reach the magma chamber, which is believed to lie beneath the central ridge crest. In other areas of volcanic activity, such as island arcs or intraplate volcanism, the lithosphere is considerably thicker and the magma is expected to collect at much greater depths. The magma that has migrated upward through the asthenosphere would be expected to form a liquid layer at the base of the lithosphere. A number of authors (i.e., Marsh and Carmichael, 1974) have proposed that the collected liquid will penetrate the lithosphere diapirically. The collected magma will certainly be lighter than the overlying lithosphere and a Rayleigh-Taylor instability will result if the lithosphere exhibits a fluid-like behavior. Laboratory studies of the diapiric penetration of a light, low-viscosity fluid have been carried out by Whitehead and Luther (1975). They found that nearly spherical bodies of the lighter fluid rose through the heavier fluid. This mechanism of magma migration is illustrated in Fig. 6.6.4a.

A model for the diapiric penetration of magma through the lithosphere is the vertical rise of a spherical magma body through a Newtonian viscous medium. The velocity at which such a body will rise has been given by Eq. (6.6.6). For lithospheric deformations a viscosity of 10^{24}–10^{26} poises is usually assumed. For these viscosities the velocities are extremely small as can be seen in Fig. 6.6.3. The transfer of heat from the magma body may significantly reduce the equivalent viscosity of the rock through which the rock is moving. The coupled problems of flow and heat transfer have been considered by Marsh (1976, 1978).

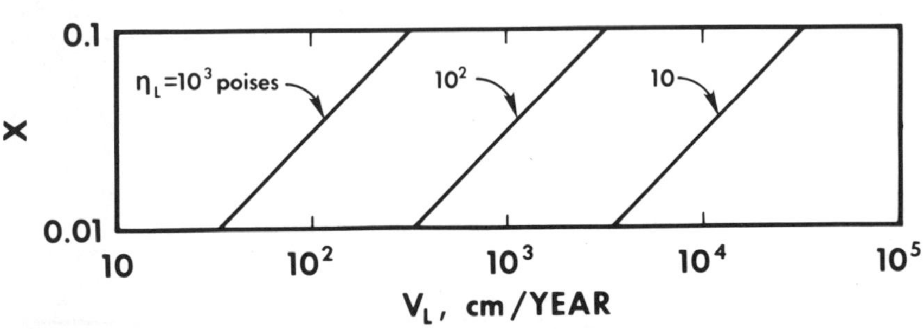

6.6.2 Migration velocity of magma (V_L) as a function of volume fraction of liquid (X).

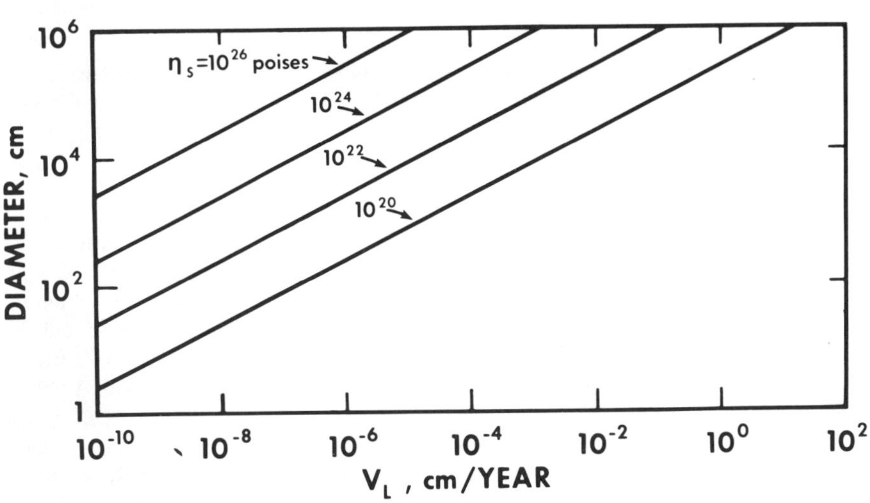

6.6.3 Velocity at which a magma body will rise (V_L) related to its diameter (d) for four values of the solid-state viscosity of the mantle.

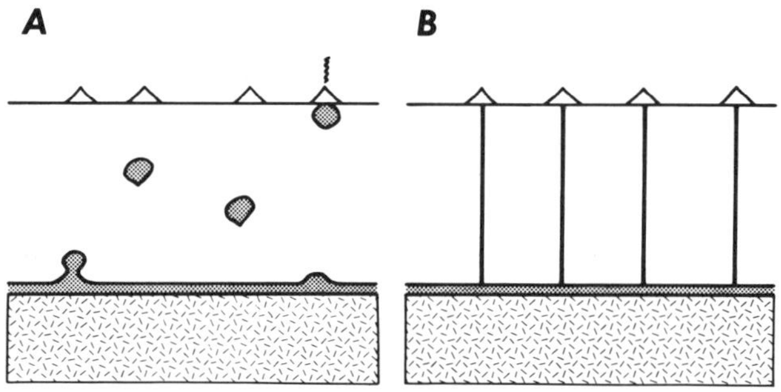

6.6.4 Magma migration. In **a** nearly spherical bodies of a lighter fluid rise through a denser fluid (as Whitehead and Luther, 1975, found experimentally) and in **b** the magma rises in continuous conduits.

An alternative model for the migration of magma through the lithosphere assumes that continuous conduits develop. These conduits act as pipes through which the magma flows. The model is illustrated in Fig. 6.6.4b. A volcanic conduit can be modelled as a circular vertical tube of diameter d. If it is assumed that the differential buoyancy is driving the flow, the vertical pressure gradient on the magma is $(\rho_s - \rho_l)g$. If the flow in the tube is laminar, the mean magma velocity is given by Schlichting (1955)

$$\bar{u}_l = \frac{d^2(\rho_s - \rho_l)g}{32\eta_l} \qquad (6.6.7)$$

If the diameter of the tube is sufficiently large the magma flow becomes turbulent. The transition occurs when $\rho_l \bar{u}_l d / \eta_l \approx 3000$. Substitution of Eq. (6.6.7) gives the critical tube diameter

$$d = \frac{45.78\eta_l^{\frac{2}{3}}}{[\rho_l(\rho_s - \rho_l)g]^{\frac{1}{3}}} \qquad (6.6.8)$$

If the diameter is greater than this, then the flow is turbulent and an empirical expression for the mean velocity is (Schlichting, 1955)

$$\bar{u}_l = 0.349\frac{(\rho_s - \rho_l)^{\frac{4}{7}} g^{\frac{4}{7}} d^{\frac{5}{7}}}{\rho_l^{\frac{3}{7}} \eta_l^{\frac{1}{7}}} \qquad (6.6.9)$$

For turbulent flow the mean velocity is a very weak function of the magma viscosity. In Fig. 6.6.5 the mean magma velocity is given as a function of the tube diameter for three values of the magma viscosity ($\rho_l = 2.7$ gm/cm^3, $\rho_s - \rho_l = 0.6$ gm/cm^3, $g = 10^3$ cm/sec^2). The mean velocities are in general quite large.

Fujii and Uyeda (1974) have suggested that the coupling between thermal dissipation and a strongly temperature-dependent viscosity will lead to flow instabilities in volcanic conduits. In a system with a constant driving force it has been shown (Gruntfest, 1963) that the decrease in viscosity caused by viscous heating can lead to flow instabilities. Hardee and Larson (1977) have questioned whether this mechanism can be responsible for the episodicity of volcanism. They doubt that the temperature dependence of the viscosity in a magma is sufficiently large over a sufficient temperature range to result in this type of instability.

It has been suggested that a fraction of the buoyancy head is required to build volcanoes to their observed height (Vogt, 1974; Edman, 1976). If this is the case, then conduits through the lithosphere are required.

Near-surface observations indicate that magma is capable of fracturing brittle rocks. A number of authors have considered this problem (Roberts, 1970; Pollard, 1973; Pollard and Muller, 1976). The wall cooling of a magma flowing through a dike has been considered by Maaloe (1973). It has been suggested by Weertman (1971) that magma may propagate through the lithosphere by hydrofracturing. Anderson and Grew (1977) have suggested a corrosion crack propagation mechanism.

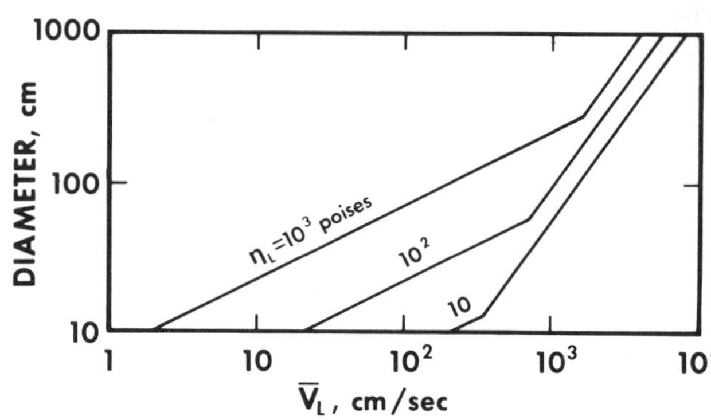

6.6.5 Mean magma velocity ($\bar{V}_L$) as a function of diameter of a conduit as illustrated in Fig. 6.6.4b for three values of magma viscosity.

ACKNOWLEDGMENTS—Discussions on Mars, with Donald Wise, proved stimulating, as did many discussions with members of other teams. The assistance of Janet Stroup, who edited much inchoate material, and of Donna Kelly, who typed most of the text, is greatly appreciated. Contributions to the various sections were as follows: 6.1—K. C. Burke; 6.2—K. C. Burke, W. S. F. Kidd, A. M. C. Şengör; 6.3—K. C. Burke, W. S. F. Kidd, J. F. Dewey and P. E. Tapponnier; 6.4—K. C. Burke and W. S. F. Kidd; 6.5—P. J. Mouginis-Mark, E. M. Parmentier, K. C. Burke and R. J. Phillips; 6.6—D. L. Turcotte.

6.7 REFERENCES

Abdulin A. A., Bespalov V. P. and Shelepova T. N. (1974) Ophiolite belts of Kazakhstan. *Doklady Akad. Nauk. S.S.S.R.* **218**, 159–162. (20–22 in translation)

Adamia S. A., Lordkipanidze M. B. and Zakariadze G. S. (1977) Evolution of an active continental margin as exemplified by the alpine history of the Caucasus. *Tectonophys.* **40**, 183–199.

Altinli E. (1966) Doğu ve güneydoğu Anadolu'nun jeolojisi. *Kisim I.M.T.A. Dergisi* **66**, 35–74.

Amm F. L. (1940) The geology of the country around Bulawayo. *Bull. Geol. Surv. S. Rhodesia* **35**, 307 pp.

Anada M. P., Sjogren W. L., Phillips R. J., Wimberly R. N. and Bills B. G. (1980) A low-order global gravity field of Venus and dynamical implications. *J. Geophys. Res.* **85**, 8303–8318.

Anderson D. L. (1975) Chemical plumes in the mantle. *Geol. Soc. Amer. Bull.* **86**, 1593–1600.

Anderson D. L. (1980) Tectonics and composition of Venus. *Geophys. Res. Lett.* **7**, 101–102.

Anderson E. M. (1951) The dynamics of faulting and dyke formation with applications to Britain, 2nd Edition. Oliver and Boyd, London. 206 pp.

Anderson O. L. and Grew P. C. (1977) Stress corrosion theory of crack propagation with applications to geophysics. *Rev. Geophys. Space Phys.* **15**, 77–104.

Anderson O. L. and Perkins P. C. (1974) Runaway temperatures in the asthenosphere resulting from viscous heating. *J. Geophys. Res.* **79**, 2136–2138.

Anderson R. N. (1972) Petrologic significance of low heat flow on the flanks of slow-spreading midocean ridges. *Geol. Soc. Amer. Bull.* **83**, 2947–2956.

Anderson R. N. (1975) Heat flow in the Mariana Marginal Basin. *J. Geophys. Res.* **80**, 4043–4048.

Anderson R. N. and Noltimier H. C. (1973) A model for the horst and graben structure of mid-ocean ridge crests based upon spreading velocity and basalt delivery to the oceanic crust. *Geophys. J. Roy. Astron. Soc.* **34**, 137–147.

Anderson R. N., Uyeda S. and Miyashiro A. (1976) Geophysical and geochemical constraints at converging plate boundaries - Part I: dehydration in the downgoing slab. *Geophys. J. Roy. Astron. Soc.* **44**, 333–357.

Anguita F. and Hernan F. (1975) A propagating fracture model versus a hot spot origin for the Canary Islands. *Earth Planet. Sci. Lett.* **27**, 11–19.

Anhaeusser C. R., Roering C., Viljoen M. J. and Viljoen R. (1968) The Barberton Mountain Land: a model of the elements and evolution of an Archean fold belt. *Trans. Geol. Soc. S. Africa* **71**, 225–253.

ARCYANA (1975) Transform fault and rift valley from bathyscaphe and diving saucer. *Science* **190**, 108–116.

Argand E. (1924) La tectonique de l'Asie. *C.R. Congr. Géol.* Intern., *13e*, **1**, 171–372.

Armstrong R. L. (1968) A model for the evolution of strontium and lead isotopes in a dynamic Earth. *Rev. Geophys. Space Phys.* **6**, 175–200.

Armstrong R. L., Leeman W. P. and Malde H. E. (1975) K-Ar dating, Quaternary and neogene volcanic rocks of the Snake River Plain, Idaho. *Amer. J. Sci.* **275**, 225–251.

Arndt N. T. (1977) Ultrabasic magmas and high-degree melting of the mantle. *Contrib. Mineral. Petrol.* **64**, 205–221.

Arpat E. and Şaroglu F. (1972) The East Anatolian Fault System: thoughts on its development. *Bull. Miner. Res. Explor. Inst. Turk.* **78** 33–39.

Atwater T. (1970) Implications of plate tectonics for the Cenozoic tectonic evolution of western North America. *Geol. Soc. Amer. Bull.* **81**, 3513–3536.

Atwater T. and Mudie J. D. (1968) Block faulting on the Gorda Rise. *Science* **159**, 729–731.

Atwater T. and Mudie J. D. (1973) Detailed near-bottom geophysical study of the Gorda Rise. *J. Geophys. Res.* **78**, 8665–8686.

Audebaud E., Capdevila R., Dalmayrac B., Debelmas J., Laubacher G., Lefevre C., Marocco R., Martinez C., Mattauer M., Megard F., Paredes J. and Thomas P. (1973) Les traits géologigues essentials des Andes Centrales (Pérou - Bolivie). *Geogr. Phys. Geol. Dyn.* **15**, 73–114.

Ave' Lallement H. (1976) Structure of the Canyon Mountain (Oregon) ophiolite complex and its implication for sea-floor spreading. *Geol. Soc. Amer. Spec. Paper 173*. 49 pp.

Backhofen W. A. (1972) *Deformation Processing*. Addison-Wesley, Reading, Mass. 326 pp.

Backstrom H. and Johanssen H. (1907) Geology. In *Scientific Results of a Journey in Central Asia 1899*–1902 (S. Hedin, ed.), **6**, pt. 2. Stockholm.

Baer A. J. (1977) Speculations on the evolution of the lithosphere. *Precambrian Res.* **5**, 249–260.

Bailey E. H., Blake M. C. and Jones D. L. (1970) On-land Mesozoic oceanic crust in California coast ranges. *U.S. Geol. Surv. Prof. Paper 700-C*, p. 70-81.

Baker B. H., Mohr P. A., Williams L. A. J. (1972) Geology of the eastern rift system of Africa. *Geol. Soc. Amer. Spec. Paper 136*. 67 pp.

Baksi A. K. and Watkins N. D. (1973) Volcanic production rates: comparison of oceanic ridges, islands and the Columbia Plateau basalts. *Science* **180**, 493–496.

Baldwin R. B. (1963) *The measure of the Moon*. Univ. Chicago Press. 488 pp.

Baldwin R. B. (1968) Lunar mascons: another interpretation. *Science* **162**, 1407–1408.

Ballard R. D. and Van Andel Tj. H. (1977) Morphology and tectonics of the inner rift valley at latitude 36° 50'N on the Mid-Atlantic Ridge. *Geol. Soc. Amer. Bull.* **88**, 507–530.

Ballard R. D., Bryan W. B., Heirtzler J. R., Keller G., Moore J. G. and Van Andel Tj. (1975) Manned submersible observations in the FAMOUS area Mid-Atlantic Ridge. *Science* **190**, 507–530.

Baragar W. R. A. (1967) Wakuach Lake map area, Quebec-Labrador (230). *Geol. Surv. Canada, Mem. 344*, 174 pp.

Barazangi M. and Dorman I. (1969) World seismicity maps compiled from ESSA, Coast and Geodetic survey, epicenter data 1961–1967. *Bull. Seismol. Soc. Amer.* **59**, 369.

Barker P. F., (1972) A spreading centre in the east Scotia Sea. *Earth Planet. Sci. Lett.* **15**, 123–132.

Bayer R., Courtillot V., Daignieres M. and Tapponnier P. (1973) Dorsales medio-oceaniques: un modele evolutif de la zone axiale. *C.R. Hebd. Seances Acad. Sci. Ser. D* **276**, 2765–2768.

Bear J. (1972) *Dynamics of Fluids in Porous Media*. Elsevier, N.Y. 764 pp.

Bellaiche G., Cheminee J. L., Francheteau J., Hekinian R., LePichon X., Needham H. D. and Ballard R. D. (1974) Inner floor of the Rift Valley: first submersible study. *Nature* **250**, 558–560.

Belt E. S. (1969) Newfoundland Carboniferous stratigraphy and its relation to the Maritimes and Ireland. In *North Atlantic-Geology and Continental Drift* (M. Kay, ed.), p. 734–753. Amer. Assoc. Petrol. Geologists Mem. 12.

Berkland J. O. (1972) Paleogene "frozen" subduction zone in the Coast Ranges of northern California. In *24th Int. Geol. Cong. Montreal 3. Tectonics*, p. 99–105.

Betz F. and Hess H. H. (1942) The floor of the north Pacific Ocean. *Geogra. Rev.* **32**, 99–116.

Bickford M. E. and Mose D. G. (1975) Geochronology of Precambrian rocks in the St. Francois Mountains, southeastern Missouri. *Geol. Soc. Amer. Spec. Paper 165*. 48 pp.

Bijlaard P. P. (1935) Beschouwingen over deknikzekerheid en de plastische vervormingen van de aardkorst in verband met de gelogie van den Oost-Indischen archipel. *Ingenieur in Nederlandsch-Indie* **11**, 135–156.

Bird J. M. (1969) Middle Ordovician gravity sliding-Taconic region. In *North Atlantic-Geology and Continental Drift* (M. Kay, ed.), p. 670–686. Amer. Assoc. Petrol. Geologists Mem. 12.

Bird J. M. and Dewey J. F. (1970) Lithosphere plate-continental margin tectonics and the evolution of the Appalachian orogen. *Geol. Soc. Amer. Bull.* **81**, 1031–1060.

Bird J. M., Dewey J. F. and Kidd W. S. F. (1971) Proto-Atlantic Ocean crust and mantle: Appalachian/Caledonian ophiolites. *Nature Phys. Sci.* **231**, 28–31.

Black R. and Girod M. (1970) Late Palaeozoic to recent igneous activity in west Africa and its relationship to basement structure. In *African Magmatism and Tectonics* (T. N. Clifford and I. G. Gass, eds.), p. 185–210. Oliver and Boyd, Edinburgh.

Blake M. C. and Jones D. L. (1973) Origin of Franciscan mélange in northern California. In *Modern and Ancient Geosynclinal Sedimentation* (R. H.

Dott and R. H. Shaver, eds.), p. 345–357. Soc. Econ. Paleontologists Mineralogists Spec. Publ. 19.

Bonatti E. (1971) Ancient continental mantle beneath oceanic ridges. *J. Geophys. Res.* **76**, 3825–3831.

Bonatti E. (1973) Origin of offsets of the Mid-Atlantic Ridge in fracture zones. *J. Geol.* **81**, 144–186.

Bonatti E., Chermak A. and Honnorez J. (1979) Tectonic and igneous emplacement of crust in oceanic transform zones. In *Deep Drilling Results in the Atlantic Ocean: Oceanic crust* (M. Talwani, C. G. Harrison and D. E. Hayes, eds.), p.239–248. Maurice Ewing Series 2. Amer. Geophys. Union, Washington, D.C.

Bonnati E., Honnorez, J. and Ferrara G. (1971) Peridotite-gabbro-basalt complex from the equatorial Mid-Atlantic ridge. *Phil. Trans. Roy. Soc. Lond.* **268**, 385–402.

Bott M. H. P. and Dean D. S. (1972) Stress at young continental margins. *Nature* **235**, 23–25.

Bottinga Y. and Allègre C. J. (1973) Thermal aspects of sea-floor spreading and the nature of the oceanic crust. *Tectonophys.* **18**, 1–17.

Bridgwater D., McGregor V. R. and Myers J. S. (1974) A horizontal tectonic regime in the Archean of Greenland and its implications for early crustal thickening. *Precambrian Res.* **1**, 179–197.

Brooks C. and Hart S. R. (1974) On the significance of komatite. *Geology* **2**, 107–110.

Brown G. F. (1978) Arabian-Nubian shield discussed on the spot. *Geotimes* **23**, 24.

Bryan W. B. (1973) Wrinkle-ridges as deformed surface crust on ponded mare lava. *Proc. Lunar Sci. Conf. 4th*, p.93–106.

Bryan W. B. and Moore J. G. (1977) Compositional variations of young basalts in the Mid-Atlantic Ridge rift valley near lat. 36°49 N. *Geol. Soc. Amer. Bull.* **88**, 556–570.

Bryan W. B., Jezek P. A. and Adams M-L. (1975) Volcanic and tectonic evolution of crater Goclenius, western Mare Fecunditatis. *Proc. Lunar Sci. Conf. 6th*, p.2563–2569.

Burke K. (1976a) Development of graben associated with the opening of the Atlantic. *Tectonophys.* **36**, 93–111.

Burke K. (1976b) The Chad basin: an active intracontinental basin. *Tectonophys.* **36**, 197–206.

Burke K. (1977) Aulacogens and continental breakup. *Ann. Rev. Earth Planet. Sci.* **5**, 371–396.

Burke K. and Dewey J. F. (1972) Orogeny in Africa. In *African Geology* (T. F. Dessauvagie and A. J. Whiteman, eds.), p.583–608. Univ. Ibadan Press, Nigeria.

Burke K. and Dewey J. F. (1973a) An outline of Precambrian plate development. In *Implications of Continental Drift to the Earth Sciences v. 2* (D. H. Tarling and S. K. Runcorn, eds.), p.1035–1045. Academic Press, London.

Burke K. and Dewey J. F. (1973b) Plume-generated triple junctions: key indicators in applying plate tectonics to old rocks. *J. Geol.* **81**, 406–433.

Burke K. and Dewey J. F. (1974) Two plates in Africa during the Cretaceous? *Nature* **249**, 313–316.

Burke K. and Kidd W. S. F. (1975) Earth, heat flow in hot-spots. In *McGraw-Hill Yearbook of Science and Technology*, p. 165–169. McGraw-Hill, N.Y. Burke K. and Kidd W. S. F. (1978) Were Archean continental geothermal gradients much steeper than those of today? *Nature* **272**, 240–241.

Burke K. and Kidd W. S. F. (1978) Were Archean continental geothermal gradients much steeper than those ot today? *Nature* **272**, 240–241.

Burke K. and Kidd W. S. F. (1980) Volcanism on earth through time. *The Continental Crust and Its Mineral Deposits.* (D. W. Strangway ed.), p. 503– 522. Geol. Assoc. Canada Spec. Paper 20, Toronto, Ontario.

Burke K. and Whiteman A. (1973) Uplift, rifting and the breakup of Africa. In *Implications of Continental Drift to the Earth Sciences 2*, p.735–755. Academic Press, London.

Burke K. and Wilson J. T. (1972) Is the African plate stationary? *Nature* **239**, 387–390.

Burke K. and Wilson J. T. (1976) Hot spots on the Earth's surface. *Sci. Amer.* **235**, 46–60.

Burke K., Dewey J. F. and Kidd W. S. F. (1974) The Tibetan Plateau: its significance for tectonics and petrology. *Geol. Soc. Amer. Abstracts with Programs* **6**, 1027–1028.

Burke K., Dewey J. F. and Kidd W. S. F. (1976) Dominance of horizontal movements, arc and microcontinental collisions during the later permobile regime. In *The Early History of the Earth* (B. F. Windley, ed.), p. 113–129. Wiley, London.

Burke K., Fox P. J. and Şengör A. M. C. (1978) Buoyant ocean floor and the evolution of the Caribbean. *J. Geophys. Res.* **83**, 3949–3954.

Burke K., Kidd W. S. F. and Wilson J. T. (1973) Relative and latitudinal motion of Atlantic hot spots. *Nature* **245**, 133–137.

Burke K. C. A., Delano L., Dewey J. F., Edelstein A., Kidd W. S. F., Nelson K. D., Şengör A. M. C. and Stroup J. (1978) *Rifts and Sutures of the World.* Contract Report NAS5-24094. Geophysics Branch, ESA Division, Goddard Space Flight Center, Greenbelt, Maryland. 238 pp.

Burnett J. L. and Jenkins C. W. (1962) Chico sheet. *Geological map of California, 1: 250,000.* Calif. Div. of Mines and Geol., Dept. of Conservation, San Francisco.

Burns J. A. (1976) Consequences of the tidal slowing of Mercury. *Icarus* **28**, 453–458.

Cady W. M. (1968) The lateral transition from the miogeosynclinal zone in northwestern New England and adjacent Quebec. In *Studies of Appalachian Geology Northern and Maritime* (E-an. Zen, W. S. White and J. B. Hadley, eds.), p. 151–161. Wiley-Interscience, N.Y.

Cady W. M. (1972) Are the Ordovician northern Appalachians and the Mesozoic cordilleran system homologous? *J. Geophys. Res.* **77**, 3806–3815.

Cann J. R. (1974) A model for oceanic crust structure developed. *Geophys. J. Roy. Astron. Soc.* **39**, 169–187.

Card K. D. (1970) Comment on "The Huronian Supergroup north of Lake Huron (Frarey and Roscoe, 1970)." In *Basins and Geosynclines of the Canadian Shield* (A. J. Baer, ed.), p. 157. Geol. Surv. Canada Paper 70–140.

Carr M. H. (1974) Tectonism and volcanism of the Tharsis region of Mars. *J. Geophys. Res.* **79**, 3943–3949.

Carr M. H. (1980) The morphology of the Martian surface. *Space Sci. Rev.* **25**, 231–284.

Carr M. and Scott D. (1978) Geologic map of Mars. *U.S.G.S. Map I-1083.*

Carr M. H., Greeley R., Blasius K., Guest J. and Murray J. (1977) Some martian volcanic features as viewed from the Viking orbiters. *J. Geophys. Res.* **82**, 3985–4015.

Casey J. F. (1979) Structure within the plutonic section, North Arm Mountain massif, Bay of Islands ophiolite complex: its bearing on tectonic processes of accreting plate margins. *International Ophiolite Symposium Abstracts*, p. 88–89. Geol. Surv. Cyprus, Nicosia.

Cathles L. M. (1975) *The Viscosity of the Earth's Mantle.* Princeton Univ. Press. 386 pp.

CAYTROUGH (1979) Geological and geophysical investigation of the mid Cayman Rise spreading center: initial results and observations. In *Deep Drilling Results in the Atlantic Ocean: Oceanic Crust* (M. Talwani, C. G. Harrison and D. E. Hayes, eds.), p. 66–93. Maurice Ewing Series 2. Amer. Geophys. Union, Washington, D.C.

Choukroune P., Francheteau J. and LePichon X. (1978) Structural observation in an oceanic transform fault from manned submersibles: transform fault "A" on the FAMOUS area. *Geol. Soc. Amer. Bull.* **89**, 1013–1029.

Christiansen R. L. and Lipman P. W. (1972) Cenozoic volcanism and plate-tectonic evolution of the western United States II—Late Cenozoic. *Phil. Trans. Roy. Soc. Lond.* **A271**, 249–284.

Church W. R. (1969) Metamorphic rocks of Burlington Peninsula and adjoining areas of Newfoundland and their bearing on continental drift in North Atlantic. *Amer. Assoc. Petrol. Geol. Mem.* **12**, 212–233.

Church W. R. (1972) Ophiolite: its definition, origin as oceanic crust and mode of emplacement in orogenic belts with special reference to the Appalachians. *Dept. of Energy, Mines and Resources of Canada Publications* **42**, 71–85.

Church W. R. and Riccio L. (1974) The sheeted dike layer of the Betts Cove ophiolite complex does not represent spreading: discussion. *Can. J. Earth Sci.* **11**, 1499–1502.

Church W. R. and Stevens R. K. (1971) Early Paleozoic ophiolite complexes of the Newfoundland Appalachians as mantle-oceanic crust sequences. *J. Geophys. Res.* **76**, 1460–1466.

Clague D. A. and Jarrard R. D. (1973) Tertiary Pacific plate motion deduced from the Hawaiian-Emperor chain. *Geol. Soc. Amer. Bull.* **84**, 1135–1154.

Clark S. P. and Ringwood A. E. (1964) Density distribution and constitution of the mantle. *Rev. Geophys. Space Phys.* **2**, 35.

Cochran J. R. (1973) Gravity and magnetic investigations of Guiana basin, western equatorial Atlantic. *Geol. Soc. Amer. Bull.* **84**, 3249–3268.

Coleman R. G. (1971) Plate tectonic emplacement of upper mantle peridotites along continental edges. *J. Geophys. Res.* **76**, 1212–1222.

Coleman R. G. (1977) *Ophiolites*. Springer-Verlag N.Y., 229 pp.

Coleman R. G. and Irwin W. P. (1977) *North American Ophiolites*. Oregon Dept. of Geol. and Mineral Ind. Bull. **95**. 183 pp.

Collette B. J. (1974) Thermal contraction joints in a spreading sea floor as origin of fracture zones. *Nature* **251**, 299–300.

Cooke D. L. and Moorhouse W. W. (1969) Timiskaming volcanism in the Kirkland Lake area, Ontario, Canada. *Can. J. Earth Sci.* **6**, 117–132.

Cooper A. K., Marlow M. S. and Scholl D. W. (1977) The Bering Sea - a multifarious marginal basin. In *Island Arcs, Deep Sea Trenches and Back-arc Basins* (M. Talwani and W. C. Pitman, eds.), p. 437–450. Maurice Ewing Series 1. Amer. Geophys. Union, Washington, D.C.

Cordell B. and Strom R. G. (1977) Global tectonics of Mercury and the Moon. *Phys. Earth Planet. Inter.* **15**, 146–155.

Courtillot V., Tapponnier P. and Varet J. (1974) Surface features associated with transform faults: a comparison between observed examples and an experimental model. *Tectonophys.* **24**, 317–329.

Courtillot V. E., Allègre C. J. and Mattauer M. (1975) On the existence of lateral relative motions on Mars. *Earth Planet. Sci. Lett.* **25**, 279–285.

Crane K. (1976) The intersection of the Sequeiros transform fault and the East Pacific Rise. *Marine Geol.* **21**, 25–46.

Curray J. R. and Moore D. G. (1974) Sedimentary and tectonic vprocesses in the Bengal deep-sea fan and geosyncline. In *Continental Margins* (C. A. Burk and C. L. Drake, eds.), p. 617–627. Springer-Verlag, N.Y.

CYAMEX (1978) First submersible study of the EPR: RITA-TAMAYO Project, 21°N. *EOS (Trans. Amer. Geophys. Union)* **59**, 1198.

Daignieres M., Courtillot V., Bayer R. and Tapponnier P. (1975) A model for the evolution of the axial zone of mid-ocean ridges as suggested by Icelandic tectonics. *Earth Planet. Sci. Lett.* **26**, 222–232.

Dalziel I. W. D., de Wit M. J. and Palmer K. F. (1974) A fossil marginal basin in the southern Andes. *Nature* **250**, 291–294.

Davies H. L. (1971) Peridotite-gabbro-basalt complex in eastern Papua: an overthrust plate of oceanic crust and mantle. *Bureau Min. Res. Geol. Geophys. Australia Bull.* **128**, 1–48.

Davis E. E. and Lister C. R. B. (1974) Fundamentals of ridge crest topography. *Earth Planet. Sci. Lett.* **21**, 405–413.

Deffeyes K. S. (1970) The Axial Valley: a steady-state feature of the terrain. In *Megatectonics of Continents and Oceans* (H. Johnson and B. L. Smith, eds.), p. 194–222. Rutgers Univ. Press, New Brunswick, N.J.

Deffeyes K. S. (1972) Plume convection with an upper-mantle temperature inversion. *Nature* **240**, 539–544.

DeLong S. E. and Fox P. J. (1977) Geological consequences of ridge subduction. In *Island Arcs, Deep-sea Trenches and Back-arc Basins*, (M. Talwani and W. C. Pitman, eds.), p. 221–229. Maurice Ewing Series 1, Amer. Geophys. Union, Washington, D.C.

DeLong S. E., Dewey J. F. and Fox P. J. (1977) Displacement history of oceanic fracture zones. *Geology* **5**, 199–202.

DeLong S. E., Fox P. J. and McDowell F. W. (1978) Subduction of the Kula Ridge at the Aleutian Trench. *Geol. Soc. Amer. Bull.* **89**, 83–95.

Deng W. -M. (1978) A preliminary study of the petrology and petrochemistry of the Quaternary volcanic rocks of northern Tibet. *Acta Geologica Sinica 2*, 148–162.

Detrick R. S. and Purdy G. M. (1978) The crustal structure of the Kane Fracture Zone from seismic refraction studies. *EOS (Trans. Amer. Geophys. Union)* **59**, 1199.

Detrick R. S., Mudie J. D., Luyendyk B. P. and MacDonald K. C. (1973) Near-bottom observations of an active transform fault (Mid-Atlantic Ridge at 37°N). *Nature Phys. Sci.* **246**, 59–61.

Dewey J. F. (1975) Finite plate evolution: implications for the evolution of rock masses at plate margins. *Amer. J. Sci.* **275A**, 260–284.

Dewey J. F. (1980) Episodicity, sequence and style at convergent plate boundaries. *The Continental Crust and Its Mineral Deposits*, (D. W. Strangway, ed.), p. 553–573. Geol. Assoc. Canada Spec. Paper 20, Waterloo, Ontario.

Dewey J. F. and Bird J. M. (1970) Mountain belts and the new global tectonics. *J. Geophys. Res.* **75**, 2625–2647.

Dewey J. F. and Bird J. M. (1971) Origin and emplacement of the ophiolite suite: Appalachian ophiolites in Newfoundland. *J. Geophys. Res.* **76**, 3179–3206.

Dewey J. F. and Burke K. (1973) Tibetan, Variscan and Precambrian basement reactivation: products of continental collision. *J. Geol.* **81**, 683–692.

Dewey J. F. and Burke K. (1974) Hot spots and continental breakup: some implications for collisional orogeny. *Geology* **2**, 57–60.

Dewey J. F. and Kidd W. S. F. (1974) Continental collisions in the Appalachian-Caledonian orogenic belt: variations related to complete and incomplete suturing. *Geology* **2**, 543–546.

Dewey J. F. and Kidd W. S. F. (1977) Geometry of plate accretion. *Geol. Soc. Amer. Bull.* **88**, 960–968.

Dewey J. F. and Şengör A. M. C. (1979) Aegean and surrounding regions: complex multiplate and continuum tectonics in a convergent zone. *Geol. Soc. Amer. Bull.* **90**, 84–92.

Dewey J. F., Fox P. J. and Kidd W. S. F. (1973) Structure and generation of oceanic crust and mantle (abstract). *Geol. Soc. Amer. Abstracts with Program*, **5**, 597–598.

Dewey J. F., Pitman W. C., Ryan W. B. F. and Bonnin J. (1973) Plate tectonics and the evolution of the Alpine system. *Geol. Soc. Amer. Bull.* **84**, 3137–3180.

Dick H. J. B. and Sclater, J. (1977) Preliminary report—ocean crust and mantle materials from the Islas Orcadas and Conrad fracture zones; southern oceans. *EOS (Trans. Amer. Geophys. Union)* **58**, 529–530.

Dickinson W. R. (1970) Relations of andesites, granites and derivative sandstones to arc-trench tectonics. *Rev. Geophys. Space Phys.* **8**, 813–859.

Dickinson W. R. (1971) Plate tectonic models of geosynclines. *Earth Planet. Sci. Lett.* **10**, 165–174.

Dickinson W. R. and Hatherton T. (1967) Andesitic volcanism and seismicity around the Pacific. *Science* **157**, 801–803.

Dickinson W. R. and Luth W. C. (1971) A model for plate tectonic evolution of mantle layers. *Science* **174**, 400–404.

Dietz R. S. (1963a) Alpine serpentinites as oceanic rind fragments. *Geol. Soc. Amer. Bull.* **74**, 947–952.

Dietz R. S. (1963b) Vredefort Ring-Bushveld Complex impact event and lunar maria (abstract). *Geol. Soc. Amer. Spec. Paper 73*, p. 35.

Dietz R. S. (1964) Sudbury structure as an astrobleme. *J. Geol.* **72**, 412–434.

Dietz R. S. (1965) Earth's original crust-lost quest? *Tectonophys.* **2**, 515–520.

Drake C. L. and Nafe J. E. (1968) The transition from ocean to continent from seismic refraction data. *Amer. Geophys. Union Monograph 12* (L. Knopoff, C. L. Drake and P. J. Hart, eds.), p. 174–185. Amer. Geophys. Union, Washington, D.C.

Dvorak J. and Phillips R. J. (1974) Gravity anomaly and structure associated with the Lamont region of the Moon. *Proc. Lunar Planet. Sci. Conf 10th*, p. 2265–2275.

Dvorak J. and Phillips R. J. (1979) Gravity anomaly and structure associated with the Lamont region of the Moon. *Proc. Lunar Planet Sci. Conf. 10th*, p. 2265–2275.

Edman J. (1976) Lithospheric thickness, subsidence of aseismic ridges and volcano height. *J. Volcan. Geotherm. Res.* **1**, 363–379.

Ellsworth W. L. and Koyanagi R. Y. (1977) Three-dimensional crust and mantle structure of Kilauea Volcano, Hawaii. *J. Geophys. Res.* **82**, 5379–5394.

Engel A. E. J. (1968) The Barberton Mountain Land: clues to the differentiation of the Earth. *Trans. Geol. Soc. S. Africa* **71**, 225–270.

Erinc S. (1953) Dogu Anadolu Cografyasi. *Istanbul Üniv. Yay.* **572**, 124 pp.

Erinc S. (1970) Kula ve Adala arasinda genc volkan reliyefi. *Istanbul Üniv. Cogr. Enst. Dergisi* **9**, 7–31.

Ewart A. (1976) Mineralogy and chemistry of modern orogenic lavas—some statistics and implications. *Earth Planet. Sci. Lett.* **31**, 417–432.

Ewing M. and Heezen B. C. (1956) Antarctica in the International Geophysical Year. *Amer. Geophys. Union Monograph 1*, 75–81.

Fahrig W. F. and Wanless R. K. (1963) Age and significance of diabase dike swarms of the Canadian Shield. *Nature* **200**, 934–937.

Fairhead J. D. (1979) A gravity link between the domally uplifted Cainozoic volcanic centers of North Africa and its similarity to the East African rift system anomaly. *Earth Planet. Sci. Lett.* **42**, 109–113.

Fairhead J. D. and Reeves C. V. (1977) Teleseismic delay times, Bouguer anomalies and inferred thickness of the African lithosphere. *Earth Planet Sci. Lett.* **36**, 63–76.

Fielder G. (1961) *Structure of the Moon's Surface*. Pergamon Press, London. 266 pp.

Fielder G. (1965) *Lunar Geology*. Lutterworth Press, London. 184 pp.

Folinsbee R. E., Baadsgaard H., Cumming G. L. and Green D. C. (1968) A very ancient island arc. *Amer. Geophys. Union Monograph 12* (L. Knopoff, C. L. Drake and P. J. Hart, eds.), p. 441–448. Amer. Geophys. Union, Washington, D. C.

Forsythe D. and Uyeda S. (1975) On the relative importance of driving forces of plate motion. *Geophys. J. Roy. Astron. Soc.* **43**, 163–200.

Fowler C. M. R. (1976) Crustal structure of the Mid-Atlantic Ridge crest at 37° N. *Geophys. J. Roy. Astron. Soc.* **47**, 459–491.

Fowler C. M. R. and Matthews D. M. (1974) A seismic refraction experiment using ocean bottom seismography and sonobuoys, on the Mid-Atlantic Ridge near 37° N. *Nature* **249**, 752–754.

Fox P. J. (1978) The effect of transform faults on the character of the oceanic crust (abstract). *Geol. Soc. Amer. Abstracts with Program* **10**, 403.

Fox P. J. and DeLong S. E. (1976) Healing of oceanic layer 2A: geophysical and geological consequences (abstract). *EOS (Trans. Amer. Geophys. Union)* **57**, 329.

Fox P. J., Lowrie A. and Heezen B. C. (1969a) Oceanographer Fracture Zone. *Deep Sea Research,* **16**, 59–66.

Fox P. J., Pitman W. C. and Shepard F. (1969b) Crustal plates in the central Atlantic: evidence for at least two poles of rotation. *Science* **165**, 487–489.

Fox P. J., Schreiber E. and Peterson J. J. (1973) The geology of the oceanic crust: compressional wave velocities of oceanic rocks. *J. Geophys. Res.* **78**, 5155–5172.

Fox P. J., Schreiber E., Rowlett H. and McCamy K. (1976) The geology of the Oceanographer Fracture Zone: a model for fracture zones. *J. Geophys. Res.* **81**, 4117–4128.

Francheteau J., Choukroune P., Hekinian R., LePichon X. and Needham H. D. (1976) Oceanic fracture zones do not provide deep sections in the crust. *Can. J. Earth Sci.* **13**, 1223–1235.

Francis T. J. G. (1974) A new interpretation of the 1968 Fernandina caldera collapse and its implications for the mid-oceanic ridges. *Geophys. J. Roy. Astron. Soc.* **39**, 301–318.

Francis T. J. G. and Porter I. T. (1973) Median valley seismology: the Mid-Atlantic Ridge near 45° N. *Geophys. J. Roy. Astron. Soc.* **34**, 279–311.

Frank F. C. (1968) Two-component flow model for convection in the Earth's upper mantle. *Nature* **220**, 350–352.

Frarey M. J. and Roscoe S. M. (1970) The Huronian supergroup north of Lake Huron. In *Basins and Geosynclines of the Canadian Shield* (A. J. Baer, ed.), p. 143–157. Geol. Surv. Canada Paper 70-40.

French B. M. (1970) Possible relations between meteorite impact and igneous petrogenesis as indicated by the Sudbury structure, Ontario, Canada. *Bull. Volc.* **34**, 466–517.

Frey H. (1979a) Thaumasia: A fossilized early forming Tharsis uplift. *J. Geophys. Res.* **84**, 1009–1023.

Frey H. (1979b) Martian canyons and African rifts: Structural comparisons and implications. *Icarus* **37**, 142–155.

Fujii N. and Uyeda S. (1974) Thermal instabilities during flow of magma in volcanic conduits. *J. Geophys. Res.* **79**, 3367–3369.

Fujita K. and Sleep N. H. (1978) Membrane stresses near mid-ocean ridge transform intersections. *Tectonophys.* **50**, 207–221.

Fyfe W. S. (1973) The granulite facies, partial melting and the Archean crust. *Phil. Trans. Roy. Soc. Lond.* **A273**, 457–462.

Gallo D. G. and Fox P. J. (1979) The influence of transform faults on the generation of oceanic lithosphere. *EOS (Trans. Amer. Geophys. Union)* **60**, 376.

Gansser A. (1964) *The Geology of the Himalayas.* Wiley-Interscience, N. Y. 289 pp.

Gansser A. (1966) *Catalogue of the Active Volcanoes of the World, Including Solfatara Fields.* Iran. Int. Volc. Assoc., Naples, Part 17. 20 pp.

Gass I. G. (1968) Is the Troodos massif of Cyprus a fragment of Mesozoic ocean floor? *Nature* **220**, 39.

Gass I. G. and Smewing J. D. (1973) Intrusion, extrusion and metamorphism at constructive margins: evidence from the Troodos Massif, Cyprus. *Nature* **242**, 26–29.

George R. P. (1978) Structural petrology of the Olympus ultramafic complex in the Troodos ophiolite, Cyprus. *Geol. Soc. Amer. Bull.* **89**, 845–865.

Gill J. B. (1976) Composition and age of Lau Basin and Ridge volcanic rocks: implications for evolution of an interarc basin and remnant arc. *Geol. Soc. Amer. Bull.* **87**, 1384–1395.

Glennie K. W., Boueff M. G. and Clarke M. W. H. (1974) Geology of the Oman Mountains. Ned. Geol. Mijnbowwkd. Genoot. Verh. **31**, Amsterdam. 423 pp.

Glennie K. W., Bouef M. G. A., Hughes-Clark M. W., Moody-Stuart M., Pilaar W. F. H. and Reinhardt B. M. (1973) Late Cretaceous nappes in Oman Mountains and their geologic evolution. *Amer. Assoc. Petrol. Geol. Bull.* **57**, 5–27.

Glikson A. Y. (1972) Petrology and geochemistry of metamorphosed Archean ophiolites, Kalgoorlie-Coolgardie, western Australia. In *Bull. 125, Bureau of Mineral Resources, Geology and Geophysics*, p. 121–189. Canberra.

Goldstein R. M., Green R. R. and Ramsey H. C., Jr. (1976) Venus Radar images. *J. Geophys Res.* **81**, 4807–4817.

Goodwin A. M. (coordinator) (1972) The Superior Province. In *Variations in Tectonic Styles in Canada* (R. A. Price and R. J. W. Douglas, eds.), p. 527–624. Geol. Assoc. Canada Spec. Paper 11.

Goodwin A. M. (1973) Plate tectonics and evolution of Precambrian crust. In *Implications of Continental Drift to the Earth Sciences v.2* (D. H. Tarling and S. K. Runcorn, eds.), p. 1047–1069. Academic Press, London.

Goodwin A. M. (1974) Precambrian belts, plumes and shield development. *Amer. J. Sci.* **274**, 987–1028.

Goodwin A. M. and Ridler R. H. (1970) The Abitibi orogenic belt. In *Basins and Geosynclines of the Canadian Shield* (A. J. Baer, ed.), p. 1–24. Geol. Surv. Canada Paper 70-40.

Goslin J., Beauzart P., Francheteau J. and LePichon X. (1972) Thickening of the oceanic layer in the Pacific Ocean. *Marine Geophys. Res.* **1**, 418–427.

Greeley R. and Spudis P. D. (1978) Volcanism in the cratered terrain hemisphere of Mars. *Geophys. Res. Lett.* **5**, 453–456.

Green A. G. (1975) On the postulated Hawaiian plume with emphasis on the limitations of seismic arrays for detecting deep mantle structure. *J. Geophys. Res.* **80**, 4028–4036.

Green D. H. (1971) Composition of basaltic magmas as indicators of conditions of origin; application to oceanic volcanism. *Phil. Trans. Roy. Soc. Lond.* **A268**, 707–725.

Green T. H. (1969) High pressure experimental studies on the origin of anorthosite. *Can. J. Earth Sci.* **6**, 427–440.

Greenbaum D. (1972) Magmatic processes at ocean ridges, evidence from the Troodos Massif, Cyprus. *Nature* **238**, 18–21.

Gregory J. W. (1921) *The Rift Valleys and Geology of East Africa.* Macmillan, London. 479 pp.

Greiner G. and Illies J. H. (1977) Central Europe: active or residual tectonic stresses. *Pure and Appl. Geophys.* **115**, 11–26.

Gruntfest I. J. (1963) Thermal feedback in liquid flow; plane shear at constant stress. *Soc. Rheo. Trans.* **7**, 195–207.

Guest J. E. and Greeley R. (1977) *Geology on the Moon.* Wykeham Pub. Ltd., London. 235 pp.

Hadley D. M., Stewart G. S. and Ebel J. E. (1976) Yellowstone: seismic evidence for a chemical mantle plume. *Science* **193**, 1237–1239.

Hall D. H. (1971) Geophysical determination of deep crust structure in Manitoba. *Geol. Assoc. Can. Spec. Paper 9*, 83–88.

Hall J. M. and Robinson P. T. (1979) Deep crustal drilling in the north Atlantic Ocean. *Science* **204**, 573–586.

Ham W. E. and Wilson J. L. (1967) Paleozoic epeirogeny and orogeny in the central United States. *Amer. J. Sci.* **265**, 332–407.

Hamilton W. (1969) Mesozoic California and the underflow of the Pacific mantle. *Geol. Soc. Amer. Bull.* **80**, 2409–2430.

Hamilton W. (1970) The Uralides and the motions of the Russian and Siberian platforms. *Geol. Soc. Amer. Bull.* **81**, 2553–2576.

Hardee H. C. and Larson D. W. (1977) Viscous dissipation effects in magma conduits. *J. Volcan. Geotherm. Res.* **2**, 299–308.

Hargraves R. B. (1976) Precambrian geologic history. *Science* **193**, 363–371.

Harris P. G., Kennedy W. Q. and Scarfe C. M. (1970) Volcanism versus plutonism—the effects of chemical composition In *Mechanisms of Igneous Intrusion*, (G. Newall and N. Rast, eds.), p. 187–200. Geol. J. Spec. Issue 2.

Harrison C. G. A. (1974) Tectonics of mid-ocean ridges. *Tectonophys.* **22**, 301–310.

Hart S. R. and Goldich S. S. (1975) Most ancient known rocks may be found in all Earth's shields. *Geotimes* **20**, 22–24.

Hart S. R., Glassley W. E. and Karig D. E. (1972) Basalts and sea-floor spreading behind the Mariana island arc. *Earth Planet. Sci. Lett.* **15**, 12–18.

Hartmann W. K. (1973) Martian surface and crust: review and synthesis. *Icarus* **19**, 550–575.

Hatherton T. and Dickinson W. R. (1968) Andesitic volcanism and seismicity in New Zealand. *J. Geophys. Res.* **73**, 4615–4619.

Hatherton T. and Dickinson W. R. (1969) The relationship between andesitic volcanism and seismicity in Indonesia, The Lesser Antilles, and other island arcs. *J. Geophys. Res.* **74**, 5301–5310.

Hawke B. R., MacLaskey D., McCord T. B., Adams J. B., Head J. W., Pieters C. M. and Zisk S. H. (1979) Multispectral mapping of the Apollo 15-Apennine region: the identification and distribution of regional pyroclastic deposits. *Proc. Lunar Planet. Sci. Conf. 10th*, p. 2995–3015.

Hawkins J. W. (1976) Petrology and geochemistry of basaltic rocks of the Lau Basin. *Earth Planet. Sci. Lett.* **28**, 283–297.

Hawkins J. W. (1977) Petrologic and geochemical characteristics of marginal basin basalts. In *Island Arcs, Deep Sea Trenches and Back-Arc Basins* (M. Talwani and W. C. Pitman, eds.), p. 355–365. Maurice Ewing Series 1. Amer. Geophys. Union, Washington, D.C.

Hayden H. (1907) Geology of the provinces of Tsang and U. in central Tibet. *Mem. Geol. Surv. India, 36* (2).

Head J. W. and McCord T. B. (1978) Imbrian-age highland volcanism on the Moon: the Gruithusen and Mairan Domes. *Science* **199**, 1433–1436.

Head J. W. and Wilson L. (1979) Alphonsus-type dark-halo craters: morphology, morphometry and eruption conditions. *Proc. Lunar Planet. Sci. Conf. 10th*, p. 2861–2897.

Heezen B. C. (1960) The rift in the ocean floor. *Sci. Amer.* **203**, 98–110.

Hennig A. (1915) Zur Petrographie und Geologie von Südwest-Tibet, Vol. 5. In *Southern Tibet* (S. Hedin, ed.), Norstedt and Söner, Stockholm. 220 pp.

Herron T. J., Ludwig W. J., Stoffa P. L., Kau T. K. and Buhl P. (1978) Structure of the East Pacific Rise crest from multichannel seismic reflection data. *J. Geophys. Res.* **83**, 798–804.

Hess H. H. (1964) The oceanic crust, the upper mantle and the Magaguez serpentinized peridotite. In *A Study of Serpentinite* (C. A. Burke, ed.), p. 169–175. *Natl. Acad. Sci.-Natl. Res. Council Publ. 1188*. Washington, D.C.

Hey R. (1977) Tectonic evolution of the Cocos-Nazca spreading center. *Geol. Soc. Amer. Bull.* **88**, 1404–1420.

Hoffman P. F. (1973) Evolution of an early Proterozoic continental margin: the Coronation geosyncline and associated aulacogens of the northwestern Canadian shield. *Phil. Trans. Roy. Soc. Lond.* **A273**, 547–581.

Hoffman P. F. (1980) Wopmay Orogen: a Wilson cycle of early Proterozoic age in the northwest of the Canadian Shield. In *The Continental Crust of the Earth and its Mineral Deposits* (D. W. Strangway, ed.), p. 523–549. Geol. Assoc. Canada Spec. Paper 20, Toronto, Ontario.

Honnorez J. and Bonatti E. (1970) Nepheline gabbro from the Mid-Atlantic Ridge. *Nature* **228**, 850–851.

Hopson C. A. and Frano C. J. (1977) Igneous history of the Point Sal ophiolite, Southern California. In *North American Ophiolites*, p.161–183. Bull. Oregon Dept. Geol. Mineral Industries 95.

Houtz R. E. and Ewing J. I. (1976) Upper crustal structure as a function of plate age. *J. Geophys. Res.* **81**, 2490–2498.

Innocenti F., Mazzuoli R., Pasquare G., Radicati di Brozolo F. and Villari L. (1976) Evolution of the volcanism in the area of interaction between the Arabian, Anatolian and Iranian plates (Lake Van, eastern Turkey). *J. Volcan. Geoth. Res.* **1**, 103–112.

Isachsen Y. W., Geraghty E. P. and Wright S. F. (1978) Investigation of holocene deformation in the Adirondacks Mountain dome. *Geol. Soc. Amer. Abstracts with Programs* **10**, 49.

Isacks B. L. and Barazangi M. (1977) Geometry of Benioff zones: lateral segmentation and downwards bending of the subducted lithosphere. In *Island Arcs, Deep Sea Trenches and Back-arc Basins* (M. Talwani and W. C. Pitman, eds.), p. 99–114. Maurice Ewing Series 1. Amer. Geophys. Union, Washington, D.C.

Isacks B. and Molnar P. (1969) Mantle earthquake mechanisms and the sinking of the lithosphere. *Nature* **223**, 1121–1124.

Jackson E. D. (1976) Linear volcanic chains on the Pacific plate. In *The Geophysics of the Pacific Ocean Basin and its Margin* (G. H. Sutton, M. H. Manghnani and R. Moberly, eds.), p. 319–335. American Geophysical Union Monograph 19. Amer. Geophys. Union, Washington, D.C.

Jackson E. D. and Wright T. L. (1970) Xenoliths in the Honolulu volcanic series, Hawaii. *J. Petrol.* **11**, 405–430.

Jackson E. D., Green H. W. II and Moores E. M. (1975) The Vourinos Ophiolite, Greece: cyclic units of lineated cumulates overlying harzburgite tectonite. *Geol. Soc. Amer. Bull.* **86**, 390–398.

Jacobs J., Russell R. D. and Wilson J. T. (1972) *Physics and Geology*, 2nd ed. McGraw-Hill, N.Y. 622 pp.

Jacobson R. R. E., Macleod W. N. and Black R. (1958) Ring complexes in the younger granite province of northern Nigeria. *Mem. Geol. Soc. London, 72 pp.*

Jakeš P. (1973) Geochemistry of continental growth. In *Implications of Continental Drift to the Earth Sciences* (D. H. Tarling and S. K. Runcorn, eds.), p. 999–1009. Academic Press, London.

Jakeš P. and Gill J. (1970) Rare-earth elements and the island arc tholeiitic series. *Earth Planet. Sci. Lett.* **9**, 17–28.

Jamieson R. A. and Strong D. F. (1978) Dynamic and chemical implications of a late syntectonic shear zone within the ophiolite aureole, St. Anthony Complex, northwestern Newfoundland. *Geol. Soc. Amer. Abstracts with Programs* **10**, 428–429.

Johnson G. L. and Vogt P. R. (1973) Mid-Atlantic Ridge from 47° to 51° north. *Geol. Soc. Amer. Bull.* **84**, 3443–3462.

Julian B. R. and Sengupta M. K. (1973) Seismic travel time evidence for lateral inhomogeneity. *Nature* **242**, 443–447.

Jung D., Kursten M. O. C. and Tarkian M. (1976) Post-Mesozoic volcanism in Iran and its relation to the subduction of the Afro-Arabian under the Eurasian plate. In *Afar Between Continental and Oceanic Rifting* (A. Pilger and A. Rösler, eds.), p. 175–190. Schweizerbart, Stuttgart.

Juteau T., Nicolas A., Dubessy J., Fruchard J. C. and Bouchez J. L. (1977) Structural relationships in the Antalya ophiolite complex, Turkey: possible model for an oceanic ridge. *Geol. Soc. Amer. Bull.* **88**, 1740–1748.

Kanasewich E. R. and Gutowski P. R. (1975) Detailed seismic analysis of a lateral mantle inhomogeneity. *Earth Planet. Sci. Lett.* **25**, 397–384.

Kanasewich E. R., Ellis R. M., Chapman C. H. and Gutowski P. R. (1972) Teleseismic array evidence for inhomogeneities in the lower mantle and the origin of the Hawaiian Islands. *Nature Phys. Sci.* **239**, 99–100.

Kanasewich E. R., Ellis R. M., Chapman C. H. and Gutowski P. R. (1973) Seismic array evidence of a core boundary source for the Hawaiian linear volcanic chain. *J. Geophys. Res.* **78**, 1361–1371.

Karig D. E. (1970) Ridges and basins of the Tonga-Kermadec Island Arc system. *J. Geophys. Res.* **75**, 239–254.

Karig D. E. (1971a) Origin and development of marginal basins in the western Pacific. *J. Geophys. Res.* **76**, 2542–2561.

Karig D. E. (1971b) Structural history of the Mariana Arc system. *Geol. Soc. Amer. Bull.* **82**, 323–344.

Karig D. E. and Mammerickx J. (1972) Tectonic framework of the New Hebrides Island Arc. *Mar. Geol.* **12**, 187–205.

Karig D. E. Caldwell J. G. and Parmentier E. M. (1976) Effects of accretion on the geometry of the descending lithosphere. *J. Geophys. Res.* **81**, 6281–6291.

Karig D. E., Ingle J. C., Bouma A. H., Ellis H., Haite N., Koizumi I., Macgregor I. D., Moore C., Ujiie H., Watanabe T., White S. M., Yasui M. and Ling H. Yi. (1973) Origin of the west Philippine basin. *Nature* **246**, 458–461.

Karson J. A. (1977) The geology of the northern Lewis Hills, western Newfoundland. Ph.D. dissertation, State University of New York, Albany. 474 pp.

Karson J. A. and Dewey J. F. (1978) Coastal complex, western Newfoundland: an early Ordovician oceanic fracture zone. *Geol. Soc. Amer. Bull.* **89**, 1037–1049.

Kaula W. M. (1975) Absolute plate motions by boundary velocity minimisations. *J. Geophys. Res.* **80**, 244–248.

Kaula W. M., Schubert G., Lingenfelter R. E. and Sjogren W. L. (1973) *Proc. Lunar Sci. Conf. 4th*, p. 2811–2819.

Keen C. and Tramontini C. (1970) A seismic refraction survey on the Mid-Atlantic Ridge. *Geophys. J. Roy. Astron. Soc.* **20**, 473–493.

Kennedy W. Q (1955) The tectonics of the Morar anticline and the problem of the northwest Caledonian front. *Quart. J. Geol. Soc. London* **110**, 357–382.

Ketin I. (1961) Über die magmatischen Erscheinungen in der Türkei. *Türk*. Jeol. Kurumu Büll. 7, 16–33.

Kidd R. G. W. (1977) A model for the process of formation of the upper oceanic crust. *Geophys. J. Roy. Astron. Soc.* **50**, 149–183.

Kidd R. G. W. and Cann J. R. (1974) Chilling statistics indicate an ocean-floor spreading origin for the Troodos Complex, Cyprus. *Earth Planet. Sci. Lett.* **24**, 151–155.

Kidd W. S. F. (1975) Widespread late Neogene and Quaternary calc-alkaline volcanism on the Tibetan Plateau. *EOS (Trans. Amer. Geophys. Union)* **56**, 453.

Kidd W. S. F. (1977) The Baie Verte Lineament, Newfoundland: ophiolite complex floor and mafic volcanic fill of a small Ordovician marginal basin. In *Island Arcs, Deep Sea Trenches and Back-arc Basins* (M. Talwani and W. C. Pitman, eds.), p.407–418. Maurice Ewing Series 1. Amer. Geophys. Union, Washington, D.C.

Kidd W. S. F., Dewey J. F. and Bird J. M. (1978) The Mings Bight Ophiolite complex, Newfoundland: Appalachian oceanic crust and mantle. *Can. J. Earth Sci.* **15**, 781–804.

Kidd W. S. F., Dewey J. F. and Nelson K. D. (1977) Medial Ordovician Ridge subduction in central Newfoundland. *Geol. Soc. Amer. Abstracts with Programs* **9**, 283–284.

Klein F. W., Einarsson P. and Wyss M. (1973) Microearthquakes on the Mid-Atlantic plate boundary on the Reykjanes Peninsula in Iceland. *J. Geophys. Res.* **78**, 5084–5099.

Klein G. de V., Kobayashi K., Chamley H., Curtis D. M., Dick H. J. B., Echols D. J., Fountain D. M., Kinoshita H., Marsh N. G., Mizumo A., Nisterenko G. V., Okada H., Sloan J. R., Waples D. M. and White S. M. (1978) Off-ridge volcanism and sea-floor spreading in the Shikoku Basin. *Nature* **273**, 746–748.

Klitin K. A. and Pavlova T. G. (1974) Ophiolite complex of the Baikal fold zone. *Doklady Akad. Nauk SSR* **215**, 413–416 (33-36 in translation).

Kobayashi K. and Isezaki N. (1976) Magnetic anomalies in the Sea of Japan and the Shikoku Basin: possible tectonic implications. In *The Geophysics of the Pacific Ocean Basin and its Margin* (G. H. Sutton, M. H. Manghnani, and R. Moberly, eds.), p. 235–251. Amer. Geophys. Union, Washington, D.C.

Koide H. and Bhattacharji S. (1975) Mechanistic interpretation of Rift Valley formation. *Science* **189**, 791–793.

Koyanagi R. Y., Endo E. T. and Ward P. L. (1976) Seismic activity on the island of Hawaii, 1970 to 1973 In *The Geophysics of the Pacific Ocean Basin and its Margin* (G. H. Sutton, M. H. Manghnani and R. Moberly, eds.), p. 169–172. Amer. Geophys. Union Monograph 19 Amer. Geophys. Union, Washington, D.C.

Koyanagi R. Y., Krivoy H. L. and Okamura A. T. (1966) The 1962 Kaoiki, Hawaii, earthquake and its aftershocks. *Bull. Seis. Soc. Amer.* **56**, 1317–1335.

Kroenke L. and Scott R. (1978) Deep Sea Drilling Project - Leg 59 - Philippine Sea. *Geotimes* **23**, 20–23.

Kuno H. (1966) Lateral variation of basalt magma type across continental margins and island arcs. *Bull. Volc.* **29**, 195–222.

Kushiro I., Yoder H. S. and Mysen B. O. (1976) Viscosities of basalt and andesitic melts at high pressures. J. Geophys. Res., **81**, 6351–6356.

Kusznir N. J. and Bott M. H. P. (1976) A thermal study of the formation of oceanic crust. *Geophys. J. Roy. Astron. Soc.* **47**, 83–95.

Lachenbruch A. H. (1973) A simple mechanical model for oceanic spreading centers. *J. Geophys. Res.* **78**, 3395–3417.

Lachenbruch A. H. (1976) Dynamics of a passive spreading center. *J. Geophys. Res.* **81**, 1883–1902.

Lamb H. (1945) Hydrodynamics. Dover, N.Y. 738 pp.

Lambeck K. (1979) Comments on the gravity and topography of Mars. *J. Geophys. Res.* **84**, 6241–6247.

Lambert R. St. J. (1976) Archean thermal regimes, crustal and upper mantle temperatures and a progressive evolutionary model for the earth. In *The Early History of the Earth.* (B. F. Windley, ed.), p. 363–373. Wiley, New York.

Lambert R. St. J., Holland J. G. and Owen P. F. (1974) Chemical petrology of a suite of calc-alkaline lavas from Mount Ararat, Turkey. *J. Geol.* **82**, 419–438.

Larson R. L. and Spiess F. N. (1969) East Pacific Rise crest: a near-bottom geophysical profile. *Science* **163**, 68–71.

Laubscher H. (1969) Mountain building. *Tectonophys.* **7**, 551–63.

Lawson A. C. (1913) The Archean geology of Rainy Lake restudied. *Mem. Geol. Survey Canada* **40**. 115 pp.

LeBlanc M. (1976) Oceanic crust at Bou Azzer. *Nature* **261**, 34 –35.

Lee W. H. K. (1967) Thermal history of the Earth. Ph.D. dissertation, Univ. of California, Los Angeles.

LeFort P. (1975) Himalayas: the collided range. Present knowledge of the continental arc. *Amer. J. Sci.* **275-A**, 1–44.

LePichon X., Francheteau J. and Bonnin J. (1973) *Plate Tectonics.* Elsevier, New York. 300 pp.

Lépine J. C. and Ruegg J. C. (1976) La seismicité du T.F.A.I., La Creation de la Lithosphere Oceanique. *Bull. Soc. Geol. France* **18**, 841–846.

Lépine J. C., Steinmetz L. and Ruegg J. C. (1972) Seismic profiles in Djibouti area, East African rifts. *Tectonophys.* **15**, 59–64.

Lister C. R. B. (1972) On the thermal balance of a midocean ridge. *Roy. Astron. Soc. Geophys. J.* **26**, 515–537.

Lister C. R. B. (1977)Qualitative models of spreading center processes including hydrothermal penetration. *Tectonophys.* **37**, 203–218.

Lliboutry L. (1976) Courants de convection et dynamique des plaques. In *Traité de Géophysique Interne,* 2 (J. Coulomb and G. Jobert, eds.), p. 501–571. Masson, Paris.

Logatchev N. A. (1977) Rift systems in East Siberia. *Nytt fra Oslofeltgruppen* **6**, 44.

Logatchev N. A. and Florensov N. A. (1978) The Baikal system of rift valleys. *Tectonophys.* **45**, 1–13.

Lonsdale P. (1978) Near bottom reconnaissance of a fast-slipping transform fault zone at the Pacific-Nazca plate boundary. *J. Geol.* **86**, 451–472.

Lotze F. (1974) *Geologie Mitteleuropas.* E. Schweizerbart'sche Verlagsbuchhandlung, Stuttgart. 491 pp.

Lowrie W. (1979) Geomagnetic reversals and ocean crust magnetisation. In *Deep Drilling Results in the Atlantic Ocean: Oceanic Crust* (M. Talwani, C. G. Harrison and D. E. Hayes, eds.), p. 135–150. Maurice Ewing Series, 2. Amer. Geophys. Union, Washington, D.C.

Lubimova E. A. (1969) Thermal history of the Earth. In *The Earth's Crust and Upper Mantle* (P. J. Hart, ed.), p. 63–77. Amer. Geophys. Union Monograph 13. Amer. Geophys. Union, Washington, D.C.

Lucchitta B. K. and Watkins J. A. (1978) Age of graben systems on the moon. *Proc. Lunar Planet. Sci. Conf. 9th* p. 3459–3472.

Ludwig W. J., Nafe J. E. and Drake C. L. (1970) Seismic refraction. In *The Sea*, v.4, (A. E. Maxwell, ed.), p. 53–84. Wiley-Interscience, N.Y.

Luyendyk B. P. and MacDonald K. C. (1977) Physiography and structure of the inner floor of the FAMOUS rift valley: observations with a deep-towed instrument package. *Geol. Soc. Amer. Bull.* **88**, 648–663.

Maaloe S. (1973) Temperature and pressure relations of ascending primary magmas. *J. Geophys. Res.* **78**, 6877–6886.

Maasha N. and Molnar P. (1972) Earthquake fault parameters and tectonics in Africa. *J. Geophys. Res.* **77**, 5731–5143.

MacDonald K. C. (1977) Near-bottom magnetic anomalies, asymmetric spreading, oblique spreading and tectonics of the accreting plate boundary on the Mid-Atlantic Ridge near 37°N. *Geol. Soc. Amer. Bull.* **88**, 541–555.

MacDonald K. C. and Luyendyk B. P. (1977) Deep-tow studies of the structure of the Mid-Atlantic Ridge crest near lat. 37° N. *Geol. Soc. Amer. Bull.* **88**, 621–636.

Makarychev G. I. and Shtreys N. A. (1973) Tectonic position of ophiolites of the southern Tien Shan. *Doklady Akad. Nauk. SSSR* **210**, 1164–1166 (92–93 in translation).

Malin M. C. (1974) Lunar red spots: possible pre-mare materials. *Earth Planet. Sci. Lett.* **21**, 331–341.

Malin M. C. (1978) Surface of Mercury and the moon: effects of resolution and lighting conditions on the discrimination of volcanic features. *Proc. Lunar Planet. Sci. Conf. 9th*, p. 3395–3409.

Malin M. C. and Saunders R. S. (1977) Surface of Venus: evidence of diverse landforms from radar observations. *Science* **196**, 987–990.

Markhinin E. K. (1968) Volcanism as an agent of formation of the earth's crust. In *The crust and Upper Mantle of the Pacific Area,* (L. Knopoff, C. L. Drake and P. J. Hart, eds.), p. 413–422. Amer. Geophys. Union Monograph 12, Amer. Geophys. Union. Washington, D.C.

Marsh B. D. (1974) Aleutian island arc magmatism. Ph.D. thesis, Univ. of California, Berkeley.

Marsh B. D. (1976) Mechanics of Benioff zone magmatism. In *The Geophysics of the Pacific Ocean Basin and its Margin* (G. H. Sutton, M. H. Manghnani and R. Moberly, eds.), p. 337–350. Amer. Geophys. Union Monograph 19. Amer. Geophys. Union, Washington, D.C.

Marsh B. D. (1978) On the cooling of andesitic magma. *Phil. Trans. Roy. Soc. London* **288A,** 611–625

Marsh B. D. and Carmichael I. S. E. (1974) Benioff zone magmatism. *J. Geophys. Res.* **79**, 1196–1206.

Masursky H., McGill G. E., Schaber G. and Schubert G. (1981a) Geomorphology of Venus, *J. Geophys. Res.*, in press.

Masursky H., Eliason E., Ford P. G., Loriot G. B. and Pettengill G. H. (1981b) Pioneer Venus radar results: a surface imaging. *J. Geophys. Res.*, in press.

Mauk F. T. and Johnston M. J. S. (1973) On the triggering of volcanic eruptions by Earth tides. *J. Geophys. Res.* **78**, 3356–3362.

Maxwell T. A. (1978) Origin of multi-ring basin ridge systems: an upper limit to elastic deformation based on a finite-element model. *Proc. Lunar Planet. Sci. Conf. 9th*, p. 3541–3559.

Maxwell T. A. and Gifford A. W. (1980) Ridge systems of Caloris: comparison with lunar basins. *Proc. Lunar Planet. Sci. Conf. 11th*, p. 2447–2462.

Maxwell T. A., El-Baz F. and Ward S. H. (1975) Distribution, morphology, and origin of ridges and arches in Mare Serenitatis. *Geol. Soc. Amer. Bull.* **86**, 1273–1278.

Maynard G. L. (1970) Crustal layer of seismic velocity of 6.9–7.6 kilometers per second under deep oceans. *Science* **168**, 120–121.

McAdoo D. C. and Burns J. A. (1975) The Coprates trough assemblage: more evidence for martian polar wander. *Earth Planet. Sci. Lett.* **25**, 347–354.

McConnell R. B. (1974) Evolution of Taphrogenic lineaments in continental platforms. *Geol. Rundsh.* **63**, 389–430.

McCulloch M. T. and Wasserburg G. T. (1978) Sm-Nd and Rb-Sr chronology of continental crust formation. *Science* **200**, 1003–1011.

McGetchin T. R. and Smyth J. R. (1978) The mantle of Mars: some possible geological implications of its high density. *Icarus* **34**, 512–536.

McGill G. E. (1980) Evidence for continental-style rifting from the Beta region of Venus. *NASA 82385*, p. 79–80.

McGregor V. R. (1973) The early Precambrian gneisses of Godthaab district, West Greenland. *Phil. Trans. Roy. Soc. Lond.* **A273**, 343–358.

McKee E. H., Noble D. C. and Silberman M. L. (1970) Middle Miocene hiatus in volcanic activity in the Great Basin area of the western United States. *Earth Planet. Sci. Lett.* **8**, 93–96.

McKenzie D. P. (1969) Speculations on the consequences and causes of plate motion. *Geophys. J. Roy. Soc.* **18**, 1–32.

McKenzie D. P. (1972) Active tectonics of the Mediterranean Region. *Geophys. J. Roy. Astron. Soc.* **30**, 109–185.

McKenzie D. and Sclater J. G. (1971) The evolution of the Indian Ocean since the late Cretaceous. *Geophys. J. Roy. Astron. Soc.* **24**, 437–528.

McKenzie D. and Weiss N. (1975) Speculations on the thermal and tectonic history of the Earth. *Geophys. J. Roy. Astron. Soc.* **42**, 131–174.

Melosh H. J. (1977) Global tectonics of a Despun Planet. *Icarus* **31** 221–243.

Melosh H. J. (1978) The tectonics of Mascon loading. *Proc. Lunar Sci. Conf. 9th*, p. 3513–3525.

Melosh H. J. (1980) Tectonic patterns on reoriented and tidally relaxed planets (abstract). In *Lunar and Planetary Science XI*, p. 726–728. Lunar and Planetary Institute, Houston.

Melosh H. J. and Dzurisin D. (1978) Tectonic implications for the gravity structure of Caloris Basin, Mercury. *Icarus* **33**, 141–144.

Melson W. G., Jarosewich E., Cifelli R. and Thompson G. (1967) Alkali olivine basalt dredged near St. Paul's rocks, Mid-Atlantic Ridge. *Nature* **215**, 381–382.

Menard H. W. (1964) *Marine Geology of the Pacific*. McGraw-Hill, N.Y. 271 pp.

Menard H. W. (1967) Sea-floor spreading, topography and the second layer. *Science* **157**, 923–924.

Menard H. W. (1973) Epeirogeny and plate tectonics. *EOS (Trans. Amer. Geophys. Union)* **54**, 1244–1255.

Menard H. W. and Atwater T. (1968) Changes in the direction of sea floor spreading. *Nature* **219**, 463–467.

Menard H. W. and Atwater T. (1969) Origin of fracture zone topography. *Nature* **222**, 1037–1040.

Merzer A. M. and Freund R. (1975) Buckling of strike-slip faults—in a model and in nature. *Geophys J. Roy. Astron. Soc.* **43**, 517–530.

Minster J. G., Jordan T. H., Molnar P. and Haines E. (1974) Numerical modelling of instantaneous plate tectonics. *Geophys. J. Roy. Astron. Soc.* **36**, 541–576.

Mitchell A. H. and Reading H. G. (1969) Continental margins, geosynclines and ocean-floor spreading. *J. Geol.* **77**, 629–646.

Miyashiro A. (1972) Metamorphism and related magmatism in plate tectonics. *Amer. J. Sci.* **272**, 629–656.

Molnar P. and Atwater T. (1973) Relative motion of hot spots in the mantle. *Nature* **246**, 288–291.

Molnar P. and Atwater T. (1978) Interarc spreading and cordilleran tectonics as alternates related to the age of subducted oceanic lithosphere. *Earth Planet. Sci. Lett.* **41**, 330–340.

Molnar P. and Francheteau J. (1975) The relative motion of "hot spots" in the Atlantic and Indian Oceans during the Cenozoic. *Geophys. J. Roy. Astron. Soc.* **43**, 763–774.

Molnar P. and Oliver J. (1969) Lateral variations of attenuation in the upper mantle and discontinuities in the lithosphere. *J. Geophys. Res.* **74**, 2648–2682.

Molnar P. and Tapponnier P. (1975) Cenozoic tectonics of Asia: effects of a continental collision. *Science* **189**, 419–426.

Molnar P. and Tapponnier P. (1978) Active tectonics of Tibet. *J. Geophys. Res.* **83**, 5361–5376.

Moores E. M. (1969) Petrology and structure of the Vourinos Ophiolite Complex of northern Greece. *Geol. Soc. Amer. Spec. Paper 118*. 74 pp.

Moores E. M. (1970) Ultramafics and orogeny, with models of the U.S. Cordillera and the Tethys. *Nature* **228**, 837–842.

Moores E. M. (1973) Plate tectonic significance of Alpine peridotite types. In *Implications of Continental Drift to the Earth Sciences* (D. H. Tarling and S. K. Runcorn, eds.), p. 963–975. Academic Press, London.

Moores E. M. (1975) The Smartville Terrain, northwestern Sierra Nevada, a major pre-late Jurassic ophiolite complex (abstract). In *Geol. Soc. Amer. Abstracts with Programs* **7**, 352.

Moores E. M. and Jackson E. D. (1974) Ophiolites and oceanic crust. *Nature* **250**, 136–138.

Moores E. M. and Vine F. J. (1971) The Troodos Massif, Cyprus and other ophiolites as oceanic crust: evaluation and implications. *Phil. Trans. Roy. Soc. Lond.* **A268**, 443–466.

Morabito L. A., Synnot S. P., Kupferman P. N. and Collins S. A. (1979) Discovery of currently active extraterrestrial volcanism. *Science* **204**, 972.

Morgan W. J. (1971) Convection plumes in the lower mantle. *Nature* **230**, 42–43.

Morgan W. J. (1972a) Deep mantle convection plumes and plate motions. *Bull. Amer. Assoc. Petrol. Geol.* **56**, 203–213.

Morgan W. J. (1972b) Deep motions and deep mantle convection. In *Studies in Earth and Space Sciences*. Mem. Geol. Soc. Amer. **132**, 7–22. Geological Society of America, Boulder, Colorado.

Morgan W. J. (1975) Heat flow and vertical movements of the crust. In *Petroleum and Global Tectonics* (A. G. Fischer and S. Judson, eds.), p. 23–43. Princeton Univ. Press, Princeton, N. J.

Muehlberger W. R. (1974) Structural history of southeastern Mare Serenitatis and adjacent highlands. *Proc. Lunar Sci. Conf. 5th*, p. 101–110.

Muehlberger W. R. (1977) Comparison of structural style between northern Mare Tranquillitatis and southern Serenitatis (abstract). In *Lunar Science VIII*, p. 694–696. The Lunar Science Institute, Houston.

Muller P. M. and Sjogren L. (1968) Mascons: lunar mass concentrations. *Science* **161**, 680–684.

Mutch T. A., Arvidson R. A., Head J. W., Jones K. L. and Saunders R. S. (1976) *The Geology of Mars*. Princeton Univ. Press, Princeton, N.J. 400 pp.

Nakamura K., Jacob K. H. and Davies J. N. (1978) Volcanoes as possible indicators of tectonic stress orientation; Aleutians and Alaska. *Pure and Appl. Geophysics* **115**, 87–112.

Needham H. D. and Francheteau J. (1974) Some characteristics of the Rift Valley in the Atlantic Ocean near 36°48′ N. *Earth Planet. Sci. Lett.* **22**, 29–43.

Needham H. D., Choukroune P., Cheminee J. L., LePichon X., Francheteau J. and Tapponnier P. (1976) The accreting plate boundary Ardoukoba rift (north east Africa) and the oceanic rift valley. *Earth Planet. Sci. Lett.* **28**, 439–453.

Neukum G. and Wise D. (1976) Mars: a standard crater curve and possible new time scale. *Science* **194**, 1381–1387.

Noble D. L. (1972) Some observations on the Cenozoic volcano-tectonic evolution of the Great Basin, western United States. *Earth Planet. Sci. Lett.* **17**, 142–150.

Norin E. (1946) Geological explorations in western Tibet. Publ. 29 of the Sino-Swedish expedition, Stockholm. 205 pp.

Oberbeck V. R., Quaide W. L., Arvidson R. E. and Aggarwal H. R. (1977) Comparative studies of lunar, martian and mercurian craters and plains. *J. Geophys. Res.* **82**, 1681–1698.

Okal E. and Anderson D. L. (1975) A study of lateral inhomogeneities in the upper mantle by multiple ScS travel-time residuals. *Geophys. Res. Lett.* **2**, 313–316.

Orcutt J., Kennett B. L. N. and Dorman L. M. (1976) Structure of the East Pacific Rise from an ocean bottom seismometer survey. *Geophys. J. Roy. Astron. Soc.* **45**, 305–320.

Osmaston M. F. (1971) Genesis of ocean ridge median valleys and continental rift valleys. *Tectonophys.* **11**, 387–405.

Oxburgh E. R. and Turcotte D. L. (1974) Membrane tectonics and the East African Rift. *Earth Planet. Sci. Lett.* **22**, 133–140.

Özelçi F. (1973a) Gravity anomalies of the eastern Mediterranean. *Bull. Mineral Res. Explor. Inst. Turk.* **80**, 54–92.

Özelçi F. (1973b) Zagros tektonik zonu ve bu zonun Anadolu'da uzanimi ile ilgili jeofizik gözlenimler. Türkiye Birinci Jeofizik Bilimsel ve Teknik Kongre Tebliğleri, p. 241–249, Ankara.

Özpeker I. (1973) Nemrut Yanardağinin Petrojenezi. Thesis. I.T.Ü. Maden Fakültesi, Istanbul. 68 pp.

Packham G. H. and Falvey D. A. (1971) An hypothesis for the formation of marginal seas in the western Pacific. *Tectonophys.* **11**, 79–109.

Palmason G. (1973) Kinematics and heat flow in a volcanic rift zone, with application to Iceland. *Geophys. J. Roy. Astron. Soc.* **33**, 451–481.

Palmason G. and Saemundsson K. (1974) Iceland in relation to the Mid-Atlantic Ridge. *Ann. Rev. Earth Planet. Sci.* **2**, 25–50.

Panayiotou A. (1980) (ed.) Ophiolites: Proceedings International Ophiolite Symposium Cyprus 1979, Geol. Survey Dept. Cyprus. 781 pp.

Pankhurst R. J., Moorbath S. and McGregor V. R. (1973) Late event in the geological evolution of the Godthaab district, West Greeland. *Nature Phys. Sci.* **243**, 24–26.

Parker R. L. and Oldenburg D. W. (1973) Thermal model of ocean ridges. *Nature Phys. Sci.* **242**, 137–139.

Parmentier E. M., Turcotte D. L. and Torrance K. E. (1975) Numerical experiments on the structure of mantle plumes. *J. Geophys. Res.* **80**, 4417–4424.

Peale S. J., Cassen P. and Reynolds R. T. (1979) Melting of Io by tidal dissipation. *Science* **203**, 892–894.

Pechmann J. B. and Melosh H. J. (1979) Global fracture patterns of a despun planet: application to Mercury. *Icarus* **38**, 243–250.

Peeples W. J., Sill W. R., May T. W. and Ward S. H. (1978) Orbital radar evidence for lunar subsurface layering in Maria Serenitatis and Crisium. *J. Geophys. Res.* **83**, 3459–3468.

Penrose field conference participants (1972) Ophiolites. *Geotimes* **17**, (12)24–25.

Peterson J. E. (1978) Volcanism in the Noachis-Hellas region of Mars, 2. *Proc. Lunar Planet. Sci. Conf. 9th*, p. 3411–3432.

Pettengill G. H., Eliason E., Ford P. G., Loriot G. B., Masursky H. and McGill G. E. (1980) Pioneer Venus radar results: altimetry and surface properties. *J. Geophys. Res.* **85**, 8261–8270.

Pettengill G. H., Ford P. G., Brown W. E., Kaula W. M., Masursky H., Eliason E. and McGill G. E. (1979) Venus: Preliminary topographic and surface imaging results from the Pioneer orbiter. *Science* **205**, 90–93.

Peyve A. V. (1969) Oceanic crust of the geologic past. *Geotectonics* **3**, 210–224.

Phillips R. J. and Ivins E. R. (1979) Geophysical observations pertaining to solid-state convection in the terrestrial planets. *Phys. Earth Planet. Inter.* **19**, 107–148.

Phillips R. J. and Lambeck K. (1980) Gravity fields of the terrestrial planets: long-wavelength anomalies and tectonics. *Rev. Geophys. Space Phys.* **18**, 27–76.

Phillips R. J. and Saunders R. S. (1975) The isostatic state of Martian topography. *J. Geophys. Res.* **80**, 2893–2898.

Phillips R. J., Banerdt W. B., Sleep N. L. and Saunders R. S. (1981a) Martian stress distributions: arguments about finite strength support of Tharsis. *J. Geophys. Res.*, in press.

Phillips R. J., Kaula W. M., McGill G. E. and Malin M. C. (1981b) Tectonics and evolution of Venus. *Science*, in press.

Phillips R. J., Conel J. E., Abbott E. A., Sjorgren W. L. and Morton J. B. (1972) Mascons: progress toward a unique solution for mass distribution. *J. Geophys. Res.* **77**, 7106–7114.

Phillips R. J., Sjogren W. L., Abbott E. A., Smith J. C., Wimberly R. N. and Wagner C. A. (1979) Gravity field of Venus: a preliminary analysis. *Science* **205**, 93–96.

Phillips R., Blasius K., Carr M., Eaton G., Gilbert C., McGetchin T., Saunders S., Sleep N., Solomon S. and Wise D. (1978) Tectonic and volcanic evolution of Tharsis province of Mars: a multidisciplinary approach (abstract). Proc. 3rd Int. Conf. on Basement Tectonics, Durango, Colorado.

Phinney W. C., Dence M. R. and Grieve R. A. F. (1978) Investigation of the Manicouagan Impact Crater: an introduction. *J. Geophys. Res.* **83**, 2729–2735.

Piper J. D. A. and Gibson I. L. (1972) Stress control of processes at extensional plate margins. *Nature* **238**, 83–86.

Pirsson L. V. (1914) Oligocene fossils from Bermuda. *Amer. J. Sci.* **35**, 331–334.

Pitman W. C. and Talwani M., eds. (1977) (eds.) *Island Arcs, Deep-Sea Trenches and Back-arc Basins*. Maurice Ewing Series 1, Amer. Geophys. Union, Washington, D.C. 470 pp.

Plescia J. B. and Saunders R. S. (1979a) The chronology of the martian volcanoes. *Proc. Lunar Planet. Sci. Conf. 10th*, p. 2841–2859.

Plescia J. B. and Saunders R. S. (1979b) Tectonism of the Tharsis region. NASA TM 80339, p. 47–49.

Poehls K. A. (1974) Seismic refraction on the Mid-Atlantic Ridge at 37° N. *J. Geophys. Res.* **79**, 3370–3373.

Pollard D. D. (1973) Derivation and evaluation of a mechanical model for sheet intrusions. *Tectonophys.* **19**, 233–269.

Pollard D. D. and Muller O. H. (1976) The effect of gradients in regional stress and magma pressure on the form of sheet intrusions in cross section. *J. Geophys. Res.* **81**, 975–984.

Pontoise B., Lépine J. C. and Ruegg J. C. (1976) Attenuation des ondes sismiques en T.F.A.I., La Creation de la Lithosphere Oceanique. *Bull. Soc. Geol. France* **18**, 847–849.

Quaide W. (1965) Rilles, ridges and domes—clues to maria history. *Icarus* **4**, 374–389.

Rabinowitz P. D. and Purdy G. M. (1976) The Kane Fracture Zone in the western central Atlantic Ocean. *Earth Planet. Sci. Lett.* **33**, 21–26.

Rabinowitz P. D. and Ryan W. B. F. (1970) Gravity anomalies and crustal shortening in the eastern Mediterranean. *Tectonophys.* **10**, 585–608.

Raitt R. W. (1963) The crustal rock. In *The Sea*, v. 3 (M. N. Hill, ed.), p. 85–102. Wiley-Interscience, N.Y.

Ramsay J. G. (1963) Structural investigations in the Barberton Mountain Land, eastern Transvaal. *Trans. Geol. Soc. S. Africa* **66**, 353–398.

Rankin D. W. (1976) Appalachian salients and recesses: late Precambrian continental breakup and the opening of the Iapetus Ocean. *J. Geophys. Res.* **81**, 5606–5619.

Ravenne C., Pascal G., Dubois J., Dugas F. and Montadert L. (1977) Model of a young intraoceanic arc: the New Hebrides island arc. In *International Symposium on Geodynamics in the Southwest Pacific*, p. 63–78. Editions Technips, Paris.

Reay A., Rooke J. M., Wallace R. C. and Whelan P. (1974) Lavas from Niuafo'ou Island, Tonga, resemble ocean floor basalts. *Geology* **2**, 605–606.

Reid I., Orcutt J. A. and Prothero W. A. (1977) Seismic evidence from a narrow zone of partial melting underlying the East Pacific Rise at 21° N. *Bull. Geol. Soc. Amer.* **88**, 678–682.

Reinhardt B. M. (1969) On the genesis and emplacement of ophiolites in the Oman Mountains geosyncline. *Schweizer. Mineralog. Petrog. Mitt.* **49**, 1–30.

Renard V., Schrumpf B., Sibuet J. C. and Carre D. (1975) Bathymetrie Detaillee d'une Partie de la Vallee du Rift et de la Faille Transformte Près de 36°50 N Dans l'Ocean Atlantique. Centre National pour l'Exploitation des Océans, Paris.

Reynolds P. and Aumento F. (1974) Deep drill 1972. Potassium-argon dating of the Bermuda drill core. *Can. J. Earth Sci.* **11**, 1269–1273.

Rhodes R. C. (1975) New evidence for the impact origin of the Bushveld Complex, South Africa. *Geology* **3**, 549–554.

Ridler R. H. (1970) Relationship of mineralization to volcanic stratigraphy in the Kirkland-Larder Lakes area, Ontario. *Proc. Geol. Assoc. Canada* **21**, p. 33–42.

Rittman A. (1953) Magmatic character and tectonic position of the Indonesian volcanoes. *Bull. Volc.* **14**, 45–58.

Robb J. M. and Kane M. F. (1975) Structure of the Vema Fracture Zone from gravity and magnetic intensity profiles. *J. Geophys. Res.* **80**, 4441–4445.

Roberts J. L. (1970) The intrusion of magma into brittle rocks. In *Mechanism of Igneous Intrusion* (G. Newall and N. Rast, eds.), p. 287–338. Galery Press, Liverpool.

Rodgers J. and Neale E. R. W. (1963) Possible "Taconic" Klippen in western Newfoundland. *Amer. J. Sci.* **261**, 713–730.

Rodolfo K. S. (1969) Bathymetry and marine geology of the Andaman Basin, and tectonic implications for southeast Asia. *Geol. Soc. Amer. Bull.* **80**, 1203–1230.

Ronca L. B. (1965) A geological model of Mare Humorum. *Icarus* **4**, 390–395.

Rosencrantz E. (1979) Basalt dolerite dike-gabbro structure, North Arm massif, Bay of Islands ophiolite complex, Newfoundland (abstract). Int. Ophiolite Sym., Abstracts, p. 100. Geol. Survey Cyprus, Nicosia.

Rosendahl B. R., Raitt R. W., Dorman L. M., Bibea L. D., Hussong D. M. and Sutton G. M. (1976) Evolution of oceanic crust: a physical model of the East Pacific Rise crust derived from seismic refraction data. *J. Geophys. Res.* **81**, 5924–5304.

Ruegg J. C. (1974) P and S velocities beneath the mid-oceanic ridges and the Poisson's ratio as a possible test of partial melting. Ph.D. thesis, Etat Univ. de Paris, Paris, France.

Ruegg J. C. (1975) Structure profonde de la croute et du manteau superieur du sud-est de l'Afar d'apres les donnees sismiques. *Ann. Geophys.* **31**, 329–360.

Runcorn S. K. (1974) On the origin of mascons and moonquakes. *Proc. Lunar Sci. Conf. 5th*, p.3115–3126.

Sapin M. (1975) Premiere campagne de sismologie experimentale aux Acores (abstract). In *3eme Reunion Annuelle des Sciences de la Terre, Montpellier*, p. 336. Societe Geologique de France, Paris.

Saunders R. S. and Malin M. (1975) Venus, geologic analysis of radar images. *Proc. Int. Colloq. Plan. Geol.*, p. 507–515. Rome, Italy.

Saunders R. S. and Malin M. C. (1977) Geologic interpretation of new observations of the surface of Venus. *Geophys. Res. Lett.* **4**, 547–550.

Saunders A. D., Tarney J., Stern C. R. and Dalziel I. W. D. (1979). Geochemistry of Mesozoic maginal basin floor igneous rocks from southern Chile. *Geol. Soc. Amer. Bull.* **90**, 237–258.

Sawkins F. J. (1976) Widespread continental rifting: some considerations of timing and mechanism. *Geology* **4**, 427–430.

Sawkins F. J. and Burke K. (1980) Extensional tectonics and mid-Palaeozoic massive sulphide occurrences in Europe. *Geol. Rundsch.* **69**, 349–360.

Schilling J.-G. (1973) Iceland mantle plume. *Nature* **246**, 141–143.

Schlichting H. (1955) *Boundary Layer Theory* McGraw-Hill, N. Y. 747 pp.

Scholz C. H., Barazangi M. and Sbar M. L. (1971) Late Cenozoic evolution of the Great Basin, western United States as an ensialic interarc basin. *Geol. Soc. Amer. Bull.* **82**, 2979–2990.

Scholz C. H., Koczynski T. A. and Hutchins D. G. (1976) Evidence for incipient rifting in southern Africa. *Geophys. J. Roy. Astron. Soc.* **44**, 135–144.

Schonfeld E. (1979) Origin of Valles Marinaris. *Proc. Lunar Planet. Sci. Conf. 10th*, p. 3031–3038.

Schouten H. and Denham C. R. (1979) Modelling the oceanic magnetic source layer. In *Deep Drilling Results in the Atlantic Ocean: Oceanic Crust.* (M. Talwani, C. G. Harrison and D. E. Hayes, eds.), p. 151–159.

Maurice Ewing Series v. 2. Amer. Geophys. Union, Washington, D.C.

Schroeder F. W. (1977) A geophysical investigation of the Oceanographer Fracture Zone and the Mid-Atlantic Ridge in the vicinity of 35° N. Ph.D. thesis, Columbia Univ. N.Y.

Schultz P. H. (1976) Floor-fractured lunar craters. *The Moon* **15**, 241–273.

Schultz P. H. and Glicken H. (1979) Impact crater and basin control of igneous processes on Mars. *J. Geophys. Res.* **84**, 8033–8047.

Sclater J. G. and Francheteau J. (1970) The implications of terrestrial heat flow observations on current tectonic and geochemical models of the crust and upper mantle of the Earth. *Geophys. J. Roy. Astron. Soc.* **20**, 509–542.

Sclater J. G., Anderson R. N. and Bell M. L. (1971) Elevation of ridges and the evolution of the central eastern Pacific. *J. Geophys. Res.* **76**, 7888–7915.

Sclater J. G., Jaupat C. and Galson D. (1980) The heat flow through oceanic and continental crust and the heat loss of the earth. *Rev. Geophys. Space Phys.* **18**, 269–311.

Scott D. H. (1978) Mars, highlands-lowlands: Viking contributions to Mariner relative age studies. *Icarus* **34**, 479–485.

Scott D. H. (1979) Geologic problems in the northern plains of Mars. *Proc. Lunar Planet Sci. Conf. 10th*, p. 3039–3054.

Scott D. H. and Carr M. H. (1978) Geologic map of Mars. U.S. Geol. Survey Misc. Inv. Map I-1047.

Scott D. H., Diaz J. M. and Watkins J. A. (1977) Lunar farside tectonics and volcanism. *Proc. Lunar Sci. Conf. 8th*, p. 1119–1130.

Scott D. H., Watkins J. A. and Diaz J. M. (1978) Regional deformation of mare surfaces. Proc. Lunar Planet. Sci. Conf. 9th, *p. 3527*–3539.

Şengör A. M. C. (1976) Collision of irregular continental margins: implications for foreland deformation of Alpine-type orogens. *Geology* **4**, 779–782.

Şengör A. M. C. and Jones I. C. (1975) A new interpretation of Martian tectonics with special reference to the Tharsis region. *Geol. Soc. Amer. Abstracts* **7**, 1264.

Şengör A. M. C. and Kidd W. S. F. (1979) Post-collisional tectonics of the Turkish-Iranian Plateau and a comparison with Tibet. *Tectonophys.* **55**, 361–376.

Şengör A. M. C., Burke K. and Dewey J. F. (1978) Rifts at high angles to orogenic belts: tests for their origin and the Upper Rhine Graben as an example. *Amer. J. Sci.* **278**, 24–40.

Seyfert C. K. and Murtaugh J. G. (1977) Terrestrial impact epochs and their relationship to orogenies and plate movements. *Geol. Soc. Amer. Abstracts with Programs* **9**, 1168–1169.

Seyfert C. K. and Sirkin L. A. (1979) Earth History and Plate Tectonics, 2nd Edition. Harper and Row, N.Y. 600 pp.

Seymen I. (1975) Kelkit vadisi kesiminde Kuzey Anadolu Fay Zonunun tektonik Seymen I. and Aydin A. (1972) Bingöl deprem fayi ve bunun Kuzey Anadolu özelliğ. I. T. U. Made Fak. Yay., Istanbul. 192 pp.

Seymen I. and Aydin A. (1972) Bingöl deprem fayi ve bunum Kuzey Anadolu Fay Zonu ile iliskisi. M.T.A. Dergisi **79**, 1–8.

Sharp R. P. (1973) Mars: Troughed terrain. *J. Geophys. Res.* **78**, 4063–4072.

Shatski N. S. (1947) Structural correlations of platforms and geosynclinal folded regions. S.S.R. Akad. Navk Izv. Geol. Ser. **5** 37–56.

Shaw H. R. and Jackson E. D. (1973) Linear island chains in the Pacific: result of thermal plumes or gravitational anchors? *J. Geophys. Res.* **78**, 8634–8652.

Sherman S. I. (1978) Faults of the Baikal Rift Zone. *Tectonophys.* **45**, 31–39.

Shibata T., DeLong S. E. and Walker D. (1979) Abyssal tholeiites from the oceanographer Fracture Zone II, phase equilibria and mixing. *Contrib. Mineral. Petrol.* **70**, 111–125.

Shor G. G., Menard H. W. and Raitt R. W. (1970) Structure of the Pacific Basin. In *The Sea*, v. 4 (A. E. Maxwell, ed.), p. 3–27. Wiley-Interscience, N. Y.

Sillitoe R. D. (1977) Mettallogeny of an Andean continental margin in South Korea: implications for opening of the Japan Sea. In *Island Arcs, Deep Sea Trenches and Back-arc Basins* (M. Talwani and W. C. Pitman, eds.), p. 303–310. Maurice Ewing Series 1. Amer. Geophys. Union, Washington, D.C.

Sjogren W. L. (1979) Mars gravity: high resolution results from Viking Orbiter 2. *Science* **203**, 1006–1010.

Sjogren W. L., Phillips R. J., Birkeland P. W. and Wimberly R. N. (1981a) Gravity anomalies on Venus. *J. Geophys. Res.*, in press.

Sleep N. H. (1969) Sensitivity of heat flow and gravity to the mechanism of sea-floor spreading. *J. Geophys. Res.* **74**, 542–549.

Sleep N. H. (1974) Segregation of magma from a mostly crystalline mushs *Geol. Soc. Am. Bull.* **85**, 1225–1232.

Sleep N. H. (1975) Formation of oceanic crust: some thermal constraints. *J. Geophys. Res.* **80**, 4037–4042.

Sleep N. H. and Biehler S. (1970) Topography and tectonics at the intersections of fracture zones with central rifts. *J. Geophys. Res.* **75** 2748–2752.

Sleep N. H. and Phillips R. J. (1979) An isostatic model for the Tharsis Province, Mars. *Geophys. Res. Lett.* **6**, 803–806.

Sleep N. H. and Rosendahl B. R. (1979) Topography and tectonics of the mid-ocean ridge axes. *J. Geophys. Res.* **84**, 6831–6834.

Sleep N. H. and Wolery T. J. (1975) Formation of oceanic crust: some physical and chemical constraints (abstract). *EOS (Trans. Amer. Geophys. Union)* **56**, p 454.

Smith A. G. (1971) Alpine deformation and the oceanic areas of the Tethys. *Geol. Soc. Amer. Bull.* **82**, 2039–2070.

Smith C. H. (1958) Bay of Islands igneous complex, western Newfoundland. *Mem. Geol. Survey Canada* **290**, 132 pp.

Smith R. B. and Sbar M. L. (1974) Contemporary tectonics and seismicity of the western United States with emphasis on the intermountain seismic belt. *Geol. Soc. Amer. Bull.* **85**, 1205–1218.

Smith R. B., Shuey R. T., Pelton J. R. and Bailey J. P. (1977) Yellowstone hot spot: contemporary tectonics and crustal properties from earthquake and aeromagnetic data. *J. Geophys. Res.* **82**, 3665–3676.

Smith R. B., Shuey R. T., Freidline R. O., Otis R. M. and Alley L. B. (1974) Yellowstone hot spot: new magnetic and seismic evidence. *Geology* **2**, 451–455.

Snydsman W. E., Lewis B. T. R. and McClain J. (1973) Upper mantle velocities on the northern Cocos plate. *Earth Planet. Sci. Lett.* **28**, 46–50.

Solomon S. C. (1973) Shear wave attenuation and melting beneath the Mid-Atlantic ridge. *J. Geophys. Res.* **78**, 6044–6059.

Solomon S. C. (1976) Some aspects of core formation in Mercury. *Icarus* **28**, 509–521.

Solomon S. C. (1978) On volcanism and thermal tectonics on one-plate planets. *Geophys. Res. Lett.* **5**, 461–464.

Solomon S C. and Chaiken J. (1976) Thermal expansion and thermal stress in the moon and terrestrial planets. *Proc. Lunar Sci. Conf. 7*, p. 3229–3243.

Solomon S. C. and Head J. W. (1979) Vertical movement in mare basins: relation to mare emplacement, basin tectonics, and lunar thermal history. *J. Geophys. Res.* **84**, 1667–1682.

Solomon S. C. and Head J. W. (1980a) Lunar mascon basins: lava filling, tectonics and evolution of the lithosphere. *Rev. Geophys. Space Phys.* **18**, 107–141.

Solomon S. C. and Head J. W. (1980b) Tharsis Province: uplift by anomalous mantle, or concentration of volcanism and tectonism in a locally thin lithosphere? (abstract). In *Lunar and Planetary Science IX*, p. 1063–1065. Lunar and Planetary Institute, Houston.

Solomon S. C. and Julian B. R. (1974) Seismic constraints on ocean-ridge mantle structure. *EOS (Trans. Amer. Geophys. Union,* **54**, p. 364.

Solomon S. C. and Sleep N. H. (1974) Some physical models for absolute plate motions. *J. Geophys. Res.* **79**, 2537–2567.

Solomon S. C., Head J. W. and Comer R. P. (1979) Thickness of the Martian lithosphere from tectonic features: evidence for lithospheric thinning beneath volcanic provinces (abstract). In *Reports of Planetary Geology Program 1978–1979*, p. 60–62. NASA TM 80339.

Spengler A. de E. and Delteil J. R. (1966) Le basin secondaire-tertiare de Cote d'Ivoire. In *Bassins sedimentaires du littoral africain le partie* (Du Reyre, ed.), p. 99–113. Assoc. Serv. Geol. Afr. Paris.

Steel R. J. (1976) Devonian basins of western Norway - sedimentary response to tectonism and to varying tectonic context. *Tectonophys.* **36**, 207–224.

Steinmetz L., Whitmarsh R. B. and Moreira V. S. (1977) Upper mantle structure beneath the mid-Atlantic ridge north of the Azores based on observations of compressional waves. *Geophys. J. Roy. Astron. Soc.* **50**, 353–380.

Steinmetz L., Hirn A., Sapin M., Whitmarsh R. B. and Moreira V. S. (1976) Zones of P wave attenuation beneath the crest of the Mid-Atlantic Ridge, La Creation de la Lithosphere Oceanique. *Bull. Soc. Geol. France* **18**, 931–936.

Stevens R. K. (1970) Cambro-Ordovician flysch sedimentation and tectonics in west Newfoundland and their possible bearing on a proto-Atlantic Ocean. *Geol. Assoc. Canada Spec. Paper 7*, 165–177.

Stevenson I. M. (1968) A geological reconnaissance of Leaf River map-area, New Quebec and Northwest Territories. *Mem. Geol. Survey Canada* 356, 112 pp.

Stöcklin J. (1968) Structural history and tectonics of Iran: a review. *Bull. Amer. Assoc. Petrol. Geol.* **52**, 1229–1258.

Stöcklin J., (1974) Possible ancient continental margins in Iran. In *The Geology of Continental Margins*, (C. A. Burk and C. L. Drake, eds.), p. 873–887 Springer-Verlag, New York.

Stoiber R. E. and Carr M. J. (1973) Quaternary volcanic and tectonic segmentation of Central America. *Bull. Volc.* **37**, 304–325.

Strom. R. G. (1979) Mercury: a post-Mariner 10 assessment. *Space Sci. Rev.* **24**, 3–70.

Strom R. G., Trask N. J. and Guest J. E. (1975a) Tectonism and volcanism on Mercury. *J. Geophys. Res.* **80**, 2478–2507.

Strom R. G., Murray B. C., Belton M. J. S., Danielson G. E., Davies M. E., Gault D. E., Hapke B., O'Leary B., Trask N., Guest J E., Anderson J. and Klaasen K. (1975b) Preliminary imaging results from the second Mercury encounter. *J. Geophys. Res.* **80**, 2345–2356.

Strong D. H. and Stevens R. K. (1974) Possible thermal explanation of contrasting Archean and Proterozoic geological regimes. *Nature* **249**, 545–546.

Stuart-Alexander D. E. (1978) Geologic map of the central farside of the Moon. U. S. Geol. Survey Misc. Inv. Map I-1047.

Sturt B. A., Thon A. and Furnes H. (1979) The Karmoy ophiolite, Southwest Norway. *Geology* **7**, 316–320.

Sugimura A. (1968) Spatial relations of basaltic magmas in island arcs. In *Basalts* v. 2, (H. H. Hess and A. Poldervaart, eds.), p. 537–571. John Wiley Interscience, N.Y.

Suppe J., Powell C. and Berry R. (1975) Regional topography, seismicity, quaternary volcanism and the present-day tectonics of the western United States. *Amer. J. Sci.* **275A**, 397–436.

Surkov Yu. A., Kirnozoo F. F., Glazov V. N., Dunchenko A. G. and Tatsil L. P. (1976) Investigations of Venusian gamma-radiation by Venera 9 and Venera 10. *Kosmicheskie Issledovania* **14**, 704.

Sutton G. M., Maynard G. L. and Hussong D. M. (1971) Widespread occurrence of a high-velocity basal layer in the Pacific crust found with repetitive sources and sonobuoys. *Amer. Geophys. Union Monograph 14,* (J. G. Heacock, ed.), p. 193–209. Amer. Geophys. Union, Washington, D.C.

Sykes L. R. (1967) Mechanism of earthquakes and nature of faulting on the mid-oceanic ridge. *J. Geophys. Res.* **72**, 2131–2153.

Sykes L. R. (1970) Earthquake swarms and sea floor spreading. *J. Geophys. Res.* **75**, 6598–6611.

Talwani M. and Eldholm O. (1972) Continental margin off Norway: a geophysical study. *Geol. Soc. Amer. Bull.* **83**, 3575–3606.

Talwani M., Windish C. C. and Langseth M. G., Jr. (1971) Reykjanes ridge crest: a detailed geophysical study. *J. Geophys. Res.* **76**, 473–517.

Tapponnier P. (1977) Evolution tectonique du système alpin en Méditerranée: poinçonnement et écrasement rigide-plastique. *Bull. Soc. Geol. France* **19**, 437–460.

Tapponnier P. and Francheteau J. (1978) Necking of the lithosphere and the mechanics of slowly accreting plate boundaries. *J. Geophys. Res.* **83**(B8), 3955–3970 and **83**(B9), 4497.

Tarney J., Dalziel I. W. D. and De Wit M. J. (1976) Marginal basin 'Rocas Verdes' complex from Chile: a model for Archean greenstone belt formation. In *The Early History of the Earth* (B. F. Windley, ed.), p. 131–146. Wiley, London.

Tarney J., Saunders A. D. and Weaver S. D. (1977) Geochemistry of volcanic rocks from the island arcs and marginal basins of the Scotia Arc region. In *Island Arcs, Deep Sea Trenches and Back-arc Basins* (M. Talwani and W. C. Pitman, eds.), p. 367–377. Maurice Ewing Series 1. Amer. Geophys. Union, Washington, D.C.

Taylor S. R. (1975) Lunar science: a post-Apollo view. Pergamon, N.Y. 372 pp.

Tchalenko J. S. (1970) Similarities between shear zones of different magnitudes. *Geol. Soc. Amer. Bull.* **81**, 1625–1640.

Temple P. G. and Zimmerman J. (1969) Tectonic significance of Alpine ophiolites in Greece and Turkey. *Geol. Soc. Amer. Abstracts with Programs* **1**, 221–222.

Thatcher W. and Brune J. N. (1971) Seismic study of an oceanic ridge earthquake swarm in the Gulf of California. *Geophys. J. Roy. Astron. Soc.* **22**, 473–489.

Thayer T. P. (1963) The Canyon Mountain complex, Oregon and the Alpine mafic magma stem. *U.S. Geol. Survey Prof. Pap. 475c*, p. C82–C85.

Thayer T. P. (1969) Peridotite-gabbro complexes as keys to petrology of mid-oceanic ridges. *Geol. Soc. Amer. Bull.* **80**, 1515–1522.

Thiessen R., Burke K. and Kidd W. S. F. (1979) African hotspots and their relation to the underlying mantle. *Geology* **7**, 263–266.

Thompson G. and Melson W. G. (1972) The petrology of oceanic crust across fracture zones in the Atlantic Ocean: evidence of a new kind of sea-floor spreading. *J. Geol.* **80**, 526–538.

Thurber C. H. and Toksöz M. N. (1978) Martian lithospheric thickness from elastic flexure theory. *Geophys. Res. Lett.* **5**, 977–980.

Tjia H. D. (1970) Lunar wrinkle ridges indicative of strike-slip faulting. *Geol. Soc. Amer. Bull.* **81**, 3095–3100.

Toksöz M. N. and Bird P. (1977) Formation and evolution of marginal basins and continental plateaus. In *Island Arcs, Deep Sea Trenches and Back-Arc Basins* (M. Talwani and W. C. Pitman, eds.), p. 379–394. Maurice Ewing Series 1. Amer. Geophys. Union, Washington, D.C.

Toksöz M. N., Arpat E. and Saroglu F. (1977) East Anatolian earthquake of 24 November 1976 - Field Observations. *Nature* **270**, 423–425.

Tozer D. C. (1973) Thermal plumes in the Earth's mantle. *Nature* **244**, 398–400.

Trask N. J. and Strom R. G. (1976) Additional evidence of mercurian volcanism. *Icarus* **28**, 559–563.

Tréhu A. M. (1975) Depth versus (age)1/2: a perspective on mid-ocean rises. *Earth Planet. Sci. Lett.* **27**, 287–304.

Trombka J. I., Arnold J. P., Adler I., Metzger A. E. and Reedy R. C. (1977) Lunar elemental analysis obtained from the Apollo gamma-ray and X-ray remote sensing experiment. In *The Soviet-American Conference on Cosmochemistry of the Moon and Planets*, p.787–796. NASA SP-370. Washington, D.C.

Trümpy R. (1969) Die helvetischen decken der Ostschweiz: versuch einer palinpastichen Korrelation und ansatze zu einer Kinematischen analyse. *Eclog. Geol. Helv.* **62**, 105–138.

Trümpy R. (1973) The timing of orogenic events in the Central Alps. In *Gravity and Tectonics*, p. 229–251. John Wiley and Sons, N.Y.

Turcotte D. L. (1974a) Membrane tectonics. *Geophys. J. Roy. Astron. Soc.* **36**, 33–42.

Turcotte D. L. (1974b) Are transform faults thermal contraction cracks? *J. Geophys. Res.* **79**, 2573–2577.

Turcotte D. L. and Ahern J. L. (1978) A porous flow model for magma migration in the asthenosphere. *J. Geophys. Res.* **83**, 767–772.

Turcotte D. L. and Burke K. C. A. (1978) Global sea-level changes and the thermal structure of the Earth. *Earth. Planet. Sci. Lett.* **41**, 341–346.

Turcotte D. L. and Oxburgh E. R. (1973) Mid-plate tectonics. *Nature* **244**, 337–339.

Turcotte D. L. and Oxburgh E. R. (1976) Stress accumulation in the lithosphere. *Tectonophys.* **35**, 183–199.

Upadhyay H. D. (1978) Phanerozoic peridotitic and pyroxenitic komatiites from Newfoundland. *Science* **202**, 1192–1195.

Upadhyay H. D., Dewey J. F. and Neale E. R. W. (1971) The Betts Cove Ophiolite Complex, Newfoundland: Appalachian oceanic crust and mantle. *Canada Geol. Assoc. Proc.* **24**, 27–34.

Urrien C. M. and Zambrano J. (1973) The geology of the basins of the Argentine continental margin. In *The Ocean Basin and Margins* (A. E. M. Nairn and F. G. Stehli, eds.), p. 135–166. Plenum Press, N.Y.

Van Andel Tj. H. and Ballard R. D. (1979) The Galapagos rift at 86° W 2. Volcanism, structure, and evolution of the rift valley. *J. Geophys. Res.* **84**, 5390–5406.

Van Andel Tj. H., von Herzen R. P. and Phillips J. D. (1971) The Vema fracture zone and the tectonics of transform shear zones in oceanic crustal plates. *Marine Geophys. Res.* **1**, 261–283.

Vening Meinesz F. A. (1950) Les 'grabens' africains, resultat de compression ou de tension dans la croute terrestre? *Inst. Roy. Colon. Belge Bull.* **21**, 539–552.

Viljoen M. J. and Viljoen R. P. (1970) Archean volcanicity and continental evolution in the Barberton region, Transvaal. In *African Magmatism and Tectonics* (T. N. Clifford and Gass I. G., eds.), p.27–39. Oliver and Boyd, Edinburgh.

Vine F. J. and Matthews D. H. (1963) Magnetic anomalies over oceanic ridges. *Nature* **199**, 947–949.

Vogt P. R. (1972) Evidence for global synchronism in mantle plume convection and possible significance for geology. *Nature* **240**, 338–342.

Vogt P. R. (1974) Volcano spacing, fractures, and thickness of the lithosphere. *Earth Planet. Sci. Lett.* **21**, 235–251.

Vogt P. R. and Avery O. E. (1974) Detailed magnetic surveys in the northeast Atlantic and Labrador Sea. *J. Geophys. Res.* **79**, 363–389.

Vogt P. R., Lowrie A., Bracey D. R. and Hey R. N. (1976) Subduction of aseismic oceanic ridges: effects on shape, seismicity and other characteristics of consuming plate boundaries. *Geol. Soc. Amer. Spec. Paper* **172**, *59 pp.*

Vogt P. T., Johnson G. L., Holcombe T. L., Gilg J. G. and Avery O. E. (1971) Episodes of sea-floor spreading recorded by the North Atlantic basement. *Tectonophys.* **12**, 211–234.

Von Zahn G. W. (1906) *Die Stellung Armeniens im Gebirgsbau von Vorderasien*. Ernst Siegfried Mittler und Sohn, Berlin. 89 pp.

Wadati K. (1935) On the activity of deep-focus earthquakes in the Japan Island and neighborhoods. *Geophys. Mag.* **8**, 305–325.

Waff H. S. (1980) Effects of the gravitational field on liquid distribution in partial melts within the upper mantle. *J. Geophys. Res.* **85**, 1815–1825.

Waff H. S. and Bulau J. R. (1977) Fluid distribution in partially molten "pyrolite" (abstract). *EOS (Trans. Amer. Geophys. Union)* **58**, 535.

Waff H. S. and Bulau J. R. (1979) Equilibrium fluid distribution in an ultramafic partial melt under hydrostatic stress conditions. *J. Geophys. Res.* **84**, 6109–6114.

Walker D., Stolper E. M. and Hays J. F. (1979) Limits on buoyancy-driven melt segregation. *Abstracts with Program, Geol. Soc. Amer.* **11**, 533.

Walker G. P. L. (1975) Intrusive sheet swarms and the identity of crustal layer 3 in Iceland. *Geol. Soc. London J.* **131**, 143–161.

Washington H. S. (1894) *The Volcanoes of the Kula Basin in Lydia*. Robert Drummond, N.Y. 65 pp.

Wasserburg G. J., Papanastassiou D. A., Tera F. and Huneke J. C. (1977) Outline of a lunar chronology. *Phil. Trans. Roy. Soc. Lond.* **A285**, 7–22.

Watson K. de P. (1943) Mafic and ultramafic rocks of the Baie Verte area, Newfoundland. *J. Geol.* **51**, 116–130.

Watts A. B. and Cochran J. R. (1974) Gravity anomalies and flexure of the lithosphere along the Hawaiian-Emperor seamount chain. *Geophys. J. Roy. Astron. Soc.* **38**, 119–141.

Watts A. B. and Weissel J. K. (1975) Tectonic history of the Shikoku marginal basin. *Earth Planet. Sci. Lett.* **25**, 239–250.

Watts A. B., Weissel J. K. and Davey F. J. (1977) Tectonic evolution of the South Fiji Basin. In *Island Arcs, Deep Sea Trenches and Back-Arc Basins*. (M. Talwani and W. C. Pitman, eds.), p.419–427. Maurice Ewing Series 1. Amer. Geophys. Union, Washington, D.C.

Weertman J. (1971) Theory of water-filled crevasses in glaciers applied to vertical magma transport beneath oceanic ridges. *J. Geophys. Res.* **76**, 1171–1183.

Weiblen P. W. and Schulz K. J. (1977) Is there any record of meteorite impact in the Archean rocks of North America? *Proc. Lunar Science Conf. 9th*, p.2749–2772.

Weidner D. J. and Aki K. (1973) Focal depth and mechanism of mid-ocean ridge earthquakes. *J. Geophys. Res.* **78**, 1818–1831.

Weissel J. K. (1977) Evolution of the Lau Basin by the growth of small plates. In *Island Arcs, Deep Sea Trenches and Back-Arc Basins* (M. Talwani and W. C. Pitman, eds.), p.429–436. Maurice Ewing Series 1. Amer. Geophys. Union, Washington, D.C.

Wellman P. and McElhinny M. W. (1970) K-Ar age of the Deccan Traps, India. *Nature* **227**, 595–596.

White G. W. and Stroup J. B. (1979) Distribution of rock types in the Mid-Cayman Rise as evidence for conjugate normal faulting in slowly spreading ridges. *Geology* **7**, 32–36.

White W. M., Schilling J-G. and Hart S. R. (1976) Evidence for the Azores mantle plume from strontium isotope geochemistry of the central North Atlantic. *Nature* **263**, 659–663.

Whitehead J. A. and Luther D. S. (1975) Dynamics of laboratory diapir and plume models. *J. Geophys. Res.* **80**, 705–717.

Whiteman A. J., Rees G., Naylor D. and Pegrum R. M. (1975) North Sea troughs. *Tectonophys.* **26**, 39–54.

Whiteside H. C. M. (1970) Volcanic rocks of the Witwatersrand Triad. In *African Magmatism and Tectonics* (T. N. Clifford and I. G. Gass, eds.), p.73–88. Oliver and Boyd, Edinburgh.

Whitford-Stark J. L. (1974) Internal origin for lunar rilled craters and the maria? *Nature* **248**, 573–576.

Whitford-Stark J. L. and Head J. W. (1980) Stratigraphy of Oceanus Procellarum basalts: sources and styles of emplacement. *J. Geophys. Res.* **85**, 6579–6609.

Whitmarsh R. B. (1973) Median valley refraction line Mid-Atlantic Ridge at 37° N. *Nature* **246**, 297–299.

Whitmarsh R. B. (1975) Axial intrusion zone beneath the median valley of the Mid-Atlantic Ridge at 37° N detected by explosion seismology. *Geophys. J. Roy. Astron. Soc.* **42**, 189–215.

Whitmarsh R. B. and Laughton A. S. (1976) A long-range sonar study of the Mid-Atlantic Ridge crest near 37° N (FAMOUS area) and its tectonic implications. *Deep-Sea Research* **23**, 1005–1023.

Wilcox R., Harding T. P. and Sealy D. R. (1973) Basic wrench tectonics. *Amer. Assoc. Petrol. Geol. Bull.* **57**, 74–96.

Wilhelms D. E. (1976) Mercury volcanism questioned. *Icarus* **28**, 551–558.

Williams H. (1967) Island of Newfoundland. Geol. Survey Canada Map 1231A.

Williams H. (1970) Red Indian Lake (east half) Newfoundland. Geol. Survey Canada Map 1196A.

Williams H. (1971) Mafic-ultramafic complexes in western Newfoundland Appalachians and the evidence for their transportation: a review and interim report. *Geol. Assoc. Canada Proc.* **24**, 9–25.

Williams H. (1973) Bay of Islands Map-Area, Newfoundland. *Canada Geol. Survey Paper* **72-34**. 7 pp.

Williams H. and Malpas J. (1972) Sheeted dikes and brecciated dike rocks within transported igneous complexes, Bay of Islands, western Newfoundland. *Can. J. Earth Sci.* **9**, 1216–1229.

Wilson H. D. B., Brisbin W. C., McRitchie W. D. and Davies J. C. (1972) Archean geology and metallogenesis of the western part of the Canadian Shield, Excursion Guidebook A33-C33. *International Geological Congress, XXIV Session*. Montreal.

Wilson J. T. (1962) Cabot Fault, an Appalachian equivalent of the San Andreas and Great Glen Faults and some implications for continental displacement. *Nature* **195**, 135–138.

Wilson J. T. (1963) Evidence from islands on the spreading of ocean floors. *Nature* **197**, 536–538.

Wilson J. T. (1965a) A new class of faults and their bearing on continental drift. *Nature* **207**, 343–347.

Wilson J. T. (1965b) Evidence from ocean islands suggesting movements in the Earth. *Phil. Trans. Roy. Soc. Lond.* **A258**, 145–167.

Wilson J. T. (1966) Did the Atlantic close and then reopen? *Nature* **211**, 676–681.

Wilson J. T. (1968a) Static or mobile Earth: the current scientific revolution. In *Gondwanaland Revisited*, Amer. Philos. Soc. Proc. 112, 309–320.

Wilson J. T. (1968b) Comparison of the Hudson Bay Arc with some other features. In *Science, History and Hudson Bay* (C. S. Beals and D. A. Shemstone, eds.), p.1015–1033. Dept. Energy, Mines and Resources, Ottawa.

Wilson J. T. (1973) Mantle plumes and plate motion. *Tectonophys.* **19**, 149–164.

Wilson R. A. M. (1959) The geology of the Xerox-Troodos area. *Mem. Cyprus Geol. Survey 1*. 135 pp.

Windley B. F. (1977) *The Evolving Continents*. Wiley, N.Y. 385 pp.

Winterer E. L. (1973) Sedimentary facies and plate tectonics of equatorial Pacific. *Bull. Amer. Assoc. Petrol. Geol.* **57**, 265–282.

Winterer E. L. (1976) Anomalies in the tectonic evolution of the Pacific. In *The Geophysics of the Pacific Ocean Basin and its Margin* (G. H. Sutton, M. H. Manghanani and R. Moberly, eds.), p. 269–278. Amer. Geophys. Union Monograph 19. Amer. Geophys. Union, Washington, D.C.

Wise D. U. (1974a) Continental margins, freeboard and the volumes of continents and oceans through time. In *The Geology of Continental Margins* (C. A. Burk and C. L. Drake, eds.), p. 45–58. Springer Verlag, N.Y.

Wise D. U. (1974b) Martian fault pattern and time sequence in relation to volcanism, northern Tharsis ridge area. *EOS (Trans. Amer. Geophys. Union)* **55**, 341.

Wise D. U., Golombek M. P. and McGill G. E. (1979a) Tharsis province of Mars: geologic sequences, geometry, and deformation mechanism. *Icarus* **38**, 456–472.

Wise D. U., Golombek M. P. and McGill G. E. (1979b) Tectonic evolution of Mars. *J. Geophys. Res.* **84**, 7934–7939.

Wood C. A. and Head J. W. (1976) Comparison of impact basins on Mercury, Mars and the Moon. *Proc. Lunar Sci. Conf. 7th*, p. 3629–3651.

Wood D. S. (1966) The Rhodesian basement. 10th Ann. Rept. Inst. African Geol. Univ. Leeds, p.18–19.

Woodland A. W. (ed.) (1975) *Petroleum and the Continental Shelf of North-west Europe*. Halsted-Wiley, London. 501 pp.

Worral D. M., Fagin S. W. and Muehlberger W. R. (1978) Ridge orientation vs. position in circular lunar maria (abstract). *EOS (Trans. Amer. Geophys. Union)* **59**, 310.

Wright C. (1975) Comments on "Seismic array evidence of a core boundary source for the Hawaiian linear volcanic chain," by E. R. Kanasewich *et al.* *J. Geophys. Res.* **80**, 1915–1919.

Wright L. A., Troxel B. W., Williams E. G., Roberts M. T. and Diehl P. E. (1974) Precambrian sedimentary units of the Death Valley Region. In *Guidebook, Death Valley Region, Shoshone California*, p.27–35. Death Valley Publishing Co.

Wyllie P. J. (1971) *The Dynamic Earth*. Wiley, N.Y. 416 pp.

Yuen D. A. and Schubert G. (1977) Asthenospheric shear flow: thermally stable or unstable? *Geophys. Res. Lett.* **4**, 503–506.

Zeschke G. (1954) Der Simav-Graben und seine Gesteine. *Tür. Jeol. Kur. Bül.* **5**.

Overleaf:

Top: The mass spectrometer on which A. O. Nier made the first precise determinations of the isotopic compositions of carbon, lead, and uranium (Harvard University, 1936–1938). The uranium and lead measurements established the ^{235}U-^{207}Pb method of geological age dating, and led to the introduction of the Pb-Pb dating method (Nier, 1939). Isotopic variations in the composition of natural carbon were first found on this instrument [Nier A. O. and Gulbransen E. A. (1939) Variations in the relative abundances of the carbon isotopes. *J. Amer. Chem. Soc.* **61,** 697–698].

Middle: ^{40}K was discovered in 1935 on this mass spectrometer at the University of Minnesota [Nier A. O. (1935) Evidence for the existence of an isotope of potassium of mass 40. *Phys. Rev.* **48,** 283–284]. Later identificaton of excess ^{40}Ar in potassium minerals at Minnesota demonstrated the potential usefulness of a geological dating method based on the decay of ^{40}K to ^{40}Ar [Aldrich L. T. and Nier A. O. (1948) Argon 40 in potassium minerals. *Phys. Rev.* **74,** 876–877].

Bottom. The first dual-collector mass spectrometer for precise measurement of carbon isotope ratios, University of Minnesota, 1941.

7. Radiogenic and stable isotopes, radiometric chronology, and basaltic volcanism

George W. Wetherill, Team Leader
—*Department of Terrestrial Magnetism*
Carnegie Institution of Washington

Claude J. Allègre
—*Institue de Physique du Globe*
Université de Paris

Christopher Brooks
—*Departemente de Géologie*
Université de Montréal

P. Eberhardt
—*Physikalisches Institut*
Universität Bern

Stanley R. Hart
—*Department of Earth and Planetary Sciences*
Massachusetts Institute of Technology

V. Rama Murthy
—*Department of Geology and Geophysics*
University of Minnesota

Fouad Tera
—*Department of Terrestrial Magnetism*
Carnegie Institution of Washington

W. Randall Van Schmus
—*Department of Geology*
University of Kansas

7.1 INTRODUCTION AND GENERAL PRINCIPLES

7.1.1 Scope of this section

Variation in the naturally occurring isotopic composition of the chemical elements can result from either chemical or nuclear processes. Chemical isotopic fractionation is a consequence of the different isotopes having slightly different chemical properties (such as free energy of chemical reaction, vapor pressure, reaction rate, or coefficient of diffusion). These effects are particularly significant for light elements which occur naturally in various chemical states. Considerable attention has been given to natural chemical isotopic fractionation of hydrogen, carbon, oxygen and sulfur. Study of the isotopic composition of these elements in terrestrial, lunar, and planetary matter is a major field of scientific inquiry. An introduction to this work is given in section 7.1.4. In addition to isotopic fractionation of chemical origin, nuclear transformations of various kinds will cause the isotopic composition of natural materials to vary.

The most prominent effect of this kind is the result of radioactive decay of naturally occurring long-lived radioactive nuclides such as ^{238}U, ^{235}U, ^{232}Th, ultimately to the stable radiogenic isotopes ^{206}Pb, ^{207}Pb, ^{208}Pb, respectively; ^{147}Sm to stable ^{143}Nd; ^{87}Rb to stable ^{87}Sr; ^{40}K to stable ^{40}Ar (as well as ^{40}Ca). A tabulation of the rate constants for radioactive decay of these isotopes may be found in Table 7.1.1.

Addition of the newly formed stable daughter isotopes, resulting from this radioactive decay, to the preexisting isotopic mixture of the daughter element will cause the isotopic composition of the daughter element in a rock or other natural system to change with time. The amount of this change will depend on both the length of this time of growth and on the relative proportion of parent to daughter elements. Therefore, the isotopic composition of the daughter element will be dependent upon natural processes which determine and alter the parent/daughter element ratios. This dependence of the isotopic abundance of the daughter element on its geochemical history introduces the possibility of using the resulting variations to infer aspects of the past history of the rock or chemical system under study. Of course many other observations, including the study of the mineralogical composition of a rock, are also relevant to its past history. However, these variations in the isotopic composition of the daughter elements have two unique characteristics:

(1) The dependence on radioactive decay introduces the possibility of referring the history to an absolute time scale.

(2) This isotopic composition is insensitive to chemical fractionation, for example as a result of fractional crystallization or equilibrium partial melting occurring during the formation of a basalt.

Decay of long-lived radioactive parent isotopes is not the only nuclear process which causes natural isotope variations. For example, short-lived isotopes such as ^{14}C and ^{26}Al are produced by cosmic rays, and their decay also provides ways to date natural events.

Table 7.1.1 Radioactive decay constants used in this chapter.

Isotope	Branch	$\lambda(yr^{-1})$	Reference
^{40}K	λ_e(to ^{40}Ar)	5.81×10^{-11}	Wetherill (1966)
	λ_β(to ^{40}Ca)	4.962×10^{-10}	Beckinsale and Gale (1969)
			Steiger and Jäger (1977)
^{87}Rb		1.42×10^{-11}	Wetherill (1966)
			Steiger and Jäger (1977)
^{129}I		4.03×10^{-8}	Katcoff *et al.* (1951)
^{147}Sm		6.54×10^{-12}	Lugmair and Marti (1978)
^{232}Th		4.948×10^{-11}	LeRoux and Glendenin (1963)
^{235}U		9.849×10^{-10}	Jaffey *et al.* (1971)
^{238}U	λ_α	1.551×10^{-10}	Jaffey *et al.* (1971)
	$\lambda_{S.F}$	7.03×10^{-17}	Roberts *et al.* (1968)
^{244}Pu	λ_α	8.47×10^{-9}	Fields *et al.* (1966)
	$\lambda_{S.F}$	1.06×10^{-11}	Fields *et al.* (1966)

These decay constants are in conformity with the conventions recommended by the I. U. G. S. subcommission on geochronology (Steiger and Jäger, 1977).

Intermediate-lived radioactive isotopes (half-life $\leq 10^8$ years, e.g., 17 m.y. ^{129}I, 82 m.y. ^{244}Pu) produced by nucleo-synthetic events prior to the formation of the solar system will have decayed to negligible levels by the present time. However, they were present during the early history of the solar system, and a quantitative record of their previous existence may be preserved in the form of their stable daughter isotopes or by the charged particle tracks produced in minerals by their radioactive decay. Measurement of these effects forms the basis of important methods for dating events in the early solar system (see references in Podosek, 1978; Wetherill, 1975c).

Natural nuclear reactions may also lead directly to the production of stable and radioactive isotopes and consequently to variations in isotopic composition. For example, cosmic ray bombardment of rocks within a few meters of the surface of an atmosphere-free planet will produce measurable quantities of stable isotopes of all the inert gases. The isotopic composition of the product will be different from the solar and primordial isotopic composition of these elements. The accumulation of the isotopes produced by this cosmic ray bombardment forms the basis for an important method of dating the "exposure age" of meteorites and lunar surface materials (Wasson, 1974, Chapter XI). In the case of meteorites, this age is usually interpreted as the time since the object under study was produced by impact fragmentation of an asteroidal parent body. This interpretation may be complicated by preexposure of the material while on the surface of the parent body or by a multiple fragmentation history subsequent to ejection from the parent body (Nishiizumi *et al.*, 1980; Wetherill, 1980). Typically, stony meteorites have exposure ages of a few million years, whereas iron meteorite exposure ages are more often several hundred million years. The ages of lunar craters have been established by analogous interpretation of accumulated cosmic ray reaction products in lunar rocks as the time of onset of cosmic ray exposure following excavation of the rock during the crater-forming impact (Eugster *et al.*, 1977).

Natural radioactivity can also cause nuclear reactions which produce similar isotopic variations. For example, alpha particles from natural uranium and thorium can produce ^{21}Ne by the reaction $^{18}O\ (\alpha, n)^{21}Ne$. The effect of this will be to enrich neon in ^{21}Ne in the Earth's interior.

A complete description of these phenomena and their application is beyond the scope of this chapter. A discussion of the general principles by which isotope effects are used to date geological events will be given in section 7.1.2. More specific applications will be described as needed in subsequent sections of the chapter.

In addition to their importance for dating events, the isotope variations resulting from natural nuclear processes can also impose an isotopic signature on a particular source and thereby be used as a tracer to learn the contribution of that source to a rock or other natural system. Trace elements can also be used as tracers in a similar way. However, as mentioned above, trace elements differ from isotopes of a single element in that their relative abundance will change during the chemical fractionation accompanying the formation of an igneous rock. This can be a disadvantage insofar as information regarding the identity of the source is decreased by this fractionation. This disadvantage is offset by the use of chemical fractionation to make inferences regarding processes such as fractional crystallization, partial melting, assimilation of wall rocks, or metasomatism, which cause the fractionation.

This similar but complementary relationship between isotope and trace element studies has led to their being used in combination, often by the same investigators. The link between these techniques is strengthened by the fact that the particular trace elements used in isotope studies have also proven to be important geochemical indicators and can be determined by the same mass spectrometric techniques required by the isotopic investigations. For these reasons, it would be unnatural to separate isotopic and trace element studies in this chapter. To facilitate understanding of the material presented in the subsequent sections, a brief introduction to the basic principles of trace element fractionation is included in section 7.1.3.

Finally, it must be recognized that use of these techniques to determine the age, source, or chemical history of a basaltic magma is essentially the inverse of the problem of predicting the isotopic and chemical composition of a basalt having a given history. The usual manner of dealing with this is to construct plausible models describing the process under consideration and to compare the prediction of the model with the experimental and observational data. Agreement between prediction and observation is interpreted as evidence supporting the validity of the model. For the most part, this chapter will follow this approach. However, at some point it will be necessary to consider the uniqueness of the conclusions obtained in this way and the extent to which it is possible to resolve small differences between possible models. In other words, the inverse problem must be solved. This is obviously more difficult, and work of this kind has only just begun. A discussion of this subject is also given in section 7.1.3.

7.1.2 Application of radioactive decay to date and otherwise characterize geological events

Basic principles

There are many and somewhat different ways in which radioactive isotopes and their stable daughter products are used to date geological events. The one common feature of these different methods is that they all measure a radioactive decay interval. If, at some time in the past (t_1), the quantity of a radioisotope was P_1, and at some earlier time $t_0 > t_1$ the quantity was P_0, the time interval between t_1 and t_0 obeys the radioactive decay law

$$P_1 = P_0 e^{-\lambda(t_0 - t_1)} \tag{7.1.1}$$

where λ is the characteristic radioactive decay constant of the isotope P. The half-life of P is simply related to the decay constant

$$t_{\frac{1}{2}} = \frac{\ln 2}{\lambda} \tag{7.1.2}$$

Equation (7.1.1) can be solved for the decay interval (t_0–t_1)

$$(t_0 - t_1) = \frac{1}{\lambda} \ln \left(\frac{P_0}{P_1}\right) \tag{7.1.3}$$

Some radiometric dating methods essentially make direct use of Eq. (7.1.3). For example, in radiocarbon (^{14}C) dating, the value of P_0 is given by the ^{14}C content of organic material of zero age, and it is assumed that the material to be dated had this ^{14}C content when it was formed. (This is not strictly true because of small variations caused by changes in the cosmic ray production rate of ^{14}C and by dilution of the relatively recent natural atmospheric carbon reservoir by CO_2 from the large-scale burning of fossil fuels—the "Suess effect.") P_1 is determined by counting the present radioactivity of the carbon in the sample. The radiocarbon age can then be calculated from Eq. (7.1.3).

More often, it is not possible to determine both P_0 and P_1 directly, and it is necessary to devise indirect ways of inferring one or both of these quantities. In most applications of long-lived radioactive isotopes to the dating of terrestrial and lunar rocks and meteorites, the present-day ($t_1 = 0$) residual quantity of the parent isotope P_1 can be measured directly in the material to be dated. However, there is no "universal" earlier concentration P_0 of the parent (for example, uranium) in a rock. In such cases, it is necessary to make use of the fact that the long-lived parent isotope decays into a stable daughter isotope. If the rock or other natural system under investigation has been "chemically closed," i.e., has undergone no exchange of the parent or daughter elements with external systems, the sum of the atomic (or molar) concentration of the parent and daughter isotopes is conserved

$$P_0 + D_0 = P_1 + D_1 \tag{7.1.4}$$

P_0 can then be eliminated from Eq. (7.1.1) by use of Eq. (7.1.4). Setting $t_1 = 0$ (the present) then gives

$$(D_1 - D_0) = P_1 (e^{\lambda t_0} - 1) \tag{7.1.5}$$

D_1 and P_1 can be measured in the present laboratory sample, but D_0 is still unknown. The original concentration D_0 of the daughter isotope will vary in a complex way from one rock to another. However, if the daughter element has at least one isotope which is not the daughter product of a naturally occurring long-lived radioisotope, the need to know the absolute value of D_0 can be eliminated. If the (atomic) concentration of this other isotope is designated as D_x, division of Eq. (7.1.5) by D_x gives the fundamental age equation

$$\left(\frac{D}{D_x}\right)_1 = \left(\frac{P}{D_x}\right)_1 (e^{\lambda t_0} - 1) + \frac{D_0}{D_{x_1}} = \left(\frac{P}{D_x}\right)_1 (e^{\lambda t_0} - 1) + \left(\frac{D}{D_x}\right)_0 \tag{7.1.6}$$

where the unknowns are the initial time (t_0) and the initial daughter isotope ratio $(D_0/D_x) = (D/D_x)_0$. The ratios $(D/D_x)_1$ and $(P/D_x)_1$ refer to ratios which can be measured in the sample at the present time.

Even though the concentration D_0 will vary from one rock to another, the isotopic composition of an initially equilibrated system will be uniform. For example, two chemically different rocks from the same fractional crystallization or partial melting sequence will have very nearly the same isotopic composition at the time they are formed. In this case, measurement of the ratios $(D/D_x)_1$ and $(P/D_x)_1$ on the two rocks will give two equations of the form of Eq. (7.1.6) which can be solved for both the unknown age t_0 and the unknown initial isotopic composition $(D/D_x)_0$. If the initial isotopic composition were homogenized in some other way, e.g., by metamorphism, the time t_0 would refer to the time of metamorphism, not to the time of initial crystallization. In general, *the "age" of a system calculated from Eq. (7.1.6) is nothing more nor less than the time since the samples analyzed had the initial daughter isotopic composition* $(D/D_x)_0$. The geological significance of the "age"—whether it is a time of crystallization, metamorphism, sedimentation, or nothing at all interesting—must be established on the basis of other evidence.

Equation (7.1.6) is also applicable to the decay of ^{40}K to ^{40}Ar by multiplying the right-hand side of Eq. (7.1.5) by a factor $\lambda_K/(\lambda_K + \lambda_{Ca})$ where λ_K is the decay constant for electron-capture and β^+ decay of ^{40}K to ^{40}Ar, and λ_{Ca} is the decay constant for β^- decay to ^{40}Ca. This same factor will appear as a multiplier of $^{40}K/^{36}Ar$ on the right-hand side of Eq. (7.1.6) and takes into account the fact that only a fraction of the decay of ^{40}K contributes to the increase of the $(D/D_x) = {}^{40}Ar/^{36}Ar$ ratio. The ^{39}Ar-^{40}Ar method is fundamentally a technical modification of the ^{40}K-^{40}Ar method in which neutron irradiation of the sample is used to transform some of the ^{39}K to ^{39}Ar by the nuclear reaction $^{39}K(n, p)^{39}Ar$. Since the ratio of $^{39}K/^{40}K$ is constant in modern potassium, the $^{40}K/^{36}Ar$ and $^{40}Ar/^{36}Ar$ ratios required for use of Eq. (7.1.6) may thus be obtained from the same sample using only argon measurements. This is accomplished by comparison with a simultaneously irradiated standard of known age. It is now the usual practice to calculate separate ^{39}Ar-^{40}Ar ages from the argon fractions released from the sample at successively higher temperatures in the laboratory. This permits checking the result for "concordance" of ^{39}Ar-^{40}Ar from different mineralogical sites in the sample. The calculated ages are then usually plotted on diagrams of age vs. fraction of argon released. A series of concordant ages, particularly at higher temperatures, thus produces a "plateau" on this diagram and is taken as an indication that the closed-system assumption was valid for those fractions corresponding to the plateau and that the age is likely to represent a geologically meaningful event. (See Merrihue and Turner, 1966; Podosek and Huneke, 1973.)

Attention must be directed toward the "closed-system" assumption used to obtain Eq. (7.1.4). Strictly speaking, this requirement can be slightly relaxed to permit loss of parent and daughter elements from the system and even gain of parent element and daughter element of the same isotopic composition, provided the *ratio* of parent to daughter concentrations remains unchanged. In such cases, Eq. (7.1.6) will still be valid. However, such "well-behaved" open-system behavior is not common. More general gains or losses of parent and daughter elements will invalidate the fundamental age Eq. (7.1.6). Unfortunately, rocks, minerals, and other natural systems we may wish to date often have had a complex open-system history, and, in such cases, use of Eq. (7.1.6) will lead to meaningless numbers with the dimension of time.

Some control over this problem can be obtained by overconstraining Eq. (7.1.6) by use of more than two cogenetic rocks to determine t_0 and $(D/D_x)_0$. It is at least plausible to suppose that the causes of open-system behavior (such as partial metamorphism, weathering, hydrothermal alteration) will affect different rocks or minerals to a different extent. This will result in different pairs of rocks giving conflicting or "discordant" ages (t_0) and different initial ratios $(D/D_x)_0$ for the solution of Eq. (7.1.6). This is a tell-tale indication that the systems have not been closed and that the calculated results are not providing the desired information. On the other hand, if a number of different samples all yield the same values of t_0 and $(D/D_x)_0$, then, at least in some general way, it seems likely the closed-system assumption is valid, and the calculated age a meaningful one.

The above qualitative discussion of open-system behavior can be expressed quantitatively by use of numerical methods for testing the "goodness of fit" of a single value of t_0 and $(D/D_x)_0$ to a set of results $(D/D_x)_1$, and $(P/D_x)_1$, obtained by measurement of a multiplicity of samples. This can be done entirely numerically, and in fact usually is, by use of well-established computational algorithms. However, a graphical expression of the "goodness of fit" is easier to visualize. This can be obtained by recognizing that Eq. (7.1.6) is the equation of a straight line for the measurable variables $(D/D_x)_1$ and $(P/D_x)_1$ with slope $(e^{\lambda t_0}-1)$ and intercept $(D/D_x)_0$ (Nicolaysen, 1961). Therefore, if the closed-system assumption is valid, the set of measured points $[(D/D_x)_1, (P/D_x)_1]$ for a group of samples which at some time t_0 all had the same value $(D/D_x)_0$, will plot on a straight line (Fig. 7.1.1). The age t_0 can be calculated from the slope of this line, $(e^{\lambda t_0}-1)$, and the initial ratio $(D/D_x)_0$ obtained from the intercept of the line on the $(D/D_x)_1$ axis. The numerical fitting procedures mentioned above are therefore equivalent to procedures, such as least-squares analysis, for fitting measured points to a straight line.

An actual example of the application of Eq. (7.1.6) in this graphical form is given in Fig. 7.1.2. The measured points fit rather well, but not exactly, on the straight line ("isochron") shown. The deviations of the measured points from the line arise in two ways:

(1) Experimental error in the measured ratios $(^{87}Sr/^{86}Sr)_1$ and $(^{87}Rb/^{86}Sr)_1$;

(2) Small but significant failure of the individual rocks to form a closed system since t_0.

One can estimate the experimental error associated with the individual measured points and apply statistical tests to determine if the deviation from the straight line can be explained entirely on the basis of such errors (Brooks *et al.*, 1972). If this turns out to be the case, it is

7.1.1 Schematic illustration of the development of an isochron. Numerical values correspond to decay of ^{238}U to ^{206}Pb in a 2000-m.y.-old system, but the principle is the same for all parent-daughter systems. Consider a sample which at time t_0 (2000 m.y. ago) had $(^{206}Pb/^{204}Pb)_o = 15.0$ and $^{238}U/^{204}Pb$ corresponding to point G. As a result of radioactive decay, the $^{238}U/^{204}Pb$ ratio has decreased by an amount FG and the $^{206}Pb/^{204}Pb$ ratio increased by an equal amount AF (note the difference in ordinate and abscissa scales). The homologous segments HI, BH and BI for a second cogenetic sample with initial ratios corresponding to point I will be larger in proportion to the initial $^{238}U/^{204}Pb$ ratio XI, and the angles AGF and BIH will be equal. Therefore, triangles such as XBI will be similar to triangle XAG, and all their angles will be equal. In particular, the angle marked θ will be the same for all such triangles and the vertices A, B, C, D, etc., will lie on a straight line, the *isochron*.

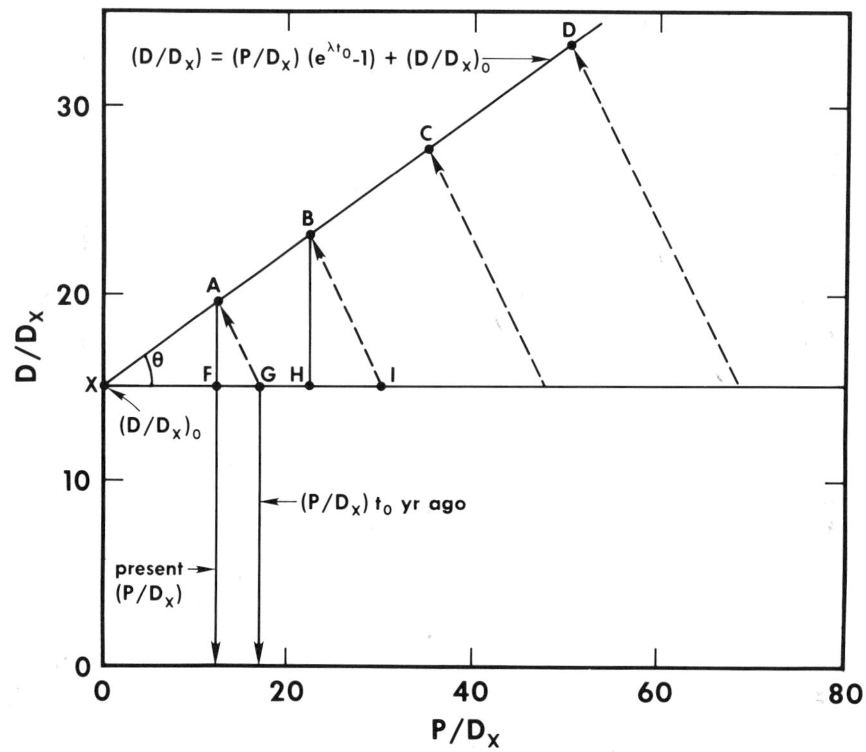

7.1.2 Plot of $^{87}Sr/^{86}Sr$ vs. $^{87}Rb/^{86}Sr$ for whole rock samples of eucrite basaltic meteorites reported by Birck and Allègre (1978a). The scatter of the points from the best fit straight line is comparable to that expected from the individual experimental errors. Therefore, *based on these data alone*, it is difficult to distinguish between the alternative hypotheses that: (1) The various rocks represent basalts which all formed at the same time 4.24 ± 0.49 b.y. ago at which time they had the same initial ratio $^{87}Sr/^{86}Sr = 0.69902$. For the most part, chemical gain or loss of Rb and Sr has not occurred since that time. (2) Although the rocks were formed simultaneously, they have subsequently been metamorphosed to varying degrees. Neither the time of original formation nor the time(s) of metamorphism are well-dated. (3) They are an unrelated group of basalts, most of which crystallized early in the solar system from sources with a possible range in initial Sr isotopic compositions.

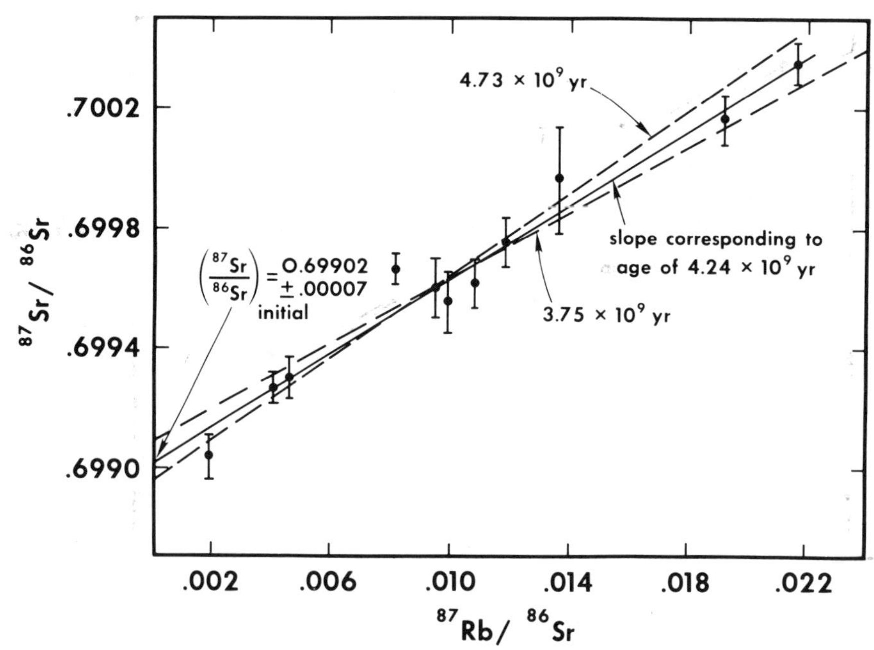

common practice to conclude that an age has been determined, and, in this case, the line on Fig. 7.1.2 is termed an *isochron*. (The term isochron will be used throughout this section and chapter to represent loci of points of the same age on various types of diagrams, including, but not restricted to, diagrams similar to Figs. 7.1.1–2.) An experimental error is assigned to the age in accordance with the standard deviation of the points from the isochron. If, on the other hand, the deviation cannot be explained entirely in terms of experimental error, the line has been called an "errorchron" by Brooks *et al.*, and it is often said that the calculated age should be rejected.

There is a certain amount of common sense in this conventional manner of treating the data, and ages meeting this criterion form the basis of most of our knowledge concerning events in the history of the Earth, Moon and meteorite parent bodies. On the other hand, there is something unsatisfactory about this approach. There definitely are many cases where the failure of the closed-system assumption causes an identifiable scatter in the points, and some of these cases are more extreme than others. It is, therefore, plausible that the degree of this geochemically-caused scatter is continuous and at some stage becomes indistinguishable from the scatter caused by errors in measurement. However, the stage at which this occurs is dependent on the accuracy of the experimental data. This leads to the paradox that as analytical procedures improve, the fraction of the measured ages deemed valid will decrease.

It must be that in the real world all apparent isochrons are covert errorchrons of small and varying degrees of severity, their true nature being camouflaged by measurement errors. The actual uncertainty in the age and initial ratios must be at least as large as that given by the measurement errors. It will be larger, however, unless it is safe to assume that the best straight-line fit to an errorchron is the best estimate of the true age, and the uncertainty in this age is limited to that determined by the scatter about this line. The criterion described above implicitly requires this assumption to be a safe one when the scatter is no more than that imposed by the present "state of the art" in analytical accuracy, but not to be safe when the scatter is larger. Neither of these implicit assumptions appears likely to be strictly correct. At the present time, however, our understanding of the effects of different geochemical processes of metamorphism, alteration, and contamination on the experimental points is too rudimentary to do much better. Even if the effects of a given alteration history were clearly predictable, the inverse problem of identifying this alteration history from the measured data would still remain. A possible approach to this problem might be to make use of the fact that under particular circumstances, some parent-daughter systems will indicate large scatter because of open-system behavior, whereas other parent-daughter systems will remain relatively closed and exhibit little scatter. For example, the $^{147}Sm/^{143}Nd$ decay system seems less susceptible to parent-daughter fractionation during metamorphism and weathering than does $^{87}Rb/^{87}Sr$. Whether the best fit to highly scattered points gives, on the average, the same age as that based on a good fit by a different method, or a systematically different value, may also be helpful in understanding the circumstances under which the age determined by an overt errorchron with a given scatter may be meaningful. It seems plausible to believe that such systematic effects will be small when the scatter is small and large only when the scatter is large, but more quantitative knowledge is required if progress is to be made in obtaining a better understanding of the extent to which age measurements are really accurate.

The situation is more complex when there is no *a priori* reason to assume that the set of points on a diagram like Fig. 7.1.1 actually represents samples which at some specific time t_0 actually had the same initial isotopic composition. A true scatter in ages and/or initial ratios could thus mimic open-system effects, but the appropriate criteria for accepting or rejecting results might well be difficult. For example, if the measured points are based on rock assemblages having two distinct ages and initial ratios, then generally the best-fit straight line to the data will yield neither the real ages nor initial values, even if such averaging of data is found to be a reasonable procedure to follow in the case of a mildly disturbed but actually cogenetic assemblage.

The conclusion of all this must be that use of isotope data for geological purposes—measuring of ages, determining the sources of basalts, or understanding the evolution of the mantle—is no different from similar use of other geological, petrological, geochemical, or geophysical data. In all cases, the link between experimental data and geological interpretation must be continually forged. There will be no cookbooks or black boxes which will permit us to grind out generally valid geological conclusions. The proper admixture of skepticism and imaginative creativity will, as has always been the case, continue to define the meaning of the appellation "a good scientist."

Mixing lines and model ages

The fundamental "age equation" (Eq. 7.1.6) is used in one way or another whenever geological conclusions are based on decay of a radioactive parent isotope to a stable daughter isotope. The type of diagram associated with Eq. (7.1.6), displayed in Figs. 7.1.1–2, also has its counterpart in the various parent-daughter systems in use.

As discussed in the previous section, the most straightforward use of this equation and diagram is to calculate the time elapsed since a set of systems with varying ratios (P/D_x) possessed a common initial isotope ratio (D/D_x). The "set of systems" can be minerals from the same rock, samples of rocks, regions of the Earth, or conceivably, entire planets. The geological mechanism responsible for achieving a uniform initial isotopic composition could result from a variety of processes including equilibrium magmatic crystallization, metamorphism, or isotopic homogenization of a primordial solar nebula. The equation simply represents a mathematical relationship; its particular meaning requires additional information.

It is often the case that this additional information is only partially available, and in such cases qualifying terms are employed to signify this. For example, if two otherwise arbitrary systems (such as rocks) are used to define a line on the diagram, it is clear that the "age" and "initial ratio" may be meaningless and may simply represent nothing more than the fact that a straight line can always be drawn through two points.

Even if more than two samples are used to define the line, this situation is not improved if the samples of intermediate composition are simply mixtures of the two end-members. This can easily be seen by consideration of two samples represented by points $[(D/D_x)_1, (P/D_x)_1]$ and $[(D/D_x)_2, (P/D_x)_2]$ These two points define a line of slope "m" and intercept "b" on the ordinate which passes through both points

$$\left(\frac{D}{D_x}\right)_1 = m\left(\frac{P}{D_x}\right)_1 + b \qquad (7.1.7)$$

$$\left(\frac{D}{D_x}\right)_2 = m\left(\frac{P}{D_x}\right)_2 + b \qquad (7.1.8)$$

Therefore

$$D_1 = mP_1 + b(D_x)_1 \qquad (7.1.9)$$

$$D_2 = mP_2 + b(D_x)_2 \qquad (7.1.10)$$

Consider a mixture of Q_1 moles of sample (1) and Q_2 moles of sample (2)

$$D_M = Q_1D_1 + Q_2D_2 \qquad (7.1.11)$$

$$P_M = Q_1P_1 + Q_2P_2 \qquad (7.1.12)$$

$$(D_x)_M = Q_1(D_x)_1 + Q_2(D_x)_2 \qquad (7.1.13)$$

where the subscript M denotes the concentration of the isotopes in the mixture. Insertion of Eqs. (7.1.9) and (7.1.10) into Eq. (7.1.11) gives

$$D_M = m[Q_1P_1 + Q_2P_2] + b[Q_1(D_x)_1 + Q_2(D_x)_2] \qquad (7.1.14)$$

which by use of Eqs. (7.1.12) and (7.1.13) and division by $(D_x)_M$ becomes

$$\left(\frac{D}{D_x}\right)_M = m\left(\frac{P}{D_x}\right)_M + b \qquad (7.1.15)$$

Therefore, the point representing the mixture $(D/D_x)_M$, $(P/D_x)_M$ lies on the same straight line (Eq. 7.1.7 and Eq. 7.1.8) as the end-members. This result is not as trivial as it may appear inasmuch as this result depends on the denominator (D_x) being the same for the ratios plotted on both ordinate and abscissa. When this condition is not realized, for example in trace element plots of (K/Rb) vs. (K/U), mixing lines will not be straight lines. This point will be discussed further below.

The terms "two point isochron" and "mixing line" are commonly used to describe these situations. Use of these terms to describe a result often has a connotation which is appropriately pejorative, in view of the fact that ages and initial ratios associated with the data may be meaningless. It should be clearly understood, however, that this is not necessarily the case, and that in some circumstances, perfectly valid conclusions may be based on such data. The burden of additional evidence or plausibility required to show that the time interval calculated is really the consequence of closed-system radioactive decay, in accordance with Eq. (7.1.6), rather than being a meaningless result, is obviously greater.

The interpretation is even more uncertain when only one sample analysis and its associated data point is available. In these circumstances, calculation of a time interval requires assuming an initial D/D_x ratio, rather than basing the value of the initial ratio upon an actual measurement. Ages calculated in this way are often called **model ages**. This term is also used to describe ages calculated in quite different ways, having only the common property that the calculation of an age value requires some additional assumptions. In the sense in which the term is used here (the assumption of an initial D/D_x ratio), whether or not the age may be considered meaningful depends on the plausibility of the particular system having actually evolved as a closed system from an initial ratio sufficiently similar to that assumed. For example, in nearly all the terrestrial ^{87}Rb-^{87}Sr geochronology carried out prior to about 1960, the initial $^{87}Sr/^{86}Sr$ ratio was simply assumed to be about 0.704, and the age calculated from the $^{87}Sr/^{86}Sr$ and $^{87}Rb/^{86}Sr$

ratios measured on a single Rb-rich mineral, e.g., mica or potassium feldspar. However, in this work, the $^{87}Sr/^{86}Sr$ ratios in the minerals studied were so high that the age calculated was essentially independent of variation of initial ratios within their usual range. The validity of the age then usually depended on whether or not the particular mineral sample studied had been isotopically disturbed only as a consequence of some geologically significant event such as magmatic crystallization or sufficiently intense metamorphism. Because this assumption was not always true, some of these earlier age determinations are no longer considered valid. On the other hand, the general distribution on the continents of the Precambrian rocks of various ages established by these measurements was nevertheless correct. Even if no further advances in the techniques of measurement and data analysis had been developed, our knowledge of Earth history is much greater than if these measurements had not been made.

Even when the assumptions discussed above are not valid, valuable information may often be expressed by these model ages. For example, suppose that a whole-rock model age of 3.6 b.y. is calculated for a meteorite from its measured $^{87}Sr/^{86}Sr$ and $^{87}Rb/^{86}Sr$ ratios by assuming an initial $^{87}Sr/^{86}Sr$ ratio in Eq. (7.1.6) equal to the lowest value found in solar system material, 0.6988 for the Allende meteorite. This age can be interpreted as showing that some isotopic disturbance occurred more recently than the time given by the model age. (This is simply a consequence of the fact that if there had been a longer final period of closed-system behavior, then Eq. (7.1.6) would predict that this closed system would have caused the development of an $^{87}Sr/^{86}Sr$ ratio greater than that observed, unless one adopts the unlikely alternative that at the time this closed system was established, the sample had a lower $^{87}Sr/^{86}Sr$ ratio than that of the most primitive solar system material presently known.) Thus, a model age can be used to distinguish between ~4.5 b.y. meteorite samples and those bearing the record of some subsequent event or series of events in the later history of the solar system. In other cases, model ages can be used to demonstrate the "memory" of geochemical processes predating a well-defined crystallization age. Many lunar basalts have ~4.5 b.y. model ages, assuming the Allende initial ratio, but are known from their $^{87}Rb/^{86}Sr$ mineral isochrons to have crystallized at specific times between 3.3 and 3.8 b.y. ago. These results show that the enrichment of the Rb/Sr ratio of the basalt at the time of its eruption was not so great that it completely altered the record of the growth of radiogenic strontium in the source region of the basalt during the interval between the formation of the solar system, and the more recent eruption and crystallization of the basalt. When combined with other data, such as major element analysis, rare-earth element abundances, and the results of laboratory petrological and trace element partition studies, these model ages contribute to the constraints on models for the generation of lunar basalts, and the premare history of the lunar mantle.

Thus, two point isochrons, mixing lines, and model ages may sometimes provide valuable knowledge concerning basaltic volcanism—and sometimes may not. Dogmatic statements regarding these matters are ill-advised, and the only general rule is that the maximum quantity of reasonably certain knowledge should be extracted from all isotopic and chemical analytical results.

Other diagrams and equations

Although Eq. (7.1.6) and the diagrams discussed (Figs. 7.1.1–2) are those used most often in work with radiogenic isotopes, there are a number of other related equations and diagrams used with sufficient frequency to warrant their discussion in this introduction. Other more specialized cases will be discussed in the subsequent sections as the need arises.

The fundamental age Eq. (7.1.6) can be used to calculate the evolution in time of the isotopic composition of a system which has a fixed ratio of (P/D_x) and a given initial ratio $(D/D_x)_o$. Thus in Fig. 7.1.3 , it is possible to use Eq. (7.1.6) to calculate the $^{143}Nd/^{144}Nd$ ratio of a planetary mantle with the $^{147}Sm/^{143}Nd$ parent-daughter ratio and the primordial initial ratio of the solar system, as inferred from data on chondritic and achondritic meteorites. It has been shown that a number of continental rocks of deep-seated origin had $^{143}Nd/^{144}Nd$ ratios falling on this line at the time of their formation; therefore, it can be hypothesized that these rocks were derived from a mantle of this kind (DePaolo and Wasserburg, 1976a). Other rocks, including oceanic basalts, do not plot along this line but deviate systematically from it. In such cases, generalizations of Eq. (7.1.6) can be derived to include the effects of variation with time of the parent-daughter ratio of the source regions, and thus define the set of mantle evolution models consistent with the observed data. Analogous (D/D_x) vs. time diagrams can be used for the other parent-daughter isotopic systems (Fig. 7.4.7).

Because of the possibly complex geochemical behavior of all the radiogenic parent-daughter pairs, much attention has been given to the simultaneous use of Eq. (7.1.6) for two or more parent-daughter systems. This permits the identification of failure of closed-system behavior, and in some cases, even permits correction for the effects of such behavior.

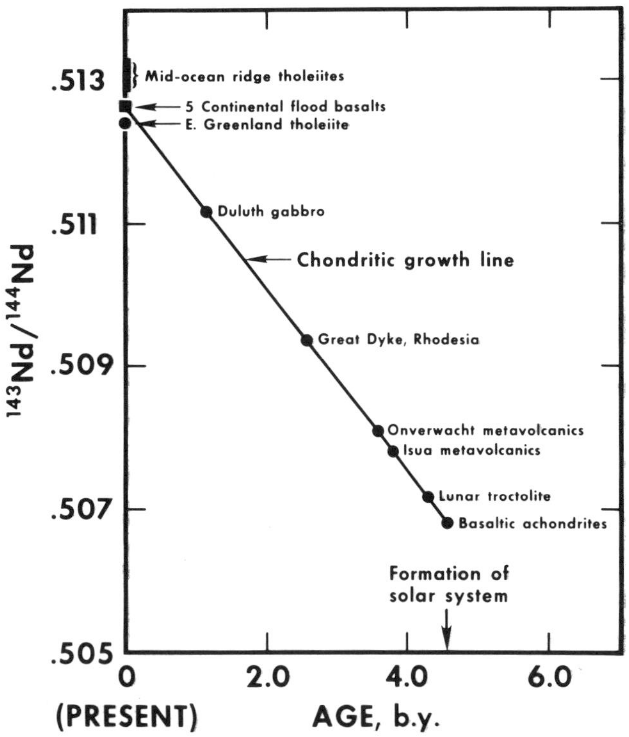

7.1.3 Straight line represents growth of $^{143}Nd/^{144}Nd$ ratio vs. time since the formation of the solar system 4.5 b.y. ago in a closed system having the $^{147}Sm/^{143}Nd$ ratio of the chondritic and achondrite meteorites (0.308), and the initial $(^{143}Nd/^{144}Nd)_0$ ratio of the achondrite Juvinas (0.5068). The experimental points represent initial Nd isotope ratios of various rocks reported by DePaolo and Wasserburg (1976a,b), Hamilton *et al.* (1977, 1978a), Richard *et al.* (1976), and O'Nions *et al.* (1977). For the most part, the data suggest that the source of these terrestrial rocks has Sm-Nd characteristics similar to this average solar system material, but that some deviations also are found, particularly for young oceanic basalts.

The most commonly used example of this simultaneous use of coupled parent-daughter systems is that of the ^{238}U-^{206}Pb and ^{235}U-^{207}Pb systems, as well as the combination of either of the U-Pb systems with ^{232}Th-^{208}Pb.

For the uranium and lead isotopes, Eq. (7.1.6) becomes

$$\left(\frac{^{206}Pb}{^{204}Pb}\right) = \left(\frac{^{238}U}{^{204}Pb}\right)(e^{\lambda_{238}t_0} - 1) + \left(\frac{^{206}Pb}{^{204}Pb}\right)_0 \qquad (7.1.16)$$

$$\left(\frac{^{207}Pb}{^{204}Pb}\right) = \left(\frac{^{235}U}{^{204}Pb}\right)(e^{\lambda_{235}t_0} - 1) + \left(\frac{^{207}Pb}{^{204}Pb}\right)_0 \qquad (7.1.17)$$

In both cases, the nonradiogenic index isotope D_x is ^{204}Pb.

By transferring the initial ratio (the second term on the right-hand side of these equations) to the left-hand side, and dividing Eq. (7.1.17) by Eq. (7.1.16), we obtain

$$\frac{\left(\frac{^{207}Pb}{^{204}Pb}\right) - \left(\frac{^{207}Pb}{^{204}Pb}\right)_0}{\left(\frac{^{206}Pb}{^{204}Pb}\right) - \left(\frac{^{206}Pb}{^{204}Pb}\right)_0} = \frac{^{235}U\,(e^{\lambda_{235}t_0} - 1)}{^{238}U\,(e^{\lambda_{238}t_0} - 1)} \qquad (7.1.18)$$

This is the equation for the so-called Pb-Pb age, introduced by Nier (1939) in his classical and pioneering discussion of uranium-lead dating of uranium-rich minerals. Use of this equation does not involve measurement of the absolute concentration of either the uranium or lead in the rock, inasmuch as only isotope ratios appear in the equation. The ratio $^{235}U/^{238}U = (1/137.8)$ is very nearly constant in natural uranium; and, as discussed below, the initial ratios can often be estimated quite well from general knowledge of the way in which the isotopic composition of lead in the Earth has evolved with time. For minerals with the high $^{207}Pb/^{204}Pb$ and $^{206}Pb/^{204}Pb$ ratios commonly found in uranium-rich minerals, uncertainties in the initial ratio do not greatly influence the value of the left-hand side of this equation. The resulting transcendental equation can then be solved for the unknown age t_0. As discussed by Nier (1939), the age calculated by this simultaneous use of the two U-Pb parent-daughter systems can, under some circumstances, be less affected by open-system behavior than ages calculated from the individual Eqs. (7.1.16–17).

This equation was used by Patterson (1956) and Houtermans (1953) to calculate the age of the Earth and the basaltic achondrite meteorites, using initial ratios inferred from meteoritic data. To some extent, the use of Pb-Pb ages has been superseded by more recent techniques for the analysis of U-Pb ages, as discussed later in this section. Nevertheless, its generalized counterpart in defining "isochrons" in these more recent treatments continues to play a central role.

The coupled Eqs. (7.1.16) and (7.1.17) can also be used as parametric equations to calculate the variation with time of the lead isotope ratios in a system having a given present-day $^{238}U/^{204}Pb$ ratio, and given present-day $^{206}Pb/^{204}Pb$ and $^{207}Pb/^{204}Pb$ ratios. These calculations define a curve passing through the present-day values on a diagram (Fig. 7.1.4) in which $^{207}Pb/^{204}Pb$ is plotted on the ordinate vs. $^{206}Pb/^{204}Pb$ on the abscissa. This type of diagram is very commonly used in Pb isotope studies. The curve plotted in Fig. 7.1.4 corresponds to a present-day $^{238}U/^{204}Pb$ ratio of 9.0, and is fitted to values of $^{207}Pb/^{204}Pb = 15.45$, and $^{206}Pb/$

^{204}Pb = 18.25 (Point E), falling in the general range of modern mid-ocean ridge basalts and Hawaiian island basalts. If these basalts are derived from a mantle source having this $^{238}U/^{204}Pb$ ratio, and the migration of uranium and lead into and out of this source has always been negligibly small, then the isotopic composition of lead from this source at any given time in the past will lie at a unique point $[(^{206}Pb/^{204}Pb)_0, (^{207}Pb/^{204}Pb)_0]$ along this curve (for example, at point "D" at 3400 m.y. ago). As will be discussed later in this chapter, many samples of ancient terrestrial lead from igneous rocks and sulfide ore deposits actually do plot in the general vicinity of this curve. This curve also passes through the point $(^{206}Pb/^{204}Pb)_0 = 9.31$ and $(^{207}Pb/^{204}Pb)_0 = 10.29$, found in the primitive unequilibrated chondritic meteorite Mezö-Madaras as well as in the differentiated iron meteorite Cañon Diablo, at point "C," corresponding to a time $t_0 = 4450$ m.y. Thus, in accordance with Eqs. (7.1.16) and (7.1.17), the curve CFDGE is the locus of points representing the evolution of lead in a closed reservoir with a present-day $^{238}U/^{204}Pb$ ratio of 9.0, which had the primitive composition "C" 4450 m.y. ago and evolved with time so as to contain Pb of isotopic composition "D" 3400 m.y. ago, and Pb of composition "E" at the present time. Insofar as the Earth's lead has actually evolved in this simple way, the $t_0 = 4450$ m.y. would be the "age of the Earth." The actual geochemical evolution of lead in the Earth is, of course, more complex than this, and calculation of the age of the Earth requires understanding the extent to which the true geochemical model for the evolution of the lead found in terrestrial rocks and minerals deviates from this oversimplified model.

Another property of the $^{207}Pb/^{204}Pb$ vs. $^{206}Pb/^{204}Pb$ diagram should be explained. Consider a rock (or more generally a "geochemical system") which at a particular time t_0 in the past contained lead of isotopic composition corresponding to some point, e.g., "D" on Fig. 7.1.4 and had a $^{238}U/^{204}Pb$ ratio designated by

$$\left(\frac{^{238}U}{^{204}Pb}\right)_D = \frac{^{238}U}{^{204}Pb}\, e^{\lambda t_0}$$

If the system remained closed between t_0 and the present time, this initial parent-daughter ratio would decay to the present-day ratio $^{238}U/^{204}Pb$. Equations (7.1.16) and (7.1.17) will then give the isotopic composition of Pb in this system at the present time (point E).

Now suppose that at time D the system under consideration undergoes geochemical fractionation, such that some portions of the system have their $^{238}U/^{204}Pb$ ratio increased, and this ratio is decreased in other portions of the system. The $^{238}U/^{204}Pb$ and $^{235}U/^{204}Pb$ ratios of each of these subsystems can then be described by the expressions

$$\left(\frac{^{238}U}{^{204}Pb}\right)_{nD} = f_n\left(\frac{^{238}U}{^{204}Pb}\right)_D ; \quad \left(\frac{^{235}U}{^{204}Pb}\right)_{nD} = f_n\left(\frac{^{235}U}{^{204}Pb}\right)_D \tag{7.1.19}$$

where f_n is the factor by which this ratio changed in subsystem n at time D as a result of the geochemical

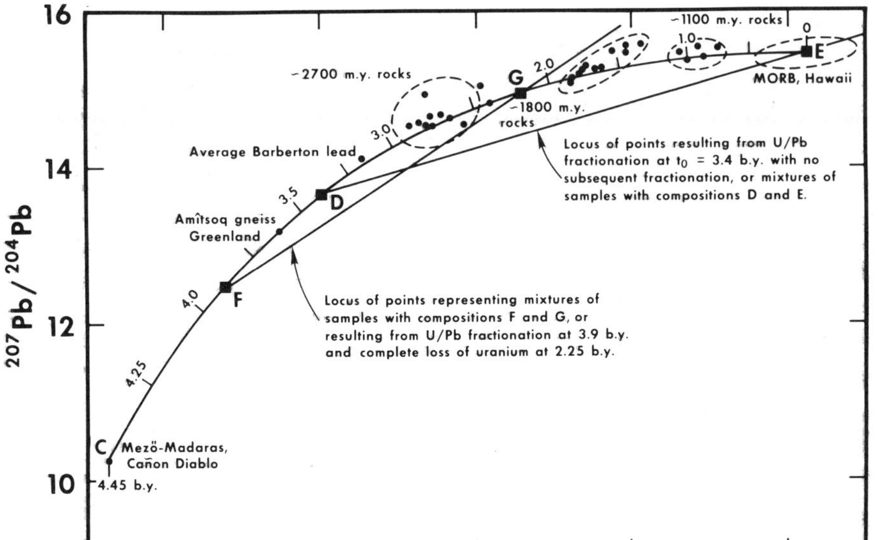

7.1.4 Plot of $^{207}Pb/^{204}Pb$ vs. $^{206}Pb/^{204}Pb$. The curve is the locus of lead isotopic compositions generated in a reservoir with a uniform $^{238}U/^{204}Pb$ ratio and an isotopic composition of lead at the present time equal to that found in MORB tholeiites. The numbers on this curve correspond to the times in the past at which this hypothetical reservoir had these lead isotopic compositions. Many samples of terrestrial lead fall near this curve at indicated times similar to the known ages of the samples, but significant deviations from this simple behavior also exist. As discussed in the text, isochrons (DE) and mixing lines (FG) are straight lines on this diagram.

fractionation. For each of the subsystems, the present-day isotopic composition of its Pb will be given by

$$\left(\frac{^{206}Pb}{^{204}Pb}\right)_{nP} = f_n \left(\frac{^{238}U}{^{204}Pb}\right)(e^{\lambda_{238}t_0} - 1) + \left(\frac{^{206}Pb}{^{204}Pb}\right)_D \quad (7.1.20)$$

$$\left(\frac{^{207}Pb}{^{204}Pb}\right)_{nP} = f_n \left(\frac{1}{137.8}\right)\left(\frac{^{238}U}{^{204}Pb}\right)(e^{\lambda_{235}t_0} - 1) + \left(\frac{^{207}Pb}{^{204}Pb}\right)_D \quad (7.1.21)$$

Because both parent and daughter isotopes are isotopes of the same elements, f_n is the same for both decay systems.

The present-day Pb isotopic compositions of the subsystem will all plot along a straight line passing through points D and E. This can be seen immediately by setting $f_n = 0$ and $f_n = 1$. The fact that the line is straight can be shown by differentiating both Eq. (7.1.20) and Eq. (7.1.21) with respect to the parameter f_n and taking the ratio of the two derivatives

$$\frac{d\left(\frac{^{207}Pb}{^{204}Pb}\right)_{nP}}{d\left(\frac{^{206}Pb}{^{204}Pb}\right)_{nP}} = \left(\frac{1}{137.8}\right)\frac{(e^{\lambda_{235}t_0} - 1)}{(e^{\lambda_{238}t_0} - 1)} \quad (7.1.22)$$

For the given value of t_0, the right-hand side of Eq. (7.1.22) is a constant. Thus the locus of the points $[(^{207}Pb/^{204}Pb)_{nP}, (^{206}Pb/^{204}Pb)_{nP}]$ has constant slope; i.e., the line is straight. This line is an isochron because it is the locus of points all of which experienced geochemical fractionation at the same time t_0 to form minerals with differing U/Pb ratios. If these minerals then remained geochemically closed up to the present time, these cogenetic minerals would all lie along the line DE. Minerals for which $f_n<1$ would plot between D and E, those with $f_n>1$ would plot on the continuation of DE beyond point E. The time in the past at which the rock was chemically fractionated can be calculated by measuring the slope of the isochron, setting it equal to the right-hand side of Eq. (7.1.22), and solving the resulting transcendental equation for t_0. Comparison with Eq. (7.1.18) shows that this procedure is a generalization of the calculation of the Pb-Pb age of a single uranium mineral. By a simple extension of the foregoing discussion, it can be shown that the locus of points representing fractionated subsystems evolving from a composition F at time t_F and sampled at a later time t_G also defines a straight line passing through the point F and the point G, the latter corresponding to a subsystem which experienced no fractionation (i.e., $f_n = 1$). The time interval $(t_G - t_F)$ can be calculated from the slope of this line if either t_G or t_F is known.

Samples characterized by points lying on segments of exactly these same straight lines DE and FG can also be produced by mixing lead from systems having the isotopic composition of the end-members of these segments. As discussed earlier in this section, this is a general property of mixing lines on plots of the form (A/B) vs. (C/D) where B = D, which is the case here with ^{204}Pb = B = D. This property therefore does *not* depend on both the parent isotopes being isotopes of the same element. Because of its relationship to the more general cases where B ≠ D, the proof of the linearity of the mixing line on the $^{207}Pb/^{204}Pb$ vs. $^{206}Pb/^{204}Pb$ diagram will be presented in a way equivalent to, but differing somewhat from, the proof of Eq. (7.1.15).

Consider a mixture of Q_1 moles of a sample (1) with Q_2 moles of a sample (2). The $^{206}Pb/^{204}Pb$ ratio of the mixture will be given by

$$\left(\frac{^{206}Pb}{^{204}Pb}\right)_M = \frac{Q_1(^{204}Pb)_1\left(\frac{^{206}Pb}{^{204}Pb}\right)_1 + Q_2(^{204}Pb)_2\left(\frac{^{206}Pb}{^{204}Pb}\right)_2}{Q_1(^{204}Pb)_1 + Q_2(^{204}Pb)_2} \quad (7.1.23)$$

where $(^{204}Pb)_1$ and $(^{204}Pb)_2$ are the molar concentrations of ^{204}Pb in the two end-members of the mixture. The ratio $(^{207}Pb/^{204}Pb)_M$ will be given by an analogous expression in terms of the ratios $(^{207}Pb/^{204}Pb)_1$ and $(^{207}Pb/^{204}Pb)_2$. This yields two equations in the variables $Q_1\,(^{204}Pb)_1$ and $Q_2\,(^{204}Pb)_2$. It will have nontrivial solutions when the determinant vanishes

$$\left[\left(\frac{^{206}Pb}{^{204}Pb}\right)_M - \left(\frac{^{206}Pb}{^{204}Pb}\right)_1\right]\left[\left(\frac{^{207}Pb}{^{204}Pb}\right)_M - \left(\frac{^{207}Pb}{^{204}Pb}\right)_2\right] - \left[\left(\frac{^{206}Pb}{^{204}Pb}\right)_M - \left(\frac{^{206}Pb}{^{204}Pb}\right)_2\right]\left[\left(\frac{^{207}Pb}{^{204}Pb}\right)_M - \left(\frac{^{207}Pb}{^{204}Pb}\right)_1\right] = 0 \quad (7.1.24)$$

Rearrangement of terms in Eq. (7.1.24) gives the linear equation

$$\left(\frac{^{207}Pb}{^{204}Pb}\right)_M = \frac{\left[\left(\frac{^{207}Pb}{^{204}Pb}\right)_2 - \left(\frac{^{207}Pb}{^{204}Pb}\right)_1\right]\left[\left(\frac{^{206}Pb}{^{204}Pb}\right)_M - \left(\frac{^{206}Pb}{^{204}Pb}\right)_1\right]}{\left[\left(\frac{^{206}Pb}{^{204}Pb}\right)_2 - \left(\frac{^{206}Pb}{^{204}Pb}\right)_1\right]} + \left(\frac{^{207}Pb}{^{204}Pb}\right)_1 \quad (7.1.25)$$

The locus of points $(^{206}Pb/^{204}Pb)_M$, $(^{207}Pb/^{204}Pb)_M$ is thus a straight line passing through the end-member isotopic compositions and with slope

$$m = \frac{\left(\frac{^{207}Pb}{^{204}Pb}\right)_2 - \left(\frac{^{207}Pb}{^{204}Pb}\right)_1}{\left(\frac{^{206}Pb}{^{204}Pb}\right)_2 - \left(\frac{^{206}Pb}{^{204}Pb}\right)_1} \qquad (7.1.26)$$

Therefore, in this diagram, mixing lines and isochrons are mathematically equivalent.

Instead of plotting $^{207}Pb/^{204}Pb$ vs. $^{206}Pb/^{204}Pb$, it is sometimes convenient to plot instead $^{207}Pb/^{206}Pb$ vs. $^{204}Pb/^{206}Pb$ as in Fig. 7.1.5 (e.g., see Fig. 7.3.5b). The reason is that ^{204}Pb is the least abundant Pb isotope and errors in its measurement may be large. This is particularly true for lunar samples and basaltic achondrites, which contain very little ^{204}Pb (see Tera and Wasserburg, 1972a). In such cases, the ^{204}Pb may primarily represent "blank": i.e., extraneous Pb introduced by the handling and preparation of the sample, rather than ^{204}Pb intrinsic to the sample. Thus, greatly disparate points in Fig. 7.1.4 may result simply from differences in these levels of contamination. A linear array may result just from analysis of the same sample with different levels of contamination, and this could be misinterpreted as an isochron. These meaningless differences are suppressed when the same data are plotted as in Fig. 7.1.5. This representation also makes it more convenient to plot those samples with greatly varying, high $^{206}Pb/^{204}Pb$ ratios on the same diagram as primordial Pb with a $^{206}Pb/^{204}Pb$ ratio of 9.31.

In Fig. 7.1.4

$$y = \left(\frac{^{207}Pb}{^{204}Pb}\right), \quad x = \left(\frac{^{206}Pb}{^{204}Pb}\right)$$

and in Fig. 7.1.5

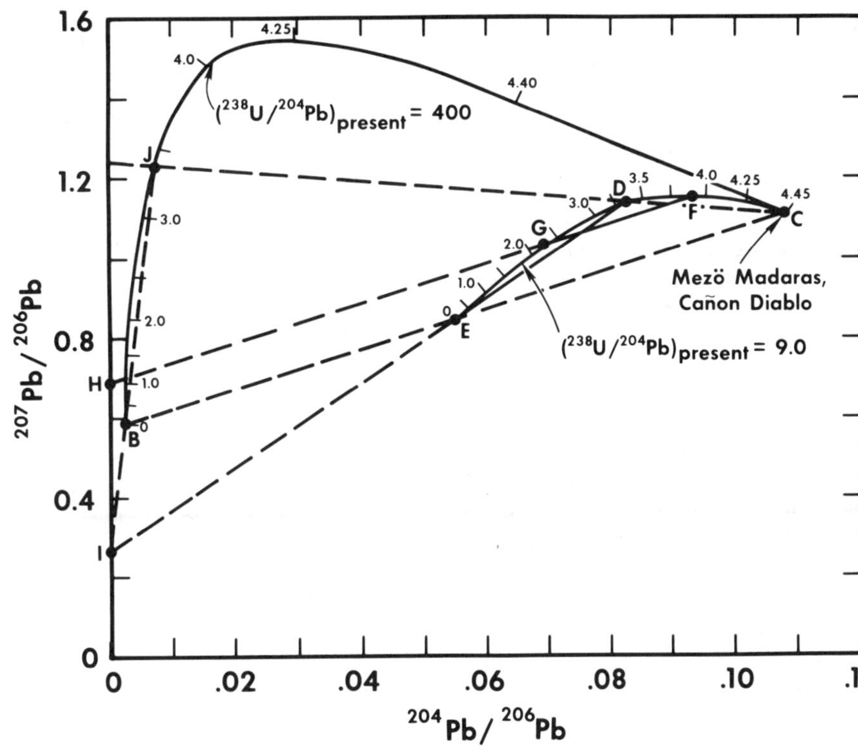

7.1.5 Diagram similar to Fig. 7.1.4, but with $^{207}Pb/^{206}Pb$ plotted vs. $^{204}Pb/^{206}Pb$. The curves represent the loci of lead isotopic compositions in reservoirs with present $^{238}U/^{204}Pb$ ratios of 400 (similar to the Moon) and 9 (similar to the Earth). The sequence of points EGDFC represent the same samples on Figs. 7.1.4 and 7.1.5. As discussed in the text, isochrons and mixing lines are also straight lines on this diagram. Points lying on the lines DE and JB represent the present isotopic composition of systems which evolved as closed systems from 4.45 b.y. until 3.4 b.y. ago and then experienced varying degrees of U/Pb fractionation at that time. The common age of the event results in their having the common intercept point I. In the format of Fig. 7.1.4 these same systems would lie on lines with the same slope but would have different intercepts. Points lying along CEB represent the present isotopic compositions of systems which evolved from the same primordial compositions C 4.45 b.y. ago in systems with constant but differing $^{238}U/^{204}Pb$ ratios. Points along CDJ represent the isotopic compositions of these systems 3.4 b.y. ago. Points lying on the line FGH correspond to the present isotopic compositions of samples which evolved from C in a closed system with $^{238}U/^{204}Pb$ of 9 until 3.9 b.y. ago. At that time they experienced varying degrees of U/Pb fractionation; in a second event at 2.1 b.y. the U/Pb ratio was reduced to zero. The intercept H represents the extreme case where the 3.9 b.y. fractionation removed all of the lead present in the sample at that time, and the corresponding $^{207}Pb/^{206}Pb$ ratio is that of the purely radiogenic Pb formed during the interval from 3.9 to 2.1 b.y. ago. This representation permits plotting data with a wider range of isotopic compositions and suppresses artifacts resulting from errors in the measurement of less abundant isotope ^{204}Pb.

$$Y = \left(\frac{^{207}Pb}{^{206}Pb}\right), \quad X = \left(\frac{^{204}Pb}{^{206}Pb}\right)$$

A straight line in Fig. 7.1.4

$$y = mx + b \qquad (7.1.27)$$

will be transformed in Fig. 7.1.5 into the straight line

$$Y = bX + m \qquad (7.1.28)$$

Thus, mixing lines and isochrons will be straight lines in both representations. An isochron which has a given slope in Fig. 7.1.4 will have the same value as the Y-intercept in Fig. 7.1.5.

Diagrams analogous to Fig. 7.1.4 are also used in connection with other parent-daughter systems such as ^{147}Sm-^{143}Nd and ^{87}Rb-^{86}Sr. For example, in Fig. 7.1.6, $^{143}Nd/^{144}Nd$ is plotted against $^{87}Sr/^{86}Sr$. Despite the similarity of this diagram to Fig. 7.1.4, some important differences should be noted. In Fig. 7.1.4 for the U-Pb decay systems, the curve representing the growth of radiogenic lead in a closed system had a marked curvature. This is a consequence of the half-life of ^{238}U (4.47 b.y.) being comparable to the age of the Earth, whereas ^{235}U is much shorter-lived (0.704 b.y.). Therefore, early in Earth history, the stable Pb daughter of ^{235}U, ^{207}Pb, was being formed at a much greater rate, relative to the ^{238}U daughter, ^{206}Pb, than is the case at present. In contrast, the half-lives of ^{147}Sm (106 b.y.) and ^{87}Rb (48.8 b.y.) are both large compared to the age of the Earth. For this reason, the analogous closed-system "growth curve" of Fig. 7.1.6 (KJH) will be nearly a straight line.

On the other hand, in general, neither isochrons nor mixing lines will be straight lines on Fig. 7.1.6. For the U-Pb systems, linear isochrons were a consequence of identical fractionation factors f_n for ($^{238}U/^{204}Pb$) and ($^{235}U/^{204}Pb$) in Eq. (7.1.19). In the case of ^{147}Sm-^{143}Nd and ^{87}Rb-^{87}Sr, the corresponding fractionation factors will depend on the actual partition coefficients for all four elements in the fractionation process, for example, during equilibrium partial melting or fractional crystallization. In fact, the f_n's for ^{147}Sm–^{143}Nd will tend to vary in the opposite direction from those of ^{87}Rb-^{87}Sr inasmuch as in solid-liquid equilibrium the daughter element neodymium will usually be enriched in the liquid relative to samarium, whereas the parent element rubidium is enriched with respect to strontium. Mixing lines will also be nonlinear. The analogues of the homogeneous linear equation of the form of Eq. (7.1.13) relating the mixing proportions Q_1 and Q_2 to the isotopic composition of the mixture can still be written in terms of the known concentrations and isotopic compositions of the end-members. However, now when the determinant of these equations is set equal to zero, a term proportional to the product ($^{143}Nd/^{144}Nd$) ($^{87}Sr/^{86}Sr$) will occur. The analogue of Eq. (7.1.25) will, therefore, also contain this nonlinear term, precluding the possibility that this equation will be linear for arbitrary end-member compositions. Only when the ratios ($^{144}Nd/^{86}Sr$) are the same for both end-members will the mixing line be a straight line.

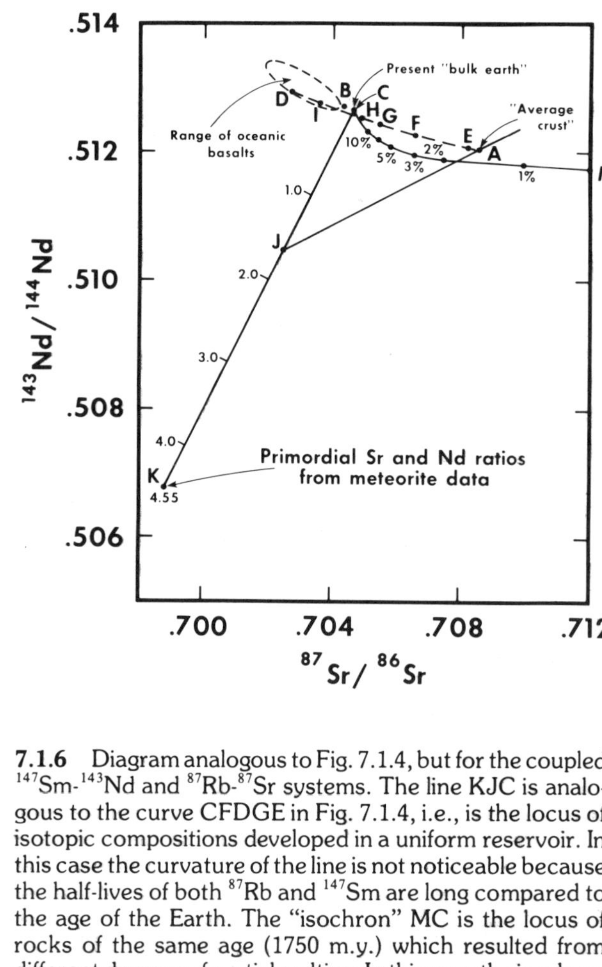

7.1.6 Diagram analogous to Fig. 7.1.4, but for the coupled ^{147}Sm-^{143}Nd and ^{87}Rb-^{87}Sr systems. The line KJC is analogous to the curve CFDGE in Fig. 7.1.4, i.e., is the locus of isotopic compositions developed in a uniform reservoir. In this case the curvature of the line is not noticeable because the half-lives of both ^{87}Rb and ^{147}Sm are long compared to the age of the Earth. The "isochron" MC is the locus of rocks of the same age (1750 m.y.) which resulted from different degrees of partial melting. In this case the isochron is a curve instead of a straight line because, unlike U-Pb systems, neither the numerators nor the denominators of the ratios plotted on the ordinate and abscissa are isotopes of the same element. Point A represents an estimate of the average composition of the Earth's crust. Points B and D are the composition of the residual mantle assuming that the crust separated from 100% and 20% of the mantle, respectively, 1750 m.y. ago, or in a sequence of events averaging 1750 m.y. in age. The curve AEFGHID is the locus of mixtures of compositions A and D. It also deviates from a straight line, as a consequence of the denominators, ^{144}Nd and ^{86}Sr, being neither identical nor proportionate to one another.

In the general case of mixtures of Q_1 and Q_2 moles of end-members with isotopic or element concentrations (A_1, B_1, C_1, D_1) and (A_2, B_2, C_2, D_2), the ratios A_1/B_1 and C_1/D_1 can be written in the form of Eq. (7.1.23). Again the nontrivial values of Q_1 and Q_2 require that the determinant of the homogeneous equations for Q_1 and Q_2 vanish. This then leads to the result

$$\frac{C_M}{D_M} = \left[\frac{A_M}{B_M} - \frac{A_1}{B_1}\right] \frac{\left[\frac{C_2 D_1}{D_2 B_1} - \frac{C_1 D_2}{D_1 B_2}\right]}{\left[\frac{A_2 D_2}{B_2 B_2} - \frac{A_1 D_1}{B_1 B_1}\right]} + \frac{C_1 D_2}{D_1 B_2} \frac{\left[\frac{A_2}{B_2} - \frac{A_1}{B_1}\right]}{\left[\frac{A_2 D_2}{B_2 B_2} - \frac{A_1 D_1}{B_1 B_1}\right]} + \frac{C_M A_M}{D_M B_M} \frac{\left[\frac{D_2}{B_2} - \frac{D_1}{B_1}\right]}{\left[\frac{A_2 D_2}{B_2 B_2} - \frac{A_1 D_1}{B_1 B_1}\right]} \tag{7.1.29}$$

This is the equation of the locus of all such mixtures, and it may be seen that it passes through the end-member compositions (A_1/B_1, C_1/D_1) and (A_2/B_2, C_2/D_2) by substitution of these values into Eq. (7.1.29). The first two terms on the right-hand side of Eq. (7.1.29) are linear in the plotted variables (C_M/D_M, A_M/B_M). However, the last term involves the product (C_M/D_M)(A_M/B_M), destroying the linearity. Only in the case (D_1/B_1) = (D_2/B_2) will this term vanish and the expression become linear.

A diagram closely related to $^{143}Nd/^{144}Nd$ vs. $^{87}Sr/^{86}Sr$ is shown in Fig. 7.1.7. Plotted in this diagram are the deviations of these isotopic ratios from a model Earth having the chondritic $^{147}Sm/^{143}Nd$ ratio, the primordial solar system ratios of $^{143}Nd/^{144}Nd$ and $^{87}Sr/^{86}Sr$ as inferred from meteorite data, and a $^{87}Rb/^{86}Sr$ ratio fitted to the best estimate of this quantity in the bulk earth at the present time (DePaolo and Wasserburg, 1977). At any time t_0 in the past, the neodymium and strontium isotopic composition of this model system can be calculated using these parameters and Eq. (7.1.6). The quantity $\epsilon_{Nd}(t)$ of a sample is then defined as

$$\epsilon_{Nd}(t) = \left\{ \frac{\left(\frac{^{143}Nd}{^{144}Nd}\right)_{smpl}}{\left(\frac{^{143}Nd}{^{144}Nd}\right)_{\text{model at time t}}} - 1 \right\} \times 10^4 \tag{7.1.30}$$

and similarly

$$\epsilon_{Sr}(t) = \left\{ \frac{\left(\frac{^{87}Sr}{^{86}Sr}\right)_{smpl}}{\left(\frac{^{87}Sr}{^{86}Sr}\right)_{\text{model at time t}}} - 1 \right\} \times 10^4 \tag{7.1.31}$$

By definition, the values of ϵ_{Nd} and ϵ_{Sr} for the model are always zero and plot at point C on Fig. 7.1.7. For young rocks in which the addition of radiogenic ^{143}Nd and ^{87}Sr has been negligible, the values of $\epsilon_{Nd}(0)$ and $\epsilon_{Sr}(0)$ are simply calculated from the measured values. For such young samples Fig. 7.1.7 can be thought of as a "blow up" of the upper portion of Fig. 7.1.6. In principle, Fig. 7.1.7 can also be used to portray the deviations

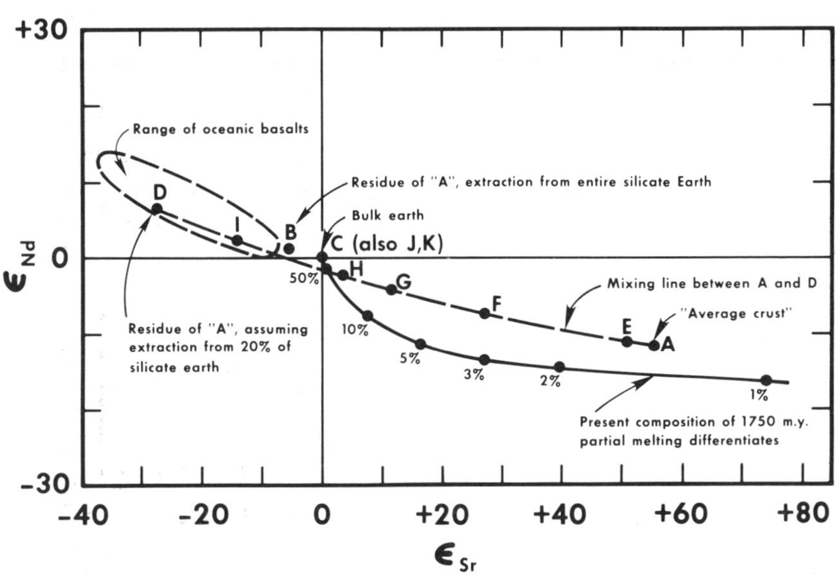

7.1.7 Diagram similar to Fig. 7.1.6, except that the deviations in parts in 10^4 of the Nd and Sr isotopic compositions relative to a uniform reservoir are plotted. The points and curves are the same as those described in Fig. 7.1.6.

of the initial Nd and Sr isotopic composition of an ancient rock of known age from the model Earth at the time of its formation "t." This is difficult in practice because of the large uncertainties in initial $^{87}Sr/^{86}Sr$ ratios of ancient rocks (see section 7.4.2). For example, the rocks of known age plotted on Fig. 7.1.3 fall along the Nd growth curve of the model Earth. If it could be shown that their initial $^{87}Sr/^{86}Sr$ ratios also corresponded to this model, all these points $[\epsilon_{Nd}(t), \epsilon_{Sr}(t)]$ would plot at point C in Fig. 7.1.7.

In all the preceding discussion, a recurring qualification was made regarding whether or not the chemical system under consideration was "closed" or "open" over the time period during which the growth of radiogenic daughter took place. This is a serious qualification, and the frequent failure of natural systems to satisfy this requirement constitutes the principal physical limitation in the application of these isotopic studies to geological problems.

Our understanding of the geochemical processes by which partial loss of parent and daughter occurs is usually insufficient to permit correction of measured parent-daughter ratios for open-system effects. Both of the parents and daughters in the ^{238}U-^{206}Pb and ^{235}U-^{207}Pb systems are isotopes of the same elements. Therefore, open-system processes will tend to fractionate both the $^{238}U/^{206}Pb$ and $^{235}U/^{207}Pb$ ratios by the same factor. For this reason, combined use of these two systems often permits identification of sufficient regularity in the open-system behavior to correct for its effects and infer the original crystallization ages. This is fortunate, inasmuch as most uranium minerals undergo partial loss of lead (and sometimes gain or loss of uranium) even in the absence of geologically significant causes.

The equations for decay of parent and growth of daughter may be modified to include open-system effects

$$\frac{d}{dt}(^{238}U) = -\lambda_{238}(^{238}U) + G_{238}(t)(^{238}U) \qquad (7.1.32)$$

$$\frac{d}{dt}(^{206}Pb) = \lambda_{238}(^{238}U) + G_{206}(t)(^{206}Pb) \qquad (7.1.33)$$

which may be combined to give

$$\frac{d}{dt}\left(\frac{^{206}Pb}{^{238}U}\right) = \lambda_{238} + \left(\frac{^{206}Pb}{^{238}U}\right)[G_8(t) + \lambda_{238}] \qquad (7.1.34)$$

where $G_8(t) = G_{206}(t) - G_{238}(t)$. An analogous expression can be written for the ^{235}U-^{207}Pb system.

The fractionation function $G_8(t)$ has the property that the ratio $^{206}Pb/^{238}U$ increases by a factor $[1+G_8(t)\Delta t]$ during the small time interval Δt purely as a consequence of open-system behavior. If $G_{238}(t)$, $G_{235}(t)$, and hence $G_8(t)$ are given as explicit functions of time, Eqs. (7.1.32–34) may be integrated immediately by standard methods. In other cases, G may be a function of the parent or daughter concentrations, or may be the result of loss of lead by diffusion. In these cases Eqs. (7.1.32–34) are primarily schematic, insofar as some other analytical or numerical procedure must first be used to find $^{206}Pb/^{238}U$ and $^{207}Pb/^{235}U$, after which the result can be cast into the form of these equations.

In one way or another, the time dependence of the U-Pb ratios have been found for a number of open-system mechanisms. These include:

A. Partial loss of Pb at some intermediate time (Wetherill, 1956)

$$\left(\frac{^{238}U}{^{206}Pb}\right) = R(e^{\lambda_{238}t_0} - e^{\lambda_{238}t_1}) + e^{\lambda_{238}t_1} - 1 \qquad (7.1.35)$$

B. Diffusion of Pb with a constant value of the diffusion coefficient $D/a^2 = R$ (Tilton, 1960)

$$\left(\frac{^{206}Pb}{^{238}U}\right) = \frac{6}{\pi^2}\sum_{n=1}^{\infty}\frac{\lambda_{238}(e^{\lambda_{238}t_0 - n^2\pi^2Rt_0} - 1)}{n^2(\lambda_{238} - n^2\pi^2R)} \qquad (7.1.36)$$

C. Loss of ^{206}Pb and ^{207}Pb proportionate to their concentration (Allègre *et al.*, 1974)

$$\left(\frac{^{206}Pb}{^{238}U}\right) = \left(\frac{\lambda_{238}}{\lambda_{238} - R}\right)(e^{\lambda_{238}t_0 - Rt_0} - 1) \qquad (7.1.37)$$

D. Other more complex results, such as time dependent diffusion (Wasserburg, 1963), and diffusion of both the parent and daughter isotopes combined with episodic loss during a short pulse of metamorphism (Wetherill, 1963).

All of these expressions have the property that they contain a parameter (R) which defines the magnitude of the open-system effects. As this parameter approaches zero, the systems approach closed-system behavior, and the ages calculated from the $^{206}Pb/^{238}U$ and $^{207}Pb/^{235}U$ ratios come into agreement with one another and are said to be *concordant*. Thus, under these circumstances, a collection of samples which have been exposed to the same geochemical history, but have responded to it in varying degrees, will provide a set of paired discordant $^{206}Pb/^{238}U$ and $^{207}Pb/^{235}U$ ages which will tend to converge toward their common true concordant age. This phenomenon of convergence was noted in the results of age measurements on uranium by Ahrens (1955). Quantitative interpretation of the phenomenon in terms of particular geochemical open systems has been given by subsequent workers (Wetherill, 1956, 1963; Tilton, 1960; Wasserburg, 1963; Allègre *et al.*, 1974; Goldich and Mudrey, 1972).

These effects are often described by use of a diagram in which the ratio of the radiogenic daughter to parent concentrations of the two U-Pb systems are plotted as $^{206}Pb/^{238}U$ on the ordinate and $^{207}Pb/^{235}U$ on the abscissa (Fig. 7.1.8). The curve labeled "concordia" is the locus of concordant ages, i.e., the two ages calculated by use of the U-Pb age (Eqs. 7.1.16 and (7.1.17) agree with one another. Discordant ages will plot in other regions in the diagram, usually within the shaded area. The arrays of points corresponding to "convergent ages" discussed above will define a curve on this diagram which will intersect concordia at the time corresponding to closed-system behavior.

For the case of episodic loss of lead and/or gain or loss of uranium at a single time in the past, the locus of points defined by the appropriate solution of Eq. (7.1.34) and its ^{235}U-^{207}Pb analog will be a straight line (AB on Fig. 7.1.8). The upper intercept with concordia defines the time of original crystallization ($R = 1$ in Eq. 7.1.35), whereas the lower intercept with concordia ($R = 0$) gives the time at which the episode of geochemical fractionation occurred. In general, other causes of open-system behavior characterized by a single parameter will lead to curved trajectories in Fig. 7.1.8. However, in many cases, the deviations of their curvature from a straight line will be small over a major portion of the trajectory, including the vicinity of the upper intercept with concordia. Because one cannot be certain of the precise mechanism which caused the open-system effects and because somewhat different upper intercepts with concordia will be found when analytical data are fit to different theoretical expressions (e.g., Eqs. 7.1.35–37), the "true age" of samples yielding highly discordant ages will have a large experimental error. When some of the ages are nearly concordant, the extrapolation to the intercept with concordia is more well-defined, and the uncertainties in the actual age will be small.

A diagram mathematically equivalent to Fig. 7.1.8 (cf. Fig. 7.3.5a) has been introduced by Tera and Wasserburg (1974c) in their discussion of U-Pb ages of lunar samples. It has some advantage in that the most accurately determined ratio, $^{207}Pb/^{206}Pb$, is plotted directly, permitting more immediate understanding of the extent to which the experimental data constrain their interpretation.

Techniques used in the dating of young volcanic rocks

^{238}U-^{230}Th (Ionium) dating. In the preceding discussion of the use of radioactive isotopes in geochronology, it was assumed that the daughter isotope was stable. This is not the case for some methods of considerable significance in the dating of young volcanic rocks, for which the daughter isotope is also radioactive. An important

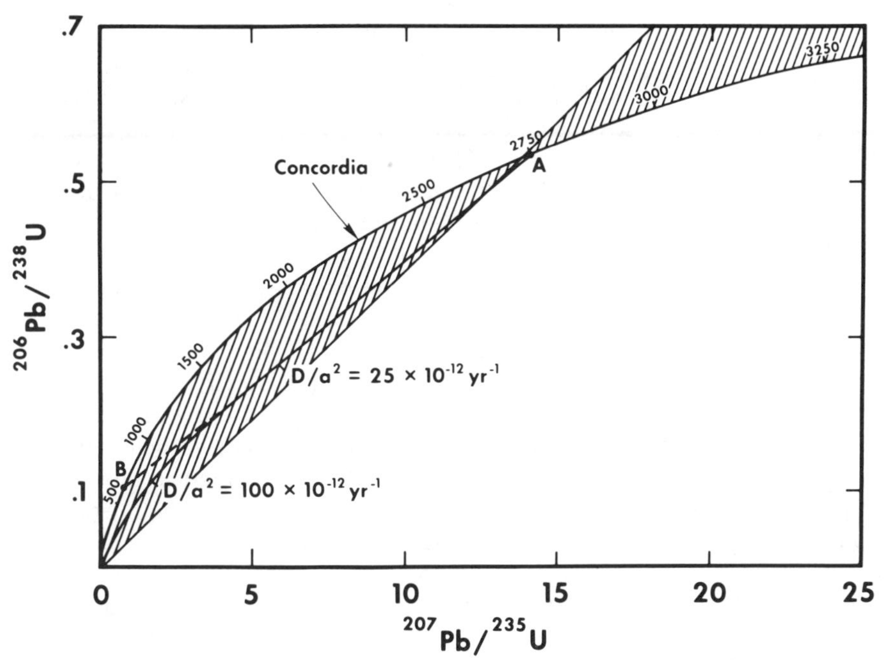

7.1.8 Plot of radiogenic $^{206}Pb/^{238}U$ vs. radiogenic $^{207}Pb/^{235}U$. The curve marked "concordia" is the locus of samples which have been closed systems since their formation. Open-system behavior is indicated by experimental points deviating from this curve. For example, the straight line AB is the locus of 2750 m.y. points corresponding to samples that experienced episodic loss of lead 600 m.y. ago. The curved line passing through Point A is the locus of 2750 m.y. samples that experienced diffusive loss of lead with various time-independent diffusion parameters D/a^2. Unless the open-system history has been unusually complex, points representing samples of true age A will plot within the shaded area.

example is ^{230}Th (also called ionium) with a half-life of 8.0×10^4 years, formed as an intermediate member of the radioactive decay chain which starts at ^{238}U and ends at stable ^{206}Pb.

The concentration of ^{230}Th produced during a short interval Δt at a time t' in the past in a rock with a present U concentration of ^{238}U will be in accordance with the basic laws of radioactive decay

$$\Delta\,(^{230}\mathrm{Th})_{t'} = \Delta t \cdot \lambda_{238}\,{}^{238}\mathrm{U}e^{\lambda_{238}t'} \qquad (7.1.38)$$

where λ_{238} is the ^{238}U decay constant and the present concentration of ^{238}U is augmented by the factor $e^{\lambda_{238}t'}$ to represent its concentration at time t' in the past.

During the interval between t' and the present, this "batch" of ^{230}Th will undergo radioactive decay, so that its present-day concentration will be

$$\Delta\,(^{230}\mathrm{Th}) = \Delta t \cdot \lambda_{238}\,{}^{238}\mathrm{U}e^{t'(\lambda_{238}-\lambda_{230})} \qquad (7.1.39)$$

where λ_{230} is the ^{230}Th decay constant.

The total accumulation from all batches produced during the interval t to the present in a rock with ^{230}Th concentration $(^{230}Th)_0$ at a time t in the past will be

$$^{230}\mathrm{Th} = (^{230}\mathrm{Th})_0\,e^{-\lambda_{230}t} + \int_0^t \lambda_{238}\,{}^{238}\mathrm{U}e^{t'(\lambda_{238}-\lambda_{230})}dt' =$$

$$(^{230}\mathrm{Th})_0\,e^{-\lambda_{230}t} + \frac{\lambda_{238}}{\lambda_{238}-\lambda_{230}}\,{}^{238}\mathrm{U}\,(e^{t(\lambda_{238}-\lambda_{230})} - 1) \qquad (7.1.40)$$

The isotopic composition of thorium, expressed as a proportion of the long-lived isotope ^{232}Th, will be

$$\left(\frac{^{230}\mathrm{Th}}{^{232}\mathrm{Th}}\right) = \frac{\lambda_{238}}{\lambda_{238}-\lambda_{230}}\left(\frac{^{238}\mathrm{U}}{^{232}\mathrm{Th}}\right)(e^{t(\lambda_{238}-\lambda_{230})} - 1) + \frac{^{230}\mathrm{Th}_0}{^{232}\mathrm{Th}}\,e^{-\lambda_{230}t} \qquad (7.1.41)$$

Because $\lambda_{230} >> \lambda_{238}$ and $\lambda_{230} >> \lambda_{232}$, this can be reduced to

$$\left(\frac{^{230}\mathrm{Th}}{^{232}\mathrm{Th}}\right) = \left(\frac{\lambda_{238}}{\lambda_{230}}\right)\left(\frac{^{238}\mathrm{U}}{^{232}\mathrm{Th}}\right)(1 - e^{-\lambda_{230}t}) + \left(\frac{^{230}\mathrm{Th}}{^{232}\mathrm{Th}}\right)_0 l^{-\lambda_{230}t} \qquad (7.1.42)$$

It can be seen from Eq. (7.1.42) that a diagram plotting $^{230}Th/^{232}Th$ vs. $^{238}U/^{232}Th$ will have properties similar to those of the familiar isochron diagram, e.g., $^{87}Sr/^{86}Sr$ vs. $^{87}Rb/^{86}Sr$. Rocks of the same age t and initial isotopic composition $(^{230}Th/^{232}Th)_0$ will plot on an isochron of slope $\lambda_{238}/\lambda_{230}(1-e^{-\lambda_{230}t})/\lambda_{230}$ and intercept $(^{230}Th/^{232}Th)_0\ e^{-\lambda_{230}t}$.

This method was first introduced for volcanic rocks by Taddeucci *et al.* (1967) and Kigoshi (1967). Soon after, Allègre (1968) developed the corresponding isochron formalism. Most of the subsequent work in this field has been carried out in Paris (Allègre and Condomines, 1976; Condomines and Tanguy, 1976; Morand *et al.*, 1978). A first important result of the technique is the precise dating of volcanic rocks with ages lying between 10,000 and 250,000 years. It is worth noting that the ^{238}U-^{230}Th method allows more accurate dating than K-Ar or ^{39}Ar-^{40}Ar methods for ages younger than 40,000 years. A sensitive test for this comparison has been the dating of the Laschamp period (Hall and York, 1978; Condomines, 1978).

Another major success of the method has been the use of the initial $^{230}Th/^{232}Th$ ratio as a tracer of magmatic evolution. For instance, the evolution of the magmatic reservoir beneath the Irazu volcano in Costa Rica has been shown to have begun 140,000 years ago, the early basaltic (tholeiitic) volcanism being rapidly superseded by the emission of andesitic lava flows. This suggests that these magma types have cogenetic affinities.

The Etna volcano in Sicily underwent a somewhat more complex evolution. The early and scattered "pre-etnean" volcanism evolved from tholeiitic to alkali lava flows before 2.1×10^5 years ago. The main volcano of alkaline nature has been subsequently erected during the last 10^5 years. Despite the preliminary character of these results, the uranium-ionium method of dating promises to permit deciphering the fine-scaled evolution of a volcanic system.

Finally, the method provides a sensitive marker of secondary processes. The $^{230}Th/^{232}Th$ activity ratio of surface water and authigenic minerals is very high (> 20). Any significant interaction of surface water with magmas ($^{230}Th/^{232}Th \sim 1$) is easily detectable. Evidence for such an interaction has been obtained from ^{238}U-^{230}Th measurements on lavas from Hawaii (Cordomines *et al.*, 1976) and more recently from mid-oceanic ridges (Morand *et al.*, pers. comm.).

Fission track dating method. In this technique radioactive processes are detected by the radiation damage "tracks" created by fission in solid minerals rather than by their isotopic products. Basically, this is a very attractive method because of its extreme sensitivity (each fission gives a track) and its simplicity (etching + microscopic examination). In practice, however, it proves to be difficult to apply with high accuracy because of experimental problems in track counting and track annealing.

Until recently, the fission track method of dating proposed by Price and Walker (1963) has been restricted to uranium-rich accessory minerals (zircon, sphene, apatite etc.) which unfortunately are not sufficiently common in volcanic rocks. In favorable cases ages as young as 500,000 years can be measured. Recently, volcanic glasses have received more attention (Störzer, 1970; Lakatos and Miller, 1972) and it has been claimed that the age limit can be lowered to about 10,000 years. The method is limited by the need for significant counting statistics and the ease with which tracks fade during subsequent thermal disturbances, even of weak intensity.

7.1.3 Trace element systematics and basaltic volcanism

Introduction

The basalts on the surfaces of the Earth, Moon, Mars, or the one or more basaltic achondrite parent bodies are complex systems which have undergone several successive chemical fractionations before reaching the surface of the planet.

A common model for a basaltic volcanic event begins with a partial melting of the source (stage 1), a transfer to the surface (stage 2), and a chemical differentiation process near or at the surface (stage 3). During stages 2 and 3, because the magma is in contact with surrounding rock of different chemical composition, chemical exchange may occur. This is defined as "contamination." To model these processes quantitatively for trace elements, following Neuman *et al.* (1954), a large number of mathematical expressions were developed and have been reviewed recently by Allègre and Minster (1978) and Shaw (1979). The processes involved in producing the observed variety of trace element effects may be complex. At present a quite simplified scheme is used to describe these processes. It has many weaknesses but seems to work fairly well in practice. These schemes assume that magmas are generated by equilibrium partial melting, that the transfer is chemically inert, and that the surface fractionation is the Rayleigh type, corresponding to closed-system fractional crystallization. Contamination is often introduced into this model by assuming a simple mixing between two end-members. Such a model will be called here "the standard model" and will be reviewed first. This model will then be applied to the trace elements which are relevant to isotope geochemistry. Assumptions and weaknesses of the standard model are discussed below and the formalism is applied to the basic problem of modelling variations in mantle chemistry and the geochemistry of basalts through geologic time.

The standard model

The partition coefficient concept. Following the approach adopted by Nernst and Berthelot in the late 1930's, a partition coefficient for a component ideally soluble in two phases α and β is defined as

$$D^{i}_{\alpha/\beta} = \frac{x^{i}_{\alpha}}{x^{i}_{\beta}}$$

where x^{i}_{α} is the molar concentration of component (i) in phase (α). Generally speaking, the difference between molar and weight concentration is ignored and

$$D^{i}_{\alpha/\beta} = \frac{C^{i}_{\alpha}}{C^{i}_{\beta}}$$

where C^{i}_{α} is the weight concentration of component (i) in phase (α).

This concept has dominated much of the progress of trace element geochemistry in recent years, with the assumption that in most cases the partition coefficient is almost constant for two well-defined phases under magmatic conditions.

In magmatic geochemistry, the reference partition coefficient is between solid and melt ($\beta = l$). Then for a mineral (α),

$$D^{i}_{\alpha/l} = \frac{C^{i}_{\alpha}}{C^{i}_{l}}$$

Since rocks are made of more than one mineral, the **bulk partition coefficient** is defined as

$$D^{i}_{w/l} = x(\alpha_1)D^{i}_{\alpha_1/l} + x(\alpha_2)D^{i}_{\alpha_2/l} + \ldots \qquad (7.1.43)$$

where $x(\alpha_1)$ and $x(\alpha_2)$,.... are the weight proportions of minerals (α_1), (α_2), ...

Equilibrium partial melting. As shown by experimental petrologists, the genesis of basalts is the result of partial melting of a peridotitic source. Since the melting behavior of complex mineral assemblages is dominated by eutectic behavior, the "virtual" mineralogical composition of the liquid varies in steps.

Assuming a complete equilibrium between solid phases and the melt, Schilling and Winchester (1966) applied the classical Berthelot-Nernst Law to these problems. The mathematical derivation for the concentration of a mineral in the liquid phase is straightforward

$$C_l = \frac{C_{o,s}}{F + (1 - F)\,\bar{D}_R} \qquad (7.1.44)$$

where $C_{o,s}$ is the concentration in the initial solid, F is the degree of partial melting (= amount of melt/ amount of initial solid), and $\overline{D}_R$ is the bulk partition coefficient for the residual solid.

It should be pointed out here that in this kind of equilibrium process, if a solid mineralogical phase is completely melted it plays no role in the partitioning. For example, consider two peridotites, one in the garnet field, the other in the plagioclase field. If 20% of both of them melt, for the usual abundances of garnet or plagioclase in these rocks both minerals will be completely melted. Assuming the other minerals (Ol, Opx, Cpx) occur in similar proportions and have the same $C_{o,s}$, the trace element composition of the two liquids will be the same.

The situation will be completely different for 5% of partial melting, for which both garnet and plagioclase may stay as residual phases. This point is made to stress the fact that $\overline{D}_R$ varies by steps and is not constant through the whole process.

Fractional crystallization. In a closed reservoir such as a magma chamber, minerals may precipitate, accumulate on the floor, stop their exchange with the magma, and become isolated. Since the partitioning between minerals and melt differs from unity, a distillation occurs to which the Rayleigh equation may be applied. Therefore, the concentration in the residual liquid is given by

$$C_l = C_{o,l}\, f^{(\overline{D}_F - 1)} \tag{7.1.45}$$

where $C_{o,l}$ is the concentration in the first liquid, which is called "primary" or "parental" magma, f is the degree of crystallization, measured by the ratio of the amount of residual magma to the amount of initial magma, and $\overline{D}_F$ is the bulk partition coefficient during fractional crystallization.

As emphasized by Presnall (1969), rejuvenating a remark originally made by Bowen, fractional crystallization proceeds by almost discontinuous steps, as for example in basaltic systems where the chemical evolution jumps from one eutectic to another quite rapidly. This behavior is not expected from a simple examination of phase diagrams.

Since the two processes (partial melting and fractional crystallization) generally occur successively, we can combine the two equations and write

$$C_l = \frac{C_{o,s}}{F + (1 - F)\,\overline{D}_R}\, f^{(\overline{D}_F - 1)} \tag{7.1.46}$$

This gives the expression for the concentration of a trace element in a lava from the point of view of the standard model.

The trace element inverse-problem. For a measurement of one trace element in one basaltic lava, we have to know five parameters ($C_{o,s}$, F, f, $\overline{D}_R$, $\overline{D}_F$). $C_{o,s}$ is of fundamental interest in problems of geodynamics because it gives the composition of the initial solid. F and f are of great petrological significance because they describe the magmatic processes. $\overline{D}_R$ and $\overline{D}_F$ are geochemical parameters which describe the partitioning of the trace elements.

One way of estimating these quantities is to choose a set of values for the parameters and use these to calculate the resulting value of C_l. In some way, the initial actual concentrations are estimated, or guessed, using indirect inferences, or "petrological intuition", and C_l (calculated) can be equated to C_l (observed). After this has been done for one element, it then has to be tried for the other elements to obtain a set of compatible values. This is in fact a trial and error approach, and consists of solving the *direct* problem. It has been excellently illustrated by Schilling and Winchester (1966) and Gast (1968a, b) in their search for the origin of basalts. This approach has the intrinsic weakness that it is not known whether the solution is unique, or only one among an infinity of possible solutions.

Another possibility is to use what in mathematics is called the inverse approach: consider the observed data as analytical results, and obtain the set of possible estimates of the unknowns. This approach is not an "intellectual inversion" but can be classified among the generalized inverse problems in mathematics. Strang (1976) is a useful reference for those unfamiliar with the subject. As a consequence of the nonuniqueness of the solutions, a range of values instead of a single set of values is obtained for the unknowns. It might be supposed that measurement of several elements in the same lava reduces the underdeterminacy of Eq. (7.1.46) because F and f are the same for all the elements. However, the problem remains underdetermined, and still cannot be solved uniquely, because each new element carries with it a new unknown $C_{o,s}$. If more lavas are now used for measurements, the $C_{o,s}$ will be the same for each element, but each lava will have its own F and f values. Mathematically far more data can be obtained than the number of unknowns, but the problem will still be underdetermined inasmuch as the equations are not independent and, moreover, their nonlinearity precludes demonstration of uniqueness.

This problem may be solved in the following way. The two processes of fractional crystallization and partial melting are considered separately, and for each of them additional external geochemical constraints are introduced in order to eliminate intrinsic indeterminacies in the problem. Examples of such constraints are the Ni content of primary magma for fractional crystallization, and smooth relationships between rare-earth partition coefficients for partial melting. Using such constraints, the problem can be solved mathematically (cf. Allègre *et al.*, 1977; Minster *et al.*, 1977; Minster and Allègre, 1978; Allègre and Minster, 1978; Provost and Allègre, 1979).

Other approaches to solutions are also possible. Consider, for example, that a sufficient number of general inversions determines a set of bulk partition coefficients for partial melting for some elements and a set of bulk partition coefficients for the crystallization of one type of volcanic suite in the basaltic sequence. Then the unknowns are reduced to three ($C_{o,s}$, F, f), and the problem becomes more simple.

In this connection, there are several simple relationships that can be used conveniently during any handling of trace element data, and that facilitate useful extraction of information:

(a) *Rayleigh fractionation.* Logarithms of Eq. (7.1.45) yield

$$\log C_l = \log C_{o,l} + (\bar{D}_F - 1) \log f \qquad (7.1.47)$$

Between two elements (a) and (b), f can be eliminated:

$$\log C_l^a = \log C_{o,l}^a + (\log C_l^b - \log C_{o,l}^b)(\bar{D}_F^a - 1)/(\bar{D}_F^b - 1) \qquad (7.1.48)$$

Then in a log-log diagram, the two elements plot on a straight line. A particular case is important: when $\bar{D}^a$ and $\bar{D}^b$ are both much smaller than one, the slope of this correlation line is equal to one. Therefore, the ratio of the two elements is constant. Such behavior is almost true for U/Th, U/Pb, and Sm/Nd in basaltic magma differentiation.

On the other hand, if only $\bar{D}^b << 1$ the slope of the log-log diagram of any element (a) vs. (b) gives ($\bar{D}^a - 1$). Such construction permits a determination of a set of bulk partition coefficients. It can be seen that for such log-log diagrams, if one initial concentration is known, the whole problem can be solved, and $C_{o,l}$ calculated for all the elements and f for all lavas. In Allègre *et al.* (1977), the constancy of the nickel content of primary basalts (200–400 ppm) was used as an external constraint and thus all $C_{o,l}$ values could be reasonably estimated. A more rigorous solution, using the generalized inverse problem approach, has been given in Minster *et al.* (1977).

(b) *Equilibrium partial melting.* Equation (7.1.44) can be written as

$$\frac{C_{o,s}}{C_l} = \bar{D}_R + F(1 - \bar{D}_R) \qquad (7.1.49)$$

and F eliminated for two elements (a) and (b)

$$\frac{1}{C_l^a} = A\frac{1}{C_l^b} + B \qquad (7.1.50)$$

where A and B are functions of $C_{o,s}^a$, $C_{o,s}^b$, $\bar{D}_R^a$ and $\bar{D}_R^b$. Then if the $\bar{D}_R$'s are constant, which is not rigorously correct but could be a good approximation, a straight line is obtained. For some elements for which $\bar{D}_R$ is small not only in comparison to 1, but also compared to F (this is more restrictive because typically $F \sim 0.1$), the ratio of two elements remains constant. This may be the case for U/Th.

In the partial melting case, external constraints are less simple to use than in the case of fractional crystallization. On the other hand, they are more general. In fact, one can use only the generally accepted range of $C_{o,s}$ values for rare-earth elements, and the observation that for these elements, partition coefficients and concentrations vary smoothly (Minster and Allègre, 1978). Of course in order to apply such a formalism, one must be certain that the series of rocks has actually been created by partial melting. If some fractional crystallization has occurred in addition to the melting process, it is necessary to correct for this secondary mechanism first. The simple graphical relationships mentioned previously may help to check whether a suite of rocks has been created by one or a combination of these two processes.

The fractionation of Rb/Sr, Sm/Nd, U/Pb, Th/Pb.

Here some practical examples of real cases can be given. It is possible to compile a set of partition coefficients for Rb, Sr, Sm, and Nd (although there are difficulties with the partition coefficient concept itself that will be discussed later). For U, Th, and Pb such values can only be tentative because little work has been done for these elements. The coefficients for Rb, Sr, Sm and Nd can then be used to calculate the possible variations of the Rb/Sr and Sm/Nd ratios during fractional crystallization and partial melting. The coefficients are given in Table 7.1.2, and the results of the calculation are shown in Fig. 7.1.9. It should be noted that for both processes the Rb/Sr ratio is more fractionated than the Sm/Nd ratio. It also can be seen that in basaltic volcanism, partial melting is a more effective source of fractionation than fractional crystallization. This is particularly true if the degree of partial melting, generally

Table 7.1.2 Mineral/melt partition coefficients.

Mineral	Rb	Sr	Sm	Nd
clinopyroxene	0.003	0.12	0.26	0.17
garnet	0.01	0.08	0.217	0.087
orthopyroxene	0.003	0.02	0.022	0.013
olivine	0.0001	0.0001	0.010	0.007
amphibole	0.045	0.188	0.34	0.19
plagioclase	0.05	1 to 2	0.05	0.05

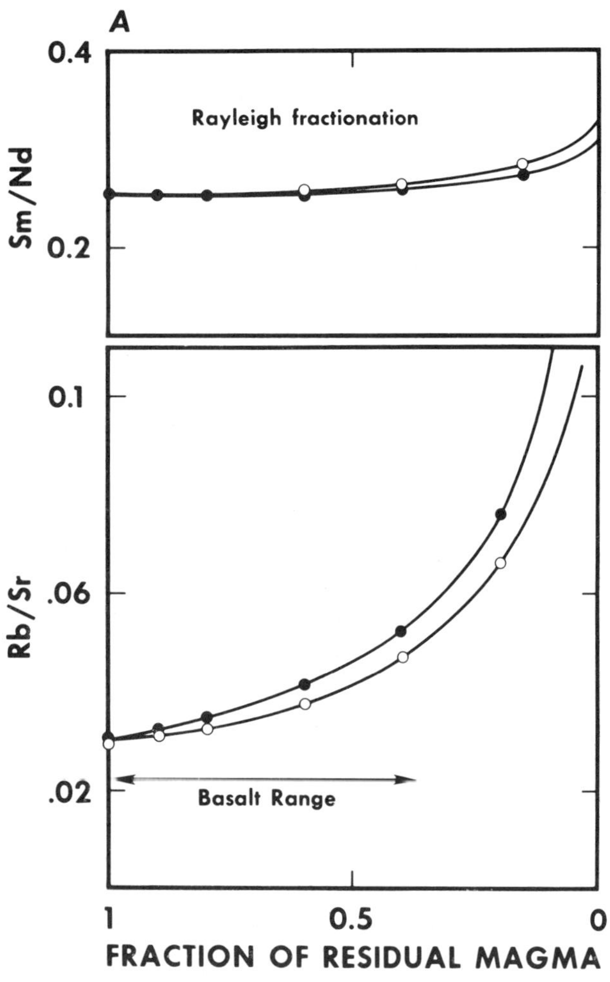

7.1.9 a. Variation of Sm/Nd and Rb/Sr ratios during a fractional crystallization of basaltic magma, assuming that a simple Rayleigh distillation law is obeyed during these processes. The partition coefficients used are given in Table 7.2. The curve (o) corresponds to a cumulate with 40% Plag, 20% Cpx, and 40% Ol + Opx. The curve (•) corresponds to a cumulate with 10% Plag, 30% Cpx, and 60% Ol + Opx. **b.** Variation of Sm/Nd and Rb/Sr ratios during a partial melting of a garnet lherzolite (•) and a plagioclase lherzolite (o). The melting is assumed to occur at the eutectic composition.

considered to be about 5% for alkali basalts and 15–20% for tholeiites, is revised downwards to 1% and 5% respectively (Yoder, pers. comm.).

A real example is given in Fig. 7.1.10, for data obtained from the Terciera Island (Azores) volcanic suite. These calculations demonstrate that agreement between the standard model and real observations can be reasonable, but they also illustrate the fact that care is needed in using the measured chemical ratios in basalts to infer the chemical ratio of their source. Corrections must be applied to the observed ratios to obtain those of the source, and these are particularly important in using observed values to determine the age of basaltic mantle source regions from Rb/Sr isochrons in which the measured initial $^{87}Sr/^{86}Sr$ ratios of basalts are plotted against the $^{87}Rb/^{86}Sr$ ratios of their source.

Stable isotopes as trace elements. Since the work of Urey (1947), it has been known that stable isotopes of the light elements can be fractionated during physical or chemical processes. Just as for trace elements, partition coefficients can be defined for isotopes. Here the concentration of a minor isotope is expressed by its ratio to that of an abundant isotope (for example, $^{18}O/^{16}O$ where ^{18}O is a trace isotope relative to the

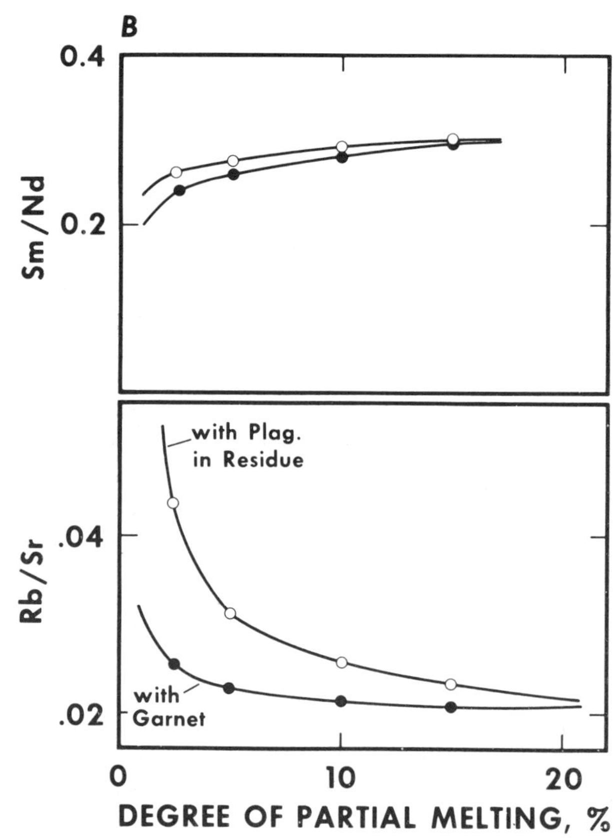

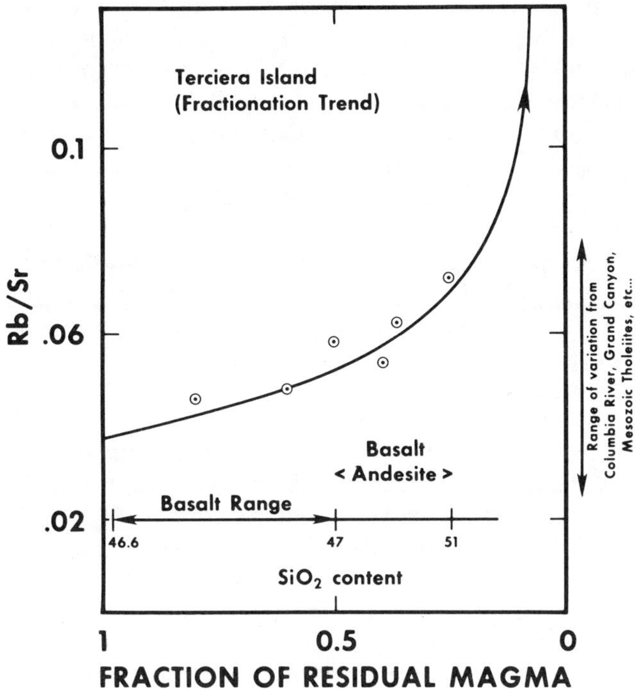

7.1.10 Rb/Sr variation within the Terceira volcanic suite (Azores). This illustrates the range of variation observed in a real fractional crystallization. The range of variation in some classic volcanic suites is illustrated for comparison.

abundant ^{16}O). The isotopic partition coefficient between two phases (A) and (B) can be written as

$$\alpha = \frac{(^{18}O/^{16}O)_A}{(^{18}O/^{16}O)_B} = \frac{D_O^{18}}{D_O^{16}} \quad (7.1.51)$$

Because the effects are very small, results are expressed not directly in ratios but in parts per thousand or per mil (‰) deviations from a standard composition

$$\delta^{18}O = \frac{[(^{18}O/^{16}O)_{smpl} - (^{18}O/^{16}O)_{std}]}{(^{18}O/^{16}O)_{std}} \times 10^3 ‰ \quad (7.1.52)$$

δD, $\delta^{13}C$, $\delta^{15}N$, $\delta^{17}O$, $\delta^{34}S$, etc., are defined in the same way. The standard for hydrogen and oxygen isotopes is mean ocean water, for carbon a specific limestone (PDB), for nitrogen atmospheric nitrogen and for sulfur the Cañon Diablo meteorite troilite.

The stable isotopes of light elements can be fractionated by equilibrium or nonequilibrium processes. Fractionation caused by equilibrium processes decreases with increasing temperature. Kinetic processes have more complex temperature dependency, but the associated fractionation may increase with temperature. The isotopic exchange rate, however, also increases with temperature, and therefore, when a kinetic process fractionates strongly at high temperature, its effect is often eliminated by fast kinetic exchange which reequilibrates the system unless one of the products is isolated from the other.

Generally speaking, at temperatures as high as magmatic temperatures, the equilibrium process is dominant, and the standard expressions for Rayleigh fractionation and equilibrium partial melting can be derived in the same way as for the trace elements.

(a) Rayleigh fractionation

Here, using the oxygen isotopes as an example, the isotopic analogs of Eq. 7.1.45 are

$$(C_{^{18}O})_l = (C_{^{18}O})_{o,l}\, f^{(D_O^{18} - 1)} \quad (7.1.53)$$

and

$$(C_{^{16}O})_l = (C_{^{16}O})_{o,l}\, f^{(D_O^{16} - 1)} \quad (7.1.54)$$

where l and o, l represent the liquid and initial liquid, respectively. Therefore

$$(^{18}O/^{16}O)_l = (^{18}O/^{16}O)_{o,l}\, f^{D_O^{16}(\alpha - 1)} \quad (7.1.55)$$

Because $D_O^{16} \cong 1$,

$$(^{18}O/^{16}O)_l = (^{18}O/^{16}O)_{o,l}\, f^{(\alpha - 1)} \quad (7.1.56)$$

However, for elements like carbon or hydrogen for which the distribution between solid and liquid is important, the partition coefficient for the elements must be considered. Therefore, the general formula for isotopic ratio R is

$$R_l = R_{o,l}\, f^{D_E(\alpha - 1)} \quad (7.1.57)$$

where D_E is the bulk partition coefficient for the chemical element and α is the isotopic partition coefficient. Then

$$\ln R_l = \ln R_{o,l} + D_E(\alpha - 1) \ln f \quad (7.1.58)$$

Using the δ definition (Eq. 7.1.52) and assuming small $\delta(10^{-3}\delta << 1)$, it follows that

$$\delta_l \cdot 10^{-3} \cong \delta_{o,l} \cdot 10^{-3} + D_E(\alpha - 1)\ln f \quad (7.1.59)$$

In the case of oxygen, $D_E \cong 1$.

(b) Equilibrium partial melting

Using Eq. (7.1.49), it can be shown that the general equation for isotope ratio R_l in equilibrium partial melting is

$$R_l = R_{o,s} \left\{ \frac{1 + \frac{F}{D_E}(1-D_E)}{\alpha + \frac{F}{D_E}(1-\alpha D_E)} \right\} \qquad (7.1.60)$$

As before, for oxygen $D_E \cong 1$ and the equation becomes

$$R_l = \frac{R_{o,s}}{\alpha + F(1-\alpha)} \qquad (7.1.61)$$

or, in δ notation for small δ,

$$\delta_l \cdot 10^{-3} = \delta_{o,s} \cdot 10^{-3} - \ln[\alpha + F(1-\alpha)] \qquad (7.1.62)$$

For elements like carbon, for which D_E is very different from 1, isotopic fractionation in partial melting can vary considerably from that given by Eq. (7.1.62). In these cases a more general version of Eq. (7.1.62), derived from Eq. (7.1.60), must be used to calculate fractionation effects.

As specific examples, take $\alpha = 0.99$ and consider the two cases for which $D_E = 1$ and $D_E = 0.2$, both for 10% of partial melting. In the first case, δ changes by +9‰, in the second by +6.5‰.

Discussion of the standard model

We will not repeat all of the comments that many people have made over the years concerning the deficiencies of the standard model, before finally using it. Some recent discussions can be found in Allègre and Minster (1978) and Shaw (1979).

Equilibrium - nonequilibrium kinetics. It is clear that fractional crystallization and equilibrium partial melting are based on exactly opposite assumptions. In the Rayleigh law, used in fractional crystallization, melt and solid do not equilibrate. In partial melting, following the Berthelot-Nernst law, they equilibrate completely. This different behavior has several consequences for the mechanism of basalt genesis. In fractional crystallization the crystals are assumed to sink, accumulate at the bottom of the magma chamber, and become isolated from the liquid. In partial melting, the liquid is assumed to stay in contact with the solid before starting to rise, or alternatively, to reequilibrate with the solid during ascent. Recent values of diffusion coefficients in liquids and solids, obtained principally by Hofmann and Giletti and their co-workers, have led to calculation and speculation concerning this assumption by Hofmann and Hart (1978), and by Allègre and Minster (1978) for the case of a magma chamber. Rough conclusions were that (1) There is no great kinetic objection to the standard model, but some second-order complication may occur; and (2) As pointed out by Albarede and Bottinga (1972), the kinetics of the process may affect the partition coefficients, and thus the relationship between experimental and actual coefficient values must be carefully evaluated.

The concept of partition coefficient. For the elements dealt with in this chapter, i.e., U, Th, Pb, Rb, Sr, K, and REE in ultramafic rocks, there is a major question concerning the validity of the partitioning concept itself as defined earlier in this section. It has been shown by ion probe imaging and by mapping of fission tracks that Rb, K, and U are distributed in cracks and grain boundary defects in most ultramafic rocks. Their behavior during a process such as partial melting may be governed more by the surface chemistry of these defects than by an equilibrium partitioning. This could invalidate all of the preceding partitioning concepts for the radioactive or radiogenic elements.

In addition, some elements may be in ultra microscopic phases, such as $<1\mu$ apatite for U, Th, and REE, which are not usually detected and which may govern the entire behavior of these elements. Their concentration in the microphases is almost constant, and the concentration of the microphases may be almost constant within a major phase. Consequently, a pseudopartitioning may be observed which is more of a statistical parameter than a thermodynamic one. The use of the ion probe will help to clarify this question in the future, but at the present time it must be considered a major problem.

The role of gas. Gases play a major role in both mantle melting and the evolution of magma chambers, as indicated by results of experimental petrology and by observation of processes and rocks in the field (for example, the explosive character of some volcanoes). Gases are important in fixing the redox conditions of magmatism and in controlling fractionation of light isotopes and certain volatile elements.

Generalized expressions for partition coefficients have been developed to include fluid (gas)-liquid (magma) partitioning. For Rayleigh fractionation, Allègre and Minster (1978) replaced the bulk partition coefficient by a new partition coefficient

$$\bar{D} = D_{S/l} = \sum_f D_{f/l} G_f \qquad (7.1.63)$$

where $D_{S/l}$ is the bulk solid-liquid partition coefficient, $D_{f/l}$ the partition coefficient between a fluid component f and the magma, and G_f the solubility of the fluid component in the magma. Shaw (1979) has derived the corresponding analogous expression for partial melting. However, at present there are insufficient data for fluid-magma partition coefficients to make these expressions useful.

The contamination problem. Contamination is certainly one of the most serious problems in the application of the standard model. In a general sense, it can be modelled in two ways: (1) as straightforward fractional mixing of the bulk chemistries of different reservoirs, or (2) as chemically and/or spacially differential mixing (which may be more common in nature) in which only parts of the reservoir of interest are contaminated, and only for certain elements. Several mathematical expressions have been derived to describe differential contamination; among these, some appear very similar to an ordinary Rayleigh law.

Contamination problems have been shown to be serious in mid-continental volcanism (Stettler and Allègre, 1979), and in orogenic volcanism (Briqueu and Lancelot, 1979). Carter *et al.* (1979) have recently demonstrated the occurrence of contamination in shallow magma chambers by a coupled Nd-Sr technique. For magma chambers embedded in granulite or in amphibolite facies rock, the resulting contaminations were shown by this technique to display the isotopic signatures of granulite or gneiss, respectively. The point must be emphasized that contamination is not necessarily equivalent to classical mixing (Briqueu and Lancelot, 1979), and consequently requires a more subtle approach to the problem than is usually taken.

Are the basic mathematical expressions appropriate? The Nernst and Rayleigh laws have been used in this discussion without explicit knowledge of the physics of the melting and crystallizing processes. Two physical assumptions are combined in the concept of Rayleigh fractionation in a magma chamber. The first, the so called "surface equilibrium" or "local equilibrium" assumption, allows the partition coefficients to be written in differential form

$$\frac{dx/dm}{x/m} = D \qquad (7.1.64)$$

where x is the quantity of a particular element and m the quantity of magma. The second assumption is that of a closed system, which is implicitly introduced when the differential equation is integrated. In recent years it has been demonstrated that in many cases the closed-system assumption is not realized in magma chambers, and that reinjection—continuous or episodic—may occur in real environments. The mathematical descriptions of these cases must include additional terms in the differential Eq. (7.1.64). At the same time, as pointed out above, exchanges between the magma and the walls of the magma chamber introduce additional complications.

The process of partial melting is described even more simplistically by the Nernst expression. When a liquid is formed in sufficient amounts it migrates upwards, and during such percolation partially exchanges with surfaces of solid grains (which may or may not themselves be affected by melting). These two components can migrate relative to each other and partially reequilibrate before finally being segregated. The effects of this behavior have never been studied, but eventually they must be.

Of course, if the mathematical expressions are not adequate, then any calculation using them (by direct or inverse methods) will fail to describe the actual behavior of the system. The fact that results obtained in many cases are quite consistent with observation is a good argument in favor of the standard model. But it is important to keep in mind that the standard model *is* a first approximation which certainly differs from a "real" model, and that it has been applied only to the behavior of certain categories of elements and may not be as applicable to other elements.

Fractionation of trace elements through geological time

One of the most important problems in terrestrial geochemistry and geochronology is that of modelling the variation of trace element concentrations and isotopic ratios in the mantle through geological time. The general formalism developed above can be used for this purpose, with the assumption that variations in the mantle are created only by extraction of magma. Time-dependent trace element fractionation results from this magma extraction.

Partitioning between mantle and magma. The bulk partition coefficient for component C between mantle (C_m) and magma (C_l) is

$$\langle D \rangle = \frac{C_m}{C_l} = \frac{(x/v)}{(dx/dv)} \qquad (7.1.65)$$

where x is the quantity of trace element C in the mantle and v is the volume of the mantle. Then

$$\frac{d}{dt}(C_m) = \frac{d}{dt}\left(\frac{x}{v}\right) = C_m\left(\frac{1}{\langle D \rangle} - 1\right)\frac{dv}{vdt} \qquad (7.1.66)$$

If the extracted liquid is created by partial melting, then from Eq. (7.1.49)

$$\langle D \rangle = \frac{C_m}{C_l} = D_R + F(1 - D_R) \qquad (7.1.67)$$

$$\frac{dC_m}{dt} = C_m \left[\frac{1}{D_R + F(1 - D_R)} - 1 \right] \frac{dv}{vdt} \qquad (7.1.68)$$

Introducing a fractionation factor ψ defined as

$$\psi = \left[\frac{1}{D_R + F(1 - D_R)} - 1 \right] \frac{dv}{vdt} \qquad (7.1.69)$$

integration of Eq. (7.1.68) yields the variation of C_m through time

$$C_m = C_{m_0} \exp \left[\int \psi (t)\, dt \right] \qquad (7.1.70)$$

Equation (7.1.70) describes the distillation of the mantle through time, and from the definition of ψ it can be seen that the variations are linked, as expected, to trace element partition coefficients. For the ratio r^{ij}_m of two elements i and j

$$r^{ij}_m = \frac{C^i_m}{C^j_m} = \frac{C^i_o}{C^j_o} \exp \left[\int \left[\frac{1}{D^i_R + F(1 - D^i_R)} - \frac{1}{D^j_R + F(1 - D^j_R)} \right] \frac{dv}{vdt}\, dt \right] \qquad (7.1.71)$$

and

$$\frac{dr^{ij}_m}{dt} = \psi^{ij} r^{ij}_m, \text{ where } \psi^{ij} = \left[\frac{1}{\langle D^i \rangle} - \frac{1}{\langle D^j \rangle} \right] \frac{dv}{vdt} \qquad (7.1.72)$$

If $\psi^{ij} > 1$, the ratio increases with time, and if $\psi^{ij} < 1$, the ratio decreases. $\psi^{ij} > 1$ when $\langle D^i \rangle < \langle D^j \rangle$ and, therefore, when element j is more incompatible than i. As examples, the Sr/Rb ratio increases with time but the Sm/Nd ratio decreases with time in this mantle model. A similar formalism can of course be applied to stable isotope fractionation.

The evolution of isotopic ratios. Equations for isotopic evolution of radiometric parent-daughter systems can be written very simply as

$$\frac{d\alpha_m}{dt} = \lambda r_m, \quad \frac{dr_m}{dt} = (\psi - \lambda)\, r_m \qquad (7.1.73\text{–}74)$$

$$\alpha_m = \frac{\text{Daughter isotope}}{\text{Reference isotope}}, \quad r_m = \frac{\text{Parent isotope}}{\text{Reference isotope}}$$

daughter and reference are isotopes of the same element, and $\psi = \psi^{ij}$ where i = parent isotope and j = reference isotope.

Integration of this system of differential equations yields parent and daughter isotopic ratios as functions of time. The most straightforward case mathematically occurs for ψ = constant. Then

$$\alpha_m = \alpha_{m_0} + \frac{\lambda r_{m_0}}{(\psi - \lambda)} [e^{(\psi - \lambda)t} - 1] \qquad (7.1.75)$$

$$r_m = r_{m_0} e^{(\psi - \lambda)t} \qquad (7.1.76)$$

This very simple model can be used to compute a rough first approximation of isotopic evolution in the mantle Sm-Nd and Rb-Sr systems, using values of partition coefficients for partial melting. But to model actual variations observed in the Earth's mantle, ψ must be a function of time (Allègre *et al.*, 1980), as shown for $^{87}Sr/^{86}Sr$ in Fig. 7.1.11 (Allègre *et al.*, 1979). In addition, the bulk fractionation factor for the mantle is more complex than just the partial melting factor used here (Allègre *et al.*, 1980), although the general approach is the same for the more complex case.

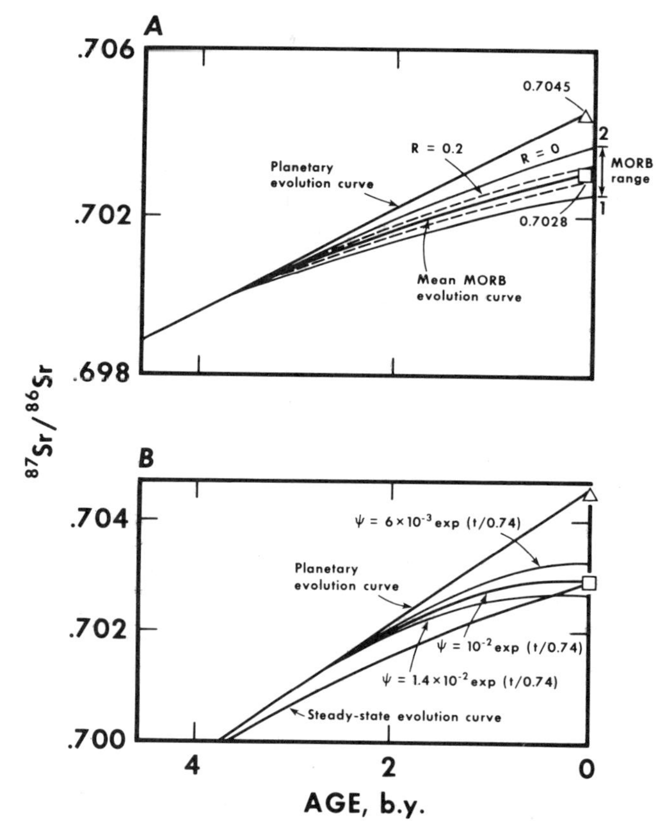

7.1.11 $^{87}Sr/^{86}Sr$ variation through time in the mantle using different assumptions for the rate constants.

7.1.4 Stable isotopes as geochemical tracers in problems of basaltic volcanism

Stable isotopes have been employed quite extensively in the study of exogenic-low temperature processes in the Earth since the classic paper by Urey *et al.* (1951) on paleothermometry. The use of stable isotopes in igneous petrology is more recent and is still not widespread, despite the fact that the measurement of these isotopes gives unique information regarding contamination with water, amount of outgassing, and contamination by sediments.

Basic principles

To first approximation, isotopes of the same element have the same chemical properties because their outer electron shells are identical. As shown theoretically by Urey (1947), however, this is only a first approximation. Because of their different masses, isotopes of the same element actually behave in slightly different ways during chemical or physical processes. Therefore, every reaction which involves change in the structural configuration of a chemical species will be characterized by a certain **isotopic fractionation**. Such fractionation is measured by comparison of an isotopic ratio before and after the process. The fractionation factor for isotopes i and j of element a is defined as the ratio

$$\alpha = \frac{(a_i/a_j)_{after}}{(a_i/a_j)_{before}} \qquad (7.1.77)$$

Generally speaking, this fractionation factor is a function of the atomic masses m_i and m_j of the two isotopes.

At the present time the only commonly measurable fractionations are for light elements, and their practical use is limited to the isotope ratios D/H, $^{13}C/^{12}C$, $^{15}N/^{14}N$, $^{17,18}O/^{16}O$ and $^{34}S/^{32}S$ for which consistent, measurable and well-recognized variations have been established in nature. It should be noted that this limitation is primarily technical. These fractionation effects are generally very small, on the order of parts per thousand, and their detection requires very accurate mass spectrometric techniques. A large number of higher mass elements would show even smaller natural isotopic fractionations if mass spectrometers were capable of detecting variations in the parts per million range.

Stable isotope variations in nature are usually expressed as δ-values (parts per thousand or per mil deviations from standard isotopic compositions), where

$$\delta = \left[\frac{(a_i/a_j)_{smpl} - (a_i/a_j)_{std}}{(a_i/a_j)_{std}}\right] \times 10^3 \text{ ‰} \qquad (7.1.78)$$

(see Eq. 7.1.52, and following, for list of common standard materials).

Natural fractionation processes are divided into two categories: *equilibrium* and *kinetic*. In equilibrium processes isotopic fractionations obey the mass action law in the case of chemical equilibrium, or more generally, the laws of thermodynamic equilibrium. For some simple cases they can be calculated by statistical thermodynamics (Urey, 1947; Bottinga, 1969a, b). More commonly, they are measured in laboratory experiments (O'Neil and Clayton, 1964; O'Neil and Taylor, 1967). The equilibrium constants depend on temperature as $1/T^2$; consequently, fractionation factors decrease with temperature and are small at magmatic temperatures. This is the reason for the more common utilization of stable isotope fractionation in the study of low temperature, rather than high temperature, geochemical processes.

Kinetic fractionation processes are less well understood, especially from the theoretical point of view. The restricted amount of work done on kinetic mechanisms seems to indicate that kinetic fractionation occurs in the direction which enriches the light isotope as the result of the reaction. In contrast to equilibrium fractionation, kinetic fractionation increases with temperature. Among the class of kinetic isotopic effects, the phenomenon of isotopic exchange is very important, both for light isotopes fractionated by physico-chemical processes and for heavy isotope compositions altered by radioactive decay. When two media (e.g., minerals) of different isotopic composition are in contact, they exchange their isotopes in the direction of isotopic equilibrium even if their chemical compositions are identical. Isotopic equilibrium may therefore occur without any chemical variation, and because the phenomenon is under kinetic control it is thermally activated. There are several geochemically important consequences of isotopic exchange:

1) Attainment of isotopic equilibrium between minerals or rocks, which sets the "radiometric clocks" used in geochronology.

2) Gradual evolution in the direction of isotopic equilibrium following a kinetic fractionation, if the species stay in contact. At high temperature a previous kinetic fractionation is often partially or completely annihilated by the isotopic exchange phenomenon.

3) Isotopic mixing without general chemical transfer.

In the use of stable isotopes for the study of magmas, it is necessary to distinguish between two types of elements. Hydrogen and oxygen are major elements in igneous rocks. Their structural localization in minerals is well-established and their study can be easily linked to petrological observations. Carbon, sulfur, and nitrogen are trace elements in igneous rocks, and are not well-localized. On the other hand, they are the major constituents of magmatic gases.

Like the trace elements discussed in section 7.1.3, stable isotopes are fractionated during the major petrogenetic processes: fractional crystallization, partial melting, contamination, and mixing. The equations describing trace element behavior can be easily adapted to describe the fractionation of the stable isotopes of both major and minor elements. Such expressions have been given above (Eqs. 7.1.51–7.1.62) and will not be repeated here.

The isotope geochemistry of oxygen and hydrogen, and then that of carbon, sulfur, and nitrogen are reviewed in the following sections. The discussion is divided in this way not only because of differences in the amount of work that has been done with these elements but also because of differences in the type of problem to be addressed. Oxygen is a dominant constituent of rocks and water. Hydrogen is the principal constituent of water but is much less abundant in rocks. Thus the coupling of H and O isotope studies represents the best approach to understanding the origin of rock in the context of exchange and interaction with water. Carbon, sulfur and nitrogen are trace elements in rocks but are abundant in volcanic gases. Thus these elements provide information concerning gas-magma interactions.

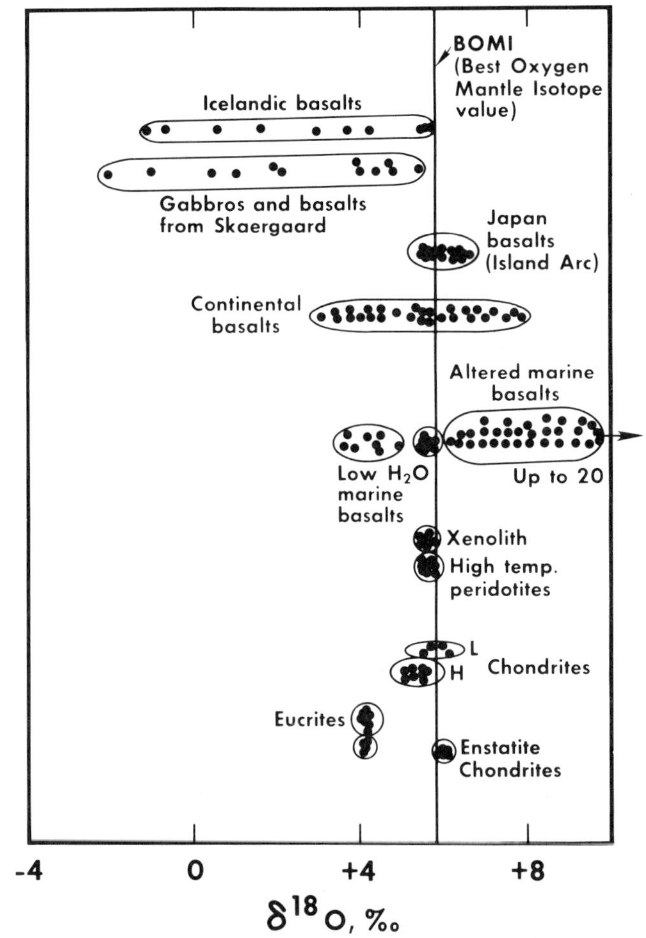

7.1.12 Values of $\delta^{18}O$ for the principal rocks involved in the discussion of basalt genesis. The continental basalt domain includes data from Massif/Central, Snake River and Deccan basalts.

Oxygen and hydrogen isotope geochemistry

Basis for O-H igneous geochemistry. Deep-seated terrestrial oxygen seems to have a fairly uniform $\delta^{18}O$ value between + 5.5 and + 5.8‰. Such values are obtained by careful analysis of both ultramafic rocks (xenolithic and alpine peridotites) and primitive basalts (such as those carrying ultramafic xenoliths). A similar range of $\delta^{18}O$ values has been found in ordinary chondrites (Taylor *et al.*, 1965; Clayton *et al.*, 1973) and lunar basalts (Taylor and Epstein, 1970; Onuma *et al.*, 1970). Chondrites, however, show distinct $\delta^{17}O$ signatures; this intriguing and significant departure from the isotopic patterns of terrestrial oxygen will be discussed later. For the present discussion, the 5.5–5.8‰ range is chosen as a reference value (BOMI: see Fig. 7.1.12) for $\delta^{18}O$.

In contrast, the acidic igneous rocks and minerals show $\delta^{18}O$ signatures significantly different from that of the mantle reservoir (Fig. 7.1.12). Rhyolitic and granitoid rocks typically have δ-values between +6 and +12‰ (Taylor and Epstein, 1963; Taylor, 1968), but examples are known with $1\delta^{18}O$ as high as + 16‰ (Taylor and Turi, 1976). Sedimentary rocks fall in the interval from + 6‰ (volcanogenic sediment) to + 22‰ (argillaceous sediments). Limestones and cherts typically range from + 20 to + 35‰ (Epstein and Taylor, 1967).

The mantle reference value for hydrogen, equivalent to BOMI for oxygen, is a subject of debate. One of the difficulties is the high abundance of hydrogen in water and its low abundance in rocks, coupled with the existence of the subduction process which brings altered oceanic crust into the mantle, thereby changing its original δD composition. As pointed out by Taylor, subduction may have "contaminated" the entire upper mantle over geological time. The δD values of such subducted materials are similar to the estimates made by Sheppard and Epstein (1970) and Craig and Lupton (1976) of present-day mantle values ($\delta D = -90$ to -50‰).

The other large oxygen reservoir is the ocean. By definition, both $\delta^{18}O$ and δD are zero since Standard

Mean Ocean Water (SMOW) is the reference standard for oxygen and hydrogen compositions. As shown by the work of Friedman, Epstein and Craig, the hydrologic cycle fractionates the oxygen and hydrogen isotopes in both the evaporation and precipitation phenomena (Friedman, 1953; Epstein and Mayeda, 1953; Craig, 1963). An important result of these processes is that fresh waters are depleted in both ^{18}O and D, and thus have negative δ values relative to the ocean reservoir. A very clear linear relationship exists between H and O isotopes for meteoric waters, which can be mapped in detail geographically as has been done for North America (see Taylor, 1974, for recent compilation of the data).

Magma-solid fractionation factors. Direct measurements of silicate melt-solid equilibrium fractionation factors for the oxygen isotopes at high temperature are extremely difficult because the fractionations are small. The only existing experimental data were obtained by Muehlenbachs and Kushiro (1974), who gave α-values (isotopic partition coefficients or fractionation factors: see Eqs. 7.1.51 and 7.1.77) for plagioclase-magma and enstatite-magma of about 1.001 to 1.0005 at "basaltic" temperatures. However, such fractionations change with temperature and in the basalt range the pyroxene-melt fractionation may be reversed, with α as low as 0.998. Another way to estimate fractionation factors is by direct measurement between glass and minerals in natural rocks. This kind of approach has been attempted by Garlick (1966), Anderson *et al.* (1971), and Muehlenbachs and Clayton (1972). Their estimates show that at high temperatures fractionations are quite small, with $\alpha \sim 0.999$ to 0.998. A more indirect method involves measurement of the total range of oxygen isotopic composition in a differentiated igneous suite, and application of the Rayleigh law (Eq. 7.1.59). If f, the degree of fractional crystallization, can be estimated for each rock type of the suite in which $\delta^{18}O$ is measured, then the fractionation factor α (assumed constant throughout the differentiation) can be calculated from the slope of a plot of $\delta^{18}O$ vs. ln f. Matsuhisa (1979) has carried out a careful study of this type on the Haihijo-jima volcanic rock suite, obtaining a $1\delta^{18}O$ range of $\sim 1‰$ for f ~ 0.1–0.8 corresponding to $\alpha = 0.9995$.

The three types of estimate described above suggest a bulk melt-solid fractionation factor α within the range 1.001–0.998 for basaltic and ultramafic systems. Taking $f \geq 0.7$ as characteristic of basalts, the Rayleigh law (Eq. 7.1.59) then yields an extreme variation in $\delta^{18}O$ for basalts of 5.3–6.4‰ (assuming a mantle value for $(\delta^{18}O)_o = 5.7‰$ in the magma source rock), as illustrated in Fig. 7.1.13. In equilibrium partial melting (Eq. 7.1.62), even for the degree of partial melting F as low as 0.01, $\delta^{18}O$ variations are restricted to the range 4.7–7.7‰ for the same α-values. Thus magmatic oxygen fractionations are small and correspondingly difficult to use in attempting to understand basalt petrogenesis in the Earth because other effects may completely obscure the picture. However, it is just this *constancy* in $\delta^{18}O$ that permits use of $^{18}O/^{16}O$ variations to monitor other processes that may have affected basaltic rocks.

Fractionation factors among minerals at igneous temperatures have been used to determine the crystallization temperatures of basalts. Such an approach by Garlick (1966) and Anderson *et al.* (1971) for terrestrial basalts, and by Onuma *et al.* (1970) and Taylor and Epstein (1970) for lunar basalts, gave temperatures around 1100°, similar to the standard petrologic estimates. More recently, Bottinga and Javoy (1975) with a more elaborate discussion of the calibration equations, redetermined these temperatures and found essentially the same result for terrestrial basalts, lunar basalts and

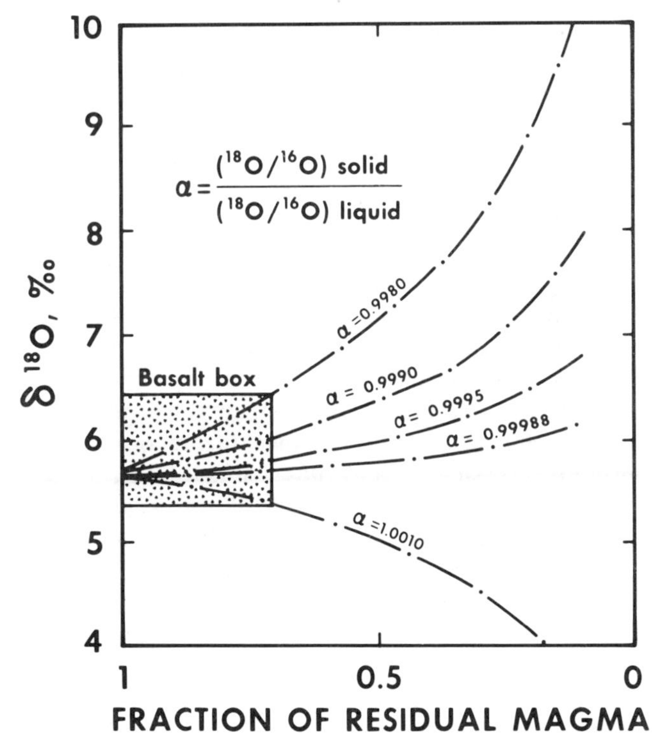

7.1.13 $\delta^{18}O$ variations during fractional crystallization, for various values of the fractionation factor α. $(\delta^{18}O)_o = 5.7$ ‰. The fraction of residual magma (equal to degree of fractional crystallization) is the parameter f in the Rayleigh equation. Basalt compositions correspond to a range of f between 0.7 and 1. The corresponding range of variation in $\delta^{18}O$ for $\alpha = 1.001$–0.998 (see text) is indicated by the shaded box.

basaltic achondrites. More important, in their analyses the minerals (and glasses) of these volcanic rocks typically seem to be in isotopic equilibrium. This is not the case for most plutonic igneous rocks, which characteristically have suffered a variety of isotopic changes during their slow cooling and interaction with fluids derived from the country rocks.

Magma-gas fractionation. Because CO_2 and H_2O are significant reservoirs for oxygen, fractionations between gases and magma are quite important. Muehlenbachs and Kushiro (1974) have carried out a set of experiments that demonstrate the possibility of complex behavior such as inversion of fractionation with temperature. However, their results were not conclusive enough to allow accurate numerical estimates of fractionation factor values.

In principle, magma-water fractionation could be estimated by extrapolating the mineral-water fractionation curves to high temperature, and by assuming the magma-mineral fractionation factor, calculating the magma-water fractionation. However, such extrapolations are difficult because in the magmatic temperature range the extrapolation may be nonlinear [as suggested by the Muehlenbachs-Kushiro (1974) experiments] and may involve "cross-overs," which means that although the minerals are enriched in ^{18}O relative to H_2O at low temperatures, above 1000° C the fluid could be enriched in ^{18}O.

Mineral-water fractionation factors. Mineral-water fractionation factors are part of the standard data base in stable isotope geochemistry. Botinga and Javoy (1975) recently reviewed the present state of knowledge of such fractionations and recalculated their temperature dependence. As is well known, the fractionation is proportional to $1/T^2$ and thus is much larger at low temperatures. More important for our problem, a number of fractionation curves have been determined or estimated for the alteration of minerals which also show quite large fractionations (some up to +20‰ for $\delta^{18}O$). Given the dependence of the partition coefficient on temperature, such effects can be understood by considering some possible scenarios of water-rock interaction sequences, shown schematically in Fig. 7.1.14a. Suppose that a water-dominant fluid interacts with minerals at high temperature. If the water/rock ratio is not very large the $\delta^{18}O$ of the water may change, moving in the direction of rock $\delta^{18}O$ values; the δD of the water remains almost constant because the rocks are relatively dry. This type of behavior is well illustrated in the thermal springs studied by Craig (1963). At the same time, the $\delta^{18}O$ of the rock shifts in the direction of the water $\delta^{18}O$ values, therefore generating low ^{18}O rocks. Assume now that the (relatively) high ^{18}O water subsequently exchanges at *low temperature* with minerals. As a result of this alteration, those minerals may acquire an extremely high $\delta^{18}O$. Isotopic changes arising from other interaction routes of this kind are illustrated in Fig. 7.1.14b. Such models have been quite systematically explored, illustrated and used by Taylor and his associates in various circumstances, and will be fundamental in the study of problems in isotopic igneous petrology.

Basaltic geochemistry of O-H isotopes

Applications of O-H isotopes. Oxygen and hydrogen isotopes have been useful in several aspects of the basalt problem. One important application is detection of the low temperature alteration processes which affect rocks. These alterations typically create new minerals such as chlorite, zeolites, and clay minerals, at low temperatures where the $\delta^{18}O$ fractionation coefficients are very large. Therefore, the $\delta^{18}O$ values of such secondary minerals are high and the $\delta^{18}O$ values of their bulk rocks will clearly fall outside the normal range. This criterion is extremely useful in detecting even minor alteration effects in ancient basalts or in submarine basalts, which will be discussed in more detail below. In continental environments Taylor (1974) has observed this kind of behavior in the Muskox and Bushveld intrusions, where it is typically associated with feldspar of the red-brick facies. The resulting $\delta^{18}O$ values may be quite high, up to +13 to +14‰.

A second process which may occur is high temperature interaction between rocks and water. Because the $^{18}O/^{16}O$ partition coefficient is very small at high temperature, this type of exchange is essentially a mixing process. Therefore, the rock becomes lower in $\delta^{18}O$ and the water becomes slightly higher. Such effects are most easily observed where the $\delta^{18}O$ is very low (e.g., in interactions with meteoric waters), but observable effects also may occur on a large scale during circulation of seawater at the oceanic ridges. In this latter case, however, the lowered $\delta^{18}O$ may be cancelled by low-temperature alteration processes that increase the $\delta^{18}O$. As a result oceanic basalts often have $\delta^{18}O$ values distinctly higher than the normal range, as shown by Garlick and Dymond (1970), Muehlenbachs and Clayton (1972), and Pineau *et al.* (1976b). Nonetheless, in some cases greenstones and altered basalts have been found with low $\delta^{18}O$ values of about +2 to +4‰ (Muehlenbachs and Clayton, 1972). Dual behavior during alteration of submarine rocks is substantiated by the ophiolite studies of Javoy (1970), Genest and Javoy

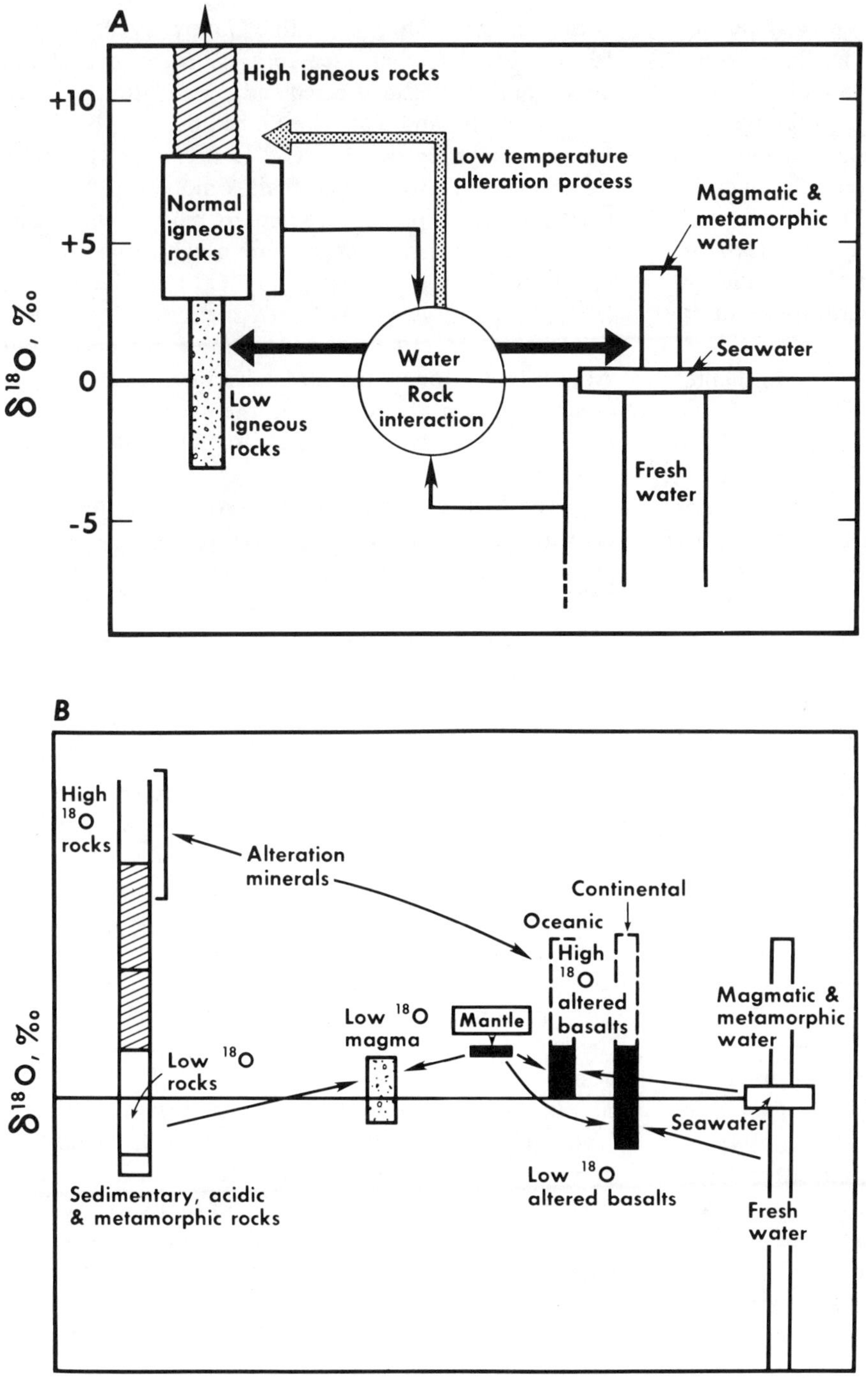

7.1.14 Water-rock interaction processes. The process is a mixture when it occurs at high temperature. When it is followed by a low-temperature process, a fractionation is superimposed onto the mixing.

(1976), Magaritz and Taylor (1974), and Heaton and Sheppard (1977). Acknowledging the fact that the alteration process may be more complex in ophiolites because several stages of alteration with different types of water may occur, these detailed studies support the concept that seawater-basalt alteration is a very important process in submarine igneous rocks. In this context, the simultaneous use of hydrogen and oxygen isotopes is very useful because they permit us to decipher the complex interactions and identify the water sources, as illustrated by Magaritz and Taylor (1974) for ophiolites and Frederichsen and Hoernes (1978) for basalts. The $\delta^{18}O$ values in oceanic basalts are correlated with H_2O contents (Taylor, 1968; Muehlenbachs and Clayton, 1972; Pineau *et al.*, 1976b); extrapolated to zero H_2O, they correspond very nicely with $\delta^{18}O = 5.5‰$. More recently, systematic studies of basalts from the Deep Sea Drilling Project (DSDP) by Frederichsen and Hoernes (1978) have clearly shown that almost all deep-sea basalts have been altered, and that such alteration occurs at moderate temperatures in the first 10 m.y. of spreading. Such an estimate is in agreement with the work of Hart *et al.* (1978) on Sr isotopes. It is extremely important to understand this phenomenon because the subduction process can be a very efficient way to transport water, and especially its hydrogen, into the mantle.

When exchange processes occur in continental areas where the fresh waters have low $\delta^{18}O$ and δD, the effects are much more clear-cut. Such phenomena have been extremely well studied, by Forester and Taylor (1977) for the Scottish Hebrides igneous complexes and also for the Skaergard instrusion in Greenland by Taylor and Epstein (1968), Taylor (1968), and Taylor and Forester (1973). This process is now known to be general around almost all volcanic centers.

The most extreme effect produced by water-magma interaction is seen in the low ^{18}O magmas. Such magmas, with a $\delta^{18}O$ of +2 to +3‰, were discovered by Forester and Taylor (1972) for acidic rocks, and by Muehlenbachs *et al.* (1974) for basaltic rocks. This phenomenon could be the result of some special basalt source composition in the mantle, but it is more probable that the magma has exchanged with a low ^{18}O source. As pointed out by Taylor, a direct dissolution of fresh water is very unlikely since the solubility of water in silicate melt at low pressure is limited and the corresponding mass balance calculation cannot explain the measured result. The most likely explanation is the roof dissolution process suggested by Taylor. Roof rocks previously altered by the meteoric-hydrothermal system that surrounds the magma chamber will have low $\delta^{18}O$ values; if these are digested or exchanged with the magma, ^{18}O-depleted silicate melts can be produced. Such a mechanism would have the consequence that the basalts which show low $\delta^{18}O$ signatures have probably resided in a shallow magma chamber for a significant time period, or at least have interacted with their upper level environment in a complex manner.

The direct assimilation of or contamination by continental crust in basaltic magma can also be detected by oxygen and possibly by deuterium isotopes. Inasmuch as granitic rocks or sedimentary rocks generally have high $\delta^{18}O$ values, mixing with normal basaltic magmas will increase the $\delta^{18}O$ in the magmas. This kind of behavior has been observed in different circumstances. In the continental environment, basaltic rocks commonly seem to have suffered some contamination because they typically show small enrichments in $\delta^{18}O$, above the primitive 5.7‰ value (Javoy, 1970; Taylor, 1968). However, when a real mixing process occurs, the bulk chemistry is also affected and a correlation between bulk chemistry and $\delta^{18}O$ should be observed, but if the process is simply due to $^{18}O/^{16}O$ exchange, then the major element variation is small compared to the $\delta^{18}O$ variation. Both processes seem to have occurred, as evidenced by the increasingly common observation of a covariance of $^{87}Sr/^{86}Sr$ with $^{18}O/^{16}O$ (Turi and Taylor, 1976; Magaritz *et al.*, 1978; Taylor and Silver, 1978; Michard *et al.*, 1979a, b). As shown by Javoy and Allègre (1969) for continental volcanism and Montigny *et al.* (1969) for arc volcanism, the existence of such a covariation is a good indication of the contamination process.

The preceding phenomena have been applied in a careful study of one of the most difficult petrogenetic problems, that of island arc volcanism (Matsuhisa, 1979). In this study of the immature Japan arc, which lies on purely oceanic basement on the Pacific side, the tholeiitic basalts have $\delta^{18}O$ of 5.5–5.7‰. On the mature arc, which is developed on the continental crust, the $\delta^{18}O$ values are higher. Matsuhisa interpreted these important observations in two ways. First, the tholeiites from the immature arc cannot be derived from altered subducted plate material with a very different value of $\delta^{18}O$. Therefore, they originate below or above the plate but not from the plate itself. Second, the basalts on the continental side are probably contaminated by the crust, by isotopic exchange or mixing.

The uniform $\delta^{18}O$ mantle: reality and cause. One of the most important conclusions of the studies quoted above is the remarkable uniformity of $\delta^{18}O$ in the pristine mantle materials. The most primitive basalts from mid-oceanic ridges, interarc basins, island and continental volcanics all have fairly uniform $\delta^{18}O$ between 5.5–5.7‰. However, Kyser and O'Neil (1978) have claimed recently that a difference of about 2‰ exists between tholeiitic and alkali basalt. Such a difference, if real, will be very important, but at the moment more data are needed before such a relationship can be definitely claimed.

The observation of the constancy in the $\delta^{18}O$ of the Earth and Moon, and its similarity with chondritic $\delta^{18}O$, is important. Muehlenbachs and Kushiro (1974) have measured diffusion coefficients for oxygen and found values on the order of 10^{-7}–10^{-8} cm^2/sec for magma and 10^{-11}–10^{-12} cm^2/sec for silicates under mantle conditions. Rough estimates of the times necessary for diffusion to homogenize oxygen isotopes in the mantle, over scale lengths of ~10 meters, are ~10^{13} seconds for magma and 10^{18} seconds for crystalline phases. As shown by

Hoffman and Hart (1977) for other elements, such slow diffusion in the solid state implies that solid-state heterogeneity may persist for quite a long time.

As mentioned above, subducted slabs are probably very different in $\delta^{18}O$ than the normal mantle. They are almost certainly characterized by a wide range of $\delta^{18}O$, from about +3‰ to values as high as +15‰. Magritz and Taylor (1976) estimate a mean value for basalts around +11 to +13‰. Therefore, the subduction process may introduce isotopic heterogeneities into the mantle. On the other hand, except for the small effects reported by Kyser and O'Neil (1978) for tholeiites vs. alkali basalts, the $\delta^{18}O$ of basalt sources seems to be quite uniform. Such uniformity may imply that the downgoing slab does not mix with mantle basalt sources, or that the subduction process cannot be extended far back in time, or that even if the basalt sources are not isotopically uniform, the basaltic magmas pass through a liquid exchange with a more homogeneous reservoir during their ascent to the surface, which thus buffers their isotopic compositions at uniform values. In themselves the oxygen isotopes cannot completely answer these questions, but they contribute to our knowledge of the processes of basalt genesis.

^{17}O anomalies or signatures. Isotopic compositions measured in mass spectrometers can be recorded in the form of ratios, for example $^{18}O/^{16}O$ and $^{17}O/^{16}O$ for the three oxygen isotopes. Such ratios normally vary together in nature (or during measurement in the mass spectrometer), since natural isotopic variations are usually the result of mass-fractionating processes. The variation of the $^{18}O/^{16}O$ ratio is twice the variation of $^{17}O/^{16}O$, because the mass difference is twice as large. In fact, if each ratio is measured and one of them chosen to *normalize* the fractionation, a constant value is obtained for the other and the normal variations thereby erased. This is exactly what is done in strontium isotope work when one normalizes the $^{87}Sr/^{86}Sr$ ratios to a constant $^{88}Sr/^{86}Sr = 8.3742$. Because of this usual covariance of the $^{18}O/^{16}O$ and $^{17}O/^{16}O$, it has not been the usual practice to measure both of these isotope ratios.

However, Clayton *et al.* (1973), in their study of oxygen in carbonaceous meteorites, showed that $^{17}O/^{16}O$ does *not* vary according to the 1/2 $\Delta(^{18}O/^{16}O)$ law but instead as if the oxygen in these samples were mixtures between a terrestrial-like component and pure ^{16}O. In further work with different types of planetary matter, from the Moon, Earth, and various classes of meteorites, Clayton and Mayeda (1975) and Clayton *et al.* (1976) found distinct $\delta^{17}O$ signatures. Samples having the same signature follow lines of slope 1/2 in a $\delta^{17}O$ vs. $\delta^{18}O$ "three isotope" diagram, but for each class of planetary material the lines are distinct, indicating different degrees of enrichment in ^{16}O. A very interesting point in this classification is the fact that lunar and terrestrial samples follow the same line and fall within roughly the same spread along this line. This relationship seems to indicate that the Moon and Earth are derived from common "nebular material" (or ancestral material) and have fractionated at approximately the same temperatures. $\delta^{18}O$ is ~5.5–5.7‰ for the Earth and ~5.6‰ for lunar samples (Clayton and Mayeda, 1975). Another interesting result, from Clayton *et al.* (1976), is the fact that basaltic achondrites are significantly lower in $\delta^{18}O$ than the Earth and Moon, and appear to have a slightly different $\delta^{17}O$ signature (see section 7.2.2).

Carbon, sulfur and nitrogen isotopes

Carbon, sulfur and nitrogen are trace elements in basalts, and in such a way that the form in which they exist (dissolved in major mineral phases, interstitial minerals, cracks, or surface coatings) is not exactly known. These elements also form important volatile compounds such as CO and CO_2 for carbon, H_2S, SO_2 and SO_3 for sulfur, and NH_3, N_2 and NO_3 for nitrogen. Isotopic fractionations between the solid and gaseous species are probably large, as has been shown for carbon (Bottinga, 1969a; Pineau *et al.*, 1973; Javoy *et al.*, 1978) and for sulphur (Sakai, 1957). The three elements differ from one another in that nitrogen may be almost inert in the mantle while both carbon and sulfur are chemically reactive and are present in different valency states. Such elements are good isotopic indicators of redox conditions as shown by Sakai (1968) and Ohmoto (1972) in their studies of ore deposits.

Carbon isotopes. Since the early work of Craig (1954), $\delta^{13}C$ in basalts has been a major problem. Basaltic $\delta^{13}C$ values range from −25 to −30‰ (relative to the PDB limestone standard), while carbonatites, diamonds and kimberlites give values around −5‰ (−3 to −8‰), as shown by Taylor *et al.* (1967), Deines and Gold (1973), and Pineau *et al.* (1973). Using the total carbon budget, Craig (1954) predicted that the deep carbon $\delta^{13}C$ should be between −7 and −12‰ but this indirect argument was based on limited data and was not completely convincing. However, as we will see, it was almost right.

A systematic approach to this problem has been carried out by Pineau *et al.* (1976a, b), Pineau (1977), and Javoy *et al.* (1978), in studies of both basalts and peridotitic material. Using step-wise heating and differential dissolution, these authors showed that basalts may contain a CO_2-rich phase with a $\delta^{13}C \cong -7‰$ very close to carbonatites and most diamonds, while residues after outgassing had $\delta^{13}C \approx -25‰$. Such experiments are possible with oceanic pillow basalts that have been only partially outgassed, but impossible with subaerial flows that are almost completely outgassed (Moore, 1965). Using a Rayleigh distillation model, Pineau *et al.* (1976) calculated the ΔCO_2-magma fractionation factors required to explain their results, and obtained values in the range $\alpha \cong 1.003$–1.005. Javoy *et al.* (1978) measured these fractionation factors in high temperature–high pressure experiments and found $\alpha = 1.0040$–1.0046 at 1180–1280°C and 1kb. Thus the laboratory experiments agree with the estimates from natural rocks. These data indicate that the phenomenon responsible for the $\delta^{13}C$ values observed in basalt is probably the following: when basaltic material starts to rise to the surface as a magma or a liquid-crystal mush, its initial composition is $^{13}C \simeq -7‰$; however, it is gradually outgassed and by CO_2-magma fractionation gradually attains a final $\delta^{13}C$ of about $-25‰$. Because these very low $\delta^{13}C$ values are typical of basalts, such a phenomenon would have to be very general. This implies that (1) a significant amount of CO_2 or carbonate must exist in the mantle, as also shown by Roedder (1965) from analysis of fluid inclusions and confirmed by experimental petrology (Eggler, 1973, 1976; Mysen *et al.*, 1975, 1976; and Wyllie and Huang, 1975, 1976); and that (2) such a mantle outgassing process, if it is going on continuously, may have consequences not only for the evolution of the atmosphere but also for other elements that may be carried in the high-CO_2 fluid. Those that form complexes with carbonates may be very important.

Studies of peridotitic materials of direct mantle origin, such as xenoliths in Alpine peridotites, show results in complete agreement with the above concept. Some of these have $\delta^{13}C$ values about $-7‰$ and are really primitive. Others have a carbonate component and a residual light component which can be separated by acid treatment or step-wise heating. Such work is the first in this direction but it shows great promise for the application of carbon isotopes in igneous petrology. For example, it raises the possibility that the contamination process could also be studied with $^{13}C/^{12}C$, in conjunction with $^{18}O/^{16}O$ and $^{87}Sr/^{86}Sr$.

Sulfur isotopes. In the early days of sulfur isotope geochemistry, Thode (1963) and his co-workers showed that sulfides associated with deep seated rocks such as gabbro-diabase-peridotites have a quite uniform $\delta^{34}S$. Very little further work has been done on the basalt problem. Sakai (1968) showed that an important redox process occurs when volcanic lavas reach the surface, and thus that it is extremely difficult to study the pristine $\delta^{34}S$ of such basalts. Recent studies of MORB by the Deep Sea Drilling Project have also shown the extreme complexity of the problem. The sulfur is present in very different forms, both reduced and oxidized. Some secondary pyrite is also found, which "interferes" in the analysis. Therefore, extremely careful treatment is required in handling these samples. The main conclusions of these studies are that when the samples are fresh and unaltered, $\delta^{34}S$ is between -0.5 and $+0.9‰$ (relative to the defined standard—troilite from the Cañon Diablo iron meteorite). Sulfur contents are quite variable and a significant fraction is in the sulfate form. These preliminary data seem to show that sulfur isotopes can be used to deduce useful information about outgassing processes and redox states in volcanic rocks. At the present stage of understanding the isotopic composition of sulfur seems to be uniform in the mantle source, and the observed results stress the importance of outgassing during transport to the surface.

Nitrogen. As for sulfur, the nitrogen isotope geochemistry of basalts is at a very primitive stage. Becker and Clayton (1978) found evidence in igneous rocks for an apparently uniform isotopic composition for nitrogen in the Earth's mantle, with $\delta^{15}N$ 17 ‰ greater than standard atmospheric nitrogen. This result, if it is confirmed, sets limitations on the nitrogen budget of the Earth and on mantle outgassing through time. For example, as speculated by Becker and Clayton (1978), a large fraction of the nitrogen can still be inside the Earth, indicating that early outgassing was less intense than previously assumed. This kind of conclusion is also in agreement with the ^3He discoveries of Craig and Lupton (1976) and Tolstikhin *et al.* (1969), and with the $\delta^{13}C$ interpretations mentioned above.

Conclusion

The stable isotopes are an extremely important tool in studying the igneous petrology of basalts. Their use, which is still in a quite primitive stage, has already demonstrated several important results:

(a) The Taylor Effect. The probability of any igneous rocks coming to the surface without at least some interaction with water is extremely small. Several types of direct or indirect interaction are possible. The oxygen isotopes are an easy and sensitive way to detect the effect of these interactions. A quite straight-forward application of this conclusion is that an ^{18}O study should be a first step in any examination of basalts where it is important to know something about the characteristics of the pristine magma.

(b) Crustal contamination, either direct or by fluid interaction, is probably an extremely general process. Such an observation reinforces the preceding point.

(c) Gas-magma processes are very important in the Earth, and the mantle may be less outgassed than we have previously thought. The role of CO_2 in the mantle is important, as demonstrated by the carbon isotope results.

7.2 CHRONOLOGIC AND ISOTOPIC STUDIES ON BASALTIC METEORITES

7.2.1 Introduction

Aside from the Earth and the Moon, the only other samples of basalt from the solar system currently available for direct analysis are those present as some of the achondritic meteorites. These meteorites are important in the context of this study because they represent the oldest known basaltic rocks in the solar system. As such, they help to define a number of the initial geochemical parameters of the solar system, particularly with regard to chronologies and isotopic systems.

Most of the known achondritic meteorites have basaltic character or could have been derived from basaltic systems; the principal exceptions are the enstatite achondrites. The principal achondrites with truly basaltic character, both mineralogically and texturally, are the *eucrites* (Duke and Silver, 1967; Stolper, 1977; section 1.2.8). Eucritic fragments also occur in the *howardites* and the silicate portions of *mesosiderites*; these, along with the apparently genetically related *diogenites*, constitute the group generally referred to as "basaltic achondrites" in recent petrologic and chemical studies (Stolper *et al.*, 1979, and sections 1.2.8 and 3.5). Eucrites consist principally of calcic plagioclase and calcium-poor pyroxene with lesser amounts of silica and minor amounts of metal. The howardites are similar to the eucrites, but although both are generally brecciated, the former are polymict breccias whereas the eucrites are monomict breccias and, in some cases (Ibitira; Wilkening and Anders, 1975), are unbrecciated. Because of their less complex structures, petrologies, and histories, the eucrites have been geochronologically studied in far more detail than most other kinds of basaltic achondrites and are the source for most of the data reviewed here.

Three smaller groups of achondrites with basaltic character are the shergottites, the nakhlites, and the chassignites, which may (or may not) come from a separate parent body than the eucrites and related meteorites (Stolper *et al.*, 1979). Achondrites that could have been derived from basaltic systems such as residual rocks, cumulates, etc., include Angra dos Reis (e.g., Prinz *et al.*, 1977; Ma *et al.*, 1977) and the ureilites for which few data are available. The genetic affinity of these meteorites is still in doubt.

This section is devoted to a review of the isotopic data that bear on the ages of basaltic achondrites, primarily the eucrites, and interpretation of their very early histories. The data will also be used to address the question of the general nature of the parent bodies of the basaltic achondrites and the processes by which the respective magmas formed. In this regard, the recent studies by Stolper (1977) and Consolmagno and Drake (1977) are particularly pertinent. They have concluded that eucrites are possibly the result of magma generation in a single parent body, and that the asteroid 4 Vesta is a likely candidate for that body. The detailed mineralogy and petrology of the basaltic achondrites are reviewed in sections 1.2.8 and 3.5.

7.2.2 Oxygen isotope results

Initial oxygen isotope studies on basaltic achondrites and lunar basalts (Taylor *et al.*, 1965; Taylor and Epstein, 1970) have shown that basaltic achondrites are isotopically lighter than their terrestrial and lunar lithologic counterparts by about 2 ‰. These results (Fig. 7.2.1) indicate that the basaltic achondrites probably formed from an oxygen isotope reservoir that was distinct from that of the Earth or Moon or chondritic metorites. Furthermore, these results would preclude chondritic meteorites, particularly the ordinary chondrites, from being the immediate precursors of the magmas which yielded the basaltic achondrites.

The recent discovery of correlated isotopic anomalies in ^{17}O and ^{18}O, unrelated to fractionation, and their application to meteorites (Clayton *et al.*, 1973, 1976) has provided further insight into the possible parentage of basaltic meteorites, while generally confirming the conclusions based on ^{18}O alone. On a three-isotope plot (Fig. 7.2.2) both lunar and terrestrial basalts lie along the same portion of the "terrestrial mass-fractionation line" (Clayton and Mayeda, 1975).

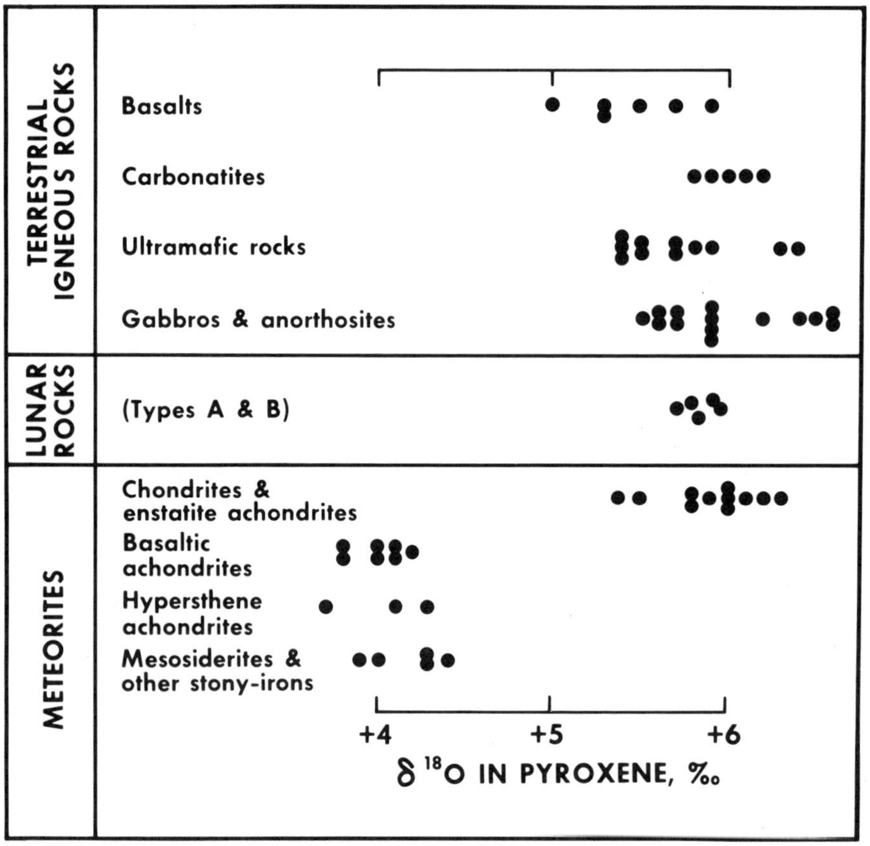

7.2.1 Comparison of $\delta^{18}O$ values of pyroxenes from the Apollo 11 lunar rocks with those from terrestrial mafic and ultramafic rocks and from meteorites (from Taylor and Epstein, 1970). More recent analyses on other lunar and meteoritic samples fall in these same groupings.

That is, the differences in oxygen isotope composition for these materials can be explained by normal isotopic fractionation processes related to oxygen isotope reservoirs of the same (or very similar) bulk compositions. The eucritic basaltic achondrites have oxygen compositions that lie near the terrestrial line (in contrast to ordinary and carbonaceous chondrites), but they are displaced along it toward ^{16}O-richer compositions (Fig. 7.2.2) and are displaced slightly below this line (Clayton, *et al.*, 1976). Thus, it appears that ordinary chondrite, eucrite, and terrestrial matter must have come from separate parts of the solar nebula or have had differing accretionary histories. Terrestrial and lunar matter appear to be closely related, but derivation of the isotopically lighter eucritic parent material apparently requires a more complex relationship. It is clear, however, that eucrites, diogenites, howardites, and mesosiderites can all be part of a single genetic system, consistent with petrologic and chemical data (sections 1.2.8 and 3.5).

Of the other achondrites with basaltic character or affinity, Angra dos Reis falls in the fractionation trend for the main group of basaltic achondrites, indicating possible genetic association. The nakhlites and shergottites, on the other hand, fall above the terrestrial frac-

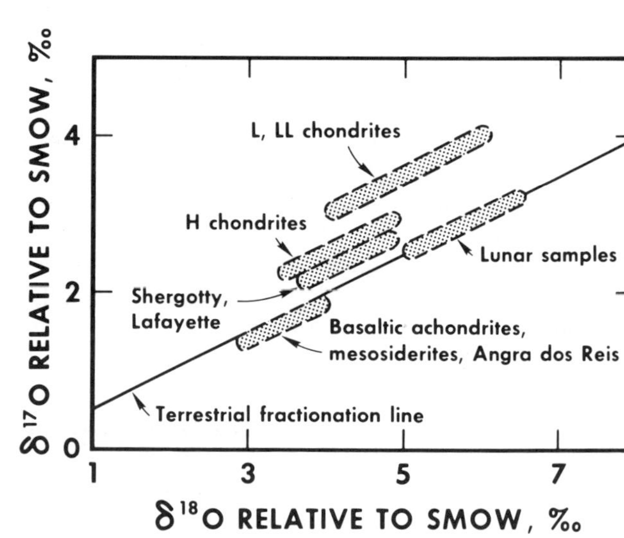

7.2.2 Three-isotope plot (essentially $^{17}O/^{16}O$ vs. $^{18}O/^{16}O$) for total rock and pyroxene samples of various meteorite types and lunar rocks and soils. H = H-Group chondrites; L, LL are L and LL-Group chondrites. Enstatite chondrites and enstatite achondrites fall within the field defined by lunar rocks and are not shown. After Clayton *et al.* (1976); difference in absolute $\delta^{18}O$ values compared to Fig. 7.2.1 is apparently due to interlaboratory bias or differences in sample preparation for the two studies.

tionation line and also are slightly isotopically heavier than the eucritic basaltic achondrites, indicating either separate genetic systems (parent bodies) or at least complex, inhomogeneous relationships in a single body (Stolper *et al.*, 1979).

7.2.3 Potential chronologies

The potential number of events pertaining to the early history of the solar system and preserved in the chronological record of meteorites is more varied than that for terrestrial or lunar samples. Recent studies have shown that most of these events have also been recorded in basaltic achondrites (Table 7.2.1). From the standpoint of early solar system basalt genesis, times t_2 through t_4 (parent body formation, differentiation, and magma generation) are of the most interest and record various stages in the evolution of a parent body. These stages may involve chemical fractionations of the parent-daughter systems, and with adequate data it is possible to put limits on the early geochemical and petrologic history of a parent body.

For isotopic systems in basaltic achondrites the four main evolutionary stages are (a) evolution in the nebula between t_0 and t_2, including the intermediate processes of condensation of solid material (t_1) and accretion of it into millimeter-sized and larger masses; (b) evolution in the parent body prior to planetary differentiation, between t_2 and t_3; (c) evolution within differentiated layers (if they existed; e.g., mantle, crust, core) of the parent body prior to magma generation, between t_3 and t_4; and (d) evolution as a rock subsequent to crystallization at t_4. In this sequence the evolution of the geochronologic systems between magma generation and crystallization as a rock is considered negligible, although a substantial portion of the chemical fractionations may take place at this time (partial melting, fractional crystallization).

Perturbations since t_4 due to cooling or later events, including remelting, shock metamorphism, and

Table 7.2.1 Chronologic events of basaltic achondrites.

Time	Event
t_0	Isolation of solar nebula and end of nucleosynthesis
t_1	Onset of condensation (Allende inclusions)
t_2	Formation of parent body (accretion)
t_3	Differentiation of parent body
t_4	Magma generation and crystallization
t_5	End of last metamorphism (if applicable)
t_6	Break-up or ejection from parent body

heterogeneous mixing, are separate problems that must be evaluated for many achondrites before their earlier history can be deduced. However, in some cases there may not be a recognizable metamorphic event, and it is from such meteorites that the best data regarding the earlier history can be obtained.

7.2.4 Review of geochronologic data

Eucrites

Absolute ages of primary crystallization, t_4, for eucrites can be obtained from a variety of geochronologic systems, namely Rb-Sr, K-Ar, Sm-Nd, U-Pb, Th-Pb and Pb-Pb. The results from K-Ar analyses will be summarized in the following discussion. The remaining ones can be divided into two categories: (a) those (U-Pb, Th-Pb, Sm-Nd, and Rb-Sr) that are dependent upon measurement of the parent/daughter ratio (Tables 7.2.2–6) and (b) that (^{207}Pb-^{206}Pb) which is only dependent upon measuring isotope ratios (Tables 7.2.7–8).

Parent/daughter crystallization ages. Rb/Sr ratios are so low for eucrites that it is not possible to calculate ages for single total-rock samples. However, with very precise measurements of Sr isotopic composition, along with ultra-clean procedures, it is possible to obtain data of sufficient quality (Tables 7.2.2–3) to determine a total-rock isochron for a suite of eucrites. This was first done by Papanastassiou and Wasserburg (1969), who reported an age of 4.30 ± 0.26 b.y. with an initial $^{87}Sr/^{86}Sr$ (I) of 0.69898 ± 0.00006. Birck and Allègre (1978a) have reported additional measurements on eucrites and separated mineral phases and obtained an isochron age of 4.47± 0.13 b.y. with I = 0.69899 ± 0.00004. Birck and Allègre (1978b) have also reported similar results for several diogenites.

Internal Rb-Sr isochrons (e.g., Fig. 7.2.3) have been determined by Allègre *et al.* (1975) and Birck and Allègre (1978a) for a number of eucrites. Papanstassiou *et al.* (1974) and Papanastassiou and Wasserburg (1976b) have reported internal isochrons for clasts from the howardite Kapoeta, and Murthy *et al.* (1977) have reported results from the mesosiderite Estherville. The results from these studies (Table 7.2.3) show that many basaltic achondrites record "young" ages relative to the total-rock isochron age of ~4.4 b.y. This poses a major problem in the interpretation of both the total-rock and the internal isochron ages. Which one(s) represents the primary crystallization age(s)?

It can be argued that the total-rock isochron for eucrites represents their primary crystallization age and that younger internal isochron ages are due to later

Table 7.2.2 Total-rock Rb-Sr analyses on basaltic achondrites.

Meteorite†	Rb (ppm)	Sr (ppm)	$^{87}Rb/^{86}Sr$ (× 100)	$^{87}Sr/^{86}Sr$	Reference
A. Eucrites					
Bereba (Be)	0.238	73.1	0.95	0.69960 ± 10	*(1)*
Ibitira (Ib)	0.137	85.4	0.463	0.69930 ± 7	*(1)*
Jonzac (Jo)	0.444	76.7	1.671	0.70006 ± 6	*(2)*
	0.459	73.1	1.913	0.70016 ± 8	*(1)*
Juvinas (Ju)	0.173	77.6	0.644	0.69938 ± 3	*(2)*
	0.111	79.1	0.407	0.69927 ± 3	*(3)*
Macibini (Ma)	0.283	75.8	1.08	0.69961 ± 8	*(1)*
Moore County (MC)	0.052	64.7	0.234	0.69908 ± 2	*(4)*
Nuevo Laredo (NL)	0.325	76.4	1.228	0.69984 ± 5	*(2)*
	0.318	77.5	1.186	0.69975 ± 9	*(1)*
Pasamonte (Pa)	0.207	77.9	0.769	0.69948 ± 11	*(2)*
	0.251	74.6	0.948	0.69962 ± 7	*(5)*
	0.331	70.6	1.355	0.69996 ± 18	*(1)*
	0.222	79.5	0.808	0.69966 ± 5	*(1)*
	0.342	70.5	1.404	0.69977 ± 5	*(6)*
	0.329	69.1	1.376	0.69972 ± 3	*(6)*
Serra de Magé (SdM)	0.072	109.2	0.191	0.69903 ± 7	*(1)*
Sioux County (SC)	0.195	72.5	0.775	0.69949 ± 4	*(2)*
	0.253	73.7	0.992	0.69955 ± 10	*(1)*
Stannern (St)	0.707	85.3	2.396	0.70048 ± 4	*(2)*
	0.776	103.7	2.16	0.70035 ± 7	*(1)*
B. Other					
Angra dos Reis *	0.056	142.6	0.112	0.69891 ± 5	*(7)*

†Abbreviations in parentheses refer to Fig. 7.2.4
*Average of four analyses
References:
(1) Birck and Allègre (1978a); *(2)* Papanastassiou and Wasserburg (1969); *(3)* Allègre *et al.* (1975); *(4)* Papanastassiou and Wasserburg (1972b); *(5)* Wetherill *et al.* (1973); *(6)* Unruh *et al.* (1977); *(7)* Wasserburg *et al.* (1977).

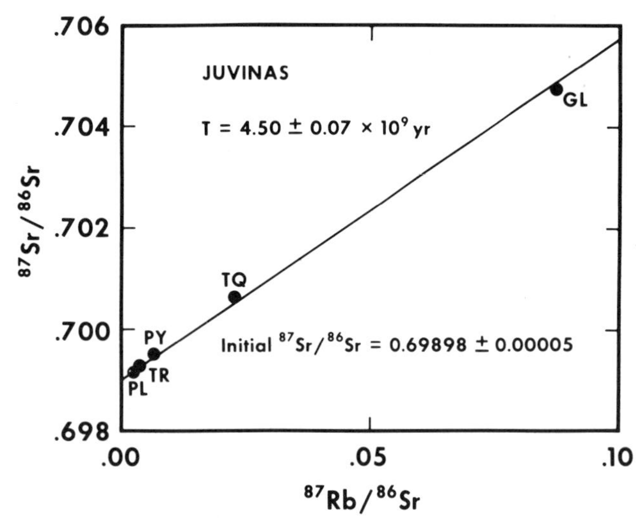

7.2.3 Rb-Sr internal isochron for the eucrite Juvinas (Allègre *et al.*, 1975). Age recalculated to $\lambda(^{87}Rb) = 1.42 \times 10^{-11}\ yr^{-1}$. PL, TR, PY, TQ and GL correspond to plagioclase, total rock, pyroxene, tridymite, and glass (quintessence), respectively.

disturbances. This assumes the basic set of assumptions for the total-rock isochrons, namely that (a) all the eucrites crystallized originally at the same time, (b) they all formed from magmas with the same I-value (or the same magma) and (c) they all have remained closed systems on the scale of the total-rock samples (a few grams or less). As discussed by Papanastassiou and Wasserburg (1969) and Birck and Allègre (1978a), the data are consistent with these assumptions.

On the other hand, it could be argued that the internal isochrons actually reflect the primary crystallization ages of the individual meteorites concerned, and that the total-rock Rb-Sr isochron age represents a parent-body "mantle isochron." That is, the individual magmas formed either from various parts of a heterogeneous, 4.4-b.y.-old mantle without significant Rb/Sr or Sr isotopic fractionation (Brooks *et al.*, 1976a, b) or by partial melting of Rb-rich phases in the source region without equilibration between them and the unfused residue (Albee and Gancarz, 1977). Thus, the Rb-Sr data alone are only sufficient to conclude that some of the eucrites (notably Juvinas and Ibitira) plus some of the other basaltic achondrites have primary crystallization ages as old as 4.4 b.y. (Table 7.2.3). For resolution of this uncertainty it is necessary to examine results from other chronologic systems.

A major weakness of the Rb-Sr internal isochron ages is that they are heavily dependent upon a single Rb-rich phase, "quintessence" (LIL-rich mesostasis), that by its very nature is readily susceptible to open-system behavior during a mild disturbance. Furthermore, since most eucrites are breccias, the possibility exists that mineral isochrons for them may be mixing lines defined by minerals from rocks of various ages and have no true age significance. (Ibitira is exempt from this uncertainty, since it is not brecciated.) However, the recently developed Sm-Nd technique (Lugmair, 1974) relies on measurements from the major mineral phases plagioclase and pyroxene, and it thus takes a major disturbance to reset this system. Recent Sm-Nd results on basaltic achondrites (Table 7.2.4), including the eucrites Pasamonte and Stannern, which show strong Rb-Sr discordance, are all in the range of 4.4–4.6 b.y. Because these ages are based on minerals they are also not subject to the "mantle isochron" uncertainty of the

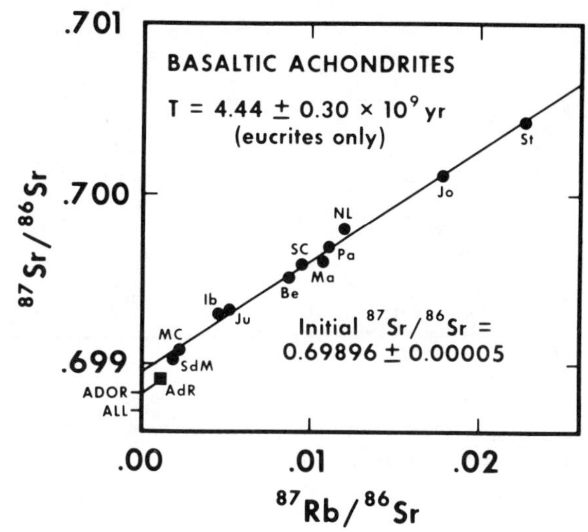

7.2.4 Averaged Rb-Sr data for total rock samples of eucrites (Table 7.2.2). Age is calculated using least-squares method of York (1966) and $\lambda(^{87}Rb) = 1.42 \times 10^{-11}$ yr^{-1}. Uncertainty quoted represents 95% confidence limits. This result is indistinguishable from the original result of Papanastassiou and Wasserburg (1969) and the recent result of Birck and Allègre (1978a).

Table 7.2.3 Rb-Sr isochron ages of basaltic achondrites and other materials.

System	Age†	$(^{87}Sr/^{86}Sr)_I$	Reference
I. Total rock isochrons			
A. Eucrites	4.30 ± 0.26	0.69898 ± 6	*(1)*
	4.47 ± 0.13	0.69899 ± 4	*(2)*
	4.44 ± 0.30	0.69896 ± 5	*(3)*
B. Chondrites	4.44 ± 0.04	0.7003 ± 4	*(4)*
	4.53 ± 0.05	0.6988 ± 4	*(5)*
II. Internal isochrons			
A. Eucrites			
Bereba	4.08 ± 0.26	0.69898 ± 7	*(3)*
Ibitira	4.42 ± 0.25	0.69895 ± 10	*(3)*
Juvinas	4.50 ± 0.07	0.69898 ± 5	*(6)*
Pasamonte	disturbed; no isochron		*(3),(7)*
Sioux County	4.10 ± 0.14	0.69897 ± 8	*(3)*
Stannern	3.2 ± 0.5	0.6993 ± 2	*(3)*
B. Howardites			
Kapoeta clast A	3.81 ± 0.05	0.69888 ± 5	*(8)*
Kapoeta clast B	3.55 ± 0.08	0.69905 ± 6	*(8)*
Kapoeta clast C	4.44 ± 0.12	0.69885 ± 4	*(9)*
C. Mesosiderites			
Estherville	4.44 ± 0.26	0.69903 ± 3	*(10)*
III. Initial $^{87}Sr/^{86}Sr$ values for selected systems			
A. Angra dos Reis *(ADOR)*		0.69883 ± 2	*(11)*
B. Allende inclusions (ALL)		0.69876 ± 2	*(12)*
C. Moon (LUNI)		0.69894 ± 3	*(9)*

†Ages in b.y.: $\lambda(^{87}Rb) = 1.42 \times 10^{-11}$ yr^{-1}.

References:
(1) Papanastassiou and Wasserburg (1969); *(2)* Birck and Allègre (1978a); *(3)* This compilation (Fig. 7.2.4); *(4)* Wetherill (1975c); *(5)* Minster *et al.* (1976); *(6)* Allègre *et al.* (1975); *(7)* Unruh *et al.* (1977); *(8)* Papanastassiou *et al.* (1974); *(9)* Papanastassiou and Wasserburg (1976b); *(10)* Murthy *et al.* (1977); *(11)* Wasserburg *et al.* (1977); *(12)* Gray *et al.* (1973).

Table 7.2.4 Sm-Nd ages of basaltic achondrites.

Meteorite	Age†	$(^{143}Nd/^{144}Nd)_I$	Reference
A. Eucrites			
Juvinas	4.56 ± 0.08	0.50677 ± 10	*(1), (2)*
Moore County	4.60 ± 0.04	0.50676 ± 7	*(3)*
Pasamonte	4.58 ± 0.12	0.50681 ± 14	*(4)*
Serra de Magé	4.41 ± 0.02	n.r.	*(5)*
Stannern	4.48 ± 0.07	n.r.*	*(6)*
B. Other			
Angra dos Reis	4.55 ± 0.04	0.50682 ± 5	*(7)*

†Ages in b.y.: $\lambda(^{147}Sm) = 6.54 \times 10^{-12}$ yr^{-1}.
*System slightly disturbed.
References:
(1) Lugmair (1974); *(2)* Lugmair *et al.* (1976); *(3)* Nakamura *et al.* (1977a); *(4)* Unruh *et al.* (1977); *(5)* Lugmair *et al.* (1977); *(6)* Lugmair and Scheinin (1975); *(7)* Lugmair and Marti (1977).

Rb-Sr total-rock system and must represent crystallization ages. The concordance of the Sm-Nd results with the Rb-Sr total-rock results indicates that (a) the Rb-Sr isochron age probably represents the age of primary crystallization, (b) mixing of minerals from several clasts in the eucrite breccias generally will not produce significantly younger ages than obtained from unbrecciated samples or single clasts, and (c) young Rb-Sr internal isochron ages are probably due to later disturbances (thermal or shock metamorphism).

It could be argued that because some of the eucrites are "disturbed," they should not be included when determining a Rb-Sr total-rock isochron for them. A plot of averaged results (Table 7.2.2) on eucrites (Fig. 7.2.4), however, shows that they all are collinear within allowable analytical uncertainty (≤1 part in 10^4 for $^{87}Sr/^{86}Sr$). A least-squares fit (York, 1966) to the data yields an age of 4.44 ± 0.30 b.y. with I = 0.69896 ± 0.00005 at the 95% confidence level, consistent with the results of Papanastassiou and Wasserburg (1969) and Birck and Allègre (1978a).

U-Pb, Th-Pb, and K-Ar results on basaltic achondrites are generally similar to Rb-Sr and Sm-Nd results. Some are concordant at ~4.5 b.y., while others show evidence of significant disturbance of the system (Tables 7.2.5–6). By and large, though, they do not add much to the general picture presented above and will not be reviewed in detail further.

In summary, then, Rb-Sr, Nd-Sm, K-Ar, U-Pb and Th-Pb measurements on basaltic achondrites indicate crystallization ages of 4.4–4.5 b.y. These are substantially older than all terrestrial rocks and most lunar rocks. Furthermore, these ages are not distinguishable from primary Rb-Sr ages on more primitive meteorites, namely chondrites (Table 7.2.3). Thus, eucritic basalts and related rocks date back to the earliest stages of planetary history.

^{207}Pb-^{206}Pb crystallization ages. Because Pb-Pb ages are primarily dependent on the measurement of isotope ratios, they are potentially much more precise than the parent/daughter systems discussed above. This is par-

Table 7.2.5 K-Ar and ^{39}Ar-^{40}Ar ages of basaltic achondrites.†

Meteorite	K-Ar Age	^{39}Ar-^{40}Ar Age	Reference
Juvinas	3.9	-	*(1)*
Kapoeta	4.32	-	*(1)*
Kapoeta (Clasts)	-	4.47, 3.4, 4.5	*(2)*
Macibini	3.5	-	*(1)*
Moore County	3.2	-	*(1)*
Nuevo Laredo	3.0–3.5	-	*(1)*
Pasamonte	3.7, 4.17, 4.07	4.02	*(1), (3)*
Petersburg	2.5, 3.95	4.32	*(1), (3)*
Sioux County	3.2	-	*(1)*
Stannern	2.9, 3.6, 3.7, 3.6	(3.4–3.7)	*(1), (3)*

†Ages in b.y.: $\lambda_e = 5.81 \times 10^{-11}$ yr^{-1}; $\lambda_\beta = 4.962 \times 10^{-10}$ yr^{-1}.
References:
(1) Anders (1963) (compilation); *(2)* Rajan *et al.* (1979); *(3)* Podosek and Huneke (1973).

Table 7.2.6 U, Th-Pb ages of basaltic achondrites.†

Meteorite	$^{206}Pb/^{238}U$	$^{207}Pb/^{235}U$	$^{208}Pb/^{232}Th$	Reference
A. Eucrites and howardites				
Bereba	4.08	4.31	3.93	*(1)*
Ibitira	4.68	4.59	4.71	*(1)*
Juvinas (breccia)	4.65	4.58	4.14	*(1)*
Juvinas (xenolith)	5.40	4.81	5.85	*(1)*
Kapoeta	4.77	4.60	5.35	*(1)*
Pasamonte	4.87 ± 0.04	4.66 ± 0.02	4.68 ± 0.09	*(2)*
Stannern	4.40	4.35	4.20	*(1)*
B. Other				
Angra dos Reis	4.63 ± 0.07		4.53 ± 0.10	*(3)*

†Ages in b.y.: $\lambda(^{238}U) = 1.551 \times 10^{-10}\ yr^{-1}$; $\lambda(^{235}U) = 9.849 \times 10^{-10}\ yr^{-1}$; $\lambda(^{232}Th) = 4.948 \times 10^{-11}\ yr^{-1}$.
References:
(1) Mahnes *et al.* (1975); *(2)* Unruh *et al.* (1977); *(3)* Wasserburg *et al.* (1977).

ticularly true for eucrites, which have relatively radiogenic Pb. Patterson (1956) used Pb-Pb data for the eucrite Nuevo Laredo to help define the age of meteorites in his pioneering work, obtaining an age of 4.49 ± 0.07 b.y. (recalculated to currently used decay constants). This was confirmed by Silver and Duke (1971) who reported an age of 4.52 ± 0.02 b.y. for the eucrites Nuevo Laredo, Juvinas, Pasamonte and Sioux County. Since then a large number of Pb-Pb measurements have been made on eucrites, as well as on other meteorite systems (Tables 7.2.7–8).

When all ^{207}Pb-^{206}Pb results from Table 7.2.7 are plotted on a $^{207}Pb/^{204}Pb$ vs. $^{206}Pb/^{204}Pb$ diagram (Fig. 7.2.5), it is apparent that they define an array that is collinear to within very small tolerances. A least-squares fit (York, 1966) to the eucrite and howardite data (omitting Bereba and Stannern and using the average result for Ibitira), plus lead from Cañon Diablo troilite, yields a slope of 0.6146 ± 0.0033 (1σ). Omitting the Cañon Diablo lead, adding Angra dos Reis or Norton County, and/or using individual analyses for Ibitira does not change the result appreciably. With all

Table 7.2.7 Total-rock Pb-Pb data on basaltic achondrites.

Meteorites†	$\frac{206}{204}$	$\frac{207}{204}$	$\left(\frac{207}{206}\right)^{*}_{R}$	Model Age	Reference
A. Eucrites and howardites					
Bereba (Be)	254.0	148.0	0.5629	4.415	*(1)*
Ibitira	148.9	96.8	0.6195	4.554	*(1)*
Ibitira	140.0	91.1	0.6184	4.552 ± 0.01	*(2)*
Ibitira (Ib) (Average)	144.5	94.0	0.6190	4.553	
Juvinas (breccia) (Ju)	37.9	27.9	0.6147	4.543	*(1)*
Juvinas (clast) (Ju)	43.1	31.3	0.6207	4.557	*(1)*
Kapoeta (Ka)	99.9	65.2	0.6066	4.524	*(1)*
Nuevo Laredo (NL)	222.4	140.1	0.6089	4.529 ± 0.005	*(3)*
Nuevo Laredo (NL)	50.3	34.9	0.5997	4.508	*(4)*
Pasamonte (Pa)	138	91.0	0.6273	4.573 ± 0.011	*(5)*
Sioux County (SC)	195.3	123.3	0.6074	4.526 ± 0.010	*(3)*
Stannern (St)	286.0	157.1	0.5307	4.329	*(1)*
B. Other achondrites					
Angra dos Reis (AdR)	214.6	137.5	0.6197	4.555 ± 0.005	*(3)*
Angra dos Reis (AdR)	495.3	385.0	0.6164	4.546 ± 0.001	*(2)*
Norton County (NC)	47.9	34.1	0.6182	4.552	*(6)*

†Abbreviations in parentheses refer to Fig. 7.2.5.
*R = radiogenic Pb, corrected for primordial Pb: $\frac{206}{204} = 9.307$; $\frac{207}{204} = 10.291$ (Tatsumoto *et al.*, 1973; Tilton, 1973).

References:
(1) Mahnes *et al.* (1975); *(2)* Wasserburg *et al.* (1977); *(3)* Tatsumoto *et al.* (1973); *(4)* Patterson (1956); *(5)* Unruh *et al.* (1977); *(6)* Huey and Kohman (1973).

Table 7.2.8 Pb-Pb isochron ages of basaltic achondrites and other materials.

System	Slope	Age†	Reference
I. Total-rock suite			
A. Achondrites	n. r.	4.52 ± 0.02	(1)
	0.6146 ± 0.0033	4.54 ± 0.02	(2)
B. Chondrites	0.5991 ± 0.0032	4.505 ± 0.008	(3)
	0.6225	4.562	(4)
II. Internal isochrons			
A. Achondrites			
Juvinas	n. r.	4.556 ± 0.012	(5)
Pasamonte	0.610 ± 0.009	4.53 ± 0.03	(6)
Angra dos Reis	n. r.	4.544 ± 0.003	(7)
B. Other			
Allende	0.6188 ± 0.0016	4.553 ± 0.004	(8)
	0.6240 ± 0.0015	4.565 ± 0.004	(9)

†Ages in b.y.
References:
(1) Silver and Duke (1971); *(2)* This work; *(3)* Huey and Kohman (1973); *(4)* Tilton (1973); *(5)* Tatsumoto and Unruh (1975); *(6)* Unruh *et al.* (1977); *(7)* Wasserburg *et al.* (1977); *(8)* Tatsumoto *et al.* (1976); *(9)* Chen and Tilton (1976).

variations considered, an age of 4.540 ± 0.020 b.y. (2σ:95% confidence limit) is the most reasonable estimate for the absolute age of eucritic basaltic achondrites based on Pb-Pb results to date.

This is consistent with, but more precise than, the results obtained using the parent/daughter systems discussed above. In addition, internal isochrons have been reported for three achondrites and yield the same ages, within limits of uncertainty, as that obtained from the total-rock isochron (Table 7.2.8). Thus, there can be little doubt that the age of 4.54 ± 0.02 b.y. represents the time of crystallization of the eucritic basaltic achondrites. The age uncertainty of ± 20 m.y. represents the uncertainty in the mean age, but it also encompasses the range of ^{207}Pb-^{206}Pb model ages obtained by correcting individual Pb compositions for nonradiogenic Pb using Cañon Diablo troilite or Mezo-Madaras total-rock Pb (Tilton, 1973; Tatsumoto *et al.*, 1973), with the possible exception of Pasamonte (Unruh *et al.*, 1977). Therefore, it must also be concluded that the time span of basalt genesis represented by most basaltic achondrites represents a brief interval in the very early history of the parent body.

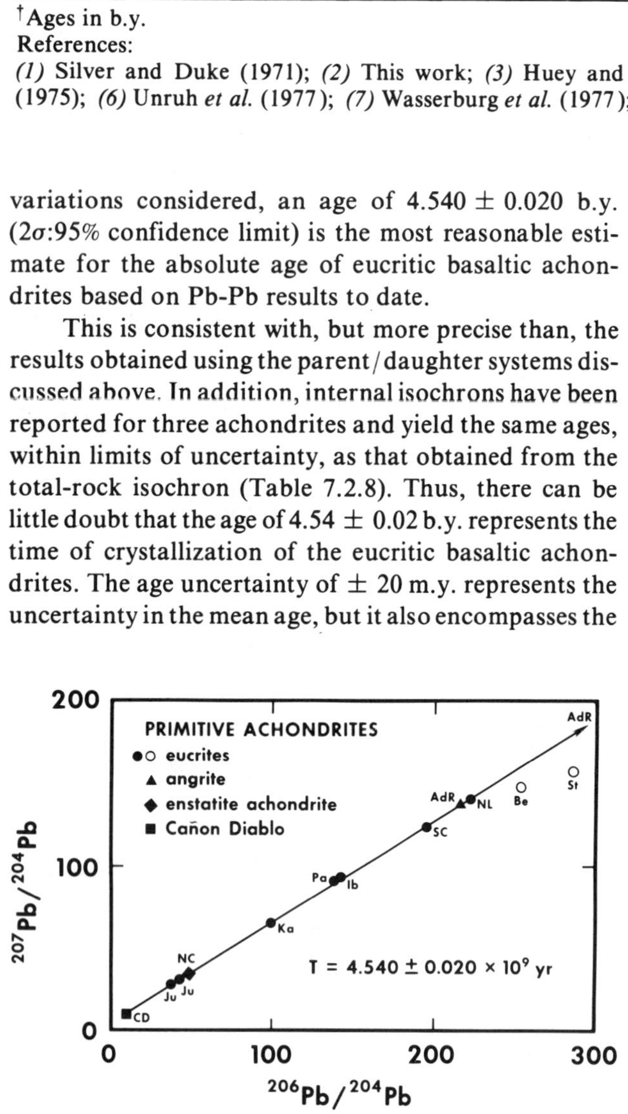

7.2.5 Total rock Pb data for basaltic achondrites (Table 7.2.7). Age quoted is estimated at 95% confidence limits from various subsets of Pb-Pb regressions that can be used (see text).

Premagmatic history of the eucrite parent body. It is not possible to uniquely define the premagmatic ($t<t_4$; Fig. 7.2.6) history of the eucrite parent body with the data currently available. Reasonable models based on evolution of Sr, Pb, and Xe isotopes have been discussed by a number of authors (e.g., Papanstassiou and Wasserburg, 1969; Gray *et al.*, 1973; Wetherill *et al.*, 1973; Wetherill, 1975; Birck and Allègré, 1978a) with the same basic conclusion: only a few tens of millions of years at most can have elapsed between initial condensation of nebular material and crystallization of eucrites.

The oldest solid-forming event in the solar system, t_1, is generally assumed to correspond to the formation of high-temperature condensates such as those present in the Allende chondrite. ^{207}Pb-^{206}Pb analyses of these materials indicate an absolute age of 4.56 ± 0.01 b.y. (Table 7.2.8) for t_1. The time, t_0, of isolation of chondritic parent material from the solar nebula can be estimated from extinct radionuclides. General interpretation of ^{129}I-^{129}Xe and ^{244}Pu-Xe results has suggested that a t_1-t_0 interval of 125–150 m.y. is the most reasonable conclusion (Wetherill, 1975c; Reynolds, 1977), but the interpretation of I-Xe and Pu-Xe results has been complicated by the discovery of extinct ^{26}Al in the

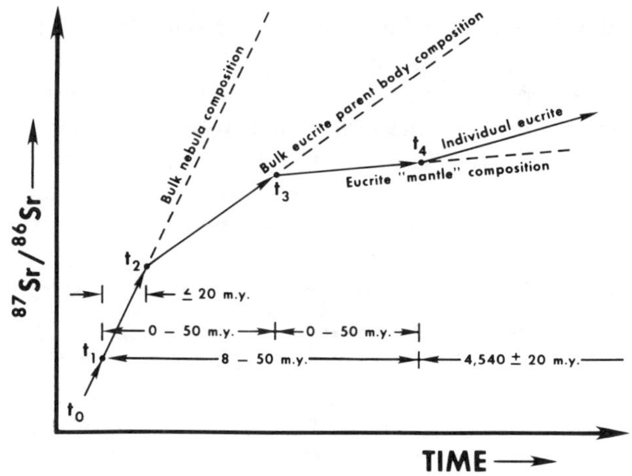

7.2.6 Diagrammatic presentation (not to scale) for growth of $^{87}Sr/^{86}Sr$ for basaltic achondrite systems at various stages in their evolution. Times indicated correspond to those in Table 7.2.1.

Allende chondrite (Lee and Papanastassiou, 1974). This problem was recently discussed by Lee *et al.* (1978) and Schramm and Clayton (1978), who conclude that there probably was a supernova in the vicinity of the solar nebula about a million years or so before formation of many of the inclusions in Allende, and which added a late "spike" of ^{26}Al. Thus, the time scale for accretion of nebular material into planetary objects may only be on the order of 10^6 years or less.

The close agreement between the absolute age of the eucrites and that of chondrites suggests that it should also be possible to use ^{129}I-^{129}Xe and ^{244}Pu-Xe systematics to examine the early chronology of the eucrites. Xenon from the fission of extinct ^{244}Pu has in fact been found in a number of basaltic achondrites (Hohenberg *et al.*, 1967; Rowe, 1970; Hohenberg, 1970; Wasserburg *et al.*, 1977). However, uncertainties in the chemical behavior of plutonium during magma generation, fractionation, and crystallization preclude calculation of precise t_4–t_0 age intervals for these meteorites (Lugmair and Marti, 1977; Wetherill, 1975c). The ^{129}I-^{129}Xe system has the potential of overcoming the chemical fractionation uncertainty, but so far no clear results have been obtained. Radiogenic ^{129}Xe has been identified in clasts from the howardite Kapoeta (Huneke *et al.*, 1977) and from the eucrite Ibitira (Wasserburg *et al.*, 1977), but no definite time intervals have been determined. Much of the trouble in dealing with Xe in the basaltic achondrites has been the presence of large amounts of trapped or spallation xenon that interferes with or, in some cases, completely masks the radiogenic or fissiogenic components (Hohenberg, 1970).

In spite of the inability to calculate relatively precise formation intervals, it can be concluded that the mere presence of radiogenic ^{129}I or ^{244}Pu fissiogenic Xe in eucrites and howardites requires that they formed very soon after the isolation of the solar nebula (intervals less than 100–200 m.y.). Within this interval the parent body had to be formed, partially melted, and the melt (magma) crystallized to form basalts.

The time interval between formation of the eucrite parent and the start of condensation, t_2–t_1, can not be fixed uniquely. However, definite limits can be placed on this interval through consideration of Sr I-values. At condensation, Sr in the nebula apparently had a $^{87}Sr/^{86}Sr$ ratio of 0.69876 ± 0.00002 (Gray *et al.*, 1973). The eucrites yield an isochron-intercept (BABI) value of 0.69896 ± 0.00005 (Table 7.2.3 and Fig. 7.2.4), and consideration of calculated I-values for individual eucrites yields a spread of only about ± 0.00008 about BABI. Assuming that there was no growth of $^{87}Sr/^{86}Sr$ between t_2 and t_4 (Rb/Sr ~0 in the source region in the parent body, or t_4–t_2 ~0), then an upper limit to t_2–t_1 can be derived based upon the Rb/Sr ratio of the nebula. If the nebula had a solar ratio of ~0.65 then t_2–t_1 ~8 m.y., but if the nebula had a chondritic ratio of ~0.25, then t_2–t_1 ~20 m.y., as pointed out initially by Papanastassiou and Wasserburg (1969). Furthermore, if each eucrite came from a separate parent body, they all would have had to form within only a few million years. Birck and Allègre (1978a) further pointed out that $^{87}Sr/^{86}Sr$ arguments can only be used to place an *upper* limit to t_2–t_1. Depending on the differentiation history (t_3) and magmatic history (t_4) of the parent body, t_2 could have occurred anytime earlier, conceivably even at or before t_1 of Allende (Fig. 7.2.6).

At present there are no data bearing on the time of differentiation, t_3, if in fact it even occurred as a discrete event. (Was there ever a mantle in the parent body?) All that can be said is that it would have had to occur sometime between t_2 and t_4. A number of recent studies (e.g., Stolper, 1977; Consolmagno and Drake, 1977; Fukuoka *et al.*, 1977; Morgan *et al.*, 1978) have concluded that the eucrites and related meteorites could have been derived from parent bodies of generally primitive composition (approximately chondritic with respect to most elements other than alkalis, which are depleted). This would be consistent with a very short t_4–t_2 interval and would indicate that a parent body mantle did not ever preexist as such, in contrast to the case for basalt genesis in the Earth and Moon.

The interval between t_2, formation of the parent body, and t_4, magma generation, can only be defined with any precision using Pb-Pb results (Table 7.2.8). Both total-rock isochron and internal isochron results indicate crystallization (t_4) of eucrites about 4.54 ± 0.02 b.y. ago, as compared to 4.56 ± 0.01 b.y. ago for the isochron defined by Allende inclusions (t_1). This yields a t_4–t_1 interval of 0–50 m.y. However, since the initial Sr in eucrites (BABI) is somewhat evolved, there must be some stage between t_1 and t_4 where this occurred. Allowing ~10 m.y. as a minimum for this, the reasonable interval for t_4–t_1 then becomes 30 ± 20 m.y., during which parent body formation (t_2), mantle formation (t_3), magma generation and crystallization (t_4) all would have to occur (Fig. 7.2.6).

A related question is whether there was (is?) more than one parent body for the basaltic achondrites. It has generally been concluded that it is not necessary to have more than one parent body for the eucrites, based on both isotopic results (above) and petrologic and chemical studies (sections 1.2.8 and 3.5). However, initial $^{87}Sr/^{86}Sr$ data for some of the other achondrites, notably Angra dos Reis and Kapoeta Clast C suggest that they were not co-magmatic with the eucrites (Fig. 7.2.4, Table 7.2.3). In order to derive all of them from a single parent body, given the time constraints discussed above, that body would have to have contained low Rb/Sr regions with different Sr isotopic compositions. This, in turn, would imply that each of these regions was isolated from higher Rb/Sr environments at different times over a span of only a few million years. The alternative, namely that these rocks were generated directly from a single high Rb/Sr source region over a span of a few million years, is untenable since meteoritic basalts with very low Rb/Sr ratios could not be generated directly from a high Rb/Sr source region. If anything, basalt petrogenesis would tend to produce a higher Rb/Sr ratio. The different source regions could have resulted from inhomogeneous accretion in a single parent body, or they could have resulted from a very early planetary differentiation. The latter alternative has the problem that Rb-rich differentiates complementary to the Rb-poor source regions are not represented among the meteorites in our collections. Thus, for the eucrites, howardites, diogenites, mesosiderites, and possibly Angra dos Reis, a single parent body seems most likely. However, Ma *et al.* (1977) concluded on the basis of trace element studies that Angra dos Reis is not simply related to the eucritic systems and could, in fact, be from a separate parent body. If so, it would be similar to the eucrite parent in being volatile-poor and in having formed basaltic magmas 4.5 b.y. ago.

Post-crystallization history of eucrites. As can be seen by the many discordant ages that have been obtained for a number of the eucritic basaltic achondrites (Tables

Table 7.2.9 Ages of noneucritic basaltic achondrites.

Meteorites	Method	Age (b.y.)	I-value	Reference
A. Nakhlites				
Nakhla	^{39}Ar-^{40}Ar	1.3		*(1)*
	Rb-Sr	1.34 ± 0.02	0.70232 ± 0.00006	*(2)*
	Rb-Sr	1.21 ± 0.01	0.70254 ± 0.00003	*(3)*
	U-Pb	1.3 ± 0.2		*(4)*
	Sm-Nd	1.27 ± 0.06		*(5)*
Lafayette	^{39}Ar-^{40}Ar	1.3		*(1)*
Governador Valadares	^{39}Ar-^{40}Ar	1.3		*(5)*
B. Shergottites				
Shergotty	^{39}Ar-^{40}Ar	0.25 (maskelynite)		*(6)*
	Rb-Sr	0.165 ± 0.011 (mineral isochron)		*(7)*
	various	~0.65 (crystallization age)		*(7)*
C. Chassignites				
Chassigny	K-Ar	1.4 ± 0.2		*(8)*
D. Angrites				
Angra dos Reis	Rb-Sr	-	0.69883 ± 0.00002	*(9)*
	Pb-Pb	4.544 ± 0.002		*(9)*
	Pb-Pb	4.555 ± 0.005		*(10)*
	Sm-Nd	4.55 ± 0.04		*(11)*

References:
(1) Podosek (1973); *(2)* Papanastassiou and Wasserburg (1974); *(3)* Gale *et al.* (1975); *(4)* Nakamura *et al.* (1977b); *(5)* Bogard *et al.* (1976); *(6)* Bogard *et al.* (1979); *(7)* Nyquist *et al.* (1979c,d); *(8)* Lancet and Lancet (1971); *(9)* Wasserburg *et al.* (1977); *(10)* Tatsumoto *et al.*, (1973); *(11)* Lugmair and Marti (1977).

7.2.3, 7.2.5, 7.2.6), these meteorites have been subjected to post-crystallization disturbances. In most cases these have been attributed to impact-produced shock-metamorphic effects, either as a direct result of the shock event and brecciation or of residual heat in an ejecta blanket. There is little evidence for any internally generated heat sources affecting these meteorites since the initial magmatism on the eucrite parent body 4.5 b.y. ago. This is in contrast to the case for the Moon and the Earth, for which basaltic magmatism extends to the present (at least for the Earth).

Other basaltic meteorites

Urey and Craig (1953) originally defined "basaltic achondrites" as those containing calcic plagioclase. This has evolved to refer commonly to eucrites and the related howardites, diogenites, and mesosiderites. However, the nakhlites, shergottites, and the chassignites have petrologic affinity to basaltic systems even though they are more volatile-rich (cf. Stolper *et al.*, 1979). The similarity to basalt is most pronounced for the two shergottites, Shergotty and Zagami (section 1.2.8). Oxygen isotope data (Fig. 7.2.2), as well as bulk chemistry and petrology suggest that the nakhlites may be closely related to the shergottites, and the chassignites are petrologically similar in that they have sodic plagioclase (Floran *et al.*, 1978). Thus, in addition to the main group of eucritic basaltic achondrites, there is also this other, perhaps genetically complex, noneucritic group.

Not only are the meteorites of this group distinctly different from the eucritic group in terms of chemistry and petrology, but they have had much more complex geochronologic histories (Table 7.2.9). Although the interpretations are not clear in all cases (e.g., resetting vs. primary crystallization), all the ages are substantially less than the 4.54 b.y. crystallization age for eucrites and Angra dos Reis. For the nakhlites, the ages cluster around 1.3 b.y. and, based on the Sm-Nd data (Nakamura *et al.*, 1977) it seems most reasonable to conclude that this represents a time of crystallization from a melt. The age of Chassigny is similar, suggesting a genetic affinity. The "ages" obtained so far for shergottites are very complex, but they could be consistent with a crystallization age of 0.6–1.2 b.y. followed by resetting of the K-Ar and Rb-Sr systems during a major shock-induced reheating 0.17–0.18 b.y. ago (Nyquist *et al.*, 1979b, c). Thus, in spite of their petrological similarity (Stolper *et al.*, 1979), it appears that the nakhlites, chassignites, and shergottites were not all formed by the same event.

Although these rocks have basaltic character, this does not mean they were produced by normal (endogenetic) basaltic volcanism 0.7–1.3 b.y. ago. If these meteorites came from an asteroidal-sized body, it is not reasonable to believe they were produced from internally generated magmas. Such a body cannot remain hot for 3–4 b.y. Thus, the melts would most likely have to have been impact generated. Total-rock Rb-Sr, Pb-Pb, and Sm-Nd data yield "model" ages on the order of 3.5–4.5 b.y., with considerable uncertainty. This suggests, however, that the melts could have been derived from a very ancient basaltic crustal surface.

It is not necessary, however, to require that the nakhlites, shergottites, or chassignites come from small bodies. If they have somehow been ejected from large bodies (e.g., Mars), then they could be the result of relatively young, internally generated basaltic magmas.

7.2.5 Chemical nature of the eucrite parent body

At present there are sufficient data to characterize the chemical composition of only the eucrite parent body with any degree of certainty. There have been several approaches used in trying to deduce the composition of the eucrite parent body, with the principal ones either based on cosmochemical arguments (Ganapathy and Anders, 1974; Morgan *et al.*, 1978; Dreibus *et al.*, 1977) or on petrogenetic arguments (Stolper, 1977; Consolmagno and Drake, 1977). These have led to generally similar results, namely that the *major* oxides (SiO_2, MgO, FeO, CaO, Al_2O_3) are present in amounts comparable to their "cosmic" (i.e., chondritic) abundances and, hence, represent primitive parent-body compositions.

The results of Consolmagno and Drake (1977) are particularly pertinent to the trace element geochemistry of geochronologic systems, especially for U, K, Rb and Sr. These authors concluded from rare-earth element considerations that eucritic magmas could be generated by 4–15% partial melting of a primitive (i.e., "cosmic") composition source region consisting of about 5% plagioclase, 10–15% pyroxene, and 80–85% "REE-free" material, namely olivine and metal. Fukuoka *et al.* (1977) concluded from more extensive trace-element studies that the eucrites Stannern and Sioux County could represent 8% and 24% partial melts, respectively, from a source having a silicate portion composed of 52–58% olivine, 40–36% orthopyroxene, and 8–6% plagioclase, with REE abundances about 1.5 times those in chondrites. Their estimate is similar to that obtained through cosmochemical arguments by Dreibus *et al.* (1977): 8% plagioclase, 40% pyroxene, 50% olivine. Stolper (1977) and Consolmagno and Drake (1977) also concluded that at least three eucrites are derivatives

Table 7.2.10 Selected trace element data for eucrites.

Eucrite	U (ppm)	K (ppm)	Rb (ppm)	Sr (ppm)	Rb/Sr	K/U	Petrogenetic Model†
Stannern	0.16	783	0.742	94.5	0.0078	5000	4% partial melt
Bereba	0.10	290	0.238	73.1	0.0033	2900	10% partial melt
Ibitira	0.09	165	0.137	85.4	0.0016	1800	10% partial melt
Jonzac		400	0.452	74.9	0.0060		10% partial melt
Juvinas	0.08	248	0.142	78.4	0.0018	3100	10% partial melt
Pasamonte	0.10	321	0.280	73.7	0.0038	3200	10% partial melt
Sioux County	0.08	295	0.224	73.1	0.0031	3700	15% partial melt
Nuevo Laredo	0.13	430	0.322	77.0	0.0042	3300	30% residual liquid
Moore County			0.052	64.7	0.0008		85% fractional crystallization
Serra de Magé		98	0.072	109.2	0.0007		50% fractional crystallization

†After Consolmagno and Drake, 1977. Residual liquid and cumulate samples assumed to have been derived from Juvinas-like melt.
References: Table 7.2.2; Birck and Allègre (1978a); Mahnes *et al.* (1976); Unruh *et al.* (1977); Tatsumoto *et al.* (1973).

from primary eucritic compositions such as Juvinas and Ibitira through fractional crystallization: Moore County and Serra de Maǵe represent cumulates from 85 and 50% fractional crystallization, respectively, and Nuevo Laredo represents a 30% residual liquid from fractional crystallization of a eucritic magma.

The K, U, Rb and Sr abundances for the eucrites modelled by Consolmagno and Drake (1977) are summarized in Table 7.2.10, from which a number of key inferences regarding the composition of the eucrite parent body may be drawn. Stannern, which represents only 4% partial melting, has a significant europium anomaly (Fig. 7.2.7), indicating that the melt was depleted in plagioclase. Because the plagioclase is enriched in Sr over Rb (Table 7.2.11), and because much of the Rb is probably present in "quintessence" or interstitial glassy phases (Fig. 7.2.3), the Rb/Sr ratio of Stannern was no doubt increased relative to that of the source material. Similar fractionations are also reflected in the relatively higher U and K abundances for Stannern (Table 7.2.10).

On the other hand, those eucrites that represent 10–15% partial melts have flat REE patterns (Fig. 7.2.7), indicating that essentially all the plagioclase is in the melt. Furthermore, given the bulk compositions estimated for the starting material (above), a 10–20% partial melt would also contain a substantial amount of the pyroxene. Hence, most of the K, U, Rb and Sr containing phases would be in the melt, with the result that the Rb/Sr, K/Rb, K/U and Th/U ratios of the "primary liquid" eucrites such as Juvinas and Ibitira are probably representative of the source region or, in effect, the eucrite parent body as a whole. The variations that do occur (e.g., Rb/Sr ratios of 0.002–0.006) could be due to heterogeneities in the parent body.

If, indeed, the eucritic liquids contain the bulk of the K, U, Rb and Sr, then they must represent about a six-fold enrichment over the original abundances in the

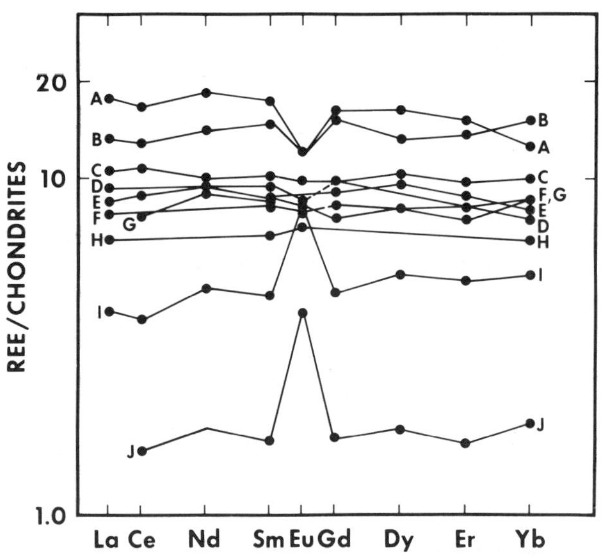

7.2.7 Summary of rare-earth data for eucrites (from Consolmagno and Drake, 1977). The main group with flat REE patterns is considered to represent 10–15% partial melting of a eucrite parent of approximately cosmic composition (with respect to REE and major elements other than alkalies). Meteorites represented are: A, Stannern; B, Nuevo Laredo; C, Pasamonte; D, Béréba; E, Juvinas; F, Haraiya; G, Jonzac; H, Sioux County; I, Moore County; J, Serra de Magé.

Table 7.2.11 Summary of selected trace element data for eucrites and the eucrite parent body.

	U (ppm)	K (ppm)	Rb (ppm)	Sr (ppm)	K/U	K/Rb	Rb/Sr	
Total rocks†	0.09	300	0.25	76	3000	1200	0.0033	10–15% p. m.†
Plagioclase	0.02	600	0.26	170	30,000	2300	0.0015	*(1),(2),(3)*
Pyroxene	(0.7)	125	0.09	23	(200)	1400	0.0039	*(1),(2),(3)*
Eucrite parent	0.015	50	0.042	13	3000	1200	0.003	This work*
	0.017	54	0.049	17	3200	1100	0.003	*(4)*

†p. m. = partial melting. Data used only for those eucrites assumed to be primary 10–15% p. m. (Consolmagno and Drake, 1977).
*Calculated assuming total rocks represent one-sixth of total bulk composition of parent and that no K, U, Rb, and Sr is present in the five-sixths residual. See text.
References: *(1)* Allègre *et al.* (1975); *(2)* Unruh *et al.* (1977); *(3)* Birck and Allègre (1978a); *(4)* Dreibus *et al.* (1977).

source region. Thus, the absolute abundances of those elements in the parent body as a whole can be estimated by about a six-fold reduction in their concentration in the "primary-liquid" eucrites (Table 7.2.11). Abundances estimated in this way agree very well with those of Dreibus *et al.* (1977).

The two meteorites examined by Consolmagno and Drake (1977) and which represent cumulate mineralogies, Moore County and Serra de Magé, have large positive europium anomalies (Fig. 7.2.7), indicating enrichment of plagioclase, with resulting enrichment of Sr over Rb. Thus, the Rb/Sr ratios of these meteorites must be less than those of the liquids from which they were derived. Furthermore, if those liquids had not fractionated Rb relative to Sr during their derivation from their respective source regions (see above), then the Rb/Sr ratios of Moore County and Serra de Magé represent lower limits to the Rb/Sr ratios of eucritic source regions, consistent with the conclusions reached above. Nuevo Laredo apparently represents a slightly more complicated case, but the slight enrichment of U, K, and Rb would also be consistent with the model outlined above.

The studies mentioned above suggest a general "cosmic" character to the composition of the basaltic achondrite parent body, but they should not be interpreted to indicate a truly "chondritic" parent material. Although the inferred abundances of U, Th, and Sr are approximately in line with the compositions of known chondrite groups, K, Rb and most other volatiles are severely depleted. Also, as Fukuoka *et al.* (1977) have shown, there is easily a 50% uncertainty in the abundances of certain elements as inferred from trace element fractionation model calculations. Furthermore, the oxygen isotope data, discussed previously, preclude any simple link between chondrites as we know them and eucritic basaltic achondrites. Thus, although the eucrite parent body may have a major element composition that is basically "cosmic," the abundances of various groups of trace elements (siderophile, chalcophile, volatile, etc.) could easily be several times more or less abundant than in chondritic meteorites.

Finally, the results of Stolper (1977) and Consolmagno and Drake (1977) could be interpreted to mean that basaltic achondritic magmas were derived by melting of "primary" material. That is, there was no earlier stage of mantle-crust-core differentiation prior to magma generation (although it would occur during magma generation). However, this is not required by their results.

7.2.6 Summary

Basaltic achondrites in general, and eucrites in particular, crystallized during a single, short period 4.54 ± 0.02 b.y. ago. They apparently represent basaltic flows and associated plutonic rocks formed on or near the surfaces of asteroidal-sized bodies such as 4 Vesta.

The magmatic events that formed the basaltic achondrites occurred 30 ± 20 m.y. after condensation of the high-temperature inclusions found in the Allende chondrite. The eucritic magmas apparently formed directly from undifferentiated primitive (but not necessarily "chondritic") material, indicating that there may have been no "mantle" or "crust" in the eucrite parent body prior to magma generation.

The eucrite parent body was approximately ($\pm$ 50%) chondritic in U, Th and Sr but strongly depleted in Na, K, and Rb. Concentrations of these elements in the parent body as a whole can be estimated to be 0.015 ppm U, 0.057 ppm Th, 50 ppm K, 0.042 ppm Rb, and 13 ppm Sr with Th/U = 3.8, K/U = 3000, K/Rb = 1200, and Rb/Sr = 0.003.

7.3 CHRONOLOGY OF LUNAR VOLCANISM

7.3.1 Introduction

The returned lunar samples from the Apollo and Luna missions can be classified into broad classes that are outlined in Fig. 7.3.1. This classification, which is based on chemistry and limited mineralogical characteristics, provides a convenient format for reviewing lunar trace elements and isotopic data in general, and helps place the geochronology of lunar basalts in context.

Isotopic investigations have shown that the major lunar geological activity, whether originating from the inside or the outside, was quite brief and it is widely accepted that the Moon has retained its present surface features for the past 3 b.y. Although highly differentiated, it, unlike the Earth, lacks the tectonic processes that sustain major magmatism for extended periods. Thus, judged by terrestrial experience, lunar magmatism is quite limited in scope and duration. To put lunar volcanism in the proper perspective, a brief outline of the lunar time scale is provided:

1. A large scale lunar differentiation occurred ~4.4 b.y. ago, resulting in the formation of alumina rich crust and sources of lunar basalts with characteristic chemical fractionation patterns. U and Th and other refractory elements were enriched in the outer layer of the Moon.

2. Excavation of giant ring basins by bombardment occurred prior to about 4 b.y. ago. The duration of the bombardment is not known. The youngest of the basins is the well preserved and only partially flooded basin of Mare Orientale on the western edge of the nearside of the Moon. The next youngest is that of Imbrium. With the exception of Orientale, all nearside giant basins are flooded.

3. Abrupt cessation of the intense bombardment occurred ~3.9 b.y. ago.

4. Extrusion of magma to the surface and flooding of the giant ring basins occurred 3.9–3.1 b.y. ago. Since then only relatively minor bombardment and magmatism have occurred.

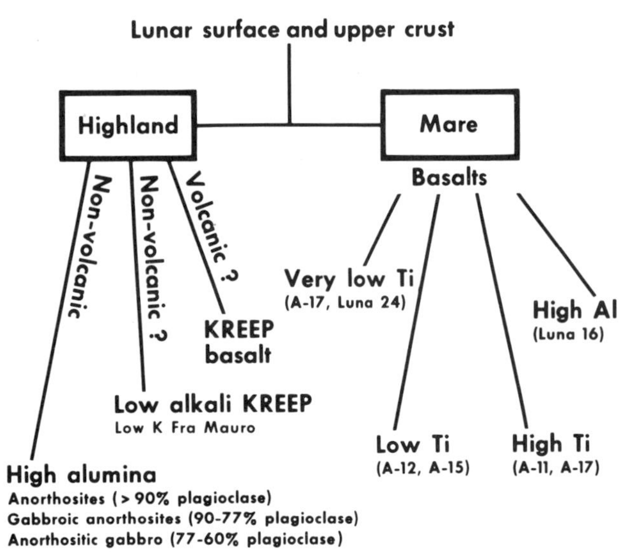

7.3.1 Schematic classification of lunar rock types.

The main aim of this chapter is to review the isotopic studies on the mare basalts. This involves a number of intimately related topics such as the ages of the rocks, duration of lunar volcanism, nature of the sources from which the basalts were derived, parent-daughter fractionation patterns associated with magma extrusion and evaluation of the current hypotheses for lunar basalt genesis. These topics in turn, together with chemical and physical parameters, contribute to the larger picture of lunar evolution as they provide important constraints that must be met by the various evolution models. These models will not be discussed in this chapter. However, an in-depth review of lunar basalts is not possible without touching on other topics such as trace element studies, cratering history and general differentiation processes.

In particular, one must keep an open mind with regard to the possibility that the so-called low-alumina highland rocks are genetically related to mare basalts. According to this view these rocks are igneous and were produced by partial melting at some depth. If so, then lunar magmatism is a continuum extending back in time beyond the age of the oldest mare basalt. In this hypothesis then, the coincidence of the end of the heavy bombardment with the apparent onset of volcanism is essentially an artifact, largely due to the severity of the bombardment that left its morphologic and isotopic imprint in the highland rocks, thus masking their immediate genetic relationship to mare basalts. [Taylor (1975) groups Apollo 14 low-alumina basalts with mare basalts, calling them "aluminous mare basalts."]

In this connection it is necessary to consider also the case of the so-called KREEP material on the Moon. The acronym was given to a class of glassy particles in Apollo 12 soil (Meyer *et al.*, 1971; Hubbard *et al.*, 1971), which were found to be rich in K, REE and P. This material, which is also rich in U and Th, was subsequently identified in samples of all Apollo missions. It is now widely accepted that KREEP must cover large areas of the lunar surface, as inferred from their high γ-ray activities (Metzger *et al.*, 1974).

Although the igneous texture of most highland rocks is either eradicated or masked by impact, it has been argued that the wide range of chemical compositions exhibited by nonmare and highland rocks would not exist if impact melting or intimate mechanical mixing and metamorphism had been very thorough. These processes can only reduce the range of chemical compositions (Hubbard *et al.*, 1973). Thus the spectrum of compositions and the common factors that relate these rocks (e.g., large Eu anomaly, REE patterns) cannot be explained entirely by impact processes. Therefore, there is considerable evidence that lunar volcanism was widespread prior to the eruption of the mare basalts.

Lunar rocks were collected from 9 different locations covering a wide area of the nearside of the Moon. Well over 100 rocks have been dated. The ages of the rocks have been determined by one or more of four well-developed radiometric clocks. The decay schemes and general characteristics of these dating methods are discussed in section 7.1. In general, the methods yield ages that are in satisfactory agreement with each other, although occasional discrepancies do occur. These discrepancies are thought to be due primarily to the different degrees of susceptibility of the methods to metamorphic events subsequent to crystallization. In general, discrepancies are much less frequent for lunar rocks than for terrestrial rocks. We have summarized in Table 7.3.1 essentially all the age data available to date on large samples of lunar basalts. When precisely determined, the mare basalt internal isochron ages are generally believed to be the true crystallization ages, dating the time of extrusion of corresponding lavas to the lunar surface. In contrast, ages of highland rocks do not have such clear cut interpretation because of the superimposed complexities resulting from the heavy bombardment of the ancient lunar crust. For this reason it is logical to treat age data in two separate sections, one dealing with mare basalts and the other with nonmare basalts and highland rocks.

7.3.2 Technical remarks about the methods

The bulk of the internal isochron crystallization ages of the lunar basalts were obtained through use of the Rb-Sr method, while the U-Pb method has provided only four isochrons of which two are two-point "isochrons." The major handicap of the U-Th-Pb method was two-fold: (a) One part was a formidable contamination problem. Mere loading of a Pb sample with a few microliters of reagents on the mass spectrometer filament could add a Pb blank of the same magnitude as the Rb or Sr blank resulting from the whole preparation and separation procedures for these two elements including the use of heavy liquids, dissolution and chemical separation on relatively large conventional cation exchange columns. (b) The second part was poor ionization efficiencies for U and Th.

In spite of all the difficulties involved, workers in the field were able, under the pressure and incentive of the Apollo program, to reduce the Pb blank by two to three orders of magnitude to the nanogram and subnanogram levels, but so far these procedures have not become routine. In establishing isochrons, the situation was often made worse by underestimation of the blank problem. However, despite the method's poor record in producing mineral isochrons, it provided extremely valuable isotopic information that shed light on the lunar history and its differentiation processes.

The Sm-Nd method is a recent arrival on the scene (Lugmair, 1974). It requires sophisticated chemical separation techniques and high resolution mass spectrometry. Both capabilities are reasonably well-established and promise to rapidly evolve the method to a rival to the Rb-Sr method as the working horse of geochronology. No blank or ionization problems are encountered. A major characteristic of the method is the chemical similarity of the parent and daughter (both are REE) making the method least susceptible to mild disturbances. In addition, while Rb resides primarily in ill-defined residual interstitial phases in lunar rocks, Sm is concentrated mainly in the pyroxene, thus making Sm-Nd isochrons (based on the major phases, pyroxene, plagioclase and ilmenite) less susceptible to doubts concerning whether or not a given isochron is possibly a mixing line of two end-members.

The ^{39}Ar-^{40}Ar method does not produce isochrons and the age of the rock is deduced from a ^{39}Ar-^{40}Ar plateau produced through stepwise thermal release of the gas. The method has been applied extensively to the dating of lunar rocks and there is good agreement between ^{39}Ar-^{40}Ar plateau ages and Rb-Sr internal isochron ages, particularly when the plateau ages are obtained from plagioclase. This is because plagioclase is significantly more retentive of ^{40}Ar than the whole rock (Lunatic Asylum, 1970a; Turner *et al.*, 1971, 1972; Podosek *et al.*, 1972). Occasionally the technique runs into difficulties when applied to rocks which have suffered complex histories resulting in significant ^{40}Ar loss. In such cases, well-defined plateaus may not be possible or, if produced, they may not be reliable (Jessberger *et al.*, 1974a). A detailed discussion of the technique is given in a recent review by Turner (1977).

Table 7.3.1 Radiometric ages of lunar mare basalts.

Sample	Location†	Description of Basalt	Age in 10^9 Years			
			^{39}Ar - ^{40}Ar	Reference	Rb - Sr	Reference
I. Mare Tranquillitatis A-11						
Low-K basalts						
10003	LM	medium grained ophitic	3.85 ± 0.03†	*(1)*	3.76 ± 0.08	*(2)*
			3.86 ± 0.07	*(3)*		
10020	LM	medium grained ophitic	~3.72	*(4,5)*		
10029	LM	medium grained ophitic	3.83 ± 0.03†	*(5,6)*		
10044	LM	medium grained ophitic	3.66 ± 0.04	*(4,5)*	3.63 ± 0.11	*(7)*
			3.68 ± 0.08	*(3)*		
			3.84 ± 0.07	*(8)*		
10045	LM	medium grained	~3.70	*(4,5)*		
10047	LM	medium grained ophitic	3.69 ± 0.03	*(1,5)*		
10050	LM		3.70 ± 0.03†	*(4,5)*		
10058	LM	medium grained ophitic	3.66 ± 0.04	*(4,5)*	3.55 ± 0.11	*(9)*
					3.55 ± 0.20	*(7)*
10062	LM	fine grained ophitic	3.79 ± 0.04†	*(5)*	3.92 ± 0.11	*(10)*
			3.77 ± 0.09	*(3)*	Sm-Nd age:	
					3.88 ± 0.06	*(10)*
High-K basalts						
10017	LM	intersertal			3.51 ± 0.05	*(7)*
					3.50 ± 0.22	*(11)*
10022	LM	intersertal	3.54 ± 0.06	*(4,5)*		
			3.54 ± 0.09	*(3)*		
10024	LM	intersertal			3.53 ± 0.07	*(12)*
10031	LM		3.55 ± 0.08	*(5)*		
10032	LM		3.53 ± 0.06	*(5)*		
10057	LM				3.55 ± 0.04	*(7,10)*
10069	LM	intersertal			3.60 ± 0.06	*(7,10)*
10071	LM	intersertal	3.47 ± 0.06†	*(13)*	3.60 ± 0.02	*(7)*
10072	LM	intersertal	3.57 ± 0.05†	*(4,5)*	3.56 ± 0.05	*(10)*
			3.48 ± 0.08	*(3)*	Sm-Nd age:	
					3.57 ± 0.03	*(10)*
II. Oceanus Procellarum A-12						
12002	N of Head	porphyritic	3.21 ± 0.05	*(14)*	3.29 ± 0.10	*(15)*
			3.18 ± 0.06	*(16,17)*		
12004	NNE of Head				3.22 ± 0.07	*(18)*
12014	NNE of Head				3.22 ± 0.11	*(51)*
12017	NNE of Head		3.16 ± 0.07	*(24)*		

Notes: †Locations are craters unless otherwise indicated. LM, Lunar Module; SEP, Surface Electrical Properties Transmitter. Ages obtained on coarse (2–4 mm) soil fragments are not included, except for Luna 16 and Luna 24. The errors given are those quoted in the original publication. Different authors may use different error definition, a fact which must be taken into account when comparing results. ^{39}Ar-^{40}Ar ages given without error and an approximate sign (~) were deduced from ill-defined (intermediate temperature) plateaus. They are included in the table since no other ages are available for these rocks. ^{39}Ar-^{40}Ar ages obtained on plagioclase are generally more reliable than whole rock ages. They are distinguished by a dagger (†). All ages have been recalculated to correspond to the following decay constants: $\lambda(^{40}K) = 5.543 \times 10^{-10} y^{-1}$; $\lambda_e(^{40}K) = 5.81 \times 10^{-11} y^{-1}$; $^{40}K/K = 1.167 \times 10^{-4}$; $\lambda(^{87}Rb) = 1.42 \times 10^{-11} y^{-1}$; $\lambda(^{147}Sm) = 6.54 \times 10^{-12} y^{-1}$ (Steiger and Jäger, 1977).

Table 7.3.1 *(Continued)*

Sample	Location†	Description of Basalt	Age in 10^9 Years: ^{39}Ar - ^{40}Ar	Reference	Rb - Sr	Reference
12020	N of Head	porphyritic	3.09 ± 0.06	*(16,17)*		
12021	NE of Head				3.26 ± 0.06	*(12)*
					3.23 ± 0.10	*(19)*
12022	N of Head	porphyritic	3.08 ± 0.06	*(16,17)*		
12031	Head				3.16 ± 0.07	*(51)*
12038	Bench	ophitic	3.08 ± 0.05	*(20)*	3.15 ± 0.21	*(21,22)*
12039	Bench				3.12 ± 0.06	*(50)*
					Sm-Nd age:	
					3.20 ± 0.05	*(51)*
12040	Bench				3.23 ± 0.04	*(12)*
					3.08 ± 0.10	*(21,22)*
12051	Surveyor	ophitic	3.23 ± 0.05	*(14)*	3.19 ± 0.10	*(15)*
			3.13 ± 0.06	*(13)*	3.09 ± 0.04	*(52)*
			3.21 ± 0.06	*(16,17)*		
12052	Head				3.15 ± 0.19	*(21,22)*
12053	Bench		3.14 ± 0.06	*(24)*		
12055	Head				3.12 ± 0.06	*(52)*
12056	Bench				Sm-Nd age:	
					3.20 ± 0.14	*(51)*
12063	Block	porphyritic	3.15 ± 0.05	*(20)*	3.27 ± 0.10	*(9)*
					3.23 ± 0.13	*(12)*
12064	Surveyor		3.15 ± 0.01†	*(23)*	3.11 ± 0.04	*(12)*
12065	Surveyor	porphyritic	3.21 ± 0.05	*(14)*	3.09 ± 0.09	*(12)*
			3.13 ± 0.06	*(16,17)*		
III. Hadley-Apennines A-15						
Quartz normative						
15065	Elbow				3.21 ± 0.04	*(25)*
15075	Elbow	porphyritic subophitic	3.45 ± 0.02	*(28)*		
15076	Elbow	porphyritic clinopyroxene	3.31 ± 0.04	*(13)*	3.26 ± 0.08	*(25)*
			3.32 ± 0.10	*(26)*		
15117	St. George				3.28 ± 0.04	*(25)*
15499	Dune	porphyritic clinopyroxene vitrophyre	3.30 ± 0.08	*(27)*		
15597	Hadley Rille	vitrophyric	3.3 ± 0.1	*(26)*		
15682	Hadley Rille				3.37 ± 0.07	*(25)*
Olivine normative						
15016	Rhysling		3.34 ± 0.08	*(26)*	3.22 ± 0.05	*(29)*
15085	Elbow				3.33 ± 0.04	*(25)*
15555	Hadley Rille	porphyritic olivine basalt	3.27 ± 0.03†	*(30)*	3.25 ± 0.04	*(25)*
			3.22 ± 0.05	*(16)*	3.23 ± 0.08	*(33)*
			3.28 ± 0.06	*(31)*	3.46 ± 0.13	*(34)*
			3.27 ± 0.06	*(32)*		

References:
(1) Stettler *et al.* (1974); *(2)* Papanastassiou and Wasserburg (1975a); *(3)* Turner (1970); *(4)* Geiss *et al.* (1977); *(5)* Guggisberg *et al.* (1979); *(6)* Guggisberg *et al.* (1977); *(7)* Papanastassiou *et al.* (1970b); *(8)* Davis *et al.* (1971); *(9)* Murthy *et al.* (1971); *(10)* Papanastassiou *et al.* (1977b); *(11)* Gopalan *et al.* (1970); *(12)* Papanastassiou and Wasserburg (1971a); *(13)* Stettler *et al.* (1973); *(14)* Turner (1971); *(15)* Papanastassiou and Wasserburg (1970b); *(16)* Alexander *et al.* (1972); *(17)* Alexander and Davis (1974); *(18)* Papanastassiou and Wasserburg (1971b); *(19)* Cliff *et al.* (1971); *(20)* Turner (1977); *(21)* Compston *et al.* (1971); *(22)* De Laeter *et al.* (1973); *(23)* Horn *et al.* (1975a); *(24)* Horn *et al.* (1975b); *(25)* Papanastassiou and Wasserburg (1973);

Table 7.3.1 *(Continued)*

Sample	Location†	Description of Basalt	Age in 10^9 Years: ^{39}Ar - ^{40}Ar	Reference	Rb - Sr	Reference
Rake Samples						
15385	Spur	15555 type	3.35 ± 0.05	*(27)*		
15607	Hadley Rille	15555 type	3.23 ± 0.12	*(27)*		
15633	Hadley Rille	15555 type	3.22 ± 0.05	*(27)*		
15659	Hadley Rille	15555 type	3.30 ± 0.04	*(27)*		
15668	Hadley Rille		3.10 ± 0.05	*(27)*		
15678	Hadley Rille		3.34 ± 0.05	*(27)*		
15683	Hadley Rille	15555 type	3.32 ± 0.03	*(27)*		
IV. Mare Serenitatis						
A-17						
70017	LM				3.59 ± 0.18	*(35)*
70035	LM	porphyritic	3.69 ± 0.07	*(13)*	3.74 ± 0.06	*(29)*
					3.65 ± 0.11	*(36)*
70135	Geophone Rock				3.67 ± 0.09	*(35)*
					Sm-Nd age:	
					3.77 ± 0.06	*(51)*
70215	LM	fine grained (devitrified) vitrophyre	3.79 ± 0.04	*(37)*		
70255	SEP		3.84 ± 0.02	*(28)*		
71055	Station 1A				3.56 ± 0.09	*(53)*
75035	Camelot	medium grained	3.72 ± 0.04†	*(39)*		
			3.71 ± 0.05	*(40)*		
75055	Camelot	coarse grained	3.73 ± 0.04†	*(41)*	3.69 ± 0.06	*(38)*
			3.71 ± 0.05	*(46)*		
			3.77 ± 0.05	*(43)*		
75075	Camelot		3.69 ± 0.02†	*(23)*	3.76 ± 0.12	*(35)*
					Sm-Nd age:	
					3.70 ± 0.07	*(44)*
79155	Van Serg		3.75 ± 0.04	*(37)*		
V. Mare Fecunditatis						
Luna 16						
L16-B1	Luna 16	fine grained ophitic	3.41 ± 0.04	*(46)*	3.35 ± 0.18	*(45)*
VI. Mare Crisium						
L24A	Luna 24		3.60 ± 0.12	*(47)*		
24170	Luna 24		3.26 ± 0.04	*(48)*	3.6 ± 0.6	*(48)*
			3.33 ± 0.20†	*(50)*	Sm-Nd age:	
					3.30 ± 0.05	*(48)*
24077,13	Luna 24	ferrogabbro	3.33 ± 0.21	*(49)*		
24077,63	Luna 24	metaferrobasalt	3.24 ± 0.06	*(49)*		
24196,9011	Luna 24	metabasalt	3.27 - 3.46	*(50)*		

References:
(26) Heidelberg, unpublished results; *(27)* Husain (1974); *(28)* Schaeffer and Schaeffer (1977); *(29)* Evensen *et al.* (1973); *(30)* Podosek *et al.* (1972); *(31)* Husain *et al.* (1972); *(32)* York *et al.* (1972); *(33)* Murthy *et al.* (1972); *(34)* Chappell *et al.* (1972); *(35)* Nyquist *et al.* (1975); *(36)* Nyquist *et al.* (1974); *(37)* Kirsten and Horn (1974); *(38)* Tera *et al.* (1974a); *(39)* Turner and Cadogan (1975); *(40)* Turner and Cadogan (1974); *(41)* Huneke *et al.* (1973); *(42)* Turner *et al.* (1973); *(43)* Kirsten *et al.* (1973); *(44)* Lugmair *et al.* (1975); *(45)* Papanastassiou and Wasserburg (1972); *(46)* Huneke *et al.* (1972); *(47)* Stettler and Albarède (1978); *(48)* The Lunatic Asylum (1978); *(49)* Schaeffer *et al.* (1978); *(50)* Hennessy and Turner (1979); *(51)* Nyquist *et al.* (1979); *(52)* Nyquist *et al.* (1977); *(53)* Tera *et al.* (1974b).

In this chapter all U-Pb ages were calculated using the more recent decay constants determined by Jaffey *et al.* (1971). These are $9.84850 \times 10^{-10} y^{-1}$ and $1.55125 \times 10^{-10} y^{-1}$ for ^{235}U and ^{238}U, respectively. The initial Pb of the planet is assumed to be identical in composition to primordial Pb from the iron meteorite Cañon Diablo as was first determined by Patterson *et al.* (1953) and later by Tatsumoto *et al.* (1972). Following the recommendation of the Subcommission on Geochronology of the I.U.G.S., we have calculated all Rb-Sr ages reported here using $\lambda = 1.42 \times 10^{-11} y^{-1}$ for ^{87}Rb. The ^{39}Ar-^{40}Ar ages have also been recalculated using the recommended new values. These are $\lambda_t = 5.543 \times 10^{-10} y^{-1}$, $\lambda_e = 5.810 \times 10^{-11} y^{-1}$ and $^{40}K/K = 1.167 \times 10^{-4}$. The effect on all three methods is a reduction of 60–80 m.y. in the age of a 4-b.y.-old system. The decay constant used for ^{147}Sm is $6.54 \times 10^{-12} y^{-1}$ (Donhoffer, 1963; Gupta and MacFarlane, 1970), and $4.9475 \times 10^{-11} y^{-1}$ for ^{232}Th (LeRoux and Glandenin, 1963).

7.3.3 Mare basalt ages

Long before the advent of the Apollo and Luna programs the maria on the Earth-facing side of the Moon were identified through telescopic observation as volcanic deposits (see for example Gilbert, 1893), although such identification was rejected by many subsequent workers. The acquisition of samples allowed for the physical, chemical and petrologic characterization of lunar basalts as well as for the unraveling of many aspects of their history and the history of the planet as a whole.

The returned lunar basalts were found to contain no water and to be very poor in volatiles (a fact dramatically illustrated by the near absence of primordial ^{204}Pb). The high fluidity of lunar lava (Murase and McBirney, 1970) coupled with the lack of volatiles, suggests that explosive volcanism is less significant on the Moon than on Earth (Head, 1976). For the most part, deposition of basaltic magma took the form of sequences of thin lava flows resulting in morphologic features resembling terrestrial basaltic flood and Hawaiian eruptions (Head, 1976).

Geochemically, lunar basalts are distinguished from terrestrial basalts on the basis of their bulk composition, trace-element content and REE fractionation patterns. Lunar basalts are generally characterized by higher Cr, Ti, Fe and lower Na, K contents. Compared with meteorites and terrestrial basalts, the lunar basalts have REE patterns that are unique. Invariably, mare basalts show a negative Eu anomaly. Such an anomaly is rare in terrestrial basalts. Terrestrial basalts generally exhibit a large enrichment in the light REE relative to the heavy ones. In contrast, lunar basalts do not show such pronounced fractionation. If anything, the opposite is the case, namely that lunar basalts show slight enrichment in the heavy REE relative to the light ones.

All the large samples of lunar basalts were returned by Apollo (missions 11, 12, 15, and 17). The Soviet's Luna program returned mostly regolithic debris. The regolith of Luna 16, however, is rich in mare basalt fragments, of which a piece has been dated. ^{39}Ar-^{40}Ar ages and Rb-Sr ages of lunar samples have been reviewed by Turner (1977) and by Nyquist (1977).

Mare Tranquillitatis (Apollo 11)

Apollo 11 landed at the southwestern periphery of Mare Tranquillitatis. The basalts collected at this landing site all have high Ti content (TiO_2 ~11 wt.%). Based on their minor- and trace-element contents they can be subdivided into two groups: high-K and low-K basalts. More comprehensive classification schemes have been discussed (Chao *et al.*, 1970; James and Jackson, 1970; Warner, 1971; James and Wright, 1972). In particular, Warner's (1971) textural classification has proven useful in the discussion of the radiometric age data (Geiss *et al.*, 1977). More recently, the classification of the Apollo 11 low-K basalts has been refined by Beaty and Albee (1978) using modal proportions of olivine and cristobalite and by Guggisberg *et al.* (1979) by introducing the normative quartz content as a further basis for classification.

Internal Rb-Sr mineral isochrons and ^{39}Ar-^{40}Ar ages for a sizable number of Apollo 11 basalts were reported by several workers at the Apollo 11 Lunar Science Conference (Compston *et al.*, 1970; Gopalan *et al.*, 1970; Lunatic Asylum, 1970; Papanastassiou and Wasserburg, 1970a; Turner, 1970). The most extensive and precise Rb-Sr internal isochron ages were those by Papanastassiou *et al.* (1970) and Papanastassiou and Wasserburg (1971a). The five high-K basalts (0.3–0.4 wt.% K_2O) investigated have, within error limits, identical ages, compatible with a time of crystallization of 3.56 ± 0.05 b.y. For the low-K basalts (0.1–0.05 wt.% K_2O), dating by the ^{39}Ar-^{40}Ar method gave much more precise ages than the Rb-Sr method, and the first results obtained by Turner (1970) established that at least one low-K basalt (10003) is distinctly older (3.86 ± 0.07

b.y.) than the high-K basalts. This result was later confirmed by an even more precise ^{39}Ar-^{40}Ar age of 3.85 ± 0.03 b.y. obtained on plagioclase separates from 10003 (Stettler *et al.*, 1974). More recently, Papanastassiou *et al.* (1977) determined the crystallization ages of two Apollo 11 mare basalts by both Rb-Sr and Sm-Nd methods. The data of the two methods are in good agreement and show an age of 3.9 b.y. for low-K basalt 10062, thus confirming the existence of older ages for some low-K Apollo 11 basalts (Fig. 7.3.2).

The initial Sr composition for the low- and high-K groups are distinctly different (Papanastassiou *et al.*, 1977). The measured low-K samples have, within error limits, identical values with an average of 0.69906 ± 0.00004. No evidence for differences between the individual subgroups has been found so far. The measured high-K rocks have distinctly higher values, with an average of 0.69930 ± 0.00006. Again, within error limits, the individual rocks have virtually identical values.

Further ^{39}Ar-^{40}Ar dating, mostly done at Bern, established that not all Apollo 11 low-K rocks were formed contemporaneously. The ages correlate with the textural and petrological properties. Geiss *et al.* (1977) discovered a correlation between age and Warner's (1971) textural classification. Also the more refined classification scheme of Beaty and Albee (1978) and Guggisberg *et al.* (1979) clearly shows a distinct correlation between crystallization age and petrological classification (Fig. 7.3.3). The two oldest rocks (10003, 10029) belong to the B2 group of Beaty and Albee (1978) and have a normative quartz content of 0.0%. The three other low-K groups (B3, B4, B1) have ages of 3.79 ± 0.04 b.y. (10062); 3.70 ± 0.03 b.y. (10050); and 3.67 ± 0.04 b.y. (10044, 10047, 10058). The normative quartz content systematically increases with decreasing age. The crystallization ages of groups B2, B3 and B4 are distinctly different, whereas B4 and B1 might be virtually contemporaneous. The best-defined ^{39}Ar-^{40}Ar age for the high-K basalts is 3.57 ± 0.05 b.y. (10072), in excellent agreement with the Rb-Sr ages. The high-K basalts are distinctly younger than any of the low-K ones. No age difference within the high-K group has yet been detected. The Apollo 11 basalts 10003 and 10029 are the oldest mare basalts known and they crystallized 3.84 ± 0.03 b.y. ago (the aluminous Apollo 14 basalt 14053 has probably an even slightly higher crystallization age of 3.89 ± 0.04 b.y.; cf. Papanastassiou and Wasserburg, 1971b; Turner *et al.*, 1972). The age of the Imbrian impact event is most likely 3.80 ± 0.04 b.y. (see section on basin chronology in this chapter). It thus appears from the available age data that the earliest dated mare basalt eruptions at Mare Tranquillitatis

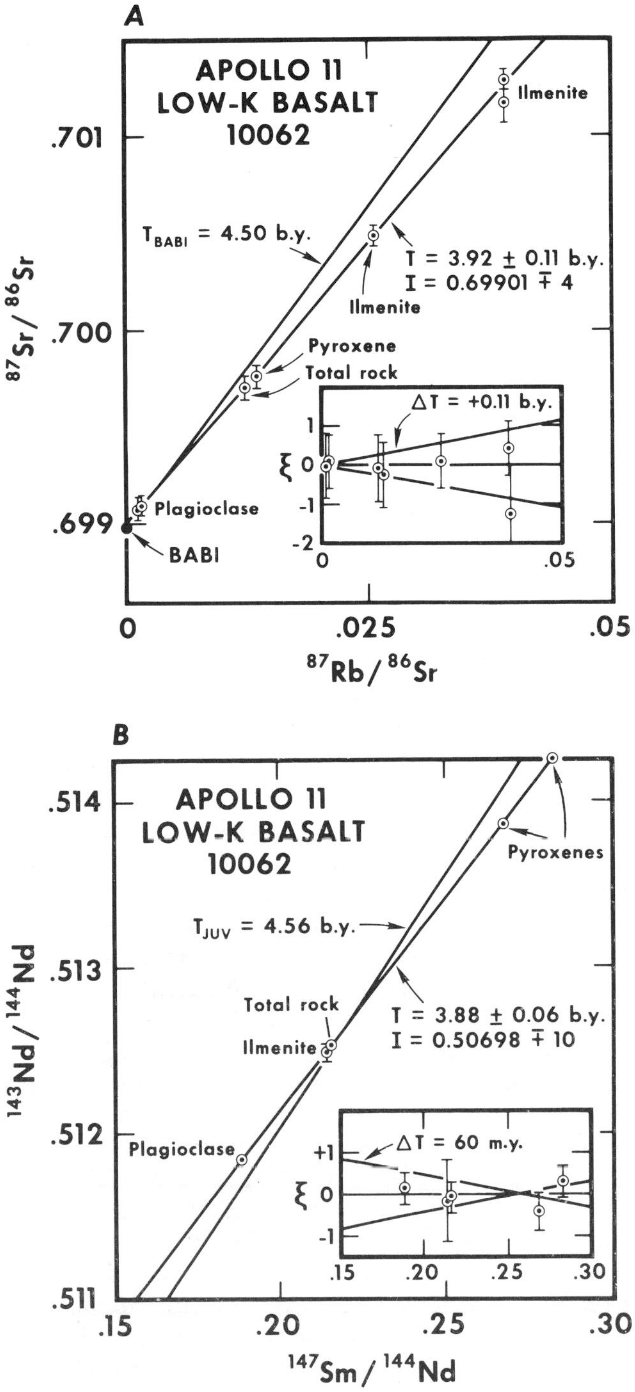

7.3.2 Two mineral isochrons for low-K mare basalt 10062 by the Rb-Sr and Sm-Nd methods (Papanastassiou *et al.*, 1977b). Insets show deviations in $^{87}Sr/^{86}Sr$ (**a**) and $^{143}Nd/^{144}Nd$ (**b**) in parts in 10^4. Results from both isochrons are in good agreement. Primarily because of the wider range of $^{143}Nd/^{144}Nd$ (5 parts in 10^3) compared with $^{87}Sr/^{86}Sr$ (3 parts in 10^3), the Sm-Nd method yields a more precise age. The age of this basalt by the ^{39}Ar-^{40}Ar method is 3.79 ± 0.04 b.y. (Guggisberg *et al.*, 1979) which, within errors, is compatible with the two internal isochron ages.

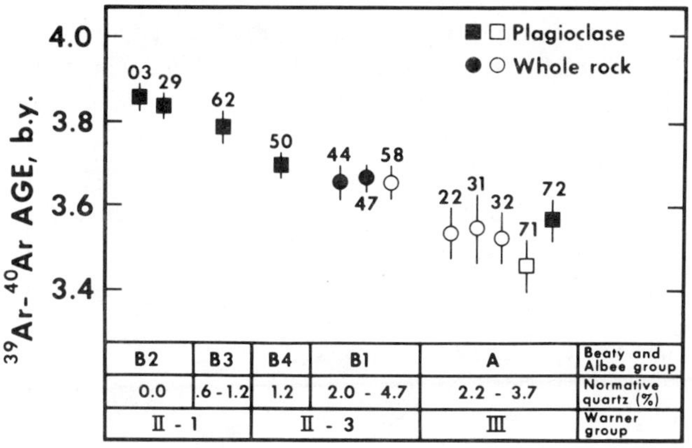

7.3.3 ^{39}Ar-^{40}Ar ages vs. petrological groupings for Apollo 11 basalts (figure from Guggisberg *et al.*, 1979). The numbers are Apollo 11 sample numbers. The petrological groups defined by Beaty and Albee (1978) and by Warner (1971) and the normative quartz content are given at the base. All ^{39}Ar-^{40}Ar ages were determined by the Bern group. Wherever a comparison is possible, there is good agreement with Turner's (1970) data. Closed symbols refer to very reliable age determinations (mostly from plagioclase). Open symbols represent somewhat less reliable determinations.

were contemporary with and perhaps triggered by the Imbrian impact event (cf. Guggisberg *et al.*, 1979). On the basis of crystallization ages of Apollo 11 basalts, it appears that the basin flooding that produced Mare Tranquillitatis continued for about 300 m.y. At least four distinct, fairly thin flows of low-K basalt must have erupted over a 200 m.y. period, followed by either one or several, as yet indistinguishable, very thin flows of high-K basalt. Not only the crystallization ages of the Apollo 11 rocks, but also their cosmic ray exposure age and history are correlated with the petrological classification (Eberhardt *et al.*, 1970; Geiss *et al.*, 1977). Low-K rocks, with the exception of the B4-type rock 10050, all have exposure ages from 65–130 m.y. (Fig. 7.3.4). They all have low $(Xe^{131}/Xe^{126})_c$ ratios, indicating that the exposure to the cosmic radiation must have occurred at or very close to the lunar surface. [The cosmogenic $(Xe^{131}/Xe^{126})_c$ ratio depends on the average shielding during exposure—low shielding corresponding to low ratios and high shielding to high ratios: cf. Eberhardt *et al.*, 1971; Burnett *et al.*, 1975.] The exposure ages of the high-K rocks vary from 30 m.y. to 500 m.y., and the systematically higher $(Xe^{131}/Xe^{126})_c$ ratios indicate that they must have received a substantial amount of their irradiation at some depth (approximately 0.5–3 m). The type B4 low-K rock 10050 has an irradiation history similar to the high-K rocks. Eberhardt *et al.* (1970) have discussed several models explaining these systematic differences in irradiation history. Their model B, proposing that the high-K basaltic bedrock resides closer to the surface at the Apollo 11 site than the low-K basaltic bedrock, is in agreement with the stratigraphic sequence obtained from the ^{39}Ar-^{40}Ar ages.

Oceanus Procellarum (Apollo 12)

The Apollo 12 basalts collected from Oceanus Procellarum are characterized by low-Ti content as compared with Apollo 11 rocks. They have been classified by Warner (1971) into two main groups: (1) porphyritic basalts, and (2) ophitic basalts. The porphyritic basalts have phenocrysts imbedded in a matrix of glass or fine-grained crystals of augite, plagioclase and opaques. Texturally, each group is further subdivided into five types. Compositionally, Apollo 12 basalts fall into four major types: Olivine (13); Pigeonite (11); Ilmenite (11); and Feldspathic (1) [James and Wright, (1972); Rhodes *et al.* (1975); Papike *et al.* (1976); Rhodes *et al.* (1977)]. The work of Rhodes *et al.* (1977) in particular has established the ilmenite basalts as a major type at the Apollo 12 site.

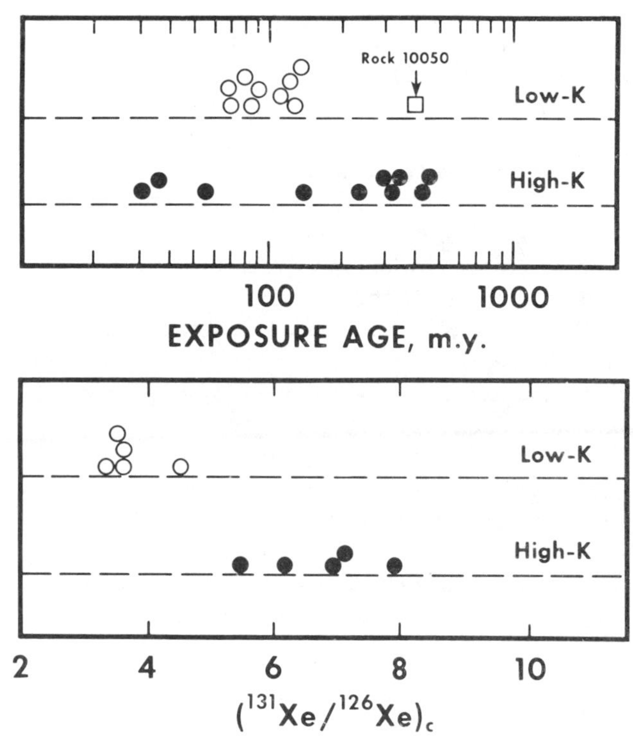

7.3.4 Apparent exposure ages and $(^{131}Xe/^{126}Xe)_c$ ratios for Apollo 11 basalts (figure adapted from Eberhardt *et al.*, 1970). Low $(^{131}Xe/^{126}Xe)_c$ ratios correspond to cosmic ray irradiation at or close to the surface; higher ratios are observed for the high-K rocks indicating irradiation at greater depth (below ~40 cm).

The ^{39}Ar-^{40}Ar ages of Turner (1971) and the Rb-Sr internal isochron ages of Papanastassiou and Wasserburg (1970b, 1971a) and Nyquist *et al.* (1977) are in accord in placing the time of crystallization of Apollo 12 basalt in a narrow time interval 3.08–3.29 b.y. ago. The Rb-Sr data of Cliff *et al.* (1971) and Compston *et al.* (1971) are in agreement with the above age range. The data clearly indicate that Apollo 12 basalts are younger than those of Apollo 11, and if the range of crystallization ages is directly correlated with the duration of magmatism, one might assume that Oceanus Procellarum was inundated in about half the time needed for flooding Mare Tranquillitatis. The range of initial ratios $(^{87}Sr/^{86}Sr)_I$ for Apollo 12 is shifted to higher values (0.6992–0.6997) than those from Apollo 11, reflecting the younger ages of Apollo 12 rocks (Papanastassiou and Wasserburg, 1971a; Nyquist *et al.*, 1977).

A two-point U-Pb internal isochron was determined for rock 12064 and was found to be compatible with the Rb-Sr age of 3.18 b.y. (Papanastassiou *et al.*, 1977 a). Recently, Sm-Nd isochrons were reported for pigeonite basalt 12039 and ilmenite basalt 12056. The two basalts have the same age of 3.2 b.y. but different initial $^{143}Nd/^{144}Nd$ indicating that these two types of basalts were derived from different sources (Nyquist *et al.* 1979a, b).

Hadley-Apennines (Apollo 15)

Apollo 15 landed at the Hadley-Apennines region on the edge of Mare Imbrium and returned basalts that are characterized by low-to-average-Ti content. On the basis of their texture and mineralogy, these basalts are divided into four types (LSPET, 1972; Brown *et al.*, 1972): (1) porphyritic clinopyroxene basalts, (2) porphyritic clinopyroxene vitrophyre, (3) porphyritic olivine basalt, and (4) vesicular basalt.

The ^{39}Ar-^{40}Ar ages on Apollo 15 basalts are generally in good agreement with Rb-Sr ages. The ages obtained by the Rb-Sr method for six rocks range from 3.21–3.37 b.y. (Papanastassiou and Wasserburg, 1973). A survey by Husain (1974), mostly on rake samples and fragments, yielded ages ranging from 3.10–3.35 b.y., obtained by the ^{39}Ar-^{40}Ar method. Thus, Apollo 15 basalts are quite similar to Apollo 12 basalts in terms of their ages and their low-Ti content. The $(^{87}Sr/^{86}Sr)_I$ values are low and slightly above BABI (0.69899). They show a very limited spread from $I = 0.69923$–0.69937 and are generally indicative of the derivation of most of these basalts from a single homogeneous source.

The porphyritic olivine basalt 15555 is one of the most widely studied lunar rocks and has been dated by four different methods. It is the first lunar basalt for which a reasonably precise U-Pb internal isochron was established (Tera and Wasserburg, 1975). The U-Pb data are shown in Fig. 7.3.5a.

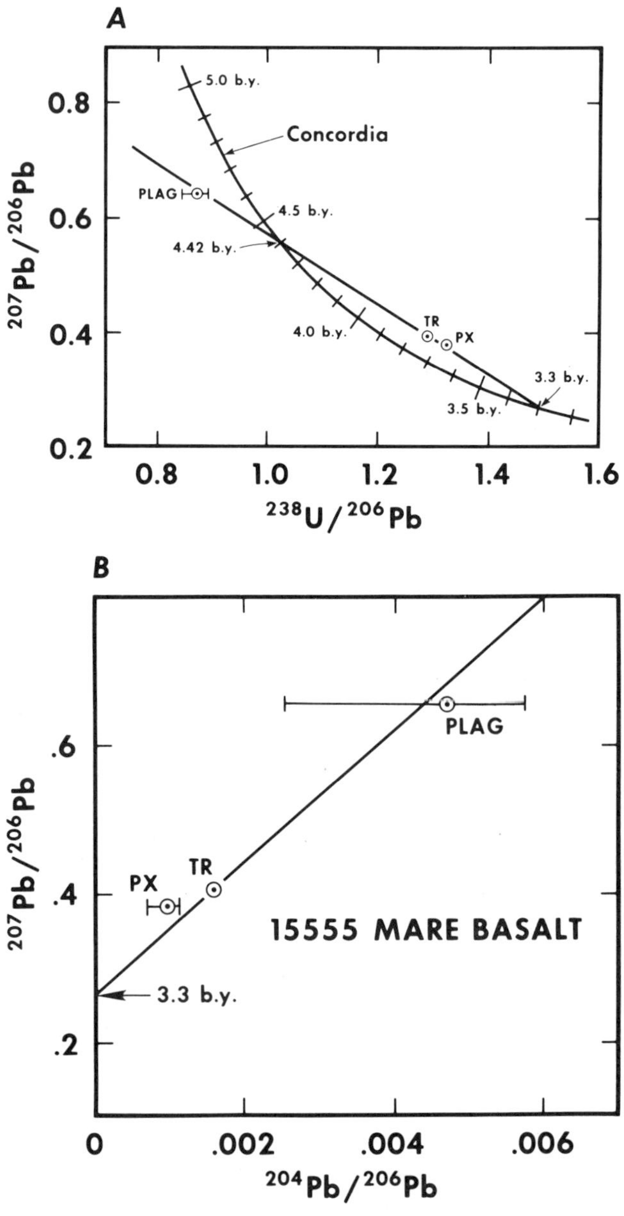

7.3.5 a. U-Pb internal isochron for lunar mare basalt 15555 (Tera and Wasserburg, 1975). The crystallization age (as given by the isochron lower intersection with concordia) is 3.30 ± 0.04 b.y., in agreement with Rb-Sr and ^{39}Ar-^{40}Ar ages (see Table 7.3.1). **b.** $^{207}Pb/^{206}Pb$–$^{204}Pb/^{206}Pb$ diagram for 15555 (Tera and Wasserburg, 1975). The "isochron" is drawn to pass through the highly precise total rock data point and the point on the $^{207}Pb/^{206}Pb$-axis corresponding to 3.30 b.y. The pyroxene datum falls off the line, indicating either analytical problems or incomplete isotopic equilibrium within the rock.

The isochron shown in Fig. 7.3.5a has a lower intersection with the concordia curve at 3.3 b.y., which is in close agreement with the Rb-Sr age (Papanastassiou and Wasserburg, 1973) as well as the ^{39}Ar-^{40}Ar age (Podosek *et al.*, 1972). The presence of radiogenic initial Pb is demonstrated by the large disparity between the Pb composition in the plagioclase and the pyroxene. From the figure, the whole rock $^{207}Pb/^{206}Pb$ model age is 3.9 b.y. (which is 600 m.y. older than the actual crystallization age) and has no time meaning. Clearly, the whole rock data point is discordant despite its proximity to the concordia curve, and the use of the near concordance of lunar basalts as evidence that U-Pb whole rock ages are really dating earlier events is not justified. Old lunar basalts plot close to concordia; this is largely a geometrical effect dictated by shallow angles of intersection of the isochron with concordia. The upper intersection with concordia does not necessarily have a time meaning, but if a two-stage model is applicable (or is a close approximation to reality), then the age of the source from which rock 15555 was derived is 4.42 b.y. old.

One of the most widely used representations of Pb data is a diagram in which $^{206}Pb/^{204}Pb$ is plotted vs. $^{207}Pb/^{204}Pb$. The application of this diagram to lunar Pb data could lead to erroneous conclusions. This is primarily due to the very low concentration of ^{204}Pb in lunar rocks, a circumstance that makes the Pb blank a potential end-member that artificially highlights linear trends. In addition, errors due to ^{204}Pb are correlated in the diagram and differences in $^{207}Pb/^{206}Pb$ are often poorly resolved. This situation may give rise to linear trends for unrelated samples. To minimize these problems the diagram has been modified (Tera and Wasserburg, 1972b) and the Pb data for rock 15555 are shown plotted on such a modified version (Fig. 7.3.5b; Tera and Wasserburg, 1975). Errors due to ^{204}Pb are not correlated in the presentation shown and the influence of blank is readily detected because it moves the data points at a distinct angle relative to the isochron (as indicated by the direction of the error bars). In addition, variations in $^{207}Pb/^{206}Pb$ ratio are clearly resolved. As discussed in section 7.1.2, the composition of Pb which decayed *in situ* (yielding the age of the rock) is obtained from the intercept and not from the slope of the isochron. It is seen that the Pb-Pb age obtained appears to be in general agreement with the U-Pb age (Fig. 7.3.5b). Such cross checks within the same method are unique to the U-Th-Pb system because of the availability of multiple radiometric clocks. These cross checks are important for verifying an age, as well as for exposing possible artifacts. However, they are rarely applied successfully because of technical difficulties, the most important of which are the Pb blank contamination and the poor ionization efficiencies of Th and U (see section 7.3.2).

In addition to mare basalts, the Apollo 15 rocks include a number of KREEP basalts. These igneous-textured rocks are thought not to be of local origin (Phinney *et al.*, 1972; Reid *et al.*, 1972; Taylor *et al.*, 1973; Carr and Meyer, 1974). This view is in accord with the results of the orbital γ-ray experiment which indicates that the Apennines are only weakly radioactive (Metzger *et al.*, 1974). It has been suggested that KREEP was transported to the Apennine front from the crater Autolycus (Phinney *et al.*, 1972; Reid *et al.*, 1972; Taylor *et al.*, 1973; Carr and Meyer, 1974). Autolycus falls on the second ring of the Imbrium Basin and it is reasonable to assume that the rings are covered by ejecta from the Imbrium Basin (Nyquist *et al.*, 1975). Rb-Sr internal isochrons for the two KREEP basalts 15434 and 15386 (Nyquist *et al.*, 1974, 1975) showed them to be distinctly older (3.83 ± 0.04 and 3.86 ± 0.01 b.y., respectively) and having initial $^{87}Sr/^{86}Sr$ values more evolved (0.70070 ± 10 and 0.70038 ± 10, respectively) than Apollo 15 mare basalts. A Sm-Nd internal isochron on 15386 yielded an age of 3.85 ± 0.08 b.y. (Carlson and Lugmair, 1979).

Mare Serenitatis (Apollo 17)

Apollo 17 landed at the Taurus-Littrow valley at the southeastern edge of the Serenitatis Basin, and returned mare basalts that are very similar to Apollo 11 basalts. Apollo 17 basalts are characterized by high-Ti content and a wide range of textures ranging from coarse-grained to vitrophyric. Papike *et al.* (1974) have divided Apollo 17 basalts into two groups on the basis of the Ti concentration. The first and least abundant group is practically identical to the Apollo 11 low-K ophitic basalts. The second group is characterized by still higher TiO_2 concentrations of ~13 wt.%. Vaniman and Papike (1977) discovered in the >0.02 mm soil fraction of Apollo 17 drill core many fragments of a new mare basalt characterized by very low titanium. This type of basalt, denoted VLT basalt, has <0.5 wt.% TiO_2 as compared with ~1.5 wt.% to ~13 wt.% for most mare basalts. Earlier, Green *et al.* (1972b) described a Luna 16 mare basalt with 0.4 wt.% TiO_2.

Apollo 17 basalts have been subject to wide compositional investigations but no general consensus regarding their genesis has been reached. Opinions range from implicitly assuming a single parental magma and ascribing the observed compositional range to near-surface crystal fractionation (Longhi *et al.*, 1974; Brown *et al.*, 1975; Kesson, 1975; Walker *et al.*, 1975), to inferring as many as five distinct parental magmas

(Laul *et al.*, 1975). On the basis of trace element investigations, however, it was concluded that the dispersion is too large to be explained by crystal fractionation alone (Warner *et al.*, 1975) and that the diversity is best explained in terms of several parental magmas (Shih *et al.*, 1975). In their trace and major element investigation, Rhodes *et al.* (1976) stressed the importance of vitrophyric, spherulitic and fine-grained textural variations on the assumption that these basalts have cooled quickly (Usselman *et al.*, 1975; Brown *et al.*, 1975) and thus should reflect magma composition more closely than the coarse-grained variants. These authors concluded that: (1) the basalts sampled at the Apollo 17 site represent three distinct types (denoted A, B and C) requiring derivation from separate and distinct parental magmas. A fourth category of coarse-grained basalts could not be confidently assigned to any group. (2) The difference in chemistry between the three types is a primary feature and is attributed to partial melting of a source that is markedly heterogeneous with respect to trace and minor elements. (3) The chemical variations within each of the three types can be explained in terms of moderate (4–22%), near-surface crystal fractionation.

A total of 18 age determinations by the ^{39}Ar-^{40}Ar method have been carried out on Apollo 17 basalts. These include dating of five large rocks and eight 2–4 mm fragments. The ^{39}Ar-^{40}Ar ages obtained range from 3.69–3.84 b.y. The actual spread might be much narrower because of possible systematic differences between different laboratories (see detailed discussion in a review by Turner, 1977). Precise dating of three rocks (75035, 75055 and 75075) by using mineral separates yielded ^{39}Ar-^{40}Ar ages ranging from 3.69–3.73 b.y. (Turner and Cadogan, 1975; Huneke *et al.*, 1973; Jessberger *et al.*, 1975; Horn *et al.*, 1975a). The Rb-Sr age data from all laboratories are in good agreement and range from 3.56–3.77 b.y. Four analyses from the same laboratory (70017, 70035, 70135, 75075) have yielded a weighted average of 3.68 ± 0.06 b.y. (Nyquist *et al.*, 1974, 1975) which is identical to the result from the precise ^{39}Ar-^{40}Ar analysis. There is also agreement among different laboratories on the value of $(^{87}Sr/^{86}Sr)_I$ for Apollo 17 basalts.

As shown in Table 2 of Murthy and Coscio (1976), the $(^{87}Sr/^{86}Sr)_I$ values are very tightly grouped and unresolvable. The values in their Table 2 are corrected for stated interlaboratory biases as given by individual laboratories (Murthy *et al.*, 1971; Evensen *et al.*, 1973; Nyquist *et al.*, 1975) relative to CIT data. The mean of all the data points is calculated to be 0.69912 ± 0.00002 (Murthy and Coscio, 1976). From the above, it appears that Apollo 17 basalts were derived from similar magma sources. This seems to indicate that the generality of the conclusion regarding the heterogeneity of the parental magmas of Apollo 17 (Rhodes *et al.*, 1976) does not apply to Rb-Sr. Nyquist *et al.* (1976) noted that the basalts of group C of Rhodes *et al.* (1976) which have essentially the same age and initial $^{87}Sr/^{86}Sr$ parameters, show a two-fold higher Rb/Sr. The authors argue that the increase was produced by small degrees of partial melting ($\lesssim$ 10%) of a parental magma containing moderate to large proportions of clinopyroxene as a residual phase while the majority of Apollo 17 basalts were probably produced by moderate partial melting ($\lesssim$ 30%). These authors (Nyquist *et al.*, 1976) reject near-surface (low pressure) crystal fractionation as the major cause of the observed Rb-Sr fractionation. Differentiation between the two alternatives is by necessity based more on deduction from the chemistry of the rocks rather than the Rb-Sr systematics as both mechanisms are compatible with the observed consistency of $(^{87}Sr/^{86}Sr)_I$ and variability of Rb/Sr.

Four U-Pb internal isochrons exist for Apollo 17 basalts and they are generally compatible with the age data from the other methods. A three-point U-Pb internal isochron for basalt 75035 yielded an age of 3.56 ± 0.40 b.y. (Nunes *et al.*, 1974). The large error quoted is a consequence of the uncertainty in the Pb blank associated with the separation of the Pb-poor plagioclase separate. A clean plagioclase separate from basalt 75055 was obtained and a two-point "isochron" of 3.8 b.y. was reported for that rock (Tera and Wasserburg, 1976). Mattinson *et al.* (1977) determined a four-point U-Pb internal isochron for basalt 70017 and obtained a 3.7 b.y. crystallization age. As usual, the plagioclase point controls the U-Pb isochron and because of the small amount of Pb analyzed ($\sim10^{-9}$g) in that phase the age obtained is imprecise. Tilton and Chen (1979) attempted to establish a U-Pb internal isochron for basalt 71055. The data obtained scatter in a manner suggestive of post-crystallization disturbance. The authors, however, suspect the contamination of some of the samples with subnanogram to nanogram amounts of Pb during preparation. Removal of the inferred contamination causes all samples to plot along a chord yielding a 3.6 b.y. age. A similar attempt on basalt 75075 by Chen *et al.* (1978) led these authors to conclude that the rock suffered a post-crystallization disturbance of the U-Pb system. The authors note that the data points most deviant from a 3.8 b.y. isochron (similar to that of 75055) are the Pb-poor plagioclase separates and that there is enough ^{204}Pb in these mineral separates to allow subtraction of a larger blank (which causes the data points to fall on the isochron). The authors argue, however, against the contamination possibility. The above examples illustrate the extreme difficulties encountered in establishing U-Pb isochrons for lunar basalts. The major stumbling block appears to be the contamination of the mineral separates with minute amounts of Pb($\leq 10^{-9}$g) prior to chemical separation.

The general consistency of geochronological data by different methods is further indicated by an Sm-Nd isochron on basalt 75075 yielding an age of 3.70 ± 0.07 b.y. (Lugmair *et al.*, 1975) and on basalt 70135 yielding an age of 3.77 ± 0.06 b.y. (Nyquist *et al.*, 1979b).

7.3.4 Interpretation of isotopic data from mare basalts

Mare basalts are the products of melts in the lunar interior. As such, and particularly in view of their relatively simple history, they are potential sources of information regarding the nature of the lunar mantle and the processes of differentiation and fractionation. The first order conclusions from the isotopic data at hand are provided below, followed by elaboration in subsequent subsections:

1. Lunar magmatism represents a short duration (3.8–3.1 b.y.) when compared with the age of the planet, which has existed as a separate body for ~4.5 b.y. In view of limited sampling, however, the existence of basalts with ages outside the established range cannot be excluded, and indeed have been inferred from crater frequency and degradation data (sections 2.2.1 and 8.6–7).

2. The duration of flooding of individual preexisting giant basins by magmas was of the order of a few hundred million years. Eruption was associated with different degrees of mild elemental fractionation.

3. The sources (reservoirs) of mare basalts are heterogeneous. The heterogeneity was established early in time (~4.4 b.y.) and was preserved to at least the time of magma eruption.

4. The sources of lunar basalts are generally old (~4.4 b.y.) approaching the age of the planet itself. This indicates that the differentiation process could not have lasted more than ~150 m.y. after accretion.

5. Despite their antiquity, these sources are not pristine primary materials but rather secondary (or higher order) differentiates.

Duration of lunar magmatism

The stated range of 3.8–3.1 b.y. for lunar magmatism is the actual range of crystallization ages as obtained largely by the Rb-Sr and the ^{39}Ar-^{40}Ar methods. In the few cases where the U-Pb and Sm-Nd methods were applied to lunar basalts, the data by all methods were found to be quite compatible. This accord between radiometric clocks of diverse characteristics, which involves parent-daughter pairs of different affinities, is a strong indication that the precise parent-daughter linear correlations established are bonafide isochrons and not mere mixing lines. The Pb literature contains occasional allegations of disagreements with other methods. With the exception of one case where the rock (basalt 12021; Tatsumoto *et al.*, 1971) showed clear evidence of post-crystallization disturbance not detected by the other methods, the alleged disagreements with the other methods can be explained by one or more of the following factors:

(a) Pb contamination problems that can produce observed trends.

(b) Internal inconsistencies which invalidate the trends upon which the alleged discrepancies rest.

(c) The amplification effects inherent in the U-Pb method due to the large disparity in the mean life of the two U parents. This is reflected in a rapid change in the Pb isotopic composition in early times and magnifies the effect of mild parent-daughter fractionation.

If most of the giant basins were excavated within a narrow time span associated with the terminal stage of heavy bombardment ~3.9 b.y. ago, then the crystallization ages of old basalts (3.8 b.y.) would seem to define the starting time of mare flooding but not necessarily the time of onset of lunar volcanism, as the latter could have preceded the last intense episode of bombardment and its morphologic features have been eradicated by it.

In 1959 the lunar surface photographs returned by the Soviet Lunik 3 revealed the striking fact that in contrast with the nearside, the farside of the Moon shows little signs of basalt plains. Since large basins are about equally distributed on both sides (Stuart-Alexander and Howard, 1970; Hartmann and Wood, 1971), the asymmetry of basalt distribution is attributed to a thicker crust on the farside rather than to different cratering histories. This deduction (Wood, 1973; Kaula *et al.*, 1974) is compatible with Apollo laser altimeter data indicating that the Moon's mass-center is offset toward Earth (Roberson and Kaula, 1972; Kaula *et al.*, 1972, 1973). Basin excavation could have triggered the initial stages of extensive lava eruption through the thinner crust of the nearside to maintain hydrostatic equilibrium. Relating lava eruption to excavation of the giant ring basins on the nearside is in harmony with the fact that the ages of old basalts (~3.8 b.y.) overlap with the terminal bombardment (3.8–4.0 b.y.; Tera *et al.*, 1973, 1974a). This mechanism would not be completely invalidated by the discovery of basalts older than 3.8 b.y. as those could have been derived from magma that was extruded to the surface by impacts preceding the terminal episode of intense bombardment. A proof for the (partial?) decoupling of lunar volcanism from giant basin excavation is the discovery of volcanoes and/or other volcanic features such as cones and domes. Such features are not lacking on the Moon and they are particularly concentrated in Oceanus Procellarum, suggesting the existence of a major center of volcanic activity in that region [see Head (1976) for literature and details]. In addition, it could be argued that the fact that magmatism lasted for at least 700 m.y. after the terminal episode of intense bombardment is an indication of lunar volcanism as a mechanism of magma extrusion independent of bombardment.

Whether magma extrusion to the surface of the Moon continued to a more recent time than ~3 b.y., is by necessity also uncertain. Although active search for basalts of ages younger than ~3 b.y. yielded negative results, it could be argued that this is a reflection of limited sampling. Studies of the lunar surface characteristics including albedo (Bryan and Adams, 1973), spectral reflectivity (Soderblom and Lebofsky, 1972; Pieters and McCord, 1975) cratering (Boyce and Dial, 1973), and morphology (Head, 1976) indicate that Apollo and Luna missions did not sample all the distinct lunar basalt units. (For a detailed review, see sections 2.2.1, 5.4, 8.6–7; Head, 1976.) In addition, Boyce *et al.* (1974) suggest, on the basis of the crater density argument, that mare basalt units as young as 1.7–2.5 b.y. exist in the Imbrium and Serenitatis basins. This younger age assignment seems quite tentative, however, because of some uncertainty associated with the calibration of crater density vs. radiometric ages, particularly in the younger age portion of the calibration curve (see sections 8.4 and 8.7).

A range of ages was found at each landing site. From these some statements were made about the duration of flooding of a number of the ring basins (section 7.3.3). These interpretations are uncertain because they are based on the assumption that the time required for filling a basin is represented by the maximum range in the ages of the surface basalts. This assumption may apply only if flooding consisted of mere thin flows covering older debris transported earlier to the basin, presumably by impacts. Assuming that such is the case and that the maximum ranges were actually established, the data indicate that the flooding of a mare basin was a slow process requiring 2–4 $\times$ 10^8 yr. Although the Apollo 17 basalts do not exhibit such a range, Papanastassiou *et al.* (1977) argue that the younger age of orange glass determined by Pb isotopes (3.48 b.y.; Tera and Wasserburg, 1976) and ^{39}Ar-^{40}Ar (3.50 $\pm$ 0.05 b.y.; Huneke *et al.*, 1973) provides strong evidence for extended volcanism in the Taurus-Littrow valley. [The actual age of the orange glass remains uncertain however. Schaeffer and Husain (1973b) report an ^{39}Ar-^{40}Ar age of 3.66 $\pm$ 0.06 b.y. for this sample. In addition, the U-Pb age is not in agreement with the Pb-Pb age (Tera and Wasserburg, 1976).]

It is reasonable to assume that the earliest lava flows that flooded a given basin are largely buried and not available for sampling. Some of the returned samples of surface mare basalts are almost as old as the basins themselves, however, which suggests the alternatives:

(a) Some of the oldest mare deposits remained exposed perhaps because the lava flooded shallow regions of the basins and escaped burial under subsequent younger flows.

(b) The rate of basin filling is not uniform, being generally fastest immediately after basin excavation. Such a mechanism would be biased toward exposing older basalts.

(c) Old basalts come from basins older than assumed.

(d) Basins exposing older basalts are shallower than basins with younger surface flows.

According to Head (1975) the older basins indeed appear to contain shallower mare fill than the younger ones. He attributes this to the possibility of decrease in the depth of older basins by approaching isostatic equilibrium due to a thinner crust in the early lunar history. This explanation appears rather tenuous, however, if one accepts the currently favored viewpoint that basin excavation took place within a narrow time span (~3.8–4.0 b.y.).

Despite the possible involvement of multiple factors governing the emplacement of mare deposits, it is interesting to note that old basalts of similar ages were returned by Apollo 11 and Apollo 17. In both cases the landing site is at the edge of a mare (Tranquillitatis and Serenitatis, respectively) where the deposits approach a feather edge less thick than the average for the basin as a whole. Apollo 15, which returned younger basalts, also landed at the margin of a mare (Imbrium) but the mare deposit is thick at the landing area because of the existence of a fault trough in the region (Carr *et al.*, 1971; Howard and Head, 1972). A 55 m thick basaltic bedrock of the Hadley Rille is exposed near the Apollo 15 landing site. The rille itself is estimated to be 350 m deep (Howard *et al.*, 1972; Swann *et al.*, 1972). Basalts of Apollo 12 are samples of thin lava flows but they do not enter in the comparison because the landing site lies in an area where a lunar basin is unrecognized (Head, 1976).

The basalt fill in Oceanus Procellarum is estimated to be generally less than 200 m thick (Pohn, 1971; Head, 1976). For Tranquillitatis Basin the average mare fill is deduced to be 50–60 m (DeHon, 1974) while the estimate for Serenitatis is 1 km (Talwani *et al.*, 1973). Mare cooling units were estimated by Brett (1975) to be uniformly thin, perhaps no thicker than 10 m. This estimate was obtained from consideration of the data of the kinetics of the chemical reactions, thermal conductivity and specific heats of lunar rocks. Later estimates by Walker *et al.* (1977), however, suggest thicker flows of several tens of meters.

Heterogeneity of magma sources

The overall isotopic data indicate that the source of mare basalt, presumably the lunar upper mantle, is heterogeneous. Like the situation on Earth, whether the heterogenity is due to layering or production of lumps (in a more or less homogeneous or layered mantle) is uncertain. The following is a brief review of the evidence of heterogeneity from various radiometric systems.

Rb-Sr data. $^{87}Sr/^{86}Sr$ has long been used in terrestrial studies as an indicator of the nature of magma sources of young basalts. The application has been extended and extensively applied to lunar rocks. The data for Apollo 11 initial Sr (I) are shown in Fig. 7.3.6 on a diagram of $^{87}Sr/^{86}Sr$ vs. age (Papanastassiou *et al.*, 1970, 1977b) and clearly show one aspect of the source's heterogeneity. The Apollo 11 high-K basalts could all have been derived at about the same time from a single reservoir, which is distinctly different from that of the low-K basalts. Similarly, the low-K basalts could all have been derived from a single reservoir but over a period of some 350 m.y. as indicated by the resolution of the data which, within the error envelopes, can be fit on a single straight-line evolution curve.

The conclusion that the two rock types of Apollo 11 basalts evolved from two different magma sources and are not the result of fractional crystallization from a single magma reservoir is in harmony with petrologic reasoning. Shih *et al.* (1975) pointed out that deriving high-K basalt by fractional crystallization from a liquid that resembles low-K basalts involves a number of implausible circumstances: enrichment of Rb, for example, by a factor of 20 or more would require 95–100% fractional crystallization, making it difficult to separate the residual liquid and to interpret the absence of accumulates among the samples.

The Apollo 12 rocks define a Rb-Sr whole rock isochron of ~4.4 b.y. (Compston *et al.*, 1971; Cliff *et al.*, 1971; Hubbard and Gast, 1971; Papanastassiou and Wasserburg, 1971a) indicating approximately closed-system history and near absence of Rb-Sr fractionation for ~4.5 b.y. (Fig. 7.3.7). In contrast with the clear-cut splitting of intial $^{87}Sr/^{86}Sr$ into two groups

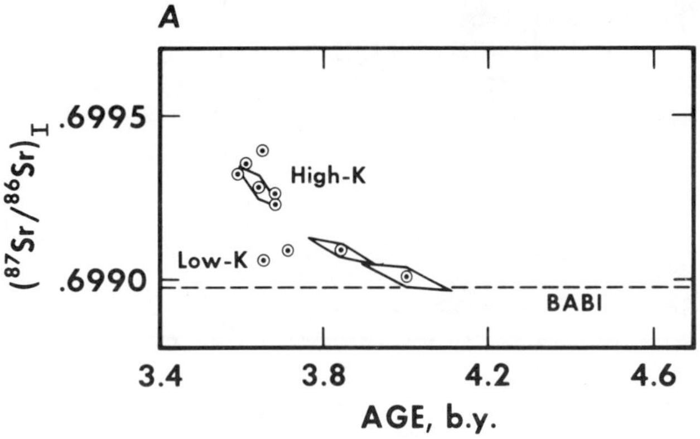

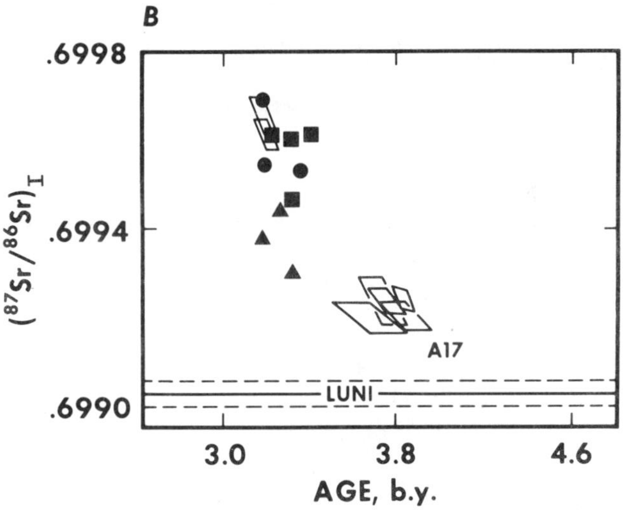

7.3.6 a. Initial $^{87}Sr/^{86}Sr$ vs. age for Apollo 11 mare basalts (from Papanastassiou *et al.*, 1977b). All high-K basalts are tightly grouped, indicating the possibility that these rocks were derived from a single source at about the same time. In contrast, low-K basalts show a distinct range. The alignment of the data in a linear trend, however, suggests that low-K basalts could also have been derived from a single source, but over an extended period . The dashed line marked "BABI" is an average initial value for basaltic achondrites (Papanastassiou and Wasserburg, 1969). **b.** Initial $^{87}Sr/^{86}Sr$ vs. age for Apollo 12 basalts. Data for olivine, pigeonite, and ilmenite basalts are shown as squares, circles and triangles, respectively. Data are from Papanastassiou and Wasserburg (1970; 1971a,b) and from Nyquist *et al.* (1977). Error polygons are shown for two pigeonite basalts. For simplicity, the rest of the Apollo 12 data are shown without error polygons. Note that initial $^{87}Sr/^{86}Sr$ values for olivine and pigeonite basalts overlap and lie within the range 0.69947–0.69969 while those of ilmenite basalts are lower and lie within the range 0.69924–0.69944. Initial $^{87}Sr/^{86}Sr$ for Apollo 17 basalts as determined by Nyquist *et al.* (1974, 1975, 1976) are shown for comparison. Lunar initial $^{87}Sr/^{86}Sr$ ratio, LUNI = 0.69903 ± 3 as determined from Apollo 16 anorthosites, is also shown. (From Nyquist *et al.*, 1977a).

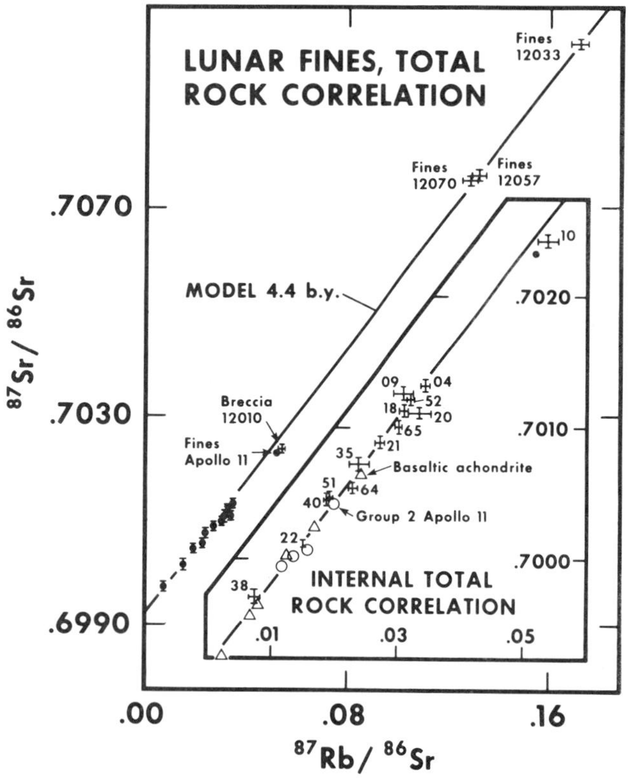

7.3.7 Rb-Sr isochron diagram showing the alignment of Apollo 12 fines and rocks along a 4.4 b.y. reference isochron. Low-K Apollo 11 basalts as well as basaltic achondrites also fall on the reference isochron. Triangles represent basaltic achondrites; filled circles, Apollo 11 fines; crosses, breccias; closed circles, Group 2 Apollo 11. (From Compston *et al.*, 1971.)

for Apollo 11 rocks, the initial ratios for Apollo 12 basalts cover an approximately continuous range and suggest the existence of multiple reservoirs. On the basis of their experimental resolution, Papanastassiou and Wasserburg (1971a) suggest that Apollo 12 basalts represent at least four magmatic sources. Further investigation by Nyquist *et al.* (1977) revealed some subtle aspects of Apollo 12 basalts (Fig. 7.3.6b): (1) Initial $^{87}Sr/^{86}Sr$ values for olivine and pigeonite basalts are indistinguishable. (2) Initial $^{87}Sr/^{86}Sr$ values for ilmenite basalts are lower. As pointed out by the authors, these observations are consistent with the interpretation of Rhodes *et al.* (1977) who on the basis of major and trace element abundances suggested the existence of two Apollo 12 types, the olivine-pigeonite magma and the ilmenite magma. The numerous lava flows inferred by Papanastassiou and Wasserburg (1970, 1971a) may be due to the possibility that more than one batch of magma contributed to each magma type (Nyquist *et al.*, 1977).

The very limited range of initial ratios for Apollo 15 rocks overlaps that of Apollo 12 rocks. The same is true of the ages of the rocks from both sites. This indicates extensive magmatism on the Moon ~3.2 b.y. ago, which tapped sources closely similar in their bulk composition and their initial $^{87}Sr/^{86}Sr$. The Rb-Sr model ages of Apollo 15 basalts are ~4.1 b.y., indicating that the basalts were enriched in Rb relative to Sr as compared with the source. Papanastassiou and Wasserburg (1973) calculate an enrichment factor of 1.4 if a 4.6 b.y. source is assumed.

The tight clustering of initial $^{87}Sr/^{86}Sr$ values for Apollo 17 basalts is suggestive of their derivation from sources with similar Rb/Sr. The whole-rock Rb-Sr data define a linear array corresponding to ~4.1 b.y. (Nyquist *et al.*, 1975; Fig. 7.3.8). Although the significance of this age is debatable it is consistent with increase of Rb/Sr in the magma relative to the source, but the exact mechanism and time of occurrence cannot be identified unambiguously from the available Rb-Sr data (see discussion above).

Assuming a two-stage model where a magma source was isolated T b.y. ago and remained closed till the time of eruption at τ (given by the crystallization age) it is possible to calculate the Rb-Sr fractionation factor F from the relationship

$$F \equiv \frac{\left(\frac{^{87}Rb}{^{86}Sr}\right)_2}{\left(\frac{^{87}Rb}{^{86}Sr}\right)_1} \cong \frac{T - \tau}{T_{BABI} - \tau} \qquad (7.3.1)$$

where the subscripts 1 and 2 refer to the first stage (the source) and the second stage (the rock), respectively. T_{BABI} is the single-stage age of the rock, i.e., the time required for the whole rock with its measured $^{87}Rb/^{86}Sr$ to evolve from $(^{87}Sr/^{86}Sr)_{BABI = 0.69898}$ to the value of $^{87}Sr/^{86}Sr$ in the rock today. Enrichment of Rb relative to Sr in the rock as compared with the source will result in $T_{BABI} < T$ (i.e., $F > 1$), while the relative depletion of Rb will cause the opposite effect. Because the decay constant of Rb is very small, the above relationship provides a good approximation to the value of F (Papanastassiou and Wasserburg, 1972a; Schonfeld and Meyer, 1972). For Apollo 11 high-K mare basalts F $\simeq$ 5. For the majority of other mare basalts from all missions F ranges from 0.7–1.5. (See review by Nyquist, 1977.)

U-Pb data. Isotopic analysis of lunar basalts showed them all to have Pb that is extremely radiogenic. $^{206}Pb/^{204}Pb$, for example, is of the order of 500 as compared with 9.307 and ~18 for primordial and terrestrial

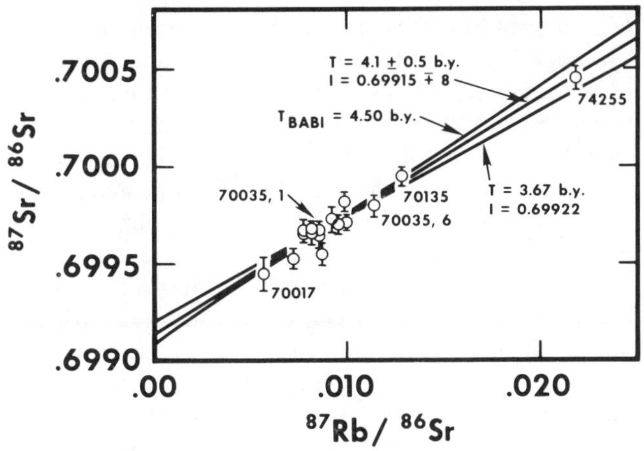

7.3.8 Rb-Sr whole rock "isochron" for sixteen Apollo 17 mare basalts. The least square fit of the data yields T = 4.1 ± 0.5 b.y. As was correctly pointed out by the authors (Nyquist *et al.*, 1975), time significance can be attached to this age only if no Rb-Sr fractionation occurred at the time of magma extrusion 3.7 b.y. ago (which is the crystallization age of these rocks). It could be argued that Rb/Sr increased at that time and that the data yield a lower limit to the age of the source.

Pb, respectively. Similarly, $\mu \equiv {}^{238}U/{}^{204}Pb$ is for most lunar basalts 500 or higher, as compared with a value of ~9 for the Earth. These very high values indicate the depletion of primordial Pb relative to U in the outer layer of the Moon.

An important parameter which describes the degree of U-Pb fractionation associated with the planet's differentiation processes is μ. The planet's content of primordial Pb as well as its distribution relative to radiogenic Pb is another parameter that might characterize the planet. In addition, the nonradiogenic isotope ^{204}Pb (which is a component of primordial Pb) is used as the reference isotope in the $^{207}Pb/^{204}Pb$ vs. $^{206}Pb/^{204}Pb$ correlation diagram. Assuming a two-stage model as has been described above and by using a slight modification of Eq. (7.1.6), it is possible to calculate the present day μ value in the reservoir from the relationship

$$\left(\frac{^{206}Pb}{^{204}Pb}\right)_I = \left(\frac{^{206}Pb}{^{204}Pb}\right)_{prim} + \mu_1(e^{\lambda_8 T} - e^{\lambda_8 \tau}) \qquad (7.3.2)$$

where I stands for initial, i.e., primordial Pb + radiogenic Pb formed in the first stage interval (T-τ). "Prim" stands for primordial Pb as was first determined by Patterson *et al.* (1953) and later by Tatsumoto *et al.* (1973) and λ_8 is the ^{238}U decay constant. In principle, initial Pb can be obtained from the U-Pb isochrons. However, if no isochron is available and the crystallization age by another method is known (as usually is the case) then the initial Pb is calculated from the relationship

$$\left(\frac{^{206}Pb}{^{204}Pb}\right)_m = \left(\frac{^{206}Pb}{^{204}Pb}\right)_I + \mu_2(e^{\lambda_8 \tau} - 1) \qquad (7.3.3)$$

where $(^{206}Pb/^{204}Pb)_m$ is a measured value and μ_2 is obtained from the present whole-rock concentrations of ^{238}U and ^{204}Pb.

In these calculations it is assumed that Pb evolved in two stages. The first stage [referred to by a subscript 1 or I, e.g., μ_1 and $(^{206}Pb/^{204}Pb)_I$] is a time interval T-τ prior to the crystallization of the rock. During the first stage Pb is assumed to have evolved in a source out of which the magma was later extruded at τ. The second (referred to by subscript 2) is from τ to the present where Pb evolved further in the crystallized rock. Both the source and the rock are supposed to be perfectly closed systems for the intervals T-τ and τ, respectively.

Unfortunately, the extreme depletion of ^{204}Pb in the lunar samples and its abundance on Earth means that all measurements involving this isotope are very sensitive to terrestrial contamination. For most basalts, the Pb isotope data obtained from internal isochrons are not reliable enough to produce precise values for initial Pb, primarily because of contamination associated with mineral separation. In most cases in the following discussion, initial Pb has been calculated from the above relationship using the relatively well-determined $(^{206}Pb/^{204}Pb)_m$ and μ_2 values for the whole rock. This initial Pb value is then used to calculate μ_1 assuming that T (the age of the source) is 4.55 b.y. Although mostly qualitative in nature, μ_1 values thus obtained provide some insight into the sources of mare basalts.

For Apollo 11 basalts, μ_2 ranges from ~300 to ~1400 and μ_1 from ~100 to ~500 (Tatsumoto, 1970). From this one calculates a fractionation factor $F \equiv \mu_2/\mu_1$ ranging from ~1.9 to ~2.7 for these basalts. The narrow range of F is reflected in the clustering of the data in a limited field in the U-Pb concordia diagram. The near constancy of F for Apollo 11 basalts contrasts with the wider range of fractionation (F = 0.8–5.0) calculated for the Rb-Sr system. This indication of decoupling of the two systems, however, could simply be an artifact and may be decided when initial leads of these rocks are reliably determined. For Apollo 12 the ranges are 450–990 and 60–300 for μ_2 and μ_1, respectively (Tatsumoto *et al.*, 1971), and correspond to fractionation factors from ~3 to ~7.5 (except for rock 12064 for which $F \simeq 1.7$). As shown schematically in Fig. 7.3.9, when $F > 1$ it means a loss of Pb relative to U in the erupting magma as compared with

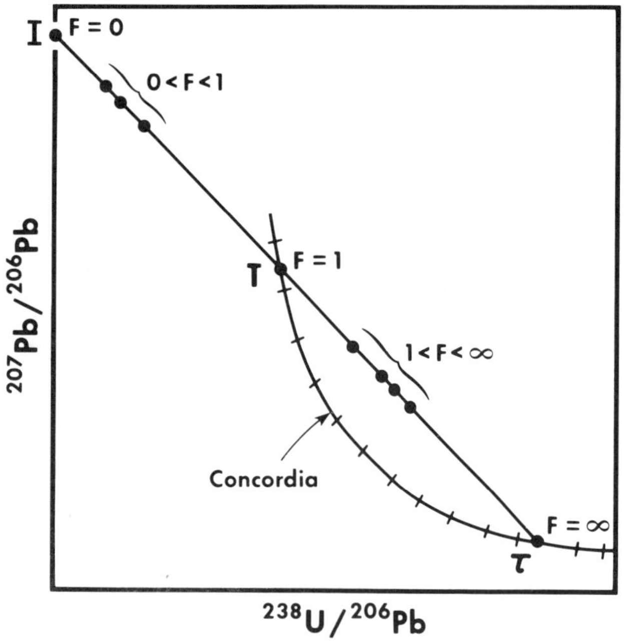

7.3.9 Schematic drawing of the U-Pb concordia diagram. The isochron intersecting concordia at T and τ can be looked upon as a mixing line of two end-members. One end point is given by the intercept I with the $^{207}Pb/^{206}Pb$ axis. This is the composition of the initial lead $(^{207}Pb/^{206}Pb)_I$ frozen unevolved since τ years ago, when its mineral phase (if it were present) was completely separated from U. Possible examples of such a phase are galenas and troilites. In a simple two-stage model, $(^{207}Pb/^{206}Pb)_I$ is assumed to have evolved in a single stage of duration T-τ. If we define a fractionation factor $F = \mu_2/\mu_1$, (see text), then for the end-member containing $(^{207}Pb/^{206}Pb)_I$, F = 0. The other end-member of the isochron is at the intersection τ and corresponds to a mineral phase that last crystallized (or metamorphosed) τ years ago and remained as a closed system since then. For such a phase, if it existed, F would be infinite. (Zircons often approximate such a phase). If a parent magma source was formed T years ago and subsequently magma was extruded from it at τ, then all the rocks crystallized from the magma at that time would plot on the isochron intersecting concordia at T and τ. The exact position of any given rock on the isochron is governed by the characteristic F value of that rock. For the special case where F = 1 (i.e., no U-Pb fractionation upon extrusion and crystallization), the rock would plot at the upper intersection. The mineral isochron of such a rock, however, would still yield its crystallization age τ.

the source. On the concordia diagram, samples of the same age derived from a single homogeneous source but experiencing different degrees of fractionation should define a line that intersects the concordia curve at two points corresponding to the crystallization age (lower intersection) and the "age" of the source (upper intersection). Apollo 12 basalts do not define such a trend (Tatsumoto *et al.*, 1971) although they are about the same age. This could in part be due to the heterogeneity of the source(s). The wide range of upper intersections suggested by the data, however, (~4.1 to > 5 b.y.) is suggestive of a complex multistage evolution not approximated by the two-stage model used in this discussion. Although by lunar standards Apollo 12 rocks define a wide range of $(^{87}Sr/^{86}Sr)_I$ values, there is no clear suggestion in the low I values of such a complex history, and the Rb-Sr model ages of ~4.5 b.y. for the rocks indicate limited Rb-Sr fractionation (F = 0.8–1.5). Thus, the Apollo 12 basalts seem to illustrate an actual decoupling of the two systems. This disparity in behavior does not represent discrepancy between the two methods as it applies to differentiation processes prior to crystallization. Indeed, one should not be surprised by conflicting whole-rock patterns from the two methods. These patterns are, except in rare special circumstances, void of strict temporal meaning and are the result of normal geochemical fractionations characteristic of various parent-daughter pairs.

If the primary material that produced the sources of mare basalts had a very low Rb/Sr ratio, then the Rb-Sr system would not be sensitive to mild Rb-Sr fractionations associated with the possibly multistage differentiation processes that produced the basalt sources. The situation is quite different for the $^{238,\,235}U$-$^{206,\,207}Pb$ system which involves a pair of parents that are isotopes of the same element, one of which is relatively short-lived, and which decay to a pair of daughters that are also isotopes of a common element. Thus even if the ratio $(U/Pb)_{prim}$ in the primary material were quite low, the resulting lead isotopic composition could be very sensitive to minor parent-daughter fractionations at early times, when the production rate of ^{207}Pb (from the relatively short-lived ^{235}U) was faster than that of ^{206}Pb. For this reason, when U-Pb internal isochrons for Apollo 12 basalts are established some day, they may reveal early multistage differentiations unrecognizable by either the Rb-Sr or Sm-Nd methods.

Apollo 15 basalts show a limited range for μ_2 (Tatsumoto *et al.*, 1972). Because of possible difficulty with ^{204}Pb these basalts are best represented by basalt 15555, which has been critically studied. For this rock μ_2 and μ_1 are 790 and 240, respectively (Tera and Wasserburg, 1974c) yielding F = 3. The whole-rock data points plot close to each other on the concordia diagram (Tatsumoto *et al.*, 1972) and imply upper intersections ~4.5 b.y. This is consistent with a limited U-Pb fractionation range and is suggestive of a simple history prior to crystallization. All basalts from the Apollo 11, 12 and 15 sites have U-Pb and Pb-Pb whole-rock model ages older than the crystallization ages, reflecting the presence of inherited initial Pb, as was actually shown by the U-Pb internal isochrons for a few rocks.

A unique feature of Apollo 17 basalts is a frequent concordance at an old age of ~4.5 b.y. (Nunes *et al.*, 1974). Again, because of apparent uncertainty in ^{204}Pb, these rocks are best represented by one basalt, 75055, which has been analyzed under stricter control (see Tera and Wasserburg, 1974). This rock is concordant at 4.42 b.y. and its measured μ_2 is 460 (Tera and Wasserburg, 1976). A two-point U-Pb "internal isochron" showed it to contain initial Pb and to have a lower intersection with concordia at 3.8 b.y. in agreement with the Rb-Sr crystallization age. By virtue of concordance at the upper intersection, $\mu_2 = \mu_1$ and F = 1 for this rock. In other words, magma extrusion was not associated with a change in U/Pb relative to the source.

In summary, the measured μ_2 values for mare basalts range from 300–1400, indicating μ_1 ranging from 60–500 for the sources of these basalts, assuming a two-stage evolution model. This suggests the existence of several distinct source regions for mare basalts. Insofar as the range in μ_1 is due to early lunar fractionations, one must conclude that the sources are not primary undifferentiated materials.

Sm-Nd data. All the available initial $^{143}Nd/^{144}Nd$ data are plotted as a function of age in Fig. 7.3.10 (Papanastassiou *et al.*, 1977b, p. 1658). The evolution of $^{143}Nd/^{144}Nd$ vs. time for a hypothetical reservoir with

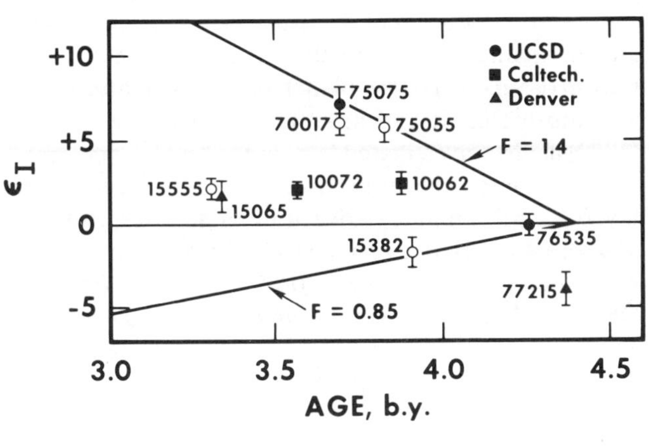

7.3.10 Deviations (in parts in 10^4) of initial $^{143}Nd/^{144}Nd$ of lunar rocks from Juvinas, plotted vs. time. Two lines are shown representing two 4.4-b.y.-old sources with Sm/Nd 1.4 and 0.85 times the Juvinas value. Filled points are samples for which Sm-Nd internal isochrons have been determined. Open points are calculated from whole rock Sm-Nd data of Lugmair (1975a) and Rb-Sr internal isochron ages. (From Papanastassiou *et al.*, 1977b).

the $^{147}Sm/^{144}Nd$ ratio of the chondrites and the primordial initial $^{143}Nd/^{144}Nd$ ratio determined for the basaltic achondrite Juvinas (Lugmair *et al.*, 1975) is shown in this figure.

The evolved nature of the initial $^{143}Nd/^{144}Nd$ ratio relative to the chondritic reservoir is evident, in agreement with the conclusion pointed out earlier—that the sources are not primary. The wide scatter of the data from any single straight evolution line is a further indication of the heterogeneity of the sources. Assuming a two-stage model, initial Nd for high-K basalt 10072 indicates that this basalt could not have evolved from the same source as low-K basalt 10062, since the two initial Nd data points in Fig. 7.3.10 do not fit on a single line intersecting the chondritic line at 4.4 b.y. This is in agreement with initial Sr data which indicate two different sources for Apollo 11 basalts; but the order is reversed showing the high-K basalt to have the less evolved initial ratio. This, however, is expected on geochemical grounds. The U-Pb system shows basalt 75055 to be among the least fractionated basalts while its initial Nd datum in Fig. 7.3.10 shows it to be among the most fractionated in the Sm-Nd system. All three systems are in agreement, however, about the coherence of the initial values for Apollo 15 basalts, but the deduced values for fractionation range from slightly above 1 (Sm-Nd) to ~3(U-Pb). Clearly, for a better understanding of the nature of the sources of mare basalts, a collective consideration of the data from various radiometric clocks is required. An extensive body of data from the Rb-Sr method is available; data from the Sm-Nd method are forthcoming; the U-Pb method still holds a mere promise.

Ages of the sources of mare basalts

All mare basalts, with the exception of Apollo 11 high-K basalts and a few Apollo 17 basalts, have model ages (T_{BABI}) clustering about 4.5 b.y., indicating that only relatively little Rb-Sr fractionation was associated with the magma extrusion and differentiation processes. Accordingly, 4.5 b.y. is an approximation to the age of the sources of mare basalts. The $^{87}Sr/^{86}Sr$ initial values of the basalts, however, do not define a trend from which a more precise age for the source can be determined (see Fig. 7.3.11). This is due to the fact that the sources are heterogeneous and the initial values are only slightly evolved above BABI. In cases where Rb and Sr are fractionated, it is not possible to uniquely define the time of fractionation. Thus the relatively young Rb-Sr model age of 3.8 b.y. for Apollo 11 high-K basalts could be due to mild enrichment of Rb relative to Sr early in the history of the Moon coupled with a marked increase in Rb/Sr ratio at the time of eruption.

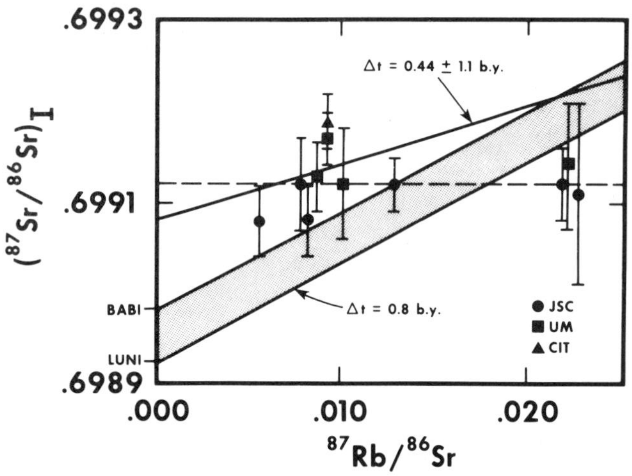

7.3.11 Initial $^{87}Sr/^{86}Sr$ (I) vs. $^{87}Rb/^{86}Sr$ for Apollo 17 basalts (from Murthy and Coscio, 1976). This figure is complementary to Fig. 7.3.8. If the 4.1 b.y. "age" of the source region is correct, then most of the data shown here could fall on a line with a slope corresponding to $\Delta t \sim 0.4$ b.y. (Δt = age of the source region—crystallization age). Because of the close similarity of the initial Sr values for the rocks (dashed line), it is not possible to resolve Δt unambiguously. The solid line is a least square fit for Apollo 17 basalts other than those with $^{87}Rb/^{86}Sr$ of 0.020 – 0.025. (For references to data see Murthy and Coscio, 1976.) Although the 4.1 b.y. age for the source regions (Fig. 7.3.8) is suggestive of a multistage history, the data shown here can be made consistent with a single-stage evolution prior to magma extrusion 3.7 b.y. ago by assuming that the source regions are 4.5 b.y. old and had initial $^{87}Sr/^{86}Sr$ values ranging for example from LUNI to BABI. The shaded region is the corresponding domain for evolution of I values for the first stage of 0.8 b.y.

It could also result from larger enrichment of Rb at more recent times close to the time of magma extrusion. If the eruption was not accompanied by Rb-Sr fractionation, then the source itself must have been enriched in Rb relative to Sr at 3.8 b.y.

One unique advantage of the U-Pb method is that if a two-stage model is assumed for the evolution of the daughter, then the U-Pb isochron would provide the age of the source irrespective of the degree of parent-daughter fractionation. This age is obtained from the upper intersection with concordia.

U-Pb internal isochrons have been determined on three basalts with crystallization ages ranging from 3.3–3.8 b.y. (Tera and Wasserburg, 1975, 1976; Papanastassiou *et al.*, 1977a). Each of the three isochrons has an upper intersection with concordia at 4.42 b.y. The striking persistence of 4.42 b.y. for three basalts from three different localities and of widely different ages is highly suggestive that 4.42 b.y. is an actual point in time. If so, it can be interpreted as the "age" of the sources from which the three basalts were derived, e.g., a time of U-Pb enrichment in the common source region of all three basalts.

U-Pb internal isochrons of reasonable precision were also determined for two highland rocks and were found to have the same upper intersection of 4.42 b.y. (Tera and Wasserburg, 1972a). Subsequently, it was found that highland samples of different lithologies define a total-rock linear array compatible with the 4.42 b.y. upper intersection (Tera *et al.*, 1974a). The data are shown in Fig. 7.3.12. The sampling sites include Apollo 14, 15, 16 and 17 and represent a wide variety of rocks including feldspathic basalts, anorthosites and breccias. The reference isochron in the figure is the internal isochron for a KREEP "basalt" from Apollo 14 (Tera and Wasserburg, 1972a). Two highland rocks (67559 and 68415) are actually concordant at the upper intersection underlining the significance of the 4.42 b.y. age.

The association of essentially all highland rocks in a total-rock isochron has been interpreted as the effect of impacts during an episode of intense bombardment that took place ~3.9 b.y. ago (lower intersection) and comminuted the ancient lunar crust of ~4.42 b.y. age (upper intersection). Because of its massiveness and the abrupt decrease in the intensity of bombardment after it, the epoch was called "The terminal lunar cataclysm" by the workers who first discovered its widespread imprint on the nearside of the Moon (Tera *et al.*, 1973, 1974a).

The simplest explanation of the 4.42 b.y. "age" is that it represents the approximate age of both the lunar crust and the sources of mare basalts. As such, it dates an early planetary process that resulted in the differentiation (layering?) of the planet and the formation of its anorthositic crust and its upper mantle. Oberli *et al.* (1979) have reinvestigated further the "cataclysm array" and considered the upper intersection to be at ~4.45 b.y. KREEP samples analyzed by these workers define a coherent linear array on the U-Pb correlation diagram with intersections at 3.86 ± 0.05 and 4.47 ± 0.03 b.y. Oberli *et al.* (1979) have also dated a highland breccia at 4.17 b.y. and calculated its upper intersection at 4.51 ± 0.04 b.y. Thus, optimistically, one may conclude that the time of lunar "primary" differentiation is 4.47 ± 0.03 b.y. The possibility that such differentiation occurred over a period of time cannot be excluded, however, in which case the upper intersection(s) represents an average value (see discussion by Tera and Wasserburg, 1974).

Sm-Nd systematics on whole rocks suggest lunar differentiation, possibly on a global scale, at ~4.36 b.y. (Lugmair and Carlson, 1978; Carlson and Lugmair,

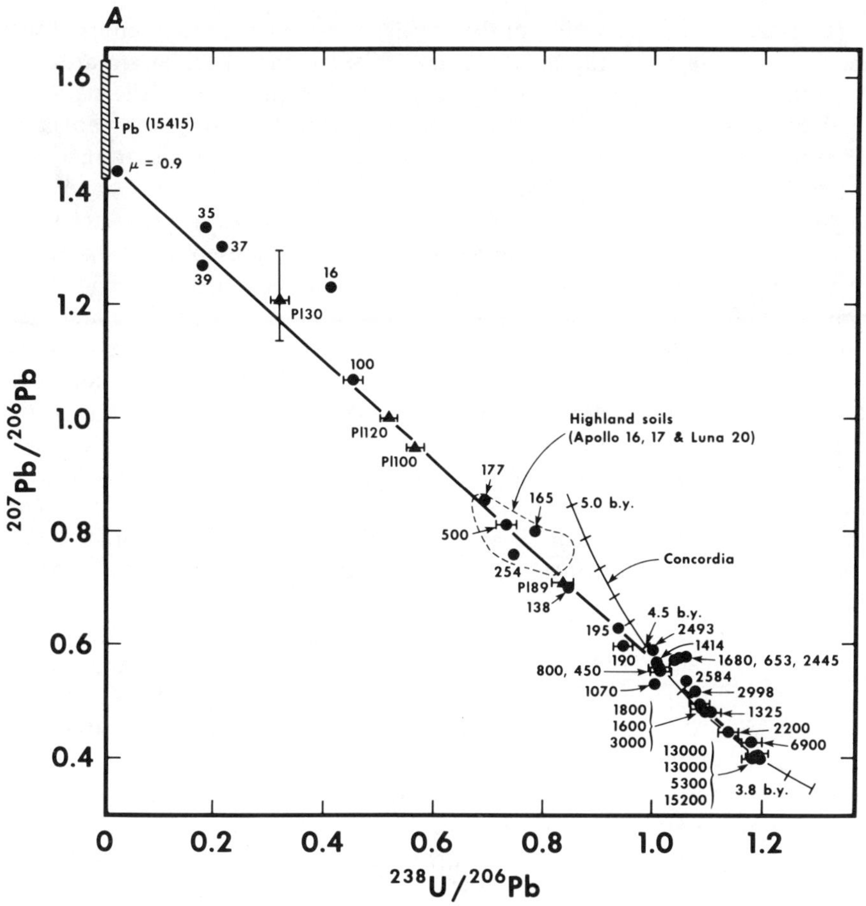

7.3.12 **a.** U-Pb data for highland rocks plotted on the concordia diagram. Data define a linear trend and cluster around an isochron intersecting concordia at 3.9 and 4.42 b.y. Plagioclase mineral separates are indicated by Pl. The μ value is given next to each datum and ranges from 15,000 at the lower intersection to <1 at the intercept with the $^{207}Pb/^{206}Pb$ axis. The general correlation of μ with the linear trend is indicative of Pb mobilization, possibly by volatilization caused by intense bombardment of the lunar surface around 3.9 b.y. ago. **b.** Shows details of data on the chord between the two intersections with concordia in Fig. 7.3.12a. Note that some samples, for example 67015 and 77017, are far off the trend. This is conceivably due to analytical difficulties (Tera and Wasserburg, 1974, Fig. 3; and Oberli *et al.*, 1978, 1979). **c.** One mare basalt (75055) and two highland rocks are concordant at 4.4 b.y. The samples fall on the reference isochron shown in Fig. 7.3.12a at its upper intersection with concordia. The Th-Pb ages of these rocks are in good agreement with their U-Pb ages. This closed-system behavior suggests the strong possibility that the upper intersection of the reference isochron at 4.42 might have a time meaning (from Tera and Wasserburg, 1974).

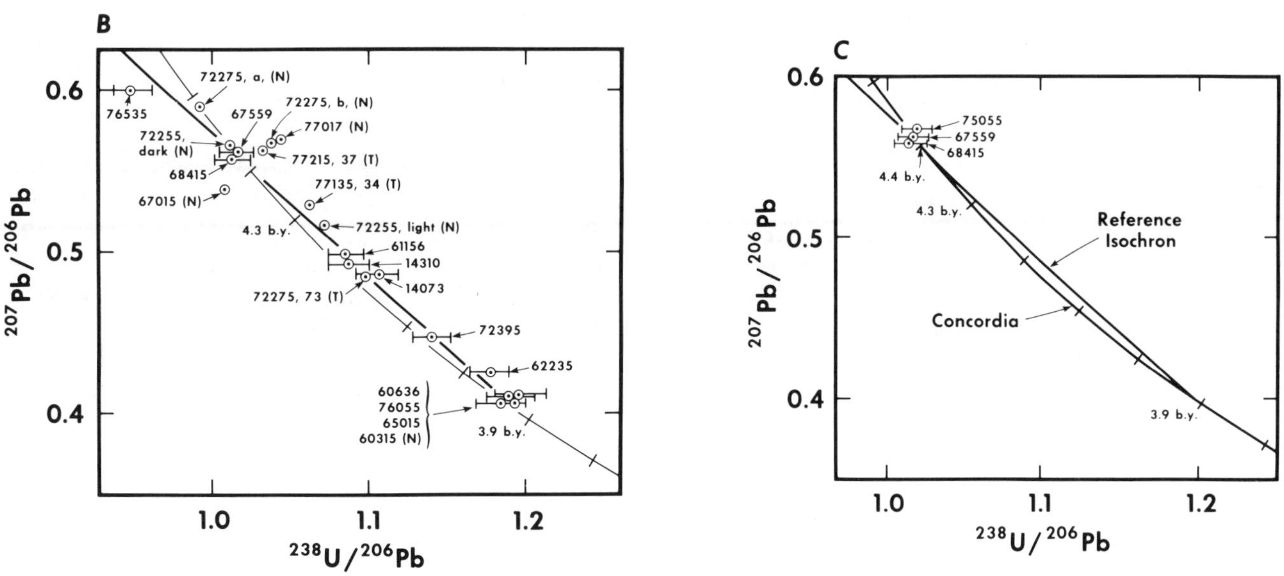

1979). More importantly, evidence for early lunar differentiation is provided by three internal isochrons older than 4.25 b.y.: dunite 72417 was dated by the Rb-Sr method at 4.45 b.y. (Papanastassiou and Wasserburg, 1975b) and norite 77215 by the Sm-Nd method at 4.37 b.y. (Nakamura *et al.*, 1976). Troctolite 76535 was dated by both the Rb-Sr and Sm-Nd methods and yielded ages of 4.51 b.y. (Papanastassiou and Wasserburg, 1976a) and 4.26 b.y. (Lugmair *et al.*, 1976), respectively (note the discrepancy). Thus, while all three methods are in qualitative agreement about early lunar differentiation, it is not possible to pinpoint an exact time through concordance of multiple radiometric methods.

Early global differentiation is compatible with the petrologic deduction that the sources of high-Ti mare basalts are late-stage differentiates formed by crystal fractionation under highly reducing conditions from a homogeneously molten outer part of the Moon (Smith *et al.*, 1970; Brett, 1973; Duncan *et al.*, 1973, 1974; Philpotts *et al.*, 1973; Taylor and Jakeš, 1974; Wakita *et al.*, 1975; Shih *et al.*, 1975). The model's requirement of rapid and extensive early melting to effect differentiation on a global scale could have been achieved by one (or a combination) of a variety of mechanisms (see Chapter 9). The very low values of initial $^{87}Sr/^{86}Sr$ for mare basalts preclude the possibility that the lunar mantle is of chondritic composition (Papanastassiou and Wasserburg, 1973). However, it could be argued that the primary primitive material was chondritic and that Rb as well as other Large Ion Lithophile elements (LIL) like K and Pb were lost by volatilization from the early melt. If the Moon is 4.55 b.y. old and the ancient radiogenic Pb was indeed lost (by evaporation or burial at depth) during early differentiation, then the process must have been completed in 50–100 m.y. so that the apparent age of the Moon would appear to be 4.45–4.50 b.y. (Tera and Wasserburg, 1974c).

7.3.5 Highlands

Introduction

In contrast with the simple history of mare basalts "highland rocks" are dominated by external forces that caused brecciation, shock metamorphism, partial or complete melting. In addition, the large basin-forming impacts were probably also a major source for the injection of mafic and KREEP material into the originally mostly anorthositic highland material. These superimposed events are manifested in the textural diversities of most highland rocks. A natural consequence of this circumstance is the ambiguity of lunar history prior to the termination of the episode of intense bombardment some 4 b.y. ago.

A pre-Apollo stratigraphic conclusion was that the smooth, lightly cratered surface of the mare was indicative of the fact that they are younger than the highlands, which are heavily cratered, often to saturation. The lava plains are also free of debris from the giant ring basins indicating that the mare filling occurred after the excavation of the basins. These observations, coupled with the fact that some of the returned mare basalts are as old as 3.8 b.y., led to the conclusion that the period of intense bombardment and excavation of the giant circular basins was restricted to a maximum duration of ~0.7 b.y. prior to 3.8 b.y. ago.

Chemically the highland rocks are characterized by high alumina content and the presence of a prominent positive Eu anomaly. They are broadly divided into four classes on the basis of their normative plagioclase content: anorthosites (>90%), gabbroic anorthosites (77.5–90%), anorthositic gabbros (65–77.5%) and high-alumina basalts (45–65%). The general aluminum-rich nature of the lunar crust is indicated by the data from the lunar orbiting experiments (Adler *et al.*, 1972a, b; Metzger *et al.*, 1974). Taylor *et al.* (1973) concluded on the basis of the data that the most dominant rock components of the highland crust are anorthositic gabbro and low-K Fra-Mauro basalt (high alumina basalt). The highest alumina group is comprised of the so-called ANT members. The acronym stands for Anorthosite-Norite-Troctolite. Rocks belonging to ANT are generally low in alkalies. The lowest alumina group consists of low-alkali and high-alkali members. The latter are known by the acronym KREEP (Hubbard *et al.*, 1971). These rocks are rich, among other things, in K, REE, P, U and Th (thus the less popular name KREEPUTH). Some KREEP rocks are classified as basalts, others are of obscure lineage. However, all KREEPs might have been created equal or near equal and the apparent dichotomy may be due to superimposed effects on a very old (4.4 b.y.?) rock type.

Lunar KREEP

The term KREEP is somewhat of a misnomer because KREEP is in fact enriched in all the incompatible trace elements including, besides K, the other radioactive parent nuclides Rb, U, Th, and Sm on which the various methods of geochronology are based. The bulk of the incompatible elements on the lunar surface is believed to be associated with a KREEP component of uncertain origin. The uncertainty stems from the fact that most KREEP rocks are polymict breccias bearing the imprints of impact metamorphism. An argument could be made to explain the few unbrecciated KREEP samples as products of impact melt. However, there seems to be a correlation between bombardment and the meteoritic content of lunar rocks (Morgan *et al.*, 1974; Hertogen *et al.*, 1977). Thus, for example, KREEP-free highland rocks have, without exception, a "high" meteoritic content as indicated by high Ir/Au. On the other extreme, Apollo 15 KREEP basalts are completely free of any detectable meteoritic component and may be bonafide igneous rocks. The crystallization ages of Apollo 15 KREEP basalts are in the range of 3.83–3.86 b.y. (Nyquist *et al.*, 1973, 1974, 1975). KREEP-enriched rocks from Apollo 14, 16, and 17 missions all show evidence of the presence of meteoritic components, and their isochron ages range from 3.82–3.97 b.y. (for details, see review by Nyquist, 1977). This is the range of ages exhibited by highland rocks in general.

Nonmare rocks which are enriched in the incompatible elements often have the chemical composition of a basalt and are often described as KREEP basalts. The term "basalt" is used loosely in this connection, and does not necessarily carry the inference of producing these rocks by partial melting in the interior of the planet.

Perhaps the most highlighted rock of a KREEP affinity is 12013 from the Apollo 12 mission. This is a breccia comprised largely of two lithologies: a white-colored component that is nearly granitic in composition, intruding a dark component with KREEP chemistry. Rb-Sr mineral isochrons on two fragments yielded ages of ~3.91 b.y. (Lunatic Asylum 1970; Bottino *et al.*, 1971) in agreement with ^{39}Ar-^{40}Ar analysis of both light and dark components (Turner, 1971). This is the age of isotope homogenization on a millimeter scale in the breccia. Whole-rock Rb-Sr "isochron," on the other hand, corresponds to 4.40 b.y., and indicates early lunar differentiation resulting in the formation of crustal rocks enriched in trace elements (Papanastassiou and Wasserburg, 1971a). The 3.9 b.y. age is assumed to indicate the time of the most recent brecciation. Subsequently, it became apparent that the majority of highland rocks experienced elemental mobilization (presumably due to bombardment) at about that time.

The genesis of KREEP and its age are difficult to ascertain. It is not even clear that a truly pristine KREEP rock has been sampled. Even in the case of the apparently pristine Apollo 15 KREEP basalts, there exists the possibility that they are impact melts from the Imbrium event (Nyquist, 1977).

Many workers (Schonfeld, 1974; Hubbard *et al.*, 1974; Palme and Wänke, 1975; and Irving, 1975) noted the uniformity of the pattern for the incompatible trace elements in KREEP-type rocks. This is highly suggestive of a single homogeneous source for the KREEP components found today in highland rocks. Whether KREEP was originally associated with the rocks of the ancient lunar crust or was mixed in them primarily as a result of the intense bombardment is not clear. In either case, however, if KREEP was produced very early in the history of the Moon (~4.4 b.y. ago as suggested by rock 12013), and the mixing with the ancient crust did not fractionate the incompatible elements to a large degree, then all KREEP-dominated rocks should have essentially the same model age. Such age would represent the approximate time of enrichment of the incompatible elements in the source of KREEP. Although the Rb-Sr model ages of highland rocks show clear clustering at ~4.3 b.y., they cover a wide range of model ages from 3.9–5.9 b.y. This scatter is attributed by Palme (1977) to Rb-Sr fractionation which was caused by the volatilization of Rb during the episode of intense bombardment. In contrast Sm-Nd data for whole-rock samples of KREEP-rich materials from four Apollo sites indicate a total range in model ages of 4.30–4.38 b.y. (Lugmair and Carlson, 1978; Carlson and Lugmair, 1979). Recognizing the refractory nature of the REE, Palme (1977) corrected the Rb/Sr ratios measured for KREEP, assuming a constant Sm/Rb in these samples, and calculated model ages for KREEP breccias from Apollo 14, 15, 16 and 17 of 4.39 ± 0.08, 4.26 ± 0.08, 4.35 ± 0.02 and 4.38 ± 0.02 b.y., respectively. The inferred age of KREEP from Palme's treatment is in general agreement with inferences from the Sm-Nd data and also the U-Pb systematics of KREEP. Oberli *et al.* (1979) have shown that KREEP samples define a tight linear array on the concordia diagram with a lower intersection with the concordia curve at 3.86 ± 0.05 b.y. (age of metamorphism of KREEP) and an upper intersection 4.47 ± 0.03 b.y. The upper intersection, in a simple two-stage model, yields the age of the source.

The genesis of KREEP is the subject of controversy between two opposed models: (1) Melting within the lunar interior, with the variable degrees of partial melting providing the mechanism for the observed variable levels of concentrations of the trace elements in the rocks; (2) Large degree of fractional crystallization of a magma ocean which is well-stirred to promote equilibrium crystallization leading to a homogeneous residual liquid which is highly enriched in the incompatibles. (For details see Warren and Wasson, 1979.)

Ages of highland rocks

The isotopic age data for highland rocks often reflect the complexities mentioned above, and a sizable age range is frequently observed for different clasts and matrix samples of the same rock. Only a detailed study combining petrological, chemical and age data can show which of these ages have a chronological significance and which only reflect partial resetting of the radiometric clock. A good example of the possible age spread that can be observed in a single highland rock is the relatively small (~1 kg) Apollo 17 light gray breccia

73215. A total of 17 different matrix and clast samples were dated with the ^{39}Ar-^{40}Ar technique, and ages between 3.86 and 4.21 b.y. were obtained (Jessberger *et al.*, 1977b). The breccia-forming event occurred at 3.86 ± 0.04 b.y. as indicted by the age of a felsite clast (James and Hammarstrom, 1977). The older ages represent the ages of individual clasts and of matrix materials, in some cases probably partially reset by the breccia formation.

The majority of the highland ages (except for Apollo 14 samples) were obtained by the ^{39}Ar-^{40}Ar technique. The reasons for this are twofold:

(1) In many highland rocks the concentration of the parent elements including Rb is very low, the Rb/Sr ratios are very unfavorable and, furthermore, no suitable minerals are present to establish internal isochrons. In addition, the matrix is frequently extremely fine-grained.

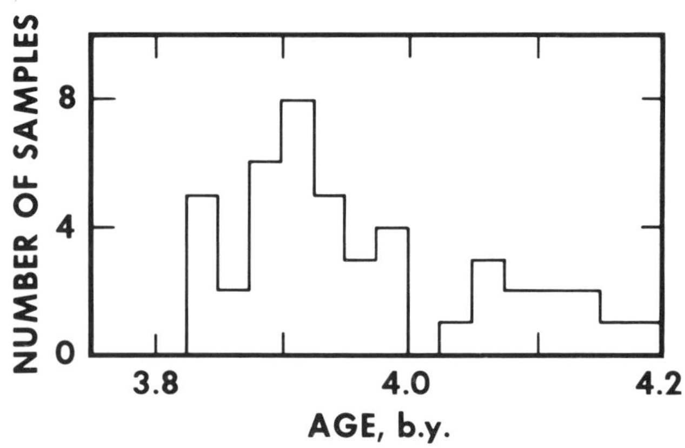

7.3.14 Histogram of high-quality ^{39}Ar-^{40}Ar plateau ages for Apollo 16 samples in the age range from 3.8 to 4.2 b.y. (Figure adopted from Maurer et al., 1978). For selection criteria see Maurer *et al.* (1978). The age distribution is similar to the one shown in Fig. 7.3.13 except for a more pronounced clustering and a gap between 4.0 and 4.02 b.y.

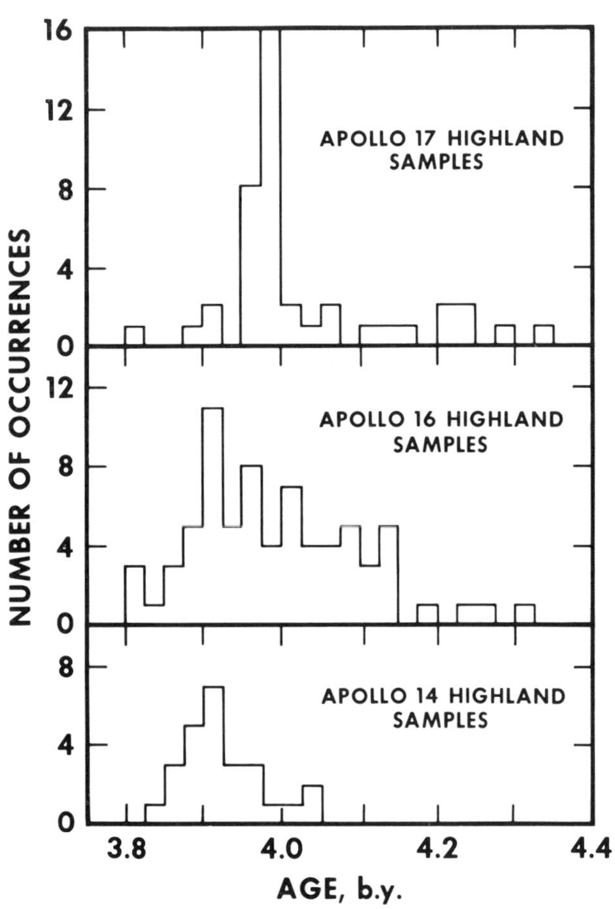

7.3.13 Histograms of ^{39}Ar-^{40}Ar ages for highland rocks from the Apollo 14, 16 and 17 landing sites. Figure from Schaeffer and Schaeffer (1977), updated by the inclusion of 23 Apollo 16 ages from Maurer *et al.* (1978); Jessberger *et al.* (1977a); Turner and Cadogan (1975).

(2) Individual clasts are very often small, and sufficient material is not available for some of the other dating methods. This is also true for the 2–4 mm fragments contained in soil samples. The individual fragments are large enough for a combined petrological, chemical and ^{39}Ar-^{40}Ar age study (Maurer *et al.*, 1978), but too small for the other age techniques.

A large number of highland ages was obtained on individual 2–4 mm fragments from the soil. This was done in order to sample the greatest possible variety of rock types from as large an area as possible. It is expected that the 2–4 mm fragments are, on the average, transported over much larger horizontal distances than larger rocks and boulders and should, at each site, extend the sampled lunar surface area considerably. In fact, the first ages substantially higher than 4.0 b.y. were measured on such fragments (Schaeffer and Husain, 1973a).

A summary of all the ^{39}Ar-^{40}Ar ages obtained on Apollo 14, 16 and 17 highland rocks is given in Fig. 7.3.13. This histogram is taken from a paper by Schaeffer and Schaeffer (1977), and is updated by the addition of 23 Apollo 16 ages (See Maurer *et al.*, (1978). The inclusion of Rb-Sr, Sm-Nd and U, Th-Pb ages will not change the overall picture, as most samples dated by these methods have also been dated by ^{39}Ar-^{40}Ar. No effort was made to evaluate the reliability of the individual ages plotted in Fig. 7.3.13. A histogram of Apollo 16 highland ages, selected on the basis of more stringent criteria, is shown in Fig. 7.3.14.

At the Fra Mauro (Apollo 14) site essentially no samples have ages larger than 4.0 b.y. (Fig. 7.3.13). This may be due to the dominance of Imbrium ejecta at that site. Most samples at the Apollo 16 and 17 sites show ages between ~3.9 and ~4.0 b.y., again reflecting the influence of major basin-forming events in obliterating the age record of the first few hundred million years. However, at both the Apollo 16 and 17 sites a fair number of samples have ages higher than 4.0 b.y. (cf. Figs. 7.3.13 and 7.3.14). There are also several samples not plotted in Fig. 7.3.13 which have complex age curves giving evidence of ages older than 4.0 b.y.—in some cases as high as 4.33 b.y. for the highest temperature steps.

The intense bombardment of the lunar crust during the first ~0.6 b.y., terminating with the formation of several of the major lunar basins, has eradicated most of the chronological information on the early differentiation of the lunar crust. So far only three larger rocks have been found which probably crystallized 4.4 b.y. ago during the initial large-scale lunar differentiation and formation of the anorthositic crust and later escaped severe metamorphism. These three oldest-dated larger lunar rocks are the dunite 72417, the troctolite 76535, and the noritic breccia 77215. A Rb-Sr internal isochron on 72417 gave an age of 4.45 ± 0.1 b.y. (Papanastassiou and Wasserburg, 1975b). The age situation for the troctolite 76535 is complex: a Rb-Sr internal isochron corresponds to an age of 4.51 ± 0.07 b.y. (Papanastassiou and Wasserburg, 1976a), whereas a Sm-Nd mineral isochron yielded an age of 4.26 ± 0.06 b.y. (Lugmair *et al.*, 1976). The Sm-Nd age is in agreement with a K-Ar age for the plagioclase (4.27 ± 0.08 b.y., Bogard *et al.*, 1975) and with two independent ^{39}Ar-^{40}Ar determinations (4.19 ± 0.02 b.y., Husain and Schaeffer, 1975; 4.16 ± 0.04 b.y., Huneke and Wasserburg, 1975). Papanastassiou and Wasserburg (1976a) have argued that the younger age dates the excavation of the rock by a major impact, whereas the Rb-Sr age is the formation age of the troctolite 76535. Internal isochrons obtained on the noritic breccia 77215 (a very large clast in the station 7 boulder) gave ages of 4.33 ± 0.04 b.y. (Rb-Sr) and 4.37 ± 0.07 b.y. (Sm-Nd) (Nakamura *et al.*, 1976). Stettler *et al.* (1978) obtained a well defined ^{39}Ar-^{40}Ar plateau age of 3.92 ± 0.03 b.y. on a plagioclase separate. The simplest explanation for these age data is a two event model: (1) formation at ~4.4 b.y.; (2) excavation at 3.92 b.y. from its original depth of >8 km with simultaneous intrusion of veinlet 77075 (Chao *et al.*, 1976; Nakamura *et al.*, 1976; Stettler *et al.*, 1978).

Individual smaller clasts in highland breccias with ages of ~4.4 b.y. have also been found. Jessberger *et al.* (1977b) obtained high temperature plateau ages of 4.35 ± 0.05 b.y. and 4.34 ± 0.04 b.y. on two strongly calcic plagioclase clasts from the polymict breccia 67435. In addition, indications for ~4.3 b.y. ages are present in the ^{39}Ar-^{40}Ar dating of other highland rocks and clasts. In some cases the apparent age is as high as 4.4 b.y. for the highest temperature steps (e.g., plagioclase from 65015, Jessberger *et al.*, 1974a; sample 78503, 7, 6, Schaeffer *et al.*, 1976; samples LB-1 and LG-1, Maurer *et al.*, 1978).

Most likely the old ages of the three rocks 72417, 76535 and 77215, and of the clasts from 67435, as well as the 4.4 b.y. age relics, reflect the initial formation of these rocks of clasts during the primary differentiation of the lunar crust.

The interpretation of the intermediate ages in the 4.0–4.4 b.y. range poses some difficulties. They could represent true crystallization ages or ages of larger impacts which either excavated the sample from considerable depth or completely reset the radiometric clock. However, they could also be artifacts due to partial resetting and mechanical mixing of older and younger minerals during larger (basin-forming) impacts. Delano and Bence (1977) have shown that the 4.19 ± 0.05 b.y. age of breccia fragment 67483, 13, 8 represents the time of excavation of this fragment. In a detailed petrological, chemical and age study of 48 Apollo 16 fragments (2–4 mm size), Maurer *et al.* (1978) present evidence that the ages in the 4.04 and 4.23 b.y. range date medium-sized impact events (crater size smaller than a few hundred kilometers). A correlation of these ages with known craters is not possible. A more detailed discussion of these results will be given below.

7.3.6 Basin chronology

The ages of the highland rocks collected at the Apollo 14, 16 and 17 sites are strongly influenced by the formation of the younger circular basins. Figure 7.3.15 shows the estimated average thickness of the ejecta layers at these three highland sites based on the work of McGetchin *et al.* (1973). At each site the ejecta layer from one specific basin is dominant: Imbrium at Apollo 14, Nectaris at Apollo 16 and South Serenitatis at Apollo 17. Many attempts have been made to assign absolute ages to the formation of these three basins from the ages observed at the Apollo 14, 16 and 17 sites (Turner *et al.*, 1973; Schaeffer and Husain, 1974; Jessberger *et al.*, 1974b; Turner and Cadogan, 1975; Schaeffer *et al.*, 1976). In the following paragraphs a short discussion of these basin ages is given.

Imbrium

Most of the KREEP-rich melt rocks at the Apollo 14 site, such as 14310, 14073, 14276, and 14001, 7, 3,

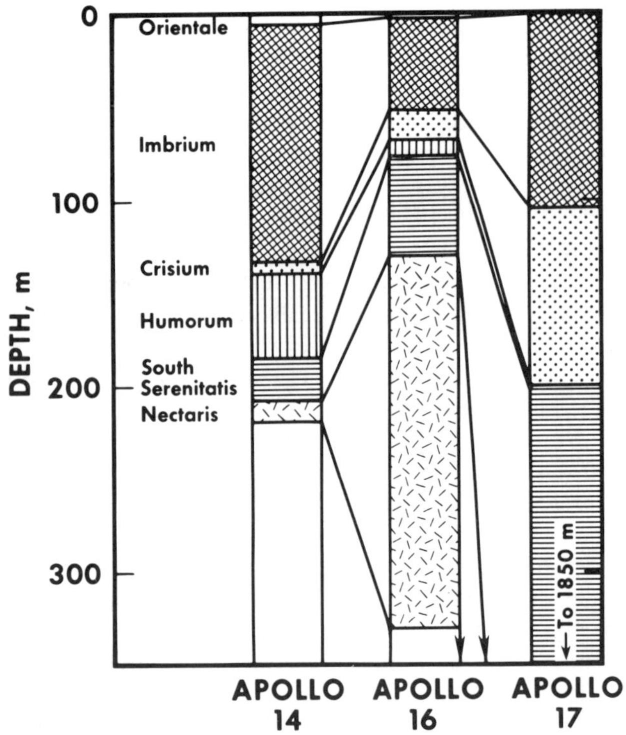

7.3.15 Average depth of ejecta, from basin forming lunar cratering events, at the Apollo 14, 16 and 17 landing sites (McGetchin *et al.*, 1973).

have well-defined Rb-Sr and ^{39}Ar-^{40}Ar ages, clustering tightly around 3.80 b.y. (for individual ages see Wasserburg and Papanastassiou, 1971, for 14276; and the compilation by Nyquist, 1977, for the other rocks). On the basis of petrology and meteoritic trace element abundance, rock 14310 has been interpreted by several authors as a possible impact melt (Dence and Plant, 1972; Green *et al.*, 1972). Hence, the age of 3.80 b.y. most likely dates the Imbrium event (Turner *et al.*, 1973; Jessberger *et al.*, 1974; Turner and Cadogan, 1975).

The low alumina ophitic Apollo 14 basalts 14053 and 14167,8,1 have distinctly higher ages of 3.86–3.89 b.y. (^{39}Ar-^{40}Ar and Rb-Sr; see Turner, 1977). They seem to predate the Imbrium impact event and represent the oldest known lunar basalts. Basaltic clasts in breccia 14321 have been extensively dated (Papanastassiou and Wasserburg, 1971b; Compston *et al.*, 1972; Mark *et al.*, 1973, 1974, 1975) at 3.9 b.y. Arguments have been presented to show that these ages also predate the Imbrium impact (Wetherill, 1975a, b).

It is reasonable to assume that Rb, being a volatile element, may have been mobilized in the Imbrium event, and thus a whole-rock isochron by the Rb-Sr method may date the time of impact. Using this approach, Nyquist *et al.* (1972) placed the Imbrium impact at 3.86 ± 0.12 b.y. This age is indistinguishable from the internal isochron ages for rocks 14053 and 14310 by U-Pb method (Tera and Wasserburg, 1972). Because of the very shallow angle at which the isochron intersects the concordia curve, the obtained U-Pb ages are not precise enough to indicate confirmation of the 3.86 b.y. age.

South Serenitatis

The age assigned to the South Serenitatis event has been reviewed by Jessberger *et al.* (1977 b). Considering the ages of different Apollo 17 breccias, they concluded that South Serenitatis formed 3.89 ± 0.05 b.y. ago. If only the 73215 felsite age is used, then the best estimate for South Serenitatis would be 3.86 ± 0.04 b.y. Within the assigned limits of error this age is compatible with the 3.90 ± 0.04 b.y. age deduced for South Serenitatis by Cadogan and Turner (1976) from the study of the Apollo 17, station 6 boulder. The ^{39}Ar-^{40}Ar chronology of the Apollo 17 station 7 boulder implies two thermal events, at 3.92 b.y. and 3.83 b.y. (Stettler *et al.*, 1978). As discussed by the authors, the correlation of these two thermal events with impacts or other geological processes poses some difficulties. Within the limits of the experimental uncertainties, either of the two events is compatible with the 3.89 ± 0.05 b.y. South Serenitatis age of Jessberger *et al.* (1977 b). If the 3.92 b.y. thermal event is ascribed to South Serenitatis, then the younger event could be related to the Imbrium or Crisium impact. If the younger event is due to South Serenitatis, then the older might represent Nectaris or North Serenitatis (for more details, see Stettler *et al.*, 1978). Nyquist *et al.* (1974) placed the time of the Serenitatis event at 3.94 ± 0.1 b.y. on the basis of Rb-Sr whole-rock isochrons of Apollo 17 noritic breccias.

Nectaris

The first ages of lunar samples distinctly higher than 4.0 b.y. were reported by Schaeffer and Husain (1973, 1974) for 2–4 mm soil fragments from the Apollo 16 landing site. They ascribed the ages of 4.13–4.23 b.y. to the formation of the Nectaris Basin. This age assignment has not remained unchallenged (Jessberger *et al.*, 1974; Turner and Cadogan, 1975). A detailed study by Maurer *et al.* (1978) on the petrology, chemistry and ^{39}Ar-^{40}Ar ages of 48 Apollo 16 soil fragments (2–4 mm size) from the rim and ejecta blanket of North Ray Crater revealed a well-defined correlation between petrology, chemistry and age of the samples (Figs. 7.3.16–17). Three distinct age groups were recognized: Group 1 with ages between 4.06 and 4.14 b.y.; Group 2 with ages from 3.83–3.96 b.y. and some young frag-

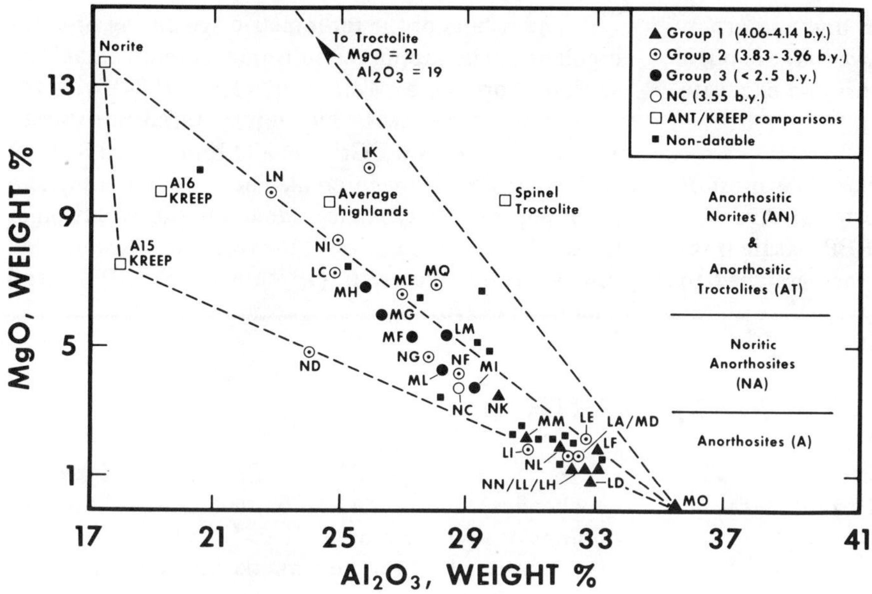

7.3.16 Al_2O_3 vs. MgO in dated 2–4 mm, coarse-fine fragments from the Apollo 16 landing site (from Maurer *et al.*, 1978). The individual fragments were dated with the ^{39}Ar-^{40}Ar method. Age Group 1 fragments (4.06–4.14 b.y.) are anorthositic, poor in mafic material. Age Group 2 fragments (3.83–3.96 b.y.) are mostly distinctly enriched in mafic material.

ments (Group 3, <2.5 b.y.). Only one fragment (3.55 b.y.) falls outside this grouping and possibly represents Theophilus ejecta.

All samples older than 4.06 b.y. are anorthositic, metamorphosed breccias of the light-matrix type, poor in KREEP and mafic components. The samples in the age Group 2 are (with one exception) more or less strongly enriched in KREEP and mafic material. No samples with ages between 3.96 and 4.06 b.y. were found. A gap remains even if the data obtained by other investigators are included (cf. Fig. 7.3.14; for selection criteria used to reject ill-defined ^{39}Ar-^{40}Ar ages see Maurer *et al.*, 1978). Based on these results the authors concluded:

(1) The ages >4.06 b.y. represent medium-sized impacts in the highland regions. The material was either excavated at this time, or the Ar clock was reset by the impact event. Possible candidates for such a medium-sized impact might be some of the unnamed craters described by Head (1974).

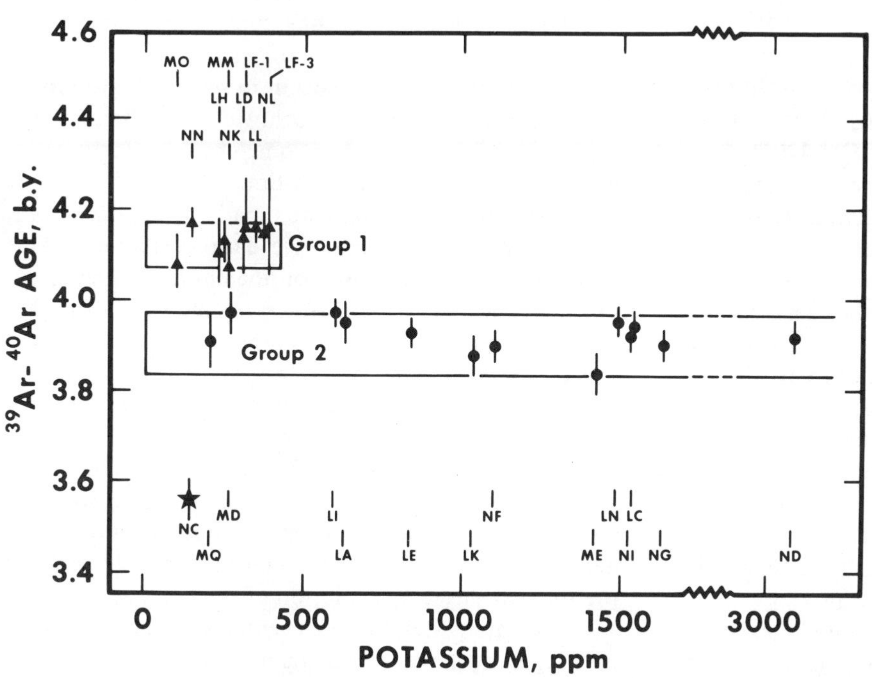

7.3.17 Potassium contents vs. ^{39}Ar-^{40}Ar plateau ages for the Apollo 16 coarse-fine fragments shown in Fig. 7.3.14 (from Maurer *et al.*, 1978). The letters (Mo, MM, etc.) are the sample designations of these authors. The fragments from age Group 1 (4.06–4.14 b.y.) cluster in a narrow range of K content between 90 and 380 ppm. The fragments from age Group 2 (3.83–3.96 b.y.) all have higher K content except for the brecciated anorthosite MD and anorthositic norite MQ. The Ar clock in these two samples was apparently reset without KREEP admixture.

(2) The fragments of the 3.83–3.96 b.y. age group come from a basin-sized impact which was large enough to penetrate the anorthositic lunar crust and excavate KREEP and mafic material.

(3) The basin-sized impact was Nectaris. As most of the ages of the Group 2 samples cluster at 3.92 b.y. (especially a completely molten KREEP basalt fragment at 3.91 ± 0.04 b.y.), this is the age assigned to Nectaris.

The results of the radiometric age dating of lunar highland rocks support a multi-impact lunar cataclysm as first proposed by Tera *et al.* (1973, 1974) with the formation of Nectaris (3.96 b.y.), Humorum, South Serenitatis (3.89 b.y.), Crisium and Imbrium (3.82 b.y.) within a 0.1 b.y. time interval. The constraints such an event imposes on dynamical models of the evolution of the small body population of the solar system have been discussed by Wetherill (1975a,b, 1976, 1977) and Chapman (1976).

7.4 THE EARTH

7.4.1 Variations in ancient terrestrial basaltic volcanism

Introduction

One inescapable fact of terrestrial evolution is that radiogenic heat production in the past was greater than it is now. Since one of the major ways heat is lost through the crust is via basaltic volcanism, especially along divergent plate boundaries, we infer that generation of basalts was always a major process in terrestrial crustal evolution. An essential question is whether this is supported by the geological record. There are at least three major aspects to the question of timing of volcanic activity:

(1) How long does a major period of volcanism last?

(2) What is the frequency of eruptions during such activity?

(3) Can individual periods of volcanism be lumped into several larger episodes of major activity with intervening episodes of lesser activity?

The answers to these questions are not necessarily easy to obtain, and in many respects definitive answers may be beyond our grasp for years to come. In part the problems are geological. A major portion of modern basalts are seafloor basalts. However, the processes that form them (e.g., plate tectonics) also tend to destroy them on a relatively short time scale, along with disruption of the continental record. Thus, for many periods in Earth history the preserved record is only fragmentary, and that which is preserved is primarily associated with the continents. On the other hand, part of the problem is analytical: how accurately and how precisely can a given lithologic unit be dated? For example, fine scale resclution of individual eruptive events is quite feasible for young basaltic assemblages using suitable techniques. For ancient assemblages, however, geochemical factors can lead to uncertainties of tens of millions of years or more.

Because of the fragmentary nature of the preserved record of basaltic volcanism, and because plutonic or felsic volcanic rocks are generally much easier to date than basalts, it is necessary to make some tectonic assumptions. By analogy with modern tectonics and basalt petrogenesis, we can assume that major orogenic events are preceded or accompanied by significant basaltic volcanism. In many cases, such as Archean greenstone belts, some of the basaltic rocks are well preserved. In other cases this may not be so. However, if many (or most) ancient orogenic belts can be attributed to some form of plate motion, then it is reasonable to assume that somewhere during a major orogenic event basalts were being produced, either along divergent or convergent plate boundaries or within plates.

Under these assumptions, the question of the timing of ancient basaltic volcanism reverts to whether orogenic events tended to occur in several well-defined periods, either locally or worldwide, or whether orogenic activity was occurring, somewhere on Earth, more or less continuously through time. One of the earliest responses to this question was made by Gastil (1960), who complied geochronologic data available at that time and concluded, based on the curve that resulted, that there were several major orogenic pulses of 100–200 m.y. duration with intervening periods of quiescence. Engel *et al.* (1965) subsequently correlated volumes of tholeiitic basalt with age, and obtained a similar picture (Fig. 7.4.1). In recent years there have been several adherents to the general idea of major worldwide orogenic pulses (Sawkins, 1976; Moorbath, 1976; Burwash, 1969) and such a model must be considered as one of the major alternatives when developing theories on the early evolution of the crust.

Assuming that discrete events, regularly (periodic) or irregularly (episodic) distributed through time, can

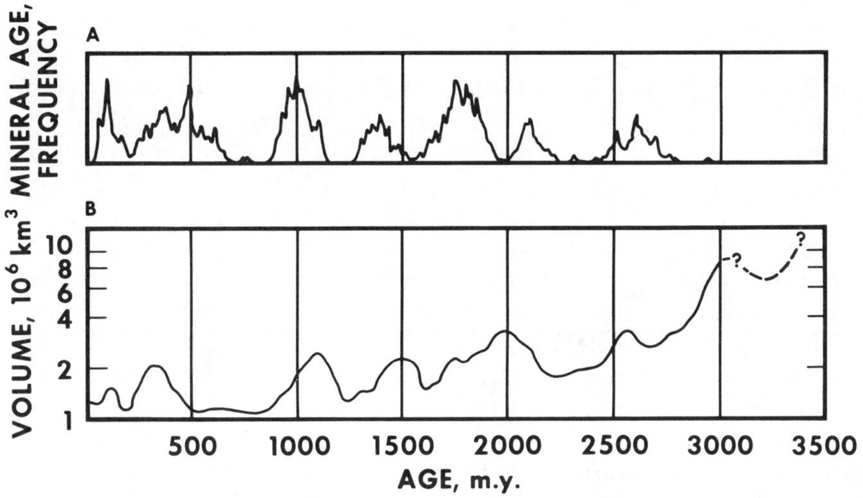

7.4.1 **a.** Frequency of mineral "age" from 0 to 3000 m.y. ago from world-wide localities, compiled by Gastil (1960, Fig. 1). **b.** Volume of continental tholeiitic basalt extruded worldwide, estimated from geologic maps by Engel *et al.* (1965, Fig. 4). From Yoder (1976a, Fig. 10-6).

be recognized, it is then essential to determine whether the chronologies are local (e.g., Canadian Shield, or North America) or global in extent. Early conclusions tended to be strongly influenced by data from only one or two continents and may not be an accurate reflection of global igneous activity.

Large amounts of geochronologic data have been generated in the last decade and more will be coming. As our worldwide coverage has increased, the additional data have tended to reinforce the general episodic model. However, it is also true that the added coverage has revealed suites of igneous rock with ages intermediate to the earlier recognized pulses, and one must still consider the basic question of completeness of the geologic record or completeness of our sampling.

Even if the episodic model stands the test of time, with much of the precise data now available it is clear that each major "pulse" may have been composed of numerous short (few million years) plutonic or volcanic episodes. Probably none of these could be considered "worldwide" individually, although broad periods of enhanced orogenesis could have been worldwide phenomena.

Resolution of the above questions is clearly important to tectonic and thermal models of the Earth. This could be particularly true if some periodicity were identified that may serve to identify the time constant, for example, of plate tectonic cycles (if they exist).

Separate from the strictly chronologic problem of occurrence of basaltic rocks in the past is that of secular variations in their compositions. We know that radiogenic systems record secular changes (e.g., initial $^{87}Sr/^{86}Sr$ ratios, section 7.4.2). However, it is not yet known to what extent certain types of basalt (e.g., island arc basalts) may have differed in the past. The differences would primarily be represented in major, minor, and trace element chemistry and could reflect (a) secular variations in the source rocks (e.g., mantle, section 7.4.3), (b) secular variations in physical conditions (e.g., geothermal gradient: Precambrian alkali basalts are conspicuously rare), or (c) different kinds of tectonic environments (different styles of plate tectonics). Furthermore, if secular variations are identified it must be determined whether they evolved more or less gradually throughout Earth history, or whether there were major periods of rapid evolution superimposed on an otherwise constant or gradually varying system.

The problems outlined above are among the most fundamental in Earth history. Unfortunately, many of them cannot be definitively resolved at present. This is not only due to the geological problems mentioned above (lack of preservation, lack of exposure, or presence of alteration), but also to an inadequate data base and an incomplete understanding of the tectonic environment during the past, especially the Precambrian. Nonetheless, we can examine these questions for several major time intervals: Phanerozoic ($<$ 600 m.y.), Proterozoic (600–2500 m.y.), and Archean (2500–3800 m.y.).

Phanerozoic

It is possible to consider variations in basaltic activity during the Phanerozoic from a variety of different viewpoints. A distinction can be made between activity in the oceans and on the continents, and a further subdivision for oceanic activity can be made between direct and indirect evidence for the amount of basaltic activity.

Direct evidence of basaltic activity in the oceans is restricted to the last 200 m.y. since there is virtually no ocean floor preserved on Earth that is older. Mapping of ridge-produced magnetic anomalies on the ocean floor and calibration of the ages of the anomalies using the results of Deep Sea Drilling have shown that virtually all the ocean floor which was made less than 10 m.y. ago is still preserved. However, progressively larger proportions of the older material have been removed from the oceans by the processes of subduction and obduction (Pitman *et al.*, 1974).

Ridge-generated ocean floor can be readily recognized because its depth below a -2.8 km datum (the average depth at which spreading ridges are generating new seafloor) is related to the square root of its age. Areas of seafloor that exist at shallower depths are mostly underlain by basaltic material produced by means other than normal ridge volcanism. Three main kinds of such abnormal seafloor can be recognized: (1) areas formed by exceptionally large amounts of volcanism at hot spots along ridges, e.g., the Iceland-Faeroes Ridge; (2) areas formed by extensive hot-spot volcanism at locations away from spreading ridges, e.g., the Hawaiian-Emperor chain plus the other numerous seamounts of the Pacific; and (3) the extensive volcanism forming such intra-oceanic plateaus as the Manihiki and the Ontong Java plateaus and the Caribbean (Burke, *et al.*, 1978).

Over the last 200 m.y. the rates of seafloor spreading and of all three processes responsible for building the superimposed volcanic edifices have varied. Pitman *et al.* (1974) have shown from the distribution of magnetic anomalies that the rate of ridge production may have been about twice as fast in the Late Cretaceous as it is today, and Larson and Pitman (1972, Fig. 7) have also indicated that the total length of mid-ocean ridges was greater at that time. Both these observations indicate that much more basalt was being erupted on Earth about 80 m.y. ago than is being erupted today or was being erupted a few tens of millions of years earlier.

The much smaller amounts of volcanism produced in other ways (currently about 10% of the whole ocean floor) have also varied with time. This can be readily seen from the varying cross-sectional area, along strike, of such features as the Hawaiian-Emperor seamount chain or the Iceland-Faeroes ridge. Vogt (1979) has made the suggestion that there are regularities in this variation so that, for example, the volcanism in all the major seamount chains was less abundant 15 m.y. ago than it is now.

Unfortunately, there is no direct way of knowing whether the older oceans that have now disappeared were occupied by the same kinds of volcanic structures, produced in the same ways, as characterize the young oceans. There is no reason to suspect that they were any different, however, and there is some indirect evidence that they were similar.

Indirect evidence that the oceans were similar in the past comes from the extent of flooding of the continents as Hallam (1977) pointed out. In general, continents lie close to sea level and ocean basins lie below sea level. When the average age of ocean floor is young the depth of floor is shallow and seawater floods the continents. Turcotte and Burke (1978) have used estimates of the extent of continental flooding throughout the Phanerozoic (from Hallam, 1977) to infer how much the amount of ridge volcanism (assuming this to dominate) has varied and have concluded that several times in the past there was about twice as much ridge volcanism as there is today. Their conclusions are firmest for the Late Cretaceous, but there is also a strong suggestion of a major transgression 400–500 m.y. ago, implying a period of enhanced ridge volcanism then as well (Fig. 7.4.2).

Because it seems unlikely that the volume of the solid earth has changed significantly in the last 100 m.y., the high seafloor spreading rate inferred for the Late Cretaceous must have been accompanied by increased subduction and/or obduction rates. Stonely (1974) has pointed out that the Late Cretaceous was a time of extensive ophiolite emplacement (by obduction) along the borders of the Tethys. Evidence of a Late Cretaceous peak in subduction has been seen (Lanphere and Reed, 1973) in the widespread circum-Pacific occurrences of granodiorites and tonalites of that age, and the argument is that the intrusives were related to underlying slabs and the episode of intrusion corresponds to a widespread intensification of subduction.

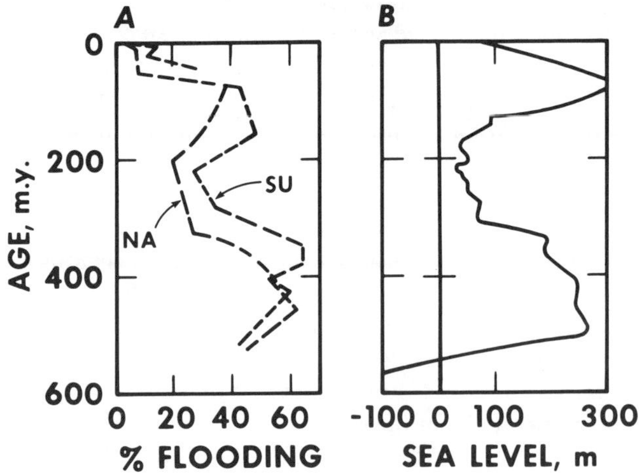

7.4.2 a. Percent flooding of North America (NA) and the Soviet Union (SU) as given by Hallam (1977) from paleogeographic maps. **b.** Relative paleo-sea level determined by Vail *et al.* (1978). Modified after Turcotte and Burke (1978, Fig. 2.)

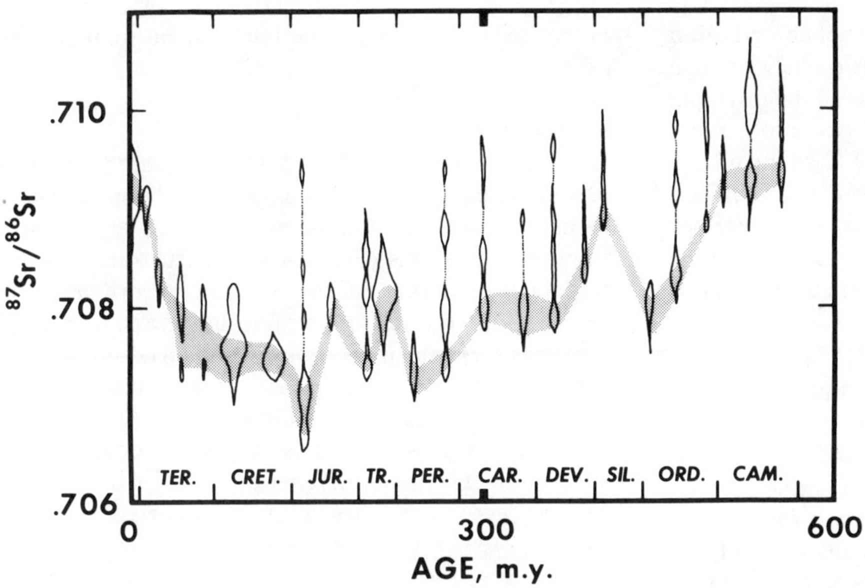

7.4.3 Variations of $^{87}Sr/^{86}Sr$ in seawater during the Phanerozoic. The minimum values for any given time probably are the best estimate of unaltered strontium compositions representative of seawater at that time. The shaded curve represents the variation of seawater composition as a function of time. Modified after Veizer and Compston (1974, Fig. 5).

Another indirect line of evidence for variations in past tectonic activity and corresponding variations in continental erosion comes from consideration of $^{87}Sr/^{86}Sr$ variations in seawater. These variations have been best documented for the Phaneozoic (Peterman *et al.*, 1970; Veizer and Compston, 1974) and are shown in Fig. 7.4.3. Veizer and Compston (1976) have also extended the data base well into the Precambrian (Fig. 7.4.4).

There have been several suggestions for causes of the $^{87}Sr/^{86}Sr$ variations in seawater. Peterman *et al.* (1970) tentatively correlated periods of low $^{87}Sr/^{86}Sr$ ratio with enhanced plate tectonic activity. Armstrong (1971), on the other hand, suggested that increases in $^{87}Sr/^{86}Sr$ ratio could be attributed to times of increased continental glaciation. Brass (1976) and Spooner (1976) have discussed the variations in seawater Sr composition in detail. Both have concluded that a major factor

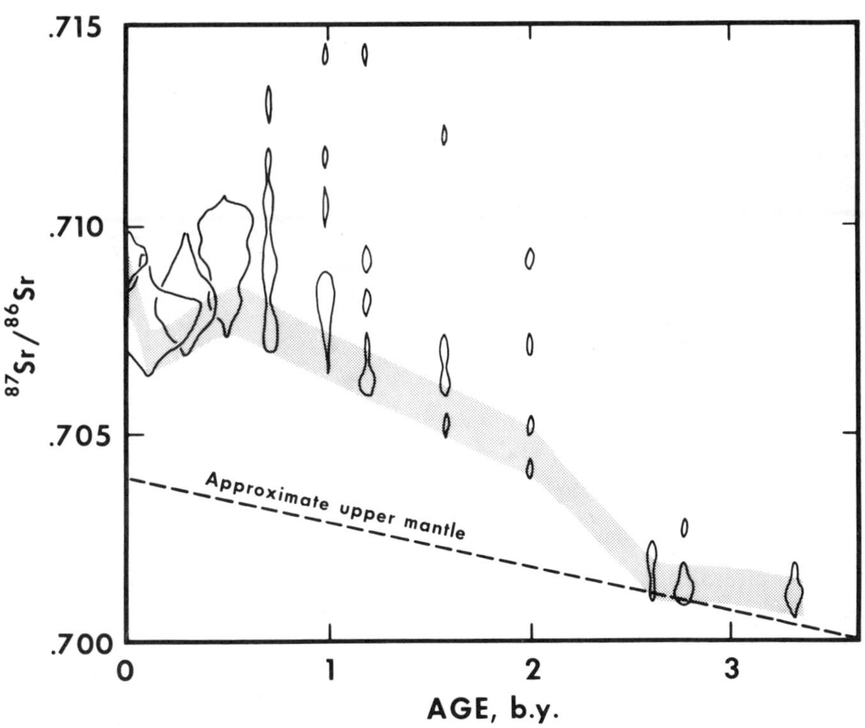

7.4.4 Variations of $^{87}Sr/^{86}Sr$ in seawater during the Precambrian. Data have been pooled for 200–300 m.y. intervals. The minimum values for any given time probably are the best estimate of unaltered compositions representative of seawater at that time. The shaded curve represents the variation of seawater strontium composition as a function of time. Modified after Veizer and Compston (1976, Fig. 2).

in causing the variations is the amount of contribution from erosion of continental basement rocks with high $^{87}Sr/^{86}Sr$ ratios, although they differ as to the principal source of strontium with low $^{87}Sr/^{86}Sr$ ratios. Brass (1976) and Brass and Turekian (1977) favor strontium from weathering of limestone as the main buffer, whereas Spooner (1976) suggested seawater-seafloor interaction. In any case, both attribute increases in the $^{87}Sr/^{86}Sr$ ratio of seawater Sr to increased contributions from continental (sialic) sources.

Increased continental erosion can come about by glaciation, as suggested by Armstrong, but Brass (1976) argues that although there is general correlation between changes in $^{87}Sr/^{86}Sr$ in seawater and periods of enhanced glaciation, especially for the last 100 m.y., the correlation is much weaker prior to that, especially for the peak in $^{87}Sr/^{86}Sr$ during the Triassic. Instead, Brass (1976) favors correlating changes in seawater Sr composition with tectonic activity and corresponding changes in inundation of the continents, a conclusion also expressed by Spooner (1976).

According to this model, the increased spreading and volcanism at ridges will cause a general rise in mean sea level, flooding large portions of continents and retarding continental erosion. This will result in a lower amount of radiogenic ^{87}Sr entering sea, with a resulting drop in the $^{87}Sr/^{86}Sr$ ratio. This effect will also be enhanced by increased subduction/obduction, for the resulting young volcanic and plutonic rocks generated, exposed, and subsequently weathered around the margins of the ocean basins will contribute essentially mantle Sr with low $^{87}Sr/^{86}Sr$ ratios.

At times of lesser ridge activity sea level will fall, causing a greater influx of radiogenic ^{87}Sr from continental sources and a corresponding rise in the $^{87}Sr/^{86}Sr$ ratio of seawater. As can be seen from Figs. 7.4.2–3, there is a crude correspondence between high sea level stands and lower (or falling) $^{87}Sr/^{86}Sr$ ratios in seawater.

The correlation is by no means excellent. This is perhaps due in part to incompleteness and lack of accuracy in some of the data; it may reflect time lags and buffering effects not fully accounted for, or it may represent an inappropriate correlation between cause and effect. In spite of these problems, the fluctuations in seawater Sr composition are regarded by most as some reflection of orogenic activity (and hence, volcanism). The fact that there are general ups and downs in both the sea level curves and the seawater Sr curves indicates a certain degree of pulsation in the orogenic behavior of the Earth during the Phanerozoic. Turcotte and Burke (1978) suggested that the variations represented cyclic behavior with a period on the order of 300 m.y., possibly related to the turnover period for mantle convection cells.

Proterozoic

For older periods the record of increased pulses of oceanic volcanism is very incomplete, and even attempts to use evidence of continental flooding or seawater $^{87}Sr/^{86}Sr$ variations (Fig. 7.4.4) are frustrated by the fragmentary nature of the geologic record and the degree of alteration for Precambrian rocks. Thus, the principal surviving indications of enhanced subduction and, by inference, enhanced spreading rates and ridge volcanism, are remnants of old island or Andean-type arcs. Even this approach is subject to considerable uncertainty, for there still remains the major question of whether tectonic regimes in the Precambrian were similar to those in the Phanerozoic. Furthermore, because of the lack of precise biostratigraphic correlation in the Precambrian, it is necessary to rely on radiometric ages and regional geology to infer major periods of basaltic volcanism.

Previous "statistical" approaches have tended to emphasize major thermal or tectonic pulses approximately 600 m.y. ago, 1000–1100 m.y. ago, 1300–1500 m.y. ago, and 1700–1800 m.y. ago (Fig. 7.4.1). In fact, the latter three of these are so well defined for the Canadian shield that they are used as a basis for defining a Precambrian time scale for that region (Stockwell, 1972). Thus it is clear that for major portions of individual continents, at least, general episodes of tectonic activity can be identified. It now remains to examine these in more detail for various continents to see if there is worldwide correlation for some or all of the events.

North America. For the Canadian shield the three major Proterozoic events are referred to as the Grenvillian (~1100 ± 100 m.y.), Elsonian (~1400 ± 100 m.y.), and Hudsonian (~1800 ± 100 m.y.) orogenies. In addition, there are significant events of intermediate age.

Although there are some basaltic rocks present in the lower Huronian (Robertson, 1976) the earliest major Proterozoic suite with basaltic nature is probably that represented by sills and dikes of the 2100-m.y.-old Nipissing Diabase (Van Schmus, 1965; Fairbairn *et al.*, 1969) and dike swarms of similar age throughout the shield (Fahrig and Wanless, 1963; Gates and Hurley, 1973; Hanson and Malhotra, 1971). As noted by Windley (1977), surprisingly few chemical studies have been carried out on the rocks of these dike swarms. The few data that are available suggest a continental tholeiite affinity, but there is little evidence to suggest that the dike swarms fed extensive continental basaltic volcanism; all of the activity may have been intrusion at depth in association with stress systems formed during early abortive attempts to break up early Proterozoic continents (Windley, 1977).

The oldest major "orogenic" events, complete with associated basaltic rocks, during the Proterozoic in North America occurred in the interval 1600–1900 m.y. ago. Furthermore, most of that activity was associated with mobile belts within and marginal to the Canadian shield 1750–1900 m.y. ago (Moore, 1977). These include the Coronation Geosyncline in the Bear Province (Hoffman, 1973; Hoffman and McGlynn, 1977), the Flin Flon–Snow Lake–Thompson belt in southeastern Churchill Province and the Circum-Ungava Geosyncline along the northeastern margin of the Superior Province (Moore, 1977), the Penokean Fold Belt south of Lake Superior (Van Schmus, 1976; Sims, 1976), and the early Proterozoic terrane of southern Wyoming and northern Colorado (Hills and Armstrong, 1979). All of these have radiometric ages in the range 1750–1900 m.y., indicating a close association in time. Much of the age control is based on Rb-Sr isochrons, which can often be subject to several percent error (Van Schmus, 1976), so that the degree of time equivalence for these areas (and similar, currently less studied areas) will have to await more precise geochronology. In the coming few years application of U-Pb on zircons and Sm-Nd total rock studies should improve the picture considerably. In any case, it is clear with the present data that the period 1750–1900 m.y. ago was one of major orogenic activity within and around the Canadian shield.

The tectonic significance of all these orogenic belts has not been fully deduced. However, in many instances the data are strongly suggestive of subduction zones active along continental margins. Chemical studies on the volcanic rocks have been rather limited until recently, so there have been few attempts to relate basalt compositions with tectonic environments. Moore (1977) has summarized much of the recent data, and other studies include those of Cudzilo (1978), Smith (1978), and Dann (1978). One common result from these studies is that early Proterozoic basalts apparently include both tholeiitic and calc-alkaline suites and bear more similarity to Archean than Phanerozoic basalt suites. In addition, attempts to use various trace-element diagrams for tectonic discrimination (Pearce and Cann, 1973; Pearce *et al.*, 1977) lead to ambiguous results. This may reflect chemical alteration in ancient volcanics that changed original trace-element patterns, or it may represent secular variations in orogenic and petrogenetic processes. Quite likely, it is a combination of both.

Assuming that these ~1800-m.y.-old mobile belts represent convergent plate-margin activity, then it would also be reasonable to conclude that there was active production of seafloor basalts along spreading ridges, analogous to the modern case. To date no remnants of such seafloor have been identified in the early Proterozoic.

In the southern portion of the continent, primarily the United States, several temporally and geographically defined events can be recognized. The oldest of these is the "Penokean Orogeny" of the Lake Superior region, which was mentioned above and occurred about 1820–1860 m.y. ago (Van Schmus, 1980). The Penokean Fold Belt includes substantial basalt as well as associated felsic volcanics and a related plutonic complex. The entire association is suggestive of a convergent plate boundary (Van Schmus, 1976; Cambray, 1978), but this is far from certain (Sims, 1976). A second pulse of volcanic and plutonic activity occurred about 1760 ± 10 m.y. ago, primarily to the south of the 1820–1860-m.y.-old complex (Van Schmus, 1978), but in this case the rocks are almost exclusively granitic to rhyolitic (Smith, 1978), and do not appear to be associated with a major local orogenic event. All of the above rocks have been affected by a metamorphic (and possible deformational) event about 1630 m.y. ago that effectively reset many of the Rb-Sr total rock systems (Van Schmus *et al.*, 1975a; Peterman, 1966).

About 1500 m.y. ago the western Great Lakes region was subjected to anorogenic plutonism of a generally quartz monzonitic character (Van Schmus *et al.*, 1975b, c) with rapakivi-anorthosite suite affinity (Silver *et al.*,1977; Emslie, 1978). The youngest significant Proterozoic event in the region was the Keweenawan rifting and basaltic volcanism (Green, 1977) that occurred 1100–1200 m.y. ago (Silver and Green, 1972).

There appear to be a number of events throughout the western part of the U.S. that were contemporaneous with or correlate with events in the Great Lakes region, although as yet no western equivalents to the ~1850-m.y.-old Penokean Orogeny have been reported. In the Colorado-Arizona-Utah-New Mexico region Silver (1968) and others (Stern *et al.*, 1971; Peterman and Hedge, 1968; Vera and Van Schmus, 1974) have demonstrated a two-stage sequence of events with older mafic volcanism (now mostly amphibolite), sedimentation, and calc-alkaline plutonism that occurred about 1750 m.y. ago, followed about 75–100 m.y. later by extensive volcanism and granitic plutonism that is especially well developed in southern Arizona (Silver, 1968). Unfortunately, the extensive Paleozoic and younger cover in the midcontinent region precludes a better understanding of the relationship between the southwestern U.S. terrane and the Great Lakes terrane. It is interesting that there is a general correspondence in the timing of various events, such as ~1750 m.y. and

~1650 m.y., even though they are represented by different results.

The ~1500-m.y.-old anorogenic plutons represent a more readily correlatable event, since similar lithologies extend in a belt from the southwestern U.S. to Labrador (Silver *et al.*, 1977; Emslie, 1978), with ages generally increasing from 1425–1450 m.y. in the west to slightly over 1500 m.y. in the east. The only known, preserved supracrustal rocks of this age are the Croteau group of Labrador, which is about 1520 m.y. old (Wanless and Loveridge, 1972) and the extensive ignimbrites of southeast Missouri, which are about 1500 m.y. old (Bickford and Mose, 1975) and extend westward into the subsurface of southern Kansas (Bickford *et al.*, 1975). In Missouri and southern Kansas basaltic rocks are nearly totally absent, although they are present in the Croteau Group (Baragar, 1977). Felsic volcanism and plutonism, also lacking mafic rocks, occurred about 1400 m.y. ago in Oklahoma (Bickford and Lewis, 1979).

Mafic sills and dikes ~1200 m.y. old occur in Arizona, California, and in the Belt Series in the northern Rocky Mountains, suggesting a correlation with similar rocks in the Canadian shield (Windley, 1977).

Thus, although the data from the Canadian shield suggest only a few major orogenic events, it is clear from studies of basement rocks in the U.S. that several events are represented, more or less uniformly from the Penokean (~1850 m.y.) to the Keweenawan and Grenville (~1100 m.y.), with recognized peaks at ~1850, ~1750, ~1650, ~1450–1500, ~1400, and ~1100 m.y. in the U.S. Within this time span there is apparently well developed regional episodicity, but it is difficult to tell which events are volumetrically more important and it is uncertain what the tectonic cause is for several.

In general, three principal types of volcanic activity can be recognized. First is that which occurs in mobile belts and which may represent subduction-zone generated magmas. The main examples of this type of volcanism are the ~1850-m.y.-old Penokean volcanics of Wisconsin and northern Michigan, the ~1750-m.y.-old terrane in Colorado and northern Arizona, the ~1850-m.y.-old volcanics in the Flin Flon–Snow Lake region, the ~1850-m.y.-old volcanics of the Circum-Ungava Geosyncline, the ~1800±100-m.y.-old volcanics of the Bear Province, and possibly the ~1300-m.y.-old volcanic rocks in the Grenville Province (Condie and Moore, 1977).

Baragar (1977) has recognized two other types of volcanic suites, a "felsite porphyry" suite and plateau basalts, both of which he considers represent stable crust volcanism. The felsite porphyries are best developed marginal to slightly older Proterozoic mobile belt assemblages and, as Baragar (1977) notes, may in these cases be genetically linked to the older orogeny. Prime examples of such assemblages are the rhyolite-granite suites of southern Wisconsin (Smith, 1978), the Great Bear batholith (Hoffman and McGlynn, 1977), the Dubawnt volcanics in the Keewatin District, and rhyolites of southern Arizona (Silver, 1968; Livingston and Damon, 1968). These are all ~1650–1750 m.y. old and have subordinate to no basalt. Three younger terranes of similar lithologies are the Croteau Group in Labrador, the rocks of the St. Francois Mountains in southeast Missouri and correlative rocks in southern Kansas, and slightly younger granite and rhyolite in Oklahoma. These are all ~1400–1500 m.y. old, but as yet are not closely associated with any known older mobile belt.

The second "stable crust" suite recognized by Baragar (1977) is the plateau basalts. The principal examples are the Keweenawan basalts (Green, 1977), the Coppermine River basalts (Baragar, 1969, 1972), the Seal Lake Group in Labrador, and the Gardar Province in southern Greenland, all of which are ~1100–1300 m.y. old, and the ~650 m.y. old Natkusiak basalts in northwestern Canada (Baragar, 1977). It is perhaps significant that the recognizable plateau basalts in North America are all late Proterozoic, but this may be mainly an artifact of preservation. There are several Proterozoic dike swarms in the Canadian shield from 2500 m.y. to 1200 m.y., especially the ~2100-m.y.-old Abitibi, Nipissing, and related swarms, for which no volcanics coexist. Thus, plateau basalts could have been much more important than indicated by the present record. Baragar (1977) has summarized the chemistry of a number of these plateau basalts and finds that they tend to have higher contents of the incompatible elements such as Ti, K, Ba, Sr, Rb, Zr, and Pb, and suggests that they are largely plume-generated, analogous to intraplate oceanic islands.

In summary, Proterozoic volcanic rocks in North America tend to fall into one of three groupings, each of which is concentrated in a particular period of time: (a) orogenic suites, ~1600–1900 m.y. old; (b) post-orogenic felsic suites ~1650–1800 m.y. and ~1400–1500 m.y. old; and (c) plateau basalts ~1100–1300 m.y. old and possibly ~2100 m.y. old. Thus, there are strong indications of episodicity in North America, especially when types of basalt terrane are considered. Furthermore, not only do the chronologies and tectonic styles indicate a mobile earth during the Proterozoic, but so do the paleomagnetic data (Burke *et al.*, 1976a; Irving and McGlynn, 1976; Windley, 1977), so that by implication there must have been significant generation of ocean floor basalts during these periods of enhanced continental activity. It remains to be seen, however, whether the pattern for North America represents local episodic response to more or less continuous seafloor spreading and continental drift, or whether there were true global pulses (fast and slow periods of plate motion).

Laurasia. Paleomagnetic studies indicate that during the Proterozoic the North American-Greenland-Scotland-Baltic shields were configured similar to more recent Pangaean reconstructions, suggesting that the Laurasian part of Pangea may have been closely related continental masses for about 1–2 b.y. prior to the Mesozoic opening of the Atlantic. The principal uncertainties in such a conclusion are the nature of the Grenville (~1100 m.y. old) and the subparallel Appalachian to Caledonian fold belts, particularly the magnitude of pre-Mesozoic oceans between North America and Europe (Roberts and Gale, 1978; see this paper also for discussion of related volcanic rocks in the 600–900 m.y. age range).

In spite of the long-term structural complexity of this block of crust, it does appear that there is a strong correlation of tectonic and petrogenetic events. This was first detailed by Burwash (1969) and later updated and expanded by Condie (1976) as shown in Fig. 7.4.5. The oldest Proterozoic events that match across the region are the ~2100–2200-m.y.-old diabase dike swarms of the Canadian shield (Van Schmus, 1965; Gates and Hurley, 1973), the Scourie dikes of northwest Scotland (Tarney, 1973), and the Jatulian diabase (and metavolcanic rocks) of Finland (Kouvo, 1976).

The most extensive period of Proterozoic orogeny throughout the Laurasian shield is that ~1750–1900 m.y. ago. This similarity was easily recognized 10 years ago (Burwash, 1969), but with the increased number of U-Pb zircon ages in the respective areas, the correlation becomes even stronger. In the Sveco-Karelian region orogenic volcanic and plutonic activity occurred ~1840 m.y. ago, followed by more alkali rich (potassic) plutonic rocks ~75 m.y. later (Kouvo, 1976). The older period is not represented in Scotland, but the Laxfordian Orogeny may correlate at ~1700–1750 m.y. (van Breeman *et al.*, 1971; Lyon and Bowes, 1977) with the younger stage. In southern Greenland rocks of the Ketilidian mobile belt yield ages of 1740 to 1800 m.y. (van Breeman *et al.*, 1974; Gulson and Krogh, 1975), very similar to those obtained to the east and west. Finally, Tugarinov and Bibikova (1976) reported U-Pb and Pb-Pb ages

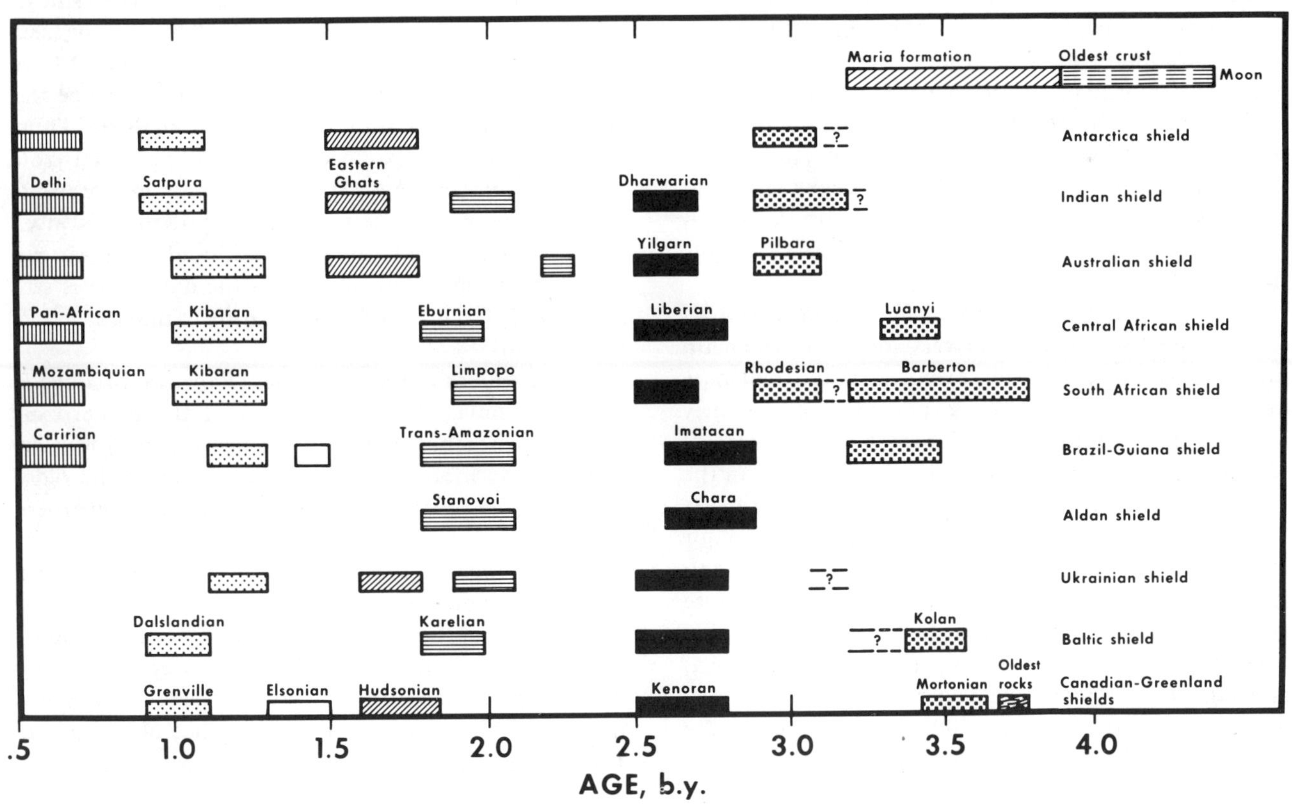

7.4.5 Major Precambrian orogenic periods as presented by Condie (1976, Fig. 5-17).

of ~1850 m.y. in the Aldan shield in Russia. Thus, modern precise chronology indicates that there was extensive, nearly contemporaneous, development of orogenic belts ~1850 m.y. ago in the Laurasian shield areas, followed ~75–100 m.y. later by more felsic stages of magmatism.

Outside North America, basaltic rocks associated with the ~1850 m.y. events are best preserved in Finland and Sweden, where Hietanen (1975) has related them to island arc development off the edge of the Archean core of the Baltic shield. The basalts occur primarily in two supracrustal suites. The older of these is the Jatulian sequence in eastern Finland, adjacent to the Archean craton, which is cut by 2100–2200-m.y.-old diabases that were apparently feeder dikes to small occurrences of mafic volcanics in the upper part of the section. This sequence also is the same age as altered ultramafic bodies (now serpentinites) associated with the Outokumpu ore deposit in Finland. The origin of these basaltic rocks is still uncertain, but they may in part have been plateau basalts formed during tensional stages on the stable platform of the Jatulian sequence, analogous to the early basalts and diabase associated with the Huronian of Canada.

The younger sections of basalt and related felsic volcanics are part of the Kalevian sequence of Finland. They occur in several places, notably in the Tampere region, are about 1870–1900 m.y. old (Kouvo, 1976), and have chemical properties suggestive of geosynclinal calc-alkaline affinity (Simonen, 1953). Hietanen (1975) suggested an island arc (subduction zone) origin for these basalts and associated plutonic rocks. Late- to post-orogenic intrusive rocks associated with the Sveco-Karelian orogeny have been dated at 1800–1850 m.y. (Kouvo, 1976).

As in North America, rapakivi granite-anorthosite suites formed in Finland, but very soon after the main orogenic and late orogenic events, ~1650–1690 m.y. ago (Kouvo, 1976). Åberg (1978) has recently completed an extensive study of the plutonic rocks of southeastern Finland and found igneous events in the 1840–1970 m.y. range for orogenic plutons and at 1650–1730 m.y. for younger plutons. He also reported a major resetting of several Rb-Sr isochrons ~1650 m.y. ago and a late metamorphic event with some granite emplacement ~1400 m.y. ago. In general, these coincide well with events to the west in North America, although there is not an exact correlation. Nonetheless, it is clear that the interval 1700–1900 m.y. ago in the Laurasian shield was one of major crustal development.

Gondwanaland. There is little strong paleomagnetic evidence to define clearly the relative positions of Laurasia and Gondwanaland in the Precambrian. It is clear, however, (Hurley and Rand, 1969) that Africa and South America were comprised of closely related continental blocks for much of the time.

Major recognized Proterozoic events occurred at ~1900 m.y. ago, ~1100 m.y. ago, and ~550 m.y. ago. These events go by various local names, but are generally interconnected with one another for a given event. Kroner (1976, 1977) concluded that they primarily represent intracontinental (ensialic) zones of remobilization of older crystalline crust rather than subduction-produced continental margin assemblages, but others (Burke and Dewey, 1972, 1973) favor continental collision models for many of the events.

One of the oldest sequences of Proterozoic basaltic rocks is contained within the Transvaal Supergroup and is about 2200–2300 m.y. old (Hunter and Hamilton, 1978). These are intruded by the Bushveld Complex, which is the world's largest known layered mafic complex. Recent studies place its main age at about 2100 m.y. (Hunter and Hamilton, 1978), with a late granitic phase at about 1950 m.y. Hunter and Hamilton (1978) conclude that the Bushveld Complex was emplaced in a tensional structural setting. It may be significant that the Bushveld was formed at about the same time as the ~2100-m.y.-old mafic dikes of the Laurasian shield.

Recent studies in the Guyana shield and in Venezuela (Gaudette *et al.*, 1977, 1978; Priem *et al.*, 1973) have revealed the existence of assemblages similar in age and lithology to many of those in the Canadian shield. Thus, this area may be more closely related to the Proterozoic of North America than of Africa. Key recognizable events are the trans-Amazonian event from ~1850–2100 m.y., rapakivi granite ~1550 m.y. old, and Pan-African events ~550 m.y. old. However, the data base is still very small, and it would be premature to make any firm correlations.

For Australia recent studies (Page, 1978; Page *et al.*, 1976) have shown that many of the Proterozoic volcanic assemblages fall in the age range 1750–1900 m.y. Furthermore, the general character of the older sequence, ~1860 m.y. old, is suggestive of a convergent plate margin assemblage (Wilson, 1978), followed by slightly younger (~1780 m.y. old) felsic volcanism and a third felsic volcanic suite ~1680 m.y. old. Thus, the timing of orogenic basaltic volcanism in Australia is very similar to that in North America, suggesting either a closer spatial relationship between the two continents in the early Proterozoic or the existence of global tectonic events in the interval 1750–1900 m.y. ago.

Summary and conclusions. Although there are large parts of the continents still poorly studied, the data that have been obtained in the last two decades still strongly indicate major pulses of orogenic activity, often accompanied by basaltic volcanism. The most extensive period involving apparent converging continental margin assemblages (island arcs or continental arcs) occurred about 1800–2000 m.y. ago. This was preceded by about 200 m.y. (~2100–2200 m.y. ago) in many regions by tensional forces and associated emplacement of diabase sills, dikes, and mafic complexes, with limited volcanism (or limited preservation of associated plateau type basalts).

The interval between 900 m.y. and 1800 m.y. appears to contain relatively few preserved convergent plate margin assemblages, although there are several plateau basalt sequences and extensive rhyolitic ignimbritic sequences preserved. Beginning about 900 m.y. ago there were apparently more convergent plate margin assemblages forming (Burke and Dewey, 1972, 1973; Windley, 1977; Roberts and Gale, 1978; Kroner, 1977), particularly in parts of the Pan-African belts and in the Appalachian-Caledonian belt.

Supportive data for the period 1800–2000 m.y. ago being one of major significance in the evolution of the crust and mantle come from numerous sources. (a) Many "mantle isochrons" yield "ages" in this range (section 7.4.2); (b) Engel *et al.* (1965) suggested there was a major change in K_2O/Na_2O ratios of igneous rocks ~2 b.y. ago (Fig. 7.4.6); (c) Veizer and Compston (1976) found an apparent rapid increase in seawater $^{87}Sr/^{86}Sr$ ratio about 2.0 b.y. ago (Fig. 7.4.4), and Perry *et al.* (1978) reported evidence for a significant change in the $\delta^{18}O$ of cherts in the early Proterozoic.

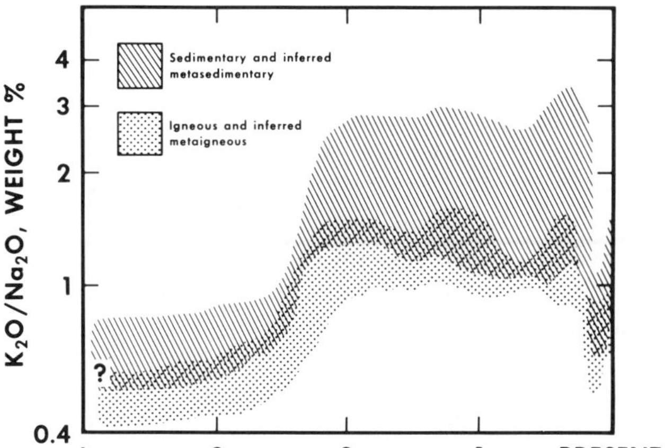

7.4.6 Variations in the petrogenetic index K_2O/Na_2O of igneous, sedimentary, and metamorphic rock complexes as a function of time. Variations in crustal environments of the constituent rocks, as indicated by the index, and by related lithologic features, rock associations, and relative abundances, suggest a crude first-order sequence of crustal evolution and global tectonics. Modified after Engel *et al.* (1965, Fig. 1).

Thus, we tentatively conclude that there have been major episodes for certain types of continental margin basaltic volcanism, several of which were essentially worldwide. Furthermore, several of these, most notably at ~1.8–2.0 b.y. ago, were probably linked to secular evolution in the mantle-crust tectonic system. On the other hand, basic thermal considerations and paleomagnetic data (Burke *et al.*, 1976a, b) suggest more or less constant mobility of plates, so that Proterozoic seafloor volcanism was probably more continuous than implied from the record preserved in continental rocks.

Archean

Archean basaltic rocks, conventionally defined as those older than 2500 m.y., are encountered in every continent. These basalts, characteristically of low metamorphic grade, form one type of Archean crustal terrain, the other being higher grade granite gneisses. Typically, as has been discussed above, age information on the supracrustal associations containing basalt comes directly from the basalts themselves, and indirectly from interbedded felsic units, sediments, or adjoining granites and granite gneisses.

The data base is extremely sparse. In Fig. 7.4.29, for instance, we have tabulated Rb-Sr age and initial-Sr data for as many mafic (not always basalt containing) associations as possible, and for the Archean there are a total of 20 individual determinations of which we would class only about 10 as being reliable. This data survey is not exhaustive and as new data are added to the compilation, the conclusion presented here may change considerably, especially as more U-Pb zircon dates appear.

Within the limitations of the available data we have reviewed the age information for the various continents and examined whether the geochronological data can answer any questions about the timing of ancient volcanism. We are somewhat hampered by the fact that not all basalt-bearing volcanic assemblages have been subjected to geochronological analysis, but for the Archean (more so than the Proterozoic) most associations are well represented. For the earliest Archean preserved, it is evident that concerted efforts have been made to date any well preserved mafic rocks, undoubtedly due to the question of whether the earliest crustal rocks were acidic or basic. A good "age" review of some continents is given in O'Nions and Pankhurst (1978). The review that follows incorporates these data and supplements them with additional data. We concentrate on discussing four continents containing at least six cratonic blocks, and these may be compared with other cratonic regions as more data emerge.

Canada and U.S.A. This Precambrian segment represents the largest known surface extent of Archean rocks in the world. It constitutes at least two major Archean cratons (Superior and Slave) as well as isolated outcrops exposed within younger cover rocks in U.S.A. (Wyoming craton). Numerous Rb-Sr total rock studies of basalts from the classic greenstone belts indicate a mean age of formation close to 2.7 b.y. with individual associations generally recording ages in the interval 2.65 to 2.75 b.y. (Folinsbee *et al.*, 1968; Hart and Davis 1969; and Fig. 7.4.1). These ages have been confirmed to a large extent by U-Pb zircon studies (Hart and Davis, 1969; Krogh and Davis, 1971; Nunes *et al.*, 1978), as well as Sm-Nd total rock studies (Zindler *et al.*, 1978). For example, in the case of the Abitibi belt, the largest surface extent of Archean volcanics in the world, the available geochronological data (Rb-Sr, Sm-Nd, U-Pb, $^{39}Ar/^{40}Ar$) indicate a remarkably short interval for the accumulation of this pile at about 2.70 ± 0.05 b.y. The zircon U-Pb data of Nunes *et al.* (1978) is particularly noteworthy in that it indicates an interval of some 25^{+13}_{-6} m.y. for the filling of a substantial stratigraphic section (Blake River Group) of the Abitibi volcanic pile. The accumulation of this pile may be interpreted (Gélinas and Brooks, 1974) as a basal section reflecting accumulation of volcanics on a constructive plate margin (Kinojévis Group) followed by volcanics characteristic of a destructive plate margin (Blake River Group).

There is evidence elsewhere in the Superior Province, however, that the volcanism and associated plutonism may have spanned an interval of some 250 m.y. (from 2700 to 2950 m.y.: Krogh and Davis, 1969; Nunes and Thurston, 1978).

A possibly older sequence of basalts may be inferred from magnesian amphibolite fragments preserved in the > 3000-m.y.-old regional tonalites of the Superior Province (Verpaelst *et al.*, 1980). These fragments are ubiquitous and commonly found with fragments of metamorphosed iron formation. However, this possibility is equivocal because in at least one instance, the tonalite masses have been remobilized at times younger than the 2.7 b.y. age of formation of greenstone belts, and the inclusions could be remnants of that volcanism. If regional plutonism is accompanied by volcanism then many of these tonalitic masses could be used as indicators of volcanism of the calc-alkaline basalt-andesite type. However, the ages of the tonalites at more than 3000 m.y. are minimum values (commonly metamorphic) and we have no firm value for their primary age of formation. Model Sr ages [i.e., time of extraction from the mantle calculated by comparing the Sr composition of the tonalite with the mantle growth curve (Fig. 7.4.30)] could be used to suggest ages in the vicinity of 3600 m.y. This, however, is conjectural for most of the Superior Province since this form of "dating" makes some gross assumptions. There is firm support for "old" volcanics in Minnesota, where inliers of very old silicic rocks are found within the 2700-m.y.-old granite-greenstone terrain. Lund (1956) reports the existence of a gabbro-diorite complex, an association reminiscent of plutonism in a convergent plate boundary. This complex is older or approximately equal in age to the Montevideo gneiss (3630 ± 60 m.y.: Goldich *et al.*, 1970; Goldich and Hedge, 1974).

Volcanism younger than the 2700 m.y. event is recorded by the presence of mafic dike swarms that cut the granite-greenstone terrains. There is at least one swarm of late Archean age (circa 2.5 b.y.) and this poses the question as to whether these dikes were feeders to surficial volcanics. However, in the case of the Abitibi metavolcanic belt where these dikes are commonly encountered, there is no firm evidence of any plateau basalt assemblages. A similar situation occurs in the Rhodesian craton (see following section) where mafic dikes (including the Great Dyke) are interpreted as a possible last abortive attempt at an Archean greenstone belt (Wilson *et al.*, 1978), although others would relate such intrusion to plutonism in a rift (Burke and Whiteman, 1973).

Thus, in résumé, when the Archean of Canada and U.S.A. is considered regionally there is evidence of volcanism at circa 3.6, ~3.0 (?) and 2.7 b.y., with the 2.7 b.y. event being the most dramatic, and evidence that some volcanism occurred in the interval 2.95 to 2.7 b.y.

Greenland and Labrador. These two regions are grouped for discussion based on the hypothesis of refitting them to form the Atlantic craton (Bridgwater *et al.*, 1973). Basaltic rocks such as typify the greenstone belts of other continents are preserved in the oldest terrain recognized in Greenland as amphibolitic schists with relic pillow structure, associated iron formation, dacites, etc. In the Isua area the volcanism and plutonism has been dated by Rb-Sr, U-Pb and Pb-Pb techniques at 3700 to 3800 m.y. (Moorbath *et al.* 1972, 1973, 1975; Baadsgaard, 1973). For Labrador this "old" period of volcanism is inferred from fragmental pieces of mafic rocks and iron formation found within gneisses that closely resemble those of Greenland (Bridgwater and Collerson, 1976). The age of these gneisses is at least 3.6 b.y. (Hurst *et al.*, 1975).

Events younger than this age are complicated due to a pervasive high-grade metamorphism at circa 2.8

b.y., evident in both Greenland and Labrador. However, volcanism younger than 3.6 b.y. is evident in both regions (Upernavik supracrustals of Labrador, Malene supracrustals of Greenland). This volcanism is not precisely dated but has been suggested in the case of Greenland to be related to the extensive 3.0 b.y. magmatism which produced the anorthosites and calc-alkaline plutonites (Bridgwater *et al.*, 1978). In Labrador there is a minimum age for the supracrustals of 3.12 ± 0.16 b.y., based on reactivation of Uivak gneisses containing interleaved Upernavik supracrustals (Bridgwater *et al.*, 1978).

The youngest Archean volcanism recognized in Greenland is that dated at circa 2.6 b.y. (e.g., Ikermit supracrustals: Pedersen *et al.*, 1974). This is not yet reported for the Labrador section. Thus like the other continents Greenland and Labrador show three possible times of basaltic volcanism, one at 3.6 to 3.8 b.y., one at 3.0 to 3.1 b.y. and one at 2.6 b.y. Of these the intermediate one is the least reliably dated, and the youngest, the least extensive.

Rhodesia and South Africa. The southern African plate contains two Archean cratonic blocks, the Rhodesian craton, and the Kaapvaal craton in South Africa.

Wilson *et al.* (1978) recognize two and possibly three distinct ages of granite-greenstone development in the Rhodesian craton. The most precisely defined age is that for the youngest volcanism and associated plutonism. Rb-Sr and Sm-Nd total-rock studies of mafic and ultramafic volcanics (termed the Upper Greenstones by Wilson *et al.*, 1978) indicate an age of formation at 2.6 to 2.7 b.y. (Hawkesworth *et al.*, 1975; Jahn and Condie, 1976; Hamilton *et al.*, 1977), synchronous with plutonism. The close agreement in age between both the Sm-Nd and Rb-Sr ages effectively nullifies the 3.08 b.y. statistical age (12 out of 50 data-points) determined by Jahn and Condie (1976) for these volcanics. The narrow range in observed ages for the Upper Greenstones suggest development in a relatively short interval (~0.1 b.y. ?).

The Upper Greenstones (previously the Upper Bulawayan Group) overlie what Wilson *et al.* (1978) refer to as the Lower Greenstones (previously Lower Bulawayan Group). The age of the latter greenstones is not precisely known, but they are tentatively correlated with tonalites and gneisses dated at 2.9 b.y. The Lower Greenstone Rb-Sr data of Jahn and Condie (1976), separated out using the map of Wilson *et al.* (1978), plot grossly scattered on the isochron diagram, with an "average" age of somewhere near 2.8 to 2.9 b.y., but the data are not at all conclusive.

The oldest greenstones (Sebakwian Group) are better positioned geochronologically. Indirect dating by Rb-Sr total rock studies of intrusive granites, and gneisses which contain the volcanics as disrupted remnants, indicate a minimum age for volcanism of 3.5 b.y. (Hickman, 1974; Moorbath *et al.*, 1977; Hawkesworth *et al.*, 1975). Hence, the Rhodesian craton records two and possibly three episodes of basaltic volcanism that occur within an interval of 1.0 b.y. The evidence suggests that the volcanism was not continuous although there is little data on the Lower Greenstones, which could range in age from 2.9 b.y. to 2.7 b.y., the age of the Upper Greenstones. The volcanism that predominates in present outcrop is clearly associated with the 2.6- to 2.7-b.y.-old event. Of interest is the Wilson *et al.* observation that the Upper Greenstones contain a lower half that is compositionally analogous with modern volcanics formed at divergent plate margins, while the upper half is more similar to volcanics formed at convergent margins. This is similar to the previously discussed Abitibi belt of Canada. Thus, the Rhodesian craton contains evidence for basaltic volcanism at ~3.5 b.y., ~2.9 b.y. (?), and 2.6 to 2.7 b.y.

In the Kaapvaal craton of South Africa, the volcanic section of the Barberton Mountain Land (Onverwacht Group) has been reliably dated at a stratigraphic level immediately overlying the Middle Marker Horizon. The age of 3.36 ± 0.10 b.y. determined by Van Niekerk and Burger (1969) by U-Pb zircon analysis agrees closely with an Rb-Sr total-rock isochron for shales from the Middle Marker Horizon (Hurley *et al.*, 1972). Unfortunately there are no reliable dates for the komatiites underlying the Middle Marker Horizon, although an "age" of 3.5 b.y. has been obtained by Jahn and Shih (1974) by means of Rb-Sr data from density fractions in which each fraction is composed of dominantly secondary minerals after primary komatiite mineralogy. The best minimum age estimate for the Onverwacht Group is provided by dated intrusive granites which generally give ages in the interval 2.9 to 3.1 b.y. (Allsopp *et al.*, 1962) although some are as old as 3.2 to 3.3 b.y. (Oosthuyzen, 1970). Thus the Kaapvaal craton of South Africa, in which the Barberton Mountain Land is found, contains evidence of basaltic (and komatiitic) volcanism in the 3.4 ± 0.1 b.y. range, with no evidence of younger (≃2.7 b.y.) volcanism as is manifest in the Rhodesian craton to the north.

Australia. There are two basalt-bearing Archean crustal blocks in Australia, the Pilbara and Yilgarn blocks. The Pilbara block, the geologically simpler of the two, comprises a layered succession of low-grade metavolcanics and metasediments within which are found large oval-shaped batholiths. The layered succession is similar to that observed in the Onverwacht of

South Africa in that it is divisible into a lower volcanic assemblage (Warrawoona Group) and an overlying (commonly unconformable) sedimentary assemblage (Gorge Creek Group overlain by the Whim Creek Group). The batholiths which intrude the Warrawoona Group are older than the Whim Creek Group which rests unconformably in at least one instance on a granite dome (Fitton *et al.*, 1975).

The Yilgarn block is more complex than the Pilbara block in that it is internally divided into three regions (Wheat Belt, Eastern Goldfields, and Murchison Provinces) differing in structural style and/or metamorphic grade. The classic Archean greenstone sequences predominate in the Eastern Goldfields and Murchison Provinces, and in the latter region it is suggested that the low-grade volcanic succession is the equivalent of cover rocks to the high-grade granites and gneisses of the Wheat Belt zone.

Within the Pilbara block, dacite flows (or sills) from the middle of the Warrawoona Group have been precisely dated at 3.45 b.y. (U-Pb zircon age: Pidgeon, 1978), and this gives a reliable time reference for the formation of the older part of the volcanic section. However, conflicting geochronological data do not allow us as yet to state the time of termination of volcanism. On the one hand a Rb-Sr total rock age of a porphyry from a stratigraphically younger section of the Warrawoona Group than that dated by Pidgeon, gives a 2.8 b.y. age (de Laeter and Trendall, 1974). On the other hand, the dated ovoid batholiths which intrude the Warrawoona Group are all about 3.0 b.y. in age [e.g., Rb-Sr (Compston and Arriens, 1968; de Laeter and Blockley, 1972; de Laeter *et al.*, 1975): Pb-Pb (Oversby, 1976)], suggesting a culminating plutonism (remobilization?) that defines a minimum age for that Group. Because of the susceptibility of silicic volcanics to record "too young" ages (Pidgeon, 1978), the most probable interpretation is that the Warrawoona Group is older than 3.0 b.y. and probably not much younger than 3.4 b.y. Nonetheless, post (?) 3.0 b.y. basaltic volcanism is suggested by the fact that the youngest Archean Group in the Pilbara block (Whim Creek) contains a basal basalt that unconformably overlies one of the ovoid batholiths (Fitton *et al.*, 1975). This post (?) 3.0 b.y. volcanism is of minor extent when compared to that of the Warrawoona Group.

The Yilgarn block, in contrast to the Pilbara block, records a dominant plutonic event at between 2.5 and 2.7 b.y. [Rb-Sr (Compston and Arriens 1968; Arriens, 1971): Pb-Pb (Oversby, 1975)]. These granites have intrusive relationships with volcanic rocks and define a minimum age for them. The geochronological data bearing directly on the age of these volcanics indicates that the volcanism was broadly synchronous with the plutonism (circa 2.6 b.y.). However, the ages are based on Rb-Sr total-rock studies, commonly of felsic volcanics, and are probably minimum-value estimators (Compston and Arriens, 1968; Hallburg *et al.*, 1976). Furthermore, there is isotopic evidence that the Wheat Belt zone contains gneisses and granites older than 2.9 b.y. (Arriens, 1971). This suggests an event coeval with the main phase of plutonism in the Pilbara block. This further implies that a volcanic sequence as old as the Warrawoona Group may exist or have existed in the Yilgarn block. As yet there is no supporting geochronological evidence. Thus while the Pilbara block records much activity (including basaltic volcanism) at $\simeq$3.0 and $\simeq$3.4 b.y., the Yilgarn block activity appears to occur at about 2.6 and 3.0 b.y. Whether the volcanism was prolonged about these age pulses awaits geochronological investigation, preferably by the U-Pb zircon technique as exemplified by Pidgeon (1978).

Table 7.4.1 Ages of Archean basaltic volcanism.

Continents	Cratonic Block	Timing of Basaltic Volcanism (b.y.) Phase I	Phase II	Phase III
North America	Superior-Slave (and extensions)	~3.6	+3.0(?)	2.7 ± 0.05
North Atlantic	Greenland-Labrador	3.7–3.8	3.0–3.1(?)	~2.6
Southern Africa	Rhodesia	+3.5	2.9(?)	2.6–2.7
	Kaapvaal	3.3–3.5	-	-
Australia	Pilbara	3.4–3.5	~3.0(?)	-
	Yilgarn	-	~3.0(?)	2.6–2.7

Summary and conclusions. In combining the data for the various continents discussed it becomes apparent that the geochronological data best supports distinct "episodes" in the Archean history of terrestrial basaltic volcanism. Three episodes may be identified and the results are summarized in Table 7.4.1. We have distinguished the three episodes as simply Phase I, II and III (oldest to youngest). Not all phases are developed in each cratonic block and some, even though identified, are of limited extent (e.g., Phase II in Pilbara). Of the three phases, Phase II is the most unreliable and future work may either eliminate it completely, or tie it by continuous volcanism to one of the other stages. For example in Canada, although Phase III is a dominantly 2.7 b.y. magmatic event, there is volcanism spanning the interval 2.95 to 2.75 b.y., which may effectively account for the volcanism and plutonism identified as Phase II. In strong support, however, of the apparently episodic nature of volcanism during the Archean is that each phase of volcanism approximately coincides with, or just precedes major plutonism.

If we are to attempt to interpret the Archean record in terms of plate tectonics, then Table 7.4.1 presents us with problems. The data suggest that volcanism and associated plutonism was episodic on a global scale. Stop-start plate tectonics with a rest period of some 500 m.y. does not satisfy simply the requirement that Earth's heat budget be continuously released via volcanism at spreading ridges. It could be argued that the actual phases as shown in Table 7.4.1 are not clearly defined. For instance for Phase I, Greenland records 3.7 to 3.8 b.y., and South Africa (Kaapvaal) 3.3 to 3.5 b.y. This is a significant difference, and it raises a possibility that can only be resolved by a vastly expanded geological/geochronological data base: namely, was Archean volcanism episodic as the current data suggest, or was it continual? We cannot answer this question with any confidence at the present time. Clearly, ongoing Archean crustal studies must respond to this question by emphasizing precise and reliable geochronology.

7.4.2 Sources of terrestrial basalts: isotopic characteristics

Introduction

In recent years, basalts have been used extensively as geochemical windows into the mantle. The initial discovery by Gast *et al.* (1964) that the mantle is chemically heterogeneous began the development of a large body of knowledge around this finding; the proof of a geochemically heterogeneous mantle is one of the major successes of modern geochemistry. This chemical heterogeneity is most clearly viewed with the radiogenic isotopes Sr, Nd and Pb, but other trace elements and trace element ratios are also useful, though their "viewing" properties involve considerably more noise than the isotopic systems. Mantle sources of basalts are not randomly heterogeneous, but each basaltic tectonic environment has characteristic trace element and isotopic fingerprints. Considerable efforts are being expended in geochemically "mapping" these various mantle regions in time and space, and in trying to understand how these geochemical signatures have originated. The following section is a review of our current understanding regarding the heterogeneous isotopic (and trace element) signatures of basalt sources, and the relationship of these signatures to tectonic processes throughout geologic time.

We begin by defining some of the terms which will be used in the following section:

LIL-elements: The large-ion-lithophile elements (also called magmatophile or incompatible elements) such as K, Rb, Cs, Sr, Ba, REE (rare-earth elements), U, Th, Pb, Zr, Nb, Y, Ta, etc.

MORB: Mid-ocean ridge basalt. Any of the basaltic rocks, usually low-K tholeiites, which are formed on spreading ridges. These may be further subdivided into "typical-MORB" or "transitional-MORB". Typical-MORBs occur on ridge segments not affected by island or triple-junction plumes or hot spots. Transitional-MORBs generally have more radiogenic isotope signatures related to proximity to islands or triple-junctions.

Heterogeneous: Compositional variations of a basalt source which exist on any scale larger than a mineralogical scale. For example, from kinetic arguments, it is likely that the minerals of a partially-molten mantle source will be in isotopic equilibrium, but isotopic heterogeneities could exist, even on a long-time scale, between partially molten mantle regions some kilometers in scale (Hofmann and Hart, 1978).

Depleted, undepleted, enriched: These terms may refer either to trace element concentration patterns, or isotopic ratio signatures. Commonly, a "depleted" trace element pattern is one in which a larger ion is depleted relative to a smaller ion of the same chemical group—i.e., Rb relative to K, Ba relative to Sr, or a light rare-earth element (La) relative to a heavy rare-earth element (Yb). Thus high K/Rb or Sr/Ba ratios, or low La/Yb ratios, are termed "depleted" ratios. Depleted may also refer simply to absolute concentrations—i.e., a basalt with low concentrations of all LIL-elements may be called "depleted." Depleted patterns or ratios frequently accompany depleted overall concentrations, but this need not necessarily be so. "Depleted" as applied to isotopic patterns generally means ratios of $^{87}Sr/^{86}Sr$,

$^{143}Nd/^{144}Nd$ or $^{206}Pb/^{204}Pb$ ($+^{207}Pb/^{204}Pb$) which are relatively low or unradiogenic. The problem is in defining *low* compared with what? For isotopes, we propose the following definition: "undepleted" is taken as a system with parent/daughter ratios and radiogenic daughter isotope ratios compatible with closed-system growth for 4.5 b.y. "Enriched" and "depleted" would then refer to systems with parent/daughter ratios (Rb/Sr, Sm/Nd or U/Pb) either higher or lower, respectively, than those required for closed-system evolution. These concepts are illustrated in Fig. 7.4.7. In the case of Sm/Nd, it is perhaps convenient to consider only one closed system, that with a Sm/Nd ratio equal to a chondritic or planetary value. Since planetary bodies such as the Earth, Moon and meteorites appear to have formed with similar Sm/Nd ratios, then this provides a single reference line against which to measure deviations. In the case of Rb/Sr and U/Pb, however, the accessible regions of the Earth appear to have parent/daughter ratios different from, say, chondrites or the Moon, and thus no unique "planetary" value can be defined. In this case, a number of closed-system growth curves are possible, and so the terms "depleted" and "enriched" need to be defined relative to the local parent/daughter ratio. It should also be noted that a "low" Nd isotope signature may accompany an "enriched" REE pattern, since for the Sm/Nd system, the parent isotope is smaller than the daughter isotope (unlike Rb/Sr and U/Pb, where enrichment of the larger ion, the parent in each case, will lead also through time to enrichment in the radiogenic daughter isotope ratios). Thus, an "enriched" REE pattern (La>Yb) will have Sm<Nd, and thus in time will generate a low ratio of $^{143}Nd/^{144}Nd$. Since no single definition of "enriched," "depleted" and "undepleted" can encompass all of these diversities, the context of the usage of these terms must be borne in mind—i.e., are they being applied to absolute trace element concentrations, trace element ratios, or radiogenic isotope ratios?

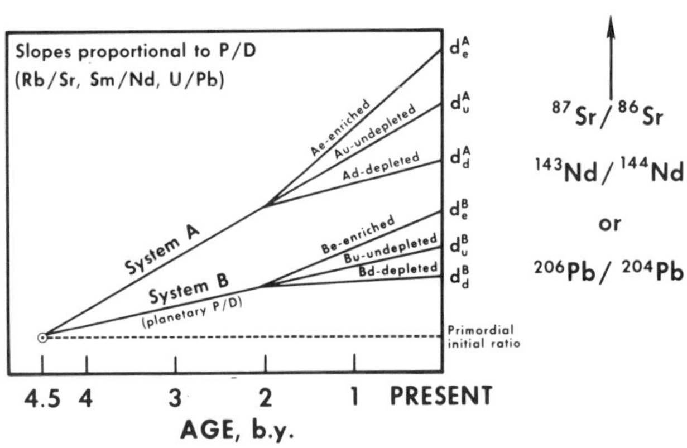

7.4.7 Schematic isotope evolution plot. Closed systems A and B, evolving with different parent/daughter ratios from the same primordial isotopic ratio, will have present-day daughter isotope ratios d^A_u, and d^B_u, and would be called "undepleted" (even though system A has a P/D ratio which is "enriched" relative to a system with planetary P/D ratio, such as B). Heterogeneous subsystems which may develop (for example, at 2 b.y.) will have P/D ratios either "enriched" (Ae, Be), or "depleted" (Ad, Bd) relative to the closed systems (Au, Bu), and will generate present day "enriched" (d^A_e, d^B_e) or "depleted" (d^A_d, d^B_d) daughter isotope ratios. Note that an "enriched" system such as Be could have an isotope ratio which is lower ("depleted") relative to some other system such as Ad, depending on the P/D ratios of their mother systems. Note also that it is theoretically possible to generate "enriched" systems from one mother, such as Be, which could have lower ("depleted") P/D ratios than a "depleted" system developed from a more "enriched" mother.

Alteration and metamorphism

Before discussing the geochemistry of basalts in detail, we will comment briefly on the problems of chemical modification during alteration and metamorphic processes.

Almost all basalts older than a few hundred million years have suffered some modification of their primary mineralogies. This modification may be accompanied by significant changes in the rock chemistry and hence the problem of alteration of basalts has been a subject of considerable study (e.g., Frey *et al.*, 1974; Hart *et al.*, 1974; Scott and Hajash, 1976), with most studies aimed at deciphering the premodification chemistry of the rock. Typically, these studies involve young basalts, although extension of the investigations to older basalts (e.g., Condie *et al.*, 1977) has also been attempted. The problem, however, of recovering an altered, ancient basalt's primary chemistry is complicated by the fact that usually that basalt has been subjected to more than one process leading to element migration. These processes may be summarized as follows:

(1) *Seawater effects*. The exchange of elements between the basalt and seawater has demonstrably caused large-scale modification of submarine basalts. The chemical differences between pillow rims and cores clearly document the chemical addition and subtractions. Typically, modern seafloor basalts show Sr-isotope compositions that are partially or totally equilibrated with seawater, i.e., between 0.7026–0.7030 and 0.709. Under submarine alteration conditions some trace elements such as the alkalies are extremely mobile while others (e.g., Nb, Y, Zr) are not. The latter have great potential for assisting in reconstructing the premodification basalt chemistry and have great application in establishing petrogenetic properties of altered basalts (discussed in more detail later).

(2) *Submarine metamorphism.* This is a process usually but not always accompanying seawater exchange effects. It leaves the basalt mineralogy generally changed to a metamorphic assemblage of minerals in the zeolite facies. This process may be a form of burial metamorphism, and may be difficult to distinguish from later low-grade regional metamorphism. Wood *et al.* (1976) studied the chemical changes accompanying zeolite grade metamorphism in eastern Iceland and showed that SiO_2, MgO, K_2O, Ba, Rb and Sr are significantly modified, the light REE may be somewhat modified, and Zr, Nb and the heavy REE are largely unmodified.

(3) *Regional metamorphism.* Whereas (1) and (2) are the normal manner in which young submarine basalts change their chemistry, ancient basalts have even more "tricks up their sleeves," i.e., regional metamorphism. Almost all ancient basalts (save some Proterozoic plateau basalts) have been subjected to a regional metamorphic imprint that has upgraded the basalt from zeolite grade to the prehnite-pumpellyite grade or higher. In fact, Archean basalts in the prehnite-pumpellyite facies are relatively rare, and the average lowest grade of metamorphism is greenschist facies. Consequently, there is even more likelihood that ancient basalt will have a modified chemistry due to these additional metamorphic effects than will younger submarine basalts.

(4) *Contact and dehydration effects.* Generally speaking, the effects of (1) to (3) above convert a pristine basalt to an assemblage of hydrated metamorphic minerals. In many of the ancient basalt occurrences (particularly the Archean) the rocks have been subjected to even further "ignominious" treatment due to the intrusion of large batholiths in the magmatic-orogenic cycle. These intrusions typically upgrade the regionally metamorphosed basalts to upper greenschist or amphibolite facies, and in the zone close to intrusions they produce dehydrated "ringing-fresh" hornfels.

Some of these effects are clearly distinguishable from others, but in the case of regional and contact metamorphism the effects may be interdependent. Nonetheless, the result of one or more of these processes could be the production of a basalt whose trace element and isotopic character was greatly different from the primary values. How do we overcome this problem?

For modern day rocks the problem is relatively simple since fresh varieties are obtainable, even in submarine situations. Study of older basalts is more complicated. Fortunately, comparison of fresh and altered modern basalts gives clues to the degree of modification that can be expected as a result of seawater exchange and submarine metamorphism. For ancient basalts at least, this component of the alteration effect can be understood by analogy with modern day basalts. There are no zeolite facies Archean basalts formed solely by burial effects on a regional scale (Archean basalts typically accumulate to thicknesses at least twice as great as modern oceanic basalts), and hence any attempt to recover their primary chemistry must take into account the "package deal" of alteration processes (1) to (3) outlined above (contact effects can generally be overcome by judicious sampling). Ways of overcoming this problem for ancient basalts include the following:

(a) comparison of trace element abundance and isotope compositions between basalt flows and less modified thick intrusive mafic equivalents;

(b) separation and analysis of relict mineral phases (e.g., clinopyroxene and plagioclase) rather than total rocks;

(c) identification and discarding of "too modified" basalts using a series of rejection criteria that include CO_2+H_2O analysis; existence of anomalous minerals in the CIPW norm (wollastonite, corundum); positioning in forbidden zones on commonly used variation diagrams; existence of limpid albite (spilite), etc. (see Gèlinas *et al.*, 1977);

(d) confirmation of broad chemical trends by means of the nonmobile elements (Y, Zr, Nb, etc.).

All of this paints a black picture for ancient basalt chemistry. However, the situation is not as bleak as it looks. Surprisingly, large thicknesses of metabasalt from Archean sections have chemical trends that are similar to those observed in unaltered modern basalts. In many instances the combined result of the alteration is to produce scatter about the original primary trends without drastically modifying the basic property of that trend. Unfortunately, there are exceptions to the observation and each case must be judged on individual merits. With judicious sampling and careful consideration of the geochemistry, meaningless scatter diagrams for Archean basalts can be avoided.

This has been restated recently in another way by Condie *et al.* (1977), who have found that 10% carbonatization and 60% epidotization of Archean tholeiite will not appreciably affect the interpretation of trace element models for magma genesis.

Modern basalt sources

This section will deal with the observed isotopic and trace element patterns and the tectonic environment in which the basalts are erupted. An enormous amount of data has been published in this area, stimulated a decade ago by the classic paper of Gast (1968a). The general results of this work have led to a fairly complete understanding of oceanic basaltic volcanism, but many inadequacies still exist, particularly in the case of island arc and continental rocks—the former due to lack of data on well-documented suites of rocks, the latter because of the difficulty of unravelling mantle vs. crustal contributions to the erupted basalts.

Oceanic mantle

Isotopic patterns. If we look first at the oceanic data, because of its relative simplicity, we see immediately that the isotopic systems Sr, Pb, and Nd all show significant geographic variations, demonstrating heterogeneity in ocean mantle sources. For example, Figs. 7.4.8–10 show the isotopic variations in Sr, Pb, and Nd in oceanic rocks. The results show that variations exist on almost all scales. Single islands, even very large ones

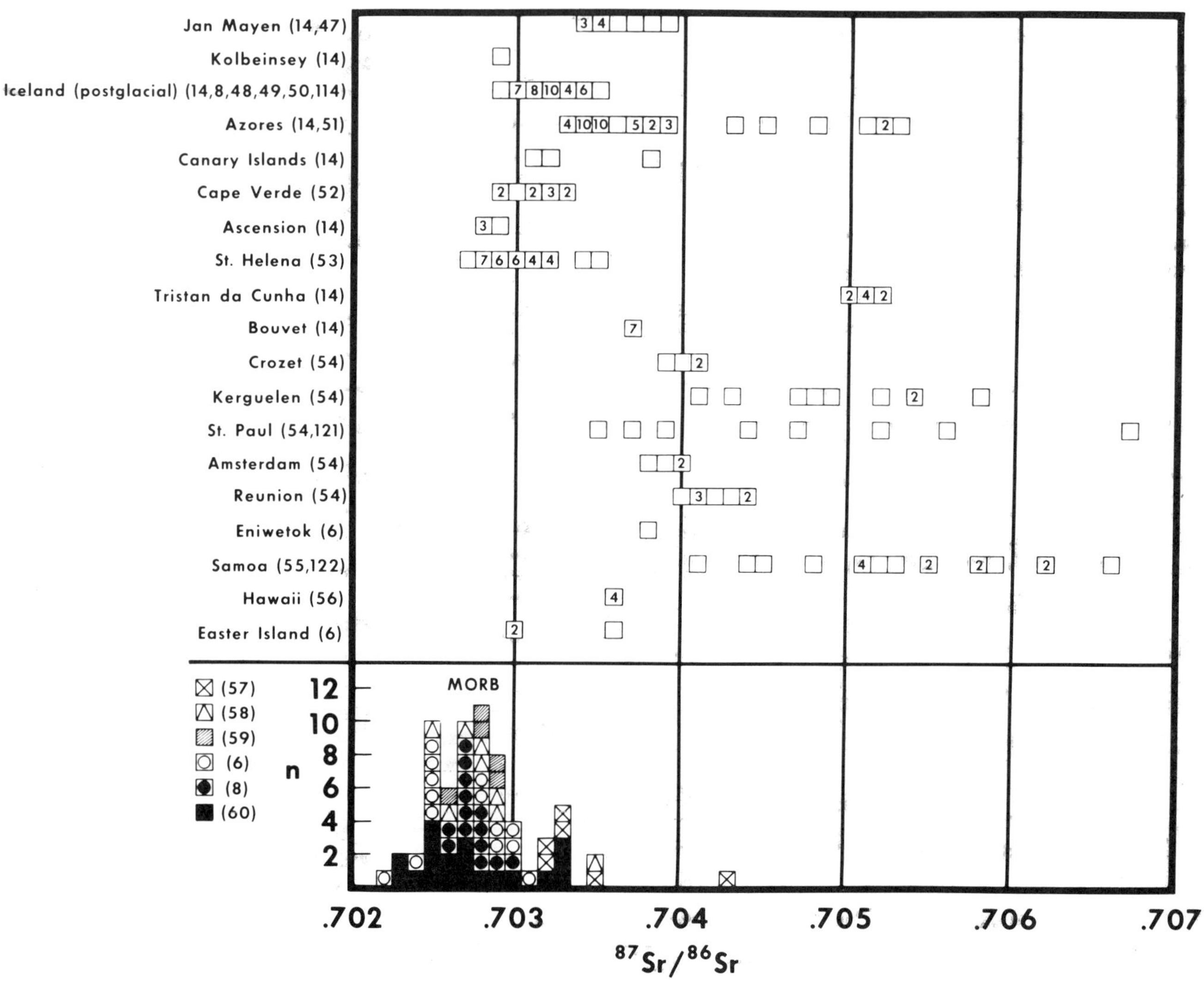

7.4.8 Comparison of $^{87}Sr/^{86}Sr$ ratios from mid-ocean ridge basalts (MORB) and from oceanic islands (from Hofmann and Hart, 1978). The MORB data are plotted as a frequency diagram (number of samples (n) vs. $^{87}Sr/^{86}Sr$) below the horizontal line. The island data are arranged geographically in the sequence North Atlantic, South Atlantic, Indian and Pacific Oceans. The number of samples for each value of $^{87}Sr/^{86}Sr$ (rounded to four significant figures) is indicated on the figure; a blank square represents one sample. The following data have been omitted: (1) all island arc data; (2) all analyses published prior to 1970 (because of less uniform analytical precison of these older analyses); (3) analyses of samples from the Reykjanes Ridge between 0 and 200 km from Iceland (because this region is similar to Iceland itself with respect to isotopic and trace element composition); (4) one anomalously low value given by Hedge and Peterman (1970), because re-analyses of the sample failed to confirm this value; (5) all data for Mid-Atlantic samples by White *et al.* (1976), because inclusion of these data would aggravate the problem inherent in this type of frequency plot, namely, weight the data excessively with a large number of analyses from one area, the Azores platform. The key on the left-hand side identifies the islands and the references in Hofmann and Hart, (1978) for islands and ocean floor samples.

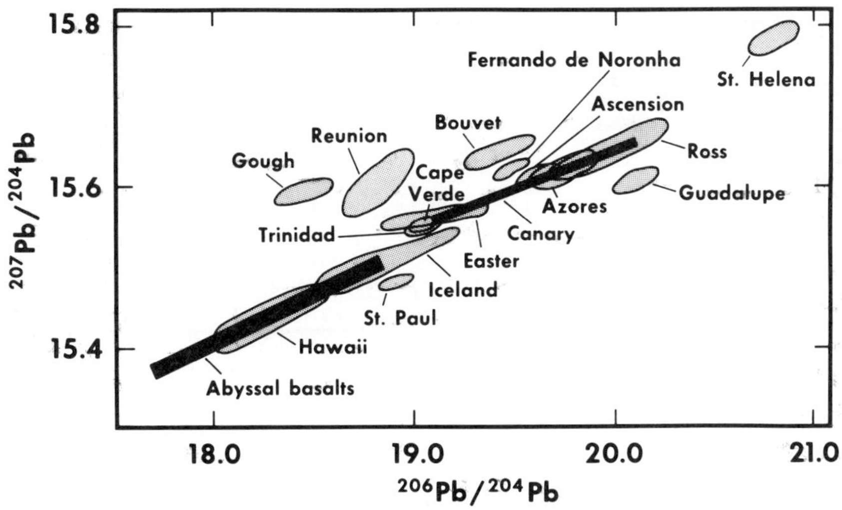

7.4.9 $^{207}Pb/^{204}Pb$ vs. $^{206}Pb/^{204}Pb$ diagram for abyssal tholeiites and rocks from oceanic islands (see Tatsumoto, 1978, this paper for references to data).

such as Iceland, may show very constant Sr isotope ratios over the whole extent of the island, but also show significant Pb and Nd variations. Islands may have uniform values on an intraisland basis but be significantly different from other islands, for example, St. Helena compared to Tristan da Cunha (Sr isotopes, Fig. 7.4.8) or St. Helena compared to Ascension (Pb isotopes, Fig. 7.4.9). Single islands may also show very

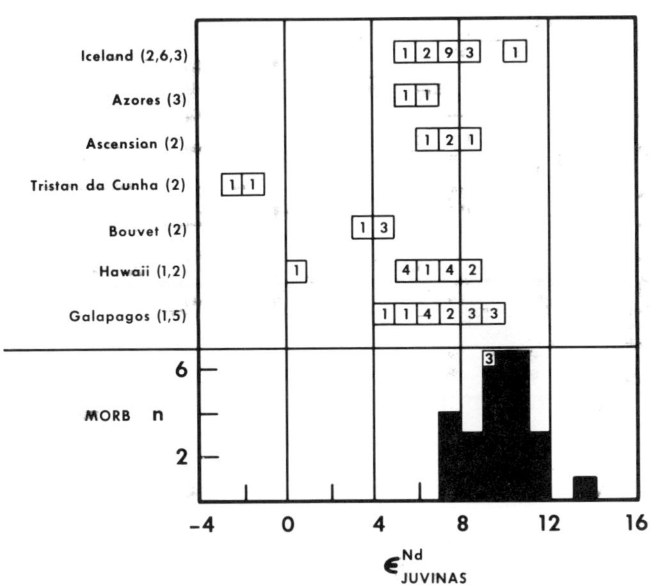

7.4.10 Nd isotope ratios from oceanic basalts. Numbers in squares are number of analyses. MORB data plotted as a histogram. References: (1) DePaolo and Wasserburg, 1976a, b; (2) O'Nions *et al.*, 1977; (3) Richard *et al.*, 1976; (4) Carlson *et al.*, 1978; (5) White *et al.*, 1978; (6) Zindler *et al.*, 1979.

large intraisland isotopic variations, as for example, Samoa (Sr, Fig. 7.4.8), the Canary Islands (Pb, Fig. 7.4.9), Iceland (Nd, Fig. 7.4.10). Isotopic variations even exist between successive lavas from a single volcano and between separate volcanoes on a given island, such as the island of Hawaii (Pb, Fig. 7.4.11).

Perhaps the most obvious and distinct finding from oceanic isotopic data is the distinction between ocean ridge-type mantle and oceanic island mantle. Ridge-type mantle shows evidence of distinct and long-lived depletions of P/D ratios for all three isotopic systems, which in most cases produce isotopic signatures almost perfectly unique relative to oceanic island mantle. This may be characterized as follows: radiogenic isotope ratios of "typical" ridge basalts (MORB) almost always show $^{87}Sr/^{86}Sr < 0.7030$, $^{206}Pb/^{204}Pb < 18.7$ and $^{143}Nd/^{144}Nd > 0.51305$. Islands, on the other hand, have $^{87}Sr/^{86}Sr > 0.7030$, $^{206}Pb/^{204}Pb > 18.5$ (except Hawaii), and $^{143}Nd/^{144}Nd < 0.51305$. Thus the Sr and Nd systems show very little overlap between MORB and island basalts; the Pb system shows partial overlap. In addition, there is a fairly regular transition from MORB to island values for these isotope systems in cases where an island straddles a spreading ridge, as in the case of Iceland or the Azores. Figure 7.4.12 shows the latitudinal variation of Sr isotope ratios along the Mid-Atlantic Ridge, with pronounced "island" type anomalies on Iceland and the Azores and the ridge segments adjacent to these islands. Note that the transition zone is considerably more abrupt near Iceland (200 km), whereas the Azores show transition zones some 500 km wide on either side of the island. In the case of Iceland, Pb isotopes show a similar transition, from typical MORB values of about 18.2 south of Iceland to values > 18.5 on and adjacent to Iceland. These variations have been

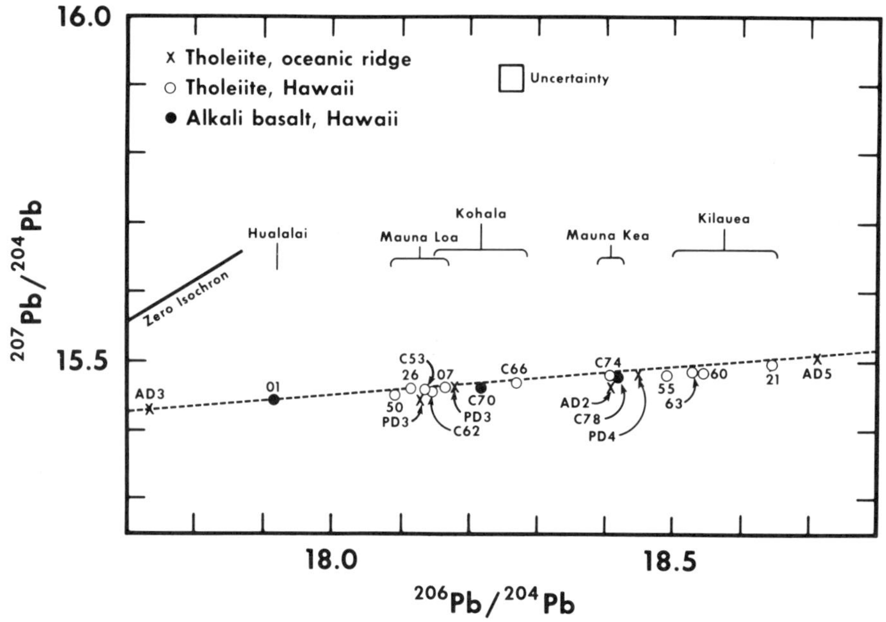

7.4.11 $^{207}Pb/^{204}Pb$ vs. $^{206}Pb/^{204}Pb$ diagram of basalts from the island of Hawaii. Shown for comparison are redetermined values of abyssal tholeiites indicated by an "X". The Pb from each volcano has a distinct isotopic composition (from Tatsumoto, 1978).

ascribed to mixing of two chemically distinct mantle sources, and a mixing model does in fact describe the Sr and Pb data quite well (Sun *et al.*, 1975), and REE data in detail (Schilling, 1973); however, when other trace elements are considered, a simple two-component mixing model does not adequately describe the data (Langmuir *et al.*, 1978).

The isotopic distinction between MORB and oceanic island basalts extends to other trace elements as well. In general, MORB are more depleted in LIL elements than island basalts, and show more depleted element ratios of, for example, Ba to Sr, Cs to Rb, Rb to K, the light REE to the heavy REE, Nb to Zr, etc. (Kay *et al.*, 1970; Hart, 1976; Kay and Hubbard, 1978; Schilling, 1975; Erlank and Kablc, 1976). These combined trace element and isotopic characteristics are quite sufficient to unequivocally separate island basalts from MORB on a chemical basis—at least if we restrict our definition of MORB to those spreading ridge locations which are distant from any islands, and the transition zones adjacent to these islands. See, for example, the variation of various elements and element ratios along the Mid-Atlantic Ridge adjacent to the Azores (Fig. 7.4.13). Where the Mid-Atlantic Ridge crosses the Azores platform there is still a distinct spreading ridge environment and yet the tholeiites erupted at this spreading center bear stronger kinship to the alkali basalts of the Azores islands than to the typical Mid-Atlantic Ridge segments further away from the Azores. This is true also of spreading ridge segments adjacent to other islands (hot spots or plumes?) such as Iceland, the Galapagos Islands, Bouvet Island and Reunion Island. In the latter three cases, where islands occur near ridge triple-junctions, it is not altogether clear whether the

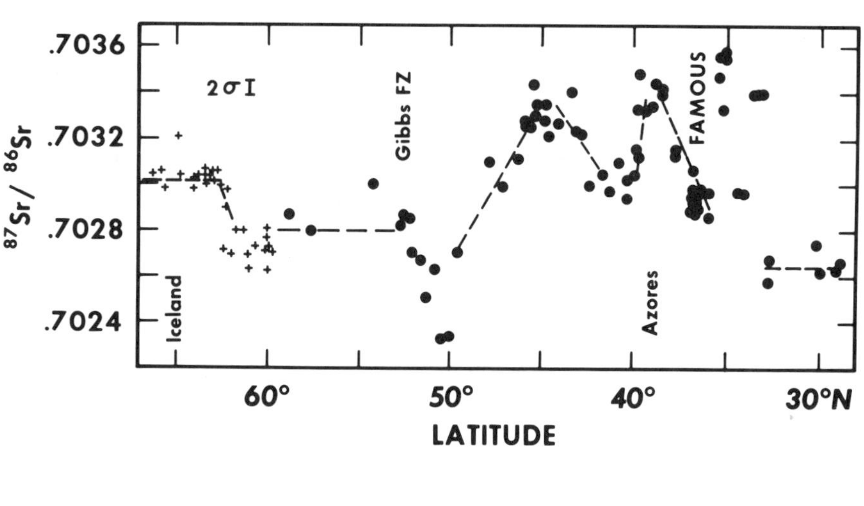

7.4.12 Variation of $^{87}Sr/^{86}Sr$ with latitude along the Mid-Atlantic Ridge. Precision is indicated by the error bar (upper left) and is $\pm$ 0.0006 ($\pm$ 2 standard deviations). Data previously reported by Hart *et al.* (1973) are shown for comparison (+). The dashed line is interpolated (from White *et al.*, 1976).

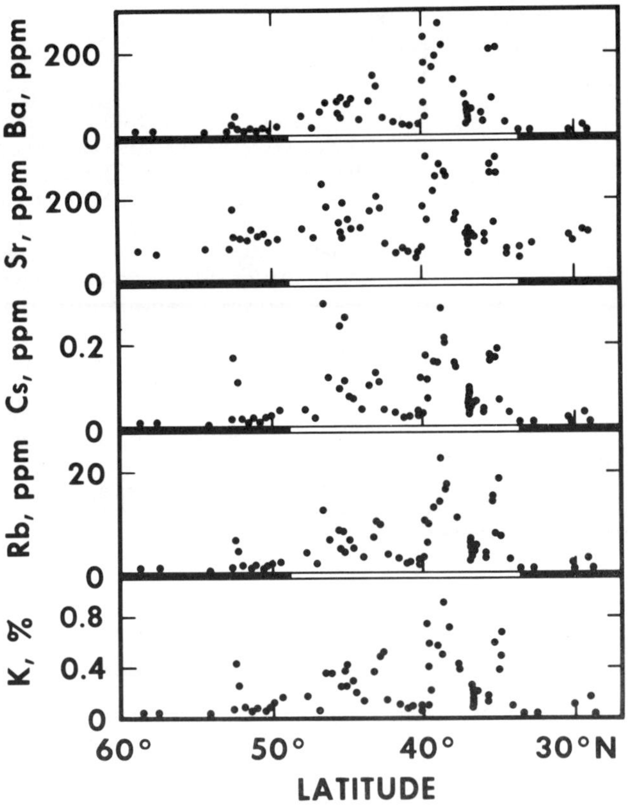

7.4.13 Variation of K, Rb, Cs, Sr and Ba with latitude along the Mid-Atlantic Ridge. Dark bars indicate normal ridge segments as inferred from Sr isotope data (from White and Schilling, 1978).

effect is due to the island, or the triple-junction, or both (if they perhaps bear causal relationship to each other).

Time duration of oceanic mantle heterogeneities. The island-ridge geochemical distinction is one of considerable age duration and this fact is of considerable importance with respect to possible kinematic arrangements of the various mantle reservoirs. Simply put, for the range of P/D ratios commonly inferred for these sources, a considerable time period is required for these sources to develop the observed differences in radiogenic daughter isotope abundances. The most unequivocal demonstration of this is derived from the Pb isotope data. (It has been argued above that basalts almost certainly erupt with isotopic ratios similar to those in the source from which they are derived. This is in contrast to P/D ratios which may be different in a derived liquid relative to the starting source, due to effects of crystal-liquid partitioning during partial melting and crystal fractionation.) The Pb data for MORB and islands are shown in Figs. 7.4.9 and 7.4.14. Basalts from these two environments lie along linear trends with positive slopes. There are at least two possible explanations for these trends:

1. A mantle source with homogeneous U/Pb ratio ($\mu = {}^{238}U/{}^{204}Pb$) separates at time t_1 into subsystems of various μ. After aging from t_1 to present, these subsystems will have Pb isotopic compositions lying along a line whose slope is proportional to the age t_1, and this line is called a secondary (or mantle) isochron (or the

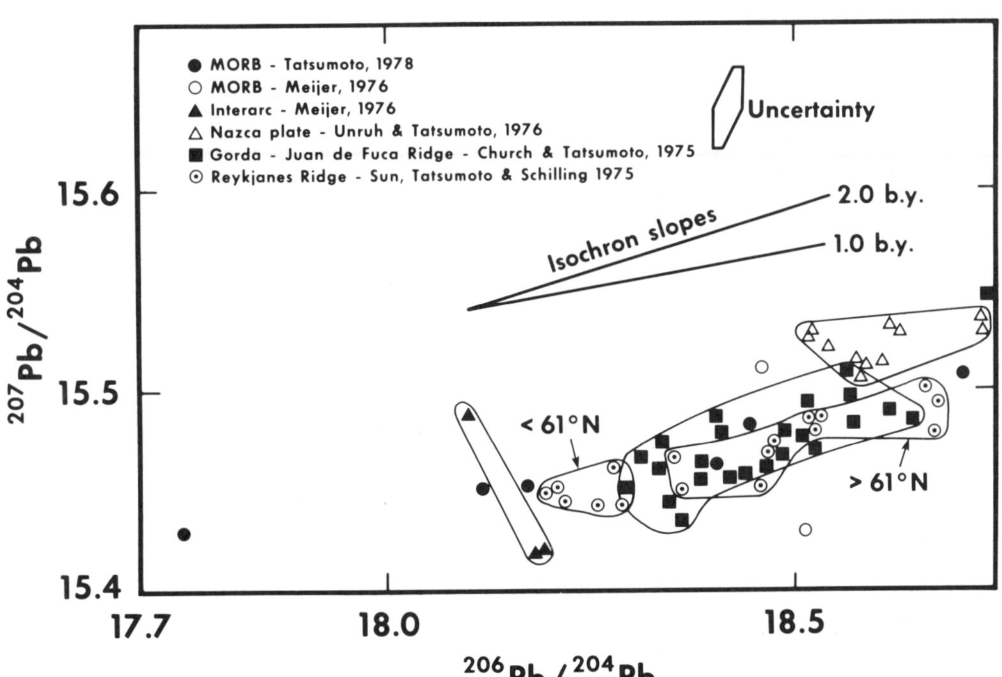

7.4.14 $^{207}Pb/^{204}Pb$ vs. $^{206}Pb/^{204}Pb$ isotope ratios for ocean floor basalts. Data from various sources as indicated. Reykjanes Ridge tholeiites are divided into two fields — those south of 61°N and those north of 61°N.

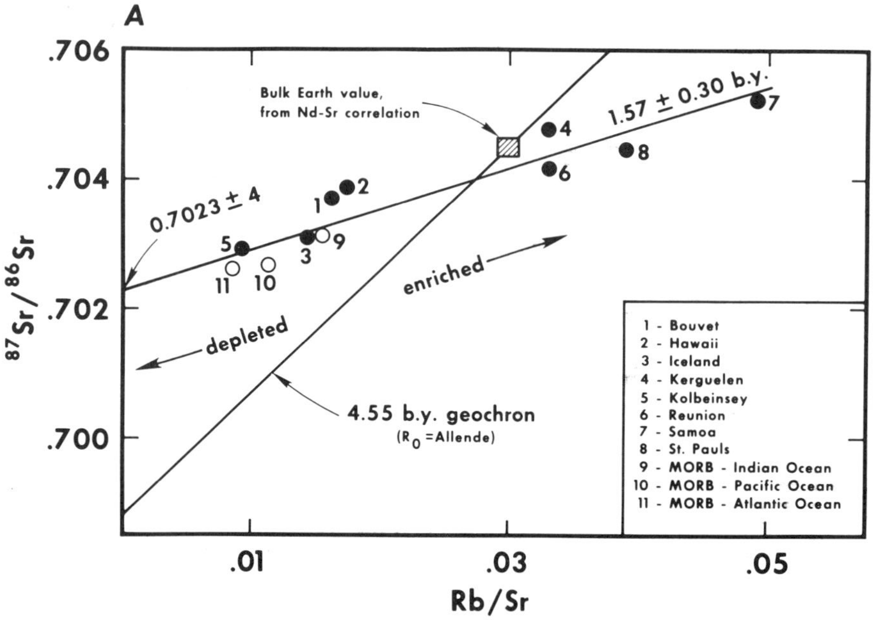

7.4.15 a. Rb-Sr isochron diagram of various oceanic basalt groups, averaged by island (solid circles) or by ridge (open circles). The data fit a linear trend which has been interpreted by Brooks *et al.* (1976a) as a "mantle" isochron of 1.6 b.y. age. A planetary or chondritic bulk earth value (square) derived from the terrestrial Sr-Nd correlation plot, is plotted on a 4.55 b.y. closed-system "geochron." Source regions for basalts plotting to the right of the geochron may be termed "enriched," those to the left may be termed "depleted" (see discussion accompanying Fig. 7.4.7).

second "stage" of a two-stage growth process). If the slopes of the trends shown in Figs. 7.4.9 and 7.4.14 are interpreted with a two-stage model of this kind, the time t_1 of development of the subsystems is in the range 1–2 b.y.

2. A mantle has developed heterogeneities in some complex way, such that two subsystems A and B exist at present. Mixing of these mantle systems (or melts derived from them) leads to a mixing-line AB. In this case the slope of this line may or may not be simply proportional to a single discrete "age" of formation of heterogeneities in U/Pb.

In the second case, even though the line is attributed to a mixing process and no formal "age" can be derived from it, some finite aging is required to originally develop the two subsystems A and B. For example, it is impossible to start with a homogeneous source at a recent time and develop the necessary "separation" of sources A and B, no matter what kind of extreme fractionations of U/Pb are envisioned. This is largely due to the fact that ^{235}U, which decays to ^{207}Pb, is of short half-life and is presently very much less abundant than ^{238}U. Thus it is difficult to produce significant variation in $^{207}Pb/^{204}Pb$ by appealing only to relatively young U/Pb fractionations. In other words, while the slope of the arrays in Figs. 7.4.9 and 7.4.14 may not correspond to an exact age, it is not devoid of age information. Even with very complex multistage U/Pb fractionation processes, it is difficult to escape the fact that Pb isotopic arrays such as this require aggregate ages for heterogeneous mantle subsystems of the order of 1 b.y. or greater. Since cycle times of oceanic litho-

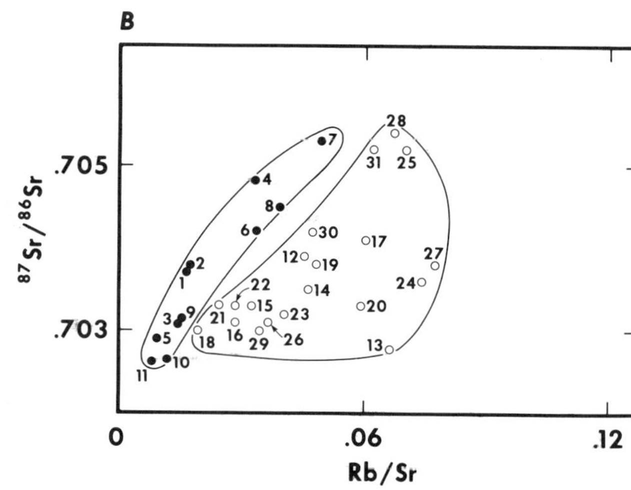

b. $^{87}Sr/^{86}Sr$ - Rb/Sr isochron plot of tholeiites (filled circles) and alkali basalts (open circles) from oceanic regions. Each point represents an average of available data on individual samples from a given island or island group. Numbers refer to the data index in Table 1, Brooks *et al.*, 1976a.

sphere are of the order of 200 m.y. or less, the observed heterogeneities *cannot* be developed by magmatic processes associated with events related to the current "cycle" of ocean crust formation (for example, extraction of island basalts cannot be said to have created a depleted mantle which is later tapped to produce ocean ridge basalts).

Returning to the Hawaiian Pb data (Fig. 7.4.11), we see a similar linear arrangement of data with a positive slope. If interpreted as a two-stage model, this slope would correspond to an age of 0.94 ± 0.42 b.y. The large error is due to the relatively small spread of points (compare variation in $^{206}Pb/^{204}Pb$ on Fig. 7.4.11 and on Fig. 7.4.14)—in fact, the Hawaiian data array is approximately collinear with the MORB data of Fig. 7.4.14, and suggests a common development for the mantle heterogeneities in both cases. Basalts from Iceland and the adjacent part of the Mid-Atlantic Ridge define a linear trend also, with an "age" of 1.5 ± 0.5 b.y. (Sun and Jahn, 1975), also collinear with the trends shown here. Reference to Fig. 7.4.9 shows that all oceanic rocks lie along this approximately linear trend, such that the trends shown above on an intraisland basis are also exhibited by the whole ensemble of oceanic rocks. Note that some islands (Ascension, etc.) show little variation and *by themselves* do not define a trend. Note also that some islands appear to lie off this "main trend," (e.g., Bouvet, Reunion, Gough, etc.); in part, this may be analytical in nature, or it may in part be related to even larger scale mantle heterogeneities (i.e., southern vs. northern hemisphere—note that most of the "anomalous" islands are southern hemisphere islands). A possible hemispherical difference for ore leads was also noted by Sinha (1971).

The Rb-Sr and Sm-Nd systems can be looked at in somewhat analogous ways. Only one more assumption is required—that the P/D ratios in the basalts are representative of the source from which they were derived. For the Rb-Sr system, this assumption is probably valid for tholeiites and melts which leave no residual mantle phase capable of fractioning Rb from Sr. This assumption thus may not be good for alkali basalts derived by low degrees of melting with possible amphibole or phlogopite as a residual phase. For the Sm-Nd system, the assumption is probably valid only for shallow-derived melts, as any melt which leaves garnet in the residue will show a fractionated Sm/Nd ratio. The oceanic Rb-Sr and Sm-Nd data are shown on isochron diagrams, Figs. 7.4.15–16. Indeed, we see a distinct difference in Fig. 7.4.15 between tholeiites and alkali basalts which, although they have similar ranges in $^{87}Sr/^{86}Sr$, are different on the average by a factor of two in Rb/Sr. (Note that this difference, if due to fractionation during partial melting, requires amphibole, not phlogopite, as a residual phase.) The tholeiites, for which the assumption of no Rb/Sr fractionation is most nearly valid, exhibit a very good linear array of points, which can be interpreted as an age (as with Pb, the array can be either interpreted as a two-stage model or as a mixing model) of 1.6 b.y. It is an interesting and perhaps very important point that this Rb-Sr "age" is very similar to the Pb isotopic "ages" discussed above. The

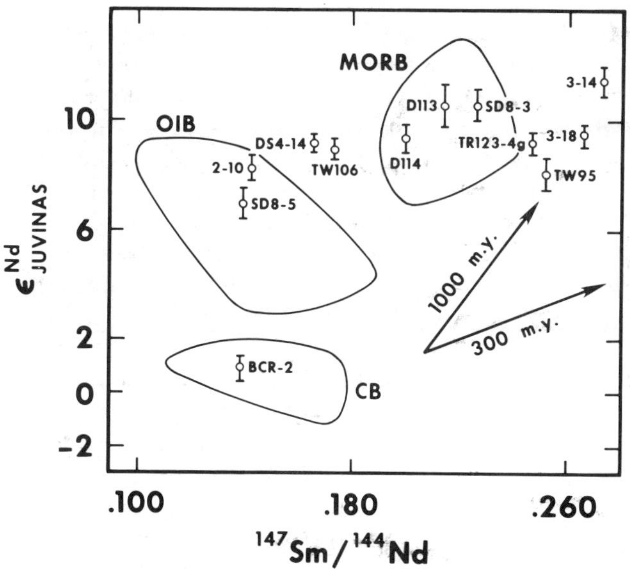

7.4.16 Nd isotopic composition vs. $^{147}Sm/^{144}Nd$. Fields for Mid-Ocean Ridge Basalts (MORB), Ocean Island Basalts (OIB), and Continental Basalts (CB) are constructed from the data of DePaolo and Wasserburg (1976a, b) and O'Nions *et al.* (1977). Isochron slopes of 0.3. and 1.0 b.y. are shown for comparison (from Carlson *et al.*, 1978).

linear array of data on the Rb-Sr diagram has been termed a *mantle isochron* (Brooks *et al.*, 1976a,b).

The Sm-Nd data in Fig. 7.4.16 are much more complex than the Rb-Sr data. There is no simple distinction between tholeiites and alkali basalts, and there are no simple linear arrays. However, if we look just at the MORB data, which presumably represent the largest degree of melting, and which are least likely to leave garnet as a residue, there is a crude linear array from which a two-stage model age of (approximately) 0.2 b.y. can be calculated. This age is very much younger than the Pb and Sr ages, and is consistent with the "cycle times" for oceanic plate circulations. The discordance of this Nd age is a very important finding and one as yet unexplained. It may be related to the existence and migration of an LIL-element-enriched (LILEN) mantle phase of the type proposed by Frey and Green (1974) and Zindler *et al.* (1979). This LILEN phase, which undoubtedly bears the strong imprint of garnet REE fractionation, may be sufficiently ubiquitous to override or overprint any longer term isotopic heterogeneity patterns of the type shown by Rb-Sr and U-Pb. Because garnet is not effective in fractionating these latter systems, the LILEN phase would not be important in controlling the isotopic imprints of Sr and Pb systems.

The ocean mantle heterogeneities so well documented by the isotope data are accompanied as well by differences in trace element character, as might be expected from the fact that the isotopic heterogeneities arise because of heterogeneities in P/D ratios (U/Pb, Rb/Sr, Sm/Nd). In general, the isotopic variations show that MORB sources are depleted in the large LIL-elements relative to the smaller LIL-elements (Rb<Sr, Nd<Sm and U<Pb). Other trace elements follow the same patterns, with the result that MORB tholeiites generally show "depleted" ratios of, for example, K/Rb, Sr/Ba, Zr/Nb, La/Yb, etc., relative to oceanic island tholeiites (Kay *et al.*, 1970; Hart, 1971, 1976; Schilling, 1971; Erlank and Kable, 1976; Kay and Hubbard, 1978). Of course certification of mantle heterogeneities based on trace element data involves more assumptions than certification based on isotopic composition, because of possible secondary trace element variations produced by the partial melting and fractional crystallization processes themselves. However, as shown initially by Gast (1968b) and amplified by many later workers (Schilling, 1975; Brooks *et al.*, 1976a, b), the observed element variations (especially trace element ratios) must in large part reflect variations in source mantle composition. This conclusion rests heavily, however, on the assumption that melting is an equilibrium process and that the degrees of melting, at least for tholeiites, are relatively large (>10%). If conventional wisdom regarding this latter point is wrong, and melting occurs in much smaller degrees (less than a few percent), then residual phases capable of fractionating many of these trace elements may be generally present after melting and the trace element data may be much less useful in delineating mantle chemistry.

Finally, we may consider the question of what mechanisms are responsible for the differences between oceanic alkali and tholeiitic basalts. We will exclude MORB tholeiites from this discussion because the evidence for a distinct mantle source chemistry for these rocks is overwhelming. Consider, however, alkali basalts and tholeiites from similar geographic regions. These basalt types occur together on many islands (Samoa, Galapagos, Iceland, Hawaii, etc.), and in many cases the isotopic evidence points to mantle sources of similar composition. For example, Pb isotopes are similar in alkali basalts and tholeiites from Kohala volcano and Mauna Kea volcano in Hawaii, even though the two volcanoes are distinctly different from each other in Pb isotopic composition (Tatsumoto, 1978). In the Hawaiian basalts which have been measured for $^{143}Nd/^{144}Nd$, there are no significant differences between tholeiites and alkali basalts (3 tholeiites—0.51292, 5 alkali basalts—0.51297; O'Nions *et al.*, 1977). Sr isotope ratios are similar in tholeiites from the Azores platform and in alkali basalts from the islands on the platform (White *et al.*, 1976), and in tholeiites and alkali basalts from Iceland (Hart *et al.*, 1973) and Samoa (Hubbard, 1971; Hedge *et al.*, 1972). Alkali basalts and tholeiites from Kerguelen show a range in Pb and Sr isotopic composition, but there are no distinct differences between the two rock types (Dosso *et al.*, 1979; Hedge *et al.*, 1973). In the Galapagos, at least one island contains both alkali basalts and tholeiites with similar Sr and Nd isotopic ratios (White *et al.*, 1978). One of the interesting things about this island is the fact that the isotopic character of the tholeiites and alkali basalts are identical to MORB, and thus distinct from all other oceanic islands. This is the first reported occurrence of island basalts with all the trace element and isotopic characteristics of MORB.

In the examples cited above the isotopic data seem to indicate that both basalt types derive from the same mantle source. Thus, the chemical differences which distinguish these alkali basalts and tholeiites from each other must relate to the peculiarities of the melting and fractionation processes involved in derivation of the basalts, not to differences in mantle composition (at least differences of long-term duration). In general, the trace element differences between the alkali basalts and tholeiites can be explained by differences of 2–4 times in the degree of melting, and by the presence of residual phases such as garnet and amphibole in the source (to produce the higher Rb/Sr and La/Yb ratios in the alkali basalts compared to the tholeiites). Ultimately, the trace element character of these alkali basalt-tholeiite combinations should prove rather definitive in specifying the differences in the petrogenetic mechanisms leading to these two basalt types. It should also be noted that small differences in the oxygen isotope composition of alkali basalts and tholeiites are being found in some areas, with the alkali basalts averaging some 0.5‰ higher than tholeiites (Kyser and O'Neil, 1978). The data base is currently insufficient to show whether this difference exists in basalts which are otherwise similar in Sr, Pb or Nd isotopic composition. Also, it is not yet clear whether the difference in $\delta^{18}O$ is primary or related to secondary effects such as magma interaction with altered wall rocks of the magma chamber, or to exchange with seawater or meteoric waters. In this connection, it is interesting to note the evidence for contamination of both alkalic and tholeiitic magmas from Hawaii based on large disequilibrium values of $^{230}Th/^{232}Th$ in these magmas (Condomines *et al.*, 1976). In this case, it was suggested that interaction of the magmas with zeolitized wall rocks was a possible mechanism for the observations, and such interaction might be expected to increase the $\delta^{18}O$ of the magmas.

The island-ridge geochemical distinction is obviously a result of different mantle compositions in the source regions of these two basalt types. Understanding of the kinematic or spatial arrangement of these different mantles, however, is under considerable controversy. Schilling (1973) initiated the idea of ridges being derived from shallow depleted asthenosphere, with the islands being supplied by plumes or blobs deriving from less depleted mantle at depths below the asthenosphere. This concept has been championed by Sun and Hanson (1975), Brooks *et al.* (1976a, b), Hofmann and Hart (1978), and White *et al.* (1976). Perhaps the most serious defect in this concept is the fact that the ocean floor is littered with seamounts, and the few of these that have been sampled and analyzed appear to be of plume-type chemistry (Houghton, pers. comm., 1978). Thus an enormous number of plumes is required and the original plume concept as proposed by Morgan begins to seem ludicrous. The alternative is to derive ridge basalts from a mantle upwelling from a depleted layer lying at depths greater than that which supplies the islands, so that islands require only relatively shallow derivation. This concept has been discussed by Allègre and Bottinga (1977), Hedge (1978), Tatsumoto (1978), and Hofmann *et al.* (1978). While this model has a certain geochemical appeal, it does appear difficult to envision in view of geophysical constraints. So-called plumes or hotspots appear to represent a fixed reference frame (Morgan, 1972; Molnar and Atwater, 1973; Minster *et al.*, 1974), with motions relative to each other which are small relative to plate velocities. On the other hand, ridges are very mobile relative to each other, as evidenced by the fact that certain plates are surrounded by ridges without intervening subduction boundaries (e.g., Africa and Antarctica). Thus it would appear difficult to support a kinematic model in which ridges represent upwelling from greater depths than islands.

Island arcs. Moving in the direction of increasing interpretative complexity, we will consider next the island and continental volcanic arc rocks. The first point to be made here is that **island arc rocks** (those where significant underlying continental crust is absent) have ranges of the isotopic ratios $^{87}Sr/^{86}Sr$, $^{206}Pb/^{204}Pb$ and $^{143}Nd/^{144}Nd$ which lie within the range of oceanic volcanic rocks (range in $^{87}Sr/^{86}Sr = 0.7027$–0.7062; $^{206}Pb/^{204}Pb = 18.0$–19.1; $^{143}Nd/^{144}Nd = 0.51297$–$0.51314$). At first glance, this might suggest simple oceanic-type mantle sources for island arc rocks. However, this simple concept disappears when the isotope systems are considered in combination. For example, on a $^{207}Pb/^{204}Pb$ vs. $^{206}Pb/^{204}Pb$ plot, some island arc rocks are displaced above the oceanic mantle trend toward higher $^{207}Pb/^{204}Pb$ ratios [e.g., the Aleutians (Kay *et al.*, 1978); Fig. 7.4.17]. On a Sr-Nd correlation plot, some island arc samples from the Marianas and the South Sandwich islands plot on the high $^{87}Sr/^{86}Sr$ side of the "main" oceanic trend as defined by DePaolo and Wasserburg (1977) and Hawkesworth *et al.* (1977); see Fig. 7.4.18. These results have been interpreted as proof that either oceanic sediment or seawater-altered oceanic crust are sometimes involved in the process of island arc magma

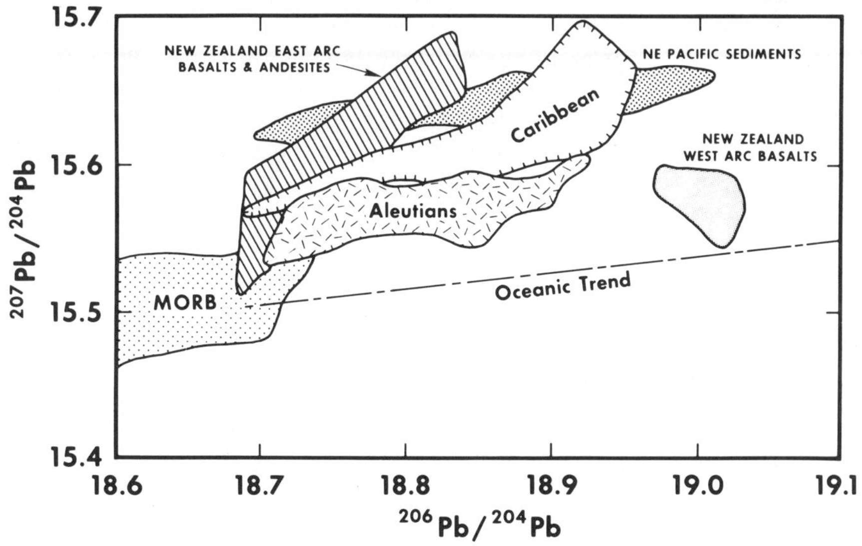

7.4.17 $^{207}Pb/^{204}Pb$ vs. $^{206}Pb/^{204}Pb$ plot for volcanic arc basalts and andesites from New Zealand, Aleutians and Caribbean compared to Pacific Ocean sediments. Line is approximate trend of oceanic basaltic rocks, as derived from Fig. 7.4.9. Field of ridge basalts, from Fig. 7.4.14, is also shown. Trends of these volcanic arc rocks strongly suggest a sedimentary-Pb component in their petrogenesis. Data from Kay *et al.* (1978), Church (1976) and Armstrong and Cooper (1971).

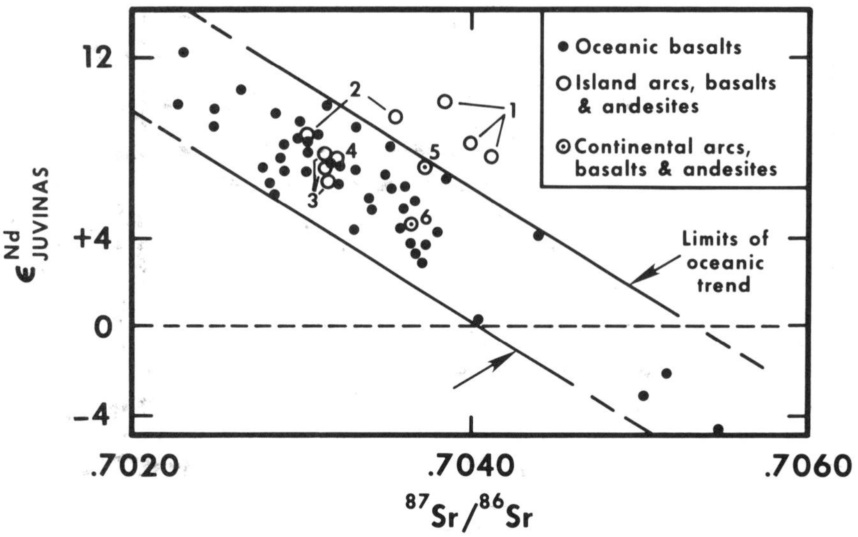

7.4.18 Sr-Nd isotope correlation plot comparing volcanic arc rocks with oceanic rocks. Data sources for oceanic rocks are given in Fig. 7.4.10. Data point numbers designate localities as follows: (1) South Sandwich; (2) Guam, Marianas; (3) Agrihan, Sarigan; (4) New Britain; (5) Costa Rica; (6) Mt. Shasta, California. Arc volcanic data from DePaolo and Wasserburg (1977), Hawkesworth *et al.* (1977) and DePaolo (1978). Only the South Sandwich arc rocks lie clearly outside the oceanic field, and this has been ascribed to mixing with sediment or oceanic crust from the subducted plate (Hawkesworth *et al.*, 1977).

genesis (Kay *et al.*, 1978; Hawkesworth *et al.*, 1977). Similarly, the position of some Mariana arc volcanic rocks on the high $^{87}Sr/^{86}Sr$ side of a $^{206}Pb/^{207}Pb$ vs. $^{87}Sr/^{86}Sr$ plot has been interpreted as a sign that altered oceanic crust has been involved in petrogenesis of these rocks (Meijer, 1976). Thus the evidence for a slab-derived component being involved in some island arc volcanism seems well established. In certain unusual tectonic areas, this process may produce very startling results, as for example the $^{87}Sr/^{86}Sr$ ratios of basalts from the Banda and Sunda arcs (Whitford *et al.*, 1977). There is evidence here for incipient underthrusting of a continental plate under an oceanic plate, and the $^{87}Sr/^{86}Sr$ ratios range from 0.704–0.705 in the arc unaffected by this process (Whitford, 1975) to values as high as 0.7091 adjacent to the underthrusting Australian plate (see Fig. 7.4.19). While Pb and Nd isotopes have not yet been studied in this area, the Sr results seem to point strongly to a sediment or continental component in the Sunda-Banda arc volcanics. Consideration of these data on a $^{87}Sr/^{86}Sr$ vs. 1/Sr-mixing plot suggests a "contaminant" with ~200 ppm Sr and $^{87}Sr/^{86}Sr$ ~0.710, a value compatible with either a crustal or sedimentary source for the contaminant.

In contrast to the above examples where a non-mantle component seems involved in some island arc petrogenesis, there are a number of areas where, at least isotopically, the volcanics do not depart significantly from typical mantle values. One well-studied example is the Tonga-Kermadec arc, where Sr isotope ratios range from 0.7033–0.7042, and where the Pb data plot essentially along the oceanic trend (Oversby and Ewart, 1972; Sinha and Hart, 1972). In this area, sediments from the ocean floor seaward of the trench plot on a trend subparallel to the volcanic line but displaced to higher $^{207}Pb/^{204}Pb$ values (Fig. 7.4.20). Consideration of the Pb content of the volcanics and sediments, and of possible mixing relationships on U-Pb isochron diagrams limits the amount of possible sediment contaminant in this area to <2–4%. It should be noted that, since sediment invariably has high $^{206}Pb/^{204}Pb$ and low μ values ($^{238}U/^{204}Pb$; Sinha and Hart, 1972), mixing of volcanic and sediment lead will tend to produce apparent U-Pb isochrons of negative slope whereas positive slopes have been observed in Tonga. [In other areas where sediment contamination has been postulated, for example the Aleutians (Kay *et al.*, 1978), μ values were not measured, unfortunately, so this same test cannot be applied.]

As with oceanic islands, island arcs appear to indicate mantle heterogeneities on a variety of scales. Some islands may be isotopically uniform (e.g., Pb in Pagan and Saipan, Marianas—Meijer, 1976; Pb in the west arc New Zealand—Armstrong and Cooper, 1971; Pb in Unimak and Sr in Adak, Aleutians—Kay *et al.*, 1978). Typically, a given arc will show large isotopic variations along its length (Pb in Marianas and Kermadecs—Meijer, 1976; Oversby and Ewart, 1972), but occasionally a single arc may show surprising isotopic uniformity along its length (Pb in Tonga—Oversby and Ewart, 1972; Sr in South Sandwich—Hawkesworth *et al.*, 1977). The situation is obviously complex, as the various isotope systems do not all behave in the same way (large variations in one isotope may be accompanied by negligible variations in another, etc.); a major hindrance to our understanding of the problem is the lack of a comprehensive data set on identical samples. A major goal for the future must be

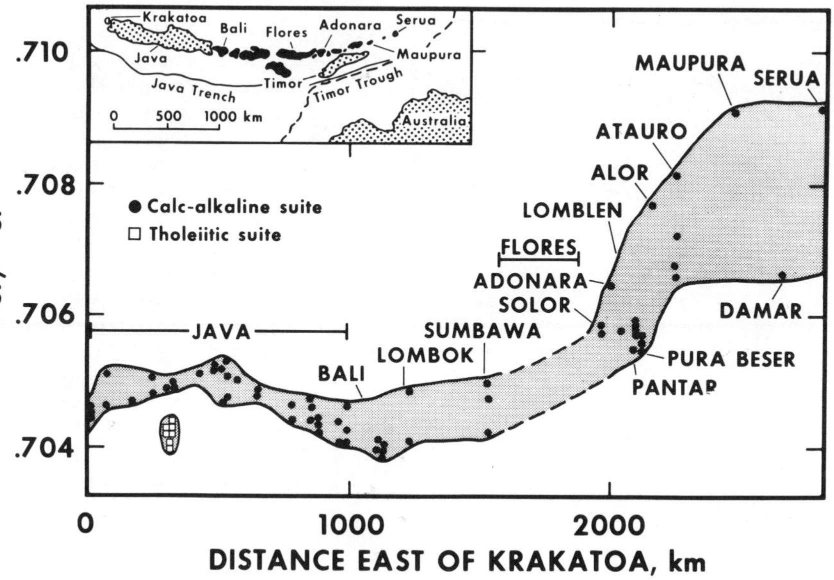

7.4.19 Sr isotope variations along the Sunda-Banda arc, Indonesia (Whitford, 1975; Whitford *et al.*, 1977). The arc segment from Krakatoa to Sumbawa has relatively normal "oceanic-type" Sr isotope ratios. The segment east of Sumbawa shows variable and enriched ratios, probably due to the presence of continental material from Australia on the subducted plate.

the application of both Sr, Pb and Nd analyses to a single geologically and geochemically well-characterized suite of samples from a number of the various arcs.

Thus, island arcs are not dissimilar from oceanic islands in their isotopic heterogeneity. The timing of development of these heterogeneities is also similar in the two tectonic environments. Pb data from the Mariana, Tonga and Kermadec arcs fall essentially along the "ocean trend," which suggests that the heterogeneities are of long duration (1–2 b.y. using two-stage models). In some arcs, where sediment or crustal contamination has been postulated, the Pb data may lie along pseudo-isochrons of even greater slope (3–4 b.y.: the Aleutians—Kay *et al.*, 1978; the Caribbean and New Zealand—Armstrong and Cooper, 1971). In these cases, where the Pb trend may be the result of mixing processes, it is obviously difficult to attach real age significance to the slope of the Pb-Pb data. In terms of Rb-Sr relationships, arc volcanics fall in the same "mantle-isochron" field as oceanic tholeiites (Fig. 7.4.21), suggesting 1–2 b.y. histories for the heterogeneities when arcs are viewed as an ensemble. On a smaller scale, some arcs show intra-arc "isochrons" (Tonga) as do some oceanic islands (Samoa, Kerguelen) but with slopes significantly lower than the arc ensemble as a whole. It may be noted that a mantle-isochron for Tonga is shown on both Rb-Sr and U-Pb isochron plots (Oversby and Ewart, 1972; Sinha and Hart, pers.

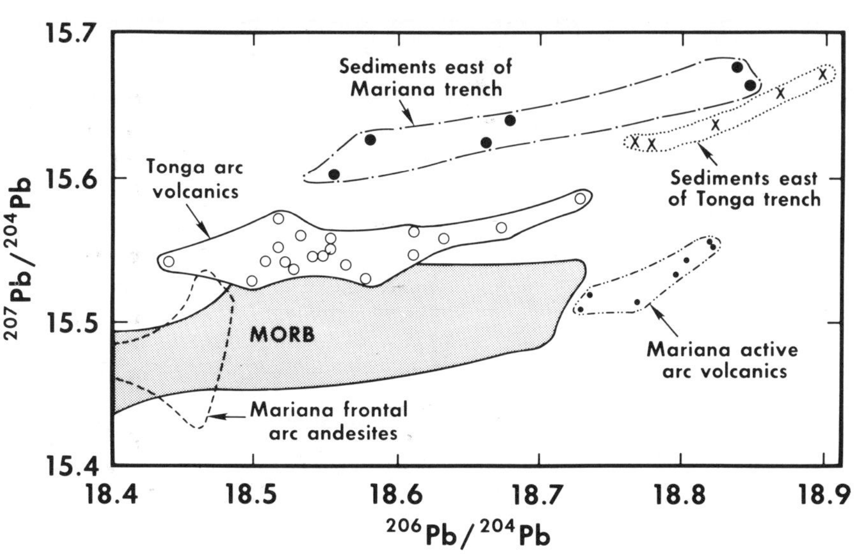

7.4.20 $^{207}Pb/^{204}Pb$ vs. $^{206}Pb/^{204}Pb$ isotope plot for volcanic rocks from the Tonga and Mariana arcs, compared to oceanic sediment on the converging oceanic plate. The Tonga trend is sub-parallel to the oceanic trend, but slightly higher. The Mariana data fall essentially along the oceanic trend. Neither arc shows a trend which is consistent with MORB —sediment mixing. Data from Sinha and Hart (1972), Oversby and Ewart (1972), Meijer (1976), and Sun (1973).

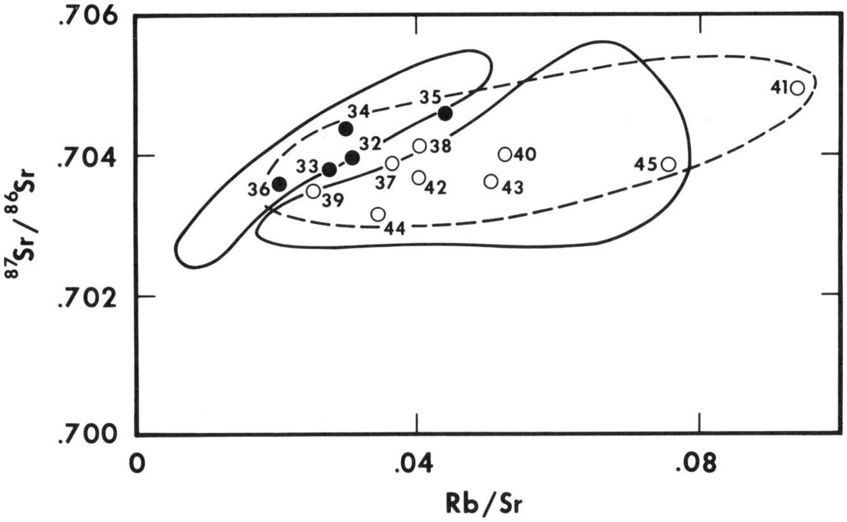

7.4.21 Rb-Sr isochron plot of island arc volcanics compared to fields of oceanic tholeiitic alkali basalts. Solid circles are tholeiitic series; open circles are transitional and calc-alkaline series. Dashed curve encloses arc rocks; solid curves enclose oceanic tholeiites (left) and oceanic alkali basalts (right). Data compiled from literature and unpublished analyses (Hart and Brooks). Average arc data given for 32-Fiji, 33-Japan, 34-Java, 35-New Zealand, 36-Tonga, 37-Aleutians, 38-Antilles, 39-Bismark, 40-East Sunda, 41-Java, 42-Marianas, 43-New Hebrides, 44-New Zealand, 45-Solomons.

comm.) with "ages" of ~500 m.y. As in the oceans, the meaning of the "age" of such mantle-isochrons is under considerable debate, ranging from interpretations of the data as the real age of two-stage history, to the "average" age of multistage or continuously evolving heterogeneities.

Continental volcanism. There is a variety of types of volcanism occurring in continental environments: subduction-related calc-alkaline volcanism (Cascades, Anes, Japan, etc.), rift-related alkaline volcanism (E. African rift, Rhine Graben, Rio Grande rift), "isolated" alkaline volcanism (most kimberlites, minettes, etc., and alkali basalts from areas such as the Southwest Cordillera of the United States; Massif Central, France; the central Sahara, Africa, etc.), and the massive continental flood basalts (Columbia River, Deccan, Paraná, Karroo, etc.).

Speaking first of continental volcanic arcs (convergent plate boundaries), it is clear that these differ fundamentally from island arc volcanics. The proportion of sialic rocks is generally much higher (as in the Andes or Sumatra; Miyashiro, 1974), and the isotopic signatures of Sr, Pb, Nd (and O and H) are consistently displaced in a direction suggesting significant crustal interaction. One of the major advances in recent years is the correlation of isotopic signatures in continental arcs with the nature of the crustal host rocks. For example, in numerous areas a regular and well-defined boundary between rocks (both volcanic and plutonic) with $^{87}Sr/^{86}Sr<0.705$ and $^{87}Sr/^{86}Sr>0.706$ can be delineated which correlates well with the presence or absence of older (cratonic) continental crust. (Following Shuto *et al.*, 1977, these rocks may be termed L-zone and H-zone, respectively.) This boundary has been defined in California by Kistler and Peterman (1973) and Early and Silver (1973), in Idaho and Washington by Armstrong *et al.* (1977), and in Japan by Shuto *et al.* (1977). In some areas, such as Idaho, the transition from L-zonc to II-zonc rocks may bc cxtrcmcly sharp (<10 km!). The Sr boundary in the western United States also agrees closely with a boundary defined on the basis of Pb isotope ratios (Zartman, 1974), where radiogenic Sr is correlated with nonradiogenic lead. In

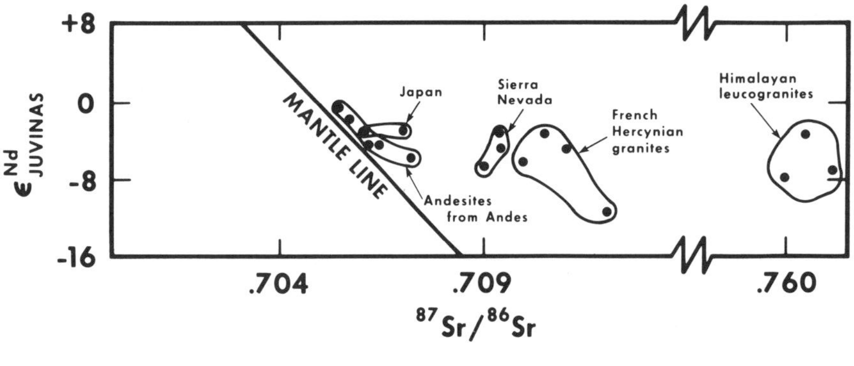

7.4.22 Sr-Nd correlation plot for young granitic rocks, mostly from convergent plate environments. Data from Ben Othman *et al.* (1979). Mantle line based on oceanic data (Fig. 7.4.18) is shown; note that most granitic rocks plot well to the right of this oceanic trend, suggesting presence of continental crustal components in their genesis.

at least one area (the Peninsular ranges, California-Mexico) the Sr boundary coincides also with an $\delta^{18}O$ boundary, with the H-zone rocks marked by high $\delta^{18}O$ values (>9.0 ‰; Taylor and Silver, 1978). It has been argued that the isotopic signature of continental rocks is determined by their derivation from a geochemically distinct subcontinental lithosphere (Brooks *et al.*, 1976b). However, in the case of the continental arc rocks described above, especially in light of the $\delta^{18}O$ data, it seems more reasonable to explain the H-zone arc rocks with crustal contamination or interaction processes. The L-zone rocks appear to have a more or less normal oceanic-type mantle signature; however, the H-zone rocks, on the craton side of the $^{87}Sr/^{86}Sr$ boundaries, probably have a combination of mantle and crustal signatures. Therefore, it is difficult at this stage to determine either the balance of mantle vs. crustal components, or the exact mechanism by which crustal contamination occurs (bulk assimilation, selective fusion, exchange through hydrothermal fluids, etc.). It is interesting to note that the larger variation in $^{87}Sr/^{86}Sr$ displayed by continental arc rocks is not accompanied by a comparable variation in $^{143}Nd/^{144}Nd$ (see Fig. 7.4.22), and it is not clear yet whether Nd shows isotopic distinctions between H- and L-zone rocks similar to those described above for Sr, Pb and O.

Continental involvement has been demonstrated convincingly in volcanism from other provinces, also. Although of uncertain tectonic regime, the calc-alkaline and alkaline volcanics of the Campanian and Roman provinces in Italy show isotopic signatures ranging from "mantle-type" to "crustal-type," with consistent correlations between, Sr, Pb, Nd and $\delta^{18}O$ (Vollmer, 1976, 1977; Turi and Taylor 1976a, b; Cox *et al.*, 1976; Carter *et al.*, 1978a, b). Here, high $^{87}Sr/^{86}Sr$ ratios correlate with high $\delta^{18}O$ and low $^{206}Pb/^{204}Pb$ and $^{143}Nd/^{144}Nd$ ratios—an isotopic signature comparable to that of the H-zone arc volcanics described above.

Turning to the intracontinental alkaline volcanism which appears distributed independent of plate boundary tectonics, we find a different situation. In the first place, many of these volcanics carry ultramafic nodules which demonstrate both a deep mantle (>100 km) origin and an uninterrupted ascent to the surface for these magmas. Thus crustal isotopic signatures should be minimal in these rocks, and they provide the best group of continental volcanic rocks on which to study the processes of basalt derivation from subcontinental mantle. For example, nodule-bearing kimberlites from South Africa have $^{87}Sr/^{86}Sr$ ~0.704–0.705 (Barrett, 1975), $^{206}Pb/^{204}Pb$ ~17.6–19.9 (Kramers, 1977), and $^{143}Nd/^{144}Nd$ ~0.51262–0.51267 (Basu and Tatsumoto, 1978). The Pb isotope data fall along the trend of oceanic basalts, and define a mantle-isochron of ~1600 m.y. (two-stage model). While the Pb and Sr isotope data are consistent with derivation from an oceanic-type mantle, the Nd isotope ratios are equal to or less than chondrites and thus lower than all but one oceanic island (Tristan da Cunha: O'Nions *et al.*, 1977). Thus the kimberlites would fall on the lower end of the oceanic Sr-Nd trend (Richard *et al.*, 1976; DePaolo and Wasserburg, 1976a, b; O'Nions *et al.*, 1977) with planetary values (Allègre *et al.*, 1979) of both $^{87}Sr/^{86}Sr$ and $^{143}Nd/^{144}Nd$. [The "least-altered" kimberlite of DePaolo (1978) has a $^{87}Sr/^{86}Sr$ ratio much higher than those normally found for fresh kimberlites (Barrett, 1975), and thus its lower $^{143}Nd/^{144}Nd$ may not be primary.] While the Sr and Nd are consistent with kimberlite derivation from a relatively unfractionated "planetary" mantle relative to the oceanic mantle, the Pb data fall to the right of the geochron and require at least a two-stage history with some period of time spent in relatively high μ ($^{238}U/^{204}Pb$) reservoirs. The kimberlite data are thus consistent with the hypothesis of Brooks *et al.* (1976b) that the subcontinental lithosphere may be relatively undepleted (relative to planetary values) with respect to Rb/Sr, Sm/Nd (and perhaps U/Pb), and that it contains chemical heterogeneities of long duration (~1600 m.y., estimated from the Pb-Pb "mantle isochron" of Kramers, 1977). Furthermore, this subcontinental lithosphere appears to have followed a planetary evolution path, perhaps since Earth formation, as judged by the fact that kimberlites with ages ranging from ~ 0.1–1.3 b.y. appear to fall on the chondritic Sm/Nd trend.

Sr, Pb and Nd studies of other intraplate continental alkaline basalts show broadly similar results, and support the general ideas derived from kimberlite studies (Peterman *et al.*, 1970; Stuckless and Irving, 1976; DePaolo, 1978; Powell and Bell, 1970, 1974), though in general, other types of alkaline basalts tend to range from planetary (undepleted) values to values more characteristic of depleted (oceanic-type) mantle. For example, $^{87}Sr/^{86}Sr$ values are frequently in the range 0.703–0.704, and $^{143}Nd/^{144}Nd$ values in the range 0.5127–0.5130, which is essentially within the range covered by oceanic islands.

Continental flood basalts, the most voluminous of continental volcanics, are also the most controversial with regard to their petrogenetic history. A classic study of Mesozoic mafic rocks from the southern hemisphere (Compston *et al.*, 1968) showed that many of them were characterized by anomalously high $^{87}Sr/^{86}Sr$ ratios (up to 0.714), and an origin involving crustal assimilation was invoked to explain these high ratios. An alternative explanation was advanced by Brooks *et al.* (1976b), who suggested that these magmas were derived from an old subcontinental lithosphere characterized by high Rb/Sr ratios (and therefore high $^{87}Sr/^{86}Sr$ ratios). Recent work has done little to resolve this controversy (Brooks and Hart, 1978; Pankhurst *et al.*, 1978; Carter *et al.*, 1978; Kyle, 1980; McDougall, 1976; Leeman, 1977). Leeman (1977) compiled the available $^{87}Sr/^{86}Sr$ data for tholeiitic flood basalts, an updated version of which is given in Table 7.4.2. Tholeiites from a few areas show Sr isotope ratios similar to oceanic basaltic rocks (for example, Columbia River Plateau, Picture Gorge, 0.7035–0.7039); tholeiites from most areas, however, are higher than the average (0.7037) of oceanic rocks, and many are higher than the highest oceanic basalts so far analyzed (0.7066). Nd isotope values range from 0.5124–0.5130, which is about the range encompassed by oceanic basalts; most Nd values, however, are consistently lower than "average" oceanic values (DePaolo, 1978; Hawkesworth *et al.*, 1979).

In many cases, samples from a given basalt complex show a positive correlation between $^{87}Sr/^{86}Sr$ and Rb/Sr ratio, with a slope which corresponds to an "age" in excess of the eruption age. This isochron correlation could be a result either of mixing of mantle magma with crustal material, or of magma derivation from an aged heterogeneous mantle source. For the latter case, the isochron correlations were termed "mantle isochrons" by Brooks *et al.* (1976b); ages calculated from these correlations range from a few hundred to a few thousand million years. When data are averaged by geographic area, the points also define a good isochron relationship, collinear with the "mantle" isochron defined by oceanic rocks (age ~1600 m.y.; Brooks *et al.*, 1976b). Similar isochronal relationships are observed for the few areas where Pb or Nd isotopes have been determined. Basalts from the Snake River Plain show a positive correlation between $^{207}Pb/^{204}Pb$ vs. $^{206}Pb/^{204}Pb$ ratios, with a slope equivalent to an "age" of 2500 m.y. (Leeman *et al.*, 1978), similar to the age of the underly-

Table 7.4.2 Initial $^{87}Sr/^{86}Sr$ ratios in representative continental tholeiitic basalts.[(a)]

Name	Age (m.y.)	$(^{87}Sr/^{86}Sr)_I$ [(b)]	N	Reference
Snake River Plain - Yellowstone				
Olivine tholeiites	0–12	0.7065 ± 5	65	*(1)*
Craters of the Moon lavas	0	0.7075 - 0.7120	21	*(1)*
King Hill lavas	3	0.7104 - 0.7176	15	*(1)*
Columbia River Plateau				
Picture Gorge	≃15	0.7035 - 0.7039	10	*(2)*
Grande Ronde	≃15	0.7045 - 0.7049	9	*(2)*
Imnaha	≃15	0.7043 - 0.7049	9	*(2)*
Lower Yakima	≃15	0.7049 - 0.7059	11	*(2)*
Middle Yakima	≃15	0.7052 - 0.7054	5	*(2)*
Upper Yakima	≃8–40	0.7077 - 0.7092	8	*(2)*
Deccan, India				
Tholeiitic basalts	≃60	0.7059 ± 3	3	*(3)*
Serra Geral, Brazil				
Dolerites, basalts	≃120–150	0.7056 ± 3	6	*(4)*
Antarctica				
Queen Maude Land basalts	≃150	0.7053 ± 20	12	*(5)*
Kirkpatrick basalt	≃150	0.7093 - 0.7143	22	*(6)*
Ferra basalts, dolerites	≃150	0.7089 - 0.7153	82	
Karroo, South Africa				
Dolerites, basalts	150–190	0.7039 - 0.7082	57	
Eastern North America				
Barndoor diabase, Connecticut	≃200	0.7061 - 0.7078	7	*(8)*
Palisades diabase	≃200	0.7151 ± 10	3	*(3)*
Brookville diabase	≃200	0.7058	1	*(7)*

[(a)] After Leeman (1971).
[(b)] Initial ratio adjusted to $^{87}Sr/^{86}Sr$ = 0.7080 in E & A Standard. Error is one sigma uncertainty in last significant digit(s).
References: *(1)* Leeman and Manton (1971); *(2)* McDougall (1976); *(3)* Faure and Hurley (1963); *(4)* Compston *et al.* (1968); *(5)* Faure *et al.* (1970); *(6)* Faure *et al.* (1974); *(7)* Barker and Long (1969); *(8)* Ragland *et al.* (pers. comm., 1972).

ing crust (and mantle?). Mafic rocks from the Tertiary Province of Northwest Scotland show isochronal relationships for all three isotopic systems, with "ages" of 1600 m.y. (Sr), 3100 m.y. (Pb) and 6200 m.y. (Nd) (Moorbath and Welke, 1969; Carter *et al.*, 1978a, b). For the Scottish rocks, it is argued that the correlations are not isochrons but rather mixing lines produced by assimilation of crustal materials into a mantle-derived magma. Carter *et al.* (1978a, b) point to the difference between the Sr and Nd "ages" (and the impossible Nd age) as evidence against their being mantle Rb/Sr isochrons (however, derivation of magma from any mantle source with garnet as a residual phase will produce melts with Sm/Nd ratios which are lower than those in the original source, thus the Sm/Nd system in basalts might not generally be expected to preserve "mantle isochrons"). Brooks and Hart (1978) pointed out that continental tholeiites lie on an extension of the oceanic "mantle isochron," and argued that the continental tholeiites need not represent mixing of continental components with mantle melts since there is no continental crust material available in the oceans to use in generating the oceanic trend.

Various other arguments have been made for and against the crustal contamination hypothesis. Leeman (1977) points out that a large range in Pb isotope ratios exists in the Snake River Plain olivine tholeiites coupled to a limited range in Sr isotope ratio, and that no correlation exists between the Sr and Pb isotopic compositions. In contrast, hybrid lavas from Craters-of-the-Moon, where crustal contamination is quite likely, show well-defined correlations between Sr and Pb isotope ratios. "Mixing" trends between $^{87}Sr/^{86}Sr$ and Sr concentration are cited in some areas as evidence for crustal contamination (McDougall, 1976), but are absent in other areas (Leeman, 1977). Correlations between $^{87}Sr/^{86}Sr$ and major elements are found in some areas (Antarctica; Faure *et al.*, 1974), with high $^{87}Sr/^{86}Sr$ corresponding to high abundances of "crustal" elements such as SiO_2, Na_2O, K_2O, etc. A mixing-type correlation also exists between $^{87}Sr/^{86}Sr$ and 1/Sr in these rocks. However, crustal contamination cannot be a bulk assimilation process in this case, since the composition of the "crustal" end-member does not correspond to any reasonable crustal material. "Selective" contamination is a viable mechanism, mainly because our limited understanding of such processes allows considerable freedom in constructing *ad hoc* contamination models. A correlation between $^{87}Sr/^{86}Sr$ and age of lavas within a given sequence was also noted in several areas and used as an argument in favor of the crustal contamination hypothesis (Faure *et al.*, 1974; McDougall, 1976). Oxygen isotope data are not available for most of these rock sequences, but the limited data available (Tasmania; Brooks and James, 1978) show values of 6.2 – 6.4‰, which are only slightly higher than pristine mantle values (5.5–6.0‰), but in the direction expected of lavas containing a crustal component.

The basic problem in these conventional flood basalt studies is lack of knowledge of possible lower crustal and "enriched" mantle compositons. One can argue that an "enriched" subcontinental mantle would be enriched in exactly those components normally termed "crustal" (i.e., SiO_2, K_2O, Na_2O, Rb>Sr, Nd>Sm and the LIL-elements in general), since these elements behave more or less similarly during most magmatic processes. On this basis, then, most chemical and isotopic arguments will be ambiguous in resolving the controversy regarding the origin of continental flood basalts. Only oxygen isotopes, for which there is no reasonable mechanism of producing high $\delta^{18}O$ mantle values, would appear to be unambiguous in differentiating between mantle and crustal signatures. Unfortunately, there are only very limited $\delta^{18}O$ data available on flood basalts at the current time.

Ancient basalt sources

Unlike modern basalts whose chemical and physical properties are generally laid against a known framework of environment of formation (oceanic or continental), ancient basalts are preserved in continental regions, and their mode of formation becomes increasingly ambiguous with antiquity. What, then, are the manners of occurrence of ancient basalts? Typically, the Proterozoic occurrences may be classed into two types: those of plateau basalt affinity, and those of mobile zone affinity. The former are the less ambiguous of the two, with field relationships to basement, volume of lava produced, dominance of tholeiitic type over minor alkaline basalts and general lack of deformation all resembling similar properties observed in Phanerozoic plateau basalt associations. By comparison, the Proterozoic mobile belt association is more complex. It is true that the long arcuate zones are difficult to interpret in any way other than a mobile belt association. However, this environment today produces a spectrum of rocks that possess the greatest compositional diversity among modern basalt associations. Linking this diversity with that previously discussed, we have a problem of characterization of ancient basalts. In the case of the basalts of Labrador, for example (Dimroth *et al.*, 1970), it is not at all clear whether the volcanic filling to the Labrador Trough is a remnant of an 1800 m.y. seafloor, a long marginal basin, or the linear accumulation of volcanic material between crustal wedges.

Nonetheless, the situation for the Proterozoic is one in which, in a general way, the basalts encountered in each continent are relatively easy to explain in terms of modern basalt-producing environments. This is not the situation at all for the Archean occurrences of basalt.

Archean basalts. The typical Archean basalt association is a synclinal one comprising a succession of ultramafic and mafic lavas, minor andesitic and silicic lavas, clastic sediments, cherts, iron formation, etc., separated and surrounded by diapiric gneissic domes. This is an association which has few direct counterparts in either the Proterozoic or Phanerozoic, although compositionally the basalts of the Archean associations resemble basalts from more than one of the modern volcanic-producing environments.

The actual mode of occurrence, tectonic setting, and broad chemical features of these ancient basalts have been dealt with at length elsewhere in this book; however, it is pertinent to keep in mind the preceding discussion in reading the following review of the knowledge that has been gained about the ancient source regions of basalts from isotope and trace element analysis of them. The underlying ambiguities stemming from uncertainty in mode of formation, compounded by the alteration of chemistries (and isotopes) by post-formational modification are real and to date have greatly inhibited our search for knowledge of ancient basalt sources. In the following subsections we will examine what progress has been made in this direction despite these underlying uncertainities.

Sr isotopes. Of the three techniques that provide information on the sources of ancient basalts, the Sr isotope technique has provided by far the most data. There is hardly a rock type or association that has not been subjected to this type of analysis and this applies equally well to felsic and mafic rocks and their sedimentary equivalents. In presenting these data it is convenient to discuss them in three categories: (1) basalt (or mafic intrusive) total rock data, (2) felsic total rock data, and (3) separated mineral phases from basalts (or intrusive equivalents).

(1) *Basalt (or mafic intrusive) total rock data.* The Sr isotope compositions for total rocks of this type are either measured directly (i.e., a suite of basalts is used to define an isochron whose intercept gives the initial Sr-isotope composition) or indirectly (the basalts do not define an isochron but individual Sr isotope ratios may be calculated for each sample using the measured Rb/Sr ratio, the Sr isotope composition, and *independent* estimate of age, e.g., a zircon U-Pb age from an inter-flow andesite or dacite).

Typically, isochrons are obtained from ancient basalts with extreme difficulty due to the alteration-modification problem. A typical set of data for Archean metavolcanics is that from the Michipicoten District of the Canadian Shield (Fig. 7.4.23) in which the data have high scatter and define an errorchron. In rare instances isochrons are encountered (e.g., Jahn and Shih, 1974, for the Komati formation) but generally speaking, data of the type given in Fig. 7.4.23 are typical. Understanding of the influence alteration has on modifying Rb-Sr systematics is underway, but few data, if any, have been published. It is apparent that in some instances H_2O and CO_2 may be used as a means of eliminating "drastically modified" basaltic rocks with the possible improvement of an errorchron to an isochron. However, in résumé, the alteration problem has effectively eliminated the simple determination of initial Sr ratios from isochrons for Archean basalts and for many Proterozoic associations.

(2) *Felsic total rock data.* The data can be discussed in two ways. First, there is the felsic volcanic associations that are genetically linked with basalts and may be used to estimate the basalt initial Sr ratios via isochrons. However, because there is the strong likelihood that the felsic volcanics may be partially or totally influenced by crustal components, and because generally the resulting initial Sr ratios have high ± values due to the position-

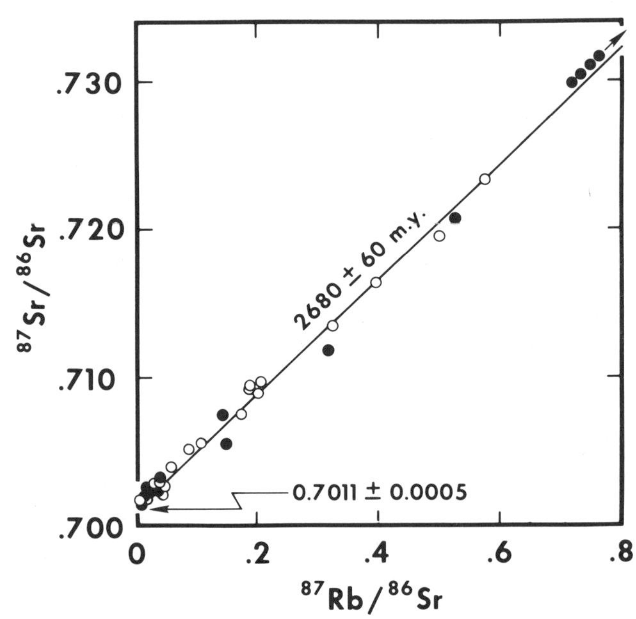

7.4.23 Rb-Sr isochron diagram for Michipicoten metavolcanics (C. Brooks, pers. comm., 1980), showing lower series (closed circles) and upper series (open circles). The data do not define an isochron. Errors are ± 2σ.

ing of data (elevated Rb/Sr), this possibility has not greatly advanced the cause for determining ancient basalt initial ratios. Gneissic and granitic rocks occur in close association with many basalts (particularly the Archean), however, and appear to be mantle-produced with respect to our knowledge of mantle growth between time of Earth formation and modern day. Because these plutonic felsic rocks have generally low volatile contents, their isotope systematics are much less affected by alteration. These associations generally define isochrons rather than errorchrons, and produce initial Sr ratios with low error limits. Granted, these rocks are compositionally a long way from basalts, but they may be used to estimate mantle compositions indirectly in much the same way as basalts.

(3) *Separated mineral phases.* In recent years, mafic and ultramafic intrusives and extrusives associated with Archean volcanic belts have been found to contain relict primary minerals such as clinopyroxene (more commonly) and plagioclase (less commonly). Clearly, separation of these primary minerals can provide a pristine sample of a primary phase of ancient mafic rocks and these can be analyzed to determine the "primary" initial Sr isotope composition of the parent rock. Like the indirect method applied to mafic total rocks, however, the successful calculation of the initial Sr composition requires an independent estimate of age.

In the following subsection we will examine what sort of data has been produced by these different approaches. The discussion will be limited to the age interval 2800–2400 m.y. because this is basically the only time slot that has sufficient data for comparative purposes. Nonetheless, the rock types, orogenies, etc., within this interval are also representative of similar 400 m.y. periods both older and younger than the selected interval. Consequently, the results, while indicating that what we know best is intermediate Earth's mantle at 2800–2400 m.y., have considerable bearing on other ages.

Results. There is only one study completed on separated mineral phases in the age range 2800–2400 m.y. (Hart and Brooks, 1977). In that study seven clinopyroxenes from mafic to ultramafic intrusions within the Abitibi Belt, Superior Province in Canada were analyzed, and they provide a reference average initial Sr isotope ratio of 0.70114 ± 13 (2 σ) for volcanic material of that belt. This value is identical with that for Michipicoten total rocks (Fig. 7.4.23). Metavolcanics of the same age, by comparison, provide age-corrected initial Sr ratios between 0.7005–0.7021 (Hart and Brooks 1977; C. Brooks, pers. comm., 1980). This difference is

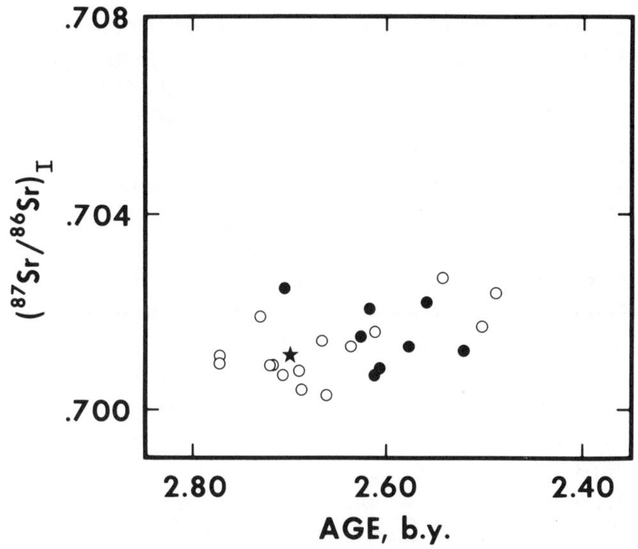

7.4.24 Initial Sr ratios for Archean rocks in the interval 2.8 to 2.4 b.y. Closed circles—mafic rocks; open circles — felsic rocks; star—Abitibi clinopyroxenes. For references to data sources see Hart and Brooks (1977).

probably the result of post-formational modification of Sr isotope compositions and does not necessarily imply that the variation in Sr initial ratios of the source regions for that interval ranged from 0.7005–0.7021. Indeed, the clinopyroxene data, which represent a surface extent of some 400 km, indicate magma source regions of relatively constant initial $^{87}Sr/^{86}Sr$. A small variation could be interpreted from the clinopyroxene data (0.7009–0.7013), although there is a scarcity of sufficiently good data to make this possibility valid.

Figure 7.4.24 graphically illustrates the results of a sampling survey of the world's mafic and felsic rocks whose Rb-Sr ages fall within the 2400–2800 m.y. interval. This graph shows the similarity of the initial Sr ratios for the mafic and felsic rocks and, in addition, emphasized the generally higher value found in the interval 2600–2400 m.y. as compared to the values obtained between 2800–2600 m.y. This was previously interpreted (Hart and Brooks, 1977) to be the result of crustal reworking and/or contamination accompanying post-orogenic igneous processes. Despite the differences in rock type, geological association, and post-formation history, the results of the survey indicate that the source region of basaltic rock in the interval 2800–2600 m.y. was relatively uniform on a global scale at that time and corresponds to an initial Sr value close to 0.7010. Most of these data involve large experimental or geologic errors, however, so that the question of mantle heterogeneity at that time cannot be firmly answered. The fact that a single value with relatively narrow error limits seems to describe Earth's mantle on a regional and local scale at that time should not be interpreted to mean that heterogeneity did not exist. Instead, it would appear to have been of a much smaller scale than the heterogeneities of the type found in modern basalt sources.

The best other major time-slot for basalt production that can be appraised in this way is the interval 2000–1600 m.y., with an orogenic "break-point" at 1800 m.y. The results of a literature survey are presented graphically in Fig. 7.4.25. As was done for the 2.8–2.4 b.y. survey, we have classed the data into felsic and mafic, and those initial Sr values derived from errorchrons rather than isochrons (Brooks *et al.*, 1972) have been purposely de-emphasized. The main features of the diagram are an increased range in observed initial Sr ratios for the interval 1.8–1.6 b.y., and a general concordance of values for both felsic and mafic rocks. The degree of scatter is so high, however, that average values calculated from these data are not likely to be meaningful. Neither can it be said that a minimum Sr isotope value for the mantle at that time was 0.7009–0.7010, since the ± values associated with the isochrons for these compositions are between 0.0010–0.0065. In fact, the mean standard error for all the data (N = 56) is 0.0027 ± 0.0025 (1σ). The main use of this type of data becomes apparent only when they are compared with similar data of other age intervals (see section below on Sr isotope evolution). Generally speaking, however,

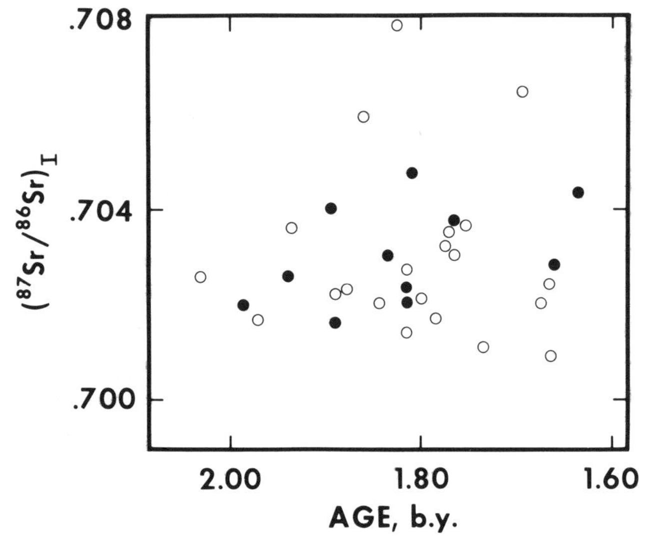

7.4.25 Initial Sr ratios for Proterozoic rocks in the interval 2.0 to 1.6 b.y. Symbols as in Fig. 7.4.24. References to data sources from Brooks and Hart (pers. comm., 1981).

there are not sufficient data to properly evaluate the "old" intervals of Precambrian history. For the younger Precambrian intervals, the likelihood of crustal contamination or anatexis influencing the results is greatly increased, and this, combined with the problem of identifying source region environment, means that even more careful selection of rocks will be required before plausible source-region compositions are obtained. Clearly, each interval must be carefully dissected, presence or absence of orogenies documented, comparison of felsic and mafic isochrons made, and in addition, time-corrected independent values determined for comparison. The 2800–2400 m.y. interval would seem to be a type case of how mantle source regions should be evaluated, and that interval is far from complete at present.

Trace elements: geochemical indicators of past tectonic environment. Trace element abundances in ancient basalts are commonly used in three ways: (1) to assess the genetic links within and between different Archean volcanic associations; (2) to determine the similarity between ancient basalts and those produced in modern, well-understood, tectonic environments; and (3) to assess the effects of post-formation modification of primary chemistries. There are many studies illustrating these and references to them may be found in O'Nions and Pankhurst (1978).

The principal modern environments can be summarized as follows (after Pearce, 1975):

Accreting Plate Margins
—ocean floor basalt series (OFB)
Converging Plate Margins
—island arc low-K tholeiitic suite (LKB)
—calc-alkaline basalt suite (CAB)
—shoshonitic suite (SHO)
Within Plate Basalt—WPB

There have been many studies in which major elements (particularly K_2O+Na_2O vs. SiO_2, and FeO–MgO vs. K_2O–Na_2O) have been used to attempt to characterize the past tectonic environment of formation and in some cases they have met with success. For instance, the volcanic pile of the Abitibi Belt can be subdivided into a basal tholeiite association followed by a superstructure of intermingled tholeiitic and calc-alkaline association (Gélinas *et al.*, 1977). However, the most reliable geochemical means of distinguishing between these types is by the use of the non- or less-mobile trace elements Ti, Zr, Y, Nb, and to a lesser extent Sr and Cr. These elements are particularly ideal for studying ancient basalts in the light of the earlier discussion on alteration. In the case of the Abitibi Belt, the non-mobile elements confirm the major element-based study. Similar findings elsewhere confirm the recognition of basalt types showing a broad range in compositions typical of that observed in modern day environments. However, it should not be forgotten that the chemical diversity in Archean basalts extends to very rare alkaline to shoshonitic basalts (Hubregtse, 1976) and to the ubiquitous komatiites. As a general rule, alkaline basalts are not common in the Archean. Komatiites, however, are very common in the Archean and early Proterozoic but almost lacking in late Proterozoic and Phanerozoic associations. One particular type of komatiite (spinifex-textured) is of mostly Archean age (Brooks and Hart, 1974). Clearly, any model used to explain the mode of formation of Archean basalts must take into account these observations. What, then, are the broad trace element characteristics of Archean basalts and how, in detail, do they compare with younger (Proterozoic and Phanerozoic) basalts?

Typically, the Archean basalts have trace element abundance levels which fall within the range of modern volcanics although commonly toward the lower part of the range. For instance, Archean tholeiites have either similar or lower contents of K, Rb, Sr, and Ba, when compared with modern tholeiites, and they typically have average K/Rb ratios between 300–400. Low abundances in these elements are also evident in the komatiites, which are viewed by some to be a continuum with the tholeiite series (e.g., O'Nions and Pankhurst, 1978). The characteristically low-K content of Archean tholeiites also applies to CAB and SHO. As noted by O'Nions and Pankhurst, Archean volcanics and their younger counterparts, both tholeiitic and calc-alkaline basalts, show no well-established compositional trend over their period of continuous generation of approximately 3000 m.y.

Because the trace element compositions (especially K, Sr, and Ba) of Archean basalts are very similar to those of tholeiitic basalts of modern island arcs that occur over shallow subduction zones, it is inviting to use the same modern day models to explain basalt genesis in the Archean. One such model advocated by Hart *et al.* (1970) involves magma generation from shallow depths (less than 50 km) by high degrees of partial melting (greater than 30%), related to gently dipping subduction zones. This type of model can be developed in many possible variations, e.g., by invoking steeper geothermal gradients, back-arc regions coupled to shallow subduction zones, etc.

An interesting application of trace element abundance levels in Archean basalts is given by Condie (1973, 1976). Using Rb/Sr distributions in Archean metabasalts, Condie estimates crustal thicknesses and depths to possible subduction zones based on analogies with modern associations, where empirical correlations have been established between the geochemical distributions and these independently determined tectonic parameters. The data are shown in Fig. 7.4.26 and from this figure it would appear that if the modern relationships are valid in the Precambrian, then the average Archean crustal thickness was 25 km; the corresponding depth to the subduction zone was 85 km. Unfortunately, the extreme mobility of Rb makes the definition of such values less than certain. To date there has been little done in attempting to use trace elements to define polarity (oceanic vs. continental) of ancient plate environments. The Sr-depth trend of Hart *et al.* (1970) has clear application for the purpose, Sr being much less mobile than K—the more commonly used polarity indicator.

There is only one trace element (Cs) that shows significantly different abundance levels in Archean

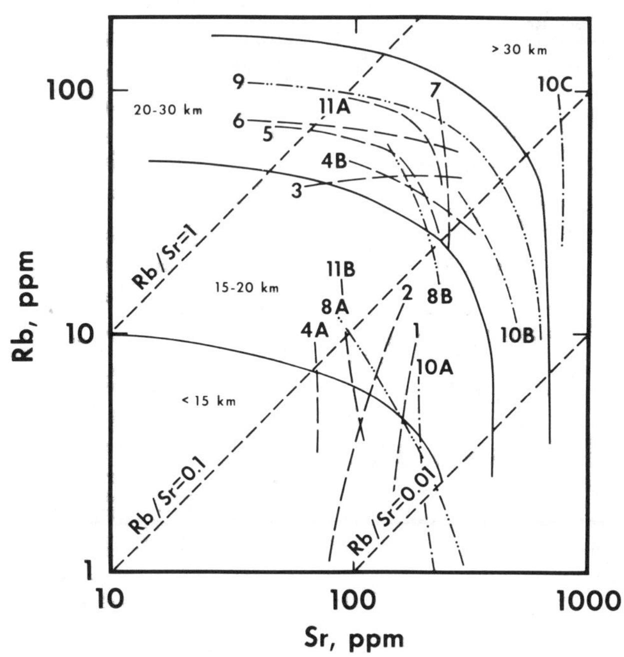

7.4.26 Rb/Sr distributions in Archean greenstone volcanic rocks and inferred crustal thickness. The crustal thickness boundaries (solid lines) have been established from modern associations between Rb/Sr trends and independently established thickness values. (After Condie, 1976; reproduced by permission of Pergamon Press). 1. Ely. 2. Chibougamau. 3. Rainy Lake. 4. Abajevis (A=Lower, B=Upper). 5. Lower series, Wawa, Ontario. 6. Clericy, Quebec. 7. Upper series, Wawa, Ontario. 8. Onverwacht (A=lower, B=Upper). 9. Schoonegezicht (Upper Fig Tree). 10. South Pass, Wyoming (A=low-K tholeiites, B=calc-alkaline volcanic rocks, C=high-K calc-alkaline volcanic rocks). 11. Yellowknife (A=second cycle, B=first cycle).

basalts when compared to modern counterparts. This observation comes from both total-rock (Hart *et al.*, 1970) and mineral separate (clinopyroxene) studies (Hart and Brooks, 1977). The conclusion advanced to explain this was that the Archean mantle may have been less depleted in Cs relative to the present-day mantle. Further studies are necessary, however, in order to support this finding, and to examine how open-system behavior of Cs in the mantle has proceeded on a temporal basis.

Evolution of basalt sources

In principle, the geochemical evolution of basalt mantle sources can be traced in considerable detail by determining the initial isotopic ratios of the available radiogenic isotopes—in the case of current usage, Sr, Pb, and Nd—in basalts of different ages. These initial ratios, provided they are representative of the mantle source (as opposed to being changed by contamination or alteration processes), can be used to determine isotopic heterogeneity in the mantle, as well as to define the ways in which the parent-daughter ratios Rb/Sr, Sm/Nd, and U/Pb have varied in the mantle sources with time. For example, Fig. 7.4.27 shows schematically a number of possible evolutionary trajectories for mantle sources. Since the slope of these trends is directly proportional to parent-daughter ratio, detailed data from past mantle trajectories can elucidate much about the relative behavior of Rb/Sr, U/Pb, and Sm/Nd in the mantle. And, of course, since these elements are coherent with other families of elements, other elements can be understood by analogy. For example, knowledge of Sm/Nd behavior can be readily extrapolated to that of the whole REE group, as these all behave coherently. Furthermore, these systems also intrinsically carry important information regarding distribution and evolution of heat sources in the Earth, as two of the parents, U and Rb, can be geochemically related to Th and K, which together comprise essentially all of the important radioactive heat sources for the Earth (excluding short-lived radioactivities such as ^{26}Al, which are only important very early in Earth history).

In principle, the data for these kinds of plots require knowledge of the *age* of a derived basalt (or other rock) and its isotopic ratio at the time of eruption. In the case of basalts, Rb/Sr ratios are relatively low, so that only minor changes are produced in $^{87}Sr/^{86}Sr$ from time of eruption to now—this also means that these materials are poor for determining age, so ages must be known from an independent approach. In the case of U/Pb and Sm/Nd, the P/D ratios in basalts are significant and a suite of co-genetic samples of varying P/D can be used to define both the age and the initial ratio.

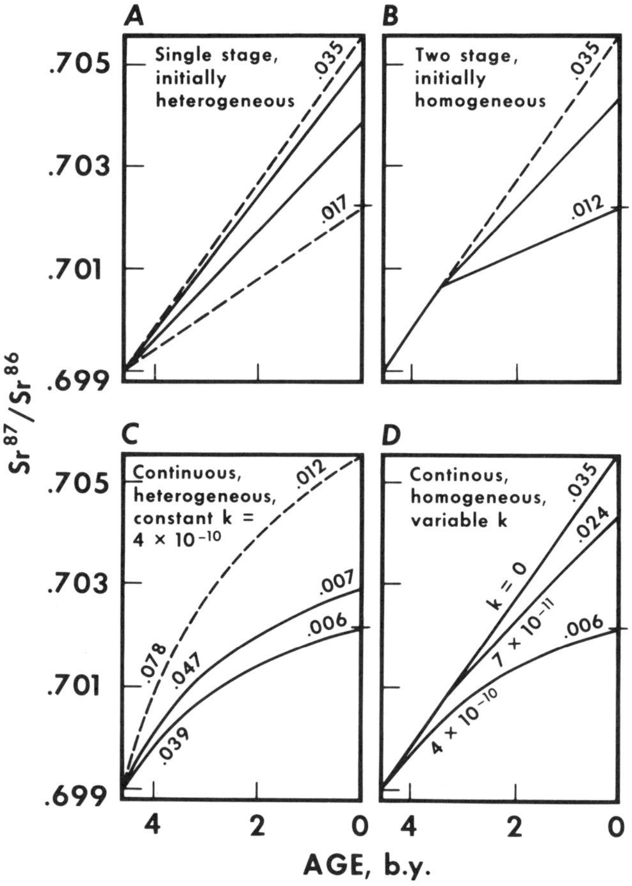

7.4.27 Diagrammatic depiction of various Rb-Sr mantle evolution models. All are shown starting at 4.5 b.y. with a uniform Sr isotope ratio of 0.699. (a) Initial heterogeneities in Rb/Sr ratio; each system evolves as a closed system. (b) System initially homogeneous in Rb/Sr ratio; heterogeneities develop at a single time (for example, 3.5 b.y., as shown). (c) Initial heterogeneities in Rb/Sr ratio; each system undergoes continuous loss of Rb and Sr with a fixed transport constant of $k = 4 \times 10^{-10}$ yr^{-1} (see Allègre, 1969 and Hart and Brooks, 1970, for description of this transport model). Numbers on curves refer to initial and final Rb/Sr ratios for several evolution trajectories. (d) Initial system homogeneous in Rb/Sr; subsystems develop which have different transport constants ranging from 0 to 4×10^{-10} yr^{-1}. Numbers on curves refer to the final Rb/Sr ratios in the systems. These models may be compared with Sr isotope data for rocks of different ages, Fig. 7.4.30b. The data do not as yet allow a definite choice of model.

Of course all of this requires samples which have been chemically closed systems from the time of crystallization to present. As discussed above, this can frequently be an insurmountable problem, as alteration and metamorphism can create major disturbances, particularly in the Rb/Sr and U/Pb systems. Th/Pb and Sm/Nd seem less susceptible to these problems, but the number of examples treated by these methods is quite limited, and thus our understanding cannot be considered very definitive. In some cases, when whole rock samples are unsuitable due to alteration effects, primary unaltered minerals may be obtainable which *have* been closed systems and these can be used with confidence. [For example, Hart and Brooks (1977) analyzed fresh clinopyroxene separated from Archean mafic rocks in Canada to provide initial Sr isotopic ratios for these samples. In this case, because of the very low P/D (Rb/Sr) ratios of clinopyroxenes, the $^{87}Sr/^{86}Sr$ ratio is virtually unchanged with time, and present-day values are very close to initial ratios.]

In the following sections, we will discuss the evolution of the three radiogenic isotopes Nd, Sr, and Pb separately, and then consider the implications of the interrelationships between the three systems.

Nd isotope evolution. We consider this system first, partly because the data are relatively sparse and the interpretation of them is therefore without many inconsistencies. As in following discussions, we will not confine ourselves solely to data from mafic rocks, as there are many cases where sialic rocks appear to show comparable results and can be used with some confidence as relatively direct derivations of mantle sources. Note that for modern rocks this may be a less valid assumption, as there is some evidence that problems of crustal contamination and derivation for sialic rocks are more severe in younger rocks than in older rocks.

All of the initial Nd isotopic ratios thus far published are plotted on a Nd evolution diagram (Fig. 7.4.28). The most striking feature of these data is the close fit of the Precambrian points to a single-stage linear evolution, starting from the primordial Nd value as determined from two achondritic meteorites. Only three mafic units diverge significantly from the planetary trend (Cape Smith, Duluth, and Bay of Islands), and these are all younger than 1.8 b.y.; they may represent the incipient development of the heterogeneities in Nd isotope ratios which are so well-documented for modern volcanic rocks. Additional data may of course reveal significant variations from this simple single-stage evolution during this time period, but the present data are alluring in their simplicity. It should be emphasized again, however, that the present mantle source regions for basalt are heterogeneous with respect to both Sm/Nd ratio and Nd isotopic ratio, and that a marked change in Nd evolution must have occurred somewhere in the time period 0.5–2.0 b.y. One of the obvious tasks for the future will be to constrain the possible trajectories in this time period with carefully documented samples.

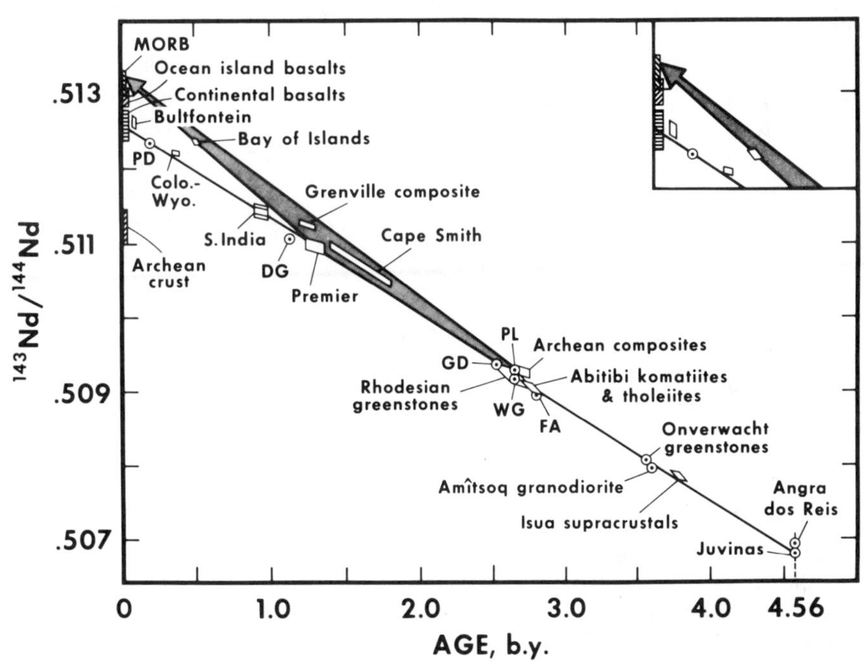

7.4.28 Nd isotope evolution plot. Initial $^{143}Nd/^{144}Nd$ ratios are derived from isochron plots or are determined from single samples by independent age correction. PD—Pallisades diabase, DG—Duluth gabbro, GD—Great dike, PL—Preissac-Lacorne granodiorite, WG—Wyoming granodiorite, FA—Fiskenaesset anorthosite. A single evolution line, with chondritic Sm/Nd ratio, fits most of the Precambrian data very satisfactorily. Most young oceanic basalts come from sources with time-integrated Sm/Nd ratios greater than chondritic (i.e., light rare-earth depleted patterns). Inset shows the ~0 age region with greater clarity. Note the contrast between this figure and the highly scattered Sr isotope evolution plot, Fig. 7.4.29. Data are from Basu and Tatsumoto, 1978; Carlson *et al.* 1978; DePaolo and Wasserburg, 1976a, b; Hamilton *et al.* 1977, 1978a, b, 1979; Hawkesworth *et al.* 1977; Jackobsen, pers. comm., 1979; Lugmair *et al.* 1976; Lugmair and Marti, 1977; McCulloch and Wasserburg, 1978; O'Nions *et al.* 1977; Richard *et al.* 1976; White *et al.* 1978; Zindler *et al.* 1978; Zindler *et al.* 1979; Zindler, pers. comm., 1979.

Sr isotope evolution. Documentation of Sr isotope evolution paths has been a goal of geochemists since the pioneering work of Hedge and Walthall (1963). The quest has been fraught with difficulties, however, largely because of the susceptibilitiy of the Rb/Sr system to post-crystallization modification by alteration, metamorphism, etc. Even now, with literally hundreds of analyses in hand, it is not possible to define the path of Sr evolution with any confidence. One is reduced either to looking at averages of large numbers of questionable data, or looking at a small handful of irreproachable data. Evidence for possible heterogeneities obviously requires a reasonably large coverage of geography and time, and thus the latter approach is not very satisfying. We will present current knowledge from both points of view below.

First we will consider all the initial Sr isotope ratios which were available as of 1976, eliminating only those where the isochron-determined error in the initial ratio is greater than ± 0.002. This is clearly not very satisfactory, as the total range of probable mantle evolution paths is not much larger than a variation of ± 0.002. Data for the age range 1200–3800 m.y. are given in Table 7.4.3 and plotted in Figs. 7.4.29–30. For reasons of clarity we have not included the individual errors in the figure; however, it may be stated that most of the scatter of the data could be accounted for by the formal regression errors in the initial ratios. Or, rephrasing the situation, it might be said that most of data are consistent, within statistical uncertainties, with a single-stage linear evolution. Thus isotopic heterogeneities may exist in the age range considered here, but demonstration of their presence is effectively masked by the statistical uncertainties. Likewise it may be said that a curved trajectory is as plausible as a linear trajectory.

Several other points may be made from these data. First, there is no discernible difference between initial ratios determined from mafic or sialic rocks. There is, however, a distinct tendency for ratios determined from isochrons where all the points are within experimental error of an isochron to be lower than those determined from errorchrons (where scatter due to geologic causes is indicated). Perhaps this is not surprising, as most geologic processes which would produce scatter on an

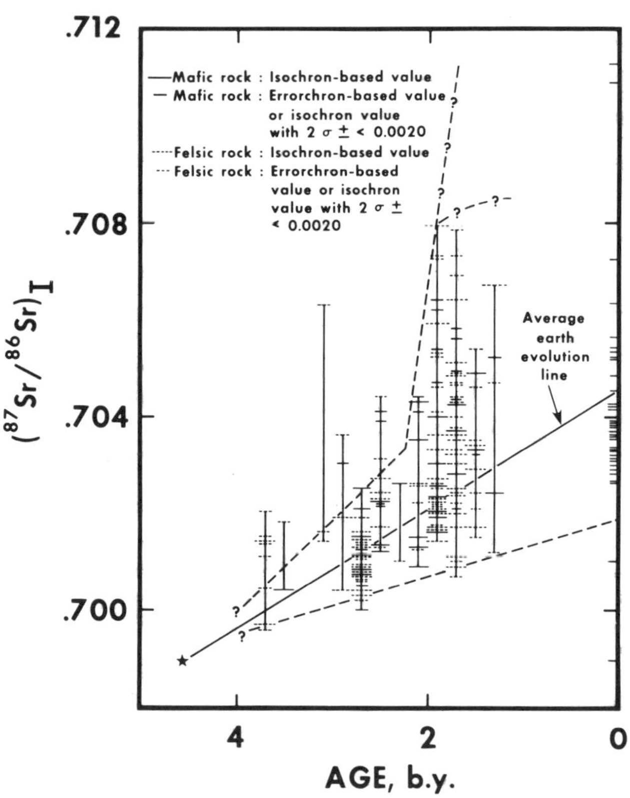

7.4.29 Strontium isotope evolution diagram. Primordial strontium (star) plotted at 4.55 b.y. is taken to be BABI, 0.69898 (Papanastassiou and Wasserburg, 1969). The heavy vertical lines emphasize the ranges for "reliable" (isochron-determined) initial Sr ratios. The dashed lines define limiting fields for these "reliable" data. Ticks along T=0 m.y. are modern oceanic volcanics (Brooks *et al.*, 1976a) and continental Mesozoic tholeiites (Brooks and Hart, 1978).

Table 7.4.3 Mean initial Sr isotopic compositions for felsic and mafic rocks in the age range 3800–1200 m.y.[a]

Age Interval (m.y.)	Felsic		Mafic	
3800 – 3600	0.7006 ± 0.0012[b]	(2/7)[c]	-	(0/0)
3600 – 3400	-	(0/1)	0.7004 ± 0.0001	(1/1)
3400 – 3200	-	(0/1)	-	(0/0)
3200 – 3000	-	(0/3)	-	(0/0)
3000 – 2800	0.7011 ± 0.0010	(2/3)	-	(0/1)
2800 – 2600	0.7010 ± 0.0005	(11/17)	0.7014 ± 0.0007	(6/11)
2600 – 2400	0.7023 ± 0.0004	(4/7)	0.7016 ± 0.0005	(3/7)
2400 – 2200	-	(0/1)	0.7026 ± 0.0009	(1/1)
2200 – 2000	0.7026 ± 0.0010	(2/4)	0.7023 ± 0.0015	(5/7)
2000 – 1800	0.7031 ± 0.0019	(12/23)	0.7024 ± 0.0008	(8/13)
1800 – 1600	0.7032 ± 0.0016	(13/30)	0.7036 ± 0.0008	(3/9)
1600 – 1400	0.7034 ± 0.0015	(3/9)	0.7041 ± 0.0011	(2/3)
1400 – 1200	0.7067 ± 0.0016	(1/3)	0.7024 ± 0.0020	(1/2)

(a) Individual data and sources from Brooks and Hart (pers. comm., 1981).

(b) Standard deviation of the mean or, in case of single values, the 1σ isochron error.

(c) (n/N) indicates number of samples averaged (n) with respect to number of values available (N). Initial ratios with 2σ error > 0.0020 or errorchron-based values are not included in the averages.

isochron diagram generally tend to rotate the isochron, thereby causing an anomalously high initial ratio.

There is also a suggestion of increasing "dispersion" of the data for the younger time periods, with data older than 2.0 b.y. showing less variation than younger data. This may be related to a real increase of mantle isotopic heterogeneity with time (as for example shown schematically in Fig. 7.4.27), or may reflect an increasing tendency toward involvement of aged sialic crust in the processes of magma generation. These processes are quite evident in igneous rocks, especially sialic rocks, emplaced along modern convergent plate boundaries wherever cratonic rocks are present (Armstrong *et al.*, 1977; Kistler and Peterman, 1973; Taylor and Silver 1978).

In terms of initial Sr isotope ratios of very high precision, the data set is very limited. The clinopyroxene data of Hart and Brooks (1977), discussed above, for samples of 2.7 b.y. mafic-ultramafic rocks from three localities from the Archean of southern Canada, averaged 0.70114 ± 13, with a total variation only slightly outside experimental error. Thus, there was no evidence for mantle heterogeneity at 2.7 b.y. for this one region (extending some 400 km only). Jahn and Shih (1974) reported a value of 0.7005 ± 10 (based on a mineral isochron) for a single sample of the 3.5 b.y. komatiitic Onverwacht Group, South Africa (this value may have uncertainties over and above the quoted errors due to the fact that the minerals are metamorphic, so the isochron relates to a metamorphic age, etc.).

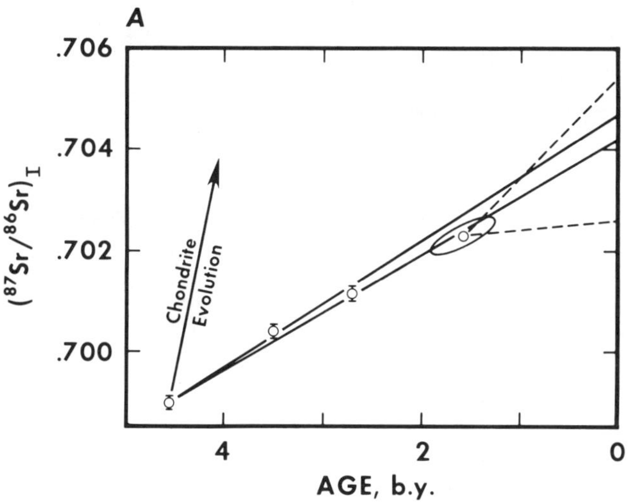

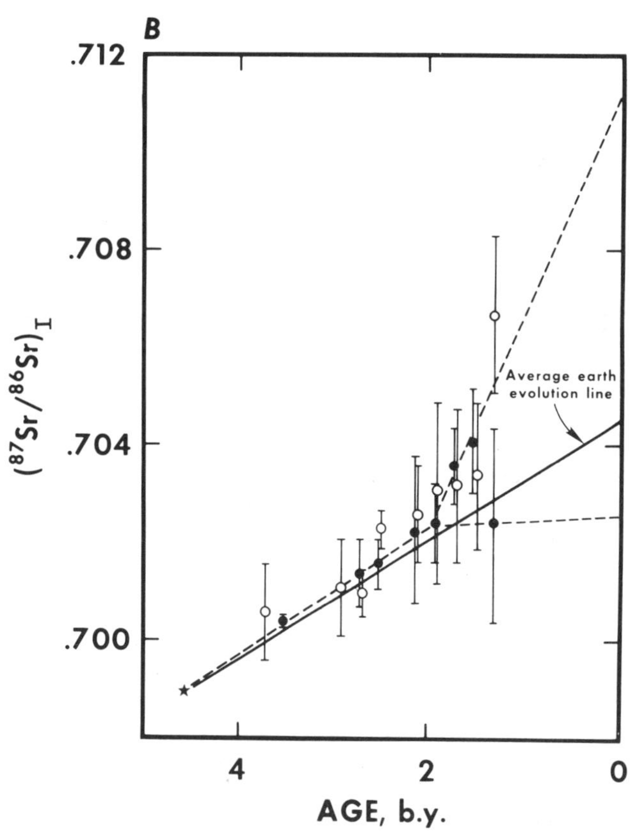

7.4.30 a. Strontium isotope evolution diagram. Primordial strontium isotope ratio, plotted at 4.55 b.y., is taken to be BABI, 0.69898 (Papanastassiou and Wasserburg, 1969); use of value from Allende, 0.69877 (Gray *et al.*, 1973) or Angra dos Reis, 0.69884 (Papanastassiou, 1970) will not affect conclusions. Value of 0.7005 at 3.5 b.y. is derived from analysis of mineral separates from a basaltic komatiite from the lower Onverwacht Group, South Africa (Jahn and Shih, 1974). It should be noted that this komatiite has been metamorphosed and the analyzed minerals are not primary; this value, therefore, cannot be considered to be of the same reliability as the other data. The value of 0.70114 at 2.7 b.y. is from Hart and Brooks (1977). Value of 0.7023 at 1.57 b.y. is the initial ratio of a mantle isochron derived from modern oceanic volcanics (Brooks *et al.*, 1976a). The uncertainty in this value is shown approximately by the error ellipse. The dashed lines drawn from the point to the $^{87}Sr/^{86}Sr$ axis represent the two limiting growth cases for this modern oceanic isochron (Samoa, upper line; Mid-Atlantic Ridge, lower line). The two solid lines represent the limiting trajectories for single-stage growth between primordial and present-day mantle Sr [latter from the Sr-Nd correlation diagram for modern oceanic basalts (O'Nions *et al.*, 1977)]. **b.** Strontium isotope evolution diagram for the averaged "reliable" data of Fig. 7.4.29. The solid circles are mafic rocks, open circles are felsic rocks. Error bars are 1σ values. The dashed lines are growth lines of the type used to describe the "mantle" isochron properties of modern oceanic volcanics in Fig. 7.4.30a, although in this figure the post-1.9 b.y. upper limiting growth line, based on mafic rocks between 1.9 and 1.2 b.y., is projected to include both the modern oceanic volcanics and the continental Mesozoic tholeiites indicated by the ticks along T = 0 in Fig. 7.4.29. The pre-1.9 b.y. growth line is the best fit line for the mafic rocks only.

When this limited set of data is plotted together with primordial Sr and the "initial" ratio determined from the "mantle" isochron for modern oceanic tholeiites (Fig. 7.4.30a), a fairly simple two-stage model is sufficient to explain the data. Starting soon after Earth formation, evolution proceeded linearly (single-stage) until (approximately) 1.6 b.y., at which time a large-scale development of heterogeneities occurred, leading to the observed dispersion in modern oceanic mantle values. While obviously based on very limited data, the model is similar to the one suggested above based on Sm-Nd data.

Pb isotope evolution. While Pb isotopes have provided us with the most detailed understanding of modern basalt genesis, because of the strength derived from having two interrelated isotopic systems, there are almost no Pb isotope data for older basalt suites. What little work has been done demonstrates very severe problems with open-system behavior in the U-Th-Pb systems—in particular, U appears to be very mobile. Sinha (1972) and Sinha *et al.* (1971) analyzed basic volcanics from Archean greenstone belts in both Canada and South Africa, and showed that U-Pb isochron plots were almost useless, with drastic depletion of U/Pb evident in South Africa and moderate enrichment of U/Pb evident in the Canadian data. While these open-system behaviors are most likely recent in origin, there was also evidence from the Pb-Pb relationships for significant earlier open-system behavior of U/Pb and Th/Pb as well. The Pb-Pb data from the Onverwacht of South Africa (Sinha, 1972) defined a good secondary isochron of 3200 m.y. age, in fairly good agreement with other age estimates for the Onverwacht. In the Canadian case, however, only 9 out of 14 analyses fell close to an isochron of the expected age (2.7 b.y.), with the other points scattered mostly toward younger ages.

In cases where U/Pb mobility is relatively recent, one can expect Pb-Pb data to still retain reasonable age relationships. In terms of identifying trends in mantle U-Pb evolution, however, one requires initial ratios for the Pb isotopes and these can only be derived from U-Pb isochron plots. Thus, any disturbance in U/Pb ratio, even of recent origin, will greatly diminish the ability to derive initial ratio information. This is the observed situation in the greenstone Pb data referred to above. Pb isotope studies of plutonic mafic rocks have been considerably more successful, with ages derived from U-Pb isochrons frequently in excellent agreement with the inferred ages of the rocks (Manhes *et al.*, 1979). As will be discussed more fully below, the data demonstrate considerable Pb isotopic heterogeneity for mantle-derived rocks, even in the earliest Archean.

The alteration problem is less severe when dealing with sialic plutonic rocks, and insofar as these types of rocks can be shown to have a mantle derivation, some information about basalt source regions may be inferred from them. While U/Pb mobility is still frequently a problem in these rocks, it is possible to derive initial Pb isotopic ratios from minerals such as K-feldspar, which have very low U/Pb ratios. Feldspars are susceptible to re-equilibration during metamorphism (Doe and Hard, 1963), so their post-crystallization histories must be carefully evaluated before they are used in tracing mantle evolution. A good review of available lead data from Archean plutonic rocks is given by Oversby (1978).

Figure 7.4.31a shows initial Pb isotopic ratios derived from K-feldspars in those rocks studied by Oversby (1975) for emplacement ages in a restricted interval of time around 2.7 b.y., for rocks from both North America and Australia. The most obvious facet of this plot is that, while the model lead ages are in reasonable agreement with the probable ages of emplacement (2600–2700 m.y.), there is a considerable dispersion of the data in terms of the U/Pb (μ) values of the sources from which these rocks were derived. Insofar as the K-feldspars have retained their primary Pb isotopic ratios (probably a good assumption in most of these cases), then the data demonstrate considerable long term heterogeneity in the Archean mantle sources prior to 2.7 b.y. ago. In terms of single-stage models, a variation in U/Pb is indicated (μ values range from 7.4–8.6). The individual suites lie very approximately along steep, almost primitive (4.5 b.y.) isochrons, suggesting a fairly long term existence of the observed heterogeneities in μ. Similar results have been evident for many years from common lead studies of galena (Robertson, 1973; Robertson and Folinsbee, 1974), only it is more difficult to relate galena source evolution to basalt mantle sources. While sulfide ore leads clearly show development in heterogeneous sources, it is quite likely that a crustal history has contributed significantly to this heterogeneous appearance.

A summary of available lead isotope data from mafic and sialic plutonic rocks is shown in Fig. 7.4.31b, using the new data representation of Manhes *et al.* (1979). This plot shows the deviation of initial isotope ratios from a single-stage closed-system evolution trend. Clearly, the Pb isotope data scatter significantly from a closed-system model, with the sialic rocks showing slightly more variation than the mafic rocks. Furthermore, in contrast to the Sr and Nd data, there seems to be no discernible tendency for the Pb data to converge and show less variation in the older rocks. This suggests that the early Archean mantle, while more or less homogeneous for Sr and Nd, was significantly heterogeneous with respect to Pb. It has been proposed that development of this very early Pb isotope heterogeneity was associated with the process of core formation (Allègre *et al.*, 1979).

Geochemical variations of mantle sources with time. It is clear from the isotopic data that substantial mantle heterogeneities exist for the elements Th, U, Pb, Rb, Sr, Sm and Nd, and these have existed at least for the past 1–2 b.y. It may be inferred from this that similar variations exist for other elements, such as the group of LIL trace elements (Zr, Cs, Y, Ba, Nb, Ta, Hf, Ga, etc.); the magnitude of these variations need not be large, however, with ratio variations (as opposed to absolute concentrations) of factors of 2–5 for Rb/Sr, 2–3 for U/Pb, and 2–3 for Sm/Nd suggested by the modern basalt data. Because of the large variation (factors of 10–100) in absolute concentration produced by short-term processes involved in basalt production (partial melting and fractional crystallization), it is difficult to trace absolute mantle source concentration variations as a function of time. For most elements, there seems to be little difference (less than a factor of 2) between Archean mantle and modern mantle sources (Hart *et al.*, 1970; Jahn *et al.*, 1974). There is a suggestion that the mantle has become depleted in Cs relative to K through time (factor of 2–3), based on the Archean clinopyroxene analyses of Hart and Brooks (1977), but this observation rests on certain assumptions relating to K/Cs partitioning between pyroxene and liquid. It has also been argued that the Archean mantle was heterogeneous with respect to the REE, based on the fact that Archean komatiites and tholeiites show an appreciable range of relative REE patterns (Sun and Nesbitt, 1977, 1978). We might point out, however, that the demonstration that these heterogeneities were of any appreciable time duration is not possible based solely on the REE data. It

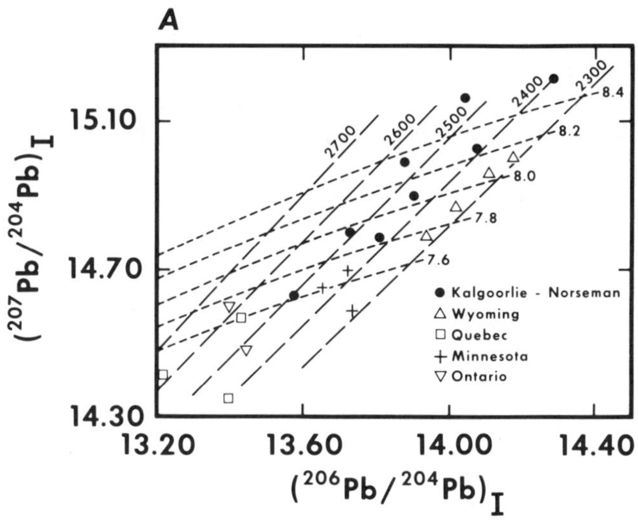

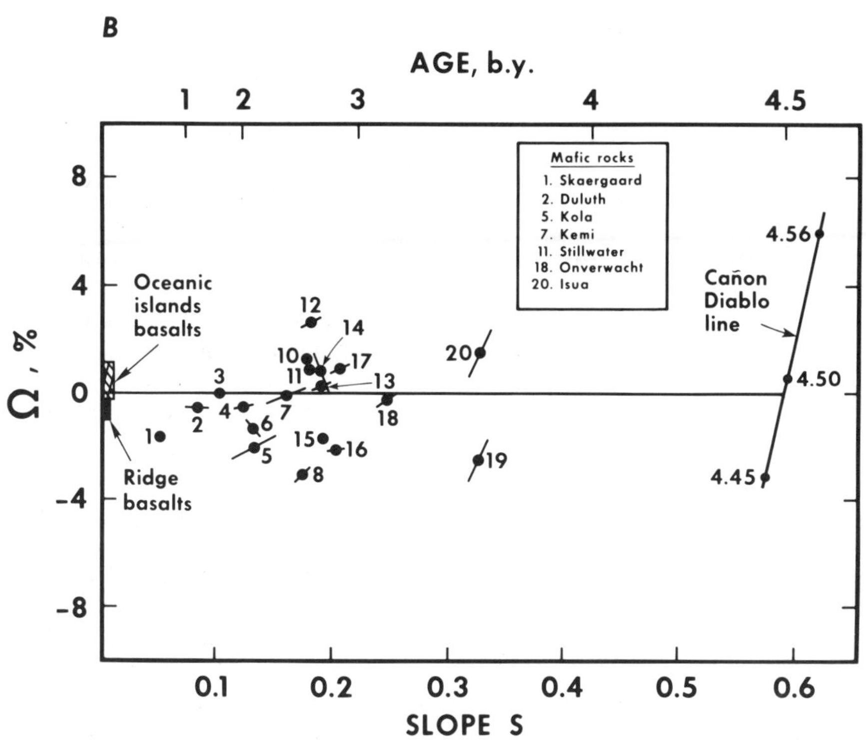

7.4.31 a. Single stage growth curve grid for μ values between 7.6 and 8.4, with isochrons for 2300–2700 m.y. Parameters used to construct the grid are $^{206}Pb/^{204}Pb = 9.307$; $^{207}Pb/^{204}Pb = 10.294$; $T_0 = 4.57 \times 10^9$ yr. Data points shown are age-corrected K-feldspar compositions for 2600 m.y. old plutonics (from Oversby, 1975). **b.** Ω-S diagram from Manhes *et al.* (1979). An "initial" ratio I and slope S are determined from data arrays on a $^{207}Pb/^{204}Pb$ vs. $^{206}Pb/^{204}Pb$ diagram. Ω is the percent deviation of I from a closed single-stage lead evolution line and S is the slope of the Pb isotope data array. The age of each rock unit data set is shown along the top. Mafic rock units are identified in the legend. All other points are sialic rocks (see Manhes *et al.*, 1979 for references). Error limits are shown by lines through the data points. Scatter of data clearly demonstrates that mafic rocks did not come from a single closed-system mantle.

is becoming quite clear from modern isotopic data that large variations in REE patterns can be produced in mantle sources on a relatively short time scale (Zindler *et al.*, 1979), and that this may be related to melt extraction episodes or migrations of LILEN-type mantle fluids, highly light-REE-enriched, during time periods concurrent with or only slightly pre-dating the episodes which produced the basaltic volcanism. Furthermore, though komatiite-tholeiite sequences show a range of REE patterns which probably requires a heterogeneous source, as pointed out by Sun and Nesbitt, these sequences show no significant dispersion of Nd isotope ratios (Zindler *et al.*, 1978, 1979), suggesting a short life time for the heterogeneities. Finally, it should be noted that the REE abundances of Archean basalts, especially tholeiites, show the same range as modern tholeiites, with little evidence for even as much as 30–50% change in REE abundances in mantle sources in the time period 3.5 b.y. to present (Sun and Nesbitt, 1977, 1978; Jahn *et al.*, 1974; Arth *et al.*, 1977; Bickle *et al.*, 1976).

It has been noted that the continental crust contains a significant portion of the Earth's complement of a number of LIL elements (Ba, U, Rb, Cs, etc.; Gast, 1968b). This implies that mantle regions must exist which are severely depleted (50–80%) in these elements. Modern ridge basalts show a highly depleted character, and it is possible that they reflect the depleted source regions from which continental crust has been derived during the past eons. Alternatively, it is possible that crust-supplying mantle regions are somehow removed from the "sampling" reservoir of the mantle, so that while some 50–80% of the mantle may be highly depleted in LIL elements, these sources are simply no longer available for production of basalt.

7.4.3 Genetic relationship between ultramafic rocks and basalts: isotopic and trace element studies

Introduction

There is a considerable body of geophysical and geochemical evidence which suggests that the Earth's upper mantle is ultramafic in composition. Basaltic liquids have been derived from such an ultramafic mantle throughout geologic time. An accurate and detailed description of the genetic relationships between the basalts and the ultramafic mantle remains a main goal of numerous petrological and geochemical studies. As discussed in section 7.1.3 and other chapters in this book (e.g., Chapters 1, 3), the composition of basalts records a complicated chemical history of processes such as melting, magma segregation, migration upward and near-surface fractionations. The major element composition of many basalts shows, however, that in general they have characteristics expected for partial melts of the ultramafic upper mantle. Thus, it is possible in principle to infer the chemical nature of the source material of basalts by the application of partial melting theory and known crystal liquid partition coefficients of various elements of interest, subject to the constraints imposed by the applicability of available models and their interpretive nature. For a planet from which samples of basalt but no direct mantle materials are available, for example the Moon, this is the principal way of estimating the chemical composition of the planetary mantle. In the case of the Earth, however, various tectonic processes have brought mantle ultramafic rocks to the surface from depths of up to 200 km. Such material can be studied directly by petrologists and geochemists to provide information on the genetic relationship between basalts and the Earth's ultramafic mantle. The purpose of this section is to review the Rb-Sr isotopic relationships and the inventory of lithophile trace elements, K, Rb, Sr and Ba in various types of ultramafic rocks in the context of basalt genesis.

It is to be noted at the outset that all ultramafic rocks observed on the Earth should bear one of the following relationships to basalts, either at the present time or some time in the prior history of the upper mantle:

1. They are possible source material of basalts.
2. They are residua left after extraction of basaltic melts.
3. They are cumulate material crystallized from basaltic melts.

It seems possible now to distinguish among these possibilities by the application of isotopic tracer methodologies discussed earlier in this chapter.

Types of ultramafic rocks available for study

Alpine peridotites and ophiolites. Ultramafic rocks composed largely of dunite, harzburgite and minor amounts of lherzolite occur throughout the major mountain belts of the Earth (Dewey and Bird, 1971; Nicolas and Jackson, 1972) either as individual massifs or as basement complexes to ophiolite sequences which are considered to be upthrust fragments of oceanic crust and mantle at convergent plate boundaries. Earlier workers (e.g., Hess, 1939, 1964) considered the alpine peridotites as direct intrusions of the upper mantle material. However, recent investigations, including detailed field and mineralogical studies, suggest a close genetic relationship between alpine peridotites and the basal ultramafic tectonites of ophiolites (e.g., Loney *et al.*, 1971; Coleman, 1971; Himmelberg and Loney, 1973).

Ultramafic rocks similar to alpine peridotites have been reported from the mid-ocean ridges, in particular where the ridges intersect major fracture zones, and in oceanic trenches. Based on studies of dredge-hauls, Bonatti (1968) concluded that these rocks bear a close resemblance in mineralogy and composition to the alpine tectonite ultramafic rocks from orogenic regions. Current ideas of plate tectonics integrate the evolution of all alpine ultramafic rocks from an initial origin at mid-ocean ridges through subsequent migration of the seafloor into orogenic environments. In plate boundary collisions, these masses and sometimes the associated oceanic crust are obducted and exposed at the surface (Dewey and Bird, 1971; Moores and Vine, 1971; Davies, 1971; Coleman, 1971).

In some instances, ultramafic complexes are known to exist with pronounced high temperature metamorphic aureoles. The classic examples are the peridotites from Lizard and Tinaquillo (Green, 1967), Mt. Albert (MacGregor 1964; Wyllie, 1970), Ronda (Dickey, 1970), and Beni Bouchera (Kornprobst, 1969). These high temperature peridotites are considered to be direct intrusions of the upper mantle material in a tensional tectonic setting (Bonini *et al.*, 1973; Moores, 1973) in contrast to that of the alpine peridotites. Experimental petrological studies clearly indicate emplacement temperatures of the order of 800–900° C and pressures of about 10 kb, indicating their origin from the mantle (MacGregor, 1964; Green, 1967).

Nodules in alkalic basalts. Ultramafic xenoliths of a wide variety of compositions and mineralogy are known to occur in alkali basalts from both the oceanic and continental regions (Forbes and Kuno, 1965). Abundant textural and mineralogic evidence shows that these xenoliths are accidental inclusions of fragments from the lithospheric plate, and represent the dominant mineralogy of the upper mantle (see for example, Kushiro and Kuno, 1963; Green and Ringwood, 1967; Wyllie, 1970 and many others).

Nodules in kimberlite pipes. Kimberlite pipes, which originate at considerable depths in the mantle, often carry a variety of xenoliths. Volumetrically, the most important are the garnet-peridotites (MacGregor and Carter, 1970). The most extensively studied are the xenoliths from the kimberlite pipes of South Africa, where studies in P-T conditions of equilibration, mineralogy and texture indicate that the material sampled may come from depths as deep as the base of the continental lithosphere, or possibly the upper part of the continental asthenosphere (Boyd, 1973; Nixon *et al.*, 1973; MacGregor and Basu, 1974). An excellent modern review of the mineralogy and petrology of these nodules is given by Meyer (1977).

Problems and approaches in isotopic and trace element studies of ultramafic rocks

Although ultramafic rocks derived from the mantle are readily available for direct studies, there are numerous problems in carrying out geochemical studies on these rocks. Foremost of these is the problem of contamination or exchange of strontium and the alkali elements in the crustal environment. Serpentinization and alteration of olivine and orthopyroxene is quite common in many ultramafic rocks, particularly the alpine peridotites. Several studies have conclusively shown that the trace element abundances and Sr isotopic compositions of mafic and ultramafic rocks can be seriously affected by alteration and exchange with seawater or groundwater (Dasch *et al.*, 1973; Hart, 1975; Pankhurst, 1968; Fenton and Faure, 1969; Jones *et al.*, 1973). In the case of ultramafic nodules in basalts, contamination from the host basalts has been noted in some cases (e.g., Burwell, 1975).

A related question concerns the location sites of the incompatible trace elements in the major mantle minerals—olivine, orthopyroxene, clinopyroxene and garnet. In the case of olivine and orthopyroxene, some

Table 7.4.4 Relative abundance data on unwashed and washed mineral separates.†

Mineral	K	Rb	Sr	$^{87}Sr/^{86}Sr$	Reference
Olivine					
unwashed	-	0.072	1.135	0.7130	*(1)*
washed	-	0.050	1.130	0.7130	*(1)*
Olivine					
unwashed	39.3	0.063	4.32	0.70684 ± 11	*(2)*
washed	21.2	0.035	0.95	0.70585 ± 18	*(2)*
Orthopyroxene					
unwashed	204.4	0.186	5.74	0.70702 ± 7	*(2)*
washed	43.6	0.057	1.51	0.70579 ± 7	*(2)*
Diopside					
unwashed	23.15	0.040	124.16	0.70244 ± 8	*(3)*
washed	17.80	0.020	124.31	0.70257 ± 6	*(3)*
Diopside					
unwashed	92.2	0.123	11.17	0.70545 ± 12	*(2)*
washed	38.7	0.054	5.29	0.70196 ± 11	*(2)*
Garnet					
unwashed	182.39	0.336	3.07	0.70657 ± 8	*(4)*
washed	8.26	0.027	1.72	-	

†Abundance data in ppm.
References: *(1)* Dasch and Green (1975); *(2)* Basu and Murthy (1977); *(3)* Menzies and Murthy (1978a); *(4)* Menzies and Murthy (1978b).

recent studies indicate that these elements reside in defect structures, cracks and intergranular films (Shimizu and Boyd, 1973; Allègre, pers. comm., 1977), and hence may not be *intrinsic* to the system, but may have been introduced into the rocks. The highly labile nature of incompatible elements in many types of ultramafic rocks and minerals (Table 7.4.4) has been thoroughly documented (Griffin and Murthy, 1968; Dasch and Green, 1975; Basu and Murthy, 1977a; Menzies and Murthy, 1978a).

In an effort to minimize the effects of some of the above problems, modern analyses of ultramafic rocks have relied exclusively on high purity mineral separates which have been surface-washed to rid them of exterior contaminant films. Data on such minerals can then be reconstituted to a "total-rock" system to provide important information on the relationship between ultramafic rocks and basalts (Brueckner, 1975; Menzies and Murthy, 1978a). Because of the extremely low abundances of these trace elements in orthopyroxene and olivine, the characteristics of these phases can be easily changed by any type of contamination. Clinopyroxene is the major mineral reservoir for these trace elements, and therefore, data on pure clinopyroxenes have been emphasized in ultramafic rock-basalt relationships (e.g., Shimizu, 1975).

Insofar as possible, we have selected the data for this review from those studies which were cognizant of these problems, and which also contain data of high analytical precision.

Isotopic and trace element studies of alpine peridotites

The first attempts to explore the genetic relationships between basalts and alpine peridotites by the use of isotopic and trace element techniques are those by Roe *et al.* (1965), Stueber (1965), and Stueber and Murthy (1966). The efforts were confined to whole-rock samples, largely of dunites and harzburgites, with a wide geographic coverage. The $^{87}Sr/^{86}Sr$ ratios exhibit a wide range from 0.7063–0.7290 and are definitely more radiogenic than the strontium in basalts from oceanic ridges and islands. Similar highly radiogenic Sr was also reported from peridotites dredged from ocean basins (Bonatti *et al.*, 1970).

The highly radiogenic strontium in alpine peridotite whole rocks seemed to preclude any contemporary relationship between alpine ultramafics and basalts in oceanic or continental regions. Although no contemporary relationship is permitted, genetic relations with basalt sometime in the past is allowed. Multistage models of the evolution of Sr in alpine ultramafic rocks (Stueber and Murthy, 1966; Hurley, 1967), shown schematically in Fig. 7.4.32, explore such possible past relationships.

Lanphere (1968) and Jones *et al.* (1973) studied the Sr isotopic composition of *cumulate* alpine peridotites from Alaska and North Carolina, respectively, and showed that these ultramafic bodies had isotopic compositions similar to those in oceanic basalts, i.e., $^{87}Sr/^{86}Sr = 0.7026–0.7047$. Since the cumulate nature of these bodies was established *a priori* from field and textural relationships, the compatibility of the isotopic evidence with the field and petrographic evidence is not surprising. But why is there no such isotopic confirmation for the vast majority of alpine peridotites where other independent evidence attests to a close genetic relationship between basalts and alpine peridotites? A reasonable answer would be that the alpine peridotites have suffered various degrees of contamination and exchange with radiogenic Sr in the crustal environment, as discussed earlier. An alternative explanation by Graham and Ringwood (1971) and Ringwood (1975) postulates that melt separation in the mantle preferen-

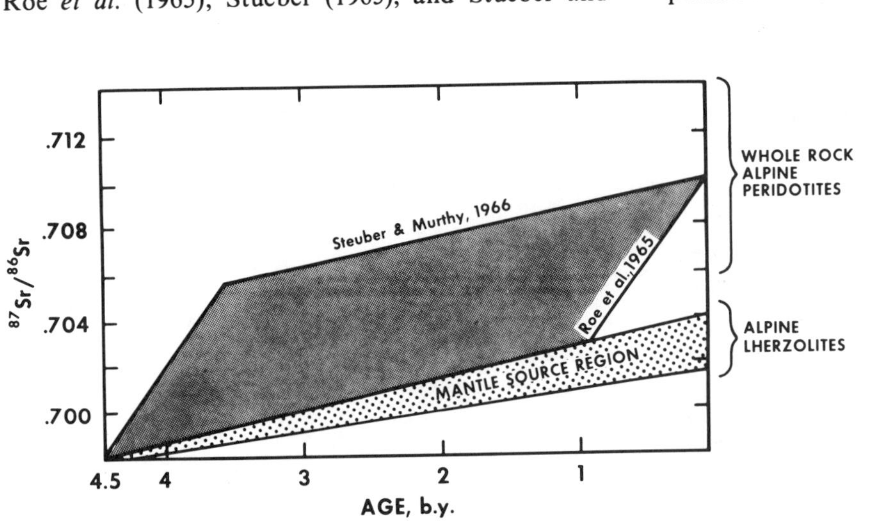

7.4.32 Sr evolution models for alpine peridotite whole rocks and lherzolites. The data of Stueber and Murthy (1966) and Roe *et al.* (1965) are schematically shown in the form of growth lines. The range of values for alpine lherzolites is computed from the modal composition and mineral data of Table 7.4.5. The field designated Mantle Source Region is a schematic single-stage growth from primordial $^{87}Sr/^{86}Sr$ to the present-day values in alpine lherzolites.

tially extracts the nonradiogenic Sr leaving behind a residue of harzburgite or dunite enriched in radiogenic Sr. Remobilization of this residual material and emplacement into a crustal environment results in an alpine peridotite in which the prior genetic relation to basaltic melt has been obliterated.

Many alpine peridotites contain minor amounts of lherzolite in addition to the harzburgite-dunite which forms the bulk of the mass. Menzies and Murthy (1976, 1978a) made a careful reexamination of four alpine peridotites, particularly concentrating on the relatively unaltered portions of the lherzolites associated with these peridotites. Hand-picked mineral separates of olivine, orthopyroxene, and diopside were analyzed, rather than whole rocks, to avoid intergranular alteration products and other contaminants which were shown to carry highly labile K, Rb, and to a lesser extent Sr (Griffin and Murthy, 1968, 1969). The data on these mineral separates for alpine lherzolites from Beni Bouchera, Morocco; Ronda, Spain; Lanzo, Italy; and Othris, Greece are given in Table 7.4.5 and shown diagrammatically in Fig. 7.4.33. The individual mineral isotopic data, when recombined to the actual modal mineralogy of these lherzolite whole rocks with a proper weighting factor for the Sr content of each phase, yield total-rock $^{87}Sr/^{86}Sr$ values from 0.7025–0.7028. This range in Sr isotopic composition is entirely compatible with present-day ocean ridge tholeiites as well as the basaltic rocks associated with the alpine peridotites themselves, as shown diagrammatically in Fig. 7.4.34.

These data are highly significant on several counts. They allow for a contemporary genetic relationship between basalts and alpine peridotites, which modern field, mineralogical and rare-earth element data strongly indicate (Bezzi and Picardo, 1971; Dickey, 1970; Menzies and Allen, 1974; Loubet *et al.*, 1975). However, it appears that the alpine lherzolites cannot be the source material of the basalts. Although the composition and the modal abundance of diopside in these alpine lherzolites is sufficient to produce basaltic melts of appropriate major element composition at reasonable degrees of partial melting, the abundance levels of K, Rb, and Sr are deficient (Menzies and Murthy, 1978a). For example, the concentrations of these lithophile elements in liquids produced by reasonable degrees of partial melting of the lherzolite (e.g., 5%, 10%, 15%, 20% and 25% melting) are shown in Table 7.4.6. In these calculations, the modal mineralogy of the mantle

Table 7.4.5 Mineral separate data from orogenic peridotites.†

Sample	Mineral	K	Rb	Sr	$^{87}Sr/^{86}Sr$	Reference
Alpine peridotites						
Lanzo, Italy	Diopside	8.00	0.011	31.10	0.70316 ± 5	*(1)*
	Diopside	31.17	0.049	14.26	0.70259 ± 13	*(1)*
	Olivine	3.71	0.005	0.12	0.70556 ± 10	*(1)*
	Orthopyroxene	16.69	0.114	0.72	0.70299 ± 10	*(1)*
Beni Bouchera, Morocco	Diopside	29.15	0.040	124.16	0.70244 ± 8	*(1)*
	Diopside	17.80	0.020	124.31	0.70257 ± 6	*(1)*
	Olivine	1.55	0.002	3.64	0.70290 ± 9	*(1)*
	Orthopyroxene	8.76	0.030	3.93	0.70265 ± 19	*(1)*
Ronda, Spain	Diopside	34.40	0.142	41.57	0.70227 ± 9	*(1)*
					0.70228 ± 6*	
	Olivine	15.81	0.030	0.37	0.70831 ± 32	*(1)*
	Orthopyroxene	11.74	0.048	0.83	0.70429 ± 62	*(1)*
Othris, Greece	Diopside	32.22	0.066	1.63	0.70370 ± 7	*(1)*
Caledonian peridotites						
Almklovden, Norway	Diopside	-	0.064	333	0.7026 ± 3	(2)
	Diopside	-	0.021	421	0.7022 ± 2	(2)
	Diopside	-	0.013	40	0.7030 ± 3	(2)
Tafjord	Diopside	-	0.046	112	0.7016 ± 2	(3)
Otroy	Diopside	-	0.008	23.5	0.7012 ± 2	(3)
	Enstatite	-	0.0013	0.142	0.7026 ± 2	(3)

†Abundance data in ppm.
*Duplicate run.
References:
(1) Menzies and Murthy (1978a); *(2)* Brueckner (1974); *(3)* Brueckner (1975).

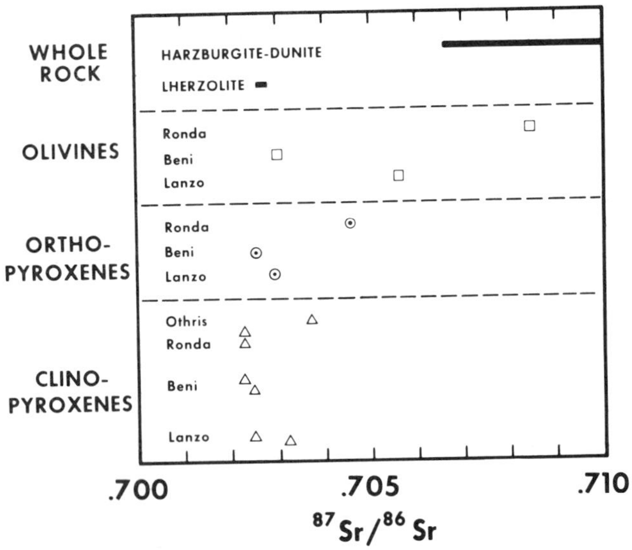

7.4.33 Ranges of Sr isotopic compositions in the constituent minerals of alpine and high temperature lherzolites. The top section of the diagram shows whole rock lherzolite data and a partial range for the whole rock dunites and harzburgites. The latter have a total range towards much more radiogenic $^{87}Sr/^{86}Sr$ values. Olivines contain variable amount of serpentine. Orthopyroxenes are crystal clear grains. Clinopyroxenes are crystal clear in the case of Ronda, Beni Bouchera, and one mineral separate from Lanzo. The separates from Othris and Lanzo are slightly clouded.

lherzolite is taken to be 10% clinopyroxene, 25% orthopyroxene and 65% olivine. For a comparison, Table 7.4.6 also lists the ranges of various trace elements in ridge tholeiites and alkalic basalts. It is clear that for any reasonable degree of partial melting (5–20%) the anhydrous lherzolitic source material cannot yield liquids whose trace element abundances resemble basalts. This conclusion rests upon the assumption that mantle melting is an equilibrium process and that basalts represent liquids produced by reasonably large degrees of partial melting in the mantle, say $\gtrsim$10%. These alpine tectonite lherzolites appear to have been subject to very small degrees of partial melting at some time in the past such that they have lost a significant fraction of their K, Rb, and Sr inventories, the degree of melting being constrained such that diopside is still largely preserved in these rocks. Menzies and Murthy (1978a) show that such liquids would have alkalic affinities and would have been removed from the lherzolites prior to their incorporation into alpine peridotite massifs. An entirely similar situation pertains to lherzolite nodules found in alkalic basalts (Basu and Murthy, 1977a).

Thus, modern isotopic and trace element studies on alpine lherzolites have established the consanguinity of alpine peridotites and basalts. Furthermore, the alpine lherzolites, which must be samples of the upper parts of the oceanic lithosphere according to modern tectonic interpretations, cannot be the source material of basalts. Rather, these lherzolites appear to be residual materials that have suffered severe depletions of K, Rb, Sr and relatively less severe depletions of the more refractory lithophile elements, such as the rare-earth elements (Menzies, 1976), in some prior fractionation episodes.

Isotopic and trace element studies of nodules in alkalic basalts

In spite of the diverse mineralogy of the accidental inclusions in alkalic basalts, Ross *et al.* (1954) were able to demonstrate that the petrology and mineral chemistry of these samples resembles that of alpine peridotites. Subsequent work (Ave'Lallemant, 1967; Den Tex, 1969; Ave'Lallemant and Carter, 1970) has demonstrated that the resemblance extends to mineral fabric, and that the deformation occurred in the upper mantle. The domi-

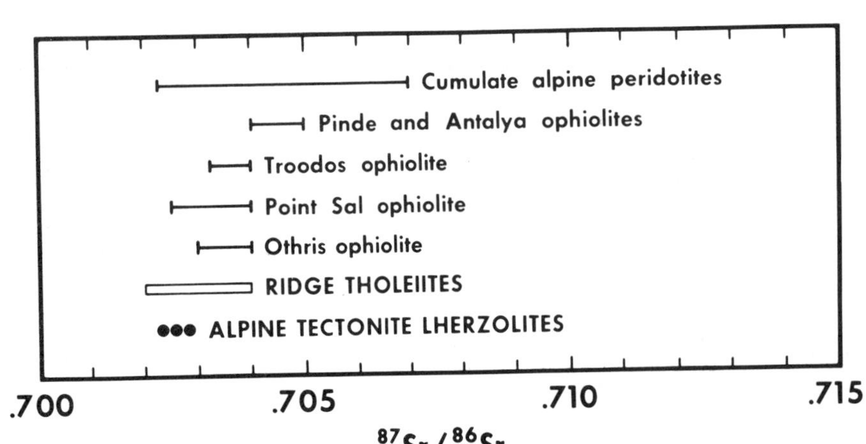

7.4.34 Ranges of Sr-isotope composition of cumulate alpine peridotites, several ophiolites and the alpine lherzolites. The lherzolite Sr isotopic compositions were calculated using mineral data.

Table 7.4.6 Concentrations of K, Rb and Sr in hypothetical partial melts of some anhydrous lherzolites.†

Sample	Degree of Melting (%)	K	Rb	Sr
Lanzo	5	194	0.73	28
	10	97	0.37	15
	15	65	0.24	11
	20	49	0.18	8
	25	39	0.15	7
Beni Bouchera	5	99	0.25	264
	10	50	0.13	144
	15	33	0.09	99
	20	25	0.06	76
	25	20	0.05	61
Victoria, Australia	5	-	2.3	93
	10	-	1.2	51
	15	-	0.78	35
	20	-	0.59	27
	25	-	0.47	22
San Quintin, Baja California	5	569	0.82	26
	10	285	0.41	14
	15	190	0.27	10
	20	143	0.20	7
	25	114	0.16	6
Ridge tholeiites		1,060–1,260	1.0–1.1	131–140
Alkali basalts		6,700–15,390	5–35	660–950

†Modal mixture for mantle lherzolite consists of 10% Cpx, 25% Opx and 65% olivine. Trace element data from Tables 7.4.5 and 7.4.7. Concentrations in ppm.

nant rock suite is spinel lherzolite (olivine 65–90%; enstatite 5–20%; diopside 5–15% and spinel 0.5–4%). Recent phase-equilibria studies (Boyd, 1973; MacGregor, 1975) show that it is possible to estimate the temperature and pressure of equilibration of the nodules, albeit with some uncertainties (Wilshire and Jackson, 1975; Meyer, 1977). Thus, the concept can be entertained that xenoliths in alkalic basalts represent a general "stratigraphy" of the mantle (MacGregor and Basu, 1974).

A number of attempts have been made to establish genetic relations between the nodules and their host basalts through Sr isotope studies. Many of these studies (Stueber and Murthy, 1966; Leggo and Hutchison, 1968; Halpern, 1969) were done before a clear understanding existed of the mineralogy and mineral chemistry in relation to mantle tectonics. With the recognition that the nodules can represent both cumulate material and true fragments of the upper mantle (White, 1966; Jackson and Wright, 1970), and with the current ideas of equilibration of the nodules at various depths in the mantle prior to their incorporation into the host lavas, it is now possible to evaluate these and other new data in the context of ultramafic rock-basalt relationships.

The earlier studies were done mainly on whole rocks and it is only possible to make some general statements from such data. The simple case is that of nodules which are isotopically similar to the host basalts. These can be interpreted as cumulate material, in agreement with many petrological studies (e.g., White, 1966; O'Hara, 1967). The highly variable $^{87}Sr/^{86}Sr$ ratios observed in nodules even from a single volcanic source (Paul, 1971; Dasch and Green, 1975; Laughlin *et al.*, 1971) must reflect a complicated history, possibly involving contamination with host basalt, crustal material, or seawater, and isotopic reequilibration with unknown reservoirs of Sr at various depths in the mantle. When nodules have $^{87}Sr/^{86}Sr$ ratios higher than the host basalt, contamination from the basalt can be excluded.

To clarify this complex situation, some recent studies have emphasized analyses of individual minerals from lherzolitic nodules (Peterman *et al.*, 1970b; Paul, 1971; Kudo *et al.*, 1972; Stueber and Ikramuddin, 1974; Dasch and Green, 1975; Shimizu, 1975; Burwell, 1975; Basu and Murthy, 1977a, and others). Some relevant mineral data are shown in Table 7.4.7. A general conclusion that emerges from these studies is that there is a considerable heterogeneity of isotopic composition of Sr in coexisting minerals; in general, olivine and orthopyroxene have significantly higher $^{87}Sr/^{86}Sr$ ratios than coexisting clinopyroxene, similar to the situation observed in some alpine lherzolites. In many cases, these variations can be attributed neither to contamination from host basalt, nor to radiogenic growth *in situ* in the minerals since the time of eruption of the host basalt. It is therefore necessary to consider that the mineral isotopic compositions reflect the prior history of the nodule in the mantle. This is in keeping with the analysis drawn from the tectonic fabric of the nodules as well as the concepts deduced from phase equilibria studies of reequilibration in the mantle.

To what extent one can see through to the time of formation or last equilibration of a given nodule depends on the extent to which the individual minerals have remained as closed systems since that time. Hofmann and Hart (1975) and Nelson and Dasch (1976) have shown that diffusion of Sr is extremely slow at subsolidus temperatures, and consequently, isotopic heterogeneities can persist over long periods of time in the mantle in the absence of coexisting melt, which would act as a diffusion "bridge." Nodules that conform to such a situation would display a linear relationship on a Rb-Sr evolution diagram (see section 7.1). The slope of the line indicates the time of formation or last complete equilibration of the nodule in the mantle. The

Table 7.4.7 Mineral separate data from spinel lherzolites from alkali basalt flows.†

Sample	Mineral	K	Rb	Sr	$^{87}Sr/^{86}Sr$	Reference
Kilbourne Hole, New Mexico	Diopside	-	0.067	43.7	0.7025	*(1)*
	Enstatite	-	0.079	1.17	0.7058	*(1)*
	Olivine	-	0.085	1.07	0.7060	*(1)*
Mt. Aldaz, Antarctica	Diopside	-	0.177	17.2	0.7022	*(1)*
	Enstatite	-	0.142	1.04	0.7058	*(1)*
	Olivine	-	0.064	0.283	0.7077	*(1)*
Victoria, Australia	Diopside	-	0.255	42.110	0.7029	*(2)*
	Enstatite	-	0.240	2.482	0.7060	*(2)*
	Olivine	-	0.050	1.130	0.7130	*(2)*
Victoria, Australia	Diopside	-	0.702	163	0.7041 ± 2	*(3)*
	Enstatite	-	0.096	2.01	0.7049 ± 2	*(3)*
	Olivine	-	0.12	1.96	0.7044 ± 4	*(3)*
	Diopside	-	0.045	140	0.7040 ± 1	*(3)*
	Enstatite	-	0.10	0.942	0.7050 ± 4	*(3)*
	Olivine	-	0.035	0.421	0.7040 ± 4	*(3)*
Baja California	Diopside	38.7	0.054	5.295	0.70196 ± 11	*(4)*
	Enstatite	43.6	0.057	1.51	0.70579 ± 7	*(4)*
	Olivine	21.2	0.035	0.953	0.70585 ± 18	*(4)*
	Diopside	23.9	0.050	2.136	0.70484 ± 10	*(5)*
	Enstatite	28.7	0.083	2.976	0.70717 ± 14	*(5)*
	Olivine	17.5	0.044	1.301	0.7055 ± 12	*(5)*

†Abundance data in ppm.
References: *(1)* Steuber and Ikramuddin (1974); *(2)* Dasch and Green (1975); *(3)* Burwell (1975); *(4)* Basu and Murthy (1977a); *(5)* Basu and Murthy (1977c).

individual $^{87}Sr/^{86}Sr$ ratios of the minerals reflect closed-system growth of radiogenic Sr appropriate to the Rb/Sr ratios of the minerals. Such relationships have been found in some lherzolitic nodules (Stueber and Ikramuddin, 1974; Dasch and Green, 1975; Basu and Murthy, 1977a) and are shown in Fig. 7.4.35. If the above interpretation is correct, these data show that the nodules in alkalic basalts are accidental inclusions of mantle material that has a complex and multi-episodic history in which melt and residue formation is a general process throughout the history of the lithosphere, both in the oceanic and continental regions. In addition, the trace element abundance levels and Sr isotopic compositions indicate that lherzolite nodules in alkalic basalts, which can be considered samples of the upper lithosphere both in continental and oceanic regions, cannot be the source material of basalts in spite of their "fertile" nature inferred from modal content of diopside, Fe/Mg ratios of olivine and orthopyroxene, and other mineral chemistry studies. This is illustrated in Table 7.4.6, which lists the concentrations of various trace elements in hypothetical liquids produced by equilibrium partial melting of source material similar to the anhydrous lherzolite nodules in alkali basalts. Comparison of the liquids to natural basalts demonstrates the inadequacy

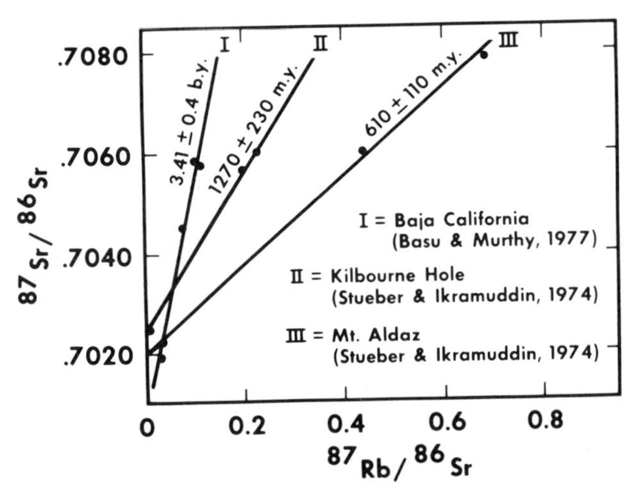

7.4.35 Apparent Rb-Sr internal isochrons for lherzolitic nodules from alkali basalts.

of these lherzolites as source material of basalts because of the trace element contents. A general conclusion that emerges from the studies of alpine lherzolite and lherzolite nodules in basalts is that the *upper* lithosphere, both in oceanic and continental regions, is barren of the lithophile trace elements needed for the production of basaltic liquids. No simple scheme of partial melting of these anhydrous lhzerolites can satisfactorily yield liquids that match the trace element abundances and characteristics of basalts.

Isotopic and trace element studies of lherzolites from kimberlite pipes

Kimberlite pipes contain a variety of crustal and mantle xenoliths of the subcontinental lithosphere. Among these xenoliths, the most pertinent in regard to basalt genesis are the garnet lherzolites. The general bulk compositional similarity between spinel lherzolite nodules from alkalic basalts and garnet lherzolites has already been noted. It is now recognized that the aluminous ultramafic composition represented in the high temperature peridotites, spinel lherzolites and garnet lherzolites corresponds to progressively higher pressure-temperature mineral assemblages of compositions relevant to basalt genesis in the lithosphere (e.g., see Yoder, 1976b).

The garnet peridotite xenoliths of South Africa have been grouped into two types on the basis of their texture and mineralogy (Nixon *et al.*, 1973). These are the "sheared" and "granular" garnet lherzolites. Estimation of the pressure-temperature conditions of these two groups of xenoliths shows that the sheared xenoliths

Table 7.4.8 Mineral separate data from garnet lherzolite from kimberlites.[(a)]

Sample	Mineral	K	Rb	Sr	$^{87}Sr/^{86}Sr$	Reference
Thaba Putsoa, Lesotho						
PHN 1611 sheared	Diopside	433.6	0.196	95.28	0.70352 ± 14	*(1)*
PHN 1596 granular	Diopside	341.2	0.015	89.3	0.70342 ± 10	*(1)*
	Garnet	1.7	0.006	0.98	-	*(1)*
PHN 1566 sheared	Diopside	163.7	0.076	89.8	0.70269 ± 11	*(1)*
PHN 1569 sheared	Diopside	110.3	0.693	503.3	0.70384 ± 12	*(1)*
JJG 352 granular	Diopside	195.4	0.463	558.0	0.70578 ± 12	*(1)*
PHN 1859 GL1 ilmdiop	Diopside	168.3	0.017	116.0	0.70289 ± 8	*(1)*
Du Toitspan						
DPA granular	Diopside	673	0.528	674.1	0.7075	*(2)*
De Beers						
	Diopside	275	1.16	551.9	0.7070	*(2)*
	Diopside	436	1.78	903.1	0.7060	*(2)*
Bultfontein						
	Diopside	157	0.654	712.7	0.7068	*(2)*
Wesselton						
	Diopside	178	0.951	504.1	0.7070	*(2)*
Bultfontein						
B10 granular	Diopside	282.46	0.478	224.98	0.70646 ± 10	*(3)*
	Garnet	182.39	0.336	3.07	0.70657 ± 8	*(3)*
CK27 granular	Diopside	296.25	0.557	141.53	0.70465 ± 8	*(3)*
	Garnet	5.37	0.018	0.246	0.70473 ± 12	*(3)*
	Mica	65710	163.91	22.05	0.72226 ± 10[(b)]	*(3)*
B69 granular	Diopside	66.17	0.127	247.65	0.70536 ± 9	*(3)*
B21 granular	Diopside	92.90	0.106	252.19	0.70537 ± 5	*(3)*
713 granular	Diopside	273.6	0.803	321.3	0.70722 ± 5	*(3)*
	Garnet	57.8	0.145	0.598	0.70871 ± 9	*(3)*
DK26 granular	Diopside	199.28	2.46	441.75	0.70583 ± 5	*(3)*
	Mica	52630	148.33	115.13	0.70569 ± 6[(b)]	*(3)*
Lesotho & Kimberley						
BD 2308 sheared	Diopside	114	0.20	686	0.7059 ± 2	*(4)*
BD 2348 sheared	Diopside	205	0.22	166	0.7045 ± 3	*(4)*
BD 2435 sheared	Diopside	153	0.22	317	0.7134 ± 2	*(4)*
BD 2458 granular	Diopside	458	0.57	189	0.7051 ± 3	*(4)*

(a) Abundance data in ppm.
(b) Corrected for radiogenic Sr growth for 75 m.y., the pipe emplacement age. $\lambda(^{87}Rb) = 1.42 \times 10^{-11} yr^{-1}$.
References:
(1) Shimizu (1975); *(2)* Barrett (1975); *(3)* Menzies and Murthy (1978b); *(4)* Kramers (1977).

were invariably equilibrated at higher T and P than the granular variety (Boyd, 1973; MacGregor, 1975: Boyd and Nixon, 1975). These studies indicate that the garnet lherzolites are samples of the subcontinental lithosphere from depths in the range of 150–175 km, and possibly from the lithosphere-asthenosphere boundary under the continents. Because of their major element compositions, the sheared nodules are considered to be more primitive and fertile enough to be potential sources of basaltic magmas (Kushiro, 1973). It has also been suggested that the granular nodules are refractory residua of partial melting and removal of liquid from material similar to the sheared nodules (Boyd and Nixon, 1972).

The eclogitic nodules in kimberlite pipes are not considered here because partial melting of such nodules does not produce basalts. Total melting is needed, which is at variance with the general state of the upper mantle from which basalts are being produced at present.

The problems of contamination discussed earlier are even more formidable in the case of nodules in kimberlites, because the kimberlitic host magma is highly enriched in the various lithophile elements that are present only in low abundance levels in the garnet lhzerolite mineral phases. In addition, the volatile-rich nature of the kimberlite magma and the mode of emplacement (Holmes, 1936; McGetchin and Ulrich, 1973) are highly conducive to contamination of the nodules by introduction of kimberlitic material as fillings in cracks, and as coatings on grain surface and intergranular spaces, and other metasomatic or alteration processes. Material balance comparisons between whole rocks and recombined mineral data as well as lability studies have clearly shown that this is the case (Griffin and Murthy, 1968, 1969; Compston and Lovering, 1969; Allsopp *et al.*, 1969; Shimizu, 1975).

Some recent data on the clinopyroxene and garnet in garnet lherzolite nodules are shown in Table 7.4.8. The $^{87}Sr/^{86}Sr$ ratios of the clinopyroxenes show a range from lows of 0.70269 in a sheared nodule and 0.70289 in an ilmenite-diopside nodule to the highest ratio of 0.7075, for a granular nodule. An anomalously high value of 0.7134 for a clinopyroxene from a sheared nodule is an exception (Kramers, 1977). Shimizu (1975) has pointed out that in the clinopyroxenes from the sheared nodules, the abundances of Rb, Cs, Ba and the ratios K/Rb, K/Cs, K/Ba and Rb/Sr vary systematically with the Fe/Mg ratio of the whole rocks, and has suggested that fractional melting processes could account for these trends. Thus, these materials are considered to be residua left after fractional melting. From the high pressure clinopyroxene-liquid partition coefficients for K, Rb, Sr and Ba (Shimizu, 1974), it is easy to show, however, that the order of magnitude variations in the residual solid phase constrain the degree of melting to a very small percentage, about 1% or less. The liquid so generated will be highly alkalic and rich in incompatible trace elements, reminiscent of the situation noted for alpine lherzolites and lherzolite nodules from alkali basalts.

The clinopyroxenes from the granular nodules appear to be different from those in the sheared nodules (Shimizu, 1975). Especially noteworthy are the relatively high concentrations of Sr and the more radiogenic Sr compositions (Table 7.4.8). The general characteristics of several trace elements and their ratios, and in particular the high concentrations of Sr, have been attributed to secondary processes involving reactions with a carbonate-rich kimberlite liquid (Shimizu, 1975).

Lherzolites as sources of basalts

The task of directly identifying the source material for basalts in the Earth requires the identification of an ultramafic rock which upon partial melting can produce melts not only with the proper *major* element compositions, but with the *minor* element and isotopic compositions observed in basalts. O'Hara and Mercy (1963) first suggested that much of the mantle may consist of aluminous peridotites, and subsequent work (O'Hara and Yoder, 1967; Ito and Kennedy, 1967; Green and Ringwood, 1967) has demonstrated that aluminous ultramafic compositions are required for basalt production. Terrestrial samples that most closely satisfy this requirement are the lherzolites found in alpine peridotite massifs, ophiolites, and nodules in basalts and kimberlites. These various types of aluminous peridotites have similar bulk chemistry (e.g., see Ringwood, 1975).

The adequacy of the aluminous peridotite compositions to satisfy simultaneously the major and the trace element characteristics of basalts was initially questioned by Griffin and Murthy (1969). They showed that the abundances of several trace elements in the major minerals of the mantle, i.e., orthopyroxene, clinopyroxene, olivine and garnet are so low that an anhydrous lherzolite mantle is incapable of production of basaltic melts with the observed trace element characteristics, in any reasonable scheme of partial melting in the mantle. More detailed and recent studies (Shimizu, 1975; Basu and Murthy, 1977a, b, c; Menzies and Murthy, 1976, 1978a) have amply confirmed this point of view (Tables 7.4.7–8). Even in situations where either the modal composition or other independent evidence from the lherzolites is suggestive of their "pristine" nature, the trace element data for K, Rb, and Sr show extreme depletions. Thus it appears that all anhydrous lhzerolite samples examined so far have lost a fraction rich in incompatible trace elements. This is certainly true for elements such as K, Rb, Sr and Ba; in some cases, as for example the high temperature lherzolites of Lizard and Tinaquillo (Frey, 1970; Frey *et al.*, 1971) and the lherzolites from alpine peridotites (Menzies, 1976), the rare-earth elements have been lost as well.

Using simple melt-solid equilibrium partitioning calculations, Menzies and Murthy (1978a) have estimated this "missing" component to be highly alkalic. The role of this component, along with the volatiles H_2O and CO_2, is discussed in the next section.

Hydrous mantle minerals and basalt genesis: possibilities and speculations

Griffin and Murthy (1969) suggested that the inadequacy of anhydrous lherzolite to produce the trace element characteristics of basalts could be overcome by the addition of amphibole and phlogopite to lherzolitic mineralogy. There is at present considerable experimental evidence to show that both these minerals are stable in relevant mantle compositions in presence of volatiles such as H_2O and CO_2 (e.g., Lambert and Wyllie, 1968; Yoder and Kushiro, 1969; Hill and Boettcher, 1970; Holloway, 1973; Merrill and Wyllie, 1975; Mysen and Boettcher, 1975; Holloway and Eggler, 1976).

Table 7.4.9 Data on mantle amphibole and phlogopite.[(a)]

Sample	K	Rb	Sr	Ba	$^{87}Sr/^{86}Sr$	Reference
Amphiboles						
Deadman Lake	18210	9.64	516	135	0.70264 ± 8	*(1)*
Hoover Dam	10350	7.64	577	292	0.70275 ± 5	*(1)*
San Carlos	12630	4.48	711	243	0.70271 ± 5	*(1)*
San Carlos	12850	5.07	155	268	-	*(2)*
Dish Hill I	15000	9.23	512	228	0.70288 ± 6	*(1)*
Dish Hill II	22010	11.23	488	137	0.70273 ± 5	*(1)*
Dish Hill III	21020	13.04	769	366	0.70266 ± 5	*(1)*
Queensland, Australia	12030	3.78	689	258	0.70285 ± 8	*(1)*
Spring Mt., New South Wales	14775	3.77	685	140	0.70246 ± 5	*(1)*
Wee Jusper, New South Wales	14240	17.70	732	108	0.70257 ± 10	*(1)*
Kakanui, New Zealand	18560	16.30	463	165	0.70290 ± 5	*(1)*
Kakanui, New Zealand	16700	14.00	425	109	-	*(2)*
Massif Centrale, France	12560	11.40	502	293	0.70311 ± 5	*(1)*
Etang de Lherz, France	13170	5.8	511	164	0.70306 ± 7	*(1)*
Nunivak, Alaska	5260	4.3	806	-	0.70337 ± 8	*(3)*
Atāq, S. Yemen I	3357	9.9	1128	-	0.70344 ± 6	*(3)*
Atāq, S. Yemen II	4003	8.4	1434	-	0.70374 ± 11	*(3)*
Phlogopites						
Kilbourne Hole, New Mexico	78160	69.0	150	2873	0.70325 ± 7	*(1)*
Wesselton, South Africa	71500	255.0	18.4	1380	-	*(2)*
Bultfontein, South Africa	65710	163.9	22.0	-	0.70226 ± 10[(b)]	*(3)*
Bultfontein, South Africa	52630	148.3	115.1	-	0.70569 ± 6[(b)]	*(3)*

[(a)] Abundance data in ppm.
[(b)] Corrected for radiogenic Sr growth for 75 m.y., the pipe emplacement age. $\lambda(^{87}Rb) = 1.42 \times 10^{-11} yr^{-1}$.
References:
(1) Basu and Murthy (1977a, c); *(2)* Griffin and Murthy (1969); *(3)* Menzies and Murthy (1978b).

These studies show that amphibole is a stable phase in the upper mantle up to temperatures of 1100° C and 30 kb pressure, whereas phlogopite is stable in the deeper mantle.

Trace element and isotopic data on these two hydrous minerals from mantle lherzolites are shown in Table 7.4.9. The amphiboles and phlogopites differ markedly in their trace element characteristics and isotopic compositions. The addition of one or the other of these phases in the source regions will have markedly different effects on basaltic melts that are produced. Basu and Murthy (1977b) have put forth the suggestion that the trace-element-rich phase needed to explain the trace element characteristics as well as the water content of the mid-ocean ridge basalts (MORB) is kaersutitic amphibole. The model envisages the production of small quantities of hydrous melt of kaersutitic composition in the oceanic low-velocity zone (LVZ) due to instability of amphibole in an otherwise anhydrous spinel-lherzolite oceanic mantle. During the asthenospheric upwelling at the mid-ocean ridges, conditions exist for larger degrees of melting of the anhydrous lherzolite (Langseth *et al.*, 1966). Thus, the final product MORB will be a mixture of kaersutitic melt of LVZ, and the melt produced from the anhydrous lherzolite. The trace element and isotopic characteristics (i.e., $^{87}Sr/^{86}Sr$, K/Rb, Ba/Sr, etc.) will be largely controlled by the kaersutitic liquid, but the absolute abundances of these lithophile elements will be substantially reduced because of the dilution by liquids produced from the anhydrous lherzolite with extremely low abundances of these lithophile elements. The model is schematically shown in Fig. 7.4.36.

Menzies and Murthy (1978b) have examined the clinopyroxene-amphibole relations in amphibole-lherzolites from the ʻAtāq volcano, South Yemen, and Nunivak Island, Alaska. The volcanism in both cases is from a tensional tectonic setting, caused by rifting of continental crust in South Arabia and spreading in the back-arc basin at Nunivak. In contrast to the situation

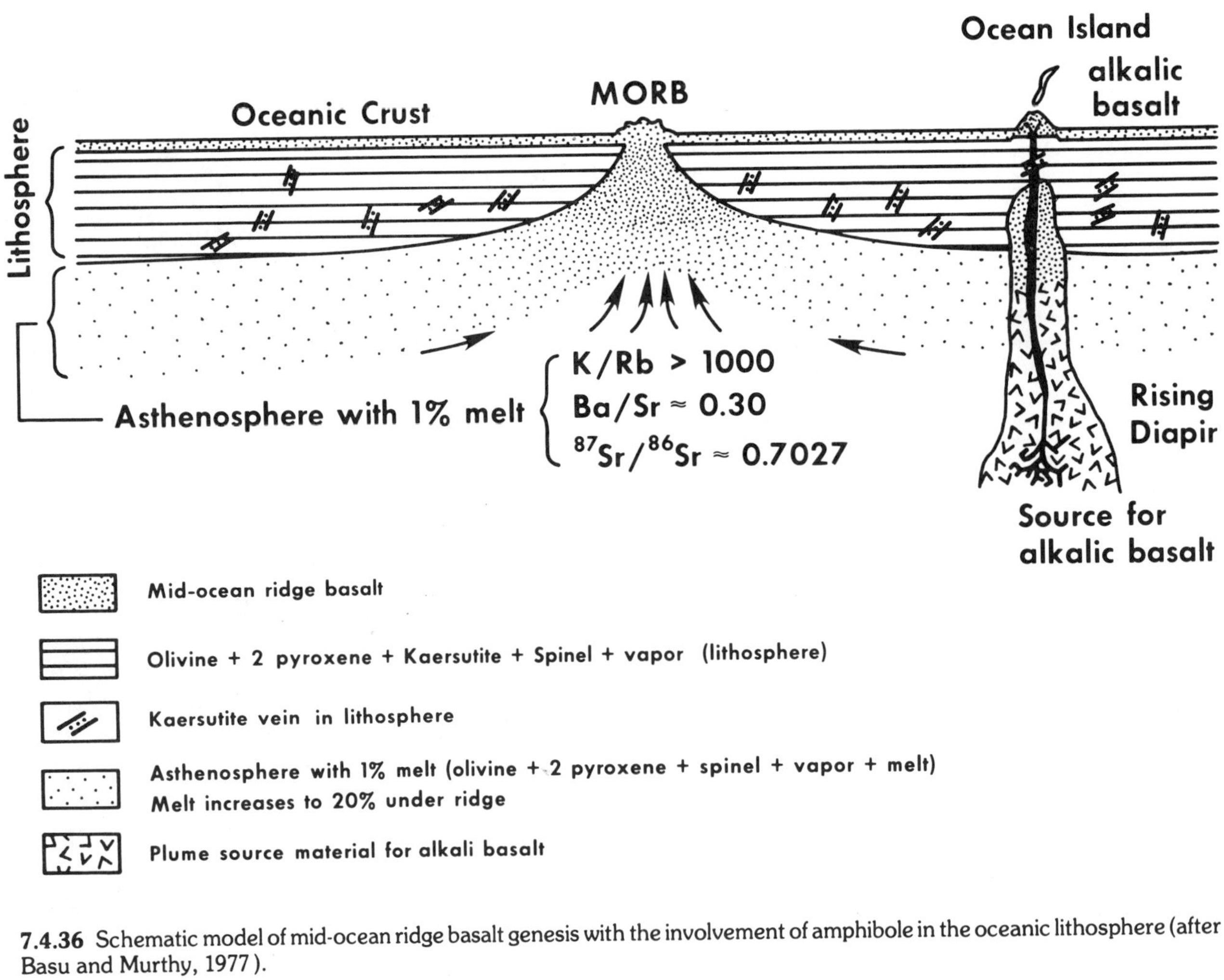

7.4.36 Schematic model of mid-ocean ridge basalt genesis with the involvement of amphibole in the oceanic lithosphere (after Basu and Murthy, 1977).

observed in kaersutites and host lavas, the amphiboles and clinopyroxenes in these lherzolites have identical $^{87}Sr/^{86}Sr$ ratios and closely match the whole-rock Sr isotopic compositions of the lherzolite and associated alkali basalts. The trace element abundances as well as the Sr isotopic compositions in these amphibole lherzolites are compatible with the suggestion that alkali basalts have formed in equilibrium with lherzolitic material containing amphibole (cf. Brooks *et al.*, 1976). Thus, it appears that the hydrous mineral amphibole plays an important role in basalt genesis both in the continental and oceanic upper mantle.

As noted earlier in this section (7.4.3), the lherzolite nodules from the deep continental lithosphere are also residual. Several workers (Shimizu and Kushiro, 1973; Carswell, 1975; Lloyd and Bailey, 1975) have noted that approximately 1% phlogopite may have to be added to such residual material to generate basaltic melts. Thus, phlogopite may also play an important role in genesis of basalts from the deep continental lithosphere.

The trace element content of all lherzolites studied so far suggests that at some stage they have lost an alkalic liquid rich in various trace elements. Volatiles such as H_2O and CO_2 would also have been removed in such a process, leaving behind a residual material that is an anhydrous assemblage of lherzolitic mineralogy devoid of the original complement of incompatible lithophile elements. Such material must make up the predominant part of the upper lithosphere and by itself cannot produce basaltic melts of reasonable trace element characteristics.

In order to generate basaltic liquids from the lithosphere, it is necessary then to augment the incompatible trace elements of the residual lherzolites. Several studies have postulated localized metasomatic or enrichment events in the mantle (Frey and Green, 1974; Lloyd and Bailey, 1975; Kay, 1977; Carter *et al.*, 1978a). If the process of removal of an alkalic and trace element rich phase from the lherzolitic upper mantle as described above is generally valid both in space and time, then an appropriate context can be provided for the process of "metasomatism" in the mantle. Transport of the trace-element-rich phase upwards, under tectonic and/or thermal control, can locally metasomatize the mantle and enrich the residual lherzolite in that locale by formation of phases such as amphibole, phlogopite and other accessories and alteration products. Subsequent partial melting in such a domain can satisfactorily generate basalts with appropriate major and trace element characteristics. Thus, the process of local metasomatism is a necessary precursor event in the mantle for basaltic magmatism.

7.4.4 Noble gases as trace elements

Introduction

Compared with other trace elements, the noble gases He, Ne, Ar, Kr and Xe have some unique properties: they are more volatile than any of the other trace elements, and radioactive decay and nuclear reactions have changed their isotopic compositions with time. Furthermore, the noble gases in the solar system were never thoroughly mixed, and different reservoirs with distinct isotopic compositions have been recognized. It is an open question whether the terrestrial noble gases were—after accretion—ever isotopically homogenized, or whether the different types of primordial noble gases can still be recognized. These properties, combined with the fact that they are chemically inert, make the noble gases potentially useful tracers for the study of the evolution of the mantle and of the history of basaltic magmas (for a recent review see Alexander and Ozima, 1978).

The elemental abundances and isotopic composition of the five noble gases have been studied in great detail in a sizable number of meteorites and lunar rocks. In contrast, most noble gas measurements in terrestrial samples have been limited to geochronological aspects (K-Ar, ^{39}Ar-^{40}Ar, and U, Th-^{4}He ages). Only in a relatively small number of terrestrial rocks have all noble gases and their isotopes been studied. Detailed investigations on the host phases of the trapped gases have so far been done only on meteorites, but not on terrestrial rocks. Furthermore, partition coefficients have not yet been determined for noble gases; the solubilities have been measured only in an enstatite melt (Kirsten, 1968), in magnetite (Lancet and Anders, 1973), and in water (Morrison and Johnstone, 1954). Data on diffusion are also very limited (Perkins and Begeal, 1971).

Sources of noble gases

The two major sources for the *terrestrial* noble gases are the *primordial* noble gases acquired during accretion of the Earth and the *nuclear* component resulting from the decay of radioactive isotopes (e.g., ^{40}K, ^{129}I, ^{238}U, ^{244}Pu) or from nuclear reactions [e.g., $^{18}O(\alpha,n)^{21}Ne$] after accretion. For the atmosphere an additional, minor source is the solar wind (Bühler *et al.*, 1976). Each of these components has a distinct isotopic signature.

Primordial noble gases. From the study of meteorites and of the solar wind we know that the noble gases in the solar system were never well mixed, and that reservoirs with significantly different elemental and isotopic abundances exist. Two major types, *solar* and *planetary*, must be distinguished (Signer and Suess, 1963). Additional, isotopically distinctive types of noble gases have been recognized (e.g., Ne-E, Xe-X; cf. Black and Pepin, 1969; Manuel *et al.*, 1972; Lewis *et al.*, 1975). Some properties of these components are listed in Table 7.4.10 and Figs. 7.4.37–39. Table 7.4.10 is not comprehensive but contains the types of primordial noble gases most relevant for the present discussion (for more details see Podosek, 1978, and references given therein).

It is important to point out that none of the known meteoritic noble gas components contain Xe with an isotopic composition matching terrestrial Xe (see Fig. 7.4.39). No satisfactory model has yet been found to explain the sizable differences.

Nuclear components. Due to the high depletion of noble gases relative to other less volatile elements, nuclear processes can contribute detectable amounts of noble gases. Most important are radioactive decay or spontaneous fission of long-lived isotopes (^{4}He from U and Th; ^{40}Ar from ^{40}K; fission Kr and Xe from ^{238}U), spontaneous fission or decay of extinct isotopes (^{129}Xe from ^{129}I; fission Kr and Xe from ^{244}Pu), and nuclear reactions such as $^{18}O(\alpha,n)^{21}Ne$ or induced fission. In extraterrestrial material (meteorites, lunar samples, interplanetary dust), spallation reactions induced by galactic or solar cosmic rays are also an important source for noble gases. Spallation reactions may have contributed a few percent of the terrestrial ^{21}Ne (Wetherill, 1954; for a recent review see Bernatowicz and Podosek, 1978).

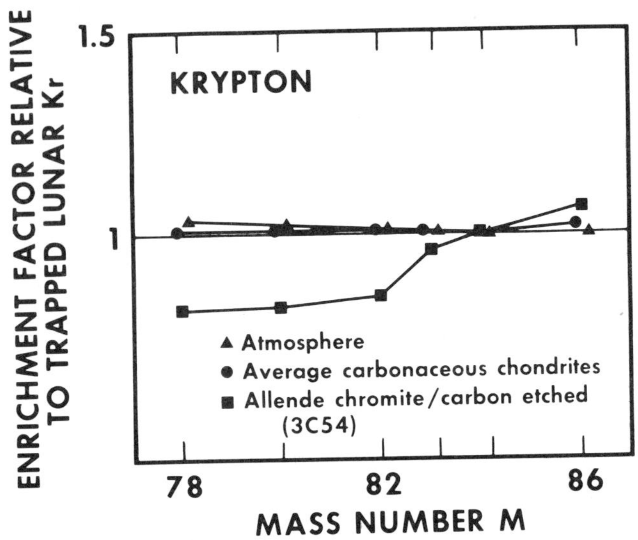

7.4.38 Isotopic composition of Kr in different solar system reservoirs. The data are normalized to the isotopic composition of trapped solar wind Kr in lunar fines, sample 12001 (BEOC 12, Eberhardt *et al.*, 1972): enrichment factor = $(^{M}Kr/^{84}Kr)_{SAMPLE} / (^{M}Kr/^{84}Kr)_{BEOC\ 12}$. The observed differences in the isotopic composition of Kr are much smaller than for Xe (cf. Fig. 7.4.39). The Allende chromite/carbon etched data (sample 3C54 in the figure; see Lewis *et al.*, 1977) represent Kr associated with the highly "anomalous" Xe (Xe-X, Fig. 7.4.39). Average composition of trapped Kr in carbonaceous chondrites (AVCC) from Eugster *et al.* (1967).

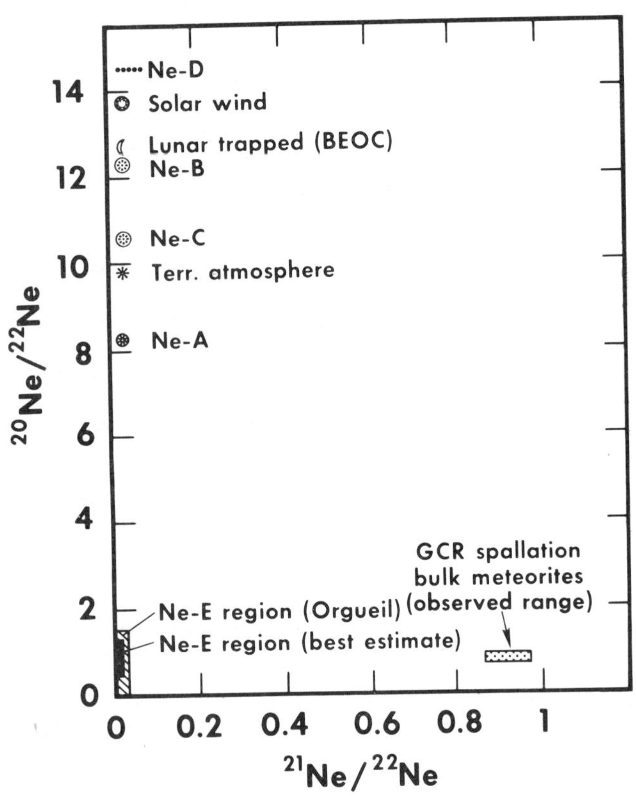

7.4.37 This figure shows the large range of neon isotopic compositions in different noble gas reservoirs within the solar system. Ne-A to Ne-E are trapped Ne components in meteorites. BEOC = Bern Oberflächen Correliert (gas emplanted in mineral grain surfaces by the solar wind); GCR = Gallactic Cosmic Ray. Data from Black (1972), Eberhardt *et al.* (1965), Eberhardt (1978), Geiss *et al.* (1972), Nyquist *et al.* (1973), and Pepin (1967).

Factors affecting noble gas concentrations

The noble gas abundances and isotopic compositions in a rock might be quite different from those in its parent magma. Factors controlling the noble gas abundance patterns of deep-sea basalts have recently been discussed by Dymond and Hogan (1978). Some of these factors are briefly reviewed here.

Table 7.4.10 Characteristics of some different types of trapped noble gases in meteorites.†

Type	Relative Elemental Abundances	Characteristic Isotopic Abundances	
Solar	Solar See Fig. 7.4.40	$^4He/^3He$	~2500
		$^{20}Ne/^{22}Ne$	~12.5
		$^{21}Ne/^{22}Ne$	~0.034
		$^{36}Ar/^{38}Ar$	~terrestrial
		Kr, Xe	(see Figs. 7.4.38 and 7.4.39)
Planetary	Similar to terrestrial atmosphere	$^4He/^3He$	~7000
		$^{20}Ne/^{22}Ne$	~8.2
		$^{21}Ne/^{22}Ne$	~0.025
		$^{36}Ar/^{38}Ar$	~terrestrial
		Kr, Xe	(see Figs. 7.4.38 and 7.4.39)
Ne-E	May be pure ^{22}Ne	$^{20}Ne/^{22}Ne$	<1.3
		$^{21}Ne/^{22}Ne$	<0.015
Xe-X			(see Fig. 7.4.39)

† Data from Pepin (1967), Black (1972), Reynolds *et al.* (1978), Eberhardt (1978).

Exchange with atmospheric gases. During crystallization and subsequent cooling, a rock might be exposed to the atmosphere or—for deep-sea basalts—to atmospheric gases dissolved in seawater. If the cooling is relatively slow, exchange and equilibration will occur. In deep-sea basalts the Ar concentration and isotopic composition change rapidly as a function of the distance from the pillow rim, with a typical scale length of the order of a few millimeters to centimeters (Dymond and Hogan, 1978). Probably only the glassy margins are affected relatively little by exchange, whereas holocrystalline basalts equilibrate completely with seawater gas (Dymond and Hogan, 1978). Hence, sample selection is very critical. For most data published so far (generally 0–1 cm depth), this may not have been done with sufficient care, seriously hampering the interpretation of the data. In other basalts the influence of exchange with atmospheric gases has not yet been studied.

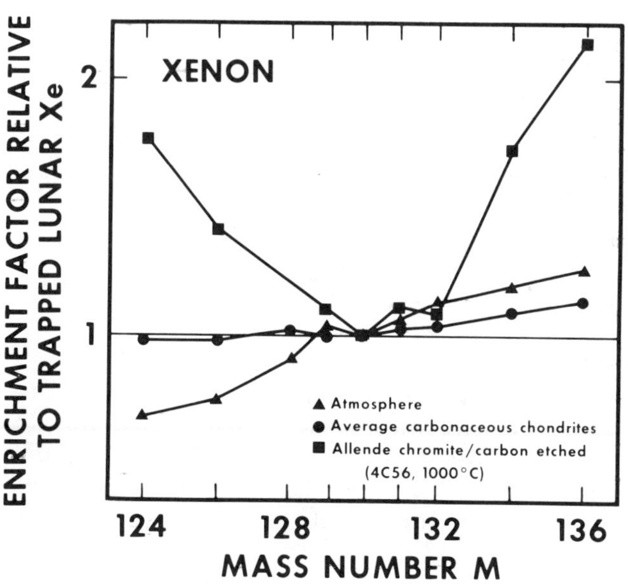

7.4.39 Isotopic composition of Xe in different solar system reservoirs. The data are normalized to the isotopic composition of trapped solar wind Xe in lunar fines, sample 12001 (BEOC 12, Eberhardt *et al.*, 1972): enrichment factor $= (^{M}Xe/^{130}Xe)_{SAMPLE} / (^{M}Xe/^{130}Xe)_{BEOC\ 12}$. The "Allende chromite/ carbon etched" composition (sample 4C56 in the figure; see Lewis *et al.*, 1977) is representative of the component designated Xe-X by Manuel *et al.* (1972) or "anomalous" by Lewis *et al.* (1975). Different interpretations for the origin of this component have been proposed by these authors (see Reynolds *et al.*, 1978 and references given therein). Average trapped Xe composition in carbonaceous chondrites (AVCC) from Eugster *et al.* (1967).

Diffusion in the magma source and during transit through the crust. The influence of diffusion on noble gas concentrations is difficult to assess since diffusion lengths, times, and constants are not well known. Isotopic compositions will be affected relatively little by diffusion, unless the diffusion occurs between two gas reservoirs containing noble gases of different isotopic compositions (e.g., between crustal material and shallow magma chambers).

In the mantle, residence times of magma melts may be of the order of 10^6 yr or more. For quartz, the typical diffusion lengths $\pi\sqrt{Dt}$ for a time period of 10^6 yr at a temperature of 1100° C are of the order of kilometers for He, 10 m for Ar, and less for Kr and Xe (diffusion constants from Perkins and Begeal, 1971). On the centimeter to meter scale equilibration between melt and solid will occur. On a larger scale the light noble gases will equilibrate preferentially to the heavier ones, and some fractionation of the noble gas composition in the basaltic liquid might result.

For a time scale of 20 days the diffusion lengths for He and Ar are of the order of 30 cm and 3 cm, respectively. Thus, extensive diffusion of noble gases in shallower magma chambers or from crustal material during eruptions is not likely.

Partial melting and crystallization. Partition coefficients of noble gases have not yet been determined; however, we may assume that the noble gases are probably concentrated in minerals with relatively open structures and low melting temperatures. Initial partial melts would then be more gas-rich than extensive partial melts. Also, fractional crystallization within a magma chamber would lead to an increase of the noble gas concentration in the residual liquid.

Partial crystallization should not change the isotopic composition except possibly for a very small mass-dependent fractionation. Partial melting could lead to large isotope effects if phases rich in radiogenic ^{40}Ar or ^{4}He (i.e., rich in K or U,Th) or rich in noble gases produced by nuclear reactions were involved.

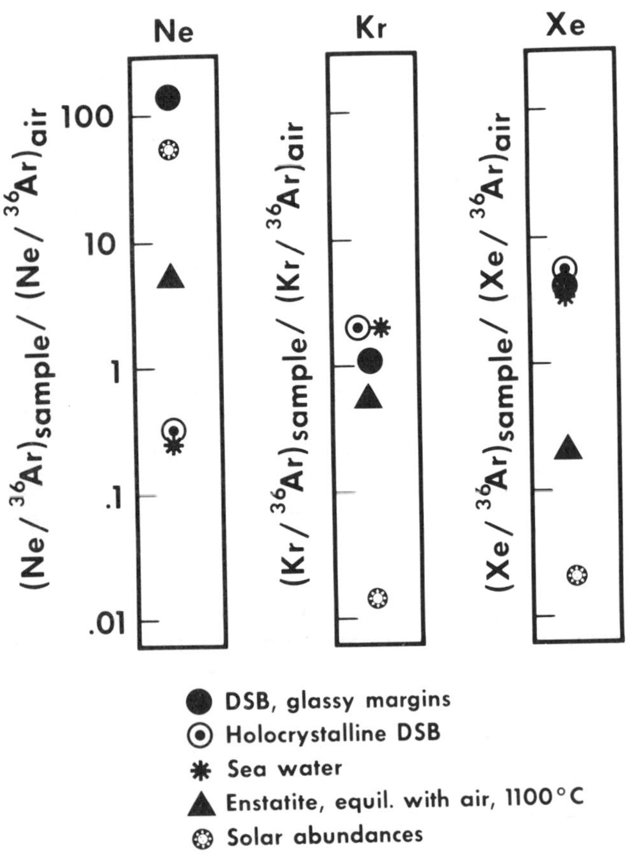

7.4.40 Abundance of noble gases relative to ^{36}Ar, normalized to the atmospheric ratios. Data for deep-sea basalts (DSB) from Dymond and Hogan (1973, 1978) and Fisher (1974, 1975). Solar abundances from Cameron (1973). Concentrations in enstatite in equilibrium with atmosphere at 1100° C calculated from data of Kirsten (1968).

Other processes. Bubble formation during eruption could remove large amounts of noble gases. No isotopic fractionation would occur and there probably would be little change in the elemental abundance ratios. Low temperature alterations caused by sea- or groundwater can lead to hydrated phases such as zeolites, which readily absorb large amounts of ambient noble gases, especially the heavy gases (Fanale and Cannon, 1971). The contamination of a sample with small amounts of such alteration products can then strongly influence the observed noble gas pattern.

Elemental abundance patterns

Only a limited number of deep-sea basalts which have not extensively exchanged with the atmosphere (using $^{40}Ar/^{36}Ar > 500$ as the criterion) have been analyzed so far (all noble gases in 4 samples; Ar, Kr and Xe in 7 additional samples). The observed abundance patterns in individual samples are reasonably similar to the average patterns shown in Fig. 7.4.40, and noble gas concentrations are of the same order of magnitude. For the samples used for Fig. 7.4.40, ^{36}Ar concentrations range from 10^{-10} to 8×10^{-10} cm^3 STP/g. Holocrystalline deep-sea basalts that have clearly exchanged with noble gases in seawater (i.e., with atmospheric $^{40}Ar/^{36}Ar \sim 295$) have at least an order of magnitude higher ^{36}Ar concentrations, and elemental abundance patterns virtually identical with that of noble gases in seawater (see Fig. 7.4.40).

The elemental abundance pattern in the glassy margins of deep-sea basalts is clearly different from the pattern in enstatite equilibrated with atmospheric noble gases (see Fig. 7.4.40). In the deep-sea basalts, Ne and Xe are 20–30 times more abundant, relative to Ar, than in the enstatite. The Xe enrichment could be due to exchange with seawater or to the presence of alteration products (see holocrystalline basalts), but the high Ne abundance seems to imply a nonatmospheric component. However, it must be pointed out that Craig and Lupton (1976) were not able to confirm the high Ne abundance observed by Dymond and Hogan (1973) in two of their four measured basalts. This discrepancy could reflect sample inhomogeneities or atmospheric Ne contamination. Hence, the strong enrichment of Ne in the glassy margins of deep-sea basalts is not yet fully established, and no conclusions on the presence of primordial gases can yet be drawn from the observed elemental abundance patterns.

The number of deep-sea basalts analyzed so far is too small to detect any regional differences, for example, between Pacific and Atlantic, in the abundance pattern. For a more detailed discussion of noble gas abundance patterns in terrestrial materials see Alexander and Ozima (1978), Ozima and Alexander (1976), Fisher (1975), Dymond and Hogan (1973), and references cited in these publications.

Isotopes

As discussed above, it is difficult to recognize non-atmospheric noble gases from the elemental abundance pattern alone. Isotopic ratios are far more powerful and less ambiguous tracers for the presence of different types of noble gas components. An excess of radiogenic isotopes (^{4}He, ^{40}Ar, etc.) proves that the sample has not completely equilibrated with atmospheric gases. Due to their distinct isotopic composition, the presence of primordial gases of the solar or planetary type can be clearly distinguished. However, a primordial component with atmospheric isotopic abundances cannot be recognized. In the following sections the isotopic results obtained so far are briefly reviewed.

Helium. Helium enriched in ^{3}He has been discovered in volcanic gases (Mamyrin *et al.*, 1969), in deep-sea water (Clarke *et al.*, 1969), in ultrabasic xenoliths (Tolstikhin *et al.*, 1974), and in the glassy rims of submarine basalts (Lupton and Craig, 1975). Tolstikhin *et al.* (1974) discussed possible nuclear reactions that could produce ^{3}He and have concluded that these sources cannot account for the observed ^{3}He excess. The high ^{3}He abundances must be due to the presence of primordial gases in the Earth's interior. The primordial ^{3}He/^{4}He ratio was $\sim 1.4 \times 10^{-4}$ (Reynolds *et al.*, 1978), much higher than the (present) atmospheric and crustal ratios of 1.4×10^{-6} and $\sim 10^{-7}$, respectively (Aldrich and Nier, 1946; Mamyrin *et al.*, 1970).

In ultrabasic xenoliths from several locations, Tolstikhin *et al.* (1974) measured ^{3}He/^{4}He ratios as high as 1.4×10^{-5} (total extraction) and 5×10^{-5} (stepwise heating results). In two Hualalai dunites, rich in CO_2 inclusions, Kaneoka and Takaoka (1978) observed ^{3}He/^{4}He ratios of $\sim 1.4 \times 10^{-5}$, similar to the submarine basalt results. Higher ratios were found in an olivine phenocryst from Kilauea (^{3}He/^{4}He $\sim 2.5 \times 10^{-5}$; Kaneoka and Takaoka, 1978); in an upper mantle-derived amphibole (kaersutite) from Kakanui, New Zealand (^{3}He/^{4}He $\sim 5 \times 10^{-5}$; Saito *et al.*, 1978); and in josephinite (^{3}He/^{4}He as high as $\sim 5 \times 10^{-4}$; Bochsler *et al.*, 1978; Downing *et al.*, 1977). Tholeiitic glasses from the Atlantic and Pacific have a rather uniform ^{3}He/^{4}He ratio of $\sim 1.4 \times 10^{-5}$ (Craig and Lupton, 1976). The He concentration is, however, very variable: samples from the Atlantic contain up to a factor of 50 more He than those from the Pacific (see Fig. 7.4.41).

Several conclusions can be drawn from these observations, despite the limited number of samples investigated so far:

(a) Primordial ^{3}He is present, and hence the Earth is not completely outgassed.

(b) The results from the submarine basalts indicate a uniform ^{3}He/^{4}He ratio in the mantle, and hence a well-mixed region of the mantle. Both Atlantic and Pacific samples show very similar ^{3}He/^{4}He values, and this well-mixed region may be of global extent. This conclusion is, however, not supported by the results from other presumably mantle-derived material. Further measurements are necessary.

(c) The variable He concentration in the tholeiitic glasses is probably due to gas loss during magma transport (e.g., by bubble formation). Hence, the different He concentrations do not necessarily imply large variations in the primordial He abundance in different magma regions.

Neon. For two glassy tholeiites from the East Pacific Rise, Craig and Lupton (1976) reported ^{20}Ne/^{22}Ne

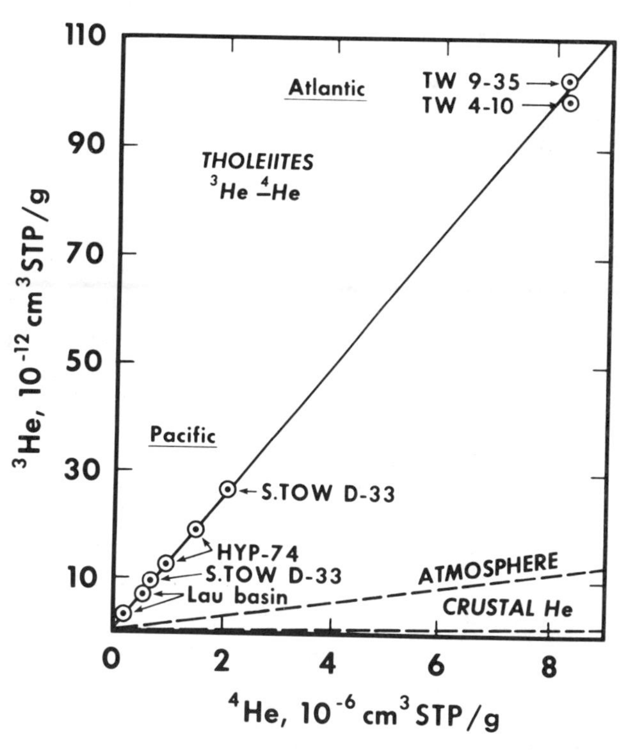

7.4.41 ^{3}He and ^{4}He concentrations in tholeiite glasses with high ^{3}He/^{4}He ratios. Figure from Craig and Lupton (1976).

values approximately 2% higher than the terrestrial ratio of 9.8. They concluded that these two deep-sea basalts contain some solar type primordial Ne ($^{20}Ne/^{22}Ne \cong 12.5$). However, further analyses by Bochsler and Stettler (1978) on an aliquot of one of these tholeiites (S. Tow D-33) did not confirm this ^{20}Ne excess.

The nonoceanic silicates which show low $^4He/^3He$ ratios (Hualalai dunites, Kilauea olivine, Kakanui amphibole) have $^{20}Ne/^{22}Ne$ ratios identical to air (Kaneoka and Takaoka, 1978; Saito *et al.*, 1978). Rison and Kyser (1977) reported ±10% variations in the $^{20}Ne/^{22}Ne$ ratio in Hawaiian xenoliths and basalts, presumably at least partially due to the addition of primordial Ne. The $^{20}Ne/^{22}Ne$ ratio in josephinite is slightly different from the atmospheric value, but not necessarily due to an addition of primordial Ne (Bochsler *et al.*, 1978; Downing *et al.*, 1977).

Neon in volcanic gases collected on the Caldera edge of Kilauea shows a 5% ^{20}Ne excess and an atmospheric $^{21}Ne/^{22}Ne$ ratio (Craig and Lupton, 1976). Phinney *et al.* (1978) measured $^{20}Ne/^{22}Ne = 11.6$ in the Harding County, New Mexico, CO_2 well. This value is midway between the atmospheric and solar values, but the 2σ error overlaps with either of these compositions. Excess ^{21}Ne has been observed in helium-bearing natural gases of crustal origin (Emerson *et al.*, 1966), in crustal rocks (Wetherill, 1954), and recently also in mantle-derived samples (Saito *et al.*, 1978; Rison and Kyser, 1977; Bochsler *et al.*, 1978). This ^{21}Ne excess is not of primordial origin but is due to the $^{18}O(\alpha, n)\ ^{21}Ne$ reaction (Wetherill, 1954).

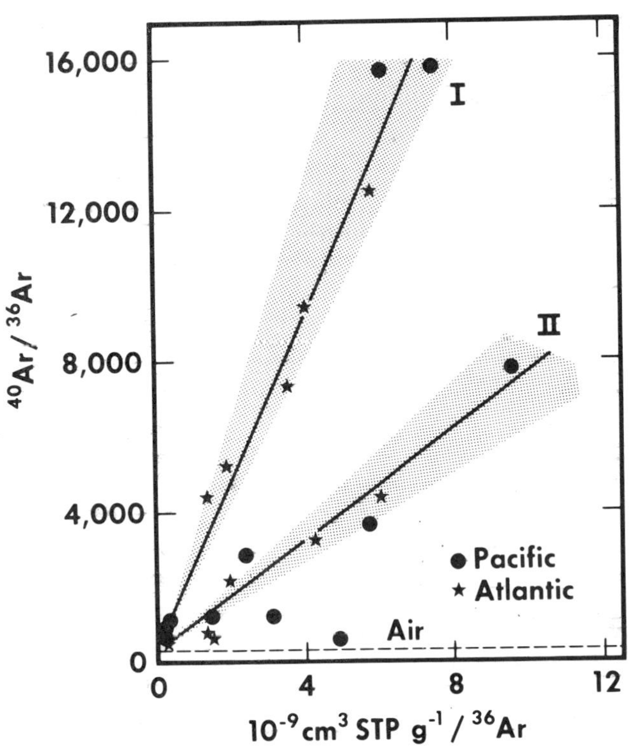

7.4.42 Ar isotopic compositions in glassy margins of deep-sea basalts plotted against reciprocal ^{36}Ar concentrations. Figure adapted from Dymond and Hogan (1978); data from Fisher (1975) and Dymond and Hogan (1973, 1978).

Argon. The $^{40}Ar/^{36}Ar$ ratio in the glassy margins of deep-sea basalts varies by more than a factor of 30. However, as shown by Dymond and Hogan (1978), most of the data points define two remarkably linear arrays in an $^{40}Ar/^{36}Ar$ vs. $1/^{36}Ar$ plot (Fig. 7.4.42). The lines can be explained by mixing of mantle-derived Ar with varying amounts of atmospheric Ar. For line I, the mantle component would contain approximately 2×10^{-6} cm^3 STP $^{40}Ar/g$ with $^{40}Ar/^{36}Ar \geq 16{,}000$. For line II, the corresponding values would be $\sim 10^{-6}$ cm^3 STP $^{40}Ar/g$ and $^{40}Ar/^{36}Ar \geq 8000$. However, partial degassing could move points from line I to line II, and it is not necessary to assume two different types of mantle gas. There seems to be no systematic distinction between Atlantic and Pacific samples in Fig. 7.4.42, implying very extended, well-mixed regions of the mantle with fairly uniform Ar concentration and isotopic composition. This would be in accordance with the $^3He/^4He$ results, but additional data are certainly required before any definite conclusions can be drawn.

Krypton and xenon. Kr and Xe isotopes have been measured in very few silicate rocks of possible mantle origin, none of which were deep-sea basalts. The isotopic compositions are essentially identical to the atmospheric values, except for a ^{129}Xe excess (Hennecke and Manuel, 1975; Kaneoka and Takaoka, 1978; Saito *et al.*, 1978). Excess ^{129}Xe from the decay of now extinct ^{129}I (half life 17×10^6 yr) has also been observed in CO_2 well gas from Harding County, New Mexico (Boulos and Manuel, 1971; Phinney *et al.*, 1978; Butler *et al.*, 1963; Wasserburg and Mazor, 1965) and in josephinite (Downing *et al.*, 1977). Two conclusions can be drawn from these observations:

(1) At least some of the Xe trapped in rocks or present in well gases must have resided in the Earth's interior essentially since the formation of the Earth.

(2) If atmospheric Xe is due to the degassing of the Earth, then this degassing probably must have occurred very early in the Earth's history, before all the ^{129}I had decayed to ^{129}Xe.

The very limited data available suggests that excess ^{129}Xe is associated with excess ^3He (Kaneoka and Takaoka, 1978; Phinney *et al.*, 1978; Downing *et al.*, 1977) and, in the case of the CO_2 well gas, also with solar type Ne (Phinney *et al.*, 1978).

In the terrestrial atmosphere Xe seems to be depleted in comparison to Kr and Ar. Canalas *et al.* (1968) and Fanale and Cannon (1971) have argued that this is due to physical adsorption on sediments. Phinney (1972) in fact observed strong Xe enrichments in shale and chert. However, the isotopic composition of the adsorbed Xe was essentially atmospheric, and thus the unique isotopic composition of terrestrial atmospheric Xe (see Fig. 7.4.39) is not a result of this adsorption. No clear evidence for the presence of solar-, AVCC-, or X-type Xe has yet been found in terrestrial rocks. Hennecke and Manuel (1975b) and Phinney *et al.* (1978) observed the possible presence of a small solar-type Xe component in a lava rock from Mount Capulin, New Mexico, and in the Harding County CO_2 well gas. However, the evidence is very marginal, and the authors themselves are reluctant to claim positive identification of a primordial solar-type Xe component in the Earth. The isotopic anomalies reported by Downing *et al.* (1977) in josephinite correspond to a mass dependent fractionation of 1% per mass unit and do not necessarily imply an exotic, nonatmospheric Xe component.

Degassing history of the Earth

Many attempts have been made to deduce the outgassing history of the Earth from noble gas isotopes, especially from the $^{40}Ar/^{36}Ar$ ratio, radiogenic ^{4}He and ^{129}Xe and fissiogenic ^{136}Xe. Such degassing models are discussed in several papers in Alexander and Ozima (1978) and in references cited in these papers. Some of the key questions are:

To what extent is the Earth degassed?
Was the degassing rate uniform with time?
Did an early catastrophic degassing occur?
Is the degassing within the mantle uniform?

At present, no unambiguous answers to these questions have been obtained. However, with a better understanding of the behavior of noble gases, and with more comprehensive and precise measurements, it should eventually be feasible to answer them.

Conclusions

Interpretation of noble gas elemental abundance patterns has not yet provided clear evidence for the presence of primordial noble gases in the Earth's mantle. In addition to experimental difficulties, too many factors not yet investigated could potentially have influenced the measured noble gas abundances. Isotopic studies have been more successful, and primordial ^{3}He and radiogenic ^{4}He, ^{40}Ar and ^{129}Xe have been unambiguously detected in mantle-derived material. Whether solar-type Ne and nonatmospheric Xe are present is still an open question.

The data so far available have clearly established the usefulness of ^{3}He and other noble gas isotopes as tracers for the origin and development of magmatic bodies (Tolstikhin, 1978; Craig *et al.*, 1978). No definite conclusions on the evolution of the mantle and the history of basaltic magmas can yet be drawn however. Extended laboratory studies and far more detailed measurements of trapped terrestrial noble gases in a large number of samples, including mineral separates, will be needed. In addition, the noble gas studies should be closely correlated with petrological and other geochemical investigations. The consortium-type approach, which has been very successful for studying lunar breccias and meteorites, would probably be advantageous. Since terrestrial noble gases are at present being studied vigorously in several laboratories, we may expect substantial progress and new developments during the next decade.

AUTHORSHIP—Section 7.1: G. W. Wetherill and C. J. Allègre. Section 7.2: W. R. Van Schmus. Section 7.3: F. Tera, P. Eberhardt, and G. W. Wetherill. Section 7.4.1: W. R. Van Schmus and C. Brooks. Section 7.4.2: S. R. Hart and C. Brooks. Section 7.4.3: V. R. Murthy. Section 7.4.4: P. Eberhardt.

7.5 REFERENCES

Åberg G. (1978) Precambrian geochronology of south-eastern Sweden. *Geologiska Föreningens i Stockholm Förhandlingar* **100**, 125–154.

Adler I., Trombka J., Gerard J., Lowman P., Schmaldebeck R., Blodgett H., Eller E., Yim L., Lamonthe R., Gorenstein P. and Bjorkholm P. (1972a) Apollo 15 geochemical x-ray fluorescence experiment: preliminary report. *Science* **175**, 436–440.

Adler I., Trombka J., Gerard J., Lowman P., Schmaldebeck R., Blodgett H., Eller E., Yim L., Lamonthe R., Osswald G., Gorenstein P., Bjorkholm P., Gursky H. and Harris B. (1972b) Apollo 16 geochemical x-ray fluorescence experiment: preliminary report. *Science* **177**, 256–259.

Ahrens L. H. (1955) The convergent lead ages of the oldest monazites and uraninites (Rhodesia, Manitoba, Madagascar and Transvaal). *Geochim. Cosmochim. Acta* **7**, 294–300.

Albarede F. and Bottinga Y. (1972) Kinetic disequilibrium in trace element partitioning between phenocrysts and host lava. *Geochim. Cosmochim. Acta* **36**, 141–146.

Albee A. L. and Gancarz A. J. (1977) Petrogenesis of lunar rocks: Rb-Sr constraints and lack of H_2O. In *The Soviet-American Conference on Cosmochemistry of the Moon and Planets*, p. 81–90. NASA SP-370. Washington, D. C.

Aldrich L. T. and Nier A. O. (1946) The abundance of He^3 in atmospheric and well helium. *Phys. Rev.* **70**, 983–984.

Alexander E. C., Jr. and Davis P. K. (1974) ^{40}Ar-^{39}Ar ages and trace element contents of Apollo 14 breccias: an interlaboratory cross-calibration of ^{40}Ar-^{39}Ar standards. *Geochim. Cosmochim. Acta* **39**, 928–981.

Alexander E. C., Jr. and Ozima M., eds. (1978) *Terrestrial Rare Gases*. Advances in Earth and Planetary Sciences, vol. 3. Japan Sci. Soc. Press, Tokyo. 229 pp.

Alexander E. C., Jr., Davis P. K. and Reynolds J. H. (1972) Rare gas analyses on neutron irradiated Apollo 12 samples. *Proc. Lunar Sci. Conf. 3rd*, p. 1787–1795.

Allègre C. J. (1968) ^{230}Th dating of volcanic rocks: a comment. *Earth Planet. Sci. Lett.* **5**, 209–210.

Allègre C. J. and Bottinga Y. (1977) The rising deep crust: an alternative to the hot spot model. *EOS (Trans. Amer. Geophys. Union)* **58**, 535–536.

Allègre C. J. and Condomines M. (1976) Fine chronology of volcanic processes using ^{238}U-^{230}Th systematics. *Earth Planet. Sci. Lett.* **28**, 395–406.

Allègre C. J. and Minster J. F. (1978) Quantitative models of trace element behavior in magmatic processes. *Earth Planet. Sci. Lett.* **38**, 1–25.

Allègre C. J., Albarede F., Grünenfelder M. and Köppel V. (1974) $^{238}U/^{206}Pb$-$^{235}U/^{207}Pb$ - $^{232}Th/^{208}Pb$ zircon geochronology in alpine and non-alpine environment. *Contrib. Mineral. Petrol.* **43**, 163–194.

Allègre C. J., Ben Othman D., Polve M. and Richard P. (1979) The Nd-Sr isotopic correlation in mantle materials and geodynamic consequences. *Phys. Earth. Planet. Inter.* **19**, 293–306.

Allègre C. J., Birck J. L., Fourcade S. and Semet M P. (1975) Rubidium-87/strontium-87 age of Juvinas basaltic achondrite and early igneous activity in the solar system. *Science* **187**, 436–438.

Allègre C. J., Brevart O., Dupre B. and Minster J. F. (1981) Isotopic and chemical effects produced in a continuously differentiating convecting earth mantle. *Proc. Roy. Soc. Lond.*, in press.

Allègre C. J., Treuil M., Minster J. F., Minster J. B. and Albarede F. (1977) Systematic use of trace elements in igneous processes. Part I. Fractional crystallization process in volcanic suites. *Contrib. Mineral. Petrol.* **60**, 57–76.

Allsopp H. L., Nicholaysen L. O. and Hahn-Weinheimer P. (1969) Rb/K ratios and Sr- isotopic compositions of minerals in eclogitic and peridotitic rocks. *Earth Planet. Sci. Lett.* **5**, 231–244.

Allsopp H. L., Roberts H. R., Schreiner G. D. L. and Hunter D. R. (1962) Rb-Sr age measurements on various Swaziland granites. *J. Geophys. Res.* **67**, 5307–5313.

Anders E. (1963) Meteorite ages. In *The Moon, Meteorites, and Comets (The Solar System IV)*, (B. Middlehurst and G. Kuiper, eds.), p. 402–495. Univ. Chicago Press, Chicago.

Anderson T., Clayton R. N. and Mayeda T. K. (1971) Oxygen isotope thermometry of mafic igneous rocks. *J. Geol.* **79**, 715–729.

Armstrong R. L. (1971) Glacial erosion and the variable isotopic composition of strontium in sea water. *Nature* **230**, 132–133.

Armstrong R. L. and Cooper J. A. (1971) Lead isotopes in island arcs. *Bull. Volc.* **35**, 27–63.

Armstrong R. L., Taubeneck W. H. and Hales P. O. (1977) Rb-Sr and geochronometry of Mesozoic granitic rocks and their Sr isotopic composition, Oregon, Washington and Idaho. *Geol. Soc. Amer. Bull.* **88**, 379–411.

Arriens P. A. (1971) The archean geochronology of Australia. *Geol. Soc. Aust., Spec. Publ.* **3**, 11–23.

Arth J. G., Arndt N. T. and Naldrett A. J. (1977) Genesis of Archean komatiites from Munro township, Ontario: trace-element evidence. *Geology* **5**, 590–594.

Ave'Lallemant H. G. (1967) Structural and petrofabric analysis of an alpine peridotite: the lherzolite of the French Pyrenees. *Leidse Geol. Mededel.* **42**, 1–57.

Ave'Lallemant H. G. and Carter N. L. (1970) Syntectonic recrystallization of olivine and modes of flow in the upper mantle. *Geol. Soc. Amer. Bull.* **81**, 2203–2220.

Baadsgaard H. (1973) U-Th-Pb dates on zircons from the early Precambrian Amitsoq gneisses Godthaab district, West Greenland. *Earth Planet. Sci. Lett.* **19**, 22–28.

Baragar W. R. A. (1969) The geochemistry of Coppermine River basalts. *Geol. Survey Can.*, Paper 69-44.

Baragar W. R. A. (1972) Coppermine River basalts, District of Mackenzie. In *Rubidium-strontium isochron age studies, Report 1* (R. K. Wanless and W. D. Loveridge, eds.), p. 21–24. Geol. Survey Can., Paper 72-23.

Baragar W. R. A. (1977) Volcanism of the stable crust. In *Volcanic Regimes in Canada* (W. R. A. Barager, L. C. Coleman and J. M. Hall, eds.), p. 377–405. Geol. Assoc. Can. Spec. Paper No. 16.

Baragar W. R. A., Plant A., Pringle G. J. and Schau M. (1977) Petrology and alteration of selected units of Mid-Atlantic Ridge basalts sampled from sites 332 and 334, DSDP. *Can. J. Earth Sci.* **14**, 837–874.

Barker D. S. and Long L. E. (1969) Feldspathoidal syenite in a quartz diabase sill, Brookville, New Jersey. *J. Petrol.* **10**, 202–221.

Barrett D. R. (1975) The genesis of kimberlites and associated rocks: strontium isotopic evidence. *Phys. Chem. Earth* **9**, 637–653.

Basu A. R. and Murthy V. R. (1977a) Ancient lithospheric lherzolite xenolith in alkali basalt from Baja California. *Earth Planet. Sci. Lett.* **35**, 246.

Basu A. R. and Murthy V. R. (1977b) Kaersutites, suboceanic low-velocity zone, and the origin of mid-oceanic ridge basalts. *Geology* **5**, 365–368.

Basu A. R. and Murthy V. R. (1977c) Trace elements and Sr-isotope geochemistry of the constituent minerals in ultramafic xenoliths from San Quentin, Baja California. In *Extended Abstracts—The 2nd International Kimberlite Conference 1977, Bishop's Lodge, Sante Fe, Oct. 3–7*. Amer. Geophys. Union., Washington, D.C.

Basu A. R. and Murthy V. R. (1977d) Significance of amphibole and phlogopite in basalt genesis - trace elemental Sr-isotopic evidence. In *Extended Abstracts—International Conference on Experimental and Trace Element Geochemistry*, p. 2–4. Sedona, Arizona.

Basu A. R. and Tatsumoto M. (1978) Nd isotopes in mantle derived rocks and minerals and the evolution of the Earth's mantle. *U.S. Geol. Surv. Open-File Rept. 78-701*, p. 23–30.

Beaty D. W. and Albee A. L. (1978) A textural, modal and chemical classification of the Apollo 11 low-K basalts (abstract). In *Lunar and Planetary Science IX*, p. 61. Lunar and Planetary Institute, Houston.

Becker R. H. and Clayton R. N. (1977) Nitrogen isotopes in igneous rocks. *EOS (Trans. Amer. Geophys. Union)* **58**, 536.

Beckinsale R. D. and Gale N. H. (1969) A reappraisal of the decay constants and branching ratio of ^{40}K. *Earth Planet. Sci. Lett.* **6**, 289–294.

Bernatowicz T. J. and Podosek F. A. (1978) Nuclear components in the atmosphere. In *Terrestrial Rare Gases* (E. C. Alexander, Jr. and M. Ozima, eds.), p. 99–136. Advances in Earth and Planetary Sciences, vol. 3. Japan Sci. Soc. Press, Tokyo.

Bezzi A. and Picardo G. B. (1971) Structural features of the Ligurian ophiolites: petrologic evidence for the oceanic floor of the N. Appennines, Italy: a contribution to the problem of the alpine-type gabbro-peridotite association. *Mem. Soc. Geol. Ital.* **10**, 53–63.

Bickford M. E. and Lewis R. D. (1979) U-Pb geochronology of exposed basement rocks in Oklahoma. *Geol. Soc. Amer. Bull.* **90**, 540–544.

Bickford M. E. and Mose D. G. (1975) Geochronology of Precambrian rocks in the St. Francois Mountains, southeast Missouri. *Geol. Soc. Amer. Spec. Paper 165*. 48 pp.

Bickford M. E., Thomas J. J. and Nelson G. R. (1975) Precambrian volcanic terrane of the mid-continent: distribution in the Kansas basement (abstract). *Abstracts with Program*, Geol. Soc. Amer. **7**, 998.

Bickle M. J., Hawkesworth C. J., Martin A., Nisbet E. G. and O'Nions R. K. (1976) Mantle composition derived from the chemistry of ultra-mafic lavas. *Nature* **263**, 577–580.

Birck J. L. and Allègre C. J. (1978a) Chronology and chemical history of the parent body of basaltic achondrites studied by the ^{87}Rb-^{87}Sr method. *Earth Planet. Sci. Lett.* **39**, 37–51.

Birck J. L. and Allègre C. J. (1978b) ^{87}Rb-^{87}Sr study of some diogenites and the Binda howardite (abstract). *Meteoritics* **13**, 394.

Black D. C. (1972) On the origins of trapped helium, neon and argon isotopic variations in meteorites—I. Gas-rich meteorites, lunar soil and breccia. *Geochim. Cosmochim. Acta* **36**, 347–375.

Black D. C. and Pepin R. O. (1969) Trapped neon in meteorites—II. *Earth Planet. Sci. Lett.* **6**, 395–405.

Bochsler P., Stettler A., Bird J. M. and Weathers M. S. (1978) Excess He^3 and Ne^{21} in josephinite. *Earth Planet. Sci. Lett.* **39**, 67–74.

Bogard D. D., Hirsch W. C. and Husain L. (1976) ^{40}Ar-^{39}Ar dating of achondritic meteorites (abstract). *Meteoritics* **11**, 251–252.

Bogard D. D., Husain L. and Nyquist L. E. (1979) ^{40}Ar-^{39}Ar age of the Shergotty achondrite and implications for its post-shock thermal history. *Geochim. Cosmochim. Acta* **43**, 1047–1055.

Bogard D. D., Nyquist L. E., Bansal B. M., Wiesmann H. and Shih C.-Y. (1975) 76535: an old lunar rock. *Earth Planet. Sci. Lett.* **26**, 69–80.

Bonatti E. (1968) Ultramafic rocks from the mid-Atlantic Ridge. *Nature* **219**, 363.

Bonatti E., Honnerez J. and Ferrara G. (1970) Equatorial Mid-Atlantic Ridge petrologic and Sr-isotope evidence for an Alpine-type rock assemblage. *Earth Planet. Sci. Lett.* **9**, 247–256.

Bonini W. E., Loomis T. P. and Robertson J. D. (1973) Gravity anomalies, ultramafic intrusions and the tectonics of the region around the Strait of Gibraltar. *J. Geophys. Res.* **78**, 1372–1382.

Bottinga Y. (1969a) Carbon isotope fractionation between graphite, diamond and carbon dioxide. *Earth Planet. Sci. Lett.* **5**, 305–307.

Bottinga Y. (1969b) Calculated fractionation factors for carbon and hydrogen isotope exchange in the system calcite-carbon dioxide-graphite-urethane-hydrogen-water vapor. *Geochim. Cosmochim. Acta* **33**, 49–64.

Bottinga Y. and Javoy M. (1975) Oxygen isotope partitioning among the minerals in igneous and metamorphic rocks. *Rev. Geophys. Space Phys.* **13**, 401–418.

Bottino M. L., Fullagar P. D., Schnetzler C. C. and Philpotts J. A. (1971) Sr isotopic measurements in Apollo 12 samples. *Proc. Lunar Sci. Conf. 2nd*, p. 1487–1491.

Boulos M. S. and Manuel O. K. (1971) The xenon record of extinct radioactivities in the Earth. *Science* **174**, 1334–1336.

Boyce J. M. and Dial A. L. (1973) Relative ages of some nearside mare units based on Apollo 17 metric photographs. *Apollo 17 Prelim. Sci. Rept.* , p. 29-26 to 29-28. NASA SP-330.

Boyce J. M., Dial A. L. and Soderblom L. A. (1974) Ages of the lunar nearside light plains and maria. *Proc. Lunar Sci. Conf. 5th*, p. 11–23.

Boyd F. R. (1973) A pyroxene geotherm. *Geochim. Cosmochim. Acta* **37**, 2533–2546.

Boyd F. R. and Nixon P. H. (1972) Ultra-mafic nodules from the Thaba Putsoa Kimberlite pipe. *Carnegie Inst. Wash., Yearb.* **71**, 362–373.

Boyd F. R. and Nixon P. H. (1975) Origins of ultramafic nodules from June Kimberlites of northern Lesotho and the Monastery Mine, South Africa. *Phys. Chem. Earth* **9**, 431–454.

Brass G. W. (1976) The variation of the marine $^{87}Sr/^{86}Sr$ ratio during Phanerozoic time: interpretation using a flux model. *Geochim. Cosmochim. Acta* **40**, 721–730.

Brass G. W. and Turekian K. K. (1977) Comment on "The strontium isotopic composition of seawater and seawater-oceanic crust interaction" by E. T. C. Spooner. *Earth Planet. Sci. Lett.* **34**, 165–166.

Brett R. (1973) The lunar crust: a product of heterogeneous accretion or differentiation of a homogeneous Moon. *Geochim. Cosmochim. Acta* **37**, 2697–2703.

Brett R. (1975) Thickness of some lunar mare basalt flows and ejecta blankets based on chemical kinetic data. *Geochim. Cosmochim. Acta* **39**, 1135–1141.

Bridgwater D. and Collerson K. D. (1976) The major petrological and geochemical characters of the 3,600 m.y. Uivak Gneisses from Labrador. *Contrib. Mineral. Petrol.* **54**, 43–59.

Bridgwater D., Collerson K. D. and Myers J. S. (1978) The development of the Archean gneiss complex of the North Atlantic region. In *Evolution of the Earth's Crust* (D. H. Tarling, ed.), p. 19–69. Academic Press, London.

Bridgwater D., Watson J. and Windley B. F. (1973) The Archean craton of the North Atlantic region. *Phil. Trans. Roy. Soc. Lond.* **A273**, 493–512.

Briqueu L. and Lancelot J. (1979) Rb-Sr systematics and crustal contamination models for calc-alkaline igneous rocks. *Earth Planet. Sci. Lett.* **43**, 385–396.

Brooks C. and Hart S. R. (1974) On the significance of komatiite. *Geology* **2**, 107–110.

Brooks C. and Hart S. R. (1978) Rb-Sr mantle isochrons and variations in the chemistry of Gondwanaland's lithosphere. *Nature* **271**, 220–223.

Brooks C. and James D. E. (1978) Strontium and oxygen isotopes in Mesozoic tholeiites and the case for heterogeneous mantle (abstract). *Abstracts with Program*, Geol. Soc. Amer. **10**, 372.

Brooks C., Hart S. R., Hofmann A. L. and James D. E. (1976a) Rb-Sr mantle isochrons from oceanic regions. *Earth Planet. Sci. Lett.* **32**, 51–61.

Brooks C., Hart S. R. and Wendt I. (1972) Realistic use of two-error regression treatment as applied to rubidium-strontium dating. *Rev. Geophys. Space Phys.* **10**, 551–577.

Brooks C., James D. E. and Hart S. R. (1976b) Ancient lithosphere: its role in young continental volcanism. *Science* **193**, 1086–1094.

Brown G. M., Emeleus C. H., Holland J. G., Peckett A. and Phillips R. J. (1972) Petrology, mineralogy and classification of Apollo 15 mare basalts. In *The Apollo 15 Lunar Samples* (J.W. Chamberlain and C. Watkins, eds.), p. 40–44. The Lunar Science Institute, Houston.

Brown G. M., Peckett A., Emeleus C. H., Phillips R. J. and Pinsent R. H. (1975) Petrology and mineralogy of Apollo 17 mare basalts. *Proc. Lunar Sci. Conf. 6th*, p. 1–13.

Brueckner H. K. (1974) Mantle Rb/Sr and $^{87}Sr/^{86}Sr$ ratios from clinopyroxenes from Norwegian garnet peridotites and pyroxenites. *Earth Planet. Sci. Lett.* **24**, 26–32.

Brueckner H. K. (1975) Contact and fracture ultramafic assemblages from Norway: Rb-Sr evidence for crustal contamination. *Contrib. Mineral. Petrol.* **49**, 39–48.

Bryan W. B. and Adams M. L. (1973) Some volcanic and structural features of Mare Serenitatis. *Apollo 17 Prelim. Sci. Rept.*, p. 30-9 to 30-12. NASA SP-330.

Buhler F., Axford W. I., Chivers H. J. A. and Marti K. (1976) Helium isotopes in an aurora. *J. Geophys. Res.* **81**, 111–115.

Burke K. C. A. and Dewey J. F. (1972) Orogeny in Africa. In *African Geology* (T. F. Dessauvagie and A. J. Whiteman, eds.), p. 583–608. Univ. Ibadan Press, Nigeria.

Burke K. C. A. and Dewey J. F. (1973) An outline of Precambrian plate development. In *Implications of Continental Drift to the Earth Sciences, vol. 2*, (D. H. Tarling and S. K. Runcorn, eds.), p. 1035–1045. Academic Press, London.

Burke K. C. A. and Whiteman A. (1973) Uplift, rifting and the breakup of Africa. In *Implications of Continental Drift to the Earth Sciences, vol. 2*, (D. H. Tarling and S. K. Runcorn, eds.), p. 735–755. Academic Press, London.

Burke K. C. A., Dewey J. F. and Kidd W. S. F. (1976a) Precambrian paleomagnetic results compatible with contemporary operation of the Wilson cycle. *Tectonophys.* **33**, 287–299.

Burke K. C. A., Dewey J. F. and Kidd W. S. F. (1976b) Dominance of horizontal movements, arc and microcontinental collisions during the later permobile regime. In *The Early History of the Earth* (B. F. Windley, ed.), p. 113–129. Wiley, London.

Burke K. C. A., Fox P. J. and Şengör A. M. C. (1978) Bouyant ocean floor and the evolution of the Caribbean. *J. Geophys. Res.* **83**, 3949–3954.

Burnett D. S., Drozd R. J., Morgan C. J. and Podosek F. A. (1975) Exposure histories of Bench Crater rocks. *Proc. Lunar Sci. Conf. 6th*, p. 2219–2240.

Burwash R. A. (1969) Comparative Precambrian geochronology of the North American, European, and Siberian shields. *Can. J. Earth Sci.* **6**, 357–365.

Burwell A. D. M. (1975) Rb-Sr isotope geochemistry of lherzolites and their constituent minerals from Victoria, Australia. *Earth Planet. Sci. Lett.* **28**, 69–78.

Butler W. A., Jeffery P. M., Reynolds J. H. and Wasserburg G. J. (1963) Isotopic variations in terrestrial xenon. *J. Geophys. Res.* **68**, 3283–3291.

Cadogan P. H. and Turner G. (1976) The chronology of the Apollo 17 Station 6 boulder. *Proc. Lunar Sci. Conf. 7th*, p. 2267–2285.

Cambray F. W. (1978) Plate tectonics as a model for the environment of deposition and deformation of the early Proterozoic (Precambrian X) of northern Michigan (abstract). *Abstracts with Program*, Geol. Soc. Amer. **10**, 367.

Cameron A. G. W. (1973) Abundances of the elements in the solar system. *Space Sci. Rev.* **15**, 121–146.

Canalas R. A., Alexander E. C., Jr. and Manuel O. K. (1968) Terrestrial abundances of noble gases. *J. Geophys. Res.* **73**, 3331–3334.

Carlson R. W. and Lugmair G. W. (1979) Sm-Nd constraints on early lunar differentiation and the evolution of KREEP. *Earth Planet. Sci. Lett.* **45**, 123–132.

Carlson R. W., Macdougall J. D. and Lugmair G. W. (1978) Differential Sm/Nd evolution in oceanic basalts. *Geophys. Res. Lett.* **5**, 229–232.

Carr M. H. and Meyer C. E. (1974) The regolith at the Apollo 15 site and its stratigraphic implications. *Geochim. Cosmochim. Acta* **38**, 1183–1197.

Carr M. H., Howard K. A. and El-Baz F. (1971) Geologic maps of the Apennine-Hadley region of the Moon. U.S. Geol. Surv. Map I-723. Washington, D.C.

Carswell D. A. (1975) Primary and secondary phlogopites and clinopyroxenes in garnet lherzolite xenoliths. *Phys. Chem. Earth* **9**, 417–429.

Carter S. R., Evensen N. M., Hamilton P. J. and O'Nions R. K. (1978a) Continental volcanics derived from enriched and depleted source regions: Nd and Sr isotope evidence. *Earth Planet. Sci. Lett.* **37**, 401–408.

Carter S. R., Evensen N. M., Hamilton P. J. and O'Nions R. K. (1978b) Neodymium and strontium isotope evidence for crustal contamination of continental volcanics. *Science* **202**, 743.

Chao E. C. T., James O. B., Minkin J. A., Boreman J. A., Jackson E. D. and Raleigh C. B. (1970) Impact metamorphic effects in lunar samples from Tranquillity Base. *Proc. Apollo 11 Lunar Sci. Conf.*, p. 287–314.

Chao E. C. T., Minkin J. A. and Thompson C. L. (1976) The petrology of 77215, a noritic impact ejecta breccia. *Proc. Lunar Sci. Conf. 7th*, p. 2287–2308.

Chapman C. R. (1976) Chronology of terrestrial planet evolution: the evidence from Mercury. *Icarus* **28**, 523–536.

Chappell B. W., Compston W., Green D. H. and Ware N. G. (1972) Chemistry, geochronology, and petrogenesis of lunar sample 15555. *Science* **175**, 415–416.

Chen J. H. and Tilton G. R. (1976) Isotopic lead investigations on the Allende carbonaceous chondrite. *Geochim. Cosmochim. Acta* **40**, 635–643.

Chen J. H., Tilton G. R., Mattinson J. M. and Vidal P. H. (1978) Lead isotope systematics of mare basalt 75075. *Proc. Lunar Planet. Sci. Conf. 9th*, p. 509–521.

Church S. E. (1976) The Cascade Mountains revisited: a re-evaluation in light of new lead isotope data. *Earth Planet. Sci. Lett.* **29**, 175–188.

Church S. E. and Tatsumoto M. (1975) Lead isotope relations in Oceanic Ridge basalts from the Juan de Fuca-Gorda Ridge area, N.E. Pacific Ocean. *Contrib. Mineral. Petrol.* **53**, 253–279.

Clarke W. B., Beg M. A. and Craig H. (1969) Excess He^3 in the sea: evidence for terrestrial primordial helium. *Earth Planet. Sci. Lett.* **6**, 213–220.

Clayton R. N. and Mayeda T. K. (1975) Genetic relations between the moon and meteorites. *Proc. Lunar Sci. Conf. 6th*, p. 1761–1769.

Clayton R. N., Grossman L. and Mayeda T. K. (1973) A component of primitive nuclear composition in carbonaceous chondrites. *Science* **182**, 485–487.

Clayton R. N., Onuma N. and Mayeda T. K. (1976) A classification of meteorites based on oxygen isotopes. *Earth Planet. Sci. Lett.* **30**, 10–18.

Cliff R. A., Lee-Hu C. and Wetherill G. W. (1971) Rb-Sr and U, Th-Pb measurements on Apollo 12 material. *Proc. Lunar Sci. Conf. 2nd*, p. 1493–1502.

Coleman R. G. (1971) Plate-tectonic emplacement of upper mantle peridotites along continental edges. *J. Geophys. Res.* **76**, 1212–1222.

Compston W. and Arriens P. A. (1968) The Precambrian geochronology of Australia. *Can. J. Earth Sci.* **5**, 561–584.

Compston W. and Lovering J. (1969) The Sr-isotope geochemistry of granulitic and eclogitic inclusions from the basic pipes at Delegate, eastern Australia. *Geochim. Cosmochim. Acta* **33**, 691–699.

Compston W., Berry H., Vernon M. J., Chappell B. W. and Kage M. J. C. (1971) Rubidium-strontium chronology and chemistry of lunar material from the Ocean of Storms. *Proc. Lunar Sci. Conf. 2nd*, p. 1471–1485.

Compston W., Chappell B. W., Arriens P. A. and Vernon M. J. (1970) The chemistry and age of Apollo 11 lunar material. *Proc. Apollo 11 Lunar Sci. Conf.*, p. 1007–1027.

Compston W., McDougall I. and Heier K. S. (1968) Geochemical comparison of the Mesozoic basaltic rocks of Antarctica, South Africa, South America and Tasmania. *Geochim. Cosmochim. Acta* **32**, 129–149.

Compston W., Vernon M. J., Berry J., Rudowiski R., Gray C. M., Ware N., Chappell B. W. and Kaye M. (1972) Apollo 14 mineral ages and thermal history of the Fra Mauro Formation. *Proc. Lunar Sci. Conf. 3rd*, p. 1487–1501.

Condie K. C. (1973) Archean magmatism and crustal thickening. *Geol. Soc. Amer. Bull.* **84**, 2981–2992.

Condie K. C. (1976) *Plate Tectonics and Crustal Evolution*. Pergamon Press, New York. 288 pp.

Condie K. C. and Moore J. M. (1977) Geochemistry of Proterozoic volcanic rocks from the Grenville Province, eastern Ontario. In *Volcanic Regimes in Canada* (W. R. A. Baragar, L. C. Coleman and J. M. Hall, eds.), p. 149–168. Geol. Assoc. of Can. Spec. Paper 16.

Condie K. C., Viljoen M. J. and Kable E. J. D. (1977) Effects of alteration on element distributions in Archean tholeiites from the Barberton Greenstone Belt, South Africa. *Contrib. Mineral. Petrol.* **64**, 75–89.

Condomines M. (1978) Age of the Olby-Laschamp geomagnetic polarity event. *Nature* **276**, 257–258.

Condomines M. and Tanguy J. C. (1976) L'âge de l'Etna determiné par le déséquilibre radioactif ^{230}Th-^{238}U. *C. R. Acad. Sci. Paris* **282**, 1661.

Condomines M., Bernat M. and Allègre C. J. (1976) Evidence for contamination of recent Hawaiian lavas from ^{230}Th-^{238}U data. *Earth Planet. Sci. Lett.* **33**, 122–125.

Consolmagno G. J. and Drake M. J. (1977) Composition and evolution of the eucrite parent body: evidence from rare earth elements. *Geochim. Cosmochim. Acta* **41**, 1271–1282.

Cox A., Doell R. and Dalrymple G. (1963) Geomagnetic Polarity Epochs and Pleistocene geochronometry. *Nature* **198**, 1049–1051.

Cox K. G., Hawkesworth C. J., O'Nions R. K. and Appleton J. D. (1976) Isotopic evidence for the derivation of some Roman region volcanics from anomalously enriched mantle. *Contrib. Mineral. Petrol.* **56**, 173–180.

Craig H. (1954) Geochemical implications of the isotopic composition of carbon in ancient rocks. *Geochim. Cosmochim. Acta* **6**, 186.

Craig H. (1963) The isotopic geochemistry of water and carbon in geothermal area. In *Conference on Nuclear Geology in Geothermal Area* (Pisa, ed.), p. 17–53. Spoleto, Italy.

Craig H. and Lupton J. E. (1976) Primordial neon, helium, and hydrogen in oceanic basalts. *Earth Planet. Sci. Lett.* **31**, 369–385.

Craig H., Lupton J. E. and Horibe Y. (1978) A mantle helium component in circum-pacific volcanic gases: Hakone, the Marianas, and Mt. Lassen. In *Terrestrial Rare Gases*. (E. C. Alexander, Jr. and M. Ozima, eds.), p. 3–16. Advances in Earth and Planetary Sciences, vol. 3. Japan Soc. Press, Tokyo.

Cudzilo T. F. (1978) Geochemistry of early Proterozoic igneous rocks, northeastern Wisconsin and upper Michigan. Unpublished Ph.D. thesis, Univ. of Kansas, Lawrence. 194 pp.

Dann J. C. (1978) Major-element variation within the Emperor igneous complex and the Hemlock and Badwater volcanic formations. Unpublished M.S. thesis, Michigan Tech. Univ., Houghton. 198 pp.

Dasch E. J. and Green D. H. (1975) Strontium isotope geochemistry of lherzolite inclusions and host basaltic rocks, Victoria, Australia. *Amer. J. Sci.* **275**, 461–469.

Dasch E. J., Hedge C. E. and Dymond J. (1973) Effect of seawater interaction on strontium isotope composition of deep-sea basalts. *Earth Planet. Sci. Lett.* **19**, 177–183.

Davies H. L. (1968) Papuan ultramafic belt. *Int'l. Geol. Congr. 23rd, Rept. 1*, p. 209–220. Prague.

Davis P. K., Lewis R. S. and Reynolds J. H. (1971) Stepwise heating analyses of rare gases from pile-irradiated rocks 10044 and 10057. *Proc. Lunar Sci. Conf. 2nd*, p. 1693–1703.

Deines P. and Gold D. P. (1973) The isotopic composition of carbonatite and kimberlite carbonates and their bearing on the isotopic composition of deep-seated carbon. *Geochim. Cosmochim. Acta* **37**, 1709–1733.

de Laeter J. R. and Blockley J. G. (1972) Granite ages within the Archean Pilbara block. *J. Geol. Soc. Aust.* **19**, 363–370.

de Laeter J. R. and Trendall A. F. (1974) The age of the Copper Hills porphyry. *Geol. Surv. W. Australia Ann. Rept. 1969*, p. 54–59.

de Laeter J. R., Lewis J. D. and Blockley J. G. (1975) Granite ages within the Shaw batholith of the Pilbara block. *Geol. Surv. W. Australia Ann. Rept. 1974*, p. 73–79.

de Laeter J. R., Vernon M. J. and Compston W. (1973) Revision of lunar Rb-Sr ages. *Geochim. Cosmochim. Acta* **37**, 700–702.

De Hon R. A. (1974) Thickness of mare material in the Tranquillitatis and Nectaris basins. *Proc. Lunar Sci. Conf. 5th*, p. 53–59.

Delano J. W. and Bence A. E. (1977) 4.2–4.3 AE anorthositic soil fragments: equilibrated or unequilibrated poly-component systems? *Proc. Lunar Sci. Conf. 8th*, p. 2029–2050.

Dence M. R. and Plant A. G. (1972) Analysis of Fra Mauro samples and the origin of the Imbrium basin. *Proc. Lunar Sci. Conf. 3rd*, p. 379–399.

Den Tex E. (1969) Origin of ultramafic rocks, their tectonic setting and history. *Tectonophys.* **7**, 457–488.

DePaolo D. J. (1978) Nd and Sr isotope systematic of young continental igneous rock. *Short Papers 4th Int. Conf. Geochr., Cosmochr., Isotope Geol.* (R. E. Zartman, ed.), p. 91–93. U.S. Geol. Survey Open-File Rept. 78–701.

DePaolo D. J. and Wasserburg G. J. (1976a) Inferences about magma sources and mantle structure from variations of $^{143}Nd/^{144}Nd$. *Geophys. Res. Lett.* **3**, 743–746.

DePaolo D. J. and Wasserburg G. J. (1976b) Nd isotopic variations and petrogenetic models. *Geophys. Res. Lett.* **3**, 249–252.

DePaolo D. J. and Wasserburg G. J. (1977) The sources of island arcs as indicated by Nd and Sr isotopic studies. *Geophys. Res. Lett.* **4**, 465–468.

Dewey J. F. and Bird J. M. (1971) Origin and emplacement of the ophiolite suite: appalachian ophiolites in Newfoundland. *J. Geophys. Res.* **76**, 3179–3206.

Dickey J. S., Jr. (1970) Partial fusion products of alpine peridotites: Serrania de la Ronda and other examples. *Mineral. Soc. Amer. Spec. Paper* **3**, 33–59.

Dimroth E., Baragar W. R. A., Bergeron R. and Jackson G. D. (1970) The filling of the Circum-Ungava Geosyncline. *Geol. Survey Can., Paper 70-40*, p. 45–142.

Doe B. R. and Hart S. R. (1963) The effect of contact metamorphism of lead in potassium feldspars near the Eldora Stock, Colorado. *J. Geophys. Res.* **68**, 3521–3530.

Donhoffer D. (1963) Determination of the half-lives of radioactive nuclides ^{147}Sm and ^{176}Lu occurring in nature by means of liquid scintillators. *Nucl. Phys.* **50**, 489–496.

Dosso L., Vidal P., Cantagrel J. M., Lameyre J., Marot A. and Zimine S. (1979) Continental fragment or oceanic island?: petrology and isotopic geochemistry evidence. *Earth Planet. Sci. Lett.* **43**, 46–60.

Downing R. G., Hennecke E. W. and Manuel O. K. (1977) Josephinite: a terrestrial alloy with radiogenic xenon-129 and the noble gas imprint of iron meteorites. *Geochem. J.* **11**, 219–229.

Dreibus G., Kruse H., Spettel B. and Wänke H. (1977) The bulk composition of the Moon and the eucrite parent body. *Proc. Lunar Sci. Conf. 8th*, p. 211–227.

Duke M. B. and Silver L. T. (1967) Petrology of eucrites, howardites, and mesosiderites. *Geochim. Cosmochim. Acta* **31**, 1637–1665.

Duncan A. R., Erlank A. J., Willis J. P. and Ahrens L. H. (1973) Composition and inter-relationships of some Apollo 16 samples. *Proc. Lunar Sci. Conf. 4th*, p. 1097–1113.

Duncan A. R., Erlank A. J., Willis J. P., Sher M. K. and Ahrens L. H. (1974) Trace element evidence for two-stage origin of some titaniferous mare basalts. *Proc. Lunar Sci. Conf. 5th*, p. 1147–1157.

Dymond J. and Hogan L. (1973) Noble gas abundance patterns in deep-sea basalts—primordial gases from mantle. *Earth Planet. Sci. Lett.* **20**, 131–139.

Dymond J. and Hogan L. (1978) Factors controlling the noble gas abundance patterns of deep-sea basalts. *Earth Planet. Sci. Lett.* **38**, 117–128.

Early T. O. and Silver L. T. (1973) Rb-Sr isotopic systematics in the Peninsular Range batholith in southern and Baja California. *EOS (Trans. Amer. Geophys. Union)* **54**, 494.

Eberhardt P. (1978) A neon-E rich phase in Orgueil: results of stepwise heating experiments. *Proc. Lunar Planet. Sci. Conf. 9th*, p. 1027–1051.

Eberhardt P., Eugster O. and Marti K. (1965) A redetermination of the isotopic composition of atmospheric neon. *Z. Naturforsch.* **20a**, 623–624.

Eberhardt P., Geiss J., Graf H., Grögler N., Krähenbühl U., Schwaller H., Schwarzmüller J. and Stettler A. (1970) Correlation between rock type and irradiation history of Apollo 11 igneous rocks. *Earth Planet. Sci. Lett.* **10**, 67–72.

Eberhardt P., Geiss J., Graf H., Grögler N., Mendia M. D., Mörgeli M., Schwaller H. and Stettler A. (1972) Trapped solar wind noble gases in Apollo 12 lunar fines 12001 and Apollo 11 breccia 10046. *Proc. Lunar Sci. Conf. 3rd*, p. 1821–1856.

Eberhardt P., Geiss J., Grögler N., Krähenbühl U., Mörgell M. and Stettler A. (1971) Potassium-argon age of Apollo 11 rock 10003. *Earth Planet. Sci. Lett.* **11**, 245–247.

Eggler D. H. (1973) Role of CO_2 on melting process in the mantle. *Carnegie Inst. Wash., Yearb. 72*, p. 245–267.

Eggler D. H. (1976) Does CO_2 cause partial melting in the low-velocity layer of the mantle? *Geology* **4**, 69–72.

Emerson D. E., Stroud L. and Meyer T. O. (1966) The isotopic abundance of neon from helium-bearing natural gases. *Geochim. Cosmochim. Acta* **30**, 847–854.

Emslie R. B. (1978) Anorthosite massifs, rapakivi granites, and late Proterozoic rifting of North America. *Precambrian Res.* **7**, 61–98.

Engel A. E. J., Engel C. G. and Havens R. G. (1965) Chemical characteristics of oceanic basalts and the upper mantle. *Geol. Soc. Amer. Bull.* **76**, 719–733.

Engel A. E. J., Itson S. P., Engel C. G., Stockney D. M. and Cray E. J., Jr. (1974) Crustal evolution and global tectonics: a petrogenic view. *Geol. Soc. Amer. Bull.* **85**, 885–891.

Epstein S. and Mayeda T. K. (1953) Variation of ^{18}O content of waters from natural sources. *Geochim. Cosmochim. Acta* **4**, 213–224.

Epstein S. and Taylor H. P. (1967) Variation of $^{18}O/^{16}O$ in minerals and rocks. In *Researches in Geochemistry* **2** (P. H. Abelson, ed.), p. 29–62. Wiley, N. Y.

Erlank A. J. and Kable E. J. D. (1976) The significance of incompatible elements in Mid-Atlantic Ridge basalts from 45° N with particular reference to Zr/Nb. *Contrib. Mineral. Petrol.* **54**, 281–291.

Eugster O., Eberhardt P. and Geiss J. (1967) Krypton and xenon isotopic composition in 3 carbonaceous chondrites. *Earth Planet. Sci. Lett.* **3**, 249–257.

Eugster O., Eberhardt P., Geiss J., Grögler N., Jungck M. and Mörgeli M. (1977) The cosmic-ray exposure history of Shorty Crater samples; the age of Shorty Crater. *Proc. Lunar Sci. Conf. 8th*, p. 3059–3082.

Evensen N. M., Murthy V. R. and Coscio M. R., Jr. (1973) Rb-Sr ages of some mare basalts and the isotopic trace element systematics in lunar fines. *Proc. Lunar Sci. Conf. 4th*, p. 1707–1724.

Fahrig W. F. and Wanless R. K. (1963) Age and significance of diabase dyke swarms of the Canadian Shield. *Nature* **200**, 934–937.

Fairbairn H. W., Hurley P. M., Card K. D. and Knight C. J. (1969) Correlation of radiometric ages of Nipissing Diabase and Huronian metasediments with Proterozoic orogenic events in Ontario. *Can. J. Earth Sci.* **6**, 489–497.

Fanale F. P. and Cannon W. A. (1971) Physical absorption of rare gas on terrigenous sediments. *Earth Planet. Sci. Lett.* **11**, 362–368.

Faure G. and Hurley P. M. (1963) The isotopic composition of strontium in oceanic and continental basalts: application to the origin of igneous rocks. *J. Petrol.* **4**, 31–50.

Faure G., Bowman J. R., Elliot D. H. and Jones L. M. (1974) Strontium isotope composition and petrogenesis of the Kirkpatrick basalt, Queen Alexandra Range, Antarctica. *Contrib. Mineral. Petrol.* **48**, 153–169.

Faure G., Hill R. L., Jones L. M. and Elliot D. H. (1970) Isotope composition and silica content of Mesozoic basalt and dolerite from Antarctica. *SCAR Symposium*, Univ. Press, Oslo.

Fenton M. D. and Faure G. (1969) The age of igneous rocks of the Stillwater complex of Montana. *Geol. Soc. Amer. Bull.* **80**, 1599–1604.

Fields P. R., Friedman A. M., Misted J., Lerner J., Stevens M., Metta D. and Sabine W. K. (1966) Decay properties of plutonium-244 and comments on its existence in nature. *Nature* **212**, 131–137.

Fisher D. E. (1974) The planetary primordial component of rare gases in the deep earth. *Geophys. Res. Lett.* **1**, 161–164.

Fisher D. E. (1975) Trapped helium and argon and the formation of the atmosphere by degassing. *Nature* **256**, 113–114.

Fitton M. J., Horwitz R. C. and Sylvester G. (1975) Stratigraphy of the earl Precambrian in the west Pilbara, W. Australia. *CSIRO (Commonwealth Scientific and Industrial Research Organization) Min. Div. Rept.* FP 11.

Floran R. J., Prinz M., Hlava P. F., Keil K., Nehru C. E. and Hinthorne J. R. (1978) The Chassigny meteorite: a cumulate with hydrous amphibole-bearing melt inclusions. *Geochim. Cosmochim. Acta* **42**, 1213–1229.

Folinsbee R. E., Baadsgaard H., Cumming G. L. and Green D. C. (1968) A very ancient island arc. *Amer. Geophys. Union Monograph* **12**, p. 441–448.

Forbes R. and Kuno H. (1965) The regional petrology of peridotite inclusions and basaltic host rocks. In *Upper Mantle Symposium Inter. Geol. Congr. 1963, New Delhi*, p. 161–179.

Forester R. W. and Taylor H. P. (1972) Oxygen and hydrogen isotopes data on the interaction of meteoric ground water with a gabbro diorite stock. San Juan Mountains (Colorado). *24th Inter. Geol. Congr., Montreal, Sect. 10*, p. 254–263.

Forester R. W. and Taylor H. P. (1977) $^{18}O/^{16}O$ D/H and $^{13}C/^{12}C$ studies of the tertiary igneous complex of Skye Scotland. *Amer. J. Sci.* **277**, 136–177.

Frederichsen H. and Hoernes S. (1978) Stable isotope exchange between oceanic crust and ocean water. *Inter. Conf. Geochron. Cosmochron., Isotope Geol.* (R. Zartman, ed.), p. 103–126. U. S. Dept. Int. Geol. Survey.

Frey F. A. (1970) Rare earth abundances in Alpine ultramafic rocks. *Phys. Earth Planet. Inter.* **3**, 323–330.

Frey F. A. and Green D. H. (1974) The mineralogy, geochemistry and origin of lherzolite inclusions in Victorian basanites. *Geochim. Cosmochim. Acta* **38**, 1023–1059.

Frey F. A., Bryan W. B. and Thompson G. (1974) Atlantic Ocean floor: geochemistry and petrology of basalts from Legs 2 and 3 of the Deep-Sea Drilling Project. *J. Geophys. Res.* **79**, 5507–5527.

Frey F. A., Haskin L. A. and Haskin M. (1971) Rare earth abundances in some ultramafic rocks. *J. Geophys. Res.* **76**, 2057–2070.

Friedmann I. (1953) Deuterium content of natural water and other substances. *Geochim. Cosmochim. Acta* **4**, 89–103.

Fukuoka T., Boynton W. V., Ma M.-S. and Schmitt R. A. (1977) Genesis of howardites, diogenites and eucrites. *Proc. Lunar Sci. Conf. 8th*, p. 187–210.

Gale N. J., Arden J. W. and Hutchinson R. (1975) The chronology of the Nakhla achondritic meteorite. *Earth Planet. Sci. Lett.* **26**, 195–206.

Ganapathy R. and Anders E. (1974) Bulk compositions of the Moon and Earth, estimated from meteorites. *Proc. Lunar Sci. Conf. 5th*, p. 1181–1206.

Garlick G. D. (1966) Oxygen isotope fractionation in igneous rocks. *Earth Planet. Sci. Lett.* **7**, 361–368.

Garlick G. D. and Dymond J. R. (1970) Oxygen isotope exchange between volcanic materials and ocean water. *Geol. Soc. Amer. Bull.* **81**, 2137–2142.

Gast P. W. (1968a) Trace element fractionation and the origin of tholeiitic and alkaline magma types. *Geochim. Cosmochim. Acta* **32**, 1057.

Gast P. W. (1968b) Upper mantle chemistry and evolution of the Earth's crust. In *The History of the Earth's Crust* (R. A. Phinney, ed.), p. 15–27. Princeton Univ. Press., Princeton, N.J.

Gast P. W., Tilton G. R. and Hedge C. (1964) Isotopic composition of and strontium from Ascension and Gough Islands. *Science* **145**, 1181.

Gastil G. (1960) The distribution of mineral dates in time and space. *Amer. J. Sci.* **258**, 1–35.

Gates T. M. and Hurley P. M. (1973) Evaluation of Rb-Sr dating methods applied to the Matachewan, Abitibi, Mackenzie and Sudbury dike swarms in Canada. *Can. J. Earth Sci.* **10**, 900–919.

Gaudette H. E., Mendoza V., Hurley P. M. and Fairbairn H. W. (1978) Geology and age of the Paraguaza rapakivi granite, Venezuela. *Geol. Soc. Amer. Bull.* **89**, 1335–1340.

Gaudette H. E., Olszewski W. J., Jr. and Mendoza V. (1977) U-Pb zircon ages of the Minicia and Macabana gneisses, Amazonas Territory, Venezuela. In *Papers Presented at V Congreso Geologico Venezolano*, p. 527–538.

Geiss J., Eberhardt P., Grögler N., Guggisberg S., Maurer P. and Stettler A. (1977) Absolute time scale of lunar mare formation and filling. *Phil. Trans. Roy. Soc. Lond.* **285A**, 151–158.

Gelinas L. and Brooks C. (1974) Archean quench-texture tholeiites. *Can. J. Earth Sci.* **11**, 324–340.

Gelinas L., Brooks C., Perrault G., Carynan R., Trundel P. and Cote R. (1977) Chemostratigraphic divisions within the Abitibi volcanic belt, Rouyn Noranda District, Quebec. In *Volcanic Regimes in Canada* (W. R. A. Baragar, L. C. Coleman and J. M. Hall, eds.), p. 265–295. Geol. Assn. Can. Spec. Paper 16.

Genest A. M. and Javoy M. (1976) Composition isotopique de l'oxygene et de l'hydrogene dans les roches spilitiques et les roches des series ophiolitiques. R.A.S.T. 4eme reunion.

Gilbert G. K. (1893) The Moon's face: a study of the origin of its features. *Bull. Phil. Soc. Wash.* **12**, 241–292.

Goldich S. S. and Hedge C. E. (1974) 3,800-Myr granitic gneiss in southwestern Minnesota. *Nature* **252**, 467–468.

Goldich S. S. and Mudrey M. G., Jr. (1972) Dilatancy model for discordant U-Pb zircon ages. In *Contributions to Recent Geochemistry and Analytical Chemistry* (A. I. Tugarinov, ed.) p. 415–418. Nauka, Moscow (Transl.). Israel Prog. for Sci. Transl., p. 466–470. Wiley, N. Y.

Goldich S. S., Hedge C. E. and Stern T. W. (1970) Age of the Morton and Montevideo gneisses and related rocks, southwestern Minnesota. *Geol. Soc. Amer. Bull.* **81**, 3671–3696.

Gopalan K., Kaushal S., Lee-Hu C. and Wetherill G. W. (1970) Rb-Sr and U, Th-Pb ages of lunar materials. *Proc. Apollo 11 Lunar Sci. Conf.*, p. 1195–1205.

Graham A. L. and Ringwood A. E. (1971) Lunar basalt genesis: the origin of the europium anomaly. *Earth Planet. Sci. Lett.* **13**, 105–115.

Gray C. M., Papanastassiou D. A. and Wasserburg G. J. (1973) The identification of early condensates from the solar nebula. *Icarus* **20**, 213–219.

Green D. H. (1967) High temperature peridotite inclusions. In *Ultramafic and Related Rocks* (P. J. Wyllie, ed.), p. 212–222. Wiley, N. Y.

Green D. H. and Ringwood A. E. (1967) The genesis of basaltic rocks. *Contrib. Mineral. Petrol.* **15**, 103–190.

Green D. H., Ringwood A. E., Ware N. G. and Hibberson W. O. (1972a) Experimental petrology and petrogenesis of Apollo 14 basalts. *Proc. Lunar Sci. Conf. 3rd*, p. 197–206.

Green J. A., Kurat G., Keil K. and Prinz M. (1972b) Electron microprobe analyses of lithic fragments, glasses, chondrites and minerals from Luna 16 fines. *Spec. Pub. 5*, p. 1–16. Univ. of New Mexico, Inst. of Meteoritics.

Green J. C. (1977) Keweenawan plateau volcanism in the Lake Superior region. In *Volcanic Regimes in Canada* (W. R. A. Baragar, L. C. Coleman and J. M. Hall, eds.), p. 407–422. Geol. Assn. of Can. Spec. Paper 16.

Griffin W. L. and Murthy V. R. (1968) K, Rb, Sr and Ba in some ultramafic rocks and minerals. *Earth Planet. Sci. Lett.* **4**, 497–501.

Griffin W. L. and Murthy V. R. (1969) Distributing K, Rb, Sr and Ba in minerals relevant to basalt genesis. *Geochim. Cosmochim. Acta* **33**, 1389–1414.

Guggisberg S., Eberhardt P., Geiss J., Grögler N., Stettler A., Brown G. M. and Peckett A. (1977) Youngest and oldest mare basalts: the temporal extent of mare filling (abstract). In *Lunar Science VIII*, p. 383–387. The Lunar Science Institute, Houston.

Guggisberg S., Eberhardt P., Geiss J., Grögler N., Stettler A., Brown G. M. and Peckett A. (1979) Classification of the Apollo-11 mare basalts according to Ar^{39}-Ar^{40} ages and petrological properties. *Proc. Lunar Planet. Sci. Conf. 10th*, p. 1–39.

Gulson B. L. and Krogh T. E. (1975) Evidence of multiple intrusion, possible resetting of U-Pb ages, and new crystallization of zircons in post-tectonic intrusions ('Rapakivi granites') and gneisses from South Greenland. *Geochim. Cosmochim. Acta* **39**, 65–82.

Gupta M. C. and MacFarlane R. D. (1970) The natural alpha radioactivity of samarium. *J. Inorg. Nucl. Chem.* **32**, 3425–3432.

Hall C. M. and York D. (1978) K-Ar and $^{40}Ar/^{39}Ar$ age of Laschamp geomagnetic polarity reversal. *Nature* **274**, 462–464.

Hallam A. (1977) Secular changes in marine inundations of USSR and North America through the Phanerozoic. *Nature* **269**, 769–772.

Hallberg J. A., Johnston C. and Bye S. M. (1976) The archean Marda Igneous Complex, Western Australia. *Precambrian Res.* **3**, 111–136.

Halpern M. (1969) $^{87}Sr/^{86}Sr$ ratios of ultramafic nodules and host basalts from the McMurdo Area and Ford Ranges, Antarctica. *Antarctic J. U.S.* **4**, 206.

Hamilton, P. J., Evensen N. M., O'Nions R. K. and Tarney J. (1979) Sm-Nd isotopic investigations of the Lewisian, N.W. Scotland. *Nature* **277**, 25–28.

Hamilton, P. J., O'Nions R. K. and Evensen N. M. (1977) Sm-Nd dating of Archean basic and ultrabasic volcanics. *Earth Planet. Sci. Lett.* **36**, 263–268.

Hamilton P. J., O'Nions R. K., Evensen N. M., Bridgewater D. and Allart J. (1978a) Sm-Nd isotopic investigations of the Isua supracrustals, West Greenland: implications for mantle evolution. *Nature* **272**, 41–43.

Hanson G. N. and Malhotra R. (1971) K-Ar ages of mafic dikes and evidence for low-grade metamorphism in northeastern Minnesota. *Geol. Soc. Amer. Bull.* **82**, 1107–1114.

Hart S. R. (1971) K, Rb, Sc, Sr and Ba contents and Sr isotope ratios of ocean floor basalts. *Phil. Trans. Roy. Soc. Lond.* **A26**, 573–587.

Hart S. R. (1975) Deep Sea Drilling Project Leg 34 Basalts and ocean ridge basalts: a geochemical comparison. *Dept. Terr. Magnet. Ann. Rept. 1974–75*, p. 218–224. Carnegie Institution, Washington.

Hart S. R. (1976) LIL-Element geochemistry, Leg 34 basalts. *Initial Reports of the Deep Sea Drilling Project XXXIV*, p. 283–288. Washington, D.C.

Hart S. R. and Brooks C. (1970) Rb-Sr mantle evolution models. *Carnegie Inst. Wash., Yearb. 68*, p. 426–429.

Hart S. R. and Brooks C. (1977) The geochemistry of early Precambrian mantle. *Contrib. Mineral. Petrol.* **61**, 109–128.

Hart S. R., Brooks C., Krogh T. E., Davis G. L. and Nava D. (1970) Ancient and modern volcanic rocks: a trace-element model. *Earth Planet. Sci. Lett.* **10**, 17–28.

Hart S. R. and Davis G. L. (1969) Zircon U-Pb and whole-rock Rb-Sr ages and early crustal development near Rainy Lake, Ontario. *Geol. Soc. Amer. Bull.* **80**, 595–616.

Hart S. R., Erlank A. J. and Kable E. J. D. (1974) Sea floor alteration: some chemical and isotopic effects. *Contrib. Mineral Petrol.* **44**, 219–230.

Hart S. R., Schilling J. G. and Powell J. L. (1973) Basalts from Iceland and along the Reykjanes Ridge: Sr isotope geochemistry. *Nature, Phys. Sci.* **246**, 104–107.

Hart S. R., Staudigel H. and Muehlenbachs K. (1978) Oceanic crust: age of alteration. *Inter. Conf. Geochron. Cosmochron., Isotope Geol.*, p. 175. U.S. Geol. Survey, Va. Open File Rept. 78–701.

Hartmann W. K. and Wood C. A. (1971) Moon: origin and evolution of multiring basins. *Moon* **3**, 3–78.

Hawkesworth C. J., Moorbath S., O'Nions R. K. and Wilson J. R. (1975) Age relationship between greenstoane belts and granites in the Rhodesian Archean craton. *Earth Planet. Sci. Lett.* **25**, 251–262.

Hawkesworth C. J., Norry M. J., Roddick J. C., Baker P. E., Francis P. W. and Thorpe R. S. (1979) $^{143}Nd/^{144}Nd$, $^{87}Sr/^{86}Sr$, and incompatible element variations in calc-alkaline andesites and plateau lavas from South America. *Earth Planet. Sci. Lett.* **42**, 45–57.

Hawkesworth C. J., O'Nions R. K., Pankhurst P. J., Hamilton P. J. and Evensen N. M. (1977) A geochemical study of island-arc and back-arc tholeiites from the Scotia Sea. *Earth Planet. Sci. Lett.* **36**, 253–262.

Head J. W. (1974) Stratigraphy of the Descartes region (Apollo 16): implications for the origin of samples. *Moon* **11**, 77–99.

Head J. W. (1975) Mode of occurrence and style of emplacement of lunar mare deposits (abstract). In *Papers Presented to the Conference on Origins of Mare Basalts and Their Implications for Lunar Evolution*, p. 61–65. The Lunar Science Institute, Houston.

Head J. W. (1976) Lunar volcanism in space and time. *Rev. Geophys. Space Phys.* **14**, 265–300.

Heaton T. H. and Sheppard S. M. (1977) Hydrogen and oxygen isotope evidence of sea-water hydrothermal alteration and ore deposition, Troodos Complex, Cyprus. In *Volcanic Processes in Ore Genesis*, p. 42–57. Spec. Publ. Geol. Soc. Lond.

Hedge C. E. (1978) Strontium isotopes in basalts from the Pacific Ocean basin. *Earth Planet. Sci. Lett.* **38**, 88–94.

Hedge C. E. and Peterman Z. E. (1970) The strontium isotopic composition of basalts from the Cordo and Juan de Fuca Rises, northeastern Pacific Ocean. *Contrib. Mineral. Petrol.* **27**, 114.

Hedge C. J. and Walthall F. G. (1963) Radiogenic strontium 87 as an index of geological processes. *Science* **140**, 1214–1217.

Hedge C. E., Peterman Z. E. and Dickinson W. R. (1972) Petrogenesis of lavas from Western Samoa. *Geol. Soc. Amer. Bull.* **83**, 2709.

Hedge C. E., Watkins N. D., Hildrett R. A. and Doering W. P. (1973) $^{87}Sr/^{86}Sr$ ratios in basalts from islands in the Indian Ocean. *Earth Planet. Sci. Lett.* **21**, 29.

Hennecke E. W. and Manuel O. K. (1975a) Noble gases in an Hawaiian xenolith. *Nature* **257**, 778–780.

Hennecke E. W. and Manuel O. K. (1975b) Noble gases in lava rock from Mount Capulin, New Mexico. *Nature* **256**, 284–287.

Hedge C. E., Watkins N. D., Hildrett R. A. and Doering W. P. (1973) $^{87}Sr/^{86}Sr$ ratios in basalts from islands in the Indian Ocean. *Earth Planet. Sci. Lett.* **21**, 29.

Hertogen J., Janssens M.-J., Takahashi H., Palme H. and Anders E. (1977) Lunar basins and craters: evidence for systematic compositional changes of bombarding population. *Proc. Lunar Sci. Conf. 8th*, p. 17–45.

Hess H. H. (1939) Island arcs, gravity anomalies and serpentinite inclusions. *Int'l Geol. Congr. 17th, Moscow. Part 2*, p. 263–283.

Hess H. H. (1964) The oceanic crust, the upper mantle and the Mayagüez serpentinized peridotite. In *A Study of Serpentinite* (C. A. Burk, ed.) p. 169–175. Nat. Acad. Sci., Nat. Res. Council Publ. 1183.

Hickman M. H. (1974) 3500-m.y.-old granite in southern Africa. *Nature* **251**, 295–296.

Hietanen A. (1975) Generation of potassium-poor magmas in the northern Sierra Nevada and the Svecofennian of Finland. *J. Res. U.S. Geol. Surv.* **3**, 631–645.

Hill R. E. T. and Boettcher A. L. (1970) Water in the Earth's mantle: melting curves of basalt-water and basalt-water-carbon dioxide. *Science* **167**, 980–982.

Hills F. A. and Armstrong R. L. (1974) Geochronology of Precambrian rocks in the Laramie Range and implications for the tectonic framework of Precambrian southern Wyoming. *Precambrian Res.* **1**, 213–225.

Himmelberg G. R. and Loney R. A. (1973) Petrology of the Vulcan Peak alpine-peridotite, S. W. Oregon. *Geol. Soc. Amer. Bull.* **84**, 1585–1600.

Hoffman P. (1973) Evolution of an early Proterozoic continental margin: the Coronation geosyncline and associated aulacogens of the northwestern Canadian shield. *Phil. Trans. Roy. Soc. Lond.* **A273**, 547–481.

Hoffman P. F. and McGlynn J. C. (1977) Great Bear Batholith: a volcanoplutonic depression. In *Volcanic Regimes in Canada* (W. R. A. Baragar, L. C. Coleman and J. M. Hall, eds.), p. 169–192. Geol. Assoc. of Can. Spec. Paper No. 16.

Hofmann A. W. and Hart S. R. (1975) An assessment of local and regional isotopic equilibrium in a partially molten mantle. *Carnegie Inst. Dept. Terr. Magnet. Ann. Rept. 1974*–75, p. 195–210. Carnegie Institution, Washington.

Hofmann A. and Hart S. R. (1978) An assessment of local and regional isotopic equilibrium in the mantle. *Earth Planet Sci. Lett.* **38**, 44–62.

Hofmann A. W., White W. M. and Whitford D. J. (1978) Geochemical constraints on mantle models: the case for a layered mantle. *Carnegie Inst. Wash., Yearb.* **77**, 548–562.

Hohenberg C. M. (1970) Xenon from the Angra dos Reis meteorite. *Geochim. Cosmochim. Acta* **34**, 185–191.

Hohenberg C. M., Munk M. N. and Reynolds J. H. (1967) Spallation and fissiogenic xenon and krypton from stepwise heating of the Pasamonte achondrite: the case for extinct plutonium-244 in meteorites; relative ages of chondrites and achondrites. *J. Geophys. Res.* **72**, 3139–3177.

Holloway J. R. (1973) The system Pargasite - H_2O-CO_2: a model for melting of a hydrous mineral with mixed volatile fluid. 1. Experimental results to 8K bar. *Geochim. Cosmochim. Acta* **37**, 651–667.

Holloway J. R. and Eggler D. H. (1976) Fluid-absent melting of peridotite containing phlogopite and dolomite. *Geophys. Lab. Ann. Rept. 1975–76*, p. 636–639.

Holmes A. (1936) A contribution to the petrology of kimberlite and its inclusions. *Trans. Geol. Soc. S. Afr.* **39**, 379–396.

Horn P., Jessberger E. K., Kirsten T. and Richter H. (1975a) ^{39}Ar-^{40}Ar dating of lunar rocks: effects of grain size neutron irradiation. *Proc. Lunar Sci. Conf. 6th*, p. 1563–1591.

Horn P., Kirsten T. and Jessberger E. K. (1975b) Are there A-12 mare basalts younger than 3.1 b.y.? Unsuccessful search for A-12 mare basalts with crystallization ages below 3.1 b.y. *Meteoritics* **10**, 417–418.

Houtermans F. G. (1953) Determination of the age of the Earth from the isotopic composition of meteoritic lead. *Nuovo Cimento* **10**, 1623–1633.

Howard K. A. and Head J. W. (1972) Regional geology of Hadley Rille. *Apollo 15 Prelim. Sci. Rept.*, p. 25–53. NASA SP-289.

Howard K. A., Head J. W. and Swann G. A. (1972) Geology of Hadley Rille. *Proc. Lunar Sci. Conf. 3rd*, p. 1–14.

Hubbard N. J. (1971) Some chemical features of lava from the Manu's Islands. *Pac. Sci.* **25**, 178.

Hubbard N. J. and Gast P. W. (1971) Chemical composition and origin of nonmare lunar basalts. *Proc. Lunar Sci. Conf. 2nd*, p. 999–1020.

Hubbard N. J., Meyer C., Gast P. W. and Wiesmann H. (1971) The composition and derivation of Apollo 12 soils. *Earth Planet. Sci. Lett.* **10**, 341–350.

Hubbard N. J., Rhodes J. M., Gast P. W., Bansal B. M., Shih C.-Y., Wiesmann H. and Nyquist L. E. (1973) Lunar rock types: the role of plagioclase in non-mare and highland rock types. *Proc. Lunar Sci. Conf. 4th*, p. 1297–1312.

Hubbard N. J., Rhodes J. M., Wiesmann H., Shih C.-Y. and Bansal B. M. (1974) The chemical definition and interpretation of rock types returned from the non-mare regions of the Moon. *Proc. Lunar Sci. Conf. 5th*, p. 1227–1246.

Hubregtse J. J. M. W. (1976) Volcanism in the Western Superior Province of Manitoba. In *The Early History of the Earth* (B. F. Windley, ed.), p. 279–287. Wiley, London.

Huey J. M. and Kohman T. P. (1973) ^{207}Pb-^{206}Pb isochron and age of chondrites. *J. Geophys. Res.* **78**, 3227–3244.

Huneke J. C. and Wasserburg G. J. (1975) Trapped Ar^{40} in troctolite 76535 and evidence for enhanced Ar^{40}-Ar^{39} age plateaus (abstract). In *Lunar Science VI*, p. 417–419. The Lunar Science Institute, Houston.

Huneke J. C., Jessberger E. K., Podosek F. A. and Wasserburg G. J. (1973) $^{40}Ar/^{39}Ar$ measurements in Apollo 16 and 17 samples and the chronology of metamorphic and volcanic activity in the Taurus-Littrow region. *Proc. Lunar Sci. Conf. 4th*, p. 1725–1756.

Huneke J. C., Podosek F. A. and Wasserburg G. J. (1972) Gas retention and cosmic ray exposure ages of a basalt fragment from Mare Fecunditatis. *Earth Planet. Sci. Lett.* **13**, 375–383.

Huneke J. C., Smith S. P., Rajan R. S., Papanastassiou D. A. and Wasserburg G. J. (1977) Comparison of the chronology of the Kapoeta parent planet and Moon (abstract). In *Lunar Science VIII*, p. 484–487. The Lunar Science Institute, Houston.

Hunter D. R. and Hamilton P. J. (1978) The Bushveld Complex. In *Evolution of the Earth's Crust* (D. H. Tarling, ed.), p. 107–173. Academic Press, N. Y.

Hurley P. M. (1967) Rb^{87}-Sr^{87} relationships in the differentiation of the mantle. In *Ultramafic and Related Rocks* (P. J. Wyllie, ed.), p. 372–375. Wiley, N. Y.

Hurley P. M. and Rand J. R. (1969) Pre-drift continental nuclei. *Science* **164**, 1229–1242.

Hurley P. M., Pinson W. H., Nagy B. and Teska T. M. (1972) Ancient age of the Middle Marker Horizon, Onverwacht Group, Swaziland Sequence, South Africa. *Earth Planet. Sci. Lett.* **14**, 360–366.

Hurst R. W., Bridgewater D., Collerson K. D. and Wetherill G. W. (1975) 3600 m.y. Rb/Sr ages in very early Archean gneisses from Saglek Bay, Labrador. *Earth Planet. Sci. Lett.* **27**, 393–403.

Husain L. (1974) ^{40}Ar-^{39}Ar chronology and cosmic ray exposure ages of the Apollo 15 samples. *J. Geophys. Res.* **79**, 2588–2606.

Husain L. and Schaeffer O. A. (1975) Lunar evolution: the first 600 million years. *Geophys. Res. Lett.* **2**, 29–32.

Husain L., Schaeffer O. A., Funkhouser J. and Sutter J. (1972) The ages of lunar material from Fra Mauro, Hadley Rille and Spur Crater. *Proc. Lunar Sci. Conf. 3rd*, p. 1557–1567.

Irving A. J. (1975) Chemical, mineralogical and textural systematics of non-mare melt rocks: implications for lunar impact and volcanic processes. *Proc. Lunar Sci. Conf. 6th*, p. 363–394.

Irving E. (1977) Drift of the major continental blocks since the Devonian. *Nature* **270**, 304–309.

Irving E. and McGlynn J. C. (1976) Proterozoic magnetostratigraphy and the tectonic evolution of Laurentia. *Phil. Trans. Roy. Soc. Lond.* **A280**, 605–628.

Ito K. and Kennedy G. C. (1967) Melting and phase relations in a natural peridotite to 40 Kilobars. *Amer. J. Sci.* **265**, 519–538.

Jackson E. D. and Wright T. L. (1970) Xenoliths in the Honolulu volcanic series, Hawaii. *J. Petrol.* **11**, 405–430.

Jaffey A. H., Flynn K. F., Glendenin L. E., Bentley W. C. and Essling A. M. (1971) Precision measurement of half-lives and specific activities of ^{235}U and ^{238}U. *Phys. Rev.* **C4**, 1889–1906.

Jahn B. M. and Condie K. C. (1976) On the age of Rhodesian greenstone belts. *Contrib. Mineral. Petrol.* **57**, 317–330.

Jahn B. J. and Shih C.-Y. (1974) On the age of the Onverwacht Group, Swaziland Sequence, South Africa. *Geochim. Cosmochim. Acta* **38**, 873–885.

Jahn B., Shih C.-Y. and Murthy V. R. (1974) Trace element geochemistry of Archean volcanic rocks. *Geochim. Cosmochim. Acta* **38**, 611–627.

James O. B. and Hammarstrom J. G. (1977) Petrology of four clasts from consortium breccia 73215. *Proc. Lunar Sci. Conf. 8th*, p. 2459–2494.

James O. B. and Jackson E. D. (1970) Petrology of the Apollo 11 ilmenite basalts. *J. Geophys. Res.* **75**, 5793–5824.

James O. B. and Wright T. L. (1972) Apollo 11 and 12 mare basalts and gabbros: classification, compositional variations and possible petrogenetic relations. *Geol. Soc. Amer. Bull.* **83**, 2357–2382.

Javoy M. (1970) Utilisation des isotopes de l'oxygène en magmatologie. Thesis Université de Paris VI.

Javoy M. and Allègre C. J. (1969) Coupled $^{18}O/^{16}O$ and $^{87}Sr/^{86}Sr$ determinations in volcanic rocks and magmatologic implications. *EOS (Trans. Amer. Geophys. Union)* **50**, 356.

Javoy M., Pineau F. and Iiyama I. (1978) Experimental determination of the isotopic fractionation between gaseous CO_2 and carbon dissolved in tholeiite magmas. *Contrib. Mineral. Petrol.* **67**, 35–39.

Jessberger E. K., Dominik B., Kirsten T. and Staudacher Th. (1977a) New Ar^{40}-Ar^{39} ages of Apollo 16 breccias and 4.42 AE old anorthosites (abstract). In *Lunar Science VIII*, p. 511–513. The Lunar Science Institute, Houston.

Jessberger E. K., Horn P. and Kirsten T. (1975) ^{39}Ar-^{40}Ar dating of lunar rocks: a methodical investigation of mare basalt 75075 (abstract). In *Lunar Science VI*, p. 441–443. The Lunar Science Institute, Houston.

Jessberger E. K., Huneke J. C., Podosek F. A. and Wasserburg G. J. (1974a) High resolution argon analysis of neutron-irradiated Apollo 16 rocks and separated minerals. *Proc. Lunar Sci. Conf. 5th*, p. 1419–1449.

Jessberger E. K., Huneke J. C. and Wasserburg G. J. (1974b) Evidence for a ~4.5 AE age of plagioclase clasts in a lunar highland breccia. *Nature* **248**, 199–202.

Jessberger E. K., Kirsten T. and Staudacher Th. (1977b) One rock and many ages—Further K-Ar data on consortium breccia 73215. *Proc. Lunar Sci. Conf. 8th*, p. 2567–2580.

Jones L. M., Hartley M. E. and Walker R. L. (1973) Strontium isotope composition of Alpine-type ultramafic rocks in the Lake Chatuge district, Georgia, N. Carolina. *Contrib. Mineral. Petrol.* **38**, 321–337.

Kaneoka I. and Takaoka N. (1978) Excess Xe^{129} and high He^3/He^4 ratios in olivine phenocrysts of Kapuho lava and xenolithic dunites from Hawaii. *Earth Planet. Sci. Lett.* **39**, 382–386.

Katcoff S., Schaeffer O. A. and Hastings J. M. (1951) Half-life of ^{129}I and the age of the elements. *Phys. Rev.* **82**, 688–690.

Kaula W. M., Schubert G., Lingenfelter R. E., Sjogren W. L. and Wollenhaupt W. R. (1972) Analysis and interpretation of lunar laser altimetry. *Proc. Lunar Sci. Conf. 3rd*, p. 2189–2204.

Kaula W. M., Schubert G., Lingenfelter R. E., Sjogren W. L. and Wollenhaupt W. R. (1973) Lunar topography from Apollo 15 and 16 laser altimetry. *Proc. Lunar Sci. Conf. 4th*, p. 2811–2819.

Kaula W. M., Schubert G., Lingenfelter R. E., Sjogren W. L. and Wollenhaupt W. R. (1974) Apollo laser altimetry and inferences as to lunar structure. *Proc. Lunar Sci. Conf. 5th*, p. 3049–3058.

Kay R. (1977) Geochemical constraints on the origin of Aleutian magmas. In *Island Arcs, Deep-Sea trenches and Back-Arc basins* (M. Talwani and W. C. Pittman, III, eds.), p. 229–242. Maurice Ewing Series 1, Amer. Geophys. Union, Washington, D. C.

Kay R. W. and Hubbard N. J. (1978) Trace elements in ocean ridge basalts. *Earth Planet. Sci. Lett.* **38**, 95–116.

Kay R. W., Hubbard N. J. and Gast P. W. (1970) Chemical characteristics and origin of oceanic ridge volcanic rocks. *J. Geophys. Res.* **75**, 1585.

Kay R. W., Sun W. W. and Lee-Hu C. N. (1978) Pb and Sr isotopes in volcanic rocks from the Aleutian Islands and Pribilof Islands, Alaska. *Geochim. Cosmochim. Acta* **42**, 263–273.

Kesson S. E. (1975) Mare basalts: melting experiments and petrogenetic interpretations. *Proc. Lunar Sci. Conf. 6th*, p. 921–944.

Kigoshi K. (1967) Ionium dating of igneous rocks. *Science* **156**, 932–934.

Kirsten T. (1968) Incorporation of rare gases in solidifying enstatite melts. *J. Geophys. Res.* **73**, 2807–2810.

Kirsten T. and Horn P. (1974) Chronology of Taurus-Littrow region-III: ages of mare basalts and highland breccias and some remarks about the interpretation of lunar highland rock ages. *Proc. Lunar Sci. Conf. 5th*, p. 1451–1475.

Kirsten T., Horn P. and Heymann D. (1973) Chronology of the Taurus-Littrow region-I: ages of two major rock types from Apollo 17 site. *Earth Planet. Sci. Lett.* **20**, 125–130.

Kistler R. W. and Peterman Z. E. (1973) Variation in Sr, Rb, K, Na and initial $^{87}Sr/^{86}Sr$ in Mesozoic granitic rocks and intruded wall rocks in central California. *Geol. Soc. Amer. Bull.* **84**, 3489–3512.

Klerkx J., Deutsch S. and DePaepe P. (1974) Rubidium, strontium content and strontium isotopic composition of strongly alkalic basalt rocks from the Cape Verde Islands. *Contrib. Mineral. Petrol.* **45**, 107.

Kornprobst J. (1969) Le massif ultrabasique des Beni Bouchera (Rif Interne Maroc). *Contrib. Mineral. Petrol.* **23**, 283–322.

Kouvo O. (1976) Chronology of orogenic events in the Svecokarelides, Finland (abstract). *Program 4th European Colloquium of Geochron. Cosmochron. and Isotope Geol.*, Amsterdam, April, p. 59.

Kramers J. D. (1977) Lead and strontium isotopes in cretaceous kimberlites and mantle derived xenoliths from S. Africa. *Earth Planet. Sci. Lett.* **34**, 419–431.

Krogh T. E. and Davis G. L. (1969) Old isotopic ages in the northwestern Grenville Province, Ontario. *Geol. Assoc. Can. Spec. Paper No. 5*, p. 189–192.

Kroner A. (1976) Proterozoic crustal evolution in parts of southern Africa and evidence for extensive sialic crust since the end of the Archaean. *Phil. Trans. Roy. Soc. Lond.* **A280**, 541–553.

Kroner A. (1977) Precambrian mobile belts of southern and eastern Africa—ancient sutures of sites of ensialic mobility? A case for crustal evolution towards plate tectonics. *Tectonophys.* **40**, 101–135.

Kudo A. M., Brookins D. G. and Laughlin A. W. (1972) Sr-isotopic disequilibrium in lherzolites from the Puerco Necks, New Mexico. *Earth Planet. Sci. Lett.* **15**, 291–295.

Kushiro I. (1973) Partial melting of garnet lherzolites in kimberlites at high pressure. In *Lesotho Kimberlites* (P. H. Nixon, ed.), p. 294–299. Lesotho National Devel. Corp., Lesotho, Africa.

Kushiro I. and Kuno H. (1963) Origin of primary basalt magmas and classification of basaltic rocks. *J. Petrol.* **4**, 75–89.

Kyle P. R. (1980) Development of heterogeneities in the subcontinental mantle: evidence from the Ferrar Group, Antarctica. *Contrib. Mineral. Petrol.* **73**, 89–104.

Kyser T. K. and O'Neil J. R. (1978) Oxygen isotope relations among oceanic tholeiites, alkali basalts and ultramafic nodules. *U. S. Geol. Surv. Open-File Report 78-701*, p. 237–240.

Lakatos S. and Miller D. S. (1972) Evidence for the effect of water content on fission-track annealing in volcanic glass. *Earth Planet. Sci. Lett.* **14**, 128–130.

Lambert I. B. and Wyllie P. J. (1968) Stability of hornblende and a model for the low velocity zone. *Nature* **219**, 1240–1241.

Lancet M. S. and Anders E. (1973) Solubilities of noble gases in magnetite: implications for planetary gases in meteorites. *Geochim. Cosmochim. Acta* **37**, 1371–1388.

Lancet M. W. and Lancet K. (1971) Cosmic-ray and gas-retention ages of the Chassigny meteorite. *Meteoritics* **6**, 81–86.

Langmuir C. H., Vocke R. D., Jr., Hanson G. N. and Hart S. R. (1978) A general mixing equation with applications to Icelandic basalts. *Earth Planet. Sci. Lett.* **18**, 380–392.

Langseth M. G., Jr., LePichon X. and Ewing M. (1966) Crustal structure of mid-ocean ridges. *J. Geophys. Res.* **71**, 5321–5355.

Lanphere M. A. (1968) Sr-Rb-K and Sr-isotopic relationships in ultramafic rocks, S.E. Alaska. *Earth Planet. Sci. Lett.* **4**, 185–190.

Lanphere M. A. and Reed B. L. (1973) Timing of Mesozoic and Cenozoic plutonic events in circum-Pacific North America. *Geol. Soc. Amer. Bull.* **84**, 3773–3782.

Larson R. L. and Pitman W. C., III (1972) World-wide correlation of Mesozoic magnetic anomalies and its implications. *Geol. Soc. Amer. Bull.* **83**, 3645–3662.

Laughlin A. W., Brookins D. G., Kudo A. M. and Causey J. D. (1971) Chemical and Sr-Isotopic investigations of ultramafic inclusions and basalt. Bandera Crater, New Mexico. *Geochim. Cosmochim. Acta* **35**, 107–113.

Laul J. C., Schmitt R. A. and Wakita H. (1975) Apollo 17 basalts and lunar evolution constraints (abstract). In *Papers Presented to the Conference on Origins of Mare Basalts and their Implications for Lunar Evolution*, p. 91–93. The Lunar Science Institute, Houston.

Lee T. and Papanastassiou D. A. (1974) ^{26}Mg isotopic anomalies in the Allende meteorite and correlation with O and Sr effects. *Geophys. Res. Lett.* **1**, 225–228.

Lee T., Schramm D. N., Welfel J. P. and Blake J. B. (1978) On the apparent conflict between the time scales inferred from the cosmochronometers ^{129}I, ^{244}Pu and ^{26}Al. In *Short Papers of the Fourth Int'l. Conf. Geochron. Cosmochron. 1978* (R. E. Zartman, ed.), p. 246–247. U.S. Geol. Surv. Open File Rept. 78-701.

Leeman W. P. (1977) Comparison of Rb/Sr, U/Pb, and rare earth characteristics of sub-continental and sub-oceanic mantle regions. In *Magma Genesis: Proc. Amer. Geophys. Union Chapman Conf.* (H. J. B. Dick, ed.), p. 149–168. Bulletin 96 of the Oregon Dept. of Geol. and Mineral Industries, Portland.

Leeman W. P. and Manton W. I. (1971) Strontium isotopic composition of basaltic lavas from the Snake River Plain, southern Idaho. *Earth Planet. Sci. Lett.* **11**, 420–434.

Leggo P. J. and Hutchison R. (1968) A Rb-Sr isotope study of ultrabasic xenoliths and their basaltic host rocks from the Massif Central, France. *Earth Planet. Sci. Lett.* **5**, 71–75.

LeRoux L. J. and Glendenin L. E. (1963) Half-life of ^{232}Th. *Proc. Nat. Meet. on Nucl. Energy*, p. 83–94. Pretoria, S. Africa.

Lewis R. S., Gros J. and Anders E. (1977) Isotopic anomalies of noble gases in meteorites and their origins 2. Separated minerals from Allende. *J. Geophys. Res.* **82**, 779–792.

Lewis R. S., Srinivasan B. and Anders E. (1975) Host phase of a strange xenon component in Allende. *Science* **190**, 1251–1262.

Livingston D. E. and Damon P. E. (1968) The ages of stratified Precambrian rock sequences in Central Arizona and Southern Sonora. *Can. J. Earth Sci.* **5**, 763–772.

Lloyd F. E. and Bailey D. K. (1975) Light element metasomatism of the continental mantle: the evidence and the consequences. *Phys. Chem. Earth* **9**, 389–416.

Loney R. A., Himmelberg G. R. and Coleman R. G. (1971) Structure and petrology of the alpine-type peridotite at Burro mountain, California. *J. Petrol.* **12**, 245–309.

Longhi J., Walker D., Grove T. L., Stolper E. M. and Hays J. F. (1974) The petrology of the Apollo 17 mare basalts. *Proc. Lunar Sci. Conf. 5th*, p. 447–469.

Loubet M., Shimizu N. and Allègre C. J. (1975) Rare-earth elements in alpine peridotites. *Contrib. Mineral. Petrol.* **53**, 1–12.

LSPET (Lunar Sample Preliminary Examination Team) (1972) The Apollo 15 lunar samples: a preliminary description. *Science* **175**, 363.

Lugmair G. W. (1974) Sm-Nd ages: a new dating method (abstract). *Meteoritics* **9**, 369.

Lugmair G. W. (1975) Sm-Nd systematics of some Apollo 17 basalts. In *Papers Presented to the Conference on Origins of Mare Basalts and their Implications for Lunar Evolution*, p. 107–110. The Lunar Science Institute, Houston.

Lugmair G. W. and Carlson R. W. (1978) The Sm-Nd history of KREEP. *Proc. Lunar Planet. Sci. Conf. 9th*, p. 689–704.

Lugmair G. W. and Marti K. (1977) Sm-Nd-Pu timepieces in the Angra dos Reis meteorite. *Earth Planet. Sci. Lett.* **35**, 273–284.

Lugmair G. W. and Marti K. (1978) Lunar initial ^{143}Nd/^{144}Nd: differential evolution of the lunar crust and mantle. *Earth Planet. Sci. Lett.* **39**, 349–357.

Lugmair G. W., Marti K., Kurtz J. P. and Scheinin N. B. (1976) History and genesis of lunar troctolite 76535 or: how old is old? *Proc. Lunar Sci. Conf. 7th*, p. 2009–2033.

Lugmair G. W. and Scheinin N. B. (1975) Sm-Nd systematics of the Stannern meteorite (abstract). *Meteoritics* **10**, 447–448.

Lugmair G. W., Scheinin N. B. and Carlson R. W. (1977) Sm-Nd systematics of the Serra de Mage eucrite (abstract). *Meteoritics* **12**, 300–301.

Lugmair G. W., Scheinin N. B. and Marti K. (1975) Sm-Nd age and history of Apollo 17 basalt 75075: evidence for early differentiation of the lunar exterior. *Proc. Lunar Sci. Conf. 6th*, p. 1419–1429.

Lunatic Asylum (1970a) Ages, irradiation history and chemical composition of lunar rocks from the Sea of Tranquility. *Science* **167**, 463–466.

Lunatic Asylum (1970b) Mineralogic and isotopic investigations on lunar rock 12013. *Earth Planet. Sci. Lett.* **9**, 137–163.

Lunatic Asylum (1978) Petrology, chemistry, age and irradiation history of Luna 24 samples. In *Mare Crisium: The View from Luna 24* (R. B. Merrill and J. J. Papike, eds.), p. 657–678. Pergamon, N. Y.

Lund R. H. (1956) Igneous and metamorphic rocks of the Minnesota River Valley. *Geol. Soc. Amer. Bull.* **67**, 1475–1490.

Lupton J. E. and Craig H. (1975) Excess He^3 in oceanic basalts: evidence for terrestrial primordial helium. *Earth Planet. Sci. Lett.* **26**, 133–139.

Lyon T. D. B. and Bowes D. R. (1977) Rb-Sr, U-Pb and K-Ar isotopic study of the Lewisian Complex between Durness and Loch Laxford, Scotland. *Krystalinikum* **13**, 53–72.

Ma M.-S., Musali A. V. and Schmitt R. A. (1977) Genesis of the Angra dos Reis and other achondritic meteorites. *Earth Planet. Sci. Lett.* **35**, 331–346.

MacGregor I. D. (1964) A study of contact metamorphic aureole surrounding the Mt. Albert ultramafic intrusion. Ph.D. Thesis, Princeton University. 195 pp.

MacGregor I. D. (1975) Petrologic and thermal structure of the upper mantle beneath South Africa in the Cretaceous. *Phys Chem. Earth.* **9**, 455–466.

MacGregor I. D. and Basu A. R. (1974) Thermal structure on the lithosphere: a petrologic model. *Science* **185**, 1007–1011.

MacGregor I. D. and Carter J. L. (1970) The chemistry of clinopyroxenes and garnets of eclogite and peridotite xenoliths from the Roberts Victor Mine, South Africa. *Phys. Earth Planet Inter.* **3**, 391–397.

Magaritz M. and Taylor H. P. (1974) Oxygen and hydrogen isotope studies of serpentinization in the Troodos ophiolite complex, Cyprus. *Earth Planet. Sci. Lett.* **23**, 8–14.

Magaritz M. and Taylor H. P. (1976) Oxygen, hydrogen, and carbon isotope studies of the Franciscan formation, Coast Ranges, California. *Geochim. Cosmochim. Acta* **40**, 215–234.

Magaritz M., Whitford D. J. and James D. E. (1978) Oxygen isotopes and the origin of high $^{87}Sr/^{86}Sr$ andesites. *Earth Planet. Sci. Lett.* **40**, 220–230.

Mamyrin B. A., Anufriev G. S., Kamenskiy I. L. and Tolstikhin I. N. (1970) Determination of the isotopic composition of atmospheric helium. *Geochim. Int.* **7**, 498.

Mamyrin B. A., Tolstikhin I. N., Anufriev G. S. and Kamenskiy I. L. (1969) Anomalous isotopic composition of helium in volcanic gases. *Dokl. Akad. nauk S.S.S.R.* **184**, 1197.

Manhes G., Allègre C. J., Dupre B. and Hamelin B. (1979) Lead-lead systematics, the age of the Earth, and the chemical evolution of our planet in a new representation space. *Earth Planet. Sci. Lett* **44**, 91–104.

Manhes G., Tatsomoto M., Unruh D., Birck J. L. and Allègre C. J. (1975) Comparative ^{238}U-^{206}Pb, ^{235}U-^{207}Pb, ^{232}Th-^{208}Pb, ^{206}Pb-^{207}Pb, and ^{87}Rb-^{87}Sr ages of basaltic achondrites and early evolution of the solar system (abstract). In *Lunar Science VI*, p. 546–547. The Lunar Science Institute, Houston.

Manuel O. K., Hennecke E. W. and Sabu D. D. (1972) Xenon in carbonaceous chondrites. *Nature* **240**, 99–101.

Mark R. K., Lee-Hu C. and Wetherill G. W. (1974) Equilibration and ages: Rb-Sr studies of breccias 14321 and 15265. *Proc. Lunar Sci. Conf. 4th*, p. 1477–1485.

Mark R. K., Lee-Hu C. and Wetherill G. W. (1975) More on Rb-Sr in lunar breccia 14321. *Proc. Lunar Sci. Conf. 6th*, p. 1501–1507.

Mark R. K., Cliff R. A., Lee-Hu C. and Wetherill G. W. (1973) Rb-Sr studies of lunar breccias and soils. *Proc. Lunar Sci. Conf. 4th*, p. 1785–1795.

Matsuhisa Y. (1979) Oxygen isotopic compositions of volcanic rocks from the East Japan Island arcs and their bearing on petrogenesis. *J. Volcanol. Geotherm. Res.* **5**, 271–296.

Mattinson J. M., Tilton G. R., Todt W. and Chen J. H. (1977) Lead isotope studies of mare basalt 70017. *Proc. Lunar Sci. Conf. 8th*, p. 1473–1487.

Maurer P., Eberhardt P., Geiss J., Grögler N., Stettler A., Brown G. M., Peckett A. and Krähenbühl U. (1978) Pre-Imbrian craters and basins: ages, compositions and excavation depths of Apollo 16 breccias. *Geochim. Cosmochim. Acta* **42**, 1687–1720.

McDougall I. (1976) Geochemistry and origin of basalt of the Columbia River Group, Oregon and Washington. *Geol. Soc. Amer. Bull.* **87**, 777–792.

McGetchin T. R., Settle M. and Head J. W. (1973) Radial thickness variation in impact crater ejecta: implications for lunar basin deposits. *Earth Planet. Sci. Lett.* **20**, 226–236.

McGetchin T. R. and Ullrich G. W. (1973) Xenoliths in maars and diatremes with inferences for the Moon, Mars and Venus. *J. Geophys. Res.* **78**, 1833–1853.

Meijer A. (1976) Pb and Sr isotopic data bearing on the origin of volcanic rocks from the Mariana Island-arc system. *Geol. Soc. Amer. Bull.* **87**, 1358–1369.

Menzies M. A. (1976) Rare-earth geochemistry of fused alpine and ophiolitic lherzolites. I. Othris, Lanzo and Troodos. *Geochim. Cosmochim. Acta* **40**, 645–656.

Menzies M. A. and Allen C. (1974) Plagioclase lherzolite-residual mantle relationships within two eastern Mediterranean ophiolites. *Contrib. Mineral. Petrol.* **45**, 197–213.

Menzies M. A. and Murthy V. R. (1976) Sr-isotopic composition of clinopyroxenes from some Mediterranean alpine lherzolites. *Geochim. Cosmochim. Acta* **40**, 1577–1581.

Menzies M. A. and Murthy V. R. (1978a) Strontium isotope geochemistry of alpine tectonite lherzolites: data compatible with a mantle origin. *Earth Planet. Sci. Lett.* **38**, 346–354.

Menzies M. A. and Murthy V. R. (1978b) Sr-isotope geochemistry of hydrous lithosphere in sub-oceanic and sub-continental regions. *4th Int'l Conf. Geochron. Cosmochron. Isotope Geology* (Extended Abstracts.) Snowmass at Aspen, Colorado.

Merrihue C. M. and Turner G. (1966) Potassium-argon dating by activation with fast neutrons. *J. Geophys. Res.* **71**, 2852–2857.

Merrill R. B. and Wyllie P. J. (1975) Kaersutite and kaersutite eclogite from Kakanvi, New Zealand-water excess and water deficient melting to 30 kilobars. *Geol. Soc. Amer. Bull.* **86**, 555–570.

Metzger A. E., Trombka J. I., Reedy R. C. and Arnold J. R. (1974) Element concentrations from lunar orbital gamma-ray measurements. *Proc. Lunar Sci. Conf. 5th*, p. 1067–1078.

Meyer C., Jr., Brett R., Hubbard N. J., Morrison D. A., McKay D. S., Aitken F. K., Takeda H. and Schonfeld E. (1971) Mineralogy, chemistry and origin of the KREEP component in soil samples from the Ocean of Storms. *Proc. Lunar Sci. Conf. 2nd*, p. 393–411.

Meyer H. O. A. (1977) Mineralogy of the upper mantle: a review of the minerals in mantle xenoliths from Kimberlite. *Earth Sci. Rev.* **13**, 251–281.

Michard-Vitrac A., Albarede F., Dupuis C. and Taylor H. P. (1979) The genesis of Varican (Hercynian) plutonic rocks. Inferences from Sr, Pb and O studies on the Maladeta igneous complex, Central Pyrenees (Spain). *Contrib. Mineral. Petrol.* **72**, 57–72.

Minster J. B., Jordan T. H., Molnar P. and Haines E. (1974) Numerical modeling of instantaneous plate tectonics. *Geophys. J. Roy. Astron. Soc.* **36**, 541–576.

Minster J. F. and Allègre C. J. (1978) Systematic use of trace elements in igneous processes III. Inverse problem of batch partial melting in volcanic suite. *Contrib. Mineral. Petrol.* **68**, 37–52.

Minster J. F., Birck J. L. and Allègre C. J. (1976) $^{87}Rb/^{87}Sr$ constraints on the primitive chronology of meteorites (abstract). *Meteoritics* **11**, 336–337.

Minster J. F., Minster J. B., Treuil M. and Allègre C. J. (1977) Systematic use of trace elements in igneous processes. Part II. Inverse problem of fractional crystallization process in volcanic suites. *Contrib. Mineral. Petrol.* **61**, 49–77.

Miyashiro A. (1974) Volcanic rock series in island arcs and active continental margins. *Amer. J. Sci.* **274**, 321–355.

Molnar P. and Atwater T. (1973) Relative motion of hot spots in the mantle. *Nature* **246**, 288–291.

Montigny R., Javoy M. and Allègre C. J. (1969) Le problème des andésites. Etude du volcanisme quaternaire du Costa Rica (Amérique Centrale) à l'aide des traceurs couplés $^{87}Sr/^{86}Sr$ et $^{18}O/^{16}O$. *Bull. Soc. Geol. France* **XI**, 794–799.

Moorbath S. (1976) Ages, isotopes, and evolution of Precambrian continental crust. *Chem. Geol.* **20**, 151–187.

Moorbath S., O'Nions R. K. and Pankhurst R. J. (1973) Early Archean age for the Isua iron formation, West Greenland. *Nature* **245**, 138–139.

Moorbath S., O'Nions R. K. and Pankhurst R. J. (1975) The evolution of early Precambrian crustal rocks at Isua, West Greenland-geochemical and isotopic evidence. *Earth Planet. Sci. Lett.* **27**, 229–239.

Moorbath S., O'Nions R. K., Pankhurst R. J., Gale N. H. and McGragor V. R. (1972) Further Rb/Sr age determinations on the very early Precambrian rocks of the Godthaab district, West Greenland. *Nat. Phys. Sci.* **240**, 78–82.

Moorbath S. and Welke H. (1969) Lead isotope studies on igneous rocks from the Isle of Skye, Northwest Scotland. *Earth Planet. Sci. Lett.* **5**, 217–230.

Moorbath S., Wilson J. F., Goodwin R. and Humm M. (1977) Further Rb-Sr age and isotope data on Early and Late Archean rocks from the Rhodesian craton. *Precambrian Res.* **5**, 229–240.

Moore J. (1965) Petrology of deep sea basalt near Hawaii. *Amer. J. Sci.* **263**, 40–52.

Moore J. M., Jr. (1977) Orogenic volcanism in the Proterozoic of Canada. In *Volcanic Regimes in Canada* (W. R. A. Baragar, L. C. Coleman and J. M. Hall, eds.), p. 127–148. Geol. Assoc. Can. Spec. Paper No. 16.

Moores E. M. (1973) Geotectonic significance of ultramafic rocks. *Earth Sci. Rev.* **9**, 241–258.

Moores E. M. and Vine J. F. (1971) The Troodos Massif, Cyprus and other ophiolites as oceanic crust: evaluation and implications. *Trans. Roy. Soc. Lond.* **A268**, 443–466.

Morand P., Condomines M., Allè C. J. (1978) Desequilibres ^{230}Th-^{238}U dans quelques lavas de la Chaîne des Puys. *C. R. Acad. Sci. Paris* **286**, 1845.

Morgan J. W., Higuchi H., Takahashi H. and Hertogen J. (1978) A "chondritic" eucrite parent body: inference from trace elements. *Geochim. Cosmochim. Acta* **42**, 27–38.

Morgan J. W., Ganapathy R., Higuchi H., Krähenbühl U. and Anders E. (1974) Lunar basins: tentative characterization of projectiles, from meteoritic elements in Apollo 17 boulders. *Proc. Lunar Sci. Conf. 5th*, p. 1703–1736.

Morrison T. J. and Johnstone N. B. (1954) Solubilities of the inert gases in water. *J. Chem. Soc.* **1954**, 3441.

Muehlenbachs K. and Clayton R. N. (1972) Oxygen isotope studies of fresh and weathered submarine basalts. *Can. J. Earth Sci.* **9**, 172–184.

Muehlenbachs K. and Kushiro I. (1974) Oxygen isotope exchange and equilibrium of silicates CO_2 or O_2. *Carnegie Inst. Wash., Yearb.* **73**, p. 232–236.

Muehlenbachs K., Anderson A. T. and Sigvaldason G. E. (1974) Low ^{18}O basalts from Iceland. *Geochim. Cosmochim. Acta* **38**, 557–558.

Murase T. and McBirney A. R. (1970) Viscosity of lunar lavas. *Science* **167**, 1491–1493.

Murthy V. R., Coscio M. R., Jr. (1976) Rb-Sr ages and isotopic systematics of some Serenitatis mare basalts. *Proc. Lunar Sci. Conf. 7th*, p. 1529–1544.

Murthy V. R., Coscio M. R., Jr. and Sabelin T. (1977) Rb-Sr internal isochron and the initial $^{87}Sr/^{86}Sr$ for the Estherville mesosiderite. *Proc. Lunar Sci. Conf. 8th*, p. 177–186.

Murthy V. R., Evensen N. M., Jahn B. M. and Coscio M. R., Jr. (1971) Rb-Sr ages and elemental abundances of K, Rb, Sr and Ba in samples from the Ocean of Storms. *Geochim. Cosmochim. Acta* **35**, 1139–1153.

Murthy V. R., Evensen N. M., Jahn B., Coscio M. R., Jr., Dragon J. C. and Pepin R. O. (1972) Rubidium-Strontium and Potassium-Argon age of lunar sample 15555. *Science* **175**, 419–421.

Mysen B. O. and Boettcher A. L. (1975) Melting of a hydrous mantle, 1 Phase relations of natural peridotites at high pressures and temperatures with controlled activities of water, carbon dioxide, and hydrogen. *J. Petrol.* **16**, 520–548.

Mysen B. O., Arculus R. J. and Eggler D. H. (1975) Solubility of carbon dioxide in melt of andesite, tholeiite and olivine nephelinite composition to 30 kbars pressure. *Contrib. Mineral. Petrol.* **53**, 227–239.

Mysen B. O., Eggler D. H., Seitz M. G. and Holloway J. R. (1976) Carbon dioxide in silicate melts and crystals. Part I: solubility measurements. *Amer. J. Sci.* **276**, 455–479.

Nakamura N., Tatsumoto M., Nunes P. D., Unruh D. M., Schwab A. P. and Wildeman T. R. (1976) 4.4 b.y.-old clast in Boulder 7, Apollo 17: a comprehensive chronological study by U-Pb, Rb-Sr and Sm-Nd methods. *Proc. Lunar Sci. Conf. 7th*, p. 2309–2333.

Nakamura N., Unruh D. M., Gensho R. and Tatsumoto M. (1977a) Evolution history of lunar mare basalts: Apollo 15 samples revisited (abstract). In *Lunar Science VIII*, p. 712–713. The Lunar Science Institute, Houston.

Nakamura N., Unruh D. M., Tatsumoto M. and Hutchison R. (1977b) Nakhla: further evidence for a young crystallization age (abstract). *Meteoritics* **12**, 324–325.

Nelson D. C. and Dasch E. J. (1976) Disequilibrium of strontium isotopes between mineral phases of parental rocks during magma genesis: a discussion. *J. Volcanol. Geotherm. Res.* **1**, 183–191.

Neuman H., Mead J. and Vitaliano C. J. (1954) Trace element variation during fractional crystallization as calculated from the distribution law. *Geochim. Cosmochim. Acta* **16**, 90–100.

Nicolas A. and Jackson E. D. (1972) Repartition en deux provinces des peridotites des chaines alpine logeant la Mediterrance: implication geotectonique. *Sweiz. Min. Petr. Mitt.* **52**, 479–495.

Nicolaysen L. O. (1961) Graphic interpretation of discordant age measurements on metamorphic rocks. *Ann. N.Y. Acad. Sci.*, 198–206.

Nier A. O. (1939) The isotopic constitution of radiogenic leads and the measurement of geological time, I and II. *Phys. Rev.* **55**, 150–163.

Nishiizumi K., Regnier S. and Marti K. (1980) Cosmic ray exposure ages of chondrites, pre-irradiation and constancy of cosmic ray flux in the past. *Earth Planet. Sci. Lett.* **50**, 156–170.

Nixon P. H., Boyd F. R. and Boulier A. M. (1973) The evidence of kimberlite and its inclusions on the constitution of the outer part of the Earth. In *Lesotho Kimberlites* (P. H. Nixon, ed.), p. 56–88. Lesotho Nat'l. Dev. Corp., Maseru.

Nunes P. D. and Thurston P. C. (1978) Evolution of a single greenstone belt over 200 m.y.—a zircon study of the Uchi Lake Area, N.W. Ontario. *U.S. Geol. Survey Open File Rept. 78*-701, p. 469–471.

Nunes P. D., Pyke D. R. and Jensen L. S. (1978) Toward an absolute age stratigraphy for the Abitibi greenstone belt, Eastern Ontario. *Geol. Assoc. Can. and Geol. Soc. Amer. Abstracts: Joint Annual Mtg. Toronto*, p. 464–465.

Nunes P. D., Tatsumoto M. and Unruh D. M. (1974) U-Th-Pb systematics of some Apollo 17 lunar samples and implications for a lunar basin excavation chronology. *Proc. Lunar Sci. Conf. 5th*, p. 1487–1514.

Nyquist L. E. (1977) Lunar Rb-Sr chronology. *Phys. Chem. Earth* **10**, 103–142.

Nyquist L. E., Bansal B. M. and Wiesmann H. (1975) Rb-Sr ages and initial $^{87}Sr/^{86}Sr$ for Apollo 17 basalts and KREEP basalt 15386. *Proc. Lunar Sci. Conf. 6th*, p. 1445–1465.

Nyquist L. E., Bansal B. M. and Wiesmann H. (1976) Sr isotopic constraints on the petrogenesis of Apollo 17 mare basalts. *Proc. Lunar Sci. Conf. 7th*, p. 1507–1528.

Nyquist L. E., Bansal B. M., Wiesmann H. and Jahn B.-M. (1974) TaurusLittrow chronology: some constraints on early lunar crustal development. *Proc. Lunar Sci. Conf. 5th*, p. 1515–1539.

Nyquist L. E., Bansal B. M., Wooden J. L. and Wiesmann H. (1977) Sr-isotopic constraints on the petrogenesis of Apollo 12 mare basalts. *Proc. Lunar Sci. Conf. 8th*, p. 1383–1415.

Nyquist L. E., Bogard D. D., Wooden J. L., Wiesmann H., Shih C.-Y., Bansal B. M. and McKay G. (1979d) Early differentiation, late magmatism, and recent bombardment of the shergottite parent planet (abstract). *Meteoritics* **14**, 502.

Nyquist L., Funk H., Schultz L. and Signer P. (1973a) He, Ne and Ar in chondritic Ni-Fe as irradiation hardness sensors. *Geochim. Cosmochim. Acta* **37**, 1655–1685.

Nyquist L. E., Hubbard N. J., Gast P. W., Bansal B. M., Wiesmann H. and Jahn B. (1973) Rb-Sr systematics for chemically defined Apollo 15 and 16 materials. *Proc. Lunar Sci. Conf. 4th*, p. 1823–1846.

Nyquist L. E., Hubbard N. J., Gast P. W., Church S. E., Bansal B. M. and Wiesmann H. (1972) Rb-Sr systematics for chemically defined Apollo 14 breccias. *Proc. Lunar Sci. Conf. 3rd*, p. 1515–1530.

Nyquist L., Shih C., Wooden J., Bansal B. and Wiesmann H. (1979a) The Sr and Nd isotopic record of Apollo 12 basalts (abstract). In *Lunar and Planetary Science* X, p. 931–933. Lunar and Planetary Institute, Houston.

Nyquist L. E., Shih C.-Y., Wooden J. L., Bansal B. M. and Wiesmann H. (1979b) The Sr and Nd isotopic record of Apollo 12 basalts: implications for lunar geochemical evolution. *Proc. Lunar Planet. Sci. Conf. 10th*, p. 77–114.

Nyquist L., Wooden J., Bansal B., Wiesmann H., McKay G. and Bogard D. (1979c) Rb-Sr age of the Shergotty achondrite and implications for metamorphic resetting of isochron ages. *Geochim. Cosmochim. Acta* **43**, 1057–1074.

Oberli F., Huneke J. C. and Wasserburg G. J. (1979) U-Pb and K-Ar systematics of cataclysm and precataclysm lunar impactites (abstract). In *Lunar and Planetary Science X*, p. 940–942. Lunar and Planetary Institute, Houston.

Oberli F., McCulloch M. T., Tera F., Papanastassiou D. A. and Wasserburg G. J. (1978) Early lunar differentiation constraints from U-Th-Pb, Sm-Nd, and Rd-Sr model gas. In *Lunar and Planetary Science IX*, p. 832–834. Lunar and Planetary Institute, Houston.

O'Hara M. J. (1967) Crystal-liquid equilibria and the origin of ultramafic nodules in basic igneous rocks. In *Ultramafic and Related Rocks* (P. J. Wyllie, ed.), p. 346–349. Wiley, N.Y.

O'Hara M. J. and Mercy E. L. P. (1963) Petrology and petrogenesis of some garnetiferous peridotites. *Trans. Roy. Soc. Edinburgh* **65**, 251–314.

O'Hara M. J. and Yoder H. S. (1967) Formation and fractionation of basic magmas at high pressures. *Scottish J. Geol.* **3**, 67–117.

Ohmoto H. (1972) Systematics of sulfur and carbon isotopes in hydrothermal ore deposits. *Econ. Geol.* **67**, 551–578.

O'Neil J. R. and Clayton R. C. (1964) Oxygen isotope geothermometry. In *Isotopic and Cosmic Chemistry* (Craig, Miller, Wasserburg, eds.), p. 157–169. North Holland.

O'Neil J. R. and Taylor H. P. (1967) The oxygen isotope and cation exchange: chemistry of feldspars. *Amer. Mineral.* **52**, 1414–1437.

O'Nions R. K. and Pankhurst R. J. (1978) Early Archean rocks and the geochemical evolution of the Earth's Crust. *Earth Planet. Sci. Lett.* **38**, 211–236.

O'Nions R. K., Hamilton P. J. and Evensen N. M. (1977) Variations in $^{143}Nd/^{144}Nd$ and $^{87}Sr/^{86}Sr$ ratios in oceanic basalts. *Earth Planet. Sci. Lett.* **34**, 13–22.

Onuma N., Clayton R. N. and Mayeda T. K. (1970) Apollo 11 rocks: oxygen isotope fractionation between minerals and an estimate of the temperature of formation. *Apollo 11 Lunar Sci. Conf.*, p. 1429–1434.

Oosthuyzen E. J. (1970) The geochronology of a suite of rocks from the granitic terrain surrounding the Barberton Mountain Land. Ph.D. Thesis, University of Witwatersrand, Johannesburg.

Oversby V. M. (1975) Lead isotopic systematics and ages of Archean acid intrusives in the Kalgoorlie-Norseman area, Western Australia. *Geochim. Cosmochim. Acta* **39**, 1107–1125.

Oversby V. M. (1976) Isotopic ages and geochemistry of Archean acid igneous rocks from the Pilbara, Western Australia. *Geochim. Cosmochim. Acta* **40**, 817–829.

Oversby V. M. (1978) Lead isotopes in Archean plutonic rocks. *Earth Planet. Sci. Lett.* **38**, 237–248.

Oversby V. M. and Ewart A. (1972) Lead isotopic compositions of Tonga-Kermadec volcanics and their petrogenetic significance. *Contrib. Mineral. Petrol.* **37**, 181–210.

Ozima M. and Alexander E. C., Jr. (1976) Rare gas fractionation patterns in terrestrial samples and the earth-atmosphere evolution model. *Rev. Geophys. Space Phys.* **14**, 385–390.

Page R. W. (1978) Response of U-Pb zircon and Rb-Sr total-rock and mineral systems to low-grade regional metamorphism in Proterozoic igneous rocks, Mount Isa, Australia. *J. Geol. Soc. Australia* **25**, 141–164.

Page R. W., Blake D. H. and Mahon M. W. (1976) Geochronology and related aspects of acid volcanics, associated granites, and other Proterozoic rocks in the Granites-Tanami region, northwestern Australia. *BMR J. of Australian Geol. and Geophys.* **1**, 1–13.

Palme H. (1977) On the age of KREEP. *Geochim. Cosmochim. Acta* **41**, 1791–1801.

Palme H. and Wänke H. (1975) A unified trace-element model for the evolution of the lunar crust and mantle. *Proc. Lunar Sci. Conf. 6th*, p. 1179–1202.

Pankhurst R. J. (1968) Strontium isotope and geochronological studies of the basic igneous province of northeast Scotland. Ph.D. Thesis, Oxford Univ., England.

Pankhurst R. J., Walsh J. N., Beckinsale R. D. and Skelhorn R. R. (1978) Isotopic and other geochemical evidence for the origin of the Loch UISG Granophyre, Isle of Mull, Scotland. *Earth Planet. Sci. Lett.* **39**, 355–363.

Papanastassiou D. A. (1970) The determination of small time differences in the formation of planetary objects. Ph.D. Thesis, Calif. Inst. of Tech., Pasadena.

Papanastassiou D. A. and Wasserburg G. J. (1969) Initial strontium isotopic abundances and the resolution of small time differences in the formation of planetary objects. *Earth Planet. Sci. Lett.* **5**, 361–376.

Papanastassiou D. A. and Wasserburg G. J. (1970) Rb-Sr ages from the Ocean of Storms. *Earth Planet. Sci. Lett.* **8**, 269–278.

Papanastassiou D. A. and Wasserburg G. J. (1971a) Lunar chronology and evolution from Rb-Sr studies of Apollo 11 and 12 samples. *Earth Planet. Sci. Lett.* **11**, 37–62.

Papanastassiou D. A. and Wasserburg G. J. (1971b) Rb-Sr ages of igneous rocks from the Apollo 14 mission and the age of Fra Mauro formation. *Earth Planet. Sci. Lett.* **12**, 36–48.

Papanastassiou D. A. and Wasserburg G. J. (1972a) Rb-Sr systematics of Luna 20 and Apollo 16 samples. *Earth Planet. Sci. Lett.* **17**, 52–63.

Papanastassiou D. A. and Wasserburg G. J. (1972b) Rb-Sr age of a Luna 16 basalt and the model age of lunar soils. *Earth Planet. Sci. Lett.* **13**, 368–374.

Papanastassiou D. A. and Wasserburg G. J. (1973) Rb-Sr ages and initial strontium in basalts from Apollo 15. *Earth Planet. Sci. Lett.* **17**, 324–337.

Papanastassiou D. A. and Wasserburg G. J. (1974) Evidence for late formation and young metamorphism in the achondrite Nakhla. *Geophys. Res. Lett.* **1**, 23–26.

Papanastassiou D. A. and Wasserburg G. J. (1975a) A Rb-Sr study of Apollo 17 boulder 3: dunite clast microclasts and matrix (abstract). In *Lunar Science VI*, p. 631–633. The Lunar Science Institute, Houston.

Papanastassiou D. A. and Wasserburg G. J. (1975b) Rb-Sr study of a lunar dunite and evidence for early lunar differentiates. *Proc Lunar Sci. Conf. 6th*, p. 1467–1489.

Papanastassiou D. A. and Wasserburg G. J. (1976a) Rb-Sr age of troctolite 76535. *Proc. Lunar Sci. Conf. 7th*, p. 2035–2054.

Papanastassiou D. A. and Wasserburg G. J. (1976b) Early lunar differentiates and lunar initial $^{87}Sr/^{86}Sr$ (abstract). In *Lunar Science VII*, p. 665–667. The Lunar Science Institute, Houston.

Papanastassiou D. A., DePaolo D. J., Tera F. and Wasserburg G. J. (1977a) An isotopic triptych on mare basalts: Rb-Sr, Sm-Nd, U-Pb (abstract). In *Lunar Science VIII*, p. 750–752. The Lunar Science Institute, Houston.

Papanastassiou D. A., DePaolo D. J. and Wasserburg G. J. (1977b) Rb-Sr and Sm-Nd chronology and genealogy of mare basalts from the Sea of Tranquillity. *Proc. Lunar Sci. Conf. 8th*, p. 1639–1672.

Papanastassiou D. A., Rajan R. S., Huneke J. C. and Wasserburg G. J. (1974) Rb-Sr ages and lunar analogs in a basaltic achondrite; implications for early solar system chronologies (abstract). In *Lunar Science V*, p. 583–585.

Papanastassiou D. A., Wasserburg G. J. and Burnett D. S. (1970) Rb-Sr ages of lunar rocks from the Sea of Tranquillity. *Earth Planet. Sci. Lett.* **8**, 1–19.

Papike J. J., Bence A. E. and Lindsley D. H. (1974) Mare basalts from the Taurus-Littrow region of the Moon. *Proc. Lunar Sci. Conf. 5th*, p. 471–504.

Papike J. J., Hodges F. N., Bence A. E., Cameron M. and Rhodes J. M. (1976) Mare basalts: crystal chemistry, mineralogy, and petrology. *Rev. Geophys. Space Phys.* **14**, 475–540.

Patterson C. C. (1956) Age of meteorites and the Earth. *Geochim. Cosmochim. Acta* **10**, 230–237.

Patterson C., Brown H. and Tilton G. (1953) Concentration of uranium and lead and the isotopic composition of lead in meteoritic material. *Phys. Rev.* **92**, 1234–1235.

Paul D. K. (1971) Strontium isotope studies in ultramafic inclusions from Dreiser Weiher, Eifel, Germany. *Contrib. Mineral. Petrol.* **34**, 22–28.

Pearce J. A. (1975) Basalt geochemistry used to investigate past tectonic environments on Cyprus. *Tectonophys.* **25**, 41–67.

Pearce J. A. and Cann J. R. (1973) Tectonic setting of basic volcanic rocks determined using trace element analysis. *Earth Planet. Sci. Lett.* **19**, 290–300.

Pearce T. H., Gorman B. E. and Birkett T. C. (1977) The relationship between major element chemistry and tectonic environment of basic and intermediate volcanic rocks. *Earth Planet. Sci. Lett.* **36**, 121–132.

Pedersen S., Larsen O., Bridgewater D. and Watterson J. (1974) Rb/Sr whole rock isochron age determinations and metamorphosed acid volcanic rocks and granitic gneisses from the Ketilidian mobile belt, S. E. Greenland. *Rapp. Grönlands Geol. Unders.* **66**, 12–20.

Pepin R. O. (1967) Trapped neon in meteorites. *Earth Planet. Sci. Lett.* **2**, 13–18.

Perkins W. G. and Begeal D. R. (1971) Diffusion and permeation of He, Ne, Ar, Kr and D_2 through silicon oxide thin films. *J. Chem. Phys.* **54**, 1683–1694.

Perry E. C., Jr., Ahmad S. N. and Swulius T. M. (1978) The oxygen isotope composition of 3,800 m.y. old metamorphosed chert and iron formation from Isukasia, West Greenland. *J. Geol.* **86**, 223–239.

Peterman Z. E. (1966) Rb-Sr dating of middle Precambrian metasedimentary rocks on Minnesota. *Geol. Soc. Amer. Bull.* **77**, 1031–1044.

Peterman Z. E. and Hedge C. E. (1968) Chronology of Precambrian events in the Front Range Colorado. *Can. J. Earth Sci.* **5**, 749–756.

Peterman Z. E., Carmichael I. S. E. and Smith A. L. (1970a) Strontium isotopes in quaternary basalts of southeastern California. *Earth Planet. Sci. Lett.* **7**, 381–384.

Philpotts J. A., Schuhmann S., Schnetzler C. C., Kouns C. W., Doan A. S., Jr., Wood F. M., Jr., Bickel A. L. and Lum R. K. L. (1973) Apollo 17: geochemical aspects of some soils, basalts, and breccia (abstract). *EOS (Trans. Amer. Geophys. Union)* **54**, 603–604.

Phinney D. (1972) ^{36}Ar, Kr, and Xe in terrestrial materials. *Earth Planet. Sci. Lett.* **16**, 413–420.

Phinney D., Tennyson J. and Frick U. (1978) Xenon in CO_2 well gas revisited. *J. Geophys. Res.* **83**, 2313–2319.

Phinney W. C., Warner J. L., Simonds C. H. and Lofgren G. E. (1972) Classification and distribution of rock types at Spur Crater. In *The Apollo 15 Lunar Samples*, (J. W. Chamberlain and C. Watkins, eds.), p. 149–153. The Lunar Science Institute, Houston.

Pidgeon R. T. (1978) 3450 m.y. old volcanics in the Archean layered greenstone succession of the Pilbara block, Western Australia. *Earth Planet. Sci. Lett.* **37**, 421–428.

Pineau F. (1977) Geochimie isotopique du carbone profond. Thèse. Paris.

Pineau F., Javoy M. and Allègre C. J. (1973) Etude systématique des isotopes de l'oxygéne du carbone et du strontium dans les carbonites. *Geochim. Cosmochim. Acta* **37**, 2363–2377.

Pineau F., Javoy M. and Bottinga Y. (1976a) $^{13}C/^{12}C$ ratios of rocks and inclusions and popping rocks in the Mid-Atlantic ridge and their bearing on the problem of isotopic composition of deep seated carbon. *Earth Planet. Sci. Lett.* **29**, 413–421.

Pineau F., Javoy M., Hawkins J. W. and Craig H. (1976b) Oxygen isotope variations in marginal basins and ocean ridge basalts. *Earth Planet. Sci. Lett.* **28**, 299–307.

Piper J. D. A. (1976) Paleomagnetic evidence for a Proterozoic supercontinent. *Phil. Trans. Roy. Soc. Lond.* **A280**, 469–490.

Pitman W. C., III, Larson R. L. and Herron E. M. (1974) The age of the ocean basins. *Geol. Soc. of Amer. Map and Chart Series* MC-6.

Podosek F. A. (1973) Thermal history of Nakhlites by the ^{40}Ar-^{39}Ar method. *Earth Planet. Sci. Lett.* **19**, 135–144.

Podosek F. A. (1978) Isotopic structures in solar system materials. *Ann. Rev. Astron. Astrophys.* **16**, 293–334.

Podosek F. A. and Huneke J. C. (1973) ^{40}Ar-^{39}Ar chronology of four calcium-rich achondrites. *Geochim. Cosmochim. Acta* **37**, 667–684.

Podosek F. A., Huneke J. C. and Wasserburg G. J. (1972) Gas retention and cosmic-ray exposure ages of lunar rock 15555. *Science* **175**, 423–425.

Pohn H. A. (1971) Geologic map of the Lansberg P region of the Moon. *U.S. Geol. Surv. Map I-627.* Washington, D.C.

Powell J. L. and Bell K. (1970) Strontium isotopic studies of alkalic rocks: localities from Australia, Spain and the Western United States. *Contrib. Mineral. Petrol.* **27**, 1–10.

Powell J. L. and Bell K. (1974) *The Alkaline Rocks.* Wiley, N.Y. 622 pp.

Presnall D. C. (1969) The geometric analysis of partial fusion. *Amer. J. Sci.* **267**, 1178–1191.

Price P. B. and Walker R. M. (1962) Observation of fossil particle tracks in natural micas. *Nature* **196**, 732.

Priem H. N. A., Boelrijk N. A. I. M., Bebeda E. H., Verdurmen E. A. Th. and Verschure R. H. (1973) Age of the Precambrian Roriama formation in northeastern South America: evidence from isotopic dating of Roriama pyroclastic volcanic rocks in Surinam. *Geol. Soc. Amer. Bull.* **84**, 1677–1684.

Prinz M., Keil K., Hlava P. F., Berkley J. L., Gomes C. B. and Curvello W. S. (1977) Studies of Brazilian meteorites, III. Origin and history of the Angra dos Reis achondrite. *Earth Planet. Sci. Lett.* **35**, 317–330.

Provost A. and Allègre C. J. (1979) Process identification and search for optimal differentiation parameters from major element data. General presentation with emphasis on the fractional crystallization processes. *Geochim. Cosmochim. Acta.* **43**, 487–501.

Rajan R. S., Huneke J. C., Smith S. P. and Wasserburg G. J. (1979) Argon-40–Argon-39 chronology of lithic clasts from the Kapoeta howardite. *Geochim. Cosmochim. Acta* **43**, 957–971.

Reid A. M., Warner J., Ridley W. I. and Brown R. W. (1972) Major element composition of glasses in three Apollo 15 soils. *Meteoritics* **7**, 395–415.

Reynolds J. (1977) Clues in the rare-gas isotopes to early solar system history. In *The Soviet-American Conference on Cosmochemistry of the Moon and Planets* (J. H. Pomeroy and N. J. Hubbard, eds.), p. 771–780. NASA SP-370.

Reynolds J. H., Frick U., Neil J. M. and Phinney D. L. (1978) Rare-gas rich separates from carbonaceous chondrites. *Geochim. Cosmochim. Acta* **42**, 1775–1797.

Rhodes J. M., Hodges F. N. and Papike J. J. (1975) Mare basalts: major element composition and classification (abstract). In *Papers Presented to the Conference on Origins of Mare Basalts and Their Implications for Lunar Evolution*, p. 135–139. The Lunar Science Institute, Houston.

Rhodes J. M., Blanchard D. P., Dungan M. A., Brannon J. C. and Rodgers K. V. (1977) Chemistry of Apollo 12 mare basalts: magma types and fractionation processes. *Proc. Lunar Sci. Conf. 8th*, p. 1305–1338.

Rhodes J. M., Hubbard N. J., Wiesmann H., Rodgers K. V., Brannon J. C. and Bansal B. M. (1976) Chemistry, classification, and petrogenesis of Apollo 17 mare basalts. *Proc. Lunar Sci. Conf. 7th*, p. 1467–1489.

Richard P., Shimizu N. and Allègre C. J. (1976) $^{143}Nd/^{146}Nd$, a natural tracer: an application to oceanic baslts. *Earth Planet. Sci. Lett.* **31**, 269–278.

Ringwood A. E. (1975) *Composition and Petrology of the Earth's Mantle.* McGraw-Hill, N.Y. 99 pp.

Rison W. and Kyser K. (1977) Rare gases, oxygen and hydrogen in Hawaiian xenoliths and basalts. *EOS (Trans. Amer. Geophys. Union)* **58**, 537.

Roberson F. I. and Kaula W. M. (1972) Apollo 15 laser altimeter. In *Apollo 15 Prelim. Sci. Rept.*, p. 48–50. NASA SP-289.

Roberts D. and Gale G. H. (1978) The Caledonian-Applachian Iapetus Ocean. In *Evolution of the Earth's Crust* (D. H. Tarling, ed.), p. 255–342. Academic Press, N.Y.

Roberts J. M., Gold R. and Armani R. J. (1968) Spontaneous fission decay of ^{238}U. *Phys. Rev.* **174**, 1482–1484.

Robertson D. K. (1973) A model discussing the early history of the Earth based on a study of lead isotope ratios from veins in some Archaen cratons of Africa. *Geochim. Cosmochim. Acta* **37**, 2099–2124.

Robertson D. K. and Folinsbee R. E. (1974) Lead isotope ratios and crustal evolution of the Slave craton at Ghost Lake, Northwest Territories. *Can. J. Earth Sci.* **11**, 819.

Robertson J. A. (1976) The Blind River uranium deposits: the ores and their setting. *Ontario Div. Mines. Misc. Paper 65*, p. 45.

Roe G. D., Pinson W. H., Jr. and Hurley P. M. (1965) Rb-Sr evidence for the origin of peridotites. *EOS (Trans. Amer. Geophys. Union)* **46**, 186.

Roedder E. (1965) Liquid CO_2 inclusions in olivine bearing nodules and phenocryst from basalts. *Amer. Mineral.* **50**, 1746–1782.

Ross C. J., Foster M. D. and Myers A. T. (1954) Origin of dunites and olivine-rich inclusions in basaltic rocks. *Amer. Mineral.* **39**, 693–737.

Rowe N. W. (1970) Evidence for decay of extinct ^{244}Pu and ^{129}I in the Kapoeta meteorite. *Geochim. Cosmochim. Acta* **34**, 1019–1025.

Saito K., Basu A. R. and Alexander E. C. (1978) Planetary-type rare gases in an upper mantle-derived amphibole. *Earth Planet. Sci. Lett.* **39**, 274–280.

Sakaï H. (1957) Fractionation of sulfur isotopes in nature. *Geochim. Cosmochim. Acta* **12**, 150–169.

Sakaï H. (1968) Isotopic properties of sulfur compounds in hydrothermal processes. *Geochem. J.* **2**, 29–49.

Sawkins F. J. (1976) Widespread continental rifting: some considerations of timing and mechanism. *Geol.* **4**, 427–430.

Schaeffer G. A. and Schaeffer O. A. (1977) ^{39}Ar-^{40}Ar ages of lunar rocks. *Proc. Lunar Sci. Conf. 8th*, p. 2253–2300.

Schaeffer O. A. and Husain L. (1973a) Early lunar history: ages of 2 to 4 mm soil fragments from the lunar highlands. *Proc. Lunar Sci. Conf. 4th*, p. 1847–1863.

Schaeffer O. A. and Husain L. (1973b) Isotopic ages of Apollo 17 lunar material. *EOS (Trans. Amer. Geophys. Union)* **54**, 614.

Schaeffer O. A. and Husain L. (1974) Chronology of lunar basin formation. *Proc. Lunar Sci. Conf. 5th*, p. 1541–1555.

Schaeffer O. A., Bence A. E., Eichhorn G., Papike J. J. and Vaniman D. T. (1978) ^{40}Ar-^{39}Ar and petrologic study of mare crisium: age and petrology of Luna 24 samples 24077, 13 and 24077, 63 (abstract). In *Lunar and Planetary Science IX*, p. 1007–1009. Lunar and Planetary Institute, Houston.

Schaeffer O. A., Husain L. and Schaeffer G. A. (1976) Ages of highland rocks: the chronology of lunar basin formation revisited. *Proc. Lunar Sci. Conf. 7th*, p. 2067–2092.

Schilling J. G. (1971) Sea-floor evolution: rare-earth evidence. *Phil. Trans. Roy. Soc. Lond.* **A268**, 663–706.

Schilling J. G. (1973) Iceland mantle plume: geochemical evidence along Reykjanes Ridge. *Nature* **242**, 565–571.

Schilling J. G. (1975) Rare-earth variations across "normal segments" of the Reykjanes Ridge, 60°–53° N, Mid-Atlantic Ridge, 29° S, and East Pacific Rise, 2°–19° S, and evidence on the composition of the underlying low-velocity layer. *J. Geophys. Res.* **80**, 1459.

Schilling J. G. and Winchester J. W. (1966) Rare earths in Hawaiian basalts. *Science* **153**, 867–869.

Schonfeld E. (1974) The contamination of lunar highland rocks by KREEP: interpretation by mixing models. *Proc. Lunar Sci. Conf. 5th*, p. 1269–1286.

Schonfeld E. and Meyer C., Jr. (1972) The abundances of components of the lunar soils by a least-squares mixing model and the formation age of KREEP. *Proc. Lunar Sci. Conf. 3rd*, p. 1397–1420.

Schramm D. N. and Clayton R. N. (1978) Did a supernova trigger the formation of the solar system? *Sci. Amer.* **239**, 124–139.

Scott R. B. and Hajash A. (1976) Initial submarine alteration of basaltic pillow lavas: a microprobe study. *Amer. J. Sci.* **276**, 480–501.

Shaw D. M. (1979) Trace element melting models. In *Origin and Distribution of the Elements*, Second Symposium, (L. H. Ahrens, ed.), p. 577–586. Pergamon Press, Oxford.

Sheppard S. M. and Epstein S. (1970) D/H and $^{18}O/^{16}O$ ratios of minerals and possible mantle or lower crustal origin. *Earth Planet. Sci. Lett.* **9**, 232–239.

Shih C., Haskin L. A., Wiesmann H., Bansal B. M. and Brannon J. C. (1975) On the origin of high-Ti mare basalts. *Proc. Lunar Sci. Conf. 6th*, p. 1255–1285.

Shimizu N. (1974) The partitioning of K, Rb, Cs and Sr and Ba between clinopyroxene and silicate liquid at high pressures. *Dept. Terr. Magnet. Ann. Rept. 1973–74*, p. 943.

Shimizu N. (1975) Geochemistry of ultramafic inclusions from Salt Lake Crater, Hawaii and from Southern African Kimberlites. *Phys. Chem. Earth.* **9**, 655–669.

Shimizu N. and Boyd F. R. (1973) Trace element contents of clinopyroxenes from garnet lherzolites in kimberlites. *Carnegie Inst. Wash., Yearb.* **72**, 272–276.

Shimizu N. and Kushiro I. (1973) Trace element content of liquid formed by partial melting of a garnet lherzolite at high pressures: a preliminary report. *Dept. Terr. Magnet. Ann. Rept. 1972–73*, p. 270–272.

Shuto K., Kagami H., Iizumi S. and Gorai M. (1977) Strontium isotopic features of the Cenozoic volcanic rocks with special reference to those of the circum-Pacific and intra-Pacific regions. *Pacific Geol.* **12**, 47.

Signer P. and Suess H. E. (1963) Rare gases in the sun, in the atmosphere and in meteorites. In *Earth Science and Meteoritics* (J. Geiss and E. D. Goldberg, eds.), p. 241–272. North Holland Publ. Co., Amsterdam.

Silver L. T. (1968) Precambrian batholiths of Arizona (abstract). Geol. Soc. Amer. Spec. Paper 121 Abstract for 1968, 558–559.

Silver L. T. and Duke M. B. (1971) U-Th-Pb isotope relations in some basaltic achondrites (abstract). *EOS (Trans. Amer. Geophys. Union)* **52**, 289.

Silver L. T. and Green J. C. (1972) Time constants for Keweenawan igneous activity (abstract). *Geol. Soc. Amer. Abstracts with Program* **4**, 665.

Silver L. T., Bickford M. E., Van Schmus W. R., Anderson J. L., Anderson T. H. and Medaris L. G., Jr. (1977) The 1.4–1.5 b.y. transcontinental anorogenic plutonic perforation of North America. *Abstracts with Program*, Geol. Soc. of Amer., **9**, 1176–1177.

Simonen A. (1953) Stratigraphy and sedimentation of the Svecofennidic, early Archean supracrustal rocks in southwestern Finland. *Bull. de la Commission Geologique de Finlande No. 60*, 64.

Sims P. K. (1976) Precambrian tectonics and mineral deposits, Lake Superior region. *Econ. Geol.* **71**, 1092–1118.

Sinha A. K. (1971) Primary differences in U/Pb ratios in two primitive crustal blocks. *Carnegie Inst. Wash., Yearb.* **69**, 405–408.

Sinha A. K. (1972) U-Th-Pb systematics and the age of the Onverwacht Series, South Africa. *Earth Planet. Sci. Lett.* **16**, 219–227.

Sinha A. K. and Hart S. R. (1972) A geochemical test of the subduction hypothesis for generation of island arc magmas. *Carnegie Inst. Wash., Yearb.* **71**, 309.

Sinha A. K., Mattinson T. M. and Hart S. R. (1971) Archean volcanics. *Carnegie Inst. Wash., Yearb.* **70**, 355–358.

Smith E. I. (1978) Precambrian rhyolites and granites in south-central Wisconsin: field relations and geochemistry. *Geol. Soc. Amer. Bull.* **89**, 875–890.

Smith J. V., Anderson A. T., Newton R. C., Olsen E. J., Wyllie P. J., Crewe A. V., Isaacson M. S. and Johnson D. (1970) Petrologic history of the Moon inferred from petrography, mineralogy, and petrogenesis of Apollo 11 rocks. *Proc. Apollo 11 Lunar Sci. Conf.*, p. 897–925.

Soderblom L. A. and Lebofsky L. A. (1972) Technique for rapid determination of relative ages of lunar areas from orbital photography. *J. Geophys. Res.* **77**, 279–296.

Spooner E. T. C. (1976) The strontium isotopic composition of seawater, and seawater-oceanic crust interaction. *Earth Planet. Sci. Lett.* **31**, 167–174.

Steiger R. H. and Jäger E. (1977) Subcommission on geochronology; convention on the use of decay constants in geo- and cosmochronology. *Earth Planet. Sci. Lett.* **36**, 359–362.

Stern T. W., Phair G. and Newell M. (1971) Boulder Creek Batholith, Colorado: Part II, Isotopic age of emplacement and morphology of zircon. *Geol. Soc. Amer. Bull.* **82**, 1615–1634.

Stettler A. and Allègre C. J. (1979) $^{87}Rb/^{87}Sr$ constraints on the genesis and evolution of the Cantal continental volcanic system (France). *Earth Planet. Sci. Lett.* **44**, 269–278.

Stettler A., Eberhardt P., Geiss J., Grögler N. and Guggisberg S. (1978) Chronology of the Apollo 17 Station 7 Boulder and the South Serenitatis impact (abstract). In *Lunar and Planetary Science IX*, p. 1113–1115. Lunar and Planetary Institute, Houston.

Stettler A., Eberhardt P., Geiss J., Grögler N. and Maurer P. (1973) ^{39}Ar-^{40}Ar ages and ^{37}Ar-^{38}Ar exposure ages of lunar rocks. *Proc. Lunar Sci. Conf. 4th*, p. 1865–1888.

Stettler A., Eberhardt P., Geiss J., Grögler N. and Maurer P. (1974) On the duration of Lava flow activity in Mare Tranquillitatis. *Proc. Lunar Sci. Conf. 5th*, p. 1557–1570.

Stockwell C. H. (1972) Revised Precambrian time-scale for the Canadian shield. *Geol. Survey of Can., Paper 72-52*, p. 4.

Stolper E. (1977) Experimental petrology of eucritic meteorites. *Geochim. Cosmochim. Acta* **41**, 587–611.

Stolper E., McSween H., Jr. and Hays J. F. (1979) A petrogenetic model of the relationships among achondritic meteorites. *Geochim. Cosmochim. Acta* **43**, 589–602.

Stonely R. (1974) Evolution of the continental margins bounding a former southern Tethys. In *The Geology of Continental Margins* (C. A. Burk and C. L. Drake, eds.), p. 889–903. Springer-Verlag, N.Y.

Störzer D. (1970) Fission track dating of volcanic glasses and the thermal history of rocks. *Earth Planet. Sci. Lett.* **8**, 55–60.

Strang, Gilbert, (1980) *Linear Algebra and Its Applications*. Academic Press. N.Y. 414 pp.

Stuart-Alexander D. E. and Howard K. A. (1970) Lunar maria and circular basins—a review. *Icarus* **2**, 440–456.

Stuckless J. S. and Irving A. J. (1976) Strontium isotope geochemistry of megacrysts and host basalts from southeastern Australia. *Geochim. Cosmochim. Acta* **40**, 209–213.

Stueber A. M. (1965) A geochemical study of ultramafic rocks. Ph.D. Thesis, Univ. of Calif., San Diego.

Stueber A. M. and Ikramuddin M. (1974) Rubidium, strontium and the isotopic composition of strontium in ultramafic nodule minerals and host basalts. *Geochim. Cosmochim. Acta* **38**, 207–216.

Stueber A. M. and Murthy V. R. (1966) Strontium isotope and alkali element abundances in ultramafic rocks. *Geochim. Cosmochim. Acta* **33**, 543–553.

Sun S. S. (1973) Lead isotope studies of young volcanic rocks from oceanic islands, mid-ocean ridges and island arcs. Ph.D. Thesis, Columbia Univ., N.Y.

Sun S. S. and Hanson G. N. (1975) Origin of Ross Island basanitoids and limitations upon the heterogeneity of mantle sources for alkali basalts and nephelinites. *Contrib. Mineral. Petrol.* **52**, 77–106.

Sun S. S. and Jahn B. (1975) Lead and strontium isotopes in post-glacial basalts from Iceland. *Nature* **255**, 527–530.

Sun S. S. and Nesbitt R. W. (1977) Chemical heterogeneity of the Archean mantle, composition of the Earth and mantle evolution. *Earth Planet. Sci. Lett.* **35**, 429–448.

Sun S. S. and Nesbitt R. W. (1978) Petrogenesis of Archean ultrabasic and basic volcanics: evidence from rare earth elements. *Contrib. Mineral. Petrol.* **65**, 301–325.

Sun S. S., Tatsumoto M. and Schilling J. G. (1975) Mantle plume mixing along the Reykjanes Ridge axis: lead isotopic evidence. *Science* **190**, 143–147.

Swann G. A., Bailey N. G., Batson R. M., Freeman V. L., Hait M. H., Head J. W., Holt H. E., Howard K. A., Irwin J. B., Larson K. B., Muehlburger W. R., Reed V. S., Rennilson J. J., Schaber G. G., Scott D. R., Silver L. T., Sutton R. L., Ulrice G. E., Wilshire H. G. and Wolfe E. W. (1972) Preliminary geologic investigation of the Apollo 15 landing site. In *Apollo 15 Prelim. Sci. Rept.*, p. 5-1 to 5-112. NASA SP-289.

Taddeuci A., Broecker W. S. and Thurber D. L. (1967) ^{230}Th dating of volcanic rocks. *Earth Planet. Sci. Lett.* **3**, 338–342.

Talwani M., Thompson G., Dent B., Kahle H. G. and Buck S. (1973) Traverse gravimeter experiment. In *Apollo 17 Prelim. Sci. Rept.*, p. 31-1 to 31-13. NASA SP-330.

Tarney J. (1973) The Scourie dyke suite and the nature of the Inverian event in Assynt. In *The Early Precambrian of Scotland and Related Rocks of Greenland* (R. G. Park and J. Tarney, eds.), p. 105–118. Dept. of Geol., Univ. of Keele, Newcastle, Eng.

Tatsumoto M. (1970) Age of the Moon: an isotopic study of U-Th-Pb systematics of Apollo 11 lunar samples. *Proc. Apollo 11 Lunar Sci. Conf.*, p. 1595–1612.

Tatsumoto M. (1978) Isotopic composition of lead in oceanic basalt and its implication to mantle evolution. *Earth Planet. Sci. Lett.* **38**, 63–87.

Tatsumoto M. and Unruh D. M. (1975) Formation and early brecciation of the Juvinas achondrite inferred from U-Th-Pb systematics (abstract). *Meteoritics* **10**, 500–501.

Tatsumoto M., Hedge C. E., Knight R. J., Unruh D. M. and Doe B. R. (1972) U-Th-Pb, Rb-Sr and K measurements on some Apollo 15 and Apollo 16 samples. In *The Apollo 15 Lunar Samples*, (J. W. Chamberlain and C. Watkins, eds.), p. 391–395. The Lunar Science Institute, Houston.

Tatsumoto M., Knight R. J. and Allègre C. J. (1973) Time differences in the formation of meteorites as determined from the ratio of lead-207 to lead-206. *Science* **180**, 1278–1283.

Tatsumoto M., Knight R. J. and Doe B. R. (1971) U-Th-Pb systematics of Apollo 12 lunar samples. *Proc. Lunar Sci. Conf. 2nd*, p. 1521–1546.

Taylor H. P. (1968) The oxygen isotope geochemistry of igneous rocks. *Contrib. Mineral. Petrol.* **19**, 1–71.

Taylor H. P. (1974) The application of oxygen and hydrogen isotope studies to problems of hydrothermal alteration and ore deposition. *Econ. Geol.* **69**, 843–883.

Taylor H. P. and Epstein S. (1963) $^{18}O/^{16}O$ ratios in rocks and coexisting minerals of the Skaergaard intrusion East Greenland. *J. Petrol.* **4**, 51–74.

Taylor H. P. and Epstein S. (1968) Hydrogen isotope evidence for influx of meteoric ground water into shallow igneous intrusions. *Geol. Soc. Amer. Spec. Publ.*, 121–294.

Taylor H. P. and Epstein S. (1970) $^{18}O/^{16}O$ ratios of Apollo 11 lunar rocks and minerals. *Proc. Apollo 11 Lunar Sci. Conf.*, p. 1613-1626.

Taylor H. P. and Forester R. W. (1973) An oxygen and hydrogen isotope study of the Skaergaard intrusion and its country rocks. *EOS (Trans. Amer. Geophys. Union)* **54**, 500.

Taylor H. P. and Silver L. T. (1978) Oxygen isotope relationships in plutonic igneous rocks of the peninsular ranges batholith, southern and Baja California. *Geol. Survey Open-File Rept. 78-701*, p. 423–426.

Taylor H. P. and Turi B. (1976) High ^{18}O igneous rocks from the Tuscan magmatic province Italy. *Contrib. Mineral. Petrol.* **55**, 33–54.

Taylor H. P., Jr., Duke M. B., Silver L. T. and Epstein S. (1965) Oxygen isotope studies of minerals in stony meteorites. *Geochim. Cosmochim. Acta* **29**, 489–512.

Taylor H. P., Frechen J. and Degens E. T. (1967) Oxygen and carbon isotope studies of carbonatites from the Leacher See District, West Germany and Alno district Sweden. *Geochim. Cosmochim. Acta* **31**, 407–430.

Taylor S. R. (1975) *Lunar Science: A Post-Apollo View*. Pergamon Press, N. Y. 372 pp.

Taylor S. R. and Jakeš P. (1974) The geochemical evolution of the Moon. *Proc. Lunar Sci. Conf. 5th*, p. 1287–1305.

Taylor S. R., Gorton M. P., Muir P., Nance W., Rudowski R. and Ware N. (1973) Lunar highlands composition: Apennine Front. *Proc. Lunar Sci. Conf. 4th*, p. 1445–1459.

Tera F. and Wasserburg G. J. (1972a) U-Th-Pb systematics in three Apollo 14 basalts and the problem of initial Pb. *Earth Planet. Sci. Lett.* **14**, 281–304.

Tera F. and Wasserburg G. J. (1972b) U-Th-Pb systematics in lunar highland samples from Luna 20 and Apollo 16 missions. *Earth Planet. Sci. Lett.* **17**, 36–51.

Tera F. and Wasserburg G. J. (1973) A response to a comment on U-Pb systematics in lunar basalts. *Earth Planet Sci. Lett.* **19**, 213–217.

Tera F. and Wasserburg G. J. (1974) U-Th-Pb systematics on lunar rocks and inferences about lunar evolution and the age of the Moon. *Proc. Lunar Sci. Conf. 5th*, p. 1571–1599.

Tera F. and Wasserburg G. J. (1975) The evolution and history of mare basalts as inferred from U-Th-Pb systematics (abstract). In *Lunar Science VI*, p. 807–809. The Lunar Science Institute, Houston.

Tera F. and Wasserburg G. J. (1976) Lunar ball games and other sports (abstract). In *Lunar Science VII*, p. 858–860. The Lunar Science Institute, Houston.

Tera F., Papanastassiou D. A. and Wasserburg G. J. (1973) A lunar cataclysm at ~3.95 AE and the structure of the lunar crust (abstract). In *Lunar Science IV*, p. 723–725. The Lunar Science Institute, Houston.

Tera F., Papanastassiou D. A. and Wasserburg G. J. (1974a) Isotopic evidence for a terminal lunar cataclysm. *Earth Planet. Sci. Lett.* **22**, 1–21.

Tera F., Papanastassiou D. A. and Wasserburg G. J. (1974b) The lunar time scale and a summary of isotopic evidence for a terminal lunar cataclysm (abstract). In *Lunar Science V*, p. 792–794. The Lunar Science Institute, Houston.

Thode H. G. (1963) Sulfur isotope geochemistry. In *Analytical Geochemistry* (D. Shaw, ed.), Univ. Toronto Press.

Tilton G. R. (1960) Volume diffusion as a mechanism for discordant lead ages. *J. Geophys. Res.* **65**, 2933–2945.

Tilton G. R. (1973) Isotopic lead ages of chondritic meteorites. *Earth Planet. Sci. Lett.* **19**, 321–329.

Tilton G. R. and Chen J. H. (1979) Lead isotope systematics of three Apollo 17 mare basalts. *Proc. Lunar Sci. Conf. 10th*, p. 259–274.

Tolstikhin I. N. (1978) A review: some recent advances in isotope geochemistry of light rare gases. In *Terrestrial Rare Gases*, (E. C. Alexander, Jr. and M. Ozima, eds.), p. 33–62. Advances in Earth and Planetary Sciences **v. 3.** Center for Academic Publications Japan/Japan Scientific Societies Press.

Tolstikhin I. N., Kamenskii I. and Mamyrin B. A. (1969) Isotopic criterion for investigation of terrestrial helium origin. *Geokhimiya* **2**, 201–204.

Tolstikhin I. N., Mamyrin B. A., Khabarin L. B. and Erlikh E. N. (1974) Isotope composition of helium in ultrabasic xenoliths from volcanic rocks of Kamchatka. *Earth Planet. Sci. Lett.* **22**, 75–84.

Tugarinov A. J. and Bibikova E. V. (1976) Age of apatite ores on the Aldan Shield (abstract). Paper presented at *4th European Colloquium of Geochron. Cosmochron. and Isotope Geol.* Amsterdam, April.

Turcotte D. L. and Burke K. (1978) Global sea-level changes and the thermal structure of the earth. *Earth Planet. Sci. Lett.* **41**, 341–346.

Turi B. and Taylor H. P. (1976a) High ^{18}O igneous rocks from the Tuscan magmatic province, Italy. *Contrib. Mineral. Petrol.* **55**, 33–54.

Turi B. and Taylor H. P. (1976b) Oxygen isotope studies of potassic volcanic rocks of the Roman Province, Central Italy. *Contrib. Mineral. Petrol.* **55**, 1–31.

Turner G. (1970) Argon-40/Argon-39 dating of lunar rock samples. *Proc. Apollo 11 Lunar Sci. Conf.*, p. 1665–1684.

Turner G. (1971) ^{40}Ar-^{39}Ar ages from the lunar maria. *Earth Planet. Sci. Lett.* **11**, 169.

Turner G. (1977) Potassium argon chronology of the Moon. *Phys. Chem. Earth* **10**, 145–195.

Turner G. and Cadogan P. H. (1974) Possible effects of ^{39}Ar recoil in ^{40}Ar-^{39}Ar dating. *Proc. Lunar Sci. Conf. 5th*, p. 1601–1615.

Turner G. and Cadogan P. H. (1975) The history of lunar bombardment inferred from ^{40}Ar-^{39}Ar dating of highland rocks. *Proc. Lunar Sci. Conf. 6th*, p. 1509–1538.

Turner G., Cadogan P. H. and Yonge C. J. (1973) Argon selenochronology. *Proc. Lunar Sci. Conf. 4th*, p. 1889–1914.

Turner G., Huneke J. C., Podosek F. A. and Wasserburg G. J. (1971) ^{40}Ar-^{39}Ar ages and cosmic ray exposure ages of Apollo 14 samples. *Earth Planet. Sci. Lett.* **12**, 19–35.

Turner G., Huneke J. C., Podosek F. A. and Wasserburg G. J. (1972) ^{40}Ar-^{39}Ar systematics in rocks and separated minerals from Apollo 14. *Proc. Lunar Sci. Conf. 3rd*, p. 1589–1612.

Unruh D. M., Nakamura N. and Tatsumoto M. (1977) History of the Pasamonte achondrite: relative susceptibility of the Sm-Nd, Rb-Sr, and U-Pb systems to metamorphic events. *Earth Planet. Sci. Lett.* **37**, 1–12.

Urey H. C. (1947) The thermodynamic properties of isotopic substances. *J. Chem. Soc.*, 562–581.

Urey H. C. and Craig H. (1953) The composition of the stone meteorites and the origin of the meteorites. *Geochim. Cosmochim. Acta* **4**, 36–82.

Urey H. C., Lowenstam H., Epstein S. and McKinney C. R. (1951) Measurements of paleotemperature and temperature of the upper Cretaceous of England, Denmark and the northern United States. *Geol. Soc. Amer. Bull.* **62**, 399–416.

Usselman T. M., Lofgren G. E., Donaldson C. H. and Williams R. J. (1975) Experimentally reproduced textures and mineral chemistries of high-titanium mare basalts. *Proc. Lunar Sci. Conf. 6th*, p. 997–1020.

Vail P. R., Mitchum R. M. and Thompson S. (1978) Seismic stratigraphy and global changes of sea level 4. Global cycles of relative changes of sea level. In *Seismic Stratigraphy-Applications to Hydrocarbon Exploration* (C. E. Payton, ed.), p. 83–97. Amer. Assoc. Pet. Geol. Mem. **26**.

Van Breemen O., Aftalion M. and Allaart J. H. (1974) Isotopic and geochronologic studies on granites from the Ketilidian mobile belt of south Greenland. *Geol. Soc. Amer. Bull.* **85**, 403–412.

Van Breemen O., Aftalion M. and Pidgeon R. T. (1971) The age of the granitic injection complex of Harris, Outer Hebrides. *Scottish J. Geol.* **7**, 139–152.

Vaniman D. T. and Papike J. J. (1977) Very low Ti(VLT) basalts: a new mare rock type from the Apollo 17 drill core. *Proc. Lunar Sci. Conf. 8th*, p. 1443–1471.

Van Niekerk C. B. and Burger A. J. (1969) A note on the minimum age of the acid lava of the Onverwacht series of the Swaziland system. *Trans. Geol. Soc. S. Africa* **72**, 9–21.

Van Schmus R. (1965) The geochronology of the Blind River—Bruce Mines area, Ontario Canada. *J. Geol.* **73**, 755–780.

Van Schmus W. R. (1976) Early and middle Proterozoic history of the Great Lakes area, North America. *Phil. Trans Roy. Soc. Lond.* **A280**, 605–628.

Van Schmus W. R. (1978) Geochronology of the southern Wisconsin rhyolites and granites. *Geoscience Wisconsin* **2**, 19–24.

Van Schmus W. R. (1980) Chronology of igneous rocks associated with the Penokean Orogeny in Wisconsin. *Geol. Soc. Amer. Spec. Pap.* **182**, 159–168.

Van Schmus W. R., Card K. D. and Harrower K. L. (1975a) Geology and ages of buried Precambrian rocks, Manitoulin Island, Ontario. *Can. J. Earth Sci.* **12**, 1175–1189.

Van Schmus W. R., Medaris L. G. and Banks P. O. (1975b) Geology and age of the Wolf River batholith, Wisconsin. *Geol. Soc. Amer. Bull.* **86**, 907–914.

Van Schmus W. R., Thurman M. E. and Peterman Z. E. (1975c) Geology and Rb-Sr chronology of middle Precambrian rocks in eastern and central Wisconsin. *Geol. Soc. Amer. Bull.* **86**, 1255–1265.

Veizer J. and Compston W. (1974) $^{87}Sr/^{86}Sr$ in Precambrian carbonates as an index of crustal evolution. *Geochim. Cosmochim. Acta* **38**, 1461–1484.

Veizer J. and Compston W. (1976) $^{87}Sr/^{86}Sr$ in Precambrian carbonates as an index of crustal evolution. *Geochim. Cosmochim. Acta* **40**, 905–914.

Vera R. H. and Van Schmus W. R. (1974) Geochronology of some Precambrian rocks in the southern Front Range, Colorado. *Geol. Soc. Amer. Bull.* **85**, 77–82.

Verpaelst P., Brooks C. and Franconi A. (1980) The 2.5 Ga Duxbury massif, Quebec: a remobilized piece of pre- 3.0 Ga sialic basement(?) *Can. J. Earth Sci.* **17**, 1–18.

Vogt P. R. (1979) Global magmatic episodes: new evidence and implications for the steady-state mid-ocean ridge. *Geology* **7**, 93–98.

Vollmer R. (1976) Rb-Sr and U-Th-Pb systematics of alkaline rocks: the alkaline rocks from Italy. *Geochim. Cosmochim. Acta* **40**, 283–295.

Vollmer R. (1977) Isotopic evidence for genetic relations between acid and alkaline rocks in Italy. *Contrib. Mineral. Petrol.* **60**, 109–118.

Wakita H., Laul J. C. and Schmitt R. A. (1975) Some thoughts on the origin of lunar ANT-KREEP and mare basalts. *Geochem. J.* **9**, 25–41.

Walker D., Longhi J., Lasaga A. C., Stolper E. M., Grove T. L. and Hays J. F. (1977) Slowly cooled microgabbros 15555 and 15065. *Proc. Lunar Sci. Conf. 8th*, p. 1521–1547.

Walker D., Longhi J., Stolper E. M., Grove T. M. and Hays J. F. (1975) Origin of titaniferous lunar basalts. *Geochim. Cosmochim. Acta* **39**, 1219–1235.

Wanless R. K. and Loveridge W. D. (1972) Rubidium-strontium isochron age studies, Rept. 1. *Geol. Surv. of Can. Paper 72-23*.

Warner J. L. (1971) Lunar crystalline rocks; petrology and geology. *Proc. Lunar Sci. Conf. 2nd* p. 469–480.

Warner R. D., Keil K., Prinz M., Laul J. C., Murali A. V. and Schmitt R. A. (1975) Mineralogy, petrology, and chemistry of mare basalts from Apollo 17 rake samples. *Proc. Lunar Sci. Conf. 6th*, p. 193–220.

Warren P. H. and Wasson J. T. (1979) The origin of KREEP. *Rev. Geophys. Space Phys.* **17**, 73–88.

Wasserburg G. J. (1963) Diffusion process in lead-uranium systems. *J. Geophys. Res.* **68**, 4823–4846.

Wasserburg G. J. and Mazor E. (1965) Spontaneous fission xenon in natural gases. In *Fluids in Subsurface Environments: A Symposium*, p. 386–398. Amer. Assn. of Petrol. Geol., Tulsa, Oklahoma.

Wasserburg G. J. and Papanastassiou D. A. (1971) Age of an Apollo 15 mare basalt: lunar crust and mantle evolution. *Earth Planet. Sci. Lett.* **13**, 97–104.

Wasserburg G. J., Tera F., Papanastassiou D. A. and Huneke J. C. (1977) Isotopic and chemical investigations on Angra dos Reis. *Earth Planet. Sci. Lett.* **35**, 294–316.

Wasson J. T. (1974) *Meteorites.* Springer-Verlag, N.Y. 316 pp.

Wetherill G. W. (1954) Variations in the isotopic abundances of neon and argon extracted from radioactive minerals. *Phys. Rev.* **96**, 679–683.

Wetherill G. W. (1956) Discordant uranium-lead ages. *EOS (Trans. Amer. Geophys. Union)* **37**, 320–326.

Wetherill G. W. (1963) Discordant uranium-lead ages 2: discordant ages resulting from diffusion of lead and uranium. *J. Geophys. Res.* **68**, 2957–2965.

Wetherill G. W. (1966) Radioactive decay constants and energies. In *Handbook of Physical Constants* (S. Clark, ed.), p.513–519. Mem., Geol. Soc. Amer. **97**.

Wetherill G. W. (1975a) Pre-mare cratering and early solar system history. In *The Soviet-American Conference on Cosmochemistry of the Moon and Planets* (J. H. Pomeroy and N. J. Hubbard, eds.), p. 553–567. NASA SP-370.

Wetherill G. W. (1975b) Late heavy bombardment of the Moon and terrestrial planets. *Proc. Lunar Sci. Conf. 6th*, p. 1539–1561.

Wetherill G. W. (1975c) Radiometric chronology of the earth solar system. *Ann. Rev. Nucl. Sci.* **25**, 283–324.

Wetherill G. W. (1976) Comments on the paper by C. R. Chapman: Chronology of terrestrial planet evolution—the evidence from Mercury. *Icarus* **28**, 537–542.

Wetherill G. W. (1977) Evolution of the Earth's planetesimal swarm subsequent to the formation of the Earth and Moon. *Proc. Lunar Sci. Conf. 8th*, p. 1–16.

Wetherill G. W. (1980) Multiple cosmic ray exposure ages of meteorites (abstract). *Meteoritics*, in press.

Wetherill G. W., Mark R. and Lee-Hu C. (1973) Chondrites: initial strontium-87/strontium-86 ratios and the early history of the solar system. *Science* **182**, 281–283.

White R. W. (1966) Ultramafic inclusions in basaltic rocks from Hawaii. *Contrib. Mineral. Petrol.* **12**, 245–314.

White W. M., Hofmann A. W. and Bow C. S. (1978) Isotope and LIL-element geochemistry of the Galapagos Island. *U.S. Geol. Surv. Open-File Rept. 78-701*, p. 449–451.

White W. M. and Schilling J. G. (1978) The nature and origin of geochemical variation of mid-Atlantic ridge basalts from the central North Atlantic. *Geochim. Cosmochim. Acta* **42**, 1501–1516.

White W. M., Schilling J. G. and Hart S. R. (1976) Strontium isotope geochemistry of the central North Atlantic: evidence for the Azores mantle plume. *Nature* **263**, 659.

Whitford D. J. (1975) Strontium isotopic studies of the volcanic rocks of the Sunda arc, Indonesia, and their petrogenetic implications. *Geochim. Cosmochim. Acta* **39**, 1287–1302.

Whitford D. J., Compston W., Nichols I. A. and Abott M. J. (1977) Geochemistry and late Cenozoic lavas from eastern Indonesia: role of subducted sediments in petrogenesis. *Geology* **5**, 571–575.

Wilkening L. L. and Anders E. (1975) Some studies of an unusual eucrite: Ibitira. *Geochim. Cosmochim. Acta* **39**, 1205–1210.

Wilshire H. G. and Jackson E. D. (1975) Problems in determining mantle geotherms from pyroxene compositions of ultramafic rocks. *J. Geol.* **83**, 313–329.

Wilson I. H. (1978) Volcanism on a Proterozoic continental margin in north western Queensland. *Precambrian Res.* **7**, 205–235.

Wilson J. F., Bickle M. J., Hawkesworth C. J., Martin A., Nisbet E. G. and Orpen J. L. (1978) Granite-greenstone terrains of the Rhodesian Archean craton. *Nature* **271**, 23–27.

Windley B. F. (1977) *The Evolving Continents.* Wiley, N.Y. 385 pp.

Wood D. A., Gibson I. L. and Thompson R. N. (1976) Elemental mobility during Zeolite facies metamorphism of the Tertiary basalts of Eastern Iceland. *Contrib. Mineral. Petrol.* **55**, 241–254.

Wood J. A. (1973) Bombardment as a cause of lunar asymmetry. *Moon* **8**, 73–103.

Wyllie P. J. (1970) Ultramafic rocks and the upper mantle. *Mineral. Soc. Amer. Spec. Paper 3*, p. 3–32.

Wyllie P. J. and Huang W. L. (1975) Influence of the mantle CO_2 in the generation of carbonatites and kimberlites. *Nature* **257**, 297–299.

Wyllie P. J. and Huang W. L. (1976) High CO_2 solubility in mantle magmas. *Geology* **4**, 21–24.

Yoder H. S., Jr. (1976a) *Generation of Basaltic Magma*. Nat. Acad. Sci., Washington, D.C. 265 pp.

Yoder H. S. (1976b) Garnet peridotites from the upper mantle. In *Generation of Basaltic Magma*, p. 30–33. Nat. Acad. Sci. Washington, D.C.

Yoder H. S. and Kushiro I. (1969) Melting of a hydrous phase: phlogopite. *Amer. J. Sci.* **267-A**, 558–582.

York D. (1966) Least squares fitting of a straight line. *Can. J. Phys.* **44**, 1079–1086.

York D., Kenyon W. J. and Doyle R. J. (1972) ^{40}Ar-^{39}Ar ages of Apollo 14 and 15 samples. *Proc. Lunar Sci. Conf. 3rd*, p. 1613–1622.

Zartman R. E. (1974) Lead isotopic provinces in the Cordillera of the Western United States and their geological significanes. *Econ. Geol.* **69**, 792–805.

Zindler A., Brooks C., Arndt N. T. and Hart S. R. (1978) Nd and Sr isotope data from komatiitic and tholeiitic rocks of Munro Township, Ontario. *U.S. Geol. Surv. Open-File Rept. 78-701*, p. 469–471.

Zindler A., Hart S. R. and Frey F. (1979) Nd and Sr isotope compositions and REE abundances in Reykjanes Peninsula basalts: evidence for mantle heterogeneity. *Earth Planet. Sci. Lett.* **45**, 249–262.

Overleaf:

Impact craters provide an important tool for clarifying age relationships among planetary provinces, such as the lightly-cratered lunar basaltic plains (~3.3 b.y. old) of Mare Imbrium which form the background of this view. The cratered upland-like terrain (foreground) is actually part of the 3.9-b.y.-old basin's multi-ring system; a fragment of the same ring is defined by the Alps mountains (horizon, right). Various nearly flooded craters, such as Archimedes (center) on the basin floor, formed in the interval between the basin-forming impact and the final lava flooding. Apollo metric photo courtesy of National Space Science Data Center.

8. Chronology of planetary volcanism by comparative studies of planetary cratering

William K. Hartmann, Team Leader
—Planetary Science Institute, Tucson

Robert G. Strom
—Lunar & PLanetary Laboratory
University of Arizona

Stuart J. Weidenschilling
—Planetary Science Institute
Tucson, Arizona

Karl R. Blasius
—Planetary Science Institute
Pasadena, California

Alex Woronow
—Lunar & Planetary Laboratory
University of Arizona

Michael R. Dence
—Department of Energy, Mines & Resources
Ottawa, Ontario, Canada

Richard A. F. Grieve
—Department of Energy, Mines & Resources
Ottawa, Ontario, Canada

Jimmy Diaz
—United States Geological Survey
Flagstaff, Arizona

Clark R. Chapman
—Planetary Science Institute
Tucson, Arizona

Eugene M. Shoemaker
—Department of Geology
California Institute of Technology

Kenneth L. Jones
—Planetary Science, Incorporated
Pasadena, California

8.1 BACKGROUND OF STUDY

The purpose of this chapter is to review and update information on the dates of volcanic events on the planets as derived from counts of craters superimposed on the volcanic landforms. The basic philosophy of this method is that recognizable primary impact craters*, formed by impacts of asteroids and comets, accumulate on planetary surfaces, and that the numbers of craters accumulated on uneroded surfaces increase with age of the surface. Crater densities (numbers/km^2) on a given planet's stratigraphic units give relative ages of the units on that planet, while crater densities divided by average crater production rates give absolute ages that can be compared from one planet to another. This general methodology is described in semi-technical reviews by McGill (1977) and Hartmann (1977a).

Such an approach to planetary surface dating has been applied in its modern form since the work of Öpik (1960), who used crater counts and impact rates to conclude that "the craters in Mare Imbrium can be accounted for by bombardment at a more or less uniform rate during 4.5 b.y. by stray bodies." Such work was also pioneered by Shoemaker *et al.* (1962) who analyzed the production rate of terrestrial fossil impact craters, and used lunar crater counts to support "the hypothesis that the lunar maria are about 4.5 billion years old, that the craters assumed to be of primary impact origin are indeed formed by impact . . . [and that] the rate of formation of craters on the Moon in the period prior to the development of the maria was much greater than during subsequent geologic history."

Similarly, early studies were performed by Baldwin (1964) who found that "the maria are of one general age, but do show differences in age [and that] . . . comparison with the flux of smaller masses in the neighborhood of the Earth leads to a probable age of the maria of 2 to 3 billion years." Hartmann (1965) reviewed the numbers of terrestrial and lunar craters and their production rates, concluding that "comparison with crater counts on the Moon gives an age of about 3.6 b.y. for the lunar maria," and also (1966a) used lunar crater counts to describe a premare period of "early intense bombardment" several hundred times more intense than post-mare cratering. Thus, the technique of crater dating can fairly be said to have converged on a reasonably accurate estimate of lunar history in pre-Apollo years. (However, a last-minute flurry of interest in data which were only partially understood, namely photo-meteor data and data on meteors detected by microbarograph records, led a few workers to revise upward their estimates of impact rates, and thus revise downward their estimates of mare ages, by factors of about ten during the last year or so before Apollo. This error reduced confidence in the whole method). Apollo and Luna spacecraft sampling of lunar rocks confirmed a spread of ages in the lunar maria Tranquillitatis, Procellarum, Fecunditatis, Imbrium, and Serenitatis, from about 3.2 – 3.8 b.y. Spacecraft data also confirmed that cratering rates were much more intense prior to about 3.8 b.y. ago, such that few rocks older than 4.0 b.y. have survived (e.g., Taylor, 1975).

Radiometric dating (Chapter 7), of course, gives vastly greater precision on the dates of specific rocks from sampled volcanic landforms than crater counts can, but is applicable only to those specific surfaces from which we have samples— namely, the Earth and a few spots on the Moon. Another limitation of radiometric dating is that it is hampered by a spread in ages of rocks from each site, due to an unknown combination of vertical and horizontal mixing of rocks at each site. For example, if a given landing sampled primarily a localized, surficial, young flow, only a meter or so thick, then generally, the rock ages would differ from those of the deeper layers. The crater counts averaging over a large region to a depth on the order of 100 m or more would give an age more characteristic of the deeper mare surface layers and having different significance to this study than the radiometric ages from this young, thin, surface flow. In practice, the rough consistency of lunar radiometric absolute chronologies and crater-derived relative chronologies serves as a check of each system against the other, indicating that less-cratered maria generally give younger rocks, but also indicating some complexity at a local (10–100 km) scale. In the absence of radiometric data for other planets, all absolute chronological conclusions currently rest on the crater chronology technique, calibrated by radiometric rock ages, both lunar and terrestrial, although important relative chronological information is added by stratigraphic studies.

Extension of the absolute crater chronologies to other planets is primarily dependent on several issues:

**For three centuries, arguments continued over whether extraterrestrial craters are really meteoritic or volcanic. In recent decades the former theory has been almost universally adopted, but primarily on grounds of photomorphology, energy considerations, and size distribution (matching that predicted for asteroid fragments), rather than on field or sample evidence. A little-recognized consequence of the Voyagers' explorations is that Io's intense volcanism has produced only various-shaped, scattered calderas, and no cratered uplands like those on the Moon and planets, and hence argues against any interpretation that most lunar or planetary craters are volcanic.*

(1) the consistency and accuracy of crater counts; (2) knowledge of the crater production rates that have applied through time to the planet in question; (3) complications due to erosional and depositional modification of craters effective on each planet; (4) the use of small (subkilometer) craters to maximize statistics, with possible confusion introduced by volcanic or secondary-impact origins of some such craters.

The first issue has been aggravated somewhat by the historical fact that different workers in the field have gathered and published crater diameter distributions (number of craters per diameter interval versus diameter) according to different systems (cumulative plots, incremental plots with logarithmic diameter increments, incremental plots with linear increments). Thus, there has been some concern whether the different counts are truly consistent. Here we address this concern by reducing many workers' counts to a single data format. We find acceptable consistency.

A much greater uncertainty applies to the second issue, the crater production rate on other planets, which is the main problem now remaining in the determination of absolute chronologies with crater counts. Various workers who have analyzed impact rates by studying orbital dynamics and asteroid/comet discovery rates have predicted crater production rates for other planets consistent within a factor of about 2 or 3. We adopt representative values and discuss uncertainties. We will also discuss sub-problems, such as (a) enlargement of craters by gravity-caused collapse of rims (with central peak formation?) during or after formation, complicating conversion of impact rates of given-sized meteorites into crater production rates of given-sized craters; and (b) additional impact velocity and gravity effects varying from planet to planet.

The third issue raises the question of precise definition of what is being measured with crater counts. From Mariner IV Mars photos, Öpik (1965, 1966) showed that there is a deficiency of small craters (D = few km), relative to the diameter distributions found on the Moon and among larger martian craters, and concluded that smaller craters were preferentially being destroyed by martian erosion, "the corrosive action of sand and dust blown around by the desert wind," a conclusion endorsed in general form by nearly all current workers (see review of "Cratering and Obliteration History of Mars", Chapman and Jones, 1977). Concurring, Hartmann (1966b) defined the age being measured by crater densities as "crater retention age," a diameter-dependent age which measures the length of time that craters of a specified diameter have been retained on a given surface. In the absence of erosion, this may be the actual age of formation of the surface (e.g., extrusion of a lava flow); but in the presence of significant erosion, it may be whatever shorter time is required to obliterate craters of the specified size being studied. Crater retention ages are thus analagous to K-Ar rock ages. The K-Ar age may be the date of solidification of the rock; however, in the presence of a later disturbance, such as a shock releasing argon from the rock, the K-Ar age may be the shorter time since the disturbance. Similarly, crater retention ages may date either the formation of the rock unit or the effective time since surface obliterative disturbance.

In the present context, we are more interested in dating volcanic rock units than in dating erosion time scales. These two effects may be distinguished by certain characteristics of the crater diameter distributions. Sparsely cratered planetary units, such as lunar maria, mercurian smooth plains, and youthful volcanic plains on Mars, where there is little evidence of massive erosive or obliterative loss processes, show similar diameter distributions. These are characterized by a power law, which apparently gives the initial creation distribution of primary impact craters, at least at $D \gtrsim 4$ km

$$N(D) = kD^b \qquad (8.1.1)$$

where

$N(D)$ = incremental number of craters in logarithmic D interval $(D, \sqrt{2}\,D)$.

D = crater diameter in km.

k = constant depending on age (crater density) of surface.

According to least-square fits to crater data presented in section 8.3, the parameter b has a value close to −1.8 for most such units (typically, −1.6 to −2.0). This parameter, b, sometimes called size index, is the slope of the diameter distribution when portrayed on a graph showing log N vs. log D.

A widely used variant of this formulation uses the cumulative number of craters

$$N_{cum}(D) = cD^b \qquad (8.1.2)$$

where

$N_{cum}(D)$ = number of craters larger than D.

c = constant different from k.

By mathematical coincidence, as long as the power law relation holds (with constant b), b is the slope on a log-log graph of both the incremental law and the cumulative law.

A large body of data confirms the b-value near −1.80 for well-preserved crater populations at $D \gtrsim 4$

km (Baldwin, 1964; Neukum *et al.*, 1975a). Other b-values occur at other sizes and for certain less well-preserved crater populations.

Figure 8.1.1 illustrates some problems that can cause varying structure in the size distribution at large sizes. Figure 8.1.1a represents the unmodified distribution approximating a power-law and plotting as a straight line in a plot of log N vs. log D. Figure 8.1.1b shows the typical downturn that results from observational incompleteness — the failure to resolve all of the smallest craters. We avoid this effect by adding counts from high resolution photos to extend the counts from low-resolution photos, and by rejecting counts below the critical diameter D_c where this downturn begins to appear. Figure 8.1.1c shows the effect of a localized group of secondary or endogenic craters: they typically have a steeper slope and cause an upturn below some diameter D_c. Gault (1970) describes a similar effect due to transition in scaling relations between smaller (strength-dominated) and larger (gravity-dominated) craters. In a few of the counts in this chapter, such as those for Olympus Mons and other shields on Mars, observers have rejected some small craters suspected of being volcanic pits. But most of the counts in this chapter include all craters seen; mostly they involve craters of $D > 2.8$ km where upturns of this type do not occur.

Obliteration processes can cause important effects. Figure 8.1.1d shows the effect of sudden blanketing of an area by an ejecta blanket or lava flow that covers all features to a given depth. This obscures craters smaller than some D_c but leaves bigger craters' rims still showing. Craters with rim heights much larger than the flow thickness would be minimally degraded, and might be counted as post-blanketing craters, even if one were trying to distinguish thc effect. In Fig. 8.1.1e, the same area has been exposed for some time, and craters have re-accumulated to some degree, leaving only a "bump" in the diameter distribution. Figure 8.1.1f shows the result of a whole series of these episodes, with flows and/or ejecta blankets of different depths: the curve becomes bumpy and has some decrease in slope due to preferential loss of small craters. Also shown in Fig. 8.1.1f is the fact that mere "cookie cutter" obliteration of small craters by accumulating large craters and their rim structures decreases the slope of the curve at lower diameters.

Figure 8.1.1g shows a different effect applicable on Mars. A constant, slow deposition of material in crater bottoms will obliterate small craters rapidly, larger craters slowly, and craters above a certain depth not at all if there has not been time for the oldest such craters to be filled in. Öpik (1966), Hartmann (1971b), Chapman (1974), and Chapman and Jones (1977) have described how this effect produces a branch with slope

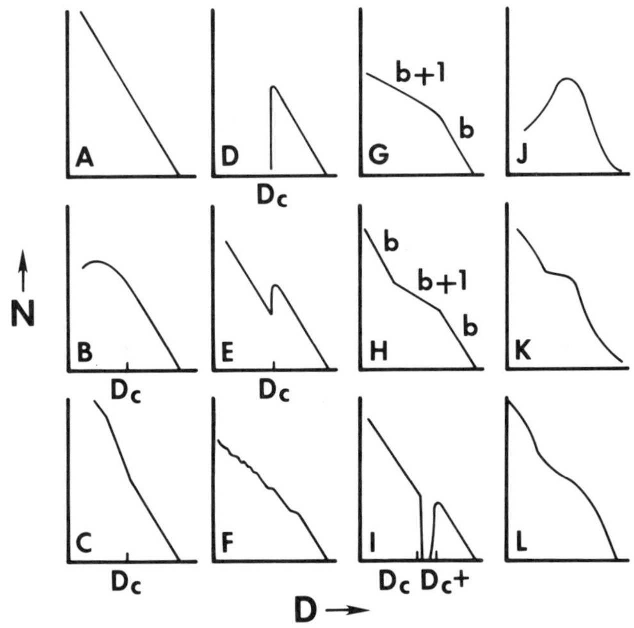

8.1.1 Schematic examples of possible variations among crater diameter frequency distributions. **a.** Normal primary impact frequency distribution found on sparsely cratered lunar and martian plains; approximates normal distribution in most provinces studied here. **b.** Turnover due to failure to resolve small craters, an observational error eliminated here by rejecting data below an incompleteness limit. **c.** Addition of secondary impacts or endogenic craters at small sizes causes upward bend. **d.** Blanketing by a single layer of depth greater than *d* erases craters below critical diameter D_c, which are shallower than *d*. **e.** As in **d**, but after a period of time, when blanket has been partially recratered. **f.** Successive episodes of ejecta blanketing and recratering, plus "cookie-cutter" removal of smaller craters builds up an irregular curve bending over due to loss of small craters. **g.** Slow steady deposition of dust builds up a steady state with old shallower craters being obliterated faster than old larger craters; magnitude of slope is one less than normal slope b; see text. **h.** As in **g**, but after slow deposition stops, allowing partial recratering. **i.** Transition from simple to complex forms, in which craters larger than D_c enlarge by wall slumping, produces gap in diameter distribution. **j.** Very primitive diameter distribution of planetesimals may have been peaked at a characteristic size, causing peak in earliest crater diameter distribution. **k.** As in **j**, but after a period of time, when primitive planetesimals begin to build up a size distribution characteristic of fragments, with a trace of the original peak remaining. **l.** Combinations of various physical effects of secondary and endogenic craters, blanketing, erosion, slumping, and early anomalous size distributions (**c-k**) could cause curvatures in size distributions on old, heavily cratered surfaces, such as actually observed on lunar, mercurian, and martian highlands.

b + 1 (smaller negative slope) as a steady state if the deposition rate is constant, as shown in Fig. 8.1.1g. After Mariner IV, Öpik and others showed that the flattened slope appeared in the Mars data from diameter about 64 km down to ~4 km. However, later missions showed that at still smaller diameters, the slope again steepened to the value characteristic of primary craters on an uneroded surface (Fig. 8.1.1h). As reviewed by Chapman and Jones (1977), this led to new models of martian erosion history. Generally, the erosion rate has decreased fairly sharply from an earlier high erosion rate to a present-day low erosion rate (Murray *et al.*, 1971; McGill and Wise, 1972; Hartmann, 1971a, b; 1973), so that small craters ($\lessapprox$ 4 km-scale) today are accumulating about as fast as they form.

A related problem, first noted by Dence (1965) and discussed by Dence *et al.* (1977), is the transition from simple craters (bowl-shaped) to large complex craters (terraced, with central peaks, etc.). This occurs by a slumping of the crater walls for craters larger than a certain size. The size is dependent on the planet's gravity and surface properties. If so, then as shown in Fig. 8.1.1i, a certain size crater (from D_c to D_c^+) would be removed from the diameter distribution, having been enlarged by slumping and counted in large diameter bins. This effect, though not yet identified in lunar and martian size distributions, may nonetheless add another complication.

Figure 8.1.1*l* represents the type of curve often found on older surfaces bearing evidence of overlapping ejecta blankets, possible erosive and obliterative effects, and the possible effects of earlier diameter distributions. These curves are often more irregular than the curves found in well-preserved younger areas, and may have segments with shallower slopes. It is thus difficult to unravel these effects and determine the history of these older, complex regions, as we will discuss in more detail in section 8.7. Judging from the lunar experience and from quantitative age estimates, these provinces usually are older than an estimated 3.5 b.y., so that while their specific histories are complex, we can often place them in the appropriate portion of planetary history, which is our objective here.

Slope variations resembling the martian curve have been reported for lunar and mercurian upland plains units (Strom and Whitaker, 1976) indicating that non-atmospheric processes probably also introduce curvatures in the crater diameter distributions. These processes might include blanketing by basin and crater ejecta (Fig. 8.1.1f). However, Strom and Whitaker (1976), Schultz (1976) and others, pointing in particular to well-preserved craters on the Orientale ejecta blanket, have suggested that primordial diameter distributions (Figs. 8.1.1j, 8.1.1k) differing from today's may be preserved in these early units. Secondary impact craters up to 25 km across, produced in basin ejecta (Wilhelms, 1976), complicate the issue. Today's meteoroid size distribution is believed to have evolved by fragmentation of crater-forming planetesimals. A different earlier size distribution may reflect dominance of accumulation over fragmentation among planetesimals.

The fourth problem, the interpretation of small craters, shows up at sizes smaller than about 2 – 3 km. Here, the slope b of the diameter distribution sharply steepens, with an excess of small craters, as shown in Fig. 8.1.1c. Their abundance makes them attractive as sources of statistics about geologic units. Most of these small craters, which virtually blanket the lunar maria at diameters smaller than a few hundred meters, have been interpreted as impact craters caused by low-velocity secondary ejecta from the larger primary craters (Shoemaker, 1965). If these small craters are produced in a uniform way by primaries or a mixture of primary and secondary impacts, they can be used to determine chronological information, with age assumed to increase with crater density. Recent examples of this relatively traditional methodology applied to small craters are given by Greeley and Gault (1970), Neukum *et al.* (1975b), Young (1975), Neukum and König (1976) and Boyce and Johnson (1977), who attempt to date a variety of maria and large primary impact craters. However, as Young notes, irregularities are found in the shape of the size distribution in some areas. Young attempts to explain these as due to crater-forming mechanics in regolith layers of different degrees of maturation and depth. On the contrary, Neukum and his co-workers find that a single distribution of small craters can be identified and applied everywhere, even on Mars, once corrected for effects such as gravity scaling. To complicate matters, Chapman *et al.* (1970) and others have argued that an unknown proportion of the small (subkilometer) craters in the lunar maria (and perhaps elsewhere) are due to endogenic processes such as collapse, not primary or secondary impact. Still more distressing, Oberbeck and Aggarwal (1977) and Wilhelms *et al.* (1978) find that secondaries, identified by morphology, are not just limited to the smallest craters. These workers identify basin secondaries reaching 20–30 km diameter; basin secondaries thus comprise a large fraction of all multikilometer craters in the highlands.

For all these reasons, some authors such as Trask (1971) and Chapman *et al.* (1970) have attempted to establish a morphological sequence of degraded states by which age could be tied to state of preservation.

Basilevsky (1976) reports support for a consistent relationship between crater degradation state and age, based on astronaut and Lunakhod crater photography and absolute exposure age measurement. Swann and Reed (1974) have attempted to use a correlation of morphology and cosmic ray exposure ages to date post-mare craters of diameter about 75 m to 1 km.

A far-reaching attempt to apply morphology of small craters to the dating problem was developed by Soderblom and Lebofsky (1972), who assumed a theoretical model of the degradation and flattening of craters under the influence of continual bombardment by fine debris. They showed that the D_L parameter, the diameter of a crater (in meters) whose rim would be eroded to a 1° slope by impacts, is theoretically related to the accumulated number of small impacts. Good correlations have been described between D_L and crystallization ages of rocks measured from Apollo and Luna sample sites (e.g., Boyce *et al.*, 1974; see also Table 8.8.2), as well as between D_L and crater density (e.g., Neukum, 1977). In principle, this utilization of small craters allows dating of quite small localized flow units, thereby leading to detailed volcanic chronologies describing emplacement of different flows within a given mare. This method has been widely applied (Boyce *et al.*, 1974; Boyce, 1976) and used to establish correlations between age and other properties such as mare chemistry (Soderblom *et al.*, 1977).

In discussing surface ages, Boyce (1976) and others have used another age parameter which makes use of similarly-sized craters. This parameter, C_S, is a limiting crater diameter where the steep size distribution of small craters intersects the estimated steady state distribution (Trask, 1966) and is believed to increase with surface age (see Shoemaker *et al.*, 1969, p. 67). Boyce derived D_L values from an empirical correlation, $D_L = 1.7\, C_S$, since the Orbiter photos he used had insufficient resolution to determine D_L directly.

In spite of the promise of these techniques, problems have been ascribed to them. Chapman *et al.* (1970) and Schultz *et al.* (1977) indicate that shallow non-impact craters might form by sagging of lava flow surfaces or other processes that could produce fresh depressions with low slopes, thus giving entirely spurious ages by the D_L criterion. Young (1975) and Schultz *et al.* (1977) argue that physical properties of the target rock or regolith (depth, maturation of regolith for example) may affect initial crater shape and degradation rate. Young (1975) also suggested seismic activity as a source of gentle slopes that might invalidate age relationships. Schultz *et al.* (1976, 1977) explicitly conclude that the erosion models in use are not valid down to slopes of 1°, though a criterion involving 20° slopes might be appropriate. Admixtures of small endogenic craters or secondary clusters would damage the D_L and C_S methods.

We will begin our report by setting aside controversies about the application of small craters and detailed chronologies of localized units. These methods are not critical, since most of this study is concerned with the *broad* development of volcanism on the other planets, not the details of stratigraphy in localized units. For example, application of crater chronologies or radiometric chronologies to constrain thermal modelling (Chapter 9), does not call for detailed stratigraphic age relationships, since thermal models of planetary evolution will, at best, show only the main trends in development of volcanic acitivity. For that purpose, it is more important to determine whether a planet's eruptive activity peaked in its first billion years, last billion years, or somewhere in the middle, than to determine if the east or west half of a certain basaltic plain is younger. Therefore, we are not putting our initial emphasis on the use of small craters because of the uncertainty that attends to their utilization, but we will return to an evaluation of small-crater methods after presenting the large-crater data. Our initial aim will be a critical understanding of the usefulness of crater count data in deriving relative and absolute dates of major periods of volcanic activity on the planets.

In summary, although we present a new review of much raw data, we do not aim at an exhaustive review of all work in the field of planetary crater studies, but rather at deriving as reliably as possible a broad-brush view of the chronology of basaltic volcanism on the planets. Therefore, represented in this chapter is a relevant cross-section of work in various fields, including planetary crater counting, terrestrial crater field work, surveying of interplanetary meteoritic bodies, and theoretical calculations of the planetary collision rate of such bodies.

8.2 METHODOLOGY AND DEFINITIONS

In order to reach the ultimate objective of comparing rates and natures of volcanic processes on different planets, it is necessary to compare crater counts by various workers and develop a standard technique for deriving crater diameter distributions. To do this we will follow several steps:

(1) Describe counts of readily recognizable impact craters on volcanic units.

(2) Present tabular and graphical summaries of these data to allow relative comparisons and future analysis by others.

(3) Develop a least-square curve fitting procedure to convert the raw counts into reliable relative crater densities meaningfully characterizing different units by their numbers of craters/km^2 (section 8.3).

(4) Review existing data and theory on the production rates of craters on the Earth-Moon system (section 8.4) and other planets (section 8.5).

(5) Attempt to derive absolute ages by dividing crater densities by crater production rates (sections 8.6 and 8.7).

(6) Compare these results with those of other techniques, such as use of small craters, and discuss remaining problems and implications (section 8.8).

(7) Consider events during the earliest planetary history for which the crater evidence is more difficult to unravel (section 8.9).

The major sources of uncertainty in the final results occur in steps (4) and (5), since the cratering rates on other planets are very poorly known. We will show in steps (1), (2) and (3) that the densities of multikilometer craters in a typical geologic province, even if determined by different observers, can easily be measured with uncertainties often less than 30%, whereas the uncertainty involved in steps (4) and (5) may be a factor 2 or 3 for the planets Mars and Mercury. While these uncertainties may sound huge to those used to radiometric dating, we point out that they allow us to meet a major goal—to determine in many cases whether a given part of a planet has been surfaced only recently by volcanic extrusions, or whether those extrusions occurred primarily before three billion years ago. Since planetary surface ages are *a priori* uncertain by several orders of magnitude, we regard even a factor 3 uncertainty as acceptable improvement!

Although the uncertainties of steps (1) and (2) are relatively small, we will critically discuss the problems and methods in each step.

Regarding step (1), we solicited raw crater counts from a number of workers, primarily from chapter authors or their associates. Since the early 1960's, various workers have independently tabulated and reported data by different systems that inhibit direct comparison of numerical counts. We solicited data primarily from those whose counts could be easily expressed in terms of numbers of craters/km^2 in $\sqrt{2}$ diameter increments (i.e., D, $\sqrt{2}$ D, 2D, . . .), so as to achieve uniform format.

The counts assembled in this way were, in general, prepared independently by different authors at different times, usually with no thought of a direct comparison of this type. Some counts in duplicate areas were made several years apart. Among some lunar and martian sets, counts have been derived from images returned by different spacecraft, using different techniques of counting, ranging from simple records on plastic overlays to semi-automated electronic devices such as those used by Greeley and Gault (1970), and stereo analysis such as that used by Neukum *et al.* (1975a). The counts were grouped by comparable geologic provinces (Fig. 8.2.1), usually with two authors' data in each province. In general, we tried to follow this definition of a **province**: On a single planet, a geographical region characterized by apparent geological homogeneity, in which craters in different regions are believed to have been affected by similar environmental conditions, and in which crater counts may be expected to characterize a meaningful crater retention age. In many cases, the provinces counted were relatively well-defined as, for example, in the case of Mare Serenitatis, or the surface of Olympus Mons. However, different authors sometimes differed in defining boundaries of less-sharply-edged provinces, such as the Tharsis region. In these cases, we had to compare slightly different areas, either by combining results into one province where we believed data could be combined meaningfully, or by treating separately each author's data.

Also, different authors use different methods to gather data within the same well-defined region. Most commonly (i.e., Jones, 1974), an author defines the boundary of a province, then publishes a catalog giving all craters above a certain diameter in that province. For example, Wood and Andersson (1980) have cataloged all craters $\geq$ 7 km on the lunar frontside. Small diameters (D) are thus represented by large numbers of craters, sometimes exceeding 100 in a given D bin. In a given province, counts usually range from at least 50 craters in the bins at small D to 1 or 2 at large D. Other authors (e.g., Hartmann in most of his work) use a "nesting-photo technique," in which photos of low resolution are used to characterize a large region, while scattered higher resolution photos may be used to provide statistics on smaller craters in local region(s) within the large area. If the province contains a landing site, nesting series of photos may be used to zero in on the specific site. The nesting-photo technique allows quicker data gathering. It may give more craters in the large-D bins (by counting larger areas), but more modest numbers of craters (typically 5 to 25) in each of the smaller-D bins. Also, this technique does not give uniform coverage throughout the broad area.

In spite of these differences, reasonably comparable crater densities, N (Eq. 8.1.1), were obtained by different authors for the same provinces. To make this comparison quantitative, Table 8.2.1 shows median ratios of N-values of one observer to the N-values of another, indicating that the median discrepancy between higher and lower N-values in individual D bins is roughly 50%. Generally, this discrepancy is attributable to factors such as normal observational errors, differences in interpretation of secondaries or morphology classes, and slight differences in defining province boundaries. But, because this is usually not a systematic bias between one observer and another, the median of the differences from one observer to another (one being

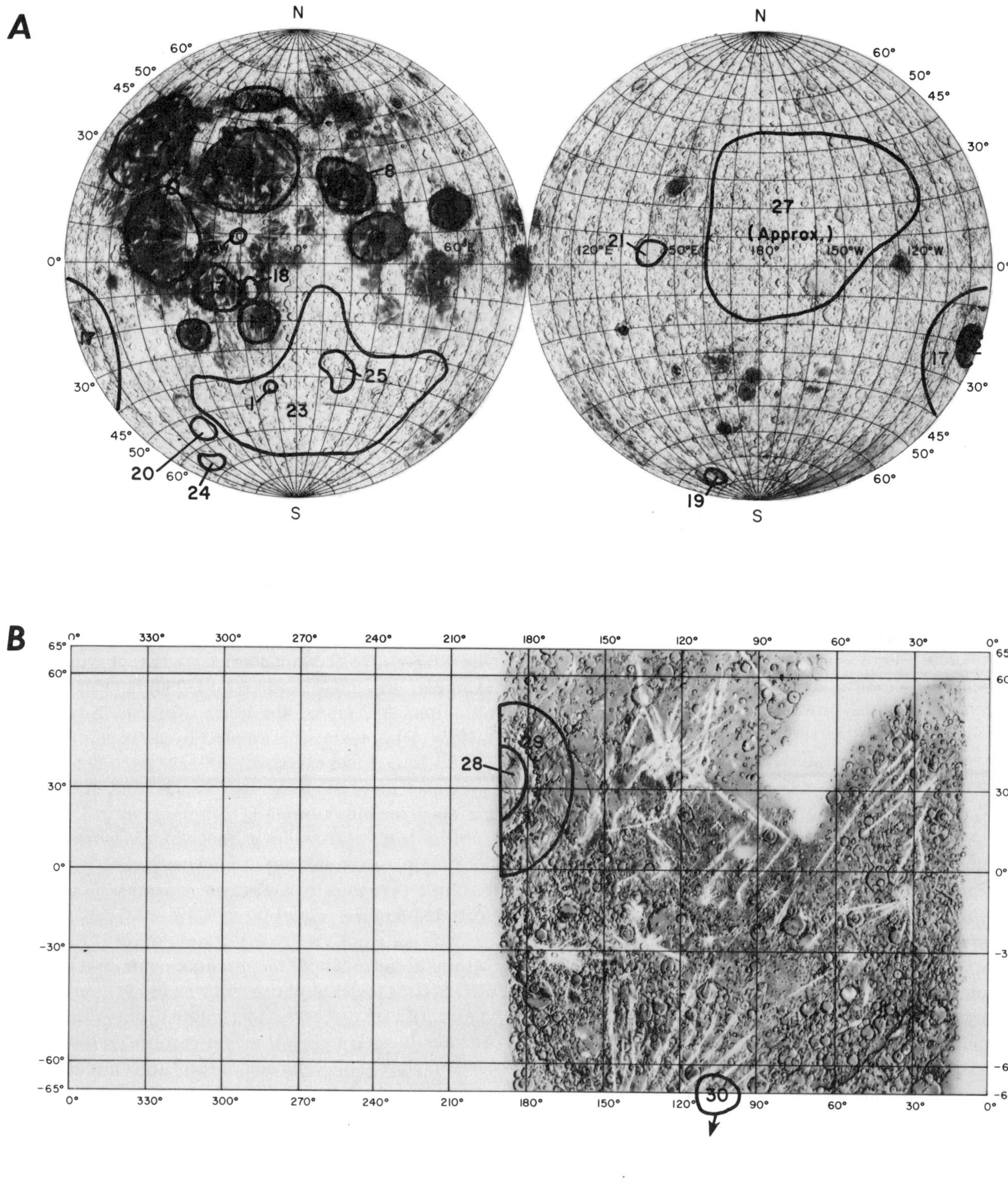

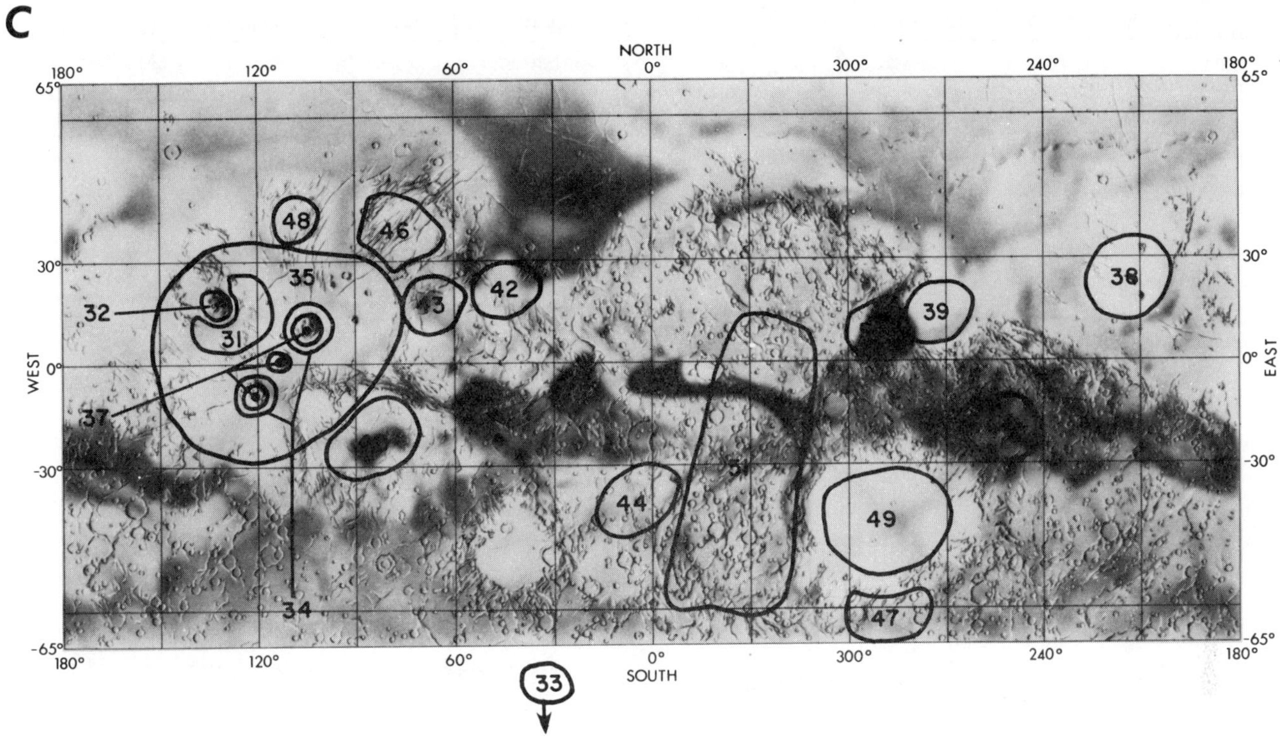

8.2.1 Maps identifying approximate province boundaries of regions studied on the Moon (**a**) Mercury (**b**)and Mars (**c**). Precise boundaries are difficult to specify, since different authors defined slightly different province boundaries, and some authors used the nesting photo technique that includes different sub-provinces at different resolutions. Province numbers are as specified in Tables 8.2.2, 8.3.2, 8.6.1, and Appendix.

sometimes higher and sometimes lower) is only about 30%. Discrepancies tend to be less among the smaller, more abundant craters of a given photo-data set and larger among the largest craters. Since we deal with least-squares fits to the whole *array* of data points, the run of crater densities in a given province is determined typically to less then 30% uncertainty. Within a given group of workers, agreeing in advance on the criteria for defining provinces and craters to be counted, even lower errors are reported (e.g., Greeley and Gault, 1970). However, we believe that this comparison of independent data sets produced in different years at different institutions establishes that uncertainties in actual counts are less than other current sources of uncertainty in the field of crater-dating (especially crater production rates on other planets), and are generally acceptable for our purposes.

The Appendix to this chapter includes plots of all the crater counts. The most straightforward utilization of impact craters is to count multikilometer craters in geologic provinces that are sparsely cratered enough to avoid overlap, but densely cratered enough to give good crater statistics. Lunar maria are good examples; however, more densely cratered provinces raise problems, whose discussion requires definition of the following terms:

- *Primary impact crater.* A crater formed by impact of an interplanetary meteoroid (i.e., asteroid, comet, or fragments thereof).
- *Secondary impact crater.* A crater formed by impact of debris ejected (usually by primary impact) from the surface of the planet being studied.
- *Endogenic crater.* A non-impact crater formed by processes internal to the planet, such as collapse of lava into a lava tube, sagging of lava in a flow, or explosively eruptive volcanism.
- *Production diameter distribution,* or *Production function.* The diameter distribution produced on a pristine surface if all craters are observed. Presumably, it would be observed on sparsely cratered (young) lava plains that were formed in a short interval and never subjected to erosion.
- *Complete geometric saturation.* The theoretical densest possible hexagonal or artificially arranged packing (defined by equal-sized craters) such that the cumulative surface area of all craters is 90.5% of the total area of the surface studied. This idealized concept was defined by Gault (1970), who found

Table 8.2.1 Comparisons of pairs of crater counts by different authors for selected provinces (see text).

Province	Original Authors of Data Sets	Number of Data Pairs	Median Higher Value / Lower Value	Median Observer 1's Value / Observer 2's Value
Moon: Average of mare	Hartmann; Strom	8	1.7	1.7
Moon: Mare Tranquillitatis	Greeley and Gault; Hartmann; Strom	7	1.6	1.4
Moon: Mare Crisium	Hartmann and Wood; Strom and Allen	6	1.5	1.3
Moon: Orientale Basin and ejecta	Hartmann, Neukum; Strom	12	1.2	1.0
Moon: Frontside Highlands	Strom; Hartmann	6	1.5	1.5
Mars: Elysium	Jones; Blasius and Sigaty	6	1.2	1.2
Mars: Solis Planum	Hartmann; Jones	8	1.4	1.1
Mars: Extended Tharsis area	Hartmann; Jones	6	1.5	1.5
Mars: Heavily cratered areas	Hartmann; Jones	12	1.3	1.3
Mars: Chryse Planitia	Cutts and Blasius; Hartmann	23	1.4	1.1
Mars: Channels	Malin; Hartmann	8	1.9	1.1
Mars: Pavonis, Ascraeus, and Arsia Mons	Blasius; Hartmann; Jones	9	2.5	1.3
Mercury: Caloris mare	Allen; Hartmann	7	1.4	1.3
Mercury: Caloris ejecta	Hartmann; Woronow	4	1.5	1.5
		Average:	1.5	1.3

that it is not reached in nature. Though not attained in nature it is often encountered in the literature as a standard relative to which real crater densities are measured.

- *Cookie-cutter saturation.* The theoretical maximum packing of impact craters from a given interplanetary population of meteoroids, in which new craters are assumed to destroy old craters only in a "cookie-cutter" mode, with no additional effects due to ejecta blankets, or other erosion or deposition. This idealized concept is adapted from definitions by various workers, including Woronow (1977), whose work suggests that it might lead to craters cumulatively covering roughly 30% to 80% of the area of the surface studied in each $\sqrt{2}$ log D increment, depending on the population index (exponent) of the impacting population's size distribution.
- *Empirical saturation.* The maximum packing of craters actually occurring in nature from a given interplanetary population of meteoroids, under the condition that the addition of new craters results in no change in the crater diameter distribution because of destruction of old craters by all naturally occurring cratering processes (excluding endogenic erosional or modifying processes originating in the planetary environment, such as wind or lava flows). In particular, this includes effects of ejecta blankets and secondary craters (Fig. 8.4.1) and may depend on the surface gravity. This concept is adapted from definitions by Gault (1970), Woronow (1977), and Chapman and Jones (1977), whose (inconsistent) results suggest that it might occur with craters cumulatively covering about 5% to 30% of the area of the surface studied in each $\sqrt{2}$ log D increment, in empirical planetary cases.
- *Equilibrium (or steady state).* The crater density maintained on a planet in the presence of environmental obliterative effects, such as deposition of windblown dust, exhumation of strata, lava flows, etc. In the absence of those effects, equilibrium should equal empirical saturation; in their presence, the equilibrium density should be less than empirical saturation densities. This concept is adapted from many workers, but is most closely related to the term as used by Woronow (1977).

Models by various workers (e.g., Öpik, 1966; Chapman *et al.*, 1970; Chapman and Jones, 1977; Schultz *et al.*, 1977; Hartmann, 1968, 1971b, 1973) show that cratering accompanied by different erosive and depositional effects can lead to widely different equilibrium crater diameter distributions with different slopes or curvatures.

- *Observed size distribution.* The size distribution of craters actually observed on a given planetary surface.

These somewhat complicated terms highlight some problems in the collection and interpretation of data. In general, we have attempted to collect data believed to reflect counts of primary impact craters. We recognize that some sets of data may include some secondary craters, especially at smaller diameters, but by working primarily at D > 2.8 km, or specifying certain procedures, we believe we minimize the effect of secondaries. The data are not homogeneous, since different workers have used their own criteria, for example, to define post-mare craters or to eliminate secondaries and endogenic craters. In *most* of the data presented here, these criteria do *not* strongly affect the data, because in most young volcanic units such as maria and martian volcanic plains, most workers believe that most craters of D > 2.8 km are relatively fresh primary impact craters. For example, of 30 craters larger than 11 km catalogued and judged classifiable by Arvidson *et al.* (1974) and Jones (1974) in the sparsely cratered Elysium volcanic plains (Jones' Region 135), 100% were classified as either "fresh" or "slightly degraded" (the least degraded of three degradation classes). Similarly, of 59 craters in the sparsely cratered volcanic plains of Tharsis (Regions 6, 7, 9), 71% fall in these classes; of 53 craters in the moderately cratered Solis Planum volcanic plains (Region 10), 87%; and of 142 craters in the most cratered plains east of Tharsis, (Regions 12, 13), 88%.

The problem becomes more severe as we work back in time toward old, heavily cratered units. In the heavily cratered Mars uplands (Region 24), the percentage of "fresh" or "slightly degraded" craters in a sample of 378 drops to 46%, due to an admixture of what are believed to be eroded or partially filled old impact craters. In addition to the morphological problems of distinguishing battered volcanic units from battered non-volcanic ancient crust (if such exists), there is a problem of interpreting what the craters tell us. Probably the most conventional view has been that the 100 km-scale craters in the cratered terrains of the Moon and other planets roughly define a common envelope creating a power-law upper bound to crater densities in the solar system, representing true empirical saturation, with craters covering about 5–12% of the surface area in each $\sqrt{2}$ log D increment. Thus, the diameter distributions in those very densely cratered areas would directly reveal neither age of the surface nor the size distribution of the cratering meteoroids; the crater diameter distribution would be altered by the effects of saturation. However, computer models by Woronow have led him to conclude that no surface in the solar system is densely enough cratered to represent empirical saturation. Diameter distributions in the uplands would thus, according to Woronow (1978) preserve their true shapes. Neukum and Wise (1976) take a similar view. Woronow (1978) and Strom and Whitaker (1976) have followed this line of thought to derive proposed changes in the diameter distributions between the era of intense cratering (4 b.y. ago) and post-mare time (last 3 b.y.), as suggested in Fig. 8.1.1a, j, and k. Still another approach to the cratered uplands comes from Oberbeck *et al.* (1977), Wilhelms *et al.* (1978), and others, who have concluded that a large fraction (possibly a majority) of highland craters even at relatively large D (up to 25 km or more), are actually old secondaries (see Fig. 8.4.1), so that the uplands of the Moon and Mercury are only sparsely cratered with primaries. Thus, again, as we move away from data on the larger craters of the obvious young volcanic units, the interpretation becomes more controversial. This shows why it is difficult to get evidence of hypothetical volcanic units much older (more cratered) than the major lunar maria, as emphasized by Fig. 8.4.1.

Regarding step (2), we have grouped the raw data for graphical presentation and analysis into geological provinces listed in Table 8.2.2. Data from these units provide the basis for our conclusions in this chapter.

As for the mode of presentation in step (2), recommendations for presentation of crater counts have been made by a Crater Analysis Techniques Working Group, which issued a report (Arvidson *et al.*, 1978) during our work. Adopting many of their recommendations, we have converted all data obtained to counts of N(D), defined in Eq. (8.1.1), in order to combine different data sets in uniform format. We present these results in the form of two types of graphs discussed by Arvidson *et al.* (1978). The "incremental number plot" gives a conventional plot of log N vs. log D. The "relative size-frequency distribution plot," developed by Robert Strom and his associates, plots the same data normalized so that a power law with a −2 exponent (which would have a slope of −2 on the first plot), appears as a horizontal line, thus emphasizing departures from this idealized power law.

The second (Strom-type) plotting technique was among two types of plots recommended by Arvidson *et al.*, 1978. In preference to the first type of plot, which we

Table 8.2.2 Groupings of data defining geologic provinces.

Province No.	Geologic Province	Usable D Interval of Original Counts(km)	Author and Publication Date, if Published	Area Represented (km^2)		Method	Notes
	Moon:						
1	Tycho crater	0.09–8	Hartmann (1968) Strom & Fielder	~1	(4)	Selected regions melts	Contains post-crater volcanics or impact.
2	Aristarchus crater	0.12–1	Hartmann (1968)	~9	(3)	Selected regions melts	Contains post-crater volcanics or impact.
3	Copernicus crater	0.25–5.6	Hartmann (1968)	~7	(3)	Selected regions melts	Contains post-crater volcanics or impact.
4	Inner (light) portion of Mare Serenitatis	2.8–16	Neukum *et al.* (1975a)	1.4	(5)		
5	Mare Crisium	1–32 4–32	Hartmann and Wood (1971) Strom and Allen	~2 1.6	(5) (5)	Selected photos LPL catalog	Hartmann data set = inner portion.
6	Mare Imbrium	0.08–16 4–64	Hartmann Strom & Allen	~2	(5)		Averages many known flows; contains Apollo 15 site.
7	Northern Oceanus Procellarum	4–32	Strom & Allen	4.3	(5)	LPL catalog	
8	Whole Mare Serenitatis	0.08–16 4–64	Hartmann Strom & Allen	~3 3	(5) (5)	Nesting photos	Average of whole mare; contains Apollo 17.
9	S. Oceanus Procellarum	0.35–45 4–45	Hartmann Strom & Allen	2 1.4	(5) (6)	Nesting photos	Contains Apollo 12 site.
10	Mare Fecunditatis	13–16 4–32	Hartmann (S. part) Strom & Allen	~3 2.8	(5) (5)	Nesting photos LPL catalog	Contains Luna 16 site.
11	Reference: Average of frontside maria	0.125–181 8–181	Hartmann Strom & Allen	4.7 6.7	(6) (6)	Catalog (ca. 1968) LPL catalog	S & A included fewer degraded craters than H. Used as Standard Reference to normalize other crater data.
12	Mare Orientale	125–8	Hartmann & Yale (1968)	~3	(4)		
13	Mare Cognitum	0.5–16 4–8	Hartmann Strom & Allen	~1.9 4.9	(5) (4)	Nesting photos LPL catalog	
14	Mare Tranquillitatis	0.07–2 0.9–32 4–45	Greeley & Gault (1970) Hartmann Strom & Allen	1 ~3 1.2	(3) (5) (5)	 Nesting photos LPL catalog	Contains Apollo 11 site.

Table 8.2.2 *(Continued)*

Province No.	Geologic Province	Usable D Interval of Original Counts(km)	Author and Publication Date, if Published	Area Represented (km^2)	Method	Notes
15	Mare Humorum	2.8–16	Hartmann & Wood (1971)	1.3 (5)	Selected photos	
		4–11	Strom & Allen	1.3 (5)	LPL catalog	
16	Mare Nubium	4–64	Strom & Allen	1.2 (5)	LPL catalog	
17	Orientale Basin and ejecta	0.35–128	Hartmann & Yale (1978)			Dates basin formation.
		8–91	Strom & Allen			
		4–45	Neukum *et al.* (1975)			
18	Fra Mauro	0.18–16	Hartmann	~3 (4)		Contains Apollo 14.
19	Schrödinger floor plains	1–11	Hartmann & Wood (1971)	~5 (4)	Nesting photos	Light, mare-like fill in a basin.
20	Basin floor near Schiller	2.8–16	Hartmann & Wood (1971)	~5 (4)	Nesting photos	Light, mare-like basin fill.
21	Plains in Mendeleev	4–32	Neukum *et al.* (1975a)	3.3 (4)		Upland crater fill.
22	Pre-mare craters	8–181	Strom		LPL catalog	Measure of interval between basin formation and lava filling.
23	Frontside uplands Intermediate D (5.7 to 64 km)	5.7–256	Hartmann	3.7 (6)	Catalog (ca. 1968)	Includes effects of probable upland flooding.
		8–512	Strom & Whitaker	~4 (6)	LPL catalog	
24	Bailly floor plains	4–91	Hartmann & Wood (1971)	~9 (4)	Nesting photos	Light, mare-like basin fill.
25	Upland plains units near Maurolycus	2–91	Hartmann	1.4 (5)	Orbiter photo counts	Probable upland volcanic flooding.
26	Heavily cratered uplands	1.0–2048	Hartmann	~2 (7)	Catalog	Relatively heavily cratered uplands, no mare, few plains; includes large basins.
			Hartmann & Wood	~3 (7)	Basin survey	
			Strom	~3 (7)	LPL catalog	
			Whitaker	~4 (7)	Basin survey	
27	"Pure" uplands	2–2048	Hartmann	~1 (5)	Various photos	Most heavily cratered upland areas. Minimal effects of upland plains.
	Mercury:					
28	Caloris Mare	5.7–91	Allen	5.1 (5)	Selected photos	Cracked mare-lake lava flows?
		2.8–64	Hartmann	~5 (5)		
29	Caloris Basin and ejecta	2.8–32	Hartmann	~3 (5)	Selected photos	Date of Caloris Basin formation.
		8–91	Woronow	~2 (6)		
30	S. Polar Quad. Highlands, Intermed. D	8–256	Allen	2.5 (6)	Selected photos Quad H-15	Includes effects of probable intervolcanic flooding.

Table 8.2.2 *(Continued)*

Province No.	Geologic Province	Usable D Interval of Original Counts(km)	Author and Publication Date, if Published	Area Represented (km^2)	Method	Notes
	Mars:					
31	Central Tharsis volcanic plains	2.8–32	Jones (1974, region 7)	2.5 (6)	Catalog	Average of plains surrounding main shield.
32	Olympus Mons	0.12–2 0.25–2.8	Blasius (1976) Hartmann (1973)	~3 (4) ~4 (4)	Selected photos, volcanic pits excluded	Last flows of largest shield volcano.
33	Polar plains S of 81°S	0.7–32	Hartmann	1.6 (5)	Selected photos	Very sparsely cratered, sedimentary layers?
34	Plains immediately around bases of Ascraeus and Arsia Mons	2.8–32	Hartmann	3 (5)	Selected photos	Volcanic units immediately predating 2 shields.
35	Extended Tharsis volcanic plains	8–256 2.8–91	Hartmann (1973) Jones (1974, regions 4, 6, 7, 9, 10)	~1 (7) 7 (6)	Selected photos Catalog region	Outer flows of Tharsis volcanics.
36	Channels emanating from localized sources or chaotic terrain	0.25–16	Hartmann Malin (1976)	~1 (5) ~5 (4)	Selected photos	Water flow from collapses?
37	3 shield volcanoes: Ascraeus, Pavonis, Arsia Mons	0.17–1.4 0.35–11 2–23	Blasius (1976) Hartmann (1978) Jones (1974, region 8)	6 (3) 2 (5)	Selected photos	Last flows, averaging over 3 shields.
38	Elysium Volcanic plains	2.8–45	Jones (1974, region 35)	1.7 (6)	Catalog	Volcanism of Elysium dome.
39	Isidis Planitia	2–32	Blasius & Sigaty	1.4 (5)	Viking photo	Sparsely cratered volcanic plain.
40	Solis Planum	0.35–32 2–32	Hartmann (1973) Jones (1974, region 10)	7.5 (5) 2.3 (6)	Selected photos Catalog	Moderately cratered volcanic plain (Mutch *et al.*, 1976).
41	Channels not associated with chaotic terrain	0.35–11	Hartmann	~1 (4)	Selected photo	Water flow from tributary systems?
42	Chryse Planitia	0.35–4 0.35–8 0.5–23	Cutts & Blasius Hartmann Blasius & Sigaty	2 (4) 4 (3) ~1 (5)	Selected Viking photos	Volcanic flows of outer Tharsis Viking I site.
43	Lunae Planum	0.35–23 2.8–23	Blasius & Sigaty Jones, region 12	~3.5 (5) 1.7 (6)	Viking photos Catalog	Volcanic plains E of Tharsis.
44	Noachis plains	2–32	Blasius & Sigaty	2.2 (5)	Viking photo	Ridges resemble lunar wrinkle ridges.

Table 8.2.2 *(Concluded)*

Province No.	Geologic Province	Usable D Interval of Original Counts(km)	Author and Publication Date, if Published	Area Represented (km^2)	Method	Notes
45	Tyrrhena Patera volcanic plains	2–16	Blasius	1.9 (5)	Selected Viking photos	Plains with "mare ridges" and "ghost craters."
46	Tempe Fossae	4–128	Jones, region 13	9.8 (5)	Catalog	Fractured plains on NE edge of Tharsis.
47	Volcanic plains on Hellas S Rim, 65° S	5.7–32	Blasius & Sigaty	8.7 (5)	Viking photos	Old volcanic plains, eroded or blanketed (?) at high latitude.
48	Alba shield	2–32	Blasius & Sigaty	1.0 (5)	Viking photo	Shield volcano, edge of Tharsis.
49	Hellas basin floor	2.8–23	Blasius & Sigaty	6.0 (5)	Viking photo	Dust-covered volcanics?
50	Syrtis Major Planitia volcanic plains	2.8–23	Blasius	5.5 (5)	Selected Viking photos	Region of flows, wrinkle ridges, and volcanic vents.
51	Heavily cratered plains	0.7–1448	Hartmann (1973)	~3 (7)	Selected photos	Evidence of erosion or obliteration process, old uplands with intercrater volcanic plains.
		4–256	Jones (1974, region 24)	1.6 (6)	Catalog	
	Earth:					
52	Eo-European Craton	23–90	Grieve & Robertson (1979)	4.5 (6)	Terrestrial mapping	Eroded continental shield.
53	N. American Craton	23–90	Grieve & Robertson (1979)	1.25 (7)	Terrestrial mapping	Eroded continental shield.

had adopted prior to this report, they recommended a cumulative plot in the format of Eq. 8.1.2, pointing out that the crater densities in the incremental plot depend on the size of the increments chosen. On the other hand, a great many data sets have been published using the $\sqrt{2}$ log increments that we have adopted, and we gathered and reduced our data in this format. Futhermore, the incremental plots we have used have one important advantage over cumulative plots in that they more clearly show changes in slope, turn-overs, or deficiencies of craters in certain diameter bins. A drawback of cumulative plots is that they never roll over or reverse slope, even if *no* craters appear in certain size bins.

In our plots, we have represented the probable error on each datum point by N/(number of craters counted in the given diameter bin)$^{1/2}$. The data points are symbolized by the initial of their author.

Regarding step (3), we have conducted experiments with several curve-fitting procedures. These procedures and the remaining steps are discussed in more detail in the remaining sections.

8.3 CURVE-FITTING PROCEDURES

Designing a curve-fitting procedure involves scientific interpretive problems, just as did tabulation of raw crater counts. The larger craters (D > 2.8 km) on most volcanic plains units on the planets tend to obey a power-law size distribution to the first order. (These are the craters most universally believed to be primary impacts.) This produces a straight line on a log-log plot of N(D) vs. D. There are theoretical and empirical indications that fragmented asteroidal bodies themselves obey such a power-law size distribution and produce a reflection of this simple distribution in the population of primary impact craters (Hartmann, 1969; Dohnanyi, 1969). The interpretation of these straight-line log-log plots should be a simple matter of least-squares curve fitting. Hartmann (1967, 1968, 1973) asserted that visual straight-line fits to these data (even without least-squares techniques) allow adequate measurements of the differences in crater density from one region to another, within the accuracy of the data, and expressed (Hartmann, 1978) some concern that formal standard deviations resulting from conventional statistical techniques could be misleading. This could occur because the traditional statistical techniques assume random distributions of the underlying population, while actual geologic processes, such as production of variable thickness ejecta blankets or degraded lava flows in certain areas, could produce subtle but significantly non-random factors affecting the form of the diameter distribution differently from one region to another. Calculated significant differences from one region to another might thus be falsely ascribed to the underlying population of planetesimals, whereas in fact they may be due to non-random geologic complexity inherent to the planet itself. Nonetheless, with increasing accuracy in crater count determinations, for example by stereo measurement techniques (Neukum *et al.*, 1975b), and with increasing attention given to crater count results, there has been increasing demand for standardized crater analysis techniques (Arvidson *et al.*, 1978).

We concluded that a comparative test of several least-squares curve-fitting procedures was thus warranted. We have fitted straight-line curves in the log N–log D plot, using those diameter intervals where this appears to be justified, but not fitting such curves to diameter distributions that appear obviously non-linear on graphical plots. Generally, we have applied our technique at the larger D's, where the straight-line (power-law) curve best approximates the primary impact crater distribution. In many of our cases, therefore, this rules out diameters below about 2.8 km., where the curve sometimes bends upward and most analysts believe that secondary impact craters or endogenic craters dominate. In a few cases, however, we will argue that we can extend below this diameter range.

Any least-squares procedure requires weighting of the data points, whether this be by equal weight or some other system. We have experimented with three different curve-fitting procedures:

- Method A. The first and simplest system, was the least-squares fit to a power law, i.e., a straight line in a log N–log D plot, with each datum point given equal weight regardless of its source or the number of craters used to determine the datum point. A problem here is that in most sets of crater counts, the largest diameter craters in the area being studied are much less abundant than the small craters, and thus are subject to large statistical uncertainty. On the other hand, this method has the advantage of not over-weighting the small craters, which are always the most suspect as to origin.
- Method B gives each data point a weight equal to the number of craters used in its determination, but with the proviso that the weight never exceed 25 even if the number of craters went higher than this. The proviso prevents exorbitant weight given to small D craters, which are of suspect origin. This is perhaps the best method.
- Method C involves somewhat more interpretation of the crater counts but may provide a result that is ultimately easiest to understand and use. It is based on the assumption that all of the primary impact craters (or at least the younger ones) on the planets were produced by a population of impacting bodies that had a single mass distribution and consequently produced crater diameter distributions with a single slope in the log N–log D diagram. It is thus assumed that if this single underlying slope could be determined from a very large population of craters in a relatively young unit where there is little chance of slope change due to secondaries or erosion effects, then this slope should be fitted to all of the diameter distributions studied on other provinces as well. This technique assumes that slopes of other value, as determined by methods A and B, are due mostly to statistical sampling variations, or, in geologically complex areas, to erosion, deposition or similar effects (Fig. 8.1.1). The technique of forcing all of the crater population to fit a single pre-determined slope has the great practical advantage of yielding parallel curves on the log N–log D diagram, thus allowing a relative density to be measured and expressed as a single ratio equally applicable at all diameters. This method is not applicable in old provinces, if these were impacted by populations with size distributions different from those in post-mare time (Fig. 8.1.1j and k), and would be questionable if gravity scaling or atmospheres changed the production function from one planet to another. The other two curve-fitting procedures have the disadvantage that the slopes vary from one province to another and thus the relative crater density varies from one diameter to another.

Method C is similar to the technique of "calibration curve" or "standard curve" fitting by Neukum *et al.* (1975a, b) and Neukum and Wise (1976), except that they have asserted that a standard production function, with a certain curvature, can be identified down to diameters of a few hundred meters or less. We have attempted to use only a "linear" (power-law) production curve derived from lunar maria down to sizes no smaller that 2.8 km. The problem of projecting to smaller sizes is discussed further in section 8.8.

Some analysts have abandoned curve-fitting and weighting procedures, representing crater density only by the cumulative number of craters larger than a certain size. However, most crater diameter distributions are so steeply sloped that this technique effectively gives nearly all the weight to the craters in the smallest one or two diameter bins. Since the smallest diameters are often suspect in terms of admixtures of secondary impact craters or endogenic craters, this method is dangerous and we reject it. Other analysts follow a curve-fitting procedure, but they report the result as a cumulative density above diameter D. The method is better, but the chosen D has often been 1 km, which we regard as dangerous because the slope of the diameter distribution steepens dramatically (and probably variably from province to province) at this small size, suggesting strong (and probably variable) effects of secondaries and perhaps endogenic craters in the 1 km size range. Extrapolations from the reported 1 km datum back up to 2 km, 4 km, or other size ranges, are thus also dangerous, since they depend strongly on the slope variations assumed.

We will now compare the results of our three curve-fitting methods. Table 8.3.1 lists the least-squares solutions for the straight-lines found in the log N–log D plot. These straight-line solutions are directly convertible to the power law shown in Eq. (8.1.1); the slope of this line is numerically identical to the exponent b. These solutions are listed for the convenience of future users of this study, but are difficult to interpret in this form. The reader can gain a much clearer impression of the crater densities by examining the log N–log D plots in the Appendix. A simplifying quantitative analysis can be carried out from the data in Table 8.3.1, as illustrated by Table 8.3.2, giving crater densities for each of the geologic provinces, calculated from the least-squares solutions, and expressed relative to the average crater densities of the front-side lunar maria. For each planet, the provinces are listed in order of increasing crater density (which means that the table starts with some of the least reliable solutions — densities calculated for the very sparsely cratered young craters, Tycho, Aristarchus and Copernicus.) Because there was some experimentation with the different least-squares methods, not all were used in all cases. Hence, solutions are not given for all three methods for each province. Except for Method C, the crater density ratio must be calculated for a specified crater diameter. We have usually given this ratio at the diameter of 4 km, which has the advantage of being usually within the diameter range of the data points, small enough to give relatively abundant statistics, but also not so small as to be in the diameter range dominated by secondaries. The result listed for Method C applies at all diameters, since the slopes of the solution and the reference curves are assumed to be parallel.

The internal consistency of our methods can be judged from the data in Table 3.3.2. Ratios of the relative crater densities calculated by the three different methods were tabulated. Based on a sample of 51 ratios from 28 different provinces, the median discrepancy between solutions by the different methods was 13%, and 76% of the solution pairs were within 40% of each other. The largest discrepancies occurred in provinces where the crater density was very small and where craters smaller than 2.8 km were involved in the solution. Examples of these cases were the craters Tycho, Aristarchus and Copernicus on the Moon, and the volcano Olympus Mons on Mars. Problems arose here because the introduction of craters smaller than 2.8 km involved complications such as (1) discrimination of non-impact craters, (2) uncertainty about a meaningful comparison curve of primary impacts in the lunar maria at these diameters (though Neukum and König, 1976, assert such a comparison curve has been identified), and (3) need for extrapolation to compare the geologic province in question to the reference data set from the lunar maria. These problems dramatize that large areas must be studied if the (more reliable) large craters are used; conversely, small craters must be understood if small areas are to be dated. These problems can also be appreciated by examining the graphs in the Appendix.

For these reasons, a fourth column labeled "D" was introduced in Table 8.3.2 to represent, in these few cases, a visual estimate of the most meaningful crater density ratio that should be applied in these few cases, weighted by craters at diameters believed most reliable, and allowing introduction of stratigraphic information that is not involved in the quantitative solutions. The final column in Table 8.3.2 gives the accepted relative crater density to no more than two significant figures, based on a comparison of all the available data. These crater densities can usually be assumed accurate within a standard error (σ) estimated at 20% based on the comparisons made above.

Table 8.3.1 Examples of least-squares solutions.[†]

Province No.	Geologic Province	D Interval	Method B Slope*	Method B Intercept	Method C (Slope Fixed at -1.80062) Intercept
	Moon:				
6	Mare Imbrium	2.8–64	-1.33 ± .13	-3.58 ± .12	-3.17
8	Mare Serenitatis	2.8–23	-1.91 ± .27	-3.02 ± .21	-3.10
9	S. Oceanus Procellarum	2.8–45	-1.91 ± .26	-2.96 ± .25	-3.06
11	Average of Maria	2.8–181	-1.80 ± .10	-2.92 ± .11	-2.92
14	Mare Tranquillitatis	2.8–45	-1.64 ± .19	-2.92 ± .17	-2.78
17	Orientale Basin	2.8–128	-1.70 ± .13	-2.65 ± .15	-2.55
21	Mare inside Mendeleev	4–32	-1.75 ± .20	-2.00 ± .17	-1.96
23	Frontside uplands	6–64	-1.32 ± .11	-2.21 ± .15	-1.98
24	Mare inside Bailly	4–91	-1.96 ± .17	-1.69 ± .16	-1.84
26B	Heavily cratered uplands	64–1448	-2.47 ± .10	-0.17 ± .21	-1.66
27	"Pure" uplands	2.8–64	-1.73 ± .04	-1.42 ± .04	-1.34
	Mercury:				
28	Caloris Mare	2.8–91	-1.79 ± .18	-2.34 ± .18	-2.32
29	Caloris Basin and ejecta	2.8–91	-2.14 ± .17	-2.05 ± .20	-2.42
30A	South Polar Quad H-15	8–64	-0.82 ± .15	-3.32 ± .21	-1.99
30B	South Polar Quad H-15	64–256	-1.59 ± .20	-2.02 ± .41	-1.60
	Mars:				
31	Central Tharsis volcanic plains	2.8–32	-1.71 ± .10	-4.05 ± .07	-3.98
32	Olympus Mons	0.25–2.8	-2.54 ± .30	-3.47 ± .14	-3.21
33	S. polar plains	4–32	-0.95 ± .13	-4.26 ± .14	-3.35
34	Plains around Ascraeus, Arsia Mons	2.8–32	-1.33 ± .19	-3.73 ± .16	-3.35
35	Extended Tharsis plains	2.8–256	-1.88 ± .07	-3.14 ± .09	-3.23
37A	Three shield volcanoes	0.18–23	-1.95 ± .10	-3.10 ± .05	-3.13
38	Elysium	2.8–45	-1.62 ± .17	-3.18 ± .15	-3.03
40	Solis Planum	2.8–32	-1.75 ± .15	-3.00 ± .13	-2.95
42	Chryse Planitia	0.35–23	-2.09 ± .10	-2.79 ± .03	-2.79
48	Alba shield volcano	2.8–32	-1.67 ± .26	-2.75 ± .19	-2.65
51	Heavily cratered plains				
A		0.7–0	-2.16 ± .18	-2.61 ± .06	-2.72
B		4–64	-0.80 ± .10	-3.41 ± .13	-2.21
C		64–1448	-2.37 ± .24	-0.90 ± .52	-2.12
	Earth:				
52	Eo-European Craton	23–91	-0.85 ± 1.02	-5.22 ± 1.66	-3.70
53	North American Craton	23–91	-1.48 ± .36	-4.35 ± .56	-3.85

[†](Highest quality data sets, or selected data sets of interest) [log N = (slope) log D + (intercept)]. Compare with plots of data, presented in Appendix.

*Note slope = exponent in power-law cumulative or log-incremental diameter distribution.

Table 8.3.2 Summary of computed and accepted values of crater densities in geologic provinces, normalized to average of lunar maria.

Province No.	Geologic Province	D Range of Data	Diameter Used in Density Ratio (if any) (km)	Crater Density (Province)/Crater Density (Lunar Maria)				
				Method				Accepted Result
				A	B	C	D[a]	
	Moon:							
1	Tycho	0.03–1.4 (A)[b]	1	0.07	0.3	NA[c]	0.1	0.1
2	Aristarchus	0.12–1.0	2.8	-	0.4	NA	0.15	0.2
3	Copernicus	0.17–1.4 (A) 0.25–6 (BC)	4	0.6	1.1	NA	0.3	0.3
4	Inner Mare Serenitatis	2.8–16	4	-	0.43	0.42	-	0.4
5	Mare Crisium	-	-	-	0.2	0.7	0.4	0.5
6	Mare Imbrium	2.8–64	4	-	0.43	0.57	-	0.50
7	N. Oceanus Procellarum	4–32	4	-	0.70	0.55	-	0.62
8	Whole Mare Serenitatis	2.8–23	4	0.48	0.70	0.67	-	0.65
9	S. Oceanus Procellarum	2.8–16 (A) 2.8–45 (B)	4	0.66	0.80	0.74	-	0.75
10	Mare Fecunditatis	2.8–181	4	-	0.81	1.05	-	0.93
11	Reference: Average of frontside maria	2.8–181	all	1.00	1.00	1.00	1.00	1.00
12	Mare Orientale	2.8–8	4	-	1.10	1.10	-	1.10
13	Mare Cognitum	2.8–16	4	-	1.28	1.25	-	1.26
14	Mare Tranquillitatis	2.8–45	4	1.05	1.28	1.40	-	1.30
15	Mare Humorum	2.8–16	4	-	1.56	1.46	-	1.50
16	Mare Nubium	4–64	4	-	3.2	2.4	2.5	2.5
17	Orientale Basin and ejecta	2.8–128	-	-	2.1	2.4	2.6	2.5
18	Fra Mauro	2.8–16	4	-	2.8	3.1	-	3.0
19	Plains in Schrödinger basin	2.8–11	-	4.7	5.6	5.1	-	5.1
20	Plains in basin near Schiller	2.8–16	4	-	6.6	7.4	-	7.0
21	Plains in Mendeleev	4–32	4	-	9.1	9.3	-	9.0
22	"Pre-mare" craters protruding in mare	8–181	100	-	9.3	NA	-	9.0
23	Frontside uplands	6–64	4	-	10.0	NA	-	10.0
24	Plains in Bailly basin	4–91	4	-	13.8	12.2	-	13.0
25	Upland plains near Maurolycus	2–64	4	-	14.0	NA	-	14.0
26	Heavily cratered basins:							
A	— all D	2–1448	4	39.0	-	NA	-	32.0
		-	64	25.0	-	-	-	
B	— large D	64–1448	64	-	35.0	NA	-	35.0

[a] D represents an eyeball fit averaged over appropriate size ranges and given consideration in cases where comparison of crater densities may require extrapolation beyond range of available data, or where stratigraphic information not available in methods A, B, and C can be used.
[b] Letters designate diameter range in analysis method A, B, or C, indicated in columns 5–8.
[c] NA = Not applicable (usually a reason to believe slope should not be force-fitted).

Table 8.3.2 *(Continued)*

Province No.	Geologic Province	D Range of Data	Diameter Used in Density Ratio (if any) (km)	Crater Density (Province)/Crater Density (Lunar Maria)				
				Method				Accepted Result
				A	B	C	D[(a)]	
27	"Pure" uplands	2.8–64	4	-	35.0	NA	-	36.0
			64	-	43.0	NA	-	
	Mercury:							
28	Caloris Mare	2–91 (A) 2.8–91 (B)	4	2.8	4.0	4.0	-	3.7
29	Caloris whole basin & ejecta	2.8–91	4	-	4.7	3.2	-	3.7
30	S. Polar uplands:							
A	— intermediate D	8–64	8	-	3.1	NA	-	5.0
			64	-	6.1	-	-	5.0
B	— large D	64–256	64	-	19.0	21.0	-	20
	Mars:							
31	Central Tharsis volcanic plains	2.8–32	4	-	0.096	0.088	-	0.1
32	Olympus Mons volcano	0.25–2 (A) 0.25–2.8 (BC)	4 4	0.25	0.10	0.52	0.15	0.15
33	S. Polar plains	4–32	4	-	0.15	0.38	0.15	0.20
34	Desert at bases of Ascraeus, Arsia Mons volcanoes	2.8–32	4	-	0.30	0.37	-	0.35
35	Extended Tharsis volcanic plains	2.8–128	4	0.45	0.54	0.49	-	0.49
36	Channels emanating from localized sources or chaotic terrain	2.0–16	4	-	0.54	0.62	-	0.58
37	3 shield volcanoes:							
A	— all D	0.18–23	4	-	0.54	0.63	-	0.63
B	— large D	2.8–23	4	-	0.68	0.63	-	0.63
38	Elysium volcanics	2.8–45	4	0.66	0.71	0.79	-	0.72
39	Isidis Planitia	2.8–32	4	-	0.72	0.81	-	0.76
40	Solis Planum volcanic plains	2.8–32	4	0.84	0.91	0.94	-	0.90
41	Channels not associated with chaotic terrain	0.35–11	4	-	1.90	0.96	1.0	1.0
42	Chryse Planitia volcanic plain	0.35–4 (A) 0.35–23 (BC)	4	1.15	0.92	1.35	1.0	1.1
43	Lunae Planum	2.8–91	4	-	1.11	1.52	1.1	1.2
44	Noachis Ridged plains	2.8–32	4	-	1.02	NA	1.5	1.3
45	Tyrrhena Patera volcanoes	2.8–16	4	-	1.36	1.53	-	1.4
46	Tempe Fossae faulted plains	4–128	4	-	1.28	1.96	-	1.6

Table 8.3.2 *(Concluded)*

Province No.	Geologic Province	D Range of Data	Diameter Used in Density Ratio (if any) (km)	Crater Density (Province)/Crater Density (Lunar Maria)				
				Method				Accepted Result
				A	B	C	D[a]	
47	Volcanic plains on Hellas S. Rim, 65°S	5.7–32	8	-	1.29	2.30	1.6	1.7
48	Alba shield volcano	2.8–32	4	-	1.81	1.86	-	1.8
49	Hellas floor	2.8–23	4	-	1.66	2.28	1.7	1.8
50	Syrtis Major Planitia volcanic plains	2.8–32	4	-	2.1	1.9	-	2.0
51	Heavily cratered plains:							
A	— small D	0.7–4	4	1.16	1.23	1.61	-	1.4
B	— intermediate D	4–32 (A)[b]	4	1.19	1.30	NA[c]	1.4	1.4
C	— large D	64–1448	64	13.0	10.0	NA	16.0	13
	Earth:							
52	Eo-European Craton	23–91	23	-	0.10	0.17	0.12	0.13
53	North American Craton	23–91	23	-	0.10	0.12	0.12	0.11

[a]D represents an eyeball fit averaged over appropriate size ranges and given consideration in cases where comparison of crater densities may require extrapolation beyond range of available data, or where stratigraphic information not available in methods A, B, and C can be used.
[b]Letters designate diameter range in analysis method A, B, or C, indicated in columns 5–8.
[c]NA = Not applicable (usually a reason to believe slope should not be force-fitted).

The precise placement of individual geologic provinces in the resulting stratigraphic column of Table 8.3.2 may be questioned in detail, but there are several indications that the general trend shown in Table 8.3.2 does meaningfully measure age differences among these provinces. For example, the inner, light-colored portion of Mare Serenitatis is universally regarded as younger than the outer dark portion based on stratigraphic information as well as visual impressions of the crater densities; our solutions agree, indicating that the inner part of Mare Serenitatis has only about 60% the crater density of the average Mare Serenitatis as a whole. Also, the range of crater densities of the various maria is quite consistent with the range of crater densities that has been reported by other workers in the past (Baldwin, 1964; Hartmann, 1967, 1972; Neukum *et al.*, 1975b). Furthermore, maria that have yielded generally younger basalts such as Imbrium and Oceanus Procellarum (Apollos 12-15) show distinctly lower crater densities than maria that have yielded old basalts, such as Mare Tranquillitatis (Apollo 11). The Fra Mauro (Apollo 14) region, which yielded still older material believed to be Imbrium ejecta, shows a still greater crater density.

8.4 EMPIRICAL DETERMINATION OF CRATER PRODUCTION RATES IN THE EARTH-MOON SYSTEM

Step (4) in our methodology (as listed in section 8.2) is to determine the crater production rates applicable to the various planets, so that the already-listed crater densities can be divided by appropriate production rates to estimate ages in years. Since craters are not observed to be forming in appreciable numbers on planets in historic time, the only empirical determination of crater production rates in the solar system comes from the Earth-Moon system where crater-bearing rock units have been dated. What is needed is a plot of crater density vs. age for each well-studied province. Shoemaker (1970), Soderblom and Lebofsky (1972), Neukum *et al.* (1975b), Neukum and König (1976), Guinness and Arvidson (1977), Hartmann (1970, 1972) and others have made such studies and established that the lunar data from rock units older than 3.2×10^9 yr reveal a lunar cratering rate that was decreasing prior to about 3.2×10^9 yr age, and slowly leveling off. The addition of terrestrial data, which come from much younger units (Shoemaker *et al.*, 1962; Hartmann, 1965; Dence, 1972), confirms that the recent rate in the Earth-Moon system has been relatively near that which was approached on the Moon about 3 to 3.2×10^9 yr ago. A similar result comes from using radiogenic or cosmic-ray ages speculatively associated with craters Tycho (~100 m.y.) and Copernicus (~850 m.y.), though these ages are questionable, since the ^{39}Ar-^{40}Ar signature once identified with supposed Copernicus ejecta has now been found in other Apollo 12 material as well (Alexander *et al.*, 1977). With the additional refinement of theoretical predictions of the terrestrial/lunar ratio expected for a given present-day meteoroid flux, it is thus possible to produce an empirical estimate of the time behavior of crater production in the Earth-Moon system from 4 b.y. ago to the present.

The present section deals mainly with the terrestrial rate and its comparison with the lunar rate to give the long-term time-dependent lunar rate, which will be used as a reference curve. A particular problem addressed here is whether the lunar rate (averaged over 3 to 4 b.y.) actually is consistent with the terrestrial rate (observed in more recent geologic time).

The average crater densities for the various lunar maria in which Apollo and Luna vehicles landed can be taken from Table 8.3.2 and compared to the radiometric ages of lunar basalts to determine the crater production rate as a function of time. As a more specific check we have tabulated (Table 8.4.1) additional counts that use the nesting technique and zero in on the actual landing sites of Apollo 11-12 and 14-16, and Luna 16. These counts are mostly from Hartmann (1972 and unpublished) and are more landing-site-specific than those in Table 8.3.2. The columns marked "Luna 24 area" and "Ave. of lunar frontside maria" are from data in Table 8.3.2, Provinces 5 and 11. Ages of rock samples for these sites were taken from Rb-Sr data in Table 7.3.1 and section 7.3.

Table 8.4.2 utilizes the data from Tables 8.3.2, 8.4.1 and 7.3.1 to establish data points for the relation of crater density vs. age for 14 dated cratered lunar and terrestrial surfaces. Figure 8.4.1 plots the data in Table 8.4.2. Error bars on the crater densities are derived from comparison of columns 2 and 3, plus the experience with our crater count solutions, described earlier. Error bars on ages are based on the range in reported solidification ages of sampled rocks (Table 7.3.1) from each site except for Apollo 16, where the standard deviation (0.17 b.y.) from 24 rocks is more representative than the total age range. In view of the great precision of individual radiometric rock age determinations, the uncertainties attached to surface ages of lunar sites may seem surprising. But the reader should note that each site yields rocks with a range of ages which may reflect rocks ejected from impacts on different surface flows, or rocks ejected from flows at different depths, or a mixture of both sources. Neukum and König (1976) discuss the effects of various lava flows in more detail.

In summary, the error bars can be viewed as approximating 1-sigma variations on either side of a most probable value of crater density and near-surface flow age. The Copernicus and Tycho entries are considered speculative and plotted with dashed lines; they are not based on firm radiogenic dates.

The terrestrial data deserve futher discussion. They are adapted from the work of Grieve and Dence (1979) and converted to lunar values. The raw terrestrial data of Grieve and Dence give crater production rates derived from two terrestrial provinces:

A. North American Craton: $0.19 \pm 0.05 \times 10^{-14}$ craters ($D > 22.6$ km)/km^2 yr.

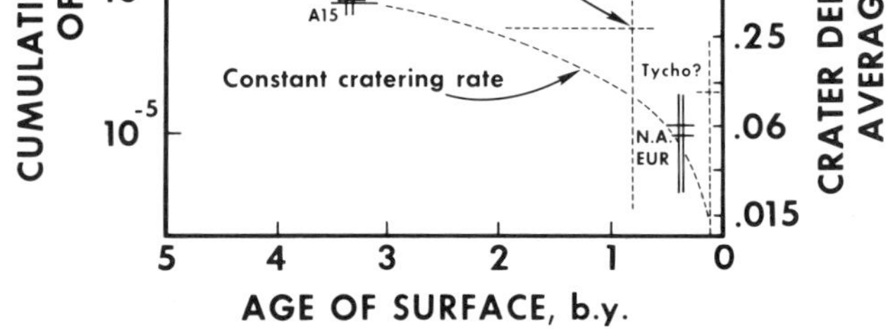

8.4.1 Plot of crater densities vs. surface unit rock ages (data and error bars from Table 8.4.2) for lunar units, plus two data points derived in text from terrestrial data. The dashed line shows the shape of a curve for constant cratering rate extending from the latest lunar mare points through the error bars for the terrestrially-derived data.

Table 8.4.1 Crater densities for areas around landing sites of Apollo 11–12, 14–16, and Luna 16 and 24

		Geological Province							
		Apollo 12 Area	Apollo 15 Area	Luna 16 Area	Apollo 11 Area	Luna 24 Area	Apollo 14 Area	Apollo 16 Area	Average of Lunar Front-side Maria
Average age of rocks (b.y.)		3.17	3.29	3.35	3.62	3.62	3.92	3.98	3.45 (average of all site averages except Apollo 14 and 16, from Table 7.3.1)
Number of rocks listed		14	7	1	10	1	26	49	
Maximum age		3.29	3.37	3.35	3.91	3.62	4.05	4.3	
Minimum age		3.09	3.22	3.35	3.50	3.62	3.82	3.8	
Standard deviation in age		.06	.05	.02	0.13	0.6	±.07	±0.16	±0.2
	D (km)								
Cumulative crater density	2.8	2.02(-4)†	1.35(-4)	2.46(-4)	4.62(-4)	1.19(-4)	8.04(-4)	6.36(-4)	3.68(-4)
	4	1.36(-4)	8.01(-5)	1.53(-4)	2.62(-4)	8.17(-5)	4.77(-4)	3.47(-4)	1.88(-4)
	5.66	8.41(-5)	5.47(-5)	1.10(-4)	1.62(-4)	6.29(-5)	2.72(-4)	1.74(-4)	9.96(-5)
	8	2.92(-5)	2.15(-5)	5.99(-5)	1.06(-4)	5.04(-5)	9.32(-5)	2.00(-4)	4.96(-5)
	11.3	7.31(-6)	4.31(-6)	2.40(-5)	7.10(-5)	3.24(-5)	3.02(-5)	1.24(-4)	2.36(-5)
	16	-	-	-	4.80(-5)	2.52(-5)	-	1.03(-4)	1.16(-5)
	22.6	-	-	-	1.60(-5)	7.21(-6)	-	6.49(-5)	7.12(-6)
Craters of diameter > D / km^2	32	-	-	-	-	-	-	3.25(-5)	3.52(-6)
	45.3	-	-	-	-	-	-	2.98(-5)	1.92(-6)
	64	-	-	-	-	-	-	-	7.2(-7)
	90.5	-	-	-	-	-	-	-	4.6(-7)
	128	-	-	-	-	-	-	-	2.0(-7)

† 2.02×10^{-4}

B. European Craton: $0.18 \pm 0.1 \times 10^{-14}$ craters ($D > 22.6$ km)/km^2 yr.

To convert from terrestrial to lunar rates, we utilize a result to be derived in the next section (Table 8.5.5), that the terrestrial/lunar crater production rate ratio is about 1.5 ± 0.6, for a fixed D, giving a multiplicative factor of 0.48–1.11. Then, to convert the above rate to a crater density for entry in Table 8.4.2, we must multiply by the average exposure age of the surface, 375 m.y. (300 m. y. for half of each province and 450 m.y. for the other half). In Table 8.4.2, these converted data are nomalized by dividing by the average crater density of lunar maria, which is $8.32\ (10^{-6})$ based on our 1.8 power law at the terrestrial reference diameter of $D = 22.6$ km.

The terrestrial cratering rates have been defined for $D > 22.6$ km, where the size index is similar to the lunar record. At smaller diameters, the terrestrial size index approaches values of −0.2 and a survey of the preservation level of terrestrial craters indicates that even on stable counting surfaces of Phanerozoic age, some

Table 8.4.2 Empirical data on lunar crater density vs. age.

Lunar Province	Crater Density for $D > 4$ km [Relative to lunar maria average = $1.88\ (10^{-4})\ km^{-2}$]: Observed[(a)]	Least Squares to Whole Province (last column, Table 8.3.2)	Adopted[(b)]	Age Range[(d)] ($\times 10^9$ yr)	Remarks
Densest cratering	-	30–36	30–36	4.3–4.55	Primeval surface, age estimated
Uplands	-	7–30	7–30	4.0–4.4	Age > oldest recovered mare lavas
Apollo 16 and 17 uplands	-	-	4–6	3.85–4.25	Fig. 7.3.13
Fra Mauro (Apollo 14)	2.53	3.0	2.2–3.3	3.85–4.1	Fig. 7.3.13
Mare Crisium (Luna 24)	0.43	0.5	.37–.56	3.2–3.6	5 rocks, Table 7.3.1
Mare Tranquillitatis (Apollo 11)	1.39	1.30	1.1–1.6	3.5–3.7	10 rocks, Rb-Sr ages, Table 7.3.1
Taurus-Littrow mare (Apollo 17)[(c)]	-	-	1.0–1.4	3.6–3.8	6 rocks, Rb-Sr ages, Table 7.3.1
Mare Fecunditatis (Luna 16)	0.81	0.93	.70–1.0	3.2–3.5	1 rock, Table 7.3.1
All frontside maria	1.00	1.00	.90–1.10	3.25–3.65	Table 8.4.1
Mare Imbrium (Apollo 15)	0.43	0.50	.37–.56	3.2–3.35	7 rocks, Rb-Sr ages, Table 7.3.1
Oceanus Procellarum (Apollo 12)	0.72	0.75	.59–.88	3.1–3.25	14 rocks, Rb-Sr ages, Table 7.3.1
Copernicus	-	0.3	.00–.6	0.4–1.7??	Inferred (see summary by Neukum & König, 1976)
Tycho	-	0.1	.00–.2	0.05–0.15??	Inferred (see summary by Neukum & König, 1976)
North American craton (lunar equivalent)	0.06±.04	-	.02–.12	0.30–0.45	Adapted from Grieve & Dence (1979)
European craton (lunar equivalent)	0.05±.04	-	.02–.14	0.30–0.45	

[(a)]From Table 8.4.1 for landing sites; adapted from Grieve and Dence (1979) for terrestrial data, using earth rate/lunar rate = 1.5±0.6, and craters of $D > 22.6$.
[(b)]For most Apollo and Luna sites, uncertainty is assumed to be ±20%. A 10% uncertainty is assumed for all "frontside maria." A ±40% uncertainty is assumed for Mare Crisium. A ±100% uncertainty is assumed for Tycho and Copernicus.
[(c)]Derived by weighting provinces 4 and 8 according to relative areas.
[(d)]Ranges are based on listings or individual ages in Table 7.3.1 and elsewhere, plus statistical standard deviations.

craters < 20 km diameter have been removed by erosion and the record is incomplete (Grieve and Robertson, 1979). Another reason for comparing directly with lunar crater densities at $D > 22.6$ km, instead of our reference maria crater diameter of 4 km, is that at 4 km, lunar craters still have a simple bowl-shaped form whereas terrestrial craters may have been modified by rim slumping and collapse. However, at diameters of 22.6 km, both lunar and terrestrial craters have undergone modification by collapse and obey energy scaling laws of the same form, differing only by a constant, which accounts for differences in the surface gravity of the two planetary bodies (Grieve and Dence, 1979).

This new analysis of the crater density-age relation, similiar to that of Neukum *et al.* (1975b), Boyce (1976), Hartmann (1972), Soderblom and Lebofsky (1972) and others, has several interesting features. It clearly reconfirms the high cratering rate that must have been present before 4 b.y. ago and indicates the rapid fall-off in cratering at that time. More uncertain is the question of how the crater production rate leveled off, and whether it reached a constant value. In a plot of crater *density* vs. age, such as Fig. 8.4.1, a constant cratering *rate* would be represented as a curve such as that shown as dashed in Fig. 8.4.1. A single such curve passes through the end of the (still rapidly decreasing) lunar data 3 b.y. ago, and through error bars on the data points in the last 1 b.y. A more literal fit to the data would require a leveling off of the cratering rate about 2.5–3 b.y. ago.

A problem in the literature on this topic is whether the cratering rate in the Earth-Moon system leveled off to a constant value, or whether there has been a slight upswing in recent time, or other minor variations. A very detailed study of the problem was given by Neukum *et al.* (1975b), Neukum and König (1976) and König *et al.* (1977). Their earlier work found possible evidence for a rise to a peak impact rate about 850 m.y. ago, followed by a slow decline with 160 m.y. half-life. This rested partly on the proposed Tycho and Copernicus ages of about 100 m.y. and 850 m.y., respectively, which they concluded were questionable. Their analysis has moved toward a more constant flux. A note added in proof to the 1976 paper refined their crater size distribution with a result "more compatible with a constant flux in the last 850 m.y.," and the 1977 paper found either a contant flux or a flux peaked slightly at ~800 m.y. ago (and decreasing since then with a half-life of 0.3–1 b.y.) would fit their data. Other work has also favored a constant crater production rate since the maria formed. Specifically investigating this problem, Guinness and Arvidson (1977) made crater counts for Tycho, Copernicus and the Apollo 12 site. Taking the ages as 109 m.y., 800 m.y. and 3.26 b.y., respectively, they found "no basis for assuming that the flux has changed with the last 3.3×10^9 yr." Young (1977) reviewed the literature and reached the same conclusion.

Probably the earlier discussions gave too much weight to the proposed Tycho and Copernicus data. Both the crater counts and rock ages are extremely uncertain. Primary impact craters are difficult to sort out from possible secondaries. The proposed Copernicus age of ~850 m.y. comes only from a degassing event of that age in material collected at the Apollo 12 site and attributed to a Copernicus ray. The proposed Tycho age of about 100 m.y. is even more shaky, based on a cluster of secondaries near the Apollo 17 landing site attributed to (remote) Tycho and a peak in Apollo 17 soil exposure ages attributed to disturbance when these secondaries were formed.

Indeed, before the Apollo 17 data were available, Wood and Hartmann (1970) compared Tycho crater statistics to the Apollo 12 crater statistics and concluded that Tycho should be 160 – 270 m.y. old if the crater production rate had been consistent with the terrestrial cratering record. Hartmann (1968) on similar grounds estimated a Copernicus age of about 2 b.y.

Other, smaller craters near Apollo landing sites (such as North Ray crater) have been more accurately dated from rock samples and can be used to calibrate crater time scales (e.g., König *et al.*, 1977). However, only very small (10's of meters) pits appear on their surfaces and they thus require considerable interpretation to tie the small-crater counts to the multikilometer crater counts — a process of interpretation we postpone until section 8.8.

Two terrestrial points, in the lower right of Fig. 8.4.1, lie marginally higher than expected for constant cratering though they have additional uncertainty due to conversion from terrestrial to lunar values and their error bars probably overlap a reasonable constant-rate curve.

Shoemaker *et al.*'s (1979) most recent reduction of asteroid orbital statistics favors the suggestion that the asteroid impact rate in the Earth-Moon system today is somewhat higher than that in the last few b.y., inferred from the average cratering rate since the maria formed. Shoemaker *et al.* (1979) have revised the calculation of collision probabilities between Earth and the known Earth-approaching asteroids as shown in Table 8.4.3, with the result that his best estimate of the modern impact rate from these objects is roughly three times the average rate evidenced from the record of lunar and terrestrial craters. A constant rate would be roughly one standard deviation down from this calculated modern impact rate. The best estimate depends on assumptions

Table 8.4.3 Earth-crossing asteroids: Crossing characteristics and collision parameters.

	Orbital Overlap with Earth	Orbital Elements [a]			P_s [b]	P_o [c]
		a (AU)	e	i (Deg.)	(10^{-9}/yr)	(10^{-9}/yr)
Aten Asteroids						
1976 UA	Continuous	0.844	0.424	6.27	14	14
2100 Ra-Shalom	Continuous	0.832	0.465	13.1	6.3	6.7
2062 Aten	Continuous	0.966	0.237	17.9	6.9	6.4
Apollo Asteroids						
1566 Icarus	Continuous	1.078	0.828	20.2	1.6	2.0
1978 SB	?	2.230	0.841	11.9	-	1
1974 MA	Part-time	1.757	0.446–0.772	53.4–32.8	0.93	2.5
2101 Adonis	Continuous	1.873	0.732	1.96	2.9	6.3
1864 Daedalus	Continuous	1.461	0.640	15.9	1.0	1.6
1865 Cerberus	Continuous	1.080	0.513	14.9	2.5	3.1
Hermes	Continuous	1.639	0.622	5.62	2.2	3.7
1981 Midas	Part-time	1.776	0.586–0.652	45.5–41.3	3.8	0.7
1862 Apollo	Continuous	1.470	0.505	6.14	3.5	4.8
2063 Bacchus	Continuous	1.077	0.320	8.78	7.4	7.7
1959 LM	?	2.155	0.674	9.2	-	2
1686 Toro	Continuous	1.368	0.438	9.15	4	4.2
1977 HA	Part-time	1.600	0.490	23.3	2.0	1.5
6743 P-L	Continuous	1.620	0.476	6.81	4.1	4.8
1620 Geographos	Continuous	1.245	0.351	14.2	3.8	3.9
1976 WA	Part-time	2.407	0.589–0.700	32.3–17.0	1.8	2.1
1950 DA	Continuous	1.683	0.535	10.7	1.8	2.4
1866 Sisyphus	Part-time	-	-	-	0?	0?
1973 NA	Part-time	2.447	0.672–0.981	67.9–51.8	0.54	0.36
1978 CA	Continuous	1.125	0.278	25.3	4.1	3.3
1863 Antinous	Part-time	-	-	-	-	-
1975 YA	Part-time	1.290	0.312–0.744	62.5–49.1	2.5	1.5
6344 P-L	Part-time	-	-	-	-	-
Amor Asteroids						
1978 DA	Part-time	-	-	-	-	-
2061 Anza	Part-time	-	-	-	-	-
1915 Quetzalcoatl	Part-time	2.49	0.60	19.7	3.5	1.5
1917 Cuyo	Part-time	-	-	-	-	-
1943 Anteros	Part-time	-	-	-	-	-
1221 Amor	Part-time	-	-	-	-	-
1580 Betulia	Part-time	2.196	0.792	26.4	0.5	1.7
1972 RA	Part-time	-	-	-	-	-
1627 Ivar	Part-time	-	-	-	-	-
6887 Alinda	Part-time	-	-	-	-	-

[a] Orbital elements shown are semi-major axis (a), eccentricity (e), and inclination (i). Most values of e and i are for the zero order state (i.e., only free oscillations of e and i are considered) derived by Shoemaker *et al.* (1979) by interpolation from computer runs of J. G. Williams, which are based on a first order theory of secular perturbation (Williams, 1969). Inclination is referred to the invariable plane. Values of e and i for ω librators, for (1685) Toro, and for (1580) Betulia are average values obtained from numerical integration of the motion of the asteroid by J. G. Williams; for Quetzalcoatl e and i are derived from a numerical integration by B. G. Marsden. Orbital elements shown are for radius vector to the node = 1.0 AU.

[b] P_s is probability of collision with Earth based on Shoemaker *et al.*'s (1979) revision of Öpik's (1951) equations. These are the best available estimates.

[c] P_o is probability of collision with Earth based on the equations of Öpik (1951), using the orbital elements listed in this table. For octuple crossers, the probabilities obtained from the two sets of e and i were summed, and, for ω librators, the two solutions for the probability were averaged. These may be compared with the new results, P_s.

such as the albedo distribution among Earth-approaching asteroids (assumed by Shoemaker to be a fifty-fifty mixture of low- and high-albedo types); higher albedos favor smaller sizes, favoring a lower estimate of today's crater production rate.

Gault (pers. comm., 1979) notes that oblique impacts reduce crater dimensions relative to impacts at normal incidence (Gault and Wedekind, 1978) and that this effect should be included in Shoemaker's calculation of the crater production rate from the observed asteroidal flux. Gault suggests this may bring the present-day predicted crater production rate down to the average level observed in post-mare time.

Certainly short-term fluctuations in production rate of small craters in the last 3 b.y. would be likely since collisions fragmenting larger bodies probably occasionally feed bursts of subkilometer asteroids into short-term resonant orbits that "dump" them into the inner solar system, consistent with clusterings of ages of some meteorite types. Ultimately, there must be some "noise" in the cratering rate records, differing from one planet to another. Such noise could affect ages of less than a few 10^8 yr derived from cratering records, but would be averaged out for older surfaces.

In summary, Fig. 8.4.1 and the recent literature suggest that the time behavior of cratering in the Earth-Moon system is known within a factor of about 2–3 for most of solar system history, and has been roughly constant within this factor since lunar mare formation. This allows crater count dating of volcanic surfaces thoughout geologic time with accuracy adequate to our purpose of establishing the general trends of volcanic activity.

8.5 EFFECTIVE CRATER PRODUCTION RATES ON THE OTHER PLANETS

The preceding section established a reference curve of crater density as a function of age for lunar surfaces, based on our moderate empirical knowledge of cratering in the Earth-Moon system. To determine absolute ages from crater counts on other planets, we need analogous curves for those bodies. Unfortunately, we have no comparable knowledge based on dated surface samples, and must infer their cratering histories by indirect means. Fortunately for our purposes, the terrestrial planets are not dynamically isolated from one another. It can be shown from considerations of celestial mechanics that crater-forming projectiles from most reasonable source populations (asteroid belt, Earth-crossing asteroids, comets, etc.) would be scattered in significant proportions among all the inner planets.

As mentioned in section 8.4, asteroid collisions could produce "noise," or short term bursts of meteoroids that might cross one planet's orbit and not another. Generally, however, near-encounters with the first planet would scatter the bodies to other planets, diluting any tendency to affect one planet without affecting others (Wetherill, 1974, 1975, 1977a). Thus we assume that, to first order, the time dependence (but not necessarily the absolute rate) of cratering on all planets was approximately that revealed by Fig. 8.4.1.

Impact histories of other planets are thus linked to that of the Moon. They can, in principle, be derived if we know the sources of the projectiles and can compute their relative probabilities of collision with the various planets. The best current source of information is the current asteroid search program of Shoemaker and his co-workers, which identifies sizes of various populations of bodies, and dynamical calculations by Wetherill and others, which give the impact rates associated with each population, planet by planet.

We must carefully distinguish between *impact* and *crater production* rates. The former refers to a fixed size of projectile and depends on the numbers of these projectiles in various types of orbits and their rates of collisions with each planet. The latter refers to a fixed size of crater and requires some assumptions about the mechanism of crater formation. Identical bodies impacting on different planets will generally produce craters of different sizes. Crater size increases with impact velocity, which depends on the projectile's orbit and the target planet's escape velocity and heliocentric distance. Surface gravity also has an effect; more energy is required to excavate a crater of given size as g increases. Appropriate scaling laws are derived from numerical simulations and extrapolations from laboratory-scale explosions and nuclear weapons tests (Gault and Wedekind, 1977; O'Keefe and Ahrens, 1978). Hartmann (1977b) has discussed their applications. He adopts scaling laws for which crater diameter is proportional to impact velocity to the power 0.6, and surface gravity to the power −0.2 (he uses a nominal slope of −2 on the diameter-frequency plot, for which the observed number of craters of a given size varies as $V^{1.2}$ and $g^{-0.4}$). Use of different exponents favored by

Table 8.5.1 Impact probabilities and velocities of impact of Earth-crossing asteroids with the terrestrial planets and the Moon.[a]

Asteroid	Impact on Mercury		Impact on Venus		Impact on Earth		Impact on Moon		Impact on Mars	
	Impact Velocity (km/sec)	Probability of Impact (10^{-9}/yr)	Impact Velocity (km/sec)	Probability of Impact (10^{-9}/yr)	Impact Velocity (km/sec)	Probability[b] of Impact (10^{-9}/yr)	Impact Velocity (km/sec)	Probability of Impact (10^{-9}/yr)	Impact Velocity (km/sec)	Probability of Impact (10^{-9}/yr)
1566 Icarus	46.8	1.37	37.7	2.72	31.8	1.6	29.9	0.11	21.1	0.24
1974 MA	42.5	0.26	36.3	1.01	32.7	0.93	30.9	0.04	25.0	0.06
2101 Adonis	22.2	0.57	27.8	7.08	27.3	2.9	25.1	0.68	21.6	0.76
1976 UA	13.7	1.33	19.5	15.78	23.2	14	12.9	0.64	-	0
1864 Daedalus	-	0	25.3	2.34	24.8	1.0	22.2	0.08	18.3	0.14
1965 Cerberus	-	0	20.6	5.28	20.0	2.5	16.7	0.16	11.0	0.39
1937 UB Hermes	-	0	19.6	6.05	21.9	2.2	19.0	0.19	17.2	0.34
1973 EA	-	0	32.4	1.53	30.1	0.64	28.1	0.04	23.3	0.07
1862 Apollo	-	0	17.5	8.33	20.3	3.5	17.1	0.23	15.1	0.41
1685 Toro	-	0	-	0	17.2	4	13.3	0.18	12.1	0.36
2063 Bacchus	-	0	-	0	15.7	7.4	11.3	0.30	-	0
PL-6743	-	0	-	0	17.0	4.1	13.1	0.18	13.8	0.31
1620 Geographos	-	0	-	0	16.1	3.8	11.8	0.18	9.8	0.41
1959 LM	-	0	-	0	15.1	15?	10.4	0.52	10.2	0.99
1950 DA	-	0	-	0	17.7	1.8	13.9	0.11	14.6	0.19
1866 Sisyphus	-	0	-	0	27.8	0.0?	25.5	0.06	22.0	0.07
1973 NA	-	0	-	0	40.3	0.54	38.8	0.04	31.4	0.04
1863 Antinous	-	0	-	0	19.5	1?	16.2	0.06	17.7	0.08
1975 YA	-	0	-	0	35.5	2.5	33.7	0.16	25.3	0.25
PL-6344	-	0	-	0	15.8	5?	11.4	0.18	16.3	0.22
rms impact velocity	34.2		27.3		24.6		21.7		19.0	
Mean probability of impact		0.18		2.51		3.88		0.21		0.27

[a]From Shoemaker (1977).
[b]From Table 8.4.3, modified slightly from Shoemaker (1977).

some workers does not seriously affect estimates of cratering rates. Corrections are added for impact velocity (for any given type of projectile orbit) and surface gravity; these vary by about a factor of 2 among all the terrestrial planets and are small. Their uncertainties should be considerably smaller than the corrections themselves.

Calculations of impact rates began with the fundamental work of Öpik (1951), who derived the collision probability per unit time for a body in a planet-crossing orbit, as a function of its orbital elements. He showed that the observed Earth-crossing asteroids have lifetimes of the order 10^7–10^8 years. Since this is much less than the age of the solar system, the present population must represent an approximate steady state, in which objects "leak" from longer-lived source orbits and are removed by collisions with the planets, or by ejection from the solar system. The only dynamically plausible sources are the main asteroid belt and the Oort comet cloud. The relative importance of these sources and the detailed mechanisms that transfer bodies into planet-crossing orbits are poorly known and are responsible for most of the uncertainty in cratering rates, and ultimately in crater-derived chronologies of planets.

If we knew the orbit of every planet-crossing body, we could apply Öpik's formulae directly, and calculate the present impact rate for each planet. In reality, our knowledge of their populations is far from complete. Statistical arguments based on the rate of their discovery indicate that there are of the order of 10^3 Earth-crossing (Apollo) asteroids larger than 1 km diameter (Wetherill, 1974b; Shoemaker and Helin, 1977; Shoemaker, 1977). As of late 1978, only 26 had been discovered; impact probabilities for 20 of these are listed in Table 8.5.1. Monte Carlo calculations by Wetherill (1979) suggest that deep Mars-crossers (Amor asteroids), with perihelia between 1 and 1.3 AU, are probably two or three times as numerous as Earth-crossers. Shoemaker (1978a, b) estimates that there are
diameter, about 1500 Amor Mars-crossers, and about
about 800 Earth-crossers larger than $1.0 \pm ^{.5}_{.3}$ km in diameter, about 1500 Amor Mars-crossers, and about 10,000 ± 5000 shallow Mars-crossers. Correcting for probable but as yet undiscovered objects, he calculates a lunar cratering rate due to Earth-crossing asteroids about a factor 2 higher than the empirical rate found in the last section.

In the same manner, Shoemaker calculates a martian crater production rate roughly 1.5 times the lunar rate. The ratio of rates (martian to lunar, etc.) does not depend on physical properties of the asteroids, and may be less model-dependent. However, Wetherill (1974a) has pointed out a potential problem with this approach. A number of shallow Mars-crossers experience secular variations of their orbital eccentricity, which can make collisions less frequent than predicted by Öpik's formulae, or prevent them entirely. Shoemaker has allowed for this effect to some extent, but a precise estimate would require calculation of the orbital evolution of every potential Mars-crosser.

Comets may cause significant cratering on the terrestrial planets. Since parabolic and short-period comets have very different orbits, we shall consider them separately. Our approach will be to estimate the relative cratering rates on the various planets for each type, and then to discuss their possible contribution to absolute cratering rates.

Parabolic comets have randomly oriented orbits. Their heliocentric velocities are always essentially equal to the local solar escape velocity. Averaging over all directions, one can show that the rms approach velocity, relative to any target planet, is $\sqrt{3}$ times the planet's orbital velocity. If comets are sent into the inner solar system by accumulation of small, random stellar perturbations, the expected number of comets per increment of perihelion distance is constant (Oort, 1950; Weissman, 1977). The number of comets per unit area is therefore inversely proportional to heliocentric distance. Corrections for impact velocity and surface gravity (Hartmann, 1977b) give the cratering rates,

Table 8.5.2 Crater production rate for parabolic comets (normalized to Moon).

Planet	rms Approach Velocity (km/sec)	rms Impact Velocity (km/sec)	Impact Rate	Crater Production Rate
Mercury	87	87	2.6	3.5
Venus	61	62	1.5	1.0
Earth	52	53	1.1	0.6
Moon	52	52	1	1
Mars	42	42	0.7	0.4

relative to the Moon, shown in Table 8.5.2. Mercury has the highest cratering rate from this source.

Periodic comets have a very different distribution of perihelion distance (q). Marsden's (1975) catalog lists 19 periodic comets with $q<1$ AU, and 37 with $1<q<1.5$ AU. Considering only these listed objects, we can use Öpik's collision probabilities and the velocity/gravity corrections to calculate relative cratering rates on the different planets. This procedure probably underestimates the martian cratering rate. The discovery rate for parabolic comets with $q \simeq 1.5$ AU is about one-fourth that for $q \simeq 1$ AU. Since Oort's theory predicts equal numbers in each interval of q, we may ascribe the difference to observational selection. If the same biases are assumed to apply to periodic comet discoveries, there are several times more Mars-crossing than Earth-crossing periodic comets. Table 8.5.3 shows impact and cratering rates for known periodic comets. The statistics are poor, especially for Mercury (two comets), but impact velocities appear to be about as high as for parabolic comets. The martian cratering rate due to *observed* objects is about equal to the lunar rate. To account for undiscovered Mars-crossing comets, we assign a weight to each known comet with $q > 1$ AU. The weight is proportional to $q^2(q-1)^2$, with a value of 4 at $q = 1.5$ AU. This procedure yields a martian cratering rate 1.4 times the lunar rate. Use of a different power of q in the weight does not affect this result significantly, as long as we constrain the weight to be 4 at 1.5 AU, by analogy with parabolic comets.

We need to combine the impact rates for comets (Tables 8.5.2–3) with estimates of asteroidal impact rates. The importance of comets relative to asteroids as crater-producers is hard to assess, as their masses are poorly known. Comets appear to follow a power-law size distribution similar to asteroids (Whipple, 1975), but parabolic comets smaller than ~1 km may be lacking (Kresák, 1978). The observed number of parabolic comets gives a very low impact rate, $\sim 10^{-8} yr^{-1}$ on all the terrestrial planets. However, their orbits indicate that most are observed only on their first passage through the inner solar system; planetary perturbations would alter their aphelion distribution, with little change in perihelia (Marsden, 1977). It seems unlikely that most parabolic comets are completely destroyed by their first perihelion passage. If they merely lose a coating of volatile material, they may persist as crater-producers, though unobserved, on later returns. They could then be responsible for a high cratering flux on Mercury.

Kresák (1978) estimates the impact rate on Earth for asteroids to be about an order of magnitude greater than for comets. In terms of crater production rates, this is partly compensated by the higher impact velocities for comets. The velocity difference is even more important in the case of Mars. Periodic comets are probably not the dominant cause of martian cratering, but they may be as important as Apollo and Amor asteroids.

Strictly speaking, the observational approach can give only present-day cratering rates, but these can be roughly extrapolated back over the last 3–4 b.y., since we believe we know the time dependence of cratering from Fig. 8.4.1.

A second approach avoids to some extent the observational problems of unobserved or extinct projectiles. Near misses are more frequent than actual collisions; each one changes the projectile's orbital parameters and the probability of subsequent encounters. It is not possible to compute the orbit of a single body deterministically through a series of encounters, but its evolution can be treated stochastically, with the

Table 8.5.3 Crater production rate for observed periodic comets (normalized to Moon).

Planet	rms Approach Velocity (km/sec)	rms Impact Velocity (km/sec)	Impact Rate	Crater Production Rate	Number of Comets[a]
Mercury	44	44	0.13	0.1	2
Venus	43	44	0.9	0.4	9
Earth	39	41	1.3	0.7	21
Moon	39	39	1	1	21
Mars[b]	24	24	2.0	0.9	56
[c]	21	21	3.6	1.4	56

[a]Listed objects from Marsden (1975).
[b]Equal weights.
[c]Weight = 1 for $q \leq 1$; $1 + 5q^2(q-1)^2$ for $q > 1$.

Table 8.5.4 Relative impact probability/unit area (normalized to Moon = 1.0).†

	Initial Orbit		Earth Impacts (%)	Mercury	Venus	Earth	Mars
	Perihelion (AU)	Aphelion (AU)					
1.	1.01	4.00	24.0	0.30	0.94	2.26	1.15
2.	1.36	4.67	0.11	0.69	1.26	1.78	0.71
3.	0.34	4.18	6.0	4.21	2.37	1.23	0.37
4.	0.19	1.97	18.5	7.50	1.97	1.08	0.45
5.	0.70	3.61	5.2	1.60	2.34	1.34	0.27
6.	1.27	1.89	27.8	1.08	2.58	2.82	11.9
7.	0.98	0.99	45.0	1.05	3.02	3.99	0.96
8.	0.67	0.82	29.6	1.08	2.44	1.36	0.18
9.	0.33	0.41	$<.01$	$>1.5 \times 10^4$	-	-	-

†From Wetherill (1975).

possible outcomes of each encounter weighted by Öpik's formulae. This Monte Carlo method was developed by Arnold (1965) and extended by Wetherill (1975; 1977a, b, c; 1979). It has the advantage that the entire distribution of orbits need not be specified, only an appropriate source orbit. Also, it can explicitly treat time-varying populations of projectiles.

This second method does not yield current absolute impact rates, but rather relative values for different planets, which can then be tied to a time dependence through Fig. 8.4.1. Absolute values can be obtained by scaling to the empirically determined lunar record. Table 8.5.4 gives Wetherill's (1975) results for a variety of starting orbits, expressed as impact probabilities per unit area, normalized to the Moon. These are also relative impact *rates* only for a steady state; Wetherill emphasizes the possibility of time variations. The relative values differ significantly for different source orbits, so the impact and cratering rates on the various planets depend on the appropriate "mix" of projectiles.

Orbits 1 to 3 represent short-period comets or related objects, derived from the long-lived reservoir of the Oort cloud. They have approximately equal impact probabilities per unit area for all the terrestrial planets. The higher values for Earth and Mercury for orbits 1 and 3, respectively, are due to the choices of initial perihelion distance, which should not be preferred. The low fraction of Earth impacts is due to the high probability of encounters with Jupiter and ejection from the solar system. Wetherill's results imply that orbits characteristic of periodic comets yield approximately equal impact rates per unit area on all the terrestrial planets. His model assumes that a projectile survives until it strikes a planet or is ejected from the solar system. The lack of observed periodic comets with small perihelia is therefore due mainly to their rapid disintegration or loss of volatiles. The active lifetime of a short-period comet ($\sim 10^4$ yr) is much shorter than its dynamical lifetime for ejection by Jupiter encounters ($\sim 10^6$ yr). If a significant fraction of comets leave substantial dead nuclei, their contribution to cratering rates will be more uniform than given in Table 8.5.3.

Orbits 4 and 5 of Table 8.5.4 are short-lived "secondary" orbits. They can be produced from either cometary or asteroidal source orbits (Wetherill, 1979).

Orbit 6 is a deep Mars-crosser (Amor asteroid) or Mars planetesimal. Its impact probability (per unit area) is an order of magnitude greater for Mars than for the Moon. The impact probabilities for Earth and Venus are about twice the lunar value, due to the relatively low approach velocities and their larger gravitational cross-sections. This type of orbit could be derived from longer-lived shallow Mars-crossers in the inner asteroid belt, or from a more eccentric cometary orbit (Wetherill, 1979). Note that Mars-crossers produce high ratios of Mars/Moon impacts, which would affect crater interpretation of martian ages. If largely dependent on shallow Mars-crossing bodies, the Mars/Moon impact ratio would be still larger. Recent calculations by Wetherill (1979), which include effects of secular resonances and Jupiter perturbations, indicate that the transition from Mars-crossing to Earth-crossing orbits is easier than indicated here. This would reduce the Mars/Moon impact ratio for asteroidal source orbits. In any case, however, the Mars-crossing orbits appear to be too stable to supply the present number of Apollo and Amor asteroids. Wetherill favors the possibility that some of them (perhaps most) are actually cores of extinct comets. They would be derived primarily from orbits like comet Encke, number 1 of Table 8.5.4. The Mars/Moon impact rates would be approximately equal, with a slightly lower martian cratering rate due to higher surface gravity. Taking both comets and asteroids into account, it seems barely possible to have a martian cratering rate less than the lunar rate.

Orbits 7 to 9 of Table 8.5.4 are planetesimal orbits that presumably were characteristic of the final stages of planetary formation. Earth- and Venus-crossers each produce a significant impact flux on the other planets. Any dynamically plausible source for the heavy lunar impact flux of ~4 b.y. ago, other than bodies in geocentric orbits, appears to require a similar flux on all of the terrestrial planets (Wetherill, 1975). We can, therefore, be fairly confident that no surviving surface in the inner solar system greatly predates the oldest lunar surface.

Mercury planetesimals (Orbit 9) present an exceptional case of a dynamically closed system. Mercury's mass is too small for it to perturb any significant number of those bodies into Venus-crossing orbits. If Mercury's accretion was delayed until after the lunar heavy bombardment, it could have had a high impact rate at a relatively later date than other planets without affecting the cratering records of the other planets. There is some suggestion in Wetherill's (1977b, c) calculations that this may have occurred. In his models for the simultaneous accretion of the terrestrial planets, Mercury requires several times 10^8 years longer to accrete than do Venus and Earth, due to capture of some Venus planetesimals into long-lived Mercury-crossing orbits. These calculations did not include the secular variation of Mercury's eccentricity. This might have extended its accretion time still further by causing some of these planetesimals to be only intermittently Mercury-crossing. Planetesimals in smaller orbits, which would occasionally encounter Mercury near its perihelion, could also have delayed accretion (Weidenschilling, 1978). High Mercury cratering rates would imply a young surface. The possibility therefore exists that Mercury's surface postdates the era of lunar basin formation, with Mercury's major impact features due to the final stages of accretion.

To summarize, the Monte Carlo calculations of Wetherill indicate relative impact rates on different planets, which can be calibrated with the known lunar/terrestrial rates. By corrections involving energy scaling and velocity/gravity effects, these can be converted to impact rates for the various planets, as reviewed by Hartmann (1977b).

Soderblom *et al.* (1974), and to some extent Neukum and his co-workers (e.g., Neukum and Wise, 1976), used a third approach in attempting to determine cratering rates and ages. They assumed that all of the

Table 8.5.5 Comparisons of estimated crater production rates (normalized to Moon) for selected populations of objects.

Type Objects	Mercury	Venus	Earth	Moon	Mars	Reference	Comments
Asteroids	0.8	1	0.9	1	2	*(1)*	Derived from Wetherill (1975).
Comets	5	2	0.7	1	0.3	*(1)*	Monte Carlo calculations.
40% comets/ 60% asteroids	2	1	1	1	2	*(1)*	
Mars-crossers favored (80%)	2	1	2	1	4	*(1)*	Derived from Wetherill (1975) Monte Carlo calculations.
Earth-crossing asteroids only	0.7	0.8	1.2	1	0.2	*(2)*	Known objects and discovery rates.
All asteroids			1.2	1	1.5	*(2)*	Known objects and discovery rates.
Parabolic comets	3.5	1	0.6	1	0.4	*(3)*	From Oort (1950) theory. Absolute rate probably low.
Periodic comets	0.1	0.4	0.7	1	0.9	*(3)*	Observed comets only.
				1	1.4	*(3)*	With unobserved Mars-crossers.
All bodies	2.5			1	1.5	*(4)*	Identification of provinces marking terminal intense bombardment.
Results accepted here:							
Maximum likely	5	2	2	1	4		
Minimum likely	0.8	0.8	0.9	1	1		
Most likely	2	1	1.5	1	2		

References:
(1) Hartmann (1977); *(2)* Shoemaker (1978); *(3)* This chapter (Weidenschilling); *(4)* Soderblom (1978).

terrestrial planets experienced heavy cratering (either a distinct episode or the tail end of declining bombardment) about 4 b.y. ago, as did the Moon, and that the highlands of Mars and Mercury have that age. The most heavily cratered surfaces on each planet are assumed to have formed at that time. Since slightly younger surfaces received substantially fewer craters, surfaces younger than 3.8–4.0 b.y. are likely to be relatively sparsely cratered. For example, while lunar uplands (4 b.y.) typically have 10 to 30 times the lunar mare average crater density, plains units (~3.5 b.y.) may have typically 0.5 to 4 times this crater density. The "benchmark" age of about 3.9 b.y. lies in the narrow gap in time. Nonetheless, the corresponding gap in cratering is large; even various lunar maria have crater densities that differ by an order of magnitude. Results based on these assumptions are consistent with other determinations, but can be placed on a firmer footing by examining crater production rates derived from asteroid and comet statistics.

Crater production rates derived by various authors, assuming different groups of cratering bodies, are summarized and synthesized in Table 8.5.5. In most cases, the results for a single planet agree within plus or minus a factor of 2. The greatest uncertainty, plus or minus a factor 2.5, is for Mercury. The "most likely" values are listed with a range that we estimate to be reasonable. These cannot be assigned rigorous uncertainties, since they depend on unknown nonstatistical factors, such as the physical structure of comets and the mechanism of planetary accretion. Undoubtedly, "worst-case" models could produce more extreme cratering rates, but we have no good evidence to support such scenarios.

The results of the preceding section (Fig. 8.4.1) gave us the absolute time dependence of the lunar cratering rate. This section (Table 8.5.5) gives us limits on the present rate on any other planet, relative to the Moon. The time dependence is assumed to be roughly the same for each planet. We are thus now ready to convert the crater densities of Table 8.3.2 into maximum and minimum ages for different planetary surfaces.

8.6 RESULTS: ESTIMATES OF AGES OF VOLCANIC FEATURES ON THE PLANETS

Because of the intense early bombardment revealed in Fig. 8.4.1 for the Moon and assumed for other planets, the relation of crater density to age is far from linear and is best determined graphically. For example, a lunar surface twice as densely cratered as a 3-b.y.-old mare is not 6 b.y. old, but only about 3.7 b.y. old.

For this reason, Fig. 8.6.1 gives a master curve through the data points of Fig. 8.4.1, defining the crater

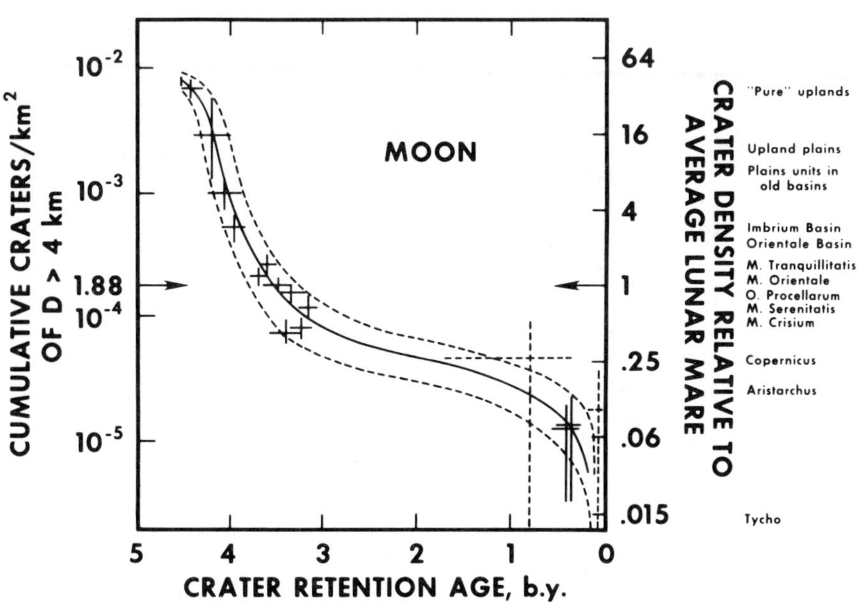

8.6.1 Crater density vs. crater retention age curve as derived for the Moon from Fig. 8.4.1. Solid line is most likely value. Dashed lines give upper and lower envelopes as discussed in the text. Observed crater densities for various provinces (from Table 8.6.1) are listed at right.

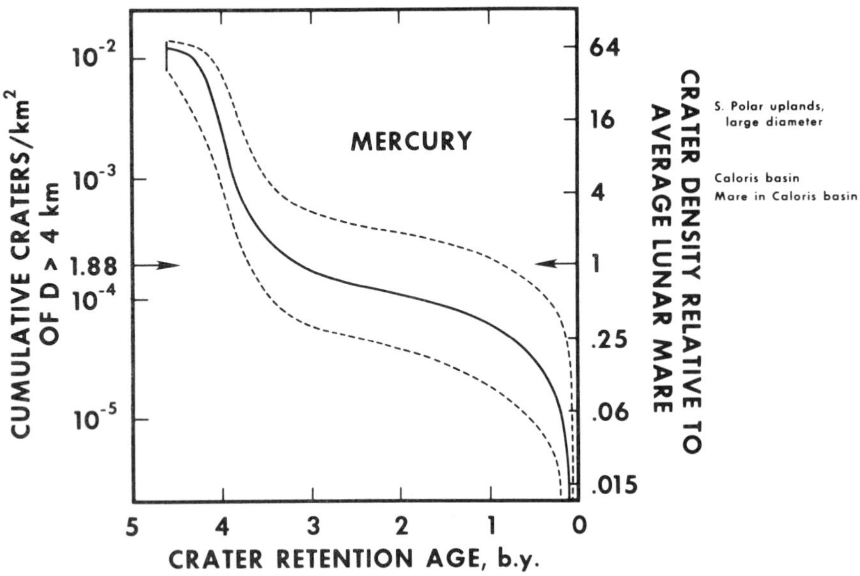

8.6.2 As in Fig. 8.6.1, for the planet Mercury. The uncertainty envelopes represent the envelopes in Fig. 8.6.1 broadened to take into account the estimated variation in converting from the lunar rate.

density-age relationship assumed here. Dashed lines give the envelopes of assumed maximum and minimum likely values (viewed as representing about one standard deviation from the most likely value, taking into account all uncertainties). The curve in Fig. 8.6.1 is then used as a base to establish the curves for the other planets.

Curves for the other planets are shown in Figs. 8.6.2–5. These are based on the data in Table 8.5.5, which gives the estimated cratering rates expected on other planets relative to the Moon. In other words, in Figs. 8.6.2–5, the lunar curve has simply been shifted up or down by factors listed in Table 8.5.5 and the uncertainty envelopes broadened appropriately to take into account the additional uncertainty in the conversion from lunar to another planetary rate. Crater densities above about 64 times the lunar mare average are not observed, and this density is assumed to correspond to an age of about 4.4 b.y. This level approaches or exceeds estimates of equilibrium steady state levels due to saturation effects.

These curves, while applied here to our own data, are designed to be generally applicable to crater counts from future researchers. Any data set that gives a crater density for craters of D $>$ 4 km (left ordinate) or a general crater density relative to the average of lunar

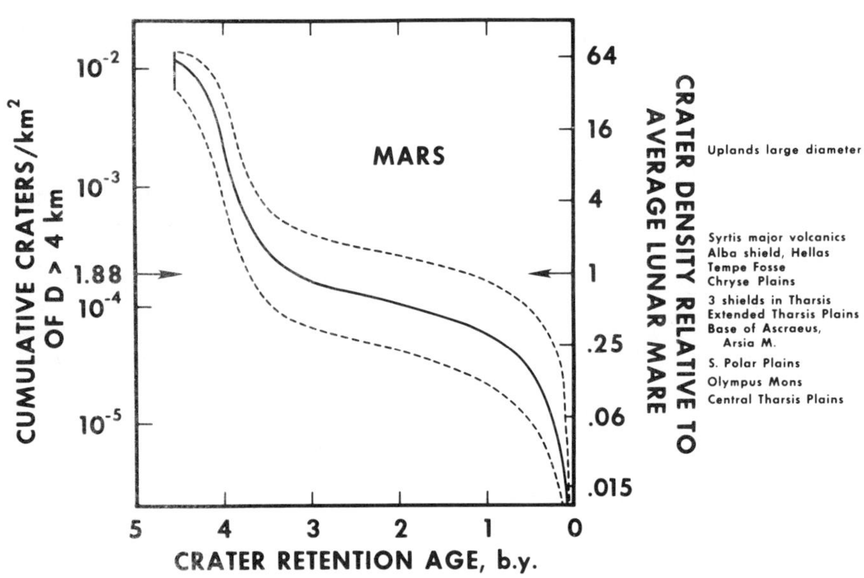

8.6.3 As in Fig. 8.6.2, for the planet Mars.

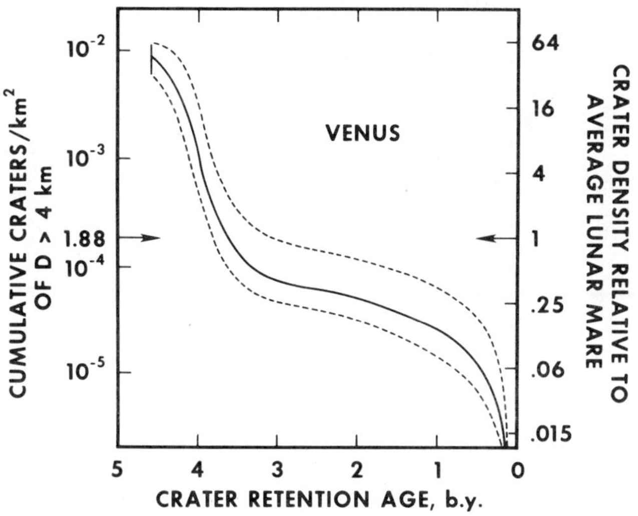

8.6.4 As in Fig. 8.6.2, for the planet Venus.

maria (right ordinate) for any terrestrial planet can be used to read an estimated age on one of these diagrams. The curves are similar in design to those given by Soderblom *et al.* (1974), Neukum *et al.* (1975b) and Neukum and Wise (1976), though we prefer the calibration at $D = 4$ km as used here to that at $D = 1$ km used by some authors, because the identification of primary impacts is much more reliable at 4 km than at 1 km, as discussed above. Conversion from crater densities at 4 km to those at larger diameters is more reliable than conversions starting from $D = 1$ km, since there is some disagreement among authors about identification of primaries as small as 1 km.

The ages derived from the figures are shown in Table 8.6.1, which lists the geologic provinces we have studied, their crater densities as derived from Table 8.3.2, and the corresponding ages read from Figs. 8.6.1–5.

In general, the results indicate that most lunar maria range in age from 2.5 to about 3.9 b.y. with evidence for some earlier resurfacing (including lava flooding) in the lunar uplands. The Caloris mare unit on Mercury may be similar in age to the lunar maria, but younger ages are also dynamically plausible. The large mare-like plains on Mars are generally found to be somewhat younger, with some activity on or near the large martian shield volcanoes almost certainly within the last 1 b.y. This finding, coupled with the March 1979 discovery of active volcanoes on Io, shows that worlds substantially smaller than Earth cannot be written off as geologically "dead."

Inherent in these curves is the result that some of the most sparsely cratered surfaces on the inner planets can fairly definitely be assigned to the last 1 b.y., or to a *late* part of planetary history, whereas heavily cratered units can be assigned to the first 1 b.y., or to an *early* period in planetary history—both findings being important to understanding of basaltic volcanism. Unfortunately, also inherent in these curves is the uncertainty attached to dates of some moderately cratered provinces, restricting their ages only to the range of about 1–3.5 b.y. While it may be interesting to learn that these provinces are definitely not 10^8 years old (an important

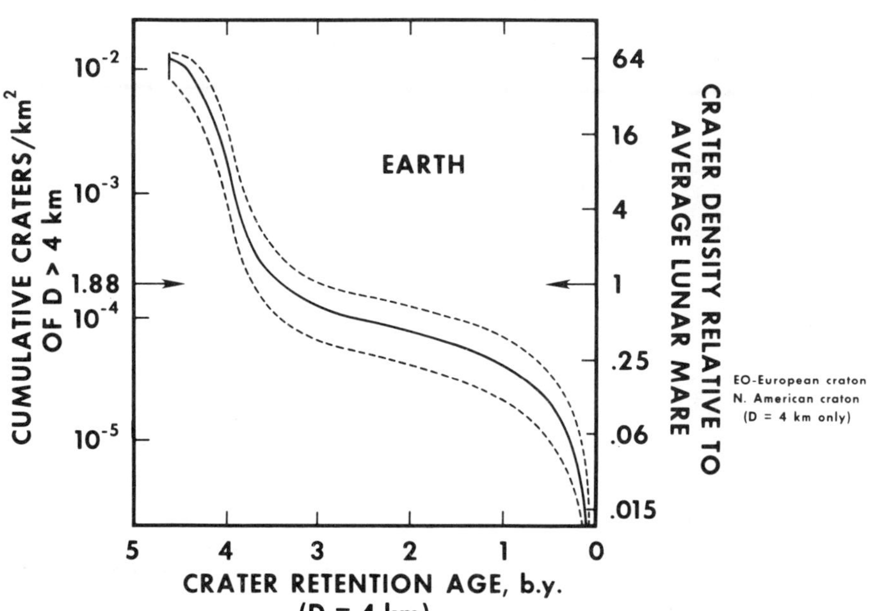

8.6.5 As in Fig. 8.6.2 for the planet Earth.

Table 8.6.1 Estimates of ages of geologic provinces on various planets.

Code No.	Geologic Province	Crater Density Relative to Average Lunar Mare	Estimated Crater Retention Age† (b.y.) Minimum Likely	Best Estimate	Maximum Likely
	Moon:				
1	Tycho	0.1	0.2	0.7	1
2	Aristarchus	0.2	0.7	1.5	2
3	Copernicus	0.3	1	2	3
4	Inner Mare Serenitatis	0.4	2.0	2.9	3.5
5	Mare Crisium	0.5	2.3	3.1	3.5
6	Mare Imbrium	0.50	2.3	3.1	3.5
7	N. Oceanus Procellarum	0.62	2.7	3.2	3.6
8	Whole Mare Serenitatis	0.65	2.8	3.2	3.7
9	S. Oceanus Procellarum	0.75	3.0	3.3	3.7
10	Mare Fecunditatis	0.93	3.2	3.5	3.7
11	Reference: Average of frontside maria	1.00	3.3	3.6	3.8
12	Mare Orientale	1.10	3.3	3.6	3.8
13	Mare Cognitum	1.26	3.4	3.6	3.9
14	Mare Tranquillitatis	1.30	3.5	3.7	3.9
15	Mare Humorum	1.50	3.6	3.7	3.9
16	Mare Nubium	2.5	3.8	3.9	4.0
17	Orientale Basin and ejecta	2.5	3.8	3.9	4.0
18	Fra Mauro	3.0	3.8	3.9	4.1
19	Fill in Schrödinger Basin	5.1	3.8	4.0	4.1
20	Fill in basin near Schiller	7.0	3.9	4.0	4.2
21	Fill in Mendeleev Crater	9	3.9	4.1	4.2
22	"Pre-mare" craters protruding in mare	9	3.9	4.1	4.2
23	Frontside uplands	10	3.9	4.1	4.2
24	Fill in Bailly Basin	13	4.0	4.1	4.2
25	Upland plains near Maurolycus	14	4.0	4.1	4.2
26	Heavily cratered uplands and basins				
A	— all D	32	4.1	4.4	4.5
B	— large D	35	4.1	4.4	4.5
27	"Pure" uplands	36	4.1	4.4	4.5
	Mercury:				
28	Caloris Mare	3.7	3.2	3.8	4.0
29	Caloris whole basin	3.7	3.2	3.8	4.0
30	S. Polar uplands				
A	— intermediate D	5	3.4	3.9	4.0
B	— large D	20	3.8	4.1	4.4

†In the cases of relatively sparsely cratered, minimally eroded volcanic plains, these are probably dates of formation of the volcanic surface units (depth $\lesssim 10^2$ m).

Table 8.6.1 *(Concluded)*

Code No.	Geologic Province	Crater Density Relative to Average Lunar Mare	Estimated Crater Retention Age† (b.y.)		
			Minimum Likely	Best Estimate	Maximum Likely
	Mars:				
31	Central Tharsis volcanic plains	0.1	0.06	0.3	1.0
32	Olympus Mons volcano	0.15	0.1	0.4	1.1
33	S. Polar plains	0.20	0.1	0.7	2.0
34	Desert at bases of Ascraeus, Arsia Mons volcanos	0.35	0.3	1.4	2.9
35	Extended Tharsis volcanic plains	0.49	0.5	1.6	3.3
36	Channels associated with chaotic terrain	0.58	0.5	2.0	3.5
37	Three shield volcanos	0.63	0.6	2.0	3.5
38	Elysium volcanics	0.68	0.7	2.6	3.5
39	Isidis Planitia	0.76	0.8	2.8	3.6
40	Solis Planum volcanic	0.90	0.9	3.0	3.7
41	Channels not associated with chaotic terrain	1.0	1.0	3.1	3.7
42	Chryse Planitia volcanic plain	1.1	1.2	3.2	3.8
43	Lunae Planum	1.2	1.3	3.2	3.8
44	Noachis ridged plains	1.3	1.7	3.3	3.8
45	Tyrrhena Patera volcanoes	1.4	1.8	3.4	3.8
46	Tempe Fossae faulted plains	1.6	2.3	3.4	3.8
47	Volcanic plains on south rim of Hellas	1.7	2.6	3.5	3.8
48	Alba shield volcano	1.8	2.6	3.5	3.8
49	Hellas floor	1.8	2.6	3.5	3.8
50	Syrtis Major Planitia volcanic plains	2.0	2.9	3.6	3.9
51	Heavily cratered plains				
A	— small $D < 4$ km	1.4	1.8	3.4	3.8
C	— large $D > 64$ km	13	3.8	4.0	4.2
	Earth:				
52	Eo-European Craton $D = 4$ km	0.13	0.2	0.4	1.0
53	North American Craton $D = 4$ km	0.11	0.3	0.5	1.1

† In the cases of relatively sparsely cratered, minimally eroded volcanic plains, these are probably dates of formation of the volcanic surface units (depth $\lesssim 10^2$ m).

result in interpreting martian climate change, for example), this result is quite frustrating. Nonetheless, we cannot change the fact that many volcanic plains have such crater density. When crater production rates become better understood, or when a few rock samples are provided for calibration of age scales, we can reduce the uncertainties.

8.7 DISCUSSION OF DATA AND INDIVIDUAL PROVINCES

Some notes are in order on interpretations of individual provinces. To clarify the discussion, consider the grouping of provinces shown in Fig. 8.7.1. Our methodology so far has been to emphasize the larger craters (generally interpreted as primary impacts) and the younger, sparsely cratered provinces (generally found to have similar size distributions approximating a power law with exponent b around −1.8 in the cumulative plot or log-incremental plot used here). In Fig. 8.7.1, we find a mean b-value to be −1.81 ± 0.44 for 36 "younger" provinces, and a median value of about −1.85. These values are nearly identical to the least-squares value of −1.80 ± 0.1 found for our reference province of average lunar frontside maria, for diameters 2.8–181 km. This agrees well with the result of a search for a standard calibration size distribution by Neukum *et al.* (1975a, b), who found that a b-value of −1.8, used by Baldwin in the diameter range 1.6–100 km, agreed well with their calibration distribution, which is nearly a power law in the range $D > 3$ km.

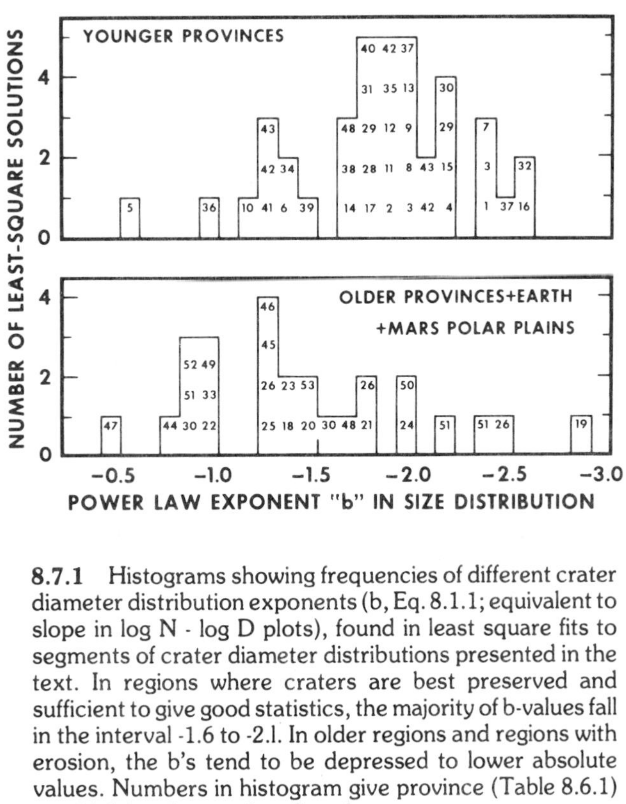

8.7.1 Histograms showing frequencies of different crater diameter distribution exponents (b, Eq. 8.1.1; equivalent to slope in log N - log D plots), found in least square fits to segments of crater diameter distributions presented in the text. In regions where craters are best preserved and sufficient to give good statistics, the majority of b-values fall in the interval -1.6 to -2.1. In older regions and regions with erosion, the b's tend to be depressed to lower absolute values. Numbers in histogram give province (Table 8.6.1) to which b-value applies. See also text.

Nonetheless, the different values shown in the upper part of Fig. 8.7.1 suggest differences from one province to another. Furthermore, the lower part of Fig. 8.7.1 shows a different situation for "older" provinces such as lunar uplands, and provinces where erosion has apparently been important, as on the Earth and the polar regions of Mars. Among the 26 least-square solutions for these provinces, the mean exponent b is only −1.45 ± 0.58, and the median is about −1.35.

What is the meaning of these different b-values? Do they rule out our assumptions that a single population of fragmental meteoroids with an essentially constant population index has been gravitationally scattered throughout the inner solar system in the last few b.y.? We would expect a common size distribution if the projectiles were scattered randomly from a common source reservoir, such as the asteroid belt. This assumption is made by most interpreters of crater counts, at least in dealing with the younger (post-mare) craters of the last 3^+ b.y.; it particularly lies at the heart of the interpretive methodology used by Neukum and his coworkers, who find a single "standard" curve that fits crater counts on all planets, after allowing for gravity and impact velocity effects.

Supporting these assumptions, we suggest that the scatter of slopes, or b-values, is not inconsistent with a single underlying meteoroid population existing during the last 3^+ b.y. This is supported by the clustering of b-values around −1.80, mentioned above, plus the fact that some of the poorer-determined values for individual geological provinces have observational uncertainties as high as ± 0.3 or more, plus the known processes of modifying initial b-values by later erosion and obliteration. Most of the scatter among younger provinces is probably due to observational uncertainty. Extreme values might also be due to observers' faulty separation of primary and secondary craters, a step required in any interpretation of these features.

Much of the scatter among older provinces must be due to erosive or obliterative mechanisms. For example, among older craters on lunar, mercurian or martian upland plains, the b-values are probably altered by any of four processes already discussed in Fig. 8.1.1: first, increases in crater density begin to cause loss of small craters by crater overlap and obliteration by ejecta blankets (Fig. 8.1.1f) (Hartmann and Wood, 1971; Chapman *et al.*, 1970; Gault, 1970), lowering the b-value. Woronow (1978) finds obliterative loss only at very high densities. Second, in the lunar and mercurian uplands, primitive heavily-cratered provinces have probably been partially covered by very old volcanic flows and then partially re-cratered. This process tends to remove smaller craters while leaving big ones, further reducing b. Depending on the sequence of flows, this can add complex curvature and structure to the diameter distribution (Fig. 8.1.1d–f). Third, Mars has apparently undergone an erosive obliteration of craters, probably involving an early episode of periods of enhanced dust transport and erosion (Arvidson, 1974; Chapman, 1974; Chapman and Jones, 1977; Hartmann, 1971b, 1973; Jones, 1974; Soderblom *et al.*, 1974). Alternatively, Oberbeck *et al.* (1977) argue that the martian craters are less affected by erosive obliteration than the above authors assert. Nonetheless, analysis of the martian plains suggests a near steady-state balance between erosive loss and replacement of craters, reducing the slope of the log N - log D curve down to near unity at $4<D<64$km (Fig. 8.1.1g–h). Subsequently, after the erosion rate declined, craters were better preserved and surfaces were re-cratered with a diameter distribution resembling that on the lunar maria (b near -1.8) seen at diameters below about 4 km. Any combination of these three processes—crater overlap and ejecta blanketing, obliteration by lava flows, or sub-areal erosion — tends to make the observed slope of the diameter distribution more shallow.

A more interesting fourth complication is that raised by Strom and Whitaker (1976) and Strom (1979), whose suggestion is that the older distributions may reflect the actual primitive production function, which differed from that today, as shown in Fig. 8.1.1 j–k. Hence, it is difficult to say whether the lower slopes found in the most ancient regions (lower part of Fig. 8.7.1) are due to obliteration/erosion effects or a remnant of a different early population of planetesimals, which was replaced by the power-law distribution characteristic of fragments as more collisions occurred in the early solar system.

The uncertainty about the underlying size distribution in ancient regions would combine with stratigraphic uncertainties in these regions to make formation ages very difficult to determine, except for the fact that these regions are restricted to the relatively narrow time interval between planet formation 4.5 b.y. ago and the end of intense accretionary bombardment about 4.0–3.8 b.y. ago.

A different problem about the shape of the diameter distribution appears at small sizes. At diameters below about 2–4 km, as discovered by Shoemaker (1965), the slope of the diameter distribution for all lunar craters (primaries, secondaries, any others) bends sharply upwards to higher absolute b-values. The behavior of the slope for primaries alone, however, is somewhat controversial, depending for its identification on correct identification of the primaries. Neukum *et al.* (1975a) and Neukum and Wise (1976) found that a single underlying curve for primaries, with a steep upward bend at these low diameters, exists on various regions of the Moon and, with a shift in D, on Mars. The upturn in their lunar "standard curve" for primaries begins at about $D = 2$ to 3 km. They clearly state that the conversion to Mars is somewhat uncertain, but find that a shift of 1.5 in D fits Mariner data well, and their martian "standard" calibration curve for primary craters bends upward at $D \simeq 1$ to 2 km. As shown in the Appendix to this chapter, lunar data sets quoted here clearly show the upward bend when all craters are counted, as seen for example, below 2–2.8 km in Province 11 (average of frontside lunar maria) or Province 17 (Orientale basin ejecta blanket). However, in some provinces where researchers attempted to identify primaries, no steep upward bend appears even at D as low as 250 m, as in lunar Provinces 1 and 2 (Tycho and Aristarchus), martian Province 32 (Olympus Mons), or Province 37 (three shields in Tharsis). The curve for Province 37 appears to extend as far as 125 m without a steep upward bend; that for Province 32 may show a hint of a bend at about $D = 250$ m. Note that the possible upward bend at about 1.4 km in Province 32 is not maintained with steep slope to smaller diameters. An upper limit to the diameter of a steep upturn in slope on Mars is set by regions where we present various authors' counts of all craters. For instance, Province 40 (Solis Planum) shows a possible slight upturn at 1–2 km, but extends down to 350 m without so sharp a bend as suggested by Neukum *et al.*; Provinces 40 (Solis Planum) and 43 (Lunae Planum) suggest gentle upturns beginning at about 1.4–2 km. One of the best counted provinces is Province 42 (Chryse), where three sets of data show a sharp upturn only below $D = 700$m, so that any sharp upturn in the underlying primary curve must be at no higher diameter. In summary, our results based on Viking counts in the relatively sparsely cratered plains of Mars suggest that the martian steep upturn analogous to that found on the Moon occurs in counts of all craters at diameters around 0.5–2 km, possibly depending on the local erosion state. In counts of craters suspected to be primaries, the upturn comes at lower size, probably 125–700 m, but possibly even less. This uncertainty about the true nature of the primary crater curve presents a serious problem in utilizing chronological data derived from fitting of "standard curves" at diameters less than 2 km.

Figures 8.7.1–2 summarize the situation. Figure 8.7.1 and the related text support the assertion that the diameter distribution of freshly formed primary craters with diameters down to about 2.8 km has fitted Eq. (8.1.1) with an exponent b, or slope of $\sim -1.8 \pm 0.3$ during the last 3.8 b.y. Neukum *et al.* (1975a) reached the same conclusion, except that they also identified a primary production function below 2.8 km as well, which has a steeper slope than we find. This is shown in the lower part of Fig. 8.7.2. Prior to about 3.8 b.y. ago, the slope may have been different, as indicated in the left part of Fig. 8.7.2. Furthermore, the largest craters and basins give evidence of a somewhat steeper slope, as shown at the top of Fig. 8.7.2. All these time- and size-dependent curvatures we avoid for the moment by concentrating on the field of Fig. 8.7.2 with relatively linear and constant size distribution, as determined by many authors.

These various results thus have little consequence to our following discussion of ages of volcanic provinces, determined from the craters of $D > 2.8$ km, but they do remain important to crater-dating methodologies that rely on small craters, to be discussed further in section 8.8.

The situation with respect to the martian chronological data is deserving of special note. Historically, since the first close-up martian photos by Mariner 4 in 1965, there have been many divergent estimates of martian crater retention ages published, leading to a widespread view that the field is divided into irreconcilable camps. What is less widely known is that the interpretations have converged in recent years, as pointed by Neukum (pers. comm., 1979) at the Second Mars Conference. If we imagine martian time divided into a very early period before 3 or 3.5 b.y., a very recent period in the last 1 b.y. and a middle period, then we can clarify the situation as follows. Even the earliest post-Mariner 4 studies disagreed in dates assigned to most features; these ranged from very early to very recent periods, depending on the method used for making lunar/martian comparison. After Mariner 9, early papers by some workers (e.g., Hartmann, 1973) assumed that the martian crater production rate would be many times higher than on the Moon due to a hypothetical high flux of Mars-crossing asteroids; they placed many of the less-cratered martian volcanoes, channels and plains in the very recent period, while the uplands stretched back to the very early period. Early analyses by other workers (e.g., Soderblom *et al.* 1974; Neukum and Wise, 1976) attempted to correlate martian ages by a more moon-like time scale, and attempted to identify provinces formed as the intense bombardment ended roughly 4 b.y. ago; these studies tended to place all or most volcanics in the very early period with the large volcanoes viewed as long inactive. Neukum and Wise (1976), in particular, concluded that the last martian volcanism on Olympus Mons ended 2.5 b.y. ago. Refinements in both methods, plus reductions in the estimated martian crater production rate as described earlier, plus new Viking counts showing very sparse cratering on Olympus Mons and elsewhere, have brought the various workers closer together. We believe that in spite of remaining differences in methodology, all current workers would agree that some old martian volcanic plains units probably have surfaces dating to the very early period, many volcanic surfaces formed in the middle period, and some volcanic activity (e.g., at least some flows on Olympus Mons) persisted to the very recent period.

For example, our results (Table 8.6.1) show volcanism on Mars ranging about 3.6–0.3 b.y. ago, with old units such as Tyrrhena Patera dated at 3.8–2.3 b.y., Elysium most likely around 3.5–0.7 b.y., and the Tharsis shield volcanoes, most likely about 2–0.4 b.y.

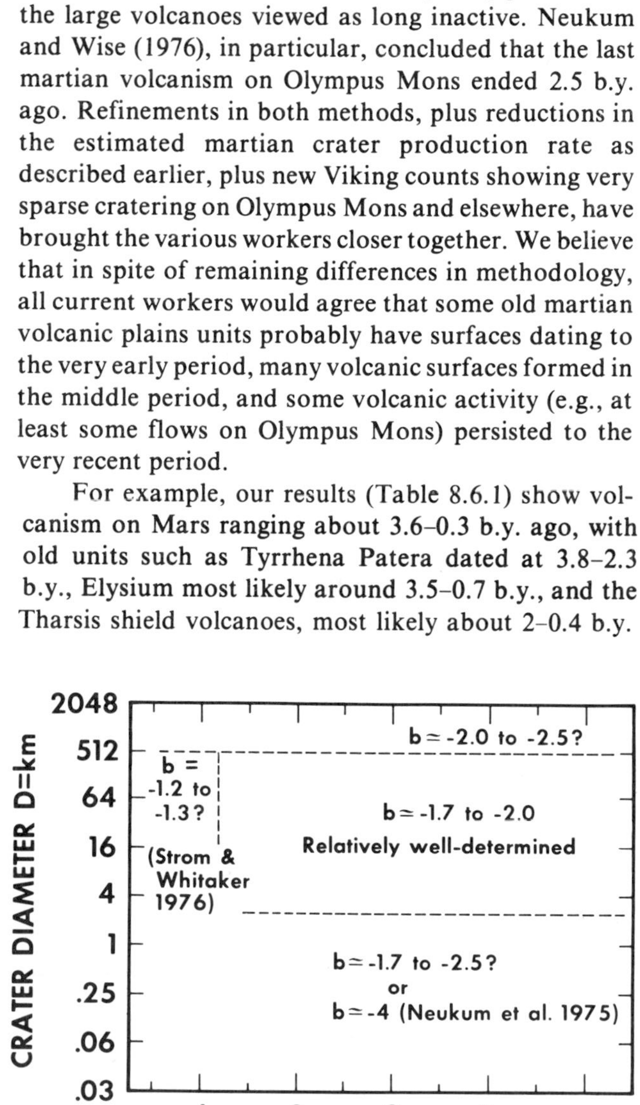

8.7.2 Schematic summary of problems dealing with diameter distribution of primary impact craters as a function of time and crater diameter. For the last 3⁺ b.y., at diameters 4 to about 512 km, behavior of the diameter distribution is relatively well-determined. Outside this zone of the diagram, different interpretations are still debated.

Neukum and Hiller (1979), who have favored old ages, give a time scale showing martian volcanism ranging from about 4.0–0.3 b.y., with Tyrrhena Patera dated about 3.9–3.8 b.y., Elysium about 3.7–3.1 b.y., and the Tharsis shields about 3.1–0.5 b.y.

The situation regarding martian erosional features is similar. We find most channel formation probably occurred about 3–2 b.y. ago, with the sediments of the martian polar plains most likely 2.0–0.1 b.y. old, and the post-erosional small craters on the heavily cratered plains most likely formed in the last 3.4 b.y. Similarly, Neukum and Hiller (1979) find intense erosion between 4 and 3.7 b.y., most channel formation during declining erosion about 3.7–3.1 b.y. ago, and "some channel erosion" continuing until 0.5 b.y. ago.

We conclude this discussion of data with a commentary on the individual provinces whose ages are estimated in Table 8.6.1, and whose crater diameter distributions relative to the lunar maria are plotted in the diagrams of the Appendix. We suggest that the reader examine the curves in the Appendix during consideration of the following comments.

1-3. Young craters Tycho, Aristarchus, Copernicus. These three data sets illustrate problems of dating young features by crater counts. In all three sets the craters counted are believed by the authors to be mostly primary impact craters. Crater counts by König (Neukum, pers. comm., 1977) agree rather well with earlier counts, as seen in the plots, although the agreement is less excellent for craters in Tycho at diameters less than 250 m.

There are not enough craters in our preferred diameter range ($D > 2.8$ km) to give relative crater densities. The counted craters are all in the diameter range corresponding to the upturned portion of the diameter distribution found in maria (dotted curve, derived from data set 11). The crater densities at these sizes approach 1–10% of the equivalent crater densities in maria, although the fraction increases somewhat as we approach diameters around 2.8 km. Relative crater densities at 1 km$<D<$4 km, derived by our least-squares-fitting techniques (Table 8.3.2) thus tend to be higher than relative densities derived by comparing the curves at, say 100 m$<D<$500 m. To claim a reliable relative or absolute age from such data, one would have to identify the diameter distribution of primaries not only in Tycho, Aristarchus and Copernicus, but also on the maria in the upturned part of the curve below $D = 2.8$ km. Neukum and his co-workers have possibly identified the latter curve, although the topic is still being debated (see section 8.8).

Schultz *et al.* (1977) raise the point that diameter distributions of craters formed in bedrock may differ from those formed by the same meteorities striking regolith. This further complicates any comparison of the small craters in Tycho, Aristarchus, etc., with small craters elsewhere. It also complicates any extrapolations of small craters in Tycho along a "standard calibration curve" to compare with larger craters elsewhere.

The ages of Tycho, Aristarchus and Copernicus and certain other similar craters are of interest to us because there appear to be volcanic features in these craters. Exceedingly rough materials on the floors of the first two and smooth "ponds" among their wall topography may be postimpact eruptive materials; apparent cinder cones and rimless volcanic pits on the floor of Copernicus also suggest history of post-impact volcanic modification (Hartmann, 1968; Wood,1979; see also section 5.4). Whether this volcanism involves only impact-created melts, or a longer-term "seepage" of deeper-seated melts through impact-induced fractures, is uncertain.

4-10. Young mare units. These units on the average are younger than the average of the frontside maria. The interior of Serenitatis has lower crater density than the dark flows around the outside border, in agreement with stratigraphic determinations of the relative ages. No flow units on the scale of a whole mare are found to be much younger than 2 b.y., although some individual smaller eruptive units within maria may have erupted more recently.

11. Average of lunar frontside maria. An important province, believed to represent a geologically well-understood and well-sampled province, thus used as a reference against which to compare the other least-square solutions.

12-16. Older lunar maria. These are the rest of the classical lunar dark maria, with higher crater densities than the average. The highest crater density in the group is about 2.5 (relative to average mare); at higher crater densities of around 3 to 6, mare-like units begin to lose their classic characteristics of smooth dark plains, due to more secondaries and primaries, overlapping ray material, or possibly in part to different compositions of earlier lavas.

17-18. Orientale and Imbrium ejecta blankets. Units apparently dating the two most recent large basin systems, Orientale and Imbrium. Figure 8.7.3 illustrates not only the Orientale blanket, but the way in which it spreads across older units, masking older small craters, but allowing some older large craters to show through more and more clearly as distance

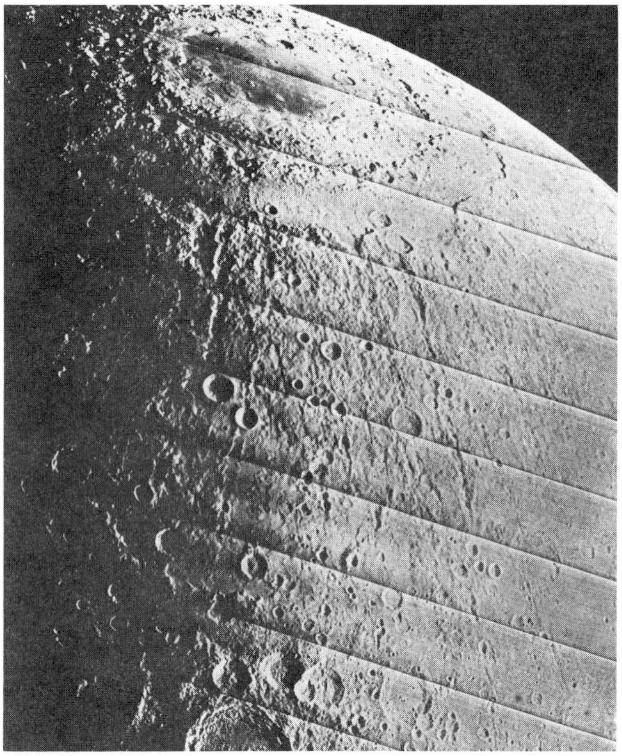

8.7.3 The effect of major basin formation and ejecta blanket formation on crater populations and early lava flows is suggested in this oblique view of the Orientale basin (top) and its ejecta blanket (central). Not only is the inner basin area entirely resurfaced and partly covered with lavas (dark mare units, top center), but older terrain far outside the basin is also resurfaced, being covered with ejecta and fractured. Large secondaries may also form in this region. Traces of early volcanism would be highly obscured. Some localized old regions (bottom edge) may approach cookie-cutter saturation, but such high levels of saturation cannot be maintained everywhere, because they would require scattered formation of large basins whose ejecta blankets would reduce crater densities locally.

from Orientale increases. This contributes to the effect shown in Fig. 8.1.1f. Counts on the surface of the Apennines also give crater densities (relative to the mare average) of around three for Imbrium, and Apennine rock samples from Apollo 15 give Imbrium dates comparable to Fra Mauro Apollo 14 samples (see sections 7.3.5–6).

19-21, 24. Mare-like fill in old basins. Good examples of older units that have the morphology of mare surfaces, but under high light appear lighter in coloration, possibly due to a thin veneer of ray material and highland debris. If this interpretation is correct, they illustrate masking of older volcanic units difficult to detect photogeologically, but important to our concept of basaltic volcanism on the planets. Figure 8.7.4 shows an especially clear case, the light-colored mare-like fill in the old double-ring basin near Schiller (Province 20). The inner ring and part of the outer ring are occupied by low crater-density mare-like material. Figure 8.7.5 shows another example, the old far-side basin Korolev, where the once-smooth fill appears heavily damaged by cratering.

22. Premare craters in mare units. These large "Archimedian" craters formed inside large basins after the basins were created, but before they filled with lava, and thus give important evidence of substantial time intervals between basin formation and the final flows covering the upper few hundred meters of the mare surfaces. The strong depletion of craters below about 32 km diameter gives good evidence of the flattening of the diameter distribution due to lava cover, as shown in Fig. 8.1.1f.

23, 25. Upland plains. Possibly analogous to upland plains on Mercury, these are broad regions between larger craters, but with lower crater density than the most heavily cratered lunar uplands. Broad regions of frontside upland (Province 23) are less

8.7.4 The double-ring lunar basin near Schiller. Schiller is the elongated crater (upper right), between the two basin rings. The basin interior to its left shows mare morphology at low light, moderately high albedo at high light, and a crater density about seven times that for average lunar maria. It is apparently an example of early mare volcanism in a small basin, later thickly covered by upland ejecta material.

8.7.5 Bailly basin, another example of an old upland basin feature with a nearly obliterated outer ring and old heavily cratered interior with a fill that has about 8 ± 2 times the average mare crater density (Hartmann and Wood, 1971). The fill may be another example of old, damaged mare-style volcanism pre-dating the more familiar smooth maria.

cratered. Are they very damaged examples of volcanism from a period around 4.2 ± 0.2 b.y. ago, but almost obliterated by the end of intense accretionary cratering?

26, 27. Heavily cratered uplands. These regions are among the most cratered provinces in the solar system, but it is uncertain whether the crater density of around 36 represents an upper limit of steady state equilibrium, or whether substantially higher densities are possible, given the size distribution of impacting meteoroids. See discussion in section 8.2.

28, 29. Caloris. The counts do not indicate a substantial difference in crater density between the basin ejecta (29) and the mare-like fill unit (28) in the central basin. They are probably more nearly the same age than, say, the Imbrium basin and its fill (Provinces 6 and 18), which can be readily distinguished by crater density. The region is illustrated in Fig. 8.7.6., which shows the ridge-laced central plain and its mare-like morphology. The Hartmann data in province 29 include only basin ejecta blanket surfaces, but the Woronow data include a small central unit as well as a larger amount of ejecta out to about one basin radius beyond the rim. Section 5.5.2 gives a further interpretation of Mercurian plains as volcanic units.

30. Mercury uplands. Note the similarity to lunar uplands (23), but with a sharper downturn in the diameter distribution below D ≅ 45 km. It is uncertain whether obliterative losses of early small craters (through early plains-forming volcanism? See section 5.5.2) or a different underlying size distri-

8.7.6 The Caloris basin region of Mercury. The ridged interior floor may be a mare-like volcanic unit. The floor and basin ejecta have about 3.7 times the crater density of average lunar maria. To the right, mercurian uplands and upland plains can be seen.

bution in the early planetesimal population is responsible. The right-hand parts of Fig. 8.7.6 show several areas of intercrater plains. Figure 8.7.7 shows a mare-like fill (lava?) in a crater in a heavily cratered upland region, again suggesting volcanism in highland areas.

31, 32. Central Tharsis and Olympus Mons. The formal solutions suggest a slightly greater density on Olympus Mons than on the surrounding plains, but this may not be significant, since the two plots involve different-sized craters, and since the Olympus Mons counts attempt to reject volcanic pits, which are hard to identify. However, there may be some very young flows in the desert plains around Tharsis; Schaber *et al.*(1978) map such units surrounding Olympus Mons, as shown in Fig. 8.7.8. In any case, these units are very important because they include flow units, similar to those in Fig. 8.7.9, so sparsely cratered that they must have erupted recently (<1 b.y. ago) according to other authors' chronological systems, including Neukum and Wise (1976), as well as our own. Thus, they establish that martian volcanism has persisted throughout most of geologic time, albeit possibly declining in intensity.

See section 2.2.4 for remote-sensing evidence that such units represent basaltic volcanism. Sections 5.6 and 6.5.2 discuss morphologic evidence of volcanic-tectonic origin of the martian plains and shields.

33. S. Polar plains of Mars. Very low densities of craters, as well as stratified deposits, have been noted on both Mariner and Viking photos. This evidence, plus the flattened slope of the curve, may indicate

8.7.7 Example of a large mercurian upland crater with a smooth floor that may be a mare-like volcanic unit.

ongoing deposition in the polar regions, depleting small craters (see Fig. 8.1.1g and h).

34. Desert surrounding Ascraeus and Arsia shields. This region gives a somewhat higher crater density. The crater counts from two sources define a fairly smooth curve (within error bars) over the range D = 350 m to 32 km, with a gentle upward bend (secondaries?) below about 2 km.

35. Extended Tharsis volcanic plains. Moving outward to regions in an arc around the central shield complex, one finds higher crater densities, with a well-defined straight curve from D = 2 to 256 km. The intermediate crater density is an example of a value giving a frustratingly wide range of plausible ages, unlike the units with substantially lower or higher crater density. Figure 8.7.10 shows sparsely cratered flows embaying craters and cratered uplands along the southwest border of the Tharsis complex. (See also section 5.6.3.)

37. Ascraeus, Pavonis and Arsia Mons shield surfaces. Averaged, these appear somewhat older than the Olympus Mons surface and central Tharsis plains around Olympus Mons. Figure 8.7.11 shows the floor of the summit caldera of Arsia Mons, revealing the very sparse cratering there.

36, 41. Channels. Two groups of channels with somewhat different morphologic features. The channels appear to have been active until slightly before the main central Tharsis volcanism, and to have been relatively inactive afterward. The channel surfaces may date the end of the period of declining erosive activity.

39, 40, 42-45. Older martian volcanic plains. Some of these units border on the outer flanks of Tharsis, indicating a long period of development there. Chapter 6, section 6.5.2 also gives evidence of an extensive period of faulting in association with the later Tharsis stages that were dominated by volcanism. Viking data show that Chryse (42) has a basalt-like composition. Figure 8.7.12 shows the moderately cratered plains of Tyrrhena Patera (45), which display wrinkle ridge systems resembling those of lunar maria, as well as a large, apparently volcanic, construct with summit caldera. These evidences support the interpretation of volcanic formation for these plains units. Tyrrhena Patera has a crater density slightly higher than that of lunar Mare Tranquillitatis.

46. Tempe Fossae. An intensely fractured province on the edge of Tharsis, with faults radiating from the Tharsis area. The fracture surface predates the central Tharsis volcanism, suggesting that this deformation is caused by older volcanic loading (section

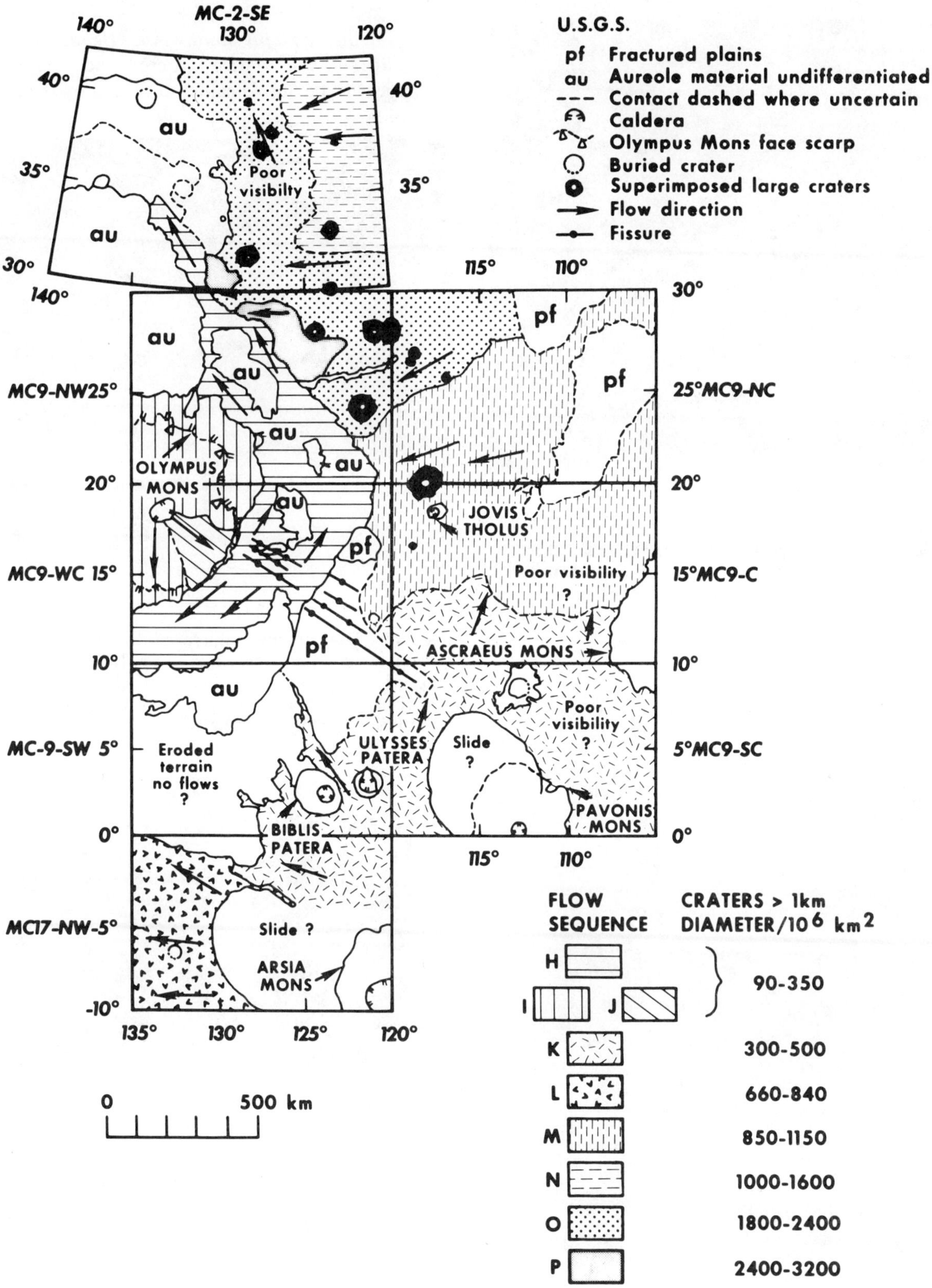

8.7.8 Complex volcanic flow units in Central Tharsis area of Mars, showing evidence for a unit (H) around Olympus Mons younger than Olympus Mons itself. This may account for the relations of ages shown in Table 8.5.5 for provinces 33-36. (From Schaber *et al.*, 1978.)

8.7.9 The southeast flank of the volcanic shield Olympus Mons is characterized by lava flows and their small scale constructs — elevated tubes and fan-shaped accumulations. A sparse population of small impact craters is also present. The area shown measures 118 × 110 km. (Viking image 222A65.)

8.7.10 At the southwest margin of the Tharsis volcanic province (near 34°S, 130°W) lava flows embay more ancient cratered terrain. Area shown measures approximately 250 km square. (Viking image 56A14.)

6.5.2).

47. Cratered plains, southwest of Hellas, 57° S (Australis Tholus). These cratered and ridged plains appear to be old volcanics degraded by the possible

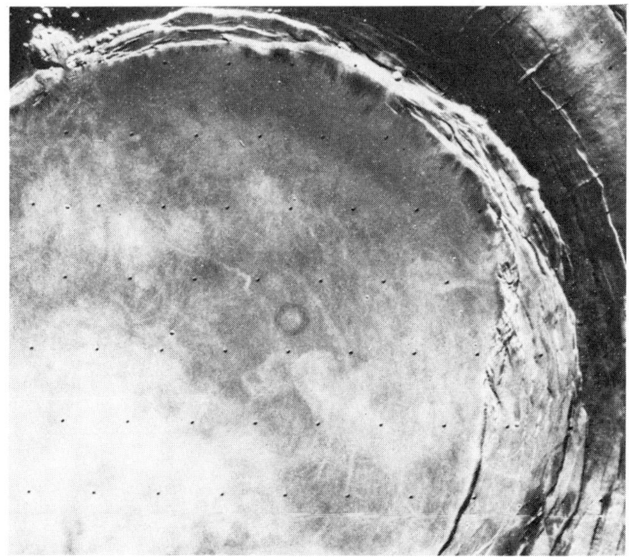

8.7.11 The summit caldera of the great shield volcano Arsia Mons measures over 120 km across. Its floor level is about 2 km below the shield summit. Both irregular (volcanic) and circular (mainly impact) small craters occur on the otherwise smooth surface (Viking image 90A06.)

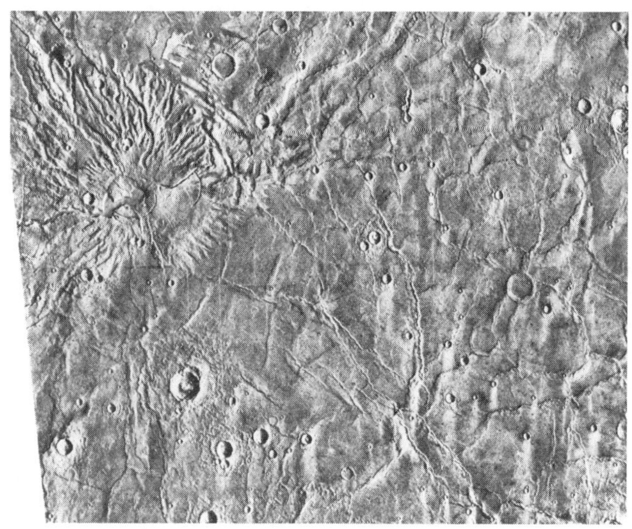

8.7.12 Plains to the east of the central volcano Tyrrhena Patera (22°S, 253°W) are characterized by prominent ridges, like those seen on lunar maria and mantled or "ghost" craters. (From VO mosaic 211-5213,87A01-24.)

8.7.13 Near the southwest rim of the Hellas Basin around 57°S, 314°W is a large area of heavily cratered ridged plains of possible early volcanic origin. The subdued character of many craters and a relative deficiency of very small craters suggests a complex history of sedimentation and erosion. (Viking image 577B01.)

sedimentation and erosion characteristic of the martian high latitudes. One portion is shown in Fig. 8.7.13. The crater density of 1.1 times the lunar mare average is high enough to remove much likelihood of such a unit being very young. Age of this volcanism is probably aroung 3.5 b.y. (certainly older than 2.6 b.y.). Other extensive upland volcanism at this age may have occurred on Mars (see 5.5.3).

48. Alba shield. The crater counts indicate that Alba, lying on the flanks of Tharsis, is older than the shields in the central Tharsis region, consistent with the general trend of older ages out from the Tharsis center. The crater-dotted flows on the shield are shown in Fig. 8.7.14. Alba has a similar age to the fractured Tempe Fossae region (46); the age difference listed in Table 8.3.2 (Alba slightly older) may be insignificant, since Alba appears superimposed on similar fractures.

49. Hellas floor. Viking photos reveal fairly dense cratering on the floor of this old basin, which was dust-obscured in the earlier Mariner photos, and regarded as possibly featureless. The basin itself is presumably still older than this measurement, which

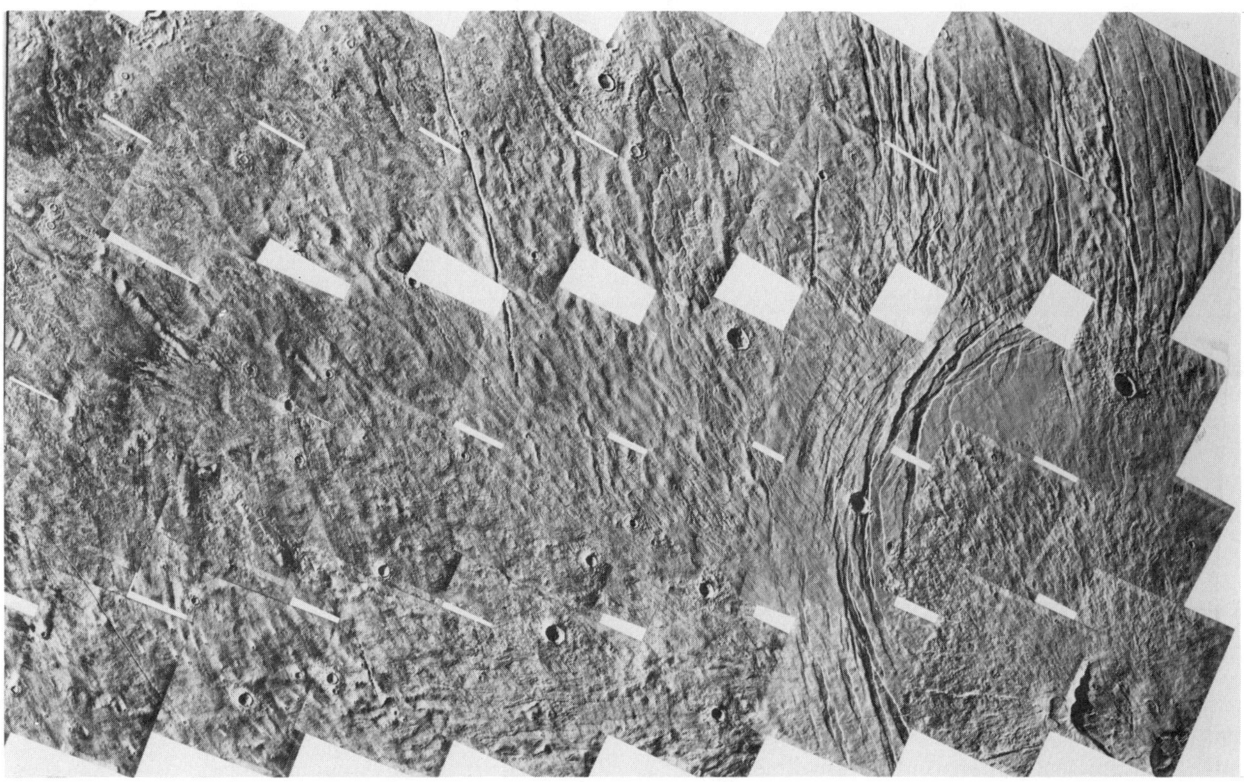

8.7.14 One of the oldest central volcanoes (as reflected by superimposed impact craters) is Alba (centered about 40°N, 110°W). Flows extend from the central caldera or the ring of faults surrounding it (lower right) for many hundreds of kilometers. Area shown measures approximately 600 × 900 km.

8.7.15 Volcanic plains showing lava flow morphology in the region Syrtis Major, around 11°N, 293°W. Wind streaks caused by veneer of wind-blown dust are prominent. (From VO mosaic 211-5601, 496A41-76.)

may reveal the crater density of volcanic fill, as in large lunar basins.

50. Syrtis Major volcanics. These are among the older volcanic plains units. As shown in Fig. 8.7.15, the cratering is moderate and volcanic landforms can be seen, although the latter are partly obscured by strong albedo features. Older volcanic units may exist in the uplands of Mars. The crater density here is approaching the value at which lunar maria begin to be obscured by possible cratering and ejecta effects.

51. Martian uplands. The remarkable crater distribution found here supports an interpretation suggested in Figs. 8.1.1g–h: at the end of early martian erosion, corresponding to crater density around 1.4 ± 0.4, surfaces that had been depleted in small craters began to reaccumulate the primary crater size distribution, which is now visible outside polar regions to sizes up to about 4 km.

52, 53. Old cratons on Earth. Severe depletion of smaller craters, presumably by known fluvial, glacial and other erosive processes, is illustrated. (See sections 5.3 and 6.4 for discussion of volcanic and tectonic histories of terrestrial cratons.)

8.8 COMPARISON OF RESULTS WITH RESULTS OF TECHNIQUES USING SMALL CRATERS

Most of our efforts have been directed to understanding the cratering chronologies of large provinces, typically the dimensions of lunar maria or larger. A major motivation for concentrating on large units is that they provide statistically significant samples of craters larger than a few kilometers in diameter whose origin by primary impact is not controversial. At the same time, they are small enough and numerous enough to give a representative sampling of volcanism through time.

Planetary basaltic volcanism often occurs on such large scale, but the spatial scales of individual volcanic events are often much smaller. A detailed understanding of the emplacement history of planetary basalts within large provinces requires some knowledge of the temporal and spatial distributions of each emplacement event (i.e., each flow). The lunar maria themselves are known to exhibit large numbers of recognizable overlapping flows (see section 5.3.2). Do these all have about the same age 3.2–3.8 b.y., or do they differ widely in age (2–4 b.y.) with only the averages being similar for mare-wide regions? It would be important to learn whether mare basins filled rapidly or whether at least the final stages of volcanism stretched over aeons of time. Crater counts averaged over an entire basin cannot hope to reveal such important details.

We must, therefore, turn to smaller craters if we are to hope for statistically significant cratering ages for small units. Aside from attempts to identify and count primaries from among the many small craters, the most widely applied age-dating technique utilizing small craters is the system developed by Soderblom, Boyce and their collaborators (e.g., Boyce *et al.*, 1974) based on a crater degradation model of Soderblom and Lebofsky (1972). Rather than the use of crater counts, the procedure involves identification on photographs of craters whose inner slopes are degraded to certain degrees by subsequent impacts. Using the Soderblom-Lebofsky model for the rate of slope degradation to a specified limit, an estimate is made of the maximum diameter (D_L) of craters degraded to that limit on each unit. The longer a surface has been exposed to cratering bombardment, the more degraded the older craters on that unit will be, and the larger will be the maximum diameter of craters degraded to the limit. The D_L ages are calibrated against radiometric ages, where available.

Because of the lack of complete photographic coverage of the Moon at high resolution, Boyce *et al.* (1974) were restricted to using Lunar Orbiter IV photography for such dating purposes; the resolution of such photography is adequate for estimating D_L, but is barely adequate for revealing other pertinent traits of the crater populations that would verify that local geological processes were not influencing the morphologies of measured craters. Two major assumptions in the Boyce-Soderblom dating methods are:

(1) All craters are primary impact craters formed with the same initial morphology, in particular with the same initial diameter/depth ratio. If secondary craters, or especially endogenic craters, are present and were formed with initially shallow shapes, the Boyce methodology might yield unrealistically old ages for the surface.

(2) The erosion process is presumed to be spatially invariant. A particular degraded morphology for a crater of a specified size is interpreted as implying a unique age. But if, for example, secondary impacts from a nearby primary augment the erosion in a locality, that unit will be judged to be unrealistically old. Further review of the Boyce-Soderblom assumptions is given by McGill (1977).

More detailed studies of crater distributions in a region can verify or refute the Boyce-Soderblom assumptions. For instance, the diameter-frequency relation of small craters should show a bend or "knee" at some size, below which the population is in empirical equilibrium. The knee would shift to larger sizes for older populations. Furthermore, a straightforward sequence of crater morphologies should be observed throughout the subkilometer size ranges, if impact erosion were in fact controlling crater morphologies (see Chapman *et al.*, 1970). Moreover, the Soderblom-Lebofsky erosional model implies a size dependence for the rate at which features degrade and are obliterated; that is, a particular "age" implies that no topographical features (of whatever origin) smaller than a limiting size would still be present.

8.8.1 Examples of small, nonimpact craters and associated interpretive problems

While tests of the Boyce age-dating scheme have not been carried out for most regions, several investigators have argued that the assumptions in fact break down in some distinct localities for which high-resolution photography is available. Chapman *et al.* (1970) argued that the size-distributions of small lunar crater morphology classes do not match the distributions predicted by the simple erosional model; Chapman *et al.* interpreted many of the smaller craters of the type used in Boyce's age-dating techniques to be of endogenic origin. Schultz *et al.* (1976) described a class of small features on some mare surfaces that must be primary volcanic features but which are so small that they should have been eroded away according to the Soderblom erosion model and the inferred age of the surfaces. Later, Schultz *et al.* (1976) interpreted a series of small-crater counts as indicating either an endogenic crater population or regional variations in the efficacy of the erosion (due, for instance, to variable strengths of target material). Either result would severely complicate dating by use of small craters.

Many researchers recognize the idealized nature of the Soderblom erosional model and of the D_L age-dating technique. But most workers have assumed that differences in regional processes and geology and secondary and endogenic cratering components add only relatively small uncertainties to the Boyce ages. In particular, they point to rough agreement between

8.8.1 Moon-like landscape in cinder field at Kapoho, Hawaii, formed in 1960 eruption and dotted by erupted volcanic boulders and pits. Foreground pit is 1-1/2 m drainage crater caused by draining of cinders into underground fissure. In the middle distance beyond it (slightly left of center) is a similar-sized volcanic impact crater. Many more numerous craters, visible in 1960's, have been destroyed by motorbike activity (see tracks in background). (Photo Nov., 1978; W.K.H.)

Boyce ages and lunar rock ages measured from several localities, relative ages determined by stratigraphic means (including counts of large craters), and regolith depths as estimated on independent criteria by Oberbeck and Quaide (1968).

Complexities associated with using simple counts of small craters, or small-crater morphometric measurements as age parameters, are illustrated by Figs. 8.8.1–5. These give examples of nonimpact craters ranging from 1.5 m–300 m diameter which, particularly if degraded by regolith formation, could be mistaken for impact pits. Those familiar with volcanic formations in the field will recognize that a wide variety of processes can cause craters in the subkilometer size range. Figure 8.8.1 is the moon-like scene in the field of regolith-like cinders caused by the 1960 volcanic eruption at Kapoho, Hawaii, onto which final eruptive stages ejected boulders. The foreground shows a 1.5 m collapse pit in which some of the cinder "regolith" has apparently drained into a subsurface fissure; to its rear is a volcanic secondary impact crater produced by one of the boulder impacts.

Figures 8.8.2–5 are features on the northeast rift of Mauna Loa, Hawaii. Figure 8.8.2 shows an 8 m pit apparently caused by collapse of a cindery surface into a lava tube, but it has a raised rim possibly constructed by spatter. With regolith cover, it could easily be taken for an impact pit. Figure 8.8.3 is another example of collapse into a tube. The inner opening is roughly 20 m across, but this is surrounded by a 40 m gentle depression in the flows, apparently caused by drainage and possible sagging into the underground cavity. If disguised by regolith that would fill the central depression, it might be interpreted as an eroded impact crater with gently sloping walls, giving a spurious D_L value. Figure 8.8.4 shows an eruptive 300 m crater filled with a lava lake floor. Again, if the walls were degraded by impact erosion and consequent regolith formation, it might be counted (for example in 100 m resolution Viking frames) as an impact pit.

8.8.2 Collapse pit with spatter rim, apparently formed by collapse into a lava tube. Diameter approximately 8 m. If slightly degraded by regolith, such a feature could be mistaken for an impact pit. Mauna Loa NE rift zone, Hawaii. (Photo, W.K.H.)

8.8.3 Collapse pit on Mauna Loa NE rift zone, showing central cavity (approximately 20 m diameter) caused by lava flowback or sagging. Such morphology, if degraded by impact erosion, could yield spurious D_L values. (Photo, W.K.H.)

8.8.4 Keanakakoi crater, Kilauea volcano, Hawaii. Diameter approximately 300 m, with flat floor formed by lava lake. Scale is comparable to smaller craters counted on martian volcanic terrains. (Photo, W.K.H.)

Figure 8.8.5 shows an example of the coalesced collapse pits forming a 20 m diameter fissure along the crest of the ridge following the northeast rift zone of Mauna Loa. Although localized eroded openings of this type might be taken as impact craters, the fissures are generally elongated enough not to be mistaken. Examples of similar fissures are seen on Olympus Mons; elongated pits of similar proportion are seen in Fig. 8.7.11, showing the summit caldera of Arsia Mons.

Although we have emphasized the larger craters and gross chronology of the larger units as most relevant to this study, we recognize the importance of identifying a remote-sensing chronometer for small units on the Moon, and have taken two approaches to evaluate the D_L ages. The first is purely empirical—to establish the degree to which D_L ages correlate with other ages. The second is to further evaluate the merit of the alternative hypothesis for small-crater origin that is most damaging to the D_L method—the endogenic hypothesis—and to see if the distributions of small-crater sizes and morphologies can be reconciled with the pure impact hypothesis for crater origin and degradation.

8.8.2 Empirical correlation of D_L and C_S ages with other ages

One useful test of the D_L methodology is to examine how well the D_L ages correlate with those established earlier in the present chapter for larger lunar regions. Soderblom (pers. comm.) used a map similar to Fig. 8.2.1a and prepared a list of D_L values averaged

8.8.5 Downhill view along crest of the NE fissure, Mauna Loa volcano, showing open rift at summit of ridge, similar to fissure and crater-chain features on martian shield volcanoes. (Photo, W.K.H.)

over these geographic provinces, most of which are classical lunar maria. These values, referenced as the "Soderblom" list, are tabulated in Table 8.8.1, along with some additional comparisons of crater densities, rock ages, and D_L values, mostly drawn from values cited in the literature. Table 8.8.1 also includes values of the additional age parameter C_S, to be discussed below. The correlation of D_L with crater density, from Table 8.8.1, is plotted in Fig. 8.8.6, where two results are apparent. First, there is a fairly good correlation of D_L and crater density. But, second, the correlation depends on including old upland sites and very young features such as Copernicus; the correlation among large-scale mare units alone is not very good.

A quite similar result was found by Neukum (1977) who compared D_L values of Soderblom and Lebofsky (1972) and Soderblom and Boyce (1972) with his own crater densities. He found a correlation $D_L \propto$ (crater density)$^{0.6}$ for units younger than the Apollo 11 site, but no correlation for older units such as upland plains and highlands.

Of course, our selection of data for comparing mare D_L and mare crater densities is poorly designed for such a comparison, because we have averaged crater counts over large regions known to contain many small flow units with somewhat different D_L values. Counts of the larger craters usually include the entire province, while counts of smaller craters and D_L values are drawn from photos dotted across the province in different patterns. Thus, the crater counts certainly are not compiled from the same distribution of units as the D_L values. Also, even if the D_L ages and crater retention ages were entirely accurate, perfect agreement is not necessarily expected. For instance, the mare D_L values reflect degradation of 200–400 m craters formed on the last lava flow, while the craters used for dating mare units range from 2.8 to about 32 km across and are probably not sensitive to the last few flows in a region. Since they penetrate further and are not modified by thin flows, they reflect ages since major build-up of lava fill ceased in basins.

Table 8.8.1 includes data on a second age parameter drawn from small craters, the C_S value, adopted by Boyce (1976) from concepts in earlier work by Shoemaker *et al.* (1969, p. 67) and Trask (1966). The C_S value is a crater diameter at the intersection between the steep branch of the crater diameter distribution, and the flatter slope at still smaller diameters, which arises because of saturation or equilibrium effects among the smaller craters. It typically occurs at about 60–300 m diameter in the maria, and at larger diameters (up to about 2 km) in the uplands. The basis for using such a parameter as an age indicator might be questioned by

Table 8.8.1 Comparison of lunar crater densities, rock ages, and age parameters D_L and C_S.

Code	Province	Crater Density Relative to Mare Average[(a)]	Median Rock Age,[(b)] b.y.	D_L Value, m: Province Average	D_L Value, m: Literature[(c)]	C_S Value, m Literature
Provinces studied here correlated with Soderblom D_L list						
4	Inner Mare Serenitatis	0.40		255		
6	Mare Imbrium	0.50		265		
8	Mare Serenitatis	0.65		250		
9	S. Oceanus Procellarum	0.75		205		
10	Mare Fecunditatis	0.93		235		
13	Mare Cognitum	1.26		235		
14	Mare Tranquillitatis	1.30		385		
15	Mare Humorum	1.50		245		
16	Mare Nubium	2.5		315		
20	Fill, basin near Schiller	7.0		550		
Landing sites and data taken from literature						
	Apollo 11	1.3	3.59		335±55 (*2*)	175±14 (*6*)
	Apollo 12	0.75	3.16		210±20 (*2*)	68±8 (*7*)
	Apollo 14	2.8	3.92		> 1000 (*3*)	
	Apollo 15	0.5	3.28		270±15 (*2*)	94±10 (*7*)
	Apollo 16	4.0	3.98		550±50 (*3*)	
	Apollo 17 (mare)	1.2	3.68		360±30 (*2*)	260±30 (*8,9*)
	Luna 16	0.9	3.4		270±30 (*2*)	
	Luna 24	0.5	3.4			138±18 (*8*)
Other sites with data from literature						
	Copernicus	0.3			100±10 (*3*)	
	Orientale ejecta	2.5			550±100 (*10*)	
	Imbrium ejecta (Apollo 14, 15)	3.0	3.92		350±30 (*4*)	730±130 (*4*)
	Humboldtianum ejecta	15 (*1*)				1100±200 (*10*)
	Nectaris ejecta	16 (*1*)				2475±1025 (*10*)
	Generalized highlands, frontside	10–35			1150±200 (*5*)	

(a)Table 8.6.1. (b)Table 7.3.1. Rb-Sr preferred; Fig. 7.3.13. (c)Soderblom list.

References:

(*1*) Hartmann and Wood (1971); (*2*) Boyce (1976); (*3*) Boyce (1975); (*4*) Soderblom and Boyce (1972); (*5*) Boyce *et al.* (1975); (*6*) Morris and Shoemaker (1969); (*7*) Boyce *et al.* (1977); (*8*) Boyce and Johnson (1977); (*9*) Lucchitta and Sanchez (1975); (*10*) Watkins *et al.* (1977).

some authors who believe that saturation or steady state effects are not evidenced, even in heavily cratered uplands, but we find a fair degree of correlation with age, as seen in Fig. 8.8.7 which shows the correlation of C_S with crater density based on Table 8.8.1. A similar result is found: a correlation is found, but it is not very strong among the extended mare provinces alone.

Table 8.8.2 summarizes the correlations among crater density, rock age, D_L, and C_S. We conclude that the D_L and C_S parameters are correlated with other age

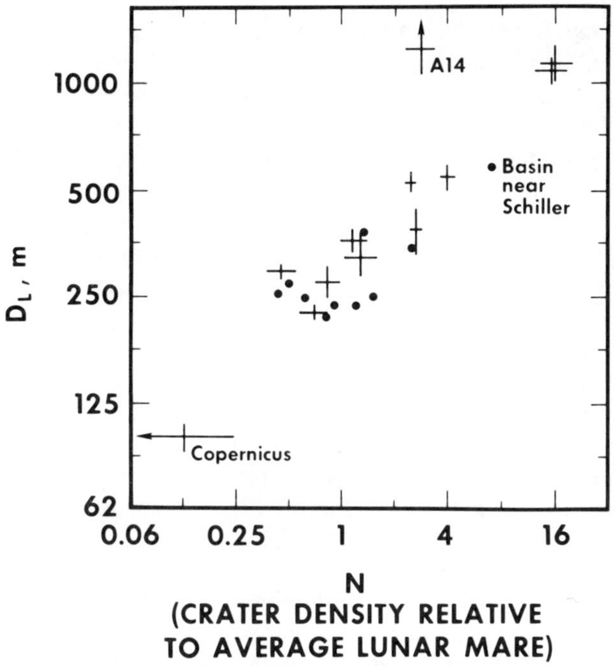

8.8.6 Correlation of D_L and crater density N values from Table 8.8.1. A correlation coefficient of 0.78 is found for these data, which include cratered uplands (two points in upper right) and other old and young units as identified.

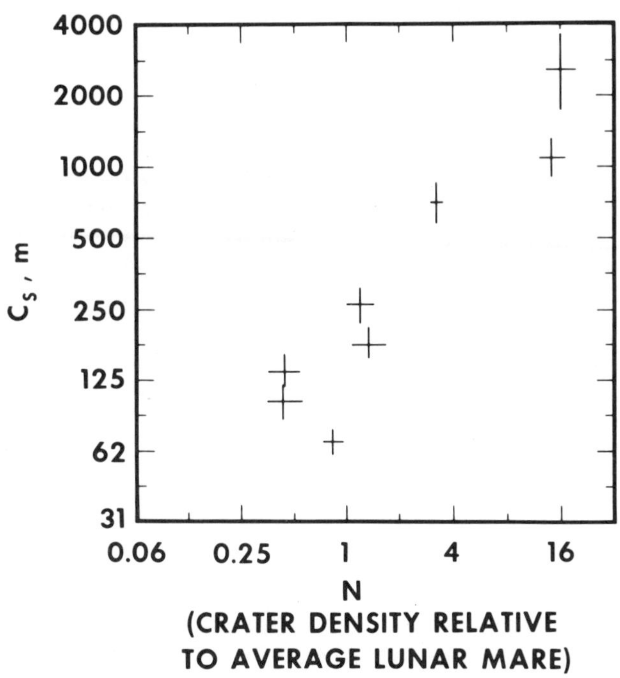

8.8.7 Correlation of C_S and crater density N values from Table 8.8.1. A correlation coefficient of 0.90 is found for these data.

Table 8.8.2 Correlation coefficients (assumed linear fits).†

	Classical Dark Mare Points Only	All Available Points
Relative crater density vs. median rock age	.82 (6 points)	.90 (9 points)
D_L vs. median rock age	.99 (5 points)	.73 (7 points)
C_S vs. median rock age	.94 (5 points)	.86 (6 points)
Relative crater density vs. D_L	.44 (14 points)	.78 (22 points)
Relative crater density vs. C_S	.70 (5 points)	.90 (8 points)
D_L vs. C_S	.94 (4 points)	.87 (6 points)

†Data drawn from Table 8.6.1. Rock age data from preliminary version of Table 8.4.2, slightly revised subsequently.

parameters. Although there is no guarantee of linear relations, we note that linear correlation coefficients typically range from 0.7 to 0.9 when provinces ranging from young craters or mare to older uplands are included. Other data in this volume also suggest that fruitful further work should be undertaken in understanding and refining these correlations. For example, section 2.2.1 includes much remote sensing data defining small units of common characteristics believed associated with younger or older ages, such as low-Ti basalts. Section 5.4 includes stratigraphic data defining similar units by relative age. It is reasonable to assume (although it may not be true) that basalt compositions have varied temporally, in which case, it is more appropriate to average large-crater counts over spectral units rather than over the units employed in this chapter. We would predict that further studies of crater counts, D_L, C_S, remote-sensing data, and stratigraphic data for selected common units will reveal better correlations between these units.

8.8.3 Evaluation of D_L in Context of Hypotheses for Small-Crater Origin

Over the past decade, evidence has continued to mount that small-crater populations on the Moon do not reflect the size and morphology distributions expected from the idealized impact model for primary impact crater origin and degradation. For instance, some mare areas show more hectometer-scale craters than some clearly older upland areas. As mentioned earlier, a minority of researchers have felt forced toward an endogenic model (such as collapse pits in lava flows; see Figs. 8.8.2–3) for the origin of a majority of small lunar craters. Despite the clear evidence for *some* small endogenic craters on the Moon (e.g., those arranged in chains) and some applicable terrestrial analogies (collapse craters on some terrestrial flows), the hypothesis has been rather *ad hoc* and has lacked aesthetic appeal. Schultz *et al.* (1977) have concluded that many of the differences in small crater characteristics from unit to unit are due not to admixtures of endogenic craters, but to differences of substrate properties in which impact craters formed.

Chapman *et al.* (1979) sought to establish whether small crater populations can be possibly reconciled with a plausible exogenic hypothesis. The conclusion is a tentative "yes." Most features of previously published crater distributions, and some new ones obtained by Chapman *et al.*, can be reconciled with impact cratering and degradation occurring on a two-layer surface, as suggested by Schultz *et al.* (1977). The two layers are a solid substrate overlain by an unbonded layer of depth the same order of magnitude as the lunar regolith. The model has obvious appeal since the two-layer model is a reasonable approximation to the real lunar surface.

Schultz *et al.* (1977) noted that a two-layer model would explain some of the peculiar features of their own crater counts. They pointed out that craters effectively formed in bedrock would be eroded and degraded at a slower rate than would a similar sized crater formed wholly in a layer of regolith. This phenomenon, applied to larger craters formed in bedrock and smaller craters formed in regolith, explained certain features of the crater size distributions that were at variance with the simple equilibrium model for crater degradation. In particular, differences in small-crater populations and morphologies between upland megaregoliths, mare deep regoliths, mare shallow regoliths, and floors of fresh craters (e.g., Tycho) are expected.

Chapman *et al.* (1979) have noted some other consequences of the two-layer model not considered by Schultz *et al.* The equilibrium numbers of craters (or craters of a particular morphological class) depend on the ratio of the crater forming rate to the degradation rate. Thus dependence of the crater production rate on surface material is as important to defining the equilibrium population as is the effect on degradation rate considered by Schultz *et al.* (1976, 1977). Well-known scaling relations between impact energy and resulting crater diameter are very different for solid materials (e.g., basalt) and unbonded materials (e.g., sand). Thus the rate of production of craters of a particular size is enhanced in regolith compared with basalt. Moreover, as noted by Oberbeck and Quaide (1968), the initial morphologies of craters having depths the order of the regolith thickness differ from usual fresh crater morphology. Similarly, craters formed in a deep upland megaregolith could degrade faster than craters penetrating a thin mare regolith. In addition to these effects in degradation rate, the *style* of crater degradation differs for craters formed in a solid substrate compared with those formed in the regolith. Together, all these effects appear to Chapman *et al.* to account for the main features of the lunar-crater size/morphology frequency relationships.

Although some puzzling observations have not yet been explained, Chapman *et al.* conclude that, on the whole, the case for predominant impact cratering as the origin of the small lunar crater populations is strengthened by these considerations. Thus, to first order, it is to be expected that the D_L parameter should correlate with the age of a surface. One the other hand, the strong dependence on surface material properties on rates and styles of crater formation and degradation has not been taken into account in the degradation model that forms a part of the D_L dating technique. Thus, the imperfect correlation of the technique with other age indicators is to be expected. Relative ages should be more reliable than absolute ages, especially those young ages extrapolated from the range results in older provinces for which the technique has been calibrated by dating of acquired samples.

A related complexity is Gault's (1970) point that some of the upturn at small diameters involves a change between smaller (strength dominated) crater forms and larger (gravity dominated) crater forms. Interpretations of small-crater "excesses" or comparison with large-crater densities need to take this effect into account.

Improving our knowledge of small craters' histories will refine crater dating techniques. However, the situation is sufficiently complex with so many competing hypotheses that it is not likely the issue will be resolved in the immediate future. If data strongly supported one major hypothesis, then ages might be decipherable. For instance, there are a number of photogeological and statistical procedures that have been proposed for separating secondary and endogenic craters from primaries; if such separation could be accomplished with certainty, the remaining variations due to differences in local geology might be fairly minor so that ages could be obtained.

8.8.4 Problems of comparing small-crater counts by different investigators

A final problem involving small craters involves some discrepancies in crater densities (and associated crater density/age relations) reported by various observers. We have established in the early part of this chapter that counts by different observers of multikilometer craters in given geologic provinces (with boundaries chosen independently by the observers) agree within about 30%. It has thus been surprising that discrepancies of factors of 3 or more have been reported between crater counts by Neukum and his associates (Neukum *et al.*, 1975b; Neukum and Wise, 1976) and others. For example, Shoemaker (1977) discovered that for craters larger than 10 km on a 3.3-b.y.-old lunar mare, the crater densities reported by Neukum *et al.* (1975b) were only 0.3 times the values found by Shoemaker himself. But, as Shoemaker (1977) noted, the result of Neukum *et al.* (1975b):

> "... is not obtained by counting craters 10 km in diameter and larger, however. Instead, they count small craters and infer the number of large craters from a 'calibration' curve of crater size-frequency distribution. ... This procedure predicts substantially fewer craters at $D \geq 10$ km than are actually found on the lunar maria."

A part of the discrepancies in counts reported is due, then, to the conversion required from actual counts at small D to equivalent reported counts expressed at large D. Indeed, Neukum and Wise (1976) find a similar explanation for seemingly gross discrepancies which they show in their plot (their Fig. 1) of lunar crater densities derived from data of Neukum, Baldwin, Hartmann and Soderblom *et al.* These data scatter by as much as factor 10 for the same lunar sites, with Neukum's lowest and Soderblom's highest. The counts are expressed on crater density scales: cumulative craters/km^2 of diameter >1 km, and incremental craters/km^2 between 4 and 10 km diameter. As Neukum and Wise (1976) point out, the explanation again involves the way in which the original data on craters of various sizes were reduced to crater densities in a selected diameter range. In the case of Neukum and Wise's use of Soderblom's data, the situation is even more confused, since the original data of Soderblom *et al.* was a measure of erosional state and was converted by them to crater density, with Neukum and Wise (1976) suggesting an error in their conversion. Neukum and Wise concluded that all these crater counts can be brought into better agreement if the required conversions from one size to another are correctly handled. However, whenever a conversion is made from counts at small diameter to densities at larger diameter, we are at the mercy of the "calibration curve," i.e., the diameter distribution assumed by the converter, not to mention any personal bias in discriminating among primaries, secondaries and endogenic craters.

Neukum (pers. comm., 1978–79), Shoemaker (1977) and Neukum *et al.* (1975b) shed further light on problem of discrepancies by pointing out that a major difference between Neukum *et al.* (1975a, b) and some other authors is in the nature of the craters being counted. Many workers have defined and counted "post-mare" craters as those whose rim crests are intact and sharp, and whose interiors are undisturbed by flows, following definitions established in lunar crater catalogs of the 1960's. Neukum *et al.*, however, have made a more detailed analysis using stereo Apollo photography resolving smaller details. In so doing, they analyzed irregularities in the size distribution which they attributed to craters whose ejecta blankets are overlapped by thin flows. Neukum's counts are thus believed to identify what might be called "post-final-flow(s)" craters while other workers have counted the more numerous "post-late-flow" craters. Both systems are reasonable, with Neukum's analysis determining crater densities since the last flows a few meters (perhaps 10 m) thick, and the other authors determining crater densities since the last flows in the upper tens or hundreds of meters. In this sense, the reported discrepancies are not "real" differences in counts, but are due to errors in equating different types of data. Both sets of crater densities can be correlated with rock ages at lunar sample sites (we are unsure to what depth these rocks sample), and thus used to predict rock ages at other unsampled sites. One must merely be sure to use "type X" crater counts only with the appropriate "type X" calibration curve of crater density vs. age.

We believe this explains much of the discrepancy between the crater density vs. age curves constructed by Neukum and his co-workers, and other workers: different types of crater densities have been plotted on the same graph of density vs. age, and thus give different curves (e.g., Fig. 1 of Neukum and Wise, 1976). Additional and less well resolved issues are: (1) Neukum favors a somewhat lower production rate of craters on Mars than we have adopted here, giving somewhat higher ages, though all workers agree that old ($\gtrsim 3$ b.y.) and young ($\lesssim 1$ b.y.) volcanic surfaces appear on Mars; (2) Neukum's "standard calibration curve," by which he projects from observed densities of sub-kilometer craters to estimated densities of large primary craters, may be somewhat different than would be adopted based on our discussion of the upturn in the diameter distribution at small sizes (section 8.7).

8.8.5 Conclusion

It would be premature to offer a final judgement on the confidence that can be attached to ages for small lunar units determined by the D_L methodology or similar procedures. But the evidence suggests that the technique could be used with fair confidence to separate lunar units of young, medium and old age. Finer age determinations, such as from one mare unit to another one at a distance, are still questionable, but are likely to be strengthened by further work. No comprehensive analyses of small crater populations, similar to the work of Boyce and his colleagues for the lunar maria, are available for Mercury, Mars, or even the highland and backside regions of the Moon, although some work is in progress. There are potential uncertainties about small-crater origin and degradation processes as well, especially for Mars.

The distribution of D_L ages across the lunar maria is shown graphically in Plate 2.6 of this book, where they may be compared with similar plots of various remote-sensing parameters. Recalling that D_L ages are presumably sensitive to the very last, relatively small-scale events in a locality, we can draw some conclusions from the D_L values concerning processes of lunar volcanism on spatial scales smaller than those treated earlier in this chapter. Mare Tranquillitatis and the northernmost mare regions are relatively old while mare units near latitude 30° (especially west of 50° longitude) are relatively young. This agrees with our finding of high crater density (1.3) in Mare Tranquillitatis and low values (0.62, 0.75) in Oceanus Procellarum (see Table 8.6.1). But individual flows on most maria have ages ranging from about 3 b.y.–3.8 b.y., a range greater than differences between the *mean* ages of the maria and other large provinces. Thus, the final stages of basaltic flooding over the front surface of the Moon occurred during an extended period of well over half a billion years during which thin flows formed the uppermost surfaces of the maria. This extended "dribble" of activity contrasts with what must have been a vastly greater rate of volcanism required during the previous 1 to 2×10^8 yr for filling the large basins with immense volumes of basalt, mostly now hidden far below the present mare surfaces. We urge further comparisons of D_L, C_S, crater densities at various D's, remote-sensing data, and stratigraphic-morphological information on various planetary surfaces to improve the understanding of small craters.

We also find that in spite of some methodological differences between use of small craters by Neukum and other workers, the actual numerical ages (estimated by using lunar crater-density/age calibrations appropriate to the actual type of crater density defined in the counting procedure) assigned to provinces in the solar system are reasonably compatible, and in the case of Mars, span the entire history.

8.9 THE PROBLEM OF UPLAND OR "PREMARE" VOLCANISM

The ideal contribution of this chapter would be to sketch the behavior of total volcanic eruptive activity (e.g., cubic km of lava/yr) on each planet. The overall history of this eruptive activity, rather than inter-mare or intra-mare details, constrains thermal models, for example. But, so far, we have determined only ages of certain surface flows. A histogram of the ages of surface flows would not indicate the history of volcanism on the planet because these flows probably cover a sequence of still earlier flows. Also, some older flows may be harder to detect than younger ones. How can we convert our cratering data and estimated chronology into the more complete picture of total volcanic eruptive activity through time?

We have presented crater counts on some early mare-like units in basins in the lunar uplands and on some inter-crater upland plains, as well as on ordinary mare units. After Apollo 16 showed that one inter-crater upland plain is a breccia unit, many researchers assumed that all upland plains units are basin ejecta, primarily from Imbrium and Orientale. However, Neukum (1977) pointed to the wide range of crater densities and ages of those plains, including values less than those for Orientale, and Neukum reaffirmed that some, especially those with more mare-like morphology, must be upland volcanic units. This section will now develop a methodology for using crater counts on these units to construct a long-term volcanic record. Since the rock dates, probable depth of lava flows, and appropriate crater statistics are best known for the Moon, we will develop this model for the Moon. The methodology described, however, is applicable in principle to all the planets. To establish the eruptive chronology, we will try to estimate the amount of lava being generated on the entire Moon in each 10^8 yr period.

Let us start with the total volumes of dark mare basalts occupying circular basins. These volumes of lava had to form in the interval between the date of basin formation and the date of the last surficial flows. The latter dates have been derived above and calibrated with known rock ages. Hartmann and Wood (1971) have counted craters on the ejecta blankets and rims of various basins and determined relative ages of the underlying basins in crater density units similar to those used here. They have also tabulated the estimated volumes of lava in several of these basins, based on depths of protruding crater rims, and have shown that these correlate fairly well with measured mascon masses. DeHon and Waskom (1976) have similarly derived depths and volumes of lavas in various basins, but find shallower depths than estimated by Hartmann and Wood. For four maria studied by both groups, the Hartmann/Wood volumes average five times that of DeHon/Waskom—a substantial discrepancy. The latter authors measured depths indicated by flooded craters with rims barely protruding; the lavas might be deeper if these craters themselves were formed on early lava flows on the basin floors. Our argument here for intense early volcanism is not qualitatively affected by the discrepancy in volume estimates. Indeed, it is strengthened if the mare volumes are reduced, relative to the volumes of earlier eruptives.

We now make the assumption that the eruption of this volume of lava (i.e., using the larger values given by Hartmann/Wood) for each basin was distributed uniformly in the time interval T, between the *basin formation* age and *surface lava flow* age. The results are shown in Table 8.9.1, which tabulates the basin age (assumed beginning of eruption), the surface mare age (assumed end of eruption), the lava volume and the resulting lava eruption rate for that particular mare system. Table 8.9.1 is not inclusive of every last bit of mare on the Moon, but is sufficient to establish the method in broad outline.

Next, we add in the obvious major flows that do not occupy circular basins but rather broad irregular depressions. These include Oceanus Procellarum, Mare Tranquillitatis and Mare Fecunditatis. Here we have no empirical data on the underlying floor age (assumed

Table 8.9.1 Volumetric flooding history of the Moon.

Basin or Province	Adopted Basin Age[(a)] (beginning of eruptions) (10^9 yr)	Adopted Lava Surface Age[(b)] (end of eruptions) (10^9 yr)	ΔT (10^8 yr)	Estimated Volume (10^5 km^3)	Eruption Rate During Activity (km^3/10^3 yr)
Basin:					
Imbrium	3.9	3.1	8	30.0	3.8
Smythii	4.2	3.8	4	5.6	1.4
Serenitatis	4.1	3.2	9	13.2	1.4
Orientale	3.9	3.6	3	2.6	0.9
Humorum	4.2	3.7	5	2.7	0.5
Nectaris	4.0	3.4	6	3.1	0.5
Crisium	4.1	3.1	10	5.6	0.6
7° S 70° E	4.2	3.5?	7	1.2	0.2
Humboldtianum	4.1	3.5	6	1.0	0.2
Grimaldi	4.1	3.2	9	0.4	0.0
Irregular maria:					
Oceanus Procellarum	4.3?	3.2	11	31.0	2.8
Tranquillitatis	4.3?	3.7	6	8.8	1.5
Fecunditatis	4.3?	3.5	8	5.6	0.7
Basins with old light "mare":					
Mendeleev	4.3?	4.1	2	1.4	0.7
Schrödinger	4.1	4.0	1	1.0	1.0
Near Schiller	4.3	4.0	3	0.4	0.1
Minimum upland flooding model: [(c)]					
Upland flooding	4.4	4.0	4	20	5.0
More plausible upland flooding model: [(d)]					
Upland flooding	4.4	4.1	3	82	27.3

[(a)]These data based on Hartmann and Wood (1971, Table VII), where results were correlated with mascon masses, and on Imbrium, Serenitatis and Nectaris ages suggested by P. Eberhardt (pers. comm., 1978). Also based on areas studied here. Typical lava depth is assumed to be 1 to 3 km, consistent with Hartmann and Wood (1971) and section 5.3.7, which suggests major mare lava expanses are <2 km thick, but approach 6–8 km in their centers. [(b)]This paper. [(c)]1/4 of uplands flooded 1/4 km deep 4.4 to 4.0 b.y. age. [(d)]1/2 of uplands flooded 1/2 km deep 4.4 to 4.1 b.y. ago.

beginning of eruptions), but we assume a date of 4.3 b.y. ago on the basis that these depressed areas were dotted by craters and cut by the rims of later basins such as the Carpathian Rim of Mare Imbrium. The eruptive activity is thus distributed uniformly in time between 4.3 b.y. and the age of the surface mare.

As can be seen in Fig. 8.9.1 (and as might be expected), the eruptive history of these classical dark mare units shows a broad peak between 4 and 3 b.y. ago. Some workers have assumed that an actual distinct surge of mare volcanism occurred. Little evidence has been found for younger rocks; any classical mare units and basins have apparently not survived in good condition from before about 4 b.y. ago.

However, as shown in Table 8.9.1, we now have firm evidence that a few basins were formed in upland areas and were flooded with mare-like material in what would normally be called "premare" time. These times correspond to crater densities above about four, relative to the average lunar maria. Are these old basin-filling materials like classical mare basalts? They have not been sampled by Apollo or Luna flights. The present surfaces have distinctly higher albedo than the classical maria, presumably either because of different chemical composition from the mare basalts, or, more likely, because of thin mantles of overlapping ray material produced by the high crater density. Hartmann and Wood (1971), who discussed this, find that while albedo extremes occur generally among the younger materials, all materials with crater density greater than about four have been reduced to an intermediate albedo characteristic of the lunar uplands, because of cumulative effects of overlapping ejecta and rays. The evident flooding of these old basins may thus very well be by ordinary mare basalts, simply disguised by a veneer of rays (see further discussion in sections 5.4.5 and 5.4.6.) Schultz and Spudis (1979) have added strong support for this by cataloging a sample of more than a hundred dark ray upland craters, which have apparently penetrated the light veneer and thrown out dark underlying material. One of the best examples lies in the old mare-like surface (crater density seven times lunar average) that floods the old multi-ring basin near Schiller shown in Fig. 8.7.4. Figure 8.7.5 shows another old basin with a damaged interior fill that appears slightly younger than the basin surroundings, and may be volcanic. Remote sensing of any basalts or other lavas with ages greater than about 3.9–4 b.y. may thus be extremely difficult since these materials may be ordinary mare basalts with substantially altered surface properties (see section 2.2.1). We lack sufficient crater statistics and depth estimates to include all probable examples from such early flooded basins in the lunar uplands, but we have included three of the best established ones in Table 8.9.1 and Fig. 8.9.1. The important point to notice is that since the flooding in these "premare" basins is constrained to a short period earlier than 4 b.y., its inclusion in Fig. 8.9.1 pushes the peak eruptive activity toward an earlier time, bending the envelope upwards to the left. This establishes schematically how the addition of small amounts of eruptive activity in "premare" time can dramatically alter the eruptive chronology curve in Fig. 8.9.1 from a model that peaks between 4 and 3 b.y. ago to a model with a rapid rate of decline in earlier times. The consequences are extremely important for thermal history models.

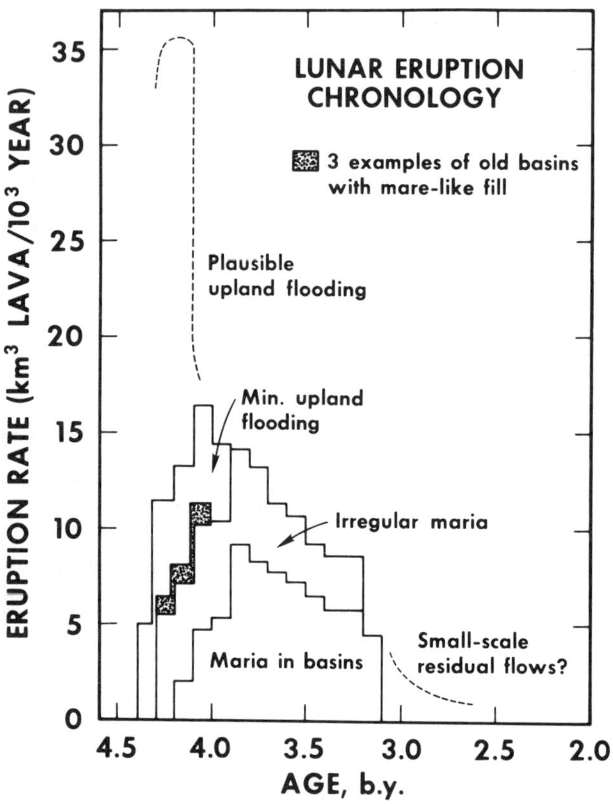

8.9.1 A reconstruction of lunar eruptive history. The classical circular maria in basin interiors, as well as classical irregular maria (volumes estimated in Table 8.9.1) define an eruptive period peaking about 3.8 b.y. ago. but addition of mare-like fill in well-documented older basins shifts activity to an earlier average time. Inclusion of minimum or plausible amounts of upland flooding, based on upland morphologies and crater counts, suggests a period of intense early eruptive volcanism declining throughout the mare period.

We thus turn, finally, to the probable lunar intercrater volcanic plains. Evidence for such flooding exists not only on the Moon, but also on Mars in eroded form, and, perhaps more dramatically, on Mercury. In large areas, for example, the lunar frontside cratered plains near Maurolycus (Province 25 in Table 8.2.2), the implication is that the final morphology seen today resulted from a competition between intense cratering and intense or sporadic volcanic flooding. The flooding had the effect of destroying the smaller craters as well as degrading some of the larger craters. Each flooding event was followed by rapid build-up of new populations of craters. Much of the evidence of the flooding was almost wiped out by cratering and was completely wiped out in some areas. Since about 85% of the Moon's surface is upland (i.e., not classical dark maria), inclusion of even small fractional areas of presumed volcanics strongly affects Fig. 8.9.1. With a minimal case, we have assumed that one-fourth of the uplands has experienced flooding to a depth of 0.25 km (a lava depth only about 0.1 that assumed for many of the classical mare-filled basins). We assume that this flooding began about 4.4 b.y. ago (the formation of the anorthositic crustal layers) and had to end by no later than 4.0 b.y. ago (a date constrained by their high crater density). The addition of even this small amount of presumed volcanic material transforms the curve in Fig. 8.9.1 from one with a peak 3.8 b.y. ago to one with a peak about 4.1 b.y. ago, smoothly declining in volcanic activity since the time represented by the earliest rock units on the Moon. A still more plausible model would call for half the lunar uplands being flooded to a depth of 0.5 km (still less depth than assumed for the mare basins) between 4.4 and 4.1 b.y. ago. In this case, as seen in the dotted line in Fig. 8.9.1, the curve is now radically transformed to show a chronology of volcanic activity exponentially declining since 4.4 b.y. ago. In summary, we find that there is evidence for volcanic activity prior to the ages of the classic dark maria and that this activity probably represents a smooth, steep decline in volcanism since the very early history of the Moon.

If there was actually a smooth decline in volcanic activity, why, then, do we seem to see two distinct types of volcanism—the light-colored upland plains, and the distinctively dark classical mare units? Two different rock chemistries is a possible, but not required, answer. As has been emphasized by Soderblom *et al.* (1974), Hartmann and Wood (1971), and others, the answer may be simply a result of the time behavior of cratering. As shown in Fig. 8.6.1., a lunar surface formed earlier than about 4 b.y. ago will have a relatively densely cratered surface, with many times the crater density of the lunar maria. These crater densities, with accompanying ejecta blankets, swarms of secondary impact pits and ray materials, are sufficient to seriously disturb the otherwise flat, dark mare morphology. At crater densities below about twice that of the average lunar maria, corresponding to about 3.8 b.y. ago, this ceases to be true. Therefore, the narrow 200 m.y. time interval from about 4 to about 3.8 b.y. ago corresponds to an important transition from the damaged high crater density surfaces to the relatively low crater density surfaces. Lava flow structures formed after about 3.8 b.y. ago retain better their original morphology and albedo characteristics. The basic reason that this time interval is so short is that the crater production rate was declining so steeply during the early history of the planets. Because it would be a rare planet that would have most of its surface formed in the narrow critical time interval, planetary surfaces tend toward bifurcation into very heavily cratered, rough "uplands" (formed >4 b.y. ago) and smooth, generally darker, lava plains (formed <3.8 b.y. ago).

In summary, there is evidence for extensive volcanism occurring well before the more obvious basaltic plains formed. This could have included basaltic eruptions with composition similar to mare basalts but modified in appearance by later cratering, or it could have included lavas of different composition. The evidence suggests that rather than having a particular pulse of mare volcanism, the Moon and other terrestrial planets began with a period of intensive volcanism (as early as 4.2–4.4 b.y. ago), which declined relatively smoothly with time.

8.10 SUMMARY AND CONCLUSIONS

All eleven planetary bodies whose surfaces have been closely studied so far (with the probable exception of overly active Io) have craters interpreted as meteorite impact craters, and these can give information on the level and chronology of geologic activity on these bodies. In the case of Io, the absence of meteorite impact craters was used correctly to infer a high level of volcanic resurfacing activity in the few hours between Voyager encounter photography and discovery of active volcanoes on Io.

We conclude that comparisons of densities of multikilometer craters on the various planets give, with the least number of assumptions, information on the gross volcanic history of each planet.

More detailed information on local age differences, age correlations with local units defined by remote sensing (Chapter 2), and age correlations with local features revealed by high resolution morphological studies (Chapter 5) require counting and examination of morphology of smaller craters of subkilometer scale. These techniques hold great promise for more detailed geologic studies, but are still in developmental stages. Among the problems still to be addressed in this field are: (1) differences in interpretation of origin of small craters and selection of primary impact craters among different workers; (2) differences in "standard" size distributions used to convert from observed numbers at small D to reported numbers at larger D; (3) further demonstrations of correlations between age parameters such as small-crater densities, large-crater densities, D_L, C_S, sample rock ages (Chapter 7), stratigraphic data, and remote-sensing compositional data; and (4) better understanding of the effective depths being sampled by rock specimens. Good correlations of D_L and C_S parameters with large-crater density are found if lunar units with a wide age spread are considered, but much poorer correlations are found among lunar maria alone, with their narrower age spread.

Starting with raw counts of multikilometer craters from various authors, we have critically reviewed and applied the steps necessary to derive relative and absolute ages. Data on 53 provinces are presented and discussed, with 25 of the provinces having data from more than one author. Good agreement among authors is found, even though different authors defined boundaries of provinces slightly differently and commonly counted craters from different photos within these boundaries. Relative crater densities from such data can be derived within about 20% standard error.

Table 8.3.2 gives the derived relative crater densities for the 53 provinces, and Table 8.6.1 the inferred absolute ages. The absolute ages depend primarily on the determination of crater production rates for the different planets, developed in section 8.5 and summarized in Table 8.5.5.

Due to the nature of the history of crater production rates in the solar system (a high rate prior to 4 b.y. ago, a rapid decline continuing to 4–3.5 b.y. ago, and a more constant rate since), planets' volcanic units tend to be highly damaged by cratering if formed before 4 b.y. ago, but relatively well-preserved if formed since about 3.8 b.y. ago. Due to the crater production rate and its uncertainties, geologic provinces older than about 3.8 b.y. or younger than about 1 b.y. can be relatively well dated, in the sense of being placed within their appropriate quarter of the planets' history with some confidence; this is a useful constraint for understanding thermal histories of planets, as discussed in Chapter 9. Unfortunately, due again to the cratering rate history and its uncertainties, provinces dated in mid-history (about 2 b.y. ago) have an uncertainty that might place them in any quarter of geologic history, making the interpretation of some important units difficult.

The largest source of uncertainty in absolute dates applies to planets outside the Earth-Moon system and comes from the need to convert observed Earth/Moon crater production rates through theoretical arguments to rates on other planets. Future cataloging of comets, Apollo and Amor asteroids, or future sampling of even a few volcanic provinces with different crater densities on other planets, could greatly reduce the current uncertainty and thus allow calibration of planet-wide chronologies by the crater-counting technique.

An ideal product of this study would be a diagram similar to Fig. 8.9.1 for each of the planets, comparing the chronologies of volcanic activity. Unfortunately, the data required to construct such diagrams are difficult to acquire. Because Mars, Earth and uplands on Mercury lack the simple basin-filling geometry of much lunar volcanism, depths and volumes of their volcanics cannot be estimated as simply as in Table 8.6.1. Also, volumes of volcanics on Earth are complicated by the sedimentary record, while ages on Mars remain uncertain. Instead, a different approach has been taken, leading to the comparison in Fig. 8.10.1. Instead of comparing the volumetric rate of activity, as in Fig. 8.9.1, Fig. 8.10.1 compares simply the percentage of surface area covered by volcanic units of various measured ages. The data from the Moon, Mars and Earth, though not precisely the same, allow some meaningful comparison. For the Moon, the areas of volcanic provinces studied here were compared as percentages of the area of the whole Moon. All of the classical dark maria together comprise only about 18% of the Moon's area; the sample of maria on which craters were counted here is somewhat less— about 11%. To this was added our minimal estimate of upland surface covered by ancient volcanic plains, about one-fourth of the uplands or 21% of the Moon placed in the 4–4.5 b.y. bin. The dashed line indicates the higher plausible amount of upland volcanic area, about one-half of the uplands. Of course, if the entire uplands are considered a solidified crust of a primeval igneous magma ocean, the dashed line would be even higher. In any case, the data show that lunar volcanism was strongly concentrated toward early lunar history.

In the case of Mars, a similar procedure was carried out, except that three histograms of surface area vs. age were constructed, one each for the most likely time

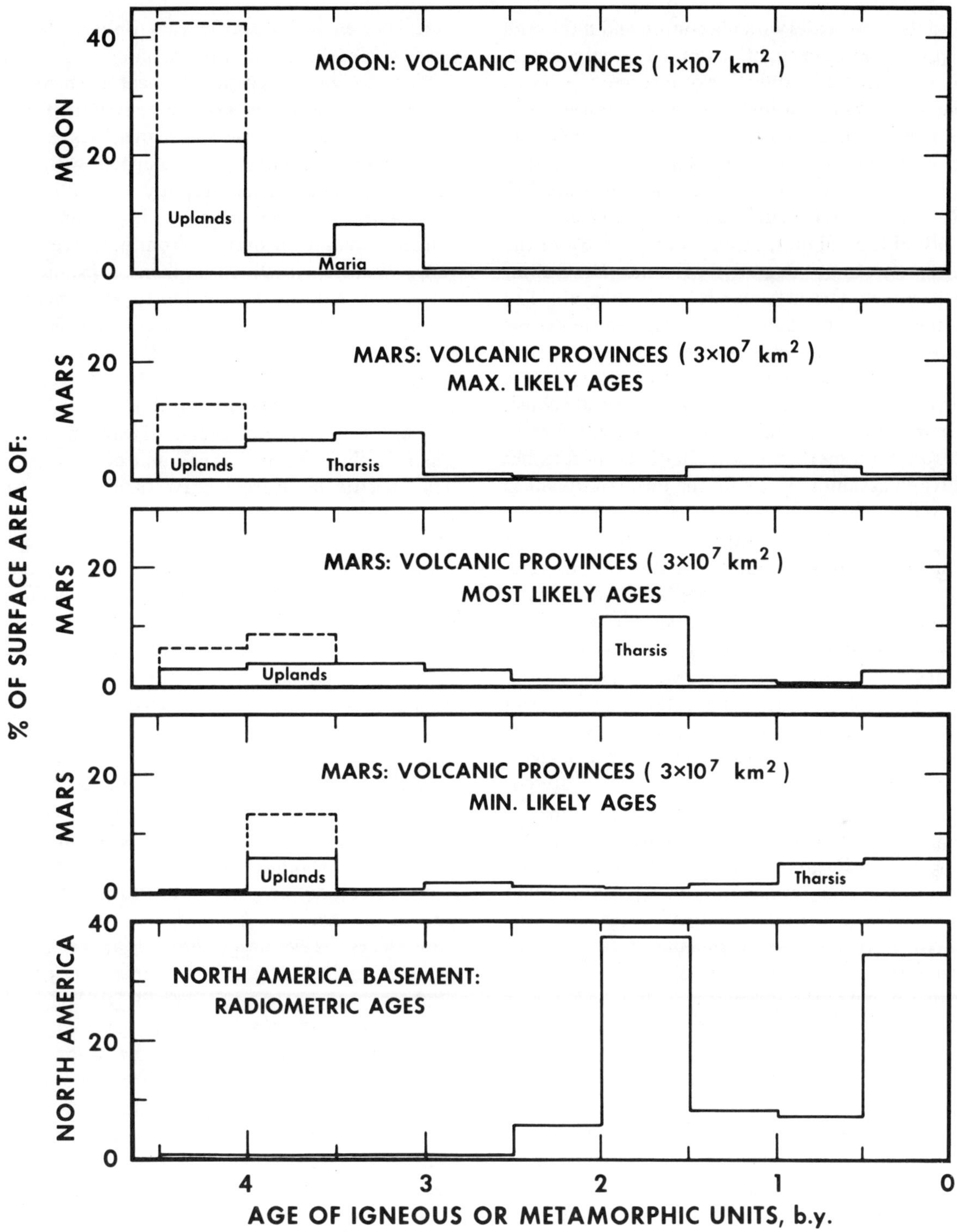

8.10.1 Comparable measures of the time distribution of igneous resurfacing on three planets: histograms of ages of volcanic units vs. percent of surface area covered by the volcanic provinces studied. See text for description of volcanic surface units included here. The graphs demonstrate that (a) any reasonable time scale for Mars has volcanic provinces spread through martian history, and (b) this places Mars between Earth and Moon in the time profile of its volcanic surface units. Note that unlike Fig. 8.9.1, this does not give the time profile of all volcanic activity on the planet since underlying volcanic units are not represented. The graphs show the age distributions of the last, still-visible surfaces created by volcanic or igneous activity.

scale, and the time scales with maximum and minimum likely ages. Areas were again computed as percentages of the whole planet. (Sediment-covered regions, such as the polar areas, were not included in the sample.) The sample of volcanic provinces studied here, including sparsely cratered plains such as Tharsis and Elysium, shields, and older plains such as Syrtis Major comprises about 15% of the planet's surface. A minimally assumed one-eighth of the uplands, or about another 6% of the planet is assumed to be degraded volcanic plains, comprising the boxes marked "uplands" in the diagram. The dashed line shows the result of including a more plausible one-fourth of the uplands as volcanics, analogous to the procedure in the lunar study (Table 8.6.1). These diagrams show that for any plausible time scale considered, martian volcanic units were being created at all times throughout martian history, a conclusion rather insensitive to assumptions about the martian uplands.

To obtain a comparable type of diagram for Earth, a terrestrial province was chosen with well-determined radiometric dates for igneous and metamorphic basement rocks, which might be viewed as the age distribution that would be found without sedimentation effects which complicate the record. An age-contoured map of ancestral North American provinces, based on radiometric dates of basement rocks (Seyfert and Sirkin, 1979), was used to derive percentages of the North American continent's basement "deformed, metamorphosed, and/or extensively intruded" at different times. While not derived from crater-dated provinces in the same way as the lunar and martian histograms, this terrestrial histogram gives some indication of the distribution of the last major igneous activity in the broad provinces defining the sub-sediment basement "surface" of North America. As shown quantitatively by the histogram, and as is qualitatively well-known, terrestrial surface units' ages are concentrated in the last part of geologic history because erosion and tectonism have obliterated earlier surfaces.

The diagrams show, then, not only the well-known fact that the Moon's exposed volcanism is old and Earth's exposed volcanism is young, but also that Mars' volcanism is more spread through time, and this places Mars clearly in an intermediate position between the lunar and terrestrial volcanic chronologies.

The present results indicate that comparable levels of volcanism, as measured by planetary surface fractions affected, either occurred later or persisted longer on more massive planets than on less massive planets. [Io does not fit this sequence because it is being heated by ongoing tidal flexing, which maintains the volcanism predicted by Peale *et al.* (1979) and subsequently discovered by Voyager I.] We also find evidence of extensive early volcanism that obliterated primitive craters, creating upland plains units and flat floors in ancient basins, which in turn have been nearly obliterated by subsequent cratering. If both findings are correct, they suggest that all planets began with intensive volcanism (or at least developed it within a few 10^8 yr), which then declined rapidly on small planets and more slowly on larger planets. This supports models of volcanic evolution derived from petrologic evolution (e.g., Walker, 1979). Hypothetical primeval intense terrestrial volcanism, similar to that proposed for Moon and Mars, has long since been obliterated by erosion and crustal unrest. Massive outpourings of large-scale mare-like flows ceased by about 2–3 b.y. ago in the case of the Moon and Mercury and by about 1 b.y. ago in the case of Mars. Volcanism continues on Earth in the form of plateau and ocean floor basalts. The specific landforms differ from planet to planet (mare plains, shield volcanoes, etc.) due to differences in mantle sources and consequent compositions, viscosities and eruption conditions on different planets. The most recent minor eruptions on each planet are a matter of conjecture. Small eruptions may have occurred recently in or near fresh impact craters on the Moon (and Mercury?) as impact-induced activity. Some activity associated with isolated martian shield volcanoes almost certainly occurred within the last 10^9 yr and may continue today. Local activity continues frequently on the Earth because of plate-tectonic disruption of the crust, presumably due to mantle convective activity.

In this study, we have not attempted to deal with Venus, because radar mapping, while revealing possible impact craters, has not yet discriminated geologic provinces that can be assigned to volcanic units. Nonetheless, evidence is good that Venus has such provinces, that they will be subject to crater-count analysis, and that they fit into the planetary sequence mentioned above. Soviet analysts have concluded that rocks at the Venera 9 and 10 sites are probably "erupted basalts," while rocks at the Venera 8 site are more granitic, based on measured U, Th and K abundances (Keldysh, 1977). Pioneer Venus radar maps indicate continent-like raised areas. These results suggest both basaltic flow activity of the type we have studied on other planets, and also a certain degree of differentiation. Soviet researchers also suggest that rocks photographed by Venera 9 and 10 indicate processes of rock erosion and fresh rock production. Finally, radar mapping has indicated that Venus has many regions about as heavily cratered ($D \gtrsim 60$ km) as the lunar uplands (Masursky *et al.*, 1979), though crater infilling and others features are indicated. Saunders and Malin (1977) interpret radar data on a large elevated feature near the Venera basaltic sites as a large volcanic construct with a summit caldera, and various other volcanic and tectonic features have been suggested (Pettingill *et al.*, 1979; Masursky *et al.*, 1979). "A great rift valley" (Saunders and Malin, 1977; Masursky *et al.*, 1979) and other possible signs of incipient plate tectonics have been found. The high proposed crater densities, however, show that Venus is not as geologically active as Earth. It may have a thicker lithosphere, shielding its surface from Earth-like tectonic turmoil. In any case, the relation of crater densities and tectonic features on Venus will be an important indicator for future analysis of volcanic activity on terrestrial planets.

In summary, crater chronologies combined with other studies such as radioisotopic dating of samples, photogeologic analysis, and theoretical thermal modelling not only reveal approximate age relationships among volcanic provinces, but also suggest histories of volcanism stretching back into the early era when surface evidence was nearly obliterated by intensive cratering. Though we are at an intermediate stage in planetary exploration, we find evidence of a sequential relation between planetary mass and the evolutionary pattern of volcanism on terrestrial planetary bodies.

8.11 APPENDIX

PLOTS OF CRATER COUNT DATA FOR 53 PLANETARY PROVINCES

This appendix includes plots of all data used in discussing crater chronologies of planetary surfaces in Chapter 8. Two formats are presented for each province. Format A plots the observed crater density in each $\sqrt{2}$ log diameter increment (1, 1.4, 2, 2.8, 4...km). Format B is a differential plot giving the difference between the observed crater density and that for a power law of form $N = k\ D^{-2}$, where N is the log incremental crater density and D is the diameter. A power-law distribution is a straight line in each plot; one with a -2 exponent is a horizontal line in format B. Format B emphasizes curvature or other departures from power laws. Plots in format A include all data points available to this study; plots in format B include just the data points used in least-square curve fitting described in the text.

Three reference curves appear in each plot. The solid line is the main reference curve, giving the fit to counts averaged over lunar maria on the Moon's front side. The middle dashed line is its extrapolation to small diameter, for purposes of reference (actual lunar mare counts curve above this dashed line as shown by dots in Figs. 1A, 2A and 3A). Upper and lower dashed curves are 10 times and 0.1 times the lunar mare reference curve, respectively, to serve as references in estimating relative crater densities in different planetary provinces.

A

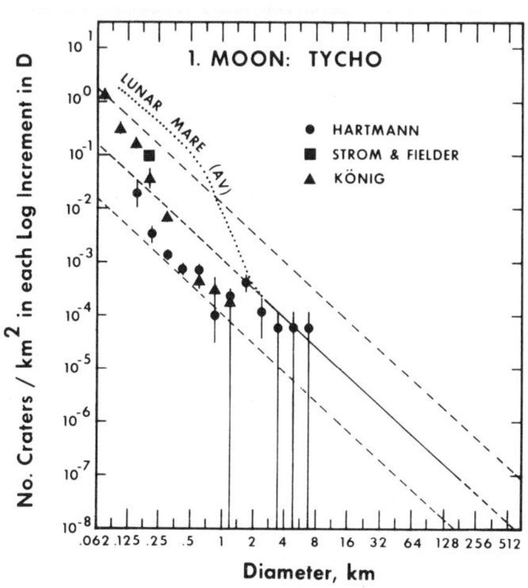

B

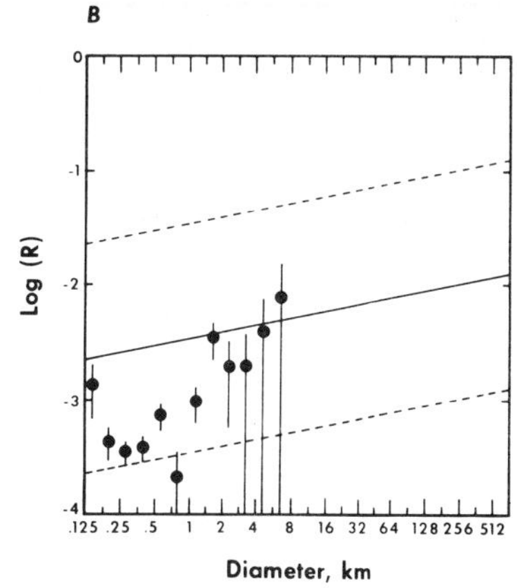

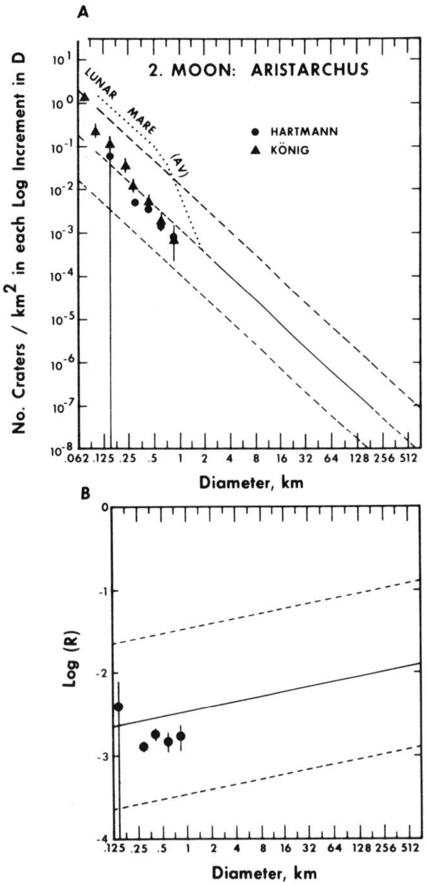
A
2. MOON: ARISTARCHUS
LUNAR MARE
(AV)
HARTMANN
KÖNIG
No. Craters / km² in each Log Increment in D
Diameter, km
B
Log (R)
Diameter, km

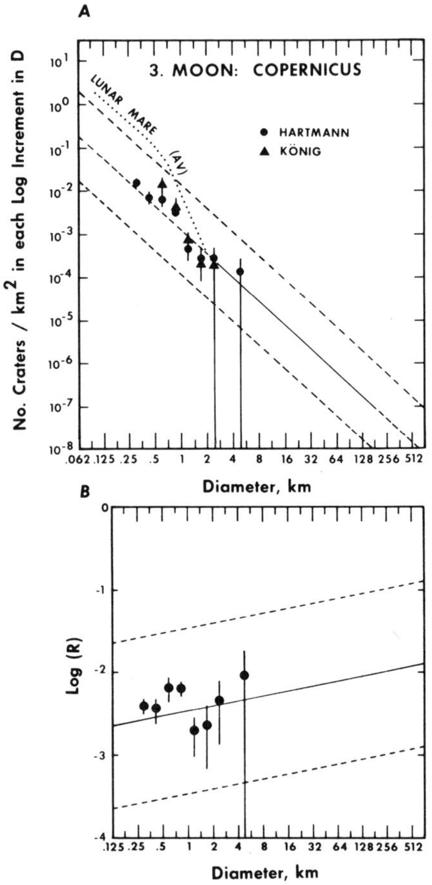
A
3. MOON: COPERNICUS
LUNAR MARE
(AV)
HARTMANN
KÖNIG
No. Craters / km² in each Log Increment in D
Diameter, km
B
Log (R)
Diameter, km

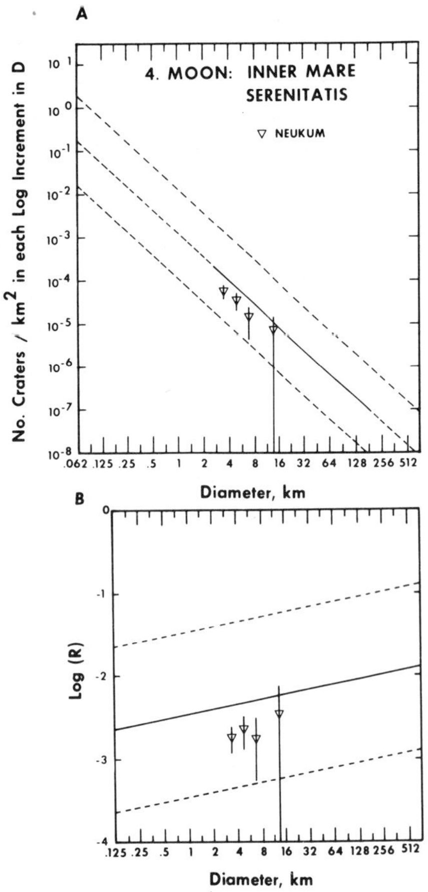
A
4. MOON: INNER MARE SERENITATIS
NEUKUM
No. Craters / km² in each Log Increment in D
Diameter, km
B
Log (R)
Diameter, km

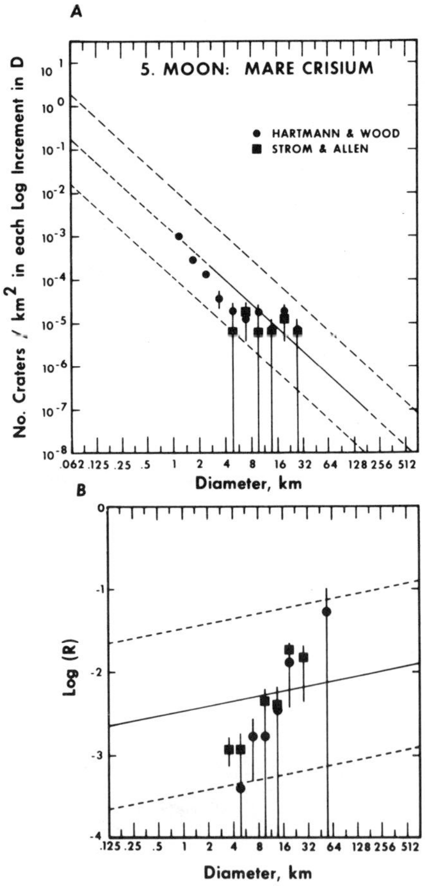
A
5. MOON: MARE CRISIUM
HARTMANN & WOOD
STROM & ALLEN
No. Craters / km² in each Log Increment in D
Diameter, km
B
Log (R)
Diameter, km

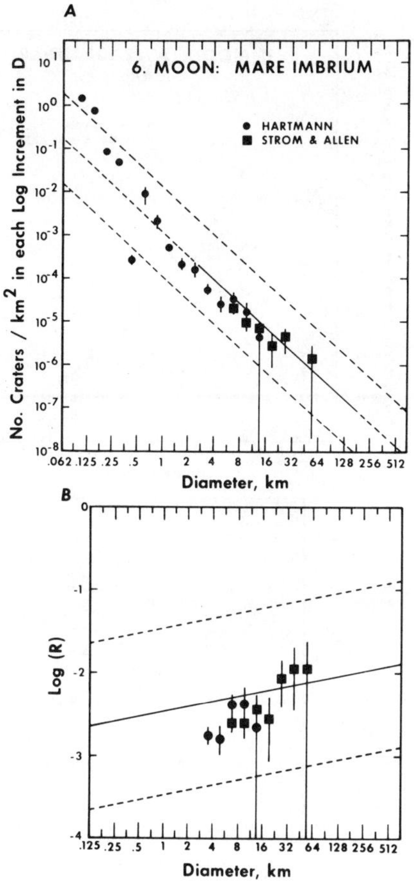
A
6. MOON: MARE IMBRIUM
HARTMANN
STROM & ALLEN
No. Craters / km2 in each Log Increment in D
Diameter, km
B
Log (R)
Diameter, km

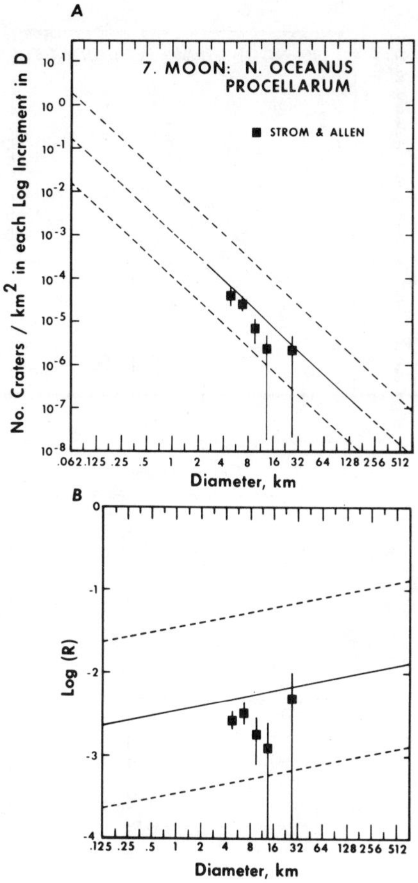
A
7. MOON: N. OCEANUS PROCELLARUM
STROM & ALLEN
No. Craters / km2 in each Log Increment in D
Diameter, km
B
Log (R)
Diameter, km

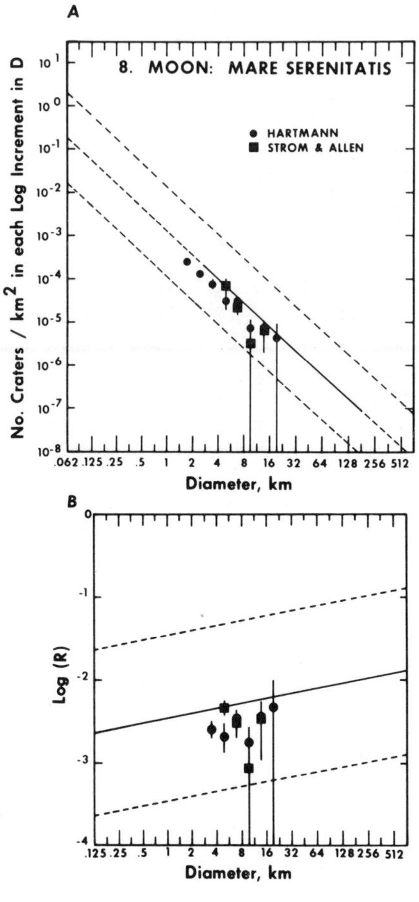
A
8. MOON: MARE SERENITATIS
HARTMANN
STROM & ALLEN
No. Craters / km2 in each Log Increment in D
Diameter, km
B
Log (R)
Diameter, km

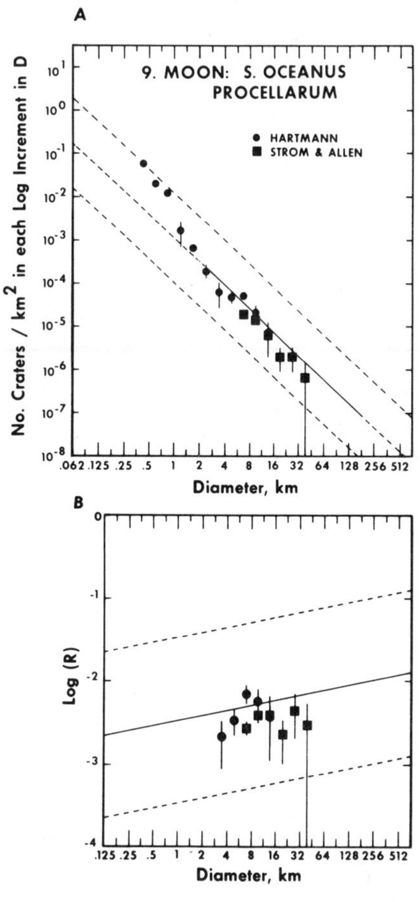
A
9. MOON: S. OCEANUS PROCELLARUM
HARTMANN
STROM & ALLEN
No. Craters / km2 in each Log Increment in D
Diameter, km
B
Log (R)
Diameter, km

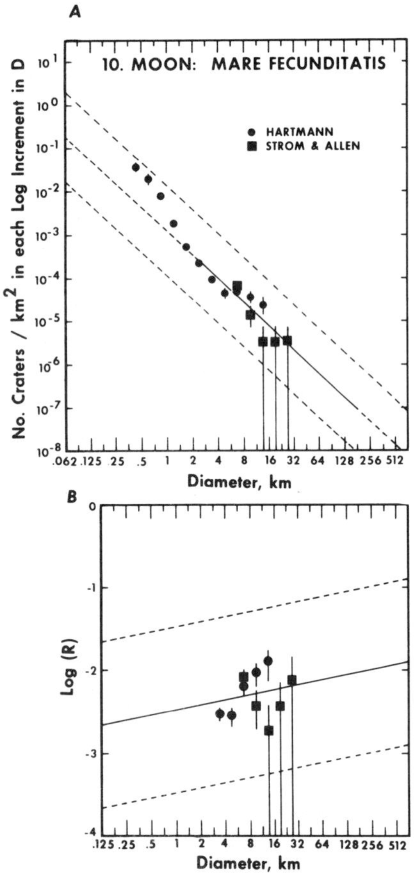
A
10. MOON: MARE FECUNDITATIS
HARTMANN
STROM & ALLEN
No. Craters / km² in each Log Increment in D
Diameter, km
B
Log (R)
Diameter, km

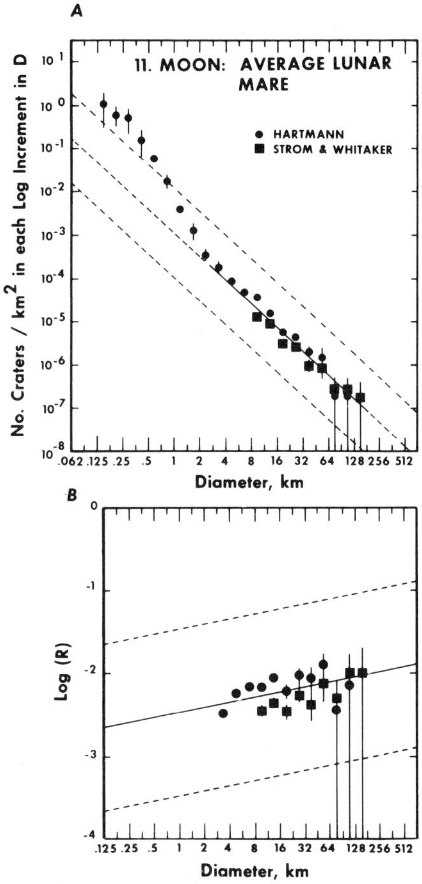
A
11. MOON: AVERAGE LUNAR MARE
HARTMANN
STROM & WHITAKER
No. Craters / km² in each Log Increment in D
Diameter, km
B
Log (R)
Diameter, km

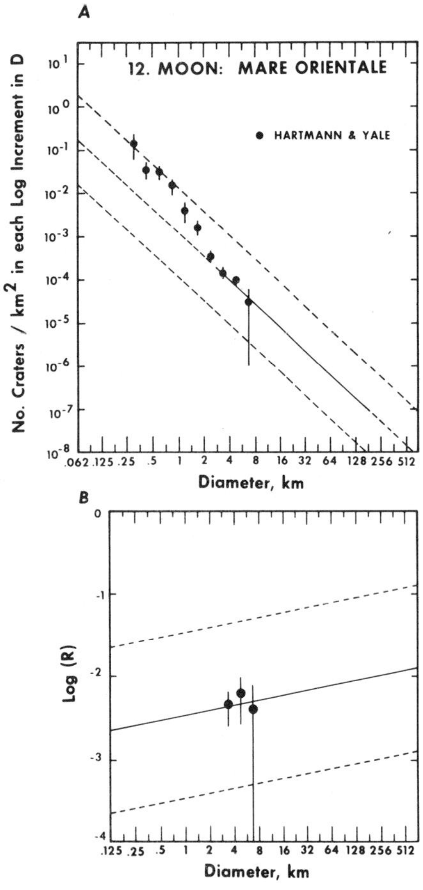
A
12. MOON: MARE ORIENTALE
HARTMANN & YALE
No. Craters / km² in each Log Increment in D
Diameter, km
B
Log (R)
Diameter, km

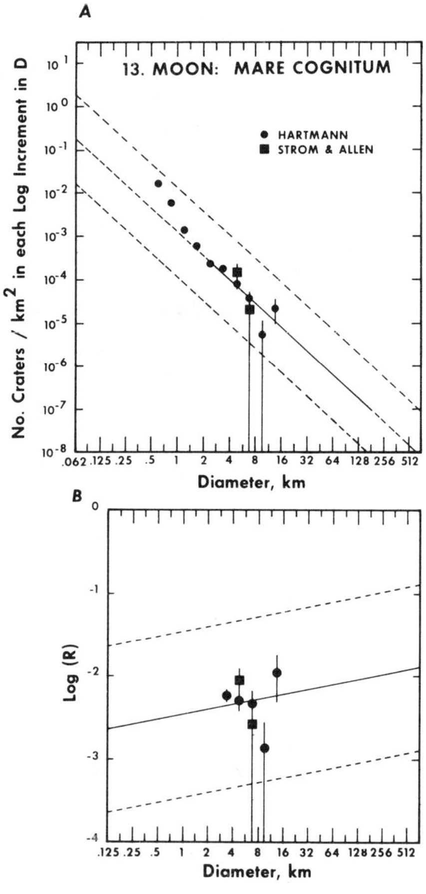
A
13. MOON: MARE COGNITUM
HARTMANN
STROM & ALLEN
No. Craters / km² in each Log Increment in D
Diameter, km
B
Log (R)
Diameter, km

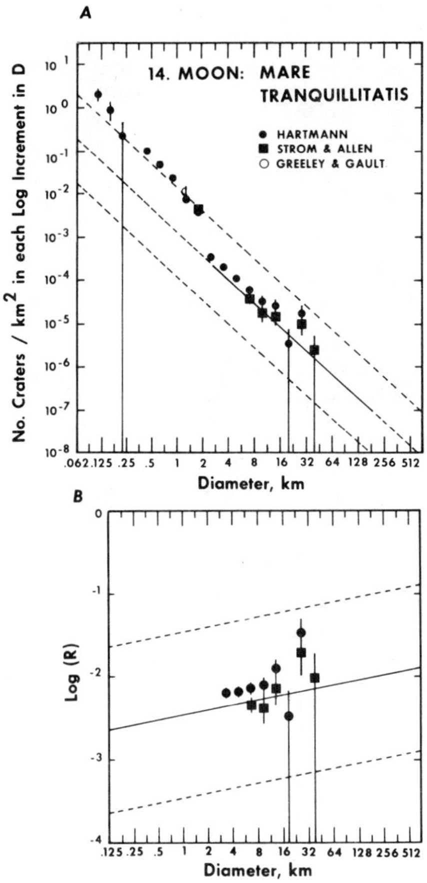
A
14. MOON: MARE TRANQUILLITATIS
HARTMANN
STROM & ALLEN
GREELEY & GAULT
No. Craters / km² in each Log Increment in D
Diameter, km
B
Log (R)
Diameter, km

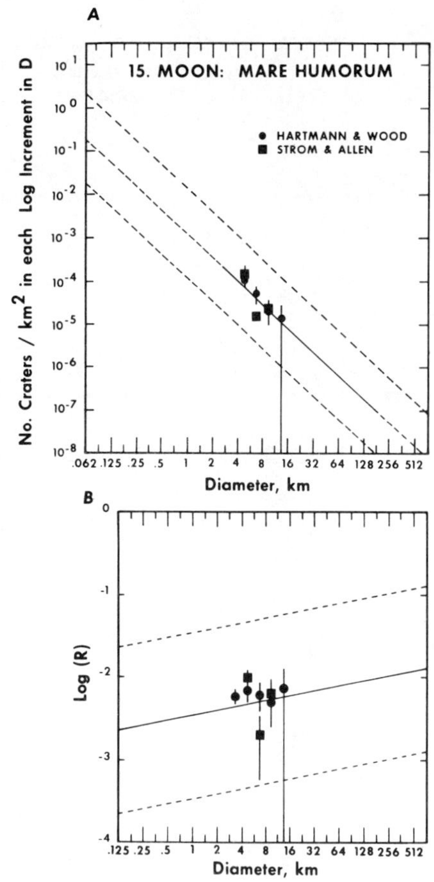
A
15. MOON: MARE HUMORUM
HARTMANN & WOOD
STROM & ALLEN
No. Craters / km² in each Log Increment in D
Diameter, km
B
Log (R)
Diameter, km

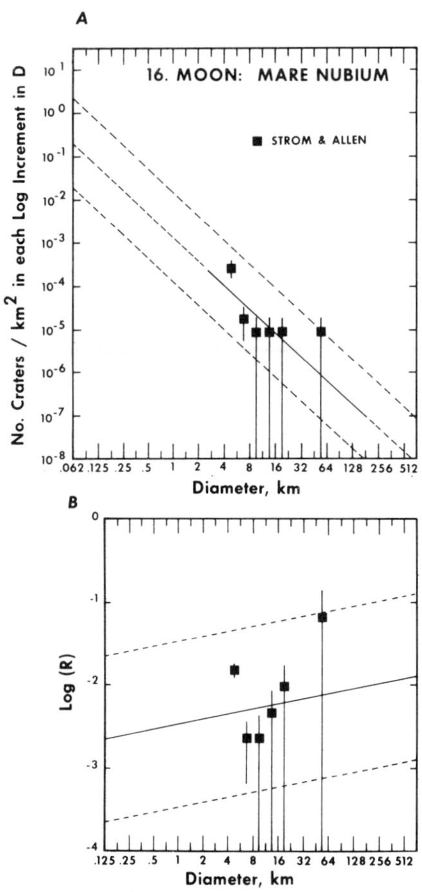
A
16. MOON: MARE NUBIUM
STROM & ALLEN
No. Craters / km² in each Log Increment in D
Diameter, km
B
Log (R)
Diameter, km

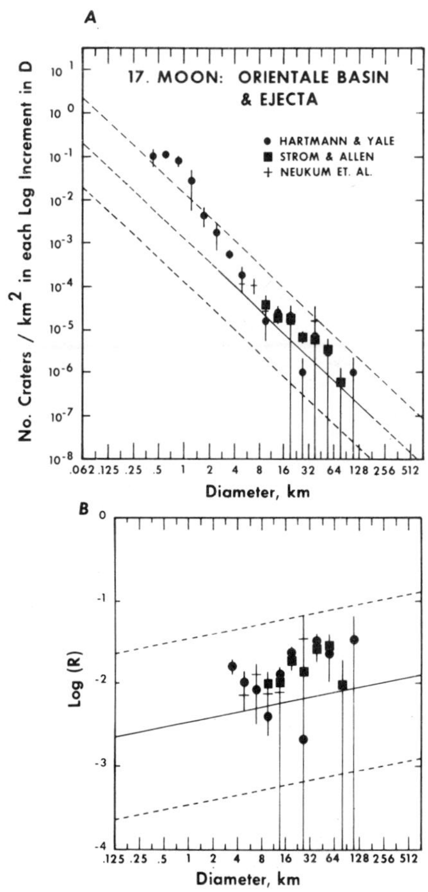
A
17. MOON: ORIENTALE BASIN & EJECTA
HARTMANN & YALE
STROM & ALLEN
NEUKUM ET. AL.
No. Craters / km² in each Log Increment in D
Diameter, km
B
Log (R)
Diameter, km

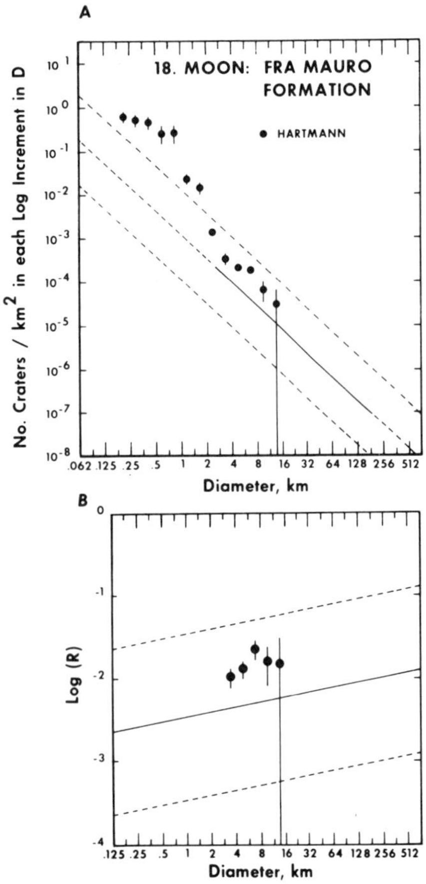
A
18. MOON: FRA MAURO FORMATION
HARTMANN
No. Craters / km^2 in each Log Increment in D
Diameter, km
B
Log (R)
Diameter, km

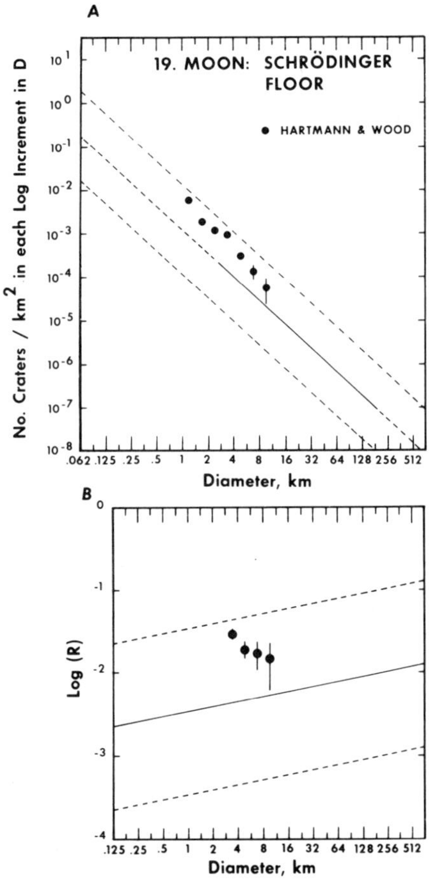
A
19. MOON: SCHRÖDINGER FLOOR
HARTMANN & WOOD
No. Craters / km^2 in each Log Increment in D
Diameter, km
B
Log (R)
Diameter, km

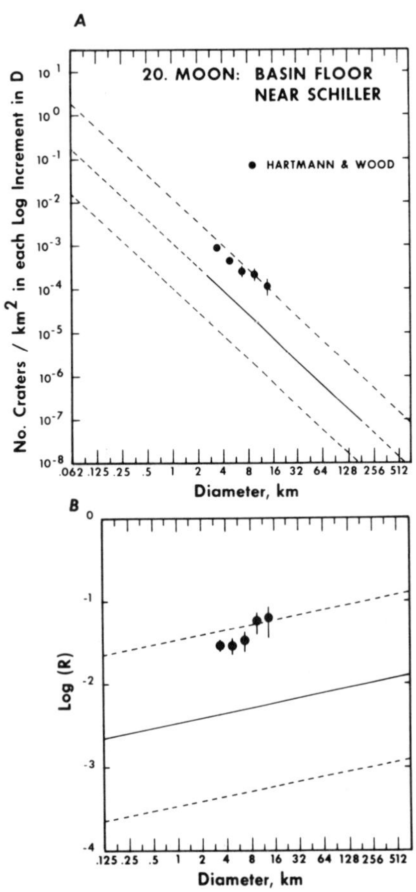
A
20. MOON: BASIN FLOOR NEAR SCHILLER
HARTMANN & WOOD
No. Craters / km^2 in each Log Increment in D
Diameter, km
B
Log (R)
Diameter, km

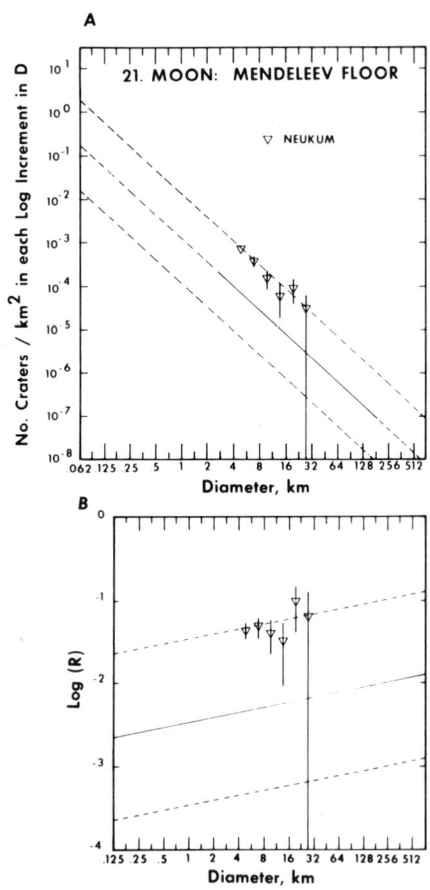
A
21. MOON: MENDELEEV FLOOR
NEUKUM
No. Craters / km^2 in each Log Increment in D
Diameter, km
B
Log (R)
Diameter, km

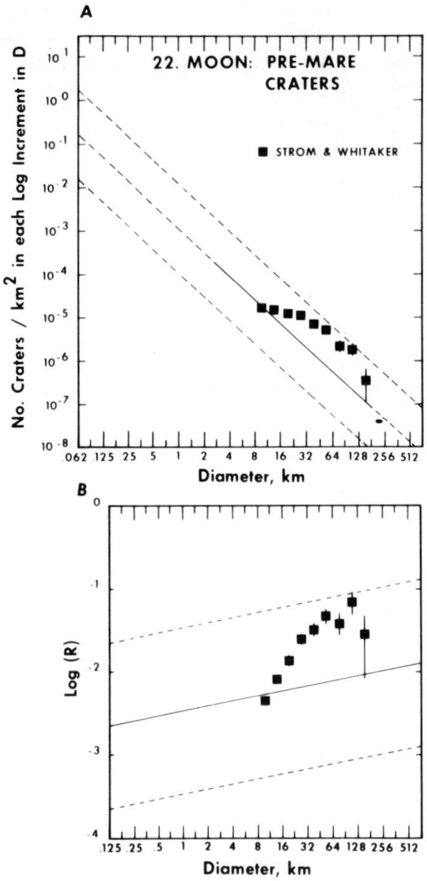
A
22. MOON: PRE-MARE CRATERS
STROM & WHITAKER
No. Craters / km² in each Log Increment in D
Diameter, km
B
Log (R)
Diameter, km

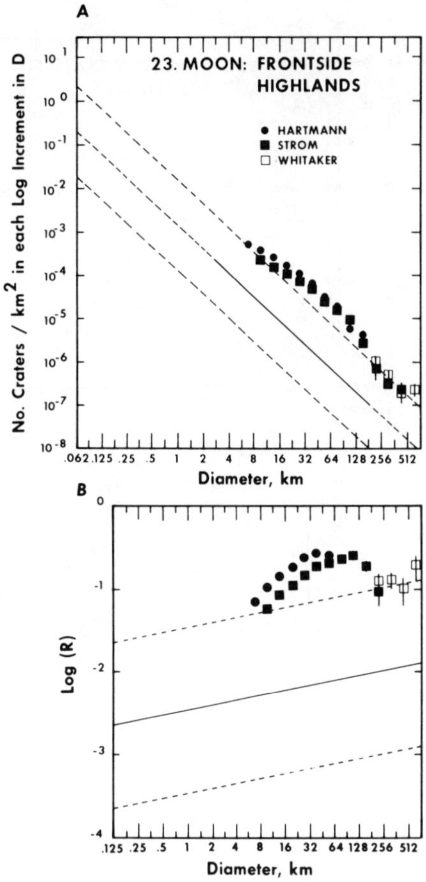
A
23. MOON: FRONTSIDE HIGHLANDS
HARTMANN
STROM
WHITAKER
No. Craters / km² in each Log Increment in D
Diameter, km
B
Log (R)
Diameter, km

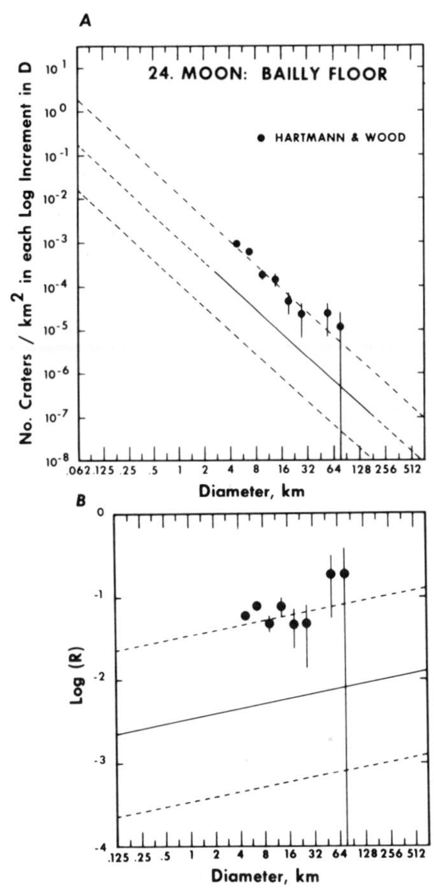
A
24. MOON: BAILLY FLOOR
HARTMANN & WOOD
No. Craters / km² in each Log Increment in D
Diameter, km
B
Log (R)
Diameter, km

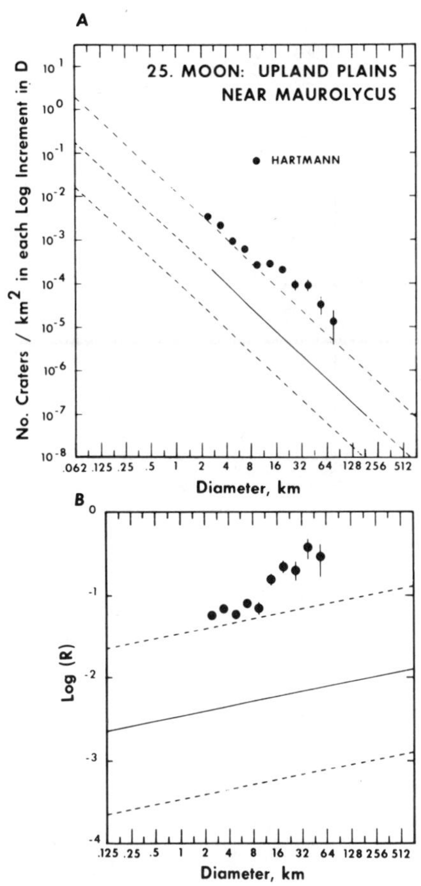
A
25. MOON: UPLAND PLAINS NEAR MAUROLYCUS
HARTMANN
No. Craters / km² in each Log Increment in D
Diameter, km
B
Log (R)
Diameter, km

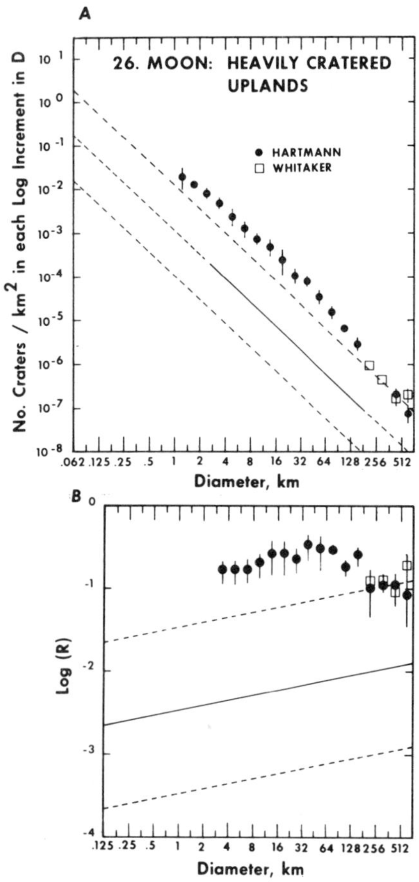
A
26. MOON: HEAVILY CRATERED UPLANDS
● HARTMANN
□ WHITAKER
No. Craters / km² in each Log Increment in D
Diameter, km
B
Log (R)
Diameter, km

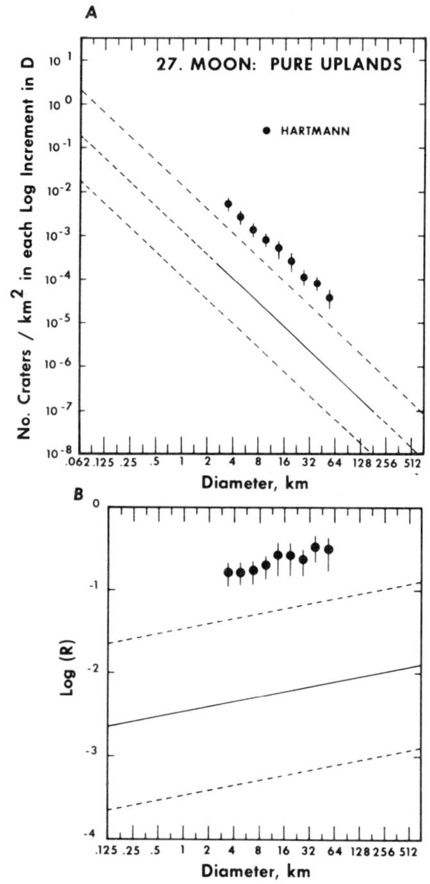
A
27. MOON: PURE UPLANDS
● HARTMANN
No. Craters / km² in each Log Increment in D
Diameter, km
B
Log (R)
Diameter, km

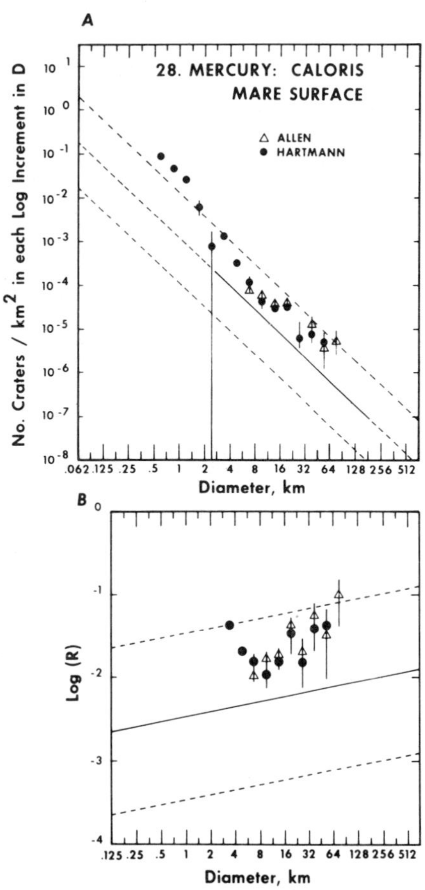
A
28. MERCURY: CALORIS MARE SURFACE
△ ALLEN
● HARTMANN
No. Craters / km² in each Log Increment in D
Diameter, km
B
Log (R)
Diameter, km

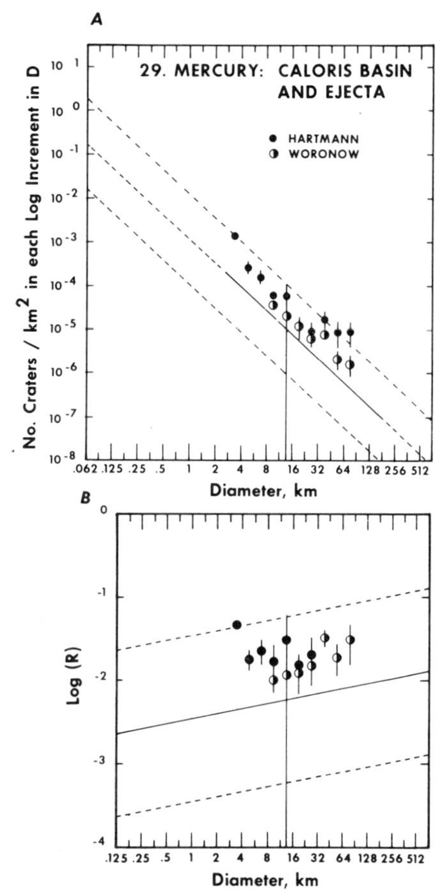
A
29. MERCURY: CALORIS BASIN AND EJECTA
● HARTMANN
◑ WORONOW
No. Craters / km² in each Log Increment in D
Diameter, km
B
Log (R)
Diameter, km

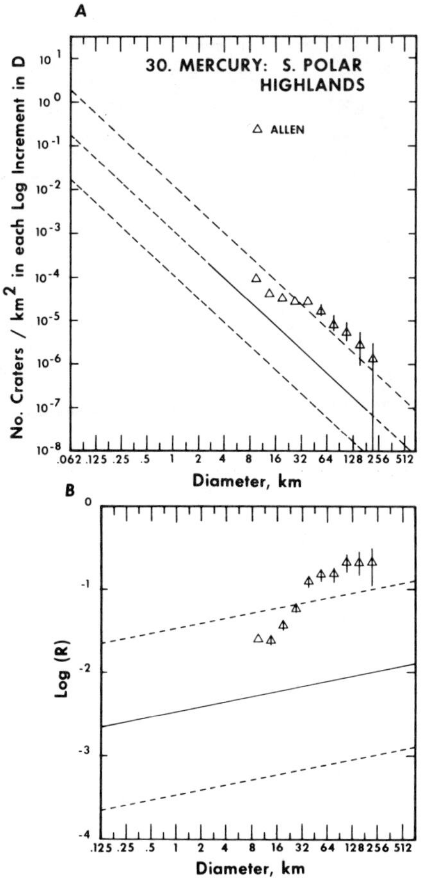
A
30. MERCURY: S. POLAR HIGHLANDS
ALLEN
No. Craters / km^2 in each Log Increment in D
Diameter, km
B
Log (R)
Diameter, km

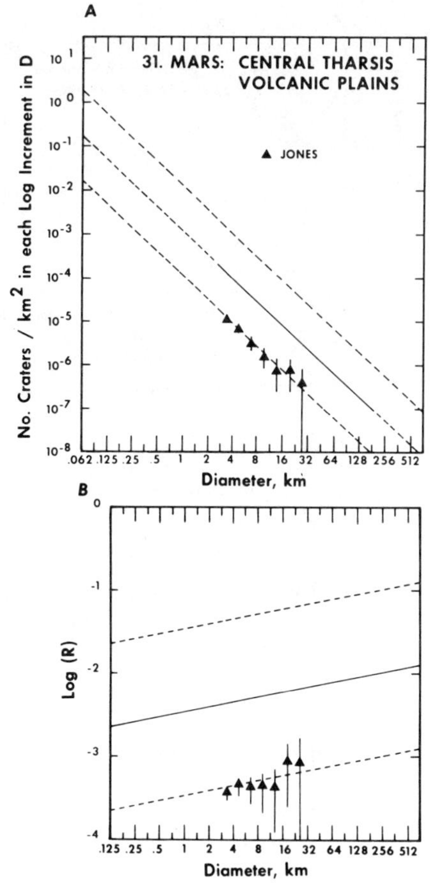
A
31. MARS: CENTRAL THARSIS VOLCANIC PLAINS
JONES
No. Craters / km^2 in each Log Increment in D
Diameter, km
B
Log (R)
Diameter, km

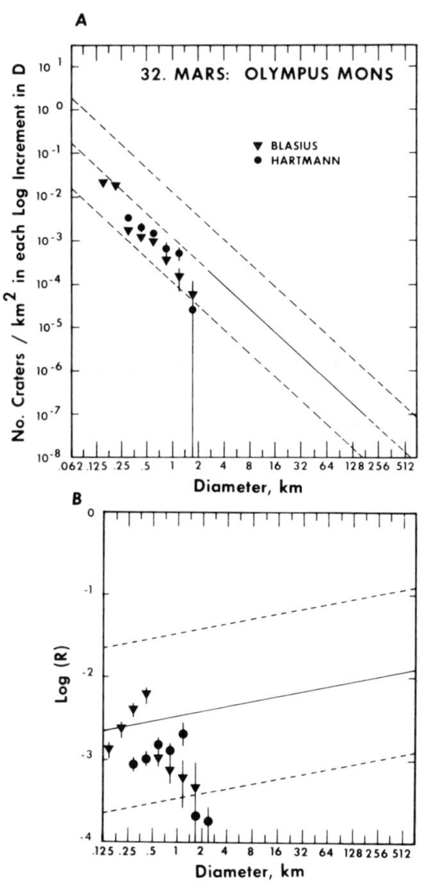
A
32. MARS: OLYMPUS MONS
BLASIUS
HARTMANN
No. Craters / km^2 in each Log Increment in D
Diameter, km
B
Log (R)
Diameter, km

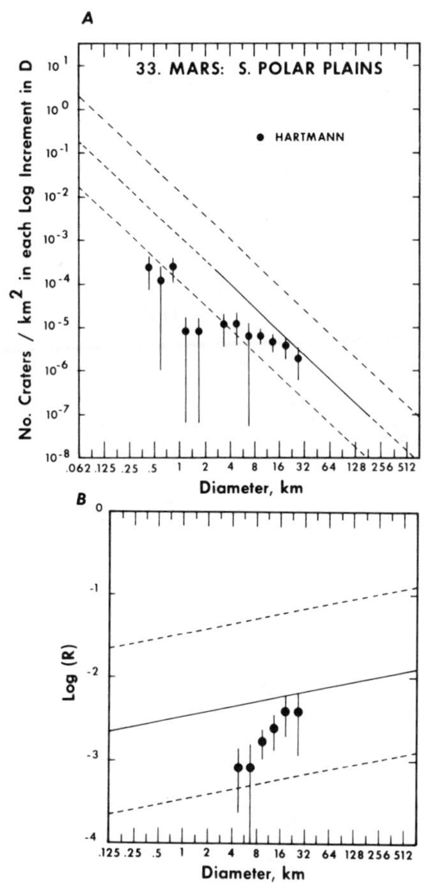
A
33. MARS: S. POLAR PLAINS
HARTMANN
No. Craters / km^2 in each Log Increment in D
Diameter, km
B
Log (R)
Diameter, km

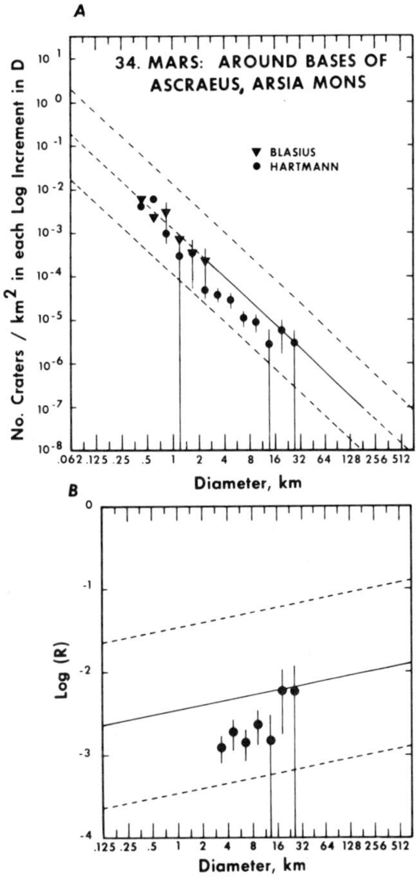
A
34. MARS: AROUND BASES OF ASCRAEUS, ARSIA MONS
BLASIUS
HARTMANN
No. Craters / km² in each Log Increment in D
Diameter, km
B
Log (R)
Diameter, km

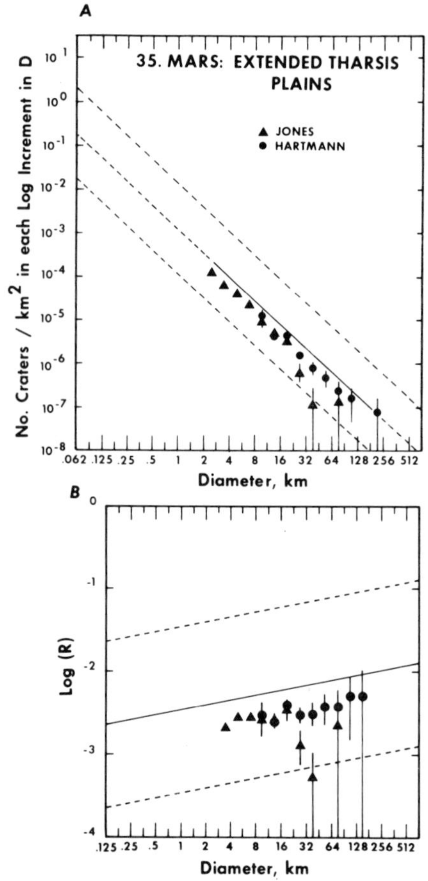
A
35. MARS: EXTENDED THARSIS PLAINS
JONES
HARTMANN
No. Craters / km² in each Log Increment in D
Diameter, km
B
Log (R)
Diameter, km

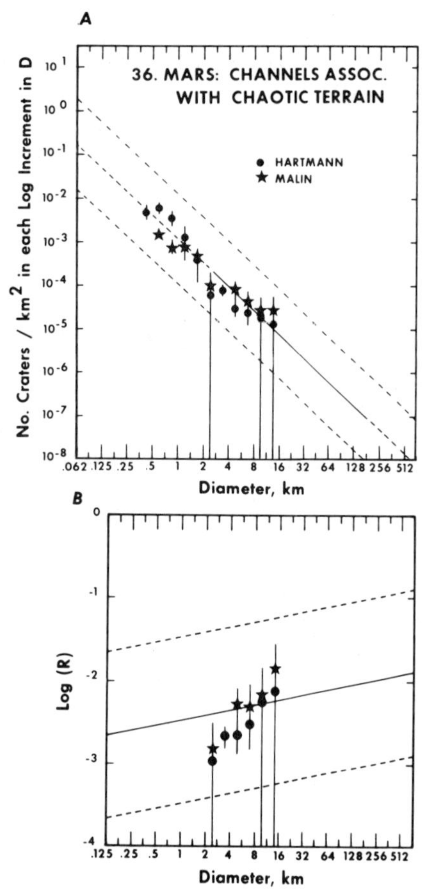
A
36. MARS: CHANNELS ASSOC. WITH CHAOTIC TERRAIN
HARTMANN
MALIN
No. Craters / km² in each Log Increment in D
Diameter, km
B
Log (R)
Diameter, km

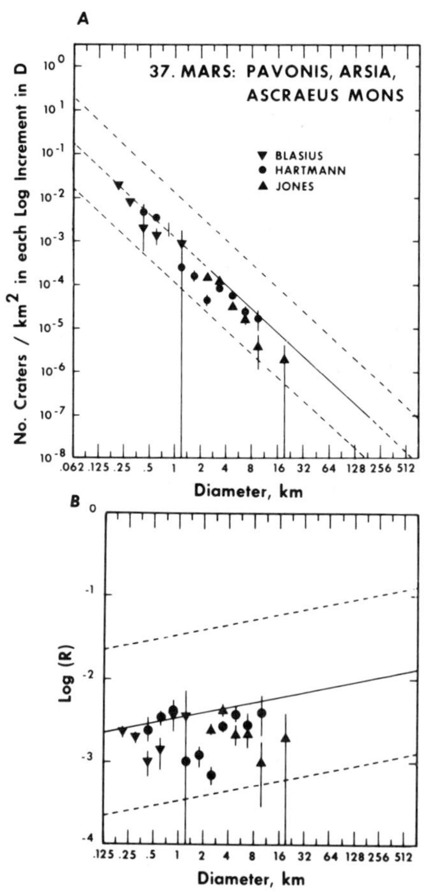
A
37. MARS: PAVONIS, ARSIA, ASCRAEUS MONS
BLASIUS
HARTMANN
JONES
No. Craters / km² in each Log Increment in D
Diameter, km
B
Log (R)
Diameter, km

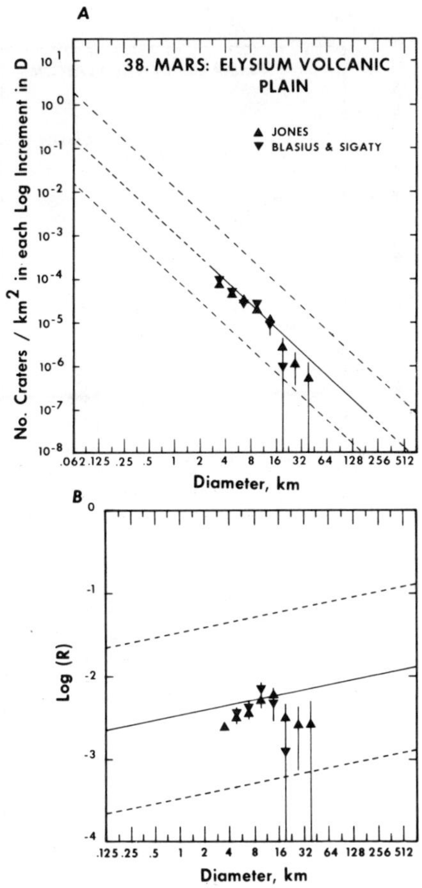
A
38. MARS: ELYSIUM VOLCANIC PLAIN
JONES
BLASIUS & SIGATY
No. Craters / km² in each Log Increment in D
Diameter, km
B
Log (R)
Diameter, km

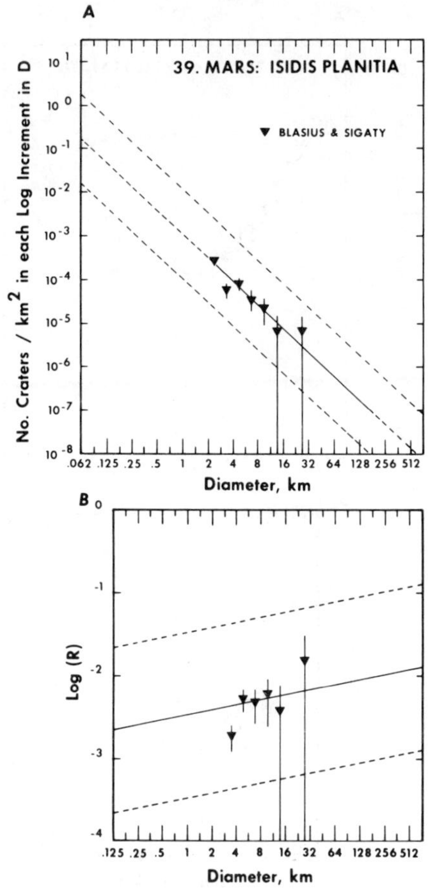
A
39. MARS: ISIDIS PLANITIA
BLASIUS & SIGATY
No. Craters / km² in each Log Increment in D
Diameter, km
B
Log (R)
Diameter, km

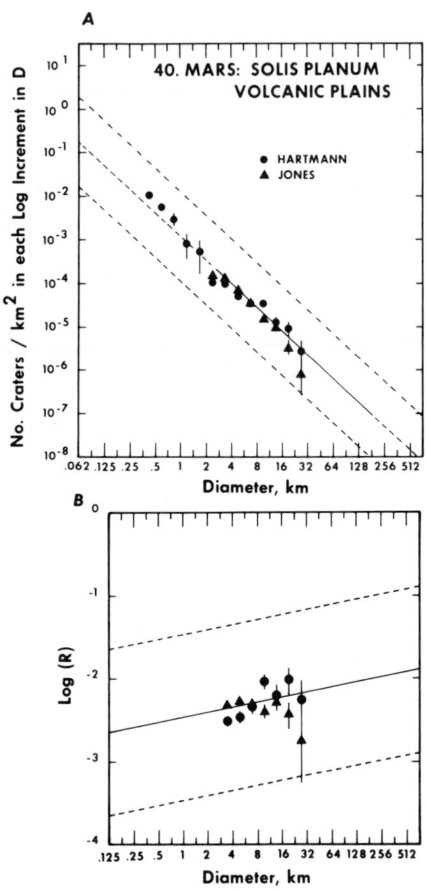
A
40. MARS: SOLIS PLANUM VOLCANIC PLAINS
HARTMANN
JONES
No. Craters / km² in each Log Increment in D
Diameter, km
B
Log (R)
Diameter, km

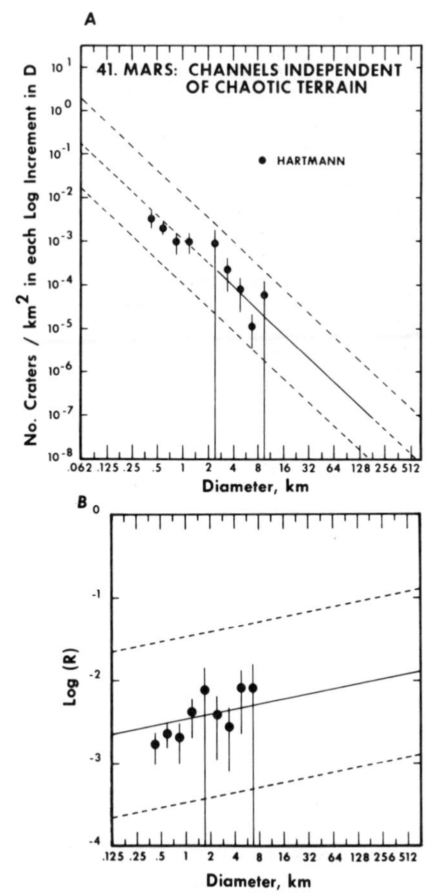
A
41. MARS: CHANNELS INDEPENDENT OF CHAOTIC TERRAIN
HARTMANN
No. Craters / km² in each Log Increment in D
Diameter, km
B
Log (R)
Diameter, km

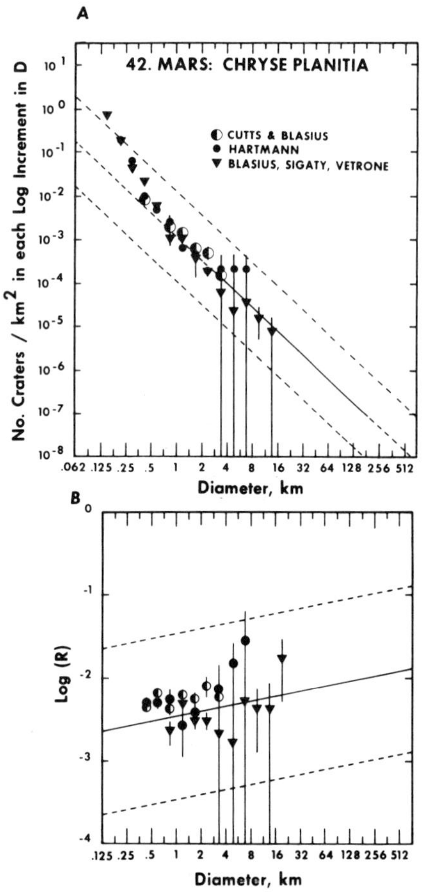
A
42. MARS: CHRYSE PLANITIA
CUTTS & BLASIUS
HARTMANN
BLASIUS, SIGATY, VETRONE
No. Craters / km^2 in each Log Increment in D
Diameter, km
.062 .125 .25 .5 1 2 4 8 16 32 64 128 256 512
B
Log (R)
Diameter, km
.125 .25 .5 1 2 4 8 16 32 64 128 256 512

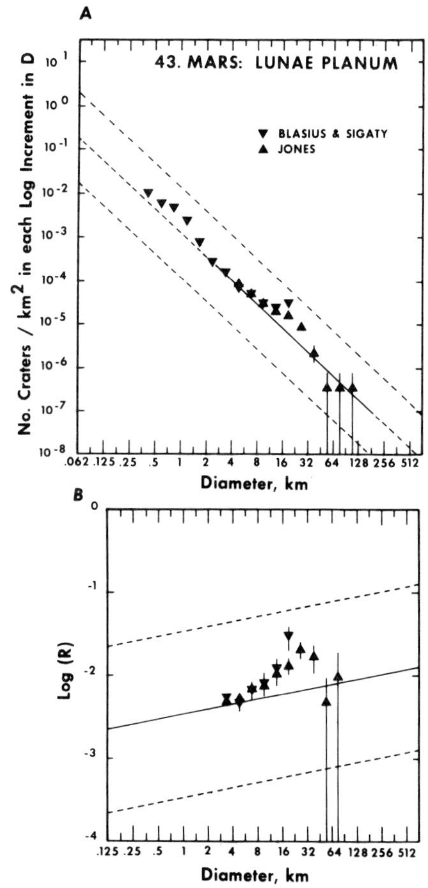
A
43. MARS: LUNAE PLANUM
BLASIUS & SIGATY
JONES
No. Craters / km^2 in each Log Increment in D
Diameter, km
.062 .125 .25 .5 1 2 4 8 16 32 64 128 256 512
B
Log (R)
Diameter, km
.125 .25 .5 1 2 4 8 16 32 64 128 256 512

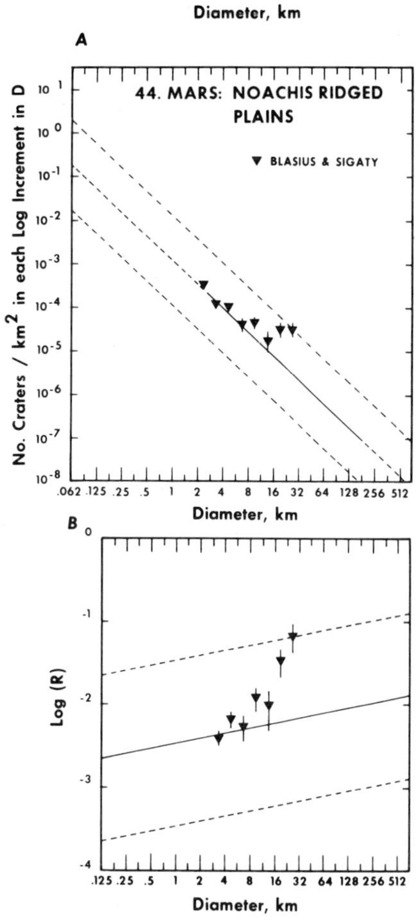
A
44. MARS: NOACHIS RIDGED PLAINS
BLASIUS & SIGATY
No. Craters / km^2 in each Log Increment in D
Diameter, km
.062 .125 .25 .5 1 2 4 8 16 32 64 128 256 512
B
Log (R)
Diameter, km
.125 .25 .5 1 2 4 8 16 32 64 128 256 512

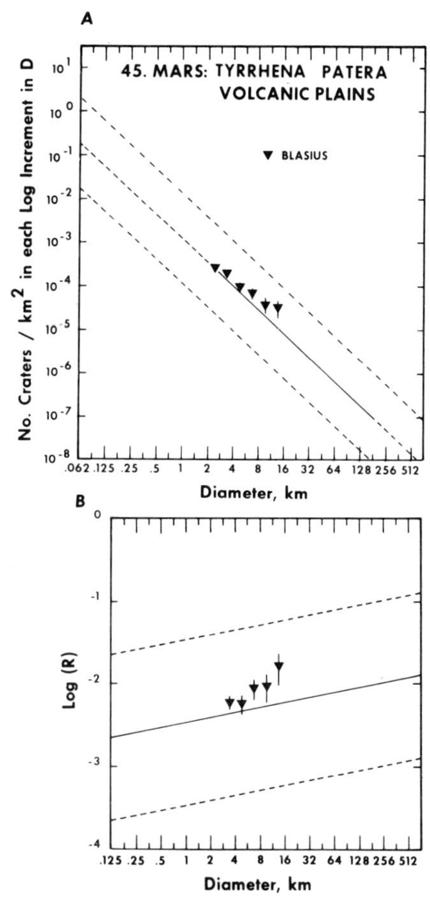
A
45. MARS: TYRRHENA PATERA VOLCANIC PLAINS
BLASIUS
No. Craters / km^2 in each Log Increment in D
Diameter, km
.062 .125 .25 .5 1 2 4 8 16 32 64 128 256 512
B
Log (R)
Diameter, km
.125 .25 .5 1 2 4 8 16 32 64 128 256 512

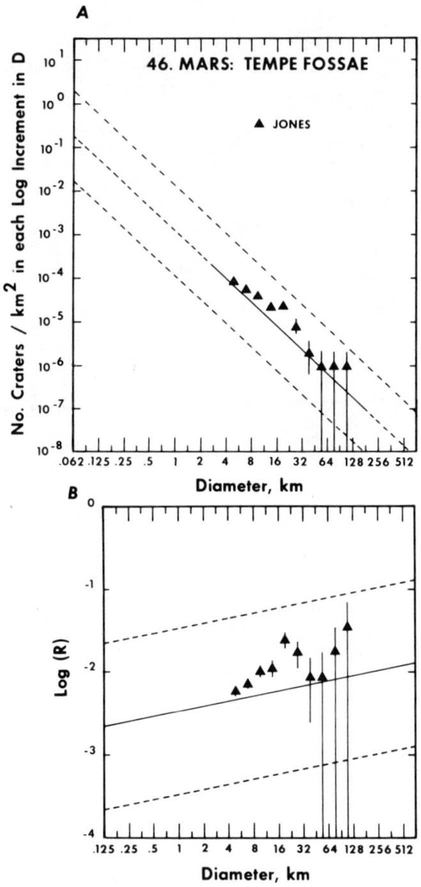
A
46. MARS: TEMPE FOSSAE
JONES
No. Craters / km2 in each Log Increment in D
Diameter, km
B
Log (R)
Diameter, km

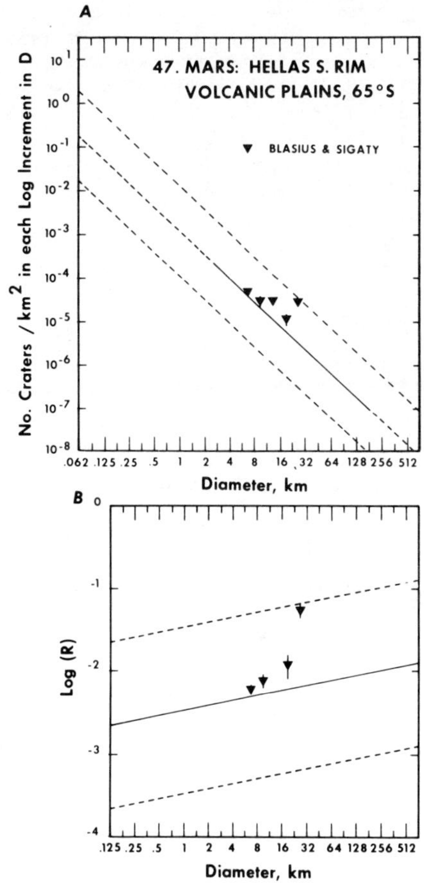
A
47. MARS: HELLAS S. RIM VOLCANIC PLAINS, 65°S
BLASIUS & SIGATY
No. Craters / km2 in each Log Increment in D
Diameter, km
B
Log (R)
Diameter, km

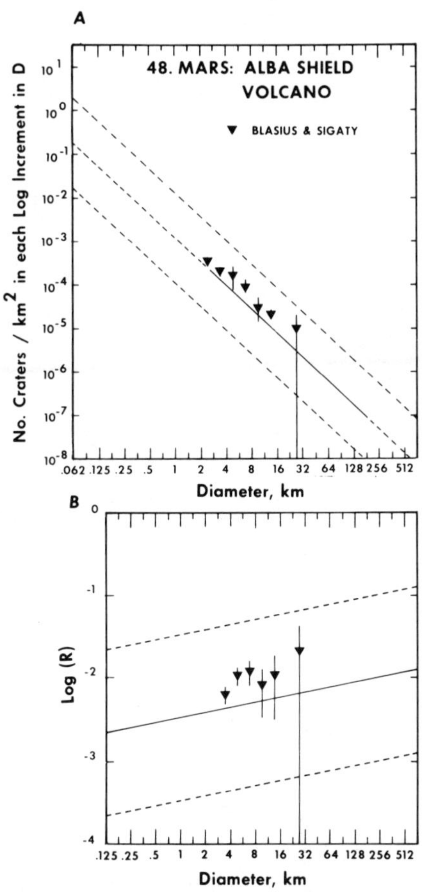
A
48. MARS: ALBA SHIELD VOLCANO
BLASIUS & SIGATY
No. Craters / km2 in each Log Increment in D
Diameter, km
B
Log (R)
Diameter, km

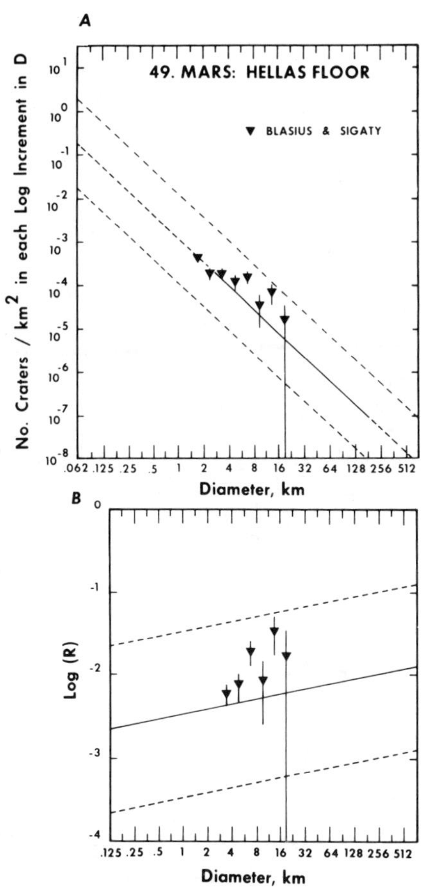
A
49. MARS: HELLAS FLOOR
BLASIUS & SIGATY
No. Craters / km2 in each Log Increment in D
Diameter, km
B
Log (R)
Diameter, km

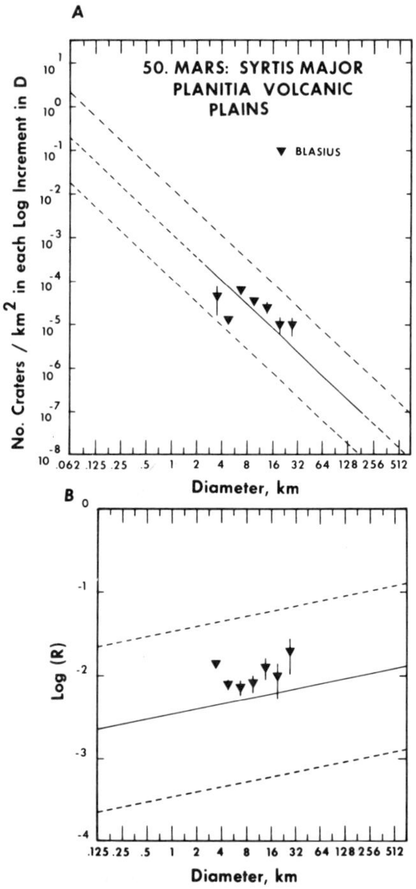
A
50. MARS: SYRTIS MAJOR PLANITIA VOLCANIC PLAINS
▼ BLASIUS
No. Craters / km² in each Log Increment in D
Diameter, km
B
Log (R)
Diameter, km

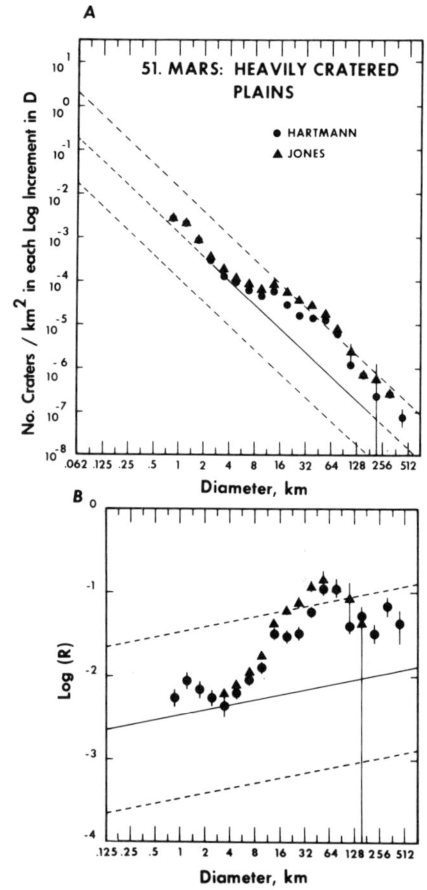
A
51. MARS: HEAVILY CRATERED PLAINS
● HARTMANN
▲ JONES
No. Craters / km² in each Log Increment in D
Diameter, km
B
Log (R)
Diameter, km

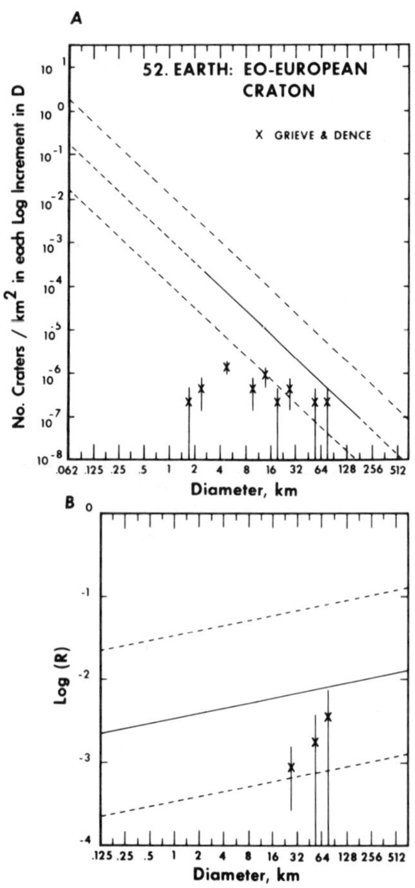
A
52. EARTH: EO-EUROPEAN CRATON
X GRIEVE & DENCE
No. Craters / km² in each Log Increment in D
Diameter, km
B
Log (R)
Diameter, km

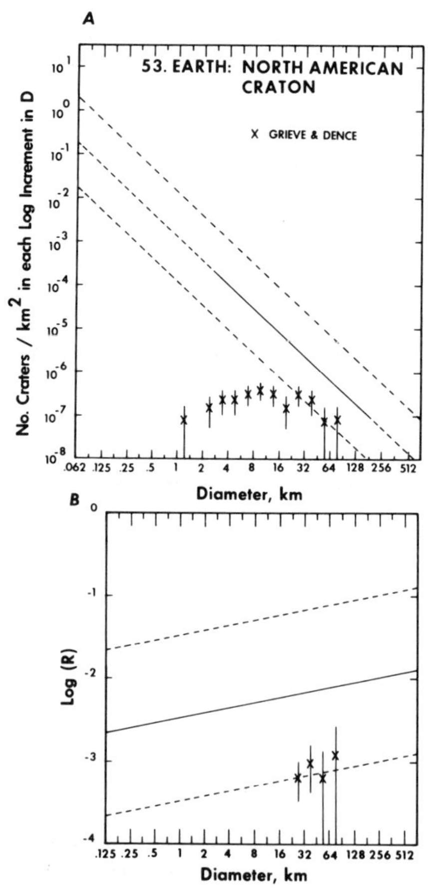
A
53. EARTH: NORTH AMERICAN CRATON
X GRIEVE & DENCE
No. Craters / km² in each Log Increment in D
Diameter, km
B
Log (R)
Diameter, km

ACKNOWLEDGMENTS—M. Dence and R. Grieve are responsible for much of the discussion of terrestrial cratering rates, section 8.4. S. Weidenschilling is responsible for much of the discussion on cratering rates on other planets, section 8.5. C. Chapman is responsible for much of the discussion of techniques using small craters, section 8.8. K. Blasius gathered most of the D_L and C_S values in Table 8.8.1. W. Hartmann is responsible for much of the discussion of "pre-mare" upland volcanism, section 8.9. J. Diaz, R. Strom, and A. Woronow assembled the graphs in Appendix 8.1. We acknowledge the help of G. Neukum, who kindly supplied information on crater counts by B. König (1977). We also thank G. Neukum, D. Wise, C. Wood, R. Pepin, and other reviewers for helpful critiques of early drafts of this chapter.

8.12 REFERENCES

Alexander E., Coscio M., Dragon J., Pepin R. and Saito K. (1977) K/Ar dating of lunar soils, III. *Proc. Lunar Sci. Conf. 8th*, p. 2725–2740.

Arkani-Hamed J. (1974) Effect of a giant impact on the thermal evolution of the Moon. *The Moon* **9,** 183–210.

Arnold J. R. (1965) The origin of meteorites as small bodies II. The model. *Astrophys. J.* **141,** 1536–1547.

Arvidson R. E. (1974) Morphologic classification of martian craters and some implications. *Icarus* **22,** 264–271.

Arvidson R., Boyce J., Chapman C., Cintala M., Fulchignoni M., Moore H., Neukum G., Schultz P., Soderblom L., Strom R., Woronow A. and Young R. (1978) Standard techniques for presentation and analysis of crater size—frequency data. *Icarus* **37,** 467–474.

Arvidson R., Mutch T. and Jones K. (1974) Craters and associated aeolian features on Mariner 9 photographs: an automated data gathering and handling system and some preliminary results. *The Moon* **9,** 104–114.

Baldwin R. B. (1964) Lunar crater counts. *Astron. J.* **69,** 377–392.

Basilevsky A. T. (1976) On the evolution rate of small lunar craters. *Proc. Lunar Sci. Conf. 7th*, p. 1005–1020.

Blasius K. (1976) The record of impact cratering on the great volcanic shields of the Tharsis region of Mars. *Icarus* **29,** 343–361.

Boyce J. M. (1975) Chronology of the major flow units in the western nearside maria (abstract). In *Papers Presented to the Conference on Origins of Mare Basalts,* p. 11–14. The Lunar Science Institute, Houston.

Boyce J. M. (1976) Ages of flow units in the lunar nearside maria based on Lunar Orbiter IV photographs. *Proc. Lunar Sci. Conf. 7th*, p. 2717–2728.

Boyce J. M. and Dial A. L. (1975) Relative ages of flow units in Mare Imbrium and Sinus Iridum. *Proc. Lunar Sci. Conf. 6th*, p. 2585–2595.

Boyce J. M., Dial A. and Soderblom L. (1974) Ages of the lunar nearside light plains and maria. *Proc. Lunar Sci. Conf. 5th*, p. 11–23.

Boyce J. M., Dial A. and Soderblom L. (1975) A summary of relative ages of lunar nearside and farside plains. U.S.G.S. Interagency Report: Astrogeology **66.**

Boyce J. M., Schaber G. G. and Dial A. (1977) Age of Luna 24 mare basalts based on crater studies. *Nature* **265,** 38–39.

Boyce J. M. and Johnson D. A. (1977) Ages of flow units in Mare Crisium, based on crater density. *Proc. Lunar Sci. Conf. 8th*, p. 3495–3502.

Chapman C. R. (1974) Cratering on Mars I. Cratering and obliteration history. *Icarus* **22,** 272–291.

Chapman C. R. (1975) Asteroid collisional evolution: evidence for a much larger early population. *Science* **190,** 553–556.

Chapman C. R. and Jones K. L. (1977) Cratering and obliteration history of Mars. *Ann. Rev. Earth Planet. Sci.* **5,** 515–540.

Chapman C. R., Aubele J., Roberts W. J. and Cutts J. A. (1979) Sub-kilometer craters: origins, ages, processes of degradation and implications for mare basalt petrogenesis (abstract). In *Lunar and Planetary Science X.* p. 190–191. Lunar and Planetary Institute, Houston.

Chapman C. R., Mosher J. and Simmons G. (1970) Lunar cratering and erosion from Orbiter 5 photographs. *J. Geophys. Res.* **75,** 1445–1466.

DeHon R. A. and Waskom J. D. (1976) Geologic structure of the eastern mare basins. *Proc. Lunar Sci. Conf. 7th*, p. 2729–2746.

Dence M. R. (1965) The extraterrestrial origin of Canadian craters. *Ann. N. Y. Acad. Sci.* **123,** 941–969.

Dence M. R. (1972) The nature and significance of terrestrial impact structures. *Proc. 24th Int. Geol. Congr., Section 15,* p. 77–89.

Dence M. R., Grieve R. A. F. and Robertson P. B. (1977) Terrestrial impact structures: principal characteristics and energy considerations. In *Impact and Explosion Cratering* (D. J. Roddy, R. O. Pepin and R. B. Merrill, eds.), 247–275. Pergamon, N.Y.

Dohnanyi J. S. (1969) Collisional model of asteroids and their debris. *J. Geophys. Res.* **74,** 2531–2554.

Eppler D. and Heiken G. (1975) Lunar crater chains of nonimpact origin. *Proc. Lunar Sci. Conf. 6th*, p. 2571–2583.

Gault D. E. (1970) Saturation and equilibrium conditions for impact cratering on the lunar surface: criteria and implications. *Radio Science* **5,** 273–291.

Gault D. E. and Wedekind J. A. (1977) Experimental hypervelocity impact into quartz sand—II, effects of gravitational acceleration. In *Impact and Explosion Cratering* (D. J. Roddy, R. O. Pepin and R. B. Merrill, eds.), p. 1231–1244. Pergamon, N.Y.

Greeley R. and Gault D. E. (1970) Precision size-frequency distributions of craters for 12 selected areas of the lunar surface. *Moon* **2,** 10–77.

Gault D. E. and Wedekind J. (1978) Experimental studies of oblique impact. *Proc. Lunar Sci. Conf. 9th*, p. 3843–3875.

Grieve R. A. F. and Dence M. (1979) The terrestrial cratering record II. The crater production rate. *Icarus* **38,** 230–242.

Grieve R. A. F. and Robertson P. B. (1979) The terrestrial cratering record I. Current status of observations. *Icarus* **38,** 212–229.

Guiness E. A. and Arvidson R. E. (1977) On the constancy of the lunar cratering flux over the past 3.3×10^9 yr. *Proc. Lunar Sci. Conf. 8th*, p. 3475–3494.

Hartmann W. K. (1965) Terrestrial and lunar flux of large meteorites in the last two billion years. *Icarus* **4,** 157–165.

Hartmann W. K. (1966a) Early lunar cratering. *Icarus* **5,** 406–418.

Hartmann W. K. (1966b) Martian cratering. *Icarus* **5,** 565–576.

Hartmann W. K. (1967) Lunar crater counts II: three lunar surface type-areas. *Comm. Lunar Planet. Lab.* **6,** 39–41.

Hartmann W. K. (1968) Lunar crater counts VI: the young craters Tycho, Aristarchus, and Copernicus. *Comm. Lunar Planet. Lab.* **7,** 145–156.

Hartmann W. K. (1969) Terrestrial, lunar, and interplanetary rock fragmentation. *Icarus* **10,** 201–213.

Hartmann W. K. (1970) Preliminary note on lunar cratering rates and absolute time-scales. *Icarus* **12,** 131–133.

Hartmann W. K. (1971a) Martian cratering 2. Asteroid impact history. *Icarus* **15,** 369–409.

Hartmann W. K. (1971b) Martian cratering 3. Theory of cratering obliteration. *Icarus* **15,** 410–428.

Hartmann W. K. (1972) Paleocratering of the Moon. Review of post-Apollo data. *Astropohys. Space Sci.* **17,** 48–64.

Hartmann W. K. (1973) Martian cratering 4. Mariner 9 initial analysis of cratering chronology. *J. Geophys. Res.* **78,** 4096–4116.

Hartmann W. K. (1977a) Cratering in the solar system. *Sci. Amer.* **236,** No. 1, 84–99.

Hartmann W. K. (1977b) Relative crater production rates on planets. *Icarus* **31,** 260–276.

Hartmann W. K. (1978) Martian cratering, V. Toward an empirical Martian chronology, and its implications. *Geophys. Res. Lett.* **5,** 450–452.

Hartmann W. K. and Hartmann A. (1968) Asteroid collisions and evolution of asteroidal mass distribution and meteoritic flux. *Icarus* **8,** 361–381.

Hartmann W. K. and Wood C. A. (1971) Moon: origin and evolution of multi-ring basins. *The Moon* **3,** 3–78.

Hartmann W. K. and Yale F. G. (1968) Lunar crater counts IV: Mare Orientale and its basin system *Commun. Lunar Planet. Lab.* **7**, 131.

Jones K. (1974) Martian obliteration history. Ph.D. thesis containing crater catalog for Mars: Brown University.

Keldysh M. V. (1977) Venus exploration with the Venera 9 and 10 spacecraft. *Icarus* **30**, 605–625.

König B. (1977) Untersuchung von Primären und Sekundären Einschlagstrukturen auf dem Mond und Laborexperimente zum Studium des Auswurfs von Sekundärteilchen, Dissertation. Ruprecht-Karl-Universität, Heidelberg.

König B., Neukum G. and Fechtig H. (1977) Recent lunar cratering: absolute ages of Kepler, Aristarchus, Tycho (abstract). In *Lunar Science VIII*, p. 555–557. Lunar Science Institute, Houston.

Kresak L. (1978) The comet and asteroid population of the Earth's environment. *Bull. Astron. Inst. Czech.* **29**, 114–125.

Lucchitta B. K. and Sanchez A. (1975) Crater studies in the Apollo 17 region. *Proc. Lunar Sci. Conf. 6th*, p. 2427–2441.

Malin M. (1976) Age of martian channels. *J. Geophys. Res.* **81**, 4825–4845.

Marsden B. G. (1975) *Catalogue of cometary orbits.* Central Bureau for Astronomical Telegrams, Cambridge, Mass.

Marsden B. G. (1977) Orbital data on the existence of Oort's cloud of comets. In *Comets, Asteroids, Meteorites: Interrelations, Evolution, and Origins*, (A. H. Delsemme, ed.) p. 79–86. University of Toledo, Toledo.

Masursky H., Eliason E., Jordan R., Pettengill G., Ford P. and McGill G. (1979) Pioneer Venus radar observations— the first year (abstract). *Bull. Amer. Astron. Soc.* **11**, 549.

McGill G. E. (1977) Craters as "fossils": the remote dating of planetary surface materials. *Geol. Soc. Amer. Bull.* **88**, 1102–1110.

McGill G. E. and Wise D. U. (1972) Regional variations in degradation and density of martian craters. *J. Geophys. Res.* **77**, 2433–2441.

Morris E. C. and Shoemaker E. (1969) Craters in Surveyor project final report, Part 2. JPL-TR-32-1265, p. 65.

Murray B. C., Soderblom L., Sharp R. and Cutts J. (1971) The surface of Mars, 1. Cratered terrains. *J. Geophys. Res.* **76**, 313–330.

Mutch T. A., Arvidson R., Head J., Jones K. and Saunders R. (1976) *The Geology of Mars.* Princeton Univ. Press, Princeton. 400 pp.

Neukum G. (1977) Different ages of lunar light plains. *Moon* **17**, 383–393.

Neukum G. and König B. (1976) Dating of individual lunar craters. *Proc. Lunar Sci. Conf. 7th*, p. 2867–2881.

Neukum G., König B. and Arkani-Hamed J. (1975a) A study of lunar impact crater size distributions. *Moon* **12**, 201–229.

Neukum G., König B., Fechtig H. and Storzer D. (1975b) Cratering in the Earth-Moon system: consequences for age determination by crater counting. *Proc. Lunar Sci. Conf. 6th*, p. 2597–2620.

Neukum G. and Wise D. U. (1976) Mars: a standard crater curve and possible new time scale. *Science* **194**, 1381–1387.

O'Keefe J. and Ahrens T. (1978) Impact flows and crater scaling on the Moon. *Phys. Earth Planet. Int.* **16**, 341–351.

Oberbeck V. and Quaide W. (1968) Genetic implications of lunar regolith thickness variations. *Icarus* **9**, 446–465.

Oberbeck V. and Aggarwal H. (1977) Topographic analysis of lunar secondaries of Copernicus and implications. *Proc. Lunar Sci. Conf. 8th*, p. 3521–3537.

Oberbeck V. R., Quaide W., Arvidson R. and Aggarwal H. (1977) Comparative studies of lunar, martian and mercurian craters and plains. *J. Geophys. Res.* **82**, 1681–1698.

Oort J. (1950) The structure of the cloud of comets surrounding the solar system, and a hypothesis concerning its origin. *Bull. Astron. Inst. Neth.* **11**, 91–110.

Öpik E. (1951) Collision probabilities with the planets and the distribution of interplanetary matter. *Proc. Roy. Irish Acad.* **54A**, 165–199.

Öpik E. (1960) The lunar surface as an impact counter. *Mon. Not. Roy. Astr. Soc.* **120**, 404–411.

Öpik E. (1965) Mariner IV and craters on Mars. *Irish Astron. J.* **7**, 92–104.

Öpik E. (1966) The Martian surface. *Science* **153**, 255–265.

Peale S. J., Cassen P. and Reynolds R. (1979) Melting of Io by tidal dissipation. *Science* **203**, 892–894.

Pettengill G., Ford P., Brown W., Kaula W., Masursky H., Eliason E. and McGill G. (1979) Venus: preliminary topographic and surface imaging results from the Pioneer Orbiter. *Science* **205**, 90–93.

Saunders R. S. and Malin M. C. (1977) Geologic interpretation of new observations of the surface of Venus. *Geophys. Res. Lett.* **4**, 547–550.

Schaber G. G., Horstman K. and Dial A. (1978) Lava flow materials in the Tharsis region of Mars. *Proc. Lunar Planet. Sci. Conf. 9th*, p. 3433–3458.

Schultz P. H. (1976) *Moon Morphology.* Univ. Texas Press, Austin. 626 pp.

Schultz P. H., Greeley R. and Gault D. E. (1976) Degradation of small mare surface features. *Proc. Lunar Sci. Conf. 7th*, p. 985–1003.

Schultz P. H., Greeley R. and Gault D. E. (1977) Interpreting statistics of small lunar craters. *Proc. Lunar Sci. Conf. 8th*, p. 3539–3564.

Schultz P. H. and Spudis P. (1979) Evidence for ancient mare volcanism. *Proc. Lunar Sci. Conf. 10th*, p. 2899–2918.

Shoemaker E. M. (1965) Preliminary analysis of the fine structure of Mare Cognitum. JPL-TR-32-700, p. 75.

Shoemaker E. M. (1970) Origin of fragmental debris on the lunar surface and the history of bombardment of the Moon. Presented at I Seminario de Geologia Lunar, Universidad de Barcelona, May 1970 (Revised January 1971).

Shoemaker E. M. (1977) Astronomically observable crater-forming projectiles. In *Impact and Explosion Cratering* (D. J. Roddy, R. O. Pepin and R. B. Merrill, eds.), p. 617–628. Pergamon, N.Y.

Shoemaker E. M., Morris E., Batson R., Holt H., Larson K., Montgomery D., Rennilson J. and Whitaker E. (1969) Television observations from Surveyor. In *Surveyor Program Results*, NASA SP-184.

Shoemaker E. M., Hackman R. and Eggleton R. (1962) Interplanetary correlation of geologic time. *Advan. Astronaut. Sci.* **8**, 70–89.

Shoemaker E. M. and Helin E. F. (1977) Populations of planet-crossing asteroids and the relationship of Apollo objects to main-belt asteroids and comets. In *Comets, Meteorites and Meteoroids* (A. H. Delsemme, ed.), p. 297–300. Univ. of Toledo Press, Toledo.

Shoemaker E. M. and Helin E. F. (1978) Earth-approaching asteroids: populations, origins, and compositional types. In *Asteroids: An Exploration Assessment*, (D. Morrison and W. C. Wells, eds.). NASA Conf. Publ. 2053, Washington.

Shoemaker E. M., Williams J. G., Helin E. F. and Wolfe R. F. (1979) Earth-crossing asteroids: orbital classes, collision rates with Earth, and origin. In *Asteroids*, (T. Gehrels, ed.), p. 253–282. Univ. Arizona Press, Tucson. 1181 pp.

Soderblom L. A., Arnold J., Boyce J. and Lin R. (1977) Regional variations in the lunar maria: age, remanent magnetism, and chemistry. *Proc. Lunar Sci. Conf. 8th*, p. 1191–1199.

Soderblom L. A. and Boyce J. M. (1972) Relative ages of some near-side and far-side terra plains based on Apollo 16 metric photography. In *Apollo 16 Prelim. Sci. Rep.*, NASA SP-315.

Soderblom L. A., Condit C., West R., Herman B. and Kreidler T. (1974) Martian planetwide crater distributions: implications for geologic history and surface processes. *Icarus* **22**, 239–263.

Soderblom L. A. and Lebofsky L. A. (1972) Technique for rapid determination of relative ages of lunar areas from orbital photography. *J. Geophys. Res.* **77**, 279–296.

Stettler A. and Albarede F. (1978) ^{39}Ar-^{40}Ar systematics of 2-mm sized rock fragments from Mare Crisium. *Earth and Planet. Sci. Lett.* **38**, 401–406.

Strom R. G. (1977) Origin and relative age of lunar and mercurian inter-crater plains. *Phys. Earth Planet. Inter.* **15**, 156–172.

Strom R. G. (1979) Mercury: a post-Mariner 10 assessment. *Space Sci. Rev.* **24**, 3–70.

Strom R. G. and Whitaker E. A. (1976) Populations of impacting bodies in the inner solar system. NASA TX-3346.

Swann G. A. and Reed V. S. (1974) A method for estimating the absolute ages of small Copernican craters and its application to the determination of Copernican meteorite flux. *Proc. Lunar Sci. Conf. 5th*, p. 151–158.

Taylor S. (1975) *Lunar Science: A Post-Apollo View.* Pergamon, N.Y. 372 pp.

Trask N. J. (1966) Size and spatial distribution of craters estimated from Ranger photographs, Ranger VIII and IX, Part 2. JPL-TR-32-800.

Trask N. J. (1971) Geologic comparison of lunar materials in the lunar equatorial belt, including Apollo 11 and 12 landing sites. U.S. Geol. Survey Prof. Paper 750D, p. D138.

Walker D., Stolper E. M. and Hays J. F. (1979) Basaltic volcanism: the importance of planet size. *Proc. Lunar Planet. Sci. Conf. 10th*, p. 1995–2015.

Wasserburg G., DiBrozolo F., Papanastassiou D., McCulloch M., Huneke J., Dymek R., DePaolo D., Chodos A. and Albee A. (1978) Petrology, chemistry, age and irradiation history of Luna 24 samples. In *Mare Crisium: The View from Luna 24,* (R. B. Merrill and J. J. Papike, eds.), p. 657–678. Pergamon Press, N.Y.

Wetherill G. W. (1976) Where do meteorites come from? A re-evaluation of the Earth-crossing Apollo objects as sources of chondrite meteorites. *Geochim. Cosmochim. Acta.* **40**, 1297–1317.

Wetherill G. W. (1977a) Evolution of the Earth's planetesimal swarm subsequent to the formation of the Earth and Moon. *Proc. Lunar Sci. Conf. 8th*, p. 1–16.

Wetherill G. W. (1977b) Dynamic accretion model for the terrestrial planets (abstract). *EOS, Trans. AGU 58*, 428.

Wetherill G. W. (1977c) Evolution of small bodies in the solar system. A. Accumulation of the terrestrial planets. *Carnegie Inst. Wash., Yearb.* **76**, 761–767.

Wetherill G. W. (1979) Steady state populations of Apollo-Amor objects. *Icarus* **37**, 96–112.

Whipple F. L. (1975) Do comets play a role in galactic chemistry and γ-ray bursts? *Astron. J.* **80**, 525–531.

Wilhelms D. E. (1976) Secondary impact craters of lunar basins. *Proc. Lunar Sci. Conf. 7th*, p. 2883–2901.

Wilhelms D., Oberbeck V. and Aggarwal H. (1978) Size-frequency distributions of primary and secondary lunar impact craters. *Proc. Lunar Planet. Sci. Conf. 9th*, p. 3735–3762.

Williams, J. G. (1969) Secular perturbations in the solar system. Ph.D. dissertation, U.C.L.A., 273 pp.

Wood C. A. (1979) Monogenetic volcanoes of the terrestrial planets. *Proc. Lunar Planet. Sci. Conf. 10th*, p. 2815–2840.

Wood C. A. and Andersson L. E. (1980) *The Lunar and Planetary Laboratory catalog of lunar craters: Part 1: Nearside.* NASA TM79328. In press.

Wood C. A. and Hartmann W. K. (1970) Appendix II: Tycho crater statistics. *Arizona Univ. Lunar Planet. Lab. Comm.* **8**, 232–233.

Woronow A. (1977) Crater saturation and equilibrium: a Monte Carlo simulation. *J. Geophys. Res.* **82**, 2447–2456.

Woronow A. (1978) A general cratering-history model and its implications for the lunar highlands. *Icarus* **34**, 76–88.

Young R. A. (1975) Mare crater size-frequency distributions: implications for relative surface ages and regolith development. *Proc. Lunar Sci. Conf. 6th*, p. 2645–2662.

Young R. A. (1977) The lunar impact flux, radiometric age correlation, and dating of specific lunar features. *Proc. Lunar Sci. Conf. 8th*, p. 3457–3473.

Young R. A., Brennen W. and Nichols D. (1974) Problems in the interpretation of lunar mare stratigraphy and relative ages indicated by ejecta from small impact craters. *Proc. Lunar Sci. Conf. 5th*, p. 159–170.

Overleaf:

Solid-state convection dominates heat transfer in the Earth's mantle and has controlled the thermal evolution of the planet. In spite of our understanding of plate motions at the Earth's surface, we do not know any of the details of the convective flow pattern at depth. Two alternative flow patterns are shown, one in which the mantle is radially zoned into separate flow regimes (top), and one in which a single convective flow system involves the entire mantle (bottom). Adapted from Richter (1979).

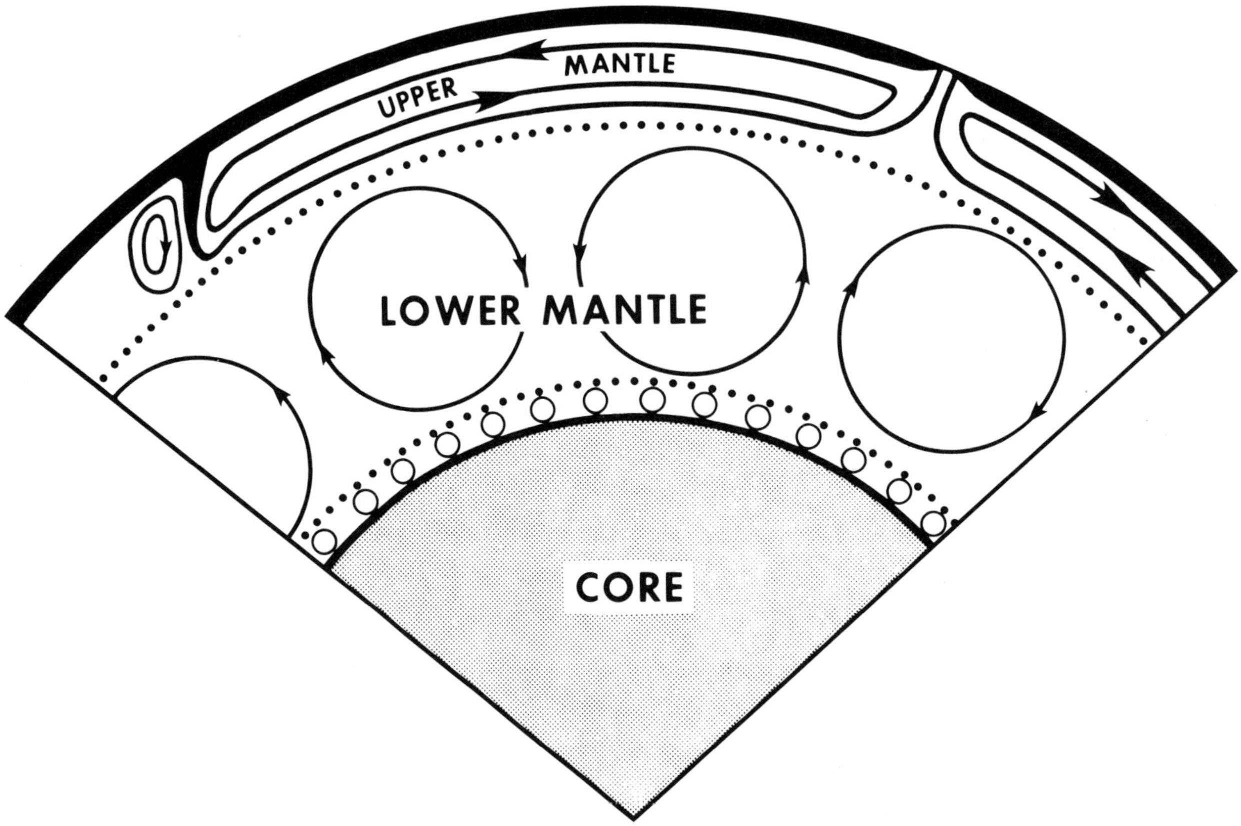
UPPER MANTLE
LOWER MANTLE
CORE

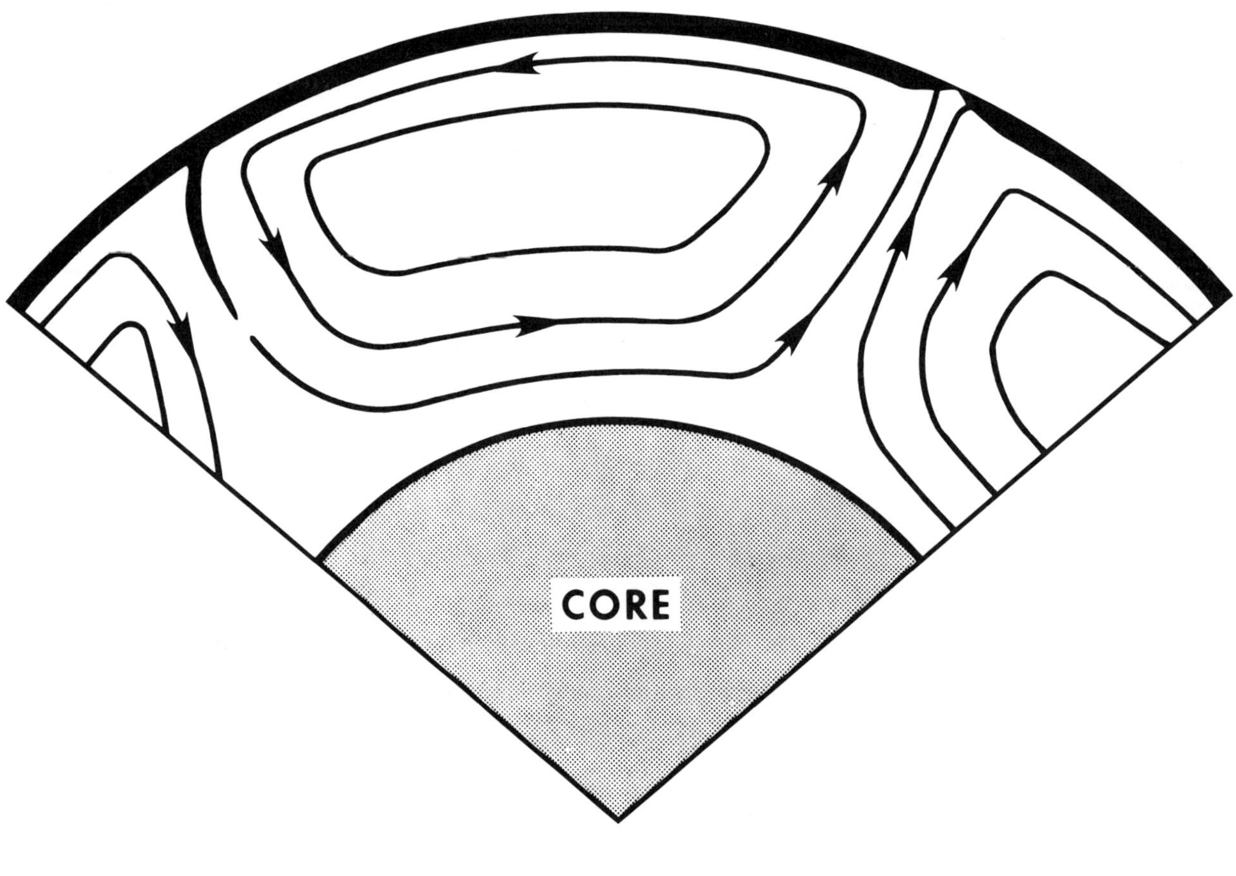
CORE

9. Thermal histories of the terrestrial planets

Sean C. Solomon, Team Leader
—*Department of Earth & Planetary Sciences*
Massachusetts Institute of Technology

Thomas J. Ahrens
—*Division of Geological & Planetary Sciences*
California Institute of Technology

Patrick M. Cassen
—*Theoretical & Planetary Studies Branch*
NASA Ames Research Center

Albert T. Hsui
—*Department of Geology*
University of Illinois

John W. Minear
—*Earth & Planetary Sciences Division*
NASA Johnson Space Center

Ray T. Reynolds
—*Theoretical & Planetary Studies Branch*
NASA Ames Research Center

Norman H. Sleep
—*Departments of Geology & Geophysics*
Stanford University

David W. Strangway
—*Department of Geology*
University of Toronto

Donald L. Turcotte
—*Department of Geological Sciences*
Cornell University

9.1 INTRODUCTION

Basaltic volcanism is both a product and a process of heat transfer in a planetary interior. Active volcanism is *prima facie* evidence that temperatures exceed the solidus over some volume at depth. Conversely, knowledge of the temperature field at depth provides a framework for understanding the occurrence and evolution of surface volcanic activity. This chapter deals with the heat engines of planets: the sources of energy, the mechanisms of heat transfer, and the evolution with time of temperature and of magma production within each of the terrestrial planets.

The development of planetary thermal histories in this chapter is based on two general and complementary goals: (1) to provide a global, physical, and quantitative context for the various observations of basaltic volcanism on the planets presented in the earlier chapters; and (2) to use thermal history models as numerical experiments to test hypotheses of planetary evolution and to derive conclusions about volcanism on the planets independent of existing observations.

In its broadest terms, a planetary thermal history is a budgetary account of energy, momentum and mass. Energy is added to planets as heat by sources both internal (radioactivity, gravitational potential energy) and external (impacts, tides, electromagnetic induction). This heat energy is lost by radiation into space, but only after the heat is transported to the surface by conduction and by the upward flux of hot material. Volcanism is the consequence of sufficient internal heat to induce at least local melting, and subsequent transfer of mass and heat by magma ascent and eruption.

Although the emphasis in this book and in this chapter is on the subject of basaltic volcanism, the thermal history of a planet cannot be divorced from a host of topics of only peripheral importance to volcanism. The global evolution of a planet is a product, for instance, of its early accretionary state, of the history of its core as may be revealed in paleomagnetism, and of processes occurring too deep within the mantle to have a direct expression in surface igneous activity. We take the philosophy in this chapter that all processes known to affect planetary thermal histories are to be considered. While this view will occasionally lead to discussions far removed from topics considered elsewhere in this book, it also provides a thorough basis for considering basaltic volcanism as a consequence of a planet's entire thermal regime.

The links between this chapter and the other chapters in this book are many. The basic structural divisions of a planet, the most conspicuous products of planetary differentiation, link this chapter to Chapter 4. The quantitative prediction of magma production requires knowledge both of the temperature and of the melting relations for probable mantle compositions, linking this chapter to the petrological and cosmochemical discussions in Chapters 1, 3 and 4. The geologic history of a planetary surface, including both volcanism and tectonism as functions of space and time, is fundamentally the product of the dynamical evolution of the planetary interior, thus linking this chapter to Chapters 2, 5, 6, 7 and 8. Finally, the quantitative thermal history models of this chapter provide a foundation for the broad discussion of the chemical and physical evolution of the terrestrial planets in Chapter 10.

9.1.1 Volcanism and thermal history

The relationship between basaltic volcanism and the thermal history of a planet is illustrated schematically in Fig. 9.1.1. Volcanism on a planet is possible only if (1) the internal heat is sufficient to at least partially melt some portion of the planetary interior and (2) the melt generated by such heating can gain access as magma to the planetary surface. Both requirements must be met. Thus, for instance, active basaltic volcanism is clear evidence for contemporaneous partial melting in the planetary mantle, but absence of volcanism does not preclude the existence of deep magmas that cannot penetrate the overlying lithosphere or that produce only deeply seated igneous intrusions.

The first condition for volcanism, melting of the interior, requires that local temperatures exceed the solidus for some portion of the planet. The melting relations for planetary mantles, as described in Chapter 3, depend on planet size and on mantle composition, particularly on the abundances of mantle volatiles. It is common practice in thermal history models to require the spherically averaged temperature profile to exceed the adopted solidus over some range of depths during the time interval over which surface observations indicate active volcanism occurred. This constraint is not strictly required; it is necessary only that the spherically averaged temperature be sufficiently close to the solidus so that local excursions from the averaged temperature yield local partial melting.

The second condition for volcanism, access of the magma to the surface, is somewhat more difficult to quantify with certainty. The subject of magma migra-

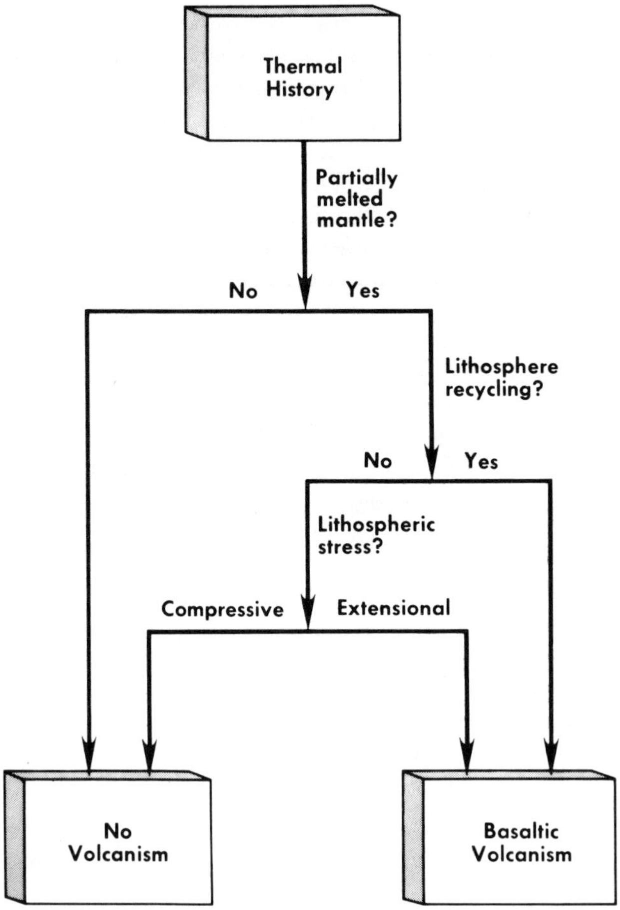

9.1.1 Schematic view of the relationship between planetary thermal history and basaltic volcanism.

tion is treated in section 6.6. If the lithosphere of a planet is continually recycled back into the mantle, then the creation of new lithosphere provides a continuous egress for the asthenospheric magma. Creation of new seafloor on Earth is, by far, the most volumetrically important mechanism by which mantle-derived magmas reach the surface to be emplaced as basalts.

For planets without lithospheric recycling, magma ascent through the lithosphere is still possible but generally occurs less readily. (Contrast the relative volumes of intraplate volcanism and ridge volcanism on the Earth.) Surface eruption of mantle-derived magmas requires both that the hydrostatic head of the magma column lie at or above the local surface and that the lithospheric stress be at least locally favorable to rise of magma through dikes or as liquid-filled, propagating cracks. The first condition is generally met except for high-density magmas generated at shallow mantle depths (perhaps the early lunar mare-type basalts rich in Fe and Ti). The second condition is favored by horizontal extensional stress, generated either locally (impact stresses, or bending stresses due to lithospheric loading) or regionally (thermal stress associated with net planetary warming). Because the early history of most terrestrial planets is one of net expansion associated with global differentiation and early internal heating, the shallow lithosphere should be in a state of generally extensional horizontal deviatoric stress, conducive to widespread volcanism. As a planet cools, the lithospheric thermal stress switches to increasing compression, a change that—in the absence of other sources of stress—signals the gradual end to surface volcanic activity.

9.1.2 Chapter outline

The subject of the thermal evolution of the Earth and the planets has a long history, dating back to Lord Kelvin's calculations on the age of the Earth and culminating in the recent widespread interest in the subject generated by the results of scientific exploration of the solar system by spacecraft. A brief review of this history, including the development of both numerical techniques and principal concepts, is given in section 9.2.

The basic theoretical formulation of the thermal history problem is presented in section 9.3. The physical equations governing thermal history are those of conservation of energy, momentum and mass. Although these differential equations can be completely prescribed, and though several terms may be readily neglected for planetary interiors, the equations cannot be solved for the general three-dimensional case over planetary length scales and geological time scales. Methods for obtaining approximate or simplified solutions, including modelling techniques necessary to represent such poorly understood phenomena as core formation and crustal differentiation, are the subjects of continuing research.

The governing equations for planetary thermal history depend on several physical properties of planetary materials, considered at length in section 9.4. Of foremost importance are the sources of heat and the parameters governing heat transfer. Both internal heat sources (radioactivity, energy of differentiation, heats of reaction) and external energy sources (impacts, tides, electromagnetic heating) must be considered. For heat transport, the values of thermal conductivity (including radiative transfer) and viscosity may be estimated for the planets from available measurements in concert with solid-state theory. It may be concluded that these values favor subsolidus convection as a dominant heat transport mechanism for the interiors of the terrestrial planets. Additional properties of planetary mantles considered in section 9.4 are adiabatic gradients, melting relations, other phase changes, and their effects on convection and on thermal history.

In section 9.5 we present a planet-by-planet discussion of thermal history for the Earth, Moon, Mars, Mercury, Venus, and for asteroids/meteorite parent bodies. For each object, the essential geological, geophysical, and geochemical constraints of planetary evolution are reviewed, one or more thermal history models consistent with those constraints are given, and the implications of successful models for basaltic volcanism are assessed.

In the final section (9.6) several common and contrasting aspects of planetary evolution are tied together, setting the stage for Chapter 10.

9.1.3 Principal conclusions

On the basis of those thermal history models for the terrestrial planets that are consistent with their volcanic and tectonic histories, their internal structures, and their available heat sources and sinks, it is clear that while basaltic volcanism has been a common feature on these bodies, the detailed history of volcanism on the terrestrial planets has varied substantially.

All of the terrestrial planets, including the Moon and some asteroids, were subject to extensive early heating sufficient to yield early core formation on the Earth and probably on Mercury, and early differentiation of a thick, low-density crust on the Moon and probably on Mars. It is likely that this early heating was accompanied by extensive early volcanism on all of these bodies. The nature of the early heat source is not clear, but for at least two of the terrestrial planets—the Moon and Mars—the early heating appears to have been confined to the outer few hundred kilometers, suggesting an exogenic (impact or electromagnetic-induction heating) rather than an endogenic (e.g., short-lived radioactivity) energy source.

Whereas early planetary differentiation appears to have been common on all of the planets, the nature and extent of that differentiation—particularly the thickness of the early crust—together with the size of the planet exerted strong controls on subsequent tectonic and volcanic evolution. It is likely that on the Earth, extensive early heating, probably augmented by the heat of core infall, led to vigorous mantle convection, rapid recycling of surface material back into the mantle, and only limited differentiation of a low-density proto-Archean crust. The Earth has probably cooled over virtually its entire history, with a consequent slow thickening of the lithosphere, a gradual reduction of typical mantle convective speeds, and the development of modern plate tectonics once the lithosphere had acquired substantial lateral rigidity. Volcanism on the Earth, at least since the end of the period of heavy bombardment of the inner solar system, has probably always been concentrated primarily at plate boundaries, particularly spreading centers, though a slow evolution of both mean plate dimensions and the chemistry of volcanic magmas might be expected as a product of the lessening of convective speeds and mean temperatures in the mantle.

Of the other terrestrial planets, only Venus may have exhibited a history broadly similar to that of the Earth, though the thick atmosphere and high surface temperatures have so far precluded the development of our understanding of the geologic history of the Venus surface to a level commensurate with that for the other bodies. For the Moon, Mars and Mercury, early crustal differentiation led to stable global lithospheres, and subsequent volcanism was more akin to intraplate volcanism on the Earth.

On the Moon, mare volcanism was fueled by remnants of the early heating that led to crustal formation as well as by later heating from radioactivity and large impacts. Thermal models provide independent support for an extended period of volcanic activity predating formation of the last few major mare basins. Cessation of mare volcanism may have been caused either by complete solidification of the lunar upper mantle or by a closing of magma ascent paths through the lunar lithosphere due to slight global contraction and consequent near-surface horizontal compressive stress.

Mars, in parallel with the Moon, experienced early heating of the outer few hundred kilometers and formation of a thick, low-density crust and of a global lithosphere prior to cessation of heavy bombardment. In contrast to the Moon, however, martian volcanism and surface extensional tectonics continued over a much longer fraction of planetary history. Thermal models can account for these features if a larger volume fraction of Mars began initially cold, compared to the Moon, so that subsequent interior heating—probably including delayed core formation—led to an extended period of global expansion and of magma generation in the martian upper mantle.

Mercury experienced more pervasive early heating than either the Moon or Mars, since the surface tectonic features indicate that the planet has been cooling and contracting since before the end of heavy bombardment. In particular, core formation and crustal differentiation could not have occurred since that time. The initial heating of Mercury and the likely heat pulse associated with core infall, assuming a grossly homogeneous initial state, should have led to widespread early volcanism on the planet, perhaps still partially preserved in the intercrater plains. That horizontal compressive stresses in Mercury's lithosphere date back to before cessation of heavy bombardment suggests that any post-bombardment volcanism was short in duration. The smooth plains, if volcanic, are therefore predicted to have been emplaced over a narrow age interval.

The parent bodies of basaltic achondritic meteorites must also have experienced early heating and melting, though the choices for early heat sources for small solar system objects are more limited than for the planets. Either short-lived radioactivity or induction heating are potential candidates. Later (1.5–3 b.y. ago) melting fueled by long-lived radioactivity is possible on the larger asteroids/meteorite parent bodies if their outer layers have low thermal conductivities appropriate to high-porosity breccias and regoliths.

9.2 HISTORICAL REVIEW

Thermal history model calculations may be thought to date from Lord Kelvin's (1864) attempt to obtain a quantitative estimate of the age of the Earth from conduction calculations of the solidification and subsequent cooling time of an initially molten silicate sphere. Since radioactivity had not yet been discovered, his effort resulted, of course, in an estimate of the age of the Earth which was much too low. In spite of much geological evidence to the contrary, Kelvin's conclusions, based as they were upon simple assumptions and correct (although incomplete) physics, became widely accepted and served as a retarding influence upon geological thought for many years (Burchfield, 1975). Hardly an auspicious beginning for thermal history calculations!

Holmes (1915) was the first to appreciate the importance of heat production of radioactive isotopes in thermal history calculations. However, calculations with a uniform distribution of radioactive elements through the mantle resulted in such high temperatures that a large fraction of the mantle would have to be molten. This result was not consistent with seismic observations that showed that the entire mantle could propagate shear waves and must therefore be solid, or at most only partially molten. This difficulty led to the hypothesis of an upward concentration of the radioactive elements in the mantle. Urey (1951, 1952), Allan and Jacobs (1956), Lubimova (1958) and MacDonald (1959), among others, constructed more detailed models of the Earth, Moon and other planetary objects by means of finite difference solutions to the equation of heat conduction for a spherically symmetric solid body with internally distributed (and time varying) heat sources.

Levin (1962) went a step further and predicted the onset of planetary melting, by employing a relation between pressure and melting point for planetary material. He also considered the effect of the heat of fusion on thermal history calculations. Anderson and Phinney (1967) extended this concept to a discussion of global differentiation of the terrestrial planets. Reynolds *et al.* (1966) and Fricker *et al.* (1967) introduced the effects of heat transfer by liquid convection and the redistribution (in space and time) of radioactive heat sources into a self-consistent model calculation. Similar numerical schemes were developed independently by Lee (1968) and Ornatskaya *et al.* (1977).

Numerous efforts have been made to include additional physical effects within the framework of spherically symmetric models of planetary thermal history. Examples of such physical effects include heat transfer by igneous processes (McConnell *et al.*, 1967), accretional energy (Hanks and Anderson, 1969; Kaula, 1979a), core formation (Siegfried and Solomon, 1974), electrical heating (Sonett *et al.*, 1975), planetary thermal expansion and thermal stress (MacDonald, 1960; Solomon and Chaiken, 1976), and crustal formation by crystal-liquid fractionation (Solomon and Longhi, 1977; Minear and Fletcher, 1978; Herbert *et al.*, 1978).

One of the early predictions of thermal history models for the Earth based on conduction theory with an upward concentration of radioactive elements was that oceanic heat flow would be low compared with continental heat flow. This prediction was based on the knowledge that the oceanic crust is thin and believed to have lower concentrations of radioactive elements than the continental crust. However, the first oceanic heat-flow measurements by Revelle and Maxwell (1952) and Bullard (1954) showed that the oceanic heat flow was approximately equal to continental values. Subsequent measurements have confirmed this result. This observation led to the conclusion that the high oceanic heat flux must be due to heat derived from the mantle.

Evidence has accumulated for many years that the Earth's mantle is not static and that flow processes affect its thermal structure. The earliest such evidence centered on the concept of continental drift. Based on geometric, geologic and paleontological arguments, Wegener (1924) proposed that continents moved across the floor of the oceans and that the motion was driven by gravitational forces. Jeffreys (1929) showed, however, that gravitational forces are many orders of magnitude too small. Even though the qualitative evidence supporting continental drift was substantive, without an alternative driving mechanism and without quantitative supporting evidence, continental drift was not accepted by a large majority of Earth scientists until the late 1960's.

Holmes (1931) was among the first to propose that the mechanism for continental drift is mantle convection. Holmes realized that the Earth's mantle behaves like a fluid on geological time scales. He carried out a Rayleigh-stability calculation for the mantle and concluded that thermal convection could be expected. Further quantitative studies of mantle convection were later given by Pekeris (1935), Hales (1936), Griggs (1939), and Tozer (1965, 1967).

Quantitative data supporting continental drift lagged behind theoretical studies of mantle convection. Using some of the first reliable paleomagnetic data, Runcorn (1956) showed that contemporaneous positions of the magnetic pole in old rocks, as measured in Europe and North America, do not coincide and that the difference between the pole positions diminishes as progressively younger rocks are considered. These results were explained by a continuous divergence between North America and Europe during the last 200 million years, i.e., continental drift.

With an improved mapping of the seafloor in the 1950's, researchers found that a mountain range, the mid-Atlantic ridge, bisects the Atlantic Ocean. This ridge is part of a world-wide ridge system. The hypothesis of seafloor spreading (Hess, 1962) was introduced to explain this symmetry. According to this hypothesis, seafloor is created by volcanism at the ocean ridge and then spreads laterally as a rigid unit, carrying along any attached continents. Striking confirmation of the seafloor spreading hypothesis was provided by Vine and Matthews (1963), who correlated the observed pattern of regular magnetic anomalies on the seafloor with periodic reversals in the Earth's magnetic field. This correlation permitted the first quantitative determination of rates of seafloor spreading.

Just as new seafloor is created at ocean ridges, it is destroyed at ocean trenches. Old seafloor bends or breaks and descends into the mantle. A natural consequence of the creation and destruction of seafloor at ocean ridges and trenches is the concept of plate tectonics (Wilson, 1965; Morgan, 1968). The Earth's surface is recognized to be composed of a set of rigid plates which are in constant relative motion with respect to each other. The relative motion of two plates is described by a rate of rotation about a pole. A large fraction of the Earth's seismicity, volcanism, and mountain building is associated with interactions between plates at their margins.

The development and confirmation of the concepts of seafloor spreading and plate tectonics led to renewed and widespread interest in studies of mantle convection. Turcotte and Oxburgh (1967) presented a boundary-layer theory for finite-amplitude thermal convection. This analysis explained the basic boundary-layer character of the surface plates and the more nearly isothermal structure of the mantle beneath the plates. A number of authors have conducted numerical calculations to model the thermal structure due to thermal convection in the mantle (Torrance and Turcotte, 1971a; Peltier, 1972: McKenzie *et al.*, 1973; Richter, 1973a; Houston and De Bremaecker, 1975; Schubert and Young, 1976).

Runcorn (1962) was the first to suggest that convection was important in the Moon. Schubert *et al.* (1969) subsequently demonstrated the likelihood that the Moon and the other terrestrial planets were unstable to subsolidus thermal convection. Turcotte and Oxburgh (1970) discussed the significance of convection for the present state of the lunar interior while Cassen and Reynolds (1973, 1974) and Cassen and Young (1975) discussed the implications for lunar thermal history. Studies of the structure of convection cells have been carried out by Turcotte *et al.* (1972) and Schubert *et al.* (1977) for the Moon and by Young and Schubert (1974) for Mars. Only recently, however, have global evolutionary models been presented that include the effects of solid convection. Toksöz *et al.* (1978) performed a restricted set of calculations by direct integration of the hydrodynamic equations while Sharpe and Peltier (1978, 1979), Cassen *et al.* (1979), Stevenson and Turner (1979), Schubert *et al.* (1979a, b), and Turcotte *et al.* (1979) conducted numerical simulations based on empirical convective heat transfer relations.

A large number of numerical experiments have been carried out using a variety of modelling techniques to test hypotheses for planetary thermal history and to explore the allowable ranges of different physical parameters. Although the computational methods used for calculating both global thermal history models and local and regional-scale thermal models have become progressively more complex and sophisticated, it is well to remember the uncertainty inherent in possibly omitted effects. The history associated with Kelvin's calculation should serve as a salutary example to those engaged in modelling calculations or in using the results of such calculations.

In spite of these uncertainties, thermal history models will continue to provide important tests for the evaluation of hypotheses of planetary origin, constitution, and surface evolution. In particular, thermal history models provide a necessary quantitative framework for the discussion of basaltic volcanism on the planets.

9.3 THERMAL MODEL CALCULATIONS

9.3.1 Equations and parameters

In principle, the thermal history of a planet is described by the solution of the energy conservation equation for the time dependence of the temperature distribution $T(r, \theta, \phi; t)$ within the body, where (r, θ, ϕ) are spherical coordinates and t is time. The energy conservation equation can be written as

$$T\,\frac{\partial S}{\partial t} + T\,\mathbf{u}\cdot\nabla S = \nabla\cdot(K\nabla T) + Q_{sources} - Q_{sinks} \qquad (9.3.1)$$

In this equation, S is the entropy per unit volume, **u** is the velocity, and K is the thermal conductivity. The left-hand side is the rate of change of enthalpy per unit volume for a moving fluid particle. At constant pressure, $TdS = \rho C_P dT$, where ρ is the density and C_P is the specific heat at constant pressure. The first term on the right in Eq. (9.3.1) is the rate of addition of heat by conduction. (Heat transfer by radiation may also be important.) In optically thick material, like rock, radiation is a diffusive process and can therefore be accounted for by an appropriate adjustment of K. (See section 9.4.2.) The source and sink terms $Q_{sources}$ and Q_{sinks} include heating due to the decay of radioactive elements, melting and other phase changes, gravitational settling of material, dissipation of mechanical strain energy (tidal heating), work done by pressure and viscous forces, and any other processes that may be important.

Some important features of planetary thermal histories can be inferred by considering typical numerical values of the terms in this equation. For instance, the importance of radioactive heating is shown by estimating its contribution to $\partial T/\partial t$. Early in its history the mean heat production per unit mass throughout the Earth's mantle due to the decay of radioactive isotopes was probably close to 3×10^{-7} erg/g-sec (see section 9.4.1). For values of $\rho = 4.5$ g/cm^3 (the mean mantle density) and $C_P = 1 \times 10^7$ erg/g-°C (typical of terrestrial silicates), radioactivity would cause a temperature rise of about 1000°C in the first billion years if heat losses were negligible.

The characteristic time for the transport of heat can be estimated by balancing the transport terms with the left-most term in Eq. (9.3.1). If only conduction and radiation were effective (no convection, or $\mathbf{u} = 0$), the time scale for energy redistribution would be $\rho C_P L^2/K$, where L is a length comparable to the planetary radius. For the typical values $K = 4 \times 10^5$ erg/cm-sec-°K and $L \geq 10^3$ km, the conduction time is greater than 3×10^{10} years—several times the age of the solar system. This means that rocks are sufficiently good insulators so that heat deposited within planets has not yet had time to escape by conduction or radiation alone. However, inspection of the convective term (second from left) in Eq. (9.3.1) reveals that velocities $u \simeq K/\rho C_P L \simeq 10^{-10}$ cm/sec could transport heat as effectively as conduction, and the time scale for energy redistribution is simply the "planet-crossing" time L/u. If the plate velocities on Earth of the order of 10^{-7} cm/sec are a measure of creep velocities within the planet, energy is effectively redistributed in less than a billion years. Thus one expects that the convection associated with plate tectonics dominates the energy transport in the Earth.

As noted in section 9.2, stability calculations have shown that solid state convection is expected to occur in all the terrestrial planets and the Moon if the viscosities of the interior of these bodies are sufficiently low (i.e., $<10^{24}$ poise). The energy conservation Eq. (9.3.1) alone is then insufficient to solve the thermal history problem, because it contains the unknown velocity, itself a function of time and space. It is therefore necessary to solve simultaneously the additional Navier-Stokes equations, which are hydrodynamic equations of mass and momentum conservation.

The basic equations for conservation of mass, momentum and energy within planetary interiors are given respectively by (e.g., Oxburgh and Turcotte, 1978)

$$\frac{D\rho}{Dt} + \rho\,\frac{\partial u_i}{\partial x_i} = 0 \qquad (9.3.2)$$

$$\rho\frac{Du_i}{Dt} = -\frac{\partial P}{\partial x_i} + \frac{\partial \tau_{ij}}{\partial x_j} + \rho\, g_i \qquad (9.3.3)$$

$$\rho C_P \frac{DT}{Dt} - T\alpha \frac{DP}{Dt} = \frac{\partial}{\partial x_i}\left(K \frac{\partial T}{\partial x_i}\right) + \tau_{ij} \frac{\partial u_i}{\partial x_j} + \rho H \qquad (9.3.4)$$

where x_i is a component of the space coordinate vector, u_i is a component of the velocity vector, $D/Dt = \partial/\partial t + u_j \partial/\partial x_j$ is the convective time derivative, P is the pressure, τ_{ij} is a component of the stress tensor, g_i is a component of the gravitational acceleration vector; α is the volume coefficient of thermal expansion, H is the heat release per unit mass, and i = 1, 2, or 3. Repeated subscripts in Eqs. (9.3.2–4) denote summation over that index. The two terms in Eq. (9.3.2) represent the time rate of change of density and the net mass flux, respectively. Equation (9.3.3) indicates the balance of the inertia force with the sum of the pressure force, the viscous force and the body force. In Eq. (9.3.4), the five terms are, respectively, the time rate of change of enthalpy, the time rate of change of compressional work, the heat transport by conduction, the viscous dissipation (or shear heating) and the internal heat release.

It should be noted that the effects of phase changes within the convective medium have been neglected in the above formulation. It has been demonstrated that most mineral assemblages will undergo phase changes as pressure and temperature are varied (see Ringwood, 1975, for a summary). Associated with each phase transformation are changes in the physical properties of the material, such as density and viscosity. Even thermal conditions may vary due to the interchange of latent heat. Studies on convection including phase changes (Schubert and Turcotte, 1971; Schubert *et al.*, 1975; Richter, 1973b) indicate that the phase changes probable in the Earth's mantle likely will not affect the global convective patterns appreciably. In modelling planetary thermal histories, therefore, the effects of phase changes have been largely ignored by current investigators.

The effects of chemical stratification, although implicitly contained in Eqs. (9.3.2–4) by the inclusion of a depth dependent ρ, have so far been largely ignored in thermal modelling. Only the obvious distinction between metal cores and silicate mantles has been taken into account. Chemical stratification could have a stabilizing effect, thereby requiring the analysis of separate, nonmixing convective layers, or perhaps preventing convection altogether (Richter and Johnson, 1974).

Following the general practice, Eqs. (9.3.2–4) are expanded about a reference adiabatic state

$$P = P_a + p, \quad T = T_a + \theta, \quad \rho = \rho_a + \rho'$$

in which the adiabatic quantities P_a, T_a and ρ_a satisfy

$$\frac{dP_a}{dx_i} = \rho_a g_i \qquad (9.3.5)$$

$$\frac{dT_a}{dP_a} = \frac{\alpha T_a}{C_P \rho_a} \qquad (9.3.6)$$

$$\rho_a = \rho_o (1 + \chi_a P_a) \qquad (9.3.7)$$

with ρ_o the zero-pressure density and χ_a the adiabatic compressibility. Furthermore, the linear equation of state can be written as $\rho' = \rho_a \alpha \theta$, since the density variation due to the compression of the perturbed pressure is small.

In order to examine the contribution of each term in the governing Eqs. (9.3.2–4), it is generally more convenient to express the equations in nondimensional forms. Therefore, we proceed to transform the expanded equations according to the nondimensional variables as given in Oxburgh and Turcotte (1978). The resultant nondimensional governing equations are

$$\frac{\partial u_i}{\partial x_i} + \frac{\Gamma}{\gamma} u_i g_i = 0 \qquad (9.3.8)$$

$$\frac{\rho_a}{Pr}\frac{Du_i}{Dt} = -\frac{\partial p}{\partial x_i} + \frac{\partial \tau_{ij}}{\partial x_j} + \frac{\Gamma}{\gamma}\rho_a g_i p - \overline{Ra}\, \rho_a g_i \theta \qquad (9.3.9)$$

$$\rho_a \frac{D\theta}{Dt} - \Gamma \rho_a g_i u_i \theta =$$

$$\frac{\partial}{\partial x_i}\left[K \frac{\partial}{\partial x_i}(\theta + \Gamma T_a)\right] + \frac{\Gamma}{\overline{Ra}} \tau_{ij} \frac{\partial u_i}{\partial x_j} + \rho_a \qquad (9.3.10)$$

In these equations, the length scale is measured in units of the planetary radius R, time in units of the conduction time $\rho_o C_P R^2/K_o$ (ρ_o and K_o are a reference density and conductivity, respectively), the pressure P and viscous stress τ_{ij} in units of $K_o \eta_o / \rho_o C_P R^2$ (η_o is a reference viscosity), and the temperature in units of $\rho_o H R^2 / K_o$.

Four dimensionless parameters appear in these equations: Pr, $\overline{Ra}$, Γ, and γ. The Prandtl number is

$$Pr \equiv \nu_o / \kappa_o \qquad (9.3.11)$$

where κ_o is the thermal diffusivity ($= K_o/\rho_o C_P$) and ν_o is the kinematic viscosity ($= \eta_o/\rho_o$). For solid planetary interiors, Pr is always very large ($\sim 10^{24}$), and therefore the acceleration terms on the left side of the momentum Eq. (9.3.9) (including the Coriolis force in a rotating planet), which are multiplied by 1/Pr, are always unimportant.

The Rayleigh number

$$\overline{Ra} \equiv \alpha \rho g H R^5 / K_o \nu_o \kappa_o \qquad (9.3.12)$$

provides a measure of the relative amplitudes of the buoyant driving force and the viscous retarding force. (Here g is the surface value of the scalar gravitational acceleration.) Stability analyses show that if $\overline{Ra}$ exceeds a critical value, which depends on boundary conditions and the precise temperature distribution but which is usually on the order of 10^3, convection will occur. The Rayleigh numbers for the Moon and terrestrial planets are given in Table 9.3.1 for the typical values of $\alpha = 3 \times 10^{-5}\,{}^\circ K^{-1}$, $\kappa_o = 10^{-2}\,cm^2/sec$, $\nu_o = 10^{22}\,cm^2/sec$, and $H = 1.7 \times 10^{-7}\,erg/cm^3$-sec, the present chondritic value. The values far exceed the critical value so that solid state convection can be expected. It should be noted that Rayleigh numbers depend critically on the viscosity of the medium. Table 9.3.1 is calculated assuming that viscosities of the interiors of the terrestrial planets are similar to that of the Earth's mantle.

Table 9.3.1 Dimensionless parameters governing convection in the terrestrial planets.

	R (km)	$\overline{Ra} = \frac{\alpha\rho g H R^5}{K_o \kappa_o \nu_o}$	$\Gamma = \frac{\alpha g R}{C_P}$
Moon	1738	3×10^6	0.08
Earth	6371	1×10^{10}	1.86
Venus	6050	9×10^9	1.60
Mars	3390	2×10^8	0.38
Mercury	2439	4×10^7	0.27

The adiabatic compression parameter

$$\Gamma \equiv \alpha g R / C_P \tag{9.3.13}$$

determines the importance of compressional heating, and together with $\overline{Ra}$ provides a measure of the viscous dissipation. This parameter is the ratio of the planet radius to the scale depth of the adiabatic temperature. Values of Γ are tabulated in Table 9.3.1 for the Moon and the terrestrial planets. From the table it may be seen that the effect of compressional heating may be important for the Earth and Venus.

The parameter

$$\gamma = \alpha / \rho_o C_P \chi_a \tag{9.3.14}$$

is the Grüneisen parameter. It is a measure of the compressibility of the material and is a physical property with values between one and two, for silicates. Equations (9.3.13–14) suggest that compressibility effects are of the same order as adiabatic gradient and viscous dissipation effects.

In order to solve Eqs. (9.3.8–10), one must specify boundary conditions (surface temperature and velocity conditions) and an "initial" temperature distribution. The latter can be taken as the temperature due to everything that happened within, say, the first 10^8 years of the planet's existence.

9.3.2 Methods of solution for thermal convection

Linear stability analysis

The convective term on the left side of the energy Eq. (9.3.4) couples the velocity and temperature fields. This nonlinearity precludes general analytic solutions for thermal convection. If the velocities are small the applicable equations can be linearized.

The linear stability problem for thermal convection constitutes an eigenvalue problem. The Rayleigh number is the eigenvalue. For a particular problem and set of boundary conditions the Rayleigh number eigenvalue is the critical value for the onset of thermal convection. For Rayleigh numbers smaller than the critical value the fluid is stable and no thermal convection occurs. The linear stability problem for a fluid layer heated from below was solved by Rayleigh (1916); with free surface boundary conditions the critical Rayleigh number is 657.5. Holmes (1931) used this analysis to show that thermal convection could be expected in the Earth's mantle.

Solutions for the critical Rayleigh number for thermal convection in a self-gravitating fluid sphere have been given by Jeffreys and Bland (1950), Chandresekhar (1952), Backus (1955), and Roberts (1965). The linearized equations are solved using an expansion in spherical harmonics, and a critical Rayleigh number is found for each harmonic by solving an eigenvalue problem. The minimum critical Rayleigh number is $Ra_c = 2214$ for a free-surface boundary condition and $Ra_c = 5758$ for a fixed-surface boundary condition. In both cases the minimum critical Rayleigh number is associated with the single convection cell, $l = 1$ mode. Schubert *et al.* (1969), Cassen and Reynolds (1973), and Cassen *et al.* (1976) have used these results to show that conduction profiles within all the terrestrial planets and the Moon are likely to be unstable and that thermal convection can be expected.

The linearized solutions are valid only near the stability limit. Also, because of the nature of linearized solutions, the amplitude of the convection is not obtained. Therefore, in order to study finite amplitude thermal convection, solutions to the nonlinear problem are required or else observed velocities must be used as constraints.

Boundary layer solutions

For large values of the Rayleigh and Prandtl numbers a boundary-layer analysis can be used to determine the structure of thermal convection cells (Turcotte and Oxburgh, 1967). Boundary-layer models are particularly useful in relating mantle convection to plate tectonics.

The boundary-layer model for thermal convection in a fluid layer heated from below is illustrated in Fig. 9.3.1. The flow is divided into cellular two-dimensional rolls; alternate rolls rotate in the clockwise and counter-clockwise directions. The entire flow field is highly viscous. On the hot and cold boundaries thin thermal boundary layers form. When the two hot boundary layers from adjacent cells meet, they separate from the boundary and form a thermal plume which rises to the upper surface. When the hot plume impinges on the upper cold boundary, a stagnation-point boundary layer is formed. As the flow splits and continues along the cool upper surface, the stagnation-point boundary layer becomes the cold thermal boundary layer on the upper surface. When two cold boundary layers from adjacent cells meet they form a cold descending plume. The buoyant forces in the ascending and descending plumes drive the flow. The core of each cell is a nearly isothermal, viscous, rotational flow.

The boundary layer structure illustrated in Fig. 9.3.1 can be interpreted directly in terms of plate tectonics. The cold surface boundary layer may be regarded as lithosphere. The conversion of the cold thermal boundary layer to a cold descending thermal plume is equivalent to the subduction of oceanic lithosphere at oceanic trenches. The gravitational body force in the descending plume drives the convection flow just as the gravitational body force on the descending lithosphere plays an important role in driving plate tectonics.

Although the boundary-layer approach is useful in understanding thermal convection in the terrestrial planets, many of the required assumptions limit its usefulness in terms of quantitative results. The planform of the convection, for instance, must be known *a priori*. In order to solve more complicated problems, numerical methods are generally required.

Numerical methods

There are three different numerical approaches that are applicable in solving the set of equations governing thermal convection. These three schemes will be briefly summarized in this section. A more detailed description can be found in Torrance (1979).

Spectral (normal mode) methods. Spectral methods are semi-analytical. Solutions are sought in the form of truncated series of smooth functions in one or more of the independent variables. The number of dimensions of the problem is effectively reduced by the number of expanded independent variables. In the case of spherical geometry, the dependent variables are generally expanded into spherical harmonics (see Young, 1974; Arkani-Hamed, 1979) which are expressed in terms of θ and ϕ. After some manipulations, a set of nonlinear ordinary differential equations in terms of the radial variable are obtained. This set of ordinary differential equations is then solved numerically to determine the thermal and flow fields. The spectral method has been demonstrated to be successful for nonlinear three-dimensional calculations. However, its application to study strong temperature and pressure dependent viscosity with adequate harmonic resolution, or non-Newtonian viscosity, has yet to be accomplished.

Finite difference numerical methods. Finite difference methods have enjoyed widespread application to solving the governing equations for thermal convection. The solution method consists of dividing the domain of interest into a prescribed grid system and approximating the partial differential equations by

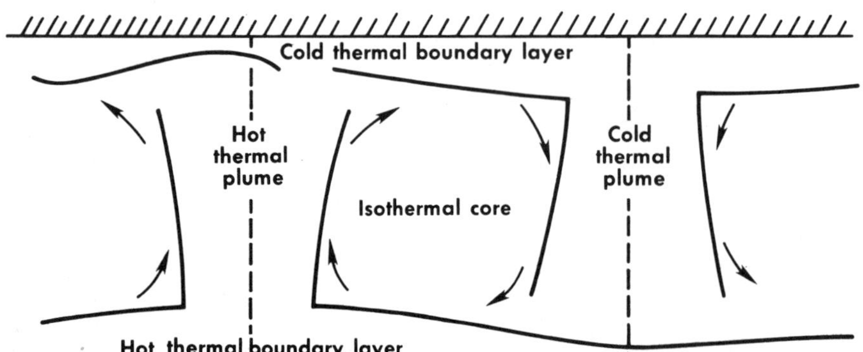

9.3.1 Boundary-layer model for two-dimensional thermal convection in a fluid layer heated from below. Lines approximately delimit boundaries of boundary layers and plumes, and arrows give schematic indication of flow directions.

equivalent finite difference equations. The finite difference equations are then solved for each nodal point of the grid system. Computations often proceed as transient problems starting from some initial state. Time steps are determined in such a way that numerical stability is preserved (Torrance and Rockett, 1969). This approach is successful in studying two-dimensional problems even for certain classes of complex rheologies. However, a large amount of computer time is required to study three-dimensional problems.

Finite element numerical methods. Application of finite element methods to solve the governing Eqs. (9.3.8–10) is a recent development and is evolving very rapidly. Successful applications to study flow problems with simple rheology have been reported by Sato and Thompson (1976), Gartling (1976) and Tabarrok and Lin (1977). The solution method is based on the principle of calculus of variations. The region of interest is first divided into a finite element grid system. The governing equations are then transformed into an integral quantity. Very often, this integral quantity represents the total energy of the system. The dependent variables at each nodal point of the finite element grid system are parameterized to form a set of algebraic equations such that the variation of the integral quantity is minimized. Solutions of the algebraic equations can be obtained using linear algebra theories. Finite element methods are extremely flexible in meeting irregular boundary conditions. Even though the stability and accuracy criteria are not yet established, it is nevertheless a very promising approach. The method has recently been applied to the problem of mantle convection in the Earth by Kopitze (1979); its application to study planetary thermal history has yet to appear in the literature.

Convection problems with numerical solutions. A wide variety of thermal convection problems in planetary interiors have been investigated to date, generally with the use of finite difference numerical techniques. Finite amplitude convection in a fluid layer heated from within has been studied by Turcotte *et al.* (1973) and by McKenzie *et al.* (1974). The structure of a thermal convection cell in a fluid layer heated from within is shown in Fig. 9.3.2. The lower boundary of the layer is insulating and the heat generated from within is lost through the upper boundary. The Rayleigh number as defined above is $\overline{Ra} = 10^5$. In the layer heated from within only a single thermal boundary layer develops on the upper boundary since there is no heat flow through the lower boundary. This boundary layer forms a single descending plume; since this plume is primarily responsible for driving the flow, the flow tends to be concentrated in this region.

In order to understand better the role of viscous dissipation in thermal convection, numerical calculations for various levels of dissipation have been carried out by varying the parameter Γ (Turcotte *et al.*, 1974; Hewitt *et al.*, 1975). The primary effect of increasing viscous dissipation is to damp out the convection. This is somewhat surprising since one might expect the heat generated by viscous dissipation to accelerate the flows. However, when Γ is increased to enlarge the effect of viscous dissipation, the depth dependence of the adiabatic temperature gradient is simultaneously enhanced. Since it is the difference between the actual temperature gradient and the adiabatic temperature gradient that drives thermal convection, an increase in the adiabatic temperature gradient inhibits thermal convection. It is concluded that viscous dissipation is unlikely to influence mantle convection significantly.

Laboratory and theoretical studies of mantle materials indicate that at the strain rates associated with mantle convection the mantle is likely to behave as a non-Newtonian fluid (section 9.4.3). In order to study the influence of a non-Newtonian rheology, Parmentier

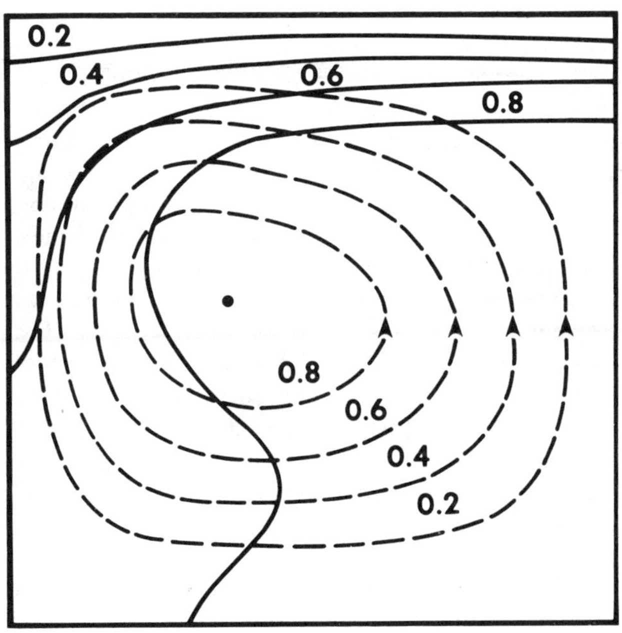

9.3.2 Thermal convection in a fluid layer heated from within and cooled from above with free surface boundaries and $\overline{Ra} = 10^5$. The solid lines are isotherms; labels give fraction of the maximum temperature difference. The dashed lines are streamlines; labels give fraction of the maximum stream function.

et al. (1976) have carried out a series of numerical calculations. It was concluded that an effective Rayleigh number could be defined for an equivalent Newtonian flow such that the stream lines and isotherms are virtually indistinguishable from the non-Newtonian case.

The mantle viscosity (or equivalent mantle viscosity for a non-Newtonian rheology) is likely to be an exponential function of temperature and pressure (see section 9.4.3). The pressure dependence is primarily through the hydrostatic pressure. The influence of viscosity variation on thermal convection in a fluid layer heated from below has been studied in a series of numerical calculations by Torrance and Turcotte (1971a, b) and by Houston and De Bremaecker (1975). These studies show that the flow is strongly concentrated near the axis of ascending flow on the upper boundary. This is to be expected because the high temperature and the small pressure contribute to a low viscosity. The rapid flow in this region tends to suppress the thermal boundary layer on the upper boundary. It is clear that a strongly temperature- and depth-dependent viscosity can significantly alter thermal convection patterns.

There are certainly striking similarities between two-dimensional thermal convection in a fluid layer and our present knowledge of mantle convection. Theoretical studies, however, show that two-dimensional thermal convection in a fluid layer with constant fluid properties and infinite Prandtl number becomes unstable at a Rayleigh number of 22,600 (Busse, 1967). Three-dimensional flow develops at higher Rayleigh numbers. Since most estimates of the Rayleigh number for the mantle are considerably larger than this value, a complex pattern of convection might be expected to occur beneath the surface plates. It has also been shown (Richter and Parsons, 1975) that thermal convection in a fluid layer heated from below with a moving, rigid upper boundary takes the form of two-dimensional rolls aligned with the direction of motion of the upper boundary, though the time scale for such roll alignment may be longer than the characteristic time for steadiness of plate motion, at least for the slow moving plates.

It is important to consider to what extent the results of constant-property two-dimensional studies must be modified in their application to mantle convection. The gravitational instabilities associated with thermal convection are generated within the cold, heavy thermal boundary layer adjacent to the upper boundary of the fluid layer. However, in the mantle the cold thermal boundary layer has considerable rigidity because of its low temperature. In order to study the influence of the rigid boundary layer on mantle convection, a series of numerical calculations have been carried out by Parmentier and Turcotte (1978). Without the removal of cold fluid from the convection cell into a thickening lithosphere, the maximum aspect ratio of convection (ratio of cell width to cell depth) that could be obtained was about 1.5. With such removal of the fluid, aspect ratios 10 and greater were obtained. These results indicate that a rigid boundary layer strongly stabilizes mantle convection.

Mantle flow associated with plate motions clearly occurs. An important question is whether mantle convection at other scales takes place. One source of evidence to examine this question is surface volcanism. There are a number of active volcanoes within plate interiors. These have been attributed by Morgan (1971, 1972a, b) to plume-like flows originating in the deep mantle.

Numerical studies of axially symmetric plumes in the mantle have been carried out by Parmentier *et al.* (1975), who concluded that because of the high viscosity of the mantle, narrowly confined plumes do not occur if the mantle is heated internally; rather, mantle flows tend to have widths and depths of equal magnitude. The inference that the degree of partial melting required for Hawaiian lavas is equal to or lower than that for mid-ocean ridges (section 3.3.4) also argues against a deep mantle origin, since rapid pressure-release from the deep mantle might be expected to lead to a much higher degree of partial melting.

Axisymmetric finite difference calculations for thermal convection in a self-gravitating, uniformly heated fluid sphere have been carried out by Hsui *et al.* (1972). As the Rayleigh number was increased it was found that the number of convection cells also increased. These calculations have been applied to the thermal structure of the Moon by Turcotte *et al.* (1972). Numerical solutions in an axisymmetric sphere with a temperature- and pressure-dependent viscosity have been given by Hsui (1978) and used for planetary thermal models by Toksöz *et al.* (1978).

9.3.3 Thermal history modelling techniques

The methods described in the previous section have mostly been applied to steady-state convection problems. Several of the methods have been used to study the full global thermal history problem for a planet, but many complex and important geophysical phenomena (e.g., the differentiation of radiogenic heat sources, the segregation of a planetary crust, and the formation of an iron-rich core) cannot readily be explicitly included.

The incorporation of these complex processes will undoubtedly prohibit a complete solution of the above equations under general circumstances. At the present time, these processes can be included in the calculations only by modelling them in such a way as to take into account their effects without specifying (or even knowing) the precise equations or parameter values that describe them. In what follows, the most commonly adopted modelling procedures for some of the complex geophysical processes will be described in more detail.

Heat transfer in the molten or partially molten state

In principle, the governing Eqs. (9.3.8–10) describe the convective heat transport in both the liquid and the solid states. In practice, however, it is rather difficult to resolve the rapid viscosity change near the solid-liquid interface. In order to have the convection in both phases taken care of automatically, a three-dimensional calculation using extremely fine grid systems and extremely small time steps would be necessary, a procedure that requires a prohibitively large amount of computer time. Therefore, simulation of heat transfer in the molten or partially molten state becomes a necessity.

A procedure for modelling convection in the molten or partially molten state was first proposed by Reynolds *et al.* (1966) and has since been adopted in almost all thermal history calculations. In their simulation, perfect convection efficiency within a discrete time interval is assumed. In other words, the excess heat beyond that necessary to keep a grid point at some specified fraction of complete melting is moved to the next higher grid point, and the temperature is set equal to the melting temperature or to a temperature at a specified position between solidus and liquidus. This modelling procedure includes latent heat and satisfies the energy conservation principle, but does not permit the occurrence of superheated melts.

Cold thermal boundary layer

At the surface of terrestrial bodies, there generally exists a cold boundary layer where viscosity increases very rapidly toward the surface. Due to the same reasons as described in the preceding section, these regions require a very large amount of computation to resolve. Therefore, in order to make the calculations feasible, these regions are generally approximated by conducting layers. The temperature at the bottom of the layer should be sufficiently low (i.e., the viscosity sufficiently high) so that solid state creep within the layer is negligible. This approximation has been adopted to calculate thermal histories of planetary objects with radii smaller than that of Mars (Toksöz *et al.*, 1978). It should be noted that the approximation is valid only if there exists no horizontal mobility of surface plates and no mixing between the surface layers and the deeper interior. For the Earth, where plate tectonics is active, a better modelling procedure for the cold surface thermal boundary layer is needed.

Radiogenic heat source differentiation

Studies of the surface heat fluxes and abundances of radioactive elements (U, Th, K) for both the Earth and the Moon imply that the surfaces of these two planets are enriched in radioactive elements while the interiors are depleted. Very little is known about the exact distribution of these heat sources in time and space. Two general procedures in use to model upwards enrichment of radioactivity are to assume an exponential decrease with depth (Lachenbruch, 1968) or to parameterize some mechanism for redistribution which depends on the local state of the evolving model. In the former method, the "skin depth" of the distribution function can be a free parameter which may vary with time. Probably the skin depth will decrease with time, because the radioactive elements are likely to be accumulated progressively near the surface by remelting and magmatic processes. Parameterization schemes for redistributing radioactive elements include that of Fricker *et al.* (1967), which uses a fixed "residual fraction" to describe the heat sources remaining in a layer following melting and a fixed "moving factor" to describe the rate of ascent of heat sources in a molten zone, and that of Solomon and Longhi (1977), which uses liquid-solid partition coefficients for U, Th, and K to redistribute heat sources between the fluid phase in any interconnected partial melt zone and the surrounding solid. For both procedures, the total abundances of heat sources must be conserved by the redistribution scheme.

Core formation

The existence of an iron-rich core inside the Earth has been confirmed by seismic studies (section 4.5.1). It is also likely that other terrestrial planets possess iron-rich cores as well. Core formation, if it occurs subsequent to planet formation, will have profound effects on the thermal history of a planetary object. The segregation of core material represents a conversion of potential energy to thermal energy. The amount of heat so derived depends on the core size, the density of core material, and increments to rotational and strain energy during differentiation. Though the actual process of core segregation is not yet understood in detail, the effect of core formation upon planetary thermal history can be incorporated approximately by estimating the energy involved. Most commonly, core formation is assumed to be a spherically symmetric process. At each time step, the interior temperature of a planet is checked against an adopted temperature at which local separation of core and mantle material can occur, say melting temperature of the core material or of the mantle material, whichever is lower. When this segregation temperature is reached, sinking of core material is assumed to take place instantaneously within the zone satisfying the condition for local separation of core and mantle material. The difference in potential energy before and after this increment of differentiation is converted to heat at some specified efficiency and is distributed in a given fashion (e.g., uniformly) within the differentiated region. A new temperature field is then calculated. By repeating this process whenever necessary, the effect of full core formation upon planetary thermal history can be roughly approximated. The thermal effects of core formation during accretion could be similarly modelled in principle, though this has not yet been done except for the trivial case of accretion of the core prior to accretion of mantle and crust.

Parameterized convection

For studies of the thermal evolution of planets it may be appropriate to parameterize the process of solid convection in order to estimate the convective heat flow without solving the Navier-Stokes equations for each thermal history calculation (Tozer, 1967; Sharpe and Peltier, 1978, 1979; Cassen *et al.*, 1979; Kaula, 1979a; Stevenson and Turner, 1979; Schubert *et al.*, 1979a, b; Turcotte *et al.*, 1979). This procedure relies on experimental data and the results of numerical calculations of convection in relatively simple configurations. On the basis of such data, convection in many cases has been shown to enhance the heat flux above that due to conduction alone by a factor proportional to Ra^β, where Ra is the Rayleigh number and β is a number between one-fifth and one-third. For a thermal history model, once it has been determined that a particular temperature distribution produces an unstable density profile, this relationship can be used to determine the convective transport term in the energy conservation Eq. (9.3.1). This simple parameterization of convective heat flux has the advantage of requiring the solution of only one partial differential equation with two independent variables. Hence extensive explorations of parameter space are feasible, and other complex phenomena associated with melting can be modelled without undue difficulty. The method has the disadvantages of providing no information on nonradial inhomogeneities, and relying on possibly incomplete studies of convection in situations essentially simpler than those encountered in planetary interiors. A strong case for the general applicability of the method has been made by Schubert *et al.* (1979a).

Usually, the parameterization of convective heat transport is expressed in terms of the Nusselt number, defined to be the ratio of convective to conductive heat flux in an equilibrium situation. For instance, the relation

$$Nu = b\ Ra^\beta \qquad (9.3.15)$$

has been found for the heat transport from a fluid layer at high Rayleigh number for the case of the fluid heated from below and transporting a steady quantity of heat. In Eq. (9.3.15), the Rayleigh number is based on the temperature difference ΔT across the layer, $Ra = \alpha g\ \Delta T\ L^3/\kappa\nu$, in contrast to the definition of $\overline{Ra}$ given in Eq. (9.3.12). Rossby (1969), Chu and Goldstein (1973) and Garon and Goldstein (1973) report the following results for the heat transport through layers of water:

Rossby (1969)

$Nu = 0.131\ Ra^{0.3},\ 3 \times 10^6 \geq Ra \geq 3.4 \times 10^4,$

Chu and Goldstein (1973)

$Nu = 0.183\ Ra^{0.278},\ 1.05 \times 10^8 \geq Ra \geq 2.76 \times 10^5,$

Garon and Goldstein (1973)

$Nu = 0.130\ Ra^{0.293},\ 3.3 \times 10^9 \geq Ra \geq 1.3 \times 10^7.$

Rossby (1969) has also measured the heat transport through layers of silicone oil and mercury. The results are

$Nu = 0.184\ Ra^{0.281},\ 3.5 \times 10^6 \geq Ra \geq 4 \times 10^3$, silicone oil;

$Nu = 0.147\ Ra^{0.257},\ 4.5 \times 10^6 \geq Ra \geq 2 \times 10^4$, mercury.

In many of these studies and some of the models that follow, the constant b in Eq. (9.3.15) is related to the critical Rayleigh number Ra_c.

A number of theories of high Rayleigh number convection provide additional support of relation (9.3.15) in the case of heating from below. Priestley (1954) argued that the heat transport, which is controlled by conduction through thin thermal boundary layers at high Ra, should be independent of layer thickness; this requires $Nu \sim Ra^{1/3}$. Kraichnan's (1962) mixing length theory also gives $Nu \sim Ra^{1/3}$, although a different relation is suggested for extremely high Rayleigh numbers. The boundary layer theory of Turcotte and Oxburgh (1967) yields $Nu = 0.167\ Ra^{1/3}$. Howard's (1966) theory of the instability of the thermal boundary layer and Long's (1976) scaling and boundary layer arguments also result in $Nu \sim Ra^{1/3}$.

Because the mantles of the planets are not simple fluid layers heated from below, it is important to extend the justification of Eq. (9.3.15) to more general situations, including internally heated convection and transient convection. First, consider the situation in which steady convection occurs in a layer of fluid that is heated uniformly throughout its interior and has an insulated lower boundary. Kulacki and Nagle (1975) and Kulacki and Emara (1977) have carried out experiments to determine the temperature difference ΔT across the layer in this circumstance. They electrically heated a dilute aqueous electrolyte to produce energy at the rate H per unit volume throughout the layer. They present their results in terms of a Nusselt number Nu′ and a Rayleigh number Ra′ defined as

$$Nu' \equiv \frac{HL}{K\Delta T/L}$$

$$Ra' \equiv \frac{\alpha g L^3}{\kappa\nu}\left[\frac{HL^2}{2K}\right] = \frac{\overline{Ra}}{2}$$

where L is the thickness of the layer, so H L is the heat flux from the top of the layer in steady state, and $K\Delta T/L$ is the conductive heat flux associated with the temperature difference ΔT across the layer. Thus, the Nusselt number Nu′ is the ratio of the actual to the conductive heat flux. $H L^2/2K$ is the temperature difference that would exist across the layer if the internally generated heat were removed by conduction.

Their steady state data are well-represented by

$$Nu' = 0.396\, Ra'^{0.227}, \qquad (9.3.16)$$
$$1.89 \times 10^3 \le Ra' \le 2.17 \times 10^{12}$$

Equation (9.3.16) can be put into the form of (9.3.15) as follows. Rewrite Ra′ as

$$Ra' = \frac{\alpha g L^3}{\kappa\nu}\frac{\Delta T}{2}\left(\frac{HL}{K\Delta T/L}\right) = \frac{Ra}{2} Nu' \qquad (9.3.17)$$

and substitute (9.3.17) into (9.3.16) to obtain

$$Nu' = 0.396\left(\frac{Ra\, Nu'}{2}\right)^{0.227} \qquad (9.3.18)$$

Solve (9.3.18) for Nu′ and find

$$Nu' = 0.228\, Ra^{0.294}$$

in approximate agreement with Rossby's (1969) data for water.

Convection in spherical shells with internal heating and insulated lower boundaries has been studied numerically by Young and Schubert (1974), Schubert and Young (1976) and Schubert *et al.* (1977). These papers presented thermal models of Mars, Earth and the Moon from which one can calculate the amounts by which the average temperatures of the convecting mantles exceed the base temperatures of the lithospheres. Log-log plots of these excess temperatures ΔT against Ra/Ra_{cr} reveal approximate power law dependences of ΔT on Ra/Ra_{cr}, which are consistent with Eq. (9.3.15) and imply $\beta \simeq 0.3$.

Schubert *et al.* (1979a) also show that Eq. (9.3.15) is consistent with the transient responses to step-function changes in the heating rate found by Kulacki and Nagle (1975) and Kulacki and Emara (1977). Finally, Cassen and Young (1975) analyzed a system (in spherical geometry) in which steady convection was driven by both internal energy sources and an imposed temperature differential. Their results also indicate that Eq. (9.3.15) gives a good estimate, if not an exact formula, for the heat flux due to convection in a constant viscosity fluid, regardless of the thermal boundary conditions. Furthermore, the experiments of Booker (1976) and Booker and Stengel (1978) with Benard convection in a variable viscosity fluid show that temperature dependent viscosity has a relatively minor effect on the Nu-Ra relation when the Rayleigh number is based on the viscosity corresponding to the mean temperature of the layer. Also, numerical calculations of convection in non-Newtonian fluids (Parmentier *et al.*, 1976; Parmentier, 1978) suggest that the relation holds even for more complicated rheologies. Thus, it appears that Eq. (9.3.15) can be applied to the thermal balance of planetary interiors with some confidence.

Discussions of thermal history calculations using parameterized convection are given in Sharpe and Peltier (1978, 1979), Cassen *et al.* (1979), and Schubert *et al.* (1979a, b). Sharpe and Peltier (1978, 1979) assume that the convective transport is given by Nu K $(\partial T/\partial r)$, which then casts the energy equation into the form of a conduction equation with enhanced thermal conductivity. In the numerical scheme of Cassen *et al.* (1979), the temperature distribution in a convecting region is reduced to an adiabat at each time step of the calculation, in a manner such that the energy transported upwards is that prescribed by the Nusselt-Rayleigh number relation applied to the region as a whole. Schubert *et al.* (1979 a, b) characterize the convecting region by a single value of the temperature. Energy balance equations are derived for the convecting mantle based on the convective transport formula, and for the lithosphere, which is purely conducting. Matching of the heat flux between the two zones couples the two energy equations. The system of equations so derived is numerically integrated with respect to time, given the initial values of temperature and lithosphere thickness.

Thermal stress

One consequence of the heating and cooling of a planet is that thermal stress is produced in the lithosphere. Such a process may be a dominant source of stress in planets lacking lithospheric recycling. Although the determination of thermal stress is not a necessary step in the description of a planetary thermal history, the surface tectonic features possibly produced by lithospheric failure in response to thermal stress provide one of the most direct tests of thermal history models (MacDonald, 1960; Solomon and Chaiken, 1976).

In the elastic portions of a spherically symmetric planet, if over some time interval the temperature changes by ΔT as a function of radius r, then the radial displacement u and the radial and tangential stress σ_r and σ_t are of the form

$$u(r) = rI(r) + Ar + B/r^2 \quad (9.3.19)$$

$$\sigma_r(r) = -\frac{2E}{1+\nu} I(r) + \frac{EA}{1-2\nu} - \frac{2EB}{(1+\nu)r^3} \quad (9.3.20)$$

$$\sigma_t(r) = \frac{E}{1+\nu} I(r) + \frac{EA}{1-2\nu} + \frac{EB}{(1+\nu)r^3} - \frac{E\alpha(r)\Delta T(r)}{3(1-\nu)} \quad (9.3.21)$$

where E, ν, and α are Young's modulus, Poisson's ratio, and the volumetric thermal expansion coefficient, respectively

$$I(r) = \frac{1}{r^3}\int_0^r \left(\frac{1+\nu}{1-\nu}\right) r^2 \frac{\alpha(r)\,\Delta T(r)}{3}\, dr$$

and A and B are constants within each layer of homogeneous E and ν, to be determined by boundary conditions. The simplest case is for a completely elastic planet of radius R and of uniform elastic constants, for which B = 0 and

$$A = \frac{2(1-2\nu)}{3(1-\nu)} \frac{1}{R^3}\int_0^R r^2\, \alpha(r)\, \Delta T(r)\, dr$$

The case of layered elastic shells overlying a fluid interior is somewhat more complicated, but straightforward.

Equations (9.3.19–21) are for an elastic medium and do not give accurate estimates of stresses after failure or in the regions of a planetary interior hot enough to deform by plastic creep. Nonetheless, the relations are a useful guide to the type and age of lithospheric failure features predicted to be present on a planetary surface for a given thermal history model.

9.4 PHYSICAL PROPERTIES

Solutions for the thermal histories of the terrestrial planets require information on many physical properties. Of fundamental importance are sources of heat, of which the decay of radioactive isotopes is perhaps the most important. Information on the spatial distribution and the time dependence of these heat sources is required. Conduction of heat depends upon the thermal conductivity and heat capacity of silicate rocks.

Solid state convection may have dominated the thermal structure and evolution of the Earth and is likely to play an important role in the other terrestrial planets. In order to determine the form and characteristic velocities of mantle convection, the viscosity of the mantle, or the equivalent viscosity for a non-Newtonian rheology, and the coefficient of thermal expansion are required.

Seismic studies have shown the presence of important discontinuities in seismic velocity and density within the Earth's upper mantle. The characteristics of some of these discontinuities have been associated with phase changes on the basis of laboratory studies. The density changes and heat release associated with such phase changes can have an important influence on convective flows. Melting relations in the mantle are important as an indicator of magma production (Chapter 3) and because the melting curve enters into one form of the expression for viscosity as a function of temperature and depth.

9.4.1 Heat sources

Radioactivity

Long-lived radioactivity. The primary source of heat produced over the span of planetary history is that due to the long-lived radioactive isotopes of U, Th, and K. The decay constants, the decay energies, and the relative and representative absolute present-day abundances for these isotopes are given in Table 9.4.1.

The spatial distribution of U, Th, and K, and hence the distribution of their released heat, are affected by episodes of melting, crystallization, magma transport, and solid-state convective mixing. Abundances of these elements in the upper crusts of the planets may be determined by direct sampling and by remote sensing, but crustal abundances for the Earth and the Moon are both much higher than can be representative of the bulk of the planet on the basis of both heat flow and internal temperature estimates.

It has been common to estimate the total current abundances of radioactive heat sources in a planet by equating the surface heat flow to the heat that would be produced by a body with the same volume as the planet and with ratios of Th/U and K/U equal to those found in crustal rocks. On this basis, Wasserburg *et al.* (1964) matched an adopted average Earth heat flow of 64 mW/m^2 and the observed ratios Th/U = 3.7 and K/U = 10,000 with a current bulk concentration of 33 ppb in the mantle and crust. (With the latest values of heat flow, decay constants and isotopic ratios, this figure is 45 ppb U; see below). Similarly, Langseth *et al.* (1976) estimated a bulk U concentration of 46 ppb for the Moon from a surface heat flux of 18 mW/m^2 and Th/U and K/U ratios of 3.7 and 2,000, respectively. On the basis of mantle abundances of other refractory elements, however, Ringwood (1975) has argued for an Earth composition closer to chondritic for U and Th (see also section 1.3.1). For a steady state thermal balance, such a composition gives a terrestrial heat production which is considerably less than the present observed heat loss. If this geochemical argument is correct, therefore, the Earth is cooling. Recent calculations by Daly and Richter (1978), Sharpe and Peltier (1978, 1979), Schubert *et al.* (1979b), Cassen *et al.* (1979), Hsui (1979), as well as models presented later, show a substantial lag between heat production and heat flow even for vigorously convecting systems and also cast doubt on the steady-state assumption.

It is worth noting that the K/U ratio may vary greatly over solar system objects, and that this variation will strongly affect the relative heat production with time because of the shorter half-life of ^{40}K than those of ^{238}U and ^{232}Th. Measured K/U ratios vary from the chondritic value of 70,000 (Mason, 1971), to 10,000 for three surface sites on Venus (Surkov, 1977) and for the crust and upper mantle of the Earth (Wasserburg *et al.*, 1964), to 3,000 to 10,000 for basaltic achondrites (Mason, 1971), to about 2,000 for returned samples from the Moon (Gast, 1972). The ratios thus range from values at which ^{40}K decay dominates radioactive heat production to values at which ^{40}K is only a relatively minor contributor. Figure 9.4.1 shows the relative heat production in terms of surface heat flow of U, Th and K with time for a steady-state Earth model, though as noted above, the steady-state assumption may yield overestimates of internal heat production.

The long-lived radioactive elements have combinations of ionic radius and valence state that prevent their being readily accommodated within the lattices of most common silicates, although K is more easily accepted than the others. In any partial melting of a silicate mineral assemblage, U, Th and K are all preferentially concentrated in the liquid phase. This preferential concentration is so efficient that the majority of the radioactive elements enters the first few percent of melt (Philpotts and Schnetzler, 1970; Dostal and Capedri, 1975), so that melting is consequently a very effective process for depleting and/or concentrating radioactive heat.

Direct comparison of crustal rocks with blocks of material from the mantle carried to the surface by lavas make it clear that processes which led to the formation of the continents from the mantle have strongly concentrated radioactive heat production within the continental crust (Table 9.4.2). Until recently, estimates of total continental radioactivity were almost entirely based on surface observations. The work of Birch *et al.*, (1968)

Table 9.4.1 Heat generation from long-lived radioactivity.

	E_i Energy Decay per Atom[(a)] (10^{-6} erg)	λ_i Decay Constant[(b)] (10^{-10} yr^{-1})	H_i Decay Energy[(c)] (erg/g-s)	Relative Isotopic Abundance[(b)]	Present Terrestrial Abundances[(d)] (10^{20} g)
^{238}U	75.9	1.55125	0.940	0.9928	1.82
^{235}U	72.4	9.8485	5.79	0.0072	0.0132
^{232}Th	63.8	0.49475	0.260	1.0	6.79
^{40}K	1.14	5.543	0.302	1.167×10^{-4}	2.14

(a)Birch (1954).
(b)Steiger and Jager (1977).
(c) $H_i = E_i \dfrac{N\lambda_i}{m_i}$, where N = Avogadro's number and m_i = atomic weight.
(d)Abundances are those which would satisfy a steady-state heat balance, with $K/U = 10^4$ and $Th/U = 3.7$ and with a mean Earth heat flow of 82 mW/m^2 (Sclater *et al.*, 1980).

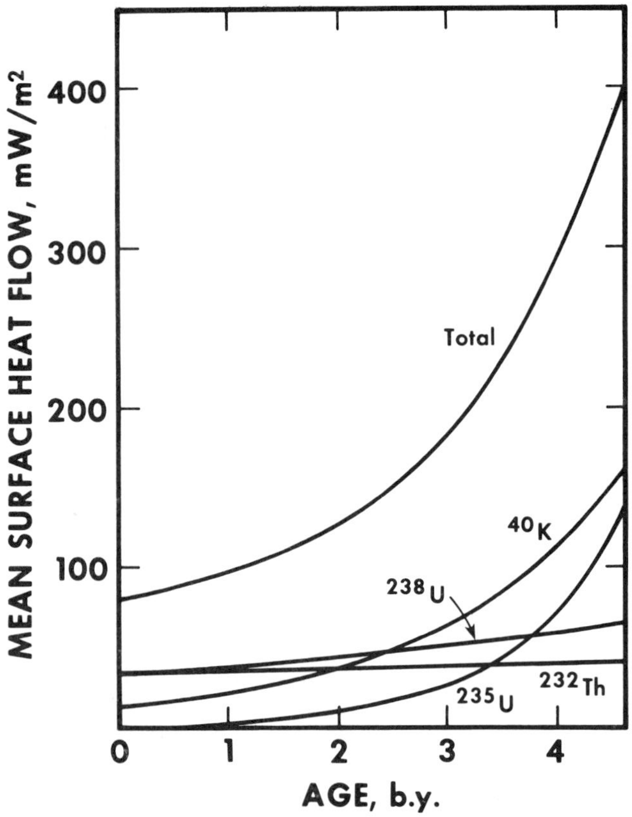

9.4.1 Heat production in the Earth during geological time assuming that the present surface flux is in equilibrium with radiogenic heat production. In the lower part of the diagram the separate contributions of U, Th and K are shown. Present day values of Th/U = 3.7 and K/U = 10,000 are assumed.

and Lachenbruch (1968), subsequently confirmed by others, has established that within crustal regions which have undergone major episodes of intrusion of igneous rock (implying mean crustal temperatures close to the melting point) there is a linear dependence of surface heat flow on the heat production in surface rocks (Fig. 9.4.2). This relation is consistent with differential erosion only if

$$A = A_0 \exp(-y/d) \qquad (9.4.1)$$

where A is the heat production per unit volume and time at the surface, y is the depth and d is the scale depth for the heat production (Lachenbruch, 1970). For most parts of the world, d is close to 8 km (Pollack and Chapman, 1977a), but there is considerable regional variation.

Although the steady-state assumption can be questioned as discussed above, it yields at least a rough estimate of the mean heat generation within the Earth. Taking the present value of 82 mW/m^2 for the average surface heat flow (Sclater *et al.*, 1980), the total Earth heat loss is about 4.2×10^{13}W; this is very much larger than the total energy released by earthquakes (given as 3×10^{10}W by Richter, 1958) and significantly larger than the energy associated with tidal dissipation (6×10^{12}W), much of which is dissipated in the oceans and shallow seas (Lambeck, 1975). If the total surface heat flow is attributed to steady state radioactive decay, the bulk Earth abundances of long lived radioactive nuclides are as given in Table 9.4.1. These values correspond to an average crust plus mantle heat generation of 1.0×10^{-11} J/kg-s (1.0×10^{-7}erg/g-s) and average U, Th and K concentrations of 45 and 168 ppb and 455 ppm, respectively, if the radioactive elements are assumed to be distributed uniformly throughout the crust and mantle. The average heat generation for the bulk Earth under the steady state assumption is 0.70×10^{-11} J/kg-s, 1.3 to 1.8 times the chondritic value (Table 9.4.2).

It is appropriate to compare these values with those measured in mantle-derived materials (Table 9.4.2). The most abundant mantle-derived magmas on Earth

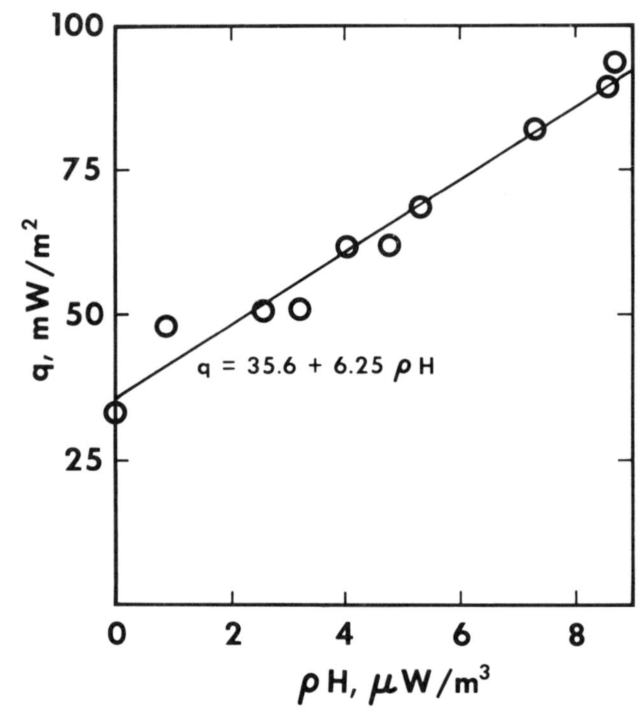

9.4.2 Dependence of surface heat flow on surface radioactive heat generation for the eastern United States (Birch *et al.*, 1968).

are the oceanic tholeiites (section 1.2.5), generated at mid-ocean ridges and forming a large fraction of the oceanic crust. Evidence from experimental petrology and geochemistry suggests that these lavas represent about 20–25% melting of the source rock (section 3.8.1). For the reasons discussed above, these basalts should contain almost all of the heat producing isotopes present in the source, which should in consequence have had an original heat production value about one-fifth to one-fourth that of the basaltic magmas. The heat production in tholeiites may be estimated from recent precise analyses (Hart, 1969; MacDougall, 1977; Ozima

Table 9.4.2 Estimates of concentrations of heat producing elements and heat production in representatives of various rock types.

	U (ppm)	Th (ppm)	K (%)	H (10^{-7} erg/g-s)	Reference
Meteorites					
Type I carbonaceous chondrites (mean)	0.012	0.038	0.053	0.040	Mason (1971)
Ordinary chondrites (mean)	0.012	0.040	0.084	0.052	Mason (1971)
Continental igneous rocks					
Diorites, quartz diorites (mean)	2.0	8.5	-	-	Clark *et al.* (1966)
Granodiorites (mean)	2.6	9.3	2.6	59	Clark *et al.* (1966)
Silicic igneous rocks (mean)	4.7	20	3.8	111	Clark *et al.* (1966)
Basaltic kimberlites, southern Africa, mean of 82 samples	2.4	-	-	-	Kresten (1974)
Oceanic tholeiites					
Mean of 8 'fresh' samples, Atlantic and Pacific	-	-	0.155	-	Hart (1969)
'Fresh' glass from chilled margins of DSDP basalts, range	0.016 - 0.450	-	-	-	MacDougall (1977)
Mean of 68 'fresh' ridge and seamount basalts	0.078	-	-	-	Fisher (1979)
Mean of 10 'weathered' basalts from Easter Island and volcanic chain	1.6	-	-	-	Fisher (1979)
'Typical' tholeiite	0.08	0.30	0.16	2.1	
Mantle-derived ultramafic nodules					
Lherzolites, Victoria, Australia					Green *et al.* (1968)
Low H	0.003	0.014	0.0012	0.070	
High H	0.114	0.457	0.011	2.3	
Lherzolites, Pacific					Wakita *et al.* (1967)
Mean of 7	0.019	0.055	0.0054	0.35	
Low H (Hualalai, Hawaii)	0.0076	0.024	0.0022	0.14	
High H (Salt Lake Crater, Hawaii)	0.040	0.124	0.0044	0.73	
Continental peridotites					Wakita *et al.* (1967)
Low H (British Columbia)	0.058	0.041	0.0033	0.079	
High H (France)	0.048	0.77	0.0362	2.6	
Dunite, Twin Sisters, Washington	0.0040	0.0097	0.00091	0.067	Wakita *et al.* (1967)
Eclogites					Heier (1963)
Salt Lake Crater, Hawaii	0.05	0.09	0.02	0.79	
Mean of 7 nodules from South African kimberlites	0.30	0.47	0.33	5.3	

et al., 1977) which have avoided material that shows evidence of sea water interaction, an effect that had given spuriously high values in earlier analyses. Using MacDougall's (1977) range in U values, a Th/U ratio of 3.7, and a K/U ratio of 2×10^4, the heat production values of tholeiites are in the range 4×10^{-8} to 12×10^{-7} erg/g-s. Based on a "typical" tholeiite with U = 0.08 ppm, Th = 0.30 ppm, and K = 0.15% (Hart, 1969; Fisher, 1979), the heat production is 2.1×10^{-7} erg/g-s (Table 9.4.2), substantially lower than 4-5 times the mantle average expected from the steady state assumption.

Many xenoliths in basalts and kimberlites may be identified as derived from the upper mantle on the evidence of their mineralogy and their bulk chemical composition. Abundances of U, Th, and K for some representative ultramafic xenoliths are given in Table 9.4.2 (see also Clark *et al.*, 1966; Wakita *et al.*, 1967). The range in heat production, a measure of the extent of depletion of the sample by prior melting events, is quite large. Even excluding the eclogite xenoliths from kimberlites, which may have been contaminated by the kimberlite matrix (Clark *et al.*, 1966), the heat production values range from as low as 7×10^{-9} erg/g-s for dunites and depleted lherzolites to at least 2×10^{-7} erg/g-s for relatively undepleted peridotites (Wakita *et al.*, 1967; Green *et al.*, 1968; Basu and Murthy, 1977).

Several conclusions are noteworthy from Table 9.4.2: (1) A substantial fraction of the Earth's radioactivity probably resides in the continental crust. (2) While some mantle-derived ultramafic nodules have heat production values comparable or even in excess of the average mantle value expected from the steady state assumption, these relatively undepleted samples are all derived from sub-continental lithosphere (or asthenosphere). (3) The heat production in the oceanic upper mantle, estimated either directly from ultramafic xenoliths or indirectly from radioactive abundances in tholeiites, is a factor of ~2 less than required to match the mean oceanic heat flow (section 9.5.1) by the steady state assumption if representative of the entire mantle (Wakita *et al.*, 1967; Fisher, 1979). This last conclusion points either to (a) a depletion of U, Th and K in the oceanic upper mantle compared to other mantle regions, and/or (b) a large fraction (~1/2) of the present-day heat flow in the oceans being supplied by heat from the core and from global cooling. As noted above and in section 9.5.1, the latter explanation is strengthened by the results of global thermal history models for the Earth.

Short-lived radioactivity. Short-lived radioactive isotopes may have provided an early heat source for the planets, depending on the time between nucleosynthesis and planetary accretion. Important short-lived radionuclides are listed in Table 9.4.3 (Fish *et al.*, 1960). Evidence for the decay of three of these parents in condensed solar nebula material has been found: ^{129}I, ^{244}Pu and ^{26}Al (Schramm and Wasserberg, 1970; Lee *et al.*, 1976). The most abundant of these was ^{26}Al with a half-life of 0.72×10^6 yrs. Lee *et al.* (1976) discovered an excess of ^{26}Mg in a Ca-Al rich chondrule from Allende, which they attribute to *in situ* decay of ^{26}Al. A solid object a few kilometers in radius or greater would have been heated to melting if it had been formed with a similar initial ^{26}Al/^{27}Al ratio.

The real unknown is whether ^{26}Al was incorporated into planetary objects quickly enough to be effective in their early melting. This depends on both the quantity of ^{26}Al generated in nucleosynthesis and its half-life compared to the accretional time of the body incorporating it. The Allende meteorite did not melt as a whole itself. The time between the end of nucleosynthesis and the accretion of planetary-sized objects, based on dynamical accretion models (the "minimum nebula" process of section 10.2.3) is on the order to 10^7 to 10^8 yr (Safronov, 1972; Weidenschilling, 1974), a time long compared to the half-life of ^{26}Al. This figure suggests that ^{26}Al may have been most important as a heat source during the early accretional phase, and hence in small planetesimals and the earliest formed central parts of larger planets.

Runcorn *et al.* (1977) have speculated on the possible heat contribution from "super heavy" transuranium elements on various time scales, but confirmation of the natural occurrence of these elements remains lacking.

Energy of differentiation

The global differentiation of a planet from a homogeneous or quasi-homogeneous initial state into dis-

Table 9.4.3 Important short-lived radionuclides.†

Nuclide	Half-life (yr)	Estimated Abundance (atoms per 10^6 Si atoms)	Initial Heat (erg/g-sec)
^{10}Be	2.5×10^6	4.5	7.0×10^{-5}
^{26}Al	7.2×10^5	100	7.4×10^{-2}
^{36}Cl	3×10^5	50	8.7×10^{-3}
^{60}Fe	$\sim 3 \times 10^5$	100	~0.16
^{129}I	1.7×10^7	0.5	5.2×10^{-7}
^{237}Np	2.2×10^6	0.03	9.5×10^{-5}
^{244}Pu	7.6×10^7	0.02	6.8×10^{-7}
^{247}Cm	$>4 \times 10^7$	0.004	$<2.8 \times 10^{-7}$

†Fish *et al.* (1960).

tinct crustal, mantle, and core regions involves a change in the gravitational energy

$$\Omega = 4\pi \int_0^R g(r)\, \rho(r)\, r^3\, dr \qquad (9.4.2)$$

where g(r) and ρ(r) are the gravitational acceleration and density at radius r. The energy is converted into heat and into much smaller contributions to strain and rotational energy (Birch, 1965). The energy released during core segregation is particularly large and may be an important early heat source for planetary thermal history. The energy $\Delta\Omega$ released by core-mantle differentiation in an originally homogenous planet, and the equivalent rise in mean planetary temperature $\Delta T = \Delta\Omega / C_P$, are given in Table 9.4.4 for the Earth, Mars, Mercury, and the Moon. The spatial distribution of this heat energy after global differentiation is not well understood; perhaps half of the energy goes toward heating the mantle (Shaw, 1978). The time scale for core differentiation is likely to be inversely related to the energy release involved (Elsasser, 1963).

Table 9.4.4 Energy released by core formation in an originally homogeneous planet.

	R Core Radius (km)	$\Delta\Omega$ Energy Release (J)	ΔT Equivalent Mean Temperature Rise (°C)
Earth	3485[(a)]	1.5×10^{31} [(b)]	2300[(b)]
Mars	1400–2100[(c)]	$(1.8\text{–}2.3) \times 10^{29}$ [(d)]	300–330[(d)]
Mercury	1840[(e)]	2×10^{29} [(f)]	680[(f)]
Moon	<400[(g)]	10^{27} [(d)]	12[(d)]

(a)Gilbert and Dziewonski (1975).
(b)Flasar and Birch (1973).
(c)Johnston and Toksöz (1977).
(d)Solomon (1979).
(e)Siegfried and Solomon (1974).
(f)Solomon (1976a).
(g)Wiskerchen and Sonett (1977); Nakamura *et al.* (1974).

The energies listed in Table 9.4.4 are for core-mantle differentiation after planetary accretion. If the terrestrial planets formed inhomogeneously with denser core material initially concentrated toward the center, the available energy of differentiation may be substantially less. It is also possible that high protoplanetary temperatures during accretion led to the separation of core and mantle material during planetary formation, particularly for Earth and Venus (Hanks and Anderson, 1969; Kaula, 1980a).

Accretional energy

The gravitational potential energy available for heating a planet during accretion is roughly GM/R erg/g, which is typically $10^4 R^2$ (with R in km) for a terrestrial-type planet. For a smaller planet (e.g., Mercury) this energy may be augmented significantly by the approach kinetic energy of impacting bodies, if they encounter the growing planet on eccentric orbits, though this additional accretional energy may be lost as extra kinetic energy of ejecta. In any event, the above gravitational potential energy is sufficient to increase the temperature of any protoplanet with $R > 10^3$ km by 10^3 °C, but heating of asteroids by this means is clearly impossible. The actual heating, of course, depends on how much of this energy is retained by the protoplanet in the face of radiation losses from its surface. This can be determined by writing an energy equation for the upper layers of the growing planet, i.e., the layers that interact with the impacting bodies. (Conduction and radiation losses are small once the layer has been buried below the interaction zone.) In the extreme limit of small impacting particles (and ignoring the effects of a primitive atmosphere), the surface would be maintained in radiative equilibrium with the energy input. The small particle assumption was the basis for many early calculations of accretional heating (e.g., Benfield, 1950; Hanks and Anderson, 1969; Mizutani *et al.*, 1972). However, Safronov (1972) has shown that the situation is quite different if impactors of kilometer scale or greater bring in most of the energy, a situation regarded as likely by Safronov (1972), Wetherill (1976) and Kaula (1979a). In this case, more energy is retained by the protoplanet than in the small particle case, the amount depending primarily on the depth of deposition, the partitioning between direct heating and kinetic energy of fragments in a given impact, the radiative loss from heated fragments at impact, and the heat transport in the impact-stirred region. The first of these factors is probably related directly to the impactor size. The second has been calculated by O'Keefe and Ahrens (1977), with the result that probably about 90% of the energy goes directly into heating for hyper-velocity impacts at 2 km/sec velocity and greater. The question of how much heat is radiated to space during the flight of small chunks of ejecta then arises. The answer depends on ejecta size distributions, temperature, and flight duration (or time before burial), factors which have yet to be rigorously analyzed. Kaula (1979a) has suggested that the stirring-mixing length is related to that fraction of the impact energy available for blasting the ejecta, a relation which allows the heat transport due to stirring to be estimated.

Kaula (1979a) has modelled the entire accumulation heating process for the Earth and Moon, allowing for a wide range of values for the uncertain parameters. He finds that the growing Earth would retain enough energy to produce large scale melting (and even vaporization). However, his Moon models do not get hot enough to melt except for implausible planetesimal mass distributions, and the melting occurs much deeper than is indicated by petrological and tectonic data. Other planets have yet to be examined in such detail, as is the case for planetary formation schemes involving gaseous protoplanets (Cameron, 1978; DeCampli and Cameron, 1979).

Compression of accreted material in the interior of a growing planet can be appreciable for the larger planets. Assuming adiabatic compression and hydrostatic pressure yields.

$$T = T_0 \exp\left[\frac{4\pi}{3}\frac{\alpha\rho G}{C_P}(R^2 - r^2)\right] \qquad (9.4.3)$$

where T is the compressed temperature (in °K) at radius r, T_0 is the uncompressed initial temperature (in °K) at radius r, and α, ρ, G and C_P are the coefficient of thermal expansion, density, gravitational constant and specific heat, respectively (assuming uniform values, for simplicity). Using $\alpha = 3\times 10^{-5}\,°K^{-1}$, $\rho = 5\,g/cm^3$, $C_P =$ 1.3 J/g-°K, and an initial temperature of 500°C throughout the planets, the additional temperature at a radius of half the planetary radius is 77, 728, 965, 119, and 25°K for Mercury, Venus, Earth, Mars, and the Moon, respectively.

Tidal heating

The amount of heating due to dissipation of the mechanical energy associated with tides depends on the frequency and magnitude of the tides and the elastic properties of the planetary material (especially the deviation from perfect elasticity). The former depends critically on the configuration and evolution of the orbit of the planetary body with respect to the tide-inducing body; however, since orbital evolution depends in turn on the dissipation of energy, the problems are coupled and generally complicated.

Naturally, the Earth-Moon system is the most thoroughly studied. Kaula (1964) showed that tidal heating of the Earth has been negligible, and that in the Moon it is presently negligible. However, because both tidal frequency and magnitude were considerably larger in the past, when the Moon was close to the Earth, early heating is not so readily ruled out. Nevertheless, calculations that rely on plausible orbital evolutions and that include the effects of nonuniform mechanical properties and large excursions in inclination (Ward, 1975) indicate that the Moon still recedes too rapidly for tidal heating to be significant (Kaula and Yoder, 1976; Peale and Cassen, 1978).

There may still be a place for tidal heating in early lunar history if the Moon was captured (Turcotte *et al.*, 1977a), or if the binary formation theory of Harris and Kaula (1975) is valid. Harris (1977) has derived some rather severe constraints on the latter possibility. Finally, Burns and Schultz (1978) have suggested that variations in Q, the dissipation factor, with time and temperature, could produce a catastrophic melting event induced by tides, but the details have yet to be worked out. It should be remarked that the proposals referred to in this paragraph generally involve global melting and therefore could only apply to very early thermal evolution.

An estimate of the tidal heating associated with the despinning of Mercury has been made by Burns (1976), who finds it to be insignificant. The same holds for the present day heating rate, in spite of the planet's large eccentricity (Peale and Cassen, 1978). There is no reason to believe that either Venus or Mars ever experienced significant tidal heating.

Perhaps the satellites of the outer planets are the only solar system objects that have experienced significant tidal heating. It has been proposed that tidal dissipation is the energy source for Io's intense volcanic activity (Peale *et al.*, 1979; Smith *et al.*, 1979), but at this writing it remains to be seen whether or not basaltic volcanism is present on that satellite. There may currently exist an ice layer on Europa that is tidally heated, and Ganymede may have been so heated in the past. But evidence for any basaltic volcanism that might have occurred on these objects probably lies beneath great depths of water or ice.

Electromagnetic heating

It is possible that the protoplanetary environment was bathed in a hydro-electromagnetic radiation field produced by the combined effects of mass loss (solar wind), solar magnetic field, and a rapid solar spin rate. In this event, protoplanetary bodies could have been heated by the ohmic losses associated with currents induced in them by the ambient field (Sonett and Colburn, 1967; Sonett *et al.*, 1968). The mechanism is attractive because it is capable, under appropriate assumptions, of melting protoplanetary material (including asteroid-sized objects) in a short time ($< 10^6$ years). The proposal suffers from our present inability to test the reality of the rather extreme conditions that must be postulated for the mechanism to work. Although it is common to associate these conditions with a T-Tauri-like primitive sun, it has not yet been definitely established that T-Tauri stars have appreciable magnetic fields, or that the sun ever went through a T-Tauri stage. The observed characteristics of T-Tauri stars and their possible relation to the planet forming stage of solar evolution are reviewed by Rydgren *et al.* (1976) and Imhoff (1978).

The mechanism may operate in either of two modes. In the first mode (unipolar induction), a current is driven through the planet by the interplanetary electric field $\mathbf{E} = \mathbf{v} \times \mathbf{B}$ ($\mathbf{v}$ is the solar wind velocity and $\mathbf{B}$ is the interplanetary magnetic field). In the second mode, time variations of the interplanetary magnetic field give rise to eddy currents within the planet. Measurement of the induced magnetic field associated with the latter mode is the basis for electromagnetic sounding of the lunar conductivity profile (although heating by this mode is miniscule in the present solar wind). Unipolar induction can be suppressed by a high crustal resistivity, which would prevent closure of the electrical circuit through the ambient plasma. Both mechanisms are inhibited by the existence of a magnetosphere around the planetary body, which could be formed by the induced magnetic field itself. It is therefore required that the solar wind have a high mass flux so that the induced fields are confined mainly within the planet. Finally, the interior of the planet must be "primed" to a modest temperature (a few hundred °C) so that the resistivity is low enough initially for induction to be effective. All of these considerations are quantified in the calculations of Sonett *et al.* (1968, 1975), wherein it is shown that, under favorable assumptions, ohmic dissipation of induced currents could produce melting in meteorite parent bodies, Mercury, and the Moon.

9.4.2 Thermal conductivity and thermodynamic parameters

Thermal conductivity

The thermal conductivity of crustal and mantle material, including both lattice conduction and radiative transport, controls heat transfer in those regions of planetary interiors in which convective heat transport is negligible. Heat is transported by conduction in planetary lithospheres, as well as in any boundary layer at the top or bottom of a distinct convection system within a planetary mantle. It is likely for instance that such a boundary layer occurs at the base of the Earth's mantle (e.g., Elsasser *et al.*, 1979), and boundary layers may also occur between nonpenetrative convection systems at intermediate mantle depths. The thermal conductivity of the dense mineralogical assemblages and at the high temperatures and pressures of the Earth's lower mantle may, in concert with future seismic data on deep boundary-layer properties, permit estimates of the heat flux across such layers.

Upper mantle. For the upper mantle of the Earth, and for possibly the entire interior of small bodies such as the Moon, pertinent thermal conductivity data for a series of mafic rocks and minerals to ~1900° K in both the lattice and radiative regime are available at ambient pressures (Kanamori *et al.*, 1968; Fujisawa *et al.*, 1968; Schatz and Simmons, 1972). Below ~400° C the thermal conductivity mechanism is believed to be largely phonon transport, which decreases with increasing temperature, whereas radiative (photon) thermal conductivity becomes increasingly important at high temperature, at least at low pressure. The total thermal conductivity K is given by the sum of K_L (lattice phonon conductivity) and K_R (radiative, or photon conductivity).

At temperatures below 400° C, the lattice thermal conductivity of most crustal and mantle rock types can be fit with an expression of the form (Birch and Clark, 1940)

$$K_L = [\alpha + \beta \quad (T - 300°K)]^{-1} \qquad (9.4.4)$$

Equation (9.4.4) has often been adopted as a basis for describing laboratory data on lattice conductivity (e.g., Horai and Simmons, 1970; Schatz and Simmons, 1972; Kanamori *et al.*, 1968; Kieffer *et al.*, 1976). A theoretical derivation of the relation $K_L \sim 1/T$ is given by Magnitsky (1967). A partial compilation of values of α and β fit to the data, up to 400°C, of Kingery *et al.* (1954) for a series of ceramics of geophysical interest are given in Table 9.4.5.

The radiative component of thermal conductivity, for semi-transparent isotropic crystals, is formally given by (Clark, 1957)

$$K_R = \frac{16\, n^2 \sigma T^3}{3 \langle e(T) \rangle} \qquad (9.4.5)$$

where n is the index of refraction, σ the Stefan-Boltzmann constant and $\langle e(T) \rangle$ the mean opacity suitably averaged over the wavelength interval prescribed by the Planck black-body distribution function

$$J(\lambda,T) = \frac{2\pi c^2 h}{\lambda^5 [e^{(hc/\lambda kT)} - 1]} \qquad (9.4.6)$$

where c is the velocity of light, h the Planck constant, λ is the wavelength, and k the Boltzmann constant. Thus

Table 9.4.5 Parameters governing the variation of lattice conductivity with temperature.

Material	α (10^{-7} cm-sec-°K/erg)	β (10^{-7} cm-sec/erg)
Al_2O_3[(a)]	2.145	0.0146
BeO[(a)]	0.261	0.0022
CaO[(a)]	5.949	0.0139
Mg_2SiO_4[(a)]	16.53	0.0308
Mg_2SiO_4[(b)]	21	0.048
C[(a)]	0.491	0.0012
MgO[(a)]	1.819	0.0112
SiO_2 (stishovite)	4.5[(c)]	(0.033–0.0012)[(d)]
$MgAl_2O_4$	5.95	0.0103
TiO_2	13.45	0.0332

(a) Kingery *et al.* (1954).
(b) Birch and Clark (1940).
(c) Soga *et al.* (1972); also Kieffer (1976).
(d) Range assumed in section 9.4.4

the mean opacity is related to the convolution of J and $\epsilon(\lambda, T)$, the measureable wavelength- and temperature-dependent opacity, by

$$\frac{1}{\langle e(T) \rangle} = \frac{\int_0^\infty \frac{1}{\epsilon(\lambda,T)} \frac{\partial J}{\partial T} d\lambda}{\int_0^\infty \frac{\partial J}{\partial T} d\lambda} \qquad (9.4.7)$$

The measurement of the opacity $\epsilon(\lambda, T)$, particularly for olivine, has been the object of a series of studies at various pressures (Faukao *et al.*, 1968; Shankland *et al.*, 1974, 1979), since the relation $K_R \sim T^3$ inferred from Eq. (9.4.5) implies that if the mineral involved does not become too opaque, radiative transport might be expected to dominate the thermal conductivity at the presumed high temperatures of terrestrial planetary mantles. Of special interest have been the studies of the temperature and pressure effects on the opacity of Fe^{2+}-bearing mafic silicates in the vicinity of 100 nm to 3000 nm wavelength. This region of the infrared spectrum is important at mantle temperatures as the J function is a maximum in this wavelength range. Moreover, Fe^{2+} in octahedral coordination with O^{2-}, as is common in virtually all rock forming silicates, gives rise to a significant absorption feature in this range (Figs. 9.4.3–4). Thus the pressure and temperature dependence of this absorption feature has provided a fundamental understanding of radiative transport in olivine in the upper mantle of the Earth (Fukao *et al.*, 1968; Shankland, 1970; Shankland *et al.*, 1974; Pitt and Tozer, 1970a, b).

Schatz and Simmons (1972) performed independent measurements of the lattice and radiative conductivities of olivine and pyroxene at room pressure and at temperatures to 1900° K. On the basis of their measurements, they proposed a general relationship for planetary mantles in which heat conduction is dominated by olivine with molar Fe/(Fe + Mg) = 0.1

$$K_L = (7.4 + 0.05\,T)^{-1} \times 10^7 \text{ erg/cm-sec-°K} \qquad (9.4.8)$$

$$K_R = \begin{cases} 0 & T \leq 500°\,K \\ 2.3\,(T - 500) \times 10^2 \text{ erg/cm-sec-°K} & T > 500°\,K \end{cases} \qquad (9.4.9)$$

where T is the absolute temperature. These relationships have been widely employed for thermal history models in the terrestrial planets.

On the basis of present-day geotherms in the Earth and perhaps in Venus, above pressures of 180 kb (corresponding to 500 km depth in the Earth), the γ-phase of stoichiometries similar to $(Mg_{0.9}, Fe_{0.1})_2\,SiO_4$ is believed to be stable (Akimoto, 1972). The optical and infrared absorption data for γ-Fe_2SiO_4 (Mao and Bell, 1972) may be used to estimate the radiative component of thermal conductivity in the transition zone of terrestrial planetary mantles over which spinel may be stable

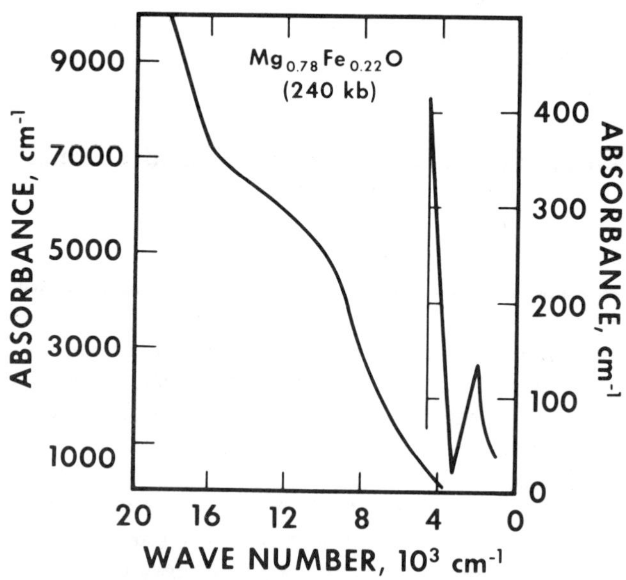

9.4.3 Absorbance versus wave number at 240 kb for $Mg_{0.78}Fe_{0.22}O$. Curve obtained by combining data of Mao (1973) and Shankland *et al.* (1979).

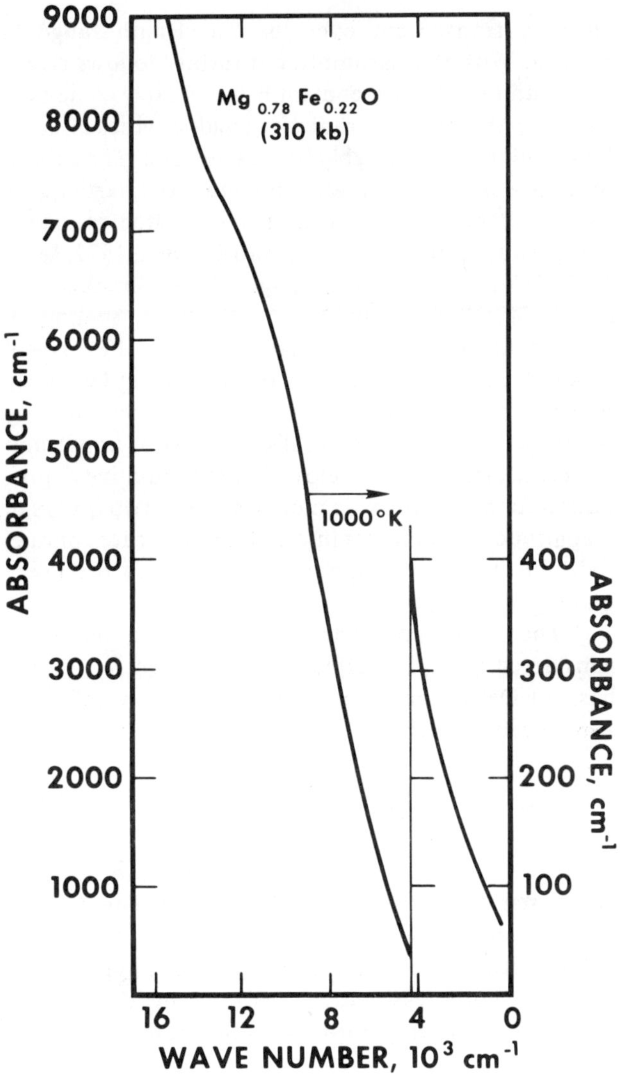

9.4.4 Absorbance versus wave number at 310 kb for $Mg_{0.78}Fe_{0.22}O$. Curve obtained by combining data of Mao (1973) and Shankland *et al.* (1979).

and compare these values to the conductivity values predicted by Eqs. (9.4.8–9) (Table 9.4.6). As can be seen by this comparison, the latter values are systematically higher than the results obtained from absorption data. This is to be expected on account of the marked pressure-induced increase in optical absorptivity observed in the 400 to 500 nm wavelength range discovered for Fe^{2+} -bearing silicates and oxides (Mao and Bell, 1972). For temperatures of 2000° K, Table 9.4.6 indicates that in spite of the increase in absorptivity in the 400 to 500 nm range, the predicted radiative conductivity is comparable to the lattice conductivity.

Lower mantle. Although our knowledge of the thermal conductivity of the upper mantle is now fairly complete, until recently little could be said definitely regarding the thermal properties below ~700 km in the so-called post-spinel (Anderson, 1967) region of the Earth. Aside from knowledge of the mineralogic species recently obtained from high pressure diamond anvil quenching experiments (e.g., Liu, 1975b), knowledge of the thermal properties at high pressure has been limited to what can be inferred from shock wave data (McQueen *et al.*, 1967; Al'tschuler and Sharapzarnov, 1971; Jeanloz and Ahrens, 1977; Jackson and Ahrens, 1979) and from extrapolating oxide properties to lower mantle conditions (Kieffer, 1976).

Recently two new studies related to the thermal conductivity of silicates have provided new constraints on the thermal conductivity of the mantles of the Earth and other planets. First, measurements of the absorption spectra of Fe^{2+} -bearing olivine at high temperatures (Shankland *et al.*, 1979), and the effect of pressure up to 310 kb on the optical and infrared absorption spectra in $Mg_{0.78}Fe_{0.22}O$ (Mao, 1973) and similar pressures on Fe_2SiO_4 (Mao and Bell, 1972) in both the olivine and spinel structures, permit calculations of the contribution to thermal conductivity by

Table 9.4.6 Radiative conductivity at 180 kb pressure in olivine.

Temperature (°K)	K_R (10^5 erg/cm-sec-°K)[a]	K_R from equation (9.4.9)[c] (10^5 erg/cm-sec-°K)	K_L from equation (9.4.8)[c] (10^5 erg/cm-sec-°K)
1000	0.682–0.665[b]	1.17	1.76
2000	2.38–2.07	3.43	0.92
3000	3.87–3.05	5.82	0.63

[a]Calculated from the spectral data of Mao and Bell (1972) and Shankland *et al.* (1979) assuming stoichiometry of $(Mg_{0.9}Fe_{0.1})_2SiO_4$.
[b]Range of values obtained from using 1 to 20 μm range of approximate sample thickness (Mao, pers. comm.).
[c]Schatz and Simmons (1972).

radiative transport in a lower mantle model of mixed oxides. Second, the work of Kieffer *et al.* (1976) reporting the thermal conductivity of quartz and fused quartz to 30 kb, together with the data for thermal diffusivity of Mg_2SiO_4, NaCl, and Fe_2SiO_4 to nearly 50 kb (Fujisawa *et al.*, 1968), provide a reasonable test of a model, based on Debye theory (Magnitsky, 1967; Kieffer, 1976), which permits prediction of the pressure effect on lattice conductivity.

Bounds on the thermal conductivity in the Earth's lower mantle are derived in this section in several steps. Radiative conductivity K_R is determined from recent optical and infrared absorption data. Taking these values of K_R into account, we then construct, based on a limited data base for the oxides, bounds on the thermal conductivity at a depth of ~700 km in the Earth (the top of the lower mantle). By calculating K_L from recent seismic velocity and density profiles, we then construct bounds on the total thermal conductivity as a function of depth in the lower mantle.

The radiative conductivity is obtained from Eqs. (9.4.5–7). The calculations are done for $Mg_{0.78}Fe_{0.22}O$ because of the availability of absorption data on such a mixture from Mao (1973). The zero pressure index of refraction, 1.8653, is obtained from (Winchell and Winchell, 1964)

$$n = 1.737 + 0.583X \qquad (9.4.10)$$

where X is the mole fraction FeO. The increase in index of refraction with compression is obtained from Drude's law (Anderson and Schreiber, 1965)

$$\frac{n^2 - 1}{\rho} = K_n \qquad (9.4.11)$$

where K_n is a constant. Assuming ideal mixing of wüstite and periclase, an initial density of 4.14 g/cm^3 and a value of $K_n = 0.5989$ is obtained from Eqs. (9.4.10–11). The values indicated in Table 9.4.7 for n at high pressure were then calculated from the density, which in turn was obtained via the Murnaghan equation. Values of 1.585 Mbar and 4.52 were assumed for the isothermal bulk modulus and its pressure derivative (Anderson *et al.*, 1968; Akimoto, 1972). The spectra shown in Figs. 9.4.2 and 9.4.3 were constructed using the $Mg_{0.78}Fe_{0.22}O$ data of Mao (1973) for wavelengths less than 2000 nm (5000 cm^{-1}) and the higher temperature (1400°C) data of Shankland *et al.* (1979) for $(Mg_{0.9}Fe_{0.1})_2SiO_4$ at longer wavelengths. The absolute absorptivity of the latter data set was multiplied by a factor of 2 to take into account the lower iron content. Calculated values of K_R are given in Table 9.4.7. It has been implicitly assumed that pure Mg_2SiO_4 or MgO are virtually transparent over the wavelength range of interest. With this assumption it further follows that if the mean mantle iron content is more likely to lie near $Fe/(Fe+Mg) \approx 0.1$, the radiative conductivities of Table 9.4.7 will increase roughly by a factor of 2. The subject of the absorptivity of mantle minerals and particularly its dependence on pressure and on Fe content is an area of vigorous current research (Mao and Bell, 1972; Mao, 1973, 1976; Abu-Eid and Langer, 1978; Shankland *et al.*, 1974, 1979; Goto *et al.*, 1979). A scaling of absorptivity directly as molar content of Fe^{2+} is not generally appropriate, and the partitioning of Fe among mineral phases and the interconnectedness of any relatively transparent mineral grains may be important. Nevertheless, we show below that the radiative contribution for a simple oxide model is one to two orders of magnitude less than that inferred for the lattice conductivity in at least the uppermost portion of the lower mantle.

The lattice conductivity in the lower mantle is considered next. The effect of pressure on K_L, on the basis of Debye theory, is given by (Mooney and Steg, 1969; Kieffer, 1976)

$$K_L \sim \frac{c^3}{\gamma^2 \rho^{2/3}} \qquad (9.4.12)$$

where c is the bulk sound velocity $1/\sqrt{\rho \chi_a}$, γ the Grüneisen parameter, and ρ the density. The data of Fujisawa *et al.* (1968) and Kieffer *et al.* (1976) are consistent with this relation, although the range of pressures, or compressions, represented by the data is narrow. For the temperature variation of K_L, Eq. (9.4.4)

Table 9.4.7 Calculated radiative conductivity of $Mg_{0.78}Fe_{0.22}O$ at two lower mantle pressures.†

Temperature (°K)	K_R (10^4 erg/cm-sec-°K)	Wavelength of Maximum Radiative Contribution (μm)
	240 kb; $n = 1.934$; $\rho = 4.575$ g/cm^3	
2000	3.50	3–4
3000	4.79	3–4
	310 kb; $n = 1.963$; $\rho = 4.763$ g/cm^3	
2000	1.72	3–4
3000	2.62	2–2.5
4000	3.61	2–2.5
5000	4.77	75

†From Eqs. (9.4.5–7) and Figs. 9.4.3–4.

may be used as an approximation. If it is assumed that in the lower mantle the temperature gradient is approximately adiabatic (section 9.4.4), Eq. (9.4.12) may be conveniently expanded about a state corresponding to a pressure and temperature appropriate to a depth of 700 km using

$$K_L(T,P)_S = K_{700} \left\{ 1 + \left[\frac{3}{c}\left(\frac{\partial c}{\partial P}\right)_S - \frac{2}{\gamma}\left(\frac{\partial \gamma}{\partial P}\right)_S - \frac{2}{3}\chi_a \right] \Delta P \right\} \quad (9.4.13)$$

The temperature variation along the adiabat is then simply given as

$$T_S = T_0 \exp \int_{\rho_0}^{\rho} \frac{\gamma}{\rho}\, d\rho \quad (9.4.14)$$

where T_o and ρ_o are evaluated at 700 km depth.

The parameters required to calculate the variation in K_L for an adiabatic lower mantle from Eqs. (9.4.13–14) can be obtained from seismic velocity and density profiles derived by inversion of free oscillation and seismic wave travel-time data. The estimation of the Grüneisen parameter and of adiabatic temperature profiles for the Earth's lower mantle is described in section 9.4.4.

We adopt the mineralogic model $2MgO+SiO_2$ (stishovite) for the lower mantle (Kieffer, 1976). Using 1900–2300°K as bounds on the mantle temperature at 700 km depth, we calculated K_L assuming the adiabatic temperature gradients of section 9.4.4. The values of β used to estimate lattice conductivity for MgO and SiO_2 are from Table 9.4.5. The rather wide but not unreasonable range of possible values of β for stishovite contributes the largest uncertainty to the present calculation.

The calculated upper and lower bounds for K in the Earth's mantle are given in Table 9.4.8 and shown in Fig. 9.4.5. The conductivity includes K_L and K_R, but $K_L >> K_R$ (compare Tables 9.4.7 and 9.4.8). Also shown are Kieffer's (1976) bounds and an extrapolation of data on dunite made by Zharkov (1958). Kieffer's (1976) bounds are lower than those given here because she assumed a linear thermal gradient in the lower mantle and adopted a higher value of β for MgO.

On the basis of these calculations it may be concluded that lattice conduction may dominate radiative transfer in the Earth's lower mantle. The total thermal conductivity of the lower mantle of the Earth is uncertain to within a factor of 4. Even within this uncertainty, however, it is clear that the Earth's lower mantle is convectively unstable (see next section) so that convection dominates conduction except in boundary layers.

Table 9.4.8 Thermal conductivity for an adiabatic lower mantle.

Depth (km)	Pressure (kb)	Thermal Conductivity (10^5 erg/cm-sec-°K)
696	248	5.0–16.3
1246	500	7.1–23.0
2071	904	11.3–36.4
2888	1357	10.5–33.9

Specific heat

Values of specific heat for a wide range of silicate minerals differ by less than about 50% over temperatures ranging from 400°K to their melting temperatures. The variation of specific heat for individual minerals over the same temperature range is less than about a factor of 2, and is even less at temperatures much greater than the Debye temperature as the specific heat approaches the Dulong-Petit limit. The relation given by Horai *et al.* (1970) for Apollo 11 mare basalt

$$C_P = 0.891 \times 10^7 + 0.206 \times 10^{-4}\, T - 0.209 \times 10^{12}\, T^{-2}\ \text{erg/g-°K}, \quad (9.4.15)$$

where T is in °K, is representative of the temperature dependence of specific heat above the Debye temperature.

Because of the small variation of C_P with temperature and with composition, most planetary thermal models are based on the assumption that C_P is constant at about 1.2 J/g-°K.

Thermal expansion

The coefficient of thermal expansion

$$\alpha = -\frac{1}{\rho}\left(\frac{\partial \rho}{\partial T}\right)_P \quad (9.4.16)$$

is an important parameter in the Rayleigh number, in the relation for the adiabatic temperature profile, and in the expressions for thermal stress.

The quantity α increases with temperature and decreases with pressure. For olivine at $P = 0$

$$\frac{\rho_0}{\rho} = 1 + a_1 T + \frac{a_2}{T} + a_3 T^2 \quad (9.4.17)$$

where T is in °K, ρ_o is at $T = 0$, $a_1 = 3.17 \times 10^{-5}$, $a_2 = 1.05$ and $a_3 = 7.2 \times 10^{-9}$ (Skinner, 1962). Thus

$$\alpha = \frac{a_1 - a_2/T^2 + 2a_3 T}{1 + a_1 T + a_2/T + a_3 T^2} \quad (9.4.18)$$

At fixed T,

$$\frac{\alpha}{\alpha_0} = 1 + \frac{P}{K_T}\left(\frac{1}{\alpha K_T}\frac{\partial K_T}{\partial T}\right)_0 \qquad (9.4.19)$$

where the subscript 0 denotes P = 0, and K_T is the isothermal bulk modulus (Birch, 1952). Alternatively, from Eq. (9.3.14)

$$\alpha = \gamma \rho_0 C_P \chi_a \qquad (9.4.20)$$

and most of the pressure dependence of α is contributed by γ. From Kieffer (1976)

$$\gamma = \gamma_0 + \gamma_0' \Delta \qquad (9.4.21)$$

where Δ is the compression or fractional volume change, and $\gamma_0' = (d\gamma/d\Delta)_{\Delta=0}$. Kieffer (1976) cites values of $\gamma_0 = 1.3$–1.5 for mantle minerals and $\gamma_0' = 1.7$ for olivine and MgO but a much higher value of γ_0' (4.5–9) for stishovite.

Because of the competing effects of pressure and temperature, α is often taken to be constant (typically 3×10^{-5} °C^{-1}) over a region of uniform chemistry and phase assemblage in planetary thermal models. This assumption is valid for α roughly to within a factor of 2 for the olivine stability field of terrestrial planetary mantles. A constant value of α may be a poor approximation, however, over the entire mantle for bodies as large as Earth or Venus. Phase changes to denser mineral forms are often accompanied by a decrease in α; α for γ-Mg_2SiO_4 (spinel) is 25% lower than for forsterite (Suzuki *et al.*, 1979). Using Eqs. (9.4.20–21), Kaula (1980b) has estimated that α decreases by a factor of 5–6 from the upper to the lowermost mantle. The effect of such a large change in α over the depth extent of a mantle convection system has not yet been investigated.

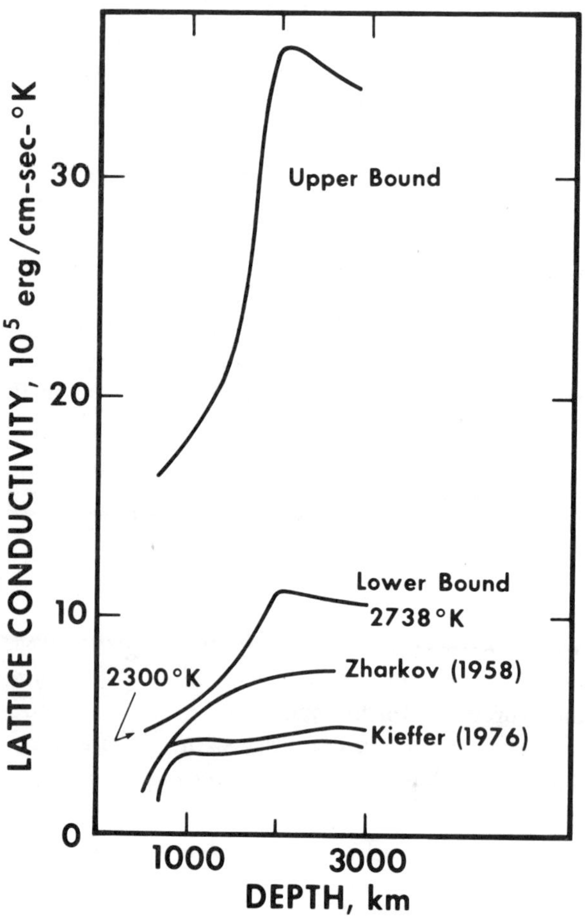

9.4.5 Upper and lower bounds of the lattice thermal conductivity versus depth for the lower mantle. Previous results of Zharkov (1958) and Kieffer (1976) are also indicated.

Heat of fusion

Heat of fusion has been measured for only a few minerals that are representative of planetary interiors. Several values at atmospheric pressure are given in Table 9.4.9. The effects of pressure on heat of fusion of individual minerals is reported to be minor, at least to 50 kb (McConnell *et al.*, 1967). However, the change in phase assemblage with pressure for a given bulk composition may cause a much more significant change in the heat of fusion of the total mineral assemblage.

An average value of the heat of fusion based on the weight percent of mineral phases assumed for the interior of a planet is usually used in thermal calculations. The heat of fusion acts effectively as a heat sink when temperatures rise above the melting point.

Table 9.4.9 Heats of fusion (J/g)[(a)].

Fayalite	452
Forsterite	950[(b)]
Albite	216
Diopside	357
Enstatite	699[(b)]
Clinoenstaite	612
Iron (δ-Fe)	271
Nickel	297
Troilite	368

[(a)]From Robie and Waldbaum (1968).
[(b)]Estimated by McConnell *et al.* (1967).

9.4.3 Rheology

As noted in section 9.3.1, convective heat transfer can dominate thermal conduction in planetary mantles if the characteristic flow velocity exceeds about 10^{-10} cm/sec or 3×10^{-3} cm/yr. The typical plate speeds on the Earth exceed this number by three orders of magnitude, and the estimated Rayleigh numbers for planetary mantles (Table 9.3.1) exceed the critical Rayleigh number for convective instability by factors of 10^4 to 10^7. Convection is therefore the primary governor of planetary thermal history.

The physical property of rocks that controls the characteristics of convection is the **rheology,** or how the rocks deform in response to stress as functions of pressure, temperature and other variables. The rheology is usually cast in terms of a viscosity or an effective viscosity, the ratio of applied shear stress to rate of shear strain. The concepts of lithosphere and asthenosphere arise naturally from a consideration of viscosity, as we first discuss below. The theoretical mechanisms governing creep in mantle minerals is next briefly reviewed, followed by a summary of proposed flow laws for mantle materials and a discussion of the effects of volatiles, partial melting, and solid-solid phase changes. Finally, we summarize the available evidence from glacial rebound for the viscosity of the Earth's mantle.

Lithosphere and asthenosphere

From the observations of glacial rebound in Fennoscandia and North America came the original concepts of lithosphere and asthenosphere. In general terms the **lithosphere** is an outer layer of finite strength, capable of supporting shear stresses for geologically long periods of time. The underlying **asthenosphere**, or sphere of weakness, behaves as a fluid over geological times and has no long term elastic strength. These definitions of lithosphere and asthenosphere are rheological in nature and are based implicitly on temperature. At low temperature, rock is elastic under low stress differences and undergoes brittle failure, or faulting, at high stress differences. At high temperature, rock creeps in response to deviatoric stress. The depth of the transition from brittle to ductile behavior in the Earth, perhaps a rather broad boundary sensitive to composition and strain rate as well as to temperature, may be loosely regarded as the base of the lithosphere as defined rheologically. For some problems in tectonics, the lithosphere may be idealized either as a perfectly elastic plate or shell or as an elastic plate with a plastic limit or yield stress. The term **elastic lithosphere** is used to refer to the elastic or elastic-plastic plate when such an approximation is appropriate.

The theory of plate tectonics and the concept of plates as thermal boundary layers has led to a second definition of lithosphere. The **thermal lithosphere** in the oceans consists of the crust and that part of the upper mantle that contributes through thermal expansion and contraction to the deepening of the seafloor with age (section 9.5.1). The thermal lithosphere is greater in thickness than the elastic lithosphere and includes material at depth without long-term strength. The thermal lithosphere may be regarded as including that part of the oceanic mantle that moves with the overlying crust at essentially the plate velocity. The concept of a thermal lithosphere extends to continents as well, though the thickness of the thermal lithosphere beneath stable continents is a matter of debate (section 9.5.1).

A third definition of lithosphere has been offered by seismologists. The seismic velocity distribution beneath oceans and many continental regions shows a pronounced **low velocity zone**, particularly for shear waves, in the uppermost 200 km of the mantle. The low velocity zones are also zones of high seismic wave dissipation, or low Q. Seismologists often identify the top of this low velocity, low Q zone with the top of the asthenosphere. While the seismic wave speeds and attenuation are controlled by temperature in the upper 200 km of the mantle, the drop in velocity and Q measured at seismic periods (10^{-1} to 10^2 sec) need not correspond to any rheological boundary for processes at geological time scales (10^4 to 10^8 yr). Thus the depth to the low velocity zone is not directly equatable to the thickness of either the elastic lithosphere or the thermal lithosphere.

Mechanisms of flow in mantle minerals

In order to understand convection in a planetary mantle, it is important to identify the dominant mechanism of deformation, since this mechanism determines how strains are related to stress and to temperature and pressure. Most of the attention of laboratory workers in the field of mantle rheology has focused on olivine, held to be the dominant mineral phase in the mantle above 350 km depth. The mechanisms governing steady-state deformation of olivine at upper mantle conditions have been treated by a number of authors (e.g., Gordon, 1965; Tozer, 1965; Weertman, 1970; Stocker and Ashby, 1973; Kohlstedt and Goetze, 1974; Weertman and Weertman, 1975; Carter, 1976; Kohlstedt *et al.*, 1976; Twiss, 1976; Ashby and Verrall, 1977). We follow the discussion of Ashby and Verrall (1977), in abbreviated form, in this section.

The important mechanisms of flow in olivine include (Ashby and Verrall, 1977): (1) cataclastic flow, or fracture; (2) low temperature plasticity, involving the gliding of dislocations; (3) diffusional flow, or deformation by the diffusive motion of ions; and (4) power-law creep, involving both climb and glide motion of dislocations. Two additional mechanisms are potentially important, but less well understood: (5) fluid phase transport, or flow in the presence of a fluid phase or partial melt (see below); and (6) creep with dynamic recrystallization. The first two mechanisms are important only at low temperature and/or at high stresses and strain rates, and are not likely to govern convective flow in planetary mantles, and are therefore not considered further below.

At very low stress levels the dominant creep mechanism is by the diffusion of ions, either through the crystal lattice (Nabarro-Herring creep) or along the grain boundary (Coble creep). This mechanism results in a linear (Newtonian) relation between stress and strain rate. The dependence of strain rate e on stress σ for uniaxial compression or simple shear can be written

$$\dot{e} = \Lambda^* \frac{D\Omega}{kTd^2} \sigma \qquad (9.4.22)$$

where D is the diffusion coefficient of the least mobile ion or ionic group, Ω is the atomic volume, k is Boltzmann's constant, T is the absolute temperature, d is the average grain diameter, and Λ^* is a constant of order 10.

Since the diffusion of ions is a thermally activated process, the diffusion coefficient is an exponential function of temperature and pressure given by

$$D = D_0 \exp\left[-\frac{E^* + PV^*}{kT}\right] \qquad (9.4.23)$$

where D_0 is a reference diffusion coefficient, E^* is an activation energy, and V^* an activation volume. Both E^* and V^* are known for olivine: $E^* = 520 \pm 20$ kJ/mole (Kohlstedt and Goetze, 1974) and $V^* = 13.4 \pm 4.0$ cm^3/mole (Ross *et al.*, 1979). This activation volume is close to the ionic volume of O^{-2} and suggests that oxygen diffusion controls the creep rate.

At higher stress levels the dominant creep mechanism is the motion of dislocations. Two possible types of dislocation motion are climb and glide. Both types can lead to a non-Newtonian power-law relationship between stress and strain rate of the form

$$\dot{e} = \Lambda \frac{D b \mu}{kT} \left(\frac{\sigma}{\mu}\right)^n \qquad (9.4.24)$$

where b is the length of the Burgers vector, μ is the shear modulus, and Λ is a constant (Weertman and Weertman, 1975; Ashby and Verrall, 1977). Note that the temperature dependence for power-law creep is also primarily through the diffusion coefficient, Eq. (9.4.23).

These scalar laws for uniaxial stress or simple shear can be combined to give a law relating the strain rate tensor $\dot{e}_{ij}$ and the deviatoric stress tensor τ_{ij} (Parmentier *et al.*, 1976)

$$e_{ij} = \left(\frac{1}{2\eta^*} + A^*\tau^{n-1}\right) \tau_{ij} \qquad (9.4.25)$$

where τ is the second invariant of the deviatoric stress tensor $\tau = (\tau_{kl}\tau_{kl})^{1/2}$, η^* is the Newtonian viscosity for diffusion creep, and A^* is a parameter equivalent to the analogous factor in Eq. (9.4.24).

Although several flow mechanisms may operate together, one mechanism dominates for a given material at a given temperature, pressure, and stress. Figure 9.4.6 shows a "deformation map" for olivine at a pressure appropriate to 250 km depth in the Earth. Within each set of heavy boundaries, one mechanism controls the steady-state creep. The effect of temperature, stress and strain rate are apparent from the figure. Maps at other pressures may be constructed from Fig. 9.4.6 through the use of Eq. (9.4.23).

An important parameter controlling the boundary between diffusional flow and power-law creep in Fig. 9.4.6 is the grain size d. As d increases, the rate of diffusional flow is lessened (Eq. 9.4.22), and the boundary separating diffusional flow and power-law creep moves to lower stress values. Figure 9.4.6 was derived for an average diameter for olivine grains of 0.1 mm, a value not likely to be too large (Stocker and Ashby, 1973). If anything, power-law creep in the mantle occupies a field larger than that shown in the figure, and is the most likely flow mechanism for a convecting upper mantle on the grounds of micromechanical data and theory (Ashby and Verrall, 1977).

Evidence from mantle-derived peridotite nodules supports the inference from deformation maps that power-law creep is a dominant flow mechanism. TEM studies on naturally deformed olivines indicates that the mineral grains have been deformed by dislocation creep (e.g., Boland *et al.*, 1971; Nicolas *et al.*, 1973).

Flow laws for mantle minerals

Steady-state creep rates have been measured under a variety of conditions for several probable upper mantle minerals. The data from a single experiment are usually fit to an expression of the form

$$\dot{e} = A\sigma^n \exp(-Q/RT) \qquad (9.4.26)$$

where R is the gas constant, T is absolute temperature, and A, n, and Q are constants determined by the fitting procedure. It may be seen that Eq. (9.4.26) is a somewhat simplified, empirical form for Eqs. (9.4.22–5).

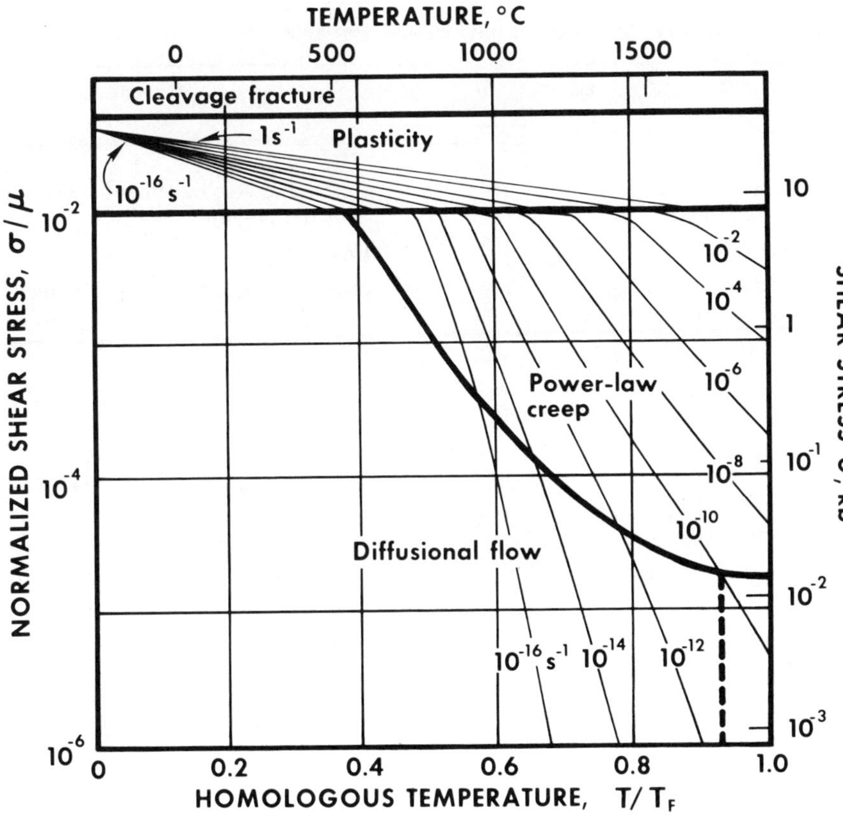

9.4.6 A deformation map for olivine at 81 kb pressure, from Ashby and Verrall (1977). Thick boundaries delimit regions in which a single deformation mechanism controls steady state creep; the dashed boundary within the diffusional flow field delimits Coble creep at lower temperatures from Nabarro-Herring creep at higher temperatures. Contours of constant strain rate are shown as thin lines. An average grain diameter of 0.1 mm and an activation volume equal to the ionic volume of O^{2-} were assumed; T_f = 2140°K.

Values of A, n, and Q for olivines and pyroxenes are given in Table 9.4.10.

Published creep data for olivine span the temperature range 900–1700° C at strain rates from 10^{-7} to $10^{-4} s^{-1}$ and at stress differences from 50 bars to 10 kb. Kohlstedt and Goetze (1974) showed that the available data on dry olivine can all be fit by a flow law of the form

$$\dot{e} = f(\sigma) \exp(-Q/RT) \tag{9.4.27}$$

where $f(\sigma)$ is shown in Fig. 9.4.7 for temperatures of 1100, 1400 and 1700° C. The relationship $f(\sigma)$ is not a single power law, but can be fit with a sum of two power laws (Ashby and Verrall, 1977) with $n = 3$ dominating at stress differences less than 1 kb and $n = 5$ dominating at higher stress differences. Ave'Lallemant's (1978) data for wet polycrystalline pyroxene show a similar behavior for $f(\sigma)$ (Table 9.4.10).

Viscosity used in solid-state convection models can also be represented in terms of T/T_f, the homologous temperature, where T_f is the melting temperature of the dominant rate-controlling mineral species (Weertman, 1970). The use of homologous temperature is a convenient and physically meaningful way of dealing with the activation energy and activation volume in Eq. (9.4.23) in those portions of the mantle where these quantities are not known from measurement. The general expression for effective viscosity is

$$\eta = F(\sigma) \exp(aT_f/T) \tag{9.4.28}$$

where $F(\sigma)$ may depend on stress σ and where a is a dimensionless constant, typically 15 to 25 for metals (Weertman, 1970). Equation (9.4.28) provides, for a given homologous temperature, an estimate of the viscosity of the Earth's lower mantle independent of that derived from analysis of glacial unloading.

Sammis *et al.* (1980) have tested Eq. (9.4.28) against measured activation volumes and melting point curves for a number of metals and nonmetals. They found that Eq. (9.4.28) works well for olivine and for metals, but less well for alkali halides. The agreement for olivine provides encouragement for the use of Eq. (9.4.28) in planetary mantles, but until more data are tested and the discrepant results explained, some caution in its general application is warranted.

Effect of water

Trace amounts of water are known to decrease the effective viscosity of silicates in the ductile regime, but the detailed physical mechanism of this weakening is not well understood for mantle minerals. The flow laws

Table 9.4.10 Constants for power-law flow laws for olivines and pyroxenes.†

Material	A ($kb^{-n}s^{-1}$)	n	Q (kJ/mole)	Reference
Mt. Burnet dunite				
Dry	5.1×10^9	3.3	464	Carter and Ave'Lallemant (1970); Carter (1976)
	4.3×10^8	3	527	Post (1977)
Wet	1.2×10^3	2.1	226	Carter and Ave'Lallemant (1970); Carter (1976)
	4.3×10^8	3.0±0.2	392±12	Post (1977)
Olivine, single crystal, dry				
Fo_{100}	7.1×10^{12}	3.5±0.5	565±63	Durham and Goetze (1977)
Fo_{92}	1.0×10^{12}	3.6±0.3	523±21	
Hedenbergite (Wo_{49}, En_{17}, Fs_{33}) single crystal	10^6	4.0±0.6	250±50	Kolle and Blacic (1978)
Enstatite (En_{89}, Fs_8, Wo_3), polycrystalline, wet	2.32×10^3	2.8±0.25	271±9	Ross and Nielson (1978)
Websterite (68% cpx, 32% opx)				
Dry, stress < 7 kb	1.6×10^2	4.3	326	Ave'Lallemant (1978)
Wet, stress < 3.5 kb	1.3×10^{10}	3.3	463	Ave'Lallemant (1978)
stress > 3.5 kb	7.9×10^5	5.3	382	Ave'Lallemant (1978)

†After Tullis (1979).

for wet and for dry olivine and pyroxene are substantially different (Table 9.4.10). The activation energy for flow in wet dunite is less than for the dry case (Carter, 1976; Post, 1977), but Ave'Lallemant (1978) found that in pyroxene the activation energy is higher for flow under hydrous conditions. Ross *et al.* (1979) attribute the lower viscosity by a factor of $\sim 10^3$ in the Carter and Ave'Lallemant (1970) experiments on "dry" dunite compared to Post's (1977) experiments on essentially the same material to 0.3% H_2O contained in serpentine in the samples used by the former workers.

An important question both for the rheology and for the solidus temperature in the Earth's upper mantle is whether any free water or other volatiles are present. The role of water-bearing phases on phase diagrams and melting relations is currently being explored (e.g., Kushiro *et al.*, 1967; Yamamoto and Akimoto, 1977), but the role of such phases on effective viscosity remains uncertain.

Effect of partial melting

For the oceanic upper mantle, comparison of model geotherms with laboratory derived melting data indicates that partial melting is likely if some free water, perhaps as little as 0.1% by weight, is present (Wyllie, 1971; Green, 1973; section 9.5.1). However, for small amounts of melting, the steady state creep rates in metal alloys (Auten *et al.*, 1974), in granite (Arzi, 1978) and in dunite and lherzolite (Goetze, 1977) are not substantially different from the creep rates at subsolidus temperatures. Arzi (1978) demonstrated that in his partial melting experiments on granites, the bulk viscosity does not decrease significantly until some 20 ± 10% melting has occurred. At lower melting volumes, creep rate is still controlled by the ability of the solid mineral grains to deform.

A consequence of these results is that in using Eq. (9.4.28) and the homologous temperature idea for multi-component systems, T_f is *not* in general the solidus temperature. Rather T_f is the temperature at which the melt fraction is sufficiently high so that deformation of the major interconnecting mineral grains no longer limits the bulk deformation. Thus T_f is closer to the temperature of first substantial melting of olivine, pyroxene and perhaps garnet (Yoder, 1976, p.59) in the upper mantle and to their high pressure equivalents (Kennedy and Higgins, 1972) in the lower mantle.

Effect of solid-solid phase changes

The effect of solid-solid phase changes on viscosity in planetary mantles is poorly understood. McKenzie and Weiss (1975), motivated by a belief in separate convection systems in the upper and lower mantle (section 9.5.1), suggested that there is an increase in the activation energy for the diffusion of oxygen ions at the 650 km seismic discontinuity. Such an increase would produce a discontinuity in viscosity at that depth, if the diffusion of oxygen controls the creep rate. The question of viscosity differences across phase changes was treated theoretically by Sammis *et al.* (1977), who concluded that changes in η due to changes in the activation energy and activation volume are probably less than one order of magnitude across any mantle phase boundary. Changes in the pre-exponential factor may still be important, however. Sammis (1976) also raised the possibility that phase boundaries in the mantle are zones of shear weakness due to the phenomenon of transformational superplasticity.

Viscosity in the Earth

The primary source of direct information on the viscosity structure of the Earth comes from studies of post-glacial rebound. Glaciers load the lithosphere and cause it to subside; when the glaciers melt the ice load is removed and the lithosphere rebounds at a rate determined by the viscosity structure of the mantle. Near coastlines and at very long wavelengths sensitive to the deep mantle, the effect of the rise in sea level during deglaciation must also be taken into account.

Using data from Fennoscandia, Haskell (1935) made the first quantitative model of glacial rebound. Approximating the mantle as a viscous half-space, Haskell derived a mantle viscosity of $\eta = 10^{22}$ poise. O'Connell (1971) first included the effects of global deglaciation, ocean loading and sphericity and concluded that a model with a uniform mantle viscosity of $\sim 10^{22}$ poise could fit his data.

Perhaps the most thorough treatment of mantle viscosity from deglaciation is that of Cathles (1975). On the basis of worldwide uplift and subsidence data on a variety of horizontal scales, Cathles concluded that (1) the (rheological) lithosphere is on average about 100 km thick; (2) the viscosity is about 4×10^{20} poise in a low viscosity channel 75 km thick, although the channel thickness and viscosity probably show substantial lateral variation; (3) the viscosity in the bulk of the mantle is $(1.0 \pm 0.1) \times 10^{22}$ poise, with no resolvable difference between upper and lower mantle; and (4) the mantle viscosity is Newtonian. Essentially the same conclusions were subsequently reached by Peltier (1976) and Peltier and Andrews (1976) using an independent analysis technique.

Several of Cathles's (1975) conclusions have been challenged, but have not yet been proven incorrect. Post and Griggs (1973) and Brennan (1974) reanalyzed Fennoscandian and North American uplift data and argued that they are better fit by non-Newtonian (power-law) creep than by a Newtonian viscosity. Cathles (1975) has shown that the effects of a low viscosity channel and of a nonlinear rheology for these data are very similar, however, and cannot be distinguished.

More recently, Peltier (1980) has asserted that a negative free air gravity anomaly whose outline generally coincides with the limit of the Laurentide ice sheet is a measure of the uplift remaining after North American deglaciation. On the basis of gravity and uplift data, Peltier (1980) concludes that the upper mantle is more viscous than the lower mantle (below 700 km depth) by a factor of perhaps as much as 50. Cathles (1975) noted that the Earth has many long wavelength free air anomalies, of comparable scale and magnitude, not associated with glacial areas; so the North American gravity anomaly need not be an indication of future rebound. The debate about lower mantle viscosity has not been settled as of this writing.

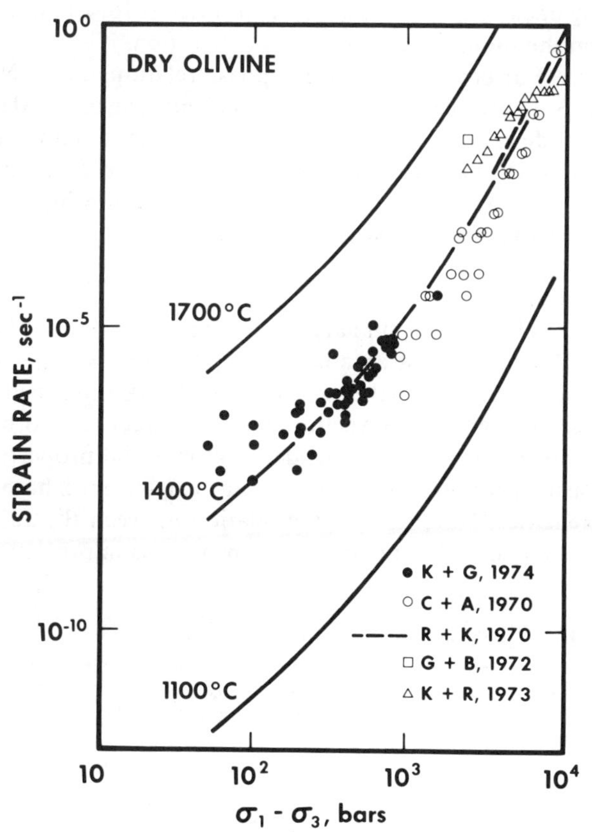

9.4.7 Strain rate versus differential stress in olivine at 1400°C and at 1100°C and 1700°C, from Kohlstedt and Goetze (1974). Data are from Carter and Ave'Lallemant (1970), Raleigh and Kirby (1970), Goetze and Brace (1972), Kirby and Raleigh (1973), and Kohlstedt and Goetze (1974). Actual data were adjusted to 1400°C using an activation energy of 522 kJ/mole.

The inference from rebound studies that mantle viscosity is Newtonian is in apparent conflict with the deduction from laboratory data and observations on mantle-derived xenoliths that deformation in the mantle occurs by power-law creep. It should be cautioned, however, that the viscosity measured by analysis of glacial unloading may not strictly apply to the mantle convection problem. Outside of the immediate vicinity of the ice sheets, the strain rates in the underlying mantle are less than in plate-tectonic associated mantle convection. Further, the total strains associated with glaciation and deglaciation are much less than in thermal convection. Finally, it has not been established that the viscosity measured in the glacial rebound problem is indicative of steady state, rather than transient, creep. Nonetheless, because linear viscous models *do* fit the glacial unloading data well, and because of their easier mathematical tractability in convection problems, a linear rheology is adopted in all thermal history models discussed in section 9.5.

Mention should be made of the promising new technique of geopiezometry that may provide an independent measure of viscosity in the upper 200 km of the mantle. In this technique, paleostress magnitudes are determined on mantle derived materials from dynamically recrystallized grain size (e.g., Mercier *et al.*, 1977) or from dislocation density (e.g., Durham *et al.*, 1977). From pyroxene geothermometry (section 9.5.1), the depth of origin and *in situ* temperature may be estimated. From these data and a flow law, the strain rate and effective viscosity may be derived (e.g., Ross *et al.*, 1979). There are several questions concerning the interpretation of these data at present, including calibrating these techniques against one another and accounting for effects of xenolith transport to the surface (Tullis, 1979), but much current effort is going toward making geopiezometry a working tool.

9.4.4 Adiabatic gradients and phase changes

Temperature changes from adiabatic heating and from phase changes modify the convection process within planets and limit the depth range over which basaltic volcanism can occur. The effect of adiabatic heating, as discussed in section 9.3.1, is important primarily for the Earth and Venus. If adiabatic heating is included in a calculation of global convection, it must be included so that energy is conserved. The relationship of adiabatic gradients to melting curves in the mantle and core of the Earth and planets are relevant to the interior state and to the radial dependence of viscosity.

Phase changes other than melting and basalt-eclogite occur at such great pressures that they are not directly relevant to production of basaltic magmas. The extent to which convection is enhanced or suppressed by phase changes, however, is of interest for questions on the radial extent of mantle convection.

For core formation the phase relations of Fe-Ni-FeS determine the range over which melting of metal or sulfide can occur. If the bulk of the core material segregates at shallow depths and then sinks in a rapidly convecting system, low pressure phase relationships will control core composition.

Adiabats

The temperature across the center of a convection cell tends toward an adiabatic profile, but whether or not adiabaticity in the mean is actually attained depends on the vigor of convection and the influence of other factors, such as the dependence of material properties on temperature, and imposed lateral temperature inhomogeneities. Nevertheless, the relation between the adiabatic and melting temperature profiles is important in

Table 9.4.11 Calculated temperatures for an adiabatic lower mantle.

Depth (km)	From Equation (9.4.14) and Seismic Velocities[a]			Assuming γ for Olivine Stoichiometry[b]	
	Model A T (°K)	Model B T (°K)	γ	Model A T (°K)	Model B T (°K)
696	1900[c]	2300[c]	1.27	1900[c]	2300[c]
1246	2043	2517	1.21	2079	2473
2071	2195	2766	1.05	2284	2657
2886	2325	2997	1.14	2476	2807

[a]From Hart (1977).
[b]From Jeanloz and Richter (1979).
[c]Assumed values, intended to bracket actual value.

determining if and where zones of melting can exist. For example, mantle material brought up along an adiabat melts if the temperature exceeds the melting temperature. This process has been used to model mid-ocean ridge basaltic melt production and to place limits on the depth of melting (section 9.5.1). Phase change effects, solid-solid and solid-liquid, must be included in estimating these profiles.

The adiabatic gradient passing through a point at radius r and absolute temperature T is

$$\frac{dT}{dr} = \left(\frac{\partial T}{\partial P}\right)_S \frac{dP}{dr} = \frac{-\alpha g T}{C_P} \qquad (9.4.29)$$

where α is the volume coefficient of thermal expansion, g is the gravitational acceleration and C_P is the specific heat at constant pressure. For $\alpha = 3\times10^{-5}$ °K^{-1}, g = 980 cm/s^2, $C_P = 1.3\times10^7$ erg/g-°K, and T ≃ 1300°K, $\frac{dT}{dr}$ ≃0.3° K/km. From Eq. (9.3.14) for the Grüneisen parameter γ, we also have

$$\frac{dT}{dr} = -\gamma\rho g \chi_a T \qquad (9.4.30)$$

where χ_a is the adiabatic compressibility. Since

$$\rho\chi_a = V_P^2 - \frac{4}{3}V_S^2 \qquad (9.4.31)$$

where V_P and V_S are the velocities of P and S waves, Eqs. (9.4.30–31) permit an estimation of adiabatic gradients in regions where seismic velocities and γ are approximately known.

Equations (9.4.30—31) may be used to derive an adiabat for the lower mantle. For this purpose, it is convenient to use an Earth model that prescribes density, pressure, and elastic velocities as functions of depth. For the present case we apply Model QM3 of Hart (1977). The Grüneisen parameter versus depth has also been derived by O. L. Anderson (1979) from the seismic velocity model of Dziewonski *et al.* (1975). The Grüneisen parameter is assumed to be given by the formula (Barron, 1955)

$$\gamma = \frac{1}{3}\left[1 + \frac{K_T}{V_P}\left(\frac{\partial V_P}{\partial P}\right)_T + \frac{2K_T}{V_S}\left(\frac{\partial V_S}{\partial P}\right)_T\right] \qquad (9.4.32)$$

which Anderson *et al.* (1968) have demonstrated to be generally valid for MgO. Here K_T is the isothermal bulk modulus. Taking the density and γ as derived by finite differences from the Earth model of Hart (1977), the temperature rise along an adiabat is given by Eq. (9.4.14). We can place some bounds on the temperature T_o and density ρ_o at the top of the lower mantle, so this depth is an appropriate reference. For the present case we have taken two values of T_o, 1900 and 2300°K, which we believe bracket current estimates of the temperature at the depth of 700 km (Ahrens, 1973), the top of the lower mantle. Values of the Grüneisen parameter and the adiabatic temperatures calculated for these two starting temperatures are given in Table 9.4.11 and Fig. 9.4.8.

Mantle melting relations

Phase diagrams describing melting relations for the upper mantle, to pressures of ~60 kb and depths approaching 200 km, are described at length in Chapter 3. In this section we confine discussion of the upper mantle to the relationship of temperatures derived from convection models (e.g., upwelling beneath a ridge) to the melting process. We also derive melting curves for the lower mantle for the estimation of viscosity through Eq. (9.4.28) and for comparison with the adiabats derived above.

Upper mantle melting. Material ascending in the mantle is thermally affected by the surface only to a characteristic depth of κ/u, where κ is the thermal diffusivity and u is the velocity of ascent. For 1 cm/yr vertical flow velocity, this depth is about 2 km. Any widespread

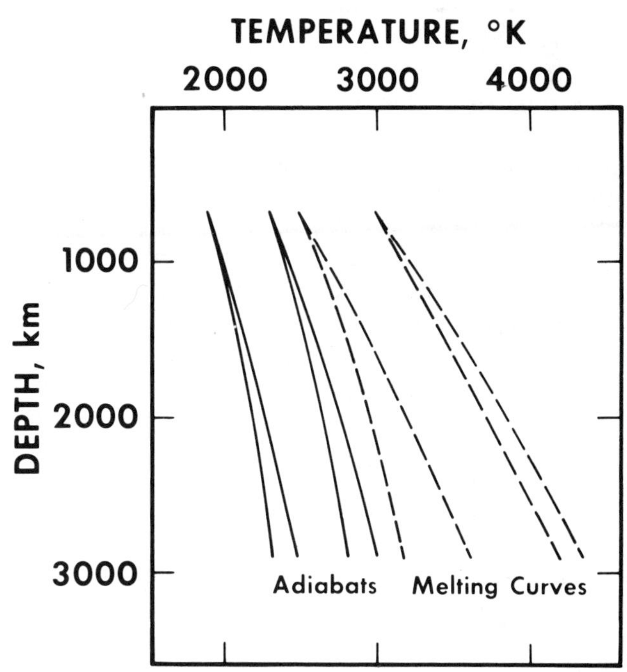

9.4.8 Adiabats and melting curves for the Earth's lower mantle. From Tables 9.4.11 and 9.4.12.

upwelling such as beneath ridges can thus be considered adiabatic.

Calculation of the temperature and the amount of partial melting in undepleted ascending mantle is somewhat simplified because the system behaves nearly as a eutectic until a harzburgite residuum is left (Kay *et al.*, 1970). Consider Fig. 9.4.9. Material ascending from depth follows the solid adiabat until the dry solidus of the mantle is intersected (5% to 30% melting range). Once the dry solidus is reached, the temperature follows the melting curve until all clinopyroxene and other aluminous, sodic, or calcic phases are used up. Some additional harzburgite dissolves into the melt as the mush continues to ascend. The temperature at 100 km depth at ridges can be estimated to be 1400° C by noting the 100° C is lost during melting of 30% basalt, 30° C is lost to adiabatic cooling of the basalt magma during ascent to the surface, and the liquidus temperature of primitive basalt is 1270° C (Usselman and Hodge, 1978). The closed system (path a, Fig. 9.4.9) and open system curves (path A, Fig. 9.4.9) are similar because the material behaves as a eutectic.

In Precambrian times ultrabasic "komatiite" lavas (section 1.2.1) were erupted at diverse places. The low temperature liquidus for a (~2.7 b.y.) Canadian komatiite studied by Arndt (1976) is 1500° C, and the liquidus for a (~3.3. b.y.) South African komatiite is 1650°C (Green *et al.*, 1975). To produce komatiite from mantle peridotite, about 60% melting is required if all the melt remains with the crystals (Arndt, 1977). This material would have a temperature of about 1730° C at 100 km depth (path b, Fig. 9.4.9).

Arndt (1977) has shown that melt is likely to segregate from komatiite source rocks as it forms. A temperature of only 1630° C at 100 km is needed to produce komatiites in this way (curve B, Fig. 9.4.9). Small amounts of komatiite could conceivably be formed from upwelling masses of already depleted mantle (curve C, Fig. 9.4.9). In this case the temperature at 100 km is 1530° C. The difference in source region temperatures between the present and ~3 b.y. ago under Arndt's (1977) hypothesis is equal to the difference between the liquidus temperatures of primitive magmas. This difference in source region temperatures may be due to a general cooling of the upper mantle by ~100°/b.y. at 100 km depth (Sleep, 1979b) or to a shoaling with time of the deepest extent of mantle upwelling beneath ridges.

Lower mantle melting. As noted above, the melting relations for mantle material are important not only for predicting magma production but also for estimating viscosity when using a relation such as Eq. (9.4.28).

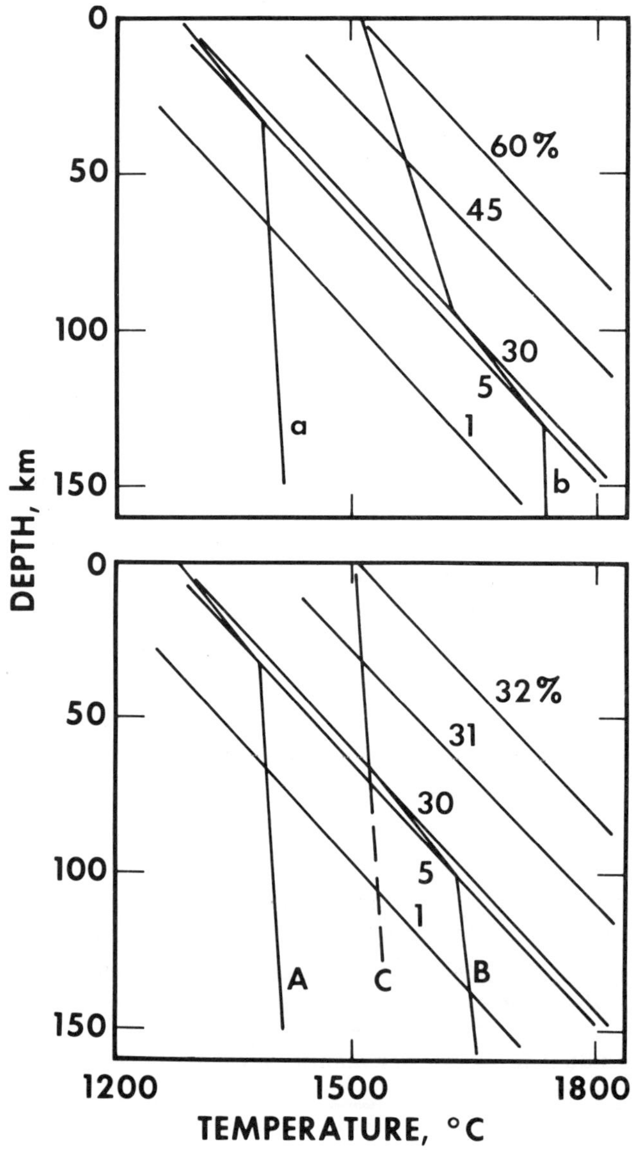

9.4.9 The temperature-depth curve of adiabatically ascending mantle material, plotted schematically with the cumulative fraction of melting. Closed system melting (melt remains in equilibrium with residuum) is assumed in the top figure, and open system melting (most of the melt segregates from residuum; only about 2% melt is retained) is assumed in the bottom figure. Rapid melting occurs between 5% and 30% where the mantle may be considered a eutectic. The temperature beneath present ridges is similar for both assumptions (curves a and A). A much higher temperature at depth is needed to produce the komatiite studied by Arndt (1976) by closed system melting (curve b) than by open system melting (curve B). Komatiites could conceivably be produced by melting adiabatically ascending residuum from previous melting events (curve C).

Such a relation is particularly useful for estimating the viscosity in high-pressure assemblages such as in the Earth's lower mantle.

An approximate melting curve T_f based on the Grüneisen parameter γ can be calculated for the lower mantle using the Lindemann law for melting

$$\frac{1}{T_f}\frac{dT_f}{dP} = \frac{2(\gamma - 1/3)}{K_f} \quad (9.4.33)$$

Here T_f is taken to be the melting point of a material which melts congruently, based on theoretical formulations relating O^{-2} vacancy diffusion to the temperature and pressure dependence of viscosity, and K_f is the bulk modulus along the melting line. The onset of significant melting is expected, on strictly mechanical grounds, to correspond to a temperature T_f somewhat above the solidus. Since MgO has a zero-pressure melting point of 3100°K and stishovite has a zero-pressure melting point of ~2800–3000°K (Jackson, 1976), we can assume two plausible trial values for T_f at a depth of 700 km of 2500°K and 3000°K. The resulting melting curves for the lower mantle are given in Table 9.4.12 and Fig. 9.4.8. Because of the decrease in free energy of formation of lower mantle phases such as $MgSiO_3$ (perovskite) with respect to oxides, such phases should have melting points lower than that of pure MgO or SiO_2. Since MgO and SiO_2 are among the higher melting point components of the mantle (Kennedy and Higgins, 1972), the Lindemann relation as used here provides an upper bound to the solidus of the lower mantle and may lie closer to the liquidus.

Solid-solid phase changes

Ordinary thermal convection is controlled by the volume coefficient of thermal expansion for mantle rock, which is about $\alpha = 3 \times 10^{-5}$ °K^{-1} in the upper mantle. For mantle convection the maximum expected temperature differences are about 1000°K, and the corresponding maximum density variation is about 3%. Inversion of seismic data has shown that there are pronounced increases (~10%) in seismic velocity and density at depths near 350 and 650 km, and perhaps lesser discontinuities at other depths.

High-pressure laboratory experiments (e.g., Ringwood, 1970a) in concert with seismological studies (Anderson, 1967) have supported the conclusion that the 350-km discontinuity is primarily associated with the transition of olivine to β-spinel structure. The thermodynamic properties of this phase change are known approximately from the laboratory studies: there is a positive density change, $\Delta\rho = 0.3$ g/cm^3, and a heat release Q = 10 J/g. The equilibrium phase boundary lies on a Clapeyron curve whose slope (dP/dT) has been reported to be between 12 bar/°K (Ma, 1974) and 64 bar/°K (Akimoto and Fujisawa, 1968) for various materials transforming between the olivine and β-phase structures. Akimoto *et al.* (1976) obtain a slope of 33 bar/°K for olivine–β spinel of mantle composition. Solomon and Paw U (1975) report an elevation of 100 ± 15 km for the olivine–β-spinel reaction in slabs using seismic travel time residuals, which corresponds to a 1000°K temperature contrast at 250 km depth for the 33 bar/°K Clapeyron slope.

The nature of the 650-km discontinuity is highly uncertain. On the basis of high pressure measurements, Liu (1976) suggested that the phase change is due to the disproportionation of the γ-spinel form of $(Mg, Fe)_2SiO_4$ to $(Mg, Fe)SiO_3$ (perovskite) plus (Mg, Fe)O (cf. Yagi *et al.*, 1979). Anderson (1977) argues, on the basis of a comparison of seismic and laboratory data, that the 650-km discontinuity does not involve γ-spinel, but may represent the reaction of garnet plus pyroxene to perovskite or mixed oxides. There may be a chemical change superposed on the phase change at 650 km (Anderson, 1977; Liu, 1979), which may pose a barrier to penetrative convection (Richter and Johnson, 1974).

Since the density changes associated with these phase changes are large it is important to examine their influence on mantle convection. Schubert *et al.* (1970) showed that the olivine–β-spinel transition in fact enhances thermal convection in the mantle. The basic mechanism is illustrated in Fig. 9.4.10. In the absence of convection the equilibrium position of the phase boundary is at depth a; the hydrostatic pressure and temperature at this point correspond to the point a on the Clapeyron curve. Assume a counterclockwise convection cell in the fluid layer heated from below. The upward flow in the fluid layer brings hotter material towards the phase change boundary; the higher temperature requires the phase boundary to lie at a greater

Table 9.4.12 Calculated melting curves for an adiabatic mantle for two melting temperatures at 700 km depth according to Lindemann Law.

Depth (km)	Pressure (kb)	T_f (°K)[(a)]		T_f (°K)[(b)]	
696	250	2500	3000	2500	3000
1246	500	2826	3391	2738	3286
2071	902	3223	3867	2989	3747
2886	1351	3603	4323	3179	4189

[(a)] γ taken from Table 9.4.12.
[(b)] γ for olivine stoichiometry taken from Jeanloz and Richter (1979).

depth (higher pressure) corresponding to point b on the Clapeyron curve. Because the elevation of heavier fluid in the region of downward flow tends to drive the fluid downwards, and the depression of the heavier fluid in the region of upward flow tends to drive the flow upwards, the density change across the phase change helps to drive the flow.

However, the phase change from olivine to β-spinel releases heat. The latent heat associated with the phase change inhibits the instability. A detailed analysis (Schubert and Turcotte, 1971) showed that for the olivine–β-spinel phase change, the distortion of the phase boundary was more important than the influence of the latent heat, and that the phase change was an important contribution to mantle convection. Numerical studies of the flow induced by the phase change have been carried out by Richter (1973a, b) and Schubert *et al.* (1975). Too little is known about the thermodynamics of candidate phase changes of the 650-km discontinuity for the effects to be predicted, but the calculations of Richter (1973a, b) indicate that convection is possible for a wide range of phase change parameters.

The above discussion of the influence of phase changes on mantle convection is based on the assumption that the phase changes are in thermodynamic equilibrium. Sung and Burns (1976) have suggested that the rate of reaction for the olivine–β-spinel phase change may be sufficiently slow to impede mantle convection. And as noted in the last section, Sammis (1976) has proposed that phase changes may significantly reduce mantle viscosities and that phase boundaries can act as zones of free sliding because small stresses will change the phase equilibrium within a grain, prompting recrystallization.

Fe-Ni-FeS phase relations

Though not directly relevant to basaltic volcanism, the phase relations of the Fe-Ni-FeS system are important for the process of core-mantle differentiation, for the physical properties of planetary cores, and for questions of the origin of planetary magnetic fields. The experimentally determined portion of the phase diagram for pure Fe (P to 200 kb, T to melting) is shown in Fig. 9.4.11. Note that the solid phase in equilibrium with liquid is δ-Fe below 53 kb, γ-Fe between 53 kb and 0.94 Mb (Liu, 1975a) and ϵ-Fe above that pressure; these three solid phases might be expected to be stable in solid portions of cores in the Moon, in Mars or Mercury, and in Venus or the Earth, respectively. Of the polymorphs of Fe, only α-Fe is ferromagnetic (see Fig. 9.4.11 for Curie point) and thus capable of carrying appreciable remanent magnetization.

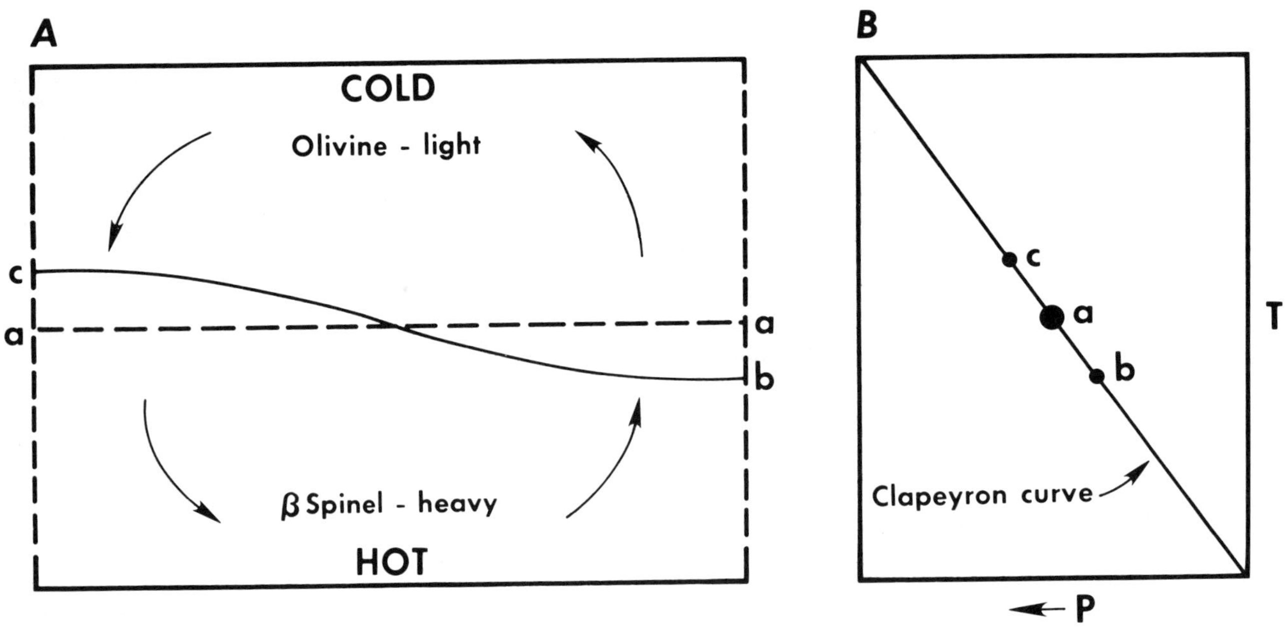

9.4.10 Illustration of the phase change instability in a fluid layer heated from below. **a.** In the region of hot upwelling flow, the phase boundary is displaced downward to the point "b". In the region of cold downwelling flow, the phase boundary is displaced upward to the point "c". The density difference across the phase change assists in driving the flow. **b.** The phase boundary, or Clapeyron curve, as a function of pressure P and temperature T.

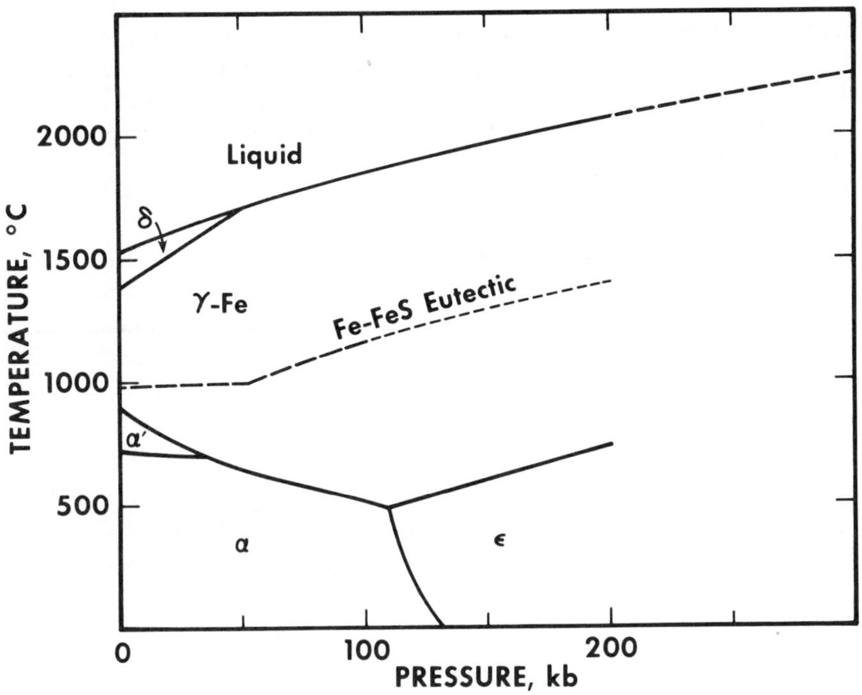

9.4.11 Approximate phase diagram for Fe. The α-γ, γ-ε, and α-ε boundaries are from Bundy (1965) and Mirwald and Kennedy (1979). The δ-γ and δ-liquid boundaries are from Strong *et al.* (1973). The γ-liquid boundary is from Liu and Bassett (1975). The α (ferromagnetic) to α′ (paramagnetic) transition—the Curie point temperature—is from Mirwald and Kennedy (1979). The eutectic temperature for the Fe-FeS system (Brett and Bell, 1969; Usselman, 1975) is shown as a dashed line.

The Earth's outer core cannot be pure Fe, but its physical properties are consistent with those of Fe plus some lighter alloying element such as S or O (e.g., Brett, 1976). The system Fe-Ni-S has been most extensively studied at planetary pressures. The Fe-FeS eutectic temperature, shown in Fig. 9.4.11, lies considerably below the melting curve for pure Fe (Brett and Bell, 1969; Usselman, 1975a). The first melt to form in a warming, homogenous planet may have the composition of the Fe-FeS eutectic: 27 weight percent S at 30 kb and 24 weight percent at 100 kb (Usselman, 1975a). Small amounts (<7 weight percent) of Ni do not affect the melting relations of the Fe-FeS system for pressures below 80 kb (Usselman, 1975a).

9.5 PLANETARY THERMAL MODELS

Possible thermal histories of the Earth, Moon, Mars, Mercury, Venus and the asteroids are presented in this section by planetary object. The thermal history models are based on one or more of the calculational schemes discussed in section 9.3 and use the physical parameters appropriate to each object (section 9.4). The goals of this section are twofold: (1) to provide a global, physical context for the observations on the form and evolution of basaltic volcanism on the planets as presented in Chapters 1–8; and (2) to look for independent conclusions on planetary history and the processes of volcanism that arise from considering the range of thermal history models consistent with these observations.

The planetary bodies are discussed in the order of decreasing level of knowledge. For the Earth, the phenomenon of plate tectonics dominates the thermal structure and volcanic processes in the outer hundred kilometers. Local thermal models of mid-ocean ridges and subduction zones are at least as important as global history models in understanding the evolution of basaltic volcanism on this planet. For the Moon, the detailed chronology of returned basalt samples and the extensive geological, geophysical, and remote sensing data permit a fairly complete understanding of the evolution of mare volcanism. This level of understanding is reflected in the large number of lunar thermal history models that have appeared in the literature since the Apollo program. For Mars, there is strong evidence for widespread volcanism but an absolute chronology is lacking; Mercury is in a similar situation, except the first-order evidence for volcanism is more equivocal. On the basis of sketchy information on global differentiation and on the relationship between the tectonic and volcanic histories for Mars and Mercury, some important conclusions may still be derived from thermal models. The history of Venus is the most poorly constrained among the terrestrial planets, and the ability of thermal models to improve on this situation is extremely limited. Finally, the thermal history of asteroids or meteorite parent bodies may be modelled, but accounting in other than an *ad hoc* fashion for the apparently very different histories of objects of generally similar mass and orbital parameters remains a major unsolved problem.

9.5.1 Earth

Most thermal models of the Earth are primarily concerned with its present thermal structure. With the general acceptance of plate tectonics, the role of thermal convection in the transport of heat within the Earth has also received general recognition. Thermal models involving convection can be applied either locally or globally. Local models have the advantage that detailed comparisons can be made with geological, geochemical, and geophysical observations, while global models are necesary to treat planetary thermal evolution.

Comprehensive studies of the thermal structure in the vicinity of ocean ridges have been carried out. Detailed comparisons with ridge topography and other observables provide confirmation of the results. A central feature of ridge thermal models is the pressure-release melting that leads to the formation of the basaltic oceanic curst, apparently the most prolific source of basaltic volcanism on the terrestrial planets. Detailed studies of the thermal structure of oceanic trench-island arc regions also have been carried out. Though a reasonable consensus has been reached concerning the thermal structure of the descending lithosphere, the heat source for the volcanism at island arcs remains a matter of debate. This uncertainty extends to the origin of the back-arc spreading which results in the formation of marginal basins.

No fully comprehensive thermal models on a global basis have been carried out. Two-dimensional calculations of the thermal structure of the upper mantle have been given by a number of authors. Such models are subject to several criticisms. Is convection confined to the upper mantle? Do two-dimensional calculations give good representations of the actual flow properties? Many questions remain with regard to the association of mantle convection and intraplate volcanism.

Approximate techniques can be used to study the effects of convection on a global basis. Calculations based on parameterized convection provide values for mean mantle temperatures, mean mantle viscosities, and other averaged quantities. This approach can also be used to study the global thermal evolution of the Earth including the effects of initial conditions and the decrease with time in the rate of radioactive heat production. Important questions such as the validity of a steady state heat balance for the Earth can be addressed.

Plate tectonics

Plate tectonics is the near-surface manifestation of the Earth's convective engine. The lithosphere may generally be regarded as the cold upper boundary layer of mantle convection cells, as described in section 9.3.2. By the creation and recycling of lithospheric plates, the Earth is able to transfer internal heat to the surface much more rapidly and efficiently than by heat conduction alone. It is likely that heat transfer by plate recycling has dominated the global thermal evolution of the Earth.

The Earth's plates are in constant relative motion, and their boundary regions are often sites of active volcanism as a consequence of that motion (see section 6.2). Plate divergence at mid-ocean ridges is closely associated with the upwelling of mantle material and with the generation of basaltic magma by pressure-release melting. Creation of new oceanic crust at mid-ocean ridges repesents the most voluminous component of basaltic volcanism on the Earth today, and probably over its entire history. Prominent lines of volcanic centers occur also over descending lithosphere at convergent plate margins, though the detailed mechanism for generation of magma at convergence zones is not well resolved. Basaltic volcanism occurs as well within plate interiors, and though intraplate volcanism is linked in some fashion with the thermal structure of the underlying mantle, a close connection to the motions of tectonic plates has not been demonstrated.

Plate tectonics provides *prima facie* evidence that mantle convection is occurring on the Earth. This is in contrast to the other terrestrial planets, for which arguments on mantle convection must be based on theoretical models and on our expectations of the rheology of mantle material. While the relative velocities of plates are fairly well known and fall in the range 1–10 cm/yr (Minster *et al.*, 1974), the relation between plate motions and the detailed pattern of mantle flow is unfortunately quite complex and has so far eluded description by all but the most simple models.

In spite of this handicap, detailed and realistic thermal models of plate boundaries can be generated simply from our knowledge of plate kinematics. These models provide a framework for understanding volcanism at ridges and at island arcs. They also serve as a basis for determining the Earth's heat budget and for extrapolating that budget into the past.

Ocean ridges and oceanic lithosphere. Ocean ridges are associated with ascending mantle convection. The rigid surface plates that make up the oceanic lithosphere are created from hot mantle rock at ocean ridges. The cooling and thickening of the oceanic lithosphere as it moves away from a ridge is an important mechanism for the transport of heat from the interior of the Earth.

In order to study the thermal structure of the oceanic-lithosphere it is appropriate to consider the equation for the one-dimensional cooling of a semi-infinite half-space

$$\frac{\partial T}{\partial t} = \kappa \frac{\partial^2 T}{\partial y^2} \tag{9.5.1}$$

where t is the age of the lithosphere and y is the depth. The necessary initial and boundary conditions are that $T = T_m$ at $t = 0$, $T = T_s$ at $y = 0$, and $T \to T_m$ as $y \to \infty$, where T_m is the temperature of hot mantle rock and T_s is the temperature of the sea floor. The solution of Eq. (9.5.1) that satisfies these boundary conditions is (Carslaw and Jaeger, 1959)

$$\frac{T - T_s}{T_m - T_s} = \text{erf}\,\frac{y}{2}\left(\frac{1}{\kappa t}\right)^{1/2} \tag{9.5.2}$$

where erf is the error function (Abramowitz and Stegun, 1964). The surface heat flow $q = \left(K\frac{dT}{dy}\right)_{y=0}$ as obtained directly from Eq. (9.5.2) is

$$q = (T_m - T_s)\left(\frac{\rho C_P K}{\pi t}\right)^{1/2} \tag{9.5.3}$$

In Fig. 9.5.1 the mean and standard deviation of heat flow measurements adjacent to the East Pacific Rise between 0° and 25° S are given as a function of crustal age (Anderson and Hobart, 1976). These results are based on 380 heat flow measurements. Also included in Fig. 9.5.1 is the predicted heat flow from Eq. (9.5.3) assuming $\rho = 3.3$ g/cm^3, $C_P = 1.0 \times 10^7$ erg/g-°K, $K = 3.3 \times 10^5$ erg/cm-sec-°K ($\kappa = 10^{-2}$cm^2/sec), and $T_m - T_s = 1200$°K.

The means of the heat flow measurements lie systematically below the theoretical curve, particularly for oceanic crust younger than about 12 m.y. This difference is attributed to the rapid convective cooling of the crust by the circulation of cold ocean water through pores and fissures. The wide scatter indicated by the large values of the standard deviations is also attributed to this process. Cracklike conduits, spaced on the scale of kilometers, are probably the loci of the exiting high temperature flow. Continuum porous flow through the bulk of the oceanic crust would redistribute heat, but produce no net heat flow anomaly. The time for the entire ocean to be circulated through the oceanic crust is a few million years (Ribando *et al.*, 1976).

The topography of the ocean floor can be directly related to the temperature distribution in the thermal boundary layer. Since the oceanic ridge is a broad feature, it is appropriate to assume that it is in hydrostatic equilibrium. Under this assumption, the increase in water depth away from the ridge, w(t), can be related to the denisty distribution in the thermal boundary layer by

$$(\rho_m - \rho_w)\, w = \int_0^\infty (\rho - \rho_m)\, dy \tag{9.5.4}$$

where ρ_m is the density of the hot mantle rock from which the lithosphere is formed and ρ_w is the density of sea water. With the assumptions that the crustal layer may be neglected and that

$$\rho - \rho_m = \alpha\,(T_m - T)\,\rho_m \tag{9.5.5}$$

and the temperature distribution given in Eq. (9.5.2), we find that

$$w = \frac{2\rho_m \alpha}{(\rho_m - \rho_w)}\left(\frac{\kappa t}{\pi}\right)^{1/2}(T_m - T_s) \tag{9.5.6}$$

The observed increase in depth away from the Mid-Atlantic Ridge and East Pacific Rise are given as functions of seafloor age in Fig. 9.5.2 (Sclater *et al.*, 1975). The observations are compared with the predicted topography given by Eq. (9.5.6) with $\alpha = 3 \times 10^{-5}$°K^{-1}

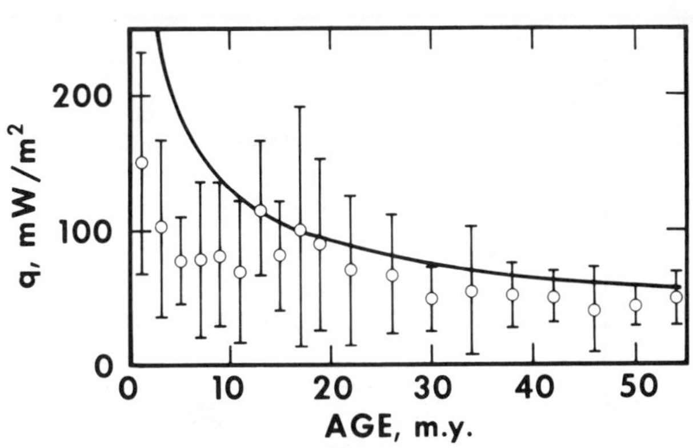

9.5.1 Mean and standard deviation of heat flow measurements adjacent to the East Pacific Rise between 0° and 25°S (Anderson and Hobart, 1976). The solid curve is the predicted heat flow from Eq. (9.5.3.)

and $\rho_m - \rho_w = 2.3\ g/cm^3$. Since the topography is related to the entire temperature distribution in the lithosphere, it is not sensitive to near-surface temperature perturbations resulting from the circulation of ocean water within the crust. The agreement shown in Fig. 9.5.2 is strong confirmation of the boundary layer theory as a general model for the thermal structure of the lithosphere.

The ocean surface geoid (gravitational equipotential surface) has been obtained directly by radar altimetry from the GEOS-3 satellite. A profile (orbit 400) that crosses the Mid-Atlantic Ridge at latitude 44.5°N is given in Fig. 9.5.3 as a function of seafloor age.

The geoid anomaly (i.e., the difference between an observation and the reference ellipsoid) in regions of hydrostatic (isostatic) equilibrium is given by

$$h = \frac{2\pi G}{g} \int_0^\infty y\,(\rho - \rho_m)\,dy \qquad (9.5.7)$$

where G is the gravitational constant (e.g., Ockendon and Turcotte, 1977). Although Eq. (9.5.7) ignores the wavelength dependence of regional compensation (plate flexure), the formula avoids the need for spherical harmonic representations of long wavelength features, provides a direct link between subsurface density anomalies and measureable ocean heights, and can be used in addition to compute lithospheric stress from geoid height (e.g., Sleep and Phillips 1979). Substitution of Eqs. (9.5.2) and (9.5.6) into Eq. (9.5.7) and integrating yields

$$\Delta h = \frac{2\pi G \rho_m \alpha (T_m - T_s)}{g} \left\{ 1 + \frac{2\rho_m \alpha (T_m - T_s)}{\pi(\rho_m - \rho_w)} \right\} \kappa t \qquad (9.5.8)$$

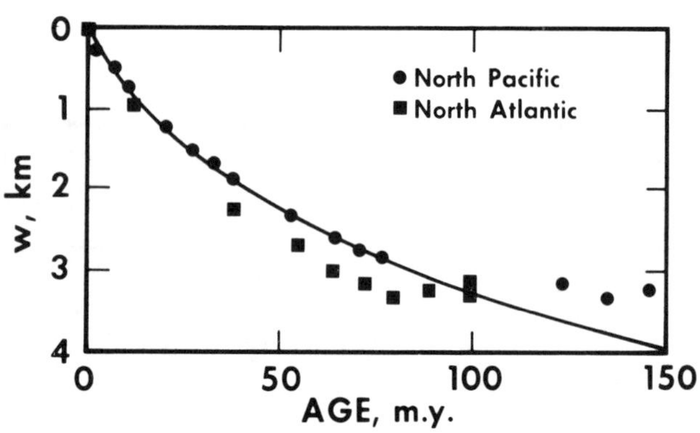

9.5.2 Depth of the seafloor relative to the ridge crest in the North Pacific and North Atlantic as a function of seafloor age (Sclater *et al.*, 1975) compared with the predictions of Eq.(9.5.6).

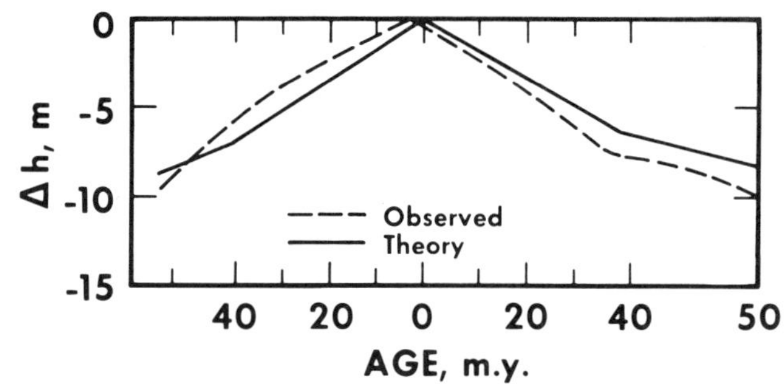

9.5.3 Profile (dashed line) of the geoid anomaly across the Mid-Atlantic ridge at latitude 44.5°N obtained from the GEOS-3 satellite (orbit 400) referenced to the ridge crest as a function of seafloor age compared with the predicted anomaly (solid line) from Eq. (9.5.8).

(Haxby and Turcotte, 1978). Taking the parameters as given above, this predicted geoid anomaly is compared with the observed anomaly in Fig. 9.5.3. The excellent agreement is further verification of the simple cooling model for the oceanic lithosphere

We have associated the oceanic lithosphere with the thermal boundary layer of mantle convection. The base of the elastic lithosphere may be defined to be an isotherm $T = T_l$, such that mantle rocks at temperatures below T_l exhibit rigidity on geological time scales and at higher temperatures exhibit a fluid-like behavior. From Eq. (9.5.2) the thickness of the oceanic lithosphere y_l is given by

$$y_l = 2(\kappa t)^{1/2}\, \mathrm{erf}^{-1} \frac{T_l - T_s}{T_m - T_s} \qquad (9.5.9)$$

Taking $T_l - T_s = 500°C$ and the other parameters as above, we find that

$$y_l = 4.3\sqrt{t} \qquad (9.5.10)$$

with t in millions of years and y_l in kilometers. For instance, for t = 100 m.y. old, y_l = 43 km by this formula.

The thickness of the elastic portion of the lithosphere may be determined from the response of the oceanic lithosphere to volcanic loads or, more generally, to the wavelength-dependent relationship between ocean floor bathymetry and free air gravity (McKenzie and Bowin, 1976; Watts, 1978; McNutt, 1979; Cochran, 1979; Detrick and Watts, 1979). On the basis of these studies, it appears that elastic plate models can explain existing oceanic data on responses to loads; that any viscous relaxation times are less than 10^6 yr or greater than 5×10^7 yr; and that the thickness of the elastic lithosphere is governed by the depth to the 450 ± 150° C isotherm in the oceanic mantle, as given approximately by Eq. (9.5.9), at the time the load was applied (Watts, 1978; Watts *et al.*, 1980).

It is evident from Fig. 9.5.2 that there is some departure of the observed topography from the predictions of the simple lithospheric cooling model at ages greater than about 80 m.y. The sea floor does not continue to subside as much as would be expected. This has led some (e.g., Langseth *et al.*, 1966; McKenzie, 1967) to model the oceanic lithosphere as a cooling, horizontal slab of fixed thickness with uniform bottom temperature. For young oceanic lithosphere this model gives the same results as the boundary layer model; however, for older lithosphere the heat flow and depth exponentially approach constant values. Parsons and Sclater (1977) give 125 ± 10 km and 1350° ± 275° C for the values of plate thickness and bottom temperature that best match heat flow and depth-age data.

There have been a number of attempts to find physical explanations for this phenomenon. Crough (1975) and Crough and Thompson (1976) have constructed models which assume that a constant input of heat from below is applied to the bottom isotherm which increases in depth with time. Schubert and Turcotte (1972) and Schubert *et al.* (1976) have suggested that viscous dissipation at the base of the lithosphere as it moves with respect to the mantle may generate sufficient heat to prevent further thickening. Forsyth (1977) suggested that the sublithospheric heat flux necessary to slow lithospheric thickening may be supplied by upper mantle radioactivity. Richter (1973c) has proposed the existence of small-scale secondary convective cells, with least horizontal dimension on the order of the thickness of the asthenosphere, immediately beneath the lithosphere and oriented with their long axes parallel to the direction of plate motion. Heat delivered by these cells prevents the downward thickening of the lithosphere. The alternation of ascending and descending limbs of such cells beneath the oceanic lithosphere might be reflected in a regular oscillation in the surface heat flow, an effect not yet observed.

Most of the Earth's volcanic activity occurs along the axes of the submarine mid-ocean ridges. Basaltic magma is more or less continuously supplied from the mantle and forms a layer about 7 km thick of extrusive and intrusive rock on top of the slowly convecting mantle. This layer is the oceanic crust, and as it moves away from the ridge axes it acquires a veneer of sediment up to 1.5 km in thickness.

Seismic studies show that the oceanic crust is remarkably uniform in its structure throughout the world's oceans, and it is clear that the generation of magma must be closely related to the spreading process. The mantle material rising under the ridge undergoes partial melting by pressure release during this ascent; the increase in the solidus temperature with depth is greater than the near-adiabatic gradient within the ascending mantle flow, and melting begins at about 50 km below the surface. The average degree of partial melting is about 20–25% (section 3.3.7), but more occurs in the center of the ascending flow and less at the side. As soon as the percentage of melt present at any point exceeds a critical value, gravitational body forces drive the lighter liquid to the vicinity of the crust through the slowly convecting mantle. Upon reaching the surface, the flow diverges and the depleted residuum after basalt segregation consitutes a volume of oceanic mantle equivalent to a layer approximately 30 km thick beneath the crust (Kay *et al.*, 1970).

The similarity solution given in Eq. (9.5.2) is not valid for the innermost few kilometers of the ridge where the finite width of the intrusion zone, lateral conduction of heat, eruption of lava, and the latent heat of basalt are important. Many early calculations following McKenzie (1967) assumed that the axial zone was maintained at the temperature of basalt intrusion. Because the model has an infinite heat loss at the ridge, the predicted shallow temperatures at the ridge axis are incorrect. Analytical models which conserve energy have been obtained by Parker and Oldenburg (1973), Oldenburg (1975), Sleep (1974, 1975a) and Lubimova and Nikitina (1975). Parker and Oldenburg (1973) assumed that the material was completely molten and that latent heat was released when material crystallizes and plates to the bottom of the lithosphere. The mathematical techniques used in that paper are given by Oldenburg (1975) along with models which assume various degrees of partial melt. Sleep (1974) ignored latent heat and assumed a constant temperature of intrusion. Sleep (1975a) balanced the latent heat of intrusion of the basaltic crust with the latent heat of melting in the underlying region.

The treatment of latent heat in any of these models is not a complete representation of the actual system because crystal settling and convection in the magma chamber permit release of latent heat without material being plated onto the top of the magma chamber. Sleep (1975a) proposed that the magma chamber at a mid-ocean ridge remained filled with mostly crystalline cumulate mush because newly added magma rose to the top of the chamber, spread laterally, and lost its latent heat to the top of the chamber. This model is contrasted with that of Oldenburg (1975), in which the latent heat of the magma chamber is lost to the sides of the chamber, which remains filled with melt. The models by

Oldenburg (1975) and Sleep (1975a) give physically relevant results at very low spreading rates. Thus the structure of the lithosphere at continental rift valleys can be approximately determined.

Ocean trenches, island arcs, and marginal basins. There is strong evidence that oceanic lithosphere descends into the mantle at oceanic trenches. Earthquake epicenters map the location of the descending plates to a depth of approximately 700 km. As the oceanic lithosphere cools with age it becomes gravitionally unstable, so that it is not surprising that it founders and sinks into the mantle. The concept of gravitational instability, however, has not yet accounted quantitatively for the wide range of ages of oceanic lithosphere currently being subducted. Continental lithosphere cannot generally participate in the subduction process because of the bouyancy of the thick, light continental crust.

Rates of subduction seem to be less variable than spreading rates, but cannot be measured directly. Current solutions for global relative plate motions (Minster *et al.*, 1974) show that the mean subduction rate is about 8.5 cm/yr. With few exceptions, most convergence rates at trenches are between 7 and 10 cm/yr. An explanation for this uniformity has been given by Forsyth and Uyeda (1975) and Chapple and Tullis (1977), who conclude that the motions of plates with extensive trench systems are dominated by a force balance on the descending plates. A balance between buoyant forces and viscous retarding forces on the descending plates within the mantle causes these plates to subduct at the observed mean speed.

The thermal structure of the descending plate has been calculated for a wide variety of situations (McKenzie, 1969; Turcotte and Oxburgh, 1969; Minear and Toksöz, 1970a, b; Hasebe *et al.*, 1970; Griggs, 1972; Toksöz *et al.*, 1971, 1973; Turcotte and Schubert, 1971, 1973; Schubert *et al.*, 1975; Hsui and Toksöz, 1979). The results of one such calculation (Schubert *et al.*, 1975) are shown in Fig. 9.5.4. The angle of dip is 45° and the convergence velocity of the plates is 8 cm/yr. The elevation of the phase boundary between olivine and β-spinel due to the cooler temperatures within the descending plate is also shown in the figure. Neglecting the heat release due to the phase change, the elevation of the phase boundary h_P is related to the temperature decrement in the slab ΔT_s by

$$h_P = \left(\frac{dP}{dT}\right)_C \frac{\Delta T_s}{\rho g} \qquad (9.5.11)$$

where $\left(\frac{dP}{dT}\right)_C$ is the slope of the Clapeyron curve. Taking this slope as 33 bar/°C, $\rho = 3.4$ g/cm^3, g = 980 cm/sec^2, and ΔT_s = 1000°C, we find that h_P = 100 km.

Because the thermodynamic properties of any phase transition at 650 km are not known, and even the nature of the 650-km discontinuity is debated (section 9.4.4), whether that discontinuity is elevated or depressed within the descending lithosphere or is unaffected by the thermal structure of the downwelling zone is unknown. Calculations of the thermal structure of the descending plate indicate that it may still be well-defined at depths below 650 km. This result would support the hypothesis that the descending lithosphere penetrates to even greater depths in the mantle, although there is little direct seismic evidence for this idea (e.g., Jordan, 1977) because of the absence of earthquakes at these depths.

From the temperature distribution, the density distribution in the plate can be obtained and the negative buoyancy due to thermal contraction can be determined. The downward body force due to the elevation of the olivine–β-spinel phase boundary can also be obtained, and is of the same magnitude as the thermal contraction term (Turcotte and Schubert, 1971). The basalt-eclogite transition may also contribute significantly to the total negative buoyancy.

The sinking of the oceanic lithosphere at trenches is in many ways analogous to the separation of the cold thermal boundary layer from the upper boundary of a convection cell as illustrated in Fig. 9.3.1. In both cases, the sinking is due to a gravitational instability. However, in the case of the oceanic lithosphere, the thermal boundary layer has a flexural rigidity associated with its elastic strength. A number of authors have treated the problem of the bending of the oceanic lithosphere at trenches as a problem involving the bending of an elastic plate with a hydrostatic restoring force. The calculated shape of the plate is in reasonably good agreement with the observed bathymetry including a slight arching seaward of the trench (e.g., Caldwell *et al.*, 1976); the apparent elastic thickness of the plate at subduction closely follows Eq. (9.5.9) as a function of sea-floor age (Caldwell and Turcotte, 1979). The outer topographic rise can also be explained as a consequence either of the bending of a viscous or an elastic-plastic plate or of the convection process (De Bremaecker, 1977; McKenzie, 1977a; Chapple and Forsyth, 1979).

The processes that lead to the initiation of subduction are poorly understood. McKenzie (1977b) has suggested that subduction is initiated by a compressional fracture of the lithosphere. Turcotte *et al.* (1977b) associated subduction with rifting of oceanic lithosphere, upwelling of asthenospheric material above the level of

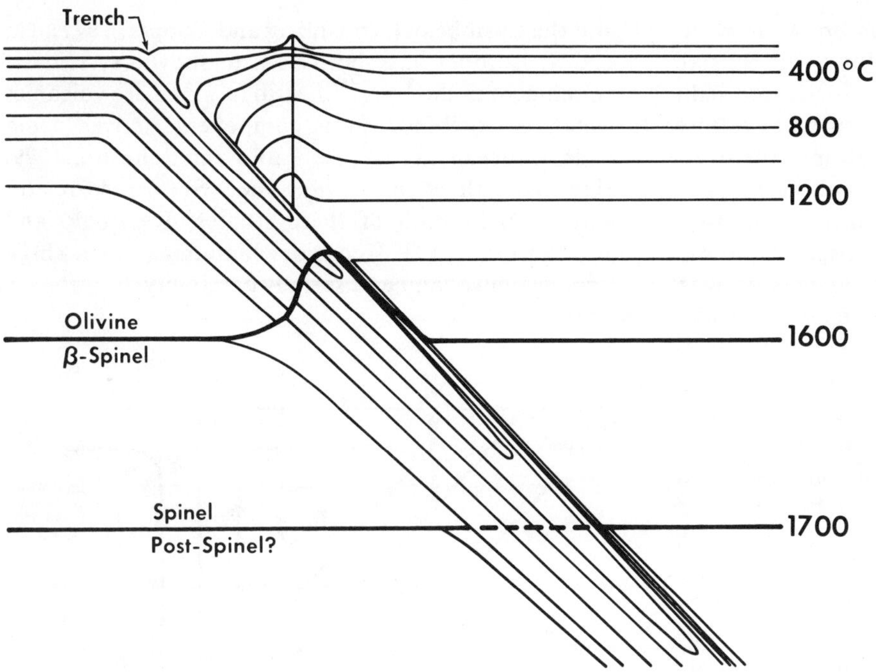

9.5.4 Thermal structure of the descending lithosphere including frictional heating on the slip zone and the structure of the olivine–β-spinel phase change (after Schubert *et al.*, 1975). The thin lines denote isotherms and the heavy lines denote phase boundaries. Lack of data on the 650-km discontinuity, including the question of whether it is due to a phase change, precludes a prediction of its structure.

the local seafloor, and subsequent foundering due to the gravitational instability.

The majority of subduction zones are situated around the margins of continents where large stress differences are expected in the transition zone between continental and oceanic lithosphere, and where there may be loading of the edge of the oceanic lithosphere with continent-derived sediment. Not all subduction zones are currently near continental margins, however, and some may have formed within entirely oceanic areas (Uyeda and Ben-Avraham, 1972).

One important aspect of oceanic trenches not predicted by thermal convection models is the associated volcanism. Well-defined lines of active volcanoes form either an island arc or create a volcanic chain on an adjacent continent. Seismic studies show that the depth to the descending plate beneath the volcanoes is 100 km or greater.

Three principal mechanisms for this volcanism have been proposed. In the first, the heat required to produce the partial melting is attributed to frictional heating on the fault zone between the descending plate and the overlying crust and mantle (McKenzie and Sclater, 1968; Oxburgh and Turcotte, 1968). For the thermal model in Fig. 9.5.4, in order to heat the descending crust to sufficiently high temperatures to produce partial melting, a shear stress of 1.35 kb is required on the slip zone. Stress models calibrated to trench depth obtained by Sleep (1975b) give stress levels of 700 bars or less. His latest models (Sleep, 1979a; Sleep and Rosendahl, 1979) give 200–250 bars stress maximum at both ridges and trenches, much less than needed for melting. It is conceivable, however, that a different rheological scheme could cause local concentrations of high stress.

In the second principal mechanism for the volcanism associated with ocean trenches, it is hypothesized that cold material from the lower lithosphere of the over-riding plate is entrained with the slab and hot material rises from the asthenosphere to take its place. The uppermost region of the recirculating flow is the source for island arc volcanoes. This mechanism is difficult to model because fluid flow is explicitly involved and because the rheology of the lowest part of the lithosphere is poorly constrained. The published computations (Andrews and Sleep, 1974) used artificial boundary conditions for the slab movement and explicitly included only material above the slab in their flow model. However, once a zone of hot material is produced above the slab, induced flow at plate velocities will tend to keep the region hot, perhaps expanding it. It should be noted that the heat source in this hypothesis is the general heat of the interior of the Earth and is not directly related to the slab. Magmas in island arcs would differ from other magmas only by differences in the depth and efficacy of segregation and by fractionation which might be more effective at island arcs where large amounts of horizontal extension are precluded.

A third class of mechanisms is a chemical and petrological one rather than a mechanical one. This class involves materials that melt or break down at modest temperatures and are carried down with the slab (e.g., Wyllie, 1971; Ringwood, 1977). Subducted sediments or hydrous minerals in oceanic crust may release water and other volatiles from the slab into the overlying mantle and may thus lower the melting point of the mantle beneath the arc. These chemical mechanisms may act with the physical mechanisms discussed above.

In general, the line of volcanoes associated with ocean trenches is very closely associated with the descending slab. Offsets in the volcanic line can be correlated with breaks in the descending plate, supporting this association. However, it has been pointed out by Isacks and Barazangi (1977) that volcanoes occur behind the trenches in northern and southern Chile, but that no active volcanoes are found in Peru (4° S to 15.5° S) or central Chile (27° S to 33° S), even though subduction is taking place off the entire western coast of South America. Since the stresses within the lithosphere are not expected to vary greatly along the plate boundary, and because the subduction velocities are similar, the volcanic line should have been continuous if fractional heating were the mechanism to generate island arc magmatism.

Location of earthquake hypocenters in this area indicates that the dip of the subducting lithosphere beneath northern and southern Chile is greater than that beneath Peru and central Chile (Isacks and Barazangi, 1977). Beneath northern and southern Chile, there is a wedge area above the slab where earthquakes are absent. This aseismic wedge is interpreted as indicating the existence of an asthenosphere. Beneath Peru and central Chile, such an asthenosphere may not be present. Therefore, the existence of an asthenosphere may be a necessary condition for island arc magma generation. It is not clear, however, whether this is a sufficient condition.

As previously discussed, the descending lithosphere has been differentiated into crustal and depleted mantle rocks by ridge processes. The crustal rocks have probably been extensively altered by interactions with seawater and some volume of sediments is probably entrained in the descending plate during subduction processes. It is conceivable that some fraction of the entrained sediments and oceanic crust melts to provide the volcanism observed behind trenches (section 7.4.2). Recognition of this component is difficult because high level contamination with hydrous oceanic crust beneath the arc and with accreted sediments is possible. Lead isotopic data indicate that the contribution of sediment is not large at many island arcs (Oversby and Ewart, 1972). Pb isotopic data do indicate a sediment contribution in the Aleutians (Kay *et al.*, 1978) and New Zealand and the Caribbean (Armstrong and Cooper, 1971). Nd and Sr isotopes have been used to show a systematic enrichment in the $^{87}Sr/^{86}Sr$ ratio in island arc volcanics which is attributed to a component derived from hydrous oceanic crust (De Paolo and Johnson, 1979; Hawkesworth *et al.*, 1979). The $^{87}Sr/^{86}Sr$ correlates with SiO_2 in both of these studies. De Paolo and Johnson (1979, p. 373) consider contamination in a high level magma chamber a serious possibility.

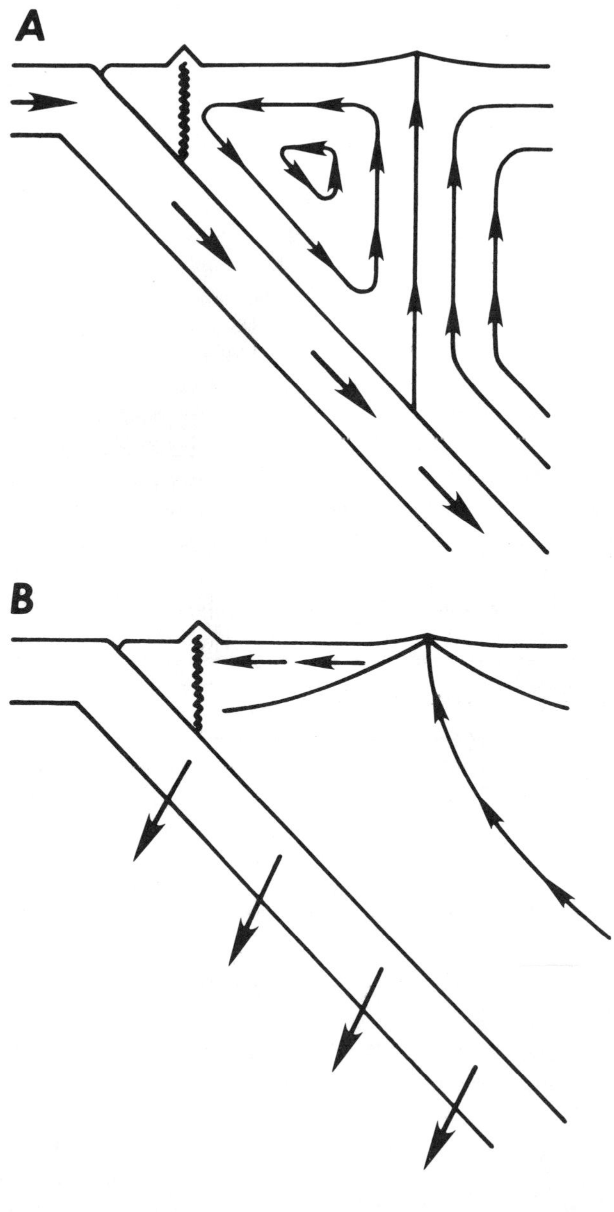

9.5.5 Models for the formation of secondary spreading centers behind island arcs. **a.** A secondary convection pattern generated by the descending plate. **b.** Ascending convection generated by the foundering of the descending plate.

Mass balances of elements such as Ar, Cl, H_2O and K may give constraints on the amount of subduction of hydrated crust and sediment and the amount of incorporation of this material in island arc volcanics (A. T. Anderson, 1975; Sleep. 1979b). The $^3He/^4He$ ratios in island arc volcanics indicate a substantial contribution from ordinary mantle (Craig *et al.*, 1978). Application of these lines of reasoning to the origin of the bulk of the volcanics is model dependent, because minor, highly mobile parts of the lavas are involved.

Behind some ocean trench systems new seafloor is generated in a manner similar to ocean ridges. For a number of years it had been known that parallel to the main island-arc there could sometimes be one, or even two, other arcs inside the first. Precision depth surveys in these regions, and the use of continous seismic profiling techniques which established the thickness and distribution of sediment from place to place, allowed Karig (1970, 1971) to demonstrate that the parallel arcs were in fact parts of the same arc which had been rifted apart. Using the hypothesis that the age of a basin on the seafloor could be determined from the age and thickness of the sediment it contained, it was possible to show that the Philippine Sea and Sea of Japan had been opened by a series of irregular rifting phases. A magnetic signature similar to that on the deep oceans was present, but it was much more irregular and could not be readily interpreted except in a few cases. A notable exception was the Scotia Sea where several hundred kilometers behind the Scotia arc a miniature ridge has become established which displays a regular pattern of magnetic stripes similar to that at the main ocean ridges (Barker, 1972).

The active parts of such marginal seas are characterized by high heat flow (80–120 mW/m^2) and they overlie the inclined zone of seismicity marking the path of the descending slab of lithosphere. The crust of back-arc basins is in many ways similar to that of the oceans (Karig, 1971), but the fill of sediment is generally much thicker and makes the volcanic part of the crust more difficult to examine. There is some indication that the basalts are similar to those of the ocean basins (e.g., Hawkins, 1976), but that the basaltic crustal layer may have lower seismic velocities or may be somewhat thinner than under the oceans (Karig, 1971).

It is clear that there is some kind of localized ascending flow above the descending slab of lithosphere which transports both heat and mass to allow the crustal extension and high surface heat flow. The basalts of the crust presumably form by the same process of pressure-release melting as that which operates at ocean ridges.

A number of explanations have been offered for the ascending flow behind arcs (Fig. 9.5.5). Oxburgh and Turcotte (1970) proposed an irregular convective flow of mantle material heated frictionally by the descent of the oceanic lithosphere. Andrews and Sleep (1974) and Toksöz and Hsui (1978a) proposed that the descending plate generated a secondary convection cell (Fig. 9.5.5a). Oxburgh and Parmentier (1978) proposed that the ascending flow may be triggered by the buoyant rise of depleted mantle separating from the warmed upper part of the descending plate. None of these mechanisms excludes any of the others and there is not at present adequate evidence to choose among them.

Mid-plate volcanism. Basaltic volcanism occurring in the Hawaiian Islands and other centers of mid-plate volcanism has many similarities to ocean ridge volcanism (sections 1.2.6, 5.3 and 6.2.2). These centers of mid-plate volcanism appear to move relatively slowly with respect to one another, compared to the relative velocity of plates. These observations led to the concept of mantle plumes (Morgan 1971, 1972a, b). According to the plume hypothesis (section 6.2.2), mid-plate volcanism is attributed to pressure-release melting within narrow columns of ascending mantle rock. The "kink" in the mantle geotherms obtained by Boyd (1973) and others from geothermometry studies of mantle xenoliths (see below) was attributed by Parmentier and Turcotte (1974) to a mantle plume impinging on the base of the lithosphere. The large amount of subsidence indicated for the Hawaiian swell and for several Pacific atolls has been ascribed by Detrick and Crough (1978) to cooling of a lithosphere thinned and heated by passage over such a plume.

Tozer (1973) attacked the concept of narrow plumes as not being applicable to the high viscosity mantle. While thermal convection at high Prandtl numbers may result in narrowly confined thermal boundary layers or thermal plumes because of the slow diffusion of heat, high viscosity generally leads to a rapid diffusion of vorticity and to a diffuse velocity structure.

With a strongly temperature-dependent viscosity, however, the high temperature region might also be a high velocity region because of the pronounced local reduction in viscosity. Analytical solutions for two-dimensional plumes in a variable viscosity fluid have been given by Yuen and Schubert (1976). They find a concentration of the flow if the temperature excess within the plume is sufficiently large. Numerical solutions for the structure of axisymmetric mantle plumes have been obtained by Parmentier *et al.* (1975). These calculations included an exponential temperature and depth dependent viscosity. Parmentier and co-workers found that relatively narrow mantle plumes are possible flow solutions if convection is confined to the upper mantle and if the heating is primarily from below.

An alternative to thermal plumes is chemical (or compositional) plumes. D. L. Anderson (1975) has proposed that mid-plate volcanism is the result of compositional plumes rising from the core-mantle boundary. Certainly the continuous supply of a lighter rock in the lower mantle could result in relatively narrow ascending plumes. The mechanism that leads to a steady or extended supply of such light material over long periods of time, however, has not been clearly defined.

Continental lithosphere. Because of its low density, continental crust is difficult to subduct. One consequence is that the average age of continental crust ($\sim 10^9$ years) is an order of magnitude greater than the average age of oceanic crust ($\sim 10^8$ years). Insofar as continents move with the surrounding oceanic lithosphere, there must be some finite thickness of mantle material which is mechanically coupled to them and moves with them—the continental lithosphere. Such lithosphere may not, however, be the thermal boundary layer of a mantle convection cell and its interaction with mantle convection must be considered separately.

The thickness of the thermal lithosphere beneath continents is a matter of current debate. One source of information on the thickness of the continental lithosphere is the surface heat flow. Surface rocks in continental areas have high concentrations of radioactive elements, which contribute significantly to measured heat flow (Fig. 9.4.2). The mantle heat flow, determined from the observed heat flow by correcting for crustal radioactivity, is about 25 mW/m^2 for many continental areas (Rao and Jessop, 1975; Sclater *et al.*, 1980). If other heat production in the continental crust and lithosphere is neglected, and if the thermal gradient corresponding to this reduced heat flow is extrapolated through the mantle, the thermal lithosphere beneath continents would have a thickness of about 200 km. More detailed determinations of the thickness of the continental lithosphere from surface heat flow data have been given by Sclater and Francheteau (1970) and by Froidevaux *et al.* (1977). Pollack and Chapman (1977a), using a somewhat different set of assumptions on the distribution of heat producing elements, concluded that the continental thermal lithosphere is somewhat thicker, reaching a maximum thickness of 300 km.

The thermal lithosphere, as noted in section 9.4.3, can in general be quite distinct from the rheological (elastic or high viscosity) lithosphere. The thickness of the latter can be determined reliably from the analysis of glacial rebound data and is about 100 km for the North American shield (Cathles, 1975; Peltier and Andrews, 1976).

Seismic travel time delays have also been used to estimate the (seismic) thickness of the continental lithosphere. Sipkin and Jordan (1975, 1976) observed a five-second delay for the vertical propagation of shear waves beneath the oceans relative to the continents, and they concluded that significant shear velocity differences between continents and oceans must extend to depths of at least 400 km. Jordan (1975, 1978) has proposed that the observed difference in travel time delays is associated with a continental "tectosphere", or the layer of crust and mantle that moves with the continental plate, which has a thickness of at least 400 km. These conclusions have been questioned, however, by Okal and Anderson (1975) and by Hales (1976), who have argued that the observed variations in travel times can be equally well explained by variations in the structure of the low velocity zone at depths shallower than 200 km.

The thickness of the continental lithosphere can also be estimated from the observed mean geoid anomaly over the continents (Turcotte and McAdoo, 1979). Using the relationship between geoid anomaly and dipole density distribution (Eq. 9.5.7), these authors found that the thickness of the continental lithosphere is roughly 180 km, a lower value than is required by Jordan's (1975) tectosphere hypothesis.

There appears also to be a gross dependence of heat flow on age for the continents, with some of the oldest parts of the crust giving values as low as 20 mW/m^2 and young, recently-active crust 100 mW/m^2 or more. This dependence is discussed by Sclater and Francheteau (1970) and has led to the suggestion that the continental lithospheric structure may also be the result of cooling from the thermal events of the plate margin which led to the addition of each increment of crust. However, for 1500 m.y.-old continental lithosphere which cooled in the same way as oceanic lithosphere, the conductive heat flux would be less than 6 mW/m^2. This value is lower than any observed, even allowing for crustal radioactivity. A more serious objection is that the same 1500 m.y.-old lithosphere should have undergone several kilometers of subsidence during the last 500 m.y. This is demonstrably not the general behavior of continental crust of this age.

If the continental thermal lithosphere is relatively thin, say 150–200 km thick, it could develop initially by conductive cooling, but it would then be necessary for further thickening and cooling to be arrested in some way, possibly by viscous heating beneath the plate while it is in motion (Froidevaux and Schubert, 1975). If, on the other hand, the estimates of Jordan (1975) are substantiated and the continental lithosphere (or tectosphere) is 400 or more kilometers thick, it would be hard to avoid the conclusion that the sub-continental mantle is at least in part different in composition from the material participating in the main upper mantle circulation. In this latter case, it would probably be necessary to invoke some process of underplating of the continents with depleted mantle (e.g., Oxburgh and Parmentier, 1978).

The subject of the deep thermal, compositional and seismic structure of the continents is clearly an important focus of ongoing research. On the basis of currently available information, we are not persuaded that systematic temperature differences between continental shields and old ocean basins persist below about 200 km depth. Lateral temperature differences below that depth are strongly indicated from seismic travel time anomalies, but such differences are more likely related to the convective flow pattern in the mantle rather than to the deep structure of continental lithosphere.

The presence of old continental lithosphere, whether 200 or 400 km thick, may nonetheless affect the pattern of underlying mantle convection cells. The boundary conditions associated with continental lithosphere differ from those associated with oceanic lithosphere. It has also been proposed that continental crust, with its relatively high rate of internal heating, can induce plate motions (Howard *et al.*, 1970; Knopoff *et al.*, 1972; Whitehead, 1972; Smith and Knopoff, 1976).

Present thermal state

The present thermal state of the Earth is an important boundary condition for models of global thermal evolution. Clearly the thermal structure in at least the upper mantle is dominated by plate tectonic phenomena, as described in the last section. In this section we deal with two aspects of the global thermal state of the present Earth: surface heat flow and geotherms.

Heat flow. The heat flow **q** in a solid body is given by Fourier's law

$$\mathbf{q} = -K\nabla T \tag{9.5.12}$$

where K is the thermal conductivity (in most general terms a tensor, but more commonly taken as a scalar) and T is the temperature field. Components of **q** are positive if heat is flowing out in the positive coordinate direction. At a planetary surface $T \simeq T(r)$, so heat flows primarily radially (vertically) and

$$\mathbf{q} = -K\, dT/dr \tag{9.5.13}$$

Measurement of heat flow requires independent measurements of the thermal gradient dT/dr and of the conductivity of the material in which the gradient was measured. Typically dT/dr is determined by measuring temperature at several depths in a drill hole, and K is determined by monitoring the transient change in temperature in response to a known heat source. A major problem in determining the thermal gradient representative of the flow of heat from the interior is to avoid the effects of time variations in surface temperature, effects which propagate to a depth z dependent on the time-scale of the variations. If the surface temperature of the planet is given by $T_0 + T_1\cos(\omega t)$, then superposed on the mean thermal gradient will be an additional temperature

$$T_1\exp\left[-(\omega/2\kappa)^{1/2} z\right]\cos\left[\omega t - (\omega/2\kappa)^{1/2} z\right]$$

(Carslaw and Jaeger, 1959). For $\kappa = 10^{-2}\,\mathrm{cm^2/sec}$, the skin depth δ (the depth at which the oscillatory term has dropped to $1/e$ of its surface value) is

$$\delta = \left(\frac{2\kappa}{\omega}\right)^{1/2} \tag{9.5.14}$$

For oscillations of 1 day (diurnal), 1 year (annual), and 10^4 yr (climatic variations), δ is 17 cm, 3.2 m, and 320 m, respectively. Thus on the continents, temperature gradient measurements should be made in holes at least ~10 m deep to avoid the effects of seasonal changes in surface temperature, and at least ~1 km deep to avoid the effects of the last ice age. These requirements are relaxed in most deep ocean environments; the ocean bottom is nearly isothermal because water has a minimum in specific volume at about 4°C. Thus fairly shallow holes (1–2 m), typically in sediments, can yield representative values of the long term conductive thermal gradient. Other potential sources of concern in the interpretation of measured thermal gradients are: (1) the effects of surface topography, so that $T \neq T(r)$; (2) the effects of heterogeneities in K, which can focus or defocus heat flux; (3) the heat transported by fluid convection, not included in Eq. (9.5.13); and (4) the effects on shallow T of recent sedimentation or erosion.

Roughly 6000 reliable measurements of surface heat flow have been made on the Earth, 75% of them in ocean floor (Sclater *et al.*, 1980). From the global point of view, the mean Earth heat flow is of greatest interest, but there are several possible routes to derive this quantity from the data. One might simply find the average of all the data, but this approach suffers from the irregular areal coverage and from the concentration of measurements in tectonically interesting areas. A more traditional approach has been to expand the data in spherical harmonics, equating the coefficient of the zero-degree term to mean Earth heat flow (e.g., Pollack and Chapman, 1977a, b). This approach, unfortunately, fails to account for heat lost in young oceanic lithosphere by hydrothermal circulation (see below).

The most careful account of terrestrial heat flow to date has been that of Sclater *et al.* (1980), who included explicitly the age dependence of heat flow for both oceans and continents as well as the heat lost by hydrothermal convection in oceanic crust and not represented in most oceanic heat flow measurements. Their results are summarized in Table 9.5.1.

The heat flux at the base of the lithosphere is of interest both because the bulk of this heat is delivered by mantle convection and because of the long-standing question of whether the heat flow into old continental and old oceanic lithosphere are substantially different. After correcting for crustal radioactivity, particularly for continents, and for lithosphere cooling, Sclater *et al.* (1980) estimate that the mantle heat flow beneath old continents lies between 19 and 31 mW/m^2, and the mantle heat flow beneath old oceanic lithosphere lies between 25 and 38 mW/m^2. Thus the heat delivered to the base of the lithosphere in old oceans and continents is not resolvably different.

The thickness of the thermal lithosphere in the geologic past may be extrapolated from the present surface heat flow if we idealized the Earth's heat flux as being partitioned into three essentially separate terms (Fig. 9.5.6): heat transfer by the processes of plate creation, cooling and subduction; heat flow delivered to the base of the lithosphere, either by upper mantle radioactivity (Forsyth, 1977), by frictional heating (Schubert *et al.*, 1976), or by small-scale mantle convection (Richter and Parsons, 1975; McKenzie and Weiss, 1975); and heat produced by lithospheric heat sources, a term dominated by radioactivity in continental crust. Several estimates of the present partitioning of heat flow have been made: plate creation and destruction currently accounts for 28–60%, mantle heat delivered to the base of the plates for 20–54%, and crustal radioactivity for 13–19% (Bickle, 1978; Turcotte and Burke, 1978 Sclater *et al.*, 1980). From sea level data and a relationship between sea level and the volume of ridges, it appears that the contribution of plate creation to total heat flow may have been larger by a factor of as much as two at earlier times in the Phanerozoic (Hays and Pitman, 1973; Turcotte and Burke, 1978). As noted in section 9.4.1 and below, the total heat production in the Earth through geologic time is probably bounded from above by the upper curve in Fig. 9.4.1, at least since ~4 b.y. ago; the actual bulk heat production in the past may lie substantially below that curve if heat flux lags heat production as indicated by convective thermal history models.

The relative fractions of heat loss in the geologic past delivered by conduction through a lithospheric layer and by recycling of surface material into the mantle is not certain. The heat loss due to plate creation is (Wolery and Sleep, 1976)

$$Q_p = \frac{2}{3} \rho C_P l \, T_l U \qquad (9.5.15)$$

where ρ and C_P are density and specific heat, l is the plate thickness, T_l is the temperature at the base of the plate and U is the rate of plate creation

$$U = vL \qquad (9.5.16)$$

where L is the total length of ridges and v is the mean half-spreading rate. Currently, U is 3 km^2/yr (Chase, 1972). The total mantle heat loss Q_m is the sum of Q_p

Table 9.5.1 Global heat loss on the Earth.[†]

	Area (10^6 km^2)	Heat Loss (10^{12}W)	Mean Heat Flow (mW/m^2)
Oceans, observed (including marginal basins)	309	20.3	66
Oceans, contribution from hydrothermal circulation	-	10.1	-
Total Oceans	309	30.4	99
Continents, observed (including continental shelves)	201	11.3	56
Continents, contribution from active volcanoes	-	0.2	-
Total Continents	201	11.5	57
Total Earth	510	42.0	82

[†]From Sclater *et al.* (1980).

and Q_c, the heat supplied to the base of the plates, say by small-scale flow,

$$Q_c \simeq \frac{KT_l}{l} A \qquad (9.5.17)$$

where A is the Earth's surface area. For a larger Q_m in the geologic past, the ratio of Q_p to Q_c need not have been the same as at present. If the ratio were, however, constant, then

$$l \sim 1/Q_m \qquad (9.5.18)$$

and

$$U = vL \sim Q_m^2 \qquad (9.5.19)$$

Equation (9.5.19) can be obtained directly from a half-space cooling model of ridges (Sleep, 1979b). It is the fundamental relationship for making calculations of the Earth's thermal history where global spreading rate is assumed. Perhaps ironically, paleomagnetic data and other information on rates of seafloor creation may thus provide better constraints on mantle convection velocities in the geological past than theoretical calculations based on fluid dynamics.

Equation (9.5.19) requires greater length of ridges, or greater spreading rates, or both. For instance, if both l and typical horizontal dimensions of plates scaled as $1/Q_m$, then $L \sim Q_m^2$ and v might be similar to the present. Burke and Kidd (1978) have argued, however, on the basis of the lack of observed products of melting of the lower crust in Archean terrain, that continental crustal geotherms in the Archean were no more than 30% steeper than present continental geotherms, so that Q_p/Q_c must have been much higher in the past than at present. The argument has been advanced (e.g., Baer, 1977) that if l is sufficiently small, oceanic lithosphere will no longer subduct and therefore Q_p will be essentially zero. Oceanic lithosphere as young as perhaps 10 m.y. old is currently subducted (Menard, 1978), however, and the lower limit on l determined by Burke and Kidd (1978) poses no problem for lithospheric recycling in the Archean. Even if young oceanic lithosphere is not gravitationally unstable *per se*, a finite-amplitude instability can result from the basalt-eclogite transition, so that once subduction of a basaltic component has been carried to a certain depth, the negative buoyancy generated by the phase change can continue to drive further subduction. Thus continental geotherms in the Archean could well have been only slightly steeper than they are today without violating our understanding of the total heat flux from the Earth at that time.

Geotherms. In our previous discussion of the thermal structure and evolution of the Earth we have emphasized that the thermal structure is not radially symmetric. Convective processes lead to strong horizontal temperature gradients. A primary example is the low temperature of the subducted lithosphere at oceanic trenches (Fig. 9.5.4). Nevertheless, mean radial profiles of temperature or geotherms are of considerable interest as a boundary condition for evolution models. We consider first the thermal structure of the upper mantle and then the thermal structure of the entire mantle.

The thermal structure of the upper mantle is dominated by the large temperature gradients in the lithosphere. Since the thermal structure of the continental lithosphere is quite different from the thermal structure

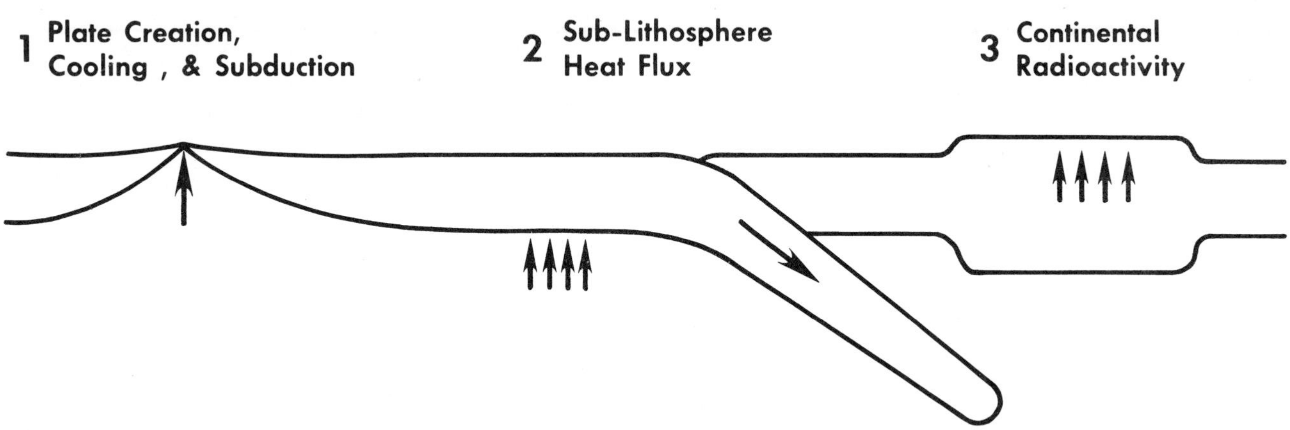

9.5.6 Schematic view of the present-day surface heat flux on the Earth, partitioned into three terms: (1) heat transferred by plate creation, cooling and subduction; (2) heat supplied to the base of the lithosphere; and (3) heat produced by radioactivity in continental crust.

of the oceanic lithosphere, these cases are considered separately. The thermal structure of the oceanic lithosphere was discussed above. It was shown (Eq. 9.5.3) that the thermal gradient decreases inversely as the square root of lithosphere age. For older lithosphere, however, the structure of the thermal lithosphere appears to approach a steady state. In order to model a mean thermal structure for the ocean basin we use the thermal structure given by Eq. (9.5.2) for an age $t = 60$m.y. This oceanic geotherm is given in Fig. 9.5.7.

Beneath the thermal boundary layer we assume that the thermal gradient is equal to the adiabatic gradient. This assumption will be discussed in some detail when we consider the whole mantle geotherm. The adiabatic temperature gradient is given by Eq. (9.4.29). For the upper mantle beneath the lithosphere, take $\alpha = 3.2 \times 10^{-5}$°$K^{-1}$, $g = 10^3$ cm/sec^2, $T = 1570$°K, and $C_p = 1.0$ J/g-°K. The resulting adiabatic temperature gradient is $(dT/dr) = -0.48$°K/km. This adiabatic gradient has been used in Fig. 9.5.7 to determine the oceanic upper mantle geotherm beneath the lithosphere.

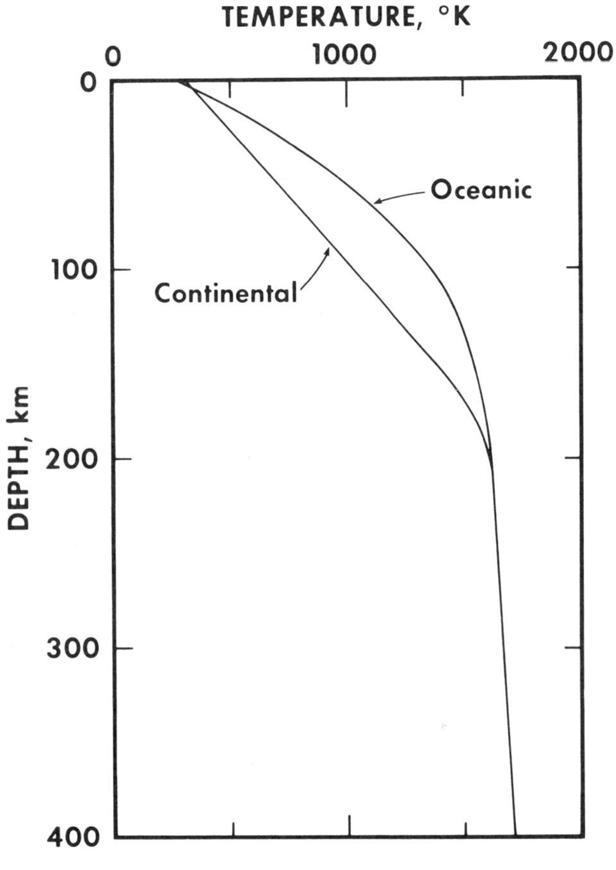

9.5.7 Upper mantle oceanic and continental geotherms (see text).

The geotherm beneath the continents is not as well understood as the geotherm beneath the oceans, because of the contribution from crustal radioactivity and the uncertainty in continental lithosphere thickness. Since there is strong evidence that the continental lithosphere is heated from below (see last section) we assume a linear temperature profile through the lithosphere. This approximation neglects the heat production in the lithosphere except in a near surface layer with a thickness of about 10 km (section 9.3.1). We further assume that the thermal lithosphere beneath continents has a thickness of 180 km (see above). The resulting continental geotherm is given in Fig. 9.5.7.

Pyroxene geothermometry (Boyd, 1973) on mantle-derived ultramafic xenoliths offers considerable promise for providing estimates of the local geotherm at the time the xenoliths were brought from deep in the lithosphere or asthenosphere to near the surface. In this technique, the temperature at which the xenolith minerals last equilibrated is estimated from the degree of crystalline solution of enstatite (orthopyroxene) in diopside (clinopyroxene) and the pressure is estimated from the Al_2O_3 content of orthopyroxenes. The procedure is not without some uncertainties and corrections yet to be worked out (e.g., Boettcher, 1974; Wilshire and Jackson, 1975), but is useful at present to corroborate other estimates of temperature.

Pyroxene geotherms for several regions designated as oceanic have been constructed by MacGregor (1974), MacGregor and Basu (1974), Mercier and Carter (1975) and Nixon and Boyd (1979). The data most readily applied to a known oceanic environment are those from xenoliths in Hawaiian alkali basalts. The Hawaiian geotherms of both Mercier and Carter (1975) and MacGregor and Basu (1974), the latter after shifting to 15% lower pressures (MacGregor and Basu, 1974) as an approximate correction for Fe and Ca to MacGregor's (1974) geobarometer based on Al_2O_3 solubility in orthopyroxene, are shown in Fig. 9.5.8. These geotherms indicate temperatures in excess of 1100°C below about 90 km depth, in adequate agreement with such other estimates of oceanic mantle temperature as oceanic heat flow and bathymetry and magma eruption temperatures (Solomon, 1976b). The geotherms intersect the solidus for pyrolite plus 0.2% H_2O (Green, 1973) at about 90 km depth, in agreement with the depth to the low velocity zone beneath 80 m.y.-old seafloor (Leeds *et al.*, 1974; Forsyth, 1977).

Several continental pyroxene geotherms are also shown in Fig. 9.5.8, including those from the Lesotho (Boyd and Nixon, 1973) and Frank Smith (Boyd, 1974) kimberlites in South Africa and the Ming Bar diatreme in Montana (Eggler and McCallum, 1974). The data of

Hearn and Boyd (1975) from Montana plot within or slightly above the field for Ming Bar. The South African kimberlite data of MacGregor and Basu (1974), after shifting to 10–15% lower pressures as a geobarometer correction, plot in the range of Boyd's kimberlite geotherms. The data from South African sheared lherzolites, which give a "kink" in the geotherms at the maximum depth shown in Fig. 9.5.8 (Boyd, 1973, 1974; Boyd and Nixon, 1973; MacGregor and Basu, 1974), are not plotted, since such a "kink" is not a steady state feature of the thermal profile.

Pyroxene geotherm data therefore support the principal conclusion from Fig. 9.5.7 that continental regions are cooler than ocean basins to depths of 150–200 km. The shield geotherms may not intersect the mantle solidus at any depth (Fig. 9.5.8), consistent with an absent or poorly developed seismic low velocity zone beneath stable continental interiors. Systematic continent-ocean differences below 200 km, however, are not required by the data.

Another check on a geotherm for the upper mantle is the temperature at the top of the seismic discontinuity regarded as the olivine–β-spinel phase change (Anderson, 1967). Taking the top of the transition at 360 km and a Clapeyron slope of 30 bar/°C, Solomon (1976b) obtained a temperature of 1300° ± 150°C at that depth. Graham (1970) obtained 1450° ± 120°C for an assumed 370 km transition depth. Within the uncertainties, dominated by the uncertainty in the depth of the proposed phase change, the geotherms in Fig. 9.5.7 fit the available data on the temperature at the olivine–β-spinel phase boundary adequately.

We now turn to the question of the geotherm throughout the mantle. It is well established that thermal convection plays a dominant role in the transport of heat in the upper mantle. The occurrence of deep focus earthquakes to depths of approximately 700 km is direct evidence that upper mantle convection is occurring to this depth.

An important, unresolved question is whether the convection associated with plate motions is confined to the upper mantle (above 700 km depth) or involves the entire mantle. If there is a barrier to thermal convection at or near 700 km depth, then separate convection systems could be present in the upper and lower mantle. Such a barrier could be posed by a phase change with

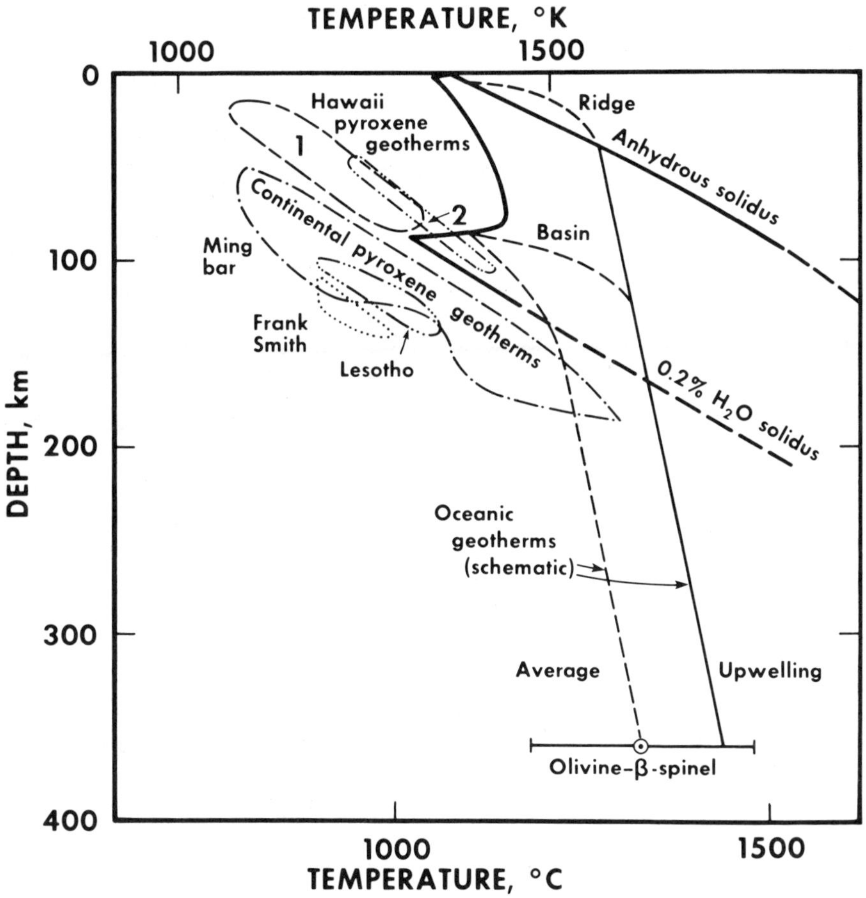

9.5.8 Pyroxene geotherms: Hawaiian geotherms 1 and 2 are respectively from MacGregor and Basu (1974) and Mercier and Carter (1975). Geotherms are shown for the Lesotho kimberlite (Boyd and Nixon, 1973), the Frank Smith kimberlite (Boyd, 1974), and the Ming Bar diatreme (Eggler and McCallum, 1974) and approximately bracket other continental geotherms. The pyrolite solidus with 0.2 percent H_2O and for anhydrous conditions is from Green (1973). Several schematic oceanic geotherms are also shown (Solomon, 1976b). The oceanic upwelling/ridge curve intersects the anhydrous solidus at the depths indicated by seismic velocity and Q data beneath ridges and follows an adiabat below. The mean oceanic curve follows an adiabat upward from the olivine–β-spinel transition at 360 km depth and intersects the solidus at 1100°C and 85 km depth. A curve for a convectively upwelling region beneath older oceanic lithosphere (perhaps Hawaii) is also sketched.

either appropriate thermodynamic properties or a rheological change, or by a chemical change involving a density increase with depth unrelated to temperature. The principal argument in favor of distinct upper mantle convection has been based on fault-plane solutions of earthquakes in subducted slabs: the mechanisms indicate that slabs which penetrate to only 200–300 km depth are under extension, but those that penetrate to 600–700 km depth are under compression (Isacks and Molnar, 1971); a substantial resistance to penetration of slabs to depths greater than 700 km is a simple, though not the only, explanation of this observation. The large energy release by earthquakes in the 500–700 km depth range and the abrupt termination of deep seismicity at 700 km depth also support the idea of a barrier to upper mantle convection near that depth, according to Richter (1979). A principal argument in favor of whole-mantle convection has been based on the finding that the viscosities of the upper and lower mantle are similar, at 10^{22} poise (Cathles, 1975; Peltier, 1976), a result recently questioned (Peltier, 1980). A further argument for whole-mantle convection is good agreement between the predicted and observed dips of subduction zones for the kinematic flow models of Hager and O'Connell (1978, 1979) if flow extends throughout the mantle. The question of a possible chemical change at 650 km depth (Anderson, 1977; Liu, 1979) is tied to the identification and thermodynamic properties, both currently uncertain, of any phase change at that depth. A more complete discussion of the arguments in favor of distinct upper mantle convection has recently been put forward by McKenzie and Weiss (1975) and by Richter (1979). the arguments in favor of whole-mantle convection have recently been summarized by O'Connel (1977) and by Davies (1977). We consider the question open as of the present.

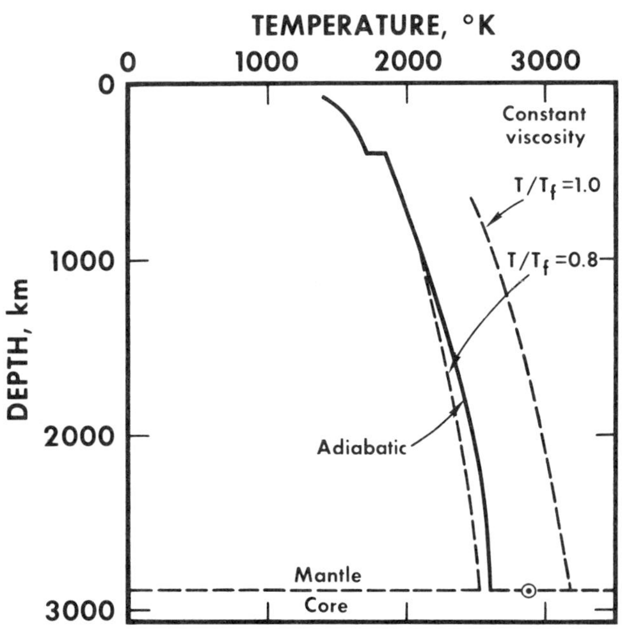

9.5.9 Mantle geotherms based on the assumptions of an adiabatic mantle and a constant viscosity mantle (T/T_f = 0.8, T_f from lowest valued melting curve in Fig. 9.4.7). The circle at the core mantle boundary is the melting point of the outer core assuming an iron-sulfur eutectic (Stacey, 1977).

In presenting a mantle geotherm we will assume for simplicity that whole mantle convection is occurring. If the 650-km seismic discontinuity acts as a barrier to convection, then an additional temperature increase would occur near this depth; there may also be additional barriers to penetrative convection above and below 650 km depth, giving rise to a number of distinct convective regimes separated by intervening boundary layers.

Rapid convection in a constant-viscosity fluid with a significant adiabatic gradient results in a vertical temperature profile that follows an adiabat (Jeffreys, 1930). We have used this result in Fig. 9.5.7 to prescribe the upper mantle geotherm. The corresponding adiabatic profile for the entire mantle has been determined by Jeanloz and Richter (1979). This profile is given in Fig. 9.5.9; it falls within the bounds given in Table 9.4.11. The increase in the adiabatic gradient due to phase changes has been included. The problem of adiabatic flow through mantle phase changes has been considered by Turcotte and Schubert (1971). For the olivine–β-spinel phase change at a mean depth of 400 km, they give a temperature increase of 160° K.

While an adiabatic temperature profile is expected for a rapidly convecting fluid with constant viscosity, the strong temperature dependence of the mantle viscosity may affect the whole mantle geotherm. From Eq. (9.4.27) or (9.4.28) it is seen that the viscosity of mantle minerals is exponentially dependent on temperature. As discussed in section 9.4.3, it is a reasonable approximation to assume that the mantle viscosity is a function only of T/T_f, where T_f is a melting curve appropriate to the major interconnecting mineral phase(s) in the mantle. If the mantle viscosity increases rapidly along a mantle adiabat, then convection may not occur at the higher viscosities. Tozer (1970) has argued that thermal convection in the mantle will result in a constant viscosity mantle rather than an adiabatic mantle. Qualitatively, he argues that the constant viscosity is required so that convection can efficiently transport heat out of the mantle.

For a constant viscosity mantle, it is appropriate to assume T/T_f equals a constant. The mantle melting curve T_f appropriate for determining homologous temperature has been discussed in section 9.4.4. The current value of T/T_f can be estimated from models of thermal evolution. The models presented below, calculated by the method described in section 9.3.3, have $T/T_f = 0.65$ to 0.80 at present (Fig. 9.5.13). Shown in Fig. 9.5.9 is $T/T_f = 0.8$ for one of the melting curves T_f in Fig. 9.4.8.

The important conclusion that emerges from Fig. 9.5.9 (and from Fig. 9.4.8) is that an adiabatic geotherm and a geotherm determined from $T/T_f =$ constant may not be significantly different, particularly considering the uncertainties in both the adiabat and the melting curve. For instance at 1000 km depth in Fig. 9.5.9, dT/dz is 0.42° K/km and 0.36° K/km for the adiabat shown and for $T/T_f = 0.8$, respectively; at 2000 km depth the two gradients are respectively 0.25 and 0.22° K/km. Thus the notions that a convecting mantle will give rise to an adiabatic thermal gradient and that a convecting system will tend toward uniform viscosity may be completely compatible. However, this later assertion remains to be rigorously demonstrated.

An important constraint on the temperature at the base of the mantle is the seismic evidence (lack of shear waves) that indicates the outer core is liquid. The melting temperature of the outer core depends strongly on its composition. Although the outer core is likely to be primarily iron, the seismic velocities indicate significant concentrations of one or more other constituents. Many authors propose sulfur as a primary secondary constituent (Brett, 1976). The melting temperature of an iron-sulfur core has been discussed by Stacey (1977). Assuming that the temperature at the outer core-mantle boundary is on the Fe-FeS eutectic, he gives a melting temperature of 2,878° K. This value, shown in Fig. 9.5.9, lies above the simple geotherms shown. This apparent conflict, though it may partly reflect ignorance of pressure effects on the Fe-FeS eutectic or of other constituents in the core (e.g., K) that may lower the melting point further, is more likely due to the neglect in Fig. 9.5.9 of a thermal boundary layer above the core-mantle boundary at the base of the lower mantle convection system (Elsasser *et al.*, 1979). The steep gradient across such a boundary layer will raise the geotherm above the melting temperature of the outer core.

Global history

The preceding discussion has emphasized the role of the plates in the terrestrial heat budget and its evolution. The importance of that role is made clear by calculations of the heat transport associated with the observed seafloor spreading and plate subduction rates. Such calculations are "kinematic" in the sense that they are not concerned with the ultimate energy source for the plate motions. It is also desirable to consider the Earth's thermal history from the point of view of a dynamical theory of convection; that is, one in which the driving force, which is derived from the Earth's internal energy including that released by the decay of radioactive elements, is taken into account. Such a theory should be applicable to the other terrestrial planets, modified only by appropriate changes in the physical parameters, so that the evolution of the Earth can be understood in the larger context of planetary evolution. The goals of the theory must necessarily be restricted to the prediction of quantities that are insensitive to the details of surface topography, plate boundaries, and other inhomogeneities that are less than planetary scale. Such quantities include the mean surface heat flow, the average mantle temperature distribution (or pressure-temperature related mantle properties), and the core temperature, all as functions of time.

Core formation. It is likely that the early thermal regime of the Earth was determined by the release of potential energy during core formation, triggered by accretional heating (Hanks and Anderson, 1969). Evidence cited below argues for early core formation, and the accretion models of Safronov (1978) and Kaula (1979a) indicate that initial core segregation could have been contemporaneous with accretion. The energy released by core formation was equivalent to a mean temperature rise of 2300° K if the equation of state of core material may be extrapolated from the properties of the present core and if accretion was homogeneous (Flasar and Birch, 1973). Inhomogeneous accretion would have provided less heating, as would the formation of an iron-sulfur eutectic core (Murthy, 1976) or the presence of endothermic reactions associated with abundant iron oxide or iron silicate. The distribution of energy depends on the details of iron segregation. Shaw (1978) has estimated that the temperature of the newly formed core would exceed 5,000° K.

The following constraints strongly support early core formation:

1. As the mantle is depleted in ^{204}Pb but not ^{207}Pb, core formation must have occurred rapidly compared with the 0.7 x 10^9 yr half life of ^{235}U (Gancarz and Wasserburg, 1977). The single stage age for core formation is 4.45 b.y. (Oversby and Ringwood, 1971). A continuous sequence of core formation over several hundred m.y. is also compatible with the lead data (Vollmer, 1977). (Chemical equilibration between iron

and silicate on the scale of grains is required to deplete the mantle of lead.)

2. Rocks thus far sampled show that the Earth's magnetic field has existed with a paleointensity of the same order as the present intensity at least back to 3.4 b.y. (McElhinny and Evans, 1968; Cisowski, 1979). If a large molten core is required to produce the Earth's magnetic field, core formation must have preceded this time.

The following weaker constraints also support early core formation, but are model-dependent:

3. Because core segregation involves a large pulse of heat (section 9.4.1), the oldest stable rock units on the Earth probably also postdate core infall (Hanks and Anderson, 1969), placing core formation prior to 3.7–3.8 b.y. ago (section 7.4.1).

4. It is inferred through data from mid-oceanic ridges that the mantle is highly depleted in $^{36}Ar/^{40}Ar$ relative to the atmosphere (Dymond and Hogan, 1978), although some rare mantle xenoliths have higher $^{36}Ar/^{40}Ar$ ratios (Saito *et al.*, 1978). The abundances of Ne, ^{36}Ar and Kr in the atmosphere are comparable to the terrestrial abundance calculated from ordinary chondrites (Fanale, 1971). Therefore, if the formation of the Earth's core caused gas depletion in the mantle, core formation must have occurred rapidly compared to the half life of ^{40}K.

5. If Rb enters the core, then a single-stage age of 4.45 b.y. can be obtained from Sr isotope systematics (Hurst, 1978).

The above considerations suggest that the Earth was hot shortly after its birth. It is safe to assume, however, that its temperature never greatly exceeded the mantle melting temperature. Any heat in excess of that needed to maintain temperatures near melting would have been rapidly removed by vigorous convection. An important result of the calculations presented in the next section is that the evolution of an initially hot Earth is insensitive to the precise initial temperature distribution. Although the chemical structure of the mantle may have been strongly affected by the details of core formation, which remain obscure because no terrestrial rocks older than ~3.8 b.y. have been found, other factors were more important in controlling the subsequent thermal evolution.

Parameterized convection models. The key to a dynamical theory of the thermal evolution of the Earth is provided by studies of convective heat transfer of the type discussed in section 9.3.3. There it was pointed out that a feature of convection systems is the surface boundary layer, which can be naturally associated with the lithospheric plates of the Earth, albeit with some important complications. It was also noted that heat transported by convection can be expressed in an approximate, but particularly simple form, in terms of the Nusselt and Rayleigh numbers. We will use this widely applicable rule to derive a simple model for the thermal evolution of the Earth that satisfies certain key observations about its present thermal state and has important implications for thermal history. The method used is that of Schubert *et al.* (1979b); it is similar to those used by Cassen *et al.* (1979) and Turcotte *et al.* (1979), and, to a lesser degree, those of Sharpe and Peltier (1978, 1979) and McKenzie and Weiss (1975).

As suggested by the discussion of the previous section, it is assumed that the Earth was initially hot, being near the melting temperature of the mantle everywhere, due to the energy of accretion and core formation. (See also McKenzie and Weiss, 1975; Sharpe and Peltier, 1979; and Kaula, 1979a, for arguments supporting this initial condition.) Let ΔT be the temperature difference between the bottom and top of the mantle (including the plates) and let ΔT_{ad} be the temperature difference if the mantle had an adiabatic thermal gradient. If compositional inhomogeneity in the mantle is ignored, then the extra-adiabatic temperature differential $(\Delta T - \Delta T_{ad})$ across the mantle determines the total buoyancy force available for driving convection, and therefore, along with the mechanical and thermal properties of the mantle, prescribes the Rayleigh number, here defined as $Ra = \alpha g (\Delta T - \Delta T_{ad}) (R - R_c)^3 / \kappa\nu$, where R and R_c are the radius of the Earth and of the Earth's core. The heat flux into and out of the mantle occurs through boundary layers, and is given by

$$q_{U,L} = \frac{K \Delta T_{U,L}}{R - R_c} Nu \qquad (9.5.20)$$

where $\Delta T_{U,L}$ is the temperature difference across the boundary layer, and subscripts U and L refer to the "upper" (surface) and "lower" (core-mantle) boundaries respectively. The Nusselt number $Nu = (Ra/Ra_{cr})^{1/3}$, where Ra_{cr} is the critical Rayleigh number, as determined both experimentally and theoretically (section 9.3.3). Implicit in Eq. (9.5.20) is the assumption that the upper and lower boundary layers have the same thickness. The heat balance equations for the mantle and core are then

$$\rho C \frac{d}{dt} \int_{R_c}^{R} 4\pi r^2 T(r) dr = 4\pi R_c^2 q_L - 4\pi R^2 q_U + \frac{4}{3}\pi (R^3 - R_c^3) H \qquad (9.5.21)$$

$$\frac{4}{3} \pi R_c^3 \rho_c C_c \frac{dT_c}{dt} = - 4\pi R_c^2 q_L$$

where subscript "c" refers to core quantities, and we have assumed for simplicity that the core is isothermal at temperature T_c. Relaxation of this assumption is important for exploring the relation between the thermal state of the core and dynamo processes (Stevenson *et al.*, 1980). Sharpe and Peltier (1978, 1979) solved a similar set of equations for the case of no mantle heat sources but with heat sources in the core. (They also assumed that the convective heat flux is everywhere proportional to the local conductive flux, so their mantle energy equation has the form of a diffusion equation.)

The viscosity of the mantle is strongly dependent on temperature and pressure and, in general, on composition and deviatoric stress; this dependence is a crucial factor in determining the evolution of the Earth. For our purposes it is sufficient to assume a linear relationship between shear stress and strain rate and to model the dependence on state variables implicitly, by using the following expressions for the kinematic viscosity, after equation (9.4.28)

$$\nu = \bar{\nu} \exp (aT_f/T) \tag{9.5.22}$$

where T_f (r) is the melting temperature of major interconnected mineral phases and $\bar{\nu}$ and a are constants. We adopt here the hypothesis that there is a strong tendency for the mantle to maintain a uniform viscosity, since large regions with low viscosity would tend to convect rapidly, which promotes cooling and hence an increase in the viscosity. Conversely, sticky regions would convect slowly, heat up, and become less sticky. Therefore, because the viscosity is such a sensitive function of temperature, there is a tendency for the temperature to be equal to a constant fraction of the melting temperature throughout the mantle: $T(r,t) = f(t)\, T_f(r)$. Equations (9.5.21) can therefore be regarded as equations for f and T_c. They are easily integrated by numerical techniques for a specified T_f (r) (which is adequately approximated by a linear function of r; see section 9.4.4), subject to a constant surface temperature T_s.

The results of two model calculations are shown in Figs. 9.5.10–13. These models were constrained by the assumptions that (1) the present average mantle heat flux, excluding crustal radioactivity, is about 57 mW/m^2 (Pollack and Chapman, 1977b; Oxburgh and Turcotte, 1978), and (2) the dynamic viscosity of the present mantle is 10^{21}–10^{22} poise (Cathles, 1975; Peltier, 1976). (Recent work by Sclater *et al.*, 1980, and Peltier, 1980, suggests that both of these values may need to be revised upwards.) The elements U, Th, and K were taken in relative abundances typical of chondritic meteorites; their ratios and absolute abundances, along with other relevant parameters are given in Table 9.5.2. Other choices of relative abundances give similar results for similar present day total heat generation rates. The main differences in the two models shown are in the values of the rheological parameters a and $\bar{\nu}$.

All solutions of Eq. (9.5.21) that meet the above requirements exhibit the following characteristics:

(1) As long as the initial condition is a "hot" Earth, the present day properties are independent of the precise initial temperatures. (2) The surface heat flow lags

Table 9.5.2 Parameters used in Earth thermal history models.

	Model 1	Model 2
Radioactive Abundances		
U, ppb	14.0	13.3
K/U	7×10^4	7×10^4
Th/U	4	4
Kinematic Viscosity (mantle)		
(10^8 cm^2/sec)	$2.6 \exp (20\, T_f/T)$	$0.73 \exp (25\, T_f/T)$

Other Constants Common to Both Models

$\rho = 4.42$ g/cm^3	$T_f(R_c) = 4000$ °K
$C = 1.2$ J/g-°K	$T_f(R) = 1600$ °K
$K = 0.042$ J/cm-sec-°K	$Ra_{cr} = 1100$
$\alpha = 3 \times 10^{-5}$ °K^{-1}	$\Delta T_{ad} = 500$° K
$g = 980$ cm/sec^2	$\rho_c = 11.04$ g/cm^3
$T_s = 300$° K	$C_c = 0.7$ J/g-°K

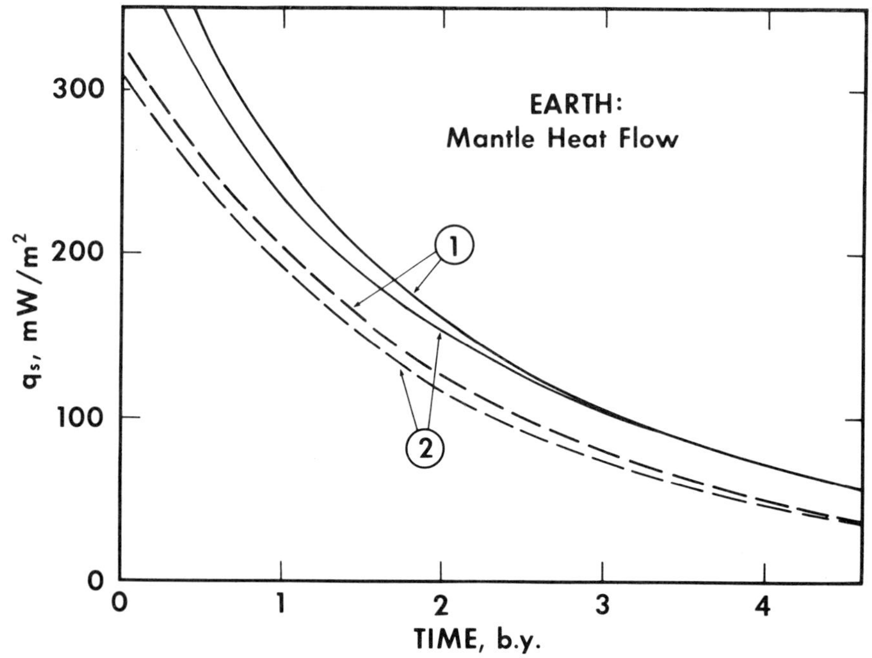

9.5.10 Heat flux q_s from the mantle versus time for two models of Earth thermal history (solid lines); and mantle heat flux predicted by a steady-state balance between radioactive heat production and Earth heat loss (dashed lines). In this figure and in all figures to follow, zero on the time axis corresponds to the formation of the planets.

that which would be produced by the internal radioactive sources alone. Thus, in these models, the cooling of the Earth provides a significant contribution to the surface flux even today. Estimates of radioactive abundances based on the steady state assumption would be too high by a factor of 20–40% (Fig. 9.5.10). A similar result has been obtained by Daly and Richter (1978), Sharpe and Peltier (1978, 1979), and Schubert *et al.* (1979a, b). Turcotte *et al.* (1979) have carried out similar calculations, but find that mantle cooling provides only about 7–12% of the surface heat flow. Part of this difference can be attributed to the fact that Turcotte and co-workers did not include core cooling (which contributes about 10% of the total heat flow in the present models), and part is apparently due to a different choice of $\bar{\nu}$ and a. Taking all results into consideration, we estimate that the mantle is currently cooling at a rate of between 20 and 80°K/b.y. (3) The calculated cooling

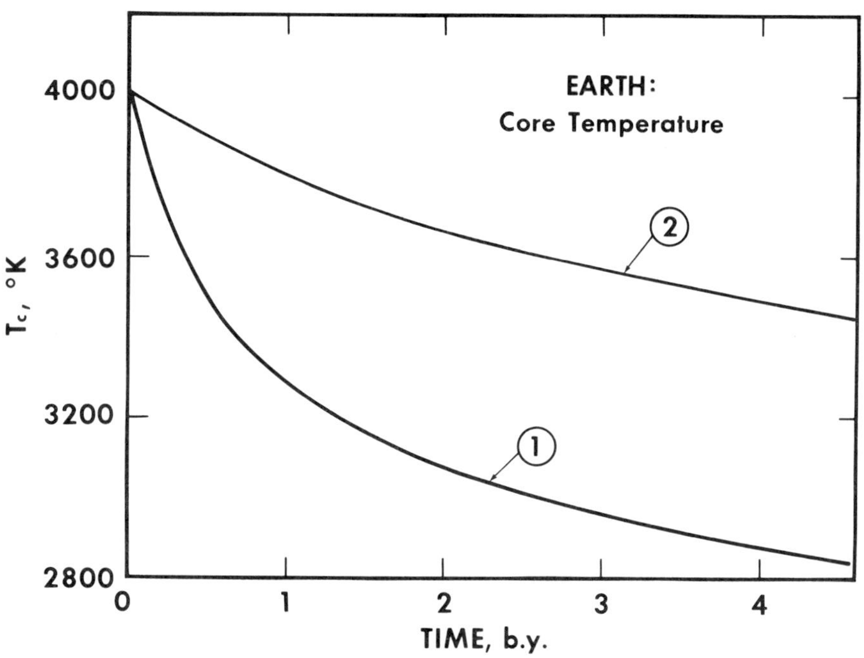

9.5.11 Core temperature T_c versus time for the two models of Earth thermal history (Table 9.5.2).

rate of the mantle is constant for the last 3 b.y. (Fig. 9.5.13), so the change in surface heat flow reflects the decay of the radioactive sources during this period. (4) The mantle and core temperatures depend on the rheological parameters (Figs. 9.5.11 and 9.5.13). In model 2, the final core temperature is near the estimated melting temperature of pure iron (Liu, 1975a). In both models, the core temperatures remain above the melting point of the Fe-FeS eutectic (Usselman, 1975b). (5) The present day heat flow from the core is approximately equal to Stacey's (1972) estimate of 20 mW/m^2 for the minimum heat flow from a thermally convecting core. The only source of core energy considered here is its heat capacity; any other sources such as latent heat of fusion, ^{40}K, or gravitational settling due to chemical differentiation, would augment the core heat flow (see Stevenson *et al.*, 1980).

These thermal history models have several geological consequences for the Archean and Proterozoic. The similarity of the composition of ancient tholeiitic lavas to recent ones, particularly with regard to trace element abundances (Hart *et al.*, 1970), can be interpreted as evidence that the upper mantle has not experienced major compositional change in the last 3 b.y. This inference is consistent with the convective thermal models, for which the thermal regime of the mantle has changed only slightly over that period. However, as noted by McKenzie and Weiss (1975) and section 9.4.4, the ultrabasic lavas produced about 2.8 b.y. ago or earlier required melting at temperatures at least 200° K higher than for the source regions of present day lavas. This, too, is consistent with the models in Figs. 9.5.10–13, for which the mean mantle temperature drop is 300° K since that time. Alternatively, the higher eruption temperature of some Archean magmas may simply be the expression of pressure-release melting during upwelling from greater depths than beneath modern ridges. The greater heat flow at ~3 b.y. ago (more than a factor of 2) and the consequently thinner thermal lithosphere may have had just as great an effect on the degree of melting experienced by magma, as would a hotter geotherm (Hart *et al.*, 1970; Turcotte *et al.*, 1979). D. L. Anderson (1979) has suggested, however, that higher upper mantle temperatures in the Archean may have expanded the stability field of basalt with respect to eclogite, inhibiting the gravitational instability of oceanic lithosphere, and perhaps leading to thicker oceanic plates than at present. Finally, the much lower viscosity (and hence higher Rayleigh number) at that time would be associated with smaller scale, but more vigorous convection, possibly resulting in smaller plates and therefore more extensive regions of igneous activity along plate boundaries. Hart *et al.* (1970), Talbot (1973), and McKenzie and Weiss (1975) have discussed this aspect of early convection, particularly with regard to the scale and setting of greenstone belts.

Certainly, it is possible to construct more complex models of the Earth's thermal evolution than those given here. For instance, the viscosity may be a more complicated function of depth than assumed above. If conjecture of McKenzie and Weiss (1975) that η increases discontinuously at 650 km depth were correct, for instance, the present models should be modified by the inclusion of another boundary layer at the phase change depth.

Also, in the models presented, no consideration was given to the possibility of two scales of convection. Richter (1973c) and McKenzie and Weiss (1975) argue that one scale is necessary in order to explain the motion

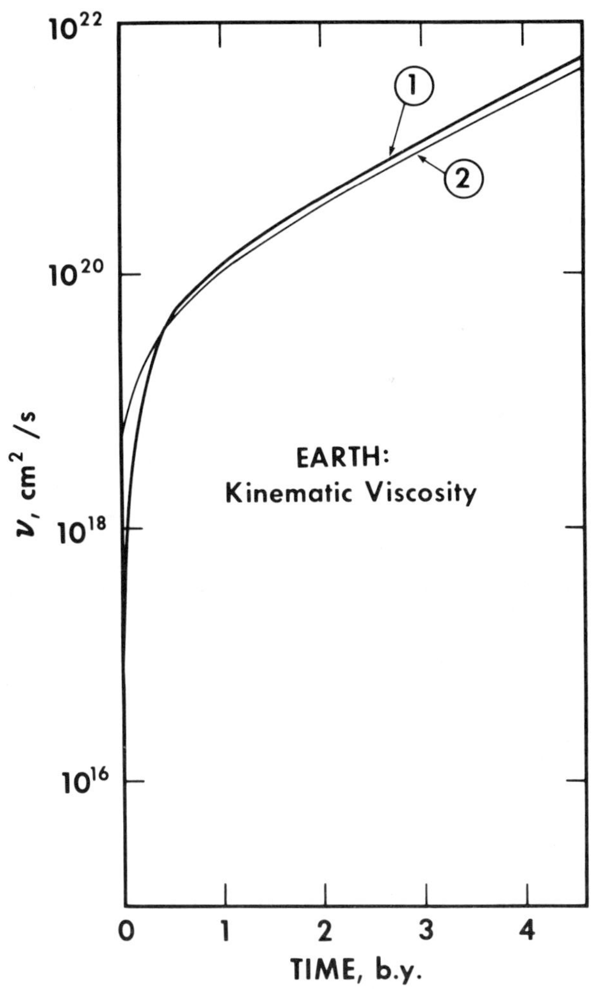

9.5.12 Kinematic viscosity ν of the mantle, exclusive of upper and lower boundary layers, versus time for the two models of Earth thermal history (Table 9.5.2).

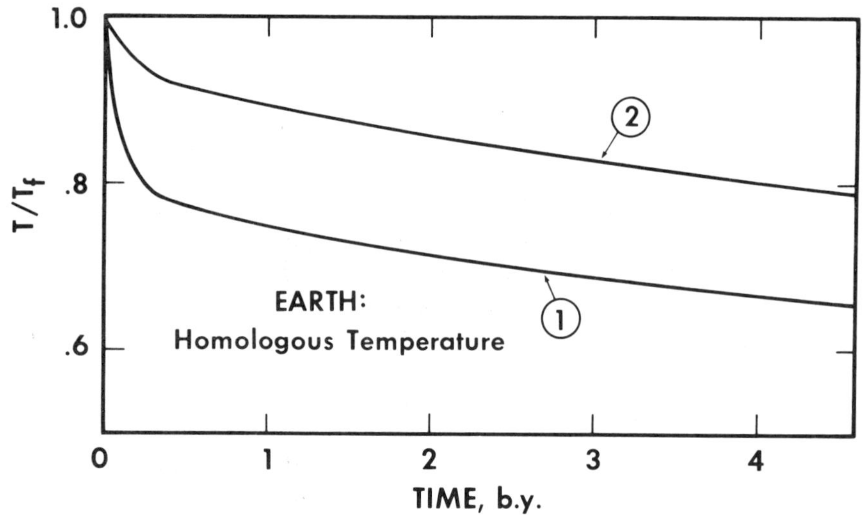

9.5.13 Homologous temperature T/T_f of the mantle, exclusive of upper and lower boundary layers, versus time for the two models of Earth thermal history (Table 9.5.2).

of the present large plates, and a smaller scale is necessary to account for most of the heat transport. The isotopic evidence (section 7.4.2) for long-lived, separate basalt regions also suggests that isolated convection systems might exist. There is no doubt that the complex interaction of plate motions with internal convection will produce cells of varying scale; this would be expected of any high Rayleigh number system. But for our purposes, the question is what are the manifestations of a bimodal distribution of convection scales on the global heat flow, average thermal regime, and rheological properties of the mantle? From calculations that alternately include and exclude the plates as thermal boundary layers, Turcotte *et al.* (1979) obtain values for the mantle viscosity that bracket the observed value. They interpret this result as support for the two-scale hypothesis. However, the short time for vorticity to be distributed over all scales greater than the smallest allowed by dissipation, in a highly viscous medium, suggests that multiple-scale convection may be a more useful concept.

Regardless of the refinements that may be warranted by these factors, the models presented in this section fulfill the requirements imposed by observations regarding the present day heat flow, mantle viscosity, and the state of the core, and clearly demonstrate the thermal effects of subsolidus convection.

Implications for basaltic volcanism

From the standpoint of basaltic volcanism, two features of present-day volcanism and of the global thermal history models are paramount: (1) the vast majority of basaltic volcanism today is at plate tectonic boundaries, principally mid-ocean ridges; and (2) the Earth has been cooling since before the time at which the oldest preserved rocks were emplaced in the crust.

The volcanism associated with mantle upwelling beneath zones of plate divergence therefore was probably a ubiquitous feature of recorded Earth history, though perhaps the characteristics of that volcanism may have evolved (e.g., ultrabasic to basic magmas) as mean convective velocities, mean mantle temperatures, or the radial extent of upper mantle convection varied with time. Spreading centers as volcanic loci may well have preceded the appearance of lithospheric plates with lateral rigidity; the earliest recycling of surface material, at least in the period since 3.8 b.y. ago following heavy bombardment of the inner solar system, may have involved easily deformable crust capping convective cells of smaller characteristic horizontal dimensions than present-day plates.

Convergence zones should have been similarly ubiquitous, though it is possible on thermal grounds that the volcanic rocks formed at the most ancient convergent boundaries may have differed from those formed at modern arcs, because island-arc volcanism probably requires the subduction of coherent units of oceanic lithosphere to depths of ~100 km or more, a feature of Earth tectonics that may not have been characteristic of the earliest history of the planet. At least the preserved Archean greenstone belts, however, exhibit chemistry very similar to modern arcs (section 7.4.2).

9.5.2 Moon

The Moon has been the focus of a large body of work on global thermal history. There have been two principal reasons for this high level of effort. First, in contrast to all of the terrestrial planets save the Earth, the Moon has—as a heritage of the Apollo program—a well-characterized history of differentiation and basaltic volcanism, and key measurements have been made of the surface heat flow and surface abundances of long-lived radionuclides. Second, in contrast to the Earth and to the other large terrestrial planets, the Moon's surface activity was concentrated into the first third of its history and the Moon therefore displays a record of events in the inner solar system no longer preserved on the Earth.

In this section, and in the sections to follow for the other terrestrial planets, we begin with a compendium of the principal constraints on global thermal evolution, including evidence for internal differentiation and constitution, the history of major surface volcanic and tectonic activity, and evidence on the present-day thermal state. We then present several thermal history models generally consistent with these constraints. The discussion closes with a brief summary of the important implications from the thermal models for the history of basaltic volcanism.

Constraints

Structure/differentiation. On the basis of seismic, gravity, topographic and chemical data, the Moon is known to have been internally differentiated into distinct crust and mantle. In the central nearside, within the seismic network of Apollo 12–16, the lunar crust is about 55 km thick (Toksöz *et al.*, 1974; section 4.5.2). The farside crust, on the basis of gravity and topographic data, is either thicker (up to at least 100 km) or of intrinsically lower density (Bills and Ferrari, 1977; Thurber and Solomon, 1978). Assuming a uniform crustal density, the average lunar crustal thickness is about 70 km.

A widely held view is that the lunar crust, exclusive of the mare basalt deposits, is on the average broadly similar in composition to the surface and near-surface rocks of the lunar highlands, known from remote sensing and direct sampling to be roughly characterizable as anorthositic gabbro (Alder *et al.*, 1973). The seismic velocity distribution in the crust is consistent with this view, but does not demand it (Toksöz *et al.*, 1974). The correlation of the abundances of several elements (e.g., Al, Th) with elevation in isostatically compensated non-mare regions offers support for the hypothesis that surface chemistry is representative of the composition of the bulk of the underlying crustal column (Metzger *et al.*, 1977; Schonfeld, 1977; Solomon, 1978b), though this interpretation is not unique.

The existence of a thick feldspar-rich lunar crust, clearly dating back to well before the end of heavy lunar bombardment ~ 3.9 b.y. ago, poses a substantial constraint on lunar thermal models. All proposed models for crustal formation are based on the concepts of early melting of the outer portions of the Moon and subsequent crystal-liquid fractionation on a global scale. In the most straightforward model for crustal differentiation, the early Moon has a surficial "magma ocean", and fractional crystallization and flotation of cumulate plagioclase during cooling of the "ocean" leads to growth of the crust (Wood *et al.*, 1970; J. V. Smith *et al.*, 1970). The complementary mafic cumulates resulting from the differentiation of such a magma ocean have been identified as the ultimate source of mare basalt liquids on the basis of rare-earth abundance patterns (Schnetzler and Philpotts, 1971) and experimental petrology studies (Walker *et al.*, 1975a). Although the details of mare basalt petrogenesis are currently debated (e.g., Walker *et al.*, 1975a, b; Hubbard and Minear, 1975, 1976; Kesson and Ringwood, 1976; Ringwood and Kesson, 1976; Shih and Schonfeld, 1976; Drake and Consolmagno, 1976), early melting and fractional crystallization of at least the outer few hundred kilometers of the Moon is the starting premise on which all models are based.

There is no consensus on the cause of this early melting. Potential candidates include short-lived radioactivity (Runcorn, 1977), induction heating (Herbert *et al.*, 1977b), and the energy of impact of large planetesimals during the later stages of accretion (Safronov, 1972, Weidenschilling, 1974, 1976; Hartmann and Davis, 1975; Wetherill, 1975, 1976; Kaula, 1979a). The last heating mechanism may lead to much greater lateral heterogeneity in heating and in efficiency of differentiation than is commonly assumed.

There are few firm constraints on the characteristics of a lunar magma ocean. Several qualitative models for a magma ocean have been offered to consider the effects on the sequence, location, and chemistry of crystallizing phases of ocean convection and of the large pressure gradient from ocean top to bottom (Walker *et al.*, 1975b; Wood, 1975; Herbert *et al.*, 1977a). The sequence and chemistry of crystallizing phases and the temperature of the magma liquid as functions of ocean fraction solidified have been quantitatively modelled by Drake (1976) and by Longhi (1977, 1978) for simple fractional crystallization in oceans of several proposed whole-moon compositions.

The isotopic systematics of mare basalts (section 7.3.4) provide an estimate of the time when the source regions were established following the early differentiation episode. Sm-Nd (Lugmair *et al.*, 1975; Lugmair, 1975) and U-Pb (Tera and Wasserburg, 1974, 1975) data indicate that mare basalt source regions, including the late-stage cumulates thought to be the ultimate source of high-Ti mare basalts (Walker *et al.*, 1975a), became closed with respect to these element pairs 4.4 b.y. ago. The simplest interpretation of these data is that differentiation of the outer portions of the Moon was essentially complete at that time. Rb-Sr ages of 4.5 to 4.6 b.y. for dunite 72417 and troctolite 76536 (Papanastassiou and Wasserburg, 1975, 1976) may record earlier solidification of parts of the lunar crust.

A major uncertainty is the fraction of the Moon that participated in the early differentiation that produced the crust. On geochemical grounds, at least the outer 300–500 km must have been involved to yield a feldspar-rich crust of the appropriate thickness (Gast, 1972; Taylor, 1978). The source regions of mare basalts, which carry a signature from that early differentiation, lie at 200–400 km depth (section 3.4.2). A major discontinuity in seismic velocity occurs in the lunar mantle at about 400 km depth (Nakamura *et al.*, 1976; Goins *et al.*, 1978), conceivably marking the base of the early differentiation zone. Finally, as discussed below, the absence of global-scale extensional or compressional tectonics on the Moon is most readily explained if early melting was confined to the upper 300 ± 100 km (Solomon, 1977). Thus the outer 300–400 km of the Moon most likely took part in the earliest lunar differentiation.

Information on the deep structure of the Moon is at best sketchy. In particular, neither seismic nor electromagnetic data have yet confirmed the presence of a dense, central core in the Moon. If a core exists, it must be less than 400 km in radius (Nakamura *et al.*, 1974; Wiskerchen and Sonett, 1977). The latest value of the dimensionless moment of inertia of the Moon, $C/MR^2=0.391 \pm 0.002$ (Blackshear and Gapcynski, 1977), requires an increase in density with depth below the crust beyond that due to self compression (Toksöz *et al.*, 1974; section 4.5.2); a dense central core is one, but not the only, possibility for providing such an increase.

Paleomagnetism. Essentially all lunar samples have been found to carry a weak remanent magnetism (e.g., Fuller, 1974). In addition, local magnetic anomalies have been detected from lunar orbit and on the lunar surface (Dyal *et al.*, 1974). If the terrestrial example were followed, this remanence would reflect a memory preserved from the time when the surface cooled in the presence of a dynamo-generated planetary magnetic field. There is essentially no global magnetic field today; an upper limit to the dipole strength is 10^{19} Gauss-cm^3 (Russell *et al.*, 1975). Further, as noted above, the evidence for a small iron core in the Moon is equivocal. The question then arises as to what is the cause of the ancient field.

One group of investigators (e.g., Runcorn, 1977) holds that the field is a memory preserved from an early dynamo. Such a dynamo must have been operating from about 4 to 3 b.y. ago when the samples became magnetized. Evidence for fields as high as 1 Oe (Stephenson *et al.*, 1974; Cisowski *et al.*, 1976) is controversial, but suggestive of an early dynamo, as is the evidence for a variation in the strength of the magnetizing field with time (Stephenson *et al.*, 1975; Lin *et al.*, 1977; Russell *et al.*, 1977). If this were the case, the bulk of the Moon probably melted early, and an early global heat source is required.

A second hypothesis is that the Moon's interior portions accreted at the temperatures below the Curie temperature and became magnetized by primordial solar, terrestrial, or interplanetary fields (Runcorn and Urey, 1973; Strangway and Sharpe, 1974). This hypothesis then requires that a large portion of the Moon remained at temperatures below the Curie point until 3 b.y. ago, so that the impact breccias and basalts could be magnetized in the presence of this field. Subsequent warming would have destroyed this interior magnetism, leaving only various surface blocks magnetized. This hypothesis requires cold initial temperatures for much of the lunar interior and relatively modest deep-seated heat sources. Remanent magnetism has been observed in chondritic meteorites (e.g., Nagata, 1979), indicating paleointensities of up to 16 Oe (Lanoix *et al.*, 1978), and suggesting that adequate external fields were present to magnetize solid bodies during the earliest days of the solar system. Time variations in field strength would be expected for this hypothesis as well as for the notion of a core dynamo.

A third hypothesis is that the samples became magnetized during shock processes. In one version of this idea, ionization would create enough of a transient field to magnetize the local surface—either by gas associated with impact release (Srnka, 1977) or by comets (Gold and Soter, 1976). Another proposal is that textural alignment of magnetic grains during shock can produce net remanent magnetization (Brecher, 1976).

There is no present consensus as to which among these hypotheses is most likely to be correct. Further work is needed to determine whether there are systematic patterns to lunar magnetic anomalies and ancient fields before the various hypotheses can be distinguished. Hood *et al.* (1978) have recently shown, for example, that the directions of magnetization associated with the orbital anomalies are not random in direction, thus tending to weaken the argument for impact hypotheses. If the issue of the origin of lunar paleomagnetism can be resolved, the characteristics of the ancient magnetic field would become an important constraint on lunar thermal history.

Volcanism. The subject of mare volcanism has been dealt with at length in each of the preceding chapters. For the purposes of modelling lunar thermal history, several of the conclusions of those chapters are noteworthy. The radiometric ages of mare basalts range from 3.1 to 3.9 b.y. (section 7.3.4). This age range is probably conservative. There is an indication from crater density and crater morphology data that mare basalt units younger than 3.1 b.y. may not yet have been sampled (Boyce *et al.*, 1974); and both the presence of basalt clasts in 3.9-b.y.-old breccias and remote sensing data suggest mare-type volcanism prior to that time (Ryder and Taylor, 1976; Conca and Hubbard, 1979). The total volume of mare basalt is relatively small, about 1% of the crustal volume (Head, 1976). In contrast to the Earth, therefore, differentiation of a permanent crust on the Moon was a far more substantial event than subsequent mare basaltic volcanism. Most of the mare basalt deposits are located in large impact basins on the lunar nearside. Within individual basins, several basalt units are discernible on the basis of remote sensing (section 2.2.3) and sample data (section 7.3.3). The basalts were emplaced as sequences of thin flows (section 5.4.2), which tended to be concentrated toward the earliest portion of volcanic activity for each basin, generally dating from the time immediately following basin formation (Solomon and Head, 1979, 1980). The source regions for mare basalts may be inferred from experimental petrology, to be located between approximately 200 and 400 km depth (section 3.4.2) and from trace element data to be chemically heterogeneous (section 7.3.4).

The question of nonmare volcanism, particularly an episode of voluminous volcanism prior to 3.9 b.y. ago, is quite important for lunar thermal history. Widespread and substantial volcanic units older than 3.9 b.y. would suggest that both this nonmare and the later mare volcanism were fueled simply by the initial heat of the Moon and by long-lived radioactivity, particularly if the volume of volcanic units decreased continuously with time as outlined in section 8.9. If mare basalts were the product of a relatively late pulse of volcanic activity—a view sometimes expressed in the literature—then an energy source (e.g., impact energy, tidal dissipation, onset of mantle convection) for that pulse must be postulated.

There is at least some evidence for an episode of premare volcanism (see section 5.4.6). Hawke and Head (1978) present evidence from photogeologic, remote sensing and sample studies for pre-Imbrian (>3.9 b.y. old) KREEP volcanism dating to at least 4.0 b.y. ago. Photogeologic evidence for a period of pre-Imbrian volcanism in the southern highlands has been given by Scott (1972) and Strom (1977). Haines *et al.* (1978) cite orbital geochemical data in support of episodes of volcanism in the eastern and western nearside highlands.

Tectonism. The Moon, in contrast to the Earth, has had only a limited tectonic history (see section 6.5.1). Subsequent to crustal formation, the Moon had no analog to terrestrial plate tectonics. The forces that shaped the lunar surface were related either to impact processes or to vertical crustal motions induced by volcanic loading.

The change in isostatic response of the Moon to surface loads shows a clear signature of the steady cooling of the Moon's outer portions. Pre-Imbrian highland topography is generally completely compensated isostatically, with at most a few limited regions with incomplete compensation (O'Keefe, 1968; Ferrari, 1977; Thurber and Solomon, 1978). By 3.8 b.y. ago, the Moon had developed a lithosphere of sufficient strength in at least some regions so that neither the Apennine Mountains nor the Orientale Basin prior to mare fill were completely compensated (Sjogren and Smith, 1976; Ferrari *et al.*, 1978).

The lunar mascons (Muller and Sjogren, 1968), excess or superisostatic masses located within major basalt-filled impact basins, indicate a lithosphere of significant strength dating from the time of mascon formation. The response of the Moon to mascon loads has been that of a plate of finite elastic strength (Kuckes, 1977; Solomon and Head, 1979). There has been lithospheric subsidence and associated tectonic failure, including the formation of mare ridges within the load region and concentric rilles outward of the ridges (Muehlberger, 1974), but a substantial fraction of the original load has remained supported until the present day.

The linear and arcuate rilles on the Moon have been interpreted as graben, indicative of crustal extension (Scott *et al.*, 1975). Most such rilles are peripheral to mare units; some rilles display larger scale regional trends. The mare ridges have most commonly been interpreted as tectonic features indicative of vertical motions within the maria and generally of horizontal surface compression (Lucchitta, 1976). The lunar rilles are restricted in age: all appear to be older than 3.6 ± 0.2 b.y. (Lucchitta and Watkins, 1978). In contrast, mare filling, mascon mare subsidence, and mare ridge formation continued until at least as recently as 3.1 b.y. ago. The lunar-wide synchronism for cessation of rille formation indicates a global change in the state of stress in the lunar lithosphere.

Both the general absence of large scale tectonic features on the Moon *unrelated* to mascon loading, features which are common on Mars and Mercury, and the timing of rille and ridge formation in mare regions are important constraints on lunar thermal history. The first observation restricts the accumulated thermal stress in the lunar lithosphere since the time of heavy bombardment to be less than that necessary to cause widespread failure. Neither this stress level nor its dependence on rate of stress accumulation are known for certain; Solomon and Chaiken (1976) and Solomon (1977) have suggested an upper bound of 1 kb stress change or, roughly equivalent, a 1 km change in lunar radius. Such a bound on accumulated stress is in accord with the strength of the lunar lithosphere estimated approximately from a fit of stress models to lunar rilles and mare ridges (Solomon and Head, 1980). This limit can be satisfied only by thermal history models starting with the outer portions of the Moon hot and the deep interior relatively cold (Solomon, 1977; Cassen *et al.*, 1979). The observation that rilles formed only prior to 3.6 ± 0.2 b.y. indicates that global horizontal surface stress was sufficiently compressive after that time so that continued mare subsidence and lithospheric flexure did not lead to further extensional failure.

The radial positions of rilles and ridges and the current gravitational and topographic data from mascon maria have been used by Solomon and Head (1979, 1980) to infer the thickness of the elastic lithosphere and its variation with space and time. At the time of graben formation, 3.6–3.8 b.y. ago, the elastic lithosphere beneath the mascon maria was 25–75 km or more thick, a range of values requiring substantial lateral heterogeneity in the temperature field within the lunar crust and upper mantle. At the time of younger mare volcanism ($\sim$ 3.0 b.y. ago) in several mascon maria, a substantially thicker ($\sim$ 100 km) elastic lithosphere is evident in the response of the Moon to the additional load.

The occurrence of moonquakes is the principal evidence for continued mild tectonic activity on the Moon today. Most of the moonquakes occur at 800–1000 km depth and appear to be causally related to tidal stresses (Lammlein *et al.*, 1974; Cheng and Toksöz, 1978). Some of the largest moonquakes, the "high-frequency teleseisms" (Nakamura, 1977), occur within the upper 100 km of the Moon and may be related to lithospheric stresses, including those associated with mascon loading (Nakamura *et al.*, 1979).

Present thermal state. An important boundary condition for the thermal history of a planetary body is the present thermal state. Heat flow has been measured at two sites on the lunar surface. In addition, the values of several physical properties of the lunar interior weakly constrain the present day selenotherm.

Heat flow. Both the *in situ* thermal conductivity and the temperature gradient have been measured in drill holes $\sim$ 2 m deep at the Apollo 15 and 17 landing sites (Langseth *et al.*, 1976). The most reliable measures of conductivity resulted from an analysis of long term periodic and transient temperature variations. With these conductivity values, the steady heat flow is 21 and 16 mW/m^2 at Apollo 15 and 17, respectively. Topographic corrections indicate that a more representative value for the Apollo 17 region is 14 mW/m^2. From the surface concentrations of radioactive elements and an assumed variation in crustal thickness, Langseth *et al.* (1976) obtained 18 mW/m^2 for the most likely value of the global lunar heat flow. If a steady state balance of heat production and heat loss were assumed, this average heat flux (with $K/U = 2000$, $Th/U = 3.7$) would give a bulk U abundance of 46 ppb at present (Langseth *et al.*, 1976). The actual U abundance is probably less than this value, however, since the gradual cooling of the Moon also contributes to the surface heat flow (section 9.4.1).

Selenotherm. The seismic and electrical properties of the lunar mantle are highly sensitive to temperature and should therefore provide constraints on the present-day temperature profile in the Moon. Unfortunately, for such constraints to be quantitatively useful, a number of debatable assumptions are first necessary. The seismic Q in the lunar crust and upper mantle (depth $<$ 500 km) is high, in excess of 3000 for both both P and S waves (Goins *et al.*, 1978), indicating that the lunar temperature is substantially cooler than the solidus temperature over that depth range. Between 500 and 1100 km depth, Q is 1500 for P and 1000 for S; and below 1100 km depth Q for S decreases by perhaps a factor of 10 or more (Nakamura *et al.*, 1976; Goins *et al.*, 1978). One interpretation of the low Q values below 1100 km depth is that the temperature makes a close approach to the solidus, although actual partial melting is not required.

The distribution of electrical conductivity with depth in the lunar mantle may be inferred from the Moon's response to transient variations in the solar wind electromagnetic field (Sonett *et al.*, 1972; Dyal *et al.*, 1974, 1976; Wiskerchen and Sonett, 1977). If the composition and oxidation state of the lunar mantle were known, the conductivity profile could be converted to a temperature profile through laboratory data on the temperature variation of conductivity on suitable minerals and rocks (Sonett and Duba, 1975; Duba *et al.*, 1976; Huebner *et al.*, 1979). Given all of the uncertainties, one may conclude only that the conductivity profiles obtained to date are consistent with a lunar temperature profile that rises steeply in the upper ~300 km of the Moon to temperatures within perhaps a few hundred degrees of the solidus, and that it may be anywhere from 0° to ~ 400°C below the solidus at greater depths.

Thermal models

The most important of the above constraints on lunar thermal history are: (1) the early differentiation of the outer portions of the Moon, (2) the chronology of volcanic activity, (3) the tectonic history, and (4) the present-day heat flow and surface abundances of radioactive elements. The first constraint leads to the requirement for initial temperatures near melting in the outer few hundred kilometers of the Moon. The second constraint requires that upper mantle temperatures be near melting until at least 3 b.y. ago. The third constraint poses a limit on lithospheric thermal stress associated with global thermal evolution. The last constraint leads to an estimate for the radiogenic heat production in the Moon and to the recognition that near-surface crustal abundances of Th, U and K are much higher than bulk abundances for the Moon as a whole.

All of the above constraints have been established as a result of measurements made during the Apollo program and on the returned lunar samples. Prior to Apollo, the subject of lunar thermal history was nonetheless one of vigorous controversy. Issues included whether the Moon is or had ever been "hot" or "cold", whether the lunar mantle had ever melted, and the nature and age of the maria (e.g., Urey, 1951, 1952, 1962; MacDonald, 1959, 1960; Levin, 1962; Anderson and Phinney, 1967; Phinney and Anderson, 1967; Fricker *et al.*, 1967; McConnell *et al.*, 1967).

There has been increasing recognition (Runcorn, 1962; Tozer, 1967; Schubert *et al.*, 1969) that solid-state convection has likely been important in controlling the thermal history of planetary bodies, including the Moon. In spite of occasional claims in the literature to the contrary, however, there are no observational data that require that solid convection has occurred in the Moon (Cassen *et al.*, 1978; Phillips and Ivins, 1979). Thermal evolution models both with and without solid convection are equally able to satisfy the constraints on lunar history and present state enumerated above. The main arguments supporting subsolidus convection stem from stability analyses for lunar temperature profiles (Schubert *et al.*, 1969; Cassen and Reynolds, 1973, 1974). These arguments are based on the assumption of a lunar mantle of uniform composition. If early lunar differentiation has resulted in stable chemical stratification of the mantle, however, then the importance of solid-state convection for lunar thermal history might be considerably lessened.

Literally dozens of thermal history models for the Moon have appeared in the literature in the decade since publication of the initial results from the Apollo program (e.g., Urey and MacDonald, 1971; Papanastassiou and Wasserburg, 1971; Hays, 1972; Wood, 1972; Reynolds *et al.*, 1972; Hanks and Anderson, 1972; Toksöz *et al.*, 1972, 1978; McConnell and Gast, 1972; Tozer, 1972, 1974; Turcotte *et al.*, 1972, 1979; Duba and Ringwood, 1973; Solomon and Toksöz, 1973; Cassen and Reynolds, 1973, 1974; Toksöz and Johnston, 1974, 1977; Strangway and Sharpe, 1974, 1975; Solomon, 1975, 1977; Hubbard and Minear, 1975, 1976; Cassen and Young, 1975; Solomon and Chaiken, 1976; Schubert *et al.*, 1977; Herbert *et al.*, 1977a, b; Solomon and Longhi, 1977; Minear and Fletcher, 1978; Solomon and Head, 1979; Cassen *et al.*, 1979). Many of the models are broadly similar, and rather than review them all, the approach taken in this section is to consider several representative models (Table 9.5.3) that fit the constraints on lunar evolution and that illustrate the principal conclusions derived from the entire body of work.

The early episode of differentiation of the outer portions of the Moon is of importance because several basic aspects of subsequent lunar evolution (upwards segregation of heat sources, establishment of mare basalt source regions) date from that time. Crustal formation and silicate differentiation are not explicitly included in standard thermal history routines for global planetary evolution, but such processes can be approximately modelled with the use of simplifying assumptions. The "magma ocean" hypothesis has received the most attention among the scenarios for early differentiation; the most detailed models for the cooling and evolution of a lunar magma ocean are those of Solomon and Longhi (1977) and Minear and Fletcher (1978).

One such model (Solomon and Longhi, 1977) is illustrated in Figs. 9.5.14–15. The initial temperature follows an adiabat in the magma ocean, intersects the liquidus at 300 km depth, and decreases quadratically with radius from the solidus immediately below the ocean to a central temperature of 0°C. Cooling of the magma ocean occurs primarily at the ocean top, and to a lesser extent through the ocean floor. At each time step, the additional crystallization due to this cooling is calculated and the proportion of the newly crystallized minerals which either sink or float in the magma ocean are determined from the crystallization sequence at low

Table 9.5.3 Parameters used in lunar thermal history models.

	Fig. 9.5.14	Fig. 9.5.16	Fig. 9.5.18	Fig. 9.5.20
Radius, km	1740	1740	1738	1740
Mean density, g/cm^3	3.34	3.34	3.34	3.34
Surface temperature, °C	-20	-20	-20	0
Present radioactive abundances				
U, ppb	40	40	35	35
K/U	2000	2000	2000	2000
Th/U	3.7	3.7	3.6	3.5
Specific heat, J/g-°C	1.2	1.2	1.2	1.2
Heat of fusion, J/g	400	400	400	400
Thermal conductivity, J/cm-s-°C	.04	.028 (crust) .04 (mantle)	Schatz & Simmons (1972)	.045
Kinematic viscosity, cm^2/s	adiabat in liquid layer	-	$(1.4 \times 10^9) \cdot \exp(25\, T_f/T)$	$(1.4 \times 10^9) \cdot \exp(E/kT)^{\dagger}$

$^{\dagger}E/k = (3.5\text{–}4.6) \times 10^4$ °K.

pressure of a magma of whole-Moon composition (Longhi, 1977). Liquidus and solidus curves for both the magma and cumulates are adjusted as the chemistry of the magma and the crystallizing minerals changes with time. The temperature profile within the remaining liquid magma is maintained at an adiabat to simulate convection, subject to conservation of energy. Heat sources are partitioned into coexisting solid and liquid phases in the proportions 1:99.

Cooling of the magma ocean is initially efficient, leading to a rapid growth of deep mafic cumulates composed of the early crystallizing minerals, principally olivine (Drake, 1976; Longhi, 1977). Once plagioclase begins to crystallize, at about 50% total crystallization, crustal growth accelerates and ocean cooling slows because of the thickening conductive surface layer. The time scale for completion of fractionation of the magma ocean, aside from a small residual liquid component (Fig. 9.5.15), is 100–200 m.y. (Solomon and Longhi, 1977; Minear and Fletcher, 1978), a figure not strongly dependent on initial ocean depth for a fixed value of final crustal thickness. This differentiation time for the magma ocean is in agreement with the time at which the mare basalt source regions were established, but does not permit an independent test of initial ocean depth.

The mafic cumulates in the model of Fig. 9.5.14–15 do not remelt following emplacement. They form with high solidus temperatures and low heat source abundances, and they rapidly lose heat to the cooling ocean and to the underlying mantle. A cumulate-remelting origin for mare basalts therefore requires the admixing of relatively low melting-point material with the mafic cumulates, either during accumulation (Walker *et al.*, 1975b) or subsequent to completion of magma ocean differentiation (Hubbard and Minear, 1975, 1976; Ringwood and Kesson, 1976). The residual liquid remaining after most of the magma ocean has differentiated (Fig. 9.5.15) can persist for some time before solidifying (4.1 to 3.8 b.y. ago for magma oceans initially 300–400 km deep, respectively) and can be readily tapped for early episodes of nonmare volcanism.

Most lunar thermal history models avoid treating early lunar differentiation in detail in favor of exploring the sensitivity of thermal models to other aspects of lunar history. One such thermal model, including upwards differentiation of heat sources and simulated convection in the liquid state but without solid-state convection, is shown in Figs. 9.5.16–17. The thermal model, from Solomon and Head (1979), is consistent with the tectonic history of the Moon, including the absence of global-scale compressional or extensional features and the cessation of rille formation at 3.6 ± 0.2 b.y. ago.

The thermal model involves initial melting of the outer 300 km of the Moon and an initially cold interior (central temperature 300° C). Because of a balance

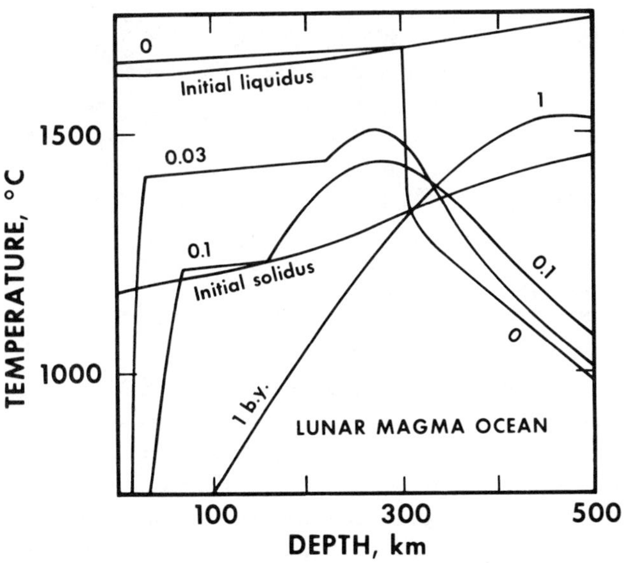

9.5.14 Thermal evolution of a lunar magma ocean (from Solomon and Longhi, 1977). Initial ocean depth is 300 km. The initial liquidus and solidus are as shown; the liquidus in the ocean as a function of ocean fraction crystallized is determined from Longhi (1977). The solidus temperature in each cumulate layer is equal to the ocean liquidus temperature at the time the layer is formed. Temperature in the liquid region follows an adiabat. Radioactive heat sources during crystallization are strongly concentrated in the liquid component.

between cooling of the lunar exterior and warming of the deep interior (Fig. 9.5.16), the lunar radius changes by less than 1 km and the horizontal thermal stress changes by less than 1 kb in the time since cessation of heavy bombardment ~ 3.9 b.y. ago (Fig. 9.5.17). The peak lunar volume is reached 3.6 b.y. ago (Fig. 9.5.16), so that horizontal stress in the lithosphere becomes increasingly compressive after that time, accounting for the ending of rille formation even though mascon mare subsidence and mare ridge formation continued to much later times. The increasingly compressive lithospheric stress may have acted to shut off access of mantle-derived magmas to the lunar surface.

A lunar thermal history model including solid-state convection (Toksöz *et al.*, 1978) is shown in Figs. 9.5.18–19. The full convective equations (9.3.8–10) are solved under the assumption of axial symmetry. The initial temperature profile follows a solidus to 500 km depth and has a central temperature of 500° C. Convection dominates heat transfer in a deepening zone that reaches lunar center at 2 b.y. after lunar formation. The present temperature profile is about 150° C below the solidus. Partial melting in the lunar mantle persists until 2 b.y. ago. Convection at present is confined to below 800 km depth.

The principal differences between thermal models incorporating solid-state convection and those that do not are the enhanced ability of convecting mantles to transfer heat to the base of the lithosphere, the generally cooler temperatures (by ~ 100–300° C) below the lithosphere, and the more rapid disappearance of partial melt in the lunar mantle (see Figs. 9.5.16 and 9.5.19).

A thermal history model for the Moon (Cassen *et al.*, 1979) based on parameterized solid-state convection (section 9.3.3) is shown in Fig. 9.5.20. The model meets the following constraints: the outer layers are initially molten; convection commences about 0.5 b.y. after formation, providing a heat pulse for mare volcanism; the temperature remains close enough to the solidus in the outer 900 km to produce magma until at least 2.5 b.y. ago; the average temperature is presently below the solidus to a depth of at least 900 km; the change in radius since 3.5 b.y. ago is less than 1 km; and the present day heat flow is between 14 and 21 mW/m^2. The initial temperature profile follows the solidus to a depth of 300 km; the initial central temperature is 650° C. The central viscosity and surface heat flow at present are, respectively, 5×10^{21} cm^2/sec and 16.5 mW/m^2.

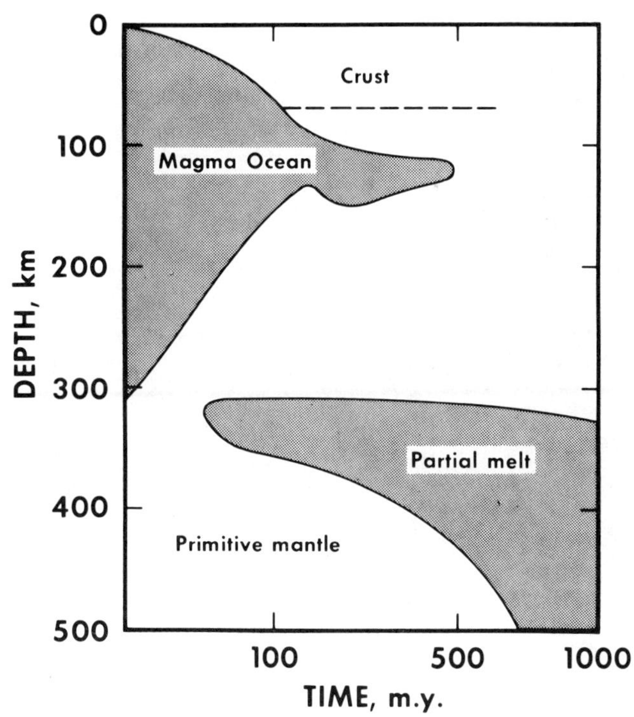

9.5.15 Evolution of magmatic liquid in the thermal model of Fig. 9.5.14. The ocean completes major fractionation by 150 m.y.; a residual liquid layer remains partially melted for an additional 350 m.y. The 'primitive' mantle beneath the ocean begins to melt at an early stage in lunar history, producing a deepening zone of partial melting. The mafic cumulates do not remelt. The scale for time t is linear in $t^{1/2}$.

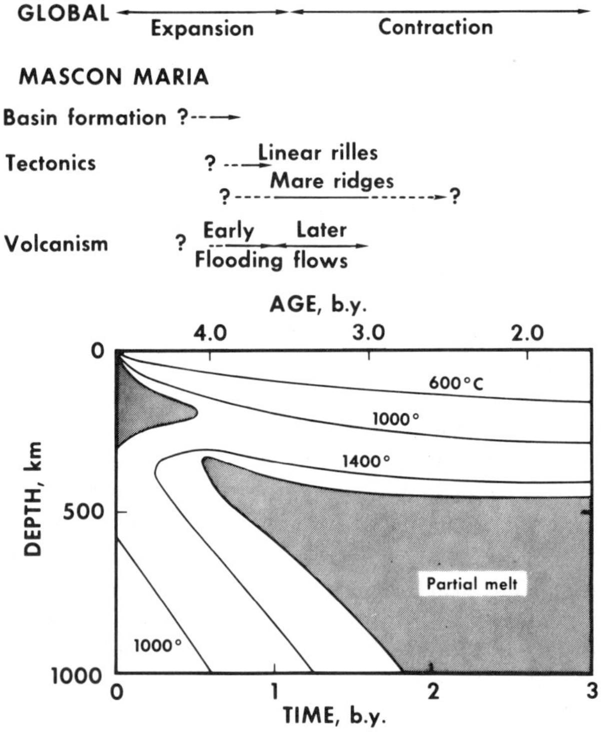

9.5.16 (Bottom) A thermal history model for the Moon with initial melting to 300 km depth and an initially cold interior (a central temperature of 300°C), from Solomon and Head (1979). (Top) Summary of temporal relationships between the global thermal expansion and contraction associated with lunar thermal history and the tectonic and volcanic histories of the mascon mare basins.

The present temperature approaches the solidus at a depth of 500 km (Fig. 9.5.20). There has been little cooling below this depth in the last 2 b.y., and the lack of surface volcanism thereafter may be attributed to the fact that the Moon, in this model, has been contracting for 2 b.y., while before that time it was expanding (Solomon, 1977). Also, while changes in the volatile distribution have not been included in these models, there would be an effect on both melting temperature and viscosity as a result of any major redistribution of volatiles which occurs from melting. Such an effect would tend to increase viscosity and melting temperature within a depleted zone, and decrease them in a zone enriched in volatiles. Magma formation would thus be less likely after the initial radioactive and volatile depletion in the 100–500 km depth region.

The parameterized convection models of Cassen *et al.* (1979) support the results of the conduction models cited above concerning the range of lunar thermal histories capable of accounting for the absence of global-scale extensional or compressional tectonics. Models with initially hot exteriors and cold interiors can yield net changes in surface lithospheric stress and in lunar radius of less than 1 kb and 1 km, respectively, since ~ 3.9 b.y. ago. Substantially hotter initial temperature profiles (e.g., total melting of the Moon) lead to the prediction of large compressive stress at the lunar surface, at variance with the absence of large-scale compressive tectonic features, for both convective and conductive lunar thermal models.

Implications for basaltic volcanism

Virtually all lunar thermal history models that have high initial temperatures in the outer few hundred kilometers of the Moon maintain temperatures in the upper mantle sufficiently close to the solidus to sustain mare volcanism until times at least as recent as 3 b.y. ago. The heat to fuel magma generation can readily be supplied by the initial heat of the Moon and by long-

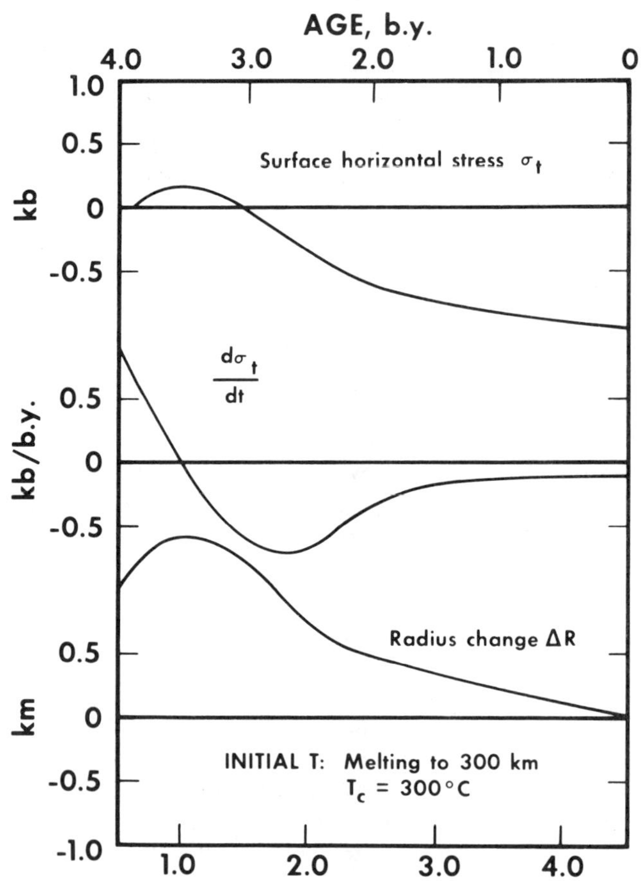

9.5.17 Changes in surface horizontal stress σ_t and lunar radius R for the thermal history model of Fig. 9.5.16. The stress σ_t is referred to an arbitrarily stress-free state at 3.9 b.y. ago.

lived radioactivity, though under special circumstances thermal models can be constructed to deliver a pulse of heat to the lunar upper mantle at ~ 3.9 b.y. ago to enhance the production of mare basaltic magma.

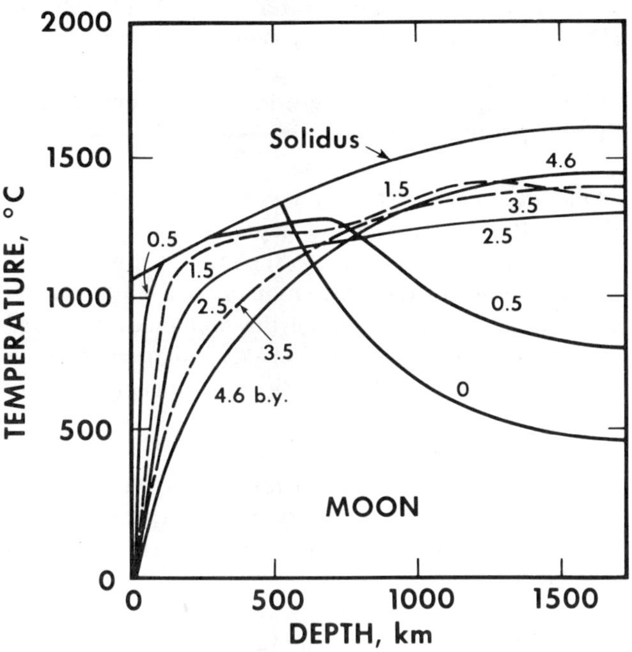

9.5.18 Angle-averaged temperature versus time for a lunar thermal history model incorporating solid-state convection (Toksöz *et al.*, 1978). Convection extends to the center of the Moon after 2 b.y.

An important implication of thermal history models in which the retained heat and the average fraction of partial melt decrease more or less continuously with time is that there should have been extensive lunar volcanism prior to 3.9 b.y. ago. The magmas need not have had close affinities to mare basalts; their chemistry would be strongly dependent on the distribution of heat sources following crust-mantle differentiation and consequently on the location and characteristics of magma source regions. The eruption sites may have reflected strong lateral heterogeneities in efficiency of heat source segregation and perhaps in the hydrostatic head of the magma columns. Nonetheless, magma generation of some sort should have been widespread in the first 600 m.y. of lunar history.

The duration of volcanism on the Moon may have been controlled as much by the state of stress in the lithosphere as by the occurrence of magmas at depth (section 9.1.1). Global expansion early in lunar history should have aided magma ascent both during early mare volcanism and during a likely earlier period of volcanism as well. Global contraction starting at least by 3.6 b.y. ago should have made magma ascent increasingly difficult, tending to localize volcanic eruptions in regions where local near-surface extensional stresses were comparable to or larger than the global surface stress. The preferred eruption sites for late-stage mare volcanism would therefore have been the edges of mare basins where extensional flexural stresses due to mare basalt loading were highest (Solomon and Head, 1979). With time, even these local extensional stresses were

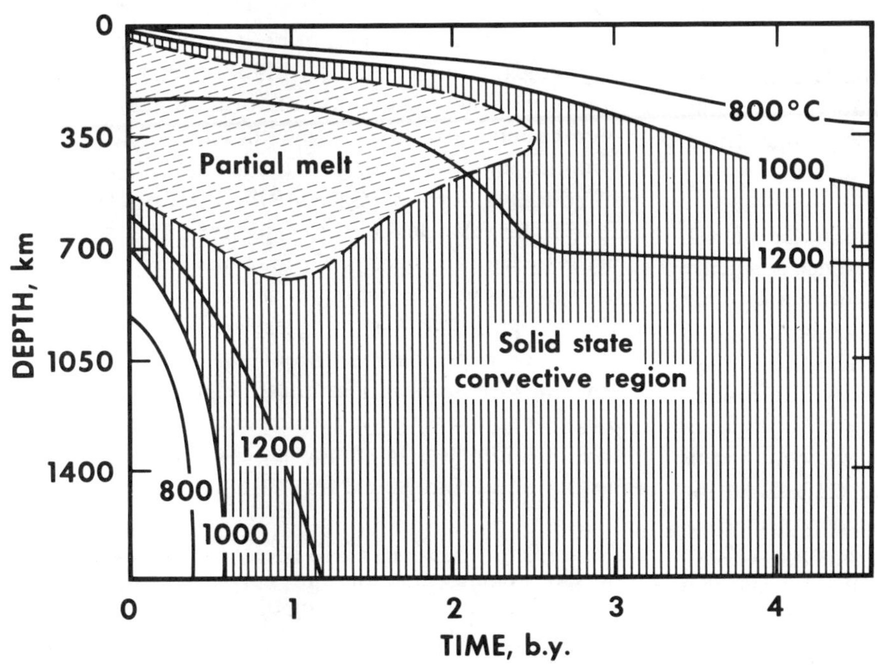

9.5.19 Isotherms and regions of melting and subsolidus convection in the thermal model of Fig. 9.5.18. Note that some partial melt can persist in upwelling regions even though the angle-averaged thermal profile lies below the solidus.

likely exceeded by the increasingly compressive global surface stress (Fig. 9.5.17), shutting off mare volcanism entirely.

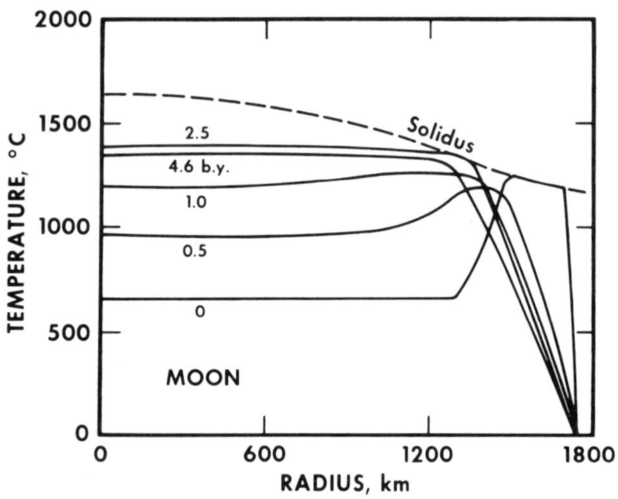

9.5.20 A thermal history model for the Moon based on parameterized subsolidus convection (Cassen *et al.*, 1979). Convection begins after 0.5 b.y.

9.5.3 Mars

Mars displays a diverse array of geological processes and shows abundant evidence of an interesting surface history, including extensive volcanic and tectonic activity. The planet has been studied with three U. S. and three Soviet fly-by missions, three U. S. orbiters and two landers. While much is currently known about Mars, many very important first-order questions remain unanswered, particularly concerning the chemistry and chronology of the planet. As a result, the thermal history of Mars is only loosely constrained.

Constraints

Structure. Mars is evidently a differentiated planet. The moment of inertia indicates the presence of a dense central core (e.g, Binder, 1969). Unfortunately, the moment of inertia is not precisely known because of a relatively large non-hydrostatic contribution to J_2, (section 4.2) which can be corrected only in a model-dependent fashion (Binder and Davis, 1973; Reasenberg, 1977; Kaula, 1979b). Neither is the core radius well-known, because it can be traded off against the assumed core density for a given crust and mantle structure. The probable core radius lies in the range 1500 to 2000 km (Johnston and Toköz, 1977); the fractional mass of the planet occupied by the core is substantially less than for the Earth (section 4.5).

Mars also has a distinct low-density crust, of quite variable thickness, as indicated by the partial to complete isostatic compensation of surface topography (Phillips *et al.*, 1973). The compensation depth for an Airy compensation mechanism, at least for ancient martian terrain, is no greater than about 100 km (Phillips and Saunders, 1975). A minimum value for the average crustal thickness is 28 ± 4 km (Bills and Ferrari, (1978), though this result depends on the adopted crust-mantle density contrast.

Magnetic field. Mars may have a small magnetic field, though this hypothesis is a matter of current dispute (Russell, 1978; Dolginov, 1978). A fit of a dipole to the available Soviet data gives a moment of 2×10^{22} Gauss-cm^3, or 3×10^{-4} times the Earth's moment, tilted 15°–20° with respect to the rotation axis and oriented in the opposite sense to the Earth's dipole moment (Dolginov *et al.*, 1973, 1976). A re-evaluation by Russell (1978) leads only to an upper bound on the moment of 2×10^{21} Gauss-cm^3, an order of magnitude less than the value of Dolginov and co-workers. Magnetic field measurements in the close vicinity of Mars are needed to resolve this issue.

Volcanism. Mars shows abundant evidence of surface volcanic activity spanning a wide range of relative geologic ages, as discussed in Chapter 5 and 8. There are numerous volcanic constructs, including the large volcanoes of the Tharsis province, and extensive volcanic plains with apparent flow fronts and other features associated with volcanic units (Carr, 1973, 1974; Malin, 1977; Carr *et al.*, 1977). There are relatively old and relatively young examples of both plains and shields.

Volcanic constructs on Mars show a general increase in height with time of formation, in that the oldest such features are a few kilometers high and the youngest shields are roughly 20 km high with respect to surrounding terrain (Carr, 1974; Blasius and Cutts, 1976). This relationship has been ascribed to an increase in the hydrostatic head of the magma with time because of a progressive deepening of the source region (Carr, 1975; Vogt, 1974). Assuming a relative density contrast of 10% between magma and average overburden and ignoring viscous head loss, these heights give depths to magma chambers varying from perhaps as little as a few tens of kilometers to somewhat over 200 km over the history of martian shield formation.

Tectonism. Mars has also been tectonically active in the past. There are numerous chasms, graben and large fractures that appear to be tectonic in origin (Sharp, 1973; Hartmann, 1973a). Almost all of these large-scale features are extensional in nature. Many, such as those radial to the Tharsis area, may represent the response to vertical uplift (Carr, 1974) or to volcanic loading (Phillips *et al.*, 1980a). There are also a number of small mare-type ridges (Lucchitta and Klockenbrink, 1979) and a proposed system of lineaments that may be related to the tectonic stress field (Binder and McCarthy, 1972).

Mars supports large non-hydrostatic stresses associated with topographic loads. Mars is the roughest of the terrestrial planets for the low harmonic order topography and gravity fields, even after scaling according to the equal stress rule (Lorell *et al.*, 1972). The isostatic response of Mars is time dependent. Geologically old terrain appears to be completely compensated (Phillips and Saunders, 1975). The long-wavelength gravity and topography over the Tharsis region of Mars, however, are inconsistent with complete isostatic compensation by a shallow Airy mechanism, i.e., by crustal thickness variations (Phillips and Saunders, 1975). Alternative explanations are dynamic support of a superisostatic topographic load (Phillips *et al.*, 1980a) and isostatic support of topography and crustal thickness variations by lateral density differences in the upper several hundred kilometers of the martian mantle (Sleep and Phillips, 1979). These last two explanations are both consistent with anomalously high mantle temperatures beneath the Tharsis region. Plescia *et al.* (1979) have argued that the tectonic response of the martian lithosphere to topographic loads has varied with time, from substantial deformation near old loads to little or no tectonic signature for young loads; they have ascribed this time variation in response to a thickening of the elastic lithosphere of Mars with time. Determination of the elastic lithosphere thickness from the radial position of circumferential graben, however, does not show a simple relationship between thickness and the surface age of the volcanic construct (Solomon *et al.*, 1979); rather the effective thickness of the elastic lithosphere appears to be controlled primarily by proximity to a major volcanic province.

The seismometer on the Viking 2 lander has detected at most one event that can be classified as a marsquake (Anderson *et al.*, 1977). If that event is a true seismic signal, and a wind-related origin for the signal cannot be ruled out, then it is a small (magnitude 3) local quake. Even though the Viking seismometer is not very sensitive by Earth or Moon standards, it may be concluded at a high level of confidence that the seismicity level on Mars is less than on Earth.

Bulk chemistry and atmosphere. The bulk chemistry of Mars is poorly known. Equilibrium condensation models for Mars (Lewis, 1972; Goettel, 1977) predict an FeS-NiS core, an olivine mantle with an Fe/(Fe+Mg) ratio higher than the Earth, and an inventory of volatiles intermediate between that of the Earth and that of carbonaceous chondrites (section 4.5.5). Such a composition is not in conflict with the limited data on internal structure (Johnston and Toksöz, 1977). Alternative compositional models for Mars have been based on an assumed higher oxidation state for Mars compared to Earth (Ringwood and Clark, 1971), on a mixture of meteorite types (Anderson, 1972), and on a terrestrial pyrolite mantle with additional FeO added to match the mean density and moment of inertia. (McGetchin and Smyth, 1978).

Rough chemical analyses have been conducted on the soil at the two Viking landing sites (Toulmin *et al.*, 1976; Clark *et al.*, 1976). The material appears to be composed dominantly of a nontronitic montmorillonite, an iron-rich clay derived by extensive chemical weathering of mafic or ultramafic rocks (Baird *et al.*, 1976).

The known volatiles on Mars include the polar caps and the atmosphere. The winter polar cap is dominantly frozen CO_2, and the residual cap during martian summer is water ice (Farmer *et al.*, 1976; Kieffer *et al.*, 1976). The abundant, apparently fluvial channels on Mars are most readily interpreted as due to earlier episodes of water flowing on the surface (Milton, 1973; Carr *et al.*, 1976).

The martian atmosphere is mainly CO_2, with traces of N_2, Ar, O, O_2, CO (Nier *et al.*, 1976a), Kr and Xe (Owen *et al.*, 1976). The ratio of Ar^{36} to Ar^{40} is $1{:}2750 \pm 500$ (Owen and Biemann, 1976), about 10 times lower than the terrestrial value. The total CO_2 and N_2 to rare gas abundances, the 75% enrichment in N^{15}/N^{14} with respect to Earth (Nier *et al.*, 1976b), and the inferred presence of liquid water on the surface all point to a past martian atmosphere more massive than the current one (McElroy *et al.*, 1976; Owen and Biemann, 1976). Long-term fluctuations in the martian climate and perhaps in the mass of the atmosphere may arise from changes in the planet's obliquity and orbital eccentricity (Ward *et al.*, 1974).

Probably the total outgassing on Mars has been less than on the Earth, and the higher $^{40}Ar/^{36}Ar$ ratio on Mars may be due to a higher crustal K abundance (Owen and Biemann, 1976) or to a lower total volatile abundance on Mars compared to the Earth.

Chronology. The relative chronology of the various large-scale surface units on Mars is reasonably well established from stratigraphic and crater density relationships (Carr *et al.*, 1973). The oldest units are the cratered terrain, an upland region as heavily cratered as the lunar highlands, and occupying approximately the hemisphere south of a great circle inclined 35° to the equator (Mutch and Saunders, 1976). Both plains deposits and volcanic constructs span a wide range of relative ages. Most, if not all, of the channels are quite old, possibly older than the oldest volcanic plains (Malin, 1976a). The surfaces of the large shield volcanoes of the Tharsis region, including Olympus Mons, are among the youngest on the planet.

The absolute chronology is a matter, however, of great debate (e.g., Hartmann, 1973b, 1977; Soderblom *et al.*, 1974; Neukum and Wise, 1976). Most workers are in agreement that the heavily cratered southern uplands probably record the terminal phase of heavy bombardment of the inner solar system, dated for the Moon at 3.9–4.0 b.y. ago. Crustal formation must have occurred prior to this time. Absolute ages of younger surface units depend upon scaling cratering flux and impacting-object size versus crater size from the Moon to Mars, a procedure that is not free of rather arbitrary assumptions (e.g., Chapman, 1974; Wetherill, 1974). Published ages for the Tharsis shields, for instance, range from a

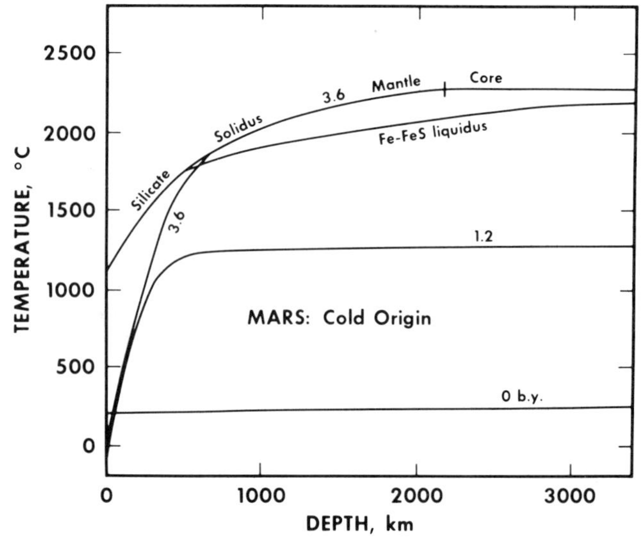

9.5.21 A thermal history model for Mars (from Solomon and Chaiken, 1976). Time in billions of years since planetary formation is shown next to the appropriate temperature profile. The termperature distribution has changed only slightly between the times 3.6 b.y. and the present.

Table 9.5.4 Parameters used in Mars thermal history models.

	Fig. 9.5.21	Fig. 9.5.23	Fig. 9.5.25
Radius, km	3390	3390	3389
Mean density, g/cm^3	3.93	3.93	3.96
Surface temperature, °C	-40	-40	-40
Present radioactive abundances			
U, ppb	11	11	15
K/U	70,000	70,000	50,000
Th/U	3.7	3.7	4
Specific heat, J/g-°C			
Silicates	1.2	1.2	1.2
Metal/sulfide	0.6	0.6	0.46
Heat of fusion, J/g			
Silicates	400	400	400
Metal/sulfide	270	270	240
Thermal conductivity			
Silicates	(*1*)	(*1*)	(*1*)
Metal/sulfide	(*2*)	(*2*)	adiabat
Kinematic viscosity (mantle), cm^2/s	-	-	$(1.4 \times 10^9)\cdot \exp(25\, T_f/T)$

References:
(*1*) From Schatz and Simmons (1972). (*2*) From Fe data of Powell *et al.* (1966).

high of 2.5 b.y. ago (Neukum and Wise, 1976) to a low on the order of a hundred million years (e.g., Hartmann, 1977). In general, the chronology developed in Chapter 8 is adopted as a working hypothesis in a later discussion.

Thermal models

Thermal history models for Mars in the era since Mariner 9 have been published by numerous authors (Hanks and Anderson, 1969; Johnston *et al.*, 1974; Young and Schubert, 1974; Fricker *et al.*, 1974; Tozer, 1974; Toksöz *et al.*, 1978; Toksöz and Hsui, 1978b). The principal features of martian history that must be described by a thermal history model are the differentiation of a distinct low-density crust and of a core, and the extended period of active planetary volcanism. Crustal differentiation must have been completed prior to termination of heavy bombardment, suggesting an early hot exterior by analogy with the Moon. The time of core formation is, however, unconstrained. The biggest uncertainty in constraining the thermal models is the lack of an absolute chronology for the surface geologic history. A second major uncertainty is the abundance and distribution of radioactive heat sources. Most workers have adopted chondritic abundances for bulk U, Th, and K, or have assumed Earth-like abundances. Both surface chemical measurements and surface heat flow observations are necessary to reduce the uncertainty in these parameters.

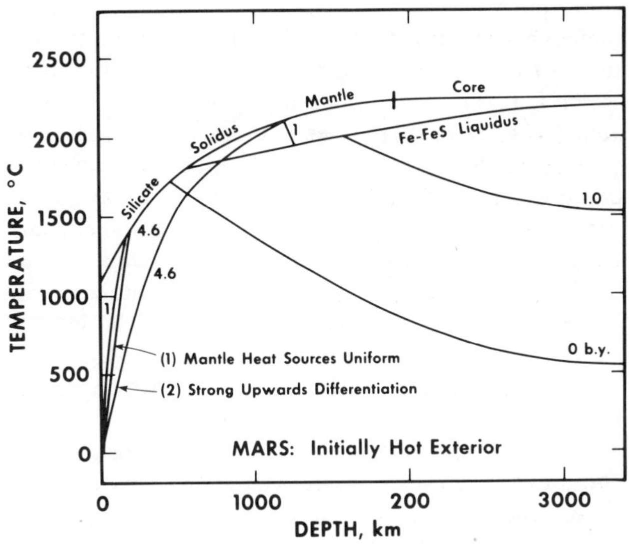

9.5.23 Two possible thermal history models for Mars (from Solomon, 1979b). After global differentiation, the sources in (1) are concentrated uniformly outside the core; in (2) they are assumed to decay exponentially with depth with a 50 km skin depth (same total abundances).

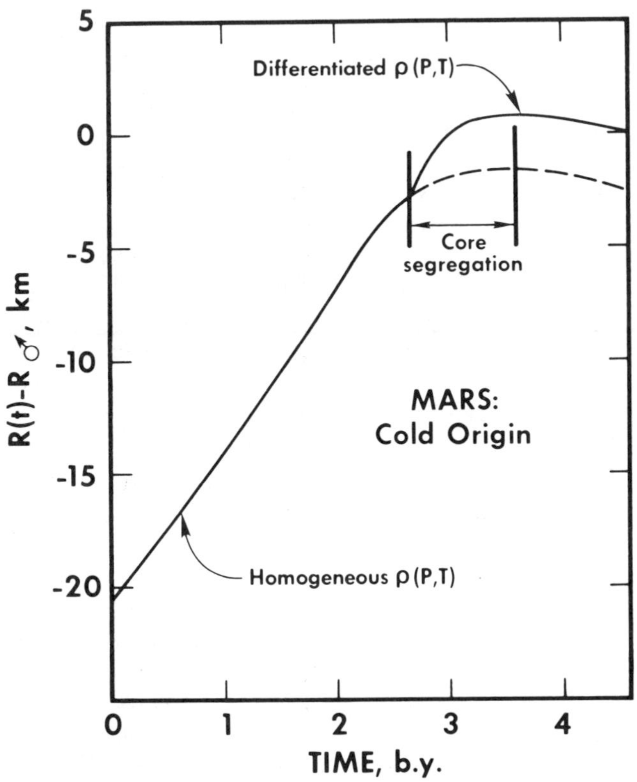

9.5.22 Radius of Mars vs. time for the thermal model of Fig. 9.5.21.

As a consequence of our lack of quantitative chronological information, many widely disparate thermal history models are possible. These models vary in the timing and magnitude of tectonic and volcanic activity to be expected. A simple such Mars history model is shown in Fig. 9.5.21. The physical parameters of this model are given in Table 9.5.4. The model starts with a uniformly cold temperature (450° K), has chondritic heat sources distributed uniformly throughout the mantle material, and includes the thermal effects of forming a (nearly pure Fe) core and of liquid-state convection. Planetary differentiation is late (2.7–3.6 b.y.) and the volume history is predominantly one of net expansion and lithospheric extension (Fig. 9.5.22). Because the outer portions of the planet start and remain cold, the model fails to account for early differentiation of a distinct crust.

A pair of similar models with an initially hot exterior, capable of segregating an early crust by crystal-liquid fractionation in the outer 400 km, are shown in Fig. 9.5.23. The models, both of which differentiate core from mantle over the first 2 b.y. of martian history, differ in the distribution of heat sources following planetary differentiation. In (1) the heat source distribution is arbitrarily taken to be uniform outside the core, which is taken to be source free (and consequently is subadiabatic). In (2) the heat sources are strongly concentrated in the crust, the effect of which is to accelerate cooling after differentiation and to yield a colder present-day temperature profile. The variation of radius with time for these two profiles is shown in Fig. 9.5.24. Model (1) is consistent with global lithospheric extension over most of martian history, with at most a short recent period of slight contraction. Model (2) predicts contraction for the last 2.5 b.y., which would have led to compressive stress (~3 kb) capable of producing lithospheric failure. The absence of large-scale compressive tectonics on Mars eliminates models such as this one.

A model which incorporates subsolidus convection and which has a generally similar initial temperature profile but somewhat different heat source abundances (K/U = 50,000; U = 15 ppb) is shown in Fig. 9.5.25. The results are rather similar to model (1) in Fig. 9.5.23, with core formation lasting 1.5 b.y. and with the mantle continuing to warm and expand until the last 1 b.y. The regions in excess of various melting temperatures are outlined in Fig. 9.5.26. Note that only angle-averaged temperature distributions are shown in the two figures. Thus where the thermal profile lies along the solidus, *mantle-wide* partial melting at that depth is indicated. Where the thermal profile lies within 200–300° C of the solidus, partial melting is still predicted to occur, but only in relatively warm regions of upwelling flow.

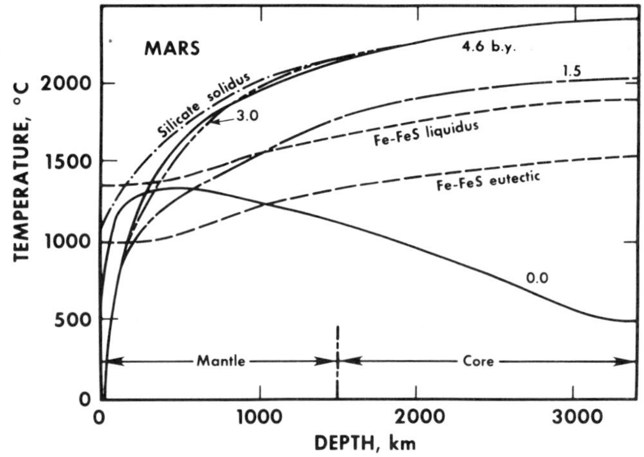

9.5.25 A thermal history model of Mars incorporating subsolidus convection (from Toksöz *et al.*, 1978).

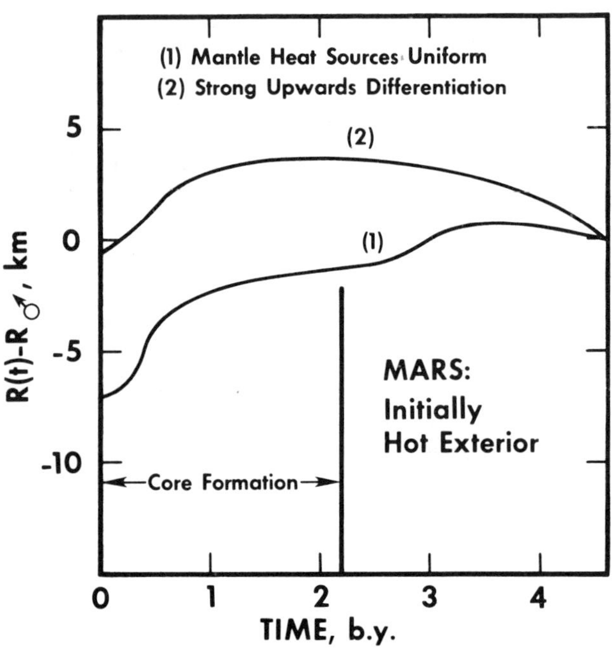

9.5.24 Radius of Mars vs. time for the two thermal models of Fig. 9.5.23.

Implications for basaltic volcanism

There is a general correspondence between expansion and widespread volcanism on the smaller terrestrial planets (Solomon, 1978a). On this basis, the widespread extensional tectonics and lack of global compressive features on Mars argue for an extended period of volcanism and probably of global expansion, though many of the extensional features may be principally the result of local rather than global sources of stress. Thermal models consistent with cessation of volcanism substantially earlier than 1 b.y. ago predict subsequent contraction and global compressive tectonics at variance with martian surface history. Thus the global thermal models lend independent support for the chronology developed in Chapter 8, in particular for a long duration of volcanism continuing from the time of heavy bombardment until the time of the most recent activity.

In view of the range of thermal models still consistent with available constraints on martian evolution, it is difficult to make specific predictions on the production with time of basaltic magmas in the martian mantle. The models in Figs. 9.5.23 and 9.5.25 give either no mantle melting or only very deep melting at present, though these results are only for a dry mantle solidus. A substantial volatile phase in the martian mantle would yield a lower solidus (see Chapter 3) and partial melting throughout martian history (Johnston *et al.*, 1974).

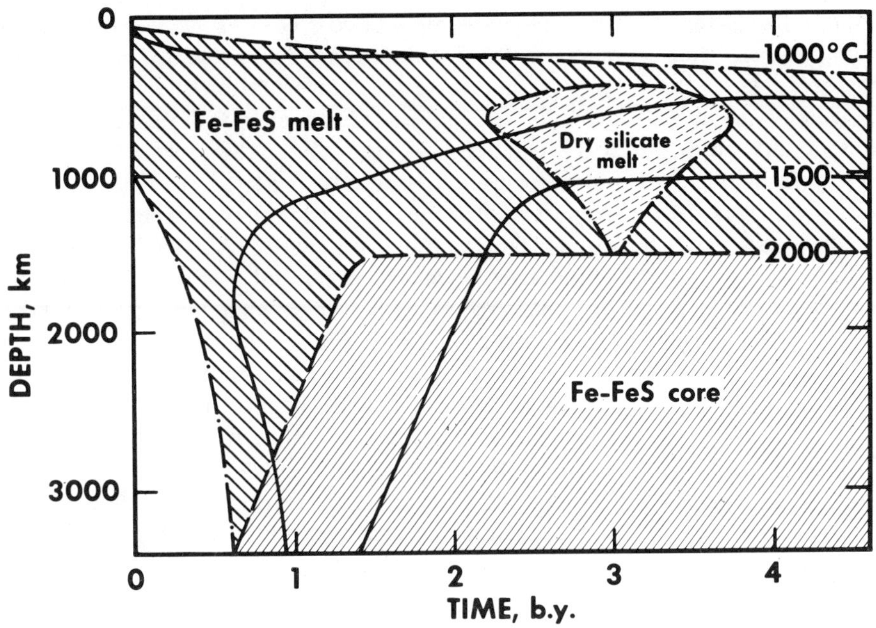

9.5.26 Evolution of isotherms and regions of melting as functions of time in the martian interior for the model of Fig. 9.5.25. Only the angle-averaged isotherms are shown.

9.5.4 Mercury

Before the flight of Mariner 10, which first encountered Mercury in March 1974, we knew Mercury's orbital parameters, spin rate, mass, and radius. It was also known that the planet's surface had electromagnetic reflection and thermal emission properties essentially like those of the Moon, which meant that the two surfaces would not be vastly different in composition and microstructure (Lyot, 1930; Harris, 1961; de Vaucouleurs, 1964; Morrison, 1970; McCord and Adams, 1972; Dollfus and Auriere, 1974; Cuzzi, 1974; among others). The Mariner 10 mission, in addition to refining values of known parameters and confirming the overall surface resemblance to the Moon, obtained photographs of about 40% of the planet to resolutions between 0.1 and 5 km, determined atmospheric constituents and put limits on the ionospheric density, discovered a steady dipolar planetary magnetic field, and measured some of the properties of the plasma environment.

Constraints

Structure. Mercury's high density implies that the planet must contain a large fraction of iron—60–70% by mass (Urey, 1952; Anderson and Kovach, 1967; Reynolds and Summers, 1969; Siegfried and Solomon, 1974). No other heavy element is sufficiently abundant to account for the density. Whether or not the iron resides primarily in a core cannot be determined by data in hand. Although J_2 was found from Mariner 10's orbit to be $(8 \pm 6) \times 10^{-5}$ (Esposito *et al.*, 1977), its value yields no direct information on the degree of central condensation because Mercury's spin is too slow to produce the observed flattening. That is, the measured J_2 must be dominated by a nonhydrostatic contribution (Solomon, 1976a) and thus the hydrostatic theory for calculating C/MR^2 from J_2 is inapplicable (see section 4.2). The presence or absence of a core could be established, in principle, by a seismometer network, or by a scheme due to Peale (1976), using measurements of the amplitude of physical librations and the obliquity (from surface instruments), and the gravitational harmonic coefficients J_2 and C_{22} (from an orbiter). There are no structural data to indicate the existence or thickness of a distinct low-density crust on Mercury.

Magnetic field. A mercurian magnetic field and magnetosphere were discovered by Mariner 10 (Ness *et al.*, 1974, 1975, 1976). The magnetosphere is similar to the Earth's in that it creates a bow shock and is bounded by a magnetopause, both of which are seen as well-defined discontinuities in the magnetic and plasma data. The field within the magnetosphere is that of a distorted

imately parallel to the planet's rotation axis, and with the same polarity sense as the Earth's present field. The field cannot be due to solar wind induction (Herbert *et al.*, 1976); it may be due to a dynamo (Ness *et al.*, 1975) or to remanence retained in the outer layers of the planet (Stephenson, 1976; Sharpe and Strangway, 1976). Mercury's slow rotation does not exclude a dynamo on theoretical grounds (Stevenson, 1975).

Volcanism. Mercury's surface, like that of the Moon, is largely the product of bombardment. It has been mapped in terms of physiographic characteristics, the three main terrain types being categorized as heavily-cratered terrain, intercrater plains, and smooth plains (Trask and Guest, 1975). No evidence for atmosphere-related erosion has been found in the Mariner 10 pictures.

There is evidence, not unequivocal, for volcanic activity on Mercury, as described in greater detail in section 5.5. Strom *et al.* (1975) and Trask and Strom (1976) have argued that morphology, distribution, albedo, and color indicate that the smooth plains have a volcanic origin. Wilhelms (1976), however, holds that the smooth plains have an origin similar to that of the lunar light plains, which are presumably basin ejecta. Schultz (1977) has described evidence for volcanism related to crater modification. A volcanic origin for the intercrater plains has also been proposed (Malin, 1976b; Strom, 1977; Cintala *et al.*, 1977).

Tectonism. Long (up to several hundred kilometers in length) and apparently globally distributed lobate scarps have been interpreted as compressional features by Strom *et al.* (1975), by Cordell and Strom (1977), and by Dzurisin (1978). The scarps were apparently produced from a time near the end of intense cratering (because many of them have been disrupted by impacts, but only by relatively fresh craters) until after the emplacement of the smooth plains (Strom, 1979). The formation of the lobate scarps apparently post-dated the emplacement of intercrater plains, because no instance of embayment of scarps by intercrater plains is observed (Strom, 1979). A simple model of thrust faulting indicates that contraction of the planet's radius by 1–2 km could account for these features (Strom *et al.*, 1975; Cordell, 1977). There is a notable absence of extensional features, except for two local regions associated with the Caloris Basin.

The scarps on Mercury have also been attributed to tidal despinning (Burns, 1976; Melosh and Dzurisin, 1978), on the basis of an apparently preferred orientation of near-equatorial scarps and of Mercury's tidally-evolved spin-orbit resonance (Peale and Gold, 1965; Colombo, 1965). The observational basis for preferred scarp orientations has been challenged by Cordell and Strom (1977), however, and the near-polar extensional features expected for a pure despinning mechanism (Melosh, 1977) are not observed. Global contraction is thus required even if despinning was a contributory source of stress.

Bulk composition and volatiles. Except for the inference on the Fe abundance from mean density noted earlier, there is no direct information on the bulk chemistry of the planet. Cosmochemical models (Lewis, 1972; Goettel, 1977) suggest that Mercury is enriched in refractories and strongly depleted in volatiles with respect to the other terrestrial planets. The equilibrium condensation model of Lewis (1972, 1973) predicts an abundance of Fe-Ni metal and a silicate component dominated by magnesian pyroxene. Several variants of this model are discussed in section 4.5.3.

Whole disc spectral reflectance measurements are consistent with a dominantly anorthositic surface with low FeO abundance (Adams and McCord, 1976; see also section 2.2.1).

Mariner 10 albedo measurements by Hapke *et al.* (1975) distinguish between inter-crater and smooth plains, but generally show less contrast than that between lunar highlands and maria. However, Schultz (1977) and Dzurisin (1977) have identified several patches of bright material, within craters, which have albedos exceeding the brightest lunar features. The nature and origin of these patches is unknown. Infrared measurements by Chase *et al.* (1974, 1976) yield lunar-like values of the surface thermal inertia and electrical skin-depth. Polarization measurements by Hapke *et al.* (1975) indicate the presence of a fine-grained regolith covering the planet, with no exposed rock surfaces discovered at 20 km resolution.

Helium and atomic hydrogen are the only unequivocally detected atmospheric constituents (Broadfoot *et al.*, 1976; Shemansky and Broadfoot, 1977). Upper limits on CO_2 and H_2O concentrations, when considered with likely loss mechanisms, imply lower than terrestrial-analog outgassing rates (Kumar, 1976). These results support the alternative possibilities either that Mercury is relatively deficient in volatiles or that it has been relatively inactive.

Chronology. A relative geologic history of Mercury was proposed by Murray *et al.* (1975) in which the heavily cratered terrain was formed by bombardment (terminal or second episode?) of an existing surface, some of which perhaps survived as intercrater plains. This bombardment was supposedly accompanied by contraction of the planet, the later stages of which produced the long scarps. The Caloris impact occurred near the end of the bombardment, and preceded the emplacement of the smooth plains, which are presumed to be volcanic. Some continued contraction postdated smooth plains formation. This history must be regarded as tentative, and revisions can be expected as analysis proceeds. For instance, Strom (1977) has since suggested that much of the intercrater plains may be younger than the heavily cratered terrain and may be associated with a period of volcanism predating smooth plains formation. Trask (1976) presents evidence that the two terrains were coeval to some extent, but still regards the intercrater plains as mainly predating the heavily cratered terrain. In any case, if the bombardment rate was similar to that in the lunar environment, the evidence indicates that the surface has undergone little modification in the last 2–3 b.y.

Thermal models

The implications of these observations for the planet's thermal history can be summarized as follows: (1) The lunar-like surface properties and the magnetic evidence for an iron-rich core require planet-wide differentiation. (2) The absence of extensional features and the relative timing of compressive faulting require that any planetary expansion was confined to times prior to the end of heavy bombardment (about 4 b.y. ago, if lunar and mercurian cratering histories are comparable). This means that core formation, which would cause an expansion of 15–20 km in radius (Solomon, 1976a), must have occurred before this time, perhaps even during planetary accretion. (3) Contraction of the planet continued after the heavy bombardment, but was limited to a radial decrease of a few kilometers, according to the estimate of Strom *et al.* (1975), based on the characteristics and extent of the global scarp system. Solomon (1977) has shown that this restriction would preclude complete core solidification, if the core was completely molten at the end of heavy bombardment.

Silicate and iron melting temperatures in Mercury rise rapidly with depth to temperatures greater than 2000° C. Therefore, early differentiation requires either that Mercury formed hot, or was heated rapidly after formation. For this to be caused by the common long-lived radioactive elements would require their abundances to be several times those of chondritic meteorites or those inferred for the Earth and Moon. Accretional energy could have been responsible for early heating if the energy deposited by impacting planetesimals was mainly buried beneath the surface of the growing planet (Safronov, 1972; Wetherill, 1976; Kaula, 1979a). The gravitational energy released by impacting objects may have been significantly augmented by the large kinetic

Table 9.5.5 Parameters used in Mercury thermal history models.

	Fig. 9.5.27	Fig. 9.5.29	Fig. 9.5.31
Radius, km	2439	2439	2439
Mean density, g/cm^3	5.43	5.43	5.43
Surface temperature, °C	107	77	27
Radioactive abundances			
U, ppb	38	44	3.2
K/U	0	0	0
Th/U	3.7	4	4.15
Specific heat, J/g-°C			
Silicates	1.2	1.2	1.2
Iron	0.6	-	-
Heat of fusion, J/g			
Silicates	400	400	400
Iron	270	240	-
Thermal conductivity, J/cm-s-°C			
Silicate	(*1*)	(*1*)	0.42
Iron	(*2*)	adiabat	-
Kinematic viscosity (mantle), cm^2/s	-	$(1.4 \times 10^9) \cdot \exp(25\, T_f/T)$	-

References:
(*1*) From Schatz and Simmons (1972). (*2*) From Powell *et al.* (1966).

energy associated with the velocity dispersion expected for planetesimals in Mercury's zone. It is also possible that Mercury was formed hot, if it accumulated from material which did not cool appreciably after condensing from a hot region of the solar nebula. Early heating mechanisms that have been proposed for other bodies may also apply: short-lived radioactive elements, such as ^{26}Al (Lee *et al.*, 1976), and electromagnetic heating (Sonett *et al.*, 1968). It is unlikely that tidal heating was significant (Burns, 1976).

The process of core formation itself was a substantial thermal event for Mercury, capable of causing a mean temperature rise of 600–700° C (Solomon, 1976a).

The energetic requirements of maintaining a liquid mercurian core, thermally convecting or otherwise, have been discussed by Gubbins (1977), Cassen (1977), and Solomon (1979). Briefly, a present day thermally convecting core apparently requires an energy source within the core in addition to latent heat of fusion and internal (thermal) energy. If solid convection has occurred in Mercury's mantle, heat sources in excess of those characteristic of meteorites, the Earth, or Moon are required in Mercury just to keep the core molten (Cassen *et al.*, 1976). These points are illustrated by the thermal history models considered below.

Stephenson (1976) has quantified the conditions under which remanent magnetization of iron-bearing rocks could account for the magnetic field measured by Mariner 10. Crustal remanence from an early external or internal source is capable of explaining the field (Stephenson, 1976; Sharpe and Strangway, 1976; Strangway, 1977), allowing the possibility of a presently solid core. However, if an early core dynamo is postulated, then according to the radius change constraint, the core would have had both to form and largely solidify before the end of heavy bombardment.

Most current thermal history models for Mercury involve extensive early heating and early global differentiation, including formation of a crust and a core. Figure 9.5.27 is an evolutionary model due to Solomon (1977) calculated under the assumption that solid-state convection in the mantle was unimportant. This assumption could be justified if, for instance, there was a systematic density increase with depth of several percent in the mantle, due to compositional stratification. Such a stratification would be necessary to offset the destabilizing thermal gradient; otherwise, mantle convection would be expected (Cassen *et al.*, 1976; Toksöz *et al.* 1978). The parameters for the model are listed in Table 9.5.5. The outer 900 km of the core remain liquid until the present; there has been a post-core-formation contraction of about 7 km (Fig. 9.5.28). The core

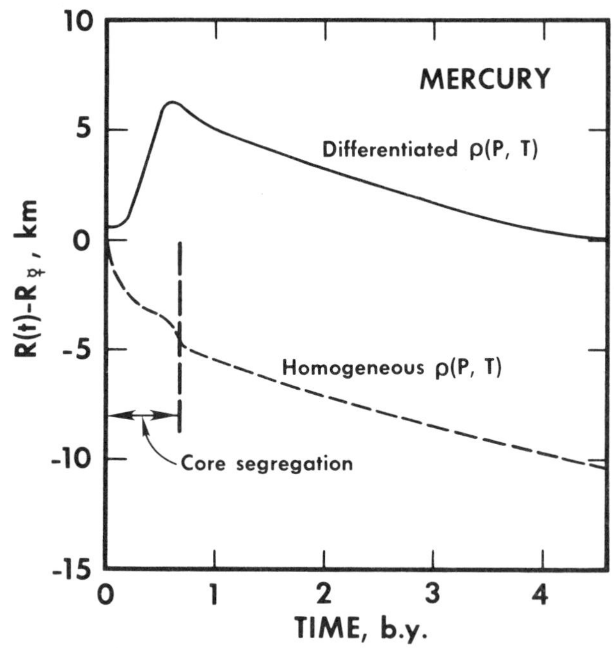

9.5.28 Radius of Mercury vs. time for the thermal model of Fig. 9.5.27 (from Solomon, 1977). Note that the planet has contracted since completion of global differentiation, with the contraction concentrated early in Mercury's history.

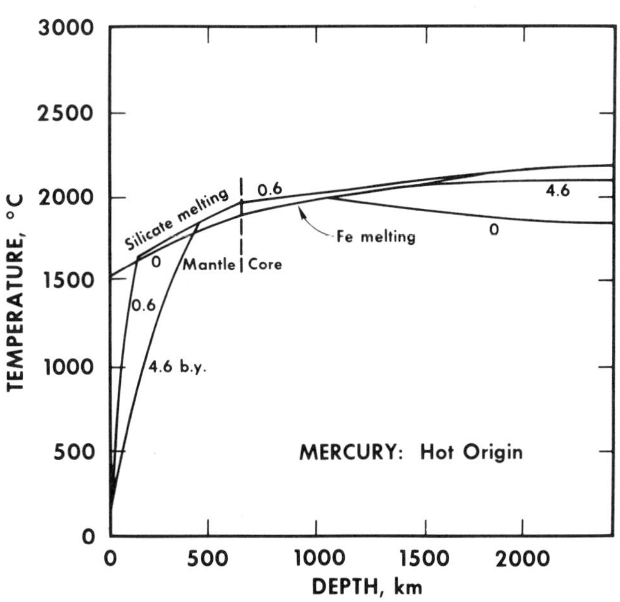

9.5.27 A thermal history model for Mercury (from Solomon, 1977). Core infall lasts from 0 to 0.6 b.y. after planetary formation. Heat sources are assumed to have been concentrated in an outer crust by the process of global differentiation. Note the partially solidified core at 4.6 b.y.

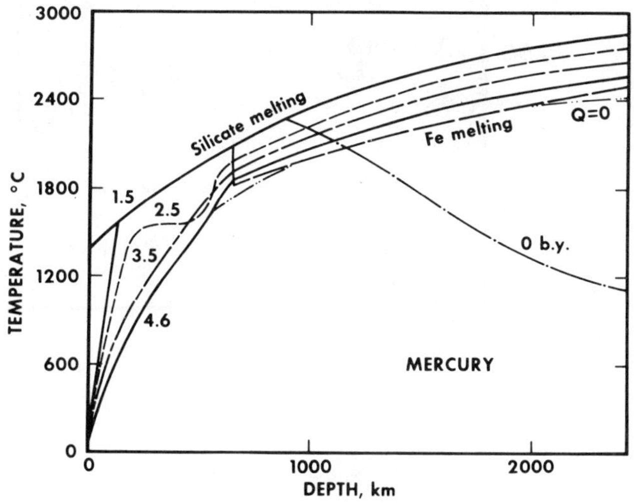

9.5.29 A thermal history model for Mercury including solid-state convection in the mantle and some radiogenic heat sources (U~2 ppb) in the core (from Toksöz *et al.*, 1978). The present-day profile if the core is without heat sources is the line marked Q = 0.

temperature gradient is sub-adiabatic (cf. Fricker *et al.*, 1976).

The (angle-averaged) temperatures for a model of Toksöz *et al.* (1978) incorporating solid-state convection is shown in Figs. 9.5.29 and 9.5.30; its parameters are also given in Table 9.5.5. Radioactive heat sources were concentrated in the mantle; some radioactivity (U~2 ppb present-day abundance) was placed in the core to prevent core solidification. A similar model without core heat sources yielded a completely solid core by the present. In these models, the heat flux from the core was constrained to be that corresponding to conduction along the core adiabat. If this condition were relaxed, solid convection in the mantle might extract heat from the core even faster, which would result in a somewhat cooler planet than is shown in Fig. 9.5.29. It is clear that additional energy sources are required to maintain a liquid core in the presence of mantle convection, if other parameters remain unchanged.

If the magnetic field of Mercury is due to crustal remanence, then evolutionary scenarios are possible that do not involve a core dynamo (Sharpe and Strangway, 1976; Strangway, 1977). One such model (Sharpe and Strangway, 1976), involving early extensive heating of only the outer portions of the planet and a primordially magnetized interior, is shown in Figs. 9.5.31 and 9.5.32. The interior remains below the Curie temperature in the model because of the very low abundances of radioactivity assumed (Table 9.5.5).

Although the observations and thermal history calculations given above define the range of thermal regimes for Mercury and some aspects of its evolution, more precise knowledge of the geophysical processes which have shaped the planet must await data on core-mantle structure, global heat flow, and the ages of major surface units.

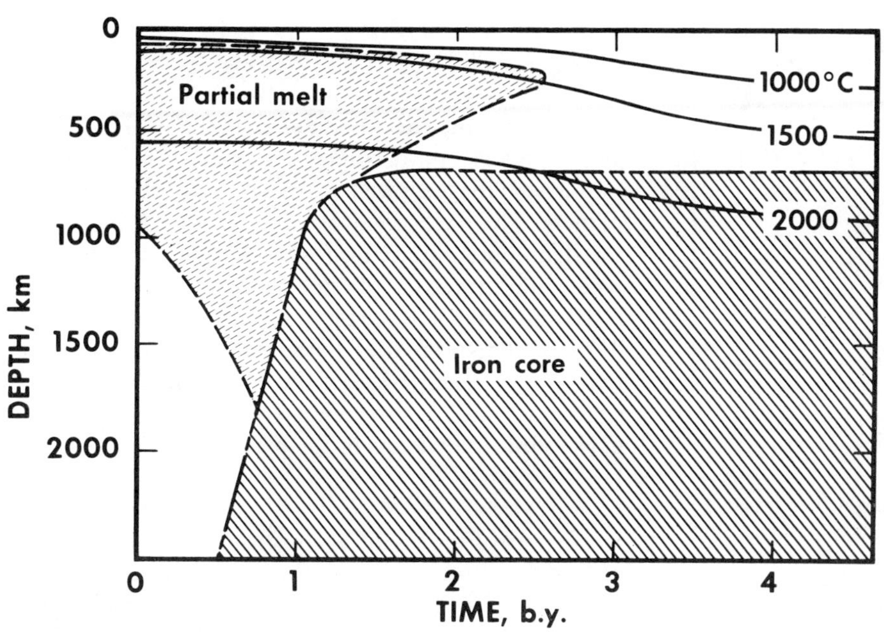

9.5.30 History of melting and core formation for the model of Fig. 9.5.29. Only the angle-averaged isotherms are shown.

Implications for basaltic volcanism

As noted above, both the smooth plains and the older intercrater plains on Mercury may be volcanic. The hypothesis of a volcanic origin for the intercrater plains (Malin, 1976b; Strom, 1977) is given independent support by the extensive mantle heating and the global extensional lithospheric stresses that would have accompanied early planetary differentiation. The association of a volcanic intercrater plains formation with global expansion requires that the formation of lobate scarps generally postdate emplacement of the intercrater plains, in agreement with observations (Strom, 1979).

The smooth plains were emplaced during the cooling and contraction stage of Mercury's history. The extensional stress conducive to eruption (Solomon, 1978a) may have been due to local sources of stress, especially impact basin formation and topography. The generally compressive global stress would have tended to shut off volcanism by over-riding the local extensional stress. Thus the duration of smooth plains volcanism on Mercury is predicted to be shorter than for the Moon and substantially shorter than for Mars. The Mercury chronology of Chapter 8 is consistent with this conclusion.

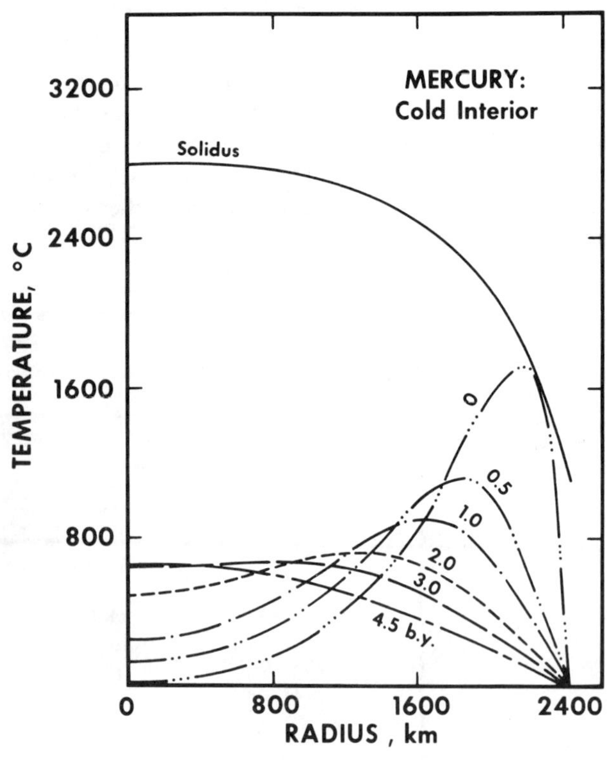

9.5.31 A thermal history model for Mercury involving an initially cold, primordially magnetized deep interior that always had very low radioactive abundances (U~3 ppb) and thus never warmed above the Curie temperature (from Sharpe and Strangway, 1976). In this model, Mercury never forms a core.

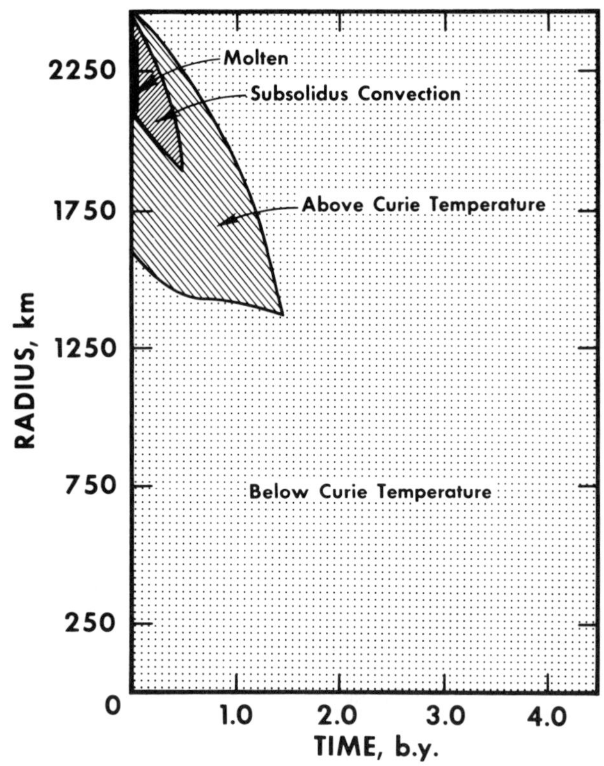

9.5.32 History of melting and the Curie temperature for the model of Fig. 9.5.31.

9.5.5 Venus

Despite the fact that Venus is the nearest planet and is also the most similar to the Earth in mass and radius, it is the least well characterized of the terrestrial planets. This lack of information is due primarily to the dense and cloudy atmosphere covering the planet which inhibits the use of remote sensing techniques, and also to the high surface temperature which makes very difficult the survival of landed instruments for more than a very brief period of time. Current and planned spacecraft missions to Venus are expected to fill large gaps in our knowledge of the planet.

Constraints

Many of the essential constraints on the evolution of Venus are not available. Review papers on the state of information available prior to the Pioneer Venus mission have been published by Hunten *et al.* (1977)

and Masursky *et al.* (1977). Preliminary results from the Pioneer Venus mission are discussed by Colin (1979a, b) and accompanying papers.

Structure. The mass, radius (and hence the mean density) and rotation rate of Venus are quite well known from Earth-based and space flight fly-by radio observations (Chapter 4). The slow rotation of the planet makes determination of the moment of inertia very difficult. The harmonic coefficients of the gravitational potential field may be dominated by near surface inhomogeneities and hence not give information on the global structure of Venus. Prior to Pioneer Venus, the second order harmonics were uncertain and the higher orders not known (Howard *et al.*, 1974). A low order gravity field will be derived from Pioneer Venus Orbiter tracking in the near future.

It is often presumed, by analogy with the Earth, that Venus should have a sizeable iron core, perhaps molten. Such an analogy is consistent with calculations from theoretical models, but is not yet required by existing data on Venus.

That Venus has differentiated a chemically distinct, low-density crust has been inferred from the offset between the equatorial centers of figure and mass (W. B. Smith *et al.*, 1970), from the isostatic compensation of large topographic features (Phillips *et al.*, 1979), and from the high radioactivity and low density at the Venera landing sites (section 2.2.2). Phillips *et al.* (1979) have determined from spacecraft gravity data that the depth of compensation of the large plateau feature at 68° N, 325° E on Venus must be less than 100 km. From the fit of a simple flexural model to the gravity and altimetry data obtained by Pioneer Venus, Phillips *et al.* (1980b) have estimated the mean thickness of a low density crust on Venus to be 100 ± 30 km if the lithosphere has no flexural rigidity and less than 100 km for a finite flexural rigidity.

Magnetic field. Magnetic field and plasma data from both Venera 4 (Dolginov *et al.*, 1969) and Mariner 5 (Bridge *et al.*, 1969) were interpreted by Russell (1976) to indicate the presence of a planetary dipole field with a strength of as much as 6.5×10^{22} Gauss-cm^3 (corresponding to a surface field of about 30 γ). The Venera 9 and 10 magnetometer experiments detected a magnetotail consistent with a moment of about 4×10^{22} Gauss-cm^3 (Dolginov *et al.*, 1977). If true, this result would support the concept of a liquid conducting core within Venus, since the volume fraction of Venus whose temperature is below the Curie temperature (and thus the volume capable of sustaining a remanent field) is very small. The rotation of Venus is sufficiently rapid as not to preclude the existence of a magnetohydrodynamic dynamo (Busse, 1976).

The Pioneer magnetometer, however, failed to detect a field of this magnitude, and early analysis indicates a limit of much less than 10^{22} Gauss-cm^3 for the present dipole magnetic moment (Russell *et al.*, 1979). The lack of an intrinsic field could imply a solid core or a liquid core in which any convective motions are insufficient to generate a magnetic field.

Volcanism. There is limited evidence for at least some volcanism in Venus' history. This evidence includes both radar data and the conclusion that the Venera 9 and 10 γ-ray (Surkov *et al.*, 1977a) and densitometer (Surkov *et al.*, 1977b) results are consistent with a basaltic surface composition (section 2.2.2).

Large volcanic features have been proposed as the cause of some regions of high radar reflectivity which appear upon the radar brightness maps of Venus. In particular, Saunders and Malin (1977) have interpreted the ground-based radar reflectivity and altimetry data for the radar bright "Beta" feature at 23° N and 84° W as indicating that the feature is a large shield volcano, 10 km high, with a summit caldera (section 5.7).

Preliminary results from the Pioneer Venus radar mapper experiment are consistent with this explanation (Pettengill *et al.*, 1979b), although somewhat lower relief is indicated for the Beta region. Two distinct shield-like structures are discernible, but the resolution is insufficient to verify the proposed summit caldera(s). The large radar-bright region Maxwell (65° N and 5° E) has also been interpreted as a volcanic construct (Campbell *et al.*, 1976), but the Pioneer data are inconclusive and more detailed information will be necessary to distinguish between volcanic and tectonic origins.

Tectonism. Earth-based radar has yielded reflectivity images and topographic height information at several to several tens of kilometer horizontal resolution over large portions of the Venus surface (Goldstein *et al.*, 1976, 1979; Campbell *et al.*, 1976, 1979). The images have been interpreted to suggest the presence of mountain belts and large rift valleys, as well as regions with numerous large circular features of a size distribution and geometry compatible with their being impact basins (Malin and Saunders, 1977; Schaber and Boyce, 1977).

Pettengill *et al.* (1979b) have described the large radar-dark area, located to the west of the Maxwell feature, as a large plateau more than 1000 km across and 3–5 km in elevation. The feature is morphologically complex, and Pettengill and co-workers conclude that it is more likely tectonic than volcanic in nature. The Pioneer imaging data show several dark circular objects resembling craters, about 250 km in diameter (Pettengill *et al.*, 1979b). The altimetry shows that these features are topographic depressions with relative depths of 500–700 meters. The shallowness is consistent with a mobile lithosphere. A rift valley mapped by the Pioneer altimetry has been described by Pettengill *et al.* (1979a).

Current interpretation of the Pioneer Venus gravity data indicates that there may be significantly more power (relative to the Earth) in the Venus gravity spectrum beyond the eighth spherical harmonic and that there is a strong gravity-topography correlation (Phillips *et al.*, 1979, 1980b). The gravity and topography data can be fit by a suite of models ranging from deep isostatic compensation and no flexural rigidity to shallow compensation and partial support of topographic variations by lithospheric strength; the maximum flexural rigidity is comparable to that for terrestrial continental shields (Phillips *et al.*, 1980b).

The high surface temperatures of Venus imply substantially higher creep rates for geological materials at crustal depths compared to Earth, if similar rheologies are assumed. On similar grounds the lithosphere on Venus might be expected to be thinner than for the Earth (Weertman, 1967; Mueller, 1969), though the high topographic relief ($\geq$ 12 km) on Venus discovered by Pioneer (Pettengill *et al.*, 1979a, b) casts such earlier conclusions in doubt. Anderson (1980) has argued that the high surface temperature expands the depth interval over which basalt is stable on Venus compared to the Earth, so that if the basalt-eclogite transition is an important element in making oceanic lithosphere on Earth unstable to subduction, a thick, hot basaltic lithosphere may be stable on Venus.

Whether there is convection in the mantle of Venus that now involves or has ever involved recycling of the lithosphere—as in plate tectonics—is an open question. If the interpretations of radar images that portions of Venus have numerous large impact craters and basins proves correct, then such areas clearly have escaped recycling for at least ~4 b.y. Weertman (1979) has estimated upper limits for the lifetimes of mountains, continental blocks and ocean-type basins to creep relaxation, ignoring sub-aerial erosion. He concluded that for appreciable elevation differences to persist over extended periods of time, the surface and near surface materials must be very dry.

Bulk composition and volatiles. Venus has a slightly smaller "uncompressed" mean density than the Earth (based upon estimates of the equation of state of planetary material). Barshay and Lewis (1976) ascribe the lower density to a lack of S in Venus due to higher temperatures of condensation and gas removal at Venus than at the Earth. Detailed models were constructed by Reynolds and Summers (1969) and by Ringwood and Anderson (1977) with the assumption of similar equation of states for Venus and Earth. Such an exercise leads to the suggestion of either less iron or a higher state of oxidation of the iron in Venus (section 4.5.4). Venus might also have a lower apparent uncompressed density than the Earth if a greater fraction of its basaltic component is near the surface within the thickened layer of basalt stability caused by the high surface temperature (Anderson, 1980).

The Venera and Mariner missions have obtained pressure and temperature profiles and estimates of the chemical composition of the Venus atmosphere (Hunten *et al.*, 1977; Hoffman *et al.*, 1977). While the surface values of 90 bar pressure and 730° K temperature are fairly well determined, as is the conclusion of a predominantly CO_2 atmosphere with about 3% N_2, the amounts of important minor constituents are still in dispute. The Pioneer mass spectrometer experimenters (Hoffman *et al.*, 1979) obtained a value of 0.85 for the $^{36}Ar/^{40}Ar$ ratio and an absolute Ar abundance some three times greater than the 19 ppm value reported by the gas chromatograph team (Oyama *et al.*, 1980). The Venera 11 and 12 results fall within this range of values (Istomin *et al.*, 1979; Gel'man *et al.*, 1979).

These argon measurements have considerable significance to the formation and subsequent evolution of the terrestrial planets (Pollack and Black, 1979). The ^{36}Ar abundance is some 20–200 times higher on Venus than on the Earth while the ^{40}Ar abundance may be somewhat lower than the terrestrial value. The assumption that planetary volatiles are more abundant as the sun-planet distance increases has been seriously called into question. Also, the hypothesis that Venus was initially formed with a low H_2O abundance has become less tenable.

The Pioneer gas chromatograph measured over 0.1% of water vapor and reported detection of O_2 and SO_2 in the lower atmosphere. The Pioneer Venus mass spectrometer team and the Venera experimenters report a more reduced atmosphere with COS and other sulfur gases, and report no detection of O_2. Detailed atmospheric modelling efforts directed toward resolving these discrepancies must also consider the composition and structure of the clouds detected by the entry probes. Lewis (1970) has pointed out the potential importance of atmospheric-surface interactions in influencing both the composition of the atmosphere and, in terms of weathering and modification of the surface material, of surface composition and morphology. Surface winds of the order of 0.5–1.0 m/sec as reported by Avduevsky

Table 9.5.6 Parameters used in Venus thermal history model.

Radioactive Abundances	
U, ppb	14
K/U	7×10^4
Th/U	4
Kinematic Viscosity (mantle)	
(10^8 cm^2/sec)	$0.73 \exp(25\, T_f/T)$
Other Constants	
$\rho = 4.36$ g/cm^3	$T_f(R_c) = 3850°$ K
$C = 1.2$ J/g-°K	$T_f(R) = 1600°$ K
$K = 0.042$ J/cm-sec-°K	$Ra_{cr} = 1100$
$\alpha = 3 \times 10^{-5}$ °K^{-1}	$\Delta T_{ad} = 470°$ K
$g = 875$ cm/sec^2	$\rho_c = 10.6$ g/cm^3
$T_s = 730°$ K	$C_c = 0.7$ J/g-°K

et al. (1976) for the Venera 9 and 10 sites could cause significant erosion and redistribution of surface materials (Iverson *et al.*, 1976). Near surface wind velocities measured by the Pioneer Venus probes are consistent with the Venera data (Counselman *et al.*, 1979).

Thermal models

Thermal history models of Venus are subject to the same major uncertainties as for models of the other planets, especially the abundances of radioactive heat sources, the initial conditions and the transfer of heat by solid-state convection. In addition, as noted above, there are even fewer constraints available to reduce the number of possible models. Fricker and Reynolds (1968) calculated thermal history models for Venus to investigate the effect of higher surface temperatures upon the temperature profile. The temperature profile was little affected below 200 km depth. Toksöz *et al.* (1978) calculated sets of thermal history models for the terrestrial planets, including Venus. Neither of those calculations included the effects of solid-state convection. If the Venus mantle has a high Rayleigh number, as in the Earth, it will be very difficult to perform adequate numerical treatments of global thermal convection.

A parameterized approach to thermal convection calculations for planetary interiors has been developed (Schubert *et al.*, 1979b) and is described in section 9.3.3. This method was used to derive a simple convection model for the Earth's thermal evolution that satisfies key observations about the present thermal state (section 9.5.1). These calculations have been extended to study models of Venus using values for the rheological and heat source parameters that provide reasonable model results for the Earth. These and other constants selected for Venus conditions are given in Table 9.5.6. Direct application of these modelling procedures, using parameters appropriate to Venus, give results very similar to those for the Earth (Figs. 9.5.33 and 9.5.34). It is clear that such models can be refined in a variety of ways, including more detailed models of the core composition and thermal structure.

Of special interest is the question of why, in spite of overall similarity to the Earth in other respects, Venus lacks a magnetic field (Russell *et al.*, 1979). In view of the strong dependence of core temperature on mantle rheology, it is possible that small compositional differences in either core or mantle could have resulted in a different thermal state for the Venus core. The slightly lower pressures within Venus (which would depress the mantle melting curve relative to that of the Earth, if the compositions were the same) and the fact that the Venus

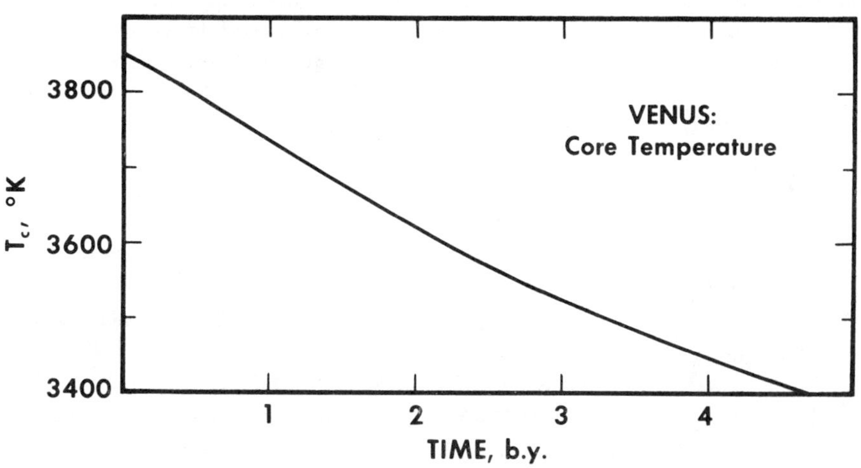

9.5.33 Core temperature, T_c, versus time for a convective model of the thermal history of Venus.

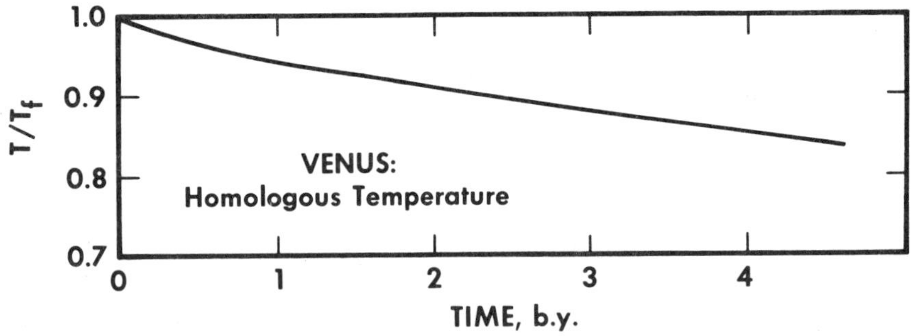

9.5.34 Homologous temperature T/T_f of the mantle, exclusive of upper and lower boundary layers, versus time for a convective model of the thermal history of Venus.

core is probably less massive than the Earth's would reduce the maximum amount of initial energy available for Venus and therefore favor earlier core freezing. On the other hand, the higher surface temperature would reduce the total buoyancy force available for driving convection, thereby reducing the Rayleigh number, and hence reducing the effectiveness of convection. But the most important factor might be the reduction of the core melting curve due to lower pressures. Stevenson *et al.* (1980) have calculated thermal histories by the methods described here, but including realistic core models. Their results indicate that, if Venus has core and mantle compositions like those of the Earth, then Venus does not yet have an inner core and hence does not have the energy released by inner core formation, which they propose is necessary for driving the Earth's dynamo.

Implications for basaltic volcanism

The lack of sufficient constraints on the thermal evolution of Venus precludes any strong conclusions about volcanism on that planet. Certainly some basaltic volcanism is to be expected on any planet that has differentiated a crust and, probably, a core. On the grounds of planetary size, surface temperature, and presumed internal heat sources, volcanism is also probable. The Venera measurements of U, Th and K and of density are consistent with basaltic chemistry at two surface sites. The type, amount and chronology of basaltic volcanism on Venus depend critically upon the extent of the interaction of the lithosphere with the mantle. The study of Venus, while less advanced than for the other terrestrial planets, is progressing rapidly and, ultimately, Venus may well prove to be the most significant among the terrestrial planets in providing information on the origin, structure and evolution of the Earth and its processes of basaltic volcanism.

9.5.6 Asteroids

Asteroids may be the parent bodies of most meteorites. Meteorite data have, therefore, been used as basis for developing interior models of asteroids (Mason, 1968; Wood, 1965, 1978, 1979) and to limit the size of asteroid parent bodies (Goles *et al.*, 1960; Wood, 1964; Goldstein and Ogilvie, 1965; Goldstein and Short, 1967; Buseck and Goldstein, 1969; Wood, 1979). These models involve melting and chemical differentiation of the asteroids. Recent telescopic observations of visible and near-infrared reflectance spectra (e.g., McCord *et al.*, 1970; Chapman, 1976) of asteroids have strengthened the asteroid-meteorite link and the case for melting and differentiation of some asteroids with attendant production of basaltic melts. It is, therefore, reasonable to constrain asteroid thermal evolution models with meteorite data, and to explore a scenario for asteroid evolution that includes very early heating, melting, chemical differentiation and the production of melts of different composition.

Constraints

Asteroid spectral reflectance. Observations of the spectral reflectance of asteroids at visible and near-infrared wavelengths have shown that the asteroids can be classified into distinct groups on the basis of their spectral curves (Bowell *et al.*, 1978) and that the spectra of many asteroids have strong similarities to the spectra of various meteorite classes, including carbonaceous and ordinary chondrites, irons, stony-irons and basaltic achondrites (Gaffey and McCord, 1977). This topic is discussed at greater length in section 2.2.7.

Only one asteroid has been demonstrated to have a spectral reflectance curve strikingly similar to that of basaltic achondrites: asteroid Vesta (McCord *et al.*, 1970; McFadden *et al.*, 1977). An apparently basaltic surface is indicated, and the argument has been advanced (Consolmagno and Drake, 1977) that Vesta is the specific parent body for certain basaltic achondrites, particularly the eucrites (section 1.2.8). Vesta, at 270 km radius, is one of the larger asteroids, but Ceres at 510 km radius has a spectral reflectance curve similar to undifferentiated carbonaceous chondrites (Chapman, 1976; Gaffey and McCord, 1977). Thus asteroid size is not a sufficient condition for internal differentiation and basaltic volcanism.

Meteorite chronologies. Meteorites constrain the time and duration of major melting events, cooling rates, and petrogenesis within their parent bodies. Chronological studies of meteorites require very early melting and crystallization of the parent bodies. Melting occurred about 4.54 ± 0.02 b.y. ago. The maximum time interval for accretion, heating, melting and crystallization is about 100–200 m.y., with a more probable interval of 30 ± 20 m.y. (see section 7.2).

The heat source for very early melting in asteroids is unresolved. Such heat sources as long-lived radionuclides (K, Th, U), accretional energy, adiabatic compression and core formation energy that are important in larger planetary objects are insignificant in the early heating of asteroid-sized objects. The most likely candiates for early heating are short-lived radionuclides, now extinct, and electromagnetic induction heating. Of several possible radionuclides (Table 9.4.3), ^{26}Al is the leading candidate. Evidence for ^{26}Al in Ca-Al rich inclusions in the Allende meteorite in sufficient concentration to melt objects as small as a few kilometers has been reported (Lee *et al.*, 1976). However, whether enough ^{26}Al was still present at the time asteroids formed so as to produce melting is unknown. Although ^{244}Pu and ^{129}I are known to have existed in the early solar system, their concentrations do not appear to have been sufficient to produce melting (Fish *et al.*, 1960).

Electromagnetic induction heating has been proposed by Sonett *et al.* (1968) and depends on the existence of a T-Tauri phase for the early sun (high spin rate, large magnetic field and large outflow of plasma); see section 9.4.1. Lack of knowledge of early solar conditions, even if the sun experienced a T-Tauri phase, make it impossible to evaluate the likelihood of significant induction heating.

Later melting has occurred on some meteorite parent bodies. The most prominent type of melting event is associated with collisional impacts that affect only the near surface of the asteroids and/or produce very little melting. Although impact generated melts and breccias are quite common among the meteorites (almost all achondrites and mesosiderites are brecciated), the effect of impacts as a heat source on global thermal evolution is probably minor. A second type of melting, postdating the very early melting and crystallization, may be represented by the nakhlite, shergottite, and chassignite meteorite classes (section 1.2.8). Isotope systematics indicate that the nakhlites crystallized from melt about 1.3 b.y. ago (see section 7.2). The lack of substantial evidence for shock (Bunch and Reid, 1975) suggests that the nakhlites were produced by internal heating at about that time.

Post-crystallization cooling rates in meteorites. Post-crystallization cooling rates have been used to estimate sizes of meteorite parent bodies and to constrain the accretional history of asteroids. Cooling rates can be estimated by two techniques. The metallographic technique is based on the composition and proportions of two nickel-iron phases, kamacite and taenite (Wood, 1964; Goldstein and Ogilvie, 1965). Metallographic cooling rates correspond to the temperature interval 400–600° C. The second technique is based on the time to cool from the crystallization temperature to the temperature at which Sr and Ar are retained (the isotopic closure temperature). Crystallization and gas retention times are datable so that the cooling time for this interval can be determined. Cooling rates determined in this manner correspond to temperatures above the isotopic closure temperature. Although closure temperatures are uncertain, it has been argued that they are $\geq$500° C (Wood, 1979). If this is true, then metallographic and isotopic closure cooling rates correspond to different temperature regimes. This and the fact that metallographic cooling rates are lower by one to two orders of magnitude than isotopic closure rates has led to consideration of asteroid models that have initial fast cooling rates that decrease significantly as the asteroid cools below 500° C (Wood, 1978, 1979).

Thermal conductivity. Several lines of evidence suggest that the thermal conductivity of asteroids after accretion should be significantly lower than that appropriate for solid silicate rocks. The porosity of carbonaceous chondrites is generally observed to be high (Ringwood, 1961); measured chondrite porosities average about 11% (Wasson, 1974). Accretion of asteroids must have involved very low collisional velocities, otherwise they could not have accreted (Kaula and Bigeleisen, 1975; Matsui and Mizutani, 1977). Low impact velocities would favor brecciation over melting and sintering. Central pressures of even the largest existing asteroid (at most a few kilobars) should have produced relatively little compaction and reduction in porosity.

Given that asteroids accreted with significant porosity, their initial thermal conductivity would have been considerably lower than that of solid rock. Lunar basalts and breccias with porosity between 15–23% under vacuum conditions have conductivities from 1.8×10^4 to 5.5×10^4 erg/cm-°C-sec at 200° C (Horai and Winkler, 1976; Fujii and Osako, 1973). Lunar fines under vacuum conditions have conductivities as low as about 3×10^2 erg/cm-°C-sec (Cremers *et al.*, 1970). Conductivity will increase with temperature because of sintering and thermally activated compaction. Once near-melting temperatures have been attained, conductivity values appropriate for solid rock can be used.

Low, porosity-dominated conductivity will be most important for long-lived radioactive heating and in the early ^{26}Al heating of asteroids that grow by continuously accreting high porosity material in which ^{26}Al concentration is declining. Long-lived radioactive heating results in a significant gradient of temperature with depth. Consequently, the outer portion of an asteroid will form a low conductivity (little sintering due to low temperature) blanket that allows the inner portion to heat up. The same effect results if asteroids accrete during the decay of ^{26}Al (see next section). Low initial conductivity has little effect on models that accrete the entire asteroid in a few half-lives of ^{26}Al. In such models, ^{26}Al heating overwhelms diffusive heat loss, even with rock thermal conductivities.

Thermal models

Numerous calculations of various aspects of the thermal evolution of asteroids exist in the literature. Models designed to relate sub-solidus cooling rates observed in meteorites to asteroid size have been based on an assumed initial temperature near melting (Allan and Jacobs, 1956; Wood, 1964; Goldstein and Short, 1967; Fricker *et al.*, 1970; Hsui and Toksöz, 1977; Toksöz *et al.*, 1978). Models based on induction heating have been used to represent the thermal history up to melting of a substantial portion of the asteroid (Sonett *et al.*, 1968, 1970; Herbert and Sonett, 1978). The most detailed computations using ^{26}Al as the early intense heat source begin at post-accretion temperatures, follow the temperature evolution through melting to solidification, and cool down to below the temperature at which nickel can diffuse in iron (~400° C). These models include core formation (iron phase separation), partitioning of radioactive elements into the silicate melt phase, melt movement, and crystal settling (Minear *et al.*, 1979).

Figure 9.5.35 shows the results of this type of model for an asteroid 100 km in radius. Radioactive abundances and other physical parameters are given in Table 9.5.7. Enough ^{26}Al (250 ppb) was assumed to produce total melting of the asteroid. This value corresponds to an initial ^{26}Al/^{27}Al ratio of 2.5×10^{-5} for an Al abundance of 0.01 by weight (Mason, 1971), or about 0.4 times the initial ^{26}Al/^{27}Al ratio inferred by Lee *et al.* (1976) for a Ca-Al rich inclusion in Allende. This concentration of ^{26}Al produces synchronous melting throughout the asteroid at about 0.6 m.y., with only a thin conduction boundary layer remaining unmelted. Electromagnetic induction heating (transverse magnetic excitation) also produces melting of the entire asteroid, except for a thin boundary layer, in about 0.5

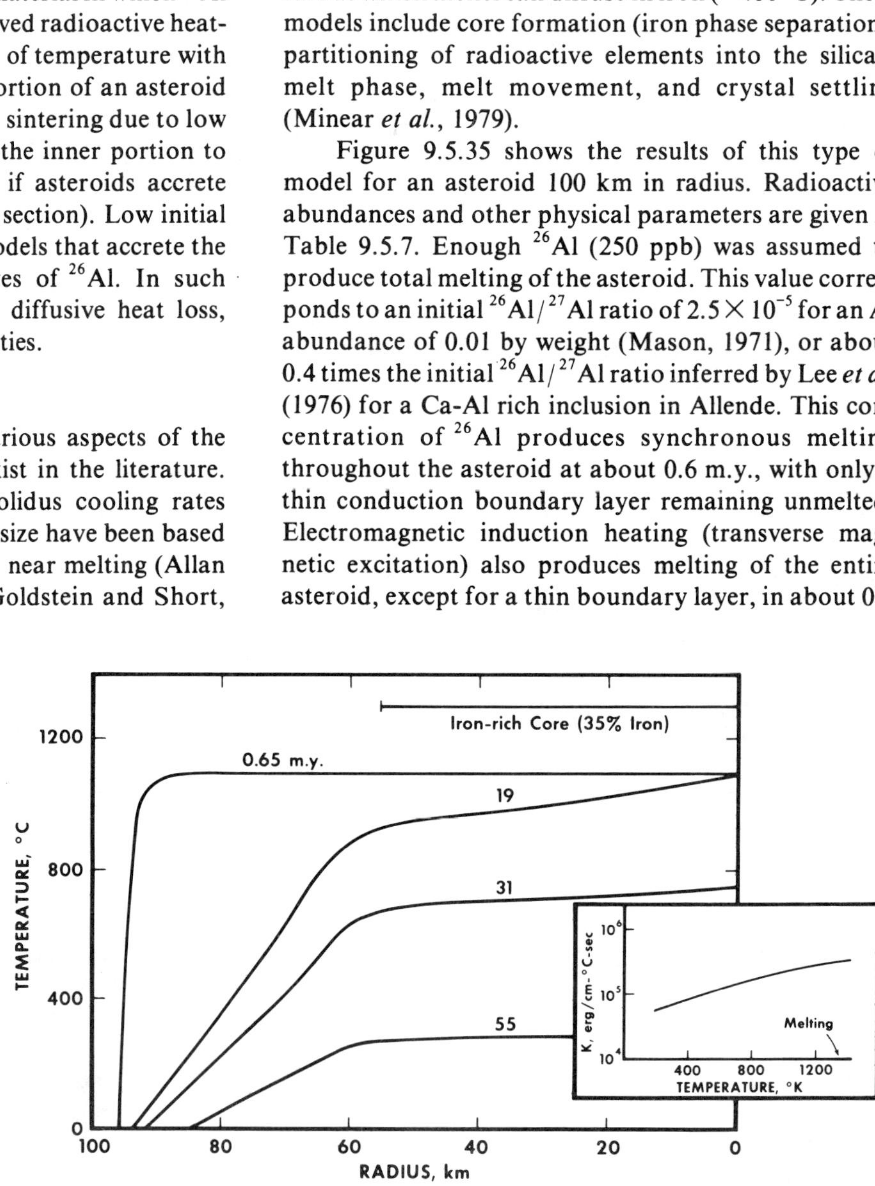

9.5.35 Temperature profiles in an asteroid heated by decay of ^{26}Al (from Minear *et al.*, 1979). Numbers on the curves indicate time after formation in 10^6 years. The initial radius of the asteroid is 100 km. Core formation is assumed to occur after a specified degree of melting (melt assumed to be of initial asteroid composition) throughout the asteroid. Core formation is modelled by assuming that the iron melt fraction from the entire asteroid volume replaces the volume occupied by melt and the core volume in the central region of the asteroid. Radioactive elements are assumed to be concentrated in the early melting silicate phase and to be removed from the core region when iron melt displaces silicate melt. Thus, on core formation, melting ceases in the core region. The core is then composed of an iron fraction and a matrix of initial asteroid composition. An initial bulk porosity of 10%, core formation at 25% melting (by volume), and an initial composition of about 13% iron were used for the temperature profiles shown. With these conditions, the core region is 35% (by volume) metallic iron, 65% iron-magnesium silicate and has a radius of about 55 km. Core formation occurred at 0.65 m.y. Thermally induced collapse of the asteroid pore structure results in a radius decrease of about 3 km. The temperature-dependent thermal conductivity used in the calculations is shown in the inset.

m.y. (Sonett, 1971; Herbert and Sonett, 1978). Cooling occurs rapidly because of convective heat transport and crystal settling (although heat loss and crystallization occur at the base of the conduction boundary layer, crystals are assumed to settle so that solidification occurs from the core outward). The asteroid therefore solidifies in about 20 m.y., a time within the interval allowed by the early chronology of meteorites.

For a given concentration of ^{26}Al, the time to melting in models of the type in Fig. 9.5.35 is essentially the same for asteroids of all sizes in which melting occurs because the intense heat source overwhelms diffusive heat loss. ^{26}Al concentrations of ≥ 250 ppb can produce total melting (except for a thin conduction boundary layer) of asteroids larger than about 10 km. Concentrations less than about 100 ppb are insufficient to produce melting in any size asteroid.

The iron-rich core in models of the type in Fig. 9.5.35 cools uniformly (Fig. 9.5.36) because it is a high conductivity (iron content) region surrounded by low conductivity material. Cooling rates in the silicate region surrounding the iron-rich core region vary by about two orders of magnitude depending on the depth. Cooling rates agree well with those reported for less detailed models (e.g., Wood, 1964; Fricker *et al.*, 1970). Metallographic cooling rates assembled by Wood (1979) are also plotted in Fig. 9.5.36. Except for the mesosiderites, cooling rates are consistent with meteorite parent bodies less than about 500 km in radius.

Table 9.5.7 Parameters used in asteroid thermal history model.

Radioactive Abundances	
U, ppb	12.5
K/U	7×10^4
Th/U	3.7
^{26}Al (initial abundance), ppb	250
Other Constants	
Initial temperature	180° K
Surface temperature	180° K
Initial radius, km	100 km
Initial density, g/cm^3	2.8
Specific heat, J/g-°C	1.2
Heat of fusion, J/g	
Iron	270
Silicate	660
Thermal conductivity	Fig. 9.5.35, inset

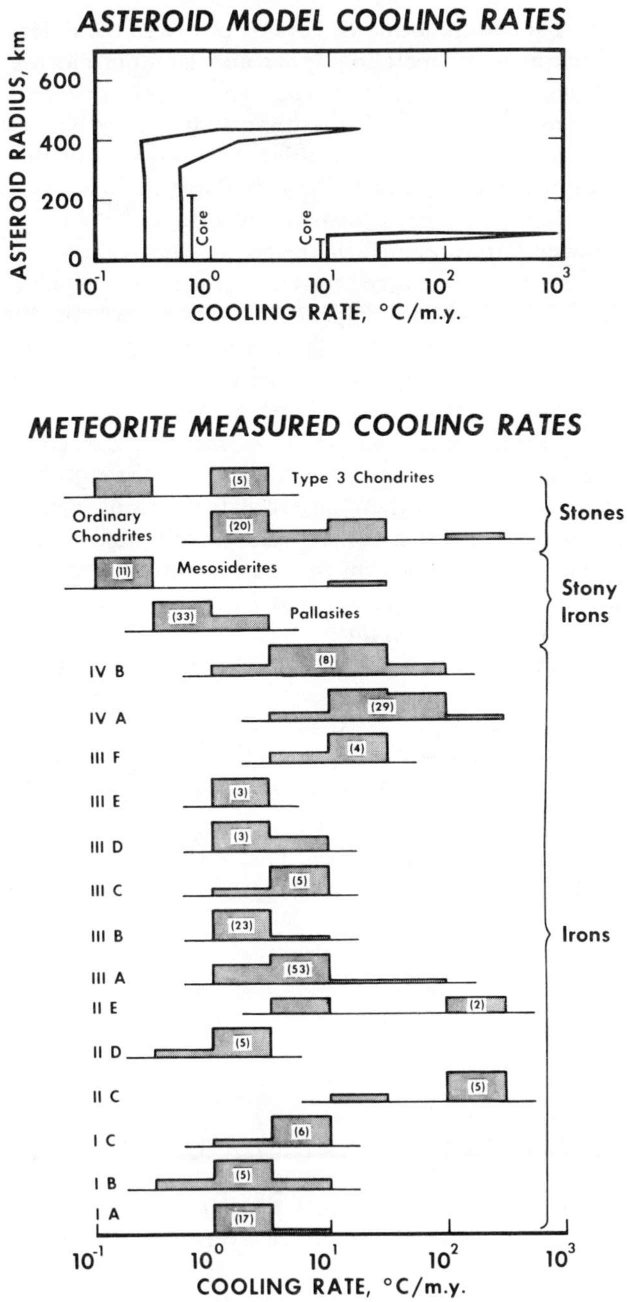

9.5.36 Cooling rates for meteorites and model asteroids. (Bottom) Histograms of meteorite metallographic cooling rates, from data compiled by Wood (1979). Histograms for each class of meteorite are normalized to the total number of measurements (indicated by the number in parentheses) reported for the class. (Top) Cooling rates within asteroids for the model of Fig. 9.5.35 and for a model with a 500 km initial radius (before collapse) but with otherwise identical physical parameters. The rates shown are for the time at each depth that the temperature passes through 700°C (upper bound) and 300°C (lower bound). Core regions formed by the sinking and aggregation of iron melt are indicated on the figure.

The low mesosiderite cooling rates and the differences between metallographic and isotopic closure cooling rates has led to the consideration of a different type of asteroid model from that shown in Fig. 9.5.35 (Wood, 1978, 1979). The primary constraint on this type of model is an early rapid cooling rate that slows dramatically when a temperature of about 500°C is reached. To accomplish this behavior Wood has modelled the meteorite parent body as a growing dust sphere heated by ^{26}Al. As temperature increases within the sphere, sintering is assumed to increase the conductivity of the dust, initially of very low conductivity. Without the growing low-conductivity dust blanket, the initial dust shpere would rapidly heat up, melt and then cool. If dust continues to accrete after the ^{26}Al has substantially decayed, the new dust will form an insulating blanket. Cooling rates that were initially high would decrease. Starting with nuclei of 5 km radius (Goldreich and Ward, 1973), melting and slow cooling through 500°C with accretion rates >1 cm/yr lasting a few million years can be attained (Wood, 1979). An interesting aspect of this model is that the lowest cooling rates observed in meteorites can be achieved in small parent bodies (~20 km radius). Very low cooling rates can be produced at shallow depth because of the thermal blanketing effect of the accreting dust layer. Planetesimal collision, rather than dust accretion, may be the mode of growth of the proto-asteroids, however (Goldreich and Ward, 1973).

It has generally been held that chondritic concentrations of U, Th and K were insufficient to produce melting in planetary bodies less than about 1000 km in radius. Calculations leading to this concept have been based on thermal conductivities appropriate for solid rock (e.g., Toksöz *et al.*, 1978). However, as noted above, the thermal conductivity of asteroids may be initially quite low. Even though sintering and compaction tend to increase conductivity in the interior of an asteroid as it warms up, surface layers of impact brecciated and dusty accretional material may provide an effective thermal blanket. Under the assumption of conductivity values that are initially equivalent to lunar breccia values (Horai and Winkler, 1976; Fujii and Osako, 1973), but that increase to rock values at the melting temperature (Fig. 9.5.35), chondritic radionuclide concentrations ($^{40}K = 100$ ppb, Th = 44 ppb, U = 12.4 ppb) can produce melting in bodies as small as about 400 km in radius. Melting in a 400 km body begins after about 1.2 b.y. and lasts for about 1.8 b.y. Melting is confined to the deep interior (about 10% by volume). Thus, it seems possible that melts could have persisted in the larger asteroids until as recently as about 1.5 b.y. ago. Such melts are sufficiently young to provide a parent melt for the nakhlites and other achondrites with apparently young crystallization ages (section 7.2).

An interesting aspect of the induction heating mechanism is that a relative heating maximum with respect to body radius exists (Herbert and Sonett, 1978), favoring heating in the radius range 25–100 km. This characteristic, together with the decrease in heating with heliocentric distance, favors the heating of intermediate-sized objects closest to the Sun. This heating bias seems consistent with the concentration of S type asteroids (assumed to be stony-iron cores of differentiated asteroids) near the inner edge of the asteroid belt as well as the observation that the largest asteroid Ceres (2.76 AU, 510 km radius) has an apparently undifferentiated surface while the smaller asteroid Vesta (2.36 AU, 270 km radius) appears to be differentiated. However, early solar characteristics, the timing of asteroid accretion and even the early heliocentric distance of present asteroids are quite uncertain.

Implications for basaltic volcanism

Thermal models for asteroids and/or meteorite parent bodies have direct implications for the character of basaltic volcanism. Early intense heat sources—either ^{26}Al or induction heating—should affect almost the entire body except for a very thin boundary layer. The expected high porosity and thin, cool boundary layer should therefore lead to efficient outgassing, with possible cold-trapping of volatiles in a very thin (a few kilometers thick) surface layer.

An important consequence of early heating by ^{26}Al is that this heat source is concentrated in the early melt phase. The potential mobility of the melt phase can thus provide an effective mechanism for redistributing the heat source, limiting melting and affecting petrogenic processes. Petrologic and geochemical arguments (Gast and Hubbard, 1970; Ringwood, 1970b; Stolper, 1977) have been advanced that certain sequences of achondrites were derived from liquids produced by increasing but small degrees (~4–20%) of melting of the source region. Most eucritic parent liquids are also the products of about 20% melting of the source region (Stolper, 1977); see section 3.5.

^{26}Al will be concentrated in the first silicate melt because Al resides principally in plagioclase which is one of the first melting phases. The amount of plagioclase present in primitive material that has been melted to produce eucrite melts has been estimated at about 5% (Consolmagno and Drake, 1977; Dreibus *et al.*, 1976). Normative mineralogies of devolatilized carbonaceous chondrites contain about 6% plagioclase. Rare earth patterns for eucrites that represent 10–15% partial melt indicate that all the plagioclase is in the melt (see section 7.2). A substantial amount of the pyroxene should also be in the melt. Essentially all of the Al is therefore in the melt phase after 10–15% partial melting.

A melt-mobile heat source provides the general mechanism for producing the variable but limited degree of partial melting supported by meteorite data. Although details of the mechanism are complicated because many factors affect the migration process, some general statements can be made. Consider melt migration to occur by fluid flow in a porous medium in which less dense melt is squeezed out as a denser solid matrix deforms (Turcotte and Ahern, 1978). The velocity of melt movement is given by Darcy's law

$$v = \frac{k}{\eta} \Delta\rho g \qquad (9.5.23)$$

where k is permeability, η is viscosity, $\Delta\rho$ is the density difference between melt and solid, and g is gravitational acceleration. From Eq. (9.5.23) it is seen that permeability, density contrast and melt viscosity influence melt velocity and hence the degree of partial melting. If melt movement is fast compared to heat generation, then melt migration is an effective way of limiting melting. Consequently, a greater density contrast, an increased permeability and a decreased melt viscosity tend to decrease the total degree of partial melting.

A basic assumption for Eq. (9.5.23) is that the solid matrix can deform to squeeze out melt. If deformation is slow compared to heat generation, then melt migration is not effective in limiting melting. Deformation rates are proportional to low powers of the shear stress (section 9.4.3). Larger bodies may have generally larger internal shear stresses and, thus, larger deformation rates. Hence, the degree of partial melting may tend to be less in larger bodies, according to this argument.

Incomplete melting does not imply that iron-nickel or iron-sulfide cores cannot be formed. Dense metal-alloy melts can form pure metal-alloy cores by displacing less dense silicate matrix. The only criterion is that the silicate matrix deform in a time short compared to solidification of the core material.

The self-limiting melt generation mechanism is a significant difference between heating by ^{26}Al decay and electrical induction heating. Geochemical and petrologic studies that bear on the degree of melting in meteorite parent bodies may provide data for evaluating the importance of these two mechanisms as early heat sources in meteorite parent bodies and in the planets.

9.6 OVERVIEW AND AFTERTHOUGHTS

It is clear from the planet-by-planet discussion of thermal history and its consequences for basaltic evolution that several important aspects of evolution are shared by the terrestrial planets. Other substantial processes such as crustal differentiation and recycling of surface material into the interior, however, have operated to very different degrees on the different bodies. The unique history of basaltic volcanism on each planet is the direct product of this different balance between thermal history events common to all planets and the variable relative importance of major global processes of heat and mass transfer. In this section, several of these thermal history events and processes that are planet-general are discussed in more detail. A broader synthesis of planetary evolution and basaltic volcanism is given in Chapter 10.

9.6.1 Early melting in the planets

There is abundant evidence for early heating and melting in each of the terrestrial planets and in some asteroid/meteorite parent bodies. In the Earth and probably in Mercury, early heating led to core formation prior to ~4 b.y. ago. In the Moon and in Mars, early heating led to formation of a thick, low-density crust within the first few hundred million years. Within the basaltic achondritic parent bodies, early melting was required to account for meteorite radiometric ages.

An important aspect of this early heating was that some of the terrestrial planets were heated to a greater volumetric extent than others. Both the Earth and Mercury were likely heated to near melting conditions essentially throughout their interiors at an early stage; probably the energy released by core formation in each planet played a significant role. For the Moon and Mars, however, early heating and melting appears to have been confined to the outer few hundred kilometers of each body, in order that the predicted lithospheric thermal stress be in accord with the observed surface tectonic history subsequent to termination of heavy bombardment. The extent of early heating on the parent bodies of basaltic achondrites is not constrained; the degree of partial melting, however, was relatively modest.

The energy for early melting could have been supplied by any or all of the several heat sources discussed in section 9.3.1 excepting long-lived radioactivity. The variation of the volumetric extent of melting among the terrestrial planets, however, places important constraints on the relative contributions of these heat sources.

The strong indication that the Moon and Mars were not initially heated throughout argues that a volumetrically uniform heat source can be excluded. This conclusion rules out ^{26}Al as an important energy source for these bodies, unless Al were somehow preferentially concentrated toward their exterior during formation, at best a remote possibility since aluminous phases are among the first to condense from a cooling solar nebula.

There is some correlation of planetary size with extent of early heating, suggesting that accretional heating was an important contributor to early melting in the larger planets. Mars, although larger than Mercury, was not as extensively heated, however, so that size alone is not the sole regulator of early history. As noted in section 9.5.6, accretional heating cannot have provided the energy to differentiate asteroids.

The volumetric extent of early heating may be a function of solar distance as well as of planetary mass. Such a relationship is supported by the decrease in extent of early heating in the sequence Mercury, Moon, Mars (and perhaps asteroids); the positions of Earth and Venus in this sequence, however, are not well constrained. An energy source dependent on solar distance, such as perhaps electromagnetic induction heating or accretional heating with total accretion time increasing with solar distance, would be consistent with these observations.

9.6.2 The role of thermal convection and plate tectonics in the planets

Applications of the theory of the stability of viscous fluids to the interiors of the terrestrial planets indicate that all of them are unstable to thermal convection if the creep viscosities of their mantles are within an order of magnitude or two of that inferred for the Earth's mantle, about 10^{22} poise. Unfortunately, the thick lithospheres of the smaller planets prevent direct confirmation of this theoretical prediction. Not only do the lithospheres of these bodies remain essentially intact, precluding earth-like plate tectonics, but they can also support substantial surface loads that mask the effects of stresses imposed by convection. Venus might not pose such difficulties, in which case observations of its surface would be very helpful in generalizing our ideas about planetary convection. But for information on the dynamic processes within the Moon, Mercury, and Mars, we must continue to rely on less direct methods, such as the modelling of the effects of convection on thermal state, magma generation, and local gravity and topography.

The implication of convection for thermal evolution is that heat transport rates can greatly exceed those due to conduction and radiation alone. It also means that the thermal states and evolutions of the planets depend strongly on their rheological properties. Tozer (1967) pointed out that there will be strong tendency for the interiors of convecting planets to tend toward a state of constant viscosity. The interplay of this tendency with other factors that affect the buoyancy and viscous forces, such as compressibility, stress level, dislocation density, and chemical stratification, has yet to be explored. These facts, together with the inhomogeneous (and perhaps inherently unsteady) nature of viscous convection preclude rigorous and detailed predictions of thermal structure, except where extensive observations can be used to justify simplifications of the theory. Nevertheless, some important gross properties of convective flows are often remarkably insensitive to the complicated details. The total heat flow through a system has this property in some circumstances, and this fact is exploited to make calculations of planetary thermal evolution.

The physical processes that drive or retard convection can also be analyzed in detail on a scale smaller than that of the entire convecting system. Such analyses have played a major role in our understanding of plate tectonics and the Earth's thermal balance. An example of this is the concept of the terrestrial plates as a thermal and rheological boundary layer.

Thermal convection in a very viscous fluid at high Rayleigh numbers results in a boundary layer structure. For a fluid which is heated from within and cooled from above, a situation applicable to the terrestrial planets, a thermal boundary layer develops adjacent to the upper boundary. The thermal structure of this boundary layer explains the thermal and mechanical structure of the lithospheric plates on the Earth. In the boundary layer theory the cold boundary layer becomes gravitationally unstable, and forms a cold descending plume which helps drive the thermal convection. This process is directly analogous to the subduction of the lithosphere at ocean trenches and the gravitational body force on the descending plate.

However, it should be noted that the mechanical rigidity of the lithosphere implies important differences between convection in a viscous fluid and that in planetary interiors. The elastic behavior of the lithosphere allows it to act as a stress guide, permitting the transmission of elastic stresses through the plate. More important, however, is the influence of the mechanical rigidity of the surface plates in terms of inhibiting their thermal instability.

In the smaller terrestrial planets, Mars, Mercury, and the Moon, this mechanical rigidity together with the positive buoyancy of a thick, low-density crust has prevented the breakup of the lithospheric shells, the initiation of subduction, and subsequent plate tectonics.

The presence of the outer rigid shells need not, however, suppress thermal convection at depth. Thermal convection can occur in these bodies with a thermal boundary layer formed in the flow beneath the rigid outer shell. Convection will be the dominant mechanism for heat transport within the planetary interior but heat transport will be by conduction through the rigid outer shell.

An important and unresolved question is the scale(s) of mantle convection within the Earth. Since the mechanical rigidity of the lithosphere suppresses surface instabilities, i.e., subduction, the horizontal extent of the surface plates may be much larger than the natural horizontal dimensions of convection cells within planetary interiors. In this case, secondary convection cells beneath the surface plates might be expected. These secondary cells might be responsible for intraplate volcanism, but there is not yet any direct evidence for their existence in the Earth.

9.6.3 Degree of global differentiation and recycling

The process of crustal differentiation has acted to a varying level of efficiency on the terrestrial planets. Known crustal volumes range from a high of 10% on the Moon to a low of half of 1% on the Earth. Such bookkeeping strictly by volume, however, can be misleading; the crust of the Earth is more differentiated chemically, for instance, than that of the Moon. As noted above, the formation of a low density crust acts to impede the recycling of lithospheric material into the underlying mantle that we associate with plate tectonics on the Earth, because the intrinsic buoyancy of the crust tends to offset any gravitational instability of cool lithosphere. Thus recycling of crust into the mantle was effectively prevented on the Moon and probably on Mars once a thick, low-density crust of global extent had formed. The dominance of either crustal differentiation or crustal recycling determined the evolutionary track and the style of basaltic volcanism on each planet.

On the Earth, the cycle of lithospheric overturn begins with crustal formation and associated mantle depletion at ridges, and ends with subduction and any associated remelting, differentiation and assimilation. At ridges, the oceanic crust forms from a 20–25% partial melting of the underlying mantle, and magma ascent leaves behind a depleted to partially depleted layer of mantle at least 30 km thick.

This differentiated mantle rock descends into the mantle when the oceanic lithosphere is subducted at ocean trenches. Several alternative hypotheses can be made with regard to its subsequent fate. One hypothesis is that the layered lithosphere is transported deep into the mantle where it is rehomogenized with the rest of the mantle and can once again participate in the plate tectonic cycle.

This hypothesis does not take account of the material removed from the subducted crust by volcanic processes at ocean trenches. Some fraction of the descending crust is partially melted and then added to the continental or island arc crust by intrusion and extrusion and is thus permanently removed from the mantle (section 7.4.2). There is considerable disagreement, however, about the fraction of the oceanic crust that is affected by this secondary differentiation process.

Because density differences associated with the oceanic crust and depleted mantle are large compared with density differences due to thermal expansion, an alternate hypothesis on the fate of subducted lithosphere is that compositional stratification of the mantle results from continued subduction. At low pressures, oceanic crust is considerably lighter than mantle rock. However, after the basaltic crustal rocks have undergone a phase change to eclogite they may be heavier than surrounding mantle material. In this form, the oceanic crust and perhaps the entire lithosphere might be expected to sink to the base of the mantle (e.g., Dickinson and Luth, 1971), or at least to the base of an upper mantle region of relatively lower density (D. L. Anderson, 1979). Depleted upper mantle is generally believed to be somewhat less dense than undepleted mantle material. If this depleted mantle layer were free to separate gravitationally from the rest of the lithosphere, it might be expected to underplate the lithosphere. Some workers believe that such underplating has occurred beneath the continents and that a substantial fraction of the continental lithosphere may be composed of depleted mantle material (see, however, section 7.4.2). Whether the crustal and the depleted and undepleted mantle components of oceanic lithosphere can in fact be gravitationally separated is an open question.

An important constraint on global recycling is the evidence for major heterogeneities in the chemistry of magma source regions in the mantle. On the basis of Rb-Sr, Sm-Nd, and U-Pb studies on mantle derived igneous rocks, the source regions of mid-ocean ridge and ocean island basalts are chemically distinct and have been separated for periods of at least 1 b.y. (section 7.4.2). Similar chemical heterogeneities have been demonstrated for sub-continental lithosphere, and source regions for continental tholeiites, alkaline basalts and kimberlites are progressively less depleted than source regions for typical oceanic basalts (section 7.4.2). The consequence of such heterogeneities for large scale convective mixing of the mantle has only recently received serious attention (McKenzie, 1979; Richter and Ribe, 1979) and is a subject of current research.

9.6.4 Magma vs. depth and time in the planets

A consequence of the early heating of the exteriors of the terrestrial planets is that the depth of shallowest partial melting in each planet should generally increase with time, as the lithosphere thickens by loss of heat from the planet to space. The predicted rate of deepening of the top of a partial melt zone depends on the rate of internal heat generation and on the surface heat loss. For comparable heat production rates, smaller planetary bodies will cool faster because of the greater surface to volume ratio.

The relationships are illustrated in several of the thermal history models of section 9.5; see, for example, Figs. 9.5.19, 9.5.26, and 9.5.30. The deepening of the isotherms and the top of the partial melt zone with time is most rapid for the Moon, less rapid for Mercury, and still less rapid for Mars.

Observational evidence supporting a progressive deepening of magma source regions with time on the planets exists only for sources beneath volcanoes, which may deepen with age beneath oceanic lithosphere on Earth (Vogt, 1974) and with time beneath large volcanic constructs on Mars (Carr, 1975). On the Moon, a progressive increase in source depth is not apparent from experimental petrological inferences on mare basalt magmas (section 3.4.2).

The predicted magma depths from simple, spherically symmetric or angle-averaged thermal history models are certainly too simple an abstraction of actual magma generation. Complexities such as heterogeneous heat sources (radioactivity, impacts) and melting relations, as well as the time-dependent lateral variations in the temperature field associated with mantle convective flow patterns, will lead to considerable variability in the timing and efficiency of magma generation as functions of position and depth.

An interesting result of geochronological studies of ancient terrestrial rocks is the evidence for "episodes" of globally enhanced volcanism at a number of times throughout recorded Earth history (section 7.4.1). On the basis of sea level curves during the Phanerozoic, global seafloor spreading rates show substantial variations over times of several hundred million years. On Mars, too, the chronology developed in Chapter 8 suggests more than one episode of substantial volcanism. An important question, if such episodicity is real, is how this time dependence in volcanic activity is related to planetary thermal history and to possible variations in convective heat transport. This question merits further attention.

9.6.5 Status of numerical modelling

Because of the complex nature of planetary interiors and the diverse processes that each planet has undergone through its history, it is inevitable that planetary evolution must be modelled using numerical techniques. Even though most of the theories of solution exist, numerical approaches are nevertheless limited by the speed and the capacity of computers available at present. In order to be computationally feasible, current models of planetary evolution necessarily employ various simplifying assumptions, such as requiring the models to be axisymmetric or to be of uniform properties. Undoubtedly these assumptions severely constrain the applicability of solutions to the study of the evolutionary history of a planetary interior. For example, axisymmetric models are probably invalid in studying the large planets such as the Earth and Venus. Therefore, a current research effort is devoted to the development of three-dimensional models with acceptable resolution and sophistication. Until such a three-dimensional model is successfully developed, the thermal evolution of large planets can be modelled only through various approximations. One such approximation currently in vogue is the parameterized convection approach described in section 9.3.3. Such models take into consideration the effects of dynamic heat transport without actually solving for the flow field. Consequently, a significant amount of numerical computation can be eliminated and an opportunity is provided to explore the effects of wider ranges in various possible parameters. While the basic idea of this approach has proved to be valuable, some of the intrinsic details of this approach can be improved, and some refinements will probably be necessary in order to generate more realistic models. One such refinement is to seek a more rigorous Nusselt number/Rayleigh number relationship for planetary evolution. The formulas adopted by most current parameterized convective models are based on steady-state, finite-Prandtl-number and constant-viscosity calculations. The validity of these formulas when applied to planetary evolution has not yet been tested and is another subject of current research. The effects of processes such as phase changes, differentiation and diapiric transport on these formulas are also currently under investigation.

As described in section 9.3.2, finite-element numerical methods are useful techniques whose power for thermal history models has not been fully realized. Because of their flexibility in approximating boundaries, the techniques are promising in modelling many surface geological processes, including those associated with magma transport and volcanism.

Finally, as described in section 9.3.3, it is apparent that many of the modelling techniques adopted for the current thermal models are somewhat arbitrary, including the partitioning of core formation energy and the dynamics of core separation and other differentiation processes. At present these problems have not yet been treated satisfactorily. Undoubtedly, refinements to improve such modelling procedures are in order so that more realistic planetary evolution models can be constructed and compared to the geological histories of the planets.

9.6.6 New measurements that might test models

For the Earth our knowledge of the details of planetary evolution, particularly for the era of plate tectonics, probably exceeds in complexity our ability to model with a thermal history calculation. For all of the remaining terrestrial bodies, it is the lack of one or more key items of knowledge that limits our ability to more tightly constrain through thermal histories the evolutionary paths for the interiors of these planets. The principal such gaps in our knowledge, from the standpoint of the thermal histories of the terrestrial planets, include the following items:

(1) Detailed information is needed on the internal structure of Mars, Mercury and Venus. At most, only bounds on mean crustal thicknesses are presently available, and there is little firm information on the dimensions of central cores. Both seismic sounding and an accurate determination of C/MR^2 for these bodies are necessary steps to achieve sufficient understanding of internal structure and its implications for planetary differentiation.

(2) The global heat flow and the surface concentrations of radioactive elements are not known for Mars, Mercury, or any asteroid. Heat flow is unknown on Venus, and the limited data on radioactive heat sources in Venus' surface materials need corroboration and extension. Also, the estimate of global lunar heat flow needs validation.

(3) Venus stands alone among the terrestrial planets in the poor level of our understanding of its surface geological history, and even of the most important processes that have shaped its surface. A global radar image of the Venus surface at a resolution comparable to the global optical images of the surfaces of Mercury and Mars would fill this major gap.

(4) An absolute chronology for surface geological history is missing for Mars, Mercury, and Venus. Our understanding of the chronology for the surface evolution of the Moon is good only for the nearside; information on the farside lags substantially.

(5) The assignment of meteorites or meteorite classes to specific parent bodies needs corroboration. The surface geology of differentiated and undifferentiated asteroids remains to be explored.

Filling these gaps in our knowledge of the terrestrial planets and of their thermal histories will be no small task. A full understanding of the processes of planetary evolution and of the role of basaltic volcanism in that evolution will require a sustained commitment to the further exploration of the solar system.

ACKNOWLEDGMENTS—We thank the many participants in the Basaltic Volcanism Study Project who contributed to this Chapter through discussions and suggestions for improvements in the text. We are also grateful to D. L. Anderson, K. Burke, N. L. Carter, K. A. Goettel, S. R. Hart, W. M. Kaula, R. J. O'Connell, E. M. Parmentier, W. R. Peltier, R. J. Phillips, C. P. Sonett andR. G. Strom for critical reviews of sections of this chapter and to G. Schubert for many helpful discussions. Finally, we thank the staff of the Lunar and Planetary Institute, particularly P. P. Jones, R. B. Merrill, R. Ridings, and S. Vass for considerable assistance with the figures and manuscript.

This work was partially supported by NASA Grants NSG-7081 and NSG-7297 and by an Alfred P. Sloan Research Fellowship at MIT (S. C. S. and A. T. H.); by NASA Grant NGR-33-070-108 at Cornell (D. L. T.); and by NSF Grant EAR 77-14479 at Northwestern (N. H. S.).

9.7 REFERENCES

Abramowitz M. and Stegun I. A. (1964) *Handbook of Mathematical Functions*. Nat. Bur. Stand. Appl. Math. Ser. **55**, Nat. Bur. Stand., Gaithersburg, Md. 1046 pp.

Abu-Eid R. M. and Langer K. (1978) Single-crystal UV-spectra of olivine (fayalite) up to 37,000 cm^{-1} at elevated pressures. *Naturwiss.* **65**, 256–257.

Adams J. B. and McCord T. B. (1976) Mercury: evidence for an anorthositic crust from reflectance spectra(abstract). In *Papers Presented to the Conference on Comparisons of Mercury and the Moon*, p. 1. The Lunar Science Institute, Houston.

Adler I., Trombka J. I., Schmadebeck R., Lowman P., Blodget H., Yin L., Eller E., Podwysocki M., Weidner J. R., Bickel A. L., Lum R. K. L., Gerard J., Gorenstein P., Bjorkholm P. and Harris B. (1973) Results of the Apollo 15 and 16 x-ray experiment. *Proc. Lunar Sci. Conf. 4th*, p. 2783–2791.

Ahrens T. J. (1973) Petrologic properties of the upper 670 km of the Earth: geophysical implications. *Phys. Earth Planet. Inter.* **7**, 167–186.

Akimoto S. I. (1972) The system MgO-FeO-SiO_2 at high pressures and temperature—phase equilibria and elastic properties. *Tectonophy.* **13**, 161–187.

Akimoto S. I., Akoaogi M., Kawada K. and Nishigawa O. (1976) Mineralogic distribution of iron in the upper half of the transition zone in the Earth's mantle. In *The Geophysics of the Pacific Ocean Basin and Its Margin*, p. 399–406. Geophys. Monograph **19**, Amer. Geophys. Union, Washington, D. C.

Akimoto S. I. and Fujisawa J. (1968) Olivine-spinel solid solution equilibria in the system Mg_2SiO_4-Fe_2SiO_4. *J. Geophys. Res.* **73**, 1467–1479.

Allan D. W. and Jacobs J. A. (1956) The melting of asteroids and the origin of meteorites. *Geochim. Cosmochim. Acta* **9**, 256–272

Al'tshuler L. V. and Sharipdzhanov I. I. (1971) Additive equations of state of silicates at high pressures. *Izv. Earth Physics*, no. 3, 167–177.

Anderson A. T. (1975) Some basaltic and andesitic gases. *Rev. Geophys. Space Phys.* **13**, 37–56.

Anderson D. L. (1967) Phase changes in the upper mantle. *Science* **157**, 1165–1173.

Anderson D. L. (1972) Internal constitution of Mars. *J. Geophys. Res.* **77**, 789–795.

Anderson D. L. (1975) Chemical plumes in the mantle. *Geol. Soc. Amer. Bull.* **86**, 1593–1600.

Anderson D. L. (1977) The 650 km discontinuity. *Geophys. Res. Lett.* **3**, 347–349.

Anderson D. L. (1979) Chemical stratification of the mantle. *J. Geophys. Res.* **84**, 6297–6298.

Anderson D. L. (1980) Tectonics and composition of Venus. *Geophys. Res. Lett.* **7**, 101–102.

Anderson D. L. and Kovach R. L. (1967) The composition of the terrestrial planets. *Earth Planet. Sci. Lett.* **3**, 19–24.

Anderson D. L., Miller W. F., Latham G. V., Nakamura Y., Toksöz M. N., Dainty A. M., Duennebier F. K.,Lazarewicz A. R., Kovach R. L. and Knight T. C. D. (1977) Seismology on Mars. *J. Geophys. Res.* **82**, 4524–4546.

Anderson D. L. and Phinney R. A. (1967) Early thermal history of the terrestrial planets. In *Mantles of the Earth and Terrestrial Planets* (S. K. Runcorn, ed.), p. 113–126. Interscience, London.

Anderson O. L. (1979) The high-temperature acoustic Grüneisen parameter in the Earth's interior. *Phys. Earth Planet. Inter.* **18**, 221–231.

Anderson O. L. and Schreiber E. (1965) The relation between refractive index and density for minerals related to the Earth's mantle. *J. Geophys. Res.* **70**, 1463–1472.

Anderson O. L., Schreiber E., Liebermann R. and Soga N. (1968) Some elastic constant data on minerals relevant to geophysics. *Rev. Geophys.* **6**, 491–524.

Anderson R. N. and Hobart M. A. (1976) The relation between heat flow, sediment thickness, and age in the eastern Pacific. *J. Geophys. Res.* **81**, 2968–2989.

Andrews D. J. and Sleep N. H. (1974) Numerical modelling of tectonic flow behind island arcs. *Geophys. J. Roy. Astron. Soc.* **38**, 237–251.

Arkani-Hamed J. (1979) Non-linear and finite amplitude thermal convection in a heterogeneous terrestrial planet. *Geophys. J. Roy. Astron. Soc.* **56**, 63–80.

Armstrong R. L. and Cooper J. A. (1971) Lead isotopes in island arcs. *Bull. Vocanol.* **35**, 27–63.

Arndt N. T. (1976) Melting relations of ultramafic lavas (komatiites) at one atmosphere and high pressure. *Carnegie Inst. Wash., Yearb.* **75**, 555–561.

Arndt N. T. (1977) Ultrabasic magmas and high-degree melting of the mantle. *Contrib. Mineral. Petrol.* **64**, 205–221.

Arzi A. A. (1978) Critical phenomena in the rheology of partially melted rocks. *Tectonophy.* **44**, 173–184.

Ashby M. F. and Verrall R. A. (1977) Micromechanisms of flow and fracture, and their relevance to the rheology of the upper mantle. *Phil. Trans. Roy. Soc. Lond.* **A288**, 59–95.

Auten T. A., Gordon R. B. and Stocker R. L. (1974) Q and creep in the mantle. *Nature* **250**, 317–318.

Avduevskii V. S., Vishnevskii S. L., Golov I. A., Karpeiskii Yu. Ya., Lavrov A. D., Likhushin V. Ya., Marov M. Y., Melnikov D. A., Pomogin N. I., Pronina N. N., Razin K. A. and Tokin V. G. (1976) Measurement of wind velocity on the surface of Venus during the operation of stations Venera 9 and Venera 10. *Cosmic Res.* **15**, 622–625.

Ave'Lallemant H. G. (1978) Experimental deformation of diopside and websterite. *Tectonophy.* **48**, 1–28.

Backus G. E. (1955) On the application of Eigenfunction expansions to the problem of the thermal instability of a fluid sphere heated within. *Phil. Mag. Ser. 7* **46**, 1310–1327.

Baer A. J. (1977) Speculations on the evolution of the lithosphere. *Precambrian Res.* **5**, 249–260.

Baird A. K., Toulmin P., III, Clark B. C., Rose H. J., Jr., Keil K., Christian R. P. and Gooding J. L. (1976) Mineralogic and petrologic implications of Viking geochemical results from Mars: interim report. *Science* **194**, 1288–1293.

Barker P. F. (1972) A spreading center in the East Scotia Sea. *Earth Planet. Sci. Lett.* **15**, 123–132.

Barron T. H. K. (1955) On the thermal expansion of solids at low temperatures. *Phil. Mag. Ser. 7* **46**, 720–734.

Basu A. R. and Murthy V. R. (1977) Ancient lithospheric lherzolite xenolith in alkali basalt from Baja California. *Earth Planet. Sci. Lett.* **35**, 239–246.

Barshay S. and Lewis J. S. (1976) Chemistry of primitive solar material. *Ann. Rev. Astron. Astrophys.* **14**, 81–93.

Benfield A. E. (1950) Temperatures of an accreting Earth. *Trans. Amer. Geophys. Union* **31**, 53–57.

Bickle M. J. (1978) Heat loss from the Earth: a constraint on Archean tectonics from the relation between geothermal gradients and the rate of plate production. *Earth Planet. Sci. Lett.* **40**, 301–315.

Bills B. G. and Ferrari A. J. (1977) A lunar density model consistent with topographic, gravitational, librational, and seismic data. *J. Geophys. Res.* **82**, 1306–1314.

Bills B. G. and Ferrari A. J. (1978) Mars topography harmonics and geophysical implications. *J. Geophys. Res.* **83**, 3497–3508.

Binder A. B. (1969) Internal structure of Mars. *J. Geophys. Res.* **74**, 3110–3118.

Binder A. B. and Davis D. R. (1973) Internal structure of Mars. *Phys. Earth Planet. Inter.* **7**, 477–485.

Binder A. B. and McCarthy D. W., Jr. (1972) Mars: the lineament systems. *Science* **176**, 279–281.

Birch F. (1952) Elasticity and constitution of the Earth's interior. *J. Geophys. Res.* **57**, 227–286.

Birch F. (1954) Heat from radioactivity. In *Nuclear Geology* (H. Faul, ed.), p. 148–174. John Wiley and Sons, N. Y.

Birch F. (1965) Energetics of core formation. *J. Geophys. Res.* **70**, 6217–6221.

Birch F. and Clark H. (1940) The thermal conductivity of rocks and its dependence upon temperature and composition, 1. *Amer. J. Sci.* **238**, 529–558.

Birch F., Roy R. F. and Decker E. R. (1968) Heat flow and thermal history in New England and New York. In *Studies of Appalachian Geology: Northern and Maritime* (E. A. Zen, W. S. White, J. B. Hadley and J. B. Thompson, eds.), p. 437–451. Interscience, N.Y.

Blackshear W. T. and Gapcynski J. P. (1977) An improved value of the lunar moment of inertia. *J. Geophys. Res.* **82**, 1699–1701.

Blasius K. R. and Cutts J. A. (1976) Shield volcanism and lithospheric structure beneath the Tharsis plateau, Mars. *Proc. Lunar Sci. Conf. 7th*, p. 3561–3573.

Boettcher A. L. (1974) Review of symposium on deep-seated rocks and geothermometry. *EOS (Trans. Amer. Geophys. Union)* **55**, 1068–1072.

Boland J. N., McLaren A. C. and Hobbs B. E. (1971) Dislocations associated with optical features in naturally deformed olivine. *Contrib. Mineral. Petrol.* **30**, 53–63.

Booker J. R. (1976) Thermal convection with strongly temperature-dependent viscosity. *J. Fluid Mech.* **76**, 741–754.

Booker J. R. and Stengel K. C. (1978) Further thoughts on convective heat transport in a variable viscosity fluid. *J. Fluid Mech.* **86**, 289–291.

Bowell E., Chapman C. R., Gradie J. C., Morrison D. and Zellner B. (1978) Taxonomy of asteroids. *Icarus* **35**, 315–335.

Boyce J. M., Dial A. L. and Soderblom L. A. (1974) Ages of the lunar nearside light plains and maria. *Proc. Lunar Sci. Conf. 5th*, p. 11–23.

Boyd F. R. (1973) A pyroxene geotherm. *Geochim. Cosmochim. Acta* **37**, 2533–2546.

Boyd F. R. (1974) Ultramafic nodules from the Frank Smith kimberlite pipe, South Africa. *Carnegie Inst. Wash., Yearb.* **73**, 285–294.

Boyd F. R. and Nixon P. H. (1973) Structure of the upper mantle beneath Lesotho. *Carnegie Inst. Wash., Yearb.* **72**, 431–445.

Brecher A. (1976) Textural remanence: a new model of lunar rock magnetism. *Earth Planet. Sci. Lett.* **29**, 131–145.

Brennan C. (1974) Isostatic recovery and the strain rate dependent viscosity of the Earth's mantle. *J. Geophys. Res.* **79**, 3993–4001.

Brett R. (1976) The current status of speculations on the composition of the core of the Earth. *Rev. Geophys. Space Phys.* **14**, 375–384.

Brett R. and Bell P. M. (1969) Melting relations in the Fe-rich portion of the system Fe-FeS at 30 kb pressure. *Earth Planet. Sci. Lett.* **6**, 479–482.

Bridge H. S., Lazarus A. J., Snyder C. W., Smith E. J., Davis L., Jr., Coleman P. J., Jr. and Jones D. E. (1969) Mariner V: plasma and magnetic fields observed near Venus. In *The Venus Atmosphere* (R. Jastrow and I. Rasool, eds.), p. 533–550. Gordon and Breach, N.Y.

Broadfoot A. L., Shemansky D. E. and Kumar S. (1976) Mariner 10: Mercury atmosphere. *Geophys. Res. Lett.* **3**, 577–580.

Bullard E. (1954) The flow of heat through the floor of the Atlantic Ocean. *Proc. Roy. Soc. Lond.* **A222**, 408–428.

Bunch T. E. and Reid A. M. (1975) The nakhlites part 1: petrography and mineral chemistry. *Meteoritics* **10**, 303–315.

Bundy F. P. (1965) Pressure-temperature phase diagram of iron to 200 kbar, 900°C. *J. Appl. Phys.* **36**, 616–620.

Burchfield J. D. (1975) *Lord Kelvin and the Age of the Earth.* Science History Publications, N.Y. 260 pp.

Burke K. and Kidd W. S. F. (1978) Were Archean continental geothermal gradients much steeper than those of today? *Nature* **272**, 240–241.

Burns J. A. (1976) Consequences of the tidal slowing of Mercury. *Icarus* **28**, 453–458.

Burns J. A. and Schultz P. (1978) Tidal heating of the Moon: a reappraisal (abstract). In *Lunar and Planetary Science IX*, p. 137–139. Lunar and Planetary Institute, Houston.

Buseck P. R. and Goldstein J. I. (1969) Olivine compositions and cooling rates of pallasitic meteorites. *Geol. Soc. Amer. Bull.* **80**, 2141–2158.

Busse F. H. (1967) On the stability of two-dimensional convection in a layer heated from below. *J. Mathematics and Physics* **46**, 140–150.

Busse F. H. (1976) Generation of planetary magnetism by convection. *Phys. Earth Planet. Inter.* **12**, 350–358.

Caldwell J. G., Haxby W. F., Karig D. E. and Turcotte D. L. (1976) On the applicability of a universal elastic trench profile. *Earth Planet. Sci. Lett.* **31**, 239–246.

Caldwell J. G. and Turcotte D. L. (1979) Dependence of the thickness of the elastic oceanic lithosphere on age. *J. Geophys. Res.* **84**, 7572–7576.

Cameron A. G. W. (1978) Physics of the primitive solar accretion disk. *The Moon and the Planets* **18**, 5–40.

Campbell D. B., Burns B. A. and Boriakoff V. (1979) Venus: further evidence of impact cratering and tectonic activity from radar observations. *Science* **204**, 1424–1427.

Campbell D. B., Dyce R. B. and Pettengill G. H. (1976) New radar image of Venus. *Science* **193**, 1123–1124.

Carr M. H. (1973) Volcanism on Mars. *J. Geophys. Res.* **78**, 4049–4062.

Carr M. H. (1974) Tectonism and volcanism of the Tharsis region of Mars. *J. Geophys. Res.* **79**, 3943–3949.

Carr M. H. (1975) Change in height of Martian volcanoes with time. *Geol. Romana* **15**, 421–422.

Carr M. H., Greeley R., Blasius K. R., Guest J. E. and Murray J. B. (1977) Some Martian volcanic features as viewed from the Viking orbiters. *J. Geophys. Res.* **82**, 3985–4015.

Carr M. H., Masursky H., Baum W. A., Blasius K. R., Briggs G. A., Cutts J. A., Duxbury T., Greeley R., Guest J. E., Smith B. A., Soderblom L. A., Veverka J. and Wellman J. B. (1976) Preliminary results from the Viking orbiter imaging experiment. *Science* **193**, 766–776.

Carr M. H., Masursky H. and Saunders R. S. (1973) A generalized geologic map of Mars. *J. Geophys. Res.* **78**, 4031–4036.

Carslaw H. S. and Jaeger J. C. (1959) *Conduction of Heat in Solids.* Oxford Univ. Press, London. 510 pp.

Carter N. L. (1976) Steady state flow of rocks. *Rev. Geophys. Space Phys.* **14**, 301–360.

Carter N. L. and Ave'Lallemant H. G. (1970) High temperature flow of dunite and peridotite. *Geol. Soc. Amer. Bull.* **81**, 2181–2202.

Cassen P. (1977) Planetary magnetism and the interiors of the Moon and Mercury. *Phy. Earth Planet. Inter.* **15**, 113–120.

Cassen P. and Reynolds R. T. (1973) Role of convection in the Moon. *J. Geophys. Res.* **78**, 3203–3215.

Cassen P. and Reynolds R. T. (1974) Convection in the Moon: effect of variable viscosity. *J. Geophys. Res.* **79**, 2937–2944.

Cassen P., Reynolds R. T., Graziani F., Summers A., McNellis J. and Blalock L. (1979) Convection and lunar thermal history. *Phys. Earth Planet. Inter.* **19**, 183–196.

Cassen P. M. and Young R. E. (1975) On the cooling of the Moon by solid convection. *The Moon* **12**, 361–368.

Cassen P., Young R. E., Schubert G. and Reynolds R. T. (1976) Implications of an internal dynamo for the thermal history of Mercury. *Icarus* **28**, 501–508.

Cassen P., Young R. E. and Schubert G. (1978) The distortion of the Moon due to convection. *Geophys. Res. Lett.* **5**, 294–296.

Cathles L. M. (1975) *The Viscosity of the Earth's Mantle.* Princeton Univ. Press, Princeton, N.J. 386 pp.

Chandrasekhar S. (1952) The thermal instability of a fluid sphere heated within. *Phil. Mag. Ser. 7* **43**, 1317–1329.

Chapman C. R. (1974) Cratering on Mars, I and II. *Icarus* **22**, 272–291.

Chapman C. R. (1976) Asteroids as meteorite parent-bodies: the astronomical perspective. *Geochim. Cosmochim. Acta* **40**, 701–749.

Chapple W. M. and Forsyth D. W. (1979) Earthquakes and bending of plates at trenches. *J. Geophys. Res.* **84**, 6729–6749.

Chapple W. M. and Tullis T. E. (1977) Evaluation of the forces that drive the plates. *J. Geophys. Res.* **82**, 1967–1984.

Chase C. G. (1972) The N-plate problem of plate tectonics. *Geophys. J. Roy. Astron. Soc.* **29**, 117–122.

Chase S. C., Jr., Miner E. D., Morrison D., Münch G., Neugebauer G. and Schroeder M. (1974) Preliminary infrared radiometry of the night side of Mercury from Mariner 10. *Science* **185**, 142–145.

Chase S. C., Jr., Miner E. D., Morrison D., Münch G. and Neugebauer G. (1976) Mariner 10 infrared radiometer results: temperatures and thermal properties of the surface of Mercury. *Icarus* **28**, 565–578.

Cheng C. H. and Toksöz M. N. (1978) Tidal stresses in the Moon. *J. Geophys. Res.* **83**, 845–854.

Chu T. Y. and Goldstein R. J. (1973) Turbulent convection in a horizontal layer of water. *J. Fluid Mech.* **60**, 141–159.

Cintala M. J., Wood C. A. and Head J. W. (1977) The effects of target characteristics on fresh crater morphology: preliminary results for the Moon and Mercury. *Proc. Lunar Sci. Conf. 8th*, p. 3409–3425.

Cisowski S. M. (1979) Paleointensity studies on 3.4 b.y. old Onverwacht basalts (abstract). *EOS (Trans. Amer. Geophys. Union)* **60**, 816.

Cisowski S. M., Fuller M. D., Wu Y. M., Rose M. F. and Wasilewski P. J. (1976) Magnetic effects of shock and their implications for magnetism of lunar samples. *Proc. Lunar Sci. Conf. 7th*, p. 3299–3320.

Clark B. C., Baird A. K., Rose H. J., Jr., Toulmin P., III, Keil K., Castro A. J., Kelliher W. C., Rowe C. D. and Evans P. H. (1976) Inorganic analyses of martian surface samples at the Viking landing sites. *Science* **194**, 1283–1288.

Clark S. P., Jr. (1957) Radiative transfer in the Earth's mantle. *Trans. Amer. Geophys. Union* **38**, 931–938.

Clark S. P., Jr., Peterman Z. E. and Heier K. S. (1966) Abundances of uranium, thorium and potassium. In *Handbook of Physical Constants* (S. P. Clark, Jr.,ed.), Geol. Soc. Amer. Mem. 97, p. 521–541. Geol. Soc. Amer., Boulder, Colo.

Cochran J. R. (1979) An analysis of isostasy in the world's oceans, 2, midocean ridge crests. *J. Geophys. Res.* **84**, 4713–4729.

Colin L. (1979a) Encounter with Venus. *Science* **203**, 743–745.

Colin L. (1979b) An encounter with Venus: an update. *Science* **205**, 44–46.

Colombo G. (1965) Rotational period of the planet Mercury. *Nature* **208**, 575.

Conca J. and Hubbard N. (1979) Evidence for early volcanism in Mare Smythii. *Proc. Lunar Planet. Sci. Conf. 10th*, p. 1727–1737.

Consolmagno G. J. and Drake M. J. (1977) Composition and evolution of the eucrite parent body: evidence from rare earth elements. *Geochim. Cosmochim. Acta* **41**, 1271–1282.

Cordell B. M. (1977) Tectonism and the interior of Mercury. Ph.D. thesis, Univ. Arizona, Tucson. 124 pp.

Cordell B. M. and Strom R. G. (1977) Global tectonics of Mercury and the Moon. *Phys. Earth Planet. Inter.* **15**, 146–155.

Counselman C. C., III, Gourevitch S. A., King R. W., Loriot G. B. and Prinn R. G. (1979) Venus winds are zonal and retrograde below the clouds. *Science* **205**, 85–87.

Craig H., Lupton J. E. and Moribe Y. (1978) A mantle helium component in circum-Pacific volcanic gases: Hakone, the Marianas, and Mt. Lassen. In *Terrestrial Rare Gases, Advances in Earth and Planetary Sciences* **3** (E. C. Alexander and M. Ozima, eds.), p. 3–16. Center for Academic Publications, Tokyo.

Cremers C. J., Birkebak R. C. and Dawson J. P. (1970) Thermal conductivity of fines from Apollo 11. *Proc. Apollo 11 Lunar Sci. Conf.*, p. 2045–2050.

Crough S. T. (1975) Thermal model of oceanic lithosphere. *Nature* **256**, 388–391.

Crough S. T. and Thompson G. A. (1976) Numerical and approximate solutions for lithospheric thickening and thinning. *Earth Planet. Sci. Lett.* **31**, 397–401.

Cuzzi J. N. (1974) The nature of the subsurface of Mercury from microwave observations at several wavelengths. *Astrophys. J.* **189**, 577–586.

Daly S. F. and Richter F. M. (1978) Convection with decaying heat sources: a simple thermal evolution model (abstract). In *Lunar and Planetary Science IX*, p. 213–214. Lunar and Planetary Institute, Houston.

Davies G. F. (1977) Whole-mantle convection and plate tectonics. *Geophys. J. Roy. Astron. Soc.* **49**, 459–486.

DeBremaecker J. C. (1977) Is the oceanic lithosphere elastic or viscous? *J. Geophys. Res.* **82**, 2001–2004.

DeCampli W. M. and Cameron A. G. W. (1979) Structure and evolution of isolated giant gaseous protoplanets. *Icarus* **38**, 367–391.

DePaolo D. J. and Johnson R. W. (1979) Magma genesis in the New Britain island-arc. *Contrib. Mineral. Petrol.* **70**, 367–379.

Detrick R. S. and Crough S. T. (1978) Island subsidence, hot spots and lithospheric thinning. *J. Geophys. Res.* **83**, 1236–1244.

Detrick R. S. and Watts A. B. (1979) An analysis of isostasy in the world's oceans, 3, aseismic ridges. *J. Geophys. Res.* **84**, 3637–3653.

deVaucoulers G. (1964) Geometric and photometric parameters of the terrestrial planets. *Icarus* **3**, 187–235.

Dickinson W. R. and Luth W. C. (1971) A model for plate tectonic evolution of mantle layers. *Science* **174**, 400–404.

Dolginov Sh. Sh. (1978) On the magnetic field of Mars: Mars 5 evidence. *Geophys. Res. Lett.* **5**, 93–95.

Dolginov Sh. Sh., Yeroshenko Ye. G. and Lewis L. (1969) Nature of the magnetic field in the neighborhood of Venus. *Cosmic Res.* **7**, 675–680.

Dolginov Sh. Sh., Yeroshenko Ye. G. and Zhuzgov L. N. (1973) The magnetic field in the very close neighborhood of Mars according to data from the Mars 2 and Mars 3 spacecraft. *J. Geophys. Res.* **78**, 4779–4786.

Dolginov Sh. Sh., Yeroshenko Ye. G. and Zhuzgov L. N. (1976) The magnetic field of Mars according to the data from the Mars 3 and Mars 5. *J. Geophys. Res.* **81**, 3353–3362.

Dolginov Sh. Sh., Zhuzgov L. N., Sharova V. A., Buzin V. B. and Yeroshenko Ye. G. (1977) Magnetosphere of the planet Venus. Izmiran preprint 19, Moscow.

Dollfus A. and Auriere M. (1974) Optical polarimetry of planet Mercury. *Icarus* **23**, 465–482.

Dostal J. and Capedri S. (1975) Partition coefficients of uranium for some rock-forming minerals. *Chem. Geol.* **15**, 285–294.

Drake M. J. (1976) Evolution of major mineral compositions and trace element abundances during fractional crystallization of a model lunar composition. *Geochim. Cosmochim. Acta* **40**, 401–411.

Drake M. J. and Consolmagno G. J. (1976) Critical review of models for the evolution of high-Ti mare basalts. *Proc. Lunar Sci. Conf. 7th*, p. 1633–1657.

Dreibus G., Spettel B. and Wänke H. (1976) Lithium as a correlated element, its condensation behavior, and its use to estimate the bulk composition of the Moon and the eucrite parent body. *Proc. Lunar Sci. Conf. 7th*, p. 3383–3396.

Duba A., Heard H. C. and Schock R. N. (1976) Electrical conductivity of orthopyroxene at 1400° C and the resulting selenotherm. *Proc. Lunar Sci. Conf. 7th*, p. 3173–3181.

Duba A. and Ringwood A. E. (1973) Electrical conductivity, internal temperatures and thermal evolution of the Moon. *The Moon* **7**, 356–376.

Durham W. B. and Goetze C. (1977) A comparison of the creep properties of pure forsterite and iron-bearing olivine. *Tectonophy.* **40**, T15–T18.

Durham W. B., Goetze C. and Blake B. (1977) Plastic flow of oriented single crystals of olivine, 2, observations and interpretations of the dislocation structures. *J. Geophys. Res.* **82**, 5755–5770.

Dyal P., Parkin C. W. and Daily W. D. (1974) Magnetism and the interior of the Moon. *Rev. Geophys. Space Phys.* **12**, 568–591.

Dyal P., Parkin C. W. and Daily W. D. (1976) Structure of the lunar interior from magnetic field measurements. *Proc. Lunar Sci. Conf. 7th*, p. 3077–3095.

Dymond J. and Hogan L. (1978) Factors controlling the noble gas abundance patterns of deep-sea basalts. *Earth Planet. Sci. Lett.* **38**, 117–128.

Dziewonski A. M., Hales A. L. and Lapwood E. R. (1975) Parametrically simple Earth models consistent with geophysical data. *Phys. Earth Planet. Inter.* **10**, 12–48.

Dzurisin D. (1977) Mercurian bright patches: evidence for physio-chemical alteration of surface material? *Geophys. Res. Lett.* **4**, 383–386.

Dzurisin D. (1978) The tectonic and volcanic history of Mercury as inferred from studies of scarps, ridges, troughs, and other lineaments. *J. Geophys. Res.* **83**, 4883–4906.

Eggler D. H. and McCallum M. E. (1974) Xenoliths in diatremes of the western United States. *Carnegie Inst. Wash., Yearb.* **73**, 294–300.

Elsasser W. M. (1963) Early history of the Earth. In *Earth Science and Meteoritics* (J. Geiss and E. Goldberg, eds.), p. 1–30. North Holland, Amsterdam.

Elsasser W. M., Olson P. and Marsh B. D. (1979) The depth of mantle convection. *J. Geophys. Res.* **84**, 147–155.

Esposito P. B., Anderson J. D. and Ng A. T. Y. (1977) Experimental determination of Mercury's mass and oblateness. *Space Res.* **17**, 639–644.

Fanale F. P. (1971) A case for catastrophic early degassing of the Earth. *Chem. Geol.* **8**, 79–105.

Farmer C. B., Davies D. W. and LaPorte D. D. (1976) Mars: northern summer ice cap—water vapor observations from Viking 2. *Science* **194**, 1339–1341.

Ferrari A. J. (1977) Lunar gravity: a harmonic analysis. *J. Geophys. Res.* **82**, 3065–3084.

Ferrari A. J., Nelson D. L., Sjogren W. L. and Phillips R. J. (1978) The isostatic state of the lunar Apennines and regional surroundings. *J. Geophys. Res.* **83**, 2863–2871.

Fish R. A., Goles G. and Anders E. (1960) The record in meteorites III. On the development of meteorites in asteroidal bodies. *Astrophys. J.* **132**, 243–258.

Fisher D. E. (1979) Uranium abundance of the oceanic crust. *Geochim. Cosmochim. Acta* **43**, 709–716.

Flasar F. M. and Birch F. (1973) Energetics of core formation: a correction. *J. Geophys. Res.* **78**, 6101–6103.

Forsyth D. W. (1977) The evolution of the upper mantle beneath mid-oceanic ridges. *Tectonophy.* **38**, 89–118.

Forsyth D. and Uyeda S. (1975) On the relative importance of the driving forces of plate motion. *Geophys. J. Roy. Astron. Soc.* **43**, 163–200.

Fricker P. E., Goldstein J. I. and Summers A. L. (1970) Cooling rates and thermal histories of iron and stony-iron meteorites. *Geochim. Cosmochim. Acta* **34**, 475–491.

Fricker P. E. and Reynolds R. T. (1968) Development of the atmosphere of Venus. *Icarus* **9**, 221–230.

Fricker P. E., Reynolds R. T. and Summers A. L. (1967) On the thermal history of the Moon. *J. Geophys. Res.* **72**, 2649–2663.

Fricker P. E., Reynolds R. T. and Summers A. L. (1974) On the thermal evolution of the terrestrial planets. *The Moon* **9**, 211–218.

Fricker P. E., Reynolds R. T., Summers A. L. and Cassen P. M. (1976) Does Mercury have a molten core? *Nature* **259**, 293–294.

Froidevaux C. and Schubert G. (1975) Plate motion and structure of the continental asthenosphere: a realistic model of the upper mantle. *J. Geophys. Res.* **80**, 2553–2564.

Froidevaux C., Schubert G. and Yuen D. A. (1977) Thermal and mechanical structure of the upper mantle: a comparison between continental and oceanic models. *Tectonophy.* **37**, 233–246.

Fujii N. and Osako M. (1973) Thermal diffusivity of lunar rocks under atmospheric and vacuum conditions. *Earth Planet. Sci. Lett.* **18**, 65–71.

Fujisawa H., Fujii N., Mitzutani H., Kanamori H. and Akimoto S. (1968) Thermal diffusivity of Mg_2SiO_4, Fe_2SiO_4 and NaCl at high pressures and temperatures. *J. Geophys. Res.* **73**, 4727–4733.

Fukao Y., Mitzutani H. and Uyeda S. (1968) Optical absorption spectra at high temperatures and radiative thermal conductivity of olivines. *Phy. Earth Planet. Inter.* **1**, 57–62.

Fuller M. (1974) Lunar magnetism. *Rev. Geophys. Space Phys.* **12**, 23–70.

Gaffey M. J. and McCord T. B. (1977) Asteroid surface materials: mineralogical characterizations and cosmological implications. *Proc. Lunar Sci. Conf. 8th*, p. 113–143.

Gancarz A. J. and Wasserburg G. J. (1977) Initial Pb of the Amitsoq gneiss, west Greenland, and implications for the age of the Earth. *Geochim. Cosmochim. Acta* **41**, 1283–1301.

Garon A. M. and Goldstein R. J. (1973) Velocity and heat transfer measurements in thermal convection. *Phys. Fluids* **16**, 1818–1825.

Gartling D. K. (1976) Recent developments in the use of finite element methods in fluid dynamics. In *Computing in Applied Mechanics*, Appl. Mechanics Div., v. 18 (R. F. Hartung, ed.), p. 65–91. Amer. Soc. Mechan. Engrs., N.Y.

Gast P. W. (1972) The chemical composition and structure of the Moon. *The Moon*, **5**, 121–148.

Gast P. W. and Hubbard N. J. (1970) Rare earth abundances in soil and rocks from the Ocean of Storms. *Earth Planet. Sci. Lett.* **10**, 94–101.

Gel'man B. F., Zolotukhin V. G., Kazakov B. V., Lamonov N. I., Lipatov A. I., Levchuk B. V., Mukhin L. M., Nenarokov D. F., Okhotnikov B. P., Sinel'nikov A. V., Rotin V. A. and Khokhlov V. N. (1979) Analysis of the chemical composition of the Venus atmosphere by the gas-chromato-graphic technique. *Pis'ma v Astronomicheskiy Zhurnal* **5**, 4.

Gilbert F. and Dziewonski A. (1975) An application of normal mode theory to the retrieval of structural parameters and source mechanisms from seismic spectra. *Phil. Trans. Roy. Soc. London* **A278**, 187–269.

Goettel K. A. (1977) Composition of the terrestrial planets (abstract). In *Reports of Planetary Geology Program, 1976–1977*, p. 7–9. NASA TMX-3511.

Goetze C. (1977) A brief summary of our present day understanding of the effect of volatiles and partial melt on the mechanical properties of the upper mantle. In *High-Pressure Research, Applications in Geophysics* (M. H. Manghnani and S. Akimoto, eds.), p. 3–24. Academic Press, N.Y.

Goetze C. and Brace W. F. (1972) Laboratory observations of high temperature rheology of rocks. *Tectonophy.* **13**, 583–600.

Goins N. R., Toksöz M. N. and Dainty A. M. (1978) Seismic structure of the lunar mantle: an overview. *Proc. Lunar Planet. Sci. Conf. 9th*, p. 3575–3588.

Gold T. and Soter S. (1976) Cometary impact and the magnetization of the Moon. *Planet. Space Sci.* **24**, 45–54.

Goldreich P. and Ward W. R. (1973) The formation of planetesimals. *Astrophys. J.* **183**, 1051–1061.

Goldstein J. I. and Ogilvie R. E. (1965) The growth of the Widmanstätten pattern in metallic meteorites. *Geochim. Cosmochim. Acta* **29**, 893–920.

Goldstein J. I. and Short J. M. (1967) Cooling rates of 27 iron and stony-iron meteorites. *Geochim. Cosmochim. Acta* **31**, 1001–1023.

Goldstein R. M., Green R. R. and Rumsey H. C. (1976) Venus radar images. *J. Geophys. Res.* **81**, 4807–4817.

Goldstein R. M., Green R. R. and Rumsey H. C. (1979) Venus radar brightness and altitude images. *Icarus* **36**, 334–353.

Goles G. G., Fish R. A. and Anders E. (1960) The record in the meteorites—I. The former environment of stone meteorites as deduced from K^{40}-Ar^{40} ages. *Geochim. Cosmochim. Acta* **19**, 177–195.

Gordon R. B. (1965) Diffusion creep in the Earth's mantle. *J. Geophys. Res.* **70**, 2413–2418.

Goto T., Ahrens T. J., Rossman G. R. and Syono Y. (1980) Absorption spectrum of shock-compressed Fe^{2+}-bearing MgO and the radiative conductivity of the lower mantle. *Phys. Earth Planet. Inter.* **22**, 277-288.

Graham E. K. (1970) Elasticity and composition of the upper mantle. *Geophys. J. Roy. Astron. Soc.* **20**, 285–302.

Green D. H. (1973) Experimental melting studies on a model upper mantle composition at high pressure under water-saturated and water-unsaturated conditions. *Earth Planet. Sci. Lett.* **19**, 37–53.

Green D. H., Morgan J. W. and Heier K. S. (1968) Thorium, uranium and potassium abundances in peridotite inclusions and their host basalts. *Earth Planet. Sci. Lett.* **4**, 155–166.

Green D. H., Nichols I. A., Viljoen M. and Viljoen R. (1975) Experimental demonstration of the existence of peridotite liquids in earliest Archean magmatism. *Geology* **3**, 11–14.

Griggs D. T. (1939) A theory of mountain building. *Amer. J. Sci.* **237**, 611–650.

Griggs D. T. (1972) The sinking lithosphere and the focal mechanism of deep earthquakes. In *The Nature of the Solid Earth* (E. C. Robertson, ed.), p. 361–384. McGraw-Hill, N.Y.

Gubbins D. (1977) Speculations on the origin of the magnetic field of Mercury. *Icarus* **30**, 186–191.

Hager B. H. and O'Connell R. J. (1978) Subduction zone dip angles and flow driven by plate motion. *Tectonophy.* **50**, 111–134.

Hager B. H. and O'Connell R. J. (1979) Kinematic models of large-scale flow in the Earth's mantle. *J. Geophys. Res.* **84**, 1031–1048.

Haines E. L., Etchegaray-Ramirez M. I. and Metzger A. E. (1978) Thorium concentrations in the lunar surface. II: deconvolution modeling and its application to the regions of Aristarchus and Mare Smythii. *Proc. Lunar Planet. Sci. Conf. 9th*, p. 2985–3013.

Hales A. (1936) Convection currents in the Earth. *Mon. Not. Roy. Astron. Soc., Geophys. Suppl.* **3**, 372–379.

Hales A. L. (1976) ScS travel residuals and the difference between continents and oceans. *Earth Planet. Sci. Lett.* **33**, 249–254.

Hanks T. C. and Anderson D. L. (1969) The early thermal history of the Earth. *Phys. Earth Planet. Inter.* **2**, 19–29.

Hanks T. C. and Anderson D. L. (1972) Origin, evolution and present thermal state of the Moon. *Phys. Earth Planet. Inter.* **5**, 409–425.

Hapke B., Danielson G. E., Jr., Klaasen K. and Wilson L. (1975) Photometric observations of Mercury from Mariner 10. *J. Geophys. Res.* **80**, 2431–2443.

Harris A. W. (1977) The effect of tidal friction on the origin and thermal evolution of the Moon (abstract). In *Lunar Science VIII*, p. 401–402. The Lunar Science Institute, Houston.

Harris A. W. and Kaula W. M. (1975) A co-accretional model of satellite formation. *Icarus* **24**, 516–524.

Harris D. L. (1961) Photometry and colorimetry of planets and satellites. In *Planets and Satellites* (G. P. Kuiper and B. M. Middlehurst, eds.), p. 272–342. Univ. Chicago Press.

Hart R. S. (1977) The distribution of seismic velocities and attenuation in the Earth. Ph.D. thesis, Calif. Inst. of Tech., Pasadena, Calif. 354 pp.

Hart S. R. (1969) K, Rb, Cs contents and K/Rb, K/Cs ratios of fresh and altered submarine basalts. *Earth Planet. Sci. Lett.* **6**, 295–303.

Hart S. R., Brooks C., Krough T. E., Davis G. L. and Nava D. (1970) Ancient and modern volcanic rocks: a trace element model. *Earth Planet. Sci. Lett.* **10**, 17–28.

Hartmann W. K. (1973a) Martian surface and crust: review and synthesis. *Icarus* **19**, 550–575.

Hartmann W. K. (1973b) Martian cratering, 4, Mariner 9 initial analysis of cratering chronology. *J. Geophys. Res.* **78**, 4096–4116.

Hartmann W. K. (1977) Relative crater production rates on planets. *Icarus* **31**, 260–276.

Hartmann W. K. and Davis D. R. (1975) Satellite-sized planetesimals and lunar origin. *Icarus* **24**, 504–515.

Hasebe K., Fujii N. and Uyeda S. (1970) Thermal processes under island arcs. *Tectonophy.* **10**, 335–355.

Haskell N. A. (1935) The motion of a viscous fluid under a surface load. *Physics* **6**, 265–269.

Hawke B. R. and Head J. W. (1978) Lunar KREEP volcanism: geologic evidence for history and mode of emplacement. *Proc. Lunar Planet. Sci. Conf. 9th.* p. 3285–3309.

Hawkesworth C. J., O'Nions R. K. and Arculus R. J. (1979) Nd and Sr isotope geochemistry of island arc volcanics, Grenada, Lesser Antilles. *Earth Planet. Sci. Lett.* **45**, 237–248.

Hawkins J. W., Jr. (1976) Petrology and geochemistry of basaltic rocks of the Lau basin. *Earth Planet. Sci. Lett.* **28**, 283–297.

Haxby W. F. and Turcotte D. L. (1978) On isostatic geoid anomalies. *J. Geophys. Res.* **83**, 5473–5478.

Hays J. D. and Pitman W. C. (1973) Lithospheric plate motion, sea-level changes and climatic and ecological consequences. *Nature* **246**, 18–22.

Hays J. F. (1972) Radioactive heat sources in the lunar interior. *Phys. Earth Planet. Inter.* **5**, 77–84.

Head J. W. (1976) Lunar volcanism in space and time. *Rev. Geophys. Space Phys.* **14**, 265–300.

Hearn B. C. and Boyd F. R. (1975) Garnet peridotite xenoliths in a Montana, U.S.A., kimberlite. *Phys. Chem. Earth* **9**, 247–256.

Heier K. S. (1963) Uranium, thorium and potassium in eclogitic rocks. *Geochim. Cosmochim. Acta* **27**, 849–860.

Herbert F., Drake M. J. and Sonett C. P. (1978) Geophysical and geochemical evolution of the lunar magma ocean. *Proc. Lunar Planet. Sci. Conf. 9th*, p. 249–262.

Herbert F., Drake M. J., Sonett C. P. and Wiskerchen M. J. (1977a) Some constraints on the thermal history of the lunar magma ocean. *Proc. Lunar Sci. Conf. 8th*, p. 573–582.

Herbert F. and Sonett C. P. (1978) Primordial metamorphism of asteroids via electrical induction in a T-Tauri-like solar wind. *Astrophys. Space Sci.* **55**, 227–239.

Herbert F., Sonett C. P. and Wiskerchen M. J. (1977b) Model "zero-age" lunar thermal profiles resulting from electrical induction. *J. Geophys. Res.* **82**, 2054–2060.

Herbert F., Wiskerchen M. J., Sonett C. P. and Chao J. K. (1976) Solar wind induction in Mercury: constraints on the formation of a magnetosphere. *Icarus* **28**, 489–500.

Hess H. H. (1962) History of ocean basins. In *Petrologic Studies, a Volume in Honor of A. F. Buddington* (A.E.J. Engel, J. L. James, and B. F. Leonard, eds.), p. 599–620. Geol. Soc. of Amer., N.Y.

Hewitt J. M., McKenzie D. P. and Weiss N. O. (1975) Dissipative heating in convective flows. *J. Fluid Mech.* **68**, 721–738.

Hoffman J. H., Keating G. M., Niemann H., Oyama V., Pollack J., Seiff A., Stewart A. I. and Von Zahn U. (1977) Composition and structure of the atmosphere of Venus. *Space Sci. Rev.* **20**, 307–328.

Hoffman J. H., Hodges R. R., Jr., McElroy M. B., Donahue T. M. and Kolpin M. (1979) The composition and structure of the Venus atmosphere: results from Pioneer Venus. *Science* **205**, 49–52.

Holmes A. (1915) Radioactivity and the Earth's thermal history. *Geol. Mag.* **62**, 102–112.

Holmes A. (1931) Radioactivity and Earth movements, XVIII. *Trans. Geol. Soc. Glasgow* **18**, 559–606.

Hood L., Russell C. T. and Coleman P. J., Jr. (1978) Evidence for a non-random magnetization of the Moon. *Geophys. Res. Lett.* **5**, 305–308.

Horai K. and Simmons G. (1970) An empirical relationship between thermal conductivity and Debye temperature for silicates. *J. Geophys. Res.* **75**, 978–982.

Horai K., Simmons G., Kanamori H. and Wones D. (1970) Thermal diffusivity, conductivity and thermal inertia of Apollo 11 lunar material. *Apollo 11 Lunar Sci. Conf.*, p. 2243–2249.

Horai K. and Winkler J. L., Jr. (1976) Thermal diffusivity of four Apollo 17 rock samples. *Proc. Lunar Sci. Conf. 7th*, p. 3183–3204.

Houston M. H., Jr. and DeBremaecker J. C. (1975) Numerical models of convection in the upper mantle. *J. Geophys. Res.* **80**, 742–751.

Howard H. T., Tyler G. L., Fielbo G., Kliore A. J., Levy G. S., Brunn D. L., Dickinson R., Edelson R. E., Martin W. L., Postal R. B., Seidel B., Sesplaukis T. T., Shirley D. L., Stelzvied C. T., Sweetnam D. N., Zygielbaum A. I., Esposito P. B., Anderson J. D., Shapiro I. I. and Reasenberg R. D. (1974) Venus: mass, gravity field, atmosphere and ionosphere as measured by the Mariner 10 dual frequency radio system. *Science* **183**, 1297–1301.

Howard L. N. (1966) Convection at high Rayleigh number. In *Proc. 11th Cong. Appl. Mech.* (H. Gortler, ed.), p. 1109–1115. Springer-Verlag, N.Y.

Howard L. N., Malkus W. V. R. and Whitehead J. A. (1970) Self-convection of floating heat sources: a model for continental drift. *Geophys. Fluid Dynamics* **1**, 123–142.

Hsui A. T. (1978) Numerical solution of finite-amplitude thermal convection with large viscosity variation in axisymmetric geometry: effect of mechanical boundary conditions. *Tectonophy.* **50**, 147–162.

Hsui A. T. (1979) The bounds of the heat production rate within the Moon. *Proc. Lunar Planet. Sci. Conf. 10th*, p. 2393–2402.

Hsui A. T. and Toksöz M. N. (1977) Thermal evolution of planetary size bodies. *Proc. Lunar Sci. Conf. 8th*, p. 447–461.

Hsui A. T. and Toksöz M. N. (1979) The evolution of thermal structures beneath a subduction zone. *Tectonophy.* **60**, 43–60.

Hsui A. T., Turcotte D. L. and Torrance K. E. (1972) Finite-amplitude thermal convection within a self-gravitating fluid sphere. *Geophys. Fluid Dyn.* **3**, 35–44.

Hubbard N. J. and Minear J. W. (1975) A physical and chemical model of early lunar history. *Proc. Lunar Sci. Conf. 6th*, p. 1057–1085.

Hubbard N. J. and Minear J. W. (1976) Petrogenesis in a modestly endowed Moon. *Proc. Lunar Sci. Conf. 7th*, p. 3421–3435.

Huebner J. S., Duba A. and Wiggins L. B. (1979) Electrical conductivity of pyroxene which contains trivalent cations: laboratory measurements and the lunar temperature profile. *J. Geophys. Res.* **84**, 4652–4656.

Hunten D. M., McGill G. E. and Nagy A. E. (1977) Current knowledge of Venus. *Space Sci. Rev.* **20**, 265–282.

Hurst R. W. (1978) Sr evolution in the west Greenland-Labrador craton: a model for early Rb depletion in the mantle. *Geochim. Cosmochim. Acta* **42**, 39–44.

Imhoff C. L. (1978) T-Tauri star evolution and evidence for planetary formation. In *Protostars and Planets* (T. Gehrels, ed.), p. 699–707. Univ. Ariz. Press, Tucson.

Isacks B. L. and Barazangi M. (1977) Geometry of Benioff zones: lateral segmentation and downwards bending of the subducted lithosphere. In *Island Arcs, Deep Sea Trenches and Back-Arc Basins* (M. Talwani and W. C. Pitman III, eds.), Maurice Ewing Ser. 1, p. 99–113. Amer. Geophys. Union, Washington, D.C.

Isacks B. and Molnar P. (1971) Distribution of stresses in the descending lithosphere from a global survey of focal-mechanism solutions of mantle earthquakes. *Rev. Geophys. Space Phys.* **9**, 103–174.

Istomin V. G., Grechnev K. V., Kochnev V. A., Pavlenko V. A., Ozerov L. N. and Klimovitskiy V. G. (1979) Measurements of the chemical and isotopic composition of an atmosphere by means of a mass-spectrometer. *Pis'ma v Astronomicheskiy Zhurnal* **5**, 3–4.

Iverson J., Greeley R. and Pollack J. B. (1976) Windblown dust on Earth, Mars and Venus. *J. Atmos. Sci.* **33**, 2425–2429.

Jackson I. (1976) Melting of the silica isotype SiO_2, BeF_2, and GeO_2 at elevated pressures. *Phys. Earth Planet. Inter.* **13**, 218–231.

Jackson I. and Ahrens T. J. (1979) Shock wave compression of single-crystal forsterite. *J. Geophys. Res.* **84**, 3039–3048.

Jeanloz R. and Ahrens T. J. (1977) Pyroxenes and olivines: structural implications of shock-wave data for high pressure phases. In *High Pressure Research: Applications in Geophysics* (M. H. Manghnani and S. Akimoto, eds.), p. 439–461. Academic Press, N. Y.

Jeanloz R. and Richter F. M. (1979) Convection, composition, and the thermal state of the lower mantle. *J. Geophys. Res.* **84**, 5497–5504.

Jeffreys H. (1929) *The Earth*, 2nd ed. MacMillan Co., N.Y. 346 pp.

Jeffreys H. (1930) The instability of a compressible fluid heated below. *Proc. Cambridge Phil. Soc.* **26**, 170–172.

Jeffreys H. and Bland M. E. M. (1950) The instability of a fluid sphere heated within. *Mon. Not. Roy. Astron. Soc., Geophys. Suppl.* **6**, 148–158.

Johnston D. H., McGetchin T. R. and Toksöz M. N. (1974) The thermal state and internal structure of Mars. *J. Geophys. Res.* **79**, 3959–3971.

Johnston D. H. and Toksöz M. N. (1977) Internal structure and properties of Mars. *Icarus* **32**, 73–84.

Jordan T. H. (1975) The continental tectosphere. *Rev. Geophys. Space Phys.* **13** **(3)**, 1–12.

Jordan T. H. (1977) Lithospheric slab penetration into the lower mantle beneath the Sea of Okhotsk. *J. Geophys.* **43**, 473–496.

Jordan T. H. (1978) Composition and development of the continental tectosphere. *Nature* **274**, 544–548.

Kanamori H., Fujii N. and Mitzutani H. (1968) Thermal diffusivity measurement of rock forming minerals. *J. Geophys. Res.* **73**, 595–605.

Karig D. E. (1970) Ridges and basins of the Tonga-Kermadec island arc system. *J. Geophys. Res.* **75**, 239–255.

Karig D. E. (1971) Origin and development of marginal basins in the western Pacific. *J. Geophys. Res.* **76**, 2542–2561.

Kaula W. M. (1964) Tidal dissipation by solid friction and the resulting orbital evolution. *Rev. Geophys.* **2**, 661–685.

Kaula W. M. (1979a) Thermal evolution of Earth and Moon growing by planetesimal impacts. *J. Geophys. Res.* **84**, 999–1008.

Kaula W. M. (1979b) Moment of inertia of Mars. *Geophys. Res. Lett.* **6**, 194–196.

Kaula W. M. (1980a) The beginning of the Earth's thermal evolution. In *"The Continental Crust and its Mineral Deposits."* (D. W. Strangway, ed.), Geol. Assoc. Canada, Special Paper 20, in press.

Kaula W. M. (1980b) Material properties for mantle convection consistent with observed surface fields. *J. Geophys. Res.* **72**, 6261-6273.

Kaula W. M. and Bigeleisen P. E. (1975) Early scattering by Jupiter and its collision effects in the terrestrial zone. *Icarus* **25**, 18–23.

Kaula W. M. and Yoder C. F. (1976) Lunar orbit evolution and heating of the Moon (abstract). In *Lunar Science VII*, p. 440–442. The Lunar Science Institute, Houston.

Kay R., Hubbard N. J. and Gast P. W. (1970) Chemical characteristics and origin of oceanic ridge volcanic rocks. *J. Geophys. Res.* **75**, 1585–1613.

Kay R. W., Sun S.-S. and Lee-Hu C.-N. (1978) Pb and Sr isotopes in volcanic rocks from the Aleutian Islands and Pribilof Islands, Alaska. *Geochim. Cosmochim. Acta* **42**, 263–273.

Kelvin Lord (1864) On the secular cooling of the Earth. *Trans. Roy. Soc. Edinburgh* **23**, 157–170.

Kennedy G. C. and Higgins G. H. (1972) Melting temperatures in the Earth's mantle. *Tectonophy.* **13**, 221–232.

Kesson S. E. and Ringwood A. E. (1976) Mare basalt petrogenesis in a dynamic Moon. *Earth Planet. Sci. Lett.* **30**, 155–163.

Kieffer H. H., Chase S. C., Jr., Martin T. Z., Miner E. D. and Palluconi F. D. (1976) Martian north pole summer temperatures: dirty water ice. *Science* **194**, 1341–1344.

Kieffer S. W. (1976) Lattice thermal conductivity within the Earth and considerations of a relationship between the pressure dependence of the thermal diffusivity and the volume dependence of the Grüneisen parameter. *J. Geophys. Res.* **81**, 3025–3030.

Kieffer S. W., Getting I. C. and Kennedy G. C. (1976) Experimental determination of the pressure dependence of the thermal diffusivity of teflon, sodium chloride, quartz, and silica. *J. Geophys. Res.* **81**, 3018–3024.

Kingery W. D., France J., Coble R. L. and Vasilos T. (1954) Thermal conductivity, 10, Data for several pure oxide materials corrected to zero porosity. *J. Amer. Ceram. Soc.* **37**, 107–111.

Kirby S. H. and Raleigh C. B. (1973) Mechanisms of high-temperature, solid-state flow in minerals and ceramics and their bearing on the creep behavior of the mantle. *Tectonophy.* **19**, 165–194.

Knopoff L., Poehls K. A. and Smith R. C. (1972) Drift of continental rafts with asymmetric heating. *Science* **176**, 1023–1024.

Kohlstedt D. L. and Goetze C. (1974) Low-stress high-temperature creep in olivine single crystals. *J. Geophys. Res.* **79**, 2045–2051.

Kohlstedt D. L., Goetze C. and Durham W. B. (1976) Experimental deformation of single crystal olivine with application to flow in the mantle. In *The Physics and Chemistry of Minerals and Rocks* (R. G. J. Strens, ed.), p. 35–49. John Wiley, London.

Kolle J. J. and Blacic J. D. (1978) Mechanical deformation of a single crystal hypersthene (abstract). *EOS (Trans. Amer. Geophys. Union)* **59**, 1185.

Kopitzke U. (1979) Finite element convection models - comparison of shallow and deep mantle convection, and temperatures in the mantle. *J. Geophys.* **46**, 97–121.

Kraichnan R. H. (1962) Mixing-length analysis of turbulent thermal convection at arbitrary Prandtl numbers. *Phys. Fluids* **5**, 1374–1389.

Kresten P. (1974) Uranium in kimberlites and associated rocks, with special reference to Lesotho occurrences. *Lithos* **7**, 171–180.

Kuckes A. F. (1977) Strength and rigidity of the elastic lunar lithosphere and implications for present-day mantle convection in the Moon. *Phys. Earth Planet. Inter.* **14**, 1–12.

Kulacki F. A. and Emara A. A. (1977) Steady and transient thermal convection in a fluid layer with uniform volumetric energy sources. *J. Fluid Mech.* **83**, 375–395.

Kulacki F. A. and Nagle M. E. (1975) Natural convection in horizontal fluid layers with volumetric energy sources. *J. Heat Transfer* **97**, 204–211.

Kumar S. (1976) Mercury's atmosphere: a perspective after Mariner 10. *Icarus* **28**, 579–591.

Kushiro I., Syono Y. and Akimoto S. (1967) Stability of phlogopite at high pressures and possible presence of phlogopite in the Earth's upper mantle. *Earth Planet. Sci. Lett.* **3**, 197–203.

Lachenbruch A. H. (1968) Preliminary geothermal model of the Sierra Nevada. *J. Geophys. Res.* **73**, 6977–6989.

Lachenbruch A. H. (1970) Crustal temperature and heat production: implications of the linear heat-flow relation. *J. Geophys. Res.* **75**, 3291–3300.

Lambeck K. (1975) Effects of tidal dissipation in the oceans on the Moon's orbit and the Earth's rotation. *J. Geophys. Res.* **80**, 2917–2925.

Lammlein D. R., Latham G. V., Dorman J., Nakamura Y. and Ewing M. (1974) Lunar seismicity, structure, and tectonics. *Rev. Geophys. Space Phys.* **12**, 1–22.

Langseth M. G., Keihm S. J. and Peters K. (1976) Revised lunar heat-flow values. *Proc. Lunar Sci. Conf. 7th*, p. 3143–3171.

Langseth M. G., LePichon X. and Ewing M. (1966) Crustal structure of the mid-ocean ridges: heat flow through the Atlantic Ocean floor and convection currents. *J. Geophys. Res.* **71**, 5321–5355.

Lanoix M., Strangway D. W. and Pearce G. W. (1978) The primordial magnetic field preserved in chondrules of the Allende meteorite. *Geophys. Res. Lett.* **5**, 73–76.

Lee T., Papanastassiou D. A and Wasserberg G. J. (1976) Demonstration of ^{26}Mg excess in Allende and evidence for ^{26}Al. *Geophys. Res. Lett.* **3**, 109–112.

Lee W. H. (1968). Effects of selective fusion on the thermal history of the Earth's mantle. *Earth Planet. Sci. Lett.* **4**, 270–276.

Leeds A. R., Knopoff L. and Kausel E. G. (1974) Variations of upper mantle structure under the Pacific Ocean. *Science* **186**, 141–143.

Levin B. J. (1962) Thermal history of the Moon. In *The Moon* (Z. Kopal and Z. K. Mikhailov, eds.), p. 157–167. Academic Press, N.Y.

Lewis J. S. (1970) Venus: atmospheric and lithospheric composition. *Earth Planet. Sci. Lett.* **10**, 73–80.

Lewis J. S. (1972) Metal/silicate fractionation in the solar system. *Earth Planet. Sci. Lett.* **15**, 286–290.

Lewis J. S. (1973) Chemistry of the planets. *Ann. Rev. Phys. Chem.* **24**, 339–351.

Lin R. P., Anderson K. A., Russell C. T., Boyce J., Masursky H., Wilhelms D. E. and Stuart-Alexander D. E. (1977) Correlations of lunar surface remanent magnetic fields to surface geologic age (abstract). In *Lunar Science VIII, Supplement A*, p. 14–15. The Lunar Science Institute, Houston.

Liu L. G. (1975a) On the (γ,ϵ,l) triple point of iron and the Earth's core. *Geophys. J. Roy. Astron. Soc.* **43**, 697–705.

Liu L. G. (1975b) Post-oxide phases of olivine and pyroxene and the mineralogy of the mantle. *Nature* **258**, 510–513.

Liu L. G. (1976) The post-spinel phase of forsterite. *Nature* **262**, 770–772.

Liu L. G. (1979) On the 650-km phase discontinuity. *Earth Planet. Sci. Lett.* **42**, 202–208.

Liu L. G. and Bassett W. A. (1975) The melting of iron up to 200 kbar. *J. Geophys. Res.* **80**, 3777–3782.

Long R. R. (1976) Relation between Nusselt number and Rayleigh number in turbulent thermal convection. *J. Fluid Mech.* **73**, 445–451.

Longhi J. (1977) Magma oceanography 2: Chemical evolution and crustal formation. *Proc. Lunar Sci. Conf. 8th*, p. 601–621.

Longhi J. (1978) Pyroxene stability and the composition of the lunar magma ocean. *Proc. Lunar Planet. Sci. Conf. 9th*, p. 285–306.

Lorell J., Born G. H., Christensen E. J., Jordon J. F., Laing P. A., Martin W. L., Sjogren W. L., Shapiro I. I., Reasenberg R. D. and Slater G. L. (1972) Mariner 9 celestial mechanics experiment: gravity field and pole direction of Mars. *Science* **175**, 317–320.

Lubimova E. A. (1958) Thermal history of the Earth with consideration of the variable thermal conductivity of its mantle. *Geophys. J. Roy. Astron. Soc.* **1**, 115–134.

Lubimova E. A. and Nikitina V. N. (1975) On heat flow singularities over mid-ocean ridges. *J. Geophys. Res.* **80**, 232–243.

Lucchitta B. K. (1976) Mare ridges and related highland scarps—result of vertical tectonism? *Proc. Lunar Sci. Conf. 7th*, p. 2761–2782.

Lucchitta B. K. and Klockenbrink J. L. (1979) Ridges and scarps in the equatorial belt of Mars (abstract). In *Lunar and Planetary Science X*, p. 750–752. Lunar and Planetary Institute, Houston.

Lucchitta B. K. and Watkins J. A. (1978) Age of graben systems on the Moon. *Proc. Lunar Planet. Sci. Conf. 9th*, p. 3459–3472.

Lugmair G. W. (1975) Sm-Nd systematics of some Apollo 17 basalts (abstract). In *Origins of Mare Basalts and Their Implications for Lunar Evolution*, p. 107–110. The Lunar Science Institute, Houston.

Lugmair G. W., Scheinin N. B. and Marti K. (1975) Sm-Nd age and history of Apollo 17 basalt 75075: evidence for early differentiation of the lunar exterior. *Proc. Lunar Sci. Conf. 6th*, p. 1419–1429.

Lyot B. (1930) La Polarisation de Mercure compareé a celle de la Lune: Resultats obtenus au Pic-du-Midi en 1930. *Comptes Rendus Acad. Sci.* 29, Oct. 1930.

Ma C. B. (1974) Reinvestigation of the olivine-spinel transformation in Ni_2SiO_4 and the incongruent melting of Ni_2SiO_4 olivine. *J. Geophys. Res.* **79**, 3321–3324.

MacDonald G. J. F. (1959) Calculations on the thermal history of the Earth. *J. Geophys. Res.* **64**, 1967–2000.

MacDonald G. J. F. (1960) Stress history of the Moon. *Planet. Space Sci.* **2**, 249–255.

MacDougall J. D. (1977) Uranium in marine basalts: concentration, distribution and implications. *Earth Planet. Sci. Lett.* **35**, 65–70.

MacGregor I. D. (1974) The system $MgO\text{-}Al_2O_3\text{-}SiO_2$: solubility of Al_2O_3 in enstatite for spinel and garnet peridotite compositions. *Amer. Mineral.* **59**, 110–119.

MacGregor I. D. and Basu A. R. (1974) Thermal structure of the lithosphere: a petrologic model. *Science* **185**, 1007–1011.

Magnitsky V. A. (1967) *The Internal Structure and Physics of the Earth*. NASA Transl. N67-23321. 447 pp.

Malin M. C. (1976a) Age of martian channels. *J. Geophys. Res.* **81**, 4825–4845.

Malin M. C. (1976b) Observations of intercrater plains on Mercury. *Geophys. Res. Lett.* **3**, 581–584.

Malin M. C. (1977) Comparison of volcanic features of Elysium (Mars) and Tibesti (Earth). *Geol. Soc. Amer. Bull.* **88**, 908–919.

Malin M. C. and Saunders R. S. (1977) Surface of Venus: evidence of diverse landforms from radar observations. *Science* **196**, 987–990.

Mao H. K. (1973) Observations of optical and electrical conductivity in magnesia wüstite at high pressures. *Carnegie Inst. Wash., Yearb.* **72**, 554–557.

Mao H. K. (1976) Charge transfer processes at high pressure. In *The Physics and Chemistry of Minerals and Rocks* (R. G. J. Strens, ed.), p. 573–581. Academic Press, N.Y.

Mao H. K. and Bell P. M. (1972) Electrical conductivity and the red shift of absorption in olivine and spinel at high pressures. *Science* **176**, 403–406.

Mason B. (1968) Composition of stony meteorites. In *Extra-terrestrial Matter* (C. A. Randall, ed.), p. 3–22. Northern Illinois Univ. Press.

Mason B. (1971) *Handbook of Elemental Abundances in Meteorites*. Gordon and Breach, N.Y. 555 pp.

Masursky H., Kaula W. M., McGill G., Pettengill G. H., Phillips R. J., Russell C. T., Schubert G. and Shapiro I. I. (1977) The surface and interior of Venus. *Space Sci. Rev.* **20**, 431–451.

Matsui T. and Mizutani H. (1977) Why is a minor planet minor? *Nature* **270**, 506–507.

McConnell R. K., Jr. and Gast P. W. (1972) Lunar thermal history revisited. *The Moon* **5**, 41–51.

McConnell R. K., Jr., McClaine L. A., Lee D. W., Aronson J. R. and Allen R. V. (1967) A model for planetary igneous differentiation. *Rev. Geophys.* **5**, 121-172.

McCord T. B. and Adams J. B. (1972) Mercury: surface composition from the reflection spectrum. *Science* **178**, 745-747.

McCord T. B., Adams J. B. and Johnson T. V. (1970) Asteroid Vesta: spectral reflectivity and compositional implications. *Science* **168**, 1445-1447.

McElhinny M. W. and Evans M. E. (1968) An investigation of the strength of the geomagnetic field in the early Precambrian. *Phys. Earth Planet. Inter.* **1**, 485-497.

McElroy M. B., Yung Y. L. and Nier A. O. (1976) Isotopic composition of nitrogen: implications for the past history of Mars' atmosphere. *Science* **194**, 70-72.

McFadden L. A., McCord T. B. and Pieters C. (1977) Vesta: the first new spectroscopic measurements. *Icarus* **31**, 439-446.

McGetchin T. R. and Smyth J. R. (1978) The mantle of Mars: some possible geological implications of its high density. *Icarus* **34**, 512-536.

McKenzie D. P. (1967) Some remarks on heat flow and gravity anomalies. *J. Geophys. Res.* **72**, 6261-6273.

McKenzie D. P. (1969) Speculations on the consequences and causes of plate motions. *Geophys. J. Roy. Astron. Soc.* **18**, 1-32.

McKenzie D. P. (1977a) Surface deformation, gravity anomalies and convection. *Geophys. J. Roy. Astron. Soc.* **48**, 211-238.

McKenzie D. P. (1977b) The initiation of trenches: a finite amplitude instability. In *Island Arcs, Deep Sea Trenches, and Back-Arc Basins* (M. Talwani and W. C. Pitman, eds.), Maurice Ewing Series 1, p. 57–61. Amer. Geophys. Union, Washington, D.C.

McKenzie D. P. (1979) Finite deformation during fluid flow. *Geophys. J. Roy. Astron. Soc.* **58**, 689–715.

McKenzie D. P. and Bowin C. (1976) The relationship between bathymetry and gravity in the Atlantic Ocean. *J. Geophys. Res.* **81**, 1903–1915.

McKenzie D. P., Roberts J. and Weiss N. (1973) Numerical models of convection in the Earth's mantle. *Tectonophy.* **19**, 89–103.

McKenzie D. P., Roberts J. M. and Weiss N. O. (1974) Convection in the Earth's mantle: towards a numerical simulation. *J. Fluid Mech.* **62**, 465–538.

McKenzie D. P. and Sclater J. G. (1968) Heat flow inside the island arcs of the northwestern Pacific. *J. Geophys. Res.* **73**, 3173–3179.

McKenzie D. P. and Weiss N. O. (1975) Speculations on the thermal and tectonic history of the Earth. *Geophys. J. Roy. Astron. Soc.* **42**, 131–174.

McNutt M. (1979) Compensation of oceanic topography: an application of the response function technique to the *Surveyor* area. *J. Geophys. Res.* **84**, 7589–7598.

McQueen R. G., Jamieson J. C. and Marsh S. P. (1967) Shock-wave compression and X-ray studies of titanium dioxide. *Science* **155**, 1401–1404.

Melosh H. J. (1977) Global tectonics of a despun planet. *Icarus* **31**, 221–243.

Melosh H. J. and Dzurisin D. (1978) Mercurian global tectonics: consequence of tidal despinning? *Icarus* **35**, 227–236.

Menard H. W. (1978) Fragmentation of the Farallon plate by pivoting subduction *J. Geol.* **86**, 99–110.

Mercier J.-C. C., Anderson D. A. and Carter N. L. (1977) Stress in the lithosphere: inferences from steady state flow of rocks. *Pure Appl. Geophys.* **115**, 199–226.

Mercier J.-C. C. and Carter N. L. (1975) Pyroxene geotherms. *J. Geophys. Res.* **80**, 3349-3362.

Metzger A. E., Haines E. L., Parker R. E. and Radocinski R. G. (1977) Thorium concentrations in the lunar surface. I: regional values and crustal content. *Proc. Lunar Sci. Conf. 8th*, p. 949–999.

Milton D. J. (1973) Water and processes of degradation in the martian landscape. *J. Geophys. Res.* **78**, 4037–4047.

Minear J. W., Clow G. and Fletcher C. R. (1979) Thermal models of asteroids (abstract). In *Lunar and Planetary Science X*, p. 842–843. Lunar and Planetary Institute, Houston.

Minear J. W. and Fletcher C. R. (1978) Crystallization of a lunar magma ocean. *Proc. Lunar Planet. Sci. Conf. 9th*, p. 263–283.

Minear J. W. and Toksöz M. N. (1970a) Thermal regime of a downgoing slab and new global tectonics. *J. Geophys. Res.* **75**, 1397–1419.

Minear J. W. and Toksöz M. N. (1970b) Thermal regime of a downgoing slab. *Tectonophy.* **10**, 367–390.

Minster J. B., Jordan T. H., Molnar P. and Haines E. (1974) Numerical modelling of instantaneous plate tectonics. *Geophys. J. Roy. Astron. Soc.* **36**, 541–576.

Mirwald P. W. and Kennedy G. C. (1979) The Curie point and the α-γ transition of iron to 53 kbar—a reexamination. *J. Geophys. Res.* **84**, 656–658.

Mooney D. L. and Steg R. G. (1969) Pressure dependence of the thermal conductivity and ultrasonic attenuation of non-metallic solids. *High Temp. High Pressure* **1**, 237–240.

Mizutani H., Matsui T. and Takeuchi K. (1972) Accretion process of the Moon. *The Moon* **8**, 476–489.

Morgan W. J. (1968) Rises, trenches, great faults, and crustal blocks. *J. Geophys. Res.* **73**, 1959–1982.

Morgan W. J. (1971) Convection plumes in the lower mantle. *Nature* **230**, 42–43.

Morgan W. J. (1972a) Deep mantle convection plumes and plate motions. *Amer. Assn. Petrol. Geol. Bull.* **56**, 203–213.

Morgan W. J. (1972b) Plate motions and deep mantle convection. In *Studies in Earth and Space Sciences*, Geol. Soc. Amer. Mem. 132, p. 7–22. Geol. Soc. Amer. Boulder, Colorado.

Morrison D. (1970) Thermophysics of the planet Mercury. *Space Sci. Rev.* **11**, 271–307.

Muehlberger W. R. (1974) Structural history of southeastern Mare Serenitatis and adjacent highlands. *Proc. Lunar Sci. Conf. 5th*, p. 101–110.

Mueller R. F. (1969) Effect of temperature on the strength and composition of the upper lithosphere of Venus. *Nature* **224**, 354–356.

Muller P. M. and Sjogren W. L. (1968) Mascons: Lunar mass concentrations. *Science* **161**, 680–684.

Murray B. C., Strom R. G., Trask N. J. and Gault D. E. (1975) Surface history of Mercury: implications for terrestrial planets. *J. Geophys. Res.* **80**, 2508–2514.

Murthy V. R. (1976) Composition of the core and the early chemical history of the Earth. In *The Early History of the Earth* (B. F. Windley, ed.), p. 21–31. Wiley, N.Y.

Mutch T. A. and Saunders R. S. (1976) The geologic development of Mars: a review. *Space Sci. Rev.* **19**, 3–57.

Nagata T. (1979) Meteorite magnetism and the early solar system magnetic field. *Phys. Earth Planet. Inter.* **20**, 324–341.

Nakamura Y. (1977) HFT events: shallow moonquakes? *Phys. Earth Planet. Inter.* **14**, 217–223.

Nakamura Y., Duennebier F. K., Latham G. V. and Dorman H. J. (1976) Structure of the lunar mantle. *J. Geophys. Res.* **81**, 4818–4824.

Nakamura Y., Latham G., Lammlein D., Ewing M., Duennebier F. and Dorman J. (1974) Deep lunar interior inferred from recent seismic data. *Geophys. Res. Lett.* **1**, 137–140.

Nakamura Y., Latham G. V., Dorman H. J., Ibrahim A. K., Koyama J. and Horvath P. (1979) Shallow moonquakes: depth, distribution and implications as to the present state of the lunar interior. *Proc. Lunar Planet. Sci. Conf. 10th*, p. 2299–2309.

Ness N. F., Behannon K. W., Lepping R. P. and Whang Y. C. (1975) The magnetic field of Mercury, 1. *J. Geophys. Res.* **80**, 2708–2716.

Ness N. F., Behannon K. W., Lepping R. P. and Whang Y. C. (1976) Observations of Mercury's magnetic field. *Icarus* **28**, 479–488.

Ness N. F., Behannon K. W. Lepping R. P., Whang Y. C. and Schatten K. H. (1974) Magnetic field observations near Mercury: preliminary results from Mariner 10. *Science* **185**, 151–160.

Neukum G. and Wise D. U. (1976) Mars: a standard crater curve and possible new time scale. *Science* **194**, 1381–1387.

Nicolas A., Boudier F. and Boullier A. M. (1973) Mechanisms of flow in naturally and experimentally deformed peridotites. *Amer. J. Sci.* **273**, 873–876.

Nier A. O., Hanson W. B., Seiff A., McElroy M. B., Spencer N. W., Duckett R. J., Knight T. C. D. and Cook W. S. (1976a) Composition and structure of the martian atmosphere: preliminary results from Viking I. *Science* **193**, 786–788.

Nier A. O., McElroy M. B. and Yung Y. L. (1976b) Isotopic composition of the martian atmosphere. *Science* **194**, 68–70.

Nixon P. H. and Boyd F. R. (1979) Garnet bearing lherzolites and discrete nodule suites from the Malaita alnoite, Solomon Islands, S.W. Pacific, and their bearing on oceanic mantle composition and geotherm. In *The Mantle Sample: Inclusions in Kimberlites and Other Volcanics, Proc. Sec. Inter. Kimberlite Conf.* **2** (F. R. Boyd and H. O. A. Meyer, eds.), p. 400–423. Amer. Geophys. Union, Washington, D.C.

Ockendon J. R. and Turcotte D. L. (1977) On the gravitational potential and field anomalies due to thin mass layers. *Geophys. J. Roy. Astron. Soc.* **48**, 479–492.

O'Connell R. J. (1971) Pleistocene glaciation and the viscosity of the lower mantle. *Geophys. J. Roy. Astron. Soc.* **23**, 299–327.

O'Connell R. J. (1977) On the scale of mantle convection. *Tectonophy.* **38**, 119–136.

Okal E. A. and Anderson D. L. (1975) A study of lateral inhomogeneities in the upper mantle by multiple ScS travel-time residuals. *Geophys. Res. Lett.* **2**, 313–316.

O'Keefe J. A. (1968) Isostasy on the Moon. *Science* **162**, 1405–1406.

O'Keefe J. D. and Ahrens T. J. (1977) Impact-induced energy partitioning, melting and vaporization on terrestrial planets. *Proc. Lunar Sci. Conf. 8th*, p. 3357–3374.

Oldenburg D. W. (1975) A physical model for the creation of the lithopshere. *Geophy. J. Roy. Astron. Soc.* **43**, 425–451.

Ornatskaya O. I., Al'ber Y. I. and Ryazantseva I. L. (1977) Calculations of the Moon's thermal history at different concentrations of radioactive elements, taking into account differentiation on melting. In *The Soviet-American Conference on Cosmochemistry of the Moon and Planets*, p. 347–366. NASA SP-300.

Oversby V. and Ewart A. (1972) Lead isotopic compositions of Tonga-Kermadec volcanics and their petrogenetic significance. *Contrib. Mineral. Petrol.* **37**, 181–210.

Oversby V. and Ringwood A. E. (1971) Time of formation of the Earth's core. *Nature* **234**, 463–465.

Owen T. and Biemann K. (1976) Composition of the atmosphere at the surface of Mars: detection of Argon-36 and preliminary analysis. *Science* **193**, 801–803.

Owen T., Biemann K., Rushneck D. R., Biller J. E., Howarth D. W and La Fleur A. L. (1976) The atmosphere of Mars: detection of Krypton and Xenon. *Science* **194**, 1293–1295.

Oxburgh E. R. and Parmentier E. M. (1978) Thermal processes in the formation of continental lithosphere. *Phil. Trans. Roy. Soc. London* **A288**, 415–429.

Oxburgh E. R. and Turcotte D. L. (1968) Problems of high heat flow and volcanism associated with zones of descending mantle convective flow. *Nature* **216**, 1041–1043.

Oxburgh E. R. and Turcotte D. L. (1970) Thermal structure of island arcs. *Geol. Soc. Amer. Bull.* **81**, 1665–1688.

Oxburgh E. R. and Turcotte D. L. (1978) Mechanisms of continental drift. *Rep. Prog. Phys.* **41**, 1249–1312.

Oyama V. I., Carle G. C., Woeller F., Pollack J. B., Reynolds R. T. and Craig R. A. (1980) Pioneer Venus gas chromatography of the lower atmosphere of Venus. *J. Geophys. Res.* **85**, 7891–7902.

Ozima M., Saito K. and Honda M. (1977) Sea water weathering effect on K-Ar age of submarine basalts. *Geochim. Cosmochim. Acta* **41**, 453–461.

Papanastassiou D. A. and Wasserburg G. J. (1971) Lunar chronology and evolution from Rb-Sr studies of Apollo 11 and 12 samples. *Earth Planet. Sci. Lett.* **11**, 37–62.

Papanastassiou D. A. and Wasserburg G. J. (1975) Rb-Sr study of a lunar dunite and evidence for early lunar differentiates. *Proc. Lunar Sci. Conf. 6th*, p. 1467–1489.

Papanastassiou D. A. and Wasserburg G. J. (1976) Rb-Sr age of troctolite 76535. *Proc. Lunar Sci. Conf. 7th*, p. 2035–2054.

Parker R. L. and Oldenburg D. W. (1973) Thermal model of ocean ridges. *Nature Phys. Sci.* **242**, 137–139.

Parmentier E. M. (1978) Study of thermal convection in non-Newtonian fluids. *J. Fluid Mech.* **84**, 1–11.

Parmentier E. M. and Turcotte D. L. (1974) An explanation of the pyroxene geotherm based on plume convection in the upper mantle. *Earth Planet. Sci. Lett.* **24**, 209–212.

Parmentier E. M. and Turcotte D. L. (1978) Two-dimensional mantle flow beneath a rigid accreting lithosphere. *Phys. Earth Planet. Inter.* **17**, 281–289.

Parmentier E. M., Turcotte D. L. and Torrance K. E. (1975) Numerical experiments on the structure of mantle plumes *J. Geophys. Res.* **80**, 4417–4424.

Parmentier E. M., Turcotte D. L. and Torrance K. E. (1976) Studies of finite amplitude non-Newtonian thermal convection with application to convection in the Earth's mantle. *J. Geophys. Res.* **81**, 1839–1846.

Parsons B. and Sclater J. G. (1977) An analysis of the variation of ocean floor bathymetry and heat flow with age. *J. Geophys. Res.* **82**, 803–827.

Peale S. J. (1976) Does Mercury have a molten core? *Nature* **262**, 765–766.

Peale S. J. and Cassen P. (1978) Contribution of tidal dissipation to lunar thermal history. *Icarus* **36**, 245–269.

Peale S. J., Cassen P. and Reynolds R. T. (1979) Melting of Io by tidal dissipation. *Science* **203**, 892–894.

Peale S. J. and Gold T. (1965) Rotation of the planet Mercury. *Nature* **206**, 1240–1241.

Pekeris C. L. (1935) Thermal convection in the interior of the Earth. *Mon. Not. Roy. Astron. Soc., Geophys. Suppl.* **3**, 343–367.

Peltier W. R. (1972) Penetrative convection in the planetary mantle. *Geophys. Fluid Dyn.* **5**, 47–88.

Peltier W. R. (1976) Glacial-isostatic adjustment—II. The inverse problem. *Geophys. J. Roy. Astron. Soc.* **46**, 669–705.

Peltier W. R. (1980) Mantle convection and viscosity. In *Proc. of the Enrico Fermi Inter. School of Physics (Course LXXVIII)* (A. Dziewonski and E. Boschi, eds.). Academic Press, N.Y., in press.

Peltier W. R. and Andrews J. T. (1976) Glacial-isostatic adjustment, I, the forward problem *Geophys. J. Roy. Astron. Soc.* **46**, 605–646.

Pettengill G. H., Ford P. G., Brown W. E., Kaula W. M., Keller C. H., Masursky H. and McGill G. E. (1979a) Pioneer Venus radar mapper experiment. *Science* **203**, 806–808.

Pettengill G. H., Ford P. G., Brown W. E., Kaula W. M., Masursky H., Eliason E. and McGill G. E. (1979b) Venus: preliminary topographic and surface imaging results from the Pioneer Orbiter. *Science* **205**, 90–93.

Phillips R. J., Banerdt W. B., Sleep N. H. and Saunders R. S. (1980a) Martian stress distributions: arguments about the finite strength support of Tharsis. *J. Geophys. Res.*, in press.

Phillips R. J. and Ivins E. R. (1979) Geophysical observations pertaining to solid state convection in the terrestrial planets. *Phys. Earth Planet. Inter.* **19**, 107–148.

Phillips R. J., Kaula W. M., McGill G. E. and Malin M. C. (1980b) Tectonics and evolution of Venus. *Science*, submitted.

Phillips R. J. and Saunders R. S. (1975) The isostatic state of martian topography. *J. Geophys. Res.* **80**, 2893–2898.

Phillips R. J., Saunders R. S. and Conel J. E. (1973) Mars: crustal structure inferred from Bouguer gravity anomalies. *J. Geophys. Res.* **78**, 4815–4820.

Phillips R. J., Sjogren W. L., Abbott E. A., Smith J. C., Wimberly R. N. and Wagner C. A. (1979) The gravity field of Venus. *Science* **205**, 93–96.

Philpotts J. A. and Schnetzler C. C. (1970) Phenocryst-matrix partition coefficients for K, Rb, Sr, and Ba, with applications to anorthosite and basalt genesis. *Geochim. Cosmochim. Acta* **34**, 307–322.

Phinney R. A. and Anderson D. L. (1967) Present knowledge about the thermal history of the Moon. In *Physics of the Moon*, p. 161–170. Science and Technology Series, v. 13, Amer. Astron. Soc., Tarzana, Calif.

Pitt G. D. and Tozer D. C. (1970a) Optical absorption measurements on natural and synthetic ferromagnesium minerals subjected to high pressures. *Phys. Earth Planet. Inter. 2*, 179–188.

Pitt G. D. and Tozer D. C. (1970b) Radiative heat transfer in dense media and its magnitude in olivines and some other ferromagnesium minerals under typical upper mantle conditions. *Phys. Earth Planet. Inter.* **2**, 189–199.

Plescia J. B., Saunders R. S. and Gregory T. (1979) Geologic evolution of the Tharsis volcanoes (abstract). In *Lunar and Planetary Science X*, p. 989–991. Lunar and Planetary Institute, Houston.

Pollack H. N. and Chapman D. S. (1977a) On the regional variation of heat flow, geotherms, and lithospheric thickness. *Tectonophy.* **38**, 279–296.

Pollack H. N. and Chapman D. S. (1977b) Mantle heat flow. *Earth Planet. Sci. Lett.* **34**, 174–184.

Pollack J. B. and Black D. C. (1979) Implications of the gas compositional measurements of Pioneer Venus for the origin of planetary atmospheres. *Science* **205**, 56–59.

Post R. L., Jr. (1977) High temperature creep of Mt. Burnett dunite. *Tectonophy.* **42**, 75–110.

Post R. L., Jr. and Griggs D. T. (1973) The Earth's mantle: evidence of non-Newtonian flow. *Science* **181**, 1242–1244.

Powell R. W., Ho C. Y. and Lilley P. E. (1966) Thermal conductivity of selected materials. Nat. Stand. Ref. Data Series, 8. Nat. Bureau of Stand., U.S. Gov't. Printing Office, Washington, D.C.

Priestly C. H. B. (1954) Convection from a large horizontal surface. *Austr. J. Phys.* **7**, 176–201.

Raleigh C. B. and Kirby S. H. (1970) Creep in the upper mantle. *Mineral. Soc. Amer. Spec. Paper* **3**, 113–121.

Rao R. U. M. and Jessop A. M. (1975) A comparison of the thermal characteristics of shields. *Can. J. Earth Sci.* **12**, 347–360.

Rayleigh Lord (1916) On convection currents in a horizontal layer of fluid when the higher temperature is on the underside. *Phil. Mag. Ser. 6* **32**, 529–546.

Reasenberg R. D. (1977) The moment of inertia and isostasy of Mars. *J. Geophys. Res.* **82**, 369–375.

Revelle R. and Maxwell A. E. (1952) Heat flow through the floor of the eastern north Pacific Ocean. *Nature* **170**, 199–200.

Reynolds R. T., Fricker P. E. and Summers A. L. (1966) Effects of melting upon thermal models of the Earth. *J. Geophys. Res.* **71**, 573–582.

Reynolds R. T., Fricker P. E. and Summers A. L. (1972) Thermal history of the Moon. In *Thermal Characteristics of the Moon* (J. W. Lucas, ed.), p. 303–337. MIT Press, Cambridge.

Reynolds R. T. and Summers A. L. (1969) Calculations on the composition of the terrestrial planets. *J. Geophys. Res.* **74**, 2494–2511.

Ribando R. J., Torrance K. E. and Turcotte D. L. (1976) Numerical models for hydrothermal circulation in the oceanic crust. *J. Geophys. Res.* **81**, 3007–3012.

Richter C. F. (1958) *Elementary Seismology*. Freeman, San Francisco. 768 pp.

Richter F. M. (1973a) Dynamical models for sea floor spreading. *Rev. Geophys. Space Phys.* **11**, 223–287.

Richter F. M. (1973b) Finite amplitude convection through a phase boundary. *Geophys. J. Roy. Astron. Soc.* **35**, 265–276.

Richter F. M. (1973c) Convection and the large-scale circulation of the mantle. *J. Geophys. Res.* **78**, 8735–8745.

Richter F. M. (1979) Focal mechanisms and seismic energy release of deep and intermediate earthquakes in the Tonga-Kermadec region and their bearing on the depth extent of mantle flow. *J. Geophys. Res.* **84**, 6783–6795.

Richter F. M. and Johnson C. E. (1974) Stability of a chemically layered mantle. *J. Geophys. Res.* **79**, 1635–1639.

Richter F. M. and Parsons B. (1975) On the interaction of two scales of convection in the mantle. *J. Geophys. Res.* **80**, 2529–2541.

Richter F. M. and Ribe N. M. (1979) Importance of advection in determining the local isotopic composition of the mantle. *Earth Planet. Sci. Lett.* **43**, 212–222.

Ringwood A. E. (1961) Chemical and genetic relationships among meteorites. *Geochim. Cosmochim. Acta* **24**, 159–197.

Ringwood A. E. (1970a) Phase transformations and the constitution of the mantle. *Phys. Earth Planet. Inter.* **3**, 109–155.

Ringwood A. E. (1970b) Petrogenesis of Apollo 11 basalts and implications of lunar origin. *J. Geophys. Res.* **75**, 6453–6479.

Ringwood A. E. (1975) *Composition and Petrology of the Earth's Mantle*. McGraw-Hill, N.Y. 618 pp.

Ringwood A. E. (1977) Petrogenesis in island arc systems. In *Island Arcs, Deep Sea Trenches and Back-Arc Basins* (M. Talwani and W. C. Pitman III, eds.), Maurice Ewing Series 1, p. 311–324. Amer. Geophys. Union, Washington, D.C.

Ringwood A. E. and Anderson D. L. (1977) Earth and Venus: a comparative study. *Icarus* **30**, 243–253.

Ringwood A. E. and Clark S. P. (1971) Internal constitution of Mars. *Nature* **234**, 89–92.

Ringwood A. E. and Kesson S. E. (1976) A dynamic model for mare basalt petrogenesis. *Proc. Lunar Sci. Conf. 7th*, p. 1697–1722.

Roberts P. H. (1965) Convection in a self-gravitating fluid sphere. *Mathematika* **12**, 128–137.

Robie R. A. and Waldbaum D. R. (1968) Thermodynamic properties of minerals and related substances at 298.1°K (25.0°C) and one atmosphere (1.013 bars) pressure and at higher temperatures. *U.S. Geol Survey Bull.* **1259**, 256 pp.

Ross J. V., Ave'Lallemant H. G. and Carter N. L. (1979) Activation volume for creep in the upper mantle. *Science* **203**, 261–263.

Ross J. V. and Nielsen K. C. (1978) High temperature flow of wet polycrystalline enstatite. *Tectonophy.* **44**, 233–262.

Rossby H. T. (1969) A study of Benard convection with and without rotation. *J. Fluid Mech.* **36**, 309–335.

Runcorn S. K. (1956) Paleomagnetic comparisons between Europe and North America. *Proc. Geol. Assn. Canada* **8**, 77–85.

Runcorn S. K. (1962) Convection in the Moon. *Nature* **195**, 1150–1151.

Runcorn S. K. (1977) Early melting of the Moon. *Proc. Lunar Sci. Conf. 8th*, p. 463–469.

Runcorn S. K., Libby L. M. and Libby W. F. (1977) Primeval melting of the Moon. *Nature* **270**, 676–681.

Runcorn S. K. and Urey H. C. (1973) A new theory of lunar magnetism. *Science* **180**, 636–638.

Russell C. T. (1976) The magnetosphere of Venus: evidence for a boundary layer and a magnetotail. *Geophys. Res. Lett.* **3**, 589–592.

Russell C. T. (1978) The magnetic field of Mars: Mars 5 evidence reexamined. *Geophys. Res. Lett.* **5**, 85–88.

Russell C. T., Coleman P. J., Jr. and Schubert G. (1975) The lunar magnetic field. *Space Res.* **15**, 621–628.

Russell C. T., Elphic R. C. and Slavin J. A. (1979) Initial Pioneer Venus magnetic field results: nightside observations. *Science* **205**, 114–116.

Russell C. T., Weiss H., Coleman P. J., Jr., Soderblom L. A., Stuart-Alexander D. E. and Wilhelms D. E. (1977) Geologic-magnetic correlations on the Moon: Apollo subsatellite results. *Proc. Lunar Sci. Conf. 8th*, p. 1171–1185.

Ryder G. and Taylor G. J. (1976) Did mare-type volcanism commence early in lunar history? *Proc. Lunar Sci. Conf. 7th*, p. 1741–1755.

Rydgren A. E., Strom S. E. and Strom K. M. (1976) The nature of the objects of Joy—A study of the T-Tauri phenomenon. *Astrophys. J. Suppl.* **30**, 307–336.

Safronov V. S. (1972) *Evolution of the Proto-planetary Cloud and Formation of the Earth and Planets.* NASA Tech. Transl. TTF-677, Washington, D.C. 206 pp.

Safronov V. S. (1978) The heating of the Earth during its formation. *Icarus* **33**, 3–12.

Saito K., Basu A. R. and Alexander E. C., Jr. (1978) Planetary-type rare gases in an upper mantle-derived amphibole. *Earth Planet. Sci. Lett.* **39**, 274–280.

Sammis C. G. (1976) The effects of polymorphic phase boundaries on vertical and horizontal motions in the Earth's mantle. *Tectonophy.* **35**, 169–182.

Sammis C. G., Smith J. C., Schubert G. and Yuen D. A. (1977) Viscosity-depth profile of the Earth's mantle: effects of polymorphic phase transitions. *J. Geophys. Res.* **82**, 3747–3761.

Sammis C. G., Smith J. C. and Schubert G. (1980) A critical assessment of estimation methods for activation volume. *J. Geophys. Res.*, in press.

Sato A. and Thompson E. G. (1976) Finite element models for creeping convection. *J. Comp. Phys.* **22**, 229–244.

Saunders R. S. and Malin M. C. (1977) Geologic interpretation of new observations of the surface of Venus. *Geophys. Res. Lett.* **4**, 542–549.

Schaber G. G. and Boyce J. M. (1977) Probable distribution of large impact basins on Venus: comparison with Mercury and the Moon. In *Impact and Explosion Cratering* (D. J. Roddy, R. O. Pepin and R. B. Merrill, eds.), p. 603–612. Pergamon., N.Y.

Schatz J. F. and Simmons G. (1972) Thermal conductivity of Earth materials at high temperatures. *J. Geophys. Res.* **77**, 6966–6983.

Schnetzler C. C. and Philpotts J. A. (1971) Alkali, alkaline earth, and rare-earth element concentrations in some Apollo 12 soils, rocks, and separated phases. *Proc. Lunar Sci. Conf. 2nd*, p. 1101–1122.

Schonfeld E. (1977) Comparison of orbital chemistry with crustal thickness and lunar sample chemistry. *Proc. Lunar Sci. Conf. 8th*, p. 1149–1162.

Schramm D. N. and Wasserburg G. J. (1970) Nucleosynthesis and the mean age of the elements. *Astrophys. J.* **162**, 57–69.

Schubert G., Cassen P. and Young R. E. (1979a) Subsolidus convective cooling histories of terrestrial planets. *Icarus* **38**, 192–211.

Schubert G., Cassen P. and Young R. E. (1979b) Core cooling by sub-solidus mantle convection. *Phys. Earth Planet. Inter.* **20**, 194–208.

Schubert G., Froidevaux C. and Yuen D. A. (1976) Oceanic lithosphere and asthenosphere: thermal and mechanical structure. *J. Geophys. Res.* **81**, 3525–3540.

Schubert G. and Turcotte D. L. (1971) Phase changes and mantle convection. *J. Geophys. Res.* **76**, 1424–1432.

Schubert G. and Turcotte D. L. (1972) One-dimensional model of shallow-mantle convection *J. Geophys. Res.* **77**, 945–951.

Schubert G., Turcotte D. L. and Oxburgh E. R. (1969) Stability of planetary interiors. *Geophys. J. Roy. Astron. Soc.* **18**, 441–460.

Schubert G., Turcotte D. L. and Oxburgh E. R. (1970) Phase change instability in the mantle. *Science* **169**, 1075–1077.

Schubert G. and Young R. E. (1976) Cooling the Earth by whole mantle sub-solidus convection: a constraint on the viscosity of the lower mantle. *Tectonophy.* **35**, 201–214.

Schubert G., Young R. E. and Cassen P. (1977) Solid state convective models of the lunar internal temperature. *Phil. Trans. Roy. Soc. Lond.* **A 285**, 523–536.

Schubert G., Yuen D. A. and Turcotte D. L. (1975) Role of phase transitions in a dynamic mantle. *Geophys. J. Roy. Astron. Soc.* **42**, 705–735.

Schultz P. H. (1977) Endogenic modification of impact craters on Mercury. *Phys. Earth Planet. Inter.* **15**, 202–219.

Sclater J. G. and Francheteau J. (1970) The implications of terrestrial heat flow observations on current tectonic and geochemical models of the crust and the upper mantle of the Earth. *Geophys. J. Roy. Astron. Soc.* **20**, 509–542.

Sclater J. G., Jaupart C. and Galson D. (1980) The heat flow through oceanic and continental crust and the heat loss of the Earth. *Rev. Geophys. Space Phys.* **18**, 269–311.

Sclater J. G., Lawver L. A. and Parsons B. (1975) Comparison of long-wavelength residual elevation and free air gravity anomalies in the North Atlantic and possible implications for the thickness of the lithospheric plate. *J. Geophys. Res.* **80**, 1031–1052.

Scott D. H. (1972) Geologic map of the Maurolycus Quadrangle of the Moon. *U.S. Geol. Surv., Misc. Geol. Inv.*, Map I-695.

Scott D. H., Diaz J. M and Watkins J. A. (1975) The geologic evaluation and regional synthesis of metric and panoramic photographs. *Proc. Lunar Sci. Conf. 6th*, p. 2531–2540.

Shankland T. J. (1970) Pressure shift of infrared absorption bands in minerals and the effect on radiative heat transport. *J. Geophys. Res.* **75**, 409–413.

Shankland T. J., Duba A. G. and Woronow A. (1974) Pressure shifts of optical absorption bands in iron-bearing garnet, spinel, olivine and pyroxene. *J. Geophys. Res.* **79**, 3273–3282.

Shankland T. J., Nitsan U. and Duba A. G. (1979) Optical absorption and radiative heat transport in olivine at high temperature. *J. Geophys. Res.* **84**, 1603–1610.

Sharp R. P. (1973) Mars: troughed terrain. *J. Geophys. Res.* **78**, 4063–4072.

Sharpe H. N. and Peltier W. R. (1978) Parameterized mantle convection and the Earth's thermal history. *Geophys. Res. Lett.* **5**, 737–740.

Sharpe H. N. and Peltier W. R. (1979) A thermal history model for the Earth with parameterized convection. *Geophys. J. Roy. Astron. Soc.* **58**, 171–203.

Sharpe H. N. and Strangway D. W. (1976) The magnetic field of Mercury and models of thermal evolution. *Geophys. Res. Lett* **3**, 285–288.

Shaw G. H. (1978) Effects of core formation. *Phys. Earth Planet. Inter.* **16**, 361–369.

Shemansky D. E. and Broadfoot A. L. (1977) Interaction of the surfaces of the Moon and Mercury with their exospheric atmospheres. *Rev. Geophys. Space Phys.* **15**, 491–499.

Shih C. and Schonfeld E. (1976) Mare basalt genesis: a cumulate-remelting model. *Proc. Lunar Sci. Conf. 7th*, p. 1757–1792.

Siegfried R. W. and Solomon S. C. (1974) Mercury: internal structure and thermal evolution. *Icarus* **23**, 192–205.

Sipkin S. A. and Jordan T. H. (1975) Lateral heterogeneity of the upper mantle determined from the travel times of ScS. *J. Geophys. Res.* **80**, 1474–1484.

Sipkin S. A. and Jordan T. H. (1976) Lateral heterogeneity of the upper mantle determined from the travel times of multiple ScS. *J. Geophys. Res.* **81**, 6307–6320.

Sjogren W. L. and Smith J. C. (1976) Quantitative mass distribution models for Mare Orientale. *Proc. Lunar Sci. Conf. 7th*, p. 2639–2648.

Skinner B. J. (1962) Thermal expansion of ten minerals. *U.S. Geol. Survey Prof. Paper 450 D*, 109–112.

Sleep N. H. (1974) Segregation of magma from a mostly crystalline mush. *Geol. Soc. Amer. Bull.* **85**, 1225–1232.

Sleep N. H. (1975a) Formation of oceanic crust: some thermal constraints. *J. Geophys. Res.* **80**, 4037–4042.

Sleep N. H. (1975b) Stress and flow beneath island arcs. *Geophys. J. Roy. Astron Soc.* **42**, 827–857.

Sleep N. H. (1979a) The double seismic zone in downgoing slabs and the viscosity of the mesosphere. *J. Geophys. Res.* **84**, 4565–4571.

Sleep N. H. (1979b) Thermal history and degassing of the Earth: some simple calculations. *J. Geol.* **87**, 671–686.

Sleep N. H. and Phillips R. J. (1979) An isostatic model for the Tharsis province, Mars. *Geophys. Res. Lett.* **6**, 803–806.

Sleep N. H. and Rosendahl B. R. (1979) Topography and tectonics of mid-oceanic ridge axes. *J. Geophys. Res.* **84**, 6831–6839.

Smith B. A., Soderblom L. A., Johnson T. V., Ingersoll A. P., Collins S. A., Shoemaker E. M., Hunt G. E., Masursky H., Carr M. H., Davies M. E., Cook A. F., II, Boyce J., Danielson G. E., Owen T., Sagan C., Beebe R. F., Veverka J., Strom R. G., McCauley J. F., Morrison D., Briggs G. A. and Suomi V. E. (1979) The Jupiter system through the eyes of Voyager I. *Science* **204**, 951–972.

Smith J. V., Anderson A. T., Newton R. C., Olsen E. J. and Wyllie P. J. (1970) A petrologic model for the Moon based on petrogenesis, experimental petrology, and physical properties. *J. Geol.* **78**, 381–405.

Smith R. C. and Knopoff L. (1976) Theory of the asymetrically-heated lithospheric plates. *Phys. Earth Planet. Inter.* **11**, 299–311.

Smith W. B., Ingals R. P., Shapiro I. I. and Ash M. E. (1970) Surface-height variations on Venus and Mercury. *Radio Sci.* **5**, 411–423.

Soderblom L. A., Condit C. D., West R. A., Herman B. M. and Kreidler T. J. (1974) Martian planetwide crater distributions: implication for geologic history and surface processes. *Icarus* **22**, 239–263.

Soga N., Klemans G. H. and Horai K. (1972) Thermal conductivity-density relationships for high and low-pressure polymorphs of silicates and germanates. *J. Geophys. Res.* **77**, 2610–2612.

Solomon S. C. (1975) Mare volcanism and lunar crustal structure. *Proc. Lunar Sci. Conf. 6th*, p. 1021–1042.

Solomon S. C. (1976a) Some aspects of core formation in Mercury. *Icarus* **28**, 509–521.

Solomon S. C. (1976b) Geophysical constraints on radial and lateral temperature variations in the upper mantle. *Amer. Mineral.* **61**, 788–803.

Solomon S. C. (1977) The relationship between crustal tectonics and internal evolution on the Moon and Mercury. *Phys. Earth Planet. Inter.* **15**, 135–145.

Solomon S. C. (1978a) On volcanism and thermal tectonics on one-plate planets. *Geophys. Res. Lett* **5**, 461–464.

Solomon S. C. (1978b) The nature of isotasy on the Moon: how big a Pratt-fall for Airy models? *Proc. Lunar Planet. Sci. Conf. 9th*, p. 3499–3511.

Solomon S. C. (1979) Formation, history and energetics of cores in the terrestrial planets. *Phys. Earth Planet. Inter.* **19**, 168–182.

Solomon S. C. and Chaiken J. (1976) Thermal expansion and thermal stress in the Moon and terrestrial planets: clues to early thermal history. *Proc. Lunar Sci. Conf. 7th*, p. 3229–3243.

Solomon S. C. and Head J. W. (1979) Vertical movement in mare basins: relation to mare emplacement, basin tectonics, and lunar thermal history. *J. Geophys. Res.* **84**, 1667–1682.

Solomon S. C. and Head J. W. (1980) Lunar mascon basins: lava filling, tectonics, and evolution of the lithosphere. *Rev. Geophys. Space Phys.* **18**, 107–141.

Solomon S. C., Head J. W. and Comer R. P. (1979) Thickness of the martian lithosphere from tectonic features: evidence for lithospheric thinning beneath volcanic provinces (abstract). In *Reports of Planetary Geology Program, 1978–1979*, p.60–62. NASA TM-80339.

Solomon S. C. and Longhi J. (1977) Magma oceanography: 1. Thermal evolution. *Proc. Lunar Sci. Conf. 8th*, p. 583–599.

Solomon S. C. and Paw U K. T. (1975) Elevation of the olivine-spinel transition in subducted lithosphere: seismic evidence. *Phys. Earth Planet. Inter.* **11**, 97–108.

Solomon S. C. and Toksöz M. N. (1973) Internal constitution and evolution of the Moon. *Phys. Earth Planet. Inter.* **7**, 15–38.

Sonett C. P. (1971) The relationship of meteoritic parent body thermal histories and electromagnetic heating by a pre-main sequence T-Tauri sun. In *Physical Studies of Minor Planets* (T. Gehrels, ed.), p. 239–245. NASA SP-267.

Sonett C. P. and Colburn D. S. (1967) Establishment of a lunar unipolar generator and assoicated shock and wake by the solar wind. *Nature* **216**, 340–343.

Sonett C. P., Colburn D. S. and Schwartz K. (1968) Electrical heating of meteorite parent bodies and planets by dynamo induction from a premain sequence T-Tauri "solar wind". *Nature* **219**, 924–926.

Sonett C. P., Colburn D. S. and Schwartz K. (1975) Formation of the lunar crust: an electrical source of heating. *Icarus* **24**, 231–255.

Sonett C. P., Colburn D. S., Schwartz K. and Keil K. (1970) The melting of asteroidal-sized bodies by unipolar dynamo induction from a primordial T-Tauri sun. *Astrophys. Space Sci.* **7**, 446–488.

Sonett C. P. and Duba A. (1975) Lunar temperature and global heat flux from laboratory electrical conductivity and lunar magnetometer data. *Nature* **258**, 118–121.

Sonett C. P., Smith B. F., Colburn D. S., Schubert G. and Schwartz K. (1972) The induced magnetic field of the Moon: conductivity profiles and inferred temperature. *Proc. Lunar Sci. Conf. 3rd*, p. 2309–2336.

Srnka L. J. (1977) Spontaneous magnetic field generation in hypervelocity impacts. *Proc. Lunar Sci. Conf. 8th*, p. 785–792.

Stacey F. D. (1972) Physical properties of the Earth's core. *Geophys. Surv.* **1**, 99–119.

Stacey F. D. (1977) A thermal model of the Earth. *Phys. Earth Planet. Inter.* **15**, 341–348.

Steiger R. H. and Jager E. (1977) Subcommission on geochronology: convention on the use of decay constants in geo- and cosmo-chronology. *Earth Planet. Sci. Lett.* **36**, 359–362.

Stephenson A. (1976) Crustal remanence and the magnetic moment of Mercury. *Earth Planet. Sci. Lett.* **28**, 454–458.

Stephenson A., Collinson D. W. and Runcorn S. K. (1974) Lunar magnetic field paleointensity determinations on Apollo 11, 16 and 17 rocks. *Proc. Lunar Sci. Conf. 5th* , p. 2859–2871.

Stephenson A., Runcorn S. K. and Collinson S. W. (1975) On changes in the intensity of the the ancient lunar magnetic field. *Proc. Lunar Sci. Conf. 6th*, p. 3049–3062.

Stevenson D. J. (1975) Dynamo generation in Mercury. *Nature* **256**, 634.

Stevenson D. J. and Turner J. S. (1979) Fluid models of mantle convection. In *The Earth, Its Origin, Evolution and Structure* (M. W. McElhinney, ed.), p. 227–263. Academic Press, London.

Stevenson D. J., Schubert G., Cassen P. and Reynolds R. T. (1980) Core evolution and magnetism of the terrestrial planets (abstract). In *Lunar and Planetary Science XI*, p. 1091–1093. Lunar and Planetary Institite, Houston.

Stocker R. L. and Ashby M. F. (1973) On the rheology of the upper mantle. *Rev. Geophys. Space Phys.* **11**, 391–426.

Stolper E. (1977) Experimental petrology of eucritic meteorites. *Geochim. Cosmochim. Acta* **41**, 587–611.

Strangway D. W. (1977) The magnetic fields of the terrestrial planets. *Phys. Earth Planet. Inter.* **15**, 121–130.

Strangway D. W. and Sharpe H. A. (1974) Lunar magnetism and an early cold Moon. *Nature* **249**, 227–230.

Strangway D. W. and Sharpe H. N. (1975) A model of lunar evolution. *The Moon* **12**, 369–397.

Strom R. G. (1977) Origin and relative age of lunar and mercurian intercrater plains. *Phys. Earth Planet. Inter.* **15**, 156–172.

Strom R. G. (1979) Mercury: a post-Mariner 10 assessment. *Space Sci. Rev.* **24**, 3–70.

Strom R. G., Trask N. J. and Guest J. E. (1975) Tectonism and volcanism on Mercury. *J. Geophys. Res.* **80**, 2478–2507.

Strong H. M., Tuft R. E. and Hanneman R. E. (1973) The iron fusion curve and γ-δ-l triple point. *Met. Trans.* **4**, 2657–2661.

Sung C. M. and Burns. R. G. (1976) Kinetics of high-pressure phase transformations: implications to the evolution of the olivine-spinel transition in the downgoing lithosphere and its consequences on the dynamics of the mantle. *Tectonophy.* **31**, 1–32.

Surkov Yu. A. (1977) Geochemical studies of Venus by Venera 9 and 10 automatic interplanetary stations. *Proc. Lunar Sci. Conf. 8th*, p. 2665–2689.

Surkov Yu. A., Kirnozov F. F., Glazov V. N. and Fedoseyev G. A. (1977a) Investigations of Venusian gamma-radiation by Venera 9 and Venera 10. *Space Res.* **17**, 659–662.

Surkov Yu. A., Kirnozov F. F., Kristianov V. K., Korchuganov E. N., Glazov V. N. and Ivanov V. F. (1977b) Investigation of the density of the Venusian surface rocks by Venera 10. *Space Res.* **17**, 651–657.

Suzuki I., Ohtani E. and Kumazawa M. (1979) Thermal expansion of γ-Mg_2SiO_4. *J. Phys. Earth* **27**, 53–62.

Tabarrok B. and Lin R. C. (1977) Finite element analysis of free convection flows. *Int. J. Heat Mass Transfer* **20**, 945–951.

Talbot C. J. (1973) A plate tectonic model for the Archean crust. *Phil. Trans. Roy. Soc. Lond.* **A273**, 413–427.

Taylor S. R. (1978) Geochemical constraints on melting and differentiation of the Moon. *Proc. Lunar Planet. Sci. Conf. 9th*, p. 15–23.

Tera F. and Wasserburg G. J. (1974) U-Th-Pb systematics on lunar rocks and inferences about lunar evolution and the age of the Moon. *Proc. Lunar Sci. Conf. 5th,* p. 1571–1599.

Tera F. and Wasserburg G. J. (1975) The evolution and history of mare basalts as inferred from U-Th-Pb systematics (abstract). In *Lunar Science VI,* p. 807–809. The Lunar Science Institute, Houston.

Thurber C. H. and Solomon S. C. (1978) An assessment of crustal thickness variations on the lunar near side: models, uncertainties, and implications for crustal differentiation. *Proc. Lunar Planet. Sci. Conf. 9th,* p. 3481–3497.

Toksöz M. N., Dainty A. M., Solomon S. C. and Anderson K. R. (1974) Structure of the Moon. *Rev. Geophys. Space Phys.* **12**, 539–567.

Toksöz M. N. and Hsui A. T. (1978a) Numerical studies of back arc convection and the formation of marginal basins. *Tectonophys.* **50**, 177–196.

Toksöz M. N. and Hsui A. T. (1978b) Thermal history and evolution of Mars. *Icarus* **34**, 537–547.

Toksöz M. N., Hsui A. T. and Johnston D. H. (1978) Thermal evolutions of the terrestrial planets. *The Moon and the Planets* **18**, 281–320.

Toksöz M. N. and Johnston D. H. (1974) The evolution of the Moon. *Icarus* **21**, 389–414.

Toksöz M. N. and Johnston D. H. (1977) The evolution of the Moon and the terrestrial planets. In *The Soviet-American Conference on Cosmochemistry of the Moon and Planets,* p. 245–328. NASA SP-370.

Toksöz M. N., Minear J. W. and Julian B. R. (1971) Temperature field and geophysical effects of a downgoing slab. *J. Geophys. Res.* **76**, 1113–1138.

Toksöz M. N., Sleep N. H. and Smith A. T. (1973) Evolution of the downgoing lithosphere and the mechanisms of deep focus earthquakes. *Geophys. J. Roy. Astron. Soc.* **35**, 285–310.

Toksöz M. N., Solomon S. C., Minear J. W. and Johnston D. H. (1972) Thermal evolution of the Moon. *The Moon* **4**, 190–213.

Torrance K. E. (1979) Numerical approaches to solid state convection in planetary bodies. *Phys. Earth Planet. Inter.* **19**, 158–167.

Torrance K. E. and Rockett J. W. (1969) Numerical study of natural convection in an enclosure with localized heating from below—creeping flow to the onset of laminar instability. *J. Fluid Mech.* **36**, 33–54.

Torrance K. E. and Turcotte D. L. (1971a) Thermal convection with large viscosity variations. *J. Fluid Mech.* **47**, 113–125.

Torrance K. E. and Turcotte D. L. (1971b) Structure of convection cells in the mantle. *J. Geophys. Res.* **76**, 1154–1161.

Toulmin P., III, Clark B. C., Baird A. K., Keil K. and Rose H. J., Jr. (1976) Preliminary results from the Viking X-ray fluorescence experiment: the first sample from Chryse Planitia, Mars. *Science* **194**, 81–84.

Tozer D. C. (1965) Heat transfer and convection currents. *Phil. Trans. Roy. Soc.* **A258**, 252–271.

Tozer D. C. (1967) Towards a theory of thermal convection in the mantle. In *The Earth's Mantle* (T. F. Gaskell, ed.), p. 325–353. Academic Press, London.

Tozer D. C. (1970) Factors determining the temperature evolution of thermally convecting Earth models. *Phys. Earth Planet. Inter.* **2**, 393–398.

Tozer D. C. (1972) The Moon's thermal state and an interpretation of the lunar electrical conductivity distribution. *The Moon* **5**, 90–105.

Tozer D. C. (1973) Thermal plumes in the Earth's mantle. *Nature* **244**, 398–400.

Tozer D. C. (1974) The internal evolution of planetary-sized objects. *The Moon* **9**, 167–182.

Trask N. J. (1976) History of basin development on Mercury (abstract). In *Papers Presented to the Conference on Comparisons of Mercury and the Moon,* p. 36. The Lunar Science Institute, Houston.

Trask N. J. and Guest J. E. (1975) Preliminary geologic terrain map of Mercury. *J. Geophys. Res.* **80**, 2461–2477.

Trask N. J. and Strom R. G. (1976) Additional evidence of mercurian volcanism. *Icarus* **28**, 559–563.

Tullis J. A. (1979) High temperature deformation of rocks and minerals. *Rev. Geophys. Space Phys.* **17**, 1137–1154.

Turcotte D. L. and Ahern J. L. (1978) Magma production and migration within the Moon. *Proc. Lunar Sci. Conf. 9th, p.* 307–318.

Turcotte D. L. and Burke K. (1978) Global sea-level changes and the thermal structure of the Earth. *Earth Planet. Sci. Lett.* **41**, 341–346.

Turcotte D. L., Cisne J. L. and Nordmann J. C. (1977a) On the evolution of the lunar orbit. *Icarus* **30**, 245–266.

Turcotte D. L., Cook F. A. and Willeman R. J. (1979) Parameterized convection within the Moon and the terrestrial planets. *Proc. Lunar Planet. Sci. Conf. 10th,* p. 2375–2392.

Turcotte D. L., Haxby W. F. and Ockendon J. R. (1977b) Lithospheric instabilities. In *Island Arcs, Deep Sea Trenches, and Back-Arc Basins* (M. Talwani and W. C. Pitman, eds.), Maurice Ewing Series 1, p. 63–69. Amer. Geophys. Union, Washington, D.C.

Turcotte D. L., Hsui A. T., Torrance K. E. and Oxburgh E. R. (1972) Thermal structure of the Moon. *J. Geophys. Res.* **77**, 6931–6939.

Turcotte D. L., Hsui A. T., Torrance K. E. and Schubert G. (1974) Influence of viscous dissipation on Benard convection. *J. Fluid Mech.* **64**, 369–374.

Turcotte D. L. and McAdoo D. C. (1979) Geoid anomalies and the thickness of the lithosphere. *J. Geophys. Res.* **84**, 2381–2387.

Turcotte D. L. and Oxburgh E. R. (1967) Finite amplitude convection cells and continental drift. *J. Fluid Mech.* **28**, 29–42.

Turcotte D. L. and Oxburgh E. R. (1969) Convection in a mantle with variable physical properties. *J. Geophys. Res.* **74**, 1458–1474.

Turcotte D. L. and Oxburgh E. R. (1970) Lunar convection. *J. Geophys. Res.* **75**, 6549–6552.

Turcotte D. L. and Schubert G. (1971) Structure of the olivine-spinel phase boundary in the descending lithosphere. *J. Geophys. Res.* **76**, 7980–7987.

Turcotte D. L. and Schubert G. (1973) Frictional heating of the descending lithosphere. *J. Geophys. Res.* **78**, 5876–5886.

Turcotte D. L., Torrance K. E. and Hsui A. E. (1973) Convection in the Earth's mantle. In *Methods of Computational Physics,* v. 13 (B. A. Bolt, ed.), p. 431–453. Academic Press, N.Y.

Twiss R. J. (1976) Structural superplastic creep and linear viscosity in the Earth's mantle. *Earth Planet. Sci. Lett.* **33**, 86–100.

Urey H. C. (1951) The origin and development of the Earth and other terrestrial planets. *Geochim. Cosmochim. Acta* **1**, 209–277.

Urey H. C. (1952) *The Planets, Their Origin and Development.* Yale Univ. Press, New Haven. 245 pp.

Urey H. C. (1962) Origin and history of the Moon. In *Physics and Astronomy of the Moon* (Z. Kopal, ed.), 1st ed., p. 481–523. Academic Press, N.Y.

Urey H. C. and MacDonald G. J. F. (1971) Origin and history of the Moon. In *Physics and Astronomy of the Moon* (Z. Kopal, ed.), 2nd ed., p. 213–289. Academic Press, N.Y.

Usselman T. M. (1975a) Experimental approach to the state of the core: Part I. The liquidus relations of the Fe-rich portion of the Fe-Ni-S system from 30 to 100 kb. *Amer. J. Sci.* **275**, 278–290.

Usselman T. M. (1975b) Experimental approach to the state of the core: Part II. Composition and thermal regime. *Amer J. Sci.* **275**, 291–303.

Usselman T. M. and Hodge D. S. (1978) Thermal control of low pressure fractionation processes. *J. Volcanol. Geotherm. Res.* **4**, 265–281.

Uyeda S. and Ben-Avraham Z. (1972) Origin and development of the Phillipine Sea. *Nature* **240**, 176–178.

Vine F. J. and Matthews D. H. (1963) Magnetic anomalies over oceanic ridges. *Nature* **199**, 947–949.

Vogt P. R. (1974) Volcano height and plate thickness. *Earth Planet Sci. Lett.* **23**, 337–348.

Vollmer R. (1977) Terrestrial lead isotopic evolution and formation time of the Earth's core. *Nature* **270**, 144–147.

Wakita H., Nagasawa H., Uyeda S. and Kuno H. (1967) Uranium, thorium, and potassium contents of possible mantle materials. *Geochem. J.* **1**, 183–198.

Walker D., Longhi J., Stolper E. M., Grove T. L. and Hays J. F. (1975a) Origin of titaniferous lunar basalts. *Geochim. Cosmochim. Acta* **39**, 1219–1235.

Walker D., Longhi J. and Hays J. F. (1975b) Differentiation of a very thick magma body and implications for the source regions of mare basalts. *Proc. Lunar Sci. Conf. 6th,* p. 1103–1120.

Ward W. R. (1975) Past orientation of the lunar spin axis. *Science* **189**, 377–379.

Ward W. R., Murray B. C. and Malin M. C. (1974) Climatic variations on Mars, 2, evolution of carbon dioxide atmosphere and polar caps. *J. Geophys. Res.* **79**, 3387–3395.

Wasserburg G. J., MacDonald G. J. F., Hoyle F. and Fowler W. A. (1964) Relative contributions of uranium, thorium, and potassium to heat production in the Earth. *Science* **143**, 465–467.

Wasson J. T. (1974) *Meteorites.* Springer-Verlag, N.Y. 316 pp.

Watts A. B. (1978) An analysis of isostasy in the world's oceans, 1, Hawaiian-Emperor seamount chain. *J. Geophys. Res.* **83**, 5989–6004.

Watts A. B., Bodine J. H. and Steckler M. S. (1980) Observations of flexure and the state of stress in the oceanic lithosphere. *J. Geophys. Res.*, in press.

Weertman J. (1967) Venus: tectonic activity. *Science* **158**, 395–396.

Weertman J. (1970) The creep strength of the Earth's mantle. *Rev. Geophys. Space Phys.* **8**, 145–168.

Weertman J. (1979) Height of mountains on Venus and the creep properties of rock. *Phys. Earth Planet. Inter.* **19**, 197–207.

Weertman J. and Weertman J. R. (1975) High temperature creep of rock and mantle viscosity. *Ann. Rev. Earth Planet. Sci.* **3**, 293–315.

Wegener A. (1924) *The Origin of Continents and Oceans*, 3rd ed. E. P. Dutton, N.Y. 246 pp.

Weidenschilling S. J. (1974) A model for accretion of the terrestrial planets. *Icarus* **22**, 426–435.

Weidenschilling S. J. (1976) Accretion of the terrestrial planets. II. *Icarus* **27**, 161–170.

Wetherill G. W. (1974) Problems associated with estimating the relative impact rates on Mars and the Moon. *The Moon* **9**, 227–231.

Wetherill G. W. (1975) Possible slow accretion of the Moon and its thermal and petrological consequences (abstract). In *Origins of Mare Basalts and Their Implications for Lunar Evolution*, p. 184–188. The Lunar Science Institute, Houston.

Wetherill G. W. (1976) The role of large bodies in the formation of the Earth and Moon. *Proc. Lunar Sci. Conf. 7th*, p. 3245–3257.

Whitehead J. A. (1972) Moving heaters as a model of continental drift. *Phys. Earth Planet. Inter.* **5**, 199–212.

Wilhelms D. E. (1976) Mercurian volcanism questioned. *Icarus* **28**, 551–558.

Wilshire H. G. and Jackson E. D. (1975) Problems in determining mantle geotherms from pyroxene compositions of ultramafic rocks. *J. Geol.* **83**, 313–329.

Wilson J. T. (1965) A new class of faults and their bearing on continental drift. *Nature* **207**, 343–347.

Winchell A. N. and Winchell H. (1964) *The Microscopic Characteristics of Artificial Inorganic Substances, Optical Properties of Artificial Minerals.* Academic Press, N.Y. 439 pp.

Wiskerchen M. J. and Sonett C. P. (1977) A lunar metal core? *Proc. Lunar Sci. Conf. 8th*, p. 515–535.

Wolery T. J. and Sleep N. H. (1976) Hydrothermal circulation and geochemical flux at mid-ocean ridges. *J. Geol.* **84**, 249–275.

Wood J. A. (1964) The cooling rates and parent planets of several iron meteorites. *Icarus* **3**, 429–459.

Wood J. A. (1965) Meteorites and asteroids. *Adv. Astronaut. Sci.* **19**, 99–118.

Wood J. A. (1972) Thermal history and early magmatism in the Moon. *Icarus* **16**, 229–240.

Wood J. A. (1975) Lunar petrogenesis in a well-stirred magma ocean. *Proc. Lunar Sci. Conf. 6th*, p. 1087–1102.

Wood J. A. (1978) Small parent planets for the meteorites: a new model (abstract). *Meteoritics* **13**, 671.

Wood J. A. (1979) Review of metallographic cooling rates of meteorites and a new model for the planetesimals in which they formed. In *Asteroids* (T. Gehrels, ed.), p. 849–891. Univ. Arizona Press, Tucson.

Wood J. A., Dickey J. S., Jr., Marvin U. B. and Powell B. N. (1970) Lunar anorthosites and a geophysical model of the Moon. *Proc. Apollo 11 Lunar Sci. Conf.*, p. 965–988.

Wyllie P. J. (1971) Role of water in magma generation and initiation of diapiric uprise in the mantle. *J. Geophys. Res.* **76**, 1328–1338.

Yagi T., Bell P. M. and Mao H. K. (1979) Phase relations in the system MgO-FeO-SiO_2 between 150 and 200 kbar at 1000°C. *Carnegie Inst. Wash., Yearb.* **78**, 614–618.

Yamamoto K. and Akimoto S. (1977) The system MgO-SiO_2-H_2O at high pressures and temperatures–stability field for hydroxylchondrodite, hydroxyl-clinohumite and 10 Å-phase. *Amer. J. Sci.* **277**, 288–312.

Yoder H. S., Jr. (1976) *Generation of Basaltic Magma.* Nat. Academy of Sci., Washington, D.C. 265 pp.

Young R. E. (1974) Finite-amplitude thermal convection in a spherical shell. *J. Fluid Mech.* **63**, 695–721.

Young R. E. and Schubert G. (1974) Temperatures inside Mars: is the core liquid or solid? *Geophys. Res. Lett.* **1**, 157–160.

Yuen D. A. and Schubert G. (1976) Mantle plumes: a boundary layer approach for Newtonian and non-Newtonian temperature-dependent rheologies. *J. Geophys. Res.* **81**, 2499–2510.

Zharakov V. N. (1958) On the thermal conductivity coefficient of the Earth's mantle. *Izv. Geophys.*, No. 11, 776–780.

Overleaf:

The surface of Venus has been found to be greatly different from the Earth's surface, as shown here in a shaded relief map generated from Pioneer Venus radar data. The differences in the surfaces of these two planets could arise because basaltic volcanism on Venus occurs on a planet which plausibly evolved with an appreciably higher surface temperature and no water. (Map generated by U.S.G.S., Flagstaff; NASA Ames Research Center; and Massachusetts Institute of Technology.)

10. Implications of basaltic volcanism for the evolution of planetary bodies

William M. Kaula, Team Leader
—*Department of Earth and Space Sciences*
University of California, Los Angeles

Fraser P. Fanale
—*Institute for Planetary Geoscience*
University of Hawaii

Don L. Anderson
—*Division of Geological and Planetary Sciences*
California Institute of Technology

10.1 INTRODUCTION

The purpose of this chapter is to place basaltic volcanism in the larger context of planetary evolution. We will attempt to synthesize the findings of the preceding chapters into a discussion of the overall evolution of terrestrial bodies. (For a summary of each chapter, see the Overview.) Although our treatment will have this broader point of view, we will avoid excessive bulk by emphasizing those aspects which are related to basaltic volcanism. The plan of this chapter is to describe scenarios of planetary evolution, incorporating other data and ideas with those from the earlier parts of the book, and then to seek heuristic patterns, to infer the leading questions remaining, and to suggest how to pursue them.

Section 10.2 is a discussion of solar system origin as it applies to the terrestrial planets in general. This phase has negligible direct connection to observable consequences of volcanism on the planets, of course. However, consideration of origin is necessary for any comparative discussion of planetary evolution, because there are significant differences in planetary properties which result in different volcanic products, but which must be attributed to circumstances of formation. We identify the boundary between "formation" and "evolution" as the time when the driving energy for major planetary changes shifted from short-lived to long-lived sources. The very early differentiation of the Moon's crust required by isotopic constraints suggests that primordial crustal separation is part of "formation" by the foregoing definition. The circumstances of initial crustal differentiation on all the planets are obscure enough that a comparative discussion is desirable. Another topic appropriately included in this section is basaltic volcanism on small bodies: in particular, the basaltic achondrite parent body and asteroids, not only because the energy sources were probably short-lived, but also because differentiation in small bodies may have a significant effect on the fractionation of protoplanetary material, affecting the bulk composition of the smaller planets (the Moon and Mercury) and the heterogeneity of the larger planets. The compositional consequences of these nebula and planetesimal processes are discussed in section 4.3.

Section 10.3 is a planet-by-planet discussion of subsequent terrestrial planet evolution, which in all cases except Io appears to be internally energized. It is our belief that the considerable differences among the planets and the present levels of understanding of terrestrial bodies are such that it is not fruitful to force a discussion of planetary evolution into a comparative framework *ab initio*. Instead, the main emphasis will be on the thermal, compositional, and tectonic evolution of the planets individually, utilizing the data and ideas presented in the other chapters to the maximum extent feasible.

Section 10.4 is an attempt to draw the leading inferences from the aforedescribed discussion. Although the main theme appears to be that the duration and complexity of a planet's evolution are both positively correlated with its size, there are significant variations from this pattern which may be inherent in planetary behavior rather than just dependent on differences in bulk properties of the planets. We include our opinions as to major problems remaining, and suggest what is needed to progress toward the solution of these problems.

10.2 ORIGIN AND EARLY EVOLUTION OF THE PLANETS

The formation of the planets is generally thought to be associated with formation of the sun, because of the paucity of interstellar material. There is, as yet, no reason to believe that the sun's formation was markedly different from the current formation of similar sized stars within the galaxy. Hypotheses of star formation are constrained to some extent by observations. It is desirable to start from this astronomical base before moving to the speculative matter of planetary formation, and then to end with aspects which are in some ways constrained by planetary, asteroidal and meteoritic properties.

10.2.1 Sun's formation and early evolution

Recently-formed stars are always observed to be close to clouds of gas and dust. This observation, together with the lack of a plausible alternative, leads to the consensus that stars are formed from interstellar clouds. These clouds are probably debris from nova and supernova explosions, because the solar system material contains heavy elements which require processes within giant stars for their synthesis (Trimble, 1975). The duration and cessation of nucleosynthesis and the number of sources affect the products of short-lived radioactive decays: ^{26}Mg, ^{129}Xe, ^{134}Xe, ^{136}Xe, and others recently identified. The **formation interval**, or the time from the end of nucleosynthesis to entrapment of the daughter isotopes in meteoritic material, is about 180 m.y. from xenon isotopes produced by ^{129}I and ^{244}Pu decay (Schramm *et al.*, 1970). However, the detection of ^{26}Mg excesses in the Allende meteorite, indicating the presence of the short-lived radioisotope ^{26}Al (Lee *et al.*, 1976; Gray and Compston, 1976), has raised the suggestion of a second source with a much shorter formation interval of 2 or 3 m.y. (Cameron and Truran, 1977). Another type of constraint is furnished by varying ratios in the stable isotopes, most notably $^{18}O/^{16}O$ and $^{17}O/^{16}O$ (Clayton and Mayeda, 1975; Clayton *et al.*, 1976). As discussed in section 7.2.2, a plot of variations within any one planetary body shows a distinct 2:1 trend of the two isotope ratios, explicable by diffusive and chemical mass fractionation effects. However, there are other differences among bodies explicable only by preservation of the consequences of incomplete mixing of materials from different nucleosynthetic sources. Isotopic differences among bodies (other than dispersion along the 2:1 line) then require that these bodies come from different parts of the solar system raw material. Conversely, agreement of isotopic ratios of two bodies suggests they came from the same part (but does not require it). Lunar rocks fall along the same 2:1 line as terrestrial rocks, and have a centroid close to that of terrestrial igneous rocks. Basaltic achondrites fall on a nearby parallel line but have a distinctly different centroid. Ordinary and carbonaceous chondritic meteorites all fall appreciably farther from the terrestrial line.

For a star to form from a gas-and-dust cloud, effects inducing contraction (gravitation, external pressure) must overcome those maintaining extension (temperature, angular momentum, turbulence, magnetic field). Considering only gravitation and temperature, a spherically symmetric isothermal cloud in equilibrium must have a density varying with radius, $\rho \propto r^{-2}$, to create a pressure gradient counterbalancing gravitational attraction. Under these extremely idealized conditions, the requirement for collapse is

$$M > 2c^2R/G \qquad (10.2.1)$$

where M is mass, R is the radius of the whole cloud, G is the gravitational constant, and c is the sound speed, proportionate to $T^{1/2}$. Some clouds are observed to have a temperature T as low as 10° K; the condition (10.2.1) for a $1M_\odot$ mass then requires a mean density ρ greater than $10^{-19}g/cm^3$ ($M_\odot$ is a solar mass). While some small dark clouds have such density and temperature, it is not known whether they can collapse, or whether stars as small as the sun must form as fragments of a much larger cloud, in which external pressures work with gravitation to collapse smaller fragments. The formation of a smallish star like the sun probably entails appreciable asymmetry of conditions, and may normally occur only within a cluster of stars.

The computed collapse of an isolated spherically symmetric star from $10^{-19}g/cm^3$ takes ~1 m.y., even without considering rotation or magnetic fields. Hence a significantly more rapid collapse requires an extraordinary external pressure.

Forming stars are identifiable because they are cooler, and hence redder, than normal stars of the same brightness. Spectral observations of forming stars indicate appreciable mass motions and heating in their vicinity. Some of the motions are interpreted as being mass outflows from the stars, in the form of T Tauri winds, on the order of 10^7 times the present solar wind, which is $\sim 10^{-14}M_\odot$/year. The heating is often interpreted as the consequence of material infalling to an **accretion disk**: an extended sheet of matter around the star. All these matters are the subject of appreciable research and debate (Strom, 1977; Ulrich, 1978).

The sun probably separated from its original cluster of stars and settled down to its long term behavior within 1 m.y. or so. Although it has undergone significant evolution (spin down due to loss of angular momentum with solar wind, heating due to increase of mean nuclear mass with hydrogen burning), the sun probably was not a major agent for moving matter or internal heating of solid bodies after this initial epoch.

In summary, there is appreciable rationale for external effects on planetary material, both from the forming sun and from outside the solar system, in the first ~30 m.y. Thereafter, conditions were essentially similar to current conditions: an isolated solar system with a sun acting mainly as a source of gravitational attraction and radiation in the visible part of the electromagnetic spectrum.

10.2.2 Solar nebular models; gravitational instability

Given the existence of a rotating accretion disk with some means of internal transfer of momentum, such as turbulent viscosity or hydromagnetic coupling, then there inevitably must be a net inward transfer of mass and a net outward transfer of angular momentum. The torque between adjacent material in the nebula must go to zero with the rotation gradient $d\Omega/dr$ at the co-rotating sun, while it must go to zero at the outer limits because the coupling effect (whether mechanical or magnetic) goes to zero (Lynden-Bell and Pringle, 1974). At an intermediate distance, there must be a maximum torque; to compensate the resulting outward flow of high angular momentum matter, conservation requires an inward flow of low angular momentum matter. Other likely properties of such an accretion disk are heating from conversion of the kinetic energy of infalling material, with consequent loss of gas, and, if of sufficient density, the formation of mass agglomerations by gravitational instability. If the gaseous disk is not sufficiently dense, these agglomerations may arise in a layer of condensed matter.

The protoplanetary material, commonly known as the **solar nebula**, plausibly was an accretion disk. The greatest uncertainties about fundamental determiners of this accretion disk are (1) the rate of infall of material into the protosolar system; (2) the relative phasing of formation of the sun and the occurrence of major gravitational instabilities in the nebula; (3) the amount and timing of "T Tauri" mass outflows from the early sun. "Minimal" theories, such as Safronov's (1972), which hypothesize a nebula a few times as massive as the planets plus their volatile solar complement ($\sim 0.015 M_\odot$), entail a slow rate of infall comparable to that calculated for formation of an isolated $1 M_\odot$ star, formation of the bulk of the sun before any significant agglomeration in the nebula, and slight outflow of mass from the sun. "Maximal" theories, such as Cameron's (1978a), which hypothesize a $\sim 2 M_\odot$ nebula, entail an infall rate ~ 30 times as great, formation of major gaseous gravitational instabilities before the existence of a large solar core, and a great subsequent mass loss from both solar outflow and accretional heating.

The gravitational instability equation (10.2.1), can be converted to a relation between a characteristic length, $\lambda \sim R$, and density, $\sim M/R^3$. For application in the solar nebula with angular rate Ω at radial distance r, two further modifications are desirable: (1) The flattened shape of the nebula makes a surface density $\sigma \sim \rho H$ appropriate, where H is a scale height normal to the plane of the nebula: $H \sim c/\Omega$. (2) The rotation of the nebula necessitates taking into account the shearing effect of a gradient in angular momentum, $d(\Omega r^2)/dr$. A consequence of this latter effect is to set an upper and a lower bound on the size of an unstable agglomeration (Safronov, 1972; Goldreich and Ward, 1973)

$$\frac{2\pi^2 G\sigma}{2\Omega^2 + r\Omega d\Omega/dr} > \lambda > \frac{c^2}{G\sigma} \qquad (10.2.2)$$

For gas, the velocity c is the thermal velocity, as used in Eq. (10.2.1). For solids, however, the velocity c is the turbulent velocity. Whether the instability which occurs is of the gas or the dust hence depends on temperature and turbulence as well as surface density σ. The minimal theories attain densities σ sufficient only for dust instability, while the maximal theories could attain gas instability. The minimal theories are more easily reconcilable with meteorite and terrestrial planet evidence. However, maximal theories explain agglomeration of the H + He rich planets, Jupiter and Saturn, more satisfactorily. Therefore, it is desirable to examine the implications and variations of both hypotheses for bodies on which basaltic volcanism has occurred.

10.2.3 Dust instability and planetesimal growth hypothesis

In this scenario, as outlined in greatest detail by Safronov (1972; see also Harris, 1978), dust grains formed by condensation settle to the central plane of the nebula in a time on the order of 1000 years. The initial planetesimals for a minimal nebula are on the order of 1 km size at 1AU distance; if gas drag or other dissipative mechanism is operating, these bodies form aggregations of ~ 10 km. This process is inexorable, so long as the time scale of reduction of mutual velocities by bumping and gas drag is shorter than the time scale of disruption due to infall of matter into the nebula or to solar activity. The relative velocities in gravitational instability are small—about 10 cm/sec. Hence the initial agglomerations were probably rather loosely put together and of low density.

Subsequent to formation of the initial planetesimals by gravitational instability, there is a growth of both planetesimal masses, due to collisions, and relative velocities, due to mutual perturbations. The rate of this growth and the velocity and mass distributions at various stages are somewhat uncertain, because the principal analytic treatment by Safronov (1972) separates mass growth from velocity growth, while computer experiments to date have been limited to extreme circumstances: uniform-sized starting populations of planetesimals, and either only the very early stages (Greenberg *et al.*, 1978) or only the very late stages (L. P. Cox, pers. comm., 1980; Wetherill, 1980a).

In any case, some plausible constraints can be stated: (1) The time scale of growth of planets apparently was long compared to the time scale of removal of volatiles from the zones of formation of proto-Earth planetesimals and meteorite parent bodies. Otherwise, the amount of volatiles swept up and held gravitationally would have been significantly greater. This constraint is most severe for Xe, for which at 500°K a planetesimal of ~ 100 km radius in a medium of solar composition, Xe:Si $\sim 10^{-6}$, would acquire more than a terrestrial or meteoritic Xe:Si ratio, $\sim 10^{-11}$. (2) The material of chondrites, once collected, cannot have been subjected to high velocity ($\gtrsim$1km/sec) impacts. Otherwise, they would evidence appreciable shock effects. (3) Either the temperature remained above $\sim$1360° K in the zone of proto-Mercury material until volatiles were removed, or subsequently some sort of mechanism acted to remove silicate-rich material in preference to iron-rich. Otherwise, the planet Mercury would have a significantly lower density. (4) Relative velocities in the later stages of planet growth were at least as great as the average noncircular components of the orbits of the Earth and Venus themselves, $\sim$1km/sec. (5) Jupiter probably had a much faster growth rate than the terrestrial planets, to account for the acquisition of H + He by Jupiter before the T Tauri wind blew it away, the absence of a planet in the asteroidal zone, the high relative velocities of asteroids, and the stunting of the growth of Mars. The dearth of matter and the amount of random motion in the asteroid-Mars zones both suggest some strong dynamical disturbance, such as might have been generated by a forming Jupiter (Weidenschilling, 1975).

Models of the evolution of a planetesimal population generally obtain relative velocities moderately smaller than the escape velocities of the average mass $\overline{m}$ of the population, $\sim(2G\overline{m}/\overline{r})^{1/2}$. The resulting capture cross sections for bodies approaching the terrestrial planets in size led to growth times well over 1 m.y. To inhibit the growth of a sizeable planet in Mercury's zone there must be either some sort of disturbing effects from the sun (Safronov, 1972) or a mechanism to remove a large portion of the planetesimals, such as a gas drag which would act more strongly on bodies of large area: mass ratio; i.e., small and silicate-rich, rather than massive and iron-rich, bodies (Weidenschilling, 1978).

Since it is a satellite, the Moon requires some special explanation. The extreme iron and refractory siderophile deficiency of the Moon (section 4.5.2) suggests previous planetary differentiation. Dynamically plausible mechanisms which minimize the energy requirements to place protolunar material in geocentric orbit indicate that this planetary differentiation took place in the Earth or in bodies in the Earth's vicinity. Oxygen isotope ratios measured in lunar and terrestrial rocks are very similar to one another, but differ markedly from those seen in meteorites, which lends support to the dynamical arguments. It is unlikely that the differentiation took place in protomoons in geocentric orbits, since known mechanisms (tidal friction, gas drag, collisions) would all act to remove the lighter, silicate-rich objects and retain the heavier, iron-rich objects in orbit around the Earth. There remains the Earth or other bodies in heliocentric orbits. Both loci involve dynamical difficulties in placing iron-deficient material in geocentric orbits. A strong argument against the differentiation having occurred in other bodies in heliocentric orbits is the absence of meteorites which appear to come from the same compositional pool as the Moon (Kaula, 1977; Ringwood, 1979; Drake, 1980).

Removal of the lunar material from the early Earth has the obvious attraction of utilizing a well established condition: a separate core and mantle at some early stage. The composition of the Moon is thus accounted for by removal of the protolunar material from the mantle after core separation. This removal may have happened well before the completion of the Earth's growth. Two mechanisms have been proposed for the removal: (1) rotational instability (O'Keefe and Sullivan, 1978); and (2) collision by another planet-sized body (Hartman and Davis, 1975; Cameron and Ward, 1976). Objections to the rotational instability hypothesis are (a) to provide sufficient energy, core formation would have to be delayed until the Earth was almost entirely formed, which in turn requires keeping the forming Earth unreasonably cool (see below); and (b) about ten lunar masses and an even greater proportion of angular momentum must be ejected from the Earth-Moon system, since the fundamental mode of rotational instability is the lowest degree non-axisymmetric distribution; i.e., a third harmonic. The main objection to the collision hypothesis is the seeming *ad hoc* character of requiring such a great impact. The terrestrial bodies are a disparate group of n = 5 bodies, however, and it should not be at all surprising if there were once six, seven, or more bodies of size comparable to Mars or Mercury (Wetherill, 1976). An additional argument in favor of the collision hypothesis is that the material is more likely to be placed in a dispersed state, so that volatiles can be lost (Kaula, 1977; Ringwood, 1979).

Acquisition of protolunar material from collisions of bodies in heliocentric orbit does not entail any severe intrinsic difficulty. Mechanisms for reducing the iron content at capture, or subsequently by additional collisions, however, are rather speculative (Ruskol, 1977; Kaula and Harris, 1975).

The minimal nebula hypothesis is a congenial working model for discussing variations in bulk composition (see Chapter 4), phasing of early phenomena (see Chapter 7), and initial thermal state (see Chapter 9) among the terrestrial bodies: so congenial it is often assumed implicitly. The hypothesis, however, does have at least two major problems: (1) How did the Jupiter and Saturn zones develop cores large enough to capture a large amount of hydrogen and helium before it was swept out? The obvious suggestion is that these zones very early had a temperature low enough for ice to condense, both H_2O and CO_2. But this turns out to be a rather marginal advantage over the terrestrial zones for any plausible density gradient (i.e., negative $d\sigma/dr$) in the nebula. (2) Why is there a steep dropoff in inert gas content in the sequence Venus-Earth-Mars? In primordial argon, $^{38}Ar + {}^{36}Ar$, the proportionate abundance ratios are 20,000:200:1. These ratios are much too great for some sort of pressure gradient in an isothermal nebula (Pollack and Black, 1979). Alternatives are (1) a steep radial gradient in solar wind implantation or (2) impact by a large body, or bodies, from closer to the sun (Wetherill, 1980b). Since the implantation gradient hypothesis requires appreciable inhibition of subsequent dynamical mixing, the large impacter seems more plausible.

10.2.4 Gas instability and giant protoplanet hypothesis

In this scenario, as described by Cameron (1978a, b), the density of the nebula becomes so great that instability of the gas can occur. For Jupiter at ~140° K, forming appreciably further out in the nebula than its present location, the required surface density σ is ~500 g-cm^{-2}. The present Jupiter with a ~15 earthmass icy and rocky core plus its solar complement spread from ~10 to 24 AU would be ~35 g-cm^{-2}. After stage (1), collapse of gas-plus-dust, evolution to the present major planets requires the following stages: (2) compositional segregation within the agglomeration, to obtain an ice- and rock-rich core; (3) stripping off most of the gaseous outer parts; and (4) removal of this excess gas from the solar nebula. Stage (2) would occur as the consequence of condensation and "rainout" within the agglomeration (Slattery, 1978; Slattery *et al.*, 1980), somewhat similar in physics to rain in the Earth's atmosphere. Stage (3) is the greatest difficulty. It is hypothesed by Cameron (1978b) to arise from a later growth of the solar core than the protoplanetary agglomeration, with a slower collapse time of the latter, so that the protosun can tidally disrupt the protoplanet. Stage (4) can be the result of the evolution of the accretion disk, in which part of the mass spirals into the sun, part is lost by "evaporation" from the outer parts, and part may be swept out by a protosolar wind.

Although the motivation for the giant protoplanet hypothesis comes from the outer solar system and astronomical considerations, its implications for the terrestrial bodies should be examined. At least three distinct scenarios can be conjectured. (1) The terrestrial planets underwent the same four stages as Jupiter and Saturn, differing mainly in having a much more drastic stripping off of their gaseous envelopes. (2) The terrestrial planets underwent stages (1) and (2), gas instability and compositional segregation, but then in addition to stripping off of gaseous envelopes, the solid cores were broken up to form planetesimals, which subsequently recombined to form planets. (3) The terrestrial planets never underwent gaseous instability, but only dust instability. In these scenarios, the planets formed considerably further from the sun than their present locations, because to lose mass from the nebula, there must be a proportionately greater loss of energy and angular momentum.

All of these scenarios require modifications to explain how Mercury, Mars, and the asteroid belt are different from the Earth and Venus. A major difficulty with scenario (1) might be the drastically greater depletions of the terrestrial bodies in inert gases than in active gases. While any segregation process would carry down with the solids more active gases than inert gases, it is difficult to imagine it being so efficiently discriminatory. Another difficulty of scenario (1) is that a mighty blast is required to remove virtually all the gaseous envelopes; however, such a blast might also account for the mercurian metal:silicate ratio. Scenario (2) is justified to explain the great depletions of all volatiles, but its energetic requirements for breaking up cores are even more extreme than those of scenario (1). Scenario (3) entails appreciably greater turbulence in the inner solar system or some sort of turbulence or accretion disk torque effect to inhibit gaseous instability, but none of these are wildly implausible. All three scenarios are difficult to reconcile with the fragile, low temperature properties of carbonaceous chondrite meteorites, which look much more like primordial than recondensed objects. In a hypothesis which requires late formation of the sun, however, it is a minor wrinkle to bring meteoritic material into the solar system at a later stage when the activity has calmed down.

The compatibility of gaseous instability hypotheses with terrestrial body properties is as yet little explored. Until a better explanation is found for the much larger scale phenomena of Jupiter and Saturn formation, however, this exploration should be continued.

10.2.5 Small body heating and differentiation

A feature of the processes leading to planetary formation is the heating of planetesimals and their resultant differentiation. Small body differentiation has produced basaltic volcanism in at least two instances: the eucrite parent body and the asteroid Vesta (which may be the same body: Consolmagno and Drake, 1977). Small body differentiation also is of interest as being an important potential contributor to differences in planetary bulk composition through collisional differentiation, as discussed in section 4.3.2; this mechanism is particularly important for any nonfission origin of the Moon. A problem for any sort of small body heating is the energy source. A source which necessarily existed is collisions. To convert a significant amount of relative kinetic energy into heat, however, requires impact velocities of well over a km/sec (O'Keefe and Ahrens, 1977). So long as planetesimal growth had not progressed beyond the creation of bodies with escape velocities of more than a km/sec (i.e., more than ~750 km in diameter), such velocities must have been rare indeed; otherwise, the planets never would have collected. While it is not to be ruled out that occasional higher impact velocities occurred as the result of injecting bodies from outside the immediate circum-solar zone (Kaula and Bigeleisen, 1975; Weidenschilling, 1975), the small portion of shocked meteorites and the exceptionality of Vesta's spectrum indicate that hypervelocity impacts between smallish planetesimals were unusual.

It has been an objection to collisional heating that the textures of the melted material would be much more brecciated and glassy than those of basaltic achondrite meteorites or other basalts. It also is difficult to imagine the classic processes of magmatic fractionation being mimicked in the few milliseconds of the high pressure phase of an impact. Experience with the lunar rocks, however, has led to the identification of more and more differentiated specimens as impact melts, so that even KREEP basalts have been hypothesized to be greatly modified by impact melting (Warren and Wasson, 1979). Collisional effects may also be important for compositional differentiation among meteorites, which in turn has implications for planetary composition and evolution, as discussed in sections 4.3.2 and 10.2.7.

The above doubts about the effectiveness of collisions have led to two other hypotheses of planetary heating, electromagnetic induction and radioactive decay. Electromagnetic induction arises from the intersection of magnetic field lines by an electrically conducting body (Sonett and Herbert, 1977). The difficulties of the hypothesis are twofold: firstly generating a sufficiently intense and rapidly changing magnetic field, and secondly raising the temperature of the planetesimal sufficiently (a few 100° C) by some other mechanism to give it an appreciable electrical conductivity. The intensity of the early solar magnetic field indicated by the remanent magnetism of meteorites, 1 gauss (Butler, 1972), may be adequate. More conjectural is whether there was a sufficiently rapid alternation of the magnetization, most plausibly arising from a sectored structure of the solar magnetic field and a rapid solar spin.

The one outstanding candidate for radiometric heating sufficient to melt planetesimals within the first few million years is ^{26}Al decay, which has a half life of 0.75 m.y. Originally proposed by Urey (1956) and elaborated by Fish *et al.* (1960), the hypothesis has been revived by discovery of a ^{26}Mg excess in inclusions of the carbonaceous chondrite Allende (Lee *et al.*, 1976; Gray and Compston, 1976). From this limited data on a few meteorite samples, it is difficult to quantify the probable significance of ^{26}Al as an early heat source for small bodies. As mentioned in section 10.2.1, the anomaly is one of those which indicate compositional heterogeneity in the solar nebula.

The intrinsic properties of eucrites, as discussed in sections 1.2.8 and 7.2, afford some appreciable constraints on the nature of their parent body. The source region of eucrites was quite dry and had metal present (Stolper, 1977; see also section 3.5). The bulk composition of the parent body could plausibly have been chondritic (Consolmagno and Drake, 1977). The spectacular depletions in refractory siderophiles (Morgan *et al.*, 1978) indicate a rather slow acting process, such as would result from a small percentage partial melt at depth, rather than a "lake" of impact melt.

The distinctly basaltic spectral signature of Vesta, discussed in section 2.2.6, may result from exceptional circumstances. Somehow, most of the original material in the asteroidal zone was cleared out. Since the escape velocity from the asteroidal belt is at least 4 km/sec, it is surprising that only a small minority of asteroids show differentiated surfaces, with Vesta's the sole basalt-like spectrum (section 2.2.6; Chapman, 1979). The differentiation of Vesta suggests either collisional heating (since Vesta is rather remote from the sun for electromagnetic induction), or a sufficiently earlier formation of Vesta compared to most other asteroids to yield appreciably greater ^{26}Al heating.

10.2.6 Initial energetics of terrestrial planets

The temperature of formation of a planet (to be distinguished from the temperature of condensation of its material) depends on, in probable order of significance: *(a)* the size of the planet, *(b)* the sizes of the bodies from which it is formed, *(c)* the heliocentric velocities of the forming bodies, *(d)* the composition of the planet, *(e)* its heterogeneity, *(f)* its epoch of formation, *(g)* its distance from the sun, and *(h)* its rate of formation.

(a) The Mass M of the planet obviously acts by its influence on the impact velocities of infalling matter, the kinetic energy per unit mass being largely GM(r)/r. Integrating over r from 0 to R gives as the mean energy/unit mass of a homogeneous body 3GM(R)/5R. This zero order result suggests that the initial mean temperatures of planets will vary as $M^{2/3}$, and that the temperature within a given planet will vary as r^2, so long as there is no convection or differentiation. The mean energy release per unit mass upon core formation would also be expected to vary proportionate to gR, or $M^{2/3}$. Core formation would reverse the temperature profile in the mantle, since the further a material element travelled in the differentiation process, the more energy it would have gained on the average, because of greater interaction with its surroundings. This consideration applies to upwardly moving, as well as downwardly moving, elements. Core separation probably got underway before the completion of planet formation for any body larger than the Moon.

(b) The principal neglect in the foregoing estimate is the loss of heat by radiation at the surface, $\sigma (T_s^4 - T_0^4)$ in energy/area/time. The strong dependence on surface temperature T_s means that normally the heat loss is governed by the rate of heat transfer from the interior to the surface, -KdT/dr. The main variable affecting the gradient dT/dr is the size of the infalling bodies: the larger the bodies, the deeper the deposition of energy. Infalling bodies also act to transfer heat outward by stirring up the planetary material. However, this effect is limited by the relatively small partitioning in hypervelocity impacts of the kinetic energy of the impacter to kinetic energy of ejecta, and attenuates more rapidly with respect to depth than heat deposition. In general, therefore, there should be a positive correlation of heating with size of body forming the planet. A difficulty in quantifying this result is the portion of energy lost by radiation from fragments created at the time of impact. The retained h energy fraction necessary to attain a prescribed temperature T at a given radius r can be obtained from, roughly,

$$h \frac{GM(r)}{r} = \rho C_p[T(r) - T_0] \qquad (10.2.3)$$

The partitioning factor h for retained heat should have a positive correlation with size of impacts, probably attaining more than 0.5 in large events (Kaula, 1980a).

As discussed in sections 10.2.2–3 above, the size of bodies forming planets depends strongly on whether the initial instability was dusty or gaseous.

(c) The heliocentric velocities of the forming bodies could have significantly contributed to the impact energy, if relative velocities were on the order of kilometers/second.

(d) The composition of a planet affects its initial heating mainly through its effect on core formation, which requires the presence of metallic iron or iron sulphide. This presence requires either a deficiency of oxygen at the time the temperature drops low enough for iron oxides to be stable (~600° K), or the introduction of reducing agent, such as carbonaceous material. In the case of Mercury, there must also be a deficiency of silica (SiO_2) when the temperature drops below the condensation point of fayalite, ~1360° K. The most marked evidence of the thermal effect of planetary composition is the different evolutions of Mercury and Mars. The initial hot state, low oxidation level, and high iron content of Mercury's material led to rapid core formation and a predominantly cooling history. The more carbonaceous-chondrite-like composition of Mars apparently led to late core formation and a mainly tensional tectonic regime (sections 9.5.3–4).

(e) The formation temperature of a planet will vary inversely with its degree of heterogeneity. A declining iron portion in time of the accreting body will accordingly decrease the amount of gravitational energy available from core formation. Any lateral heterogeneity, such as large planetesimal infall or instabilities associated with core formation, will induce convection and volcanism, leading to more rapid heat loss.

(f) The epoch of formation of the planet is critical not only because of solar-dependent processes, such as the ambient temperature T_0 and the removal of volatiles alluded to in section 10.3.4 above, but also because of the possible presence of short-lived radionuclides, such as ^{26}Al, discussed in section 10.2.5.

(g) The distance from the sun has both thermal and dynamical effects. The thermal effects are manifested mainly in planetary compositions, as discussed in section 4.5. Only in the case of Mercury might solar insolation have a significant effect on a planetary thermal state. Probably more important were dynamical effects, since orbital velocities vary inversely with solar distance. In addition, there were interaction of a strong solar wind with the nebula (Elmegreen, 1978), and drag effects of the nebula gas (Weidenschilling, 1978).

(h) The rate of formation of a planet probably was of comparatively small import, since, as discussed above, the mass was probably acquired mainly as large

planetesimals which bury their heat. In fact, there may be some reason to expect a mild inverse correlation of heating with rate of formation, since a slower rate would occur with larger planetesimals, inhibiting growth due to the enhanced relative velocities (Kaula, 1979). The body most likely to have an initial thermal state enhanced by rapid growth is the Moon, if it were made from debris around the Earth resulting from a major collision.

10.2.7 Early planet differentiation

As discussed in section 4.5, all the terrestrial bodies have significant indication of crustal separation, while three, the Earth, Mars, and Mercury, appear to require a core from either moment of inertia or magnetic field. However, any planet which has been hot enough to differentiate a crust, has a plausible content of radioactive heat sources, and has reduced iron (or sufficient reducing agent present) will differentiate a core. The Moon is anomalous in having very little iron. The Moon has not enough iron in its reduced state to give evidence of a metallic core other than some puzzling remanent magnetism. Also affecting core formation is compositional heterogeneity of infall. The slow cooling rates inferred for iron meteorites (Wood, 1964; Goldstein and Short, 1967) indicate that there existed large heterogeneous planetesimals. An unknown of terrestrial planet formation is the degree of heterogeneity: in particular, the sizes of predominantly iron planetesimals within a mainly silicate population. However, the temperature dependence of viscosity, together with the large stresses induced in any silicate center of a protoplanet, appear sufficient for rapid early core formation starting from only 10 km iron concentrations (Stevenson, 1980).

Only for the Earth, Moon, and the eucrite parent body is there radiometric evidence that crustal differentiation occurred very early, and even in the case of the Earth it is an indirect inference (though rather inescapable), as discussed in section 7.4. But these constitute a sufficient range of conditions that it seems plausible to assume that all terrestrial bodies differentiated crusts early. For the Earth and Moon there is also evidence that the amount of crust created early is comparable to the present bulk. Again, this evidence is more clear cut for the Moon, where the terrae lead isotopes point rather conclusively to a general differentiation 4.4 b.y. ago (see section 7.3). Furthermore, given that as small a body as the Moon differentiated a thick crust more than ~ 4.0 b.y. ago, it is thermomechanically most plausible that the energy for the differentiation was delivered within the first ~0.1 b.y. after solar system origin. Hence the energy source could not have been any of the long-lived radioactivities (U, Th, and K) but must have been one of the short-lived processes (collisional, electromagnetic, or radioactive: see sections 10.2.5–6.)

Whether these early crustal differentiations had some similarities to basaltic volcanism as discussed in this book is problematical. The eucritic meteorites evidently came from a body (or bodies) not only appreciably smaller than the planets, but also drier than the Earth and probably more reduced than either the Moon or the outer parts of the Earth. The short time scale of planetary differentiation indicates that it was a much more violent process than the subsequent tectonics and volcanism indicated by the crater count stratigraphy in Fig. 8.9.1. The ~70 km thickness of the lunar crust (or other compositional variation accounting for isostasy) indicates that the differentiation was a global phenomenon, involving a major part of the planet. On the other hand, the ~14 km average thickness of the Earth's crust and the rates of crustal formation in accessible geological time, as discussed in section 6.2, indicate that the Earth has recycled an appreciable amount of crust to the interior. Even a recycling of radioisotopes (Sm/Nd, Rb/Sr) in the first ~2.0 b.y. do not conflict with any of the isotopic indicators of long-term separation of oceanic and continental source regions discussed in section 7.4. The indication of inverse correlation of crustal thickness with planetary size, Table 4.2.1, may be related to either the compressibility of melts (E. Stolper, pers. comm., 1980) or to the basalt-eclogite phase transition and the floating or sinking of anorthite-rich "rockbergs" in a magma ocean. The crusts of both Earth and Moon are about one-fourth as thick as the depth at which the basalt-eclogite phase transition intersects the solidus (Fig. 3.3.27), and hence the depth at which sinking, rather than rising, of plagioclase would occur. The same could be true of Mars, but Venus appears to have a greater ratio of crustal thickness to phase transition depth.

A question closely interacting with early planetary differentiation is that of the degree of compositional heterogeneity in formation. Heterogeneity is most strongly indicated for the Earth, in which a reduced iron core and hydrous outer parts co-exist. However, heterogeneity cannot be ruled out for the other terrestrial bodies. In the dust instability and planetesimal growth hypotheses discussed in section 10.2.3 heterogeneities would arise from cooling of the nebula, differential settling of condensate grains in the nebula, and interzone mixing, such as a greater admixture of volatile material from further out in the solar system in the later stages of growth, as suggested by Anders and Owen

(1977). Dynamical rationales can be contrived to explain the dearth of active volatiles in Mars and the Moon, but not for the aforementioned drastic differences in primordial inert gas abundances. In the gas instability and giant protoplanet hypotheses, heterogeneity would be a plausible consequence of the "rainout" of the solid planet from its gaseous envelope due to differences in condensation temperatures, droplet densities, and other factors (Slattery, 1978; Slattery *et al.*, 1980). Such processes could lead to an iron-enriched center and volatile-enriched outer parts, since the rainout rate likely was more rapid than the cooling rate (in part due to the energy resulting from the rainout itself). Where calcaluminous and ferromagnesian silicates would become located relative to each other is more conjectural. Why Mars is small and lacking in volatiles is not clear in this hypothesis, and the Moon must be formed by a separate mechanism, such as a major collision with the Earth.

The relatively small amount of evidence available suggests that the planets were as different in the characters of their initial differentiations as they have been in subsequent compositional evolution. We do not know whether the very early differentiations of the Earth produced material like Archean basalts: 0.8 b.y. is too large an extrapolation range. But the greater heating and volatile content of the Earth suggest that, if the Moon had a "magma ocean," the Earth had an outer layer which was highly disrupted for a much longer time, with appreciable ferromagnesian content and a great deal of recycling: quite different from what the Moon settled down to within 0.1 or 0.2 b.y., if the collected samples of highland rocks are at all representative.

In conclusion, about all that can be suggested is: (1) the outer parts of the terrestrial planets were quite hot in their first few 0.1 b.y., (2) the degree of heating had a strong positive dependence on planetary size, (3) the ability to stabilize an initial differentiated crust plausibly had an inverse correlation with size, and (4) there could have been an initial gradient in composition from more refractory at the center to more volatile in the outer parts in each planet. More uncertain is whether volatile correlation with solar distance or planetary size dominated. Certainly the correlation is predominantly with size out to Mars, and the resulting volatile contents probably had significant influence on crustal stabilization.

10.3 SUBSEQUENT EVOLUTION OF THE PLANETS

As mentioned in the introduction, for the stages of planetary evolution subsequent to the time when short-lived radioactivities, material infalls, and solar processes (such as electromagnetic induction) could have had important influences our knowledge of the differences in planetary character makes separate discussion of each body appropriate.

10.3.1 The Earth

Long after the early heating and differentiation discussed in sections 10.2.6–7, this largest of the terrestrial planets is still convecting vigorously. As reviewed in section 9.5.1, this mantle convection and the resulting plate tectonics are plausibly the consequence of heat which is to a minor (but not to a negligible) fraction primordial and to a major fraction radioactive, aided by a temperature dependent rheology and a thermal expansivity enhanced by phase transitions. The major current debate is over the extent of material exchange between the upper and the lower mantles.

Despite the ample material brought to the surface by this vigorous convection, the present basaltic crust is but a thin veneer. Currently, the Earth is subducting oceanic crust about as fast as it is being formed. If this return is at a rate proportionate to the energy provided to convection, then there must be a large amount of basaltic material in the mantle, as has been pointed out by Dickinson and Luth (1971), Ringwood (1975), and others.

While plate tectonics requires this appreciable circulation of matter, isotopic data indicate that there are several reservoirs in the mantle which have remained separate for the order of 1 to 2 b.y. (or have equivalently slow mixing rates). One of these reservoirs provides the mid-ocean ridge basalts and the other provides the continental flood basalts, and possibly one component of the blend that supplies the oceanic islands.

The mantle is heterogeneous in chemistry, isotopic composition and seismic velocities. But it is not clear whether or how the seismological and geochemical heterogeneities are related. The largest lateral variations in seismic velocities occur in the outer 200–250 km of the Earth and are related to such surface tectonic features as shields, trenches, rises, and volcanic belts, as discussed in section 6.2. The mantle also has radially inhomogeneous rheology, with the lithosphere, asthenosphere, and transition zone being the main subdivisions of the upper mantle. It also is not clear whether the chemical differences, particularly evident in the trace element and isotope geochemistry, are related to the lateral or radial variations in mantle structure.

Ocean floor basalts have comparatively uniform and low large-ion radioisotope content: $^{87}Sr/^{86}Sr < 0.703$, $^{206}Pb/^{204}Pb < 18.7$, and $^{144}Nd/^{143}Nd > 0.51305$, whereas continental basalts and basalts from ocean islands not associated with island arcs have less uniform and higher ratios (see section 7.4.2). In some cases there is close geographic proximity of lavas derived from distinctly different source regions. Radioisotopes impose a time constraint. Reservoirs with different element ratios—Rb/Sr, U/Pb, Th/Pb, and Sm/Nd have existed for the order of 1 to 2 b.y.—ten times older than present oceanic plates. The source region for mid-ocean ridge basalts (MORB) provides rather uniform composition lavas throughout the world. It must therefore be immense in size and global in nature. On the other hand, continental and ocean island basalts, although having a wide geographic distribution, are much less voluminous and have variations in $^{206}Pb/^{207}Pb$, suggesting separation of sources for 1.5 b.y. or more (Sun and Hanson, 1975). These source regions are enriched in large-ion lithophile (LIL) elements and either are more variable in composition than MORB source regions or are mixed with variable amounts of MORB before they reach the surface.

The leading question for long term evolution of the mantle might be said to be its heterogeneity; in particular, the extent to which it is radial: a chemically layered mantle; or lateral: isolated blobs and regional inhomogeneities. Particularly pertinent to basalts is the history of their immediate parent materials. In the case of oceanic tholeiites, the question is whether the picritic material from which it is generated at 30–70 kilometers depth (see section 3.3.7) is itself a recent differentiate from a more ultramafic source, or a much earlier differentiate. Recent differentiation requires either remixing of previous differentiates or retention of primordial material (despite the amount of previous nearsurface differentiation probably being sufficient to utilize an entire mantle volume of rock). Differentiation long ago requires an intervening storage place for this picritic eclogite, and therefore, dense material.

Schilling (1971) and Sun and Hanson (1975) proposed that the low-velocity zone, or asthenosphere, is the depleted reservoir and the source of mid-ocean ridge basalts. Plume basalts, i.e., magmas from the nondepleted reservoir, are then attributed to deeper sources. A difficulty of this hypothesis is that the volume of MORB, integrated over time, is greater than the volume of the low velocity zone. As an alternative to deep sources of plume basalts, Tatsumoto (1978), Hedge (1978) and Frey *et al.* (1978) proposed that the enriched reservoir is shallow and the depleted MORB reservoir is deeper. The requirement of basal, rather than internal, heating for convective plumes also suggests a shallow origin for plume basalts.

Another hypothesis is that differentiation during Archean led to two layers in the upper mantle (Anderson, 1979). As the Earth cooled, the base of this original crust transformed to eclogite and sank through the upper mantle, resulting in a layer of peridotite overlying a layer of picritic eclogite. Melting in the eclogite layer then leads to ascent of the MORB material, plus a possibly pyroxenite residue underplating of older oceanic lithosphere. The basalt and eclogite portions of the oceanic lithosphere return to the eclogite layer by subduction. The shallow refractory peridotite layer, enriched by metasomatism, is hypothesized to be the locus of the source regions for the less abundant continental and ocean island basalts. In this hypothesis the picritic eclogite diapirs become buoyant upon partial melting and melt extensively upon ascent. This evolution is possible because eclogite has a small melting interval. Nearly complete melting of the eclogite source material is required to satisfy the bulk chemistry and trace element uniformity of MORB. The feasible degree of melting upon pressure release and the rate of rise of a diapir relative to its heat loss rate both need better quantification.

Composition of the mantle to 670 km depth

It is usually inferred that olivine is the main constituent of the mantle. This inference is based primarily on petrological evidence but it is also consistent with the seismic data. This evidence applies most strongly to the region of the mantle above about 220 km. Between 220 and 670 km deep, the seismic properties are consistent with either garnet-rich peridotite or eclogite.

Ringwood (1975) and others hypothesize that the refractory peridotite in the uppermost mantle grades into a mantle material that could be the parent (or grandparent) material of MORB: pyrolite, compositionally equivalent to a garnet-rich peridotite. Such material would not yield a sharp discontinuity at depth 220 km; instead, the olivine and pyroxene would undergo transitions to spinel + stishovite + periclase forms at depths of about 350–450 km. Pyroxene and garnet react between 200 and 400 km to form a complex solid solution. As discussed in section 3.3, oceanic tholeiites have a composition which agrees experimentally with a source region that is mainly olivine and pyroxene, with a minor portion of garnet material. Xenoliths found in alkali basalts and in kimberlite pipes which tap ~200 km deep in the mantle are predominantly peridotitic, with only a minor eclogitic portion. These xenoliths are enriched in LIL and H_2O and are the potential source materials for continental tholeiites. MORB is less enriched in LIL and appears to come from a relatively dry source. Continental and ocean island basalts have a range of compositions indicating varying degrees of partial melting, but MORB is invariably tholeiitic.

The Anderson (1979) proposal that the region of the mantle between 220 and 670 km is eclogitic is based mainly on the sharpness of these discontinuities and the changes in seismicity patterns associated with them. The 220-km discontinuity is close to the base of the low-velocity zone. It is near the maximum depth of earthquakes in continental regions and in regions where young, <50 m.y.-old oceanic lithosphere has subducted. The 670-km discontinuity is also sharp and global and is close to the maximum depth of deep focus earthquakes, which are restricted to regions where old oceanic lithosphere has subducted. Below 670 km the olivine-pyroxene material goes to ilmenite and perovskite structures which are denser than the garnet phases.

Composition of the mantle below 670 km

Olivine in the denser ilmenite and perovskite phases as the main constituent below 670 km is consistent with seismic velocities. The principal debate in interpreting seismological and high pressure laboratory data is whether the Fe:Mg ratio must increase with depth. Part of this problem is the temperature gradient. If the gradient is adiabatic, then the increase falls within the ~ ± 5% uncertainty of shock wave experiments for density of olivine material as a function of pressure and Fe:Mg ratio.

In any case, the lower mantle composition is of interest relevant to the basalt source problem because any plausible calculation of the crustal recycling rate extrapolated back in time requires a repository for returned material appreciably larger than the mantle above 670 km depth. *If* returned eclogite can separate from peridotite, then the mantle below 670 km would be largely residual, refractory peridotite. If this separation does not occur, then the lower mantle would have a pyrolitic composition depleted in some large-ion lithophiles.

While the above-stated bulk calculation applies to the problem of mantle interchange over the entire 4.5 b.y. history of the Earth, the evolution of isotopes—particularly ^{87}Sr and ^{143}Nd—constrains more strongly the evolution over the last 2.5 b.y. or so. Both isotopes require that the MORB be drawn from a reservoir depleted in these lithophiles. *If* this depletion can occur *only* by crustal differentiation, then over this time not more than one-third of the mantle can be depleted and contributing to MORB (section 7.4.2; Wasserburg and DePaolo, 1979; O'Nions *et al.*, 1979). The obvious hypothesis is that this depleted reservoir is the mantle above 670 km depth, about one-fourth of the total mass of the mantle. But to decide whether this region is the only plausible reservoir requires more careful considerations of the scenarios by which the present mantle came to be.

Evolution of the mantle

As discussed in section 9.5.1, the thermal evolution of the Earth has been most influenced by a decline in heat sources. We wish to explore what this history implies for mantle composition. As estimated in section 4.5.1, the primitive mantle had roughly the composition of an "undepleted" garnet peridotite or a peridotitic komatiite (see Tables 4.5.2–3). Early differentiation processes plausibly could have led to the development of a thick mafic-felsic crust, too buoyant to be subducted, leaving a peridotite mantle still containing an appreciable component of crustal material. As the outer layer cooled, some of the crustal mafics would have converted to eclogite. If there were large enough agglomerations, this eclogite would have settled through the upper mantle and have been halted near 670 km by the phase changes of the peridotite mantle. Partial melting would have concentrated lithophile elements into a protocontinental crust. Whether this process would have been more efficient in the Earth's history when the geothermal gradient was high is debatable. Certainly mantle heterogeneities in the Rb/Sr and Sm/Nd systems developed more strongly after Archean times (see section 7.4.2).

As discussed in section 3.3, the final fractionation leading to mid-ocean ridge basalts occurs at shallow depth, less than 80 km, enriched in plagioclase and probably depleted in olivine relative to a "pyrolite" peridotite. The parent magma is picritic (sections 3.3.7–8; O'Hara, 1968; Green *et al.*, 1979). Whether this source could be modified eclogite rather than modified peridotite depends on mechanisms of eclogite ascent, as well as the efficient global scale mixing mentioned above. If there were an eclogite layer, the highest temperatures in the mantle, relative to melting temperatures, would be toward the bottom of the thermal boundary layer at about 220 km deep. For this depth, estimates of the melting temperature of eclogite are about 200–300°C less than peridotite.

In summary, as the Earth's mantle differentiates and cools it is evolving from a more homogeneous to a more heterogeneous system. If basalts are classified according to age, then the variety of characteristics in isotopic, major, and trace element chemistry becomes more and more marked with time. For at least the last ~2.0 b.y. the isotopes demand separation of sources among continental and ocean island basalts. On the other hand, they suggest the sources of ocean ridge basalts could be globally homogeneous and well-mixed. The central question is the degree to which lateral heterogeneity has evolved into a layered mantle. Since Birch (1952), the main (but not unanimous) push has been to reconcile the seismological and equation-of-state data with a uniform composition, accounting for discontinuities with phase transitions. If the 220 km discontinuity is sharp, however, there must be some compositional change, since an appropriate phase transition of abundant material is lacking. As high pressure experimental and seismological data are refined, similar difficulties may evolve in reconciling the seismic velocity curve below 600 km with compositional homogeneity.

10.3.2 The Moon

The principal relevant properties of the Moon are conveniently characterized as geological, compositional and geophysical. In the following discussion, we place emphasis on differences from the Earth.

Geological properties

The dominant shaper of the surface is exogenic: the great basins and their consequent throwout and stress patterns, as discussed in sections 5.4 and 6.5. Some of the manifestations of internally generated activity, such as sinuous rilles, appear to be influenced in their occurrence by the basins. Probably the most important geological constraints in lunar evolution over the last ~4.4 b.y. are: (1) the net amount of expansion or contraction since a single coherent surface was established is very slight, less than ~0.1%; (2) morphometry and remote sensing of ancient basins and upland plains indicate voluminous volcanism from shallow sources; (3) the post-Imbrian mare lava flows constitute a very minor proportion of the total crust, perhaps 0.2% of the total volume (Hörz, 1978). These properties indicate that the main course of lunar history has been one of activity declining to negligible levels. The lack of distinct overall expansion or contraction requires compensatory changes within the volume integral of temperature over the applicable time interval (Solomon and Chaiken, 1976; see also section 9.5.2). The only way such a balance could plausibly occur would be for the outer part of the Moon to be cooling as the center heats up, implying an ever thickening lithosphere. There is some uncertainty as to the effective time of establishment of a "single coherent surface," but it is not more recent than 3.9 b.y. ago; i.e., not since the Imbrium and Orientale impacts. An auxiliary indicator of a predominantly cooling regime is younger ages for compressive features (wrinkle ridges) than tensile (circum-maria rilles) in ringed maria (Solomon and Head, 1979).

The small volume of mare volcanism, its severe partitioning of trace elements, and its occurrence 0.5–1.2 b.y. after differentiation of the bulk of the crust imply rather deep source regions. The low viscosity implied by the great extent and low scarp heights of Imbrium flows (Schaber, 1973) also suggests melting of deep source regions undersaturated in silica.

Compositional properties

Lunar rocks, as discussed in sections 1.2.9–10, 2.2.1 and 7.3 have several major element characteristics unmistakenly indicative of a constitution and history quite different from the Earth's:

A dearth of volatiles: no perceptible H_2O, and lower levels of K_2O and Na_2O.

Lower oxidation level: no Fe_2O_3, in addition to several lesser indicators, such as Cr^{2+}.

Lower silica content: SiO_2 often less than 48%.

Higher content of refractory lithophiles, such as Al_2O_3.

Higher ultramafic components: FeO, MgO, and (in most) TiO_2.

These characteristics require the source regions (and hence probably the bulk of the outer half of the Moon) to be low in volatiles and possibly high in refractory silicates compared to the Earth, and the crustal differentiation to be less thorough: less reworking or hydration to obtain highly siliceous rock.

Trace element abundances and isotopic ratios also require a dry, oxygen-poor initial constitution subjected to very early major differentiation, followed by a history of markedly dwindling activity. The early differentiation extended rather deep—perhaps 300 km—since features of the mare basalts which came to the surface 3.1–3.9 b.y. ago require that either their source regions were subject to differentiation, or that the magmas had major interaction with differentiated material in their ascents. Most restrictive on lunar history is the isotopic evidence discussed in section 7.3. The Rb-Sr model age from lunar soils indicates a formation time of 4.60–4.65 b.y. ago, with an initial strontium intercept of 0.6989–0.6990. The ^{207}Pb-^{206}Pb isochron for terra rocks appears to require a differentiation which is early—4.42 b.y. ago—but distinctly later than formation. The enrichment factors of Rb-Sr for a variety of rocks and other considerations, such as the composition of KREEP basalts, indicate significant intracrustal layering at this early stage: a highly anorthositic surface layer, a more basic lower layer, and intervening zones of residual melt greatly enriched in incompatible elements. Subsequent differentiation was at a considerably lower rate, but not completely negligible, as indicated by Rb-Sr enrichment factors.

Geophysical properties

The Moon appears to be closer to isostatic equilibrium than the Earth, in the sense of the stresses implied: the topographic and gravitational irregularities of the Moon are appreciably less than six times the Earth's. Assuming isostatic compensation by variation in crustal thickness and using altimetry and gravimetry to extrapolate from the seismologically determined crustal thickness in the northeast Procellarum region, one obtains ~70 km for the global mean crustal thickness. It is entirely plausible, however, that a significant part of the compensation is by lateral variations in the upper mantle Mg:Fe ratio (Wasson and Warren, 1980; section 4.2).

Travel times in the eastern Procellarum area indicate a gradual increase in seismic velocity with depth to ~6.8 km/sec at 20 km, relative constancy to 56 km depth and then a jump to more than 8 km/sec. The increase in the top 20 km is most plausibly attributed to a decrease in porosity. The 6.8 km/sec velocity in the lower crust is the principal seismological puzzle; it is difficult to reconcile with a specific anorthositic-gabbroic mix (Liebermann and Ringwood, 1976). The upper mantle velocity below 56 km is consistent with plausible olivine-pyroxene compositions.

A significant constraint on the nature of basaltic deposits on the Moon is the nature of the gravity profiles from orbits crossing ringed maria "mascons." These profiles have appreciable short wavelength content, necessarily the "edge" effects of a shallow high density layer about 1 km thick.

Summary

Basaltic volcanism on the Moon thus can be characterized as two main types: "primordial" and "evolutionary," the former associated with the early crustal differentiation $\gtrsim$ 4.4 b.y. ago and the latter with subsequent development.

No pristine samples certain to be primordial basalts have been obtained. There is much evidence, however, of an early differentiation of large enough scale to have had significant extrusions. While there are few rock specimens as well preserved as the ancient meteorites, the Moon constitutes the best constrained instance of very early circumstances in the inner solar system. The highland basalts indicate that, even on a satellite, conditions in the first 0.2 b.y. were hot enough to lead to major differentiations and redifferentiations. The thickness of the lunar crust requires that this heating and differentiation extend some hundreds of kilometers deep.

In contrast, the evolutionary basalts of the Moon—the mare basalts—are, on the one hand, the best documented series of rocks extant, but, on the other, representative of a quantitatively minor process. The mare basalts are only a fraction of a percent of the bulk of the lunar crust. However, they indicate that partial melting and the resultant volcanism are rather inexorable processes. Even on a very dry smallish planet which has already undergone a significant differentiation, magmatism at appreciable depth can produce basalts with enough pressure behind them to come to the surface, so long as the planet is not strongly cooling overall. Furthermore, such basalts can have some variety of compositions, with trends which are not entirely associated with temporal evolution, indicating heterogeneity of source regions.

The Moon is undoubtedly the best understood planet in its evolution over the last ~4.4 b.y. However, the Moon's evolution (as distinguished from origin) is still not without mysteries. In particular, the source of the fields creating the remanent magnetism as young as 3.5 b.y. is obscure (but see Stevenson, 1980).

10.3.3 Mars

Spectacular volcanoes, up to 27 km high with flows up to 800 km in length, together with extensive plains flows, provide clear evidence of the importance of volcanism on Mars. Low impact crater counts have been found for some areas, requiring activity within the last 1 b.y. on even the longest time scales, and within the last ~200 m.y. on the shortest. Estimates of the predominant ages on the Tharsis shield volcanoes range from about 500 m.y. to more than 3.0 b.y. (see Fig. 8.10.1). However, while the period of peak activity is uncertain, the duration of martian volcanism and associated tectonics is very long.

There is abundant and varied evidence suggesting that the volcanism is dominantly basaltic. As described in section 2.2.5, Earth-based reflectance spectra of the dark areas have long been interpreted as being consistent with a basaltic mineralogy. This interpretation is not based upon curve shape alone but upon the observation of at least one absorption band at ~1.0 μm, suggestive of octahedrally coordinated Fe^{2+} as in olivine and/or pyroxene. The bright area spectra also have sometimes been interpreted as containing a "surviving" pyroxene component and magnetite, although the spectra of these areas are generally thought to be dominated by weathered materials such as clays, hydrated oxides of iron and possibly salts. Bright region spectra are nearly identical over the planet, while significant differences occur among dark region spectra. A weathered "bright spectrum" assemblage is generally derivable from mafic rocks by processes thought to occur under Mars's surface (and possibly subsurface) conditions, but it is difficult to "work back" to a primary igneous rock composition by consideration of weathering products alone (see below).

The Viking lander sites, which may be more boulder-rich than the average of Mars, show an abundance of extensively vesiculated boulders which have a different reflectance spectra from the dust. On the basis of spectral reflectance measurements by the lander imagery system, the soil appears to be an assemblage of weathering products—especially Fe-rich clays. However, the boulders seem to have low albedos and a visible reflectance compatible with a possible basaltic composition. The lander X-ray fluorescence analysis experiment (XRF), a partial elemental analysis, was found to be compatible with an assemblage of weathering products dominated by Fe-rich clay, such as is suggested by the lander spectral results. One question is whether the weathering products must represent a "closed-system" alteration of the source igneous rocks, or whether the surface weathering products could have been fractionated and segregated by winds or other agents. On Earth, weathering products tend to become chemically and mechanically segregated into sandstones, limestones, shales and evaporites, but in the absence of liquid water, the same may not be true for Mars. If so, the XRF experiment results on the weathered soil could correspond to that of the original igneous rock, which would be unlike any meteorite (except possibly shergottites) or terrestrial rock, but more similar to primitive basalts or some ultramafics than to other types. For example, the SiO_2 content of the soils is $\leq$50%, which suggests a basaltic or even more mafic rock. Thus the bulk of the Earth-based and Viking lander data favor a basaltic or ultramafic mineralogy.

Also, the great length of lava flows (up to 800 km) and the low slopes (4–6°) suggest not only high effusion rates, but also low viscosity, low yield strength magmas; i.e., iron-rich basaltic or ultramafic magmas. There seem to be fewer spatter cones or side vents on Mars relative to the Earth, consistent with a fluid magma. There is some complexity involved in such interpretations, however, owing to the importance of eruption *rates* in determining flow lengths, slopes, etc., but even a high rate suggests low viscosity. Therefore, morphological observations do support a mafic/ultramafic composition for eruptive rocks on Mars, but are slightly more ambiguous than spectroscopic observations. In summary, volcanic activity has been a dominant force in supplying material to the surface of Mars and creating the surface morphology as well.

There evidently has been extensive and prolonged mafic/ultramafic volcanism on Mars. Two factors are key to rationalizing such activity: (1) bulk planetary (or bulk mantle) chemical composition, and (2) planetary energy history, as reflected in time-temperature curves for each depth, especially those depths which might participate in supplying surface magmas. Mars either was hot at one time or accreted in a partially inhomogeneous way because it definitely has a substantial (1500–2000 km) core much denser than Mars's average density. Knowledge of the radial density distribution is too rudimentary to allow us to say what the proportion of Fe^0 to FeS to FeO might be in Mars's core, but at least one is likely to be an important constituent. A core would satisfy the density and moment-of-inertia constraints while yielding a thermal evolution consistent with the tectonic history. Most of the tectonic features on Mars appear to be extensional (see section 6.5), which argues against large scale cooling over the last 1–2 b.y. of Mars's history. The relative youth of Olympus Mons—some almost uncratered areas on its flanks—also suggests that Mars stayed warm rather late.

The absence of marked cooling in the last 2.0 b.y. precludes drastic upward differentiation of lithophilic radioactive elements. If there had been such differentiation, then the planet would have been cooling considerably during the last 2 b.y. and the extensional features, coupled with the apparent absence of compressional features, would not be evident.

The evidence is sketchy as to whether Mars was significantly heated early in its evolution. Erosional effects, such as dust deposition, have modified or obliterated many features (see section 8.7). There is an indication that formation of a chemically distinct martian crust occurred prior to 3.5 b.y. ago, in that old terrain features, such as Hellas, appear compensated. Therefore, fairly shallow vertical density differences were set in below at some time. If they were set in *after* Hellas formed, then Hellas's shape would have been destroyed at the time of differentiation. On the other hand, it is possible that compensation could have been achieved without much surface distortion. Also much smaller bodies—Mercury, the Moon and the eucrite parent body—differentiated crusts very early, and hence to assume that Mars experienced a similar early burst of energy seems less *ad hoc* than to assume it somehow escaped (see section 9.5.3).

As discussed in sections 4.3 and 10.2, the dryness of Mars poses problems for nebula models and the resulting planetary compositions. In any case, the apparent dearth of volatiles, the modest degree of central concentration required by the moment-of-inertia, the mean density, and theoretically deduced temperature-depth curves produce a Mars mantle model with a high FeO content (~21%). Despite uncertainties in FeO/FeS/Fe^0 ratios and core composition, Mars's mantle must be denser, uncompressed, than the Earth's. Almost certainly this is due to a higher FeO content, but the uncertainties alluded to above allow FeO to range from 14% to 30%.

The only major point of convergence of all the models is that Mars's mantle is more FeO rich by a factor of, say, two or three than that of Earth. The question is whether martian volcanism as discussed above could be reconstructed with a martian mantle that was exactly like that of the Earth in every way *except* for a much higher FeO content. Plausible Earth mantle compositions generate normative mantle mineralogies which produce cotectic-like melts like olivine tholeiites. As discussed in section 3.6, changing the composition only by adding FeO produces a different normative mantle composition for Mars. Melting of that composition produces iron-rich and ultrabasic lavas. These melts would have a composition similar to that observed by the Viking lander X-ray-fluorescence experiment. This brings us back to a question we raised earlier: "What does an assemblage of weathering products tell you about the primary igneous source from which they were derived?" On Mars, it probably tells us a great deal. It seems as if at least recent alteration of Mars took place without oceans. Ultraviolet-activated H_2O vapor-rock interaction or perhaps local subsurface liquids appear to have been responsible. In either case, the ion mobility on Mars was probably low. Moreover, the general agreement between analyses of Viking 1 and 2 lander material supports the notion that alteration on Mars was of the closed-system variety, in contrast to most sedimentary processes on Earth. Similarly, fractionations associated with the processes that produced the duricrust, or caused by aeolean transport, seem slight.

Hence there is good coupling between the observed characteristics of igneous activity on Mars, global geophysical constraints, and the high FeO content inferred from formation models. This consistent story concerning Mars volcanism can be put together without reference to the bulk volatile content of Mars, its degassing history and its atmospheric history, because, as emphasized in Chapter 3, volatiles have little influence on abundant basalts. Mars's probable accretion history, its subsequent thermal history, zonal structure, and obvious widespread volcanism all suggest that there has been ample opportunity for at least the outer portion of Mars to outgas. Hence the small Mars surface and atmospheric inventories of ^{36}Ar and other volatiles suggest that Mars is a volatile poor planet. If there was an early atmospheric loss, it is uncertain whether it would produce (an unobserved) elemental fractionation in the remaining nonradiogenic rare gas inventory. Another problem is potassium. If the Soviet Mars 5 orbital γ-ray data and interpretations are correct, then the *global* surface K/U ratio for Mars is only one-third that of the Earth's crust and one-tenth that of chondrites (Anders and Owen, 1977). Hence for any models but those for a volatile-poor Mars, it seems necessary to lose the early atmosphere, to bury much of the K in the interior, and to preserve any remaining volatiles from anything like efficient outgassing despite continuing volcanism. A volatile-poor Mars is difficult to reconcile with systematic patterns of planetary properties as the result of origin. Aside from the nonmonotonic trend in volatiles with solar distance, Mars seems to break the coupling between volatile content and degree of oxidation—a key feature in some planetary formation models.

Many of the controversies could probably be resolved by a combination of global geochemical (γ-ray) and mineralogical mapping from orbit and analysis of carefully selected samples from several sites. In the meantime, an Fe-rich martian mantle and the apparent occurrence of widespread basaltic/ultramafic volcanism on Mars are two of the very few things about the planet on which almost all investigators agree.

10.3.4 Mercury

As discussed in sections 2.2.3 and 5.5, direct evidence of basaltic composition on Mercury's surface is limited to reflection spectrophotometry indicating pyroxene, and a morphology of some units, such as the broad plains around the Caloris Basin, suggestive of low viscosity lava flows.

Indirect indicators of volcanism are those features of Mercury requiring a fairly warm thermal history. The predominantly compressive tectonics of the surface, marked by lobate scarps, suggests an early differentiation, bringing most of the radioactive lithophiles near the surface, so as to produce a net cooling over most of Mercury's evolution. The high density and the magnetic field of Mercury require a core of metallic iron, in contrast to the silicate surface. The initial high temperature of Mercury's material plus the energy released by core formation would have been sufficient for an early crustal differentiation.

There are indications of early intensive volcanism in the extensive upland "incrater" plains of Mercury (sections 8.5–7). However, for most of its history Mercury has been the most compressive in its tectonics of all the terrestrial planets (now that significant rifts have been found on Venus: Pettengill *et al.*, 1979a). As discussed in section 9.5, the onset of the net cooling to produce a compressive regime cuts off volcanism in a one-plate planet. Hence Mercury's volcanism probably terminated earlier than that of the other terrestrial planets or the Moon, perhaps not long after the "cataclysm" of 3.9 b.y. ago, of which the Caloris Basin was plausibly a product. This conjecture is not inconsistent with the cratering data: Mercury's surface is cratered everywhere, but the absolute time scale is rather uncertain (see section 8.10).

The greatest problems of explaining Mercury's constitution go back to origin circumstances (see section 4.5.3): in particular, the depletion of magnesian silicates relative to iron necessary to explain its high mean density. It is likely that the oxidation level on Mercury is sufficiently low that despite the low bulk content of magnesium, the comparatively small mantle probably has a higher Mg:Fe ratio than that of the Earth or the Moon.

The Moon is probably the best guide for the nature of basaltic rocks which may eventually be sampled on the mercurian surface. We should expect to find a highly brecciated surface with specimens that, while not identifiable as basalts petrographically, will have compositions corresponding to the basic component of a primordial crustal differentiation. The mercurian analogue of mare basalts should be scarce or entirely absent. The failure to observe dark plains is consistent with the modelling hypothesis that volcanism gets cut off early in a one-plate planet of predominantly cooling thermal history. Whether the broad plains around Caloris are some sort of lava flow—perhaps impact-triggered—rather than relatively fine ejecta, is rather conjectural. All these inferences are, of course, derived from spacecraft observations of only half the planet.

10.3.5 Venus

Radar measurements of the near equatorial topography of Venus showed variations in height of ± 3 km and wavelengths of 3000 km (Campbell *et al.*, 1972). Such large variations imply isostasy, and hence a crustal differentiation. The first positive indication of differentiation was the high content of K, U, Th inferred from the Venera 8 gamma-ray experiment (Vinogradov *et al.*, 1973). Now the Pioneer Venus radar mapper experiment shows that while most of Venus's surface is gently undulating, it has two sizeable plateaus, both of ~7 km relief (Pettengill *et al.*, 1980). Hence Venus must have been tectonically active to the extent that granite-like radioactive content, at some locations, is not implausible. The most unmistakable difference between the tectonics of Venus and the Earth is the absence of any feature like an ocean rise. Hence, if plate tectonics ever existed on Venus, it must have terminated, and the surface must be entirely choked with crust (Phillips *et al.*, 1981).

As described in section 4.5, the venusian mantle would have MgO/(MgO + FeO) ratio like the Earth's, if its lower mean density were the consequence of an iron or sulfur deficiency, but it would have a somewhat lower ratio if the lower density came from an oxygen excess. Since volatiles apparently have a negligible role in volumetrically major basaltic differentiations (see section 3.6), the abundant basalts on Venus should be similar to the abundant basalts on Earth. However, this conjecture is based upon the assumption that Venus's density has a simple explanation. At our present state of ignorance, it cannot be ruled out that Venus is both appreciably depleted in iron and has a low oxidation level: i.e., that it is somewhere between the Earth and the Moon in mantle constitution. Hence, it could conceivably have basalts intermediate between lunar and terrestrial in major element chemistry and in characteristics dependent on redox state. An intriguing question will be the degree of siderophile depletion in venusian surface rocks, since such depletion is an indicator of whether material has been in a planet which has undergone core differentiation (Ringwood, 1979).

While volatiles may not influence abundant basalt character—terrestrial tholeiites and their venusian analogues, whatever they may be—they could be significant in the more complex and later stages of crustal differentiation, such as those associated with subduction zones, continental rifting, "hot spots," etc., on Earth. It is likely that Venus's interior has a lower volatile content than the Earth. An outgassing efficiency comparable to that of the Earth probably existed in the first 2 or 3 b.y. Once the surface became hot enough for any water to boil off, it would no longer be recycled, as it is on Earth. The high surface temperature keeps the venusian lithosphere from becoming sufficiently more dense than the underlying mantle to subduct.

While Venus appears to be tectonically dead now, it could have had a rich history, of which the two plateaus are remnants. Hence a balanced exploration program requires mapping by radar satellites as well as landers to obtain compositional information directly. It would be a pity to have several landers all fall on venusian analogues of ocean floor and Precambrian shield, and miss the critical one percent of surface area which contains analogues of island arcs, rift volcanism and other important features of limited extent.

10.3.6 Galilean satellites

Asteroids and achondrite parent bodies can plausibly be passed over as nonevolving bodies (see section 10.4.2 for possible exceptions), and the outer Galilean satellites as icy rather than terrestrial planets. The same cannot be done with the dense inner Galilean satellites, Io and Europa, which now have been photographed by the Voyager probes.

Io manifestly must have an energy source different from the other terrestrial bodies. This source, probably tidal (Peale *et al.*, 1979), appears to be rather shallow, because of the widespread occurrence of volcanism: more than 300 possible volcanic vents have been found. The explosiveness of the volcanism probably arises from its volatile content; in fact, the volcanism is probably not basaltic at all, but constituted of more volatile sulfur compounds, which would drastically lower the energy requirements. Although only sulfur compounds have been identified spectroscopically, the morphology of some calderas suggests that silicate volcanism exists.

Europa also appears to be nonbasaltic in its near surface evolution. The mildness of the topographic variations observed by Voyager 2 suggests a predominantly icy surface layer, consistent with theoretical models.

10.4 LEADING INFERENCES

In this section, we will attempt to set up a general model of terrestrial planet evolution, then go back to reexamine the planets with a view to identifying variations and exceptions. Next, we will discuss the consequences of the general model and variations for basaltic volcanism. Finally, we will suggest the outstanding problems and directions for future research. Aside from earlier chapters of this book, principal references are Kaula (1975) and Walker *et al.* (1979).

10.4.1 A general model

The most obvious independent variable is planetary mass. As discussed in section 10.2.6, the more massive a planet, the more heating it is likely to have in the formation process. The absolute magnitude of this heating will depend on the efficiency of heat retention upon impact. The relative magnitude for planets formed from planetesimals appreciably smaller than themselves (but still big enough to bury significant heat upon impact) will be at radius r approximately proportional to r^2, up to a radius at which solidus temperatures are attained. Above this radius, convection should keep the temperature curve close to the solidus.

A priori, there is no clear rule as to the minimum-sized planet to approach melting temperature in its formation and thus to attain a convective regime in its outer parts. This uncertainty is primarily due to ignorance as to the efficiency of heat retention in large impacts. If we adopt h = 50% for the retention factor (Kaula, 1980a), then the minimum-sized planet to attain melting temperatures in its outer parts, determined from (cf. section 10.2.3)

$$\max(\Delta T) \approx h \frac{GM}{RC_p}(1 + \frac{1}{2\theta}) \approx 1700° K \qquad (10.4.1)$$

will be R ~ 1900 km, using 3.7 g-cm^{-3} for density and 4 for θ, the mean ratio GM/Rv^2, where v is approach velocity. The most important consequence of attaining such temperatures is obviously core formation, which adds an extra pulse of energy yielding an average temperature change

$$\Delta T \approx (1-P)\, P\, \Delta\rho \frac{GM}{3R\bar{\rho}C_p} \qquad (10.4.2)$$

where P is the fraction of the mass in the core, $\Delta\rho$ is the density difference between core and mantle, $\bar{\rho}$ is the mean density, and the other symbols have their usual meanings. For a large planet, core formation probably started well before the completion of planetary growth, but the thermal result will be roughly the same. The temperature rise in the later stages may be great enough that, with the help of instability of a lighter center (Stevenson, 1980), core formation is a rapidly self-accelerating process. An important feature of core formation is that the energy released tends to be maximized near the core-mantle boundary, thus further inducing mantle-wide convection. Core formation also has compositional implications: it would remove siderophiles and chalcophiles from the mantle, plus, in the Earth and Venus, a portion of elements which may undergo a change in character at higher pressures, such as has been suggested for potassium. For a small planet of relatively high oxidation level, core formation may be delayed until temperatures are raised radioactively, and may be a slow process because the temperature rises are modest.

Other effects of early heating will be outgassing and crustal differentiation. Outgassing will evidently go more rapidly the greater the heating. Crustal separation is a different matter, since it involves rocks of varying ratios of calc-aluminous to ferromagnesian silicates, and hence of small density differentials. The vigorous convection induced by heating and outgassing may bring about remixing rather than separation, particularly if more than a small amount of melting is involved. However, the strong inverse correlation of crustal thickness with planetary size (see Table 4.2.1) may be related to the depth at which the basalt-eclogite phase transition intersects the solidus, and hence below which calc-aluminous material is more likely to sink than rise: about 50 km on the Earth, but 300 km on the Moon. Positively correlated with planet size must be the degree of recycling of crustal material to the mantle. This process affects thermal evolution because the degree of upward differentiation of U, Th, K will be proportionate to the degree of crustal differentiation. This proportionality should be nonlinear, because large-ion lithophiles are upwardly enriched more severely and are less likely to be recycled than the major crustal constituents. In any case, it seems that an additional factor contributing to the prolongation of activity in larger planets is the return, or retention, of more radioactivity in their mantles.

Radioactive heat should become the dominant energy source within a time that the heat would melt its host rocks if it were not removed, since formation heat in excess of this amount would have largely been removed by convection. For chondritic composition this time is 1.2 b.y. The time is somewhat longer for planets with deficiencies in potassium. Therefore, it is useful to discuss idealizations of planetary evolution whose heat sources are entirely radiogenic as, indeed, was customary until rather recently.

While the ultimate asymptote of planetary evolution (ignoring solar evolution) is complete upward differentiation of heat sources which meanwhile have dwindled to negligible rates, a more useful quasi-asymptote is a planet whose heat loss rate is equal to its radioactive heat generation rate. Given a distribution of heat generation Q(r) with radius r, from the equation of heat transfer (Eq. 9.3.1), the temperature at depth $D \ll R$, the planetary radius, will be $T(D) \approx$

$$T(0) + \frac{D}{4\pi R^2 K}\int_0^R Q(r)\, r^2 dr - \frac{1}{K}\int_0^D\int_0^y Q(z)dz\, dy \quad (10.4.3)$$

where K is the thermal conductivity. With uniform heat sources, Eq. (10.4.3) predicts that the temperature rise T(D)-T(0) will vary proportionate to R. Given further the observation that small planets appear to have been more effective at upwardly differentiating lithophiles and their associated heat sources, then the proportionality will have a higher than linear dependence on R. Hence the smaller a planet, the thicker the lithosphere—the layer in which radial heat transfer is dependent on conduction rather than on convection. This condition is true for any planet which has cooled off enough that the outer layers are representable by a single continuous lithosphere, such as the Moon, Mercury, and Mars. It is partly true for planets characterized by plate tectonics, in which some heat is brought essentially all the way to the surface by convection.

The foregoing discussion of lithospheric thickness brought up the temperature dependence of viscosity in an implicit manner. It is observed that silicate viscosities tend to be proportionated to $\exp(T_m/T)$, where T_m is the melting temperature. Since the melting gradient is steeper than the adiabatic gradient, as discussed in section 9.4.4 and 9.5.1, a convecting planet may tend toward a state where the viscosity in the convecting zone will increase with depth. To some extent this increase would be counteracted by the "governor" tendencies of temperature-dependent viscosity: the higher the temperature in a region, the quicker convection can cool off the region by removing heat, and vice-versa. But it seems likely a planet will develop a zone of high viscosity toward the bottom of the mantle. An important unknown affecting this inference is the pressure dependence of viscosity (Kaula, 1980b).

The transition of a planet from the formation-dominated early hot stages to the radioactivity-dominated late quasi-equilibrium stages will depend on a variety of factors. A body which is small, and therefore of high area:mass ratio, can get rid of its heat quickly and hence approach equilibrium more rapidly. For bodies of less than 500 km radius, such as the asteroids and the meteorite parent bodies, the characteristic time of this temperature decay is short compared to the ~1.2 b.y. of the radioactive heating time. Therefore, these bodies are dismissed in section 10.3 as "nonevolving." For a body of ~1700 km radius, such as the Moon, the situation is marginal; consequently, there is an era of interaction between cooling of the outer parts and heating of the inner parts by radioactivity before the cooling prevails to the extent that an impenetrably thick lithospheric layer dominates the transfer of both heat and magma. For a body of large radius, say more than 5000 km, the much greater store of initial energy, the low area:mass ratio, and the relative inefficiency of upward differentiation of large-ion lithophiles keep it active until current times.

The preceding discussion is based almost entirely on one independent variable: the mass. Next in significance is bulk composition. Composition in turn depends primarily on distance from the sun and secondarily on planetary mass: there appear to be positive correlations of volatile content and oxidation level with distance from the sun and planet size. The correlation with solar distance is obviously a consequence of a negative temperature gradient in the nebula until gases were expelled, as well as planets being made from material in their respective zones (see section 10.2). The correlation with planetary mass does not have as clear an explanation, the most plausible being an addition of more volatile material to the inner nebula late during planetary growth which the larger planets acquired more easily. This hypothesis does not, however, explain the remarkable differences in inert gas abundances among Venus, the Earth and Mars.

The most evident consequence of compositional difference is the contrast in evolution of Mercury and Mars. The planet of low oxidation and volatile level, Mercury, had a large component of reduced iron and differentiated a core early, thus attaining a net cooling and a compressive tectonic regime for most of its history. The planet of high oxidation level, Mars, had a marginally small amount of reduced iron, and hence apparently required radiogenic heat to attain core separation circumstances relatively late in its evolution. However, these thermal circumstances led to longer persistence of a tensile tectonic regime, so that the major volcanic constructs of Tharsis could occur.

A third variable affecting planetary evolution is occurrence as a satellite rather than as a planet. In general, a satellite should reflect hotter circumstances of formation than an isolated body of the same size, due to the influence of its planetary companion. The Moon and Galilean satellites indicate that composition is greatly affected. Planetary influence may also be felt in the course of evolution, as is most evident for Io.

A fourth variable is surface temperature. While it may have little influence on bulk evolution because of the temperature dependence of viscosity, it may be of significance for near surface differentiations through its influence on magma thermodynamics, buoyancy of the lithosphere, and tectonic settings.

Finally, volatile content as a variable independent from mass is important for crustal evolution for its influence on the character of volcanism and subduction (and hence recycling), and on more advanced differentiations, andesitic to granitic.

In summary, the larger a planet, the hotter it runs; however, the increase in heat transport effectiveness with temperature is such that a terrestrial planet should rarely get much above melting for realistic growth times. The larger a planet, the longer it takes to attain a quasi-equilibrium state, in which the heat loss is not much more than contemporaneous radioactive heat generation. Also positively correlated with planet size are complexity and variety, arising from, first, the domination of convection in heat transfer, leading to a plate tectonic surface situation; second, the greater degree of recycling of crustal material; third, the attainment of pressures sufficient for more phase transitions, which have strong bearing on both heat transport and differentiation; and fourth, the higher volatile content.

10.4.2 Variations and exceptions

We list here phenomena unexplained by, or difficult to reconcile with, the general picture we have presented:

(1) The low Mg:Fe ratio in Mercury. As discussed in section 4.5.3, there does exist a narrow temperature band in which iron condenses but magnesian silicates do not; however, this seems too weak an effect to account for the large departure of Mercury from the chondritic ratio in a nebula of realistically evolving temperature. Consequently, resort must be made to effects dependent on dynamical evolution, as discussed in section 4.5.

(2) The lower density of Venus compared to the Earth. Explanation of the difference by a lower sulfur content in Venus but a solar complement in the Earth is inconsistent with depletion of other volatiles in the Earth. The alternate explanations by higher oxidation or lower iron content in Venus are inconsistent with monotonic trends in nebula properties, and require resort to dynamical effects, perhaps associated with Venus's lack of a satellite, as discussed in section 4.5. The explanation by a thick basaltic, rather than eclogitic, layer requires strong upward differentiation of lithophiles and higher temperatures at depth (to keep

basalt above the solidus at pressures greater than ~15kb) to be quantitatively sufficient.

(3) The higher inert gas content of Venus. Hypothesis dependent on steep radial gradients at, or soon after, condensation in the nebula (e.g., solar wind implantation or pressure effects) severely limit the allowable subsequent interzonal mixing to account for the order-of-magnitude differences among Venus, Earth and Mars in abundance of primordial argon. Hence acquisition of primordial argon by large body impact in the late stages of formation seems more plausible.

(4) The drastic differences in evolution of Venus and the Earth: a water content smaller by a factor of more than 10^{-2}, a magnetic field smaller by a factor of more than 10^{-4}. Neat explanations exist for both of these. For the water content, Venus is close enough to the sun for the water to be boiled off and dissociated (Walker, 1975; Pollack, 1979). For the magnetic field, Venus lacks an energy source from either precessional torques—no Moon—or core differentiation, since its central pressure is too low (Stevenson *et al.*, 1980). However, both explanations are only marginally satisfying, leading to speculation as to whether Venus was different in the past.

(5) Evidence of mantle heterogeneity. Departures from spherical symmetry and compositional homogeneity must exist to some degree in all the terrestrial planet mantles. However, the Earth dominates our attention not only because of accessibility, but also because of the greater range of conditions. The most marked heterogeneity in the Earth is the simultaneous presence of a reduced iron core and volatiles in the outer parts. Also rather inescapable is the need to keep the source regions of certain categories of rocks, such as basalts from continents and ocean rises separated for more than ~2.0 b.y. as discussed in section 7.4. Geophysical data, such as heat flow, gravity variations, seismic velocities and the tectonic plate motions, indicate that there are significant lateral variations to a few hundred kilometers deep which must be at least partly thermal, but may also be compositional. More in debate is the need for compositional differences with depth to satisfy seismological models associated with both the 220-km and 670-km discontinuities and the deeper mantle. As discussed in section 4.5, there are differing interpretations of equations-of-state and phase transitions as to whether there is a distinct increase in Fe/Mg and/or Si/Mg with depth below 670 km (Ringwood, 1975; Anderson, 1977). As discussed in section 10.3.1, if the 220-km discontinuity is sharp, there are no phase changes in the abundant Earth materials to match it. While it is the current general opinion that there is heat transfer throughout the mantle by convection, it is debatable whether there is transfer of matter as well as heat between lower and upper mantles (section 9.5.1; Stevenson and Turner, 1979). The principal need for all-mantle convection cells is mainly to get flow patterns consistent with plate velocities and subduction zone dip angles (Hager and O'Connell, 1979). Finally, the implications of plausible cosmogonies for mantle heterogeneity are not at all clear. While classical heterogeneous accretion, in which the entire core is formed before the mantle is added, is virtually impossible to reconcile with physically realistic models, the mantle still could be significantly influenced by smaller scale heterogeneities associated with the planetesimals that formed the Earth.

(6) The Moon's bulk composition. As discussed in section 4.5.2, the Moon definitely has drastic depletions in iron and volatiles; it may also have an enrichment in calc-aluminous silicates relative to ferromagnesian silicates. This pattern is quite inconsistent with a monotonic trend with respect to condensation temperature, thus suggesting a combination of differentiation in an earlier body or bodies together with dynamical effects, such as hitting the Earth with an extra large planetesimal (Kaula, 1977; Ringwood, 1979).

(7) The low volatile content of Mars compared to the Earth and Venus. It is difficult to account for the present atmospheric consititution of Mars, particularly the isotope ratios, without inferring a bulk composition deficient in volatiles (Anders and Owen, 1977). If Mars has a cosmic Fe:Mg ratio, the lower mean density requires a higher oxygen abundance, resulting in a smaller core and a lower Mg/(Mg + Fe) ratio in the mantle. In combining a small core and dryness, Mars resembles the Moon, and both bodies reinforce the hypothesis that the planets acquired their chemically active volatiles from an additional component similar in composition to carbonaceous chondrites rather late in the formation process, so that planet size was a significant factor in determining the amount acquired (Anders, 1977; Ringwood, 1979). The findings as to venusian composition emphasize that the explanation for the abundances of inert gases is not necessarily the same as for the abundances of active gases.

(8) The differences in composition within the asteroid belt. Most asteroids have spectra similar to carbonaceous chondrites (section 2.2.6; Chapman, 1979), which reinforces the idea that volatile content increases with distance from the sun. The minority of asteroids with spectra suggesting differentiation are moderately concentrated toward the inner edge of the

belt, and have a weak correlation with size. These statistics suggest that effects which would be somewhat random in their influence (most obviously, collisions) have some influence on the degree of differentiation of asteroids, as well as original heliocentric distance.

(9) The energy source for melting in the parent bodies of differentiated meteorites. This problem overlaps, and could be combined with, item (8) above. As discussed in section 10.2, currently it is quite speculative as to whether this source is collisional, electromagnetic, or short-lived radioactivity.

(10) The presence of relatively young achondrites. As discussed in Chapter 3 and Walker *et al.* (1979), there exist igneous meteorites of rather young age: the shergottites, about 0.6 b.y. old, and the nakhlites and chassignites, about 1.3 b.y. old. Both are rather volatile-rich cumulates. Their ages and petrology suggest they are products of differentiation on a sizeable planet. Given the limited availability of transfer mechanisms over the last ~4 b.y., the obvious scenario to produce these meteorites involves an asteroid impact on Mars, to which the obvious objection is that the resulting textures should be highly shocked, and that there should occur abundant high pressure phases. Shergottites contain maskelynite: shock-vitrified plagioclase. The physics of large impacts is imperfectly known; the scaling up from small events is not straightforward. Furthermore, the presence of volatiles (such as permafrost) greatly enhances ejection capability with little shock (Kieffer and Simonds, 1980). Hence it cannot be ruled out that moderately shocked meteorites could be ejecta from Mars (Wasson and Wetherill, 1979). Otherwise, we are left with a similar problem to item (9) above, only worse (see also section 9.5.6).

Other variations and exceptions could be selected, of course. The preceding list is to some extent a mixture of real variations from the model based on planetary size and distance from the sun, items (2), (3), (4), (6) and (7); related matters which involve more detailed consideration, items (1) and (5); or which affect only a minor portion of terrestrial matter, items (8), (9) and (10). Most of them involve solar system origin, which must have entailed a complex intermediate stage of relatively few sizeable bodies, leading to an appreciable variance of the actual outcome from the average of all possible planetary systems. For the purposes of interpreting basaltic volcanism, the terrestrial bodies must therefore be taken as having some systematic patterns, but also in some ways as having been formed differently.

10.4.3 Consequences for, and constraints from, basaltic volcanism

Ideally we should take models of planetary composition (Chapter 4) and evolution (Chapter 9) to predict basaltic volcanism properties for comparison with observation. Most important is to infer source characteristics (Chapter 3) from compositions (Chapters 1 and 2), abundance and morphology (Chapter 5), locus and manner of occurrence (Chapter 6), and chronology (Chapters 7 and 8). The discrepancies between observation and prediction should then be used to decide the next steps of data acquisition, experimentation, and modelling.

Practically, any discussion becomes shaped by those bodies from which samples have been obtained: the Earth, the Moon, and the basaltic achondrite parent body (or bodies), since the strongest constraints come from the petrological experiments discussed in Chapter 3 and the isotopic measurements discussed in Chapter 7. Hence a more sensible course is to take a combination of the strong observational constraints with the modelling considerations to predict circumstances which are less well known: surface compositions of unsampled planets; the constitution and thermal states of planetary interiors; and the past evolutions of the planets, with emphasis on their surface manifestations.

From the terrestrial, lunar, and achondritic samples we have evidence of two major phases of basaltic volcanism, which may be described as **primordial** and **evolutionary**.

Primordial basalts

Achondritic meteorites and lunar highland rocks indicate that differentiations producing basaltic compositions occurred more than 4.4 b.y. ago. This is early enough that we should expect the energy sources to be other than long-lived radioactive, and we should hope the source material to be truly primary. The eucritic meteorites show most clearly a primary source which could be chondritic in major element composition (section 3.5 and Stolper, 1977). They also suggest that magmatic differentiation in the presence of metal efficiently removes siderophiles, that the eucrite parent body (EPB) was very dry, and that heat sources much greater than accretionary existed in some small bodies.

The lunar evidence is more indirect. The Fra Mauro basalts are highly-processed fragments which apparently were a minor but significant part of the general differentiation of a crust of predominantly feldspathic composition. This differentiation was sufficiently complex that a primordial source composition is not inferrable. More important, the petrology and isotopic systematics of highland rocks, together with geophysical data, indicate that the Moon may have differentiated a primordial crust thick enough that either at least half of the Moon's bulk was involved or that the Moon is enriched in CaO and Al_2O_3 relative to chondrites. In either case, the outer part of the early Moon was extremely hot as well as extremely dry. Lateral variations in highland petrology—e.g., the distribution of KREEP—are consistent with the hypothesis that the heating came from infall of sizeable bodies.

Since the Moon is special because it is a satellite and the EPB may differ because it accreted early enough to be heated by Al^{26} or electromagnetic effects, it is difficult to extrapolate to primordial crustal formation on the terrestrial planets proper. As indicated in Table 4.2.1, there appears to be an inverse correlation of crustal thickness with size. As mentioned in section 10.2.7, this may have some relation to the basalt-eclogite phase transition. A further inference is that the larger a planet, the more it recycled its crustal material. This is obviously true for the Earth, on which no rock ages older than 3.8 b.y. have been found, but rather unsure for Venus. Whether it is true for the smaller planets, Mars and Mercury, is uncertain to the extent that the Moon may be anomalous in its early history because it is a satellite. The implication for subsequent planetary evolution is that "primordial" (i.e., resulting from formation processes) mantle material is likely to be exceptional, since any sort of crustal differentiation and recycling is bound to have different effects on some constituents than on others. This is even more true for planets which underwent early core differentiation.

While it is highly desirable to find primordial basalts on Mars or Mercury, we should not expect to find them on any body larger than Ceres (which may not have any basalt at all).

Evolutionary basalts

Terrestrial basalts and lunar mare basalts further indicate that basaltic volcanism can originate from appreciably modified sources over a long period of planetary evolution. The contrasts between lunar mare basalts and the most abundant terrestrial basalts, ocean rise tholeiites, emphasize the effects of planet size. Mare basalts are generally considered to be partial melts from depths greater than 200 km; they are either from cumulate sources or have interacted significantly with cumulates, and have severe trace element fractionations suggesting small percent partial melts or other complexities. On the other hand, ocean rise tholeiites are large percentage (~20%) partial melts which have differentiated at shallow depths, 30–50 km, either from mantle material depleted in the aluminous phases or from a more basic magma (e.g., a picrite), which has risen from a deeper differentiation. The depths of sources, inferred from high pressure phase equilibria studies, are suggestive of lithospheric thickness in the cases of both lunar mare basalts 3.1–3.9 b.y. ago and recent oceanic tholeiites. Both also show little influence of volatiles, although for different reasons: mare basalts, because of a bulk deficiency; tholeiites, because they result from large percentages of partial melting.

Both lunar and terrestrial evolutionary basalts have variations in composition with location which must at least in part be attributed to variations in source composition. The lunar variations are comparatively modest. At one time, the principal trend appeared to be an inverse correlation of Ti content with age. More recently, exceptions to this trend have been discovered. The terrestrial variations described in Chapters 1 and 3 are considerably greater, and of more interest as furnishing examples of varying conditions of basalt formation: pressure, percent melt, H_2O, CO_2, Fe/Mg, etc. Furthermore, some of these petrogenetic variations have correlations with age or with trace element constraints on source isolation. As discussed in section 7.4, some continental basalts have nearly chondritic $^{143}Nd/^{144}Nd$, which requires separation of their source from the depleted ocean rise basalt source for $\gtrsim$2.0 b.y. Although the nontholeiitic varieties are of minor abundance on Earth, some of them may be of closer resemblance to basalts on planets which are one-plate or of higher volatile content.

Combining the trends of the general model set up in section 10.4.1 with the associated basaltic characteristics outlined above, and taking into account the variations and exceptions described in section 10.4.2, we can conjecture the pattern in time of volcanic activity on the terrestrial bodies.

Mercury, being formed hot, probably differentiated both its core and its crust fairly early. Analogy with the Moon suggests that although it took longer to cool in the outer parts, the differentiation was fairly thorough. Whether it resulted in a mainly olivine mantle and a thick crust or an olivine + spinel + diopside (or melilite) mantle with a thin crust depends on a prime unknown: the degree to which Mercury's bulk composition is depleted in SiO_2 relative to MgO (Smith, 1979; Stolper, 1980). Subsequent volcanism was slight, and probably terminated before 3.5 b.y. ago because of the early onset of a compressive regime. Sampling of the mercurian surface should obtain few pristine specimens; mostly brecciated rocks of olivine tholeiite, or perhaps even komatiite-like composition with low Fe/Mg and no alkalis. Volumetrically minor lavas of less than 4.0 b.y. age may be found, with characteristics indicating formation as small percentage partial melts at considerable pressure.

Venus appears to have been tectonically active, from its altimetry (Pettengill *et al.*, 1980) and gravity field (Sjogren *et al.*, 1980). Since Venus no doubt has had as long a history as the Earth, evidence of a primordial differentiation will be obliterated and a comparable complexity should prevail. There should be about the same abundance of volcanism, and the abundant basalts should be similar to those on Earth, olivine tholeiites consistent with near surface crystal-liquid equilibrium (at, if anything, lower pressures than of Earth, because of higher surface temperature), with perhaps a higher Fe:Mg ratio on the average. Trace element contents in abundant basalts may reflect the lower pressure or differences in compositional evolution. More marked differences between Venus and the Earth in major element chemistry should occur in its less abundant basalts, such as the venusian analogues of island arc basalts, if they exist: differences in near surface temperatures and volatile content could be much more critical here. At this point it is rather speculative as to whether venusian basalts will reflect lower volatile or alkali content. A respect in which Venus may differ significantly from the Earth is the extent to which the record of the past has been obliterated by erosion.

There is still ample room for speculation about the Earth and Moon, in the attempts to extrapolate into the interior or into the past. However, the extrapolation has a different character pertaining much more to scientific inference than to what an exploring spacecraft may find. Hence we have put these questions in the next section on future directions.

Mars supports the highest pile of basalt in the solar system, possibly because it still maintained a tensional tectonic state after it had already developed a rather thick lithosphere. This later-stage tensional state was plausibly induced by later core formation, in turn the consequence of the relatively small portion of free iron. The height and spacing of the volcanoes are dependent on both the thickness and the depth of sources. Source depths are in turn proportionate to lithospheric thickness, and hence to mechanical strength. Therefore, the pressures reflected by the younger martian basalt compositions may be even higher than those indicated by mare basalts, a consequence of the higher gravity and the longer persistence of tensional tectonics. The principal effect of the higher oxidation level should be a higher Fe:Mg ratio (McGetchin and Smyth, 1978). The low viscosity suggested by the low slopes of lava flows is consistent with this higher iron content. More uncertain is whether there is a higher abundance than in the Earth of elements of intermediate volatility: such as Si, and K, leading to higher proportion of quartz-normative basalts on the one hand and alkalinity on the other. However, the Soviet Mars 5 gamma-ray experiment indicates a K/U lower than terrestrial by one-third (Anders and Owen, 1977). An important volatile effect on Mars discussed in section 10.3.3 is the weathering of martian volcanics, complicating interpretation of remote-sensing data.

The differentiated asteroids should afford the best opportunity to view pristine primordial basalts and their contexts. If the chondritic meteorites are any guide, the range of oxidation level and volatile content among asteroids should be wide.

As discussed in section 10.3.6, Europa and Io manifest the effects of high volatile content, resulting in volcanic activity not easily compared to basaltic volcanism.

10.4.4 Outstanding problems and directions for the future

The obvious way to answer many of the questions about basaltic volcanism on the other planets is to go there and look, meanwhile having improved techniques on how to look. However, other questions require deeper understanding to answer. Significantly, most of these questions pertain to basaltic volcanism on the Earth itself. The following list represents a consensus of the authors of this volume as to these problems and, in some cases, their possible solutions.

(1) What is the relative quantitative significance of different basalt types on Earth? How do they vary with tectonic context, upper mantle composition, and other factors? The suites assembled and described in Chapter 1 are still a sampling, and much work remains to be done to achieve a comprehensive quantitative mapping of basalts by composition, volume, and epoch. The diversity as well as the accessibility of the Earth makes it the fundamental basis for understanding igneous differentiations on other planets.

(2) What is the evolution of igneous rocks with time and how does it constrain the Earth's evolution? So far, the only distinct change in composition determined is the shift from Archean to post-Archean some 2.5 b.y. ago. Surely there should be subtler trends inferrable from suites generated at intermediate epochs.

(3) How can the different effects on basaltic composition—source composition, percent melting, crystallization, mixing, fluid separation—be better discriminated? Most of the improvements needed are fairly evident: phase equilibria experiments at higher pres-

sures; experiments on five or more components, and better means to represent the results; better trace element partitioning coefficients, and their application to samples; better physical models of melting, magma fractionation, etc. The problem seems more to decide among these many possibilities.

(4) How can basaltic volcanism on Earth, and its relationship to other (intermediate and acidic) volcanism be used to predict conditions on other planets? Are some of the less abundant types more sensitive indicators of differences in volatile content, temperature, etc.?

(5) How are isotopic ratios affected by the various processes affecting basalts, including post-eruptive alterations as well as magmatic differentiations? The significance of isotopic measurements—the events discriminated, etc.—is often uncertain, because of complex local circumstances: contamination, stability of mineral sites, metamorphic effects, etc.

(6) What is the relationship of convection to magmatism, and of magmatism to volcanism? There is a crude positive correlation of volcanism with indicators of convective upstreams, such as positive gravity anomalies. However, within a broad "hot spot" area, volcanism seems to shift sporadically from one site to another, and from one composition to another. Obviously, local heterogeneities in composition and temperature play some role, but how these interact with changing physical circumstances [see (7) below] is not clear. Also quite uncertain is the portion of each magmatic event which reaches the surface.

(7) How does magma penetrate the lithosphere? To what extent is it dependent on larger scale tectonic setting? Various aspects of the physics of magma ascent have been modelled, but all such models were severely hypothesis-dependent. Observation, as well as thermal considerations, suggest that volatiles may be significant in enabling a magma to penetrate to the surface. The volumetric domination of ocean rise basalts on the Earth, the predominantly tensional regime of Mars, and the shutting off of volcanism on the Moon and Mercury (since entering their compressive phases) make it evident that macroscale tectonics are important.

(8) How can isotopic indicators of mantle inhomogeneity be improved, and how can they be used to constrain convective and evolutionary models? Several problems are entangled here, such as the refinement of the temporal restrictions of some isotopic systems, the definition of the volumes represented by isotopic samples, the implication of macroscale fluid dynamical modellings for local elements of matter, and differing migration rates of solid and gaseous components. Most of these difficulties will probably be solved in a one-by-one way, but a couple of decades hence the combination of isotopic measurements with thermal convective modellings should give a much better picture of mantle evolution: the degree of interchange between reservoirs, the relationship of heat transfer to material transfer, the principal differentiations, the rate of outgassing, etc.

(9) What are the physical properties of the deeper mantle, and the resulting thermomechanical state? Not only are the effects of phase transitions on viscosity, thermal expansion, and other properties below ~400 km imperfectly known or unknown, the implications of these changes in physical properties are not well understood; as, for example, the relationship of temperature dependence of viscosity to the convective tendency to drive the temperature curve to an adiabat.

(10) How can the interplanetary correlation of volcanic chronologies be improved? The relationship between the cratering rates on different planets is an analytic problem which is model-dependent: it appears to require definition of the reservoirs in which small bodies have been kept, and identification of the mechanism(s) by which these bodies are transferred to trajectories which cross planetary orbits. Possibly, ways will be found to short-circuit some of these dynamical difficulties to give a more effective interplanetary chronology.

(11) What are the thermodynamics and hydrodynamics of great impacts? Better understanding of these processes is important to several problems, such as early heating of the planets and planetesimals, and thence their differentiation; and the evolution of the surfaces of the Moon, Mercury and Mars. Fairly simple considerations indicate that the scaling up from small impacts currently feasible for experiment or computation is by no means simple, and that the energy partitionings, fragment sizes, and material effects may change appreciably.

ACKNOWLEDGMENTS—We are grateful to R. J. Arculus, P. M. Cassen, W. K. Hartmann, R. J. Phillips, S. C. Solomon, D. M. Walker, G. W. Wetherill, D. E. Wilhelms, and J. A. Wood for detailed reviews. The following support for work on Chapter 10 was received: W. M. Kaula, NASA grant NGL 05-007-002; D. L. Anderson, NASA grants NSG-7610 and NGL 05-007-069.

10.5 REFERENCES

Anders E. (1976) Chemical compositions of the Moon, Earth and eucrite parent body. *Phil. Trans. Roy. Soc. Lond.* **A285**, 23–40.

Anders E. and Owen T. (1977) Mars and Earth: origin and abundance of volatiles. *Science* **198**, 453–465.

Anderson D. L. (1975) Chemical plumes in the mantle. *Geol. Soc. Amer. Bull.* **86**, 1593–1600.

Anderson D. L. (1977) Composition of the mantle and core. *Ann. Rev. Earth Planet. Sci.* **5**, 179–202.

Anderson D. L. (1979) The upper mantle transition region: eclogite? *Geophys. Res. Lett.* **6**, 433–435.

Birch F. (1972) Elasticity and constitution of the Earth's interior. *J. Geophys. Res.* **52**, 277–286.

Butler R. F. (1972) Natural remanent magnetization and thermomagnetic properties of the Allende meteorite. *Earth Planet. Sci. Lett.* **17**, 120–128.

Cameron A. G. W. (1978a) Physics of the primitive solar disk. *Moon and Planets* **18**, 5–40.

Cameron A. G. W. (1978b) The primitive solar accretion disk and the formation of the planets. In *The Origin of the Solar System* (S. F. Dermott, ed.), p. 49–73. John Wiley, N.Y.

Cameron A. G. W. and Truran J. W. (1977) The supernova trigger for formation of the solar system. *Icarus* **30**, 447–461.

Cameron A. G. W. and Ward W. R. (1976) The origin of the Moon (abstract). *Lunar Science VII*, p. 120–122. The Lunar Science Institute, Houston.

Campbell D. B., Dyce D. B., Ingalls R. P., Pettengill G. H. and Shapiro I. I. (1972) Venus: topography revealed by radar data. *Science* **175**, 514–516.

Chapman C. R. (1979) The asteroids: nature, interrelations, origin, and evolution. In *Asteroids* (T. Gehrels, ed.), p. 25–60. Univ. Arizona Press, Tucson.

Clayton R. N. and Mayeda T. K. (1975) Genetic relations between the Moon and meteorites. *Proc. Lunar Sci. Conf. 6th*, p. 1761–1769.

Clayton R. N., Onuma N. and Mayeda T. K. (1976) A classification of meteorites based on oxygen isotopes. *Earth Planet. Sci. Lett.* **30**, 10–18.

Consolmagno G. J. and Drake M. J. (1977) Composition and evolution of the eucrite parent body: evidence from rare-earth elements. *Geochim. Cosmochim. Acta* **41**, 1271–1282.

Dickinson W. R. and Luth W. C. (1971) A model for plate tectonic evolution of mantle layers. *Science* **174**, 400–404.

Drake M. J. (1980) Trace elements as quantitative probes of differentiation processes in planetary interiors. *Rev. Geophys. Space Phys.* **18**, 11–26.

Elmegreen B. C. (1978) On the interaction between a strong stellar wind and a surrounding disk nebula. *Moon and Planets* **19**, 261–278.

Fish R. A., Goles G. G. and Anders E. (1960) The record in the meteorites. III On the development of meteorites in asteroidal bodies. *Astrophys. J.* **132**, 243–258.

Frey F. A., Green D. H. and Roy S. D. (1978) Integrated models of basalt petrogenesis: a study of quartz tholeiites to olivine melilites from S.E. Australia utilizing geochemical and experimental petrological data. *J. Petrol.* **19**, 463–573.

Goldreich P. and Ward W. R. (1973) The formation of planetesimals. *Astrophys. J.* **183**, 1051–1061.

Goldstein J. R. and J. M. Short (1967) The iron meteorites, their thermal history and parent bodies. *Geochim. Cosmochim. Acta* **31**, 1733–1770.

Gray C. M. and Compston W. (1976) Excess ^{26}Mg in the Allende meteorite. *Nature* **251**, 495–497.

Green D. H., Hibberson W. O. and Jaques A. L. (1979) Petrogenesis of mid-ocean ridge basalts. In *The Earth: its Origin, Structure and Evolution* (M. W. McElhinny, ed.), p. 265–299. Academic Press, N.Y.

Greenberg R., Wacker J. F., Hartmann W. K. and Chapman C. R. (1978) Planetesimals to planets: numerical simulation of collisional evolution. *Icarus* **35**, 1–26.

Hager B. H. and O'Connell R. J. (1979) Kinematic models of large-scale flow in the Earth's mantle. *J. Geophys. Res.* **84**, 1031–1048.

Harris A. W. (1978) Dynamics of planetesimal formation and planetary accretion. In *The Origin of the Solar System* (S. F. Dermott, ed.), p. 469–492. John Wiley, N.Y.

Hartmann W. K. and Davis D. (1975) Satellite-sized planetesimals and lunar origin. *Icarus* **24**, 504–515.

Hedge C. F. (1978) Strontium isotopes in basalts from the Pacific Ocean basin. *Earth. Planet. Sci. Lett.* **38**, 88–94.

Hoffman J. H., Hodges R. R., McElroy M. B., Donahue T. M. and Kolpin M. A. (1979) Venus lower atmospheric composition: preliminary results from Pioneer Venus. *Science* **203**, 800–802.

Hörz F. (1978) How thick are mare basalts? *Proc. Lunar Sci. Conf. 9th*, p. 3311–3331.

Kaula W. M. (1975) The seven ages of a planet. *Icarus* **26**, 1–15.

Kaula W. M. (1977) On the origin of the Moon, with emphasis on bulk composition. *Proc. Lunar Sci. Conf. 8th*, p. 321–331.

Kaula W. M. (1979) Thermal evolution of Earth and Moon growing by planetesimal impacts. *J. Geophys. Res.* **84**, 999–1008.

Kaula W. M. (1980a) The beginning of Earth's thermal evolution. In *The Continental Crust and its Mineral Deposits* (D. Strangway, ed.), Geol. Assoc. Canada, Spec. Paper 20, p. 25–34.

Kaula W. M. (1980b) Material properties for mantle convection consistent with observed surface fields. *J. Geophys. Res.* **85**, 7031–7044.

Kaula W. M. and Bigeleisen P. E. (1975) Early scattering by Jupiter and its collision effects in the terrestrial zone. *Icarus* **25**, 18–33.

Kaula W. M. and Harris A. W. (1975) Dynamics of lunar origin and orbital evolution. *Rev. Geophys. Space Phys.* **13**, 363–371.

Kieffer S. W. and Simonds C. H. (1980) The role of volatiles and lithology in the impact cratering process. *Rev. Geophys. Space Phys.* **18**, 143–181.

Lee T., Papanastassiou D. A. and Wasserburg G. J. (1976) Demonstration of ^{26}Mg excess in Allende and evidence for ^{26}Al. *Geophys. Res. Lett.* **3**, 109–112.

Liebermann R. C. and Ringwood A. E. (1976) Elastic properties of anorthite and the nature of the lunar crust. *Earth. Planet. Sci. Lett.* **31**, 69–74.

Lynden-Bell D. and Pringle J. E. (1974) The evolution of viscous discs and the origin of the nebular variables. *Mon. Not. Roy. Astron. Soc.* **168**, 603–637.

McGetchin T. R. and Smyth J. R. (1978) The mantle of Mars: some possible geological implications of its high density. *Icarus* **34**, 512–536.

Morgan J. W., Higuchi H., Takakashi H. and Hertogen J. (1978) A "chondritic" eucrite parent body: inference from trace elements. *Geochim. Cosmochim. Acta* **42**, 27–38.

O'Hara M. J. (1968) Are ocean-floor basalts primary magmas? *Nature* **220**, 683–686.

O'Keefe J. A. and Sullivan F. (1978) Fission origin of the Moon: cause and timing. *Icarus* **35**, 272–283.

O'Keefe J. D. and Ahrens T. J. (1977) Impact-induced energy partitioning, melting, and vaporization on terrestrial planets. *Proc. Lunar Sci. Conf. 8th*, **p. 3357**–3374

O'Nions R. K., Evensen N. M. and Hamilton P. J. (1979) Geochemical modeling of mantle differentiation and crustal growth. *J. Geophy. Res.* **84**, 6091–6101.

Peale S. J., Cassen P. and Reynolds R. T. (1979) Melting of Io by tidal dissipation. *Science* **203**, 892–894.

Pettengill G. H., Eliason E., Ford P. G., Loriot G. B., Masursky H. and McGill G. E. (1980) Pioneer Venus radar results: altimetry and surface properties. *J. Geophys. Res.* **85**, 8261–8270.

Pettengill G. H., Ford P. G., Brown W. E., Kaula W. M., Keller C. H., Masursky H. and McGill G. E. (1979a) Pioneer Venus Radar mapper experiment. *Science* **203**, 806–808.

Pettengill G. H., Ford P. G., Brown W. E., Kaula W. M., Keller C. H., Masursky H. and McGill G. E. (1979b) Venus: preliminary topographic and surface imaging results from the Pioneer Orbiter. *Science* **205**, 90–93.

Phillips R. J., Kaula W. M., McGill G. E. and Malin M. C. (1981) Tectonics and evolution of Venus. *Science*, in press.

Pollack J. B. (1979) Climatic change on the terrestrial planets. *Icarus* **37**, 479–553.

Ringwood A. E. (1975) *Composition and Petrology of the Earth's Mantle.* McGraw-Hill, N.Y. 618 pp.

Ringwood A. E. (1979) *Origin of the Earth and Moon.* Springer-Verlag, Berlin. 295 pp.

Ruskol E. L. (1977) The origin of the Moon. In *The Soviet-American Conference on Cosmochemistry of the Moon and Planets*, p. 815–822. NASA, SP-370. Washington, D.C.

Safronov V. S. (1972) *Evolution of the Protoplanetary Cloud and Formation of the Earth and Planets.* NASA TTF-667. Washington, D.C. 206 pp.

Schaber G. G. (1973) Lava flows from Mare Imbrium: geologic evolution from Apollo orbital photography. *Proc. Lunar Sci. Conf. 4th*, p. 73–92.

Schilling J. G. (1971) Sea-floor evolution: rare-earth evidence. *Phil. Trans. Roy. Soc. Lond.* **A286**, 663–706.

Schramm D. N. and Wasserburg G. J. (1970) Nucleochronologies and the mean age of the elements. *Astrophys. J.* **162**, 57–69.

Sjogren W. L., Phillips R. J., Birkeland P. W. and Wimberly R. N. (1980) Gravity anomalies on Venus. *J. Geophys. Res.* **85**, 8295–8302.

Slattery W. L. (1978) Protoplanetary Core formation by rainout of iron drops. *Moon and Planets* **19**, 443–456.

Slattery W. L., DeCampli W. M. and Cameron A. G. W. (1980) Protoplanetary core formation by rain-out of minerals. *Harvard Cen. Astrophys. Preprint 1280.* 9 pp.

Smith J. V. (1979) Mineralogy of the planets: a voyage in space and time. *Mineral. Mag.* **43**, 1–89.

Solomon S. C. and Chaiken J. (1976) Thermal expansion and thermal stress in the Moon and terrestrial planets: clues to early thermal history. *Proc. Lunar Sci. Conf. 7th*, p. 3229–3243.

Solomon S. C. and Head J. W. (1979) Vertical movement in mare basins: relation to mare emplacement, basin tectonics, and lunar thermal history. *J. Geophys. Res.* **84**, 1667–1682.

Sonett C. P. and Herbert F. L. (1977) Pre-main sequence heating of planetoids. In *Comets, Asteroids, and Meteorites* (A. Delsemme, ed.), p. 429–437. Univ. of Toledo Press.

Stevenson D. J. (1980) Lunar asymmetry and paleomagnetism. *Nature* **287**, 520–521.

Stevenson D. J. and Turner J. S. (1979) Fluid models of mantle convection. In *The Earth: Origin, Evolution, and Structure* (M. McElhinny, ed.), p. 227–263. John Wiley, N.Y.

Stevenson D. J., Schubert G., Cassen P. and Reynolds R. J. (1980) Core evolution and magnetism of the terrestrial planets (abstract). In *Lunar and Planetary Science XI*, p. 1091–1093. Lunar and Planetary Institute, Houston.

Stolper E. (1977) Experimental petrology of eucritic meteorites. *Geochim. Cosmochim. Acta* **41**, 587–611.

Stolper E. (1980) Predictions of mineral assemblages in planetary interiors. *Proc. Lunar Sci. Conf. 11th*, p. 235–250.

Strom S. E. (1977) Early stages of stellar evolution. In *Star Formation* (T. DeJong and A. Maeder, eds.), p. 179–197. Reidel, Dordrecht.

Sun S. -S. and Hanson G. N. (1975) Evolution of the mantle: geochemical evidence from alkali basalts. *Geology* **3**, 297–302.

Tatsumoto M. (1978) Isotopic composition of lead in oceanic basalt and its implication to mantle evolution. *Earth Planet. Sci. Lett.* **38**, 63–87.

Trimble V. (1975) The origin and abundances of the chemical elements. *Rev. Mod. Phys.* **47**, 877–976.

Ulrich R. K. (1978) The status of T. Tauri models. In *Protostars and Planets* (T. Gehrels, ed.), p. 718–773. Univ. of Arizona Press, Tucson.

Urey H. C. (1956) Diamonds, meteorites, and the origin of the solar system. *Astrophys. J.* **124**, 623–637.

Vinogradov A. P., Surkov Y. A. and Kimozov F. F. (1973) The contents of uranium, thorium, and potassium in the rocks of Venus as measured by Venera 8. *Icarus* **20**, 253–259.

Walker D., Stolper E. M. and Hays J. F. (1979) Basaltic volcanism: the importance of planet size. *Proc. Lunar Sci. Conf. 10th*, p. 1995–2015.

Walker J. C. G. (1975) Evolution of the atmosphere of Venus. *J. Atmos. Sci.* **32**, 1248–1256.

Warren P. H. and Wasson J. T. (1979) The origin of KREEP. *Rev. Geophys. Space Phys.* **17**, 73–88.

Wasserburg G. J. and DePaolo D. J. (1979) Models of earth structure inferred from neodymium and strontium isotopic abundances. *Proc. Natl. Acad. Sci. USA* **76**, 3594–3598.

Wasson J. T. and Warren P. H. (1980) Contribution of the mantle to the lunar asymmetry. *Icarus*, in press.

Wasson J. T. and Wetherill G. W. (1979) Dynamical, chemical and isotopic evidence regarding the formation locations of asteroids and meteorites. In *Asteroids* (T. Gehrels, ed.), p. 926–974. Univ. of Arizona Press, Tucson.

Weidenschilling S. J. (1975) Mass loss from the region of Mars and the asteroid belt. *Icarus* **26**, 361–366.

Weidenschilling S. J. (1978) Iron/silicate fractionation and the origin of Mercury. *Icarus* **35**, 99–111.

Wetherill G. W. (1976) The role of large bodies in the formation of the Earth and Moon. *Proc. Lunar Sci. Conf. 7th*, p. 3245–3257.

Wetherill G. W. (1980a) Numerical calculations relevant to the accumulation of the terrestrial planets. In *The Continental Crust and its Mineral Deposits.* (D. Strangway, ed.), Geol. Assoc. Canada, Spec. Paper 20, in press.

Wetherill G. W. (1980b) Could the solar wind have been the source of the high concentration of ^{36}Ar in the atmosphere of Venus? (abstract). In *Lunar and Planetary Science XII*, p. 1239–1241. Lunar and Planetary Institute, Houston.

Wood J. A. (1964) The cooling rates and parent planets of several iron meteorites. *Icarus* **3**, 429–459.

Locality Index

Entries refer to pages on which planetary bodies or features on those bodies are discussed. Boldface entries identify extensive discussions, descriptions, definitions or photographs of the object referenced. Names of localities or features appear in the subject index, as well, if they denote the archetype of a class of features or objects (e.g., Fra Mauro basalts) or if physical properties are discussed.

Subject Index

Boldface entries denote extensive discussions, descriptions, definitions or photographs of the object or term referenced.